BERGEY'S MANUAL OF

Systematic Bacteriology

Second Edition

Volume Four

The *Bacteroidetes*, *Spirochaetes*, *Tenericutes* (*Mollicutes*), *Acidobacteria*, *Fibrobacteres*, *Fusobacteria*, *Dictyoglomi*, *Gemmatimonadetes*, *Lentisphaerae*, *Verrucomicrobia*, *Chlamydiae*, and *Planctomycetes*

BERGEY'S MANUAL OF

Systematic Bacteriology

Second Edition

Volume Four

The *Bacteroidetes*, *Spirochaetes*, *Tenericutes* (*Mollicutes*), *Acidobacteria*, *Fibrobacteres*, *Fusobacteria*, *Dictyoglomi*, *Gemmatimonadetes*, *Lentisphaerae*, *Verrucomicrobia*, *Chlamydiae*, and *Planctomycetes*

Noel R. Krieg, James T. Staley, Daniel R. Brown, Brian P. Hedlund, Bruce J. Paster, Naomi L. Ward, Wolfgang Ludwig and William B. Whitman
EDITORS, VOLUME FOUR

William B. Whitman
DIRECTOR OF THE EDITORIAL OFFICE

Aidan C. Parte
MANAGING EDITOR

Springer

William B. Whitman
Bergey's Manual Trust
Department of Microbiology
527 Biological Sciences Building
University of Georgia
Athens, GA 30602-2605
USA

ISBN: 978-0-387-95042-6 e-ISBN: 978-0-387-68572-4
DOI: 10.1007/978-0-387-68572-4
Springer New York Dordrecht Heidelberg London

Library of Congress Control Number: 2010936277

Printed on acid-free paper.

Springer is part of Springer Science+Business Media (www.springer.com)

This volume is dedicated to our colleague
Karl-Heinz Schleifer,
who retired from the Board of Trustees of Bergey's Manual as this volume was in preparation.
We deeply appreciate his efforts as an editor, author and officer of the Trust. He has devoted many years to helping the Trust meet its objectives.

Preface to volume 4 of the second edition of *Bergey's Manual of Systematic Bacteriology*

Prokaryotic systematics has remained a vibrant and exciting field of study, one of challenges and opportunities, great discoveries and gradual advances. To honor the leaders of our field, the Trust presented the 2010 Bergey's Award in recognition of outstanding contributions to the taxonomy of prokaryotes to Antonio Ventosa.

We expect that this will be the last volume to be edited by Jim Staley and Noel Krieg, both of whom served on the Trust for many years and continued to be active after their retirements. Noel contributed to *Bergey's Manual of Determinative Bacteriology*, in both the 8th edition as an author and the 9th edition as an editor. Moreover, he also edited volume 1 of the 1st edition of *Bergey's Manual of Systematic Bacteriology*. This was a massive achievement for this volume included the "Gram-negatives" and comprised more than one-third of that edition. In the 2nd edition of *Bergey's Manual of Systematic Bacteriology*, Noel and Jim edited volume 2 along with Don Brenner. This three-part volume comprised the *Proteobacteria* and was also a major portion of the current edition. In the current volume, Noel edited the phylum *Bacteroidetes*, which is the largest phylum in this work. In addition to his passion for prokaryotic systematics, Noel is well known by his colleagues and students at Virginia Tech as a dedicated and passionate teacher.

Jim Staley's service to the Trust paralleled that of Noel's. Jim also contributed to *Bergey's Manual of Determinative Bacteriology*, in both the 8th edition as an author and the 9th edition as an editor. He edited volume 3 of the 1st edition of *Bergey's Manual of Systematic Bacteriology* along with Marvin Bryant and Norbert Pfennig and volume 2 of the 2nd edition. In the current volume, he edited the phyla *Acidobacteria*, *Chlamydiae*, *Dictyoglomi*, *Fibrobacteres*, *Fusobacteria*, and *Gemmatimonadetes*. More importantly, he coached and mentored the rest of us throughout the entire editorial process. Most recently, Jim has led the efforts of the Trust to form Bergey's International Society for Microbial Systematics (BISMiS), whose purpose is to promote excellent research in microbial systematics as well as enhance global communication among taxonomists who study the *Bacteria* and *Archaea*. The society will also serve internationally as an advocate for research efforts on microbial systematics and diversity. We wish Jim the best in this new adventure.

Acknowledgements

The Trust is indebted to all of the contributors and reviewers, without whom this work would not be possible. The Editors are grateful for the time and effort that each has expended on behalf of the entire scientific community. We also thank the authors for their good grace in accepting comments, criticisms, and editing of their manuscripts.

The Trust recognizes its enormous debt to Aidan Parte, whose enthusiasm and professionalism have made this work possible. His expertise and good judgment have been extremely valued.

We also recognize the special efforts of Jean Euzéby in checking and correcting where necessary the nomenclature and etymology of every described taxon in this volume.

The Trust also thanks its Springer colleagues, Editorial Director Andrea Macaluso and Production Manager Susan Westendorf. In addition, we thank Amina Ravi, our manager at our typesetters, SPi, for her work in the proofing and production of the book.

We thank our current copyeditors, proofreaders and other staff, including Susan Andrews, Joanne Auger, Francis Brenner, MaryAnn Brickner, Travis Dean, Robert Gutman, Judy Leventhal and Linda Sanders, without whose hard work and attention to detail the production of this volume would be impossible. Lastly, we thank the Department of Microbiology at the University of Georgia for its assistance and encouragement in thousands of ways.

William B. (Barny) Whitman

Contents

Preface to volume 4 ix
Contributors xix
On using the *Manual* xxv
Road map of volume 4 phyla 1
Taxonomic outline of volume 4 21

Phylum XIV. *Bacteroidetes* phyl. nov. 25
Class I. *Bacteroidia* class. nov. 25
Order I. *Bacteroidales* ord. nov. 25
Family I. *Bacteroidaceae* 25
Genus I. *Bacteroides* 27
Genus II. *Acetofilamentum* 41
Genus III. *Acetomicrobium* 42
Genus IV. *Acetothermus* 44
Genus V. *Anaerorhabdus* 45
Family II. *Marinilabiliaceae* fam. nov. 49
Genus I. *Marinilabilia* 49
Genus II. *Anaerophaga* 51
Genus III. *Alkaliflexus* 53
Family III. *Rikenellaceae* fam. nov. 54
Genus I. *Rikenella* 55
Genus II. *Alistipes* 56
Family IV. *Porphyromonadaceae* fam. nov. 61
Genus I. *Porphyromonas* 62
Genus II. *Barnesiella* 70
Genus III. *Dysgonomonas* 71
Genus IV. *Paludibacter* 76
Genus V. *Petrimonas* 77
Genus VI. *Proteiniphilum* 77
Genus VII. *Tannerella* 78
Family V. *Prevotellaceae* fam. nov. 85
Genus I. *Prevotella* 86
Genus II. *Xylanibacter* 102
Class II. *Flavobacteriia* class. nov. 105
Order I. *Flavobacteriales* ord. nov. 105
Family I. *Flavobacteriaceae* 106
Genus I. *Flavobacterium* 112
Genus II. *Aequorivita* 155
Genus III. *Algibacter* 157
Genus IV. *Aquimarina* 158
Genus V. *Arenibacter* 161
Genus VI. *Bergeyella* 165

Genus VII. *Bizionia* . . . 166
Genus VIII. *Capnocytophaga* . . . 168
Genus IX. *Cellulophaga* . . . 176
Genus X. *Chryseobacterium* . . . 180
Genus XI. *Cloacibacterium* . . . 197
Genus XII. *Coenonia* . . . 198
Genus XIII. *Costertonia* . . . 199
Genus XIV. *Croceibacter* . . . 199
Genus XV. *Dokdonia* . . . 201
Genus XVI. *Donghaeana* . . . 201
Genus XVII. *Elizabethkingia* . . . 202
Genus XVIII. *Empedobacter* . . . 210
Genus XIX. *Epilithonimonas* . . . 212
Genus XX. *Flaviramulus* . . . 213
Genus XXI. *Formosa* . . . 214
Genus XXII. *Gaetbulibacter* . . . 218
Genus XXIII. *Gelidibacter* . . . 219
Genus XXIV. *Gillisia* . . . 221
Genus XXV. *Gramella* . . . 226
Genus XXVI. *Kaistella* . . . 227
Genus XXVII. *Kordia* . . . 228
Genus XXVIII. *Krokinobacter* . . . 230
Genus XXIX. *Lacinutrix* . . . 231
Genus XXX. *Leeuwenhoekiella* . . . 232
Genus XXXI. *Lutibacter* . . . 234
Genus XXXII. *Maribacter* . . . 235
Genus XXXIII. *Mariniflexile* . . . 238
Genus XXXIV. *Mesonia* . . . 239
Genus XXXV. *Muricauda* . . . 240
Genus XXXVI. *Myroides* . . . 245
Genus XXXVII. *Nonlabens* . . . 248
Genus XXXVIII. *Olleya* . . . 249
Genus XXXIX. *Ornithobacterium* . . . 250
Genus XL. *Persicivirga* . . . 254
Genus XLI. *Polaribacter* . . . 255
Genus XLII. *Psychroflexus* . . . 258
Genus XLIII. *Psychroserpens* . . . 261
Genus XLIV. *Riemerella* . . . 262
Genus XLV. *Robiginitalea* . . . 264
Genus XLVI. *Salegentibacter* . . . 266
Genus XLVII. *Sandarakinotalea* . . . 269
Genus XLVIII. *Sediminicola* . . . 270
Genus XLIX. *Sejongia* . . . 271
Genus L. *Stenothermobacter* . . . 275
Genus LI. *Subsaxibacter* . . . 275
Genus LII. *Subsaximicrobium* . . . 277
Genus LIII. *Tenacibaculum* . . . 279
Genus LIV. *Ulvibacter* . . . 283
Genus LV. *Vitellibacter* . . . 284
Genus LVI. *Wautersiella* . . . 285
Genus LVII. *Weeksella* . . . 286
Genus LVIII. *Winogradskyella* . . . 288
Genus LIX. *Yeosuana* . . . 291
Genus LX. *Zhouia* . . . 292
Genus LXI. *Zobellia* . . . 292

Family II. *Blattabacteriaceae* fam. nov. 315
Genus I. *Blattabacterium* 315
Family III. *Cryomorphaceae* 322
Genus I. *Cryomorpha* 323
Genus II. *Brumimicrobium* 323
Genus III. *Crocinitomix* 326
Genus IV. *Fluviicola* 327
Genus V. *Lishizhenia* 327
Genus VI. *Owenweeksia* 328
Class III. *Sphingobacteriia* class. nov. 330
Order I. *Sphingobacteriales* ord. nov. 330
Family I. *Sphingobacteriaceae* 331
Genus I. *Sphingobacterium* 331
Genus II. *Pedobacter* 339
Family II. *Chitinophagaceae* fam. nov. 351
Genus I. *Chitinophaga* 351
Genus II. *Terrimonas* 356
Family III. *Saprospiraceae* fam. nov. 358
Genus I. *Saprospira* 359
Genus II. *Aureispira* 361
Genus III. *Haliscomenobacter* 363
Genus IV. *Lewinella* 366
Class IV. *Cytophagia* class. nov. 370
Order I. *Cytophagales* 370
Family I. *Cytophagaceae* 371
Genus I. Cytophaga 371
Genus II. *Adhaeribacter* 375
Genus III. *Arcicella* 377
Genus IV. *Dyadobacter* 380
Genus V. *Effluviibacter* 387
Genus VI. *Emticicia* 388
Genus VII. *Flectobacillus* 389
Genus VIII. *Flexibacter* 392
Genus IX. *Hymenobacter* 397
Genus X. *Larkinella* 404
Genus XI. *Leadbetterella* 405
Genus XII. *Meniscus* 406
Genus XIII. *Microscilla* 408
Genus XIV. *Pontibacter* 410
Genus XV. *Runella* 412
Genus XVI. *Spirosoma* 415
Genus XVII. *Sporocytophaga* 418
Family II. *Cyclobacteriaceae* fam. nov. 423
Genus I. *Cyclobacterium* 423
Genus II. *Algoriphagus* 426
Genus III. *Aquiflexum* 433
Genus IV. *Belliella* 434
Genus V. *Echinicola* 437
Genus VI. *Rhodonellum* 440
Family III. *Flammeovirgaceae* fam. nov. 442
Genus I. *Flammeovirga* 442
Genus II. *Fabibacter* 447
Genus III. *Flexithrix* 448
Genus IV. *Persicobacter* 450
Genus V. *Reichenbachiella* 452

Genus VI. *Roseivirga* . . . 453
Order II. *Incertae sedis* . . . 457
Family I. *Rhodothermaceae* fam. nov. . . . 457
Genus I. *Rhodothermus* . . . 458
Genus II. *Salinibacter* . . . 460
Order III. *Incertae sedis* . . . 465
Genus I. *Thermonema* . . . 465
Order IV. *Incertae sedis* . . . 467
Genus I. *Toxothrix* . . . 467

Phylum XV. *Spirochaetes* phyl. nov. . . . 471
Class I. *Spirochaetia* class. nov. . . . 471
Order I. *Spirochaetales* . . . 471
Family I. *Spirochaetaceae* . . . 473
Genus I. *Spirochaeta* . . . 473
Genus II. *Borrelia* . . . 484
Genus III. *Cristispira* . . . 498
Genus IV. *Treponema* . . . 501
Family II. *Brachyspiraceae* . . . 531
Genus I. *Brachyspira* . . . 531
Family III. *Brevinemataceae* fam. nov. . . . 545
Genus I. *Brevinema* . . . 545
Family IV. *Leptospiraceae* . . . 546
Genus I. *Leptospira* . . . 546
Genus II. *Leptonema* . . . 556
Genus III. *Turneriella* . . . 558
Hindgut spirochetes of termites and *Cryptocercus punctulatus* . . . 563

Phylum XVI. *Tenericutes* . . . 567
Class I. *Mollicutes* . . . 568
Order I. *Mycoplasmatales* . . . 574
Family I. *Mycoplasmataceae* . . . 575
Genus I. *Mycoplasma* . . . 575
Genus II. *Ureaplasma* . . . 613
Family II. *Incertae sedis* . . . 639
Genus I. *Eperythrozoon* . . . 640
Genus II. *Haemobartonella* . . . 642
Order II. *Entomoplasmatales* . . . 644
Family I. *Entomoplasmataceae* . . . 645
Genus I. *Entomoplasma* . . . 646
Genus II. *Mesoplasma* . . . 649
Family II. *Spiroplasmataceae* . . . 654
Genus I. *Spiroplasma* . . . 654
Order III. *Acholeplasmatales* . . . 687
Family I. *Acholeplasmataceae* . . . 687
Genus I. *Acholeplasma* . . . 688
Family II. *Incertae sedis* . . . 696
Genus I. "*Candidatus* Phytoplasma" gen. nov. . . . 696
Order IV. *Anaeroplasmatales* . . . 719
Family I. *Anaeroplasmataceae* . . . 720
Genus I. *Anaeroplasma* . . . 720
Genus II. *Asteroleplasma* . . . 722

Phylum XVII. *Acidobacteria* phyl. nov. 725
Class I. *Acidobacteriia*. 727
Order I. *Acidobacteriales* 727
Family I. *Acidobacteriaceae* fam. nov. 728
Genus I. *Acidobacterium* 728
Genus II. *Edaphobacter*. 729
Genus III. *Terriglobus* 730
Class II. *Holophagae*. 731
Order I. *Holophagales*. 731
Family I. *Holophagaceae* 732
Genus I. *Holophaga* 732
Genus II. *Geothrix* 732
Order II. *Acanthopleuribacterales* 734
Family I. *Acanthopleuribacteraceae* 734
Genus I. *Acanthopleuribacter*. 734

Phylum XVIII. "*Fibrobacteres*" 737
Class I. *Fibrobacteria* class. nov. 739
Order I. *Fibrobacterales* ord. nov. 739
Family I. *Fibrobacteraceae* fam. nov. 739
Genus I. *Fibrobacter*. 740

Phylum XIX. "*Fusobacteria*". 747
Class I. *Fusobacteriia* class. nov. 747
Order I. *Fusobacteriales* ord. nov. 747
Family I. *Fusobacteriaceae* fam. nov. 748
Genus I. *Fusobacterium*. 748
Genus II. *Cetobacterium* 758
Genus III. *Ilyobacter*. 759
Genus IV. *Propionigenium* 761
Family II. *Leptotrichiaceae* fam. nov. 766
Genus I. *Leptotrichia* 766
Genus II. *Sebaldella*. 769
Genus III. *Sneathia*. 770
Genus IV. *Streptobacillus* 771

Phylum XX. *Dictyoglomi* phyl. nov. 775
Class I. *Dictyoglomia* class. nov. 776
Order I. *Dictyoglomales* ord. nov. 776
Family I. *Dictyoglomaceae* fam. nov. 776
Genus I. *Dictyoglomus* 776

Phylum XXI. *Gemmatimonadetes* 781
Class I. *Gemmatimonadetes*. 781
Order I. *Gemmatimonadales*. 781
Family I. *Gemmatimonadaceae* 782
Genus I. *Gemmatimonas* 782

Phylum XXII. *Lentisphaerae* 785
Class I. *Lentisphaeria* class. nov. 787
Order I. *Lentisphaerales* 788
Family I. *Lentisphaeraceae* fam. nov. 788
Genus I. *Lentisphaera* 788

Order II. *Victivallales* . . . 791
Family I. *Victivallaceae* fam. nov. . . . 791
Genus I. *Victivallis* . . . 791

Phylum XXIII. *Verrucomicrobia* phyl. nov. . . . 795
Class I. *Verrucomicrobiae* . . . 799
Order I. *Verrucomicrobiales* . . . 802
Family I. *Verrucomicrobiaceae* . . . 803
Genus I. *Verrucomicrobium* . . . 803
Genus II. *Prosthecobacter* . . . 805
Family II. *Akkermansiaceae* fam. nov. . . . 809
Genus I. *Akkermansia* . . . 809
Family III. *Rubritaleaceae* fam. nov. . . . 812
Genus I. *Rubritalea* . . . 812
Class II. *Opitutae* . . . 817
Order I. *Opitutales* . . . 820
Family I. *Opitutaceae* . . . 820
Genus I. *Opitutus* . . . 820
Genus II. *Alterococcus* . . . 821
Order II. *Puniceicoccales* . . . 823
Family I. *Puniceicoccaceae* . . . 824
Genus I. *Puniceicoccus* . . . 824
Genus II. *Cerasicoccus* . . . 825
Genus III. *Coraliomargarita* . . . 827
Genus IV. *Pelagicoccus* . . . 829
Class III. *Spartobacteria* class. nov. . . . 834
Order I. *Chthoniobacterales* ord. nov. . . . 836
Family I. *Chthoniobacteraceae* fam. nov. . . . 837
Genus I. *Chthoniobacter* . . . 837
Genus II. "*Candidatus* Xiphinematobacter" . . . 838

Phylum XXIV. "*Chlamydiae*" . . . 843
Class I. *Chlamydiia* class. nov. . . . 844
Order I. *Chlamydiales* . . . 844
Family I. *Chlamydiaceae* . . . 845
Genus I. *Chlamydia* . . . 846
Family II. "*Candidatus Clavichlamydiaceae*" . . . 865
Genus I. "*Candidatus Clavichlamydia*" . . . 865
Family III. *Criblamydiaceae* . . . 867
Genus I. *Criblamydia* . . . 867
Family IV. *Parachlamydiaceae* . . . 867
Genus I. *Parachlamydia* . . . 868
Genus II. *Neochlamydia* . . . 869
Genus III. *Protochlamydia* gen. nov. . . . 870
Family V. "*Candidatus* Piscichlamydiaceae" . . . 872
Genus I. "*Candidatus* Piscichlamydia" . . . 872
Family VI. *Rhabdochlamydiaceae* fam. nov. . . . 873
Genus I. *Rhabdochlamydia* gen. nov. . . . 873
Family VII. *Simkaniaceae* . . . 874
Genus I. *Simkania* . . . 875
Genus II. "*Candidatus* Fritschea" . . . 875
Family VIII. *Waddliaceae* . . . 876
Genus I. *Waddlia* . . . 877

Phylum XXV. "*Planctomycetes*" 879
Class I. *Planctomycetia* class. nov. 879
Order I. *Planctomycetales*. 879
Family I. *Planctomycetaceae* 880
Genus I. *Planctomyces*. 881
Genus II. *Blastopirellula* 895
Genus III. *Gemmata* 897
Genus IV. *Isosphaera* 900
Genus V. *Pirellula* 903
Genus VI. *Rhodopirellula* 906
Genus VII. *Schlesneria*. 910
Genus VIII. *Singulisphaera*. 913
Order II. "*Candidatus* Brocadiales" ord. nov. 918
Family I. "*Candidatus* Brocadiaceae" fam. nov. 918

Author index. 927

Index of scientific names of *Archaea* and *Bacteria* 929

Contributors

Wolf-Rainer Abraham
Helmholtz Center for Infection Research, 125, Chemical Microbiology, Inhoffenstrasse 7, D-38124 Braunschweig, Germany
wolf-rainer.abraham@helmholtz-hzi.de

Rudolf Amann
Department of Molecular Ecology, Max Planck Institute for Marine Microbiology, Celsiusstrasse 1, D-28359 Bremen, Germany
ramann@mpi-bremen.de

Josefa Antón
Department of Physiology, Genetics and Microbiology, University of Alicante, Apartado 99, 03080 Alicante, Spain
anton@ua.es

Mitchell F. Balish
Miami University, Department of Microbiology, 80 Pearson Hall, Oxford, OH, USA
balishmf@muohio.edu

Patrik M. Bavoil
Department of Microbial Pathogenesis, University of Maryland Dental School, 650 West Baltimore Street, Baltimore, MD 21201, USA
pbavoil@umaryland.edu

Yoshimi Benno
Benno Laboratory, Center for Intellectual Property Strategies, RIKEN, Wako, Saitama 351-0198, Japan
benno@jcm.riken.jp, benno828@riken.jp

Jean-François Bernardet
Unité de Virologie et Immunologie Moléculaires, Institut National de la Recherche Agronomique, Domaine de Vilvert, 78352 Jouy-en-Josas cedex, France
jean-francois.bernardet@jouy.inra.fr

Luise Berthe-Corti
Institute for Chemistry and Biology of the Marine Environment (ICBM), University of Oldenburg, P.O. Box 2503, D-26111, Oldenburg, Germany
luise.berthe.corti@uni-oldenburg.de

Joseph M. Bové
18 chemin Feyteau, 33650 La Brède, France
joseph.bove@wanadoo.fr

John P. Bowman
Tasmanian Institute of Agricultural Research, Private Bag 54, University of Tasmania, Hobart, TAS 7001, Australia
john.bowman@utas.edu.au

Janet M. Bradbury
Infectious Diseases Group, Department of Veterinary Pathology, Jordan Building, University of Liverpool, Leahurst Neston CH64 7TE, UK
jmb41@liverpool.ac.uk

Ingrid Brettar
Helmholtz-Zentrum für Infektionsforschung GmbH (HZI), Helmholtz Centre for Infection Research, Department of Vaccinology & Applied Microbiology, Inhoffenstrasse 7, D-38124 Braunschweig, Germany
inb@gbf.de

John A. Breznak
Department of Microbiology and Molecular Genetics, Michigan State University, East Lansing, MI 48824-4320, USA
breznak@msu.edu

Daniel R. Brown
Infectious Diseases and Pathology, Box 110880, University of Florida, Basic Sciences Building BSB3-31/BSB3-52, 1600 SW Archer Road, Gainesville, FL 32610, USA
brownd@vetmed.ufl.edu

Andreas Brune
Max Planck Institute for Terrestrial Microbiology, Karl-von-Frisch-Straße, 35043 Marburg, Germany
brune@mpi-marburg.mpg.de

Alke Bruns
Lohmann Animal Health GmbH & Co. KG Zeppelinstrasse 2, D-27472 Cuxhaven, Germany
alke.bruns@lohmann.de

Brita Bruun
Department of Clinical Microbiology, Hillerød Hospital, 3400 Hillerød, Denmark
bgb@noh.regionh.dk

Sandra Buczolits
Institut fur Bakteriologie, Mykologie und Hygiene, Veterinarmedizinische Universitat, Veterinarplatz 1, A-1210 Wien, Austria
sandra.buczolits@vu-wien.ac.at

Hans-Jürgen Busse
Institut für Bakteriologie, Mykologie und Hygiene, Veterinarmedizinische Universitat, Veterinarplatz 1, A-1210 Wien, Austria
Austriahans-juergen.busse@vu-wien.ac.at

Margaret K. Butler
Australian Institute for Bioengineering and Nanotechnology, The University of Queensland, Building 75, St Lucia, QLD 4072, Australia
m.butler1@uq.edu.au

Michael J. Calcutt
201 Connaway Hall, University Missouri-Columbia, Columbia, MO, USA
calcuttm@missouri.edu

Richard W. Castenholz
Center for Ecology & Evolutionary Biology, Department of Biology, University of Oregon, Eugene, OR 97403-1210, USA
rcasten@darkwing.uoregon.edu

Marie Anne Chattaway
Molecular Identification Services Unit, Department for Bioanalysis and Horizon Technologies, Health Protection Agency Centre for Infections, 61 Colindale Avenue, London NW9 5EQ, UK

Jang-Cheon Cho
Department of Biological Sciences, Division of Biology and Ocean Sciences, Inha University, Incheon 402-751, Republic of Korea
chojc@inha.ac.kr

Richard Christen
UMR 6543 CNRS – Université Nice Sophia Antipolis, Centre de Biochimie, Parc Valrose, F-06108 Nice, France
christen@unice.fr

Jongsik Chun
School of Biological Sciences, Seoul National University, 56-1 Shillim-dong, Kwanak-gu, Seoul 151-742, Republic of Korea
jchun@snu.ac.kr

John D. Coates
Department of Plant and Microbial Biology, 271 Koshland Hall, University of California, Berkeley, Berkeley, CA 94720, USA
jcoates@nature.berkeley.edu

August Coomans
Zoological Institute, Department of Biology, Faculty of Sciences, Ghent University, K.L. Ledeganckstraat 35, B-9000 Ghent, Belgium
august.coomans@ugent.be

Kyle C. Costa
Department of Microbiology and Astrobiology Program, Box Number 357242, University of Washington, Seattle, WA 98195, USA
kccosta@u.washington.edu

Robert E. Davis
Molecular Plant Pathology Laboratory, USDA-Agricultural Research Service, Beltsville, MD 20705, USA
robert.davis@ars.usda.gov

Willem M. de Vos
Laboratory of Microbiology, Wageningen University, 6703 CT Wageningen, The Netherlands
willem.devos@wur.nl

Svetlana N. Dedysh
S.N. Winogradsky Institute of Microbiology RAS, Prospect 60-Letya Oktyabrya 7/2, Moscow 117312, Russia
dedysh@mail.ru

Muriel Derrien
Laboratory of Microbiology, Wageningen University, Hesselink van Suchtelenweg 4, 6703 CT Wageningen, The Netherlands
muriel.derrien@wur.nl

Kirstin J. Edwards
Applied and Functional Genomics, Centre for Infections, Health Protection Agency, 6l Colindale Avenue, London NW9 5EQ, UK
kirstin.edwards@hpa.org.uk

Jean P. Euzéby
Ecole Nationale Veterinaire, 23 chemin des Capelles, B.P. 87614, 31076 Toulouse cedex 3, France
euzeby@bacterio.org

Mark Fegan
Biosciences Research Division, Department of Primary Industries, Attwood, VIC 3049, Australia
mark.fegan@dpi.vic.gov.au

Sydney M. Finegold
Infectious Diseases Section (111 F), VA Medical Center West Los Angeles, Los Angeles, CA 90073, USA
sidfinegol@aol.com

Cecil W. Forsberg
Department of Molecular and Cellular Biology, Science Complex, University of Guelph, Guelph, ON, Canada N1G 2W1
cforsber@uoguelph.ca

John A. Fuerst
School of Chemistry and Molecular Biosciences, The University of Queensland, Brisbane St Lucia, QLD 4072, Australia
j.fuerst@uq.edu.au

Ferran Garcia-Pichel
School of Life Sciences, Arizona State University, Tempe, AZ 85287-4501, USA
ferran@asu.edu

George M. Garrity
Department of Microbiology, Michigan State University, 6162 Biomedical and Physical Sciences Building, East Lansing, MI 48824-4320, USA
garrity@msu.edu

Gail E. Gasparich
Department of Biological Sciences, 8000 York Road, Towson University, Towson, MD 21252, USA
ggasparich@towson.edu

Saheer E. Gharbia
Applied and Functional Genomics, Centre for Infections, Health Protection Agency, 6l Colindale Avenue, London NW9 5EQ, UK
saheer.gharbia@hpa.org.uk

Peter Gilbert (Deceased)
School of Pharmacy and Pharmaceutical Sciences, University of Manchester, Manchester, UK

Stephen J. Giovannoni
Department of Microbiology, Oregon State University, Corvallis, OR 97331-3804, USA
steve.giovannoni@orst.edu

John I. Glass
The J. Craig Venter Institute, 9704 Medical Center Drive, Rockville, MD, USA
jglass@jcvi.org

Dawn Gundersen-Rindal
Insect Biocontrol Laboratory, U.S. Department of Agriculture, Agricultural Research Service, Henry A. Wallace Beltsville Agricultural Research Center, Plant Sciences Institute, Beltsville, MD 20705, USA
gundersd@ba.ars.usda.gov

Hafez M. Hafez
Free University Berlin, Königsweg 63, 14163 Berlin, Germany
hafez@vetmed.fu-berlin.de

Nigel A. Harrison
Research and Education Center, 3205 College Avenue, University of Florida, Fort Lauderdale, FL 33314-7799, USA
naha@ufl.edu

Brian P. Hedlund
School of Life Sciences, University of Nevada Las Vegas, Box 4004, 4505 Maryland Parkway, Las Vegas, NV 89154-4004, USA
brian.hedlund@unlv.edu

Peter Hirsch
Institut für Allgemeine Mikrobiologie der Biozentrum, Universität Kiel, Am Botanischen Garten 1-9, D-24118 Kiel, Germany
phirsch@ifam.uni-kiel.de

Manfred G. Höfle
Helmholtz-Zentrum für Infektionsforschung GmbH (HZI), Helmholtz Centre for Infection Research, Department of Vaccinology & Applied Microbiology, Inhoffenstrasse 7, D-38124 Braunschweig, Germany
mho@gbf.de

Stanley C. Holt
The Forsyth Institute, Boston, MA, USA
stanglous@yahoo.com

Matthias Horn
Department of Microbial Ecology, University of Vienna, Althanstrasse 14, 1090 Vienna, Austria
horn@microbial-ecology.net

Celia J. Hugo
Department of Microbial, Biochemical and Food Biotechnology, University of the Free State, Bloemfontein 9300, South Africa
hugocj.sci@ufs.ac.za

Roar L. Irgens
715 Lilac Drive, Mount Vernon, WA 98273-6613, USA
alpinehiker@comcast.net

Elena P. Ivanova
Faculty of Life and Social Science, Swinburne University of Technology, H31, P.O. Box 218, Hawthorn, VIC 3122, Australia
eivanova@groupwise.swin.edu.au

Peter H. Janssen
Grasslands Research Centre, AgResearch, Private Bag 11008, Palmerston North 4442, New Zealand
peter.janssen@agresearch.co.nz

Mike S. M. Jetten
Radboud University, Institute for Water and Wetland Research, Department of Microbiology, Huygens Building Room HG02.339, Heyendaalseweg 135, 6525 AJ Nijmegen, The Netherlands
m.jetten@science.ru.nl

Karl-Erik Johansson
Department of Bacteriology, Ulls Vaeg 2A-2C, National Veterinary Institute (SVA), SE-751 89 Uppsala, Sweden
kaggen@sva.se

Bernhard Kaltenboeck
Department of Pathobiology, College of Veterinary Medicine, Auburn University, 270 Greene Hall, Auburn, AL 36849-5519, USA
kaltebe@auburn.edu

Yoichi Kamagata
Bioproduction Research Institute, National Institute of Advanced Industrial Science and Technology (AIST), Sapporo, Hokkaido 062-8517, Japan
y.kamagata@aist.go.jp

Srinivas Kambhampati
Professor of Insect Genetics and Evolution, Department of Entomology, Kansas State University, Manhattan, KS 66506, USA
srini@ksu.edu

Peter Kämpfer
Institut für Angewandte Mikrobiologie, Justus-Liebig-Universität Giessen, Heinrich-Buff-Ring 26-32 (IFZ), D-35392 Giessen, Germany
peter.kaempfer@umwelt.uni-giessen.de

Hiroaki Kasai
Marine Biotechnology Institute, 3-75-1, Heita, Kamaishi, Iwate 026-0001, Japan
hkasai@kitasato-u.ac.jp

Sang-Jin Kim
Marine Biotechnology Research Centre, Korea Ocean Research & Development Institute, Department of Marine Biotechnology, University of Science and Technology, P.O. Box 29, Ansan 425-600, Republic of Korea
s-jkim@kordi.re.kr

Seung Bum Kim
Department of Microbiology, Chungnam National University, 220 Gung-dong, Yusong, Daejon 305-764, Republic of Korea
sbk01@cnu.ac.kr

Eija Könönen
Laboratory Head, Anaerobe Reference Laboratory, Department of Bacterial and Inflammatory Diseases, National Public Health Institute (KTL), Mannerheimintie 166, FIN-00300, Helsinki, Finland
eija.kononen@ktl.fi

Noel R. Krieg
617 Broce Drive, Blacksburg, VA 24060-2801, USA
nrk@vt.edu

Lee R. Krumholz
Department of Botany and Microbiology and Institute for Energy and the Environment, 770 Van Vleet Oval, The University of Oklahoma, Norman, OK 73019, USA
krumholz@ou.edu

J. Gijs Kuenen
Delft University of Technology, Department of Biotechnology, Julianalaan 67, 2628 BC Delft, The Netherlands
j.g.kuenen@tnw.tudelft.nl

Irina S. Kulichevskaya
S.N. Winogradsky Institute of Microbiology RAS, Prospect 60-Letya Oktyabrya 7/2, Moscow 117312, Russia
kulich2@mail.ru

Cho-chou Kuo
Department of Pathobiology, University of Washington, Seattle, WA 98195, USA
cckuo@u.washington.edu

Edward R. Leadbetter
Department of Marine Chemistry and Geochemistry, Woods Hole Oceanographic Institution, Mail Stop 52, Woods Hole, MA 02543, USA
eleadbetter@whoi.edu

Kuo-Chang Lee
School of Chemistry and Molecular Biosciences, The University of Queensland, Brisbane St Lucia, QLD 4072, Australia
kuochang.lee@uqconnect.edu.au

Susan Leschine
University of Massachusetts Amherst, Department of Microbiology, 639 North Pleasant Street, Amherst, MA 01003, USA
suel@microbio.umass.edu

Ralph A. Lewin (Deceased)
Scripps Institution of Oceanography, University of California, La Jolla, CA 92093, USA

Chengxu Liu
11511 Reed Hartman Hwy, Cincinnati, OH 45241, USA
chengxu66@yahoo.com, liu.c.34@pg.com

Julie M. J. Logan
Molecular Identification Services, Centre for Infections, Health Protection Agency, 61 Colindale Avenue, London NW9 5EQ, UK
julie.logan@hpa.org.uk

Wolfgang Ludwig
Lehrstuhl für Mikrobiologie, Technische Universität München, Am Hochanger 4, D-85350 Freising, Germany
ludwig@mikro.biologie.tu-muenchen.de

Heinrich Lünsdorf
Helmholtz Center for Infection Research, Vaccinology Department, Inhoffenstrasse 7, D-38124 Braunschweig, Germany
heinrich.luensdorf@helmholtz-hzi.de, lhu@gbf.de

Alexandre J. Macedo
Universidade Federal do Rio Grande do Sul, Faculdade de Farmácia and Centro de Biotecnologia, Porto Alegre, RS, Brazil
alexandre.macedo@ufrgs.br

Rosa Margesin
Institut für Mikrobiologie, Universität Innsbruck, Technikerstrasse 25, A-6020 Innsbruck, Austria
rosa.margesin@uibk.ac.at

Meghan May
Department of Infectious Diseases and Pathology, College of Veterinary Medicine, University of Florida, Gainesville, FL 32611-0880, USA
mmay@ufl.edu

Joanne B. Messick
Comparative Pathobiology, School of Veterinary Medicine, Purdue University, West Lafayette, IN 47907, USA
jmessic@purdue.edu

Valery V. Mikhailov
Pacific Institute of Bioorganic Chemistry, Far-Eastern Branch of Russian Academy of Sciences, Prospect 100 Let Vladivostoku 159, Vladivostok 690022, Russia
mikhailov@piboc.dvo.ru

Yasuyoshi Nakagawa
Biological Resource Center (NBRC), Department of Biotechnology, National Institute of Technology and Evaluation, 2-5-8, Kazusakamatari, Kisarazu, Chiba 292-0818, Japan
nakagawa-yasuyoshi@nite.go.jp

Jason B. Navarro
School of Life Sciences, University of Nevada, Las Vegas, 4505 Maryland Parkway, Las Vegas, NV 89154-4004, USA
jason.navarro@unlv.edu

Olga I. Nedashkovskaya
Pacific Institute of Bioorganic Chemistry, of the Far-Eastern Branch of the Russian Academy of Sciences, Pr. 100 let Vladivostoku 159, Vladivostok 690022, Russia
olganedashkovska@piboc.dvo.ru, olganedashkovska@yahoo.com

Harold Neimark
Department of Microbiology & Immunology, Box 44, 450 Clarkson Avenue, State University of New York, College of Medicine, Brooklyn, NY 11203, USA
neimah25@bmec.hscbklyn.edu

Denis I. Nikitin
Russian Academy of Sciences, Institute of Microbiology, Prosp. 60 let Oktyabrya, 7a, 117811 Moscow, Russia
nikitin@inmi.host.ru

Steven J. Norris
Greer Professor and Vice Chair for Research, Department of Pathology and Laboratory Medicine, University of Texas Medical School at Houston, MSB 2.120, P.O. Box 20708, Houston, TX 77030, USA
steven.j.norris@uth.tmc.edu

Ingar Olsen
Institute of Oral Biology, Faculty of Dentistry, P.B. 1052 Blindern, N-0316 Oslo, Norway
ingaro@odont.uio.no

Huub J. M. Op den Camp
Radboud University, Institute for Water and Wetland Research, Department of Microbiology, Huygens Building Room HG02.340, Heyendaalseweg 135, 6525 AJ Nijmegen, The Netherlands
h.opdencamp@science.ru.nl

Bruce J. Paster
The Forsyth Institute, 245 First Street, Cambridge, MA 02142-1200, USA
bpaster@forsyth.org

Bharat K. C. Patel
Microbial Discovery Research Unit, School of Biomolecular and Physical Sciences, Griffith University, Nathan Campus, Kessels Road, Brisbane, QLD 4111, Australia
b.patel@griffith.edu.au

Caroline M. Plugge
Laboratory of Microbiology, Wageningen University, Dreijenplein 10, 6703 HB Wageningen, The Netherlands
caroline.plugge@wur.nl

Lakshani Rajakurana
Molecular Identification Services Unit, Department for Bioanalysis and Horizon Technologies, Health Protection Agency Centre for Infections, 61 Colindale Avenue, London NW9 5EQ, UK

Merja Rautio
Division of Clinical Microbiology, Huslab, Jorvi Hospital, Espoo, Finland

Gundlapally S.N. Reddy
Centre for Cellular and Molecular Biology (CCMB), Uppal Road, Hyderabad 500 007, India
gsnr@ccmb.res.in, sarithagsn@yahoo.com

Laura B. Regassa
Georgia Southern University, 202 Georgia Avenue, Statesboro, GA, USA
lregassa@georgiasouthern.edu

Joël Renaudin
UMR Génomique Diversité et Pouvoir Pathogène, INRA-Université de Bordeaux 2, IBVM, 71 Avenue Edouard Bourlaux, BP 81, F-33883 Villenave d'Ornon, France
renaudin@bordeaux.inra.fr

Alexander H. Rickard
Department of Epidemiology, School of Public Health, University of Michigan, 1415 Washington Heights, 4647 SPH Tower, Ann Arbor, MI, USA
alexhr@umich.edu

Janet A. Robertson
Department of Medical Microbiology and, Immunology, University of Alberta, 1-49 Medical Sciences Building, Edmonton, AB, Canada T6G 2H7
janet.robertson@ualberta.ca, janrob2@shaw.ca

Ramon Rosselló-Mora
Institut Mediterrani d'Estudis Avançats (CSIC-UIB), C/Miquel Marque's 21, E-07290 Esporles, Mallorca, Spain
rossello-mora@uib.es

Collette Saillard
UMR 1090 Génomique Diversité Pouvoir Pathogène, INRA, Université Victor Ségalen Bordeaux 2, 71 Avenue Edouard Bourlaux BP 81, F-33883 Villenave d'Ornon, France
saillard@bordeaux.inra.fr

Mitsuo Sakamoto
Microbe Division/Japan Collection of Microorganisms, RIKEN BioResource Center, Wako, Saitama 351-0198, Japan
sakamoto@jcm.riken.jp

Bernhard Schink
Lehrstuhl für Mikrobielle Ökologie, Fakultät für Biologie, Universität Konstanz, Fach M 654, D-78457 Konstanz, Germany
bernhard.schink@uni-konstanz.de

Heinz Schlesner
Institut für Allgemeine Mikrobiologie, Christian-Albrechts-Universität, Am Botanischen Garten 1-9, D-24118 Kiel, Germany
hschlesner@t-online.de

Jean M. Schmidt
Arizona State University, Department of Microbiology, Box 2701, Tempe, AZ 85287, USA
jean.schmidt@asu.edu

Ira Schwartz
Department of Microbiology & Immunology, New York Medical College, Valhalla, NY 10595, USA
schwartz@nymc.edu

Haroun N. Shah
Molecular Identification Services Unit, Department for Bioanalysis and Horizon Technologies, Health Protection Agency Centre for Infections, 61 Colindale Avenue, London NW9 5EQ, UK
haroun.shah@hpa.org.uk

Wung Yang Shieh
Institute of Oceanography, National Taiwan University, Taipei, Taiwan
winyang@ntu.edu.tw

Sisinthy Shivaji
Centre for Cellular and Molecular Biology (CCMB), Uppal Road, Hyderabad 500 007, India
shivas@ccmb.res.in

Lindsay I. Sly
Department of Microbiology and Parasitology, University of Queensland, Brisbane, QLD, Australia
l.sly@uq.edu.au

Robert M. Smibert
Department of Anaerobic Microbiology, Virginia Tech, Blacksburg, VA, USA

Yuli Song
8700 Mason Montgomery Road, MBC DV2-5K1, 533; Mason, OH 45040-9462, USA
song.y.7@pg.com

Anne M. Spain
Department of Botany and Microbiology, 770 Van Vleet Oval, The University of Oklahoma, Norman, OK 73019, USA
aspain@ou.ed

James T. Staley
Department of Microbiology, University of Washington, Seattle, WA 98195-2700, USA
jtstaley@u.washington.edu

Thaddeus B. Stanton
Agricultural Research Service – Midwest Area, National Animal Disease Center, United States Department of Agriculture, P.O. Box 70, 1920 Dayton Ave, Building 24, Ames, IA 50010-0070, USA
thad.stanton@ars.usda.gov

Richard S. Stephens
Program in Infectious Diseases and Immunity, School of Public Health, University of California at Berkeley, 235 Warren Hall, Berkeley, CA 94720, USA
rss@berkeley.edu

Marc Strous
Max Planck Institute for Marine Microbiology, Celsiusstrasse 1, 28359 Bremen, Germany
mstrous@mpi-bremen.de

Paula Summanen
Anaerobe Laboratory, VA Medical Center West Los Angeles, 11301 Wilshire Boulevard, Building 304, Room E3-237, Los Angeles, CA 90073, USA
carlsonph@aol.com

Makoto Suzuki
Kyowa Hakko Bio Co., Ltd, 1-6-1 Ohtemachi, Chiyoda-ku, Tokyo 100-8185, Japan
makoto.suzuki@kyowa-kirin.co.jp

Anne C. R. Tanner
Department of Molecular Genetics, The Forsyth Institute, 140 The Fenway, Boston, MA 02115, USA
annetanner@forsyth.org

Séverine Tasker
School of Clinical Veterinary Science, University of Bristol, Langford, Bristol BS40 5DU, UK
s.tasker@bristol.ac.uk

David Taylor-Robinson
6 Vache Mews, Vache Lane, Chalfont St Giles, Buckingham HP8 4UT, UK
dtr@vache99.freeserve.co.uk

J. Cameron Thrash
Department of Microbiology, Oregon State University, Corvallis, OR, USA
thrashc@onid.orst.edu

Stefanie Van Trappen
BCCM/LMG Bacteria Collection, Laboratorium voor Microbiologie, Universiteit Gent, K.L. Ledeganckstraat 35, B-9000 Gent, Belgium
stefanie.vantrappen@ugent.be

Marc Vancanneyt
BCCM/LMG Bacteria Collection, Faculty of Sciences, Ghent University, K.L. Ledeganckstraat 35, B-9000 Ghent, Belgium

Tom T. M. Vandekerckhove
BioBix: Laboratory for Bioinformatics and Computational Genomics, Department of Molecular Biotechnology, Faculty of Bioscience Engineering, Ghent University, Coupure Links 653, B-9000 Ghent, Belgium
tom.vandekerckhove@ugent.be

Guiqing Wang
Department of Pathology, New York Medical College, Valhalla, NY 10595, USA
guiqing_wang@nymc.edu

Naomi L. Ward
Department of Molecular Biology, University of Wyoming, Department 3944, 1000 E. University Ave, Laramie, WY 82071, USA
nlward@uwyo.edu

Robert Whitcomb (Deceased)
Patagonia, AZ 85624, USA

William B. Whitman
Department of Microbiology, University of Georgia, 527 Biological Sciences Building, Cedar Street, Athens, GA 30602-2605, USA
whitman@uga.edu

David L. Williamson
4 Galahad Lane, Nesconset, NY 11767-2220, USA
dwmson11767@yahoo.com

Hana Yi
School of Biological Sciences, Seoul National University, NS70, 56-1 Shillim-dong, Kwanak-gu, Seoul 151-742, Korea
dimono@snu.ac.kr

Jaewoo Yoon
Institute of Molecular and Cellular Biosciences, The University of Tokyo, 1-1-1 Yayoi Bunkyo-Ku, Tokyo 113-0032, Japan
yjw222@hotmail.com

Erwin G. Zoetendal
Laboratory of Microbiology, Wageningen University, Hesselink van Suchtelenweg 4, 6703 CT Wageningen, The Netherlands
egzoetendal@hotmail.com

Richard L. Zuerner
Infectious Bacterial Diseases Research Unit, USDA-ARS-NADC, P.O. Box 70, 1920 Dayton Avenue, Building 24, Ames, IA 50010, USA
richard.zuerner@ars.usda.gov

On using the *Manual*

NOEL R. KRIEG AND GEORGE M. GARRITY

Citation

The *Systematics* is a peer-reviewed collection of chapters, contributed by authors who were invited by the Trust to share their knowledge and expertise of specific taxa. Citations should refer to the author, the chapter title, and inclusive pages rather than to the editors.

Arrangement of the Manual

As in the previous volumes of this edition, the *Manual* is arranged in phylogenetic groups based upon the analyses of the 16S rRNA presented in the introductory chapter "Road map of the phyla *Bacteroidetes*, *Spirochaetes*, *Tenericutes* (*Mollicutes*), *Acidobacteria*, *Fibrobacteres*, *Fusobacteria*, *Dictyoglomi*, *Gemmatimonadetes*, *Lentisphaerae*, *Verrucomicrobia*, *Chlamydiae* and *Planctomycetes*". These groups have been substantially modified since the publication of volume 1 in 2001, reflecting both the availability of more experimental data and a different method of analysis. Since volume 4 includes only the phylum *Firmicutes*, taxa are arranged by class, order, family, genus and species. Within each taxon, the nomenclatural type is presented first and indicated by a superscript T. Other taxa are presented in alphabetical order without consideration of degrees of relatedness.

Articles

Each article dealing with a bacterial genus is presented wherever possible in a definite sequence as follows:

a. Name of the genus. Accepted names are in boldface, followed by "defining publication(s)", i.e. the authority for the name, the year of the original description, and the page on which the taxon was named and described. The superscript AL indicates that the name was included on the Approved Lists of Bacterial Names, published in January 1980. The superscript VP indicates that the name, although not on the Approved Lists of Bacterial Names, was subsequently validly published in the International Journal of Systematic and Evolutionary Microbiology (or the International Journal of Systematic Bacteriology). Names given within quotation marks have no standing in nomenclature; as of the date of preparation of the Manual they had not been validly published in the International Journal of Systematic and Evolutionary Microbiology, although they may have been "effectively published" elsewhere. Names followed by the term "nov." are newly proposed but will not be validly published until they appear in a Validation List in the International Journal of Systematic and Evolutionary Microbiology. Their proposal in the Manual constitutes only "effective publication", not valid publication.

b. Name of author(s). The person or persons who prepared the Bergey's article are indicated. The address of each author can be found in the list of Contributors at the beginning of the *Manual*.

c. Synonyms. In some instances a list of some synonyms used in the past for the same genus is given. Other synonyms can be found in the *Index Bergeyana* or the *Supplement to the Index Bergeyana*.

d. Etymology of the name. Etymologies are provided as in previous editions, and many (but undoubtedly not all) errors have been corrected. It is often difficult, however, to determine why a particular name was chosen, or the nuance intended, if the details were not provided in the original publication. Those authors who propose new names are urged to consult a Greek and Latin authority before publishing in order to ensure grammatical correctness and also to ensure that the meaning of the name is as intended.

e. Salient features. This is a brief resume of the salient features of the taxon. The most important characteristics are given in **boldface**. The DNA G+C content is given.

f. Type species. The name of the type species of the genus is also indicated along with the defining publication(s).

g. Further descriptive information. This portion elaborates on the various features of the genus, particularly those features having significance for systematic bacteriology. The treatment serves to acquaint the reader with the overall biology of the organisms but is not meant to be a comprehensive review. The information is normally presented in the following sequence:

Colonial morphology and pigmentation
Growth conditions and nutrition
Physiology and metabolism
Genetics, plasmids, and bacteriophages
Phylogenetic treatment
Antigenic structure
Pathogenicity
Ecology

h. Enrichment and isolation. A few selected methods are presented, together with the pertinent media formulations.

i. Maintenance procedures. Methods used for maintenance of stock cultures and preservation of strains are given.

j. Procedures for testing special characters. This portion provides methodology for testing for unusual characteristics or performing tests of special importance.

k. Differentiation of the genus from other genera. Those characteristics that are especially useful for distinguishing the genus from similar or related organisms are indicated here, usually in a tabular form.

l. Taxonomic comments. This summarizes the available information related to taxonomic placement of the genus and indicates the justification for considering the genus a distinct taxon. Particular emphasis is given to the methods of molecular biology used to estimate the relatedness of the genus to other taxa, where such information is available. Taxonomic information regarding the arrangement and status of the various species within the genus follows. Where taxonomic controversy exists, the problems are delineated and the various alternative viewpoints are discussed.

m. Further reading. A list of selected references, usually of a general nature, is given to enable the reader to gain access to additional sources of information about the genus.

n. Differentiation of the species of the genus. Those characteristics that are important for distinguishing the various species within the genus are presented, usually with reference to a table summarizing the information.

o. List of species of the genus. The citation of each species is given, followed in some instances by a brief list of objective synonyms. The etymology of the specific epithet is indicated. Descriptive information for the species is usually presented in tabular form, but special information may be given in the text. Because of the emphasis on tabular data, the species descriptions are usually brief. The type strain of each species is indicated, together with the collection(s) in which it can be found. (Addresses of the various culture collections are given in the article in volume 1 entitled *Culture Collections: An Essential Resource for Microbiology*.) The 16S rRNA gene sequence used in phylogenetic analysis and placement of the species into the taxonomic framework is given, along with the GenBank (or other database) accession number. Additional comments may be provided to point the reader to other well-characterized strains of the species and any other known DNA sequences that may be relevant.

p. Species *incertae sedis*. The List of Species may be followed in some instances by a listing of additional species under the heading "Species *Incertae sedis*" or "Other organisms", etc. The taxonomic placement or status of such species is questionable, and the reasons for the uncertainty are presented.

q. References. All references given in the article are listed alphabetically at the end of the family chapter.

Tables

In each article dealing with a genus, there are generally three kinds of table: (a) those that differentiate the genus from similar or related genera, (b) those that differentiate the species within the genus, and (c) those that provide additional information about the species (such information not being particularly useful for differentiation). The meanings of symbols are as follows:

+, 90% or more of the strains are positive
d, 11–89% of the strains are positive
–, 90% or more of the strains are negative
D, different reactions occur in different taxa (e.g., species of a genus or genera of a family)
v, strain instability (NOT equivalent to "d")
w, weak reaction.
nd, not determined or no data.
nr, not reported.

These symbols, and exceptions to their use, as well as the meaning of additional symbols, are given in footnotes to the tables.

Use of the *Manual* for determinative purposes

Many chapters have keys or tables for differentiation of the various taxa contained therein. For identification of species, it is important to read both the generic and species descriptions because characteristics listed in the generic descriptions are not usually repeated in the species descriptions.

The index is useful for locating the articles on unfamiliar taxa or in discovering the current classification of a particular taxon. Every bacterial name mentioned in the *Manual* is listed in the index. In addition, an up-to-date outline of the taxonomic framework is provided in the introductory chapter "Road map of the phyla *Bacteroidetes*, *Spirochaetes*, *Tenericutes* (*Mollicutes*), *Acidobacteria*, *Fibrobacteres*, *Fusobacteria*, *Dictyoglomi*, *Gemmatimonadetes*, *Lentisphaerae*, *Verrucomicrobia*, *Chlamydiae* and *Planctomycetes*".

Errors, comments, and suggestions

As in previous volumes, the editors and authors earnestly solicit the assistance of all microbiologists in the correction of possible errors in *Bergey's Manual of Systematic Bacteriology*. Comments on the presentation will also be welcomed as well as suggestions for future editions. Correspondence should be addressed to:

Editorial Office
Bergey's Manual Trust
Department of Microbiology
University of Georgia
Athens, GA 30602-2605, USA
Tel: +1-706-542-4219; fax +1-706-542-6599
e-mail: bergeys@uga.edu

Road map of the phyla *Bacteroidetes*, *Spirochaetes*, *Tenericutes* (*Mollicutes*), *Acidobacteria*, *Fibrobacteres*, *Fusobacteria*, *Dictyoglomi*, *Gemmatimonadetes*, *Lentisphaerae*, *Verrucomicrobia*, *Chlamydiae*, and *Planctomycetes*

Wolfgang Ludwig, Jean Euzéby and William B. Whitman

This revised road map updates previous outlines of Garrity and Holt (2001) and Garrity et al. (2005) with the description of additional taxa and new phylogenetic analyses. While the outline/road map seeks to be complete for all taxa validated prior to July 1, 2006, some taxa described after that date are included.

The new phylogenetic trees are strict consensus trees based on various maximum-likelihood and maximum-parsimony analyses and corrected according to results obtained when applying alternative treeing methods. Multifurcations indicate that a common branching order was not significantly supported after applying alternative treeing approaches. Detailed branching orders are shown if supported by at least 50% of the "treeings" performed in addition to the maximum-likelihood approach.

Given that the focus is on the higher taxonomic ranks, rather restrictive variability filters were applied. Consequently, resolution power is lost for lower levels. Of special importance, relationships within genera lack the resolution that would be obtained with genus–family level analyses. Furthermore, the type strain tree, which is available online at www.bergeys.org, is an extract of comprehensive trees comprising some thousand sequences. Thus, trees for the specific groups in subsequent chapters, which are based upon smaller datasets and include the variable sequence positions, may differ with respect to detailed topology, especially at levels of closer relationships within and between genera. In the trees shown here, branch lengths – in first instance – indicate significance and only approximate estimated number of substitutions.

Starting with the second edition of *Bergey's Manual of Systematic Bacteriology*, the arrangement of content follows a phylogenetic framework or road map based largely on analyses of the nucleotide sequences of the ribosomal small-subunit RNA rather than on phenotypic data (Garrity et al., 2005). Implicit in the use of the road map are the convictions that prokaryotes have a phylogeny and that phylogeny matters. However, phylogenies, like other experimentally derived hypotheses, are not static but may change whenever new data and/or improved methods of analysis become available (Ludwig and Klenk, 2005). Thus, the large increases in data since the publication of the taxonomic outlines in the preceding volumes have led to a re-evaluation of the road map. Not surprisingly, the taxonomic hierarchy has been modified or newly interpreted for a number of taxonomic units. These changes are described in the following paragraphs.

The taxonomic road map proposed in volume 1 and updated and emended in volume 2 was derived from phylogenetic and principal-component analyses of comprehensive datasets of small-subunit rRNA gene sequences. A similar approach is continued here. Since the introduction of comparative rRNA sequencing (Ludwig and Klenk, 2005; Ludwig and Schleifer, 2005), there has been a continuous debate concerning the justification and power of a single marker molecule for elucidating phylogeny and establishing taxonomy of organisms. Although generally well established in taxonomy, the polyphasic approach cannot currently be applied for sequence-based analyses due to the lack of adequate comprehensive datasets for alternative marker molecules. Even in the age of genomics, the datasets for non-rRNA markers are poor in comparison to more than 400,000 rRNA primary structures available in general and special databases (Cole et al., 2007; Pruesse et al., 2007). Nevertheless, the data provided by the full genome-sequencing projects allow the definiton of a small set of genes representing the conserved core of prokaryotic genomes (Cicarelli et al., 2006; Ludwig and Schleifer, 2005). Furthermore, comparative analyses of the core gene sequences globally support the small-subunit rRNA derived view of prokaryotic evolution. Although the tree topologies reconstructed from alternative markers differ in detail, the major groups (and taxa) are verified or at least not disproved (Ludwig and Schleifer, 2005). Consequently, the structuring of this volume is based on updated and curated (http://www.arb-silva.de; Pruesse et al., 2007) databases of processed small-subunit rRNA primary structures.

Data analysis

The current release of the integrated small-subunit rRNA database of the SILVA project (Pruesse et al., 2007) provides the basis for these phylogenetic analyses. The tools of the ARB software package (Ludwig et al., 2004) were used for data evaluation, optimization, and phylogenetic inference. A subset of about 33,000 high-quality sequences from *Bacteria* was extracted from the current SILVA SSU Ref database. Among the criteria for restrictive quality analyses and data selection were coverage of at least positions 18–1509 (*Escherichia coli* 16S rRNA numbering), no ambiguities or missing sequence stretches, no chimeric primary structures, low deviation from overall and group-specific consensus and conservation profiles, and good agreement of tree topologies and branch length with processed sequence data. Unfortunately, only some of the type strain sequences successfully passed this restrictive quality check. The alignment of the sequences of this subset, as well as all type strain sequences initially excluded given incompleteness or lower quality, was manually evaluated and optimized. Phylogenetic treeing was first based on the high-quality dataset and performed applying phylum specific position filters (50% positional identity). The partial or lower quality type strain sequences were subsequently added using a special ARB-tool allowing the optimal positioning of branches to the reference tree without admitting topology changes (Ludwig and Klenk, 2005). The consensus trees used for evaluating or modifying the taxonomic outline were based on maximum-likelihood analyses (RAXML, implemented in the ARB package; Stamatakis et al., 2005) and further evaluated by maximum-parsimony and distance matrix analyses with the respective ARB tools (Ludwig et al., 2004).

Taxonomic interpretation

The phylogenetic conclusions were used for evaluating and modifying the taxonomic outline of the phyla "*Bacteroidetes*", "*Spirochaetes*", *Tenericutes* (*Mollicutes*), "*Acidobacteria*", "*Fibrobacteres*", "*Fusobacteria*", "*Dictyoglomi*", *Gemmatimonadetes*, *Lentisphaerae*, "*Verrucomicrobia*", "*Chlamydiae*", and "*Planctomycetes*". These include all the phyla not described in earlier volumes with the exception of the *Actinobacteria*, which will be included in the fifth and last volume of this edition. There is no particular rationale for inclusion in this volume. Although some of the phyla may be related in a kingdom or superphylum (i.e., "*Chlamydiae*", *Lentisphaerae*, "*Planctomycetes*", and "*Verrucomicrobia*") (Griffiths and Gupta, 2007; Lee et al., 2009; Pilhofer et al., 2008; Wagner and Horn, 2006), most are unrelated to each other (Figure 1). Some are major pathogens of humans, other animals, and plants. Some are exotic and only described in the last decade.

In order to ensure applicability and promote acceptance, the proposed taxonomic modifications were made following a conservative procedure. The overall organization follows the type "taxon" principle as applied in the previous volumes. Taxa defined in the outline of the preceding volumes were only unified, dissected, or transferred in the cases of strong phylogenetic support. This approach is justified by the well-known low significance of local tree topologies (also called "range of unsharpness" around the nodes; Ludwig and Klenk, 2005). Thus, many of the cases of paraphyletic taxa found were maintained in the current road map if the respective (sub)-clusters rooted closely together, even if they were separated by intervening clusters representing other taxa. While reorganization of these taxa may be warranted, it was not performed in the absence of confirmatory evidence. The names of validly published, but phylogenetically misplaced, type strains are also generally maintained. These strains are mentioned in the context of the respective phylogenetic groups. In cases of paraphyly, all concerned species or higher taxa are assigned to the respective (sub)-groups. New higher taxonomic ranks are only proposed if species or genera – previously assigned to different higher taxonomic units – are significantly unified in a monophyletic branch.

Upon the recommendation of the Judicial Commission (De Vos et al., 2005), many of the names and classifications previously proposed by Cavalier-Smith (2002) are not used in this work. The classification used categories not covered by the Rules of the Code and priority and proposed types without standing in nomenclature. For these reasons, the following phylum (or

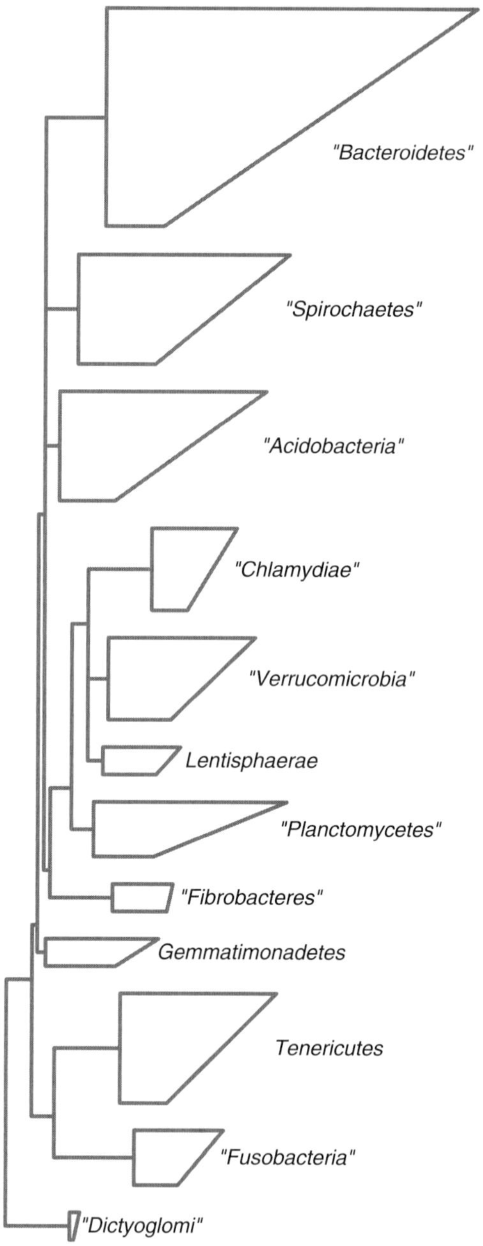

FIGURE 1. Phyla *Bacteroidetes, Spirochaetes, Tenericutes, Acidobacteria, Fibrobacteres, Fusobacteria, Dictyoglomi, Gemmatimonadetes, Lentisphaerae, Verrucomicrobia, Chlamydiae,* and *Planctomycetes.* While the phyla *Lentisphaerae, Verrucomicrobia, Chlamydiae,* and *Planctomycetes* may be specifically related to each other, the other phyla included in volume 4 are not related.

division) names are not used: *Planctobacteria*, *Sphingobacteria*, and *Spirochaetae*. Likewise, the following class names are not used: *Acidobacteria*, *Chlamydiae*, *Flavobacteria*, *Planctomycea*, and *Spirochaetes*. Lastly, priority for the order name *Acidobacteriales* is no longer attributed to Cavalier-Smith (2002).

Phylum "*Bacteroidetes*"

In previous classifications, the phylum "*Bacteroidetes*" was proposed to comprise three classes, "*Bacteroidia*", "*Flavobacteriia*", and "*Sphingobacteriia*" (Garrity et al., 2005). While the analyses performed here, which were based upon many more sequences and differ-

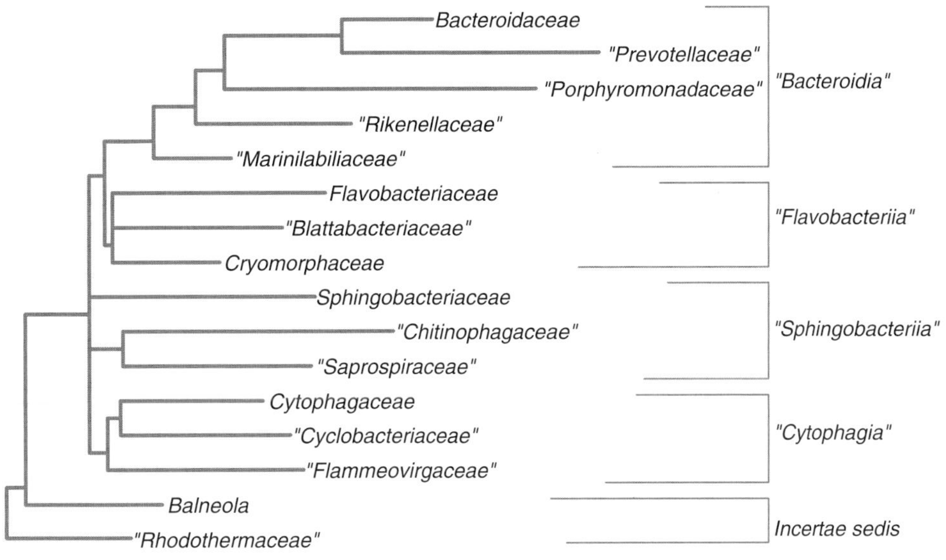

FIGURE 2. Overview of the phylum "*Bacteroidetes*". This phylum contains 15 families classified within four classes. Currently, the *incertae sedis* taxa *Balneola* and "*Rhodothermaceae*" are classified within the class "*Cytophagia*".

ent methods, generally support this conclusion, they also justify formation of a fourth class within this phylum, the "*Cytophagia*" (Figure 2). This new class comprises many genera previously classified within the "*Flexibacteraceae*", "*Flammeovirgaceae*", and *Crenotrichaceae* (see below). Thus, the phylum "*Bacteroidetes*" comprises at least four phylogenetic groups that are well delineated on the basis of their 16S rRNA gene sequences. In addition, two groups are affiliated with the phylum but could not be readily assigned to one of these classes. While additional evidence may warrant classification with one of the known or novel classes, these organisms were grouped within *Incertae sedis* of the "*Cytophagia*" for the present time (Figure 2).

Class "*Bacterioidia*" and order "*Bacteroidales*"

The class "*Bacterioidia*" contains five families, all classified within the order "*Bacteroidales*". These families include the four families proposed previously (Garrity et al., 2005), *Bacteroidaceae*, "*Rikenellaceae*", "*Porphyromonadaceae*", and "*Prevotellaceae*", as well as a new family proposed here, "*Marinilabiliaceae*" (Figure 3). In addition, on the basis of the dissimilarity of its 16S rRNA gene sequence to other members of the order, *Odoribacter* (*Bacteroides*) *splanchnicus* may represent an additional undescribed family or a member of the "*Marinilabiliaceae*". However, chemotaxonomic characteristics and analyses of the *fimA* gene imply a close relationship to the family "*Porphyromonadaceae*" (Hardham et al., 2008). Therefore, its reclassification is not proposed at this time. Lastly, the recently described marine organism, *Prolixibacter bellariivorans*, appears to represent a deep lineage in this class but whose affiliation with these families is ambiguous (Holmes et al., 2007).

Family *Bacteroidaceae*

In addition to the type genus, *Bacteroides*, this family comprises three monospecific genera, *Acetofilamentum*, *Acetothermus*, and *Anaerorhabdus*, and one genus, *Acetomicrobium*, comprising two species. Because complete 16S rRNA gene sequences are not available for representatives of these four genera, these assignments are tentative. Two genera previously assigned to this family have also been reassigned. As recommended by Morotomi et al. (2007), *Megamonas* has been transferred to the *Firmicutes*. Based on its rRNA gene sequence, *Anaerophaga* has been transferred to the new family "*Marinilabiliaceae*".

The genus *Bacteroides* comprises at least six lineages or clades. The type species, *Bacteroides fragilis*, together with *Bacteroides acidifaciens, caccae, finegoldii, nordii, ovatus, salyersiae, thetaiotaomicron*, and *xylanisolvens*, represent a cluster slightly separated from the other members of the genus. If supported by other evidence, each of the other lineages could be classified as new genera within this family. The other lineages are represented by *Bacteroides cellulosilyticus* and *intestinalis*; *Bacteroides coprosuis* and *propionifaciens*; *Bacteroides pyogenes, suis*, and *tectus*; *Bacteroides barnesiae, coprocola, coprophilus, dorei, helcogenes, massiliensis, plebeius, salanitronis, uniformis*, and *vulgatus*. The species *Bacteroides eggerthii, gallinarum*, and *stercoris* cannot clearly be assigned to one of the lineages.

In addition to these clades within the genus, the following validly published species are probably misclassified. *Bacteroides splanchnicus* was recently reclassified as *Odoribacter splanchnicus* (Hardham et al., 2008); this genus may represent a novel member of family "*Porphyromonadaceae*" (see above). *Bacteroides capillosus* and *cellulosolvens* are probably members of the phylum *Firmicutes*. In addition, rRNA gene sequences are not available for *Bacteroides capillus, forsythus, furcosus, polypragmatus*, and *salivosus*, so their assignment is uncertain.

Lastly, the family *Bacteroidaceae* appears to be paraphyletic, and the family "*Prevotellaceae*" falls within the radiation of *Bacteroides* clades. Because the members of the "*Prevotellaceae*" are generally closely related and the branch length to the *Bacteroidaceae* is fairly long, this conclusion is tentative. While these families were not combined at this time, this classification may warrant further investigation.

Family "*Marinilabiliaceae*"

This family represents a group of sister but not clearly monophyletic branches within the "*Bacteroidales*" and comprises three genera.

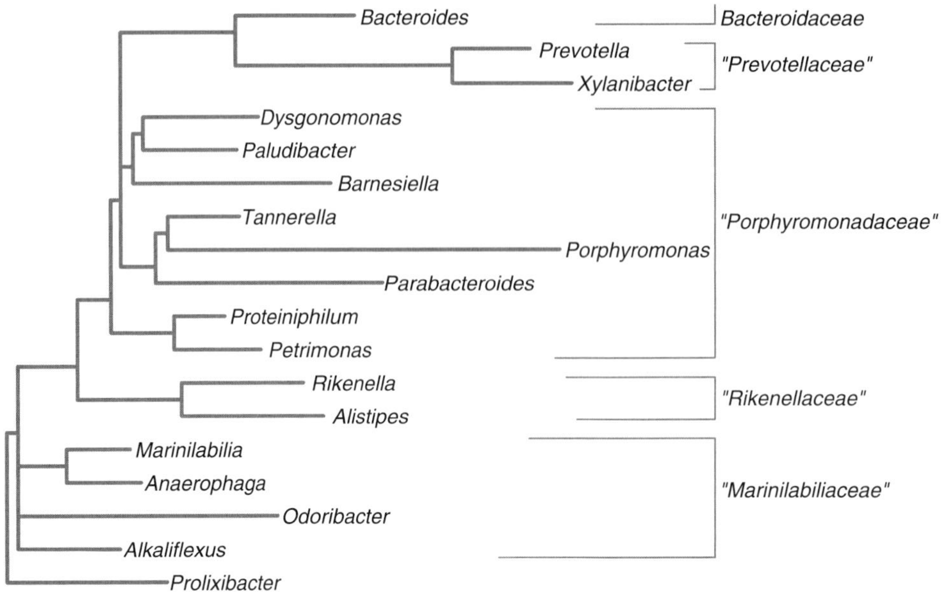

FIGURE 3. Genera of the class "*Bacteroidia*". This class comprises five families and the genus *Prolixibacter*, which has not yet been assigned to a family.

The type genus, *Marinilabilia*, contains two species, *Marinilabilia salmonicolor* and *agarovorans*, and was previously classified within the "*Rikenellaceae*" (Garrity et al., 2005). The remaining taxa include *Alkaliflexus imshenetskii* and *Anaerophaga thermohalophila*, the latter of which was formerly classified within the *Bacteroidaceae* (Garrity et al., 2005). The current analysis suggests that *Cytophaga fermentans* should be reclassified as a novel genus that is associated with this family. Lastly, rRNA analyses suggest that *Odoribacter* (*Bacteroides*) *splanchnicus*, which was proposed after the deadline for inclusion in this volume, may represent an additional member of the "*Marinilabiliaceae*". However, chemotaxonomic characteristics and analyses of the *fimA* gene imply a close relationship to the family "*Porphyromonadaceae*" (Hardham et al., 2008), so this classification is not proposed at this time.

Family "*Rikenellaceae*"

The family comprises the monospecific genus *Rikenella microfusus* and the closely related genus *Alistipes*. The latter genus comprises the type species *Alistipes putredinis* and *Alistipes finegoldii*, *onderdonkii*, and *shahii*. *Marinilabilia*, which was classified within this family by Garrity et al. (2005), is now classified within the family "*Marinilabiliaceae*".

Family "*Porphyromonadaceae*"

The genus *Porphyromonas*, which was formed by reclassification of various species of *Bacteroides* (Shah and Collins, 1988), is the type for this family. Originally, this family comprised the genera *Porphyromonas*, *Dysgonomonas*, and *Tannerella* (Garrity et al., 2005).

The genus *Porphyromonas* comprises five subclusters: (1) the type species *Porphyromonas asaccharolytica* and *Porphyromonas circumdentaria*, *endodontalis*, *gingivicanis*, and *uenonis*; (2) *Porphyromonas cangingivalis*, *canoris*, *levii*, and *somerae*; (3) *Porphyromonas creviorIcanis*, *gingivalis*, and *gulae*; (4) *Porphyromonas catoniae* and *macacae*; and (5) *Porphyromonas cansulci*.

The genus *Dysgonomonas* comprises the type *Dysgonomonas gadei* and the closely related species *Dysgonomonas capnocytophagoides* and *mossii*.

The last genus is monospecific, *Tannerella forsythia*.

The current analyses add five other genera to this family. These include three monospecific genera represented by *Paludibacter propionicigenes*, *Petrimonas sulfuriphila*, and *Proteiniphilum acetatigenes*. Also included is the recently described genus comprising *Barnesiella viscericola* and *intestinihominis* (Morotomi et al., 2008; Sakamoto et al., 2007). Lastly, the genus *Parabacteroides* comprises the type species *Parabacteroides distasonis* and three closely related species *Parabacteroides goldsteinii*, *johnsonii*, and *merdae* (Sakamoto and Benno, 2006). This last genus was also described after the deadline for inclusion in this volume.

Family "*Prevotellaceae*"

Although the family "*Prevotellaceae*" appears within the cluster of species of the family *Bacteroidaceae*, the genera representing the "*Prevotellaceae*" are well separated from the *Bacteroidaceae*. Therefore, both families are continued in the current classification. The genus *Prevotella*, which was formed by reclassification of various species of *Bacteroides* (Shah and Collins, 1990), is the type for this family. It comprises a number of phylogenetic groups, each of which may warrant reclassification into one or more genera if supported by additional evidence: (1) the type species *Prevotella melaninogenica* and *Prevotella histolytica* and *veroralis*; (2) *Prevotella denticola* and *multiformis*; (3) *Prevotella corporis*, *disiens*, *falsenii*, *intermedia*, *nigrescens*, and *pallens*; (4) *Prevotella maculosa*, *oris*, and *salivae*; (5) *Prevotella bryantii*, and *multisaccharivorax*; (6) *Prevotella baroniae*, *buccae* and *dentalis*; (7) *Prevotella enoeca* and *pleuritidis*; (8) *Prevotella buccalis* and *timonensis*; (9) *Prevotella loescheii* and *shahii*; (10) *Prevotella brevis* and *ruminicola*; and (11) *Prevotella amnii* and *bivia*.

The species *Prevotella albensis, bergensis, copri, marshii, oralis, oulorum, paludivivens,* and *stercorea* cannot be clearly assigned to one of the lineages. *Xylanibacter oryzae* is also found within the radiation of the described *Prevotella* clusters. *Prevotella tannerae* represents a more distant branch of the family. In contrast, *Prevotella heparinolytica* and *zoogleoformans* are clearly separated from the other members of this family and may warrant reclassification. Lastly, *Hallella seregens* is closely related to *Prevotella dentalis,* which has priority (Willems and Collins, 1995). Therefore, *Hallella seregens* is not used.

Class "*Flavobacteriia*" and order "*Flavobacteriales*"

This class comprises a single order, "*Flavobacteriales*", and is essentially unchanged from the original proposal of Garrity et al. (2005). The order comprises three families, *Flavobacteriaceae*, "*Blattabacteriaceae*", and *Cryomorphaceae* (Figures 4 and 5) (Bowman et al., 2003). The family "*Myroidaceae*" proposed by Garrity et al. (2005) was judged to be insufficiently resolved from the *Flavobacteriaceae* and was not used.

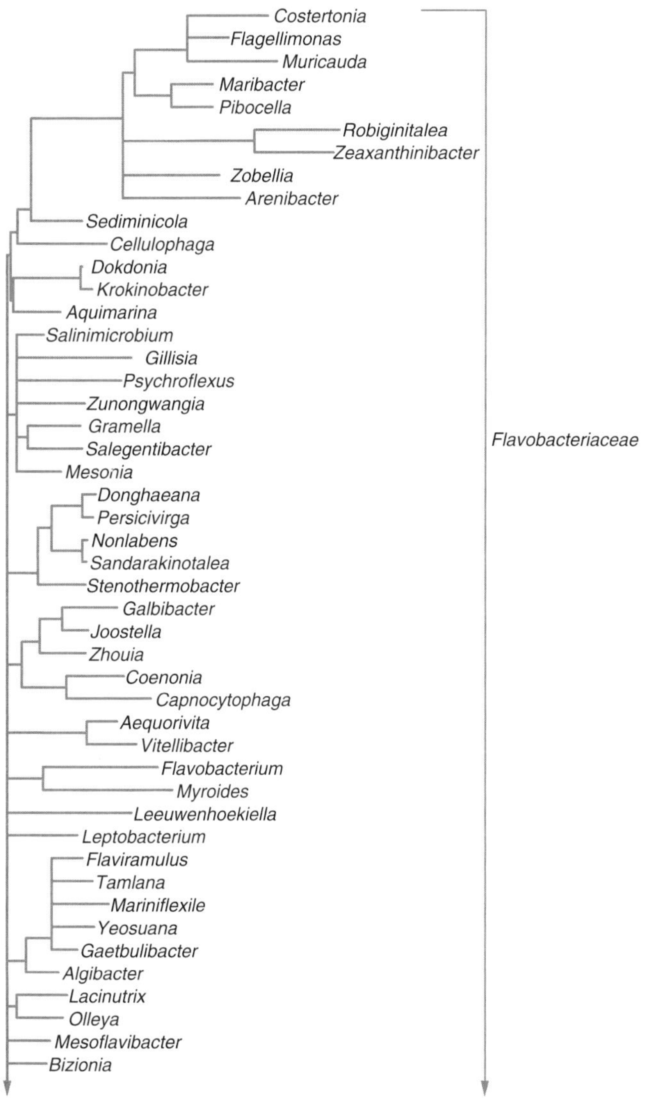

FIGURE 4. Genera of the class "*Flavobacteriia*". This class comprises three families. The first part of the family *Flavobacteriaceae* is shown here.

Family *Flavobacteriaceae*

This extraordinarily diverse family comprises over 70 genera. The rRNA analyses indicate the presence of many phylogenetic clusters that may warrant separation into novel families if supported by additional evidence. Many of the clusters described here are identical to those found by Bernardet and Nakagawa (2006) or include mostly taxa described after their work. Cluster (1) includes the type genus *Flavobacterium* and *Myroides*. This latter genus includes three closely related species, *Myroides odoratus, odoratimimus,* and *pelagicus*. Although the genus *Flavobacterium* is very diverse, the rRNA phylogeny lacks clear indication of clades that might serve as the basis for further subdivision. Species included in this genus include the type species *Flavobacterium aquatile* and, in alphabetical order, *Flavobacterium antarcticum, aquidurense, branchiophilum, columnare, croceum, cucumis, daejeonense, defluvii, degerlachei, denitrificans, flevense, frigidarium, frigidimaris, frigoris, fryxellicola, gelidilacus, gillisiae, granuli, hercynium, hibernum, hydatis, johnsoniae, limicola, micromati, omnivorum, pectinovorum, psychrolimnae, psychrophilum, saccharophilum, saliperosum, segetis, soli, succinicans, suncheonense, tegetincola, terrae, terrigena, weaverense, xanthum,* and *xinjiangense*. In addition, there are some species for which sequences are not available, including *Flavobacterium acidificum, acidurans, oceanosedimentum,* and *thermophilum*.

(2) *Capnocytophaga ochracea* (type species), *canimorsus, cynodegmi, gingivalis, granulosa, haemolytica, ochracea,* and *sputigena*; and *Coenonia anatina*. Although *Galbibacter mesophilus* (Khan et al., 2007c), *Joostella marina* (Quan et al., 2008), and *Zhouia amylolytica* are associated with this cluster, this relationship is not strong.

(3) *Actibacter sediminis* (Kim et al., 2008a); *Aestuariicola saemankumensis* (Yoon et al., 2008d); *Lutibacter litoralis*; *Lutimonas vermicola* (Yang et al., 2007); *Polaribacter filamentus* (type species), *butkevichii, franzmannii,* and *glomeratus*; *Polaribacter dokdonensis* (which forms a separate clade from the other species of this genus); *Tenacibaculum maritimum* (type species), *adriaticum, aestuarii, aiptasiae, amylolyticum, galleicum, litopenaei, litoreum, lutimaris, mesophilum, ovolyticum, skagerrakense,* and *soleae*.

(4) *Chryseobacterium gleum* (type species), *aquaticum, aquifrigidense, arothri, balustinum, bovis, caeni, daecheongense, daeguense, defluvii, flavum, formosense, gambrini, gregarium, haifense, hispanicum, hominis, hungaricum, indologenes, indoltheticum, jejuense, joostei, luteum, marina, molle, oranimense, pallidum, piscium, scophthalmum, shigense, soldanellicola, soli, taeanense, taichungense, taiwanense, ureilyticum, vrystaatense,* and *wanjuense*. In addition to these species, the following taxa appear within the radiation of *Chryseobacterium,* including *Epilithonimonas tenax, Kaistella koreensis, Sejongia antarctica* (type species) and *jeonii*. Other taxa within this cluster include *Bergeyella zoohelcum, Cloacibacterium normanense, Elizabethkingia meningoseptica* (type species) and *miricola, Empedobacter brevis, Ornithobacterium rhinotracheale, Riemerella anatipestifer* (type species) and *columbina, Wautersiella falsenii,* and *Weeksella virosa*.

(5) *Arenibacter latericius* (type species), *certesii, echinorum, palladensis,* and *troitsensis*; *Cellulophaga algicola, baltica,* and *pacifica* (a clade which does not include the type species); *Costertonia aggregata*; *Flagellimonas eckloniae* (Bae et al., 2007); *Maribacter*

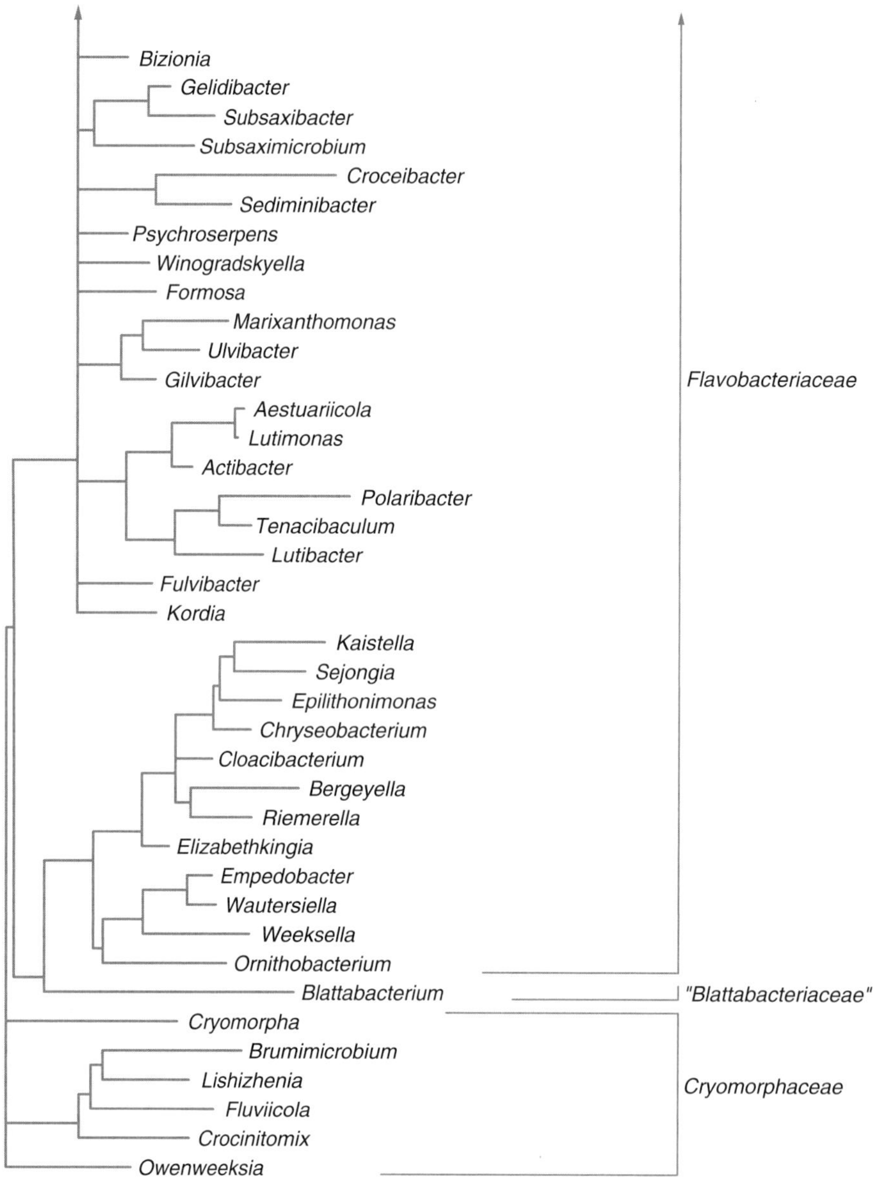

FIGURE 5. Genera of the class "*Flavobacteriia*". This class comprises three families. The second part of the family *Flavobacteriaceae* and the remaining two families are shown here.

sedimenticola (type species), *aquivivus, arcticus, dokdonensis, forsetii, orientalis, polysiphoniae*, and *ulvicola*; *Muricauda ruestringensis* (type species), *aquimarina, lutimaris*, and *flavescens*; *Pibocella ponti* (which appears within the cluster of *Maribacter* species); *Robiginitalea biformata* and *myxolifaciens*; *Sediminicola luteus*; *Zeaxanthinibacter enoshimensis* (Asker et al., 2007); and *Zobellia galactanivorans* (type species), *amurskyensis, laminariae, russellii*, and *uliginosa*. In addition, *Cellulophaga lytica* (the type species of this genus) and *fucicola* appear as either a deep branch of this cluster (Bernardet and Nakagawa, 2006) or as an associated but independent group (this analysis). In either case, the reclassification of *Cellulophaga algicola, baltica*, and *pacifica* to a new genus would appears to be warranted. Lastly, the type strain of *Pibocella ponti* has been lost. If available, this strain would be reclassified within *Maribacter*. For that reason, this genus is not included in the outline.

(6) *Algibacter lectus* and *mikhailovii*; *Flaviramulus basaltis*; *Gaetbulibacter saemankumensis* and *marinus*; *Mariniflexile gromovii* and *fucanivorans*; *Tamlana crocina* (Lee, 2007); and *Yeosuana aromativorans*.

(7) *Croceibacter atlanticus* and *Sediminibacter furfurosus* (Khan et al., 2007a).

(8) *Gelidibacter algens* (type species), *gilvus, mesophilus*, and *salicanalis*; *Subsaxibacter broadyi*; *Subsaximicrobium wynnwilliamsii* (type species) and *saxinquilinus*.

(9) *Lacinutrix copepodicola, algicola*, and *mariniflava*; and *Olleya marilimosa*.

(10) *Gilvibacter sediminis* (Khan et al., 2007a); *Marixanthomonas ophiurae* (Romanenko et al., 2007). *Ulvibacter litoralis* and *antarcticus*.

(11) *Dokdonia donghaensis*; *Krokinobacter genikus* (type species), *diaphorus*, *eikastus*, and *genicus*.

(12) *Donghaena dokdonensis*; *Nonlabens tegetincola*; *Persicivirga xylanidelens*; *Sandarakinotalea sediminis*; and *Stenothermobacter spongiae*.

(13) *Gillisia limnaea* (type species), *hiemivivida*, *illustrilutea*, *mitskevichiae*, *myxillae*, and *sandarakina*; *Gramella echinicola* (type species) and *portivictoriae*; *Mesonia algae* (type species) and *mobilis*; *Psychroflexus torques* (type species), *gondwanensis*, and *tropicus*; *Salegentibacter salegens* (type species), *agarivorans*, *flavus*, *holothuriorum*, *mishustinae*, *salaries*, and *salinarum*; *Salinimicrobium catena*, *terrae*, and *xinjiangense* (Chen et al., 2008; Lim et al., 2008); and *Zunongwangia profunda* (Qin et al., 2007). Among these taxa, *Salinimicrobium catena* was previously classified as *Salegentibacter catena* (Lim et al., 2008).

(14) *Aequorivita antarctica* (type species), *crocea*, *lipolytica*, and *sublithincola* and *Vitellibacter vladivostokensis*.

In addition to these well delineated clusters, a large number of taxa were not closely associated with any of these clusters or each other. These include: *Aquimarina muelleri* (type species), *brevivitae*, *intermedia*, and *latercula*; *Bizionia paragorgiae* (type species), *gelidisalsuginis*, and *saleffrena*; a second clade of *Bizionia* species including *Bizionia algoritergicola* and *myxarmorum*; *Formosa algae* (type species) and *agariphila*; *Fulvibacter tottoriensis*; *Kordia algicida*; *Leeuwenhoekiella marinoflava* (type species), *aequorea*, and *blandensis*; *Leptobacterium flavescens*; *Mesoflavibacter zeaxanthinifaciens*; *Psychroserpens burtonensis* (type species) and *mesophilus*; and *Winogradskyella thalassocola* (type species), *epiphytica*, *eximia*, and *poriferorum*. The rRNA gene sequences of the following pairs of genera are closely related, which may justify combining them: *Sandarakinotalea–Nonlabens*; *Dokdonia–Krokinobacter*.

Family "*Blattabacteriaceae*"

This family comprises *Blattabacterium cuenoti*, which is an endosymbiont of insects that has not been grown in pure culture.

Family *Cryomorphaceae*

Proposed by Bowman et al. (2003) to include novel genera of cold-tolerant marine bacteria isolated from sea ice and other polar environments, this family comprises six monospecific genera: *Cryomorpha ignava*, *Brumimicrobium glaciale*, *Crocinitomix catalasitica*, *Fluviicola taffensis*, *Lishizhenia caseinilytica*, and *Owenweeksia hongkongensis*. The phylogenetic analyses conducted here suggest that this family is polyphenetic and contains three lineages that cluster together at the base of the phylogenetic tree for the *Flavobacteriales*. *Cryomorpha* and *Owenweeksia* each comprise one monogeneric lineage, with the remaining four genera comprising the third lineage. However, in the absence of additional evidence, these lineages were not separated at this time.

Class "*Sphingobacteriia*" and order "*Sphingobacteriales*"

This class comprises a single order, the "*Sphingobacteriales*". It is more circumscribed than the original proposal (2005) and excludes many taxa previously classified within the "*Flexibacteraceae*". The order comprises three families: *Sphingobacteriaceae*, "*Chitinophagaceae*", and "*Saprospiraceae*" (Figure 6). The family *Crenotrichaceae* was removed because the type genus *Crenothrix* was transferred to the *Proteobacteria* (Stoecker et al., 2006). The genus *Chitinophaga* then became the type for a new family within the order. Based upon their rRNA gene sequence similarities, the genera *Rhodothermus* and *Salinibacter*, which were also previously classified within the *Crenotrichaceae*, were transferred to the class "*Cytophagia*" as an order *incertae sedis* (see below). Similarly, *Balneola*, which was described after the deadline for

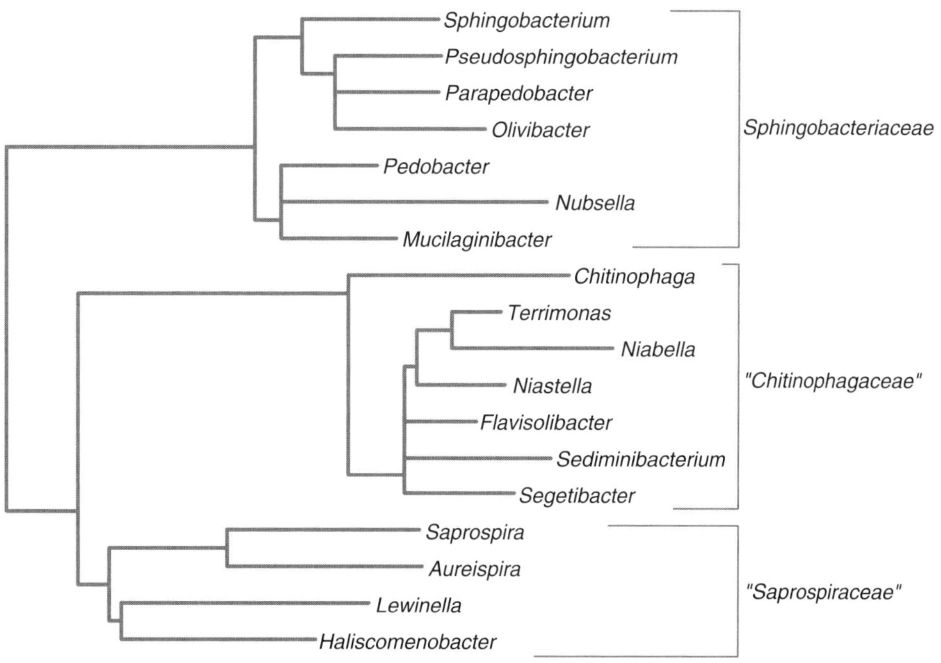

FIGURE 6. Genera of the class "*Sphingobacteriia*".

inclusion in this volume, was classified within the *Crenotrichaceae* based in part upon its similarity to *Rhodothermus* (Urios et al., 2006). Analyses performed here suggest that it may also be a deep lineage of the "*Cytophagia*". Lastly, the sequence of the 16S rRNA gene for *Toxothrix trichogenes* is not available, so this genus was transferred to *incertae sedis*. Even with these changes, this class is not clearly monophyletic (Figure 2). The families "*Chitinophagaceae*" and "*Saprospiraceae*" may represent a sister lineage to the family *Sphingobacteriaceae*. However, it the absence of confirmatory evidence, the grouping of the three families into one class was retained in the current outline.

Family *Sphingobacteriaceae*

At the time these analyses were performed, seven genera were identified within this family. The genera *Sphingobacterium*, *Olivibacter*, *Parapedobacter*, and *Pseudosphingobacterium* form one phylogenetic cluster. The genera *Pedobacter*, *Mucilaginibacter*, and *Nubsella* form a second cluster. In addition, the species *Flexibacter canadensis* represents an additional deep phylogenetic group within this family that may warrant classification as a novel genus.

The genus *Sphingobacterium* comprises *Sphingobacterium spiritivorum* (type species), *anhuiense*, *canadense*, *composti*, *daejeonense*, *faecium*, *kitahiroshimense*, *mizutaii*, *multivorum*, *siyangense*, and *thalpophilum*. Interestingly, the species epithet *Sphingobacterium composti* was independently proposed for two different organisms by Ten et al. (2006, 2007) and Yoo et al. (2007). Because *Sphingobacterium composti* Ten et al. (2007) has priority, the species of Yoo et al. (2007) warrants renaming. In addition, this genus contains *Sphingobacterium antarcticum*, whose rRNA gene sequence is not available. Related to the genus *Sphingobacterium* are the taxa *Olivibacter sitiensis* (type species), *ginsengisoli*, *soli*, and *terrae* (Ntougias et al., 2007; Wang et al., 2008), *Parapedobacter koreensis* (type species) and *soli* (Kim et al., 2007b, 2008b), and *Pseudosphingobacterium domesticum* (Kim et al., 2007b; Vaz-Moreira et al., 2007). These genera were described after the deadline for inclusion in this volume.

The second cluster is composed of *Pedobacter* species, which itself comprises four subclusters. The first subcluster contains *Pedobacter heparinus* (type species), *africanus*, *caeni*, *cryoconitis*, *duraquae*, *ginsengisoli*, *himalayensis*, *metabolipauper*, *panaciterrrae*, *piscium*, *steynii*, and *westerhofensis*. The second subcluster comprises *Pedobacter insulae* and *koreeensis*. The third subcluster comprises *Pedobacter daechungensis*, *lentus*, *saltans*, and *terricola*, which may warrant reclassification into a novel genus if supported by additional evidence. A fourth subcluster is represented by *Mucilaginibacter gracilis*, *kameinonesis*, and *paludis* (Pankratov et al., 2007; Urai et al., 2008). A number of species were not closely associated with any of these clusters or each other: *Nubsella zeaxanthinifaciens* (Asker et al., 2008); *Pedobacter agri*, *aquatilis*, *composti*, *roseus*, *sandarokinus*, *suwonensis*, and *terrae*.

Family "*Chitinophagaceae*"

This family contains two phylogenetic clusters. The first cluster includes the genus *Chitinophaga*. This genus comprises *Chitinophaga pinensis* (type species), *arvensicola*, *filiformis*, *ginsengisegetis*, *ginsengisoli*, *japonensis*, *sancti*, *skermani*, and *terraei*. The second cluster includes six related genera with ten species: *Flavisolibacter ginsengisoli* and *ginsengiterrae* (Yoon and Im, 2007); *Niabella aurantiaca* and *soli* (Kim et al., 2007a; Weon et al., 2008a); *Niastella koreensis* (type species) and *yeongjuensis* (Weon et al., 2006); *Sediminibacterium salmoneum* (Qu and Yuan, 2008); *Segetibacter koreensis* (An et al., 2007); and *Terrimonas ferruginea* (type species) and *lutea*. *Flavisolibacter*, *Niabella*, *Niastella*, *Sediminibacterium*, and *Segetibacter* were described after the deadline for inclusion in this volume (Weon et al., 2006).

Family "*Saprospiraceae*"

As originally proposed by Garrity et al. (2005), the family comprises three related genera and nine species. These include: *Saprospira grandis*; *Haliscomenobacter hydrossis*, and *Lewinella cohaerens* (type species), *agarilytica*, *antarctica*, *lutea*, *marina*, *nigricans*, and *persica*. Recently, the newly discovered genus *Aureispira* (*marina* and *maritime*) has also been classified within this family (Hosoya et al., 2006, 2007).

Class "*Cytophagia*" and order *Cytophagales*

Analyses performed here of the rRNA genes indicate that many of the genera previously classified within the families "*Flexibacteraceae*" and "*Flammeovirgaceae*" are not closely related to the "*Sphingobacteriia*" and should be transferred to a novel class (Figure 7). The order *Cytophagales* is designated the type for the new class. The genus *Cytophaga* is the type for the order and family *Cytophagaceae*. Because the family *Cytophagaceae* includes the type genera of the families "*Flexibacteraceae*" and *Spirosomaceae*, these classifications are not used.

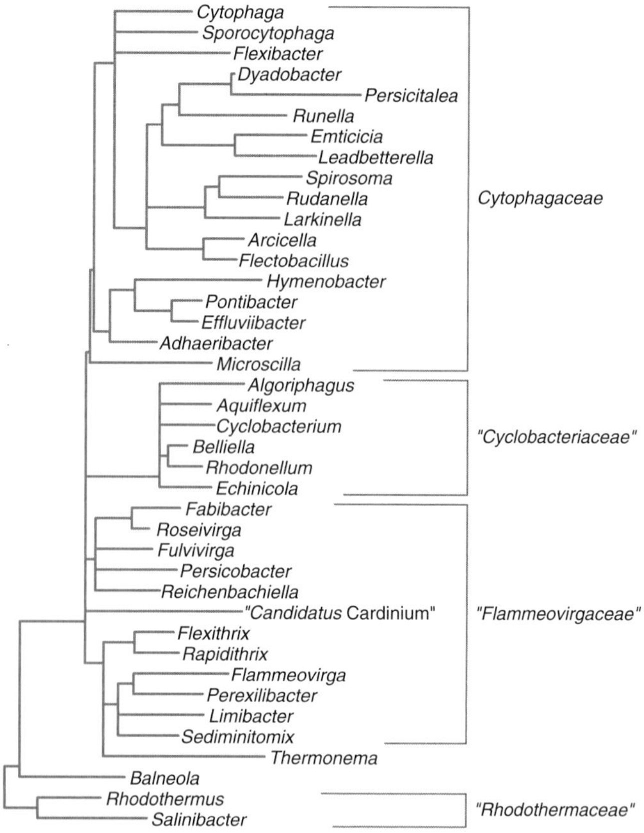

FIGURE 7. Genera of the class "*Cytophagia*". This class comprises three families and four orders *incertae sedis*.

Family *Cytophagaceae*

This family comprises 19 genera distributed within seven phylogenetic clusters: the first cluster includes *Cytophaga hutchinsonii* (type species) and *aurantiaca*; the second cluster includes *Sporocytophaga myxococcoides* (type and only species); the third cluster includes *Effluviibacter roseus*; *Hymenobacter roseosalivarius* (type species), *actinosclerus*, *aerophilus*, *chitinivorans*, *deserti*, *gelipurpurascens*, *norwichensis*, *ocellatus*, *psychrotolerans*, *rigui*, *soli*, and *xinjiangensis*; and *Pontibacter actiniarum* (type species), *akesuensis*, and *korlensis*. In addition, *Adhaeribacter aquaticus* appears to represent a deep lineage in this cluster.

The fourth cluster includes *Arcicella aquatica* and *rosea*; *Dyadobacter fermentans* (type species), *alkalitolerans*, *beijingensis*, *crusticola*, *ginsengisoli*, *hamtensis*, and *koreensis*; *Emticicia ginsengisoli* and *oligotrophica*; *Flectobacillus major* (type species) and *lacus*; *Larkinella insperata*; *Leadbetterella byssophila*; *Persicitalea jodogahamensis* (Yoon et al., 2007b); *Rudanella lutea* (Weon et al., 2008b); *Runella slithyformis* (type species), *defluvii*, *limosa*, and *zeae*; and *Spirosoma linguale*, *panaciterrae*, and *rigui*. The fifth cluster includes *Flexibacter roseolus*, *elegans*, and *Microscilla marina*. The sixth cluster comprises *Flexibacter flexilis* (type species). The seventh cluster comprises *Flexibacter ruber*.

Cyclobacterium and *Reichenbachiella*, two genera previously classified with this group (Garrity et al., 2005), have been transferred to the "*Cyclobacteriaceae*" and "*Flammeovirgaceae*", respectively. In addition, *Meniscus glaucopis* is retained within the *Cytophagaceae* even though the sequence of its rRNA gene is not available.

Family "*Cyclobacteriaceae*"

This family includes the genus *Cyclobacterium*, which was previously classified within the "*Flexibacteraceae*", and five related genera: *Cyclobacterium marinum* (type species), *amurskyense*, and *lianum*; *Aquiflexum balticum*; *Algoriphagus ratkowskyi* (type species), *alkaliphilus*, *antarcticus*, *aquimarinus*, *boritolerans*, *chordae*, *halophilus*, *locisalis*, *mannitolivorans*, *marincola*, *ornithinivorans*, *terrigena*, *vanfongensis*, *winogradskyi*, and *yeomjeoni*. This cluster includes *Chimaereicella* and *Hongiella* species that were transferred to *Algoriphagus* (Nedashkovskaya et al., 2007b); *Belliella baltica*; *Echinicola pacifica* (type species) and *vietnamensis*; and *Rhodonellum psychrophilum* represent further genera.

Family "*Flammeovirgaceae*"

This family includes the genus *Flammeovirga* and at least seven related genera and one *Candidatus* taxon. This family comprises two phylogenetic groups which are neighbors in all trees but not clearly monophyletic. In addition, *Thermonema*, which was previously classified in this family (Garrity et al., 2005), possesses only low similarity to the other genera and was reclassified to an order *incertae sedis*. Subsequently, it was found that this reassignment was equivocal, and analyses with more representatives of this family are ambiguous (Figure 7). For the purposes of this road map, this genus was retained in an order *incertae sedis*. As a result, this family comprises two phylogenetic groups: *Flammeovirga aprica* (type species), *arenaria*, *kamogawensis*, and *yaeyamensis*; *Flexibacter aggregans*, *litoralis*, and *polymorphus*, which appear to be misclassified; *Flexithrix dorotheae*; *Limibacter armeniacum* (Yoon et al., 2008b); *Perexilibacter aurantiacus* (Yoon et al., 2007a); *Rapidithrix thailandica* (Srisukchayakul et al., 2007); and *Sediminitomix flava* (Khan et al., 2007b).

The second group comprises *Fabibacter halotolerans*; *Fulvivirga kasyanovii* (Nedashkovskaya et al., 2007a); *Reichenbachiella agariperforans*; *Roseivirga ehrenbergii* (type species), *echinicomitans*, *seohaensis*, and *spongicola*; and *Persicobacter diffluens*.

In addition, *Flexibacter tractuosus*, which appears to be misclassified, and "*Candidatus* Cardinium hertigii", a symbiont of parasitoid wasps (Zchori-Fein et al., 2004), are neighboring lineages.

Class "*Cytophagia*" orders *incertae sedis*

In addition to the members of these families whose taxonomic position is relatively well defined, three deep lineages are classified within "*Cytophagia*" as separate orders *incertae sedis*. These lineages include (1) the family "*Rhodothermaceae*", comprising *Rhodothermus marinus* and *Salinibacter ruber*; (2) the genus *Balneola*, with species *Balneola vulgaris* (type) and *alkaliphila*, which were described after the deadline for inclusion in the volume (Urios et al., 2006, 2008); and (3) *Thermonema lapsum* (type species) and *rossianum* (which may also be assigned to the "*Flammeovirgaceae*").

The assignment of the first two lineages to this class is ambiguous, and their reclassification may be warranted with additional evidence. *Toxothrix trichogenes*, for which the rRNA gene sequence is not available, is also included as *incertae sedis* within this class.

Phylum "*Spirochaetes*"

As a result of the current analyses of 16S rRNA gene sequences, a single class and order are recognized within the phylum "*Spirochaetes*". Members of the "*Spirochaetes*" possess a cellular ultrastructure unique to bacteria with internal organelles of motility, namely periplasmic flagella.

Class "*Spirochaetia*" and order *Spirochaetales*

The class comprises a single order. The order *Spirochaetales* comprises four families that are well delineated by 16S rRNA gene sequences (Figure 8). Compared to the previous outline (Garrity et al., 2005), the families *Spirochaetaceae* and *Leptospiraceae* are retained in the current classification. However, the genus *Serpulina* was judged to be a subjective synonym of *Brachyspira* (Ochiai et al., 1997). As a consequence, the family "*Serpulinaceae*" was replaced with "*Brachyspiraceae*". The genus *Brevinema* was also transferred from the family *Spirochaetaceae* to a novel family "*Brevinemataceae*" in recognition of the differences in 16S rRNA gene sequences. Lastly, four genera of arthropod symbionts for which no sequences are available were transferred from the *Spirochaetaceae* to a fifth family, *incertae sedis*.

Family *Spirochaetaceae*

This family comprises four genera that are well delineated on the basis of their 16S rRNA gene sequences. Compared to previous classifications, the genus *Brevinema* was transferred to a new family on the basis of substantial differences in its 16S rRNA gene sequence. Likewise, the genera *Clevelandina*, *Diplocalyx*, *Hollandina*, and *Pillotina* were transferred to a family *incertae sedis* in the absence of rRNA gene sequences.

The culture for the type species of the genus *Spirochaeta*, *Spirochaeta plicatilis*, is not available, and its rRNA gene has not

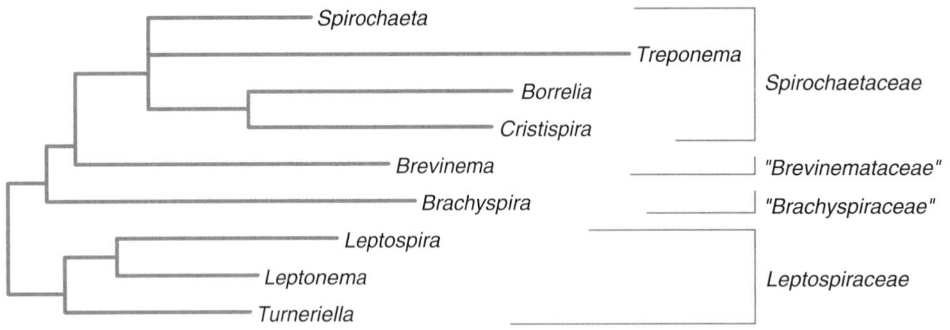

FIGURE 8. Genera of the phylum "*Spirochaetes*".

been sequenced. Of the remaining species, three are more closely related to *Treponema* and should probably be reclassified within that group (see below). The remaining species of the genus *Spirochaeta* comprise at least seven phylogenetic groups: (1) *Spirochaeta africana* and *asiatica*; (2) *Spirochaeta alkalica*, *americana*, and *halophila*; (3) *Spirochaeta aurantia*; (4) *Spirochaeta bajacaliforniensis* and *smaragdinae*; (5) *Spirochaeta coccoides*; (6) *Spirochaeta isovalerica* and *litoralis*; and (7) *Spirochaeta thermophila*.

The genus *Borrelia* comprises three phylogenetic groups. One group contains the type species, *Borrelia anserina*, and the causative agents of relapsing fever, *Borrelia coriaceae*, *crocidurae*, *duttonii*, *hermsii*, *hispanica*, *miyamotoi*, *parkeri*, *persica*, *recurrentis*, *theileri*, and *turicatae*.. Many of these species are transmitted by soft-bodied ticks. The second group includes the causative agent of Lyme disease, *Borrelia burgdorferi*, and species transmitted by hard-bodied ticks, *Borrelia afzelii*, *burgdorferi*, *garinii*, *japonica*, *lusitaniae*, *sinica*, *spielmanii*, *tanukii*, *turdi*, and *valaisiana*. The third group consists solely of *Borrelia turcica*.

In addition, sequences are not available for some named species, including *Borrelia baltazardii*, *brasiliensis*, *caucasica*, *dugesii*, *graingeri*, *harveyi*, *latyschewii*, *mazzottii*, *tillae*, and *venezuelensis*.

The genus *Cristispira* is represented by a single species, *Cristispira pectinis*, which is related to *Borrelia*. This microorganism has been identified in the crystalline styles of oysters (Paster et al., 1996).

The genus *Treponema* comprises three phylogenetic groups: *Treponema pallidum* (type species), "*calligyrum*", *denticola*, *medium*, *phagedenis*, *putidum*, "*refringens*", and "*vincentii*"; in addition, *Spirochaeta zuelzerae* is associated with this group. The second group comprises *Treponema amylovorum*, *berlinense*, *bryantii*, *brennaborense*, *lecithinolyticum*, *maltophilum*, *parvum*, *pectinovorum*, *porcinum*, *saccharophilum*, *socranskii*, and *succinifaciens*. The third group comprises *Treponema azotonutricium* and *primitia*; in addition, *Spirochaeta caldaria* and *stenostrepta* are associated with this group. Lastly, no sequence is available for *Treponema minutum*, so its placement is ambiguous.

Family "*Brachyspiraceae*"

This family comprises a single genus of closely related species: *Brachyspira aalborgi* (type species), *alvinipulli*, *hyodysenteriae*, *innocens*, *intermedia*, *murdochii*, and *pilosicoli*. Many of the species in this genus were previously classified in the genus *Serpulina*, which is not used in the current classification (Ochiai et al., 1997).

Family "*Brevinemataceae*"

This family is represented by a single genus and species, *Brevinema andersonii*, isolated from rodents.

Family *Leptospiraceae*

This family comprises the large genus *Leptospira* and two monospecies genera, *Leptonema illini* and *Turneriella parva*. These latter genera were previously classified within the *Leptospira*. However, on the basis of differences in their 16S rRNA gene sequences, they were transferred to novel genera.

The genus *Leptospira* comprises three phylogenetic groups: (1) *Leptospira interrogans* (type species), *alexanderi*, *borgpetersenii*, *kirschneri*, *noguchii*, *santarosai*, and *weilii*; (2) *Leptospira broomii*, *fainei*, *inadai*, *licerasiae*, and *wolffii*; and (3) *Leptospira biflexa*, *meyeri*, and *wolbachii*.

Spirochaetales family *incertae sedis*

This family includes four genera of symbionts of arthropod invertebrates. Although their morphologies have been described in detail (Bermudes et al., 1988), their 16S rRNA genes have not been sequenced, and their phylogenetic placements are unknown. They are *Clevelandina reticulitermitidis*, *Diplocalyx calotermitidis*, *Hollandina pterotermitidis*, and *Pillotina calotermitidis*.

Phylum *Tenericutes*

This phylum comprises a single class, *Mollicutes*, which was previously classified within the *Firmicutes* (Garrity et al., 2005). Elevation of these organisms to a separate phylum is justified in part by analyses of a number of conserved phylogenetic markers such as the elongation factor Tu and RNA polymerase (Ludwig and Schleifer, 2005). This classification is further supported by the presence of a wall-less cytoplasmic membrane which is a distinctive cellular structure of this group.

Class *Mollicutes*

This class comprises four orders, *Mycoplasmatales*, *Entomoplasmatales*, *Acholeplasmatales*, and *Anaeroplasmatales*. While these orders do not agree well with the 16S rRNA gene phylogeny, efforts to reorganize the taxonomy are confounded by the presence of many human and animal pathogens within the group and the priority of some genus names that are seldom used (Brown et al., 2010). A major difficulty is the polyphyletic nature of the genus *Mycoplasma*, species of which are found in 13 distinct clusters distributed over three deep lineages. A fuller discussion of the complexities of this group along with rRNA gene trees is found in the chapter on *Mycoplasmatales*.

Order *Mycoplasmatales*

This order is the type for the class and comprises two families and four genera. The genera *Mycoplasma* and *Ureaplasma* are classified within the family *Mycoplasmataceae*. The other two genera, *Eperythrozoon* and *Haemobartonella*, contain many blood parasites that have not been cultivated. Although some of the species have been transferred to the genus *Mycoplasma*, the genera are classified within a family *incertae sedis* in recognition of the remaining uncertainties in their classification.

Family *Mycoplasmataceae*

This family contains the genera *Mycoplasma* and *Ureaplasma*. While *Ureaplasma* is well defined on the basis of its rRNA gene sequence phylogeny, *Mycoplasma* is found in at least three deep phylogenetic lineages or groups. The first group contains the type species, *Mycoplasma mycoides*, which is actually more closely related to *Entomoplasma*, the type genus of the order *Entomoplasmatales*, than to most other species of *Mycoplasma* and *Ureaplasma*. A second lineage, called the "*pneumoniae* group", includes the genus *Ureaplasma* as well as four *Mycoplasma* clusters. The third lineage, called the "*hominis* group", includes the remaining eight *Mycoplasma* clusters.

The group containing the type species includes: *Mycoplasma mycoides* (type species), *capricolum*, *cottewii*, *putrefaciens*, and *yeatsii*.

The "*hominis* group" includes eight clusters of *Mycoplasma* species. (1) The "*bovis*" cluster comprises *Mycoplasma adleri*, *agalactiae*, *bovigenitalium*, *bovis*, *californicum*, *caviae*, *columbinasale*, *columbinum*, *felifaucium*, *fermentans*, *gallinarum*, *iners*, *leopharyngis*, *lipofaciens*, *maculosum*, *meleagridis*, *opalescens*, *phocirhinis*, *primatum*, *simbae*, and *spermatophilum*. (2) The "*equigenitalium*" cluster comprises *Mycoplasma elephantis* and *equigenitalium*. (3) The "*hominis*" cluster comprises *Mycoplasma alkalescens*, *anseris*, *arginini*, *arthritidis*, *auris*, *buccale*, *canadense*, *cloacale*, *equirhinis*, *falconis*, *faucium*, *gateae*, *gypis*, *hominis*, *hyosynoviae*, *indiense*, *orale*, *phocicerebrale*, *phocidae*, *salivarium*, *spumans*, and *subdolum*. (4) The "*lipophilum*" cluster comprises *Mycoplasma hyopharyngis* and *lipophilum*. (5) The "*neurolyticum*" cluster comprises *Mycoplasma bovoculi*, *collis*, *cricetuli*, *conjunctivae*, *dispar*, *flocculare*, *hyopneumoniae*, *hyorhinis*, *iguanae*, *lagogenitalium*, *molare*, *neurolyticum*, and *ovipneumoniae*. (6) The "*pulmonis*" cluster comprises *Mycoplasma agassizii*, *pulmonis*, and *testudineum*. (7) The "*sualvi* " cluster comprises *Mycoplasma moatsii*, *mobile*, and *sualvi*. (8) The "*synoviae*" cluster comprises *Mycoplasma alligatoris*, *anatis*, *bovirhinis*, *buteonis*, *canis*, *citelli*, *columborale*, *corogypsi*, *crocodyli*, *cynos*, *edwardii*, *felis*, *gallinaceum*, *gallopavonis*, *glycophilum*, *leonicaptivi*, *mustelae*, *oxoniensis*, *pullorum*, *sturni*, *synoviae*, and *verecundum*.

The "*pneumoniae* group" includes four clusters of *Mycoplasma* species and *Ureaplasma*. (1) The "*fastidiosum*" cluster comprises *Mycoplasma cavipharyngis* and *fastidiosum*. (2) The "hemotrophic" cluster comprises many species that were formerly classified within the genera *Eperythrozoon* and *Haemobartonella* (see below), including *Mycoplasma coccoides*, *haemocanis*, *haemofelis*, *haemomuris*, *ovis*, *suis*, and *wenyonii*. (3) The "*muris*" cluster comprises *Mycoplasma iowae*, *microti*, *muris*, and *penetrans*. (4) The "*pneumoniae*" cluster comprises *Mycoplasma alvi*, *amphoriforme*, *gallisepticum*, *genitalium*, *imitans*, *pirum*, *pneumoniae*, and *testudinis*. The genus *Ureaplasma* comprises *Ureaplasma urealyticum* (type species), *canigenitalium*, *cati*, *diversum*, *felinum*, *gallorale*, and *parvum*.

Mycoplasmatales family *incertae sedis*

This family includes the genera of blood parasites *Eperythrozoon* and *Haemobartonella*. Species whose 16S rRNA genes have been sequenced are also classified within the *Mycoplasma* hemotrophic cluster. On the basis of their 16S rRNA gene sequences, the species of these genera are intermixed in two groups. The first group comprises *Eperythrozoon coccoides* (type species) and *Haemobartonella canis* and *felis*. *Haemobartonella muris*, which is the type species of its genus, is a deep lineage in this group. Upon reclassification to *Mycoplasma*, the *Haemobartonella* species were renamed *haemocanis*, *haemofelis*, and *haemomuris*, respectively, to distinguish them from previously named *Mycoplasma* species. The second group comprises *Eperythrozoon ovis*, *suis*, and *wenyonii*.

Order *Entomoplasmatales*

This order contains two families, *Entomoplasmataceae* and *Spiroplasmataceae*. The order is paraphyletic because it includes the type species of the genus *Mycoplasma*, most species of which are classified in the *Mycoplasmatales*.

Family *Entomoplasmataceae*

This family comprises the genera *Entomoplasma* and *Mesoplasma*. However, on the basis of their 16S rRNA gene sequences, some species of *Acholeplasma* appear to be misclassified within this group. The family comprises four phylogenetic lineages: (1) *Entomoplasma ellychniae* (type species), *Mesoplasma florum* (type species), and *Mesoplasma chauliocola*, *coleopterae*, *corruscae*, *entomophilum*, *grammopterae*, and *tabanidae*; (2) *Mesoplasma photuris*, *seiffertii*, and *syrphidae*; *Entomoplasma lucivorax*, *luminosum*, and *somnilux*; and *Acholeplasma multilocale*; (3) *Mesoplasma lactucae*; and (4) the group containing the type species of *Mycoplasma*, *Mycoplasma mycoides* (see above).

Family *Spiroplasmataceae*

This family comprises the single genus *Spiroplasma*, which itself comprises three relatively deep phylogenetic lineages. In fact, these lineages are no more closely related to each other than to some *Mycoplasma* species. These lineages include (1) *Spiroplasma citri* (type species), *chrysopicola*, *insolitum*, *melliferum*, *penaei*, *phoeniceum*, *poulsonii*, and *syrphidicola*; (2) *Spiroplasma alleghenense*, *cantharicola*, *chinense*, *corruscae*, *culicicola*, *diabroticae*, *diminutum*, *gladiatoris*, *helicoides*, *lampyridicola*, *leptinotarsae*, *lineolae*, *litorale*, *montanense*, *sabaudiense*, *turonicum*, and *velocicrescens*; and (3) *Spiroplasma ixodetis* and *platyhelix*.

Order *Acholeplasmatales* and family *Acholeplasmataceae*

This order comprises the family *Acholeplasmataceae* and a family *incertae sedis* of uncultured plant pathogens classified within "*Candidatus* Phytoplasma". On the basis of their 16S rRNA gene sequences, both of these groups are well defined phylogenetically. The family *Acholeplasmataceae* comprises four closely related lineages that are all classified with the genus *Acholeplasma*: (1) *Acholeplasma laidlawii* (type species), *equifetale*, *granularum*, *oculi*, and *pleciae*; (2) *Acholeplasma axanthum*, *cavigenitalium*, and *modicum*; *Mycoplasma feliminutum*; (3) *Acholeplasma brassicae*, *morum*, and *vituli*; and (4) *Acholeplasma palmae* and *parvum*.

Order *Anaeroplasmatales* and family *Anaeroplasmataceae*

This order and family comprises two genera which, on the basis of 16S rRNA gene sequence similarity, are not closely related. *Anaeroplasma* is related to members of the order *Acholeplasmatales*. The second genus, *Asteroleplasma*, appears to represent a very deep lineage within the phylum. The genus *Anaeroplasma* comprises three closely related species: *Anaeroplasma abactoclasticum* (type species), *bactoclasticum*, and *varium*. In addition, the species *Anaeroplasma intermedium* has been described for which no sequence is available. *Asteroleplasma anaerobium* is the sole species in the genus *Asteroleplasma*.

Phylum "*Acidobacteria*"

With only seven species, this phylum of mostly oligotrophic heterotrophs comprises two classes of validly published bacteria (Figure 9). However, surveys of environmental DNA indicate that this is one of the most abundant groups of bacteria in soil and many other habitats.

Class "*Acidobacteriia*", order "*Acidobacteriales*", and family "*Acidobacteriaceae*"

These taxa comprise two monospecific genera, represented by *Acidobacterium capsulatum* and *Terriglobus roseus*, and *Edaphobacter modestus* (type species) and *aggregans*.

Class *Holophagae*, order *Holophagales*, family *Holophagaceae*, order *Acanthopleuribacterales*, and family *Acanthopleuribacteraceae*

The family *Holophagaceae* comprises two monospecific genera, represented by *Holophaga foetida* and *Geothrix fermentans*. The family *Acanthopleuribacteraceae* comprises one monospecific genus, *Acanthopleuribacter*.

Phylum "*Fibrobacteres*"

This phylum comprises the class "*Fibrobacteria*", the order "*Fibrobacterales*", the family "*Fibrobacteraceae*", and the genus *Fibrobacter*. This genus contains two species, *Fibrobacter succinogenes* (type species) and *intestinalis*.

Phylum "*Fusobacteria*"

This phylum comprises a single class, "*Fusobacteriia*", and order "*Fusobacteriales*". Two families are currently described (Figure 10). While the family "*Leptotrichiaceae*" is well defined on the basis of 16S rRNA gene sequences, the family "*Fusobacteriaceae*" is more complicated. It comprises five genera. The genus *Fusobacterium* is paraphyletic and includes the lineage containing the genus *Cetobacterium*. The genera *Ilyobacter* and *Propionigenium* are also intermixed. If additional evidence supports these conclusions, reclassification within this family would be warranted.

The phylogenetic groups within the family "*Fusobacteriaceae*" are (1) *Fusobacterium nucleatum* (type species), *canifelinum*, *equinum*, *gonidiaformans*, *mortiferum*, *necrogenes*, *necrophorum*, *perfoetens*, *periodonticum*, *russii*, *simiae*, *ulcerans*, and *varium*; and *Cetobacterium ceti* (type species) and *somerae*, representing a deeper branch; (2) *Ilyobacter polytropus* (type species), *insuetus*, and *tartaricus*; *Propionigenium modestum* (type species) and *maris*; and (3) *Psychrilyobacter atlanticus*, which was described after the deadline for inclusion in this volume, but it appears to be a deep lineage of this family (Zhao et al., 2009).

The phylogenetic groups within the family "*Leptotrichiaceae*" are (1) *Leptotrichia buccalis* (type species), *hofstadii*, *shahii*, *trevisanii*, and *wadei*; (2) *Leptotrichia goodfellowii*; (3) *Sebaldella termitidis*; (4) *Sneathia sanguinegens*; and (5) *Streptobacillus moniliformis*.

Phylum "*Dictyoglomi*"

This phylum comprises the class "*Dictyoglomia*", the order "*Dictyoglomales*", the family "*Dictyoglomaceae*", and the genus

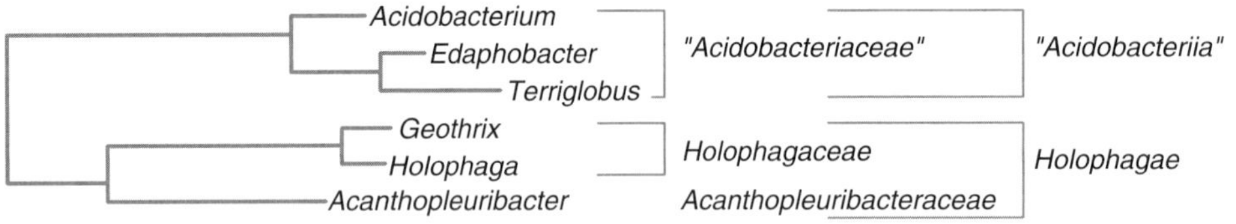

FIGURE 9. Genera of the phylum "*Acidobacteria*".

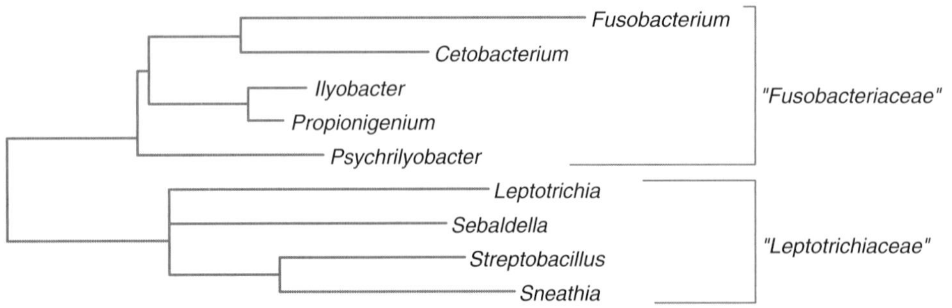

FIGURE 10. Genera of the phylum "*Fusobacteria*".

Dictyoglomus. This genus contains two species, *Dictyoglomus thermophilum* (type species) and *turgidum*.

Phylum *Gemmatimonadetes*

This phylum comprises the class *Gemmatimonadetes*, the order *Gemmatimonadales*, the family *Gemmatimonadaceae*, and the genus *Gemmatimonas*. This genus contains one species, *Gemmatimonas aurantiaca*.

Phylum *Lentisphaerae*

On the basis of their 16S rRNA gene sequences and other molecular markers, this phylum is related to the phyla "*Verrucomicrobia*", "*Chlamydiae*", and "*Planctomycetes*", which form a deep group within the *Bacteria*. The phylum *Lentisphaerae* comprises the class "*Lentisphaeria*" and two orders. The order *Lentisphaerales* comprises the family "*Lentisphaeraceae*" and the monospecific genus *Lentisphaera*, the type of which is *Lentisphaera araneosa*. The order *Victivallales* comprises the family "*Victivallaceae*"and the monospecific genus *Victivallis*, the type of which is *Victivallis vadensis*.

Phylum "*Verrucomicrobia*"

On the basis of their 16S rRNA gene sequences and other molecular markers, this phylum is related to the phyla "*Chlamydiae*", *Lentisphaerae*, and "*Planctomycetes*", which form a deep group within the bacteria. "*Verrucomicrobia*" comprises three classes, *Verrucomicrobiae*, *Opitutae*, and "*Spartobacteria*" (Figure 11).

Currently, the class *Verrucomicrobiae* comprises the order *Verrucomicrobiales*, which comprises the families *Verrucomicrobiaceae*, "*Akkermansiaceae*", and "*Rubritaleaceae*". The family *Verrucomicrobiaceae* comprises *Verrucomicrobium spinosum*, *Prosthecobacter fusiformis* (type species), *debontii*, *dejongeii*, and *vanneervenii*. In addition, *Prosthecobacter fluviatilis*, which was described after the deadline for inclusion in this volume, is a member of this family (Takeda et al., 2008). The family "*Akkermansiaceae*" comprises the monospecific genus *Akkermansia*, the type of which is *Akkermansia muciniphila*. The family "*Rubritaleaceae*" comprises *Rubritalea marina* (type species), *sabuli*, *spongiae*, *squalenifaciens*, and *tangerina*.

In addition to these genera, four genera were described after the deadline for inclusion in this volume. *Persicirhabdus sediminis*; and *Roseibacillus ishigakijimensis* (type species), *persicicus*, and *ponti* (Yoon et al., 2008a), are affiliated with the family "*Rubritaleaceae*". The remaining two genera, *Haloferula rosea* (type species), *harenae*, *helveola*, *phyci*, *rosea*, and *sargassicola* (Yoon et al., 2008c) and *Luteolibacter pohnpeiensis* (type species) and *algae* (Yoon et al., 2008a), appear to be members of the order *Verrucomicrobiales*, but their affiliation with a particular family is more ambiguous. For this reason, they have not been included in the Taxonomic Outline.

The class *Opitutae* comprises the orders *Opitutales* and *Puniceicoccales*. The order *Opitutales* comprises a single family, *Opitutaceae*, and two monospecific genera, the type species of which are *Opitutus terrae* and *Alterococcus agarolyticus*. The order *Puniceicoccales* comprises a single family, *Puniceicoccaceae*, and four genera. The genera form two clusters. The first cluster includes three monospecific genera, the type species of which are *Puniceicoccus vermicola*, *Cerasicoccus arenae*, and *Coraliomargarita akajimensis*. The second cluster includes *Pelagicoccus mobilis* (type species), *albus*, *croceus*, and *litoralis*. In addition, the genus "*Fucophilus*", which has been described but whose name has never been validly published, is a member of this family.

The class "*Spartobacteria*" comprises the order "*Chthoniobacterales*", which includes the family "*Chthoniobacteraceae*". This family comprises "*Chthoniobacter flavus*" and the nematode symbionts "*Candidatus* Xiphinematobacter brevicolli" (type species), "americani", and "rivesi".

Phylum "*Chlamydiae*"

On the basis of their 16S rRNA gene sequences and other molecular markers, this phylum is related to the phyla *Lentisphaerae*, "*Planctomycetes*", and "*Verrucomicrobia*", which form a deep group within the bacteria. All known members of the phylum "*Chlamydiae*" are obligate intracellular bacteria and multiply in eukaryotic hosts, including humans and other animals and protozoa. They also possess a developmental cycle that is characterized by morphologically and physiologically distinct stages. The intracellular lifestyle of chlamydiae is thus thought to be an ancient trait of this phylum (Everett et al., 1999). As a consequence of the intracellular lifestyle, no species has ever been grown in axenic culture. Because of changes to the Bacteriological Code beginning in 1997, only the species described before that time have been validly published, and many of the newer taxa are limited to *Candidatus* status (Labeda, 1997; Murray and Stackebrandt, 1995). In addition, even though some species have been cultivated in the free-living amoebae *Acanthamoeba castellanii* and *Acanthamoeba polyphaga*, they have not been deposited in two public culture collections, and thus their names have not been validly published (Heyrman et al., 2005).

The phylum "*Chlamydiae*" comprises a single class, "*Chlamydiia*", and order, *Chlamydiales*. The order comprises eight families of varying relatedness based upon 16S rRNA gene sequence similarities (Figure 12). The family *Chlamydiaceae* contains the type genus for the order. Two taxonomies are in widespread use for this family. One taxonomy assigns all species within this family to the genus *Chlamydia*. The second taxonomy classifies many of these species within a second genus, *Chlamydophila*, in recognition of their differences in a variety of molecular markers including the 16S rRNA gene and some phenotypic markers (Everett et al., 1999). The merits of these approaches have been discussed (Everett and Andersen, 2001; Schachter et al., 2001). While this taxonomic outline uses the taxonomy of Everett et al. (1999), the first taxonomy is used by the authors of the chapter *Chlamydiaceae* (Kuo and Stephens, 2010). On the basis of the taxonomy of Everett et al. (1999), the genus *Chlamydia* comprises *Chlamydia trachomatis* (type species), *muridarum*, and *suis*. The genus *Chlamydophila* comprises *Chlamydophila psittaci* (type species), *abortus*, *caviae*, *felis*, *pecorum*, and *pneumoniae*.

The remaining families in the order are: "*Candidatus* Clavichlamydiaceae" comprising "*Candidatus* Clavichlamydia salmonicola"; "*Criblamydiaceae*" comprising "*Criblamydia sequanensis*"; *Parachlamydiaceae* comprising *Parachlamydia acanthamoebae* (type species and genus), *Neochlamydia hartmannellae*, and "*Protochlamydia amoebophila*"; "*Candidatus* Piscichlamydiaceae" comprising "*Candidatus* Piscichlamydia salmonis"; "*Rhabdochlamydiaceae*" comprising "*Candidatus* Rhabdochlamydia porcellionis" and "*Candidatus* Rhabdochlamydia crassificans"; *Simkaniaceae* comprising *Simkania negevensis* (type species and genus) and "*Candidatus* Fritschea

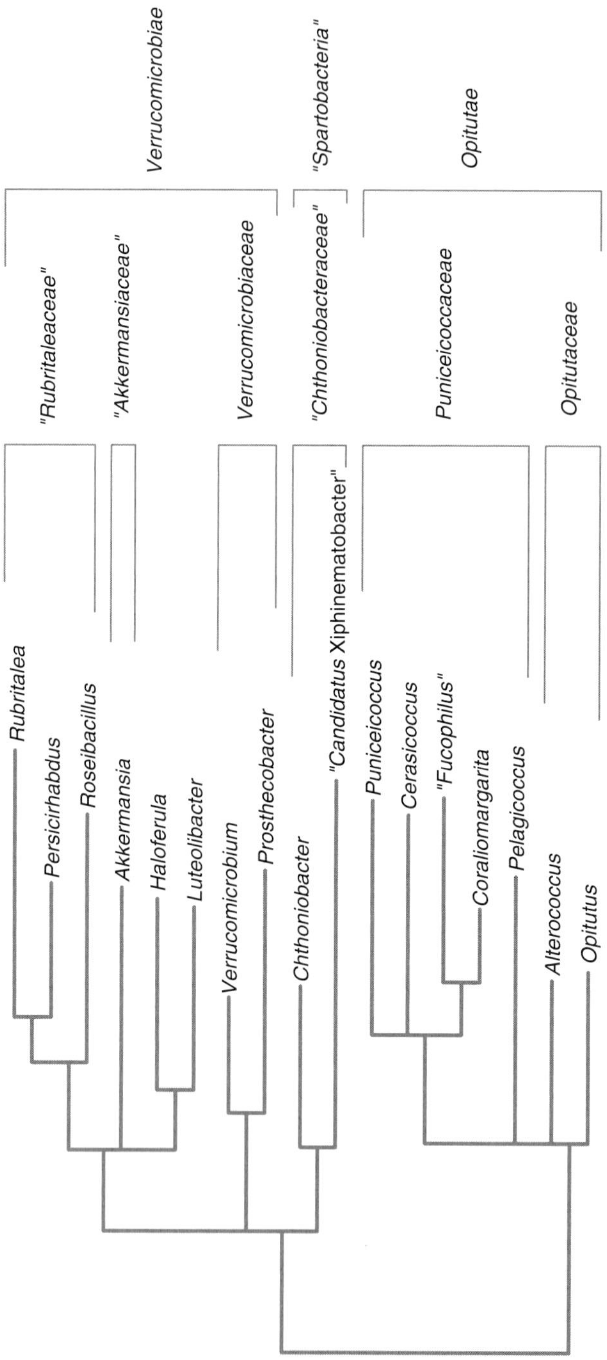

FIGURE 11. Genera of the phylum "*Verrucomicrobia*".

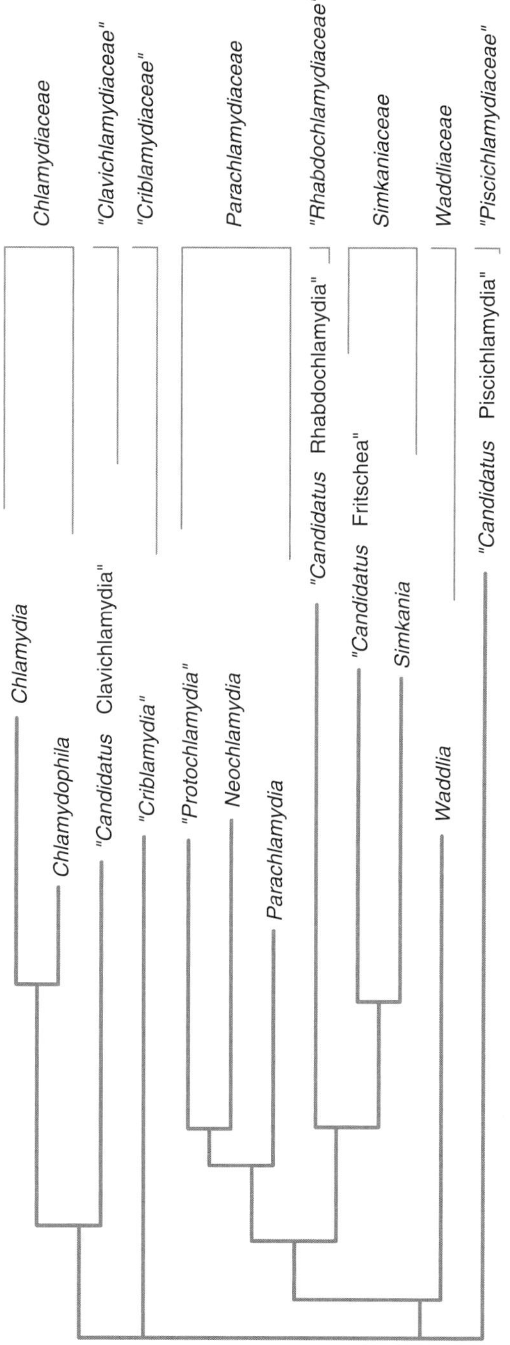

FIGURE 12. Genera of the phylum "*Chlamydiae*".

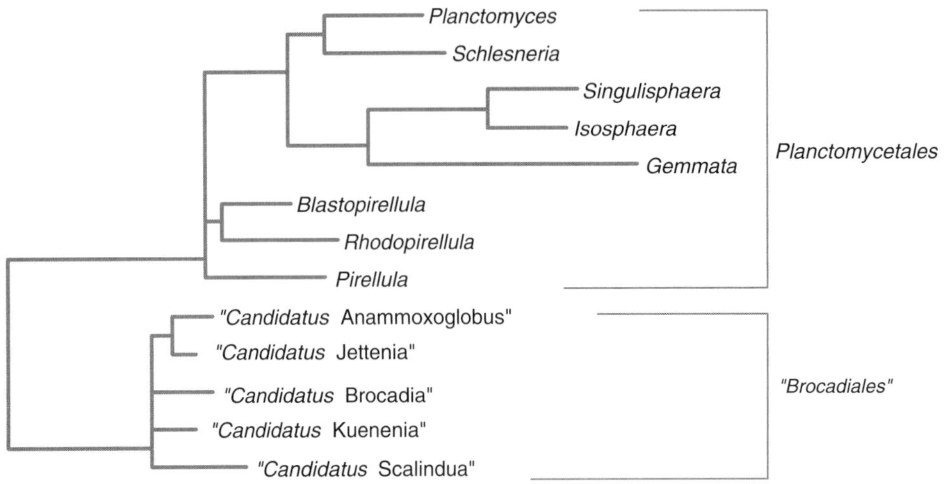

FIGURE 13. Genera of the phylum "*Planctomycetes*".

bemisiae" and "*Candidatus* Fritschea eriococci"; and *Waddliaceae* comprising *Waddlia chondrophila* and "*Waddlia malaysiensis*" (Chua et al., 2005).

Phylum "*Planctomycetes*"

This phylum comprises a single class, "*Planctomycetia*", and two orders, *Planctomycetales* and "*Brocadiales*" (Figure 13). The order *Planctomycetales* comprises the family *Planctomycetaceae*, containing eight diverse genera. The type genus is *Planctomyces*. However, a strain and 16S rRNA gene sequence are not available for the type species, *Planctomyces bekefii*, or for two other validly published species in this genus, *Planctomyces guttaeformis* and *stranskae*. Therefore, the taxonomy of this group is based upon the properties of the species that are available: *Planctomyces brasiliensis*, *limnophilus*, and *maris*. Most of the other genera in this family are monospecific and represented by *Blastopirellula marina*, *Gemmata obscuriglobus*, *Isosphaera pallida*, *Pirellula staleyi* (type species) and *marina*, *Rhodopirellula baltica*, *Schlesneria paludicola*, and *Singulisphaera acidiphila*. In addition to these, *Zavarzinella formosa* was described after the deadline for this volume but could be classified within this family (Kulichevskaya et al., 2009).

The order "*Brocadiales*" and family "*Brocadiaceae*" comprises *Candidatus* species. They include "*Candidatus* Brocadia anammoxidans" and "fulgida", "*Candidatus* Anammoxoglobus propionicus", "*Candidatus* Jettenia asiatica", "*Candidatus* Kuenenia stuttgartiensis", and "*Candidatus* Scalindua brodiae", "sorokinii", and "wagneri".

References

An, D.S., H.G. Lee, W.T. Im, Q.M. Liu and S.T. Lee. 2007. *Segetibacter koreensis* gen. nov., sp. nov., a novel member of the phylum *Bacteroidetes*, isolated from the soil of a ginseng field in South Korea. Int. J. Syst. Evol. Microbiol. *57*: 1828–1833.

Asker, D., T. Beppu and K. Ueda. 2007. *Zeaxanthinibacter enoshimensis* gen. nov., sp. nov., a novel zeaxanthin-producing marine bacterium of the family *Flavobacteriaceae*, isolated from seawater off Enoshima Island, Japan. Int. J. Syst. Evol. Microbiol. *57*: 837–843.

Asker, D., T. Beppu and K. Ueda. 2008. *Nubsella zeaxanthinifaciens* gen. nov., sp. nov., a zeaxanthin-producing bacterium of the family *Sphingobacteriaceae* isolated from freshwater. Int. J. Syst. Evol. Microbiol. *58*: 601–606.

Bae, S.S., K.K. Kwon, S.H. Yang, H.S. Lee, S.J. Kim and J.H. Lee. 2007. *Flagellimonas eckloniae* gen. nov., sp. nov., a mesophilic marine bacterium of the family *Flavobacteriaceae*, isolated from the rhizosphere of *Ecklonia kurome*. Int. J. Syst. Evol. Microbiol. *57*: 1050–1054.

Bermudes, D., D. Chase and L. Margulis. 1988. Morphology as a basis for taxonomy of large spirochetes symbiotic in wood-eating cockroaches and termites: *Pillotina* gen. nov., nom. rev., *Pillotina calotermitidis* sp. nov., nom. rev., *Diplocalyx* gen. nov., nom. rev., *Diplocalyx calotermitidis* sp. nov., nom. rev., *Hollandina* gen. nov., nom. rev., *Hollandina pterotermitidis* sp. nov., nom. rev., and *Clevelandina reticulitermitidis* gen. nov., sp. nov. Int. J. Syst. Bacteriol. *38*: 291–302.

Bernardet, J.F. and Y. Nakagawa. 2006. An introduction to the family *Flavobacteriaceae*. *In* The Prokaryotes: a Handbook on the Biology of Bacteria, 3rd edn, vol. 7, *Proteobacteria: Delta and Epsilon Subclasses. Deeply Rooting Bacteria* (edited by Dworkin, Falkow, Rosenberg, Schleifer and Stackebrandt). Springer, New York, pp. 455–480.

Bowman, J.P., C. Mancuso, C.M. Nichols and J.A.E. Gibson. 2003. *Algoriphagus ratkowskyi* gen. nov., sp. nov., *Brumimicrobium glaciale* gen. nov., sp. nov., *Cryomorpha ignava* gen. nov., sp. nov. and *Crocinitomix catalasitica* gen. nov., sp. nov., novel *flavobacteria* isolated from various polar habitats. Int. J. Syst. Evol. Microbiol. *53*: 1343–1355.

Brown, D.R., M. May, J.M. Bradbury, K.-E. Johansson and H. Neimark. 2010. Order I. *Mycoplastamales*. *In* Bergeys Manual of Systematic Bacteriology, 2nd edn, vol. 4, The *Bacteroidetes*, *Spirochaetes*, *Tenericutes* (*Mollicutes*), *Acidobacteria*, *Fibrobacteres*, *Fusobacteria*, *Dictyoglomi*, *Gemmatimonadetes*, *Lentisphaerae*, *Verrucomicrobia*, *Chlamydiae*, and *Planctomycetes* (edited by Krieg, Staley, Brown, Hedlund, Paster, Ward, Ludwig and Whitman). Springer, New York, pp. 574–644.

Cavalier-Smith, T. 2002. The neomuran origin of *archaebacteria*, the negibacterial root of the universal tree and bacterial megaclassification. Int. J. Syst. Evol. Microbiol. *52*: 7–76.

Chen, Y.G., X.L. Cui, Y.Q. Zhang, W.J. Li, Y.X. Wang, C.J. Kim, J.M. Lim, L.H. Xu and C.L. Jiang. 2008. *Salinimicrobium terrae* sp. nov., isolated from saline soil, and emended description of the genus *Salinimicrobium*. Int. J. Syst. Evol. Microbiol. *58*: 2501–2504.

Chua, P.K., J.E. Corkill, P.S. Hooi, S.C. Cheng, C. Winstanley and C.A. Hart. 2005. Isolation of *Waddlia malaysiensis*, a novel intracellular bacterium, from fruit bat (*Eonycteris spelaea*). Emerg. Infect. Dis. *11*: 271–277.

Cicarelli, F.D., T. Doerks, C. von Mering, C.J. Creevey, B. Snel and P. Bork. 2006. Toward automatic reconstruction of a highly resolved tree of life. Science *311*: 1283–1287.

Cole, J.R., B. Chai, R.J. Farris, Q. Wang, A.S. Kulam-Syed-Mohideen, D.M. McGarrell, A.M. Bandela, E. Cardenas, G.M. Garrity and J.M. Tiedje. 2007. The ribosomal database project (RDP-II): introducing myRDP space and quality controlled public data. Nucleic Acids Res. *35*: D169–D172.

De Vos, P., H.G. Truper and B.J. Tindall. 2005. Judicial Commission of the International Committee on Systematics of Prokaryotes Minutes. Xth International (IUMS) Congress of Bacteriology and Applied Microbiology. Int. J. Syst. Bacteriol. *55*: 525–532.

Everett, K.D., R.M. Bush and A.A. Andersen. 1999. Emended description of the order *Chlamydiales*, proposal of *Parachlamydiaceae* fam. nov. and *Simkaniaceae* fam. nov., each containing one monotypic genus, revised taxonomy of the family *Chlamydiaceae*, including a new genus and five new species, and standards for the identification of organisms. Int. J. Syst. Bacteriol. *49*: 415–440.

Everett, K.D.E. and A.A. Andersen. 2001. Radical changes to chlamydial taxonomy are not necessary just yet: reply. Int. J. Syst. Evol. Microbiol. *51*: 251–253.

Garrity, G.M. and J.G. Holt. 2001. The road map to the manual. *In* Bergey's Manual of Systematic Bacteriology, 2nd edn, vol. 1, *The Archaea and the Deeply Branching and Phototrophic Bacteria* (edited by Boone, Castenholz and Garrity). Springer, New York, pp. 119–166.

Garrity, G.M., J.A. Bell and T. Lilburn. 2005. The revised road map to the manual. *In* Bergey's Manual of Systematic Bacteriology, 2nd edn, vol. 2, *The Proteobacteria, Part A, Introductory Essays* (edited by Brenner, Krieg, Staley and Garrity). Springer, New York, pp. 159–206.

Griffiths, E. and R.S. Gupta. 2007. Phylogeny and shared conserved inserts in proteins provide evidence that *Verrucomicrobia* are the closest known free-living relatives of chlamydiae. Microbiology *153*: 2648–2654.

Hardham, J.M., K.W. King, K. Dreier, J. Wong, C. Strietzel, R.R. Eversole, C. Sfintescu and R.T. Evans. 2008. Transfer of *Bacteroides splanchnicus* to Odoribacter gen. nov. as *Odoribacter splanchnicus* comb. nov., and description of *Odoribacter denticanis* sp. nov., isolated from the crevicular spaces of canine periodontitis patients. Int. J. Syst. Evol. Microbiol. *58*: 103–109.

Heyrman, J., M. Rodriguez-Diaz, J. Devos, A. Felske, N.A. Logan and P. De Vos. 2005. *Bacillus arenosi* sp. nov., *Bacillus arvi* sp. nov. and *Bacillus humi* sp. nov., isolated from soil. Int. J. Syst. Evol. Microbiol. *55*: 111–117.

Holmes, D.E., K.P. Nevin, T.L. Woodard, A.D. Peacock and D.R. Lovley. 2007. *Prolixibacter bellariivorans* gen. nov., sp. nov., a sugar-fermenting, psychrotolerant anaerobe of the phylum *Bacteroidetes*, isolated from a marine-sediment fuel cell. Int. J. Syst. Evol. Microbiol. *57*: 701–707.

Hosoya, S., V. Arunpairojana, C. Suwannachart, A. Kanjana-Opas and A. Yokota. 2006. *Aureispira marina* gen. nov., sp. nov., a gliding, arachidonic acid-containing bacterium isolated from the southern coastline of Thailand. Int. J. Syst. Evol. Microbiol. *56*: 2931–2935.

Hosoya, S., V. Arunpairojana, C. Suwannachart, A. Kanjana-Opas and A. Yokota. 2007. *Aureispira maritima* sp. nov., isolated from marine barnacle debris. Int. J. Syst. Evol. Microbiol. *57*: 1948–1951.

Khan, S.T., Y. Nakagawa and S. Harayama. 2007a. *Sediminibacter furfurosus* gen. nov., sp. nov. and *Gilvibacter sediminis* gen. nov., sp. nov., novel members of the family *Flavobacteriaceae*. Int. J. Syst. Evol. Microbiol. *57*: 265–269.

Khan, S.T., Y. Nakagawa and S. Harayama. 2007b. *Sediminitomix flava* gen. nov., sp. nov., of the phylum *Bacteroidetes*, isolated from marine sediment. Int. J. Syst. Evol. Microbiol. *57*: 1689–1693.

Khan, S.T., Y. Nakagawa and S. Harayama. 2007c. *Galbibacter mesophilus* gen. nov., sp. nov., a novel member of the family *Flavobacteriaceae*. Int. J. Syst. Evol. Microbiol. *57*: 969–973.

Kim, B.Y., H.Y. Weon, S.H. Yoo, S.B. Hong, S.W. Kwon, E. Stackebrandt and S.J. Go. 2007a. *Niabella aurantiaca* gen. nov., sp. nov., isolated from a greenhouse soil in Korea. Int. J. Syst. Evol. Microbiol. *57*: 538–541.

Kim, J.H., K.Y. Kim, Y.T. Hahm, B.S. Kim, J. Chun and C.J. Cha. 2008a. *Actibacter sediminis* gen. nov., sp. nov., a marine bacterium of the family *Flavobacteriaceae* isolated from tidal flat sediment. Int. J. Syst. Evol. Microbiol. *58*: 139–143.

Kim, M.K., J.R. Na, D.H. Cho, N.K. Soung and D.C. Yang. 2007b. *Parapedobacter koreensis* gen. nov., sp. nov. Int. J. Syst. Evol. Microbiol. *57*: 1336–1341.

Kim, M.K., Y.A. Kim, Y.J. Kim, N.K. Soung, T.H. Yi, S.Y. Kim and D.C. Yang. 2008b. *Parapedobacter soli* sp. nov., isolated from soil of a ginseng field. Int. J. Syst. Evol. Microbiol. *58*: 337–340.

Kulichevskaya, I.S., O.I. Baulina, P.L. Bodelier, W.I. Rijpstra, J.S. Damste and S.N. Dedysh. 2009. *Zavarzinella formosa* gen. nov., sp. nov., a novel stalked, Gemmata-like planctomycete from a Siberian peat bog. Int. J. Syst. Evol. Microbiol. *59*: 357–364.

Kuo, C.C. and R.S. Stephens. 2010. Family I. *Chlamydiaceae*. *In* Bergey's Manual of Systematic Bacteriology, 2nd edn, vol. 4, The *Bacteroidetes, Spirochaetes, Tenericutes* (*Mollicutes*), *Acidobacteria, Fibrobacteres, Fusobacteria, Dictyoglomi, Gemmatimonadetes, Lentisphaerae, Verrucomicrobia, Chlamydiae,* and *Planctomycetes* (edited by Krieg, Staley, Brown, Hedlund, Paster, Ward, Ludwig and Whitman). Springer, New York, p. 845.

Labeda, D.P. 1997. International Committee on Systematic Bacteriology, VIIth International Congress of Microbiology and Applied Bacteriology Minutes. Int. J. Syst. Bacteriol. *47*: 597–600.

Lee, K.C., R.I. Webb, P.H. Janssen, P. Sangwan, T. Romeo, J.T. Staley and J.A. Fuerst. 2009. Phylum *Verrucomicrobia* representatives share a compartmentalized cell plan with members of bacterial phylum *Planctomycetes*. BMC Microbiol. *9*: 5.

Lee, S.D. 2007. *Tamlana crocina* gen. nov., sp. nov., a marine bacterium of the family *Flavobacteriaceae*, isolated from beach sediment in Korea. Int. J. Syst. Evol. Microbiol. *57*: 764–769.

Lim, J.M., C.O. Jeon, S.S. Lee, D.J. Park, L.H. Xu, C.L. Jiang and C.J. Kim. 2008. Reclassification of *Salegentibacter catena* Ying *et al.* 2007 as *Salinimicrobium catena* gen. nov., comb. nov. and description of *Salinimicrobium xinjiangense* sp. nov., a halophilic bacterium isolated from Xinjiang province in China. Int. J. Syst. Evol. Microbiol. *58*: 438–442.

Ludwig, W., O. Strunk, R. Westram, L. Richter, H. Meier, Yadhukumar, A. Buchner, T. Lai, S. Steppi, G. Jobb, W. Forster, I. Brettske, S. Gerber, A.W. Ginhart, O. Gross, S. Grumann, S. Hermann, R. Jost, A. Konig, T. Liss, R. Lussmann, M. May, B. Nonhoff, B. Reichel, R. Strehlow, A. Stamatakis, N. Stuckmann, A. Vilbig, M. Lenke, T. Ludwig, A. Bode and K.H. Schleifer. 2004. ARB: a software environment for sequence data. Nucleic Acids Res. *32*: 1363–1371.

Ludwig, W. and H.P. Klenk. 2005. Overview: a phylogenetic backbone and taxonomic framework for procaryotic systematics. *In* Bergey's Manual of Systematic Bacteriology, 2nd edn, vol. 2, *The Proteobacteria, Part A, Introductory Essays* (edited by Brenner, Krieg, Staley and Garrity). Springer, New York, pp. 49–65.

Ludwig, W. and K.H. Schleifer. 2005. Molecular phylogeny of bacteria based on comparative sequence analysis of conserved genes. *In* Microbial Phylogeny and Evolution, Concepts and Controversies (edited by Sapp). Oxford University Press, New York, pp. 70–98.

Morotomi, M., F. Nagai and H. Sakon. 2007. Genus *Megamonas* should be placed in the lineage of *Firmicutes*; *Clostridia*; *Clostridiales*; '*Acidaminococcaceae*'; *Megamonas*. Int. J. Syst. Evol. Microbiol. *57*: 1673–1674.

Morotomi, M., F. Nagai, H. Sakon and R. Tanaka. 2008. *Dialister succinatiphilus* sp. nov. and *Barnesiella intestinihominis* sp. nov., isolated from human faeces. Int. J. Syst. Evol. Microbiol. *58*: 2716–2720.

Murray, R.G. and E. Stackebrandt. 1995. Taxonomic note: implementation of the provisional status *Candidatus* for incompletely described procaryotes. Int. J. Syst. Bacteriol. *45*: 186–187.

Nedashkovskaya, O.I., S.B. Kim, D.S. Shin, I.A. Beleneva and V.V. Mikhailov. 2007a. *Fulvivirga kasyanovii* gen. nov., sp. nov., a novel member of the phylum *Bacteroidetes* isolated from seawater in a mussel farm. Int. J. Syst. Evol. Microbiol. *57*: 1046–1049.

Nedashkovskaya, O.I., M. Vancanneyt, S.B. Kim, B. Hoste and K.S. Bae. 2007b. *Algibacter mikhailovii* sp. nov., a novel marine bacterium of the family *Flavobacteriaceae*, and emended description of the genus *Algibacter*. Int. J. Syst. Evol. Microbiol. *57*: 2147–2150.

Ntougias, S., C. Fasseas and G.I. Zervakis. 2007. *Olivibacter sitiensis* gen. nov., sp. nov., isolated from alkaline olive-oil mill wastes in the region of Sitia, Crete. Int. J. Syst. Evol. Microbiol. *57*: 398–404.

Ochiai, S., Y. Adachi and K. Mori. 1997. Unification of the genera *Serpulina* and *Brachyspira*, and proposals of *Brachyspira hyodysenteriae* comb. nov., *Brachyspira innocens* comb. nov. and *Brachyspira pilosicoli* comb. nov. Microbiol. Immunol. *41*: 445–452.

Pankratov, T.A., B.J. Tindall, W. Liesack and S.N. Dedysh. 2007. *Mucilaginibacter paludis* gen. nov., sp. nov. and *Mucilaginibacter gracilis* sp. nov., pectin-, xylan- and laminarin-degrading members of the family *Sphingobacteriaceae* from acidic Sphagnum peat bog. Int. J. Syst. Evol. Microbiol. *57*: 2349–2354.

Paster, B.J., D.A. Pelletier, F.E. Dewhirst, W.G. Weisburg, V. Fussing, L.K. Poulsen, S. Dannenberg and I. Schroeder. 1996. Phylogenetic position of the spirochetal genus *Cristispira*. Appl. Environ. Microbiol. *62*: 942–946.

Pilhofer, M., K. Rappl, C. Eckl, A.P. Bauer, W. Ludwig, K.H. Schleifer and G. Petroni. 2008. Characterization and evolution of cell division and cell wall synthesis genes in the bacterial phyla *Verrucomicrobia*, *Lentisphaerae*, *Chlamydiae*, and *Planctomycetes* and phylogenetic comparison with rRNA genes. J. Bacteriol. *190*: 3192–3202.

Pruesse, E., C. Quast, K. Knittel, B. Fuchs, W. Ludwig, J. Peplies and F.O. Glöckner. 2007. SILVA: a comprehensive online resource for quality checked and aligned rRNA sequence data compatible with ARB. Nucleic Acids Res. *35*: 7188–7196.

Qin, Q.L., D.L. Zhao, J. Wang, X.L. Chen, H.Y. Dang, T.G. Li, Y.Z. Zhang and P.J. Gao. 2007. *Wangia profunda* gen. nov., sp. nov., a novel marine bacterium of the family *Flavobacteriaceae* isolated from southern Okinawa Trough deep-sea sediment. FEMS Microbiol. Lett. *271*: 53–58.

Qu, J.H. and H.L. Yuan. 2008. *Sediminibacterium salmoneum* gen. nov., sp. nov., a member of the phylum *Bacteroidetes* isolated from sediment of a eutrophic reservoir. Int. J. Syst. Evol. Microbiol. *58*: 2191–2194.

Quan, Z.X., Y.P. Xiao, S.W. Roh, Y.D. Nam, H.W. Chang, K.S. Shin, S.K. Rhee, Y.H. Park and J.W. Bae. 2008. *Joostella marina* gen. nov., sp. nov., a novel member of the family *Flavobacteriaceae* isolated from the East Sea. Int. J. Syst. Evol. Microbiol. *58*: 1388–1392.

Romanenko, L.A., M. Uchino, G.M. Frolova and V.V. Mikhailov. 2007. *Marixanthomonas ophiurae* gen. nov., sp. nov., a marine bacterium of the family *Flavobacteriaceae* isolated from a deep-sea brittle star. Int. J. Syst. Evol. Microbiol. *57*: 457–462.

Sakamoto, M. and Y. Benno. 2006. Reclassification of *Bacteroides distasonis*, *Bacteroides goldsteinii* and *Bacteroides merdae* as *Parabacteroides distasonis* gen. nov., comb. nov., *Parabacteroides goldsteinii* comb. nov. and *Parabacteroides merdae* comb. nov. Int. J. Syst. Evol. Microbiol. *56*: 1599–1605.

Sakamoto, M., P.T. Lan and Y. Benno. 2007. *Barnesiella viscericola* gen. nov., sp. nov., a novel member of the family *Porphyromonadaceae* isolated from chicken caecum. Int. J. Syst. Evol. Microbiol. *57*: 342–346.

Schachter, J., R.S. Stephens, P. Timms, C. Kuo, P.M. Bavoil, S. Birkelund, J. Boman, H. Caldwell, L.A. Campbell, M. Chernesky, G. Christiansen, I.N. Clarke, C. Gaydos, J.T. Grayston, T. Hackstadt, R. Hsia, B. Kaltenboeck, M. Leinonnen, D. Ojcius, G. McClarty, J. Orfila, R. Peeling, M. Puolakkainen, T.C. Quinn, R.G. Rank, J. Raulston, G.L. Ridgeway, P. Saikku, W.E. Stamm, D.T. Taylor-Robinson, S.P. Wang and P.B. Wyrick. 2001. Radical changes to chlamydial taxonomy are not necessary just yet. Int. J. Syst. Evol. Microbiol. *51*: 249; author reply 251–253.

Shah, H.N. and M.D. Collins. 1988. Proposal for reclassification of *Bacteroides asaccharolyticus*, *Bacteroides gingivalis*, and *Bacteroides endodontalis* in a new genus, *Porphyromonas*. Int. J. Syst. Bacteriol. *38*: 128–131.

Shah, H.N. and D.M. Collins. 1990. *Prevotella*, a new genus to include *Bacteroides melaninogenicus* and related species formerly classified in the genus *Bacteroides*. Int. J. Syst. Bacteriol. *40*: 205–208.

Srisukchayakul, P., C. Suwanachart, Y. Sangnoi, A. Kanjana-Opas, S. Hosoya, A. Yokota and V. Arunpairojana. 2007. *Rapidithrix thailandica* gen. nov., sp. nov., a marine gliding bacterium isolated from samples collected from the Andaman sea, along the southern coastline of Thailand. Int. J. Syst. Evol. Microbiol. *57*: 2275–2279.

Stamatakis, A.P., T. Ludwig and H. Meier. 2005. RAxML-II: a program for sequential, parallel & distributed inference of large phylogenetic trees. Concurrency Comput. Pract. Exp. *17*: 1705–1723.

Stoecker, K., B. Bendinger, B. Schoning, P.H. Nielsen, J.L. Nielsen, C. Baranyi, E.R. Toenshoff, H. Daims and M. Wagner. 2006. Cohn's *Crenothrix* is a filamentous methane oxidizer with an unusual methane monooxygenase. Proc. Natl. Acad. Sci. U.S.A. *103*: 2363–2367.

Takeda, M., A. Yoneya, Y. Miyazaki, K. Kondo, H. Makita, M. Kondoh, I. Suzuki and J. Koizumi. 2008. *Prosthecobacter fluviatilis* sp. nov., which lacks the bacterial tubulin *btubA* and *btubB* genes. Int. J. Syst. Evol. Microbiol. *58*: 1561–1565.

Ten, L.N., O.M Liu, W.T. Im, Z. Aslam and S.T. Lee. 2006. *Sphingobacterium composti* sp. nov., a novel DNase-producing bacterium isolated from compost. J. Microbiol. Biotechnol. *16*: 1728–1733.

Ten, L.N., Q.-M. Liu, W.-T. Im, Z. Aslam and S.-T. Lee. 2007. List of new names and new combinations previously effectively, but not validly, published. Validation List no. 116 Int. J. Syst. Evol. Microbiol *57*: 1372.

Urai, M., T. Aizawa, Y. Nakagawa, M. Nakajima and M. Sunairi. 2008. Mucilaginibacter kameinonensis sp., nov., isolated from garden soil. Int. J. Syst. Evol. Microbiol. *58*: 2046–2050.

Urios, L., H. Agogue, F. Lesongeur, E. Stackebrandt and P. Lebaron. 2006. *Balneola vulgaris* gen. nov., sp. nov., a member of the phylum *Bacteroidetes* from the north-western Mediterranean Sea. Int. J. Syst. Evol. Microbiol. *56*: 1883–1887.

Urios, L., L. Intertaglia, F. Lesongeur and P. Lebaron. 2008. *Balneola alkaliphila* sp. nov., a marine bacterium isolated from the Mediterranean Sea. Int. J. Syst. Evol. Microbiol. *58*: 1288–1291.

Vaz-Moreira, I., M.F. Nobre, O.C. Nunes and C.M. Manaia. 2007. *Pseudosphingobacterium domesticum* gen. nov., sp. nov., isolated from home-made compost. Int. J. Syst. Evol. Microbiol. *57*: 1535–1538.

Wagner, M. and M. Horn. 2006. The *Planctomycetes*, *Verrucomicrobia*, *Chlamydiae* and sister phyla comprise a superphylum with biotechnological and medical relevance. Curr. Opin. Biotechnol. *17*: 241–249.

Wang, L., L.N. Ten, H.G. Lee, W.T. Im and S.T. Lee. 2008. *Olivibacter soli* sp. nov., *Olivibacter ginsengisoli* sp. nov. and *Olivibacter terrae* sp. nov., from soil of a ginseng field and compost in South Korea. Int. J. Syst. Evol. Microbiol. *58*: 1123–1127.

Weon, H.Y., B.Y. Kim, S.H. Yoo, S.Y. Lee, S.W. Kwon, S.J. Go and E. Stackebrandt. 2006. *Niastella koreensis* gen. nov., sp. nov. and *Niastella yeongjuensis* sp. nov., novel members of the phylum *Bacteroidetes*, isolated from soil cultivated with Korean ginseng. Int. J. Syst. Evol. Microbiol. *56*: 1777–1782.

Weon, H.Y., B.Y. Kim, J.H. Joa, S.W. Kwon, W.G. Kim and B.S. Koo. 2008a. *Niabella soli* sp. nov., isolated from soil from Jeju Island, Korea. Int. J. Syst. Evol. Microbiol. *58*: 467–469.

Weon, H.Y., H.J. Noh, J.A. Son, H.B. Jang, B.Y. Kim, S.W. Kwon and E. Stackebrandt. 2008b. *Rudanella lutea* gen. nov., sp. nov., isolated from an air sample in Korea. Int. J. Syst. Evol. Microbiol. *58*: 474–478.

Willems, A. and M.D. Collins. 1995. 16S ribosomal RNA gene similarities indicate that *Hallella seregens* (Moore and Moore) and *Mitsuokella dentalis* (Haapasalo *et al.*) are genealogically highly related and are members of the genus *Prevotella*: emended description of the genus *Prevotella* (Shah and Collins) and description of *Prevotella dentalis* comb. nov. Int. J. Syst. Bacteriol. *45*: 832–836.

Yang, S.J., Y.J. Choo and J.C. Cho. 2007. *Lutimonas vermicola* gen. nov., sp. nov., a member of the family *Flavobacteriaceae* isolated from the marine polychaete *Periserrula leucophryna*. Int. J. Syst. Evol. Microbiol. *57*: 1679–1684.

Yoo, S.H., H.Y. Weon, H.B. Jang, B.Y. Kim, S.W. Kwon, S.J. Go and E. Stackebrandt. 2007. *Sphingobacterium composti* sp. nov., isolated from cotton-waste composts. Int. J. Syst. Evol. Microbiol. *57*: 1590–1593.

Yoon, J., S. Ishikawa, H. Kasai and A. Yokota. 2007a. *Perexilibacter aurantiacus* gen. nov., sp. nov., a novel member of the family '*Flammeovirgaceae*' isolated from sediment. Int. J. Syst. Evol. Microbiol. *57*: 964–968.

Yoon, J., S. Ishikawa, H. Kasai and A. Yokota. 2007b. *Persicitalea jodogahamensis* gen. nov., sp. nov., a marine bacterium of the family '*Flexibacteraceae*', isolated from seawater in Japan. Int. J. Syst. Evol. Microbiol. *57*: 1014–1017.

Yoon, J., Y. Matsuo, K. Adachi, M. Nozawa, S. Matsuda, H. Kasai and A. Yokota. 2008a. Description of *Persicirhabdus sediminis* gen. nov., sp. nov., *Roseibacillus ishigakijimensis* gen. nov., sp. nov., *Roseibacillus ponti* sp. nov., *Roseibacillus persicicus* sp. nov., *Luteolibacter pohnpeiensis* gen. nov., sp. nov. and *Luteolibacter algae* sp. nov., six marine members of the phylum '*Verrucomicrobia*', and emended descriptions of the class *Verrucomicrobiae*, the order *Verrucomicrobiales* and the family *Verrucomicrobiaceae*. Int. J. Syst. Evol. Microbiol. *58*: 998–1007.

Yoon, J., Y. Matsuo, H. Kasai and A. Yokota. 2008b. *Limibacter armeniacum* gen. nov., sp. nov., a novel representative of the family '*Flammeovirgaceae*' isolated from marine sediment. Int. J. Syst. Evol. Microbiol. *58*: 982–986.

Yoon, J., Y. Matsuo, A. Katsuta, J.H. Jang, S. Matsuda, K. Adachi, H. Kasai and A. Yokota. 2008c. *Haloferula rosea* gen. nov., sp. nov., *Haloferula harenae* sp. nov., *Haloferula phyci* sp. nov., *Haloferula helveola* sp. nov. and *Haloferula sargassicola* sp. nov., five marine representatives of the family *Verrucomicrobiaceae* within the phylum '*Verrucomicrobia*'. Int. J. Syst. Evol. Microbiol. *58*: 2491–2500.

Yoon, J.H., S.J. Kang, Y.T. Jung and T.K. Oh. 2008d. *Aestuariicola saemankumensis* gen. nov., sp. nov., a member of the family *Flavobacteriaceae*, isolated from tidal flat sediment. Int. J. Syst. Evol. Microbiol. *58*: 2126–2131.

Yoon, M.H. and W.T. Im. 2007. *Flavisolibacter ginsengiterrae* gen. nov., sp. nov. and *Flavisolibacter ginsengisoli* sp. nov., isolated from ginseng cultivating soil. Int. J. Syst. Evol. Microbiol. *57*: 1834–1839.

Zchori-Fein, E., S.J. Perlman, S.E. Kelly, N. Katzir and M.S. Hunter. 2004. Characterization of a '*Bacteroidetes*' symbiont in *Encarsia* wasps (Hymenoptera: Aphelinidae): proposal of '*Candidatus* Cardinium hertigii'. Int. J. Syst. Evol. Microbiol. *54*: 961–968.

Zhao, J.S., D. Manno and J. Hawari. 2009. *Psychrilyobacter atlanticus* gen. nov., sp. nov., a marine member of the phylum *Fusobacteria* that produces H_2 and degrades nitramine explosives under low temperature conditions. Int. J. Syst. Evol. Microbiol. *59*: 491–497.

Taxonomic outlines of the phyla *Bacteroidetes*, *Spirochaetes*, *Tenericutes* (*Mollicutes*), *Acidobacteria*, *Fibrobacteres*, *Fusobacteria*, *Dictyoglomi*, *Gemmatimonadetes*, *Lentisphaerae*, *Verrucomicrobia*, *Chlamydiae*, and *Planctomycetes*

Wolfgang Ludwig, Jean Euzéby and William B. Whitman

All taxa recognized within this volume of the rank of genus and above are listed below. Within each classification, the nomenclatural type is listed first followed by the remaining taxa in alphabetical order. Taxa appearing on the Approved Lists are denoted by the superscript AL. Taxa that were otherwise validly published are denoted by the superscript VP. Taxa that have not been validly published are presented in quotations. Taxa which were described after the deadline of July 1, 2006, and are therefore not included in this volume are indicated by an asterisk (*).

- Phylum XIV. "*Bacteroidetes*"
  - Class I. "*Bacteroidia*"
    - Order I. "*Bacteroidales*"
      - Family I. *Bacteroidaceae*^{AL}
        - Genus I. *Bacteroides*$^{AL(T)}$
        - Genus II. *Acetofilamentum*^{VP}
        - Genus III. *Acetomicrobium*^{VP}
        - Genus IV. *Acetothermus*^{VP}
        - Genus V. *Anaerorhabdus*^{VP}
      - Family II. "*Marinilabiliaceae*"
        - Genus I. *Marinilabilia*$^{VP(T)}$
        - Genus II. *Alkaliflexus*^{VP}
        - Genus III. *Anaerophaga*^{VP}
      - Family III. "*Rikenellaceae*"
        - Genus I. *Rikenella*$^{VP(T)}$
        - Genus II. *Alistipes*^{VP}
      - Family IV. "*Porphyromonadaceae*"
        - Genus I. *Porphyromonas*$^{VP(T)}$
        - Genus II. *Barnesiella*^{VP}
        - Genus III. *Dysgonomonas*^{VP}
        - Genus IV. *Paludibacter*^{VP}
        - Genus V. *Parabacteroides*^{VP}*
        - Genus VI. *Petrimonas*^{VP}
        - Genus VII. *Proteiniphilum*^{VP}
        - Genus VIII. *Tannerella*^{VP}
      - Family V. "*Prevotellaceae*"
        - Genus I. *Prevotella*$^{VP(T)}$
        - Genus II. *Xylanibacter*^{VP}
  - Class II. "*Flavobacteriia*"
    - Order I. "*Flavobacteriales*"
      - Family I. *Flavobacteriaceae*^{VP}
        - Genus I. *Flavobacterium*$^{AL(T)}$
        - Genus II. *Actibacter*^{VP}*
        - Genus III. *Aequorivita*^{VP}
        - Genus IV. *Aestuariicola*^{VP}*
        - Genus V. *Algibacter*^{VP}
        - Genus VI. *Aquimarina*^{VP}
        - Genus VII. *Arenibacter*^{VP}
        - Genus VIII. *Bergeyella*^{VP}
        - Genus IX. *Bizionia*^{VP}
        - Genus X. *Capnocytophaga*^{VP}
        - Genus XI. *Cellulophaga*^{VP}
        - Genus XII. *Chryseobacterium*^{VP}
        - Genus XIII. *Cloacibacterium*^{VP}
        - Genus XIV. *Coenonia*^{VP}
        - Genus XV. *Costertonia*^{VP}
        - Genus XVI. *Croceibacter*^{VP}
        - Genus XVII. *Dokdonia*^{VP}
        - Genus XVIII. *Donghaeana*^{VP}
        - Genus XIX. *Elizabethkingia*^{VP}
        - Genus XX. *Empedobacter*^{VP}
        - Genus XXI. *Epilithonimonas*^{VP}
        - Genus XXII. *Flagellimonas*^{VP}*
        - Genus XXIII. *Flaviramulus*^{VP}
        - Genus XXIV. *Formosa*^{VP}
        - Genus XXV. *Fulvibacter*^{VP}*
        - Genus XXVI. *Gaetbulibacter*^{VP}
        - Genus XXVII. *Galbibacter*^{VP}*
        - Genus XXVIII. *Gelidibacter*^{VP}
        - Genus XXIX. *Gillisia*^{VP}
        - Genus XXX. *Gilvibacter*^{VP}*
        - Genus XXXI. *Gramella*^{VP}
        - Genus XXXII. *Joostella*^{VP}*
        - Genus XXXIII. *Kaistella*^{VP}
        - Genus XXXIV. *Kordia*^{VP}
        - Genus XXXV. *Krokinobacter*^{VP}
        - Genus XXXVI. *Lacinutrix*^{VP}
        - Genus XXXVII. *Leeuwenhoekiella*^{VP}
        - Genus XXXVIII. *Leptobacterium*^{VP}*
        - Genus XXXIX. *Lutibacter*^{VP}

Genus XL. *Lutimonas*[VP*]
Genus XLI. *Maribacter*[VP]
Genus XLII. *Mariniflexile*[VP]
Genus XLIII. *Marixanthomonas*[VP*]
Genus XLIV. *Mesoflavibacter*[VP*]
Genus XLV. *Mesonia*[VP]
Genus XLVI. *Muricauda*[VP]
Genus XLVII. *Myroides*[VP]
Genus XLVIII. *Nonlabens*[VP]
Genus XLIX. *Olleya*[VP]
Genus L. *Ornithobacterium*[VP]
Genus LI. *Persicivirga*[VP]
Genus LII. *Polaribacter*[VP]
Genus LIII. *Psychroflexus*[VP]
Genus LIV. *Psychroserpens*[VP]
Genus LV. *Riemerella*[VP]
Genus LVI. *Robiginitalea*[VP]
Genus LVII. *Salegentibacter*[VP]
Genus LVIII. *Salinimicrobium*[VP*]
Genus LIX. *Sandarakinotalea*[VP]
Genus LX. *Sediminibacter*[VP*]
Genus LXI. *Sediminicola*[VP]
Genus LXII. *Sejongia*[VP]
Genus LXIII. *Stenothermobacter*[VP]
Genus LXIV. *Subsaxibacter*[VP]
Genus LXV. *Subsaximicrobium*[VP]
Genus LXVI. *Tamlana*[VP*]
Genus LXVII. *Tenacibaculum*[VP]
Genus LXVIII. *Ulvibacter*[VP]
Genus LXIX. *Vitellibacter*[VP]
Genus LXX. *Wautersiella*[VP]
Genus LXXI. *Weeksella*[VP]
Genus LXXII. *Winogradskyella*[VP]
Genus LXXIII. *Yeosuana*[VP]
Genus LXXIV. *Zeaxanthinibacter*[VP*]
Genus LXXV. *Zhouia*[VP]
Genus LXXVI. *Zobellia*[VP]
Genus LXXVII. *Zunongwangia*[VP*]
Family II. "*Blattabacteriaceae*"
Genus I. *Blattabacterium*[AL(T)]
Family III. *Cryomorphaceae*[VP]
Genus I. *Cryomorpha*[VP(T)]
Genus II. *Brumimicrobium*[VP]
Genus III. *Crocinitomix*[VP]
Genus IV. *Fluviicola*[VP]
Genus V. *Lishizhenia*[VP]
Genus VI. *Owenweeksia*[VP]
Class III. "*Sphingobacteriia*"
Order I. "*Sphingobacteriales*"
Family I. *Sphingobacteriaceae*[VP]
Genus I. *Sphingobacterium*[VP(T)]
Genus II. *Mucilaginibacter*[VP*]
Genus III. *Nubsella*[VP*]
Genus IV. *Olivibacter*[VP*]
Genus V. *Parapedobacter*[VP*]
Genus VI. *Pedobacter*[VP]
Genus VII. *Pseudosphingobacterium*[VP*]
Family II. "*Chitinophagaceae*"
Genus I. *Chitinophaga*[VP(T)]
Genus II. *Flavisolibacter*[VP*]
Genus III. *Niabella*[VP*]
Genus IV. *Niastella*[VP*]
Genus V. *Sediminibacterium*[VP*]
Genus VI. *Segetibacter*[VP*]
Genus VII. *Terrimonas*[VP]
Family III. "*Saprospiraceae*"
Genus I. *Saprospira*[AL(T)]
Genus II. *Aureispira*[VP]
Genus III. *Haliscomenobacter*[AL]
Genus IV. *Lewinella*[VP]
Class IV. "*Cytophagia*"
Order I. *Cytophagales*[AL(T)]
Family I. *Cytophagaceae*[AL]
Genus I. *Cytophaga*[AL(T)]
Genus II. *Adhaeribacter*[VP]
Genus III. *Arcicella*[VP]
Genus IV. *Dyadobacter*[VP]
Genus V. *Effluviibacter*[VP]
Genus VI. *Emticicia*[VP]
Genus VII. *Flectobacillus*[AL]
Genus VIII. *Flexibacter*[AL]
Genus IX. *Hymenobacter*[VP]
Genus X. *Larkinella*[VP]
Genus XI. *Leadbetterella*[VP]
Genus XII. *Meniscus*[AL]
Genus XIII. *Microscilla*[AL]
Genus XIV. *Persicitalea*[VP*]
Genus XV. *Pontibacter*[VP]
Genus XVI. *Rudanella*[VP*]
Genus XVII. *Runella*[AL]
Genus XVIII. *Spirosoma*[AL]
Genus XIX. *Sporocytophaga*[AL]
Family II. "*Cyclobacteriaceae*"
Genus I. *Cyclobacterium*[VP(T)]
Genus II. *Algoriphagus*[VP]
Genus III. *Aquiflexum*[VP]
Genus IV. *Belliella*[VP]
Genus V. *Echinicola*[VP]
Genus VI. *Rhodonellum*[VP]
Family III. "*Flammeovirgaceae*"
Genus I. *Flammeovirga*[VP(T)]
Genus II. *Fabibacter*[VP]
Genus III. *Flexithrix*[AL]
Genus IV. *Fulvivirga*[VP*]
Genus V. *Limibacter*[VP*]
Genus VI. *Perexilibacter*[VP*]
Genus VII. *Persicobacter*[VP]
Genus VIII. *Rapidithrix*[VP*]
Genus IX. *Reichenbachiella*[VP]
Genus X. *Roseivirga*[VP]
Genus XI. *Sediminitomix*[VP*]
Order II. *Incertae sedis*
Family I. "*Rhodothermaceae*"
Genus I. *Rhodothermus*[VP(T)]
Genus II. *Salinibacter*[VP]
Order III. *Incertae sedis*
Genus I. *Balneola*[VP*]
Order IV. *Incertae sedis*

- Genus I. *Thermonema*[VP]
- Order V. *Incertae sedis*
  - Genus I. *Toxothrix*[AL]

Phylum XV. "*Spirochaetes*"
- Class I. "*Spirochaetia*"
  - Order I. *Spirochaetales*[AL(T)]
    - Family I. *Spirochaetaceae*[AL]
      - Genus I. *Spirochaeta*[AL(T)]
      - Genus II. *Borrelia*[AL]
      - Genus III. *Cristispira*[AL]
      - Genus IV. *Treponema*[AL]
    - Family II. "*Brachyspiraceae*"
      - Genus I. *Brachyspira*[VP(T)]
    - Family III. "*Brevinemataceae*"
      - Genus I. *Brevinema*[VP(T)]
    - Family IV. *Leptospiraceae*[AL]
      - Genus I. *Leptospira*[AL(T)]
      - Genus II. *Leptonema*[VP]
      - Genus III. *Turneriella*[VP]
    - Family V. *Incertae sedis*
      - Genus I. *Clevelandina*[VP]
      - Genus II. *Diplocalyx*[VP]
      - Genus III. *Hollandina*[VP]
      - Genus IV. *Pillotina*[VP]

Phylum XVI. *Tenericutes*[VP]
- Class I. *Mollicutes*[AL]
  - Order I. *Mycoplasmatales*[AL(T)]
    - Family I. *Mycoplasmataceae*[AL]
      - Genus I. *Mycoplasma*[AL(T)]
      - Genus II. *Ureaplasma*[VP]
    - Family II. *Incertae sedis*
      - Genus I. *Eperythrozoon*[AL]
      - Genus II. *Haemobartonella*[AL]
  - Order II. *Entomoplasmatales*[VP]
    - Family I. *Entomoplasmataceae*[VP]
      - Genus I. *Entomoplasma*[VP(T)]
      - Genus II. Mesoplasma[VP]
    - Family II. *Spiroplasmataceae*[VP]
      - Genus I. *Spiroplasma*[AL(T)]
  - Order III. *Acholeplasmatales*[VP]
    - Family I. *Acholeplasmataceae*[AL]
      - Genus I. *Acholeplasma*[AL(T)]
    - Family II. *Incertae sedis*
      - Genus I. "*Candidatus* Phytoplasma"
  - Order IV. *Anaeroplasmatales*[VP]
    - Family I. *Anaeroplasmataceae*[VP]
      - Genus I. *Anaeroplasma*[AL(T)]
      - Genus II. *Asteroleplasma*[VP]

Phylum XVII. "*Acidobacteria*"
- Class I. "*Acidobacteriia*"
  - Order I. "*Acidobacteriales*"[(T)]
    - Family I. "*Acidobacteriaceae*"
      - Genus I. *Acidobacterium*[VP(T)]
      - Genus II. *Edaphobacter*[VP]
      - Genus III. *Terriglobus*[VP]
- Class II. *Holophagae*[VP]
  - Order I. *Holophagales*[VP(T)]
    - Family I. *Holophagaceae*[VP]
      - Genus I. *Holophaga*[VP(T)]
      - Genus II. *Geothrix*[VP]
  - Order II. *Acanthopleuribacterales*[VP]
    - Family I. *Acanthopleuribacteraceae*[VP]
      - Genus I. *Acanthopleuribacter*[VP(T)]

Phylum XVIII. "*Fibrobacteres*"
- Class I. "*Fibrobacteria*"
  - Order I. "*Fibrobacterales*"[(T)]
    - Family I. "*Fibrobacteraceae*"
      - Genus I. *Fibrobacter*[VP(T)]

Phylum XIX. "*Fusobacteria*"
- Class I. "*Fusobacteriia*"
  - Order I. "*Fusobacteriales*"[(T)]
    - Family I. "*Fusobacteriaceae*"
      - Genus I. *Fusobacterium*[AL(T)]
      - Genus II. *Cetobacterium*[VP]
      - Genus III. *Ilyobacter*[VP]
      - Genus IV. *Propionigenium*[VP]
      - Genus V. *Psychrilyobacter*[VP*]
    - Family II. "*Leptotrichiaceae*"
      - Genus I. *Leptotrichia*[AL(T)]
      - Genus II. *Sebaldella*[VP]
      - Genus III. *Sneathia*[VP]
      - Genus IV. *Streptobacillus*[AL]

Phylum XX. "*Dictyoglomi*"
- Class I. "*Dictyoglomia*"
  - Order I. "*Dictyoglomales*"[(T)]
    - Family I. "*Dictyoglomaceae*"
      - Genus I. *Dictyoglomus*[VP(T)]

Phylum XXI. *Gemmatimonadetes*
- Class I. *Gemmatimonadetes*[VP]
  - Order I. *Gemmatimonadales*[VP(T)]
    - Family I. *Gemmatimonadaceae*[VP]
      - Genus I. *Gemmatimonas*[VP(T)]

Phylum XXII. *Lentisphaerae*
- Class I. "*Lentisphaeria*"
  - Order I. *Lentisphaerales*[VP(T)]
    - Family I. "*Lentisphaeraceae*"
      - Genus I. *Lentisphaera*[VP(T)]
  - Order II. *Victivallales*[VP]
    - Family I. "*Victivallaceae*"
      - Genus I. *Victivallis*[VP(T)]

Phylum XXIII. "*Verrucomicrobia*"
- Class I. *Verrucomicrobiae*[VP]
  - Order I. *Verrucomicrobiales*[VP(T)]
    - Family I. *Verrucomicrobiaceae*[VP]
      - Genus I. *Verrucomicrobium*[VP(T)]
      - Genus II. *Prosthecobacter*[VP]
    - Family II. "*Akkermansiaceae*"
      - Genus I. *Akkermansia*[VP(T)]
    - Family III. "*Rubritaleaceae*"
      - Genus I. *Rubritalea*[VP(T)]
      - Genus II. *Persicirhabdus*[VP*]
      - Genus III. *Roseibacillus*[VP*]
- Class II. *Opitutae*[VP]
  - Order I. *Opitutales*[VP(T)]
    - Family I. *Opitutaceae*[VP]
      - Genus I. *Opitutus*[VP(T)]
      - Genus II. *Alterococcus*[VP]

Order II. *Puniceicoccales*VP
Family I. *Puniceicoccaceae*VP
Genus I. *Puniceicoccus*VP(T)
Genus II. *Cerasicoccus*VP
Genus III. *Coraliomargarita*VP
Genus IV. *Pelagicoccus*VP
Class III. "*Spartobacteria*"
Order I. "*Chthoniobacterales*"(T)
Family I. "*Chthoniobacteraceae*"
Genus I. "*Chthoniobacter*"(T)
Genus II. "*Candidatus* Xiphinematobacter"

Phylum XXIV. "*Chlamydiae*"
Class I. "*Chlamydiia*"
Order I. *Chlamydiales*AL(T)
Family I. *Chlamydiaceae*AL
Genus I. *Chlamydia*AL(T)
Genus II. *Chlamydophila*VP
Family II. "*Clavichlamydiaceae*"
Genus I. "*Candidatus* Clavichlamydia"
Family III. "*Criblamydiaceae*"
Genus I. "*Criblamydia*"(T)
Family IV. *Parachlamydiaceae*VP
Genus I. *Parachlamydia*VP(T)
Genus II. *Neochlamydia*VP
Genus III. "*Protochlamydia*"
Family V. "*Piscichlamydiaceae*"
Genus I. "*Candidatus* Piscichlamydia"(T)
Family VI. "*Rhabdochlamydiaceae*"
Genus I. "*Candidatus* Rhabdochlamydia"VP(T)
Family VII. *Simkaniaceae*VP
Genus I. *Simkania*VP(T)
Genus II. "*Candidatus* Fritschea"VP
Family VIII. *Waddliaceae*VP
Genus I. *Waddlia*VP(T)

Phylum XXV. "*Planctomycetes*"
Class I. "*Planctomycetia*"
Order I. *Planctomycetales*VP(T)
Family I. *Planctomycetaceae*VP
Genus I. *Planctomyces*AL(T)
Genus II. *Blastopirellula*VP
Genus III. *Gemmata*VP
Genus IV. *Isosphaera*VP
Genus V. *Pirellula*VP
Genus VI. *Rhodopirellula*VP
Genus VII. *Schlesneria*VP
Genus VIII. *Singulisphaera*VP
Order II. "*Brocadiales*"
Family I. "*Brocadiaceae*"
Genus I. "*Candidatus* Brocadia"VP(T)
Genus II. "*Candidatus* Anammoxoglobus"
Genus III. "*Candidatus* Jettenia"
Genus IV. "*Candidatus* Kuenenia"
Genus V. "*Candidatus* Scalindua"

Acknowledgements

Jean-François Bernardet, Brian Hedlund, Noel Krieg, Olga Nedashkovskaya, Bruce Paster, Bernhard Schink, Jim Staley and Naomi Ward provided valuable advice concerning the preparation of this outline.

Phylum XIV. Bacteroidetes phyl. nov.

Noel R. Krieg, Wolfgang Ludwig, Jean Euzéby and William B. Whitman

Bac.te.ro.i.de′tes. N.L. fem. pl. n. *Bacteroidales* type order of the phylum; N.L. fem. pl. n. *Bacteroidetes* the phylum of *Bacteroidales*.

The phylum *Bacteroidetes* is a phenotypically diverse group of Gram-stain-negative rods that do not form endospores. They are circumscribed for this volume on the basis of phylogenetic analysis of 16S rRNA gene sequences. The phylum contains four classes, *Bacteroidia*, *Cytophagia*, *Flavobacteriia*, and *Sphingobacteriia*. In addition, the genera *Rhodothermus*, *Salinibacter*, and *Thermonema* appear to represent deep groups of the phylum that can not be readily assigned to any of the four classes.

Type order: **Bacteroidales** ord. nov.

Class I. **Bacteroidia** class. nov.

Noel R. Krieg

Bac.te.ro.i.di′a. N.L. masc. n. *Bacteroides* type genus of the type order *Bacteroidales*; suff. *-ia* ending proposed by Gibbons and Murray and by Stackebrandt et al. to denote a class; N.L. neut. pl. n. *Bacteroidia* the *Bacteroidales* class.

The class presently contains one order, *Bacteroidales*. The description of the class is the same as that given for the order.

Type order: **Bacteroidales** ord. nov.

Order I. **Bacteroidales** ord. nov.

Noel R. Krieg

Bac.te.ro.i.da′les. N.L. masc. n. *Bacteroides* type genus of the order; suff. *-ales* ending to denote an order; N.L. fem. pl. n. *Bacteroidales* the *Bacteroides* order.

The order presently includes the families *Bacteroidaceae*, *Marinilabiliaceae*, *Porphyromonadaceae*, *Prevotellaceae*, and *Rikenellaceae*. Straight, fusiform or thin **rods and coccobacilli** that **stain Gram-negative**. **Nonsporeforming**. **Mostly anaerobic**, although some are facultatively anaerobic, and saccharolytic, although proteins and other substrates may be utilized. **Nonmotile or motile by gliding**.

Type genus: **Bacteroides** Castellani and Chalmers 1919, 959[AL].

Reference

Castellani, A. and A.J. Chalmers. 1919. Manual of Tropical Medicine, 3rd edn. Williams Wood and Co., New York, pp. 959–960.

Family I. **Bacteroidaceae** Pribam 1933, 10[AL]

The Editorial Board

Bac.te.ro.i.da.ce′a.e. N.L. masc. n. *Bacteroides* type genus of the family; suff. *-aceae* ending to denote a family; N.L. fem. pl. n. *Bacteroidaceae* the *Bacteroides* family.

The family *Bacteroidaceae* was circumscribed for this volume on the basis of phylogenetic analysis of 16S rRNA gene sequences; the family contains the genera *Acetofilamentum*, *Acetomicrobium*, *Acetothermus*, *Anaerorhabdus*, and *Bacteroides*. The genera *Anaerophaga* and *Megamonas*, which were previously classified with this family, have been transferred to the family *Marinilabiliaceae* and the *Firmicutes*, respectively (Morotomi et al., 2007). **Straight rods** that **stain Gram-negative**. **Anaerobic**. **Nonsporeforming**. Some characteristics that differentiate the genera are given in Table 1.

Type genus: **Bacteroides** Castellani and Chalmers 1919, 959[AL].

TABLE 1. Some characteristics that differentiate the genera of the family *Bacteroidaceae*[a]

Characteristic	*Bacteroides*	*Acetofilamentum*	*Acetomicrobium*	*Acetothermus*	*Anaerorhabdus*
Cell size (μm)	0.5–2.0 × 1.6–12	0.18–0.3 × 2–8	0.6–0.8 × 2–7	0.5 × 5–8	0.3–1.5 × 1–3
Motility	–	–	+	–	–
Saccharolytic	+	+	+	+	–
Growth occurs below 50°C	+	+	+	–	+
Major products of glucose fermentation	Succinate and acetate; trace to moderate amounts of isobutyrate and isovalerate	Acetate and H_2 (in a molar ratio of 1:2) and CO_2	Acetate and H_2 or acetate, lactate, ethanol, and H_2	Acetate and H_2 in a molar ratio of 1:1:2.	Acetate and lactate
Isolated from	Major constituents of the normal human colonic flora; also found in smaller numbers in the female genital tract; not common in the mouth or upper respiratory tract	Secondary sewage sludge	Sewage sludge	Thermophilically fermenting sewage sludge	Infected appendix, lung abscesses and abdominal abscesses; infrequently from human and pig feces
DNA G+C content (mol%)	39–49	~47	47	~62	34

[a]Symbols: +, >85% positive; –, 0–15% positive.

Genus I. **Bacteroides** Castellani and Chalmers 1919, 959^AL emend. Shah and Collins 1989, 85

YULI SONG, CHENGXU LIU AND SYDNEY M. FINEGOLD

Bac.te.ro.i'des. N.L. n. *bacter* rod; L. suff. *-oides* (from Gr. suff. *-eides*, from Gr. n. *eidos* that which is seen, form, shape, figure), resembling, similar; N.L. masc. n. *Bacteroides* rodlike.

Rod-shaped cells with rounded ends. Gram-stain-negative. Cells are fairly uniform if smears are prepared from young cultures on blood agar. Nonmotile. **Anaerobic.** Colonies are 1–3 mm in diameter, smooth, white to gray, and nonhemolytic on blood agar. Chemo-organotrophic. **Saccharolytic.** Weakly proteolytic. Most species grow in the presence of 20% bile, but are not always stimulated. Esculin is usually hydrolyzed. Nitrate is not reduced to nitrite. Indole variable. **Major fermentation products are succinate and acetate.** Trace to moderate amounts of isobutyrate and isovalerate may be produced. **Predominant cellular fatty acid is $C_{15:0}$ anteiso.**

DNA G+C content (mol%): 39–49.

Type species: **Bacteroides fragilis** (Veillon and Zuber 1898) Castellani and Chalmers 1919, 959^AL.

Further descriptive information

As defined in *Bergey's Manual of Systematic Bacteriology*, 1st edition, the genus *Bacteroides* comprised more than 60 species (Holdeman et al., 1984). These species exhibited a variety of cellular morphologies and were biochemically and physiologically extremely heterogeneous. Therefore, Shah and Collins (1989) proposed that this genus should be restricted to *Bacteroides fragilis* and closely related organisms. The application of molecular biology techniques such as DNA–DNA hybridization and 16S rRNA gene sequencing has done much to clarify the inter- and intrageneric structure of *Bacteroides*. 16S rRNA gene sequence analysis indicates that the genus *Bacteroides* is equivalent to the *Bacteroides* cluster within the *Bacteroides* subgroup of the "*Bacteroidetes*" (previously referred as the phylum *Cytophaga–Flavobacteria–Bacteroides*) (Paster et al., 1994; Woese, 1987). The currently described species that conform to the emended description of the genus *Bacteroides* based on biochemical, chemical, and genetic criteria include the members of the "*Bacteroides fragilis* group" [not a formal taxonomic group, but the ten species conforming to the 1989 proposal (Shah and Collins) to restrict the genus] and several other later described species (Figure 14). Other *Bacteroides* species with validly published names that do not conform to the emended generic description require further study (Table 5).

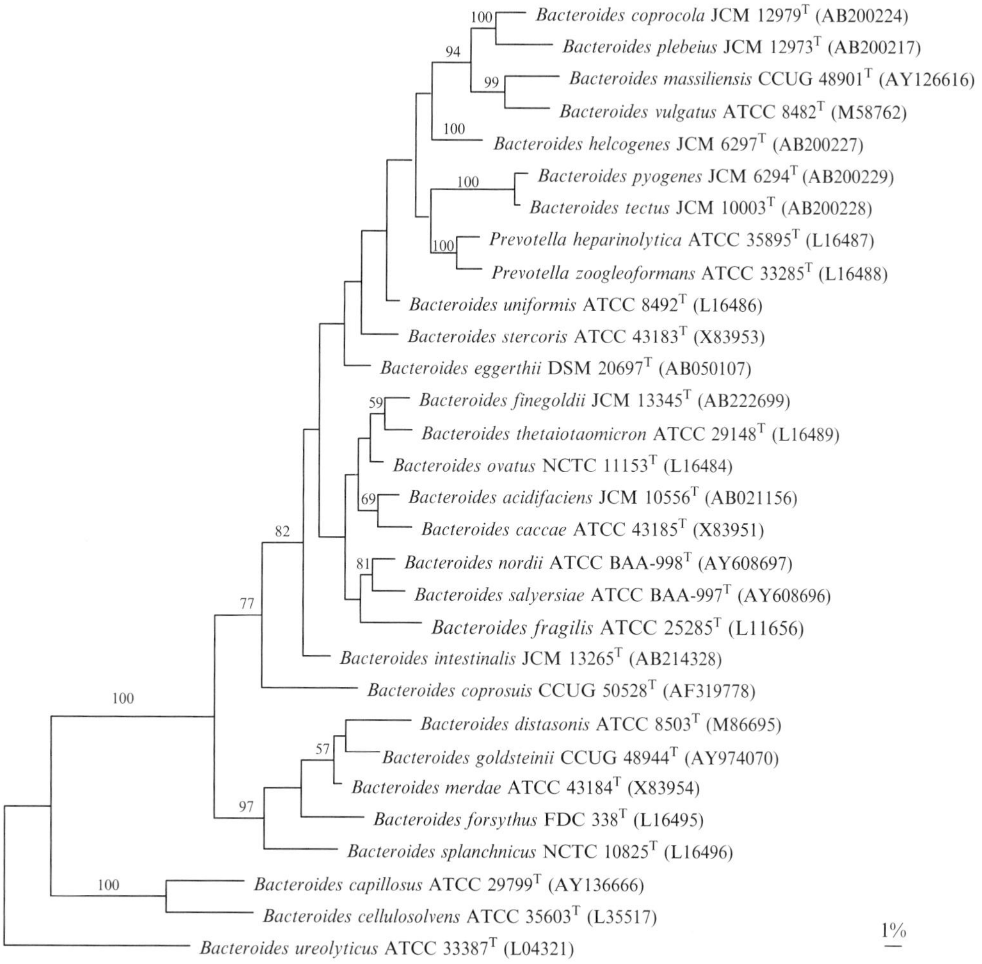

FIGURE 14. Phylogenetic tree based on 16S rRNA gene sequence comparisons (>1300 aligned bases) showing the phylogenetic relationships of *Bacteroides* species. The tree, constructed using the neighbor joining method, was based on a comparison of approximately 1400 nt. Bootstrap values, expressed as percentages of 1000 replications, are given at branching points. Bar = 1% sequence divergence.

TABLE 2. Descriptive characteristics of different *Bacteroides* species[a,b]

Characteristic	1. *B. fragilis*	2. *B. acidifaciens*	3. *B. caccae*	4. *B. coprocola*	5. *B. coprosuis*	6. *B. distasonis*	7. *B. dorei*	8. *B. eggerthii*	9. *B. finegoldii*	10. *B. goldsteinii*	11. *B. helcogenes*
Cell size (μm)	0.8–1.3 × 1.6–8.0	0.8–1.3 × 1.6–8.0	1.4–1.6 × 2.5–12	0.8–1.0 × 4.0	0.8–3.0 × 0.5–1.5	1.3–1.9 × 1.6–5.0	1.6–4.2 × 0.8–1.2	0.4–1.0 × 1.0–6.0	0.8 × 1.5–4.5	0.9–1.5 × 1.2–10	0.5–0.6 × 0.8–4.0
Growth in 20% bile	+	+	+	+	+	+	+	+	+	+	w
Production of:											
Indole	–	v	–	+	–	–	–	+	–	–	–
Catalase	+	nr	d	–	–	d	nr	–	nr	nr	–
Starch hydrolyzed	+	d	w	nr	+	+	nr	w	nr	nr	+
Gelatin digested	w	nr	w	–	–	w	nr	–	–	nr	–
G6PDH and 6PGDH	+	+	+	nr	nr	+	nr	+	nr	nr	nr
Acid produced from:											
Arabinose	–	nr	+	–	–	d	+	+	+	–	–
Cellobiose	d	nr	d	+	–	+	–	d	+	+	+
Xylan	–	nr	–	nr	nr	–	nr	+	nr	–	+
Glucose	+	nr	+	+	+	+	+	+	+	+	+
Glycogen	+	nr	w	nr	nr	–	nr	+	nr	nr	+
Inositol	–	nr	nr	nr	nr	–	nr	–	nr	nr	nr
Lactose	+	nr	nr	+	nr	+	+	+	+	nr	+
Maltose	+	nr	nr	+	+	+	+	+	+	nr	+
Mannitol	–	nr	nr	–	nr	–	–	–	–	nr	nr
Mannose	+	nr	nr	+	nr	+	+	+	+	+	+
Melezitose	–	nr	d	–	nr	+	–	–	–	nr	–
Melibiose	w	nr	+	nr	nr	+	nr	–	nr	nr	+
Raffinose	+	nr	+	+	nr	+	+	–	+	+	+
Rhamnose	–	nr	d	+	nr	d	+	w	+	+	nr
Ribose	–	nr	+	nr	nr	d	nr	–	nr	nr	nr
Salicin	–	nr	d	+	nr	+	–	–	+	–	+
Sucrose	+	nr	+	+	nr	+	+	–	nr	+	+
Trehalose	–	nr	+	–	nr	+	–	–	–	–	–
Xylose	+	nr	+	+	–	+	+	+	+	+	+
Enzyme activity:[d]											
α-Galactosidase	+	nr	+	+	w	+	+	+	+	+	nr
β-Galactosidase	+	nr	+	+	+	+	+	+	+	+	nr
α-Glucosidase	+	nr	w	+	+	+	+	+	+	+	nr
β-Glucosidase	+	nr	w	+	+	+	+	+	+	–	nr
α-Arabinosidase	–	nr	+	+	–	+	+	+	+	–	nr
β-Glucuronidase	–	nr	–	–	–	–	+	–	–	–	+
N-Acetyl-β-glucosaminidase	+	nr	+	+	+	+	+	+	+	+	nr
Glutamic acid decarboxylase	+	nr	w	+	nr	+	+	+	+	–	v
α-Fucosidase	+	nr	+	+	+	–	+	–	–	–	nr
Alkaline phosphatase	+	nr	+	+	+	+	+	+	+	+	nr
Arginine arylamidase	+	nr	+	–	+	+	+	–	–	+	–
Leucine arylamidase	+	nr	+	–	–	+	+	–	–	+	–
Glutamyl glutamic acid arylamidase	+	nr	+	+	+	+	+	+	+	+	nr
Glycine arylamidase	–	nr	–	–	–	+	+	–	–	+	–
Leucyl glycine arylamidase	+	nr	+	+	+	+	+	+	+	+	nr
Alanine arylamidase	+	nr	+	+	+	+	+	+	+	+	nr
Major metabolic end product(s)[e]	A, S, p, pa (ib, iv, l)	A, S	A, S, p, (iv)	nr	A, S, p	A, S, p (pa, ib, iv, l)	nr	A, S, p (pa, ib, iv, l)	nr	A, S, p (iv, f)	A, S (p, iv)

[a]Symbols: +, >85% positive; d, different strains give different reactions (16–84% positive); –, 0–15% positive; $+^{-}$, usually positive, sometimes negative; w, weak reaction; nr, not reported.

[b]Esculin is usually hydrolyzed and nitrate is not reduced to nitrite.

[c]Type strain is negative.

[d]Enzyme activity tested using API kits.

[e]Fermentation products: A, acetic acid; B, butyric acid; P, propionic acid; PA, phenylacetic acid; S, succinic acid; ib, isobutyric acid; iv, isovaleric acid; f, formic acid; l, lactic acid. Capital letters indicate >1 meq/100 ml of broth; small letters, <1 meq/100 ml; products in parentheses may or may not be detected.

12. *B. intestinalis*	13. *B. massiliensis*	14. *B. merdae*	15. *B. nordii*	16. *B. ovatus*	17. *B. plebeius*	18. *B. pyogenes*	19. *B. salyersiae*	20. *B. stercoris*	21. *B. suis*	22. *B. tectus*	23. *B. thetaiotaomicron*	24. *B. uniformis*	25. *B. vulgatus*
0.8 × 1.0–5.0	0.8–1.4 × 2.1–3.9	1.0 × 3.1–12	0.8–1.5 × 0.5–5.0	1.3–2.0 × 1.6–5.0	0.8 × 1.0–5.0	0.3–0.5 × 0.5–5.0	0.8–1.5 × 0.5–5.0	1.6 × 2.4–12.6	0.4–0.6 × 0.5–2.0	0.6–0.8 × 1.2–5.0	0.7–2.0 × 1.0–4.0	0.8 × 1.5	0.5–0.8 × 1.5–7.7
+	+	+	+	+	+	w	+	+	w	+	+	d	+
+	–	–	+	+	–	–	+	+	–		+	+	–
nr	–	d	–	d	–	nr	–	–	nr	–	+	d	d
nr	+	+	nr	+	nr	d	nr	+	+	d	+	d	+
–	–	–	nr	d	–	–	nr	–	nr	–	–	–	+
nr	nr	+	nr	nr	nr	nr	nr	+	nr	nr	+	+	+
+	–	d	–	+	+	–	+	d	+	–	+	+	+
+	–	v	+	+	+	d	+	d	+	d	d	+	–
nr	nr	–	–	+	nr	–	–	d	+	nr	–	d	d
+	+	+	+	+	+	+	+	+	+	d	+	+	+
nr	nr	–	nr	d	nr	nr	nr	+	+	nr	+	d	+
nr	nr	nr	nr	–	nr	nr	nr	nr	–	nr	–	–	–
+	+	nr	nr	+	+	+	nr	nr	+	$+^{c}$	+	+	+
+	+	nr	nr	+	+	$+^{-}$	nr	nr	+	$+^{c}$	+	+	+
nr	nr	nr	nr	d	nr	nr	nr	nr	nr	nr	–	–	–
+	+	nr	+	+	+	+	+	nr	nr	–	+	+	+
nr	nr	+	+	$+^{-}$	nr	nr	nr	–	nr	nr	d	–	–
nr	nr	+	+	+	nr	nr	nr	–	+	–	+	w	w
+	+	+	d	+	+	d	+	+	+	–	+	+	+
nr	nr	+	d	w	nr	nr	nr	+	nr	–	+	d	+
nr	nr	v	nr	+	nr	nr	nr	d	nr	nr	w	d	d
nr	nr	+	nr	+	nr	nr	nr	–	+	–	d	d	–
+	+	+	+	+	+	v	+	+	nr	–	+	+	+
nr	nr	+	nr	+	nr	nr	nr	–	–	–	+	–	–
+	–	+	+	+	+	–	+	+	+	–	+	+	+
+	+	+	w	+	+	–	+	–	nr	–	+	+	+
+	+	+	+	+	+	+	+	+	nr	+	+	+	+
+	+	v	+	+	+	w	+	+	nr	+	+	+	+
+	–	v	+	+	+	–	+	w	nr	–	+	+	–
+	–	+	–	+	+	nr	–	–	nr	–	+	+	+
–	–	–	–	–	+	–	–	–	nr	–	–	–	–
+	+	+	+	+	+	w	+	+	nr	+	+	+	+
+	–	–	+	+	+	nr	d	–	nr	nr	+	v	d
+	+	–	–	–	+	w	–	–	nr	+	+	+	d
+	+	+	+	+	+	+	+	+	nr	+	+	+	+
–	–	+	–	–	d	nr	–	–	nr	d	+	–	–
–	+	+	–	–	d	–	–	–	nr	–	+	–	–
+	–	+	+	+	+	nr	+	+	nr	nr	d	+	+
–	–	+	–	–	+	nr	–	–	nr	d	–	–	+
+	+	+	+	+	+	nr	+	+	nr	+	+	+	+
+	+	+	+	+	+	nr	+	+	nr	nr	+	+	+
nr	nr	A, S, p (ib, iv)	A, S (iv, p, f)	A, S, p, pa (ib, iv, l)	nr	A, S (p, ib)	A, S (iv, p, f)	A, S, p, f (iv, ib)	A, S (p, ib)	A, p, S (ib, iv)	A, p, S (ib, iv, l)	a, p, l, S (ib, iv)	A, p, S

Morphology and general characteristics. Microscopically, members of the genus *Bacteroides* are pale-staining, Gram-stain-negative bacilli with rounded ends, occurring singly or in pairs. Cells are approximately 0.5–2.0 × 1.6–12 µm. Cells are uniform, pleomorphic, or vacuolized, traits that are medium- and age-dependent. Colonies on blood agar are 1–3 mm in diameter, smooth, circular, entire, convex, white to gray, and nonhemolytic. According to the characteristic special potency antibiotic identification disk profile of *Bacteroides* species, they are resistant to vancomycin (5 µg), kanamycin (1000 µg), and colistin (10 µg). Good growth on bacteroides bile esculin (BBE) medium and other 20% bile (2% oxgall)-containing media is characteristic of species of the genus *Bacteroides*, with the exception of *Bacteroides helcogenes*, *Bacteroides pyogenes*, and some strains of *Bacteroides uniformis*. On BBE agar, *Bacteroides* species hydrolyze esculin, blackening the agar except for most strains of *Bacteroides vulgatus* and some strains of *Bacteroides pyogenes* and *Bacteroides stercoris* that are esculin-negative. β-Galactosidase, alkaline phosphatase, leucyl glycine arylamidase, and alanine arylamidase activities are detected. Both nonhydroxylated and 3-hydroxylated long-chain fatty acids are present. The nonhydroxylated acids are predominantly of the straight-chain saturated, anteiso- and iso-methyl branched-chain types. Characteristics of the *Bacteroides* species are presented in Table 2.

Nutrition and growth conditions. Carbon dioxide is utilized or required and incorporated into succinic acid (Caldwell et al., 1969). Hemin and vitamin K_1 are required or highly stimulatory for growth and are added routinely to media for *Bacteroides*. Growth of described species is most rapid at 37°C and pH near 7.0.

Antibiotics and drug resistance. The *Bacteroides* species are among the most antibiotic resistant of the anaerobes. They are usually resistant to penicillins and not uncommonly resistant to expanded and broad-spectrum cephalosporins (including relatively β-lactamase-resistant drugs such as cefoxitin) and clindamycin (Wexler and Finegold, 1998). Antimicrobial agents that are active against >99% of clinical isolates of *Bacteroides* include metronidazole, chloramphenicol, and carbapenems; however, strains of the *Bacteroides fragilis* group with resistance to imipenem and metronidazole are also encountered. Agents active against 95–99% of *Bacteroides fragilis* isolates include the β-lactam/β-lactamase inhibitor combinations. *Bacteroides fragilis* group species other than *Bacteroides fragilis* are more likely to be resistant to β-lactam/β-lactamase inhibitor combinations than *Bacteroides fragilis*. The level of chloramphenicol susceptibility remains quite high, whereas almost uniform resistance to aminoglycosides is observed and resistance to quinolones is common. Several multicenter surveys have documented an alarming gradual increase in resistance rates of *Bacteroides* species worldwide (Snydman et al., 2002). Multiresistant *Bacteroides fragilis* group strains, some of which are capable of transferring resistance genes, have been isolated in several countries. Although the rate of *Bacteroides fragilis* resistance to carbapenems or β-lactam/β-lactamase inhibitor combination drugs remains low, some investigators have noted a gradual increase in minimum inhibitory concentrations. Agents such as clindamycin and some cephalosporins, which have traditionally been considered good choices against *Bacteroides fragilis* group strains, are losing activity against these clinically important pathogens. The classical mechanisms of resistance to β-lactams are: (a) production of β-lactamases; (b) alteration of penicillin-binding proteins; and (c) changes in outer membrane permeability to β-lactams. Resistance to clindamycin is mediated by modification of the ribosome. Tetracycline resistance is mediated by both tetracycline efflux and ribosomal protection. 5-Nitroimidazole resistance appears to be caused by a combination of decreased antibiotic uptake and decreased nitroreductase activity (Fang et al., 2002). Two main mechanisms – alteration of target enzymes (gyrase) caused by chromosomal mutations in encoding genes and reduced intracellular accumulation due to increased efflux of the drug – are associated with quinolone resistance in the *Bacteroides fragilis* group (Oh and Edlund, 2003).

Natural habitats and clinical significance. Members of the "*Bacteroides fragilis* group" are major constituents of the normal human colonic flora and are also found in smaller numbers in the female genital tract, but are not common in the mouth or upper respiratory tract. The "*Bacteroides fragilis* group" makes up one-third of all clinical isolates of anaerobes. *Bacteroides fragilis* is the most frequently encountered, with *Bacteroides thetaiotaomicron* second. They are recovered from most intra-abdominal infections and may occur in infections at other sites. *Bacteroides fragilis* strains producing a potent zinc-dependent metalloprotease or enterotoxin with a variety of pathological effects on intestinal mucosal cells have been identified. Enterotoxin-producing *Bacteroides fragilis* strains have been isolated from the intestinal tracts of young farm animals and small children with diarrhea and in cases of extra-abdominal infections including bacteremia, but they have also been isolated from fecal samples from healthy children and adults (Sears et al., 1995). To date, three types of the enterotoxin, each having different virulence and geographical distribution, have been characterized. Other members of the *Bacteroides fragilis* group are present less often in infection.

Enrichment and isolation procedures

A complex medium containing peptone, yeast extract, vitamin K_1, and hemin is recommended for isolation of most species. The use of selective media along with nonselective media will increase yield and save time in terms of isolation of *Bacteroides*. Fresh or pre-reduced media (commercially available from Anaerobe Systems) are recommended, as both increase the initial isolation efficiency. The recommended minimum medium setup for culture of clinical specimens includes: (a) a nonselective, enriched, Brucella base sheep blood agar plate supplemented with vitamin K_1 and hemin; and (b) a BBE agar plate for specimens from below the diaphragm for the selection and presumptive identification of the "*Bacteroides fragilis* group" (and *Bilophila* spp.). When indicated, a phenylethyl alcohol-sheep blood agar plate to prevent overgrowth by aerobic Gram-stain-negative rods and swarming of some clostridia may also be used. After inoculation, the anaerobic plates should be placed immediately in an anaerobic environment containing at least 5% carbon dioxide. Carbon dioxide is required for optimal growth of most of the saccharolytic species. The plates are examined after incubation for 48 h at 35 or 37°C.

Maintenance procedures

The following procedures are those of Jousimies-Somer et al. (2002). Cultures of *Bacteroides* species can be maintained by regularly subculturing. A young, actively growing culture can be put into stock from liquid or solid media by freezing at −70°C or lyophilizing in a well-buffered medium containing no fermentable carbohydrate. Supplemented (vitamin K_1 and hemin) thioglycolate medium or chopped meat broth incubated for 24–48 h, depending on the growth rate of the isolate, can be used to prepare stock cultures. Broth culture (0.5 ml) is added to an equal volume of sterile 20% skim milk or 15% glycerol prepared in an unbreakable screw-capped vial. For toxin-producing organisms, storage under anaerobic conditions at −70°C after a minimal number of subcultures may be important.

Differentiation of the genus *Bacteroides* from other genera

Bacteroides can be differentiated from other known related genera by 16S rRNA gene sequence analysis because the members form a distinct cluster within the *Bacteroides* subgroup with mean interspecies similarities of approximately 93% (Paster et al., 1994). The stability of this cluster has also been verified by bootstrap analysis with a confidence level of 100%. Species of the genus *Bacteroides* can also be distinguished from species of other closely related genera by bile resistance, cellular fatty acid content, fermentation, and pigmentation (Table 3). Members of the genus can be distinguished from those of the most closely related genera *Prevotella* and *Porphyromonas* by their resistance to 20% bile. In addition, *Bacteroides* species are highly fermentative and generally produce acetic and succinic acids as the major end products of glucose metabolism, in contrast to asaccharolytic species in the genus *Porphyromonas* and moderately saccharolytic species in *Prevotella. Bacteroides* species can also be readily distinguished from *Porphyromonas* species by not being pigmented and containing predominantly 12-methyl-tetradecanoic acid ($C_{15:0}$ anteiso) as the long-chain fatty acid, in contrast to pigmented *Porphyromonas* species, which contain mainly 13-methyl-tetradecanoic acid ($C_{15:0}$ iso). All *Bacteroides* species examined so far possess the enzymes glucose-6-phosphate dehydrogenase (G6PDH) and 6-phosphogluconate dehydrogenase (6PGDH), which differs from other genera. Furthermore, the dibasic amino acid of their cell wall peptidoglycan is *meso*-diaminopimelic acid. The principal respiratory quinones are menaquinones with 10 or 11 isoprene units or both. Sphingolipids are produced (Shah and Collins, 1983). Unfortunately, these features have not been tested in many of the recently described *Bacteroides* species. Characteristics that differentiate *Bacteroides* from related taxa within the phylum "*Bacteroidetes*" are presented in Table 3.

Taxonomic comments

The taxonomy of the genus *Bacteroides* has been in a state of great change in recent years. On the basis of 16S rRNA gene sequence analysis, *Bacteroides* forms the "CFB group" with *Cytophaga* and *Flavobacterium* (Woese, 1987). The "CFB group" is divided into the *Cytophaga, Flavobacterium,* and *Bacteroides* subgroups. The *Bacteroides* subgroup is further divided into the *Prevotella* cluster, *Porphyromonas* cluster, *Bacteroides* cluster, and two new unnamed clusters. Accordingly, it is now generally

TABLE 3. Descriptive characteristics of *Bacteroides* and related genera[a]

Characteristic	*Bacteroides*[b]	*Porphyromonas*[c]	*Prevotella*	*Alistipes*[d]	*Tannerella*
Pigment production	−	$+^{-}$	+/−	$+^{-}$	−
Growth in 20% bile	+	−	−	$+^{-}$	−
Susceptibility to:[e]					
Vancomycin (5 μg)	R	S/R	R	R	R
Kanamycin (1 mg)	R	R	S/R	R	S
Colistin (10 μg)	R	R	S/R	R	S
Catalase production	−	−	−	$-^{+}$	−
Indole production	+/−	+/−	+/−	$+^{-}$	−
Nitrate reduction	−	−	−	−	−
G6PDH and 6PGDH dehydrogenase production	+	−	−	−	−
Proteolytic activity	−	+/−	+/−	+	+
Carbohydrate fermentation	+	$-^{+}$	+	$+^{-}$	−
Major metabolic end product(s)[f]	A, S	A, B	A, S	S	A, S, PA
Major cellular fatty acids	$C_{15:0}$ anteiso	$C_{15:0}$ iso	$C_{15:0}$ anteiso	$C_{15:0}$ iso	$C_{15:0}$ anteiso
DNA G+C content (mol%)	40–48	40–55	39–60	55–58	44–48
Type species	*B. fragilis*	*P. asaccharolytica*	*P. melaninogenica*	*A. putredinis*	*T. forsythia*

[a]Symbols: +, positive; −, negative; $-^{+}$, usually negative, sometimes positive; +/−, positive or negative; $+^{-}$, usually positive, sometimes negative.

[b]*Bacteroides sensu stricto.*

[c]Unlike other *Porphyromonas* spp., *P. catoniae* does not produce pigment and is moderately saccharolytic.

[d]Unlike other *Alistipes* spp., *A. putredinis* does not produce pigment, is susceptible to bile, catalase-positive, and nonsaccharolytic.

[e]Special potency antimicrobial identification disks; R, resistant; S, susceptible; S/R, either sensitive or resistant.

[f]A, acetic acid; B, butyric acid; PA, phenylacetic acid; S, succinic acid.

recognized that the majority of the original *Bacteroides* species fall into three genera: *Prevotella* (bile-sensitive, moderately saccharolytic, pigmented and nonpigmented species), *Porphyromonas* (bile-sensitive, pigmented, asaccharolytic species), and *Bacteroides* (bile-resistant, nonpigmented, saccharolytic species) (Shah and Collins, 1988, 1989, 1990). Several other genera have been described subsequently for those clearly unrelated *Bacteroides* taxa that do not conform to these three major groups (e.g., *Alistipes, Anaerorhabdus, Dichelobacter, Fibrobacter, Megamonas, Mitsuokella, Rikenella, Sebaldella, Tannerella,* and *Tissierella*) (Table 4). However, a large number of taxa still remain unclassified and several, such as *Bacteroides capillosus, Bacteroides cellulosolvens, Bacteroides ureolyticus,* and *Bacteroides splanchnicus,* are retained in *Bacteroides* until valid new genera are proposed. The genus *Bacteroides* now contains mainly the species that were formerly described as the "*Bacteroides fragilis* group" (including *Bacteroides eggerthii*), as well as several later described species. It is evident from the phylogenetic data (Figure 14) that three

TABLE 4. Taxonomy changes in the genus *Bacteroides*

Previous name	References	Current name	Reference(s)
Species reclassified into the genus Porphyromonas:			
B. asaccharolyticus	Holdeman and Moore (1970)	*Porphyromonas asaccharolytica*	Shah and Collins (1988)
B. endodontalis	van Steenbergen et al. (1984)	*Porphyromonas endodontalis*	Shah and Collins (1988)
B. gingivalis	Coykendall et al. (1980)	*Porphyromonas gingivalis*	Shah and Collins (1988)
B. levii	Johnson and Holdeman (1983)	*Porphyromonas levii*	Shah et al. (1995)
B. macacae	Slots and Genco (1980)	*Porphyromonas macacae*	Love (1995)
B. salivosus	Love et al. (1987)	*Porphyromonas macacae*	Love (1995)
Species reclassified into the genus Prevotella:			
B. bivius	Holdeman and Johnson (1977)	*Prevotella bivia*	Shah and Collins (1990)
B. ruminicola subsp. *brevis*	Bryant et al. (1958)	*Prevotella brevis*	Avgustin et al. (1997)
B. buccae	Holdeman and Johnson (1982)	*Prevotella buccae*	Shah and Collins (1990)
B. capillus	Kornman and Holt (1981)	*Prevotella buccae*	Shah and Collins (1990)
B. pentosaceus	Shah and Collins (1981)	*Prevotella buccae*	Shah and Collins (1990)
B. buccalis	Shah and Collins (1981)	*Prevotella buccalis*	Shah and Collins (1990)
B. corporis	Johnson and Holdeman (1983)	*Prevotella corporis*	Shah and Collins (1990)
B. denticola	Shah and Collins (1981)	*Prevotella denticola*	Shah and Collins (1990) emend. Wu et al. (1992)
B. disiens	Holdeman and Johnson (1977)	*Prevotella disiens*	Shah and Collins (1990)
B. heparinolyticus	Okuda et al. (1985)	*Prevotella heparinolytica*	Shah and Collins (1990)
B. intermedius	Holdeman and Moore (1970)	*Prevotella intermedia*	Shah and Collins (1990)
B. loescheii	Holdeman and Johnson (1982)	*Prevotella loescheii*	Shah and Collins (1990) emend. Wu et al. (1992)
B. melaninogenicus	Oliver and Wherry (1921) emend. Roy and Kelly (1939)	*Prevotella melaninogenica*	Shah and Collins (1990) emend. Wu et al. (1992)
B. oralis	Loesche et al. (1964)	*Prevotella oralis*	Shah and Collins (1990)
B. oris	Holdeman and Johnson (1982)	*Prevotella oris*	Shah and Collins (1990)
B. oulorum	Shah et al. (1985)	*Prevotella oulorum*	Shah and Collins (1990)
B. ruminicola subsp. *ruminicola*	Bryant et al. (1958)	*Prevotella ruminicola*	Shah and Collins (1990)
B. veroralis	Watabe et al. (1983)	*Prevotella veroralis*	Shah and Collins (1990) emend. Wu et al. (1992)
B. zoogleoformans	Weinberg et al. (1937) emend. Cato et al. (1982)	*Prevotella zoogleoformans*	Shah and Collins (1990) emend. Moore et al. (1994)
Species reclassified into other genera:			
B. amylophilus	Hamlin and Hungate (1956)	*Ruminobacter amylophilus*	Stackebrandt and Hippe (1986)
B. forsythus	Tanner et al. (1986)	*Tannerella forsythia*	Sakamoto et al. (2002)
B. furcosus	Veillon and Zuber (1898)	*Anaerorhabdus furcosa*	Shah and Collins (1986)
B. gracilis	Tanner et al. (1981)	*Campylobacter gracilis*	Vandamme et al. (1995)
B. hypermegas	Harrison and Hansen (1963)	*Megamonas hypermegale*	Shah and Collins (1982b)
B. microfusus	Kaneuchi and Mitsuoka (1978)	*Rikenella microfusus*	Collins et al. (1985)
B. multiacidus	Mitsuoka et al. (1974)	*Mitsuokella multacida*	Shah and Collins (1982a)
B. nodosus	Beveridge (1941)	*Dichelobacter nodosus*	Dewhirst et al. (1990)
B. ochraceus	Prévot et al. (1956)	*Capnocytophaga ochracea*	Leadbetter et al. (1979)
B. pneumosintes	Olitsky and Gates (1921)	*Dialister pneumosintes*	Moore and Moore (1994)
B. praeacutus	Tissier (1908)	*Tissierella praeacuta*	Collins and Shah (1986b)
B. putredinis	Weinberg et al. (1937)	*Alistipes putredinis*	Rautio et al. (2003)
B. succinogenes	Hungate (1950)	*Fibrobacter succinogenes*	Montgomery et al. (1988)
B. termitidis	Sebald (1962)	*Sebaldella termitidis*	Collins and Shah (1986a)

bile-resistant "*Bacteroides fragilis* group" species – *Bacteroides distasonis*, *Bacteroides merdae*, and the recently described *Bacteroides goldsteinii* – cluster close to the bile-sensitive *Tannerella forsythia* (formerly *Bacteroides forsythus*) and display a loose affinity with the first subcluster (the genus *Porphyromonas*), suggesting that a novel genus should be established to accommodate these three species. Sakamoto and Benno (2006) created the genus *Parabacteroides* to encompass three species that were previously classified in *Bacteroides*, viz., *Parabacteroides distasonis* (previously *Bacteroides distasonis*), *Parabacteroides goldsteinii* (previously *Bacteroides goldsteinii*), and *Parabacteroides merdae* (previously *Bacteroides merdae*). A fourth species, *Parabacteroides johnsonii*, was subsequently added (Sakamoto et al., 2007). The description of *Parabacteroides* is as follows (Sakamoto and Benno, 2006). Rods (0.8–1.6 × 1.2–12 μm). Nonmotile. Non-spore-forming. Gram-stain-negative. Obligately anaerobic. Colonies on Eggerth–Gagnon (EG) agar plates are 1–2 mm in diameter, gray to off-white gray, circular, entire, slightly convex, and smooth. Saccharolytic. Major end products are acetic and succinic acids; lower levels of other acids may be produced. Growth occurs in the presence of 20% bile. Esculin is hydrolyzed. Indole is not produced. G6PDH, 6PGDH, malate dehydrogenase, and glutamate dehydrogenase are present, but not α-fucosidase. The principal respiratory quinones are menaquinones MK-9 and MK-10. Both nonhydroxylated and 3-hydroxylated long-chain fatty acids are present. The nonhydroxylated acids are predominantly of the saturated straight chain and anteiso-methyl branched chain types. The DNA G+C content is 43–46 mol%. The type species is *Parabacteroides distasonis* (Eggerth and Gagnon, 1933) Sakamoto and Benno 2006, 1602[VP].

Interestingly, the bile-sensitive oral species *Prevotella heparinolytica* and *Prevotella zoogleoformans* fall, phylogenetically, within the *Bacteroides* cluster. The Subcommittee on Gram-negative Rods of the International Committee on Systematics of Prokaryotes recommended designating them *Bacteroides heparinolytica* and *Bacteroides zoogleoformans*, respectively, but studies other than 16S rRNA gene sequence analysis should be performed before the status of these taxa is finalized.

List of species of the genus *Bacteroides*

1. **Bacteroides fragilis** (Veillon and Zuber 1898) Castellani and Chalmers 1919, 959[AL] [*Bacillus fragilis* Veillon and Zuber 1898, 536; *Fusiformis fragilis* (Veillon and Zuber 1898) Topley and Wilson 1929, 393; *Ristella fragilis* (Veillon and Zuber 1898) Prévot 1938; *Bacteroides fragilis* subsp. *fragilis* (Veillon and Zuber 1898) Castellani and Chalmers 1919, 959]

fra'gi.lis. L. masc. adj. *fragilis* fragile (relating to the brittle colonies that may form under some culture conditions).

Characteristics are as described for the genus and as given in Table 2, with the following additional characteristics. Cells occur singly or in pairs and have rounded ends. Vacuoles are often present, particularly in broth media containing a fermentable carbohydrate. Surface colonies on blood agar are 1–3 mm in diameter, circular, entire, low convex, and translucent to semi-opaque. They often have an internal structure of concentric rings when reviewed by obliquely transmitted light. In general, strains produce no hemolysis on blood agar, although a few strains may be slightly hemolytic, particularly in the area of confluent growth. Growth is often enhanced by 20% bile. Growth may occur at 25 or 45°C. Most strains grow at pH 8.5, but they grow more slowly and less luxuriantly than at pH 7.0. Cells can survive exposure to air for at least 6–8 h. Hemin is either required for growth or markedly stimulates growth. The major fatty acids are $C_{15:0}$ anteiso, $C_{15:0}$ iso, $C_{17:0}$ iso 3-OH, and $C_{16:0}$. Strains have been isolated from various types of human clinical specimens or conditions including appendicitis, peritonitis, endocarditis, bacteremia, perirectal abscess, infected pilonidal cyst, postsurgical wound infections, and lesions of the urogenital tract; occasionally isolated from the mouth and vagina. It is the major obligately anaerobic, Gram-stain-negative bacterium isolated from various infections.

Although *Bacteroides fragilis* strains show little phenotypic variability, the species comprises genetically heterogeneous strains (Johnson, 1978). A DNA–DNA hybridization study by Johnson and Ault (1978) led to the distinction of two DNA homology groups, I and II, with about 80% of strains isolated in clinical studies assigned to homology group I. Similarly, two genotypically distinct *Bacteroides fragilis* groups have been identified on the basis of ribotyping, restriction fragment length polymorphism analysis, PCR-generated fingerprinting, insertion sequence content, 16S rRNA gene sequence alignments, multilocus enzyme electrophoresis, and genetic marker sequence analysis (Gutacker et al., 2000, 2002). One group is characterized by the presence of the *cfiA* gene (encoding a metallo-β-lactamase of Ambler's class B) and the absence of the *cepA* gene (encoding a β-lactamase of Ambler's class A). The second group is characterized by the absence of the *cfiA* gene and of the associated insertion sequences, the frequent presence of the *cepA* gene, and a higher genetic heterogeneity (Podglajen et al., 1995; Ruimy et al., 1996). By including strains obtained from Johnson (1978) in their 16S rRNA gene sequence comparison, Ruimy et al. (1996) showed that the two groups described above could be related to DNA homology groups II and I, respectively. However, data obtained so far are still insufficient to clarify definitively whether the two groups may be considered as two distinct genospecies that have diverged recently, or if they represent two *Bacteroides fragilis* groups not yet separated at the species level, but evolving in this direction.

Putative virulence factors of *Bacteroides fragilis* include attachment mechanisms, relative aerotolerance, extracellular enzyme production, and resistance to complement-mediated killing and phagocytosis. A polysaccharide capsule contributes to this resistance and resistance to T-cell activity (Patrick, 2002). An enterotoxin termed *Bacteroides fragilis* toxin, or BFT, is a recognized virulence factor. BFT has been characterized as a 20-kDa zinc-dependent metalloprotease (Moncrief et al., 1995) that mediates the cleavage of E-cadherin, resulting in an altered morphology of certain human intestinal carcinoma cell lines (particularly

HT29/C1cells), fluid accumulation in ligated lamb ileal loops, and intestinal epithelial cell proliferation. It has been reported that the *bft* gene is contained in a 6-kb pathogenicity island termed the *Bacteroides fragilis* pathogenicity island or BfPAI (Franco et al., 1999).

The genomes of the *Bacteroides fragilis* type strain, NCTC 9343^T, and a clinical strain, YCH46, have been sequenced. The genome of *Bacteroides fragilis* NCTC 9343^T contains a single circular chromosome of 5,205,140 bp that is predicted to encode 4274 genes and a plasmid, pBF9343. It shows considerable variation, even within the same strain, and "invertable promoters" appear to regulate much of this variance in many surface-exposed and secreted proteins (Cerdeno-Tarraga et al., 2005). The complete genome sequence has revealed an unusual breadth (in number and in effect) of DNA inversion events that potentially control expression of many different components, including surface and secreted components, regulatory molecules, and restriction-modification proteins. This may be related to its niche as a commensal and opportunistic pathogen, because the resulting diversity in surface structures could increase both immune evasion and the ability to colonize novel sites.

Source: human clinical specimens.

DNA G+C content (mol%): 41–44 (HPLC).

Type strain: ATCC 25285, CCUG 4856, CIP 77.16, DSM 2151, JCM 11019, LMG 10263, NCTC 9343.

Sequence accession no. (16S rRNA gene): AB050106, X83935, M11656; genome sequences of strain NCTC 9343^T, NC_003228, CR626927; genome sequences of strain YCH46, NC_006347, AP006841.

2. **Bacteroides acidifaciens** Miyamoto and Itoh 2000, 148^{VP}

a.ci.di'fa.ci.ens. N.L. n. *acidum* (from L. adj. *acidus* sour) acid; L. v. *facio* produce; N.L. part. adj. *acidifaciens* acid-producing.

Characteristics are as described for the genus and as given in Table 2, with the following additional characteristics. After 48 h incubation, surface colonies on EG agar (Mitsuoka et al., 1965) are 1–3 mm in diameter, circular, entire, raised, convex, smooth, and grayish. The specific character of this species is reduction of the pH of pre-reduced anaerobically sterilized peptone-yeast extract (PY) broth with Fildes' digest (PYF) (Fildes et al., 1936) broth without carbohydrate.

Source: mouse cecum.

DNA G+C content (mol%): 39.4–42.4 (HPLC).

Type strain: strain A40, JCM 10556.

Sequence accession no. (16S rRNA gene): AB021164.

3. **Bacteroides caccae** Johnson, Moore and Moore 1986, 499^{VP}

cac'cae. Gr. n. *kakkê* feces; N.L. gen. n. *caccae* of feces, referring to the source of isolate.

Characteristics are as described for the genus and as given in Table 2, with the following additional characteristics. Cells of the type strain from peptone-yeast-glucose (PYG) broth cultures occur singly or in pairs; cells may appear vacuolated or beaded in strains from broth cultures in media with a fermentable carbohydrate. Surface colonies on supplemented brain heart infusion (BHI) blood agar plates (Jousimies-Somer et al., 2002) incubated for 48 h are 0.5–1 mm in diameter, circular, entire, convex, gray, translucent, shiny, and smooth. Rabbit blood may be slightly hemolyzed. Strains grow equally well at 30 and 37°C, but less well at 25 and 45°C. The type strain reduces neutral red and does not produce H_2S. Pyruvate is converted to acetate. Lactate and threonine are not utilized. The major fatty acids are $C_{15:0}$ anteiso, $C_{15:0}$ iso, $C_{17:0}$ iso 3-OH, and $C_{16:0}$.

Source: human feces and blood.

DNA G+C content (mol%): 40–42 (HPLC).

Type strain: ATCC 43185, CCUG 38735, CIP 104201, JCM 9498, NCTC 13051, VPI 3452A.

Sequence accession no. (16S rRNA gene): X83951.

Additional remarks: Previously referred to as "3542A" DNA homology group (Johnson, 1978; Johnson and Ault, 1978).

4. **Bacteroides coprocola** Kitahara, Sakamoto, Ike, Sakata and Benno 2005, 2146^{VP}

co.pro.co'la. Gr. n. *kopros* feces; L. suff. *-cola* (from L. n. *incola*) inhabitant; N.L. n. *coprocola* inhabitant of feces.

Characteristics are as described for the genus and as given in Table 2, with the following additional characteristics. Surface colonies on EG blood agar incubated for 48 h are 1.0 to approximately 3.0 mm in diameter, disc shaped, and grayish white. The major fatty acids are $C_{15:0}$ anteiso, $C_{16:0}$ 3-OH, $C_{16:0}$, $C_{17:0}$ iso 3-OH, and $C_{18:1}$ ω9*c*.

Source: feces of healthy humans.

DNA G+C content (mol%): 42.4 (HPLC).

Type strain: strain M16, DSM 17136, JCM 12979.

Sequence accession no. (16S rRNA gene): AB200224.

5. **Bacteroides coprosuis** Whitehead, Cotta, Collins, Falsen and Lawson 2005, 2517^{VP}

co.pro.su'is. Gr. n. *kopros* feces; L. gen. n. *suis* of a pig; N.L. gen. n. *coprosuis* of pig feces, from which the organism was isolated.

Characteristics are as described for the genus and as given in Table 2, with the following additional characteristics. Cells usually occur singly, but occasionally in pairs in PYG. Colonies grown on BHI agar plates after 48 h incubation are cream-colored, circular, convex, entire, opaque, and reach a diameter of 1 mm. Strains grow at 25–37°C, but not at 42 or 45°C, with an optimum of 37°C. The major fatty acids are $C_{15:0}$ anteiso and $C_{17:0}$ iso 3-OH. Significant amounts of $C_{17:0}$ iso and $C_{15:0}$ iso are also present.

Source: swine feces.

DNA G+C content (mol%): 36.4 (HPLC).

Type strain: PC139, CCUG 50528, NRRL B-41113.

Sequence accession no. (16S rRNA gene): AJ514258, AF319778.

6. **Bacteroides distasonis** Eggerth and Gagnon 1933, 403^{AL} [*Ristella distasonis* (Eggerth and Gagnon 1933) Prévot 1938, 291; *Bacteroides fragilis* subsp. *distasonis* (Eggerth and Gagnon 1933) Holdeman and Moore 1970, 35]

dis.ta.so'nis. N.L. gen. masc. n. *distasonis* of Distaso, named after A. Distaso, a Romanian bacteriologist.

Characteristics are as described for the genus and as given in Table 2, with the following additional characteristics.

In PYG, cells usually occur singly, but occasionally in pairs. Colonies on sheep-blood agar are pinpoint to 0.5 mm in diameter, circular, entire, convex, translucent to opaque, gray-white, soft, smooth, and α-hemolytic (sheep blood). Hemin is required or is highly stimulatory for growth. Resazurin, but not neutral red, is reduced. Propionate is not formed from lactate or threonine. Neither pyruvate nor gluconate is converted to products other than those found after growth in PY. The major fatty acids are $C_{15:0}$ anteiso, $C_{15:0}$ iso, $C_{17:0}$ iso 3-OH, and $C_{16:0}$. Isolated primarily from human feces, where it is one of the most common species; also isolated from human clinical specimens.

This species has been recently reclassified as *Parabacteroides distasonis* Sakamoto and Benno (2006); see *Taxonomic comments*, above).

Source: human feces, human clinical specimens.

DNA G+C content (mol%): 43–45 (HPLC).

Type strain: ATCC 8503, CCUG 4941, CIP 104284, DSM 20701, JCM 5825, NCTC 11152.

Sequence accession no. (16S rRNA gene): M86695.

7. **Bacteroides dorei** Bakir, Sakamoto, Kitahara, Matsumoto and Benno 2006c, 1642[VP]

do.re'i. N.L. gen. masc. n. *dorei* of Doré, in honor of the French microbiologist Joel Doré, in recognition of his many contributions to intestinal (gut) microbiology.

Cells are Gram-stain-negative rods, anaerobic, nonmotile, and non-spore-forming. Typical cells are 1.6–4.2 × 0.8–1.2 µm and occur singly. Colonies on EG agar plates after 48 h incubation at 37°C under 100% CO_2 are circular, whitish, raised, and convex, and attain a diameter of 2.0 mm. Optimum temperature for growth is 37°C. Growth occurs in the presence of bile. Esculin is not hydrolyzed. Nitrate is not reduced. No activity is detected for urease and gelatin. Acid is produced from glucose, sucrose, xylose, rhamnose, lactose, maltose, arabinose, mannose, and raffinose. Acid is not produced from cellobiose, salicin, trehalose, mannitol, glycerol, melezitose, or sorbitol. Positive reactions are obtained using API rapid ID 32A for α-fucosidase, α-galactosidase, β-galactosidase, 6-phospho-β-galactosidase, α-glucosidase, β-glucosidase, α-arabinosidase, β-glucuronidase, *N*-acetyl-β-glucosaminidase, glutamic acid, decarboxylase, alkaline phosphatase, arginine arylamidase, leucyl glycine arylamidase, phenylalanine arylamidase, leucine arylamidase, tyrosine arylamidase, alanine arylamidase, glycine arylamidase, histidine arylamidase, glutamyl glutamic acid arylamidase, and serine arylamidase. Negative reactions are obtained for arginine dihydrolase, and proline arylamidase. Major fatty acids are $C_{15:0}$ anteiso (26–32%), $C_{17:0}$ iso 3-OH (17–19%) and $C_{18:1}$ ω*9c* (9–12%).

Source: human feces.

DNA G+C content (mol%): 43 (HPLC).

Type strain: 175, JCM 13471, DSM 17855 [strain 219 (JCM 13472) is included in this species].

Sequence accession no. (16S rRNA gene): AB242142.

8. **Bacteroides eggerthii** Holdeman and Moore 1974, 260[AL]

eg.gerth'i.i. N.L. gen. masc. n. *eggerthii* of Eggerth, named after Arnold H. Eggerth, an American bacteriologist.

Characteristics are as described for the genus and as given in Table 2, with the following additional characteristics. In PYG broth, cells are pleomorphic rods, ranging from coccoid to large rods with vacuoles or swellings, occurring singly or in pairs. Colonies on blood agar are punctiform, circular, entire convex, translucent, gray-white, shiny, smooth, and nonhemolytic. Hemin markedly stimulates growth, but is not required. Without hemin, malate and lactate are produced; with hemin, succinate and acetate are produced. Vitamin B_{12} is required for production of propionate from succinate. The optimum temperature for growth is 37°C. There is good growth at 30 and 45°C, moderate growth at 25°C. Neutral red is reduced. The major fatty acids are $C_{15:0}$ anteiso, $C_{15:0}$ iso, $C_{17:0}$ iso 3-OH, and $C_{16:0}$.

Source: human feces and occasionally from clinical specimens.

DNA G+C content (mol%): 44–46 (HPLC).

Type strain: ATCC 27754, CCUG 9559, CIP 104285, DSM 20697, NCTC 11155.

Sequence accession no. (16S rRNA gene): AB050107, L16485.

9. **Bacteroides finegoldii** Bakir, Kitahara, Sakamoto, Matsumoto and Benno 2006b, 934[VP]

fine.gold'i.i. N.L. gen. masc. n. *finegoldii* of Finegold, in honor of Sydney M. Finegold, a contemporary researcher in anaerobic bacteriology and infectious diseases.

Characteristics are as described for the genus and as given in Table 2, with the following additional characteristics. Cells occur singly. Surface colonies on EG blood agar plates after 48 h are 1–2 mm in diameter, circular, translucent-whitish, raised, and convex. The major fatty acids are $C_{15:0}$ anteiso and $C_{17:0}$ iso 3-OH.

Source: feces of healthy humans.

DNA G+C content (mol%): 42.4–43 (HPLC).

Type strain: strain 199, JCM 13345, DSM 17565.

Sequence accession no. (16S rRNA gene): AB222699.

10. **Bacteroides goldsteinii** Song, Liu, Lee, Bolanos, Vaisanen and Finegold 2006, 499[VP] (Effective publication: Song, Liu, Lee, Bolanos, Vaisanen and Finegold 2005a, 4526.)

gold.stein'i.i. N.L. gen. masc. n. *goldsteinii* of Goldstein, in honor of an infectious disease clinician who has done much work with anaerobes, Ellie J.C. Goldstein.

Characteristics are as described for the genus and as given in Table 2, with the following additional characteristics. Cells occur singly, occasionally in pairs. Colonies on Brucella blood agar plates at 48 h are gray, circular, convex, entire, opaque, and attain a diameter of 1–2 mm. The major fatty acids are $C_{15:0}$ anteiso and $C_{17:0}$ iso 3-OH. Significant amounts of $C_{18:1}$ω9*c* and $C_{17:0}$ anteiso 3-OH are also present. Habitat is probably the human gut.

This species has been recently reclassified as *Parabacteroides goldsteinii* (Sakamoto and Benno, 2006; see *Taxonomic comments*, above).

Source: human clinical specimens of intestinal origin.

DNA G+C content (mol%): 43 (HPLC).

Type strain: WAL 12034, CCUG 48944, ATCC BAA-1180.

Sequence accession no. (16S rRNA gene): AY974070.

11. **Bacteroides helcogenes** Benno, Watabe and Mitsuoka 1983a, 896[VP] (Effective publication: Benno, Watabe and Mitsuoka 1983b, 404.)

hel.co.ge'nes. Gr. n. *helkos* abscess; N.L. suff. *-genes* (from Gr. v. *gennaô* to produce), producing; N.L. adj. *helcogenes* abscess-producing.

Characteristics are as described for the genus and as given in Table 2, with the following additional characteristics. Cells are single or in pairs. After 48 h of incubation on EG agar, surface colonies are 1.0–2.0 mm in diameter, circular with entire edges, convex, translucent to opaque, gray to gray-white, shiny, and smooth. Most strains do not grow in peptone-yeast-Fildes-glucose (PYFG) broth containing 20% bile. The major fatty acids are $C_{15:0}$ anteiso and $C_{17:0}$ iso 3-OH.

Source: swine abscesses and feces.

DNA G+C content (mol%): 45.3–46.3 (HPLC).

Type strain: strain P 36-108, ATCC 35417, CCUG 15421, DSM 20613, JCM 6297.

Sequence accession no. (16S rRNA gene): AB200227.

12. **Bacteroides intestinalis** Bakir, Kitahara, Sakamoto, Matsumoto and Benno 2006a, 153[VP]

in.tes.ti.na'lis. L. n. *intestinum* gut, intestine; L. masc. suff. *-alis* suffix denoting pertaining to; N.L. masc. adj. *intestinalis* pertaining to the intestine.

Characteristics are as described for the genus and as given in Table 2, with the following additional characteristics. Cells occur singly. Surface colonies on EG blood agar plates after 2 d are 1–3 mm in diameter, circular, translucent-whitish, raised, and convex. The major fatty acids are $C_{15:0}$ anteiso and $C_{17:0}$ iso 3-OH.

Source: feces of healthy humans.

DNA G+C content (mol%): 44 (HPLC).

Type strain: strain 341, DSM 17393, JCM 13265.

Sequence accession no. (16S rRNA gene): AB214328.

13. **Bacteroides massiliensis** Fenner, Roux, Mallet and Raoult 2005, 55[VP]

mas.si'li.en.sis. L. masc. adj. *massiliensis* of Massilia, the ancient Roman name for Marseille, France, where the type strain was isolated.

Characteristics are as described for the genus and as given in Table 2, with the following additional characteristics. Cells usually occur singly. Surface colonies on sheep blood agar plates after 2 d are 1–2 mm in diameter, circular, white-grayish, translucent, raised, and convex. No hemolysis on sheep blood agar. Optimum growth temperature is 37°C, but growth is observed at 25–42°C. The major fatty acid is $C_{15:0}$ anteiso.

Source: human blood culture.

DNA G+C content (mol%): 49 (HPLC).

Type strain: strain B84634, CCUG 48901, CIP 107942, JCM 13223.

Sequence accession no. (16S rRNA gene): AY126616, AB200226.

14. **Bacteroides merdae** Johnson, Moore and Moore 1986, 499[VP]

mer'dae. L. gen. n. *merdae* of feces, referring to the source of the isolate.

Characteristics are as described for the genus and as given in Table 2, with the following additional characteristics. Cells occur singly or in pairs or short chains. Surface colonies on BHI blood agar plates supplemented with vitamin K and hemin are 0.5–1.0 mm in diameter, circular to slightly irregular, entire, convex, white, shiny, and smooth. Rabbit blood is slightly hemolyzed. The type strain reduces neutral red and resazurin and produces hydrogen sulfide after incubation for 5 d. Pyruvate is converted to acetate and propionate. Lactate, threonine, and gluconate are not utilized. The major fatty acids are $C_{15:0}$ anteiso, $C_{15:0}$ iso, $C_{17:0}$ iso 3-OH, and $C_{16:0}$.

This species has been recently reclassified as *Parabacteroides merdae* (Sakamoto and Benno, 2006; see *Taxonomic comments*, above).

Source: human feces and occasionally clinical specimens.

DNA G+C content (mol%): 43–46 (HPLC).

Type strain: strain ATCC 43184, CCUG 38734, CIP 104202, JCM 9497, NCTC 13052, VPI T4-1.

Sequence accession no. (16S rRNA gene): AY169416, X83954, U50416.

Additional remarks: Previously referred to as "T4-1" DNA homology group (Johnson, 1978; Johnson and Ault, 1978).

15. **Bacteroides nordii** Song, Liu, McTeague and Finegold 2005b, 983[VP] (Effective publication: Song, Liu, McTeague and Finegold 2004, 5569.)

nor.di.i. N.L. gen. masc. n. *nordii* of Nord, to honor Carl Erik Nord, who has contributed much to our knowledge of anaerobic bacteriology and intestinal bacteriology.

Characteristics are as described for the genus and as given in Table 2, with the following additional characteristics. Cells occur singly or in pairs. Colonies on Brucella blood agar plates at 48 h are gray, circular, convex, entire, opaque, and attain a diameter of 1–2 mm. The major fatty acids are $C_{15:0}$ anteiso, $C_{15:0}$ iso, $C_{16:0}$ 3-OH, and $C_{18:1}$ ω9*c*. Probably of intestinal origin.

Source: clinical sources such as peritoneal fluid, appendix tissue, and intra-abdominal abscesses.

DNA G+C content (mol%): 41.4 (HPLC).

Type strain: strain WAL 11050, ATCC BAA-998, CCUG 48943.

Sequence accession no. (16S rRNA gene): AY608697.

16. **Bacteroides ovatus** Eggerth and Gagnon 1933, 405[AL] [*Pasteurella ovata* Eggerth and Gagnon 1933) Prévot 1938, 292; *Bacteroides fragilis* subsp. *ovatus* (Eggerth and Gagnon 1933) Holdeman and Moore 1970, 35]

o.va'tus. L. masc. adj. *ovatus* ovate, egg-shaped (relating to the cellular shape).

Characteristics are as described for the genus and as given in Table 2, with the following additional characteristics. Cells are oval, occurring singly, but occasionally in pairs. Capsules are detected in some strains. Surface colonies on blood agar are 0.5–1.0 mm, circular, entire, convex, pale buff, semi-opaque, and may have a mottled appearance. There is no hemolysis of sheep blood. Hemin is required or is highly stimulatory for growth. The major fatty acids are $C_{15:0}$ anteiso, $C_{17:0}$ iso 3-OH, $C_{18:1}$ ω9*c*, and $C_{15:0}$ iso.

Source: human feces and occasionally human clinical specimens.

DNA G+C content (mol%): 39–43 (HPLC).

Type strain: ATCC 8483, BCRC 10623, CCUG 4943, CIP 103756, DSM 1896, JCM 5824, NCTC 11153.

Sequence accession no. (16S rRNA gene): AB050108, L16484, X83952.

17. **Bacteroides plebeius** Kitahara, Sakamoto, Ike, Sakata and Benno 2005, 2146[VP]

ple.bei'us. L. masc. adj. *plebeius* common, of low class.

Characteristics are as described for the genus and as given in Table 2, with the following additional characteristics. Colonies grown on EG blood agar plates are 1–3 mm in diameter, disc-shaped, grayish-white, and translucent. The major fatty acids are $C_{15:0}$ anteiso, $C_{17:0}$ iso 3-OH, $C_{16:0}$ 3-OH, $C_{18:1}$ ω9*c*, and $C_{16:0}$.

Source: feces of healthy humans.

DNA G+C content (mol%): 42.4–43.9 (HPLC).

Type strain: strain M12, DSM 17135, JCM 12973.

Sequence accession no. (16S rRNA gene): AB200217–AB200222.

18. **Bacteroides pyogenes** Benno, Watabe and Mitsuoka 1983a, 896[VP] (Effective publication: Benno, Watabe and Mitsuoka 1983b, 402.)

py.o'ge.nes. Gr. n. *pyum* pus; N.L. suff. *-genes* (from Gr. v. *gennaô* to produce) producing; N.L. adj. *pyogenes* pus-producing.

Characteristics are as described for the genus and as given in Table 2, with the following additional characteristics. Cells occur singly or occasionally in pairs with longer rods. After 2 d of incubation on EG agar, surface colonies are 0.5–1.5 mm in diameter, circular with entire edges, low convex, translucent to opaque, cream to light gray, shiny, and smooth. Usually, no hemolysis occurs on blood agar. Strains do not grow well in PYFG–20% bile broth. Some strains hydrolyze esculin.

Source: swine abscess and feces.

DNA G+C content (mol%): 46.1–47.6 (HPLC).

Type strain: strain P 39-88, ATCC 35418, CCUG 15419, DSM 20611, JCM 6294.

Sequence accession no. (16S rRNA gene): AB200229.

19. **Bacteroides salyersiae** corrig. Song, Liu, McTeague and Finegold 2005b, 983[VP] (Effective publication: Song, Liu, McTeague and Finegold 2004, 5569.) [*Bacteroides salyersae* (*sic*) Song, Liu, McTeague and Finegold 2004, 5569]

sal'yer.si.ae. N.L. gen. fem. n. *salyersiae* of Salyers, named after Abigail Salyers, an American bacteriologist who has contributed so much to our knowledge of intestinal bacteriology and anaerobic bacteriology in general.

Characteristics are as described for the genus and as given in Table 2, with the following additional characteristics. Cells occur singly, occasionally in pairs. Colonies on Brucella blood agar plates at 24 h are gray, circular, convex, entire, opaque, and attain a diameter of 1–2 mm. The major fatty acids are $C_{15:0}$ anteiso, $C_{15:0}$ iso, $C_{16:0}$ 3-OH, $C_{16:0}$, and $C_{18:1}$ ω9*c*. Probably of intestinal origin.

Source: clinical sources such as peritoneal fluid, appendix tissue, and intra-abdominal abscesses.

DNA G+C content (mol%): 42.0 (HPLC).

Type strain: strain WAL 10018, ATCC BAA-997, CCUG 48945.

Sequence accession no. (16S rRNA gene): AY608696.

20. **Bacteroides stercoris** Johnson, Moore and Moore 1986, 501[VP]

ster'co.ris. L. n. *stercus* feces; L. gen. n. *stercoris* of feces, referring to source of isolate.

Characteristics are as described for the genus and as given in Table 2, with the following additional characteristics. Cells occur singly and in pairs. Vacuolated cells are seen sometimes in cultures in broths that contain a fermentable carbohydrate. Surface colonies on supplemented BHI blood agar with rabbit blood are 0.5–1.0 mm in diameter, circular, entire, convex, transparent to translucent, shiny, smooth, and β-hemolytic. Resazurin is reduced, neutral red is not reduced. Pyruvate is converted to acetate. Lactate, threonine, and gluconate are not used. The major fatty acids are $C_{15:0}$ anteiso, $C_{15:0}$ iso, $C_{17:0}$ iso 3-OH, and $C_{16:0}$.

Source: human feces and occasionally human clinical specimens.

DNA G+C content (mol%): 43–47 (HPLC).

Type strain: ATCC 43183, CCUG 38733, CIP 104203, JCM 9496, NCTC 13053, VPI B5-21.

Sequence accession no. (16S rRNA gene): X83953, U50417.

Additional remarks: previously referred to as the "subsp. a" DNA homology group (Johnson, 1978; Johnson and Ault, 1978).

21. **Bacteroides suis** Benno, Watabe and Mitsuoka 1983a, 896[VP] (Effective publication: Benno, Watabe and Mitsuoka 1983b, 403.)

su'is. L. n. *sus* swine; L. gen. n. *suis* of the pig.

Characteristics are as described for the genus and as given in Table 2, with the following additional characteristics. Cells are short rods, single or in pairs; longer rods occur only occasionally. After 2 d of incubation on EG agar, surface colonies are 0.5–1.5 mm in diameter, circular with entire edge, low convex, translucent to opaque, gray to gray-white, shiny, and smooth. Slight greening or indefinite hemolysis often apparent on blood agar plates. Either no growth or markedly inhibited and delayed growth occurs in PYFG broth containing 20% bile.

Source: swine abscess and feces.

DNA G+C content (mol%): 42.3–45.7 (HPLC).

Type strain: P 38024, ATCC 35419, DSM 20612, CCUG 15420, JCM 6292.

Sequence accession no. (16S rRNA gene): DQ497991.

22. **Bacteroides tectus** corrig. Love, Johnson, Jones, Bailey and Calverley 1986, 126[VP] [*Bacteroides tectum* (*sic*) Love, Johnson, Jones, Bailey and Calverley 1986, 126]

tec'tus. L. masc. adj. *tectus* (from L. v. *tego*) secret, concealed, hidden, referring to difficulty in species identification.

Characteristics are as described for the genus and as given in Table 2, with the following additional characteristics. Cells occur singly, in pairs, and in short chains. On sheep blood agar, surface colonies after 48 h are 2–3 mm in diameter, circular, entire, dome shaped, grayish, and opaque.

Source: subcutaneous fight wound abscesses from cats and dogs, pyothorax of cats, and mouths of normal cats.

DNA G+C content (mol%): 47–48 (HPLC).

Type strain: strain 160, ATCC 43331, CCUG 25929, JCM 10003, NCTC 11853.

Sequence accession no. (16S rRNA gene): AB200228.

23. **Bacteroides thetaiotaomicron** (Distaso 1912) Castellani and Chalmers 1919, 960[AL] [*Bacillus thetaiotaomicron* Distaso 1912, 444; *Sphaerocillus thetaiotaomicron* (Distaso 1912) Prévot 1938, 300; *Bacteroides fragilis* subsp. *thetaiotaomicron* (Distaso 1912) Holdeman and Moore 1970, 35]

the.ta.i.o.ta.o'mi.cron. N.L. n. *thetaiotaomicron* a combination of the Greek letters *theta, iota* and *omicron* (relating to the morphology of vacuolated forms).

Characteristics are as described for the genus and as given in Table 2, with the following additional characteristics. Cells are pleomorphic and occur singly or in pairs. Cells from blood agar plates or PY broth are smaller and more homogeneous in size and shape than cells from media with a fermentable carbohydrate. Capsules are present in some strains. Surface colonies on blood agar are punctiform, circular, entire, convex, semiopaque, whitish, soft, and shiny. Sheep blood is not hemolyzed. Hemin is either required or is highly stimulatory for growth. The major fatty acids are $C_{15:0}$ anteiso, $C_{17:0}$ iso 3-OH, $C_{18:1}$ ω9*c*, and $C_{16:0}$ 3-OH.

The genome of *Bacteroides thetaiotaomicron* VPI-5482 has been sequenced. The genome of *Bacteroides fragilis* NCTC 9343^T contains a single circular chromosome of 6,260,361 bp. The genome has large expansions of many paralogous groups of genes that encode products essential to the organism's ability to successfully compete in the human colonic microbiota. Most notable of these is the organism's abundant machinery for utilizing a large variety of complex polysaccharides as a source of carbon and energy. The proteome also reveals the organism's extensive ability to adapt and regulate expression of its genes in response to the changing ecosystem. These factors suggest a highly flexible and adaptable organism that is exquisitely equipped to dominate in its challenging and competitive niche (Xu et al., 2003).

Source: human feces, frequently found in human clinical specimens.

DNA G+C content (mol%): 40–43 (HPLC).

Type strain: ATCC 29148, CCUG 10774, CIP 104206, DSM 2079, JCM 5827, NCTC 10582, VPI 5482.

Sequence accession no. (16S rRNA gene): M58763, AB050109, L16489; genome sequence for strain VPI 5482, NC_004663, AE015928.

24. **Bacteroides uniformis** Eggerth and Gagnon 1933, 400[AL]

u.ni.form'is. L. masc. adj. *uniformis* having only one form, uniform.

Characteristics are as described for the genus and as given in Table 2, with the following additional characteristics. Cells occur singly or in pairs; an occasional filament may be seen. Vacuoles that do not greatly swell the cell are often present in cells grown in media containing a fermentable carbohydrate. Surface colonies on blood agar are 0.5–2.0 mm in diameter, circular, entire, low convex, gray-to-white, and translucent to slightly opaque. There usually is no hemolysis of blood agar, but some strains may produce a slight greening of the agar. The optimum temperature for growth is 35–37°C. Growth may occur at 25 or 45°C. The major fatty acids are $C_{15:0}$ anteiso, $C_{15:0}$ iso, $C_{17:0}$ iso 3-OH, and $C_{16:0}$. Strains of *Bacteroides uniformis* and *Bacteroides thetaiotaomicron* are very similar phenotypically, but they differ in their DNA G+C content and are not related by DNA–DNA hybridization analysis. In general, strains of *Bacteroides uniformis* do not grow as well in 20% bile as strains of *Bacteroides thetaiotaomicron.*

Source: part of human and swine fecal flora; also isolated from various human clinical specimens.

DNA G+C content (mol%): 45–48 (HPLC).

Type strain: ATCC 8492, CCUG 4942, CIP 103695, DSM 6597, JCM 5828, NCTC 13054.

Sequence accession no. (16S rRNA gene): L16486, AB050110.

25. **Bacteroides vulgatus** Eggerth and Gagnon 1933, 401[AL] [*Pasteurella vulgatus* (Eggerth and Gagnon, 1933) Prévot 1938, 292; *Bacteroides fragilis* subsp. *vulgatus* (Eggerth and Gagnon 1933) Holdeman and Moore 1970, 35]

vul.ga'tus. L. masc. adj. *vulgatus* common (referring to the frequent occurrence of the species in fecal flora).

Characteristics are as described for the genus and as given in Table 2, with the following additional characteristics. Cells may be pleomorphic with swellings or vacuoles, but less so than strains of *Bacteroides fragilis.* In broth cultures, cells usually occur singly, occasionally in pairs or short chains. Capsules are detected in some strains. Surface colonies on blood agar are 1–2 mm in diameter, circular, entire, convex, and semi-opaque. Sheep blood is not hemolyzed. Hemin is required or is highly stimulatory for growth. The major fatty acids are $C_{15:0}$ anteiso, $C_{15:0}$ iso, $C_{17:0}$ iso 3-OH, and $C_{16:0}$.

Source: human feces, occasionally isolated from human infections.

DNA G+C content (mol%): 40–42 (HPLC).

Type strain: ATCC 8482, BCRC (formerly CCRC) 12903, CCUG 4940, CIP 103714, DSM 1447, IFO (now NBRC) 14291, JCM 5826, LMG 7956, LMG 17767, NCTC 11154.

Sequence accession no. (16S rRNA gene): AJ867050, AB050111.

Species *incertae sedis*

1. **Bacteroides capillosus** (Tissier 1908) Kelly 1957, 433[AL] [*Bacillus capillosus* Tissier 1908, 193; *Ristella capillosa* (Tissier 1908) Prévot 1938, 292]

ca.pil.lo'sus. L. masc. adj. *capillosus* very hairy.

Comparative 16S rRNA gene sequence analysis demonstrates that *Bacteroides capillosus* is phylogenetically far removed from the genus *Bacteroides* and, in fact, has a closer affinity to clostridia in the *Firmicutes* and is closely related genealogically to the type strain of *Clostridium orbiscindens* in *Clostridium* cluster IV.

Cells are straight or curved rods, 0.7–1.1 × 1.6–7.0 µm, occurring singly, in pairs, or in short chains after 24 h incubation in glucose broth. Vacuoles, swellings, and filaments with tapered ends are observed. Surface colonies are minute to 1 mm, circular, entire, convex, translucent, and smooth. Growth of most strains is enhanced by hemin, rumen fluid, or Tween 80. Strains are generally nonfermentative unless Tween 80 is added to the medium, in which case they may be slightly fermentative. Good growth occurs at 37 and 45°C, slight growth at 30°C, and no growth at 25°C. Other characteristics are given in Table 5.

Source: cysts and wounds, human mouth, human infant and adult feces, intestinal tracts of dogs, mice, and termites, and sewage sludge.

DNA G+C content (mol%): 60 (HPLC).

Type strain: ATCC 29799, CCUG 15402.

Sequence accession no. (16S rRNA gene): AY136666.

2. **Bacteroides cellulosolvens** Murray, Sowden and Colvin 1984, 186[VP]

cell.u.lo.sol'vens. N.L. n. *cellulosum* cellulose; L. v. *solvere* to dissolve; N.L. part. adj. *cellulosolvens* cellulose dissolving, so named because of its ability to ferment cellulosic substrates.

Comparative 16S rRNA gene sequence analysis has demonstrated that *Bacteroides cellulosolvens* is phylogenetically far removed from the genus *Bacteroides* and, in fact, has a closer affinity to the clostridia of the *Firmicutes* and is closely related genealogically to the type strain of *Acetivibrio cellulolyticus* in *Clostridium* cluster III (Lin et al., 1994).

Cells are straight, rod-shaped, approximately 0.8 × 6 µm, and occur singly. They are nonmotile and no flagella are detected by staining or electron microscopy. Non-spore-forming. Bright

TABLE 5. Characteristics of misplaced *Bacteroides* species[a,b]

Characteristic	1. *B. capillosus*	2. *B. cellulosolvens*	3. *B. coagulans*	4. *B. galacturonicus*	5. *B. pectinophilus*	6. *B. polypragmatus*	7. *B. splanchnicus*	8. *B. xylanolyticus*
Growth in 20% bile	–	–	nr	+	+	nr	+	nr
Esculin hydrolyzed	+	nr	–	–	–	+	+	nr
Indole produced	–	–	+	nr	nr	+	+	nr
Nitrate reduced	–	–	–	nr	nr	–	–	nr
Catalase produced	nr	–	nr	–	–	–	nr	nr
Starch hydrolyzed	–	–	nr	–	nr	+	–	nr
Gelatin digested	w	–	+	–	+	–	d	nr
Acid produced from:								
Arabinose	–	–	–	–	–	+	+	+
Cellobiose	–	–	–	–	–	+	–	+
Xylan	–	–	–	–	–	nr	nr	+
Sorbitol	–	–	–	nr	nr	nr	nr	nr
Fructose	–	–	–	d	–	+	d	nr
Glucose	w	–	–	–	–	+	+	+
Glycogen	–	–	–	nr	nr	+	–	nr
Inositol	–	–	–	nr	nr	–	–	nr
Lactose	–	–	–	–	–	+	+	nr
Maltose	–	–	–	–	–	+	–	nr
Mannitol	–	–	–	nr	nr	+	–	nr
Mannose	–	–	–	–	–	+	+	+
Melezitose	–	–	–	nr	nr	+	–	nr
Melibiose	–	–	–	nr	nr	+	d	nr
Raffinose	–	–	–	nr	nr	+	–	nr
Rhamnose	–	–	–	–	–	+	–	nr
Ribose	–	–	–	–	–	+	–	nr
Salicin	–	–	–	nr	nr	+	–	nr
Sucrose	–	–	–	–	–	–	–	nr
Trehalose	–	–	–	nr	nr	+	–	nr
Xylose	–	–	–	–	–	+	–	+
Fermentation products[c]	s, a (l, f, p)	H_2, CO_2, a, ethanol, l[d]	a (f, s, l)	A, F (l, ethanol)[e]	A, F (l)[e]	H_2, CO_2, ethanol, a (b)	S, A, P (b, ib, iv, l)	H_2, CO_2, ethanol, a

[a]Symbols: +, >85% positive; d, different strains give different reactions (16–84% positive); –, 0–15% positive; w, weak reaction; nr, not reported. Fermentation products: A, acetic acid; B, butyric acid; F, formic acid; ib, isobutyric acid; iv, isobutyric acid; l, lactic acid; p, propionic acid; S, succinic acid. Upper-case letters indicate >1 meq/100 ml of broth; lower-case letters, <1 meq/100 ml; products in parentheses may or may not be detected.

[b]*Bacteroides ureolyticus* is not included because it was described under the genus *Campylobacter* in *Bergey's Manual of Systematic Bacteriology*, 2nd edition, vol. 2, part C.

[c]End products from glucose fermentation unless indicated.

[d]End products from cellobiose fermentation.

[e]End products from polygalacturonate or pectin.

yellow, circular, convex colonies with rough margins are produced on cellobiose agar. Colonies reach a maximum diameter of 1.0–1.25 mm after 5 d of growth at 35°C. On Solka Floc agar, colonies reach their maximum size after 10 d of growth and show zones of cellulose clearing around the periphery of the colonies and in the agar below. Anaerobic medium is required for growth. Of the substrates tested, only cellulose and cellobiose support growth. The addition of bile (2% oxgall) to cellobiose broth inhibits growth, whereas Tween 80, hemin, and vitamins B_{12} and K_1 have no effect. Other characteristics are given in Table 5.

Source: sewage sludge (type strain).

DNA G+C content (mol%): 43 (HPLC).

Type strain: WM2, ATCC 35603, DSM 2933, NRCC 2944.

Sequence accession no. (16S rRNA gene): L35517.

3. **Bacteroides coagulans** Eggerth and Gagnon 1933, 409[AL] [*Pasteurella coagulans* (Eggerth and Gagnon 1933) Prévot 1938, 292]

co.a'gu.lans. L. part. adj. *coagulans* curdling, coagulating (milk).

Cells of the type strain are small ovoid rods, 0.6 × 0.8–2.0 µm, occurring singly, in pairs, or in short chains. Surface colonies on horse blood agar are punctate, circular, entire, slightly raised, translucent, and nonhemolytic. Glucose broth cultures are only slightly turbid; no acid is produced. Growth is inhibited by Tween 80. Grows poorly, if at all, at 25 and 45°C; does not grow at pH 8.5. Other characteristics are given in Table 5.

Source: human feces, urogenital tract, and occasionally from human clinical specimens.

DNA G+C content (mol%): 37 (HPLC).

Type strain: ATCC 29798, DSM 20705, JCM 12528, LMG 8206.

Sequence accession no. (16S rRNA gene): DQ497990.

4. **Bacteroides galacturonicus** Jensen and Canale-Parola 1987, 179[VP] (Effective publication: Jensen and Canale-Parola 1986, 886.)

ga.lac.tu.ro'ni.cus. N.L. n. *acidum galacturonicum* galacturonic acid; L. masc. suff. *-icus* suffix of various meanings, but signifying in general made of or belonging to; N.L. masc. adj. *galacturonicus* pertaining to galacturonate, with reference to the ability to ferment this compound or polygalacturonate.

Cells grown in broth containing polygalacturonate as the fermentable substrate (PF broth; Jensen and Canale-Parola, 1986) are rod-shaped, 0.6 × 3.5–6 µm, arranged singly or in pairs, and rarely in short chains. Cells up to 12 µm in length are present in cultures. Non-spore-forming, Gram-stain-negative. Six to eight peritrichous flagella per cell are present; however, motility has not been observed under the culture conditions used. Surface colonies on PF agar are white with a tan tinge, irregularly shaped with uneven margins, flat, opaque, rough, and measure 4–5 mm in diameter. Optimal growth is between 35 and 40C, with no growth at 20 or 45C. Obligate anaerobe. Pectin, polygalacturonate, D-galacturonate, and D-glucuronate are fermented. Folic acid, pantothenic acid, and biotin are required as growth factors. Neither hemin nor vitamin K is required for growth. Other characteristics are given in Table 5.

Source: human feces.

DNA G+C content (mol%): 36 (T_m).

Type strain: strain N6, ATCC 43244, DSM 3978.

Sequence accession no. (16S rRNA gene): DQ497994.

5. **Bacteroides pectinophilus** Jensen and Canale-Parola 1987, 179[VP] (Effective publication: Jensen and Canale-Parola 1986, 886.)

pec.ti.no'phil.us. N.L. n. *pectinum* pectin; Gr. adj. *philos* loving; N.L. masc. adj. *philus* loving; N.L. masc. adj. *pectinophilus* pectin-loving.

Cells grown in PF broth are rod-shaped, 0.5 × 2.5–5.0 µm, arranged singly or in pairs, and rarely in short chains. Non-spore-forming, Gram-stain-negative, nonmotile. Surface colonies on PF agar are cream, circular, flat, opaque, smooth, and measure 2–3 mm in diameter. Optimal growth is between 35 and 40°C, with no growth at 20 or 45°C. Obligate anaerobe. Pectin, polygalacturonate, D-galacturonate, and D-glucuronate are fermented. Other characteristics are given in Table 5.

Source: human feces.

DNA G+C content (mol%): 45 (T_m).

Type strain: strain N3, ATCC 43243.

Sequence accession no. (16S rRNA gene): DQ497993.

6. **Bacteroides polypragmatus** Patel and Breuil 1982, 266[VP] (Effective publication: Patel and Breuil 1981, 292.)

po.ly.prag'ma.tus. Gr. *polypragmâtos* busy about many things; N.L. masc. adj. *polypragmatus* versatile, reflecting its activity on many carbohydrates.

Cells from Syt broth (Patel and Breuil, 1981) are 0.5–1.0 × 3–5 µm and have pointed ends. Most of the cells are single or in pairs and a few longer chains (50 µm) are also present. Young cells (18 h) exhibit motility and electron microscopic observation indicates the presence of peritrichous flagella. Staining cells with aqueous nigrosin dye demonstrates the presence of capsular material. Surface colonies are glistening, cream, have serrated margins, and are 2–3 mm in diameter after 1 week of incubation. Saccharoclastic. Grows between pH 5.6 and 9.2 and at temperatures ranging from 10 to 43°C. Optimum growth is at pH 7.0–7.8 and 30–35°C. The presence of vitamin K, hemin, or Tween 80 does not affect growth. Tests for acetylmethylcarbinol, ammonia, nitrate reduction, and digestion of gelatin, milk and meat are negative. Major fermentation products are CO_2, H_2, ethanol, and acetic acid. Other characteristics are given in Table 5.

Source: sewage sludge.

DNA G+C content (mol %): 61 (T_m).

Type strain: strain GP4, NRC 2288.

Sequence accession no. (16S rRNA gene): none available.

7. **Bacteroides splanchnicus** Werner, Rintelen and Kunstek-Santos 1975, 133[AL]

splanch'ni.cus. Gr. pl. n. *splanchna* the "innards"; L. masc. suff. *-icus*, suffix used with the sense of pertaining to; N.L. masc. adj. *splanchnicus* pertaining to the internal organs (referring to the source of isolation).

Phylogenetic analysis of *Bacteroides splanchnicus* based on 16S rRNA gene sequences indicates that *Bacteroides splanchnicus* branches off separately from the three major clusters mentioned previously, but belongs within the *Bacteroides* subgroup of the CFB phylum (Paster et al., 1994). It most likely represents a novel genus.

Cells from glucose broth are 0.7 × 1.0–5.0 µm. Colonies on anaerobic blood agar are punctiform to 1 mm in diameter, circular to slightly irregular, entire, convex, translucent, opalescent to yellowish, smooth, and shiny. Hemin is highly stimulatory, but not required. Without hemin, malate, acetate, and hydrogen are produced; with hemin, succinate, acetate, butyrate, and decreased amounts of hydrogen are produced. The presence of vitamin B_{12} results in propionate production from succinate and decreased butyrate production. Cells contain sphingolipids. Other characteristics are given in Table 5.

Source: human feces and vagina and occasionally from intra-abdominal infections.

DNA G+C content (mol%): not reported.

Type strain: ATCC 29572, CCUG 21054, CIP 104287, DSM 20712, LMG 8202, NCTC 10825.

Sequence accession no. (16S rRNA gene): L16496.

8. **Bacteroides xylanolyticus** Scholten-Koerselman, Houwaard, Janssen and Zehnder 1988, 136[VP] (Effective publication: Scholten-Koerselman, Houwaard, Janssen and Zehnder 1986, 543.)

xy.lan.o.lyt'i.cus. N.L. neut. n. *xylanum* xylan; Gr. masc. adj. *lutikos* able to loose, able to dissolve; N.L. masc. adj. *xylanolyticus* xylan-dissolving.

An anaerobic non-spore-forming, Gram-stain-negative rod that is motile with peritrichous flagella. This organism ferments xylan and many soluble sugars (glucose, cellobiose, mannose, xylose, and arabinose). Other hemicelluloses such as gum xanthan, laminaran, locust bean gum, and gum arabic are not utilized. Does not use cellulose. Fermentation products are carbon dioxide, hydrogen, acetate, and ethanol. Produces carboxymethyl cellulase and xylanase, especially when growing on xylan. Growth is optimal between 25 and 40°C and between pH 6.5 and 7.5.

DNA G+C content (mol%): 34.8 ± 0.8 (UV absorbance-temperature profile).

Type strain: strain X5-1, DSM 3808.

Sequence accession no. (16S rRNA gene): none available.

Genus II. **Acetofilamentum** Dietrich, Weiss, Fiedler and Winter 1989, 93[VP] (Effective publication: Dietrich, Weiss, Fiedler and Winter 1988b, 273.)

THE EDITORIAL BOARD

A.ce'to.fil.a.men'tum. L. n. *acetum* vinegar; L. neut. n. *filamentum* a spun thread; N.L. neut. n. *Acetofilamentum* an acetate-producing, filamentous, threadlike bacterium.

Straight rods. Single cells are 0.18–0.3 µm × 2–8 µm. Filaments of more than 100 µm in length may be formed. Gram-stain-negative. Nonsporeforming. **Nonmotile**. Temperature range, 25–40°C; optimum, 35–38°C. pH range, 6.5–9.0; optimum, 7.3–8.0. **Yeast extract is required**. Chemo-organotrophic. **Anaerobic, having a strictly fermentative type of metabolism. Hexoses are fermented by glycolysis to acetate and H_2 (in a molar ratio of 1:2) and CO_2.** Isolated from sewage sludge fermenters.

DNA G+C content (mol%): ~47 (T_m).

Type species: **Acetofilamentum rigidum** Dietrich, Weiss, Fiedler and Winter 1989, 93[VP] (Effective publication: Dietrich, Weiss, Fiedler and Winter 1988b, 273.).

Further descriptive information

The straight, stiff filaments return to the straight form after being bent. In the filaments, a spacerlike septum is formed, reminiscent of the spacers found in the filaments of *Methanobacterum hungatei.*

The cell-wall peptidoglycan contains ornithine as the diamino acid, unlike most other Gram-stain-negative bacteria, which usually contain diaminopimelic acid. A glycine bridge occurs between the ornithine on one peptide chain and the D-alanine on another. The peptidoglycan structure may be a variation of the A3β type described by Schleifer and Kandler (1972).

Small, round, whitish to transparent colonies with a smooth surface are formed on agar media.

H_2 and acetate production is directly proportional to the concentration of yeast extract present, up to 0.4%. Increasing H_2 partial pressures inhibit growth, but elevated acetate concentrations do not affect growth.

Carbon sources include glucose, fructose, galactose, ribose, arginine, and alanine. Scant growth occurs on glycine. The following carbon sources are not used: cellulose, starch, glycogen, maltose, sucrose, mannose, xylose, mannitol, glutamate, glycerol, formate, methanol, and phenol.

Enrichment and isolation procedures

For isolation, serially diluted samples of secondary sewage sludge are inoculated onto solidified Medium 1* supplemented with 10% raw sewage sludge supernatant fluid (sterilized by filtration) and incubated under N_2/CO_2 (80:20%, 300 kPa) at 37°C. Colonies of other organisms appear after 2 weeks, but colonies of *Acetofilamentum rigidum* have a diameter <1 mm and develop after 4 weeks of incubation. When subcultured into serum bottles containing liquid Medium 1, the organisms form a cotton-like growth at the bottom after 6 weeks. Pure cultures of these filamentous organisms are obtained by repeated streaking onto solidified Medium 1.

Substrate specificity is tested by omitting glucose from Medium 1 and adding other potential carbon sources from sterile stock solutions.

---

*Medium 1 (Dietrich et al., 1988b) consists of the following ingredients (g/l): K_2HPO_4, 0.255; KH_2PO_4, 0.255; $(NH_4)_2SO_4$, 0.255; NaCl, 0.5; $MgSO_4 \cdot 7H_2O$, 0.1; $CaCl_2 \cdot 2H_2O$, 0.07; yeast extract, 2; peptone, 2; glucose, 2; $FeSO_4 \cdot 7H_2O$, 0.002; $NaHCO_3$, 6; vitamin solution (Wolin et al., 1963), 10 ml; trace mineral solution (Wolin et al., 1963), 10 ml. After boiling under N_2, 0.1 mg resazurin and 50 ml of a reducing agent (containing 2 g NaOH, 0.62 g $Na_2S \cdot 9H_2O$, and 0.62 g cysteine–HCl) are added per liter of medium. The media are dispensed in 10- or 20-ml portions into 25-ml serum tubes or 120-ml serum bottles in an anaerobic chamber, sealed with a butyl rubber stopper and aluminum cap, gassed with 300 kPa N_2/CO_2 (80:20%), H_2/CO_2 (80:20%), or with N_2, and autoclaved. A solidified medium 1 is obtained by addition of agar (20 g/l); the medium is dispensed into Petri dishes in an anaerobic chamber.

Taxonomic comments

The taxonomic placement of *Acetofilamentum* as a member of the family *Bacteroidaceae* is uncertain because of the lack of rRNA gene sequence data.

Differentiation of the genus *Acetofilamentum* from other genera

Characteristics differentiating the genus *Acetofilamentum* from other genera are listed in Table 6.

TABLE 6. Characteristics differentiating the genus *Acetofilamentum* from other genera[a]

Characteristic	*Acetofilamentum*	*Acetomicrobium*	*Acetothermus*	*Acetivibrio*
Shape of single cells	Straight tapered rods	Straight tapered rods	Straight rods	Curved rods
Long filaments formed	+	–	–[b]	–
Peptidoglycan contains ornithine	+	–	+	nr
Dimensions of single cells (μm):				
Width	0.18–0.3	0.6–0.8	0.5	0.4–0.8
Length	2–8	2.0–7.0	5.0–8.0	4.0–10.0
Motility	–	+	–	+
Flagellar arrangement	–	Single subpolar or 1 or 2 lateral	–	Single attached one-third of the way along the concave side, or by a bundle attached linearly on the concave side
Optimum temperature (°C)	35–38	58–73	58	35
Ethanol formed from glucose	–	D	–	+
Substrates fermented:				
Cellulose	–	nr	–	D
Galactose	+	nr	–	D
Glucose	+	+	+	D
Glycogen	–	–	D	nr
Maltose	–	+	–	D
Mannitol	–	–	–	D
Mannose	–	nr	–	D
Ribose	+	D	–	–
Starch	–	D	–	nr
Sucrose	–	D	–	–
Xylose	–	D	–	–
Glycerol	nr	D	nr	nr
DNA G+C content (mol%)	47	45–47	62	37–44

[a]Symbols: +, >85% positive; –, 0–15% positive; D, different reactions occur in different taxa (species of genus or genera of a family); w, weak reaction; nr, not reported.

[b]Except in the absence of Vitamin B_{12}, when filaments up to 50 μm long may be formed.

List of species of the genus *Acetofilamentum*

1. **Acetofilamentum rigidum** Dietrich, Weiss, Fiedler and Winter 1989, 93[VP] (Effective publication: Dietrich, Weiss, Fiedler and Winter 1988b, 273.)

ri′gi.dum. L. neut. adj. *rigidum* stiff.

The characteristics are as described for the genus and as listed in Table 6.

DNA G+C content (mol%): ~47 (T_m).

Type strain: MN, DSM 20769.

Genus III. **Acetomicrobium** Soutschek, Winter, Schindler and Kandler 1985, 223[VP] (Effective publication: Soutschek, Winter, Schindler and Kandler 1984, 388.)

The Editorial Board

A.ce′to.mi.cro′bi.um. L. n. *acetum* vinegar; Gr. adj. *mikros* small; Gr. n. *bios* life; N.L. neut. n. *microbium* a small living being, a microbe; N.L. neut. n. *Acetomicrobium* a microorganism producing acetic acid.

Straight tapered rods, 0.6–0.8 × 2–7 μm, occurring mostly in pairs or short chains. Gram-stain-negative. Nonsporeforming. The murein contains *meso*-diaminopimelic acid. **Motile by means of a single subpolar flagellum or a few lateral flagella. Optimum temperature, 58–73°C.** pH range for growth, 6.2–8.0. Catalase-negative. Cytochromes are not present. Chemoorganotrophic. **Anaerobic, having a strictly fermentative type of metabolism.** One species ferments glucose to acetate, CO_2, and H_2; the other ferments glucose to acetate, lactate, ethanol, CO_2, and H_2. One species ferments pentoses. Good growth depends on the presence of yeast extract or peptone and carbohydrate. Isolated from sewage sludge.

DNA G+C content (mol%): 45–47.

Type species: **Acetomicrobium flavidum** Soutschek, Winter, Schindler and Kandler 1985, 223[VP] (Effective publication: Soutschek, Winter, Schindler and Kandler 1984, 388.).

Further descriptive information

Colonies on solidified RCM medium* are 3-4 mm in diameter, convex, smooth, yellowish, circular, with an entire margin.

The peptidoglycan contains *meso*-diaminopimelic acid and is of the m-A_2pm direct type, variation A1γ, as designated by the nomenclature of Schleifer and Kandler (1972).

Acetomicrobium flavidum ferments glucose by glycolysis to acetate, CO_2, and H_2 in a 1:1:2 ratio. No ethanol is formed. *Acetomicrobium faecale* ferments glucose by glycolysis to acetate, lactate, ethanol, CO_2, and H_2.

Small amounts of volatile acids, e.g., propionic acid, are only formed from complex substrates, e.g., yeast extract and peptone.

Enrichment and isolation procedures

Acetomicrobium flavidum was isolated from sludge samples that were diluted anaerobically in growth vials containing 10 ml of RCM. The diluted samples were streaked onto RCA plates in the anaerobic chamber and incubated anaerobically. After repeated plating, single colonies were transferred into Medium 1.† Cultures were incubated without shaking at 58°C.

For isolation of *Acetomicrobium faecale*, 400 ml of mesophilically digested sewage sludge and 50 ml of fresh sewage sludge were incubated at 72°C. Then 50 ml portions of digested sludge were replaced daily for 10 d by fresh sewage sludge. The sludge addition was later replaced by the continuous addition of sterile culture medium‡ containing 5 g/l glucose. After 4 weeks of continuous operation, samples were plated and, after incubation, single colonies were picked from the agar plates in an anaerobic chamber.

Taxonomic comments

No significant DNA–DNA hybridization occurs between *Acetomicrobium* and *Acetothermus* (Dietrich et al., 1988c). *Acetomicrobium faecale* exhibits a DNA–DNA hybridization value of 64% with *Acetomicrobium flavidum*, indicating that the two species belong in the same genus (Dietrich et al., 1988c)

The taxonomic placement of *Acetomicrobium* as a member of the family *Bacteroidaceae* is not certain because of the lack of rRNA gene sequence data.

Differentiation of the genus *Acetomicrobium* from other genera

Characteristics differentiating *Acetomicrobium* from *Acetofilamentum*, *Acetothermus*, and *Acetivibrio* are listed in Table 6 in the chapter on *Acetofilamentum*.

Differentiation of species of the genus *Acetomicrobium*

Table 7 lists the differential features of the species of *Acetomicrobium*.

List of species of the genus *Acetomicrobium*

1. **Acetomicrobium flavidum** Soutschek, Winter, Schindler and Kandler 1985, 223[VP] (Effective publication: Soutschek, Winter, Schindler and Kandler 1984, 388.)

   fla'vi.dum. L. neut. adj. *flavidum* yellowish.

   The characteristics are as described for the genus and as listed in Table 7. Glucose, fructose, maltose, cellobiose, glycerol, and starch are fermented. Sucrose, melibiose, mannose, ribose, arabinose, xylose, sorbitol, and mannitol are not fermented.

   Source: a thermophilic (60°C) sewage sludge fermentor.

   DNA G+C content (mol%): 47 (T_m).

   Type strain: TN, ATCC 43122, DSM 20664, LMG 6941.

   Sequence accession no.: not reported.

2. **Acetomicrobium faecale** corrig. Winter, Braun and Zabel 1988, 136[VP] (Effective publication: Winter, Braun and Zabel 1987, 75.)

   fae.ca'le. L. n. *faex faecis* dregs, faeces; L. neut. suff. *-ale* suffix denoting pertaining to; N.L. neut. adj. *faecale* pertaining to faeces, fecal. (Note: the original specific epithet *faecalis* was corrected to *faecale* by Euzéby (1998).)

   The characteristics are as described for the genus and as listed in Table 7. Motile by means of 1 or 2 lateral flagella. Growth occurs best at temperatures 70–73°C and at a pH of 6.5-7.0. Colonies are whitish to translucent, 1-2 mm in diameter, convex, and round with an entire edge. Yeast extract is required for growth. Increasing the NaCl concentration from 0 to 3% causes a decrease in growth; no growth occurs with 4% NaCl. Unlike *Acetomicrobium flavidum*, *Acetomicrobium faecale* ferments pentoses. Carbohydrates used for growth include arabinose, xylose, ribose, fructose, glucose, galactose, mannose, maltose, rhamnose, sucrose, lactose, cellobiose, melibiose, and salicin. Sorbitol, starch, xylan, mannitol, glycogen, glycerol, and cellulose are not used.

   Source: a thermophilic (70°C) continuous culture initially inoculated with sewage sludge.

   DNA G+C content (mol%): 45 (T_m).

   Type strain: DSM 20678.

   Sequence accession no.: not reported.

---

*RCM medium contains the following ingredients (g/l of distilled water): yeast extract, 3; meat extract, 10; peptone from casein, 10; sodium acetate, 3; NaCl, 5; glucose, 5; starch, 1; resazurin, 0.001; cysteine–HCl, 0.25; sodium sulfide, 0.25; and NaOH, 0.66. Agar (20 g) was added for a solid medium. For primary isolation, the supernatant solution of centrifuged digested sludge is used instead of distilled water to prepare the medium. The final pH is adjusted to 7 under an N_2 atmosphere.

†Medium 1 contains (per liter of distilled water): K_2HPO_4, 0.255 g; KH_2PO_4, 0.255 g; $(NH_4)_2SO_4$, 0.255 g; NaCl, 0.5 g; $MgSO_4 \cdot 7H_2O$, 0.1 g; $CaCl_2 \cdot 2H_2O$, 0.07 g; yeast extract, 10 g; $FeSO_4 \cdot 7H_2O$, 1 mg; resazurin, 0.5 mg; $NaHCO_3$, 6 g; cysteine–HCl, 0.625 g; $Na_2S \cdot 9H_2O$, 0.625 g; glucose, 0.15 g; vitamin solution (Wolin et al., 1963), 10 ml; and mineral solution (Wolin et al., 1963), 10 ml.

‡Medium for isolation of *Acetomicrobium faecale* contained (per liter): K_2HPO_4, 0.255 g; KH_2PO_4, 0.255 g; $(NH_4)_2SO_4$, 0.255 g; NaCl, 0.5 g; $MgSO_4 \cdot 7H_2O$, 0.1 g; $CaCl_2 \cdot 2H_2O$, 0.07 g; yeast extract, 2 g; trypticase, 2 g; sodium acetate, 5 g; $FeSO_4 \cdot 7H_2O$, 2 mg; vitamin solution (Wolin et al., 1963), 10 ml; trace mineral solution (Wolin et al., 1963), 10 ml; and $NaHCO_3$, 6 g. After boiling under nitrogen, 0.1 mg resazurin and 50 ml of reducing agent containing 2 g NaOH, 0.62 g $Na_2S \cdot 9H_2O$, and 0.62 g cysteine–HCl were added. The reduced medium was dispensed and autoclaved under anaerobic conditions.

TABLE 7. Characteristics differentiating species of the genus *Acetomicrobium*[a]

Characteristic	*A. faecale*	*A. flavidum*
Source	Sewage sludge incubated at 72°C	Sewage sludge incubated at 60°C
Morphology:		
Cell width (μm)	0.6	0.8
Cell length (μm)	3–7	2–3
Flagellar arrangement	1–2 subpolar	Single polar, or a few lateral
Optimum temperature (°C)	70–73	58
Temperature range (°C)	60–75	35–65
Yellow pigmented colonies	–	+
Fermentation of arabinose, ribose, xylose	+	–
End products from glucose fermentation	Acetate, lactate, ethanol, CO_2, H_2	Acetate, CO_2, H_2

[a]Symbols: +, >85% positive; –, 0–15% positive.

Genus IV. **Acetothermus** Dietrich, Weiss and Winter 1988a, 328[VP] (Effective publication: Dietrich, Weiss and Winter 1988c, 179.)

The Editorial Board

A.ce'to.ther'mus. L.n. *acetum* vinegar; Gr. adj. *thermos* hot; N.L. masc. n. *Acetothermus* a thermophilic microorganism producing acetic acid.

Straight rods 0.5 × 5–8 μm, occurring singly or in pairs. Gram-stain-negative. Nonsporeforming. **Nonmotile.**

Chemoorganotrophic. **Anaerobic, having a strictly fermentative type of metabolism. Optimum temperature, 58°C;** no growth occurs below 50°C or above 60°C. Optimum pH, 7-8. Yeast extract, peptone, and **vitamin B_{12}** are required. **Only glucose and fructose are used as energy sources. Glucose is fermented by glycolysis to acetate, CO_2, and H_2 in a molar ratio of 1:1:2.** Increasing H_2 partial pressures inhibit growth, but elevated acetate concentrations do not affect growth. Isolated from thermophilically fermenting sewage sludge.

DNA G+ C content (mol%): ~62.

Type species: **Acetothermus paucivorans** Dietrich, Weiss and Winter 1988a, 328[VP] (Effective publication: Dietrich, Weiss and Winter 1988c, 179.).

Further descriptive information

In the absence of vitamin B_{12} the organisms form long filaments 0.5 μm in diameter and up to 50 μm long. Optimal growth occurs in a medium containing 0.2% yeast extract and 0.2% peptone. Growth is directly proportional to the concentration of vitamin B_{12} up to 1 μg/l. NaCl is not required; >1 g/l is inhibitory.

The cell-wall peptidoglycan contains ornithine as the diamino acid, unlike most other Gram-stain-negative bacteria, which usually contain diaminopimelic acid as the diamino acid. The peptidoglycan structure corresponds to the directly cross-linked A1β type described by Schleifer and Kandler (1972).

Glucose and fructose are used as energy sources. The following carbon sources are not used: arabinose, mannose, ribose, xylose, galactose, mannitol, sorbitol, sucrose, maltose, lactose, melibiose, raffinose, starch, cellulose, glycerol, Casamino acids, meat extract, glycogen, pyruvate, glycine, alanine, glutamate, formate, and methanol.

Enrichment and isolation procedures

Enrichment and isolation procedures from sewage sludge digesting at 60°C were done using the medium described for isolation of *Acetomicrobium faecale* (see footnote in the chapter on *Acetomicrobium* for formula) modified by the omission of acetate.

Taxonomic comments

No significant DNA–DNA hybridization occurs between *Acetothermus* and *Acetomicrobium* (Dietrich et al., 1988c). The taxonomic placement of *Acetothermus* as a member of the family *Bacteroidaceae* is not certain because of the lack of rRNA gene sequence data.

Differentiation of the genus *Acetothermus* from other genera

Characteristics differentiating *Acetothermus* from *Acetofilamentum, Acetivibrio*, and *Acetomicrobium* are listed in Table 6 in the chapter on *Acetofilamentum.*

List of species of the genus *Acetothermus*

1. **Acetothermus paucivorans** Dietrich, Weiss and Winter 1988a, 328[VP] (Effective publication: Dietrich, Weiss and Winter 1988c, 179.)

pau.ci.vo'rans. L. masc. adj. *paucus* few, little; L. v. *vorare* to eat, swallow; L. part. adj. *vorans* eating, swallowing; N.L. part. adj. *paucivorans* utilizing only a very restricted number of the supplied substrates.

The characteristics are as described for the genus.

Source: thermophilically fermenting sewage sludge.

DNA G+C content (mol%): 62 (T_m).

Type strain: TN, DSM 20768. (Note: this strain is no longer listed in the DSMZ catalog.)

Sequence accession no.: not reported.

Genus V. **Anaerorhabdus** Shah and Collins 1986, 573[VP] (Effective publication: Shah and Collins 1986, 86.)

Haroun N. Shah

An.ae.ro.rhab′dus. Gr. pref. *an* not; Gr. n. *aer* air: *anaero* (not living) in air; Gr. fem. n. *rhabdos* rod; N.L. fem. n. *Anaerohabdus* rod-shaped bacterium not living in air.

Short rods. Nonspore-forming. Nonmotile. Gram-negative. Anaerobic. **Nonsaccharolytic,** although a few carbohydrates may be fermented weakly. Acetic and lactic acids are the major metabolic end products in peptone-yeast extract-glucose broth (PYG). **Glucose-6-phosphate dehydrogenase is produced, but 6-phosphogluconate dehydrogenase, malate dehydrogenase, and glutamate dehydrogenase are absent. Sphingolipids and menaquinones are not produced.** The nonhydroxylated long-chain fatty acids are primarily of the straight-chain saturated and **monounsaturated types. Methyl branched fatty acids are either absent or present in small amounts.** Isolated from infected appendix, lung abscesses, and abdominal abscesses. Infrequently isolated from human and pig feces.

DNA G+C content (mol%): 34.

Type species: **Anaerorhabdus furcosa** (Veillon and Zuber 1898) Shah and Collins 1986, 573[VP] [Effective publication: *Anaerorhabdus furcosus* (*sic*) (Veillon and Zuber 1898) Shah and Collins 1986, 86.] [*Bacillus furcosus* Veillon and Zuber 1898, 541; *Fusiformis furcosus* (Veillon and Zuber 1898) Topley and Wilson 1929, 302; *Bacteroides furcosus* (Veillon and Zuber 1898) Hauduroy, Ehringer, Urbain, Guillot and Magrou 1937, 61[AL]; *Ristella furcosa* (Veillon and Zuber 1898) Prévot 1938, 291].

Enrichment and isolation procedures

Anaerorhabdus isolates grow on blood agar producing small colonies (0.5 mm in diameter) but is improved by culturing on fastidious anaerobic agar. Growth is improved by the addition of rumen fluid.

Differentiation of the genus *Anaerorhabdus* from other genera

Anaerorhabdus differs from *Bacteroides fragilis* and related species in having a significantly lower DNA base composition, in being nonfermentative or weakly fermentative, and in lacking 6-phosphogluconate dehydrogenase, malate dehydrogenase, and glutamate dehydrogenase. *Anaerorhabdus* primarily synthesizes fatty acids of the straight-chain saturated and monounsaturated type, and methyl-branched acids are present in only small amounts. In contrast, the fatty acids of *Bacteroides fragilis* and related species are mainly of the straight-chain saturated, *anteiso*-, and *iso*-methyl branched chain types, with monounsaturated acids either absent or present in only trace amounts. Moreover, *Anaerorhabdus* lacks menaquinones and sphingolipids, whereas *Bacteroides fragilis* and related species synthesize menaquinones and sphingolipids.

List of species of the genus *Anaerorhabdus*

1. **Anaerorhabdus furcosa** (Veillon and Zuber 1898) Shah and Collins 1986, 573 [Effective publication: *Anaerorhabdus furcosus* (*sic*) (Veillon and Zuber 1898) Shah and Collins 1986, 86.] [*Bacillus furcosus* Veillon and Zuber 1898, 541; *Fusiformis furcosus* (Veillon and Zuber 1898) Topley and Wilson 1929, 302; *Bacteroides furcosus* (Veillon and Zuber 1898) Hauduroy, Ehringer, Urbain, Guillot and Magrou 1937, 61[AL]; *Ristella furcosa* (Veillon and Zuber 1898) Prévot 1938, 291].

fur.co′sa. L. fem. adj. *furcosa* forked (pertaining to cell shape). (Note: the original epithet *furcosus* was corrected by Euzéby and Boemare, 2000.)

The description is the one given for the genus, plus the following additional features. Pleomorphic rods, 0.3–1.5 × 1–3 µm, occurring singly, in pairs, or short chains. Some cells appear to be forked or Y-shaped. Colonies on blood agar are 0.5 mm in diameter, circular, entire, low convex, translucent, colorless to gray, smooth, and shiny. Glucose broth cultures are turbid with smooth sediment; growth is stimulated by rumen fluid. Optimum growth temperature, 30–37°C; slight growth occurs at 25°C but none at 45°C. Acetic and lactic acids are the major metabolic end products in PYG broth; trace amounts of formic and succinic acids may also be produced. Most strains hydrolyze esculin. Phosphate positive. Indole, acetylmethylcarbinol, urease, and H_2S are not produced. Gelatin liquefaction is usually negative although a few strains may be weakly positive. Starch and hippurate are not hydrolyzed. Nitrate is not reduced.

The cell walls do not contains diaminopimelic acid, heptose, or ketodeoxyoctonic acid (KDO). Non-hydroxylated long chain fatty acids are primarily of the straight-chain saturated and mono-unsaturated types with hexadecanoic, octadecanoic, and octadecenoic acids predominating.

DNA G+C content (mol%): 34 (T_m).

Type strain: ATCC 25662, VPI 3253.

Sequence accession no. (16S rRNA gene): none available.

References

Avgustin, G., R.J. Wallace and H.J. Flint. 1997. Phenotypic diversity among ruminal isolates of *Prevotella ruminicola*: proposal of *Prevotella brevis* sp. nov., *Prevotella bryantii* sp. nov., and *Prevotella albensis* sp. nov. and redefinition of *Prevotella ruminicola*. Int. J. Syst. Bacteriol. *47*: 284–288.

Bakir, M.A., M. Kitahara, M. Sakamoto, M. Matsumoto and Y. Benno. 2006a. *Bacteroides intestinalis* sp. nov., isolated from human faeces. Int. J. Syst. Evol. Microbiol. *56*: 151–154.

Bakir, M.A., M. Kitahara, M. Sakamoto, M. Matsumoto and Y. Benno. 2006b. *Bacteroides finegoldii* sp. nov., isolated from human faeces. Int. J. Syst. Evol. Microbiol. *56*: 931–935.

Bakir, M.A., M. Sakamoto, M. Kitahara, M. Matsumoto and Y. Benno. 2006c. *Bacteroides dorei* sp. nov., isolated from human faeces. Int. J. Syst. Evol. Microbiol. *56*: 1639–1643.

Benno, Y., J. Watabe and T. Mitsuoka. 1983a. *In* Validation of the publication of new names and new combinations previously effectively published outside the IJSB. List no. 12. Int. J. Syst. Bacteriol. *33*: 896–897.

Benno, Y., J. Watabe and T. Mitsuoka. 1983b. *Bacteroides pyogenes* sp. nov., *Bacteroides suis* sp. nov., and *Bacteroides helcogenes* sp. nov., new species from abscesses and feces of pigs. Syst. Appl. Microbiol. *4*: 396–407.

Beveridge, W.I.B. 1941. Foot-rot in sheep: a transmissible disease due to infection with *Fusiformis nodosus* (n. sp.). Studies on its causes, epidemiology and control. Counc. Sci. Ind. Res. Aust. Bull. *140*: 1–56.

Bryant, M.P., N. Small, C. Bouma and H. Chu. 1958. *Bacteroides ruminicola* n. sp. and *Succinimonas amylolytica*; the new genus and species; species of succinic acid-producing anaerobic bacteria of the bovine rumen. J. Bacteriol. *76*: 15–23.

Caldwell, D.R., M. Keeney and P.J. Van Soest. 1969. Effects of carbon dioxide on growth and maltose fermentation by *Bacteroides amylophilus*. J. Bacteriol. *98*: 668–676.

Castellani, A. and A.J. Chalmers. 1919. Manual of Tropical Medicine, 3rd edn. Williams Wood, New York, pp. 959–960.

Cato, E.P., R.W. Kelley, W.E.C. Moore and L.V. Holdeman. 1982. *Bacteroides zoogleoformans* (Weinberg, Nativelle, and Prevot 1937) corrig., comb. nov.: emended description. Int. J. Syst. Bacteriol. *32*: 271–274.

Cerdeno-Tarraga, A.M., S. Patrick, L.C. Crossman, G. Blakely, V. Abratt, N. Lennard, I. Poxton, B. Duerden, B. Harris, M.A. Quail, A. Barron, L. Clark, C. Corton, J. Doggett, M.T. Holden, N. Larke, A. Line, A. Lord, H. Norbertczak, D. Ormond, C. Price, E. Rabbinowitsch, J. Woodward, B. Barrell and J. Parkhill. 2005. Extensive DNA inversions in the *B. fragilis* genome control variable gene expression. Science *307*: 1463–1465.

Collins, M.D., H.N. Shah and T. Mitsuoka. 1985. Reclassification of *Bacteroides microfusus* (Kaneuchi and Mitsuoka) in a new genus *Rikenella*, as *Rikenella microfusus* comb. nov. Syst. Appl. Microbiol. *6*: 79–81.

Collins, M.D. and H.N. Shah. 1986a. Reclassification of *Bacteroides termitidis* Sebald (Holdeman and Moore) in a new genus *Sebaldella termitidis*, as *Sebaldella termitidis* comb. nov. Int. J. Syst. Bacteriol. *36*: 349–350.

Collins, M.D. and H.N. Shah. 1986b. Reclassification of *Bacteroides praeacutus* Tissier (Holdeman and Moore) in a new genus, *Tissierella*, as *Tissierella praeacuta* comb. nov. Int. J. Syst. Bacteriol. *36*: 461–463.

Coykendall, A.L., F.S. Kaczmarek and J. Slots. 1980. Genetic heterogeneity in *Bacteroides asaccharolyticus* (Holdeman and Moore 1970) Finegold and Barnes 1977 (Approved Lists, 1980) and proposal of *Bacteroides gingivalis* sp. nov. and *Bacteroides macacae* (Slots and Genco) comb. nov. Int. J. Syst. Bacteriol. *30*: 559–564.

Dewhirst, F.E., B.J. Paster, S. Lafontaine and J.I. Rood. 1990. Transfer of *Kingella indologenes* (Snell and Lapage 1976) to the genus *Suttonella* gen. nov. as *Suttonella indologenes* comb. nov., transfer of *Bacteroides nodosus* (Beveridge 1941) to the genus *Dichelobacter* gen. nov. as *Dichelobacter nodosus* comb. nov., and assignment of the genera *Cardiobacterium*, *Dichelobacter*, and *Suttonella* to *Cardiobacteriaceae* fam. nov. in the gamma division of *Proteobacteria* on the basis of 16S ribosomal RNA sequence comparisons. Int. J. Syst. Bacteriol. *40*: 426–433.

Dietrich, G., N. Weiss and J. Winter. 1988a. Validation of the publication of new names and new combinations previously effectively published outside the IJSB. List no 26. Int. J. Syst. Bacteriol. *38*: 328–329.

Dietrich, G., N. Weiss, F. Fiedler and J. Winter. 1988b. *Acetofilamentum rigidum* gen. nov., sp. nov., a strictly anaerobic bacterium from sewage sludge. Syst. Appl. Microbiol. *10*: 273–278.

Dietrich, G., N. Weiss and J. Winter. 1988c. *Acetothermus paucivorans*, gen. nov., sp. nov., a strictly anaerobic, thermophilic bacterium from sewage sludge, fermenting hexoses to acetate, CO_2 and H_2. Syst. Appl. Microbiol. *10*: 174–179.

Dietrich, G., N. Weiss, F. Fiedler and J. Winter. 1989. Validation of the publication of new names and new combinations previously effectively published outside the IJSB. List no28. Int. J. Syst. Bacteriol. *39*: 93–94.

Distaso, A. 1912. Contribution à l'étude sur l'intoxication intestinale. Zentralbl. Bakteriol. Parasitenkd. Infektionskr. Hyg. Abt. I Orig. *62*: 433–468.

Eggerth, A.H. and B.H. Gagnon. 1933. The *Bacteroides* of Human Feces. J. Bacteriol. *25*: 389–413.

Euzéby, J.P. 1998. Taxonomic note: necessary correction of specific and subspecific epithets according to Rules 12c and 13b of the International Code of Nomenclature of *Bacteria* (1990 Revision). Int. J. Syst. Bacteriol. *48*: 1073–1075.

Euzéby, J.P. and N.E. Boemare. 2000. The modern Latin word *rhabdus* belongs to the feminine gender, inducing necessary corrections according to Rules 65(2), 12c(1) and 13b of the *Bacteriological Code* (1990 Revision). Int. J. Syst. Evol. Microbiol. *50*: 1691–1692.

Fang, H., C. Edlund, M. Hedberg and C.E. Nord. 2002. New findings in beta-lactam and metronidazole resistant *Bacteroides fragilis* group. Int. J. Antimicrob. Agents *19*: 361–370.

Fenner, L., V. Roux, M.N. Mallet and D. Raoult. 2005. *Bacteroides massiliensis* sp. nov., isolated from blood culture of a newborn. Int. J. Syst. Evol. Microbiol. *55*: 1335–1337.

Fildes, P., G.M. Richardson, B.C.G.E. Knight and G.P. Gladstone. 1936. A nutrient mixture suitable for the growth of *Staphylococcus aureus*. Br. J. Exp. Pathol. *17*: 481–484.

Franco, A.A., R.K. Cheng, G.T. Chung, S. Wu, H.B. Oh and C.L. Sears. 1999. Molecular evolution of the pathogenicity island of enterotoxigenic *Bacteroides fragilis* strains. J. Bacteriol. *181*: 6623–6633.

Gutacker, M., C. Valsangiacomo and J.-C. Piffaretti. 2000. Identification of two genetic groups in *Bacteroides fragilis* by multilocus enzyme electrophoresis: distribution of antibiotic resistance (*cfiA*, *cepA*) and enterotoxin (*bft*) encoding genes. Microbiology *146*: 1241–1254.

Gutacker, M., C. Valsangiacomo, M.V. Bernasconi and J.C. Piffaretti. 2002. *recA* and *glnA* sequences separate the *Bacteroides fragilis* population into two genetic divisions associated with the antibiotic resistance genotypes *cepA* and *cfiA*. J. Med. Microbiol. *51*: 123–130.

Hamlin, L.J. and R.E. Hungate. 1956. Culture and physiology of a starch-digesting bacterium (*Bacteroides amylophilus* n. sp.) from the bovine rumen. J. Bacteriol. *72*: 548–554.

Harrison, A.P., Jr. and P.A. Hansen. 1963. *Bacteroides hypermegas* nov. spec. Antonie van Leeuwenhoek *29*: 22–28.

Hauduroy, P., G. Ehringer, A. Urbain, G. Guillot and J. Magrou. 1937. Dictionnaire des bactéries pathogènes. Masson et Cie, Paris.

Holdeman, L.V. and W.E.C. Moore. 1970. *Bacteroides*. Outline of Clinical Methods in Anaerobic Bacteriology, 2nd revn (edited by Cato, Cummins, Holdeman, Johnson, Moore, Smibert and Smith). Virginia Polytechnic Institute Anaerobe Laboratory, Blacksburg, VA, pp. 57–66.

Holdeman, L.V. and W.E.C. Moore. 1974. New genus, *Coprococcus*, twelve new species, and emended descriptions of four previously described species of bacteria from human feces. Int. J. Syst. Bacteriol. *24*: 260–277.

Holdeman, L.V. and J.L. Johnson. 1977. *Bacteroides disiens* sp. nov. and *Bacteroides bivius* sp. nov. from human clinical infections. Int. J. Syst. Bacteriol. *27*: 337–345.

Holdeman, L.V. and J.L. Johnson. 1982. Description of *Bacteroides loescheii* sp. nov. and emendation of the descriptions of *Bacteroides melaninogenicus* (Oliver and Wherry) Roy and Kelly 1939 and *Bacteroides denticola* Shah and Collins 1981. Int. J. Syst. Bacteriol. *32*: 399–409.

Holdeman, L.V., R.W. Kelly and W.E.C. Moore. 1984. Genus I. *Bacteroides*. *In* Bergey's Manual of Systematic Bacteriology, vol. 1 (edited by Krieg and Holt). Williams & Wilkins, Baltimore, pp. 604–631.

Hungate, R.E. 1950. The anaerobic mesophilic cellulolytic bacteria. Bacteriol. Rev. *14*: 1–49.

Jensen, N.S. and E. Canale-Parola. 1986. *Bacteroides pectinophilus* sp. nov. and *Bacteroides galacturonicus* sp. nov., two pectinolytic bacteria from the human intestinal tract. Appl. Environ. Microbiol. *52*: 880–887.

Jensen, N.S. and E. Canale-Parola. 1987. *In* Validation of the publication of new names and new combinations previously effectively published outside the IJSB. List no. 23. Int. J. Syst. Bacteriol. *37*: 179–180.

Johnson, J.L. 1978. Taxonomy of the bacteroides. I. Deoxyribonucleic acid homologies among *Bacteroides fragilis* and other saccharolytic *Bacteroides* species. Int. J. Syst. Bacteriol. *28*: 245–256.

Johnson, J.L. and D.A. Ault. 1978. Taxonomy of the bacteroides. II. Correlation of phenotypic characteristics with deoxyribonucleic acid homology groupings for *Bacteroides fragilis* and other saccharolytic *Bacteroides* species. Int. J. Syst. Bacteriol. *28*: 257–268.

Johnson, J.L. and L.V. Holdeman. 1983. *Bacteroides intermedius* comb. nov. and descriptions of *Bacteroides corporis* sp. nov. and *Bacteroides levii* sp. nov. Int. J. Syst. Bacteriol. *33*: 15–25.

Johnson, J.L., W.E.C. Moore and L.V.H. Moore. 1986. *Bacteroides caccae* sp. nov., *Bacteroides*-merdae sp. nov., and *Bacteroides stercoris* sp. nov. isolated from human feces. Int. J. Syst. Bacteriol. *36*: 499–501.

Jousimies-Somer, H.R., P. Summanen, D.M. Citron, E.J. Baron, H.M. Wexler and S.M. Finegold. 2002. Wadsworth-KTL anaerobic bacteriology manual. Star Publishing Company, Belmont, CA.

Kaneuchi, C. and T. Mitsuoka. 1978. *Bacteroides microfusus*, a new species from intestines of calves, chickens, and Japanese quails. Int. J. Syst. Bacteriol. *28*: 478–481.

Kelly, C.D. 1957. Genus I. *Bacteroides* Castellani and Chalmers 1919. *In* Bergey's Manual of Determinative Bacteriology, 7th edn (edited by Breed, Murray and Smith). Williams & Wilkins, Baltimore, pp. 424–436.

Kitahara, M., M. Sakamoto, M. Ike, S. Sakata and Y. Benno. 2005. *Bacteroides plebeius* sp. nov. and *Bacteroides coprocola* sp. nov., isolated from human faeces. Int. J. Syst. Evol. Microbiol. *55*: 2143–2147.

Kornman, K.S. and S.C. Holt. 1981. Physiological and ultrastructural characterization of a new *Bacteroides* species (*Bacteroides capillus*) isolated from severe localized periodontitis. J. Periodont. Res. *16*: 542–555.

Leadbetter, E.R., S.C. Holt and S.S. Socransky. 1979. *Capnocytophaga*: new genus of Gram-negative gliding bacteria. 1. General characteristics, taxonomic considerations and significance. Arch. Microbiol. *122*: 9–16.

Lin, C., J.W. Urbance and D.A. Stahl. 1994. *Acetivibrio cellulolyticus* and *Bacteroides cellulosolvens* are members of the greater clostridial assemblage. FEMS Microbiol. Lett. *124*: 151–155.

Loesche, W.J., S.S. Socransky and R.J. Gibbons. 1964. *Bacteroides oralis*, proposed new species isolated from the oral cavity of man. J. Bacteriol. *88*: 1329–1337.

Love, D.N., J.L. Johnson, R.F. Jones, M. Bailey and A. Calverley. 1986. *Bacteroides tectum* sp. nov. and characteristics of other nonpigmented *Bacteroides* isolates from soft-tissue infections from cats and dogs. Int. J. Syst. Bacteriol. *36*: 123–128.

Love, D.N., J.L. Johnson, R.F. Jones and A. Calverley. 1987. *Bacteroides salivosus* sp. nov., an asaccharolytic, black-pigmented species from cats. Int. J. Syst. Bacteriol. *37*: 307–309.

Love, D.N. 1995. *Porphyromonas macacae* comb. nov., a consequence of *Bacteroides macacae* being a senior synonym of *Porphyromonas salivosa*. Int. J. Syst. Bacteriol. *45*: 90–92.

Mitsuoka, T., T. Sega and S. Yamamoto. 1965. [Improved methodology of qualitative and quantitative analysis of the intestinal flora of man and animals]. Zentralbl. Bakteriol. [Orig.] *195*: 455–469.

Mitsuoka, T., A. Terada, K. Watanabe and K. Uchida. 1974. *Bacteroides multiacidus*, a new species from feces of humans and pigs. Int. J. Syst. Bacteriol. *24*: 35–41.

Miyamoto, Y. and K. Itoh. 2000. *Bacteroides acidifaciens* sp. nov., isolated from the caecum of mice. Int. J. Syst. Evol. Microbiol. *50*: 145–148.

Moncrief, J.S., R. Obiso, Jr., L.A. Barroso, J.J. Kling, R.L. Wright, R.L. Van Tassell, D.M. Lyerly and T.D. Wilkins. 1995. The enterotoxin of Bacteroides fragilis is a metalloprotease. Infect. Immun. 63: 175–181.

Montgomery, L., B. Flesher and D. Stahl. 1988. Transfer of *Bacteroides succinogenes* (Hungate) to *Fibrobacter* gen. nov. as *Fibrobacter succinogenes* comb. nov. and description of *Fibrobacter intestinalis* sp. nov. Int. J. Syst. Bacteriol. *38*: 430–435.

Moore, L.V.H., J.L. Johnson and W.E.C. Moore. 1994. Descriptions of *Prevotella tannerae* sp. nov. and *Prevotella enoeca* sp. nov. from the human gingival crevice and emendation of the description of *Prevotella zoogleoformans*. Int. J. Syst. Bacteriol. *44*: 599–602.

Moore, L.V.H. and W.E.C. Moore. 1994. *Oribaculum catoniae* gen. nov., sp. nov., *Catonella morbi* gen. nov., sp. nov., *Hallella seregens* gen. nov., sp. nov., *Johnsonella ignava* gen. nov., sp. nov., and *Dialister pneumosintes* gen. nov., comb. nov., nom. rev., anaerobic Gram-negative bacilli from the human gingival crevice. Int. J. Syst. Bacteriol. *44*: 187–192.

Morotomi, M., F. Nagai and H. Sakon. 2007. Genus *Megamonas* should be placed in the lineage of *Firmicutes*; *Clostridia*; *Clostridiales*; 'Acidaminococcaceae'; *Megamonas*. Int. J. Syst. Evol. Microbiol. *57*: 1673–1674.

Murray, W.D., L.C. Sowden and J.R. Colvin. 1984. *Bacteroides cellulosolvens* sp. nov., a cellulolytic species from sewage sludge. Int. J. Syst. Bacteriol. *34*: 185–187.

Oh, H. and C. Edlund. 2003. Mechanism of quinolone resistance in anaerobic bacteria. Clin. Microbiol. Infect. *9*: 512–517.

Okuda, K., T. Kato, J. Shiozu, I. Takazoe and T. Nakamura. 1985. *Bacteroides heparinolyticus* sp. nov. isolated from humans with periodontitis. Int. J. Syst. Bacteriol. *35*: 438–442.

Olitsky, P.K. and F.L. Gates. 1921. Experimental studies of the naso-pharyngeal secretions from influenza patients. J. Exp. Med. *33*: 713–729.

Oliver, W.W. and W.B. Wherry. 1921. Notes on some bacterial parasites of the human mucous membranesq. J. Infect. Dis. *28*: 341–344.

Paster, B.J., F.E. Dewhirst, I. Olsen and G.J. Fraser. 1994. Phylogeny of *Bacteroides*, *Prevotella*, and *Porphyromonas* spp. and related bacteria. J. Bacteriol. *176*: 725–732.

Patel, G.B. and C. Breuil. 1981. Isolation and characterization of *Bacteroides polypragmatus* sp. nov., an isolate which produces carbon dioxide, hydrogen and acetic acid during growth on various organic substrates. *In* Advances in Biotechnology (edited by Moo-Young and Robinson). Pergamon Press, Toronto, pp. 291–296.

Patel, G.B. and C. Breuil. 1982. *In* Validation of the publication of new names and new combinations previously effectively published outside the IJSB. List no. 8. Int. J. Syst. Bacteriol. *32*: 266–268.

Patrick, S. 2002. *Bacteroides*. *In* Molecular Medical Microbiology (edited by Sussman). Academic Press, London, pp. 1921–1948.

Podglajen, I., J. Breuil, I. Casin and E. Collatz. 1995. Genotypic identification of two groups within the species *Bacteroides fragilis* by ribotyping and by analysis of PCR-generated fragment patterns and insertion sequence content. J. Bacteriol. *177*: 5270–5275.

Prévot, A.R. 1938. Etudes de systematique bacterienne. III. Invalidite du genre *Bacteroides* Castellani et Chalmers demembrement et reclassification. Ann. Inst. Pasteur *20*: 285–307.

Prévot, A.R., P. Ardieux, L. Joubert and F. De Cadore. 1956. Recherches sur Fusiformis nucleatus (Knorr) et son pouvoir pathogène pour l'homme et les animaux. Ann. Inst. Pasteur (Paris) *91*: 787–798.

Pribram, E. 1933. Klassification der Schizomyceten. F. Deuticke, Leipzig, pp. 1–143.

Rautio, M., E. Eerola, M.L. Vaisanen-Tunkelrott, D. Molitoris, P. Lawson, M.D. Collins and H. Jousimies-Somer. 2003. Reclassification of *Bacteroides putredinis* (Weinberg et al. 1937) in a new genus *Alistipes* gen. nov., as *Alistipes putredinis* comb. nov., and description of *Alistipes finegoldii* sp. nov., from human sources. Syst. Appl. Microbiol. *26*: 182–188.

Roy, T.E. and C.D. Kelly. 1939. Genus VIII. *Bacteroides* Castellani and Charmers. *In* Bergey's Manual of Determinative Bacteriology, 5th edn (edited by Bergey, Breed, Murray and Hitchens). Williams & Wilkins, Baltimore, pp. 569–570.

Ruimy, R., I. Podglajen, J. Breuil, R. Christen and E. Collatz. 1996. A recent fixation of *cfiA* genes in a monophyletic cluster of *Bacteroides fragilis* is correlated with the presence of multiple insertion elements. J. Bacteriol. *178*: 1914–1918.

Sakamoto, M., M. Suzuki, M. Umeda, I. Ishikawa and Y. Benno. 2002. Reclassification of *Bacteroides forsythus* (Tanner et al. 1986) as *Tannerella forsythensis* corrig., gen. nov., comb. nov. Int. J. Syst. Evol. Microbiol. 52: 841–849.

Sakamoto, M. and Y. Benno. 2006. Reclassification of *Bacteroides distasonis*, *Bacteroides goldsteinii* and *Bacteroides merdae* as *Parabacteroides distasonis* gen. nov., comb. nov., *Parabacteroides goldsteinii* comb. nov. and *Parabacteroides merdae* comb. nov. Int. J. Syst. Evol. Microbiol. *56*: 1599–1605.

Sakamoto, M., M. Kitahara and Y. Benno. 2007. *Parabacteroides johnsonii* sp. nov., isolated from human faeces. Int. J. Syst. Evol. Microbiol. *57*: 293–296.

Schleifer, K.H. and O. Kandler. 1972. Peptidoglycan types of bacterial cell walls and their taxonomic implications. Bacteriol. Rev. *36*: 407–477,

Scholten-Koerselman, I., F. Houwaard, P. Janssen and A.J.B. Zehnder. 1986. *Bacteroides xylanolyticus* sp. nov., a xylanolytic bacterium from methane producing cattle manure. Antonie van Leeuwenhoek *52*: 543–554.

Scholten-Koerselman, I., F. Houwaard, P. Janssen and A.J.B. Zehnder. 1988. *In* Validation of the publication of new names and new combinations previously effectively published outside the IJSB. List no. 24. Int. J. Syst. Bacteriol. *38*: 136–137.

Sears, C.L., L.L. Myers, A. Lazenby and R.L. Van Tassell. 1995. Enterotoxigenic *Bacteroides fragilis*. Clin. Infect. Dis. *20 Suppl 2*: S142–S148.

Sebald, M. 1962. Étude sur les bactéries anaérobies gram-négatives asporulées. Thèses de l'Université Paris, Imprimerie Barnéoud S.A., Laval, France.

Shah, H.N. and M. Collins. 1981. *Bacteroides buccalis*, sp. nov., *Bacteroides denticola*, sp. nov., and *Bacteroides pentosaceus*, sp. nov., new species of the genus *Bacteroides* from the oral cavity. Zentralbl. Bakteriol. Parasitenkd. Infektionskr. Hyg. I Abt. Orig. C. *2*: 235–241.

Shah, H.N. and M.D. Collins. 1982a. Reclassification of *Bacteroides multiacidus* (Mitsuoka, Terada, Watanabe and Uchida) in a new genus *Mitsuokella*, as *Mitsuokella* multiacidus comb. nov. Zentralbl. Bakteriol. Parasitenkd. Infektionskr. Hyg. I Abt. Orig. C. *3*: 491–494.

Shah, H.N. and M.D. Collins. 1982b. Reclassification of *Bacteroides hypermegas* (Harrison and Hansen) in a new genus *Megamonas*, as *Megamonas* hypermegas comb. nov. Zentralbl. Bakteriol. Parasitenkd. Infektionskr. Hyg. Abt. 1 Orig. *C3*: 394–398.

Shah, H.N. and M.D. Collins. 1983. Genus *Bacteroides*. A chemotaxonomical perspective. J. Appl. Bacteriol. *55*: 403–416.

Shah, H.N., M.D. Collins, J. Watabe and T. Mitsuoka. 1985. *Bacteroides oulorum* sp. nov., a nonpigmented saccharolytic species from the oral cavity. Int. J. Syst. Bacteriol. *35*: 193–197.

Shah, H.N. and M.D. Collins. 1986. Reclassification of *Bacteroides furcosus* Veillon and Zuber (Hauduroy, Ehringer, Urbain, Guillot and Magrou) in a new genus *Anaerorhabdus*, as *Anaerorhabdus* furcosus comb. nov. Syst. Appl. Microbiol. *8*: 86–88.

Shah, H.N. and M.D. Collins. 1986. *In* Validation of the publication of new names and new combinations previously effectively published outside the IJSB. List no. 22. Int. J. Syst. Bacteriol. *36*: 573–576.

Shah, H.N. and M.D. Collins. 1988. Proposal for reclassification of *Bacteroides asaccharolyticus*, *Bacteroides gingivalis*, and *Bacteroides endodontalis* in a new genus, *Porphyromonas*. Int. J. Syst. Bacteriol. *38*: 128–131.

Shah, H.N. and M.D. Collins. 1989. Proposal to restrict the genus *Bacteroides* (Castellani and Chalmers) to *Bacteroides fragilis* and closely related species. Int. J. Syst. Bacteriol. *39*: 85–87.

Shah, H.N. and D.M. Collins. 1990. *Prevotella*, a new genus to include *Bacteroides melaninogenicus* and related species formerly classified in the genus *Bacteroides*. Int. J. Syst. Bacteriol. *40*: 205–208.

Shah, H.N., M.D. Collins, I. Olsen, B.J. Paster and F.E. Dewhirst. 1995. Reclassification of *Bacteroides levii* (Holdeman, Cato, and Moore) in the genus *Porphyromonas*, as *Porphyromonas levii* comb. nov. Int. J. Syst. Bacteriol. *45*: 586–588.

Slots, J. and R.J. Genco. 1980. *Bacteroides melaninogenicus* subsp. *macacae*: new subspecies from monkey periodontopathic indigenous microflora. Int. J. Syst. Bacteriol. *30*: 82–85.

Snydman, D.R., N.V. Jacobus, L.A. McDermott, R. Ruthazer, E.J. Goldstein, S.M. Finegold, L.J. Harrell, D.W. Hecht, S.G. Jenkins, C. Pierson, R. Venezia, J. Rihs and S.L. Gorbach. 2002. National survey on the susceptibility of *Bacteroides fragilis* Group: report and analysis of trends for 1997–2000. Clin. Infect. Dis. *35*: S126–134.

Song, Y., C. Liu, J. Lee, M. Bolanos, M.L. Vaisanen and S.M. Finegold. 2005a. "*Bacteroides goldsteinii* sp. nov." isolated from clinical specimens of human intestinal origin. J. Clin. Microbiol. *43*: 4522–4527.

Song, Y., C. Liu, J. Lee, M. Bolaños, M.-L. Vaisanen and S.M. Finegold. 2006. *In* List of new names and new combinations previously effectively, but not validly, published. List no. 108. Int. J. Syst. Evol. Microbiol. *56*: 499–500.

Song, Y.L., C.X. Liu, M. McTeague and S.M. Finegold. 2004. "*Bacteroides nordii*" sp. nov. and "*Bacteroides* salyersae" sp. nov. isolated from clinical specimens of human intestinal origin. J. Clin. Microbiol. *42*: 5565–5570.

Song, Y.L., C.X. Liu, M. McTeague and S.M. Finegold. 2005b. *In* Validation of publication of new names and new combinations previously effectively published outside the IJSEM. List no. 103. Int. J. Syst. Evol. Microbiol. *55*: 983–985.

Soutschek, E., J. Winter, F. Schindler and O. Kandler. 1984. *Acetomicrobium flavidum*, gen. nov., sp. nov., a thermophilic, anaerobic bacterium from sewage-sludge, forming acetate, CO_2 and H_2 from glucose. Syst. Appl. Microbiol. *5*: 377–390.

Soutschek, E., J. Winter, F. Schindler and O. Kandler. 1985. *In* Validation of the publication of new names and new combinations previously effectively published outside the IJSB. List no.17. Int. J. Syst. Bacteriol. *35*: 223–225.

Stackebrandt, E. and H. Hippe. 1986. Transfer of *Bacteroides amylophilus* to a new genus *Ruminobacter* gen. nov., nom. rev. as *Ruminobacter amylophilus* comb. nov. Syst. Appl. Microbiol. *8*: 204–207.

Tanner, A.C.R., S. Badger, C.H. Lai, M.A. Listgarten, R.A. Visconti and S.S. Socransky. 1981. *Wolinella* gen-nov, *Wolinella succinogenes* (*Vibrio succinogenes* Wolin et al.) comb. nov., and description of *Bacteroides gracilis* sp. nov., *Wolinella recta* sp. nov., *Campylobacter concisus* sp. nov., and *Eikenella corrodens* from humans with periodontal disease. Int. J. Syst. Bacteriol. *31*: 432–445.

Tanner, A.C.R., M.A. Listgarten, J.L. Ebersole and M.N. Strezempko. 1986. *Bacteroides forsythus* sp. nov., a slow-growing, fusiform *Bacteroides* sp. from the human oral cavity. Int. J. Syst. Bacteriol. *36*: 213–221.

Tissier, H. 1908. Recherches sur la flore intestinale normale des enfants agés d'un an à cinq ans. Ann. Inst. Pasteur (Paris) *22*: 189–208.

Topley, W.W.C. and G.S. Wilson. 1929. The Principles of Bacteriology and Immunity, vol. 1. Edward Arnold, London.

Van Steenbergen, T.J.M., A.J. Van Winkelhoff, D. Mayrand, D. Grenier and J. De Graaff. 1984. *Bacteroides endodontalis* sp. nov., an asaccharolytic black-pigmented bacteriodes species from infected dental root canals. Int. J. Syst. Bacteriol. *34*: 118–120.

Vandamme, P., M.I. Daneshvar, F.E. Dewhirst, B.J. Paster, K. Kersters, H. Goossens and C.W. Moss. 1995. Chemotaxonomic analyses of *Bacteroides gracilis* and *Bacteroides ureolyticus* and reclassification of B. gracilis as *Campylobacter gracilis* comb. nov. Int. J. Syst. Bacteriol. *45*: 145–152.

Veillon, A. and A. Zuber. 1898. Recherches sur quelques microbes strictement anaérobies et leur rôle en pathologie. Arch. Med. Exp. *10*: 517–545.

Watabe, J., Y. Benno and T. Mitsuoka. 1983. Taxonomic study of *Bacteroides oralis* and related organisms and proposal of *Bacteroides veroralis* sp. nov. Int. J. Syst. Bacteriol. *33*: 57–64.

Weinberg, M., R. Nativelle and A.R. Prévot. 1937. Les microbes anaérobies. Masson et Cie, Paris.

Werner, H., G. Rintelen and H. Kunstek-Santos. 1975. A new butyric acid-producing *Bacteroides* species: *B. splanchnicus* n. sp. Zentralbl. Bakteriol. Parasitenkd. Infektionskr. Hyg. Abt. I *231*: 133–144.

Wexler, H.M. and S.M. Finegold. 1998. Current susceptibility patterns of anaerobic bacteria. Yonsei Med. J. *39*: 495–501.

Whitehead, T.R., M.A. Cotta, M.D. Collins, E. Falsen and P.A. Lawson. 2005. *Bacteroides coprosuis* sp. nov., isolated from swine-manure storage pits. Int. J. Syst. Evol. Microbiol. *55*: 2515–2518.

Winter, J., E. Braun and H.P. Zabel. 1987. *Acetomicrobium faecalis* spec. nov., a strictly anaerobic bacterium from sewage sludge, producing ethanol from pentoses. Syst. Appl. Microbiol. *9*: 71–76.

Winter, J., E. Braun and H.P. Zabel. 1988. *In* Validation of the publication of new names and new combinations previously effectively published outside the IJSB. List no. 24. Int. J. Syst. Bacteriol. *38*: 136–137.

Woese, C.R. 1987. Bacterial evolution. Microbiol. Rev. *51*: 221–271.

Wolin, E.A., M.G. Wolin and R.S. Wolfe. 1963. Formation of methane by bacterial extracts. J. Biol. Chem. *238*: 2882–2886.

Wu, C.C., J.L. Johnson, W.E.C. Moore and L.V.H. Moore. 1992. Emended descriptions of *Prevotella denticola*, *Prevotella loescheii*, *Prevotella veroralis*, and *Prevotella melaninogenica*. Int. J. Syst. Bacteriol. *42*: 536–541.

Xu, J., M.K. Bjursell, J. Himrod, S. Deng, L.K. Carmichael, H.C. Chiang, L.V. Hooper and J.I. Gordon. 2003. A genomic view of the human–*Bacteroides thetaiotaomicron* symbiosis. Science *299*: 2074–2076.

Family II. **Marinilabiliaceae** fam. nov.

Wolfgang Ludwig, Jean Euzéby and William B. Whitman

Ma.ri.ni.la.bi.li.a.ce′ae. N.L. fem. n. *Marinilabilia* type genus of the family; suff. *-aceae* ending to denote family; N.L. fem. pl. n. *Marinilabiliaceae* the *Marinilabilia* family.

The family *Marinilabiliaceae* was circumscribed for this volume on the basis of phylogenetic analysis of 16S rRNA gene sequences and includes the type genus *Marinilabilia*, *Alkaliflexus*, and *Anaerophaga*. All genera are composed of **slender, flexible rods**. Except for *Anaerophaga*, which does not grow on solid surfaces in culture, cells are **motile by gliding**. Cells **stain Gram-negative**. **Saccharolytic** and **require NaCl** for growth. Some characteristics that differentiate the three genera are given in Table 8.

Type genus: **Marinilabilia** Nakagawa and Yamasato 1996, 600[VP].

TABLE 8. Some characteristics that differentiate the genera of the family *Marinilabiliaceae*[a]

Characteristic	*Marinilabilia*	*Alkaliflexus*	*Anaerophaga*
Motile by gliding	+	+	nt
Can grow aerobically	+	–	–
Major products of glucose fermentation	Acetate, succinate, propionate, lactate, formate and H_2	Propionate, acetate, succinate and formate	Propionate, acetate, and succinate
Growth above 50°C	–	–	+
Isolated from	Marine mud containing decayed algae	Mud from an alkaline soda lake	Oil contaminated sediment in an oil separation tank
DNA G+C content (mol%)	37–41	44	41

[a]nt, Not tested.

Genus I. **Marinilabilia** Nakagawa and Yamasato 1996, 600[VP]

Makoto Suzuki

Ma.ri.ni.la.bil′i.a. L. adj. *marinus* marine, pertaining to the sea; L. adj. *labilis* gliding; N.L. fem. n. *Marinilabilia* marine gliding organisms.

Short to very long flexible rods, 0.3–0.5 × 2–50 µm. Resting stage is absent. **Motile by gliding. Gram-negative.** Colonies are pink to salmon in color. **Chemo-organotrophic. Facultatively anaerobic**, having both respiratory and fermentative types of metabolism. Catalase-positive. Decompose several kinds of biomacromolecules. **Marine organisms** requiring elevated salt concentrations. Optimum temperature: 28–37°C. Optimum pH: around 7. The major polyamine is spermidine. **The major respiratory quinone is MK-7.**

DNA G+C content (mol%): 41 (HPLC).

Type species: **Marinilabilia salmonicolor** (Veldkamp 1961) Nakagawa and Yamasato 1996, 600[VP] emend. Suzuki, Nakagawa, Harayama and Yamamoto 1999, 1555 (*Cytophaga salmonicolor* Veldkamp 1961, 339).

Further descriptive information

Phylogenetic analysis based on 16S rRNA gene sequences indicates that the genus *Marinilabilia* belongs to the family "*Marinilabiaceae*". The phylogenetic relationships of the genus *Marinilabilia* and related organisms are shown in Figure 15. The nearest neighbor is the genus *Anaerophaga* and the similarities between the species of *Marinilabilia* and *Anaerophaga* are 90.3–92.4%. Two misclassified facultatively anaerobic *Cytophaga* species, *Cytophaga fermantans* and *Cytophaga xylanolytica*, are more distantly related to members of the genus *Marinilabilia*. There is 86.8–89.6% similarity between these two organisms and members of the genus *Marinilabilia*.

Cells of *Marinilabilia salmonicolor* are slender rods with rounded or slightly tapered ends. They are extremely flexible and motile by gliding. The cells usually vary from 2 to 30 µm in length, but occasionally, very long elements of up to 50 µm long can be seen. These long cells occur even in young cultures and often show flexing and gliding movements. The mean length of cells grown aerobically on agar plates is usually greater than that of cells from stab or anaerobic liquid cultures. In the stationary phase, spherical cells are always found. They vary in diameter from 1.0 to 3.5 µm.

Spermidine is the dominant cellular polyamine when cells are cultivated in marine broth (Hosoya and Hamana, 2004).

Strains of the genus *Marinilabilia* can grow well anaerobically in glucose-mineral medium supplemented with a vitamin mixture (Veldkamp, 1961). Under anaerobic conditions, growth is poor when bicarbonate is omitted from the medium. Under aerobic conditions, *Marinilabilia* strains can grow well when the medium contains 0.1% (w/v) yeast extract, corn steep liquor, or nutrient broth.

Products of fermentation are acetic acid, succinic acid, propionic acid, lactic acid, formic acid, CO_2, H_2, and trace amounts of ethanol (Veldkamp, 1961).

Marinilabilia strains inhabit coastal marine sediments. They have been isolated from marine mud containing decayed algae. One biovar can actively degrade agar.

Enrichment and isolation procedures

Organisms of this group have been recognized as incidentally occurring organisms in enrichment cultures of green sulfur bacteria (Veldkamp, 1961). They were originally enriched by using anaerobic cultures in which agar was the carbon and energy source. The anaerobic enrichment cultures were incubated in the dark at 30°C for 3–5 d. Anaerobic plate cultures were used for the isolation of these organisms. The medium used for these cultures contained (in g/l): NaCl, 30; KH_2PO_4, 1; NH_4Cl, 1; $MgCl_2·6H_2O$, 0.5; $CaCl_2$, 0.04; $NaHCO_3$, 5; $Na_2S·9H_2O$, 0.1; ferric citrate, 0.03; agar, 5; yeast extract, 0.3; pH 7. The medium was supplemented with a trace element mixture (Veldkamp, 1961).

Maintenance procedures

Cultures are maintained as paraffin-covered stab cultures at 4°C; transfers are made monthly. Cultures can be kept in broth containing 10% glycerol at −80°C or in the gas phase of liquid nitrogen. They can be preserved by a liquid-drying method using 75% bovine serum in marine broth as a protective reagent.

Differentiation of the genus *Marinilabilia* from other genera

Marinilabilia can be differentiated from related organisms by its relation to oxygen, salt requirement, some physiological characters, and cell mass color. Differential features are given in Table 9.

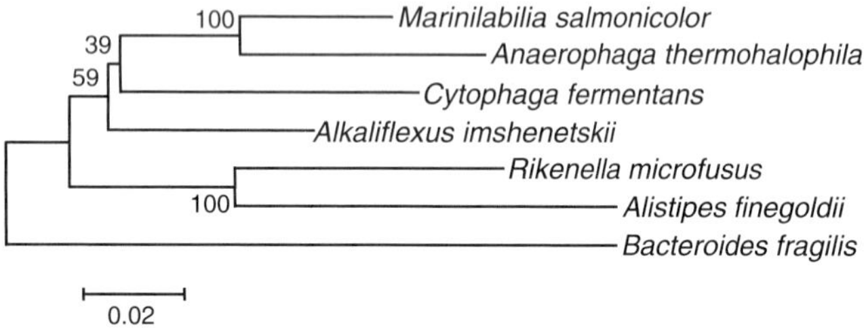

FIGURE 15. 16S rRNA gene sequence-based phylogenetic tree of the genus *Marinilabilia* and related organisms. The tree was constructed by the neighbor-joining method. The numbers at nodes show the percentage bootstrap value of 500 replicates. Bar = evolutionary distance of 0.02 (K_{nuc}).

Marinilabilia can be differentiated from *Anaerophaga, Alkaliflexus, Alistipes*, and *Rikenella* by its relation to oxygen. It can also be differentiated from *Cytophaga fermentans* by its cell mass color and from *Cytophaga xylanolytica* by its salt requirement.

Taxonomic comments

There are two known isolates of the genus *Marinilabilia.* Strain NBRC 15948^T (=ATCC 19041^T), the type strain of *Marinilabilia salmonicolor*, was isolated from marine mud off the coast of California. The second strain, NBRC 14957 (=ATCC 19043), which was isolated from the same environment, has the same characteristics as the type strain and, in addition, is able to degrade agar. These two strains were originally reported as *Cytophaga salmonicolor* and "*Cytophaga salmonicolor* var. *agarovorans*" by Veldkamp (1961). Reichenbach (1989) described *Cytophaga agarovorans* to accommodate "*Cytophaga salmonicolor* var. *agarovorans*", based on its remarkable agar-degrading ability and higher DNA G+C content. Later, Nakagawa and Yamasato (1996) included these two organisms in the genus *Marinilabilia* as *Marinilabilia salmonicolor* and *Marinilabilia agarovorans.* Phylogenetic analysis of these two organisms using 16S rRNA gene and *gyrB* sequences indicated their close relatedness (Nakagawa and Yamasato, 1996; Suzuki et al., 1999). In particular, the *gyrB* sequences of strains NBRC 15948^T and NBRC 14957 are identical, as determined for a 1.2 kb region. DNA–DNA hybridization also confirmed their close relatedness: over 70% DNA relatedness was found between the two strains (Suzuki et al., 1999). Reichenbach (1989) stated that there was a small percentage difference in the G+C content of the DNA between *Marinilabilia agarovorans* and *Marinilabilia salmonicolor*, namely, 41 vs. 37 mol%, respectively (buoyant density method). A second determination of the G+C content of *Marinilabilia* strains was performed by the HPLC method (Suzuki et al., 1999) and indicated that the G+C content was almost the same for the two strains; by this method,

TABLE 9. Differential characteristics of *Marinilabilia* and related organisms[a]

Characteristic	*Marinilabilia*	*Anaerophaga*	*Alkaliflexus*	*Cytophaga fermentans*	*Cytophaga xylanolytica*	*Alistipes*	*Rikenella*
Cell size, μm	0.3–0.5 × 2–20	0.3 × 4–8	0.3–0.4 × 4–10	0.5–0.7 × 2–10	0.4 × 3–24	0.3–0.5 × 0.9–3.0	0.2–0.3 × 0.3–1.5
Gliding motility	+	−	+	+	+	−	−
Cell mass color	Pink to salmon	Orange to red	Pink	Yellow	Orange to salmon	Gray, brown to black	Gray
Relation to oxygen							
Facultative anaerobe	+	−	−	+	−	−	−
Strict anaerobe	−	+	+	−	−	+	+
Aerotolerant anaerobe	−	−	−	−	+	−	−
Catalase	+	−	+	+	w	trace	nd
Salt requirement	+	+	+	+	−	nd	−
DNA G+C content (mol%)	37–41	42	44	39	46	55–58	60–61

[a]Symbols: +, >85% positive; −, 0–15% positive; w, weak reaction; nd, not determined.

the DNA G+C contents of strains NBRC 14957 (*Marinilabilia agarovorans*) and NBRC 15948[T] (*Marinilabilia salmonicolor*) were 41.2 and 41.5 mol%, respectively. All of the above information indicates that these two strains belong to a single species. Since *Marinilabilia salmonicolor* was first described as *Cytophaga salmonicolor*, *Marinilabilia salmonicolor* is a senior subjective synonym of *Marinilabilia agarovorans*. However, the agar-degrading ability of strain NBRC 14957 is a prominent biochemical characteristic; therefore, this strain is considered as *Marinilabilia salmonicolor* biovar Agarovorans.

List of species of the genus *Marinilabilia*

1. **Marinilabilia salmonicolor** (Veldkamp 1961) Nakagawa and Yamasato 1996, 600[VP] emend. Suzuki, Nakagawa, Harayama and Yamamoto 1999, 1555 (*Cytophaga salmonicolor* Veldkamp 1961, 339)

sal.mo.ni′co.lor. L. masc. n. *salmo -onis* salmon; L. masc. n. *color* color; N.L. adj. *salmonicolor* intended to mean salmon-colored.

Slender flexible cylindrical rods with rounded or slightly tapering end, 0.3–0.5 × 2–50 µm, usually around 10–20 µm in length. Resting stages are not observed. Motile by gliding. Gram-negative. The color of the cell mass is salmon to pink. Chemo-organotrophic. Facultative anaerobe. Good growth is obtained in a normal aerobic atmosphere. Marine strains require elevated salt concentrations. Peptones, Casamino acids, yeast extract, ammonium, and nitrate are suitable nitrogen sources. Arabinose, xylose, glucose, galactose, mannose, fructose, sucrose, lactose, maltose, cellobiose, trehalose, raffinose, inulin, and starch are fermented. Gelatin is liquefied slowly. Catalase is produced. The optimum temperature is 28–37°C. The optimum pH is around 7. The major respiratory quinone is MK-7. The major polyamine is spermidine.

Source: marine mud containing decayed algae.

DNA G+C content (mol%): 41.2–41.5 (HPLC).

Type strain: ATCC 19041, CIP 104809, DSM 6480, LMG 1346, NBRC 15948.

Note: strain NBRC 14957 can degrade agar and is biovar Agarovorans.

Genus II. **Anaerophaga** Denger, Warthmann, Ludwig and Schink 2002, 177[VP]

BERNHARD SCHINK

An.a.e.ro.pha′ga. an Gr. pref. *an* not; Gr. n. *aer aeros* air; Gr. v. *phagein* to devour, to eat up; N.L. fem. n. *Anaerophaga* an anaerobic eater.

Slender flexible rods with rounded ends. Strictly anaerobic. Chemo-organotrophic. Nonphotosynthetic. Have a fermentative type of metabolism, using organic compounds as substrates. Inorganic electron acceptors are not used. Media containing a reductant, e.g., cysteine, are necessary for growth. Catalase-negative. Isolated from oil-contaminated sediment in an oil separation tank.

Type species: **Anaerophaga thermohalophila** Denger, Warthmann, Ludwig and Schink 2002, 177[VP].

DNA G+C content (mol%): 42.

Further descriptive information

The only described species so far is *Anaerophaga thermohalophila*. The strains Fru22[T] and Glc12 were isolated as possible agents for tenside (surfactant) production in microbially improved oil recovery (MIOR). Therefore, the organism was enriched and selected under conditions of elevated temperature (50°C) and increased salinity (7.5%, w/v).

The isolated strains ferment hexoses and pentoses to equal molar amounts of acetate, propionate, and succinate (Denger and Schink, 1995; Denger et al., 2002), according to the following equations:

$$3\,C_6H_{12}O_6 \rightarrow 2\,C_2H_3O_2^- + 2\,C_3H_5O_2^- + 2\,C_4H_4O_4^{2-} + 8\,H^+ + 2\,H_2O, \text{ or}$$
$$9\,C_5H_{10}O_5 \rightarrow 5\,C_2H_3O_2^- + 5\,C_3H_5O_2^- + 5\,C_4H_4O_4^{2-} + 20\,H^+ + 5\,H_2O$$

No CO_2 is released in this type of mixed-acid fermentation.

Upon prolonged exposure of fully grown cultures to daylight, an orange-red pigment was produced, which could be extracted from cell pellets by acetone or hexane, indicating that it was a lipophilic component. Absorption spectra of these extracts showed maxima at 488 and 518 nm and a further shoulder around wavelength 460 nm, as is typical of carotenoids (Reichenbach et al., 1974). Carotenoids are known to be produced by several representatives of aerobic gliding bacteria, including *Flexibacter* sp., *Cytophaga* sp., and several myxobacteria (Reichenbach and Dworkin, 1981). Strain Fru22[T] is the first strict anaerobe producing such pigments, although the moderately oxygen-tolerant anaerobe *Cytophaga xylanolytica* also produces carotenoids (Haack and Breznak, 1993). When pigmented cell material after centrifugation was treated with 10% KOH, the color turned dark-red to brownish, similar to the flexirubins of *Cytophaga*, *Sporocytophaga*, and *Flexibacter* spp. (Achenbach et al., 1978; Reichenbach et al., 1974).

Unfortunately, these strains do not grow on agar surfaces, even if the medium is poured and stored in an oxygen-free glove box under a N_2/CO_2 atmosphere (90/10, v/v). Thus, whether gliding motility occurs could not be determined. Spore-like structures or spheres are observed in ageing cultures.

Surface-active compounds were produced by strain Fru22[T] optimally at the end of exponential growth (Denger and Schink, 1995). This tenside efficiently stabilized hexadecane/water emulsions. Production was enhanced in the presence of hexadeca ne, which provides a lipophilic surface in the culture. The surface-active compound(s) was associated partly with the cells and cell surfaces, but was also released into the culture medium.

Enrichment and isolation procedures

Anaerophaga thermohalophila strain Fru22^T and a similar strain, Glc12, were enriched originally from blackish-oily sedimentary residues in an oil separation tank near Hannover, Germany. The mineral salts medium for enrichment and cultivation was bicarbonate-buffered (50 mM), cysteine-reduced (1 mM), and contained, together with other minerals, 75 g of NaCl and 4.0 g of $MgCl_2 \cdot 7H_2O$ per liter (Denger and Schink, 1995). The pH was 6.7–6.8. During the enrichment, the medium received a few drops (50 µl per 25 ml of medium) of hexadecane to provide a lipophilic boundary layer. Subcultures were inoculated with oily drops from the surface of the preculture. All details concerning cultivation and physiological characterization have been described before (Denger and Schink, 1995). Growth rates and yields were best at 20–70 g of total salt per liter. At 80 g of salt per liter, growth was partly inhibited, and there was no growth at <20 g of salt per liter.

Maintenance procedures

Cultures are maintained either by repeated transfer at intervals of 2–3 months or by freezing in liquid nitrogen using techniques common for strictly anaerobic bacteria. No information exists about survival upon lyophilization.

Differentiation of the genus *Anaerophaga* from other genera

There are substantial differences between *Anaerophaga thermohalophila* strain Fru22^T, *Cytophaga fermentans*, and the described species of the genus *Marinilabilia*. First, strain Fru22 is strictly anaerobic. Moreover, the growth parameters of strain Fru22, especially its growth up to 55°C and its salt tolerance up to 12% (w/v) salt with an optimum around 6% salt, clearly separate this strain from all described *Cytophaga* and *Marinilabilia* species. So far, only few *Cytophaga* species show temperature maxima at 40–45°C (*Cytophaga aprica*, *Cytophaga lytica*) or 45°C (*Cytophaga diffluens*); all three species also tolerate NaCl concentrations up to 6%, but, unlike *Anaerophaga thermohalophila*, all are strict aerobes. *Cytophaga aprica* and *Cytophaga diffluens* have been reclassified recently as *Flammeovirga aprica* and *Persicobacter diffluens* (Nakagawa et al., 1997).

The genus *Capnocytophaga* has been separated from other *Cytophaga* species as a genus of facultatively anaerobic bacteria that need CO_2 at enhanced concentrations for efficient growth. In addition, *Capnocytophaga* species produce acetate, propionate, and succinate during sugar fermentation; however, all have been described as facultatively aerobic organisms, whereas *Anaerophaga thermohalophila* is strictly anaerobic. Moreover, all species of *Capnocytophaga* described so far are associated with higher animals, especially the oral cavity of man (Leadbetter et al., 1979). No thermophilic or halophilic representatives of *Capnocytophaga* are known.

Taxonomic comments

As indicated by phylogenetic analysis of 16S rDNA sequences, strain Fru22^T is a member of the phylum *Bacteroidetes*, comprising *Cytophaga*, *Flavobacterium*, and *Bacteroides* as representative genera. Based on the similarities of 16S rRNA gene sequences, the closest relatives are representatives of the genus *Marinilabilia* (formerly *Cytophaga*; Suzuki et al., 1999). However, overall sequence similarities of around 91.6% indicate only a moderate relationship and justify the establishment of a new genus, *Anaerophaga*. These genera share a common origin with *Cytophaga fermentans* (Bachmann, 1955); the corresponding overall 16S rRNA gene sequence similarities are 87.9–89.1%. Strain Fru22 shares several properties with *Cytophaga* and *Marinilabilia* species, e.g., its morphology (thin, slender rods), production of a flexirubin-like pigment, and production of sphere-like structures in ageing cultures (Reichenbach, 1992; Staley et al., 1989). Moreover, the facultatively anaerobic species *Cytophaga fermentans*, *Cytophaga xylanolytica*, *Marinilabilia salmonicolor*, and *Marinilabilia agarovorans* all have been described to produce acetate together with propionate and succinate as main products of sugar fermentation (Haack and Breznak, 1993; Staley et al., 1989). The G+C content of the DNA of *Cytophaga fermentans* and *Marinilabilia* sp. is in the range of 30–42 mol%; strain Fru22 (41.8 mol%) would be at the upper limit of this range.

List of species of the genus *Anaerophaga*

1. **Anaerophaga thermohalophila** Denger, Warthmann, Ludwig and Schink 2002, 177^{VP}

ther.mo.ha.lo'phi.la. Gr. n. *thermê* heat; Gr. n. *hals*, *halos* salt; N.L. adj. *philus -a -um* (from Gr. adj. *philos -ê -on*) friend, loving; N.L. fem. adj. *thermohalophila* heat and salt loving.

Slender flexible rods (0.3 × 3–8 µm) with rounded ends. Formation of spheres and spore-like structures occur in ageing cultures. Strictly anaerobic. Catalase- and oxidase-negative. Cytochromes of the *b*-type present. Glucose, fructose, arabinose, xylose, cellobiose, mannose, trehalose, raffinose, galactose, starch, and lactose are used for growth. Hexoses and pentoses are fermented to equimolar amounts of acetate, propionate, and succinate. No growth occurs with ribose, sorbose, rhamnose, dulcitol, mannitol, glycerol, glycerate, tartrate, malate, glycolate, lactate, pyruvate, succinate, fumarate, methanol, ethanol, ethylene glycol, acetoin, alanine, serine, threonine, glutamate, aspartate, proline, cellulose, arabinogalactan, chitin, yeast extract, or peptone. Growth requires media with an enhanced CO_2/bicarbonate content and salt concentrations of at least 2% (w/v). Growth occurs with 2–12% (w/v) salt (optimum, 2–6%). Growth occurs at 37–55°C (optimum, 50°C); no growth occurs at 30 and 60°C. Ageing cultures exposed to daylight form an orange-red carotenoid pigment similar to flexirubin. Habitat: anoxic subsurface sites of enhanced temperature and salt content.

Source: contaminated oil tank sediment.

DNA G+C content (mol%): 41.8 ± 0.7 (HPLC).

Type strain: Fru 22, DSM 12881, OCM 798.

Sequence accession no. (16S rRNA gene): AJ418048.

Genus III. **Alkaliflexus** Zhilina, Appel, Probian, Llobet Brossa, Harder, Widdel and Zavarzin 2005, 1395[VP] (Effective publication: Zhilina, Appel, Probian, Llobet Brossa, Harder, Widdel and Zavarzin 2004, 251.)

The Editorial Board

Al.ka.li.flex'us. N.L. n. *alkali* (from Ar. *al-qalyi* the ashes of saltwort) soda ash; L. masc. part. adj. *flexus* bent; N.L. masc. n. *Alkaliflexus* referring to life in basic surroundings and to bending/flexible cells.

Thin, flexible rods with slightly tapered ends. A nonrefractile, spherical, cyst-like structure develops at one pole during late growth phase. **Gliding motility is present.** Endospores are not formed. Fruiting bodies are not formed. Gram-negative. **Anaerobic**, with low to modest tolerance to oxygen. Organisms have a strictly **fermentative** type of metabolism; exogenous electron acceptors are not used. Mesophilic. **Alkaliphilic**, grows at pH 7.5–10.2 (optimum, approx. pH 8.5). Sodium ions are required for growth (optimum, 2%; range, 0.08–5.3%). Carbonate/bicarbonate ions (but not chloride) are also required for growth. Cell mass is yellow to pink due to carotenoids. The temperature range for growth is 10–44°C (optimum, 35°C); no growth occurs at 5 or 55°C. Chemoorganotrophic. Several carbohydrates are utilized, particularly those that may result from the hydrolysis of cellulose, hemicelluloses, and other natural polysaccharides. Isolated from a cellulose-degrading enrichment culture originating from an alkaline soda lake.

DNA G+C content (mol%): 43–44.

Type species: **Alkaliflexus imshenetskii** Zhilina, Appel, Probian, Llobet Brossa, Harder, Widdel and Zavarzin 2005, 1395[VP] (Effective publication: Zhilina, Appel, Probian, Llobet Brossa, Harder, Widdel and Zavarzin 2004, 251.).

Enrichment and isolation procedures

Anaerobic cellulose-decomposing alkaliphilic communities were initially enriched in the alkaline (pH 10) soda medium described by Kevbrin et al. (1999). (*Note*: *Alkaliflexus* does not utilize cellulose and was probably stimulated by cellulose degradation products such as cellobiose that were produced by other bacteria.) After isolation by the methods described by Widdel and Bak (1992), cultures were propagated in the medium of Nedashkovskaya et al. (2006)*.

Differentiation of the genus *Alkaliflexus* from related genera

The adaptation to high pH (optimum, pH 8.5) is a physiological feature that distinguishes the present isolates from all *Marinilabilia*, *Cytophaga*, and *Anaerophaga* species, as well as from any hitherto known member of the phylum *Bacteroidetes*. Moreover, *Marinilabilia* and *Cytophaga* are facultative anaerobes, whereas *Alkaliflexus* is a strict anaerobe. *Anaerophaga* is catalase-negative, whereas *Alkaliflexus* is catalase-positive. Other distinguishing features have been indicated by Nedashkovskaya et al. (2006).

Taxonomic comments

Analysis of 16S rRNA gene sequences indicates that the closest phylogenetic relatives of *Alkaliflexus* are *Marinilabilia salmonicolor*, *Cytophaga fermentans*, and *Anaerophaga thermophila*.

List of species of the genus *Alkaliflexus*

1. **Alkaliflexus imshenetskii** Zhilina, Appel, Probian, Llobet Brossa, Harder, Widdel and Zavarzin 2005, 1395[VP] (Effective publication: Zhilina, Appel, Probian, Llobet Brossa, Harder, Widdel and Zavarzin 2004, 251.)

im.she.net'ski.i. N.L. gen. masc. n. *imshenetskii* of Imshenetskii, named after Aleksandr A. Imshenetskii (1905–1992), a microbiologist who devoted much of his research to gliding bacteria and the microbial degradation of cellulose.

Characteristics are as described for the genus, with the following additional features. Cell dimensions are 0.25–0.4 × 4–10 μm. Cells usually occur singly or in pairs. Cells survive desiccation. Cells are yellow to salmon-colored or pink due to the presence of a carotenoid or carotenoids. No dependence of color on the presence of light during growth has been observed. The temperature range for growth is 10–44°C, with the optimum being around 35°C; no growth occurs at 5 or 55°C. Catalase-positive. Best growth occurs under strictly anaerobic conditions with a reducing agent such as sulfide. Growth occurs on arabinose, xylose, glucose, galactose, mannose, cellobiose, maltose, sucrose, trehalose, xylan, starch, glycogen, dextran, dextrin, pectin, pullulan, and *N*-acetylglucosamine. Growth is most rapid on cellobiose or starch. No growth occurs on methanol, ethanol, pyruvate, succinate, fumarate, malate, glycerol, mannitol, inositol, ribose, fructose, lactose, agarose, cellulose, alginate, agar, xanthan, chitin, chitosan, Casamino acids, peptone, gelatin, or yeast extract. The main products of sugar fermentation are propionate, acetate, and succinate; minor products are formate, fumarate, and hydrogen. Vitamins or yeast extract are needed for growth. The predominant cellular fatty acids are $C_{15:0}$ isomers. Habitat: anoxic soda lake sediments rich in organic nutrients. The type strain was isolated as a component of an anaerobic cellulose-degrading enrichment culture originating from Lake Verkhneye Beloye (south-eastern Transbaikal region, Russia).

DNA G+C content (mol%): 43.0–44.4 (T_m).

Type strain: Z-7010, DSM 15055, VKM B-2311.

Sequence accession no. (16S rRNA gene): AJ784993.

*The basal medium of Nedashkovskaya et al. (2006) is composed of (g/l): NH_4Cl, 0.2; $MgCl_2 \cdot 6H_2O$, 0.05; KH_2PO_4, 0.2; Na_2CO_3, 7.4; $NaHCO_3$, 18.5; $Na_2S \cdot 9H_2O$, 0.5; and yeast extract, 0.2. The latter three ingredients are added from separately autoclaved stock solutions (sulfide was autoclaved under N_2) to the anoxically cooled medium. Non-chelated trace elements and vitamins are added as described by Widdel and Bak (1992). The medium is dispensed into bottles or tubes with a headspace (30% of total volume) of N_2 and sealed with butyl rubber stoppers. Organic substrates are added before inoculation from separately sterilized stock solutions. Sugars are sterilized by filtration.

References

Achenbach, H., W. Kohl and H. Reichenbach. 1978. The flexirubin-type pigments – a novel class of natural pigments from gliding bacteria. Rev. Latinoam. Quim. *9*: 111–124.

Bachmann, B.J. 1955. Studies on *Cytophaga fermentans*, n.sp., a facultatively anaerobic lower myxobacterium. J. Gen. Microbiol. *13*: 541–551.

Denger, K. and B. Schink. 1995. New halo- and thermotolerant fermenting bacteria producing surface-active compounds. Appl. Microbiol. Biotechnol. *44*: 161–166.

Denger, K., R. Warthmann, W. Ludwig and B. Schink. 2002. *Anaerophaga thermohalophila* gen. nov., sp. nov., a moderately thermohalophilic, strictly anaerobic fermentative bacterium. Int. J. Syst. Evol. Microbiol. *52*: 173–178.

Haack, S.K. and J.A. Breznak. 1993. *Cytophaga xylanolytica* sp. nov., a xylan-degrading, anaerobic gliding bacterium. Arch. Microbiol. *159*: 6–15.

Hosoya, R. and K. Hamana. 2004. Distribution of two triamines, spermidine and homospermidine, and an aromatic amine, 2-phenylethylamine, within the phylum *Bacteroidetes*. J. Gen. Appl. Microbiol. *50*: 255–260.

Kevbrin, V.V., T.N. Zhilina and G.A. Zavarzin. 1999. Decomposition of cellulose by the anaerobic alkaliphilic microbial community. Microbiology (En. transl. from Mikrobiologiya) *68*: 601–609.

Leadbetter, E.R., S.C. Holt and S.S. Socransky. 1979. *Capnocytophaga*: new genus of Gram-negative gliding bacteria. 1. General characteristics, taxonomic considerations and significance. Arch. Microbiol. *122*: 9–16.

Nakagawa, Y. and K. Yamasato. 1996. Emendation of the genus *Cytophaga* and transfer of *Cytophaga agarovorans* and *Cytophaga salmonicolor* to *Marinilabilia* gen. nov: phylogenetic analysis of the *Flavobacterium cytophaga* complex. Int. J. Syst. Bacteriol. *46*: 599–603.

Nakagawa, Y., K. Hamana, T. Sakane and K. Yamasato. 1997. Reclassification of *Cytophaga aprica* (Lewin 1969) Reichenbach 1989 in *Flammeovirga* gen. nov. as *Flammeovirga aprica* comb. nov. and of *Cytophaga diffluens* (*ex* Stanier 1940; emend. Lewin 1969) Reichenbach 1989 in *Persicobacter* gen. nov. as *Persicobacter diffluens* comb. nov. Int. J. Syst. Bacteriol. *47*: 220–223.

Nedashkovskaya, O.I., S.B. Kim, M. Vancanneyt, C. Snauwaert, A.M. Lysenko, M. Rohde, G.M. Frolova, N.V. Zhukova, V.V. Mikhailov, K.S. Bae, H.W. Oh and J. Swings. 2006. *Formosa agariphila* sp. nov., a budding bacterium of the family *Flavobacteriaceae* isolated from marine environments, and emended description of the genus *Formosa*. Int. J. Syst. Evol. Microbiol. *56*: 161–167.

Reichenbach, H., H. Kleinig and H. Achenbach. 1974. The pigments of *Flexibacter elegans*: novel and chemosystematically useful compounds. Arch. Microbiol. *101*: 131–144.

Reichenbach, H. and M. Dworkin. 1981. The order *Cytophagales*. In The Prokaryotes: a Handbook on Habitats, Isolation, and Identification of *Bacteria* (edited by Starr, Stolp, Trüper, Balows and Schlegel). Springer, Berlin, pp. 356–379.

Reichenbach, H. 1989. Genus I. *Cytophaga*. *In* Bergey's Manual of Systematic Bacteriology, vol. 1 (edited by Staley, Bryant, Pfennig and Holt). Williams & Wilkins, Baltimore, pp. 2015–2050.

Reichenbach, H. 1992. The order *Cytophagales*. *In* The Prokaryotes: a Handbook on the Biology of Bacteria: Ecophysiology, Isolation, Identification, Applications, 2nd edn, vol. 4 (edited by Balows, Trüper, Dworkin, Harder and Schleifer). Springer, New York, pp. 3631–3675.

Staley, J.T., M.P. Bryant, N. Pfennig and J.G. Holt. 1989. Bergey's Manual of Systematic Bacteriology, vol. 3. Williams & Wilkins, Baltimore.

Suzuki, M., Y. Nakagawa, S. Harayama and S. Yamamoto. 1999. Phylogenetic analysis of genus *Marinilabilia* and related bacteria based on the amino acid sequences of GyrB and emended description of *Marinilabilia salmonicolor* with *Marinilabilia agarovorans* as its subjective synonym. Int. J. Syst. Bacteriol. *49*: 1551–1557.

Veldkamp, H. 1961. A study of two marine agar-decomposing, facultatively anaerobic myxobacteria. J. Gen. Microbiol. *26*: 331–342.

Widdel, F. and F. Bak. 1992. Gram-negative mesophilic sulfate-reducing bacteria. *In* The Prokaryotes: a Handbook on the Biology of Bacteria: Ecophysiology, Isolation, Identification, Applications, 2nd edn, vol. 4 (edited by Balows, Trüper, Dworkin, Harder and Schleifer). Springer, New York, pp. 3352–3378.

Zhilina, T.N., R. Appel, C. Probian, E.L. Brossa, J. Harder, F. Widdel and G.A. Zavarzin. 2004. *Alkaliflexus imshenetskii* gen. nov. sp. nov., a new alkaliphilic gliding carbohydrate-fermenting bacterium with propionate formation from a soda lake. Arch. Microbiol. *182*: 244–253.

Zhilina, T.N., R. Appel, C. Probian, E. Llobet Brossa, J. Harder, F. Widdel and G.A. Zavarzin. 2005. *In* Validation of publication of new names and new combinations previously effectively published outside the IJSEM. List no. 104. Int. J. Syst. Evol. Microbiol. *55*: 1395–1397.

Family III. **Rikenellaceae** fam. nov.

The Editorial Board

Ri.ke.nel.la.ce.ae. N.L. fem. n. *Rikenella* type genus of the family; *-aceae* ending to denote family; N.L. fem. pl. n. *Rikenellaceae* the *Rikenella* family.

The family *Rikenellaceae* is a phenotypically diverse assemblage of genera that was circumscribed for this volume on the basis of phylogenetic analysis of 16S rRNA gene sequences. The family contains the genera *Rikenella* (type genus) and *Alistipes*. Both are Gram-negative rods that lack swimming motility. Some characteristics that differentiate the two genera are given in Table 10.

Type genus: **Rikenella** Collins, Shah and Mitsuoka 1985a, 375[VP] (Effective publication: Collins, Shah and Mitsuoka 1985b, 80.).

TABLE 10. Some characteristics that differentiate the genera of the family *Rikenellaceae*[a]

Characteristic	*Rikenella*	*Alistipes*
Major products of glucose fermentation	Propionic and succinic acids, with moderate amounts of acetic acid and trace amounts of alcohols	Succinic acid with minor amounts of acetic acid
Major quinone	MK-8	nr
Source	Fecal or cecal specimens from calves, chickens, and Japanese quails	Human and animal specimens of intestinal origin
DNA G+C content (mol%)	59.5-60.7	55–58

[a]nr, Not reported.

Genus I. **Rikenella** Collins, Shah and Mitsuoka 1985a, 375[VP] (Effective publication: Collins, Shah and Mitsuoka 1985b, 80.)

THE EDITORIAL BOARD

Ri.ke.nel'la. L. fem. dim. ending *-ella*; N.L. fem. dim. n. *Rikenella* named after the RIKEN-Institute, Japan, where the organisms were originally isolated.

Rods with pointed ends occurring singly, in pairs, or short chains. Short, filamentous, or swollen forms may occur in media with a high concentration of fermentable carbohydrate. Nonmotile. Nonsporeforming; strains do not survive 70°C for 10 min. Gram negative. **Anaerobic**. Growth enhanced by a fermentable carbohydrate (e.g., glucose). Optimum temperature ~37°C. Poor growth at 45°C and usually no growth below 25°C. **Relatively few sugars are fermented,** but glucose, lactose, mannose, and melibiose are fermented. **Major end products of glucose fermentation are propionic and succinic acids, with moderate amounts of acetic acid,** and trace amounts of alcohols. **Glucose-6-phosphate dehydrogenase, 6-phosphogluconate dehydrogenase, and glutamate dehydrogenase are absent**. Esculin is not hydrolyzed. Nitrate is not reduced. Nonhydroxylated and 3-hydroxylated long-chain fatty acids present. **The fatty acids are primarily of the iso-methyl branched-chain types**; moderate amounts of straight-chain saturated acids are also present. **Menaquinones are produced**. Isolated from fecal or cecal specimens from calves, chickens, and Japanese quails.

DNA G+C content (mol%): 59.5-60.7.

Type species: **Rikenella microfusus** (Kaneuchi and Mitsuoka 1978) Collins, Shah and Mitsuoka 1985a, 375[VP] (Effective publication: Collins, Shah and Mitsuoka 1985b, 80) (*Bacteroides microfusus* Kaneuchi and Mitsuoka 1978, 478.).

Further descriptive information

Cells grown on Eggerth–Gagnon agar* for 2 d are 0.15-0.3 × 0.3-1.5 μm (Collins et al., 1985b). Cells from glucose broth are 0.5-0.9 × 1.0-5.0 μm (Holdeman et al., 1984).

Surface colonies on Eggerth–Gagnon agar are 0.5-1 mm in diameter, circular, convex, entire, smooth, translucent, grayish, and β-hemolytic. Some strains form raised, opaque, gray colonies that are 1.0-1.5 mm in diameter.

Growth in peptone-yeast extract-glucose (PYG) broth is enhanced by 20% bile or by 0.0005% hemin plus 0.00005% menadione, but not by 10% rumen fluid or 0.1% Tween 80. Cultures in peptone-yeast extract-Fildes enrichment† - glucose (PYFG) broth are turbid with smooth sediment and have a final pH of 5.5-5.8. Only trace amounts of gas are produced.

The major fatty acid is 13-methyltetradecanoic acid. An unsaturated menaquinone of the MK-8 type is present.

Strains have been isolated from the feces of healthy 2-month-old calves fed an artificial diet, from the feces of adult Japanese quails, and from the cecal contents of chickens. The type strain occurred at a concentration of 10^9/g (wet wt) in Japanese quail feces. Analysis of partial 16S rRNA gene sequences obtained from guts of various termite species by Ohkuma et al. (2002) indicated that one phylogenetic cluster, composed of the phylotypes from a single termite species, was related to the genus *Rikenella*.

Pathogenicity has not been reported.

Enrichment and isolation procedures

In the method of Kaneuchi and Mitsuoka (1978), samples are emulsified in an anaerobic buffer, plated onto modified Eggerth–Gagnon agar‡ and glucose-blood-liver agar, and incubated at 37°C anaerobically under 100% CO_2. Grayish to slightly reddish mottled colonies are selected after 2-3 d.

Strains can be maintained on prereduced Eggerth–Gagnon liver slants with H_2CO_3/CO_2 buffer and stored at 4°C, with serial transfer every 3 months.

Differentiation of the genus *Rikenella* from other genera

Unlike *Bacteroides fragilis* and related species, *Rikenella microfusus* lacks glucose-6-phosphate dehydrogenase and 6-phosphogluconate dehydrogenase (which are characteristic of the hexose monophosphate shunt-pentose pathway) and glutamate dehydrogenase. Unlike *Mitsuokella multiacida*, *Rikenella microfusus* cells are smaller in diameter and attack relatively few carbohydrates. They also contain predominantly long-chain, iso-methyl-branched fatty acids, whereas *Mitsuokella multiacida* contains mainly straight-chain saturated and unsaturated fatty acids and only small amounts of methyl branched-chain fatty acids. In addition, *Rikenella microfusus* contains unsaturated menaquinones of the MK-8 type, whereas *Mitsuokella multiacida* does not contain menaquinones.

Taxonomic comments

Comparative analysis of 16S rRNA sequences of *Bacteroides* and related organisms by Paster et al. (1994) indicated that *Rikenella microfusus*, together with *Bacteroides putredinis* and two species of *Cytophaga*, belong to a subgroup separate from the bacteroides subgroup (the latter being composed of prevotella, bacteroides, and porphyromonas clusters). However, analysis of base signatures indicated that these species do belong within the bacteroides subgroup.

*The medium of Eggerth and Gagnon (1933) consists of beef infusion broth containing 1.5% agar, 1% peptone, and 0.4% disodium phosphate. The pH is adjusted to 7.6–7.8. Before dispensing the medium into plates, 5% (v/v) of sterile blood and 0.15% sterile glucose (in the form of a 10% solution) are added.

†Fildes enrichment (Willis, 1960) is prepared by adding 50 ml of defibrinated sheep blood, 6 ml of concentrated HCl, and 2 g of pepsin to 150 ml of sterile normal saline. This solution is mixed thoroughly in stoppered bottle and incubated at 55°C overnight. Sufficient 20% NaOH solution is added (usually about 12 ml) to make the digest strongly alkaline to cresol red indicator (about pH 8). The pH is then restored to 7.2 with concentrated HCl. After adding 0.25% chloroform as a preservative, the digest is stored at 4°C. The enrichment digest is usually added to a final concentration of 3–5% (v/v). After addition to a sterile medium, the medium is steamed for 30 min to drive off residual chloroform.

‡Modified Eggerth–Gagnon agar (Mitsuoka et al., 1965) consists of 930 ml of horsemeat infusion broth, 10 g Proteose peptone No. 3 (Difco), 5 g yeast extract, 4 g Na_2HPO_4, 0.5 g soluble starch, 1.5 g glucose, 0.2 g cystine (dissolved in HCl), 10 ml Antifoam B (Dow Corning, in 10% solution), 15 g agar; 50 ml horse blood, and 10 ml cysteine–HCl (in 5% solution). The pH is 7.6–7.8.

List of species of the genus *Rikenella*

1. **Rikenella microfusus** (Kaneuchi and Mitsuoka 1978) Collins, Shah and Mitsuoka 1985a, 375[VP] (Effective publication: Collins, Shah and Mitsuoka 1985b, 80.) (*Bacteroides microfusus* Kaneuchi and Mitsuoka 1978, 478)

mi.cro.fus'us. Gr.adj. *mikros* small; L. n. *fusus* a spindle; N.L. n. *microfusus* (nominative in apposition) a small spindle (referring to cellular morphology).

The description is as given for the genus, with the following additional information. Acid is produced from glucose, lactose, mannose, and melibiose. Most strains (75%) produce acid from galactose. No acid is produced from amygdalin, arabinose, cellobiose, dextrin, dulcitol, erythritol, fructose, glycerol, glycogen, inositol, inulin, maltose, mannitol, α-methylglucoside, α-methylmannoside, melezitose, raffinose, rhamnose, ribose, salicin, sorbose, sorbitol, starch, sucrose, trehalose, and xylose.

β-Glucuronidase, glutamic acid decarboxylase, and malate dehydrogenase are produced. Gelatin is weakly hydrolyzed. Casein, starch, and urea are not hydrolyzed. Indole and acetylmethylcarbinol are not produced. No lecithinase activity occurs on egg yolk agar. H_2S production is not evident on sulfide-indole-motility medium.

Growth is inhibited by 0.001% brilliant green, 0.005% crystal violet, neomycin (1600 μg/ml; slight growth may occur sometimes with a few strains), and rifampin (15 μg/ml). Strains are resistant to bacitracin (3 U/ml), cephalothin (300 μg/ml), colistin 10 μg/ml), polymyxin B (10 μg/ml), and vancomycin (10 μg/ml). There is slight to moderate growth in the presence of kanamycin (1000 μg/ml), penicillin (15 μg/ml), and erythromycin (60 μg/ml). Most strains are inhibited by neomycin (1600 μg/ml) and rifampin (15 μg/ml).

DNA G+C content (mol%): 59.5-60.7 (T_m).

Type strain: ATCC 29728, CCUG 54772, DSM 15922, JCM 2053, NCTC 11190.

Sequence accession no. (16S rRNA gene): L16498.

Genus II. **Alistipes** Rautio, Eerola, Väisänen-Tunkelrott, Molitoris, Lawson, Collins and Jousimies-Somer 2003b, 1701[VP] (Effective publication: Rautio, Eerola, Väisänen-Tunkelrott, Molitoris, Lawson, Collins and Jousimies-Somer 2003a, 186.)

Eija Könönen, Yuli Song, Merja Rautio and Sydney M. Finegold

A.li.sti'pes. L. adj. *alius* other; L. masc. n. *stipes* a log, stock, post, trunk of a tree, stick; N.L. masc. n. *alistipes* the other stick.

Straight or slightly curved rods 0.2–0.9 μm in diameter × 0.5–4 μm in length, with rounded ends. Swellings may occur, but spores are not formed. Cells usually occur singly or in pairs, occasionally in longer filaments. Nonmotile. Gram negative. **Obligately anaerobic**. Indole is produced. Gelatin is digested. Nitrate is not reduced to nitrite. Arginine and urea are not hydrolyzed. Produce **succinic acid as the major glucose metabolic end product**, with minor amounts of acetic acid. The major long-chain fatty acid produced is $C_{15:0}$ iso.

Source: human and animal specimens of intestinal origin.

DNA G+C content (mol%): 55–58.

Type species: **Alistipes putredinis** (Weinberg, Nativelle and Prévot 1937) Rautio, Eerola, Väisänen-Tunkelrott, Molitoris, Lawson, Collins and Jousimies-Somer 2003b, 1701(Effective publication: Rautio, Eerola, Väisänen-Tunkelrott, Molitoris, Lawson, Collins and Jousimies-Somer 2003a, 186.) (*Bacillus putredinis* Weinberg, Nativelle and Prévot 1937, 755; *Ristella putredinis* Prévot 1938, 291; *Bacteroides putredinis* Kelly 1957, 420).

Further descriptive information

Phylogenetic analysis of the 16S rRNA gene sequence indicates that the genus *Alistipes* belongs to the family *Rikenellaceae* within the phylum *Bacteroidetes* (Figure 16).

Colonies are pinpoint to 1 mm in diameter, circular, convex, gray, and opaque or translucent on supplemented (5% sheep blood, 1 μg/ml vitamin K_1, 5 μg/ml hemin) Brucella agar after anaerobic incubation for 4 d. In case of pigment-producing *Alistipes* species, strains are (weakly) β-hemolytic. A light brown or brown pigmentation appears on rabbit-laked-blood (RLB) agar within 4–5 d and on kanamycin-vancomycin-laked blood (KVLB; containing 5% lysed sheep blood) agar plates, but it also appears on Brucella blood agar after extended incubation (10 d).

Strains grow well on solid media under anaerobic conditions. The optimal growth temperature is 37°C. Growth is poor in liquid media even in the presence of supplements. For the demonstration of carbohydrate fermentation, addition of 0.5% Tween 80 may stimulate the growth, but even with this supplementation, the growth can be poor.

According to their characteristic special potency antibiotic identification disk profile, *Alistipes* species are resistant to vancomycin (5 μg), kanamycin (1000 μg), and colistin (10 μg). Bile resistance, which can be tested on bacteroides-bile-esculin (BBE) agar with or without gentamicin, is typical of pigment-producing *Alistipes* species. As with other strict anaerobes, *Alistipes* is susceptible to metronidazole. Susceptibility to penicillin varies due to β-lactamase production by some strains.

Pigment-producing, bile-resistant *Alistipes* species have been isolated from various human clinical specimens, such as inflamed appendix tissue, peritoneal fluid of patients with appendicitis, antibiotic-associated diarrheal feces, and urine (Rautio et al., 2003a; Song et al., 2006). *Alistipes putredinis* has been isolated from inflamed appendix tissue, abdominal and peri-rectal abscesses, pilonidal and Bartholin cysts, human blood, and sheep foot rot (Cato et al., 1979; Holdeman et al., 1984). The clinical significance in these conditions remains unclear, however, since the organisms appear in mixed cultures with other anaerobes and/or aerobes. The isolation of *Alistipes* species from appendix tissue, both inflamed and noninflamed (Rautio et al., 2000), and from the feces of healthy individuals

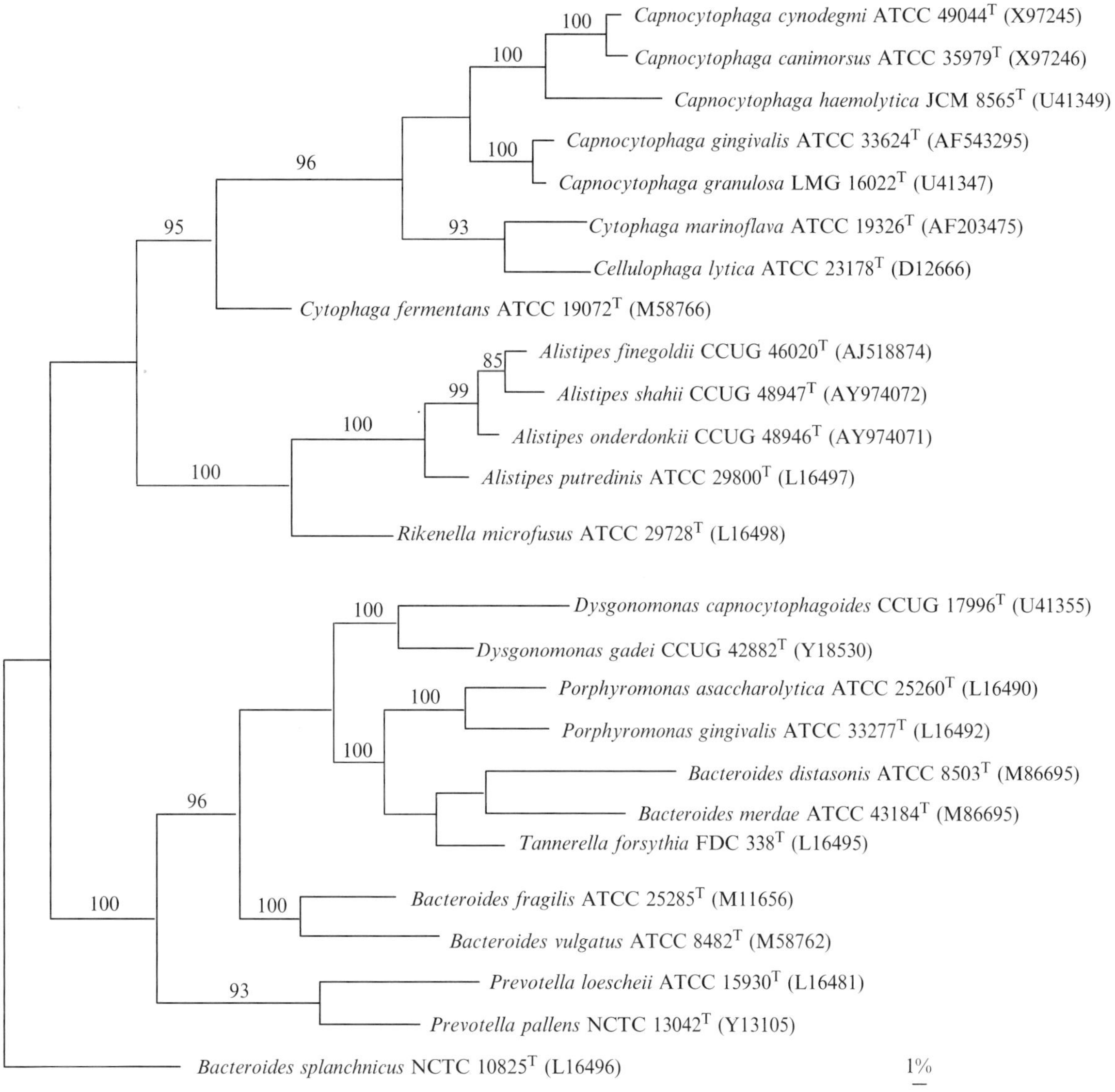

FIGURE 16. Phylogenetic tree based on 16S rRNA gene sequence comparisons (>1300 aligned bases) showing the phylogenetic relationship of the *Alistipes* species to related taxa. The tree was constructed using the neighbor-joining method, following distance analysis of aligned sequences. Bootstrap analysis was used, with 1000 repetitions.

(Rautio et al., 2003a; Song et al., 2006) as well as from farm soil (Cato et al., 1979) indicates that the habitat is the human and animal gut.

Enrichment and isolation procedures

Alistipes strains can be isolated on solid culture media appropriate for anaerobes, including nonselective Brucella blood agar enriched with vitamin K_1 and hemin, and fastidious anaerobe agar (FAA; Lab M). In addition, pigment-producing, bile-resistant species are also isolated from selective KVLB and BBE agar plates (Rautio et al., 1997).

Growth in liquid media is especially poor. Tween 80 (0.5%) has been used to enhance growth of *Alistipes putredinis* (Holdeman et al., 1984). Of various supplements (bile, formate-fumarate, hemin, horse serum, pyruvate, sodium bicarbonate, and Tween) tested for enhancement of growth of pigment-producing *Alistipes* species in thioglycolate broth medium, bile and horse serum showed a weak stimulation (Rautio et al., 1997).

Maintenance procedures

For long term storage, young (2–3 d) cultures are transferred into vials containing sterilized 20% skim milk and kept frozen at –70°C. Twenty-year-old strains have been revived successfully from stocks by scraping frozen bacterial suspension onto fresh or prereduced Brucella blood agar and incubating the plates in anaerobic jars or in an anaerobic chamber for 3–5 d before subcultivation for verifying the purity of strains.

Methods for characterization tests

In general, the anaerobic methods described in the Wadsworth-KTL Anaerobic Bacteriology Manual (Jousimies-Somer et al., 2002) are suitable for the study of members of this genus. The spot indole test may give false-negative results, therefore, the

TABLE 11. Characteristics differentiating *Alistipes* from related genera[a]

Characteristic	*Alistipes*[b]	*Bacteroides*[c]	*Porphyromonas*[d]	*Prevotella*	*Rikenella*	*Tannerella*
Growth in air and CO_2	–	–	–	–	–	–
Gram reaction	–	–	–	–	–	–
Pigment production	+	–	+	D	–	–
Growth in 20% bile	+	+	–	–	+	–
Susceptibility to:[e]						
Vancomycin (5 μg)	R	R	D	R	R	R
Kanamycin (1 mg)	R	R	R	D	S	S
Colistin (10 μg)	R	R	R	D	R	S
Catalase production	–	–	–	–	–	–
Indole production	+	D	D	D	–	–
Nitrate reduction	–	–	–	–	–	–
Proteolytic activity	+	–	D	D	D	+
Carbohydrate fermentation	+	+	–	+	–	–
Major metabolic end product(s)[f]	S	A, S	A, B	A, S	P, S	A, B, IV, P, PA
Major cellular fatty acid	$C_{15:0}$ iso	$C_{15:0}$ anteiso	$C_{15:0}$ iso	$C_{15:0}$ anteiso	$C_{15:0}$ iso	$C_{15:0}$ anteiso
DNA G+C content (mol%)	55–58	40–48	40–55	39–60	60–61	44–48
Type species	*A. putredinis*	*B. fragilis*	*P. asaccharolytica*	*P. melaninogenica*	*R. microfusus*	*T. forsythia*

[a]Symbols: +, positive; –, negative; D, different reactions in different taxa (species).
[b]Unlike other *Alistipes* spp., *Alistipes putredinis* does not produce pigment, and is susceptible to bile, catalase positive, and asaccharolytic.
[c]*Bacteroides sensu stricto.*
[d]Unlike other *Porphyromonas* spp., *Porphyromonas catoniae* does not produce pigment, and is moderately saccharolytic.
[e]Special potency antimicrobial identification disks. Symbols: R, resistant; S, susceptible.
[f]Symbols: A, acetic acid; B, butyric acid; IV, isovaleric acid; P, propionic acid; PA, phenylacetic acid; S, succinic acid.

tube indole test is recommended. Commercial API ZYM and API rapid ID 32 A test kits (bioMérieux) and individual Rosco diagnostic tablets (Rosco) can be useful for examining biochemical characteristics. A heavy inoculum from young (2–3 d) cultures should be used for biochemical testing to avoid poor reproducibility of reactions. The enzyme profiles generated by the API ZYM test kit (bioMérieux) proved to be most useful in distinguishing the *Alistipes* species from each other. For demonstration of carbohydrate fermentation, prereduced, anaerobically sterilized (PRAS) peptone-yeast-sugar broth tubes, with or without additional supplements, are used, but results can be affected by poor growth of *Alistipes* strains in liquid media.

Differentiation of the genus *Alistipes* from other genera

Characteristics that differentiate *Alistipes* from related taxa within the phylum *Bacteroidetes* (previously referred to as the "*Cytophaga–Flavobacteria–Bacteroides*" phylum) are presented in Table 11.

Taxonomic comments

The genus *Alistipes* currently includes four validly published species, *Alistipes putredinis* and *Alistipes finegoldii* (Rautio et al., 2003a, b), and *Alistipes onderdonkii* and *Alistipes shahii* (Song et al., 2006). Phylogenetic analyses of the 16S rRNA gene sequence reveal further heterogeneity within the genus, since occasional strains that are phenotypically similar to but phylogenetically diverse from *Alistipes* species described in the current literature have been reported (Song et al., 2005, 2006). Also, preliminary results on bile-resistant, pigment-producing *Alistipes*-like organisms in ongoing studies by the authors (unpublished) revealed (among approximately 150 strains examined) several groups having a 16S rRNA gene sequence divergence of more than 3% compared to any known *Alistipes* species.

List of species of the genus *Alistipes*

1. **Alistipes putredinis** (Weinberg, Nativelle and Prévot 1937) Rautio, Eerola, Väisänen-Tunkelrott, Molitoris, Lawson, Collins and Jousimies-Somer 2003b, 1701 (Effective publication: Rautio, Eerola, Väisänen-Tunkelrott, Molitoris, Lawson, Collins and Jousimies-Somer 2003a, 186.) (*Bacillus putredinis* Weinberg, Nativelle and Prévot 1937, 755; *Ristella putredinis* Prévot 1938, 291; *Bacteroides putredinis* Kelly 1957, 420)

put.re'di.nis. L. n. *putredo -inis* rottenness, putridity; L. gen. n. *putredinis* of putridity.

The description is based on previous literature (Cato et al., 1979; Holdeman et al., 1984; Rautio et al., 2003a; Song et al., 2006). Surface colonies on supplemented Brucella sheep blood agar after incubation for 4 d are pinpoint to 0.5 × mm in diameter, circular, entire or slightly irregular, low convex, translucent, gray, dull, and smooth. Colonies do not produce pigment. No growth occurs in 20% bile. Although only traces of acid products can be detected in a 6-d-old PYG broth (pH > 6), in a 24-h-old meat chopped carbohydrate broth, the main acid detected is succinate. Catalase and indole are produced. The type strain, examined with the API ZYM and API rapid ID 32 A test kits, is positive for alkaline and acid phosphatases, esterase, esterase lipase, naphthol-AS-BI-phosphohydrolase, α-glucosidase, leucyl glycine, alanine, and serine arylamidases, glutamic decarboxylase, and indole. Some characteristics of the species are listed in Table 12.

DNA G+C content (mol%): 55 (HPLC).

Type strain: ATCC 29800, CCUG 45780, CIP 104286, DSM 17216.

Sequence accession no. (16S rRNA gene): L16497.

TABLE 12. Characteristics differentiating species of the genus *Alistipes*[a].

Characteristic	*Alistipes putredinis*	*Alistipes finegoldii*	*Alistipes onderdonkii*	*Alistipes shahii*
Cell morphology	Straight or slightly curved rods, 0.3–0.5 × 0.9–3 µm, singly or in pairs	Straight rods, 0.2 × 0.8–2 µm, singly, long filaments occur	Straight rods, 0.3–0.9 × 0.5–3 µm, singly or in pairs	Straight rods, 0.5–0.8 × 0.6–4 µm, singly or in pairs
Colony morphology	Pinpoint to 0.5 mm, circular, slightly irregular, low convex, translucent	0.3–1.0 mm, circular, entire, translucent or opaque	0.5–0.8 mm, circular, entire, opaque	0.5–1 mm, circular, entire, opaque
Hemolysis	–	β	β	β
Pigment production	–	+	+	+
Growth in 20% bile	–	+	+	+
Catalase production	+	–	–	–
Enzyme activities:[b]				
α-Chymotrypsinase	–	+	d	d
α-Fucosidase	–	+	–	+
α-Glucosidase	+	+	+	+
β-Glucosidase	–	–	–	+
Glucose fermentation	–	+	+	+
Metabolic end products:				
Major	S	S	S	S
Minor (trace) amounts[c]	a, iv, p	a, p, (iv, l)	a, p, (iv, l)	a, iv
Major cellular fatty acid	$C_{15:0}$ iso	$C_{15:0}$ iso	$C_{15:0}$ iso	$C_{15:0}$ iso
DNA G+C content (mol%)	55	57	56	58
Type strain	ATCC 29800	CCUG 46020	CCUG 48946	CCUG 48947
Isolation sit	Human and animal feces, appendicitis and, related infections, intraabdominal and, perianal infections, sheep foot rot, farm soil	Human feces, appendix tissue	Human feces, appendix tissue, abdominal abscess, urine	Human feces, appendix tissue, intraabdominal fluid

[a]Symbols: +, >85% positive; d, different strains give different reactions (16–84% positive); –, 0–15% positive.
[b]Based on the reactions generated by the API ZYM test kit.
[c]Symbols: a, acetic acid; iv, isovaleric acid; l, lactic acid; p, propionic acid; S, major amount of succinic acid.

2. **Alistipes finegoldii** Rautio, Eerola, Väisänen-Tunkelrott, Molitoris, Lawson, Collins and Jousimies-Somer 2003b, 1701[VP] (Effective publication: Rautio, Eerola, Väisänen-Tunkelrott, Molitoris, Lawson, Collins and Jousimies-Somer 2003a, 186.)

fine.gold'i.i. N.L. gen. masc. n. *finegoldii* of Finegold; named after Sydney M. Finegold, an American contemporary researcher and clinician in recognition of his contribution to anaerobic bacteriology and infectious diseases.

The description is based on previous literature (Rautio et al., 2003a, 1997; Song et al., 2006). Surface colonies on supplemented Brucella sheep blood agar after 4 d of incubation are pinpoint to 1.0 mm in diameter, circular, entire, raised, gray, translucent or opaque, and (weakly) β-hemolytic. Colonies are light brown to brown. No fluorescence is observed under long-wave UV light (365 nm), but the colonies appear black. Esculin hydrolysis differs among strains. Acid is produced from glucose. The strains, examined with the API ZYM and API rapid ID 32 A test kits , are positive for alkaline and acid phosphatases, esterase, esterase lipase, α-chymotrypsin, naphthol-AS-BI-phosphohydrolase, α-galactosidase, β-galactosidase, α-glucosidase, *N*-acetyl-β-glucosaminidase, α-fucosidase, leucyl glycine, alanine, and glutamyl glutamic arylamidases, and indole. Some characteristics of the species are listed in Table 12.

DNA G+C content (mol%): 57 (HPLC).

Type strain: AHN 2437, CCUG 46020, CIP 107999, DSM 17242.

Sequence accession no. (16S rRNA gene): AJ518874.

3. **Alistipes onderdonkii** Song, Könönen, Rautio, Liu, Bryk, Eerola and Finegold 2006, 1988[VP]

on.der.don'ki.i. N.L. gen. masc. n. *onderdonkii* of Onderdonk; named after Andrew B. Onderdonk, a contemporary American microbiologist, for his contribution to increased knowledge about the intestinal microbiota and anaerobic bacteria.

The description is based on the investigation of 15 strains (Song et al., 2006). Cells are 0.2–0.5 μm × 0.5–3 μm. Surface colonies on supplemented Brucella sheep blood agar after 4 d are pinpoint to 0.8 mm in diameter, circular, entire, convex, opaque, gray, and (weakly) β-hemolytic. Colonies are light brown to brown. No fluorescence is observed under long-wave UV light (365 nm), but the colonies appear black. Grows in the presence of 20% bile. Indole positive. Catalase negative. Esculin hydrolysis differs among strains. Acid is produced from glucose. When tested by the API rapid ID 32 A system (bioMeriéux), mannose and raffinose are fermented. The strains, examined with the API ZYM and API rapid ID 32A test kits, are positive for alkaline and acid phosphatases, esterase, esterase lipase, naphthol-AS-BI-phosphohydrolase, α-galactosidase, β-galactosidase, α-glucosidase, *N*-acetyl-β-glucosaminidase, leucyl glycine, alanine, and glutamyl glutamic arylamidases, and indole. Some characteristics of the species are listed in Table 12.

DNA G+C content (mol%): 56 (HPLC).

Type strain: WAL 8169, ATCC BAA-1178, CCUG 48946.

Sequence accession no. (16S rRNA gene): AY974071.

4. **Alistipes shahii** Song, Könönen, Rautio, Liu, Bryk, Eerola and Finegold 2006, 1999[VP]

sha'hi.i. N.L. gen. masc. n. *shahii* of Shah; to honor Haroun N. Shah, a contemporary British microbiologist for his contributions to anaerobic bacteriology.

The description is based on the investigation of six strains (Song et al., 2006). Cells are 0.1–0.2 μm × 0.6–4 μm. Surface colonies on supplemented Brucella sheep blood agar after 4 d are pinpoint to 1.0 mm in diameter, circular, entire, convex, opaque, gray, and (weakly) β-hemolytic. Colonies are light brown to brown. No fluorescence is observed under long-wave UV light (365 nm), but the colonies appear black. Grows in the presence of 20% bile. Indole positive. Catalase negative. Esculin is hydrolyzed. Acid is produced from glucose. When tested by the API rapid ID 32 A system (bioMeriéux), mannose and raffinose are fermented. The strains, examined with the API ZYM and API rapid ID 32 A test kits (bioMeriéux), are positive for alkaline and acid phosphatases, esterase, esterase lipase, naphthol-AS-BI-phosphohydrolase, α-galactosidase, β-galactosidase, α-glucosidase, β-glucosidase, *N*-acetyl-β-glucosaminidase, α-fucosidase, leucyl glycine, alanine, and glutamyl glutamic arylamidases, and indole. Some characteristics of the species are listed in Table 12.

DNA G+C content (mol%): 58 (HPLC).

Type strain: WAL 8301, ATCC BAA-1179, CCUG 48947.

Sequence accession no. (16S rRNA gene): AY974072.

References

Cato, E.P., L.V. Holdeman and W.E.C. Moore. 1979. Proposal of neotype strains for seven non-saccharolytic *Bacteroides* species. Int. J. Syst. Bacteriol. *29*: 427–434.

Collins, M.D., H.N. Shah and T. Mitsuoka. 1985a. *In* Validation of the publication of new names and new combinations previously effectively published outside the IJSB. List no. 18. Int. J. Syst. Bacteriol. *35*: 375–376.

Collins, M.D., H.N. Shah and T. Mitsuoka. 1985b. Reclassification of *Bacteroides microfusus* (Kaneuchi and Mitsuoka) in a new genus *Rikenella*, as *Rikenella microfusus* comb. nov. Syst. Appl. Microbiol. *6*: 79–81.

Eggerth, A.H. and B.H. Gagnon. 1933. The *Bacteroides* of human feces. J. Bacteriol. *25*: 389–413.

Holdeman, L.V., R.W. Kelly and W.E.C. Moore. 1984. Genus I. *Bacteroides*. *In* Bergey's Manual of Systematic Bacteriology, vol. 1 (edited by Krieg and Holt). Williams & Wilkins, Baltimore, pp. 604–631.

Jousimies-Somer, H.R., P. Summanen, D.M. Citron, E.J. Baron, H.M. Wexler and S.M. Finegold. 2002. Wadsworth-KTL Anaerobic Bacteriology Manual. Star Publishing Company, Belmont, CA.

Kaneuchi, C. and T. Mitsuoka. 1978. *Bacteroides microfusus*, a new species from intestines of calves, chickens, and Japanese quails. Int. J. Syst. Bacteriol. *28*: 478–481.

Kelly, C.D. 1957. Genus I. *Bacteroides* Castellani and Chalmers 1919. *In* Bergey's Manual of Determinative Bacteriology, 7th edn (edited by Breed, Murray and Smith). Williams & Wilkins, Baltimore, pp. 424–436.

Mitsuoka, T., T. Sega and S. Yamamoto. 1965. Improved methodology of qualitative and quantitative analysis of the intestinal flora of man and animals. Zentralbl. Bakteriol. [Orig.] *195*: 455–469.

Ohkuma, M., S. Noda, Y. Hongoh and T. Kudo. 2002. Diverse bacteria related to the bacteroides subgroup of the CFB phylum within the gut symbiotic communities of various termites. Biosci. Biotechnol. Biochem. *66*: 78–84.

Paster, B.J., F.E. Dewhirst, I. Olsen and G.J. Fraser. 1994. Phylogeny of *Bacteroides*, *Prevotella*, and *Porphyromonas* spp. and related bacteria. J. Bacteriol. *176*: 725–732.

Prévot, A.R. 1938. Etudes de systematique bacterienne. III. Invalidite du genre *Bacteroides* Castellani et Chalmers demembrement et reclassification. Ann. Inst. Pasteur *20*: 285–307.

Rautio, M., M. Lonnroth, H. Saxen, R. Nikku, M.L. Vaisanen, S.M. Finegold and H. Jousimies-Somer. 1997. Characteristics of an unusual anaerobic pigmented gram-negative rod isolated from normal and inflamed appendices. Clin. Infect. Dis. *25 Suppl* 2: S107–S110.
Rautio, M., H. Saxen, A. Siitonen, R. Nikku and H. Jousimies-Somer. 2000. Bacteriology of histopathologically defined appendicitis in children. Pediatr. Infect. Dis. J. *19*: 1078–1083.
Rautio, M., E. Eerola, M.L. Vaisanen-Tunkelrott, D. Molitoris, P. Lawson, M.D. Collins and H. Jousimies-Somer. 2003a. Reclassification of *Bacteroides putredinis* (Weinberg et al., 1937) in a new genus *Alistipes* gen. nov., as *Alistipes putredinis* comb. nov., and description of *Alistipes finegoldii* sp. nov., from human sources. Syst. Appl. Microbiol. *26*: 182–188.
Rautio, M., E. Eerola, M.L. Väisänen-Tunkelrott, D. Molitoris, P. Lawson, M.D. Collins and H.R. Jousimies-Somer. 2003b. *In* Validation of the publication of new names and new combinations previously effectively published outside the IJSEM. List no. 94. Int. J. Syst. Evol. Microbiol. *53*: 1701–1702.
Song, Y., C. Liu, M. Bolanos, J. Lee, M. McTeague and S.M. Finegold. 2005. Evaluation of 16S rRNA sequencing and reevaluation of a short biochemical scheme for identification of clinically significant *Bacteroides* species. J. Clin. Microbiol. *43*: 1531–1537.
Song, Y., E. Kononen, M. Rautio, C. Liu, A. Bryk, E. Eerola and S.M. Finegold. 2006. *Alistipes onderdonkii* sp. nov. and *Alistipes shahii* sp. nov., of human origin. Int. J. Syst. Evol. Microbiol. *56*: 1985–1990.
Weinberg, M., R. Nativelle and A.R. Prévot. 1937. Les microbes anaérobies. Masson et Cie, Paris.
Willis, A.T. 1960. Anaerobic Bacteriology in Clinical Medicine. Butterworths, London.

Family IV. **Porphyromonadaceae** fam. nov.

Noel R. Krieg

Por.phy.ro.mo.na.da.ce′a.e. N.L. fem. n. *Porphyromonas* type genus of the family; suff. *-aceae* ending to denote a family; N.L. fem. pl. n. *Porphyromonadaceae* the *Porphyromonas* family.

The family *Porphyromonadaceae* is a phenotypically diverse group of genera that was circumscribed for this volume on the basis of phylogenetic analysis of 16S rRNA gene sequences. The family contains the genera *Porphyromonas* (type genus), *Barnesiella*, *Dysgonomonas*, *Paludibacter*, *Petrimonas*, *Proteiniphilum*, and *Tannerella*. In addition, *Parabacteroides*, which was described after the deadline for this volume should be classified within this family (Sakamoto and Benno, 2006; Sakamoto et al., 2007a). All are **nonmotile rods** that **stain Gram-negative**. Except as noted below, **strict anaerobes** and **saccharolytic**. Some characteristics that differentiate the genera are given in Table 13.

Type genus: **Porphyromonas** Shah and Collins 1988, 129[VP] emend. Willems and Collins 1995, 580.

TABLE 13. Some characteristics that differentiate the genera of the family *Porpyromonadaceae*[a]

Characteristic	*Porphyromonas*	*Barnesiella*	*Dysgonomonas*	*Paludibacter*	*Petrimonas*	*Proteiniphilum*	*Tannerella*
Cell shape	Short rods or coccobacilli	Rods	Coccobacilli to short rods	Rods with ends usually round to slightly tapered	Rods	Rods	Fusiform cells
Growth in the presence of bile	–	–	+	–	nt	–	–
Can grow aerobically	–	–	+	–	–	–	–
N-Acetylglucosamine required for growth	–	–	–	–	–	–	+, except bite wound isolates
Saccharolytic	–; Some species are weakly positive	+	+	+	+	–	+
Products of glucose fermentation:	Butyric and acetic acids; propionic, isovaleric, isobutyric, and phenylacetic acids may also be produced	Acetic and succinic acids	Propionic, lactic, and succinic acids	Acetic and propionic acids; succinic acid is a minor product	Acetic acid and H_2	na	Acetic, butyric, isovaleric, propionic, and phenylacetic acids; smaller amounts of isobutyric and succinic acids may be produced
Predominant menaquinone	MK-9, MK-10	MK-11, MK-12	nt	MK-8	MK-8	nt	MK-10, MK-11
Isolated from:	Oral infections and various other clinical specimens of human and animal origin	Chicken cecum	Human clinical specimens and stools	Rice plant residue (rice straw) collected from irrigated rice-field soil	Oilfield well head	UASB reactor treating brewery wastewater	Human subgingival, gingival, and periodontal pockets, in dental root canals, and around infected dental implants
DNA G+C content (mol%)	44–55	52	38	39	41	47–49	44–48

[a]Symbols: +, >85% positive; –, 0–15% positive; w, weak reaction; na, not available; nt, not tested.

Genus I. **Porphyromonas** Shah and Collins 1988, 129[VP] emend. Willems and Collins 1995, 580.

Paula Summanen and Sydney M. Finegold

Por.phy.ro.mo′nas. Gr. adj. *porphyreos* purple; Gr. n. *monas* unit; N.L. fem. n. *Porphyromonas* porphyrin cell.

Short rods or coccobacilli, 0.3–1 × 0.8–3.5 µm. Gram-negative, non-sporeforming, and nonmotile. **Obligately anaerobic**. Generally cells form **brown to black colonies on blood agar due to protoheme production**. Most species are asaccharolytic: growth is not significantly affected by carbohydrates but is enhanced by protein hydrolysates such as proteose peptone or yeast extract. **Major fermentation products are usually n-butyric acid and acetic acid**; propionic, isovaleric, isobutyric, and phenylacetic acid may also be produced. The major cellular fatty acid is 13-methyltetradecanoic acid ($C_{15:0}$ iso). Indole is produced by most strains. Nitrate is not reduced to nitrite. Esculin is not hydrolyzed. Most species do not hydrolyze starch. Isolated from oral infections and various other clinical specimens of human and animal origin.

DNA G+C content (mol%): 40–55 (T_m).

Type species: **Porphyromonas asaccharolytica** (Holdeman and Moore 1970) Shah and Collins 1988, 128[VP] [*Bacteroides asaccharolyticus* (Holdeman and Moore 1970) Finegold and Barnes 1977, 390; *Bacteroides melaninogenicus* subsp. *asaccharolyticus* Holdeman and Moore 1970, 33].

Further descriptive information

Cell morphology. Most cells in broth cultures are small (0.3–1.0 × 0.8–3.5 µm); however, occasionally longer cells and filaments (≥5 µm) may be formed. Cells from growth on a solid medium are commonly shorter and can appear spherical.

Cell-wall composition. The cell-wall peptidoglycan contains lysine as the diamino acid. 2-Keto-3-deoxyoctulosonic acid is absent. The principal respiratory quinones are unsaturated menaquinones with 9 or 10 isoprene units. Both nonhydroxylated and 3-hydroxylated fatty acids are present. The nonhydroxylated fatty acids are predominantly methyl-branched-chain fatty acids. The predominant fatty acid is $C_{15:0}$ iso; a few species contain comparable amounts of $C_{15:0}$ iso and $C_{15:0}$ anteiso acids. The 3-hydroxylated fatty acids are generally straight-chain saturated fatty acids. Information on the fatty acid content and cell wall composition can be found in Collins et al. (1994) and Brondz and Olsen (1991).

Colony morphology. Colonies on blood agar plates are usually round, entire, smooth (occasionally rough), shiny, convex, and 0.5–3 mm in diameter. The colonies of all but one species (*Porphyromonas catoniae*) are pigmented. The black pigmentation of *Porphyromonas* is caused by the accumulation of hemin used as an iron source for bacterial growth. The species vary in the degree and rapidity of pigment production depending primarily on the type of blood used in the growth medium. Laked rabbit blood agar is considered the most reliable medium for detecting the pigment. The pigmentation ranges from tan to black and may take several days to develop.

Growth conditions. The optimum temperature for growth is 37°C. *Porphyromonas* species favor a slightly alkaline environmental pH and 100% humidity. Hemin and vitamin K_1 are either required or greatly stimulate the growth of most species. Although some species are weakly saccharolytic, their growth is not significantly affected by carbohydrates. Nitrogenous substances, such as proteose peptone, trypticase, and yeast extract markedly enhance growth.

Metabolism and metabolic pathways. Malate dehydrogenase and glutamate dehydrogenase are present; glucose-6-phosphate dehydrogenase and 6-phosphogluconate dehydrogenase are absent from most species. Proteolytic activity is variable. The strains have a limited ability to ferment amino acids such as aspartate and asparagine.

Genetics. *Porphyromonas* forms a distinct phylogenetic group, and 16S rRNA gene sequencing can be reliably used to differentiate *Porphyromonas* from other genera and from each other, with the exception of *Porphyromonas asaccharolytica* and *Porphyromonas uenonis*, where DNA–DNA reassociation studies are required to distinguish these two species genetically.

The genome of *Porphyromonas gingivalis* has been sequenced and studied by the Forsyth Institute and The Institute for Genomic Research (TIGR) (Nelson et al., 2003). The genome size of this species was determined to be 2176 kb from *Xba*I restriction enzyme digests and 2250 kb from *Spe*I digests. Information and a schematic representation of the *Porphyromonas gingivalis* W83 genome can be found at www.tigr.org. Comparative analysis of the whole-genome sequence with other available complete genome sequences confirmed the close relationship between the phylum *Bacteroidetes* [*Cytophaga–Flavobacteria–Bacteroides* (CFB)] and the green sulfur bacteria. Within the phylum "*Bacteroidetes*" the genomes of *Bacteroides thetaiotaomicron* and *Bacteroides fragilis* were most similar to that of *Porphyromonas gingivalis*. The genome analysis revealed a range of virulence determinants that relate to the novel biology of this bacterium. It also revealed that *Porphyromonas gingivalis* can metabolize a range of amino acids and generate end products that are toxic to the human host.

Antibiotic susceptibility. *Porphyromonas* species are generally very susceptible to most of the antimicrobial agents commonly used for treatment of anaerobic infections, such as amoxicillin-clavulanate, piperacillin-tazobactam, ampicillin-sulbactam, imipenem, cephalosporins, and metronidazole. β-Lactamase production has been described in *Porphyromonas asaccharolytica* (Aldridge et al., 2001), *Porphyromonas catoniae* (Kononen et al., 1996), *Porphyromonas somerae* (Summanen et al., 2005), and *Porphyromonas uenonis* (Finegold et al., 2004). The frequency of β-lactamase production has been reported at approximately 20%. Animal-derived *Porphyromonas* species are more often β-lactamase producers than those derived from humans. Occasional resistance to clindamycin and ciprofloxacin may occur in *Porphyromonas asaccharolytica* and *Porphyromonas somerae*; also, ciprofloxacin resistance of *Porphyromonas gingivalis* has been reported (Lakhssassi et al., 2005).

Pathogenicity. Some species are considered true pathogens and are associated with human or animal infections. In particular, *Porphyromonas gingivalis* is a major causative agent in the initiation and progression of severe forms of periodontal disease. *Porphyromonas gingivalis* possesses a multitude of cell-surface associated and extracellular activities that contribute to its virulence potential. Several of these factors are adhesins that interact with other bacteria, epithelial cells, and extracellular matrix proteins. Secreted or cell-bound enzymes, toxins, and hemolysins play a significant role in the spread of the organism through tissue, in tissue destruction, and in evasion of host defenses.

Ecology. Several of the members of *Porphyromonas* are indigenous bacterial flora in the oral cavity of humans and animals. Many species are also found in the urogenital and intestinal tracts. *Porphyromonas* species have been isolated from oral infections, and from many infections throughout the body, e.g., blood, amniotic fluid, umbilical cord, pleural empyema, peritoneal and pelvic abscesses, inflamed endometrium, and other infected tissues. *Porphyromonas* species of animal origin have been encountered in humans with animal bite infections.

Enrichment and isolation procedures

A complex medium containing peptone, yeast extract, vitamin K_1, and hemin, and supplemented with 5% blood is recommended for isolation of *Porphyromonas* from body sites. Fresh or prereduced media (commercially available from Anaerobe Systems, Morgan Hill, California) are recommended, as they increase isolation efficacy. For determination of pigmentation, rabbit blood (laked) is preferable to blood from other animals. Kanamycin-vancomycin laked-blood agar with reduced vancomycin concentration (2 µg/ml) may be used as a selective medium when *Porphyromonas* species are sought (Jousimies-Somer et al. 2002). Formulas for basal media are given in Jousimies-Somer et al. (2002).

Inoculated media should immediately be placed in an anaerobic environment, such as an anaerobic pouch, jar, or chamber. If anaerobic systems utilizing palladium catalysts are used, the anaerobic gas mixture must contain hydrogen to enable the palladium to reduce oxygen to water. The plates may be examined after a 48-h incubation. However, a total incubation period of at least 7 d is recommended, because not all *Porphyromonas* species may be detected with shorter incubation times.

Maintenance procedures

Isolates can be put into stock from broth or plate cultures. Freezing at –70°C or lyophilization of young cultures grown in a well-buffered liquid or solid medium is satisfactory for storage of *Porphyromonas* species. Storage of lyophilized cultures at 4°C is recommended. Even with the best storage conditions, only a portion of the original cell population survives. Therefore, a large inoculum in a supportive medium and minimal exposure to oxygen are recommended for the recovery of viable cultures from stored material.

To maintain stock strains in the laboratory, it is advisable to transfer them weekly in chopped meat medium or other suitable medium that does not contain a fermentable carbohydrate.

Differentiation of the genus *Porphyromonas* from other genera

The pigmented *Porphyromonas* species can be differentiated from other anaerobic, Gram-stain-negative genera with relative ease (Table 14). The special-potency antibiotic disks (vancomycin, 5 µg; kanamycin, 1000 µg; and colistin, 10 µg) can be used to separate the Gram-stain-negative genera: *Porphyromonas* species are generally sensitive to vancomycin and resistant to kanamycin and colistin, whereas the other Gram-stain-negative genera are resistant to vancomycin and vary in their resistance to kanamycin and colistin.

A distinctive feature of the porphyromonads is that the predominant fatty acid is 13-methyltetradecanoic acid ($C_{15:0}$ iso); by contrast, *Bacteroides* and *Prevotella* contain 12-methyl tetradecanoic acid ($C_{15:0}$ anteiso) as their major cellular fatty acid. *Bacteroides* and *Prevotella* also produce a simpler metabolic end product profile of mainly acetic and succinic acids (Table 14).

Taxonomic comments

The taxonomy of pigmented Gram-negative bacilli has changed greatly since the first edition of *Bergey's Manual of Systematic Bacteriology*. The genus *Porphyromonas* was created in 1988 (Shah and Collins, 1988), and since then some species

TABLE 14. Differentiation of the genus *Porphyromonas* from other anaerobic Gram-negative rods[a]

Characteristic	*Porphyromonas*[b]	*Alistipes*[c]	*Bacteroides*[d]	*Prevotella*	*Rikenella*	*Tannerella*
Susceptibility to:[e]						
Vancomycin (5 µg)	S	R	R	R	R	R
Kanamycin (1000 µg)	R	R	R	R	S	S
Colistin (10 µg)	R	R	R	D	R	S
Pigment	+	+	–	D	–	–
Growth in 20% bile	–	+	+	–	+	–
Proteolytic activity	D	+	–	D	–	+
Major metabolic end products from PYG[f]	A, B, iV	S	A, S	A, S	P, S	A, S, PA
Major long-chain fatty acids	$C_{15:0}$ iso	$C_{15:0}$ iso	$C_{15:0}$ anteiso	$C_{15:0}$ anteiso	$C_{15:0}$ iso	$C_{15:0}$ anteiso
DNA G+C (mol%)	40–55	55–58	40–48	39–60	60–61	44–48

[a]+, 90% or more of the strains are positive; –, 10% or more of the strains are negative; D, different reaction in different species.
[b]*Porphyromonas catoniae* does not produce pigment, is vancomycin-resistant, and fermentative.
[c]*Alistipes putredinis* does not produce pigment, is susceptible to bile, and nonfermentative.
[d]*Bacteroides sensu stricto.*
[e]Special potency antimicrobial identification disks; R, resistant; S, susceptible; D, differs among species.
[f]A, acetic acid; B, butyric acid; iV, isovaleric acid; P, propionic acid; PA, phenylacetic acid; S, succinic acid.

previously included in the genus *Bacteroides* (*Porphyromonas gingivalis*, *Porphyromonas asaccharolytica*, *Porphyromonas endodontalis*, *Porphyromonas levii*, and *Porphyromonas macacae*) were reclassified as *Porphyromonas* species, and several new species described. *Porphyromonas* now includes 16 validly published species. *Porphyromonas salivosa* is a later heterotypic synonym of *Porphyromonas macacae* (Love, 1995).

The porphyromonads form a natural, but deep, phylogenetic group (Figure 17). The species exhibit levels of 16S rRNA sequence divergence of up to 15%. Phylogenetically, the closest related species to *Porphyromonas* are found in the genera *Bacteroides* and *Prevotella* (approx. 82–89% similarity).

Differentiation of species of the genus *Porphyromonas*

With the exception of *Porphyromonas catoniae*, *Porphyromonas* species form a phenotypically homogeneous group: they are pigmented, sensitive to the special-potency vancomycin disk, and produce butyric acid as the major metabolic end product. *Porphyromonas catoniae* differs in all these aspects. The characteristics useful in differentiating the species are given in Table 15.

Numerous PCR-based identification or detection systems have been described for *Porphyromonas gingivalis* and *Porphyromonas endodontalis* (de Lillo et al., 2004; Fouad et al., 2002; Gomes et al., 2005; Jervoe-Storm et al., 2005; Kuboniwa et al., 2004; Kumar et al., 2003; Noguchi et al., 2005; Seol et al., 2006).

List of species of the genus *Porphyromonas*

1. **Porphyromonas asaccharolytica** (Holdeman and Moore 1970) Shah and Collins 1988, 129[VP] (*Bacteroides asaccharolyticus* (Holdeman and Moore 1970) Finegold and Barnes 1977, 390[AL]; *Bacteroides melaninogenicus* subsp. *asaccharolyticus* Holdeman and Moore 1970, 33)

a.sac.cha.ro.ly′ti.cus. Gr. pref. *a* not; Gr. n. *sakchâr* sugar; N.L. fem. adj. *lytica* (from Gr. fem. adj. *lutikê*) able to loosen, able to dissolve; N.L. fem. adj. *asaccharolytica* not digesting sugar.

The description is from Shah and Collins (1988) and Holdeman et al. (1984). Proteolytic activity is weak, but gelatin liquefaction is positive and fibrinolytic activity is present. Starch is not hydrolyzed. Growth is stimulated by 0.5% NaCl. Cellular and colonial morphology and other characteristics are as described for the genus and as given in Table 15. *Porphyromonas asaccharolytica* can be distinguished from *Porphyromonas uenonis* by phenotypic tests (Table 15); however, DNA–DNA reassociation studies are required to distinguish these two species genetically. Susceptible to piperacillin-tazobactam, ampicillin-sulbactam, imipenem, meropenem, trovafloxacin, and metronidazole. Most strains (≥90%) are susceptible to cefoxitin, ciprofloxacin, and clindamycin. β-Lactamase production has been described in *Porphyromonas asaccharolytica* at the rate of 21% (Aldridge et al., 2001).

Source: various human clinical infections.

DNA G+C content (mol%): 52–54 (T_m)

Type strain: ATCC 25260, CCUG 7834, DSM 20707, JCM 6326, LMG 13178, L16490, VPI 4198.

Sequence accession no. (16S rRNA gene): L16490.

2. **Porphyromonas cangingivalis** Collins, Love, Karjalainen, Kanervo, Forsblom, Willems, Stubbs, Sarkiala, Bailey, Wigney and Jousimies-Somer 1994, 676[VP]

can.gin.gi.val′is. L. n. *canis* dog; L. n. *gingiva* gum; L. fem. suff. *-alis* suffix denoting pertaining to; N.L. fem. adj. *cangingivalis* pertaining to the gums of dogs.

The description is from Collins et al. (1994). In cooked-meat carbohydrate medium and on sheep blood agar plates cells are 0.3–0.6 × 0.8–1.5 µm and occur singly and in clumps; occasionally filaments up to 16 µm long are observed. Some strains indent agar, exhibit peripheral flattening, and have a roughened and dry surface appearance on sheep blood agar. On egg yolk agar, colonies are yellow or orange. After incubation of 5 d, the pH range in media containing carbohydrates generally is 6.3–6.5. Ammonia is produced in cooked meat medium. Neither lactate nor threonine is converted to propionate, and pyruvate is not utilized. Other characteristics are as described for the genus and as given in Table 15. Some strains are susceptible to penicillin, amoxicillin, carbenicillin, and erythromycin; 35% of the strains produce β-lactamase.

Source: diseased or healthy periodontal pockets of dogs with naturally occurring periodontitis.

DNA G+C content (mol%): 55 (T_m).

Type strain: ATCC 700135, CCUG 47700, NCTC 12856, VPB 4874, X76259.

Sequence accession no. (16S rRNA gene): X76259.

3. **Porphyromonas canoris** Love, Karjalainen, Kanervo, Forsblom, Willems, Stubbs, Sarkiala, Bailey, Wigney and Jousimies-Somer 1994, 207[VP]

can′or.is. L. n. *canis* dog; L. gen. n. *oris* of the mouth; N.L. gen. n. *canoris* of a dog's mouth.

The description is from Love et al. (1994). In cooked-meat carbohydrate medium and on sheep blood agar plates cells are 0.3–0.6 × 0.8–1.5 µm and occur singly and in clumps; occasionally filaments up to 16 µm long are observed. On sheep blood agar, colonies at 48 h are circular and rough, and have an orange pigmentation. After incubation of 5 d, the pH range in media containing carbohydrates is 6.3–6.5. Ammonia is produced in cooked meat medium. Lactate is converted to propionate, but pyruvate is not utilized and threonine is not converted to propionate. Other characteristics are as described for the genus and as given in Table 15. Strains are susceptible to penicillin, amoxicillin, carbenicillin, and erythromycin.

Source: subgingival pockets of dogs with naturally occurring periodontitis.

DNA G+C content (mol%): 49–51 (T_m).

Type strain: CCUG 36550, CIP 104881, JCM 11138, NCTC 12835, VPB 4878.

Sequence accession no. (16S rRNA gene): X76261.

4. **Porphyromonas cansulci** Collins, Love, Karjalainen, Kanervo, Forsblom, Willems, Stubbs, Sarkiala, Bailey, Wigney and Jousimies-Somer 1994, 678[VP]

can.sul′ci. L. n. *canis* dog; L. gen. n. *sulci* of a furrow; L. gen. n. *cansulci* of a dog's furrow, referring to the habitat in the mouths of dogs.

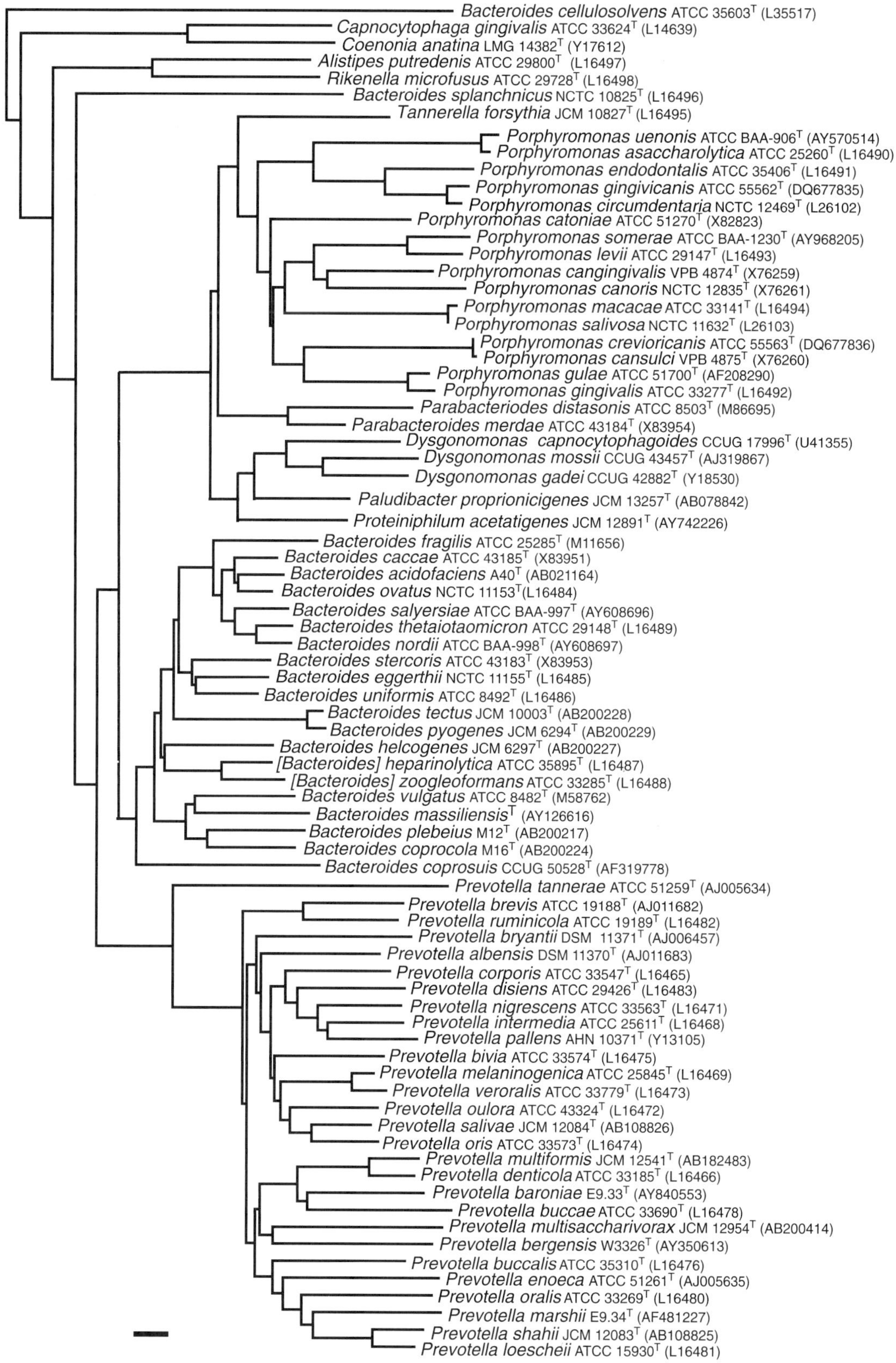

FIGURE 17. Unrooted tree showing the phylogenetic position of *Porphyromonas* within the *Bacteroides* subgroup of the *Bacteroidetes* (*Cytophaga–Flavobacter–Bacteroides*) phylum. The tree was constructed by the maximum-parsimony method and is based on a comparison of approximately 1400 nt. Bootstrap values, expressed as a percentage of 1000 replications, are given at the branching points. The scale bar indicates 1% sequence divergence. Courtesy of Paul A. Lawson.

TABLE 15. Differentiation of the species of the genus *Porphyromonas*[a]

Characteristic	*P. asaccharolytica*	*P. cangingivalis*	*P. canoris*	*P. cansulci*	*P. catoniae*	*P. circumdentaria*	*P. crevioricanis*	*P. endodontalis*	*P. gingivalis*	*P. gingivicanis*	*P. gulae*	*P. levii*	*P. macacae*[e]	*P. somerae*	*P. uenonis*
Pigment production	+	+	+	+	–	+	+	+	+	+	+	+	+	+	+
Fluorescence	+	–	+	+	–	+	+	+	–	+	–	d	–	d	+
Hemagglutinin activity	–	–	–	–	na	–	+	–	+	–	+	–	–	na	na
Indole	+	+	+	+	–	+	+	+	+	+	+	–	+	–	+
Catalase	–	+	+	+	–	+	–	–	–	+	+	–	+	–	–
Lipase	–	–	–	–	–	–	–	–	–	–	–	–	$+^{-}$	–	–
Preformed enzyme activity:[b]															
α-Fucosidase	+	–	–	–	+	–	–	–	–	–	–	–	–	–	–
α-Galactosidase	–	–	–	–	d	–	–	–	–	–	–	–	+	–	–
β-Galactosidase	–	–	+	–	+	–	–	–	–[f]	–	+	+	–[f]	+	–
N-Acetyl-β-glucosaminidase	–	–	+	–	+	–	–	–	+	–	+	+	+	+	d
Chymotrypsin	–	+	+	–	d	+	–	–	–	–	na	+	+	+	–
Trypsin	–	–	–	–	d	–	–	–	+	–	+	–	+	–	–
Fermentation of:[c]															
Glucose	–	–	–	–	+	–	–	–	–	–	–	w	w	w	–
Lactose	–	–	–	–	+	–	–	–	–	–	–	w	w	w	–
Maltose	–	–	–	–	+	–	–	–	–	–	–	w	–	w	–
Glucose-6-phosphate and 6-phosphogluconate dehydrogenases present	–	+	+	–	+	–	–	–	–	–	na	d	d	na	na
Major long-chain fatty acids	$C_{15:0}$ iso	$C_{15:0}$ iso	$C_{15:0}$ iso	$C_{15:0}$ iso	$C_{15:0}$ iso, $C_{15:0}$ anteiso	$C_{15:0}$ iso	$C_{15:0}$ iso	$C_{15:0}$ iso	$C_{15:0}$ iso	$C_{15:0}$ iso	na	$C_{15:0}$ iso, $C_{15:0}$ anteiso	$C_{15:0}$ iso	$C_{15:0}$ iso, $C_{15:0}$ anteiso	$C_{15:0}$ iso
Metabolic end products from PYG[d]	A, P, ib, B, IV, s	A, p, ib, B, IV	A, P, ib, b, IV, s	A, P, ib, B, IV, S, pa	a, P.iv, l, S	A, P, ib, b, IV, s, pa	A, p, ib, B, IV, s, pa	A, P, ib, B, IV, s	A, P, ib, B, IV, s, pa	A, p, ib, B, IV, s	A, P, ib, B, IV, s, pa	A, P, ib, B, IV, s	A, P, ib, B, IV, s, pa	A, P, ib, B, IV, s	A, P, ib, B, IV, s

[a]+, 90% or more of the strains are positive; –, 10% or more of the strains are negative; d, 11–89% of the strains are positive; w, weak positive reaction; na, data not available.

[b]Reaction by API ZYM System or Rosco Diagnostic tablets. Reactivity in these systems is not always identical (see footnote f).

[c]Fermentation of most other carbohydrates have been reported negative.

[d]Upper-case letters indicate major metabolic products from peptone-yeast-glucose (PYG), lower-case letters indicate minor products, and parentheses indicate a variable reaction for the following acids: A, acetic; P, propionic; IB, isobutyric; B, butyric; IV, isovaleric; V, valeric; L, lactic; S, succinic; PA, phenylacetic.

[e]*Porphyromonas salivosa* is a later heterotypic synonym. The cat biovar (*Porphyromonas salivosa*) is lipase-positive, does not ferment sorbitol, and may fluoresce under UV light.

[f]Negative by the API ZYM System; positive with the Rosco *o*-nitrophenyl-β-D-galactopyranoside test.

The description is from Collins et al. (1994). On egg yolk agar, colonies may be yellow or orange. After incubation of 5 d, the pH range in media containing carbohydrates is 6.3–6.5. Ammonia is produced in cooked-meat medium. Neither lactate nor threonine is converted to propionate and pyruvate is not utilized. Other characteristics are as described for the genus and as given in Table 15. Strains are susceptible to penicillin, amoxicillin, carbenicillin, and erythromycin. β-Lactamase producing strains have not been detected.

Source: periodontal pockets of dogs with naturally occurring periodontitis.

DNA G+C content (mol%): 49–51 (T_m).

Type strain: CCUG 47702, NCTC 12858, VPB 4875.

Sequence accession no. (16S rRNA gene): X76260.

Taxonomic note: Strain ATCC 55563[T] of *Porphyromonas crevioricanis* Hirasawa and Takada 1994, 640[VP] exhibits 99.9% rRNA gene sequence homology with *Porphyromonas cansulci* 12858[T]. The taxonomic standing of these two species remains to be determined.

5. **Porphyromonas catoniae** (Moore and Moore 1994) Willems and Collins 1995, 581[VP] (*Oribaculum catoniae* Moore and Moore 1994, 189[VP])

ca.to′ni.ae. N.L. gen. fem. n. *catoniae* of Cato, named in honor of Elizabeth P. Cato, an American microbiologist.

The description is from Willems and Collins (1995) and Moore and Moore (1994). Cells grown in PYG broth are 0.6 × 0.8–1.7 μm and occur in pairs and short chains. Cells grown in media containing fermentable carbohydrate may be highly vacuolated. Surface colonies on blood agar plates incubated for 2 d are 0.2–2 mm in diameter, circular, entire, flat to low convex, and transparent. Most strains are not hemolytic on rabbit blood agar; an occasional strain may be beta-hemolytic. No colonies with a dark pigment are produced. Saccharolytic. Abundant growth occurs in peptone-yeast extract or PYG broth. Broth cultures are turbid with a smooth to fine granular sediment. The pH range of PYG broth cultures is 5.0–5.5. Gelatin is hydrolyzed. Starch is hydrolyzed by most strains; some strains hydrolyze milk and meat. H_2S is produced by some strains. The major cellular fatty acids are $C_{15:0}$ iso and $C_{15:0}$ anteiso; substantial amounts of $C_{13:0}$ iso are also present. Sensitive to clindamycin and chloramphenicol. Two of 85 strains were resistant to erythromycin, one strain to tetracycline, and six strains to penicillin. β-Lactamase producing strains have been described (Kononen et al., 1996).

Source: gingival crevices of humans with gingivitis or periodontitis and from humans with healthy gingivae.

DNA G+C content (mol%): 49 (T_m).

Type strain: ATCC 51270, CCUG 41358, NCTC 13056, VPI N3B-3.

Sequence accession no. (16S rRNA gene): X82823.

6. **Porphyromonas circumdentaria** Love, Bailey, Collings and Briscoe 1992, 435^VP^

cir.cum.den.ta′ri.a. L. adv. and prep. *circum* around, about; L. fem. adj. *dentaria* pertaining to teeth; N.L. fem. adj. *circumdentaria* referring to the isolation of the organism from the vicinity of the teeth.

The description is from Love et al. (1992). In cooked-meat carbohydrate medium and on sheep blood agar plates cells are 0.3–0.6 × 0.8–1.5 μm and occur singly, in clumps, and occasionally as filaments up to 10 μm long. On sheep blood agar, surface colonies at 72 h are 1–2 mm in diameter, and greenish brown. Both colony size and pigment formation are enhanced by growth in the presence of *Staphylococcus epidermidis*. After incubation for 5 d, the pH range in media containing carbohydrates generally is 6.3–6.5. Ammonia is produced in cooked meat medium. Gelatin is hydrolyzed. Neither lactate nor threonine is converted to propionate, and pyruvate is not utilized. Other characteristics are as described for the genus and as given in Table 15.

Source: soft tissue infections (abscesses and empyemas), gingival margins, and subgingival plaque of felines.

DNA G+C content (mol%): 40–42 (T_m).

Type strain: ATCC 51356, CCUG 41934, JCM 13864, NCTC 12469, VPB 3329.

Sequence accession no. (16S rRNA gene): L26102.

7. **Porphyromonas crevioricanis** Hirasawa and Takada 1994, 640^VP^

cre.vi.o.ri.ca′nis. N.L. gen. n. *crevi* (*sic*) of a crevice; L. gen. n. *oris* of the mouth; L. gen. n. *canis* of a dog; N.L. gen. n. *crevioricanis* of the crevice of a dog's mouth.

The description is from Hirasawa and Takada (1994). Cellular and colonial morphology and other characteristics are as described for the genus and as given in Table 15. Susceptible to penicillin, amoxicillin, sulbenicillin, and erythromycin.

Source: gingival crevicular fluid obtained from beagles.

DNA G+C content (mol%): 44–45 (T_m).

Type strain: NUM 402, ATCC 55563.

Sequence accession no. (16S rRNA gene): DQ677836.

Taxonomic note: Strain NCTC 12858^T^ of *Porphyromonas cansulci* Collins et al. 1994, 678^VP^ exhibits 99.9% rRNA gene sequence homology with *Porphyromonas crevioricanis* ATCC 55563^T^. The taxonomic standing of these two species remains to be determined.

8. **Porphyromonas endodontalis** (van Steenbergen, VanWinkelhoff, Mayrand, Grenier and de Graaff 1984) Shah and Collins 1988, 129^VP^ (*Bacteroides endodontalis* van Steenbergen, VanWinkelhoff, Mayrand, Grenier and de Graaff 1984, 119^VP^)

en.do.don′ta.lis. Gr. adv. *endon* within; Gr. n. *odous -ontos* tooth; L. fem. suff. *-alis* suffix denoting pertaining to; N.L. fem. adj. *endodontalis* pertaining to the inside of a tooth, within teeth.

The description is from Shah and Collins (1988) van Steenbergen et al. (1984), and Holdeman et al. (1984). Protoheme is the major porphyrin produced, but protoporphyrin is also present. Most strains produce colonies that adhere strongly to the agar plates, and growth in liquid media is slow.

Gelatin is hydrolyzed. No proteolytic or collagenolytic activity is detected. Arginine is hydrolyzed; starch is not hydrolyzed. H_2S is produced. Cellular and colonial morphology and other characteristics are as described for the genus and as given in Table 15.

Specific primers based on 16S rRNA sequencing have been described for culture-independent identification of *Porphyromonas endodontalis* (de Lillo et al., 2004). Similarly, the use of PCR amplification of 16S rDNA and the downstream intergenic spacer region (ISR) for culture-independent detection of periodontal pathogens revealed that *Porphyromonas endodontalis* was as strongly associated with periodontitis as *Porphyromonas gingivalis* (Kumar et al., 2003).

Source: infected dental root canals, periodontal pockets, and other oral sites.

DNA G+C content (mol%): 49–51 (T_m).

Type strain: HG370, ATCC 35406, NCTC 13058.

Sequence accession no. (16S rRNA gene): L16491.

9. **Porphyromonas gingivalis** (Coykendall, Kaczmarek and Slots 1980) Shah and Collins 1988, 129^VP^ (*Bacteroides gingivalis* Coykendall, Kaczmarek and Slots 1980, 559^VP^)

gin.gi.val′is. L. n. *gingiva* gum; L. fem. suff. *-alis* suffix denoting pertaining to; N.L. fem. adj. *gingivalis* pertaining to the gums, gingival.

The description is from Shah and Collins (1988) and Holdeman et al. (1984). Protoheme is the major porphyrin produced, but traces of protoporphyrin also occur. Several amino acids, such as aspartate, arginine, cystine, histidine, serine, tryptophan, leucine, methionine, phenylalanine, and isoleucine, are utilized. Proteases are present. Starch is not hydrolyzed. Cellular and colonial morphology and other characteristics are as described for the genus and as given in Table 15.

Porphyromonas gingivalis possesses a multitude of cell-surface-associated and extracellular activities, such as adhesins, secreted or cell-bound enzymes, toxins, and hemolysins that contribute to its virulence potential. Differences in the virulence exist, but the mechanisms underlying these differences are not yet fully understood. Multilocus sequence typing of *Porphyromonas gingivalis* strains from different geographic origins showed high genetic diversity of the species (Enersen et al., 2006).

Porphyromonas gingivalis is generally very susceptible to most of the antimicrobial agents commonly used for treatment of anaerobic infections, such as amoxicillin-clavulanate, piperacillin-tazobactam, ampicillin-sulbactam, imipenem, cephalosporins, and metronidazole. Ciprofloxacin resistance of *Porphyromonas gingivalis* has been reported (Lakhssassi et al., 2005).

Porphyromonas gingivalis is considered a major periodontal pathogen and reported to be a cause of extraoral infections, such as lung abscesses and pulmonary infections. It has also been suggested that *Porphyromonas gingivalis* may contribute to the development of atheromas in cardiovascular disease.

Source: infected dental root canals, periodontal pockets, and other oral sites.

DNA G+C content (mol%): 46–48 (T_m).

Type strain: 2561, ATCC 33277, CCUG 25893, CCUG 25928, CIP 103683, DSM 20709, JCM 12257, NCTC 11834.

Sequence accession no. (16S rRNA gene): L16492.

10. **Porphyromonas gingivicanis** Hirasawa and Takada 1994, 639[VP]

gin.gi.vi.ca'nis. L. fem. n. *gingiva* gum; L. gen. n. *canis* of a dog; N.L. gen. n. *gingivicanis* of the gums of a dog.

The description is from Hirasawa and Takada (1994). Cellular and colonial morphology and other characteristics are as described for the genus and as given in Table 15. Susceptible to penicillin, amoxicillin, sulbenicillin, and erythromycin.

Source: gingival crevicular fluid obtained from beagles.

DNA G+C content (mol%): 41–42 (T_m).

Type strain: NUM 301, ATCC 55562.

Sequence accession no. (16S rRNA gene): DQ677835.

11. **Porphyromonas gulae** Fournier, Mouton, Lapierre, Kato, Okuda and Menard. 2001, 1187[VP]

gu'lae. L. n. *gula* [animal] mouth; L. gen. n. *gulae* from the animal mouth, referring to its isolation from subgingival plaque of various animal hosts.

The description is from Fournier et al. (2001). *Porphyromonas gulae* encompasses the "animal *Porphyromonas gingivalis*". Cellular and colonial morphology and other characteristics are as described for the genus and as given in Table 15. Twenty strains were susceptible to amoxicillin and amoxycillin/clavulanate.

Source: subgingival plaque samples of various mammals, including cat, dog, coyote, wolf, bear, and non-human primate, such as squirrel monkey and spider monkey. *Porphyromonas gulae* is the most prominent species of the genus *Porphyromonas* to be found in the oral cavity of mammals.

DNA G+C content (mol%): 51 (T_m).

Type strain: Loup 1, ATCC 51700, CCUG 47701, NCTC 13180.

Sequence accession no. (16S rRNA gene): AF208290.

12. **Porphyromonas levii** (Johnson and Holdeman 1983) Shah, Collins, Olsen, Paster and Dewhirst 1995, 586[VP] (*Bacteroides levii* Johnson and Holdeman 1983, 15[VP])

lev'i.i. N.L. gen. masc. n. *levii* of Lev, named after Meir Lev, the American-English microbiologist, who first isolated this organism.

The description is from Shah et al. (1995), Holdeman et al. (1984), and Summanen et al. (2005). Cellular and colonial morphology and other characteristics are as described for the genus and as given in Table 15. Protoheme is the major porphyrin produced, but traces of protoporphyrin also occur. Succinate stimulates growth and can replace the requirement for heme. Few sugars, such as glucose and lactose, are weakly fermented. Glucose broth cultures are turbid with smooth sediment and a final pH of 5.5. Most other commonly occurring sugars, such as arabinose, cellobiose, melezitose, melibiose, raffinose, rhamnose, ribose, salicin, sucrose, trehalose, and xylose, are not fermented. Some amino acids, such as asparagine, tryptophan, phenylalanine, and glutamine, are utilized. Proteases are present. Starch is not hydrolyzed.

Porphyromonas levii is genetically distant from *Porphyromonas somerae*; however, these two species cannot readily be differentiated by phenotypic tests. Although both species contain major $C_{15:0}$ iso, $C_{15:0}$ anteiso acids, the cluster analysis of the cellular fatty acid profiles (Euclidian distance of principal components accounting for the greatest variance of the organisms) differentiates the two species.

Source: bovine rumen, cattle horn abscess, bovine summer mastitis, and bovine necrotizing vulvovaginitis.

DNA G+C content (mol%): 46–48 (T_m).

Type strain: LEV, ATCC 29147, CCUG 21027, CCUG 34320, HAMBI 467, NCTC 11028, VPI 10450, VPI 3300.

Sequence accession no. (16S rRNA gene): L16493.

13. **Porphyromonas macacae** (Slots and Genco 1980) Love 1995, 91[VP] [*Bacteroides melaninogenicus* subsp. *macacae* Slots and Genco 1980, 84[VP]; *Bacteroides macacae* (Slots and Genco 1980) Coykendall, Kaczmarek and Slots 1980, 563[VP]; *Porphyromonas salivosa* Love, Bailey, Collings and Briscoe 1992, 438[VP]]

ma.ca'cae. N.L. fem. n. *Macaca* genus name of the macaque; N.L. gen. n. *macacae* of the macaque.

The description is from Love et al. (1995) and Holdeman et al. (1984). After 6 d on blood agar plates, colonies are 0.1–0.2 mm, entire, dome shaped, and creamy brown. After 9 d in the presence of *Staphylococcus epidermidis*, however, the colonies are 1.0–1.5 mm in diameter, entire, umbonate with a central depression. As incubation progresses, the surfaces of colonies may become wrinkled with multiple central depressions and peripheral ridging. Gelatin is hydrolyzed. Lipase is not produced. Other characteristics are as described for the genus and as given in Table 15.

Porphyromonas macacae ATCC 33141 can be distinguished by phenotypic criteria from cat strains, suggesting that cat and monkey biovars exist. (See discussion of *Porphyromonas salivosa* under *Taxonomic note*, below.)

Source: oral cavities, subcutaneous abscesses, and pyothoraxes of animals, including cats and monkeys. Important pathogen in animal bite infections in humans.

DNA G+C content (mol%): 43–44 (T_m).

Type strain: 7728-L6C, Slots' strain 7728-L6C, ATCC 33141, CCUG 47703, DSM 20710, NCTC 13100.

Sequence accession no. (16S rRNA gene): L16494.

Taxonomic note: Strain NCTC 11632^T of *Porphyromonas salivosa* (Love et al., 1987) Love et al., 1992, 438^{VP} exhibits 99.3% rRNA gene sequence homology (Paster et al., 1994) and a mean level of DNA–DNA hybridization of 81% (Love, 1995) with *Porphyromonas macacae* ATCC 33141^T; therefore, the species *Porphyromonas salivosa* was included in the species *Porphyromonas macacae* by Love (1995). However, *Porphyromonas macacae* ATCC 33141^T can be distinguished by phenotypic criteria from cat strains (*Porphyromonas salivosa*), suggesting that cat and monkey biovars of *Porphyromonas macacae* exist. The members of the cat biovar have different colonial morphologies on blood agar: the colonies are 0.5–1.5 mm in diameter, circular, entire, dome-shaped, and brown–black at 72 h on blood agar plates. Furthermore, unlike *Porphyromonas macacae* ATCC 33141^T, the cat biovars produce different whole-cell protein and proteinase profiles on SDS-PAGE gels, and are lipase-positive and sorbitol-negative.

14. **Porphyromonas salivosa** (Love, Johnson, Jones, and Calverley 1987) Love, Bailey, Collings and Briscoe 1992, 438^{VP} = *Porphyromonas macacae* (senior heterotypic synonym) (*Bacteroides salivosus* Love, Johnson, Jones and Calverley 1987, 308)

sal.i.vo′sa. L. fem. adj. *salivosa* resembling saliva, slimy.

The description is from Love et al. (1995), Love et al. (1992), and Love et al. (1987). After 72 h on blood agar plates, colonies are 0.5–1.5 mm, entire, dome shaped, and brown–black. Gelatin is hydrolyzed. Lipase is produced.

Porphyromonas salivosa NCTC 11632^T can be distinguished by phenotypic criteria from monkey strains (*Porphyromonas macacae*), suggesting that cat and monkey biovars exist. (See *Taxonomic note*, above, under *Porphyromonas macacae*.)

DNA G+C content (mol%): 42–44 (T_m).

Type strain: ATCC 49407, NCTC 11632, VPB 157, CCUG 33478.

Sequence accession no. (16S rRNA gene): L26103.

15. **Porphyromonas somerae** Summanen, Durmaz, Vaisanen, Liu, Molitoris, Eerola, Helander and Finegold 2006, 925^{VP} (Effective publication: Summanen, Durmaz, Vaisanen, Liu, Molitoris, Eerola, Helander and Finegold 2005, 4458.)

so′mer.ae. N.L. gen. fem. n. *somerae*, of Somer, named in honor of the late Finnish microbiologist Hannele Jousimies-Somer.

The description is from Summanen et al. (2005). Colonies incubated on blood agar for 2 d often exhibit a "patchy" growth pattern, with larger colonies surrounded by smaller colonies; they are circular, entire, and convex. The colonies on laked rabbit blood agar are white–yellow to tan; after 4 d of incubation, the colonies are pigmented (light brown to dark brown) and show no or occasionally weak red fluorescence under long-wave UV light. Weakly saccharolytic, the pH of glucose, lactose, and maltose cultures is 5.3–5.4 after 5 d of incubation. Most other commonly occurring sugars, such as arabinose, cellobiose, melezitose, melibiose, raffinose, rhamnose, ribose, salicin, sucrose, trehalose, and xylose are not fermented. β-lactamase is produced by 21% of the strains. Some strains are resistant to clindamycin. Other characteristics are as described for the genus and as given in Table 15.

Porphyromonas somerae is genetically distant from *Porphyromonas levii*; however, these two species cannot readily be differentiated by phenotypic tests. Although both species contain major $C_{15:0}$ iso, $C_{15:0}$ anteiso acids, the cluster analysis of the cellular fatty acid profiles (Euclidian distance of principal components accounting for the greatest variance of the organisms) differentiates the two species.

Source: various clinical specimens of non-oral origin, mainly from chronic foot infections of diabetics or other patients with vascular insufficiency.

DNA G+C content (mol%): 47.8 (T_m).

Type strain: WAL 6690, ATCC BAA-1230, CCUG 51464.

Sequence accession no. (16S rRNA gene): AY968205.

16. **Porphyromonas uenonis** Finegold, Vaisanen, Rautio, Eerola, Summanen, Molitoris, Song, Liu and Jousimies-Somer 2005, 547^{VP} (Effective publication: Finegold, Vaisanen, Rautio, Eerola, Summanen, Molitoris, Song, Liu and Jousimies-Somer 2004, 5301.)

ue.no′nis. N.L. gen. masc. n. *uenonis* of Ueno, in honor of the late Japanese microbiologist Kazue Ueno.

The description is from Finegold et al. (2004). Growth is stimulated by 5% horse serum and similar additives. Cellular and colonial morphology and other characteristics are as described for the genus and as given in Table 15. *Porphyromonas uenonis* can be distinguished from *Porphyromonas asaccharolytica* by phenotypic tests (Table 15); however, DNA–DNA reassociation studies are required to distinguish these two species genetically. Susceptible to most antimicrobial agents. Some strains produce β-lactamase.

Source: part of a mixed flora in various infections, which apparently have their origin in the intestinal tract. The habitat is probably the human gut.

DNA G+C content (mol%): 52.5 (T_m).

Type strain: WAL 9902, ATCC BAA-906, CCUG 48615.

Sequence accession no. (16S rRNA gene): AY570514.

Other organisms

Porphyromonas bennonis, a novel species isolated from human clinical specimens, was recently published in IJSEM (Summanen et al., 2009).

DNA G+C content (mol%): 58 (T_m).

Type strain: WAL 1926C, ATCC BAA-1629, CCUG 55979.

Sequence accession no. (16S rRNA gene): EU414673.

The 16S rRNA gene sequence of an organism called "*Porphyromonas canis*" was published in GenBank with accession number AB0034799. "*Porphyromonas canis* sp. nov. isolated from dog". Unpublished. Strain ATCC 55562^T of *Porphyromonas gingivicanis* Hirasawa and Takada 1994, 640^{VP} exhibits 99.9% rRNA gene sequence homology with "*Porphyromonas canis*"

JCM 10100, suggesting that *Porphyromonas canis* is a later synonym of *Porphyromonas gingivicanis*.

No valid description of the species "*Porphyromonas canis*" can be found in the literature. However, the following information is available.

DNA G+C content (mol%): 52.5 (T_m).

Type strain: JCM 10100.

Sequence accession no. (16S rRNA gene): AB034799.

Taxonomic note: Hardham, Dreier, Wong, Sfintescu and Evans (2005) mentioned an organism named "*Porphyromonas denticanis*". They indicated that *Porphyromonas salivosa*, "*Porphyromonas denticanis*" (a novel species), and *Porphyromonas gulae* were the most frequently isolated black-pigmented anaerobic bacteria associated with canine periodontitis. However, no valid description of "*Porphyromonas denticanis*" can be found in the literature, and there are no GenBank entries for it.

Genus II. **Barnesiella** Sakamoto, Lan and Benno 2007b, 344[VP]

The Editorial Board

Bar.ne.si.el'la. N.L. dim. fem. n. *Barnesiella* named after the British microbiologist Ella M. Barnes, who contributed much to knowledge of intestinal bacteriology and to anaerobic bacteriology in general.

Rods (0.8–1.6 × 1.7–11 μm). Nonsporeforming. Nonmotile. Gram-negative. Obligately anaerobic. On Eggerth–Gagnon agar, colonies are 1–2 mm in diameter, gray to off-white–gray, circular, entire, slightly convex, and smooth. Saccharolytic, with a strictly fermentative type of metabolism. Acetic and succinic acids are the main fermentation products. Growth is inhibited on a medium containing 20% bile. Esculin is hydrolyzed. Indole-negative. The predominant menaquinones are MK-11 and MK-12. Isolated from the chicken cecum.

DNA G+C content (mol%): 52.

Type species: **Barnesiella viscericola** Sakamoto, Lan and Benno 2007b, 345[VP].

Further descriptive information

Phylogenetic analyses of chicken cecal microbiota, based on 16S rRNA gene clone library analyses, have revealed a large number of novel phylotypes (Lan et al., 2002; Zhu et al., 2002). Using a special anaerobic culture technique, Sakamoto et al. (2007b) obtained unusual strains. Like *Porphyromonas* strains, these obligately anaerobic, nonsporeforming, nonmotile, Gram-negative rods were inhibited by 20% bile. However, their major menaquinones (MK-11 and MK-12) differed from those of most other genera of the family *Porphyromonadaceae*, which had menaquinones 8, 9, or 10 (except *Tannerella*, which has MK-10 and MK-11). Phylogenetic analysis has shown that these strains represent a new genus in the family.

Enrichment and isolation procedures

Barnesiella strains were isolated from the chicken cecum under strictly anaerobic conditions by the method described by Lan et al. (2002). They were maintained on Eggerth–Gagnon (EG) agar (Merck) supplemented with 5% (v/v) horse blood, with incubation for 2 d at 37°C in an atmosphere of 100% CO_2.

Differentiation of the genus *Barnesiella* from other closely related genera

Some characteristics differentiating the genus *Barnesiella* from other genera are shown in Table 16.

Taxonomic comments

Analyses of the 16S rRNA gene sequences of two cecal isolates that failed to grow in the presence of 20% bile indicated that the strains were related to *Parabacteroides distasonis* (86% sequence similarity). The two strains exhibited 100% 16S rRNA gene sequence similarity with each other.

TABLE 16. Some characteristics differentiating the genus *Barnesiella* from other related genera[a,b]

Characteristic	*Barnsiella*	*Dysgonmonas*	*Paludibacter*	*Parabacteroides*	*Porphyomonas*	*Proteiniphilum*	*Tannerella*
Growth in presence of 20% bile	–	+	–	+	–	–	–
Saccharolytic	+	+	+	+	–	–	–
Major end products of glucose fermentation	Acetic, succinic acids	Propionic, lactic, succinic acids	Acetic, propionic acids	Acetic, succinic acids	na	na	na
Major menaquinones	MK-10, MK-11	nr	MK-8	MK-9, MK-10	MK-9, MK-10	nr	MK-10, MK-11
Brown to black colonies on blood agar	–	–	–	–	+	–	–
DNA G+C content (mol%)	52	38–39	39	43–46	40–55	47	44–48

[a]Symbols: +, >85% positive; –, 0–15% positive; na, not applicable; nr, not reported.

[b]Data taken from Chen and Dong (2005); Hofstad et al. (2000); Lawson et al. (2002a); Sakamoto and Benno (2006); Sakamoto et al. (2002); Ueki et al. (2006); Sakamoto et al. (2007b).

List of species of the genus *Barnesiella*

1. **Barnesiella viscericola** Sakamoto, Lan and Benno 2007b, 345[VP]

vis.ce.ri'co.la. L. n. *viscus, visceris* intestine; L. suff. n. *-cola* (from L. n. *incola*) inhabitant; N.L. fem. n. *viscericola* inhabitant of the intestine.

The characteristics are as given for the genus, with the following additional features. Urease and catalase-negative. Gelatin is digested. Acid is produced from D-cellobiose, glucose, maltose, D-mannose, and sucrose, but not from L-arabinose, glycerol, lactose, D-mannitol, D-melezitose, D-raffinose, L-rhamnose, salicin, D-sorbitol, D-trehalose, or D-xylose. Using the Rapid ID 32A tests, all strains are positive for α-galactosidase, β-galactosidase, α-glucosidase, β-glucosidase, *N*-acetyl-β-glucosaminidase, glutamic acid decarboxylase, α-fucosidase, alkaline phosphatase, leucylglycine arylamidase, and alanine arylamidase. Raffinose is fermented. All of the other tests with the Rapid ID 32A system give negative results. The major end products are acetic acid and succinic acid; lower levels of other acids may be produced. Both non-hydroxylated and 3-hydroxylated long-chain fatty acids are present. The major cellular fatty acids are $C_{15:0}$ anteiso and $C_{15:0}$ iso. The predominant respiratory quinones are MK-11 (65–66%) and MK-12 (21–24%). MK-10 is present as a minor menaquinone (10–11%).

Source: the chicken cecum.

DNA G+C content (mol%): 52 (HPLC).

Type strain: C46, DSM 18177, JCM 13660.

Sequence accession no. (16S rRNA gene): AB267809.

Genus III. **Dysgonomonas** Hofstad, Olsen, Eribe, Falsen, Collins and Lawson 2000, 2194[VP]

Ingar Olsen

Dys.go.no.mo'nas. Gr. pref. *dys-* with notion of hard, bad, unlucky; Gr. n. *gonos* that which is begotten, reproduction; Gr. fem. n. *monas* a monad, unit; N.L. fem. n. *Dysgonomonas* intended to mean a weakly growing monad.

Coccobacilli to short rods. Nonmotile. Gram-negative. Facultatively anaerobic. Colonies are 1–2 mm in diameter, nonadherent, entire, gray–white, smooth, and nonhemolytic and have a slight aromatic odor. Growth is not observed on MacConkey agar. Requires X factor for growth. May be catalase-positive or -negative. Oxidase-negative. Glucose is fermented, producing acid but no gas. Alkaline phosphatase is generated but not arginine dihydrolase. Nitrate is not reduced. H_2S and acetoin are not produced. Esculin may not be hydrolyzed; gelatin and urea are not hydrolyzed. Indole may be produced. Long-chain cellular fatty acids include straight-chain saturated, anteiso- and iso-methyl branched and 3-hydroxy types. Isolated from human clinical specimens and stools.

DNA G+C content (mol%): 38 (T_m).

Type species: **Dysgonomonas gadei** Hofstad, Olsen, Eribe, Falsen and Lawson 2000, 2194[VP].

Further descriptive information

Dysgonomonas capnocytophagoides (Hofstad et al., 2000) and *Dysgonomonas mossii* (Lawson et al., 2002a, b) are members of the CDC DF (dysgonic fermenter)-3 group (Daneshvar et al., 1991; Wallace et al., 1989). *Dysgonomonas gadei* (Hofstad et al., 2000) was isolated at the Gade Institute, Bergen, Norway. Comparative 16S rRNA sequence analysis indicates that *Dysgonomonas* is a distinct genus in Family III. "*Porphyromonadaceae*" in the phylum "*Bacteroidetes*". See *Taxonomic comments* for details.

Cells can be coccobacilli to short rods (Figures 18 and 19). Nitrate is not reduced and the oxidase reaction is negative. The catalase reaction has been reported as negative (Koneman et al., 1997) and as negative or positive (Hofstad et al., 2000). The organisms produce acid by fermentation of glucose (Hofstad et al., 2000; Koneman et al., 1997), xylose and maltose; most strains produce acid from sucrose and lactose,

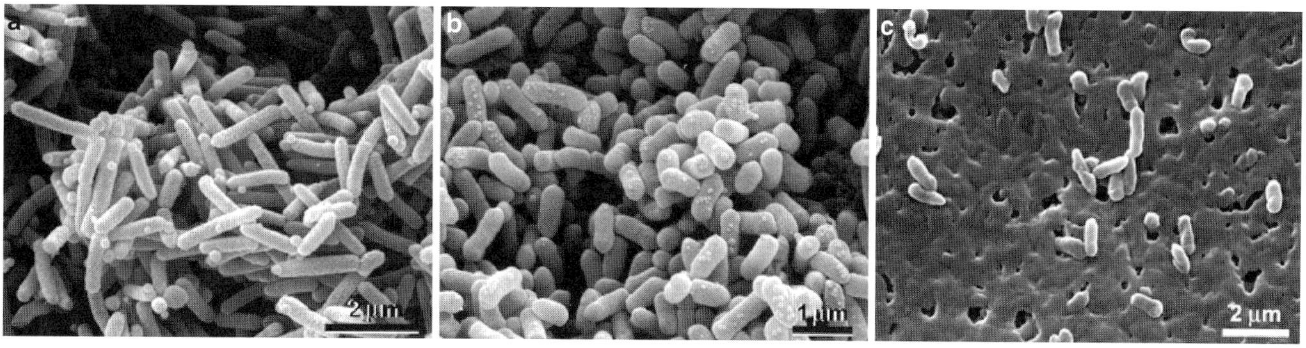

FIGURE 18. Scanning electron microscopy of cells from colonies of *Dysgonomonas gadei* CCUG 42886[T] (a), *Dysgonomonas mossii* CCUG 43457[T] (b), and *Dysgonomonas capnocytophagoides* CCUG 17996[T] (c). Cells were cultured anaerobically for 48 h at 37°C on human blood agar supplemented with hemin and vitamin K.

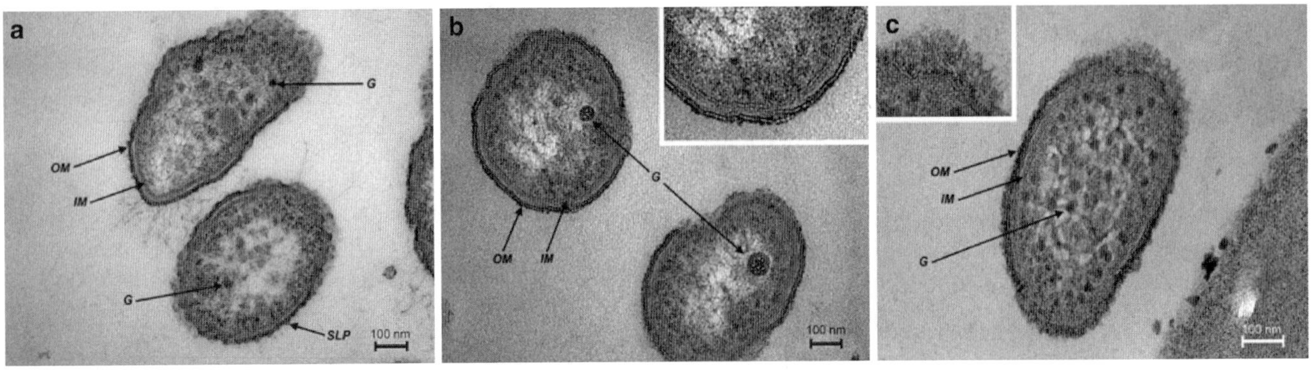

FIGURE 19. Transmission electron microscopy of *Dysgonomonas gadei* CCUG 42886ᵀ (**a**), *Dysgonomonas mossii* CCUG 43457ᵀ (**b**), and *Dysgonomonas capnocytophagoides* CCUG 17996ᵀ (**c**). Cells were cultured anaerobically for 48 h at 37°C on human blood agar supplemented with hemin and vitamin K. *OM*, outer membrane; *IM*, inner cytoplasmic membrane; *G*, granule; *SLPS*, scale-like protrusion.

but not from mannitol (Koneman et al., 1997). Esculin may be hydrolyzed (Hofstad et al., 2000; Koneman et al., 1997). Furthermore, alkaline phosphatase is produced but not arginine dihydrolase. H_2S and acetoin are not produced. Indole may be generated (Hofstad et al., 2000).

Isolates of *Dysgonomonas* species are rare (Koneman et al., 1997; Martínez-Sánchez et al., 1998). They have been recovered from clinical sources such as blood, wounds, urine, peritoneal fluid, umbilicus, stools, and gallbladder. Asymptomatic carriers have been found. The first report on these organisms was made by Wagner et al. (1988), who made multiple isolations in pure culture from the stools of an elderly woman with common variable hypogammaglobulinemia of long standing. DF-3 was also found in stool specimens by Gill et al. (1991) and Grob et al. (1999); several of the patients suffered from prolonged diarrhea. In another study with immunocompromised patients or patients with severe underlying disease, including HIV or inflammatory bowel disease, DF-3 was isolated from stool specimens during a 1-year period (Blum et al., 1992); the clinical spectrum associated with DF-3 ranged from asymptomatic carrier state to symptomatic with chronic diarrhea. Heiner et al. (1992) described an association of enteric DF-3 infection with HIV coinfection and common variable hypogammaglobulinemia. Further, Aronson and Zbick (1988) isolated DF-3 from a 24-year-old man with relapse of acute lymphocytic leukemia, and Grob et al. (1999) detected DF-3 in bacteremia from a patient with acute myelocytic leukemia during aplasia. The former patient became granulocytopenic during intensive chemotherapy and DF-3 was isolated from blood cultures. Recently, a case of *Dysgonomonas capnocytophagoides* in blood culture from a severely neutropenic patient treated for acute myeloid leukemia was reported (Hansen et al., 2005). A soft-tissue abscess in a diabetic patient and a postoperative urinary tract infection in an 81-year-old man yielded DF-3, and in the latter case, DF-3 was recovered together with *Escherichia coli* (Bangsborg et al., 1990; Schonheyder et al., 1991). DF-3 has also been isolated together with *Candida albicans*, *Candida glabrata*, *Staphylococcus aureus*, and enterococci (Lawson et al., 2002a). Melhus (1997) recovered DF-3 from a decubitous ulcer of a subfebrile patient with diarrhea.

Disk diffusion and broth dilution have been used to assess antimicrobial susceptibility of DF-3 isolates (Aronson and Zbick, 1988; Blum et al., 1992; Gill et al., 1991; Heiner et al., 1992; Wagner et al., 1988). The DF-3 strains were resistant to penicillin, ampicillin, ampicillin-sulbactam, aztreonam, aminoglycosides, cephalosporins (including cephalotin, cefoxitin, ceftriaxone, cefoperazone, and ceftazidime), erythromycin, ciprofloxacin, and vancomycin (Koneman et al., 1997). They are usually susceptible to trimethoprim-sulfamethoxazole and chloramphenicol, and variably susceptible to piperacillin, clindamycin, tetracycline, and imipenem. *Dysgonomonas gadei* was sensitive to metronidazole, clindamycin, doxycycline, imipenem, meropenem, and trimethoprim/sulfamethoxazole (Hofstad et al., 2000). The organism was resistant to cefoxitin and other cephalosporins (cefotaxime, cefpirome, ceftazidime, ceftriaxone, cefuroxime, and cephalotin), aminoglycosides (gentamicin, netilmicin, sulfadiazine), fluoroquinolones (ciprofloxacin, oxafloxacin), vancomycin, and teicoplanin. Similarly, *Dysgonomonas capnocytophagoides* was susceptible to ampicillin, tetracycline, chloramphenicol, clindamycin, and trimethoprim-sulfamethoxazole, while it was resistant to penicillin, cephalosporins, meropenem, aminoglycosides, and ciprofloxacin (Hansen et al., 2005).

Enrichment and isolation procedures

Culture from stool specimens is best performed on cefoperazone-vancomycin-amphotericin blood agar incubated at 35°C in 5–7% CO_2 (Koneman et al., 1997). Growth is relatively slow with pinpoint colonies visible after 24 h. After 48–72 h the colonies turn gray-white, smooth, and are nonhemolytic. A sweet odor may be produced by the organism on agar media (Bernard et al., 1991; Blum et al., 1992; Gill et al., 1991; Wagner et al., 1988).

Taxonomic comments

In 2000, Hofstad et al. isolated an organism from a human gallbladder that resembled CDC Group DF-3 organisms and named it *Dysgonomonas gadei*, thereby creating a new genus *Dysgonomonas*. The authors simultaneously reclassified the organisms previously designated CDC group DF-3 as *Dysgonomonas capnocytophagoides*. Lawson et al. (2002a, b) described another *Dysgonomonas* species, *Dysgonomonas mossii*, from human clinical specimens.

The CDC Group DF-3 has been considered closely related to *Capnocytophaga* species. After whole-cell protein electrophoresis, a separate position was occupied by a DF-3 strain when compared to well-characterized reference strains representing seven *Capnocytophaga* species (Vandamme et al., 1996). The whole-cell protein pattern of *Dysgonomonas gadei* was separate from those of two DF-3 strains and reference strains of *Capnocytophaga*, *Bacteroides*, and *Prevotella* (Hofstad et al., 2000). The nearest correlation was seen with *Bacteroides uniformis* at approximately 58%.

The cell walls of DF-3 strains had large amounts (24%) of anteiso-branched-chain fatty acid ($C_{15:0}$ anteiso), moderate amounts of saturated iso-branched-chain acids ($C_{14:0}$ iso and $C_{15:0}$ iso), and small to moderate amounts of both branched- and straight-chain hydroxy acids ($C_{15:0}$ 3-OH, $C_{16:0}$ iso 3-OH, $C_{16:0}$ 3-OH, and $C_{17:0}$ iso 3-OH) (Wallace et al., 1989). The dominating cellular fatty acid content of *Dysgonomonas gadei* CCUG 42882^T is $C_{15:0}$ anteiso, $C_{16:0}$, $C_{14:0}$ iso, and $C_{16:0}$ iso 3-OH, which is similar to that of *Dysgonomonas capnocytophagoides* CCUG 17996^T and CCUG 42515 (Hofstad et al., 2000). For comparison, Moore et al. (1994) found that $C_{15:0}$ anteiso, $C_{15:0}$ iso, $C_{17:0}$ iso 3-OH, and $C_{16:0}$ are the major cellular fatty acids in *Bacteroides* and *Prevotella*, and $C_{15:0}$ iso the major cellular fatty acid of *Porphyromonas*. Bernard et al. (1991) detected $C_{15:0}$ iso and $C_{17:0}$ iso 3-OH as the major cellular fatty acids in *Capnocytophaga*. The overall fatty acid composition of DF-3 organisms and *Capnocytophaga* species is clearly different from that of CDC group DF-3-like organisms (Daneshvar et al., 1991). The isoprenoid quinone content of four group DF-3-like strains was similar with ubiquinone-9 (Q-9) and Q-10 as the major quinone, whereas two other group DF-3-like strains had Q-7 as their major quinones, with smaller amounts of Q-8 and Q-9. Notably, CDC group DF-3 strains F9489 and G4990 did not contain quinones.

Bernard et al. (1991) found that CDC group DF-3 organisms produced significant quantities of propionic acid as metabolic product. Also lactic and succinic acids were detected, but the quantities varied among strains.

Comparative 16S rRNA sequence analysis has indicated that CDC group DF-3 is phylogenetically related to but different from *Bacteroides*, *Porphyromonas*, *Prevotella*, and kindred organisms (Paster et al., 1994; Vandamme et al., 1996). Actually, *Dysgonomonas gadei* and *Dysgonomonas capnocytophagoides* formed a distinct phylogenic cluster within the *Bacteroides–Prevotella–Porphyromonas* group (Hofstad et al., 2000). Sequence analysis showed that *Dysgonomonas mossii* differed from *Dysgonomonas gadei* (91.7%) and *Dysgonomonas capnocytophagoides* (94.3%) (Lawson et al., 2002a). Other taxa were related to these species (Figure 20) but showed much lower levels of sequence similarity, including

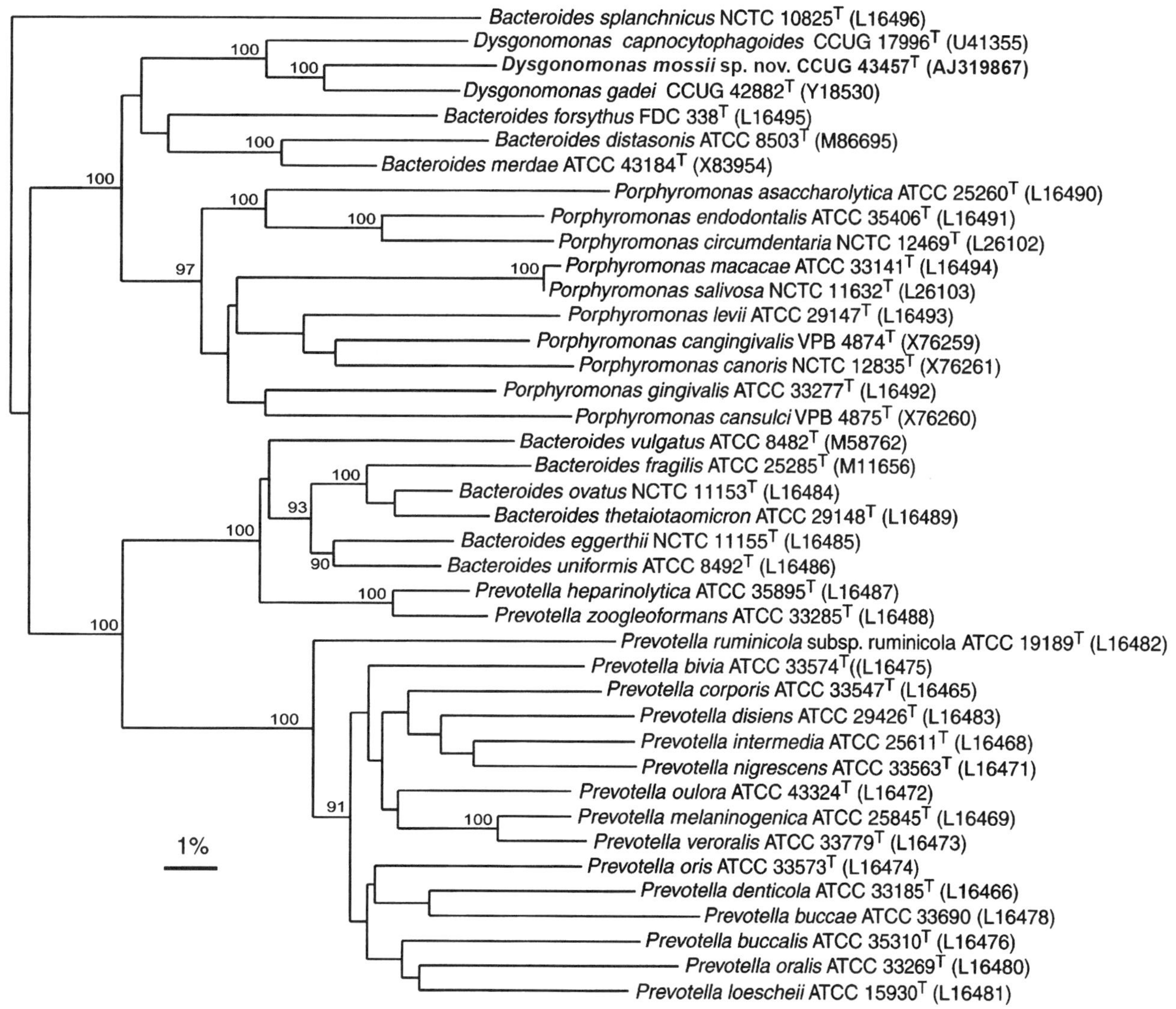

FIGURE 20. Phylogenetic tree of *Dysgonomonas* species and related organisms. (Reproduced with permission from Lawson et al., 2002a. Syst. Appl. Microbiol. *25:* 194–197.)

Bacteroides (85–87%), *Porphyromonas* (84–88%), and *Prevotella* (79–85%) and the misclassified strict anaerobic species *Bacteroides distasonis* (87%), *Bacteroides forsythus* (88%), *Bacteroides merdae* (89%), and *Bacteroides splanchnicus* (83%) (Hofstad et al., 2000; Lawson et al., 2002a). Comparative sequence 16S rRNA analysis unequivocally demonstrated three strains in a hitherto unrecognized subline within the *Dysgonomonas* clade (Lawson et al., 2002a) Divergence values of >5% with *Dysgonomonas capnocytophagoides* and *Dysgonomonas gadei* indicated that they may merit classification as a distinct species. Dewhirst et al. (1999) found that strain ASF 519 of *Bacteroides distasonis* fell into an unnamed genus containing *Bacteroides distasonis, Bacteroides merdae, Bacteroides forsythus,* and CDC group DF-3. *Capnocytophaga* species were clearly distinct from *Dysgonomonas* (Hofstad et al., 2000).

Differentiation of the species of the genus *Dysgonomonas*

Similarities and differences between the three *Dysgonomonas* spp. are listed in Tables 17–19. Contrary to the other species *Dysgonomonas gadei* produces catalase. *Dysgonomonas gadei* can further be distinguished from *Dysgonomonas capnocytophagoides* and *Dysgonomonas mossii* by the absence of β-glucuronidase production by the latter two. Unlike the other two species, *Dysgonomonas capnocytophagoides* ferments trehalose but does not produce *N*-acetyl-β-glucosaminidase and α-fucosidase. *Dysgonomonas mossii* does not generate glutamyl glutamic acid arylamidase.

TABLE 17. Salient characteristics of the species of the genus *Dysgonomonas*[a]

Characteristic	*D. gadei*	*D. capnocytophagoides*	*D. mossii*
Colony pigmentation	Gray–white	Gray–white	Gray–white
Aromatic odor	Slight	Slight	Slight
Cells	Coccobacilli	Coccobacilli and short rods	Coccobacilli and short rods
Growth on blood agar:			
CO_2 required	+	+	+
at 25°C	+	+	+
at 35–37°C	+	+	+
at 42–43°C	–	nd	–
Microaerophilic	+	+	nd
Anaerobic	+	+	+
Growth on MacConkey agar	–	–	–
Catalase	+	–	–
Indole	+	d	+
Acetoin	–	–	–
Esculin hydrolysis	+	d	+
Gelatin hydrolysis	–	–	–
H_2S production	–	–	nd
Nitrate reduction	–	–	–
Oxidase	–	–	–
Resistant to ox bile	+	+	+
Starch hydrolysis	nd	nd	+
Urea hydrolysis	–	–	–

[a]Symbols: +, >85% positive; d, different strains give different reactions (16–84% positive); –, 0–15% positive; nd, not determined.

TABLE 18. Fermentation profiles of species of the genus *Dysgonomonas*[a]

Acid produced from	*D. gadei*	*D. capnocytophagoides*	*D. mossii*
Adonitol, dulcitol	–	–	nd
L-Arabinose, lactose, D-mannose, sucrose, D-xylose	+	+	+
D-Arabitol, L-arabitol	nd	–	nd
Cellobiose, fructose, salicin	+	nd	+
Erythritol, glycogen	–	nd	nd
Inositol, D-mannitol	–	–	w
Maltose	nd	+	+
Melibiose, melezitose	+	+	nd
Raffinose	+	+	d
L-Rhamnose	+	nd	w
D-Ribose	w	nd	nd
D-Sorbitol	–	–	nd
Starch	+	nd	nd
Trehalose	+	–	+

[a]Symbols: +, >85% positive; –, 0–15% positive; w, weak reaction; nd, not determined.

List of species of the genus *Dysgonomonas*

1. **Dysgonomonas gadei** Hofstad, Olsen, Eribe, Falsen, Collins and Lawson 2000, 2194[VP]

ga'de.i. N.L. gen. masc. n. *gadei* of the Gade Institute, Bergen, Norway, where the organism was first isolated.

The following description is according to Hofstad et al. (2000). Nonmotile, Gram-negative coccobacilli that grow relatively slowly on blood agar. After 48 h of aerobic incubation at 35°C in a CO_2-enriched atmosphere, the colonies are 1–2 mm in diameter, nonadherent, entire, gray–white, smooth, and nonhemolytic, with a slight aromatic odor. After incubation for a few more d the colonies become coalesced, butyrous, and α-hemolytic. Growth can be seen at 25°C but not at 43°C. Growth occurs under microaerophilic and strictly anaerobic conditions. No growth occurs on MacConkey agar but does occur on nutrient agar around X and XV discs, suggesting a growth dependence for heme. Catalase-positive and oxidase-negative. No nitrate reduction

TABLE 19. Enzymic profiles of species of the genus *Dysgonomonas*[a,b]

Enzyme	*D. gadei*	*D. capnocytophagoides*	*D. mossii*
α-Arabinosidase	+	+	d
Acid phosphatase, phosphoamidase	+	+	nd
α-Fucosidase	+	–	+
α-Galactosidase, α-glucosidase, alanine, arylamidase, alkaline phosphatase, β-glucosidase, leucyl glycine arylamidase	+	+	+
α-Mannosidase	w	–	w
Arginine arylamidase, arginine, dihydrolase, glutamic acid, decarboxylase, glycine arylamidase, histidine arylamidase, leucine arylamidase, proline arylamidase, phenylalanine arylamidase, pyroglutamic acid arylamidase, serine arylamidase, tyrosine arylamidase	–	–	–
β-Galactosidase	w	+	+
β-Galactosidase 6-phosphate	–	+	w
β-Glucuronidase	+[a]	–	–
Chymotrypsin	+	–	nd
Cystine arylamidase, lipase C-14, valine arylamidase	–	–	nd
Esterase C-4	–	d	nd
Ester lipase C-8	w	–	nd
Glutamyl glutamic acid arylamidase	+	+	–
Lysine decarboxylase, ornithine decarboxylase	nd	–	nd
N-Acetyl-β-glucosaminidase	+	–	+
Trypsin	+	–	nd
Urease	–	nd	nd

[a]Symbols: +, >85% positive; d, different strains give different reactions (16–84% positive); –, 0–15% positive; nd, not determined.
[b]Positive with API ID32A and rapid ID32E, negative with API ZYM.

or production of H_2S or acetoin. Esculin is hydrolyzed, but gelatin and urea are not. Indole-positive. Resistant to ox bile. Glucose is fermented with production of acid but no gas. Acid is also produced from L-arabinose, cellobiose, fructose, lactose, D-mannose, melezitose, melibiose, raffinose, L-rhamnose, D-ribose (weak reaction), salicin, starch, sucrose, trehalose, and xylose. Positive reactions occur for *N*-acetyl-β-glucosaminidase, acid phosphatase, alanine arylamidase, alkaline phosphatase, α-arabinosidase, ester lipase C8 (weak), α-galactosidase, β-galactosidase (weak reaction), α-glucosidase, β-glucosidase, glutamyl glutamic acid arylamidase, α-mannosidase (weak), α-fucosidase, chymotrypsin, alanine arylamidase, leucyl glycine arylamidase, phosphoamidase, and trypsin. Positive and negative reactions are summarized in Tables 17–19.

Source: an infected human gallbladder, but the habitat is unknown.

DNA G+C content (mol%): not determined.

Type strain: ATCC BAA-286, CCUG 42882, CIP 106420.

Sequence accession no. (16S rRNA gene): Y18530.

2. **Dysgonomonas capnocytophagoides** Hofstad, Olsen, Eribe, Falsen, Collins and Lawson 2000, 2194[VP] (CDC Group DF-3, Wallace, Hollis, Weaver and Moss 1989, 735)

cap.no.cy.to.pha.goi′des. N.L. n. *Capnocytophaga* a genus of CO_2-requiring bacteria; L. suff. *-oides* (from Gr. suff. *-eides*, from Gr. n. *eidos* that which is seen, form, shape, figure) resembling, similar; N.L. adj. *capnocytophagoides* like *Capnocytophaga*, referring to some properties shared between these organisms.

The following description is based on results obtained by Wallace et al. (1989) and Hofstad et al. (2000). Nonmotile, Gram-negative coccobacilli to short rods. Colonies are 1–2 mm in diameter, nonadherent, entire, gray–white, smooth, and non-hemolytic with a slight aromatic odor after 48 h of aerobic incubation on blood agar at 35°C in a CO_2-enriched atmosphere (7.5%). No growth on MacConkey agar. Facultatively anaerobic. Catalase- and oxidase-negative. Does not produce H_2S or acetoin. Esculin may be hydrolyzed, but gelatin and urea are not. Indole may be produced. The organism is resistant to ox bile and does not reduce nitrate. Major products from glucose fermentation are propionic, lactic, and succinic acids. Glucose fermentation does not produce gas. Acid is produced from L-arabinose, lactose, maltose, D-mannose, melibiose, raffinose, sucrose, and D-xylose. Positive reactions occur for acid phosphatase, alanine arylamidase, alkaline phosphatase, α-arabinosidase, α-galactosidase, β-galactosidase, β-galactosidase 6-phosphate, α-glucosidase, β-glucosidase, glutamyl glutamic acid arylamidase, leucyl glycine arylamidase, and phosphoamidase. Esterase C-4 production is variable between strains. Positive and negative reactions are listed in Tables 17–19.

Source: human clinical specimens, but the habitat is unknown.

DNA G+C content (mol%): 38 (HPLC).

Type strain: CCUG 17996, CIP 107043, LMG 11519.

Sequence accession no. (16S rRNA gene): U41355.

3. **Dysgonomonas mossii** Lawson, Falsen, Inganäs, Weyant and Collins 2002a, 1915[VP] (Effective publication: Lawson, Falsen, Inganäs, Weyant and Collins 2002a, 194.)

moss′i.i. N.L. gen. masc. n. *mossii* of Moss, to honor Claude Wayne Moss, an American microbiologist who has contributed much to microbial taxonomy.

The following description is according to Lawson et al. (2002a). Nonmotile, Gram-negative coccobacilli to short rods. After anaerobic incubation for 48 h at 37°C on blood agar, colonies are 1–2 mm in diameter, nonadherent, entire, gray–white, smooth, and nonhemolytic, and produce a slightly aromatic odor. Growth occurs at 25°C but not at 42°C. No growth occurs on MacConkey agar. The organisms grow on nutrient agar around X and XV disks, indicating that they have a requirement for heme. Catalase- and oxidase-negative. Nitrate is not reduced to nitrite. Indole is produced, but not acetoin. The organisms are resistant to ox bile. Esculin and starch are hydrolyzed, but gelatin and urea are not. Glucose is fermented with production of acid but no gas. Acid is produced (conventional methods) from L-arabinose, cellobiose, fructose, inositol (weak reaction), lactose, mannitol (weak reaction), maltose, L-rhamnose (weak reaction), salicin, sucrose, trehalose, and D-xylose. With API rapid ID 32A, acid is produced from mannose, and D-raffinose may be fermented. Positive reactions are obtained for *N*-acetyl-β-glucosaminidase, alanine arylamidase, alkaline phosphatase, α-galactosidase, β-galactosidase, β-galactosidase 6-phosphate (weak), α-glucosidase, β-glucosidase, leucyl glycine arylamidase, α-mannosidase (weak reaction), and α-fucosidase. Arabinosidase may be produced. Positive and negative reactions are summarized in Tables 17–19.

Source: clinical sources, but the habitat is not known.

DNA G+C content (mol%): 38.5 (HPLC).

Type strain: CDC F9489, CCUG 43457, CIP 107079.

Sequence accession no. (16S rRNA gene): AJ319867.

Genus IV. **Paludibacter** Ueki, Akasaka, Suzuki and Ueki 2006, 43[VP]

The Editorial Board

Pa.lu.di.bac'ter. L. n. *palus -udis* a swamp, marsh; N.L. masc. n. *bacter* a rod; N.L. masc. n. *Paludibacter* rod living in swamps.

Rods (0.5–0.6 µm × 1.3–1.7 µm), with the ends usually round to slightly tapered. Nonmotile. Nonsporeforming. Gram-negative. Strictly anaerobic. Chemo-organotrophic. Optimum growth temperature, 30°C. No growth occurs at 37°C. Oxidase and catalase-negative. Nitrate is not reduced. Various sugars are fermented, and acetate and propionate are the major fermentation end products with succinate as a minor product. Major cellular fatty acids are $C_{15:0}$ anteiso, $C_{15:0}$, and $C_{17:0}$ anteiso 3-OH. The major respiratory quinone is MK-8(H_4).

DNA G+C content (mol%): 39.3 (HPLC).

Type species: Paludibacter **propionicigenes** Ueki, Akasaka, Suzuki and Ueki 2006, 43[VP].

Further descriptive information

Paludibacter propionicigenes was isolated from a rice plant residue (rice straw) sample collected from irrigated rice-field soil in the Shonai Branch of the Yamagata Agricultural Experimental Station (Fujishima-machi, Yamagata, Japan) during the flooding period of the field.

Cells can be cultivated at 30°C under an atmosphere of 95% N_2/5% CO_2 in peptone-yeast (PY) broth supplemented with 1% glucose (PYG broth). Colonies on PY4S* agar are grayish white, translucent, circular with a smooth surface, and 1.0–1.5 mm in diameter after 48 h. Propionate and acetate are produced from fermentation of glucose at a ratio of 2:1. The predominant cellular fatty acids are $C_{15:0}$ anteiso (30.8%), $C_{15:0}$ (19.0%), $C_{17:0}$ anteiso 3-OH (17.9%), and $C_{17:0}$ iso 3-OH (6.2%).

Enrichment, isolation, and maintenance procedures

The type strain was isolated using the anaerobic roll tube method. PY4S agar can be used for maintenance of the strain on agar slants.

Differentiation of the genus *Paludibacter* from related genera

Paludibacter propionicigenes is a strict anaerobe isolated from plant residue, whereas *Dysgonomonas capnocytophagoides* and *Dysgonomonas mossii* are facultative anaerobes isolated from human clinical specimens. *Paludibacter propionicigenes* cannot grow at 37°C and is inhibited by bile, whereas *Parabacteroides merdae*, *Dysgonomonas capnocytophagoides*, and *Dysgonomonas mossii* can grow at 37°C and are bile tolerant.

Taxonomic comments

By analysis of 16S rRNA gene sequences, the closest neighboring species of *Paludibacter propionicigenes* are *Dysgonomonas capnocytophagoides* (90.9% similarity), *Dysgonomonas mossii* (89.8%), and *Parabacteroides merdae* (88.7%).

List of species of the genus *Paludibacter*

1. **Paludibacter propionicigenes** Ueki, Akasaka, Suzuki and Ueki 2006, 43[VP]

pro.pi.on.i.ci'ge.nes. N.L. n. *acidum propionicum* propionic acid; N.L. suff. *-genes* (from Gr. v. *gennaô* to produce) producing; N.L. part. adj. *propionicigenes* producing propionic acid.

The characteristics are as described for the genus, with the following additional features. Elongated cells are occasionally formed both singly and in chains of the short cells. Spherical cells sometimes develop after storage of slant cultures at 4°C. Growth occurs at pH 5.0–7.6 (optimum, 6.6) and at 15–35°C. Growth at 33°C is much slower than at 30°C. The NaCl concentration range for growth is 0–0.5% in PYG medium. The following compounds are used for growth and acid production: arabinose, cellobiose, fructose, galactose, glucose, mannose, maltose, melibiose, glycogen, soluble starch, and xylose. The following compounds are not used: cellulose, dulcitol, ethanol, fumarate, glycerol, inositol, lactate, lactose, malate, mannitol, melezitose, pyruvate, rhamnose, raffinose, ribose, salicin, sorbitol, sorbose, succinate, sucrose, trehalose, and xylan. Esculin is hydrolyzed, but gelatin and urea are not. H_2S and indole are not produced. No growth occurs in the presence of 0.01% bile salts.

Source: rice plant residue in anoxic rice-field soil in Japan.

DNA G+C content (mol%): 39.3 (HPLC).

Type strain: WB4, DSM 17365, JCM 13257.

Sequence accession no. (16S rRNA gene): AB078842.

---

*PY4S agar is peptone-yeast extract (PY) broth supplemented with (g/l): glucose, 0.25; cellobiose, 0.25; maltose, 0.25; soluble starch, 0.25; and agar, 15.0.

Genus V. **Petrimonas** Grabowski, Tindall, Bardin, Blanchet and Jeanthon 2005, 1118[VP]

The Editorial Board

Pe.tri.mo′nas. L. fem. n. *petra* rock, stone; L. fem. n. *monas* a unit, monad; N.L. fem. n. *Petrimonas* stone monad.

Straight rods (0.7–1 × 1.5–2.0 μm) during exponential growth; some longer cells (0.5 × 4.0 μm) may be observed in old cultures. Nonsporeforming. Gram-stain-negative. Chemo-organotrophic. Mesophilic. **Strictly anaerobic, having a fermentative type of metabolism. End products of glucose fermentation are acetate, H_2, and CO_2.** Carbohydrates and some organic acids are fermented. Tryptone is required for growth. **The predominant menaquinone is MK-8**, with smaller amounts of MK-7 and MK-9. The major polar lipids are phosphatidylethanolamine, an unidentified phospholipid, two unidentified aminophospholipids, three unidentified phosphoglycolipids, a glycolipid, an aminolipid, and two additional uncharacterized lipids. The fatty acids include straight chain and branched fatty acids; in addition, 2-OH and 3-OH fatty acids are present. **The major cellular fatty acids are $C_{15:0}$ anteiso**, $C_{13:0}$ anteiso, $C_{15:0}$ iso, and $C_{15:0}$. The major hydroxy fatty acids are C3-$OH_{-16:0}$ iso, C3-$OH_{17:0}$ iso, and $C_{2\text{-OH }17:0}$; about one-third of each of all three appear to be amide-linked. Isolated from a biodegraded oil reservoir.

DNA G+C content (mol%): 40.8 (HPLC).

Type species: **Petrimonas sulfuriphila** Grabowski, Tindall, Bardin, Blanchet and Jeanthon 2005, 1119[VP].

Further descriptive information

To date, *Petrimonas sulfuriphila* is the only member of the phylum *Bacteroidetes* to be isolated from a producing well of a biodegraded oil reservoir.

Enrichment and isolation procedures

Enrichment was performed in nitrate broth medium (Difco). Isolation of single colonies was accomplished using the anaerobic roll-tube technique (Hungate, 1969) with nitrate broth solidified with 2% (w/v) purified agar (Difco).

Differentiation of the genus *Petrimonas* from other genera

The predominance of MK-8 as menaquinone is a feature that distinguishes the novel isolate from all its phylogenetic relatives, including *Tannerella* (MK-10, MK-11), *Bacteroides* (MK-10, MK-11), *Parabacteroides distasonis* (MK-10); *Parabacteroides merdae* (MK-9, MK10), and *Porphyromonas* (MK-9, MK-10).

Taxonomic comments

According to 16S rRNA gene sequence analysis by Grabowski et al. (2005), the closest cultivated relatives of the type strain of *Petrimonas sulfuriphila* are *Tannerella forsythia* (88% similarity) and *Parabacteroides merdae* (87% similarity). However, very high rRNA gene sequence similarities (99.6%) were found between *Petrimonas sulfuriphila* and certain environmental clones retrieved from a dechlorinating consortium (GenBank accession no. AJ488088) and the bovine rumen (AB003390).

List of species of the genus *Petrimonas*

1. **Petrimonas sulfuriphila** Grabowski, Tindall, Bardin, Blanchet and Jeanthon 2005, 1119[VP]

sul.fu.ri.phi′la. L. n. *sulfur* sulfur; Gr. adj. *philos -ê -on* loving; N.L. fem. adj. *sulfuriphila* sulfur-loving, indicating that sulfur stimulates growth.

The characteristics are as described for the genus, with the following additional features. Cells occur singly or in pairs. The temperature range for growth at pH 7.2 is 15–40°C; optimum, 37–40°C. Growth occurs in the presence of 0–4% NaCl; optimum, 0%. Yeast extract and elemental sulfur stimulate growth. The following substrates support growth when yeast extract, tryptone, and elemental sulfur are present: glucose, arabinose, galactose, maltose, mannose, rhamnose, lactose, ribose, fructose, sucrose, lactate, mannitol, glycerol, and cellobiose. Fumarate, pyruvate, and Casamino acids support weak growth. Acetate, formate, butyrate, propionate, methanol, peptone, ethanol, propanol, butanol, toluene, sorbose, and cellulose are not used. Elemental sulfur is reduced to sulfide. Nitrate is reduced to NH_4^+.

Source: isolated from an oilfield well head in the Western Canadian Sedimentary Basin (Canada).

DNA G+C content (mol%): 40.8 (HPLC).

Type strain: BN3, DSM 16547, JCM 12565.

Sequence accession no. (16S rRNA gene): AY570690.

Genus VI. **Proteiniphilum** Chen and Dong 2005, 2259[VP]

The Editorial Board

Pro.tei′ni.phi.lum. N.L. neut. n. *proteinum* protein; Gr. adj. *philos -ê -on* loving; N.L. neut. n. *Proteiniphilum* protein-loving.

Rods 0.6–0.9 × 1.9–2.2 μm. **Motile by means of lateral flagella**. Nonsporeforming. Gram-stain-negative. **Obligately anaerobic.** Growth occurs at 20–45°C. Oxidase and catalase-negative. Chemo-organotrophic. Proteolytic. Yeast extract and peptone can be used as energy sources. **Nonsaccharolytic. Carbohydrates, alcohols, and organic acids (except pyruvate) are not used.** Gelatin is not hydrolyzed. **Not resistant to 20% bile. The major fermentation products from peptone-yeast extract (PY) medium are acetic acid and propionic acid.** Nitrate is not reduced. Cellular fatty acids mainly consist of iso-branched fatty acids, predominantly $C_{15:0}$ anteiso.

DNA G+C content (mol %): 46.6–48.9 (T_m).

Type species: **Proteiniphilum acetatigenes** Chen and Dong, 2005, 2261[VP].

Further descriptive information

During studies of syntrophic propionate-degrading bacteria from methanogenic environments, Chen and Dong (2005) reported the isolation of mixed cultures consisting of three different organisms that, in combination, were capable of degrading propionate to acetate and methane. The three organisms were: a syntrophic, propionate-degrading bacterium; an organism resembling *Methanobacterium formicicum*; and a rod-shaped bacterium which did not consume propionate or synthesize methane but did accelerate the degradation of propionate by the triculture. Chen and Dong classified the latter organism as the representative of a new genus, *Proteiniphilum*, and a new species, *Proteiniphilum acetatigenes*.

Colonies of *Proteiniphilum acetatigenes* on (PY) agar are circular, slightly convex, white, translucent, and reach 1.5 mm in diameter after 3 d at 37°C. Optimum growth occurs at 37°C. The pH range for growth is 6.0–9.7; optimum is 7.5–8.0.

In addition to yeast extract and peptone, pyruvate, glycine and L-arginine can be used as carbon and energy sources. Tryptone, L-serine, L-threonine, and L-alanine support weak growth. Pyruvate is converted to acetic acid and CO_2.

Enrichment and isolation procedures

Proteiniphilum was isolated from methanogenic propionate-degrading mixtures obtained from the granule sludge of an upflow anaerobic sludge blanket reactor for treatment of brewery wastewater. Isolation was accomplished by initial serial dilution in anaerobic PY medium followed by selection of colonies on anaerobic roll tubes. Repeated subculturing of colonies in PY broth followed by colony isolation is performed until purity is assured.

Maintenance procedures

Routine cultivation can be accomplished in pre-reduced PY broth in anaerobic tubes under an atmosphere of 100% oxygen-free N_2 with incubation at 37°C.

Differentiation of the genus *Proteiniphilum* from related genera

The rod shape of *Proteiniphilum* differs from the coccobacillary shape of *Dysgonomonas* species; moreover, unlike *Proteiniphilum*, *Dysgonomonas* is nonmotile, facultatively anaerobic, ferments glucose, and has a higher DNA G+C content (47–49 vs 38 mol%). The motility of *Proteiniphilum* and failure to form succinic acid as a major end product from glucose differentiate *Proteiniphilum* from *Parabacteroides distasonis*, *Parabacteroides merdae*, and *Tannerella forsythia*. Unlike *Tannerella forsythia*, *Proteiniphilum* does not produce phenylacetic acid and butyric acid from glucose.

Taxonomic comments

Analysis of 16S rRNA gene sequences indicates that the nearest neighbors of the type strain are *Dysgonomonas* species (89.6–90.6% sequence similarity). The type strain is more distantly related to *Parabacteroides* (85–87% sequence similarity), *Porphyromonas* (84–88%), *Prevotella* (79–85%), and *Tannerella forsythia* (89.3%).

List of species of the genus *Proteiniphilum*

1. **Proteiniphilum acetatigenes** Chen and Dong 2005, 2261^{VP}

a.ce′ta.ti.gen.es. N.L. *acetas -atis* acetate; N.L. suff. *-genes* (from Gr. v. *gennaô* to produce) producing; N.L. part. adj. *acetatigenes* acetate-producing.

The characteristics are as described for the genus, with the following additional features. Milk is not curdled. Starch and esculin are hydrolyzed. Indole is not produced. Urease, lecithinase and lipase are not produced. Methyl red and Voges–Proskauer tests are negative. H_2S is not produced from peptone or thiosulfate. NH_3 is produced from yeast extract, peptone and L-arginine. Acetic acid is the main product from fermentation of yeast extract, peptone, pyruvate and L-arginine; propionic acid is also produced. The following substrates are not used: adonitol, amygdalin, L-arabinose, L-aspartate, butanedioic acid, cellobiose, cellulose, citrate, L-cysteine, dulcitol, erythritol, esculin, ethanol, D-fructose, fumarate, D-galactose, gluconate, D-glucose, L-glutamine, glycogen, hippurate, L-histidine, β-hydroxybutyric acid, inositol, inulin, L-isoleucine, D-lactose, L-leucine, L-lysine, malate, malonate, D-maltose, mannitol, mannose, melibiose, methanol, L-methionine, phenylacetic acid, L-phenylalanine, L-proline, 1-propanol, raffinose, rhamnose, ribitol, ribose, salicin, D-sorbitol, sorbose, succinate, sucrose, starch, trehalose, tryptophan, L-tyrosine, L-valine, xylan, and xylose. The major cellular fatty acids are $C_{15:0}$ anteiso (46.21%), $C_{15:0}$ (8.90%), $C_{17:0}$ iso 3-OH (5.93%), and $C_{17:0}$ anteiso (5.15%).

Source: the type strain was isolated from the granule sludge of an Upflow Anaerobic Sludge Blanket (UASB) reactor treating brewery wastewater.

DNA G+C content (mol%): 46.6–48.9 (T_m).

Type strain: TB107, AS 1.5024, JCM 12891.

Sequence accession no. (16S rRNA gene): AY742226.

Genus VII. **Tannerella** Sakamoto, Suzuki, Umeda, Ishikawa and Benno 2002, 848^{VP}

Mitsuo Sakamoto, Anne C. R. Tanner and Yoshimi Benno

Tan.ne.rel′la. L. dim. suffix *-ella*; N.L. fem. dim. n. *Tannerella* named after the American microbiologist Anne C. R. Tanner, for her contributions to research on periodontal disease.

Fusiform cells, generally 0.3–0.5 × 1–30 μm. Gram-negative. Nonmotile. **Obligately anaerobic. *N*-acetylmuramic acid (NAM) is required for growth** (some strains do not require NAM). **Growth is inhibited in the presence of 20% bile**. The major end products are acetic acid, butyric acid, isovaleric acid, propionic acid, and phenylacetic acid; smaller amounts of isobutyric acid and succinic acid may be produced. Esculin is hydrolyzed. Indole variable. **Trypsin activity is positive**. Glucose-6-phosphate dehydrogenase (G6PDH), 6-phosphogluconate dehydrogenase (6PGDH), malate dehydrogenase, and glutamate dehydrogenase are present. The principal respiratory quinones are menaquinones MK-10 and MK-11. Both nonhydroxylated and

3-hydroxylated long-chain fatty acids are present. The nonhydroxylated acids are predominantly of the saturated straight-chain and anteiso methyl-branched-chain types. **The ratio of $C_{15:0}$ anteiso to $C_{15:0}$ iso is high (>20). Isolated originally and principally from human oral cavity. *Tannerella forsythia* has been associated with periodontal disease, root canal infections, and peri-implantitis.**

DNA G+C content (mol%): 44–48.

Type species: **Tannerella forsythia** corrig. (Tanner, Listgarten, Ebersole and Strzempko 1986) Sakamoto, Suzuki, Umeda, Ishikawa and Benno 2002, 848[VP] (*Bacteroides forsythus* Tanner, Listgarten, Ebersole and Strzempko 1986, 216).

Sequence accession no. (16S rRNA gene): L16495.

Further descriptive information

The genus *Tannerella* is a member of the *Bacteroides* subgroup of the *Cytophaga–Flexibacter–Bacteroides* (CFB) phylum (Paster et al., 1994) that is referred to as the phylum *Bacteroidetes* in this edition of the *Manual.*

In the absence of NAM, cells of *Tannerella forsythia* are often pleomorphic and occur as rods with tapered (fusiform) or rounded ends, long filaments, and spheroids. The spheroids frequently occur in the center of cells and enlarge the diameter of the cell to approximately 5 μm. The spheroids are observed free from the rod forms (Tanner et al., 1979); this cell morphology has been attributed to the presence of a weakened cell wall (Wyss, 1989). In the presence of NAM, cells are regular with rounded or tapered ends (Braham and Moncla, 1992).

In cell cross-sections stained with uranyl acetate and lead citrate and observed by electron microscopy, the outer membrane consists of an outer layer that is approximately 4 nm thick, an inner layer that is 1 nm thick, and a 2 nm electron-lucent space in between. The inner membrane consists of two electron-dense layers, each approximately 2 nm wide, with a 2 nm intervening electron-lucent layer. The inner and outer membranes are separated by a moderately electron-dense zone that is about 16 nm wide and appears to lack a distinct peptidoglycan layer. A regular structured external layer surrounds the outer membrane and consists of subunits, which in some cross-sections resemble adjacent arches that are 10 nm wide and 10 nm high. These subunits are separated from the outer membrane by a 12 nm electron-lucent zone. Electron-dense radial connections appear to extend from the subunits to the outer surface of the outer membrane. Negative staining of tangential sections through the cell periphery indicates that the subunits in the outermost layer are packed in an orthogonal array (Tanner et al., 1986).

The predominant cellular fatty acids are $C_{15:0}$ anteiso and $C_{16:0}$ 3-OH. In contrast, *Bacteroides* and *Parabacteroides* species possess significant levels of $C_{15:0}$ anteiso and $C_{17:0}$ iso 3-OH. Moreover, the ratio of $C_{15:0}$ anteiso to $C_{15:0}$ iso in whole-cell methanolysates of *Tannerella forsythia* is very much higher than those of *Bacteroides* and *Parabacteroides* species (Sakamoto and Benno, 2002, 2006): the ratios range from 22.8 to 95.2% in *Tannerella forsythia* strains but only from 1.9 to 10.3% in *Bacteroides* and *Parabacteroides* species. This is an important feature that differentiates *Tannerella forsythia* from *Bacteroides* and *Parabacteroides* species.

Colonies are pale, speckled pink, circular, entire, and slightly convex on Trypticase soy agar supplemented with 5% (v/v) sheep blood (TSBA) and a NAM disk (Braham and Moncla, 1992). The colonies are generally small with a diameter of 0.5–2 mm after 7–10 d incubation at 37°C. *Tannerella forsythia* is a strict anaerobe and requires CO_2 and H_2 in the atmosphere for growth on agar surfaces. Growth on agar is stimulated by the presence of *Fusobacterium nucleatum* (and strains of *Streptococcus sanguinis, Bacteroides fragilis, Campylobacter rectus, Prevotella intermedia,* and *Veillonella parvula*) in a satellite pattern (Tanner et al., 1986).

The growth of *Tannerella forsythia* strains is inhibited on bacteroides bile esculin agar. In contrast, the growth of *Bacteroides* and *Parabacteroides* species is not inhibited on the media containing 20% bile. In addition, *Tannerella forsythia* requires exogenous NAM as a growth factor (Wyss, 1989). These characteristics are important features that differentiate *Tannerella forsythia* from the *Bacteroides fragilis* group and the genus *Parabacteroides.* Braham and Moncla (1992) reported that *Tannerella forsythia* strains isolated from subgingival plaque samples from monkeys (*Macaca fascicularis*) also require NAM for growth. However, *Tannerella forsythia* strains recovered from cat and dog bite wound infections in humans do not require NAM for growth (Hudspeth et al., 1999). These results suggest that there are host-specific biotypes within the species of *Tannerella forsythia.*

Tannerella forsythia does not react in the API 20A or API 20E test series but does react in other resting cells tests, including API ZYM, API AN-IDENT (Tanner et al., 1985, 1986), and fluorogenic substrate tests (Maiden et al., 1996). *Tannerella forsythia* strains give positive results for α-glucosidase, β-glucosidase, α-fucosidase, and β-glucuronidase. In the absence of NAM there is no detectable pH decrease in media supplemented with carbohydrates (Tanner et al., 1985, 1986). These results suggest that *Tannerella forsythia* may grow too poorly in that medium to ferment the carbohydrate, that cells may have trouble transporting the sugars across their membranes for use in the glycolytic pathway, or may overproduce methyl glyoxyl, thereby leading to toxicity and growth inhibition (Maiden et al., 2004).

The genome-sequencing project of *Tannerella forsythia* ATCC 43037[T] has been completed by The Institute for Genomic Research (TIGR). A total of 3034 open reading frames are predicted from the genomic sequence of 3,405,543 bp in length based on the annotation provided by the Oral Pathogen Sequences Databases (ORALGEN, http://www.oralgen.lanl.gov). The preliminary genomic annotations as well as a variety of bioinformatics tools useful for studying the *Tannerella forsythia* genome are available at both the ORALGEN and Bioinformatics Resource for Oral Pathogens (BROP, http://www.brop.org) (Chen et al., 2005) project web sites.

The minimum inhibitory concentrations (MICs) of antimicrobial agents have been reported by Takemoto et al. (1997). *Tannerella forsythia* strains are most sensitive to clindamycin and metronidazole. *Tannerella forsythia* shows a comparatively lower susceptibility to ciprofloxacin. Most strains are sensitive to penicillin G, ampicillin, amoxycillin, tetracycline, doxycycline, and erythromycin.

Tannerella forsythia has been associated with advanced forms of periodontal disease, including severe and refractory periodontitis. Unlike other periodontopathic bacteria, such as *Porphyromonas gingivalis* and *Prevotella intermedia, Tannerella forsythia* is difficult to culture, and its prevalence in periodontal disease may be underestimated. *Tannerella forsythia* is synergistic with *Porphyromonas gingivalis,* and the presence of both accelerates progression of the disease (Yoneda et al., 2001). Pathogenic factors include a trypsin-like protease (PrtH) (Saito et al., 1997) and a variety of glycosidases, such as α-D-glucosidase, *N*-acetyl-β-D-glucosaminidase (Hughes et al., 2003), and sialidase (Ishikura et al., 2003). Tan et al. (2001) determined the prevalence and the association of the *prtH* gene of *Tannerella forsythia* in adult periodontitis and periodontally healthy patients. Among the 86 diseased sites examined, 73 (85%) were colonized by *Tannerella forsythia* with the *prtH* genotype. In sites of the periodontally healthy, only 7 of 73 (10%) possessed *Tannerella forsythia* with the *prtH* genotype. These findings suggest a strong association of the *prtH* gene of *Tannerella forsythia* with adult periodontitis.

The outer surface of *Tannerella forsythia* is covered by an S-layer (surface layer) composed of two protein subunits (200 and 210 kDa) that mediate adherence and invasiveness (Sabet et al., 2003). Although mice immunized with purified S-layer and *Tannerella forsythia* whole cells did not develop any abscesses when challenged with viable *Tannerella forsythia* cells, unimmunized mice developed abscesses (Sabet et al., 2003). Other virulence factors include the cell-surface-associated protein (BspA) that mediates adherence to fibronectin (an extracellular matrix component), and fibrinogen (a clotting factor) (Sharma et al., 1998). Honma et al. (2001) constructed a BspA-defective mutant of *Tannerella forsythia* that showed a reduced ability to adhere to fibronectin and fibrinogen compared with the wild-type strain (ATCC 43037^T). Sharma et al. (2005) demonstrated alveolar bone loss in mice infected with the *Tannerella forsythia* wild-type strain, whereas this effect was impaired when the BspA mutant was used.

Huang et al. (2003) investigated the distribution of *Tannerella forsythia* genotypes in a Japanese periodontitis population, and also the relationship between different genotypes and the periodontal status, using the arbitrarily primed polymerase chain reaction (AP-PCR) method. *Tannerella forsythia* ATCC 43037^T and 137 clinical bacterial isolates were separated into 11 distinct AP-PCR genotypes. The majority of *Tannerella forsythia* strains examined belonged to AP-PCR genotypes I (including ATCC 43037^T), II, III, and IV (accounting for 39.7, 20.6, 10.3, and 10.3%, respectively). The strains of Types I and III were mainly isolated from chronic periodontitis subjects, whereas the strains of Types II and IV were mainly isolated from aggressive periodontitis subjects.

A *Tannerella* phylotype (oral clone BU063: GenBank accession no. for 16S rRNA gene is AY008308) was associated with periodontal health, in contrast to *Tannerella forsythia* strains that were associated with periodontal disease (Kumar et al., 2003; Leys et al., 2002).

Tannerella forsythia is found in the subgingival, gingival, and periodontal pockets, in dental root canals (Conrads et al., 1997), and around infected dental implants (Tanner et al., 1997) of humans. *Tannerella forsythia* has been also found in monkey (*Macaca fascicularis*) subgingival plaque samples (Braham and Moncla, 1992). In addition, *Tannerella forsythia* strains have been recovered from cat- and dog-bite wound infections in humans (Hudspeth et al., 1999). *Tannerella forsythia* has also been detected in atheromatous plaques (Haraszthy et al., 2000) and in buccal epithelial cells (Rudney et al., 2005) of humans. From clone libraries of 16S rRNA gene sequences from termite guts, phylotypes related to *Tannerella forsythia* have been found (Hongoh et al., 2005, 2003).

Enrichment and isolation procedures

Isolation and identification of *Tannerella* species has been described (Braham and Moncla, 1992; Tanner et al., 1998; Umeda et al., 1996). Rapid presumptive identification of clinical isolates of *Tannerella forsythia* is based on the following eight criteria (Braham and Moncla, 1992): (1) positive activity for α-glucosidase, (2) positive activity for β-glucosidase, (3) positive activity for sialidase, as it has been reported that sialidase activity is useful for identification of *Tannerella forsythia* (Moncla et al., 1990), (4) positive activity for trypsin-like enzyme, (5) negative indole production, (6) requirement for NAM, (7) colonial morphology, and (8) Gram stain morphology from blood agar medium deficient in NAM. The trypsin-like activity is measured by the benzoyl-DL-arginine-naphthylamide (BANA) test (Loesche et al., 1990). BANA-positive non-black-pigmented bacteria are subcultured with the plates with or without the NAM disk. Subsequently, bacterial colonies that grow only on the plates with NAM are screened as *Tannerella forsythia*. Finally, five filter paper fluorescent spot tests [for indole (Sutter and Carter, 1972), α-glucosidase, β-glucosidase, sialidase, and trypsin-like enzyme] are used for identification of *Tannerella forsythia* (Braham and Moncla, 1992). For the trypsin-like enzyme assay, *N*-α-carbobenzoxy-l-arginine-7-amino-4-methylcoumarin hydrochloride (CAMM), fluorogenic substrate, is used.

An agar-free broth medium for *Tannerella forsythia* has been developed by Wyss (1989). This medium consists of BHI broth supplemented with fetal calf serum (FCS) and NAM. Some strains grew without the addition of FCS.

Maintenance procedures

The organism can be maintained by transfer on the same media used for isolation. Recommended procedures for long-term preservation are lyophilization, freezing at −80°C, or storage in liquid nitrogen. Cryoprotective agents such as 10% glycerol or DMSO should be added to cultures before freezing, and heavy cell concentrations should be used.

Differentiation of the genus *Tannerella* from other genera

Differential characteristics of the genus *Tannerella* and some related taxa are shown in Table 20. Although *Tannerella forsythia* is phylogenetically related to *Parabacteroides* species, the ratio of $C_{15:0}$ anteiso to $C_{15:0}$ iso in whole-cell methanolysates of *Tannerella forsythia* is different from those of *Parabacteroides* species. While the ratios of $C_{15:0}$ anteiso to $C_{15:0}$ iso range from 22.8 to 95.2 in *Tannerella forsythia* strains (Sakamoto et al., 2002), those of *Parabacteroides* species range from 3.1 to 10.3. In addition, although the major menaquinones of *Tannerella forsythia* are MK-10 and MK-11, the major menaquinones of *Parabacteroides* species are MK-9 and MK-10 (Sakamoto and Benno, 2006).

Taxonomic comments

As shown in Figure 21, *Tannerella* exhibits a close phylogenetic association with the genus *Parabacteroides* (Sakamoto and Benno, 2006).

The name *Bacteroides forsythus* was published by Tanner et al. (1986) but did not appear on a Validation List in the IJSEM.

TABLE 20. Differential characteristics of the genus *Tannerella* and some related taxa[a]

Characteristic	*Tannerella*	*Parabacteroides*	*Bacteroides*	*Dysgonomonas*	*Paludibacter*	*Porphyromonas*	*Prevotella*	*Proteiniphilum*
Growth in bile	–	+	+	+	–	–	–	–
Aerobic growth	–	–	–	+	–	–	–	–
NAM required	+[b]	–	–	–	–	–	–	–
α-Fucosidase produced	+	–	D	D	nt	–[c]	D	nt
Catalase produced	D	D	D	D	–	D	D	–
Indole produced	D	–	D	D	–	D	D	–
Esculin hydrolyzed	+	+	D	D	+	–	D	+
Pigment produced	–	–	–	–	–	+[d]	D	–
Metabolism	NF	F	F	F	F	NF[e]	MF	NF
Major end products	A, B, IV, P, PA	A, S	A, S	P, L, S	A, P	A, B, IV, P, PA, S	A, S	A, P
Presence of:								
G6PDH	+	+	+	nt	nt	D	–	nt
6PGDH	+	+	+	nt	nt	D	–	nt
Proteolytic activity	+	–	–	D	nt	D	D	+
Major cellular fatty acids	$C_{15:0}$ anteiso	$C_{15:0}$ anteiso	$C_{15:0}$ anteiso	$C_{14:0}$ iso, $C_{15:0}$ anteiso and $C_{16:0}$ iso 3-OH	$C_{15:0}$ anteiso, $C_{15:0}$ and $C_{17:0}$ anteiso 3-OH	$C_{15:0}$ iso[f]	$C_{15:0}$ anteiso	$C_{15:0}$ anteiso
Ratio of $C_{15:0}$ anteiso to $C_{15:0}$ iso	22.8–95.2	3.1–10.3	1.9–8.2	6.0–8.8	28	<1	1.2–11.3	12.3
Predominant menaquinones	MK-10, MK-11	MK-9, MK-10	MK-10, MK-11	nt	MK-8	MK-9, MK-10	MK-10, MK-11, MK-12, MK-13[g]	nt
Growth at 37°C	+	+	+	+	–	+	+	+
DNA G+C content (mol%)	44–48	43–46	40–49	38–38.5	39.3	40–55	40–60	46.6
Principal habitat	Periodontal pockets	Feces	Feces	Human clinical specimen	Irrigated rice field soil	Oral cavities	Oral cavities	UASB sludge

[a]Data from Chen and Dong (2005), Hofstad et al. (2000), Lawson et al. (2002a), Sakamoto et al. (2002, 2006), Song et al. (2005), and Ueki et al. (2006). Symbols and abbreviations: +, positive; –, negative; D, different reactions in different species; nt, not tested; NF, nonfermentative; F, fermentative; MF, moderately fermentative; A, acetic acid; B, butyric acid; IV, isovaleric acid; L, lactic acid; P, propionic acid; PA, phenylacetic acid; S, succinic acid; G6PDH, glucose-6-phosphate dehydrogenase; 6PDGH, 6-phosphogluconate dehydrogenase; UASB, upflow anaerobic sludge blanket.
[b]The bite wound isolates do not require *N*-acetylmuramic acid (NAM) for growth.
[c]*Porphyromonas asaccharolytica* produces α-fucosidase.
[d]*Porphyromonas catoniae* does not produce a black pigment on blood agar
[e]Some species are weakly saccharolytic.
[f]*Porphyromonas catoniae* contains approximately equal amounts of $C_{15:0}$ anteiso and $C_{15:0}$ iso as the predominant fatty acids.
[g]*Prevotella dentalis* lacks menaquinones.

Thus, the name was illegitimate when Sakamoto et al. (2002) created the genus *Tannerella* for the organism in and renamed *Bacteroides forsythus* as *Tannerella forsythensis*. Although this name is legitimate, Maiden et al. (2003) requested an Opinion from the Judicial Commission that the original adjectival form of the specific epithet be conserved, i.e., "*Tannerella forsythia*". In Opinion 85, the Judicial Commission (2008) agreed with the proposal of Maiden et al. (2003).

Further reading

Holt, S.C. and J.L. Ebersole. 2005. *Porphyromonas gingivalis*, *Treponema denticola*, and *Tannerella forsythia*: the "red complex", a prototype polybacterial pathogenic consortium in periodontitis. Periodontol. 2000 *38*: 72–122.

Socransky, S.S. and A.D. Haffajee. 2005. Periodontal microbial ecology. Periodontol. 2000 *38*: 135–187.

Tanner, A.C. and J. Izard. 2006. *Tannerella forsythia*, a periodontal pathogen entering the genomic era. Periodontol. 2000 *42*: 88–113.

List of species of the genus *Tannerella*

1. **Tannerella forsythia** corrig. (Tanner, Listgarten, Ebersole and Strzempko 1986) Sakamoto, Suzuki, Umeda, Ishikawa and Benno 2002, 848[VP] (*Bacteroides forsythus* Tanner, Listgarten, Ebersole and Strzempko 1986, 216)

for.sy.then'sis. N.L. fem. adj. *forsythia* pertaining to the Forsyth Dental Center, where the species was first isolated.

The characteristics are as given for the genus and as listed in Table 20.

DNA G+C content (mol%): 46 (T_m). It is currently reported at 46.8 mol% based on the genome sequence (http://www.oralgen.lanl.gov).

Type strain: JCM 10827, ATCC 43037, CCUG 21028 A, CCUG 33064, CCUG 33226, CIP 105219, FDC 338.

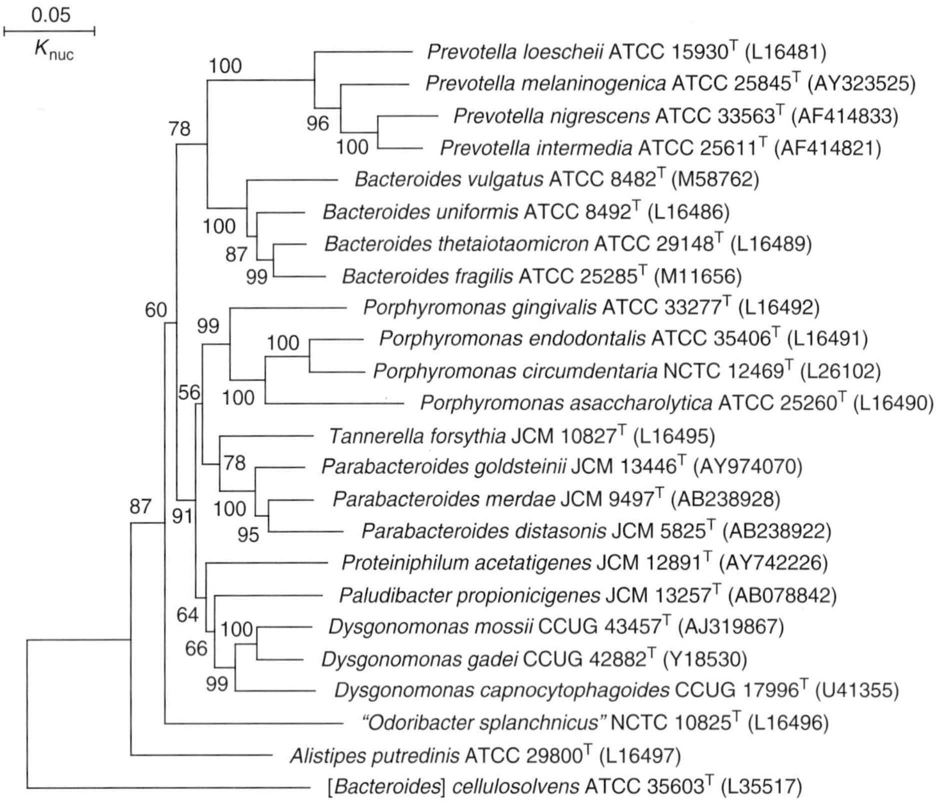

FIGURE 21. Phylogenetic tree showing the relationship between the genus *Tannerella* and some related taxa. The tree was constructed by the neighbor-joining method based on 16S rRNA gene sequences. Numbers at nodes indicate the percentage bootstrap values of 1000 replicates. Bar = 0.05 substitutions per nucleotide position. Accession numbers for 16S rRNA gene sequences are given for each strain.

Sequence accession no. (16S rRNA gene): L16495.

Genome: GenBank accession number withheld until publication of the original paper. Meanwhile, the sequence is available at ORALGEN.

References

Aldridge, K.E., D. Ashcraft, K. Cambre, C.L. Pierson, S.G. Jenkins and J.E. Rosenblatt. 2001. Multicenter survey of the changing *in vitro* antimicrobial susceptibilities of clinical isolates of *Bacteroides fragilis* group, *Prevotella*, *Fusobacterium*, *Porphyromonas*, and *Peptostreptococcus* species. Antimicrob. Agents Chemother. *45*: 1238–1243.

Aronson, N.E. and C.J. Zbick. 1988. Dysgonic fermenter 3 bacteremia in a neutropenic patient with acute lymphocytic leukemia. J. Clin. Microbiol. *26*: 2213–2215.

Bangsborg, J.M., W. Frederiksen and B. Bruun. 1990. Dysgonic fermenter 3-associated abscess in a diabetic patient. J. Infect. *20*: 237–240.

Bernard, K., C. Cooper, S. Tessier and E.P. Ewan. 1991. Use of chemotaxonomy as an aid to differentiate among *Capnocytophaga* species, CDC group DF-3, and aerotolerant strains of *Leptotrichia buccalis*. J. Clin. Microbiol. *29*: 2263–2265.

Blum, R.N., C.D. Berry, M.G. Phillips, D.L. Hamilos and E.W. Koneman. 1992. Clinical illnesses associated with isolation of dysgonic fermenter 3 from stool samples. J. Clin. Microbiol. *30*: 396–400.

Braham, P.H. and B.J. Moncla. 1992. Rapid presumptive identification and further characterization of *Bacteroides forsythus*. J. Clin. Microbiol. *30*: 649–654.

Brondz, I. and I. Olsen. 1991. Multivariate analyses of cellular fatty acids in *Bacteroides*, *Prevotella*, *Porphyromonas*, *Wolinella*, and *Campylobacter* spp. J. Clin. Microbiol. *29*: 183–189.

Chen, S. and X. Dong. 2005. *Proteiniphilum acetatigenes* gen. nov., sp. nov., from a UASB reactor treating brewery wastewater. Int. J. Syst. Evol. Microbiol. *55*: 2257–2261.

Chen, T., K. Abbey, W.J. Deng and M.C. Cheng. 2005. The bioinformatics resource for oral pathogens. Nucleic Acids Res. *33*: W734–740.

Collins, M.D., D.N. Love, J. Karjalainen, A. Kanervo, B. Forsblom, A. Willems, S. Stubbs, E. Sarkiala, G.D. Bailey, D.I. Wigney and H. Jousimies-Somer. 1994. Phylogenetic analysis of members of the genus *Porphyromonas* and description of *Porphyromonas cangingivalis* sp. nov. and *Porphyromonas cansulci* sp. nov. Int. J. Syst. Bacteriol. *44*: 674–679.

Conrads, G., S.E. Gharbia, K. Gulabivala, F. Lampert and H.N. Shah. 1997. The use of a 16S rDNA directed PCR for the detection of endodontopathogenic bacteria. J. Endod. *23*: 433–438.

Coykendall, A.L., F.S. Kaczmarek and J. Slots. 1980. Genetic heterogeneity in *Bacteroides asaccharolyticus* (Holdeman and Moore 1970) Finegold and Barnes 1977 (Approved Lists, 1980) and proposal of *Bacteroides gingivalis* sp. nov. and *Bacteroides macacae* (Slots and Genco) comb. nov. Int. J. Syst. Bacteriol. *30*: 559–564.

Daneshvar, M.I., D.G. Hollis and C.W. Moss. 1991. Chemical characterization of clinical isolates which are similar to CDC group DF-3 bacteria. J. Clin. Microbiol. *29*: 2351–2353.

de Lillo, A., V. Booth, L. Kyriacou, A.J. Weightman and W.G. Wade. 2004. Culture-independent identification of periodontitis-associated

Porphyromonas and *Tannerella* populations by targeted molecular analysis. J. Clin. Microbiol. *42*: 5523–5527.

Dewhirst, F.E., C.C. Chien, B.J. Paster, R.L. Ericson, R.P. Orcutt, D.B. Schauer and J.G. Fox. 1999. Phylogeny of the defined murine microbiota: altered Schaedler flora. Appl. Environ. Microbiol. *65*: 3287–3292.

Enersen, M., I. Olsen, A.J. van Winkelhoff and D.A. Caugant. 2006. Multilocus sequence typing of *Porphyromonas gingivalis* strains from different geographic origins. J. Clin. Microbiol. *44*: 35–41.

Finegold, S.M. and E.M. Barnes. 1977. Report of the ICSB Taxonomic subcommittee on Gram-negative anaerobic rods. Proposal that the saccharolytic and asaccharolytic strains at present classified in the species *Bacteroides melaninogenicus* (Oliver and Wherry) be reclassified into two species as *Bacteroides melaninogenicus* and *Bacteroides asaccharolyticus*. Int. J. Syst. Bacteriol. *27*: 388–391.

Finegold, S.M., M.L. Vaisanen, M. Rautio, E. Eerola, P. Summanen, D. Molitoris, Y.L. Song, C.X. Liu and H. Jousimies-Somer. 2004. *Porphyromonas uenonis* sp. nov., a pathogen for humans distinct from *P. asaccharolytica* and *P. endodontalis*. J. Clin. Microbiol. *42*: 5298–5301.

Finegold, S.M., M.L. Vaisanen, M. Rautio, E. Eerola, P. Summanen, D. Molitoris, Y.L. Song, C. Liu and H. Jousimies-Somer. 2005. *In* Validation of publication of new names and new combinations previously effectively published outside the IJSEM. List no. 102. Int. J. Syst. Bacteriol. *55*: 547–549.

Fouad, A.F., J. Barry, M. Caimano, M. Clawson, Q. Zhu, R. Carver, K. Hazlett and J.D. Radolf. 2002. PCR-based identification of bacteria associated with endodontic infections. J. Clin. Microbiol. *40*: 3223–3231.

Fournier, D., C. Mouton, P. Lapierre, T. Kato, K. Okuda and C. Menard. 2001. *Porphyromonas gulae* sp. nov., an anaerobic, Gram-negative coccobacillus from the gingival sulcus of various animal hosts. Int. J. Syst. Evol. Microbiol. *51*: 1179–1189.

Gill, V.J., L.B. Travis and D.Y. Williams. 1991. Clinical and microbiological observations on CDC group DF-3, a gram-negative coccobacillus. J. Clin. Microbiol. *29*: 1589–1592.

Gomes, B.P., R.C. Jacinto, E.T. Pinheiro, E.L. Sousa, A.A. Zaia, C.C. Ferraz and F.J. Souza-Filho. 2005. *Porphyromonas gingivalis*, *Porphyromonas endodontalis*, *Prevotella intermedia* and *Prevotella nigrescens* in endodontic lesions detected by culture and by PCR. Oral Microbiol. Immunol. *20*: 211–215.

Grabowski, A., B.J. Tindall, V. Bardin, D. Blanchet and C. Jeanthon. 2005. *Petrimonas sulfuriphila* gen. nov., sp. nov., a mesophilic fermentative bacterium isolated from a biodegraded oil reservoir. Int. J. Syst. Evol. Microbiol. *55*: 1113–1121.

Grob, R., R. Zbinden, C. Ruef, M. Hackenthal, I. Diesterweg, M. Altwegg and A. van Graevenitz. 1999. Septicemia caused by *Dysgonomonas* fermenter 3 in a severely immunocompromised patient and isolation of the same micro-organism from a stool specimen. J. Clin. Microbiol. *37*: 1617–1618.

Hansen, P.S., T.G. Jensen and B. Gahrn-Hansen. 2005. *Dysgonomonas capnocytophagoides* bacteraemia in a neutropenic patient treated for acute myeloid leukaemia. APMIS *113*: 229–231.

Haraszthy, V.I., J.J. Zambon, M. Trevisan, M. Zeid and R.J. Genco. 2000. Identification of periodontal pathogens in atheromatous plaques. J. Periodontol. *71*: 1554–1560.

Hardham, J., K. Dreier, J. Wong, C. Sfintescu and R.T. Evans. 2005. Pigmented-anaerobic bacteria associated with canine periodontitis. Vet. Microbiol. *106*: 119–128.

Heiner, A.M., J.A. DiSario, K. Carroll, S. Cohen, T.G. Evans and A.O. Shigeoka. 1992. Dysgonic fermenter-3: a bacterium associated with diarrhea in immunocompromised hosts. Am. J. Gastroenterol. *87*: 1629–1630.

Hirasawa, M. and K. Takada. 1994. *Porphyromonas gingivicanis* sp. nov. and *Porphyromonas crevioricanis* sp. nov., isolated from beagles. Int. J. Syst. Bacteriol. *44*: 637–640.

Hofstad, T., I. Olsen, E.R. Eribe, E. Falsen, M.D. Collins and P.A. Lawson. 2000. *Dysgonomonas* gen. nov. to accommodate *Dysgonomonas gadei* sp. nov., an organism isolated from a human gall bladder, and *Dysgonomonas capnocytophagoides* (formerly CDC group DF-3). Int. J. Syst. Evol. Microbiol. *50*: 2189–2195.

Holdeman, L.V. and W.E.C. Moore. 1970. *Bacteroides*. Outline of Clinical Methods in Anaerobic Bacteriology, 2nd revn (edited by Cato, Cummins, Holdeman, Johnson, Moore, Smibert and Smith). Virginia Polytechnic Institute Anaerobe Laboratory, Blacksburg, VA, pp. 57–66.

Holdeman, L.V., R.W. Kelley and W.E.C. Moore. 1984. Family I. *Bacteroidaceae* Pribram 1933, 10[AL]. *In* Bergey's Manual of Systematic Bacteriology, vol. 1 (edited by Krieg and Holt). Williams & Wilkins, Baltimore, pp. 602–603.

Hongoh, Y., M. Ohkuma and T. Kudo. 2003. Molecular analysis of bacterial microbiota in the gut of the termite *Reticulitermes speratus* (Isoptera; Rhinotermitidae). FEMS Microbiol. Ecol. *44*: 231–242.

Hongoh, Y., P. Deevong, T. Inoue, S. Moriya, S. Trakulnaleamsai, M. Ohkuma, C. Vongkaluang, N. Noparatnaraporn and T. Kudo. 2005. Intra- and interspecific comparisons of bacterial diversity and community structure support coevolution of gut microbiota and termite host. Appl. Environ. Microbiol. *71*: 6590–6599.

Honma, K., H.K. Kuramitsu, R.J. Genco and A. Sharma. 2001. Development of a gene inactivation system for *Bacteroides forsythus*: construction and characterization of a BspA mutant. Infect. Immun. *69*: 4686–4690.

Huang, Y., M. Umeda, Y. Takeuchi, M. Ishizuka, K. Yano-Higuchi and I. Ishikawa. 2003. Distribution of *Bacteroides forsythus* genotypes in a Japanese periodontitis population. Oral Microbiol. Immunol. *18*: 208–214.

Hudspeth, M.K., S. Hunt Gerardo, M.F. Maiden, D.M. Citron and E.J. Goldstein. 1999. Characterization of *Bacteroides forsythus* strains from cat and dog bite wounds in humans and comparison with monkey and human oral strains. J. Clin. Microbiol. *37*: 2003–2006.

Hughes, C.V., G. Malki, C.Y. Loo, A.C. Tanner and N. Ganeshkumar. 2003. Cloning and expression of α-D-glucosidase and *N*-acetyl-β-glucosaminidase from the periodontal pathogen, *Tannerella forsythensis* (*Bacteroides forsythus*). Oral Microbiol. Immunol. *18*: 309–312.

Hungate, R.E. 1969. A roll tube method for cultivation of strict anaerobes. *In* Methods in Microbiology, vol. 3B (edited by Norris and Ribbons). Academic Press, London, pp. 117–132.

Ishikura, H., S. Arakawa, T. Nakajima, N. Tsuchida and I. Ishikawa. 2003. Cloning of the *Tannerella forsythensis* (*Bacteroides forsythus*) *siaHI* gene and purification of the sialidase enzyme. J. Med. Microbiol. *52*: 1101–1107.

Jervoe-Storm, P.M., M. Koltzscher, W. Falk, A. Dorfler and S. Jepsen. 2005. Comparison of culture and real-time PCR for detection and quantification of five putative periodontopathogenic bacteria in subgingival plaque samples. J. Clin. Periodontol. *32*: 778–783.

Johnson, J.L. and L.V. Holdeman. 1983. *Bacteroides intermedius* comb. nov. and descriptions of *Bacteroides corporis* sp. nov. and *Bacteroides levii* sp. nov. Int. J. Syst. Bacteriol. *33*: 15–25.

Jousimies-Somer, H.R., P. Summanen, D.M. Citron, E.J. Baron, H.M. Wexler and S.M. Finegold. 2002. Wadsworth-KTL anaerobic bacteriology manual. Star Publishing Company, Belmont, CA.

Judicial Commission of the International Committee on Systematics of Prokaryotes. 2008. The adjectival form of the epithet in *Tannerella forsythensis* Sakamoto *et al.* 2002 is to be retained and the name is to be corrected to *Tannerella forsythia* Sakamoto *et al.* 2002. Opinion 85. Int. J. Syst. Evol. Microbiol. *58*: 1974.

Koneman, E.W., S.D. Allen, W.M. Janda, P.C. Shreckenberger and W.C. Winn Jr. 1997. CDC group DF-3. *In* Color Atlas and Textbook of Diagnostic Microbiology. Lippincott, Philadelphia, pp. 413–414.

Kononen, E., M.-L. Vaisanen, S.M. Finegold, R. Heine and H. Jousimies-Somer. 1996. Cellular fatty acid analysis and enzyme profiles of *Porphyromonas catoniae* – a frequent colonizer of the oral cavity in children. Anaerobe *2*: 329–335.

Kuboniwa, M., A. Amano, K.R. Kimura, S. Sekine, S. Kato, Y. Yamamoto, N. Okahashi, I.T. and S. Shizukuishi. 2004. Quantitative detection of periodontal pathogens using real-time polymerase chain reaction with TaqMan probes. Oral Microbiol. Immunol. *19*: 168–176.

Kumar, P.S., A.L. Griffen, J.A. Barton, B.J. Paster, M.L. Moeschberger and E.J. Leys. 2003. New bacterial species associated with chronic periodontitis. J. Dent. Res. *82*: 338–344.

Lakhssassi, N., N. Elhajoui, J.P. Lodter, J.L. Pineill and M. Sixou. 2005. Antimicrobial susceptibility variation of 50 anaerobic periopathogens in aggressive periodontitis: an interindividual variability study. Oral Microbiol. Immunol. *20*: 244–252.

Lan, P.T., H. Hayashi, M. Sakamoto and Y. Benno. 2002. Phylogenetic analysis of cecal microbiota in chicken by the use of 16S rDNA clone libraries. Microbiol. Immunol. *46*: 371–382.

Lawson, P.A., E. Falsen, E. Inganas, R.S. Weyant and M.D. Collins. 2002a. *Dysgonomonas mossii* sp. nov., from human sources. Syst. Appl. Microbiol. *25*: 194–197.

Lawson, P.A., E. Falsen, E. Inganas, R.S. Weyant and M.D. Collins. 2002b. *In* Validation of the publication of new names and new combinations previously effectively published in the IJSEM. List no. 88. Int. J. Syst. Evol. Microbiol. *52*: 1915–1916.

Leys, E.J., S.R. Lyons, M.L. Moeschberger, R.W. Rumpf and A.L. Griffen. 2002. Association of *Bacteroides forsythus* and a novel *Bacteroides* phylotype with periodontitis. J. Clin. Microbiol. *40*: 821–825.

Loesche, W.J., W.A. Bretz, D. Kerschensteiner, J. Stoll, S.S. Socransky, P. Hujoel and D.E. Lopatin. 1990. Development of a diagnostic test for anaerobic periodontal infections based on plaque hydrolysis of benzoyl-DL-arginine-naphthylamide. J. Clin. Microbiol. *28*: 1551–1559.

Love, D.N., J.L. Johnson, R.F. Jones and A. Calverley. 1987. *Bacteroides salivosus* sp. nov., an asaccharolytic, black-pigmented species from cats. Int. J. Syst. Bacteriol. *37*: 307–309.

Love, D.N., G.D. Bailey, S. Collings and D.A. Briscoe. 1992. Description of *Porphyromonas circumdentaria* sp. nov. and reassignment of *Bacteroides salivosus* (Love, Johnson, Jones and Calverley 1987) as *Porphyromonas* (Shah and Collins 1988) *salivosa* comb. nov. Int. J. Syst. Bacteriol. *42*: 434–438.

Love, D.N., J. Karjalainen, A. Kanervo, B. Forsblom, E. Sarkiala, G.D. Bailey, D.I. Wigney and H. Jousimies-Somer. 1994. *Porphyromonas canoris* sp. nov., an asaccharolytic, black-pigmented species from the gingival sulcus of dogs. Int. J. Syst. Bacteriol. *44*: 204–208.

Love, D.N. 1995. *Porphyromonas macacae* comb. nov., a consequence of *Bacteroides macacae* being a senior synonym of *Porphyromonas salivosa*. Int. J. Syst. Bacteriol. *45*: 90–92.

Maiden, M.F., C. Pham and S. Kashket. 2004. Glucose toxicity effect and accumulation of methylglyoxal by the periodontal anaerobe *Bacteroides forsythus*. Anaerobe *10*: 27–32.

Maiden, M.F.J., A. Tanner and P.J. Macuch. 1996. Rapid characterization of periodontal bacterial isolates by using fluorogenic substrate tests. J. Clin. Microbiol. *34*: 376–384.

Maiden, M.F.J., P. Cohee and A.C.R. Tanner. 2003. Proposal to conserve the adjectival form of the specific epithet in the reclassification of *Bacteroides forsythus* Tanner *et al.* 1986 to the genus *Tannerella* Sakamoto *et al.* 2002 as *Tannerella forsythia* corrig., gen. nov., comb. nov. Request for an Opinion. Int. J. Syst. Evol. Microbiol. *53*: 2111–2112.

Martínez-Sánchez, L., F.J. Vasallo, F. García-Garrote, L. Alcalá, M. Rodríguez-Créíxems and E. Bouza. 1998. Clinical isolation of a DF-3 micro-organism and review of the literature. Clin. Microbiol. Infect. *4*: 344–346.

Melhus, A. 1997. Isolation of dysgonic fermenter 3, a rare isolate associated with diarrhoea in immunocompromised patients. Scand. J. Infect. Dis. *29*: 195–196.

Moncla, B.J., P. Braham and S.L. Hillier. 1990. Sialidase (neuraminidase) activity among gram-negative anaerobic and capnophilic bacteria. J. Clin. Microbiol. *28*: 422–425.

Moore, L.V.H., D.M. Bourne and W.E.C. Moore. 1994. Comparative distribution and taxonomic value of cellular fatty acids in 33 genera of anaerobic gram-negative bacilli. Int. J. Syst. Bacteriol. *44*: 338–347.

Moore, L.V.H. and W.E.C. Moore. 1994. *Oribaculum catoniae* gen. nov., sp. nov., *Catonella morbi* gen. nov., sp. nov., *Hallella seregens* gen. nov., sp. nov., *Johnsonella ignava* gen. nov., sp. nov., and *Dialister pneumosintes* gen. nov., comb. nov., nom. rev., anaerobic Gram-negative bacilli from the human gingival crevice. Int. J. Syst. Bacteriol. *44*: 187–192.

Nelson, K.E., R.D. Fleischmann, R.T. DeBoy, I.T. Paulsen, D.E. Fouts, J.A. Eisen, S.C. Daugherty, R.J. Dodson, A.S. Durkin, M. Gwinn, D.H. Haft, J.F. Kolonay, W.C. Nelson, T. Mason, L. Tallon, J. Gray, D. Granger, H. Tettelin, H. Dong, J.L. Galvin, M.J. Duncan, F.E. Dewhirst and C.M. Fraser. 2003. Complete genome sequence of the oral pathogenic bacterium *Porphyromonas gingivalis* strain W83. J. Bacteriol. *185*: 5591–5601.

Noguchi, N., Y. Noiri, M. Narimatsu and S. Ebisu. 2005. Identification and localization of extraradicular biofilm-forming bacteria associated with refractory endodontic pathogens. Appl. Environ. Microbiol. *71*: 8738–8743.

Paster, B.J., F.E. Dewhirst, I. Olsen and G.J. Fraser. 1994. Phylogeny of *Bacteroides*, *Prevotella*, and *Porphyromonas* spp. and related bacteria. J. Bacteriol. *176*: 725–732.

Rudney, J.D., R. Chen and G.J. Sedgewick. 2005. *Actinobacillus actinomycetemcomitans*, *Porphyromonas gingivalis*, and *Tannerella forsythensis* are components of a polymicrobial intracellular flora within human buccal cells. J. Dent. Res. *84*: 59–63.

Sabet, M., S.W. Lee, R.K. Nauman, T. Sims and H.S. Um. 2003. The surface (S-) layer is a virulence factor of *Bacteroides forsythus*. Microbiology *149*: 3617–3627.

Saito, T., K. Ishihara, T. Kato and K. Okuda. 1997. Cloning, expression, and sequencing of a protease gene from *Bacteroides forsythus* ATCC 43037 in *Escherichia coli*. Infect. Immun. *65*: 4888–4891.

Sakamoto, M., M. Suzuki, M. Umeda, I. Ishikawa and Y. Benno. 2002. Reclassification of *Bacteroides forsythus* (Tanner *et al.* 1986) as *Tannerella forsythensis* corrig., gen. nov., comb. nov. Int. J. Syst. Evol. Microbiol. *52*: 841–849.

Sakamoto, M. and Y. Benno. 2006. Reclassification of *Bacteroides distasonis*, *Bacteroides goldsteinii* and *Bacteroides merdae* as *Parabacteroides distasonis* gen. nov., comb. nov., *Parabacteroides goldsteinii* comb. nov. and *Parabacteroides merdae* comb. nov. Int. J. Syst. Evol. Microbiol. *56*: 1599–1605.

Sakamoto, M., M. Kitahara and Y. Benno. 2007a. *Parabacteroides johnsonii* sp. nov., isolated from human faeces. Int. J. Syst. Evol. Microbiol. *57*: 293–296.

Sakamoto, M., P.T. Lan and Y. Benno. 2007b. *Barnesiella viscericola* gen. nov., sp. nov., a novel member of the family *Porphyromonadaceae* isolated from chicken caecum. Int. J. Syst. Evol. Microbiol. *57*: 342–346.

Schonheyder, H., T. Ejlertsen and W. Frederiksen. 1991. Isolation of a dysgonic fermenter (DF-3) from urine of a patient. Eur. J. Clin. Microbiol. Infect. Dis. *10*: 530–531.

Seol, J.H., B.H. Cho, C.P. Chung and K.S. Bae. 2006. Multiplex polymerase chain reaction detection of black-pigmented bacteria in infections of endodontic origin. J. Endod. *32*: 110–114.

Shah, H.N. and M.D. Collins. 1988. Proposal for reclassification of *Bacteroides asaccharolyticus*, *Bacteroides gingivalis*, and *Bacteroides endodontalis* in a new genus, *Porphyromonas*. Int. J. Syst. Bacteriol. *38*: 128–131.

Shah, H.N., M.D. Collins, I. Olsen, B.J. Paster and F.E. Dewhirst. 1995. Reclassification of *Bacteroides levii* (Holdeman, Cato and Moore) in the genus *Porphyromonas*, as *Porphyromonas levii* comb. nov. Int. J. Syst. Bacteriol. *45*: 586–588.

Sharma, A., H.T. Sojar, I. Glurich, K. Honma, H.K. Kuramitsu and R.J. Genco. 1998. Cloning, expression, and sequencing of a cell surface antigen containing a leucine-rich repeat motif from *Bacteroides forsythus* ATCC 43037. Infect. Immun. *66*: 5703–5710.

Sharma, A., S. Inagaki, K. Honma, C. Sfintescu, P.J. Baker and R.T. Evans. 2005. *Tannerella forsythia*-induced alveolar bone loss in mice involves leucine-rich-repeat BspA protein. J. Dent. Res. *84*: 462–467.

Slots, J. and R.J. Genco. 1980. *Bacteroides melaninogenicus* subsp. *macacae*: new subspecies from monkey periodontopathic indigenous microflora. Int. J. Syst. Bacteriol. *30*: 82–85.

Song, Y., C. Liu, J. Lee, M. Bolanos, M.L. Vaisanen and S.M. Finegold. 2005. "*Bacteroides goldsteinii* sp. nov." isolated from clinical specimens of human intestinal origin. J. Clin. Microbiol. *43*: 4522–4527.

Summanen, P.H., B. Durmaz, M.L. Vaisanen, C. Liu, D. Molitoris, E. Eerola, I.M. Helander and S.M. Finegold. 2005. *Porphyromonas somerae* sp. nov., a pathogen isolated from humans and distinct from *Porphyromonas levii*. J. Clin. Microbiol. *43*: 4455–4459.

Summanen, P.H., B. Durmaz, M.L. Vaisanen, C. Liu, D. Molitoris, E. Eerola, I.M. Helander and S.M. Finegold. 2006. *In* List of new names and new combinations previously effectively, but not validly, published. Validation List no. 109. Int. J. Syst. Evol. Microbiol. *56*: 925–927.

Summanen, P.H., P.A. Lawson and S.M. Finegold. 2009. *Porphyromonas bennonis* sp. nov., isolated from human clinical specimens. Int. J. Syst. Evol. Microbiol. *59*: 1727–1732.

Sutter, V.L. and W.T. Carter. 1972. Evaluation of media and reagents for indole-spot tests in anaerobic bacteriology. Am. J. Clin. Pathol. *58*: 335–338.

Takemoto, T., H. Kurihara and G. Dahlen. 1997. Characterization of *Bacteroides forsythus* isolates. J. Clin. Microbiol. *35*: 1378–1381.

Tan, K.S., K.P. Song and G. Ong. 2001. *Bacteroides forsythus prtH* genotype in periodontitis patients: occurrence and association with periodontal disease. J. Periodont. Res. *36*: 398–403.

Tanner, A., M.F. Maiden, K. Lee, L.B. Shulman and H.P. Weber. 1997. Dental implant infections. Clin. Infect. Dis. 25 Suppl. 2: S213–217.

Tanner, A.C., C. Haffer, G.T. Bratthall, R.A. Visconti and S.S. Socransky. 1979. A study of the bacteria associated with advancing periodontitis in man. J. Clin. Periodontol. *6*: 278–307.

Tanner, A.C., M.N. Strzempko, C.A. Belsky and G.A. McKinley. 1985. API ZYM and API An-Ident reactions of fastidious oral gram-negative species. J. Clin. Microbiol. *22*: 333–335.

Tanner, A.C., M.F. Maiden, J.J. Zambon, G.S. Thoren and R.L. Kent, Jr. 1998. Rapid chair-side DNA probe assay of *Bacteroides forsythus* and *Porphyromonas gingivalis*. J. Periodont. Res. *33*: 105–117.

Tanner, A.C.R., M.A. Listgarten, J.L. Ebersole and M.N. Strezempko. 1986. *Bacteroides forsythus* sp. nov, a slow-growing, fusiform *Bacteroides* sp. from the human oral cavity. Int. J. Syst. Bacteriol. *36*: 213–221.

Ueki, A., H. Akasaka, D. Suzuki and K. Ueki. 2006. *Paludibacter propionicigenes* gen. nov., sp. nov., a novel strictly anaerobic, Gram-negative, propionate-producing bacterium isolated from plant residue in irrigated rice-field soil in Japan. Int. J. Syst. Evol. Microbiol. *56*: 39–44.

Umeda, M., Y. Tominaga, T. He, K. Yano, H. Watanabe and I. Ishikawa. 1996. Microbial flora in the acute phase of periodontitis and the effect of local administration of minocycline. J. Periodontol. *67*: 422–427.

Van Steenbergen, T.J.M., A.J. Van Winkelhoff, D. Mayrand, D. Grenier and J. De Graaff. 1984. *Bacteroides endodontalis* sp. nov., an asaccharolytic black-pigmented bacteriodes species from infected dental root canals. Int. J. Syst. Bacteriol. *34*: 118–120.

Vandamme, P., M. Vancanneyt, A. van Belkum, P. Segers, W.G. Quint, K. Kersters, B.J. Paster and F.E. Dewhirst. 1996. Polyphasic analysis of strains of the genus *Capnocytophaga* and Centers for Disease Control group DF-3. Int. J. Syst. Bacteriol. *46*: 782–791.

Wagner, D.K., J.J. Wright, A.F. Ansher and V.J. Gill. 1988. Dysgonic fermenter 3-associated gastrointestinal disease in a patient with common variable hypogammaglobulinemia. Am.J. Med. *84*: 315–318.

Wallace, P.L., D.G. Hollis, R.E. Weaver and C.W. Moss. 1989. Characterization of CDC group DF-3 by cellular fatty acid analysis. J. Clin. Microbiol. *27*: 735–737.

Willems, A. and M.D. Collins. 1995. Reclassification of *Oribaculum catoniae* (Moore and Moore 1994) as *Porphyromonas catoniae* comb. nov. and emendation of the genus *Porphyromonas*. Int. J. Syst. Bacteriol. *45*: 578–581.

Wyss, C. 1989. Dependence of proliferation of *Bacteroides forsythus* on exogenous N-acetylmuramic acid. Infect. Immun. *57*: 1757–1759.

Yoneda, M., T. Hirofuji, H. Anan, A. Matsumoto, T. Hamachi, K. Nakayama and K. Maeda. 2001. Mixed infection of *Porphyromonas gingivalis* and *Bacteroides forsythus* in a murine abscess model: involvement of gingipains in a synergistic effect. J. Periodont. Res. *36*: 237–243.

Zhu, X.Y., T. Zhong, Y. Pandya and R.D. Joerger. 2002. 16S rRNA-based analysis of microbiota from the cecum of broiler chickens. Appl. Environ. Microbiol. *68*: 124–137.

Family V. **Prevotellaceae** fam. nov.

Noel R. Krieg

Pre.vo.tell.a.ce′a.e. N.L. fem. n. *Prevotella* type genus of the family; suff. *-aceae* ending to denote a family; N.L. fem. pl. n. *Prevotellaceae* the *Prevotella* family.

At present, the family is comprised of the type genus *Prevotella* and *Xylanibacter*. The organisms are **anaerobic**, **nonsporeforming**, **nonmotile**, rods that **stain Gram-negative**. **Saccharolytic**. The principal respiratory quinones are unsaturated **menaquinones** with 8–13 isoprene units, depending upon the species. Some characteristics that differentiate the genera are given in Table 21.

Type genus: **Prevotella** Shah and Collins 1990, 205^{VP} emend. Willems and Collins 1995, 834^{VP}.

TABLE 21. Some characteristics that differentiate the genera of the family *Prevotellaceae*

Characteristic	*Prevotella*	*Xylanibacter*
Cell shape	Short rods	Short to filamentous rods
Major products of glucose fermentation	Succinate and acetate	Acetate, propionate, and succinate
Isolated from	Primarily isolated from oral cavity, but also urogenital and intestinal tracts of animals	Anoxic rice-plant residue
DNA G+C content (mol%)	40–52	44

Genus I. **Prevotella** Shah and Collins 1990, 205[VP] emend. Willems and Collins 1995, 834[VP]

Haroun N. Shah, Marie Anne Chattaway, Lakshani Rajakurana and Saheer E. Gharbia

Pre.vo.tel'la. N.L. dim. ending *ella*; N.L. fem. n. *Prevotella*, named after the French microbiologist, A. R. Prévot, a pioneer in anaerobic microbiology.

Short rods. Nonsporeforming. Nonmotile. Gram-stain-negative. Anaerobic. **Moderately saccharolytic. Growth is inhibited by 20% bile. Succinic and acetic acids are the major metabolic end products** in peptone-yeast extract-glucose (PYG) broth, but lower levels of other short-chain acids may be produced. **Malate dehydrogenase and glutamate dehydrogenase are present, but glucose-6-phosphate dehydrogenase and 6-phosphogluconate dehydrogenase are absent.** Sphingolipids are produced. The nonhydroxylated long-chain fatty acids are primarily of the **straight-chain saturated and monounsaturated types;** methyl branched fatty acids are either absent or present in small amounts. **Porphyrins** are produced by pigmented species, while **menaquinones are the sole respiratory quinones found in all species so far studied. Diaminopimelic acid is the only dibasic amino acid in the peptidoglycan** (Miyagawa et al., 1981). Neither heptose nor 2-keto-3-deoxyoctulosonic acid has so far been reported (Hofstad, 1974). The primary site of isolation is the oral cavity, but more recently species have been reported from the intestinal tract of man and animals.

DNA G+C content (mol%) of most species: 40–52.

Type species: **Prevotella melaninogenica** (Oliver and Wherry 1921) Shah and Collins 1990, 206[VP] emend. Wu, Johnson, Moore and Moore 1992, 536[VP] [*Fusiformis nigrescens* Schwabacher, Lucas and Rimington 1947, 109; *Bacteroides melaninogenicus* (Oliver and Wherry 1921) Roy and Kelly 1939, 569[AL]; *Ristella melaninogenica* (Oliver and Wherry 1921) Prévot 1938, 290; *Hemophilus melaninogenicus* (*sic*) (Oliver and Wherry 1921) Bergey, Harrison, Breed, Hammer and Huntoon 1930, 314; *Bacterium melaninogenicum* Oliver and Wherry 1921, 341].

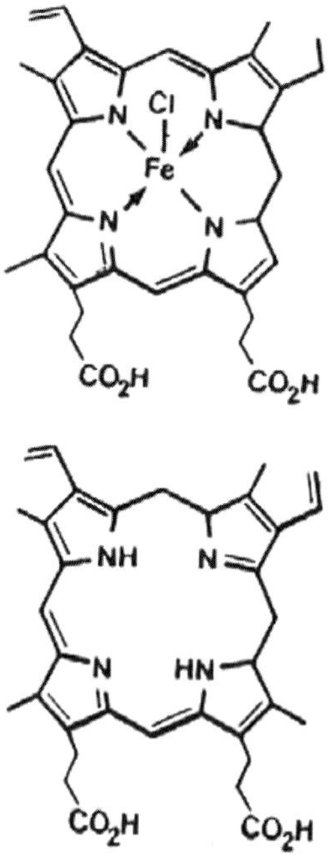

FIGURE 22. Major porphyrins found in pigmented species of *Prevotella* that give the colonies their characteristic appearance on blood agar plates. (Top) The non-UV fluorescent protohemin showing its stabilization by iron in the ring structure. (Bottom) Protoporphyrin; the absence of iron from the porphyrin system enables excitation under long-wavelength UV radiation (365 nm) and the brilliant red fluorescence observed in some species (see text).

Further descriptive information

The genus *Prevotella* includes species that produce black, brown, beige or non-pigmented colonies on blood agar plates. However, all appear to have the capacity to produce protoporphyrin and protoheme (Figure 22), which form the prosthetic group of cytochromes. Prior to colonies becoming black (due to over production of protoheme), the brown/beige centers of the colonies show brilliant red fluorescence under long wave ultraviolet radiation (365 nm) due to the production of protoporphyrin (Shah et al., 1979). Once black, (~3–7 d) the UV fluorescence is quenched by virtue of the stabilization of the ring structure of the iron-porphyrin complex (Shah et al., 1979) and colonies appear consistently black. Factors such as the type of blood, hemolysis, temperature, pH, the presence of other isolates, etc., determine the depth and rate of pigmentation. Other characteristic electron carriers present in *Prevotella* species are respiratory quinones. However, only one type, *viz.* menaquinones, is produced; ubiquinones have not been reported. The former are detectable by scanning cell extracts in iso-octane solution between 240 and 360 nm. Absorption maxima at 244, 248, 260, 269 and 328 nm are characteristic of such compounds (Figure 23). Reduction (e.g., with potassium borohydrate) results in bleaching of absorption at 260– 269 nm and a concomitant increase in absorption at 246 nm. These spectral characteristics are consistent with the presence of menaquinones that may be used to confirm members of the genus. However, the number of isoprene units is characteristic for several species and can be ascertained by chromatography or mass spectrometry (Shah and Collins, 1980) (Figure 24). Both the porphyrins and menaquinones form part of an anaerobic electron transport system that is essential to the physiology of this group and therefore they are likely to be stable characters.

Enrichment and isolation procedures

Because energy conservation in *Prevotella* species resolves around porphyrins and menaquinones, their respective precursors, *viz.* heme and menadione, are required for growth. On blood agar plates, it is unnecessary to add these compounds to the medium, but their inclusion in liquid culture is essential for growth. Hemin is added at 0.5 mg%, while menadione is added at 0.05 mg%. Peptides, rather than free amino acids, are required as sources of nitrogen for growth. This requirement may be met by protein hydrolysates such as trypticase, proteose peptone or yeast extract that have a pronounced effect upon growth. However, these species are also moderately saccharolytic; hence, the inclusion of glucose also helps to stimulate growth. There are many published methods for enrichment and isolation of these species, but all media contain the above components with a thiol reducing agent. A typical enrichment

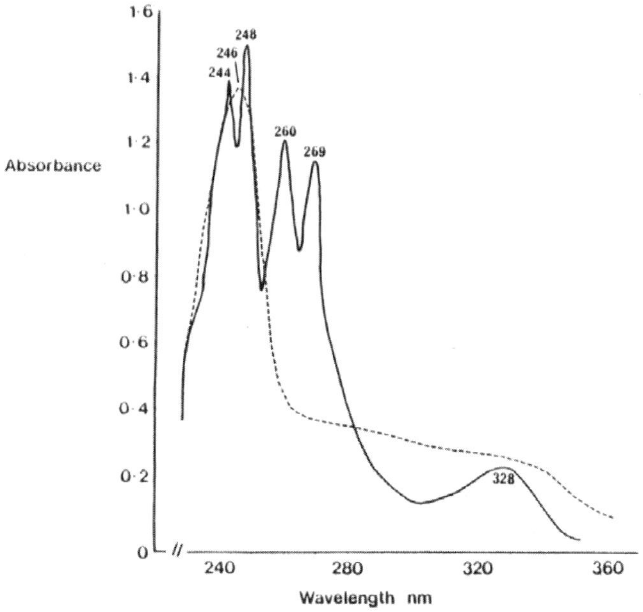

FIGURE 23. Demonstration of the presence of menaquinones in *Prevotella* species. Dried cells are taken up in isooctane solution and the absorption spectrum between 240 and 350 nm recorded. Characteristic maxima at 244, 248, 260, 269, and 328 nm that are reduced with potassium borohydride (see above, dotted lines) demonstrate the presence of menaquinones. Confirmation of the number of isoprene units is achieved using mass spectrometry (see Figure 24).

medium, often referred to as "BM" (Shah et al., 1976) consists of the following (g/l): trypticase (10), proteose peptone (10), yeast extract (5), glucose (5), sodium chloride (5) and cysteine hydrochloride (0.75), together with hemin and menadione at the above concentrations. The medium is adjusted to pH 7.4 and autoclaved for the standard 121°C for 15 min. The medium needs to be prepared fresh, especially for isolation. Many of these species are particularly difficult to isolate initially, but once cultured they adapt readily.

Maintenance procedures

Prevotella species possess an electron transport system in which the anaerobic reduction of fumarate takes place. The redox potential of the fumarate/succinate couple at pH 7.0 is about -200 mV greater that the redox couples used by more exacting anaerobes such as methane producing or nitrogen-fixing bacteria (-400 mV). Consequently, the very rigid procedures used for the latter are unnecessary for *Prevotella* species. Once isolated, most standard blood agar media (with 0.5%, v/v, horse blood) may be used to maintain the organisms. Broth cultures are useful for obtaining large quantities of biomass, but are not useful for maintaining cells. On blood agar, cells will grow rapidly (1–2 d) butmay be lost if care is not taken to subculture every few days. This is particularly so for species such as *Prevotella intermedia* or *Prevotella nigrescens* that grow rapidly and produce a jet black pigment on blood agar plates (Shah and Gharbia, 1992). If the entire colony becomes black and often "dry", the organisms rapidly loose their viability. When subculturing from such colonies, a large number of colonies should be taken over to the fresh plate to ensure their survival. It is essential therefore, for long-term storage and weekly subculture, that cells are taken from the early exponential phase prior to pigmentation or from central parts of the colony that are non-pigmented/beige during the first few days of growth. Dense cells suspensions (~10^8 per ml) may be lyophilized or kept frozen on beads at -80°C. These should be prepared with minimum expose to air or, if available, prepared in an anaerobic chamber. Maintenance on standard laboratory slope media or growth and storage in Robertson Cooked Meat broth will result in a loss of viability.

Differentiation of the genus *Prevotella* from other genera

Members of the genera *Prevotella* and *Porphyromonas* previously belonged to the genus *Bacteroides* (Holdeman et al., 1984; Olsen and Shah, 2001; Werner, 1991). However, they differed so markedly in biochemical and chemical properties from the type species, *Bacteroides fragilis*, that they were reclassified into separate genera. *Porphyromonas* encompasses species that are nonfermentative or very weakly fermentative, whereas *Prevotella* species are moderately fermentative. These proposals for reclassification were made well before the acceptance of 16S rRNA gene sequencing, but have now been substantiated by comparative sequence analysis of this gene (see Figure 25), and the number of new species added to both genera continue to increase significantly.

Biochemically, *Prevotella* species appear to occupy an ecological niche where there is a limited range of carbohydrates and free amino acids. Thus, unlike *Bacteroides* species that thrive in the intestinal tract and are able to ferment a vast array of available carbohydrates, the natural habitat of *Prevotella* and *Porphyromonas* species appears to be mainly the oral cavity (especially supra- and sub- gingival dental plaque), where there is a more restricted range of fermentable substrates. *Porphyromonas* species are generally nonfermentative and flourish in more inaccessible, anoxic environments as part of diverse bacterial communities where nitrogenous substrates rather than carbohydrates are available. Some species such as *Porphyromonas gingivalis* produce potent proteinases such as α- and β-gingivain (Shah et al., 1993) that release amino acids and peptides for growth (Shah and Williams, 1987a, b). *Prevotella* species appear to occupy an "intermediate niche" and their biochemical properties are reflected in their function. Some species have limited proteolytic activities, and all species, except the recently proposed "*Prevotella massiliensis*" (Berger et al., 2005), ferment a limited range of carbohydrates such as sucrose or lactose (see Table 22). They differ markedly from the genus *Bacteroides* (*sensu stricto*) in lacking enzymes of the pentose phosphate pathway-hexose monophosphate shunt (*viz.* glucose-6-phosphate dehydrogenase and 6-phosphogluconate dehydrogenase), and because their major habitat is the oral cavity they are unable to grow in the presence of bile. Glutamate and aspartate have been identified among the amino acids metabolized, (Shah and Williams, 1987a, b). Consequently, the enzymes glutamate dehydrogenase and malate dehydrogenase are present and are key markers of the genus. Aspartate catabolism is inextricably linked to the utilization of an electron transport system involving both cytochromes (e.g., cytochrome *b*) and the respiratory quinones (e.g., menaquinones), in which fumarate acts as an electron sink in accept-

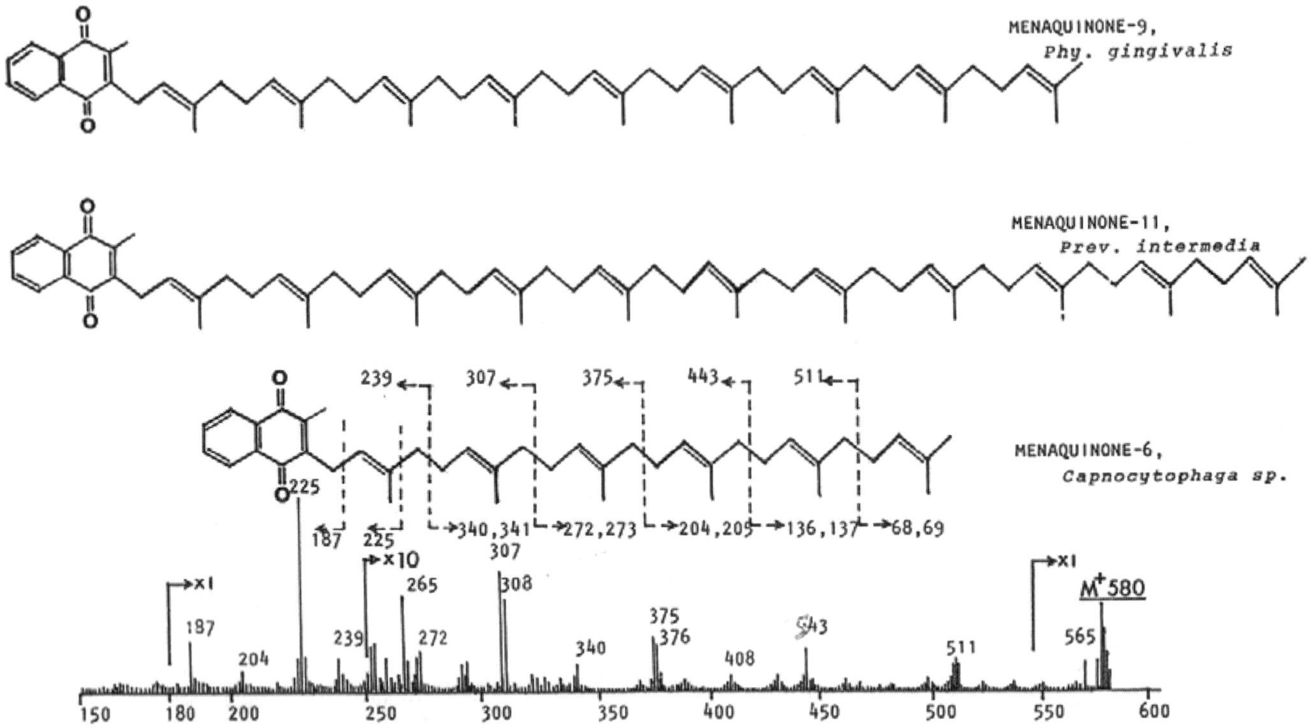

FIGURE 24. Mass spectra of the unsaturated menaquinones of three species. Below. *Bacteroides ochraceus* (later *Capnocytophaga*) that previously belonged to this group. It was excluded initially on the basis that it contained six isoprene units (MK-6). Similarly (Top) *Porphyromonas* was separated on the basis of major levels of MK-9. By contrast, most *Prevotella* species analyzed to date possess significant levels of MK-10, MK-11, or MK-12 with lower levels of MK-8, MK-9, and MK-13. The mass spectrum (MK-6, Bottom) shows the characteristic fragmentation pattern of such a compound. The major mass ion at 580 is its molecular weight, 565 (M^+ −15) loss of the ring methyl group (CH_3), while mass ions at 239, 307, 375, 443, and 511 are attributable to the loss of isoprene (68 mass units). Intense peaks at m/e 187 and 225 are derived from the 1–4 naphthoquinone nucleus and are characteristic of menaquinones. This property, which was first introduced for the characterization of the *Bacteroidaceae* (Shah and Collins, (1980, 1983), is now increasingly being reintroduced to support new diversity revealed by16S rRNA sequence analysis (see e.g., Sakamoto et al., 2005a, b).

ing reducing equivalents from reduced coenzymes generated during catabolism. Components of this system are therefore key features of this genus and have been used as exclusionary criteria in delineating species of this genus. Species are therefore characterized by the presence of unsaturated menaquinones with predominantly 10–13 isoprene units. Both nonhydroxylated and 3-hydroxylated fatty acids are present, with the former being predominantly straight-chain saturated, anteiso- and iso-methyl and branched-chain types. The DNA base compositions are within the approximate range of 40–52 mol% G+C.

Differentiation of the species of the genus *Prevotella*

The genus *Prevotella* includes 34 validly published species and one that is not validly published. Table 22 gives the phsiological characteristics helpful in differentiating the species. The list of species is given below.

List of species of the genus *Prevotella*

1. **Prevotella melaninogenica** (Oliver and Wherry 1921) Shah and Collins 1990, 206[VP] emend. Wu, Johnson, Moore and Moore 1992, 536[VP] [*Fusiformis nigrescens* Schwabacher, Lucas and Rimington 1947, 109; *Bacteroides melaninogenicus* (Oliver and Wherry 1921) Roy and Kelly 1939, 569[AL]; *Ristella melaninogenica* (Oliver and Wherry 1921) Prévot 1938, 290; *Hemophilus melaninogenicus* (*sic*) (Oliver and Wherry 1921) Bergey, Harrison, Breed, Hammer and Huntoon 1930, 314; *Bacterium melaninogenicum* Oliver and Wherry 1921, 341].

me.la.ni.no.ge'ni.ca. N.L. n. *melaninum* melanin; N.L. adj. *genicus -a -um* producing, probably derived from Gr. n. *genetes* a begetter; N.L. fem. adj. *melaninogenica* melanin producing. [Note: the black pigment produced by this organism is due to protoheme (Shah et al., 1979) and not to melanin, as originally thought.]

The description is from Holdeman and Johnson (1982), Holdeman et al. (1984), Miyagawa et al. (1978, 1979, 1981), Miyagawa and Suto (1980), Shah et al. (1979, 1976), Shah and Collins, (1990, 1980, 1983), Shah and Williams (1982), Gregory et al. (1978), Lambe et al. (1974) and van Steenbergen et al. (1979). Cells are rods 0.5–0.8 × 0.9–2.5 µm; occasional cells are 10 µm long in glucose broth cultures. Colonies grown on blood agar are 0.5–2 mm in diameter, circular and entire, convex and shiny; cells are usually

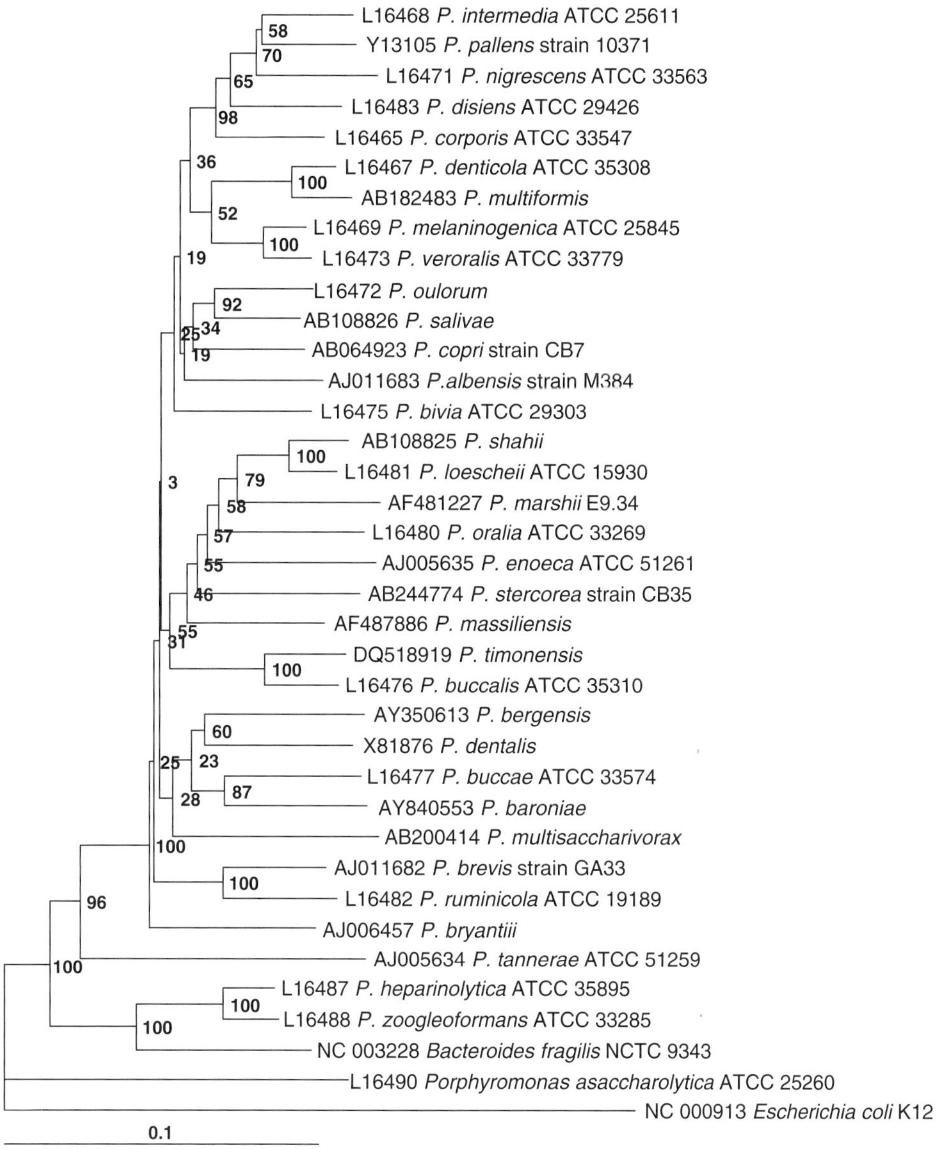

FIGURE 25. Neighbor-joining phylogenetic tree of 16S rRNA gene sequences of the type strains of *Prevotella* species deposited in GenBank to show the relationships of strains within the genus. Bootstrap values (expressed as percentages of 1000 replicates) are shown at the nodes. The numbers that precede each strain are the GenBank accession numbers, while those that follow the species names are the Culture Collection numbers. The type species of *Bacteroides* (*Bacteroides fragilis*, NCTC 9343) and *Porphyromonas* (*Porphyromonas asaccharolytica*, ATCC 25260), both former members of the same genus, and *Escherichia coli* (K12) are included for comparison.

darker in the center of the colony; edges are gray to light brown and become darker upon continued incubation (5–14 d). Before the pigmentation develops, the lighter areas of the colony exhibit bright red fluorescence under long wave ultraviolet radiation (365 nm), which is helpful in the presumptive identification of this species. Cells grown on hemolyzed laked blood agar develop pigment more rapidly; some strains do not produce black pigment on horse blood agar, but do on rabbit blood agar; some strains produce beta-hemolysis on rabbit blood agar. Glucose broth cultures are usually turbid with a smooth or stringy sediment and pH of 4.6–5.0 and growth is stimulated with the addition of hemin (1 µg/ml is the requirement for the type strain) or vitamin K_1 (0.1 µg/ml). The majority of strains grow at pH 8.5 and 25°C; some grow at 45°C.

Cells ferment dextrin, fructose, glucose, glycogen, inulin, lactose, maltose, mannose, raffinose, starch, sucrose, but not amygdalin, arabinose, cellobiose, erythritol, inositol, mannitol, melibiose, melezitose (variable), rhamnose, ribose, salcin, sorbitol, trehalose or xylan. Esculin (variable) and starch are hydrolyzed and gelatin is digested; nitrate is not reduced; catalase, indole and lecithinase are not produced.

Succinate and acetate are the major fermentation end products from glucose broth culture (1% peptone, 1% yeast extract, vitamin K_1 and hemin); minor products include formate, isobutyrate and isovalerate. H_2 is not produced. Lactate and pyruvate are not utilized; threonine is not converted to propionate. Cells have superoxide dismutase activity. The long-chain fatty acid composition is mainly anteiso- and iso-methyl branched acids with small amounts of straight-

TABLE 22. Distinguishing physiological characteristics of species of the genus *Prevotella*[a,b]

Characteristic	*P. albensis*	*P. baroniae*	*P. bergensis*	*P. bivia*	*P. brevis*	*P. bryantii*	*P. buccae*	*P. buccalis*	*P. corpri*	*P. corporis*	*P. dentalis*	*P. denticola*	*P. disiens*	*P. enoeca*	*P. heparinolytica*	*P. intermedia*	*P. loescheii*
Pigment production on blood agar	−	−	−	−	−	−	−	−	−	+	−	±	−	−	−	+	+
Arabinose	nd	−	+	−	nd	nd	+	−	nd	−	+	−	−	−	+	−	−
Cellobiose	nd	+	+	−	nd	nd	+	+	nd	−	+	−	−	−	+	−	+
Indole	nd	−	−	−	nd	nd	−	−	nd	−	−	−	−	−	+	+	+(−)−
Lactose	nd	+	nd	+	nd	nd	+	+	+	−	+	+	−	+	+	−	+
Mannose	−	+	+	nd	nd	nd	+	+	nd	+	+	+	−	+	+	v	+
Raffinose	−	+	−	nd	nd	nd	+	+	+	−	+	+	−	−	−	v	+
Rhamnose	nd	−	+	−	−	+	+ (−)	−	nd	−	nd	−	−	nd	nd	−	−
Salicin	nd	+	+	−	−	+	+	−	+	−	−	−	−	−	+	−	−
Sucrose	−	+	−	−	nd	nd	+	+	+	−	w	+	−	−	+	+	+
Xylose	+	−	+	−	−	+	+	−	nd	−	−	−	−	−	+	−	−
Esculin hydrolysis	nd	+	+	−	nd	nd	+	+	nd	−	+	+	−	v	+	−	+(−)+
Gelatin liquefaction	nd	w	−	nd	nd	nd	+	−	−	+	−	+	+	+	−	+	+
β-*N*-Acetyl-glucosaminidase	nd	nd	nd	+	nd	nd	−	+	nd	+	nd	+	−	nd	nd	−	+
α-Fucosidase	nd	nd	nd	+	nd	nd	−	+	−	−	nd	+	−	nd	+	+(−)	+
β-Xylosidase	nd	nd	nd	−	nd	nd	+	−	nd	nd	nd	nd	nd	nd	+	−	nd
β-Glucosidase	nd	nd	nd	−	nd	nd	+	+	+	−	nd	−	−	nd	+	−	+(−)
Glycine aminopeptidase	nd	nd	nd	+	nd	nd	−	−	nd	−	nd	−	+	nd	+	nd	−

Characteristic	*P. marshii*	*P. massiliensis*	*P. melaninogenica*	*P. multiformis*	*P. multisaccharivorax*	*P. nigrescens*	*P. oralis*	*P. oris*	*P. oulora*	*P. pallens*	*P. ruminicola*	*P. salivae*	*P. shahii*	*P. stercorea*	*P. tannerae*	*P. timonensis*	*P. veroralis*	*P. zoogleoformans*
Pigment production on blood agar	−	−	+	−	−	+	−	−	−	+	−	−	+	−	+	−	−	−
Arabinose	−	−	−	−	+	−	−	+	−	−	+/w	+	−	nd	−	−	−	v
Cellobiose	−	−	−	+	+	−	+	+	−	−	+/w	+	−	nd	−	−	+	+
Indole	−	+	−	−	−	+	−	−	−	+	−	−	−	−	−	−	−	−(+)
Lactose	−	−	+	+	+	−	+	+	+	−	+/w	+	+	−	v	+	+	+
Mannose	v	−	+	+	+	+	+	+	+	−	nd	+	+	−	v	−	+	nd
Raffinose	−	−	+	+	+	+	+	+	+	+	nd	+	+	−	v	−	+	nd
Rhamnose	−	−	−	−	+	nd	+	− (+)	−	nd	+	−	−	nd	nd	−	−	−
Salicin	−	−	−	−	+	−	+	+	−	−	+/w	+	−	nd	−	−	−	v
Sucrose	−	nd	+	+	+	+	+	+	−	+	+	+	+	−	v	−	+	+
Xylose	−	−	−	−	+	−	−	+	−	−	±	+	−	nd	−	−	−	v
Esculin hydrolysis	−	−	− v	−	+	−	+	+	−	−	+	+	−	−	−	−	+	+
Gelatin liquefaction	+	−	+	+	+	+	v	v	−	+	nd	−	+	−	+	+	v	nd
β-*N*-Acetyl-glucosaminidase	nd	nd	+	nd	nd	nd	+	+	+	nd	+	nd	nd	nd	nd	+	+	+
α-Fucosidase	nd	nd	+(−)	v	nd	nd	+	+	+	nd	+	nd	nd	+	nd	+	+	nd
β-Xylosidase	nd	nd	−	nd	nd	nd	−	+	−	nd	nd	nd	nd	nd	nd	nd	nd	nd
β-Glucosidase	nd	nd	−	−	nd	nd	+	+	+	nd	nd	nd	nd	−	nd	−	v	nd
Glycine aminopeptidase	nd	nd	nd	−	nd	nd	+	−	−	nd	−	nd	nd	nd	nd	nd	−	−

[a]Symbols: +, Acid produced from sugar or enzyme activity; −, negative; v, variable; w, weak; (−), some strains are negative; nd, no data available.

chain acids; the major fatty acid is 12-methyltetradecanoic acid (C_{15} anteiso). The principal respiratory quinones are unsaturated menaquinones with 10 and 11 isoprene units.

Source: isolated from human gingival crevices and from human clinical specimens.

DNA G+C content (mol%): 36–40 (T_m, Bd).

Type strain: ATCC 25845, CCUG 4944 B, CIP 105346, DSM 7089, JCM 6325, NCTC 12963, B282VPI 4196, VPI 15087.

Sequence accession no. (16S rRNA gene): L16469.

2. **Prevotella albensis** Avguštin, Wallace and Flint 1997, 286[VP]

al.ben'sis. N.L. fem. adj. *albensis* referring to Alba, the ancient name for Scotland north of the Forth and Clyde, where the type strain was isolated.

The species was distinguished from *Prevotella ruminicola* and reported to contain several genotypes. The description is based on those given by Bryant et al. (1958), Bryant and Robinson (1962), Pittman and Bryant (1964), Holdeman et al. (1984), Augustin et al. (1997) and on our own observations.Cells are 0.8–1.0 × 1.0–8.0 µm. They have tapered to rounded ends and are encapsulated. Cells contain inclusion bodies after several days growth. It grows well in a carbohydrate medium producing slime and sediment, with a terminal pH of between 4.6 and 5.7. Most strains grow well in a defined medium containing glucose, CO_2, mineral salts, heme, B vitamins, volatile fatty acids (especially, acetate, 2-methylbutyric or isobutyric acid), methionine or cysteine and ammonia or peptides as a nitrogen source. Surface colonies on blood agar plates are 0.5–1 mm in diameter, circular with an entire edge, low convex, translucent to semi-opaque, white, shiny and smooth.

Carbohydrate fermentation reactions are variable among isolates but most ferment glucose, xylose and salicin, but not melibiose, sucrose, mannose, *N*-acetylglucosamine, raffinose or inulin. Much emphasis has been placed on carboxymethylcellulose activities among species isolated from the rumen; however, this species shows only weak activity.

The major metabolic end products from glucose broths are acetic and succinic acids, with low to trace amounts of formic, propionic, isobutyric and isovaleric acids. The long-chain fatty acid composition is mainly anteiso- and iso-methyl branched acids with small amounts of straight-chain acids; the major fatty acid is 12-methyltetradecanoic acid ($C_{15:0}$ anteiso) and pentadecanoic ($C_{15:0}$). The principal respiratory quinones are menaquinones MK-11 and MK-12 with low levels (~1–5%) of MK-8, MK-9 and MK-13.

Source: the type strain was isolated from sheep rumen.

DNA G+C content (mol%): 39–43 (T_m).

Type strain: M384, CIP 105472, DSM 11370, JCM 12258.

Sequence accession no. (16S rRNA gene): AJ011683.

3. **Prevotella baroniae** Downes, Sutcliffe, Tanner and Wade 2005, 1554[VP]

bar'on.i.ae. N.L. gen. fem. n. *baroniae* of Baron, named in honor of Ellen Jo Baron, the American microbiologist, for her contributions to clinical microbiology.

The description is from Downes et al. (2005). Most cells are coccobacillary, 0.6 × 0.6–2.0 µm. Rod-shaped cells 3–8 µm long may be observed. Colonies on Fastidious Anaerobic Agar after 5 d growth are 1.2–3.8 mm in diameter, circular and high convex; colony morphology can vary from entire to undulate. They are opaque with a shiny gray periphery and an off-white center, which appears matt in some strains. Growth in peptone-yeast extract (PY) broth is good and is stimulated by the addition of 1% (w/v) fermentable carbohydrates. Optimum growth temperature is 37°C.

Cells are saccharolytic and ferment cellobiose, fructose, glucose, lactose, maltose, mannose, melibiose, raffinose, salicin and sucrose. Acid is not produced from arabinose, mannitol, melezitose, rhamnose, ribose, sorbitol or trehalose. Arginine and urea are not hydrolyzed, gelatin is weakly hydrolyzed and esculin is hydrolyzed. Catalase and indole are not produced and nitrate is not reduced.

Acetic and succinic acids are the major end products of fermentation in PYG broth; minor products include isovaleric acid, and minor amounts of isobutyric acid may be produced in PY broth. The non-hydroxylated fatty acid profile consists predominantly of $C_{15:0}$ iso, $C_{15:0}$ anteiso, $C_{16:0}$ and $C_{17:0}$ anteiso.

Source: isolated from the human oral cavity in patients with endodontic and periodontal infections or dentoalveolar abscesses and from the dental plaque of healthy subjects.

DNA G+C content (mol%): 52 (HPLC).

Type strain: E9.33, CCUG 50418, DSM 16972, JCM 13447.

Sequence accession no. (16S rRNA gene): AY840553.

4. **Prevotella bergensis** Downes, Sutcliffe, Hofstad and Wade 2006, 611[VP]

berg.en'sis. N.L. fem. adj. *bergensis* referring to Bergen, the Norwegian city where the first strains were isolated.

The description is from Downes et al. (2006). Most cells are bacilli (0.7–0.8 × 0.8–6 µm), but short rod/cocci are often seen. Colonies on Fastidious Anaerobic Agar after 4 d growth are 0.6–0.8 mm in diameter, circular and entire, convex, opaque, and gray to off-white. Growth in broth media produces moderate turbidity and is stimulated by the addition of fermentable carbohydrates. Cells are saccharolytic and ferment arabinose, cellobiose, fructose, glucose, lactose, maltose, mannose, rhamnose, salicin and xylose. Mannitol and trehalose are fermented variably. Melezitose, melibiose, raffinose, ribose, sorbitol and sucrose are not fermented. Arginine, gelatin and urea are not hydrolyzed. Esculin is hydrolyzed. Catalase and indole are not produced and nitrate is not reduced.

Acetic and succinic acids are the major end products of fermentation in PYG broth; trace products include isovaleric acid. The non-hydroxylated fatty acid profile is predominately $C_{14:0}$ iso, $C_{15:0}$ iso and $C_{15:0}$ anteiso. The menaquinone composition is unknown.

Source: isolated from infections of the skin and soft-tissue abscesses.

DNA G+C content (mol%): 48 (HPLC).

Type strain: 94067913, CCUG 51224, DSM 17361, JCM 13869.

Sequence accession no. (16S rRNA gene): AY350613

5. **Prevotella bivia** (Holdeman and Johnson 1977) Shah and Collins 1990, 206[VP] (*Bacteroides bivius* Holdeman and Johnson 1977, 341[VP])

bi'vi.us L. fem. adj. *bivia* having two ways, pertaining to the saccharolytic and proteolytic activities of the species.

The description is taken from Holdeman et al. (1977), Hammann and Werner (1981) and our own observations. Cells are 0.7–1.0 × 1.3–4.6 μm and occur in pairs or short chains. On blood agar, colonies are 0.5–2.0 mm, circular, entire to slightly erose, convex, translucent to semi-opaque, smooth and glistening. Some strains show orange to pink fluorescence under longwave ultraviolet radiation (365 nm). Isolates typically ferment dextrin, glucose, lactose, maltose and mannose, but not amygdalin, cellobiose, inositol, inulin, mannitol, melezitose, melibiose, raffinose, rhamnose, ribose, salicin, sucrose, trehalose or xylose. Starch is hydrolyzed but not esculin.

Prevotella bivia produces acetate and succinate as major metabolic ends from a glucose broth. The cell wall contains *meso*-diaminopimelic acid. The long-chain fatty acid composition is mainly anteiso- and iso-methyl branched acids with lower levels of straight-chain acids. The major fatty acid is 12-methyltetradecanoic acid ($C_{15:0}$ anteiso). The principal respiratory quinones are unsaturated menaquinones with 10 and 11 isoprene units.

Source: usually isolated from infections in the urogenital or abdominal region, but now generally considered as part of the normal flora of the vagina.

DNA G+C content (mol%): 40 (T_m).

Type strain: ATCC 29303, CCUG 9557, CIP 105105, DSM 20514, JCM 6331, LMG 6452, NCTC 11156.

Sequence accession no. (16S rRNA gene): L16475.

6. **Prevotella brevis** (Bryant, Small, Bouma and Chu 1958) Avguštin, Wallace and Flint 1997, 286[VP] (*Bacteroides ruminicola* subsp. *brevis* Bryant, Small, Bouma and Chu 1958, 18[AL])

bre'vis. L. fem. adj. *brevis* short.

The morphological and biochemical characteristics of this species are based on those of Bryant et al. (1958) and Holdeman et al. (1984) and on chemotaxonomic characters reported by Shah and Collins (1980, 1983). The species was originally part of a heterogeneous collection of strains that were assigned to *Bacteroides ruminicola* as *Bacteroides ruminicola* subspecies *brevis* (biovars 1 and 2). Even though DNA–DNA hybridization values (Shah et al., 1982) [see Holdeman et al. (1984)] between the various subspecies of *Bacteroides ruminicola* indicated they were distinct species, their taxonomic status remained unchanged because of the lack of reliable phenotypic characters for delineating each subspecies. They were reassigned to the genus *Prevotella* (Shah and Collins, 1990) and elevated to species rank (along with *Prevotella ruminicola* subsp. *ruminicola* and *Prevotella ruminicola* subsp. *brevis* (biovar 3) now *Prevotella bryantii*) by Avguštin et al. (1997). Clear phenotypic circumscription of these closely related species are still lacking, but it is evident that they represent distinct lineages on the basis of 16S rRNA sequences. However, many of the ill-defined taxa remain *incertae sedis*.

Cells are 0.8–1.0 × 1.0–5.0 μm and are often coccoid to oval. Surface colonies on blood agar plates are 0.5–1 mm in diameter, circular with an entire edge, low convex, translucent to semi-opaque, white, shiny, and smooth. It grows well in a carbohydrate medium producing slime and sediment with a terminal pH of between 4.6 and 5.7. Most strains grow well in a defined medium containing glucose, CO_2, mineral salts, heme, B vitamins, volatile fatty acids (especially, acetate, 2-methylbutyric or isobutyric acid), methionine or cysteine and ammonia or peptides as a nitrogen source.

Carbohydrate fermentation reactions differ among isolates. Most isolates ferment arabinose, cellobiose, dextrin, fructose, glucose, inulin, lactose, maltose, mannose, melibiose, raffinose and sucrose, but not inositol, mannitol, rhamnose, ribose, trehalose or xylose. Strains do not exhibit carboxymethylcellulose activity. The major metabolic end products from glucose broths are acetic and succinic acids, with low to trace amounts of propionic, iso-butyric, formic and isovaleric acids.

The long-chain fatty acid composition is mainly anteiso- and iso-methyl branched acids with small amounts of straight-chain acids; the major fatty acid is 12-methyltetradecanoic acid ($C_{15:0}$ anteiso) and pentadecanoic ($C_{15:0}$). The principal respiratory quinones are menaquinones MK-13 and MK-12 with low levels (~1–5%) of MK-11, MK-10 and MK-9.

Source: the type strain was isolated from bovine rumen.

DNA G+C content (mol%): 45–52 (T_m).

Type strain: GA33, ATCC 19188, CIP 105473).

Sequence accession no. (16S rRNA gene): AJ011682.

7. **Prevotella bryantii** Avguštin, Wallace and Flint 1997, 287[VP]

bry.an'ti.i. N.L. gen. masc. n. *bryantii* of Bryant, named after Marvin Bryant, an American microbiologist.

The morphological and biochemical characteristics of this species are based on that of Bryant et al. (1958) and Holdeman et al. (1984), and on chemotaxonomic characters reported by Shah and Collins (1980, 1983). The species was formerly *Bacteroides ruminicola* subspecies *brevis* (biovar 3) (see above, *Prevotella brevis*) that was reassigned to the genus *Prevotella* (Shah and Collins, 1990) and elevated to species rank as *Prevotella bryantii* by Avguštin et al. (1997).

Cells are 0.8–1.0 × 1.0–5.0 μm and are often coccoid to oval. Surface colonies on blood agar plates are 0.5–1 mm in diameter, circular with an entire edge, low convex, translucent to semi-opaque, white, shiny, and smooth. It grows well in a carbohydrate medium producing slime and sediment, with a terminal pH of between 4.6 and 5.7. Most strains grow well in a defined medium containing glucose, CO_2, mineral salts, heme, B vitamins, volatile fatty acids (especially, acetate, 2-methylbutyric or isobutyric acid), methionine or cysteine and ammonia or peptides as a nitrogen source. Carbohydrate fermentation reactions differ among isolates. Most isolates ferment arabinose, dextrin, fructose, glucose, inulin, lactose, maltose, mannose, melibiose, raffinose, sucrose, xylose and xylan, but not *N*-acetylglucosamine, inositol, mannitol, rhamnose, ribose or trehalose. Strains exhibit carboxymethylcellulose activity.

The major metabolic end products from glucose broths are acetic and succinic acids with low to trace amounts of formic, propionic, isobutyric and isovaleric acids. The long-chain fatty acid composition is mainly anteiso- and iso-methyl branched acids with small amounts of straight-chain acids; the major fatty acid is 12-methyltetradecanoic acid ($C_{15:0}$ anteiso) and pentadecanoic ($C_{15:0}$). The principal respiratory quinones are menaquinones MK-13 and MK-12 with low levels (~1–5%) of MK-11, MK-10 and MK-9.

Source: isolated from the reticulo-rumen of cattle, sheep and elk; it is the predominant species of the rumen of most ruminants. Phenotypically similar strains have been reported from man, but these have been assigned to either *Prevotella oris* or *Prevotella buccae.*

DNA G+C content (mol%): 40–43 (T_m).

Type strain: B14, CIP 105474, DSM 11371.

Sequence accession no. (16S rRNA gene): AJ006457.

8. **Prevotella buccae** (Holdeman, Moore, Churn and Johnson 1982) Shah and Collins 1990, 207[VP] (*Bacteroides buccae* Holdeman, Moore, Churn and Johnson 1982, 128[VP]; *Bacteroides capillus* Kornman and Holt, 1982, 266 *Bacteroides pentosaceus* Shah and Collins 1982, 266)

buc'cae. L. gen. n. *buccae* of the mouth, referring to a major natural habitat of the species.

The species was reported by Holdeman et al. (1982) as *Bacteroides buccae* because of its site of isolation. It was named *Bacteroides capillus* by Kornman and Holt (1981) because of its pronounced capsule, and it was also named *Bacteroides pentosaceus* by Shah and Collins (1981) because of its characteristic capacity to ferment pentoses. However, the names *Bacteroides capillus* and *Bacteroides pentosaceus* are regarded as later heterotypic synonyms of *Bacteroides buccae* (Johnson and Holdeman, 1985). The species was later reclassified as *Prevotella buccae* by Shah and Collins (1990). It was delineated from among the *Bacteroides oralis-Bacteroides ruminicola* complex using long-chain cellular fatty acid and menaquinone profiles (Shah and Collins, 1980), DNA–DNA hybridization (Shah et al., 1982) and multilocus enzyme electrophoresis (Shah and Williams, 1982). The description is based on these papers and Johnson and Holdeman (1985).

Cells occur in pairs or short chains, are rod-shaped and are 0.5–0.6 × 0.8–6.5 µm in PYG or BM broth cultures. Cells on brain heart infusion agar (BHIA) roll tubes (Holdeman et al., 1980) produce subsurface colonies 1–2 mm in diameter, lenticular and translucent. Colonies grown on blood agar plates are 0.5–3 mm in diameter, circular with an entire edge, low convex, translucent to semi-opaque, white or buff, shiny, smooth and nonhemolytic on rabbit blood. There is no action on rabbit blood or egg yolk agar (Holdeman et al., 1977). Most cells do not grow in PYG containing 10% bile; cells that do grow are markedly inhibited and delayed. Growth is stimulated by heme and for some strains this is an absolute requirement for cytochrome biosynthesis. Pre-reduced broth cultures after incubation at 37°C for 24 h are uniformly turbid with a smooth sediment; the terminal pH after 3–5 d is 4.5–5.0 in broths with carbohydrates or 5.1–5.6 in the absence of sugars.

Cells typically ferment the pentose sugars arabinose and xylose, in addition to cellobiose, dextrin, fructose, glucose, glycogen, lactose, maltose, mannose, melibiose, raffinose, salicin, starch and sucrose. Acid is not produced from melezitose, trehalose, inositol or mannitol. Nitrate is not reduced; indole and catalase are not produced. Gelatin is digested by most strains.

Succinic and acetic acids are the major fermentation end products from glucose broth, with minor amounts of formic, isobutyric and isovaleric acids; no H_2 is produced. The long-chain fatty acid composition is mainly anteiso- and iso-methyl branched acids, with lower levels of straight-chain acids. The major fatty acid is 12-methyltetradecanoic acid $C_{15:0}$ anteiso). The principal respiratory quinones are unsaturated menaquinones with 13, 12 and 11 isoprene units.

Source: type strain was isolated from gingival crevice with moderate periodontitis. The normal habitat appears to be the oral cavity.

DNA G+C content (mol%): 52 (T_m, Bd).

Type strain: ATCC 33574, CCUG 15401, CIP 105106, JCM 12245, VPI D3A-6.

Sequence accession no. (16S rRNA gene): L16477.

9. **Prevotella buccalis** (Shah and Collins 1982) Shah and Collins 1990, 207[VP] (*Bacteroides buccalis* Shah and Collins 1982, 266[VP])

buc.ca'lis. L. n. *bucca* the mouth; L. fem. suff. *-alis* suffix denoting pertaining to; N.L. fem. adj. *buccalis* buccal, pertaining to the mouth.

The description is from Shah et al. (1981) and was distinguished from the *Bacteroides oralis-Bacteroides ruminicola* complex by using long-chain cellular fatty acid and menaquinone profiles (Shah and Collins, 1980, 1983), DNA–DNA hybridization (Shah et al., 1982) and multilocus enzyme electrophoresis (Shah and Williams, 1982).

Cells are short rods on agar; cells grown in broth culture are elongated (0.5–0.8 × 0.9–6.0 µm) and pleomorphic. Colonies grown on agar after 3 d are 0.2–0.5 mm in diameter, circular and entire, convex, shiny, smooth and translucent cream to light gray; they are non-hemolytic on horse blood agar. The organisms ferment cellobiose, fructose, glucose, lactose, maltose, mannose, raffinose, salicin and sucrose, but not arabinose, glycerol, mannitol, melezitose, rhamnose, sorbitol, trehalose or xylose. Starch is not hydrolyzed; indole is not produced; gelatin is not digested; esculin is hydrolyzed.

Acetic and succinic acids are the major metabolic end products in glucose broth, with minor amounts of lactic acid; terminal pH is 5.5. The long-chain fatty acid composition is mainly anteiso- and iso-methyl branched acids with small amounts of straight-chain acids; the major fatty acid is 12-methyltetradecanoic acid ($C_{15:0}$ anteiso). The principal respiratory quinones are unsaturated menaquinones with 11, 12 and 13 isoprene units.

Source: isolated initially from human dental plaque, but has since been reported from other sites.

DNA G+C content (mol%): 45–46 (T_m, Bd).

Type strain: ATCC 35310, CCUG 15557, DSM 20616, JCM 12246, NCDO 2354, NCTC 13064.

Sequence accession no. (16S rRNA gene): L16476.

10. **Prevotella copri** Hayashi, Shibata, Sakamoto, Tomita and Benno 2007, 943[VP]

cop'ri. N.L. gen. n. *copri* from Gr. gen. n. *kopron* of/from feces.

The description is from Hayashi et al. (2007). Unlike most *Prevotella* species, which are of oral origin, these isolates are from human feces, but unusually, growth is inhibited on Bacteroides bile esculin agar.

The cell dimensions have not been reported. Colonies on Eggerth–Gagnon blood agar after 48 h incubation at 37°C under 100% CO_2 are white, circular and convex. Optimum temperature for growth is 37°C. Esculin is hydrolyzed. Indole is not produced. No activity is detected for urease. Gelatin is not hydrolyzed. Acid is produced from glucose, lactose, sucrose, maltose, raffinose, salicin, xylose, arabinose, cellobiose and rhamnose. Positive reactions are obtained with the API ZYM system for alkaline phosphatase, acid phosphatase, naphthol-AS-BI-phosphohydrolase, α-galactosidase, β-galactosidase, α-glucosidase and β-glucosidase; negative reactions are obtained for lipase (C4), leucine arylamidase, valine arylamidase, cystine arylamidase, trypsin, β-gluconidase, *N*-acetyl-β-D-glucosaminidase, α-mannosidase and α-fucosidase. Positive reactions are obtained with the An-Ident system for a glucosidase, α-arabinofuranosidase, β-gluconidase, alkaline phosphatase, α-galactosidase, indoxylacetate hydrolase and arginine and alanine aminopeptidases; negative reactions are obtained for *N*-acetyl-β-D-glucosaminidase, α-L-fucosidase, β-galactosidase and pyroglutamic acid arylamidase, and for leucine, proline, tyrosine, arginine, histidine, phenylalanine and glycine aminopeptidases. Catalase is negative.

The major metabolic end products are succinic and acetic acids. The predominant long-chain fatty acids are $C_{16:0}$, $C_{18:1}$ ω9*c* and $C_{15:0}$ anteiso. The principal respiratory quinones are menaquinones are MK-12 and MK-11 with lower levels of MK-8 to -10 and MK-13 (<20%).

Source: The source is human feces. The habitat is the intestinal tract.

DNA G+C content (mol%): 44–45 (HPLC).

Type strain: CB7, JCM 13464, DSM 18205.

Sequence accession no. (16S rRNA gene): AB064923.

11. **Prevotella corporis** (Johnson and Holdeman 1983) Shah and Collins 1990, 207[VP] (*Bacteroides corporis* Johnson and Holdeman 1983, 19[VP])

cor'po.ris. L. gen. n. *corporis* of the body; pertaining to the isolation of this organism from human clinical specimens.

The description is from Johnson and Holdeman (1983) and our own observations. Cells occur singly, in pairs and in short chains and are 0.9–1.6 × 1.6–4.0 μm long in PYG broth cultures. Growth and fermentation are stimulated by addition of 10% (v/v) serum; hemin and vitamin K are required for growth. Cells mainly appear coccoid, but occasionally cells up to 11 μm long can be observed. At 48–72 h colonies on anaerobic blood agar are minute to 1 mm in diameter, circular and entire, convex and buff with brown edges; dark brown colonies develop after 4–7 d incubation. Glucose broth cultures are turbid with smooth or ropey sediment that tends to adhere to the bottom of the tube; the terminal pH value is 4.8–5.1.

Cells ferment dextrin and glucose; the majority of strains ferment fructose (except the type strain), glycogen, maltose, mannose and starch. Cells do not ferment amygdalin, arabinose, cellobiose, erythritol, esculin, inositol, lactose, mannitol, melezitose, melibiose, rhamnose, ribose, salicin, sorbitol or xylose. Indole and catalase are not produced. Starch is hydrolyzed. Nitrate is not reduced. Gelatin is digested.

Succinic and acetic acids are the major fermentation end products from PYG broth cultures with vitamin K and hemin; minor acids produced may include isobutyric and isovaleric acids; no H_2 is produced. Neither lactate nor pyruvate is used; threonine is not converted to propionate. The long-chain fatty acid composition is mainly anteiso- and iso-methyl branched acids with small amounts of straight-chain acids; the major fatty acid is 12-methyltetradecanoic acid ($C_{15:0}$ anteiso). The principal respiratory quinones are unsaturated menaquinones with 11 and 12 isoprene units.

Source: the type strain was isolated from a cervical swab.

DNA G+C content (mol%): 43–46 (T_m).

Type strain: Lambe 532-70A, ATCC 33547, CIP 105107, JCM 8529, NCTC 13065, VPI 9342.

Sequence accession no. (16S rRNA gene): L16465.

12. **Prevotella dentalis** (Haapasalo, Ranta, Shah, Ranta, Lounatmaa and Kroppenstedt 1986a) Willems and Collins 1995, 834[VP] (*Mitsuokella dentalis* Haapasalo, Ranta, Shah, Ranta, Lounatmaa and Kroppenstedt 1986a, 566[VP])

den.ta'lis. L. n. *dens dentis* a tooth; L. fem. suff. *-alis* suffix denoting pertaining to; L. fem. adj. *dentalis* pertaining to the teeth.

The species was originally placed in the genus *Mitsuokella* because of its high mol% G+C content (~ 56–60) (Haapasalo et al., 1986a), which is characteristic of the genus *Mitsuokella.* However, by other criteria, this taxon differed so significantly from *Mitsuokella* that it was reclassified as *Prevotella dentalis* by Willems et al. (Willems and Collins, 1995). This species may be synonymous with *Hallella seregens* (Moore and Moore, 1994).

The description is from Haapasalo et al. (1986a, b) and subsequent observations in our laboratory. Cells occur singly and are blunt-ended oval rods 0.7 × 1–2 μm. Peritrichous fimbriae and a thick capsule-like structure are present. Poor growth occurs in liquid media, growth is enhanced on blood agar media supplemented with hemolyzed blood. No growth occurs on kanamycin-vancomycin laked blood agar. Colonies grown on enriched horse blood agar after 3 d are 1–2 mm in diameter, convex, irregular, translucent, wet and mucoid, with a characteristic water drop appearance. Alpha-hemolysis usually occurs on horse and sheep blood agar after 7 d. Using pre-reduced anaerobically sterilized PY broth as a base, cells ferment arabinose, cellobiose, fructose, galactose, glucose, lactose, maltose, mannose and raffinose; melibiose and sucrose are weakly fermented (pH 5.5–5.7). Cells do not ferment erythritol, mannitol, melezitose, rhamnose, salicin or xylose; Starch and gelatin are not hydrolyzed. Esculin is not hydrolyzed except when a chromogenic substrate for the detection of a constitutive enzyme is used with cells grown on agar plates.

The major metabolic end products are succinate and acetate. Hydroxylated and nonhydroxylated long-chain fatty acid methyl esters are present in whole-cell methanol lysates. The major fatty acids are 3-hydroxyhexadecanoic acid ($C_{16:0}$ 3-OH), hexadecanoic acid ($C_{16:0}$) and 12-methyl tridecanoic acid ($C_{14:0}$ iso).

Source: isolated from human dental root canals.

DNA G+C content (mol%): 56–60 (T_m, Bd).

Type strain: ES2772, ATCC 49559, CCUG 48288, DSM 3688, JCM 13448, NCTC 12043.

Sequence accession no. (16S rRNA gene): X81876.

13. **Prevotella denticola** (Shah and Collins 1982) Shah and Collins 1990, 207[VP] emend. Wu, Johnson and Moore 1992, 536[VP] (*Bacteroides denticola* Shah and Collins 1982, 266[VP] emend. Holdeman and Johnson 1982, 404)

den.ti'co.la. L. n. *dens, dentis* tooth; L. suff. *-cola* (from L. n. *incola*) inhabitant, dweller; N.L. n. *denticola* tooth dweller.

The description is from Shah and Collins (1981) and Holdeman et al. (1984). The species was distinguished from the *Bacteroides oralis-Bacteroides ruminicola* complex by using long-chain cellular fatty acid and menaquinone profiles (Shah and Collins, 1980, 1983), DNA–DNA hybridization (Shah and Collins, 1982) and multilocus enzyme electrophoresis (Shah and Williams, 1982).

Cells are rods and occur in pairs and short chains 0.5–0.7 × 0.7–6.0 μm in glucose broth cultures. Colonies grown on blood agar are 1–2 mm in diameter, circular and entire, low convex, semi-opaque, shiny, smooth and often appear to have white or buff concentric rings. Colonies grown on rabbit blood agar develop pigment more quickly and are darker. Most cells grown on hemolyzed rabbit blood agar after 7 d produce dark brown or black colonies; occasional strains do not develop colonies with a definite pigment. Most colonies show red fluorescence under long wavelength ultraviolet radiation (365 nm). Glucose broth cultures are turbid, have smooth sediment and have a pH of 4.5–4.9. Growth is stimulated by the addition of hemin; vitamin K_1 is not often required. There is no growth in PYG broth with 20% bile.

Cells ferment dextrin, fructose, glucose, glycogen, inulin, lactose, maltose, mannose, raffinose, sucrose and starch. Cells do not ferment amygdalin, arabinose, cellobiose, erythritol, gum arabic, inositol, larch arabinogalactan, mannitol, melibiose, melezitose, rhamnose, salicin, sorbitol, trehalose, xylan or xylose. Esculin and starch are hydrolyzed. Gelatin digestion is variable. Nitrate is not reduced. Catalase, indole and lecithinase are not produced.

Succinate and acetate are the major fermentation end products from glucose broth cultures (1% glucose and peptone, 0.5% yeast extract, vitamin K_1 and hemin); minor production of lactate, isobutyrate and isovalerate may be detected; no H_2 is produced. The long-chain fatty acid composition is mainly anteiso- and iso-methyl branched acids with lower levels of straight-chain acids. The major fatty acid is 12-methyltetradecanoic acid ($C_{15:0}$ anteiso). The principal respiratory quinones are unsaturated menaquinones with 11 and 12 isoprene units.

Source: the normal habitat is the oral cavity but the organisms may be recovered from human clinical specimens.

DNA G+C content (mol%): 49–51 (T_m, Bd).

Type strain: Socransky 1210, Shah and Collins 1210, ATCC 35308, CCUG 29542, CIP 104478, DSM 20614, JCM 13449, NCDO 2352, NCTC 13067.

Sequence accession no. (16S rRNA gene): L16467.

14. **Prevotella disiens** (Holdeman and Johnson 1977) Shah and Collins 1990, 207[VP] (*Bacteroides disiens* Holdeman and Johnson 1977, 337[VP])

di'si.ens. N.L. part. adj. *disiens* (*sic*) going in two different directions; intended to refer to the fact that the organism is both saccharolytic and proteolytic.

The description is from Holdeman and Johnson (1977) and our own observations. Cells are non-pigmented rods and occur predominantly in pairs or occasionally short chains with longer rods; the longer rods occur both singly and in chains with the short rods. Cells are 0.6–0.9 × 2.0–8.2 μm. Colonies grown on Brucella-laked blood agar plates after 2 d are minute to 2 mm in diameter, circular and entire, convex, translucent to opaque, smooth, shiny and white. There is no hemolysis on blood agar plates; slight greening maybe observed in areas of confluent growth. No brown or black pigmentation is observed on BHI-laked blood agar; some strains show light orange to pink fluorescence on blood agar plates. Cultures in pre-reduced PYG broth are moderately turbid; slight smooth sediment may be observed. Growth is enhanced in broth with fermentable carbohydrate and is turbid and smooth, with a granular or flocculent sediment; growth occurs well in BM broth. Optimal growth occurs at 37°C, but growth can occur at 25 and 45°C. The terminal pH after 1 d is 4.9–5.2 with fermentable carbohydrates and 5.9–6.1 without fermentable carbohydrates. Growth is inhibited in 6.5% NaCl broth.

Strains ferment dextrin, fructose, glucose, glycogen, maltose and starch, but not amygdalin, arabinose, cellobiose, erythritol, esculin, galactose, inositol, lactose, mannitol, mannose, melezitose, melibiose, raffinose, rhamnose, salicin, sorbitol, sucrose, trehalose or xylose. Gelatin and starch are hydrolysed. Nitrate is not reduced. Indole and catalase are not produced. Esculin is not hydrolysed.

Succinic and acetic acids are the major fermentation end products in PYG broth, with minor levels of isovaleric, isobutyric, formic, propionic and lactic acids. The long-chain fatty acid composition is mainly anteiso- and iso-methyl branched acids with lower levels of straight-chain acids. The major fatty acid is 12-methyltetradecanoic acid ($C_{15:0}$ anteiso). The principal respiratory quinones are unsaturated menaquinones with 10 and 11 isoprene units.

Source: recovered from a wide range of human clinical specimens.

DNA G+C content (mol%): 40–42 (T_m).

Type strain: ATCC 29426, CCUG 9558, CIP 105108, DSM 20516, JCM 6334, LMG 6453, NCTC 11157, VPI 8057.

Sequence accession no. (16S rRNA gene): L16483.

15. **Prevotella enoeca** Moore, Johnson and Moore 1994, 601[VP]

e.noe'ca. Gr. n. *enoikos* inhabitant, dweller in a place; N.L. fem. adj. *enoeca* inhabiting, because the organism is an inhabitant of the gingival crevice.

The description is from Moore et al. (1994). Cells occur in pairs and short chains and are 0.5 × 2.2–4.5 μm; filaments up to 8 μm long are observed. Colonies grown on rabbit blood agar for 2 d are 1–2 mm in diameter, circular and entire, convex, transparent to translucent and not hemolytic or pigmented. Broth cultures are turbid with a smooth sediment. Terminal pH values of glucose broth cultures incubated for 5 d are 4.8–5.4. Cells ferment fructose, glycogen, lactose, maltose and mannose, but not cellobiose, raffinose, salicin, sucrose, trehalose or xylose. Esculin in not hydrolyzed by the majority of strains. Gelatin may be liquefied.

Succinic and acetic acids are the major end products of fermentation; trace products include formic acid. No H_2 is produced. The major cellular fatty acids are $C_{15:0}$ anteiso, $C_{16:0}$, $C_{16:0}$ 3-OH and $C_{15:0}$ iso with low levels of $C_{17:0}$ iso 3-OH and $C_{18:2}$ dimethyl acetyl, $C_{14:0}$, $C_{14:0}$ iso and $C_{17:0}$ iso.

Source: isolated from the gingival crevices of humans with healthy gingiva or with periodontitis.

DNA G+C content (mol%): 47 (T_m).

Type strain: ATCC 51261, CIP 104472, JCM 12259, NCTC 13068, VPI D194A-25A.

Sequence accession no. (16S rRNA gene): AJ005635.

16. **Prevotella heparinolyticus** (Okuda, Kato, Shiozu, Takazoe and Nakamura 1985) Shah and Collins 1990, 207[VP] (*Bacteroides heparinolyticus* Okuda, Kato, Shiozu, Takazoe and Nakamura 1985, 439[VP])

he.pa'ri.no.ly.ti.cus. N.L. n. *heparinum* (from Gr. n. *hêpar* liver) heparin; N.L. adj. *lyticus -a -um* (from Gr. adj. *lutikos -ê -on*) able to loosen, able to dissolve; N.L. fem. adj. *heparinolytica* heparin dissolving.

The description is from Okuda et al. (1985). Colonies grown on blood agar plates are 0.5–0.7 × 0.8–2.0 μm; occasional cells are up to 10 μm long or longer in broth. The cytoplasmic and outer membranes exhibit a trilamellar structure; the outer wall is separated from the outer membrane by an electron-lucent layer; a distinctive capsular structure surrounding the cell is detected by phase-contrast microscopy and India ink negative staining. Colonies on blood agar after 4 d are circular, entire, convex, translucent, smooth and gray. No dark brown or black colonies are produced on hemolyzed rabbit blood agar; sheep blood is not hemolyzed. Broth cultures are turbid and have a dense sediment; harvested cells are markedly viscous. The terminal pH after 5 d in 1% glucose broth is 5.4–5.6. Strains ferment cellobiose, esculin, fructose, glucose, lactose, maltose, mannose, sucrose, salicin, starch and xylose, but not raffinose, mannitol or sorbitol. Nitrate is not reduced. Indole, H_2S, catalase and gelatinase are not produced. Starch and esculin are hydrolyzed. The reliable property that distinguishes this taxon from other non-pigmented *Prevotella* species is its heparin-degrading activity. In addition to heparinase, alkaline phosphatase, esterase, lipase, phosphatase, trypsin, chemotrypsin, glucosidases and galactosidases are produced.

Succinic acid is the major end product of fermentation in PYG broth; low levels of acetic, propionic and isovaleric acids are also produced.

Source: isolated from human periodontal lesions.

DNA G+C content (mol%): 47–49 (Bd).

Type strain: HEP, ATCC 35895, CCUG 27827, LMG 10142.

Sequence accession no. (16S rRNA gene): L16487.

17. **Prevotella intermedia** (Holdeman and Moore 1970) Shah and Collins 1990, 207[VP] [*Bacteroides intermedius* (Holdeman and Moore 1970) Johnson and Holdeman 1983, 18[VP]; *Bacteroides melaninogenicus* subspecies *intermedius* Holdeman and Moore 1970, 33]

in.ter.me'di.us. L. fem. adj. *intermedia* intermediate.

The description is from Johnson and Holdeman (1983), Miyagawa et al. (1978, 1979, 1981), Miyagawa and Suto (1980), Shah et al. (1976, 1979), Shah and Collins (1980, 1983, 1990) and Shah and Williams (1982). Cells are 0.4–0.7 × 1.5–2.0 μm; occasional cells are up to 12 μm long. Colonies on rabbit blood agar after 2 d are 0.5–2.0 mm in diameter, circular and entire, low convex, translucent, smooth and beta-hemolytic; older or larger colonies may be opaque; colonies are tan, gray, reddish brown or black. Cells grown on hemolyzed blood agar develop pigment more rapidly (~1–2 d) and are darker. Glucose broth cultures are turbid with a smooth (occasionally ropey or slightly mucoid) sediment with a final pH of 4.9–5.4. Hemin is required for growth, while vitamin K_1 is highly stimulatory or required for fermentation. Growth is inhibited by 6.5% NaCl. Strains grow well between 25 and 45°C. Acid is produced by the majority of strains from fructose, glycogen, inulin, maltose, mannose, raffinose, starch and sucrose, but not from cellobiose, esculin, gum arabic, lactose, ribose, trehalose or xylan. Indole is produced, Starch is hydrolyzed. No H_2 is produced. Neither lactate nor pyruvate is used. Threonine is not converted to propionate.

Succinic and acetic acids are the major end products of fermentation in cultures grown in PYG broth with vitamin K_1 and hemin; minor products include isobutyric and isovaleric acids. No heptose or 2-keto-3-deoxyoctulosonic acid has been detected. The long-chain fatty acid composition is mainly anteiso- and iso-methyl branched acids with small amounts of straight-chain acids; the major fatty acid is 12-methyltetradecanoic acid ($C_{15:0}$ anteiso). The principal respiratory quinones are unsaturated menaquinones with 11 isoprene units with lower levels of MK-10 and MK-12.

Source: isolated initially from human dental plaque but has since been reported from head, neck and pleural infections, from blood, and from abdominal and pelvic sites.

DNA G+C content (mol%): 41–44 (T_m, Bd).

Type strain: ATCC 25611, NCTC 13070, Finegold B422, CCUG 24041, CIP 101222, CIP 103682, DSM 20706, JCM 12248, JCM 11150, VPI 4197.

Sequence accession no. (16S rRNA gene): L16468.

18. **Prevotella loescheii** (Holdeman and Johnson 1982) Shah and Collins 1990, 207[VP] (*Bacteroides loescheii* Holdeman and Johnson 1982, 406[VP])

loesche'i.i. N.L. gen. masc. n. *loescheii* of Loesche, named after Walter J. Loesche, an American oral microbiologist.

The description is from Holdeman and Johnson (1982) Miyagawa et al. (1978, 1979, 1981), Miyagawa and Suto (Miyagawa and Suto, 1980), Shah et al. (1979, 1976), Shah and Collins, (1990, 1980, 1983) and Shah and Williams (1982). Cells are rods 0.4–0.6 × 0.8–15.0 μm and occur singly, in pairs and in short chains in glucose broth cultures. Colonies grown on anaerobic blood agar are 1–2 mm in diameter, circular and entire, low convex, translucent, shiny and smooth. Colonies grown on whole blood agar after 48 h are white or buff; up to 14 d, colonies will usually become light brown. Colonies grown on hemolyzed rabbit blood agar up to 21 d develop pigment more rapidly, although definite black or brown colonies will not develop with some strains; most strains will produce a slight beta-hemolysis on rabbit blood agar. Among the pigmented species, colonies of *Prevotella loescheii* strains exhibit the most pronounced fluorescence under ultraviolet radiation (365 nm), due to overproduction of protoporphyrin (Shah et al., 1979). Glucose broth cultures are turbid and have a smooth sediment and a pH of 4.9–5.4. Growth is stimulated by the addition of hemin and 10% serum. Cells ferment cellobiose, dextrin, fructose, glucose, glycogen, inulin (majority), lactose, maltose, mannose, raffinose, sucrose and starch, but not amygdalin, arabinose, erythritol, inositol, mannitol, melezitose, rhamnose, ribose, salicin, sorbitol, trehalose or xylose. Esculin and starch are hydrolyzed and gelatin is digested. Nitrate is not reduced. Catalase, indole and lecithinase are not produced.

Succinate and acetate are the major fermentation end products from glucose broth culture (1% peptone, 1% yeast extract, vitamin K_1 and hemin); trace amounts of lactate and formate may be detected; no H_2 is produced. No heptose or 2-keto-3-deoxyoctulosonic acid has been reported. The long-chain fatty acid composition is mainly anteiso- and iso-methyl branched acids with small amounts of straight-chain acids; the major fatty acid is 12-methyltetradecanoic acid ($C_{15:0}$ anteiso). The principal respiratory quinones are unsaturated menaquinones with 11 isoprene units with lower levels of MK-10 and MK-12.

Source: the normal habitat appears to be the oral cavity, but isolates have been reported from clinical specimens.

DNA G+C content (mol%): 46–48 (T_m).

Type strain: Loesche 8B, ATCC 15930, NCTC 11321, CCUG 5914, JCM 8530, JCM 12249, VPI 9085.

Sequence accession no. (16S rRNA gene): L16481.

19. **Prevotella marshii** Downes, Sutcliffe, Tanner and Wade 2005, 1554[VP]

mar'shi.i. N.L. gen. masc. n. *marshii* of Marsh, named in honor of British microbiologist Philip Marsh, for his contributions to oral microbiology.

The description is from Downes et al. (2005). Cells are short rods (~0.4 × 0.9–3 μm), but occasionally elongated cells are seen that are up to 6 μm long, occurring singly or in pairs. Colonies on Fastidious Anaerobic Agar after 5 d are 1.8–3.5 mm in diameter, circular and entire, convex, watery, opaque and gray to off-white/gray to greenish gray; heavier growth appears to be cream in color. Under plate microscopy, colonies have a striated appearance with concentric rings of varying opacity and iridescent hues of pink, green and yellow. Broth media cultures are moderately turbid; growth is stimulated by the addition of glucose, fructose or maltose. Cells are moderately saccharolytic and ferment glucose and maltose strongly, fructose and mannose are fermented variably and weakly; arabinose, cellobiose, lactose, mannitol, melezitose, melibiose, raffinose, rhamnose, ribose, salicin, sorbitol, sucrose and trehalose are not fermented. Esculin, arginine and urea are not hydrolyzed; gelatin is hydrolyzed. Catalase and indole are not produced. Nitrate is not reduced.

Acetic and succinic acids are the major end products of metabolism in PYG broth; propionic acid is moderately produced. The long-chain fatty acid profile predominantly comprises $C_{15:0}$ anteiso, $C_{15:0}$ iso, $C_{15:0}$ and $C_{14:0}$ iso.

Source: isolated from the human oral cavity in subjects with endodontic and periodontal infections and from subgingival dental plaque in healthy subjects.

DNA G+C content (mol%): 51 (HPLC).

Type strain: E9.34, CCUG 50419, DSM 16973, JCM 13450.

Sequence accession no. (16S rRNA gene): AF481227.

20. **Prevotella multiformis** Sakamoto, Huang, Umeda, Ishikawa and Benno 2005a, 818[VP]

mul.ti.for'mis. L. fem. adj. *multiformis* many-shaped, multiform.

The description is from Sakamoto et al. (2005a). Cells are short rods (~0.5–0.8 × 1.6–6.6 μm) or cocci (coccobacilli) (0.8 × 0.9–1.0 μm) and occur singly. Colonies on Eggerth–Gagnon blood agar after 2 d are 1–2 mm in diameter, circular and entire, slightly convex, gray to light brown and smooth. Cells forming small colonies are cocci and rods, whereas large colonies consist only of cocci. There is no growth in 20% bile. Acid is produced from cellobiose, glucose, glycerol, lactose, maltose, mannose, raffinose and sucrose, but not arabinose, mannitol, melezitose, rhamnose, salicin, sorbitol, trehalose or xylose. Esculin is not hydrolyzed. Gelatin is digested. Catalase, indole and urease are not produced.

Acetic and succinic acids are the major end products of fermentation from glucose broth (composed of peptone, 1%; yeast extract, 1%; and glucose, 1%); minor products include isovaleric acid. Both non-hydroxylated and 3-hydroxylated long-chain fatty acids are present. The major cellular fatty acids are $C_{15:0}$ anteiso, $C_{15:0}$ iso, $C_{17:0}$ iso 3-OH and $C_{18:1}$ ω9*c*. The principal respiratory quinones are menaquinones MK-11 and MK-12 with lower levels of MK-8, MK-9, MK-10 and MK-13.

Source: isolated from sub gingival plaque from a patient with chronic periodontitis.

DNA G+C content (mol%): 51 (HPLC).

Type strain: PPPA21, DSM 16608, JCM 12541.

Sequence accession no. (16S rRNA gene): AB182483.

21. **Prevotella multisaccharivorax** Sakamoto, Umeda, Ishikawa and Benno 2005b, 1842[VP]

mul.ti.sac.cha.ri.vo'rax. L. adj. *multus* many/much; L. n. *saccharum* sugar; L. adj. *vorax* devouring, ravenous, voracious; N.L. fem. adj. *multisaccharivorax* liking to eat many sugars.

The description is from Sakamoto et al. (2005b). Cells are 0.8 × 2.5–8.3 μm. Colonies are 0.5–0.7 mm in diameter, gray to off-white–gray, circular, entire, slightly convex and smooth on Eggerth–Gagnon blood agar. Acid is produced from arabinose (variable), cellobiose, glucose, glycerol, lactose, maltose, mannitol, mannose, melezitose, raffinose, rhamnose, salicin (variable), sorbitol, sucrose, trehalose and xylose. Esculin is hydrolyzed. Indole is not produced. Gelatin is digested. Catalase and urease are not produced.

The major end products from glucose broth (composed of peptone, 1%; yeast extract, 1%; and glucose, 1%) are succinic and acetic acids, with lower levels of isovaleric acid. Both nonhydroxylated and 3-hydroxylated long-chain fatty acids are present. The major fatty acids are $C_{18:1}$ ω9*c* and $C_{16:0}$. The principal respiratory quinones are menaquinones MK-12 and MK-13 with lower levels of (<10%) MK-10 and MK-11.

Source: isolated from subgingival plaque from patients with chronic periodontitis.

DNA G+C content (mol%): 50 (HPLC).

Type strain: PPPA20. JCM12954, DSM 17128.

Sequence accession no. (16S rRNA gene): AB200414.

22. **Prevotella nigrescens** Shah and Gharbia 1992, 545[VP]

ni.gres'cens. L. part. adj. *nigrescens* becoming black, referring to the characteristic black colonies formed on blood agar plates.

The description is from Shah and Gharbia (1992), Miyagawa et al. (1978, 1979, 1981), Miyagawa and Suto (Miyagawa and Suto, 1980), Shah et al. (1976, 1979), Shah and Collins, (1980, 1983, 1990) and Shah and Williams (1982). Cells are 0.4–0.7 × 1.5–2.0 µm; occasional cells are up to 12 µm long. Colonies on rabbit blood agar after 2 d are 0.5–2.0 mm in diameter, circular and entire, low convex, translucent, smooth and beta-hemolytic; older or larger colonies may be opaque; colonies are reddish brown or black. Cells grown on hemolyzed blood agar develop pigment more rapidly (~1–2 d) and are darker. Glucose broth cultures are turbid with a smooth (occasionally ropey or slightly mucoid) sediment and a final pH of 4.9–5.4. Hemin and vitamin K_1 are required for growth. Growth is inhibited by 6.5% NaCl and strains grow well between 25 and 45°C. Acid is produced from fructose, glycogen, inulin, maltose, mannose, raffinose, starch and sucrose, but not from cellobiose, esculin, gum arabic, lactose, ribose, trehalose or xylan. Indole is produced. Starch is hydrolyzed. No H_2 is produced.

The major fermentation products from a glucose-containing broth (e.g., BM) are acetic, isobutyric, isovaleric and succinic acids. The long-chain fatty acid composition is mainly anteiso- and iso-methyl branched acids with small amounts of straight-chain acids; the major fatty acid is 12-methyltetradecanoic acid ($C_{15:0}$ anteiso). The principal respiratory quinones are unsaturated menaquinones with 11 isoprene units with lower levels of MK-10 and MK-12.

Sources: isolated initially from human dental plaque but has since been reported from head, neck and pleural infections, from blood, and from abdominal and pelvic sites.

DNA G+C content (mol%): 40–44 (T_m).

Type strain: Lambe 729-74, ATCC 33563, CCUG 9560, CIP 105552, DSM 13386, JCM 6322, JCM 12250, NCTC 9336, VPI 8944.

Sequence accession no. (16S rRNA gene): L16471.

23. **Prevotella oralis** (Loesche, Socransky and Gibbons 1964) Shah and Collins 1990, 207[VP] (*Ristella* oralis (Loesche, Socransky and Gibbons 1964, 1334) Prévot, Turpin and Kaiser 1967, 264

o.ra'lis. L. n. *os oris* the mouth; L. fem. suff. *-alis* suffix denoting pertaining to; N.L. fem. adj. *oralis* pertaining to the mouth.

The description is from Loesche et al. (1964), Holdeman et al. (1984), Shah et al. (1979, 1976), Shah and Collins (1990, 1980, 1983) and Shah and Williams (1982). Cells from glucose broth are 0.5–0.8 × 1.0–5.0 µm, arranged in pairs and in chains. Surface colonies on hemolyzed rabbit blood agar are 1.0 mm, circular, entire, convex, shiny, smooth, semi-opaque and buff. Strains are nonhemolytic on horse blood infusion agar. In 24 h, glucose broth cultures are moderately turbid with a smooth sediment and pH of 4.5–4.9. Strains grow in up to 3% O_2 and survive exposure to air for up to 80 min. No H_2S is detected in peptone iron agar or in semisolid agar with 0.02% ferrous sulfate and 0.03% sodium hyposulfite, but there is a slight blackening of lead acetate paper suspended over cultures in trypticase broth. The following sugars are fermented: amygdalin, fructose, cellobiose, glucose, glycogen, lactose, maltose, mannose, melibiose, raffinose, salicin, starch and sucrose. Acid is not produced from arabinose, inositol, mannitol, melezitose, trehalose or xylose.

The major fermentation products from a glucose-containing broth (e.g., BM) are acetic, isobutyric, isovaleric and succinic acids. The long-chain fatty acid composition is mainly anteiso- and iso-methyl branched acids with small amounts of straight-chain acids; the major fatty acid is 12-methyltetradecanoic acid ($C_{15:0}$ anteiso). The principal respiratory quinones are unsaturated menaquinones with 11 isoprene units with lower levels of MK-10 and MK-12.

Source: isolated from the gingival crevice area of man and from infections, usually of the oral cavity and upper respiratory and genital tracts.

DNA G+C content (mol%): 43 (T_m).

Type strain: ATCC 33269, CCUG 15408, DSM 20702, JCM 12251, NCTC 11459, VPI D27B-24.

Sequence accession no. (16S rRNA gene): L16480.

24. **Prevotella oris** (Holdeman, Moore, Churn and Johnson 1982) Shah and Collins 1990, 207[VP] (*Bacteroides oris* Holdeman, Moore, Churn and Johnson 1982, 126[VP])

or'is. L. gen. n. *oris* of the mouth, referring to a major natural habitat of the species.

The description is from Holdeman et al. (1982) and our own observations. The rod-shaped cells occur in pairs or short chains and are 0.5–0.8 × 0.8–2.6 µm in PYG broth. Cells on BHIA roll tubes (Holdeman et al., 1980) produce subsurface colonies 1–2 mm in diameter and are lenticular and translucent. Colonies grown on blood agar plates are 0.5–1 mm in diameter, circular with an entire edge, low convex, translucent to semi-opaque, white to buff, shiny and smooth. There is no action on rabbit blood or egg yolk agar (Holdeman et al., 1977). Most cells do not grow in PYG containing 10% bile; cells that do grow are markedly inhibited and delayed. Growth is stimulated and required by some strains by the addition of heme. Pre-reduced broth cultures after incubation at 37°C for 24 h are uniformly turbid with a smooth sediment. The terminal pH value after 3–5 d is 4.5–4.9 in media with carbohydrates that are fermented by all strains, and 5.0–5.6 in media with sugars that are not fermented by all strains. All strains ferment dextrin, fructose, glucose, lactose, maltose, mannose, raffinose, starch and sucrose; most strains ferment amygdalin, arabinose, cellobiose, glycogen, melibiose, rhamnose, ribose and salicin. Cells do not ferment erythritol, inositol, mannitol, melezitose, sorbitol or trehalose. Esculin and starch are hydrolyzed. Nitrate is not reduced. Indole and catalase are not produced. Gelatin is digested by most strains.

Succinic and acetic acids are the major fermentation end products from PYG broth culture; minor amounts produced may include isobutyric and isovaleric acids; no H_2 is produced. The long-chain fatty acid composition is mainly anteiso- and iso-methyl branched acids with small amounts of straight-chain acids; the major fatty acid is 12-methyltetradecanoic acid ($C_{15:0}$ anteiso). The principal respiratory quinones are unsaturated menaquinones with 11 isoprene units (MK-11) with lower levels of MK-10 and MK-12.

Source: isolated from human periodontal flora, systemic human infections, and from large intestines of chickens.

DNA G+C content (mol%): 42–46 (T_m).

Type strain: ATCC 33573, CCUG 15405, CIP 104480, JCM 8540, JCM 12252, NCTC 13071, VPI D1A-1A.

Sequence accession no. (16S rRNA gene): L16472.

25. **Prevotella oulorum** corrig. (Shah, Collins, Watabe and Mitsuoka 1985) Shah and Collins 1990, 207[VP] (*Bacteroides oulorum* Shah, Collins, Watabe and Mitsuoka 1985, 195[VP])

ou.lo'rum. Gr. n. *oulon* the gums; N.L. gen. pl. n. *oulorum* of the gums.

The description is from Shah et al. (1985) and previous reports: Shah et al. (1976, 1979), Shah and Collins (1980, 1983, 1990) and Shah and Williams (1982). Cells are short rods to cocci, 0.5 × 1.0–1.5 μm, occurring singly, in pairs or in chains of rods. Colonies on Eggerth–Gagnon blood agar after 2 d are 1 mm in diameter, circular and entire, convex, translucent, smooth and shiny, while colonies on blood agar after 2 d show a similar morphology but are low, convex and opaque. Cells grow well in either pre-reduced PYG or BM medium supplemented with hemin and menadione. The terminal pH in glucose-containing broth media is 5.0; the terminal pH in media without glucose is pH 6.2. Cells are saccharolytic, producing acid from glucose, fructose, glycogen, inulin, lactose, maltose, mannose, raffinose, starch and sucrose, but not from arabinose, cellobiose, glycerol, mannitol, melezitose, rhamnose, salicin, sorbitol, trehalose, xylan or xylose. Esculin, but not gelatin, is hydrolyzed. Indole is not produced.

Acetate and succinate are the major end products of fermentation. Both nonhydroxylated long-chain fatty acids (~90%) and 3-hydroxy long-chain fatty acids are present. The major cellular fatty acids are $C_{15:0}$ anteiso, $C_{15:0}$ iso and $C_{17:0}$ iso; minor cellular fatty acids are $C_{17:0}$ iso 3-OH, $C_{17:0}$ anteiso, $C_{16:0}$, $C_{18:0}$, $C_{16:0}$ iso and $C_{14:0}$ iso; trace amounts include $C_{14:0}$ iso, $C_{15:0}$, $C_{13:0}$ anteiso and $C_{18:1}$ ω9*c*. The major respiratory quinones are menaquinones with 10 isoprene units (MK-10, >75%) with lower levels of MK-9 (<5%) and MK-11 (<20%).

Source: isolated from the gingival crevice.

DNA G+C content (mol%): 45–46 (T_m, Bd).

Type strain: WPH 179, ATCC 43324, CCUG 20177, CIP 104477, NCTC 11871.

Sequence accession no. (16S rRNA gene): L16472.

26. **Prevotella pallens** Könönen, Eerola, Frandsen, Jalava, Mättö, Salmenlinna and Jousimies-Somer 1998a, 49[VP]

pal'lens. L. part. adj. *pallens* being or looking or growing pale, referring to the weak pigmentation of colonies on blood agar plates.

This species was first referred to as a *Prevotella intermedia*/*nigrescens*-like organism by Könönen, Mättö, et al. (1998b). The description is from Könönen, Eerola, et al. (1998a) and our own observations. Cells grown in broth are short rods or coccobacilli, 0.5–0.8 × 1.2–2 μm. Colonies on supplemented Brucella sheep agar after 3 d are 0.8–1.4 mm diameter, circular and entire, raised convex, smooth and glistening, with a distinct halo of metallic sheen around the colonies and with weak cream pigmentation. Colonies on rabbit laked blood agar after 3–5 d are tan to light brown and exhibit red fluorescence under long-wavelength UV light (365 nm) due to over production of protoporphyrin. Acid is produced from glucose, maltose and sucrose in PYG broth, but not from arabinose, cellobiose, esculin, lactose (except one strain) or xylose. Gelatin is liquefied. Indole is positive. Lipase is negative.

Acetic and succinic acids are the major end products of fermentation; another frequent product is isovaleric acid. The major cellular fatty acids are $C_{15:0}$ anteiso, $C_{15:0}$ iso, $C_{17:0}$ iso 3-OH and an unknown cellular fatty acid with equivalent chain-length at 13.570. The principal respiratory quinones are menaquinones with 11 isoprene units (MK-11) with lower levels of MK-10 and MK-12.

Source: the type strain was isolated from saliva of a young child.

DNA G+C content (mol%): unknown.

Type strain: AHN 10371, ATCC 700821, CCUG 39484, CIP 105551, JCM 11140, NCTC 13042.

Sequence accession no. (16S rRNA gene): Y13105.

27. **Prevotella ruminicola** (Bryant, Small, Bouma and Chu 1958) Shah and Collins 1990, 207[VP] emend. Avguštin, Wallace and Flint 1997, 286[VP] (*Bacteroides ruminicola* Bryant, Small, Bouma and Chu 1958, 18[AL]; *Ruminobacter ruminicola* (Bryant, Small, Bouma and Chu 1958) Prévot 1966, 121).

ru.mi.ni'co.la. L. n. *rumen -inis* first stomach of ruminants, rumen; L. suff. *-cola* (from L. n. *incola*) inhabitant, deweller; N.L. n. *ruminicola* inhabitant of the rumen.

The morphological and biochemical characteristics of this species are based on the reports of Bryant et al. (1958) and Holdeman et al. (1984) and on the chemotaxonomic characters reported by Shah and Collins (1980, 1983).

Cells are 0.8–1 μm × 0.8–30 μm, with slightly tapered and with rounded ends. Most cells are 1.2–6 μm long. Irregular granules and bipolar staining are seen in some cells. Cells are often encapsulated and are either single or in pairs, with some chains containing long to coccoid cells. Surface colonies are smooth, entire, convex, opaque and light buff in color. Growth occurs at 30 and 37°C. Most strains produce clear zones that are visible after staining with Congo red. The following carbohydrates are fermented: arbutin, salicin, raffinose, mannose, sucrose, melibiose, xylose, arabinose, galactose, fructose, cellobiose, maltose, sucrose, inulin, sucrose, xylan, pectin and esculin. Lactate, trehalose, cellulose, glycerol, mannitol, inositol and gum arabic are not fermented. The species is also proteolytic and plays a key role in ruminal protein degradation. Many strains have activities that suggest that they play important roles in the utilization of polysaccharides of plant origin, including xylans, pectins and starch, and in the metabolism of peptides and proteins. Starch is hydrolyzed and nitrate is not reduced. Acetylmethylcarbinol, H_2S, catalase and indole are not produced.

The major metabolic end products from glucose broth are acetic, formic and succinic acids, with low to trace amounts of propionic, isobutyric and isovaleric acids. The long-chain fatty acid composition is mainly anteiso- and iso-methyl branched acids, with small amounts of straight-chain acids; the major fatty acid is 12-methyltetradecanoic acid ($C_{15:0}$ anteiso) and pentadecanoic ($C_{15:0}$), but other profiles may exist (see Shah and Collins, 1983). The principal respiratory quinones are menaquinones MK-12 and MK-11 with low levels (~1–5%) of MK-10 and MK-13.

Source: reticulo-rumen of cattle, sheep and elk and presumed to be among the more numerous bacteria in the rumen of most ruminants. Also isolated from the intestinal contents of chickens.

DNA G+C content (mol%): 45–52 (T_m).

Type strain: Bryant 23, ATCC 19189, CIP 105475, JCM 8958.

Sequence accession no. (16S rRNA gene): L16482.

28. **Prevotella salivae** Sakamoto, Suzuki, Huang, Umeda, Ishikawa and Benno 2004, 882[VP]

sa.li'vae. L. fem. gen. n. *salivae* from/of saliva from which the micro-organism was isolated.

The description is from Sakamoto et al. (2004). Cells are ~0.5–0.8 × 0.8–1.7 µm. Colonies are 1–2 mm in diameter, gray to light brown, circular, entire, slightly convex and smooth on Eggerth–Gagnon agar plates. Acid is produced from arabinose, cellobiose, glucose, lactose, maltose, mannose, raffinose, salicin, sucrose and xylose. Acid is not produced from glycerol, mannitol, melezitose, rhamnose, sorbitol or trehalose. Esculin is hydrolyzed. Indole is not produced. Gelatin is not digested. Catalase and urease are not produced.

Major end products (from a (w/v) peptone (1%), yeast extract 1%) glucose (1%) broth culture) are succinic and acetic acids; small amounts of isovaleric acid are also produced. Malate dehydrogenase and glutamate dehydrogenase (low activity) are present. Both nonhydroxylated and 3-hydroxylated long-chain fatty acids are present. Major fatty acids are $C_{15:0}$ anteiso, $C_{17:0}$ iso 3-OH and $C_{18:1}$ ω9c. The predominant respiratory quinones are menaquinones MK-11 (>67%), MK-12 (20%) and MK-10 (13%).

Source: isolated from the saliva of a patient with chronic periodontitis.

DNA G+C content (mol%): 41–42 (HPLC).

Type strain: EPSA11, DSM 15606, JCM 12084.

Sequence accession no. (16S rRNA gene): AB108826.

29. **Prevotella shahii** Sakamoto, Suzuki, Huang, Umeda, Ishikawa and Benno 2004, 881[VP]

sha'hi.i. N.L. gen. masc. n. *shahii* of Shah, named after the Trinidadian-born British microbiologist Haroun N. Shah, for his contributions to the taxonomy of the genus *Bacteroides* and related taxa.

The description is from Sakamoto et al. (2004). Most cells are ~0.5–0.8 × 8.3–10.0 µm. Colonies are 1–2 mm in diameter, tan to light brown, circular, entire, slightly convex and smooth on Eggerth–Gagnon blood agar plates. Acid is produced from glucose, lactose, maltose, mannose, raffinose and sucrose, but not from arabinose, cellobiose, glycerol, mannitol, melezitose, rhamnose, salicin, sorbitol, trehalose or xylose. Esculin is not hydrolyzed. Indole is not produced. Gelatin is digested. Catalase and urease are not produced.

Succinic and acetic acids are major end products from glucose broth containing peptone (1%), yeast extract (1%) and glucose (1%). Both non-hydroxylated long-chain fatty acids (~90%) and 3-hydroxy long-chain fatty acids are present. The major cellular fatty acids are $C_{18:1}$ ω9*c*, $C_{16:0}$ and $C_{16:0}$ 3-OH. The major respiratory quinones are menaquinones with 11 isoprene units (MK-11, >50%); MK-10 (>15%) and MK-12 (>20%) are also present, with lower levels of MK-8 (<1%) and MK-9 (<1%).

Source: isolated from the normal oral flora.

DNA G+C content (mol%): 44–45 (HPLC).

Type strain: EHS11, DSM 15611, JCM 12083.

Sequence accession no. (16S rRNA gene): AB108825.

30. **Prevotella stercorea** Hayashi, Shibata, Sakamoto, Tomita and Benno 2007, 943[VP]

ster.co're.a. L. fem. adj. *stercorea* pertaining to feces.

The description is from Hayashi et al. (2007). Colonies on Eggerth–Gagnon blood agar plates after 48 h incubation at 37°C under 100% CO_2 gas are translucent, circular, entire, slightly convex and smooth. Optimum temperature for growth is 37°C.

Acid is produced from glucose, lactose, sucrose, maltose, mannose and raffinose, but not from arabinose, melezitose, rhamnose, salicin or xylose. Growth is inhibited on *Bacteroides* bile esculin agar. Indole is not produced, esculin is not hydrolyzed, nor is gelatinase or urease activity detected. With the API ZYM system, positive reactions are obtained for alkaline phosphatase, acid phosphatase, naphthol-AS-BI-phosphohydrolase, β-galactosidase, α-glucosidase, *N*-acetyl-β-glucosaminidase and α-L-fucosidase; negative reactions are obtained for esterase (C4), esterase lipase (C8), lipase (C4), leucine arylamidase, valine arylamidase, cystine arylamidase, trypsin, chymotrypsin, β-gluconidase and β-glucosidase.

The major metabolic end products are succinic and acetic acids with small amounts of isovaleric acid. Both non-hydroxylated long-chain fatty acids and 3-hydroxy long-chain fatty acids are present. The major fatty acids are $C_{18:1}$ ω9c, $C_{15:0}$ iso and $C_{15:0}$ anteiso. The principal respiratory quinones are menaquinones MK-13 (>50%) and MK-12 (>25%) with lower levels of MK-9 (1%), MK-10 (5%) and MK-11 (<10%).

Source: the type strain was isolated from human feces. The habitat appears to be the human intestinal tract.

DNA G+C content (mol%): 48–49 (HPLC).

Type strain: CB35, DSM 18206, JCM 13469.

Sequence accession no. (16S rRNA gene): AB244774.

31. **Prevotella tannerae** Moore, Johnson and Moore 1994, 600[VP]

tan'ne.rae. N.L. gen. fem. n. *tannerae* of Tanner, in honor of Anne C. R. Tanner, an American microbiologist.

The description is from Moore et al. (1994). Cells occur in short chains grown in PYG broth and are 0.3 × 0.7–8 µm; filaments up to 14 µm long are observed. Colonies on blood agar plates after 2 d are 1–2 mm in diameter, circular, low convex and translucent to transparent. Colonies grown on rabbit blood agar after 5 d may be black, tan to brown to red, or colorless; occasional colonies are surrounded by small clear zones of hemolysis. Broth cultures are cloudy with a smooth sediment. Strains require addition of 10% sterile serum for optimum growth and acid production. Terminal pH values of glucose broth after 5 d are 4.8–5.3. Acid is produced by most strains from dextrin, fructose, glycogen, lactose and maltose, and less reliably (20–50% of isolates) from raffinose, sucrose and trehalose and mannose. Acid is not produced from amygdalin, cellobiose, melibiose, salicin or xylose. Esculin is not hydrolyzed by the majority of strains. Gelatin is liquefied and starch is hydrolyzed by most strains. Succinic and acetic acids are the major end products of fermentation; minor products include formic and isovaleric acids and trace amounts of isobutyric acid. No H_2 is produced. The major cellular fatty acids are $C_{15:0}$ anteiso, $C_{15:0}$ iso, $C_{17:0}$ iso 3-OH and $C_{18:2}$ dimethyl acetyl; minor cellular fatty acids are $C_{16:0}$, $C_{14:0}$, $C_{16:0}$ 3-OH, $C_{14:0}$ iso and $C_{17:0}$ iso.

Source: the type strain was isolated from gingival crevice of an adult with gingivitis. The normal habitat of this species is the human gingival crevice.

DNA G+C content (mol%): 45 (T_m).

Type strain: ATCC 51259, CCUG 34292, NCTC 13073, VPI N14B-15.

Sequence accession no. (16S rRNA gene): AJ005634.

32. **Prevotella timonensis** Glazunova, Launay, Raoult and Roux 2007, 885[VP]

ti.mo.n.en'sis. N.L. fem. adj. *timonensis* pertaining to the Hospital de la Timone, the hospital in Marseille, France, from where the type strain was isolated.

The description is from Glazunova et al. (2007). Cells are straight rods. After 48 h growth in Trypticase Soy Broth (TSB) medium, rods are 0.3–0.5 × 0.8–1.4 μm and occur singly. Growth occurs on sheep blood agar and in TSB liquid medium. After 72 h growth on blood sheep agar, surface colonies are circular, white-grayish, smooth, shiny and 1–2 mm in diameter. The temperature range for growth is 25–37°C, with an optimum at 37°C. Acid is produced from glucose, lactose and maltose. With the API 20A system, gelatin hydrolysis is positive; negative results are obtained for indole production, urease, esculin hydrolysis and fermentation of mannitol, sucrose, salicin, xylose, arabinose, glycerol, cellobiose, mannose, melezitose, raffinose, sorption, rhamnose and trehalose. With the API ID 32A system, positive enzyme reactions occur for alkaline phosphatase, β-galactosidase, α-glucosidase, N-acetyl-β-glucosaminidase, α-fucosidase, arginine arylamidase, leucyl glycine arylamidase and alanine arylamidase. No activity is detected for glutamic acid decarboxylase, arginine dihydrolase, β-glucosidase, α-arabinosidase, β-glucuronidase, proline arylamidase, phenylalanine arylamidase, leucine arylamidase, pyroglutamic acid arylamidase, tyrosine arylamidase, glycine arylamidase, histidine arylamidase, glutamyl glutamic acid arylamidase or serine arylamidase. The major cellular fatty acids are $C_{14:0}$, $C_{16:0}$, $C_{14:0}$ iso and a mixture of $C_{18:2}$ ω6,9*c* and $C_{18:0}$.

Source: the type strain was isolated from a human breast abscess.

DNA G+C content (mol%): unknown.

Type strain: 4401737, CCUG 50105, CIP 108522.

Sequence accession no. (16S rRNA gene): DQ518919.

33. **Prevotella veroralis** (Watabe, Benno and Mitsuoka 1983) Shah and Collins 1990, 207[VP] emend. Wu, Johnson, Moore and Moore 1992, 536[VP] (*Bacteroides veroralis* Watabe, Benno and Mitsuoka 1983, 62[VP]).

ver'o.ral.is. L. adj. *verus* true; N.L. adj. *oralis* pertaining to the mouth and also a specific epithet; N.L. fem. adj. *veroralis* the true (*Prevotella*) *oralis*.

The description is from Watabe et al. (1983), Wu et al. (1992) and our own observations. The rod-shaped cells are 0.5 × 1.0–1.5 μm and occur in pairs of short rods or short chains of long rods. Colonies grown on Eggerth–Gagnon blood agar plates for 2 d are 1.0–2.0 mm in diameter, circular, entire, convex, translucent, smooth and shiny. Growth is not stimulated by bile. Cultures in pre-reduced PY broth are moderately turbid and may have a slight smooth sediment. Growth is enhanced in broth supplemented with fermentable carbohydrate. The optimal temperature is 37°C; the terminal pH in PYG broth is 4.7, and the pH without fermentable carbohydrate is 6.1. Strains ferment cellobiose, esculin, fructose, glycogen, inulin, melibiose, raffinose, starch, sucrose and xylan, but not amygdalin, glycerol, mannitol, rhamnose, ribose, sorbitol or salicin. Starch and esculin are hydrolyzed. Gelatin is liquefied. Nitrate is not reduced. Indole is not produced.

The major fermentation products from a glucose-containing broth (e.g., BM) are acetic, isobutyric, isovaleric and succinic acids. The long-chain fatty acid composition is mainly anteiso- and iso-methyl branched acids, with small amounts of straight-chain acids. The major fatty acids are $C_{18:1}$ ω9c and $C_{15:0}$ anteiso. The principal respiratory quinones are unsaturated menaquinones with approximately equivalent levels (~40–50%) of MK-10 and MK-11 with lower levels of MK-12 (<10%).

Source: isolated from the human oral cavity.

DNA G+C content (mol%): 42 (T_m).

Type strain: ATCC 33779, CCUG 15422, JCM 6290, VPI D22A-7.

Sequence accession no. (16S rRNA gene): L16473.

34. **Prevotella zoogleoformans** (Weinberg, Nativelle and Prévot 1937) Cato, Kelley, Moore and Holdeman 1982, 273[VP] [*Bacterium zoogleoformans* (*sic*) Weinberg, Nativelle and Prévot 1937, 725; *Capsularis zoogleoformans* (*sic*) Prévot 1938, 293]

zo.o.gle'o.for'mans. Gr. adj. *zôos* alive, living; Gr. masc. noun *gloios* gum, glue; N.L. fem. n. *zoogleoea* inhabitant of glue; L. part. adj. *formans* forming; N.L. part. adj. *zoogleoformans* forming zoogloea (pertaining to the glutinous mass produced in broth cultures).

The description is from Cato et al. (1982), Holdeman et al. (1984) and Moore et al. (1994). Cells grown in glucose broth are 0.6–1.0 × 0.8–8.0 μm, have rounded ends, and many stain irregularly. They usually occur singly, occasionally in pairs. No capsules are detected. Cells form a viscous, glutinous mass in broth cultures and produce distinctive towers of "zoogloeal" slime on meat particles. Surface colonies on blood agar plates are 0.5–2 mm, circular, entire, pulvinate, opaque, buff colored, shiny, smooth and butyrous. Hemin is required for growth. No growth occurs at 25°C; variable growth occurs at 45°C. In pre-reduced media, the most reliable fermentation reactions are obtained when 10% (v/v) sterile serum is added and cultures are inoculated and incubated in a gaseous atmosphere containing 90% N_2 and only 10% CO_2. Acid is produced from salicin, raffinose, mannose, sucrose, melibiose, xylose, arabinose, galactose, fructose, cellobiose, maltose, sucrose, inulin, sucrose and xylan, but not from melezitose, rhamnose, ribose, trehalose, mannitol or inositol. Pectin and esculin are hydrolyzed. The species is proteolytic and plays a key role in ruminal protein degradation. Many strains have activities that suggest that they play important roles in the utilization of polysaccharides of plant origin, including xylans, pectins and starch, and in the metabolism of peptides and proteins. Starch is hydrolyzed and nitrate is not reduced. Acetylmethylcarbinol, H_2S, catalase, and indole are not produced.

The major fermentation products from a glucose-containing broth (e.g., BM) are acetic, isobutyric, isovaleric and succinic acids. The long-chain fatty acid composition is mainly anteiso- and iso-methyl branched acids with small amounts of straight-chain acids; the major fatty acids are $C_{18:1}$ ω9*c* and $C_{15:0}$ anteiso.

Source: isolated from the oral cavity and intestinal tract of man and animals.

DNA G+C content (mol%): 47 (T_m).

Type strain: ATCC 33285, CCUG 20495, CIP 104479, NCTC 13075, VPI D28K-1.

Sequence accession no. (16S rRNA gene): L16488.

Other organisms

"*Prevotella massiliensis*" has been reported on the basis of comparative 16S rRNA sequence analysis. The species clusters with other *Prevotella* but differs markedly from members of the genus by its inability to ferment any sugars tested while its long-chain fatty acid profile is more compatible with *Clostridium botulinum* or *Bifidobacterium bifidum* (see below). The following description is from Berger et al. (2005):

"Prevotella massiliensis" Berger, Adekambi, Mallet and Drancourt 2005, 973

mas.si.li.en′sis. L. fem. adj. *massiliensis* pertaining to Massilia, the ancient Roman name of Marseille, France, where the organism was isolated and characterized.

The description is from Berger et al. (2005). Cells are 1–2 µm which when viewed under electron microscopy exhibit a trilamellar membrane. They are nonmotile and possess no flagella. Colonies are seen on blood agar after 72 h incubation and are non-pigmented, circular, convex, shiny with a smooth surface, approximately 1 mm diameter and non-hemolytic. Growth occurs at 37°C, but not at ambient temperature nor at 44°C; better growth is achieved on 5% sheep blood agar than on trypticase soy or chocolate agar. Using the API 20A identification strip for anaerobes, it is indole-positive. The physiological reaction spectrum of this species is incomplete, but unlike other *Prevotella* species it appears to be nonfermentative. Thus acid is not produced from glucose, mannitol, sucrose, maltose, lactose, salicin, xylose, arabinose, glycerol, cellobiose, mannose, melezitose, raffinose, sorbitol, rhamnose or trehalose. Gelatin and esculin are not hydrolyzed. Other differential tests include its susceptibility to penicillin, imipenem, clarithromycin, clindamycin, metronidazole and kanamycin and its resistance to vancomycin.

The short-chain metabolic end products are unknown, but the long-chain fatty acid profiles are at variance with *Prevotella* species and are more compatible with those of *Clostridium botulinum* or *Bifidobacterium bifidum.* Thus, the major fatty acids are: $C_{16:0}$, $C_{16:0}$ 3-OH, $C_{14:0}$ and $C_{18:2}$.

Source: isolated from the blood of a patient hospitalized in an intensive care unit. Habitat is not known.

DNA G+C content (mol%): unknown.

Type strain: CIP 107630.

Sequence accession no. (16S rRNA gene): AF487886.

Genus II. **Xylanibacter** Ueki, Akasaka, Suzuki, Hattori and Ueki 2006, 2220[VP]

The Editorial Board

Xy.la.ni.bac′ter. N.L. n. *xylanum* xylan; N.L. masc. n. *bacter* a rod; N.L. masc. n. *Xylanibacter* rod decomposing xylan.

Short to filamentous rods. Nonmotile. Nonsporeforming. Gram-stain-negative. **Strictly anaerobic.** Chemo-organotrophic. Optimum growth temperature, 30°C. Oxidase and catalase-negative. Nitrate is not reduced. Hemin markedly stimulates growth. **Various sugars are utilized, including xylan. In the presence of cyanocobalamin and hemin, acetate, propionate, and succinate are produced as major fermentation products.** The major cellular fatty acids are $C_{15:0}$ anteiso and $C_{17:0}$ iso 3-OH. The major respiratory quinones are **MK-12(H_2) and MK-13(H_2)**.

DNA G+C content (mol%): 43.6 (HPLC).

Type species: **Xylanibacter oryzae** Ueki, Akasaka, Suzuki, Hattori and Ueki 2006, 2220[VP].

Further descriptive information

Xylanibacter oryzae was isolated from a sample of rice-plant residue (rice stubble and roots) collected from irrigated rice-field soil at the Shonai Branch of the Yamagata Agricultural Experimental Station (Tsuruoka, Yamagata, Japan) during the flooding period of the field.

The addition of hemin to the medium greatly enhances growth, which otherwise occurs very slowly. Cells grown in peptone-yeast extract-glucose (PYG) broth with hemin are mainly short rods (0.6–0.7 × 2–2.6 µm) but also some longer rods (4–10 µm), whereas cells grown in the absence of hemin are often filamentous rods (20–50 µm long), sometimes in chains. Addition of a B-vitamin mixture or vitamin K does not affect the growth rate.

The end products of glucose fermentation are acetate, succinate, and a small amount of propionate; however, the level of propionate is markedly increased by the addition of either the B-vitamin mixture or cyanocobalamin.

Enrichment, isolation, and maintenance procedures

Rice plant residue samples were homogenized under N_2 gas using a Waring blender. Isolation of the type strain was accomplished using the anaerobic roll-tube method.

For maintenance, PY4S agar* (which has no hemin) is preferred.

*PY4S agar is peptone-yeast extract (PY) broth supplemented with (g/l): glucose, 0.25; cellobiose, 0.25; maltose, 0.25; soluble starch, 0.25; and agar, 15.0.

Differentiation of the genus *Xylanibacter* from other genera

Unlike *Xylanibacter oryzae*, *Prevotella bivia* hydrolyzes gelatin but not esculin, forms no propionate, does not form acid from arabinose, cellobiose, salicin, sucrose, trehalose, and xylose, and has MK-9, MK-10, and MK-11 as its predominant menaquinones. *Prevotella albensis* differs from *Xylanibacter oryzae* by its failure to form acid from carboxymethylcellulose, mannose, sucrose, and trehalose, and by the presence of MK-11 and MK-12 as its predominant menaquinones. *Prevotella oulorum* differs from *Xylanibacter oryzae* by its failure to form propionate, formation of acid from inulin but not arabinose, cellobiose, salicin, trehalose, and xylose, and by the presence of MK-10 as its predominant menaquinone.

Taxonomic comments

By 16S rRNA gene sequence analysis, the nearest neighbor of *Xylanibacter oryzae* is *Prevotella bivia* with a sequence similarity of 89.5%. Other neighbors are *Prevotella albensis* (sequence similarity of 89.1%) and *Prevotella oulorum* (sequence similarity of 89.1%).

List of species of the genus *Xylanibacter*

1. **Xylanibacter oryzae** Ueki, Akasaka, Suzuki, Hattori and Ueki 2006, 2220[VP]

o'ry.zae. L. fem. n. *oryza* rice and the genus name of rice; L. gen. n. *oryzae* from/of rice or rice plants, referring to rice-plant residue from which the strain was isolated.

The characteristics are as described for the genus, with the following additional features. pH range, 4.7–7.3; optimum, 5.7–6.2. Temperature range, 15–35°C; optimum, 30°C; growth at 37°C is considerably delayed. The NaCl concentration range for growth is 0–0.5% in PYG medium containing hemin and B-vitamin mixture. The following compounds are growth substrates: arabinose, carboxymethylcellulose, cellobiose, fructose, galactose, glucose, lactose, maltose, mannose, pectin, pyruvate, rhamnose, ribose, salicin, soluble starch, sucrose, trehalose, xylose, and xylan. Acids but not gas are produced from these substrates. The following substrates are not used: cellulose powder, filter paper, fumarate, glycerol, inositol, inulin, lactate, malate, mannitol, melezitose, sorbose, and succinate. Esculin is hydrolyzed but gelatin and urea are not. H_2S and indole are not produced. No growth occurs in the presence of bile salts.

Source: stubble and roots of rice-plant residue in anoxic rice-field soil in Japan.

DNA G+C content (mol%): 43.6 (HPLC).

Type strain: KB3, DSM 17970, JCM 13648.

Sequence accession no. (16S rRNA gene): AB078826.

References

Avguštin, G., R.J. Wallace and H.J. Flint. 1997. Phenotypic diversity among ruminal isolates of *Prevotella ruminicola*: proposal of *Prevotella brevis* sp. nov., *Prevotella bryantii* sp. nov., and *Prevotella albensis* sp. nov. and redefinition of *Prevotella ruminicola*. Int. J. Syst. Bacteriol. *47*: 284–288.

Berger, P., T. Adekambi, M.N. Mallet and M. Drancourt. 2005. *Prevotella massiliensis* sp. nov. isolated from human blood. Res. Microbiol. *156*: 967–973.

Bergey, D.H., F.C. Harrison, R.S. Breed, B.W. Hammer and F.M. Huntoon. 1930. Bergey's Manual of Determinative Bacteriology, 3rd edn. Williams & Wilkins, Baltimore.

Bryant, M.P., N. Small, C. Bouma and H. Chu. 1958. *Bacteroides ruminicola* n. sp. and *Succinimonas amylolytica*; the new genus and species; species of succinic acid-producing anaerobic bacteria of the bovine rumen. J. Bacteriol. *76*: 15–23.

Bryant, M.P. and I.M. Robinson. 1962. Some nutritional characteristics of predominant culturable ruminal bacteria. J. Bacteriol. *84*: 605–614.

Cato, E.P., R.W. Kelley, W.E.C. Moore and L.V. Holdeman. 1982. *Bacteroides zoogleoformans* (Weinberg, Nativelle and Prevot 1937) corrig., comb. nov.: emended description. Int. J. Syst. Bacteriol. *32*: 271–274.

Downes, J., I. Sutcliffe, A.C. Tanner and W.G. Wade. 2005. *Prevotella marshii* sp. nov. and *Prevotella baroniae* sp. nov., isolated from the human oral cavity. Int. J. Syst. Evol. Microbiol. *55*: 1551–1555.

Downes, J., I.C. Sutcliffe, T. Hofstad and W.G. Wade. 2006. *Prevotella bergensis* sp. nov., isolated from human infections. Int. J. Syst. Evol. Microbiol. *56*: 609–612.

Glazunova, O.O., T. Launay, D. Raoult and V. Roux. 2007. *Prevotella timonensis* sp. nov., isolated from a human breast abscess. Int. J. Syst. Evol. Microbiol. *57*: 883–886.

Gregory, E.M., W.E. Moore and L.V. Holdeman. 1978. Superoxide dismutase in anaerobes: survey. Appl. Environ. Microbiol. *35*: 988–991.

Haapasalo, M., H. Ranta, H. Shah, K. Ranta, K. Lounatmaa and R.M. Kroppenstedt. 1986a. *Mitsuokella dentalis* sp. nov. from dental root canals. Int. J. Syst. Bacteriol. *36*: 566–568.

Haapasalo, M., H. Ranta, H. Shah, K. Ranta, K. Lounatmaa and R.M. Kroppenstedt. 1986b. Biochemical and structural characterization of an unusual group of Gram-negative, anaerobic rods from human periapical osteitis. J. Gen. Microbiol. *132*: 417–426.

Hammann, R. and H. Werner. 1981. Presence of diaminopimelic acid in propionate-negative *Bacteroides* species and in some butyric acid-producing strains. J. Med. Microbiol. *14*: 205–212.

Hayashi, H., K. Shibata, M. Sakamoto, S. Tomita and Y. Benno. 2007. *Prevotella copri* sp. nov. and *Prevotella stercorea* sp. nov., isolated from human faeces. Int. J. Syst. Evol. Microbiol. *57*: 941–946.

Hofstad, T. 1974. The distribution of heptose and 2-keto-3-deoxyoctonate in *Bacteroidaceae*. J. Gen. Microbiol. *85*: 314–320.

Holdeman, L.V. and W.E.C. Moore. 1970. *Bacteroides*. Outline of Clinical Methods in Anaerobic Bacteriology, 2nd revn (edited by E.P. Cato, C.S. Cummins, L.V. Holdeman, J.L. Johnson, W.E.C. Moore, R.M. Smibert and L.D.S. Smith). Virginia Polytechnic Institute Anaerobe Laboratory, Blacksburg, VA, pp. 57–66.

Holdeman, L.V., E.P. Cato and W.E.C. Moore (editors). 1977. Anaerobe Laboratory Manual, 4th edn. Anaerobe Laboratory, Virginia Polytechnic Institute and State University, Blacksburg, VA.

Holdeman, L.V. and J.L. Johnson. 1977. *Bacteroides disiens* sp. nov. and *Bacteroides bivius* sp. nov. from human clinical infections. Int. J. Syst. Bacteriol. *27*: 337–345.

Holdeman, L.V., E.P. Cato, J.A. Burmeister and W.E.C. Moore. 1980. Descriptions of *Eubacterium timidum* sp. nov. *Eubacterium brachy* sp. nov. and *Eubacterium nodatum* sp. nov. isolated from human periodontitis. Int. J. Syst. Bacteriol. *30*: 163–169.

Holdeman, L.V. and J.L. Johnson. 1982. Description of *Bacteroides loescheii* sp. nov. and emendation of the descriptions of *Bacteroides melaninogenicus* (Oliver and Wherry) Roy and Kelly 1939 and *Bacteroides denticola* Shah and Collins 1981. Int. J. Syst. Bacteriol. *32*: 399–409.

Holdeman, L.V., W.E.C. Moore, P.J. Churn and J.L. Johnson. 1982. *Bacteroides oris* and *Bacteroides buccae*, new species from human periodontitis and other human infections. Int. J. Syst. Bacteriol. *32*: 125–131.

Holdeman, L.V., R.W. Kelly and W.E.C. Moore. 1984. Genus I. *Bacteroides*. *In* Bergey's Manual of Systematic Bacteriology, vol. 1 (edited by Krieg and Holt). Williams & Wilkins, Baltimore, pp. 604–631.

Johnson, J.L. and L.V. Holdeman. 1983. *Bacteroides intermedius* comb. nov. and descriptions of *Bacteroides corporis* sp. nov. and *Bacteroides levii* sp. nov. Int. J. Syst. Bacteriol. *33*: 15–25.

Johnson, J.L. and L.V. Holdeman. 1985. Bacteriodes capillus Kornman and Holt and *Bacteroides pentosaceus* Shah and Collins, later synonyms of *Bacteroides buccae* Holdeman et al. Int. J. Syst. Bacteriol. *35*: 114.

Könönen, E., E. Eerola, E.V. Frandsen, J. Jalava, J. Mättö, S. Salmenlinna and H. Jousimies-Somer. 1998a. Phylogenetic characterization and proposal of a new pigmented species to the genus *Prevotella*: *Prevotella pallens* sp. nov. Int. J. Syst. Bacteriol. *48*: 47–51.

Könönen, E., J. Mättö, M.L. Väisänen-Tunkelrott, E.V. Frandsen, I. Helander, S. Asikainen, S.M. Finegold and H. Jousimies-Somer. 1998b. Biochemical and genetic characterization of a *Prevotella intermedia/ nigrescens*-like organism. Int. J. Syst. Bacteriol. *48*: 39–46.

Kornman, K.S. and S.C. Holt. 1981. Physiological and ultrastructural characterization of a new *Bacteroides* species (*Bacteroides capillus*) isolated from severe localized periodontitis. J. Periodont. Res. *16*: 542–555.

Kornman, K.S. and S.C. Holt. 1982. *In* Validation of the publication of new names and new combinations previously effectively published outside the IJSB. List no. 8. Int. J. Syst. Bacteriol. *32*: 266–268.

Lambe, D.W., Jr. 1974. Determination of *Bacteroides melaninogenicus* serogroups by fluorescent antibody staining. Appl. Microbiol. *28*: 561–567.

Loesche, W.J., S.S. Socransky and R.J. Gibbons. 1964. *Bacteroides oralis*, proposed new species isolated from the oral cavity of man. J. Bacteriol. *88*: 1329–1337.

Miyagawa, E., R. Azuma and T. Suto. 1978. Distribution of sphingolipids in *Bacteroides* species. J. Gen. Appl. Microbiol. *24*: 341–348.

Miyagawa, E., R. Azuma and T. Suto. 1979. Cellular fatty acid composition in Gram-negative obligately anaerobic rods. J. Gen. Microbiol. *25*: 41–51.

Miyagawa, E. and T. Suto. 1980. Cellular fatty acid composition in *Bacteroides oralis* and *Bacteroides ruminicola*. J. Gen. Appl. Microbiol. *26*: 331–343.

Miyagawa, E., R. Azuma and T. Suto. 1981. Peptidoglycan composition of gram-negative obligately anaerobic rods. J. Gen. Appl. Microbiol. *22*: 199–208.

Moore, L.V.H., J.L. Johnson and W.E.C. Moore. 1994. Descriptions of *Prevotella tannerae* sp. nov. and *Prevotella enoeca* sp. nov. from the human gingival crevice and emendation of the description of *Prevotella zoogleoformans*. Int. J. Syst. Bacteriol. *44*: 599–602.

Moore, L.V.H. and W.E.C. Moore. 1994. *Oribaculum catoniae* gen. nov., sp. nov., *Catonella morbi* gen. nov., sp. nov., *Hallella seregens* gen. nov., sp. nov., *Johnsonella ignava* gen. nov., sp. nov., and *Dialister pneumosintes* gen. nov., comb. nov., nom. rev., anaerobic Gram-negative bacilli from the human gingival crevice. Int. J. Syst. Bacteriol. *44*: 187–192.

Okuda, K., T. Kato, J. Shiozu, I. Takazoe and T. Nakamura. 1985. *Bacteroides heparinolyticus* sp. nov. isolated from humans with periodontitis. Int. J. Syst. Bacteriol. *35*: 438–442.

Oliver, W.W. and W.B. Wherry. 1921. Notes on some bacterial parasites of the human mucous membranesq. J. Infect. Dis. *28*: 341–344.

Olsen, I. and H.N. Shah. 2001. International Committee on Systematics of Prokaryotes Subcommittee on the taxonomy of Gram-negative anaerobic rods. Minutes of the meetings, 9 and 10 July 2000, Manchester, UK. Int. J. Syst. Evol. Microbiol. *51*: 1943–1944.

Pittman, K.A. and M.P. Bryant. 1964. Peptides and other nitrogen sources for growth of *Bacteroides ruminicola*. J. Bacteriol. *88*: 401–410.

Prévot, A.R. 1938. Etudes de systematique bacterienne. III. Invalidite du genre *Bacteroides* Castellani et Chalmers demembrement et reclassification. Ann. Inst. Pasteur *20*: 285–307.

Prévot, A.R. 1966. Manual for the classification and determination of the anaerobic bacteria, 1st Am. Ed. edn. Lea and Febiger, Philadelphia.

Prévot, A. R., A. Turpin and P. Kaiser. 1967. Les Bactéries Anaérobies. Dunod, Paris.

Roy, T.E. and C.D. Kelly. 1939. Genus VIII. *Bacteroides* Castellani and Charmers. *In* Bergey's Manual of Determinative Bacteriology, 5th edn (edited by Bergey, Breed, Murray and Hitchens). Williams & Wilkins, Baltimore, pp. 569–570.

Sakamoto, M., M. Suzuki, Y. Huang, M. Umeda, I. Ishikawa and Y. Benno. 2004. *Prevotella shahii* sp. nov. and *Prevotella salivae* sp. nov., isolated from the human oral cavity. Int. J. Syst. Evol. Microbiol. *54*: 877–883.

Sakamoto, M., Y. Huang, M. Umeda, I. Ishikawa and Y. Benno. 2005a. *Prevotella multiformis* sp. nov., isolated from human subgingival plaque. Int. J. Syst. Evol. Microbiol. *55*: 815–819.

Sakamoto, M., M. Umeda, I. Ishikawa and Y. Benno. 2005b. *Prevotella multisaccharivorax* sp. nov., isolated from human subgingival plaque. Int. J. Syst. Evol. Microbiol. *55*: 1839–1843.

Schwabacher, H., D.R. Lucas and C. Rimington. 1947. *Bacterium melaninogenicum* – a misnomer. J. Gen. Microbiol. *1*: 109–120.

Shah, H.N., R.A. Williams, G.H. Bowden and J.M. Hardie. 1976. Comparison of the biochemical properties of *Bacteroides melaninogenicus* from human dental plaque and other sites. J. Appl. Bacteriol. *41*: 473–495.

Shah, H.N., R. Bonnett, B. Mateen and R.A. Williams. 1979. The porphyrin pigmentation of subspecies of *Bacteroides melaninogenicus*. Biochem. J. *180*: 45–50.

Shah, H.N. and M.D. Collins. 1980. Fatty acid and isoprenoid quinone composition in the classification of *Bacteroides melaninogenicus* and related taxa. J. Appl. Bacteriol. *48*: 75–87.

Shah, H.N. and M. Collins. 1981. *Bacteroides buccalis*, sp. nov., *Bacteroides denticola*, sp. nov., and *Bacteroides pentosaceus*, sp. nov., new species of the genus *Bacteroides* from the oral cavity. Zentralbl. Bakteriol. Parasitenkd. Infektionskr. Hyg. Abt. 1 Orig. C. *2*: 235–241.

Shah, H.N. and M.D. Collins. 1982. *In* Validation of new names and new combinations not previously published in the IJSB. List no. 8. Int. J. Syst. Bacteriol. *32*: 266–268.

Shah, H.N., T.J.M. Vansteenbergen, J.M. Hardie and J. Degraaff. 1982. DNA-base composition, DNA–DNA reassociation and isoelectric-focusing of proteins of strains designated *Bacteroides oralis*. FEMS Microbiol. Lett. *13*: 125–130.

Shah, H.N. and R.A.D. Williams. 1982. Dehyrogenase patterns in the taxonomy of *Bacteroides*. J. Gen. Microbiol. *128*: 2955–2965.

Shah, H.N. and M.D. Collins. 1983. Genus *Bacteroides*. a chemotaxonomical perspective. J. Appl. Bacteriol. *55*: 403–416.

Shah, H.N., M.D. Collins, J. Watabe and T. Mitsuoka. 1985. *Bacteroides oulorum* sp. nov., a nonpigmented saccharolytic species from the oral cavity. Int. J. Syst. Bacteriol. *35*: 193–197.

Shah, H.N. and R.A.D. Williams. 1987a. Catabolism of aspartate and asparagine by *Bacteroides intermedius* and *Bacteroides gingivalis*. Curr. Microbiol. *15*: 313–318.

Shah, H.N. and R.A.D. Williams. 1987b. Utilization of glucose and amino acids by *Bacteroides intermedius* and *Bacteroides gingivalis*. Curr. Microbiol. *15*: 241–246.

Shah, H.N. and D.M. Collins. 1990. *Prevotella*, a new genus to include *Bacteroides melaninogenicus* and related species formerly classified in the genus *Bacteroides*. Int. J. Syst. Bacteriol. *40*: 205–208.

Shah, H.N. and S.E. Gharbia. 1992. Biochemical and chemical studies on strains designated *Prevotella intermedia* and proposal of a new pigmented species, *Prevotella nigrescens* sp. nov. Int. J. Syst. Bacteriol. *42*: 542–546.

Shah, H.N., S.E. Gharbia and K. Brocklehurst. 1993. Isolation and characterization of gingivain, a cysteine proteinase from *Porphyromonas gingivalis* W83 using covalent chromatography by thiol-disulfide interchange and 2,2′-dipyridyl disulfide as a two-hydronic state thiol-specific catalytic site titrant inhibitor and reactivity probe. *In* Biology of the species of *Porphyromonas gingivalis* (edited by Shah, Mayrand and Genco). CRC Press, Boca Raton, FL, pp. 245–258.

Ueki, A., H. Akasaka, D. Suzuki, S. Hattori and K. Ueki. 2006. *Xylanibacter oryzae* gen. nov., sp. nov., a novel strictly anaerobic, Gram-negative, xylanolytic bacterium isolated from rice-plant residue in flooded rice-field soil in Japan. Int. J. Syst. Evol. Microbiol. *56*: 2215–2221.

Van Steenbergen, J.J.M., J.J. De Soet and M. De Graff. 1979. DNA base composition of various strains of *Bacteroides melaninogenicus*. FEMS Microbiol. Lett. *5*: 127–130.

Watabe, J., Y. Benno and T. Mitsuoka. 1983. Taxonomic study of *Bacteroides oralis* and related organisms and proposal of *Bacteroides veroralis* sp. nov. Int. J. Syst. Bacteriol. *33*: 57–64.

Weinberg, M., R. Nativelle and A.R. Prévot. 1937. Les Microbes Anaérobies. Masson et Cie, Paris.

Werner, H. 1991. International Committee on Systematic Bacteriology Subcommittee on Gram-negative anaerobic rods. Minutes of the meeting, 13 and 14 September 1990, Osaka, Japan. Int. J. Syst. Bacteriol. *41*: 590–591.

Willems, A. and M.D. Collins. 1995. 16S ribosomal RNA gene similarities indicate that *Hallella seregens* (Moore and Moore) and *Mitsuokella dentalis* (Haapasalo et al.) are genealogically highly related and are members of the genus *Prevotella*: emended description of the genus *Prevotella* (Shah and Collins) and description of *Prevotella dentalis* comb. nov. Int. J. Syst. Bacteriol. *45*: 832–836.

Wu, C.C., J.L. Johnson, W.E.C. Moore and L.V.H. Moore. 1992. Emended descriptions of *Prevotella denticola*, *Prevotella loescheii*, *Prevotella veroralis*, and *Prevotella melaninogenica*. Int. J. Syst. Bacteriol. *42*: 536–541.

Class II. **Flavobacteriia** class. nov.

JEAN-FRANÇOIS BERNARDET

Fla.vo.bac.te′ri.i.a. N.L. neut. n. *Flavobacterium* type genus of the type order *Flavobacteriales*; suff. *-ia*, ending proposal by Gibbons and Murray and by Stackebrandt et al. to denote class; N.L. neut. pl. n. *Flavobacteriia*, the *Flavobacterium* class.

The class *Flavobacteriia* is circumscribed for this volume on the basis of the phylogenetic analyses of the 16S rRNA sequences in contrast to the proposal by Cavalier-Smith (2002), which was based on a few basic morphological, chemotaxonomic and physiological features. This class includes only the order *Flavobacteriales* (ord. nov., this volume). The description of the class is the same as the order.

Type order: **Flavobacteriales** ord. nov.

Reference

Cavalier-Smith, T. 2002. The neomuran origin of archaebacteria, the negibacterial root of the universal tree and bacterial megaclassification. Int. J. Syst. Evol. Microbiol. *52*: 7–76.

Order I. **Flavobacteriales** ord. nov.

JEAN-FRANÇOIS BERNARDET

Fla.vo.bac.te.ri′a.les. N.L. neut. n. *Flavobacterium* type genus of the order; suff. *-ales* ending to denote an order; N.L. fem. pl. n. *Flavobacteriales* the *Flavobacterium* order.

The order *Flavobacteriales* is circumscribed for this volume on the basis of the phylogenetic analyses of the 16S rRNA sequences. This order contains the families *Flavobacteriaceae*, *Blattabacteriaceae*, and *Cryomorphaceae*. It consists of Gram-stain-negative, nonspore-forming rods or filaments devoid of gas vesicles and intracellular granules of poly-β-hydroxybutyrate, which usually multiply by binary fission; ring-shaped cells are not formed. Members of the family *Blattabacteriaceae* are nonmotile, nonflagellated, nonculturable and poorly characterized intracellular symbionts of insects. The families *Flavobacteriaceae* and *Cryomorphaceae* consist of strictly aerobic or facultatively anaerobic chemo-organotrophs with respiratory metabolism (fermentative metabolism occurs in a few members of the family *Flavobacteriaceae*), nonmotile or motile by gliding, with yellow or orange colonies due to the production of carotenoid and/or flexirubin type pigments.

Many members of these families require NaCl or seawater salts for growth and most of them require organic compounds present in yeast extract, peptone and/or Casamino acids. Members of the family *Flavobacteriaceae* are widespread in soil or fresh, brackish or sea water in temperate, tropical or polar areas, while cultured members of the family *Cryomorphaceae* are so far restricted to low temperature marine habitats. Some members of the family *Flavobacteriaceae* are pathogenic for humans, fish, or amphibians. Menaquinone 6 (MK-6) is the major respiratory quinone in all members of the family *Flavobacteriaceae*.

DNA G+C content (mol%): 27–56.

Type genus: **Flavobacterium** Bergey, Harrison, Breed, Hammer and Huntoon 1923, 97^{AL} emend. Bernardet, Segers, Vancanneyt, Berthe, Kersters and Vandamme 1996, 139^{VP}.

Further descriptive information

The three families that were grouped into the order *Flavobacteriales* on the basis of phylogenetic analyses represent very different numbers of organisms. While the family *Blattabacteriaceae* is so far limited to a single nonculturable species and the family *Cryomorphaceae* comprises four genera each only represented by a single species, the family *Flavobacteriaceae* currently includes more than 61 genera, some of them comprising more than 40 species. The only *Blattabacterium* species known so far is characterized by its very special habitat. On the other hand, differentiation of the two other families is hindered by the variability of most key characteristics between and within their genera and species. Hence, it is not possible to propose a key to the families of the order *Flavobacteriales* or a table of differentiating characteristics and 16S rRNA sequence analysis remains the best method to allocate a new isolate to a family.

References

Bergey, D.H., F.C. Harrison, R.S. Breed, B.W. Hammer and F.M. Huntoon. 1923. Bergey's Manual of Determinative Bacteriology. Williams & Wilkins, Baltimore.

Bernardet, J.-F., P. Segers, M. Vancanneyt, F. Berthe, K. Kersters and P. Vandamme. 1996. Cutting a Gordian knot: emended classification and description of the genus *Flavobacterium*, emended description of the family *Flavobacteriaceae*, and proposal of *Flavobacterium hydatis* nom. nov. (basonym, *Cytophaga aquatilis* Strohl and Tait 1978). Int. J. Syst. Bacteriol. *46*: 128–148.

Family I. **Flavobacteriaceae** Reichenbach 1992b, 327[VP] (Effective publication: Reichenbach 1989b, 2013.) emend. Bernardet, Segers, Vancanneyt, Berthe, Kersters and Vandamme 1996, 145 emend. Bernardet, Nakagawa and Holmes 2002, 1057

Jean-François Bernardet

Fla.vo.bac.te'ri.a.ce'a.e. N.L. neut. n. *Flavobacterium* type genus of the family; *-aceae* ending to denote a family; N.L. fem. pl. n. *Flavobacteriaceae* the family *Flavobacterium*.

Gram-stain-negative, nonsporeforming, short or moderately long to filamentous, rigid to flexible rods. Coiled and helical cells are formed by members of some taxa, but ring-shaped cells are not formed. Cells of some species may form coccoid bodies in old cultures. Cells usually divide by binary fission, but a budding process has also been reported in *Formosa agariphila*. Nonmotile or motile by gliding. Flagella are absent; an exception is *Polaribacter irgensii*, although flagellar motility could not be observed in the only strain available. Cells of the three *Muricauda* species produce long and relatively thick appendages. Colonies are round or uneven to rhizoid, convex to flat, and may be sunken into or adherent to the agar. Nondiffusible, light to bright yellow to orange pigments are usually produced, but members of some genera form nonpigmented colonies.

Chemo-organotrophic. Unsaturated menaquinone with six isoprene units (MK-6) is the only or major respiratory isoprenoid quinone. Growth is usually aerobic, but microaerobic to anaerobic growth occurs in members of some genera. The type of metabolism is respiratory or (rarely) fermentative with oxygen as electron acceptor, except for a few species that may use nitrates or nitrites as electron acceptors. Nitrates are usually not reduced. Members of a few genera are able to reduce nitrates, but they cannot reduce nitrites; an exception is *Flavobacterium denitrificans*, a facultative anaerobe that grows by denitrification in mineral medium. Oxidase and catalase activities are usually present. Growth occurs on organic nitrogen compounds (peptones, yeast extract, or Casamino acids) as the sole source of nitrogen and often also of carbon and energy (although carbohydrates are the preferred sources of carbon and energy); inorganic nitrogen is also used by most members of the family. Vitamins and amino acids are not required. Members of most species are able to degrade a range of organic macromolecules such as proteins (e.g., casein, gelatin, etc.), lipids (e.g., lecithin, Tweens, etc.), and simple or complex carbohydrates (e.g., esculin, starch, pectin, agar, chitin, carboxymethylcellulose, etc.); however, crystalline cellulose (i.e., filter paper) is not degraded. Most members of the family are halophilic to varying degrees, requiring NaCl or seawater for growth; some others are merely halotolerant. Most species are mesophilic, but a number of species and genera are psychrophilic. Pigments belong to the carotenoid or flexirubin type or both. Intracellular granules of poly-β-hydroxybutyrate are absent. Sphingophospholipids are absent. Homospermidine is usually the major polyamine. Members of the family typically contain high levels of branched saturated, branched monounsaturated, and branched hydroxy C_{15}–C_{17} fatty acids.

The organisms are most frequently isolated from soil or from fresh, brackish, or seawater, in temperate, tropical, or polar areas. Some species are frequently isolated from food and dairy products. The organisms also occur in biofilms in various environments. Several species are pathogenic for humans and other warm-blooded animals; other species are pathogenic for freshwater and marine fish, amphibians, and other aquatic animals, or marine microalgae. Unidentified members of the family have also been found to occur in the guts of insects and intracellularly in amoebae. Analysis of 16S rRNA gene sequences show that the family belongs to the phylum *Bacteroidetes* phyl. nov., class *Flavobacteriia*, order *Flavobacteriales* ord. nov.. The most closely related organisms are members of the families "*Blattabacteriaceae*" and *Cryomorphaceae* (Bowman et al., 2003).

DNA G+C content (mol%): 27–56.

Type genus: **Flavobacterium** Bergey, Harrison, Breed, Hammer and Huntoon 1923, 97[AL] emend. Bernardet, Segers, Vancanneyt, Berthe, Kersters and Vandamme 1996, 139.

Historical, phylogenetic, and taxonomic overview

The history and evolution of the family *Flavobacteriaceae* and of the genera it comprises have already been extensively presented (Bernardet and Nakagawa, 2006; Bernardet et al., 2002; Bernardet et al., 1996). In the previous edition of *Bergey's Manual of Systematic Bacteriology*, the organisms currently grouped in the family (as well as many organisms now removed from the family) were divided among the chapters "Genus *Flavobacterium*" (Holmes et al., 1984c) and "Order I. *Cytophagales*" (Reichenbach, 1989b). The first mention of the family *Flavobacteriaceae* appeared in Jooste's PhD thesis (Jooste, 1985). The brief description of the family subsequently included in *Bergey's Manual* (Reichenbach, 1989b) was based on the features of the genus *Flavobacterium* as it was defined at that time, i.e., yellow-orange to nonpigmented, nongliding, strictly aerobic organisms retrieved from various environments and from clinical specimens, which may become pathogenic. Extensively revised descriptions of the genus *Flavobacterium* and of the family *Flavobacteriaceae* were published following a thorough polyphasic study; at that time, the family comprised eight genera and the organisms that would later become the genera *Myroides* and *Tenacibaculum* (Bernardet et al., 1996). The publication of minimal standards for the description of new taxa in the family was the opportunity to further emend the description of the family, then comprising 18 genera and two generically misclassified organisms (Bernardet et al., 2002). Following improvement and increased accessibility of molecular sequencing techniques and extensive investigations of various remote, mostly polar and/or marine environments, the number of new genera attributed to the family has been growing at an exponential pace. At the time of writing (February 2006), 54 genera belong to the family (including one with a name that has not been validly published, "*Fucobacter*") and several more are "in press" or accepted for publication in the *International Journal of Systematic and Evolutionary Microbiology*. See *Recently published genera*, below, for more information. The description of *Pibocella ponti* (Nedashkovskaya et al., 2005b) was actually based on an erroneous 16S rRNA gene sequence (O. Nedashkovskaya, personal communication); consequently, this organism is not considered in this chapter.

The 54 genera are listed in Table 23 and their phylogenetic relationships are shown in Figure 26. Phylogenetic analyses

TABLE 23. Differential characteristics of genera in the family *Flavobacteriaceae*[a,b]

Characteristic	*Flavobacterium* (35)	*Aequorivita* (4)	*Algibacter* (1)	*Aquimarina* (1)	*Arenibacter* (4)	*Bergeyella* (1)	*Bizionia* (5)	*Capnocytophaga* (7)	*Cellulophaga* (5)	*Chryseobacterium* (19)	*Coenonia* (1)	*Croceibacter* (1)	*Dokdonia* (1)	*Donghaeana* (1)	*Elizabethkingia* (2)	*Empedobacter* (1)	*Epilithonimonas* (1)	*Formosa* (2)
Habitat	FL (fe, me, te), S or P	FL (te, me) or S	S (me)	FL (me)	FL (me) or S	P or S	FL (me) or S	P or S	FL (me) or S	FL (fe, me, te), S or P	P	FL (me) or S	FL (me)	FL (me)	FL (fe, te), S or P	P or S	FL (fe) or S	FL (me) or S
Cold or polar environment	d	+	–	–	–	–	d	–	d	–	–	–	–	–	–	–	–	–
Cell morphology	Rods or filaments	Rods	Rods	Rods	Rods	Rods	Rods	Rods	Rods	Rods	Rods	Rods	Rods and elongated rods	Rods and elongated rods	Rods	Rods	Rods	Rods; buds may be produced
Spherical cells in stationary phase	d	–	–	–	–	–	–	–	–	–	–	–	–	–	–	–	–	–
Production of yellow pigment	+ (F and/or C)	+ (C)	+ (F–)	+ (F)	+ (C)	–	+ (C)	+ (F/nd)	+ (C)	+ F	–	+ C	+ (F–)	+ (F–)	–/(+) (F)	(+) (F)	+ (F)	+ (C)
Gliding motility	d	–	+	+	d	–	–	+	+	–	–	–	–	–	–	–	–	+
Na⁺ or seawater requirement	–	d	+	+	d	–	+	–	+	–	–	+	+	+	–	–	–	d
Optimal growth conditions	A/FA (31/4)	A	FA	A	A	A	A	A/ME/AN	A	A	ME/AN	A	A	A	A	A	A	FA
Growth at (°C):																		
25	d	+	+	+	+	+	+/(+)	nd	+	+	(+)	+	+	+	+	+	+	+
37	–	–	–	–	+	+	–	+	d	d	+	–	–	–	+	d	–	–
42	–	–	–	–	d	–	–	nd	–	–	–	–	–	–	–	–	–	–
Acid production from:																		
Glucose	d	–	+	–	d	–	–	+	d	d	+	nd	–	–	+	d	–	+
Sucrose	d	–	+	–	+	–	–	d	d	–	–	nd	–	–	–	–	nd	–
Production of:																		
DNase	d	d	–	+	–	–	d	nd	d	d	nd	+	nd	nd	+	+	–	–/nd
Urease	d	d	nd	–	d	+	d	d	d	d *	–	+	–	–	d	V*	–	–/V
Oxidase	d	–	+	+	+	+	+	–	+	+	+	–	+	+	+	+	+	+/V
Catalase	+/(+)	+	+	+	+	+	+	d	+	+/nd	+	+	+	+	+	+	+	+
H_2S	d	–	–	–	d	–	+	–	–	–	nd	nd	–	–	d	–	nd	–
Indole	–	nd	–	–	–	+	–	–	–	d	–	–	–	–	d	+	–	–
β-Galactosidase	d	–	+	nd	+	–	–	d	nd	d	+	–	–	–	+	–	–	d
Nitrate reduction	d	–	–	–	d	–	–	d	d	d	–	–	–	–	–	–	+	–/V
Carbohydrate utilization	d	d	+	–	+	–	–	d	nd	+	+	+	–	–	nd	d	+	+
Degradation of:																		
Agar	d	–	+	–	–	–	–	–	+	–	–	–	–	–	–	–	–	d
Starch	d	d	+	+	–	–	–	d	+	d	nd	+	–	+	–	V	(+)	d
Esculin	d	d	nd	nd	+	–	d	d	nd	+	+	–	+	+	+	–	+	+
Gelatin	d	+	+	+	d	+	+	d	+	+	–	+	–	–	+	+	–	+
Resistance to penicillin G	d/nd	nd	+	+	+	–	nd	–	+	+/nd	nd	nd	–	–	+/nd	+	–	+
DNA G+C content (mol%)	30–41[c]	33–39	31–33	32–33	37–40	35–37	38–45	36–44	32–36	29–39	35–36	35	38	37	35–38	31–33	38	34–36

[a]Symbols: +, 90% or more of the strains or species are positive; –, 10% or less of the strains or species are positive; (+), 90% or more of the strains or species give a weak or delayed positive response; d, 11–89% of the strains or species are positive; V, varies between references; and *, data in these two boxes were erroneously inverted in Bernardet et al. (2002). Where optimal growth conditions vary between members of the genus, the different conditions are listed, separated by slashes (the number of species that can grow under the corresponding conditions is shown in parentheses for the type genus). Abbreviations: FL, free-living; P, parasitic; S, saprophytic; fe, freshwater environment; me, marine environment; te, terrestrial environment; nd, not determined; F, flexirubin-type pigments; F–, pigments that do not belong to the flexirubin-type but that have not been further characterized; C, carotenoid-type pigments; A, aerobic growth; ME, microaerobic growth or growth in a CO_2-enriched atmosphere; FA, facultatively anaerobic growth; and AN, anaerobic growth.

[b]The number of species is specified in parentheses after the name of the genus.

[c]A DNA G+C content of 41 mol% has been reported for *Flavobacterium saliperosum* (Wang et al., 2006).

[d]The transfer of *Stanierella latercula* to the genus *Aquimarina* has been proposed recently (O. Nedashkovskaya, personal communication).

Data from Colwell et al. (1966), Lewin (1969), Lewin and Lounsbery (1969), Holmes et al. (1977, 1978, 1984c, 1986b, a), Oyaizu and Komagata (1981), Yabuuchi et al. (1983), London et al. (1985), Dees et al. (1986), Bruun and Ursing (1987), McGuire et al. (1987), Bernardet and Grimont (1989), Reichenbach (1989b), Ursing and Bruun (1991), Holmes (1992), Dobson et al. (1993), Segers et al. (1993b), Ostland et al. (1994), Vandamme et al. (1994a, b, 1996a, 1999), Yamamoto et al. (1994), Bernardet et al. (1996), Vancanneyt et al. (1996, 1999), Bowman et al. (1997, 1998), Hugo (1997), Gosink et al. (1998), McCammon et al. (1998), Johansen et al. (1999), Bowman (2000), McCammon and Bowman (2000), Yamaguchi and Yokoe (2000), Barbeyron et al. (2001), Bruns et al. (2001), Humphry et al. (2001), Ivanova et al. (2004, 2001), Suzuki et al. (2001), Bowman and Nichols (2002), Macián et al. (2002), Sakai et al. (2002), Cho and Giovannoni (2003, 2004), Hugo et al. (2003), Kämpfer et al. (2003), Li et al. (2003a), Nedashkovskaya et al. (2003a–c, 2004b, c, 2005a, c–f, h, i, 2006d, g), Tamaki et al. (2003), Zhu et al. (2003), Kim et al. (2004, 2005a, b), Sohn et al. (2004),Van Trappen et al. (2004a, b, 2005), Aslam et al. (2005), de Beer et al. (2005), Bowman and Nichols (2005), Jung et al. (2005), Lau et al. (2005a–c, 2006), Mancuso Nichols et al. (2005), Shen et al. (2005), Shimomura et al. (2005), Yi et al. (2005a, b), Yoon et al. (2005a–d, 2006b, c), Young et al. (2005), Khan et al. (2006a), O'Sullivan et al. (2006), Park et al. (2006b), Wang et al. (2006), J.-F. Bernardet (unpublished results), J.P. Bowman (personal communication), and O. Nedashkovskaya (personal communication).

TABLE 23. (Continued)

Characteristic	*"Fucobacter"* (1)	*Gaetbulibacter* (1)	*Gaetbulimicrobium* (1)	*Gelidibacter* (4)	*Gillisia* (5)	*Gramella* (2)	*Kaistella* (1)	*Kordia* (1)	*Krokinobacter* (3)	*Lacinutrix* (1)	*Leeuwenhoekiella* (2)	*Maribacter* (5)	*Mesonia* (1)	*Muricauda* (3)	*Myroides* (2)	*Nonlabens* (1)	*Olleya* (1)	*Ornithobacterium* (1)
Habitat	FL (me) or S	FL (me)	FL (me)	FL (me) or S	FL (me) or S	FL (me) or S	FL (fe)	FL (me) S, or P	FL (me) or S	S (me)	FL (me) or S	FL (me) or S	FL (me) or S	FL (me)	P or S	S (me)	S (me)	P
Cold or polar environment	–	–	–	d	d	–	–	–	–	+	d	–	–	–	–	–	+	–
Cell morphology	Rods	Rods	Rods	Rods or filaments	Rods	Rods	Rods	Rods	Rods	Straight or slightly curved rods	Rods	Rods	Rods	Rods with append-ages	Rods	Rods	Rods with tapered ends	Plump rods
Spherical cells in stationary phase	–	–	–	+	–	–	–	–	–	–	–	–	–	–	–	–	–	–
Production of yellow pigment	+	+ (F–)	+ (C)	+ (C)	+ (F–)	+ (C)	+ (nd)	+ (C)	+ (C)	+ (F–)	+ (F–)	+ (F–)	d (F–)	+ (F–)	+ (F)	+ (F–)	+ (F–)	–
Gliding motility	–	+	+	+	–	+	–	–	nd	–	+	+	–	–	–	–	+	–
Na^+ or seawater requirement	+	+	+	d	d	+	–	+	+	+	–	+	+	+	–	+	+	–
Optimal growth conditions	A	FA	A	A	A	A	A	A	A	A	A	A	A	A/FA	A	A	A	A/ME/AN
Growth at (°C):																		
25	+	+	+	+	d	+	+	+	+	+	+	+	+	+	+	+	+	–
37	nd	+	+	d	–	d	+	+	–	–	+/d	–	–	+	+	+	–	+
42	nd	–	–	–	–	–	d	–	–	–	–	–	–	–	–	+	–	+
Acid production from:																		
Glucose	+	d	nd	+	d	d	–	–	–	–	–	d	–	+	–	–	+	d
Sucrose	–	d	–	+/nd	+/nd	d	nd	nd	nd	–	d	d	–	+	–	+	nd	–
Production of:																		
DNase	nd	nd	nd	d	d	d	nd	nd	+	–	–	d	–	nd	+	+	–	–
Urease	+	–	–	d	d	–	d	–	–	–	–	–	–	–	+	–	–	+
Oxidase	+	+	+	–	+	+	+	+	+	–	+	+	+	+	+	+	+	+
Catalase	+	+	+	+	+	+	+	–	+	+	+	+	+	+	+	+	+	–
H_2S	nd	–	–	–	–	–	–	–	nd	–	–	–	+	–	nd	–	–	–
Indole	–	nd	–	–	–	–	+	–	–	–	–	–	–	–	–	–	–	–
β-Galactosidase	–	–	–	d	–	d	–	–	+	–	+	d	–	nd	–	–	–	+
Nitrate reduction	–	+	–	–	d	–	d	+	–	–	–/V	d	–	–	–	–	–	–
Carbohydrate utilization	nd	+	+	+	d	+	+	+	+	+	+	+	–	+	–	+	+	+
Degradation of:																		
Agar	nd	–	–	–	–	–	–	–	–	–	–/V	d	–	–	–	–	–	–
Starch	–	+	+	d	d	+	–	+	–	–	+	d	–	–	–	+	–	nd
Esculin	+	+	+	+	d	+/nd	+	–	+	–	+	+/nd	nd	+	–	–	–	–
Gelatin	+	–	nd	d	d	+	+	+	+	+	+	d	+	–	+	+	+	–
Resistance to penicillin G	nd	–	–	nd	+/nd	–/nd	nd	nd	nd	nd	d/nd	+	+	nd	nd	–	nd	d
DNA G+C content (mol%)	32	35	36	37–42	32–39	40	41–42	34	33–39	37	35–38	35–39	32–34	41–45	30–38	34	49	37–39

comparing the sequences of the 16S rRNA and DNA gyrase B subunit (*gyrB*) genes (Bernardet and Nakagawa, 2006; Suzuki et al., 2001) have resulted in very similar trees showing a clear delineation of the family within the phylum *Bacteroidetes*. The family comprises two separate, well-defined clades; one clade includes the genera *Weeksella, Empedobacter, Elizabethkingia, Chryseobacterium, Riemerella,* and several allied genera, whereas all other genera are grouped in the other clade (Figure 26). This dual structure could be the basis for a future splitting of the family, providing clear differential features are found between members of the two clades. When the distribution of significant traits (e.g., pathogenicity for humans and animals; ability to grow under microaerobic, facultatively anaerobic or anaerobic conditions; halophilic and/or psychrophilic growth conditions; presence of yellow pigment) among the phylogenetic tree is evaluated, it appears that all members of the smaller clade (i.e., *Weeksella, Riemerella,* etc.) are devoid of gliding motility and unable to grow under halophilic (except some *Chryseobacterium* species) or psychrophilic (except members of the genus *Sejongia*) conditions. This clade also comprises most of the nonpigmented and pathogenic members of the family *Flavobacteriaceae*. Conversely, most members of the other, larger "marine clade" (Bowman, 2006; Bowman and Nichols, 2005) are halotolerant to halophilic, about half of them display gliding motility, and a third are psychrophilic.

Although most members of the family do comply with the theoretical threshold value of 97% 16S rRNA gene sequence similarity below which two bacterial strains belong to different species (i.e., they share less than 70% DNA similarity) (Stackebrandt and Goebel, 1994), there are examples where members of different species actually share up to 98–99% sequence similarity (Bowman, 2000; Li et al., 2003a; Vandamme et al., 1996a).

TABLE 23. (Continued)

Characteristic	*Persicivirga* (1)	*Polaribacter* (4)	*Psychroflexus* (3)	*Psychroserpens* (1)	*Riemerella* (2)	*Robiginitalea* (1)	*Salegentibacter* (3)	*Sejongia* (2)	*Stanierella* (1)[d]	*Stenothermobacter* (1)	*Subsaxibacter* (1)	*Subsaximicrobium* (2)	*Tenacibaculum* (6)	*Ulvibacter* (1)	*Vitellibacter* (1)	*Weeksella* (1)	*Winogradskyella* (4)	*Zobellia* (5)
Habitat	FL (me)	FL (me) or S	FL (me) or S	FL (me) or S	P	FL (me)	FL (me) or S	FL (te)	FL (me) or S	S (me)	FL (te) or S	FL (te) or S	FL (me), S or P	FL (me) or S	FL (me) or S	P or S	S (me)	FL (me) or S
Cold or polar environment	–	+	d	+	–	–	d	+	–	–	+	+	–	–	–	–	–	–
Cell morphology	Rods	Rods, filaments, or coils; gas vesicles	Rods, filaments, or coils	Ring shaped, helical or coiled cells	Rods	Rods	Rods	Rods	Rods	Rods in chains	Cocco-bacilli	Rods	Rods or filaments	Rods	Rods	Rods	Rods	Rods
Spherical cells in stationary phase	–	+	d	+	–	+	–	–	–	–	–	+	+/(+)	–	–	–	–	–
Production of yellow pigment	+ (F)	+ (C/nd)	+ (C)	+ (C)	–	+ (C)	+ (F–)	+ (F–)	+ (C)	+ (F–)	+ (F–)	+ (F–)	+/(+) (C/F–)	+ (F)	+ (F)	–	+ (F–)	+ (F)
Gliding motility	–	–	d	–	–	–	–	–	–	+	+	+	+	–	–	–	+	+
Na^+ or seawater requirement	+	+	+	+	–	+	d	–	+	+	+	+	d	–	–	–	+	+
Optimal growth conditions	A	A	A	A	A/ME/AN	A	A	A	A	A	A	A	A	A	A	A	A	A
Growth at (°C):																		
25	nd	–	d	–	d	+	+	+	+	+	–	(+)/–	+	+	+	+	+	+
37	–	–	d	–	+	+	d	–	–	–	–	–	d	–	+	+	d	d
42	–	–	d	–	+	+	–	–	–	–	–	–	–	–	+	+	d	d
Acid production from:																		
Glucose	–	+	d	–	+	–	d	+	–	–	–	–	–/nd	nd	–	–	d	d
Sucrose	nd	d	d	–	–	nd	d	–	–	–	nd	nd	–/nd	nd	–	–	d	d
Production of:																		
DNase	nd	nd	d	–	nd	–	d	+	+	–	–	d	+	–	+	–	d	d
Urease	–	–	d	–	d	–	–	–	–	–	–	–	–/nd	–	–	–	–	–
Oxidase	–	d	d	–	+	+	+	+	+	+	nd	nd	+	+	+	+	+	+
Catalase	–	+/(+)	+	+	+	+	+	+	–	(+)	+	+	+	+	+	+	+	+
H_2S	nd	–	–/nd	–	–/nd	nd	+	–	+	–	–	–	–/nd	–	–	–	d	–
Indole	–	–	–/nd	–	d	–	–	+	nd	–	–	–	nd	–	–	+	–	–
β-Galactosidase	nd	d	–	d	–	+	+	–	V	+	–	–	nd	–	–	–	–	+
Nitrate reduction	–	–	d	–	–	–	d	–	V	–	–	–	d	+	–	–	–	+
Carbohydrate utilization	–	+	+	–	+	+	+	–	+	+	–	+	d	–	+	–	d	+
Degradation of:																		
Agar	–	–	–	–	–	–	–	–	+	–	–	–	–	–	–	–	d	+
Starch	–	+/(+)	+	–	nd	+	+	+	+	+	–	d	d	–	–	–	d	d
Esculin	–	d	d	–	d	+	+	+	nd	nd	d	+	–	nd	nd	–	nd	+
Gelatin	+	d	d	d	+	–	+	+	+	+	+	+	+	+	+	+	+	+
Resistance to penicillin G	+	nd	nd	nd	–	–	–	nd	+	–	nd	nd	–/nd	+	+	–	d	d
DNA G+C content (mol%)	35	31–34	32–39	27–29	29–37	55–56	37–38	34–36	34	41	35	39–40	30–35	36–38	41	37–38	33–36	36–43

Habitat

As shown in Table 23, flavobacteria may be free-living (in freshwater, marine, or terrestrial environments), saprophytic, or parasitic. Recent surveys of bacterial communities in various environments [reviewed by Jooste and Hugo (1999) and Bernardet and Nakagawa (2006)] using culture-dependent or -independent methods have revealed many members of the family *Flavobacteriaceae*. They seem to be particularly abundant in marine, mostly polar environments (Abell and Bowman, 2005; Bano and Hollibaugh, 2002; Bowman, 2006; Kirchman et al., 2003; Köpke et al., 2005) where they play an important role in the degradation of organic matter (Barbeyron et al., 2001; Descamps et al., 2006; Elifantz et al., 2005; Humphry et al., 2001; Sakai et al., 2002) and in the nutrient turnover in oceans (Cottrell and Kirchman, 2000; Pinhassi et al., 1999, 2004). Many of them are associated with marine phytoplankton (Nicolas et al., 2004) and diatoms (Grossart et al., 2005) or their detritus (Abell and Bowman, 2005; Pinhassi et al., 2005); they represent one of the dominant bacterial groups in the sediments of sea-cage salmon farms (Bissett et al., 2006). As a consequence, most of the recently described members of the family have been retrieved from marine plants and animals, seawater, sea ice, or lake water with marine salinity (Barbeyron et al., 2001; Bowman, 2005; Bowman and Nichols, 2002, 2005; Bruns et al., 2001; Cho and Giovannoni, 2003, 2004; Ivanova et al., 2004; Jung et al., 2005; Khan et al., 2006a; Lau et al., 2005b, c, 2006; Mancuso Nichols et al., 2005; McCammon and Bowman, 2000; Nedashkovskaya et al., 2003a, b, c, 2004a, b, c, d, 2005a–i; Sakai et al., 2002; Sohn et al., 2004; Van Trappen et al., 2003, 2004a, b; Yoon et al., 2005a, 2006b, c). Other habitats in which flavobacteria occur abundantly and from which new taxa have been described include: freshwater and river biofilms

(Crump et al., 1999; Manz et al., 1999; Van Trappen et al., 2002) and glaciers (Xiang et al., 2005; Zhu et al., 2003); soil (Bodour et al., 2003; Bowman and Nichols, 2002; Radianingtyas et al., 2003; Rosado and Govind, 2003); and food and dairy products (Bernardet et al., 2006; Hugo et al., 1999, 2004a, b). A few species are also able to infect a variety of plants and animals, as well as humans (Bernardet et al., 2006, 2005; Hugo et al., 2004a, b); others are harbored by insects (Campbell et al., 2004; Dugas et al., 2001; Lysyk et al., 1999) or amoebae (Horn et al., 2001).

Identification

Table 23 lists the main differential features of the 54 genera currently included in the family *Flavobacteriaceae*. Additional information is provided in the chapters dealing with individual genera in this edition. Key characteristics for differentiating members of the family are the presence of gliding motility, the production and type of yellow pigment, the requirement of NaCl or seawater, the temperature range for growth, the ability to grow under microaerobic, facultatively anaerobic or anaerobic conditions, the presence of a number of enzymic activities, and the DNA G+C content (mol%). The emended description of the family (Bernardet et al., 2002) is still relevant since most of the 29 new taxa described since then comply with it, although the proportion of halophilic and/or psychrophilic organisms has increased and a few facultatively anaerobic organisms have been described (i.e., *Algibacter, Formosa, Gaetbulibacter,* and *Muricauda*). In contrast to members of the family *Flavobacteriaceae*, those of the closely related family "*Blattabacteriaceae*" are unculturable endosymbionts of insects (Clark and Kambhampati, 2003). Although the family *Cryomorphaceae* shares a number of phenotypic traits with the family *Flavobacteriaceae*, the two families are distinctly separate in phylogenetic trees (Bowman et al., 2003). The features that distinguish members of the family *Flavobacteriaceae* from those of related taxa in the phylum *Bacteroidetes* are mostly chemotaxonomic, such as the absence of sphingophospholipids (unlike members of the family *Sphingobacteriaceae*; Steyn et al., 1998) and the presence of MK-6 as the only or major respiratory quinone (unlike all tested members of other families; Hanzawa et al., 1995). The conditions in which some of the above-mentioned differential properties should preferably be determined have been specified previously (Bernardet and Nakagawa, 2006; Bernardet et al., 2002). Various commercial identification galleries, strips, or plates may be used for the identification of members of the family *Flavobacteriaceae*. Many of them, however, cannot grow at incubation temperatures recommended by the manufacturers. Therefore, growth conditions must be adapted to the organisms studied and results must be interpreted carefully.

Candidatus genera

An intraerythrocytic bacterium infecting several frog species and transmitted by a leech feeding on frogs has recently been allocated to the family *Flavobacteriaceae* following analysis of its 16S rRNA and *gyrB* gene sequences (Zhang and Rikihisa, 2004). Initially assumed to be related to *Aegyptianella pullorum* (family *Ehrlichiaceae*, order *Rickettsiales*), which causes similar infections in bird species, the new organism was designated "*Aegyptianella ranarum*". However, BLAST searches showed that its closest relative was *Elizabethkingia meningoseptica* (previously *Chryseobacterium meningosepticum*) and that it made a clade with *Weeksella virosa* and *Bergeyella zoohelcum* in phylogenetic trees based on 16S rRNA and *gyrB* genes, respectively. Interestingly, *Elizabethkingia meningoseptica* has also been reported to be responsible for outbreaks of hemorrhagic septicemia in frogs (Bernardet et al., 2005 and references therein). Since no cultivation of the organism could be achieved, it was described as "*Candidatus* Hemobacterium ranarum" and brief genus and species descriptions were published, although not in the *International Journal of Systematic and Evolutionary Microbiology* (Zhang and Rikihisa, 2004). Members of the genus "*Hemobacterium*" are nonmotile and nonculturable Gram-stain-negative rods that grow inside red blood cells. Cells of "*Candidatus* Hemobacterium ranarum" grow in membrane-bound inclusions in frog erythrocytes; the type strain is "Toronto" and accession numbers for the 16S rRNA and *gyrB* gene sequences are AY208995 and AY208996, respectively.

Recently published genera

Since the submission and review of the family chapter, the following new genera have been added to the family *Flavobacteriaceae* and their names have been validly published in the *International Journal of Systematic and Evolutionary Microbiology*: *Actibacter* (Kim et al., 2008), *Aestuariicola* (Yoon et al., 2008), *Cloacibacterium* (Allen et al., 2006), *Costertonia* (Kwon et al., 2006b), *Croceitalea* (Lee et al., 2008), *Eudoraea* (Alain et al., 2008), *Flagellimonas* (Bae et al., 2007), *Flaviramulus* (Einen and Øvreås, 2006), *Fulvibacter* (Khan et al., 2008), *Galbibacter* (Khan et al., 2007b), *Gilvibacter* (Khan et al., 2007a), *Jejuia* (Lee et al., 2009), *Joostella* (Quan et al., 2008), *Kriegella* (Nedashkovskaya et al., 2008), *Leptobacterium* (Mitra et al., 2009), *Lutaonella* (Arun et al., 2009), *Lutibacter* (Choi and Cho, 2006), *Lutimonas* (Yang et al., 2007), *Mariniflexile* (Nedashkovskaya et al., 2006a), *Marixanthomonas* (Romanenko et al., 2007), *Mesoflavibacter* (Asker et al., 2007b), *Planobacterium* (Peng et al., 2009), *Pseudozobellia* (Nedashkovskaya et al., 2009), *Salinimicrobium* (Lim et al., 2008), *Sandarakinotalea* (Khan et al., 2006c), *Sediminibacter* (Khan et al., 2007a), *Sediminicola* (Khan et al., 2006b), *Tamlana* (Lee, 2007), *Wautersiella* (Kämpfer et al., 2006), *Yeosuana* (Kwon et al., 2006a), *Zeaxanthinibacter* (Asker et al., 2007a), *Zhouia* (Liu et al., 2006), and *Zunongwangia* corrig (Qin et al., 2007).

Descriptions of *Cloacibacterium, Costertonia, Flaviramulus, Lutibacter, Mariniflexile, Sandarakinotalea, Sediminicola, Wautersiella, Yeosuana,* and *Zhouia* are given below. In addition, the genera *Gaetbulimicrobium* and *Stanierella* have been transeferred to *Aquimarina* (Nedashkovskaya et al., 2006f).

Acknowledgements

The author is indebted to Y. Nakagawa (National Institute of Technology and Evaluation, Biological Resource Center, Kisarazu, Japan) who kindly provided the phylogenetic tree (Figure 26) and to J.P. Bowman (University of Tasmania, Hobart, Australia) and O.I. Nedashkovskaya (Pacific Institute of Bioorganic Chemistry, Vladivostok, Russia) for useful comments on the manuscript.

Further reading

Website of the International Committee on Systematics of Prokaryotes (ICSP) Subcommittee on the Taxonomy of *Flavobacterium* and *Cytophaga*-like Bacteria (http://www.the-icsp.org/subcoms/Flavobacterium_Cytophaga.htm).

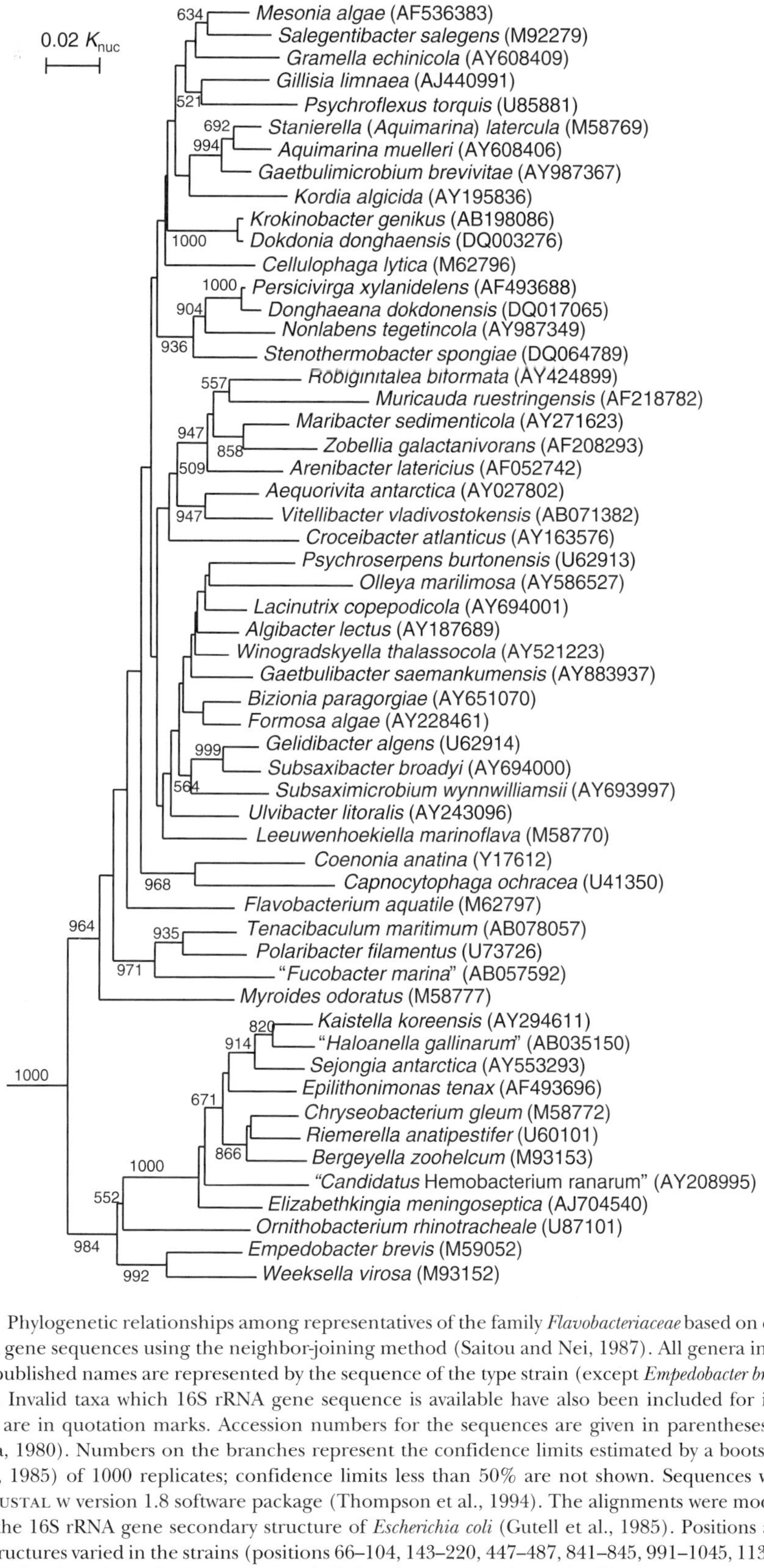

FIGURE 26. Phylogenetic relationships among representatives of the family *Flavobacteriaceae* based on comparisons of 16S rRNA gene sequences using the neighbor-joining method (Saitou and Nei, 1987). All genera in the phylum with validly published names are represented by the sequence of the type strain (except *Empedobacter brevis*) of their type species. Invalid taxa which 16S rRNA gene sequence is available have also been included for information; their names are in quotation marks. Accession numbers for the sequences are given in parentheses. Bar = 0.02 K_{nuc} (Kimura, 1980). Numbers on the branches represent the confidence limits estimated by a bootstrap analysis (Felsenstein, 1985) of 1000 replicates; confidence limits less than 50% are not shown. Sequences were aligned using the CLUSTAL W version 1.8 software package (Thompson et al., 1994). The alignments were modified manually against the 16S rRNA gene secondary structure of *Escherichia coli* (Gutell et al., 1985). Positions at which the secondary structures varied in the strains (positions 66–104, 143–220, 447–487, 841–845, 991–1045, 1134–1140, and 1446–1456; *Escherichia coli* numbering system) and all sites that were not determined in any sequence were excluded from the analysis. *Agrobacterium tumefaciens*, *Bacillus subtilis*, and *Escherichia coli* were used as outgroups.

Genus I. **Flavobacterium** Bergey, Harrison, Breed, Hammer and Huntoon 1923, 97[AL] emend. Bernardet, Segers, Vancanneyt, Berthe, Kersters and Vandamme 1996, 139

Jean-François Bernardet and John P. Bowman

Fla.vo.bac.te′ri.um. L. adj. *flavus* yellow; L. neut. n. *bacterium* a rod or staff and, in biology, a bacterium (so called because the first ones observed were rod-shaped); N.L. neut. n. *Flavobacterium* a yellow bacterium.

Straight or slightly curved, single rods with rounded or slightly tapered ends typically about 0.3–0.5 µm in diameter and variable in length, often 2–5 µm; shorter (1 µm) or filamentous (10–40 µm) cells and pleomorphism also occur. The longer rods are flexible. Do not form endospores. Several species produce spherical degenerative forms (spheroplasts) in stationary growth phase. **Nonmotile or motile by gliding. Flagella have not been reported. Colonies are pale to bright yellow due to the production of nondiffusible, nonfluorescent carotenoid or flexirubin types of pigments or both.** Gram-stain-negative. Chemoorganotrophic. **Most species are obligately aerobic, having a strictly respiratory type of metabolism** with oxygen as the terminal electron acceptor. A few species may also grow weakly under microaerobic to anaerobic conditions. **About half of the species are able to reduce nitrate to nitrite, but only one species is able to carry out complete denitrification.** The optimum growth temperature range is 20–30°C for most temperate species and 15–20°C for most cold-living species. Most species grow readily on nutrient and tryptic soy agars; no growth factors are required. Most species also grow on media containing up to 2–4% NaCl. **Usually positive for catalase and oxidase.** About half of the species are able to oxidize carbohydrates. Esculin and starch are hydrolyzed by most species, but agar and carboxymethylcellulose are hydrolyzed by only a few species. Strong proteolytic activity occurs. Menaquinone MK-6 is the only or predominant respiratory quinone. Predominant cellular fatty acids are $C_{15:0}$, $C_{15:0}$ iso, $C_{15:0}$ iso 3-OH, $C_{15:0}$ anteiso, $C_{15:1}$ ω6*c*, $C_{15:1}$ iso $_{G}$, $C_{16:0}$ iso 3-OH, and $C_{17:0}$ iso 3-OH, as well as $C_{15:0}$ iso 2-OH and/or $C_{16:1}$ ω7*c* and/or $C_{16:1}$ ω7*t*. Sphingophospholipids are absent in all tested species. Homospermidine is the major polyamine in all tested species. **Occur in soil and in freshwater, marine, or saline environments in warm, temperate, or polar locations. Three species are pathogenic for freshwater fish** and three others have occasionally been isolated from diseased freshwater fish.

DNA G+C content (mol%): 30–41.

Type species: **Flavobacterium aquatile** (Frankland and Frankland 1889) Bergey, Harrison, Breed, Hammer and Huntoon 1923, 100 (*Bacillus aquatilis* Frankland and Frankland 1889, 381) (not *Flavobacterium antarcticum* as erroneously printed in Bernardet and Bowman, 2006).

Further descriptive information

Phylogenetic position. Sequence analysis of 16S rRNA and DNA gyrase large subunit genes has shown that the genus *Flavobacterium* is a member (and the type genus) of the family *Flavobacteriaceae*, order *Flavobacteriales* ord. nov., class *Flavobacteriia*, phylum *Bacteroidetes* [previously the *Cytophaga–Flexibacter–Bacteroides* or *Cytophaga–Flavobacterium–Bacteroides* (CFB) group]. Together with about 50 other genera, it belongs to the larger of the two well-defined clades in the phylogenetic tree of the family, where it occupies a rather separate and deep position (see Figure 27 in the treatment of the family *Flavobacteriaceae*). This clade contains most of the halophilic/halotolerant members of the family and most of its psychrophilic/psychrotolerant members. Most organisms in the clade are yellow-pigmented and about half of them display gliding motility.

Data on 16S rRNA gene sequencing, DNA–DNA hybridization, and DNA G+C content in the genus *Flavobacterium* have been reviewed recently (Bernardet and Bowman, 2006) and will not be expounded further here. It is, however, necessary to stress the fact that high levels of 16S rRNA gene sequence similarity (i.e., 97.5–98.9%) have been reported between some *Flavobacterium* species (Cousin et al., 2007; Park et al., 2007; Van Trappen et al., 2004b, 2005; Yi and Chun, 2006; Zdanowski et al., 2004). Hence, DNA–DNA hybridization experiments may reveal novel species even when such similarity values are observed. For instance, two novel Antarctic isolates sharing nearly 99% 16S rRNA gene sequence similarity were shown to display only 34% DNA relatedness and the presence of several differentiating phenotypic features allowed the description of two novel species, *Flavobacterium weaverense* and *Flavobacterium segetis* (Yi and Chun, 2006). Similarly, the type (and so far only) strain of *Flavobacterium glaciei* shared 97.2–97.9% 16S rRNA gene sequence similarity with the type strains of *Flavobacterium succinicans*, *Flavobacterium granuli*, and *Flavobacterium hydatis*, but its DNA relatedness values with these three organisms were only 48, 44, and 42%, respectively (Zhang et al., 2006). At the time of writing, the genus *Flavobacterium* comprised 40 species with validly published names (Tables 24). Four additional species have unequivocally been shown not to belong to the genus, although they have not yet been assigned to other or new genera (see the *List of species*, below).

Cell morphology. Cells of *Flavobacterium* strains are usually single, straight, or slightly curved (e.g., *Flavobacterium croceum*; Park et al., 2006a) rods with rounded or slightly tapered ends. Chains of 3–4 cells, sometimes connected by cellular bridges, have been reported in *Flavobacterium denitrificans*. Cells are typically about 0.3–0.5 µm in diameter and 2–5 µm in length, but pleomorphism occurs frequently. Rods of about 1 µm long have been reported in some species and filamentous, flexible cells of 10–40 µm also occur. The size of bacterial cells may decrease during the stationary phase, as reported in *Flavobacterium weaverense* (Yi and Chun, 2006). Rods of *Flavobacterium defluvii* form characteristic knars (Park et al., 2007). Endospores are not formed. The presence of spherical degenerative forms (spheroplasts) in ageing liquid cultures has been reported for *Flavobacterium aquidurense*, *Flavobacterium branchiophilum*, *Flavobacterium columnare*, *Flavobacterium denitrificans*, *Flavobacterium gelidilacus*, *Flavobacterium hercynium*, *Flavobacterium hydatis*, *Flavobacterium indicum*, *Flavobacterium johnsoniae*, *Flavobacterium omnivorum*, *Flavobacterium psychrolimnae*, *Flavobacterium psychrophilum*, *Flavobacterium saliperosum*, and *Flavobacterium succinicans* (Reichenbach, 1989b; Bernardet and Bowman, 2006 and references therein; Cousin et al., 2007; J.F. Bernardet, unpublished data) (Table 24). Although accurate drawings of *Flavobacterium columnare* spheroplasts had

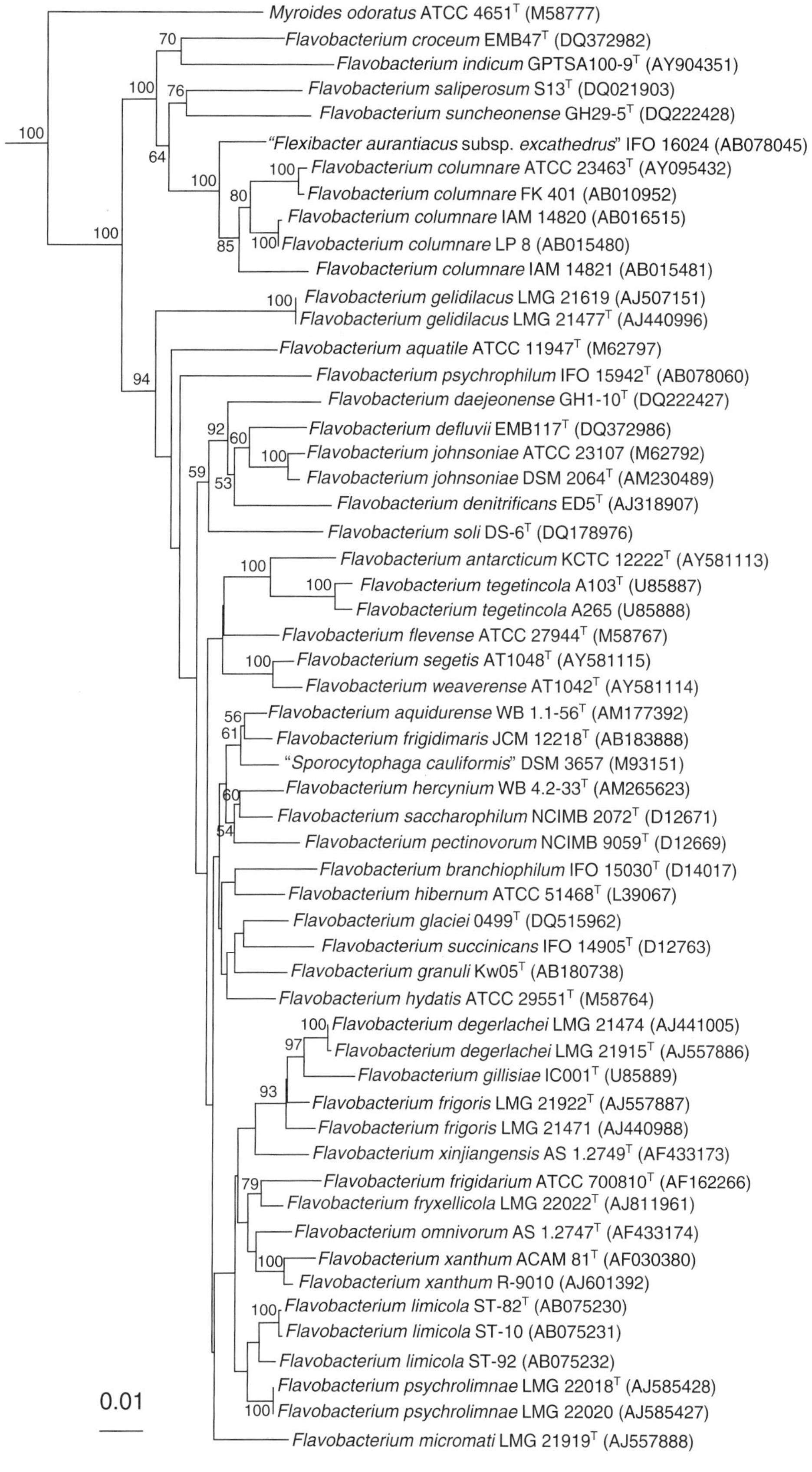

FIGURE 27. Maximum-likelihood distance phylogenetic tree created by the neighbor-joining procedure (Saitou and Nei, 1987) based on nearly complete 16S rRNA gene sequences of members of the genus *Flavobacterium. Myroides odoratus* was used as the outgroup. Bootstrap values >50% are indicated at nodes. Bar = maximum-likelihood distance.

TABLE 24. Characteristics that differentiate species in the genus *Flavobacterium* with validly published names[a,b]

	F. aquatile	*F. aquidurense*	*F. antarcticum*	*F. branchiophilum*	*F. columnare*	*F. croceum*	*F. daejeonense*
Source	Temperate freshwater	Temperate freshwater	Polar soil	Temperate freshwater and tissues of freshwater fish	Temperate freshwater and tissues of freshwater fish	Activated sludge	Temperate soil
Colony morphology on Anacker and Ordal's agar	Low convex, circular, with entire margins	Flat, spreading, with undulated margins	Convex, circular, with entire margins	Low convex, circular, with entire margins	Flat, adherent to the agar, with rhizoid margins	nd	nd
Adsorption of Congo red	−	nd	−	−	+	−	−
Gliding motility	V	−	(+)[c], −[d]	−	+	−	−
Presence of spheroplasts	−	+	−	+	+	nd	nd
Flexirubin type pigments	−	+	−	−	+	−	−
Growth on:							
Marine agar 2216	−	nd	+	−	−	nd	nd
Nutrient agar	+	+	+	−	V	−	+
Tryptic soy agar	(+)	+	+	−	−	(+)	+
Temperature range (°C)[e]	<15 to >30	13–30	5–24	5–30	15–37	10–45	5–35
Optimum temperature (°C)	nd	19–28	21	18–25	25–30	25–35	nd
Tolerance to NaCl (%, w/v)[e]	0	0–1	0–4	0–0.2	0–0.5	0–1	0–3
Optimum NaCl concentration (%, w/v)	0	nd	0	0	0	0	nd
Facultative anaerobe	−	−	−	−	−	(+)	−
Glucose utilization	nd	+	−	nd	−	nd	+
Acid produced from carbohydrates	+	nd	+	+	−	+	+
Degradation of:							
Gelatin	+	−	+	+	+	+	−
Casein	+	+	+	+	+	+	−
Starch	+	+	−	+	+[f]	−	+
Carboxymethylcellulose	−	nd	−	−	−	−	−
Agar	−	−	−	−	−	−	−
Alginate	nd	nd	−	nd	nd	nd	−
Pectin	nd	nd	−	nd	nd	nd	−
Chitin	−	nd	−	−	−	nd	−
Esculin	+	+	−	−	−	−	+
DNA	−	−	+	−	+	nd	−
Urea	−	−	−	−	−	−	−
Tween compounds	+	nd	+	+	+	−	nd
Tyrosine	+	+	−	+	−	−	−
Brown diffusible pigment on tyrosine agar	−	+	(+)	−	d	−	nd
Precipitate on egg yolk agar	+	nd	−	+	+	nd	−
Production of:							
Cytochrome oxidase	+	+	+	+	+	+	+
β-Galactosidase	+	+	−	+	−	−	+
H_2S	−	nd	−	−	+	−	nd
Nitrate reduction	+	D	−	−	d	−	+
DNA G+C content (mol%)[g]	32 (33)	34	38	29–31 (33–34)	30 (32–33)	41	35

(Continued)

	F. defluvii	*F. degerlachei*	*F. denitrificans*	*F. flevense*	*F. frigidarium*	*F. frigidimaris*	*F. frigoris*
Source	Activated sludge	Polar saline lakes	Temperate soil	Temperate freshwater	Polar marine sediment	Polar seawater	Polar freshwater and saline lakes
Colony morphology on Anacker and Ordal's agar	nd	Flat, circular, with entire margins	Flat, spreading, with lobate to rhizoid margins[c]	Low convex, circular, sunk into the agar	Flat, circular, with entire margins	Flat, spreading, with filamentous margins[c]	Flat, circular, with entire or spreading margins[c]
Adsorption of Congo red	–	–	–[c]	–	–[c], +[d]	–[c]	–
Gliding motility	+	+[c], –[d]	+[c]	+	–	+	+[c], –[d]
Presence of spheroplasts	nd	–	(+)[c]	–	–	–	–
Flexirubin type pigments	+	–	+	–	–	+	–
Growth on:							
Marine agar 2216	nd	+	–[c]	+	+	+	+
Nutrient agar	+	+	+[c]	+	+	+	–
Tryptic soy agar	+	+	+[c]	+	+	+	+
Temperature range (°C)[e]	10–40	5–30	10–30	0–30	0–24	2–26	5–25
Optimum temperature (°C)	25–30	20	25	20–25	15	18	nd
Tolerance to NaCl (%, w/v)[e]	0–2	0–5	?–2	0–1	0–9	?–3	0–5
Optimum NaCl concentration (%, w/v)	0–1	nd	nd	nd	nd	nd	nd
Facultative anaerobe	(+)	–	+	–	–	–	–
Glucose utilization	nd	+	+	+	+	+	+
Acid produced from carbohydrates	+	–	nd	+	–	+	–
Degradation of:							
Gelatin	+	–	+	–	+	+	–
Casein	+	–	nd	–	+	+	+
Starch	–	+	+	V	–	+	+
Carboxymethylcellulose	+	–	nd	–	–	–	–
Agar	–	–	–	+	–	–	–
Alginate	nd	–	nd	–	nd	nd	–
Pectin	nd	–	+	+	–	+	–
Chitin	nd	–	–	–	–	+	–
Esculin	+	+	nd	+	+	+	+
DNA	nd	–	nd	–	–	nd	–
Urea	+	–	–	–	–	–	–
Tween compounds	+	nd	nd	+	nd	nd	nd
Tyrosine	–	–	nd	–	–	nd	+
Brown diffusible pigment on tyrosine agar	–	–	nd	–	–	nd	–
Precipitate on egg yolk agar	nd	–	nd	–	–	nd	–
Production of:							
Cytochrome oxidase	–	+	–	+	+	–	+
β-Galactosidase	(+)	–	+[c]	+	–	+	–
H_2S	–	–	nd	–	–	–	–
Nitrate reduction	–	–	+	V	–	–	d
DNA G+C content (mol%)[g]	34	34	35	33–35	35	35	34–35

(Continued)

TABLE 24. (Continued)

	F. fryxellicola	*F. gelidilacus*	*F. gillisiae*	*F. glaciei*	*F. granuli*	*F. hercynium*	*F. hibernum*
Source	Polar freshwater lakes	Polar freshwater and saline lakes	Sea ice	Frozen soil	Wastewater treatment plant	Temperate freshwater	Polar freshwater lake
Colony morphology on Anacker and Ordal's agar	Flat, circular, with entire margins	Flat, circular, with entire or rhizoid margins[c]	nd	nd	Flat, circular, with entire margins[c]	Flat, translucent, no margin visible	Flat, circular, with erose to rhizoid margins[c]
Adsorption of Congo red	–	–	nd	–	–	nd	–
Gliding motility	–	+	–	–	–	+	+
Presence of spheroplasts	–	(+)[c]	–	nd	–	+	–
Flexirubin type pigments	–	–	–	–	–	+	+
Growth on:							
Marine agar 2216	–	+	+	–	–[c]	nd	–
Nutrient agar	+	+	+	+	+	+	+
Tryptic soy agar	(+)	+	+	+	+	+	+
Temperature range (°C)[e]	5–25	5–25	0–27	4–25	15–37	12–29	−7 to 30
Optimum temperature (°C)	20	20	20	21	25–30	20–27	26
Tolerance to NaCl (%, w/v)[e]	0–2	0–5	0–5	0–1	0–2	0–1	0–2
Optimum NaCl concentration (%, w/v)	nd	nd	nd	nd	nd	nd	nd
Facultative anaerobe	–	–	–	–	–	–	–
Glucose utilization	+	–	+	+	+	+	+
Acid produced from carbohydrates	–	–	+	–	–	nd	+
Degradation of:							
Gelatin	–	D	–	+	–	d	+
Casein	–	+	+	+	–	+	+
Starch	–	+	+	+	–	+	+
Carboxymethylcellulose	–	–	–	–	–	nd	–
Agar	–	–	–	–	–	–	–
Alginate	–	–	nd	–	nd	nd	–
Pectin	–	–	–	–	nd	nd	–
Chitin	–	–	+	–	–	nd	–
Esculin	+	–	+	+	nd	+	+
DNA	–	–	–	–	–	–	+
Urea	–	–	–	–	+	–	–
Tween compounds	nd	nd	+	nd	nd	nd	+
Tyrosine	–	–	–	–	–	+	+
Brown diffusible pigment on tyrosine agar	–	–	–	–	nd	+	–
Precipitate on egg yolk agar	–	–	–	nd	–	nd	–
Production of:							
Cytochrome oxidase	+	+	–	+	+	+	+[c]
β-Galactosidase	–	–	–	–	+	+	+
H_2S	–	–	–	–	–	nd	–
Nitrate reduction	–	–	–	+	–	+	+
DNA G+C content (mol%)[g]	35–36	30	32	37	36	38	34

	F. hydatis	*F. indicum*	*F. johnsoniae*	*F. limicola*	*F. micromati*	*F. omnivorum*	*F. pectinovorum*
Source	Temperate freshwater	Warm spring water	Temperate soil and freshwater	Cold freshwater	Polar freshwater and saline lakes	Frozen soil	Temperate soil
Colony morphology on Anacker and Ordal's agar	Flat, spreading, with filamentous margins	Low convex, circular, with entire margins[c]	Flat, spreading, with filamentous margins	Flat, circular, with entire margins	Flat, circular, with rhizoid margins[c]	Convex, circular, with entire margins[c]	Low convex, circular, with entire margins
Adsorption of Congo red	–	–[c]	–	+[d], –[c]	–	–[c]	–
Gliding motility	+	(+)[c], –[d]	+	–	–	–	+
Presence of spheroplasts	+	+[c]	+	–	–	(+)[c]	–
Flexirubin type pigments	+	–	+	–	–	–	+
Growth on:							
Marine agar 2216	–	+	–	–	(+)	–	–
Nutrient agar	+	(+)	+	+	(+)	+	+
Tryptic soy agar	+	–	+	+	(+)	+	+
Temperature range (°C)[e]	5–35	15–42	10–30	0–25	5–25	0 to <20	?–30
Optimum temperature (°C)	20–25	37	25–30	15–20	20	11	20–25
Tolerance to NaCl (%, w/v)[e]	0–2	0–2	0–1	0–1.5	0–2	?–3.5	?–1
Optimum NaCl concentration (%, w/v)	0	nd	nd	nd	nd	nd	nd
Facultative anaerobe	+	–	d	–	–	–	–
Glucose utilization	+	(+)	+	+	–	+	+
Acid produced from carbohydrates	+	+	+	–	–	–	+
Degradation of:							
Gelatin	+	+	+	+	–	–	+
Casein	+	+	+	+	–	+	+
Starch	+	(+)	+	+	–	+	+
Carboxymethylcellulose	V	–	+	–	–	+	+
Agar	–	–	–	+	–	–	–
Alginate	–	–	+	–	–	+	+
Pectin	+	–	+	–	–	+	+
Chitin	(+)	–	+	–	–	+	+
Esculin	+	–	+	+	+	+	+
DNA	+	–	+	–	–	–	+
Urea	–	–	+	+	–	–	–
Tween compounds	+	+	+	–	nd	–	+
Tyrosine	+	–	+	+	–	–	+
Brown diffusible pigment on tyrosine agar	–	–	d	+	–	–	–
Precipitate on egg yolk agar	–	–	–	–	–	–	–
Production of:							
Cytochrome oxidase	V	+	+	+	+	+	V
β-Galactosidase	+	–	+	–	–	+	+
H_2S	–	–	–	–	–	–	–
Nitrate reduction	+	–	d	–	–	+	+
DNA G+C content (mol%)[g]	32–34	31	33 (35)	34–35	33–34	35	34 (36)

(Continued)

TABLE 24. (Continued)

	F. psychrolimnae	*F. psychrophilum*	*F. saccharophilum*	*F. saliperosum*	*F. segetis*	*F. soli*	*F. succinicans*
Source	Polar freshwater lakes	Temperate freshwater and tissues of freshwater fish	Temperate freshwater	Temperate freshwater	Polar soil	Temperate soil	Temperate freshwater
Colony morphology on Anacker and Ordal's agar	Convex, circular, with entire margins	Low convex, circular, with entire or uneven margins	Flat, spreading, sunk into the agar	Low convex, circular, with entire margins[c]	Convex, circular to spreading, with entire to uneven margins[c]	Convex, circular, with entire margins[c]	Flat, spreading, with filamentous margins
Adsorption of Congo red	–	–	–	–	–	–	–
Gliding motility	–	(+)	+	–	–	–	+
Presence of spheroplasts	nd	+	–	+[c]	–[c]	–[c]	+
Flexirubin type pigments	–	+	+	+	–	–	–
Growth on:							
Marine agar 2216	–	–	–	–	+	+	–
Nutrient agar	+	+	+	+	+	+	+
Tryptic soy agar	+	–	+	+	+	+	+
Temperature range (°C)[e]	5–30	5–23	4–30	20–34	5–22	4–33	2 to <37
Optimum temperature (°C)	20	15–20	25	28	14	25	25
Tolerance to NaCl (%, w/v)[e]	0–2	0–1	0–2	0–1	0–3	0–2	nd
Optimum NaCl concentration (%, w/v)	nd	0	nd	0.1	0	0–0.5	nd
Facultative anaerobe	–	–	–	–	–	–	+
Glucose utilization	+	–	+	nd	+	+	+
Acid produced from carbohydrates	–	–	nd	–	+	–	+
Degradation of:							
Gelatin	–	+	+	+	–	+	(+)
Casein	+	+	+	+	+	–	+
Starch	+	–	+	–	+	–	+
Carboxymethylcellulose	–	–	(+)	–	–	–	(+)
Agar	–	–	+	–	–	–	–
Alginate	–	–	nd	–	–	nd	nd
Pectin	–	–	+	–	–	nd	nd
Chitin	–	–	–	–	–	nd	–
Esculin	+	–	+	–	+	+	+
DNA	–	(+)	–	nd	+	–	+
Urea	–	–	–	nd	–	–	–
Tween compounds	nd	+	+	nd	+	+	+
Tyrosine	–	D	+	+	(+)	–	–
Brown diffusible pigment on tyrosine agar	–	–	–	+	(+)	–	–
Precipitate on egg yolk agar	–	+	–	+	–	nd	–
Production of:							
Cytochrome oxidase	+	(+)	V	–	+	+	V
β-Galactosidase	+	–	+	–[c]	+	–	+
H_2S	–	–	+	–	–	–	–
Nitrate reduction	–	–	+	–	–	–	d
DNA G+C content (mol%)[g]	34–35	32 (33–34)	32–36 (36)	41	35	37	38 (34–37)

	F. suncheonense	*F. tegetincola*	*F. weaverense*	*F. xanthum*	*F. xinjiangense*
Source	Temperate soil	Polar saline lake	Polar soil	Polar soil and polar freshwater lakes	Frozen soil
Colony morphology on Anacker and Ordal's agar	nd	Convex, circular, with entire margins	Convex, circular, with entire margins	nd	Convex, circular, with entire margins[c]
Adsorption of Congo red	–	–	–	nd	–[c]
Gliding motility	–	+	–	–	–
Presence of spheroplasts	nd	–	–[c]	–	–
Flexirubin type pigments	–	–	–	–	–
Growth on:					
Marine agar 2216	nd	+	+	+	–
Nutrient agar	+	+	+	+	+
Tryptic soy agar	(+)	(+)	+	+	–
Temperature range (°C)[e]	15–37	0–27	5–20	<0 to 25	0 to <20
Optimum temperature (°C)	nd	20	15	15–20	11
Tolerance to NaCl (%, w/v)[e]	0–1	0–5	0–3	<3	?–3.5
Optimum NaCl concentration (%, w/v)	nd	nd	1	nd	nd
Facultative anaerobe	–	–	–	–	–
Glucose utilization	–	+	+	+	+
Acid produced from carbohydrates	–	+	+	+	–
Degradation of:					
Gelatin	+	–	–	+	+
Casein	+	–	+	+	+
Starch	–	–	+	+	–
Carboxymethylcellulose	–	–	–	–	–
Agar	–	–	–	–	–
Alginate	–	nd	–	–	–
Pectin	–	–	–	–	–
Chitin	–	–	–	–	+
Esculin	–	–	–	+	+
DNA	–	–	–	–	–
Urea	–	–	+	–	–
Tween compounds	nd	+	+	+	–
Tyrosine	+	–	(+)	–	–
Brown diffusible pigment on tyrosine agar	nd	–	(+)	–	–
Precipitate on egg yolk agar	+	–	–	–	–
Production of:					
Cytochrome oxidase	+	–	+	+	+
β-Galactosidase	–	–	–	–	–
H_2S	nd	–	–	+	+
Nitrate reduction	–	–	–	V	–
DNA G+C content (mol%)[g]	39	32–34	37	36	34

(Continued)

TABLE 24. (Continued)

[a]Symbols: +, 90% or more of the strains are positive; –, 10% or less of the strains are positive; (+), 90% or more of the strains give a weak or delayed positive response; d, 11–89% of the strains are positive; V, varies between authors; nd, no data available.

[b]Data from: Anderson and Ordal, 1961; Van der Meulen et al., 1974; Inoue and Komagata, 1976; Christensen, 1977; Strohl and Tait, 1978; Agbo and Moss, 1979; Oyaizu and Komagata, 1981; Holmes et al., 1984c; Holt, 1988; Bernardet, 1989; Bernardet and Grimont, 1989; Reichenbach, 1989b; Wakabayashi et al., 1989; Carson et al., 1993; Ostland et al., 1994; Bernardet et al., 1996; Shamsudin and Plumb, 1996; McCammon et al., 1998; Triyanto and Wakabayashi, 1999a; McCammon and Bowman, 2000; Humphry et al., 2001; Tamaki et al., 2003; Van Trappen et al. (2003, 2004b, 2005); Zhu et al., 2003; Aslam et al., 2005; Horn et al., 2005; Nogi et al., 2005b; Shoemaker et al., 2005; Yi et al., 2005a; Kim et al., 2006a; Park et al., 2006a; Saha and Chakrabarti, 2006; Wang et al., 2006; Yi and Chun, 2006; Yoon et al., 2006d; Zhang et al., 2006; Cousin et al., 2007; Park et al., 2007; and J.-F. Bernardet (unpublished data). Since some of the characteristics listed were tested by several authors using different methods, the original publications should be consulted for direct comparison.

[c]J.-F. Bernardet, unpublished data.

[d]Data from the original description.

[e]Extreme values of growth temperature and NaCl concentration ranges must be interpreted with care: bacterial growth may still be possible between them and the next lower or higher temperatures or NaCl concentrations tested.

[f]Degradation of starch by *Flavobacterium columnare* strains is medium-dependent (see comments on this species in the text).

[g]Values in parentheses are those determined by Bernardet et al. (1996) when different from previously published values.

TABLE 25. Fatty acid composition (%) of *Flavobacterium* species[a,b]

	F. aquatile	*F. aquidurense*	*F. antarcticum*	*F. branchiophilum*	*F. columnare*	*F. croceum*	*F. daejeonense*
No. of strains tested	1	5	1	7	5	1	1
Culture conditions[c]	mShieh, 48 h, 25°C	NA	R2A, 5 d, 15°C	mShieh, 48 h, 25°C	mShieh, 48 h, 25°C	R2A	R2A, 48 h, 28°C
Temperate/polar/warm environment	T	T	P	T	T	T	T
Procedure	GLC	MIDI	MIDI	GLC	GLC	MIDI	MIDI
Saturated:							
$C_{14:0}$			2				1
$C_{15:0}$	13	7	8	11	4	11	
$C_{16:0}$		1		1		1	5
$C_{18:0}$							
Unsaturated:							
$C_{15:1}$[d]							
$C_{15:1}$ ω6*c*[d]	9	7	2	8			3
$C_{10.1}$ ω5*c*							
$C_{16:1}$ ω7*c*							
$C_{16:1}$ ω9*c*							
$C_{16:1}$ *cis*[d]							
$C_{17:1}$							
$C_{17:1}$ ω6*c*	4	6		5			2
$C_{17:1}$ ω8*c*		1					1
$C_{18:1}$ ω5*c*			1				
Branched-chain:							
$C_{13:0}$ iso				1	2	1	
$C_{14:0}$ iso	1		3			9	1
$C_{14:0}$ iso 3-OH						3	
$C_{15:0}$ iso	22	15	16	22	39	9	23
$C_{15:0}$ iso 2-OH							
$C_{15:0}$ iso 3-OH	7	8	6	14	9	6	6
$C_{15:0}$ anteiso	2	3	7	3	1	5	6
$C_{15:0}$ anteiso 3-OH							
$C_{15:1}$ iso[d]							
$C_{15:1}$ anteiso							
$C_{15:1}$ anteiso A			3			2	
$C_{15:1}$ iso G[d]	9	6	15	11	13	12	4
$C_{15:1}$ iso ω10*c*[d]							
$C_{15:1}$ anteiso ω10*c*							
$C_{16:0}$ iso	2	1	4		2	9	1
$C_{16:0}$ anteiso							
$C_{16:0}$ iso 3-OH	5	2	5	2	3	17	2
$C_{16:1}$ iso[d]							
$C_{16:1}$ iso G[d]						3	
$C_{16:1}$ iso H[d]	2		3				
$C_{16:1}$ iso ω6*c*							
$C_{17:0}$ iso							
$C_{17:0}$ iso 3-OH	7	12	3	6	12	3	11
$C_{17:0}$ iso ω9*c*							
$C_{17:0}$ anteiso							
$C_{17:0}$ anteiso 3-OH							
cyclo-$C_{17:0}$ ω7,8*c*							
$C_{17:1}$ iso							
$C_{17:1}$ iso ω5*c*							
$C_{17:1}$ iso ω7*c*							
$C_{17:1}$ iso ω9*c*	5	8	2		8		2
$C_{17:1}$ anteiso							
$C_{17:1}$ anteiso ω5*c*							
$C_{17:1}$ anteiso ω9*c*			1				
$C_{17:1}$ anteiso B/I							
Hydroxy:							
$C_{14:0}$ 3-OH							
$C_{15:0}$ 2-OH		1				2	
$C_{15:0}$ 3-OH	2	3		2		1	2
$C_{16:0}$ 2-OH							
$C_{16:0}$ 3-OH		2	2	2		1	5
$C_{17:0}$ 2-OH		1					
$C_{17:0}$ 3-OH		1					
Summed features:[e]							
$C_{16:1}$ iso I and/or $C_{14:0}$ 3-OH[d]							
$C_{15:0}$ iso 2-OH and/or $C_{16:1}$ ω7*c* and/or $C_{16:1}$ ω7*t*[d]	4	12	11	5		4	21
$C_{17:1}$ iso I and/or $C_{17:1}$ anteiso B[d]							
Unknown:[f]							
ECL 13.566		1					1
ECL 16.582		1					1
ECL unspecified							

(Continued)

TABLE 25. (Continued)

	F. defluvii	*F. degerlachei*	*F. denitrificans*	*F. flevense*	*F. frigidarium*	*F. frigidimaris*	*F. frigoris*
No. of strains tested	1	14	1	1	1	1	19
Culture conditions[c]	NA	R2A, 20°C	NA	mShieh, 48 h, 25°C	XMM, 15°C	CY, 24 h, 20°C	R2A, 20°C
Temperate/polar/warm environment	T	P	T	T	P	P	P
Procedure	MIDI	MIDI	MIDI	GLC	GC-MS	GC-MS	MIDI
Saturated:							
$C_{14:0}$					2		
$C_{15:0}$	6	7	7	8	5	10	7
$C_{16:0}$	2	1	1	1	3	1	2
$C_{18:0}$							
Unsaturated:							
$C_{15:1}$[d]					1		
$C_{15:1}$ ω6*c*[d]		11	4	9		5	11
$C_{16:1}$ ω5*c*							
$C_{16:1}$ ω7*c*						14	
$C_{16:1}$ ω9*c*							
$C_{16:1}$ *cis*[d]					44		
$C_{17:1}$							
$C_{17:1}$ ω6*c*	2	7	4	3		6	4
$C_{17:1}$ ω8*c*		1	1				1
$C_{18:1}$ ω5*c*							
Branched-chain:							
$C_{13:0}$ iso							
$C_{14:0}$ iso		3		2	4		3
$C_{14:0}$ iso 3-OH							
$C_{15:0}$ iso	20	6	22	15	9	27	7
$C_{15:0}$ iso 2-OH							
$C_{15:0}$ iso 3-OH	11	5	7	5		8	4
$C_{15:0}$ anteiso	4	8	2	10	15	3	10
$C_{15:0}$ anteiso 3-OH							
$C_{15:1}$ iso[d]		5			2		3
$C_{15:1}$ anteiso		1					
$C_{15:1}$ anteiso A							
$C_{15:1}$ iso G[d]	8		5	6			
$C_{15:1}$ iso ω10*c*[d]						9	
$C_{15:1}$ anteiso ω10*c*							
$C_{16:0}$ iso	2	4	1	1	9		7
$C_{16:0}$ anteiso					2		
$C_{16:0}$ iso 3-OH	6	10	1	4		1	8
$C_{16:1}$ iso[d]		4			1		4
$C_{16:1}$ iso G[d]							
$C_{16:1}$ iso H[d]			1				
$C_{16:1}$ iso ω6*c*							
$C_{17:0}$ iso			1				
$C_{17:0}$ iso 3-OH	13	5	11	4		6	3
$C_{17:0}$ iso ω9*c*	2						
$C_{17:0}$ anteiso							
$C_{17:0}$ anteiso 3-OH							
cyclo-$C_{17:0}$ ω7,8*c*							
$C_{17:1}$ iso							
$C_{17:1}$ iso ω5*c*							
$C_{17:1}$ iso ω7*c*						7	
$C_{17:1}$ iso ω9*c*		1	14				2
$C_{17:1}$ anteiso							
$C_{17:1}$ anteiso ω5*c*							
$C_{17:1}$ anteiso ω9*c*							
$C_{17:1}$ anteiso B/I							
Hydroxy:							
$C_{14:0}$ 3-OH							
$C_{15:0}$ 2-OH							
$C_{15:0}$ 3-OH	2	2		2		2	1
$C_{16:0}$ 2-OH					1		
$C_{16:0}$ 3-OH	5	2	1	4	2	2	2
$C_{17:0}$ 2-OH							
$C_{17:0}$ 3-OH	1						
Summed features:[e]							
$C_{16:1}$ iso I and/or $C_{14:0}$ 3-OH[d]	1						
$C_{15:0}$ iso 2-OH and/or $C_{16:1}$ ω7*c* and/or $C_{16:1}$ ω7*t*[d]	10	13	11	18			15
$C_{17:1}$ iso I and/or $C_{17:1}$ anteiso B[d]			2				
Unknown:[d]							
ECL 13.566			1				
ECL 16.582			1				
ECL unspecified					3		

(Continued)

TABLE 25. (Continued)

	F. fryxellicola	*F. gelidilacus*	*F. gillisiae*	*F. glaciei*	*F. granuli*	*F. hercynium*	*F. hibernum*
No. of strains tested	3	22	1	1	1	5	2
Culture conditions[c]	R2A, 20°C	R2A, 20°C	TSA, 2 d, 20°C	mPYG, 3 d, 21°C	TSA, 2 d, 30°C	NA	MM, 25°C
Temperate/polar/warm environment	P	P	P	P	T	T	P
Procedure	MIDI	MIDI	GC-MS	MIDI	MIDI	MIDI	GC-MS
Saturated:							
$C_{14:0}$						2	
$C_{15:0}$	7	10	7	14	5	2	8
$C_{16:0}$	3		3	2	1	3	3
$C_{18:0}$			1		1		
Unsaturated:							
$C_{15:1}$[d]							
$C_{15:1}$ ω6*c*[d]	8	6	22	5	6	1	4
$C_{16:1}$ ω5*c*						1	
$C_{16:1}$ ω7*c*			22				19
$C_{16:1}$ ω9*c*							1
$C_{16:1}$ *cis*[d]							
$C_{17:1}$							
$C_{17:1}$ ω6*c*	5	3	1	10	4	1	
$C_{17:1}$ ω8*c*	1			3	1		
$C_{18:1}$ ω5*c*		1					
Branched-chain:							
$C_{13:0}$ iso					3		
$C_{14:0}$ iso	3	4	1		1		
$C_{14:0}$ iso 3-OH							
$C_{15:0}$ iso	8	12	5	8	28	23	20
$C_{15:0}$ iso 2-OH							
$C_{15:0}$ iso 3-OH	5	6	10	5	10	7	11
$C_{15:0}$ anteiso	4	8	6	8	3	3	8
$C_{15:0}$ anteiso 3-OH							1
$C_{15:1}$ iso[d]	3	10		6			
$C_{15:1}$ anteiso		1		1			
$C_{15:1}$ anteiso A							
$C_{15:1}$ iso G[d]					5	6	
$C_{15:1}$ iso ω10*c*[d]			5				5
$C_{15:1}$ anteiso ω10*c*			1				1
$C_{16:0}$ iso	10	8	4	1		1	1
$C_{16:0}$ anteiso							
$C_{16:0}$ iso 3-OH	11	10	1	3	2	1	3
$C_{16:1}$ iso[d]	5	4					
$C_{16:1}$ iso G[d]							
$C_{16:1}$ iso H[d]							
$C_{16:1}$ iso ω6*c*			3				2
$C_{17:0}$ iso							
$C_{17:0}$ iso 3-OH	5	6	1	4	11	8	3
$C_{17:0}$ iso ω9*c*							
$C_{17:0}$ anteiso			2				1
$C_{17:0}$ anteiso 3-OH							1
cyclo-$C_{17:0}$ ω7,8*c*							
$C_{17:1}$ iso							
$C_{17:1}$ iso ω5*c*			1				
$C_{17:1}$ iso ω7*c*			4				3
$C_{17:1}$ iso ω9*c*	2	2		3	8	3	
$C_{17:1}$ anteiso							
$C_{17:1}$ anteiso ω5*c*							2
$C_{17:1}$ anteiso ω9*c*							
$C_{17:1}$ anteiso B/I					2		
Hydroxy:							
$C_{14:0}$ 3-OH							1
$C_{15:0}$ 2-OH							
$C_{15:0}$ 3-OH	1	1		2			
$C_{16:0}$ 2-OH							
$C_{16:0}$ 3-OH	2			2		7	2
$C_{17:0}$ 2-OH						1	
$C_{17:0}$ 3-OH				1			
Summed features:[e]							
$C_{16:1}$ iso I and/or $C_{14:0}$ 3-OH[d]						1	
$C_{15:0}$ iso 2-OH and/or $C_{16:1}$ ω7*c* and/or $C_{16:1}$ ω7*t*[d]	14	2		10	3	26	
$C_{17:1}$ iso I and/or $C_{17:1}$ anteiso B[d]						1	
Unknown:[f]							
ECL 13.566					4		
ECL 16.582						1	
ECL unspecified							

(Continued)

TABLE 25. (Continued)

	F. hydatis	*F. indicum*	*F. johnsoniae*	*F. limicola*	*F. micromati*	*F. omnivorum*	*F. pectinovorum*
No. of strains tested	1	1	1	1	3	1	1
Culture conditions[c]	mShieh, 48 h, 25°C	mR2A, 24 h, 30°C	mShieh, 48 h, 25°C	23°C	R2A, 20°C	PYG, 2 d, 11°C	mShieh, 48 h, 25°C
Temperate/polar/warm environment	T	W	T	T	P	P	T
Procedure	GLC	MIDI	GLC	GC-MS	MIDI	MIDI	GLC
Saturated:							
$C_{14:0}$							
$C_{15:0}$	10		6	13	8	4	7
$C_{16:0}$	1		3	1	5		
$C_{18:0}$							
Unsaturated:							
$C_{15:1}$[d]				7			
$C_{15:1}$ ω6*c*[d]	5		1		6	4	6
$C_{16:1}$ ω5*c*							
$C_{16:1}$ ω7*c*				5		18	
$C_{16:1}$ ω9*c*							
$C_{16:1}$ *cis*[d]							
$C_{17:1}$				1			
$C_{17:1}$ ω6*c*	4		2		5	5	5
$C_{17:1}$ ω8*c*					1		
$C_{18:1}$ ω5*c*							
Branched-chain:							
$C_{13:0}$ iso							
$C_{14:0}$ iso	1		2	2	2		1
$C_{14:0}$ iso 3-OH							
$C_{15:0}$ iso	18	19	25	21	7	9	24
$C_{15:0}$ iso 2-OH							
$C_{15:0}$ iso 3-OH	9	5	7	4	4	7	8
$C_{15:0}$ anteiso			3	9	6	10	2
$C_{15:0}$ anteiso 3-OH							
$C_{15:1}$ iso[d]				7	4		
$C_{15:1}$ anteiso							
$C_{15:1}$ anteiso A							
$C_{15:1}$ iso G[d]	4	18	5				8
$C_{15:1}$ iso ω10*c*[d]						5	
$C_{15:1}$ anteiso ω10*c*						2	
$C_{16:0}$ iso	2	5	3	2	9	5	2
$C_{16:0}$ anteiso							
$C_{16:0}$ iso 3-OH	7	5	4	2	11	9	5
$C_{16:1}$ iso[d]				6	4		
$C_{16:1}$ iso G[d]							
$C_{16:1}$ iso H[d]	1						
$C_{16:1}$ iso ω6*c*						6	
$C_{17:0}$ iso							
$C_{17:0}$ iso 3-OH	8	9	9	4	5	9	12
$C_{17:0}$ iso ω9*c*							
$C_{17:0}$ anteiso							
$C_{17:0}$ anteiso 3-OH							
cyclo-$C_{17:0}$ ω7,8*c*				5			
$C_{17:1}$ iso				1	3		
$C_{17:1}$ iso ω5*c*				1	3		
$C_{17:1}$ iso ω7*c*				1	3		
$C_{17:1}$ iso ω9*c*	3		2	1	3	6	5
$C_{17:1}$ anteiso				1	3		
$C_{17:1}$ anteiso ω5*c*				1	3		
$C_{17:1}$ anteiso ω9*c*				1	3		
$C_{17:1}$ iso				6			
$C_{17:1}$ iso ω5*c*							
$C_{17:1}$ iso ω7*c*							
$C_{17:1}$ iso ω9*c*	3		2		1	6	5
$C_{17:1}$ anteiso				3			
$C_{17:1}$ anteiso ω5*c*							
$C_{17:1}$ anteiso ω9*c*							
$C_{17:1}$ anteiso B/I							
Hydroxy:							
$C_{14:0}$ 3-OH							
$C_{15:0}$ 2-OH							
$C_{15:0}$ 3-OH	2		1				2
$C_{16:0}$ 2-OH							
$C_{16:0}$ 3-OH	5		5		3		
$C_{17:0}$ 2-OH							
$C_{17:0}$ 3-OH							
Summed features:[e]							
$C_{16:1}$ iso I and/or $C_{14:0}$ 3-OH[d]							
$C_{15:0}$ iso 2-OH and/or $C_{16:1}$ ω7*c* and/or $C_{16:1}$ ω7*t*[d]	13	17	13		16		5
$C_{17:1}$ iso I and/or $C_{17:1}$ anteiso B[d]							
Unknown:[f]							
ECL 13.566							
ECL 16.582							
ECL unspecified				1			

(Continued)

TABLE 25. (Continued)

	F. psychrolimnae	*F. psychrophilum*	*F. saccharophilum*	*F. saliperosum*	*F. segetis*	*F. soli*	*F. succinicans*
No. of strains tested	4	5	1	1	1	1	2
Culture conditions[c]	R2A, 20°C	mShieh, 48 h, 19°C	mShieh, 48 h, 25°C	mM1, 24 h, 25°C	R2A, 5 d, 15°C	TSA, 3 d, 25°C	mShieh, 48 h, 25°C
Temperate/polar/warm environment	P	T	T	T	P	T	T
Procedure	MIDI	GLC	GLC	MIDI	MIDI	MIDI	GLC
Saturated:							
$C_{14:0}$							
$C_{15:0}$	5	6	9	5	5	8	12
$C_{16:0}$	2				1	2	
$C_{18:0}$							
Unsaturated:							
$C_{15:1}$[d]							
$C_{15:1}$ ω6*c*[d]	8	6	7		9	2	11
$C_{16:1}$ ω5*c*							
$C_{16:1}$ ω7*c*				3			
$C_{16:1}$ ω9*c*							
$C_{16:1}$ *cis*[d]							
$C_{17:1}$							
$C_{17:1}$ ω6*c*	5	2	8		7	1	4
$C_{17:1}$ ω8*c*						1	
$C_{18:1}$ ω5*c*	1						
Branched-chain:							
$C_{13:0}$ iso		1		1			
$C_{14:0}$ iso	4	2	1	1	4		2
$C_{14:0}$ iso 3-OH					1		
$C_{15:0}$ iso	7	20	10	28	5	27	17
$C_{15:0}$ iso 2-OH				1			
$C_{15:0}$ iso 3-OH	4	9	6	5	5	6	9
$C_{15:0}$ anteiso	5	4	1	4	9	3	1
$C_{15:0}$ anteiso 3-OH							
$C_{15:1}$ iso[d]	4					6	
$C_{15:1}$ anteiso							
$C_{15:1}$ anteiso A					2		
$C_{15:1}$ iso G[d]		12	7	1	4		9
$C_{15:1}$ iso ω10*c*[d]							
$C_{15:1}$ anteiso ω10*c*							
$C_{16:0}$ iso	10	3	4	7	5	4	1
$C_{16:0}$ anteiso							
$C_{16:0}$ iso 3-OH	10	3	5	2	11	3	5
$C_{16:1}$ iso[d]	8					2	
$C_{16:1}$ iso G[d]							
$C_{16:1}$ iso H[d]		4	2	2	7		2
$C_{16:1}$ iso ω6*c*							
$C_{17:0}$ iso							
$C_{17:0}$ iso 3-OH	4	9	10	9	4	12	6
$C_{17:0}$ iso ω9*c*							
$C_{17:0}$ anteiso							
$C_{17:0}$ anteiso 3-OH							
cyclo-$C_{17:0}$ ω7,8*c*							
$C_{17:1}$ iso							
$C_{17:1}$ iso ω5*c*							
$C_{17:1}$ iso ω7*c*							
$C_{17:1}$ iso ω9*c*	4	12	6	19	2	10	2
$C_{17:1}$ anteiso							
$C_{17:1}$ anteiso ω5*c*							
$C_{17:1}$ anteiso ω9*c*							
$C_{17:1}$ anteiso B/I							
Hydroxy:							
$C_{14:0}$ 3-OH							
$C_{15:0}$ 2-OH							
$C_{15:0}$ 3-OH	1		3		2		3
$C_{16:0}$ 2-OH							
$C_{16:0}$ 3-OH			1		1		2
$C_{17:0}$ 2-OH				1			
$C_{17:0}$ 3-OH							
Summed features:[e]							
$C_{16:1}$ iso I and/or $C_{14:0}$ 3-OH[d]							
$C_{15:0}$ iso 2-OH and/or $C_{16:1}$ ω7*c* and/or $C_{16:1}$ ω7*t*[d]	13	2	12		14	6	8
$C_{17:1}$ iso I and/or $C_{17:1}$ anteiso B[d]							
Unknown:[f]							
ECL 13.566							
ECL 16.582							
ECL unspecified							

(Continued)

TABLE 25. (Continued)

	F. suncheonense	*F. tegetincola*	*F. weaverense*	*F. xanthum*	*F. xinjiangense*
No. of strains tested	1	1	1	1	1
Culture conditions[c]	R2A, 48 h, 28°C	TSA, 2 d, 20°C	R2A, 5 d, 15°C	TSA, 2 d, 20°C	PYG, 2 d, 11°C
Temperate/polar/warm environment	T	P	P	P	P
Procedure	MIDI	GC-MS	MIDI	GC-MS	MIDI
Saturated:					
$C_{14:0}$	1	1			
$C_{15:0}$		7	8	7	9
$C_{16:0}$	1	2	1	2	
$C_{18:0}$					
Unsaturated:					
$C_{15:1}$[d]					
$C_{15:1}$ ω6*c*[d]	1	2	14	8	13
$C_{16:1}$ ω5*c*					
$C_{16:1}$ ω7*c*		18		23	18
$C_{16:1}$ ω9*c*					
$C_{16:1}$ *cis*[d]					
$C_{17:1}$					
$C_{17:1}$ ω6*c*		2	8		11
$C_{17:1}$ ω8*c*					
$C_{18:1}$ ω5*c*	1				
Branched-chain:					
$C_{13:0}$ iso	2				
$C_{14:0}$ iso		1	5		2
$C_{14:0}$ iso 3-OH			1		
$C_{15:0}$ iso	30	8	3	11	12
$C_{15:0}$ iso 2-OH					
$C_{15:0}$ iso 3-OH	11	6	3	13	7
$C_{15:0}$ anteiso	2	15	4	11	6
$C_{15:0}$ anteiso 3-OH		1		1	
$C_{15:1}$ iso[d]					
$C_{15:1}$ anteiso					
$C_{15:1}$ anteiso A			1		
$C_{15:1}$ iso G[d]	12		3		
$C_{15:1}$ iso ω10*c*[d]		9		7	4
$C_{15:1}$ anteiso ω10*c*		6		2	
$C_{16:0}$ iso	1	5	9	1	
$C_{16:0}$ anteiso					
$C_{16:0}$ iso 3-OH	1	3	15	2	6
$C_{16:1}$ iso[d]					
$C_{16:1}$ iso G[d]					
$C_{16:1}$ iso H[d]			12		
$C_{16:1}$ iso ω6*c*		4		1	2
$C_{17:0}$ iso					
$C_{17:0}$ iso 3-OH	18	3	2		5
$C_{17:0}$ iso ω9*c*					
$C_{17:0}$ anteiso					
$C_{17:0}$ anteiso 3-OH					
cyclo-$C_{17:0}$ ω7,8*c*					
$C_{17:1}$ iso					
$C_{17:1}$ iso ω5*c*		4		3	
$C_{17:1}$ iso ω7*c*		3		4	
$C_{17:1}$ iso ω9*c*	8		1		
$C_{17:1}$ anteiso					
$C_{17:1}$ anteiso ω5*c*					
$C_{17:1}$ anteiso ω9*c*					
$C_{17:1}$ anteiso B/I					
Hydroxy:					
$C_{14:0}$ 3-OH					
$C_{15:0}$ 2-OH					
$C_{15:0}$ 3-OH	1	1	2	1	2
$C_{16:0}$ 2-OH					
$C_{16:0}$ 3-OH	1	1	1	2	4
$C_{17:0}$ 2-OH					
$C_{17:0}$ 3-OH					
Summed features:[e]					
$C_{16:1}$ iso I and/or $C_{14:0}$ 3-OH[d]					
$C_{15:0}$ iso 2-OH and/or $C_{16:1}$ ω7*c* and/or $C_{16:1}$ ω7*t*[d]	10		6		
$C_{17:1}$ iso I and/or $C_{17:1}$ anteiso B[d]					
Unknown:[f]					
ECL 13.566					
ECL 16.582	1				
ECL unspecified					

[a]Values are given as percentages of total fatty acids. Fatty acids amounting to less than 1% of the total fatty acids in all strains tested are not included; therefore, the percentages do not total 100%. All values are rounded up. When several strains were analyzed, rounded up means are given, but standard deviations are not given.

TABLE 25. (Continued)

[b]Data from: Bernardet et al. (1996), McCammon et al. (1998), McCammon and Bowman (2000), Humphry et al. (2001), Tamaki et al. (2003), Van Trappen et al. (2003, 2004b, 2005), Zhu et al. (2003), Aslam et al. (2005), Nogi et al. (2005b), Yi et al. (2005a), Kim et al. (2006a), Park et al. (2006a, 2007), Saha and Chakrabarti (2006), Wang et al. (2006), Yi and Chun (2006), Yoon et al. (2006d), Zhang et al. (2006), and Cousin et al. (2007). The fatty acid composition of *Flavobacterium denitrificans* was determined by Cousin et al. (2007).

[c]Growth medium, duration, and temperature are specified when available. Abbreviations and compositions of growth media: NA, not available (growth conditions unspecified); mShieh, modified Shieh agar (per l: 5.0 g peptone, 1.0 g yeast extract, 10 mg sodium acetate, 10 mg $BaCl_2{\cdot}H_2O$, 0.1 g K_2HPO_4, 50 mg KH_2PO_4, 0.3 g $MgSO_4{\cdot}7H_2O$, 6.7 mg $CaCl_2{\cdot}2H_2O$, 1.0 mg $FeSO_4{\cdot}7H_2O$, 50 mg $NaHCO_3$, 15.0 g agar; Song et al., 1988); R2A (per l: 0.5 g yeast extract, 0.5 g proteose peptone, 0.5 g Casamino acids, 0.5 g glucose, 0.5 g soluble starch, 0.3 g sodium pyruvate, 0.3 g K_2HPO_4, 0.05 g $MgSO_4{\cdot}7H_2O$, 15.0 g agar; Reasoner and Geldreich, 1985); mR2A, modified R2A agar (i.e., containing 0.75 g/l proteose peptone and 0.25 g/l Casamino acids; Saha and Chakrabarti, 2006); XMM, X minimal medium (per l: 0.02 g $FeSO_4$, 0.2 g $MgSO_4$, 0.75 g KNO_3, 0.5 g K_2HPO_4, 0.04 g $CaCl_2$, 5.0 g soluble xylan, 15.0 g agar; Humphry et al., 2001); CY (per l: 3.0 g casitone, 1.36 g $CaCl_2{\cdot}2H_2O$, 1.0 g yeast extract, 15.0 g agar; Nogi et al., 2005b); TSA, tryptic soy agar (per l: 15.0 g pancreatic digest of casein, 5.0 g enzymatic digest of soybean meal, 5.0 g NaCl, 15.0 g agar); PYG, peptone-yeast extract-glucose agar [per l: 5.0 g polypeptone, 5.0 g tryptone, 10.0 g yeast extract, 10.0 g glucose, 40 ml salt solution, 15.0–20.0 g agar. Salt solution (per l): 0.2 g $CaCl_2$, 0.4 g $MgSO_4{\cdot}7H_2O$, 1.0 g K_2HPO_4, 1.0 g KH_2PO_4, 10.0 g $NaHCO_3$, 2.0 g NaCl; Zhu et al., 2003]; mPYG, modified peptone-yeast extract-glucose agar (per l: 5.0 g peptone, 0.2 g yeast extract, 3.0 g beef extract, 5.0 g glucose, 0.5 g NaCl, and 1.5 g $MgSO_4{\cdot}7H_2O$; amount of agar unspecified; Zhang et al., 2006); MM, maintenance medium [per l: 1.0 g yeast extract, 5.0 g lactose (filter-sterilized), 0.5 g K_2HPO_4, 0.5 g $(NH_4)_2SO_4$, 0.55 g NaCl, 15.0 g agar; McCammon et al., 1998]; and mM1, modified medium M1 [per l: 5.0 g peptone, 0.2 g yeast extract, 2.0 beef extract, 1.0 g NaCl, 12.0 g agar (modified from Weeks, 1955; S.-J. Liu, personal communication)].

[d]Fatty acids that are not fully identified, making their significance unclear.

[e]Fatty acids that could not be separated by GC using the Microbial Identification System (Microbial ID).

[f]The identity of the fatty acid is unknown. The equivalent chain length (ECL) is specified when available.

been published by Garnjobst (1945), these forms have been mistaken for microcysts by several authors. Consequently, *Flavobacterium columnare* has long been considered a myxobacterium (Ordal and Rucker, 1944).

Flagella have not been reported (see *Fine structure*, below), but gliding motility over wet surfaces (such as agar or glass in wet mounts) occurs in 18 of the 40 *Flavobacterium* species (Table 24). A speed of approximately 5 μm/s has been reported for cells of *Flavobacterium johnsoniae* (Nelson et al., 2007). Cells typically also display the movement accurately described by Reichenbach (1989b): "Often a cell oscillates or rotates in a conical orbit with one pole attached and the rest of the cell pointing away from the surface." Because this type of bacterial motility is only evidenced when special procedures are followed (Bernardet et al., 2002; Reichenbach, 1989b), it may actually have been overlooked in some species. For instance, the presence of gliding motility in *Flavobacterium aquatile* has been a matter of debate, as it seems that it only occurs in rather special culture conditions [see *Taxonomic comments* and the *List of Species*, below, as well as Holmes and Owen (1979) and Holmes et al. (1984c)]. *Flavobacterium denitrificans* was described as a motile organism, but the type of motility was not specified (Horn et al., 2005) and no motility has been reported in *Flavobacterium antarcticum* (Yi et al., 2005a), *Flavobacterium degerlachei* or *Flavobacterium frigoris* (Van Trappen et al., 2004b), or *Flavobacterium indicum* (Saha and Chakrabarti, 2006). However, cells from fresh broth cultures of these five organisms possess distinct gliding motility (J.-F. Bernardet, personal observation).

As repeatedly mentioned, gliding motility seems to be favored by growth on nutrient-poor media, such as Anacker and Ordal's (Anacker and Ordal, 1955), and by a high amount of moisture on the surface of the agar (Bernardet et al., 2002; McCammon et al., 1998; Nogi et al., 2005b; Reichenbach, 1989b; Van Trappen et al., 2003). In order to explain the mechanisms of gliding motility, a number of non-gliding mutants of *Flavobacterium* strains, obtained through repeated subcultivation on agar plates (Glaser and Pate, 1973; Gorski et al., 1992; Pate, 1988) or treatment of cultures with nitrosoguanidine (Abbanat et al., 1986; Godchaux et al., 1990), has been studied extensively. The latter procedure produced mutants of the *Flavobacterium johnsoniae* type strain that were deficient in both gliding motility and sulfonolipid synthesis. Restoration of the sulfonolipids (through provision of a specific biosynthetic precursor) also resulted in recovery of the ability to glide, demonstrating that sulfonolipids are specifically required for gliding motility (Abbanat et al., 1986). The existence of non-gliding mutants with normal lipid content indicated, however, that these are not the only molecules involved. Godchaux et al. (1990) showed that two non-gliding mutants of *Flavobacterium johnsoniae* were deficient in the high-molecular-weight (H) fraction of the outer membrane polysaccharide. One of the mutants was also deficient in sulfonolipids, but could be cured by provision of the specific precursor, a process that also resulted in the return of both the H fraction and gliding motility. Hence, the polysaccharide may be the component that is directly involved in gliding and the presence of sulfonolipids in the outer membrane is necessary for the synthesis or accumulation of the polysaccharide (as also suggested by the fact that the second non-gliding, polysaccharide-deficient mutant had normal sulfonolipid content). Recently, a mutant of the fish pathogen *Flavobacterium psychrophilum* that was deficient in gliding motility, growth on iron-depleted media, and extracellular proteolytic activity has been shown to exhibit enhanced biofilm formation and decreased virulence and cytotoxicity; hence, gliding motility and biofilm formation appear to be antagonistic properties (Álvarez et al., 2006). Over the last few years, a number of non-gliding mutants of *Flavobacterium johnsoniae* obtained through transposon mutagenesis have been used to identify several genes involved in gliding motility (see *Genetics*, below). Although some clues have been obtained, the molecular mechanisms of gliding in *Flavobacterium* strains have not yet been fully elucidated.

Cell-wall composition. Although there is clear and unambiguous evidence that lipids (i.e., polar lipids, quinones, and fatty acids) are very useful chemotaxonomic markers in flavobacteria, difficulties in interpreting published data have hindered their exploitation. Table 25 presents the fatty acid composition data of members of the genus *Flavobacterium*. Although some genera in the family *Flavobacteriaceae* display rather distinct fatty acid compositions (Bernardet et al., 2006; Bernardet and Nakagawa, 2006; Hugo et al., 1999), it is currently impossible to assess whether all *Flavobacterium* species display overall similar fatty acid profiles or species-specific profiles. The main reason is the considerable heterogeneity in the conditions used to grow the different species for fatty acid analysis (Table 25). *Flavobacterium limicola* is a striking example of the influence of growth temperature on fatty acid composition (Tamaki et al., 2003), but growth medium and duration also affect the fatty acid profile considerably, as shown by the significantly different profiles of the *Flavobacterium xanthum* type strain cultivated on tryptic soy agar (McCammon and Bowman, 2000) or modified Shieh agar (Bernardet et al., 1996). Moreover, different analytical procedures have been used to extract and identify fatty acids and the resolution of some of them (e.g., the MIDI system) has changed over the years, affecting the naming of some overlapping peaks and hindering the exact identification of some compounds. Hence, a number of the fatty acids listed in Table 25 (e.g., $C_{15:1}$ iso, $C_{16:1}$ iso, and $C_{16:1}$ *cis*) are not fully identified and their significance cannot be evaluated. The ten species included in the genus *Flavobacterium* at the time when its description was emended (Bernardet et al., 1996) were grown under the same conditions and their fatty acid compositions were determined using the same procedure. The very similar profile found in all ten species was one of the arguments for delineation of the genus. The various growth conditions used for the subsequently described *Flavobacterium* species now make this kind of interpretation impossible. In order to fully exploit the chemotaxonomic significance of fatty acids in the genus *Flavobacterium* in the future, all strains compared should be cultivated under the same conditions and their fatty acid composition should be determined using the same procedure. The only possible general comment regarding the fatty acid composition of *Flavobacterium* species so far is that all of them contain rather high amounts of $C_{15:0}$ iso and most of them also contain significant amounts of $C_{15:0}$, $C_{15:0}$ iso 3-OH, $C_{15:0}$ anteiso, $C_{15:1}$ ω6*c*, $C_{15:1}$ iso G, $C_{16:0}$ iso 3-OH, and $C_{17:0}$ iso 3-OH, as well as $C_{15:0}$ iso 2-OH and/or $C_{16:1}$ ω7*c* and/or $C_{16:1}$ ω7*t* (Table 25). This confirms and extends to other *Flavobacterium* species the opinion of Fautz et al. (1981) that the fatty acid profiles of *Flavobacterium johnsoniae*, *Flavobacterium hydatis*, and *Flavobacterium aquatile* are dominated by branched-chain compounds, both non-hydroxylated (mainly $C_{15:0}$ iso) and hydroxylated ($C_{15:0}$ iso 3-OH and $C_{17:0}$ iso 3-OH). Flavobacteria are interesting in that they produce both saturated and unsaturated straight-chain fatty acids and iso/anteiso-branched-chain fatty acids (B. Tindall, personal communication).

Divergent opinions have been published regarding the specificity of fatty acid profiles in the genus *Flavobacterium*. Recently, fatty acid analysis has been evaluated as an identification method for the fish pathogen *Flavobacterium columnare*. The fatty acid profile shared by all 31 strains studied differed significantly from those of the type strains of *Flavobacterium aquatile*, *Flavobacterium johnsoniae*, *Flavobacterium hydatis*, and *Flavobacterium psychrophilum* cultivated under the same conditions (Shoemaker et al., 2005). This study also revealed that one of the isolates received as *Flavobacterium columnare* had been misidentified and actually clustered with the *Flavobacterium johnsoniae* type strain. When the fatty acid compositions of ten *Flavobacterium* strains isolated from creek water and their closest phylogenetic neighbors were determined in parallel and presented as a dendrogram, the novel isolates formed two very coherent clusters that could be readily separated from the reference strains; these clusters were described as the two novel species *Flavobacterium aquidurense* and *Flavobacterium hercynium* (Cousin et al., 2007). Another study, however, showed that the fatty acid clusters delineated within a collection of polar *Flavobacterium* isolates did not correspond to the species: for instance, fatty acid cluster 5 included all *Flavobacterium micromati* and *Flavobacterium frigoris* strains as well as some *Flavobacterium degerlachei* strains, whereas the latter species also comprised strains belonging to fatty acid cluster 6 (Van Trappen et al., 2004b). This confirmed that members of the genus can differ significantly in their fatty acid composition, even when grown under the same conditions.

Godchaux and Leadbetter (1983) reported on unusual sulfonolipids (capnine and *N*-acylated versions of capnine, collectively termed capnoids) as significant (ca. 4%) outer membrane lipids in *Flavobacterium johnsoniae*; *N*-acylated capnines represented >99% of the total capnoids. These compounds are involved in gliding motility (see *Cell morphology*, above), but they cannot be used as chemotaxonomic markers as they also occur in representatives of several other genera of gliding bacteria in the phylum *Bacteroidetes* and in other bacterial groups. The same authors later reported that ornithine amino lipids represent another type of unusual lipids in the outer membrane of *Flavobacterium johnsoniae* (Pitta et al., 1989). The sulfonolipids and ornithine lipids are apparently co-regulated, as their total amount remains constant at 40% of total cellular lipids regardless of mutations or growth conditions.

The type (and so far only) strains of *Flavobacterium croceum* and *Flavobacterium defluvii* are the only new members of the genus in which the polar lipid composition has been studied (Park et al., 2006a, 2007). Although small amounts of phosphatidylglycerol and phosphatidylcholine were present, phosphatidylethanolamine was the predominant polar lipid of the inner membrane, as already reported for *Flavobacterium johnsoniae* (Pitta et al., 1989). Other interesting aspects of the lipids in members of the phylum are the significant proportions of amino-acid-based lipids and the large amounts of hydroxylated fatty acids that are amide-linked rather than ester-linked, confirming the peculiar nature of the polar lipids (B. Tindall, personal communication).

Published data on lipopolysaccharide, capsular polysaccharide, outer-membrane proteins, and glycopeptides of the fish-pathogenic *Flavobacterium* species, as well as on their antigenic properties and role in virulence, have been reviewed by Bernardet and Bowman (2006). Some of these components induce a high titer of protective antibodies and could represent targets for vaccines in the future (Dumetz et al., 2006).

Fine structure. Little information is available on the fine structure of members of the genus *Flavobacterium*. The transmission electron microscopy studies performed on some species (Horn et al., 2005; Møller et al., 2005; Bernardet and Bowman, 2006 and references therein; Park et al., 2006a, 2007; Wang et al., 2006; Yoon et al., 2006d; Liu et al., 2007) revealed that the structure of the cell wall was typical of Gram-stain-negative bacteria. Non-flagellar appendages and fimbriae-like structures were observed in *Flavobacterium aquatile*, *Flavobacterium branchiophilum*, and *Flavobacterium frigidarium*. Recently, cryoelectron tomography revealed tufts of 5-nm-wide filaments extending from and distributed unevenly on the outer membrane of *Flavobacterium johnsoniae* (Liu et al., 2007). As these filaments were absent in cells of a nonmotile mutant, but restored after the mutant was complemented, they may represent the adhesive surface organelles of the gliding motility machinery. In broth cultures, *Flavobacterium psychrophilum* produces long, tubular blebs that release membrane vesicles into the supernatant. These vesicles display a typical membrane bilayer, contain various proteases, and express several antigenic proteins (Møller et al., 2005). A regular capsule was reported in *Flavobacterium columnare*, *Flavobacterium frigidarium*, and *Flavobacterium hibernum* (Bernardet and Bowman, 2006 and references therein), but the capsule layer observed in *Flavobacterium psychrophilum* was thin and irregular (Møller et al., 2005). A capsule also seems to exist in *Flavobacterium johnsoniae*, since "*Flavobacterium columnare*" strain ATCC 43622 studied by MacLean et al. (2003) was actually later identified as *Flavobacterium johnsoniae* (Darwish et al., 2004; Shoemaker et al., 2005).

Colonial or cultural characteristics. The composition of the agar medium strongly affects colonial morphology of the *Flavobacterium* strains that display gliding motility because low nutrient content promotes this type of motility (see above) (Reichenbach, 1989b). Hence, colonies on nutrient-rich agars are usually circular, low convex to convex, translucent to opaque, smooth and shiny, with entire or uneven to undulate edges, whereas they may be flat and spreading with irregular to rhizoid or filamentous margins on low-nutrient agars (Bernardet and Bowman, 2006). Colonies of *Flavobacterium pectinovorum* merely display a sticky consistency, but the flat, rhizoid colonies of *Flavobacterium columnare* usually strongly adhere to the agar, a trait that may be lost after repeated subcultures. They also frequently display raised, warty to nodular centers that were originally mistaken for the fruiting bodies produced by myxobacteria (Ordal and Rucker, 1944). Different colony types may occur on the same agar plate, as reported for *Flavobacterium columnare* and *Flavobacterium psychrophilum* (Bernardet, 1989; Bernardet and Kerouault, 1989). After prolonged incubation, colonies of *Flavobacterium weaverense* and *Flavobacterium segetis* display a mucoid consistency (Yi and Chun, 2006). The consistency of colonies may also be affected by growth temperature: for instance, colonies of *Flavobacterium hibernum* are mucoid when grown at 25°C and gelatinous at 4°C (McCammon et al., 1998). Colonies of the agar-liquefying species *Flavobacterium flevense*, *Flavobacterium saccharophilum*, and *Flavobacterium tegetincola* are typically sunk into the agar; in *Flavobacterium limicola*, no liquefaction occurs and agar degradation is only visualized using potassium iodide solution (Tamaki et al., 2003).

Colonies of *Flavobacterium* strains are usually pale to bright yellow or orange due to the production of non-diffusible, non-fluorescent carotenoid or flexirubin types of pigments or both (Bernardet and Bowman, 2006; Bernardet et al., 1996, 2002; Reichenbach, 1989b). Ten of the 40 *Flavobacterium* species have been shown to produce predominantly flexirubin type pigments using the simple KOH test (Table 24): the color of the biomass shifts from yellow to dark pink, red or brown when covered with 20% KOH (Bernardet et al., 2002; Reichenbach, 1989b). Some of these species may in addition produce minor amounts of carotenoids, but the more sophisticated technique necessary to detect these pigments has not been performed. Carotenoid pigments have also not been looked for in a number of *Flavobacterium* species in spite of negative KOH tests (e.g., *Flavobacterium croceum*, *Flavobacterium soli*, *Flavobacterium daejeonense*, *Flavobacterium suncheonense*, *Flavobacterium segetis*, and *Flavobacterium weaverense*). Several *Flavobacterium* species also produce a pink to dark brown, diffusible melanin-like pigment when grown on tyrosine-containing agar (Table 24).

Liquid cultures of *Flavobacterium* strains are usually uniformly turbid. However, cells of *Flavobacterium columnare* frequently adhere to each others (resulting in numerous, tiny aggregates in suspension in the medium) and to the surface of the glass flask (producing a yellow filamentous ring at the upper level of the broth).

Additional information on the colonial and cultural characteristics of *Flavobacterium* strains, as well as the most suitable methods to study them, can be found in Reichenbach (1989b), Bernardet et al. (2002), and Bernardet and Bowman (2006).

Nutrition and growth conditions. Procedures for isolating and cultivating *Flavobacterium* strains have been reviewed extensively (Bernardet and Bowman, 2006; Hugo and Jooste, 2003; Jooste and Hugo, 1999; Reichenbach, 1989b, 1992a). With the exception of some freshwater, NaCl-sensitive species (including the rather fastidious fish-pathogenic species), most *Flavobacterium* species grow readily on commercial media such as nutrient and tryptic soy agars, as well as in the corresponding broths (Table 24). No growth factors are required. The fish pathogens grow well on media containing no NaCl and low nutrient concentrations such as Anacker and Ordal's medium (Anacker and Ordal, 1955), R2A (Reasoner and Geldreich, 1985), and modified Shieh's medium (Song et al., 1988). Many other media, including some selective ones, have been proposed for these economically significant fish pathogens; they have been reviewed extensively by Bernardet and Bowman (2006). All other *Flavobacterium* species tested also grow well on Anacker and Ordal's agar on which gliding strains display their typical spreading colonial morphology optimally (Table 24). Gliding motility is also most active in Anacker and Ordal's broth (Reichenbach, 1989b; Bernardet et al., 2002; J.-F. Bernardet, unpublished data). A glucose-containing medium has been used to grow *Flavobacterium omnivorum* and *Flavobacterium xinjiangense* (Zhu et al., 2003), as well as *Flavobacterium glaciei* (Zhang et al., 2006). Although not essential for growth, glucose has also been included in the growth media for *Flavobacterium columnare* (Song et al., 1988) and *Flavobacterium psychrophilum* (Cepeda et al., 2004; Daskalov et al., 1999). Under optimal conditions, the minimum doubling time is 7.3 h for *Flavobacterium denitrificans* (Horn et al., 2005), 6.8 h for *Flavobacterium segetis*, and 2.9 h for *Flavobacterium weaverense* (Yi and Chun, 2006).

The *Flavobacterium* strains that are able to grow on marine media (e.g., Difco marine 2216E agar) are those that have been retrieved from marine or saline environments, as well as a few strains isolated from temperate (*Flavobacterium soli*) or polar (*Flavobacterium antarcticum*, *Flavobacterium segetis*, *Flavobacterium weaverense*, and *Flavobacterium xanthum*) soil. These species, however, do not require NaCl for growth and are therefore halotolerant, not halophilic. The range of tolerated and optimum

NaCl concentrations for growth are listed in Table 24. Among described species, the highest resistance to NaCl (9%) was found in *Flavobacterium frigidarium* (Humphry et al., 2001), but isolates from saline environments tentatively identified as *Flavobacterium* sp. were recently shown to grow optimally in the presence of 10% NaCl and to tolerate up to 20% (Ghozlan et al., 2006). Although isolated from a freshwater lake, *Flavobacterium flevense* also grows well on marine agar; this fact, together with the ability of *Flavobacterium flevense* to degrade agar, could be a reminiscence of the time when the lake still received an inflow of seawater from the North Sea (Van der Meulen et al., 1974). Other soil and freshwater species, including the above-mentioned fish pathogens, do not grow on marine agar. Growth of *Flavobacterium croceum*, *Flavobacterium limicola*, and *Flavobacterium saliperosum* is severely inhibited by NaCl concentrations of 1–1.5% (Park et al., 2006a; Tamaki et al., 2003; Wang et al., 2006). The fish-pathogenic species are even more NaCl-sensitive: although *Flavobacterium columnare* may tolerate up to 0.5% NaCl (Bernardet and Grimont, 1989), growth is already reduced in the presence of only 0.3% NaCl (Chowdhury and Wakabayashi, 1988). The maximum NaCl concentrations tolerated by *Flavobacterium branchiophilum* and *Flavobacterium psychrophilum* are 0.2% and 0.5–1%, respectively (Bernardet and Kerouault, 1989; Holt et al., 1989; Ostland et al., 1994). Studying the in vitro survival of *Flavobacterium psychrophilum* in water, Madetoja et al. (2003) found that a salinity of ≥6‰ drastically reduced the number of culturable cells.

Despite the rather chilling specific epithets (e.g., *antarcticum, frigidarium, frigidimaris, frigoris, gelidilacus, glaciei, hibernum, psychrolimnae, psychrophilum*, etc.) given to the cold-living *Flavobacterium* species, most of them are actually psychrotolerant rather than true psychrophilic organisms. The optimum temperature for growth of the 19 *Flavobacterium* species that have been retrieved from cold or polar environments falls within the 11–21°C range, except for *Flavobacterium hibernum*, which grows optimally at 26°C (Table 24). Intriguingly, the latter species is also the one that is able to grow at the lowest temperature, i.e., −7°C (McCammon et al., 1998). Other psychrotolerant and psychrophilic *Flavobacterium* species are usually able to grow down to 0–5°C. The temperature range for growth of *Flavobacterium* strains that occur in temperate environments is usually 10–25°C, but some of them are able to grow at temperatures as low as 0°C or as high as 37°C (Table 24). *Flavobacterium croceum* is even able to grow up to 45°C (Park et al., 2006a). *Flavobacterium indicum* is the only *Flavobacterium* species that has been retrieved from a warm environment, i.e., spring water at 37–38°C, corresponding to its optimal growth temperature; its growth temperature range is 15–42°C (Saha and Chakrabarti, 2006). The most serious outbreaks of fish disease caused by *Flavobacterium columnare* usually occur in water at 20–30°C (Wakabayashi, 1991), whereas diseases caused by the other major fish pathogen, *Flavobacterium psychrophilum*, typically occur at 3–15°C (Borg, 1960; Holt et al., 1989).

Metabolism and metabolic pathways. Except for the general facts listed at the beginning of this chapter, little is known about these aspects for individual species. Almost all *Flavobacterium* strains are strictly aerobic chemoorganotrophs with a respiratory type of metabolism (Bernardet and Bowman, 2002, 2006; Bernardet et al., 1996). Facultatively anaerobic growth, however, has been reported in *Flavobacterium hydatis* and *Flavobacterium succinicans* when peptone or yeast extract is provided (Anderson and Ordal, 1961; Bernardet and Bowman, 2006; Chase, 1965; Reichenbach, 1989b; Strohl and Tait, 1978). Recently, weak and delayed anaerobic growth was also reported in *Flavobacterium antarcticum* (Yi et al., 2005a), *Flavobacterium croceum* (Park et al., 2006a), *Flavobacterium weaverense* and *Flavobacterium segetis* (Yi and Chun, 2006), and *Flavobacterium defluvii* (Park et al., 2007). Weak anaerobic growth of the latter organism occurred after 16 d, presumably by fermentation of carbohydrates (Park et al., 2007). The most intriguing *Flavobacterium* species described recently is *Flavobacterium denitrificans*, a facultative anaerobe that grows by carrying out complete denitrification in mineral medium, producing N_2O as a transient intermediate during the reduction of nitrate to nitrite (Horn et al., 2005). This non-fermentative organism uses O_2, NO_3^-, and NO_2^- as electron acceptors, but not sulfate or Fe^{3+}. Fifteen of the 40 *Flavobacterium* species are able to supplement their energy metabolism by reducing nitrate to nitrite, but *Flavobacterium denitrificans* is the first truly denitrifying member of the genus described so far. Some *Flavobacterium johnsoniae* strains are able to grow by anaerobic respiration using nitrate as electron acceptor (Reichenbach, 1989b; Stanier, 1947). A "var. *denitrificans*" was proposed by Stanier (1947) for one of these strains as it was an active denitrifier, but its inclusion in the species *Flavobacterium johnsoniae* has been subsequently questioned (Reichenbach, 1989b).

Catalase is produced by all *Flavobacterium* species, cytochrome oxidase by all but a few species, and β-galactosidase by 18 species of the 38 that have been tested. Indole production has not been tested in *Flavobacterium frigidimaris, Flavobacterium glaciei, Flavobacterium indicum, Flavobacterium saliperosum*, or *Flavobacterium soli*; it is negative in all other *Flavobacterium* species. The production of hydrogen sulfide has only been reported in *Flavobacterium columnare, Flavobacterium saccharophilum, Flavobacterium xanthum*, and *Flavobacterium xinjiangense*.

Most *Flavobacterium* species degrade gelatin and casein and several species also hydrolyze starch and esculin. Other polysaccharides such as chitin, pectin, and carboxymethylcellulose are only hydrolyzed by a few *Flavobacterium* species and the only agarolytic species are *Flavobacterium flevense, Flavobacterium limicola*, and *Flavobacterium saccharophilum* (Table 24). Crystalline cellulose (i.e., filter paper) is not decomposed (Bernardet and Bowman, 2006). Saccharolytic capacities vary widely among members of the genus. Some species are able to utilize a wide range of carbohydrates (e.g., *Flavobacterium flevense, Flavobacterium frigidimaris, Flavobacterium hydatis, Flavobacterium johnsoniae, Flavobacterium pectinovorum*, and *Flavobacterium saccharophilum*), whereas others utilize few or no carbohydrates and prefer amino acids and proteins (e.g., *Flavobacterium antarcticum, Flavobacterium gelidilacus, Flavobacterium glaciei, Flavobacterium indicum, Flavobacterium micromati, Flavobacterium saliperosum, Flavobacterium suncheonense, Flavobacterium tegetincola, Flavobacterium weaverense*, and the three fish-pathogenic species). About half of the *Flavobacterium* species produce acid from carbohydrates and degrade Tween compounds and tyrosine, whereas only a few species are able to degrade DNA and urea. The ability of many *Flavobacterium* strains to degrade a variety of biomacromolecules explains their role in mineralizing organic matter in soil and aquatic environments. The combination of polysaccharide-degrading enzymes and extracellular proteases produced by the fish-pathogenic *Flavobacterium* species is responsible for the extensive skin and muscular necrotic lesions observed in infected fish (Bernardet and Bowman, 2006).

The recently published whole-genome sequence of an *Flavobacterium psychrophilum* strain has yielded some interesting data on its metabolism (Duchaud et al., 2007). Although most genes encoding the enzymes of the glycolytic and pentose phosphate pathways have been identified, the sugar kinase and phosphotransferase systems that allow carbohydrate uptake were found to be missing. This may explain why *Flavobacterium psychrophilum* is unable to use carbohydrates as sources of carbon and energy. Host proteins are first degraded to oligopeptides by secreted proteases; oligopeptides are transported inside the bacterium by extensive uptake systems and degraded to amino acids by numerous peptidases. Amino acids are then processed by various catabolic pathways. Interestingly, *Flavobacterium psychrophilum* also produces cyanophycin synthetase and cyanophycinase, two enzymes also found in cyanobacteria and a few other bacteria. The former synthesizes cyanophycin, a non-ribosomally synthesized polypeptide occurring as storage inclusions in the cytoplasm, whereas the latter degrades cyanophycin to release carbon, nitrogen, and energy in nutrient-poor conditions (Krehenbrink et al., 2002). Hence, this unusual process may also occur in *Flavobacterium psychrophilum*, although cyanophycin inclusions have not been reported so far. Genes encoding the enzymes that degrade lipids to fatty acids have also been identified in the *Flavobacterium psychrophilum* genome. All enzymes of the tricarboxylic acid cycle are present for processing the degradation products of host proteins and lipids. Although *Flavobacterium psychrophilum* is a strict aerobe whose genome contains the genes encoding all components of the aerobic respiratory chain, mechanisms allowing the bacterium to deal with microaerobic conditions have also been identified. They may allow *Flavobacterium psychrophilum* to colonize oxygen-poor environments. The genome has also revealed a number of stress-response mechanisms that probably allow the bacterium to resist iron starvation, oxidative stress, and the reactive oxygen radicals produced in fish macrophages.

The psychrophilic and psychrotolerant *Flavobacterium* species have to overcome the physical constraints resulting from low temperature. For instance, *Flavobacterium frigidimaris* produces a large amount of various $NAD(P)^+$-dependent, cold-active dehydrogenases, some of which are thermolabile, whereas others are unexpectedly thermostable (Nogi et al., 2005b). Only the genes orthologous to the *Flavobacterium frigidimaris* genes encoding psychrophilic and thermolabile enzymes have been identified in the genome of *Flavobacterium psychrophilum*. It also contains genes involved in the regulation of membrane fluidity, in the maintenance of protein synthesis, and in the production of proteins with antioxidant properties (Duchaud et al., 2007).

Enzymic activities of *Flavobacterium* strains that may have biotechnological applications in the future have been reviewed recently (Bernardet and Bowman, 2006). In addition, a cold-active, thermolabile xylanase has been described from a creek bed soil isolate sharing 99% full-length 16S rRNA gene sequence similarity with the type strain of *Flavobacterium hibernum* (Lee et al., 2006a), and two *Flavobacterium* strains isolated from lake sediment in Finland have been shown to degrade casein, starch, and carboxymethylcellulose at 5°C (Männistö and Haggblom, 2006). A soil isolate that was able to rapidly metabolize the organophosphorus pesticide cadusafos has also been definitely allocated to the genus *Flavobacterium* on the basis of the high similarity of its full-length 16S rRNA gene sequence with those of several *Flavobacterium* species (Karpouzas et al., 2005). However, many "*Flavobacterium* sp." strains have actually been tentatively identified on the basis of partial 16S rRNA gene sequence and/or poor phenotypic data. For instance, "*Flavobacterium* sp." ATCC 27551, whose parathion hydrolase activity has been the subject of many publications (e.g., Mulbry and Karns, 1989; Manavathi et al., 2005; Khajamohiddin et al., 2006) dealing with the microbial detoxification of hazardous organophosphate contaminants (such as the insecticides diazinon and parathion), has actually been shown to be a facultatively anaerobic and flagellated strain (Sethunathan and Yoshida, 1973). Degrading activities of the following contaminants with possible applications in bioremediation have been reported in other poorly (probably erroneously) identified "*Flavobacterium* sp.": the solid propellant perchlorate (Okeke and Frankenberger, 2005), gasoline (Lu et al., 2006), various aromatic and chlorinated aromatic hydrocarbons (Plotnikova et al., 2006), the herbicide atrazine (Smith and Crowley, 2006), and the pesticide pentachlorophenol (Pu and Cutright, 2007). The β-galactosidase secreted by a *Flavobacterium* strain isolated from post-processing water in a shrimp and fish processing plant in Greenland met a number of requirements that are necessary for industrial dairy processes (Sørensen et al., 2006).

Genetics. The *Bacteroides thetaiotaomicron* transposable element Tn*4351* has been the first tool for mutagenesis in members of the phylum *Bacteroidetes* (Cooper et al., 1997). This transposon has also been shown to function in *Flavobacterium johnsoniae* and cloning vectors based on the cryptic plasmid pCP1 of *Flavobacterium psychrophilum* have been developed to genetically manipulate *Flavobacterium* species and related organisms (McBride and Baker, 1996; McBride and Kempf, 1996). Gliding motility mutants of *Flavobacterium columnare* and *Flavobacterium johnsoniae* had already been obtained through conventional methods (see *Cell-wall composition*, above), but Tn*4351* mutagenesis has been used to obtain non-gliding mutants of *Flavobacterium johnsoniae* to elucidate the genetic mechanisms of gliding motility. To date, 15 different genes (*gldA* to *gldN* and *ftsX*) required for gliding motility have been identified. Cells with mutations in any of these genes are completely nonmotile, deficient in chitin utilization, and resistant to infection by bacteriophages (Braun et al., 2005; Kempf and McBride, 2000; McBride, 2001; McBride and Braun, 2004; McBride et al., 2003). Conversely, some gliding activity is retained in *secDF* and *sprA* mutants (Nelson and McBride, 2006; Nelson et al., 2007). Unlike all other proteins involved in gliding motility, sprA is at least partially exposed on the cell surface (Nelson et al., 2007). The same group of scientists has sequenced the pCP1 plasmid harbored by most *Flavobacterium psychrophilum* strains (M.J. McBride, sequence no. NC_004811, http://www.ncbi.nlm.nih.gov; Chakroun et al., 1998; Madsen and Dalsgaard, 2000; Izumi and Aranishi, 2004). Genetic tools have also been developed for *Flavobacterium psychrophilum*, including selectable markers, plasmid cloning vectors, a reporter system, and a transposon (Álvarez et al., 2004).

The whole-genome sequence of *Flavobacterium psychrophilum* JIP 02/86 (=CIP 103535=ATCC 49511) was recently determined (Duchaud et al., 2007) and consists of a 2,861,988 bp circular chromosome and one copy of the above-mentioned pCP1 cryptic plasmid. The mean G+C content of the whole chromosome is 32.54 mol%. The chromosome contains six rRNA genes, 49 tRNA genes, and 2,432 predicted protein-coding genes. No function could be assigned to 45% of the genes and little conservation was observed when the genome of JIP 02/86 was compared to other published complete genome sequences. The closest to *Flavobacterium psychrophilum* among these was *Bacteroides fragilis* YCH46, but only 50% orthologous genes and very limited synteny were observed between the two genomes. No counterpart was found in other published genomes for 401 putative proteins (16%). The ongoing whole-genome sequence of other members of the genus (*Flavobacterium psychrophilum* CSF-259-93, *Flavobacterium columnare* ATCC 49512, and *Flavobacterium johnsoniae* ATCC 17061^T) and of members of the family *Flavobacteriaceae* ("*Gramella forsetii*" KT0803 and *Leeuwenhoekiella blandensis* MED 217^T) will allow interesting comparisons between the genomes. For instance, the genome size of the pathogen *Flavobacterium psychrophilum* (2.9 Mb) is low compared to that of related environmental organisms such as *Flavobacterium johnsoniae* (6 Mb) and *Leeuwenhoekiella blandensis* (4.2 Mb). The NCBI website (http://www.ncbi.nlm.nih.gov/genomes/lproks.cgi) provides information on all these genomes, except that of *Flavobacterium columnare* for which a separate website is available (http://microgen.ouhsc.edu/f_columnare/f_columnare_home.htm).

The genotyping techniques that have been applied to collections of *Flavobacterium* strains have been reviewed recently (Bernardet and Bowman, 2006 and references therein). Random amplified polymorphic DNA, rRNA gene restriction pattern analysis (ribotyping), restriction fragment length polymorphism analysis, repetitive extragenic palindromic DNA-PCR (rep-PCR), and plasmid profiles have resulted either in the differentiation of *Flavobacterium* species or in intra-specific typing with taxonomic or epidemiological applications. The two major fish pathogens *Flavobacterium columnare* and *Flavobacterium psychrophilum*, of which large collections of strains are available and which are economically significant, have been the subjects of most molecular typing studies.

Early reports of phages in *Flavobacterium columnare* and *Flavobacterium johnsoniae* have been mentioned by Reichenbach (1989b). The online directory of phage names (Ackermann and Abedon 2001) lists a number of other phages reportedly isolated from *Flavobacterium* strains. They include: T-ϕD1B, a member of the family *Myoviridae*, morphotype A1; ϕCB38, a member of the family *Podoviridae*, morphotype C1; and CMF-1-F, O6N-12P, O6N-24P, and ϕMT5, four different virus species of the family *Siphoviridae*, morphotype B1. A new flavobacteria-infecting siphonophage, phage 11b, has been isolated recently from strain "*Bacteroidetes* bacterium 11B", an Arctic sea-ice isolate that shares 97–99% 16S rRNA gene sequence similarity with *Flavobacterium frigoris*, *Flavobacterium degerlachei*, and other polar *Flavobacterium* isolates (Borriss et al., 2003, 2007). Flavobacteriophage 11b has been cultivated and its genome and proteome have been characterized. Its host-specificity is reflected by a low DNA G+C content similar to that of *Flavobacterium* species and by the significant similarity of five of its coding sequences with those of closely related bacteria of the phylum *Bacteroidetes* (Borriss et al., 2007).

Antibiotic sensitivity. Overall, the environmental *Flavobacterium* strains that have been tested for their susceptibility to antimicrobial drugs showed no outstanding resistance (Aslam et al., 2005; Humphry et al., 2001; Park et al., 2006a, 2007; Saha and Chakrabarti, 2006; Strohl and Tait, 1978; Wang et al., 2006; Yoon et al., 2006d). Drug susceptibility of the fish-pathogenic *Flavobacterium* species has been studied extensively to evaluate antibiotics for selective isolation media and for treatment of infected fish. Intrinsic resistance to tobramycin was found in *Flavobacterium columnare* and *Flavobacterium psychrophilum*; the former species was also resistant to neomycin and polymyxin B. Bath or oral antibiotic treatments are still the most effective control method for *Flavobacterium* infections in fish, but resistance to the widely used oxytetracycline, oxolonic acid, and amoxycillin has progressively developed in fish farm isolates; no resistance to florfenicol has been reported yet (Bernardet and Bowman, 2006 and references therein).

Pathogenicity. Strains of *Flavobacterium johnsoniae* have occasionally been isolated from cases of "soft rot" in various plants; it was probably an opportunistic pathogen (Lelliott and Stead, 1987; Liao and Wells, 1986). An organism named "*Cytophaga allerginae*" (later shown to share significant DNA relatedness with *Flavobacterium hydatis* and *Flavobacterium johnsoniae*; Bernardet and Bowman, 2006) was retrieved from the air humidification system of a textile facility and identified as the source of the endotoxin that caused severe cases of hypersensitivity pneumonitis in several workers (Flaherty et al., 1984; Liebert et al., 1984). No other cases have been reported since then.

Freshwater fish are the animals most prone to flavobacterial infections, resulting in considerable economic losses worldwide. Fish diseases caused by *Flavobacterium columnare*, *Flavobacterium psychrophilum*, and *Flavobacterium branchiophilum* have been reviewed extensively (Austin and Austin, 1999; Bernardet and Bowman, 2006; Holt et al., 1993; Nematollahi et al., 2003; Shotts and Starliper, 1999) and basic information is also provided below in the list of species. Although mainly a free-living organism, *Flavobacterium johnsoniae* has been isolated frequently from diseased fish (Christensen, 1977; J.-F. Bernardet, unpublished data) and is considered to be an opportunistic pathogen in a few cases (Carson et al., 1993; Rintamäki-Kinnunen et al., 1997). Originally isolated from diseased fish, *Flavobacterium hydatis* (previously [*Cytophaga*] *aquatilis*; Strohl and Tait, 1978) and *Flavobacterium succinicans* (previously [*Cytophaga*] *succinicans*; Anderson and Ordal, 1961) may also act as opportunistic pathogens. Several other *Flavobacterium*-like organisms have been reported from diseased fish, but they have subsequently been transferred to other or new genera (Bernardet and Bowman, 2006).

Virulence mechanisms of fish-pathogenic *Flavobacterium* species are as yet poorly understood. Adhesive properties, agglutination, and lysis of fish erythrocytes, the ability to escape the bactericidal mechanisms of phagocytes, and production of extracellular proteases and chondroitin AC lyase have been suggested as important virulence factors of *Flavobacterium psychrophilum*, *Flavobacterium columnare*, and *Flavobacterium branchiophilum*, but virulence varies considerably among strains (Bernardet and Bowman, 2006 and references therein). New insights have resulted from the recent whole-genome sequencing of *Flavobacterium psychrophilum* strain JIP 20/86 (Duchaud et al., 2007; see above). Genome analysis has revealed a number of mechanisms that confer on *Flavobacterium psychrophilum* the ability to colonize and degrade fish tissues: adhesion, biofilm formation, gliding motility, iron acquisition, production of proteases able to hydrolyze the components of skin and muscular tissues, production of cell surface proteins important for bacteria–host interactions, and secretion systems transporting toxins to the bacterial surface. Additional data on virulence will likely result from comparison of this genome to those of other pathogenic (*Flavobacterium branchiophilum*, *Flavobacterium columnare*) and environmental (*Flavobacterium johnsoniae*) *Flavobacterium* species when available (see *Genetics*, above).

Ecology. Members of the genus *Flavobacterium* occur in a variety of soil and freshwater, marine, or saline habitats in warm, temperate, or polar environments (Table 24) (Bernardet and Bowman, 2006). Of the 40 *Flavobacterium* species with validly published names, only one (*Flavobacterium indicum*) has been isolated from a warm habitat (i.e., spring water at 37–38°C), 18 are from cold or polar habitats, and eight are from habitats characterized by moderate salinity such as seawater, sea ice, marine sediment, and saline lake water. *Flavobacterium* strains have not been isolated from hypersaline environments so far. Ghozlan et al. (2006) investigated bacterial communities in water and soil samples from various habitats in Egypt displaying a wide range of salinity. Strains allocated to the genus *Flavobacterium* on the basis of 155 phenotypic tests were only retrieved from salt lakes and solar salterns with non-saturating brines (salinity 13.5–25%) and shown to grow optimally in the presence of 10% NaCl. Several of the recently described *Flavobacterium* species have been isolated following extensive investigation of bacterial communities in Antarctic soil (*Flavobacterium weaverense* and *Flavobacterium segetis*; Yi and Chun (2006) and seawater (*Flavobacterium frigidimaris*; Nogi et al., 2005b). Temperate or frozen soils have also yielded a number of novel *Flavobacterium* species, such as *Flavobacterium omnivorum* and *Flavobacterium xinjiangense* (Zhu et al., 2003), *Flavobacterium soli* (Yoon et al., 2006d), *Flavobacterium daejeonense* and *Flavobacterium suncheonense* (Kim et al., 2006a), and *Flavobacterium glaciei* (Zhang et al., 2006). *Flavobacterium denitrificans* (Horn et al., 2005), isolated from the gut of an earthworm, is also most likely a soil organism (H.L. Drake, personal communication); it participates in the production of large amounts of nitrous oxide, a greenhouse gas, by earthworms. Among the NaCl-sensitive species, *Flavobacterium glaciei* has been isolated from frozen soil (Zhang et al., 2006), *Flavobacterium aquidurense* and *Flavobacterium hercynium* are from calcium-rich creek water (Cousin et al., 2007), *Flavobacterium saliperosum* is from freshwater lake sediment (Wang et al., 2006), and *Flavobacterium croceum* and *Flavobacterium defluvii* are from activated sludge (Park et al., 2006a, 2007). The fish-pathogenic species (see below) are also very sensitive to NaCl concentration.

Among the few investigations of cultivable bacterial communities using complete or nearly complete 16S rRNA gene sequences are studies from river epilithon (i.e., the biofilm attached to rock surfaces comprising algal and bacterial communities embedded in organic matter) in England (O'Sullivan et al., 2006) and from the calcium-rich water of a German creek (Brambilla et al., 2007). Both studies revealed the same breadth of diversity among *Flavobacterium* strains, but only a few isolates could be affiliated with described species, with the majority of isolates representing putative novel species. Descriptions of novel *Flavobacterium* species (*Flavobacterium aquidurense* and *Flavobacterium hercynium*; Cousin et al., 2007) have already resulted from one of these studies. Kisand et al. (2005) studied the distribution and seasonal dynamics of five *Flavobacterium* strains along a Northern Baltic Sea river-marine transect. Two strains were the most abundant bacteria in the river water (mean >10,000 cells/ml), but they decreased in number from the river to the estuary and the open sea. The remaining three strains were only occasionally detectable. Authentic *Flavobacterium* isolates were also recovered from another estuarine environment (Selje et al., 2005) and from water beneath a Canadian glacier (Cheng and Foght, 2007). Interestingly, freeze-tolerant bacteria producing ice-active macromolecules are not restricted to polar areas. A *Flavobacterium* strain exhibiting ice-nucleating activity, but unable to inhibit recrystallization was isolated from soil in a temperate Canadian location (Wilson et al., 2006). New *Flavobacterium* species will most likely emerge from investigations of these and other environments.

Most studies of bacterial communities and phylogenetic diversity in various environments, however, have been based on partial 16S rRNA gene sequences, cloned 16S rRNA genes, or denaturing gradient gel electrophoresis bands. Consequently, the resulting affiliation of the corresponding organisms to the genus *Flavobacterium* is only tentative. For instance, *Flavobacterium* sp. were among the surface and endophytic bacteria that colonize orchid roots (Tsavkelova et al., 2007), and Kim et al. (2006b) reported that 87% of the bacterial community associated with the roots of tomato plants belonged to undescribed members of the genus *Flavobacterium*. Some of the isolates recovered from permafrost in Siberia (Vishnivetskaya et al., 2006) and China (Bai et al., 2006) were identified as *Flavobacterium* sp. *Flavobacterium* strains were also found in three different habitats of a paddy field in Japan: among the epibiotic and intestinal microbial community of several microcrustacean species (Niswati et al., 2005), among the microbial community responsible for the decomposition of buried rice straw compost (Tanahashi et al., 2005), and in the floodwater (Shibagaki-Shimizu et al., 2006). The bacterioplankton communities of water from oligotrophic, eutrophic, and mesotrophic lakes (i.e., lakes with low concentrations of nutrients and low overall productivity, high concentrations of nutrients and high productivity, and intermediate characteristics, respectively) in Germany all contained *Flavobacterium* strains (Gich et al., 2005). A study of the abundance and seasonal dynamics of members of the class *Flavobacteriia* in four eutrophic lakes in Sweden revealed that *Flavobacterium* strains formed one of the two dominant *Flavobacteriia* lineages responsible for the observed *Flavobacteriia* blooms (i.e., episodes where *Flavobacteriia* contributed in excess of 30% of total bacterioplankton). *Flavobacterium* spp. alone or in combination with members of the other lineages also contributed more than 50% of the total 16S rRNA gene pool during some cyanobacterial blooms (Eiler and Bertilsson, 2007). The two *Flavobacterium* strains identified on the basis of their fatty acid profile and partial 16S rRNA sequence from lake and pond sediment in Finland were able to grow at 0°C (Männistö and Haggblom, 2006). Larvae of some butterflies species (Xiang et al., 2006) have also been shown to harbor *Flavobacterium* strains in the gut. In the search for potential bacterial pathogens to control the hazelnut pest beetle *Oberea linearis*, Bahar and Demirbag (2007) characterized a number of strains isolated from diseased or dead larvae. Conventional phenotypic tests, commercial galleries, and near-full-length 16S rRNA gene sequences have clearly identified one of the isolates as a *Flavobacterium* sp., but bioassays on beetle larvae have failed to reveal any insecticidal effect.

The fish-pathogenic *Flavobacterium* species (e.g., *Flavobacterium branchiophilum*, *Flavobacterium columnare*, and *Flavobacterium psychrophilum*) have mainly been isolated from superficial lesions or internal organs of infected fish, and only rarely from water (Bernardet and Bowman, 2006). Although *Flavobacterium psychrophilum* cells are able to survive in water for several months, they rapidly lose their ability to multiply (Madetoja et al., 2003). Hence, these species seem to be mainly parasitic, as also suggested by their rather fastidious growth on artificial media (especially *Flavobacterium branchiophilum*). The few strains of *Flavobacterium hydatis* and *Flavobacterium succinicans* available and the fish isolates of *Flavobacterium johnsoniae* have been isolated exclusively from external lesions of fish. This suggests that these three species, which grow well on artificial media, are probably saprophytic or commensal organisms that may colonize fish lesions initiated by other bacteria or parasites. Other saprophytic or commensal *Flavobacterium* strains have been reported from the surface of fish eggs and from fish skin, gills, and gut (Huber et al., 2004; Bernardet and Bowman, 2006 and references therein; Hu et al., 2007), but the identification of some of these strains was only based on scant data.

Flavobacterium strains have also been found in air humidification systems (see *Pathogenicity*, above), in amoebae (Horn et al., 2001; Müller et al., 1999), in the water and sediment of intensive fish culture systems (Bernardet and Bowman, 2006 and references therein), and in the wastewater treatment plant of a beer brewery (*Flavobacterium granuli*; Aslam et al., 2005). Many "*Flavobacterium*" strains have reportedly been involved in the spoilage or degradation of various food products, but they have either been subsequently allocated to other genera or shown to play only a minor role (Bernardet and Bowman, 2006 and references therein). Recently, some of the spoilage pectinolytic organisms isolated from various minimally processed frozen vegetables sold in supermarkets in Botswana were tentatively identified as *Flavobacterium* sp. on the basis of commercial identification systems and additional phenotypic tests (Manani et al., 2006).

In the different habitats they colonize, *Flavobacterium* strains participate in the degradation of various organic compounds such as polysaccharides and proteins. For instance, five *Flavobacterium* sp. represented important metabolizers of dissolved organic matter in river estuaries of the Northern Baltic area (Kisand et al., 2002, 2005). The proteases and chondroitin AC lyase secreted by the fish-pathogenic *Flavobacterium* species are responsible for severe necrotic lesions by destroying fish tissues (see *Pathogenicity*, above; Bernardet and Bowman, 2006). Some of the enzymes, especially the cold-active ones, produced by *Flavobacterium* strains may have biotechnological applications in the future (see *Metabolism and metabolic pathways*, above; Bernardet and Bowman, 2006).

Enrichment and isolation procedures

Most *Flavobacterium* strains are easily isolated and cultivated, usually requiring no enrichment or selective isolation procedure and growing in an ambient air atmosphere. Serial dilutions of the source material, such as water, soil, or fish tissue, are spread directly onto agar media and incubated at a temperature corresponding to that of the original habitat. The media used for isolation and cultivation of *Flavobacterium* strains are presented in the section on nutrition and growth conditions. Relevant information on the isolation and cultivation of some species now included in the genus *Flavobacterium* has also been provided by Reichenbach (1989b, 1992a).

The enrichment procedures that have been used occasionally have taken advantage of the special physiological characteristics displayed by some *Flavobacterium* strains. For instance, a pre-enrichment step at 2–4°C has been performed to promote the isolation of psychrophilic/psychrotolerant *Flavobacterium* species from cold environments (e.g., McCammon and Bowman, 2000; Zhu et al., 2003; Yi et al., 2005a; Yi and Chun, 2006; Zhang et al., 2006). Similarly, isolation of chitin-, pectin-, or xylan-degrading *Flavobacterium* strains has been achieved by using an enrichment step on media containing the specific substrate (Agbo and Moss, 1979; Reichenbach (1989b) and references therein; Humphry et al. (2001).

Maintenance procedures

Most cold-living *Flavobacterium* strains can be stored for a few weeks or months at 2–4°C as live cultures on agar plates or slants, whereas agar deeps (e.g., Anacker and Ordal's medium containing 0.4% agar; Anacker and Ordal, 1955) should preferably be used for short-term preservation of other *Flavobacterium* strains (J.-F. Bernardet, unpublished data). For long-term preservation, *Flavobacterium* strains can be lyophilized or maintained at −70°C or −80°C using glycerol (10–30%, v/v) as a cryoprotectant (Bernardet and Bowman, 2006; Reichenbach, 1989b; Yi and Chun, 2006). Commercial media for cryopreservation have also been used successfully (Cepeda et al., 2004).

Procedures for testing special characters

The phenotypic characteristics that allow differentiation of *Flavobacterium* species are listed in Table 24 and those that distinguish *Flavobacterium* species from members of related genera are detailed in the next section. The procedures recommended for testing these characteristics have been specified previously (Bernardet et al., 2002). Additional details have been provided for those features that are unusual or require some experience, such as production of spherical degenerative cells in the stationary phase, colony morphology and consistency, adsorption of Congo red, gliding motility, ability to degrade cellulose or its derivatives, yellow pigment type (flexirubin and/or carotenoid), production of cytochrome oxidase and β-galactosidase, salinity requirement, capnophilic metabolism, and use of commercial identification systems (Bernardet and Bowman, 2006; Bernardet et al., 2002).

Differentiation of the genus *Flavobacterium* from closely related taxa

Nowadays, differentiation of *Flavobacterium* strains from members of other genera in the family *Flavobacteriaceae* is usually achieved initially through a comparison of 16S rRNA gene sequences with those deposited in databases and the determination of phylogenetic distances.

The features that should be included in the description of novel *Flavobacterium* species have been specified in the minimal standards for describing new taxa in the family *Flavobacteriaceae* (Bernardet et al., 2002). The phenotypic characteristics that differentiate members of genera in the family *Flavobacteriaceae* are listed in Table 23 in the treatment of the family *Flavobacteriaceae* in this volume. However, because the genus *Flavobacterium* contains by far the largest number of species in the family, many of these characteristics also vary within the genus and consequently cannot be used to circumscribe the genus *Flavobacterium* from other genera. The following characteristics vary within the genus: production of spherical degenerative cells in stationary phase; gliding motility; ability to grow at 25°C; acid production from glucose and sucrose; presence of DNase, urease, cytochrome oxidase, H_2S, and β-galactosidase activities; nitrate reduction; carbohydrate utilization; and degradation of agar, starch, esculin, and gelatin. Hence, the only characteristics common to all *Flavobacterium* species so far are: production of yellow pigments; the presence of catalase activity; the absence of a requirement for Na^+ or seawater; and the inability to grow under anaerobic conditions (except for five of the 40 species), to produce indole, or to grow at 37 and 42°C (except for *Flavobacterium croceum* and *Flavobacterium indicum*).

Information on the relevance of fatty acid composition to identify *Flavobacterium* strains at the genus and species level has been given above (see *Cell-wall composition*). Application of whole-cell protein analysis by SDS-PAGE to members of the genus *Flavobacterium* has been reviewed recently (Bernardet and Bowman, 2006). Another procedure, MALDI-TOF (Matrix-Assisted Laser Desorption/Ionization Time-of-Flight) mass spectrometry analysis, has been used to rapidly group 60 freshwater isolates according to their whole-cell protein pattern; more than half of the isolates grouped around the single *bona fide Flavobacterium* strain contained in the database (Brambilla et al., 2007).

The DNA G+C content of *Flavobacterium* strains (i.e., 30–41 mol%) is similar to that of most other members of the family *Flavobacteriaceae* (see Table 23 in the treatment of the family *Flavobacteriaceae*). DNA G+C composition only allows differentiation of *Flavobacterium* strains from *Olleya* (49 mol%) and *Robiginitalea* (55–56 mol%) strains, and from some strains of *Bizionia* (38–45 mol%), *Capnocytophaga* (36–44 mol%), *Muricauda* (41–45 mol%), *Psychroserpens* (27–29 mol%), and *Zobellia* (36–43 mol%).

Taxonomic comments

The early history of the genus *Flavobacterium* and of its type species (*Flavobacterium aquatile*) has been reviewed several times (especially in Holmes and Owen (1979), but also: Weeks, 1955; Holmes and Owen, 1981; Holmes, 1992; Holmes et al., 1984c). Following the description of "*Bacillus aquatilis*" by Frankland and Frankland (1889), the species was designated the type species of the genus *Flavobacterium* when it was created by Bergey et al. (1923). As no type strain had been designated by the Franklands and no original isolate was still available, strain Taylor F36, reputed to be a re-isolate of the original organism from the same source, has become generally accepted (although not formally proposed) as the type strain of *Flavobacterium aquatile* (Holmes and Owen, 1979; Weeks, 1955). However, this strain displayed serious phenotypic discrepancies compared to the meagre original description of "*Bacillus aquatilis*" and it shared some characteristics with members of the genus *Cytophaga*, such as the structure of the cell wall (Follett and Webley, 1965) and movements that were considered to be gliding by some investigators (Perry, 1973). Moreover, its place and date of isolation were not known with certainty. Since Holmes and Owen (1979) felt that "*Flavobacterium aquatile* was an unfortunate choice as the type species of *Flavobacterium* because the species is currently represented by only one strain which is of uncertain taxonomic affinities", they proposed to substitute [*Flavobacterium*] *breve* for *Flavobacterium aquatile* as the type species of the genus. They also proposed an emended description of the genus that took into account its successive restrictions. Indeed, because the only original criteria for including an organism in the genus *Flavobacterium* were the production of yellow to orange pigments and the ability to produce acid weakly from carbohydrates (Bergey et al., 1923), the genus had grown to include Gram-stain-positive, Gram-stain-negative, flagellated, gliding, and nonmotile bacteria with high or low DNA G+C content (Holmes and Owen, 1979). In the successive editions of *Bergey's Manual*, flagellated, Gram-stain-positive, and gliding strains were excluded, but species with high and low DNA G+C content were still included (Weeks, 1974).

The genus *Flavobacterium* as emended by Holmes and Owen (1979) was restricted to clinical, environmental, or food bacteria that were aerobic, Gram-stain-negative, chemoorganotrophic, oxidase- and catalase-positive, and nonmotile, usually producing yellow to orange pigments, having a strictly respiratory type of metabolism, actively proteolytic but poorly saccharolytic, with a DNA G+C content of 31–40 mol%. Because the Judicial Commission (Wayne, 1982) had declined to issue the opinion requested by Holmes and Owen (1979) for rejecting the name *Flavobacterium aquatile* as a *nomen dubium* and for changing the type species of the genus *Flavobacterium*, *Flavobacterium aquatile* had to remain the type species of the genus. The description of the genus in the following edition of the *Manual* (Holmes et al., 1984c) was essentially the same as the one these authors had proposed in 1979. However, although all seven species shared the above-mentioned characteristics, the following species or groups of species could already be distinguished on the basis of phenotypic features: (i), *Flavobacterium aquatile*; (ii), [*Flavobacterium*] *breve*, [*Flavobacterium*] *balustinum*, and [*Flavobacterium*] *meningosepticum*; (iii), [*Flavobacterium*] *odoratum*; and (iv), [*Flavobacterium*] *multivorum* and [*Flavobacterium*] *spiritivorum*. The same "natural groups", some of them expanded by the description of novel species, appeared in the next review on the genus *Flavobacterium* (Holmes, 1992).

In 1996, the results of an extensive polyphasic study of more than 100 bacterial strains representing a variety of flavobacteria (mainly attributed to the genera *Cytophaga*, *Flavobacterium*, *Flexibacter*, *Microscilla*, and *Sphingobacterium*) were published (Bernardet et al., 1996). The main result of the 16S rRNA analysis included in this study was the distinct delineation of a "*Flavobacterium aquatile* rRNA cluster" that comprised *Flavobacterium aquatile*, *Flavobacterium branchiophilum*, [*Cytophaga*] *flevensis*, [*Cytophaga*] *aquatilis*, [*Cytophaga*] *johnsonae* (*sic*), [*Cytophaga*] *pectinovora*, [*Cytophaga*] *saccharophila*, [*Cytophaga*] *succinicans*,

"*Cytophaga allerginae*", "*Cytophaga xantha*", [*Flexibacter*] *columnaris*, [*Flavobacterium*] *psychrophilus*, [*Flavobacterium*] *aurantiacus*, "*Flavobacterium aurantiacus* subsp. *excathedrus*", "*Promyxobacterium flavum*", and "*Sporocytophaga cauliformis*". Due to the low resolution of the 16S rRNA analysis used at that time, [*Flavobacterium*] *odoratum* was also included in this cluster. Strikingly, all other species included in the genus *Flavobacterium* as delineated by Holmes et al. (1984c) and Holmes (1992) appeared only remotely related to the "*Flavobacterium aquatile* rRNA cluster" and had to be reclassified in other or new genera (i.e., *Chryseobacterium*, *Empedobacter*, and *Bergeyella*; see Vandamme et al. (1994a) and the treatment of these genera and the family *Flavobacteriaceae*). The organisms grouped in the "*Flavobacterium aquatile* rRNA cluster" shared similar fatty acid and whole-cell protein profiles, a number of phenotypic and chemotaxonomic characteristics, and a rather narrow range of DNA G+C content, as well as similar habitats. Hence, both phenotypic and phylogenetic criteria supported inclusion of members of the cluster into an emended genus *Flavobacterium* with *Flavobacterium aquatile* as the type species (Bernardet et al., 1996). An emended description of the genus was proposed that took into account characteristics that were common to all its members. Conversely, as [*Flavobacterium*] *odoratum* differed from the other species by several characteristics and by its clinical origin, it was decided to reclassify this species in a separate genus, *Myroides*, as *Myroides odoratus* (Vancanneyt et al., 1996). The reclassification of [*Cytophaga*] *aquatilis* in the genus *Flavobacterium* made it necessary to propose a new species name for this taxon, *Flavobacterium hydatis*, to avoid the new combination becoming a junior homonym of *Flavobacterium aquatile*. The reclassification of [*Cytophaga*] *johnsonae* was also the opportunity for correcting its specific epithet to *johnsoniae* (Bernardet et al., 1996). As the two [*Flexibacter*] *aurantiacus* strains shared many phenotypic features, similar fatty acid profiles, and high levels of DNA–DNA relatedness with the type strain of *Flavobacterium johnsoniae*, it was proposed to include them in this species (Bernardet et al., 1996). Although clearly attributed to the emended genus *Flavobacterium*, the taxa "*Cytophaga allerginae*", "*Cytophaga xantha*", "*Flexibacter aurantiacus* subsp. *excathedrus*", "*Promyxobacterium flavum*", and "*Sporocytophaga cauliformis*" were all, except the latter, represented by a single, poorly investigated strain and their names had not been validly published. Hence, pending further research and isolation of additional strains, it was decided not to formally describe novel *Flavobacterium* species for these organisms and to refer to them merely as "*Flavobacterium* sp.". The only strain of "*Cytophaga xantha*" was later included in a study of Antarctic isolates and formally described as *Flavobacterium xanthum* (McCammon and Bowman, 2000); since then, novel strains have been isolated (Van Trappen et al., 2005).

Hence, the genus *Flavobacterium* as it was defined in 1996 comprised soil or freshwater, Gram-stain-negative, aerobic, gliding (except *Flavobacterium branchiophilum*) organisms that contained menaquinone MK-6 as the major respiratory quinone, produced yellow non-diffusible pigments and catalase, and had very similar fatty acid profiles; *Flavobacterium* strains were unable to grow on seawater-containing media (except *Flavobacterium flevense*) or hydrolyze crystalline cellulose; the DNA G+C content was 32–37 mol% and the optimum temperature range was 20–30°C (except *Flavobacterium psychrophilum*). The genus contained *Flavobacterium aquatile* and the following nine other species: *Flavobacterium branchiophilum*, *Flavobacterium columnare*, *Flavobacterium flevense*, *Flavobacterium hydatis*, *Flavobacterium johnsoniae*, *Flavobacterium pectinovorum*, *Flavobacterium psychrophilum*, *Flavobacterium saccharophilum*, and *Flavobacterium succinicans* (Bernardet et al., 1996). Sixteen additional species had been described at the time of publication of the most recent review on the genus (Bernardet and Bowman, 2006). Description of the following 14 novel species have been published since then: *Flavobacterium denitrificans* (Horn et al., 2005), *Flavobacterium frigidimaris* (Nogi et al., 2005b), *Flavobacterium croceum* (Park et al., 2006a), *Flavobacterium daejeonense* and *Flavobacterium suncheonense* (Kim et al., 2006a), *Flavobacterium glaciei* (Zhang et al., 2006), *Flavobacterium indicum* (Saha and Chakrabarti, 2006), *Flavobacterium saliperosum* (Wang et al., 2006), *Flavobacterium segetis* and *Flavobacterium weaverense* (Yi and Chun, 2006), *Flavobacterium soli* (Yoon et al., 2006d), *Flavobacterium defluvii* (Park et al., 2007), and *Flavobacterium aquidurense* and *Flavobacterium hercynium* (Cousin et al., 2007). Hence, the genus currently comprises 40 species with validly published names. The description of several other *Flavobacterium* species is pending and it is likely that some of the organisms affiliated to the genus *Flavobacterium* reported from investigations of various environments (e.g., Kisand et al., 2002; Zdanowski et al., 2004; Gich et al., 2005; Green et al., 2006; Kim et al., 2006b; Männistö and Haggblom, 2006; O'Sullivan et al., 2006; Vishnivetskaya et al., 2006; Cheng and Foght, 2007) will ultimately be formally described.

As the genus *Flavobacterium* progressively expanded, some of the recently described species have proved to share only moderate 16S rRNA gene sequence similarity with other members of the genus. For instance, *Flavobacterium croceum*, *Flavobacterium saliperosum*, and *Flavobacterium indicum* share 94, 93.3, and 92.1% 16S rRNA gene sequence similarity, respectively, with the type and only strain of the type species, *Flavobacterium aquatile* (Park et al., 2006a; Saha and Chakrabarti, 2006; Wang et al., 2006). Sequence similarity values as low as 89–92.5% with other *Flavobacterium* species have been reported for *Flavobacterium saliperosum* and *Flavobacterium suncheonense* (Kim et al., 2006a; Wang et al., 2006). As a consequence, *Flavobacterium columnare*, *Flavobacterium croceum*, *Flavobacterium indicum*, *Flavobacterium saliperosum*, *Flavobacterium suncheonense*, and "*Flexibacter aurantiacus* subsp. *excathedrus*" form a distinct, smaller branch in the phylogenetic tree compared to all other *Flavobacterium* species (Figure 27) and this branching is supported by very high bootstrap values. However, 16S rRNA gene sequence similarity values within this smaller branch are moderate to low. For instance, *Flavobacterium columnare*, *Flavobacterium croceum*, *Flavobacterium saliperosum*, and *Flavobacterium suncheonense* share about 93–94% sequence similarity and the sequence similarity between *Flavobacterium indicum* and *Flavobacterium columnare* is as low as 90% (Kim et al., 2006a; Park et al., 2006a; Saha and Chakrabarti, 2006; Wang et al., 2006). Consequently, it would most likely be impossible to include all these species into a single new genus if it was decided to remove them from the genus *Flavobacterium*. Moreover, it has not been possible to identify common features shared by these organisms that would support their differentiation from members of the main branch. Because fatty acid profiles of *Flavobacterium* species have been determined using different procedures and different cultural conditions (see *Cell-wall composition*, above), it has unfortunately been impossible to compare the fatty acid profiles of the different clusters and branches currently grouped in the genus. Hence, members of the smaller branch should preferably be maintained in the genus *Flavobacterium* pending further studies.

As shown in Figure 27, the branching pattern of the various 16S rRNA gene sequences available for *Flavobacterium columnare* strains clearly shows the three genomic groups (genomovars) repeatedly reported in this species (Arias et al., 2004; Darwish and Ismaiel, 2005; Schneck and Caslake, 2006; Triyanto and Wakabayashi, 1999a, b). Distinct 16S rRNA restriction patterns and particular signatures in the 16S rRNA gene sequence further support delineation of these groups (Triyanto and Wakabayashi, 1999a). The type strain and strain FK 401 belong to genomic group I, strains IAM 14820 and LP 8 to genomic group II, and strain IAM 14821 to genomic group III. 16S rRNA gene sequence similarities and DNA–DNA relatedness among members of genomic group I were 99.4–99.8% and 83–93%, respectively, whereas sequence similarities and DNA–DNA relatedness between representatives of groups I and II were 98.0–98.2% and 44–73%, respectively. The only representative of group III included in the study shared 96.5–97.2% sequence similarity and 45–70% DNA relatedness with members of the other groups (Triyanto and Wakabayashi, 1999a). Slightly lower values were reported by Schneck and Caslake (2006), with 97.9–98.7% 16S rRNA gene sequence similarity among members of genomic group I and 95.8–96.6% sequence similarity between representatives of groups I and III. However, these studies have failed to uncover a correlation between the genomic groups and either the temperature of isolation, the fish host, or the geographic origin. Furthermore, no phenotypic evidence has yet been found to support the splitting of *Flavobacterium columnare* into three different species or subspecies.

Taxonomic and nomenclatural issues in the genus *Flavobacterium* are dealt with by the Subcommittee on the Taxonomy of *Flavobacterium* and *Cytophaga*-like Bacteria of the International Committee on Systematics of Prokaryotes. The subcommittee has issued minimal standards for describing new taxa of the family *Flavobacteriaceae* (Bernardet et al., 2002).

It should be noted that the descriptions of each of 21 *Flavobacterium* species (*Flavobacterium aquatile*, *Flavobacterium antarcticum*, *Flavobacterium croceum*, *Flavobacterium daejeonense*, *Flavobacterium defluvii*, *Flavobacterium denitrificans*, *Flavobacterium flevense*, *Flavobacterium frigidarium*, *Flavobacterium frigidimaris*, *Flavobacterium gillisiae*, *Flavobacterium glaciei*, *Flavobacterium granuli*, *Flavobacterium indicum*, *Flavobacterium omnivorum*, *Flavobacterium saliperosum*, *Flavobacterium segetis*, *Flavobacterium soli*, *Flavobacterium suncheonense*, *Flavobacterium weaverense*, *Flavobacterium xanthum* and *Flavobacterium xinjiangense*) of the 40 species with validly published names was based on a single isolate, even though additional strains of *Flavobacterium xanthum* were subsequently isolated (Van Trappen et al., 2005). Descriptions of all these species, apart from *Flavobacterium aquatile* and *Flavobacterium flevense*, were published between 2001 and 2007, although the practice of single-strain description had been discouraged repeatedly (Bernardet et al., 2002; Christensen et al., 2001).

Acknowledgements

The authors wish to acknowledge Jean Euzéby (Ecole Nationale Vétérinaire, Toulouse, France) for correcting the etymology of a number of specific epithets in the List of Species, Brita Bruun (Hillerød Hospital, Denmark) for critically reviewing the manuscript, Richard Holt (Oregon State University) for checking Borg's PhD thesis in search of the first mention of [*Cytophaga*] *psychrophila*, Brian Tindall (DSMZ, Germany) for sharing his expertise on fatty acid analysis and chemotaxonomy, and Gregory D. Wiens (USDA/ARS) for tracing the problematic reference Chester (1897).

Further reading

Website of the International Committee on Systematics of Prokaryotes (ICSP) Subcommittee on the Taxonomy of *Flavobacterium* and *Cytophaga*-like Bacteria (http://www.the-icsp.org/subcoms/Flavobacterium_Cytophaga.htm)

Differentiation of species of the genus *Flavobacterium*

Characteristics that differentiate species of the genus are listed in Table 24. Chemotaxonomic and molecular methods have also been evaluated for their ability to differentiate *Flavobacterium* species. Provided identical culture conditions are used, fatty acid profiles may be used to characterize some *Flavobacterium* species, e.g., *Flavobacterium columnare* (see *Cell-wall composition*, above). However, most of the fatty acid clusters delineated in a collection of polar *Flavobacterium* isolates actually contained multiple taxa, as shown by rep-PCR fingerprinting (Van Trappen et al., 2003, 2002). For instance, all strains of *Flavobacterium frigoris*, *Flavobacterium fryxellicola*, *Flavobacterium micromati*, *Flavobacterium psychrolimnae*, and *Flavobacterium xanthum* strains and most *Flavobacterium degerlachei* strains belonged to the same fatty acid cluster, whereas all 22 strains of *Flavobacterium gelidilacus* belonged to a different cluster (Van Trappen et al., 2002, 2003, 2004b, 2005). Strains sharing identical rep-PCR profiles were closely related, although not always at the species level. This technique allowed either differentiation of *Flavobacterium* species from each other or intraspecies typing. For instance, species-specific rep-PCR profiles were found in *Flavobacterium degerlachei*, *Flavobacterium micromati*, *Flavobacterium fryxellicola*, and *Flavobacterium psychrolimnae*, whereas two or three different rep-PCR profiles occurred in *Flavobacterium frigoris* and *Flavobacterium gelidilacus* (Van Trappen et al., 2003, 2004b, 2005).

List of species of the genus *Flavobacterium*

Preliminary note. All species with validly published names included in the genus *Flavobacterium* in the previous edition of the *Manual* (Holmes et al., 1984c) and in Holmes (1992), except *Flavobacterium aquatile*, have been reclassified in other or new genera. They are currently known under the following names: *Chryseobacterium balustinum*, *Chryseobacterium gleum*, *Chryseobacterium indologenes*, *Chryseobacterium indoltheticum*, *Elizabethkingia meningoseptica*, *Empedobacter brevis*, *Myroides odoratus*, *Sphingobacterium multivorum*, *Sphingobacterium spritivorum*, and *Sphingobacterium thalpophilum* (see the corresponding chapters in this volume; Takeuchi and Yokota, 1992; Vandamme et al., 1994a; Vancanneyt et al., 1996; Yabuuchi et al., 1983; Kim et al., 2005b). According to Takeuchi and Yokota (1992), *Flavobacterium yabuuchiae* (Holmes et al., 1988) is a later heterotypic synonym of *Sphingobacterium spiritivorum* (Holmes et al., 1982; Yabuuchi et al., 1983). Described by Yabuuchi et al. (1983), *Sphingobacterium mizutae* was transferred to the genus *Flavobacterium* by Holmes et al. (1988) under the name *Flavobacterium mizutaii*; however, this organism has been clearly assigned to the genus *Sphingobacterium* in recent publications (see treatment of the genus *Sphingobacterium*; Gherna and Woese, 1992; Kim et al., 2006a).

Most bacterial species included in the *Flavobacterium* chapters of the 7th and 8th editions of the *Manual* (Weeks, 1974; Weeks and Breed, 1957) and considered *species incertae sedis* by Holmes et al. (1984c) have also been reclassified. They are currently known under the following names: *Halomonas halmophila*, *Microbacterium esteraromaticum*, *Microbacterium maritypicum*, *Novosphingobium capsulatum*, *Novosphingobium resinovorum*, *Pedobacter heparinus*, *Planomicrobium okeanokoites*, *Psychroflexus gondwanensis*, *Salegentibacter salegens*, *Terrimonas ferrigunea*, and *Zobellia uliginosa* (see the corresponding chapters in this edition of the *Manual*). According to Yabuuchi et al. (1979) and to the catalogs of the major culture collections, the type and only strain of [*Flavobacterium*] *devorans* should be assigned to *Sphingomonas paucimobilis*. Among the *species incertae sedis* listed in the previous edition of the *Manual* (Holmes et al., 1984c) [*Flavobacterium*] *acidificum*, [*Flavobacterium*] *acidurans*, [*Flavobacterium*] *oceanosedimentum*, and [*Flavobacterium*] *thermophilum* have still not been assigned to other genera. Their names have been either included in the Approved Lists of Bacterial Names (Skerman et al., 1980) or validly published. Strains are available in culture collections, but 16S rRNA sequences are not. The reasons for excluding these poorly characterized species from the genus *Flavobacterium* have been given by Holmes et al. (1984c) and Holmes (1993).

The name "*Flavobacterium aureum*" had been proposed, although not validly published, for *Flavobacterium* strains of CDC group IIb by Price and Pickett (1981). These strains have later been transferred to the genus *Chryseobacterium* (see the corresponding chapter in this volume). A bacterium responsible for hemorrhagic septicemia in farmed bullfrog in Taiwan had received the provisional name "*Flavobacterium ranacida*" (Faung et al., 1996 and references therein); this organism and similar isolates from other frog species were later identified as *Elizabethkingia meningoseptica* (previously *Chryseobacterium meningosepticum*; Bernardet et al., 2005). Strain NRRL B-184, tentatively named "*Flavobacterium aurantiacum*", is able to remove aflatoxin from peanut milk (Diarra et al., 2005). According to the online catalog of the Agricultural Research Service Culture Collection (USA), it is actually a strain of *Rhodococcus corynebacteroides* (previously *Nocardia corynebacteroides*). "*Flavobacterium chlorophenolica*" ATCC 39723, isolated from pentachlorophenol-contaminated soil in Minnesota, can mineralize pentachlorophenol at concentrations of 100–200 mg/l. This organism, which has been the subject of many publications since 1985, has recently been reclassified in the genus *Sphingomonas* on the basis of its 16S rRNA gene sequence (Nohynek et al., 1995).

Editorial note: the following novel species have been added to the genus *Flavobacterium* after this chapter was completed and their names have been validly published in the *International Journal of Systematic* and *Evolutionary Microbiology*: *Flavobacterium terrigena* (Yoon et al., 2007a), *Flavobacterium terrae* and *Flavobacterium cucumis* (Weon et al., 2007), *Flavobacterium filum* (Ryu et al., 2007), *Flavobacterium ceti* (Vela et al., 2007), *Flavobacterium lindanitolerans* (Jit et al., 2008), *Flavobacterium cheniae* (Qu et al., 2008) and *Flavobacterium anhuiense* (Liu et al., 2008).

A complete list of organisms currently belonging to the genus or that were previously attributed to it are available in the List of Prokaryotic Names with Standing in Nomenclature (LPSN) (Euzéby, 1997; http://www.bacterio.cict.fr/). This list also provides the current taxonomic status and standing in nomenclature of the taxa that were subsequently excluded from the genus *Flavobacterium*. For instance, according to the LPSN, the name "*Flavobacterium farinofermentans*" Bai (1983) has no standing in nomenclature.

1. **Flavobacterium aquatile** (Frankland and Frankland 1889) Bergey, Harrison, Breed, Hammer and Huntoon 1923, 100^{AL} emend. Bernardet, Segers, Vancanneyt, Berthe, Kersters and Vandamme 1996, 140^{VP} [*Bacillus aquatilis* Frankland and Frankland 1889, 381; *Bacterium aquatilis* (Frankland and Frankland 1889) Chester 1897, 96; *Flavobacterium aquatilis* (Frankland and Frankland 1889) Bergey, Harrison, Breed, Hammer and Huntoon 1923, 100; *Chromobacterium aquatilis* (Frankland and Frankland 1889) Topley and Wilson 1929, 404; *Empedobacter aquatile* (Frankland and Frankland 1889) Brisou, Tysset and Jacob 1960, 359]

a.qua'ti.le. L. neut. adj. *aquatile* living in water.

Rods, approximately 0.5–0.7 × 1.0–3.0 µm; longer rods and filaments also occur. Colonies on agar media at 25–30°C are light yellow in color, circular with entire edges, even on Anacker and Ordal's agar on which gliding bacteria usually produce spreading colonies. At 15–20°C, colonies are bright orange with a mucoid consistency. Growth does not occur at 37°C. Catalase-positive. Peptones are used as nitrogen sources, but urea and Casamino acids are not. Tributyrin and lecithin are hydrolyzed. Indole is not produced. Acid but no gas is produced aerobically from glucose, lactose, maltose, and sucrose, but not from arabinose, cellobiose, ethanol, fructose, mannitol, raffinose, rhamnose, salicin, trehalose, or xylose (Holmes et al., 1984c; Wakabayashi et al., 1989). Resistant to the vibriostatic compound O/129 (Bernardet et al., 1996). Other characteristics are as given in the genus description and in Tables 24 and 25.

The long and complex history of this organism, its status as type species of the genus, the unique strain, and its much debated motility and status as neotype strain have been reviewed extensively (see *Taxonomic comments*, above, as well as: Weeks, 1955, 1981; Holmes and Owen, 1979; Holmes et al., 1984c; Holmes, 1992). Perry (1973) observed gliding motility in hanging drop preparations and on agar plates on which he had added small glass beads (although the strain did not produce spreading growth on agar, as most gliding bacteria do), but his view was not shared by other authors (Holmes et al., 1984c and references therein). Consequently, the ability of *Flavobacterium aquatile* to glide still has to be unequivocally established. Using recommended techniques (Bernardet et al., 2002), gliding motility was not observed (J.-F. Bernardet, personal observation). The pseudoflagella and nonfunctional flagella initially reported were later shown by electron microscopy to be non-flagellar appendages similar to those found on several gliding bacteria (Holmes and Owen, 1979; Holmes et al., 1984c and references therein). Although originally reported to contain ubiquinones (Callies and Mannheim, 1978), *Flavobacterium aquatile* was later shown to contain MK-6 as its major respiratory quinone (Oyaizu and Komagata, 1981), as do all members of the family *Flavobacteriaceae*.

Source: the only available strain has purportedly been isolated from the water of the same deep well in the chalk region of Kent, England, as the original organism.

DNA G+C content (mol%): 32–33 (T_m).

(Neo)type strain: Taylor F36, ATCC 11947, CCUG 29304, CIP 103744, DSM 1132, IAM 12316, IFO (now NBRC) 15052, LMG 4008, NCIMB 8694, NCTC 9758.

Sequence accession no. (16S rRNA gene): M62797.

Note: the erroneous page number in Chester (1897) used in previous publications has been corrected thanks to G.D. Wiens who traced and provided the original paper.

2. **Flavobacterium antarcticum** Yi, Oh, Lee, Kim and Chun 2005a, 640[VP]

ant.arc'ti.cum. L. neut. adj. *antarcticum* southern and, by extension, pertaining to Antarctica.

Rods, approximately 0.3–0.4 × 0.5–1.3 μm, frequently forming pairs. Although gliding motility was not reported in the original description, cells distinctly gliding and rotating have been observed in 48-h-old cultures in Anacker and Ordal's broth and tryptic soy broth (J.-F. Bernardet, personal observation). Movements are special, however, being shorter but sharper than those of other gliding *Flavobacterium* strains. Colonies on R2A agar are circular, convex, glistening, and yellow with entire margins. The consistency is butyrous, becoming mucoid after prolonged incubation. In addition to the media listed in Table 24, growth occurs on R2A agar, but not on cetrimide or MacConkey agars. Growth is aerobic (minimum doubling time, 4.9 h), but weak and poor growth was also observed under microaerobic and anaerobic conditions, respectively.

Catalase-positive. Arginine dihydrolase activity is present, but phenylalanine deaminase activity is absent. Indole is not produced and alkalinization does not occur on Christensen's citrate agar. Acid is produced from glucose and maltose, but not from arabinose, cellobiose, fructose, lactose, mannitol, raffinose, rhamnose, salicin, sucrose, trehalose, or xylose. The list of enzymic activities detected in the API ZYM gallery is given in the original description. None of the compounds contained in the API 20NE gallery is assimilated. No precipitate is formed on egg-yolk agar, but the colonies are surrounded by clear zones. Elastin is not degraded. The maximum adsorption peak of the pigment is at 452 nm and the next shoulder peak is at 479 nm. Other characteristics are as given in the genus and original species descriptions and in Tables 24 and 25.

Source: the type and, so far, only strain was isolated from a soil sample of a penguin habitat near the King Sejong Station on King George Island, Antarctica.

DNA G+C content (mol%): 38.0 (HPLC).

Type strain: AT1026, CIP 108750, IMSNU 14042, JCM 12383, KCTC 12222.

Sequence accession no. (16S rRNA gene): AY581113.

3. **Flavobacterium aquidurense** Cousin, Päuker and Stackebrandt 2007, 247[VP]

a.qui.du.ren'se. L. fem. n. *aqua* water; L. adj. *durus* hard; L. suff. *-ensis*, *-e* suffix used with the sense of belonging to or coming from; N.L. neut. adj. *aquidurense* pertaining to hard water.

Rods, approximately 0.9 × 2.8 μm. Colonies on R2A, nutrient, and tryptic soy agars are 1–2.2 mm in diameter, yellow, circular, convex, and spreading with filamentous margins (R2A agar), non-spreading with undulate margins (tryptic soy agar), or umbonate with undulate margins (nutrient agar). Larger (3–8 mm), flat, spreading colonies with filamentous margins are formed on DSMZ medium 67. No growth occurs on MacConkey agar. Growth is aerobic and microaerobic.

Catalase-positive. Indole is not produced. Characteristics included in API ZYM, API 20NE, and GN MicroPlate galleries are given in the original description. The following biochemical characteristics are strain-dependent: nitrate reduction and utilization of *N*-acetyl-D-galactosamine, L-arabinose, D-cellobiose, D-fructose, D-galactose, α-D-lactose, melibiose, sucrose, trehalose, turanose, methyl pyruvate, L-alaninamide, L-alanine, and glycyl-L-aspartic acid. Other characteristics are as given in the genus and original species descriptions and in Tables 24 and 25.

Source: the spring of the Westerhöfer Bach, a hard-water creek from the western slopes of the Harz Mountains, in Westerhof, 40 km north of Göttingen, Germany. Five strains were isolated and characterized, but only the type strain was deposited in culture collections.

DNA G+C content (mol%): 33.5 (T_m).

Type strain: Cousin WB 1.1-56, CIP 109242, DSM 18293.

Sequence accession no. (16S rRNA gene): AM177392.

4. **Flavobacterium branchiophilum** corrig. Wakabayashi, Huh and Kimura 1989, 215[VP] emend. Bernardet, Segers, Vancanneyt, Berthe, Kersters and Vandamme 1996, 140.

bran.chi.o.phi'lum. Gr. n. *branchion* gill; Gr. adj. *philos* loving; N.L. neut. adj. *branchiophilum* gill loving.

The original spelling of the specific epithet, *branchiophila* (*sic*), was corrected by von Graevenitz (1990).

Rods, approximately 0.5 × 5.0–8.0 μm. Filamentous cells 15–40 μm long also occur. *Flavobacterium branchiophilum* is one of the most fastidious *Flavobacterium* species (J.-F. Bernardet, personal observation). No growth occurs on tryptic soy, nutrient, or Mueller–Hinton agars. Growth is distinctly better on plain Anacker and Ordal's agar than on the enriched version that contains 5 g/l tryptone instead of 0.5 g/l and is commonly used for growing other fish pathogens. Other convenient media are modified Shieh's agar (Song et al., 1988), *Microcyclus–Spirosoma* agar (NCIMB medium 81), and medium for freshwater flexibacteria (NCIMB medium 218; Lewin, 1969). The best growth is achieved on casitone yeast extract agar (NCIMB medium 101) at 20–25°C (J.-F. Bernardet, unpublished data).

Catalase-positive. Acid but no gas is produced aerobically from cellobiose, fructose, glucose, maltose, raffinose, sucrose, and trehalose, but not from arabinose, lactose, mannitol, rhamnose, salicin, or xylose. Tributyrin and lecithin are degraded. Indole is not produced. The list of substrates hydrolyzed in the API ZYM gallery is given in the original description and in Bernardet et al. (1996). Susceptible to the vibriostatic compound O/129 (Bernardet et al., 1996). Other characteristics are as given in the genus and original species descriptions and in Tables 24 and 25. A DNA G+C content range of 29–31 mol% was originally reported (Wakabayashi et al., 1989), but subsequent investigation of six strains yielded a range of approximately 33–34 mol% (Bernardet et al., 1996; J.-F. Bernardet, unpublished data).

Source: diseased gills of various freshwater fish species (mainly salmonids). This organism is the main causal agent of bacterial gill disease of fish in Japan, Republic of Korea, Hungary, Oregon (USA), and in Ontario (Canada) (Bernardet and Bowman, 2006). The type strain was isolated in 1977 from the gills of a diseased fingerling yamame (*Oncorhynchus masou*) from a hatchery in Gumma Prefecture, Japan.

DNA G+C content (mol%): 32.5–34.2 (T_m).

Type strain: Wakabayashi BGD-7721, ATCC 35035, CCUG 33442, CIP 103527, IFO (now NBRC) 15030, LMG 13707, NCIMB 12904.

Sequence accession no. (16S rRNA gene): D14017.

5. **Flavobacterium columnare** (Bernardet and Grimont 1989) Bernardet, Segers, Vancanneyt, Berthe, Kersters and Vandamme 1996, 140[VP] [*Bacillus columnaris* Davis 1922, 263; *Chondrococcus columnaris* (Davis 1922) Ordal and Rucker 1944, 18; *Cytophaga columnaris* (Davis 1922) Garnjobst 1945, 127 and Reichenbach 1989b, 2046; *Flexibacter columnaris* (Davis 1922) Leadbetter 1974, 107 and Bernardet and Grimont 1989, 352]

co.lum.na're. L. neut. adj. *columnare* rising as a pillar, referring to the shape of the aggregates formed by the bacteria on external lesions of infected fish.

Rods, approximately 0.3–0.5 × 3.0–10.0 µm in 48-h-old broth cultures; filamentous cells up to 25 µm occur. Colonies best display their typical flat, spreading, and rhizoid morphology on Anacker and Ordal's agar; they are greenish-yellow and adhere strongly to the agar (see *Colonial or cultural characteristics*, above). Some colonies have a raised, globular, or warty center. Different colony types (i.e., more or less rhizoid) may coexist on the same plate. In agitated Anacker and Ordal's broth cultures, numerous filamentous tufts of bacteria adhere to the glass and/or to the magnet. The particular odor of *Flavobacterium columnare* cultures has been repeatedly reported. Growth does not occur on nutrient agar (J.-F. Bernardet, personal observation), contrary to the report by Reichenbach (1989b). Growth at 37°C is strain-dependent. *Flavobacterium columnare* is the only *Flavobacterium* species whose colonies unequivocally adsorb Congo red; the Congo red-staining material has been shown to be an extracellular galactosamine glycan in the slime (Johnson and Chilton, 1966). Electron microscopy and biochemical studies have revealed the presence of a capsule that is thicker and denser in highly virulent strains than in low virulence ones (Decostere et al., 1999).

Catalase-positive. Arginine dihydrolase and lysine and ornithine decarboxylase activities are absent. Actively proteolytic. Different types of extracellular proteases have been identified; in combination with enzymes degrading the complex acidic polysaccharides of connective tissue (e.g., chondroitin sulfate and hyaluronic acid), they play a major role in the extensive skin and muscular necrotic lesions displayed by diseased fish (see comments below for *Flavobacterium psychrophilum*). Peptones and casein hydrolysate serve as nitrogen sources, but NH_4^+ and NO_3^- do not. No acid is produced from carbohydrates. Tributyrin and lecithin are hydrolyzed. Starch degradation is medium-dependent: it is negative on 0.2% starch Anacker and Ordal's agar (Bernardet and Grimont, 1989), but positive on 0.2% starch Hsu–Shotts agar (Shamsudin and Plumb, 1996; Shoemaker et al., 2005). This point has been confirmed recently by testing ten strains in parallel on both media (C. Shoemaker, personal communication). The antibiotics to which *Flavobacterium columnare* are resistant and the substrates it hydrolyzes in the API ZYM gallery are listed in Bernardet and Grimont (1989) and Bernardet et al. (1996). Susceptible to the vibriostatic compound O/129 (Bernardet et al., 1996). Other characteristics are as given in the genus description and in Tables 24 and 25. Additional information can be found in Reichenbach (1989b), Bernardet and Grimont (1989), and Bernardet et al. (1996).

Three different genomic groups (genomovars) have been reported in *Flavobacterium columnare* (see *Taxonomic comments*, above), but the species has not been split due to the lack of phenotypic characteristics to differentiate them. Strain ATCC 43622 was erroneously considered a *Flavobacterium columnare* strain as shown by 16S rRNA and fatty acid data (McCammon et al., 1998; Shoemaker et al., 2005); it may actually belong to *Flavobacterium johnsoniae* (Darwish et al., 2004). Unfortunately, this strain has been used to characterize the glycopeptides of *Flavobacterium columnare* that may provide target molecules for fish vaccines (Vinogradov et al., 2003) and its 16S rRNA gene sequence has been used to infer phylogenetic relationships and to design PCR primers aimed at detecting and identifying *Flavobacterium columnare* strains (Bader and Shotts, 1998; Bader et al., 2003; Toyama et al., 1996).

Flavobacterium columnare is the causal agent of "columnaris disease" and is recognized as an important fish pathogen in temperate to warm freshwater worldwide, causing serious economic losses in a considerable variety of farmed and wild, food and ornamental fish. Virulence for fish has been demonstrated experimentally. For recent reviews on the bacterium and the infection it causes, see Austin and Austin (1999), Shotts and Starliper (1999), and Bernardet and Bowman (2006). No type strain had been designated and only strain NCIMB 1038[T] was available in culture collections at the time when Reichenbach was writing the *Flavobacterium* chapter for the first edition of *Bergey's Manual of Systematic Bacteriology*. This strain was isolated in 1955 from the kidney of a diseased chinook salmon (*Oncorhynchus tschawytcha*) from the Snake River, WA, USA. However, because strain NCIMB 1038[T] had lost the typical rhizoid and adherent colony morphology through repeated subcultures, Reichenbach (1989b) proposed to use as a neotype a substrain (NCIMB 2248[NT]) that still displayed these characteristics.

Source: external lesions (and less frequently from internal organs) of diseased fish.

DNA G+C content (mol%): 32–33 (T_m, Bd).

(Neo)type strain: 1-S-2cl (Anacker and Ordal, 1955), ATCC 23463, CIP 103531, IAM 14301, LMG 10406, LMG 13035, NBRC 100251, NCIMB 1038, NCIMB 2248. The type strain belongs to genomovar 1.

Sequence accession no. (16S rRNA gene): Genomovar 1: AB078047 (type strain), AB010952 (strain FK 401). Genomovar 2: AB016515 (strain IAM 14820), AB015480 (strain LP 8). Genomovar 3: AB015481 (strain IAM 14821).

Reference strains: Genomovar 1: JIP 39/87 (ATCC 49513, CIP 103532), JIP 44/87 (ATCC 49512, CIP 103533). Genomovar 2: Wakabayashi EK28 (IAM 14820). Genomovar 3: Wakabayashi PH 97028 (IAM 14821, JCM 21327).

6. **Flavobacterium croceum** Park, Lu, Ryu, Chung, Park, Kim and Jeon 2006a, 2445[VP]

cro'ce.um. L. neut. adj. *croceum* saffron-colored, yellow.

Straight or slightly curved rods, approximately 0.3–0.5 × 1.0–3.2 µm. Colonies on R2A agar are yellow, slightly raised, and circular with entire margins. Growth is severely inhibited by NaCl concentrations above 1%. Weak anaerobic growth occurs only after 16 d on R2A agar.

Catalase-negative. Arginine dihydrolase and lysine and ornithine decarboxylase activities are absent. Xanthine and hypoxanthine are not hydrolyzed. Indole is not produced. Citrate is not utilized. Acid is produced from L-arabinose, D-glucose, *myo*-inositol, lactose, melibiose, and raffinose, but not from D-fructose, D-galactose, D-mannitol, D-mannose, arbutin, or salicin. Enzymic activities detected in the API ZYM gallery and antibiotic resistances are listed in the original description. The major polar lipid is phosphatidylethanolamine; minor amounts of phosphatidylglycerol and phosphatidylcholine are also present. Other characteristics are as given in the genus and original species descriptions and in Tables 24 and 25.

Source: sludge that performed enhanced biological phosphorus removal in a laboratory-scale sequencing batch reactor (Republic of Korea); sodium acetate was supplied as sole carbon source.

DNA G+C content (mol%): 40.8 (HPLC).

Type strain: EMB47, DSM 17960, KCTC 12611.

Sequence accession no. (16S rRNA gene): DQ372982.

7. **Flavobacterium daejeonense** Kim, Weon, Cousin, Yoo, Kwon, Go and Stackebrandt 2006a, 1647[VP]

da.e.je.o.nen'se. N.L. neut. adj. *daejeonense* of or belonging to Daejeon, a city in the Republic of Korea, from where the type strain was isolated.

Rods, approximately 0.5 × 2.0–3.0 µm. Colonies on R2A agar are yellow, convex, and circular with entire margins. No growth occurs on MacConkey agar.

Catalase-positive. Arginine dihydrolase activity is absent. Indole is not produced. D-Glucose, L-arabinose, D-mannose, and maltose are assimilated, but D-mannitol, *N*-acetylglucosamine, potassium gluconate, capric acid, adipic acid, malic acid, trisodium citrate, and phenylacetic acid are not. Enzymic activities detected in the API ZYM gallery are listed in the original description. In the API 50CH gallery, acid is produced from L-arabinose, D-xylose, D-galactose, D-glucose, D-fructose, D-mannose, esculin, maltose, D-lactose, melibiose, sucrose, inulin, and raffinose. Other characteristics are as given in the genus and species descriptions and in Tables 24 and 25.

Source: greenhouse soil, Daejeon city, Republic of Korea.

DNA G+C content (mol%): 35.0 (HPLC).

Type strain: GH1-10, DSM 17708, KACC 11422.

Sequence accession no. (16S rRNA gene): DQ222427.

8. **Flavobacterium defluvii** Park, Ryu, Vu, Ro, Yun and Jeon 2007, 235[VP]

de.flu'vi.i. L. gen. n. *defluvii* of sewage.

Rods, approximately 0.4–0.5 × 2.3–6.5 µm. Some cells show a knar. Colonies on R2A agar are pale yellow, glistening, translucent, irregular, slightly raised with curled margins, and slightly sticky. Growth in R2A broth is severely inhibited by the addition of more than 2% NaCl. Good growth occurs on Luria–Bertani agar. Anaerobic growth is not observed after 7 d at 30°C on R2A agar, but weak growth occurs after 16 d. Because the strain is unable to reduce nitrate to nitrite and consequently grow by anaerobic respiration using nitrate or nitrite as an electron acceptor, it probably grows anaerobically by fermenting carbohydrates.

Catalase-positive. Tyrosine is not hydrolyzed, contrary to the statement in Table 1 of the original description (C.O. Jeon, personal communication). Indole and acetoin are not produced. Citrate is not utilized. Acid is produced from L-arabinose, arbutin, D-fructose, D-galactose, D-glucose, lactose, D-mannitol, D-mannose, melibiose, *myo*-inositol, and raffinose, but not from salicin. Enzymic activities detected in the API ZYM gallery and antibiotic resistances are listed in the original description. The major polar lipid is phosphatidylethanolamine. Other characteristics are as given in the genus and original species descriptions and in Tables 24 and 25.

Source: sludge that performed enhanced biological phosphorus removal in a laboratory-scale sequencing batch reactor (Pohang, Republic of Korea).

DNA G+C content (mol%): 33.5 (HPLC).

Type strain: Park EMB117, DSM 17963, KCTC 12612.

Sequence accession no. (16S rRNA gene): DQ372986.

9. **Flavobacterium degerlachei** Van Trappen, Vandecandelaere, Mergaert and Swings 2004b, 89[VP]

de.ger.lache'.i. N.L. masc. gen. n. *degerlachei* of de Gerlache, named after Adrien de Gerlache, the Belgian pioneer who conducted the first scientific expedition in Antarctica in 1897–1899.

Rods, approximately 1 × 3–4 µm, frequently forming pairs or short chains. Although gliding was not reported in the original description, gliding and rotating cells have been observed distinctly in a 48-h-old R2A broth culture (J.-F. Bernardet, personal observation). Colonies on R2A agar are yellow, convex, translucent with entire margins, non-adherent, and 1–3 mm in diameter after 6 d.

Catalase-positive. Arginine dihydrolase, lysine and ornithine decarboxylase, and tryptophan deaminase activities are absent. Enzymic activities detected in the API ZYM gallery are listed in the original description. Glucose, mannose, and maltose are assimilated in the API 20NE gallery. Acid is not produced from carbohydrates in the API 20E gallery. Indole and acetoin are not produced.

Citrate is not utilized. Other characteristics are as given in the genus and original species descriptions and in Tables 24 and 25.

Source: microbial mats from hyposaline to saline lakes in the Vestfold Hills and Larsemann Hills, eastern Antarctica. The 14 original isolates all belong to the same rep-PCR cluster, but are divided into two different fatty acid clusters.

DNA G+C content (mol%): 33.8–34.2 (HPLC).

Type strain: Mergaert R-9106, AC23, CIP 108386, DSM 15718, LMG 21915.

Sequence accession nos (16S rRNA gene): AJ557886 (type strain), AJ441005 (strain LMG 21474).

10. **Flavobacterium denitrificans** Horn, Ihssen, Matthies, Schramm, Acker and Drake 2005, 1263[VP]

de.ni.tri'fi.cans. N.L. part. adj. *denitrificans* denitrifying.

Rods, approximately 0.3–0.9 × 0.8–5.0 µm, frequently forming chains of 3–14 cells. Cells may be connected by cellular bridges. Although motility was mentioned in the original description, the type of motility was not specified. Distinct though slow gliding and rotating movements, as well as spheroplasts, were observed in a 48-h-old Anacker and Ordal's broth culture (J.-F. Bernardet, personal observation). Colonies on tryptic soy agar are yellow to dark yellow, flat, circular to irregular, with entire margins. The doubling time under optimal conditions is 7.3 h. Facultatively anaerobic growth occurs in mineral medium by complete denitrification. Membranes contain *c*-type cytochromes.

Catalase-positive. Arginine dihydrolase activity is present. Lysine decarboxylase and phenylalanine deaminase activities are absent. Indole is not produced. Arabinose, cellobiose, fructose, fumarate, gelatin, glucose, glutamate, inulin, lactose, maltose, mannitol, mannose, *N*-acetylglucosamine, pectin, starch, succinate, and xylose are utilized. 1-Butanol, 1-propanol, acetate, butyrate, chitin, citrate, ethanol, ethanolamine, glycerol, glycolate, isobutyrate, inositol, isovalerate, lactate, oxalate, propionate, raffinose, sucrose, sorbitol, and tartrate are not utilized. Ammonium is used as a nitrogen source. O_2, NO_3^-, and NO_2^- are used as electron acceptors. N_2O is produced as an intermediate during the reduction of NO_3^- to N_2. SO_4^{2-} and Fe^{3+} are not used as electron acceptors. Does not grow by fermentation. Other characteristics are as given in the genus and original species descriptions and in Tables 24 and 25. The fatty acid composition has been determined by Cousin et al. (2007).

Source: gut of the earthworm *Aporrectodea caliginosa* collected from garden soil in Bayreuth, Germany. Because denitrifying bacteria retrieved from the gut are usually derived from the soil, the authors considered *Flavobacterium denitrificans* to be a soil organism (H.L. Drake, personal communication).

DNA G+C content (mol%): 34.6 (HPLC).

Type strain: ED5, ATCC BAA-842, CIP 109214, DSM 15936.

Sequence accession no. (16S rRNA gene): AJ318907.

11. **Flavobacterium flevense** (van der Meulen, Harder and Veldkamp 1974) Bernardet, Segers, Vancanneyt, Berthe, Kersters and Vandamme 1996, 141[VP] (*Cytophaga flevensis* van der Meulen, Harder and Veldkamp 1974, 340 and Reichenbach 1989c, 2038)

fle.ven'se. L. n. *Flevum* a former inner sea in the central part of The Netherlands which is now a freshwater lake called IJsselmeer; N.L. neut. adj. *flevense* of or belonging to *Flevum*.

Rods, approximately 0.5–0.7 × 2.0–5.0 µm; almost coccoid in old cultures. Colonies are circular, low convex, sunk in the agar, and yellow-colored due to the production of carotenoid pigments. Spreading is observed occasionally after 1–2 weeks on nutrient agar.

Catalase activity was originally reported as present. Reichenbach (1989b) considered it absent, whereas others have observed a distinctly positive reaction (J.-F. Bernardet, personal observation). Peptones, Casamino acids, NH_4^+, and NO_3^- are used as N sources. Grows on peptone alone, but the presence of carbohydrates enhances growth. Tributyrin is hydrolyzed, but lecithin is not. Agar, agarose, arabinose, galactose, glucose, inulin, lactose, maltose, mannose, pectin, raffinose, rhamnose, sucrose, and xylose are assimilated. Cellobiose, fucose, ribose, trehalose, and alcohols are not assimilated. van der Meulen et al. (1974) and Reichenbach (1989b) found that starch was not degraded and that nitrate was reduced; opposite results, however, have been obtained by others (J.-F. Bernardet, personal observation). Indole is not produced. Resistant to novobiocin, penicillin G, polymyxin B, and streptomycin; sensitive to chloramphenicol, tetracycline, erythromycin, nalidixic acid, nitrofurantoin, and sulfafurazole. Susceptible to the vibriostatic compound O/129 (Bernardet et al., 1996). Other characteristics are as given in the genus and original species descriptions and in Tables 24 and 25. Additional information can be found in Christensen (1977), Reichenbach (1989b), and Bernardet et al. (1996).

Its ability to grow on media containing full-strength seawater and to degrade agar may suggest that *Flavobacterium flevense* was originally a marine organism that has progressively adapted to freshwater conditions following the separation of its habitat (polders) from the sea (Reichenbach, 1989b; Van der Meulen et al., 1974). A total of 25 Gram-stain-negative, yellow-pigmented, agarolytic strains have been isolated from the IJsselmeer and habitats in contact with this lake, but only strain A-34[T] has been studied in detail and deposited in culture collections.

Source: freshwater and wet soil collected in the IJsselmeer (an old bay in the North Sea) area, The Netherlands.

DNA G+C content (mol%): 33–34.9 (Bd, T_m).

Type strain: van der Meulen A-34, ACAM 579, ATCC 27944, CCUG 14832, CIP 104740, DSM 1076, IFO (now NBRC) 14960, LMG 8328, NCIMB 12056.

Sequence accession no. (16S rRNA gene): M58767.

12. **Flavobacterium frigidarium** Humphry, George, Black and Cummings 2001, 1242[VP]

fri.gi.da'ri.um. L. neut. adj. *frigidarium* of or for cooling, intended to mean belonging to the cold.

Rods, approximately 0.5–0.7 × 0.8–2 µm, occurring singly or in pairs, with a thick capsule and pili. Colonies on XMM agar (per l: 0.02 g $FeSO_4$, 0.2 g $MgSO_4$, 0.75 g KNO_3, 0.5 g K_2HPO_4, 0.04 g $CaCl_2$, 5.0 g soluble xylan, 15.0 g agar) and on tryptic soy agar are bright yellow, circular, and raised with entire margins. In the original description, the authors carefully stated that "there seemed to be adsorption of Congo red by colonies, indicating the presence of extracellular galactosamine glycan; this may be the capsular substance around the cell". Adsorption of Congo red could not be confirmed, however (J.-F. Bernardet, personal observation). The bright yellow carotenoid pigment has been tentatively identified as zeaxanthin. Doubling times in XMM broth at 0, 5, 10, 15, and 20°C are 34.8, 28.1, 10.9, 9.6, and 13.1 h (±10%), respectively.

Catalase-positive. Arginine dihydrolase activity is absent. Indole and acetoin are not produced. Neither acid nor gas are produced from glucose, fructose, xylose, mannitol, and maltose. Enzymic activities detected in the API ZYM gallery and antibiotic resistances are listed in the original description. Growth occurs on casein, fructose, gelatin, glucose, glycerol, laminarin, maltose, mannitol, mannose, succinate, tryptone, xylan, and xylose as sole carbon and energy sources. No growth occurs on acetate, adipate, agar, arabinose, caprate, carboxymethylcellulose, chitin, citrate, galactose, gluconate, inositol, lactose, malate, *N*-acetylglucosamine, pectin, phenylacetate, pyruvate, raffinose, ribose, starch, sucrose, trehalose, or tyrosine. Other characteristics are as given in the genus and original species descriptions and in Tables 24 and 25.

Source: shallow-water Southern Ocean sediment recovered near Adelaide Island, Antarctica, during a systematic search for xylanolytic, laminarinolytic, pectinolytic, and cellulolytic bacteria.

DNA G+C content (mol%): 35 (T_m).

Type strain: A2i, ATCC 700810, CIP 107124, LMG 21010, NCIMB 13737.

Sequence accession no. (16S rRNA gene): AF162266.

Note: the strain deposited in the ATCC was originally named "*Flavobacterium xylanivorum*".

13. **Flavobacterium frigidimaris** Nogi, Soda and Oikawa 2005a, 1743[VP] (Effective publication: Nogi, Soda and Oikawa 2005b, 314.)

fri.gi.di.ma'ris. L. adj. *frigidus* cold; L. gen. n. *maris* of the sea; N.L. gen. n. *frigidimaris* of a cold sea.

Rods, approximately 0.6–0.8 × 1.5–2.0 µm. Possibly capsulate (J.-F. Bernardet, personal phase-contrast microscopy observation). Colonies on tryptic soy agar are orange-yellow, circular, and low convex with entire to slightly irregular margins. Good growth also occurs on Luria–Bertani agar and on CY agar (per l: 3.0 g casitone, 1.36 g $CaCl_2 \cdot 2H_2O$, 1.0 g yeast extract, 15 g agar; pH 7.2).

Catalase-positive. Large amounts of $NAD(P)^+$-dependent alcohol, aldehyde, threonine, valine, and glutamate dehydrogenases are produced. These enzymes are all active at low temperature; the latter three are thermolabile, whereas the two former are unexpectedly thermostable. Acid is produced from arabinose, cellobiose, fructose, galactose, glucose, maltose, mannitol, mannose, raffinose, sucrose, trehalose, and xylose. No acid is produced from glycerol, inositol, lactose, rhamnose, or sorbitol. Other characteristics are as given in the genus and original species descriptions and in Tables 24 and 25.

Source: Antarctic seawater.

DNA G+C content (mol%): 34.5 (HPLC).

Type strain: KUC-1, DSM 15937, JCM 12218.

Sequence accession no. (16S rRNA gene): AB183888.

14. **Flavobacterium frigoris** Van Trappen, Vandecandelaere, Mergaert and Swings 2004b, 90[VP]

fri.go'ris. L. masc. n. *frigor* cold; L. gen. n. *frigoris* of the cold.

Rods, <1 × 4–6 µm. Gliding motility was not reported in the original description, but gliding and rotating cells have been observed in 48-h-old Anacker and Ordal's and R2A broth cultures (J.-F. Bernardet, personal observation). Colonies on R2A agar are yellow, convex, translucent with entire margins, non-adherent, and 2–5 mm in diameter after 6 d.

Catalase-positive. Arginine dihydrolase, lysine and ornithine decarboxylase, and tryptophan deaminase activities are absent. Enzymic activities detected in the API ZYM gallery are listed in the original description. Glucose, mannose, and maltose are assimilated in the API 20NE gallery. No acid is produced from carbohydrates in the API 20E gallery. Acetoin and indole are not produced. Citrate is not utilized. Nitrate reduction is strain-dependent; the type strain does not reduce nitrate. Other characteristics are as given in the genus and original species descriptions and in Tables 24 and 25.

Source: microbial mats from freshwater, hyposaline, and saline lakes in the Vestfold Hills, Larsemann Hills, and McMurdo Dry Valleys, eastern Antarctica. The 19 original isolates all belong to the same fatty acid cluster but are divided into two different rep-PCR clusters. Four additional strains originating from the same locations were identified during a later study (Van Trappen et al., 2005).

DNA G+C content (mol%): 33.8–34.5 (HPLC).

Type strain: Mergaert R-9014, WA11, CIP 108385, DSM 15719, LMG 21922.

Sequence accession no. (16S rRNA gene): AJ557887 (type strain), AJ440988 (strain LMG 21471).

15. **Flavobacterium fryxellicola** Van Trappen, Vandecandelaere, Mergaert and Swings 2005, 771[VP]

fry.xel.li'co.la. N.L. n. *Fryxellum* Lake Fryxell; L. suff. *-cola* from L. n. *incola* dweller; N.L. n. (nominative in apposition) *fryxellicola* a dweller of Lake Fryxell.

Rods, approximately 1.0–1.5 × 3.0–4.0 µm. Colonies on R2A agar are yellow-orange, convex, and translucent with entire margins after 6 d.

Catalase-positive. Arginine dihydrolase, lysine and ornithine decarboxylase, and tryptophan deaminase activities are absent (API 20E). Enzymic activities detected in the API ZYM gallery are listed in the original description. Glucose and maltose are assimilated, but arabinose, mannitol, mannose, *N*-acetylglucosamine, gluconate, caprate, adipate, malate, citrate, and phenylacetate are not (API 20NE). Acid is not produced from carbohydrates (API 20E). Acetoin is produced, but indole is not produced. Citrate is not utilized. Other characteristics are as given in the genus and original species descriptions and in Tables 24 and 25.

Flavobacterium fryxellicola belongs to the same fatty acid cluster as all strains of *Flavobacterium frigoris*, *Flavobacterium micromati*, *Flavobacterium psychrolimnae*, and *Flavobacterium xanthum*, plus some *Flavobacterium degerlachei* strains, but it displays a distinct rep-PCR profile (Van Trappen et al., 2002, 2003, 2004b, 2005).

Source: microbial mats in the fresh/brackish water of Lake Fryxell, McMurdo Dry Valleys, eastern Antarctica.

DNA G+C content (mol%): 35.2–35.9 (HPLC).

Type strain: Mergaert R-7548, FR 64, CIP 108325, DSM 16209, LMG 22022.

Sequence accession no. (16S rRNA gene): AJ811961.

16. **Flavobacterium gelidilacus** Van Trappen, Mergaert and Swings 2003, 1243[VP]

ge.li.di.la′cus. L. adj. *gelidus* ice-cold; L. n. *lacus* lake; N.L. gen. n. *gelidilacus* of an ice-cold lake, referring to the isolation source, microbial mats in Antarctic lakes.

Rods, approximately <1 × 2–4 μm; some longer cells also occur. Gliding was reported as strain-dependent in the original description, the type strain and strain LMG 21619 being negative for this characteristic. However, gliding and rotating cells, as well as spheroplasts, have been observed in 4-d-old Anacker and Ordal's broth culture of the type strain (J.-F. Bernardet, personal observation). Colonies on R2A agar are yellow to orange, circular, convex, and translucent, 1–4 mm in diameter with entire margins after 6 d at 20°C. Colonies do not adhere to the agar.

Catalase-positive. Arginine dihydrolase, lysine and ornithine decarboxylase, and tryptophan deaminase activities are absent (API 20E). Gelatin is degraded, except by strain LMG 21619. Enzymic activities detected in the API ZYM gallery are listed in the original description. Glucose, arabinose, mannose, mannitol, *N*-acetylglucosamine, maltose, gluconate, caprate, adipate, malate, citrate, and phenylacetate are not assimilated (API 20NE). Acid is not produced from glucose, mannitol, inositol, sorbitol, rhamnose, sucrose, melibiose, amygdalin, or arabinose (API 20E). Acetoin and indole are not produced. Citrate is not utilized. Other characteristics are as given in the genus and original species descriptions and in Tables 24 and 25.

The 22 original isolates all belong to the same fatty acid cluster (that also contains all strains of *Flavobacterium frigoris*, *Flavobacterium micromati*, *Flavobacterium psychrolimnae*, and *Flavobacterium xanthum*, and some *Flavobacterium degerlachei* strains), but are divided into three different rep-PCR clusters.

Source: microbial mats from freshwater and saline lakes in the Vestfold Hills and Larsemann Hills, eastern Antarctica.

DNA G+C content (mol%): 30.0–30.4 (HPLC).

Type strain: Mergaert R-8899, AC 35, CIP 108171, DSM 15343, LMG 21477.

Sequence accession no. (16S rRNA gene): AJ440996 (type strain), AJ507151 (strain LMG 21619).

17. **Flavobacterium gillisiae** McCammon and Bowman 2000, 1059[VP]

gil.lis′i.ae. N.L. gen. fem. n. *gillisiae* of Gillis, named after Monique Gillis, a Belgian microbiologist who pioneered new techniques for bacterial taxonomy.

Rods, approximately 0.4–0.5 × 2–5 μm. Colonies on R2A agar are orange, convex, circular with entire margins, and have a butyrous consistency. No growth occurs fermentatively or by anaerobic respiration using ferric iron, nitrate, nitrite, or trimethylamine *N*-oxide as electron acceptors.

Catalase-positive. Arginine dihydrolase activity is present, but lysine and ornithine decarboxylase and tryptophan deaminase activities are absent. Acetoin and indole are not produced. Citrate is not utilized. Sodium nitrate, ammonium chloride, L-glutamate, peptone, and Casamino acids serve as nitrogen sources. Vitamins are not required for growth but are stimulatory. Acid is produced from glucose, mannose, galactose, fructose, sucrose, trehalose, cellobiose, maltose, mannitol, and glycerol, but not from arabinose, rhamnose, xylose, lactose, melibiose, raffinose, adonitol, sorbitol, or inositol. Uric acid and xanthine are not degraded. Other characteristics are as given in the genus and original species descriptions and in Tables 24 and 25.

Source: coastal sea ice from Prydz Bay, Antarctica.

DNA G+C content (mol%): 32 (T_m).

Type strain: ACAM 601, CIP 107449, LMG 21422.

Sequence accession no. (16S rRNA gene): U85889.

18. **Flavobacterium glaciei** Zhang, Wang, Liu, Dong and Zhou 2006, 2924[VP]

gla.ci′e.i. L. gen. n. *glaciei* of ice, referring to the isolation source, the China No. 1 glacier.

Rods, approximately 0.45–0.55 × 2.7–6.3 μm. Colonies on PYG agar (per l: 5.0 g peptone, 0.2 g yeast extract, 3.0 g beef extract, 5.0 g glucose, 0.5 g NaCl, and 1.5 g $MgSO_4 \cdot 7H_2O$; amount of agar unspecified) are yellow, smooth, convex, and circular with entire margins.

Catalase-positive. Arginine dihydrolase, lysine and ornithine decarboxylase, and tryptophan deaminase activities are absent. Acid is not produced from carbohydrates. The following substrates are assimilated: glucose, maltose, sucrose, trehalose, mannose, dextrin, asparagine, L-glutamic acid, and glucose 1-phosphate. The following substrates are not assimilated: fructose, fucose, galactose, arabinose, cellobiose, rhamnose, lactose, raffinose, melibiose, sorbitol, turanose, xylitol, glycerol, inositol, erythritol, serine, threonine, alanine, histidine, leucine, aspartic acid, malonic acid, lactic acid, acetic acid, citric acid, pyruvate, succinate, and uridine. Acetoin is not produced. Citrate is not utilized. Other characteristics are as given in the genus and original species descriptions and in Tables 24 and 25.

Source: frozen soil collected from the China No. 1 glacier, Xinjiang Uygur Autonomous Region, China.

DNA G+C content (mol%): 36.5 (T_m).

Type strain: 0499, CGMCC 1.5380, CIP 109489, JCM 13953.

Sequence accession no. (16S rRNA gene): DQ515962.

19. **Flavobacterium granuli** Aslam, Im, Kim and Lee 2005, 750[VP]

gra.nu′li. L. gen. n. *granuli* of a small grain, pertaining to a granule, from which the type strain was isolated.

Rods, approximately 0.3–0.5 × 2.0–5.0 μm. Colonies on R2A agar are yellow, smooth, non-glossy, convex, circular with entire margins, and 2–4 mm in diameter after 2 d. Colonies on Columbia and brain-heart infusion agars are bright yellow to orange-yellow.

Catalase-positive. Arginine dihydrolase, lysine and ornithine decarboxylase, and tryptophan deaminase activities are absent. Acid is not produced from glucose. Xylan is not degraded. The following substrates are utilized as sole carbon sources: glucose, mannose, *N*-acetylglucosamine, maltose, propionate, and L-proline. The following substrates are not utilized as sole carbon sources: L-arabinose, mannitol, gluconate, caprate, adipate, malate, citrate, phenylacetate, salicin, D-melibiose, L-fucose, D-sorbitol, valerate, histidine, 2-ketogluconate, 3-hydroxybutyrate, 4-hydroxybutyrate, rhamnose, D-ribose, inositol, D-sucrose, itaconate, suberate, malonate, acetate, lactate, L-alanine, 5-ketogluconate, glycogen, 3-hydroxybenzoate, and L-serine. Resistant to ampicillin, tetracycline, and streptomycin; sensitive to kanamycin. Acetoin is produced. Citrate is not utilized. Other characteristics are as given in the genus and original species descriptions and in Tables 24 and 25.

Source: granules used in the wastewater treatment plant of a beer-brewing factory in Kwang-Ju, Republic of Korea. These granules contained aerobic bacteria even though they had been kept under anaerobic conditions for 2 years.

DNA G+C content (mol%): 36.2 (HPLC).

Type strain: Lee Kw05, CIP 108709, KCTC 12201, IAM 15099.

Sequence accession no. (16S rRNA gene): AB180738.

20. **Flavobacterium hercynium** Cousin, Päuker and Stackebrandt 2007, 248[VP]

her.cy'ni.um. L. neut. adj. *hercynium* of or belonging to *Hercynia*, the Roman name for a mountain range in Germany where the type strain was isolated.

Rods, approximately 0.7–0.8 × 5.8–6.3 µm. Gliding motility is observed on DSMZ medium 67 and on Anacker and Ordal's, nutrient, and R2A agars, but not on DNA or tryptic soy agars. Colonies are brown-beige on medium 67 and Anacker and Ordal's agars, and bright yellow to dark orange on R2A, DNA, nutrient, and tryptic soy agars. They are flat, spreading, with irregular lobate margins or no visible margins on medium 67, Anacker and Ordal's, and R2A agars and circular, non-spreading, umbonate with undulate margins on DNA and tryptic soy agars. No growth occurs on MacConkey agar. Growth is aerobic and microaerobic.

Catalase-positive. Indole is not produced. Gelatin hydrolysis and assimilation of L-arabinose and trisodium citrate are strain-dependent. Characteristics included in API ZYM, API 20NE, and GN MicroPlate galleries are given in the original description. Other characteristics are as given in the genus and original species descriptions and in Tables 24 and 25.

Source: isolated from a site located about 320 m downstream of the spring of the Westerhöfer Bach, a hard-water creek from the western slopes of the Harz Mountains, in Westerhof, 40 km north of Göttingen, Germany. Five strains were isolated and characterized, but only the type strain was deposited in culture collections.

DNA G+C content (mol%): 37.5 (T_m).

Type strain: WB 4.2-33, CIP 109241, DSM 18292.

Sequence accession no. (16S rRNA gene): AM265623 (type strain), AM177627 (strain WB 4.2-78).

21. **Flavobacterium hibernum** McCammon, Innes, Bowman, Franzmann, Dobson, Holloway, Skerratt, Nichols and Rankin 1998, 1411[VP]

hi.ber'num. L. neut. adj. *hibernum* of winter.

Rods, approximately 0.7 × 1.8–13 µm. A thick capsule is produced when cultures are incubated at 4°C. Colonies on "maintenance medium" [per l: 1.0 g yeast extract, 5.0 g lactose (filter-sterilized), 0.5 g K_2HPO_4, 0.5 g $(NH_4)_2SO_4$, 0.55 g NaCl, 15.0 g agar] are discrete, opaque, shiny, convex, smooth, and yellow. Consistency is mucoid when plates are incubated at 25°C, but gelatinous at 4°C. Colonies on nutrient-poor media are flat, translucent, and spreading. Colonies on tryptic soy agar are orange-yellow. The theoretical minimum temperature for growth is −7°C.

Catalase-positive. Arginine dihydrolase, lysine and ornithine decarboxylase, and tryptophan deaminase activities are absent (API 20E). Oxidase activity was not found in the original study, but retesting of the type strain using dimethyl-*p*-phenylenediamine resulted in a distinctly positive oxidase reaction (J.-F. Bernardet, personal observation). Acid is produced from glucose, rhamnose, sucrose, and arabinose. Citrate is not utilized. Indole and acetoin are not produced (API 20E). Glucose, lactose, galactose, fructose, and sucrose are assimilated, but raffinose is not. The list of substrates utilized in the Biolog system is given in the original description. Utilization of formic acid, glucuronamide, and saccharic acid is strain-dependent. The optimum temperature for β-galactosidase activity is 38.6°C. Other characteristics are as given in the genus and original species descriptions and in Tables 24 and 25.

Source: surface freshwater of Crooked Lake, Vestfold Hills, Antarctica. Recently, a strain sharing 99% full-length 16S rRNA gene sequence similarity with the type strain of *Flavobacterium hibernum* and producing a cold-active xylanase has been isolated from a stream in California (Lee et al., 2006a).

DNA G+C content (mol%): 34 (type strain; HPLC).

Type strain: ACAM 376, ATCC 51468, CIP 105745, DSM 12611, LMG 21424.

Sequence accession no. (16S rRNA gene): L39067.

Note: the strains deposited in the Australian Collection of Antarctic Micro-organisms were originally named "*Flavobacterium ameridies*".

22. **Flavobacterium hydatis** (Strohl and Tait 1978) Bernardet, Segers, Vancanneyt, Berthe, Kersters and Vandamme 1996, 141[VP] (*Cytophaga aquatilis* Strohl and Tait 1978, 302 and Reichenbach 1989c, 2043)

hy'da.tis. Gr. n. *hûdor* water; N.L. gen. n. *hydatis* from water.

Following reclassification of [*Cytophaga*] *aquatilis* in the genus *Flavobacterium*, the original specific epithet had to be changed because the new combination would have become a junior homonym of *Flavobacterium aquatile*. Hence, the new specific epithet *hydatis*, whose meaning is similar to that of the original epithet, was proposed for this taxon (Bernardet et al., 1996).

Rods, approximately 0.4–0.7 × 1.5–15.0 µm. Flexible. Coccoid forms and long filaments may occur. Colonies on relatively rich media are circular with entire margins, convex, and yellow-pigmented. Anaerobic growth occurs in the presence of fermentable carbohydrates (see below) and yeast extract (or peptone), or during nitrate reduction. Colonies from anaerobically grown cultures are a light cream color or non-pigmented. Growth may occur up to 35°C, but the cultures degenerate rapidly at 30°C. Colonies do not adsorb Congo red, but the slime layer reportedly stains with ruthenium red.

Catalase-positive. In contrast to the original description, no hydrolysis of carboxymethylcellulose and a positive oxidase reaction have been found (J.-F. Bernardet, personal observation). Oxidase activity was also found by Oyaizu and Komagata (1981), and a weak reaction was reported by Reichenbach (1989b). Acid is produced fermentatively from arabinose, cellobiose, fructose, galactose, glucose, lactose, maltose, mannose, raffinose, sucrose, xylose, glycerol, dulcitol, and mannitol. The same compounds are also metabolized aerobically with acid production. Strong proteolytic activity. Tributyrin is hydrolyzed, but not lecithin. Indole is not produced. Citrate is not utilized. Peptones, Casamino acids, NH_4^+, and NO_3^- can be used as N sources. Resistant to ampicillin, penicillin G, and polymyxin B; sensitive to chloramphenicol, erythromycin, kanamycin, nalidixic acid, neomycin, novobiocin, and tetracycline. Resistant to the vibriostatic compound O/129 (Bernardet et al., 1996). Other characteristics are as given in the genus and original species descriptions and in Tables 24 and 25. Additional information can be found in Reichenbach (1989b) and in Bernardet et al. (1996).

Source: isolated in 1974 from the gills of a salmon (species unrecorded) suffering from "bacterial gill disease", Platte River Fish Hatchery, Michigan, USA. Four of the five original strains have been lost. Pathogenicity has not been demonstrated experimentally.

DNA G+C content (mol%): 32.0–33.7 (Bd).

Type strain: Strohl strain N, Reichenbach Cyaq1, ATCC 29551, CCUG, 35201, CIP 104741, DSM 2063, IFO (now NBRC) 14958, LMG 8385, NRRL B-14732.

Sequence accession no. (16S rRNA gene): M58764.

23. **Flavobacterium indicum** Saha and Chakrabarti 2006, 2620[VP]

in'di.cum. L. neut. adj. *indicum* of or belonging to India, from where the type strain was isolated.

Rods, approximately 0.1–0.2 × 1–3 µm, occurring singly and occasionally in pairs. Filamentous cells appear in ageing cultures (Saha and Chakrabarti, 2006), followed by numerous small spheroplasts (J.-F. Bernardet, personal observation). Colonies on modified R2A agar [containing 0.025% (w/v) Casamino acids instead of 0.05% and 0.075% (w/v) peptone instead of 0.05%] are yellowish orange, circular, and convex to slightly umbonate with irregular margins. Although no gliding motility was originally observed when a broth culture was examined under phase-contrast microscopy, a thin film of translucent growth with finger-like projections appeared at the margins of 5–7-D-old colonies, suggesting that gliding motility did occur on modified R2A agar (Saha and Chakrabarti, 2006). Indeed, moderate gliding and rotating was observed in Anacker and Ordal's and R2A broth cultures (J.-F. Bernardet, personal observation). Good growth also occurs on 100× diluted tryptic soy agar (colonies on this medium are smaller and more adherent than on modified R2A) and on plain R2A, R3A, Anacker and Ordal's, full- and quarter-strength ZoBell, yeast extract, and NaCl-free nutrient agars. Weak growth occurs on plain nutrient, Luria–Bertani, and LY (Reichenbach, 1989b) agars. No growth occurs on plain tryptic soy, Simmons' citrate, or MacConkey agars.

Catalase activity is very weak. Hypoxanthine and xanthine are not hydrolyzed. No precipitate is formed on egg-yolk agar, but colonies are surrounded by clear zones. D-Fructose and (to a lesser extent) inulin are utilized. L-Arabinose, D-amygdalin, D-galactose, D-glucose, glycerol, glycogen, D-mannose, L-rhamnose, D-sorbitol, and L-xylose are very weakly utilized. Adonitol, D-arabinose, L-arabitol, arbutin, D-cellobiose, dulcitol, *myo*-inositol, D-lactose, D-mannitol, maltose, melibiose, melezitose, raffinose, D-ribose, L-sorbose, sucrose, trehalose, xylitol, and D-xylose are not utilized. Acid is produced from D-fructose and (very weakly) from D-glucose, glycogen, inulin, L-xylose, and D-mannose, but not from the other carbohydrates tested. Antibiotic resistances are listed in the original description. Other characteristics are as given in the genus and original species descriptions and in Tables 24 and 25.

Source: water of a warm (37–38°C) spring in a forest reserve in Assam, India. The spring was undisturbed by human activities but frequented by wild animals.

DNA G+C content (mol%): 31.0 (T_m).

Type strain: Saha and Chakrabarti GPTSA100-9, CIP 109464, DSM 17447, MTCC 6936.

Sequence accession no. (16S rRNA gene): AY904351.

24. **Flavobacterium johnsoniae** (Stanier 1947) Bernardet, Segers, Vancanneyt, Berthe, Kersters and Vandamme 1996, 141[VP] [*Cytophaga johnsonae* Stanier 1947, 306 and Reichenbach 1989b, 2041; *Cytophaga johnsonii* (*sic*) (Stanier 1947) Stanier 1957, 860]

john.so.ni'ae. N.L. gen. fem. n. *johnsoniae* of Johnson, named after the American microbiologist Delia E. Johnson, who made an early study of chitinolytic gliding soil bacteria.

Following reclassification of [*Cytophaga*] *johnsonae* in the genus *Flavobacterium*, the original specific epithet has been corrected to *johnsoniae*.

Flexible rods, approximately 0.3–0.4 × 1.5–15 µm. Cells tend to become almost coccoid in ageing cultures, preceding the development of spheroplasts. Colonies on relatively rich media are circular with entire margins, convex, and bright yellow, whereas they are flat, spreading with rhizoid to filamentous margins, and pale yellow on media containing low nutrient concentration (e.g., Anacker and Ordal). Deep orange and pale yellow variants have been reported for the type strain. Zeaxanthin is produced in addition to flexirubin-type pigments (Reichenbach, 1989b).

Catalase activity is present, though moderate. Peptones, Casamino acids, glutamate, proline, urea, and NH_4^+ are good N sources. Growth occurs on mineral salts media containing NH_4^+ and a carbohydrate, such as arabinose, cellobiose, chitin, galactose, glucose, inulin, lactose, maltose, mannose, raffinose, starch, sucrose, and xylose. Acid is produced from sugars and various polysaccharides are degraded. Strongly proteolytic. Indole is not produced. Tributyrin is degraded, but not lecithin. The organism has a strictly aerobic metabolism. Reichenbach and other authors (Reichenbach, 1989b and references therein) considered the report of Christensen (1977) that the type strain was able to ferment glucose and to produce gas from NO_2^- to be erroneous. This strain is unable to use NO_3^- or fumarate as electron acceptor, but one of Stanier's five original [*Cytophaga*] *johnsonae* isolates grew anaerobically "to a slight extent", reducing NO_3^- to NO_2^-. It is uncertain, however, whether all five strains belonged to the same species. This is even more the case of Stanier's sixth strain, labeled "var. *denitrificans*", which grew abundantly in anaerobic conditions by denitrification and produced copious amounts of N_2. Resistant to the vibriostatic compound O/129 (Bernardet et al., 1996). Other characteristics are as given in the genus and species descriptions and in Tables 24 and 25. Additional information can be found in Christensen (1977), Reichenbach (1989b), and Bernardet et al. (1996).

Overall, *Flavobacterium johnsoniae* shows considerable phenotypic and genomic diversity (Reichenbach, 1989b). Many strains isolated from soil and freshwater environments and resembling Stanier's original strains have been labeled [*Cytophaga*] *johnsonae* or *Flavobacterium johnsoniae*, but it is likely that many of them actually belong to different species. Many *Flavobacterium johnsoniae*-like organisms have also been isolated from superficial lesions of diseased fish and considered opportunistic pathogens. There have been, however, cases of fish disease in which *Flavobacterium johnsoniae* indeed seemed to be the causative agent, but DNA–DNA hybridization between the isolates and the type strain has not been performed to confirm their identification (Bernardet and Bowman, 2006 and references therein). Strains ATCC 29585 and 29586, isolated from diseased fish in Manitoba, Canada, as well as the soil isolate NCIMB 11391 have shown insignificant DNA relatedness with the type strain and consequently cannot be considered *bona fide Flavobacterium johnsoniae* strains (Bernardet et al., 1996). Conversely, the two [*Flexibacter*] *aurantiacus* (Lewin, 1969) strains Lewin DWO and Lewin PSY included in the same study were shown to share significant DNA relatedness with the type strain of *Flavobacterium johnsoniae*; the proposed transfer of these two strains to the species *Flavobacterium johnsoniae* was supported by similarities in phenotypic features and fatty acid composition (Bernardet et al., 1996). The *Flavobacterium johnsoniae* type strain and the two [*Flexibacter*] *aurantiacus* strains, however, showed rather different whole-cell protein profiles. Strain DSM 425 and the *Flavobacterium johnsoniae* type strain share only 22% DNA relatedness (Reichenbach, 1989b) and occupied rather separate positions in phylogenetic trees (data not shown). Another strain, Hayes S4/1 (=LMG 13161=NCIMB 11391) is available in culture collections, but DNA–DNA hybridization with the type strain has not been performed. Extensive studies are needed to properly delineate this species and to reject misidentified strains; novel *Flavobacterium* species may emerge from this process.

Source: as no type strain had been designated by Stanier (1947), Reichenbach (1989b) designated strain ATCC 17061 the cotype strain because it seemed to be one of the authentic strains coming from Stanier's original collection and it had been extensively studied. This strain had been isolated from a soil sample from Rothamsted Experimental Station, UK. Of the only two strains that have been shown to share high DNA relatedness with the type strain, viz., NCIMB 1382 (=Lewin DWO=ATCC 23107=LMG 3987) and NCIMB 1455 (=Lewin PSY=ATCC 23108=LMG 10404), only the origin of the former strain is known: garden soil, Minneapolis, Minnesota. Most other strains available in culture collections have been retrieved from soil and freshwater habitats. Recently, three strains sharing very high near full-length *(16S rRNA gene):* rRNA gene sequence similarity with the type strain of *Flavobacterium johnsoniae* have been isolated from river epilithon in England (O'Sullivan et al., 2006).

DNA G+C content (mol%): 33–35.2 (T_m).

(Co)type strain: van Niel MYX 1.1.1, Reichenbach Cy j3, ATCC 17061, CCUG 35202, CFBP 3036, CIP 100931, DSM 2064, IFO (now NBRC) 14942, LMG 1340, LMG 1341, JCM 8514, NCIMB 11054.

Sequence accession no. (16S rRNA gene): AM230489 (type strain), M62792 (strain ATCC 23107).

25. **Flavobacterium limicola** Tamaki, Hanada, Kamagata, Nakamura, Nomura, Nakano and Matsumura 2003, 523[VP]

li.mi.co'la. L. n. *limus* mud; L. suff. *-cola* from L. gen. n. *incola* dweller; N.L. n. (nominative in apposition) *limicola* a mud-dweller.

Rods, approximately 0.3–0.6 × 1.1–3.2 µm. Chains of 5–20 cells occur frequently and filamentous cells occasionally appear. Colonies on tryptic soy agar are bright orange-yellow, circular, and convex with entire margins. Adsorption of Congo red by colonies was reported in the original description, but others have not been able to observe it (J.-F. Bernardet, personal observation). Although agar is degraded, no liquefaction occurs and colonies are not sunk in the agar.

Catalase-positive. Arginine dihydrolase activity is present, but lysine and ornithine decarboxylase and tryptophan deaminase activities are absent. Other enzymic activities determined using API 20E, 20NE, ID32E, and ZYM galleries are listed in the original description. The production of extracellular proteases (investigated using hide powder azure as the substrate) by the three isolates is higher at 5°C than at 15 and 23°C. Indole and acetoin are not produced. Citrate is not utilized. Acid is not produced from any of the carbohydrates included in API 50CH strips and GN MicroPlates. In the latter, amino acids such as L-alanine, L-alanylglycine, L-asparagine, L-aspartic acid, L-glutamic acid, glycyl-L-aspartic acid, glycyl-L-glutamic acid, L-ornithine, L-proline, L-serine, and L-threonine allow for better growth than carbohydrates such as D-glucose, mannose, maltose, sucrose, starch, glycogen, and dextrin. Yeast extract stimulates growth. Uric acid is hydrolyzed, but xanthine is not. Agar degradation is revealed by flooding colonies with an I/KI solution. Other characteristics are as given in the genus and original species descriptions and in Tables 24 and 25.

Source: freshwater river sediments in Ibaraki Prefecture, Japan.

DNA G+C content (mol%): 34.0–34.8 (method not specified).

Type strain: Tamaki ST-82, CIP 107957, DSM 15094, JCM 11473, LMG 21930.

Sequence accession no. (16S rRNA gene): AB075230 (type strain), AB075231 (strain ST-10), AB075232 (strain ST-92).

26. **Flavobacterium micromati** Van Trappen, Vandecandelaere, Mergaert and Swings 2004b, 89[VP]

mi.cro.ma'ti. N.L. gen. n. *micromati* arbitrary name referring to the MICROMAT project ("Biodiversity of microbial mats in Antarctic") in connection with which the strains were isolated.

Rods, <1 × 3–4 µm. Colonies on R2A agar are orange-red, convex, translucent with entire margins, non-adherent, and 1–3 mm in diameter after 6 d.

Catalase-positive. Arginine dihydrolase, lysine and ornithine decarboxylase, and tryptophan deaminase activities are absent. Enzymic activities detected in the API ZYM gallery are listed in the original description. No growth occurs on carbohydrates in the API 20NE gallery and no acid is produced from carbohydrates in the API 20E gallery. Acetoin is produced. Indole is not produced. Citrate is not utilized. Other characteristics are as given in the genus and original species descriptions and in Tables 24 and 25.

The three original isolates all belong to the same rep-PCR and fatty acid clusters.

Source: microbial mats from freshwater, hyposaline, and saline lakes in the Vestfold Hills and Dry Valleys, eastern Antarctica

DNA G+C content (mol%): 33.1–34.4 (HPLC).

Type strain: Mergaert R-9192, GR 28, CIP 108161, LMG 21919.

Sequence accession no. (16S rRNA gene): AJ557888.

27. **Flavobacterium omnivorum** Zhu, Wang and Zhou 2003, 856[VP]

om.ni.vo'rum. L. n. *omne -is* everything; L. v. *vorare* to devour; N.L. neut. adj. *omnivorum* eating everything, referring to the ability of the strain to degrade a wide range of macromolecules.

Rods, approximately 0.8 × 2–5 µm; filamentous cells also occur. A limited number of spheroplasts was observed in 4-d-old Anacker and Ordal's broth culture (J.-F. Bernardet, personal observation). Colonies on PYG agar [per l: 5.0 g polypeptone, 5.0 g tryptone, 10.0 g yeast extract, 10.0 g glucose, 40 ml salt solution, 15.0–20.0 g agar. Salt solution (per l): 0.2 g $CaCl_2$, 0.4 g $MgSO_4 \cdot 7H_2O$, 1.0 g K_2HPO_4, 1.0 g KH_2PO_4, 10.0 g $NaHCO_3$, 2.0 g NaCl] are orange-yellow, circular, convex, and smooth with entire margins.

Catalase-positive. Arginine dihydrolase, lysine and ornithine decarboxylase, and tryptophan deaminase activities are absent. Indole and acetoin are not produced. Citrate is not utilized. Peptone, Casamino acids, sodium nitrate, and ammonium chloride serve as nitrogen sources, but L-glutamate does not. Acid is not produced from glucose, fructose, xylose, mannitol, or maltose. Uric acid and xanthine are not degraded. Glucose, cellobiose, maltose, fructose, sorbose, sucrose, lactose, rhamnose, mannose, trehalose, melibiose, and arabinose are assimilated, but sorbitol, galactose, D-mannitol, raffinose, glycerol, inositol, and xylose are not. Other characteristics are as given in the genus and original species descriptions and in Tables 24 and 25.

Source: frozen soil enclosed within the China No. 1 glacier, Xinjiang Uygur Autonomous Region, China.

DNA G+C content (mol%): 35.2 (T_m).

Type strain: Zhou ZF-8, AS 1.2747, CIP 108050, JCM 11313, LMG 21986.

Sequence accession no. (16S rRNA gene): AF433174.

28. **Flavobacterium pectinovorum** (Dorey 1959) Bernardet, Segers, Vancanneyt, Berthe, Kersters and Vandamme 1996, 141[VP] [*Flavobacterium pectinovorum* Dorey 1959, 94; *Empedobacter pectinovorum* (Dorey 1959) Kaiser 1961, 210; *Cytophaga pectinovora* (Dorey 1959) Reichenbach 1989c, 2042]

pec.ti.no.vo'rum. N.L. n. *pectinum* pectin, methylated polygalacturonic acids in plant cell walls; L. v. *vorare* to devour; N.L. neut. adj. *pectinovorum* pectin-devouring.

Rods, approximately 0.4–0.5 × 1.0–5.0 µm; longer forms (15–25 µm) may appear in rich media and in older cultures. Gliding motility, though not observed by Dorey (1959), has been reported by all other authors. Colonies on all media are bright yellow, circular, and low convex, with entire to lobate margins when incubated at 30°C, whereas pale yellow, flat, spreading swarms are produced at 22°C. The bacterial mass collected on agar is difficult to remove from the loop due to the production of a dry, sticky slime.

Catalase activity is present but moderate. Reichenbach (1989b) reported the presence of oxidase activity, whereas others have found a negative oxidase reaction (J.-F. Bernardet, personal observation). Peptones, Casamino acids, NH_4^+, and NO_3^- are used as N sources. Grows on mineral media containing arabinose, cellobiose, glucose, inulin, lactose, maltose, pectin, sucrose, or xylose as the sole carbon and energy source, but glycerol, mannitol, and salicin do not allow growth. Acid is produced aerobically from carbohydrates. An inducible, extracellular pectinase (polygalacturonase) is produced; it creates shallow depressions around colonies on pectate gels before liquefying the whole gel. Strongly proteolytic. Indole is not produced. Susceptible to the vibriostatic compound O/129 (Bernardet et al., 1996). Other characteristics are as given in the genus and original species descriptions and in Tables 24 and 25. Additional information can be found in Christensen (1977), Reichenbach (1989b), and Bernardet et al. (1996).

The species was named *Flavobacterium pectinovorum* by Dorey (1959) but was not included in the *Approved Lists of Bacterial Names* (Skerman et al., 1980). Christensen (1977) proposed to reclassify the species as [*Cytophaga*] *johnsonae*, but Reichenbach (1989b) restored it as an independent species under the combination [*Cytophaga*] *pectinovora*, a decision supported by phenotypic differences and very low DNA relatedness between the two type strains.

Source: soil in Southern England. The seven other original strains differed only slightly from the type strain, but they have subsequently been lost. Recently, six strains sharing more than 99% near full-length 16S rRNA gene sequence similarity with the type strain of *Flavobacterium pectinovorum* have been isolated from a hard-water creek in Germany (Brambilla et al., 2007).

DNA G+C content (mol%): 34.0–35.5 (T_m, Bd).

Type strain: Dorey 81, Reichenbach Cy p1, ATCC 19366,

CIP 104742, DSM 6368, IFO (now NBRC) 15945, JCM 8518, LMG 4031, NCIMB 9059, USCC 1365.

Sequence accession no. (16S rRNA gene): D12669, AM230490.

29. **Flavobacterium psychrolimnae** Van Trappen, Vandecandelaere, Mergaert and Swings 2005, 771[VP]

psy.chro.lim'nae. Gr. adj. *psychros* cold; Gr. fem. N. *limna* lake; N.L. gen. n. *psychrolimnae* of a cold lake.

Rods, approximately 0.5 × 2.0 µm. Colonies on R2A agar are yellow, convex, and translucent with entire margins after 6 d.

Catalase-positive. Arginine dihydrolase, lysine and ornithine decarboxylase, and tryptophan deaminase activities are absent (API 20E). Enzymic activities detected in the API ZYM gallery are listed in the original description. β-Galactosidase activity is detected in API 20E and API 20NE galleries, but not in the API ZYM gallery (S. Van Trappen, personal communication). Glucose, maltose, and mannose are assimilated, but arabinose, mannitol, *N*-acetylglucosamine, gluconate, caprate, adipate, malate, citrate, and phenylacetate are not (API 20NE). Acid is not produced from carbohydrates (API 20E). Acetoin and indole are not produced. Citrate is not utilized. Other characteristics are as given in the genus and original species descriptions and in Tables 24 and 25.

Flavobacterium psychrolimnae belongs to the same fatty acid cluster as all strains of *Flavobacterium frigoris, Flavobacterium fryxellicola, Flavobacterium micromati,* and *Flavobacterium xanthum* and some strains of *Flavobacterium degerlachei,* but it displays a distinct rep-PCR profile (Van Trappen et al., 2002, 2003, 2004b, 2005).

Source: microbial mats in fresh/brackish water lakes in the McMurdo Dry Valleys, eastern Antarctica.

DNA G+C content (mol%): 33.8–34.5 (HPLC).

Type strain: Mergaert R-7582, FR 57, CIP 108326, DSM 16141, LMG 22018.

Sequence accession no. (16S rRNA gene): AJ585428 (type strain), AJ585427 (LMG 22020).

30. **Flavobacterium psychrophilum** (Bernardet and Grimont 1989) Bernardet, Segers, Vancanneyt, Berthe, Kersters and Vandamme 1996, 142[VP] [*Cytophaga psychrophila* Borg 1948, 120 and Reichenbach 1989c, 2044; *Flexibacter psychrophilus* (Borg 1948) Bernardet and Grimont 1989, 353]

psy.chro.phi'lum. Gr. adj. *psychros* cold; Gr. adj. *philos* loving; N.L. neut. adj. *psychrophilum* cold-loving.

Rods, approximately 0.4–0.5 × 1.0–5.0 µm; a few cells are longer (8–12 µm). Flexible cells may appear. Gliding motility is slow and readily discernable in some strains only. Narrow and uneven zones of spreading growth may appear at the margins of colonies on Anacker and Ordal's agar. Different colony types may occur on the same plate. The optimum growth temperature is 19–20°C for Japanese isolates (Uddin and Wakabayashi, 1997), but only 15°C for most US isolates (Holt et al., 1989). Growth at 23 and 25°C is strain-dependent and usually weak.

Catalase activity is present, though moderate. Arginine dihydrolase and lysine and ornithine decarboxylase activities are absent. Actively proteolytic. The different types of extracellular proteases have been considered important virulence factors, playing a significant role in the extensive skin and muscular necrotic lesions displayed by diseased fish. These proteases degrade components of muscle, cartilage, and connective tissue such as elastin, type IV collagen, fibrinogen, gelatin, laminin, fibronectin, actin, and myosin. The two metalloproteases recently purified from *Flavobacterium psychrophilum* exhibit a broad range of hydrolytic activity and are adapted to low temperatures (Secades et al., 2001, 2003). Although unable to hydrolyze any of the commonly tested carbohydrates, *Flavobacterium psychrophilum* strains can degrade the complex acidic polysaccharides of connective tissue such as chondroitin sulfate and hyaluronic acid (Bernardet and Bowman, 2006 and references therein). In combination with the extracellular proteases, these polysaccharide-degrading enzymes most likely participate in tissue necrosis. Peptones and Casamino acids are suitable nitrogen sources. Acid is not formed from carbohydrates. Indole is not produced. Tributyrin and lecithin are hydrolyzed. The antibiotics to which *Flavobacterium psychrophilum* is resistant and the substrates it hydrolyzes in the API ZYM gallery are listed in Bernardet and Grimont (1989) and Bernardet et al. (1996). Susceptible to the vibriostatic compound O/129. Other characteristics are as given in the genus description and in Tables 24 and 25. Additional information and references can be found in Bernardet and Grimont (1989), Bernardet et al. (1996), and Bernardet and Bowman (2006). The whole-genome sequence of strain JIP 02/86 (=TG 02/86=ATCC 49511=CIP 103535) has been published (Duchaud et al., 2007).

Flavobacterium psychrophilum is the causal agent of severe infections ("bacterial cold-water disease" and "rainbow trout fry syndrome") in salmonid fish and has been isolated from external lesions and internal organs of the diseased fish. It is a significant pathogen that causes considerable economic losses in the salmonid aquaculture worldwide. Virulence for fish has been demonstrated experimentally. For reviews on the bacterium and on the diseases it is responsible for, see Holt et al. (1993), Austin and Austin (1999), Shotts and Starliper (1999), Nematollahi et al. (2003), and Bernardet and Bowman (2006). Because no type strain had been designated previously, Reichenbach (1989b) proposed strain NCIMB 1947, deposited by E.J. Ordal, as the cotype strain. This strain was isolated from the kidney of a diseased juvenile coho salmon (*Oncorhynchus kisutch*) in the state of Washington, but the year of isolation is unknown.

Source: external lesions and internal organs of infected salmonid fish.

DNA G+C content (mol%): 32.3–33.8 (T_m, Bd).

(Co)type strain: Ordal 3068, ATCC 49418, CCUG 35200, CIP 103534, DSM 3660, JCM 8519, LMG 13179, IAM 14308, IFO (now NBRC) 15942, NCIMB 1947.

Sequence accession no. (16S rRNA gene): AB078060 (type strain). The other sequence available for this strain, AF090991, is significantly different.

Note: the first mention of the name [*Cytophaga*] *psychrophila* has long been considered to be in a paper by Borg (1960), but an earlier mention was found in his PhD thesis (Borg, 1948) (R.A. Holt, personal communication), thus allowing the correction to the citation in the present description.

31. **Flavobacterium saccharophilum** (Agbo and Moss 1979) Bernardet, Segers, Vancanneyt, Berthe, Kersters and

Vandamme 1996, 142[VP] (*Cytophaga saccharophila* Agbo and Moss 1979, 363 and Reichenbach 1989c, 2039)

sac.cha.ro.phi'lum. Gr. n. *saccharon* sugar; Gr. adj. *philos* loving; N.L. neut. adj. *saccharophilum* sugar-loving.

Rods, approximately 0.5–0.7 × 2.5–6.0 µm in length. Pleomorphism is common. Almost coccoid cells occur in ageing agar cultures, whereas slender flexible cells may appear in pure peptone liquid media. Gliding motility is slow, but active. Colonies on nutrient and tryptic soy agars are small, convex, circular, yellow-colored, slimy, and sit in shallow depressions. Large haloes of softened (not liquefied) agar surrounding the colonies are revealed by flooding with Lugol's reagent.

The presence of catalase and oxidase activities originally reported by Agbo and Moss (1979) was not confirmed by Reichenbach (1989b). Catalase activity was observed, but no oxidase activity was found using both a disc and a liquid oxidase reagent [dimethyl-*p*-phenylenediamine and tetramethyl-*p*-phenylenediamine, respectively (bioMérieux); J.-F. Bernardet, personal observation]. Peptones and NH_4^+ serve as sole source of nitrogen. Growth is possible on media containing only peptone, but is considerably improved by the addition of carbohydrates. Arabinose, cellobiose, fructose, galactose, glucose, lactose, maltose, mannose, melibiose, raffinose, rhamnose, sucrose, trehalose, and xylose, as well as agar, agarose, arabinogalactan, carrageenan, pectin, starch, and a few other polysaccharides are used as sole carbon and energy sources. The production of extracellular agarase is induced by the presence of agar or other galactans and polysaccharides associated with plants (e.g., arabinogalactan, starch, and pectin) and repressed by the presence of galactose or glucose. Following repeated subcultivation, two of the original isolates lost their ability to degrade agar. Oxidative metabolism of glucose was reported in the original description. Strongly proteolytic. Indole is not produced. The antibiotics to which *Flavobacterium saccharophilum* is resistant are listed in Agbo and Moss (1979) and Reichenbach (1989b). The substrates hydrolyzed in the API ZYM gallery are listed in Bernardet et al. (1996). Susceptible to the vibriostatic compound O/129. Other characteristics are as given in the genus and original species descriptions and in Tables 24 and 25. Additional information can be found in Reichenbach (1989b) and Bernardet et al. (1996). Although a DNA G+C content of 32 mol% was found using buoyant density (Reichenbach, 1989b), values close to 36 mol% were reported by Agbo and Moss (1979) and by Bernardet et al. (1996) using T_m.

Source: six strains were originally isolated during the summer of 1976 from the River Wey, Guildford, Surrey, UK. Although not formally designated the type strain in the original description, strain 024 is the only currently available strain. Recently, several strains sharing high near full-length 16S rRNA gene sequence similarity with the type strain of *Flavobacterium saccharophilum* have been isolated from river epilithon in England (O'Sullivan et al., 2006) and from a hard-water creek in Germany (Brambilla et al., 2007).

DNA G+C content (mol%): 35.7 (T_m).

Type strain: 024 Agbo and Moss, ACAM 581, ATCC 49530, CIP 104743, DSM 1811, IFO (now NBRC) 15944, JCM 8520, LMG 8384, NCIMB 2072.

Sequence accession no. (16S rRNA gene): D12671, AM230491.

32. **Flavobacterium saliperosum** Wang, Liu, Dai, Wang, Jiang and Liu 2006, 441[VP]

sa.li.pe.ro'sum. L. n. *sal, salis* salt; L. neut. adj. *perosum* detesting, hating; N.L. neut. adj. *saliperosum* salt-hating.

Rods, approximately 0.35–0.55 × 1.5–2.5 µm. Small spheroplasts have been observed in 3-d-old Anacker and Ordal's broth culture (J.-F. Bernardet, personal observation). Colonies on modified ATCC medium M1 are yellow, circular, and smooth with entire margins. On tryptic soy agar, colonies are orange-yellow and circular with irregular margins.

Catalase-positive. Acetoin is not produced. Acid is not produced from lactose, arabinose, rhamnose, raffinose, ribose, galactose, melibiose, melezitose, sucrose, xylose, mannose, fucose, fructose, glucose, cellobiose, maltose, salicin, mannitol, or sorbitol. Antibiotic resistances are listed in the original description. Other characteristics are as given in the genus and original species descriptions and in Tables 24 and 25.

Source: sediment of the freshwater Taihu Lake, Jiangsu Province, China. The water quality of the lake has been declining in recent years, with algal blooms occurring regularly.

DNA G+C content (mol%): 41 (T_m).

Type strain: S13, CGMCC 1.3801, CIP 109164, JCM 13331.

Sequence accession no. (16S rRNA gene): DQ021903.

Note: the 16S rRNA sequence was originally deposited in GenBank under the name "*Flavobacterium saliodium*".

33. **Flavobacterium segetis** Yi and Chun 2006, 1243[VP]

se.ge'tis. L. gen. n. *segetis* of the soil.

Rods, approximately 0.2–0.3 × 1.1–2.3 µm. Colonies on R2A agar are orange, convex, translucent, glistening, and circular with entire margins. Consistency is butyrous, becoming mucoid after prolonged incubation. Colonies do not adhere to the agar. No growth occurs on cetrimide or MacConkey agars. Grows well under aerobic conditions (minimum doubling time, 6.8 h), weakly under microaerobic conditions (i.e., with 5–15% O_2 and 5–12% CO_2), and poorly under anaerobic conditions (i.e., with 4–10% CO_2).

Catalase-positive. Arginine dihydrolase and L-phenylalanine deaminase activities are absent. The list of enzymic activities detected in API ZYM galleries is given in the original description. In this commercial system, β-galactosidase activity is not detected, whereas a positive result is observed in API 20NE galleries. Indole is not produced and citrate is not utilized. Acid is produced from D-cellobiose, D-glucose, trehalose, maltose, and sucrose, but not from D-fructose, D-mannitol, raffinose, D-salicin, D-xylose, L-arabinose, or L-rhamnose. Acid production from lactose only occurs after 4 weeks. Glucose, mannose, *N*-acetylglucosamine, and maltose are utilized as sole carbon sources in API 20NE galleries, but arabinose, mannitol, gluconate, caprate, adipate, malate, citrate, and phenylacetate are not. The maximum adsorption peak of pigment is at 472 nm and the next peak is at 452 nm. Other characteristics are as given in the genus and original species descriptions and in Tables 24 and 25.

Source: a soil sample of a penguin habitat near the King Sejong Station on King George Island, Antarctica.

DNA G+C content (mol%): 35 (HPLC).

Type strain: AT1048, CIP 109332, IMSNU 14050, JCM 12385, KCTC 12224.

Sequence accession no. (16S rRNA gene): AY581115.

34. **Flavobacterium soli** Yoon, Kang and Oh 2006d, 999[VP]

so'li. L. gen. n. *soli* of the soil.

Rods, approximately 0.3–0.6 × 1.0–3.0 μm; filamentous cells also occur. Colonies on tryptic soy agar are cream-yellow, slightly convex, smooth, glistening, circular with entire margins, with a mucous to viscid consistency, and 2.0–3.0 mm in diameter after 3 d at 25°C. No anaerobic growth occurs on plain tryptic soy agar or on tryptic soy agar supplemented with nitrate.

Catalase-positive. Arginine dihydrolase, lysine and ornithine decarboxylase, and tryptophan deaminase activities are absent. Xanthine and hypoxanthine are not hydrolyzed. Indole is not produced. Antibiotic resistances and the list of enzymic activities detected with the API ZYM gallery are given in the original description. The only substrates utilized as sole carbon sources in the API 50CH gallery are L-arabinose, glucose, mannose, esculin, cellobiose, maltose, starch, gentiobiose, and 5-ketogluconate. Other characteristics are as given in the genus and original species descriptions and in Tables 24 and 25.

Source: soil samples collected on the island of Dokdo, East Sea (disputed between the Republic of Korea and Japan).

DNA G+C content (mol%): 36.9 (HPLC).

Type strain: DS-6, CIP 108840, KCTC 12542.

Sequence accession no. (16S rRNA gene): DQ178976.

35. **Flavobacterium succinicans** (Anderson and Ordal 1961) Bernardet, Segers, Vancanneyt, Berthe, Kersters and Vandamme 1996, 142[VP] [*Cytophaga succinicans* Anderson and Ordal 1961, 136 and Reichenbach 1989c, 2045; *Flexibacter succinicans* (Anderson and Ordal 1961) Leadbetter 1974, 106]

suc.ci'ni.cans. L. n. *succinum* amber; N.L. n. *acidum succinicum* succinic acid (derived from amber); N.L. part. adj. *succinicans* intended to mean forming succinic acid.

Rods, approximately 0.5 × 4.0–6.0 μm; longer (up to 40 μm) cells may occur. In ageing cultures, cell length decreases and diameter increases slightly, preceding the appearance of spheroplasts. Colonies on media with high nutrient concentration are low convex and circular with irregular margins, whereas pale yellow, thin, spreading colonies with rhizoid to filamentous margins appear on low-nutrient media. Non-spreading variants occur and different colony types may coexist on the same agar plate. Growth is usually aerobic, but facultatively anaerobic growth may occur if a fermentable carbohydrate and CO_2 (supplied as 15–25 mM $NaHCO_3$) are available. When grown anaerobically, the cell mass is unpigmented, whereas yellow-orange carotenoid pigments are produced during aerobic growth.

Catalase activity is absent or weak. Peptones, Casamino acids, glutamate, NH_4^+, and NO_3^- serve as N sources. All strains ferment glucose, galactose, lactose (with a lag period of 6–38 d), maltose, mannose, and starch; some strains also ferment other carbohydrates. Fermentation products are succinate, acetate, and formate. Only the type strain reduces NO_3^- to NO_2^-. Susceptible to the vibriostatic compound O/129. Slight differences between the three strains in the list of fermentable carbohydrates, the catalase reaction, and the reduction of nitrate have been noticed by Anderson and Ordal (1961) and Reichenbach (1989b). Strain NCIMB 2278 has also been found to differ from the two other strains in its fatty acid composition and strain NCIMB 2279 displays a different whole-cell protein profile (Bernardet et al., 1996). No DNA hybridization experiments have been performed to verify whether the three strains belong to the same species. Other characteristics are as given in the genus and original species descriptions and in Tables 24 and 25. Additional information can be found in Reichenbach (1989b) and Bernardet et al. (1996).

Source: the type strain (strain Anderson and Ordal 8) was isolated in 1954 from the eroded caudal fin of a fingerling chinook salmon (*Oncorhynchus tschawytscha*) at the University of Washington hatchery. This strain was not formally designated the type strain in the original description; it was proposed as the cotype by Reichenbach (1989b). The two other strains were isolated in 1957 from the water of a tank containing salmonid fish in the same hatchery (strain Anderson and Ordal 16) and from a lesion on an adult chinook salmon taken from the Snake River at Brownlee Dam (Idaho) (strain Anderson and Ordal 14). The pathogenicity of the isolates for fish has not been tested experimentally; they were considered nonpathogenic by Anderson and Ordal (1961). No other strain has been isolated since then.

DNA G+C content (mol%): 34.0–36.7 (Bd, T_m).

(Co)type strain: Anderson and Ordal 8, Reichenbach Cysu3, CIP 104744, DSM 4002, IFO (now NBRC) 14905, LMG 10402, NCIMB 2277.

Sequence accession no. (16S rRNA gene): D12673.

36. **Flavobacterium suncheonense** Kim, Weon, Cousin, Yoo, Kwon, Go and Stackebrandt 2006a, 1648[VP]

sun.che.o.nen'se. N.L. neut. adj. *suncheonense* of or belonging to Suncheon, a city in the Republic of Korea, from where the type strain was isolated.

Rods, approximately 0.3 × 1.5–2.5 μm. Colonies on R2A agar are yellow, convex, and circular with entire margins. No growth occurs on MacConkey agar.

Catalase-positive. Arginine dihydrolase activity is absent. Indole is not produced. None of the carbohydrates in the API 20NE gallery is assimilated. Acid is not produced from any of the substrates in the API 50CH gallery. Enzymic activities detected in the API ZYM gallery are listed in the species description. Other characteristics are as given in the genus and original species descriptions and in Tables 24 and 25.

Source: greenhouse soil, Suncheon city, Republic of Korea.

DNA G+C content (mol%): 39.0 (HPLC).

Type strain: GH29-5, DSM 17707, KACC 11423.

Sequence accession no. (16S rRNA gene): DQ222428.

37. **Flavobacterium tegetincola** McCammon and Bowman 2000, 1060[VP]

te.ge.tin'co.la. L. n. *teges -etis* a mat or covering; L. n. *incola* dweller; N.L. n. *tegetincola* (nominative in apposition) a mat-dweller, pertaining to the cyanobacterial mat habitat.

Rods, approximately 0.4–0.5 × 2–5 µm. Colonies on R2A agar are yellow, convex, circular with entire margins, and have a butyrous consistency. No growth occurs fermentatively or by anaerobic respiration using ferric iron, nitrate, nitrite, or trimethylamine *N*-oxide as electron acceptors.

Catalase-positive. Arginine dihydrolase, lysine and ornithine decarboxylase, and tryptophan deaminase activities are absent. Acetoin and indole are not produced. Citrate is not utilized. Casamino acids serve as nitrogen sources, but not sodium nitrate, ammonium chloride, or L-glutamate. Vitamins are not required for growth. Acid is produced from glucose, fructose, and mannitol, but not from arabinose, galactose, rhamnose, xylose, sucrose, trehalose, cellobiose, maltose, melibiose, raffinose, adonitol, sorbitol, inositol, or glycerol. Uric acid and xanthine are not degraded. Other characteristics are as given in the genus and original species descriptions and in Tables 24 and 25.

Source: the type strain was isolated from cyanobacterial mat material collected from Ace Lake, a marine salinity meromictic lake located in the Vestfold Hills area of Antarctica. Only one other strain is known.

DNA G+C content (mol%): 32.0–34.0 (T_m).

Type strain: A103, ACAM 602, CIP 107447, LMG 21423.

Sequence accession no. (16S rRNA gene): U85887 (type strain), U85888 (ACAM 603).

38. **Flavobacterium weaverense** Yi and Chun 2006, 1242[VP]

wea.ve.ren'se. N.L. neut. adj. *weaverense* of or belonging to the Weaver Peninsula in the Antarctic, the geographical origin of the type strain.

Rods, approximately 0.3–0.5 × 1.6–12.5 µm during the exponential growth phase and 0.3–0.4 × 0.7–1.5 µm during the stationary phase. Cells occur singly or in chains in R2A and tryptic soy broths; filamentous cells occur in Anacker and Ordal's broth. Colonies on R2A agar are bright yellow, convex, translucent, glistening, and circular with entire margins. The colony consistency is viscous, becoming mucoid after prolonged incubation. Colonies are orange on tryptic soy agar. Colonies do not adhere to the agar. No growth occurs on cetrimide or MacConkey agars. Grows well under aerobic conditions (minimum doubling time, 2.9 h), weakly under microaerobic conditions (i.e., with 5–15% O_2 and 5–12% CO_2), and poorly under anaerobic conditions (i.e., with 4–10% CO_2).

Catalase-positive. L-Phenylalanine deaminase activity is absent. Arginine dihydrolase activity is present using Thornley's semi-solid medium, but absent with the API 20NE gallery. The list of enzymic activities detected in the API ZYM gallery is given in the original description. Indole is not produced and citrate is not utilized. Acid is produced from D-glucose and maltose, but not from D-cellobiose, D-fructose, D-mannitol, raffinose, D-salicin, trehalose, D-xylose, lactose, L-arabinose, L-rhamnose, or sucrose. Glucose, mannose, *N*-acetylglucosamine, and maltose are assimilated (API 20NE), but arabinose, mannitol, gluconate, caprate, adipate, malate, citrate, and phenylacetate are not. The maximum adsorption peak of pigment is at 451 nm and the next peak is at 479 nm. Other characteristics are as given in the genus and original species descriptions and in Tables 24 and 25.

Source: a soil sample collected on the Weaver Peninsula, King George Island, Antarctica.

DNA G+C content (mol%): 37 (HPLC).

Type strain: AT1042, CIP 109334, IMSNU 14048, JCM 12384, KCTC 12223.

Sequence accession no. (16S rRNA gene): AY581114.

39. **Flavobacterium xanthum** (Inoue and Komagata 1976) McCammon and Bowman 2000, 1060[VP] (*Cytophaga xantha* Inoue and Komagata 1976, 169 and Reichenbach 1989c, 2049)

xan'thum. Gr. adj. *xanthos* yellow; N.L. neut. adj. *xanthum* yellow.

The following description of the type strain combines the data of Inoue and Komagata (1976), Reichenbach (1989b), and McCammon and Bowman (2000). Rods, approximately 0.4–0.5 × 0.7–2 µm. Longer, filamentous cells appear in ageing cultures. Colonies on PYG agar (1% peptone, 0.5% yeast extract, 0.3% glucose, pH 7.2; amount of agar not specified) are bright yellow to orange-yellow, smooth, opaque, and circular with entire margins. No fermentation occurs and there is no growth by anaerobic respiration using ferric iron, nitrate, nitrite, or trimethylamine *N*-oxide as electron acceptors. Casamino acids, peptone, ammonium chloride, and L-glutamate serve as nitrogen sources. Vitamins are not required for growth.

Catalase-positive. Arginine dihydrolase, lysine and ornithine decarboxylase, and tryptophan deaminase activities are absent. Acid is produced from cellobiose, fructose, glucose, maltose, mannitol, mannose, sucrose, trehalose, and xylose, but not from adonitol, arabinose, galactose, glycerol, inositol, lactose, melibiose, rhamnose, raffinose, or sorbitol. Formate, acetate, lactate, and hippurate are assimilated, but succinate, fumarate, citrate, propionate, protocatechuate, and *p*-hydroxybenzoate are not. Uric acid and xanthine are not degraded. Acetoin and indole are not produced. Citrate is not utilized. Other characteristics are as given in the genus and original species description and in Tables 24 and 25. Although a G+C content of 39.3 mol% was originally reported (Inoue and Komagata, 1976), values of approximately 36 mol% were subsequently published (Bernardet et al., 1996; McCammon and Bowman, 2000).

Source: the type strain was isolated from soil (a mud pool according to McCammon and Bowman, 2000) in Showa Station, Antarctica, in February 1967. Eleven additional strains were isolated from microbial mats in Antarctic freshwater lakes by Van Trappen et al. (2005).

DNA G+C content (mol%): 36.0 (T_m).

Type strain: Inoue 5-O-c, ACAM 81, CIP 107448, DSM 3661, IAM 12026, IFO (now NBRC) 14972, LMG 8372, NCIMB 2069.

Sequence accession no. (16S rRNA gene): AF030380 (type strain), AJ601392 (R-9010).

40. **Flavobacterium xinjiangense** Zhu, Wang and Zhou 2003, 856[VP]

xin.ji.ang.en'se. N.L. neut. adj. *xinjiangense* pertaining to Xinjiang, an autonomous region in north-west China from where the type strain was isolated.

Rods, approximately 0.8 × 2.5–5 µm. Colonies on PYG agar [per l: 5.0 g polypeptone, 5.0 g tryptone, 10.0 g yeast extract, 10.0 g glucose, 40 ml salt solution, 15.0–20.0 g agar. Salt solution (per l): 0.2 g $CaCl_2$, 0.4 g $MgSO_4 \cdot 7H_2O$, 1.0 g K_2HPO_4, 1.0 g KH_2PO_4, 10.0 g $NaHCO_3$, 2.0 g NaCl] are pale yellow, circular, convex, and smooth, with entire margins. Good growth also occurs on CIP medium 469 (per l: 10 g peptone, 10 g yeast extract, 5 g sodium chloride, 20 g agar).

Catalase-positive. Arginine dihydrolase, lysine and ornithine decarboxylase, and tryptophan deaminase activities are absent. Indole and acetoin are not produced. Citrate is not utilized. Peptone and Casamino acids serve as N sources, but sodium nitrate, ammonium chloride, and L-glutamate do not. Acid is not produced from glucose, fructose, xylose, mannitol, or maltose. Uric acid and xanthine are not degraded. Glucose, cellobiose, maltose, fructose, sucrose, arabinose, and xylose are assimilated, but sorbitol, sorbose, galactose, D-mannitol, raffinose, lactose, mannose, trehalose, glycerol, rhamnose, melibiose, and inositol are not. Other characteristics are as given in the genus and original species descriptions and in Tables 24 and 25.

Source: frozen soil enclosed within the China No. 1 glacier, Xinjiang Uygur Autonomous Region, China.

DNA G+C content (mol%): 34.4 (T_m).

Type strain: Zhou ZF-6, AS 1.2749, CIP 108285, JCM 11314, LMG 21985.

Sequence accession no. (16S rRNA gene): AF433173.

Other organisms

These organisms, whose names have not been validly published, have been assigned to the genus *Flavobacterium* by DNA–rRNA hybridization (Bernardet et al., 1996) and, for two of them, by 16S rRNA gene sequencing (Figure 27). Except for "*Sporocytophaga cauliformis*", only one strain is available in culture collections (see *Taxonomic comments*, above).

1. "**Cytophaga allerginae**" Liebert, Hood, Deck, Bishop and Flaherty 1984, 941

al.ler.gi'nae. Gr. pl. n. *alloi* strangers; Gr. n. *ergon* work, act, deed; N.L. gen. n. *allerginae* intended to mean connected with, causing an allergy.

The following description is a combination of data from the original description (Liebert et al., 1984) and unpublished data (J.-F. Bernardet, unpublished observations). Rods, approximately 0.3–0.5 × 3.5–9.0 µm; cells up to 30 µm occur occasionally and spheroplasts appear in ageing broth cultures. Cells display gliding motility, pivoting, and flexing movements. Colonies on nutrient agar are bright yellow, and circular with entire margins. Colonies on Anacker and Ordal's agar and on other media with low nutrient concentration are yellow, flat, and irregular with spreading rhizoid margins. Colonies do not adsorb Congo red and do not adhere to the agar. The positive KOH test and the maximum adsorption peak at 450 nm suggest that the yellow pigments belong to the flexirubin type. Grows at 10–30°C, but not at 4 or 37°C. Grows at pH 5.5–9.5 and in nutrient broth containing 1.5% NaCl, but not in the presence of 3.0% NaCl. Good growth occurs under aerobic conditions; weak growth also occurs in half-strength nutrient broth and on plate count agar after 48 h under anaerobic conditions. Contrary to Reichenbach's statement (1989b), anaerobic growth seems to occur even in the absence of nitrate. In addition to the above-mentioned media, growth also occurs on nutrient, plate count, Cook's (per l: 2 g tryptone, 10 g agar), and vy/2 (per l: 5 g bakers' yeast, 1 g $CaCl_2 \cdot 2H_2O$, 1 mg vitamin B_{12}, 15 g agar) agars, but not on MacConkey agar.

Catalase-positive. β-Galactosidase activity is present, but arginine dihydrolase, lysine and ornithine decarboxylase, phenylalanine deaminase, urease, and DNase activities are absent. Casein, gelatin, tyrosine, Tweens 20 and 80, glycerol tributyrate, starch, chitin, and carboxymethylcellulose are degraded, but agar, alginate, and cellulose are not. A dark brown diffusible melanic pigment is produced on tyrosine agar and a precipitate is produced on egg-yolk agar. Acid is produced from glucose, maltose, cellobiose, and arabinose, but not from lactose or sucrose. Glucose is not fermented. Indole and H_2S are not produced. Nitrate is not reduced. Antibiotic resistances are listed in the original description. A biologically active endotoxin is produced. According to Liebert et al. (1984), the only available strain of "*Cytophaga allerginae*" shares approximately 78% DNA relatedness with the *Flavobacterium hydatis* type strain. However, J.-F. Bernardet (unpublished data) found only 34% DNA relatedness between these two strains, and 16% between "*Cytophaga allerginae*" and the *Flavobacterium johnsoniae* type strain. Fatty acid and whole-cell protein analyses as well as DNA–rRNA hybridization have confirmed that "*Cytophaga allerginae*" belongs to the genus *Flavobacterium* (Bernardet et al., 1996), but 16S rRNA studies and additional DNA–DNA hybridizations are needed to allocate it to a new or a described species. Additional information can be found in Liebert et al. (1984), Reichenbach (1989b), and Bernardet et al. (1996).

Source: an industrial water spray air humidification system; isolated following several cases of lung disease with hypersensitivity pneumonitis-like symptoms that occurred in the late 1970s in a United States textile facility. The strain was identified as the source of the endotoxin responsible for the symptoms (Flaherty et al., 1984; Liebert et al., 1984). This was the first report of a *Flavobacterium* strain implicated in an industrial health-related disease; no other case has been reported since then.

DNA G+C content (mol%): 34–34.8 (T_m).

Type strain: WF-164, ATCC 35408.

2. "**Flexibacter aurantiacus var. excathedrus**" Lewin 1969, 200

au.ran.ti'a.cus. N.L. masc. adj. *aurantiacus* orange-colored. ex.cath.e'drus. L. prep. *ex* out of; L. n. *cathedra* a chair, a stool, an arm-chair and, by extension, a cathedral; N.L. masc. adj. *excathedrus* from a cathedral, referring to the origin of the strain, a pool in a cathedral.

Rods, approximately 0.3–0.5 × 5.0–10.0 µm. Active gliding occurs. Colonies on modified Shieh agar (Song et al., 1988) are yellow, low convex, and circular with regular to lobate margins (J.-F. Bernardet, unpublished data). No growth occurs on media with half-strength sea water. Good growth occurs at 25°C and the highest growth temperature is 40°C.

Catalase-negative. Gelatin, tyrosine, and starch are degraded, but agar, alginate, and carboxymethylcellulose are not. A dark brown diffusible melanic pigment is produced on tyrosine agar. H_2S is not produced and nitrate is not reduced. Tryptone, Casamino acids, and sodium glutamate are used as nitrogen sources. Additional information can be found in Lewin (1969), Lewin and Lounsbery (1969), and Reichenbach (1989b). Fatty acid and whole-cell protein analyses have shown that "*Flexibacter aurantiacus* var. *excathedrus*" belongs to the genus *Flavobacterium* (Bernardet et al., 1996).

Flexibacter aurantiacus was proposed by Lewin (1969) and Lewin and Lounsbery (1969) to accommodate two strains received as "*Cytophaga aurantiaca*" and "*Cytophaga psychrophila*", but DNA–DNA hybridization later demonstrated that they actually represented two strains of *Flavobacterium johnsoniae* (Bernardet et al., 1996) (see above). Two variants had also been proposed by Lewin (1969) and Lewin and Lounsbery (1969): the marine organism "*Flexibacter aurantiacus* var. *copepodarum*" and the freshwater isolate "*Flexibacter aurantiacus* var. *excathedrus*". DNA–rRNA hybridization experiments have allocated these strains to the marine clade of the family *Flavobacteriaceae* and to the close vicinity of *Flavobacterium columnare*, respectively (Bernardet et al., 1996). The position of "*Flexibacter aurantiacus* var. *excathedrus*" was later defined more accurately by 16S rRNA studies (Figure 27), but DNA–DNA hybridizations are needed to determine whether it can be considered a *Flavobacterium columnare* strain or if the description of a novel species is necessary. Any further study will have to take into consideration the taxonomic status of the three genomovars of *Flavobacterium columnare* (see *Taxonomic comments*, above).

Source: a freshwater pool in (or near?) a cathedral, Cartago, Costa Rica.

DNA G+C content (mol%): 32.5–34.5 (T_m).

Type strain: Lewin CR-134, ATCC 23086, IFO 16024, LMG 3986.

Sequence accession no. (16S rRNA gene): AB078045.

3. "**Promyxobacterium flavum**" Imshenetski and Solntseva 1945, 224

fla'vum. L. neut. adj. *flavum* golden yellow.

This organism is remarkably pleomorphic. Rods may be stout, almost coccoid (approx. 0.6–0.7 × 1.0–2.5 μm) to slender and flexible (approx. 0.4–0.5 × 2.5–6.0 μm). The stout forms are dominant in ageing broth cultures. Gliding motility is present. Colonies are bright yellow and compact with more or less lobate to spreading margins on rich media, whereas yellow, thin, spreading swarms tend to occupy all the surface of media with low nutrient concentration. The yellow pigments belong to the flexirubin type (positive KOH test). Good growth occurs on nutrient and modified Shieh agars (Song et al., 1988) at 30°C (Bernardet et al., 1996). No growth occurs on media prepared with full-strength seawater.

Metabolism is strictly aerobic. Catalase-negative. Weakly oxidase-positive. Proteins, peptones, and nitrate may serve as sole nitrogen sources. Glucose may serve as the sole source of carbon and energy. Reichenbach (1989b) reported reasonable growth on a mineral agar containing KNO_3 and glucose. Acid is produced from glucose. Gelatin, starch, hemicellulose, and pectin are degraded, but agar, chitin, and cellulose are not. Additional information can be found in Reichenbach (1989b). Fatty acid and whole-cell protein analyzes as well as DNA–rRNA hybridization have shown that "*Promyxobacterium flavum*" belongs to the genus *Flavobacterium* (Bernardet et al., 1996), but 16S rRNA studies and DNA–DNA hybridizations are needed to allocate it to a new or a described species.

Source: the original strains isolated from soil in Russia by Imshenetski and Solntseva (1945) have been lost. A strain was isolated by Vozniakovskaia and Rybakova (1969) from the rhizosphere of a tomato plant in Russia. As its scant description was in accordance with the original species description, this strain has been accepted as the neotype strain by Reichenbach (1989b). The above-mentioned characteristics are those reported by this author for the neotype strain.

DNA G+C content (mol%): 33.0–34.5 (Bd, T_m).

(Neo)type strain: Vozniakovskaya 19, VKM B-1553, DSM 3577, LMG 10389.

4. "**Sporocytophaga cauliformis**" Gräf 1962, 124

cau.li.for'mis. L. n. *caulis* stalk, stem; L. adj. suff. *-formis* shaped like; N.L. fem. adj. *cauliformis* stalk-shaped, stalked.

Rods, approximately 0.4 × 2.0–5.0 μm, flexible, and gliding. Colonies on relatively rich media such as nutrient agar are compact with more or less spreading margins, whereas thin, large, spreading swarms appear on media with low nutrient concentration such as CY agar (per l: 3.0 g casitone, 1.36 g $CaCl_2 \cdot 2H_2O$, 1.0 g yeast extract, 15.0 g agar, pH 7.2). The yellow to orange pigments belong to the flexirubin type (positive KOH test). Grows at 5–30°C and optimally at 22°C.

Catalase- and urease-negative. Oxidase-positive. Peptones may serve as sole N source; NH_3 is released from peptones. Acid is not produced from carbohydrates. Starch is degraded, but cellulose and agar are not. Indole and H_2S are not produced. Nitrate is not reduced. According to Reichenbach (1989b), the classification of this organism in the genus *Sporocytophaga* resulted from the misinterpretation of several degenerative phenomena. Some confusion also occurred with other freshwater isolates and two types were described, each of them represented by a single strain. "*Sporocytophaga cauliformis*" type 2 was considered the type strain of the species; it is catalase-negative and does not produce acid from carbohydrates. Conversely, "*Sporocytophaga cauliformis*" type 1 (Gräf Z2=NCIMB 9487=DSM 3656=LMG 8362) is positive for these two traits. Also, the DSMZ Catalog of Strains states that type 2 degrades chitin and pectin, whereas no such mention is made for type 1.

Fatty acid and whole-cell protein analyses, as well as DNA–rRNA hybridization, have unequivocally shown that the two "*Sporocytophaga cauliformis*" strains belong to the genus *Flavobacterium* (Bernardet et al., 1996). A 16S rRNA gene sequence is available for the type strain (type 2) (Gherna and Woese, 1992). In phylogenetic trees, its closest neighbors are *Flavobacterium saccharophilum*, *Flavobacterium pectinovorum*, and several recently described *Flavobacterium* species (Figure 27), but DNA–DNA hybridization data are needed to elucidate their relationship at the species level. Moreover, phenotypic differences, discrepancies in the respective proportions of some fatty acids, distinct positions in the dendrogram derived from whole-cell protein profiles, and rather different DNA G+C contents suggest that the two "*Sporocytophaga cauliformis*" strains probably belong to different species (Bernardet et al., 1996).

Source: water, Lake Constance (Bodensee), Germany.

DNA G+C content (mol%): 35.0–36.2 [Bd, T_m; Type 2 (type strain)]; 31.0–33.5 (Bd, T_m; Type 1).

Type strain: ("*Sporocytophaga cauliformis*" type 2) Gräf Z6, NCIMB 9488, DSM 3657, LMG 8363.

Sequence accession no. (16S rRNA gene): M93151.

Genus II. **Aequorivita** Bowman and Nichols 2002, 1538[VP]

John P. Bowman

Ae.quo.ri.vi'ta. L. n. *aequor -oris* the even surface of the sea in its quiet state; L. fem. n. *vita* life; N.L. fem. n. *Aequorivita* life (living being) at the sea surface.

Cells are straight or slightly curved **rods or filaments**, 0.5–20 × 0.2–0.5 µm. **Gram-stain-negative**. Cells occur singly or in pairs. Spores and resting cells are not present. Gas vesicles and helical or ring-shaped cells are not formed. **Nonmotile. Strictly aerobic, having an oxidative metabolism**. Catalase-positive. Oxidase test is weakly positive. **Chemoheterotrophic. Non-agarolytic. Colonies are orange or yellow** due to production of carotenoids. **Flexirubin pigments are not produced**. Some species **require Na^+** for growth. Tolerates up to 6% NaCl; growth with 10% NaCl is either weak or absent. Best growth occurs in organic media containing 1–3% NaCl or seawater salts. All species **produce alkaline phosphatase and esterase and hydrolyze L-tyrosine and gelatin**. Some species are strongly proteolytic and/or lipolytic. **Psychroactive.** Growth occurs between −2 and 25°C. **Optimal growth occurs at 20°C. Neutrophilic,** with optimal growth occurring at pH 7.0–7.5. The major fatty acids are $C_{15:1}$ ω10*c* iso, $C_{15:1}$ ω10*c* anteiso, $C_{15:0}$ iso, $C_{15:0}$ anteiso, $C_{16:1}$ ω6*c* iso, $C_{17:1}$ ω5*c* iso, $C_{17:1}$ ω7*c* anteiso, and $C_{16:0}$ 3-OH iso. **Known habitats include polar ocean seawater, sea-ice, and quartz stone sublithic communities** located in polar maritime ice-free zones. Also reported to be associated with deep-sea invertebrates.

DNA G+C content (mol%): 33–39.

Type species: **Aequorivita antarctica** Bowman and Nichols 2002, 1539[VP].

Further descriptive information

The genus *Aequorivita* belongs to the family *Flavobacteriaceae* and is most closely related to the seawater species *Vitellibacter vladivostokensis* (Nedashkovskaya et al., 2003c). With *Vitellibacter*, it forms a relatively distinct phylogenetic lineage in the family *Flavobacteriaceae.*

Phenotypic traits differentiating the species of *Aequorivita* are shown in Table 26. All species possess catalase, oxidase (weakly), alkaline phosphatase, and esterase (Tween 80 as substrate) activities. L-Tyrosine is hydrolyzed and, thus, presumably metabolized. Gelatin is also liquefied by all species. The species possesses a range of DNA G+C values (Table 26) that are typical of members of the family *Flavobacteriaceae.* Fatty acid profiles are quite similar between the species and are composed of almost entirely branched chain C_{15}–C_{17} fatty acids (Table 27).

The species exhibit only a limited capacity to utilize sole carbon and energy sources, showing only poor, equivocal growth on defined media containing a variety of different substrates even when supplemented with yeast extract and a vitamin solution. This suggests other unknown growth factors, present in complex media used for routine growth, are required by strains of this genus.

Aequorivita species have been found to be negative for the following phenotypic characteristics: anaerobic growth (including growth by ferric iron reduction as well as carbohydrate fermentation in the presence and absence of nitrate); oxidative and fermentative acid production from sugars and sugar alcohols; tolerance to NaCl levels of 12% or greater; tolerance to 1% ox bile salts; hydrolysis of polysaccharides (however, *Aequorivita antarctica* can attack starch); decomposition of uric acid and xanthine; lysine decarboxylase, arginine dihydrolase, ornithine decarboxylase, α-arabinosidase, α-fucosidase, β-glucosidase, β-galactosidase, 6-phospho-β-galactosidase, and β-glucuronidase activities; indole production and Voges–Proskauer tests; and the bathochromic shift assay to detect the presence of flexirubin type pigments.

Although the presently described species were obtained from Antarctic marine and marine-influenced ecosystems (Bowman and Nichols, 2002; Smith et al., 2000), other studies indicate that members of the genus may be relatively widespread, present in the Arctic and in association with deep-sea invertebrates (Sfanos et al., 2005; unpublished sources). Of the four known species in the genus *Aequorivita,* all except *Aequorivita crocea,* have been isolated from quartz stone sublithic samples. These constitute cyanobacterial-dominated biofilm-like communities existing on the bottom surfaces of semi-translucent quartz stones, protected from surrounding extreme weather conditions, particularly freezing and desiccation. The location of the quartz stones studied so far has been entirely confined to coastal ice-free regions of Antarctica. *Aequorivita* strains were not isolated from soil underlying quartz stone subliths or the immediately adjacent soils, suggesting that quartz stone sublithic communities represented a form of refuge or oasis for these bacteria. *Aequorivita antarctica, Aequorivita crocea,* and *Aequorivita lipolytica* have also been found in coastal Antarctic sea-ice and under-ice seawater, suggesting that sublithic environments may be inoculated by windborne marine particulate deposition (Smith et al., 2000).

Enrichment and isolation procedures

Marine agar (Difco, Oxoid) is used for isolation of *Aequorivita* species from sea-ice and seawater samples. Potentially, isolates can be obtained from sea-ice samples, pre-melted at 2–4°C in sterile seawater, and polar seawater samples, diluted in artificial seawater and directly plated onto marine agar. Incubation proceeds at 2–4°C. For isolation of strains from quartz stone subliths, sublithic biomass is scraped off stones using a sterile scalpel and suspended, vortexed, and then diluted in trypticase soy broth containing 1% NaCl. The suspensions are serially diluted and spread onto agar plates and incubated at 4–10°C. *Aequorivita* strains can be grown routinely either on marine agar or trypticase soy agar containing 1% NaCl. No selective media are available for this genus at present and so strains must be identified empirically from amongst the orange- or yellow-pigmented colonies arising from primary isolation plates.

Maintenance procedures

Strains can be maintained for many years cryopreserved in marine broth containing 20–30% glycerol. Strains survive for many months, potentially years, on thick agar plates or on slants, stored at 2–4°C. Strains may also be maintained by lyophilization, although the survival level is presently unknown.

TABLE 26. Differential characteristics of *Aequorivita* species[a]

Characteristic	*A. antarctica*	*A. crocea*	*A. lipolytica*	*A. sublithincola*
Morphology	Rods, filaments	Rods	Rods	Rods, filaments
Dimensions (μm)	0.5–20 × 0.2–0.3	0.5–3.0 × 0.4–0.5	0.5–3.0 × 0.4–0.5	0.5–20 × 0.4–0.5
Pigment	Orange	Yellow	Yellow	Orange
Na^+ requirement for growth	+	–	+	+
D-Glucose utilization	+	–	–	–
Urease	d	–	d	+
α-Galactosidase	d	–	–	–
α-Glucosidase	d	–	–	–
N-Acetyl-β-glucosaminidase	–	–	–	+
Lipase (olive oil)	–	–	+	–
Extracellular DNase	–	d	–	–
Egg yolk reaction	–	–	+	–
Hydrolysis of:				
Esculin	+	+	+	–
Tributyrin	–	–	+	–
Starch	+	–	–	–
Casein	–	+	+	–
Elastin	d	+	+	–
DNA G+C content (mol%)	38–39	33–34	35–36	36–37

[a]Symbols: +, >85% positive; d, different strains give different reactions (16–84% positive); –, 0–15% positive.

TABLE 27. Whole-cell fatty acids present in *Aequorivita* species[a]

Fatty acid	*A. antarctica*	*A. crocea*	*A. lipolytica*	*A. sublithincola*
Saturated fatty acids:				
$C_{15:0}$	tr	tr	tr	3.1
$C_{16:0}$	1.9	2.9	1.2	tr
Branched-chain fatty acids:				
$C_{14:0}$ iso	tr	tr	tr	tr
$C_{15:1}$ ω10*c* iso	8.5	8.9	10.5	12.9
$C_{15:1}$ ω10*c* anteiso	16.0	13.8	23.0	8.5
$C_{15:0}$ iso	7.6	15.9	16.3	16.8
$C_{15:0}$ anteiso	15.7	19.2	20.7	17.8
$C_{16:1}$ ω6*c* iso	3.0	6.2	7.4	6.7
$C_{16:0}$ iso	1.4	4.7	1.2	5.3
$C_{17:1}$ ω5*c* iso	5.8	8.6	4.8	9.7
$C_{17:1}$ ω7*c* anteiso	5.0	3.2	4.2	2.5
Monounsaturated fatty acids:				
$C_{15:1}$ ω6*c*	tr	nd	nd	tr
$C_{16:1}$ ω7*c*	7.8	2.9	1.2	tr
Hydroxy fatty acids:				
$C_{15:0}$ 3-OH iso	5.4	2.9	1.7	2.1
$C_{15:0}$ 3-OH anteiso	5.8	1.6	1.6	tr
$C_{16:0}$ 3-OH iso	9.2	4.4	2.1	5.6
$C_{17:0}$ 3-OH iso	2.0	nd	2.3	4.5
$C_{17:0}$ 3-OH anteiso	2.9	4.2	nd	4.0
Other	2.7	2.5	tr	tr

[a]Values are given as percentages of total fatty acids; tr, trace level detected (<1% of total fatty acids); nd, not detected.

Differentiation of the genus *Aequorivita* from other genera

The most salient traits useful for distinguishing *Aequorivita* from other members of the *Flavobacteriaceae* include a strictly oxidative metabolism, the inability to form acid from carbohydrates, the inability to grow on defined mineral salts media, a lack of flexirubin production, a lack of motility, low growth temperatures (e.g., growth at 4°C, no growth at 30°C or more), a requirement for Na^+ for growth, and a lack of agarolytic activity. *Aequorivita* differs most significantly from its closest relative, *Vitellibacter vladivostokensis* (Nedashkovskaya et al., 2003c), in terms of flexirubin production (positive for *Vitellibacter vladivostokensis*) and growth temperature (*Vitellibacter vladivostokensis* can grow at 43°C). The DNA G+C content of *Vitellibacter vladivostokensis* is 41 mol%, which is higher than that found for all *Aequorivita* species. The fatty acid profile of *Vitellibacter vladivostokensis* also differs from that of *Aequorivita* spp. by having a very high $C_{15:0}$ iso content.

Taxonomic comments

The genus *Aequorivita* includes four species, *Aequorivita antarctica*, *Aequorivita crocea*, *Aequorivita lipolytica*, and *Aequorivita sublithincola*. Genomic DNA–DNA hybridization levels were highest between *Aequorivita lipolytica* and *Aequorivita crocea*, ranging from 44–57%. Both of these species occur as rod-like cells (0.5–3.0 × 0.4–0.5 μm) and form yellow colonies. This corresponds clearly with 16S rRNA gene sequence data, which indicates that these species are closely related, sharing a gene sequence similarity of 98%. By comparison, *Aequorivita antarctica* and *Aequorivita sublithincola* strains appear as rod-like to filamentous cells (0.5–20 × 0.2–0.5 μm) and form orange colonies. These species are more genetically distinct, especially *Aequorivita sublithincola*, which occurs as an outlier in the *Aequorivita* 16S rRNA gene cluster. *Aequorivita antarctica* shares 32–39% DNA–DNA hybridization with *Aequorivita crocea* and *Aequorivita lipolytica*. *Aequorivita sublithincola* genomic DNA hybridization values with the other species are below background hybridization levels.

List of species of the genus *Aequorivita*

1. **Aequorivita antarctica** Bowman and Nichols 2002, 1539[VP]

ant.arc'ti.ca. L. fem. adj. *antarctica* southern; here of or belonging to Antarctica.

Characteristics are as given for the genus and as listed in Table 26.

Source: seawater (ice-covered polar regions), sea-ice, and quartz stone sublithic material.

DNA G+C content (mol%): 38–39 (T_m).

Type strain: SW49, ACAM 640, DSM 14231.

Sequence accession no. (16S rRNA gene): AY027802.

2. **Aequorivita crocea** Bowman and Nichols 2002, 1540[VP]

cro'ce.a. L. fem. adj. *crocea* saffron-colored, yellow.

Characteristics are as given for the genus and as listed in Table 26.

Source: seawater (ice-covered polar regions) and sea-ice.

DNA G+C content (mol%): 33–34 (T_m).

Type strain: Y12-2, ACAM 642, DSM 14293.

Sequence accession no. (16S rRNA gene): AY027806.

3. **Aequorivita lipolytica** Bowman and Nichols 2002, 1539[VP]

li.po.ly'ti.ca. Gr. n. *lipos* fat; Gr. adj. *lytikos* dissolving; N.L. fem. adj. *lipolytica* fat-dissolving.

Characteristics are as given for the genus and as listed in Table 26.

Source: seawater (ice-covered polar regions) and quartz stone sublithic material.

DNA G+C content (mol%): 35–36 (T_m).

Type strain: Y10-2, ACAM 641, DSM 14236.

Sequence accession no. (16S rRNA gene): AY027805.

4. **Aequorivita sublithincola** Bowman and Nichols 2002, 1540[VP]

sub.lith.in'co.la. L. pref. *sub* below; Gr. n. *lithos* stone; L. nom. n. *incola* an inhabitant; N.L. n. *sublithincola* an inhabitant below stone.

Characteristics are as given for the genus and as listed in Table 26. The description is based on a single isolate.

Source: quartz stone sublithic material.

DNA G+C content (mol%): 36–37 (T_m).

Type strain: 9-3, ACAM 643, DSM 14238.

Sequence accession no. (16S rRNA gene): AY170749.

Genus III. **Algibacter** Nedashkovskaya, Kim, Han, Rhee, Lysenko, Rohde, Zhukova, Frolova, Mikhailov and Bae 2004d, 1260[VP]

Olga I. Nedashkovskaya and Seung Bum Kim

Al.gi.bac'ter. L. fem. n. *alga* seaweed; N.L. masc. n. *bacter* from Gr. n. *bacterion* rod; N.L. masc. n. *Algibacter* rod isolated from seaweed.

Thin rods usually measuring 0.4–0.5 × 2–3 µm. **Motile by gliding.** Produce non-diffusible orange pigments. No flexirubin types of pigments are formed. **Chemoorganotrophs. Can ferment D-glucose.** Positive for oxidase, catalase, alkaline phosphatase, and β-galactosidase. Chitin and cellulose (CM-cellulose and filter paper) are not degraded, but other polysaccharides including agar, alginate, starch, gelatin, and Tweens may be decomposed. Do not grow without seawater or sodium ions. **The major respiratory quinone is MK-6.** Phosphatidylethanolamine is the main polar lipid compound.

DNA G+C content (mol%): 31–33 (T_m).

Type species: **Algibacter lectus** Nedashkovskaya, Kim, Han, Rhee, Lysenko, Rohde, Zhukova, Frolova, Mikhailov and Bae 2004d, 1260[VP].

Further descriptive information

Phylogenetic analysis of nearly complete 16S rRNA gene sequences of strains of the genus *Algibacter* has revealed that its close relatives are members of the genera *Bizionia* and *Formosa* with a similarity range of 94.2–95.3%.

The main cellular fatty acids are straight-chain unsaturated and branched-chain unsaturated fatty acids ($C_{15:0}$ iso, $C_{15:0}$ anteiso, $C_{15:1}$ iso, $C_{15:0}$, $C_{15:1}$ ω6*c*, $C_{15:0}$ iso 3-OH, and $C_{17:0}$ iso 3-OH).

On marine agar 2216 (Difco), the algibacters form regular, round, shiny colonies that are sunk in the agar, transparent with entire edges, and bright orange with diameters of 3–4 mm after 48 h at 28°C.

All isolated strains have been grown on media containing 0.5% peptone and 0.1–0.2% yeast extract (Difco), prepared with native or artificial seawater or supplemented with 2–3% NaCl. Growth occurs at 4–35°C, at pH 5.5–10.0, and with 1–6% NaCl. The optimal temperature for growth is 21–23°C and the optimal pH is between 7.5 and 8.3.

Strains of the genus *Algibacter* are susceptible to carbenicillin, lincomycin, and oleandomycin, but resistant to ampicillin, benzylpenicillin, gentamicin, kanamycin, neomycin, polymyxin B, streptomycin, and tetracycline.

The algibacters are dwellers of coastal marine environments and strains have been isolated from seaweeds collected in the temperate latitudes.

Enrichment and isolation procedures

The algibacters were isolated from seaweeds using the direct plating technique on marine agar (Difco). Natural or artificial seawater is sufficient for their cultivation. All isolated strains have been grown on media containing 0.5% peptone and

0.1–0.2% yeast extract (Difco). The algibacters remain viable on marine agar (Difco) or other rich media based on natural or artificial seawater for several weeks. They can survive storage in marine broth or artificial seawater supplemented with 20% glycerol (v/v) at −80°C for at least 5 years.

Differentiation of the genus *Algibacter* from other genera

Algibacters are able to move by means of gliding, to hydrolyze agar, and to produce acid from carbohydrates, in contrast with their nearest neighbors, *Bizionia* species (Bowman and Nichols, 2005; Nedashkovskaya et al., 2005a). DNA G+C content may also be used for differentiation of these two genera (31–33 mol% for *Algibacter* and 37–45 mol% for *Bizionia*). The combination of phylogenetic distances, the distinctiveness of the cellular fatty acid composition, and growth only with seawater or NaCl enable algibacters to be differentiated from their other nearest neighbor, *Formosa* (Nedashkovskaya et al., 2006d). *Algibacter* strains clearly differ from those of the close relative *Lacinutrix* in the ability to move by gliding, to produce acid from carbohydrates, to hydrolyze agar, and by the lower DNA G+C content (Bowman and Nichols, 2005; Nedashkovskaya et al., 2004d). In comparison with species of the other closest relative, *Gaetbulibacter*, algibacters produce gelatinase, but not nitrate reductase (Jung et al., 2005). In addition, they can be distinguished by their temperature range for growth (31–33°C for *Algibacter* and 13–40°C for *Gaetbulibacter*). D-Glucose fermentation is a helpful phenotypic feature that enables differentiation between algibacters and their close relatives *Bizionia*, *Lacinutrix*, *Olleya*, *Psychroserpens*, and *Winogradskyella* (Bowman et al., 1997; Bowman and Nichols, 2005; Lau et al., 2005a; Mancuso Nichols et al., 2005; Nedashkovskaya et al., 2005a). Algibacters produce agarase and amylase, in contrast to the genus *Olleya*, and the DNA G+C content is lower than the value determined for *Olleya* (31–33 and 49 mol%, respectively) (Mancuso Nichols et al., 2005). Several phenotypic features, e.g. gliding motility and production of oxidase, agarase, and amylase, that are found in algibacters are not observed among *Psychroserpens* strains (Bowman et al., 1997). Species of the genus *Psychroserpens* hydrolyze casein and Tween 80 and have a lower DNA G+C content (27–29 mol%), in contrast to the algibacters (31–33 mol%).

List of species of the genus *Algibacter*

1. **Algibacter lectus** Nedashkovskaya, Kim, Han, Rhee, Lysenko, Rohde, Zhukova, Frolova, Mikhailov and Bae 2004d, 1260[VP]

lec′tus. L. masc. adj. *lectus* chosen, select, referring to a bacterium that forms select, beautiful colonies.

Cells are rod-shaped, 0.4–0.5 µm in width and 2–3 µm in length. Colonies are circular, 3–4 mm in diameter, convex, shiny, sunk in the agar, bright orange in color, and translucent on solid media containing high nutrient components. Na^+ ions are required for growth. Growth occurs at 4–35°C. The optimal temperature for growth is 21–23°C. Growth occurs with 1–6% NaCl. The pH range for growth is 5.5–10.0, with optimum growth occurring between pH 7.5 and 8.3. Agar, gelatin, alginate, starch, and Tweens 20 and 40 are hydrolyzed, but casein, Tween 80, and DNA are not. Acid is produced from D-cellobiose, L-fucose, D-galactose, D-glucose, maltose, sucrose, and DL-xylose, but not from L-arabinose, D-lactose, melibiose, raffinose, glycerol, inositol, or mannitol. L-Rhamnose and *N*-acetylglucosamine are oxidized. D-Lactose and D-mannose are utilized, but not L-arabinose, adonitol, dulcitol, mannitol, inositol, sorbitol, malonate, or citrate. Nitrate is not reduced to nitrite. Indole, H_2S, and acetoin (Voges–Proskauer reaction) are not produced. The organisms are susceptible to carbenicillin, lincomycin, and oleandomycin, but they are resistant to ampicillin, benzylpenicillin, gentamicin, kanamycin, neomycin, polymyxin B, streptomycin, and tetracycline.

Source: three strains were isolated from the green algae *Acrosiphonia sonderi* and *Ulva fenestrata*, Troitsa Bay, Sea of Japan.

DNA G+C content (mol%): 31–33 (T_m).

Type strain: KMM 3902, KCTC 12103, DSM 15365.

Sequence accession no. (16S rRNA gene): AY187689.

Genus IV. **Aquimarina** Nedashkovskaya, Kim, Lysenko, Frolova, Mikhailov, Lee and Bae 2005f, 227[VP] emend. Nedashkovskaya, Vancanneyt, Christiaens, Kalinovskaya, Mikhailov and Swings 2006f, 2039

Olga I. Nedashkovskaya, Marc Vancanneyt and Seung Bum Kim

Aqui.ma.ri′na. L. fem. n. *aqua* water; L. adj. *marinus -a -um* marine; N.L. fem. n. *Aquimarina* an organism of seawater.

Thin rods usually measuring **0.2–0.5 µm in width and 1.0–15.0 µm in length. Cells of some species can move by means of gliding.** Produce non-diffusible carotenoid pigments. Flexirubin type of pigments can be produced. **Chemoorganotrophs. Strictly aerobic.** Halophilic. Positive for oxidase, catalase, and alkaline phosphatase. Arginine dihydrolase, lysine decarboxylase, ornithine decarboxylase, and tryptophan deaminase are not produced. **Casein, gelatin, and Tweens 20, 40 and 80 are hydrolyzed.** Urea and cellulose (CM-cellulose and filter paper) are not attacked, but agar, starch, DNA, and chitin may be decomposed. Nitrate reductase is not produced. **Require seawater or sodium ions for growth.** Indole and acetoin production is negative. **The major respiratory quinone is MK-6. Marine**, from coastal habitats.

DNA G+C content (mol%): 31–38 (T_m).

Type species: **Aquimarina muelleri** Nedashkovskaya, Kim, Lysenko, Frolova, Mikhailov, Lee and Bae 2005f, 227[VP].

Further descriptive information

The predominant cellular fatty acids are straight-chain unsaturated and branched-chain unsaturated fatty acids ($C_{15:1}$ iso, $C_{15:0}$ iso, $C_{15:0}$ iso 3-OH, $C_{17:1}$ iso ω9*c*, $C_{17:0}$ iso 3-OH) and summed feature 3, comprising iso-$C_{15:0}$ iso 2-OH and/or $C_{16:1}$ ω7*c* (Table 28).

TABLE 28. Fatty acid content of species of the genus *Aquimarina*[a,b]

Fatty acid	*A. muelleri* LMG 22569[T]	*A. brevivitae* SMK-19[T]	*A. intermedia* KMM 6258[T]	*A. latercula* LMG 1343[T]
$C_{13:0}$ iso	1.5		tr	tr
$C_{14:0}$ iso		1.3		
$C_{15:0}$	1.0	4.5	1.4	5.3
$C_{13:0}$ iso ω6			tr	1.6
$C_{15:0}$ 2-OH				1.1
$C_{15:0}$ iso	22.2	20.2	25.9	18.4
$C_{15:1}$ iso	5.2	5.2	7.8	7.3
$C_{15:0}$ iso 3-OH	9.7	6.9	7.0	7.0
$C_{16:0}$		1.2		
$C_{16:0}$ 3-OH	1.5		tr	2.0
$C_{16:0}$ iso	tr	4.0	tr	2.7
$C_{16:1}$ iso H		1.4	tr	1.4
$C_{16:0}$ iso 3-OH	tr	2.9	tr	2.5
$C_{17:0}$ 3-OH		1.0		1.1
$C_{17:1}$ ω6*c*		1.4	tr	1.6
$C_{17:1}$ iso ω9*c*	9.9	3.7	12.7	4.7
$C_{17:0}$ iso 3-OH	37.6	31.9	32.6	30.5
$C_{18:1}$ ω5*c*			1.0	
Summed feature 3[c]	6.8	3.7	5.0	6.4

[a]Values refer to the percentage of total fatty acids. Fatty acids amounting to less than 1% in all taxa are not given. Data from Yoon et al. (2006b) and Nedashkovskaya et al. (2006f).

[b]tr, less than 1%.

[c]Summed features consist of one or more fatty acids that could not be separated by the Microbial Identification System. Summed feature 3: $C_{15:0}$ iso 2-OH and/or $C_{16:1}$ ω7*c*.

On marine agar 2216 (Difco), strains of *Aquimarina* form regular or rhizoid, round, usually shiny colonies that are sunk into the agar or flat, red, dark-red, orange or yellow-brownish colonies with a diameter of 1–4 mm after 48 h at 28°C.

All isolated strains have been grown on media containing 0.5% peptone and 0.1–0.2% yeast extract (Difco), prepared with natural or artificial seawater or supplemented with 2–3% NaCl. All strains grow at 10–33°C and with 1–5% NaCl.

Strains of *Aquimarina* are susceptible to ampicillin, chloramphenicol, doxycycline, erythromycin, lincomycin, and oleandomycin, but resistant to gentamicin, kanamycin, neomycin, and polymyxin B. The susceptibility to benzylpenicillin, carbenicillin, streptomycin, and tetracycline varies depending on the strain.

The aquimarinas inhabit coastal marine environments. Strains have been isolated from seawater and echinoderms in the temperate latitudes.

Enrichment and isolation procedures

The aquimarinas were isolated by direct plating on marine agar (Difco). Natural or artificial seawater is satisfactory for their cultivation. All isolated strains have been grown on media containing 0.5% peptone and 0.1–0.2% yeast extract (Difco). The strains may remain viable on marine agar (Difco) or other rich media based on natural or artificial seawater from 3 d to several weeks. They can survive storage in marine broth or artificial seawater supplemented with 20% (v/v) glycerol at −80°C for at least 5 years.

Differentiation of the genus *Aquimarina* from other genera

Aquimarina species clearly differ from their closest relatives, members of the genus *Kordia*, by production of flexirubin type pigments and the absence of acid formation from carbohydrates (Nedashkovskaya et al., 2005f, 2006f; Sohn et al., 2004). Moreover, aquimarinas produce catalase and grow with Na^+ ions, in contrast to *Kordia algicida*, which requires not only Na^+, but also Ca^{2+} and Mg^{2+} ions for growth.

Taxonomic comments

The 16S rRNA gene sequence similarities of *Aquimarina intermedia* with *Aquimarina muelleri* and *Aquimarina latercula* are 96.1 and 96.3%, respectively. The sequence similarity between *Aquimarina muelleri* and *Aquimarina latercula* is 95.4%.

These phylogenetic divergences, taken together with a number of differences in phenotypic characteristics and fatty acid composition, enable separation of the representatives of *Aquimarina muelleri* from their nearest neighbor, [*Cytophaga*] *latercula*. The latter was reclassified as [*Stanierella*] *latercula* in the same study. Furthermore, 16S rRNA gene sequence similarities between KMM 6258[T], a novel strain, and its nearest neighbors *Aquimarina muelleri* and [*Stanierella*] *latercula* were 96.3 and 96.4%, respectively. The intermediate phylogenetic position of this isolate supported the placement of KMM 6258[T], *Aquimarina muelleri*, and [*Stanierella*] *latercula* in the same genus as three distinct species, according to criteria published by Stackebrandt and Goebel (1994). Fatty acid analysis of KMM 6258[T] and the type strains of *Aquimarina muelleri* and [*Stanierella*] *latercula*, carried out according to the standard protocol of the Microbial Identification System (Microbial ID) under the same conditions, showed that the cellular fatty acid profiles of all strains tested were consistent with the description of the genus *Aquimarina* (Table 28). The fatty acid composition and the phylogenetic position of [*Stanierella*] *latercula* justified its transfer to the genus *Aquimarina* as *Aquimarina latercula* (Nedashkovskaya et al., 2006f). Because gliding motility and catalase production were not observed among cells of [*Stanierella*] *latercula* in this and previous studies (Nedashkovskaya et al., 2005f; Reichenbach, 1989d), in contrast to other members of the genus *Aquimarina*, an emended description of the genus *Aquimarina* was proposed by Nedashkovskaya et al. (2006f).

Phylogenetic analysis based on 16S rRNA gene sequences indicates that the genus *Aquimarina* forms a cluster with species of the genera [*Gaetbulimicrobium*], *Gelidibacter* and *Algibacter* (Figure 28). The nearest neighbor of *Aquimarina* is the genus [*Gaetbulimicrobium*], consisting of a single species, [*Gaetbulimicrobium*] *brevivitae*, described by Yoon et al. (2006b). A close phylogenetic distance (16S rRNA gene sequence similarity of 96.2%) between the genera *Aquimarina* and [*Gaetbulimicrobium*] justified inclusion of [*Gaetbulimicrobium*] *brevivitae* in the genus *Aquimarina* as *Aquimarina brevivitae*. Similar fatty acid compositions and many common phenotypic traits also supported inclusion of [*Gaetbulimicrobium*] in the genus *Aquimarina* (Table 28).

Differentiation of species of the genus *Aquimarina*

Although species of the genus *Aquimarina* are phylogenetically close to each other and have similar fatty acid compositions (Table 28) and phenotypic characteristics, they can be differentiated by phenotypic traits presented in Table 29.

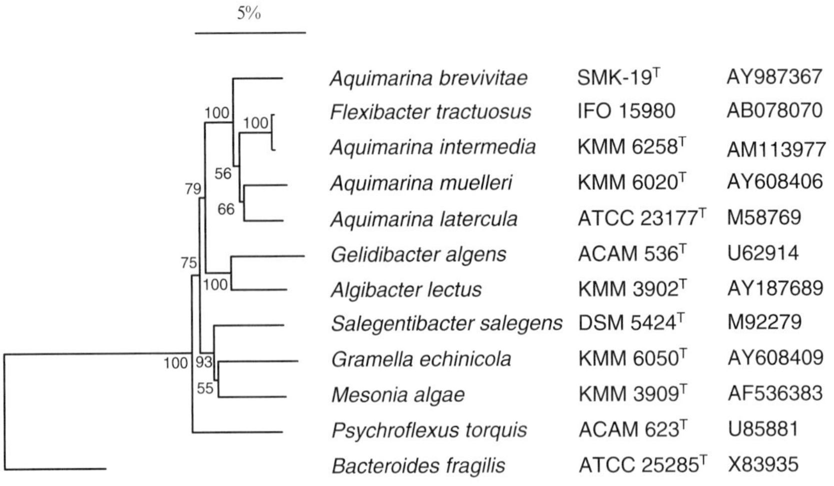

FIGURE 28. Phylogenetic tree of type strains of *Aquimarina* and related species based on 16S rRNA gene sequences.

TABLE 29. Differential characteristics of species in the genus *Aquimarina*[a,b]

Characteristic	*A. muelleri*	*A. brevivitae*	*A. intermedia*	*A. latercula*
Gliding motility	+	+	+	–
Catalase	+	+	+	–
β-Galactosidase	–	–	+	+
Flexirubin type pigments	+	–	+	+
Temperature range for growth (°C)	4–34	10–41	4–36	4–33
Salinity range for growth (%)	1–8	1–10	1–10	1–5
Acid formation from carbohydrates	–	+	–	–
Hydrolysis of:				
Agar	–	–	–	+
Starch	+	+	+	–
Chitin	+	nd	–	+
Utilization of:				
L-Arabinose, D-mannose, sucrose	–	–	+	–
D-Glucose	–	+	+	–
H_2S production	–	–	+	+
Susceptibility to:				
Benzylpenicillin	–	+	+	+
Carbenicillin	+	+	+	–
Streptomycin	–	+	–	+
Tetracycline	+	–	+	–
Maximum absorption of pigments (nm)	445.8	474, 505	469.8	465.0
DNA G+C content (mol%)	31–33	36	37.1	34

[a]Symbols: +, >85% positive; –, 0–15% positive; nd, not detected.

[b]Data from Reichenbach (1989d), Nedashkovskaya et al. (2005f, 2006f), and Yoon et al. (2006b).

List of species of the genus *Aquimarina*

1. **Aquimarina muelleri** Nedashkovskaya, Kim, Lysenko, Frolova, Mikhailov, Lee and Bae 2005f, 227[VP]

mu.el'le.ri. N.L. gen. masc. n. *muelleri* of Müller, named after Otto Friedrich Müller (1730–1784), the famous Danish naturalist, for his contributions to the development of marine microbiology.

Cells are 0.3–0.5 × 5–7 µm. Colonies are irregular shaped, flat with non-entire edges, 3–5 mm in diameter, and dark-yellow- or brown-colored on marine agar. Alginate is not degraded. No acid is produced from arabinose, cellobiose, fucose, galactose, glucose, lactose, maltose, melibiose, raffinose, rhamnose, sucrose, xylose, adonitol, dulcitol, glycerol, inositol, mannitol, or *N*-acetylglucosamine.

Source: seawater collected in Amursky Bay, Gulf of Peter the Great, Sea of Japan.

DNA G+C content (mol%): 31.6–32.5 (T_m).

Type strain: KMM 6020, KCTC 12285, LMG 22569.

Sequence accession no. (16S rRNA gene): AY608406.

2. **Aquimarina brevivitae** (Yoon, Kang, Jung, Oh and Oh 2006b) Nedashkovskaya, Vancanneyt, Christiaens, Kalinovskaya, Mikhailov and Swings 2006f, 2040^{VP} (*Gaetbulimicrobium brevivitae* Yoon, Kang, Jung, Oh and Oh 2006b, 118)

bre.vi.vi′tae. L. adj. *brevis* short; L. gen. n. *vitae* of life; N.L. gen. n. *brevivitae* of a short life, referring to the short-lived cultures of the type strain.

According to Yoon et al. (2006b), cells are 0.2–0.3 × 1.0–15.0 μm. Colonies are circular to irregular, slightly raised, smooth, orange-colored, and 1–2 mm in diameter after cultivation for 48 h at 37°C on marine agar. Carotenoid pigments are produced. Optimal growth is at 37°C, pH 7.0–8.0, and with 2–3% NaCl. Esculin, hypoxanthine, Tween 60, and tyrosine are hydrolyzed, but xanthine is not. Maltose and pyruvate are utilized, but D-cellobiose, D-fructose, D-galactose, trehalose, D-xylose, acetate, citrate, succinate, benzoate, L-malate, salicin, formate, and L-glutamate are not. No acid is produced from L-arabinose, D-fructose, D-galactose, lactose, D-mannose, D-melezitose, melibiose, raffinose, L-rhamnose, D-ribose, sucrose, trehalose, D-xylose, *myo*-inositol, D-mannitol, or D-sorbitol. API ZYM testing indicated the presence of esterase (C4), esterase lipase (C8), leucine arylamidase, valine arylamidase, acid phosphatase, naphthol-AS-BI-phosphohydrolase, and β-glucosidase activities, and the absence of lipase (C14), cystine arylamidase, trypsin, α-chymotrypsin, α-galactosidase, β-glucuronidase, α-glucosidase, *N*-acetyl-β-glucosaminidase, α-mannosidase, and α-fucosidase activities. Susceptible to cephalothin and novobiocin. The main polar lipids are phosphatidylethanolamine, unidentified phospholipids, and an amino-group-containing lipid that is ninhydrin-positive.

Source: tidal flat sediment at Saemankum, Pyunsan, Yellow Sea, Korea.

DNA G+C content (mol%): 36.0 (HPLC).

Type strain: SMK-19, KCTC 12390, DSM 17196.

Sequence accession no. (16S rRNA gene): AY987367.

3. **Aquimarina intermedia** Nedashkovskaya, Vancanneyt, Christiaens, Kalinovskaya, Mikhailov and Swings 2006f, 2040^{VP}

in.ter.me′di.a. L. fem. adj. *intermedia* intermediate, pertaining to the phylogenetic position of the strain studied.

Cells are 0.4–0.5 × 2.1–3.2 μm. Gliding motility occurs. On marine agar, colonies are 2–3 mm in diameter, circular, shiny with entire edges, and reddish. Optimal growth occurs at 25–28°C and with 2–5% NaCl. No acid is produced from arabinose, cellobiose, fucose, galactose, glucose, lactose, maltose, melibiose, raffinose, rhamnose, sorbose, sucrose, xylose, adonitol, dulcitol, glycerol, inositol, mannitol, citrate or *N*-acetylglucosamine.

Source: a single strain was isolated from the sea urchin *Strongylocentrotus intermedius* collected in Troitsa Bay, Gulf of Peter the Great, Sea of Japan.

DNA G+C content (mol%): 37.1 (T_m).

Type strain: KMM 6258, DSM 17527, JCM 13506, LMG 23204.

Sequence accession no. (16S rRNA gene): AM113977.

4. **Aquimarina latercula** (Lewin 1969) Nedashkovskaya, Vancanneyt, Christiaens, Kalinovskaya, Mikhailov and Swings 2006f, 2040^{VP} (*Stanierella latercula* Nedashkovskaya, Kim, Lysenko, Frolova, Mikhailov, Lee and Bae 2005f, 228; *Cytophaga latercula* Lewin 1969, 200)

la.ter′cu.la. L. masc. dim. n. *laterculus* a small brick; N.L. fem. adj. *latercula* brick-like, brick-red color.

Cells are 0.4–0.5 × 2.1–3.2 μm. Gliding movement does not occur. On marine agar, colonies are 1–2 mm in diameter, circular, shiny with entire edges, dark-red-pigmented, and sunk into the agar. Optimal growth is observed at 23–25°C and with about 1.5% NaCl. No acid is produced from arabinose, cellobiose, fucose, galactose, glucose, lactose, maltose, melibiose, raffinose, rhamnose, sorbose, sucrose, xylose, adonitol, dulcitol, glycerol, inositol, mannitol, citrate, or *N*-acetylglucosamine.

Source: a single strain was isolated from the outflow of a marine aquarium, La Jolla, California, USA.

DNA G+C content (mol%): 34.0 (T_m).

Type strain: ATCC 23177, CIP 104806, DSM 2041, JCM 8515, LMG 1343, NBRC 15938, NCIMB 1399.

Sequence accession no. (16S rRNA gene): M58769.

Genus V. **Arenibacter** Ivanova, Nedashkovskaya, Chun, Lysenko, Frolova, Svetashev, Vysotskii, Mikhailov, Huq and Colwell 2001, 1992^{VP} emend. Nedashkovskaya, Vancanneyt, Cleenwerck, Snauwaert, Kim, Lysenko, Shevchenko, Lee, Park, Frolova, Mikhailov, Bae and Swings 2006g, 159

Olga I. Nedashkovskaya and Marc Vancanneyt

Are.ni.bac′ter. L. fem. n. *arena* sand; N.L. masc. n. *bacter* from Gr. n. *baktron* rod; N.L. masc. n. *Arenibacter* a sand-dwelling rod.

Rod-shaped cells, 0.4–0.7 × 1.6–5.0 μm, with slightly irregular sides and rounded ends. **Cells of some species have gliding motility.** On marine agar 2216 (Difco), strains of the genus *Arenibacter* form regular, round, shiny, convex colonies with a diameter of 1–4 mm after 48 h at 28°C. Produce non-diffusible, dark-orange carotenoid pigments. **No flexirubin type pigments are produced. Chemoorganotrophs. Strictly aerobic.** Positive for oxidase, catalase, and alkaline phosphatase. The following enzyme activities are present: acid and alkaline phosphatases, α- and β-galactosidases, α- and β-glucosidases, *N*-acetylglucosaminidase, esterase lipase (C8), leucine-, valine- and cystine-arylamidases, trypsin, α-chymotrypsin, naphthol-AS-BI-phosphohydrolase, α-mannosidase, and α-fucosidase. Arginine dihydrolase, lysine decarboxylase, ornithine decarboxylase, and tryptophan deaminase are not produced. Agar, casein, starch, alginic acids, cellulose (CM-cellulose and filter paper), chitin, and Tween 80 are not attacked, but gelatin, urea, DNA, and Tweens 20 and 40 can be hydrolyzed by strains of some species. Some species require seawater or Na^+ for growth. Acetoin is produced; indole and H_2S are not produced. **The major respiratory quinone is MK-6.**

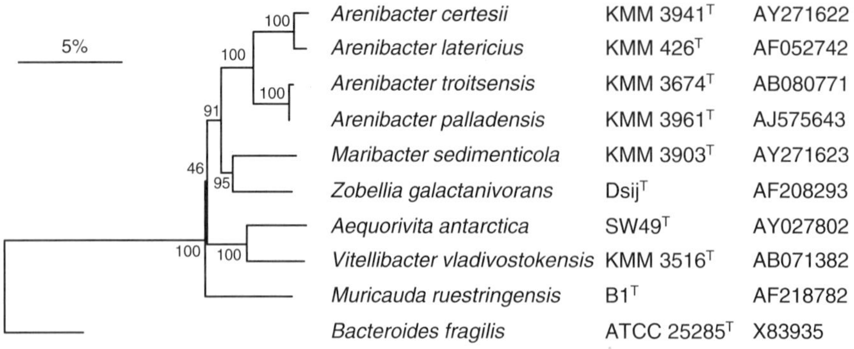

FIGURE 29. Phylogenetic tree of type strains of *Arenibacter* and related species based on 16S rRNA gene sequences.

The main cellular fatty acids are $C_{15:1}$ iso, $C_{15:0}$ iso, $C_{15:0}$, $C_{17:0}$ iso 3-OH, and summed feature 3 comprising $C_{15:0}$ iso 2-OH and/or $C_{16:1}$ ω7*c*. The main polar lipid is phosphatidylethanolamine. Marine, isolated from coastal habitats.

DNA G+C content (mol%): 37–41 (T_m).

Type species: **Arenibacter latericius** Ivanova, Nedashkovskaya, Chun, Lysenko, Frolova, Svetashev, Vysotskii, Mikhailov, Huq and Colwell 2001, 1994[VP] emend. Nedashkovskaya, Vancanneyt, Cleenwerck, Snauwaert, Kim, Lysenko, Shevchenko, Lee, Park, Frolova, Mikhailov, Bae and Swings 2006g, 159.

Further descriptive information

Phylogenetic analysis based on 16S rRNA gene sequences indicated that the genus *Arenibacter* forms a cluster with species of the genera *Maribacter*, *Muricauda*, *Robiginitalea*, and *Zobellia* (Figure 29). The 16S rRNA gene sequence similarities between *Arenibacter* species were 94.8–99.7%. A DNA–DNA hybridization level of 62% was found between *Arenibacter palladensis* KMM 3961[T] and *Arenibacter troitsensis* KMM 3674[T]. The latter two strains had low DNA–DNA hybridization values (6–20%) with *Arenibacter certesii* KMM 3941[T] and *Arenibacter latericius* KMM 426[T].

The lipopolysaccharide of *Arenibacter certesii* KMM 3941[T] consists of α-D-rhamnose, α-mannose, and α-galactosyluronic acid phosphate. Lipid A is made up of a bis-phosphorylated disaccharide unit composed of 2,3-diamino-2,3-dideoxy-β-D-glucopyranose and glucosamine (Silipo et al., 2005).

The cellular fatty acid compositions of *Arenibacter* species are given in Table 30.

TABLE 30. Cellular fatty acid compositions (%) of *Arenibacter* species[a]

Fatty acid	*A. latericius* (*n* = 4)[b]	*A. certesii* KMM 3941[T]	*A. palladensis* KMM 3961[T]	*A. troitsensis* KMM 3674[T]
$C_{15:0}$ iso	6.9–15.8 (8.1)	7.7	8.7	6.8
$C_{15:0}$ anteiso	4.8–13.5 (9.3)	6.3	3.3	3.2
$C_{15:1}$ iso	4.9–14.0 (14.0)	7.2	12.7	12.2
$C_{15:1}$ anteiso	0.6–2.9 (2.9)	0.8	0.5	0.6
$C_{15:0}$	4.2–16.0 (14.2)	11.5	15.0	13.6
$C_{15:1}$ ω6*c*	1.0–2.3 (2.3)	1.9	2.6	1.2
$C_{16:0}$ iso	0.5–1.3 (0.5)	1.7	0.2	0.3
$C_{16:0}$	1.1–2.7 (1.9)	1.0	0.6	1.5
$C_{17:1}$ iso ω9*c*	2.2–4.6 (2.9)	4.7	4.0	5.3
$C_{17:1}$ ω8*c*	0.5–2.0 (1.3)	2.4	0.5	0.9
$C_{17:1}$ ω6*c*	0.7–2.9 (2.4)	3.0	1.4	1.1
$C_{15:0}$ 2-OH	0.7–1.0 (0.6)	0.6	0.4	0.4
$C_{15:0}$ iso 3-OH	4.6–5.7 (5.6)	3.5	5.3	5.1
$C_{15:0}$ 3-OH	0.0–1.4 (0)	0.6	2.2	1.6
$C_{16:0}$ iso 3-OH	2.1–5.7 (2.1)	7.2	1.6	2.2
$C_{16:0}$ 3-OH	0.6–1.3 (1.3)	0.8	2.0	2.2
$C_{17:0}$ iso 3-OH	6.9–14.4 (6.9)	13.3	17.4	21.9
$C_{17:0}$ 2-OH	2.1–5.1 (2.1)	3.8	1.0	1.7
Summed feature 3	9.8–11.9 (9.8)	13.5	11.1	9.6

[a]Only fatty acids accounting for more than 1.0% for one of the strains are indicated. Summed feature 3 consists of one or more of the following fatty acids which could not be separated by the Microbial Identification System: $C_{15:0}$ iso 2-OH, $C_{16:1}$ ω7*c*, and $C_{16:1}$ ω7*t*.

[b]The range of values for four strains is given. Values for the type strain, *Arenibacter latericius* KMM 426[T], are indicated in parentheses.

All isolated strains have been grown on media containing 0.5% peptone and 0.1–0.2% yeast extract (Difco), prepared with natural or artificial seawater or supplemented with 2–3% NaCl. All species grow with 1–6% NaCl and at 10–32°C. Optimal temperature for growth is 23–30°C.

Acid is produced from maltose, D-cellobiose, and sucrose. No acid is produced from adonitol, L-arabinose, dulcitol, inositol, mannitol, or L-sorbose. L-Arabinose, D-glucose, D-lactose, and D-mannose are utilized, but inositol, mannitol, sorbitol, malonate, and citrate are not.

Based on the Biolog GN2 system of testing substrate utilization, the type strains of the four *Arenibacter* species utilize the following substrates: dextrin, D-galactose, gentiobiose, α-D-lactose, maltose, raffinose, sucrose, trehalose, and turanose. Substrates not utilized include α-cyclodextrin, Tween 80, adonitol, D-arabitol, i-erythritol, *myo*-inositol, psicose, xylitol, monomethyl succinate, acetic acid, *cis*-aconitic acid, citric acid, formic acid, D-galactonic acid lactone, D-gluconic acid, D-glucosaminic acid, D-glucuronic acid, α-hydroxybutyric acid, β-hydroxybutyric acid, γ-hydroxybutyric acid, *p*-hydroxyphenylacetic acid, itaconic acid, α-ketoglutaric acid, α-ketovaleric acid, malonic acid, propionic acid, quinic acid, D-saccharic acid, sebacic acid, succinic acid, bromosuccinic acid, succinamic acid, glucuronamide, alaninamide, D-alanine, L-alanyl-glycine, L-asparagine, glycyl-L-glutamic acid, L-histidine, hydroxy-L-proline, L-leucine, L-ornithine, L-phenylalanine, L-pyroglutamic acid, D-serine, D-serine, DL-carnitine, γ-aminobutyric acid, urocanic acid, inosine, uridine, thymidine, phenylethylamine, putrescine, 2-aminoethanol, 2,3-butanediol, and glucose 6-phosphate.

Strains of the genus *Arenibacter* are susceptible to lincomycin, but are resistant to kanamycin, benzylpenicillin, neomycin, streptomycin, gentamicin, and polymyxin B. Susceptibility to ampicillin, erythromycin, carbenicillin, cephaloridin, oleandomycin, tetracycline, and oxacillin differs among species.

The arenibacters inhabit coastal marine environments. They have been isolated from seaweeds and sediments in the temperate latitudes. The strains are accumulators of iodide (Amachi et al., 2005). An ability to produce α-*N*-acetylgalactosaminidase occurs among strains of *Arenibacter latericius* (Ivanova et al., 2001). One of the forms of this enzyme, α-*N*-acetylgalactosaminidase IV, isolated from cells of the type strain KMM 426^T, is able to remove the blood type specificity of human A (II)-erythrocytes, transforming them to O (I)-erythrocytes. The enzyme is active at pH 7–8 and has molecular mass of 84 kDa (Bakunina et al., 2002).

Enrichment and isolation procedures

The arenibacters were isolated by direct platting on marine agar or on medium B containing 0.5% (w/v) Bacto peptone (Difco), 0.2% (w/v) casein hydrolysate (Merck), 0.2% (w/v) Bacto yeast extract (Difco), 0.1% (w/v) glucose, 0.002% (w/v) KH_2PO_4, 0.005% (w/v) $MgSO_4 \cdot 7H_2O$ and 1.5% (w/v) Bacto agar (Difco) in 50% (v/v) natural seawater and 50% (v/v) distilled water. Natural or artificial seawater is suitable for cultivation. All isolated strains have been grown on media containing 0.5% peptone and 0.1–0.2% yeast extract (Difco). Strains of the genus *Arenibacter* may remain viable on marine agar or other rich media based on natural or artificial seawater for several weeks. They have survived storage at −80°C for at least 5 years.

Differentiation of the genus *Arenibacter* from other genera

Species of the genus *Arenibacter* grow at 37°C and form dark-orange colonies on marine agar, in contrast with the yellow-pigmented *Maribacter* species (Ivanova et al., 2001; Nedashkovskaya et al., 2003b) (Nedashkovskaya et al., 2004a, 2004b, 2006g; Yoon et al., 2005b). *Arenibacter* species can be differentiated from those of *Robiginitalea* by their inability to hydrolyze starch and lower DNA G+C content (37–41 mol% vs 55–56 mol%, respectively) (Cho and Giovannoni, 2004). An inability to produce agarase and lack of flexirubin pigments distinguish *Arenibacter* species from those of *Zobellia* (Barbeyron et al., 2001; Nedashkovskaya et al., 2004f; Reichenbach, 1989d). *Arenibacter*, whose cells are regular rods, differs from its close relative, *Muricauda*, which can form appendages on the cell walls (Bruns et al., 2001).

TABLE 31. Differential characteristics of *Arenibacter* species[a]

Characteristic	*A. latericius*	*A. certesii*	*A. palladensis*	*A. troitsensis*
Gliding motility	–	–	+	–
Na^+ requirement for growth	+	+	–	+
Nitrate reduction	+	+	–	+
Degradation of:				
Gelatin	–	–	–	+
DNA	d	–	–	–
Urea	+	+	–	–
Tween 20	d	–	–	–
Tween 40	d	–	d	+
Tween 80	–	–	–	–
Growth with:				
8% NaCl	+	+	+	–
10% NaCl	–	+	+	–
Growth at 42°C	+	–	–	+
Acid production from:				
D-Galactose, D-glucose	+	+	+	–
D-Lactose, L-raffinose	+	+	–	+
Melibiose	d	+	+	–
L-Fucose	d	+	+	+
L-Rhamnose	d	–	+	–
DL-Xylose	–	–	+	–
Glycerol	+	–	–	–
N-Acetylglucosamine	d	+	+	–
Production of:				
Esterase (C4), lipase (C14)	–	–	+	–
β-Glucuronidase	–	–	+	+
Susceptibility to:				
Ampicillin	+	+	–	–
Carbenicillin	+	–	–	–
Oleandomycin	+	+	–	+
Tetracycline	–	–	–	+
DNA G+C content (mol%)	37–38	38	40–41	40

[a]Symbols: +, positive; –, negative; d, differs among strains.

Taxonomic comments

Originally the genus *Arenibacter* was described as consisting of non-gliding bacteria that required Na^+ for growth and were unable to decompose gelatin (Ivanova et al., 2001). However, studies of novel algal isolates and *Arenibacter troitsensis* KMM 3674^T (Nedashkovskaya et al., 2003b, 2006g) indicate that these and other phenotypic traits differ among species and are helpful in differentiating the *Arenibacter* species from one another (Tables 31 and 32).

List of species of the genus *Arenibacter*

1. **Arenibacter latericius** Ivanova, Nedashkovskaya, Chun, Lysenko, Frolova, Svetashev, Vysotskii, Mikhailov, Huq and Colwell 2001, 1994[VP] emend. Nedashkovskaya, Vancanneyt, Cleenwerck, Snauwaert, Kim, Lysenko, Shevchenko, Lee, Park, Frolova, Mikhailov, Bae and Swings 2006g, 159

la.te.ri′ci.us. L. masc. adj. n. *latericius* made or consisting of bricks, here pertaining to the dark-orange pigmentation of the colonies.

Cells are 0.4–0.6 × 2.1–5.0 µm. Growth occurs at 10–42°C and with 1–8% NaCl. According to the Biolog system, L-orni-

TABLE 32. Differences in substrate utilization by the type strains of *Arenibacter* species using the Biolog GN2 testing system[a]

Substrate	*A. latericius* KMM 426[T]	*A. certesii* KMM 3941[T]	*A. palladensis* KMM 3961[T]	*A. troitsensis* KMM 3674[T]
Glycogen, D-galacturonic acid, glycyl-L-aspartic acid, DL-α-glycerol phosphate, D-sorbitol	–	+	–	–
N-Acetyl-D-galactosamine, L-aspartic acid	+	–	–	–
N-Acetyl-D-glucosamine, L-glutamic acid	+	–	+	–
Cellobiose, D-mannose, α-ketobutyric acid	–	–	+	+
D-Fructose, DL-lactic acid	–	+	+	+
L-Fucose	–	–	+	+
Melibiose, methyl β-D-glucoside, methylpyruvate, α-D-glucose, α-lactose	+	–	+	+
L-Alanine	+	+	–	–
L-Proline	–	–	–	+
Glucose 1-phosphate, L-threonine	–	–	+	–

[a]Symbols: +, positive; –, negative.

thine, uridine, glycerol, and glucose 6-phosphate are utilized. D-Fructose, α-D-lactose, lactulose, D-mannitol, succinic acid, glucuronamide, alaninamide, L-alanyl-glycine, L-asparagine, glycyl L-glutamic acid, and D- and L-serine are utilized weakly. Susceptible to erythromycin and cephaloridin. Not susceptible to oxacillin. The acetoin test is positive in API galleries (bioMérieux). Sphingophospholipids are absent.

Source: a sediment sample collected in the South-China Sea and from the holothurian *Apostichopus japonicus* and the brown alga *Chorda filum*, which inhabit Troitsa Bay, Sea of Japan, and the Sea of Okhotsk (The Pacific Ocean), respectively.

DNA G+C content (mol%): 37–38 (T_m).

Type strain: KMM 426, VKM B-2137D, LMG 19693, CIP 106861.

Sequence accession no. (16S rRNA gene): AF052742.

2. **Arenibacter certesii** Nedashkovskaya, Kim, Han, Lysenko, Mikhailov and Bae 2004a, 1174[VP]

cer.te'si.i. N.L. gen. masc. n. *certesii* of Certes, named after A. Certes for his contributions to the development of marine microbiology.

Cells are nonmotile rods, 0.4–0.7 µm wide and 3–5 µm long. Colonies are circular, low convex, shiny with entire edges, and 1–3 mm in diameter on marine agar 2216. No growth is observed without Na^+. Growth occurs with 1–10% NaCl and at 4–38°C. Fumarate is not utilized. Nitrate is reduced.

Source: the single available strain was isolated from the green alga *Ulva fenestrata* collected in Troitsa Bay, Gulf of Peter the Great, Sea of Japan.

DNA G+C content (mol%): 37.7 (T_m).

Type strain: KMM 3941, KCTC 12113, CCUG 48006.

Sequence accession no. (16S rRNA gene): AY271622.

3. **Arenibacter palladensis** Nedashkovskaya, Vancanneyt, Cleenwerck, Snauwaert, Kim, Lysenko, Shevchenko, Lee, Park, Frolova, Mikhailov, Bae and Swings 2006g, 159[VP]

pal.la.den'sis. N.L. masc. adj. *palladensis* pertaining to Pallada Bay, where the strains were isolated.

Cells are 0.4–0.5 × 1.6–2.3 µm. On marine agar, colonies are 2–4 mm in diameter and circular with entire edges. Growth is observed at 4–38°C (optimum, 23–25°C). Growth occurs with 0–10% NaCl. According to the results obtained with the Biolog GN2 MicroPlate system, the type strain utilizes D-fructose, lactulose, and D-mannose. Does not utilize α-cyclodextrin, Tweens 40 and 80, adonitol, L-arabinose, D-arabitol, i-erythritol, *myo*-inositol, D-mannitol, psicose, L-rhamnose, xylitol, monomethyl succinate, acetic acid, citric acid, formic acid, D-galactonic acid lactone, D-gluconic acid, D-glucosaminic acid, D-glucuronic acid, α-and β-hydroxybutyric acids, *p*-hydroxyphenylacetic acid, itaconic acid, α-ketoglutaric acid, α-ketovaleric acid, malonic acid, propionic acid, quinic acid, D-saccharic acid, sebacic acid, succinic acid, bromosuccinic acid, succinamic acid, glucuronamide, alaninamide, L-alanyl-glycine, L-asparagine, glycyl-L-glutamic acid, L-histidine, hydroxy-L-proline, L-leucine, L-ornithine, L-phenylalanine, L-pyroglutamic acid, D- and L-serine, DL-carnitine, γ-aminobutyric acid, uronic acid, inosine, uridine, thymidine, phenylethylamine, putrescine, 2-aminoethanol, 2,3-butanediol, glycerol, or glucose 6-phosphate.

Source: the green alga *Ulva fenestrata* in Pallada Bay, Sea of Japan.

DNA G+C content (mol%): 39–41 (T_m).

Type strain: KMM 3961, LMG 21972, CIP 108849.

Sequence accession no. (16S rRNA gene): AJ575643.

4. **Arenibacter troitsensis** Nedashkovskaya, Suzuki, Vysotskii and Mikhailov 2003b, 1289[VP]

tro.it.sen'sis. N.L. masc. adj. *troitsensis* of or belonging to Troitsa Bay, from where the organism was isolated.

Cells are nonmotile rods, 0.4–0.7 µm wide and 3–5 µm long. Colonies are circular, low convex, shiny with entire edges, 1–3 mm in diameter on marine agar. No growth is observed without Na^+. Grows with 1–6% NaCl and at 10–42°C.

Source: a sediment sample from Troitsa Bay, Gulf of Peter the Great, Sea of Japan.

DNA G+C content (mol%): 40.0 (T_m).

Type strain: KMM 3674, JCM 11736.

Sequence accession no. (16S rRNA gene): AB080771.

Genus VI. **Bergeyella** Vandamme, Bernardet, Segers, Kersters and Holmes 1994a, 830[VP]

The Editorial Board

Ber.gey.el'la. L. dim. suff. *-ella*; N.L. fem. dim. n. *Bergeyella* named in honor of David H. Bergey.

Rods with parallel sides and rounded ends, typically 0.6 × 2–3 µm. **Intracellular granules of poly-β-hydroxybutyrate are absent.** Endospores are not formed. Gram-stain-negative. **Nonmotile.** Do not glide or spread. **Aerobic**, having a strictly respiratory type of metabolism. Grows at 18–42°C. Growth on solid media is not pigmented. Colonies are circular (0.5–2 mm in diameter), low convex, smooth, and shiny with entire edges. **Catalase- and oxidase-positive.** Agar is not digested. Chemo-organotrophic. Nonsaccharolytic. **Indole-positive.** No growth at 42°C, on MacConkey agar, or on β-hydroxybutyrate agar. Grows on blood and chocolate agar but poorly on nutrient agar. Branched-chain fatty acids ($C_{15:0}$ iso, $C_{15:0}$ iso 2-OH, $C_{17:1}$ iso, $C_{17:0}$ iso 3-OH) are predominant. Sphingophospholipids are absent. The major polyamine is homospermidine. *Bergeyella* appears to be part of the oral and/or nasal microbiota of dogs and cats; rare but severe human wound infections can be caused by bites from these animals.

DNA G+C content (mol%): 35–37 (T_m).

Type species: **Bergeyella zoohelcum** (Holmes et al., 1987) Vandamme, Bernardet, Segers, Kersters and Holmes 1994a, 830[VP] (*Weeksella zoohelcum* Holmes, Steigerwalt, Weaver and Brenner 1987, 179; CDC Group IIj).

Further descriptive information

Circular, entire, nonhemolytic, nonpigmented colonies having a butyrous consistency develop on 5% horse blood agar (Holmes et al., 1986b). Only scant growth occurs on nutrient agar.

Gelatin is hydrolyzed by the plate method, but not by the stab method. No change occurs in glucose O/F medium. No growth occurs in the presence of 0.0075% (w/v) KCN. No growth occurs at 5°C, on MacConkey agar, or on cetrimide agar. Nitrate, nitrite, and selenite are not reduced. Esculin, Tween 80, and starch are not hydrolyzed.

The following tests give a negative reaction: oxidation of gluconate; phenylalanine deaminase; H_2S production; alkalinization of Christensen's citrate medium; arginine deamidase; arginine dihydrolase; lysine decarboxylase; ornithine decarboxylase; 3-ketolactose; ONPG hydrolysis; acid from 10% glucose and lactose; acid or gas from glucose; and acid (in ammonium salts medium under aerobic conditions) from glucose, adonitol, arabinose, cellobiose, dulcitol, ethanol, fructose, glycerol, inositol, lactose, maltose, mannitol, raffinose, rhamnose, salicin, sorbitol, sucrose, trehalose, and xylose.

Bergeyella zoohelcum has been found in 38–90% of nasal and oral fluids and gingival scrapings of dogs (Reina and Borrell, 1992). It can be an opportunistic pathogen of humans, acquired from dog or cat bites, causing rare but serious infections in humans such as leg abscesses (Reina and Borrell, 1992), polymicrobial tenosynovitis (Isotalo et al., 2000), septicemia (Kivinen et al., 2003), acute cellulitis (Shukla et al., 2004), and pneumonia (Grimault et al., 1996). *Bergeyella zoohelcum* has usually been thought not to be pathogenic for its animal hosts; however, it has been associated with respiratory disease in a cat (Decostere et al., 2002). *Bergeyella*-like strains have also been isolated from a variety of food sources (Botha et al., 1998).

Goldstein et al. (1999) reported that ten *Bergeyella zoohelcum* isolates were sensitive to gatifloxacin, levofloxacin, sparfloxacin, ciprofloxacin, and grepafloxacin. Kivinen et al. (2003) reported that their *Bergeyella zoohelcum* strain was sensitive to all β-lactam antibiotics, including penicillin and cefuroxim, as well as to aminoglycosides and quinolones; however, it was resistant to cotrimoxazole. A strain isolated from a respiratory infection in a cat was susceptible to ampicillin, amoxicillin-clavulanic acid, enrofloxacin, gentamicin, apramycin, ceftiofur, spectinomycin, neomycin, and trimethoprim-sulfamethoxazole, but was resistant to tetracycline, flumequine, and colistin (Decostere et al., 2002).

Enrichment and isolation procedures

There is no specific medium for the selective isolation of *Bergeyella* strains. Growth occurs on blood and chocolate agar, but may be scanty. The organism is best identified by amplifying and sequencing the 16S rRNA gene and comparing it with that of the type strain.

Differentiation of the genus *Bergeyella* from other genera

The following features differentiate the genus *Bergeyella* from the genus *Weeksella*: positive urease activity, failure to grow at 42°C, and failure to grow on MacConkey agar and β-hydroxybutyrate agar.

Taxonomic comments

Bergeyella zoohelcum strains initially belonged Centers for Disease Control and Prevention group IIj. Holmes et al. (1986b) classified the strains of this group as members of a novel species of the genus *Weeksella*, *Weeksella zoohelcum*. The genus *Weeksella* contained two species: *Weeksella zoohelcum* and the type species *Weeksella virosa*. No significant DNA–DNA hybridization occurred between the two species (Holmes et al., 1986b). rRNA gene analysis later showed that *Weeksella* was heterogeneous and *Weeksella zoohelcum* was placed in a new genus, *Bergeyella*, as *Bergeyella zoohelcum* (Vandamme et al., 1994a). *Bergeyella zoohelcum* is currently the only species in the genus. The nearest related genera based rRNA cistron analysis are *Riemerella* and *Chryseobacterium* (Vandamme et al., 1994a).

List of species of the genus *Bergeyella*

1. **Bergeyella zoohelcum** (Holmes, Steigerwalt, Weaver and Brenner 1987) Vandamme, Bernardet, Segers, Kersters and Holmes 1994a, 830[VP] (*Weeksella zoohelcum* Holmes, Steigerwalt, Weaver and Brenner 1987, 179; CDC Group IIj)

zoo.hel'cum. Gr. n *zoon* an animal; Gr. neut. n. *helkos* a wound; N.L. gen. pl. n. *zoohelcum* of animal wounds, because strains are isolated from cat and dog bites and scratches.

Characteristics are as described for the genus.

Source: oral fluids and gingival scrapings of dogs and human wounds caused by cat and dog bites/scratches.

DNA G+C content (mol%): 35–37 (T_m).

Type strain: D658, ATCC 43767, CCUG 12568, CCUG 30535, CIP 103041, NBRC 16014, LMG 8351, LMG 12996, NCTC 11660.

Sequence accession no. (16S rRNA gene): M93153.

Genus VII. **Bizionia** Nedashkovskaya, Kim, Lysenko, Frolova, Mikhailov and Bae 2005d, 377[VP]

Olga I. Nedashkovskaya, Seung Bum Kim and Valery V. Mikhailov

Bi.zi.o'ni.a. N.L. fem. n. *Bizionia* named after the Italian naturalist Bartolomeo Bizio, for his important contribution to the development of microbiology.

Thin rods usually measuring **0.3–0.5 × 1.0–5.0 µm**. No gliding motility. Produce non-diffusible carotenoid pigments. Flexirubin type pigments are not produced. **Chemo-organotrophs. Strictly aerobic.** Positive for oxidase, catalase, and alkaline phosphatase. Agar, starch, and chitin are not hydrolyzed, but casein and gelatin are hydrolyzed. Nitrate reductase is not produced. No acid is formed from carbohydrates. **Marine**, isolated from coastal habitats. Require seawater or sodium ions for growth. **The major respiratory quinone is MK-6.**

DNA G+C content (mol%): 37–45 (T_m).

Type species: **Bizionia paragorgiae** Nedashkovskaya, Kim, Lysenko, Frolova, Mikhailov and Bae 2005d, 377[VP].

Further descriptive information

16S rRNA gene sequence analysis indicates that *Bizionia* forms a cluster with species of the genera *Formosa* (Figure 30).

Major cellular fatty acids are $C_{15:0}$ iso, $C_{15:0}$ anteiso $C_{15:1}$ iso, $C_{15:1}$ anteiso, $C_{16:1}$ iso, $C_{16:0}$ iso, $C_{16:0}$ iso 3-OH, and summed feature 3, comprising $C_{15:0}$ iso 2-OH and/or $C_{16:1}$ ω7*c*. It should be noted that cells of strain *Bizionia paragorgiae* KMM 6029^T contain 77.1% branched fatty acids.

On marine agar, colonies are 2–4 mm in diameter, circular, shiny, yellow or golden-yellow, and have a butyrous consistency and entire edges after 48 h at 28°C.

Bizionia strains require peptone or yeast extract for growth and cannot use inorganic nitrogen sources such as sodium nitrate or ammonium chloride (Bowman and Nichols, 2005). All isolated strains have been grown on media containing 0.5% peptone and 0.1–0.2% yeast extract (Difco), prepared with natural or artificial seawater or supplemented with 2–3% NaCl. Growth occurs at −2 to 36°C (optimum at 23–25°C). Growth occurs with 1–17% NaCl (optimum at 1.5–6% NaCl).

Strains of the genus are susceptible to ampicillin, carbenicillin, lincomycin, oleandomycin, streptomycin, and tetracycline. They are resistant to benzylpenicillin, gentamicin, kanamycin, neomycin, and polymyxin B.

Representatives of the genus *Bizionia* inhabit coastal marine environments. Some have been isolated from corals in the temperate latitudes. However, most bizionias are cold-adaptive organisms, isolated from Antarctic sea-ice brine and from exoskeletal slime on coastal amphipods.

Enrichment and isolation procedures

Bizionia strains have been isolated by direct plating on marine agar (Difco). Either natural or artificial seawater is suitable for their cultivation. All isolated strains have been grown on media containing 0.5% peptone and 0.1–0.2% yeast extract (Difco). *Bizionia* strains may remain viable on marine agar (Difco) or other rich media based on natural or artificial seawater for several weeks. They have survived storage at −80°C for at least 5 years.

Taxonomic comments

Bizionia species show 96–98% 16S rRNA gene sequence similarity to each other. DNA–DNA hybridization experiments indicate that the species are distinct from one another (Bowman and Nichols, 2005).

Differentiation of the genus *Bizionia* from other genera

Bizionia can be differentiated clearly from its closest relative, *Formosa*, by its absence of gliding motility, its inability to produce acid from carbohydrates, its ability to ferment D-glucose and hydrolyze casein, its requirement for NaCl for growth, and the higher G+C content of its DNA.

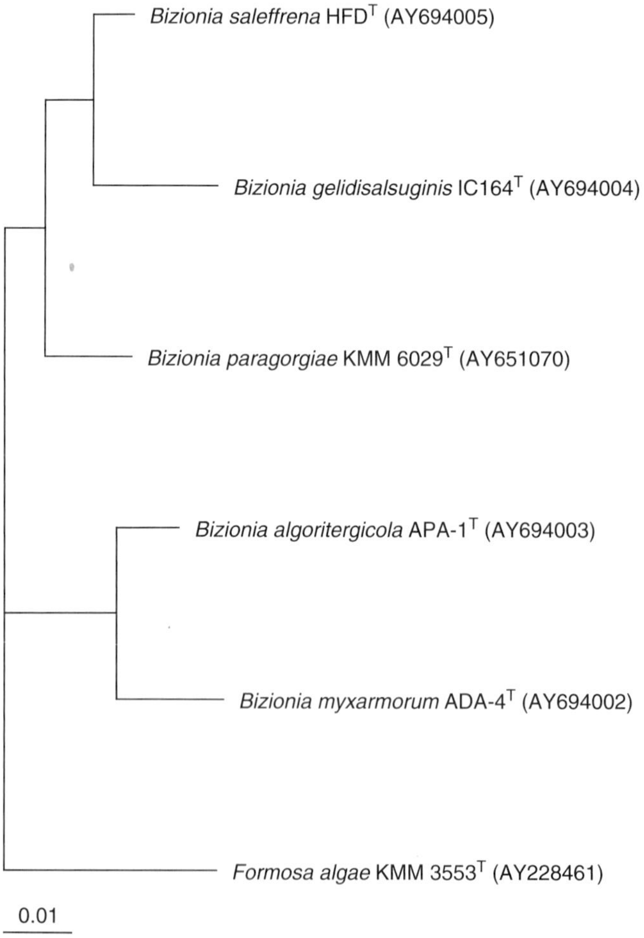

FIGURE 30. Phylogenetic tree based on 16S rRNA gene sequences of the type strains of the genus *Bizionia*. The tree was generated by the neighbor-joining method (Saitou and Nei, 1987). Bar = 0.01 substitutions per nucleotide position.

TABLE 33. Differential characteristics of species of the genus *Bizionia*[a,b]

Characteristic	*B. paragorgiae*	*B. algoritergicola*	*B. gelidisalsuginis*	*B. myxarmorum*	*B. saleffrena*
Arginine dihydrolase	–	+	+	+	–
Salinity range for growth (%)	1–8	1–10	1–15	1–10	1–15
Temperature range (°C)	4 to 36	–2 to 25	–2 to 25	–2 to 30	–2 to 25
Hydrolysis of:					
Esculin	+	–	–	–	–
Elastin	nd	–	–	–	+
DNA	–	+	–	+	–
Tween 80	+	+	–	+	+
Urea	–	+	–	+	+
L-Tyrosine	nd	+	+	+	+
Egg yolk reaction	nd	–	–	+	+
Utilization of:					
Acetate	–	d[c]	+	+	–
Propionate	–	–	+	–	–
Valerate	–	–	–	+	–
L-Alanine	–	–	+	+	–
L-Histidine	–	+	+	+	+
L-Proline	–	–	–	+	–
L-Serine	–	–	–	+	–

[a]Symbols: +, >85% positive; d, different strains give different reactions (16–84% positive); –, 0–15% positive; nd, not determined.

[b]Data from Bowman and Nichols (2005) and Nedashkovskaya et al. (2005d).

[c]Only the type strain is negative.

Differentiation of species of the genus *Bizionia*

Although *Bizionia* species are phylogenetically close to each other and have similar fatty acid compositions and phenotypic characteristics, they can be distinguished by the set of phenotypic traits shown in Table 33.

List of species of the genus *Bizionia*

1. **Bizionia paragorgiae** Nedashkovskaya, Kim, Lysenko, Frolova, Mikhailov and Bae 2005d, 377[VP]

pa.ra.gor'gi.ae. N.L. gen. n. *paragorgiae* of *Paragorgia*, the generic name of the soft coral *Paragorgia arborea*, from which the type strain was isolated.

Cells are 0.4–0.5 × 1.9–2.3 µm. On marine agar, colonies are 2–4 mm in diameter, circular, shiny, and yellow with entire edges. The optimal growth temperature is 23–25°C. Tween 40 is hydrolyzed, but not Tween 20 or agar. H_2S is produced, but not indole or acetoin (Voges–Proskauer reaction). Acid is not formed from L-arabinose, D-cellobiose, L-fucose, D-galactose, D-glucose, D-lactose, maltose, melibiose, raffinose, L-rhamnose, D-sucrose, DL-xylose, citrate, adonitol, dulcitol, glycerol, inositol, or mannitol. Does not utilize L-arabinose, D-glucose, D-lactose, D-mannose, sucrose, mannitol, inositol, sorbitol, malonate, or citrate.

Source: isolated from the soft coral *Paragorgia arborea* collected in the Makarov Bay, Iturup Island, Sea of Okhotsk.

DNA G+C content (mol%): 37.6 (T_m).

Type strain: KCTC 12304, KMM 6029, LMG 22571.

Sequence accession no. (16S rRNA gene): AY651070.

2. **Bizionia algoritergicola** Bowman and Nichols 2005, 1483[VP]

al.go.ri.ter.gi'co.la. L. n. *algor* the cold; L. n. *tergum* outer covering or surface; L. fem. or masc. suff. *-cola* (from L. n. *incola*) the dweller, inhabitant; N.L. fem. or masc. n. *algoritergicola* the inhabitant of a cold surface/covering.

Cells are 0.3–0.5 × 1–3 µm. On marine agar, colonies are circular, shiny, convex, and golden-yellow with a butyrous consistency and entire edges. Optimal growth is observed with 1.7–2.3% NaCl.

Source: isolated from an unidentified marine amphipod collected in East Antarctica.

DNA G+C content (mol%): 45.0 (T_m).

Type strain: APA-1, ACAM 1056, CIP 108533.

Sequence accession no. (16S rRNA gene): AY694003.

3. **Bizionia gelidisalsuginis** Bowman and Nichols 2005, 1482[VP]

gel.id.i.sal.su'gin.is. L. adj. *gelidus* icy; L. fem. n. *salsugo -inis* the brine; N.L. gen. n. *gelidisalsuginis* of icy brine.

Cells are 0.4–0.5 × 1.5–3.5 µm. On marine agar, colonies are circular, convex, butyrous, and golden-yellow with entire edges. Optimum growth occurs with 5.7% NaCl.

Source: isolated from sea-ice brine in East Antarctica.

DNA G+C content (mol%): 39.0 (T_m).

Type strain: IC164, ACAM 1057, CIP 108536.

Sequence accession no. (16S rRNA gene): AY694004.

4. **Bizionia myxarmorum** Bowman and Nichols 2005, 1483[VP]

myx.ar.mor'um. Gr. n. *myxa* slime; L. gen. pl. n. *armorum* of defensive armors; N.L. pl. gen. n. *myxarmorum* of armors slime (of the slime on the carapaces of crustacean hosts).

Cells are 0.3–0.5 × 1.5–3.5 µm. On marine agar, colonies are circular, convex, butyrous, and golden-yellow with entire edges. Optimal growth occurs with 1.5–2.3% NaCl.

Source: isolated from an unidentified marine amphipod collected in East Antarctica.

DNA G+C content (mol%): 43.0 (T_m).

Type strain: ADA-4, ACAM 1058, CIP 108535.

Sequence accession no. (16S rRNA gene): AY694002.

5. **Bizionia saleffrena** Bowman and Nichols 2005, 1482[VP]

sal.ef.fre'na. L. masc. n. *sal salis* salt; L. fem. adj. *effrena* unbridled; N.L. fem. adj. *saleffrena* unbridled by salt (referring to the species' good growth on salt-containing media).

Cells are 0.4–0.5 × 1.5–5 µm. On marine agar, colonies are circular, convex, butyrous, and golden-yellow with entire edges. Optimum growth occurs with 5.7% NaCl.

Source: isolated from sea-ice brine in East Antarctica.

DNA G+C content (mol%): 40.0 (T_m).

Type strain: HFD, ACAM 1059, CIP 108534.

Sequence accession no. (16S rRNA gene): AY694005.

Genus VIII. **Capnocytophaga** Leadbetter, Holt and Socransky 1982, 266[VP] (Effective publication: Leadbetter, Holt and Socransky 1979, 13.)

Stanley C. Holt

Gr. n. *capnos* smoke; N.L. fem. n. *Cytophaga* a bacterial genus name; N.L. fem. n. *Capnocytophaga* bacteria requiring carbon dioxide and related to the cytophagas.

Short to elongate flexible fusiform to rod-shaped cells, 0.42–0.6 µm in diameter and 2.5–5.7 µm in length. Ends of cells are usually round to tapered. Cells can be pleomorphic. Capsules and sheaths not formed. Gram-stain-negative. Resting stages not known. No flagella. Motility when it occurs is by "gliding" on solid culture media (surface translocation). Facultatively anaerobic. Growth occurs in air with 5% CO_2. Some strains are reported to grow aerobically without CO_2; however, **primary isolation and initial growth require CO_2**. Optimum temperature: 35–37°C. Chemo-organotrophic, with fermentative type of metabolism. **Carbohydrates are used as fermentable substrates and energy sources. Fermentation of glucose yields chiefly acetate and succinate as major acidic end products**; trace amounts of isovalerate are formed. Complex polysaccharides such as dextran, glycogen, inulin, and starch may be fermented. Catalase and oxidase reactions differ among species. Positive for *o*-nitrophenyl-β-D-galactoside (ONPG) and benzidine. Found in association with both animal and human hosts. Originally isolated from the human oral cavity where the cells occupied periodontal pockets. Pathogenicity is uncertain, but members of the genus have been implicated as pathogens in several human and animal infections.

DNA G+C content (mol%): 34–44.

Type species: **Capnocytophaga ochracea** Leadbetter, Holt and Socransky 1982, 266[VP] (Effective publication: Leadbetter, Holt and Socransky 1979, 14.).

Further descriptive information

The genus consists of five human oral species: *Capnocytophaga gingivalis, Capnocytophaga ochracea, Capnocytophaga sputigena, Capnocytophaga granulosa,* and *Capnocytophaga haemolytica.* Two additional species, *Capnocytophaga canimorsus* and *Capnocytophaga cynodegmi,* form part of the canine and feline oral microbiota. The genus belongs to the family *Flavobacteriaceae,* class *Flavobacteriia.*

Cells can be single, short or elongate flexible rods, or long filaments (Figure 31). Pleomorphic cells have also been observed when these organisms have been grown in liquid. In liquid culture, cells also grow in characteristic tight masses (Figure 31). When grown on trypticase soy agar (TSA) containing 5% (v/v) sheep's blood (TS-blood agar) prepared by BBL, cells form flat, thin colonies that, when viewed by oblique illumination, have uneven edges (Figures 32–34) that can form fingerlike projections – characteristic of the cells' gliding motility. Gliding motility (surface translocation; Altschul et al., 1997) is consistent with strains of the genus being sensitive to actinomycin D (Dworkin, 1969). Laboratory-prepared TS-blood agar did not support this typical spreading or gliding motility reproducibly; hence, spreading is medium-dependent. Macroscopically, colonies (after incubation for 24 h at 35–37°C) are very small to pinpoint; between 48 and 96 h, colonies are convex and 2–4 mm in diameter. Some colonies appear to pit the agar (Figure 34). Blood is not required for cell growth or gliding motility; the harder the agar surface, the more characteristic the cell spreading or gliding motility [see Leadbetter et al. (1979) for a discussion of the effects of medium and agar concentration on gliding motility]. Cultures have a characteristic sour, bitter almond odor.

The salient staining, growth, genetic and physiological characteristics of the genus *Capnocytophaga* are listed in Table 34. Pigmentation of colonies on an agar surface varies from white–gray to pink and orange–yellow. However, when cells are centrifuged into pellets, all strains are yellow to orange. Pigmentation is not inhibited by the carotenoid inhibitor diphenylamine. The cells are positive for ONPG and benzidine (Collins et al., 1982). Menaquinone-6 and trace amounts of menaquinone-5 constitute the sole respiratory quinones (Collins et al., 1982; Speck et al., 1987).

A detailed study of the fine structure of the genus *Capnocytophaga* was carried out by Holt et al. (1979), Poirier et al. (1979), and Yamamoto et al. (1994). The cell envelope is typical of Gram-stain-negative bacteria (Figure 35). The outer membrane has a tendency to slough material from its surface (Figure 36), which is also typical of Gram-stain-negative

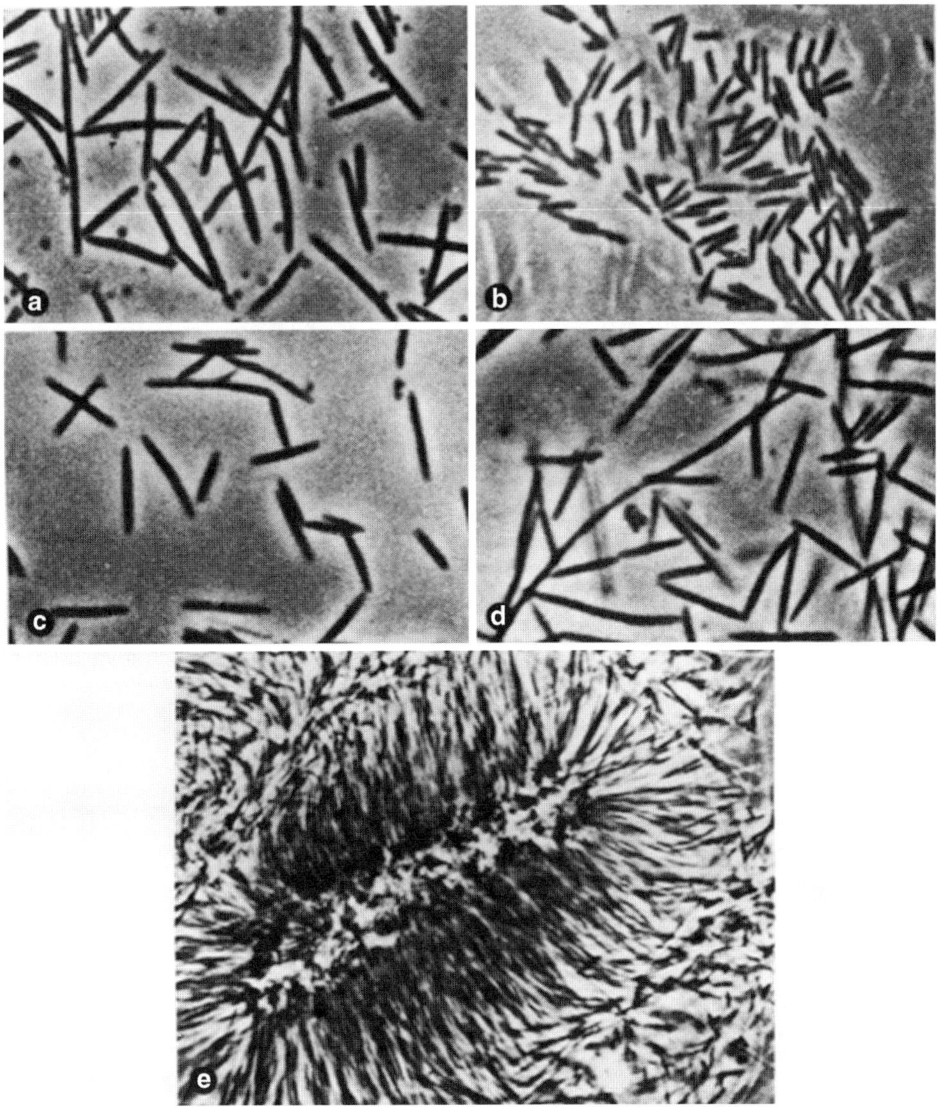

FIGURE 31. Phase-contrast photomicrographs of representative strains of *Capnocytophaga* species. Strains in (a–d) were grown on trypticase soy agar + 5% sheep's blood and that in (e) was grown in trypticase soy broth. (a) *Capnocytophaga sputigena* strain 4^T (=ATCC 33612^T); (b) *Capnocytophaga ochracea* strain 25^T (=ATCC 33596^T); (c) *Capnocytophaga gingivalis* strain 27^T (=ATCC 33624^T); (d) and (e), *Capnocytophaga* strain 60-38.1 grown in liquid and on a plate, respectively. [Reproduced with permission from Holt et al. (1979). Arch. Microbiol. *122:* 17–27, Springer, Berlin.]

micro-organisms. When cells are removed from agar surfaces and prepared for electron microscopy, they are held together by extracellular material (Figure 37), with the outer-most layer of *Capnocytophaga* species staining positively with ruthenium red – presumptive of an acidic mucopolysaccharide layer (Figure 38). Cell division resembles that seen in other Gram-stain-negative bacteria, being characterized by a constriction of the central region of the cell.

There have been several studies reported on the enzymes of *Capnocytophaga* (see Table 35; also Laughon et al., 1982; Nakamura and Slots, 1982; Poirier and Holt, 1983; Lillich and Calmes, 1979; Takeshita et al., 1983). Recently, Ohishi et al. (2005) studied the activity of various inhibitors on the aminopeptidases of *Capnocytophaga granulosa*. *Capnocytophaga* species display significant aminopeptidase activity, exhibiting leucine, valine, and cystine aminopeptidases by the API ZYM system (Laughon et al., 1982; Slots, 1981), as well as being positive for α-glucosidase. Although the API ZYM pattern is distinct for the aminopeptidases, the *Capnocytophaga* species vary enough in the other enzymic reactions tested to make these chromogenic assays of only limited value for correlating API ZYM reactions with key phenotypic characteristics. *Capnocytophaga ochracea* possesses high levels of phosphoenolpyruvate (PEP) carboxykinase, which is the only CO_2 (or HCO_3^-)-fixing enzyme in the genus (Kapke et al., 1980). Cell-free extracts of *Capnocytophaga ochracea* also contain an NAD-specific glutamate dehydrogenase, as well as high levels of acid and alkaline phosphatase. Detectable levels of the malic enzyme, pyruvic acid carboxylase, PEP carboxylase, or PEP carboxytransphosphorylase have not been observed. Glucose is transported into *Capnocytophaga ochracea*

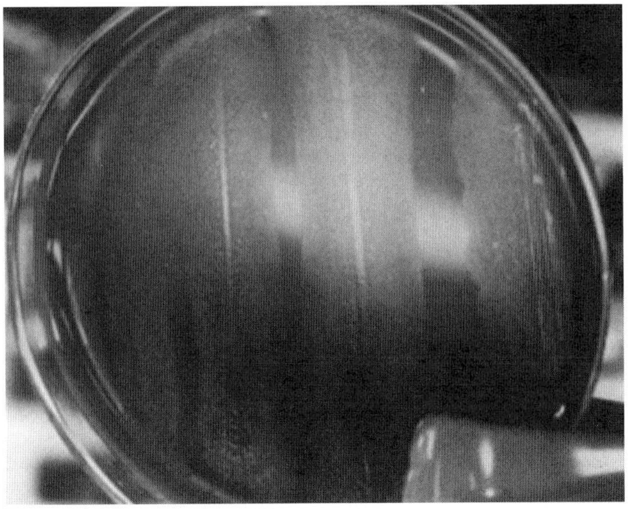

FIGURE 32. *Capnocytophaga* strain grown on blood agar plate and photographed by oblique illumination. Cells spread from a central streak line in opposite directions (courtesy of M. Newman, University of CA School of Dentistry, Los Angeles).

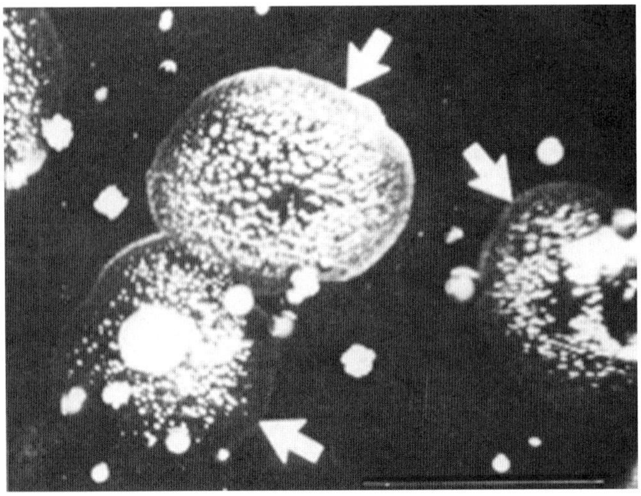

FIGURE 33. Diluted sample of subgingival plaque after 28 h of growth on TS-blood agar prepared by BBL. Arrows: *Capnocytophaga* species displaying characteristic gliding morphology. Bar = 1 cm. (Reproduced with permission from M.G. Newman, V.L. Sutter, M.J. Pickett, U. Blachman, J.R. Greenwood, V. Grinenko and D. Citron. J. Clin. Microbiol. *10:* 557–562 (1977), American Society for Microbiology, Washington, DC.)

by a PEP/phosphotransferase system and then catabolized to pyruvic acid by the Embden–Meyerhof–Parnas pathway (Calmes et al., 1980). Although key enzymes of the Entner–Doudoroff, hexose phosphoketolase, or Warburg–Dickens pathways have not been found, definitive proof of their absence from *Capnocytophaga* will require tracing the distribution of specifically labeled glucose carbons to their specific end products.

Information has been published on the chemical composition of the lipopolysaccharide of *Capnocytophaga* after phenol/water and butanol/water preparative procedures, as well as the biological activity of these molecules (Poirier and Holt, 1983). The lipid chemistry of the genus has also been reported

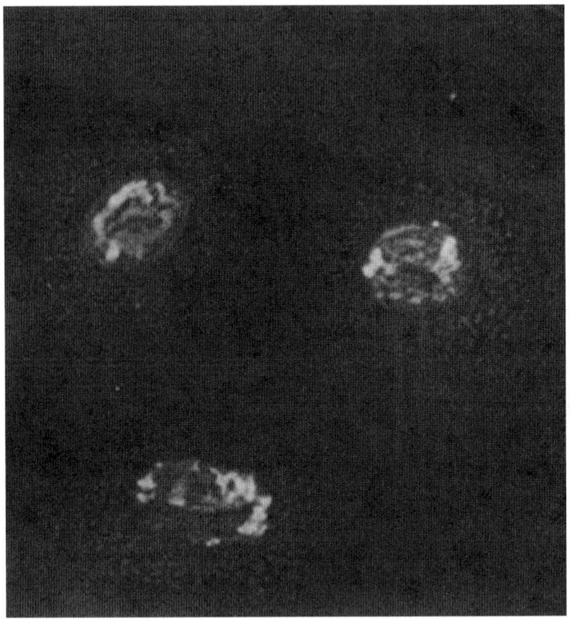

FIGURE 34. *Capnocytophaga* strain grown on blood agar plate. Colonies are flat with concentrically spreading growth from a central point of inoculation. Magnification, ×3.5. [Reproduced with permission from R.E. Weaver, D.G. Hollis and E.J. Bottone (1985). *In* Manual of Clinical Microbiology, 4th edn (edited by Lennette, Balows, Hausler and Shadomy), American Society for Microbiology, Washington, DC.]

TABLE 34. Key characteristics of the genus *Capnocytophaga*[a]

Characteristic	*Capnocytophaga*
Yellow/orange pigmentation	+
Gliding motility	+
Gram reaction	–
Cells fusiform to rod-shaped	+
CO_2 required	+
Bitter almond odor	+
Catalase activity	d
Oxidase activity	d
Nitrate reduction to nitrite	d
Growth at 35–37°C on/in:	+
Blood agar	+
Air	–[b]
Air + 5% CO_2	+
7.5% H_2 + 80% N_2	+
80% N_2 + 10% H_2 + 10 CO_2	+
Thayer–Martin agar	+
Chocolate agar	+
Blood agar + 50 μg/ml bacitracin + 100 μg/ml polymyxin B	+
MacConkey agar	–
Actinomycin D inhibition	+[c]

[a]Symbols: +, >85% positive; d, different strains give different reactions (16–84% positive); –, 0–15% positive.

[b]Some strains are positive after initial cultivation; *Capnocytophaga granulosa* can grow aerobically.

[c]Some strains are resistant.

(Dees et al., 1982; Holt et al., 1979). Cellular fatty acid analysis of strains of *Capnocytophaga* (Dees et al., 1982; Yamamoto et al., 1994; Vandamme et al., 1996b; Table 36) has revealed the predominant fatty acids to be a saturated, iso-branched-chain,

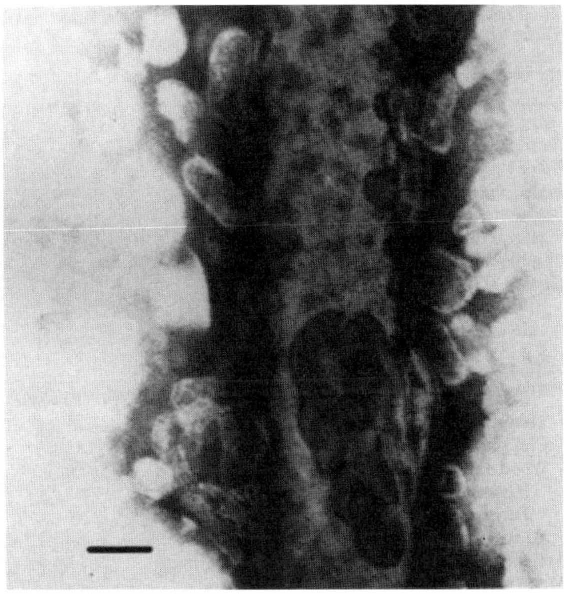

FIGURE 35. Thin section of *Capnocytophaga ochracea* strain 25^T (=ATCC 33596^T). The outer membrane encloses a periplasmic space, with the cytoplasmic membrane limiting the electron-opaque ribosomes in the cytoplasmic region. The peptidoglycan is closely adherent to the outer aspect of the cytoplasmic membrane. An electron-opaque "fuzz" covers the surface of the outer membrane. Stained with uranyl acetate-lead citrate. Bar = 0.25 µm.

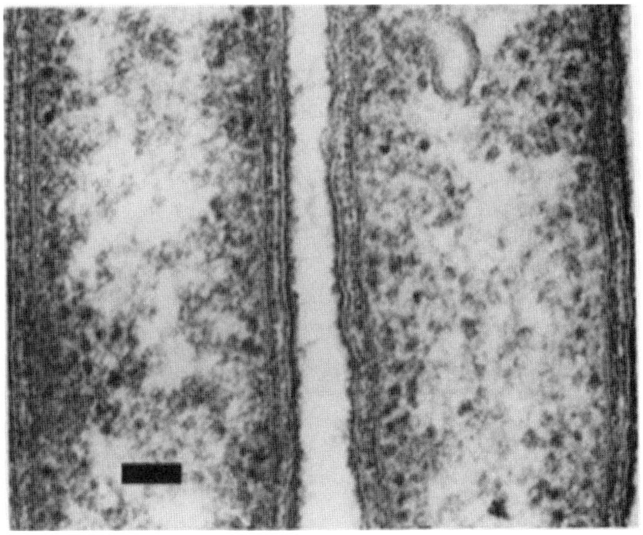

FIGURE 36. Negatively stained (ammonium molybdate at pH 7.2) *Capnocytophaga ochracea* strain 25^T (=ATCC 33596^T). Numerous outer membrane fragments were in the process of being removed from the cell surface. A large mesosome is apparent. Bar = 0.5 µm.

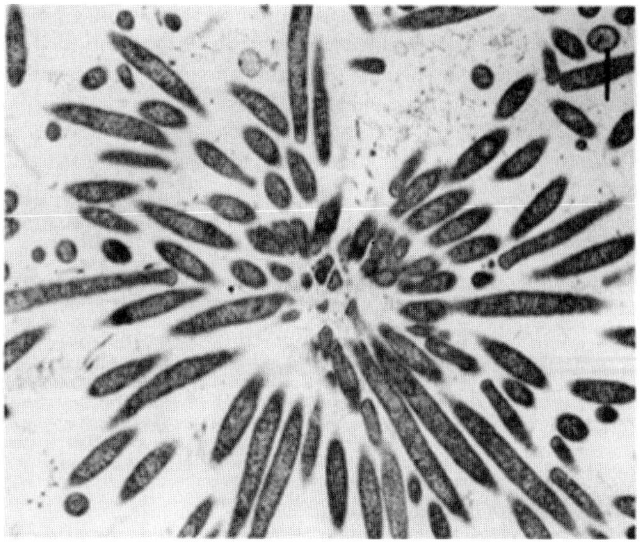

FIGURE 37. Thin section of *Capnocytophaga sputigena* strain 4^T (=ATCC 33612^T). Cells were grown on BBL TS-blood agar at 37°C for 48 h. An agar square was cut from the plate and fixed in gluataldehyde/osmium tetroxide by standard techniques. Cells were thin-sectioned parallel to the cell-covered agar surface. Bar = 2.5 µm.

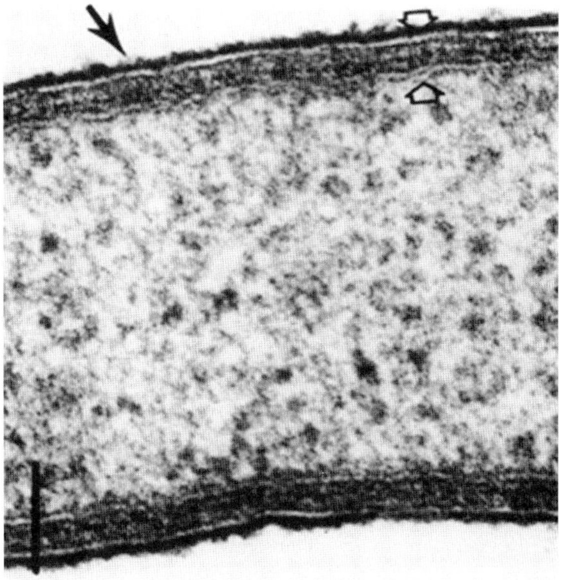

FIGURE 38. Thin section of ruthenium red-fixed *Capnocytophaga ochracea* strain 25^T (=ATCC 33596^T). The ruthenium red has stained the extracellular polysaccharide (arrow). The large periplasmic space (open arrows) encloses the outer membrane, the electron-opaque periplasm, and the inner cytoplasmic membrane. Bar = 0.25 µm.

15 carbon acid (13-methyltetradecanoate) and a saturated, iso-branched-chain 3-hydroxy acid (13-methyl-3-hydroxy-3-hydroxytetradecanoate). Small amounts of 15-methyl-3-hydroxy-hexadecanoate also occur. The major phospholipids are phosphatidylethanolamine and an ornithine-amino lipid, as well as lesser amounts of lysophosphatidylethanolamine. *Capnocytophaga* species also contain the unique sulfonolipid, 1-deoxy-15-methylhexadecasphingamine-l-sulfonic acid, or capnine, which provides a unique characteristic for the genus (Godchaux and Leadbetter, 1980, 1983, 1984).

In vitro agar dilution testing of the sensitivity of *Capnocytophaga* to a variety of antimicrobial agents (Forlenza et al., 1981; Sutter et al., 1981) has been reported (Tables 37 and 38). Essentially all isolates are sensitive at a minimum inhibitory concentration (MIC_{90}) of at least 1 μg/ml to penicillin, ampicillin, carbenicillin, erythromycin, clindamycin, and tetracycline. Doses of tetracycline, metronidazole, cefoxitin, and chloramphenicol of between 1 and 3.12 μg/ml are effective bactericidal agents, killing 90% of strains examined. For *Capnocytophaga* strains, MIC_{90} values for cephalothin and cefazolin were only 25 and 50 μg/ml, respectively, whereas for cefamandole, the MIC_{90} was 3.12 μg/ml. The aminoglycoside antibiotics (i.e., streptomycin, kanamycin, gentamicin, and neomycin) do not inhibit growth of the *Capnocytophaga* species tested, even at concentrations of 50 μg/ml.

Immunological analysis of the antigenic components of *Capnocytophaga* species has revealed the presence of two cell-envelope-associated antigens: a group-specific antigen, which has been found in all species examined so far; and a type-specific antigen, which has been found only in clinical isolates (Murayama et al., 1982; Stevens et al., 1980). An exopolysaccharide has been recovered from the surface of *Capnocytophaga ochracea* strain 25[T], which exhibits immunoregulatory properties upon DEAE-Sepharose CL-SB column purification (Bolton and Dyer, 1983; Dyer and Bolton, 1985).

Capnocytophaga was originally isolated from the human oral cavity; cells occupied periodontal pockets (Leadbetter et al., 1979; Slots and Genco, 1984). Recently, two additional species have been recovered from dog and cat mouths (Brenner et al., 1989; Butler et al., 1977; Ciantar et al., 2001). Although the pathogenicity of these organisms is not yet clear, capnocytophagas have been implicated as pathogens in several human and animal infections (Fung et al., 1983; Gill, 2000; Kristensen et al., 1995; Lion et al., 1996). Apart from their involvement in human oropharyngeal infections, members of the genus have also been implicated in multiple infections causing keratitis, tonsillitis, and stomatitis. Hematogenous spread from the oropharynx has implicated them in osteomyelitis and endocarditis (Kamma et al., 1995). The organisms are frequently recovered from neutropenic patients as endogenous infections or wound infections as a result of animal bites (Lion et al., 1996; Shurin et al., 1979). *Capnocytophaga* infections due to animal bite are usually a result of transmission of *Capnocytophaga canimorsus*; the bacteria can cause clinical manifestations including septicemia, disseminated intravascular coagulopathy, tissue destruction, meningitis, and even death (Fischer et al., 1995; Greene, 1998; Hawkey et al., 1984; Mossad et al., 1997; Valtonen et al., 1995). Fatal complications because of *Capnocytophaga* infection are found in immunocompromised patients or patients suffering from a chronic disease (Parenti and Snydman, 1985). In addition to their being frequent isolates from oral sites, members of the genus are also recovered from pulmonary lesions and abscesses, as well as from healthy oral and non-oral sites in their hosts (van Palenstein Helderman, 1981), and from patients with septic arthritis (Winn et al., 1984).

Capnocytophaga species may be found as part of the normal microbiota of humans and primates. They are routinely recovered from periodontal lesions (Newman and Sims, 1979; Newman and Socransky, 1977), as well as from soft-tissue infections and bacteremias. They are frequently recovered from individuals diagnosed with juvenile periodontitis; although previous

TABLE 35. Enzymic activities of human oral *Capnocytophaga* strains[a,b]

Enzyme	*C. ochracea*	*C. gingivalis*	*C. sputigena*
C4 esterase	1	1	1
C8 esterase lipase	2	1	2
C14 lipase	1	0, 1	0, 1
Leucine aminopeptidase	2	2	2
Valine aminopeptidase	2	2	2
Cystine aminopeptidase	2	2	2
Trypsin	0, 1	0	0, 1
Chymotrypsin	1	0, 1	1
Acid phosphatase	2	2	2
Alkaline phosphatase	2	2	2
Phosphoamidase	1	1	1
α-Galactosidase	0	0	0
β-Galactosidase	1	1	1
β-Glucuronidase	0	0	0
α-Glucosidase	2	2	2
β-Glucosidase	1	1	1
N-Acetyl-β-glucosaminidase	1	0	1
α-Mannosidase	0	0	0
α-Fucosidase	0	0	0
ONPG	2	0	2

[a]Data have been modified from Kristiansen et al. (1984) and Laughon et al. (1982).

[b]Numbers indicate color intensities of chromogenic reactions: 0, no activity; 0, 1, no activity to weak activity; 1, weak activity; and 2, strong activity.

TABLE 36. Cellular fatty acid composition (%) of *Capnocytophaga* species[a,b]

Fatty acid	*C. ochracea*	*C. canimorsus*	*C. cynodegmi*	*C. haemolytica*	*C. gingivalis*	*C. granulose*	*C. sputigena*
$C_{13:0}$ iso	tr–2.1	1.3	1.0	tr	3.5–3.9	1.5–2.1	1.5–2.9
$C_{14:0}$	1.1–1.5	1.7	2.2	tr–21.2	tr–1.0	1.1–1.9	1.1–2.2
$C_{15:0}$ iso	51.5–63.5	66.7	66.5	60.7–69.1	63.5–68.4	72.8–75.5	55.3–68.0
$C_{15:0}$	tr–1.9	nd	nd	tr	1.2–1.6	tr	tr
$C_{16:0}$	3.6–12.6	2.3	2.8	4.9–9.8	1.9–4.1	4.2–5.6	4.2–4.7
$C_{16:0}$ 3-OH	10.0	4.3	4.6	4.9	4.4	5.4	4.4
$C_{17:0}$ iso	1.2–8.1	nd	nd	1.6–8.4	tr–5.7	tr–5.0	2.3–5.9
$C_{18:0}$	12.4	nd	nd	15.3	14.4	10.0	25.8

[a]Data from Yamamoto et al. (1994) and Vandamme et al. (1996b).

[b]Values listed are percentages of the total fatty acids; nd, not determined; tr, <1%.

TABLE 37. Antimicrobial susceptibility of *Capnocytophaga* strains[a]

	MIC (μg/ml)[b]			MBC (μg/ml)[c]		
		For % of strains			For % of strains	
Antimicrobial agent	Range	50	90	Range	50	90
Penicillin	<0.20–0.39	<0.20	<0.20	<0.20–0.78	<0.20	0.20
Ampicillin	<0.20–0.39	<0.20	<0.20	<0.20–0.78	<0.20	0.39
Oxacillin	0.78–25.00	6.25	25.00	6.25–25.00	25.00	25.00
Carbenicillin	<0.50–2.00	<0.50	<0.50	<0.50–2.00	<0.50	1.00
Cephalothin	1.56–>50.00	12.50	25.00	3.12–>50.00	12.50	50.00
Cefazolin	0.78–>50.00	6.25	50.00	3.12–>50.00	12.50	>50.00
Cefamandole	<0.20–25.00	0.78	3.12	0.78–>50.00	3.12	50.00
Cefoxitin	<0.20–6.25	0.39	1.56	<0.20–6.25	0.78	3.12
Erythromycin	<0.20–50.00	<0.20	0.78	<0.20–50.00	<0.20	0.78
Tetracycline	<0.20–1.56	<0.20	0.78	<0.20–1.56	0.20	1.56
Chloramphenicol	<0.20–6.25	0.39	6.25	0.78–12.50	6.25	6.25
Clindamycin	<0.20–0.39	<0.20	<0.20	<0.20–0.39	0.20	<0.20
Metronidazole	<0.25–8.00	<0.25	2.00	<0.25–8.00	0.50	2.00
Vancomycin	<0.20–50.00	3.12	25.00	12.50–>50.00	50.00	50.00
Amikacin	12.50–>50.00	<50.00	>50.00	All–>50.00	>50.00	>50.00
Gentamicin	25.00–>50.00	<50.00	>50.00	All–>50.00	>50.00	>50.00
Tobramycin	All–>50.00	<50.00	>50.00	All–>50.00	>50.00	>50.00

[a]Data are from Forlenza et al. (1981). Thirteen isolates were tested.

[b]MIC, Lowest antibiotic concentration at which turbidity was visible after 48 h (anaerobic incubation).

[c]MBC, Value calculated from places inoculated with 0.01 ml of broth from each clear MIC tube onto BBL TS-blood agar plates. Designated lowest antibiotic concentration at which these plates were devoid of bacterial growth after 48 h (anaerobic incubation).

TABLE 38. Antimicrobial susceptibility of *Capnocytophaga* strains[a]

Antimicrobial agent	Range[b]	MIC_{50}[c]	MIC_{90}[c]
Penicillin G	0.5–2	1	2
Cephalexin	1–128	4	64
Cefaclor	0.5–8	2	8
Cephradine	2–128	4	128
Cefamandole	0.5–64	16	32
Cefoxitin	1–8	2	8
Cefoperazone	0.5–32	8	32
Moxalactam	0.125–8	1	8
Erythromycin	0.125–4	0.5	1
Clindamycin	≤0.062–4	0.062	0.125
Chloramphenicol	2–8	4	8
Metronidazole	2–128	4	16
Tetracycline	0.25–2	0.5	1
Bacitracin	2–16	8	16
Colistin	64–>128	>128	>128
Kanamycin	128–>128	>128	>128
Nalidixic acid	32–>128	>128	>128
Vancomycin	4–128	32	64

[a]Data are from Sutter et al. (1981). Twenty-seven isolates were tested.

[b]Concentrations are in μg/ml, except for penicillin G and bacitracin, which are in units/ml.

[c]MIC_{50} and MIC_{90}, Minimal concentrations required to inhibit 50 and 90% respectively, of the tested strains. Units are as described in footnote b.

reports have indicated some clinical association between the genus and the clinical outcome of the disease in young adults, this has not been confirmed. Members of the genus are also recovered from a variety of lesions and/or abscesses from compromised hosts. The genus has been isolated from patients with hematological malignancy and profound neutropenia. These patients routinely have oral mucosal ulcerations, which may be the portal of entry of *Capnocytophaga* into the bloodstream. *Capnocytophaga* species have also been recovered from sputum and throat samples, spinal fluid, vagina, cervix, amniotic fluid, trachea, and eyes.

Enrichment and isolation procedures

Capnocytophagas can be isolated selectively by streaking onto Thayer–Martin selective medium or on TS-blood agar containing bacitracin (50 μg/ml) and polymyxin B (100 μg/ml) (Mashimo et al., 1983). *Capnocytophaga* species have also been isolated by standard laboratory procedures employing TS-blood agar-based media (Forlenza et al., 1980; Gilligan et al., 1981; Leadbetter et al., 1979), blood agar, and chocolate agar. *Capnocytophaga* species do not grow on MacConkey agar. Rummens et al. (1985, 1986) have reported a selective medium ("CAP") composed of a GC agar supplemented with 1% hemoglobin, 1% Polyvitex, and an antibiotic mixture of polymyxin B (15 U/ml), vancomycin (5 μg/ml), trimethoprim (2.5 μg/ml), and amphotericin (2.5 μg/ml) to be an excellent medium for the recovery of *Capnocytophaga* species from a variety of oral clinical specimens. The plates are incubated for 3–5 d at 35–37°C in an anaerobic chamber (5% CO_2 + 10% H_2 + 85% N_2) or in Brewer jars (under 5% CO_2 + 95% air atmosphere). *Capnocytophaga* species grow as flat, concentrically spreading films from a central colony or point of inoculation (Figures 32–34).

Colonies are removed easily from the agar surface, diluted in sterile phosphate-buffered saline or in a reduced transport fluid (Syed and Loesche, 1972), and restreaked to Thayer–Martin or TS-blood agar. Pure isolates can be obtained by employing this procedure.

Maintenance procedures

Capnocytophaga species can be maintained on TS-blood agar or TS agar. The characteristic spreading morphology is best observed and maintained on BBL TS-blood agar plates or on TS agar [3% (w/v) agar concentration]. After suitable incubation in an anaerobic environment with at least 5% CO_2 or in Brewer jars (5% CO_2/95% air), stock cultures can be maintained on agar slants or Petri dish agar surfaces in Brewer jars at room temperature or in the refrigerator. Cultures can also be frozen in liquid nitrogen or freeze-dried by standard bacteriological procedures.

Taxonomic comments

Capnocytophaga ochracea ATCC 33596^T [strain 25^T of Leadbetter et al. (1979)] was compared to *Bacteroides ochraceus* (VPI 5567, VPI 5568, VPI 5569), and the Centers for Disease Control (CDC) biogroup DF-1 by DNA–DNA hybridization and phenotypic characteristics (Newman and Sims, 1979; Newman and Socransky, 1977; Williams et al., 1979) and most of these strains were designated *Capnocytophaga ochracea* (Leadbetter et al., 1979); one DF-1 strain was designated *Capnocytophaga gingivalis* by Williams et al. (1979). The original isolate of Loesche, *Bacteroides oralis* var. *elongatus* (strain SS31), is extant as *Bacteroides ochraceus* (originally ATCC 27872^T). DNA–DNA hybridization (Williams et al., 1979), serological cross-reactivity (Stevens et al., 1980; B.F. Hammond, personal communication), and phenotypic characteristics (Socransky et al., 1979) have identified it as *Capnocytophaga ochracea.* Yamamoto et al. (1994) isolated two additional *Capnocytophaga* species from human supragingival dental plaque. *Capnocytophaga haemolytica* [Group A of Yamamoto et al. (1994)] was strongly α-hemolytic and aminopeptidase-negative. *Capnocytophaga granulosa* was capable of aerobic growth, was non-hemolytic, and contained granular inclusions.

Acknowledgements

This presentation of the genus *Capnocytophaga* is dedicated to Professor Edward R. Leadbetter. Ed was one of the original discoverers of this genus along with Dr Sigmund Socransky and me. I first met Ed Leadbetter in 1964, after submitting my PhD thesis to Professor Roger Stanier at the University of California for his comments. After the work had been read and approved by Leadbetter, Stanier, John Ingraham, and Gerry Marr (UC-Davis), I was pronounced fit to go out into the cruel world of academics. Through these past 40 years, it has been my fortune to have been associated with Ed Leadbetter. Ed taught me the importance of understanding the microbiology of our environment, was instrumental in introducing me to the Delft School of Microbiology, as well as to the genius that was Kees van Niel, and all those other renowned microbe hunters who found their way to Pacific Grove. Ed taught me to look beyond the surface of biology and to attempt to understand the interrelations of microbes to each other and to us in our complex environment.

Louis Pasteur said: "...chance favors the prepared mind..." – it was Ed Leadbetter who made me realize the true meaning of this phrase.

Differentiation of the species of the genus *Capnocytophaga*

The differential characteristics of *Capnocytophaga* species are listed in Table 39.

TABLE 39. Differentiation of *Capnocytophaga* species[a,b]

Characteristic	*C. ochracea*	*C. canimorsus*	*C. cynodegmi*	*C. gingivalis*	*C. granulosa*	*C. haemolytica*	*C. sputigena*
Aerobic growth	–	–	–	–	–	+	–
Hemolysis	w	–	–	–	–	+	–
Esculin hydrolysis	+	d	(+)	–	–	+	+
Nitrate reduction to nitrite	d	–	d	–	–	+	+
Oxidase	–	+	+	–	–	–	–
Catalase	–	+	+	–	–	–	–
Acid production from:							
Fructose	(+)	d	+	(+)			(+)
Glycogen	d	d	(+)	+	–	d	–
Inulin	+	–	d	+	–	d	+
Melibiose	–	–	+	–			–
Raffinose	+	–	d	+			+
Sucrose	+	–	d	+	+	+	+

[a]Symbols: +, >90% positive at 48 h; (+), >90% positive at 7 d; –, <10% positive at 7 d; w, weak reaction; blank space, not reported.

[b]Data from Brenner et al. (1989) and Yamamoto et al. (1994).

List of species of the genus *Capnocytophaga*

1. **Capnocytophaga ochracea** Leadbetter, Holt and Socransky 1982, 266[VP] (Effective publication: Leadbetter, Holt and Socransky 1979, 14.)

o.chra'ce.a. L. n. *ochra* yellow ochre; N.L. fem. adj. *ochracea* of the color of ochre, yellow.

Characteristics are as described for the genus. The majority (92%) of strains of *Capnocytophaga ochracea* produce acid from lactose. A variable number (11–89%) are capable of fermenting galactose with acid production. Only 8% of the strains are capable of reducing nitrate. The majority of strains analyzed so far produce acid from amygdalin, esculin, and glycogen, as well as hydrolyze starch and dextran.

Source: human supragingival dental plaque.

DNA G+C content (mol%): 38–39 (T_m).

Type strain: 25, ATCC 33596.

Sequence accession no. (16S rRNA gene): X67610, L14635.

Editorial note: in the effective publication of the species by Leadbetter, Holt and Socransky (1979), the type strain is designated ATCC 27872. However, in Validation List no. 8 (IJSB 32: 266–267, 1980), the type strain is designated "strain 25 (=ATCC 33596)". Footnote (c) in Validation List no. 8 states: "Type designated by the author(s) in a personal communication to the editor, IJSB; this announcement establishes the type and satisfies the requirement for valid publication". Footnote (e) states: "Through a clerical error, the wrong strain was cited as the type strain in the original publication; the strain cited here is the actual type (S. Holt, personal communication to the editor, IJSB)".

2. **Capnocytophaga canimorsus** Brenner, Hollis, Fanning and Weaver 1990, 105[VP] (Effective publication: Brenner, Hollis, Fanning and Weaver 1989, 233; CDC group DF-2.)

ca.ni.mor′sus. L. n. *canis* dog; L. n. *morsus* bite, biting; N.L. gen. n. *canimorsus* of a dog bite.

This species was proposed by Brenner et al. (1989) for *Capnocytophaga* strains belonging to CDC Dysgonic Fermenter group 2 (DF-2). The organisms were first described by Bobo and Newton (1976) from a human case of septicemia with meningitis. Butler et al. (1977) extended the original observations by Bobo and Newton and reported a high frequency of association between dog bites in the antecedents and the importance of splenectomy as a predisposing condition. The organism was given the epithet CDC group DF-2 (see above). Brenner et al. (1989) were able to describe similarities between DF-2 and DF-2-like micro-organisms with another closely related bacterium responsible for localized infections due to dog bites and *Capnocytophaga* (originally DF-1) species described by Leadbetter et al. (1979). DF-2 was later found to reside in the oral cavity of dogs and cats. Transmission to humans by bite, scratch, or mere exposure to animals, and their common characteristics with members of the genus *Capnocytophaga* led Brenner et al. (1989) to propose the name *Capnocytophaga canimorsus* for members of group DF-2 and *Capnocytophaga cynodegmi* for the DF-2-like species.

Cells are thin, Gram-stain-negative, nonsporeforming rods, 1–3 µm in length. *Capnocytophaga canimorsus* differs from the other *Capnocytophaga* species by its positive oxidase and catalase reactions, but it does resemble *Capnocytophaga* in other characteristics. For biochemical characteristics of *Capnocytophaga canimorsus* (DF-2 strains), see Brenner et al. (1989).

Source: the type strain was isolated in 1961 from the blood of a male subject. Other strains have been isolated from human cerebrospinal fluid and wounds and the mouths of healthy dogs.

DNA G+C content (mol%): 36–38 (T_m) (type strain).

Type strain: ATCC 35979, CDC 7120, CIP 103936, NCTC 12242.

Sequence accession no. (16S rRNA gene): X97246, L14637.

3. **Capnocytophaga cynodegmi** Brenner, Hollis, Fanning and Weaver 1990, 105[VP] (Effective publication: Brenner, Hollis, Fanning and Weaver 1989, 235.)

cy.no.deg′mi. Gr. *kyon, kynos* dog; Gr. n. *degmos* a bite; N.L. gen. n. *cynodegmi* of a dog bite.

This species was proposed for CDC group DF-2-like organisms. Strains are similar to those of *Capnocytophaga canimorsus* in that they are positive for oxidase and catalase, but resemble *Capnocytophaga* species in other characteristics. For biochemical characteristics of DF-2-like strains, see Brenner et al. (1989).

Source: the type strain was isolated from the mouth of a dog. Other strains have been isolated from human dog-bite wounds.

DNA G+C content (mol%): 34–36 (T_m) (type strain).

Type strain: E6447, ATCC 49044, CCUG 24742, CIP 103937, LMG 11513, NCTC 12243.

Sequence accession no. (16S rRNA gene): X97245, L14638.

4. **Capnocytophaga gingivalis** Leadbetter, Holt and Socransky 1982, 266[VP] (Effective publication: Leadbetter, Holt and Socransky 1979, 14.)

gin.gi.va′lis. N.L. fem. adj. *gingivalis* of the gum.

The physiological and morphological characteristics are as described for the genus (Tables 34, 36, and 39). It does not produce acid end products from lactose or galactose. Only 8% of the *Capnocytophaga gingivalis* strains examined reduced nitrate. *Capnocytophaga gingivalis* is physiologically inactive, compared with *Capnocytophaga ochracea*; it does not ferment amygdalin, salicin, cellobiose, esculin, or glycogen and it does not hydrolyze starch.

Source: human supragingival dental plaque.

DNA G+C content (mol%): 40 (T_m) (type strain).

Type strain: 27, ATCC 33624, CCUG 9715, CIP 102945, DSM 3290, JCM 12953, LMG 11514, NCTC 12372.

Sequence accession no. (16S rRNA gene): X67608, L14639.

5. **Capnocytophaga granulosa** Yamamoto, Kajiura, Hirai and Watanabe 1994, 327[VP]

gra.nu.lo′sa. N.L. fem. adj. *granulosa* granular.

Although sharing the major characteristics of the other *Capnocytophaga* species, this species exhibits high aminopeptidase activity (Table 35) and granular inclusions in its cells (Figure 39). The aminopeptidase enzyme was purified and classified as a metallopeptidase and endopeptidase because it released N-terminal amino acid residues exclusively, but did not hydrolyze benzoyl-Arg-*p*-nitroanilide. The enzyme also differs from that found in the other *Capnocytophaga* species in that it was isolated from the envelope fraction and found to hydrolyze a synthetic substrate for chymotrypsin or trypsin. The molecular mass of the *Capnocytophaga gingivalis* aminopeptidase has been reported to be 64 kDa, whereas

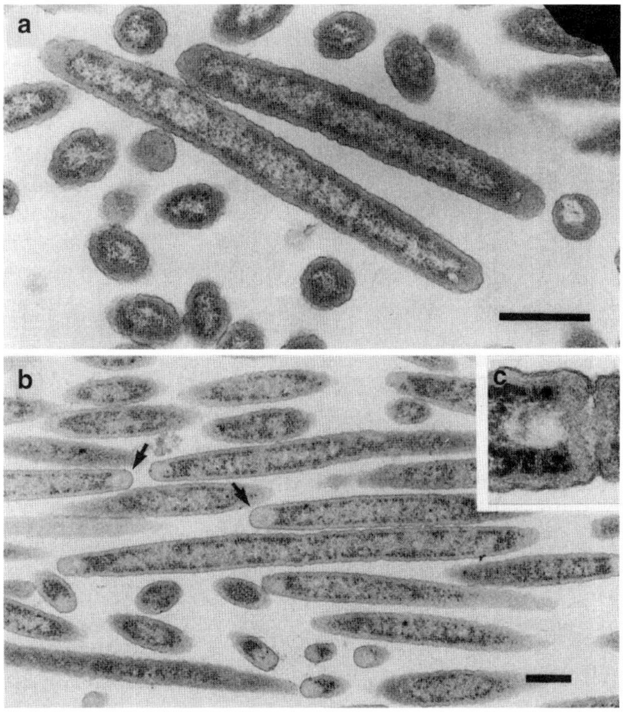

FIGURE 39. Thin section of *Capnocytophaga haemolytica* (a) and *Capnocytophaga granulosa* (b) grown anaerobically in GAM broth at 37°C for 18 h. Note the clear areas representative of inclusion bodies at the cell poles. Bar = 0.5 µm. (Reprinted with permission from Yamamoto et al., 1994, Int. J. Syst. Bacteriol. *44:* 324–329.)

that isolated and purified from *Capnocytophaga granulosa* was larger (86 kDa). The *Capnocytophaga granulosa* aminopeptidase also shared several characteristics with aminopeptidase B (arginyl aminopeptidase), one of the major aminopeptidases in mammalian organs.

Both *Capnocytophaga granulosa* and *Capnocytophaga haemolytica* contain electron-transparent inclusion bodies when grown in GAM broth (Figure 39). The role of these in the physiology of this species is to be determined.

Source: human dental plaque.

DNA G+C content (mol%): 42–44 (T_m) (type strain).

Type strain: B0611, ATCC 51502, CCUG 32991, CIP 104128, DSM 11449, JCM 8566, LMG 16022, NCTC 12948.

Sequence accession no. (16S rRNA gene): X97248.

6. **Capnocytophaga haemolytica** Yamamoto, Kajiura, Hirai and Watanabe 1994, 327[VP]

hae.mo.ly′ti.ca. Gr. n. *haema* blood; Gr. adj. *lyticus -a -um* able to loosen, able to dissolve; N.L. fem. adj. *haemolytica* blood dissolving.

The phenotypic characteristics of this species are as described for the genus *Capnocytophaga*. However, the levels of DNA relatedness between *Capnocytophaga haemolytica* and the other *Capnocytophaga* species (excluding *Capnocytophaga granulosa*) are less than 20%. Characteristics of *Capnocytophaga haemolytica* are its hemolytic activity and the absence of aminopeptidase activity.

Source: human dental plaque.

Type strain: A0404, ATCC 51501, CCUG 32990, CIP 104125, DSM 11385, JCM 8565, LMG 16021, NCTC 12947.

DNA G+C content (mol%): 44 (T_m) (type strain).

Sequence accession no. (16S rRNA gene): X97247.

7. **Capnocytophaga sputigena** Leadbetter, Holt and Socransky 1982, 266[VP] (Effective publication: Leadbetter, Holt and Socransky 1979, 14.)

spu.ti.ge′na. N.L. fem. adj. *sputigena* sputum-produced.

Physiological and morphological characteristics are as described for the genus (Tables 34 and 39). *Capnocytophaga sputigena* possesses a variable ability to ferment lactose, fructose, and dextran. The strains examined were unable to ferment galactose. Of the *Capnocytophaga sputigena* strains examined, 80% hydrolyzed esculin; none of the examined strains were capable of utilizing cellobiose, glycogen, or starch.

Source: human supragingival dental plaque.

DNA G+C content (mol%): 38–40 (T_m) (type strain).

Type strain: 4, ATCC 33612, CCUG 9714, CIP 104301, DSM 7273, LMG 11518, NCTC 11653.

Sequence accession no. (16S rRNA gene): X67609, L14636.

Genus IX. **Cellulophaga** Johansen, Nielsen and Sjøholm 1999, 1238[VP]

John P. Bowman

Cel.lu.lo.pha′ga. N.L. n. *cellulosum* cellulose; Gr. v. *phagein* to eat; N.L. fem. n. *Cellulophaga* eater of cellulose.

Rod-shaped cells with rounded ends, 1.5–5 × 0.4–0.8 μm. **Gram-stain-negative**. Cells occur singly or in pairs, occasionally in chains. Spores and resting cells are not present. Gas vesicles and helical or ring-shaped cells are not formed. **Motile by gliding. Strictly aerobic, with an oxidative type of metabolism. Catalase-positive. Chemoheterotrophic. Colonies are yellow to orange** due to production of carotenoids. **Flexirubin pigments are not produced.** Strains **require Na^+** for good growth. Good growth occurs between 1 and 5% NaCl; growth does not occur in 10% NaCl. Optimum growth occurs in organic media containing seawater salts. All species: **hydrolyze agar, carrageenan, gelatin, starch, and esculin**; **produce esterase** (with Tweens as substrates) **and possess β-galactosidase activity; and decompose L-tyrosine and form dark brown pigments on media containing L-tyrosine.** All species **form acid from carbohydrates** including D-glucose, D-galactose, sucrose, trehalose, and cellobiose. Some species are strongly proteolytic, degrading elastin and fibrinogen. Some species **can attack and lyse living and dead eukaryotic cells**, including yeasts and unicellular algal species. All species grow between 10 and 25°C, with temperature optima depending on the species. Neutrophilic, with optimal growth occurring at pH 7.0–7.5. The major fatty acids are $C_{15:0}$, $C_{15:1}$ ω10*c* iso, $C_{15:0}$ iso, $C_{16:1}$ ω7*c*, $C_{17:1}$ ω7*c* iso, $C_{15:0}$ 3-OH iso, $C_{16:0}$ 3-OH iso, and $C_{17:0}$ 3-OH iso. The major polyamine is **homospermidine**. The major isoprenoid quinone is **menaquinone-6**. The organisms are cosmopolitan inhabitants of marine ecosystems including marine coastal mud and sand, tide pools, seawater (brackish and open ocean), and surfaces of marine benthic, pelagic and sea-ice-associated macroalgae and microalgae.

DNA G+C content (mol%): 32–38 (HPLC, T_m).

Type species: **Cellulophaga lytica** (Lewin 1969) Johansen, Nielsen and Sjøholm 1999, 1239[VP] (*Cytophaga lytica* Lewin 1969, 199).

Further descriptive information

The genus *Cellulophaga* is a member of the family *Flavobacteriaceae*, making up a relatively distinct if somewhat broad phylogenetic sublineage (see *Taxonomic comments*). The genus includes five species, *Cellulophaga lytica*, *Cellulophaga algicola*, *Cellulophaga baltica*, *Cellulophaga fucicola*, and *Cellulophaga pacifica*. *Cellulophaga* species form a single 16S rRNA gene cluster that is composed of two distinct subgroups. *Cellulophaga lytica* and *Cellulophaga fucicola* form one subgroup, while *Cellulophaga algicola*, *Cellulophaga baltica*, and *Cellulophaga pacifica* form the second subgroup. These subgroups differ in terms of their 16S rRNA gene sequences by approximately 5–6%. It would seem that there could be grounds to split the genus into two separate sections; however, at present, this is very difficult to justify owing to the many distinct phenotypic and chemotaxonomic features that *Cellulophaga* species share.

The species form characteristically spreading yellow–orange colonies on marine media. These colonies soften and depress the agar immediately around them due to the secretion of agarolytic exoenzymes. Colonial pigmentation can occur with a greenish metallic iridescence or opalescence and appears to be mainly due to the production of the carotenoid zeaxanthin (Lewin and Lounsbery, 1969). Colonies may appear compact if grown on media with a high nutrient content. Spreading growth and agarolytic activity are more evident on nutrient-poor media, e.g., 0.1% tryptone, 0.1% peptone, and 1.2% agar prepared in seawater. *Cellulophaga* species possess a relatively rapid gliding motility; again, this property is most evident in nutrient-poor situations.

Characteristics that differentiate species of the genus are shown in Table 40. *Cellulophaga* species are generally strongly saccharolytic and are notably active against marine-type polysaccharides, especially agars and alginates, substances commonly produced by various marine macroalgae. The name of the genus is unfortunately not exactly apt because no species in the genus is actually able to degrade cellulose, whether provided in a crystalline powdered form or in the form of paper. However, most species can degrade carboxymethylcellulose, with the exception of *Cellulophaga pacifica*. Nevertheless, *Cellulophaga* species produce numerous exoenzymes that remain, to date, nearly completely unstudied.

Cellulophaga species, like other members of the family *Flavobacteriaceae*, have menaquinone-6 as the major isoprenoid quinone, accumulate primarily the polyamine homospermidine (Hamana and Nakagawa, 2001), and possess primarily branched chain C_{15}–C_{17} fatty acids (Table 41). Strains also tend to accumulate the biogenic amine 2-phenylethylamine (Hosoya and Hamana, 2004), presumably due to the decarboxylation of phenylalanine.

Strains of *Cellulophaga lytica* as well as undescribed *Cellulophaga* isolates have been shown to exhibit predatory behavior on pelagic unicellular algae (Skerratt et al., 2002). This activity appears to be due to the release of lytic enzymes in the vicinity of the host cells. *Cellulophaga baltica* and *Cellulophaga fucicola* also exhibit pronounced lytic activity against autoclaved yeast cells and are strongly proteolytic (Johansen et al., 1999).

Cellulophaga lytica-like bacteria are considered, somewhat anecdotally, to be very common in various marine settings including tidal pools and on the surfaces of macroalgae (especially red algae) species (Lewin, 1969). The temperature profile of *Cellulophaga lytica* (growth occurs between >4°C and 35–40°C, optimum 22–30°C) suggests that it favors temperate to tropical marine ecosystems. *Cellulophaga baltica* and *Cellulophaga fucicola* have a preference for brackish seawater (optimal growth occurs with 20 g/l sea salts) and are at least indigenous to the Baltic Sea, occurring on benthic macroalgae (*Fucus serratus* L. in particular) and in surrounding seawater (Johansen et al., 1999). *Cellulophaga algicola* represents a cold-adapted species in the genus, preferring temperatures between 10 and 20°C, and has been isolated mainly from the surfaces of algae associated with coastal Antarctic sea-ice (Bowman, 2000) *Cellulophaga* species forming biofilms on macroalgae may influence the reproductive colonization success of macroalgal populations (Patel et al., 2003). *Cellulophaga algicola* is noted as producing several cold-adapted exoenzymes (Nichols et al., 1999). Molecular studies suggest *Cellulophaga* is widespread in Antarctic sea-ice (Brown and Bowman, 2001). *Cellulophaga pacifica* was isolated from seawater collected from the Sea of Japan in the north-west Pacific Ocean (Nedashkovskaya et al., 2004e). Several other studies have also identified *Cellulophaga* isolates from diverse marine locations, including toxic dinoflagellate blooms, salmonid gill tissue, and deep-sea sediment (Green et al., 2004; various unpublished sources).

TABLE 40. Differential phenotypic characteristics of *Cellulophaga* species[a]

Characteristic	*C. lytica*	*C. algicola*	*C. baltica*	*C. fucicola*	*C. pacifica*
Growth on:					
Nutrient agar	d (w)	d (w)	–	–	–
Nutrient agar + 8% NaCl	+	+	–	–	+
Growth at:					
–2°C	–	+	–	–	–
4°C	–	+	+	+	+
30°C	+	–	+	+	+
37°C	+	–	–	–	–
Hydrolysis of:					
Chitin	d	–	–	–	
Alginate	–	+	+	+	nd
Carboxymethylcellulose	+	d	+	+	–
Casein	d	+	+	–	–
Elastin	–	–	+	+	nd
Fibrinogen	–	–	+	–	nd
Oxidase activity	+	+	–	–	+
Extracellular DNase	–	d	+	+	–
Lipase (olive oil)	d	d	–	–	nd
Lysis of autoclaved yeast cells	–	–	+	+	–
Nitrate reduction to nitrite	–	+	–	+	+
Urease activity	–	–	–	+	nd
Urate decomposition	–	–	+	+	nd
H_2S from L-cysteine	d	–	nd	nd	–
Acid production from:					
D-Fructose	+	d	+	+	nd
D-Mannose	+	+	+	–	nd
L-Arabinose	d	–	–	+	+
D-Xylose	d	–	+	–	D
N-Acetyl-D-glucosamine	–	d	–	–	–
D-Mannitol	+	–	–	–	–
Maltose	+	+	d	–	+
Lactose	+	+	–	–	+
Glycerol	d	–	–	–	–
L-Rhamnose	d	–	–	+	–
DNA G+C content (mol%)	32–34	36–38	33	32	32–34

[a]Symbols: +, >85% positive; d, different strains give different reactions (16–84% positive); –, 0–15% positive; w, weak reaction; nd, not determined.

Enrichment and isolation procedures

No specific procedure has been established for specific isolation of *Cellulophaga* species. Macroalgal dwelling species can be obtained by homogenizing pieces of seaweed material in sea salts solutions and, after sufficient dilution, spread onto CYT agar (1 g casein, 0.5 g yeast extract, 0.5 g $CaCl_2 \cdot H_2O$, 0.5 g $MgSO_4 \cdot H_2O$, 15 g agar, 1000 ml deionized water, pH 7.3; Johansen et al., 1999) or marine agar (Difco, Oxoid) (Bowman, 2000). Yellow–orange colonies, containing rod-like cells and surrounded by softened agar due to agarolysis, are presumptive for *Cellulophaga* species. Subsequent confirmation using 16S rRNA gene sequencing is generally required because several other members of the family *Flavobacteriaceae* are also capable of agarolytic activity.

TABLE 41. Whole-cell fatty acid contents of members of the genus *Cellulophaga*[a]

Fatty acid	*C. lytica*	*C. algicola*	*C. baltica*	*C. fucicola*	*C. pacifica*[b]
Saturated fatty acids:					
$C_{14:0}$	tr	tr	tr	tr	
$C_{15:0}$	9.3 (2.0)	14.3 (2.4)	12.2	9.8	15.2 (2.7)
$C_{16:0}$	1.7 (1.0)	1.3 (0.5)	3.4	1.2	
Branched-chain fatty acids:					
$C_{15:1}$ ω10*c* iso	10.3 (1.7)	7.5 (2.0)	9.8	9.0	17.9 (4.2)
$C_{15:1}$ ω10*c* anteiso	tr	tr	nd	tr	
$C_{15:0}$ iso	18.9 (3.6)	7.5 (1.8)	13.6	21.4	8.3 (1.9)
$C_{15:0}$ anteiso	1.0 (0.6)	2.5 (1.0)	2.6	1.2	
$C_{16:1}$ ω6*c* iso	1.4 (1.0)	2.2 (1.1)	1.1	1.3	
$C_{16:0}$ iso	tr	2.7 (1.3)	tr	1.4	
$C_{17:1}$ ω7*c* iso	5.1 (0.9)	6.1 (1.5)	5.2	6.7	
$C_{17:1}$ ω7*c* anteiso	1.5 (0.5)	tr	tr	tr	
Monounsaturated fatty acids:					
$C_{15:1}$ ω6*c*	2.5 (0.8)	2.6 (0.6)	2.3	1.4	
$C_{16:1}$ ω7*c*	9.0 (1.5)	19.2 (3.0)	16.9	13.8	13.3 (4.1)
$C_{17:1}$	tr	nd	tr	tr	
Hydroxy fatty acids:					
$C_{15:0}$ 3-OH iso	6.2 (1.9)	8.6 (2.3)	6.1	3.2	
$C_{15:0}$ 3-OH anteiso	1.1 (0.7)	1.5 (1.0)	tr	1.8	
$C_{15:0}$ 3-OH	tr	tr	tr	tr	
$C_{16:0}$ 3-OH iso	5.2 (3.4)	6.5 (1.9)	7.3	5.0	
$C_{16:0}$ 3-OH	tr	1.8 (0.9)	1.1	1.9	
$C_{17:0}$ 3-OH iso	20.8 (4.0)	4.5 (2.0)	14.0	16.9	4.8 (2.7)
$C_{17:0}$ 3-OH anteiso	tr	tr	tr	tr	
Other	1.5 (0.8)	1.3 (0.5)	0.8	1.7	

[a]Values represent percentages of total fatty acids (standard deviations based on results from three experiments are given in parentheses); tr, trace level detected (<1% of total fatty acids); nd, not detected.

[b]Only the major fatty acid constituents were reported in the description of *Cellulophaga pacifica* (Nedashkovskaya et al., 2004e).

Maintenance procedures

Strains can be maintained for many years cryopreserved in marine broth containing 20–30% glycerol. Strains may also be maintained by lyophilization using 20% skim milk as a cryoprotectant.

Differentiation of the genus *Cellulophaga* from other genera

The most salient traits useful for differentiation of *Cellulophaga* spp. from other members of the *Flavobacteriaceae* include a strictly oxidative metabolism, the ability to form acid from carbohydrates, the ability to grow in defined mineral salts media, the lack of flexirubin pigment production, gliding motility, a requirement for Na^+ for growth, and agarolytic activity. *Cellulophaga* species have several similarities to other members of the family *Flavobacteriaceae*, in particular members of the genus *Zobellia* (Barbeyron et al., 2001); however, *Cellulophaga* species can be differentiated from these species primarily by lack of budding cells, lack of flexirubin-type pigments, and strictly aerobic growth (Table 42).

TABLE 42. Differentiation of genus *Cellulophaga* from other Na^+-requiring, agarolytic, yellow-pigmented members of the family *Flavobacteriaceae*[a]

Characteristic	*Cellulophaga*	*Algibacter*	*Formosa*	*Maribacter*	*Winogradskyella*	*Zobellia*
Budding cells	–	–	+	–	–	–
Flexirubins	–	+	–	+	+	+
Anaerobic growth	–	+	+	–	–	–
Growth at 42°C	–	–	–	–	D	D
Nitrate reduction	D	–	D	D	–	+
DNA G+C content (mol%)	32–38	32–36	34–36	35–39	33–36	36–43

[a]Symbols: +, >85% positive; –, 0–15% positive; D, different results occur in different taxa (species of a genus).

Taxonomic comments

As part of his study of marine bacteria, Lewin (1969) described *Cytophaga lytica*. Reichenbach (1989c) opined that there was considerable heterogeneity in the genus *Cytophaga* and *Cytophaga lytica* was identified, along with many other *Cytophaga* species, as being a part of a distinct group of bacteria, probably constituting a separate genus. Indeed, phylogenetic and chemotaxonomic studies (Nakagawa and Yamasato, 1993) showed that *Cytophaga lytica* was quite distinct from other *Cytophaga* species, in particular the *Cytophaga* species *sensu stricto* such as *Cytophaga hutchinsonii*, which belong to a completely separate section of phylum *Bacteroidetes* and are non-marine in terms of their ecology. There was for a while some confusion that *Cytophaga lytica* was possibly closely related to the marine species *Cytophaga marinoflava*. Different 16S rRNA gene sequences present in the GenBank database were identified as giving conflicting phylogenetic relationships for *Cytophaga lytica* and *Cytophaga marinoflava* (Clayton et al., 1995). This was probably due to a strain mix-up. Reappraisal of the data indicated that *Cytophaga marinoflava* was distinct from *Cytophaga lytica* (and other *Cellulophaga* species) phylogenetically, as well as in terms of phenotypic characteristics. *Cytophaga marinoflava* was renamed *Leeuwenhoekiella marinoflava* (Nedashkovskaya et al., 2005i). Johansen et al. (1999) placed *Cytophaga lytica* in the genus *Cellulophaga* along with novel isolates from Baltic Sea samples, which were described as *Cellulophaga baltica* and *Cellulophaga fucicola*. Subsequently, *Cellulophaga algicola* and *Cellulophaga pacifica* were described from other marine samples. At one point, *Cytophaga uliginosa* (ZoBell and Upham, 1944) Reichenbach (1989c) was also transferred to the genus *Cellulophaga* (Bowman, 2000), mainly on the basis of various phenotypic similarities (e.g., cellular and colony morphology, gliding motility, agarolytic activity, and fatty acid content). However, with the subsequent description of several other genera in the family *Flavobacteriaceae*, it quickly became evident that *Cellulophaga uliginosa* belonged to a separate genus and was thus renamed *Zobellia uliginosa* (Barbeyron et al., 2001).

List of species of the genus *Cellulophaga*

1. **Cellulophaga lytica** (Lewin 1969) Johansen, Nielsen and Sjøholm 1999, 1239[VP] (*Cytophaga lytica* Lewin 1969, 199)

ly'ti.ca. L. fem. adj. *lytica* (from Gr. fem. adj. *lutikê* able to loosen, able to dissolve), loosening, dissolving.

Characteristics are as given for the genus and as listed in Table 40. In addition, *Cellulophaga lytica* strains can use peptones, Casamino acids, and L-glutamate as nitrogen sources and can produce ammonia from peptones. Growth can occur on peptone alone.

Source: isolated from coastal sand and mud, tidal pools, and macroalgal samples.

DNA G+C content (mol%): 32–34 (T_m).

Type strain: ATCC 23178, CIP 103822, DSM 7489, IFO 14961, JCM 8516, LMG 1344, VKM B-1433.

Sequence accession no. (16S rRNA gene): M62796.

Additional remarks: 16S rRNA and GyrB gene sequences for several *Cellulophaga lytica* strains have been published by Suzuki et al. (2001) under GenBank accession numbers AB032509–AB032513 (16S rRNA gene sequences) and AB034213–AB034218 (GyrB gene sequences).

2. **Cellulophaga algicola** Bowman 2000, 1866[VP]

al.gi.co'la. L. n. *alga* alga; L. suff. *-cola* (from L. n. *incola*) inhabitant, dweller; N.L. fem. n. *algicola* algae-dweller.

Characteristics are as given for the genus and as listed in Table 40.

Source: isolated from sea-ice algal material and polar macroalgae.

DNA G+C content (mol%): 36–38 (T_m).

Type strain: IC166, ACAM 630.

Sequence accession no. (16S rRNA gene): AF001366.

3. **Cellulophaga baltica** Johansen, Nielsen and Sjøholm 1999, 1238[VP]

bal'ti.ca. N.L. fem. adj. *baltica* of or belonging to the Baltic Sea.

Characteristics are as given for the genus and as listed in Table 40. In addition, on Biolog GN MicroPlates, most strains (>80%) are positive for utilization of D-fructose, L-fucose, D-galactose, α-D-glucose, sucrose, trehalose, methyl pyruvate, α-ketobutyrate, and L-glutamate; some strains tested (20–70%) are positive for L-arabinose, gentiobiose, maltose, D-mannose, psicose, L-rhamnose, formate, γ-hydroxybutyrate, 2-oxoglutarate, glucuronamide, L-proline, and L-threonine. Negative results (<10% of strains) were recorded for the following substrates: dextrin, α-cyclodextrin, glycogen, Tween 40, Tween 80, *N*-acetyl-D-galactosamine, *N*-acetyl-D-glucosamine, adonitol, D-arabitol, cellobiose, i-erythritol, *myo*-inositol, α-D-lactose, lactulose, D-mannitol, melibiose, methyl-β-D-glucoside, raffinose, D-sorbitol, turanose, xylitol, acetate, *cis*-aconitate, citrate, D-galactonate lactone, D-glucosaminate, D-galacturonate, D-gluconate, D-glucuronate, α-hydroxybutyrate, β-hydroxybutyrate, *p*-hydroxyphenylacetate, itaconate, α-ketovalerate, DL-lactate, malonate, monomethylsuccinate, propionate, quinate, D-saccharate, sebacate, bromosuccinate, succinate, succinamate, alaninamide, D-alanine, L-alanine, L-alanylglycine, L-asparagine, L-aspartate, glycyl-L-aspartate, glcyl-L-glutamate, L-histidine, hydroxy-L-proline, L-leucine, L-ornithine, L-phenylalanine, L-pyroglutamate, D-serine, L-serine, DL-carnitine, γ-aminobutyrate, urocanate, inosine, uridine, thymidine, phenylethylamine, putrescine, 2-aminoethanol, 2,3-butanediol, glycerol, glucose 1-phosphate, glucose 6-phosphate, and DL-α-glycerol phosphate.

Source: isolated from the brown alga *Fucus serratus* L. and surrounding seawater of the Baltic Sea.

DNA G+C content (mol%): 33 (HPLC).

Type strain: NN015840, ATCC 700862, CIP 106307, LMG 18535.

Sequence accession no. (16S rRNA gene): AJ005972.

4. **Cellulophaga fucicola** Johansen, Nielsen and Sjøholm 1999, 1238[VP]

fu.ci.co'la. N.L. n. *Fucus* seaweed genus; L. suff. *-cola* (from L. n. *incola*) inhabitant, dweller; N.L. fem. n. *fucicola Fucus*-dweller.

Data are based on a single isolate. Characteristics are as given for the genus and as listed in Table 40. In addition, on

Biolog GN MicroPlates the type strain is positive for utilization of Tween 40, L-arabinose, D-fructose, L-fucose, L-rhamnose, D-sorbitol, methyl pyruvate, and α-ketobutyrate. All other substrates tested were negative (see description of *Cellulophaga baltica* for substrate list).

Source: isolated from the brown alga *Fucus serratus* L. from the Baltic Sea.

DNA G+C content (mol%): 32 (HPLC).

Type strain: NN015860, ATCC 700863, CIP 106308, LMG 18536.

Sequence accession no. (16S rRNA gene): AJ005973.

5. **Cellulophaga pacifica** Nedashkovskaya, Suzuki, Lysenko, Snauwaert, Vancanneyt, Swings, Vysotskii and Mikhailov 2004e, 611[VP]

pa.ci'fi.ca. L. fem. adj. *pacifica* peaceful; referring to the Pacific Ocean, from which the organism was isolated.

Characteristics are as given for the genus and as listed in Table 40.

Source: isolated from seawater collected from the Sea of Japan, north-west Pacific Ocean.

DNA G+C content (mol%): 32 (HPLC), 33 (T_m).

Type strain: JCM 11735, KMM 3664, LMG 21938.

Sequence accession no. (16S rRNA gene): AB100840.

Additional remarks: GenBank accession numbers for the 16S rRNA gene sequence of other *Cellulophaga pacifica* isolates are AB100842 (strain KMM 3915) and AB100841 (strain KMM 3669).

Genus X. **Chryseobacterium** Vandamme, Bernardet, Segers, Kersters and Holmes 1994a, 829[VP]

Jean-François Bernardet, Celia J. Hugo and Brita Bruun

Chry.se.o.bac.te'ri.um. Gr. adj. *chryseos* golden; L. neut. n. *bacterium* a small rod; N.L. neut. n. *Chryseobacterium* a yellow rod.

Straight, single rods with rounded ends typically about 0.5 µm in width and variable in length, often 1–3 µm; filamentous cells and pleomorphism also occur. Do not form endospores. Spherical degenerative forms do not appear in ageing liquid cultures. **Colonies are pale to bright yellow due to the production of non-diffusible, non-fluorescent flexirubin-type pigments.** A strong, aromatic odor is produced. Gram-stain-negative. **Nonmotile. Flagella, gliding movement, or swarming growth have not been reported. Obligately aerobic, having a strictly respiratory type of metabolism** with oxygen as the terminal electron acceptor. **No denitrification occurs and most species do not reduce nitrate or nitrite.** Chemo-organotrophic. Most environmental isolates grow at 5°C, all grow at 15–30°C, and several grow up to 37°C; clinical isolates do not grow at 5°C, but grow at 15–37°C, and some grow at up to 42°C. Grow readily on the usual commercial media; no growth factors are required. Most species also grow in media containing up to 3–5% NaCl. No growth or weak growth on cetrimide agar, but members of a few species are able to grow on MacConkey agar. **Catalase- and oxidase-positive.** Several carbohydrates are oxidized. Esculin is hydrolyzed, but agar is not. Strong proteolytic activity occurs. **Most strains are resistant to a wide range of antimicrobial agents.** Menaquinone-6 (MK-6) is the only or predominant respiratory quinone. Predominant cellular fatty acids are $C_{15:0}$ iso, $C_{17:0}$ iso 3-OH, $C_{17:1}$ iso ω9*c*, and summed feature 4 (comprising $C_{15:0}$ iso 2-OH and/or $C_{16:1}$ ω7*c*/*t*). **Members of most species occur in soil and in fresh or seawater habitats. Two species and unclassified strains also occur in human clinical specimens** and hospital environments, but their pathogenicity to man is not well-documented. Two species and unclassified strains have been retrieved from diseased fish. Four species occur in food or dairy products; they may be involved in spoilage.

DNA G+C content (mol%): 29–39.

Type species: **Chryseobacterium gleum** (Holmes, Owen, Steigerwalt and Brenner 1984b) Vandamme, Bernardet, Segers, Kersters and Holmes 1994a, 830[VP] (*Flavobacterium gleum* Holmes, Owen, Steigerwalt and Brenner 1984b, 23).

Further descriptive information

Phylogenetic position. The genus *Chryseobacterium* is a member of the family *Flavobacteriaceae*, phylum *Bacteroidetes*. It belongs to the smaller of the two well-defined clades in the family comprising non-gliding, non-halophilic, and mostly non-psychrophilic organisms, as well as most unpigmented and several pathogenic members of the family (see the Family *Flavobacteriaceae* in this volume of the *Manual*). Besides *Chryseobacterium*, this clade contains the genera *Elizabethkingia*, *Riemerella*, *Bergeyella*, *Kaistella*, *Sejongia*, *Epilithonimonas*, *Ornithobacterium*, *Empedobacter*, and *Weeksella*, as well as some poorly characterized organisms (Figure 40). Except for the three latter genera that are distinctly separate, all genera appear to be highly related phylogenetically and share a number of phenotypic features. Because of this overall very high 16S rRNA gene sequence similarity among members of the clade, some branches (e.g., the *Sejongia* branch) occupy different positions in published phylogenetic trees. *Chryseobacterium hispanicum* is distinctly separate from the other *Chryseobacterium* species. The two *Elizabethkingia* species were only recently separated from the genus *Chryseobacterium* (Kim et al., 2005b). The uncharacterized organisms "*Haloanella gallinarum*" and "*Candidatus* Amoebinatus massiliae" are located on the *Kaistella* branch (Figure 40); given their nearly identical 16S rRNA gene sequences, they may actually belong to the same species.

Several novel species have been described during the short period of time that has elapsed since the writing of the last review on the genus *Chryseobacterium* (Bernardet and Nakagawa, 2006). At the time of writing (July 2006), the genus comprised 18 species with validly published names; the name of one additional species, "*Chryseobacterium proteolyticum*", has not been validly published. The number of species will likely keep growing since many organisms phylogenetically related to members of the genus *Chryseobacterium* have already been reported from investigations of various environments using both culture-dependent and -independent methods (e.g., Dugas et al., 2001; O'Sullivan et al., 2002; Wery et al., 2003; Drancourt et al., 2004; Greub et al., 2004; Bernardet et al., 2005).

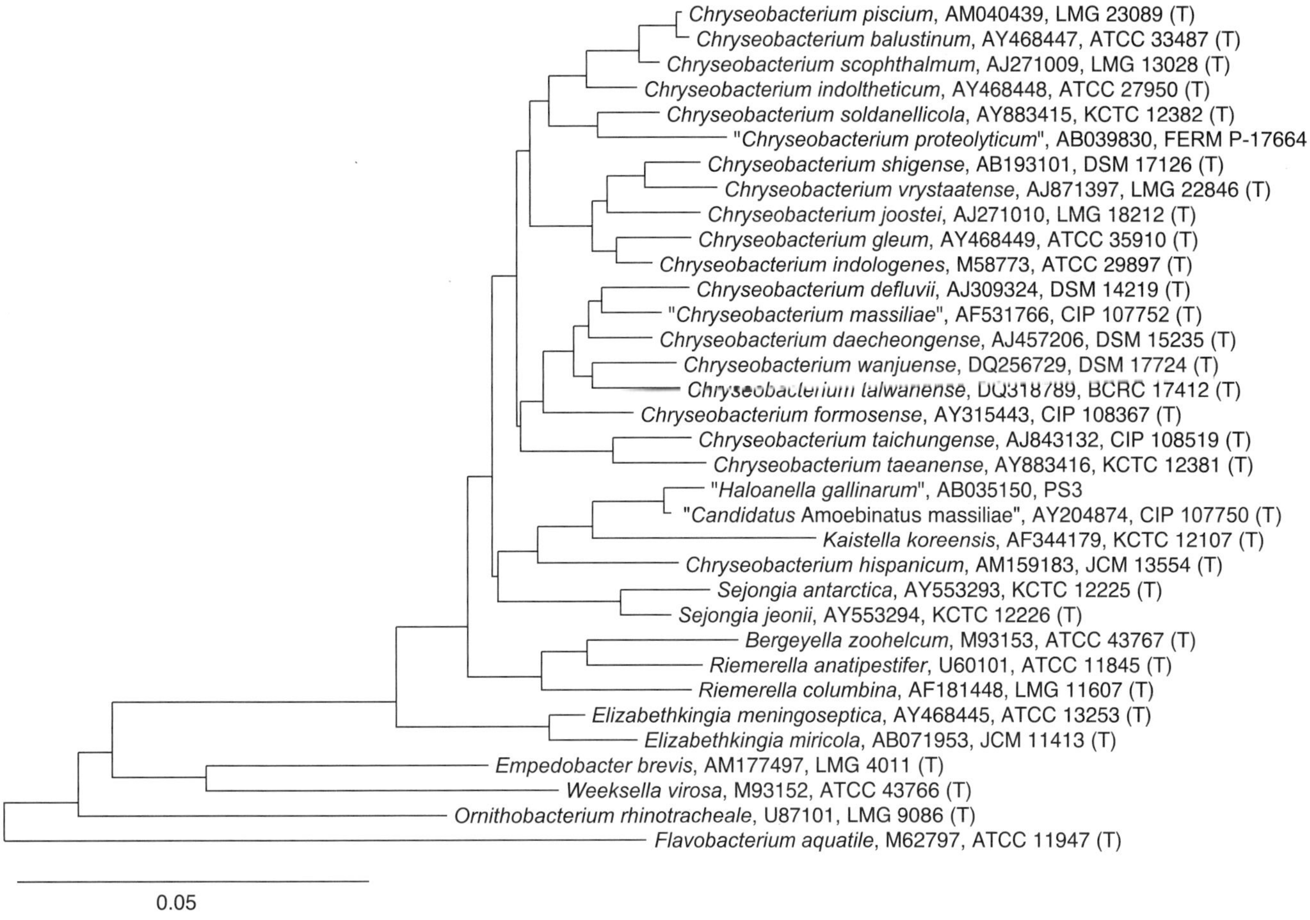

FIGURE 40. Relationships between *Chryseobacterium* species and relatives based on 16S rRNA gene sequences. Only the type strain of each species was included. The sequences of "*Candidatus* Chryseobacterium timonae" (Drancourt et al., 2004) and *Epilithonimonas tenax* (O'Sullivan et al., 2006) were not included in the analysis. GenBank accession numbers are shown. Distances in the tree were calculated using 1164 positions, the ARB neighbor-joining algorithm and a Jukes–Cantor correction. The 16S rRNA gene sequence of *Flavobacterium aquatile* was used as the outgroup. Bar = 0.05 substitutions per nucleotide position. (Courtesy T. Lilburn of the American Type Culture Collection.)

Cell morphology. Members of the genus *Chryseobacterium* commonly appear as small, straight rods, about 0.5 µm wide and 1–3 µm long, with rounded ends and parallel sides. Slightly curved rods have been reported in *Chryseobacterium shigense* (Shimomura et al., 2005) and filamentous forms have been observed in broth cultures of members of several species and in clinical specimens. Interestingly, strains of three *Chryseobacterium* sp. consisting of ultrasmall cells (about 0.2–0.4 µm wide and 0.4–0.7 µm long) have been retrieved recently from the ice core of a Greenland glacier (Miteva and Brenchley, 2005). They may represent intrinsically small organisms or starved, minute forms of normal-sized microbes resulting from extreme environmental conditions (low nutrient concentrations and subzero temperatures). Contrary to the situation in members of some other genera in the family *Flavobacteriaceae* (e.g., *Flavobacterium, Gelidibacter, Tenacibaculum,* and *Polaribacter,* see the family *Flavobacteriaceae* in this volume), *Chryseobacterium* strains do not produce spherical degenerative forms in ageing liquid cultures. Also, in contrast to many members of the family, cells of *Chryseobacterium* species are devoid of gliding motility.

Cell-wall composition. Table 43 presents the fatty acid composition of members of the genus *Chryseobacterium* and related genera. "*Chryseobacterium proteolyticum*" was not analyzed in this respect (Yamaguchi and Yokoe, 2000) and only the dominant fatty acids in *Chryseobacterium shigense* were specified (Shimomura et al., 2005). Although all *Chryseobacterium* species contain the same fatty acids in rather similar proportions, they can be differentiated from other members of the clade (e.g., *Elizabethkingia, Empedobacter, Riemerella,* and *Bergeyella*), which all display rather distinct fatty acid compositions (Bernardet et al., 2006; Hugo et al., 1999). Hence, as for other members of the family *Flavobacteriaceae,* the fatty acid profile is mostly a chemotaxonomic marker at the genus level (Bernardet and Nakagawa, 2006).

Fine structure. Since electron microscopy studies have not been widely performed on members of the genus, little information is available on the fine structure of the cells. Transmission electron microscopy has revealed a thick cell wall (i.e., about 50 nm) in *Chryseobacterium scophthalmum* (Mudarris et al., 1994) and *Chryseobacterium joostei* (Hugo et al., 2003). The scanning electron microscopy studies performed on cells of "*Chryseobacterium proteolyticum*" (Yamaguchi and Yokoe, 2000), *Chryseobacterium daecheongense* (Kim et al., 2005a), *Chryseobacterium shigense* (Shimomura et al., 2005), and the ultrasmall *Chryseobacterium*

TABLE 43. Fatty acid composition (%) of *Chryseobacterium* species and closely related taxa[a]

Fatty acid	*C. gleum* (5)	*C. balustinum* (1)	*C. daecheongense* (1)	*C. defluvii* (1)	*C. formosense* (1)	*C. hispanicum* (1)	*C. indologenes* (45)	*C. indoltheticum* (1)	*C. joostei* (11)	*C. piscium* (4)	*C. scophthalmum* (2)	*C. shigense* (1)	*C. soldanellicola* (1)	*C. taeanense* (1)	*C. taichungense* (1)	*C. taiwanense* (1)	*C. vrystaatense* (7)	*C. wanjuense* (1)	*Elizabethkingia meningoseptica* (5)	*Elizabethkingia miricola* (2)	*Bergeyella zoohelcum* (1)	*Riemerella anatipestifer* (16)	*Riemerella columbina* (13)	*Kaistella koreensis* (3)	*Sejongia antarctica* (1)	*Sejongia jeonii* (1)	*Epilithonimonas tenax* (1)
$C_{12:0}$ iso	nd	nd	nd	nd	nd	nd	nd	nd	nd	nd	nd	na	nd	nd	nd	nd	nd	nd	nd	nd	nd	nd	nd	nd	tr	1	nd
$C_{13:0}$ iso	tr	tr	2	3	4	3	tr	nd	tr	1	tr	na	tr	tr	tr	1	1	tr	1	2	2	15	10	10	3	3	tr
$C_{13:0}$ anteiso	nd	nd	nd	nd	nd	nd	nd	nd	nd	nd	nd	na	nd	nd	nd	nd	nd	nd	nd	nd	nd	nd	2	nd	3	4	nd
$C_{13:1}$	nd	nd	nd	nd	nd	nd	nd	nd	nd	nd	nd	nd	nd	nd	nd	nd	nd	nd	nd	nd	nd	nd	nd	nd	nd	nd	2
$C_{14:0}$ iso	nd	nd	nd	nd	nd	nd	nd	nd	nd	nd	nd	na	nd	nd	nd	nd	nd	nd	nd	nd	nd	nd	1	2	2	5	4
$C_{14:0}$ anteiso	nd	nd	nd	nd	nd	nd	nd	nd	nd	nd	nd	nd	nd	nd	nd	nd	nd	nd	nd	nd	nd	nd	nd	nd	nd	nd	1
ECL 13.566[b]	1	2	2	tr	tr	1	2	2	1	tr	3	na	2	3	7	3	1	3	2	2	2	2	nd	nd	tr	tr	nd
$C_{15:0}$	nd	nd	nd	nd	nd	nd	nd	nd	nd	nd	nd	na	nd	nd	nd	nd	nd	nd	nd	nd	nd	nd	nd	nd	3	2	nd
$C_{15:0}$ 2-OH	nd	nd	nd	nd	nd	nd	nd	nd	nd	nd	nd	na	nd	nd	nd	nd	nd	nd	nd	nd	nd	nd	1	nd	2	2	nd
$C_{15:0}$ iso	35	32	51	59	52	26	34	29	35	38	35	40	42	36	35	43	42	40	44	46	48	53	45	50	14	12	20
$C_{15:0}$ iso 3-OH	3	3	2	3	2	4	3	2	3	2	3	na	3	3	4	4	3	4	3	3	4	8	4	3	1	1	2
$C_{15:0}$ anteiso	tr	tr	1	3	2	4	tr	6	tr	3	tr	na	2	1	tr	1	2	tr	1	1	nd	6	22	13	15	24	15
$C_{15:1}$ anteiso A	nd	nd	nd	nd	nd	nd	nd	nd	nd	nd	nd	na	nd	nd	nd	nd	nd	nd	nd	nd	nd	nd	nd	nd	7	nd	nd
$C_{16:0}$	1	2	2	1	2	2	tr	1	tr	1	1	na	1	2	1	1	1	nd	tr	1	tr	tr	nd	tr	nd	nd	8
$C_{16:0}$ 3-OH	1	1	tr	tr	tr	4	1	tr	1	1	1	na	1	tr	3	tr	1	tr	3	3	nd	nd	nd	nd	nd	nd	6
$C_{16:0}$ iso	nd	nd	nd	nd	nd	nd	nd	nd	nd	nd	nd	na	nd	nd	nd	nd	nd	nd	nd	nd	nd	nd	nd	nd	3	6	nd
$C_{16:0}$ iso 3-OH	nd	tr	tr	tr	1	nd	nd	1	nd	tr	nd	na	tr	tr	1	nd	tr	nd	tr	tr	nd	1	nd	2	5	9	2
$C_{16:1}$ iso H	nd	nd	nd	nd	nd	nd	nd	nd	nd	nd	nd	na	nd	nd	nd	nd	nd	nd	nd	nd	nd	nd	nd	nd	4	9	nd
$C_{16:1}$ ω5*c*	nd	nd	nd	nd	nd	5	nd	nd	nd	nd	nd	nd	nd	nd	nd	nd	nd	nd	nd	nd	nd	nd	nd	nd	nd	nd	nd
ECL 16.580[b]	2	1	1	tr	1	2	2	1	2	1	2	na	2	1	2	tr	1	nd	2	1	1	tr	nd	nd	tr	tr	nd
$C_{17:0}$ 2-OH	nd	tr	nd	tr	nd	nd	nd	3	nd	nd	tr	na	tr	tr	nd	tr	nd	nd	nd	nd	nd	nd	1	nd	3	2	nd
$C_{17:0}$ iso	2	1	3	2	2	nd	tr	tr	tr	1	tr	na	tr	1	tr	1	tr	3	tr	tr	nd	nd	nd	nd	nd	nd	tr
$C_{17:0}$ iso 3-OH	22	17	16	14	11	18	19	14	20	16	16	20	18	19	22	17	15	22	15	15	14	13	7	9	6	4	10
$C_{17:0}$ anteiso 3-OH	nd	nd	nd	nd	nd	nd	nd	nd	nd	nd	nd	nd	nd	nd	nd	nd	nd	nd	nd	nd	nd	nd	nd	nd	nd	nd	3
$C_{17:1}$ iso ω9*c*	20	27	8	5	4	1	24	26	23	19	25	+[c]	15	16	9	18	20	12	8	7	18	nd	nd	6	21	9	1
$C_{17:1}$ anteiso ω9*c*	nd	nd	nd	nd	nd	nd	nd	nd	nd	nd	nd	na	nd	nd	nd	nd	nd	nd	nd	nd	nd	nd	nd	nd	3	2	nd
$C_{18:1}$ ω5*c*	tr	tr	nd	nd	nd	nd	tr	tr	tr	nd	tr	na	tr	tr	nd	nd	tr	nd	tr	tr	2	nd	nd	nd	2	tr	nd
Summed feature 3[d]	na	na	na	na	na	27	na	na	na	11	na	na	na	na	na	8	9	nd	na	na	na	na	na	na	na	na	21
Summed feature 4[d]	12	9	10	8	7	nd	11	11	12	nd	12	+[c]	10	11	14	nd	tr	11	20	17	9	nd	3	3	3	3	nd
Summed feature 5[d]	tr	tr	nd	na	nd	nd	tr	nd	tr	nd	tr	na	tr	tr	nd	tr	na	nd	tr	na	2	nd	nd	nd	nd	nd	nd

[a]The fatty acid composition of "*Chryseobacterium proteolyticum*" has not been determined (Yamaguchi and Yokoe, 2000) and is therefore not included in this table. Fatty acid percentages amounting to less than 1% of the total fatty acids in all species are not included; therefore, the percentages do not total 100%. When several strains were analyzed, rounded up means are given (standard deviations are not given); the number of strains analyzed is given in parentheses after the name of each species. Data are taken from studies using different growth times, media, and temperatures. Abbreviations: tr, trace (less than 1%); na, not available; and nd, not detected. Data from Segers et al. (1993a), Mudarris et al. (1994), Vancanneyt et al. (1999), Hugo et al. (2003), Kämpfer et al. (2003), Li et al. (2003b), Kim et al. (2004), de Beer et al. (2005), Kim et al. (2005a, b), Shen et al. (2005), Shimomura et al. (2005), Yi et al. (2005b), Young et al. (2005), de Beer et al. (2006), Gallego et al. (2006), O'Sullivan et al. (2006), Park et al. (2006b), Tai et al. (2006), and Weon et al. (2006).

[b]ECL, equivalent chain-length (i.e., the identity of the fatty acids is unknown).

[c]According to Shimomura et al. (2005), "Considerable proportions of summed feature 4 and $C_{17:1}$ iso ω9*c* are also present."

[d]Fatty acids that could not be separated by GC using the Microbial Identification System (Microbial ID) software were considered summed features. Summed feature 3 contained $C_{15:0}$ iso 2-OH and/or $C_{16:1}$ ω7*c*. Summed feature 4 contained $C_{15:0}$ iso 2-OH and/or $C_{16:1}$ ω7*c*/*t*. Summed feature 5 contained $C_{17:1}$ iso I and/or $C_{17:1}$ anteiso B.

strains retrieved from a Greenland glacier (Miteva and Brenchley, 2005) have not revealed any special structure.

Colonial or cultural characteristics. Colonies of *Chryseobacterium* strains are circular, convex, translucent to opaque, smooth, and shiny with entire edges. On commonly used agar media (blood, tryptic soy, R2A, and nutrient), colonies are usually bright-yellow-pigmented owing to the production of flexirubin-type pigments (Bernardet et al., 2002). The type of yellow pigments in *Chryseobacterium defluvii*, *Chryseobacterium soldanellicola*, and *Chryseobacterium taeanense* has not been determined, but they most likely belong to the flexirubin type. Colonial pigmentation of most of the recently described species (i.e., *Chryseobacterium defluvii*, *Chryseobacterium daecheongense*, *Chryseobacterium formosense*, *Chryseobacterium taichungense*, *Chryseobacterium taiwanense*, *Chryseobacterium soldanellicola*, and *Chryseobacterium taeanense*) has been described as "yellowish" or "white–yellow". Indeed, the production of pigment is highly dependent on growth temperature, the presence of daylight, and composition of the culture medium (Bruun, 1982; Holmes et al., 1984a; Hugo et al., 2003). A sentence in the original description of the genus (Vandamme et al., 1994a) stated that non-pigmented strains occurred; this actually referred to the strains of *Elizabethinkingia meningoseptica* that have recently been removed from the genus *Chryseobacterium* (Kim et al., 2005b). When grown on tyrosine-containing agar, strains of *Chryseobacterium balustinum*, *Chryseobacterium gleum*, *Chryseobacterium indoltheticum*, and *Chryseobacterium joostei* and some strains of *Chryseobacterium scophthalmum* also produce a pinkish-brown to dark brown, diffusible melanic pigment (Hugo et al., 2003; Mudarris et al., 1994); J.-F. Bernardet, unpublished data); this trait was not investigated for recently described species. After prolonged incubation, colonies of some *Chryseobacterium* species (i.e., *Chryseobacterium balustinum*, *Chryseobacterium daecheongense*, *Chryseobacterium formosense*, *Chryseobacterium joostei*, *Chryseobacterium soldanellicola*, and *Chryseobacterium taeanense*) display a mucoid consistency; this may be related to the production of extracellular slimy substances, as reported for *Chryseobacterium defluvii* and *Chryseobacterium formosense*. Broth cultures are usually uniformly turbid. *Chryseobacterium* strains give off a strong aromatic odor.

Nutrition and growth conditions. *Chryseobacterium* strains are not fastidious. They grow readily on common commercial media such as blood, tryptic soy, and nutrient agars, and the R2A medium of Reasoner and Geldreich (1985). Growth in corresponding broths is also good. The original description of the genus *Chryseobacterium*, containing only six species at the time, stated that all strains grew at 30°C, and that most strains grew at 37°C (Vandamme et al., 1994a). However, several of the novel species described subsequently showed psychrotolerant growth at 5°C: *Chryseobacterium joostei* (Hugo et al., 2003); *Chryseobacterium vrystaatense* (de Beer et al., 2005); *Chryseobacterium daecheongense* (Kim et al., 2005a); *Chryseobacterium soldanellicola* and *Chryseobacterium taeanense* (Park et al., 2005); *Chryseobacterium shigense* (Shimomura et al., 2005); *Chryseobacterium piscium* (de Beer et al., 2006); *Chryseobacterium hispanicum* (Gallego et al., 2006); *Chryseobacterium taiwanense* (Tai et al., 2006); and *Chryseobacterium wanjuense* (Weon et al., 2006). A *Chryseobacterium* strain isolated from forest soil in Northern Finland was able to grow at 0°C and to degrade Tween 80 at 5°C (Männistö and Haggblom, 2006). Possible novel *Chryseobacterium* species have also been isolated from Antarctic soil using a packed-column bioreactor maintained at 10°C (Wery et al., 2003). When soil samples from Canada were subjected to 48 freeze–thaw cycles in a cryocycler, the surviving fraction of the bacterial community displayed a considerable increase in freeze–thaw tolerance; among these viable bacteria, a *Chryseobacterium* sp. (identified by 16S rRNA gene sequencing) was able to inhibit ice recrystallization (a characteristic of antifreeze proteins) and the supernatant of its culture enhanced the freeze–thaw tolerance of an *Enterococcus* strain (Walker et al., 2006).

Although only three of the 19 *Chryseobacterium* species have been retrieved from marine environments (i.e., *Chryseobacterium indoltheticum*, *Chryseobacterium piscium*, and *Chryseobacterium scophthalmum*), most other species are also halotolerant and able to grow in the presence of 3–5% NaCl (w/v), for instance on marine agar 2216E (Difco). This is even true for many strains of the clinical organisms *Chryseobacterium gleum* and *Chryseobacterium indologenes*, which are also environmental species. Conversely, strains of the three marine species are able to grow on NaCl-free media such as tryptic soy agar (de Beer et al., 2006; Mudarris et al., 1994) J.-F. Bernardet, unpublished data).

Metabolism and metabolic pathways. These aspects have not been investigated extensively in members of the genus *Chryseobacterium* and little is known besides the details listed at the beginning of this chapter and in Table 44. *Chryseobacterium* strains are chemo-organotrophs with a strictly respiratory type of metabolism (Bernardet et al., 2002; Vandamme et al., 1994a). The original description of *Chryseobacterium scophthalmum*, however, stated that members of this species display both respiratory and fermentative metabolisms (Mudarris et al., 1994). Holmes et al. (1984a, b) also reported that some strains of *Chryseobacterium gleum*, *Chryseobacterium indologenes*, and other strains of CDC group IIb (see below) exhibited anaerobic respiration with nitrate as the terminal electron acceptor. Fumarate can also be used as a terminal electron acceptor by *Chryseobacterium indologenes* (Yabuuchi et al., 1983). These organisms are able to grow in the presence of 5–10% CO_2 (B. Bruun, unpublished data). The three above-mentioned *Chryseobacterium* glacier strains were isolated under anaerobic conditions (Miteva and Brenchley, 2005). Strains of *Chryseobacterium daecheongense* (Kim et al., 2005a) and *Chryseobacterium piscium* (de Beer et al., 2006), as well as some strains of *Chryseobacterium gleum* and *Chryseobacterium indologenes* and other unnamed CDC group IIb strains (Bruun, 1982; Holmes et al., 1984b; Weyant et al., 1996; Yabuuchi et al., 1983), reduce nitrate as well as nitrite (Table 44). Members of *Chryseobacterium* species show catalase, oxidase, phosphatase, and strong proteolytic activities. Several carbohydrates are oxidized, but glycerol and trehalose are not oxidized by members of any species (contrary to the statement in the original description of the genus). Esculin is hydrolyzed, but not agar (Vandamme et al., 1994a).

A few studies have been performed on the enzymic activities of some *Chryseobacterium* strains that may have practical or industrial applications in the future. A study on the biodegradation of the insecticide, herbicide, and disinfectant pentachlorophenol (PCP), whose residues represent a health hazard, has shown that a *Chryseobacterium gleum* strain had a higher PCP degradation capacity than that of a *Pseudomonas* strain and an *Agrobacterium radiobacter* strain (Yu and Ward, 1996). Other studies (cited by Yu and Ward, 1996) have reported degradation of chlorinated phenols by *Chryseobacterium gleum* and suggested

TABLE 44. Differential characteristics of *Chryseobacterium* species and closely related taxa[a]

Characteristic	*C. gleum*	*C. balustinum*	*C. daecheongense*	*C. defluvii*	*C. formosense*	*C. hispanicum*	*C. indologenes*	*C. indoltheticum*	*C. joostei*	*C. piscium*	"*C. proteolyticum*"	*C. scophthalmum*	*C. shigense*
Source	Diseased humans, water, soil and hospital environments	Diseased fresh-water fish	Sediment, freshwater lake	Activated sludge	Rhizosphere of lettuce	Drinking water	Diseased humans, water, soil and hospital environments	Seawater, marine mud	Raw cow's milk	Healthy marine fish	Soil	Seawater, healthy and diseased marine fish	Lactic acid beverage
Production of yellow flexirubin-type pigments	+	+	+	+[b]	+	(+)	+	+	+	+	+	+	+
Growth on:													
Marine agar 2216	+	–	nd	nd	nd	nd	(+)	+	(+)	nd	nd	+	nd
Cetrimide agar	–	–	nd	nd	nd	nd	(+)	–	(+)	–	nd	–	nd
MacConkey agar	d	+	–	–	–	–	d	+	+	–	–	–	–
Growth with 3% NaCl	d	–	nd	nd	nd	–	d	+	+	+	nd	+	nd
Growth at (°C):													
5	–	(+)	–	–	–	(+)	–	+	+	+	nd	d	+
25	+	+	nd	+	+	+	+	+	+	+	nd	+	+
30	+	+	+	+	+	+	+	+	+	+	nd	nd	+
37	+	–	+	+	–	–	+	–	–	–	+	–	–
42	d[f]	–	–	nd	–	–	–[f]	–	–	–	–	–	–
Degradation of:													
Esculin	+[f]	+	+	+	+	+	+[f]	+	+	+	+	+	nd
DNA	+	+	nd	nd	nd	–	+	+	+	(+)	+	+	nd
Starch	d	(+)	+	nd	nd	+	+	–	+	–	+	–	+
Tween 80	+	+	+	nd	–	–	+	+	+	d	+	+	nd
Tyrosine	d	(+)	nd	nd	nd	nd	–	+	+	–	+	+	nd
Urea	V[f]	–	–	–	–	–	V[f]	–	d	+	–	+	–
Production of:													
Indole	+	+	–	+	+	(+)	+	+	+	+	+	–	+
β-Galactosidase	d	–	–	nd	–	–	–	–	–	nd	+	+	nd
L-Phenylalanine deaminase	V[f]	–	nd	nd	nd	nd	–[f]	d	–	+	–	d	nd
Precipitate on 10% egg yolk agar	+	–	nd	nd	nd	nd	+	+	+	+	nd	+	nd
Reduction of:													
Nitrate	d	V	+	–	–	+	d	–	–	+	–	–	–
Nitrite	d	–	–	nd	nd	nd	d	–	–	–	–	–	nd
Acid production from:													
L-Arabinose	V[f]	–	–	–	–	+	–[f]	–	–	–	+	–	–
Cellobiose	–	–	+	(+)	–	nd	–	–	–	d	–	–	nd
Ethanol	V[f]	+	–	nd	nd	nd	–[f]	–	–	nd	–	+	nd
D-Fructose	+	+	+	nd	nd	+	(+)	–	+	–	nd	–	+
D-Glucose	+	+	–	+	(+)	+	+	+	+	d	+	–	+
Glycerol	d	–	+	nd	nd	+	(+)	–	d	d	(+)	–	–
Lactose	–	–	–	–	nd	–	–	–	–	–	–	–	–
Maltose	+	–	–	+	–	+	+	+	+	d	+	–	–
D-Mannitol	–	–	–	–	–	–	–	–	d	d	(+)	–	–
Salicin	V[f]	–	–	–	nd	nd	–[f]	–	–	nd	–	–	nd
Sucrose	–	V	–	–	–	nd	–	–	–	–	+	–	nd
Trehalose	+	–	+	+	+	–	+	–	+	d	+	+	–
D-Xylose	V[f]	–	+	–	(+)	+	–[f]	–	–	nd	nd	–	–
Utilization of malonate	–	–	–	–	–	nd	+	nd	–	nd	–	nd	nd
Alkaline reaction on Christensen's citrate	d	–	nd	nd	–	–	+	V	d	nd	nd	+	–
DNA G+C content (mol%)	36–39	35	37	39[g]	nd	34	37–39	34	36–37	34	37	33–35	37

[a]Symbols and abbreviations: +, 90% or more of the strains are positive; –,10% or less of the strains are positive; (+) 90% or more of the strains give a weak or delayed positive reaction; d, 11–89% of the strains are positive; V, varies between references; and nd, not determined. Since some of the characteristics listed were tested by several authors using different methods, the original publications should be consulted for direct comparison. Data from Bruun (1982), Yabuuchi et al. (1983, 1990), Holmes et al. (1984b, 1986a), Bruun and Ursing (1987), Ursing and Bruun (1991), Segers et al. (1993a), Mudarris et al. (1994), Vandamme et al. (1994a), Weyant et al. (1996), Vancanneyt et al. (1999), Yamaguchi and Yokoe (2000), Hugo et al. (2003), Kämpfer et al. (2003), Li et al. (2003b), Kim et al. (2004), Bernardet et al. (2005), de Beer et al. (2005), Kim et al. (2005a, b), Shen et al. (2005), Shimomura et al. (2005), Yi et al. (2005b), Young et al. (2005), de Beer et al. (2006), Gallego et al. (2006), O'Sullivan et al. (2006), Park et al. (2006b), Tai et al. (2006), Weon et al. (2006), J.-F. Bernardet (unpublished data), and C.J. Hugo (unpublished data).

[b]The type of yellow pigment in *Chryseobacterium defluvii*, *Chryseobacterium soldanellicola*, *Chryseobacterium taeanense*, and *Kaistella koreensis* has not been determined (Kämpfer et al., 2003; Kim et al., 2004; Park et al., 2006b). The yellow pigment produced by members of the genus *Sejongia* is not of the flexirubin type (Yi et al., 2005b).

[c]Strains of *Elizabethkingia meningoseptica* are either not pigmented or produce a weak yellow pigment (e.g., the type strain; Bruun and Ursing, 1987) that seems to belong to the flexirubin type (J.-F. Bernardet, unpublished data).

C. soldanellicola	*C. taeanense*	*C. taichungense*	*C. taiwanense*	*C. vrystaatense*	*C. wanjuense*	*Elizabethkingia meningoseptica*	*Elizabethkingia miricola*	*Bergeyella zoohelcum*	*Riemerella anatipestifer*	*Riemerella columbina*	*Kaistella koreensis*	*Sejongia antarctica*	*Sejongia jeonii*	*Epilithonimonas tenax*
Root of sand dune plant	Root of sand dune plant	Soil	Soil	Raw chicken meat	Soil	Diseased humans and animals, water, soil and hospital environments	Condensation water, contaminated commercial enzyme	Oral and nasal cavities of dogs and cats; diseased humans, dogs, and cats	Respiratory tract of various healthy and diseased bird species (except pigeon)	Respiratory tract of diseased pigeons	Fresh-water stream	Soil	Soil	Epilithon-covered stone in a river
+[b]	+[b]	+	+	+	+	d[c]	–	–	–[d]	–[d]	+[b]	+[b]	+[b]	+
nd	nd	nd	nd	nd	nd	+	nd[e]	–	–	(+)	nd	+	+	–
nd	nd	nd	nd	+	nd	(+)	(+)	–	–	–	nd	–	–	–
–	–	–	nd	–	(+)	d	+	–	–	–	–	–	–	+
+	+	nd	+	d	–	d	nd	–	–	+	nd	+	+	–
+	+	–	+	+	+	–	–	–	–	–	–	+	+	+
+	+	+	+	+	+	+	+	+	+	+	+	+	+	+
+	+	+	+	+	+	+	+	+	+	+	+	–	+	+
+	+	+	+	–	+	+	+	+	+	+	+	–	–	–
–	–	–	+	–	–	–	–	–	+	+	d	–	–	–
+	+	+	+	+	+	+	+	–	–	+	+	+	+	+
–	–	nd	nd	+	(+)	+	+	–	–	–	nd	+	+	–
nd	nd	nd	+	–	+	–	V	–	+	–	–	+	+	(+)
–	+	nd	+	+	nd	V	+	–	nd	nd	nd	+	+	–
nd	nd	nd	nd	–	+	+	–	V	–	–	nd	(+)	–	nd
–	–	–	–	+	–	–	+	+	d	d	d	–	–	–
–	–	(+)	+	+	–	+	d	+	d	–	+	+	+	–
nd	nd	(+)	–	nd	+	+	+	–	–	–	–	–	–	–
nd	nd	nd	nd	–	nd	(+)	–	–	+	+	nd	+	+	nd
nd	nd	nd	nd	+	nd	d	+	–	–	–	nd	+	+	nd
–	–	–	–	–	–	–	–	–	–	–	d	–	–	+
–	–	nd	–	+	nd	d	–	–	nd	nd	nd	–	–	–
–	–	–	+	nd	nd	–	–	–	–	–	nd	–	–	nd
nd	nd	–	+	nd	nd	V	–	–	–	–	nd	–	–	nd
nd	nd	nd	nd	nd	nd	d	nd	–	–	–	nd	nd	nd	nd
–	–	nd	d	nd	nd	+	+	–	(+)	(+)	nd	–	–	nd
nd	nd	(+)	nd	nd	+	+	+	–	+	+	–	+	+	–
–	–	nd	–	nd	nd	d	d	–	–	–	nd	nd	nd	nd
–	–	–	–	nd	nd	V	+	–	–	–	nd	–	–	nd
–	–	(+)	nd	nd	–	+	+	–	+	+	nd	+	+	nd
–	–	–	–	nd	nd	+	+	–	–	–	nd	–	–	nd
nd	nd	–	nd	nd	nd	–	–	–	–	–	nd	nd	nd	nd
nd	nd	–	–	nd	nd	–	–	–	–	–	nd	–	–	nd
–	–	(+)	–	nd	+	+	+	–	–	–	nd	–	–	nd
–	–	(+)	nd	nd	nd	V	–	–	–	–	nd	–	–	nd
nd	nd	–	nd	nd	+	–	–	–	–	–	nd	nd	nd	nd
nd	nd	–	–	nd	nd	d	+	–	–	–	nd	–	–	nd
29	32	nd	37	37–38	38	36–38	35	35–37	33–35	36–37	41–42	34	36	38

[d]Strains of *Riemerella anatipestifer* and *Riemerella columbina* grown on trypticase soy agar produce a diffusible, light yellow-brown non-flexirubin pigment that gives a beige color to the colonies (J.-F. Bernardet, unpublished data). On Columbia blood agar, colonies of *Riemerella columbina* are grayish-white to beige, whereas those of *Riemerella anatipestifer* are nonpigmented (Vancanneyt et al., 1999).

[e]Although its growth on marine agar has not been tested, the type strain of *Elizabethkingia miricola* has been shown to tolerate seawater (Li et al., 2003b).

[f]According to Holmes et al. (1984b), strains of *Chryseobacterium gleum* varied in acid production from arabinose and xylose, in growth at 42°C, and in urease production, and were negative for acid production from salicin. According to Yabuuchi et al. (1990), strains of *Chryseobacterium gleum* and *Chryseobacterium indologenes* may be distinguished by their positive and negative reaction, respectively, for the following tests: growth at 41°C; production of phenylalanine deaminase; acid production from L-arabinose, ethanol, salicin, and D-xylose; hydrolysis of esculin after 4 h (strains of both species hydrolyze esculin after 24 h); and production of urease on Christensen's urea agar slant after 40 h. According to Ursing and Bruun (1991), members of *Chryseobacterium gleum* and of one allied genomic group could be differentiated from other CDC group IIb strains by their ability to grow at 40°C and to produce acid from arabinose, xylose, and salicin.

[g]Data from Kim et al. (2005a, b).

that a reductive dechlorination pathway is involved. Organophosphorus tri-esterases, involved in hydrolysis of the tri-ester bonds of pesticides such as diazinon and parathion, have been reported from a *Chryseobacterium balustinum*-like organism (D. Siddavattam, personal communication). Other chemical contaminants, such as aniline (Radianingtyas et al., 2003), and furan and phenolic compounds (López et al., 2004), have been degraded by a mixed culture that included a *Chryseobacterium indologenes* strain. A novel protein-glutaminase purified from "*Chryseobacterium proteolyticum*" deamidates glutaminyl residues in proteins and is regarded by the authors as a promising method of improving protein functionality in food systems (Yamaguchi et al., 2001). A *Chryseobacterium* strain has been shown to produce an extracellular keratinase, which may be used for hydrolysis of poultry feathers and depilation of bovine pelts (Brandelli and Riffel, 2005). Screening for strains able to degrade chitosan (a deacetylated derivative of chitin) yielded a *Chryseobacterium* strain whose 16S rRNA gene sequence shared more than 97% similarity with that of *Chryseobacterium taichungense* (Yu and Ward, 1996). Its chitosanase was highly similar to the ChoA chitosanase of the betaproteobacterium *Mitsuaria chitosanitabida*. Oligosaccharides produced by enzymic hydrolysis of chitosan have many industrial applications. Other potentially useful enzymes discovered in *Chryseobacterium* strains include an endopeptidase that cleaves human plasminogen (Lijnen et al., 2000) and various cold-active proteases (Morita et al., 1997). Conversely, enzymes produced by some *Chryseobacterium* strains may have a negative impact; for instance, heat-stable metalloproteases may play a role in the spoilage of milk and milk products (Venter et al., 1999).

Genetics. The use of 16S rRNA gene sequencing and DNA–DNA hybridization in the delineation of the genus *Chryseobacterium* and of *Chryseobacterium* species has been reviewed recently (Bernardet et al., 2006). Importantly, levels of 16S rRNA gene sequence similarity may be very high (i.e., 98–99%) between *Chryseobacterium* species (de Beer et al., 2006; Li et al., 2003b). Consequently, DNA–DNA hybridization studies, or possibly sequencing of a number of household genes (Kuhnert and Korczak, 2006), are necessary even when two strains share high levels of sequence similarity. The DNA G+C contents of members of the genus were considered to be in the range 33–39 mol% (Bernardet et al., 2006) until lower values were reported for the novel species *Chryseobacterium taeanense* (32.1 mol%) and *Chryseobacterium soldanellicola* (28.8 mol%) (Park et al., 2006b). The usefulness of randomly amplified polymorphic DNA (RAPD) typing has been evaluated recently on a collection of *Chryseobacterium* fish isolates. The clusters of strains delineated by this technique matched some of the clusters yielded by SDS-PAGE analysis of whole-cell protein profiles, although RAPD analysis differentiated strains within other SDS-PAGE clusters (Bernardet et al., 2005).

Antibiotic sensitivity. Most studies on the antibiotic sensitivity of *Chryseobacterium* strains have naturally enough focused on [*Chryseobacterium*] *meningosepticum* (now *Elizabethkingia meningoseptica*) strains because of this species' pathogenicity to humans. In some recent studies (Fraser and Jorgensen, 1997; Hsueh et al., 1997), results of *Chryseobacterium* strains have been pooled with those of *Elizabethkingia meningoseptica* strains, whereas, in other studies, results are given separately for strains of *Chryseobacterium indologenes*, *Chryseobacterium gleum*, and unnamed members of CDC group IIb (see below) (Chang et al., 1997; Kirby et al., 2004; Spangler et al., 1996). Based on these studies, it appears that these taxa, like *Elizabethkingia meningoseptica*, are naturally resistant to polymyxins, aminoglycosides (gentamicin, streptomycin, kanamycin), chloramphenicol, and most β-lactams (penicillins, cephalosporins, carbapenems).

Chryseobacterium indologenes and *Chryseobacterium gleum* have, like *Elizabethkingia meningoseptica*, been found to produce carbapenem-hydrolyzing β-lactamases (Bellais et al., 1999, 2000a, b, 2002b), which means that they are resistant to extended-spectrum cephalosporins and carbapenems, but not necessarily to ureidopenicillins, i.e., piperacillin. An extended-spectrum β-lactamase has also been described in a strain of *Chryseobacterium gleum* (Bellais et al., 2002c). Members of *Chryseobacterium* species have been found, for the most part, to be resistant to tetracyclines, erythromycin, and linezolid, whereas they display intermediate sensitivity or resistance to clindamycin and vancomycin. Sensitivity to trimethoprim-sulfamethoxazole also varies, some strains being sensitive. The most active antibiotics are rifampicin and the newer quinolones (gatifloxacin, levofloxacin, sparfloxacin) (Fraser and Jorgensen, 1997; Kirby et al., 2004; Spangler et al., 1996).

A constitutive resistance to ampicillin, polymyxin B, chloramphenicol, and oxytetracycline was also found in fish-pathogenic *Chryseobacterium* strains; the antibiotic treatment of diseased fish was successful only when oxolonic acid or sulfamethoxazole-trimethoprim were used (Michel et al., 2005). The 50 *Chryseobacterium scophthalmum* isolates tested *in vitro* were all resistant to tetracyclines, amidoglycosides, lincomycin, oleandomycin, penicillin, and sulfadiazine, but susceptible to chloramphenicol, sulfamethoxazole-trimethoprim, furazolidone, fusidic acid, and novobiocin. Infection caused by *Chryseobacterium scophthalmum* in turbot (*Scophthalmus maximus*) was successfully controlled by furazolidone administered by injection or bath (Mudarris and Austin, 1989).

Pathogenicity. Among past and present *Chryseobacterium* species, [*Chryseobacterium*] *meningosepticum* is by far the most important from a clinical point of view. This species, now reclassified as *Elizabethkingia meningoseptica*, has joined [*Chryseobacterium*] *miricola* in the genus *Elizabethkingia* (Kim et al., 2005b) and is described elsewhere in this volume.

The pathogenicity for man is less well-documented for the two *Chryseobacterium* species of clinical interest, *Chryseobacterium indologenes* and *Chryseobacterium gleum*, taxa which were grouped previously, together with allied strains, as CDC group IIb (King, 1959) or *Flavobacterium* species group IIb (Holmes, 1992). However, there are approximately 35 cases of bacteremia caused by *Chryseobacterium indologenes*/group IIb reported in the literature and most of these cases are described in three published series (Hsueh et al., 1996b, c; Stamm et al., 1975). Bacteremias have been associated with indwelling devices, such as arterial and central venous catheters, endotracheal tubes, and drains, in seriously immunocompromised patients, who often have been placed under intensive care. The exact contribution of group IIb strains to the infections has been difficult to determine due to the often polymicrobial nature of the infections and to the severe underlying conditions in the patients. In some of the intubated patients with bacteremia, the same group IIb strain has also been isolated from lower tract specimens and the clinical diagnosis has been ventilator-associated pneumonia (Hsueh et al., 1996b; Stamm et al., 1975). There are also sporadic case reports of infections in the literature, including bacteremias (Akay et al., 2006; Lin et al., 2003; Nulens et al., 2001; Siegman-Igra et al., 1987), a case of meningitis following surgery and irrigation with tap water of a paranasal sinus (Bagley et al., 1976), urinary tract infections associated with nephrostomy drainage or urinary catheters, pyomyositis in intravenous drug abusers, and infected burn wounds (Hsueh et al., 1996a–c, 1997).

No infections seem definitely to have been ascribed to *Chryseobacterium gleum* during the past 20 years, which may be due to the fact that automatic identification systems dominating the market do not include this species in their databases. Prior to the extensive use of these systems, the bacterium had mainly been isolated from the vagina, but strains have also been found on catheters and in various body fluids (Holmes et al., 1984b).

Although not yet formally described, the two possible novel species "*Chryseobacterium massiliae*" and "*Candidatus* Chryseobacterium timonae" have been isolated from a human nasal swab and from blood, respectively (Drancourt et al., 2004; Greub et al., 2004); however, their pathogenicity to man is doubtful.

Members of some *Chryseobacterium* species have been associated with diseases in various animal species and their significance in veterinary medicine has been reviewed recently (Bernardet et al., 2006). The only *bona fide Chryseobacterium balustinum* strain currently available in culture collections was isolated from a diseased freshwater fish (Brisou et al., 1959); since no other cases have been reported since, it most likely acted as an opportunistic pathogen. Extensive investigations of cases of gill hyperplasia and hemorrhagic septicemia in turbot led to the description of *Chryseobacterium scophthalmum* (Bernardet et al., 2006; Mudarris and Austin, 1989; Mudarris et al., 1994). Because strains were also retrieved from healthy turbot and seawater and since no other cases have been reported since the original study, this species is also most likely an opportunistic pathogen. A collection of *Chryseobacterium* strains isolated from a variety of fish species and geographical areas was recently subjected to a polyphasic taxonomic study. Although all isolates originated from external lesions or internal organs of diseased fish, their virulence was not assessed experimentally. None of the isolates belonged to *Chryseobacterium balustinum* or *Chryseobacterium scophthalmum*, but two of them were attributed to *Chryseobacterium joostei*; other strains, grouped in three clusters, likely constitute the core of novel *Chryseobacterium* species (Bernardet et al., 2005). Finally, diseased leopard frogs (*Rana pipiens*) reared in laboratory facilities have been found to harbor *Chryseobacterium indologenes*-like strains (Olson et al., 1992).

Ecology. *Chryseobacterium* species have been isolated from a variety of environments, as reviewed recently by Bernardet et al. (2006). The presence of members of the family *Flavobacteriaceae* in various complex bacterial communities has been revealed by studies using culture-independent techniques, but only a few of them specifically mention *Chryseobacterium* strains (Gich et al., 2005; Green et al., 2006; Männistö and Haggblom, 2006; O'Sullivan et al., 2002). Even when bacterial strains were actually isolated, they were seldom studied in detail; the vague denomination "*Flavobacterium* sp." used by many authors may also very likely cover *Chryseobacterium* strains.

Environmental *Chryseobacterium* species have been isolated from: freshwater and lake sediments (Gich et al., 2005; Hayes, 1977; Kim et al., 2005a; Leff et al., 1998; Tatum et al., 1974); the rhizosphere of various plants (Domenech et al., 2006; Park et al., 2006b; Young et al., 2005); soil from temperate (Hayes, 1977; Shen et al., 2005; Tai et al., 2006; Tatum et al., 1974; Weon et al., 2006; Yamaguchi and Yokoe, 2000), tropical (Radianingtyas et al., 2003), and polar (Männistö and Haggblom, 2006; Walker et al., 2006; Wery et al., 2003) regions; glacier ice (Miteva and Brenchley, 2005); and drinking water (Gallego et al., 2006; Pavlov et al., 2004; Ultee et al., 2004).

The significance of *Chryseobacterium* in these habitats has only been investigated in a few instances. *Chryseobacterium* spp. have been reported from composted materials, soil environments, and plant surfaces (Green et al., 2006 and references therein). They were shown to form the largest group after *Pseudomonas* sp. among the culturable bacterial communities of the plant rhizosphere (McSpadden Gardener and Weller, 2001; Park et al., 2005), to colonize seed surfaces, and to be among the most persistent bacteria in root samples (Green et al., 2006). Some *Chryseobacterium* species have been investigated for their ability to adhere to plant roots and to inhibit plant pathogens (Albareda et al., 2006). For instance, *Chryseobacterium soldanellicola*, one of the novel bacterial species described from the rhizosphere of sand-dune plants, inhibited growth of the plant-pathogenic fungus *Fusarium oxysporum* (Park et al., 2006b). A similar result was obtained with a *Chryseobacterium* strain tentatively identified as *Chryseobacterium balustinum* isolated from the rhizosphere of lupins (*Lupinus albus*); this strain, alone or in combination with other bacterial strains, also displayed a growth-promoting effect on pepper and tomato plants (Domenech et al., 2006). Conversely, although the number of *Chryseobacterium* strains in the rhizosphere of wheat increased significantly during the course of a fungal disease, they were unable to inhibit the fungus (McSpadden Gardener and Weller, 2001).

Chryseobacterium species are also present in freshwater and marine environments (Campbell and Williams, 1951; de Beer et al., 2006; Engelbrecht et al., 1996a, b; Gallego et al., 2006; Gennari and Cozzolino, 1989; González et al., 2000; Harrison, 1929; Mudarris and Austin, 1989; Mudarris et al., 1994). Some of these bacteria may be agents of fish diseases (Bernardet et al., 2005; Mudarris and Austin, 1989; Mudarris et al., 1994) (see above). *Chryseobacterium* strains have also been reported from the epibiotic bacterial community of freshwater microcrustaceans (Niswati et al., 2005), from the gut of various insect species, and from the endosymbiotic microflora of amoebae (Bernardet et al., 2006 and references therein).

Industrial environments such as activated sewage sludge (Kämpfer et al., 2003) and paper mill slimes (Oppong et al., 2003) also harbor *Chryseobacterium* strains. Any relationship between *Chryseobacterium taichungense* and the tar that contaminated the soil from which it was isolated was not investigated (Shen et al., 2005).

Chryseobacterium species are frequently isolated from food environments. These sources include the dairy environment, such as fresh cow milk in South Africa (Hugo et al., 2003; Jooste, 1985; Jooste et al., 1985, 1986; Welthagen and Jooste, 1992) and a lactic acid beverage in Japan (Shimomura et al., 2005). They have also been isolated from meat and poultry products (García-López et al., 1998; Hayes, 1977). An extensive study of *Chryseobacterium* strains retrieved from chicken portions in a poultry abattoir in South Africa (de Beer, 2005) led to the description of *Chryseobacterium vrystaatense* (de Beer et al., 2005). *Chryseobacterium* strains also occur on the surface or in the gut of apparently healthy marine and freshwater fish (de Beer et al., 2006; Gennari and Cozzolino, 1989; González et al., 2000; Lijnen et al., 2000; Morita et al., 1997); J.-F. Bernardet, unpublished data). The role of these bacteria in food is still unclear, but some studies indicate a role in the spoilage of milk and milk products because of the presence of metalloproteases (Venter et al., 1999). *Chryseobacterium shigense*, however, has been considered as a part of the normal bacterial community in the lactic acid beverage from which it was isolated (Shimomura et al., 2005). *Chryseobacterium* strains retrieved from fish may also be involved with spoilage (Engelbrecht et al., 1996a, b; Gennari and Cozzolino, 1989; Harrison, 1929), as suggested by the production of urea and phenylalanine deaminase by *Chryseobacterium piscium* (de Beer et al., 2006).

Members of CDC group IIb, including *Chryseobacterium indologenes* and *Chryseobacterium gleum*, are also found widely in soil and water. Investigations on the microflora of sink drains revealed the presence of several *Chryseobacterium* strains (as well as strains of the closely related species "*Haloanella gallinarum*", see Figure 40) (Bruun et al., 1989; McBain et al., 2003). Because of their natural habitat in water, *Chryseobacterium* strains are also found in the hospital environment and consequently on patients' inner and outer body surfaces, where they can colonize indwelling devices, e.g., endotracheal tubes and intravascular catheters. From here, they can spread and occasionally cause bacteremia, having the potential to colonize other indwelling devices. However, *Chryseobacterium* strains are not part of the normal flora of humans. *Chryseobacterium* and allied unnamed bacteria are the most frequently isolated flavobacteria in the clinical microbiology laboratory (Holmes and Owen, 1981).

Enrichment and isolation procedures

Chryseobacterium species are not difficult to isolate and cultivate; they usually do not require enrichment or selective isolation procedures and they grow in ambient air (Bernardet et al., 1996). The isolation and preservation media used for *Chryseobacterium* species have been discussed in detail by Jooste and Hugo (1999), Hugo and Jooste (2003), and Bernardet et al. (2006).

Cultivation of environmental *Chryseobacterium* strains is usually done on nutrient, tryptic soy, or brain-heart infusion (BHI) agars (Kim et al., 2005a; Shen et al., 2005; Tai et al., 2006; Young et al., 2005), but Luria–Bertani and plate count agars have also been used (Gallego et al., 2006; Tai et al., 2006). Another medium that has also been used for isolation of these bacteria is R2A agar (Gallego et al., 2006; Kämpfer et al., 2003; Miteva and Brenchley, 2005; Park et al., 2005; Weon et al., 2006). A new cultivation approach using a microdispenser and microtiter plates and resulting in highly enriched or pure cultures recently enabled several members of the phylum *Bacteroidetes*, including a *Chryseobacterium* strain, to be isolated from the water of German lakes (Gich et al., 2005).

Strains of some environmental *Chryseobacterium* species have been isolated initially by using enrichment media in screening studies aimed at isolating bacteria exhibiting special metabolic activities. For instance, the two original strains of *Chryseobacterium indoltheticum* were isolated on a chitin-containing enrichment medium during a search for chitin-decomposing microorganisms in samples of marine mud (Campbell and Williams, 1951). Subsequently, this species was shown to grow well on media devoid of chitin. "*Chryseobacterium proteolyticum*" was isolated from soil samples during a search for protein-deaminating bacteria by using an enrichment medium containing 0.1% carboxybenzoxy-Gln-Gly (Yamaguchi and Yokoe, 2000). *Chryseobacterium defluvii* was isolated using a selective enrichment medium for a phosphorus-removing mixed bacterial culture from activated sludge; it was subsequently cultivated on R2A and nutrient agars (Kämpfer et al., 2003). *Chryseobacterium formosense* was retrieved during a screening of rhizobacteria for proteolytic activity on skimmed milk agar and subcultivated on BHI agar (Young et al., 2005). By successive rounds of filtration (pore size 0.2 µm) and recultivation at 5°C, a number of ultrasmall psychrophiles was isolated from the deep core of a Greenland glacier and three of them were attributed to the genus *Chryseobacterium* by 16S rRNA gene sequence analysis (Miteva and Brenchley, 2005).

Chryseobacterium strains from freshwater fish are best isolated on tryptic soy or nutrient agars (Bernardet et al., 2005; Michel et al., 2005). For the isolation of *Chryseobacterium* species from marine fish, BHI, glucose-yeast extract, marine 2216E, or nutrient agars may be used (de Beer et al., 2006). Other isolation media include proteose (2%) medium and plate count agar containing 0.5% NaCl (*Chryseobacterium balustinum*; Harrison, 1929; Holmes et al., 1984a; Engelbrecht et al., 1996a, b). *Chryseobacterium scophthalmum* was isolated initially from diseased turbot on medium K (0.1% yeast extract, 0.5% beef extract, 0.6% casein, 0.2% tryptone, 0.1% $CaCl_2$, 1.5% agar, and 750 ml aged seawater per liter, pH 7.2) (Mudarris and Austin, 1989; Mudarris et al., 1994); it also grows well on marine 2216E, nutrient, and tryptic soy agars.

Nutrient agar (Shimomura et al., 2005) and plate count agar (Jooste, 1985; Jooste et al., 1985, 1986; Welthagen and Jooste, 1992) are the preferred isolation media for food-associated *Chryseobacterium* species. Examples are *Chryseobacterium shigense* (Shimomura et al., 2005), *Chryseobacterium joostei* (Hugo et al., 2003), and *Chryseobacterium vrystaatense* (de Beer et al., 2005). Incubation is at 25–30°C for 48–96 h; however, *Chryseobacterium vrystaatense* was first incubated at 4°C for 24 h and then at 25°C for 48 h in order to slow the strong growth of pseudomonads and give the *Chryseobacterium* strains an opportunity to compete (de Beer, 2005).

Isolation of *Chryseobacterium* species from clinical specimens is done on blood, nutrient, or heart infusion agars (Bruun, 1982; Holmes et al., 1984b; Yabuuchi et al., 1983). An amoebal co-culture was used to recover a novel, as yet undescribed *Chryseobacterium* species from polymicrobial human nasal swabs; the novel species, "*Chryseobacterium massiliae*", was among the few amoeba-resistant bacteria in the swabs (Drancourt et al., 2004; Greub et al., 2004).

Maintenance procedures

The preservation media used for *Chryseobacterium* species have been discussed in detail by Jooste and Hugo (1999), Hugo and Jooste (2003), and Bernardet et al. (2006).

For short-term preservation of *Chryseobacterium* species, nutrient agar slants in metal screw-capped bijou bottles stored at

4°C were used for "*Chryseobacterium proteolyticum*" (Yamaguchi and Yokoe, 2000) and *Chryseobacterium taichungense* (Shen et al., 2005). BHI agar was used in the same way for *Chryseobacterium formosense* (Young et al., 2005). Cultures of *Chryseobacterium scophthalmum* were maintained on slopes of medium K (see above) at 4°C with transfer every 6–8 weeks (Mudarris and Austin, 1989).

For longer-term preservation, *Chryseobacterium* strains may be freeze-dried on filter paper discs (AA discs) and stored in screw-capped Wassermann tubes at −20°C (e.g., *Chryseobacterium joostei*; Hugo et al., 2003). Preservation at −80°C may employ BHI broth with 20% (v/v) glycerol (Brandelli and Riffel, 2005; Park et al., 2006b) or CAS broth (1% casitone, 0.1% $MgSO_4 \cdot 7H_2O$, pH 6.8 unadjusted; Reichenbach, 1989b), ox broth, or tryptic soy broth with 10% (v/v) glycerol (Tai et al., 2006; J.-F. Bernardet, unpublished data; B. Bruun, unpublished data).

The preferred long-term preservation method is, however, lyophilization [Mudarris and Austin, 1989; Bernardet et al., 2002, 2006; Tai et al., 2006; BCCM-LMG Bacteria Collection (http://bccm.belspo.be/lmg.htm); American Type Culture Collection (http://www.atcc.org/)].

Procedures for testing special characters

A number of phenotypic features are of special importance for differentiation between *Chryseobacterium* species (Tables 43 and 44) and related genera in the family *Flavobacteriaceae* (see below). The performance of these phenotypic tests requires some experience and the conditions under which they are performed are critical for reliability and reproducibility; these conditions have been extensively specified previously (Bernardet et al., 2006; Bernardet and Nakagawa, 2002, 2006).

Differentiation of the genus *Chryseobacterium* from closely related taxa

The characteristics differentiating members of the genus *Chryseobacterium* from those of the most closely related taxa (i.e., *Elizabethkingia, Epilithonimonas, Bergeyella, Riemerella, Kaistella,* and *Sejongia*) are listed in Tables 43 and 44. They include the production of flexirubin-type pigments, growth on various media and at different temperatures, DNA G+C content, a number of biochemical traits, and fatty acid composition. In addition, members of the genus *Riemerella* have a distinctly capnophilic metabolism (Segers et al., 1993a; Vancanneyt et al., 1999). Some additional features have been proposed by Laffineur et al. (2002); all *Chryseobacterium indologenes, Elizabethkingia meningoseptica,* and *Empedobacter brevis* strains studied shared the same characteristics, but *Chryseobacterium indologenes* strains differ from *Bergeyella zoohelcum* and *Weeksella virosa* strains by the production of pyrrolidonyl arylamidase and resistance to desferrioxamine and colistin. The features differentiating *Chryseobacterium* from other, more distant, members of the family *Flavobacteriaceae* are listed in Table 23 of the family *Flavobacteriaceae* in this volume. In addition to the above-mentioned features, they include cell morphology, the absence of gliding motility, halotolerance, and respiratory metabolism. Features that should be included in the description of novel *Chryseobacterium* species have been specified in the minimal standards for describing new taxa in the family *Flavobacteriaceae* (Bernardet et al., 2002). The use of commercial identification galleries and strips as well as the relevance of fatty acid and whole-cell protein profiles, respiratory quinones, polyamines, and polar lipids for the differentiation between *Chryseobacterium* species and from related taxa have been discussed recently (Bernardet et al., 2006). In an ongoing study of food spoilage caused by *Chryseobacterium* strains, Biolog GN2 MicroPlates yielded differentiating profiles for the type strains of the seven *Chryseobacterium* and two *Elizabethkingia* species tested (C.J. Hugo, personal communication).

Taxonomic comments

The genus *Chryseobacterium* was proposed in 1994 to accommodate six bacterial species previously included in the genus *Flavobacterium*, but which had been clearly separated from the type species of the genus, *Flavobacterium aquatile*, as a result of an extensive polyphasic study (Vandamme et al., 1994a). These six original *Chryseobacterium* species were *Chryseobacterium balustinum, Chryseobacterium gleum, Chryseobacterium indologenes, Chryseobacterium indoltheticum, Chryseobacterium meningosepticum,* and *Chryseobacterium scophthalmum.* Although *Chryseobacterium balustinum* and *Chryseobacterium indoltheticum* were the two oldest species (described in 1929 and 1951, respectively), they were not chosen as the type species of the new genus since they were rather poorly characterized and they were each represented by a single strain. The human pathogen *Chryseobacterium meningosepticum* was also not selected, although it had been most extensively studied and many strains were available. This was because 16S rRNA studies had already shown that it occupied a rather separate position compared to the five other *Chryseobacterium* species. *Chryseobacterium gleum* was chosen as the type species since an extensive genomic and phenotypic study had shown that the 12 strains available were rather homogeneous (Holmes et al., 1984b), contrary to the *Chryseobacterium indologenes* strains (Yabuuchi et al., 1983). Novel species progressively joined the genus; conversely, *Chryseobacterium meningosepticum* and *Chryseobacterium miricola* (Li et al., 2003b) were subsequently moved to a new genus, *Elizabethkingia* (Kim et al., 2005b).

A rather confusing issue concerns the group of strains initially named "*Flavobacterium* CDC group IIb" by King (1959). The descriptions of *Chryseobacterium indologenes* and *Chryseobacterium gleum* were based on some of its members, but since many CDC group IIb strains still cannot be assigned to any named species, it was proposed to continue referring to them as "*Chryseobacterium* sp. CDC group IIb" (Bernardet et al., 2006; Ursing and Bruun, 1991).

There are, at the present time, a few other taxonomic or nomenclatural problems in the genus *Chryseobacterium.* "*Chryseobacterium proteolyticum*", although well characterized (Yamaguchi and Yokoe, 2000), was not published in the *International Journal of Systematic and Evolutionary Microbiology* or included in a validation list in this journal; hence, its name is not valid. "*Chryseobacterium massiliae*" has not been described formally and "*Candidatus* Chryseobacterium timonae" is only known through its 16S rRNA gene sequence (Drancourt et al., 2004; Greub et al., 2004).

Taxonomic and nomenclatural issues in the genus *Chryseobacterium* are dealt with by the Subcommittee on the Taxonomy of *Flavobacterium* and *Cytophaga*-like Bacteria of the International Committee on Systematics of Prokaryotes. The subcommittee has issued the above-mentioned minimal standards (Bernardet et al., 2002).

Attention might be drawn to the fact that of the 18 *Chryseobacterium* species with validly published names, 11 (*Chryseobacterium balustinum, Chryseobacterium daecheongense, Chryseobacterium defluvii, Chryseobacterium formosense, Chryseobacterium hispanicum, Chryseobacterium shigense, Chryseobacterium soldanellicola, Chryseobacterium taeanense, Chryseobacterium taichungense, Chryseobacterium taiwanense,* and *Chryseobacterium wanjuense*) have been described on the basis of only one isolated strain and that descriptions of all these species, except *Chryseobacterium balustinum,* were published from 2003 to 2006. In this regard, it should be noted that a proposal was made in 2001 for not publishing species consisting of less than five strains or species for which easily performed phenotypic tests for differentiation from other species in the same genus were not available (Christensen et al., 2001). Tests that differentiate *Chryseobacterium* species are often sparse, as seen in Table 44. Bearing in mind the facts that different studies employ different variations of tests and media bearing the same name and that a certain level of phenotypic variation is to be expected within a species, this underscores the need for a polyphasic study of a large number of *Chryseobacterium* strains including all species using the same defined methods for all strains.

Editorial note: The following novel species have been added to the genus *Chryseobacterium* after this chapter was completed and their names have been validly published in the *International Journal of Systematic and Evolutionary Microbiology*: *Chryseobacterium caeni* (Quan et al., 2007; this paper includes the detailed fatty acid composition of *Chryseobacterium shigense*); *Chryseobacterium daeguense* (Yoon et al., 2007b); *Chryseobacterium flavum* (Zhou et al., 2007); *Chryseobacterium luteum* (Behrendt et al., 2007); *Chryseobacterium haifense* (Hantsis-Zacharov and Halpern, 2007); *Chryseobacterium hominis* (Vaneechoutte et al., 2007); *Chryseobacterium ureilyticum, Chryseobacterium gambrini, Chryseobacterium pallidum,* and *Chryseobacterium molle* (Herzog et al., 2008); *Chryseobacterium arothri* (Campbell et al., 2008); and *Chryseobacterium soli* and *Chryseobacterium jejuense* (Weon et al., 2008).

Acknowledgements

The authors wish to acknowledge Jean Euzéby (Ecole Nationale Vétérinaire, Toulouse, France) for correcting the etymology of a number of specific epithets in the List of Species and Tim Lilburn (American Type Culture Collection, Manassas, VA) for providing the phylogenetic tree (Figure 40).

Further reading

Website of the International Committee on Systematics of Prokaryotes (ICSP) Subcommittee on the Taxonomy of *Flavobacterium* and *Cytophaga*-like Bacteria (http://www.the-icsp.org/subcoms/Flavobacterium_Cytophaga.htm).

Differentiation of species of the genus *Chryseobacterium*

Characteristics that differentiate the various species of the genus are listed in Table 44.

List of species of the genus *Chryseobacterium*

1. **Chryseobacterium gleum** (Holmes, Owen, Steigerwalt and Brenner 1984b) Vandamme, Bernardet, Segers, Kersters and Holmes 1994a, 830[VP] (*Flavobacterium gleum* Holmes, Owen, Steigerwalt and Brenner 1984b, 23)

gle'um. Gr. neut. adj. *gloion* slippery, sticky; N.L. neut. adj. *gleum* (*sic*) sticky.

Rods, approximately 0.5 µm in diameter with varying lengths, 1.0–3.0 µm. Filaments are common. Good growth occurs on tryptic soy and nutrient agars. Colonies on nutrient agar at 35°C are circular, entire and smooth, bright-yellow-pigmented, and up to 2 mm in diameter, becoming mucoid after incubation for 5 d. Cultures have a strong aromatic odor. There is no hemolysis on blood agar. Nitrate reduction and anaerobic growth in the presence of nitrate have been reported for some *Chryseobacterium gleum* strains (Holmes et al., 1984b). A dark brown pigment is produced on tyrosine agar. Tween 20 is hydrolyzed. Growth occurs on β-hydroxybutyrate, but lipid inclusion granules are not formed. Gelatin, casein, and tributyrin are hydrolyzed. Arginine dihydrolase, and lysine and ornithine decarboxylases are not produced. 3-Ketolactose is not produced. In addition to the data listed in Table 44, acid is not produced from adonitol, dulcitol, inositol, raffinose, rhamnose, or sorbitol. H_2S production is negative when tested using the triple-sugar iron test and positive when the lead acetate test is used. The list of substrates hydrolyzed in API ZYM galleries is given in the original species description. Other characteristics are as given in the genus description and in Tables 43 and 44. The description of *Chryseobacterium gleum* was based on the thorough study of a homogeneous group of 12 CDC group IIb strains (Holmes et al., 1984b); however, phenotypic differentiation of *Chryseobacterium indologenes, Chryseobacterium gleum,* and allied members of CDC group IIb is problematic. As acid production from the differentiating carbohydrates (arabinose, xylose, and salicin) is weak and dependent on the type of carbohydrate medium used, the most reliable differentiating test for *Chryseobacterium gleum* is probably its ability to grow at 40°C. This characteristic is only shared with a few allied group IIb strains and not with the type and other strains of *Chryseobacterium indologenes,* according to Ursing and Bruun (1991). This test is, however, not widely used and not applicable in modern automated identification systems. These systems do not contain *Chryseobacterium gleum* in their databases with the result that strains of *Chryseobacterium gleum* and other group IIb strains are identified as *Chryseobacterium indologenes.* Consequently, only publications of infections with *Chryseobacterium indologenes* appear in recent medical literature. Ironically, this means that the original intention of the proposal by Yabuuchi et al. (1983), i.e., for *Chryseobacterium indologenes* to encompass all CDC group IIb strains, is now what is taking place.

Source: ubiquitous in soil and water and, therefore, the hospital environment. Not a member of normal human flora, but occasionally found in clinical specimens due to its occurrence in hospital water. Isolated mainly from vaginal specimens, but also from body fluids.

DNA G+C content (mol%): 36.6–39.0 (type strain: 37.0–38.0) (T_m) (Holmes et al., 1984b; Hugo et al., 1999).

Type strain: Owen F93, Holmes CL 4/79, ATCC 35910, CCUG 14555, NCTC 11432, LMG 8334, LMG 12447, CIP 103039, IFO (now NBRC) 15054, JCM 2410, NCIMB 13462.

Sequence accession nos (16S rRNA gene): M58772, AY468449.

Reference strains: 11 other strains are included in the species description (Holmes et al., 1984b).

2. **Chryseobacterium balustinum** (Harrison 1929) Vandamme, Bernardet, Segers, Kersters and Holmes 1994a, 830[VP] [*Empedobacter balustinum* (Harrison 1929) Brisou, Tysset and Vacher 1959, 690; *Flavobacterium balustinum* Harrison 1929, 233]

ba.lus.ti'num. Etymology uncertain but possibly derived from L. n. *balux* gold sand or gold dust, probably in reference to the bright yellow color (Holmes et al., 1984a).

Rods, 0.5–0.7 µm in diameter and up to 3.5 µm in length; filaments (10–20 µm) occur. Colonies on tryptic soy and nutrient agars at 25–30°C are circular, convex, entire, smooth, 2 mm in diameter, and bright-yellow-pigmented. The consistency, originally butyrous, becomes mucoid after 2–3 d. There is no hemolysis on blood agar. A brownish pink pigment is produced on tyrosine agar. Other characteristics are as given in the genus description and in Tables 43 and 44. No strain dating back to the original description has been preserved; hence, the only authentic strain available in culture collections is the one isolated from the blood of a diseased freshwater fish (dace, *Leuciscus leuciscus*) in France (Brisou et al., 1959). Harrison (1929) first isolated the bacterium from the scales of halibut (*Hippoglossus hippoglossus*) caught in the Pacific Ocean. Brisou et al. (1959) considered their own isolate as a freshwater variant of [*Flavobacterium*] *balustinum*, renamed it "*Empedobacter balustinum*", and demonstrated that it was able to kill various fish species when injected intraperitoneally, but not rats or mice. It is actually not certain whether Brisou's strain indeed belongs to the same species as the strain described by Harrison. Samples of intestinal content, skin, and muscle from healthy freshwater fish have recently yielded bacterial isolates phenotypically similar to *Chryseobacterium balustinum* (González et al., 2000; Morita et al., 1997), but they have not been definitely identified and their significance has not been elucidated. Polyphasic studies of collections of fish or dairy *Chryseobacterium* isolates have failed to identify novel *Chryseobacterium balustinum* strains (Bernardet et al., 2005; Hugo et al., 1999).

Source: fish.

DNA G+C content (mol%): 33 (Holmes et al., 1984a, b), 34.7 (Yabuuchi et al., 1983), or 35.0 (Hugo et al., 1999) (all by T_m).

(Neo) type strain: CCTM 724, NCTC 11212, ATCC 33487, LMG 8329, LMG 4010, CIP 103103, IFO (now NBRC) 15053, CCUG 13228, NCIMB 2270.

Sequence accession no. (16S rRNA gene): AY468447.

Reference strains: the strain LMG 12856 (=Holmes 42/78=NCTC 11409=CCUG 22401) (16S rRNA gene sequence accession no. AY468481) should not be considered a *Chryseobacterium balustinum* strain; a polyphasic study has demonstrated it is actually highly related to the *Chryseobacterium indologenes* type strain (Bernardet et al., 2005).

3. **Chryseobacterium daecheongense** Kim, Bae, Schumann and Lee 2005a, 136[VP]

dae.che.ong.en'se. N.L. neut. adj. *daecheongense* pertaining to Lake Daechong, Korea, from where the type strain was recovered.

Rods, approximately 0.5 µm in diameter and 0.8–2.0 µm in length. Good growth occurs on R2A, tryptic soy, and nutrient agars at 28–37°C. No growth occurs on β-hydroxybutyrate. Colonies are yellow-pigmented and become mucoid after 3 d of incubation. *sym*-Homospermidine is the major polyamine and phosphatidylethanolamine is the major polar lipid. Nitrate is reduced, but not nitrite. Casein and gelatin are hydrolyzed. Lists of compounds that are utilized as carbon sources and those that show positive reactions in API ZYM galleries are given in the original species description. Other characteristics are as given in the genus description and in Tables 43 and 44. Highest 16S rRNA gene sequence similarity is with the type strain of *Chryseobacterium defluvii* (97.9%), but DNA–DNA relatedness between the two species is only 34%.

Source: the type (and so far only) strain was isolated from the sediment of a freshwater lake in Korea.

DNA G+C content (mol%): 36.6 (HPLC).

Type strain: CPW406, DSM 15235, KCTC 12088.

Sequence accession no. (16S rRNA gene): AJ457206.

4. **Chryseobacterium defluvii** Kämpfer, Dreyer, Neef, Dott and Busse 2003, 95[VP]

de.flu'vi.i. L. gen. n. *defluvii* of sewage.

Rods, approximately 2.0 µm long; diameter unspecified. Good growth occurs on R2A, tryptic soy, and nutrient agars at 25–30°C. Colonies are yellow-pigmented and circular with entire edges; after prolonged incubation, they merge owing to the production of extracellular slimy substances. In addition to the data listed in Table 44, no acid is produced from adonitol, D-arabitol, dulcitol, erythritol, i-inositol, melibiose, methyl α-D-glucoside, raffinose, L-rhamnose, or D-sorbitol. Lists of compounds that are utilized as carbon sources and chromogenic substrates that are hydrolyzed are given in the original species description. *sym*-Homospermidine is the major polyamine and phosphatidylethanolamine is the major polar lipid. Other characteristics are as given in the genus description and in Tables 43 and 44. Contains the highest proportion of $C_{15:0}$ iso fatty acid (58.5%) of all *Chryseobacterium* species. The highest 16S rRNA gene sequence similarity is with *Chryseobacterium indoltheticum* (96.2%).

Source: the type (and so far only) strain was isolated from activated sewage sludge in Germany.

DNA G+C content (mol%): 38.8 (HPLC) [not determined originally by Kämpfer et al. (2003), but by Kim et al. (2005a, b)].

Type strain: Kämpfer B2, DSM 14219, LMG 22469, CIP 107207.

Sequence accession no. (16S rRNA gene): AJ309324.

Note: Chryseobacterium defluvium is the name given for the organism originally deposited in 16S rRNA sequence databases.

5. **Chryseobacterium formosense** Young, Kämpfer, Shen, Lai and Arun 2005, 426[VP]

for.mo.sen'se. N.L. neut. adj. *formosense* pertaining to Formosa (Taiwan).

Rods, approximately 1.0 µm in diameter and 2.0 µm in length. Good growth occurs on nutrient, tryptic soy, and BHI agars at 20–30°C. Colonies are yellow-pigmented and circular with entire edges; after prolonged incubation, they become mucoid and merge owing to the production of extracellular slimy substances. Negative for arginine dihydrolase, lysine decarboxylase, and tryptophan deaminase. Lists of compounds that are utilized as carbon sources and that show positive reactions in API ZYM galleries are given in the species description. In API ZYM galleries, *Chryseobacterium formosense* showed β-glucosidase activity, but no α-glucosidase activity, in contrast to *Chryseobacterium defluvii* and *Chryseobacterium scophthalmum*. Other characteristics are as given in the genus description and in Tables 43 and 44. The highest 16S rRNA gene sequence similarity is with *Chryseobacterium indoltheticum* (97.5%), but the two type strains share only 27% DNA–DNA relatedness.

Source: the type (and so far only) strain was isolated from the rhizosphere of garden lettuce (*Lactuca sativa* L.) from Kuohsing, Taiwan.

DNA G+C content (mol%): not determined.

Type strain: CC-H3-2, CCUG 49271, CIP 108367.

Sequence accession no. (16S rRNA gene): AY315443.

6. **Chryseobacterium hispanicum** Gallego, García and Ventosa 2006, 1591[VP]

his.pa'ni.cum. L. neut. adj. *hispanicum* from Spain.

Rods, approximately 0.9–1.5 µm in diameter and 4.0 µm in length. Good growth occurs on R2A, plate count, and tryptic soy agars at 25–28°C. Colonies on tryptic soy agar are white–yellow, opaque, round, convex, and 2–3 mm in diameter after 48 h at 28°C. Growth occurs with 0–2% NaCl (optimal growth without NaCl). The pH range for growth is 5.0–10.0 (optimal growth at pH 7.0). Methyl red test is negative. Acetoin and H_2S are not produced. Casein and gelatin are hydrolyzed. Glucose is not fermented. D-Glucose, L-arabinose, D-mannose, and maltose are used as sole carbon sources. Arginine dihydrolase activity is absent. The list of substrates that show positive reactions in API ZYM galleries is given in the original species description. In addition to the data listed in Table 44, acid is produced from D-arabinose and D-mannose, but not from D-galactose. Other characteristics are as given in the genus description and in Tables 43 and 44. The highest 16S rRNA gene sequence similarity (96.0%) is with *Chryseobacterium indoltheticum* and *Chryseobacterium scophthalmum*; it occupies a separate position in the phylogenetic tree compared to the other *Chryseobacterium* species.

Source: the type (and so far only) strain was isolated from drinking water in Seville, Spain.

DNA G+C content (mol%): 34.3 (T_m).

Type strain: VP48, CECT 7129, CCM 7359, JCM 13554.

Sequence accession no. (16S rRNA gene): AM159183.

7. **Chryseobacterium indologenes** (Yabuuchi, Kaneko, Yano, Moss and Miyoshi 1983) Vandamme, Bernardet, Segers, Kersters and Holmes 1994a, 830[VP] (*Flavobacterium indologenes* Yabuuchi, Kaneko, Yano, Moss and Miyoshi 1983, 595)

in.do.lo'ge.nes. N.L. neut. n. *indolum* indole; N.L. suff. *-genes* (from Gr. v. *gennaô* to produce), producing; N.L. neut. adj. *indologenes* (*sic*), indole-producing.

Rods, approximately 0.5 µm in diameter and 1.0–3.0 µm in length. Filaments are common. Good growth occurs on tryptic soy and nutrient agars. Bright-yellow-pigmented. Colonies on nutrient agar at 35°C are circular, convex, entire, smooth, and up to 2 mm in diameter. Cultures have a strong aromatic odor. There is no hemolysis on blood agar. Yabuuchi et al. (1983) reported that this organism can grow under anaerobic conditions in the presence of fumarate and that some strains are able to reduce nitrate to N_2, whereas others reduce neither nitrate nor nitrite. No pigment is produced on tyrosine agar. Gelatin is hydrolyzed. In addition to the data listed in Table 44, acid is produced from glycogen and mannose, but not from rhamnose, ribose, melibiose, raffinose, melezitose, adonitol, dulcitol, or inulin. Lysine and ornithine decarboxylases, and arginine dihydrolase are not produced. H_2S production is negative when tested using the triple-sugar iron test and positive when the lead acetate test is used. Other characteristics are as given in the genus description and in Tables 43 and 44. See above under *Chryseobacterium gleum* for the differentiation from other *Chryseobacterium* species of clinical interest. The highest 16S rRNA gene sequence similarity (99.0%) is with *Chryseobacterium gleum*, but the two type strains share only 22% DNA–DNA relatedness.

Source: ubiqitous in soil and water and therefore in the hospital environment. Not a member of normal human flora, but together with allied members of CDC group IIb the most commonly isolated flavobacteria in the clinical microbiology laboratory. Associated with device-related infections in immunocompromised patients, but its role in these infections is unclear.

DNA G+C content (mol%): 37.6–38.3 (type strain: 37.7) (Yabuuchi et al., 1983) or 38.0–39.0 (type strain: 38.0) (Hugo et al., 1999) (all using T_m).

Type strain: RH 542, ATCC 29897, NCTC 10796, LMG 8337, LMG 12453, LMG 12454, CIP 101026, CCUG 14483, CCUG 14556, IFO (now NBRC) 14944, GIFU 1347, CDC 3716.

Sequence accession nos (16S rRNA gene): M58773, AY468450 (type strains).

Reference strains: besides the other strains included in the description of the species, LMG 12452 (=CCUG 12408=CDC B998) shares high DNA–DNA relatedness (Ursing and Bruun, 1991) and similar protein and RAPD profiles (Bernardet et al., 2005) with the type strain.

8. **Chryseobacterium indoltheticum** (Campbell and Williams 1951) Vandamme, Bernardet, Segers, Kersters and Holmes 1994a, 830[VP] [*Beneckea indolthetica* (Campbell and Williams 1951) Campbell 1957, 331; *Flavobacterium indoltheticum* Campbell and Williams 1951, 903]

in.dol.the'ti.cum. N.L. n. *indolum* indole; Gr. adj. *thetikos* positive; N.L. neut. adj. *indoltheticum* (*sic*), indole-positive.

Of the two strains isolated by Campbell and Williams (1951) during a screening for chitin-decomposing bacteria in marine mud samples of unknown origin, only one, the type strain, remains in culture collections. This strain is nonmotile, although it was originally described as motile by means of peritrichous flagella. This discrepancy casts a doubt on the whole original description; it is also possible that the available strain is not one of the original ones. Hence, the features presented in Table 44 were taken from Holmes et al. (1984a), Holmes (1992), Hugo et al. (2003), and Bernardet et al. (2005). Furthermore, another reference strain (LMG 13342) displays a clear positive reaction for the phenylalanine deaminase test, contrary to the type strain. Both strains grow well on tryptic soy agar and produce a dark brown, diffusible pigment on tyrosine agar (J.-F. Bernardet, unpublished data). Other characteristics are as given in the genus description and in Tables 43 and 44. Holmes et al. (1984a) considered this taxon a *species incertae sedis*, but recent polyphasic studies including 16S rRNA gene sequence data have clearly demonstrated that *Chryseobacterium indoltheticum* is a member of the genus *Chryseobacterium* (Bernardet et al., 2005; Hugo et al., 1999, 2003; Vandamme et al., 1994a). These investigators, however, did not succeed in finding new strains.

Source: marine mud samples of unknown origin.

DNA G+C content (mol%): about 34.0 in most studies (T_m); a surprisingly low value (30.0) was found by Hugo et al. (1999) (T_m).

Type strain: Weeks F37, ATCC 27950, CCUG 33445, CIP 103168, LMG 4025, NCIMB 2220.

Sequence accession no. (16S rRNA gene): AY468448 (type strain).

Reference strains: strain LMG 13342 (=PHLS F32=PHLS A 78/68) (16S rRNA gene sequence accession no. AY468444) exhibits very similar phenotypic features, whole-cell protein profiles and 16S rRNA gene sequence (more than 99%) to those of the type strain (Bernardet et al., 2005).

9. **Chryseobacterium joostei** Hugo, Segers, Hoste, Vancanneyt and Kersters 2003, 775[VP]

joos'te.i. N.L. gen. masc. n. *joostei* of Jooste, named after P.J. Jooste, who isolated the first strains of this species from South African raw milk.

Rods, approximately 0.5 µm in diameter and 1.0 µm in length. The cell wall is approximately 50 nm thick. Colonies on nutrient agar are smooth, shiny, bright yellow, circular with entire edges, and butyrous in consistency, becoming mucoid after incubation for 5 d. A brown, diffusible pigment is produced on tyrosine agar. Growth is observed on β-hydroxybutyrate (without production of β-hydroxybutyrate inclusion granules), but not on Simmons' citrate agar. All strains show phosphatase activity, but not gluconate oxidation, or arginine dihydrolase, lysine decarboxylase, or ornithine decarboxylase activities. Produces indole when using Kovac's reagent, but not when using Ehrlich's reagent. 3-Ketolactose is not produced. Hydrolyzes casein, gelatin, and Tween 20. Negative for the methyl red test; does not produce acetoin, does not tolerate 0.0075% (w/v) KCN, and cannot reduce 0.4% (w/v) selenite. In addition to the data listed in Table 44, acid is not produced from 1% (w/v) dulcitol, inositol, raffinose, rhamnose, or sorbitol. Lists of compounds that are utilized/not utilized as sole carbon sources and substrates that are hydrolyzed/not hydrolyzed in API ZYM galleries are given in the original species description. Other characteristics are as given in the genus description and in Tables 43 and 44.

Source: strains were isolated originally from raw milk samples taken from different farms and from tankers arriving at dairy factories across South Africa; they were distinguished from reference strains of the other *Chryseobacterium* species by their specific whole-cell protein profile (Hugo et al., 1999). More recently, two additional strains were retrieved from ulcerative lesions of Atlantic salmon (*Salmo salar*) in Finland (Bernardet et al., 2005).

DNA G+C content (mol%): 36–37 (type strain: 36.7) (T_m).

Type strain: Hugo Ix5a, LMG 18212, CCUG 46665, CIP 105533.

Sequence accession no. (16S rRNA gene): AJ271010 (type strain); AY468479 (LMG 18208).

Reference strains: LMG 18207 [strain Eh 9 of Jooste (1985)], LMG 18208 [strain Hb 7a of Jooste (1985)], LMG 18209 [strain Hb 7b of Jooste (1985)], LMG 18210 [strain Hbg 12b of Jooste (1985)], LMG 18211 [strain Ip 8b of Jooste (1985)], LMG 18213 [strain Ix 11 of Jooste (1985)], LMG 18214 [strain Sp 6a of Jooste (1985), LMG 18215 [strain J 68 of Welthagen and Jooste (1992)], LMG 18216 [strain J 75 of Welthagen and Jooste (1992)], LMG 18217 [strain J 79 of Welthagen and Jooste (1992)], LMG 22906 [=CIP 108621; strain UOF CR694 of Bernardet et al. (2005)]; and LMG 22907 [=CIP 108622; strain UOF CR1094 of Bernardet et al. (2005)].

Note: Chryseobacterium joosteii is the name given for the organism originally deposited in 16S rRNA sequence databases.

10. **Chryseobacterium piscium** de Beer, Hugo, Jooste, Vancanneyt, Coenye and Vandamme 2006, 1321[VP]

pis'ci.um. L. pl. gen. n. *piscium* of fish.

Rods, approximately 0.5 µm in diameter and 1.0 µm in length. Colonies on nutrient agar are shiny, yellow, and translucent with entire edges. Growth occurs in nutrient broth containing 5% NaCl. Strains test positive for gelatin and casein hydrolysis. The reaction on triple-sugar iron agar and 10% lactose is alkaline. All strains show positive reactions with the Biolog GN2 MicroPlate system for only four substrates, namely gentiobiose, D-mannose, succinic acid monomethyl ester, and acetic acid. The highest 16S rRNA gene sequence similarities are with *Chryseobacterium balustinum*, *Chryseobacterium scophthalmum*, and *Chryseobacterium indoltheticum* (99.3, 98.9, and 97.4%, respectively); DNA–DNA relatedness values with these three species are 57, 51, and 52%, respectively. The four *Chryseobacterium piscium* strains were originally distinguished from reference strains of the other *Chryseobacterium* species by their specific whole-cell protein profile, although *Chryseobacterium balustinum* and *Chryseobacterium indoltheticum* display very similar profiles. Other characteristics are as given in the genus description and in Tables 43 and 44.

Source: all four currently known strains have been isolated from fresh fish from the South Atlantic ocean off the South African coast during 1996 (Lingalo, 1997). Production of urea and phenylalanine deaminase suggests that *Chryseobacterium piscium* may be involved in spoilage of the fish.

DNA G+C content (mol%): 33.6 (HPLC; type strain).

Type strain: strain R-23621, LMG 23089, CCUG 51923.

Sequence accession no. (16S rRNA gene): AM040439.

Reference strains: LMG 23086 (=strain R-23611), LMG 23087 (=strain R-23616) and LMG 23088 (=strain R-23620).

11. **Chryseobacterium scophthalmum** (Mudarris, Austin, Segers, Vancanneyt, Hoste and Bernardet 1994) Vandamme, Bernardet, Segers, Kersters and Holmes 1994a, 830^{VP} (*Flavobacterium scophthalmum* Mudarris, Austin, Segers, Vancanneyt, Hoste and Bernardet 1994, 450)

scoph.thal'mum. N.L. n. *Scophthalmus* scientific name of a genus that encompasses the turbot; N.L. neut. adj. *scophthalmum (sic)* pertaining to turbot.

Rods, 0.5 µm in diameter and 1.0–2.0 µm in length; filaments occur. The cell wall is 50 nm thick. The presence of gliding motility initially reported (Mudarris and Austin, 1989) has not been confirmed during further investigations (Mudarris et al., 1994); J.-F. Bernardet, unpublished data). Good growth occurs on marine 2216E, nutrient, and tryptic soy agars at 25–30°C. Colonies are circular, smooth, convex with entire edges, and deep golden-yellow-pigmented. On tyrosine agar, a pink, diffusible pigment is produced by some strains (e.g., the type strain), progressively turning brown. The type strain yields a strong positive reaction in the phenylalanine deaminase test; the six other strains tested are negative. Metabolism is reportedly respiratory and fermentative. Arginine dihydrolase, and lysine and ornithine decarboxylases are not produced. The methyl red test is negative; acetoin is not produced. Blood (β-hemolysis occurs around the areas of growth), casein, gelatin, tributyrin, and Tweens 20, 40, 60, and 85 are degraded, but agar, cellulose, and chitin are not degraded. Lists of compounds that are utilized as carbon sources and substrates that are hydrolyzed in API ZYM galleries are given in the original species description. Sodium nitrate, vitamin-free Casamino acids, and yeast extract are utilized as nitrogen sources. Growth occurs in the presence of 0–4% (w/v) NaCl, but not in the presence of 5% NaCl. Other characteristics are as given in the genus description and in Tables 43 and 44. The highest 16S rRNA gene sequence similarities (98.9 and 97.8%) are with *Chryseobacterium balustinum* and *Chryseobacterium indoltheticum*, respectively (Li et al., 2003b), but the corresponding levels of DNA–DNA relatedness are only 44 and 37% (Mudarris et al., 1994).

Source: the type strain was isolated in 1987 from the gills of a diseased turbot (*Scophthalmus maximus*) in Scotland. Forty-nine other strains were isolated at the same time from healthy and diseased turbot and from seawater. Diseased fish suffered from gill hyperplasia and hemorrhagic septicemia. They displayed hemorrhages in the eyes, skin, and jaw; several internal organs also exhibited necrosis and hemorrhages (Mudarris and Austin, 1989; Mudarris et al., 1994). No other strains of *Chryseobacterium scophthalmum* have been found subsequently, even when collections of *Chryseobacterium* strains isolated from diseased fish have been extensively investigated (Bernardet et al., 2005).

DNA G+C content (mol%): 33.5–34.7 (type strain: 34.1) (T_m).

Type strain: Austin MM1, ATCC 700039, LMG 13028, CIP 104199, CCM 4109, NCIMB 13463, CCUG 33454.

Sequence accession no. (16S rRNA gene): AJ271009 (type strain).

Reference strains: Austin MM1A (=LMG 13029), Austin MM1B (=LMG 13030), Austin MM1D (=LMG 13031), Austin MM2B (=LMG 13032), Austin MM2C (=LMG 13033), and Austin MM4 (=LMG 13034).

Note: the synonym *Cytophaga scophthalmis* is listed in 1989 in the catalog of strains of the Czech Collection of Microorganisms.

12. **Chryseobacterium shigense** Shimomura, Kaji and Hiraishi 2005, 1905^{VP}

shi.gen'se. N.L. neut. adj. *shigense* pertaining to Shiga prefecture in Japan, the geographical area of isolation of the type strain.

Straight or slightly curved rods, approximately 0.6 µm in diameter; the published scanning electron micrograph shows rods approximately 2.0–4.0 µm in length. Good growth occurs on nutrient agar at 20–30°C. Colonies are deep yellow, circular, and shiny. Casein and gelatin are hydrolyzed. The highest 16S rRNA gene sequence similarity (95.7%) is with *Chryseobacterium joostei*. However, unlike *Chryseobacterium joostei*, *Chryseobacterium shigense* does not grow on MacConkey agar and does not produce acid from maltose or trehalose. The fatty acid composition is not given in detail in the original description, only the amounts of the two major fatty acids ($C_{15:0}$, 39.7%; and $C_{17:0}$ iso 3-OH, 19.6%) are specified. Other characteristics are as given in the genus description and in Tables 43 and 44.

Source: the type (and so far only) strain was isolated in Japan from a commercial fresh lactic acid beverage; it was considered as a part of the normal bacterial community of the preparation.

DNA G+C content (mol%): 36.6 (HPLC).

Type strain: GUM-Kaji, BAMY 1001, DSM 17126, NCIMB 14047.

Sequence accession no. (16S rRNA gene): AB193101.

13. **Chryseobacterium soldanellicola** Park, Jung, Lee, Lee, Do, Kim and Bae 2006b, 436^{VP}

sol.da.nel'li.co.la. N.L. n. *soldanella* the species epithet of a plant belonging to the genus *Calystegia*; L. suff. *-cola* (from L. n. *incola*) dweller; N.L. masc. n. *soldanellicola* dweller of *Calystegia soldanella.*

Rods of unspecified size. Good growth occurs on R2A, tryptic soy, and nutrient agars at 25–30°C. Colonies are circular with entire edges, shiny, and yellow, becoming mucoid after prolonged incubation. Growth occurs in the presence of 0–4% (w/v) NaCl within 14 d. The pH range for growth is 5–7 and the optimal pH for growth is 5. Acetoin is not produced. Negative for arginine dihydrolase, lysine decarboxylase, and tryptophan deaminase. Gelatin is hydrolyzed. Glucose, arabinose, mannose, and maltose are weakly assimilated. Compounds that are utilized as carbon sources are given in the original species description. *Chryseobacterium soldanellicola* has the lowest G+C content in the genus. Other characteristics are as given in the genus description and in Tables 43 and 44.

The highest 16S rRNA gene sequence similarity (97.2%) is with *Chryseobacterium indoltheticum*. The type strain displays antifungal activity against the plant pathogen *Fusarium oxysporum*.

Source: the type (and so far only) strain was isolated from the roots of beach morning glory (*Calystegia soldanella*) growing on a sand dune in the coastal area of Tae-an, Chungnam, Korea.

DNA G+C content (mol%): 28.8 (unspecified method).

Type strain: PSD 1-4, KCTC 12382, NBRC 100864.

Sequence accession no. (16S rRNA gene): AY883415.

14. **Chryseobacterium taeanense** Park, Jung, Lee, Lee, Do, Kim and Bae 2006b, 436[VP]

tae.an.en′se. N.L. neut. adj. *taeanense* pertaining to Tae-an, a geographical region in Chungnam, Korea, where the type strain was isolated.

Rods of unspecified size. Good growth on R2A, tryptic soy, and nutrient agars at 25–30°C. Colonies are circular with entire edges, shiny, and yellow, becoming mucoid after prolonged incubation. Growth occurs in the presence of 0–6% (w/v) NaCl within 14 d. The pH range for growth is 5–9 and the optimal pH for growth is 5. Acetoin is not produced. Negative for arginine dihydrolase, lysine decarboxylase, and tryptophan deaminase. Gelatin is hydrolyzed. Weakly assimilates glucose, arabinose, mannose, and maltose. Compounds that are utilized as carbon sources are given in the original species description. Other characteristics are as given in the genus description and in Tables 43 and 44. The highest 16S rRNA gene sequence similarity (97.8%) is with *Chryseobacterium taichungense*.

Source: the type (and so far only) strain was isolated from the roots of *Elymus mollis* (wild rye) growing on a sand dune in the coastal area of Tae-an, Chungnam, Korea.

DNA G+C content (mol%): 32.1 (unspecified method).

Type strain: PHA3-4, KCTC 12381, NBRC 100863.

Sequence accession no. (16S rRNA gene): AY883416.

15. **Chryseobacterium taichungense** Shen, Kämpfer, Young, Lai and Arun 2005, 1302[VP]

tai.chung′en.se. N.L. neut. adj. *taichungense* pertaining to Taichung, a province in Taiwan.

Rods, approximately 2.0 µm in length; diameter unspecified. Good growth occurs on nutrient, BHI, and tryptic soy agars at 25–30°C. Colonies are yellow, shiny, and circular with entire edges, becoming mucoid after prolonged incubation due to the production of extracellular substances. Grows in the pH range 6.0–9.0; optimal pH range for growth is 7.0–8.0. Gelatin is hydrolyzed. In addition to the data listed in Table 44, acid is produced weakly from adonitol, i-inositol, and L-rhamnose; no acid is produced from D-arabitol, dulcitol, erythritol, melibiose, methyl-α-D-glucoside, raffinose, or D-sorbitol. Lists of compounds that are utilized as carbon sources, substrates that show positive reactions in API ZYM galleries, and chromogenic substrates that are hydrolyzed are given in the original species description. Other characteristics are as given in the genus description and in Tables 43 and 44. The highest 16S rRNA gene sequence similarity (96.8%) is with *Chryseobacterium indologenes* and *Chryseobacterium gleum*.

Source: the type (and so far only) strain was isolated from tar-contaminated soil in Taichung, Taiwan.

DNA G+C content (mol%): not determined.

Type strain: Shen CC-TWGS1-8, CCUG 50001, CIP 108519.

Sequence accession no. (16S rRNA gene): AJ843132.

16. **Chryseobacterium taiwanense** Tai, Kuo, Lee, Chen, Yokota and Lo 2006, 1774[VP]

tai.wan.en′se. N.L. neut. adj. *taiwanense* pertaining to Taiwan, from where the type strain was recovered.

Rods, approximately 0.6–0.9 µm in diameter and 1.0–2.2 µm in length. Good growth on nutrient and tryptic soy agars at 30°C. Colonies on tryptic soy agar are yellowish, translucent, and shiny with entire edges. Growth occurs on tryptic soy agar containing 4% NaCl. The pH range for growth is 5.0–10.0 (optimum 6.0–8.0). Casein and gelatin are hydrolyzed. Lists of substrates that show positive reactions in API ZYM galleries and on GN2 MicroPlates are given in the original species description. Other characteristics are as given in the genus description and in Tables 43 and 44. The highest 16S rRNA gene sequence similarities (96.7–97.2%) are with *Chryseobacterium daecheongense*, *Chryseobacterium defluvii*, and *Chryseobacterium taichungense*, but DNA–DNA relatedness between the four type strains is only 8.5–24.2%.

Source: the type (and so far only) strain was isolated from farmland soil in Taiwan.

DNA G+C content (mol%): 36.8 (HPLC).

Type strain: Soil-3-27, BCRC 17412, IAM 15317, LMG 23355.

Sequence accession no. (16S rRNA gene): DQ318789.

17. **Chryseobacterium vrystaatense** de Beer, Hugo, Jooste, Willems, Vancanneyt, Coenye and Vandamme 2005, 2151[VP]

vry.staa.ten′se. N.L. neut. adj. *vrystaatense* pertaining to Vrystaat (Free State), the South African province where these bacteria were isolated.

Rods, approximately 0.5 µm in diameter and 1.0 µm in length. Colonies on nutrient agar are shiny, deep golden yellow, and translucent as single colonies, with entire edges. It should be noted that, although the table of differentiating characteristics in the original description of the species (de Beer et al., 2005) indicated no growth on MacConkey agar, the formal description stated that the species did grow on this medium; after personal communication with the author, it was confirmed that no growth was observed on MacConkey agar. All strains grow in nutrient broth containing 2% (w/v) NaCl; growth in 3 and 4% NaCl broth is strain-dependent (negative for the type strain). Reactions on triple-sugar iron agar and 10% lactose are alkaline. The ability to produce H_2S varies depending on incubation time on triple-sugar iron agar and sulfide indole motility medium. All strains show positive reactions using the Biolog GN2 MicroPlate system for Tweens 40 and 80, gentiobiose, α-D-glucose, D-mannose, trehalose, succinic acid monomethyl ester, acetic acid, and L-asparagine. Variable reactions are observed for L-aspartic acid, L-glutamic acid, glycyl-L-glutamic acid, L-serine, glycerol, dextrin, α-ketovaleric acid, L-alanyl-glycine, L-threonine, D-mannitol, uridine, thymidine, and inosine.

Other characteristics are as given in the genus description and in Tables 43 and 44. Whole-cell protein analysis delineated three subgroups in the collection of 36 isolates studied, but DNA–DNA relatedness values obtained among representative isolates of these subgroups were above 90%, demonstrating that they actually form a homogeneous genospecies (de Beer et al., 2005). The highest 16S rRNA gene sequence similarities are with *Chryseobacterium joostei*, *Chryseobacterium indologenes*, and *Chryseobacterium gleum* (96.9, 97.1, and 97.1%, respectively); the *Chryseobacterium vrystaatense* type strain shares 46% DNA–DNA relatedness with the *Chryseobacterium joostei* type strain.

Source: all *Chryseobacterium vrystaatense* strains were isolated from raw chicken-portion samples collected at different processing stages from a chicken processing plant in 2003 in the Free State province of South Africa (de Beer et al., 2005).

DNA G+C content (mol%): 37.0–37.6 (type strain: 37.1) (HPLC).

Type strain: de Beer 161, LMG 22846, CCUG 50970.

Sequence accession nos (16S rRNA gene): AJ871397 (type strain), AJ871398 (LMG 22954).

Reference strains: LMG 22847 (=strain R-23600), LMG 22848 (=strain R-23500), and LMG 22954 (=strain R-23533).

18. **Chryseobacterium wanjuense** Weon, Kim, Yoo, Kwon, Cho, Go and Stackebrandt 2006, 1502[VP]

wan.ju.en′se. N.L. neut. adj. *wanjuense* pertaining to Wanju, the Korean province where the type strain was isolated.

Rods, approximately 0.7–0.8 μm in diameter and 2.0–3.5 μm in length. Good growth occurs on R2A, nutrient, and tryptic soy agars at 28°C. Colonies on R2A agar are yellow with entire edges. Growth occurs with 0–2% NaCl (optimum 0–1% NaCl). The pH range for growth is 5.0–9.0 (optimum pH 7.0). Casein and gelatin are hydrolyzed. Chitin and carboxymethylcellulose are not hydrolyzed. Glucose is not fermented. Arginine dihydrolase activity is absent. Lists of substrates that show positive reactions in the API ZYM, 50 CH, 20NE, and ID 32 GN galleries are given in the original species description. Other characteristics are as given in the genus description and in Tables 43 and 44. The highest 16S rRNA gene sequence similarity (97.7%) is with *Chryseobacterium daecheongense*, but the two type strains share only 28% DNA–DNA relatedness.

Source: the type (and so far only) strain was isolated from greenhouse soil cultivated with lettuce in Korea.

DNA G+C content (mol%): 37.8 (HPLC).

Type strain: R2A10-2, KACC 11468, DSM 17724.

Sequence accession no. (16S rRNA gene): DQ256729.

Other organisms

1. **"Chryseobacterium proteolyticum"** Yamaguchi and Yokoe 2000, 3342

pro.te.o.ly′ti.cum. N.L. neut. n. *proteinum* protein; Gr. adj. *lutikos -ê -on* dissolving; N.L. neut. adj. *proteolyticum* (*sic*) protein-dissolving, proteolytic.

Rods, approximately 0.4–0.5 μm in diameter and 0.8–2.0 μm in length. Good growth occurs on nutrient agar at 30°C. Colonies are circular, and orange or light pinkish cream. The pH range for growth is 5–9; optimal range for growth is pH 6–8. Negative for 3-ketolactose and acetoin production. Lysine and ornithine decarboxylases, and arginine dihydrolase are negative. Casein and gelatin are hydrolyzed. In addition to the data listed in Table 44, acid is produced from soluble starch, but not from adonitol, inositol, or inulin. Other characteristics are as given in the genus description and in Tables 43 and 44. The fatty acid profile was not determined. The highest 16S rRNA gene sequence similarities are with *Chryseobacterium gleum* and *Chryseobacterium indologenes* (96.0 and 95.9%, respectively). The name of this well-described species has not been validly published and the strains have not been deposited in a recognized culture collection or been made available for research. "*Chryseobacterium proteolyticum*" has been shown to produce an enzyme with a great potential for industrial applications, the first protein-deaminating enzyme of microbial origin.

Source: the two strains were isolated from damp soil samples in a rice field (type strain) and on the bank of a brook from Tsukuba, Ibaraki, Japan.

DNA G+C content (mol%): 37.1 for both strains (HPLC).

Type strain: Yamaguchi and Yokoe 9670, FERM P-17664 (Patent Micro-organism Depository, National Institute of Bioscience and Human Technology, Tsukuba, Japan).

Sequence accession no. (16S rRNA gene): AB039830 (type strain).

Other reference strain: Yamaguchi and Yokoe 9671.

2. **"Chryseobacterium massiliae"** Greub, La Scola and Raoult 2004, 474

mas.si′li.ae. L. gen. n. *massiliae* of *Massilia*, the Latin name of Marseille.

A strain provisionally named "*Chryseobacterium massiliae*" was recovered from a nasal swab from a homeless woman suffering from rhinitis in a hospital in Marseille, France, using amoebal co-culture; it was one of the amoeba-resistant bacteria in the sample (Drancourt et al., 2004). The strain was deposited in culture collections, but no detailed study was published. It grows on medium used for growing strains of *Afipia* and *Legionella* (medium 23; CIP catalog of strains) at 30°C in the presence of 5% CO_2.

Source: human nasal swab.

Type strain: Greub 90B, CIP 107752, CCUG 51329.

Sequence accession no. (16S rRNA gene): AF531766.

3. "***Candidatus* Chryseobacterium timonae"** Drancourt, Berger and Raoult 2004, 2198

ti′mo.nae. N.L. gen. n. *timonae* of the hospital, La Timone.

Only the 16S rRNA gene sequence of this strain is available.

Source: the strain was isolated from the blood of a 73-year-old man suffering from ulcerative colitis in a hospital in Marseille, France.

Sequence accession no. (16S rRNA gene): AY244770.

Genus XI. **Cloacibacterium** Allen, Lawson, Collins, Falsen and Tanner 2006, 1314^{VP}

THE EDITORIAL BOARD

Clo.a.ci.bac.te'ri.um. L. fem. n. *cloaca* a sewer, canal; L. neut. n. *bacterium* a small rod; N.L. neut. n. *Cloacibacterium* a sewer rod.

Pleomorphic rods. Nonmotile. Gram-stain-negative. A flexirubin-type pigment is not produced. **Yellow to orange carotenoid-type pigments are produced. Facultatively anaerobic.** Catalase- and oxidase-positive. The major end product of glucose fermentation under both aerobic and anaerobic conditions is pyruvate. Cellular fatty acids consist mainly of branched-chain fatty acids, with $C_{13:0}$ iso, $C_{15:0}$ iso, $C_{15:1}$ iso, and $C_{17:0}$ iso 3-OH predominating. **The predominant respiratory quinone is MK-6.**

DNA G+C content (mol%): 31.

Type species: **Cloacibacterium normanense** Allen, Lawson, Collins, Falsen and Tanner 2006, 1314^{VP}.

Further descriptive information

In broth cultures, cells are up to 27 µm long, but cells from agar plate cultures are only 5–9 µm long. Colonies grown on tryptic soy agar for 48 h are 1.0–1.5 mm in diameter, round, entire, and waxy. The temperature range for growth is 18–36°C (optimum, 30°C). No growth occurs at 4°C or at 40°C or above. All strains grow at pH 7 and 8 (optimum, pH 7), and some grow at pH 6 and 9.

Numbers of *Cloacibacterium normanense* in raw sewage from two different treatment plants were 1.4×10^5 and 1.4×10^4 cells/ml, as estimated by a most probable number-PCR method (Allen et al., 2006). The organisms, however, were not detected by this method in treated effluent or in human stool specimens, indicating that *Cloacibacterium normanense* is unlikely to be a dominant member of the human gastrointestinal tract.

Most isolates tested are resistant to erythromycin (15 µg) and kanamycin (30 µg). All isolates so far tested are sensitive to carbenicillin (100 µg), chloramphenicol (30 µg), ciprofloxacin (5 µg), doxycycline (30 µg), gentamicin (10 µg), oxytetracycline (30 µg), sulfathiazole (0.25 mg), and tetracycline (30 µg).

Taxonomic comments

Analysis of 16S rRNA gene sequences indicates that the bacterial species most closely related to *Cloacibacterium normanense* are *Riemerella anatipestifer* (88.8% sequence similarity), *Riemerella columbina* (94.6% sequence similarity), and *Bergeyella zoohelcum* (94.4% sequence similarity).

Differentiation of the genus *Cloacibacterium* from related genera. Unlike *Cloacibacterium*, *Bergeyella* and *Riemerella* do not produce a yellow pigment. *Bergeyella* and *Riemerella* also have a higher DNA G+C content (35–37 and 33–37 mol%, respectively, vs 31 mol% for *Cloacibacterium*). Unlike *Cloacibacterium*, *Bergeyella* hydrolyzes urea, but not DNA or esculin; also, *Bergeyella* is non-saccharolytic and fails to form acid from glucose. In contrast to *Cloacibacterium*, the genus *Riemerella* is capnophilic and most strains of *Riemerella* grow at 42°C.

List of species of the genus *Cloacibacterium*

1. **Cloacibacterium normanense** Allen, Lawson, Collins, Falsen and Tanner 2006, 1314^{VP}

nor.man.en'se. N.L. neut. adj. *normanense* pertaining to the city of Norman, Oklahoma, USA, where the organism was first isolated.

Characteristics are as described for the genus, with the following additional features. No growth occurs on MacConkey agar. Indole-positive. Casein, esculin, gelatin, and starch are hydrolyzed, but alginate, agar, cellulose, chitin, hypoxanthine, pectin, urea, uric acid, and xanthine are not hydrolyzed. DNA is weakly hydrolyzed. Methyl red- and Voges–Proskauer-negative. Nitrate is not reduced. With the API system, acid is produced from α-cyclodextrin and mannose, but not from alanine, ribose, mannitol, sorbitol, lactose, trehalose, raffinose, hippurate, glycogen, melibiose, melezitose, sucrose, L-arabinose, D-arabitol, tagatose, or raffinose. Positive for the following enzyme activities: alkaline phosphatase, acid phosphatase, alanine arylamidase, arginine dihydrolase (weak reaction), arginine arylamidase, chymotrypsin, esterase C4 (weak reaction), ester lipase C8, α-glucosidase, β-glucosidase, glycine arylamidase, glycyl tryptophan arylamidase, proline arylamidase, leucine arylamidase, leucyl glycine arylamidase, phenylalanine arylamidase, alanine phenylalanine proline arylamidase, leucine arylamidase, naphthol-AS-BI-phosphohydrolase, pyroglutamic acid arylamidase, tyrosine arylamidase, histidine arylamidase, glutamyl glutamic acid arylamidase, serine arylamidase, and valine arylamidase. The following enzyme activities are absent: *N*-acetyl-β-glucosaminidase, α-arabinosidase, α-fucosidase, α-galactosidase, β-galactosidase, β-glucuronidase, glutamic acid arylamidase, α-mannosidase, β-mannosidase, methyl β-D-glucopyranoside, pyroglutamic acid arylamidase, lipase C14, and trypsin. The major fatty acids are $C_{13:0}$ iso (8–9%), $C_{15:0}$ iso (40–46%), $C_{15:1}$ iso (7–10%), and $C_{17:0}$ iso 3-OH (5–9%).

Source: the type strain was isolated from untreated human wastewater.

DNA G+C content (mol%): 31 (HPLC).

Type strain: NRS1, ATCC BAA-825, CCUG 46293, CIP 108613, DSM 15886.

Sequence accession no. (16S rRNA gene): AJ575430.

Genus XII. **Coenonia** Vandamme, Vancanneyt, Segers, Ryll, Köhler, Ludwig and Hinz 1999, 873[VP]

THE EDITORIAL BOARD

Coe.no'ni.a. Gr. n. *koinônia* community, association; N.L. fem. n. *Coenonia* refers to the association between these bacteria and a host.

Rods, 0.2–0.4 × 1.25–2.5 µm. Nonsporeforming. Nonmotile. Gram-stain-negative. **Microaerophilic**. Growth is optimal at 37°C in a microaerobic and CO_2-enriched atmosphere; it is weak under anaerobic conditions and absent in aerobic conditions. Catalase- and oxidase-positive. **Hyaluronidase- and chondroitin sulfatase-positive**. The major fatty acid components of all strains examined are branched-chain fatty acids, including $C_{13:0}$ iso, $C_{15:0}$ iso, $C_{15:0}$ anteiso, $C_{15:0}$ iso 3-OH, $C_{16:0}$ 3-OH, and $C_{17:0}$ iso 3-OH.

DNA G+C content (mol%): 35–36.

Type species: **Coenonia anatina** Vandamme, Vancanneyt, Segers, Ryll, Köhler, Ludwig and Hinz 1999, 873[VP].

Further descriptive information

Coenonia anatina causes an exudative septicemic disease in ducks and geese. Strains have been isolated as pure cultures under microaerobic conditions from samples of the respiratory tract (lungs, air sac fibrin, pericard) and brain of ducks and geese, always associated with signs similar to those of *Riemerella anatipestifer*-associated exudative septicemia.

In the northern part of Germany, *Coenonia anatina* strains have been recovered from over 30% of the cases of septicemia-related losses in commercial duck flocks in the period 1994–1995. Vaccination with an inactivated whole culture of *Coenonia anatina* strains has induced protection of ducklings against *Coenonia anatina*-associated disease. These data suggest its pathogenic role (Vandamme et al., 1999).

Differentiation of the genus *Coenonia* from other genera

Coenonia is positive for hyaluronidase and chondroitin sulfatase activities, whereas *Capnocytophaga* species are negative (Vandamme et al., 1999).

Taxonomic comments

Based on 16S rRNA sequence analysis, Vandamme et al. (1999) reported that *Coenonia anatina* exhibited a moderate but distinct relationship to *Capnocytophaga* species, with an overall 16S rRNA sequence similarity of 88.8–90.2%. Similarity values to other representatives of the phylum *Bacteroidetes* were 88.6% and lower.

List of species of the genus *Coenonia*

1. **Coenonia anatina** Vandamme, Vancanneyt, Segers, Ryll, Köhler, Ludwig and Hinz 1999, 873[VP]

a.na.ti'na. L. fem. adj. *anatina* of a duck.

In 16-h-old cultures on blood agar, the mean cell size is 0.2–0.4 × 1.25–2.5 µm. Some cells are spherically swollen, spindle or lemon-shaped, and arranged in short chains. Colonies are flat to convex, circular, nonpigmented to white, with entire edges and a smooth surface on blood agar. Nonhemolytic. Special growth factors are not required. Growth occurs on conventional media. No growth occurs on MacConkey agar or on Simmons citrate agar. Hyaluronidase and chondroitin sulfatase activity is present, but not urease, gelatinase, arginine dihydrolase, phenylalanine deaminase, ornithine decarboxylase, or lysine decarboxylase activities. Nitrate is not reduced. Esculin is hydrolyzed. Acetylmethylcarbinol is produced, but not indole. The methyl red reaction is negative. Citrate is not utilized. No alkali is formed from malonate. Acid is produced in a buffered single-substrate test (BSS medium) from D-glucose, D-fructose, maltose, D-mannose, lactose, D-galactose, *N*-acetylglucosamine, lactulose, and dextrin, but not from trehalose, adonitol, L-arabinose, dulcitol, *myo*-inositol, D-mannitol, D-sorbitol, sucrose, salicin, sorbose, or D-xylose. The following activities are present (API ZYM system): alkaline phosphatase, acid phosphatase, esterase C4, ester lipase C8, leucine, valine arylamidase, cystine arylamidase, naphthol-AS-BI-phosphohydrolase, β-galactosidase, α-glucosidase, *N*-acetyl-β-glucosaminidase, α-fucosidase, and trypsin. The following are not detected: lipase C14, chymotrypsin, β-glucuronidase, α-galactosidase, β-glucosidase, and α-mannosidase. The following reactions are positive (API ID 32E system): acid from glucose, α-glucosidase, β-glucosidase, α-galactosidase, β-galactosidase, α-maltosidase, lipase, *N*-acetyl-β-glucosaminidase, and L-aspartic acid arylamidase. The following reactions are negative: arginine dihydrolase, ornithine and lysine decarboxylase, urease, β-glucuronidase, indole production, and acidification of L-arabitol, galacturonate, 5-ketogluconate, mannitol, L-arabinose, D-arabitol, trehalose, rhamnose, inositol, maltose, adonitol, palatinose, sucrose, cellobiose, and sorbitol. The major fatty acid components of all strains examined are $C_{13:0}$ iso (~34%), $C_{14:0}$ (~3%), $C_{15:0}$ iso (~41%), $C_{15:0}$ anteiso (~3%), $C_{15:0}$ iso 3-OH (~6.5%), $C_{16:0}$ 3-OH (~4%) and $C_{17:0}$ iso 3-OH (~6.5%).

Source: the type strain was isolated from a Pekin duck in Germany.

DNA G+C content (mol%): 35–36 (T_m).

Type strain: 1502-91, CCUG 46148, CIP 106119, LMG 14382.

Sequence accession no. (16S rRNA gene): Y17612.

Additional remarks: *Coenonia anatina* was previously known as *Riemerella anatipestifer*-like taxon 1502 (Hinz et al., 1998).

Genus XIII. **Costertonia** Kwon, Lee and Lee 2006b, 1352VP

THE EDITORIAL BOARD

Cos.ter.to′ni.a. N.L. fem. n. *Costertonia* named after J.W. Costerton, a famous American biofilm microbiologist.

Rods, 0.35–0.41 × 0.50–0.57 µm, but sometimes they can sometimes be longer than 4 µm. Cells form irregular aggregates during growth in liquid media. **Motile.** Gliding motility is absent. Gram-stain-negative. Aerobic. **Colonies are orange** on marine agar due to non-diffusible carotenoid pigments. Flexirubin-type pigments are absent. Optimal growth occurs at pH 7.5–8.0 and 26–32°C. **NaCl, Ca^{2+}, and K^+ are all required for growth. Nitrate is reduced** to N_2. Oxidase- and catalase-positive. **The major respiratory quinone is MK-6.** The major cellular fatty acids are $C_{15:0}$ iso, $C_{15:1}$ iso, and $C_{15:0}$.

DNA G+C content (mol%): 35.8 (T_m).

Type species: **Costertonia aggregata** Kwon, Lee and Lee 2006b, 1352VP.

Enrichment and isolation procedures

Costertonia aggregata was isolated from a mature marine biofilm, including various marine algae, covering a rock-bed of the East Sea (Sea of Japan), Korea. A sample of biofilm was obtained with a razor blade and dispersed in sterile seawater. The dispersed biofilm was serially diluted in sterile seawater and the dilutions were spread onto marine agar 2216 (Difco). After incubation for 1 week at 25°C, a small orange colony was selected and propagated on marine agar.

Preservation can be accomplished in 20% glycerol solution at −80°C.

Differentiation of the genus *Costertonia* from related genera

Unlike *Maribacter* (91.2–92.4% 16S rRNA gene sequence similarity), *Zobellia* (90.7–91.5%), and *Muricauda* species, *Costertonia* has a requirement not only for Na^+, but also for other components of seawater, i.e., Ca^{2+} and K^+. It also contains a larger amount of $C_{15:0}$ iso fatty acid. The DNA G+C content of *Costertonia* is lower than that of *Muricauda* (35.8 mol% vs 41–45.4 mol%), and *Muricauda* species fail to hydrolyze gelatin and reduce nitrate. Unlike *Costertonia* species, *Zobellia* species hydrolyze agar and, in contrast to both *Maribacter* and *Zobellia*, the single *Costertonia* species does not exhibit gliding motility.

Taxonomic comments

16S rRNA gene sequence analysis has shown that *Costertonia aggregata* is related to members of the genera *Maribacter* (91.2–92.4% similarity), *Zobellia* (90.7–91.5%), and *Muricauda* (90.7–91.4%).

List of species of the genus *Costertonia*

1. **Costertonia aggregata** Kwon, Lee and Lee 2006b, 1352VP

ag.gre.ga′ta. L. fem. adj. *aggregata* joined together, referring to the formation of aggregates during cultivation in liquid media.

Characteristics are as described for the genus, with the following additional features. Growth occurs at 10–35°C, pH 6.5–9.0, and with 1.5–12.0% sea salts. Nitrate is reduced to N_2 (API 20 E test strip). Positive for β-glucosidase, β-galactosidase, urease, arginine dihydrolase, and protease. The following substrates can be used as carbon sources: acetic acid, *N*-acetyl-D-galactosamine, *N*-acetyl-D-glucosamine, *cis*-aconitic acid, adonitol, alaninamide, D-alanine, L-alanine, L-alanyl-glycine, γ-aminobutyric acid, 2-aminoethanol, L-arabinose, D-arabitol, L-asparagine, L-aspartic acid, bromosuccinic acid, cellobiose, citric acid, dextrin, L-erythritol, D-fructose, L-fucose, D-galactonic acid lactone, D-galactose, D-galacturonic acid, gentiobiose, D-gluconic acid, α-D-glucose, glucose 1-phosphate, glucose 6-phosphate, glucuronamide, D-glucuronic acid, L-glutamic acid, glycerol, DL-α-glycerol phosphate, glycogen, glycyl L-aspartic acid, glycyl L-glutamic acid, β-hydroxybutyric acid, L-histidine, γ-hydroxybutyric acid, *p*-hydroxyphenylacetic acid, hydroxy-L-proline, inosine, *myo*-inositol, α-ketoglutaric acid, α-ketovaleric acid, DL-lactic acid, α-D-lactose, lactulose, L-leucine, malonic acid, maltose, D-mannitol, D-mannose, melibiose, methyl β-D-glucoside, methyl pyruvate, L-ornithine, phenylethylamine, L-proline, propionic acid, putrescine, L-pyroglutamic acid, quinic acid, raffinose, L-rhamnose, D-saccharic acid, D-serine, L-serine, D-sorbitol, succinic acid, sucrose, L-threonine, trehalose, turanose, Tweens 40 and 80, uridine, urocanic acid, and xylitol as sole carbon sources. The dominant fatty acids are $C_{15:0}$ iso (39.7%), $C_{15:1}$ iso ω10 (22.4%), $C_{15:0}$ (7.8%), and $C_{16:1}$ ω9 (4.6%).

Source: isolated from a mature marine biofilm covering a rock-bed of the East Sea (Sea of Japan), Korea.

DNA G+C content (mol%): 35.8 (T_m).

Type strain: KOPRI 13342, JCM 13411, KCCM 42265.

Sequence accession no. (16S rRNA gene): DQ167246.

Genus XIV. **Croceibacter** Cho and Giovannoni 2003, 935VP (Effective publication: Cho and Giovannoni 2003, 82.)

JANG-CHEON CHO AND STEPHEN J. GIOVANNONI

Cro.cei.bac′ter. L. adj. *croceus* saffron-colored; N.L. masc. n. *bacter* rod; N.L. masc. n. *Croceibacter* saffron-colored rod shaped bacterium.

Rod-shaped cells with rounded ends. Gram-stain-negative. **Non-motile and no gliding motility**. Aerobic, mesophilic, and chemoheterotrophic. Endospores are not produced. **Carotenoid pigments are produced, but not flexirubin pigments**. NaCl is required for growth. **Gelatin, DNA, starch, casein, and elastin are degraded**. The major fatty acid types are branched acids and hydroxy acids, including $C_{17:0}$ 3-OH iso, $C_{15:0}$ iso, $C_{15:1}$ iso, and $C_{17:1}$ ω9*c* iso.

DNA G+C content (mol%): 34.8.

Type species: **Croceibacter atlanticus** Cho and Giovannoni 2003, 935VP (Effective publication: Cho and Giovannoni 2003, 82.).

Further descriptive information

Phylogenetic analyses based on 16S rRNA gene sequences show that the genus forms a distinct monophyletic clade with several uncultured environmental clones and belongs to the family *Flavobacteriaceae*.

Cells of *Croceibacter* are nonmotile, straight rod-shaped cells, approximately 1.9 µm long and 0.4 µm wide, dividing by binary fission. Transmission electron microscopy has shown that the cells lack flagella, endospores, intracellular granules, and vacuoles.

Colonies of *Croceibacter* can be seen easily on marine agar 2216 after 2 d culture at 25°C. They are 1.8–3.0 mm in diameter, bright orange or saffron-colored, uniformly circular, convex, opaque, and with a smooth surface after 5 d at 25°C. The major fatty acid types in the cell membrane are branched acids (38.9%) and hydroxy acids (41.1%). The predominant fatty acids are $C_{17:0}$ 3-OH iso (28.0%), $C_{15:0}$ iso (13.3%), $C_{15:1}$ iso (9.2%), and $C_{17:1}$ ω9*c* iso (9.4%).

Organisms are obligately aerobic, mesophilic, NaCl-requiring chemoheterotrophs. The temperature, pH, and NaCl concentration ranges for growth are 10–28°C (optimum, 20–23°C), pH 6.0–10.0 (optimum, pH 7.5–8.0), and 0.5–15% (w/v) NaCl (optimum, 3.0%). The type strain of *Croceibacter atlanticus* is catalase-positive and oxidase-negative. Urease and arginine dehydrolase activities are present. It is negative for indole production and denitrification activity and does not produce acid from glucose. It produces carotenoid pigments with wavelength absorbance spectral peaks at 318 and 483 nm. Photosynthetic gene clusters, bacteriochlorophyll *a*, and photosynthetic activity are not found in the *Croceibacter atlanticus* type strain. The only *Croceibacter* species identified thus far can degrade several high molecular mass compounds, including gelatin, DNA, starch, casein, and elastin after prolonged incubation for 2 weeks, but cannot degrade esculin, dextran, cellulose, alginate, chitin, or carrageenan. According to sole carbon source utilization tests, *Croceibacter* strains utilize a variety of carbohydrates, sugar alcohols, organic acids, and amino acids as sole carbon sources. The type strain of *Croceibacter atlanticus*, the type species of the genus, utilizes DL-glyceraldehyde, D-arabinose, D-galactose, D-fructose, trehalose, maltose, melezitose, D-mannose, D-mannitol, D-sorbitol, pyruvic acid, succinic acid, gluconic acid, L-glutamic acid, L-ornithine, L-proline, L-alanine, L-serine, and L-leucine as sole carbon sources.

The type strain of *Croceibacter atlanticus* is susceptible to chloramphenicol, nalidixic acid, tetracycline, erythromycin, vancomycin, rifampin, and benzylpenicillin, but resistant to kanamycin, carbenicillin, streptomycin, ampicillin, puromycin, gentamicin, and cycloheximide.

Enrichment and isolation procedures

The original liquid culture of *Croceibacter atlanticus* HTCC2559^T was obtained at a depth of 250 m from the Sargasso Sea by high throughput culturing techniques (Connon and Giovannoni, 2002) using a low nutrient heterotrophic medium [LNHM, consisting of filtered (0.2 µm pore diameter) and autoclaved seawater amended with 1.0 µM NH_4Cl and 0.1 µM KH_2PO_4] supplemented with 0.001% (w/v) of each of D-glucose, D-ribose, succinic acid, pyruvic acid, glycerol, and *N*-acetyl-D-glucosamine, and 0.002% (v/v) ethanol (Rappé et al., 2002). Single colonies of the strain were obtained easily by spreading the liquid culture on marine agar 2216 after incubation for 5 d at 25°C. Therefore, it is likely that *Croceibacter* can be obtained by spreading seawater on marine agar using a serial dilution plating method.

Maintenance procedures

Frozen stocks as glycerol suspensions (10% glycerol) in liquid nitrogen or at −70°C are routinely used to start a new culture. Either colonies (scraped from the surface of agar media) or broth cultures can be used for preparing glycerol stocks. Lyophilization of liquid cultures is also recommended for long-term storage. Frozen stocks and lyophilization stocks are stable for at least 2 years (confirmed by survival tests). Working stock cultures can be maintained at 4°C only for 3 weeks, so frequent subculturing is required for short-term storage.

Differentiation of the genus *Croceibacter* from other genera

The physiological property of the genus that best differentiates it from related genera is its growth temperature range. The type strain of *Croceibacter atlanticus* does not grow at either 4°C or 30°C, which differentiates it from other genera in the family *Flavobacteriaceae*. Psychrophilic members of the *Flavobacteriaceae* grow at 4°C, but not 30°C, and other members of the family cannot grow at 4°C, but do grow at 30°C or higher. The genus *Croceibacter* can also be differentiated from the genera *Aequorivita* and *Arenibacter* by its different cellular fatty acid profiles and macromolecule degradation patterns.

Taxonomic comments

The genus *Croceibacter* currently contains only one species, *Croceibacter atlanticus*. Phylogenetic analyses based on 16S rRNA gene sequences using different treeing algorithms (distance analyses, maximum-parsimony, and maximum-likelihood) have shown that the genus belongs to the family *Flavobacteriaceae* in the phylum *Bacteroidetes*. 16S rRNA gene sequence similarities to members of the genera *Leeuwenhoekiella*, *Aequorivita*, *Psychroserpens*, and *Arenibacter* in the family *Flavobacteriaceae* are only 88–92%. Because of these low sequence similarities, the genus *Croceibacter* forms a very distinct clade within the family. Classification of *Croceibacter* as a genus of the family is based not only on 16S rRNA gene sequence analysis, but also on phenotypic and chemotaxonomic properties.

The type strain of *Croceibacter atlanticus* is most closely related to several as-yet-uncultured environmental clones retrieved from sea-ice, arctic deep-sea water, and coastal mud by high similarity values (98.5–99.5%), indicating that the major habitat of the genus *Croceibacter* is marine environments.

List of species of the genus *Croceibacter*

1. **Croceibacter atlanticus** Cho and Giovannoni 2003, 935^{VP} (Effective publication: Cho and Giovannoni 2003, 82.)

at.lan′ti.cus. L. masc. adj. *atlanticus* of the Atlantic Ocean, referring to the isolation of the species from the Atlantic Ocean.

Characteristics of the species are as described for the genus.
Source: marine environments.
DNA G+C content (mol%): 34.8 ± 0.1 (HPLC).
Type strain: HTCC2559, ATCC BAA-628, KCTC 12090.
Sequence accession no. (16S rRNA gene): AY163576.

Genus XV. **Dokdonia** Yoon, Kang, Lee and Oh 2005a, 2326[VP]

The Editorial Board

Dok.do'ni.a. N.L. fem. n. *Dokdonia* named after Dokdo, an island located on the East Sea in Korea, from where the organisms were isolated.

Rods, 0.3–0.6 × 1.5–5.0 µm. Nonmotile. **Gliding motility does not occur.** Nonsporeforming. Gram-stain-negative. **Strictly aerobic.** Colonies are yellow. Catalase- and oxidase-positive. **Growth does not occur in the absence of NaCl.** Nitrate is not reduced. **Casein is hydrolyzed, but not agar and starch. Acid is not formed from a variety of carbohydrates.** The major fatty acids are $C_{15:0}$ iso, $C_{17:0}$ 3-OH iso, and $C_{15:1}$ iso. The predominant menaquinone is MK-6. Isolated from seawater.

DNA G+C content (mol%): 38 (HPLC).

Type species: **Dokdonia donghaensis** Yoon, Kang, Lee and Oh 2005a, 2326[VP].

Enrichment and isolation procedures

Dokdonia donghaensis was isolated from seawater by standard dilution plating techniques on marine agar 2216 (MA; Difco) with incubation at 25°C.

Differentiation of the genus *Dokdonia* from related genera

Dokdonia differs from *Cellulophaga* by its lack of gliding motility, its failure to hydrolyze agar and starch, and its inability to form acid from any of a large variety of carbohydrates.

Taxonomic comments

As indicated by 16S rRNA gene sequence analysis by Yoon et al. (2005a), the nearest neighbors of *Dokdonia* are species of the genus *Cellulophaga* (90.4–92.2% sequence similarity). *Dokdonia* showed less than 91.7% similarity to the other species of the *Flavobacteriaceae*.

Yoon et al. (2005a) reported that the two available strains of *Dokdonia donghaensis* exhibited a mean DNA–DNA relatedness level of 78%.

List of species of the genus *Dokdonia*

1. **Dokdonia donghaensis** Yoon, Kang, Lee and Oh 2005a, 2326[VP]

dong.ha.en'sis. N.L. fem. adj. *donghaensis* of or belonging to Donghae, the Korean name of the East Sea of Korea, where Dokdo is located and from where the organism was isolated.

The description is as given for the genus, with the following additional features. Colonies on MA are circular, slightly convex, glistening, smooth, yellow, and 1.0–2.0 mm in diameter after 3 d at 30°C. Growth occurs at 4 and 35°C and optimum growth is at 30°C; no growth occurs at 36°C. Optimum pH is 7.0–8.0; growth occurs at pH 5.5, but not at 5.0. Optimal growth occurs in the presence of 2% (w/v) NaCl; no growth occurs in the presence of 7% NaCl. Anaerobic growth does not occur on MA or on MA supplemented with nitrate. Esculin and Tween 60 are hydrolyzed, but not hypoxanthine or xanthine. Indole and H_2S are not produced. Arginine dihydrolase, lysine decarboxylase, ornithine decarboxylase, and tryptophan deaminase are absent. The following enzyme activities are present (API ZYM system): alkaline phosphatase, esterase (C4), esterase lipase (C8), leucine arylamidase, and valine arylamidase. Negative for *N*-acetyl-β-glucosaminidase, acid phosphatase, α-chymotrypsin, cystine arylamidase, α-fucosidase, α-galactosidase, β-galactosidase, α-glucosidase, β-glucosidase, β-glucuronidase, lipase (C14), α-mannosidase, naphthol-AS-BI-phosphohydrolase, and trypsin activities. Growth occurs on peptone and tryptone as sole carbon and nitrogen sources, but not on DL-aspartate, Casamino acids, D-cellobiose, D-fructose, D-galactose, D-glucose, L-glutamate, L-leucine, L-proline, D-ribose, sucrose, or trehalose. No acid is produced from L-arabinose, D-cellobiose, D-fructose, D-galactose, D-glucose, *myo*-inositol, lactose, maltose, D-mannitol, D-mannose, melibiose, raffinose, L-rhamnose, D-ribose, sucrose, trehalose, D-sorbitol, or D-xylose. Susceptible to carbenicillin, cephalothin, chloramphenicol, lincomycin, oleandomycin, penicillin G, and tetracycline, but not to ampicillin, polymyxin B, gentamicin, novobiocin, kanamycin, neomycin, and streptomycin. The major fatty acids (>10% of total fatty acids) are $C_{15:0}$ iso, $C_{17:0}$ 3-OH iso, and $C_{15:1}$ iso. The major polar lipids are phosphatidylethanolamine, unidentified phospholipids, an unidentified glycolipid, and an amino-group-containing lipid that is ninhydrin-positive.

Source: the type strain was isolated from seawater collected near Dokdo Island, Korea.

DNA G+C content (mol%): 38 (HPLC).

Type strain: DSW-1, DSM 17200, KCTC 12391.

Sequence accession no. (16S rRNA gene): DQ003276.

Genus XVI. **Donghaeana** Yoon, Kang, Lee and Oh 2006c, 190[VP]

The Editorial Board

Dong.hae.a'na. N.L. n. *Donghae* the Korean name of the East Sea in Korea; L. fem. suff. *-ana* suffix used with the sense of belonging to; N.L. fem. adj. used as a substantive *Donghaeana* pertaining to the East Sea of Korea, where the island of Dokdo is located.

Rods or elongated rods, 0.4–0.6 × 1.0–30.0 µm. Nonmotile. Gliding motility is not present. Nonsporeforming. Gram-stain-negative. Strictly aerobic. **Growth does not occur in the absence of NaCl.** Nitrate is not reduced. Anaerobic growth does not occur with nitrate. **Casein, starch, and Tweens 20 and 40 are hydrolyzed, but not agar, gelatin, or urea.** β-Galactosidase and α-glucosidase are not produced. Acid is not produced from glucose or maltose. **Glucose and sucrose are not utilized.** The predominant menaquinone is MK-6. The major polar lipids are phosphatidylethanolamine, unidentified phospholipids, and amino-group-containing lipids that are ninhydrin-positive.

DNA G+C content (mol%): 36.9.

Type species: **Donghaeana dokdonensis** Yoon, Kang, Lee and Oh 2006c, 190[VP].

Enrichment and isolation procedures

Donghaeana dokdonensis was isolated from seawater by a standard dilution plating technique on marine agar 2216 (MA; Difco) with incubation at 20°C.

Differentiation of the genus *Donghaeana* from related genera

Donghaeana is similar to *Nonlabens* in many characteristics, but differs in its inability to grow at 37°C, its hydrolysis of casein, but not gelatin, its inability to utilize glucose and sucrose, and a higher DNA G+C content (36.9 mol% vs 33.6 mol%). *Donghaeana* differs from *Stenothermobacter* by its lack of gliding motility, the absence of β-galactosidase and α-glucosidase activities, its hydrolysis of casein, but not gelatin, its inability to utilize glucose and sucrose, and a slightly lower DNA G+C content (36.9 mol% vs 41.0 mol%).

Taxonomic comments

Using 16S rRNA gene sequence analysis, the nearest neighbors of *Donghaeana dokdonensis* are *Nonlabens tegetincola* and *Stenothermobacter spongiae*, with sequence similarity values of 93.3–98.2% (Yoon et al., 2006c).

List of species of the genus *Donghaeana*

1. **Donghaeana dokdonensis** Yoon, Kang, Lee and Oh 2006c, 190[VP]

dok.do.nen'sis. N.L. fem. adj. *dokdonensis* of or belonging to Dokdo, a Korean island, where the type strain was isolated.

Characteristics are as described for the genus, with the following additional features. After 3 d at 25°C, colonies on MA are 0.8–1.4 mm in diameter, circular, convex, glistening, smooth, and orange. Temperature range for growth is 4–32°C; optimum is 25°C. Optimum pH is 7.0–8.0; growth occurs at pH 5.5, but not pH 5.0. Optimal NaCl concentration is 2%; no growth occurs at >8% NaCl. Esculin and Tween 60 are hydrolyzed, but hypoxanthine, xanthine, and L-tyrosine are not. Indole is not produced. Arginine dihydrolase, lysine decarboxylase, ornithine decarboxylase, and tryptophan deaminase are absent. The following enzymes are present (API ZYM system): esterase (C4), esterase lipase (C8), leucine arylamidase, valine arylamidase, acid phosphatase, and naphthol-AS-BI-phosphohydrolase. The following are not present: *N*-acetyl-β-glucosaminidase, α-chymotrypsin, cystine arylamidase, α-fucosidase, α-galactosidase, β-glucuronidase, α-glucosidase, β-glucosidase, lipase (C14), α-mannosidase, and trypsin. Growth occurs on peptone and tryptone as sole carbon and nitrogen sources, but no growth occurs on DL-aspartate, Casamino acids, cellobiose, D-fructose, D-galactose, L-glutamate, L-leucine, L-proline, D-ribose, or trehalose. No acid is produced from L-arabinose, D-fructose, D-galactose, *myo*-inositol, lactose, D-mannitol, D-mannose, melezitose, melibiose, raffinose, L-rhamnose, D-ribose, D-sorbitol, sucrose, trehalose, or D-xylose. Susceptible to chloramphenicol, cephalothin, novobiocin, and lincomycin, but not to polymyxin B, ampicillin, gentamicin, kanamycin, or neomycin. The major fatty acids (>10% of total fatty acids) are $C_{15:0}$ iso (19.2%), $C_{15:0}$ anteiso (11.1%), $C_{17:0}$ 3-OH iso (10.5%), and $C_{15:0}$ (10.2%).

Source: the type strain was isolated from seawater.

DNA G+C content (mol%): 36.9 (HPLC).

Type strain: DSW-6, DSM 17205, KCTC 12402.

Sequence accession no. (16S rRNA gene): DQ017065.

Genus XVII. **Elizabethkingia** Kim, Kim, Lim, Park and Lee 2005b, 1291[VP]

Brita Bruun and Jean-François Bernardet

E.liz.a.beth.kin'gi.a. N.L. fem. n. *Elizabethkingia* named in honor of Elizabeth O. King, who first described bacteria associated with infant meningitis, notably [*Flavobacterium*] *meningosepticum* in 1959.

Straight, single rods with rounded ends, typically 0.5 µm wide and variable in length, often forming filaments. Do not form endospores. Spherical degenerative forms do not appear in ageing liquid cultures. **Nonmotile.** Flagellar motility, gliding movement, and swarming growth have not been reported. **Colonies are typically nonpigmented or weakly yellow pigmented**, circular, convex, smooth and shiny, with entire edges, and up to 2 mm in diameter. A strong aromatic odor is produced. Gram-stain-negative. **Obligately aerobic, having a strictly respiratory type of metabolism** with oxygen as the terminal electron acceptor. **Nitrate is not reduced.** Chemoorganotrophic. Growth occurs at 22–37°C, but not at 5 or 42°C. Growth occurs readily on the usual commercial media. No growth factors are required. Growth on seawater media is possible. Growth on cetrimide and MacConkey agars is slow and weak, but is also strain-dependent. **Catalase, oxidase, phosphatase, and β-galactosidase activities are present.** Acid, but no gas, is produced from a number of carbohydrates in media with low peptone concentrations. Strong proteolytic activity occurs. Esculin, gelatin, and casein are hydrolyzed, but agar is not. Malonate is not utilized. **Resistant to a wide range of antimicrobial agents.** Menaquinone MK-6 is the predominant respiratory quinone. Predominant cellular fatty acids are $C_{15:0}$ iso, $C_{17:0}$ iso 3-OH, $C_{17:1}$ iso ω9*c*, and summed feature 4 (comprising $C_{15:0}$ iso 2-OH and/or $C_{16:1}$ ω7*c*/*t*). As they are ubiquitous in soil and freshwater, **strains of *Elizabethkingia meningoseptica* occasionally occur in the hospital environment and clinical specimens; they are opportunistic pathogens of humans and various animals.**

DNA G+C content (mol%): 35.0–38.2.

Type species: **Elizabethkingia meningoseptica** (King 1959) Kim, Kim, Lim, Park and Lee 2005b, 1291[VP] (*Flavobacterium meningosepticum* King 1959, 247; *Chryseobacterium meningosepticum* Vandamme, Bernardet, Segers, Kersters and Holmes 1994a, 830).

Further descriptive information

The genus *Elizabethkingia*, which contains *Elizabethkingia meningoseptica*, the most important clinical species of the family *Flavobacteriaceae*, and *Elizabethkingia miricola* (Li et al., 2003a), was recently separated from the genus *Chryseobacterium* (Kim et al., 2005b). However, members of the two genera still share a very considerable number of phenotypic features, as well as most of their taxonomic history (Bernardet et al., 2006). Hence, this chapter will focus on the few differential characteristics between *Elizabethkingia* and *Chryseobacterium* and readers are invited to refer to the *Chryseobacterium* chapter in this volume of the *Manual*. In particular, the phylogenetic tree (Figure 40) and the tables showing fatty acid composition (Table 43) and phenotypic characteristics (Table 44) that appear in the *Chryseobacterium* chapter all include the two *Elizabethkingia* species as well as the closest phylogenetic relatives of the two genera; they will consequently not be repeated here.

Phylogenetic position. Analysis of 16S rRNA gene sequences shows that the genus *Elizabethkingia* belongs to the family *Flavobacteriaceae*, phylum *Bacteroidetes*. Interspecific 16S rRNA gene sequence similarity values may be higher than 98%. Together with the genera *Chryseobacterium*, *Riemerella*, *Bergeyella*, *Kaistella*, *Sejongia*, *Ornithobacterium*, *Empedobacter*, and *Weeksella*, *Elizabethkingia* belongs to the smaller of the two well-defined clades in the family (see Figure 40 in the treatment of *Chryseobacterium* in this volume). In contrast to most members of the larger clade, it shares with these genera a number of significant traits, such as halophilic and psychrophilic characteristics and the absence of gliding motility. Most nonpigmented and pathogenic members of the family belong to the smaller clade.

The separate position of [*Chryseobacterium*] *meningosepticum* compared to the other *Chryseobacterium* species was established in 1994 when the genus *Chryseobacterium* was described (Vandamme et al., 1994a) following extensive phylogenetic investigations (Bauwens and De Ley, 1981; Mudarris et al., 1994; Segers et al., 1993a) of members of the phylogenetic branch that would eventually become the family *Flavobacteriaceae* (Bernardet et al., 1996). However, because of the low resolution of the technique used at that time (i.e., DNA–rRNA hybridization) and because the number of shared characteristics far exceeded that of differential ones, it was decided to include the organism known as [*Flavobacterium*] *meningosepticum* (which grouped strains previously known as CDC group IIa) in the new genus *Chryseobacterium*, while acknowledging that it was the most aberrant member of the genus and consequently not a good candidate for the type species (Vandamme et al., 1994a). In 2003, a novel species [*Chryseobacterium*] *miricola* (Li et al., 2003a) was shown to share the separate position of [*Chryseobacterium*] *meningosepticum* that had been confirmed and determined more precisely in the meantime using 16S rRNA gene sequence analysis. As shown in Figure 40 in the *Chryseobacterium* chapter and in other published phylogenetic trees (Bernardet et al., 2006; de Beer et al., 2005; Kim et al., 2005b; Li et al., 2003a), [*Chryseobacterium*] *meningosepticum* and [*Chryseobacterium*] *miricola* are not the closest relatives of the genus *Chryseobacterium* according to 16S rRNA gene sequence analysis as the three branches comprising (i) the genera *Riemerella* and *Bergeyella*, (ii) the two *Sejongia* species, and (iii) *Kaistella koreensis* fall in between.

Polyphasic investigation of six strains received as [*Chryseobacterium*] *meningosepticum* and of the [*Chryseobacterium*] *miricola* type strain by Kim et al. (2005b) showed that members of the two species share 97.8–98.4% 16S rRNA gene sequence similarity with each other and only approximately 91–94% sequence similarity with the other *Chryseobacterium* species; these values are similar to those reported in the description of [*Chryseobacterium*] *miricola* (Li et al., 2003a). The proposal of the new genus *Elizabethkingia* to accommodate [*Chryseobacterium*] *meningosepticum* and [*Chryseobacterium*] *miricola* was based on their phylogenetic distance to other *Chryseobacterium* species, as well as on differences in fatty acid profiles and a few phenotypic features (Kim et al., 2005b). The results of DNA–DNA hybridization experiments between and among members of the genera *Elizabethkingia* and *Chryseobacterium* have been reviewed recently (Bernardet et al., 2006).

Cell morphology. Cells of *Elizabethkingia* occur as straight rods of variable length, often 1.0–3.0 μm long and about 0.5 μm wide, with rounded ends and parallel sides. Filament formation is environmentally influenced and occurs in broth cultures, old cultures, and clinical specimens, such as cerebrospinal fluid. The presence of a capsule has been reported in some *Elizabethkingia meningoseptica* strains following their passage on mice (Holmes et al., 1984a). In contrast to members of some other genera in the family *Flavobacteriaceae* (e.g., *Flavobacterium*, *Gelidibacter*, *Tenacibaculum*, and *Polaribacter*, see the chapter Family *Flavobacteriaceae* in this volume), *Elizabethkingia* strains do not produce spherical degenerative forms in ageing liquid cultures. Also, in contrast to many members of the family, cells of *Elizabethkingia* species are devoid of gliding motility.

Cell-wall composition. Table 43 of the *Chryseobacterium* chapter in this volume presents the fatty acid composition of members of the genus *Elizabethkingia* and of related genera. The fatty acid compositions of the two *Elizabethkingia* species as determined by Kim et al. (2005b) are similar. It is important to note that the fatty acid profiles of the two *Elizabethkingia miricola* strains reported in this study differ significantly from the composition given previously for the type strain (Li et al., 2003a). Whether these discrepancies result from different growth conditions cannot be demonstrated since conditions are not specified or incompletely specified in the corresponding publications. Although *Elizabethkingia* strains contain mainly the same fatty acids as other members of the clade, differences in the proportions of some components allow strains to be differentiated from *Chryseobacterium*, *Empedobacter*, *Riemerella*, *Sejongia*, *Kaistella*, and *Bergeyella* strains, which all display rather distinct fatty acid compositions (Bernardet et al., 2006; Hugo et al., 1999). Compared with these organisms, strains of *Elizabethkingia* contain rather low amounts of $C_{17:1}$ iso ω9*c* and a high proportion of summed feature 4 (comprising $C_{15:0}$ iso 2-OH and/or $C_{16:1}$ ω7*c*/*t*). Branched-chain fatty acids account for 80% and 88% of all fatty acids in *Elizabethkingia meningoseptica* and *Elizabethkingia miricola*, respectively (Li et al., 2003a; Yabuuchi et al., 1983). As for other members of the family *Flavobacteriaceae*, the fatty acid profile is mainly a chemotaxonomic marker at the genus level (Bernardet and Nakagawa, 2006).

Fine structure. No study specifically investigating the fine structure of *Elizabethkingia meningoseptica* cells has been published. Electrón micrographs of the bacterium in samples of infected mice lungs were obtained recently in the course of a study evaluating the virulence mechanisms of *Elizabethkingia meningoseptica* on a murine pulmonary infection model (Lin et al., 2006) (see below, section Pathogenicity). Micrographs revealed the usual structure common to all Gram-stain-negative bacteria, but it was not possible to assess whether cells of *Elizabethkingia meningoseptica* displayed the thick cell wall reported in some *Chryseobacterium* species (see the corresponding chapter in this volume).

Colonial or cultural characteristics. Colonies of *Elizabethkingia* strains are circular, convex, translucent to opaque, smooth, and shiny, with entire edges. Similar to the colonies of many *Chryseobacterium* strains, their butyrous consistency turns mucoid after a few days of incubation (Kim et al., 2005b). Colonies of *Elizabethkingia miricola* are very sticky on solid media (Kim et al., 2005b). Broth cultures are usually uniformly turbid.

On commonly used media such as tryptic soy, blood, and nutrient agars, colonies are usually nonpigmented (i.e., white to creamy-white) or very pale yellow. Colonies of *Elizabethkingia miricola* are "white–yellow" according to Li et al. (2003a) and Kim et al. (2005b). The latter authors also described as white–yellow the colonies of the five *Elizabethkingia meningoseptica* strains they studied. Production of pigment by *Elizabethkingia meningoseptica* is strain-dependent; however, a weak yellow pigment occurs in the DNA groups containing the neonatal meningitis and bacteremia strains, whereas the other DNA groups are nonpigmented (see below; Bruun and Ursing, 1987). The type of pigment produced weakly by some *Elizabethkingia* strains is discussed below. The production of a diffusible, pinkish-brown melanic pigment on tyrosine agar by *Elizabethkingia meningoseptica* is strain-dependent (Bernardet et al., 2005). No hemolysis occurs on blood agar, but the medium may show a green discoloration (Holmes et al., 1984a).

Although no particular odor was reported for strains of *Elizabethkingia meningoseptica* (Bruun and Ursing, 1987; Holmes et al., 1984a; Kim et al., 2005b; Li et al., 2003a) and *Elizabethkingia miricola* (Kim et al., 2005b; Li et al., 2003a), all *Elizabethkingia meningoseptica* strains studied were found to give off a rather strong aromatic odor similar to that of *Chryseobacterium* strains (J.-F. Bernardet, unpublished data; B. Bruun, unpublished data).

Nutrition and growth conditions. *Elizabethkingia* strains grow readily on various commercially available media such as blood, tryptic soy, brain-heart infusion (BHI), and nutrient agars. Visible colonies (1.0–1.5 mm in diameter) appear within 24 h on nutrient agar. Growth of *Elizabethkingia meningoseptica* on MacConkey agar varies among strains (Holmes et al., 1984b; Weyant et al., 1996). Growth of the *Elizabethkingia miricola* type strain on MacConkey agar was described as slow by Li et al. (2003a); however, Kim et al. (2005b) reported that the two strains of this species grow well on this medium. Very slow growth of *Elizabethkingia miricola* occurs on cetrimide agar (Li et al., 2003a). Growth on β-hydroxybutyrate has been reported for *Elizabethkingia meningoseptica* (Holmes et al., 1984a).

Both species grow in ambient air and at temperatures of 22–37°C. Growth is more limited at room temperature (18–20°C) and no growth is observed at 5 or 42°C. The growth rate of the *Elizabethkingia miricola* type strain on BHI agar plates was similar at 30 and 37°C (Li et al., 2003a). The growth curve of several *Elizabethkingia meningoseptica* isolates in Luria broth at 37°C was determined by Lin et al. (2006); all strains reached exponential phase within 6 h. Strains of the *Elizabethkingia meningoseptica* DNA groups comprising the neonatal meningitis and bacteremia isolates could grow at 40°C, in contrast to the other DNA groups (Bruun and Ursing, 1987); interestingly, these strains were the same as those that were weakly pigmented (see above), the other strains being unable to grow at 40°C and devoid of pigment. Some of the *Elizabethkingia meningoseptica* isolates studied by Olsen (1966) grew at up to 41°C, whereas others ceased growing at 38°C; the author suggested that this difference could be related to the climate in the country of isolation.

Both *Elizabethkingia* species are halotolerant – a feature observed in members of most *Chryseobacterium* species. Strains of *Elizabethkingia meningoseptica* are able to grow on marine agar 2216E (Difco) (Bernardet et al., 2005). The ability to grow in the presence of 3% NaCl in broth culture varies among strains. The type strain of *Elizabethkingia miricola* can tolerate seawater (Li et al., 2003a).

Metabolism and metabolic pathways. Available data on the metabolism of *Elizabethkingia meningoseptica* mainly result from three sources that have included rather extensive collections of strains (B. Holmes and R.J. Owen, unpublished data cited by Holmes et al., 1984a, b; Bruun and Ursing, 1987; Weyant et al., 1996). Recent studies have focused on a few *Elizabethkingia meningoseptica* strains and on the only two available strains of *Elizabethkingia miricola* (Bernardet et al., 2005; Kim et al., 2005b; Li et al., 2003a). *Elizabethkingia* strains are chemo-organotrophs with a strictly respiratory type of metabolism (Bernardet et al., 2002, 2006; Vandamme et al., 1994a). The type strain of *Elizabethkingia miricola* and all *Elizabethkingia meningoseptica* strains grow well in the presence of 5–10% (v/v) CO_2 (B. Bruun, unpublished data; Li et al., 2003a). Members of *Elizabethkingia* species show catalase, oxidase, and phosphatase activities. They also produce β-galactosidase, but discrepancies have been reported between the different techniques used to test this characteristic in *Elizabethkingia meningoseptica* strains (Bernardet et al., 2005; Bruun and Ursing, 1987).

Table 44 of the *Chryseobacterium* chapter in this volume presents the differential characteristics of members of *Elizabethkingia* and related genera. *Elizabethkingia* strains demonstrate strong proteolytic activity, hydrolyzing casein and gelatin, and produce indole from tryptophan (Bernardet et al., 2005; Bruun and Ursing, 1987; Weyant et al., 1996). All 149 *Elizabethkingia meningoseptica* strains studied by the CDC (Weyant et al., 1996) and all 52 strains studied by Bruun and Ursing (1987) produced indole. In the latter study, no strains produced urease, in contrast to the study of 49 *Elizabethkingia meningoseptica* strains by Holmes et al. (1984b) and to the CDC's study (Weyant et al., 1996), where the trait varied among strains. The authors attributed the variable results reported for urease production to methodological differences (direct enzymic method versus Christensen's urea). Disagreement exists between Li et al. (2003a) and Kim et al. (2005b) as to whether the type strain of *Elizabethkingia miricola* produces indole or not, but they agree on the production of urease by the two strains (Kim et al., 2005b; Li et al., 2003a). In contrast to the *Elizabethkingia miricola* type strain (Li et al.,

2003a), the only three strains of *Elizabethkingia meningoseptica* tested also degrade tyrosine (Bernardet et al., 2005).

The production of a precipitate on egg yolk agar is strain-dependent in *Elizabethkingia meningoseptica* (Bernardet et al., 2005); it occurs in the two *Elizabethkingia miricola* strains (Li et al., 2003a).

Several carbohydrates including D-glucose, D-fructose, D-mannitol, mannose, lactose, and maltose are oxidized, but cellobiose, rhamnose, raffinose, galactose, sucrose, melezitose, salicin, adonitol, dulcitol, sorbitol, and inositol are not oxidized (Bruun and Ursing, 1987; Holmes et al., 1984b; Kim et al., 2005b; Li et al., 2003a; Weyant et al., 1996). Acid production from trehalose (number of positive strains/number of strains tested is 51/52), melibiose (25/52), D-xylose (9/52), and L-arabinose (3/52) is strain-dependent in *Elizabethkingia meningoseptica* (Bruun and Ursing, 1987). The two *Elizabethkingia miricola* strains produce acid from trehalose, but not from arabinose or xylose (Kim et al., 2005b; Li et al., 2003a). Esculin is hydrolyzed by *Elizabethkingia* strains, but chitin and agar are not. The description of the genus *Elizabethkingia* states that starch is not hydrolyzed (Kim et al., 2005b). Starch hydrolysis was indeed listed as negative in the table of phenotypic characteristics of *Elizabethkingia miricola*; yet, it was described as positive in the species description (Li et al., 2003a).

Production of H_2S was considered negative for all members of the genus *Elizabethkingia* by Kim et al. (2005b), although a positive result was reported for the type strain of *Elizabethkingia miricola* by Li et al. (2003a). None of the *Elizabethkingia meningoseptica* strains studied (Bruun and Ursing, 1987; Holmes et al., 1984a; Weyant et al., 1996) reduce nitrate; this trait is also negative in the two *Elizabethkingia miricola* strains (Kim et al., 2005b; Li et al., 2003a).

In accordance with other members of the family *Flavobacteriaceae*, *Elizabethkingia* strains contain no sphingophospholipids and menaquinone MK-6 is their only or major respiratory quinone (Bernardet et al., 2002; Kim et al., 2005b). Homospermidine is the major polyamine in *Elizabethkingia meningoseptica* (Hamana and Matsuzaki, 1990), as it is in several *Chryseobacterium* species and in members of the genera *Bergeyella*, *Ornithobacterium*, *Riemerella*, and *Weeksella* (Hamana and Nakagawa, 2001).

Very few enzymic activities of *Elizabethkingia* strains that may have practical or industrial applications have been reported to date. An endoglycosidase activity has been detected in an *Elizabethkingia meningoseptica* strain; the availability of the cleavage site of this enzyme, named endo F, was extremely variable in the different *N*-linked glycoproteins tested, making endo F a possible tool for studying glycans and protein backbones and for defining the nature of the glycan/protein interface (Elder and Alexander, 1982). This bacterial strain was later shown to actually represent a second strain of *Elizabethkingia miricola* (Kim et al., 2005b). The proline-specific endopeptidase produced by an *Elizabethkingia meningoseptica* strain was heterologously expressed in *Escherichia coli* cells and efficient methods were devised to purify the enzyme that accumulated in the periplasmic space (Diefenthal and Dargatz, 1995).

Genetics. After early DNA–DNA hybridization studies of [*Flavobacterium*] *meningosepticum* strains had shown that the type strain shared only 29–65% DNA relatedness with the other strains (Owen and Snell, 1976), further studies have confirmed the considerable intraspecific heterogeneity. In an investigation of 52 [*Flavobacterium*] *meningosepticum* strains (Ursing and Bruun, 1987), two main DNA groups were found that were about 40–55% interrelated and comprising 4 and 48 strains, respectively. The type strain of the species was found in the smaller group. Biochemical or physiological characteristics that could differentiate the two DNA groups were not found (Bruun and Ursing, 1987); therefore, a subdivision of the species was not proposed. The larger DNA group was further divided into four subgroups when ΔT_m values were taken into consideration; 25 of the 28 strains in the largest subgroup were invasive neonatal isolates, indicating that the DNA groups defined within the species might differ with regard to pathogenic significance (Ursing and Bruun, 1987).

The recent study by Kim et al. (2005b) further confirmed the isolated position of the type strain compared to other [*Chryseobacterium*] *meningosepticum* strains: although four strains shared nearly identical 16S rRNA gene sequences and 90–100% DNA–DNA relatedness, the type strain shared about 98% 16S rRNA gene sequence similarity with them and a DNA–DNA relatedness value of only 31–35%. Again, no phenotypic clue was found to support a division of the species. The sixth "[*Chryseobacterium*] *meningosepticum*" strain studied proved to share very high 16S rRNA gene sequence similarity and DNA–DNA relatedness values with the [*Chryseobacterium*] *miricola* type strain, indicating that it actually represents a second strain of this species.

The fact that strains of *Elizabethkingia meningoseptica* and *Elizabethkingia miricola* show only 23–54% DNA relatedness, while sharing 97.8–98.4% 16S rRNA gene sequence similarity (see above; Kim et al., 2005b) should be emphasized. The same situation occurs between members of the genus *Chryseobacterium* (see the corresponding chapter in this volume). Consequently, DNA–DNA hybridization experiments or possibly sequencing of a number of household genes (Kuhnert and Korczak, 2006) are necessary even when 16S rRNA gene sequence similarities between different strains are above the theoretical threshold value of 97%.

The DNA G+C contents of members of the genus are in the range 35.0–38.2 mol% (Kim et al., 2005b). In this study, five *Elizabethkingia meningoseptica* strains showed DNA G+C contents of 37.2 ± 0.6 mol%, in accordance with previous data (Holmes et al., 1984a; Ursing and Bruun, 1987). Thanks to the identification of a second strain of *Elizabethkingia miricola*, the G+C content for the species was re-evaluated as 35.3 ± 0.3 mol%, compared to 34.6 mol% as reported previously for the type strain (Li et al., 2003a).

Plasmids have not been detected in *Elizabethkingia meningoseptica* and allied taxa using the single colony lysis technique (Owen and Holmes, 1981).

An invasive neonatal strain and an environmental strain of *Elizabethkingia meningoseptica* selected among the DNA groups delineated by Ursing and Bruun (1987) were recently compared using suppressive subtraction hybridization (Lin et al., 2006). The genes present specifically in the invasive strain were identified, but their function remains unknown as their identity to known references and their deduced amino acid sequences were low.

McBride and Baker (1996) and McBride and Kempf (1996) demonstrated that genetic manipulation of *Elizabethkingia meningoseptica* strains is possible by using the genetic tools (selectable marker, suicide vector, and transposon) they developed to manipulate *Flavobacterium johnsoniae*.

Three different molecular typing methods have been used so far to type *Elizabethkingia meningoseptica* clinical isolates for epidemiological and taxonomic studies. Ribotyping was applied to the collection of 52 strains previously characterized using extensive phenotypic analysis (Bruun and Ursing, 1987) and DNA–DNA hybridization experiments (Ursing and Bruun, 1987) (see above). Ribotyping proved to be a useful epidemiological tool because it allowed the differentiation of strains within genomic groups and because the similar profiles found in isolates originating from the same outbreak (Bruun et al., 1989) pointed to a common source of infection (Colding et al., 1994). The technique was also found to have taxonomic relevance because it was able to identify the DNA groups of Ursing and Bruun. In particular, the only three isolates sharing high DNA relatedness with the type strain also shared with it a unique *Hin*dIII ribotype. Ribotyping was also applied to another collection of 92 isolates of *Elizabethkingia meningoseptica* whose serotypes had been determined previously (Quilici and Bizet, 1996). Again, ribotyping clustered the isolates belonging to some serotypes (e.g., C, G, and N) and differentiated strains within other serotypes; however, the different banding patterns found in isolates originating from the same outbreak suggested different sources of infection. Hence, as reported for many other bacterial species, the type of information yielded by ribotyping depends on the restriction enzyme (or combination of enzymes) used.

Pulsed-field gel electrophoresis of bacterial DNAs previously subjected to enzymic macrorestriction demonstrated the identity of environmental and clinical isolates (Hoque et al., 2001), the relapse of a catheter-related bacteremia (Sader et al., 1995), and the identity of isolates involved in different outbreaks (Green et al., 1999).

Randomly amplified polymorphic DNA (RAPD) analysis was recently evaluated on a collection of *Chryseobacterium* and *Elizabethkingia* fish isolates and reference strains. In contrast to the results of 111 phenotypic characteristics and of whole-cell protein profiles (SDS-PAGE) that clustered the three included *Elizabethkingia meningoseptica* strains, RAPD yielded three different profiles, demonstrating that it could probably be used to differentiate and type larger collections of strains (Bernardet et al., 2005). On the other hand, multilocus enzyme electrophoresis was not able to differentiate between all *Chryseobacterium*, *Elizabethkingia*, and *Empedobacter* species (Hugo and Jooste, 1997).

Antigenic structure. Six serovars of *Elizabethkingia meningoseptica*, A–F, were initially reported from human infections, mainly invasive neonatal infections (King, 1959). The most commonly found type is serovar C, seen in about two-thirds of typed isolates from neonatal infections (Bloch et al., 1997). Richard et al. (1979a, b) have described additional serovars, G–N, derived from clinical specimens, but not invasive neonatal infections. Some of these new serovars (i.e., I, J, and L) actually appear to be strains of *Empedobacter brevis* (Holmes et al., 1984a). Other investigators have found cross-reactions between strains of *Elizabethkingia meningoseptica* and strains of CDC group IIb (containing *Chryseobacterium gleum*, *Chryseobacterium indologenes*, and allied strains) and CDC group IIf (King, 1959; Owen and Lapage, 1974; Price and Pickett, 1981).

Antibiotic sensitivity. As observed in *Chryseobacterium gleum*, *Chryseobacterium indologenes* and other, unnamed members of CDC group IIb (see above), *Elizabethkingia meningoseptica* strains are naturally resistant to polymyxins, aminoglycosides (e.g., gentamicin, streptomycin), chloramphenicol, and most β-lactam antibiotics, including penicillin and ampicillin (Aber et al., 1978; Bruun, 1987; Chang et al., 1997; Fraser and Jorgensen, 1997; Kirby et al., 2004; Olsen, 1967). *Elizabethkingia meningoseptica* strains produce at least three β-lactamases, two unrelated carbapenem-hydrolyzing class B metallo-β-lactamases [BlaB (Rossolini et al., 1998) and GOB (Bellais et al., 2000a)], plus a non-inducible extended-spectrum β-lactamase, Ambler class A, CME-2 (Bellais et al., 2000b). Consequently, *Elizabethkingia meningoseptica* is resistant to carbapenems (imipenem, meropenem) and extended spectrum cephalosporins (cefotaxime, ceftazidime, cefepime). Some strains are, however, susceptible to ureidopenicillins and piperacillin-tazobactam susceptibility ranged from 39 to 70% in two studies (Fraser and Jorgensen, 1997; Kirby et al., 2004). Recently, the high levels and rates of imipenem resistance in *Elizabethkingia meningoseptica* isolates from Hangzhou, China, were shown to result from the presence of heterogeneous BlaB and/or GOB metallo-β-lactamases, as well as from other, undefined mechanisms of carbapenem resistance (Chen et al., 2006).

Elizabethkingia meningoseptica strains are mainly resistant to tetracyclines, erythromycin, and linezolid, and resistant or intermediately susceptible to clindamycin and vancomycin (Bruun, 1987; Fraser and Jorgensen, 1997; Kirby et al., 2004). Sensitivity to trimethoprim-sulfamethoxazole varies, ranging from 33 to 80% of strains (Fraser and Jorgensen, 1997; Kirby et al., 2004). The antibiotics that are most active against *Elizabethkingia meningoseptica* are minocycline, rifampicin, and the newer quinolones (levofloxacin, gatifloxacin, sparfloxacin, moxifloxacin), whereas susceptibility to ciprofloxacin varies (Fraser and Jorgensen, 1997; Kirby et al., 2004; Spangler et al., 1996).

No data exist on the antibiotic sensitivity of the two strains of *Elizabethkingia miricola*, but it is presumed that susceptibilities are about the same as described above for *Elizabethkingia meningoseptica*.

Pathogenicity. The pathogenicity of *Elizabethkingia meningoseptica* has been reviewed recently by Bernardet et al. (2006). Among past and present *Chryseobacterium* species, *Elizabethkingia meningoseptica* is by far the most important species found in human infections. Invasive neonatal and infant disease, notably meningitis, has been reported from all over the world since the disease was first reported in 1944 (although bacterial etiology was first established retrospectively by King, 1959). The infection is not common. Von Graevenitz (1985) estimated that about 120 cases of invasive neonatal disease had been reported in the literature by 1985, while Bloch et al. (1997) found 308 patient reports in the English language literature up to 1994 of cultures positive for *Elizabethkingia meningoseptica*, of which about 60% were judged to represent true infections (Bloch et al., 1997 and references therein). About half of these occurred in infants <3 months of age, of which 85% represented cases of meningitis, sometimes with concomitant bacteremia and pneumonia. The mortality was 57%, with post-meningitis sequelae, mainly hydrocephalus, occurring in the majority of the survivors. Prematurity occurred in more than half of the cases and accounted for 66% of the mortality. In the non-English language literature,

15 cases, mainly of neonatal meningitis, have been published (see Bernardet et al. (2006) for references). Since 1994, 13 cases of neonatal infection have been reported (Chiu et al., 2000; Güngör et al., 2003; Hoque et al., 2001; Tekerekoglu et al., 2003; Tizer et al., 1995).

Elizabethkingia meningoseptica infections in infants typically occur as small epidemics in neonatal wards, including neonatal intensive care units. Surveillance cultures from uninfected infants hospitalized with the cases and from the hospital environment often reveal extensive colonization with *Elizabethkingia meningoseptica* (see Bernardet et al. (2006).

Disease after the age of 1 year is rare, with less than 100 cases reported in the English language literature up to 1994 (Bloch et al., 1997 and references therein). Since then, few cases have been published (Chiu et al., 2000; Gunnarsson et al., 2002; Lin et al., 2004; Manfredi et al., 1999; Ozkalay et al., 2006; Sader et al., 1995). Adult cases occur as outbreaks in hospitals or as sporadic infections after trauma/surgery or in patients severely immunocompromised because of malignancy, end-stage hepatic and renal disease, and extensive burns. Outbreaks have taken place after administration of contaminated medicine, such as parenteral anesthetics and aerosolized polymyxin B, or ingestion of contaminated water or ice cubes. Clinically, the most common manifestations have been bacteremia and pneumonia, but colonization without signs of infection also occurs, with *Elizabethkingia meningoseptica* being found in wounds, urine, and mucous membranes.

Strains of *Elizabethkingia meningoseptica* have been isolated from a variety of diseased animals, such as cats, dogs, turtles, snakes, and several bird species, as well as various frogs and fish species (for a recent review, see Bernardet et al. (2006), and references therein). Infected birds displayed various pathological signs: pericarditis or joint infection in chicken, inflamed hemorrhagic traumatic lesion in pigeon, and liver infection in zebra finch (Vancanneyt et al., 1994). Frogs infected by *Elizabethkingia meningoseptica* exhibited signs of hemorrhagic septicemia, leg and abdominal swelling, extreme lethargy, and severe enlargement of liver, spleen, and kidney (Chung, 1990; Green et al., 1999; Mauel et al., 2002). Severe outbreaks caused mass mortality in all developmental stages of frogs in intensive frog aquaculture systems; recurrence of outbreaks was observed frequently (Chung, 1990). The uncultured organism successively named "*Aegyptianella ranarum*" and "*Candidatus* Hemobacterium ranarum" that replicates in the red blood cells of several frog species and that is phylogenetically related to *Elizabethkingia meningoseptica* (Zhang and Rikihisa, 2004) is discussed in the Family *Flavobacteriaceae* in this volume.

Infections in fish imported to France from Africa and Asia have been reported recently. Samples of internal organs of diseased Koi carp (*Cyprinus carpio*), snake fish (*Erpetoichthys calabaricus*), and gourami (*Colisia lalia*) yielded pure cultures of *Elizabethkingia meningoseptica*; all fish suffered from hemorrhagic septicemia and carps also displayed skin lesions (Bernardet et al., 2005; N. Keck and J.-F. Bernardet, unpublished data). An *Elizabethkingia meningoseptica* strain retrieved from *Sinperca chuatsi* was one of the bacterial fish pathogens used by Yin et al. (2006) to test the bactericidal activity of an antimicrobial peptide from cDNA previously isolated from a fish leukocyte cDNA library. The minimal bactericidal concentration of the synthetic mature peptide, epinecidin-1, on *Elizabethkingia meningoseptica* was the lowest of all Gram-stain-negative bacteria tested.

Although various animal species have been reported to harbor *Elizabethkingia meningoseptica*, its transfer from infected animals to humans is unlikely as no transfer from human to human has ever been reported.

Only scant information is currently available on the virulence mechanisms of *Elizabethkingia meningoseptica*. A possible clue was found in a strain isolated from a diseased bullfrog (see above), tentatively named "[*Flavobacterium*] *ranacida*" (Chung, 1990; Faung et al., 1996) but later shown to actually represent an *Elizabethkingia meningoseptica* strain (Bernardet et al., 2005). As the concentration of its major fatty acid, 13-methyl myristate, was increased it was able to induce changes in morphology and ultimately lysis of human platelets. At non-lytic concentrations, the fatty acid inhibited platelet response to various agents (Faung et al., 1996). The strong lytic activity of 13-methyl myristate on platelets and its ability to perturb membrane function could explain, at least partly, the hemorrhagic septicemia caused by *Elizabethkingia meningoseptica* in infected frogs.

In an attempt to evaluate the virulence mechanisms of *Elizabethkingia meningoseptica*, environmental strains and isolates from cases of invasive infection were selected from those studied by Ursing and Bruun (1987) and tested in penetration and cytotoxicity assays on cell cultures, in serum susceptibility tests, and in a murine pulmonary infection model (Lin et al., 2006). All strains were resistant to the bactericidal activity of normal human sera, but were unable to penetrate cultured cells and devoid of cytotoxic activity. Electron microscopic studies of samples of infected lung following intratracheal challenge of mice revealed that a strain originating from an invasive infection was able to invade the epithelial cells of the respiratory tract in contrast to the environmental isolate. Both strains, however, were cleared from the lung within 7 d; they did not spread to the liver or spleen and the mice survived the infection (Lin et al., 2006).

Ecology. Although the majority of *Elizabethkingia meningoseptica* strains have been isolated from clinical specimens, strains have also been found in soil, river water, and water reservoirs, as well as in various humid environments in hospitals, such as apparatus (bronchoscopes, respirators), medicine, catheters, sinks, and tap water (Bruun, 1982; Bruun et al., 1989; du Moulin, 1979; Heeg et al., 1994; Holmes, 1987; Hoque et al., 2001; Olsen, 1969; Owen and Holmes, 1981; Ursing and Bruun, 1987). An *Elizabethkingia meningoseptica* strain was part of the bacterial community in the water of a spent nuclear fuel pool in a nuclear power plant (Chicote et al., 2005). The environmental distribution of *Elizabethkingia meningoseptica* is also reflected by the strains identified from diseased birds, mammals, and freshwater animals (see above) and from lake amoebae (Hadas et al., 2004). Although *Elizabethkingia meningoseptica* appears to be rather resistant to chlorine and other disinfectants (Green et al., 1999 and references therein), the implementation of strict hygienic procedures such as isolation, plus disinfection and autoclaving of equipment, usually succeeds in stopping outbreaks (Bruun et al., 1989; Chung, 1990; Green et al., 1999; Heeg et al., 1994). However, in one neonatal intensive care unit, screening cultures of water tanks and taps remained positive even after sink taps were repaired and chlorinated and after the water temperature was raised to 43°C; only when taps were replaced did cultures become negative (Hoque et al., 2001). Strains of *Elizabethkingia meningoseptica* usually belong to biofilm bacterial communities, as was shown during an outbreak in a frog-housing facility (Green et al., 1999).

So far, no clinical isolate of *Elizabethkingia miricola* is known. The type strain was retrieved from the condensation water of the space station Mir (Li et al., 2003a) and, although previously considered to be a strain of *Elizabethkingia meningoseptica*, the second strain was isolated from a contaminated commercial preparation of carboxypeptidase A (Elder and Alexander, 1982; Kim et al., 2005b).

Contrary to some *Chryseobacterium* species (see the corresponding chapter in this volume and Bernardet et al. (2006), strains of *Elizabethkingia* have not been reported from food products so far. Two dairy isolates grouped with the *Elizabethkingia meningoseptica* type strain as a result of multilocus enzyme electrophoresis (Hugo and Jooste, 1997), but a later comparison of fatty acid and whole-cell protein profiles did not confirm this identification (Hugo et al., 1999).

Enrichment and isolation procedures

The two species of *Elizabethkingia* are not fastidious and do not require enrichment or special procedures for isolation and cultivation. They grow well at 37°C (also at room temperature) on ordinary media, such as blood, tryptic soy, BHI, and nutrient agars, and are often recognized in the clinical microbiology laboratory by their resistance to antibiotics to which most other Gram-stain-negative rods are susceptible.

Maintenance procedures

Preservation methods used for *Elizabethkingia* strains do not differ from those used for strains of *Chryseobacterium* and related organisms (see the corresponding chapter in this volume and Bernardet et al. (2006). Commercial agar deeps of bacterial strain storage medium (e.g., Bio-Rad) are convenient for short-term preservation. For long-term preservation, *Elizabethkingia* strains may be frozen at −80°C in casitone broth with 10% (v/v) glycerol, BHI broth with 20% (v/v) glycerol, or ox broth with 10% (v/v) glycerol (Reichenbach, 1989b; Park et al., 2006b; B. Bruun, unpublished data). The last method has successfully preserved strains of *Elizabethkingia meningoseptica* for at least 25 years. The preferred method is, however, lyophilization (Bernardet et al., 2002, 2006).

Procedures for testing special characters

Phenotypic differentiation between the two *Elizabethkingia* species and from related *Chryseobacterium* species and *Empedobacter brevis* is difficult and requires strict adherence to the described methods and the conditions under which they are performed. These methods and conditions have been thoroughly reviewed previously (Bernardet et al., 2002, 2006; Bernardet and Nakagawa, 2006).

One of the most significant differential characteristics between members of the family *Flavobacteriaceae* is the production of yellow pigments. The pigments belong either to the carotenoid- or flexirubin-types, but members of some *Flavobacterium* species may produce both types of pigments (see the family *Flavobacteriaceae* and genus *Flavobacterium* chapters in this volume). An easy test based on the color shift induced by a 20% KOH solution is available to differentiate between the two types (Reichenbach, 1989b), but drops of the solution must be deposited on a rather thick mass of bacterial cells collected with a loop (as recommended in the minimal standards for describing new taxa in the family *Flavobacteriaceae*; Bernardet et al., 2002) rather than directly on colonies. This requirement is especially important for weakly pigmented colonies such as those of *Elizabethkingia miricola* and some *Elizabethkingia meningoseptica* strains, as shown by the negative result obtained by Kim et al. (2005b) when they directly flooded *Elizabethkingia* colonies with KOH. When the KOH solution was deposited on bacterial masses, however, those *Elizabethkingia meningoseptica* strains that produced pale yellow colonies turned light brown, whereas no color shift occurred for nonpigmented strains (Bernardet et al., 2006). Hence, small amounts of flexirubin-type pigments seem to be produced by pigmented *Elizabethkingia meningoseptica* strains. It is important to note that pigment production is highly dependent on growth temperature, the presence of daylight, and the composition of the culture medium (Bruun, 1982; Holmes et al., 1984a; Hugo and Jooste, 2003).

Differentiation of the genus *Elizabethkingia* from other genera

The characteristics that differentiate members of the genus *Elizabethkingia* from those of the most closely related taxa (especially *Chryseobacterium*, but also *Bergeyella*, *Riemerella*, *Kaistella*, and *Sejongia*) are listed in Tables 43 and 44 of the *Chryseobacterium* chapter in this volume. They include various biochemical traits, production of yellow pigments, growth at different temperatures, G+C contents, and fatty acid composition. For characteristics that differentiate *Elizabethkingia* from more distantly related members of the family *Flavobacteriaceae*, see the corresponding chapter in this volume.

Although restricted to well-equipped research laboratories, the determination of fatty acid and whole-cell protein profiles allows a clear differentiation between strains of *Elizabethkingia*, *Chryseobacterium*, and related genera (Bernardet et al., 2005, 2006; de Beer et al., 2005, 2006; Hugo et al., 1999; Kim et al., 2005b). The fatty acid profile of members of the genus *Elizabethkingia* mainly differs from that of *Chryseobacterium* strains in the proportions of $C_{17:1}$ iso ω9*c* and summed feature 4 (see Table 43 of the *Chryseobacterium* chapter). Although only one strain of each reference species was analyzed, the table of fatty acid compositions published by Kim et al. (2005b) is interesting because all strains were grown under the same conditions, thus allowing direct comparison.

Differentiation of *Elizabethkingia meningoseptica* strains from isolates of *Chryseobacterium gleum*, *Chryseobacterium indologenes*, and *Empedobacter brevis* in the clinical microbiology laboratory is based on a limited number of phenotypic tests, including the absence of or slow production of yellow pigment, acid from trehalose, and β-galactosidase (ONPG test) (Bernardet et al., 2006; Ursing and Bruun, 1991). Differentiation between strains of the two *Elizabethkingia* species rests only on the sticky colony consistency and the ability of *Elizabethkingia miricola* to hydrolyze 2-naphthyl butyrate (API ZYM galleries) in contrast to *Elizabethkingia meningoseptica* (Kim et al., 2005b).

Commercial identification galleries and kits are mostly adapted to clinical organisms that grow well at 37°C. They have been used to study collections of *Elizabethkingia meningoseptica* isolates, although discrepancies between the different commercial galleries and conventional tests have been reported (Bernardet et al., 2006, 2005; Bruun and Ursing, 1987; Kim et al., 2005b). Most commercial identification systems include *Elizabethkingia meningoseptica* in their analytical profile index. This species can be differentiated from *Chryseobacterium* species using the API 20E, 20NE, ID 32E, ID 32GN, and Biotype 100 galleries, the Vitek GNI+ and Vitek2 ID-GNB systems (Bernardet et al., 2006 and references therein), and the Biolog GN2 MicroPlate (C.J. Hugo, personal communication). Although not a clinical organism and not included in databases, *Elizabethkingia miricola* was also tested successfully on various commercial galleries (Kim et al., 2005b; Li et al., 2003a). Fourteen of the 95 tests in the Biolog GN2 MicroPlate system differentiated the type strain of *Elizabethkingia miricola* from the type strain of *Elizabethkingia meningoseptica* (C.J. Hugo and A. Mielmann, personal communication) but, because *Elizabethkingia meningoseptica* is a phenotypically heterogeneous species, the importance of this finding is unclear.

When the minimal standards for describing new taxa in the family *Flavobacteriaceae* were published (Bernardet et al., 2002), the genus *Elizabethkingia* had not yet been separated from *Chryseobacterium* and [*Chryseobacterium*] *miricola* had not been described. However, the general principles for the family and the minimal standards for describing new *Chryseobacterium* species should be followed when novel *Elizabethkingia* species are described.

Taxonomic comments

The genus *Elizabethkingia* has only recently been proposed to accommodate two former *Chryseobacterium* species, [*Chryseobacterium*] *meningosepticum* and [*Chryseobacterium*] *miricola*, that shared a separate position compared to all other *Chryseobacterium* species in phylogenetic trees (Kim et al., 2005b). Physiological and biochemical tests differentiating the two species are sparse (see above; also see Table 44 of the *Chryseobacterium* chapter in this volume). The two *Elizabethkingia miricola* strains show DNA–DNA hybridization values of 23–54% to the five *Elizabethkingia meningoseptica* strains included in the study of Kim et al. (2005b). The argument for creating the novel species *Elizabethkingia miricola* is, however, confounded by the fact that *Elizabethkingia meningoseptica* is known to be a heterogeneous species divided into two main DNA groups about 40–55% related to each other (Ursing and Bruun, 1987). The type strain of the species is in the small DNA group consisting of only four strains versus a total of 48 strains in the larger DNA group. This heterogeneity was confirmed by Kim et al. (2005b) (see above). In spite of its genomic heterogeneity, *Elizabethkingia meningoseptica* was chosen as the type species of the new genus because of its long standing description, its clinical significance, and because the numerous strains available had been subjected to extensive phenotypic and genomic studies (B. Holmes and R.J. Owen, unpublished data cited in Holmes et al., 1984b; Bruun and Ursing, 1987; Ursing and Bruun, 1987; Kim et al., 2005b). Hopefully, the new status of *Elizabethkingia meningoseptica* as the type species of a genus should prevent further changes in its nomenclature. For the time being, however, we are left with the genomic heterogeneity of this species. Hence, it is possible that the species will be split in the future, provided sound and convenient differential characteristics between the resulting species are found.

Taxonomic and nomenclatural issues in the genus *Elizabethkingia* are dealt with by the Subcommittee on the Taxonomy of *Flavobacterium* and *Cytophaga*-like Bacteria of the International Committee on Systematics of Prokaryotes. The subcommittee has issued the above-mentioned minimal standards (Bernardet et al., 2002).

Acknowledgements

The authors wish to acknowledge Tim Lilburn (American Type Culture Collection) for supplying the phylogenetic tree (Figure 40 in this volume of the *Manual*).

Further reading

Website of the International Committee on Systematics of Prokaryotes (ICSP) Subcommittee on the Taxonomy of *Flavobacterium* and *Cytophaga*-like Bacteria (http://www.the-icsp.org/subcoms/Flavobacterium_Cytophaga.htm).

List of species of the genus *Elizabethkingia*

1. **Elizabethkingia meningoseptica** (King 1959) Kim, Kim, Lim, Park and Lee 2005b, 1291^{VP} (*Flavobacterium meningosepticum* King 1959, 247; *Chryseobacterium meningosepticum* Vandamme, Bernardet, Segers, Kersters and Holmes 1994a, 830)

me.nin.go.sep'ti.ca. Gr. n. *meninx, meningos* meninges, membranes covering the brain; Gr. adj. *septikos* putrefactive; N.L. fem. adj. *meningoseptica* presumably referring to the association of the bacterium with both meningitis and septicemia.

Rods, approximately 0.5 µm in diameter and various lengths (1–3 µm). Filaments are common. Good growth occurs on tryptic soy, blood, and nutrient agars, while growth on MacConkey agar is strain-dependent. Growth occurs at 22–37°C; strains from invasive neonatal disease grow at 40°C. Colonies on nutrient agar at 35°C are circular, entire, and smooth, 1–2 mm in diameter. Cultures have a characteristic aromatic odor. Strains deriving from invasive neonatal disease (e.g., the type strain) are weakly yellow-pigmented, while others are nonpigmented. Oxidase- and catalase-positive. Urease is sometimes produced using Christensen's urea. Indole and β-galactosidase are produced. Gelatin is hydrolyzed. 2-Naphthyl butyrate is not hydrolyzed in the API ZYM gallery. Production of a precipitate on egg yolk agar is strain-dependent. Tyrosine is degraded by some strains. Other biochemical and physiological characteristics are given in Tables 43 and 44 of the *Chryseobacterium* chapter in this volume. Several serovars have been described. DNA–DNA hybridization studies have delineated two main DNA groups that cannot be differentiated phenotypically.

Source: the type strain was isolated from cerebrospinal fluid in Massachusetts in 1949. The species is not a member of normal human flora, but it causes invasive disease in neonates and occasionally in immunocompromised patients.

DNA G+C content (mol%): 36.4–37.9 (type strain: 37.9 ± 0.3) (T_m; Holmes et al., 1984a; Ursing and Bruun, 1987); 36.6–37.8 (type strain: 37.1) (HPLC; Kim et al., 2005b).

Type strain: strain 14 of King (1959), ATCC 13253, NCTC 10016, LMG 12279, CCUG 214 (serovar A).

Sequence accession nos (16S rRNA gene): AY468445, AJ704540.

Reference strains: the three strains that shared high DNA relatedness with the type strain in the study of Ursing and Bruun (1987) are: Richard's strain 3.83 (no longer available), GTC 10941 (Gifu 1606) and NCTC 13393 (Holmes' strain E 847, Greaves' strain F2, NCIB 7-61). Other reference strains are: NCTC 10585 (strain 422 of King, 1959; serovar B); NCTC 10586 (strain 3375 of King, 1959; serovar C); NCTC 10587 (strain 6925 of King, 1959; serovar D); NCTC 10588 (strain 8388 of King, 1959; serovar E); NCTC 10589 (strain 8707 of King, 1959; serovar F); NCTC 11305 (CIP 78.30; Richard et al., 1979b; serovar G); NCTC 11306 (CIP 79.5; Richard et al., 1979a; serovar H); and NCTC 11309 (CIP 79.29; Richard et al., 1979a; serovar K).

2. **Elizabethkingia miricola** (Li, Kawamura, Fujiwara, Naka, Liu, Huang, Kobayashi and Ezaki 2003a) Kim, Kim, Lim, Park and Lee 2005b, 1292[VP] (*Chryseobacterium miricola* Li, Kawamura, Fujiwara, Naka, Liu, Huang, Kobayashi and Ezaki 2003a, 527)

mi.ri'co.la. N.L. neut. n. *mirum* derived from *mir* (peace in Russian, name of the Russian space station); L. suff. *-cola* from L. masc. or fem. n. *incola* inhabitant; N.L. masc. or fem. n. *miricola* inhabitant of the Mir space station.

Rods, approximately 0.5 μm in diameter and 1.0–2.5 μm in length. Good growth occurs on tryptic soy, blood, BHI, nutrient, and MacConkey agars. Very slow growth occurs on cetrimide agar. Growth occurs at 22–37°C, but not at 5 or 42°C. Colonies are white–yellow, circular, entire, smooth, and very sticky on solid media. Oxidase- and catalase-positive. Indole and β-galactosidase are produced. Gelatin and urea are hydrolyzed. 2-Naphthyl butyrate is hydrolyzed in the API ZYM gallery. The type strain forms a precipitate when grown on egg yolk agar. D-Mannitol, sodium citrate, and *N*-acetyl-D-glucosamine are assimilated, but tyrosine is not degraded. Other biochemical and physiological characteristics are given in Tables 43 and 44 of the *Chryseobacterium* chapter in this volume.

Source: the type strain was isolated from condensation water in the space station Mir; a second strain was recently identified from the contaminated commercial preparation of an enzyme.

DNA G+C content (mol%): 35.3 ± 0.3 (type strain: 35.0) (HPLC; Kim et al., 2005b).

Type strain: strain W3-B1 of Li et al. (2003a), DSM 14571, JCM 11413, GTC 862.

Sequence accession no. (16S rRNA gene): AB071953.

Reference strain: ATCC 33958.

Genus XVIII. **Empedobacter** (*ex* Prévot 1961) Vandamme, Bernardet, Segers, Kersters and Holmes 1994a, 830[VP]

The Editorial Board

Em.pe.do.bac'ter. Gr. adj. *empedos* fixed, immovable; N.L. masc. n. *bacter* a small rod; N.L. masc. n. *Empedobacter* nonmotile rod

Rods, typically 0.5 μm wide and 1–2 μm long, but some longer rods may be present. Gram-stain-negative. **Nonmotile**. Nonsporeforming. Intracellular poly-β-hydroxybutyrate granules are absent. **Aerobic**, having a strictly respiratory type of metabolism. Chemo-organotrophic. All strains grow at 30°C; some strains grow at 37°C. **Growth on solid media is light yellow**. Growth occurs on MacConkey agar and β-hydroxybutyrate agar. **Positive for catalase, oxidase, and phosphatase. Indole-positive**. Several carbohydrates are oxidized, but glycerol and trehalose are not. **Casein, gelatin, tributyrin, DNA, and Tween 20 are hydrolyzed, but not esculin, urea, or agar.** Fatty acids $C_{15:0}$ iso, $C_{16:1}$ ω7*c*, $C_{16:1}$ ω5*c*, $C_{16:0}$, $C_{16:0}$ 3-OH, and $C_{17:0}$ iso 3-OH are predominant. Sphingophospholipids are absent. The only respiratory quinone is menaquinone-6. Resistant to a wide range of antimicrobial agents.

DNA G+C content (mol%): 31–33 (T_m).

Type species: **Empedobacter brevis** (Holmes and Owen 1982) Vandamme, Bernardet, Segers, Kersters and Holmes 1994a, 830[VP] [*Flavobacterium breve* (*ex* Lustig 1890) Holmes and Owen 1982, 233; *Bacillus brevis* Lustig 1890, 521; *Bacillus canalicolis brevis* Cornil and Babes 1890, 292; *Bacillus canalis parvus* Eisenberg 1891, 362; *Bacterium canalis parvis* (*sic*) (Eisenberg 1891) Chester 1897, 130; *Bacterium canale* Mez 1898, 55; *Bacterium breve* (Lustig 1890) Chester 1901, 69; *Flavobacterium brevis* (*sic*) (Lustig 1890) Bergey, Harrison, Breed, Hammer and Huntoon 1923, 116; *Pseudobacterium brevis* (*sic*) (Lustig 1890) Krasil'nikov 1949, 239; *Empedobacter breve* (*sic*) (Lustig 1890) Prévot 1961, 181].

Further descriptive information

Good growth occurs on nutrient agar. Colonies are circular, low convex, smooth, and shiny, with entire edges. The colony size is pinpoint to 2.0 mm in diameter on nutrient agar; colonies on blood agar are pinpoint to 2.5 mm in diameter and nonhemolytic. The light yellow pigmentation does not change with variation of medium and temperature. Growth occurs at room temperature (18–22°C); some strains grow at 37°C; no strains grow at 5 or 42°C. Although some strains grow at 37°C, the majority will only produce acid from carbohydrates when incubated at 30°C and prolonged incubation may be necessary.

The cellular polyamine homospermidine, characteristic of members of the *Flavobacterium–Cytophaga–Sphingobacterium* complex, is present (Hamana and Nakagawa, 2001).

The pathogenicity of *Empedobacter brevis* for humans and other animals is not clear, but at least in some instances it seems to be an opportunistic pathogen. Mori (1888) inoculated laboratory animals with canal water and, when the animals died, he isolated *Empedobacter brevis* from their blood and internal organs. Whether the organism caused death of the animals is uncertain, but the cultures were pathogenic for guinea pigs, mice, and rabbits. Holmes et al. (1978) reviewed the history of isolation of *Empedobacter brevis* strains from various clinical specimens and indicated that there was no definite evidence that these strains were the causative agents of the infections, since some were obtained from mixed cultures, others were possible contaminants, and still others were isolated from cases where the clinical details were not well described. *Empedobacter brevis* has been isolated from the anterior chamber and the vitreous in a series of patients who had developed endophthalmitis after cataract extraction by the same surgeon on the same day; the source of infection was unknown (Janknecht et al., 2002). *Empedobacter brevis* has been isolated from the cerebrospinal fluid of a dog, where it occurred at high levels (Haburjak and Schubert, 1997). Various species of flavobacteria, including *Empedobacter brevis*, were isolated from damaged gills, pins, and skin ulcers of the Atlantic salmon (*Salmo salar*), rainbow trout (*Oncorhynchus mykiss*), and cod (*Gadus morhua*) and from exoskeleton lesions of king crab (*Paralithodes camtschaticus*) (Serdyuk, 1998). *Empedobacter brevis* was one of the species isolated from moribund Koi carp and goldfish, which often had severe ulceration; however, the primary causative agent was "atypical" *Aeromonas salmonicida*, with aggravation by the secondary invaders *Aeromonas hydrophila* and *Pseudomonas fluorescens* (Robertson and Austin, 1994).

Empedobacter brevis is also a parasite of certain protists. Córdova et al. (2002) found *Empedobacter brevis* and several other bacterial species to be present in the cytoplasm of *Alexandrium catenella* – a chain-forming dinoflagellate that is toxic and causes paralytic shellfish poisoning; however, *Empedobacter brevis* was not associated with production of the algal saxitoxin. In several instances, the *Alexandrium catenella* clones were simultaneously infected by different species of bacteria. The intracellular bacteria were alive and appeared to be dividing. Antibiotic treatment of *Alexandrium catenella* did not generate bacteria-free cells and led to the killing of the host cells. Bottone et al. (1992) showed that when the amoebae *Acanthamoeba castellanii* and *Acanthamoeba polyphaga* were cocultivated with *Empedobacter brevis*, *Xanthomonas maltophilia*, and *Pseudomonas paucimobilis*, enhancement of amoebal growth was better than that obtained in the presence of *Staphylococcus aureus*, *Staphylococcus epidermidis*, and *Escherichia coli*, the standard cocultivation species used for isolation of amoebae from clinical specimens. The authors concluded that contamination of contact lens care systems with *Acanthamoeba* spp. and a bacterium capable of supporting amoebic growth might be involved in amoeba-induced keratitis by providing large inocula of amoebae. Wadowsky and Yee (1983) found that satellite growth of *Legionella pneumophila* occurred with an environmental isolate of *Flavobacterium breve* on a medium deficient in cysteine – an amino acid required by *Legionella pneumophila* for growth; presumably, *Empedobacter brevis* secreted cysteine or a related compound into the medium.

Empedobacter brevis is resistant to a wide range of antimicrobial agents (Vandamme et al., 1994a). Janknecht et al. (2002) found an isolate of *Empedobacter brevis* to be resistant to aminoglycosides, cephalosporins (cefotiam, ceftriaxone, ceftazidim), and fosfomycin, but sensitive to ampicillin, piperacillin, imipenem, quinolones, tetracycline, and chloramphenicol. The occurrence of intrinsic metallo-β-lactamases in *Empedobacter brevis* may partially explain the resistance of the organism to penicillin. Bellais et al. (2002a) reported that the Ambler class B β-lactamase gene bla_{EBR-1} occurred in a strain of *Empedobacter brevis* isolated from a rectal swab. When cloned and expressed in *Escherichia coli*, the enzyme hydrolyzed penicillins, cephalosporins, and carbapenems efficiently, but not aztreonam.

Empedobacter brevis can be found free living in water and in food (Hugo and Jooste, 1997). Turtura et al. (2000) reported that *Empedobacter brevis* was one of several species isolated from black crusts taken from open-air stone monuments in Bologna, Italy, that dated back to between the twelfth and nineteenth centuries. The presence of these species in the black crusts was not occasional, but instead the result of an ecologically significant process of colonization.

Enrichment and isolation procedures

Special procedures are not normally required for the isolation of *Empedobacter brevis*. However, the ability of this species to grow on MacConkey agar and β-hydroxybutyrate agar may provide some selectivity.

Differentiation of the genus *Empedobacter* from other genera

The genus *Empedobacter* contains a single species, *Empedobacter brevis*. The nearest relative of *Empedobacter* is *Weeksella*, which also includes a single species, *Weeksella virosa* (Vandamme et al., 1994a). *Empedobacter brevis* differs from *Weeksella virosa* in that it forms yellow colonies, has a saccharolytic metabolism, exhibits DNase activity, is resistant to penicillin, fails to grow at 42°C, and has a lower DNA G+C content (31–33 mol% vs 35–38 mol%).

Features differentiating *Empedobacter* from other genera of rRNA superfamily V are listed by Vandamme et al. (1994a). In a study of 130 food isolates of *Chryseobacterium* and *Empedobacter*, Hugo and Jooste (1997) reported that use of the multilocus enzyme electrophoresis technique was able to correctly identify the *Empedobacter* group.

Taxonomic comments

The taxonomic history of *Empedobacter brevis* prior to 1984 has been reviewed by Holmes et al. (1984a), who included *Empedobacter brevis* in the genus *Flavobacterium* as *Flavobacterium breve*. In a later review by Holmes (1992), members of *Flavobacterium* were divided into four groups, of which the first group contained *Flavobacterium breve* and several other *Flavobacterium* species. Vandamme et al. (1994a) indicated that, based on chemotaxonomic and phenotypic features and on DNA–rRNA hybridization experiments, *Flavobacterium breve* and *Weeksella virosa* belonged to a separate position in rRNA superfamily V. However, the level of genotypic divergence between the two species was about 6°C [$T_{m(e)}$] and the two species differed in many biochemical features, including fatty acid composition. Consequently, Vandamme et al. (1994a) concluded that the two taxa should be considered as members of different genera and, accordingly, they revived the genus name *Empedobacter* for *Flavobacterium breve*, a name that had been applied previously to the same species by Prévot (1961).

List of species of the genus *Empedobacter*

1. **Empedobacter brevis** (Holmes and Owen 1982) Vandamme, Bernardet, Segers, Kersters and Holmes 1994a, 830[VP] [*Flavobacterium breve* (*ex* Lustig 1890) Holmes and Owen 1982, 233; *Bacillus brevis* Lustig 1890, 521; *Bacillus canalicolis brevis* Cornil and Babes 1890, 292; *Bacillus canalis parvus* Eisenberg 1891, 362; *Bacterium canalis parvis* (*sic*) (Eisenberg 1891) Chester 1897, 130; *Bacterium canale* Mez 1898, 55; *Bacterium breve* (Lustig 1890) Chester 1901, 69; *Flavobacterium brevis*

(*sic*) (Lustig 1890) Bergey, Harrison, Breed, Hammer and Huntoon 1923, 116; *Pseudobacterium brevis* (*sic*) (Lustig 1890) Krasil'nikov 1949, 239; *Empedobacter breve* (*sic*) (Lustig 1890) Prévot 1961, 181]

bre'vis. L. masc. adj. *brevis* short.

The following characteristics are taken from Holmes et al. (1984a). No acid is produced from adonitol, arabinose, cellobiose, dulcitol, ethanol, fructose, glycerol, inositol, lactose, mannitol, raffinose, rhamnose, salicin, sorbitol, sucrose, trehalose, or xylose. Some strains hydrolyze Tween 80. The following tests are all negative: acid and/or gas production from glucose in a peptone/water medium, oxidation of gluconate, 3-ketolactose production, malonate utilization, arginine deamidase, arginine dihydrolase, lysine decarboxylase, ornithine decarboxylase, phenylalanine deaminase, β-galactosidase (ONPG), alkali production on Christensen's citrate, fluorescent pigment production on King's medium B, pigment production on tyrosine agar, growth on cetrimide agar and Simmons' citrate agar, reduction of 0.4% (w/v) selenite, H_2S production, KCN tolerance, nitrate reduction, and production of opalescence on lecithovitellin agar.

Source: various sources including canal water, clinical specimens, food, fish and marine animals, dogs, and protists.

DNA G+C content (mol%): 31–33 (T_m).

Type strain: CL88/76 (Holmes et al., 1978), NCTC 11099, ATCC 43319, CCUG 7320, CIP 103104, IFO (NBRC) 14943, LMG 4011.

Sequence accession nos: M33888 (5S rRNA), D14022 (16S rRNA).

Genus XIX. **Epilithonimonas** O'Sullivan, Rinna, Humphreys, Weightman and Fry 2006, 177[VP]

THE EDITORIAL BOARD

Ep.i.lith.on.i.mo'nas. N.L. n. *epilithon -is* (or *epilithonum -i*) epilithon; L. fem. n. *monas* a unit, monad; N.L. fem. n. *Epilithonimonas* a monad isolated from epilithon.

Rounded rods, 0.6–0.7 × 1.0–2.5 µm, that do not form chains or filaments. No gliding motility. Gram-stain-negative. Strictly aerobic. Colonies are orange. **Flexirubin pigments are present.** Growth occurs between 4 and 30°C. **No growth occurs on marine agar.** Oxidase- and catalase-positive. **Nitrate is reduced.** A few carbohydrates, including glucose, are utilized for growth. **Acid is not produced from glucose. Indole-negative.** Fatty acids contain a large proportion of saturated branched-chain, monosaturated and hydroxy fatty acids. The most abundant individual fatty acids are $C_{15:0}$ iso and summed feature 3 ($C_{16:1}$ ω7*c*/$C_{15:0}$ iso 2-OH). Na^+ is not required. Isolated from freshwater environments.

DNA G+C content (mol%): 37.5 (HPLC).

Type species: **Epilithonimonas tenax** O'Sullivan, Rinna, Humphreys, Weightman and Fry 2006, 177[VP].

Enrichment and isolation procedures

Epilithonimonas tenax was isolated from the surface film on river stones. Samples of the epilithon were serially diluted in sterile deionized water containing 1 µl/l Tween 20. The spread-plate technique was used with various common media such as plate count agar (PCA) containing 20 µg/ml cycloheximide to inhibit eukaryotic micro-organisms. Plates were incubated at 20°C for 10 d and orange, spreading colonies were selected, subcultured twice on their original isolation medium, and purified on PCA.

Differentiation of the genus *Epilithonimonas* from related genera

Epilithonimonas differs from members of the genera *Bergeyella*, *Elizabethkingia*, *Riemerella*, and *Sejongia* by possessing flexirubin-type pigments. An ability to reduce nitrate distinguishes it from *Bergeyella*, *Elizabethkingia*, *Kaistella*, *Riemerella*, *Sejongia*, and most species of *Chryseobacterium*, except *Chryseobacterium balustinum* and *Chryseobacterium daecheongense*. Lack of indole production distinguishes it from *Bergeyella*, *Elizabethkingia*, *Kaistella*, *Sejongia*, and most species of *Chryseobacterium*, except *Chryseobacterium daecheongense* and *Chryseobacterium scophthalmum*. The inability of *Epilithonimonas* to form acid from glucose differentiates it from *Elizabethkingia*, *Riemerella*, *Sejongia*, and most species of *Chryseobacterium*, except *Chryseobacterium daecheongense* and *Chryseobacterium scophthalmum*.

Taxonomic comments

16S rRNA gene sequence analysis has shown that *Epilithonimonas tenax* belongs to a monophyletic cluster containing the genera *Bergeyella*, *Chryseobacterium*, *Elizabethkingia*, *Kaistella*, *Riemerella*, and *Sejongia*. It forms a discrete lineage positioned between the genus *Elizabethkingia* and the other genera (O'Sullivan et al., 2006).

List of species of the genus *Epilithonimonas*

1. **Epilithonimonas tenax** O'Sullivan, Rinna, Humphreys, Weightman and Fry 2006, 177[VP]

ten'ax. L. fem. adj. *tenax* sticky, holding firm, referring to the organism's viscous colonies.

Characteristics are as described for the genus, with the following additional features. Colonies on PCA after 1 week at 20°C are 3–6 mm in diameter, bright orange, opaque, and smooth, with an entire edge and a viscous consistency. Growth occurs on trypticase soy agar, nutrient agar, MacConkey agar, and DNase agar, but not on marine agar 2216 or cetrimide agar. Esculin and starch are hydrolyzed, but not agar, arginine, DNA, Tween 80, or gelatin. No β-galactosidase, urease, or xylanase activities. Growth occurs on the following substrates: acetic acid, L-alanyl-glycine, α-cyclodextrin, dextrin, gentiobiose, α-D-glucose, L-glutamic acid, glycogen, glycyl-L-aspartic acid, glycyl-L-glutamic acid, inosine, α-ketovaleric acid, mannose, maltose, L-ornithine, L-proline, L-serine, Tween 40, sucrose, L-threonine, thymidine, trehalose, and uridine. No growth occurs on the following substrates: *N*-acetyl-D-galactosamine, *N*-acetylglucosamine, *cis*-aconitic acid, adipate, adonitol, alaninamide, D-alanine, L-alanine, γ-aminobutyric acid, 2-aminoethanol, arabinose, D-arabitol, L-asparagine, L-aspartic acid, bromosuccinic acid, 2,3-butanediol, caprate, DL-carnitine, cellobiose, citric acid, i-erythritol, formic acid, D-fructose, L-fucose, D-galactonic acid, D-galactose, D-galacturonic acid, D-gluconic acid, D-glucosaminic acid, glucuronamide, D-glucuronic acid, glucose 1-phosphate, glucose 6-phosphate, glycerol, DL-α-glycerol phosphate, L-histidine, α-hydroxybutyric acid, β-hydroxybutyric acid, γ-hydroxybutyric acid, *p*-hydroxyphenylacetic acid, hydroxy-L-proline, *myo*-inositol, itaconic acid, α-ketobutyric acid, α-ketoglutaric acid, DL-lactic acid, α-D-lactose, lactulose, L-leucine, malate, malonic acid, mannitol, melibiose, methyl-β-D-glucoside, methylpyruvate, monomethyl succinate, phenylacetate, L-phenylalanine, phenylethylamine, propionic acid, D-psicose, putrescine, L-pyroglutamic acid, quinic acid, raffinose, L-rhamnose, D-saccharic acid, sebacic acid, D-serine, D-sorbitol, succinamic acid, succinic acid, turanose, Tween 80, urocanic acid, or xylitol. Susceptible to penicillin G, ampicillin, and rifampin, but not to chloramphenicol, streptomycin, kanamycin, or tetracycline.

Source: the type strain was isolated from epilithon-covered stones from the River Taff in Cardiff, UK.

DNA G+C content (mol%): 37.5 (HPLC).

Type strain: EP105, DSM 16811, NCIMB 14026.

Sequence accession no. (16S rRNA gene): AF493696.

Genus XX. **Flaviramulus** Einen and Øvreås 2006, 2460[VP]

THE EDITORIAL BOARD

Fla.vi.ra′mu.lus. L. adj. *flavus* yellow; L. masc. dim. n. *ramulus* small branch; N.L. masc. n. *Flaviramulus* small yellow branch.

Cells in stationary phase are rods, 0.2–0.3 × 1–3 µm. Exponential phase cells are rods, but stationary phase cells are pleomorphic, 0.2–0.3 µm in diameter and 1–30 µm long. Branched and curled cells are also seen in stationary phase cultures. As cultures age, cells degenerate into spheroplasts ranging from 0.15 to 1 µm in diameter. **Cells are motile by polar adhesion to a glass surface, with a rotational movement**; cells grip the surface at the opposite pole and then are released at the first pole. Gram-stain-negative. **Aerobic. Oxidase-negative. Cells contain yellow and orange carotenoids.** Flexirubin-type pigments are not found. **Seawater, yeast extract or thiamine, and amino acids are required for growth.** Menaquinone-6 is the major respiratory quinone.

DNA G+C content (mol%): 31.4 (T_m).

Type species: **Flaviramulus basaltis** Einen and Øvreås 2006, 2460[VP].

Further descriptive information

Einen and Øvreås (2006) reported the isolation of four yellow-pigmented, Gram-stain-negative, motile strains of bacteria from the glassy rind of basaltic pillow lavas collected from the seafloor in the Norwegian/Greenland Sea at a depth of 1,300 m below sea level. The seafloor temperature at the sampling site was −0.7°C.

Motility is best observed in fresh cultures grown in FM broth* containing <0.05% yeast extract. Cells adherent to a glass surface have a rotational movement of about three rotations per second. Polar appendages occur on many of the cells that are motile, but not on nonmotile cells. No spreading growth occurs on marine agar (MA) or FM agar.

The *Flaviramulus* strains were able to produce acid from the following compounds: *N*-acetyl-D-glucosamine, D-arabinose, D-cellobiose, D-fructose, D-glucose, lactose, maltose, D-mannitol, L-rhamnose, D-ribose, sucrose, and xylan. Alcohols did not support growth. The following compounds were used in the Biolog system: L-aspartic acid, cellobiose, cyclodextrin, dextrin, D-fructose, D-galactose, D-galacturonic acid, gentiobiose, D-glucose, D-glucose 1-phosphate, D-glucose 6-phosphate, D-glucuronic acid, L-glutamic acid, glycogen, glycyl-L-aspartic acid, glycyl-L-glutamic acid, ketobutyric acid, ketoglutaric acid, DL-lactic acid, D-lactose, lactulose, maltose, D-mannose, methyl-D-glucoside, L-ornithine, L-proline, pyruvic acid methyl ester, raffinose, L-rhamnose, succinamic acid, succinic acid, succinic acid monomethyl ester, sucrose, trehalose, turanose, and Tweens 40 and 80. The following Biolog substrates were not utilized: acetic acid, *N*-acetyl-D-galactosamine, *cis*-aconitic acid, adonitol, L-alaninamide, D-alanine, L-alanine, L-alanyl-glycine, L-arabinose, D-arabitol, L-asparagine, bromosuccinic acid, citric acid, i-erythritol, formic acid, L-fucose, D-galactonic acid lactone, D-gluconic acid, D-glucosaminic acid, glucuronamide, L-histidine, α-, β-, and γ-hydroxybutyric acid, *p*-hydroxyphenylacetic acid, hydroxy-L-proline, *myo*-inositol, itaconic acid, ketovaleric acid, L-leucine, malonic acid, D-mannitol, melibiose, L-phenylalanine, propionic acid, D-psicose, L-pyroglutamic acid, quinic acid, D-saccharic acid, sebacic acid, D-sorbitol, and xylitol.

Enrichment and isolation procedures

The outer glassy rim of the basalt was chipped off and the glass was placed in sterile seawater and stored in closed flasks in the dark for 3 years at 10°C. Portions of the seawater were plated onto MA [marine broth 2216 (Difco) solidified with 1.5% agar] and incubated at 10°C. After incubation for 14 d, small yellow colonies were selected and purified.

Cultures can be grown in FM broth at 22°C. Preservation can be accomplished at −80°C in marine broth supplemented with 15% glycerol.

*The Flavobacteriaceae Medium (FM) of Einen and Øvreås (2006) contains (g/l distilled water): sea salts (Sigma), 30; D-glucose, 5; and yeast extract, 0.5; pH adjusted to 8.0.

Differentiation of the genus *Flaviramulus* from related genera

Several of the phenotypic characteristics of *Flaviramulus* differ from those of members of the genera *Gaetbulibacter* and *Algibacter*. The most important difference between *Flaviramulus* and the genera *Gaetbulibacter* and *Algibacter* is that *Flaviramulus* is obligately aerobic, whereas *Gaetbulibacter* and *Algibacter* are capable of fermentation/anaerobic respiration. In addition, *Flaviramulus* is oxidase-negative, grows at 0°C, grows in media with 0.5% NaCl, and hydrolyzes Tween 80. Phenotypic differences between *Flaviramulus* and other genera of *Flavobacteriaceae* have been described by Einen and Øvreås (2006).

Taxonomic comments

16S rRNA gene sequence analysis showed that the type strain of *Flaviramulus basaltis* is closely related to the type strains of *Gaetbulibacter saemankumensis* and *Algibacter lectus* (94.3 and 94.8% similarities, respectively).

List of species of the genus *Flaviramulus*

1. **Flaviramulus basaltis** Einen and Øvreås 2006, 2460[VP]

ba.sal'tis. L. masc. gen. n. *basaltis* of basalt, pertaining to the source of isolation.

The following properties are in addition to those given for the genus. Colonies grown on MA are shiny, dark yellow, circular, and convex, with entire margins. Cells are catalase-positive and alkaline phosphatase-positive. Growth occurs from −2.0 to 34.0°C, with optimum growth at 17.5–22.8°C. The salinity range for growth is 3–60 g/l sea salts, with an optimum of 24–60 g/l. Growth is supported at pH 6.5–8.6, with optimum growth at pH 6.5–8.2. Utilizes and produces acids from several sugars. Utilizes organic acids and amino acids. Does not utilize alcohols. Hydrolyzes L-tyrosine, esculin, carrageenan, gelatin, starch, DNA, urea, and Tweens 20, 40 and 80, but not agar, cellulose, or chitin. Produces H_2S from cysteine. Branched fatty acids are predominant.

Source: the type strain was isolated from seafloor basalt offshore of Jan Mayen in the Norwegian/Greenland Sea at a depth of 1300 m.

DNA G+C content (mol%): 31.4 ± 0.6 (T_m).

Type strain: H35, CIP 109091, DSM 18180.

Sequence accession no. (16S rRNA gene): DQ361033.

Genus XXI. **Formosa** Ivanova, Alexeeva, Flavier, Wright, Zhukova, Gorshkova, Mikhailov, Nicolau and Christen 2004, 707[VP] emend. Nedashkovskaya, Kim, Vancanneyt, Snauwaert, Lysenko, Rohde, Frolova, Zhukova, Mikhailov, Bae, Oh and Swings 2006d, 166

Elena P. Ivanova

For.mo'sa. L. fem. adj. (used as a substantive) *Formosa* beautiful, finely formed.

Rod-shaped cells with slightly irregular sides and pointed ends, approx. 0.8–1.8 × 0.4–0.9 μm. **Cells have gliding motility. Buds may be produced. Do not form endospores or resting stages. Gram-stain-negative**. Poly-β-hydroxybutyrate is not accumulated as an intracellular reserve product. An arginine dihydrolase system is not present. Aerobic; however, anaerobic growth can occur on D-glucose by anaerobic respiration with nitrate. Chemo-organotrophic. Positive for oxidase, catalase, and alkaline phosphatase. May require Na^+ for growth. Agar hydrolysis differs among species. Produce non-diffusible carotenoid pigments. Flexirubin pigments are not detected. Isolated from marine habitats of the North Pacific.

DNA G+C content (mol%): 34–36.

Type species: **Formosa algae** Ivanova, Alexeeva, Flavier, Wright, Zhukova, Gorshkova, Mikhailov, Nicolau and Christen 2004, 707[VP] emend. Nedashkovskaya, Kim, Vancanneyt, Snauwaert, Lysenko, Rohde, Frolova, Zhukova, Mikhailov, Bae, Oh and Swings 2006d, 161.

Further descriptive information

Phylogenetically, *Formosa* is a member of the family *Flavobacteriaceae*. Species of the genus form a distinct lineage within the family. Similar topologies of phylogenetic trees are retrieved by neighbor-joining, maximum-likelihood, and maximum-parsimony algorithms and supported by high bootstrap values. The level of 16S rRNA gene sequence similarity with the two phylogenetically most closely related genera, *Psychroserpens* and *Gelidibacter*, is almost equidistant, reaching 94.2% sequence similarity and 82 sequence differences with, for example, *Psychroserpens burtonensis*. The closest uncultured sequence, from bacteria from the toxic dinoflagellate *Alexandrium tamarense* (Groben et al., 2000), had 97.2% sequence similarity and 27 differences.

Phosphatidylethanolamine is the only phospholipid identified in bacteria of this genus. Amino-containing lipids that were detected on TLC plates are ninhydrin-positive (Ivanova et al., 2004).

MK-6 was the major lipoquinone among isoprenoid quinones, as detected using HPLC (Nedashkovskaya et al., 2006d).

The predominant cellular fatty acids are branched-chain saturated and unsaturated, namely, $C_{15:0}$, $C_{15:0}$ iso, $C_{15:1}$ ω6c, $C_{15:0}$ iso, $C_{15:0}$ iso 3-OH, $C_{17:0}$ iso 3-OH, and summed feature 3 (comprising any combination of $C_{16:1}$ ω7*c*, $C_{16:1}$ ω7*t*, and $C_{15:0}$ iso 2-OH) (Table 45). In spite of the intraspecific heterogeneity of the fatty acids content of micro-organisms belonging to the family *Flavobacteriaceae* (Bernardet et al., 1996, 2002; Holmes, 1984a, 1993), bacteria of the genus *Formosa* retain a high proportion of branched-chain saturated and unsaturated cellular fatty acids, a characteristic feature of the family *Flavobacteriaceae*.

The cell morphology of *Formosa algae* KMM 3553[T] has been observed by atomic force microscopy (Ivanova et al., 2004) and that of strain KMM 3901[T] has been observed by scanning and transmission electron microscopy as described by Bruns et al. (2001). Cells of strain KMM 3901[T] showed budding fission (Figure 41). The buds are usually formed on mature cells of strain KMM 3901[T]. In addition to producing buds, cells of KMM 3901[T] formed a fibrillar network that connected them to one another. The formation of a fibrillar network by flavobacteria

TABLE 45. Fatty acid content (%) of whole-cell hydrolysates of *Formosa* strains[a]

Fatty acid	*F. algae* KMM 3553[T]	*F. agariphila* KMM 3901[T]	*F. agariphila* KMM 3962
$C_{15:0}$	15.5	8.7	11.4
$C_{15:0}$ 2-OH	1.5	1.8	1.6
$C_{15:0}$ 3-OH	2.3	2.4	4.0
$C_{15:1}$ ω6*c*	8.5	6.0	11.8
$C_{15:0}$ iso	17.1	12.7	17.2
$C_{15:1}$ iso G	9.9	6.5	11.4
$C_{15:0}$ iso 3-OH	6.7	7.7	10.5
$C_{15:0}$ anteiso	4.7	3.4	1.6
$C_{15:1}$ anteiso A	1.0	tr	–
$C_{16:0}$	1.1	1.6	–
$C_{16:0}$ 3-OH	tr	2.1	tr
$C_{16:0}$ iso	1.1	2.1	–
$C_{16:1}$ iso H	tr	2.5	–
$C_{16:0}$ iso 3-OH	4.0	8.9	3.1
$C_{17:0}$ 2-OH	1.5	tr	–
$C_{17:1}$ ω8*c*	1.2	tr	–
$C_{17:1}$ ω6*c*	3.5	3.2	1.4
$C_{17:1}$ iso ω9*c*	1.3	1.4	tr
$C_{17:0}$ iso 3-OH	9.6	8.5	10.7
Summed feature 3	5.9	15.8	12.1

[a]Values of less than 1% for all strains are not shown. Summed feature 3, which could not be separated by the Microbial Identification System, consists of one or more of the following fatty acids: $C_{15:0}$ iso 2-OH, $C_{16:1}$ ω7*c*, and $C_{16:1}$ ω7*t*. tr, Trace amount (<1% of total); –, not detected.

was first reported by Reichenbach (1989b). Similar features have also been found in strains of *Winogradskyella* (Nedashkovskaya et al., 2005a). The extracellular fibrils may facilitate attachment of cells to suitable substrates and to other cells, followed by biofilm formation on the surface of algae in natural habitats (Nedashkovskaya et al., 2005a).

On solid media containing high nutrient concentrations (e.g., marine agar), colonies formed by *Formosa algae* KMM 3553[T] are circular, 1–3 mm in diameter, and low-convex. Colonies formed by *Formosa agariphila* KMM 3901[T] are round, flat, 3–4 mm in diameter, translucent, and sunk into the agar, whereas colonies of *Formosa agariphila* KMM 3962 are 2–3 mm in diameter and slightly sunk into the agar after 24–48 h. Colonies are pale to light yellow depending on the strain. The yellow color of the colonies is due to production of carotenoid pigments (with absorbance peaks at 455 and 480 nm). Flexirubin pigments are not detected.

Under laboratory conditions, growth of *Formosa agariphila* KMM 3901[T] in broth results in the formation of small aggregates. Cells of strains KMM 3553[T] and KMM 3901[T] produce a dense slime on solid media.

There are no specific growth requirements for strains of *Formosa*. Some strains may require Na^{+} ions for growth. Natural seawater [50% (v/v)] may need to be added to the base medium when testing for acid production from carbohydrates. The temperature range for growth is 4–34°C, with optimum growth at 21–23°C.

All strains of *Formosa algae* and *Formosa agariphila* KMM 3962 show remarkable resistance to benzylpenicillin, gentamicin, kanamycin, neomycin, polymyxin B, streptomycin ampicillin, carbenicillin, lincomycin, and oleandomycin. *Formosa agariphila* KMM 3901[T] is susceptible to ampicillin, carbenicillin, lincomycin, and oleandomycin.

Formosa strains have been isolated from marine habitats of the North Pacific. These organisms may be free-living or colonizers of the surfaces of marine algae. *Formosa algae* represents one of the numerically dominant groups of the brown alga (*Fucus evanescens*)-degrading enrichment community. *Formosa agariphila* KMM 3901[T] was isolated from the green alga *Acrosiphonia sonderi* in Troitsa Bay, Gulf of Peter the Great, East Sea (also known as the Sea of Japan). *Formosa agariphila* KMM 3962 was isolated from seawater in Amursky Bay (Gulf of Peter the Great), East Sea.

Enrichment and isolation procedures

Species of *Formosa* can usually be isolated by direct inoculation of 0.1 ml algal homogenate or 0.1 ml seawater onto marine agar 2216 (Difco) or medium B.* Strains of *Formosa algae* were isolated from enrichment experiments during the degradation of the thallus of brown algae (*Fucus evanescens*) (Ivanova et al., 2002a, b), with addition of a protein inhibitor for the endo-(1 → 3)-β-D-glucanases (Yermakova et al., 2002) to the enrichment culture.

Short-term survival of *Formosa* can be achieved for approximately 2 weeks on marine agar 2216 plates stored at 4°C. Long-term preservation of cultures can be accomplished in marine broth supplemented with 30% glycerol at −80°C.

Differentiation of the genus *Formosa* from other genera

Differential phenotypic features for the genus *Formosa* and its close relatives within the family *Flavobacteriaceae* are presented in Table 46. For most flavobacteria, determining generic affiliation based on phenotypic traits is difficult because of the great variation among species. A polyphasic taxonomic approach is recommended to differentiate different members of *Flavobacteriaceae*. Species of *Formosa* share many similar phenotypic features with species of *Algibacter* and *Gaetbulibacter*. In contrast to *Formosa*, *Psychroserpens burtonensis* (Bowman et al., 1997) has a characteristic cell morphology, a non-saccharolytic metabolism, and a low DNA G+C content. Strains of *Gelidibacter algens* have different cell morphology and also contain significant levels of characteristic $C_{15:1}$ anteiso ω10*c* and $C_{17:1}$ anteiso ω7*c* fatty acids.

Taxonomic comments

The level of rRNA gene sequence similarity between *Formosa algae* KMM 3553[T] and *Formosa agariphila* KMM 3901[T] is 98.9–99.1%.

Differentiation of species of the genus *Formosa*

The identification of novel isolates as existing species of *Formosa* may be problematic due to high interspecies variation in phenotypic characteristics, as occurs, for example, in *Formosa agariphila*. The correct allocation of newly discovered taxa will require partial and/or full sequencing of 16S rRNA genes and DNA–DNA hybridization. The differential characteristics of species of *Formosa* are given in Table 47.

*Medium B contains: Bacto peptone (Difco), 0.02 g; casein hydrolysate (Merck), 0.02 g; Bacto yeast extract (Difco), 0.02 g; 0.1% glucose, 0.01 g; KH_2PO_4, 0.002 g; $MgSO_4 \cdot 7H_2O$, 0.0005 g; Bacto agar (Difco), 15 g; 500 ml natural seawater and 500 ml distilled water; pH 7.5–7.8.

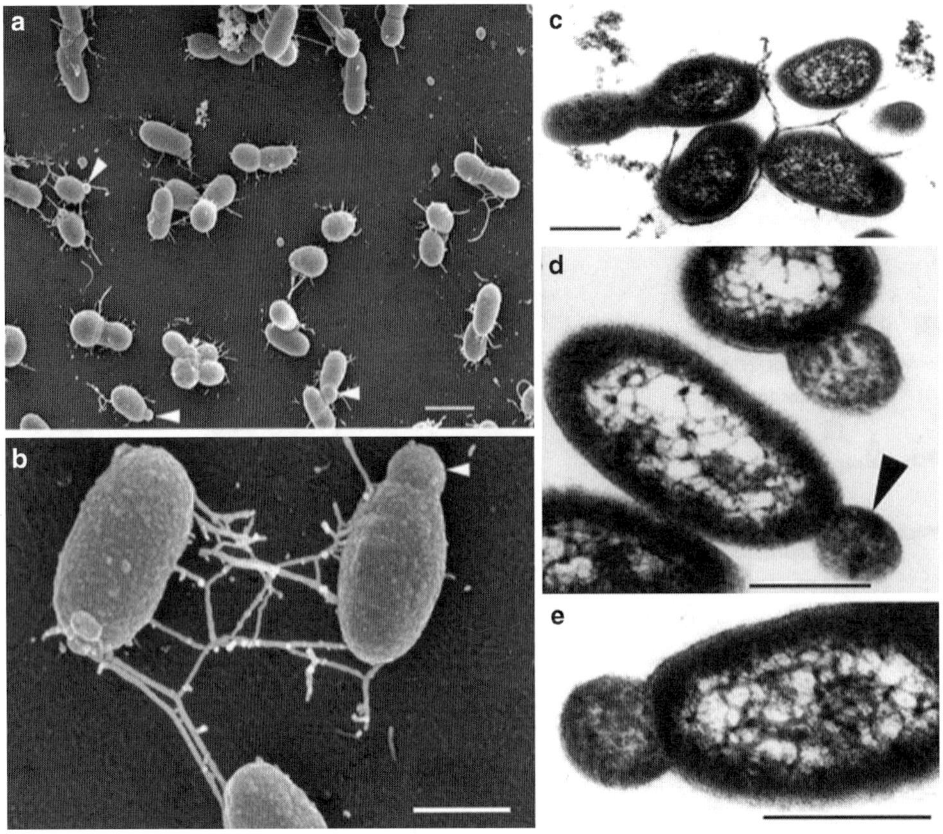

FIGURE 41. (a, b) Scanning electron micrographs of cells of *Formosa agariphila* strain KMM 3901^T, showing the rod-shaped morphology, budding (arrowheads), and thread-like structures formed by bacterial cells. (c–e) Transmission electron micrographs of ultrathin sections of cells of strain KMM 3901^T, showing thread-like structures (c, compare with (b)) and bud formation on mature cells (indicated by an arrowhead in d). Bars = 1 µm (a) and 0.5 µm (b–e). Reproduced from Nedashkovskaya et al. (2006d).

TABLE 46. Differential characteristics of the genus *Formosa* and other allied genera of the family *Flavobacteriaceae*[a,b]

Characteristic	*Formosa*	*Algibacter*	*Gaetbulibacter*	*Gelidibacter*	*Lacinutrix*	*Olleya*	*Psychroserpens*	*Subsaxibacter*	*Subsaximicrobium*
Fermentation of glucose	+	+	+	–	–	–	–	–	–
Gliding motility	+	+	+	+	–	+	–	+	+
Presence of budding	+	–	–	–	–	–	–	–	–
Oxidase activity	+	+	+	–	nd	+	–	nd	nd
Catalase activity	+	+	+	+	+	+	+	+	+
Acid formation from carbohydrates	+	+	+	+	–	+	–	–	–
Temperature range for growth, °C	4–34	4–35	13–40	-2 to 37	-2 to 29	4–30	0–19	-2 to 20	-2 to 25
Salinity range for growth (%)	0–8	1–6	1–7	0–8	0.6–12	1.2–5	2.4–8	0.6–7	1.5–12
Hydrolysis of:									
Agar	D	+	–	–	–	–	–	–	–
Casein	–	–	–	D	–	–	+	–	D
Gelatin	+	+	–	D	+	+	D	+	+
Starch	D	+	+	D	–	–	–	–	D
Tween 80	–	–	–	D	+	+	+	+	+
Nitrate reduction	D	–	+	–	–	–	–	–	–
DNA G+C content (mol%)	34–36	31–33	34–35	36–42	37	49	27–29	35–36	38–41

[a]Symbols: +, 90% or more of the strains are positive; –, 10% or less of the strains are positive; D, different reactions in different taxa (species of a genus or genera of a family); nd, not determined.

[b]Data are from Bowman et al. (1997); Macián et al. (2002); Ivanova et al. (2004); Jung et al. (2005); Nedashkovskaya et al. (2004d, 2006d); Bowman and Nichols (2005); Mancuso Nichols et al. (2005); Jung et al. (2005).

TABLE 47. Phenotypic characteristics of *Formosa* strains[a,b]

Characteristic	*F. algae* KMM 3553[T]	*F. agariphila* (*n* = 2)
β-Galactosidase activity	–	+
Hydrolysis of agar	–	+
Acid production from:		
D-Galactose	–	+
Glycerol	+	–
Utilization of L-arabinose and D-lactose	–	+
Susceptibility to tetracycline	+	–
DNA G+C content (mol%)	34.0	35–36

[a]Symbols: +, 90% or more of the strains are positive; –, 10% or less of the strains are positive; *n*, number of strains.

[b]All strains are positive for the following characteristics: gliding motility; oxidase, catalase, and alkaline phosphatase activities; glucose fermentation; growth in 1–6% NaCl and at 4–30°C; hydrolysis of gelatin and esculin; acid formation from L-fucose, D-glucose, maltose, DL-xylose, and mannitol; and utilization of D-mannose. All strains are negative for the following characteristics: flexirubin pigments; urease activity; nitrate reduction; H_2S, indole, and acetoin production; hydrolysis of casein, cellulose (CM-cellulose and filter paper), chitin, Tween 80, and DNA; acid formation from L-arabinose, cellobiose, D-lactose, melibiose, raffinose, L-rhamnose, L-sorbose, sucrose, acetate, citrate, fumarate, malate, adonitol, inositol, and sorbitol; utilization of sucrose, inositol, sorbitol, citrate, and malonate; and susceptibility to benzylpenicillin, gentamicin, kanamycin, neomycin, polymyxin B, and streptomycin.

List of species of the genus *Formosa*

1. **Formosa algae** Ivanova, Alexeeva, Flavier, Wright, Zhukova, Gorshkova, Mikhailov, Nicolau and Christen 2004, 707[VP] emend. Nedashkovskaya, Kim, Vancanneyt, Snauwaert, Lysenko, Rohde, Frolova, Zhukova, Mikhailov, Bae, Oh and Swings 2006d, 166

al′gae. L. fem. gen. n. *algae* of an alga, pertaining to the source of isolation, brown algae.

Characteristics are as described for the genus plus the following features. Exhibits gliding motility. Colonies are light yellow, circular, 1–3 mm in diameter, and low-convex on solid media containing high nutrient concentrations. Produces carotenoid pigments with absorbance peaks at 455 and 480 nm. Flexirubin pigments are absent. No growth is detected at 2 or 37°C; optimum growth is at 23°C. Grows at pH 5.0–10.0; optimum pH is 8.0–8.5. Alkalitolerant. Grows in 0–6% NaCl. Does not hydrolyze agar or chitin. Starch, gelatin, and urea are weakly hydrolyzed. Hydrolyzes esculin and Tween 40. Acid is produced from L-fucose, D-glucose, maltose, DL-xylose, and mannitol, but not from L-arabinose, cellobiose, D-lactose, melibiose, raffinose, L-rhamnose, L-sorbose, sucrose, acetate, citrate, fumarate, malate, adonitol, inositol, or sorbitol. Utilizes D-mannose, but not sucrose, inositol, sorbitol, citrate, or malonate (API identification system; bioMérieux). H_2S, indole, and acetoin (Voges–Proskauer reaction) are not produced. Nitrate can be reduced to nitrite. According to the Biolog system, utilizes Tween 40, D-galactose, gentiobiose, α-D-glucose, monosuccinate, citric acid, D-glucuronic acid, succinamic acid, succinic acid, alaninamide, glycyl-L-aspartic acid, hydroxy-L-proline, L-ornithine, L-pyroglutamic acid, urocanic acid, thymidine, 2-aminoethanol, and glycerol. Does not utilize α-cyclodextrin, dextrin, glycogen, Tween 80, *N*-acetyl D-glucosamine, adonitol, cellobiose, L-arabinose, D-arabitol, i-erythritol, D-fructose, L-fucose, *myo*-inositol, α-D-lactose, lactulose, maltose, D-mannitol, D-mannose, melibiose, methyl α-D-glucoside, D-psicose, raffinose, L-rhamnose, D-sorbitol, sucrose, trehalose, turanose, xylitol, acetic acid, *cis*-aconitic acid, formic acid, D-galactonic acid lactone, D-galacturonic acid, D-glucosaminic acid, D-glucuronic acid, α-hydroxybutyric acid, β-hydroxybutyric acid, γ-hydroxybutyric acid, *p*-hydroxyphenylacetic acid, itaconic acid, α-ketoglutaric acid, α-ketobutyric acid, α-ketovaleric acid, DL-lactic acid, malonic acid, propionic acid, quinic acid, D-saccharic acid, sebacic acid, bromosuccinic acid, glucuronamide, D-alanine, L-alanine, L-alanylglycine, L-asparagine, L-aspartic acid, L-glutamic acid, L-histidine, L-leucine, L-ornithine, L-phenylalanine, L-proline, D-serine, L-serine, L-threonine, DL-carnitine, γ-aminobutyric acid, inosine, uridine, phenylethylamine, 2,3-butanediol, DL-α-glycerol phosphate, α-D-glucose 1-phosphate, or D-glucose 6-phosphate. Not susceptible to ampicillin, lincomycin, benzylpenicillin, kanamycin, oleandomycin, tetracycline, neomycin, streptomycin, gentamicin, or polymyxin B. Phosphatidylethanolamine is the only phospholipid identified. Sphingophospholipids are absent.

Source: brown algae.

DNA G+C content (mol%): 34–35 (T_m).

Type strain: KMM 3553, CIP 107684, KCTC 12364.

Sequence accession no. (16S rRNA gene): AY228461.

2. **Formosa agariphila** Nedashkovskaya, Kim, Vancanneyt, Snauwaert, Lysenko, Rohde, Frolova, Zhukova, Mikhailov, Bae, Oh and Swings 2006d, 166[VP]

a.ga.ri.phi′la. N.L. neut. n. *agarum* agar; Gr. adj. *philos* loving; N.L. fem. adj. *agariphila* agar-loving.

Cells are 0.4–0.6 μm in width, 0.8–1.2 μm in length, and can be connected by thread-like structures. Budding morphology may be observed. On marine agar, colonies are 2–4 mm in diameter, yellow, circular, flat or convex, opaque or translucent, shiny with entire edges, and sunk into the agar. Growth occurs at 4–33°C; optimal temperature for growth is 21–23°C. Growth occurs in 1–8% NaCl. Agar, gelatin, and esculin are hydrolyzed, but not casein, DNA, Tween 80, cellulose (CM-cellulose and filter paper), or chitin. Acid is formed from L-fucose, D-galactose, D-glucose, maltose, DL-xylose, and mannitol, but not from L-arabinose, cellobiose, D-lactose, melibiose, L-rhamnose, raffinose, L-sorbose, sucrose, adonitol, glycerol, dulcitol, inositol, or sorbitol. Utilizes L-arabinose, D-lactose, D-mannose, and sucrose, but not inositol, sorbitol, malonate, or citrate. Produces β-galactosidase. Nitrate is not reduced to nitrite. H_2S, indole, and acetoin (Voges–Proskauer reaction) production are not produced. Some strains are susceptible to ampicillin, carbenicillin, lincomycin, and oleandomycin. Resistant to benzylpenicillin, gentamicin, kanamycin, neomycin, polymyxin B, tetracycline, and streptomycin.

Source: algae and seawater.

DNA G+C content (mol%): 35–36 (T_m).

Type strain: KMM 3901, KCTC 12365, LMG 23005, DSM 15362.

Sequence accession no. (16S rRNA gene): AY187688.

Genus XXII. **Gaetbulibacter** Jung, Kang, Lee, Lee, Oh and Yoon 2005, 1848[VP]

THE EDITORIAL BOARD

Gaet.bu.li.bac′ter. N.L. n. *gaetbulum -i* gaetbul, the Korean name for a tidal flat; N.L. masc. n. *bacter* from Gr. neut. n. *baktron* rod; N.L. masc. n. *Gaetbulibacter* rod isolated from a tidal flat.

Rods, 0.4–0.5 × 3.0–4.5 µm. Non-flagellated. **Gliding motility is present.** Nonsporeforming. Gram-stain-negative. Aerobic. Growth also occurs under anaerobic conditions on marine agar and on marine agar with nitrate. Colonies are yellow. Flexirubin pigments are absent. Optimal pH is 7.0–8.0; weak growth at pH 5.5; no growth at pH 5.0. **NaCl range for growth is 0–7%**; optimum is 2–5%, but no growth is seen at >8%. Catalase- and oxidase-positive. Urease-negative. Nitrate is reduced. Starch is hydrolyzed, but not casein or gelatin. Acid is produced from a variety of carbohydrates. The predominant menaquinone is MK-6. Isolated from tidal flat sediments.

DNA G+C content (mol%): 34.7–34.9 (HPLC).

Type species: **Gaetbulibacter saemankumensis** Jung, Kang, Lee, Lee, Oh and Yoon 2005, 1848[VP].

Further descriptive information

Jung et al. (2005) reported that growth occurred under anaerobic conditions after incubation in an anaerobic chamber on both marine agar (Difco) and on marine agar supplemented with nitrate, both of which had been prepared anaerobically using nitrogen.

Gaetbulibacter strains are sensitive to the following antibiotics (concentration per disc is given in parentheses): lincomycin (15 µg), benzylpenicillin (10 U), and carbenicillin (25 µg). They are resistant to ampicillin (10 µg), gentamicin (10 µg), polymyxin B (300 U), streptomycin (30 µg), tetracycline (30 µg), and neomycin (15 µg).

Enrichment and isolation procedures

The three available strains of *Gaetbulibacter saemankumensis* were isolated by the dilution plating technique on marine agar 2216 (Difco), with incubation at 30°C.

Differentiation of the genus *Gaetbulibacter* from related genera

The ability of *Gaetbulibacter* strains to reduce nitrate distinguishes them from members of the genera *Algibacter*, *Bizionia*, and *Formosa*. Unlike members of the genus *Bizionia*, *Gaetbulibacter* strains have gliding motility, form acid from carbohydrates, hydrolyze starch, do not hydrolyze casein or gelatin, and fail to produce H_2S. *Gaetbulibacter* strains differ from those of *Algibacter* by growing in the presence of 7% NaCl and by their failure to hydrolyze gelatin. They differ from *Formosa* strains by growing in the presence of 7% NaCl and by lacking urease activity.

Taxonomic comments

By 16S rRNA gene sequence analysis, the closest neighbors of *Gaetbulibacter saemankumensis* are *Algibacter lectus* (94.9% sequence similarity), *Bizionia paragorgiae* (94.5%), and *Formosa algae* (93.8%).

DNA–DNA relatedness values among the three available strains of *Gaetbulibacter saemankumensis* ranged from 83 to 106%, indicating that the strains are members of the same species (Jung et al., 2005).

List of species of the genus *Gaetbulibacter*

1. **Gaetbulibacter saemankumensis** Jung, Kang, Lee, Lee, Oh and Yoon 2005, 1848[VP]

sae.man.kum.en′sis. N.L. masc. adj. *saemankumensis* of or belonging to Saemankum, from where the organism was originally isolated.

Characteristics are as given for the genus, with the following additional features. Esculin, tyrosine, and Tween 20 are hydrolyzed, but not hypoxanthine, xanthine, or Tweens 40, 60, and 80. The following enzyme activities are present (API ZYM system): *N*-acetyl-β-glucosaminidase, acid phosphatase, alkaline phosphatase, cystine arylamidase, esterase (C4), esterase lipase (C8), α-glucosidase, lipase (C14), leucine arylamidase, and valine arylamidase. Activities of the following enzymes are not found: α-galactosidase, β-galactosidase, β-glucuronidase, α-mannosidase, α-fucosidase, naphthol-AS-BI-phosphohydrolase, and trypsin. α-Chymotrypsin and β-glucosidase activities differ among strains (absent in the type strain). The following compounds are utilized as sole carbon and energy sources: cellobiose, D-fructose, D-galactose, D-glucose, L-rhamnose, and D-xylose. The following compounds are not utilized: L-alanine, L-asparagine, D-gluconic acid, glycerol, L-malic acid, melibiose, propionic acid, pyruvic acid, raffinose, L-serine, D-sorbitol, succinic acid, and trehalose. Utilization of maltose, L-proline, and sucrose differs among strains (positive for the type strain). Acid is produced from cellobiose, D-galactose, lactose, and maltose, but not from L-arabinose, D-fructose, D-mannitol, melezitose, melibiose, *myo*-inositol, raffinose, D-ribose, D-sorbitol, trehalose, or D-xylose. Acid production differs among strains for D-glucose and sucrose (positive for the type strain) and for D-mannose and L-rhamnose (negative for the type strain). The major cellular fatty acids are $C_{15:0}$ iso, $C_{17:0}$ iso 3-OH, $C_{15:1}$ iso, $C_{15:0}$ anteiso, $C_{15:0}$ iso 3-OH, and $C_{16:1}$ ω7*c* and/or $C_{15:0}$ iso 2-OH.

Source: the type strain was isolated from tidal flat sediment at Saemankum, Pyunsan, Korea.

DNA G+C content (mol%): 34.7–34.9 (HPLC).

Type strain: SMK-12, DSM 17032, KCTC 12379.

Sequence accession no. (16S rRNA gene): AY883937 (type strain).

Genus XXIII. **Gelidibacter** Bowman, McCammon, Brown, Nichols and McMeekin 1997, 675[VP]

John P. Bowman

Ge.li.di.bac′ter. L. adj. *gelidus* ice-cold or icy; N.L. masc. n. *bacter* rod; N.L. masc. n. *Gelidibacter* ice-cold or icy rod.

Cells are **rod-shaped or filamentous with rounded ends**, 0.4–0.6 × 1–15 µm. **Gram-stain-negative**. Cells occur singly, in pairs, and occasionally in chains. Spores and resting cells are not present. Gas vesicles and helical or ring-shaped cells are not formed. **Motile by gliding. Strictly aerobic, with an oxidative type of metabolism. Catalase-positive. Oxidase-negative. Chemoheterotrophic. Colonies are yellow** due to production of carotenoids. **Flexirubin pigments are not produced**. Strains **require Na^+** for good growth. Best growth occurs in organic media containing seawater salts. Hydrolyze esculin, produce glutamate decarboxylase, and utilize D-glucose, glycogen and L-proline. Growth occurs between <4 and 25°C (optimum, 20–25°C). Neutrophilic, with optimal growth occurring at pH 7.0–7.5. The major fatty acids include $C_{15:1}$ ω10*c* iso, $C_{15:1}$ ω10*c* anteiso, $C_{15:0}$ iso, $C_{15:0}$ anteiso, $C_{16:1}$ ω7*c*, and $C_{16:0}$ 3-OH iso. The major polyamine is **homospermidine** and the major isoprenoid quinone is **menaquinone-6.** The organisms inhabit seawater, polar sea-ice, polar saline lakes, and quartz stone sublithic communities from maritime, ice-free areas of Antarctica.

DNA G+C content (mol%): 37–42.

Type species: **Gelidibacter algens** Bowman, McCammon, Brown, Nichols and McMeekin 1997, 675[VP].

Further descriptive information

The genus *Gelidibacter* is a member of the family *Flavobacteriaceae*. The first strains of the genus to be isolated were from several Antarctic marine samples and the genus is most closely related to other Antarctic genera, including *Subsaxibacter* and *Subsaximicrobium* (Bowman and Nichols, 2005). The genus contains four species: *Gelidibacter algens*, *Gelidibacter gilvus*, *Gelidibacter mesophilus*, and *Gelidibacter salicanalis*.

Gelidibacter species typically form pale to bright yellow colonies with a mucoid consistency and with a spreading margin. The species are all active gliders. Gliding motility is best seen on nutrient-poor media or media containing utilizable carbohydrates. Cells appear usually as rods of highly variant lengths from coccobacilli to filamentous cells up to 15–20 µm long. In older cultures, cells may degenerate into spheroplast-like bodies approximately 1–2 µm in diameter. This feature is particularly obvious in some strains of *Gelidibacter algens*.

Species in the genus *Gelidibacter* are strictly aerobic, saccharolytic chemoheterotrophs, but are also able to utilize a range of amino acids and organic acids. Amino acids are required for growth of some species; however, the specific amino acids required are currently unknown. Thus, for growth on single carbon source media, yeast extract (approx. 0.1%) must be added. The ability to degrade macromolecules is present, although it varies considerably between species and even between strains. Acid is formed oxidatively from carbohydrates, although this can sometimes be weak. Acid production is best observed on a phenol red-containing oxidation/fermentation medium developed for marine bacteria (Leifson, 1963). Phenotypic tests that differentiate the species are shown in Table 48. All species synthesize several peptide arylamidases (Bowman and Nichols, 2005). *Gelidibacter* species do not denitrify, produce lipase (with olive oil as the substrate), or tolerate ox bile salts (1% concentration added to marine agar). Strains are also unable to degrade agar, alginate, carboxymethylcellulose, cellulose (crystalline power, filter paper), chitin, xylan, xanthine, or uric acid. *Gelidibacter* strains are usually negative for dextranase activity; however, equivocal results have been obtained on marine agar that contains dextran blue (Bowman et al., 1997). The following enzymic activities are not present among known *Gelidibacter* species: α-arabinosidase, β-glucuronidase, pyroglutamate arylamidase, lysine decarboxylase, and ornithine decarboxylase. H_2S is not formed from L-cysteine or thiosulfate, and indole is not produced. *Gelidibacter* species are unable to utilize the following carbon sources: L-fucose, salicin, valerate, citrate, itaconate, phenylacetate, caprate, adipate, suberate, DL-3-hydroxybutyrate, DL-lactate (some strains of *Gelidibacter algens* may exhibit weak growth on this substrate), or L-histidine. Due to weak growth on substrates and the need to add yeast extract to the medium for many strains, carbon source data should be considered with caution.

Gelidibacter species, like other members of the family *Flavobacteriaceae*, have menaquinone-6 as the major isoprenoid quinone, accumulate primarily the polyamine homospermidine (Hamana and Nakagawa, 2001), and possess primarily branched chain C_{15}–C_{17} fatty acids (Table 49).

Gelidibacter species inhabit various polar marine and marine-derived ecosystems, including seawater, sea-ice (especially in samples rich in algae) (Bowman et al., 1997), saline polar lakes, and quartz stone sublithic communities (Bowman and Nichols, 2005; Smith et al., 2000). 16S rRNA gene clones of this genus are frequently recovered from these ecosystems, which suggests that the genus comprises numerically significant populations in various permanently cold, aerobic, marine econiches (Brinkmeyer et al., 2003; Brown and Bowman, 2001; Groudieva

TABLE 48. Differential characteristics of *Gelidibacter* species[a]

Characteristic	*G. algens*	*G. gilvus*	*G. mesophilus*	*G. salicanalis*
Growth on marine agar at 25°C	d (w)	+	+	+
Yeast extract requirement	−	+	+	−
Tolerance to NaCl at:				
2.0 M	d (w)	−	−	+
2.5 M	−	−	−	+ (w)
Starch hydrolysis	+	−	+	−
Egg yolk reaction	−	−	+	−
Gelatin hydrolysis	d	−	+	+
Casein hydrolysis	d	−	−	+
Nitrate reduction to nitrate	d	−	−	−
Tyrosine decomposition	+	−	+ (w)	+
Enzyme activities:				
Arginine dihydrolase	−	−	−	+
Esterase (Tween 80)	d	+	−	−
Extracellular DNase	+	−	−	−
Urease	−	−	−	+ (w)
α-Galactosidase	−	+	−	+
β-Galactosidase	−	+	−	+
6-Phospho-β-galactosidase	−	+	−	+
α-Glucosidase	+	+	+	−
β-Glucosidase	+	+	+	−
α-Fucosidase	−	+	−	+
N-Acetyl-β-glucosaminidase	d	+	+	+
Alkaline phosphatase	+	+	−	+
Proline arylamidase	−	+	−	−
Acid production from D-glucose	+	+ (w)	+	+ (w)
Utilization of:				
N-Acetyl-D-glucosamine	d	+	−	+
DL-Arabinose	−	+	−	−
D-Mannose	−	−	+	−
Maltose	d	+	+	+
Sucrose	d	+	+	−
Acetate	−	+	−	−
Propionate	−	d	−	−
L-Alanine	−	+	−	+ (w)
L-Serine	−	d	−	+
DNA G+C content (mol%)	37	39	39–40	42

[a]Symbols: +, >85% positive; d, different strains give different reactions (16–84% positive); −, 0–15% positive; w, weak reaction.

TABLE 49. Whole-cell fatty acids of *Gelidibacter* species[a]

Fatty acid	*G. algens*	*G. gilvus*	*G. mesophilus*	*G. salicanalis*
Saturated fatty acids:				
$C_{13:0}$	nd	nd	nd	tr
$C_{14:0}$	tr	tr	tr	tr
$C_{15:0}$	2.4	3.2	5.3	4.1
$C_{16:0}$	1.6	2.2	1.9	1.2
$C_{18:0}$	tr	1.2	tr	tr
Branched chain fatty acids:				
$C_{13:0}$ iso/anteiso	nd	nd	tr	tr
$C_{14:1}$ iso/anteiso	1.7	tr	1.0	tr
$C_{14:0}$ iso	2.8	tr	tr	tr
$C_{15:1}$ ω10*c* iso	11.4	5.3	5.7	6.7
$C_{15:1}$ ω10*c* anteiso	15.9	14.6	11.8	16.6
$C_{15:0}$ iso	8.8	5.0	7.7	3.4
$C_{15:0}$ anteiso	17.7	13.7	10.5	11.5
$C_{16:1}$ iso/anteiso	10.3	1.6	1.4	5.1
$C_{16:0}$ iso	4.4	1.4	3.5	3.5
$C_{17:1}$ iso	2.3	1.1	2.1	1.9
$C_{17:1}$ anteiso	3.1	3.4	1.9	2.5
Monounsaturated fatty acids:				
$C_{15:1}$	2.7	2.8	3.0	4.2
$C_{16:1}$ ω7*c*	4.3	6.8	7.4	9.5
$C_{16:1}$ ω5*c*	1.0	1.1	tr	nd
Hydroxy fatty acids:				
$C_{15:0}$ 3-OH iso	2.2	6.2	4.0	3.5
$C_{15:0}$ 3-OH anteiso	tr	9.8	7.9	10.6
$C_{15:0}$ 3-OH	tr	1.5	1.3	0.9
$C_{16:0}$ 3-OH iso	4.1	12.2	11.5	10.7
$C_{17:0}$ 3-OH iso	tr	3.1	2.8	nd
$C_{17:0}$ 3-OH anteiso	1.6	11.3	10.5	1.0
Other	nd	2.3	1.0	nd

[a]Values represent percentages of total fatty acids; tr, trace level detected (<1% of total fatty acids); nd, not detected.

et al., 2004; Junge et al., 2002). Molecular analysis and cultural studies, however, indicate that this genus has a widespread distribution in the world's oceans. Members of the genus have been either isolated or detected in seawater, benthic and salt marsh sediment, and on the surfaces of microalgae, macroalgae and salmonid fish, in several (mainly temperate) locations (Madrid et al., 2001; Kelly and Chistoserdov, 2001; Bowman and Nowak, 2004; Macián et al., 2002; Pinhassi et al., 2003; various unpublished sources).

Enrichment and isolation procedures

No selective methods are available for direct isolation of *Gelidibacter* species. Empirical approaches for isolation can be achieved most simply by suspension and dilution of marine samples (as indicated above) followed by serial dilution onto marine agar (Difco Laboratories, Oxoid). Plates are incubated aerobically in the dark between 4 and 20°C. The appearance of golden-yellow and orange colonies can be indicative of *Gelidibacter* strains, but identification requires further confirmation with phenotypic characterization and 16S rRNA gene sequence analysis.

Maintenance procedures

Strains can be maintained for many years cryopreserved in marine broth containing 20–30% glycerol. Strains can survive for many months, potentially years, on thick agar plates or on slants when stored at 2–4°C. Strains may also be maintained by lyophilization, although the survival level is somewhat variable if 20% skim milk is used as a cryoprotectant.

Differentiation of the genus *Gelidibacter* from other genera

The most useful traits for differentiating *Gelidibacter* from other members of the *Flavobacteriaceae* include yellow pigmentation, lack of flexirubin pigments, gliding motility, strictly oxidative metabolism, ability to form acid from carbohydrates, ability to grow at low temperatures (e.g., 4°C), requirement for Na^+ for growth (or good growth on media containing sea salts), and lack of agarolytic activity. Several genera of the family *Flavobacteriaceae* possess at least some proportion of the above traits; thus, any phenotypic-based identifications are at best tentative and 16S rRNA gene sequencing is required for any definitive identification.

List of species of the genus *Gelidibacter*

1. **Gelidibacter algens** Bowman, McCammon, Brown, Nichols and McMeekin 1997, 675[VP]

al'gens. L. part. adj. *algens* feeling cold, referring to the low temperature of the typical habitats of this species.

Characteristics are as given for the genus, as listed in Table 48, and as indicated in the text. In addition, peptones alone can support growth (1% peptone in seawater). Acid is formed oxidatively from D-mannose, starch, and (weakly) from dextran.

Source: sea-ice core, Ellis Fjord, Vestfold Hills, Antarctica (type strain).

DNA G+C content (mol%): 36–38 (T_m).

Type strain: ACAM 536, ATCC 700364.

Sequence accession no. (16S rRNA gene): U62914.

Additional remarks: sequence accession numbers (16S rRNA gene sequences) for *Gelidibacter algens* strains ACAM 550 and ACAM 551 are U62916 and U62915, respectively.

2. **Gelidibacter gilvus** Bowman and Nichols 2005, 1483[VP]

gil'vus. L. masc. adj. *gilvus* pale yellow.

Characteristics are as given for the genus, as listed in Table 48, and as indicated in the text. In addition, *Gelidibacter gilvus* strains form pale yellow colonies, unlike other *Gelidibacter* species, which produce colonies with more intense pigmentation. Also, unlike other *Gelidibacter* species, colonies of *Gelidibacter gilvus* are very sticky, viscid, and gummy in consistency due to the copious production of a so-far uncharacterized, cell-attached exopolysaccharide. Acid is produced oxidatively from L-arabinose, D-mannose, D-galactose, D-fructose, L-rhamnose, D-xylose, D-mannitol, *N*-acetyl-D-glucosamine, cellobiose, lactose, maltose, sucrose, trehalose, melibiose, inositol, and glycerol, but not from melezitose, raffinose, dextran, adonitol, or D-sorbitol.

Source: sea-ice algae from fast sea-ice cores obtained from the coastal areas of the Vestfold Hills, an ice-free region of East Antarctica (type strain).

DNA G+C content (mol%): 39 (T_m).

Type strain: IC158, ACAM 1054, CIP 108531.

Sequence accession no. (16S rRNA gene): AF001369.

3. **Gelidibacter mesophilus** Macián, Pujalte, Márquez, Ludwig, Ventosa, Garay and Schleifer 2002, 1328[VP]

me.so.phi'lus. Gr. n. *mesos* middle; Gr. adj. *philus* loving; N.L. masc. adj. *mesophilus* middle (temperature)-loving, mesophilic.

Characteristics are as given for the genus, as listed in Table 48, and as indicated in the text. Gliding motility is best observed on seawater mineral salts media containing carbohydrates, including L-rhamnose, raffinose, sucrose, cellobiose, or melibiose. In addition to those mentioned in Table 48, the following carbon sources are utilized: D-galactose, trehalose, L-rhamnose, sucrose, melibiose, cellobiose, lactose, and amygdalin.

Source: seawater from the Mediterranean coast of Valencia, Spain (type strain).

DNA G+C content (mol%): 39–40 (T_m).

Type strain: 2SM29, CECT 5103, DSM 14095.

Sequence accession no. (16S rRNA gene): AJ344133.

Additional remarks: the GenBank accession number for the 16S rRNA gene sequence of *Gelidibacter mesophilus* CECT 5104 (=2SM28) is AJ344134.

4. **Gelidibacter salicanalis** Bowman and Nichols 2005, 1484[VP]

sal.i.can.a'lis. L. n. *sal salis* salt; L. n. *canalis -is* channel; N.L. gen. n. *salicanalis* of the salt channel, referring to the sea-ice brine channel habitat of this species.

Characteristics are as given for the genus, as listed in Table 48, and as indicated in the text. Colonies are golden with a smooth edge on nutrient-rich media. On nutrient-poor media (such as 0.1× marine agar), colonies have a more pronounced spreading margin indicative of gliding cells. Acid is produced from L-arabinose, D-mannose, D-galactose, *N*-acetyl-D-glucosamine, cellobiose, lactose, maltose, sucrose, and trehalose.

Source: sea-ice algae collected from pack-ice brine, Southern Ocean (type strain).

DNA G+C content (mol%): 42 (T_m).

Type strain: IC162, ACAM 1053, CIP 108532.

Sequence accession no. (16S rRNA gene): AY694009.

Genus XXIV. **Gillisia** Van Trappen, Vandecandelaere, Mergaert and Swings 2004a, 446[VP]

Stefanie Van Trappen

Gil.lis'i.a. N.L. fem. n. *Gillisia* named after Monique Gillis, a Belgian bacteriologist, who has made major contributions to bacterial taxonomy.

Rod-shaped cells, generally 0.5–0.7 × 3.0–4.0 µm (*Gillisia limnaea*, *Gillisia mitskevichiae*, and *Gillisia myxillae*) or 0.3–0.6 × 1.0–10.0 µm (*Gillisia illustrilutea*, *Gillisia sandarakina*, and *Gillisia hiemivivida*). **Gram-stain-negative**. Endospores are not produced. **Nonmotile** by gliding. Strictly **aerobic** with respiratory metabolism. Chemo-organotrophic. Catalase-positive. Produce **yellow** or **orange pigments**. Flexirubin type pigments are not formed. **Halotolerant** to **halophilic**. Optimal growth occurs at 20–25°C, with the species *Gillisia illustrilutea* and *Gillisia sandarakina* being true **psychrophiles**. Strains are characterized by a **unique fatty acid composition**, with large amounts of branched fatty acids ($C_{15:0}$ anteiso, $C_{15:1}$ iso, $C_{16:0}$ iso, and $C_{16:0}$ iso 3-OH are the main constituents).

DNA G+C content (mol%): 32–39.

Type species: **Gillisia limnaea** Van Trappen, Vandecandelaere, Mergaert and Swings 2004a, 447[VP].

Further descriptive information

16S rRNA gene sequence analysis demonstrates that the genus belongs to the family *Flavobacteriaceae*, with the genera *Algibacter*, *Aquimarina*, *Bizionia*, *Formosa*, *Mesonia*, *Psychroflexus*, *Salegentibacter*, *Stanierella*, and *Ulvibacter* as nearest neighbors.

Fatty acids. In the genus *Gillisia*, the most dominant fatty acids are branched fatty acids with $C_{15:0}$ anteiso, $C_{15:1}$ iso, $C_{16:0}$ iso, and $C_{16:0}$ iso 3-OH as the main constituents. However, the clear differences observed between *Gillisia limnaea* and *Gillisia mitskevichiae* on the one hand and *Gillisia illustrilutea*, *Gillisia sandarakina*, and *Gillisia hiemivivida* on the other (see Table 50) could be due to differences in growth conditions of the cultures that were used for fatty acid analysis. Strains of *Gillisia limnaea* and *Gillisia mitskevichiae* were grown on marine agar for 48 h at 20 or 25°C, whereas strains of *Gillisia illustrilutea*, *Gillisia sandarakina*, and *Gillisia hiemivivida* were cultivated on marine agar at 15 or 20°C for 3–5 d to yield sufficient cells. For *Gillisia myxillae*, the exact growth conditions of the culture that was used to determine the fatty acid composition are not known.

Colonial and cultural characters. On marine agar, species have yellow (light orange for *Gillisia sandarakina* and *Gillisia hiemivivida*) colonies that are circular and convex with diameters of 1–4 mm and entire margins. The pigments are non-diffusible. *Gillisia illustrilutea*, *Gillisia sandarakina*, and *Gillisia hiemivivida* form colonies with a butyrous consistency. For *Gillisia limnaea*, colonies on the AOA medium of Anacker and Ordal (1955) are flat, round with entire margins, and 0.7–0.9 mm in diameter after 14 d incubation.

Nutrition and growth conditions. *Gillisia* species are cold-adapted. For *Gillisia limnaea*, growth occurs at 5–25°C with optimal growth at 20°C. The type strain of *Gillisia mitskevichiae* grows at 4–31°C with optimal growth at 25°C. The type strain of *Gillisia myxillae* grows at 4–28°C with optimal growth at 12–20°C; no growth occurs at 36°C or higher. Strains of *Gillisia illustrilutea* and *Gillisia sandarakina* grow at 1–20°C, but not at 25°C or higher, indicating that these species are true psychrophiles. For *Gillisia hiemivivida*, growth occurs at 1–25°C with no growth at 30°C or higher. *Gillisia illustrilutea*, *Gillisia sandarakina*, and *Gillisia hiemivivida* can also grow at −2°C in marine broth.

All species of the genus *Gillisia* can tolerate high salt concentrations with *Gillisia limnaea* and *Gillisia illustrilutea* being moderately halotolerant (growth occurs in up to 1.5 M NaCl), whereas the other species are halophilic (for *Gillisia mitskevichiae*, growth is dependent on 0.2–2.5 M NaCl; for *Gillisia myxillae*, growth occurs in 2.0–10.0% NaCl with an optimum of 4.0–6.0% NaCl; for *Gillisia sandarakina* and *Gillisia hiemivivida*, growth is dependent on 0.2–1.5 M NaCl).

TABLE 50. Whole-cell fatty acid profiles of species of the genus *Gillisia*[a]

Fatty acid	*G. limnaea*	*G. mitskevichiae*	*G. illustrilutea*	*G. sandarakina*	*G. hiemivivida*	*G. myxillae*
$C_{15:0}$	–	4.4	4.3	4.3	4.3	–
$C_{16:0}$	tr	–	1.7	tr	1.8	1.6
$C_{14:0}$ iso	tr	1.2	tr	tr	1.3	–
$C_{15:1}$ iso	9.2	11.8	14.6	13.2	3.0	8.9
$C_{15:1}$ anteiso	2.5	2.4	17.6	12.9	13.1	2.3
$C_{15:0}$ iso	7.2	7.5	3.2	3.2	3.5	16.6
$C_{15:0}$ anteiso	9.7	5.1	10.2	9.1	19.5	12.5
$C_{16:1}$ iso	3.0	6.0	3.8	6.2	4.2	–
$C_{16:0}$ iso	7.4	9.3	14.4	15.8	2.7	6.2
$C_{17:1}$ iso	7.3	4.0	2.2	3.4	1.8	3.6
$C_{17:1}$ anteiso	7.7	2.1	3.3	3.0	4.7	3.5
$C_{15:1}$	1.2	1.9	1.4	1.7	2.3	tr
$C_{17:1}$	2.0	4.7	–	–	–	2.5
$C_{15:0}$ iso 3-OH	tr	1.1	1.4	1.5	3.8	1.9
$C_{15:0}$ anteiso 3-OH	–	–	9.9	8.6	14.6	–
$C_{15:0}$ 2-OH	4.0	2.4	–	–	–	3.8
$C_{15:0}$ 3-OH	–	–	–	tr	1.1	–
$C_{16:0}$ iso 3-OH	4.3	7.2	4.9	7.0	8.9	4.9
$C_{17:0}$ iso 3-OH	9.8	6.7	1.4	1.7	1.4	9.4
$C_{17:0}$ 2-OH	13.6	3.8	–	–	–	7.3
$C_{17:0}$ 3-OH	–	–	6.0	5.5	1.9	–
$C_{17:0}$ anteiso 3-OH	–	–	1.6	1.2	tr	–
Summed feature 3[b]	8.4	11.1	2.8	4.4	4.0	7.7

[a]Data from: Bowman and Nichols (2005), Nedashkovskaya et al. (2005c), Van Trappen et al. (2004a), Lee et al. (2006b). Values are percentages of total fatty acids. tr, Trace amounts (<1.0% of total); –, not detected.

[b]Summed feature 3 comprises $C_{15:0}$ iso 2-OH and/or $C_{16:1}$ ω7*c*.

Enrichment and isolation procedures

Species of the genus *Gillisia* have been isolated from several aquatic habitats and the surface of a marine sponge. The type species, *Gillisia limnaea*, was isolated from a microbial mat from Lake Fryxell in the McMurdo Dry Valleys in Antarctica. *Gillisia mitskevichiae* was derived from a seawater sample that was collected in Amursky Bay, Gulf of Peter the Great, Sea of Japan. The type strain of *Gillisia myxillae* was isolated from the surface of the marine sponge *Myxilla incrustans*. *Gillisia illustrilutea*, *Gillisia sandarakina*, and *Gillisia hiemivivida* were isolated from an Antarctic sea-ice algal assemblage. All strains were isolated on marine agar at relatively low temperatures (4–25°C) and with incubation for up to 14 d. For *Gillisia limnaea* and *Gillisia illustrilutea*, growth can also be obtained on trypticase soy agar (TSA), nutrient agar (NA), and the R2A medium of Reasoner and Geldreich (1985).

Maintenance procedures

Cultures can be maintained in the laboratory by transfer on the same media used for isolation. The recommended conditions for long-term storage are cryopreservation at −80°C or lyophilization.

Differentiation from closely related taxa

Table 51 indicates characteristics that are useful for distinguishing members of the genus *Gillisia* from related genera within the family *Flavobacteriaceae*.

Taxonomic comments

Based on 16S rRNA gene sequences of the type strains, the genus *Gillisia* forms a distinct lineage within the family *Flavobacteriaceae*, which is supported by a very high bootstrap value (100%; see Figure 42). The highest sequence similarities (89.5–91.3%) are found with members of the genera *Algibacter*, *Aquimarina*, *Bizionia*, *Formosa*, *Mesonia*, *Psychroflexus*, *Salegentibacter*, [*Stanierella*], and *Ulvibacter*. Sequence similarities within the genus *Gillisia* are clearly higher (around 93.0–98.0%).

Differentiation of species of the genus *Gillisia*

The differential characteristics of species of the genus *Gillisia* are presented in Table 52.

TABLE 51. Differential characteristics of the genus *Gillisia* and other related taxa[a]

Characteristic	*Gillisia*	*Algibacter*	*Aquimarina*	*Bizionia*	*Formosa*	*Mesonia*	*Psychroflexus*	*Salegentibacter*	*Stanierella*	*Ulvibacter*
Pigments	Y/O	O	Y	Y	Y	Y	O	Y	O/R	Y/O
Flexirubins	–	–	+	–	–	–	–	–	–	+
Gliding motility	–	+	+	–	+	–	D	–	–	+
Catalase	+	+	+	+	+	+	+	+	–	+
Growth at 25°C	D	+	+	+	+	+	D	+	+	+
Requirement for sea salts	D	+	+	–	–	nd	D	–	+	+
Nitrate reduction	D	–	–	–	+	–	D	D	–	+
H_2S production	–	–	–	nd	nd	+	–	+	+	–
Acid from D-glucose	D	+	–	–	nd	–	+	+	–	–
Hydrolysis of:										
Agar	D	+	–	–	–	–	–	–	+	–
Gelatin	D	+	+	+	D	+	D	+	+	+
Casein	D	–	+	+	nd	+	–	–	+	–
Starch	D	+	+	+	D	–	+	+	–	–
Esculin	D	nd	nd	–	nd	nd	D	+	nd	nd
DNA	D	–	+	D	nd	–	+	+	+	–
DNA G+C content (mol%)	32–39	31–33	31–33	38–45	34	32–34	31–39	37–38	34	36–38

[a]Data from: Bowman and Nichols (2005), Ivanova et al. (2004), Nedashkovskaya et al. (2004c, d, 2005c, d, f), Van Trappen et al. (2004a), and Lee et al. (2006b). Symbols: +, 90% or more of the strains are positive; –, 10% or less of the strains are negative; D, different reactions in different species of the genus; Y, yellow; O, orange; R, red; nd, not determined.

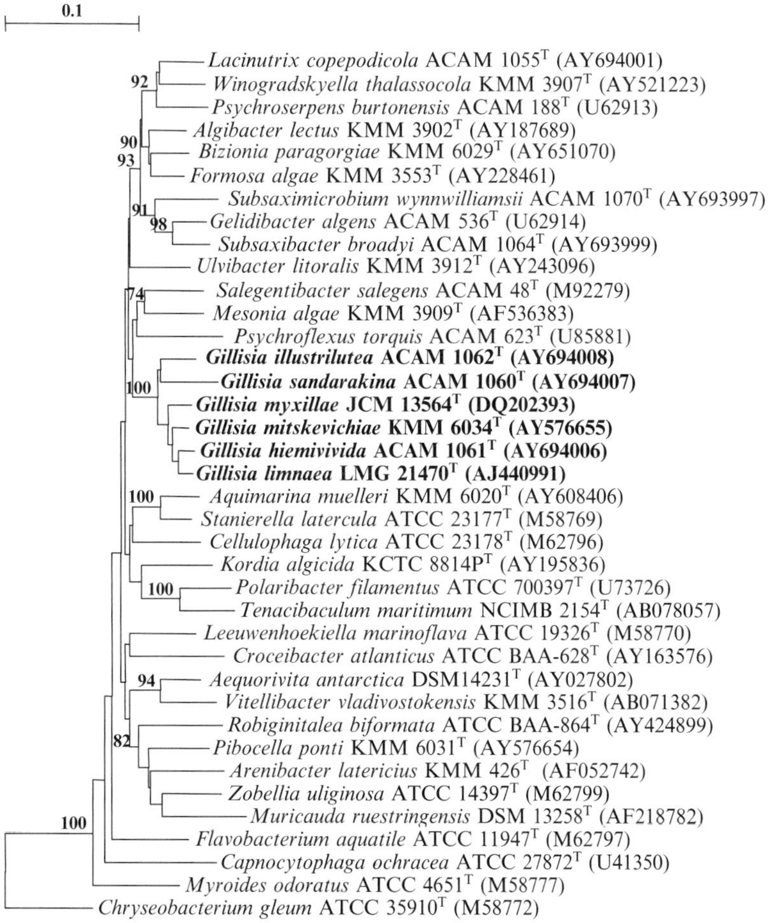

FIGURE 42. Neighbor-joining dendrogram showing the estimated phylogenetic relationships of the genus *Gillisia* within the family *Flavobacteriaceae* on the basis of 16S rRNA gene sequences. *Chryseobacterium gleum* was chosen as outgroup. Bootstrap values over 70% are shown (percentages of 500 replicates). Bar, 1 nt substitution per 10 nt. Sequence database accession numbers for reference strains are shown in parentheses.

TABLE 52. Differential characteristics of species of the genus *Gillisia*[a]

Characteristic	*G. limnaea*	*G. hiemivivida*	*G. illustrilutea*	*G. mitskevichiae*	*G. sandarakina*	*G. myxillae*
Pigments	Y	O	Y	Y	O	Y
Growth at 25°C	+	+	–	+	–	+
Growth without salt	+	–	+	–	–	–
Requirement for divalent cations	–	+	–	–	+	nd
Tolerance to NaCl:						
2.0 M	–	–	–	+	–	+
2.5 M	–	–	–	+	–	–
Arginine dihydrolase	–	–	–	nd	–	–
Nitrate reduction to nitrite	–	–	–	–	+	–
Hydrolysis of:						
Gelatin	+	+	d	+	+	–
Casein	–	–	–	+	–	+
Starch	–	+	–	–	+	+
Tween 80	–	+	–	+	+	–
L-Tyrosine	–	+	d	nd	+	nd
Esculin	+	–	–	nd	–	nd
DNA	–	–	+	+	–	+
Urea	–	+	–	+	–	–
Glutamate decarboxylase	+	+	d	nd	+	nd
α-Galactosidase	–	–	+	nd	–	–
Alkaline phosphatase	+	+	d	+	+	+
Sole carbon and energy sources:						
Glycogen	(+)	+	d	nd	–	–
N-Acetyl-D-glucosamine	–	+	d	nd	–	–
D-Arabinose	–	+	d	–	+	+
D-Mannose	–	+	+	–	+	–
Maltose	–	+	+	nd	–	–
Sucrose	–	+	+	+	–	+
L-Alanine	–	(+)	+	nd	–	–
DNA G+C content (mol%)	39	34	32	36	36	34

[a]Data from: Bowman and Nichols (2005), Nedashkovskaya et al. (2005c), Van Trappen et al. (2004a), and Lee et al. (2006b). Symbols: +, 90% or more of the strains are positive; (+), weak positive reaction for 90% or more of the strains; –, 10% or less of the strains are negative; d, 11–89% of the strains are positive; Y, yellow; O, orange; nd, not determined.

List of species of the genus *Gillisia*

1. **Gillisia limnaea** Van Trappen, Vandecandelaere, Mergaert and Swings 2004a, 447[VP]

lim.nae'a. Gr. adj. *limnaios* of or from the marsh; N.L. fem. adj. *limnaea* living in the water, referring to the isolation source, microbial mats in Lake Fryxell.

The morphological characteristics are as described for the genus. Cells are 0.7 µm wide and 3.0 µm long. Grows at 5–25°C; optimal growth at 20°C. Weak growth is observed at 30°C and no growth occurs at 37°C. Convex, translucent colonies with diameters of 1–3 mm and entire margins are formed on marine agar plates after 6 d. Colonies on AOA are flat, round with entire margins, and 0.7–0.9 mm in diameter after 14 d. Growth also occurs on nutrient agar and R2A and colonies do not adhere to the agar. No growth on TSA. Growth occurs in up to 5% NaCl, but not in 10% NaCl. The physiological and nutritional characteristics of *Gillisia limnaea* are presented in Table 52. Positive for cytochrome oxidase, catalase, and β-galactosidase. Esculin and gelatin are hydrolyzed. Acids are not produced from carbohydrates.

Source: the type strain was isolated from microbial mats, Lake Fryxell, Antarctica.

DNA G+C content (mol%): 37.8–38.9 (HPLC; Mesbah et al., 1989).

Type strain: LMG 21470, DSM 15749.

Sequence accession no. (16S rRNA gene): AJ440991.

2. **Gillisia hiemivivida** Bowman and Nichols 2005, 1485[VP]

hi.em.i.vi.vi'da. L. fem. n. *hiems -emis* the cold (of winter); L. fem. adj. *vivida* lively; N.L. fem. adj. *hiemivivida* lively in the cold.

The morphological characteristics are as described for the genus. Cells are 0.4–0.6 µm wide and 1.0–7.0 µm long. Growth occurs at 22°C in marine broth. Good growth occurs on marine agar at 1–25°C. No growth occurs at 30°C or higher. Colonies are circular and convex, with entire edges and butyrous consistency on marine agar. Requires sea salts for growth. Media supplemented only with Na^+ ions do not support growth. Grows in media containing 0.2–1.5 M NaCl, with optimal growth occurring in approximately

0.3–0.5 M NaCl. Grows in media prepared with 2× strength sea salts, but no growth occurs in 4× strength sea salts. Does not grow on TSA containing 1% (w/v) NaCl, TSA, or NA. Does not require yeast extract for growth. Can use inorganic nitrogen sources such as sodium nitrate and ammonium chloride. The physiological and nutritional characteristics of *Gillisia hiemivivida* are presented in Table 52. In mineral salts growth media, can utilize D-glucose, glycogen, *N*-acetylglucosamine, DL-arabinose, D-mannose, maltose, sucrose, sodium propionate, L-alanine (weak growth), and L-proline as sole carbon and energy sources. Hydrolyzes gelatin, starch, Tween 80, L-tyrosine, and urea. Produces α-glucosidase, β-glucosidase, and alkaline phosphatase.

Source: the type strain was isolated from sea-ice algae collected from fast sea cores obtained from the coastal areas of the Vestfold Hills, an ice-free region of East Antarctica.

DNA G+C content (mol%): 34.0 (T_m; Sly et al., 1986).

Type strain: IC154, ACAM 1061, CIP 108528.

Sequence accession no. (16S rRNA gene): AY694006.

3. **Gillisia illustrilutea** Bowman and Nichols 2005, 1484[VP]

il.lus.tri.lu.te'a. L. adj. *illustris* bright; L. fem. adj. *lutea* yellow; N.L. fem. adj. *illustrilutea* bright yellow.

The morphological characteristics are as described for the genus. Cells are 0.3–0.5 µm wide and 1.0–10.0 µm long. Growth occurs at 22°C in marine broth. Good growth occurs on marine agar at 1–20°C. No growth occurs at 25°C or higher. Colonies are circular and convex, with entire edges and a butyrous consistency on marine agar. Not halophilic. Grows in 0–1.25 M NaCl, with optimal growth occurring in approximately 0.2–0.3 M NaCl. Grows in media prepared with 2× strength sea salts, whereas no growth occurs in 4× strength sea salts. Grows on TSA containing 1% (w/v) NaCl, TSA, and NA. Does not require yeast extract for growth and can use inorganic nitrogen sources such as sodium nitrate and ammonium chloride. The physiological and nutritional characteristics of *Gillisia illustrilutea* are presented in Table 52. In mineral salts growth media, can utilize D-glucose, D-mannose, maltose, sucrose, sodium propionate, and L-proline as sole carbon and energy sources. Hydrolyzes DNA and produces α-galactosidase, α-glucosidase, and β-glucosidase.

Source: the type strain was isolated from sea-ice algae collected from fast sea cores obtained from the coastal areas of the Vestfold Hills, an ice-free region of East Antarctica.

DNA G+C content (mol%): 32.0 (T_m; Sly et al., 1986).

Type strain: IC157, ACAM 1062, CIP 108530.

Sequence accession no. (16S rRNA gene): AY694008.

4. **Gillisia mitskevichiae** Nedashkovskaya, Kim, Lee, Mikhailov and Bae 2005c, 322[VP]

mit.ske.vi'chi.ae. N.L. gen. fem. n. *mitskevichae* of Mitskevich, named in honor of Irina N. Mitskevich, Russian marine microbiologist, for her contributions to the development of marine microbiology.

The morphological characteristics are as described for the genus. Cells are 0.5–0.7 µm wide and 3.0–4.0 µm long. Growth occurs at 4–31°C. Colonies are circular, convex, and shiny with entire edges, 1–3 mm in diameter on marine agar 2216. No growth is observed without Na^+; grows in 1–12% NaCl. The pH range for growth is 5.5–10.0, with optimum growth occurring between pH 7.6 and 8.3. The physiological and nutritional characteristics of *Gillisia mitskevichiae* are presented in Table 52. Positive for oxidase, catalase, urease, and alkaline phosphatase. Casein, gelatin, DNA, urea, and Tweens 40 and 80 are degraded. Forms acid from glucose, sucrose, and *N*-acetylglucosamine. Susceptible to ampicillin, carbenicillin, oleandomycin, lincomycin, streptomycin, and tetracycline; resistant to kanamycin, benzylpenicillin, neomycin, gentamicin, and polymyxin B.

Source: the type strain was isolated from seawater collected in Amursky Bay, Gulf of Peter the Great, Sea of Japan.

DNA G+C content (mol%): 36.4 (T_m; Marmur and Doty, 1962).

Type strain: KMM 6034, KCTC 12261, NBRC 100590, LMG 22575.

Sequence accession no. (16S rRNA gene): AY576655.

5. **Gillisia myxillae** Lee, Lau, Tsoi, Li, Plakhotnikova, Dobretsov, Wu, Wong and Qian 2006b, 1797[VP]

my.xil'lae. N.L. fem. n. *Myxilla* systematic name of a genus of sponges; N.L. gen. fem. n. *myxillae* of/from *Myxilla*, referring to the isolation of the type strain from the sponge *Myxilla incrustans*.

The morphological characteristics are as described for the genus. Cells are 0.5 µm wide and 1.3–2.0 µm long. Growth occurs at 4–28°C, with optimum growth occurring between 12.0 and 20.0°C. Colonies are circular, convex with a smooth surface and entire edges, and 2–4 mm in diameter on marine agar. No growth is observed without Na^+; grows in 2–10% NaCl. The pH range for growth is 5.0–10.0, with optimum growth occurring between pH 7.0 and 9.0. The physiological and nutritional characteristics of *Gillisia myxillae* are presented in Table 52. Positive for oxidase, catalase, lipase (C14), and tryptophan deaminase. Casein, starch, DNA, and Tweens 20 and 40 are degraded. Susceptible to benzylpenicillin, chloramphenicol, ampicillin, and tetracycline; resistant to streptomycin and kanamycin.

Source: the type strain was isolated from the surface of the marine sponge, *Myxilla incrustans*, Friday Harbor, San Juan Island, WA.

DNA G+C content (mol%): 34.6 (HPLC; Mesbah et al., 1989).

Type strain: JCM 13564, NRRL B-41416.

Sequence accession no. (16S rRNA gene): DQ202393.

6. **Gillisia sandarakina** Bowman and Nichols 2005, 1484[VP]

san.da.ra.kin'a. N.L. fem. adj. *sandarakina* (from Gr. fem. adj. *sandarakinê*) of orange color.

The morphological characteristics are as described for the genus. Cells are 0.4–0.5 µm wide and 1.0–10.0 µm long. Growth occurs at 22°C in marine broth. Good growth occurs on marine agar at 1–20°C. No growth occurs at 25°C or higher. Colonies are small (1 mm diameter), circular, and convex, with entire edges and butyrous consistency on marine agar. Requires sea salts for growth. Media supplemented only with Na^+ ions do not support growth. Grows in 0.2–1.5 M NaCl, with optimal growth occurring in approximately 0.3–0.5 M NaCl. Grows in media prepared with 2× strength sea salts, but no growth occurs with 4× strength sea salts. Does not grow on TSA, TSA containing 1% (w/v)

NaCl, or NA. Does not require yeast extract for growth and can use inorganic nitrogen sources such as sodium nitrate and ammonium chloride. The physiological and nutritional characteristics of *Gillisia sandarakina* are presented in Table 52. In mineral salts growth media, can utilize D-glucose, DL-arabinose, D-mannose, sodium propionate (weak growth), and L-proline as sole carbon and energy sources. Hydrolyzes gelatin, starch, Tween 80, and L-tyrosine. Produces α-glucosidase, β-glucosidase, and alkaline phosphatase. Reduces nitrate to nitrite.

Source: the type strain was isolated from sea-ice algae collected from fast sea cores obtained from the coastal areas of the Vestfold Hills, an ice-free region of East Antarctica

DNA G+C content (mol%): 36.0 (T_m; Sly et al., 1986).

Type strain: IC148, ACAM 1060, CIP 108529.

Sequence accession no. (16S rRNA gene): AY694007.

Genus XXV. **Gramella** Nedashkovskaya, Kim, Lysenko, Frolova, Mikhailov, Bae, Lee and Kim 2005e, 393[VP]

Olga I. Nedashkovskaya, Seung Bum Kim and Valery V. Mikhailov

Gra.mel'la. N.L. dim. fem. n. *Gramella* named after of the Danish pharmacologist and pathologist Hans Christian Gram (1853–1938), who proposed the differential staining of bacteria.

Thin rods usually measuring **0.5–0.7 × 2.1–3.6 µm. Motile by means of gliding.** Produce non-diffusible yellow–orange pigments. No flexirubin type of pigments are produced. **Chemo-organotrophs. Strictly aerobic.** Halophilic. Positive for oxidase, catalase, and alkaline phosphatase. Arginine dihydrolase, and lysine and ornithine decarboxylases are not produced. Agar, urea, and chitin are not hydrolyzed. **The major respiratory quinone is MK-6. Marine**, from coastal habitats. Require seawater or sodium ions for growth.

DNA G+C content (mol%): 39–40.

Type species: **Gramella echinicola** Nedashkovskaya, Kim, Lysenko, Frolova, Mikhailov, Bae, Lee and Kim 2005e, 393[VP].

Further descriptive information

Phylogenetic analysis based on 16S rRNA gene sequences indicated that the genus *Gramella* forms a cluster with species of the genera *Salegentibacter* and *Mesonia* (Lau et al., 2005c; Nedashkovskaya et al., 2005e). The 16S rRNA gene sequence similarity between *Gramella echinicola* KMM 6050[T] and *Gramella portivictoriae* UST040801-001[T] is 98%. The level of DNA–DNA reassociation between strains KMM 6050[T] and UST040801-001[T] is 13% (Lau et al., 2005c).

The main cellular fatty acids are straight-chain unsaturated and branched-chain unsaturated fatty acids ($C_{15:0}$ anteiso, $C_{15:0}$ iso, $C_{15:0}$, $C_{16:0}$ iso, $C_{16:0}$ iso 3-OH, $C_{17:0}$ iso 3-OH, and summed feature 3 comprising $C_{15:0}$ iso 2-OH and/or $C_{16:1}$ ω7*c*).

On marine agar 2216 (Difco), colonies are regular, round, convex, smooth, nontransparent, and bright orange or yellow, with entire edges and diameters of 2–4 mm after 48 h at 28–30°C.

All isolated strains have been grown on media containing 0.5% peptone and 0.1–0.2% yeast extract (Difco), prepared with native or artificial seawater or supplemented with 2–3% NaCl. Good growth is observed on marine agar 2216 (Difco). Growth is observed at 4–37°C. The optimum temperature for growth is 23–30°C. Growth occurs in 1–15% NaCl and at pH 6.0–10.0 (optimum pH 7.0–8.0).

Gramella strains are susceptible to tetracycline, but resistant to kanamycin and streptomycin. Some strains may be susceptible to ampicillin, chloramphenicol, doxycycline, erythromycin, neomycin, and penicillin, but resistant to gentamicin, carbenicillin, lincomycin, oleandomycin, and polymyxin B.

The gramellas inhabit coastal marine environments and have been isolated from the sea urchin (*Strongylocentrotus intermedius*) and sediments collected in the temperate latitudes.

Enrichment and isolation procedures

Gramella strains were isolated from echinoderms and sediment samples by direct plating on marine agar (Difco). Either natural or artificial seawater is suitable for their cultivation. All isolated strains have been grown on media containing 0.5% peptone and 0.1–0.2% yeast extract (Difco). Cultures may stay alive on marine agar or other rich medium based on natural or artificial seawater for several weeks. They have survived storage at −80°C for at least 5 years.

Differentiation of the genus *Gramella* from other genera

There are many common traits between the genus *Gramella* and its closest relatives, *Mesonia algae* and *Salegentibacter* species. However, *Gramella* strains differ from *Mesonia* strains by their ability to hydrolyze starch and by a higher DNA G+C content (Nedashkovskaya et al., 2003a). Members of the genus *Gramella* may be distinguished from those of the genus *Salegentibacter* by production of H_2S and by a higher DNA G+C content (Dobson et al., 1993; McCammon and Bowman, 2000; Nedashkovskaya et al., 2004g, 2005e, h, 2006c).

Acknowledgements

We thank Dr Pei-Yuan Qian from Coastal Marine Laboratory/Department of Biology, The Hong Kong University of Science and Technology for kind providing us with the micrograph of strain *Gramella portivictoriae* UST040801-001[T].

List of species of the genus *Gramella*

1. **Gramella echinicola** Nedashkovskaya, Kim, Lysenko, Frolova, Mikhailov, Bae, Lee and Kim 2005e, 393[VP]

   e.chi.ni.co'la. L. n. *echinus* a hedgehog, a sea urchin; L. suff. *-cola* (from L. n. incola) dweller; N.L. n. *echinicola* a sea-urchin-dweller.

   Cells are 0.5–0.7 × 2.1–2.7 µm. On marine agar, colonies are 2–4 mm in diameter, circular, shiny with entire edges, and yellow–orange in color. Growth occurs at 4–37°C. Hydrolyzes gelatin, starch, and Tweens 40 and 80, but not agar, urea, or chitin. No acid is formed from L-arabinose, cellobiose,

L-fucose, D-lactose, melibiose, L-rhamnose, DL-xylose, citrate, adonitol, dulcitol, glycerol, inositol, or mannitol. Does not utilize D-lactose, D-mannose, inositol, malonate, or citrate. Nitrate is not reduced. H_2S and indole are not produced.

Source: the only available strain was isolated from the sea urchin *Strongylocentrotus intermedius* in Troitsa Bay, Sea of Japan.

DNA G+C content (mol%): 39.6 (T_m).

Type strain: KMM 6050, KCTC 12278, NBRC 100593, LMG 22585.

Sequence accession no. (16S rRNA gene): AY608409.

2. **Gramella portivictoriae** Lau, Tsoi, Li, Plakhotnikova, Dobretsov, Wong and Qian 2005c, 2499[VP]

por.ti.vic.to'ri.ae. L. n. *portus* harbor; L. gen. fem. n. *victoriae* of victory; N.L. gen. fem. n. *portivictoriae* of Victoria Harbor, Hong Kong, the source of isolation of the type strain.

According to Lau et al. (2005c), cells of the type strain are 0.6 µm wide and 3.6 µm long Figure 43. On marine agar, colonies are 2–4 mm in diameter, circular, convex with smooth surfaces, and yellow. Growth is observed at 4–36°C. Growth occurs at pH 6.0–10.0 (optimum, 7.0–8.0). Esculin, gelatin, starch, and Tweens 40 and 80 are hydrolyzed, but not agar, urea, or chitin. D-Galactose, D-glucose, sucrose, and glycerol, but not citrate, are used as sole carbon sources on agar media supplemented with 4% (w/v) carbon source. Utilization of γ-hydroxybutyric acid is observed with the MicroLog 3 system. No growth or acid production occurs with carbon sources in the API 50CH, API 20E, and API 20NE test systems. Positive for the following enzyme activities: acid phosphatase; α-chymotrypsin; cystine, leucine, and valine arylamidases; esterase (C4); esterase lipase (C8); lipase (C14); α-galactosidase; α- and β-glucosidases; trypsin; and naphthol-AS-BI-phosphohydrolase. Negative for *N*-acetyl-β-glucosaminidase, α-fucosidase, β-glucuronidase, α-mannosidase, and tryptophan deaminase activities. Nitrate is not reduced. H_2S and indole are not produced. Susceptible to chloramphenicol and tetracycline. Resistant to kanamycin and streptomycin.

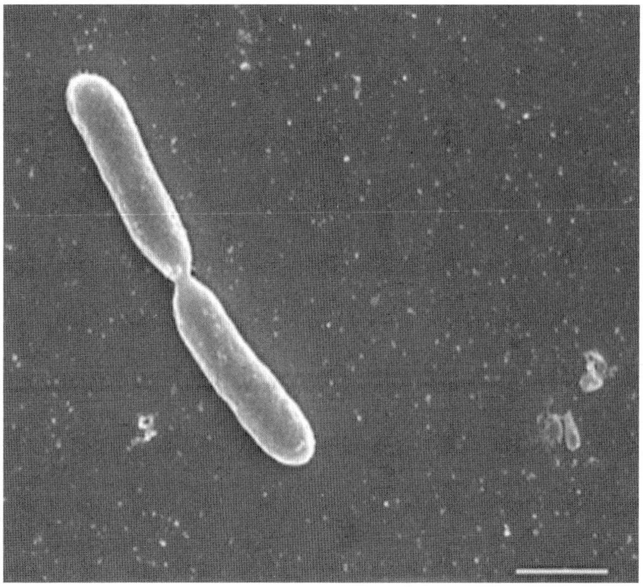

FIGURE 43. Micrograph of strain *Gramella portivictoriae* UST040801-001[T] showing the rod-shaped morphology; bar = 1 µm. (Printed with permission of P.-Y. Qian.)

Source: the only available strain was isolated from sediment in Victoria Harbor, Hong Kong, South-China Sea.

DNA G+C content (mol%): 39.9 (T_m).

Type strain: UST040801-001, JCM 13192, NRRL B-41137.

Sequence accession no. (16S rRNA gene): DQ002871.

Genus XXVI. **Kaistella** Kim, Im, Shin, Lim, Kim, Lee, Park, Lee and Lee 2004, 2323[VP]

THE EDITORIAL BOARD

Ka.is.tel'la. L. dim. suff. *-ella*; N.L. fem. dim. n. *Kaistella* arbitrary name after KAIST, Korea Advanced Institute of Science and Technology.

Rod-shaped cells. Nonsporeforming. Gram-stain-negative. Flagella not present. Aerobic. Colonies are yellow. Catalase- and oxidase-positive. **No growth occurs on MacConkey agar.** Nitrate is not reduced to N_2. **Indole-positive. β-Glucosidase-positive.** Negative for arginine dihydrolase and β-galactosidase. **No acid is produced from glucose.** Esculin and gelatin are hydrolyzed, but not starch. Glucose, arabinose, and β-hydroxybutyrate are not assimilated. The major isoprenoid quinone is MK-6. Isolated from freshwater.

DNA G+C content (mol%): 41–42.

Type species: **Kaistella koreensis** Kim, Im, Shin, Lim, Kim, Lee, Park, Lee and Lee 2004, 2323[VP].

Further descriptive information

The genus description is based on data for three strains. Analysis based on 16S rRNA gene sequences indicated that the genus *Kaistella* belongs to the *Chryseobacterium–Bergeyella–Riemerella* branch of the family *Flavobacteriaceae.*

Regarding the long-chain fatty acid content of *Kaistella*, $C_{15:0}$ iso was dominant, with significant levels of $C_{15:0}$ anteiso, $C_{17:0}$ iso 3-OH, and $C_{17:1}$ iso ω9*c*. *Kaistella* strains possessed the distinctive fatty acids $C_{13:0}$ iso and $C_{15:0}$ anteiso, similar to those found in members of the genera *Bergeyella* and *Riemerella.*

The following compounds are assimilated: acetate, adipate, L-alanine, caprate, citrate, glycogen, 4-hydroxybenzoate, inositol, itaconate, DL-lactate, malate, malonate, maltose, mannitol, phenylacetate, L-proline, propionate, rhamnose, and suberate. No assimilation of the following occurs: *N*-acetylglucosamine, arabinose, fucose, glucose, histidine, 3-hydroxybenzoate, β-hydroxybutyrate, 2-ketogluconate, 5-ketogluconate, mannose, melibiose, D-ribose, salicin, L-serine, sorbitol, D-sucrose, and valerate.

Differentiation of the genus *Kaistella* from other genera

Kaistella strains differ from those of *Chryseobacterium* in that they do not produce acid from glucose, lack β-galactosidase activity, cannot hydrolyze starch or assimilate glucose, arabinose or 3-hydroxybutyrate, and cannot grow on MacConkey agar. *Kaistella* strains differ from those of *Bergeyella* and *Riemerella* in that they are indole-positive, have β-glucosidase activity, and grow aerobically on nutrient agar. Also, the DNA G+C contents of *Kaistella* strains are higher than those of *Chryseobacterium*, *Bergeyella*, and *Riemerella* species.

Taxonomic comments

Kaistella koreensis Chj707[T] was most closely related to *Chryseobacterium balustinum* ATCC 33487[T] and *Chryseobacterium scophthalmum* LMG 13028[T] (94.3 and 94.1% 16S rRNA gene sequence similarities, respectively). Additional information can be found in Kim et al. (2004).

List of species of the genus *Kaistella*

1. **Kaistella koreensis** Kim, Im, Shin, Lim, Kim, Lee, Park, Lee and Lee 2004, 2323[VP]

ko.re.en′sis. N.L. fem. adj. *koreensis* pertaining to Korea, where the organisms were first isolated.

The description is the same as that of the genus.

Source: the available strains were isolated from natural mineral water from Daejeon City, Korea.

DNA G+C content (mol%): 41.2–41.6 (HPLC).

Type strain: Chj707, IAM 15050, KCTC 12107.

Sequence accession no. (16S rRNA gene): AF344179 (type strain, partial sequence).

Additional remarks: the GenBank accession numbers for the 16S rRNA gene sequences of *Kaistella koreensis* strains Ko2 (=KCTC 12108=IAM 15051) and Ko10 (=KCTC 12109=IAM 15052) are AY294611 (complete sequence) and AY299974 (partial sequence), respectively.

Genus XXVII. **Kordia** Sohn, Lee, Yi, Chun, Bae, Ahn and Kim 2004, 678[VP]

Sang-Jin Kim

Kor′di.a. N.L. fem. n. *Kordia* arbitrary name derived from the acronym KORDI, which stands for Korea Ocean Research and Development Institute.

Straight rods. Nonmotile and nongliding. Gram-stain-negative. Strictly aerobic. Oxidase-positive, but catalase-negative. Produce **extracellular polysaccharides and carotenoid-type pigments**, but not flexirubin-type pigments. Several carbohydrates are used as sole carbon sources. **Unable to grow in the absence of any of the ions Na^+, Ca^{2+}, and Mg^{2+}. Require 1–5% (w/v) NaCl for growth**, with optimal growth in 3% NaCl. Degrade gelatin, skimmed milk, and starch. Major cellular fatty acids are saturated iso-branched and 3-hydroxy iso-branched fatty acids. The respiratory quinone is menaquinone-6.

DNA G+C content (mol%): 34.

Type species: **Kordia algicida** Sohn, Lee, Yi, Chun, Bae, Ahn and Kim 2004, 678[VP].

Further descriptive information

Cells are 0.3–0.5 × 2–5 μm when grown in ZoBell 2216e medium at 25°C for 24 h. Resting stages are absent.

The colony color is yellow on ZoBell 2216e and tryptic soy agar (TSA) containing seawater, but white on peptone-seawater agar medium. After 3 d, colonies are about 1.2 mm in diameter, slightly convex (elevation), entire (margin), and round (configuration). Growth occurs at 5–40°C, pH 6–10, and 1–5% NaCl. Grows optimally at 30°C, pH 7–8 and 3% NaCl. Sea salts (Na^+, Ca^{2+}, and Mg^{2+}) are required for growth. H_2S and indole are not produced. Nitrate is reduced to nitrite. Acid is not produced from glucose. The O/F test is negative with glucose. No growth occurs on urea, NH_4, or NO_3 as inorganic nitrogen sources; growth does occur when Casamino acids, sodium glutamate, peptone, tryptone, and yeast extract are supplied instead of an inorganic nitrogen source. Gelatin, skimmed milk, and starch are degraded, but not cellulose, carboxymethylcellulose, agar, chitin, alginate, pectin, or inulin.

The following substrates are oxidized: α-cyclodextrin, dextrin, glycogen, *N*-acetyl-D-glucosamine, adonitol, i-erythritol, gentiobiose, α-D-glucose, maltose, D-mannitol, D-mannose, raffinose, sucrose, citric acid, D-glucuronic acid, α-ketoglutaric acid, DL-lactic acid, quinic acid, succinic acid, alaninamide, L-aspartic acid, L-glutamic acid, glycyl-L-aspartic acid, glycyl-L-glutamic acid, L-ornithine, L-proline, L-threonine, DL-carnitine, uridine, glucose 1-phosphate, and glucose 6-phosphate.

The following substrates are not oxidized: Tween 40, Tween 80, *N*-acetyl-D-galactosamine, L-arabinose, D-arabitol, cellobiose, D-fructose, L-fucose, D-galactose, *myo*-inositol, α-D-lactose, lactulose, melibiose, methyl β-D-glucoside, D-psicose, L-rhamnose, D-sorbitol, trehalose, turanose, xylitol, methyl pyruvate, monomethyl succinate, acetic acid, *cis*-aconitic acid, formic acid, D-galactonic acid lactone, D-galacturonic acid, D-gluconic acid, D-glucosaminic acid, α-hydroxybutyric acid, β-hydroxybutyric acid, γ-hydroxybutyric acid, *p*-hydroxyphenylacetic acid, itaconic acid, α-ketobutyric acid, α-ketovaleric acid, malonic acid, propionic acid, D-saccharic acid, sebacic acid, bromosuccinic acid, succinamic acid, glucuronamide, D-alanine, L-alanine, L-alanylglycine, L-asparagine, L-histidine, hydroxyl-L-proline, L-leucine, L-phenylalanine, L-pyroglutamic acid, D-serine, L-serine, γ-aminobutyric acid, urocanic acid, inosine, thymidine, phenylethylamine, putrescine, 2-aminoethanol, 2,3-butanediol, glycerol, and DL-α-glycerol phosphate.

Distinctive features of the fatty acid profile are the presence of a high level of $C_{15:0}$ iso (41.2%) and the presence of $C_{15:0}$ iso 3-OH, $C_{16:0}$ iso 3-OH, and $C_{17:0}$ iso 3-OH.

The type strain was isolated from a seawater sample associated with a red tide in Masan Bay, Republic of Korea.

Enrichment and isolation procedures

Kordia algicida was isolated by enrichment culture of seawater samples collected during an outbreak of red tides caused by a microalga. A seawater sample was collected at a depth of 1 m from Masan Bay, Republic of Korea, during an algal bloom caused by a marine microalga, *Skeletonema costatum*. The sample was filtered through a 1.2-μm pore-size membrane filter, co-cultured with *Skeletonema costatum*, and incubated at 20°C under a cycle of light (approx. 5,000 lux for 14 h) and dark (10 h) periods. The cells of *Skeletonema costatum* in a co-culture tube were completely killed after 3 d. The bacterium responsible for killing *Skeletonema costatum* was isolated from the co-culture tube using ZoBell 2216e agar medium (Sohn et al., 2004). *Kordia algicida* also showed algicidal activity against other algal species, namely, *Thalasiossira* sp., *Heterosigma akashiwo*, and *Cochlodinium polykrikoides*.

Maintenance procedures

Kordia algicida can be maintained as a glycerol suspension (20%, w/v) at −80°C.

Differentiation of the genus *Kordia* from other genera

Characteristics that differentiate the genus from other phylogenetically related genera in the family *Flavobacteriaceae* are given in Table 53.

Taxonomic comments

Kordia algicida is the only member of the genus *Kordia* and, based on 16S rRNA gene sequence analysis, is not closely related to any known genera. The highest similarities were to the type strains of *Cytophaga latercula* (92.5%), *Flexibacter tractuosus* (92.1%), *Salegentibacter salegens* (91.0%), *Cellulophaga lytica* (90.9%), *Cellulophaga fucicola* (90.7%), *Tenacibaculum maritimum* (90.5%), and *Polaribacter franzmannii* (90.1%). The position of *Kordia algicida* is not stable within the *Cytophaga–Flavobacterium–Bacteroides* complex, because branching positions in the trees varied depending on phylogenetic methods employed (Sohn et al., 2004). The estimated genome size of *Kordia algicida* is about 5.0 Mbp (https://moore.jcvi.org/moore/SingleOrganism.do?speciesTag=KAOT1&pageAttr=pageMain). The partial genome sequencing is now finished and the further annotation work is in progress (http://www.moore.org/program_areas/science/initiatives/marine_microbiology/initiative_marine_microbiology.asp).

TABLE 53. Differential characteristics of *Kordia algicida* and phylogenetically most closely related genera in the family *Flavobacteriaceae*[a,b]

Characteristic	*Kordia*	*Aquimarina*	*Krokinobacter*	*Tenacibaculum*	*Polaribacter*	*Lutibacter*	*Dokdonia*[c]
Pigment	Y	V	Y	Y	V	Y	Y
Flexirubin reaction	−	D	−	−[d]	−	−	−
Cell shape:							
Rod	+	+	+	+	D	+	+
Filament	−	D	nd	nd	D	−	+
Gliding motility	−	D	nd	D	−	−	−
Sea salts requirement	+	+	+	+	D	+	+
Growth at 4°C	+	D	−	D	+	+[e]	+
Growth at 25°C	+	+	+	+	D	+	+
Nitrate reduction	+	−	−	D	D	−	−
Acid from carbohydrates	−	D	−	nd	+	−	−
Enzyme activity:							
Catalase	−	D	+	+	+	+	+
Oxidase	+	+	+	+	D	−	+
Agarase	−	D	−	nd	nd	nd	−
Amylase	+	D	−	D	+	+	−
β-Galactosidase	−	D	+	nd	D	+	−
Caseinase	+	+	+	+[d]	D	nd	+
Esculinase	−	nd	nd	−	D[f]	+	+
Gelatinase	+	+	+	D	D	+	−
DNA G+C content (mol%)	34	31–37.1	33–39	30–33.6	30–33.2	33.9	38
Isoprenoid quinone	MK-6	MK-6[g]	nd	MK-6	MK-6[f]	MK-6	MK-6

[a]Symbols: +, >85% positive; −, 0–15% positive; D, different reactions occur in different taxa (species of a genus); w, weak reaction; nd, not determined; Y, yellow; V, variable color.

[b]Data from Sohn et al. (2004), Nedashkovskaya et al. (2005g, 2006f), Khan et al. (2006a), Jung et al. (2006), Gosink et al. (1998), Choi and Cho (2006), and Yoon et al. (2005a).

[c]Based on phylogenetic analysis, this genus should be reclassified in the same clade as the genus *Krokinobacter*.

[d]Except *Tenacibaculum litoreum* which has not been determined.

[e]Grows at 5°C.

[f]Except *Polaribacter butkevichii* which has not been determined.

[g]Data not available for *Aquimarina latercula*.

List of species of the genus *Kordia*

1. **Kordia algicida** Sohn, Lee, Yi, Chun, Bae, Ahn and Kim 2004, 678[VP]

   al.gi'ci.da. L. fem. n. *alga -ae* alga; L. suff. *-cida* (from L. v. *caedere* to cut or to kill), killer; N.L. n. *algicida* alga-killer.

   The description of this species is the same as that of the genus.

   Source: the type strain was isolated from a seawater sample associated with a red tide in Masan Bay, Republic of Korea.

   DNA G+C content (mol%): 34 (HPLC).

   Type strain: OT-1, KCTC 8814P, NBRC 1000336.

   Sequence accession no. (16S rRNA gene): AY195836.

Genus XXVIII. **Krokinobacter** Khan, Nakagawa and Harayama 2006a, 326[VP]

The Editorial Board

Kro.ki'no.bac.ter. Gr. adj. *krokinos* yellow; N.L. masc. n. *bacter* from Gr. n. *bakterion* rod; N.L. masc. n. *Krokinobacter* a yellow, rod-like bacterium.

Rods, 0.5–0.7 × 2.5–4.0 µm. Gram-stain-negative. **Aerobic. Lemon-yellow** carotenoid-type pigments are present. **Flexirubin-negative.** Catalase- and oxidase-positive. **Gelatin, casein, and DNA are hydrolyzed, but not agar, carrageenan, cellulose, chitin, starch, or urea.** Very weak growth, or no growth, occurs at temperatures below 10°C and above 30°C. **No growth occurs in the presence of 0 or 10% NaCl.** No acid is produced from glucose. **Nitrate is not reduced.** Denitrification does not occur. The following substrates are utilized: aspartic acid, dextrin, cellobiose, D-fructose, gentiobiose, α-D-glucose, glutamic acid, glycine, glycyl-L-aspartic acid, lactose, lactulose, maltose, methyl-β-D-glucoside, ornithine, raffinose, sucrose, trehalose, turanose, proline, threonine, and uridine. Do not utilize the following substrates: aconitic acid, adonitol, alaninamide, alanine, aminobutyric acid, aminoethanol, arabitol, asparagine, bromosuccinic acid, butanediol, carnitine, citric acid, erythritol, formic acid, fucose, galactonic acid lactone, galactose, galacturonic acid, gluconic acid, glucosaminic acid, glucose phosphate, glucuronamide, glucuronic acid, glycerol, glycerol phosphate, glycyl-L-glutamic acid, histidine, α-, β- and γ-hydroxybutyric acids, *p*-hydroxyphenylacetic acid, inosine, inositol, itaconic acid, α-ketobutyric acid, α-ketoglutaric acid, α-ketovaleric acid, leucine, malonic acid, mannose, melibiose, phenylalanine, phenylethylamine, propionic acid, psicose, putrescine, pyroglutamic acid, pyruvate, quinic acid, rhamnose, saccharic acid, sebacic acid, serine, sorbitol, succinamic acid, succinic acid, succinate, thymidine, urocanic acid, and xylitol. The major cellular fatty acids are $C_{15:0}$ iso, $C_{15:1}$ iso, $C_{17:0}$ iso 3-OH, and summed feature A ($C_{16:1}$ ω7*c* and/or $C_{15:0}$ iso 12-OH; fatty acids that could not be separated by GC). Isolated from marine sediment samples.

DNA G+C content (mol%): 33–39.

Type species: **Krokinobacter genikus** Khan, Nakagawa and Harayama 2006a, 326[VP].

Enrichment and isolation procedures

Five *Krokinobacter* strains were isolated from marine sediment samples collected from the Pacific coastline of Japan. Three-day-old colonies on marine agar were lemon-yellow, slightly convex, and 1–2 mm in diameter. The organisms were Gram-stain-negative rods, 0.5–0.7 × 2.5–4.0 µm, and were negative for the presence of flexirubin-type pigments. Their optimum growth temperature was 15–25°C.

Differentiation of the genus *Krokinobacter* from related genera

Unlike members of the genus *Cellulophaga*, *Krokinobacter* strains are unable to hydrolyze agar and starch. Their ability to hydrolyze casein and DNA, but not agar or starch distinguishes them from strains of *Algibacter*. Unlike *Psychroserpens* strains, *Krokinobacter* strains hydrolyze DNA and have a higher DNA G+C content (33–39 mol% vs 27–29 mol%). *Krokinobacter* strains differs from members of the genera *Psychroflexus* and *Salegentibacter* by their ability to hydrolyze casein, but not starch, and also differ from *Salegentibacter* strains by reducing nitrate.

Taxonomic comments

Based on 16S rRNA gene sequence analysis, the five strains formed a distinct monophyletic cluster within the family *Flavobacteriaceae*. The most closely related species with a validly published name was *Cellulophaga lytica*, with sequence similarities of 91.2–91.7%.

DNA–DNA hybridization experiments with the five strains indicated that the strains should be classified as representatives of three species: *Krokinobacter genikus* (three strains), *Krokinobacter diaphorus* (one strain), and *Krokinobacter eikastus* (one strain).

Differentiation of species of the genus *Krokinobacter*

Krokinobacter genikus can be differentiated from *Krokinobacter diaphorus* and *Krokinobacter eikastus* by its ability to utilize acetic acid, but not mannitol, and by the absence of arachidonic acid ($C_{20:4}$ ω6*c*) in its fatty acid profile. *Krokinobacter diaphorus* differs from the other two species in being unable to use both acetic acid and mannitol and in having a lower DNA G+C content (33 mol% vs 37–38 mol%). *Krokinobacter eikastus* can be differentiated from the other two species by its ability to use mannitol, but not acetic acid.

List of species of the genus *Krokinobacter*

1. **Krokinobacter genikus** Khan, Nakagawa and Harayama 2006a, 326[VP]

   ge'ni.kus. N.L. masc. adj. *genikus* from Gr. masc. adj. *genikos* principal, typical.

   The characteristics are as given for the genus, with the following additional features. Colonies on marine agar 2216 are yellow and slightly convex. Grows optimally at 20°C and with 3% NaCl. Acetic acid is utilized, but not mannitol. Unlike other *Krokinobacter* species, does not contain arachidonic acid as part of the cellular fatty acid composition.

Source: the type strain was isolated from marine sediment at Odawara, Japan.

DNA G+C content (mol%): 37–39 (HPLC).

Type strain: Cos-13, CIP 108744, NBRC 100811.

Sequence accession no. (16S rRNA gene): AB198086.

2. **Krokinobacter diaphorus** Khan, Nakagawa and Harayama 2006a, 327[VP]

di.aph'or.us. N.L. masc. adj. *diaphorus* from Gr. masc. adj. *diaphoros* different, unlike.

The characteristics are as given for the genus, with the following additional features. Colonies on marine agar 2216 are slightly convex and yellowish. Grows optimally at 20°C and with 3% NaCl. Neither acetic acid nor mannitol is utilized. The cellular fatty acids include arachidonic acid.

Source: the type strain was isolated from marine sediment at Kisarazu, Japan.

DNA G+C content (mol%): 33 (HPLC).

Type strain: MSKK-32, CIP 108745, NBRC 100817.

Sequence accession no. (16S rRNA gene): AB198089.

3. **Krokinobacter eikastus** Khan, Nakagawa and Harayama 2006a, 327[VP]

ei.kas'tus. N.L. masc. adj. *eikastus* from Gr. masc. adj. *eikastos* similar, comparable.

The characteristics are as given for the genus, with the following additional features. The optimal growth temperature is 20°C and the optimal salt concentration is 3%. Colonies are slightly convex and yellowish on marine agar 2216. Mannitol is utilized, but acetate is not utilized. The cellular fatty acids include arachidonic acid.

Source: the type strain was isolated from marine sediment collected at Kisarazu, Japan.

DNA G+C content (mol%): 38 (HPLC).

Type strain: PMA-26, CIP 108743, NBRC 100814.

Sequence accession no. (16S rRNA gene): AB198088.

Genus XXIX. **Lacinutrix** Bowman and Nichols 2005, 1482[VP]

John P. Bowman

La.ci.nu'trix. L. n. *lacus* lake; L. fem. n. *nutrix* feeder; N.L. fem. n. *Lacinutrix* lake feeder (in the sense of contributing to lake food chains).

Straight or slightly curved **rod-shaped cells**, 0.4–0.5 × 1–2 µm. Cells occur singly or in pairs. Spores and resting cells are not present. Gas vesicles and helical or ring-shaped cells are not formed. **Nonmotile. Gram-stain-negative. Strictly aerobic, with an oxidative type of metabolism. Catalase- and oxidase-positive. Chemoheterotrophic. Colonies are golden-yellow** due to production of carotenoids. **Flexirubin pigments are not produced. Sodium ions are required** for growth. Best growth occurs in organic media containing seawater salts. Grows between −2 and 25°C, with temperature optima of 15–20°C in seawater-containing media. Neutrophilic, with optimal growth occurring at about pH 7.5. The major fatty acids include $C_{15:1}$ ω10*c* iso, $C_{15:1}$ ω10*c* anteiso, $C_{15:0}$ iso, $C_{15:0}$ anteiso, and $C_{16:0}$ iso. The major isoprenoid quinone is **menaquinone-6.** Habitat: Antarctic saline lakes in association with the calanoid copepod species *Paralabidocera antarctica*.

DNA G+C content (mol%): 37.

Type species: **Lacinutrix copepodicola** Bowman and Nichols 2005, 1482[VP].

Further descriptive information

The genus *Lacinutrix* is a member of the family *Flavobacteriaceae*, represented at the time of writing by a single species, *Lacinutrix copepodicola*, which was isolated directly from copepod samples collected from an Antarctic lake with marine salinity (Bowman and Nichols, 2005). Physiologically, the species has features that are consistent with the majority of members of the family *Flavobacteriaceae*, including Gram-stain-negative cell walls, an aerobic chemoheterotrophic metabolism, the ability to synthesize carotenoids, the formation of mainly branched chain fatty acids, and the possession of menaquinones as the primary respiratory isoprenoid quinones. Based on available cultivation and molecular data (unpublished data from the National Center of Biotechnology Information; http://www.ncbi.nlm.nih.gov), members of the genus *Lacinutrix* appear to be widely distributed in marine habitats, including polar sea-ice, and in association with marine algae and animals collected from high latitude oceanic regions. Members of the marine genus *Psychroserpens*, which are the closest relatives of *Lacinutrix*, also have a similar distribution pattern in the marine environment.

Isolation and maintenance procedures

Strains can be isolated directly from source material, typically cold water marine samples (<10°C), as described above by serial dilution plating directly onto marine agar 2216 (Difco) and incubation at 2–15°C. Routine cultivation should be performed aerobically on marine agar 2216 at 15–20°C. Strains can be maintained as frozen suspensions in marine broth 2216 containing 20–30% (v/v) glycerol at −70°C. Strains may survive for several months, potentially years, on thick-agar (marine agar 2216) plates or on slants, stored at 2–4°C.

Differentiation of the genus *Lacinutrix* from other genera

The most useful traits for differentiating *Lacinutrix* from other members of the *Flavobacteriaceae* include yellow pigmentation, lack of flexirubin pigments, lack of motility, a strictly oxidative metabolism, formation of acid from carbohydrates, growth at low temperatures (e.g., 4°C), requirement for sodium ions for growth (or good growth on sea-salts-containing media), and lack of agarolytic activity. Several genera of the family *Flavobacteriaceae* possess at least some proportion of the above traits; thus, any phenotypic-based identifications are at best tentative and 16S rRNA gene sequencing is required for any definitive identification.

List of species of the genus *Lacinutrix*

1. **Lacinutrix copepodicola** Bowman and Nichols 2005, 1482[VP]

co.pe.pod.i'col.a. N.L. neut. pl. n. *copepoda* copepods (small type of crustacean); L. fem. or masc. suff. *-cola* (from L. n. *incola*) the dweller, inhabitant; N.L. masc. or fem. n. *copepodicola* the inhabitant of copepods.

Characteristics are as described for the genus, with the following additional features. Colonies are golden-yellow, circular, convex, with an entire edge and a butyrous consistency on marine agar 2216. Growth occurs in 0.1–2.0 M NaCl, with optimal growth at seawater salinities (0.3–0.4 M NaCl). Grows well in media containing 2× strength sea salts; no growth occurs with 4× strength sea salts. Does not require divalent cations found in seawater for growth; all salinity requirements can be met with NaCl. Grows poorly on trypticase soy agar (Oxoid) containing 1% (w/v) NaCl. Does not grow on trypticase soy agar or nutrient agar (Oxoid). Can grow on a mineral salts medium prepared with sea salts and containing ammonium chloride or sodium nitrate as the sole nitrogen source and D-glucose as the sole carbon and energy source. Other sole carbon and energy sources include glycogen, D-mannose, maltose, sucrose, and L-proline, but not *N*-acetyl-D-glucosamine, L-rhamnose, DL-arabinose, L-fucose, D-melibiose, salicin, acetate, propionate, valerate, caprate, adipate, suberate, citrate, DL-3-hydroxybutyrate, DL-lactate, itaconate, phenylacetate, L-alanine, L-histidine, or L-serine. Acid production from carbohydrates is undetectable in Leifson's oxidation/fermentation medium (Leifson, 1963). Gelatin, Tween 80, and L-tyrosine are hydrolyzed, but not agar, chitin, carboxymethylcellulose, dextran, xylan, xanthine, uric acid, casein, elastin, starch, esculin, DNA, or urea. Produces alkaline phosphatase; produces *N*-acetyl-β-glucosaminidase weakly. Negative for arginine dihydrolase, lysine decarboxylase, ornithine decarboxylase, and glutamate decarboxylase. H_2S is not produced from L-cysteine. Indole is not produced. No growth occurs in the Simmons' citrate test. Lipase (olive oil substrate) and lecithinase (egg yolk reaction) activities are not observed. Nitrate is not reduced. Negative for α-arabinosidase, α-galactosidase, β-galactosidase, 6-phospho-β-galactosidase, α-glucosidase, β-glucosidase, β-glucuronidase, and α-fucosidase. Positive for arginine arylamidase, leucyl glycine arylamidase, phenylalanine arylamidase, leucine arylamidase, tyrosine arylamidase, glycine arylamidase, histidine arylamidase, serine arylamidase, alanine arylamidase, and gutamyl glutamate arylamidase. Negative for proline arylamidase and pyroglutamate arylamidase. Does not tolerate 1% (w/v) ox bile salts.

Source: isolated directly from the calanoid copepod species *Paralabidocera antarctica* dwelling in Ace Lake in the Vestfold Hills, an ice-free region of East Antarctica.

DNA G+C content (mol%): 37 (T_m).

Type strain: DJ3, ACAM 1055, CIP 108538.

Sequence accession no. (16S rRNA gene): AY649001.

Additional remarks: the GenBank accession number (16S rRNA gene sequence) for *Lacinutrix copepodicola* KMM 3838 is AB261015.

Genus XXX. **Leeuwenhoekiella** Nedashkovskaya, Vancanneyt, Dawyndt, Engelbeen, Vandemeulebroecke, Cleenwerck, Hoste, Mergaert, Tan, Frolova, Mikhailov and Swings 2005i, 1035[VP]

Olga I. Nedashkovskaya, Marc Vancanneyt and Valery V. Mikhailov

Leeu.wen.hoe.ki.el'la. N.L. fem. dim. n. *Leeuwenhoekiella* named in honor of the famous Dutchman Antonie van Leeuwenhoek (1632–1723), discoverer of micro-organisms.

Thin rods usually measuring 0.4–0.7 × 1.5–4.0 µm. **Motile by gliding.** Produce non-diffusible yellow pigments. No flexirubin-type pigments are formed. **Chemo-organotrophs. Strictly aerobic.** Positive for oxidase, catalase, alkaline phosphatase, and β-galactosidase. **Halotolerant.** Agar, DNA, urea, cellulose (CM-cellulose and filter paper), and chitin are not attacked. Casein, gelatin, starch, and Tweens 20, 40, and 80 are hydrolyzed. Nitrate is not reduced. H_2S, indole, and acetoin are not produced. **Marine**, from coastal habitats. Can grow without seawater or Na^+. **The major respiratory quinone is MK-6.**

DNA G+C content (mol%): 35–43 (T_m).

Type species: **Leeuwenhoekiella marinoflava** (Reichenbach 1989c) Nedashkovskaya, Vancanneyt, Dawyndt, Engelbeen, Vandemeulebroecke, Cleenwerck, Hoste, Mergaert, Tan, Frolova, Mikhailov and Swings 2005i, 1035[VP].

Further descriptive information

Phylogenetic analysis based on 16S rRNA gene sequences revealed that the genus *Leeuwenhoekiella* forms a distinct lineage within the family *Flavobacteriaceae* with sequence similarity levels below 92.2% with members of the following genera: *Vitellibacter*, *Aequorivita*, *Arenibacter*, *Muricauda*, *Zobellia*, and *Maribacter* (Figure 44).

Dominant cellular fatty acids are straight-chain unsaturated and branched-chain unsaturated fatty acids $C_{15:1}$ iso, $C_{15:0}$ iso, $C_{17:1}$ iso ω9*c*, $C_{17:0}$ iso 3-OH, and summed feature 3 ($C_{15:0}$ iso 2-OH and/or $C_{16:1}$ ω7*c*).

Phenotypic properties of the type strains are given in Table 54. On marine agar 2216 (Difco), colonies are regular, round, usually shiny, and yellow, with a diameter of 2–4 mm after 48 h at 28°C.

All isolated strains have been grown on media containing 0.5% peptone and 0.1–0.2% yeast extract (Difco), prepared with natural or artificial seawater or supplemented with 2–3% NaCl. Growth is observed at 4–41°C (optimum, 23–30°C). Growth occurs in 0–17% NaCl; optimum growth varies from 0 to 5% NaCl. Utilizes L-arabinose, D-glucose, D-lactose, D-mannose, and sucrose. Inositol, sorbitol, malonate, and citrate are not utilized.

Strains are susceptible to carbenicillin, lincomycin, doxycycline, erythromycin, and chloramphenicol. They are resistant to ampicillin, gentamicin, kanamycin, neomycin, oleandomycin, polymyxin B, and streptomycin.

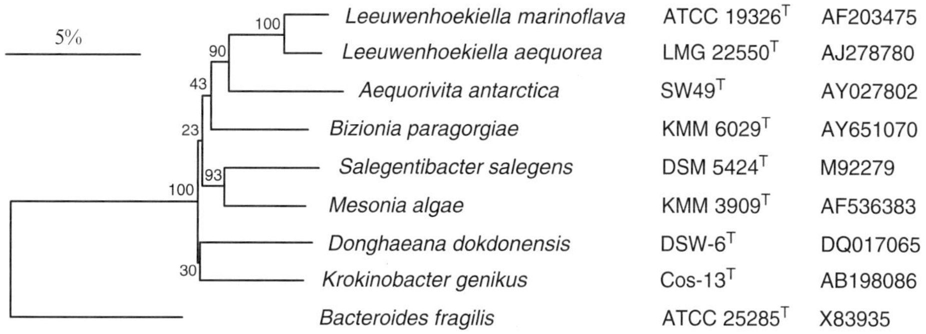

FIGURE 44. Phylogenetic tree based on the 16S rRNA gene sequences of *Leeuwenhoekiella marinoflava* ATCC 19326^T and *Leeuwenhoekiella aequorea* LMG 22550^T and representative members of related genera in the family *Flavobacteriaceae*. The tree was generated by the neighbor-joining method (Saitou and Nei, 1987). The *numbers* at nodes indicate bootstrap values (%). Bar = 0.05 substitutions per nucleotide position.

TABLE 54. Phenotypic properties of the type strains of *Leeuwenhoekiella* species[a,b]

Characteristic	*L. marinoflava* LMG 1345^T	*L. aequorea* LMG 22550^T	*L. blandensis* MED 217^T
Salinity range for growth (%)	0–15	0–15	0–17
Optimum salinity (%)	1–3	0–5	2–4
Temperature range for growth (°C)	4–37	4–37	10–41
Optimum temperature (°C)	21–23	23–25	28–30
Acid production from:			
L-Arabinose, cellobiose, L-fucose, D-lactose, maltose, melibiose, raffinose, L-rhamnose, L-sorbose	–	–	–
D-Glucose, DL-xylose	–	–	+
D-Galactose	+	+	–
Sucrose	–	+	–
N-Acetylglucosamine	–	–	–
Glycerol	+	+	+
Mannitol	–	+	–
Adonitol, dulcitol, inositol, sorbitol	–	–	–
Utilization of mannitol	–	+	–
Susceptibility to:			
Benzylpenicillin	+	v	–
Tetracycline	–	+	+

[a]Symbols: +, positive; –, negative; v, variable.

[b]Data are from Nedashkovskaya et al. (2005i) and Pinhassi et al. (2006).

Leeuwenhoekiella species inhabit the coastal marine environment. They have been isolated from seawater and echinoderms in the temperate and Antarctic latitudes.

Enrichment and isolation procedures

Leeuwenhoekiella species were isolated from seawater and the sea urchin *Strongylocentrotus intermedius* by direct platting on marine agar. *Leeuwenhoekiella blandensis* MED 217^T was isolated from a surface seawater sample enriched with Na_2HPO_4. For strain isolation, 0.1 ml of a 100× dilution of a water sample was spread onto Zobell agar plates prepared with natural seawater. Either natural or artificial seawater is suitable for cultivation. All isolated strains can grow on media containing 0.5% peptone and 0.1–0.2% yeast extract (Difco). The strains remain viable on marine agar or other rich media based on natural or artificial seawater from 3 d to several weeks. They can survive storage at −80°C for at least 5 years.

Differentiation of the genus *Leeuwenhoekiella* from other genera

An ability to grow in 15% NaCl differentiates *Leeuwenhoekiella* species from all their closest relatives. *Leeuwenhoekiella* species may be distinguished from members of the genus *Arenibacter* by their caseinase and amylase activities. A lack of flexirubin-type pigments, failure to hydrolyze agar, lack of a requirement for Na^+ ions or seawater for growth, and failure to reduce nitrates to nitrites clearly separate *Leeuwenhoekiella* strains from representatives of the genus *Zobellia*. Phenotypic features such as production of caseinase, gelatinase, and amylase, and the ability to grow without Na^+ ions or seawater may be helpful in differentiating *Leeuwenhoekiella* from *Muricauda* species. *Leeuwenhoekiella* species move by gliding and form acid from carbohydrates and decompose starch, in contrast to members of the genera *Vitellibacter* and *Aequorivita*. *Leeuwenhoekiella* strains can be separated from those of the genus *Maribacter* by growth without Na^+ ions or seawater, growth at 37°C, and by hydrolysis of casein.

Taxonomic comments

16S rRNA gene sequence similarity between *Leeuwenhoekiella aequorea* LMG 22550^T and *Leeuwenhoekiella marinoflava* ATCC 19326^T was 97.1%. DNA–DNA binding values between *Leeuwenhoekiella aequorea* LMG 22550^T and *Leeuwenhoekiella marinoflava* LMG 1345^T ranged from 9 to 14%. Sequence similarities between the two copies of the 16S rRNA gene of strain *Leeuwenhoekiella blandensis* MED 217^T and sequences of *Leeuwenhoekiella aequorea* LMG 22550^T and *Leeuwenhoekiella marinoflava* ATCC 19326^T were 96.1–96.7 and 95.7–96.7%, respectively. DNA–DNA hybridization experiments carried out between strains MED 217^T and ATCC 19326^T show a level of DNA–DNA reassociation of 21%. These data indicate that strains LMG 22550^T and ATCC 19326^T represent two separate species within the genus *Leeuwenhoekiella* (Wayne et al., 1987). The genome of strain *Leeuwenhoekiella blandensis* MED 217^T was sequenced and its size was equal to 4.24 Mbp. This value is consistent with that reported for *Leeuwenhoekiella marinoflava* (Callies and Mannheim, 1980). The size of the genome of *Leeuwenhoekiella marinoflava* ATCC 19326^T was determined to be 2.26×10^9 Da or 4.20 Mbp.

List of species of the genus *Leeuwenhoekiella*

1. **Leeuwenhoekiella marinoflava** (Reichenbach 1989c) Nedashkovskaya, Vancanneyt, Dawyndt, Engelbeen, Vandemeulebroecke, Cleenwerck, Hoste, Mergaert, Tan, Frolova, Mikhailov and Swings 2005i, 1035^{VP} [*Cytophaga marinoflava* (*ex* Colwell, Citarella and Chen 1966) Reichenbach 1989c, 2036; *Cytophaga marinoflava* Colwell, Citarella and Chen 1966, 1102]

ma.ri.no.fla′va. L. adj. *marinus* marine; L. adj. *flavus* golden yellow; N.L. fem. adj. *marinoflava* (*sic*) marine and yellow-pigmented.

Rod-shaped cells, 0.5–0.6 × 1.6–2.3 µm. On marine agar 2216, colonies are 2–4 mm in diameter, circular with entire edges, and bright yellow. No acid is formed from trehalose, citrate, acetate, fumarate, or malate. Susceptible to carbenicillin, doxycycline, erythromycin, and chloramphenicol. Resistant to ampicillin and streptomycin.

Source: the single available strain was isolated from seawater collected in the North Sea off Aberdeen, Scotland.

DNA G+C content (mol%): 38 (T_m).

Type strain: ATCC 19326, DSM 3653, JCM 8517, LMG 1345, NBRC 14170, NCIMB 397.

Sequence accession no. (16S rRNA gene): M58770.

2. **Leeuwenhoekiella aequorea** Nedashkovskaya, Vancanneyt, Dawyndt, Engelbeen, Vandemeulebroecke, Cleenwerck, Hoste, Mergaert, Tan, Frolova, Mikhailov and Swings 2005i, 1036^{VP}

ae.quo.re′a. L. fem. adj. *aequorea* of the sea, marine.

Rod-shaped cells, 0.5–0.6 × 1.6–2.3 µm. On marine agar 2216, colonies are 2–4 mm in diameter, circular with entire edges, and bright yellow. No acid is formed from acetate, citrate, fumarate, or malate.

Source: the type strain and several other strains were isolated from Antarctic seawater. Strain LMG 22555 was isolated from the sea urchin *Strongylocentrotus intermedius* in the Troitsa Bay, Sea of Japan.

DNA G+C content (mol%): 35–36 (T_m).

Type strain: LMG 22550, CCUG 50091.

Sequence accession no. (16S rRNA gene): AJ278780.

3. **Leeuwenhoekiella blandensis** Pinhassi, Bowman, Nedashkovskaya, Lekunberri, Gomez-Consarnau and Pedrós-Alió 2006, 1492^{VP}

bla.den′sis. L. fem. adj. *blandensis* pertaining to *Blande* or *Blanda*, the name the Romans used for the city of Blanes, which has given its name to the Bay of Blanes, where the type strain was isolated.

Rod-shaped cells 0.4–0.7 × 1.5–4.0 µm; short chains were also observed. On marine agar 2216, colonies are 2–3 mm in diameter, round with entire edges, shiny, and bright yellow. Forms acid from D-glucose and DL-xylose. No acid is produced from D-galactose, sucrose, or mannitol.

Source: the type strain was isolated from a surface sea water sample enriched with inorganic phosphate, from the Bay of Blanes in the northwestern Mediterranean Sea on the coast of Spain.

DNA G+C content (mol%): 42.5 (T_m).

Type strain: MED 217, CECT 7118, CCUG 51940.

Sequence accession nos (16S rRNA gene): DQ294290 and DQ294291 (two copies).

GenBank accession number (genome sequence): AANC 00000000.

Genus XXXI. **Lutibacter** Choi and Cho 2006, 773^{VP}

The Editorial Board

Lu.ti.bac′ter. L. n. *lutum* mud; N.L. masc. n. *bacter* rod; N.L. masc. n. *Lutibacter* rod from mud.

Rod-shaped cells. Nonmotile. Gram-stain-negative. Heterotrophic. Aerobic. **Catalase-positive. Oxidase-negative.** The predominant menaquinone is MK-6. Dominant fatty acids are $C_{15:0}$ iso 3-OH, $C_{15:0}$ iso, $C_{15:0}$ anteiso, and $C_{16:0}$ iso 3-OH. Cells contain carotenoids, but **no flexirubin-type pigments**. The genus is a member of the family *Flavobacteriaceae.*

DNA G+C content (mol%): 33.9 (HPLC).

Type species: **Lutibacter litoralis** Choi and Cho 2006, 775^{VP}.

Enrichment and isolation procedures

A slurry of tidal flat sediment was spread on a plate of marine agar 2216 (Difco) and incubated at 30°C for 1 week. A colony was selected and subsequently purified four times on marine agar.

Maintenance procedures

The organism can be maintained on marine agar at 4°C. It can be preserved in marine broth containing 30% (v/v) glycerol at −80°C.

Differentiation of the genus *Lutibacter* from other genera

For characteristics that differentiate the genus *Lutibacter* from related genera, see Table 70 in the treatment on *Tenacibaculum* in this volume. In particular, *Lutibacter* can be differentiated from *Polaribacter* (Gosink et al., 1998) by its colony color (yellow vs salmon pink to orange), ability to grow at 25°C, and utilization of citrate, L-leucine, tartrate, pyruvate, and succinate. *Lutibacter* can be distinguished from *Tenacibaculum* (Suzuki et al., 2001) by its hydrolysis of esculin, negative oxidase reaction, failure to grow at pH 6, and growth on pyruvate and succinate. *Lutibacter* can also be differentiated from *Polaribacter* and *Tenacibaculum* by its large proportion of $C_{15:0}$ anteiso and by the proportions of several fatty acids, including $C_{16:0}$ iso 3-OH, $C_{15:1}$ iso $_{G}$, and summed feature 3 ($C_{16:1}$ ω7*c* and/or $C_{15:0}$ iso 2-OH).

Taxonomic comments

Phylogenetic analyses based on 16S rRNA gene sequences revealed that the type strain formed a very robust clade with *Tenacibaculum* and *Polaribacter* species. The closest relatives of the type strain were members of the genera *Tenacibaculum* (90.6–91.8%) and *Polaribacter* (91.0–91.5%) (Choi and Cho, 2006).

List of species of the genus *Lutibacter*

1. **Lutibacter litoralis** Choi and Cho 2006, 775[VP]

li.to.ra'lis. L. masc. adj. *litoralis* of the shore.

Characteristics are as described for the genus, with the following additional features. Cells are 0.3–0.8 × 1.0–5.7 µm. Spherical cells occur in old cultures. On marine agar 2216 (Difco), colonies are circular, entire, convex, shining, opaque, and yellow. Absorption maxima of the pigments are 450 and 475 nm. Growth occurs at 5–30°C (optimum, 25–30°C) and pH 7–8. Grows in sea salt concentrations of 1–5% (w/v). Positive for catalase, amylase, gelatinase, and DNase. Negative for oxidase, nitrate reductase, and Tween 80 hydrolysis. With the API 20NE system, esculin hydrolysis and gelatinase activities are positive, but nitrate reductase, indole production, and acid production from glucose, arginine dihydrolase and urease are negative. Grows on acetone, citrate, D-fructose, raffinose, D-salicin, D-sorbitol, glycine, glycogen, *myo*-inositol, L-arginine, L-lysine, L-ornithine, pyruvic acid, succinate, tartrate, urea, Casamino acids, L-leucine, peptone, tryptone, and yeast extract. No growth occurs on acetate, acetamide, α-ketobutyric acid, benzoate, DL-cysteine, cellobiose, D-galactose, D-glucose, D-mannitol, D-mannose, D-ribose, trehalose, D-xylose, ethanol, formic acid, glycerol, inulin, 2-propanol, lactose, L-arabinose, L-ascorbate, L-asparagine, L-rhamnose, maleic acid, *N*-acetylglucosamine, polyethylene glycol, salicylate, sucrose, thiamine, DL-aspartate, L-proline, or L-glutamate.

Source: the type strain was isolated from tidal flat sediment in Ganghwa, Korea.

DNA G+C content (mol%): 33.9 (HPLC).

Type strain: CL-TF09, JCM 13034, KCCM 42118.

Sequence accession no. (16S rRNA gene): AY962293.

Genus XXXII. **Maribacter** Nedashkovskaya, Kim, Han, Lysenko, Rohde, Rhee, Frolova, Falsen, Mikhailov and Bae 2004b, 1021[VP]

Olga I. Nedashkovskaya and Seung Bum Kim

Ma.ri.bac'ter. L. neut. n. *mare* the sea; N.L. masc. n. *bacter* from Gr. n. *baktron* rod; N.L. masc. n. *Maribacter* rod inhabiting marine environments.

Rod-shaped cells, 0.2–0.7 × 1.2–10 µm, with rounded ends. Motile by gliding. Produce non-diffusible yellow carotenoid pigments. No flexirubin pigments are produced. **Chemo-organotrophs. Strictly aerobic.** Positive for oxidase, catalase, and alkaline phosphatase. Arginine dihydrolase, and lysine and ornithine decarboxylases are not produced. Casein, urea, and chitin are not attacked. Tween 40 is hydrolyzed. Hydrolysis of agar, gelatin, starch, Tweens 20 and 80, and DNA varies among species. Indole, H_2S, and acetoin are not produced. **Marine,** from coastal habitats. May require seawater or sodium ions for growth. **The major respiratory quinone is MK-6. The main polar lipid is phosphatidylethanolamine.**

DNA G+C content (mol%): 35–39 (T_m).

Type species: **Maribacter sedimenticola** Nedashkovskaya, Kim, Han, Lysenko, Rohde, Rhee, Frolova, Falsen, Mikhailov and Bae 2004b, 1021[VP].

Further descriptive information

Phylogenetic analysis based on 16S rRNA gene sequences indicated that the genus *Maribacter* created a cluster with *Zobellia* species (Figure 45). The 16S rRNA gene sequence similarities of *Maribacter* and its nearest neighbor *Zobellia uliginosa* were 92.9–94.3%, but their relationship did not have significant bootstrap support. Species of the genus *Maribacter* possess 16S rRNA gene sequence similarities to one another that range from 95.1 to 97.4%. Only the 16S rRNA gene sequence similarity value between strains *Maribacter aquivivus* KMM 3949^T and *Maribacter ulvicola* KMM 3951^T is higher than 97%. The DNA–DNA hybridization value between strains KMM 3949^T and KMM 3951^T was 51%.

The main cellular fatty acids are $C_{15:1}$ iso, $C_{15:0}$ iso, $C_{15:0}$, $C_{17:0}$ iso 3-OH, and summed feature 3 comprising $C_{15:0}$ iso 2-OH and/or $C_{16:1}$ ω7*c*. Further fatty acid data can be found in Table 55.

On marine agar 2216 (Difco), maribacters form regular, round, shiny, yellow–orange colonies with entire or rhizoid edges, and are 1–4 mm in diameter after 48 h at 28°C. Some are sunken into the agar.

All isolated strains have been grown on media containing 0.5% peptone and 0.1–0.2% yeast extract (Difco), prepared with natural or artificial seawater or supplemented with 2–3%

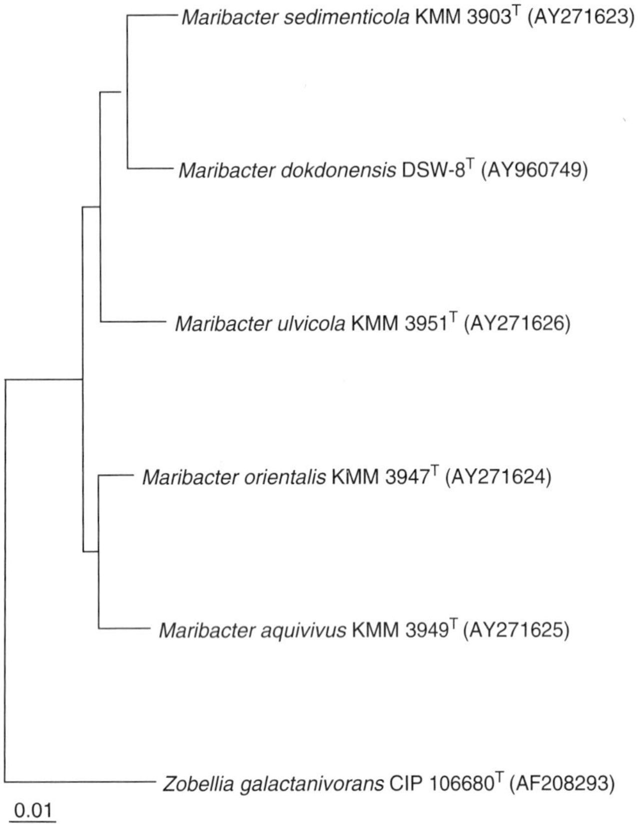

FIGURE 45. Phylogenetic tree based on the 16S rRNA gene sequences of the type strains of species of the genus *Maribacter* and the related species *Zobellia galactanivorans*. Bar = 0.01 substitutions per nucleotide position.

NaCl. Growth occurs at 4–35°C (optimum, 21–24°C) and in 1–7% NaCl (optimum, 1–3% NaCl). Growth does not occur in the absence of NaCl or in the presence of greater than 10% (w/v) NaCl. The optimal pH for growth is 7.0–8.0; growth occurs at pH 5.5, but not at pH 5.0.

No acid is produced from citrate, adonitol, *N*-acetylglucosmine, glycerol, inositol, sorbitol, or sorbose. Mannitol, inositol, sorbitol, malonate, and citrate are not utilized.

Maribacter strains are susceptible to carbenicillin, lincomycin, and oleandomycin, but are resistant to benzylpenicillin, gentamicin, kanamycin, neomycin, polymyxin B, and streptomycin. Some species may be susceptible to novobiocin and tetracycline and resistant to cephalothin.

Species of the genus *Maribacter* inhabit coastal marine environments. They have been isolated from seawater, sediments, and seaweeds collected in temperate latitudes.

TABLE 55. Whole-cell fatty acid profiles (%) of *Maribacter* species[a]

Fatty acid	*M. sedimenticola* KMM 3903T	*M. aquivivus* KMM 3949T	*M. dokdonensis* DSW-8T	*M. orientalis* KMM 3947T	*M. ulvicola* KMM 3951T
$C_{13:1}$	0.7	1.1		0.3	1.1
$C_{14:0}$	0.5	1.0	0.9	0.5	0.9
$C_{14:1}$ ω5		1.0			0.9
$C_{15:0}$ iso	20.5	12.3	19.7	10.6	13.6
$C_{15:0}$ anteiso	1.2	1.3	0.3	2.3	1.9
$C_{15:1}$ iso	16.9	13.6	16.1	10.1	18.9
$C_{15:0}$	6.3	14.5	3.8	12.3	8.1
$C_{15:1}$ ω6*c*	1.7	4.8	0.6	2.5	1.6
$C_{16:0}$ iso	1.1	0.7		0.3	0.3
$C_{16:0}$	1.0	0.5	1.3–1.4	1.2	1.0
$C_{16:1}$ ω7/$C_{15:0}$ iso 2-OH	5.8	12.9	8.6	11.4	12.2
$C_{17:1}$ iso ω5*c*		1.4			1.2
$C_{17:1}$ iso ω9*c*	2.3	2.2	2.0	4.0	2.2
$C_{17:1}$ ω6*c*	0.5	1.7	0.2	1.3	0.5
$C_{15:0}$ iso 3-OH	5.4	3.2	5.0	2.9	4.1
$C_{15:0}$ 3-OH	2.4	2.3	1.4	1.5	1.5
$C_{16:0}$ iso 3-OH	1.7	2.5	1.1	2.1	1.7
$C_{16:0}$ 3-OH	2.2	2.9	5.4	3.0	3.7
$C_{17:0}$ iso 3-OH	20.4	11.6	28.6	18.8	14.5
$C_{18:1}$ iso				2.4	

[a]Data from Nedashkovskaya et al. (2004b) and Yoon et al. (2005b).

Enrichment and isolation procedures

The maribacters inhabit coastal marine environments and were isolated from seawater, sediments, and seaweeds by direct plating on marine agar (Difco). Natural or artificial seawater is sufficient for their cultivation. All isolated strains have been grown on media containing 0.5% peptone and 0.1–0.2% yeast extract (Difco). Strains of *Maribacter* may remain viable on marine agar (Difco) or other rich media based on natural or artificial seawater for several weeks. They have survived storage at −80°C for at least 5 years.

Differentiation of the genus *Maribacter* from other genera

Maribacter strains differ from those of its closest relative, the genus *Zobellia*, by their inability to produce flexirubin-type pigments and lack of caseinase activity (Barbeyron et al., 2001; Nedashkovskaya et al., 2004b, f; Yoon et al., 2005b).

Differentiation of species of the genus *Maribacter*

Species of the genus *Maribacter* have many common phenotypic traits; however, there are enough variable phenotypic properties to differentiate them clearly (Table 56).

List of species of the genus *Maribacter*

1. **Maribacter sedimenticola** Nedashkovskaya, Kim, Han, Lysenko, Rohde, Rhee, Frolova, Falsen, Mikhailov and Bae 2004b, 1021[VP]

se.di.men.ti.co'la. L. masc. n. *sedimentum* sediment, L. suff. *-cola* (from L. n. *incola*) dweller; N.L. masc. n. *sedimenticola* sediment dweller.

Cells are 0.5–0.7 × 2–10 μm (Figure 46). On marine agar, colonies are 2–4 mm in diameter, circular, shiny with entire edges, yellow, and slightly sunken into the agar. Growth occurs at 4–33°C (optimum, 22–24°C) and in 1–6% NaCl.

Source: isolated from sediment in Troitsa Bay, Gulf of Peter the Great, Sea of Japan.

TABLE 56. Differential characteristics of *Maribacter* species[a,b]

Characteristic	*M. sedimenticola*	*M. aquivivus*	*M. dokdonensis*	*M. orientalis*	*M. ulvicola*
Nitrate reduction	+	+	−	−	−
β-Galactosidase	−	+	−	+	+
Growth in/at:					
6% NaCl	+	+	+	−	−
32°C	+	−	+	+	+
Hydrolysis of:					
Agar	+	+	+	−	+
Gelatin	+	+	−	+	−
Starch	+	−	−	−	−
Alginate	−	+	d	+	−
DNA	−	+	d	−	−
Tween 20	−	−	+	+	+
Tween 80	+	+	+	−	+
Acid production from:					
Arabinose	+	−	+	+	−
L-Fucose	+	−	nd	−	+
Galactose, melibiose, xylose	−	−	+	+	−
Cellobiose, glucose, lactose, maltose, sucrose	−	−	+	+	+
Raffinose, D-mannitol	−	−	+	−	−
Rhamnose	−	−	−	−	+
Utilization of:					
Arabinose	+	−	+	+	−
Glucose, lactose, mannose, sucrose	−	+	+	+	+
Susceptibility to:					
Ampicillin	−	−	−	−	+
Tetracycline	+	−	+	−	−

[a]Symbols: +, >85% positive; d, different strains give different reactions (16–84% positive); −, 0–15% positive; nd, not determined.

[b]Data from Nedashkovskaya et al. (2004b) and Yoon et al. (2005b).

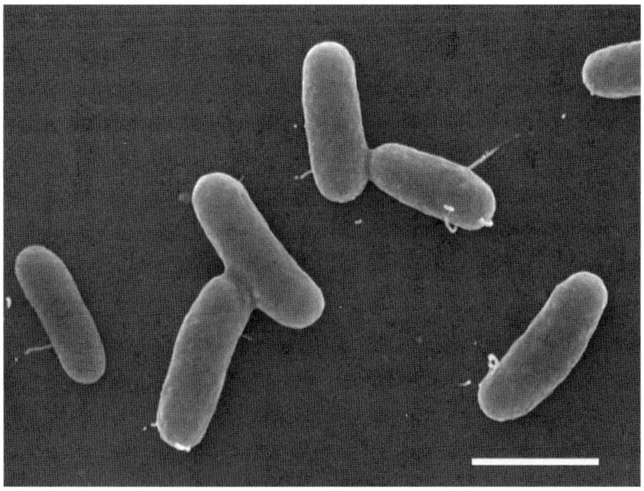

FIGURE 46. Micrograph of *Maribacter sedimenticola* KMM 3903[T] showing rod-shaped cells. Bar = 1 µm.

DNA G+C content (mol%): 37.0 (T_m).
Type strain: KMM 3903, KCTC 12966, CCUG 47098.
Sequence accession no. (16S rRNA gene): AY271623.

2. **Maribacter aquivivus** Nedashkovskaya, Kim, Han, Lysenko, Rohde, Rhee, Frolova, Falsen, Mikhailov and Bae 2004b, 1022[VP]

a.qui.vi'vus. L. fem. n. *aqua* water; L. adj. *vivus* alive; N.L. masc. adj. *aquivivus* living in water.

Cells are 0.4–0.5 × 1.2–1.4 µm. On marine agar, colonies are 2–4 mm in diameter, circular, shiny with entire edges, yellow, and sunken into the agar. Growth occurs at 4–30°C (optimum, 21–23°C) and in 1–7% NaCl.

Source: isolated from seawater in Amursky Bay, Gulf of Peter the Great, Sea of Japan.
DNA G+C content (mol%): 35.0 (T_m).
Type strain: KMM 3949, KCTC 12968, CCUG 48009.
Sequence accession no. (16S rRNA gene): AY271625.

3. **Maribacter dokdonensis** Yoon, Kang, Lee, Lee and Oh 2005b, 2054[VP]

dok.do.nen'sis. N.L. masc. adj. *dokdonensis* pertaining to Dokdo, the Korean island from where the strains were isolated.

Cells are rods, 0.3–0.4 × 0.8–4.0 µm. Colonies are circular or rhizoid, glistening, slightly convex, yellow, and 1.0–2.0 mm in diameter after incubation for 3 d on marine agar at 30°C. Growth occurs at 4–35°C. Optimal pH for growth is 7.0–8.0; growth occurs at pH 5.5, but not at pH 5.0. Optimal growth occurs in the presence of 2–3% (w/v) NaCl; growth does not occur in the absence of NaCl or in the presence of greater than 10% (w/v) NaCl. Esculin, tyrosine, and Tween 60 are hydrolyzed, but hypoxanthine and xanthine are not. Anaerobic growth does not occur on marine agar or on marine agar supplemented with nitrate. Tests for arginine dihydrolase, lysine and ornithine decarboxylases, and tryptophan deaminase are negative. In assays with the API ZYM system, esterase (C4), esterase lipase (C8), leucine arylamidase, acid phosphatase, naphthol-AS-BI-phosphohydrolase, β-glucosidase, *N*-acetyl-β-glucosaminidase, cystine arylamidase, trypsin, α-chymotrypsin, α- and β-galactosidases, β-glucuronidase, α-glucosidase, α-mannosidase, and α-fucosidase are absent. D-Fructose and salicin are utilized, but acetate, succinate, benzoate, L-malate, pyruvate, formate, and L-glutamate are not utilized. Acid is produced from D-mannose, melezitose, and trehalose, but not from D-ribose. Acid production from D-fructose differs among strains (positive for the type strain). Susceptible to novobiocin (5 µg), but not to cephalothin (30 µg). The major polar lipids are phosphatidylethanolamine, two unidentified phospholipids, an unidentified glycolipid and an amino group-containing lipid that is ninhydrin-positive.

Source: isolated from seawater off Dokdo, an island located at the edge of East Sea, Korea.
DNA G+C content (mol%): 35.9–36.1 (HPLC).
Type strain: DSW-8, KCTC 12393, DSM 17201.
Sequence accession no. (16S rRNA gene): AY960749.

4. **Maribacter orientalis** Nedashkovskaya, Kim, Han, Lysenko, Rohde, Rhee, Frolova, Falsen, Mikhailov and Bae 2004b, 1022[VP]

o.ri.en.ta'lis. L. masc. adj. *orientalis* eastern, bacterium inhabiting the East.

Cells are 0.5–0.7 × 2–10 μm. On marine agar, colonies are 2–4 mm in diameter, circular, shiny with entire edges, and yellow–orange. Growth occurs at 4–32°C (optimum, 21–23°C) and in 1–5% NaCl.

Source: isolated from seawater in Amursky Bay, Gulf of Peter the Great, Sea of Japan.

DNA G+C content (mol%): 39.0 (T_m).

Type strain: KMM 3947, KCTC 12967, CCUG 48008.

Sequence accession no. (16S rRNA gene): AY271624.

5. **Maribacter ulvicola** Nedashkovskaya, Kim, Han, Lysenko, Rohde, Rhee, Frolova, Falsen, Mikhailov and Bae 2004b, 1022[VP]

ul.vi.co'la. N.L. fem. n. *Ulva* generic name of the green alga *Ulva fenestrata*; L. suff. *-cola* (from L. n. *incola*) dweller; N.L. n. *ulvicola Ulva fenestrata* dweller.

Cells are flexible rods, 0.25–0.30 × 4–6 μm (Figure 47). On marine agar, colonies are 2–4 mm in diameter, circular, shiny with entire edges, yellow, and slightly sunken into the agar. Growth occurs at 4–32°C (optimum, 21–23°C) and in 1–4% NaCl. Agar is weakly hydrolyzed.

Source: isolated from the green alga *Ulva fenestrata*, collected in Troitsa Bay, Gulf of Peter the Great, Sea of Japan.

DNA G+C content (mol%): 35–37 (T_m).

Type strain: KMM 3951, KCTC 12969, DSM 15366.

Sequence accession no. (16S rRNA gene): AY271626.

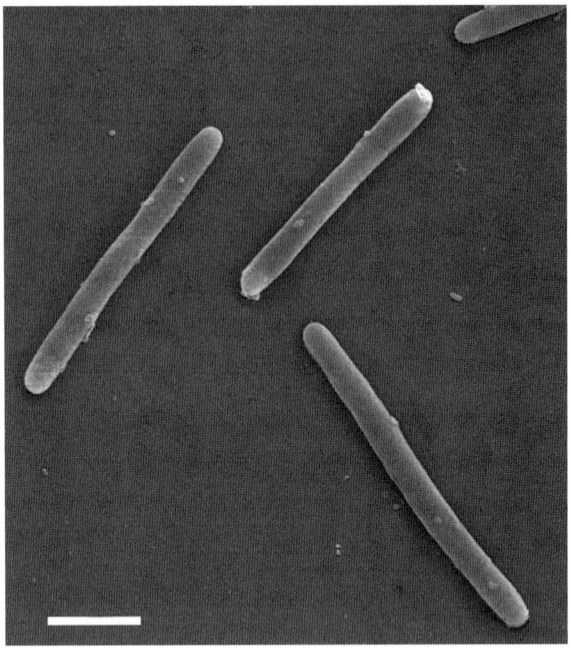

FIGURE 47. Micrograph of *Maribacter ulvicola* KMM 3951[T] showing rod-shaped cells. Bar = 1 μm.

Genus XXXIII. **Mariniflexile** Nedashkovskaya, Kim, Kwak, Mikhailov and Bae 2006a, 1636[VP]

The Editorial Board

Ma.ri.ni.fle'xi.le. L. adj. *marinus -a -um* marine; L. part. adj. *flexilis -e* pliant, pliable, flexible; N.L. neut. n. *Mariniflexile* a flexible marine bacterium.

Rods, 0.4–0.5 × 2–3 μm. Motile by gliding. Nonsporeforming. Gram-stain-negative. **Colonies are orange.** Chemo-organotrophic. **Do not require Na^+ ions for growth,** but grow best with 1–2% NaCl. **Oxidase- and catalase-positive.** Alkaline phosphatase-positive. **D-Glucose is fermented in the O-F test. Agar and starch are not hydrolyzed.** The predominant cellular fatty acids are straight-chain saturated, branched-chain saturated and unsaturated fatty acids, $C_{15:0}$ iso, $C_{15:0}$ anteiso, $C_{15:1}$ iso, $C_{15:0}$, $C_{17:0}$ iso 3-OH, and summed feature 3 (comprising $C_{16:1}$ ω7 and/or $C_{15:0}$ iso 2-OH). The main menaquinone is MK-6. Isolated from the sea urchin *Strongylocentrotus intermedius*.

DNA G+C content (mol%): 35.7 (T_m).

Type species: **Mariniflexile gromovii** Nedashkovskaya, Kim, Kwak, Mikhailov and Bae 2006a, 1636[VP].

Further descriptive information

Mariniflexile gromovii was isolated from the sea urchin *Strongylocentrotus intermedius* in Troitsa Bay, Gulf of Peter the Great, East Sea (Sea of Japan).

Colonies of *Mariniflexile gromovii* are circular, 2–3 mm in diameter, convex, shiny, orange, and translucent on marine agar 2216 (Difco). Growth occurs at 4–37°C; optimum, 23–25°C. The salinity range for growth is 0–6% NaCl.

Mariniflexile gromovii is susceptible to ampicillin, carbenicillin, lincomycin, and tetracycline. It is resistant to benzylpenicillin, gentamicin, kanamycin, neomycin, oleandomycin, polymyxin B, and streptomycin.

Enrichment and isolation procedures

Portions (0.1 ml) of homogenates of sea urchin tissues were inoculated onto plates of marine agar. After primary isolation and purification, strains were cultivated at 28°C on marine agar. The organism can be preserved at −80°C in marine broth containing 20% (v/v) glycerol.

Differentiation of the genus *Mariniflexile* from related genera

Mariniflexile gromovii differs from *Yeosuana aromativorans* by: its ability to form acid from L-fucose and DL-xylose; its ability to ferment glucose in the O-F test; positive oxidase reaction; gliding motility; a lower G+C content (35–36 mol% vs 51–52 mol%); its ability to grow in the absence of Ca^{2+}, Mg^{2+}, and NaCl; and its broader temperature range for growth (4–37°C vs 23–39°C). *Mariniflexile gromovii* can be differentiated from *Algibacter lectus* by: its ability to grow in the absence of NaCl; its inability to hydrolyze agar and starch; and a higher G+C content (35–36 mol% vs 31–33 mol%).

Taxonomic comments

On the basis of 16S rRNA gene sequence analysis, the nearest neighbors of *Mariniflexile gromovii* are the species *Algibacter lectus* and *Yeosuana aromativorans*, with sequence similarities of 93.8 and 93.6%, respectively. The rRNA gene sequence of *Mariniflexile gromovii* has less than 93.1% similarity to those of other members of the *Flavobacteriaceae*.

List of species of the genus *Mariniflexile*

1. **Mariniflexile gromovii** Nedashkovskaya, Kim, Kwak, Mikhailov and Bae 2006a, 1636[VP]

gro.mo'vi.i. N.L. gen. masc. n. *gromovii* of Gromov, named in honor of B.V. Gromov, the Russian aquatic and marine microbiologist.

Characteristics are as described for the genus, with the following additional features. Decomposes gelatin. The following compounds are not hydrolyzed: casein, Tweens 20, 40 and 80, urea, cellulose (carboxymethylcellulose and filter paper), chitin, and DNA. Acid is produced from L-fucose and DL-xylose, but not from L-arabinose, cellobiose, D-galactose, D-glucose, glycerol, inositol, D-lactose, maltose, mannitol, melibiose, raffinose, or sucrose. L-Rhamnose and *N*-acetylglucosamine are oxidized. D-Lactose, D-mannose, and sucrose are utilized, but not L-arabinose, adonitol, citrate, dulcitol, inositol, malonate, mannitol, or sorbitol. Nitrate is not reduced. Indole, H_2S, and acetoin (Voges–Proskauer reaction) production are negative. Cellular fatty acids accounting for more than 1.0% of the total are $C_{15:1}$ iso (16.9%), $C_{15:1}$ anteiso (1.6%), $C_{15:0}$ iso (15.0%), $C_{15:0}$ anteiso (5.4%), $C_{15:0}$ (13.8%), $C_{15:1}$ ω6*c* (3.1%), $C_{15:0}$ iso 3-OH (4.3%), $C_{15:0}$ iso 2-OH (1.4%), $C_{15:0}$ 3-OH (1.6%), $C_{16:1}$ iso (1.5%), $C_{16:0}$ iso (1.0%), $C_{16:0}$ (1.4%), $C_{16:0}$ iso 3-OH (2.5%), $C_{16:0}$ 3-OH (1.1%), $C_{17:1}$ ω6*c* (1.1%), $C_{17:0}$ 3-OH (8.9%), and summed feature 3 (8.4%; comprising $C_{16:1}$ ω7 and/or $C_{15:0}$ iso 2-OH).

Source: isolated from the sea urchin *Strongylocentrotus intermedius* in Troitsa Bay, Gulf of Peter the Great, East Sea (Sea of Japan).

DNA G+C content (mol%): 35.7 (T_m).

Type strain: KMM 6038, KCTC 12570, LMG 22578.

Sequence accession no. (16S rRNA gene): DQ312294.

Genus XXXIV. **Mesonia** Nedashkovskaya, Kim, Han, Lysenko, Rohde, Zhukova, Falsen, Frolova, Mikhailov and Bae 2003a, 1970[VP] emend. Nedashkovskaya, Kim, Zhukova, Kwak, Mikhailov and Bae 2006e, 2435

Olga I. Nedashkovskaya and Seung Bum Kim

Me.so'ni.a. N.L. fem. n. *Mesonia* arbitrary name, derived from the abbreviation MES (Marine Experimental Station of Pacific Institute of Bioorganic Chemistry, FEB RAS) near the site where the bacteria were isolated.

Thin rods usually measuring 0.4–0.5 × 1.0–2.3 µm. **Cells may be motile by means of gliding.** Produce non-diffusible yellow pigments. No flexirubin pigments are formed. **Chemo-organotrophs. Strictly aerobic.** Positive for oxidase, catalase, and alkaline phosphatase and negative for β-galactosidase. Agar, alginate, starch, chitin, and cellulose (carboxymethylcellulose and filter paper) are not hydrolyzed, but gelatin and Tween 20 are hydrolyzed. Do not grow without seawater or Na^+. **The major respiratory quinone is MK-6.** Phosphatidylethanolamine is the main polar lipid. **Marine**, isolated from coastal seaweeds.

DNA G+C content (mol%): 32–37.

Type species: **Mesonia algae** Nedashkovskaya, Kim, Han, Lysenko, Rohde, Zhukova, Falsen, Frolova, Mikhailov and Bae 2003a, 1970[VP].

Further descriptive information

Phylogenetic analysis of nearly complete 16S rRNA gene sequences indicate that *Mesonia* forms a cluster group with the genera *Salegentibacter* and *Gramella*, with similarities of 92.0–92.5% and 92.1–92.2%, respectively (see the phylogenetic tree in the treatment of the family *Flavobacteriaceae* in this volume).

Dominant fatty acids of the genus *Mesonia* are straight-chain unsaturated and branched-chain unsaturated fatty acids. The fatty acid profile includes $C_{15:0}$ iso, $C_{15:1}$ iso, $C_{15:0}$, $C_{16:0}$ iso, $C_{15:0}$ iso 3-OH, $C_{17:1}$ iso ω9*c*, $C_{17:0}$ iso 3-OH, and summed feature 3 ($C_{16:1}$ ω7 and/or $C_{15:0}$ iso 2-OH).

On marine agar 2216 (Difco), *Mesonia* strains form regular, round, convex, smooth, and opaque colonies with entire edges and diameters of 1–4 mm after 48 h at 28°C. Usually, the colonies are yellow, but strains that form white colored colonies also occur.

All isolated strains have been grown on media containing 0.5% peptone and 0.1–0.2% yeast extract (Difco), prepared with natural or artificial seawater or supplemented with 2–3% NaCl. Good growth is observed on marine agar 2216. Growth occurs at 4–39°C (optimum, 21–30°C) and with 1–15% NaCl.

Mesonia strains are susceptible to lincomycin and resistant to gentamicin, kanamycin, neomycin, polymyxin B, and streptomycin. Susceptibility to ampicillin, benzylpenicillin, carbenicillin, oleandomycin, and tetracycline differs among strains.

Strains of the genus *Mesonia* have been isolated from the green alga *Acrosiphonia sonderi*, a common dweller of coastal marine environment, and from seawater samples in the temperate latitudes.

Enrichment and isolation procedures

Mesonia strains were isolated from seaweed and seawater samples by direct plating on marine agar. Natural or artificial seawater is suitable for their cultivation. All isolated strains have been grown on media containing 0.5% peptone and 0.1–0.2% yeast extract (Difco). Strains of the genus may remain viable for several weeks on marine agar or other rich medium based on natural or artificial seawater. They have survived storage at −80°C for at least 5 years.

Differentiation of the genus *Mesonia* from other genera

The results of phenotypic examination indicate that strains of the genus *Mesonia* can be differentiated from those of its closest relative, *Gramella*, by their inability to hydrolyze starch and their lower DNA G+C content. The absence of nitrate reductase allows *Mesonia* strains to be differentiated from those of *Salegentibacter*.

Differentiation of species of the genus *Mesonia*

Table 57 lists characteristics that differentiate the two species of *Mesonia*.

TABLE 57. Phenotypic characteristics of *Mesonia* species[a]

Characteristic	*M. mobilis* KMM 6059[T]	*M. algae*
Gliding motility	+	–
Growth at 39°C	+	–
H_2S production	–	+
Hydrolysis of casein and Tween 40	–	+
Acid production from D-glucose and maltose	+	–
Utilization of L-arabinose, D-glucose and D-mannose	+	–
Susceptibility to:		
Benzylpenicillin	+	–
Carbenicillin and oleandomycin	–	+
DNA G+C content (mol%)	36.1	32–34

[a]Data are from Nedashkovskaya et al. (2003a) (for four strains of *Mesonia algae*) and Nedashkovskaya et al. (2006e).

List of species of the genus *Mesonia*

1. **Mesonia algae** Nedashkovskaya, Kim, Han, Lysenko, Rohde, Zhukova, Falsen, Frolova, Mikhailov and Bae 2003a, 1970[VP]

al'gae. L. n. *alga* alga, seaweed; L. gen. n. *algae* of alga.

Cells are 0.4–0.5 × 1.6–2.3 µm. On marine agar, colonies are 2–4 mm in diameter, circular, shiny with entire edges, and yellow, although white strains occur. Growth occurs at 4–34°C (optimum, 21–23°C) and with 1–15% NaCl. Gelatin, casein, and Tweens 20, 40 and 80 are hydrolyzed, but not agar, alginate, starch, cellulose (carboxymethylcellulose and filter paper), or chitin. No acid is formed from L-arabinose, cellobiose, L-fucose, D-galactose, D-glucose, D-lactose, maltose, D-mannose, melibiose, raffinose, L-rhamnose, sucrose, DL-xylose, citrate, adonitol, dulcitol, glycerol, inositol, or mannitol. Does not utilize L-arabinose, D-glucose, D-lactose, D-mannose, sucrose, mannitol, inositol, sorbitol, *N*-acetylglucosamine, gluconate, caprate, adipate, malate, malonate, phenylacetate, or citrate. Nitrate is not reduced. H_2S is produced, but indole and acetoin (Voges–Proskauer reaction) production are negative.

Source: isolated from the green alga *Acrosiphonia sonderi* collected in Troitsa Bay, Gulf of Peter the Great, Sea of Japan.

DNA G+C content (mol%): 32.7–34.0 (T_m).

Type strain: KMM 3909, KCTC 12089, CCUG 47092.

Sequence accession no. (16S rRNA gene): AF536383.

2. **Mesonia mobilis** Nedashkovskaya, Kim, Zhukova, Kwak, Mikhailov and Bae 2006e, 2435[VP]

mo'bi.lis. L. fem. adj. *mobilis* movable, mobile, referring to the ability to move by gliding.

Cells are 0.4–0.5 × 1.0–2.1 µm. On marine agar, colonies are 1–3 mm in diameter, circular, convex, and shiny with entire edges. Produces non-diffusible yellow pigments. Flexirubin-type pigments are absent. Growth occurs at 4–39°C (optimum, 28–30°C), with 1–12% NaCl (optimum, 3–4% NaCl) and at pH 6.0–9.5 (optimum, pH 7.5). Gelatin and Tween 20 are hydrolyzed, but not agar, alginate, casein, starch, Tweens 40 and 80, DNA, urea, cellulose (carboxymethylcellulose and filter paper), or chitin. Forms acid from D-glucose and maltose, but not from L-arabinose, cellobiose, L-fucose, D-galactose, D-lactose, melibiose, raffinose, L-rhamnose, sucrose, DL-xylose, citrate, adonitol, dulcitol, glycerol, inositol, mannitol, malate, or fumarate. L-Arabinose and D-mannose are utilized, but D-lactose, sucrose, mannitol, inositol, and sorbitol are not. Nitrate is not reduced. H_2S, indole, and acetoin (Voges–Proskauer reaction) production are negative. Susceptible to ampicillin and benzylpenicillin; resistant to carbenicillin, oleandomycin, and tetracycline.

Source: isolated from seawater collected in Troitsa Bay, Gulf of Peter the Great, Sea of Japan.

DNA G+C content (mol%): 36.1 (T_m).

Type strain: KMM 6059, KCTC 12708, LMG 23670.

Sequence accession no. (16S rRNA gene): DQ367409.

Genus XXXV. **Muricauda** Bruns, Rohde and Berthe-Corti 2001, 2005[VP] emend. Yoon, Lee, Oh and Park 2005d, 1018

Alke Bruns and Luise Berthe-Corti

Mu'ri.cau.da. L. masc. n. *mus* mouse (gen. *muris* of the mouse); N.L. fem. n. *cauda* tail; N.L. fem. n. *Muricauda* tail of the mouse, referring to electron micrographs in which the bacterial appendages appeared like mouse tails.

Rod-shaped cells. Size varies between 0.2–0.6 × 1.1–6.0 µm depending on the species. Cells of older cultures (3–4 d) of *Muricauda ruestringensis* and *Muricauda aquimarina* are characterized by long (up to 3 µm) and relatively thick **appendages** (*Muricauda ruestringensis*: 20–30 nm). Gram-stain-negative. **Nonmotile** in hanging drop preparations. **Facultatively anaerobic or strictly aerobic.** Chemo-organotrophic, with both an oxidative and a fermentative type of metabolism. Endospores are not formed. Colonies are circular, slightly convex, 0.5–1.2 mm in diameter and **yellow in color** (*Muricauda ruestringensis*) or golden yellow (*Muricauda aquimarina* and *Muricauda flavescens*) after 3 d. No flexirubin-type pigment is produced. Catalase-positive. Oxidase-positive. Nitrate is not reduced. H_2S is not produced. Optimal growth occurs at 20–37°C, at pH 6.5–7.5, and in the presence of 2% NaCl (*Muricauda ruestringensis*) to 3% NaCl (*Muricauda aquimarina* and *Muricauda flavescens*). No hydrolysis of high-molecular-mass carbohydrates such as starch, agar, or crystalline cellulose occurs. The fatty acid profile is characterized by high amounts of branched fatty acids. The genus is closely related to other *Flavobacteriaceae*, which have been isolated from marine and marine-derived habitats.

DNA G+C content (mol%): 41–45.

Type species: **Muricauda ruestringensis** Bruns, Rohde and Berthe-Corti 2001, 2005[VP].

Further descriptive information

Phylogeny. Several members of the *Flavobacteriaceae* have been isolated from marine or marine-derived habitats. A comparison of 16S rRNA gene sequence data of all genera that have been isolated mainly from these habitats clearly indicates that the genus *Muricauda* is closely related to the genera *Arenibacter*, *Maribacter*, *Pibocella*, *Robiginitalea*, and *Zobellia* (Figure 48). The genus *Muricauda* comprises three species with validly published names: *Muricauda ruestringensis*, which is represented by only one type strain, strain B1[T] (Bruns et al., 2001); *Muricauda aquimarina*, represented by two strains (SW-63[T] and SW-72); and *Muricauda flavescens*, represented by strains SW-62[T] and SW-74 (Yoon et al., 2005d).

Morphology. When grown on marine agar, all species produce yellow to golden yellow colonies of 0.5–1.5 mm in diameter after 3 d. Colonies are circular, convex in shape with a glistening surface. Young cells from suspension cultures are rod-shaped with rounded edges. However, with increasing age of the cultures the populations differentiate into larger coccus-shaped cells and into rod-shaped cells. *Muricauda ruestringensis* and *Muricauda aquimarina* have been described as exhibiting appendages when grown in liquid cultures. *Muricauda flavescens* has not yet been described as exhibiting appendages; however, the formation of vesicles has been observed clearly by electron microscopy. The production of appendages and vesicles is a distinctive characteristic of the genus *Muricauda*, although some other species within the *Cytophaga–Flavobacteria–Bacteroides* group are known to have cell structures that look very much like appendages, i.e., *Flavobacterium aquatile* strain ATCC 11947[T] (Thomson et al., 1981), *Flexibacter* strain BH3, *Cytophaga johnsonae*, *Polaribacter irgensii*, and *Winogradskyella thalassocola*. The appendages of *Flexibacter* consist of extruded material derived from the lipopolysaccharide membrane and are probably related to the gliding motility of this species (Humphrey et al., 1979). The nature of the appendages of *Flavobacterium aquatile* has not yet been characterized. Gosink et al. (1998) showed that the appendages of *Polaribacter irgensii* strain 23-P[T] are true flagella, although motility in this genus has never been observed. The recently isolated *Winogradskyella thalassocola* has been described to form network-like structures whose composition and function is also not known (Nedashkovskaya et al., 2005a).

Fine structure. Ultrathin sections of *Muricauda ruestringensis* show the typical ultrastructure of a Gram-stain-negative bacterium, with the exception that there are holes in the outer membrane of older cultures (Figure 49). In addition, older cultures (3–4 d) are characterized by long and relatively thick appendages, which can be observed by light microscopy at 1,000× when stained with Nile Red. This staining indicates that they are lipid-containing structures (Müller et al., 2001). Bruns et al. (2001) have characterized the appendages of the type strain *Muricauda ruestringensis* strain B1[T]. They are 20–30 μm long and often located in a polar position. The appendages exhibit a maturation process that begins with the formation of a vesicle at the cell surface (Figure 49), after which a stalk develops between vesicle and cell surface (Figure 49). The vesicle-like structures at the end of the stalks exhibit fibrillar-like structures on their surface (Figure 49). Ultrathin sections have revealed that the appendages are formed as a continuum of the outer membrane and are covered by a membrane (Figure 49). Because older cultures of *Muricauda ruestringensis* often form flocks, Bruns et al. (2001) suggested that the appendages might enable the cells to connect to each other.

Life cycle. Müller et al. (2001) suggest that *Muricauda ruestringensis* exhibits a complex life cycle in which the appendages of the cells are involved as a DNA-containing structure. By dual staining with DAPI and Nile Red, the authors showed that DNA was transferred from cells into the vesicles during the process of vesicle formation (Müller et al., 2001). They also demonstrated that fully developed vesicles of *Muricauda ruestringensis* B1[T] are liberated from the progenitor cell and contain a double set of chromosomal DNA. The authors suggested an asymmetric life cycle for *Muricauda ruestringensis* strain B1[T] (Müller et al., 2001). This type of multiplication seems to be unique among the *Flavobacteriaceae*. *Muricauda aquimarina* also shows relative long and thick appendages (Yoon et al., 2005d). The character of these appendages, however, has not been analyzed.

Chemotaxonomically significant components. All described strains of the genus are yellow- to golden yellow-pigmented when grown on the complex medium marine agar 2216 (Difco). The strains produce no flexirubin-type pigment (Bruns et al., 2001; unpublished data from the authors). All members of the genus exhibit an unsaturated menaquinone with six isoprene units (MK-6) as the respiratory pigment.

The genus *Muricauda* is characterized by cellular fatty acid profiles containing high amounts of straight-chain, branched and hydroxy fatty acids. The major cellular fatty acids of all three species are $C_{15:1}$ iso, $C_{15:0}$ iso, and $C_{17:0}$ 3-OH iso. The growth medium strongly affects the proportion of the straight-chain fatty acid $C_{15:0}$. The proportion of $C_{15:0}$ in *Muricauda ruestringensis* is about 1% when cells are cultivated with mineral PAC* medium containing sodium pyruvate (0.1%, w/v), sodium acetate (0.4%, w/v), and Casamino acids (0.1%, w/v) as the carbon source. In contrast, when cells are cultivated on the complex medium marine agar 2216 (Difco), this fatty acid amounts to 13%.

Growth conditions. All described *Muricauda* species grow well in liquid cultures or on solid media composed of artificial

*PAC medium contains (per liter of artificial seawater of 2.25% salinity): sodium pyruvate, 1.0 g; sodium acetate, 4.0 g; Casamino acids, 1.0 g; vitamin solution, 3.0 ml; mineral salt solution (DSMZ medium no. 124; DSMZ Catalogue of Strains, 1983), 1.0 ml; $NaNO_3$, 1.46 g; and Na_2HPO_4, 1.4 g; pH 7.2.

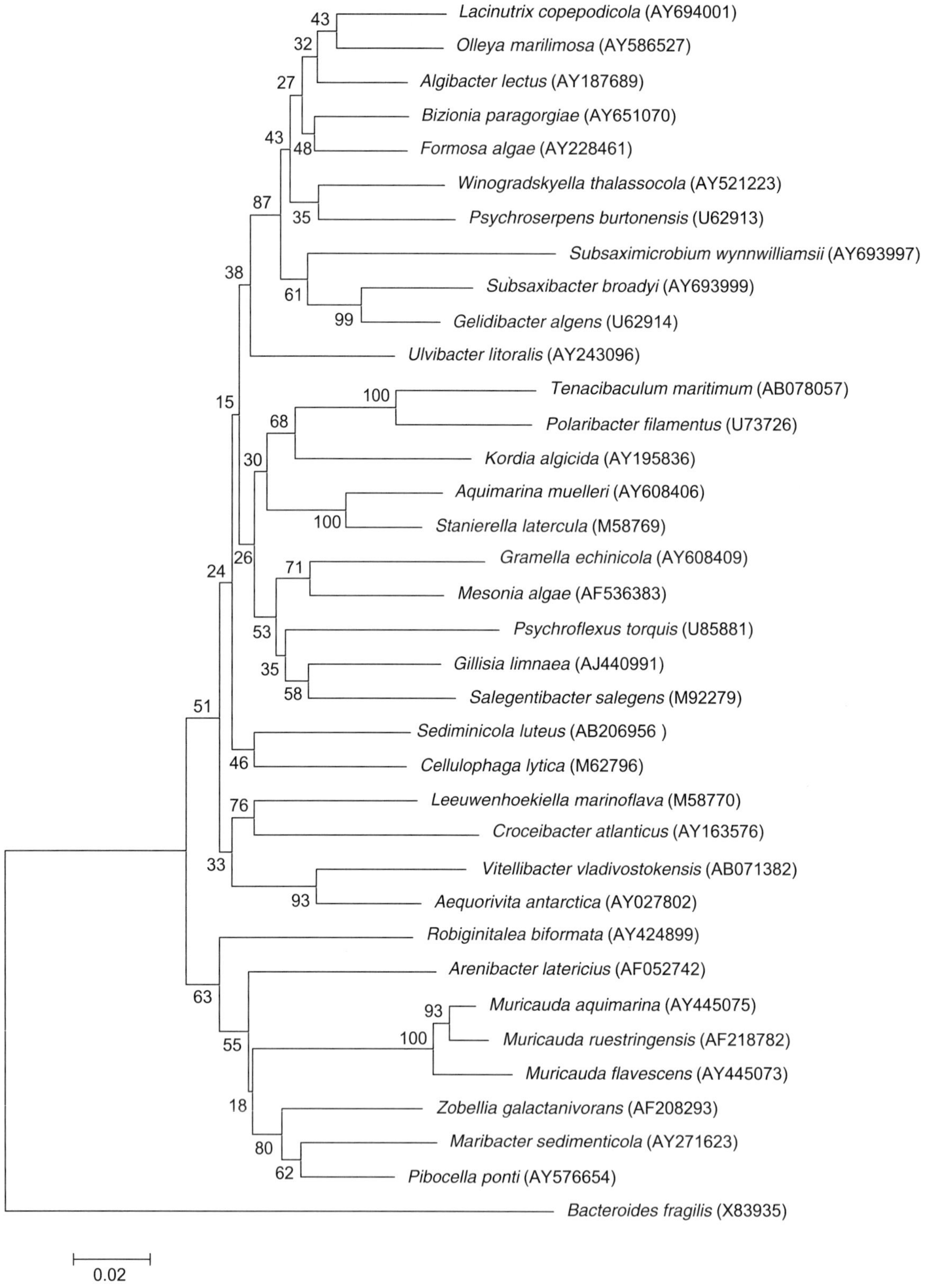

FIGURE 48. Neighbor-joining tree based on 16S rRNA gene sequences of marine members of the family *Flavobacteriaceae*. The tree shows the position of the *Muricauda* species *Muricauda aquimarina*, *Muricauda flavescens*, and *Muricauda ruestringensis*. Bootstrap values (expressed as percentages of 1,000 replications) are shown at the branching points. *Bacteroides fragilis* was used as the outgroup. Bar = 0.02 substitutions per nucleotide position.

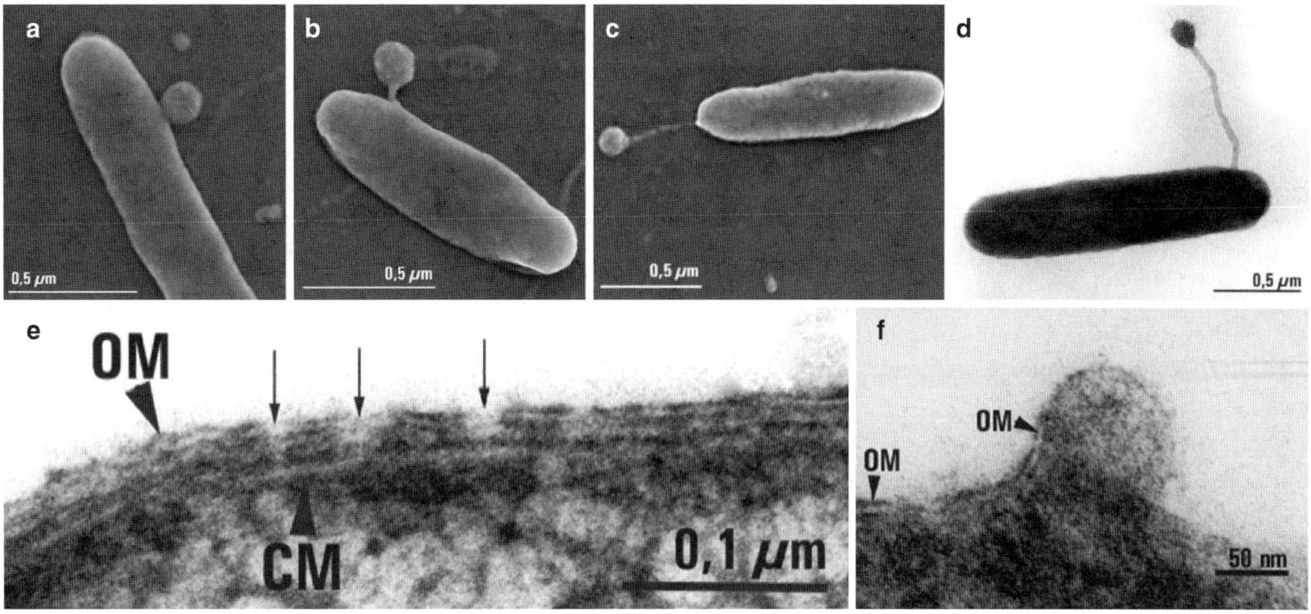

FIGURE 49. Electron micrographs of the cell morphology and ultrastructure of *Muricauda ruestringensis*. Scanning electron micrographs: (a) formation of cell appendages starts with a vesicle-like structure on the bacterial surface (bar = 0.5 μm); (b, c) formation and growth of a stalk up to a length of approximately 5 μm (bars = 0.5 μm); (d) single cell with an appendage that has a vesicle-like structure which exhibits fibrillar-like structures (bar = 0.5 μm). Ultrathin sections of ultrastructural features: (e) older cells show holes (arrows) in the cell membrane which otherwise show a typical Gram-negative arrangement (bar = 0.1 μm); (f) appendages are formed as a continuum of the outer membrane (bar = 50 nm). OM, Outer membrane; CM, cytoplasmic membrane.

seawater supplemented with amino acids and trace elements or in marine broth. *Muricauda* species show good growth in media containing 0.5–9.0% NaCl, but not in media with an NaCl concentration of more than 10%. The optimal NaCl concentration varies between *Muricauda* species: 3% NaCl is optimal for *Muricauda ruestringensis*, whereas 2% is optimal for *Muricauda aquimarina* and *Muricauda flavescens*. The pH tolerance for growth varies between 6.0 and 8.0, with optimal growth at pH 6.5–7.5. All species have a wide temperature range showing good growth between 8 and 40°C. Bruns et al. (2001) report 30°C as optimal for *Muricauda ruestringensis* and Yoon et al. (2005d) report 37°C to be optimal for *Muricauda aquimarina* and *Muricauda flavescens*. All species grow under oxic conditions. *Muricauda ruestringensis* also shows good growth at dissolved oxygen concentrations as low as 0.8 μmol/l and weak growth under anoxic conditions. Anoxic growth has not been observed for *Muricauda aquimarina* and *Muricauda flavescens*.

Metabolism and metabolic pathways. All members of the genus *Muricauda* are chemo-organotrophic. *Muricauda aquimarina* and *Muricauda flavescens* are described as having an exclusively respiratory-type metabolism. *Muricauda ruestringensis* is described as being facultatively anaerobic, exhibiting an anaerobic metabolism when peptone and yeast extract are supplied as the carbon source. All *Muricauda* species are able to hydrolyze esculin, tyrosine, and Tweens 20, 40, and 60. They grow with the sugars cellobiose, fructose, lactose, mannose, raffinose, and sucrose, or with succinate. Acid is produced from cellobiose, D-fructose, D-glucose, lactose, maltose, D-mannose, melibiose, sucrose, trehalose, and raffinose. All species are unable to reduce nitrate or nitrite and do not produce H_2S. They are also unable to degrade high-molecular-mass carbohydrates such as starch or agar. Birchwood xylan is not degraded when tested on solid marine salts basal medium using the method of Baumann and Baumann (1981). *Muricauda* species do not hydrolyze casein, gelatin, hypoxanthine, urea, or xanthine. No growth occurs with acetate, pyruvate, glutamate, butyrate, citrate, lactate, formate, benzoate, methanol, ethanol, gelatin, starch, urea, xanthine, serine, mannitol, or hexadecane when these are supplied as the sole carbon source. No acids are produced from D-sorbitol, *myo*-inositol, D-ribose, D-mannitol, adonitol, or L-rhamnose.

Ecology. All known *Muricauda* species have been isolated from salt water habitats. *Muricauda ruestringensis* strain B1^T was isolated from a continuous culture with a suspension composed of artificial seawater and intertidal sediment taken from the German North Sea coast and which contained an autochthonous community. The culture was run at a dissolved oxygen concentration of 0.8 μmol/l (Berthe-Corti and Bruns, 1999). The intertidal Wadden sediment from the German North Sea coast is a nutrient-rich habitat characterized by strong shifts in temperature, salinity, and oxygen concentration. It is rich in species belonging to the *Cytophaga–Flavobacteria–Bacteroides* group (Llobet-Brossa et al., 1998). *Muricauda aquimarina* and *Muricauda flavescens* have been taken from the water column of a salt lake near Hwajinpo Beach of the East Sea in Korea. This lake is fed by a river and has a direct connection to the East Sea in Korea.

Enrichment and isolation procedures

All *Muricauda* species were isolated using the standard dilution plating technique. Before isolation, *Muricauda ruestringensis* was enriched in a sediment-suspension culture (as described above). Isolation was performed at 25°C on solid modified nutrient agar* *Muricauda aquimarina* and *Muricauda flavescens* have been isolated from the water column of a salt lake using marine agar 2216 (Difco). Cultures were incubated at 30°C.

Maintenance procedures

Muricauda can be maintained in the laboratory by transfer onto the isolation medium described above. Long-term storage of *Muricauda ruestringensis* has only been tested in the liquid PAC medium containing 20–25% glycerol. J.-H. Yoon recommends using nutrient broth from Difco with the addition of 20% glycerol for long-term storage of *Muricauda aquimarina* and *Muricauda flavescens* (personal communication).

Differentiation of the genus *Muricauda* from other closely related taxa

Phylogenetically, the genus *Muricauda* is clearly affiliated to the family of *Flavobacteriaceae.* It shares the inability to degrade crystalline cellulose (filter paper) with the other members of the *Flavobacteriaceae.* Among the complex organic substrates that some members of the phylum *Bacteroidetes* are able to degrade, cellulose has a particular taxonomic significance. The inability to degrade crystalline cellulose is one major characteristic used to distinguish members of the *Flavobacteriaceae* from those of the genus *Cytophaga.* Moreover, the genus *Muricauda* produces yellow to golden yellow pigments, which is a typical characteristic of most *Flavobacteriaceae.* Production of flexirubin-type pigment is a further distinguishing characteristic within the *Flavobacteriaceae. Muricauda* was negative for the flexirubin reaction, whereas other members of the *Flavobacteriaceae* such as *Zobellia* produce this pigment (Table 58). While all closely related members of the *Flavobacteriaceae* have a strictly aerobic respiratory metabolism, *Muricauda ruestringensis* is a facultatively anaerobic organism. All other differential characteristics of closely related *Flavobacteriaceae* isolated from marine or marine-derived habitats are summarized in Table 58.

Finally, the most striking differential characteristic making *Muricauda* an extraordinary member of the *Flavobacteriaceae* is its phenotypic appearance. The appendages formed as a continuum of the outer membrane have never been described for other marine *Flavobacteriaceae* (Table 58).

Acknowledgements

The authors wish to express their appreciation to J.-H. Yoon for providing the strains *Muricauda aquimarina* SW-63^T and *Muricauda flavescens* SW-62^T.

List of species of the genus *Muricauda*

1. **Muricauda ruestringensis** Bruns, Rohde and Berthe-Corti 2001, 2005^{VP}

rue.strin′gen.sis. N.L. fem. adj. *ruestringensis* of or belonging to to the former village of Rüstringen, which was destroyed by a tidal wave in 1362.

Cells are 0.3–0.6 × 1.1–2.7 µm. Cells from older cultures (3–4 d) exhibit long and relatively thick (20–30 nm) appendages. On mineral agar, colonies are yellow, circular, slightly convex, and 0.5–1.0 mm in diameter after 3 d at 30°C. The temperature range for growth is 8–40°C, with optimal growth at 20–30°C. The pH tolerance ranges between pH 6.0 and 8.0. Optimal pH for growth is 6.5–7.5. Slightly halophilic, showing good growth in media containing 0.5–9.0% (w/v) NaCl, with optimal growth at 3% (w/v) NaCl. Facultatively anaerobic. Growth on marine agar occurs under oxic and under anoxic conditions. The predominant respiratory lipoquinone is MK-6. The major fatty acids are $C_{15:0}$ iso, $C_{15:1}$ iso, and $C_{17:0}$ 3-OH iso.

Source: strain B1^T was isolated from an enrichment culture containing intertidal sediment of the North Sea coast near the former village of Ruestringen.

DNA G+C content (mol%): 41.4 (HPLC).

Type strain: B1, DSM 13258, LMG 19739.

Sequence accession no. (16S rRNA gene): AF218782.

2. **Muricauda aquimarina** Yoon, Lee, Oh and Park 2005d, 1019^{VP}

a.qui.ma.ri′na. L. n. *aqua* water; L. adj. *marinus* of the sea; N.L. fem. adj. *aquimarina* pertaining to seawater.

Cells are 0.2–0.5 × 2.5–6.0 µm. On marine agar, colonies are golden yellow, circular, slightly convex, and 0.8–1.2 mm in diameter. Growth occurs at 10°C, but not at 4°C; maximum growth temperature is 44°C. Optimal pH for growth is around 7.0. Growth occurs in media containing up to 9% (w/v) NaCl, but does not occur in the absence of NaCl. No growth under anoxic conditions. The major fatty acids are $C_{15:0}$ iso, $C_{15:1}$ iso, and $C_{17:0}$ 3-OH iso.

Source: the type strain was isolated from a salt lake near Hwajinpo Beach of the East Sea in Korea.

DNA G+C content (mol%): 44.1–44.2 (HPLC).

Type strain: SW-63, JCM 11811, KCCM 41646.

Sequence accession no. (16S rRNA gene): AY445075.

3. **Muricauda flavescens** Yoon, Lee, Oh and Park 2005d, 1018^{VP}

fla.ves′cens. L. v. *flavesco* to become golden yellow; L. part. adj. *flavescens* becoming golden yellow.

Most characteristics of relevance are identical to those of the species *Muricauda aquimarina.* Cells are 0.2–0.5 × 2.5–6.0 µm. On marine agar, colonies are golden yellow, circular, slightly convex, and 0.8–1.2 mm in diameter. Growth occurs at 10°C, but not at 4°C; the maximum growth temperature is 44°C. The optimal pH for growth is around 7.0. Growth occurs up to 9% (w/v) NaCl (Yoon et al., 2005d), but also occurs in a medium without the addition of NaCl and containing (per liter) 17.0 g casein peptone, 3.0 g soya peptone, 2.5 g D-glucose, and 2.5 g K_2HPO_4 (unpublished data from the authors). No growth occurs under anoxic conditions. The major fatty acids are $C_{15:0}$ iso, $C_{15:1}$ iso, and $C_{17:0}$ 3-OH iso.

Source: the type strain was isolated from a salt lake near Hwajinpo Beach of the East Sea in Korea.

DNA G+C content (mol%): 45.2–45.4 (HPLC).

Type strain: SW-62, JCM 11812, KCCM 41645.

Sequence accession no. (16S rRNA gene): AY445073.

*Solid modified nutrient agar contains (g/l of artificial seawater): Bacto Peptone (Difco), 2.5; yeast extract, 1.5; and agar, 15.0. The artificial seawater contains (g/l of distilled water): NaCl, 23.6; KCl, 0.64; $MgCl_2 \cdot 6H_2O$, 4.53; $MgSO_4 \cdot 7H_2O$, 5.94; $CaCl_2 \cdot 2H_2O$, 1.3. To avoid precipitation, the $CaCl_2 \cdot 2H_2O$ should be sterilized separately. The pH of the medium must be adjusted to 7.2.

TABLE 58. Differential characteristics of the genus *Muricauda* and other closely related genera of the family *Flavobacteriaceae* isolated from marine or marine-derived habitats[a]

Character	*Muricauda*	*Aequorivita*	*Arenibacter*	*Maribacter*	*Pibocella*	*Robiginitalea*	*Vitellibacter*	*Zobellia*
Appendages	+	–	–	–	–	–	–	–
Cell length (μm)	1.1–2.7	0.5–20.0	3.0–5.0	0.8–10.0	1.6–2.3	1.6–5.6	3.0–10.0	1.2–8.0
Cell width (μm)	0.3–0.6	0.2–0.5	0.4–0.7	0.3–0.7	0.4–0.5	0.3–0.7[b]	0.3–0.5	0.3–0.5
Pigment color	Yellow to golden yellow	Yellow–orange	Orange	Yellow–orange	Yellow	Orange	Yellow–orange	Yellow–orange–red
Flexirubin reaction	–	–	–	–	–	–	+	+
Gliding motility	–	–	–	+	+	–		+
Hydrolysis of:								
Gelatin	–	+	D	D	+	–	+	+
Starch	–	D	–	D	+	+	–	D
Esculin	+	D	na	na	na	+	na	na
Cellulose	–[c]	–	–	na	–	–	–	–
Agar	–[c]	–	–	D	–	na	–	+
DNA	na	–	–	D	–	–	+	D
Production of:								
Indole	–	na	–	–	–	–	–	–
H_2S	–	–	D	–	–	na	–	–
Nitrate reduction	–	–	+	D	–	–	–	+
Strictly aerobic	D	+	+	+	+	+	+	+
Facultatively anaerobic	D	–	–	–	–	–	–	D
Optimum temperature (°C)	20–37	20	28–30	21–30	21–23	30	28	20–35
Highest NaCl concentration tolerated (%)	9	6–10	6–10	4–10	13	10	6	6–10[d]
Major fatty acids (%):								
$C_{15:0}$ iso	14.7–23.7	7.6–17	8.5–17	12–21	8.7	24–28	69	17–23
$C_{15:1}$ iso	19.5–21.6	8.9–13	14–19	10–19	12	14–21	2.4	8.8–15
$C_{17:1}$ iso 3-OH	17.3–20.9	0–4.5	0–6	11–29	5.6	25–27	0.8	15–26
DNA G+C content (mol%)	41–45	33–39	37–40	35–39	36	55–56	41	36–43

[a]+, Positive reaction; –, negative reaction; na, no data available; D, different reactions in different species. All genera are positive for catalase and oxidase.

[b]Cells in stationary phase are coccoid, 0.6–1.2 μm in diameter.

[c]L. Berthe-Corti, unpublished data.

[d]No data for *Zobellia uliginosa* available.

Genus XXXVI. **Myroides** Vancanneyt, Segers, Torck, Hoste, Bernardet, Vandamme and Kersters 1996, 930[VP]

THE EDITORIAL BOARD

My.roi′des. Gr. n. *myron* perfume; L. suff. *-oides* (from Gr. suff. *-eides*, from Gr. n. *eidos* that which is seen, form, shape, figure), resembling, similar; N.L. masc. n. *Myroides* resembling perfume.

Rods, 0.5 × 1–2 μm. Longer rods and long chains (containing 4–10 cells) may occur in broth medium. Nonmotile**. No gliding motility** or swarming occurs. Gram-stain-negative. **Yellow or yellow to orange colonies are usually produced**. A characteristic fruity odor is produced by most strains. **Aerobic**, having a strictly respiratory type of metabolism with oxygen as the terminal electron acceptor. Good growth occurs on nutrient agar and MacConkey agar. No hemolysis occurs on blood agar. Growth occurs at 18–22 and 37°C, but not at 5 or 42°C. NaCl is not required, but growth occurs in the presence of at least 5% NaCl. Positive for oxidase and gelatinase; catalase-positive or (in one species) weakly positive. Indole-negative. Nitrate is not reduced. Negative for arginine dihydrolase and β-galactosidase. Esculin is not hydrolyzed. **Nonsaccharolytic**. The major isoprenoid quinone is menaquinone-6. The dominant polyamine component is homospermidine. The dominant fatty acids are $C_{15:0}$ iso, $C_{15:0}$ iso 3-OH, $C_{16:0}$ 3-OH, $C_{17:0}$ iso 3-OH, and $C_{17:1}$ iso ω9*c*. Two species are sources of nosocomial infections in humans; a third has been isolated from seawater.

DNA G+C content (mol%): 30–38.

Type species: **Myroides odoratus** (Stutzer and Kwaschnina 1929) Vancanneyt, Segers, Torck, Hoste, Bernardet, Vandamme and Kersters 1996, 931[VP] (*Flavobacterium odoratum* Stutzer *in* Stutzer and Kwaschnina 1929, 221).

Further descriptive information

Vancanneyt et al. (1996) reported that *Myroides odoratus* and *Myroides odoratimimus* hydrolyzed the following substrates, as determined by API ZYM galleries: 2-naphthyl-phosphate, 2-naphthyl-butyrate, 2-naphthyl-caprylate, L-leucyl-2-naphthylamide, and naphthol-AS-BI-phosphate. *N*-Benzoyl-DL-arginine-2-naphthylamide, 6-Br-2-naphthyl-β-D-galactopyranoside, 2-naphthyl-β-D-galactopyranoside, naphthol-AS-BI-β-D-glucuronide, 2-naphthyl-α-D-glucopyranoside, 6-Br-2-naphthyl-α-D-glucopyranoside, 1-naphthyl-*N*-acetyl-β-D-glucosaminide, 6-Br-2-naphthyl-α-D-mannopyranoside, and 2-naphthyl-α-L-fucopyranoside are not hydrolyzed. Very weak or no hydrolysis occurs with the substrates 2-naphthylmyristate, L-valyl-2-naphthylamide, L-cystyl-2-naphthylamide, and *N*-glutaryl-phenylalanine-2-naphthylamide.

Myroides odoratus is most commonly recovered from human urine specimens, sometimes in significant numbers, and from colonization of amputation sites (Davis et al., 1979; Holmes et al., 1979; Vancanneyt et al., 1996). Strains of *Myroides odoratimimus* have been isolated from human wounds, urine, human leg slough, pus, and a water sample (Vancanneyt et al., 1996). The role of *Myroides odoratus* and *Myroides odoratimimus* as causative agents of infection is ill-defined and data are still emerging. Yağci et al. (2000) reported an outbreak of *Myroides* odoratimimus that occurred in a urology ward of a hospital over a 3-year period and described the use of DNA fingerprinting for characterization of the isolates. Green et al. (2001) reported a case of bacteremia and recurrent cellulitis caused by *Myroides odoratus* in an evidently immunocompetent male; this appears to be the first documented life-threatening infection in an immunocompetent host. Motwani et al. (2004) reported a case of cellulitis caused by *Myroides odoratus* that progressed to septic shock.

Some strains of *Myroides odoratus* have been found in insect guts (Spiteller et al., 2000) and freshwater fishes (the bacterial niche being unidentified; González et al., 2000).

The type strain of *Myroides pelagicus* was obtained during the isolation of crude oil-utilizing and -emulsifying bacteria from seawater by Maneerat et al. (2005). *Myroides pelagicus* (and also *Myroides odoratus* and *Myroides odoratimimus*) produced surface-active compounds in marine broth 2216 (Difco), which were identified as cholic acid, deoxycholic acid, and their glycine conjugates. These findings were the first indication that bile acids could be produced by prokaryotic cells. Maneerat et al. (2006) reported that the type strain of *Myroides pelagicus* could grow on weathered crude oil and was capable of emulsifying it. Bile acids cannot emulsify crude oil and Maneerat et al. (2006) showed that the biosurfactant able to emulsify crude oil was excreted in the culture supernatant and was identified as being a mixture of L-ornithine lipids.

Enrichment and isolation procedures

Special procedures are not normally required for the isolation of *Myroides odoratus* and *Myroides* odoratimimus. Myroides *pelagicus* was isolated from seawater and can be cultured in marine broth 2216 (Difco).

Differentiation of the genus *Myroides* from other genera

The genus differs from *Flavobacterium* species by its lack of gliding motility, its ability to grow well at 37 °C, its salt tolerance, and differences in its fatty acid composition. Whereas *Flavobacterium* species are mainly soil and freshwater organisms, *Myroides odoratus* and *Myroides odoratimimus* are associated with nosocomial infections in humans. The characteristics of being non-saccharolytic and indole-negative differentiate *Myroides odoratus* from *Elizabethkingia meningosepticum* and other similar medically important organisms.

Taxonomic comments

Bernardet et al. (1996) emended the description of the genus *Flavobacterium* and restricted members of this genus to aerobic, chemoheterotrophic, Gram-stain-negative rods that occur in soil and freshwater, have gliding motility, form cream to yellow colonies, and decompose several polysaccharides but not cellulose. *Flavobacterium odoratum* (now *Myroides odoratum*) did not have gliding motility and occurred in clinical specimens, so it was removed from the genus *Flavobacterium*. Analysis of 16S rRNA gene sequences had also indicated that *Flavobacterium odoratum* was phylogenetically distinct from the other members of the *Flavobacterium* (Gherna and Woese, 1992; Nakagawa and Yamasato, 1993, 1996) (Vandamme et al., 1994a). In a polyphasic taxonomic study that included DNA–rRNA hybridizations and an analysis of whole-cell protein patterns, fatty acid compositions, and phenotypic characteristics, Vancanneyt et al. (1996) concluded that *Flavobacterium odoratum* should be placed in a new genus, *Myroides*. Moreover, they also gave species status, *Myroides odoratimimus*, to a genetically homogeneous subgroup within *Flavobacterium odoratum* based on the lack of significant DNA–DNA hybridization with *Myroides odoratum* strains.

Yoon et al. (2006a) reported that the type strain of *Myroides pelagicus* shared 93–95% 16S rRNA gene sequence similarity with the type strains of *Myroides odoratus* and *Myroides odoratimimus*. Moreover, the DNA–DNA relatedness values of the strain with the type strains of *Myroides odoratus* and *Myroides odoratimimus* were less than 70%, indicating that the strain belonged to a novel species.

Differentiation of species of the genus *Myroides*

Table 59 lists some features that differentiate the three species of the genus.

List of species of the genus *Myroides*

1. **Myroides odoratus** (Stutzer and Kwaschnina 1929) Vancanneyt, Segers, Torck, Hoste, Bernardet, Vandamme and Kersters 1996, 931[VP] (*Flavobacterium odoratum* Stutzer *in* Stutzer and Kwaschnina 1929, 221)

   o.do.ra'tus. L. part. masc. adj. *odoratus* perfumed.

   Characteristics are as given for the genus and as listed in Table 59, with the following additional features given by Vancanneyt et al. (1996). No positive reaction is observed in Biotype 100 galleries. In the Biolog GN MicroPlate assay, all strains, even the negative control, oxidize all wells after a few hours of incubation. DNA–DNA hybridization values among the strains range from 55 to 100%. The major fatty acids are $C_{15:0}$ iso, $C_{15:0}$ iso 3-OH, $C_{15:0}$ anteiso, $C_{16:0}$ 3-OH, $C_{17:0}$ iso 3-OH, and $C_{17:1}$ iso ω9*c*.

   Source: commonly recovered from human urine specimens, sometimes in significant numbers, and from colonization of

TABLE 59. Features that differentiate the type strains of the species of the genus *Myroides*[a]

Characteristic	*M. odoratus*	*M. odoratimimus*	*M. pelagicus*
Cell length (μm)	11–12	3.5–4.0	0.5–1.0
Colony color	Yellow	Pale yellow	Yellow to orange
NaCl tolerance for growth (%, w/v)	0–5	0–6	0–9
pH range for growth	6–9	6–9	5–9
Catalase	+	+	Weak
Urease	+	+	−
Esterase (C4), esterase lipase (C8)	+	+	−
Utilization of:[b]			
L-Histidine	+	+	−
α-Hydroxybutyric acid	+	−	+
Succinamic acid	+	Weak	+
Urocanic acid	+	+	−
Nitrite reduction	+[c]	+[c]	−

[a]Data taken from Yoon et al. (2006a).

[b]Using the Biolog system.

[c]Nitrite is characteristically reduced, but not nitrate (Vancanneyt et al., 1996).

amputation sites. Some strains have been found in insect guts and freshwater fishes.

DNA G+C content (mol%): 35–38 (T_m).

Type strain: ATCC 4651, CCUG 7321, CIP 103105, DSM 2801, IFO (now NBRC) 14945, JCM 7458, LMG 1233, NCTC 11036.

Sequence accession no. (16S rRNA gene): M58777.

2. **Myroides odoratimimus** Vancanneyt, Segers, Torck, Hoste, Bernardet, Vandamme and Kersters 1996, 931[VP]

o.do.ra.ti'mi.mus. L. part. masc. adj. *odoratus* a specific epithet; L. n. *mimus* a mimic, an imitator; N.L. n. *odoratimimus* imitator of (*Myroides*) *odoratus*.

Characteristics are as given for the genus and as listed in Table 59, with the following additional features given by Vancanneyt et al. (1996). As determined by Biotype 100 galleries, growth occurs on L-aspartate, L-glutamate, and L-proline. Growth on L-malate, succinate, and fumarate differs depending on the strain studied. When the Biolog GN MicroPlate assay is used, all strains oxidize Tween 40, Tween 80, methylpyruvate, monomethylsuccinate, acetic acid, α-hydroxybutyric acid, α-ketovaleric acid, DL-lactic acid, propionic acid, succinic acid, bromosuccinic acid, succinamic acid, alaninamide, L-alanine, L-alanylglycine, L-asparagine, L-aspartic acid, L-glutamic acid, glycyl-L-aspartic acid, glycyl-L-glutamic acid, L-leucine, L-ornithine, L-proline, L-serine, L-threonine, inosine, uridine, and thymidine. Reactions differ among strains for oxidation of L-fucose, formic acid, β-hydroxybutyric acid, γ-hydroxybutyric acid, α-ketobutyric acid, α-ketoglutaric acid, L-histidine, L-phenylalanine, D-serine, urocanic acid, phenylethylamine, 2,3-butanediol, and glycerol.

The following substrates are not oxidized: α-cyclodextrin, dextrin, glycogen, *N*-acetyl-D-galactosamine, *N*-acetyl-D-glucosamine, adonitol, L-arabinose, D-arabitol, cellobiose, *meso*-erythritol, D-fructose, D-galactose, gentiobiose, α-D-glucose, *myo*-inositol, α-D-lactose, lactulose, maltose, D-mannitol, D-mannose, melibiose, methyl β-D-glucoside, D-psicose, raffinose, L-rhamnose, D-sorbitol, sucrose, trehalose, turanose, xylitol, *cis*-aconitic acid, citric acid, D-galactonic acid lactone, D-galacturonic acid, D-gluconic acid, D-glucosaminic acid, D-glucuronic acid, *p*-hydroxyphenylacetic acid, itaconic acid, malonic acid, quinic acid, D-saccharic acid, sebacic acid, glucuronamide, D-alanine, hydroxy-L-proline, L-pyroglutamic acid, DL-carnitine, γ-aminobutyric acid, putrescine, 2-aminoethanol, DL-α-glycerolphosphate, glucose 1-phosphate, and glucose 6-phosphate.

DNA–DNA hybridization values among the strains range from 80 to 100%. The major fatty acids are $C_{13:0}$ iso, $C_{15:0}$, $C_{15:0}$ iso, $C_{15:0}$ iso 3-OH, $C_{16:0}$ 3-OH, $C_{17:0}$ iso 3-OH, $C_{17:1}$ iso ω9*c*, and summed feature 4 ($C_{15:0}$ iso 2-OH and/or $C_{16:1}$ ω7*c*/*t*).

Source: human wounds, urine, human leg slough, pus, and a water sample.

DNA G+C content (mol%): 30–35 (T_m).

Type strain: CCUG 39352, CIP 105170, JCM 7460, LMG 4029, NCTC 11180.

Sequence accession no. (16S rRNA gene): AJ854059.

3. **Myroides pelagicus** Yoon, Maneerat, Kawai and Yokota 2006a, 1919[VP]

pe.la'gi.cus. L. masc. adj. *pelagicus* belonging to the sea.

Characteristics are as given for the genus and as listed in Table 59, with the following additional features given by Yoon et al. (2006a). Colonies on LB agar are circular, convex, and yellow to orange. Temperature range for growth is 10–37°C; no growth occurs at 4 or 45°C. The pH range for growth is 5–9. NaCl is not required for growth, but the organism can tolerate up to 9% NaCl (w/v). Oxidase-positive. Catalase is weakly positive.

The following reactions are positive: gelatin hydrolysis, Voges–Proskauer test, citrate utilization, alkaline phosphatase, leucine arylamidase, acid phosphatase, and naphthol-AS-BI-phosphohydrolase. The following substrates are utilized (Biolog MicroPlate system): glycogen, methyl pyruvate, monomethyl succinate, acetic acid, α-ketoglutaric acid, α-ketovaleric acid, DL-lactic acid, succinic acid, bromosuccinic acid, succinamic acid, L-alanine, L-alanyl glycine, L-asparagine, L-aspartic acid, L-glutamic acid, glycyl L-aspartic acid, glycyl L-glutamic acid, L-leucine, L-ornithine, L-proline, L-serine, L-threonine, inosine, uridine, and thymidine.

The following reactions are negative: ONPG, arginine dihydrolase, lysine decarboxylase, ornithine decarboxylase, urease, production of H_2S and indole, nitrate and nitrite reduction, esterase (C4), esterase lipase (C8), lipase (C4), valine arylamidase, cystine arylamidase, trypsin, chymotrypsin, α-galactosidase, β-galactosidase, α-glucosidase, β-glucosidase, and tryptophan deaminase. The following substrates are not utilized (Biolog MicroPlate system): cyclodextrin, dextrin, Tween 40, Tween 80, *N*-acetyl-D-glucosamine, *N*-acetyl-D-galactosamine, adonitol, L-arabinose, D-arabitol, cellobiose, i-erythritol, D-fructose, L-fucose, D-galactose, gentiobiose, α-D-glucose, *myo*-inositol, α-D-lactose, lactulose, maltose, D-mannitol, D-mannose, melibiose, methyl β-D-glucoside, D-psicose, raffinose, L-rhamnose,

D-sorbitol, sucrose, trehalose, furanose, xylitol, *cis*-aconitic acid, citric acid, formic acid, D-galactonic acid lactone, D-galacturonic acid, D-gluconic acid, D-glucosaminic acid, D-glucuronic acid, α-hydroxybutyric acid, γ-hydroxybutyric acid, *p*-hydroxyphenylacetic acid, itaconic acid, α-ketobutyric acid, malonic acid, propionic acid, quinic acid, D-saccharic acid, sebacic acid, glucuronamide, alaninamide, D-alanine, L-histidine, hydroxyl-L-proline, L-phenylalanine, L-pyroglutamic acid, D-serine, DL-carnitine, γ-aminobutyric acid, urocanic acid, phenylethylamine, putrescine, 2-aminoethanol, 2,3-butanediol, glycerol, DL-α-glycerol phosphate, glucose 1-phosphate, and D-glucose 6-phosphate.

The major quinone system is menaquinone-6. The major cellular fatty acids are $C_{15:0}$ iso, $C_{17:1}$ iso ω9*c*, and $C_{17:0}$ iso 3-OH.

Source: the type strain was isolated from seawater, off the coast of Thailand.

DNA G+C content (mol%): 33.6 (HPLC).

Type strain: SM1, IAM 15337, KCTC 12661.

Sequence accession no. (16S rRNA gene): AB176662.

Genus XXXVII. **Nonlabens** Lau, Tsoi, Li, Plakhotnikova, Dobretsov, Wong, Pawlik and Qian 2005b, 2281[VP]

The Editorial Board

Non.la'bens. L. adv. *non* not; L. part. adj. *labens* gliding; N.L. fem. n. (N.L. part. adj. used as a substantive) *Nonlabens* non-gliding.

Short rods. Nonmotile. Nonsporeforming. Gram-stain-negative. Strictly aerobic. MK-6 is the only respiratory quinone. **Colonies are orange. NaCl** (2.0–4.0%) **is required for growth.** Flexirubin-type pigments are not produced. Catalase- and oxidase-positive. **Acetoin is produced. Gelatin is hydrolyzed. No α- and β-glucosidase and β-galactosidase activities are present.** Isolated from a microbial mat in a subtropical estuary.

DNA G+C content (mol%): 33.6 (HPLC).

Type species: **Nonlabens tegetincola** Lau, Tsoi, Li, Plakhotnikova, Dobretsov, Wong, Pawlik and Qian 2005b, 2281[VP] emend. Lau, Tsoi, Li, Plakhotnikova, Dobretsov, Wu, Wong, Pawlik and Qian 2006, 184.

Enrichment and isolation procedures

Nonlabens tegetincola was isolated from a microbial mat covering a polystyrene surface retrieved from an estuarine mangrove habitat in the Bahamas. The isolation medium was a special marine agar (SMA) consisting of 5.0 g peptone and 3.0 g yeast extract dissolved in seawater. The medium was sterilized by filtration (0.22 µm pore diameter) and solidified with agar. Orange colonies, 2–4 mm in diameter, were selected after 48 h at 30°C.

Differentiation of the genus *Nonlabens* from related genera

Members of the genus *Nonlabens* differ from those of *Psychroflexus* in that they require NaCl, are able to hydrolyze gelatin, lack α- and β-glucosidase activities, and are unable to grow at temperatures below 12°C. They differ from members of the genus *Salegentibacter* in their requirement for NaCl, orange pigmentation (rather than yellow), production of acetoin, lack of β-galactosidase activity, and inability to grow at temperatures below 12°C.

Taxonomic comments

Phylogenetic analysis based on 16S rRNA gene sequences indicates that the nearest neighbors of *Nonlabens* are members of the genera *Psychroflexus* and *Salegentibacter*, with sequence similarity values of 88.5–90.7% (Lau et al., 2005b).

List of species of the genus *Nonlabens*

1. **Nonlabens tegetincola** Lau, Tsoi, Li, Plakhotnikova, Dobretsov, Wong, Pawlik and Qian 2005b, 2281[VP] emend. Lau, Tsoi, Li, Plakhotnikova, Dobretsov, Wu, Wong, Pawlik and Qian 2006, 184

te.get.in'col.a. L. n. *teges -etis* mat; L. n. *incola* an inhabitant; N.L. n. *tegetincola* mat-inhabitant, pertaining to its microbial mat habitat.

Characteristics are as described for the genus, with the following additional features. Colonies on SMA agar are orange, circular, 2.0–4.0 mm in diameter, and convex, with smooth surfaces and entire translucent margins. No diffusible pigment is produced. Grows at 12.0–44.0°C (optimum 28.0–36.0°C) and pH 5.0–10.0. Acetoin is produced, but not indole or H_2S. Nitrate is not reduced. DNA, gelatin, starch, and Tweens 20, 40 and 80 are hydrolyzed, but not agar, casein, cellulose, or chitin. The following enzymes are present: acid phosphatase, alkaline phosphatase, α-chymotrypsin, cystine arylamidase, leucine arylamidase, valine arylamidase, esterase (C4), esterase lipase (C8), lipase (C14), trypsin, and naphthol-AS-BI-phosphohydrolase. The following enzymes are not present: *N*-acetyl-β-glucosaminidase, arginine dihydrolase, α-fucosidase, α-galactosidase, β-galactosidase, α-glucosidase, β-glucosidase, β-glucuronidase, α-mannosidase, lysine decarboxylase, ornithine decarboxylase, tryptophan deaminase, and urease. Except for acid production from sucrose, no growth or acid production is observed from carbon sources included in the API 50CH, API 20E, and API 20NE test systems. Utilization of 57 carbon sources occurs when tested by the MicroLog 3 system; the compounds used include acetic acid, adonitol, L-alaninamide, L-alanine, L-alanylglycine, D-arabitol, L-aspartic acid, citric acid, α-cyclodextrin, dextrin, L-erythritol, D-fructose, L-fucose, D-galactonic acid lactone, D-galactose, D-galacturonic acid, D-gluconic acid, D-glucosaminic acid, α-D-glucose, glucuronamide, D-glucuronic acid, L-glutamic acid, glycogen, glycyl-L-aspartic acid, glycyl-L-glutamic acid, α-hydroxybutyric acid, *myo*-inositol, α-ketobutyric acid, α-ketoglutaric acid, α-ketovaleric acid, DL-lactic acid, α-D-lactose, lactulose, malonic acid, maltose, D-mannitol, D-mannose, melibiose, methyl β-D-glucoside, methylpyruvate, monomethyl succinate, L-ornithine, L-proline, propionic acid, quinic acid, D-psicose, raffinose, L-rhamnose, D-saccharic acid, L-serine, D-sorbitol, succinamic acid, sucrose, L-threonine, trehalose, turanose, and uridine. The predominant fatty acids are $C_{15:0}$ iso, $C_{15:0}$ iso 3-OH, $C_{16:0}$ iso, $C_{16:0}$ iso 3-OH, $C_{17:0}$ iso 3-OH, summed feature 3 (comprising $C_{15:0}$ iso 2-OH and/or $C_{16:1}$ ω7*c*), and an unknown fatty acid with a carbon chain length equivalent of 13.6 (altogether representing 78.3% of the total). Susceptible to ampicillin, chloramphenicol, penicillin, streptomycin, and tetracycline, but not to kanamycin.

Source: the type strain was isolated from a microbial mat in an estuarine mangrove habitat in the Bahamas.

DNA G+C content (mol%): 33.6 (HPLC).

Type strain: UST030701-324, JCM 12886, NRRL B-41136.

Sequence accession no. (16S rRNA gene): AY987349.

Genus XXXVIII. **Olleya** Mancuso Nichols, Bowman and Guezennec 2005, 1560[VP]

The Editorial Board

Ol.ley'a. N.L. fem. n. *Olleya* named in honor of June Olley, who has made significant contributions to the area of predictive microbiology.

Rods, approximately 0.3–0.5 × 2.0–2.5 µm. Motile by gliding. Endospores are not formed. Gram-stain-negative. Cell mass is orange/yellow. Flexirubin pigments are absent. Growth occurs at 4–30°C, but not at 37°C. Strictly aerobic. Chemoheterotrophic. Catalase-positive. **Oxidase-positive. NaCl or sea salts are required for growth. Acid is produced from carbohydrates.** β-Galactosidase activity is not present. **Agar and starch are not hydrolyzed.** The major fatty acids include $C_{15:1}$ iso ω10*c*, $C_{15:0}$ iso, $C_{15:0}$ iso β-OH, $C_{15:1}$ anteiso ω10*c*, $C_{15:0}$, and $C_{15:0}$ iso α-OH.

DNA G+C content (mol%): 49 (T_m).

Type species: **Olleya marilimosa** Mancuso Nichols, Bowman and Guezennec 2005, 1560[VP].

Enrichment and isolation procedures

Olleya marilimosa was isolated from material sampled from the cod end of a plankton net (20 µm pore size) trawled through the Southern Ocean; the sea temperature was 4°C and the salinity was 3.5%. Aliquots of 20 µl from the net were spread onto SNA (nutrient broth made with seawater plus 12 g/l agar) and SNA + 2% glucose and incubated at 2°C. After initial isolation, colonies were subcultured onto marine agar (MA)* and MA-glucose.

Differentiation of the genus Olleya from related genera

Members of the genus *Olleya* can be differentiated from those of the genus *Lacinutrix* by their gliding motility, oxidase activity, growth at 30°C, acid formation from carbohydrates, and higher DNA G+C content (49 mol% vs 37 mol%). They differ from members of the genus *Bizionia* by their gliding motility, acid production from carbohydrates, utilization of carbohydrates, inability to produce H_2S and hydrolyze casein, and higher DNA G+C content (49 mol% vs 38–45 mol%). They differ from members of the genus *Algibacter* by their aerobic instead of fermentative metabolism, lack of β-galactosidase, inability to hydrolyze agar and starch, and higher DNA G+C content (49 mol% vs 31–33 mol%).

Taxonomic comments

By 16S rRNA gene sequence analysis, the taxa most closely related to *Olleya marilimosa* are *Lacinutrix copepodicola, Bizionia saleffrena, Bizionia paragorgiae,* and *Algibacter lectus,* having sequence similarities of 94.0, 94.1, 94.2, and 94.5%, respectively (Mancuso Nichols et al., 2005).

List of species of the genus *Olleya*

1. **Olleya marilimosa** Mancuso Nichols, Bowman and Guezennec 2005, 1560[VP]

mar.i.lim.o'sa. L. gen. neut. n. *maris* of the sea; L. adj. *limosus -a -um* full of slime, slimy; N.L. fem. adj. *marilimosa* of the sea and slimy.

Characteristics are as described for the genus, with the following additional features. After 1 week at 20°C on MA, colonies are orange/yellow, translucent, 1–2 mm in diameter, circular, and convex, with an entire edge and a butyrous consistency. Colonies exhibit a spreading margin on dilute agar and an enhanced mucoid morphology when grown on MA-glucose. The pH range for growth is 5–9. Growth occurs between 1.2 and 5.2% NaCl, with optimal growth occurring in approximately 1.2–2.9% NaCl. Yeast extract or peptone is required for growth. Nitrate is not reduced to nitrite. Indole, DNase, β-galactosidase, lipase, urease, and acetoin (Voges–Proskauer reaction) are not produced. Degrades Tween 80, elastin, gelatin, and tyrosine, but not agar, starch, esculin, casein, cellulose, or xanthine. Citrate is utilized as a sole carbon source, but uric acid is not. Acid is produced from glucose. Glucose, maltose, and mannose are assimilated, but arabinose, mannitol, D-gluconate, capric acid, adipic acid, malate, and trisodium citrate are not. The following enzyme activities are present: *N*-acetyl-β-glucosaminidase, alanine arylamidase, alkaline phosphatase, arginine arylamidase, glutamyl glutamic acid arylamidase, glycine arylamidase, histidine arylamidase, leucine arylamidase, leucyl glycine arylamidase, phenylalanine arylamidase, serine arylamidase, and tyrosine arylamidase. The following enzyme activities are not present: α-arabinosidase, arginine

*MA contains: yeast extract (Oxoid L21), 1 g; bacteriological peptone (Oxoid L375), 5 g; artificial sea salts (Sigma S9883), 32 g; agar, 15 g; and distilled water, 1,000 ml. For MA-glucose, a glucose stock solution is autoclaved separately before being combined with MA to give a final concentration of 3% glucose.

dihydrolase, α-fucosidase, α-galactosidase, β-galactosidase, β-galactosidase-6-phosphate, α-glucosidase, β-glucosidase, β-glucuronidase, glutamic acid decarboxylase, lysine decarboxylase, ornithine decarboxylase, proline arylamidase, pyroglutamic acid arylamidase, and tryptophan deaminase. The exopolysaccharide produced by the type strain is composed of the following monosaccharides (percentage of total sugars): arabinose, 5; mannose, 74; galactose, 3; glucose, 8; glucuronic acid, 8; and *N*-acetylglucosamine, 1 (Nichols et al., 2005).

Source: the type strain was isolated from particulate material from the Southern Ocean at approximately 65° 32′ 06 S 143° 10′ 16 E.

DNA G+C content (mol%): 49 (T_m).

Type strain: CAM030, ACAM 1065, CIP 108537.

Sequence accession no. (16S rRNA gene): AY586527.

Genus XXXIX. **Ornithobacterium** Vandamme, Segers, Vancanneyt, Van Hove, Mutters, Hommez, Dewhirst, Paster, Kersters, Falsen, Devriese, Bisgaard, Hinz and Mannheim 1994b, 35^VP^

Hafez M. Hafez and Peter Vandamme

Or.ni.tho.bac.te′ri.um. Gr. n. *ornis -ithos* bird; L. neut. n. *bacterium* rod; N.L. neut. n. *Ornithobacterium* bird bacterium, because it was first isolated from birds.

Cells are pleomorphic rods, 0.2–0.4 × 1.3 µm. Gram-stain-negative. **Nonsporeforming. Nonmotile.** Metabolism is **chemoorganotrophic** and **mesophilic**. Growth occurs at 30–42°C. *Ornithobacterium* strains grow in various atmospheric conditions, but preferentially **microaerobically**. *Ornithobacterium* strains exhibit various levels of capnophilic metabolism. Colonies are not pigmented on common growth media. **Oxidase-positive. Catalase-negative. β-Galactosidase (ONPG)-positive.** Most strains are urease-positive. **Indole-negative. Nitrate is not reduced to nitrite.** This genus has been referred to as *Pasteurella*-like, *Kingella*-like, or pleomorphic Gram-stain-negative rod (PGNR); the name TAXON 28 has also been used. *Ornithobacterium* strains cause respiratory disease in avian species worldwide.

DNA G+C content (mol%): 37–39.

Type species: **Ornithobacterium rhinotracheale** Vandamme, Segers, Vancanneyt, Van Hove, Mutters, Hommez, Dewhirst, Paster, Kersters, Falsen, Devriese, Bisgaard, Hinz and Mannheim 1994b, 35^VP^.

Further descriptive information

The genus *Ornithobacterium* is a member of the family *Flavobacteriaceae* and represents a distinct line of descent within this family. Related bacteria are the bird pathogens *Riemerella anatipestifer* and *Coenonia anatina* (Vandamme et al., 1994a). The family includes *Flavobacterium*, the type genus, and the genera *Bergeyella*, *Capnocytophaga*, *Chryseobacterium*, *Ornithobacterium*, *Riemerella*, and *Weeksella* (Bernardet et al., 2002).

Morphology and physiology. Growth in broth media reveals more cellular pleomorphism than growth on agar surfaces. In addition, cells have a more variable length (0.6–5 µm).

Most strains grow aerobically, microaerobically, anaerobically, and in a CO_2-enriched atmosphere. Growth occurs at 30, 35, and 42°C; weak or no growth occurs at 24°C. Strains produce convex, circular, and grayish white to yellowish colonies, with an entire edge and smooth surface on blood agar. Colonial adherence, spreading, or corrosion is not observed. No hemolysis occurs on blood agar. Smooth, nonpigmented colonies develop within 2 d of incubation on rich peptone, peptone-blood, or chocolate agar at 36°C. No growth occurs on MacConkey agar, Endo agar, Drigalski agar, or Simmons' citrate medium. No growth factors are required.

Ornithobacterium is oxidase-positive; however, Ryll et al. (2002) were able to isolate and identify an oxidase-negative strain of *Ornithobacterium rhinotracheale* from turkey flocks in Germany. Alkaline phosphatase activity is present. Urease production may test positive either on Lautrop's non-proliferative test or after a prolonged incubation time (up to 7 d) and may yield variable results in both Christensen urea agar slants and urea broth. Nitrates are not reduced; some strains reduce nitrites, but no denitrification occurs. Arginine dihydrolase is usually present if strains are grown in Möller's medium (Difco) within 3 d of incubation at 36°C. Lysine and ornithine decarboxylases, phenylalanine deaminase, lecithinase, DNase, and gelatinase activities are absent. Hydrogen sulfide is not detected in Kligler's agar, triple-sugar iron agar, or SIM agar. Esculin is not hydrolyzed. Acetylmethylcarbinol is produced in the Voges–Proskauer test. The methyl red test is negative. Hyaluronidase and chondroitin sulfatase activities are present.

Ethanol is not oxidized. D-Glucose in OF medium is either weak and slowly oxidized or not detected. Most strains use D-galactose, D-glucose, D-mannose, lactose, and sucrose as carbon sources. D-Xylose, D-mannitol, D-sorbitol, and malonate are usually not used as carbon sources. In general, carbohydrates are better catabolized in media supplemented with 2% chicken serum. Wheat starch is acidified by most strains within 1–5 d of incubation at 36°C.

Due to the variable ability of *Ornithobacterium rhinotracheale* strains to grow in the liquid media that are normally used for identification purposes, the biochemical reactions of *Ornithobacterium rhinotracheale* may appear inconsistent. However, under optimal conditions, the biochemical properties of the organism are fairly constant.

Ecology and pathogenicity. Infection with *Ornithobacterium rhinotracheale* has been recognized in many countries worldwide and incriminated as a possible additional causative agent in respiratory disease complexes in poultry (Allymehr, 2006; Arns et al., 1998; Back et al., 1998a; Canal et al., 2005; Charlton et al., 1993; Dudouyt et al., 1995; El-Gohary, 1998; El-Sukhon et al., 2002; Hafez et al., 1993; Hafez and Friedrich, 1998; Hinz et al., 1994; Koga and Zavaleta, 2005; Naeem et al., 2003; Sakai et al., 2000; Soriano et al., 2002; Tanyi et al., 1995; Travers et al., 1996; Turan and Ak, 2002; van Beek et al., 1994; J. DuPreez, personal communication). Retrospective examination of culture collections has revealed that *Ornithobacterium rhinotracheale* had already been isolated from turkeys in 1981 and from rooks in 1983 in Germany. Also, examination of bacterial isolates collected before 1990 from chickens, partridge, and turkeys in the United Kingdom, Israel, Belgium, and France confirmed that they were strains of *Ornithobacterium rhinotracheale* (Hinz and Hafez, 1997; Vandamme et al., 1994b).

Typing. Although several investigators have described microbiological isolation and identification of *Ornithobacterium rhinotracheale*, there are few reports using molecular typing techniques. Amonsin et al. (1997) characterized 55 *Ornithobacterium rhinotracheale* isolates from eight countries of four continents by multilocus enzyme electrophoresis (MLEE), repetitive sequence based-PCR (rep-PCR), and 16S rRNA gene sequencing. Using MLEE, the *Ornithobacterium rhinotracheale* isolates could be discriminated into six electrophoretic types (ETs), of which only three were recovered from domesticated poultry. A total of 50 of the 55 isolates (90.9%) examined were assigned to one of two closely related clones (ET 1 and ET 2) that comprise the ET 1 complex. Furthermore, the data suggested a geographic component to the variation typical of a clonal population structure. In addition, the results suggested host specificity among clones of *Ornithobacterium rhinotracheale*. Virtually all the *Ornithobacterium rhinotracheale* isolates recovered from domesticated poultry were assigned to the ET 1 complex, whereas only a single non-ET 1 complex isolate was recovered from domesticated poultry populations. In contrast, none of the four isolates from rooks or guinea fowl were assigned to the common ET 1 complex. Moreover, the clustering obtained by analysis of rep-PCR banding patterns predicted that *Ornithobacterium rhinotracheale* clones infecting passeriform birds (rooks) are genetically distinct from clones infecting galliform birds (chickens, turkeys, and guinea fowl).

In another study, Leroy-Setrin et al. (1998) compared 23 strains of *Ornithobacterium rhinotracheale* which were isolated in France between 1994 and 1995 from 17 geographical regions using plasmid profiles, ribotyping, and random amplified polymorphic DNA (RAPD) analysis. All isolates were poorly discriminated by ribotyping although different enzymes were used. The RAPD method gave a good level of discrimination and demonstrated the genetic diversity of the *Ornithobacterium rhinotracheale* strains isolated among several species in France.

An endogenous plasmid has been reported in one strain of *Ornithobacterium rhinotracheale* (Back et al., 1997, cited by Jansen et al., 2004). Jansen et al. (2004) too were able to detect the plasmid pOR1 in only two isolates of the numerous *Ornithobacterium rhinotracheale* strains that were analyzed. These two isolates might have represented the same strain, since both isolates originated from the same geographical region and were the rare serotype K. The apparent rare occurrence of pOR1 suggests that the plasmid was introduced into *Ornithobacterium rhinotracheale* on a single recent occasion and has not spread in the population.

Within the species *Ornithobacterium rhinotracheale*, several serotypes seem to exist. Serological typing can be carried out using serological examination with known positive antisera in agar gel precipitation (AGP), or ELISA. In AGP, monospecific reactions can be obtained using heat-extracted antigen. However, some cross-reactions have been observed (Hafez and Sting, 1999; van Empel et al., 1997). Currently 18 serotypes, designated A to R, seem to exist. Most of the chicken isolates belong to serotype A. The turkey isolates are more heterogeneous and belong to serotypes A, B, and D. Serotype C could only be isolated from chickens and turkeys in South Africa and USA (van Empel and Hafez, 1999).

van Empel et al. (1998) (from van Empel, 1998) characterized 56 isolates belonging to different serotypes, independently isolated from bird species from various countries, by amplified fragment length polymorphism (AFLP) analysis. They were able to group *Ornithobacterium rhinotracheale* strains into five minor or three major clusters that showed some association with serotyping.

Popp and Hafez (2003) investigated several *Ornithobacterium rhinotracheale* isolates from turkeys and chickens originating from Germany, Hungary, and Spain by pulsed-field gel electrophoresis (PFGE) of genomic macro-restriction fragments using the enzyme *Sal*I. In general, most isolates showed differences in DNA fingerprints, although the overall profiles were very similar and a correlation between geographic origin, serotype, and DNA fingerprint pattern was observed. In contrast, Koga and Zavaleta (2005) recently investigated 25 *Ornithobacterium rhinotracheale* isolates from broilers, breeders, and layers from several geographic zones of Peru using PCR and rep-PCR techniques. All 25 isolates tested had a genetic profile similar to that of the *Ornithobacterium rhinotracheale* type strain, which was isolated from a turkey in the UK.

Finally, typing of *Ornithobacterium rhinotracheale* isolates has also been performed using the primers M13 (5′-TATGTAAAACGACGGCCAGT-3′) and ERIC 1R (5′-ATGTAAGCTCCTGGGGATTCAC-3′); variations were found between all tested serotypes (Hafez and Beyer, 1997; Hung and Alvarado, 2001).

Hemagglutination. Fitzgerald et al. (1998) tested 25 *Ornithobacterium rhinotracheale* isolates for their ability to hemagglutinate chicken red blood cells. Ten of the 25 isolates, which were sensitive to fosfomycin (MIC values below 128 μg/ml), were able to agglutinate red blood cells. The remaining 15 isolates were resistant to fosfomycin (MIC values above 128 μg/ml). Only five of these isolates had the ability to agglutinate red blood cells. The ability of certain isolates of *Ornithobacterium rhinotracheale* to agglutinate red blood cells raises the issue of differences in virulence. The ability to agglutinate red blood cells could be used as an alternative method for serotyping *Ornithobacterium rhinotracheale*. Similar results were obtained by testing some isolates from Mexico by Soriano et al. (2002). They found that all isolates tested showed hemagglutination activity with glutaraldehyde-fixed erythrocytes.

Antimicrobial susceptibility. Most published reports on antimicrobial susceptibility are based on disc diffusion tests and are difficult to compare. Using this method, all tested *Ornithobacterium rhinotracheale* isolates from Germany showed high susceptibility to amoxycillin, chloramphenicol, and chlortetracycline. Of these isolates, 90 and 36% were susceptible to erythromycin and furazolidone, respectively. In addition, only 6% of tested isolates were susceptible to enrofloxacin. None of isolates were susceptible to apramycin, neomycin, gentamicin, or sulfonamide/trimethoprim (Hafez et al., 1993), or to lincospectin/spectinomycin, cortimoxazal (sulfonamide) or lincospectin. The susceptibility to enrofloxacin seems to be location-related, since most turkey isolates from Germany and the Netherlands are resistant, whereas 98% of isolates from France (Dudouyt et al., 1995) and 71% of isolates from Belgium (Devriese et al., 1995) are susceptible to enrofloxacin.

Nagaraja et al. (1998) reported on the antimicrobial susceptibility of *Ornithobacterium rhinotracheale* isolates from the USA. All tested isolates were susceptible to tylosin, ampicillin, penicillin, erythromycin, and spectinomycin. Of 68 samples, 54 were susceptible to tetracycline, neomycin, and sarafloxacin and a small number of isolates were susceptible to streptomycin, sulfa-trimethoprim, and gentamicin. These isolates differed significantly from German isolates in their pattern of susceptibility for at least two antibiotics, erythromycin and sarafloxacin. In Canada, *Ornithobacterium rhinotracheale* could be isolated from enrofloxacin-treated birds in mono-cultures (Joubert et al., 1999). van Veen et al. (2001) tested *Ornithobacterium rhinotracheale* isolates collected in the Netherlands between 1996 and 1999 in the agar gel diffusion test for their sensitivity to amoxycillin, tetracycline, enrofloxacin, and trimethoprim/sulfonamide. The percentages of strains susceptible to amoxycillin and tetracycline decreased in successive years from approximately 62 to 14% and four strains were resistant to enrofloxacin or sulfonamide/trimethoprim. Twelve multiresistant strains were tested against seven alternative antibiotics; they were resistant to all antibiotics tested except clavulanic acid-potentiated amoxycillin. Varga et al. (2001) examined *Ornithobacterium rhinotracheale* isolates from Hungary using MIC determinations. Among the 16 drugs examined, penicillin G, ampicillin, ceftazidim (MICs from ≤0.06 to 0.12 μg/ml), erythromycin, tylosin, tilmicosin (with some exceptions, MICs were from ≤0.06 to 1 μg/ml) and tiamulin (MICs from ≤0.06 to 2 μg/ml) were the most effective. Lincomycin, oxytetracycline, and enrofloxacin also gave good inhibition, but at a higher concentration with most strains (MICs ranged in most cases from 2 to 8 μg/ml). The other antibiotics inhibited the growth of *Ornithobacterium rhinotracheale* only in very high concentrations (colistin) or not at all (apramycin, spectinomycin, polymyxin B). Popp and Hafez (2002) investigated the susceptibility profiles of *Ornithobacterium rhinotracheale* isolates from several countries using the MIC method. The obtained results showed that 84–88% of the isolates were susceptible to amoxycillin, cetiofur, and tiamulin. Of the isolates, 45% were susceptible to chlortetracycline and 30% to tetracycline and penicillin, whereas 25% were susceptible to tiamulin. In addition, only 6–12% of tested isolates were susceptible to enrofloxacin.

Malik et al. (2003) examined in vitro antibiotic resistance profiles of 125 isolates of *Ornithobacterium rhinotracheale* strains isolated from turkeys in Minnesota in 1996–2002. The majority of isolates were sensitive to clindamycin, erythromycin, spectinomycin, and ampicillin. Resistance against sulfachloropyridiazine decreased from 1996 to 2002, but an increase in resistance was seen against gentamicin, ampicillin, sulfa-trimethoprim, and tetracycline. The resistance against penicillin remained constant from year to year. Soriano et al. (2003) determined the MIC of ten antimicrobial drugs for Mexican isolates and found a marked resistance trend. The susceptibility of *Ornithobacterium rhinotracheale* to amoxycillin, enrofloxacin, and oxytetracycline was variable. However, consistently higher MIC values were obtained for gentamicin, fosfomycin, trimethoprim, sulfamethazine, sulfamerazine, sulfaquinoxaline, and sulfachloropyridazine.

Host range. Strains have been isolated from respiratory tracts of turkey, chicken, chukar partridge, duck, goose, guinea fowl, gull, ostrich, partridge, pheasant, pigeon, quail, and rook, mostly associated with infections such as tracheitis, pericarditis, sinusitis, airsacculitis, and pneumonia (Charlton et al., 1993; Hafez et al., 1993; Leroy-Setrin et al., 1998; van Empel et al., 1997; Vandamme et al., 1994b).

Transmission. *Ornithobacterium rhinotracheale* is transmitted horizontally by direct and indirect contact. *Ornithobacterium rhinotracheale* infection appears to have become endemic and can affect every new restocking even in previously cleaned and disinfected poultry houses, especially in areas with intensive poultry production, as well as in multiple age farms. Such infection can occur despite the fact that *Ornithobacterium rhinotracheale* is highly sensitive to different chemical disinfectants and preparations based on different organic acids (formic and glyoxyl acids) and that preparations containing different aldehydes can inactivate *Ornithobacterium rhinotracheale* in vitro at concentrations of 0.5% within 15 min (Hafez and Schulze, 2003). Studies of the ability of *Ornithobacterium rhinotracheale* to remain viable in poultry litter indicate that *Ornithobacterium rhinotracheale* can survive for 1 d at 37°C, 6 d at 22°C, 40 d at 4°C, and at least 150 d at −12°C; however, it does not survive for 24 h at 42°C. Survival at the lower temperatures may be associated with the higher incidence of *Ornithobacterium rhinotracheale* infection in poultry during winter months (Lopes et al., 2002b). Surveillance of exposure to *Ornithobacterium rhinotracheale* infection in the field has shown that the prevalence of infection is higher during winter months. In addition, Amonsin et al. (1997) suggested that *Ornithobacterium rhinotracheale* might be introduced to domesticated poultry flocks from wild bird populations.

Vertical transmission is suspected based on some reports of the isolation of *Ornithobacterium rhinotracheale* at a very low incidence from reproductive organs and hatching eggs, infertile eggs, and dead embryos (Back et al., 1998b; El-Gohary, 1998; Tanyi et al., 1995). This might be the reason for the worldwide prevalence of the disease (Tanyi et al., 1995). On the other hand, Varga et al. (2001) found that *Ornithobacterium rhinotracheale* did not survive on eggshell at 37°C for more than 24 h, whereas upon inoculation into embryonated chicken eggs, it killed embryos by the 9th day and, from the 14th day post-inoculation, no *Ornithobacterium rhinotracheale* could be cultured from the eggs at all. This suggested that *Ornithobacterium rhinotracheale* was not transmitted via eggs during hatching. van Veen et al. (2004) observed that specific pathogen-free broiler chickens that were placed in hatching incubators at a commercial turkey hatchery during hatch showed respiratory tract lesions at post-mortem examination that were positive for *Ornithobacterium rhinotracheale* by bacteriological and immunohistological examination.

Pathogenicity. The severity of clinical signs, duration of the disease, and mortality are extremely variable and are influenced by many environmental factors such as poor management, inadequate ventilation, high stocking density, poor litter conditions, poor hygiene, high ammonia level, concurrent diseases, and the type of secondary infection. Further information concerning the clinical aspects of *Ornithobacterium* infections can be found in the following references: DeRosa et al., 1997; Droual and Chin, 1997; Hafez, 1998, 2002; Heeder et al., 2001; Hafez et al., 1993; Jirjis et al., 2004; Marien et al., 2005; Odor et al., 1997; Ryll et al., 1996; Sprenger et al., 1998; Szalay et al., 2002; Travers, 1996; van Beek et al., 1994; Vandekerchove et al., 2004; van Empel et al., 1996; van Empel and Hafez, 1999; Van Loock et al., 2005; van Veen et al., 2000b; Zorman-Rojs et al., 2000.

Determination of pathogenicity by the embryo lethality test has been done by Hafez and Popp (2003) using specific pathogen-free hatching eggs as described for *Escherichia coli* by Wooley et al. (2000).

Several attempts to combat infection by using vaccines have been carried out with various results (Bisschop et al., 2004; Bock et al., 1997; Cauwerts et al., 2002; Hafez and Sting, 1999; Lopes et al., 2002a; Schuijffel et al., 2005, 2006; van Empel and van den Bosch, 1998).

Isolation and detection procedures

Isolation. Samples for bacterial culture should be collected in the early stages of the disease. Ornithobacterium rhinotracheale can usually be isolated from trachea, tracheal swabs, lungs, and air sacs. Culture of heart, blood, and liver tissue under field conditions has yielded negative results. However, the bacteria could be isolated from these organs, as well as from joints, brains, ovary, and oviduct, after experimental infections. Primary isolation and initial in vitro growth should be performed on 5–10% sheep blood agar plates incubated under microaerobic conditions at 37°C for 24–48 h. These conditions provide the highest isolation rates and optimal growth. After several subcultivations, some strains may be adapted to grow under aerobic conditions, although growth is always significantly better under microaerobic conditions (van Empel and Hafez, 1999).

Serological detection of *Ornithobacterium*. Serological examination can be carried out using ELISA tests or other methods. The following references should be consulted for details: Bock et al. (1997), Back et al. (1998b), Erganis et al. (2002), Hafez and Sting (1996), Hafez et al. (2000), Lopes et al. (2000), and van Empel et al. (1996, 1997). The advantage of serological tests over bacteriological examination is that antibodies persist for several weeks after infection, whereas bacterial shedding is briefer. However, *Ornithobacterium rhinotracheale* excretion and antibody response may also be affected by a number of factors such as antibiotic therapy and vaccination. The influence of antibiotic therapy on the serological response to *Ornithobacterium rhinotracheale* remains unclear. Popp and Hafez (2002) investigated the effect of drug therapy using amoxycillin on antibody kinetics after experimental infection. The results showed that the immediate treatment did not influence the antibody response, whereas the treatment that started at the seventh day post-infection resulted in lower antibody response compared to infected controls.

Other detection methods. In field trials using a sensitive immunohistochemical staining method, *Ornithobacterium rhinotracheale* was the cause of 70% of the cases of respiratory symptoms in broiler chickens, whereas through bacteriology and/or serology only 30% of the cases could be connected to *Ornithobacterium rhinotracheale* (van Empel et al., 1999; van Veen et al., 2000a).

Finally, molecular detection of *Ornithobacterium rhinotracheale* infection can be performed through a specific PCR using the primer combination OR16S-F1 (5′-GAGAATTAATTTACCGATTAAG) and OR16S-R1 (5′-TTCGCTTGGTCTCCGAAGAT). This combination amplifies a 784 bp fragment on the 16S rRNA gene of *Ornithobacterium rhinotracheale*, but not that of other closely related bacteria with which *Ornithobacterium rhinotracheale* can be confused (Hung and Alvarado, 2001; van Empel et al., 1998).

Maintenance procedures

Cultures may be stored for many years by lyophilization, freezing at −80°C, or in liquid nitrogen. Cryoprotective agents such as 10% glycerol or dimethylsulfoxide should be added to cultures before freezing. After a storage period of 6 weeks at 6–8°C, blood agar cultures are subcultivable on blood agar.

Differentiation of the genus *Ornithobacterium* and other bird pathogens of the family *Flavobacteriaceae*

Ornithobacterium rhinotracheale, like its phylogenetic neighbors *Riemerella columbina* and *Coenonia anatina*, was initially recognized during the course of long-term studies on the etiology of respiratory tract infections in birds as phenotypically unusual isolates. Table 60 lists the differential diagnostic characteristics that can be used to distinguish *Ornithobacterium rhinotracheale* from other bird pathogens of the family *Flavobacteriaceae*, namely, *Riemerella anatipestifer*, *Riemerella columbina*, *Coenonia anatina*, and *Chryseobacterium meningosepticum*. Species differentiation can also be achieved through whole-cell protein electrophoresis and whole-cell fatty acid analyses (Segers et al., 1993a; Vancanneyt et al., 1999; Vandamme et al., 1994b, 1999). Analysis of the outer-membrane-proteins of 56 *Ornithobacterium rhinotracheale* strains from various countries and belonging to different serotypes confirmed the high protein electrophoretic homogeneity within this species (van Empel et al., 1998).

List of species of the genus *Ornithobacterium*

1. **Ornithobacterium rhinotracheale** Vandamme, Segers, Vancanneyt, Van Hove, Mutters, Hommez, Dewhirst, Paster, Kersters, Falsen, Devriese, Bisgaard, Hinz and Mannheim 1994b, 35[VP]

rhi.no.tra.che.a′le. Gr. n. *rhis rhinos* nose, nostril; L. n. *trachia* windpipe; L. neut. adj. suff. *-ale* pertaining to; N.L. neut. adj. *rhinotracheale* relating to the nostrils and windpipe, because the organism was first isolated there.

The description is as given for the genus, with the following additional information. The following enzyme activities are always present: alkaline and acid phosphatase, ester lipase C8, leucine arylamidase, phosphoamidase, α-glucosidase, β-glucosaminidase, phosphodiesterase, alanine arylamidase, glycine arylamidase, lysine arylamidase, proline arylamidase, α-glutamyl-α-glutamic acid arylamidase, glycyl-phenylalanine arylamidase, phenylalanyl-arginine arylamidase, prolyl-arginine arylamidase, seryl-methionine arylamidase, 2-glycyl-glycyl-arginine arylamidase, and alanyl-phenylalanyl-prolyl-alanine arylamidase. All strains exhibit strong or weak α- and β-galactosidase, esterase C4, valine and cysteine arylamidase, and

TABLE 60. Main differential characteristics of *Ornithobacterium rhinotracheale* and other bird pathogens of the family *Flavobacteriaceae*[a]

Characteristic	*Ornithobacterium rhinotracheale*	*Chryseobacterium meningosepticum*	*Coenonia anatina*	*Riemerella anatipestifer*	*Riemerella columbina*
Aerobic growth on blood agar	–	+	–	+	+
Growth on MacConkey agar	–	+	–	–	–
Colony pigmentation	–	+	–	–	+
Catalase	–	+	+	+	+
Urease	d	d	–	d	d
Indole production	–	d	–	d	–
Esculin hydrolysis (β-D-glucosidase)[b]	–	+	+	–	+
Gelatinase	–	+	–	+	+
Hyaluronidase[c]	+	nd	+	–	–
β-D-Galactosidase[b]	+	–	+	–	–
α-D-Glucosidase[d]	+	+	+	+	+
N-Acetyl-β-glucosaminidase[c]	+	–	+	–	–
α-L-Fucosidase[d]	–	–	+	–	–
Utilization of carbon sources for growth[b]	–	+	–	–	–
Host spectrum	Birds	Humans and birds	Anatine birds	Birds and pigs	Pigeon

[a]Symbols: +, >85% positive; d, different strains give different reactions (16–84% positive); –, 0–15% positive; nd, not determined.

[b]API 20NE test results.

[c]Identical to the chondroitin sulfatase test results.

[d]API ZYM test results.

trypsin activities. The following activities are always absent: β-glucuronidase, β-glucosidase, α-mannosidase, α-fucosidase, lipase C14, phenylalanine deaminase, hippurate hydrolysis, γ-glutamyl arylamidase, and phospholipase. No acid is produced from D-glucosaminic acid, D-saccharic acid, L-fucose, D-mannitol, D-sorbitol, trehalose, or D-xylose. Lyxose, ribose, glucose, galactose, and mannose are the principal carbohydrate components in whole-cell extracts; altrose, sorbose, and heptose do not occur. The branched fatty acids $C_{15:0}$ iso, $C_{15:0}$ iso 3-OH, and $C_{17:0}$ iso 3-OH are the major fatty acid components. Additional fatty acids present in small quantities in all strains are two unidentified fatty acids with equivalent chain-length values of 13.566 and 16.580, $C_{16:0}$, $C_{17:0}$ iso, and $C_{16:0}$ iso 3-OH.

Source: the type strain was isolated from a turkey in the UK.

DNA G+C content (mol%): 37–39 (T_m).

Type strain: ATCC 51463, CCUG 23171, CIP 104009, LMG 9086, MCCM 01774.

Sequence accession nos (16S rRNA gene): L19156 and U87101 (type strain). The accession numbers of additional reference strains included in a polyphasic taxonomic study (Vandamme et al., 1994b) are U87102 (LMG 11553), U87103 (LMG 15870), U87104 (LMG 11554), and U87105 (LMG 14578).

Genus XL. **Persicivirga** O'Sullivan, Rinna, Humphreys, Weightman and Fry 2006, 177[VP]

The Editorial Board

Per.si.ci.vir'ga. L. neut. n. *persicum* peach; L. fem. n. *virga* rod; N.L. fem. n. *Persicivirga* peach-colored rod.

Irregularly straight to curved rods, 0.5–0.6 × 2.2–7.5 µm. No visible flagella. **No gliding motility.** Gram-stain-negative. **Growth occurs at 4–20°C;** no growth occurs at 30°C. Colonies are peach–orange colored. **Flexirubin pigments are present.** Strictly aerobic. **Oxidase- and catalase-negative. Growth requires Na^+ions or natural seawater. Nitrate is not reduced. Do not utilize carbohydrates for growth.** The most abundant fatty acids are saturated branched-chain, unsaturated branched-chain and hydroxy fatty acids; $C_{15:0}$ iso and $C_{15:0}$ anteiso are the most abundant individual fatty acids. Isolated from coastal seawater.

DNA G+C content (mol%): 34.7.

Type species: **Persicivirga xylanidelens** O'Sullivan, Rinna, Humphreys, Weightman and Fry 2006, 178[VP].

Enrichment and isolation procedures

Coastal seawater samples were serially diluted in aged seawater from a circulating marine aquarium that had been filter-sterilized (0.2 µm pore diameter) after dilution to 36‰ salinity with distilled water (O'Sullivan et al., 2006). Plating of dilutions was done on three different agar media that were made with this sterilized seawater: plate count agar (PCA; Oxoid) plus 50 µg/ml kanamycin; casein-yeast-tryptone (Holmes, 1992); and R2A (Difco). Plates were incubated for 10 d at 20°C. Suspected colonies were subcultured twice and then purified on PCA containing aged seawater.

Viability can be maintained at 4°C on a low-nutrient medium (SAP2 agar; Holmes, 1992). Strains can also be preserved at –80°C in SP5 liquid medium (Holmes, 1992) containing 30% glycerol.

Differentiation of the genus *Persicivirga* from related genera

Unlike *Cellulophaga* strains, *Persicivirga xylanidelens* lacks gliding motility, is catalase-negative, does not hydrolyze agar or starch, does not grow at 30°C, and does not produce acid from glucose. In contrast to *Salegentibacter* strains, *Persicivirga xylanidelens* requires NaCl, but does not grow with 15% NaCl, does not produce acid from glucose, does not reduce nitrate, does not grow at 30°C, does not hydrolyze esculin or starch, and is catalase- and oxidase-negative. Unlike *Psychroflexus* strains, it is catalase- and oxidase-negative, grows at 20°C, does not hydrolyze starch, and does not produce acid from glucose. Unlike *Gillisia* strains, *Persicivirga xylanidelens* is catalase- and oxidase-negative, grows at 4°C, and hydrolyzes Tween 80 but not esculin. *Persicivirga xylanidelens* differs from *Mesonia* strains by failing to grow at 30°C, being catalase- and oxidase-negative, and failing to grow in the presence of 15% NaCl.

Taxonomic comments

O'Sullivan et al. (2006) indicated that, based on 16S rRNA gene sequence analyses, the type strain of *Persicivirga xylanidelens* was related to members of the large marine *Flavobacteriaceae* branch and formed a low bootstrap grouping with members of the genus *Cellulophaga*. Its position varied according to other sequences used in the analysis and, in some trees, it was more closely associated with the genera *Psychroflexus*, *Gillisia*, *Mesonia*, and *Salegentibacter*, or formed a separate branch on its own. However, the strain exhibited <90% 16S rRNA gene sequence similarity to any recognized species within the *Flavobacteriaceae*.

List of species of the genus *Persicivirga*

1. **Persicivirga xylanidelens** O'Sullivan, Rinna, Humphreys, Weightman and Fry 2006, 178[VP]

xy.lan.i.del'ens. N.L. n. *xylanum* xylan; L. part. adj. *delens* destroying; N.L. part. adj. *xylanidelens* xylan-destroying.

Characteristics are as described for the genus, with the following additional features. Colonies on PCA plus artificial seawater for 1 week at 20°C are 1–2 mm in diameter, peach-orange, opaque, smooth with an entire edge, and have a creamy consistency. Cell mass is orange-pigmented and flexirubin pigments are synthesized. Growth occurs at 4 and 20°C, but not at 30 or 37°C. Growth occurs in the presence of 5% NaCl, but not in 10% NaCl. Growth occurs on marine agar 2216, but not on tryptone soy agar, nutrient agar, MacConkey agar, cetrimide agar, or DNase agar (all containing 2.5% NaCl). Xylanase activity is present, but urease activity is absent. Tween 80 and gelatin are hydrolyzed, but agar, arginine, esculin, and starch are not hydrolyzed. Nitrate and nitrite are not reduced. Indole is not produced. Acid is not produced from glucose. Resistant to chloramphenicol, penicillin G, streptomycin, kanamycin, ampicillin, and tetracycline, but sensitive to rifampin. Growth is not detected for any substrates included in the API 20NE and Biolog GN2 commercial arrays.

Source: the type strain was isolated from coastal sea water, Hope Cove, near Plymouth, UK.

DNA G+C content (mol%): 34.7 (HPLC).

Type strain: SW256, DSM 16809, NCIMB 14027.

Sequence accession no. (16S rRNA gene): AF493688.

Genus XLI. **Polaribacter** Gosink, Woese and Staley 1998, 231[VP]

James T. Staley

Po.lar.i.bac'ter. N.L. adj. *polaris* pertaining to the geographic poles; N.L. masc. n. *bacter* from Gr. n. *baktron* rod or staff; N.L. masc. n. *Polaribacter* rod-shaped bacteria from polar habitats.

Rods, curved rods, or filaments. Cell size varies between 0.8 and 48 μm in length and 0.25 and 1.6 μm in diameter depending on species, growth medium, temperature, and physiological state of the culture. Nonmotile; some cells may have flagella, but motility has not been observed. Coccoid bodies often seen in ageing cultures. **Gas vesicles are produced by some species.** Gram-stain-negative. Heterotrophic. **Aerobic. Psychrophilic or mesophilic.** Organisms grow well in marine media or media that have been supplemented with NaCl. Organisms produce acid from a variety of carbohydrates. Starch is hydrolyzed. Cells can grow on yeast extract. Colonies produce yellow, orange, salmon, or pink non-diffusible pigments that are not flexirubins. **Some strains produce rhodopsin pigments** that stimulate growth in presence of light. Some strains have been isolated from polar marine and sea ice environments and grow at temperatures of 10°C or lower. **All strains have been isolated from marine habitats.**

DNA G+C content (mol%): 30–33.

Type species: **Polaribacter filamentus** Gosink, Woese and Staley 1998, 232[VP].

Further descriptive information

The genus *Polaribacter* was named because the initial strains of this genus that were isolated from polar sea ice communities formed a novel clade in the phylum *Bacteroidetes* based on 16S rRNA gene sequence analyses. Similarities were noted between the three novel polar species and a previously described species, *Flectobacillus glomeratus*. Therefore, the phylogenetic, phenotypic, and genotypic properties of this latter species were compared to those of members of the new genus *Polaribacter* (Gosink et al., 1998). In particular, DNA–DNA hybridization values between *Flectobacillus glomeratus* and the three polar species of *Polaribacter* were from 14 to 31%, which is within the expected range of values for members of a genus. Based on this information, *Flectobacillus glomeratus* was reclassified as an additional member of the genus *Polaribacter* and renamed *Polaribacter glomeratus*. Two novel species have been reported recently from marine habitats in temperate zone waters.

Recent reports indicate that at least some *Polaribacter* species produce bacterial rhodopsins. These have been shown to stimulate growth in the presence of light (Gómez-Consarnau et al., 2007).

Enrichment, isolation and maintenance

These bacteria were first isolated from the sea ice microbial community using Ordal's seawater cytophaga (SWCm) agar (Irgens et al., 1989) with succinate as a carbon source. However, strains also grow well on marine agar 2216 (Difco). The organisms are maintained best by freezing cultures in growth medium with 15% glycerol at −80°C or in liquid nitrogen. Many strains do not survive well using lyophilization.

Differentiation of the genus *Polaribacter* from related genera

The closest relatives of members of the genus *Polaribacter* are found in the genus *Tenacibaculum*. However, *Polaribacter* species differ from *Tenacibaculum* species by 16S rRNA gene sequence similarities which range, for example, from 92.4–93.1% for *Tenacibaculum lutimaris* (Yoon et al., 2005c). *Tenacibaculum* species move by gliding motility and in this regard also differ from *Polaribacter* species in which motility has not yet been demonstrated. *Tenacibaculum maritimum* is a fish pathogen that lives in association with black and red sea bream. Other members of the genus *Tenacibaculum* have been isolated from various other marine environments including pelagic marine habitats, tidal flats, macroalgae, sponges, and halibut eggs.

Differentiation of the species of the genus *Polaribacter*

The species can be differentiated on the basis of nutritional characteristics including carbon source utilization, growth on various media, biochemical properties, and temperature range for growth (Table 61). Fatty acid composition differs among those strains that have been tested (Table 62).

List of species of the genus *Polaribacter*

1. **Polaribacter filamentus** Gosink, Woese and Staley 1998, 232[VP]

fil.a.men'tus. L. n. *filum* thread; N.L. masc. adj. *filamentus* threadlike, filamentous.

Characteristics are as described for the genus and as listed in Tables 61 and 62, with the following additional features. Colonies are orange- to salmon-colored, circular, flat to convex, entire, smooth, opaque, and butyrous. Organisms grow from 4 to 19°C, but not at 21°C. L-Glutamate and glycerol are used as carbon sources. Gelatin is hydrolyzed. Acid is produced from DL-arabinose, rhamnose, D-galactose, D-glucose, mannose, and glycerol. Principal fatty acids include $C_{15:0}$ iso 3-OH (22%), $C_{15:0}$ iso (22%), and $C_{15:1}$ iso $_{G}$ (12%) when grown on SWCm agar (Irgens et al., 1989) at 10°C.

Source: isolated from surface seawater in pack ice, 350 km north of Deadhorse, Alaska.

DNA G+C content (mol%): 32 ± 1 (T_m).

Type strain: 215, ATCC 700397, CIP 106479.

Sequence accession no. (16S rRNA gene): U73726.

2. **Polaribacter butkevichii** Nedashkovskaya, Kim, Lysenko, Kalinovskaya, Mikhailov, Kim and Bae 2006b, 1459[VP] (Effective publication: Nedashkovskaya, Kim, Lysenko, Kalinovskaya, Mikhailov, Kim and Bae 2005g, 411.)

but.ke.vi'chi.i. N.L. gen. masc. n. *butkevichii* of Butkevich, named in honor of V.S. Butkevich, a Russian marine microbiologist.

Characteristics are as described for the genus and as listed in Tables 61 and 62, with the following additional features. Colonies are orange, circular, convex, viscous entire, smooth, translucent, and butyrous. Sodium ions are required for growth. Casein, gelatin, starch, DNA, Tween 40, and Tween 80 are hydrolyzed. Acid is produced from D-glucose, D-galactose, D-maltose, DL-xylose, and glycerol. A yellow non-diffusible pigment is produced. Growth occurs at 4–32°C.

Source: the type strain was isolated from seawater from Amursky Bay, Gulf of Peter the Great, Sea of Japan, Russia.

DNA G+C content (mol%): 32.4 (T_m).

Type strain: KMM 3938, CCUG 48005, KCTC 12100.

Sequence accession no. (16S rRNA gene): AY189722.

3. **Polaribacter dokdonensis** Yoon, Kang and Oh 2006e, 1252[VP]

dok.do.nen'sis. N.L. masc. adj. *dokdonensis* of or belonging to Dokdo, an island of Korea where the type strain was isolated.

Characteristics are as described for the genus and as listed in Tables 61 and 62, with the following additional features. Colonies are orange, circular, convex, and smooth. Optimal growth occurs at 25–28°C. NaCl is required for growth. Tweens 20, 40, 60, and 80 are hydrolyzed, but casein, xanthine, and L-tyrosine are not. Rhodopsin pigments are produced.

Source: the type strain was isolated from seawater off Dokdo in the East Sea of Korea (also known as the Sea of Japan).

DNA G+C content (mol%): 30 (HPLC).

Type strain: DSW-5, DSM 17204, KCTC 12392.

Sequence accession no. (16S rRNA gene): DQ004686.

4. **Polaribacter franzmannii** Gosink, Woese and Staley 1998, 233[VP]

franz.man'ni.i. N.L. gen. masc. n. *franzmannii* of Franzmann, named in honor of Peter D. Franzmann, an Australian microbiologist and polar researcher.

Characteristics are as described for the genus and as listed in Tables 61 and 62, with the following additional features. Colonies are orange, circular, flat to convex, entire, smooth, opaque, and viscous. Cells can grow on glycerol and *N*-acetylglucosamine. Acid is produced from D-fructose, D-galactose, D-glucose, mannose, lactose, maltose, sucrose, and trehalose. Esculin and gelatin are hydrolyzed. Cells can grow at temperatures of <4 to 10°C, but not at 15°C. Principal fatty acids include $C_{15:0}$ iso 3-OH (17%), $C_{15:1}$ iso G (11%), and 9% each of $C_{15:0}$ iso, $C_{15:1}$ ω6*c*, and $C_{16:1}$ ω7*c* when grown on SWCm agar (Irgens et al., 1989) at 10°C.

Source: isolated from fast sea ice in McMurdo Sound, Antarctica.

DNA G+C content (mol%): 32 ± 1 (T_m).

Type strain: 301, ATCC 700399, CIP 106480.

Sequence accession no. (16S rRNA gene): U14586.

TABLE 61. Nutritional, physiological, and genotypic properties of *Polaribacter* species[a,b]

Characteristic	*P. filamentus*	*P. butkevichii*	*P. dokdonensis*	*P. franzmannii*	*P. glomeratus*	*P. irgensii*
Cell morphology	Filamentous	Rods	Rods	Irregular rods	Curved rods	Filamentous
Cell diameter (μm)	0.5–1.2	0.4–0.6	0.4–0.6	0.8–1.6	0.4–0.6	0.25–0.5
Cell length (μm)	1.6–32	0.6–2.0	0.8–5.0	4–16	0.6–2.0	0.8–48
Coil formation	–	+	–	–	+	–
Coccoid cells in ageing cultures	+	–	+	+	–	+
Motility	–	–	–	–	–	–
Gas vesicles	+	–	–	+	–	+
Utilization as a carbon source:						
Yeast extract	+	+	na	+	+	+
Casamino acids	+	+	–	+	+	+
DL-Malate	–	+	–	–	–	+
Sucrose	–	+	–	–	+	–
L-Glutamate	+	+	–	–	+	–
Succinate	–	+	na	–	–	–
Glycerol	+	+	–	+	–	–
N-Acetylglucosamine	–	–	na	+	–	–
Growth on various solid media:						
Nutrient agar + 2.5% NaCl	+	na	–	–	–	–
Trypticase soy agar	–	+	–	–	+	–
Marine agar 2216 (Difco)	+	–[c]	na	+	+	+
Modified seawater cytophaga medium	+	na	na	+	+	+
Acid production from:						
DL-Arabinose	+	–	–	–	–	–
D-Ribose	ng	na	na	ng	–	–
D-Xylose	–	+	–	ng	–[c]	–
L-Rhamnose	+	na	–	ng	+[c]	ng
D-Fructose	–	na	+	+	–	ng
D-Glucose, D-mannose	+	+	na	+	+[c]	+
Macromolecule hydrolysis:						
Esculin	+	na	–	+	–	–
Starch	w	+	+	w	+	w
Gelatin	w	+	–	+	+	–
Biochemical tests:						
Urease	–	–	na	–	–	–
β-Galactosidase	–	+	na	+	–	–
Oxidase	–	+	na	+	+	w
Catalase	+	w	na	+	w	+
Nitrate reduction	–	–	na	–	+[c]	–
Growth at:						
4°C, 10°C	+	na	+	+	+	+
15°C	+	na	+	–	+	–
19°C	w	na	+	–	+	–
21°C	–	+	+	–	+	–
25°C	–	+	+	–	–	–
Absorbance wavelength of ethanolic extracts (nm)	451, 475, 506	na	448–449, 472	451, 506	451, 476, 505	450, 475, 506
DNA G+C content (mol%)	32 ± 1	32.4	30	32 ± 1	33.2	31

[a]Symbols: +, >85% positive; –, 0–15% positive; w, weak positive reaction; na, not available; ng, no growth.

[b]Data from Gosink et al. (1998); Larkin and Borrall (1984); McGuire et al. (1987); Irgens et al. (1989); Nedashkovskaya et al. (2005g); Yoon et al. (2006e).

[c]Differs among various studies.

5. **Polaribacter glomeratus** (McGuire, Franzmann and McMeekin 1988) Gosink, Woese and Staley 1998, 233[VP] (*Flectobacillus glomeratus* McGuire, Franzmann and McMeekin 1988, 136)

glo.mer.a'tus. L. masc. part. adj. *glomeratus* (from L. v. *glomerare* to form into ball, glomerate), formed into a ball, glomerated.

Characteristics are as described for the genus and as listed in Tables 61 and 62, with the following additional features. Colonies are yellow or tan, circular, convex, entire, smooth, opaque, and butyrous. Grows on sucrose and L-glutamate. Acid is produced from rhamnose, D-galactose, D-glucose, mannose, maltose, sucrose, and dextrin. Gelatin

TABLE 62. Fatty acid composition (%) of type strains of selected *Polaribacter* species[a,b]

Fatty acid	*P. filamentus*	*P. butkevichii*	*P. dokdonensis*	*P. franzmannii*	*P. irgensii*
13:0 iso	5	16	5	5	2
14:0 iso	nd	8	1	nd	nd
15:1 iso G	12	nd	15	11	6
15:1 iso I/H/13:0 3-OH	6	nd	nd	nd	nd
15:0	nd	8	2	nd	nd
15:0 iso	22	13	17	9	12
15:0 anteiso	6	nd	nd	4	6
15:1 ω6*c*	9	nd	5	9	3
16:1 ω7*c*	4	nd	7	9	2
15:0 iso 3-OH	22	12	14	17	38
15:0 3-OH	2	7	4	7	10
15:0 Δ6	nd	8	nd	nd	nd

[a]Growth conditions varied among strains; fatty acid values less than 1% are not reported.

[b]nd, Not detected.

is hydrolyzed. Nitrate is reduced to nitrite. Positive for oxidase and weakly positive for catalase. Cells can grow at temperatures of <4 to 21°C. Phylogenetic analyses based on 16S rRNA gene sequences show that this species is a member of the genus *Polaribacter* and is very distantly related to *Flectobacillus major*, the type species of *Flectobacillus*, the genus to which the bacterium was originally ascribed.

Source: Antarctic marine environments.

DNA G+C content (mol%): 33–33.2 (T_m).

Type strain: ACAM 171, ATCC 43844, CIP 103112, LMG 13858, UQM 3055.

Sequence accession no. (16S rRNA gene): M58775.

6. **Polaribacter irgensii** Gosink, Woese and Staley 1998, 233[VP]

ir.gen'si.i. L. gen. masc. n. *irgensii* of Irgens, named in honor of Roar L. Irgens, the first microbiologist to observe polar marine gas vacuolate bacteria.

Characteristics are as described for the genus and as listed in Tables 61 and 62, with the following additional features. Colonies are orange, circular, convex, entire, smooth, translucent, and butyrous. Rhodopsin has been reported in this species. The type strain can grow on DL-malate. Acid is produced from a number of carbohydrates including D-galactose, D-glucose, mannose, and glycerol. Growth occurs from at least as low as −1.5 to 12°C (Irgens et al., 1989). Principal fatty acids include $C_{15:0}$ iso 3-OH (38%), $C_{15:0}$ iso (12%), and $C_{15:0}$ 3-OH (10%) when grown on SWCm agar (Irgens et al., 1989) at 10°C.

Source: isolated from seawater near the pack ice edge in Penola Strait, Palmer Peninsula, Antarctica.

DNA G+C content (mol%): 31 (T_m).

Type strain: 23-P, ATCC 700398, CIP 106478.

Sequence accession no. (16S rRNA gene): M61002.

Genus XLII. **Psychroflexus** Bowman, McCammon, Lewis, Skerratt, Brown, Nichols and McMeekin 1999, 2[VP] (Effective publication: Bowman, McCammon, Lewis, Skerratt, Brown, Nichols and McMeekin 1998, 1606.)

John P. Bowman

Psy.chro.flex'us. Gr. adj. *psychros* cold; L. masc. n. *flexus* bend, curve; N.L. masc. n. *Psychroflexus* cold bend.

Cells are straight or slightly curved **rods**, 0.4–1.5 × 0.5–3 µm. One species forms partially coiled filaments up to 50 µm in length in liquid media, whereas on solid surfaces the filaments formed are of indeterminant length. Coccoid-like cells are formed in old cultures. Cells occur singly or in pairs. Spores and resting cells are not present. Gas vesicles and helical or ring-shaped cells are not formed. Nonmotile or motile by gliding. **Gram-stain-negative. Strictly aerobic, having an oxidative type of metabolism.** Catalase-positive. Oxidase-positive. **Chemoheterotrophic. Non-agarolytic. Colonies are bright orange** due to production of carotenoids, including β-carotene. **Flexirubin pigments are not produced**. Produce alkaline phosphatase, α-glucosidase, β-glucosidase, and α-amylase. Strains **usually require Na+** for growth. Strains are **stenohaline or halotolerant**. Stenohaline species grow over a narrow range of seawater salinity. **Moderate halophiles** grow best with elevated NaCl concentrations (optimal levels 5–10%). Yeast extract may be required for growth by some strains. Species range from psychrophilic to mesophilic in terms of the response of their growth rates to temperature. **Neutrophilic**, with optimal growth occurring at pH 7.0–7.5. The major fatty acids are $C_{15:1}$ ω10*c* anteiso, $C_{15:0}$ anteiso, $C_{16:0}$ iso, $C_{16:0}$ 3-OH iso, and $C_{17:0}$ 3-OH iso/anteiso. One species produces **polyunsaturated fatty acids**. The major respiratory quinone is menaquinone-6. The major polyamine is homospermidine. **Habitats, depending on the species, include polar sea-ice algal assemblages or hypersaline lakes.**

DNA G+C content (mol%): 33–36.

Type species: **Psychroflexus torquis** Bowman, McCammon, Lewis, Skerratt, Brown, Nichols and McMeekin 1999, 1[VP] (Effective publication: Bowman, McCammon, Lewis, Skerratt, Brown, Nichols and McMeekin 1998, 1607.).

Further descriptive information

The genus *Psychroflexus* is a member of the family *Flavobacteriaceae* and comprises three species, *Psychroflexus torquis*, *Psychroflexus gondwanensis*, and *Psychroflexus tropicus*, which exhibit evidence of considerable econiche specialization.

Psychroflexus torquis includes rod-like to filamentous cells that are capable of gliding motility and have an absolute requirement for seawater and an associated stenohaline growth pattern. The species is slow-growing (doubling time 20–24 h at 10°C) and highly psychrophilic. In liquid media, partly coiled filaments are observed frequently (Bowman et al., 1998). On agar media, *Psychroflexus torquis* tends to form thin, filamentous cells that are often hundreds of micrometers long.

All *Psychroflexus* species produce catalase, cytochrome *c* oxidase, alkaline phosphatase, α-glucosidase, β-glucosidase, and acid from mannose. All produce α-amylases. Properties known to be absent from *Psychroflexus* species include: hydrolysis of casein, agar, chitin, uric acid, and xanthine; nitrate reduction; production of H_2S (from L-cysteine and thiosulfate); and lysine decarboxylase, ornithine decarboxylase, arginine dihydrolase, tryptophan deaminase, phenylalanine deaminase, α-arabinosidase, α-fucosidase, α-galactosidase, β-galactosidase, 6-phospho-β-galactosidase, β-glucuronidase, and α-mannosidase activities.

Other traits differentiating the species are shown in Table 63.

Psychroflexus species form homospermidine and menaquinone-6 as the major polyamine and quinone, respectively (Hamana and Nakagawa, 2001). In addition, *Psychroflexus torquis* accumulates high levels of 2-phenylethylamine (Hosoya and Hamana, 2004). The major fatty acids amongst *Psychroflexus* species are branched-chain fatty acids with chain lengths of 15 or 16 carbon atoms (Table 64).

A draft genome sequence is available for the type strain of *Psychroflexus torquis* ATCC 700755^T. This strain contains a single genome of 4.29 Mbp, contains no extrachromosomal DNA, and has an overall G+C content of 35 mol%. A total of 3862 predicted open reading frames is present. The genome includes three *rrna* operons (coding rRNA subunit genes) and 38 tRNA genes. The genome has a relatively low gene coding level, only 82%. This is explained by the unusually heavy infiltration of lateral gene transferred (LGT) genetic elements into the genome, forming a series of "islands" ranging in size from 9 to 142 kb in length. These genomic islands are demarcated by lower than average coding density and the presence of phage-type integrases, retron-type reverse transcriptases, and various transposase-like insertional elements. Indeed, more than 200 of these gene types are present in the genome. Based on annotation data and database comparisons, as much as 25% of the *Psychroflexus torquis* genome comprises these LGT islands. Genes coding polyunsaturated fatty acid synthesis genes (*pfa* operon) (Allen and Bartlett, 2002) and a large exopolysaccharide genetic locus are located on genomic islands, suggesting that LGT elements may have been critical for the acquisition of some of the more unusual traits of this species. The remainder of the genome contains, at higher coding density, genes conserved amongst the phylum *Bacteroidetes* that code for central and intermediary metabolism and housekeeping cellular activities. The *Psychroflexus torquis* genome has a metabolome comparable to that of other members of the *Flavobacteriaceae* that have genome sequence data (e.g., *Croceibacter atlanticus*, *Robiginitalea biformata*, *Polaribacter irgensii* and several others), with a preference for the oxidative catabolism of carbohydrates. Numerous response regulatory genes and transporters are present suggesting that the micro-organism is particularly sensitive to the surrounding environment. Many genes are present for the import and export of high molecular mass substances, inorganic ions, solutes, and carbohydrates. Two *BetT* high affinity choline/glycine betaine/carnitine transporters and eight putative proline:sodium ion symporters are present with possible protective roles, conferring osmotolerance and/or cryotolerance. Overall, peptide sequences have the greatest similarities with those of the genome of *Croceibacter atlanticus*, a pelagic member of the *Flavobacteriaceae*. The lack of various genes or pathways for synthesis of pantothenic acid and cobalamin, as well as specific receptor proteins for these compounds, indicates that these vitamins are essential nutrients for growth. This is borne out by the need to supplement defined mineral salts media with yeast extract before any growth of *Psychroflexus torquis* can be achieved. Highly similar orthologs to genes found necessary for gliding (*gld* genes) in *Flavobacterium johnsoniae* (Braun and

TABLE 63. Differential characteristics of *Psychroflexus* species[a]

Characteristic	*P. torquis*	*P. gondwanensis*	*P. tropicus*
Morphology	Rods, coiled filaments	Rods	Rods
Gliding motility	+	–	–
Salinity range for growth (%)	1.5–6	0–15	1–20
Optimum salinity for growth (%)	3	5	7.5–10
Requirement for divalent cations in seawater	+	–	–
Temperature growth range (°C)	<0 to 15–20	0–30	4–43
Esculin hydrolysis	–	+	–
Extracellular Dnase	+	+	–
Gelatin hydrolysis	–	d	–
Esterase (Tween 80)	+	+	–
Urease	–	d	nd
Oxidative acid production from:			
L-Arabinose	–	+	–
D-Fructose	–	–	+
D-Glucose	d (w)	+	–
D-Mannose	+	+	–
D-Xylose	–	+	–
Cellobiose	–	+	–
Maltose	d (w)	+	–
D-Arabitol	–	–	+
D-Mannitol	–	–	+
D-Sorbitol	–	–	+
Glycerol	d	–	–
DNA G+C content (mol%)	33 (T_m), 35 (genome)	36	35

[a]Symbols: +, >85% positive; d, different strains give different reactions (16–84% positive); –, 0–15% positive; w, weak reaction; nd, not determined.

TABLE 64. Whole-cell fatty acid composition (%) present in *Psychroflexus* species cultivated on marine agar 2216 at 15°C[a]

Fatty acid	*P. torquis*	*P. gondwanensis*	*P. tropicus*
Saturated fatty acids:			
$C_{14:0}$	tr	tr	–
$C_{15:0}$	4.2 (0.6)	1.9 (0.8)	tr
$C_{16:0}$	tr	tr	–
$C_{17:0}$	–	tr	–
$C_{18:0}$	tr	tr	–
Branched chain fatty acids:			
$C_{13:0}$ iso	tr	1.4 (0.4)	tr
$C_{14:1}$ iso	–	2.8 (0.3)	–
$C_{14:0}$ iso	1.0 (0.2)	4.8 (0.8)	2.7
$C_{15:1}$ ω10*c* iso	tr	2.2 (1.0)	11.8
$C_{15:1}$ ω10*c* anteiso	16.9 (1.8)	18.4 (2.5)	12.9
$C_{15:0}$ iso	1.1 (0.3)	2.1 (0.5)	16.7
$C_{15:0}$ anteiso	35.2 (4.4)	23.0 (2.7)	19.3
$C_{16:1}$ iso	tr	tr	–
$C_{16:0}$ iso	6.0 (1.2)	10.9 (0.3)	3.4
$C_{17:0}$ iso	tr	tr	–
Hydroxy fatty acids:			
$C_{14:0}$ 3-OH iso	–	tr	tr
$C_{15:0}$ 3-OH iso	tr	tr	2.9
$C_{15:0}$ 3-OH anteiso	tr	tr	–
$C_{15:0}$ 3-OH	2.5 (0.7)	tr	2.8
$C_{16:0}$ 3-OH iso	15.4 (1.0)	18.5 (1.6)	10.1
$C_{16:0}$ 3-OH	1.2 (1.2)	tr	tr
$C_{17:0}$ 3-OH iso	tr	tr	10.0
$C_{17:0}$ 3-OH anteiso	10.9 (3.4)	6.6 (0.8)	–
$C_{17:0}$ 3-OH	tr	tr	–
$C_{18:0}$ 3-OH iso	–	tr	–
Polyunsaturated fatty acids:			
$C_{20:4}$ ω6*c*	2.1 (1.2)	–	–
$C_{20:5}$ ω3*c*	4.9 (3.0)	–	–

[a]Values represent the mean percentages of the total fatty acids (values in parentheses are standard deviations); tr, trace level detected (<1% of total fatty acids); –, not detected.

McBride, 2005) are present in the *Psychroflexus torquis* genome. The bright orange pigment of *Psychroflexus torquis* colonies is indicative of carotenoid biosynthesis and genes encoding the synthesis of β-carotene and zeaxanthin are present. The species also appears to be able to synthesize proteorhodopsin (Friedrich et al., 2002) and possesses a blue light receptor gene ortholog, although at this stage it is unknown if light provides an enhancement to the growth of this species.

Psychroflexus torquis has the unusual ability to produce the polyunsaturated fatty acids eicosapentaenoic acid (EPA) and arachidonic acid (Nichols et al., 1997). As the incubation temperature is reduced from 15 to 2°C, the production of EPA increases more than fourfold, indicating the importance of this lipid to *Psychroflexus torquis* for low temperature growth. No other member of the family *Flavobacteriaceae* has been shown to be able to form polyunsaturated fatty acids, including the two other known *Psychroflexus* species. Extreme psychrophily is also relatively uncommon in the family, although psychroactive growth is very prevalent. So far only isolated from dense algal assemblages within sea-ice, *Psychroflexus torquis* has a pronounced epiphytic lifestyle.

In comparison, both *Psychroflexus gondwanensis* and *Psychroflexus tropicus* are inhabitants of hypersaline lakes, including those of Antarctica and the Hawaiian Archipelago, respectively. Neither form filaments nor are they capable of gliding; they appear to be mainly planktonic bacteria. The growth rates of these two species are much faster than that of *Psychroflexus torquis*. These species are halotolerant and have relatively high salinity optima for growth, suggesting a preference for moderate but not extremely hypersaline ecosystems. Both *Psychroflexus gondwanensis* and *Psychroflexus tropicus* exhibit growth at low temperatures, but have growth rate temperature optima of 25–35°C, as compared to 10–15°C found for *Psychroflexus torquis*, when grown in liquid media.

Enrichment and isolation procedures

Currently, no specific method of enrichment is available for isolation of *Psychroflexus* species. Plating melted sea-ice material or hypersaline lake water onto marine agar media or organic media containing high levels of salt (10–12%) can be used to directly isolate *Psychroflexus* species, with bright orange colonies being indicative of the genus. *Psychroflexus torquis* colonies are typically very viscid and elastic due to production of an exopolysaccharide and are best isolated from sea-ice samples rich in algae. Incubation temperatures depend on the species (see above).

Maintenance procedures

Strains can be maintained for many years cryopreserved in marine broth containing 20–30% glycerol. Strains survive for many months, potentially years, on thick agar plates or on slants stored at 2–4°C. Strains may also be maintained by lyophilization, although the procedure is generally not useful for *Psychroflexus torquis*.

Differentiation of the genus *Psychroflexus* from other genera

The most salient traits useful for distinguishing *Psychroflexus* spp. from other members of the *Flavobacteriaceae* include orange pigmentation, lack of flexirubin pigments, strictly aerobic oxidative metabolism, the ability to use carbohydrates such as starch, and the ability to grow at low temperatures.

Taxonomic comments

Psychroflexus gondwanensis was originally described as *Flavobacterium gondwanense* (Dobson et al., 1993).

List of species of the genus *Psychroflexus*

1. **Psychroflexus torquis** Bowman, McCammon, Lewis, Skerratt, Brown, Nichols and McMeekin 1999, 2[VP] (Effective publication: Bowman, McCammon, Lewis, Skerratt, Brown, Nichols and McMeekin 1998, 1607.)

tor′quis. L. masc. n. *torquis* a twisted neck-chain, referring to coiling of cellular filaments.

Characteristics are as given for the genus, as listed in Table 63, and as given in the text. In addition, *Psychroflexus torquis* forms orange, circular, mucoid, viscid to elastic, convex or dome-like, opaque, entire, smooth, glistening colonies up to 10 mm in diameter.

Source: inhabits sea-ice algal assemblages.

DNA G+C content (mol%): 33 (T_m), 35 (genome sequence).

Type strain: 651, ACAM 623, ATCC 700755, CIP 106069.

Sequence accession no. (16S rRNA gene): U85881.

Sequence accession no. (genome): NZ_AAPR00000000.

2. **Psychroflexus gondwanensis** corrig. (Dobson, Colwell, Franzmann and McMeekin 1993) Bowman, McCammon, Lewis, Skerratt, Brown and Nichols and McMeekin 1999, 2[VP] (Effective publication: Bowman, McCammon, Lewis, Skerratt, Brown, Nichols and McMeekin 1998, 1608.) (*Flavobacterium gondwanense* Dobson, Colwell, Franzmann and McMeekin 1993, 81)

gond.wa.nen'sis. N.L. masc. adj. *gondwanensis* of or belonging to Gondwanaland or Gondwana, one of the two ancient supercontinents, which originally included Antarctica and other, separate continental landmasses.

Characteristics are as given for the genus, as listed in Table 63, and as given in the text. In addition, *Psychroflexus gondwanensis* forms orange, circular, butyrous or viscid, convex, opaque, entire, smooth, glistening colonies up to 5 mm in diameter.

Source: inhabits various hypersaline lakes of Antarctica (James et al., 1994).

DNA G+C content (mol%): 36 (T_m).

Type strain: ACAM 44, ATCC 51278, CCUG 33444, CIP 104040, DSM 5423, LMG 13192.

Sequence accession no. (16S rRNA gene): M92278.

Additional note: the original spelling, *gondwanense*, did not agree in gender with the generic name *Psychroflexus* and thus was corrected on Validation List no. 68, Int. J. Syst. Bacteriol. *49:* 1–3, 1999).

3. **Psychroflexus tropicus** Donachie, Bowman and Alam 2004, 937[VP]

trop'i.cus. L. masc. adj. *tropicus* tropical, relating to the species tropical lake habitat.

Characteristics are as given for the genus, as listed in Table 63, and as given in the text. In addition, *Psychroflexus tropicus* forms orange, circular, butyrous, convex, opaque, entire, smooth, glistening colonies up to 4 mm in diameter. Esterases (C_4, C_8, and C_{14}), and leucine, valine, and cystine arylamidases are produced. Trypsin, chymotrypsin, acid phosphatase, and phosphohydrolase activities are present. In Biolog GN trays, the following carbon sources test positively in the presence of 7.5% NaCl, but not with 2 or 4% NaCl: L-alanine, L-alanyl-glycine, L-asparagine, L-aspartate, L-glutamate, L-leucine, L-ornithine, L-proline, L-serine, L-threonine, monomethyl succinate, and L-alaninamide. Glycerol, D-glucose, D-fructose, D-mannose, D-sorbitol, trehalose, starch, and D-arabitol are utilized. Acid is produced oxidatively from sucrose and 5-ketogluconate.

Source: inhabits hypersaline lakes of the Hawaiian Archipelago.

DNA G+C content (mol%): 35 (T_m).

Type strain: LA1, ATCC BAA-734, DSM 15496.

Sequence accession no. (16S rRNA gene): AF513434.

Genus XLIII. **Psychroserpens** Bowman, McCammon, Brown, Nichols and McMeekin 1997, 674[VP]

John P. Bowman

Psy.chro.ser'pens. Gr. adj. *psychros* cold; L. masc. or fem. n. *serpens* serpent; N.L. masc. n. *Psychroserpens* cold serpent.

Cells are **flexible rods with rounded ends**, 0.5–0.6 × 2–6 µm, occurring singly or in pairs. Cultures also frequently include **coiled or roughly helical filamentous cells** up to 20 µm in length. Spores, resting cells, and gas vesicles are not present. **Nonmotile. Gram-stain-negative. Colonies are yellow due to accumulation of carotenoids. Flexirubin pigments are not produced. Strictly aerobic, having an oxidative type of metabolism.** Catalase-positive. Oxidase-negative. **Chemoheterotrophic. Non-agarolytic. Carbohydrates are not utilized, except for glycogen.** Acid is not formed from carbohydrates. **Seawater salts and yeast extract are required** for growth. **Psychrophilic.** Grows between −2 and 18°C, with optimal growth at 10–12°C and pH 6–8. The major fatty acids are $C_{15:0}$, $C_{15:0}$ iso, $C_{15:0}$ anteiso, $C_{15:1}$ ω10*c* iso, $C_{15:1}$ ω10*c* anteiso, $C_{16:1}$ ω11*c* iso, $C_{15:1}$ ω11*c*, $C_{15:1}$ ω4*c*, and $C_{16:1}$ ω5*c*. **Isolated from saline lakes of Antarctica.**

DNA G+C content (mol%): 27–29.

Type species: **Psychroserpens burtonensis** Bowman, McCammon, Brown, Nichols and McMeekin 1997, 674[VP].

Further descriptive information

Psychroserpens is a genus of the family *Flavobacteriaceae* and currently includes one species, *Psychroserpens burtonensis*. *Psychroserpens burtonensis* was isolated from the waters of Burton Lake, a saline lake located in the Vestfold Hills ice-free region of Eastern Antarctica (68° S 78° E). Burton Lake has salinity levels and composition that are nearly the same as seawater and receives an inflow of seawater from a closely adjacent fjord in the summer when surface ice melts sufficiently. The species is characterized by psychrophilic growth and a preference for organic acids as carbon sources.

Psychroserpens burtonensis occasionally forms ring-shaped cells and more frequently helical or coiled cells. This is most obvious in stationary growth phase cultures. In young, exponentially growing cultures, cells tend to appear as flexible rods. The organism requires vitamins and other unknown substances in yeast extract for growth. Growth is also markedly stimulated by the addition of Tween 20 or Tween 40. The species can hydrolyze casein, Tween 20, and Tween 40. Some strains can also attack gelatin and Tween 80. The species forms alkaline phosphatase. The species cannot degrade most polysaccharides (starch, chitin, dextran, agar), esculin, DNA, L-tyrosine, uric acid, or xanthine. Other standard tests (indole production, Voges–Proskauer test, H_2S production, nitrate reduction, etc.) are negative (Bowman et al., 1997).

Psychroserpens burtonensis does not in general utilize carbohydrates, alcohols, aromatics, or amino acids. Organic acids and some other compounds are preferred for growth and include: glycogen, α-glycerol phosphate, butyrate, malonate, glutarate, pimelate, azelate, citrate, oxaloacetate, citrate, and 2-aminobutyrate. Some strains can also utilize isobutyrate, *trans*-aconitate, and L-serine. A detailed phenotypic description of the species has been published (Bowman et al., 1997).

The major fatty acids are $C_{15:0}$, $C_{15:0}$ iso, $C_{15:0}$ anteiso, $C_{15:1}$ ω10*c* iso, $C_{15:1}$ ω10*c* anteiso, $C_{16:1}$ ω11*c* iso, $C_{15:1}$ ω11*c*, $C_{15:1}$ ω4*c*, and $C_{16:1}$ ω5*c* (Bowman et al., 1997). The major isoprenoid quinone is menaquinone-6 and the major polyamine is homospermidine (Hamana and Nakagawa, 2001).

Undescribed strains and environmental clones closely related to *Psychroserpens burtonensis* but probably representing cold- and non-cold-adapted novel *Psychroserpens* species have been obtained or detected from a variety of marine habitats, suggesting the genus is widely distributed. Habitats include Antarctic and Arctic sea-ice and seawater, gills of salmon cultured in marine waters, sea surface microlayer, marine biofilms, and the surfaces of macroalgae including kelp (Abell and Bowman, 2005; Agogué et al., 2005; Bowman and Nowak, 2004; Brinkmeyer et al., 2003; Brown and Bowman, 2001).

Enrichment and isolation procedures

Psychroserpens burtonensis can be isolated by direct plating of sample material onto marine agar followed by incubation at 4–10°C. The species grows slowly, with colonies taking at least 5–10 d to appear and grow to full size.

Maintenance procedures

For long-term storage, *Psychroserpens burtonensis* can be cryopreserved at −80°C in marine broth 2216 supplemented with 30% glycerol. Strains survive well on marine agar 2216 plates or slants stored at 2–4°C. The species shows generally poor survival following lyophilization.

Differentiation of the genus *Psychroserpens* from other genera

The most salient traits useful for distinguishing *Psychroserpens* from other members of the *Flavobacteriaceae* include coiled/helical cellular morphology, yellow pigmentation, lack of motility, strictly aerobic oxidative metabolism, inability to use carbohydrates, lack of reactivity in most standard phenotypic tests, and a requirement for low temperatures. The species also has a distinctive fatty acid profile, although various components can only be accurately identified by GC-MS analysis.

List of species of the genus *Psychroserpens*

1. **Psychroserpens burtonensis** Bowman, McCammon, Brown, Nichols and McMeekin 1997, 675[VP]

bur.ton.en′sis. N.L. masc. adj. *burtonensis* pertaining to Burton Lake, Antarctica, the body of water from where the organism was first isolated.

Characteristics are as given for the genus and as given in the text.

Source: saline lakes of Antarctica.

DNA G+C content (mol%): 27–29 (T_m).

Type strain: ACAM 188, ATCC 700359, CIP 105822.

Sequence accession no. (16S rRNA gene): U62913.

Genus XLIV. **Riemerella** Segers, Mannheim, Vancanneyt, De Brandt, Hinz, Kersters and Vandamme 1993a, 774[VP] emend. Vancanneyt, Vandamme, Segers, Torck, Coopman, Kersters and Hinz 1999, 293

The Editorial Board

Rie.me.rel′la. N.L. fem. n. *Riemerella* named in honor of O.V. Riemer, who first described *Riemerella anatipestifer* infections in geese in 1904 and referred to the disease as *septicemia anserum exsudativa* (Riemer, 1904).

The following description is based on those given by Segers et al. (1993a) and Vancanneyt et al. (1999). **Rods, 0.2–0.5 × 1.0–2.5 µm.** Nonmotile. No gliding motility. Nonsporeforming. Gram-stain-negative. **All strains grow microaerobically and most of them grow aerobically on blood agar.** Some strains grow anaerobically at 37°C. Colonies are smooth and nonpigmented or grayish-white to beige. Growth on litmus lactose agar is strain-dependent. No growth occurs on MacConkey agar. Most strains show a positive Voges–Proskauer reaction. Nitrates are not reduced. **Acid production from glucose is frequently negative in peptone-containing media.** The following enzymes are present: **oxidase, catalase**, gelatinase, α-glucosidase, α-maltosidase, alkaline phosphatase, acid phosphatase, esterase lipase C8, esterase C4, naphthol-AS-Bl-phosphohydrolase, leucine arylamidase, valine arylamidase, cystine arylamidase, and L-aspartic acid arylamidase. The following reactions are strain-dependent: urease, chymotrypsin, trypsin, arginine dihydrolase, indole production, hemolysis on blood, and esculin hydrolysis. Depending on the micro-test kit used, the following enzyme reactions give variable results: β-glucosidase, α-galactosidase, and lipase (these three reactions are positive in the API ID32E and negative in the API ZYM system); and *N*-acetyl-β-glucosaminidase [negative in API ZYM and API ID32E, positive in LRA-ZYM-Oxidase (Hinz et al., 1998); only some representative *Riemerella anatipestifer* strains have been tested using the latter system]. The following enzyme activities are absent: chondroitin sulfatase, hyaluronidase, β-galactosidase, β-glucuronidase, α-mannosidase, α-fucosidase, trypsin, ornithine decarboxylase, and lysine decarboxylase. None of the strains use malonate as a carbon source or assimilate the following compounds (API 20NE system): D-glucose, L-arabinose, D-mannose, D-mannitol, *N*-acetylglucosamine, maltose, D-gluconate, caprate, adipate, L-malate, citrate, or phenylacetate. No acid production occurs with the following (API ID32E system): D- and L-arabitol, galacturonate, 5-ketogluconate, phenol red, D-mannitol, maltose, adonitol, palatinose, sucrose, L-arabinose, trehalose, rhamnose, inositol, sorbitol, and cellobiose. Using the buffered single substrate (BSS) test, however, acidification of the following carbohydrates may be detected: D-glucose, maltose, D-mannose, and dextrin and, to a lesser extent, D-fructose, L-sorbose, and trehalose, but no acid is produced from lactose, D-galactose, *N*-acetyl-D-glucosamine, lactulose, trehalose, sucrose, D-mannitol, L-arabinose, *myo*-inositol, D-sorbitol, D-xylose, dulcitol, salicin, or adonitol. Menaquinone-6 is the major respiratory quinone detected in the type species. The dominant fatty acids are the branched-chain fatty acids $C_{13:0}$ iso, $C_{15:0}$ iso, $C_{15:0}$ anteiso, $C_{15:0}$ iso 3-OH, and $C_{17:0}$ iso 3-OH. Isolated mainly from diseased birds and, in a few cases, from pigs.

DNA G+C content (mol%): 29–37.

Type species: **Riemerella anatipestifer** (Hendrickson and Hilbert 1932) Segers, Mannheim, Vancanneyt, De Brandt, Hinz, Kersters and Vandamme 1993a, 775^VP [*Moraxella anatipestifer* (Hendrickson and Hilbert 1932) Bruner and Fabricant 1954, 461; *Hemophilus anatipestifer* Hauduroy, Ehringer, Urbain, Guillot and Magrou 1937, 247; *Pasteurella anapestifer* (*sic*) Hauduroy, Ehringer, Guillot, Magrou, Prévot, Rosset and Urbain 1953, 367; *Pasteurella anatipestifer* (Hendrickson and Hilbert 1932) Breed, Murray and Smith 1957, 397; *Pfeifferella anatipestifer* Hendrickson and Hilbert 1932, 249].

Further descriptive information

The illness caused by *Riemerella anatipestifer* is an exudative septicemia in ducks, pigeons, and other domestic and wild birds. *Riemerella anatipestifer* has been isolated from ducks, geese, turkeys, chickens, pheasants, quails, and wild free-living waterfowl (Asplin, 1955; Bangun et al., 1987; Bruner et al., 1970; Graham et al., 1938; Hendrickson and Hilbert, 1932; Loh et al., 1992; Pierce and Vorhies, 1973; Sandhu and Rimlet, 1997; Singh et al., 1983; Smith et al., 1987; Zehr and Ostendorf, 1970). The infection can be peracute, acute, or chronic. The organism is thought to be transmitted vertically by transovarian passage. Lateral transmission occurs through injuries, such as toenail scratches of the webbed foot in ducklings, or by entrance through the respiratory epithelium. The disease is of major economic importance in the duck industry, since the mortality in ducklings is usually 2–30% and can be as high as 95%. Anapestifer disease can be treated with a range of antibiotics.

The CAMP cohemolysin has been implicated as a potential virulence factor of *Riemerella anatipestifer* and the CAMP cohemolysin gene, *cam*, of reference strain 30/90 of serotype 19 has been cloned and expressed in *Escherichia coli* (Crasta et al., 2002). The amino acid sequence of the Cam protein has high homology with sequences of *O*-sialoglycoprotein endopeptidases and the Cam protein has been shown to be a sialoglycoprotease by its ability to hydrolyze radioiodinated glycophorin A (Crasta et al., 2002).

There are at least 21 serotypes of *Riemerella anatipestifer* that have been identified by slide and tube agglutination tests with antisera (Loh et al., 1992; Pathanasophon et al., 1994). Inactivated bacterins and live vaccines are available that can offer protection against homologous strains or serotypes of *Riemerella anatipestifer*, but are unable to protect against heterologous serotypes (Pathanasophon et al., 1996; Sandhu, 1991).

An ELISA based on a recombinant fragment of an *Riemerella anatipestifer* surface protein has been developed to aid early detection of the infection in ducks (Huang et al., 2002); also, a direct and indirect antibody ELISA has been developed for detection of duck yolk IgY and duck serum IgY. Kardos et al. (2007) have described a PCR assay that can rapidly identify *Riemerella anatipestifer* from bacterial cultures.

Riemerella columbina seems to be uniquely associated with disease in pigeons (Vancanneyt et al., 1999).

Enrichment and isolation procedures

A definitive diagnosis of anatipestifer disease can best be done by isolating and identifying the organism. *Riemerella* strains can be isolated from the organs of infected birds (liver, lung, spleen, heart, blood, kidney, brain) at necropsy by plating onto chocolate agar or 5% sheep blood agar, with incubation for 24 h at 37°C in a microaerobic CO_2-containing atmosphere, e.g., composed of 5% O_2/10% CO_2/85% N_2 or of 5% O_2/3.5% CO_2/7.5% H_2/84% N_2.

Differentiation of the genus *Riemerella* from related genera

Table 44 in the chapter on the genus *Chryseobacterium* lists phenotypic characteristics that are helpful in differentiating *Riemerella* from *Chryseobacterium* and *Bergeyella*.

Taxonomic comments

DNA–rRNA hybridization studies by Segers et al. (1993a) showed that *Riemerella anatipestifer* belongs to the family *Flavobacteriaceae* and that it is related particularly to the genera *Chryseobacterium* and *Bergeyella*. 16S rRNA gene sequence analysis by Subramaniam et al. (1997) supported the conclusion that the closest phylogenetic neighbors of *Riemerella anatipestifer* were members of the genera *Chryseobacterium* and *Bergeyella*.

Vancanneyt et al. (1999) found that *Riemerella columbina* strains were highly related to those of *Riemerella anatipestifer* based on DNA–rRNA hybridization experiments; however, there was no significant DNA–DNA hybridization with the type strain of that species, thereby, establishing the separate species status of *Riemerella columbina*.

Differentiation of the species of the genus *Riemerella*

Riemerella columbina can be differentiated from *Riemerella anatipestifer* by its grayish-white to beige pigmentation on Columbia blood agar and by its hydrolysis of esculin. Also, see the species description of *Riemerella columbina* for additional features that may help to differentiate the two species.

List of species of the genus *Riemerella*

1. **Riemerella anatipestifer** (Hendrickson and Hilbert 1932) Segers, Mannheim, Vancanneyt, De Brandt, Hinz, Kersters and Vandamme 1993a, 775^VP [*Moraxella anatipestifer* (Hendrickson and Hilbert 1932) Bruner and Fabricant 1954, 461; *Hemophilus anatipestifer* Hauduroy, Ehringer, Urbain, Guillot and Magrou 1937, 247; *Pasteurella anapestifer* (*sic*) Hauduroy, Ehringer, Guillot, Magrou, Prévot, Rosset and Urbain 1953, 367; *Pasteurella anatipestifer* (Hendrickson and Hilbert 1932) Breed, Murray and Smith 1957, 397; *Pfeifferella anatipestifer* Hendrickson and Hilbert 1932, 249]

a.na.ti.pes'ti.fer. L. n. *anas -atis* duck; L. adj. *pestifer* that brings destruction, noxious, pernicious; N.L. n. *anatipestifer* one who brings destruction of ducks.

The description is as given for the genus, with the following additional features (Segers et al., 1993a). Cells occur singly, in pairs, or in short chains. Smooth, nonpigmented colonies develop within 2 d during microaerobic incubation on rich peptone, peptone-blood, or chocolate agar at 36°C. Optimum temperature, 37°C; most strains grow at 45°C, but not at 4°C. Thiamine is required, but low concentrations of pyrithiamine and amprolium are inhibitory. The litmus milk reaction is strain-dependent. Most strains liquefy gelatin, Löffler's blood serum, and coagulated egg medium. Growth occurs in Huddleson's thionine medium, in Huddleson's basic fuchsin medium, and on agar containing 10% bile in serum. No growth occurs on agar containing 40% bile in serum, citrate agar, in KCN broth or in glycerol phosphate medium. H_2S is not produced. The following enzyme activities are present: cystine arylamidase, phosphoamidase, α-glucosidase, and esterase C4. The following enzyme activities are absent: α-galactosidase, β-glucosidase, β-glucosaminidase, lipase C14, and phenylalanine deaminase. Highly susceptible to penicillin, but highly resistant to polymyxin B and kanamycin. The fatty acids include branched fatty acids $C_{13:0}$ iso, $C_{15:0}$ iso, $C_{15:0}$ anteiso, $C_{15:0}$ iso 3-OH, $C_{17:0}$ iso 3-OH, and an unidentified fatty acid with an equivalent chain-length of 13.566. Segers et al. (1993a) reported that menaquinone-7 is the sole respiratory quinone in the type strain of *Riemerella anatipestifer*; however, Vancanneyt et al. (1999) reported that menaquinone-6 is the major respiratory quinone detected in the type species.

Source: the type strain was isolated from a duck in the United States.

DNA G+C content (mol%): 35 (T_m).

Type strain: LMG 11054, ATCC 11845, CCUG 14215, CCUG 21370, CIP 82.28, JCM 9532, LMG 11606, MCCM 00568, NCTC 11014.

Sequence accession no. (16S rRNA gene): U10877, U60101.

Additional note: the illness caused by this organism is not restricted to ducks.

2. **Riemerella columbina** Vancanneyt, Vandamme, Segers, Torck, Coopman, Kersters and Hinz 1999, 294[VP]

co.lum.bi'na. L. fem. adj. *columbina* pertaining to pigeons.

The following description is based on that given by Vancanneyt et al. (1999). The description of *Riemerella columbina* is as for the genus. On Columbia blood agar, all strains show good growth when incubated aerobically at 37°C and microaerobically at 24, 37, and 42°C and produce a gray–white or beige pigment (*Riemerella anatipestifer* strains are nonpigmented). No growth occurs anaerobically at 37°C. No growth occurs on litmus lactose agar (variable for *Riemerella anatipestifer* strains). All strains tested hydrolyze esculin (negative for *Riemerella anatipestifer* strains). When using the API ID32E system, all strains are positive for β-glucosidase activity (negative for *Riemerella anatipestifer* strains; this reaction is, however, negative for *Riemerella columbina* strains in the API ZYM system). Indole is not produced. All strains exhibit chymotrypsin activity (variable reactions for *Riemerella anatipestifer* strains). Using the BSS test, all strains produce acid from D-glucose, maltose, D-mannose, and dextrin (variable reactions for *Riemerella anatipestifer*). A negative or weakly positive reaction occurs for acid production from D-fructose and L-sorbose. The dominant fatty acids are the branched-chain fatty acids $C_{13:0}$ iso, $C_{13:0}$ anteiso, $C_{15:0}$ iso, $C_{15:0}$ anteiso, $C_{15:0}$ iso 3-OH, $C_{17:0}$ iso 3-OH, and "summed feature 4".

Source: isolated from pigeons with respiratory disease. The type strain was isolated from a pigeon palatine cleft in Germany.

DNA G+C content (mol%): 36 (T_m).

Type strain: Hinz x183-89, CCUG 47689, CIP 106288, LMG 11607.

Sequence accession no. (16S rRNA gene): AF181448.

Genus XLV. **Robiginitalea** Cho and Giovannoni 2004, 1104[VP]

Jang-Cheon Cho and Stephen J. Giovannoni

Ro.bi.gi.ni.tal'e.a. L. gen. n. *robiginis* rust; L. fem. n. *talea* a rod; N.L. fem. n. *Robiginitalea* a rust-colored rod.

Cell shape varies from straight rods in exponential phase to coccoid cells in stationary phase. Gram-stain-negative. **Nonmotile and no gliding motility.** Nonsporeforming. Poly-β-hydroxybutyrate granules are not formed. Obligately aerobic and chemoheterotrophic. **Produce carotenoid pigments, but not flexirubin pigments.** Require NaCl for growth. Oxidase- and catalase-positive. **Degrade starch and esculin.** Utilize a variety of carbon compounds as sole carbon sources. The major fatty acid types are branched acids and hydroxy acids.

DNA G+C content (mol%): 55–56.

Type species: **Robiginitalea biformata** Cho and Giovannoni 2004, 1105[VP].

Further descriptive information

Phylogenetic analysis based on 16S rRNA gene sequences has shown that the genus forms a distant lineage within the family *Flavobacteriaceae*. The genus *Robiginitalea* currently contains only one species, *Robiginitalea biformata*, with HTCC 2501[T] as the type strain.

Cell morphology varies with the growth stage of the cells. Cells in early exponential phase are straight rods, approximately 3.6 μm long and 0.5 μm wide, whereas cells in stationary phase become coccoid, 0.6–1.2 μm in diameter. Transmission electron microscopy has shown that the cells lack flagella, endospores, intracellular granules, and vacuoles. Gliding motility and flagella are not observed.

Colonies on marine agar 2216 grown at 30°C for 4 d are 1.0–2.0 mm in diameter, rusty orange-colored, uniformly circular, opaque, and pulvinate.

The predominant fatty acids are $C_{15:0}$ iso (24–28%), $C_{15:1}$ iso (14–21%), and $C_{17:0}$ 3-OH iso (25–27%). The major fatty acid types are branched acids and hydroxy acids, comprising 75.7–77.9% total fatty acids.

The organisms are obligately aerobic, mesophilic, NaCl-requiring, and chemoheterotrophic. The temperature, pH, and NaCl concentration ranges for growth are 10–44°C (optimum 30°C), pH 6.0–9.0 (optimum pH 8.0–8.5), and 0.25–10% (w/v) NaCl (optimum 2.5%). Tests for esculin hydrolysis and β-galactosidase activity are positive. Tests for denitrification, indole production, glucose acidification, arginine dihydrolase, urease, and gelatin hydrolysis are negative. Carotenoid pigments are produced with wavelength absorbance peaks at 339 and 457 nm, with a major peak at 457 nm. Photosynthetic gene clusters, bacteriochlorophyll *a*, and photosynthetic activity are not found. Starch is degraded after 3 weeks of incubation, but not gelatin, DNA, casein, elastin, dextran, cellulose, alginate, chitin, or carrageenan. Sole carbon sources include D-ribose, D-xylose, D-glucose, D-fructose, L-sorbose, D-mannose, sucrose, β-lactose, trehalose, cellobiose, maltose, melibiose, raffinose, *N*-acetyl-D-glucosamine, succinic acid, propionic acid, lactic acid, L-glutamate, L-ornithine, L-proline, and L-serine. The organisms are susceptible to nalidixic acid, tetracycline, erythromycin, rifampin, and benzylpenicillin, but resistant to chloramphenicol, kanamycin, carbenicillin, streptomycin, ampicillin, puromycin, vancomycin, gentamicin, and cycloheximide.

The type strain of the single species of the genus was isolated from the oceanic surface, indicating that its major habitat is the marine environment.

Enrichment and isolation procedures

The original liquid cultures of *Robiginitalea biformata* were obtained from the surface of the Sargasso Sea by a dilution-to-extinction method using low nutrient heterotrophic medium consisting of filtered (0.2 µm pore size) and autoclaved seawater supplemented with 1.0 µM NH_4Cl and 0.1 µM KH_2PO_4 amended with 0.001% (w/v) each of D-glucose, D-ribose, succinic acid, pyruvic acid, glycerol, and *N*-acetyl D-glucosamine, and 0.002% (v/v) ethanol (Connon and Giovannoni, 2002). Single colonies were obtained easily by spreading the liquid culture on marine agar 2216 (Difco) after incubation for 5 d at 25°C. Therefore, it is likely that *Robiginitalea* can be obtained by spreading seawater on marine agar using a serial dilution plating method. Currently, there is no special method to enrich *Robiginitalea* from the mixed marine microbial community.

Maintenance procedures

Frozen stocks as a glycerol suspension (10–30%) stored in liquid nitrogen or at −70°C are routinely used to start a new culture. Lyophilization of liquid cultures is also recommended for long-term storage. Frozen stocks and lyophilization stocks are stable for at least 1 year (confirmed by survival tests). Working stock cultures can be maintained on slants at 4°C for 2 months.

Differentiation of the genus *Robiginitalea* from related genera

16S rRNA gene sequence analysis is the most convincing evidence for the uniqueness of the genus *Robiginitalea*. All the currently classified genera in the family *Flavobacteriaceae* show less than 90% of 16S rRNA gene sequence similarity to strains of the genus *Robiginitalea*. One of major differentiating characteristics of the genus compared to other members of the family *Flavobacteriaceae* is its DNA G+C content, 54.7–56.4 mol%, which is more than 10 mol% higher than that of the other genera within the family. Other differential properties of the genus include pleomorphic properties and the upper limit of the growth temperature (44°C). Although the fatty acid profile of *Robiginitalea biformata* is similar to that of *Muricauda ruestringensis*, it differs significantly from those of members of the genera *Zobellia*, *Arenibacter*, *Aequorivita*, *Vitellibacter*, and *Croceibacter*.

Taxonomic comments

The only species currently included in the genus is *Robiginitalea biformata*. Classification of *Robiginitalea* as a distinct genus of the family *Flavobacteriaceae* is based on 16S rRNA gene sequence analyses as well as phenotypic and chemotaxonomic properties. The 16S rRNA gene sequence similarities of *Robiginitalea biformata* to members of other genera in the family *Flavobacteriaceae* are only 88–89%. None of the other taxa with validly published names shows more than 90% 16S rRNA gene sequence similarity to *Robiginitalea biformata*. Strains that are most closely related to *Robiginitalea biformata* belong to the genera *Zobellia* (88.6–89.7%), *Arenibacter* (88.3–88.4%), *Aequorivita* (88.3–89.2%), and *Vitellibacter* (87.8–88.0%). Phylogenetic analysis has shown that the genus forms a distinct linage within the *Flavobacteriaceae*, and its sister polyphyletic clade contains the genera *Muricauda*, *Zobellia*, *Arenibacter*, *Aequorivita*, and *Vitellibacter*.

List of species of the genus *Robiginitalea*

1. **Robiginitalea biformata** Cho and Giovannoni 2004, 1105^{VP}

bi.for.ma'ta. L. fem. adj. *biformata* double-formed, two-shaped, pertaining to different cell morphology in different growth phases.

Characteristics of the species are as described for the genus.

Source: seawater from the surface of the western Sargasso Sea, Atlantic Ocean.

DNA G+C content (mol%): 55–56 (HPLC).

Type strain: HTCC2501, ATCC BAA-864, KCTC 12146.

Sequence accession no. (16S rRNA gene): AY424899.

Genus XLVI. **Salegentibacter** McCammon and Bowman 2000, 1062^{VP} emend. Ying, Liu, Wang, Dai, Yang and Liu 2007, 221

John P. Bowman

Sal.e.gent'i.bac.ter. L. n. *salis* salt; L. part. adj. *egentis* needy; Gr. n. *bakterion* rod; L. masc. n. *Salegentibacter* salt-needy rod.

Cells are **rod-shaped**, 0.5–0.8 × 1.2–11.5 μm. Cells may occur singly, in pairs, or in short chains. Spores and resting cells are not present. Gas vesicles and helical or ring-shaped cells are not formed. **Nonmotile or motile by gliding.** Gram-stain-negative. **Strictly aerobic, having an oxidative type of metabolism.** Catalase-positive. Most species are oxidase-positive. **Chemoheterotrophic. Colonies are bright-yellow-pigmented** due to production of carotenoids. **Flexirubin pigments are not produced.** Most strains form alkaline phosphatase, amylase, Tween esterases (Tweens 20, 40, and 80), gelatinase, elastinase, and alginase, and produce H_2S from L-cysteine. Strains do not form cellulases, chitinases, produce indole from L-tryptophan, or form acetoin (Voges–Proskauer test). **Some species require Na^+ ions for growth. Halotolerant.** Tolerates 8% or more NaCl. The major fatty acids are $C_{15:0}$, $C_{15:1}$ iso, $C_{15:0}$ iso, $C_{15:0}$ anteiso, summed feature 3 ($C_{16:1}$ ω7*c*/$C_{15:0}$ 2-OH iso), $C_{16:0}$ 3-OH iso, $C_{17:0}$ 3-OH iso, and $C_{17:0}$ 2-OH. The major respiratory quinone is **menaquinone-6**. The major polyamine is homospermidine. Habitats include the surfaces of marine fauna including holothurians, sea urchins and sponges, hypersaline lake surface waters, and marine sediment.

DNA G+C content (mol%): 37–44 (T_m, HPLC).

Type species: **Salegentibacter salegens** (Dobson, Colwell, McMeekin and Franzmann 1993) McCammon and Bowman 2000, 1062^{VP} (*Flavobacterium salegens* Dobson, Colwell, McMeekin and Franzmann 1993, 81).

Further descriptive information

The genus *Salegentibacter* is a member of the family *Flavobacteriaceae* and is composed of six known species: *Salegentibacter salegens*, *Salegentibacter holothuriorum*, *Salegentibacter mishustinae*, *Salegentibacter flavus*, *Salegentibacter agarivorans*, and *Salegentibacter catena*. All species produce distinctive yellow-pigmented colonies on complex marine media such as marine agar 2216 (Difco). Pigmentation of *Salegentibacter salegens* is suppressed when grown on Tween 20-containing growth media (Dobson et al., 1993). *Salegentibacter* species are strictly aerobic chemoheterotrophs with a saprophytic or epiphytic lifestyle consuming naturally occurring proteins, polysaccharides, and carbohydrates. *Salegentibacter salegens*, detected by indirect immunofluorescence, has been observed to colonize diatoms and chlorophytes of various saline, meromictic Antarctic lakes (James et al., 1994). *Salegentibacter* species have been recovered from various moderately hypersaline ecosystems (Ghozlan et al., 2006).

Phenotypic characteristics that enable differentiation of *Salegentibacter* species are shown in Table 65. All species of the genus are considerably halotolerant, with *Salegentibacter salegens*, *Salegentibacter mishustinae*, and *Salegentibacter agarivorans* able to grow in the presence of up to 20% (w/v) NaCl. Regarding temperature responses, *Salegentibacter* species are mesophilic, with some species able to grow at 0–4°C or up to 42°C. Optimal growth for most species occurs at the salinity and pH of seawater (1–4% salinity and around pH 8.0). Although most species are nonmotile, *Salegentibacter agarivorans* has observable gliding motility and, also unlike the other species of the genus, is agarolytic (Nedashkovskaya et al., 2006c). Although *Salegentibacter* species appear as relatively featureless rod-shaped cells, cells of *Salegentibacter catena* have been shown to form appendage-like structures in the stationary growth phase (Ying et al., 2007). The appendages may be large bleb-like structures (Zhou et al., 1998). Fatty acid compositions of some *Salegentibacter* species are given in Table 66.

Enrichment and isolation procedures

For isolation of *Salegentibacter* strains from water or sediment, samples are serially diluted on an organic, salt-containing medium. Isolation from marine fauna usually involves homogenization of tissue samples in seawater. For the isolation of most *Salegentibacter* species, marine agar 2216 (Difco) can be used for primary isolation and subsequent purification. For *Salegentibacter catena*, a low-nutrient agar medium can be used (0.05% peptone, 0.01% yeast, and 1.5% agar, dissolved in seawater) for primary isolation. Marine agar 2216 can be used for subsequent purification.

Maintenance procedures

The most convenient form of storage of *Salegentibacter* species is cryopreservation in marine broth 2216 containing 20% (v/v) glycerol at −80°C. *Salegentibacter* species can also be readily lyophilized.

Differentiation of the genus *Salegentibacter* from other genera

The most salient traits useful for distinguishing *Salegentibacter* species from other members of the *Flavobacteriaceae* include yellow pigmentation, lack of flexirubin pigments, lack of motility of most species, strictly aerobic oxidative metabolism, ability to use carbohydrates such as starch, proteolytic activity, and, in particular, the ability to grow at NaCl concentrations of 8% or more.

Taxonomic comments

Salegentibacter salegens was originally described as *Flavobacterium salegens* (Dobson et al., 1993), but later phylogenetic and fatty acid analyses indicated that the species was misclassified at the genus level (McCammon and Bowman, 2000).

List of species of the genus *Salegentibacter*

1. **Salegentibacter salegens** (Dobson, Colwell, McMeekin and Franzmann 1993) McCammon and Bowman 2000, 1062^{VP} (*Flavobacterium salegens* Dobson, Colwell, McMeekin and Franzmann 1993, 81)

TABLE 65. Distinguishing characteristics of *Salegentibacter* species[a]

Characteristic	*S. salegens*	*S. holothuriorum*	*S. mishustinae*	*S. flavus*	*S. agarivorans*	*S. catena*
Cell diameter (μm)	0.5–0.8	0.5–0.7	0.5–0.7	0.5–0.7	0.5–0.7	0.5–0.8
Cell length (μm)	1.2–11.5	2.7–5.3	2.5–5.1	2.5–4.0	2.0–4.7	2.0–6.0
Gliding motility	–	–	–	–	+	–
NaCl growth range (%)	0–20	1–8	1–18	1–10	1–18	1–10
Temperature growth range (°C)	2–34	4–37	4–36	10–34	4–41	15–42
Oxidase	+	+	+	+	+	–
DNase	+	+	–	nd	+	–
β-Galactosidase	+	+	+	nd	+	–
Nitrate reduction to nitrite	+	–	–	–	+	–
Agar hydrolysis	–	–	–	–	+	–
Casein hydrolysis	–	–	+	–	–	+
Urease activity	d	–	–	+	–	–
Utilization of:						
Lactose	d	+	+	–	+	–
Sucrose	d	–	+	–	+	–
D-Glucose	+	+	+	–	+	+
D-Mannose	d	+	+	nd	+	–
L-Arabinose	–	–	–	–	+	–
Maltose	+	+	+	nd	+	–
Production of acid from:						
L-Arabinose	d	–	–	–	+	–
Lactose	d	+	–	nd	+	–
D-Glucose	d	+	–	nd	+	+
Sucrose	d	–	+	nd	+	–
D-Galactose	+	+	–	nd	+	–
Cellobiose	–	–	–	nd	+	
DL-Xylose	–	–	–	–	+	–
L-Fucose	–	+	–	nd	+	–
Raffinose	–	–	+	nd	+	–
N-Acetylglucosamine	–	+	–	nd	+	
DNA G+C content (mol%)	37–38	37	37–38	40	39	44

[a]Symbols: +, >85% positive; d, different strains give different reactions (16–84% positive); –, 0–15% positive; nd, not determined.

sal′e.gens. L. n. *sal* salt; L. adj. *egens* needy; *salegens* needing salt.

The characteristics are the same as given for the genus, as listed in Tables 65 and 66, and as described in the text. In addition, nitrate and ammonia are utilized as inorganic nitrogen sources. Esculin is hydrolyzed. Negative for ornithine and lysine decarboxylases. L-Arginine, D-gluconate, and L-ornithine are either utilized or are stimulatory. D-Fructose, cellobiose, maltose, raffinose, pyruvate, fumarate, succinate, L-alanine, L-proline, and L-tryptophan are either utilized or are stimulatory for some strains. The following compounds are neither utilized nor are stimulatory: L-rhamnose, DL-xylose, trehalose, salicin, adonitol, D-mannitol, D-sorbitol, citrate, malonate, L-asparagine, L-cysteine, L-glycine, L-histidine, L-isoleucine, L-leucine, L-lysine, L-serine, L-threonine, L-tyrosine, L-valine, and hydroxy-L-proline. Acid is produced from mannitol by some strains. No acid is produced from L-rhamnose, L-sorbose, melibiose, adonitol, dulcitol, D-sorbitol, inositol, or glycerol. The optimal salinity for growth is ~5%. Growth occurs between pH 5 and 10 (optimum pH 8.0–8.5). Colonies are bright yellow, circular, convex, have an entire edge, and are 1–3 mm in diameter after 3–5 d on marine agar 2216 at 25°C.

Source: isolated from saline, meromictic lakes of Antarctica (James et al., 1994).

DNA G+C content (mol%): 37–38 (T_m, HPLC). The DNA G+C content was reported originally as 39–41 mol% (T_m) by Dobson et al. (1993); however, subsequent reanalysis by spectrophotometry and by HPLC indicated that the type strain has mean G+C values of 37 mol% (McCammon and Bowman, 2000; Nedashkovskaya et al., 2004, 2005).

Type strain: ACAM 48, ATCC 51522, CCUG 33447, CIP 104041, DSM 5424, LMG 13193.

Sequence accession no. (16S rRNA gene): M92279.

2. **Salegentibacter agarivorans** Nedashkovskaya, Kim, Vancanneyt, Shin, Lysenko, Shevchenko, Krasokhin, Mikhailov, Swings and Bae 2006c, 884[VP]

a.gar.i.vo′rans. N.L. n. *agarum* agar, algal polysaccharide; L. v. *vorare* to devour, to digest; N.L. part. adj. *agarivorans* agar-digesting.

TABLE 66. Whole-cell fatty acid composition (%) of some *Salegentibacter* species cultivated on marine 2216 agar at 20–30°C[a]

Fatty acid	*S. salegens*	*S. agarivorans*	*S. catena*	*S. mishustinae*
Saturated fatty acids:				
$C_{15:0}$	7.6	5.6	10.4	6.7
$C_{16:0}$	–	–	4.3	1.8
$C_{18:0}$	–	–	2.6	1.3
Branched-chain fatty acids:				
$C_{14:0}$ iso	–	–	2.0	–
$C_{15:1}$ iso	17.7	12.1	5.0	12.3
$C_{15:1}$ anteiso	3.5	1.9	1.3	1.3
$C_{15:0}$ iso	8.5	12.3	6.8	12.1
$C_{15:0}$ anteiso	8.5	7.1	6.4	7.9
$C_{16:1}$ iso	2.6	1.5	2.8	1.9
$C_{16:0}$ iso	3.1	2.8	13.5	4.6
$C_{17:1}$ iso	2.0	2.5	2.3	2.2
$C_{17:1}$ anteiso	–	–	2.3	–
Monounsaturated straight-chain fatty acids:				
$C_{15:1}$ ω6*c*	5.7	2.3	2.6	2.4
Summed feature 3 ($C_{16:1}$ ω7*c*/$C_{15:0}$ 2-OH iso)	6.1	12.2	6.3	7.9
$C_{17:1}$ ω6*c*	4.1	3.8	2.4	3.5
$C_{17:1}$ ω8*c*	–	–	1.5	–
$C_{18:1}$ ω7*c*	–	–	1.3	–
Hydroxy fatty acids:				
$C_{15:0}$ 3-OH iso	2.5	3.6	1.2	2.9
$C_{15:0}$ 3-OH	3.9	2.3	1.2	2.1
$C_{15:0}$ 2-OH	2.4	3.5	1.8	2.6
$C_{16:0}$ 3-OH iso	5.9	3.9	3.6	4.7
$C_{16:0}$ 3-OH	–	1.3	–	0.5
$C_{17:0}$ 3-OH iso	5.9	9.8	5.2	8.6
$C_{17:0}$ 2-OH	4.7	4.5	5.0	3.1

[a]–, Not detected or at a level below 1% total fatty acid content.

The characteristics are as given for the genus, as listed in Tables 65 and 66, and as described in the text, with the following additional features. Gliding motility is present. Inositol, D-mannitol, D-sorbitol, citrate, and malonate are not utilized. No acid is produced from L-rhamnose, L-sorbose, adonitol, dulcitol, inositol, D-mannitol, or D-sorbitol. From the Microlog GN2 plate system, the following substrates produce positive results: α-D-glucose, α-lactose, sucrose, methylpyruvate, monomethylsuccinate, D-galactonate, D-gluconate, β-hydroxybutyrate, *p*-hydroxyphenylacetate, itaconate, 2-oxoglutarate, DL-lactate, propionate, succinate, succinamate, alaninamide, L-alanylglycine, L-asparagine, L-aspartate, L-glutamate, glycyl-L-aspartate, L-phenylalanine, L-proline, L-pyroglutamate, L-threonine, and urocanate. Based on the API ZYM system, positive for the following enzyme activities: α-galactosidase, β-galactosidase, α-glucosidase, β-glucosidase, *N*-acetyl-β-glucosaminidase, alkaline phosphatase, acid phosphatase, naphthol-AS-BI-phosphohydrolase, C8 esterase/lipase, leucine arylamidase, valine arylamidase, trypsin, and α-chymotrypsin. Colonies are bright yellow, shiny, slimy, circular, convex, sunken into the agar surface, and are 1–3 mm in diameter after 3–5 d incubation on marine agar 2216 at 25°C. The optimal salinity for growth is 2–4%. Growth occurs between pH 5.7 and 10.0; optimum pH is 7.5–8.3.

Source: isolated from the tissues of a sponge (*Artemisina* sp.) collected near Onecotan Island, Kuril Islands, Sea of Okhotsk.

DNA G+C content (mol%): 39 (HPLC).

Type strain: KCTC 12560, KMM 7019, LMG 23205.

Sequence accession no. (16S rRNA gene): DQ191176.

3. **Salegentibacter catena** Ying, Liu, Wang, Dai, Yang and Liu 2007, 221[VP]

ca.te'na. L. n. *catena* chain, referring to the fact that cells frequently occur in chains.

Cells frequently occur in chains. Cells produce appendage-like structures when in stationary growth phase. The characteristics are otherwise the same as given for the genus, as listed in Tables 65 and 66, and as described in the text, with the following additional features. Esculin is hydrolyzed, but not carboxymethylcellulose. The following compounds are not utilized: L-arabinose, L-fucose, L-rhamnose, D-galactose, DL-xylose, maltose, melibiose, raffinose, adonitol, D-mannitol, inositol, D-sorbitol, D-gluconate, acetate, propionate, pyruvate, DL-lactate, citrate, L-malate, succinate, malonate, L-alanine, and L-proline. No acid is produced from D-mannose, L-rhamnose, melibiose, adonitol, dulcitol, inositol, D-mannitol, or D-sorbitol. Based on the API ZYM system, strong activity is exhibited for the following enzymes: α-glucosidase, alkaline phosphatase, acid phosphatase, naphthol-AS-BI-phosphohydrolase, C8 esterase/lipase, leucine arylamidase, and valine arylamidase. Weak activity occurs for C4 esterase, C14 lipase, cystine arylamidase, and trypsin. Colonies are saffron yellow, shiny, circular, convex, have an entire edge, and are 3–5 mm in diameter after 5 d on marine agar 2216 at 25°C. The optimal salinity for growth is 1–4%. Grows between pH 6.5 and 9.0, with optimum growth at pH 7.5–8.2.

Source: isolated from marine sediment collected from the Xijiang oilfield located in the South China Sea.

DNA G+C content (mol%): 44 (T_m).

Type strain: HY1, CGMCC 1.6101, JCM 14015.

Sequence accession no. (16S rRNA gene): DQ640642.

4. **Salegentibacter flavus** Ivanova, Bowman, Christen, Zhukova, Lysenko, Gorshkova, Mitik-Dineva, Sergeev and Mikhailov 2006, 585[VP]

fla'vus. L. masc. adj. *flavus* golden yellow.

The characteristics are the same as given for the genus, as listed in Table 65, and as described in the text. In addition, DL-xylose, raffinose, salicin, adonitol, inositol, D-mannitol, D-sorbitol, and citrate are not utilized. No acid is produced from L-rhamnose, inositol, D-mannitol, or D-sorbitol. Colonies are bright yellow, shiny, circular, convex, and are 1–3 mm in diameter after 3–5 d on marine agar 2216 at 25°C. The optimal salinity for growth is ~3%. Growth occurs between pH 5 and 10. Optimum growth occurs at pH 8.0–8.5.

Source: isolated from marine sediment collected from the Chazhma Sea, Sea of Japan.

DNA G+C content (mol%): 40 (HPLC).

Type strain: Fg 69, CIP 107843, KMM 6000.

Sequence accession no. (16S rRNA gene): AY6822200.

5. **Salegentibacter holothuriorum** Nedashkovskaya, Suzuki, Vancanneyt, Cleenwerck, Zhukova, Vysotskii, Mikhailov and Swings 2004g, 1109^VP^

ho.lo.thu.ri.o'rum. N.L. gen. pl. n. *holothuriorum* of holothurians (sea cucumbers).

The characteristics are the same as given for the genus, as listed in Table 65, and as described in the text. In addition, D-mannitol, inositol, D-sorbitol, citrate, succinate, and malonate are not utilized. Acid is not produced from L-rhamnose, L-sorbose, melibiose, adonitol, dulcitol, inositol, D-mannitol, D-sorbitol, or glycerol. The optimal salinity for growth is ~3%. Colonies are bright yellow, shiny, circular, convex, have an entire edge, and are 1–3 mm in diameter after 3–5 d on marine agar 2216 at 25°C. The major phospholipid present in cells is phosphatidylethanolamine.

Source: isolated from the tissues of a holothurian (*Apostichopus japonicus*) collected from the Sea of Japan.

DNA G+C content (mol%): 37 (HPLC).

Type strain: KMM 3524, LMG 21968, NBRC 100249.

Sequence accession no. (16S rRNA gene): AB116148.

6. **Salegentibacter mishustinae** Nedashkovskaya, Kim, Lysenko, Mikhailov, Bae and Kim 2005h, 237^VP^

mi.shu'sti.nae. N.L. gen. n. *mishustinae* of Mishustina, in honor of Irina E. Mishustina, a Russian marine microbiologist.

The characteristics are the same as given for the genus, as listed in Tables 65 and 66, and as described in the text. In addition, D-mannitol, inositol, D-sorbitol, citrate, succinate, and malonate are not utilized. Acid is not produced from L-rhamnose, L-sorbose, melibiose, adonitol, dulcitol, inositol, D-mannitol, D-sorbitol, or glycerol. Colonies are bright yellow, shiny, circular, convex, have an entire edge, and are 13 mm in diameter after 3–5 d on marine agar 2216 at 25°C.

Source: isolated from the tissues of a sea urchin (*Strongylocentrotus intermedius*) collected from Troitsa Bay, Gulf of Peter the Great, Sea of Japan.

DNA G+C content (mol%): 37–38 (HPLC).

Type strain: KCTC 12263, KMM 6049, LMG 22584, NBRC 100592.

Sequence accession no. (16S rRNA gene): AY576653.

Genus XLVII. **Sandarakinotalea** Khan, Nakagawa and Harayama 2006c, 960^VP^

THE EDITORIAL BOARD

San.da.ra.ki.no.ta'le.a. Gr. adj. *sandarakinos -e -on* of orange color; L. fem. n. *talea* a slender staff, a rod; N.L. fem. n. *Sandarakinotalea* an orange-colored rod.

Rods, 0.5–0.7 × 3–5 µm. Nonmotile. Gliding motility is not present. Gram-stain-negative. **Aerobic. Colonies are light orange** due to a carotenoid pigment. Flexirubin-type pigments are not formed. Catalase- and oxidase-positive. Growth occurs at 10–40°C; optimum, 15–20°C; no growth occurs at 4°C. **Growth requires NaCl and either a potassium or magnesium salt. Sea salts are required for growth. Starch, DNA, and gelatin are hydrolyzed. Nitrate is not reduced.** The predominant fatty acids are $C_{15:0}$ iso, $C_{15:0}$ anteiso, $C_{16:0}$ iso, and $C_{17:0}$ iso 3-OH. **Pentadecenoic acid ($C_{15:1}$) is absent.** The only respiratory quinone present is MK-6. Isolated from marine sediment.

DNA G+C content (mol%): 35–37.

Type species: **Sandarakinotalea sediminis** Khan, Nakagawa and Harayama 2006c, 962^VP^.

Further descriptive information

Khan et al. (2006c) reported that the mean percentage values for cellular fatty acids assayed in four isolates of *Sandarakinotalea sediminis* were as follows: $C_{15:0}$ iso (27%), $C_{17:0}$ iso 3-OH (13%), $C_{15:0}$ anteiso (10%), $C_{16:0}$ iso (10%), $C_{17:1}$ iso ω9*c* (7%), $C_{16:0}$ iso 3-OH (6%), summed feature 4 ($C_{16:1}$ ω7*c*/$C_{15:0}$ iso 2-OH) (6%), $C_{16:0}$ iso (3%), $C_{15:0}$ iso 3-OH (3%), $C_{17:1}$ ω6*c* (3%), $C_{17:1}$ anteiso ω9*c* (2%), $C_{14:0}$ iso (2%), and $C_{17:0}$ 2-OH (2%). Fatty acids amounting to less than 1% of the total fatty acid content in all strains are not listed.

Enrichment and isolation procedures

Sandarakinotalea sediminis has been isolated from marine sediments. Strains are cultured at 20°C on 1.2% agar plates or slants prepared with 0.5× marine agar 2216 (Difco) diluted in artificial seawater (Naigai Chemicals).

For long-term preservation, cells can be stored at −80°C in artificial seawater supplemented with 20% (v/v) glycerol.

Differentiation of the genus *Sandarakinotalea* from related genera

Sandarakinotalea strains differ from those of the genera *Algibacter, Aquimarina, Kordia, Mesonia, Psychroflexus,* and *Salegentibacter* by being unable to grow at 4°C. Unlike *Kordia* strains, *Sandarakinotalea* strains are catalase-positive and unable to reduce nitrate. In contrast to *Algibacter* and *Aquimarina* strains, they do not exhibit gliding motility and have higher G+C content (35–37 mol% vs 31–33 mol%). Unlike *Psychroflexus* strains, they can hydrolyze gelatin. *Sandarakinotalea* strains differ from those of *Mesonia* by their ability to hydrolyze DNA and starch. They can be distinguished from both *Mesonia* and *Aquimarina* strains by their ability to grow at 37°C and by their inability to grow in the presence of 8–10% NaCl.

Taxonomic comments

Based on 16S rRNA gene sequence analysis, the species most closely related to *Sandarakinotalea sediminis* are *Psychroflexus torquis* and *Algibacter lectus* (90.7 and 90.9% similarities, respectively). *Sandarakinotalea sediminis* has 88–90.1% similarity to members of related genera in the family *Flavobacteriaceae* such as *Aquimarina, Kordia, Mesonia,* and *Salegentibacter.*

The four strains of *Sandarakinotalea sediminis* isolated so far share DNA–DNA reassociation values of 67–99% with each other.

List of species of the genus *Sandarakinotalea*

1. **Sandarakinotalea sediminis** Khan, Nakagawa and Harayama 2006c, 962[VP]

se.di.mi'nis. L. gen. n. *sediminis* of sediment.

The characteristics are as given for the genus, with the following additional features. Cells may form filaments 5–10 μm in length after incubation for 6 or 7 d. Colonies are circular or irregular. Growth is supported by 10, 30, 50, and 70% (v/v) artificial seawater. Growth in 0.2× LB medium [2 g tryptone (Difco) and 1 g yeast extract (Difco) dissolved in 1,000 ml water] requires addition of NaCl and either a potassium or magnesium salt. Agar, cellulose, carboxymethylcellulose, chitin, esculin, and urea are not hydrolyzed. Indole is not produced from tryptophan. Acid is not produced from glucose. The following substrates are oxidized (Biolog GN2 MicroPlate system): L-alanyl glycine, L-aspartic acid, α-cyclodextrin, dextrin, L-glutamic acid, glycogen, glycyl L-aspartic acid, glycyl L-glutamic acid, α-ketobutyric acid, α-ketovaleric acid, maltose, mannitol, monomethyl succinate, L-ornithine, L-proline, L-serine, and L-threonine. The following compounds are not oxidized: acetic acid, *N*-acetyl-D-galactosamine, *N*-acetyl-D-glucosamine, *cis*-aconitic acid, adonitol, alaninamide, D-alanine, L-alanine, γ-aminobutyric acid, 2-aminoethanol, L-arabinose, D-arabitol, L-asparagine, bromosuccinic acid, 2,3-butanediol, DL-carnitine, cellobiose, citric acid, i-erythritol, D-fructose, formic acid, L-fucose, D-galactose, D-galactonic acid lactone, D-galacturonic acid, gentiobiose, D-gluconic acid, D-glucosaminic acid, α-D-glucose, glucose 1-phosphate, glucose 6-phosphate, glucuronamide, D-glucuronic acid, glycerol, DL-α-glycerol phosphate, glycyl L-aspartic acid, glycyl L-glutamic acid, L-histidine, α-, β- and γ-hydroxybutyric acids, *p*-hydroxyphenylacetic acid, hydroxy-L-proline, *myo*-inositol, itaconic acid, α-ketoglutaric acid, DL-lactic acid, α-D-lactose, lactulose, L-leucine, malonic acid, melibiose, methyl β-D-glucoside, methylpyruvic acid, phenylethylamine, L-phenylalanine, propionic acid, D-psicose, putrescine, L-pyroglutamic acid, quinic acid, raffinose, L-rhamnose, D-saccharic acid, sebacic acid, D-serine, D-sorbitol, succinamic acid, succinic acid, sucrose, trehalose, turanose, Tweens 40 and 80, urocanic acid, uridine, thymidine, and xylitol. Mannose is oxidized only by the type strain CKA-5^T and inosine is oxidized only by reference strain CKA-36.

Source: the type strain was isolated from marine sediment collected on the Pacific Ocean coast at Katsuura, Japan.

DNA G+C content (mol%): 35–37 (HPLC).

Type strain: CKA-5, LMG 23247, NBRC 100970.

Sequence accession no. (16S rRNA gene): AB206954.

Genus XLVIII. **Sediminicola** Khan, Nakagawa and Harayama 2006b, 843[VP]

THE EDITORIAL BOARD

Se.di.mi.ni.co'la. L. n. *sedimen -inis* sediment; L. masc. suff. *-cola* (from L. n. *incola*) an inhabitant; N.L. masc. n. *Sediminicola* an inhabitant of sediment, referring to the source of the strains.

Rods, 0.5–0.7 × 3–5 μm. Nonmotile. Gram-stain-negative. Colonies are golden-yellow. Flexirubin-type pigments are not produced. **Aerobic.** Catalase-positive. Weakly positive for oxidase. Growth occurs at 10–30°C; optimum, 20°C. **Growth occurs in 10–100% strength artificial seawater; NaCl alone is not sufficient to support growth. Nitrate is reduced. Gelatin, casein, and starch are hydrolyzed.** Predominant cellular fatty acids are $C_{15:0}$ iso, $C_{15:0}$ anteiso, $C_{15:1}$ iso, $C_{17:1}$ iso ω9*c*, $C_{17:0}$ iso 3-OH, and summed feature 3 ($C_{16:1}$ iso ω7*c*/i-$C_{15:0}$ 2-OH). **Pentadecanoic acid ($C_{15:0}$) is absent.** Isolated from sediments collected on the shores of the Pacific Ocean and the Sea of Japan.

DNA G+C content (mol%): 38–40.

Type species: **Sediminicola luteus** Khan, Nakagawa and Harayama 2006b, 843[VP].

Enrichment and isolation procedures

Isolation can be accomplished on marine agar 2216 (Difco). Cultures are grown at 20°C. Long-term preservation can be achieved by storing the organisms at −80°C in artificial seawater containing 20% (v/v) glycerol.

Differentiation of the genus *Sediminicola* from other genera

Sediminicola strains can be differentiated phenotypically from those of other genera by the following combination of features: the ability to reduce nitrate; the ability to degrade gelatin, casein, and starch; and a lack of pentadecanoic acid ($C_{15:0}$).

Taxonomic comments

Analysis of 16S rRNA gene sequences indicates that members of the genus *Sediminicola* have 87.9–90.5% sequence similarity to those of the following genera: *Arenibacter*, *Maribacter*, *Muricauda*, *Robiginitalea*, *Ulvibacter*, *Vitellibacter*, and *Zobellia*.

The four *Sediminicola luteus* strains isolated so far exhibit 99.3–99.7% rRNA gene sequence similarity to one another. They also show DNA–DNA reassociation values of 93–104% with each other.

List of species of the genus *Sediminicola*

1. **Sediminicola luteus** Khan, Nakagawa and Harayama 2006b, 843[VP]

lu'te.us. L. masc. adj. *luteus* golden yellow, because the colony color is golden yellow.

The characteristics are as described for the genus, with the following additional features. Seven-day-old cells form filaments of 5–10 μm. Gelatin, casein, and starch are hydrolyzed, but not urea, chitin, cellulose, carboxymethylcellulose, or DNA. Nitrate is reduced. Indole is not produced from tryptophan. Acid is not produced from glucose. The following compounds are oxidized (Biolog GN2 MicroPlate system): *N*-acetyl-D-galactosamine, L-aspartic acid, cellobiose, α-cyclodextrin, dextrin, D-fructose, D-galactose, gentiobiose, α-D-glucose, glucose 1-phosphate, L-glutamic acid, glycogen, glycyl L-aspartic acid, glycyl L-glutamic acid, DL-lactic acid, α-D-lactose, lactulose, maltose, D-mannitol, D-mannose, melibiose, methyl β-D-glucoside, L-ornithine, L-proline, raffinose, sucrose, L-threonine, trehalose, and turanose. Does not oxidize the following compounds: acetic acid, *N*-acetyl D-glucosamine, *cis*-aconitic acid, adonitol, alaninamide, D-alanine, L-alanine, L-alanyl glycine, γ-aminobutyric acid, 2-aminoethanol, L-asparagine, bromosuccinic acid, 2,3-butanediol, D-arabitol, DL-carnitine, citric acid, D-galactonic acid lactone, D-galacturonic acid, D-gluconic acid, D-glucosaminic acid, D-glucuronic acid, glycerol, DL-α-glycerol phosphate, α-, β- and γ-hydroxybutyric acid, *p*-hydroxyphenylacetic acid, i-erythritol, formic acid, L-fucose, glucose 6-phosphate, glucuronamide, inosine, *myo*-inositol, itaconic acid, L-histidine, hydroxy-L-leucine, α-ketoglutaric acid, α-ketovaleric acid, malonic acid, methyl pyruvic acid, monomethyl succinic acid, phenylethylamine, L-phenylalanine, propionic acid, putrescine, L-pyroglutamic acid, quinic acid, L-rhamnose, D-serine, L-serine, D-sorbitol, xylitol, D-saccharic acid, sebacic acid, succinamic acid, succinic acid, thymidine, Tween 40, and Tween 80.

Source: the type strain was isolated from marine sediment on the shore of the Sea of Japan.

DNA G+C content (mol%): 38–40 (HPLC).

Type strain: CNI-3, LMG 23246, NBRC 100966.

Sequence accession no. (16S rRNA gene): AB206957.

Genus XLIX. **Sejongia** Yi, Yoon and Chun 2005b, 414[VP]

Hana Yi and Jongsik Chun

Se.jong'i.a. N.L. fem. n. *Sejongia* named after the King Sejong Station, where the type species was isolated.

Rods with parallel sides and rounded ends, generally 0.4–0.5 × 1.0–3.1 µm. Gram-stain-negative. Spores are not formed. Nonmotile. **Do not glide** or spread. **Aerobic**, exhibiting only weak growth under microaerobic and anaerobic conditions. **Pychrotolerant**, with optimum growth temperatures of 19–22°C. Optimum growth is observed at pH 7–8 and in 0% NaCl. Growth occurs on R2A, Anacker and Ordal's agar (AOA), nutrient agar (NA) and tryptic soy agar (TSA), but not on cetrimide or MacConkey agars. Colonies are convex, translucent, circular, glistening, butyrous, and **yellow** with entire margins, becoming mucoid after prolonged incubation on R2A. Do not adhere to agar plates. Catalase- and oxidase-positive. Congo red is not adsorbed. The maximum absorption peak of the yellow pigment in ethanolic extracts is at 452 nm and the next shoulder peak is at 480 nm. **No flexirubin-type pigments are produced.** Acid is produced aerobically or fermentatively from some carbohydrates. The major isoprenoid quinone is MK-6. The predominant cellular fatty acids are $C_{15:0}$ iso (12.2–13.6%), $C_{15:0}$ anteiso (15.2–24.2%), and $C_{17:1}$ iso ω9*c* (8.6–21.3%).

DNA G+C content (mol%): 34–36.

Type species: **Sejongia antarctica** Yi, Yoon and Chun 2005b, 414[VP].

Further descriptive information

Based on 16S rRNA gene sequence analysis, the genus is phylogenetically related to the genera *Bergeyella*, *Chryseobacterium*, *Kaistella*, and *Riemerella* in the family *Flavobacteriaceae*. See *Taxonomic comments* section for detailed information. Differential characteristics for the species are given in Table 67 and other characteristics are given in Table 68.

Fatty acids. The dominant fatty acids in *Sejongia antarctica* are $C_{15:0}$ iso, $C_{15:0}$ anteiso, and $C_{17:1}$ iso ω9*c*; those in *Sejongia jeonii* are $C_{15:0}$ iso and $C_{15:0}$ anteiso. Fairly large differences in the amount of $C_{15:0}$ anteiso (15.2% vs 24.2%) and $C_{17:1}$ iso ω9c (21.3% vs 8.6%) are found between *Sejongia antarctica* and *Sejongia jeonii.* Overall, the fatty acid compositions of *Sejongia* species are similar to those of phylogenetically related species, but they differ from them somewhat in quantities. This may result from different growth media and the low growth temperature of *Sejongia* spp. compared with other mesophilic species. The full fatty acid composition of *Sejongia antarctica* is $C_{17:1}$ iso ω9*c* (21.3%), $C_{15:0}$ anteiso (15.2%), $C_{15:0}$ iso (13.6%), $C_{15:1}$ anteiso A (6.6%), $C_{17:0}$ iso 3-OH (5.6%), $C_{16:0}$ iso 3-OH (5.1%), $C_{16:1}$ iso $_H$ (3.6%), $C_{17:0}$ 2-OH (3.3%), $C_{13:0}$ anteiso (3.2%), $C_{16:0}$ iso (2.8%), summed feature 3 ($C_{15:0}$ iso 2-OH and/or $C_{16:1}$ ω7*c*/*t*) (2.7%), $C_{15:0}$ (2.6%), $C_{13:0}$ iso (2.5%), $C_{17:1}$ anteiso ω9*c* (2.5%), $C_{15:0}$ 2-OH (1.9%), $C_{14:0}$ iso (1.5%), and $C_{18:1}$ ω5*c* (1.5%); trace amounts (less than 1% of total fatty acids) of $C_{15:0}$ iso 3-OH, $C_{12:0}$ 3-OH, ECL (equivalent chain-length, i.e., the identity of the fatty acids is unknown) 13.566, $C_{15:1}$ iso F, $C_{12:0}$ iso, $C_{14:0}$ iso 3-OH, ECL 16.580, and $C_{17:1}$ ω8*c* are also present. The full fatty acid composition of *Sejongia jeonii* is $C_{15:0}$ anteiso (24.2%), $C_{15:0}$ iso (12.2%), $C_{16:1}$ iso $_H$ (9.1%), $C_{16:0}$ iso 3-OH (9.0%), $C_{17:1}$ iso ω9*c* (8.6%), $C_{16:0}$ iso (5.7%), $C_{14:0}$ iso (5.0%), $C_{17:0}$ iso 3-OH (4.4%), $C_{13:0}$ anteiso (3.6%), $C_{13:0}$ iso (2.9%), summed feature 3 (2.6%), $C_{17:0}$ 2-OH (2.3%), $C_{17:1}$ anteiso ω9*c* (1.9%), $C_{15:0}$ 2-OH (1.9%), $C_{15:0}$ (1.5%), $C_{15:0}$ iso 3-OH (1.3%), and $C_{12:0}$ iso (1.0%); trace amounts of $C_{18:1}$ ω5*c*, $C_{14:0}$ iso 3-OH, ECL 16.580, $C_{17:1}$ ω6*c*, ECL 13.566, and $C_{14:0}$ are also present.

TABLE 67. Characters useful for differentiation of the species of *Sejongia*[a]

Characteristic	*S. antarctica*	*S. jeonii*
Growth at 30°C	–	+
Elastinase	–	+
Lecithinase	w	+
L-Tyrosine hydrolysis	del	–
Fermentative acid production from D-mannose	–	+
$C_{15:0}$ anteiso (% of total fatty acids)	15	24
$C_{17:1}$ iso ω9*c* (% of total fatty acids)	21	9
DNA G+C content (mol%)	34	36

[a]Symbols: +, positive; –, negative; w, weak; del, delayed.

TABLE 68. Other characteristics of *Sejongia* species[a]

Characteristic	*S. antarctica*	*S. jeonii*
Yellow pigment	+	+
Flexirubin production, Congo red adsorption, NaCl requirement, growth on cetrimide/ MacConkey agars, nitrate and nitrite reduction, H_2S production, alkaline reaction on Christensen's citrate	–	–
Relation to oxygen	Aerobic	Aerobic
Temperature range (optimum) (°C)	4–28 (19)	4–31 (22)
Growth on R2A/AOA/NA/TSA	+	+
Catalase, oxidase	+	+
Indole production	+	+
Amylase, caseinase, esculinase, DNase, gelatinase, L-phenylalanine deaminase, Tweenase	+	+
Agarase, alginase, arginine dihydrolase, β-galactosidase, cellulase, chitinase, pectinase, urease	–	–
API ZYM results:		
Alkaline phosphatase, esterase lipase (C8), leucine arylamidase, valine arylamidase, acid phosphatase, naphthol-AS-BI-phosphohydrolase, α-glucosidase	+	+
Esterase (C4), lipase (C14), cystine arylamidase, trypsin, α-chymotrypsin, α-galactosidase, β-galactosidase, β-glucuronidase, β-glucosidase, *N*-acetyl-β-glucosaminidase, α-mannosidase, α-fucosidase	–	–
Oxidative acid production from:		
D-Glucose, maltose	+	+
L-Arabinose, D-cellobiose, D-fructose, lactose, D-mannitol, D-raffinose, L-rhamnose, D-salicin, sucrose, D-trehalose, D-xylose	–	–
Fermentative acid production from:		
D-Glucose, maltose, starch, glycogen	+	+
Glycerol, erythritol, D-arabinose, L-arabinose, D-ribose, D-xylose, L-xylose, D-adonitol, methyl β-D-xylopyranoside, D-galactose, D-fructose, L-sorbose, L-rhamnose, dulcitol, inositol, D-mannitol, D-sorbitol, methyl α-D-mannopyranoside, methyl α-D-glucopyranoside, *N*-acetylglucosamine, amygdalin, arbutin, esculin ferric citrate, salicin, cellobiose, D-lactose (bovine origin), melibiose, sucrose, trehalose, inulin, melezitose, raffinose, xylitol, gentiobiose, turanose, D-lyxose, tagatose, D-fucose, L-fucose, D-arabitol, L-arabitol, potassium gluconate, potassium 2-ketogluconate, potassium 5-ketogluconate	–	–
Major fatty acids	$C_{15:0}$ iso, $C_{15:0}$ anteiso, $C_{17:1}$ iso ω9*c*	$C_{15:0}$ iso, $C_{15:0}$ anteiso
Isoprenoid quinone	MK-6	MK-6

[a]Symbols: +, positive; –, negative.

Colonial characteristics. Freshly grown cells are catalase- and oxidase-positive. Colonies of *Sejongia* appear convex, translucent, circular, glistening, and butyrous with entire margins; they then become mucoid after prolonged incubation and do not adhere to the agar. The colonial color is bright to pale yellow depending on growth media. The maximum absorption peak of the yellow pigment in ethanolic extracts is at 452 nm and the next shoulder peak is at 480 nm. Flexirubins are not detected from a bathychromatic shift test of the absorption peak or from a color change test of cell mass by addition of KOH solution.

Motility. Gliding motility is not observed by direct microscopic examination of the edge of colonies in exponential phase on AOA agar (Anacker and Ordal, 1955), R2A medium (Reasoner and Geldreich, 1985), and CY agar (casitone, 3 g; yeast extract, 1 g; $CaCl_2 \cdot 2H_2O$, 1 g; sea salts, 40 g; agar, 15 g; distilled water, 1 l) plates. Cellular motility is not observed by the hanging drop technique for cells in exponential phase in R2A and CY broth. Congo red is not adsorbed by directly flooding colonies on agar plates with 0.01% aqueous Congo red solution.

Nutrition and growth conditions. Maximum growth is observed on R2A and abundant growth is observed on AOA, NA, and TSA. No growth is observed on cetrimide or MacConkey agars. The organisms are aerobic and grow poorly under microaerobic conditions [with 5–15% O_2 and 5–12% CO_2 created by the Campy Pak Plus system (BBL)] and anaerobic conditions [with 4–10% CO_2 created by the GasPak Plus system (BBL)]. Microaerobic growth is better than anaerobic growth. Growth is observed at pH 6–11 (optimum pH 7–8) and 0–3% NaCl (optimum 0%). The temperature range for growth determined by a temperature-gradient incubator (TVS 126MA; Advantec) in R2A broth is 5–30°C. Cardinal temperature plots based on the Ratkowsky temperature growth model (Ratkowsky et al., 1983) are given in Figure 50. The notional minimum, optimum, and maximum growth temperatures for *Sejongia antarctica* are –16.6, 18.9, and 28.2°C and those for *Sejongia jeonii* are –10.9, 21.5, 30.9°C, respectively. According to Isaksen and Jorgensen (1996) who defined psychrotolerant bacteria as those that are able to grow at ≤0°C, but not at >35°C and have temperature optima at ≤25°C, the two *Sejongia* species are psychrotolerant.

Metabolism and metabolic pathways. Neither nitrate nor nitrite is reduced. Indole is produced, but H_2S is not. There is no alkaline reaction on Christensen's citrate agar (Christensen, 1949). L-Phenylalanine deaminase activity is present, but arginine dihydrolase, β-galactosidase, and urease activities

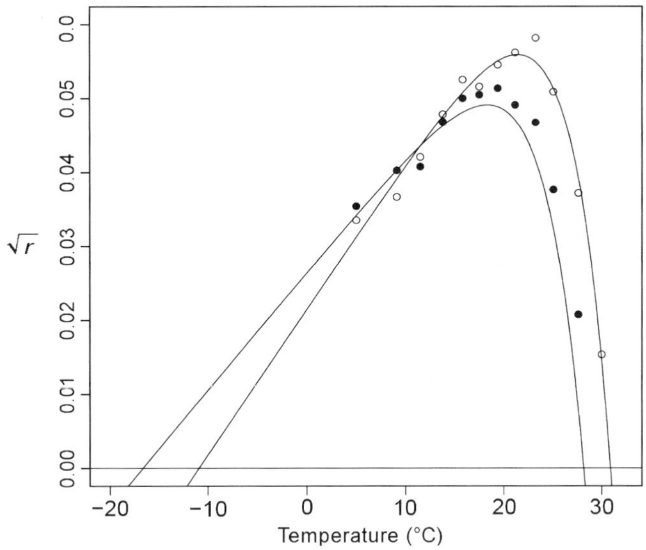

FIGURE 50. Fitted Ratkowsky model of growth versus temperature data for *Sejongia antarctica* (filled circles) and *Sejongia jeonii* (open circles). √*r* is the square root of the growth rate.

are absent. Casein, esculin, DNA, gelatin, egg yolk, starch, and Tween 80 are decomposed, but agar, alginate, carboxymethylcellulose, chitin, and pectin are not. Positive reactions for both lecithinase and Tween esterase are delayed in *Sejongia antarctica*. L-Tyrosine is hydrolyzed weakly by *Sejongia antarctica*, but not by *Sejongia jeonii*. Elastin is decomposed by *Sejongia jeonii*, but not by *Sejongia antarctica*. None of the compounds contained in API 20NE kits (glucose, arabinose, mannose, mannitol, *N*-acetylglucosamine, maltose, gluconate, caprate, adipate, malate, citrate, and phenylacetate) are assimilated as a sole carbon source. Alkaline phosphatase, esterase lipase (C8), leucine arylamidase, valine arylamidase, acid phosphatase, naphthol-AS-BI-phosphohydrolase, and α-glucosidase are positive. Esterase (C4), lipase (C14), cystine arylamidase, trypsin, α-chymotrypsin, α-galactosidase, β-galactosidase, β-glucuronidase, β-glucosidase, *N*-acetyl-β-glucosaminidase, α-mannosidase, and α-fucosidase are negative with the API ZYM kit. Acid is produced aerobically from D-glucose and maltose, but not from L-arabinose, cellobiose, D-fructose, lactose, D-mannitol, raffinose, L-rhamnose, D-salicin, sucrose, trehalose, or D-xylose on the modified O/F agar of Leifson (1963) (casitone, 1.0 g; yeast extract, 0.1 g; ammonium sulfate, 0.5 g; Tris buffer, 0.5 g; phenol red, 0.01 g; Bacto agar, 15 g; distilled water, 1 l; adjusted to pH 7.0). Acid is produced fermentatively from D-glucose, maltose, starch, and glycogen, but not from glycerol, erythritol, D-arabinose, L-arabinose, D-ribose, D-xylose, L-xylose, D-adonitol, methyl β-D-xylopyranoside, D-galactose, D-fructose, D-mannose, L-sorbose, L-rhamnose, dulcitol, inositol, D-mannitol, D-sorbitol, methyl α-D-mannopyranoside, methyl α-D-glucopyranoside, *N*-acetylglucosamine, amygdalin, arbutin, esculin ferric citrate, salicin, cellobiose, D-lactose (bovine origin), melibiose, sucrose, trehalose, inulin, melezitose, raffinose, xylitol, gentiobiose, turanose, D-lyxose, tagatose, D-fucose, L-fucose, D-arabitol, L-arabitol, potassium gluconate, potassium 2-ketogluconate, or potassium 5-ketogluconate with the API 50 CH kit using API 50 CHB/E medium. *Sejongia jeonii* produces acid fermentatively from D-mannose, but *Sejongia antarctica* does not.

Ecology. The two *Sejongia* species with validly published names were cold-adapted psychrotolerant bacteria isolated from Antarctica (Yi et al., 2005b). *Sejongia antarctica* and *Sejongia jeonii* were isolated from an eutrophic terrestrial environment, i.e., a soil and a moss sample from a penguin habitat, near the King Sejong Station on King George Island, Antarctica.

Enrichment and isolation procedures

Terrestrial samples for isolation of psychrotolerant *Sejongia* strains should be kept at a low temperature. Initial enrichment of samples in refrigerated liquid media at 4°C for 1–2 d may enhance the isolation. The enriched samples can be applied directly to isolation media and incubated at low temperatures (4–10°C) for up to 1 month.

Maintenance procedures

Freshly grown cultures may be maintained on agar plates at 4°C by monthly transfer. For long-term storage, a dense suspension of cells in 20% glycerol may be maintained at −80°C or may also be freeze-dried.

Differentiation from closely related taxa

Members of the genus *Sejongia* can be differentiated from those of the genus *Chryseobacterium* by their lack of flexirubin-type pigments, and from those of the genera *Bergeyella* and *Riemerella* by their yellow pigmentation.

Taxonomic comments

The two *Sejongia* species, i.e., *Sejongia antarctica* and *Sejongia jeonii*, show 97.7% 16S rRNA gene sequence similarity, corresponding to 31 nucleotide differences. They belong to separate genomic species, sharing a low DNA–DNA relatedness value of 27% based on slot-blot DNA–DNA hybridization (Yi et al., 2005b). Based on 16S rRNA gene sequence similarities, the closest bacterial relatives of the genus *Sejongia* are *Chryseobacterium* (91.9–95.7% for *Sejongia antarctica*; 93.4–96.3% for *Sejongia jeonii*), *Elizabethkingia* (92.5–93.3%; 93.3–93.9%), *Riemerella* (92.6–93.0%; 92.3–93.5%), *Bergeyella* (92.6%; 92.5%), and *Kaistella* (92.5%; 93.3%). This relationship between *Sejongia* spp. and other members of the above-mentioned genera is also highlighted in the phylogenetic tree (Figure 51), a neighbor-joining tree of *Sejongia* and related taxa based on 16S rRNA gene sequences. The genera *Sejongia*, *Bergeyella*, *Chryseobacterium*, *Elizabethkingia*, *Kaistella*, and *Riemerella* form a monophyletic suprageneric clade in the family *Flavobacteriaceae*. The branching order within this suprageneric group and the position of *Sejongia* vary depending on the tree-making algorithm used. The two *Sejongia* species were consistently recovered as a robust monophyletic clade that could be differentiated from *Chryseobacterium* spp. at the time of its description (Yi et al., 2005b). Young et al. (2005) simultaneously described a novel *Chryseobacterium* species named *Chryseobacterium formosense*. Based on comprehensive phylogenetic analysis, *Chryseobacterium formosense* is phylogenetically placed between *Sejongia* and *Chryseobacterium* spp., although its position is not stable (Figure 51). *Chryseobacterium formosense* shows high 16S rRNA gene sequence similarities to both *Chryseobacterium* (95.3–97.4%) and *Sejongia* (95.7–96.3%) strains. The generic status of *Chryseobacterium*

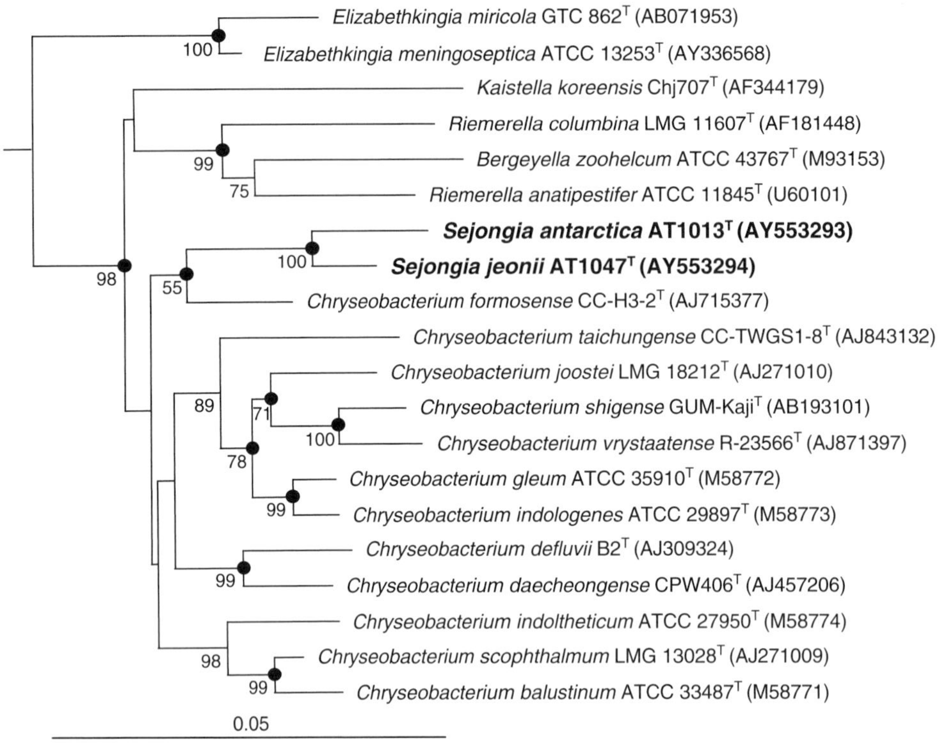

FIGURE 51. Neighbor-joining tree of the genus *Sejongia* and related taxa based on 16S rRNA gene sequences. *Numbers* at nodes are percentages of bootstrap support (>50%) from 1000 resampled datasets. *Filled circles* indicate that the corresponding nodes (groupings) were also recovered in maximum-likelihood and maximum-parsimony trees. *Flavobacterium aquatile* (M62797), *Ornithobacterium rhinotracheale* (U87101), *Empedobacter brevis* (D14022), and *Weeksella virosa* (M93152) were used as outgroups (not shown). The manual sequence alignment based on bacterial 16S rRNA secondary structure and phylogenetic analysis was performed using the program JPHYDIT (Jeon et al., 2005). Bar = 0.05 nt substitution per position.

formosense should be re-evaluated using additional physiological and chemotaxonomic properties.

Although the genus *Sejongia* forms a distinct monophyletic clade within the suprageneric group containing the genera *Bergeyella, Chryseobacterium, Elizabethkingia, Kaistella*, and *Riemerella*, the distinctness of the generic status of *Sejongia* has become less clear by the addition of *Chryseobacterium formosense*. Because *Sejongia* and *Chryseobacterium formosense* were validly published simultaneously (Yi et al., 2005b; Young et al., 2005), the relationship between these two taxa was not reported at the time of their description. However, *Sejongia* spp. and *Chryseobacterium formosense* can be easily differentiated at the genus level by the presence of flexirubin. *Chryseobacterium formosense*, like all chryseobacteria, has flexirubin, but *Sejongia* species do not. The taxonomic status of *Chryseobacterium formosense* should be re-evaluated with additional phenotypic and genotypic methods in future.

List of species of the genus *Sejongia*

1. **Sejongia antarctica** Yi, Yoon and Chun 2005b, 414[VP]

ant.arc'ti.ca. L. fem. adj. *antarctica* southern, and by extension of or belonging to Antarctic, the geographical origin of the type strain.

Cells are approximately 1.0–3.1 × 0.4–0.5 μm. Grows at 4–28°C with notional optimum of 18.9°C. Does not decompose elastin. Hydrolyzes L-tyrosine weakly. Does not produce acid fermentatively from D-mannose.

Source: the type strain was isolated from a soil sample of penguin habitats near the King Sejong Station on King George Island, Antarctica.

DNA G+C content (mol%): 34 (HPLC).

Type strain: AT1013, IMSNU 14040, KCTC 12225, JCM 12381.

Sequence accession no. (16S rRNA gene): AY553293.

2. **Sejongia jeonii** Yi, Yoon and Chun 2005b, 414[VP]

je.on'i.i. N.L. gen. masc. n. *jeonii* of Jeon, named in honor of the late Jae Gyu Jeon, who devoted his life to polar research.

Cells are approximately 1.0–3.1 × 0.4–0.5 μm. Grows at 4–31°C with notional optimum of 21.5°C. Decomposes elastin. Does not hydrolyze L-tyrosine. Produces acid fermentatively from D-mannose.

Source: the type strain was isolated from a moss sample of penguin habitats near the King Sejong Station on King George Island, Antarctica.

DNA G+C content (mol%): 36 (HPLC).

Type strain: AT1047, IMSNU 14049, KCTC 12226, JCM 12382.

Sequence accession no. (16S rRNA gene): AY553294.

Genus L. **Stenothermobacter** Lau, Tsoi, Li, Plakhotnikova, Dobretsov, Wu, Wong, Pawlik and Qian 2006, 183[VP]

THE EDITORIAL BOARD

Ste.no.ther.mo.bac'ter. Gr. adj. *stenos* narrow; Gr. adj. *thermos* hot; N.L. masc. n. *bacter* rod; N.L. masc. n. *Stenothermobacter* a rod with narrow temperature range, pertaining to the narrow temperature range that supports growth of the type strain.

Rods, >2.5 µm in length with tapered ends, forming chains of up to four cells. Gram-stain-negative. **Slow gliding motility is present.** Colonies are orange. Flexirubin-type pigments are not produced. Temperature range for growth, 20.0–36.0°C; optimum, 28.0–30.0°C. **Require 2.0–6.0% NaCl for growth. Strictly aerobic.** Chemo-organotrophic. Oxidase-positive. Catalase activity is very weak. Acetoin is produced, but not H_2S. **DNA and casein are not hydrolyzed. β-Galactosidase and α-glucosidase are produced, but not β-glucosidase.** MK-6 is the only respiratory quinone. Isolated from the marine sponge *Lissodendoryx isodictyalis.*

DNA G+C content (mol%): 41.0

Type species: **Stenothermobacter spongiae** Lau, Tsoi, Li, Plakhotnikova, Dobretsov, Wu, Wong, Pawlik and Qian 2006, 184[VP].

Enrichment and isolation procedures

Stenothermobacter spongiae was isolated from tissue of the marine sponge *Lissodendoryx isodictyalis* on a special marine agar (SMA) consisting of 5.0 g peptone and 3.0 g yeast extract dissolved in seawater. The medium was sterilized by filtration (0.22 µm pore diameter) and solidified with agar. Orange colonies 2–4 mm in diameter were selected after 48 h at 30°C.

Differentiation of the genus *Stenothermobacter* from related genera

Stenothermobacter spongiae can be differentiated from *Nonlabens tegetincola* by its gliding motility, higher DNA G+C content (41.0 mol% vs 33.6 mol%), inability to hydrolyze DNA, narrower temperature range for growth (20–36°C vs 12–44°C), and by its α-glucosidase and β-galactosidase activities. It can be differentiated from *Donghaena dokdonensis* by its gliding motility, β-galactosidase activity, inability to hydrolyze casein, and higher DNA G+C content (41.0 mol% vs 36.9 mol%).

Taxonomic comments

Analysis of 16S rRNA gene sequences by Lau et al. (2006) indicated that the nearest neighbors of *Stenothermobacter spongiae* were the type strains of *Nonlabens tegetincola* (93.3% sequence similarity) and *Donghaena dokdonensis* (93.6%). The sequence similarity to strains of other species of the *Flavobacteriaceae* was 90.9% or less.

List of species of the genus *Stenothermobacter*

1. **Stenothermobacter spongiae** Lau, Tsoi, Li, Plakhotnikova, Dobretsov, Wu, Wong, Pawlik and Qian 2006, 184[VP]

spon'gi.ae. L. gen. n. *spongiae* of a sponge, pertaining to the isolation source of the type strain.

The description is as given for the genus, with the following additional features. Colonies on SMA are orange, circular, 2.0–4.0 mm in diameter, and convex with smooth surfaces and entire margins. No diffusible pigment is produced. pH range for growth is 6.0–10.0. Acetoin is produced, but not indole or H_2S. Nitrate is not reduced. Gelatin, starch, and Tweens 20, 40 and 80 are hydrolyzed, but not agar, casein, carboxymethylcellulose, chitin, or DNA. The following enzyme activities are present: acid phosphatase, alkaline phosphatase, β-galactosidase, α-glucosidase, α-chymotrypsin, cystine arylamidase, leucine arylamidase, valine arylamidase, esterase (C4), esterase lipase (C8), lipase (C14), trypsin, and naphthol-AS-BI-phosphohydrolase. No activity is found for *N*-acetyl-β-glucosaminidase, arginine dihydrolase, α-fucosidase, α-galactosidase, β-glucosidase, β-glucuronidase, α-mannosidase, lysine decarboxylase, ornithine decarboxylase, tryptophan deaminase, or urease. Growth, but not acid production, occurs with the following sole carbon sources (API 20E, API 20NE, and API 50CH systems): D-arabinose, D-galactose, D-glucose, glycerol, D-mannitol, melibiose, D-sorbitol, starch, and sucrose. The following carbon sources are utilized (MicroLog 3 system): acetic acid, L-alaninamide, L-alanine, L-alanylglycine, L-aspartic acid, L-glutamic acid, glycyl-L-aspartic acid, glycyl-L-glutamic acid, α-ketoglutaric acid, α-ketovaleric acid, maltose, melibiose, monomethyl succinate, L-ornithine, propionic acid, L-proline, raffinose, and sucrose. Susceptibility is exhibited toward ampicillin, chloramphenicol, penicillin, streptomycin, and tetracycline, but not to kanamycin. The predominant fatty acids (>5%) are $C_{15:0}$ anteiso, $C_{15:0}$ iso, $C_{15:0}$ iso 3-OH, $C_{17:0}$ iso 3-OH, $C_{17:1}$ iso ω9*c*, and summed feature 3 (comprising $C_{15:0}$ iso 2-OH and/or $C_{16:1}$ ω7*c*) (altogether representing 76.2% of the total fatty acids).

Source: the type strain was isolated from tissue of the marine sponge *Lissodendoryx isodictyalis* in the Bahamas.

DNA G+C content (mol%): 41 (HPLC).

Type strain: UST030701-156, JCM 13191, NRRL B-41138.

Sequence accession no. (16S rRNA gene): DQ064789.

Genus LI. **Subsaxibacter** Bowman and Nichols 2005, 1481[VP]

JOHN P. BOWMAN

Sub.sa.xi.bac'ter. L. pref. *sub* below; L. neut. n. *saxum* stone; N.L. masc. n. *bacter* rod; N.L. masc. n. *Subsaxibacter* bacterial rod living below stone.

Small coccibacilli, 0.4–1.0 × 0.3–0.5 µm. **Gram-stain-negative.** Cells occur singly or in pairs. Spores and resting cells are not present. Gas vesicles and helical or ring-shaped cells are not formed. **Gliding motility is present. Strictly aerobic. Oxidative metabolism. Catalase-positive. Oxidase weakly positive. Chemoheterotrophic. Colonies are orange** due to production of carotenoids. **Flexirubin pigments are not produced. Sodium ions** are required for growth. Best growth occurs in organic media containing seawater salts. **Psychrophilic.** Grows between −2 and 20°C, with temperature optima at 10–15°C in seawater or NaCl-containing media. Neutrophilic, with optimal growth occurring at about pH 7.5–8.0. The major fatty acids include $C_{15:1}$ iso ω10*c*, $C_{15:1}$ anteiso ω10*c*, $C_{15:0}$ iso, $C_{15:0}$ anteiso, $C_{16:1}$ ω7*c*, $C_{17:1}$, and $C_{16:0}$ 3-OH iso. The major isoprenoid quinone is **menaquinone-6.** Habitat: Antarctic quartz stone sublithic cyanobacterial biofilms.

DNA G+C content (mol%): 35.

Type species: **Subsaxibacter broadyi** Bowman and Nichols 2005, 1481[VP].

Further descriptive information

The genus *Subsaxibacter* is a member of the family *Flavobacteriaceae* and is represented currently by a single species, *Subsaxibacter broadyi* (Bowman and Nichols, 2005). This species was isolated from cyanobacteria-dominated biofilm material obtained from the undersides of quartz stones, partially buried in feldfield soils of an ice-free coastal area of Antarctica (see the treatment of the genera *Gelidibacter* and *Subsaximicrobium* for more details) (Smith et al., 2000). Physiologically, the genus has many features that are typical of the family *Flavobacteriaceae* (Bernardet et al., 2002). Based on currently available cultivation and molecular data (published and unpublished data from the National Center of Biotechnology Information; http://www.ncbi.nlm.nih.gov), the genus *Subsaxibacter* has yet to be isolated from other saline habitats; however, a cultivation-independent survey of corroding concrete samples from a sewer system (Okabe et al., 2007) detected 16S rRNA gene sequences that grouped with those of *Subsaxibacter*.

Isolation and maintenance procedures

Strains of *Subsaxibacter* can be potentially isolated directly from quartz stone biofilm scrapings thoroughly suspended in a 1:10 dilution of trypticase soy broth (Oxoid) supplemented with 1% (w/v) NaCl. The suspension is then serially diluted onto the same medium containing 1.5% agar and incubated aerobically in the dark at 2–10°C for several weeks. Bright orange viscid colonies that form slowly may be indicative of *Subsaxibacter* strains. Routine cultivation should be performed aerobically on marine agar 2216 at 10–15°C.The species is slow-growing compared to related species of the genus *Subsaximicrobium* and colonies may take up to 2 weeks to form. Strains can be maintained as frozen suspensions in marine broth 2216 containing 20–30% (v/v) glycerol at −70°C. Strains may survive for several months, potentially years, on thick agar marine agar 2216 plates or on slants, stored at 2–4°C.

Differentiation of the genus *Subsaxibacter* from other genera

The most useful traits for differentiating *Subsaxibacter* from other members of the *Flavobacteriaceae* include: orange pigmentation; lack of flexirubin pigments; ability to glide; strictly oxidative metabolism; inability to form acid from carbohydrates; requirement for low temperatures for growth (e.g., <15–20°C) and no growth at 25°C; requirement for sodium ions for growth (good growth on sea salts-containing media); and lack of agarolytic activity. These basic traits can be used to readily differentiate *Subsaxibacter* from most of its relatives in the family *Flavobacteriaceae. Subsaxibacter* strains differ from those of *Subsaximicrobium*, which share the same quartz stone habitat, in being much more slow-growing, forming comparatively viscid colonies, possessing a relatively narrow salinity tolerance, requiring yeast extract for growth, being unable to hydrolyze most macromolecules, and having an overall general lack of bioactivity in standard biochemical tests. Several other genera of the family *Flavobacteriaceae* possess at least some proportion of the traits possessed by *Subsaxibacter* and, thus, any phenotypic-based identification may require 16S rRNA gene sequencing to achieve a completely definitive identification.

List of species of the genus *Subsaxibacter*

1. **Subsaxibacter broadyi** Bowman and Nichols 2005, 1481[VP]

broa.dy'i. N.L. gen. masc. n. *broadyi* of Broady, named in honor of P.A. Broady, an Antarctic microbiologist from New Zealand who pioneered the study of many Antarctic terrestrial biomes.

Characteristics are as described for the genus, with the following additional features. Colonies are bright orange, circular, convex, with an entire edge and a viscid consistency on marine agar 2216. No growth occurs at 25°C; optimal temperature for growth, 10–15°C. Growth occurs in 0.1–1.25 M NaCl, with optimal growth at seawater salinities (0.3–0.4 M NaCl). Grows well in media containing double-strength sea salts (70 g/l dehydrated sea salts), but no growth occurs with quadruple-strength sea salts. Does not require divalent cations found in seawater for growth; all salinity requirements can be met with NaCl. Grows well on trypticase soy agar (Oxoid) containing 1% (w/v) NaCl, but poor or no growth occurs on trypticase soy agar and nutrient agar (Oxoid). Requires yeast extract for growth. The following compounds are not utilized in a basal mineral salts medium containing yeast extract, a vitamin solution, and ammonium chloride: D-glucose, glycogen, maltose, sucrose, *N*-acetyl-D-glucosamine, L-rhamnose, DL-arabinose, L-fucose, D-mannose, melibiose, salicin, acetate, propionate, valerate, caprate, adipate, suberate, citrate, DL-3-hydroxybutyrate, DL-lactate, itaconate, phenylacetate, L-alanine, L-proline, L-histidine, and L-serine. Acid production from carbohydrates is undetectable in Leifson's oxidation/fermentation medium (Leifson, 1963). Gelatin and Tween 80 are hydrolyzed. Strains may also hydrolyze esculin. L-Tyrosine may be hydrolyzed weakly. Casein, elastin, starch, agar, chitin, carboxymethylcellulose, dextran, xylan, xanthine, uric acid, DNA, and urea are not hydrolyzed. Alkaline phosphatase is produced. Some strains are positive for glutamate decarboxylase activity. Activity not observed for arginine dihydrolase, lysine decarboxylase, or ornithine decarboxylase. H_2S is not produced from L-cysteine. Indole is not produced. No growth occurs in the Simmons' citrate test. Lipase (olive oil as the substrate) and lecithinase (egg yolk reaction) activity is not observed. Nitrate is not reduced. Negative for α-arabinosidase, α-galactosidase, α-glucosidase, β-glucosidase, β-galactosidase, 6-phospho-β-galactosidase, β-glucuronidase, *N*-acetyl-β-D-glucosaminidase, and α-fucosidase activity. Positive for arginine arylamidase, leucyl glycine arylamidase, phenylalanine arylamidase, leucine arylamidase, tyrosine arylamidase, glycine arylamidase, histidine arylamidase, serine arylamidase, alanine arylamidase, and glutamyl glutamate arylamidase. Negative for pyroglutamate arylamidase and proline arylamidase. Does not tolerate 1% (w/v) ox bile salts.

Source: the type strain was isolated from cyanobacterial biofilms attached to the undersides of partially buried quartz stones found in the Vestfold Hills, an ice-free region of East Antarctica.

DNA G+C content (mol%): 35 (T_m).

Type strain: P7, ACAM 1064, CIP 108527.

Sequence accession no. (16S rRNA gene): AY693999.

Genus LII. **Subsaximicrobium** Bowman and Nichols 2005, 1480[VP]

John P. Bowman

Sub.sa.xi.mi.cro'bi.um. L. pref. *sub* below; L. neut. n. *saxum* stone; N.L. neut. n. *microbium* microbe; N.L. neut n. *Subsaximicrobium* microbe living below stone.

Straight **rod-shaped cells,** 0.3–0.5 × 1–6 µm. In stationary growth phase cultures, cells may transform into spheroplasts, 1–3 µm in diameter. **Gram-stain-negative.** Cells occur singly or in pairs. Spores and resting cells are not present. Gas vesicles and helical or ring-shaped cells are not formed. **Gliding motility** is present. **Strictly aerobic. Oxidative metabolism. Catalase-positive. Oxidase-positive. Chemoheterotrophic. Colonies are orange** due to production of carotenoids. **Flexirubin pigments are not produced. Sodium ions are required** for growth. Best growth occurs in organic media containing seawater salts. **Psychrophilic.** Grows between −2 and 20°C, with temperature optima at 15–20°C in seawater or NaCl-containing media. Neutrophilic, with optimal growth occurring at about pH 7.5–8.0. The major fatty acids include $C_{15:1}$ iso ω10*c*, $C_{15:1}$ anteiso ω10*c*, $C_{15:0}$ iso, $C_{15:0}$ anteiso, $C_{16:1}$ ω7*c*, $C_{15:0}$ anteiso 3-OH, and $C_{16:0}$ iso 3-OH. The major isoprenoid quinone is **menaquinone-6**. Inhabits Antarctic quartz stone sublithic cyanobacterial biofilms.

DNA G+C content (mol%): 39–40.

Type species: **Subsaximicrobium wynnwilliamsii** Bowman and Nichols 2005, 1481[VP].

Further descriptive information

The genus *Subsaximicrobium* is a member of the family *Flavobacteriaceae*, represented currently by two closely related species, *Subsaximicrobium wynnwilliamsii* and *Subsaximicrobium saxinquilinus* (Bowman and Nichols, 2005). Both species were isolated from cyanobacteria-dominated biofilm material obtained from the undersides of quartz stones, partially buried in feldfield soils of an ice-free coastal area of Antarctica called the Vestfold Hills (latitude 78°S, longitude 68°E). The quartz stone sublithic habitat is an unusual oasis-like zone in which marine-type bacteria are found. Within the sublithic biofilms, these bacteria may be partially protected from the surrounding harsh soil conditions, i.e., regular diurnal freezing, continuous desiccation, and, during summer, relatively intense ultraviolet light radiation. The microbial community of quartz stone subliths, which includes *Subsaximicrobium* and species of its immediate relatives, *Subsaxibacter* and *Gelidibacter*, has been described using cultivation-dependent and -independent methods (Smith et al., 2000). Physiologically, the genus has features that are typical of the family *Flavobacteriaceae* (Bernardet et al., 2002). The two species of *Subsaximicrobium* are quite closely related sharing up to 60% DNA–DNA similarity (mean 50–55%). Phenotypic tests differentiating the species are shown in Table 69 and are based on available data only. Most strains that were isolated from Vestfold Hills quartz stone material belonged to *Subsaximicrobium wynnwilliamsii* (Bowman and Nichols, 2005). Based on currently available cultivation and molecular data (published and unpublished data from the National Center of Biotechnology Information; http://www.ncbi.nlm.nih.gov), members of the genus *Subsaximicrobium* have yet to be isolated from other saline habitats. The species require sodium ions for growth and the sublithic habitats in which they are found are located in a maritime locale, which suggests that they may also occur in cold marine waters or sea-ice.

Isolation and maintenance procedures

Strains can be isolated directly from quartz stone biofilm scrapings thoroughly suspended in 1:10 dilution of trypticase soy broth (Oxoid) supplemented with 1% NaCl. The suspension is then serially diluted onto the same medium containing 1.5% agar and incubated aerobically in the dark at 2–15°C for 1–4 weeks. Bright orange colonies that form within 1–2 weeks are usually indicative of *Subsaximicrobium*. Routine cultivation and storage should be performed aerobically on marine agar 2216 at 15–20°C. Strains can be maintained as frozen suspensions in

TABLE 69. Phenotypic characteristics that distinguish the species of *Susbsaximicrobium*[a]

Phenotypic characteristic	*S. wynnwilliamsii*	*S. saxinquilinus*
Pigment of cell mass	Orange or yellow-gold	Orange
Growth on marine agar 2216 at 25°C	d	−
Arginine dihydrolase activity	−	+
Casein hydrolysis	d (w)	+
L-Tyrosine hydrolysis	+	−
DNA hydrolysis	−	+
Glutamate decarboxylase activity	d	−
β-Glucosidase activity	d	+
Growth on D-mannose	−	+
Growth on sucrose	+	−

[a]Symbols: +, >85% positive; d, different strains give different reactions (16–84% positive); −, 0–15% positive; w, weak reaction.

marine broth 2216 containing 20–30% (v/v) glycerol at −70°C. Strains may survive for several months, potentially years, on thick agar marine agar 2216 plates or on slants, stored at 2–4°C.

Differentiation of the genus *Subsaximicrobium* from other genera

The most useful traits for differentiating *Subsaximicrobium* from other members of the *Flavobacteriaceae* include orange pigmentation, lack of flexirubin pigments, ability to glide, strictly oxidative metabolism, ability to form acid from carbohydrates, relatively rapid growth at low temperatures (e.g., 4°C) and at 25°C, requirement for sodium ions for growth (or good growth on sea salts-containing media), and lack of agarolytic activity. These basic traits can be used to differentiate *Subsaximicrobium* from its closest relatives *Subsaxibacter* and *Gelidibacter*. *Subsaxibacter* differs in being much more slow-growing, forming crumbling and/or viscid colonies, having a lower salinity tolerance, possessing a requirement for yeast extract, and being unable to hydrolyze casein or starch. *Gelidibacter* differs in that its species have a yellow pigment, mainly possess a lower salt tolerance (an exception is *Gelidibacter salicanalis*), and are able to form detectable acid production from D-glucose. Several other genera of the family *Flavobacteriaceae* possess at least some proportion of the traits possessed by *Subsaximicrobium*; thus, any phenotypic-based identification is at best tentative and 16S rRNA gene sequencing is required for any definitive identification.

List of species of the genus *Subsaximicrobium*

1. **Subsaximicrobium wynnwilliamsii** Bowman and Nichols 2005, 1481[VP]

wynn.wil.li.ams'i.i. N.L. gen. masc. n. *wynnwilliams* of Wynn-Williams, named in honor of the late David Donaldson Wynn-Williams, British Antarctic microbiologist.

Characteristics are as described for the genus, with the following additional features. Colonies are usually bright orange, circular, and convex, with an entire edge and a butyrous consistency on marine agar 2216. A minority of strains may possess a golden-yellow pigmentation. Most strains grow poorly at 25°C; the optimal temperature for growth is 15–20°C. Growth occurs in 0.25–2.0 M NaCl, with optimal growth at seawater salinities (0.3–0.4 M NaCl). Poor growth occurs in media containing double-strength sea salts (70 g/l dehydrated sea salts), and no growth occurs with quadruple-strength sea salts. Does not require divalent cations found in seawater for growth; all salinity requirements can be met with NaCl. Grows well on trypticase soy agar (Oxoid) containing 1% (w/v) NaCl, but poor or no growth occurs on trypticase soy agar and nutrient agar (Oxoid). Growth can occur on a mineral salts medium prepared with sea salts and containing ammonium chloride or sodium nitrate as the sole nitrogen source and D-glucose as the sole carbon and energy source. Other sole carbon and energy sources include glycogen, maltose, sucrose, and L-proline, but not *N*-acetyl-D-glucosamine, L-rhamnose, DL-arabinose, L-fucose, D-mannose, melibiose, salicin, propionate, valerate, caprate, adipate, suberate, citrate, DL-3-hydroxybutyrate, DL-lactate, itaconate, phenylacetate, L-histidine, or L-serine. Some strains (but not the type strain) can grow on acetate and/or L-alanine. Acid production from carbohydrates is undetectable in Leifson's oxidation/fermentation medium (Leifson, 1963). Gelatin, Tween 80, esculin, starch, and L-tyrosine are hydrolyzed, but not agar, chitin, carboxymethylcellulose, dextran, xylan, xanthine, uric acid, DNA, or urea. Most strains can decompose starch and weakly decompose casein. Some strains can also degrade elastin. Produces alkaline phosphatase and α-glucosidase. Most strains possess glutamate decarboxylase activity; some strains are positive for β-glucosidase and weakly positive for *N*-acetyl-β-D-glucosaminidase. No activity is observed for arginine dihydrolase, lysine decarboxylase, or ornithine decarboxylase. H_2S is not produced from L-cysteine. Indole is not produced. No growth occurs in the Simmons' citrate test. Lipase (olive oil as the substrate) or lecithinase (egg yolk reaction) activity is not observed. Nitrate is not reduced. Negative for α-arabinosidase, α-galactosidase, β-galactosidase, 6-phospho-β-galactosidase, β-glucuronidase, and α-fucosidase activity. Positive for arginine arylamidase, leucyl glycine arylamidase, phenylalanine arylamidase, leucine arylamidase, tyrosine arylamidase, glycine arylamidase, histidine arylamidase, serine arylamidase, alanine arylamidase, and glutamyl glutamate arylamidase. Negative for pyroglutamate arylamidase. Some strains may produce proline arylamidase. Does not tolerate 1% (w/v) ox bile salts.

Source: the type strain was isolated from cyanobacterial biofilms attached to the undersides of partially buried quartz stones found in the Vestfold Hills, an ice-free region of East Antarctica.

DNA G+C content (mol%): 40 (T_m).

Type strain: G#7, ACAM 1070, CIP 108525.

Sequence accession no. (16S rRNA gene): AY693997.

2. **Subsaximicrobium saxinquilinus** Bowman and Nichols 2005, 1481[VP]

sax.in.qui.li'nus. L. n. *saxum* stone; L. masc. n. *inquilinus* the denizen; N.L. masc. n. *saxinquilinus* the denizen of stone.

Characteristics are as described for the genus, with the following additional features. Colonies are usually bright orange, circular, convex, with an entire edge and a butyrous consistency on marine agar 2216. No growth occurs at 25°C or higher temperatures on marine agar 2216. Grows in 0.1–2.0 M NaCl, with optimal growth at seawater salinities (0.3–0.4 M NaCl). Grows well in media containing double-strength sea salts (70 g/l dehydrated sea salts), but no growth occurs with quadruple-strength sea salts. Does not require divalent cations found in seawater for growth; all salinity requirements can be met with NaCl. Grows well on trypticase soy agar (Oxoid) containing 1% (w/v) NaCl, but poor or no growth occurs on trypticase soy agar and nutrient agar (Oxoid). Growth can occur on a mineral salts medium prepared with sea salts and containing ammonium chloride or sodium nitrate as the sole nitrogen source and D-glucose as the sole carbon and energy source. Other sole carbon and energy sources include glycogen, D-mannose, maltose, and L-proline, but not *N*-acetyl-D-glucosamine, L-rhamnose,

DL-arabinose, L-fucose, melibiose, sucrose, salicin, acetate, propionate, valerate, caprate, adipate, suberate, citrate, DL-3-hydroxybutyrate, DL-lactate, itaconate, phenylacetate, L-alanine, L-histidine, or L-serine. Acid production from carbohydrates is undetectable in Leifson's oxidation/fermentation medium (Leifson, 1963). Casein, gelatin, Tween 80, esculin, starch, and DNA are hydrolyzed, but not agar, chitin, carboxymethylcellulose, dextran, xylan, xanthine, uric acid, L-tyrosine, elastin, or urea. Most strains can decompose starch and weakly decompose casein. Some strains can also degrade elastin. The following activities occur: alkaline phosphatase, arginine dihydrolase, α-glucosidase, and β-glucosidase. No activity is observed for glutamate decarboxylase, lysine decarboxylase, or ornithine decarboxylase. H_2S is not produced from L-cysteine. Indole is not produced. No growth occurs in the Simmons' citrate test. Lipase (olive oil as the substrate) or lecithinase (egg yolk reaction) activity is not observed. Nitrate is not reduced. Negative for α-arabinosidase, α-galactosidase, β-galactosidase, 6-phospho-β-galactosidase, β-glucuronidase, *N*-acetyl-β-D-glucosaminidase, and α-fucosidase. Positive for arginine arylamidase, leucyl glycine arylamidase, phenylalanine arylamidase, leucine arylamidase, tyrosine arylamidase, glycine arylamidase, histidine arylamidase, serine arylamidase, alanine arylamidase, and glutamyl glutamate arylamidase. Negative for proline arylamidase and pyroglutamate arylamidase. Does not tolerate 1% (w/v) ox bile salts.

Source: the type strain was isolated from cyanobacterial biofilms attached to the undersides of partially buried quartz stones found in the Vestfold Hills, an ice-free region of East Antarctica.

DNA G+C content (mol%): 39 (T_m).

Type strain: Y4-5, ACAM 1063, CIP 108526.

Sequence accession no. (16S rRNA gene): AY693998.

Genus LIII. **Tenacibaculum** Suzuki, Nakagawa, Harayama and Yamamoto 2001, 1650^{VP}

Makoto Suzuki

Te.na.ci.ba'cu.lum. L. adj. n. *tenax -acis* holding fast; L. neut. n. *baculum* stick; N.L. neut. n. *Tenacibaculum* rod-shaped bacterium that adheres to the surface of marine organisms.

Rods, 1.5–30 × 0.4–0.5 µm. Ring-shaped cells and gas vesicles are not formed. Spores are not formed. Cells are **nonflagellated. Cells are motile by gliding.** Gram-stain-negative. Cells produce a yellow pigment that is mainly zeaxanthin. **Flexirubin-type pigments are absent. Chemo-organotrophic. Aerobic. Catalase- and oxidase-positive.** Grow well on media containing seawater. **The major respiratory quinone is menaquinone-6.** All strains have been isolated from marine environments.

DNA G+C content (mol%): 31–33.

Type species: **Tenacibaculum maritimum** (Wakabayashi, Hikida and Masumura 1986) Suzuki, Nakagawa, Harayama and Yamamoto 2001, 1650^{VP} [*Flexibacter maritimus* Wakabayashi, Hikida and Masumura 1986, 398 emend. Bernardet and Grimont 1989, 353; *Cytophaga marina* (*ex* Hikida, Wakabayashi, Egusa and Masumura 1979) Reichenbach 1989a, 2044; *Flexibacter marinus* Hikida, Wakabayashi, Egusa and Masumura 1979, 427].

Further descriptive information

The genus *Tenacibaculum* belongs to the family *Flavobacteriaceae* and is closely related to the genus *Polaribacter* with 16S rRNA gene sequence similarities between strains of these species of 91.0–93.3% (Figure 52). 16S rRNA gene sequence similarities between strains of *Tenacibaculum* species are 93.3–98.5%.

The cells are rod-shaped and often elongate to 10–30 µm and sometimes up to 100 µm. Pleomorphism is not usually observed, but spherical cells are observed in aged cultures.

The dominant cellular fatty acids are the branched-chain unsaturated fatty acid $C_{15:0}$ iso and hydroxylated fatty acids $C_{16:0}$ iso 3-OH, $C_{17:0}$ iso 3-OH, and summed feature 3 (comprising $C_{15:0}$ iso 2-OH and/or $C_{16:1}$ ω7*c*).

All species grow on media containing peptone and yeast extract prepared with natural or artificial seawater, such as marine agar 2216 (Difco). Some species are able to grow on media supplemented with 1–3% NaCl only. Growth is observed at 4–34°C. Some species can grow at 40°C, but *Tenacibaculum ovolyticum* cannot grow at 28°C or higher.

Members of the genus *Tenacibaculum* are commonly found in marine environments. They usually attach to or associate with the surface of marine organisms such as fishes, macroalgae or microalgae, and invertebrates.

Enrichment and isolation procedures

Strains of the genus *Tenacibaculum* have been isolated from marine samples, particularly from the surfaces of marine organisms. Two fish-pathogenic species, *Tenacibaculum maritimum* and *Tenacibaculum ovolyticum*, have been isolated from the surface of diseased fish or fish eggs (Hansen et al., 1992; Wakabayashi et al., 1986). These species were isolated by direct streaking of external lesions, kidney tissues, or adherent epiflora of eggs onto agar plates. *Tenacibaculum mesophilum* and *Tenacibaculum amylolyticum* were isolated from homogenates of marine macroalgae or sponges (Suzuki et al., 2001). *Tenacibaculum skagerrakense* was isolated from filtered (0.8 µm pore diameter) seawater (Frette et al., 2004). *Tenacibaculum lutimaris*, *Tenacibaculum litoreum*, and *Tenacibaculum aestuarii* were isolated from tidal flat sediment by the dilution plating technique or by direct streaking onto agar plates (Choi et al., 2006; Jung et al., 2006; Yoon et al., 2005c). The media for isolation of these bacteria were standard or diluted marine agar or cytophaga agar made with 70–100% seawater.

Maintenance procedures

For temporary preservation, cultures may be stored at 4°C for several weeks. For long-term preservation, cultures can be kept in broth containing 10% glycerol at −80°C or in the gas phase of liquid nitrogen. Lyophilization may give satisfactory results.

Differentiation of the genus *Tenacibaculum* from other genera

The differentiating features of *Tenacibaculum* and closely related genera are listed in Table 70. *Tenacibaculum* strains can be differentiated from those of *Polaribacter* by cell morphology, cell mass

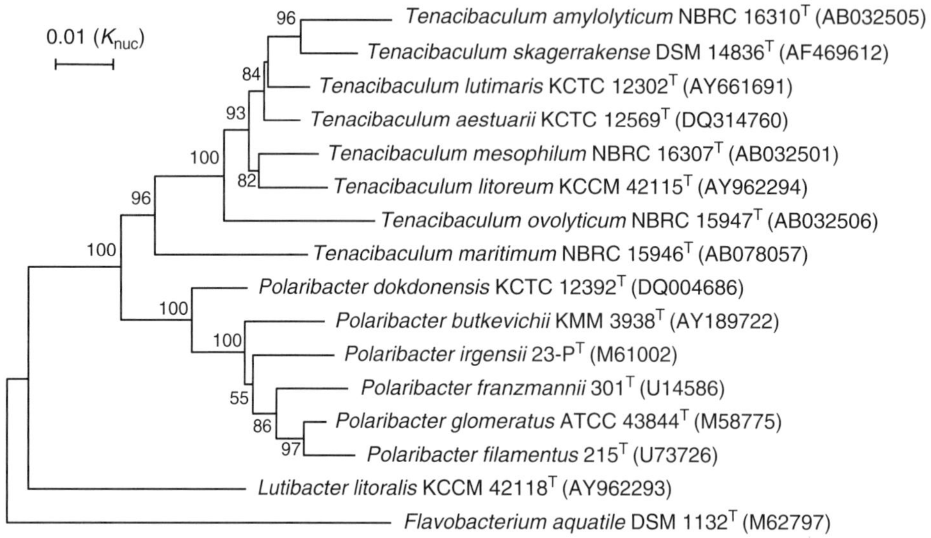

FIGURE 52. Neighbor-joining tree obtained with the 16S rRNA gene sequences of strains of *Tenacibaculum*, *Polaribacter*, and *Lutibacter* species. Bootstrap values (500 resamplings) greater than 50% are indicated at the nodes. Bar = genetic distance of 0.01 (K_{nuc}).

TABLE 70. Differential characteristics of the genus *Tenacibaculum* and related genera[a]

Characteristic	*Tenacibaculum*	*Polaribacter*	*Lutibacter*
Cell morphology	Rods	Pleomorphic rods	Rods
Cell mass color	Pale to bright yellow	Salmon to orange	Yellow
Motility by gliding	+	–	–
Oxidase	+	d	–
Growth at 25°C	+	–	+
Growth at pH 6	+	nd	–
Degradation of casein	+	–	nd
Growth on pyruvate	–	–	+
Growth on succinate	–	–	+

[a]Symbols: +, >85% positive; d, different strains give different reactions (16–84% positive); –, 0–15% positive; nd, not determined.

color, and growth temperature. They can also be differentiated from those of *Lutibacter* by oxidase activity and growth pH range.

Taxonomic comments

Tenacibaculum maritimum is a fish-pathogenic bacterium and some intraspecific varieties have been studied. Two major O-serotypes (O1 and O2) have been described in *Tenacibaculum maritimum* strains isolated from marine fish in Spain (Avendaño-Herrera et al., 2004a). These serotypes were associated mainly with the host fish species (sole or turbot). Serotype O1 contained the majority of the sole isolates, whereas serotype O2 consisted of turbot isolates. There was an intermediate minor serotype, O1/O2, for Japanese strains, which were isolated from sea bream. In 2005, a novel O-serotype, O3, was reported from strains isolated from cultured sole from Portugal and south of Spain (Avendaño-Herrera et al., 2005).

Intraspecific diversity of *Tenacibaculum maritimum* was also studied by the randomly amplified polymorphic DNA (RAPD) method (Avendaño-Herrera et al., 2004c). The results of RAPD analysis indicated there were two major groups that correlated strongly with the O-serotype grouping.

Species-specific PCR primers using 16S rRNA gene sequences for *Tenacibaculum maritimum* were reported by Toyama et al. (1996). A nested PCR method for the specific diagnosis of *Tenacibaculum maritimum* in fish samples was reported by Avendaño-Herrera et al. (2004b).

Differentiation of species of the genus *Tenacibaculum*

The differential characteristics of species of the genus *Tenacibaculum* are listed in Table 71.

TABLE 71. Differential characteristics of species of the genus *Tenacibaculum*[a,b]

Characteristic	*T. maritimum*	*T. aestuarii*	*T. amylolyticum*	*T. litoreum*	*T. lutimaris*	*T. mesophilum*	*T. ovolyticum*	*T. skagerrakense*
Cell size (µm)	2–30 × 0.5	2–3.5 × 0.3	2–5 × 0.4	2–35 × 0.3–0.5	2–10 × 0.5	1.5–10 × 0.5	2–20 × 0.5	2–15 × 0.5
Gliding motility	+	+	+	+	+	+	+	–
Growth temperature (°C)	15–34	9–41	20–35	5–40	10–39	15–40	4–25	10–40
Sea salt requirement	+	–	+	–	–	–	+	+
Degradation of:								
Starch	–	–	+	+	–	–	–	+
Gelatin	–	+	+	+	+	+	+	nd
Tween 80	+	+	+	+	–	+	+	–
Nitrate reduction	+	–	w	+	v	–	+	+
Growth on:								
Aspartate	–	nd	–	–	–	+	–	+
Proline	–	–	+	+	–	+	–	+
Glutamate	w	–	+	–	–	+	–	+
L-Leucine	–	nd	–	–	–	–	–	w
Sucrose	–	–	–	–	nd	–	–	+
DNA G+C content (mol%)	31.3–32.5	33.6	30.9	30	32.3–32.8	31.6–32.0	30.3–32.0	35.2

[a]Symbols: +, >85% positive; –, 0–15% positive; v, variable; w, weak reaction; nd, not determined.

[b]Data from Wakabayashi et al. (1986), Hansen et al. (1992), Suzuki et al. (2001), Frette et al. (2004), Yoon et al. (2005c), Choi et al. (2006), and Jung et al. (2006).

List of species of the genus *Tenacibaculum*

1. **Tenacibaculum maritimum** (Wakabayashi, Hikida and Masumura 1986) Suzuki, Nakagawa, Harayama and Yamamoto 2001, 1650[VP] [*Flexibacter maritimus* Wakabayashi, Hikida and Masumura 1986, 398 emend. Bernardet and Grimont 1989, 353; *Cytophaga marina* (*ex* Hikida, Wakabayashi, Egusa and Masumura 1979) Reichenbach 1989a, 2044; *Flexibacter marinus* Hikida, Wakabayashi, Egusa and Masumura 1979, 427]

mar.i.ti′mum. L. neut. adj. *maritimum* of the sea, maritime.

The characteristics are as described for the species and as listed in Tables 70 and 71, with the following additional features. Rods, 0.5 × 2–30 µm. Spherical cells are often present. Cell mass color is pale yellow. Colonies have an uneven edge and diameters of less than 5 mm at 5 d. Microcysts are not formed. Growth occurs at 15–34°C, with optimum growth at 30°C. The pH range for growth is 5.9–8.6; no growth occurs at pH 5. Seawater, at least 1/3 strength, is required for growth. Growth occurs on Casamino acids. Nitrate is reduced. Gelatin, Tween 80, and DNA are hydrolyzed, but not starch. Esterase lipase (C8) is present (API ZYM system); lipase (C14), trypsin, and naphthol-AS-BI-phosphohydrolase activities are weakly positive. Lysis of dead cells of *Escherichia coli* is negative. L-Glutamate is utilized weakly as a carbon source, but citrate, L-leucine, sucrose, L-proline, and DL-aspartate are not utilized.

Source: isolated from diseased red sea bream (*Pagrus major*), black sea bream (*Acanthopagrus schlegeli*), rock bream (*Oplegnathus fasciatus*), sole (*Solea* spp.), turbot (*Scophthalmus maximus*), and gilthead (*Sparus aurata*). The type strain was isolated from a diseased red sea bream fingerling (*Pagrus major*) reared in a floating net cage offshore Ondo, Hiroshima Prefecture, Japan.

DNA G+C content (mol%): 31.3–32.5 (T_m).

Type strain: R2, ATCC 43398, CIP 103528, CCUG 35198, LMG 11612, NBRC 15946, NCIMB 2154.

Sequence accession no. (16S rRNA gene): AB078057.

2. **Tenacibaculum aestuarii** Jung, Oh and Yoon 2006, 1580[VP]

aes.tu′a.ri.i. L. gen. n. *aestuarii* of the tidal flat, from where the organism was isolated.

The characteristics are as described for the species and as listed in Tables 70 and 71, with the following additional features. Cells are 0.3 × 2.0–3.5 µm. Cell mass color is pale yellow. Colonies are smooth, irregular with spreading edges, greenish, and glistening. No growth occurs under anaerobic conditions on marine agar 2216 (Difco) or on marine agar with nitrate. Optimum pH is 7.5–8.5; weak growth occurs at pH 5.5; no growth at pH 5.0. No growth occurs in the presence of more than 7% (w/v) NaCl. Growth does not occur in the absence of NaCl. Tyrosine and Tweens 20, 40, 60 and 80 are hydrolyzed, but esculin, urea, hypoxanthine, and xanthine are not. With the API ZYM system (bioMérieux), alkaline phosphatase, esterase (C4), leucine arylamidase, valine arylamidase, α-chymotrypsin, acid phosphatase, phosphohydrolase, and β-glucosidase are present, but cystine arylamidase, α-galactosidase, β-galactosidase, β-glucuronidase, α-glucosidase, *N*-acetyl-β-glucosaminidase, α-mannosidase, and α-fucosidase are absent. Peptone and tryptone are utilized as sole carbon and energy sources. D-Glucose, D-galactose, D-fructose, cellobiose, trehalose, and L-leucine are not utilized. Acid is not produced from D-sorbitol, *myo*-inositol, D-xylose, D-ribose, D-fructose, D-mannitol, melibiose, L-arabinose, melezitose, D-glucose, D-galactose, L-rhamnose, D-mannose, cellobiose, lactose, sucrose, maltose, trehalose, or raffinose. Susceptible to cephalothin, lincomycin, oleandomycin, and carbenicillin, but not to polymyxin B, streptomycin, penicillin G, ampicillin, gentamicin, novobiocin, tetracycline, kanamycin, or neomycin. The major cellular fatty acids are $C_{15:0}$ iso, $C_{16:0}$ iso 3-OH, and summed feature 3 (comprising $C_{15:0}$ iso 2-OH and/or $C_{16:1}$ ω7*c*).

Source: the type strain was isolated from tidal flat sediment at Saemankum, Pyunsan, Korea.

DNA G+C content (mol%): 33.6 (HPLC).

Type strain: SMK-4, JCM 13491, KCTC 12569.

Sequence accession no. (16S rRNA gene): DQ314760.

3. **Tenacibaculum amylolyticum** Suzuki, Nakagawa, Harayama and Yamamoto 2001, 1650[VP]

am.y.lo.ly'ti.cum. Gr. n. *amulon* starch; N.L. adj. *lyticus -a -um* (from Gr. adj. *lutikos -ê -on*) able to loosen, able to dissolve; N.L. neut. adj. *amylolyticum* starch-dissolving.

The characteristics are as described for the species and as listed in Tables 70 and 71, with the following additional features. Rods, 0.4 × 2–5 µm. Cell mass color is bright yellow. Colonies on 1/5× LBM agar (tryptone, 2.0 g; yeast extract, 1.0 g; agar, 15 g, 1000 ml Jamarin S synthetic seawater at pH 7.2; Suzuki et al., 2001) are circular with a spreading and undulating margin, a flat elevation, and 20–30 mm diameter after 5 d incubation. At least half-strength seawater is required for sufficient growth. Growth occurs at 19–35°C, with optimum growth at 30°C in liquid media. No growth occurs at temperatures at or below 15°C and at or above 40°C. pH range for growth is 5.3–8.3 in liquid media. Nitrate is reduced weakly. Casein, starch, tyrosine, gelatin, Tween 80, and DNA are hydrolyzed. Growth occurs on peptone, tryptone, Casamino acids, L-proline, and L-glutamate as the sole carbon and nitrogen sources.

Source: the type strain was isolated from the green alga *Avrainvillea riukiuensisi*, which was collected in Palau.

DNA G+C content (mol%): 31 (HPLC).

Type strain: CIP 107214, DSM 13766, MBIC4355, NBRC 16310.

Sequence accession no. (16S rRNA gene): AB032505.

4. **Tenacibaculum litoreum** Choi, Kim, Hwang, Yi, Chun and Cho 2006, 639[VP]

li.to.re'um. L. neut. adj. *litoreum* of the shore.

The characteristics are as described for the species and as listed in Tables 70 and 71, with the following additional features. Straight rods, 0.3–0.5 × 2–35 µm. Spherical cells are rare. Cell mass is pale yellow. On marine agar 2216, colonies are irregular with spreading edges, greenish, glistening, and 5–10 mm in diameter after 5 d incubation at 30°C. Growth occurs at 5–40°C, with optimum growth at 35–40°C. pH range for growth is 6–10; no growth occurs at pH 5. Growth occurs in NaCl concentrations of 3–5% (w/v) and in sea salt concentrations of 1–10% (w/v). Starch, gelatin, DNA, and Tween 80 are hydrolyzed. Positive for nitrate reductase and gelatinase, but negative for indole production, acid production from glucose, arginine dihydrolase, urease, hydrolysis of esculin, and β-galactosidase (API 20NE system). Alkaline phosphatase, esterase (C4 and C8), leucine arylamidase, valine arylamidase, cystine arylamidase, trypsin, α-chymotrypsin, acid phosphatase, and naphthol-AS-BI-phosphohydrolase are present, but lipase (C14), α-galactosidase, β-galactosidase, β-glucuronidase, α-glucosidase, β-glucosidase, *N*-acetyl-β-glucosaminidase, α-mannosidase and α-fucosidase are absent (API ZYM system). Growth occurs on Casamino acids, tryptone, yeast extract, peptone, L-arginine, L-ornithine, L-proline, and L-lysine. No growth occurs on acetate, benzoate, citrate, maleic acid, ethanol, glycerol, L-leucine, tartrate, pyruvic acid, succinate, sucrose, L-glutamate, D-ribose, DL-aspartate, *N*-acetylglucosamine, L-arabinose, D-xylose, D-fructose, D-glucose, D-mannose, trehalose, inulin, D-mannitol, D-sorbitol, D-salicin, raffinose, D-galactose, urea, or lactose. Major fatty acids are summed feature 3 (comprising $C_{15:0}$ iso 2-OH and/or $C_{16:1}$ ω7*c*; 19.6%), $C_{15:0}$ iso (18.8%), and $C_{17:0}$ iso 3-OH (13.6%). Also contains minor amounts of $C_{18:3}$ ω6*c*(6,9,12) (1.5%) and summed feature 4 ($C_{17:1}$ iso $_I$ and/or $C_{17:1}$ anteiso B; 1.3%).

Source: the type strain was isolated from tidal flat sediment in Ganghwa, Korea.

DNA G+C content (mol%): 30 (T_m).

Type strain: CL-TF13, KCCM 42115, JCM 13039.

Sequence accession no. (16S rRNA gene): AY962294.

5. **Tenacibaculum lutimaris** Yoon, Kang and Oh 2005c, 797[VP]

lu.ti.ma'ris. L. neut. n. *lutum* mud; L. neut. n. *mare -is* the sea, marine; N.L. gen. n. *lutimaris* of a mud of the sea.

The characteristics are as described for the species and as listed in Tables 70 and 71, with the following additional features. Rods, 0.5 × 2–10 µm. Cell mass color is pale yellow. Colonies are irregular, smooth, and glistening on marine agar 2216 at 30°C. Colonies are adherent to the agar surface. Growth occurs at 10–39°C, with optimum growth at 30–37°C. No growth occurs at 4°C or above 41°C. Optimum pH is 7–8; no growth occurs at pH 4.5. Optimal growth occurs in the presence of 2–3% (w/v) NaCl; no growth occurs in the absence of NaCl. Tyrosine is hydrolyzed, but esculin, hypoxanthine, Tweens 20, 40 and 60, xanthine, and urea are not hydrolyzed. H_2S is not produced. Growth under anaerobic conditions does not occur on marine agar. The ability to grow under anaerobic conditions on marine agar supplemented with nitrate differs among strains (negative for type strain). Growth occurs on peptone, tryptone, and Casamino acids as sole carbon and nitrogen sources. No growth occurs on D-glucose, sucrose, D-ribose, DL-aspartate, L-proline, L-glutamate, or L-leucine. No acid is produced from L-arabinose, cellobiose, D-fructose, D-galactose, D-glucose, lactose, maltose, D-mannose, melibiose, melezitose, raffinose, L-rhamnose, D-ribose, sucrose, trehalose, D-xylose, adonitol, D-sorbitol, *myo*-inositol, or D-mannitol. Major fatty acids are summed feature 3 (comprising $C_{15:0}$ iso 2-OH and/or $C_{16:1}$ ω7*c*), $C_{15:0}$ iso, $C_{16:0}$ iso 3-OH, $C_{15:0}$, and $C_{17:0}$ iso -OH.

Source: the type strain was isolated from a tidal flat on Daepo Beach in the Yellow Sea, 3Korea.

DNA G+C content (mol%): 32.3–32.8 (HPLC).

Type strain: TF-26, KCTC 12302, DSM 16505.

Sequence accession no. (16S rRNA gene): AY661691.

6. **Tenacibaculum mesophilum** Suzuki, Nakagawa, Harayama and Yamamoto 2001, 1650[VP]

me.so.phi'lum. Gr. n. *mesos* middle; Gr. adj. *philos* loving; N.L. neut. adj. *mesophilum* middle (temperature)-loving, i.e., mesophilic.

The characteristics are as described for the species and as listed in Tables 70 and 71, with the following additional features. Rods, 0.5 × 1.5–10 µm. Spherical cells are rare. Cell mass color is yellow. Colonies on 1/5× LBM medium are circular with a spreading, undulating margin and a flat elevation, and have diameters of 40–50 mm after 5 d. Growth requires 1/10 strength seawater or 1% (w/v) NaCl. Growth occurs at 15–40°C, with optimum growth at 28–35°C in liquid media. No growth occurs at temperatures at or below 10°C and at or above 45°C. Growth occurs at pH 5.3–9.0 in liquid media; no growth occurs at pH 5.0. Nitrate is not reduced. Casein, tyrosine, gelatin, Tween 80, and DNA are hydrolyzed. Growth occurs on peptone, tryptone, Casamino acids, DL-aspartate, L-proline, and L-glutamate as sole carbon and nitrogen sources. Citrate, L-leucine, and sucrose are not used as carbon sources.

Source: the type strain was isolated from the sponge *Halichondria okadai*, which was collected in Numazu, Japan.

DNA G+C content (mol%): 31–32 (HPLC).

Type strain: CIP 107215, DSM 13764, MBIC1140, NBRC 16307.

Sequence accession no. (16S rRNA gene): AB032501.

7. **Tenacibaculum ovolyticum** (Hansen, Bergh, Michaelsen and Knappskog 1992) Suzuki, Nakagawa, Harayama and Yamamoto 2001, 1650[VP] (*Flexibacter ovolyticus* Hansen, Bergh, Michaelsen and Knappskog 1992, 457)

o.vo.lyt'ic.um. L. n. *ovum* egg; N.L. adj. *lyticus -a -um* (from Gr. adj. *lutikos -ê -on*) able to loosen, able to dissolve; N.L. neut. adj. *ovolyticum* egg-damaging.

The characteristics are as described for the species and as listed in Tables 70 and 71, with the following additional features. Rods, 0.4 × 2–20 µm that occasionally grow to filaments 70–100 µm long. Cell mass color is pale yellow. Microcysts are not formed. The cells exhibit gliding motility. Oxidase-positive. Congo red is not adsorbed. Acid is not produced from carbohydrates. Gelatin, tyrosine, DNA, and Tween 80 are hydrolyzed, but not agar, starch, carboxymethylcellulose, cellulose, chitin, or urea. Nitrate reductase is present. H_2S is not produced. Pigment is produced on tyrosine. Tryptone, Casamino acids, yeast extract, and sodium glutamate can serve as nitrogen sources in a medium based on artificial seawater; KNO_3 cannot be used. Ammonia is not produced. Arginine dihydrolase, lysine decarboxylase, and ornithine decarboxylase activities are negative. Citrate is not used as a sole carbon source. Indole is not produced. Requires 50% seawater for growth; alternatively, media based on artificial seawater containing 1–3% NaCl may be used. No high-viscosity extracellular polysaccharide is produced in liquid cultures. Growth is not stimulated by halibut egg homogenate. Growth occurs at 4°C, but not at 30°C. The following enzymic activities are present (API ZYM system): acid phosphatase, alkaline phosphatase, esterase (C4), esterase lipase (C8), leucine arylamidase, valine arylamidase, and naphthol-AS-BI-phosphohydrolase. The following activities are absent: lipase (C4), cystine arylamidase, trypsin, chymotrypsin, α-galactosidase, β-galactosidase, β-glucuronidase, α-glucosidase, β-glucosidase, *N*-acetyl-β-glucosaminidase, α-mannosidase, and α-fucosidase.

Source: the type strain was isolated from the adherent epiflora of halibut eggs in Western Norway.

DNA G+C content (mol%): 30–32 (T_m).

Type strain: EKD002, ATCC 51887, CCUG 35199, CIP 106403, IAM 14318, LMG 13026, NBRC 15947, NCIMB 13127.

Sequence accession no. (16S rRNA gene): AB032506.

8. **Tenacibaculum skagerrakense** Frette, Jørgensen, Irming and Kroer 2004, 523[VP]

ska.ger.rak.en'se. N.L. neut. adj. *skagerrakense* of or belonging to Skagerrak, Denmark, referring to the place of isolation.

The characteristics are as described for the species and as listed in Tables 70 and 71, with the following additional features. Rods, 0.5 × 2–15 µm, during exponential growth. Spherical cells occur often in the stationary phase. Gliding motility is not observed. Cell mass color is yellow. Colonies are circular and have a spreading edge. At least 1/4-strength seawater is required for growth. Growth occurs in up to 150% strength seawater. Growth occurs at 10–40°C, with optimum growth at 25–37°C. Resistant to temperatures up to 50°C. pH range, 6–9. Casein, collagen, hydroxyethylcellulose, starch, barley β-glucan, and pullulan are hydrolyzed. Growth occurs on sucrose, aspartate, and (weakly) L-leucine. Nitrate is reduced.

Source: the type strain was isolated from the <0.8 µm fraction of a seawater sample taken from Skagerrak, Denmark, at a depth of 30 m

DNA G+C content (mol%): 35.2 (HPLC).

Type strain: D30, ATCC BAA-458, DSM 14836.

Sequence accession no. (16S rRNA gene): AF469612.

Genus LIV. **Ulvibacter** Nedashkovskaya, Kim, Han, Rhee, Lysenko, Falsen, Frolova, Mikhailov and Bae 2004c, 121[VP]

Olga I. Nedashkovskaya and Seung Bum Kim

Ul.vi.bac'ter. N.L. fem. n. *Ulva* generic name of the green alga *Ulva fenestrata*; N.L. masc. n. *bacter* from Gr. n. *bakterion* rod; N.L. masc. n. *Ulvibacter* rod isolated from *Ulva fenestrata*.

Thin rods, usually measuring 0.4–0.5 × 2.5–7.3 µm. **No gliding motility occurs.** Produce non-diffusible yellow–orange pigments. **Flexirubin pigments are produced. Chemo-organotrophs. Strictly aerobic.** Positive for oxidase, catalase, and alkaline phosphatase. Gelatin, Tweens 20 and 40, and DNA are hydrolyzed, but agar, alginate, casein, starch, chitin, and cellulose (carboxymethylcellulose and filter paper) are not hydrolyzed. Carbohydrates are not oxidized or utilized. Growth does not occur without seawater or Na^+. **The major respiratory quinone is MK-6. Marine**, isolated from coastal seaweeds.

DNA G+C content (mol%): 36–38.

Type species: **Ulvibacter litoralis** Nedashkovskaya, Kim, Han, Rhee, Lysenko, Falsen, Frolova, Mikhailov and Bae 2004c, 121[VP].

Further descriptive information

Phylogenetic analysis of almost-complete 16S rRNA gene sequences indicate that the genus *Ulvibacter* forms a cluster that includes members of the family *Flavobacteriaceae* such as *Cellulophaga, Arenibacter, Zobellia, Muricauda, Aequorivita, Vitellibacter,* and *Leeuwenhoekiella marinoflava* (Nedashkovskaya et al., 2004c). The closest relatives of *Ulvibacter* are *Cellulophaga* and *Leeuwenhoekiella.*

The predominant cellular fatty acids are the straight-chain unsaturated and branched-chain unsaturated fatty acids $C_{15:0}$ iso, $C_{15:1}$ iso, $C_{16:0}$ iso, $C_{16:0}$ iso 3-OH, and $C_{17:0}$ iso 3-OH.

On marine agar 2216 (Difco), ulvibacters form regular, round, convex, smooth, shiny, yellow–orange colonies with entire edges and diameters of 2–4 mm after 48 h at 28°C.

Ulvibacter strains grow on media containing 0.5% peptone and 0.1–0.2% yeast extract (Difco), prepared with natural or artificial seawater or supplemented with 2–3% NaCl. Good growth is observed on marine agar 2216. Growth occurs at 4–36°C (optimum, 21–23°C) and with 1–6% NaCl.

Strains of *Ulvibacter* are susceptible to carbenicillin, lincomycin, oleandomycin, and tetracycline, and resistant to ampicillins, gentamicin, kanamycin, neomycin, polymyxin B, and streptomycin.

The ulvibacters are inhabitants of coastal marine environments. Strains have been isolated from seaweeds collected in the temperate latitudes.

Enrichment and isolation procedures

Strains of *Ulvibacter* have been isolated from seaweeds by direct plating on marine agar. Natural or artificial seawater is suitable for their cultivation. All isolated strains have been grown on media containing 0.5% peptone and 0.1–0.2% yeast extract (Difco). Growth occurs at 4–36°C (optimum, 21–23°C) and with 1–6% NaCl. The organisms may remain viable for several weeks on marine agar or other rich medium based on natural or artificial seawater. They have survived storage at −80°C for at least 5 years.

Differentiation of the genus *Ulvibacter* from other genera

Phenotypic analysis indicates that *Ulvibacter* strains can be clearly differentiated from those of its closest relatives, *Cellulophaga* and *Leeuwenhoekiella,* by production of flexirubin-type pigments and by an inability to oxidize carbohydrates and hydrolyze starch (Johansen et al., 1999; Nedashkovskaya et al., 2004c, e, 2005i). Moreover, the absence of agarase helps to separate *Ulvibacter* from *Cellulophaga,* and the NaCl requirement differentiates the ulvibacters from *Leeuwenhoekiella.*

List of species of the genus *Ulvibacter*

1. **Ulvibacter litoralis** Nedashkovskaya, Kim, Han, Rhee, Lysenko, Falsen, Frolova, Mikhailov and Bae 2004c, 121^VP^

li.to.ra'lis. L. masc. adj. *litoralis* of or belonging to the seashore.

Rod-shaped cells 0.4–0.5 × 2.5–7.3 µm. On marine agar, colonies are 2–4 mm in diameter, circular, convex, shiny with entire edges, viscous, and yellow–orange in color. Growth is observed at 4–36°C (optimum, 21–23°C). Growth occurs at 1–6% NaCl. Gelatin, Tweens 20 and 40, and DNA are hydrolyzed. Does not hydrolyze agar, casein, alginate, starch, Tween 80, cellulose (carboxymethylcellulose and filter paper), or chitin. No acid is produced from cellobiose, L-fucose, D-galactose, melibiose, raffinose, L-rhamnose, DL-xylose, adonitol, dulcitol, or glycerol. Does not utilize L-arabinose, D-glucose, D-lactose, D-mannose, maltose, sucrose, mannitol, inositol, sorbitol, malonate, citrate, *N*-acetylglucosamine, gluconate, caprate, adipate, malate, or phenylacetate. Nitrate is reduced. Indole, H_2S, and acetoin (Voges–Proskauer reaction) are not produced.

Source: two strains were isolated from the green alga *Ulva fenestrata* collected in Troitsa Bay, Sea of Japan.

DNA G+C content (mol%): 36–38 (T_m).

Type strain: KMM 3912, KCTC 12104, CCUG 47093.

Sequence accession no. (16S rRNA gene): AY243096.

Genus LV. **Vitellibacter** Nedashkovskaya, Suzuki, Vysotskii and Mikhailov 2003c, 1285^VP^

Olga I. Nedashkovskaya, Makoto Suzuki and Valery V. Mikhailov

Vi.tel.li.bac'ter. L. n. *vitellus* egg yolk; N.L. masc. n. *bacter* rod; N.L. masc. n. *Vitellibacter* egg-yolk-colored rod.

Thin rods, usually measuring 0.3–0.5 × 3–10 µm. **No gliding motility.** Produce nondiffusible, bright orange pigments. Flexirubin-type pigments are formed. **Chemo-organotrophs. Strictly aerobic.** Positive for oxidase, catalase, and alkaline phosphatase. Gelatin, casein, DNA, and Tweens 20 and 40 are hydrolyzed, but agar, alginate, starch, urea, chitin, and cellulose (carboxymethylcellulose and filter paper) are not attacked. **Marine,** from coastal habitats. Growth does not occur without seawater or Na^+. **The major respiratory quinone is MK-6. The main cellular fatty acids** are branched-chain saturated fatty acids $C_{15:0}$ iso and $C_{15:0}$ anteiso.

DNA G+C content (mol%): 41–42.

Type species: **Vitellibacter vladivostokensis** Nedashkovskaya, Suzuki, Vysotskii and Mikhailov 2003c, 1285^VP^.

Further descriptive information

According to phylogenetic analysis based on 16S rRNA gene sequences, members of the genus *Vitellibacter* form a coherent cluster with those of the genus *Aequorivita* with sequence similarities of 92.9–93.6%. 16S rRNA gene sequence similarities of *Vitellibacter vladivostokensis* KMM 3516^T^ with other closely related members of the family *Flavobacteriaceae,* including members of the genera *Ulvibacter, Leeuwenhoekiella, Robiginitalea, Arenibacter, Zobellia, Muricauda,* and *Maribacter,* are less than 90.5%.

Dominant fatty acids of the genus *Vitellibacter* are branched-chain saturated fatty acids $C_{15:0}$ iso and $C_{15:0}$ anteiso. The use of fatty acid composition to differentiate the genus *Vitellibacter* from other members of the family *Flavobacteriaceae* has been discussed by Nedashkovskaya et al. (2003c).

On marine agar 2216 (Difco), vitellibacters form regular, round, convex, smooth, nontransparent, bright orange colonies with entire edges and diameters of 2–4 mm after 48 h at 28°C.

Strains have been grown on media containing 0.5% peptone and 0.1–0.2% yeast extract (Difco), prepared with natural or artificial seawater or supplemented by 2–3% NaCl. Good growth is observed on marine agar 2216 (Difco). Growth occurs at 4–43°C (optimum, 28°C) and with 1–6% NaCl.

The single available strain of the genus *Vitellibacter* is susceptible to carbenicillin, lincomycin, and oleandomycin and resistant to ampicillins, gentamicin, kanamycin, neomycin, polymyxin B, streptomycin, and tetracycline.

Vitellibacter vladivostokensis inhabits coastal marine environments. It was isolated from the edible holothurian *Apostichopus japonicus* collected in the temperate latitudes.

Enrichment and isolation procedures

Vitellibacter vladivostokensis was isolated from *Apostichopus japonicus* by direct plating on marine agar. Natural or artificial seawater is suitable for its cultivation. Vitellibacters may remain viable on marine agar or other rich media based on natural or artificial seawater for several weeks. They have survived storage at −80°C for at least 5 years.

Differentiation of the genus *Vitellibacter* from other genera

Vitellibacter can be differentiated from its closest relative, *Aequorivita*, by its ability to produce flexirubin-type pigments, grow at 43°C, and hydrolyze DNA (Bowman and Nichols, 2002; Nedashkovskaya et al., 2003c).

List of species of the genus *Vitellibacter*

1. **Vitellibacter vladivostokensis** Nedashkovskaya, Suzuki, Vysotskii and Mikhailov 2003c, 1285^{VP}

vla.di.vo.sto.ken′sis. N.L. adj. *vladivostokensis* pertaining to Vladivostok, a city in Asian Russia, where the organism was first isolated.

Cells are 0.3–0.5 × 3–10 µm. Flexirubin-type pigments are produced. Gelatin, casein, DNA, and Tweens 20 and 40 are hydrolyzed, but not agar, starch, alginate, urea, or Tweens 60 and 80. No acid is formed from L-arabinose, cellobiose, L-fucose, D-galactose, D-glucose, D-lactose, maltose, melibiose, raffinose, L-rhamnose, L-sorbose, sucrose, DL-xylose, adonitol, dulcitol, glycerol, inositol, mannitol, sorbitol, *N*-acetylglucosamine, or citrate. Glycerol, *N*-acetylglucosamine, acetate, citrate, malonate, tartrate, and alanine are utilized, but L-arabinose, D-glucose, D-lactose, D-mannose, sucrose, inositol, mannitol, and sorbitol are not utilized. Nitrate is not reduced to nitrite. H_2S and indole are not produced. The predominant fatty acids are $C_{15:0}$ iso (68.8%) and $C_{15:0}$ anteiso (8.4%).

Source: the single available strain was isolated from the holothurian *Apostichopus japonicus* collected in Troitsa Bay, Gulf of Peter the Great, Sea of Japan.

DNA G+C content (mol%): 41.3 (T_m).

Type strain: KMM 3516, IFO 16718.

Sequence accession no. (16S rRNA gene): AB071382.

Genus LVI. **Wautersiella** Kämpfer, Avesani, Janssens, Charlier, De Baere and Vaneechoutte 2006, 2328^{VP}

PETER KÄMPFER

Wau.ter.si.el′la. N.L. fem. dim. n. *Wautersiella* named after Georges Wauters, a Belgian microbiologist, who first recognized this group of organisms as a separate entity.

Rods, 0.5–1.0 × 2.0–3.0 µm, with rounded ends. **Nonmotile.** Gram-stain-negative. **Aerobic**, having a strictly respiratory type of metabolism with oxygen as the terminal electron acceptor. **Oxidase- and catalase-positive.** Colonies on blood agar at 37°C are circular, entire, slightly convex, smooth, glistening, and pale beige. Some strains display yellow-pigmented colonies. The major cellular fatty acids are $C_{15:0}$ iso and summed feature 4 ($C_{15:0}$ iso 2-OH and/or $C_{16:1}$ ω7*t*). The hydroxy acids include the hydroxylated fatty acids $C_{15:0}$ iso 3-OH and $C_{17:0}$ iso 3-OH. Isolated from clinical specimens.

DNA G+C content (mol%): 33.8–34.4.

Type species: **Wautersiella falsenii** Kämpfer, Avesani, Janssens, Charlier, De Baere and Vaneechoutte 2006, 2328^{VP}.

Further descriptive information

Wautersiella falsenii is described as a Gram-stain-negative nonfermenting species with phenotypic characters resembling those of members of the genera *Empedobacter* and *Weeksella*. A total of 26 isolates were collected between 1980 and 2004 by at least ten different Belgian clinical laboratories, originating from various human clinical samples including blood, wounds, pus, respiratory tract, ear discharge, vaginal swab, and pleural fluid.

Growth occurs on blood agar, nutrient agar (Oxoid), tryptone soy agar (Oxoid), trypticase soy broth (BBL) supplemented with 1.5% agar (BBL), and R2A agar (Oxoid) at 20–37°C.

Differentiation of the genus *Wautersiella* from related genera

Tables 72 and 73 list the main differential characteristics of the genera belonging to this rRNA group.

Taxonomic comments

16S rRNA gene sequence analyses of the type strain of *Wautersiella falsenii* clearly place the genus in the family *Flavobacteri-*

TABLE 72. Differential biochemical characteristics of *Wautersiella falsenii* genomovars 1 and 2, *Empedobacter brevis*, and *Weeksella virosa*[a,b]

Characteristic	*Wautersiella falsenii* genomovar 1	*Wautersiella falsenii* genomovar 2	*Empedobacter brevis*	*Weeksella virosa*
Esculin hydrolysis, 4 h	+	–	–	–
Galacturonate alkalinization	+	–	–	–
Gelatin hydrolysis, 24 h	+	–	+	+
β-Galactosidase (ONPG)	–	d	–	–
Casein hydrolysis within 8 h	–	–	+	–
Urease	+	+	–	–
Colistin resistance	+	+	+	–
Acid from glucose and maltose	+	+	+	–

[a]Symbols: +, >85% positive; d, different strains give different reactions (16–84% positive); –, 0–15% positive.

[b]Data from Kämpfer et al. (2006).

aceae. In phylogenetic trees, the highest similarities occur with strains of *Empedobacter* (94–95%) and *Weeksella virosa*. The 16S rRNA gene sequences of the *Wautersiella falsenii* isolates were similar to one another, but two clusters (genomovars) could be distinguished. The sequence similarities for the 14 isolates of genomovar 1 were 99.5–100% and the similarities for the 12 isolates of genomovar 2 were 99.4–100%. The similarity between both clusters was 98.3–99.5%. The presence of two clearly different groups was also shown by tRNA-intergenic length polymorphism (tDNA-PCR) analysis, which enabled differentiation of the species from all other species that have been studied thus far with this technique (Kämpfer et al., 2006).

List of species of the genus *Wautersiella*

1. **Wautersiella falsenii** Kämpfer, Avesani, Janssens, Charlier, De Baere and Vaneechoutte 2006, 2328[VP]

fal.sen′i.i. N.L. gen. n. *falsenii* of Falsen, in honor of the contemporary Norwegian microbiologist Enevold Falsen.

The characteristics are as described for the genus and as listed in Tables 72 and 73, with the following additional features. The main cellular fatty acids are $C_{15:0}$ iso, $C_{17:0}$ iso 3-OH, summed feature 4 ($C_{15:0}$ iso 2-OH and/or $C_{16:1}$ ω7*t*), and $C_{15:0}$ iso 3-OH. Acid is produced from glucose and maltose oxidatively. Tests for the following are positive: urease, indole production, alkaline phosphatase, trypsin (benzyl-arginine arylamidase), and pyrrolidonylaminopeptidase. The following are negative: citrate utilization, nitrate reduction (nitrite reduction is variable), lysine decarboxylase, ornithine decarboxylase, and arginine dihydrolase. Gelatin hydrolysis is positive or weakly and delayed positive. Casein hydrolysis is negative or weakly and delayed positive.

The species contains two genomovars that differ in their 16S rRNA gene sequences and tDNA-PCR patterns. All genomovar 1 isolates (n = 14) display rapid hydrolysis of gelatin and esculin, are β-galactosidase (ONPG)-negative, and all except one alkalinize galacturonate. All genomovar 2 isolates (n = 12) except one are negative for esculin hydrolysis and galacturonate alkanization and most isolates are β-galactosidase-positive and weakly positive for gelatin hydrolysis. The fatty acid profile is as described for the genus, with only minor differences between both genomovars (Table 73).

Source: the type strain of the species and of genomovar 1, NF 993[T], was isolated from a surgical wound. The reference strain of genomovar 2, NF 770 (=CCUG 51537=CIP 108860), was isolated from blood.

DNA G+C content (mol%): 33.8–34.4 (HPLC); values for the type strain of genomovar 1 and the reference strain of genomovar 2 were 33.8 ± 0.4 and 34.4 ± 0.2 mol%, respectively.

Type strain: NF 993, CCUG 51537, CIP 108861.

Sequence accession no. (16S rRNA gene): AM084341.

Genus LVII. **Weeksella** Holmes, Steigerwalt, Weaver and Brenner 1987, 179[VP] (Effective publication: Holmes, Steigerwalt, Weaver and Brenner 1986b, 185.)

The Editorial Board

Weeks.el′la. N.L. dim. ending *-ella*; N.L. fem. n. *Weeksella* named after Owen B. Weeks for his contributions to the taxonomy of the genus *Flavobacterium.*

Rods with parallel sides and rounded ends, typically 0.6 μm wide and 2–3 μm long. Intracellular granules of poly-β-hydroxybutyrate are absent. Endospores are not formed. Gram-stain-negative. **Nonmotile.** Do not glide or spread. **Aerobic**, having a strictly respiratory type of metabolism. **Grow at temperatures from 18 to 42°C. Growth on solid media is not pigmented. Catalase- and oxidase-positive. Urease-negative. Growth occurs on MacConkey agar and β-hydroxybutyrate agar.** Casein is digested. Agar is not digested. Chemo-organotrophic. **Nonsaccharolytic. Indole is produced.** Not found in the general environment, but seem to be parasites, saprophytes, or commensals of the internal surfaces of man and other warm-blooded animals.

DNA G+C content (mol%): 35–38 (T_m).

Type species: **Weeksella virosa** Holmes, Steigerwalt, Weaver and Brenner 1987, 179[VP] (Effective publication: Holmes, Steigerwalt, Weaver and Brenner 1986b, 185.).

Further descriptive information

Colonies are circular (diameter of 0.5–2 mm), low convex, smooth and shiny, with entire edges. On nutrient agar, colonies

TABLE 73. Long-chain fatty acid composition of *Wautersiella falsenii* genomovars 1 and 2, *Empedobacter brevis*, and *Weeksella virosa*[a,b]

Fatty acid	*Wautersiella falsenii* genomovar 1	*Wautersiella falsenii* genomovar 2	*Empedobacter brevis*	*Weeksella virosa*
Unknown 11.543[c]	1.25 ± 0.15	0.88 ± 0.45	nd	nd
$C_{13:0}$ iso	tr	nd	tr	1
$C_{13:1}$ AT 12–13	1.16 ± 0.31	1.4 ± 0.43	nd	nd
$C_{14:1}$ ω5*c*	tr	tr	nd	nd
$C_{14:0}$	1.38 ± 0.21	1.1 ± 0.1	nd	tr
$C_{14:0}$ 2-OH	tr	tr	nd	nd
Unknown 13.566[c]	4.24 ± 1.4	6.55 ± 2.21	1.9 ± 1.4	2
$C_{15:0}$ iso	31.62 ± 0.86	25.83 ± 1.8	23.6 ± 2.2	46
$C_{15:0}$ iso 2-OH	nd	nd	nd	10
$C_{15:0}$ iso 3-OH	8.2 ± 2.0	5.58 ± 0.58	4.8 ± 0.7	5
$C_{15:0}$ anteiso	nd	nd	tr	nd
$C_{15:0}$	tr	tr	nd	nd
$C_{15:1}$ ω6*c*	tr	tr	nd	nd
$C_{16:0}$	3.2 ± 1.36	4.98 ± 0.73	3.8 ± 0.4	4
$C_{16:0}$ iso	nd	nd	nd	1
$C_{16:0}$ 3-OH	4.58 ± 0.95	3.05 ± 0.5	3.6 ± 0.7	1
$C_{16:0}$ iso 3-OH	0.96 ± 0.16	1.23 ± 0.38	tr	nd
$C_{16:1}$ ω5*c*	4.74 ± 1.03	5.55 ± 0.77	nd	1
Unknown 16.580[c]	1.24 ± 0.24	1.5 ± 0.07	1.4 ± 0.1	nd
$C_{17:0}$ 2-OH	tr	tr	tr	nd
$C_{17:0}$ iso	1.1 ± 0.45	1.23 ± 0.48	tr	3
$C_{17:0}$ iso 3-OH	13.32 ± 1.67	13.85 ± 1.86	17.1 ± 2.1	7
$C_{17:1}$ iso ω8*c*	nd	nd	nd	5
$C_{17:1}$ iso ω9*c*	1.22 ± 0.31	2.3 ± 0.47	3.9 ± 0.5	nd
$C_{17:1}$ iso ω12*t*	nd	nd	nd	8
$C_{18:0}$	nd	nd	nd	1
$C_{18:1}$ ω9*c*	nd	nd	nd	1
$C_{18:2}$	nd	nd	nd	2
Summed features:[d]	nd	nd	nd	nd
1	1.78 ± 0.19	0.83 ± 0.28	nd	nd
2	tr	tr	nd	nd
4	13.96 ± 2.65	15.35 ± 1.99	19.4 ± 1.6	nd
5	3.44 ± 0.44	6.1 ± 0.39	6.7 ± 0.8	nd

[a]Fatty acid percentages amounting to less than 1% of the total fatty acids in all strains were not included. Means ± SD are given; tr, trace (less than 1.0%); nd, not detected.

[b]Data from Kämpfer et al. (2006).

[c]Unknown fatty acid; numbers indicate equivalent chain-length.

[d]Fatty acids that could not be separated by GC using the Microbial Identification System (Microbial ID) software were considered summed features. Summed feature 1 contains $C_{15:1}$ iso H and/or $C_{13:0}$ 3-OH; summed feature 2 contains $C_{16:1}$ iso I and/or $C_{14:0}$ 3-OH; summed feature 4 contains $C_{15:0}$ iso 2-OH and/or $C_{16:1}$ ω7*t*; and summed feature 5 contains $C_{17:1}$ iso I and/or $C_{17:1}$ anteiso B.

are circular, entire, mucoid, and nonpigmented after incubation for 24 h. Nonhemolytic on 5% horse blood agar.

Susceptible to penicillin and to most antimicrobial agents (Pedersen et al., 1970; Von Graevenitz, 1981). Tween 20 is hydrolyzed. Gelatin is hydrolyzed (by both stab and plate methods).

The following tests all give a negative reaction: acidity in glucose O-F medium (alkaline reaction); growth in the presence of 0.0075% KCN; growth at 5°C; growth on cetrimide agar; reduction of nitrate, nitrite, and selenite; hydrolysis of esculin and starch; citrate utilization (Simmons' medium); malonate utilization; gluconate oxidation; phenylalanine deaminase; urease (Christensen's medium); opalescence on lecithovitellin agar; extracellular DNase; indole production (as detected by Kovacs' reagent); H_2S production (lead acetate paper and triple-sugar iron agar methods); lipid inclusion granules (following growth on β-hydroxybutyrate); alkaline reaction on Christensen's citrate; arginine desimidase; arginine dihydrolase; lysine decarboxylase; ornithine decarboxylase; 3-ketolactose formation; ONPG hydrolysis; acid or gas from glucose in peptone-water medium; acid from 10% (w/v) glucose and lactose; and acid production (in ammonium salt medium under aerobic conditions) from glucose, adonitol, arabinose, cellobiose, dulcitol, ethanol, fructose, glycerol, inositol, lactose, maltose, mannitol, raffinose, rhamnose, salicin, sorbitol, sucrose, trehalose, and xylose.

Hydrolysis of the following substrates occurs (using API ZYM galleries): 2-naphthyl phosphate at pH 8.5 and at pH 5.4; naphthol-AS-BI-phosphodiamide; bis-(*p*-nitrophenyl)-phosphate; L-lysyl-β-naphthylamide; L-aspartyl-β-naphthylamide; L-alanyl-β-naphthylamide; DL-methionyl-β-naphthylamide; glycyl-glycyl-β-naphthylamide hydrobromide; glycyl-L-phenylalanyl-β-naphthylamide; glycyl-L-prolyl-β-naphthylamide; L-leucyl-glycyl-β-naphthylamide; α-L-glutamyl-β-naphthylamide; and *N*-carbobenzoxy-glycyl-glycyl-L-arginine-β-naphthylamide.

No hydrolysis of the following substrates occurs (using API ZYM galleries): 2-naphthyl butyrate; 2-naphthyl myristate; L-cystyl-2-naphthylamide; *N*-glutaryl-phenylalanine-2-naphthylamide; 6-bromo-2-naphthyl-α-D-galactopyranoside; 2-naphthyl-β-D-galactopyranoside; naphthol-AS-BI-β-D-glucuronic acid; 2-naphthyl-α-D-glucopyranoside; 6-bromo-2-naphthyl-β-D-glucopyranoside; 1-naphthyl-*N*-acetyl-β-D-glucosaminide; 6-bromo-2-naphthyl-α-D-mannopyranoside; 2-naphthyl-α-L-fucopyranoside; 6-bromo-2-naphthyl-β-D-xylopyranoside; *p*-nitrophenyl-α-D-xylopyranoside; *p*-nitrophenyl-β-D-fucopyranoside; *p*-nitrophenyl-β-L-fucopyranoside; *o*-nitrophenyl-*N*-acetyl-α-D-glucosaminide; *p*-nitrophenyl lactoside; *p*-nitrocatechol-sulfate; 4-methylumbelliferyl-arabinopyranoside; 4-methyl umbelliferyl-cellobiopyranoside; L-hydroxyprolyl-β-naphthylamide; γ-L-glutamyl-β-naphthylamide; *N*-benzoyl-L-leucyl-naphthylamide; *N*-carbobenzoxy-L-arginine-4-methoxy-β-naphthylamide hydrochloride; and L-prolyl-β-naphthylamide hydrochloride.

The pathogenicity of *Weeksella virosa* is not known. Most strains have been isolated from women (urine, cervix, vagina, Bartholin's gland cyst, blood, umbilical stump, and ear), but some have been isolated from the urethra, spinal fluid, and blood of men (Mardy and Holmes, 1988; Tatum et al., 1974).

Enrichment and isolation procedures

There is no specific medium for the selective isolation of *Weeksella* strains. Growth occurs on nutrient agar and blood agar. The organism is best identified by amplifying and sequencing the 16S rRNA gene and comparing it with that of the type strain.

Differentiation of the genus *Weeksella* from other genera

The following features differentiate members of the genus *Weeksella* from those of the genus *Bergeyella*: ability to grow at 42°C, ability to grow on MacConkey agar and β-hydroxybutyrate agar, and negative urease activity.

Taxonomic comments

Phenotypic characterization by Holmes et al. (1986b) of 29 strains of aerobic Gram-stain-negative rods previously assigned to CDC group IIf (Tatum et al., 1974) indicated that the strains form a homogeneous species; it had been suggested previously by Holmes and Owen (1981) that they belong to a prospective new genus. In addition, ten of the strains chosen for DNA–DNA hybridization experiments yielded reassociation values of 96–100% (and, under stringent conditions, 77–100%) to strain NCTC 11634[T] of group IIf, indicating that the strains belonged to a single genospecies. There was no significant DNA–DNA hybridization to any other species tested (including CDC group IIj; see below). Holmes et al. (1986b) compared the phenotypic characteristics of group IIf to those of other genera of Gram-stain-negative nonfermenters that had similar DNA G+C contents to those of group IIf and they concluded that, based on both scientific grounds and practical considerations, the species should be placed in a new genus, *Weeksella*, as *Weeksella virosa*.

Holmes et al. (1986a) also placed the strains of another CDC group, group IIj, as a second species in the genus *Weeksella*, viz., *Weeksella zoohelcum*. However, rRNA gene sequence analysis later showed that the genus *Weeksella* was heterogeneous (Vandamme et al., 1994a) and *Weeksella zoohelcum* was placed in a new genus, *Bergeyella, as Bergeyella zoohelcum*. Thus, *Weeksella virosa* currently remains the only species in the genus *Weeksella*.

List of species of the genus *Weeksella*

1. **Weeksella virosa** Holmes, Steigerwalt, Weaver and Brenner 1987, 179[VP] (Effective publication: Holmes, Steigerwalt, Weaver and Brenner 1986b, 185.)

vi.ro'sa. L. fem. adj. *virosa* slimy.

Characteristics are as described for the genus.

Source: most strains have been isolated from clinical specimens.

DNA G+C content (mol%): 35–38 (T_m).

Type strain: NCTC 11634, ATCC 43766, CCUG 30538, CIP 103040, IFO (now NBRC) 16016, LMG 12995.

Sequence accession no. (16S rRNA gene): M93152.

Genus LVIII. **Winogradskyella** Nedashkovskaya, Kim, Han, Snauwaert, Vancanneyt, Swings, Kim, Lysenko, Rohde, Frolova, Mikhailov and Bae 2005a, 51[VP]

Olga I. Nedashkovskaya, Seung Bum Kim and Valery V. Mikhailov

Wi.no.grad.sky.el'la. N.L. fem. n. *Winogradskyella* named after Sergey Winogradskyi, Russian soil microbiologist who made great contributions to the taxonomy of bacteria of the *Cytophaga–Flavobacterium–Bacteroides* phylum.

Rod-shaped cells, 0.4–0.6 × 1.0–1.3 µm, with rounded ends. **Motile by gliding.** Produce non-diffusible yellow carotenoid pigments. No flexirubin pigments are produced. Cells can form network-like structures. **Chemo-organotrophs. Strictly aerobic.** Oxidase-, catalase-, and alkaline phosphatase-positive. Arginine dihydrolase, lysine decarboxylase, ornithine decarboxylase, and β-galactosidase are not produced. Gelatin and Tween 40 are hydrolyzed, but casein, chitin, and urea are not attacked. Hydrolysis of agar, gelatin, starch, DNA, and Tweens 20 and 80 varies among species. Indole and H_2S are not produced. Nitrate is not reduced. **Marine**, from coastal habitats. Require seawater or Na^+ for growth. **The major respiratory quinone is MK-6. The main polar lipid is phosphatidylethanolamine.**

DNA G+C content (mol%): 32–37.

Type species: **Winogradskyella thalassocola** Nedashkovskaya, Kim, Han, Snauwaert, Vancanneyt, Swings, Kim, Lysenko, Rohde, Frolova, Mikhailov and Bae 2005a, 52[VP].

Further descriptive information

Phylogenetic analysis based on 16S rRNA gene sequences indicates that the genus *Winogradskyella* forms a cluster with members of the family *Flavobacteriaceae* such as *Psychroserpens*, *Olleya*, *Lacinutrix*, *Algibacter*, and *Gaetbulibacter*. *Psychroserpens burtonensis* was found to be the nearest neighbor (Figure 53). These relationships with *Winogradskyella* are supported by a high bootstrap value and also by different tree-making algorithms. The low sequence similarity values between strains of *Winogradskyella* and other *Cytophaga*-like bacteria described to date are 85.6–92.1%. *Winogradskyella* species possess 16S rRNA gene sequence similarities to one another that range from 94.7 to 97.1%. The DNA–DNA relatedness values between the type strains of *Winogradskyella* species are 34–45%.

The predominant cellular fatty acids are $C_{15:1}$ iso, $C_{15:0}$ iso, $C_{16:0}$ iso 3-OH, $C_{17:0}$ iso 3-OH, and summed feature 3 ($C_{15:0}$ iso 2-OH and/or $C_{16:1}$ ω7*c*).

Cells of *Winogradskyella* can form network-like structures or aggregates (Figure 54). On marine agar 2216 (Difco), *Winogradskyella* species form regular, round, shiny, yellow colonies with diameters of 2–4 mm after 48 h at 28°C. The colonies are usually slightly sunk in the agar.

All isolated strains have been grown on media containing 0.5% peptone and 0.1–0.2% yeast extract (Difco), prepared with natural or artificial seawater or supplemented with 2–3% NaCl. Growth occurs in 1–8% NaCl, at pH 6.0–10.0, and at 4–44°C. The optimal growth temperature is 21–25°C.

No acid is formed from L-arabinose, D-galactose, D-lactose, melibiose, L-rhamnose, DL-xylose, adonitol, dulcitol, inositol, sorbitol, or citrate. Does not utilize L-arabinose, D-lactose, D-sucrose, inositol, mannitol, sorbitol, malonate, or citrate.

Strains of the genus *Winogradskyella* are susceptible to chloramphenicol and resistant to kanamycin.

Winogradskyella species inhabit coastal marine environments. They have been isolated from seaweeds and sponges.

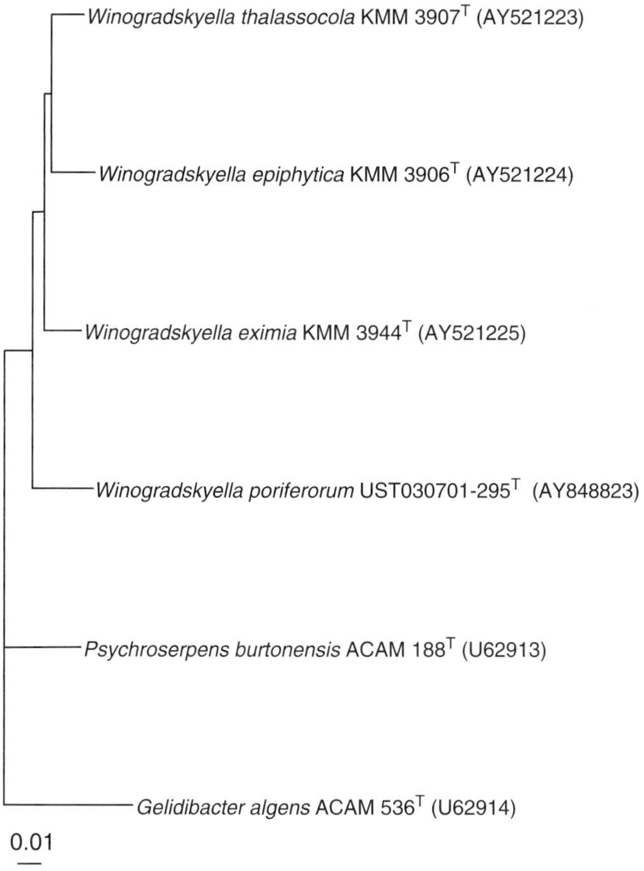

FIGURE 53. Phylogenetic tree based on the 16S rRNA gene sequences of the type strains of the genus *Winogradskyella* and representative members of the family *Flavobacteriaceae*. Bar = 0.01 substitutions per nucleotide position.

Enrichment and isolation procedures

Winogradskyella species were isolated from samples of seaweeds and sponges by direct plating on marine agar (Difco) and modified marine agar prepared with filtered natural seawater (Lau et al., 2005a; Nedashkovskaya et al., 2005a). Either natural or artificial seawater is suitable for their cultivation. All isolated strains have been grown on media containing 0.5% peptone and 0.1–0.2% yeast extract (Difco). *Winogradskyella* strains may remain viable on marine agar or other rich media based on natural or artificial seawater for several weeks. They have survived storage at −80°C for at least 5 years.

Differentiation of the genus *Winogradskyella* from other genera

Members of the genus *Winogradskyella* can be differentiated from those of its closest relative, *Psychroserpens*, by the presence of gliding motility, oxidase activity, and higher DNA G+C content (32–37 mol% vs 27–29 mol%) (Bowman et al., 1997; Lau et al., 2005a; Nedashkovskaya et al., 2005a). The latter feature clearly distinguishes *Winogradskyella* from another relative, *Olleya*, which has a DNA G+C content of 49 mol% (Mancuso Nichols et al., 2005). *Winogradskyella* species are able to move by means of gliding and cannot grow with 12% NaCl, in contrast to members of the genus *Lacinutrix* (Bowman and Nichols, 2005). An inability to ferment D-glucose differentiates *Winogradskyella* strains from those of the genera *Algibacter* and *Gaetbulibacter* (Jung et al., 2005; Nedashkovskaya et al., 2004d). Moreover, *Winogradskyella* strains may be separated from those of *Gaetbulibacter* by the presence of gliding motility and by the absence of nitrate reduction.

Differentiation of the species of the genus *Winogradskyella*

Differential features of *Winogradskyella* species are given in Table 74.

List of species of the genus *Winogradskyella*

1. **Winogradskyella thalassocola** Nedashkovskaya, Kim, Han, Snauwaert, Vancanneyt, Swings, Kim, Lysenko, Rohde, Frolova, Mikhailov and Bae 2005a, 52[VP]

tha.las.so.co'la. Gr. n. *thalassa* the sea; L. suffix *-cola* dweller; N.L. n. *thalassocola* a sea dweller

Cells are 0.4–0.6 × 1.0–1.3 µm and can form network-like structures or aggregates. On marine agar, colonies are 2–4 mm in diameter, circular, shiny with entire edges, yellow, and viscous. Growth occurs at 4–33°C. Optimal temperature for growth is 21–23°C. Growth is observed in 1–8% NaCl. No acid is produced from L-arabinose, D-galactose, D-lactose, melibiose, L-rhamnose, DL-xylose, citrate, adonitol, dulcitol, or inositol. L-Arabinose, D-lactose, sucrose, mannitol, inositol, sorbitol, malonate, and citrate are not utilized. β-Galactosidase activity is negative. Susceptible to carbenicillin and lincomycin; resistant to gentamicin, neomycin, polymyxin B, and streptomycin.

Source: isolated from brown alga *Chorda filum*, collected in Troitsa Bay in the Gulf of Peter the Great, Sea of Japan.

DNA G+C content (mol%): 34.6 (T_m).

Type strain: KMM 3907, KCTC 12221, LMG 22492, DSM 15363.

Sequence accession no. (16S rRNA gene): AY521223.

2. **Winogradskyella epiphytica** Nedashkovskaya, Kim, Han, Snauwaert, Vancanneyt, Swings, Kim, Lysenko, Rohde, Frolova, Mikhailov and Bae 2005a, 53[VP]

e.pi.phy'ti.ca. N.L. fem. adj. *epiphytica* (derived from Gr. *epi* on and *phyt-* relating to plants) onto plant, pertaining to the original isolation from the surface of the algal fronds.

Cells are 0.4–0.6 × 1.0–1.3 µm and can form network-like structures or aggregates. On marine agar, colonies are 2–4 mm in diameter, circular, shiny with entire edges, yellow, and viscous. Growth occurs at 4–37°C (optimum, 23–25°C). Growth occurs in 1–8% NaCl. No acid is formed from L-arabinose, D-galactose, D-lactose, melibiose, L-rhamnose, DL-xylose, citrate, adonitol, dulcitol, or inositol. Does not utilize L-arabinose, D-lactose, sucrose, mannitol, inositol, sorbitol, malonate, or citrate. β-Galactosidase activity is negative. Nitrate is not reduced. Susceptible to carbenicillin and lincomycin. Resistant to gentamicin, neomycin, and polymyxin B.

Source: isolated from the green alga *Acrosiphonia sonderi*, collected in Troitsa Bay in the Gulf of Peter the Great, Sea of Japan.

DNA G+C content (mol%): 35.2 (T_m).

Type strain: KMM 3906, KCTC 12220, LMG 22491, CCUG 47091.

Sequence accession no. (16S rRNA gene): AY521224.

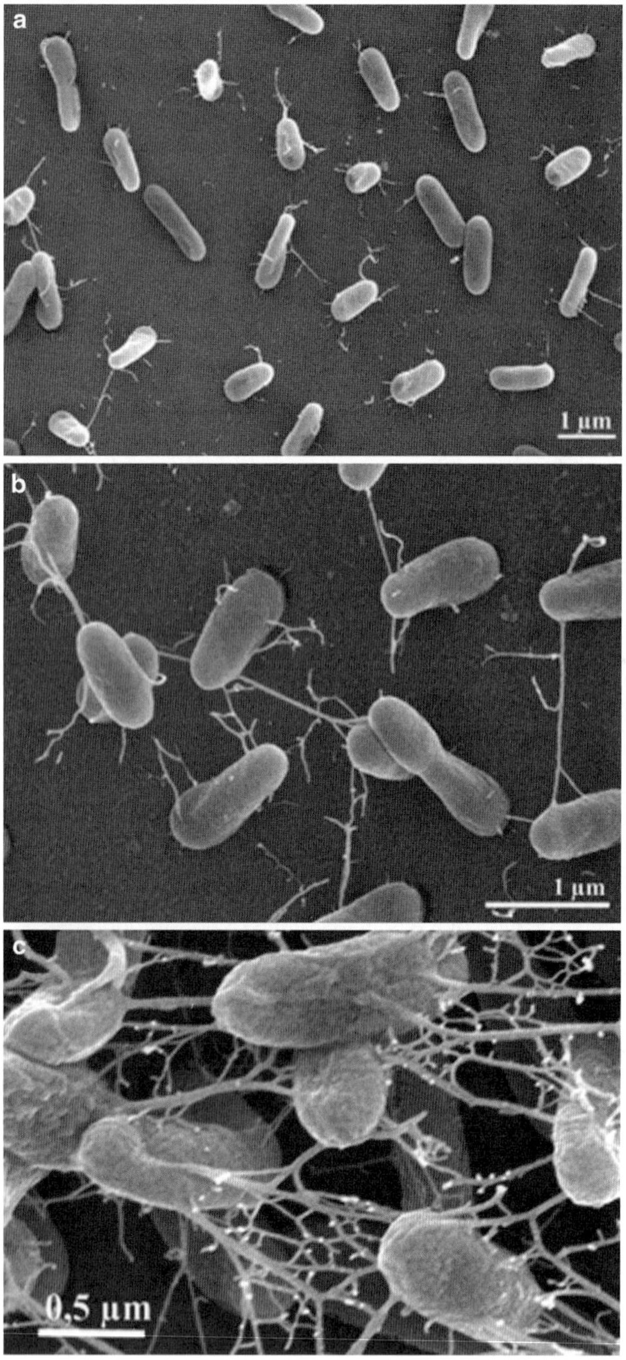

FIGURE 54. Scanning electron micrographs of *Winogradskyella thalassocola* KMM 3907[T] showing the rod-shaped morphology and network-like structures formed by cells.

TABLE 74. Differential features of *Winogradskyella* species[a,b]

Characteristic	*W. thalassocola*	*W. epiphytica*	*W. eximia*	*W. poriferorum*
H_2S production	–	–	+	–
Acetoin production	–	–	–	+
Degradation of:				
Agar	+	+	+	–
Casein, starch	–	–	+	–
Tween 20	–	+	+	+
Tween 80	–	+	–	+
DNA	–	+	–	+
Growth at/in:				
37°C	–	+	–	+
44°C	–	–	–	+
8% NaCl	+	+	–	–
Acid formation from:				
D-Glucose, maltose	+	–	+	–
Cellobiose	+	–	–	–
Sucrose, mannitol	–	–	+	–
Utilization of:				
D-Glucose, D-mannose	+	–	+	–
Susceptibility to:				
Ampicillin	–	+	–	+
Benzylpenicillin, streptomycin	–	–	–	+
Oleandomycin	+	+	–	nd
Tetracycline	–	+	+	+
DNA G+C content (mol%)	34.6	35.2	36.1	32.8

[a]Symbols: +, >85% positive; –, 0–15% positive; nd, not determined.
[b]Data are taken from Nedashkovskaya et al. (2005a) and Lau et al. (2005a).

3. **Winogradskyella eximia** Nedashkovskaya, Kim, Han, Snauwaert, Vancanneyt, Swings, Kim, Lysenko, Rohde, Frolova, Mikhailov and Bae 2005a, 54[VP]

e.xi'mi.a. L. fem. adj. *eximia* excellent.

Cells are 0.4–0.6 × 1.0–1.3 µm and can form network-like structures or aggregates. On marine agar, colonies are 2–4 mm in diameter, circular, shiny with entire edges, yellow, and viscous. Growth occurs in 1–5% NaCl and at 4–33°C. The optimal temperature for growth is 21–23°C. Does not form acid from L-arabinose, D-galactose, D-lactose, melibiose, L-rhamnose, DL-xylose, citrate, adonitol, dulcitol, or inositol. L-Arabinose, D-lactose, D-sucrose, mannitol, inositol, sorbitol, malonate and citrate are not utilized. β-Galactosidase activity is negative. Susceptible to lincomycin; resistant to carbenicillin, gentamicin, neomycin, and polymyxin B.

Source: isolated from the brown alga *Laminaria japonica*, collected in the Gulf of Peter the Great, Sea of Japan.

DNA G+C content (mol%): 36.1 (T_m).

Type strain: KMM 3944, KCTC 12219, LMG 22474.

Sequence accession no. (16S rRNA gene): AY521225.

4. **Winogradskyella poriferorum** Lau, Tsoi, Li, Plakhotnikova, Dobretsov, Lau, Wu, Wong, Pawlik and Qian 2005a, 1591[VP]

por.if.er.or'um. N.L. gen. pl. n. *poriferorum* of the phylum *Porifera*, referring to the isolation source sponge, of the phylum *Porifera*.

On marine agar, colonies are 2–4 mm in diameter, convex, smooth, and circular with entire edges. Growth is observed at 12–44°C, at pH 6.0–10.0, and in 1–4% NaCl. Esculin is utilized as a sole carbon source, but none of the other substrates tested in the API 50CH system are utilized. In the API galleries (bioMérieux), the following enzyme activities are exhibited: α-chymotrypsin, cystine arylamidase, leucine arylamidase, valine arylamidase, esterase (C4), esterase lipase (C8), lipase (C14), acid phosphatase, naphthol-AS-BI-phosphohydrolase, and trypsin. Tryptophan deaminase, *N*-acetyl-β-glucosaminidase, lysine decarboxylase, ornithine decarboxylase, arginine dihydrolase, α- and β-galactosidases, α- and β-glucosidases, β-glucuronidase, α-fucosidase, α-mannosidase, and urease activities are not observed.

Source: isolated from the surface of the sponge *Lissodendoryx isodictyalis* in the Bahamas.

DNA G+C content (mol%): 32.8 (T_m).

Type strain: UST030701-295, NRRL B-41101, JCM 12885.

Sequence accession no. (16S rRNA gene): AY848823.

Genus LIX. **Yeosuana** Kwon, Lee, Jung, Kang and Kim 2006a, 731[VP]

The Editorial Board

Yeo.su.a'na. N.L. fem. n. *Yeosuana* named after Yeosu City, where the type strain of the type species was isolated.

Rods, 0.2–0.3 × 0.7–1.7 µm. Nonmotile. **No gliding motility.** Gram-stain-negative. Colonies are yellowish-brown due to carotenoid pigments. Flexirubin type pigments are not produced. Temperature range for growth, 23–39°C; optimum, 33–46°C. **Requires 0.5–4.0% NaCl; optimum, 2.0%. Growth does not occur without supplementation with either $CaCl_2$ or $MgCl_2$, even in the presence of NaCl.** Yeast extract is not required for growth. **Aerobic. Oxidase-negative.** Catalase is weakly positive. **Nitrate is not reduced. Indole and H_2S are not produced. Gelatin is hydrolyzed, but not urea, agar, casein, or starch. Acid is not produced from carbohydrates.** Major cellular fatty acids are straight-chain and branched-chain unsaturated fatty acids. Major respiratory quinones are MK-5 and MK-6.

DNA G+C content (mol%): 51.4 (HPLC).

Type species: **Yeosuana aromativorans** Kwon, Lee, Jung, Kang and Kim 2006a, 731[VP].

Enrichment and isolation procedures

One gram of estuarine sediment (from Gwangyang Bay, Yeosu City, Republic of Korea) was enriched for 2 years at 10–30°C in MM2 medium* supplemented with 1 mg/l each of pyrene and benzo(a)pyrene (BaP). After serial dilution, a sample of the slurry was plated onto marine agar 2216 (MA). The type strain of *Yeosuana aromativorans* was one of isolates able to grow on MA.

Differentiation of the genus *Yeosuana* from related genera

The DNA G+C content of *Yeosuana* (51.4 mol%) is much higher than that of members of the genera *Algibacter*, *Bizionia*, *Formosa*, *Gaetbulibacter*, *Gelidibacter*, *Subsaxibacter*, and *Subsaximicrobium*, whose values range from 31 to 45 mol%; moreover, unlike these genera, *Yeosuana* requires sea salts (not merely NaCl) for growth. *Yeosuana* differs from members of the genera *Gelidibacter*, *Algibacter*, *Formosa*, and *Gaetbulibacter* by its failure to form acid from carbohydrates. Gliding motility is present in *Algibacter*, *Formosa*, *Gaetbulibacter*, *Gelidibacter*, *Subsaxibacter*, and *Subsaximicrobium*, but does not occur in *Yeosuana*. *Yeosuana* does not hydrolyze agar (unlike *Algibacter* species), casein (unlike *Bizionia* and *Subsaximicrobium* species), or starch (unlike *Bizionia*, *Formosa*, *Gaetbulibacter*, and *Subsaximicrobium* species). It differs from *Formosa* and *Gaetbulibacter* species by its failure to reduce nitrate.

Taxonomic comments

Kwon et al. (2006a) reported that, by 16S rRNA gene sequence analysis, the closest neighbors of *Yeosuana aromativorans* are *Gelidibacter algens* (94.7% gene sequence similarity), *Gaetbulibacter saemankumensis* (94.5%), and *Gelidibacter gilvus* (94.4%). *Yeosuana* also has high sequence similarities to members of the genera *Subsaximicrobium* (93.3%), *Subsaxibacter* (93.9%), *Gaetbulibacter* (94.5%), *Algibacter* (94.2%), *Bizionia* (93.6–94.3%), and *Formosa* (93.2%).

List of species of the genus *Yeosuana*

1. **Yeosuana aromativorans** Kwon, Lee, Jung, Kang and Kim 2006a, 731[VP]

a.ro.ma.ti.vo'rans. L. n. *aroma -atis* spice; L. part. adj. *vorans* devouring; N.L. part. adj. *aromativorans* degrading aromatic compounds.

Characteristics are as described for the genus, with the following additional features. Forms yellowish-brown colonies, 1.0–1.5 mm in diameter on MA. β-Glucosidase, β-galactosidase, and protease activities are present. The following compounds are utilized: α-cyclodextrin, dextrin, cellobiose, D-fructose, gentiobiose, α-D-glucose, α-D-lactose, maltose, D-mannose, sucrose, methyl pyruvate, α-ketobutyric acid, L-proline, glucose 1-phosphate, and glucose 6-phosphate. The following are utilized weakly: D-galactose, α-ketoglutaric acid, L-alanine, L-alanyl glycine, L-asparagine, L-glutamic acid, glycyl L-aspartic acid, glycyl L-glutamic acid, hydroxy-L-proline, uridine, and L-threonine. Polycyclic aromatic hydrocarbons are degraded, including pyrene and BaP. The predominant fatty acids are $C_{15:0}$ iso (21.7%), $C_{15:0}$ anteiso (14.9%), $C_{15:1}$ ω10*c* iso (14.8%), and $C_{16:1}$ (10.9%).

Source: the type strain was isolated from estuarine sediment of Gwangyang Bay, Yeosu City, Korea.

DNA G+C content (mol%): 51.4 (HPLC).

Type strain: GW1-1, JCM 12862, KCCM 42019.

Sequence accession no. (16S rRNA gene): AY682382.

*MM2 contains 18 mM $(NH_4)_2SO_4$, 1 µM $FeSO_4·7H_2O$, and 100 µl of 1 M KH_2PO_4/Na_2HPO_4 buffer solution in 1 l aged seawater, pH 7.2.

Genus LX. **Zhouia** Liu, Wang, Dai, Liu and Liu 2006, 2826[VP]

The Editorial Board

Zhou'i.a. N.L. fem. n. *Zhouia* named after Pei-Jin Zhou, a pioneer of environmental microbiology in China.

Rods, 0.25–0.3 × 1.3–3.0 µm. Nonmotile. **No gliding motility.** Nonsporeforming. Gram-stain-negative. **Strictly aerobic. NaCl is required for growth. Growth occurs at 42°C.** Colonies are yellow to pale yellow. Catalase-positive. Oxidase-negative. Urease-negative. Flexirubin pigments are not produced. Nitrate is not reduced. Starch and gelatin are hydrolyzed, but not agar, casein, or Tweens 40 and 80. **Acid is produced from cellobiose, but not glucose.** Glucose and sucrose are utilized. The menaquinone is MK-6. The major fatty acids are $C_{15:1\,G}$ iso, $C_{15:0}$ iso, summed feature 4 ($C_{15:0}$ iso 2-OH and/or $C_{16:1}$ ω7*c*/*t*), and $C_{15:0}$.

DNA G+C content (mol%): 34.5.

Type species: **Zhouia amylolytica** Liu, Wang, Dai, Liu and Liu 2006, 2827[VP].

Enrichment and isolation procedures

Strains were isolated by spreading dilutions of marine sediment samples onto marine agar 2216 (MA; Difco), with subsequent incubation at 30°C. Colonies on MA are circular, slightly raised, smooth, yellow to pale yellow, and 2.0–3.0 mm in diameter after incubation for 2 d.

Differentiation of the genus *Zhouia* from related genera

Zhouia amylolytica, the only species currently recognized, can be differentiated from members of the genus *Coenonia* by its requirement for NaCl, hydrolysis of gelatin, inability to hydrolyze Tween 40, production of acid from cellobiose, and its lack of $C_{13:0}$ as a major cellular fatty acid. It differs from other members of the *Flavobacteriaceae* such as *Aquimarina*, *Cellulophaga*, *Gaetbulimicrobium*, *Psychroflexus*, and *Winogradskyella* by the presence of summed feature 4 ($C_{15:0}$ iso 2-OH and/or $C_{16:1}$ ω7*c*/*t*) as a characteristic fatty acid.

Taxonomic comments

Based on 16S rRNA gene sequence analysis, the nearest neighbor of the type strain of *Zhouia amylolytica* is *Coenonia anatina* LMG 14382[T], with 90.2% similarity.

List of species of the genus *Zhouia*

1. **Zhouia amylolytica** Liu, Wang, Dai, Liu and Liu 2006, 2827[VP]

a.my.lo.ly'ti.ca. Gr. n. *amylos* starch; N.L. adj. *lyticus* from Gr. adj. *lutikos* dissolving; N.L. fem. adj. *amylolytica* dissolving starch, pertaining to the ability of the bacterium to hydrolyze starch.

The description is as given for genus, with the following additional features. Growth occurs at 7–42°C, with optimum growth at 30°C; growth does not occur at 4 or 45°C. Temperature range for growth, 6.0–8.0; optimum, 7.2–7.4; no growth at pH 5.8 or 8.2. NaCl is required for growth; optimum concentration, 4.5–5%; no growth occurs in >9% NaCl. The following compounds are utilized: acetic acid, *N*-acetyl-D-galactosamine, *N*-acetyl-D-glucosamine, L-alanine, L-alanyl glycine, L-asparagine, L-aspartic acid, cellobiose, α-cyclodextrin, dextrin, D-fructose, L-fucose, D-galactonolactone, D-galactose, gentiobiose, D-glucose 1-phosphate, D-glucose 6-phosphate, L-glutamic acid, glycogen, glycyl-L-aspartic acid, glycyl-L-glutamic acid, inosine, α-ketobutyric acid, α-ketoglutaric acid, α-ketovaleric acid, DL-lactic acid, α-D-lactose, L-leucine, maltose, melibiose, methyl-β-D-glucoside, L-ornithine, L-proline, propionic acid, D-psicose, raffinose, L-rhamnose, D-sorbitol, succinic acid monomethyl ester, L-threonine, trehalose, turanose, urocanic acid, and uridine. The following compounds are not utilized: *cis*-aconitic acid, adonitol, L-alaninamide, D-alanine, γ-aminobutyric acid, 2-aminoethanol, L-arabinose, D-arabitol, bromosuccinic acid, 2,3-butanediol, DL-carnitine, citric acid, i-erythritol, formic acid, D-galactonic acid, D-gluconic acid, D-glucosaminic acid, glucuronamide, D-glucuronic acid, glycerol, DL-α-glycerol phosphate, L-histidine, α-hydroxybutyric acid, β-hydroxybutyric acid, γ-hydroxybutyric acid, *myo*-inositol, itaconic acid, malonic acid, D-mannitol, methyl pyruvate, L-phenylalanine, phenylethylamine, putrescine, L-pyroglutamic acid, quinic acid, D-saccharic acid, sebacic acid, D-serine, L-serine, succinamic acid, succinic acid, thymidine, Tweens 40 and 80, and xylitol. The major fatty acids are $C_{15:1\,G}$ iso (24.2%), $C_{15:0}$ iso (14.9%), summed feature 4 ($C_{15:0}$ iso 2-OH and/or $C_{16:1}$ ω7*c*/*t*; 10.7%), and $C_{15:0}$ (9.4%).

Source: the type strain was isolated from sediment from the South China Sea, China.

DNA G+C content (mol%): 34.5 (T_m).

Type strain: HN-171, CGMCC 1.6114, JCM 14016.

Sequence accession no. (16S rRNA gene): DQ423479.

Genus LXI. **Zobellia** Barbeyron, L'Haridon, Corre, Kloareg and Potin 2001, 993[VP]

Olga I. Nedashkovskaya and Makoto Suzuki

Zo.bel'li.a. N.L. fem. n. *Zobellia* named after C.E. ZoBell, who isolated and characterized numerous marine bacteria, notably [*Cytophaga*] *uliginosa* in 1944, and for his general contribution to the taxonomy of marine bacteria.

Rod-shaped cells with rounded ends, **0.3–0.5 × 1.2–8.0 µm. Motile by gliding.** Produce non-diffusible yellow–orange carotenoid pigments. Flexirubin-type pigments are produced. **Chemoorganotrophs. Strictly aerobic. The major respiratory quinone is MK-6. The main polar lipid is phosphatidylethanolamine.** Positive for oxidase, catalase, β-galactosidase, and alkaline phosphatase. Arginine dihydrolase, lysine decarboxylase, ornithine decarboxylase, and urease are not produced. H_2S, indole, and acetoin are not produced. Agar and gelatin are hydrolyzed; hydrolysis of casein, gelatin, starch, alginate, chitin, DNA, and Tweens varies among species. **Marine**, isolated from coastal habitats. Require seawater or Na^+ for growth.

DNA G+C content (mol%): 36–44 (T_m).

Type species: **Zobellia galactanivorans** (Barbeyron, Gerard, Potin, Henrissat and Kloareg 1998) corrig. Barbeyron, L'Haridon, Corre, Kloareg and Potin 2001, 994[VP] (*Cytophaga drobachiensis* Barbeyron, Gerard, Potin, Henrissat and Kloareg 1998, 528).

Further descriptive information

Phylogenetic analysis based on 16S rRNA gene sequences indicates that the genus *Zobellia* forms a cluster with *Arenibacter*, *Maribacter*, *Muricauda*, and *Robiginitalea* (Figure 55). The closest relatives of *Zobellia* strains are members of the genus *Maribacter*, with similarities of 92.1–94.1%. 16S rRNA gene similarities between members of the genus *Zobellia* and other close neighbors range from 88.9 to 91%. 16S rRNA gene sequence similarities between the various *Zobellia* species are 96.7–99.3%.

Dominant cellular fatty acids are straight-chain unsaturated and branched-chain unsaturated fatty acids $C_{15:1}$ iso, $C_{15:0}$ iso, $C_{17:0}$ iso 3-OH, and summed feature 3, comprising $C_{15:0}$ iso 2-OH and/or $C_{16:1}$ ω7*c*/*t* (see Table 75).

Differential phenotypic characteristics are given in Table 76. On marine agar 2216 (Difco), strains of *Zobellia* form regular, round, shiny colonies, usually with entire edges, which are sunk in the agar. Colonies are yellow–orange to dark red or dark orange with a diameter of 2–4 mm after 48 h at 28°C. The colonies usually spread rapidly on the surface of the agar.

All isolated strains have been grown on media containing 0.5% peptone and 0.1–0.2% yeast extract (Difco), either prepared with natural or artificial seawater, or supplemented with 2–3% NaCl. Growth is observed at 4–45°C (optimum, 21–35°C). Growth occurs in 1–10% NaCl (optimum, 2–3% NaCl). Zobellias grow between pH 5.8 and 9.4 (optimum of about 7.0).

Acid is produced from L-fucose, D-galactose, D-lactose, melibiose, and L-sorbose, but not from *N*-acetylglucosamine, citrate, adonitol, dulcitol, glycerol, or inositol. All species utilize L-arabinose, D-glucose, D-lactose, D-mannose, sucrose, and mannitol, but not inositol, sorbitol, malonate, or citrate.

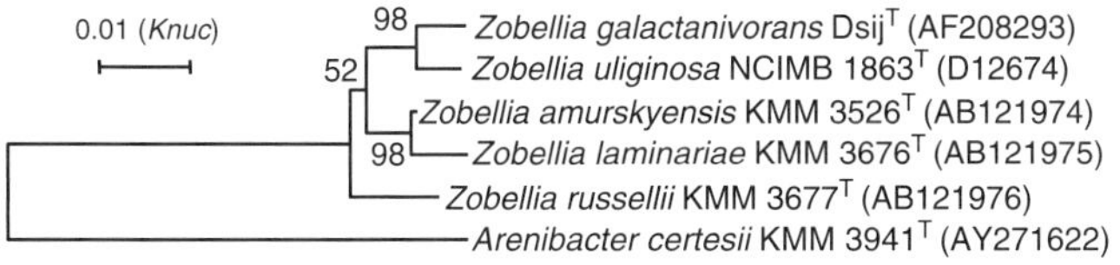

FIGURE 55. Phylogenetic tree showing relationship among the species of the genus *Zobellia*, The tree was generated by the neighbor-joining method (1987). Numbers at the nodes indicate bootstrap values (%) of 1000 replicates (only values ≥50% are cited). Bar = genetic distance of 0.01 (K_{nuc}).

Zobellia species are susceptible to carbenicillin, lincomycin, and oleandomycin. They are resistant to gentamicin, kanamycin, neomycin, and polymyxin B.

Zobellia species are common inhabitants of the marine environment. They have been isolated from seawater, sediments, and seaweeds in the temperate latitudes.

Enrichment and isolation procedures

Zobellia species were isolated by direct plating on marine agar or ZoBell agar (Nedashkovskaya et al., 2004f; ZoBell and Upham, 1944), except *Zobellia galactanivorans*, formerly [*Cytophaga*] *drobachiensis* (Barbeyron et al., 1998), which was isolated on a basal salts medium (Quatrano and Caldwell, 1978) supplemented with 2% (w/v) ι-carrageenan. Either natural or artificial seawater is suitable for their cultivation. All isolated strains have been grown on media containing 0.5% peptone and 0.1–0.2% yeast extract (Difco). Strains may remain viable on marine agar or other rich medium based on natural or artificial seawater for several weeks. They have survived storage at −80°C for at least 5 years.

Differentiation of the genus *Zobellia* from other genera

Zobellia strains differ from those of all their close relatives, i.e., members of the genera *Arenibacter*, *Maribacter*, *Muricauda* and *Robiginitalea*, by their production of flexirubin-type pigments (Bruns et al., 2001; Cho and Giovannoni, 2004; Nedashkovskaya et al., 2003b, 2004a, b, f, 2006g; Yoon et al., 2005b). Also, *Zobellia* strains can be differentiated from those of *Arenibacter*, *Muricauda*, and *Robiginitalea* by their ability to hydrolyze agar. Moreover, *Zobellia* strains liquefy gelatin, move by gliding, and reduce nitrate to nitrite, in contrast to strains of *Muricauda* and *Robiginitalea*.

Taxonomic comments

The genus *Zobellia* was created by Barbeyron et al. (2001) and contains Gram-stain-negative, aerobic, gliding, agarolytic, flexirubin-producing bacteria of marine origin. One of the species, *Zobellia uliginosa*, was isolated from beach sand in California and was originally classified as *Flavobacterium uliginosum* (ZoBell and Upham, 1944). Later, it was transferred to the genus *Cytophaga* as *Cytophaga uliginosa* (Reichenbach, 1989a), reclassified in the genus *Cellulophaga* as *Cellulophaga uliginosa* (Bowman, 2000) and finally assigned to the new genus *Zobellia* based on its higher DNA G+C content, its maximum growth temperature, the presence of flexirubin pigments, and its unique phylogenetic position (Barbeyron et al., 2001).

List of species of the genus *Zobellia*

1. **Zobellia galactanivorans** (Barbeyron, Gerard, Potin, Henrissat and Kloareg 1998) corrig. Barbeyron, L'Haridon, Corre, Kloareg and Potin 2001, 994[VP] (*Cytophaga drobachiensis* Barbeyron, Gerard, Potin, Henrissat and Kloareg 1998, 528)

ga.lac.ta.ni.vo'rans. N.L. n. *galactan* polygalactose; L. v. *vorare* to devour; N.L. fem. adj. *galactanivorans* galactan-devouring.

Cells are 0.3–0.4 × 3.0–8.0 μm. On marine agar, colonies are 2–4 mm in diameter, circular, shiny with entire or irregular edges, yellow–orange, and sunk into the agar; they spread very fast on agar plates. Growth occurs between 10 and 45°C (optimum, about 35°C) and in a salt concentration of 1.0–8% NaCl (optimum, 2%). The following substrates are utilized: *N*-acetylglucosamine, DL-glycerol phosphate, mannitol, L-rhamnose, sucrose, phosphate, and L-serine, but not adipate, caprate, gluconate, α-ketobutyric acid, 2-ketogluconate, D-lyxose, malate, phenylacetate, D-tagatose, L-threonine, or urocanic acid. Carrageenans (κ and ι) are decomposed.

TABLE 75. Whole-cell fatty acid composition of *Zobellia* species[a]

Fatty acid	*Z. galactanivorans* Dsij^T	*Z. amurskyensis* KMM 3526^T	*Z. laminariae* KMM 3676^T	*Z. russellii* KMM 3677^T	*Z. uliginosa* CIP 104808^T
Straight-chain:					
$C_{14:0}$	tr	1.0	tr	tr	nd
$C_{15:0}$	7.5	14.4	12.5	11.0	10.2
$C_{15:0}$ 3-OH	nd	tr	nd	nd	nd
$C_{15:1}$ ω6*c*	1.1	3.2	2.7	1.7	1.4
$C_{16:0}$	2.2	1.0	tr	2.4	2.6
$C_{16:0}$ 3-OH	3.0	2.4	2.6	4.9	2.9
$C_{17:1}$ ω6*c*	nd	1.2	1.0	nd	nd
$C_{18:1}$ ω6*c*	nd	0.7	1.1	nd	nd
Branched:					
$C_{15:0}$ iso	21.1	22.5	16.8	20.1	21.9
$C_{15:0}$ iso 3-OH	8.3	4.6	6.1	5.9	6.7
$C_{15:0}$ anteiso	1.8	1.0	1.0	nd	1.4
$C_{15:1}$ iso	8.8	10.4	12.3	14.9	12.0
$C_{17:0}$ iso 3-OH	23.7	15.1	22.4	19.7	25.9
$C_{17:1}$ iso ω9	5.1	3.8	3.1	2.4	3.6
Summed feature 3[b]	14.5	15.5	14.9	14.3	9.9

[a]Data are taken from Nedashkovskaya et al. (2004f). Those fatty acids for which the mean amount for all taxa was less than 1% are not given; nd, not detected; tr, trace.

[b]Summed feature 3 consists of one or more of the following fatty acids which could not be separated by the Microbial Identification system: $C_{16:1}$ ω7*c*, $C_{16:1}$ ω7*t*, and $C_{15:0}$ iso 2-OH.

Source: the single strain was isolated from the red alga *Delesseria sanguinea*, collected in the English Channel, France.

DNA G+C content (mol%): 43.4 (T_m), 43.0 (HPLC).

Type strain: Dsij, CIP 106680, DSM 12802.

Sequence accession no. (16S rRNA gene): AF208293.

Note: the original spelling *galactanovorans* (*sic*) has been corrected (Int. J. Syst. Evol. Microbiol. Notification that new names and new combinations have appeared in volume 51, part 3, of the IJSEM, 2001, *51:* 1231–1233).

2. **Zobellia amurskyensis** Nedashkovskaya, Suzuki, Vancanneyt, Cleenwerck, Lysenko, Mikhailov and Swings 2004f, 1647[VP]

a.mur.sky.en′sis. N.L. fem. adj. *amurskyensis* of Amursky Bay, where the bacterium was isolated.

Cells are 0.4–0.5 × 1.2–1.4 μm. On marine agar, colonies are 2–4 mm in diameter, circular, shiny with entire edges, dark orange, and sunk into the agar. Growth occurs at 4–32°C (optimum, 23–25°C) and in 1–6% NaCl (optimum, 2% NaCl).

Source: isolated from seawater, collected in Amursky Bay, Gulf of Peter the Great, Sea of Japan.

DNA G+C content (mol%): 37.1 (T_m).

Type strain: KMM 3526, LMG 22069, CCUG 47080.

Sequence accession no. (16S rRNA gene): AB121974.

3. **Zobellia laminariae** Nedashkovskaya, Suzuki, Vancanneyt, Cleenwerck, Lysenko, Mikhailov and Swings 2004f, 1647[VP]

la.mi.na′ri.ae. N.L. gen. n. *laminariae* of *Laminaria*, the generic name of the brown alga *Laminaria japonica*, from which the bacterium was isolated.

TABLE 76. Differential phenotypic characteristics of *Zobellia* species[a,b]

Characteristic	Z. galactanivorans	Z. amurskyensis	Z. laminariae	Z. russellii	Z. uliginosa
Growth in/at:					
8% NaCl	+	–	–	+	–
10% NaCl	–	–	–	+	–
32°C	+	+	–	+	+
37°C	+	–	–	+	+
42°C	+	–	–	–	+
Hydrolysis of:					
Casein	+	–	–	–	+
Starch	+	+	–	+	+
Alginate	+	+	–	+	+
Chitin	–	–	–	–	+
DNA	–	+	–	+	+
Tween 20	+	+	–	+	–
Tween 40	–	–	+	+	–
Tween 80	–	+	–	+	+
Acid from:					
L-Arabinose, D-cellobiose	+	–	+	+	–
D-Glucose, L-rhamnose	–	+	+	+	–
Maltose	+	+	+	+	–
Raffinose	–	–	+	–	–
Sucrose	–	+	+	+	+
L-Xylose	–	–	–	+	–
Mannitol	–	–	+	+	–
Susceptibility to:					
Ampicillin, benzylpenicillin	–	–	–	–	+
Streptomycin	–	+	–	–	–
Tetracycline	–	–	–	+	–

[a]Symbols: +, >85% positive; –, 0–15% positive.

[b]Data are taken from: Barbeyron et al. (2001), Bowman (2000), Bruns et al. (2001), Cho and Giovannoni (2004), Nedashkovskaya et al. (2004f), Reichenbach (1989a), Yoon et al. (2005b), and ZoBell and Upham (1944).

Cells are 0.4–0.5 × 1.2–1.4 μm. On marine agar, colonies are 2–4 mm in diameter, circular, shiny with entire edges, dark red, and sunk into the agar. Growth occurs at 4–30°C (optimum, 21–23°C), and in salt concentrations from 1.5 to 6% NaCl (optimum, 2%).

Source: isolated from the brown alga *Laminaria japonica*, collected in Troitsa Bay, Gulf of Peter the Great, Sea of Japan.

DNA G+C content (mol%): 36–37 (T_m).

Type strain: KMM 3676, LMG 22070, CCUG 47083.

Sequence accession no. (16S rRNA gene): AB121975.

4. **Zobellia russellii** Nedashkovskaya, Suzuki, Vancanneyt, Cleenwerck, Lysenko, Mikhailov and Swings 2004f, 1647[VP]

rus.sel′li.i. N.L. gen. n. *russellii* of H.L. Russell, an American scientist, for his contribution to the development of marine microbiology.

Cells are 0.4–0.5 × 1.2–1.4 μm. On marine agar, colonies are 2–4 mm in diameter, circular, shiny with entire edges, dark orange, and sunk into the agar. Growth occurs at 4–38°C (optimum, 25–28°C) and with NaCl concentrations between 1 and 10% NaCl (optimum, 2–3%).

Source: isolated from the green alga *Acrosiphonia sonderi*, collected in Troitsa Bay, Gulf of Peter the Great, Sea of Japan.

DNA G+C content (mol%): 38.6 (T_m).

Type strain: KMM 3677, LMG 22071, CCUG 47084.

Sequence accession no. (16S rRNA gene): AB121976.

5. **Zobellia uliginosa** Barbeyron, L'Haridon, Corre, Kloareg and Potin 2001, 995[VP] [*Cellulophaga uliginosa* Bowman 2000, 1867; *Cytophaga uliginosa* (ZoBell and Upham 1944) Reichenbach 1989a, 2037; *Agarbacterium uliginosum* (ZoBell and Upham 1944) Breed 1957, 326; *Flavobacterium uliginosum* ZoBell and Upham, 1944, 263]

u.li.gi.no'sa. L. fem. adj. *uliginosa* moist, marshy.

Cells are 0.4–0.5 × 1.2–1.4 µm. On marine agar, colonies are 2–4 mm in diameter, circular, shiny with entire edges, dark orange, sunk into the agar, and have very tenacious slime. Growth occurs at 4–37°C (optimum, 25–28°C), and at NaCl concentrations from 1 to 6% (optimum, 2%). β-Glucosidase (esculin test) is positive. No brown pigment is formed with tyrosine. The following substrates are utilized: *N*-acetylglucosamine, gluconate, 2-ketogluconate, α-ketobutyric acid, mannitol, L-threonine, and urocanic acid. DL-Glycerol phosphate, L-rhamnose, and L-serine are not utilized.

Source: isolated from a marine mud, Lemon, Costa Rica.

DNA G+C content (mol%): 42.9 (HPLC, T_m).

Type strain: ZoBell 553, ACAM 538, ATCC 14397, CCUG 33448, CECT 4277, CIP 104808, DSM 2061, IFO (now NBRC) 14962, LMG 3809, NCIMB 1863.

Sequence accession no. (16S rRNA gene): M62799.

References

Abbanat, D.R., E.R. Leadbetter, I.W. Godchaux and A. Escher. 1986. Sulphonolipids are molecular determinants of gliding motility. Nature *324*: 367–369.

Abell, G.C. and J.P. Bowman. 2005. Ecological and biogeographic relationships of class *Flavobacteria* in the Southern Ocean. FEMS Microbiol. Ecol. *51*: 265–277.

Aber, R.C., C. Wennersten and R.C. Moellering, Jr. 1978. Antimicrobial susceptibility of flavobacteria. Antimicrob. Agents Chemother. *14*: 483–487.

Ackermann, H.W. and S.T. Abedon. 2001. Bacteriophage Names 2000 The Bacteriophage Ecology Group. (http://www.phage.org/names.htm).

Agbo, J.A.C. and M.O. Moss. 1979. The isolation and characterization of agarolytic bacteria from a lowland river. J. Gen. Microbiol. *115*: 355–368.

Agogué, H., E.O. Casamayor, M. Bourrain, I. Obernosterer, F. Joux, G.J. Herndl and P. Lebaron. 2005. A survey on bacteria inhabiting the sea surface microlayer of coastal ecosystems. FEMS Microbiol. Ecol. *54*: 269–280.

Akay, M., E. Gunduz and Z. Gulbas. 2006. Catheter-related bacteremia due to *Chryseobacterium indologenes* in a bone marrow transplant recipient. Bone Marrow Transplant. *37*: 435–436.

Alain, K., L. Intertaglia, P. Catala and P. Lebaron. 2008. *Eudoraea adriatica* gen. nov., sp. nov., a novel marine bacterium of the family *Flavobacteriaceae*. Int. J. Syst. Evol. Microbiol. *58*: 2275–2281.

Albareda, M., M.S. Dardanelli, C. Sousa, M. Megias, F. Temprano and D.N. Rodriguez-Navarro. 2006. Factors affecting the attachment of rhizospheric bacteria to bean and soybean roots. FEMS Microbiol. Lett. *259*: 67–73.

Allen, E.E. and D.H. Bartlett. 2002. Structure and regulation of the omega-3 polyunsaturated fatty acid synthase genes from the deep-sea bacterium *Photobacterium profundum* strain SS9. Microbiology *148*: 1903–1913.

Allen, T.D., P.A. Lawson, M.D. Collins, E. Falsen and R.S. Tanner. 2006. *Cloacibacterium normanense* gen. nov., sp. nov., a novel bacterium in the family *Flavobacteriaceae* isolated from municipal wastewater. Int. J. Syst. Evol. Microbiol. *56*: 1311–1316.

Allymehr, M. 2006. Seroprevalence of *Ornithobacterium rhinotracheale* infection in broiler and broiler breeder chickens in West Azerbaijan Province, Iran. J. Vet. Med. A Physiol. Pathol. Clin. Med. *53*: 40–42.

Altschul, S., T. Madden, A. Schaffer, J. Zhang, Z. Zhang, W. Miller and D. Lipman. 1997. Gapped BLAST and PSI-BLAST: a new generation of protein database search programs. Nucleic Acids Res. *25*: 3389–3402.

Álvarez, B., P. Secades, M.J. McBride and J.A. Guijarro. 2004. Development of genetic techniques for the psychrotrophic fish pathogen *Flavobacterium psychrophilum*. Appl. Environ. Microbiol. *70*: 581–587.

Álvarez, B., P. Secades, M. Prieto, M.J. McBride and J.A. Guijarro. 2006. A mutation in *Flavobacterium psychrophilum tlpB* inhibits gliding motility and induces biofilm formation. Appl. Environ. Microbiol. *72*: 4044–4053.

Amachi, S., Y. Mishima, H. Shinoyama, Y. Muramatsu and T. Fujii. 2005. Active transport and accumulation of iodide by newly isolated marine bacteria. Appl. Environ. Microbiol. *71*: 741–745.

Amonsin, A., J.F. Wellehan, L.L. Li, P. Vandamme, C. Lindeman, M. Edman, R.A. Robinson and V. Kapur. 1997. Molecular epidemiology of *Ornithobacterium rhinotracheale*. J. Clin. Microbiol. *35*: 2894–2898.

Anacker, R.L. and E.J. Ordal. 1955. Study of a bacteriophage infecting the myxobacterium *Chondrococcus columnaris*. J. Bacteriol. *70*: 738–741.

Anderson, R.L. and E.J. Ordal. 1961. *Cytophaga succinicans* sp. n., a facultatively anaerobic, aquatic myxobacterium. J. Bacteriol. *81*: 130–138.

Arias, C.R., T.L. Welker, C.A. Shoemaker, J.W. Abernathy and P.H. Klesius. 2004. Genetic fingerprinting of *Flavobacterium columnare* isolates from cultured fish. J. Appl. Microbiol. *97*: 421–428.

Arns, C., H.M. Hafez, T. Yano, M. Monterio, M. Alves, H. Domingues and L. Coswig. 1998. *Ornithobacterium rhinotracheale*: Detecto serologica em aves matrzes e Fragos de Corte. Proc. Assoc. Broiler Producers. Presented at the APINCO`98, Campinas, Brazil.

Arun, A.B., W.M. Chen, W.A. Lai, J.H. Chou, F.T. Shen, P.D. Rekha and C.C. Young. 2009. Lutaonella thermophila gen. nov., sp. nov., a moderately thermophilic member of the family *Flavobacteriaceae* isolated from a coastal hot spring. Int. J. Syst. Evol. Microbiol. *59*: 2069–2073.

Asker, D., T. Beppu and K. Ueda. 2007a. *Zeaxanthinibacter enoshimensis* gen. nov., sp. nov., a novel zeaxanthin-producing marine bacterium of the family *Flavobacteriaceae*, isolated from seawater off Enoshima Island, Japan. Int. J. Syst. Evol. Microbiol. *57*: 837–843.

Asker, D., T. Beppu and K. Ueda. 2007b. *Mesoflavibacter zeaxanthinifaciens* gen. nov., sp. nov., a novel zeaxanthin-producing marine bacterium of the family *Flavobacteriaceae*. Syst. Appl. Microbiol. *30*: 291–296.

Aslam, Z., W.T. Im, M.K. Kim and S.T. Lee. 2005. *Flavobacterium granuli* sp. nov., isolated from granules used in a wastewater treatment plant. Int. J. Syst. Evol. Microbiol. *55*: 747–751.

Asplin, F.D. 1955. A septicemic disease of ducklings. Vet. Rec. *67*: 854–858.

Austin, B. and D.A. Austin. 1999. Bacterial Fish Pathogens: Disease of Farmed and Wild Fish, 3rd edn. Springer/Praxis Publishing, Chichester.

Avendaño-Herrera, R., B. Magarinos, S. Lopez-Romalde, J.L. Romalde and A.E. Toranzo. 2004a. Phenotypic characterization and description of two major O-serotypes in *Tenacibaculum maritimum* strains from marine fishes. Dis. Aquat. Org. *58*: 1–8.

Avendaño-Herrera, R., B. Magarinos, A.E. Toranzo, R. Beaz and J.L. Romalde. 2004b. Species-specific polymerase chain reaction primer sets for the diagnosis of *Tenacibaculum maritimum* infection. Dis. Aquat. Org. *62*: 75–83.

Avendaño-Herrera, R., J. Rodriguez, B. Magarinos, J.L. Romalde and A.E. Toranzo. 2004c. Intraspecific diversity of the marine fish pathogen *Tenacibaculum maritimum* as determined by randomly amplified polymorphic DNA-PCR. J. Appl. Microbiol. *96*: 871–877.

Avendaño-Herrera, R., B. Magariños, M.A. Morinigo, J.L. Romalde and A.E. Toranzo. 2005. A novel O-serotype in *Tenacibaculum maritimum* strains isolated from cultured sole (*Solea senegalensis*). Bull. Eur. Assoc. Fish. Pathol. *25*: 70–74.

Back, A., S. Sprenger, G. Rajashekara, D.A. Halvorson and K.V. Nagaraja. 1997. Antimicrobial sensitivity of *Ornithobacterium rhinotracheale* isolated from different geographic location. Proceedings of the 48th North Central Avian Disease Conference, Des Moines, IA, pp. 15–18.

Back, A., D. Halvorson, G. Rajashekara and K.V. Nagaraja. 1998a. Development of a serum plate agglutination test to detect antibodies to *Ornithobacterium rhinotracheale*. J. Vet. Diagn. Invest. *10*: 84–86.

Back, A., G. Rajashekara, R.B. Jeremiah, D.A. Halvorson and K.V. Nagaraja. 1998b. Tissue distribution of *Ornithobacterium rhinotracheale* in experimentally infected turkeys. Vet. Rec. *143*: 52–53.

Bader, J.A. and J.E.B. Shotts. 1998. Identification of *Flavobacterium* and *Flexibacter* species by species-specific polymerase chain reaction primers to the 16S rRNA gene. J. Aquat. Anim. Health *10*: 311–319.

Bader, J.A., C.A. Shoemaker and P.H. Klesius. 2003. Rapid detection of columnaris disease in channel catfish (*Ictalurus punctatus*) with a new species-specific 16-S rRNA gene-based PCR primer for *Flavobacterium columnare*. J. Microbiol. Methods *52*: 209–220.

Bae, S.S., K.K. Kwon, S.H. Yang, H.S. Lee, S.J. Kim and J.H. Lee. 2007. *Flagellimonas eckloniae* gen. nov., sp. nov., a mesophilic marine bacterium of the family *Flavobacteriaceae*, isolated from the rhizosphere of *Ecklonia kurome*. Int. J. Syst. Evol. Microbiol. *57*: 1050–1054.

Bagley, D.H., Jr, J.C. Alexander Jr, V.J. Gill, R. Dolin and A.S. Ketcham. 1976. Late *Flavobacterium* species meningitis after craniofacial exenteration. Arch. Intern. Med. *136*: 229–231.

Bahar, A.A. and Z. Demirbag. 2007. Isolation of pathogenic bacteria from *Oberea linearis* (*Coleoptera: Cerambycidae*). Biologia *62*: 13–18.

Bai, J.Y. 1983. [Serological studies of *Flavobacterium farinofermentans* nov. sp. – anti-O serotyping and distribution]. Zhonghua Yu Fang Yi Xue Za Zhi *17*: 138–140.

Bai, Y., D. Yang, J. Wang, S. Xu, X. Wang and L. An. 2006. Phylogenetic diversity of culturable bacteria from alpine permafrost in the Tianshan Mountains, northwestern China. Res. Microbiol. *157*: 741–751.

Bakunina, I.Y., R.A. Kuhlmann, L.M. Likhosherstov, M.D. Martynova, O.I. Nedashkovskaya, V.V. Mikhailov and L.A. Elyakova. 2002. Alpha-N-acetylgalactosaminidase from marine bacterium *Arenibacter latericius* KMM 426^T removing blood type specificity of A-erythrocytes. Biochemistry (Mosc.) *67*: 689–695.

Bangun, A., J.L. Johnson and D.N. Tripathy. 1987. Taxonomy of *Pasteurella anatipestifer*. I. DNA base composition and DNA–DNA hybridization analysis. Avian Dis. *31*: 43–45.

Bano, N. and J.T. Hollibaugh. 2002. Phylogenetic composition of bacterioplankton assemblages from the Arctic Ocean. Appl. Environ. Microbiol. *68*: 505–518.

Barbeyron, T., A. Gerard, P. Potin, B. Henrissat and B. Kloareg. 1998. The kappa-carrageenase of the marine bacterium *Cytophaga drobachiensis*. Structural and phylogenetic relationships within family-16 glycoside hydrolases. Mol. Biol. Evol. *15*: 528–537.

Barbeyron, T., S. L'Haridon, E. Corre, B. Kloareg and P. Potin. 2001. *Zobellia galactanovorans* gen. nov., sp. nov., a marine species of *Flavobacteriaceae* isolated from a red alga, and classification of *Cytophaga uliginosa* (ZoBell and Upham 1944) Reichenbach 1989 as *Zobellia uliginosa* gen. nov., comb. nov. Int. J. Syst. Evol. Microbiol. *51*: 985–997.

Baumann, P. and L. Baumann. 1981. The marine Gram-negative eubacteria; genera *Photobacterium, Beneckea, Alteromonas, Pseudomonas*, and *Alcaligenes*. *In* The Prokaryotes: A Handbook on Habitats, Isolation, and Identification of *Bacteria* (edited by Starr, Stolp, Trüper, Balows and Schlegel). Springer, New York, pp. 1302–1330.

Bauwens, M. and J. De Ley. 1981. Improvements in the taxonomy of *Flavobacterium* by DNA:rRNA hybridization. *In* The *Flavobacterium-Cytophaga* group, Gesellschaft für Biotechnologische Forschung Monograph Series No. 5 (edited by Reichenbach and Weeks). Verlag Chemie, Weinheim, pp. 27–31.

Behrendt, U., A. Ulrich, C. Sproer and P. Schumann. 2007. *Chryseobacterium luteum* sp. nov., associated with the phyllosphere of grasses. Int. J. Syst. Evol. Microbiol. *57*: 1881–1885.

Bellais, S., S. Leotard, L. Poirel, T. Naas and P. Nordmann. 1999. Molecular characterization of a carbapenem-hydrolyzing beta-lactamase from *Chryseobacterium* (*Flavobacterium*) *indologenes*. FEMS Microbiol. Lett. *171*: 127–132.

Bellais, S., D. Aubert, T. Naas and P. Nordmann. 2000a. Molecular and biochemical heterogeneity of class B carbapenem-hydrolyzing beta-lactamases in *Chryseobacterium meningosepticum*. Antimicrob. Agents Chemother. *44*: 1878–1886.

Bellais, S., L. Poirel, T. Naas, D. Girlich and P. Nordmann. 2000b. Genetic-biochemical analysis and distribution of the Ambler class A beta-lactamase CME-2, responsible for extended-spectrum cephalosporin resistance in *Chryseobacterium* (*Flavobacterium*) *meningosepticum*. Antimicrob. Agents Chemother. *44*: 1–9.

Bellais, S., D. Girlich, A. Karim and P. Nordmann. 2002a. EBR-1, a novel Ambler subclass B1 beta-lactamase from *Empedobacter brevis*. Antimicrob. Agents Chemother. *46*: 3223–3227.

Bellais, S., T. Naas and P. Nordmann. 2002b. Genetic and biochemical characterization of CGB-1, an Ambler class B carbapenem-hydrolyzing beta-lactamase from *Chryseobacterium gleum*. Antimicrob. Agents Chemother. *46*: 2791–2796.

Bellais, S., T. Naas and P. Nordmann. 2002c. Molecular and biochemical characterization of Ambler class A extended-spectrum beta-lactamase CGA-1 from *Chryseobacterium gleum*. Antimicrob. Agents Chemother. *46*: 966–970.

Bergey, D.H., F.C. Harrison, R.S. Breed, B.W. Hammer and F.M. Huntoon. 1923. Bergey's Manual of Determinative Bacteriology. Williams & Wilkins, Baltimore.

Bernardet, J.-F. 1989. '*Flexibacter columnaris*': first description in France and comparison with bacterial strains from other origins. Dis. Aquat. Org. *6*: 37–44.

Bernardet, J.-F. and P.A.D. Grimont. 1989. Deoxyribonucleic acid relatedness and phenotypic characterization of *Flexibacter columnaris* sp. nov., nom. rev., *Flexibacter psychrophilus* sp. nov., nom. rev., and *Flexibacter maritimus* Wakabayashi, Hikida, and Masumura 1986. Int. J. Syst. Bacteriol. *39*: 346–354.

Bernardet, J.-F. and B. Kerouault. 1989. Phenotypic and genomic studies of "*Cytophaga psychrophila*" isolated from diseased rainbow trout (*Oncorhynchus mykiss*) in France. Appl. Environ. Microbiol. *55*: 1796–1800.

Bernardet, J.-F. and J.P. Bowman. 2006. The genus *Flavobacterium*. *In* The Prokaryotes: A Handbook on the Biology of Bacteria, 3rd edn, vol. 7, *Proteobacteria*: Delta and Epsilon Subclasses. Deeply Rooting *Bacteria* (edited by Dworkin, Falkow, Rosenberg, Schleifer and Stackebrandt). Springer, New York, pp. 481–531.

Bernardet, J.-F. and Y. Nakagawa. 2006. An introduction to the family *Flavobacteriaceae*. In The Prokaryotes: A Handbook on the Biology of Bacteria, 3rd edn, vol. 7, *Proteobacteria*: Delta and Epsilon Subclasses. Deeply Rooting *Bacteria* (edited by Dworkin, Falkow, Rosenberg, Schleifer and Stackebrandt). Springer, New York, pp. 455–480.

Bernardet, J.-F., P. Segers, M. Vancanneyt, F. Berthe, K. Kersters and P. Vandamme. 1996. Cutting a gordian knot: emended classification and description of the genus *Flavobacterium*, emended description of the family *Flavobacteriaceae*, and proposal of *Flavobacterium hydatis* nom. nov. (basonym, *Cytophaga aquatilis* Strohl and Tait 1978). Int. J. Syst. Bacteriol. *46*: 128–148.

Bernardet, J.-F., Y. Nakagawa and B. Holmes. 2002. Proposed minimal standards for describing new taxa of the family *Flavobacteriaceae* and emended description of the family. Int. J. Syst. Evol. Microbiol. *52*: 1049–1070.

Bernardet, J.-F., M. Vancanneyt, O. Matte-Tailliez, L. Grisez, P. Tailliez, C. Bizet, M. Nowakowski, B. Kerouault and J. Swings. 2005. Polyphasic study of *Chryseobacterium* strains isolated from diseased aquatic animals. Syst. Appl. Microbiol. *28*: 640–660.

Bernardet, J.-F., C. Hugo and B. Bruun. 2006. The genera *Chryseobacterium* and *Elizabethkingia*. In The Prokaryotes: A Handbook on the Biology of

Bacteria, 3rd edn, vol. 7, *Proteobacteria*: Delta and Epsilon Subclasses. Deeply Rooting *Bacteria* (edited by Dworkin, Falkow, Rosenberg, Schleifer and Stackebrandt). Springer, New York, pp. 638–676.

Berthe-Corti, L. and A. Bruns. 1999. The impact of oxygen tension on cell density and metabolic diversity of microbial communities in alkane degrading continuous-flow cultures. Microb. Ecol. *37*: 70–77.

Bisschop, S.P., M. Van Vuuren and B. Gummow. 2004. The use of a bacterin vaccine in broiler breeders for the control of *Ornithobacterium rhinotracheale* in commercial broilers. J. S. Afr. Vet. Assoc. *75*: 125–128.

Bissett, A., J. Bowman and C. Burke. 2006. Bacterial diversity in organically-enriched fish farm sediments. FEMS Microbiol. Ecol. *55*: 48–56.

Bloch, K.C., R. Nadarajah and R. Jacobs. 1997. *Chryseobacterium meningosepticum:* an emerging pathogen among immunocompromised adults. Report of 6 cases and literature review. Medicine (Baltimore) *76*: 30–41.

Bobo, R.A. and E.J. Newton. 1976. A previously undescribed gram-negative bacillus causing septicemia and meningitis. Am. J. Clin. Pathol. *65*: 564–569.

Bock, R., P. Freidlin, M. Manoim, A. Inbar, A. Frommer, P. Vandamme and P. Wilding. 1997. *Ornithobacterium rhinotracheale* (ORT) associated with a new turkey respiratory tract infectious agent in Israel. Proceedings of the 11th International Congress of the World Veterinary Poultry Association, Budapest, p. 120.

Bodour, A.A., K.P. Drees and R.M. Maier. 2003. Distribution of biosurfactant-producing bacteria in undisturbed and contaminated arid Southwestern soils. Appl. Environ. Microbiol. *69*: 3280–3287.

Bolton, R.W. and J.K. Dyer. 1983. Suppression of murine lymphocyte mitogen responses by exopolysaccharide from *Capnocytophaga ochracea*. Infect. Immun. *39*: 476–479.

Borg, A.F. 1948. Studies on myxobacteria associated with diseases in salmonid fishes. PhD thesis, University of Washington, Seattle.

Borg, A.F. 1960. Studies on myxobacteria associated with diseases in salmonid fishes. Wildl. Dis. *8*: 85.

Borriss, M., E. Helmke, R. Hanschke and T. Schweder. 2003. Isolation and characterization of marine psychrophilic phage-host systems from Arctic sea ice. Extremophiles *7*: 377–384.

Borriss, M., T. Lombardot, F.O. Glockner, D. Becher, D. Albrecht and T. Schweder. 2007. Genome and proteome characterization of the psychrophilic *Flavobacterium* bacteriophage 11b. Extremophiles *11*: 95–104.

Botha, W.C., P.J. Jooste and C.J. Hugo. 1998. The incidence of *Weeksella*- and *Bergeyella*-like bacteria in the food environment. J. Appl. Microbiol. *84*: 349–356.

Bottone, E.J., R.M. Madayag and M.N. Qureshi. 1992. *Acanthamoeba keratitis*: synergy between amebic and bacterial cocontaminants in contact lens care systems as a prelude to infection. J. Clin. Microbiol. *30*: 2447–2450.

Bowman, J.P. 2000. Description of *Cellulophaga algicola* sp. nov., isolated from the surfaces of Antarctic algae, and reclassification of *Cytophaga uliginosa* (ZoBell and Upham 1944) Reichenbach 1989 as *Cellulophaga uliginosa* comb. nov. Int. J. Syst. Evol. Microbiol. *50*: 1861–1868.

Bowman, J.P. 2006. The marine clade of the family *Flavobacteriaceae*: the genera *Aequorivita, Arenibacter, Cellulophaga, Croceibacter, Formosa, Gelidibacter, Gillisia, Maribacter, Mesonia, Muricauda, Polaribacter, Psychroflexus, Psychroserpens, Robiginitalea, Salegentibacter, Tenacibaculum, Ulvibacter, Vitellibacter* and *Zobellia*. In The Prokaryotes: A Handbook on the Biology of Bacteria, 3rd edn, vol. 7, *Proteobacteria*: Delta and Epsilon Subclasses. Deeply Rooting *Bacteria* (edited by Dworkin, Falkow, Rosenberg, Schleifer and Stackebrandt). Springer, New York, pp. 677–694.

Bowman, J.P. and D.S. Nichols. 2002. *Aequorivita* gen. nov., a member of the family *Flavobacteriaceae* isolated from terrestrial and marine Antarctic habitats. Int. J. Syst. Evol. Microbiol. *52*: 1533–1541.

Bowman, J.P. and D.S. Nichols. 2005. Novel members of the family *Flavobacteriaceae* from Antarctic maritime habitats including *Subsaximicrobium wynnwilliamsii* gen. nov., sp. nov., *Subsaximicrobium saxinquilinus* sp. nov., *Subsaxibacter broadyi* gen. nov., sp. nov., *Lacinutrix copepodicola* gen. nov., sp. nov., and novel species of the genera *Bizionia, Gelidibacter* and *Gillisia*. Int. J. Syst. Evol. Microbiol. *55*: 1471–1486.

Bowman, J.P. and B. Nowak. 2004. Salmonid gill bacteria and their relationship to amoebic gill disease. J. Fish Dis. *27*: 483–492.

Bowman, J.P., S.A. McCammon, J.L. Brown, P.D. Nichols and T.A. McMeekin. 1997. *Psychroserpens burtonensis* gen. nov., sp. nov., and *Gelidibacter algens* gen. nov., sp. nov., psychrophilic bacteria isolated from Antarctic lacustrine and sea ice habitats. Int. J. Syst. Bacteriol. *47*: 670–677.

Bowman, J.P., S.A. McCammon, T. Lewis, J.H. Skerratt, J.L. Brown, D.S. Nichols and T.A. McMeekin. 1998. *Psychroflexus torquis* gen. nov., sp. nov., a psychrophilic species from Antarctic sea ice, and reclassification of *Flavobacterium gondwanense* (Dobson *et al.* 1993) as *Psychroflexus gondwanense* gen. nov., comb. nov. Microbiology *144*: 1601–1609.

Bowman, J.P., S.A. McCammon, T.E. Lewis, J.H. Skerratt, J.L. Brown, D.S. Nichols and T.A. McMeekin. 1999. *In* Validation of publication of new names and new combinations previously effectively published outside the IJSB, List no. 68. Int. J. Syst. Bacteriol. *49*: 1–3.

Bowman, J.P., C. Mancuso, C.M. Nichols and J.A.E. Gibson. 2003. *Algoriphagus ratkowskyi* gen. nov., sp. nov., *Brumimicrobium glaciale* gen. nov., sp. nov., *Cryomorpha ignava* gen. nov., sp. nov. and *Crocinitomix catalasitica* gen. nov., sp. nov., novel *flavobacteria* isolated from various polar habitats. Int. J. Syst. Evol. Microbiol. *53*: 1343–1355.

Brambilla, E., O. Päuker, S. Cousin, U. Steiner, A. Reimer and E. Stackebrandt. 2007. High phylogenetic diversity of *Flavobacterium* spp. isolated from a hardwater creek, Harz Mountains, Germany. Org. Divers. Evol. *7*: 145–154.

Brandelli, A. and A. Riffel. 2005. Production of an extracellular keratinase from *Chryseobacterium* sp. growing on raw feathers. Electr. J. Biotechnol. *8*: 35–42.

Braun, T.F. and M.J. McBride. 2005. *Flavobacterium johnsoniae* GldJ is a lipoprotein that is required for gliding motility. J. Bacteriol. *187*: 2628–2637.

Braun, T.F., M.K. Khubbar, D.A. Saffarini and M.J. McBride. 2005. *Flavobacterium johnsoniae* gliding motility genes identified by mariner mutagenesis. J. Bacteriol. *187*: 6943–6952.

Breed, R.S. 1957. Genus IV. *Agarbacterium* Angst 1929. *In* Bergey's Manual of Determinative Bacteriology, 7th edn (edited by Breed, Murray and Smith). Williams & Wilkins, Baltimore, pp. 322–328.

Breed, R.S., E.G.D. Murray and N.R. Smith (editors). 1957. Bergey's Manual of Determinative Bacteriology. Williams & Wilkins, Baltimore.

Brenner, D.J., D.G. Hollis, G.R. Fanning and R.E. Weaver. 1989. *Capnocytophaga canimorsus* sp. nov. (formerly CDC Group Df-2), a cause of septicemia following dog bite, and *C. cynodegmi* sp. nov., a cause of localized wound-infection following dog bite. J. Clin. Microbiol. *27*: 231–235.

Brenner, D.J., D.G. Hollis, G.R. Fanning and R.E. Weaver. 1990. *In* Validation of the publication of new names and new combinations previously effectively published outside the IJSB. List no. 32. Int. J. Syst. Bacteriol. *40*: 105–106.

Brinkmeyer, R., K. Knittel, J. Jurgens, H. Weyland, R. Amann and E. Helmke. 2003. Diversity and structure of bacterial communities in arctic versus antarctic pack ice. Appl. Environ. Microbiol. *69*: 6610–6619.

Brisou, J., C. Tysset and B. Vacher. 1959. Etude de trois souches microbiennes, famille des *Pseudomonadaceae*, dont la synergie provoque une maladie de caractère septicémique chez les poissons blancs de la Dordogne, du Lot et de leurs affluents. Ann. Inst. Pasteur *96*: 689–696.

Brisou, J., C. Tysset and A. Jacob. 1960. Etude d'un germe de la famille des *Pseudomonadaceae* (Tribu des Chromobactereae) *Empedobacter* aquatile isolé d'un produit frais de charcuterie. Arch. Inst. Pasteur Algérie *38*: 353–360.

Brown, M.V. and J.P. Bowman. 2001. A molecular phylogenetic survey of sea-ice microbial communities (SIMCO). FEMS Microbiol. Ecol. *35*: 267–275.

Bruner, D.W. and J. Fabricant. 1954. A strain of *Moraxella anatipestifer* (*Pfeifferella anatipestifer*) isolated from ducks. Cornell Vet. *44*: 461–464.

Bruner, D.W., C.I. Angstrom and J.I. Price. 1970. *Pasteurella anatipestifer* infection in pheasants. A case report. Cornell Vet. *60*: 491–494.

Bruns, A., M. Rohde and L. Berthe-Corti. 2001. *Muricauda ruestringensis* gen. nov., sp. nov., a facultatively anaerobic, appendaged bacterium from German North Sea intertidal sediment. Int. J. Syst. Evol. Microbiol. *51*: 1997–2006.

Bruun, B. 1982. Studies on a collection of strains of the genus *Flavobacterium*. 1. Biochemical studies. Acta Pathol. Microbiol. Immunol. Scand. [B] *90*: 415–421.

Bruun, B. 1987. Antimicrobial susceptibility of *Flavobacterium meningosepticum* strains identified by DNA–DNA hybridization. Acta Pathol. Microbiol. Immunol. Scand. [B] *95*: 95–101.

Bruun, B. and J. Ursing. 1987. Phenotypic characterization of *Flavobacterium meningosepticum* strains identified by DNA–DNA hybridization. Acta Pathol. Microbiol. Immunol. Scand. Sect. B *95*: 41–47.

Bruun, B., E.T. Jensen, K. Lundstrom and G.E. Andersen. 1989. *Flavobacterium meningosepticum* infection in a neonatal ward. Eur. J. Clin. Microbiol. Infect. Dis. *8*: 509–514.

Butler, T., R.E. Weaver, T.K. Ramani, C.T. Uyeda, R.A. Bobo, J.S. Ryu and R.B. Kohler. 1977. Unidentified gram-negative rod infection. A new disease of man. Ann. Intern. Med. *86*: 1–5.

Callies, E. and W. Mannheim. 1978. Classification of the *Flavobacterium-Cytophaga* complex on the basis of respiratory quinones and fumarate respiration. Int. J. Syst. Bacteriol. *28*: 14–19.

Callies, E. and W. Mannheim. 1980. Deoxyribonucleic acid relatedness of some menaquinone-producing *Flavobacterium* and *Cytophaga* strains. Antonie van Leeuwenhoek *46*: 41–49.

Calmes, R., G.W. Rambicure, W. Gorman and T.T. Lillich. 1980. Energy metabolism in *Capnocytophaga ochracea*. Infect. Immun. *29*: 551–560.

Campbell, L.L. 1957. Genus *Beneckea* Campbell. *In* Bergey's Manual of Determinative Bacteriology, 7th edn (edited by Breed, Murray and Smith). Williams & Wilkins, Baltimore, pp. 328–332.

Campbell, L.L. and O.B. Williams. 1951. A study of chitin-decomposing micro-organisms of marine origin. J. Gen. Microbiol. *5*: 894–905.

Campbell, C.L., D.L. Mummey, E.T. Schmidtmann and W.C. Wilson. 2004. Culture-independent analysis of midgut microbiota in the arbovirus vector *Culicoides sonorensis* (Diptera: Ceratopogonidae). J. Med. Entomol. *41*: 340–348.

Campbell, S., R.M. Harada and Q.X. Li. 2008. *Chryseobacterium arothri* sp. nov., isolated from the kidneys of a pufferfish. Int. J. Syst. Evol. Microbiol. *58*: 290–293.

Canal, C.W., J.A. Leao, S.L. Rocha, M. Macagnan, C.A. Lima-Rosa, S.D. Oliveira and A. Back. 2005. Isolation and characterization of *Ornithobacterium rhinotracheale* from chickens in Brazil. Res. Vet. Sci. *78*: 225–230.

Carson, J., L.M. Schmidtke and B.L. Munday. 1993. *Cytophaga johnsonae*: a putative skin pathogen of juvenile farmed barramundi, *Lates calcarifer* Bloch. J. Fish Dis. *16*: 209–218.

Cauwerts, K., P. De Herdt, F. Haesebrouck, J. Vervloesem and R. Ducatelle. 2002. The effect of *Ornithobacterium rhinotracheale* vaccination of broiler breeder chickens on the performance of their progeny. Avian. Pathol. *31*: 619–624.

Cepeda, C., S. García-Márquez and Y. Santos. 2004. Improved growth of *Flavobacterium psychrophilum* using a new culture medium. Aquaculture *238*: 75–82.

Chakroun, C., F. Grimont, M.C. Urdaci and J.-F. Bernardet. 1998. Fingerprinting of *Flavobacterium psychrophilum* isolates by ribotyping and plasmid profiling. Dis. Aquat. Org. *33*: 167–177.

Chang, J.C., P.R. Hsueh, J.J. Wu, S.W. Ho, W.C. Hsieh and K.T. Luh. 1997. Antimicrobial susceptibility of flavobacteria as determined by agar dilution and disk diffusion methods. Antimicrob. Agents Chemother. *41*: 1301–1306.

Charlton, B.R., S.E. Channing-Santiago, A.A. Bickford, C.J. Cardona, R.P. Chin, G.L. Cooper, R. Droual, J.S. Jeffrey, C.U. Meteyer, H.L. Shivaprasad and R.L. Walker. 1993. Preliminary characterization of a pleomorphic Gram-negative rod associated with avian respiratory disease. J. Vet. Diagn. Invest. *5*: 47–51.

Chase, J.M. 1965. Nutrition of some aquatic myxobacteria. MSc thesis, University of Washington, Seattle.

Chen, G.X., R. Zhang and H.W. Zhou. 2006. Heterogeneity of metallo-beta-lactamases in clinical isolates of *Chryseobacterium meningosepticum* from Hangzhou, China. J. Antimicrob. Chemother. *57*: 750–752.

Cheng, S.M. and J.M. Foght. 2007. Cultivation-independent and -dependent characterization of *Bacteria* resident beneath John Evans Glacier. FEMS Microbiol. Ecol. *59*: 318–330.

Chester, F.D. 1897. Report of the mycologist: bacteriological work. Del. Coll. Agric. Exp. Sta. 9th Annual Report. Mercantile Printing Co, Wilmington, DE: 38–145.

Chester, F.D. 1901. A Manual of Determinative Bacteriology. Macmillan, New York.

Chicote, E., A.M. Garcia, D.A. Moreno, M.I. Sarro, P.I. Lorenzo and F. Montero. 2005. Isolation and identification of bacteria from spent nuclear fuel pools. J. Ind. Microbiol. Biotechnol. *32*: 155–162.

Chiu, C.H., M. Waddingdon, W.-S. Hsieh, D. Greenberg, P.C. Schreckenberger and A.M. Carnahan. 2000. Atypical *Chryseobacterium meningosepticum* and meningitis and sepsis in newborns and the immunocompromised, Taiwan. Emerg. Infect. Dis. *6*: 481–486.

Cho, J.-C. and S.J. Giovannoni. 2003. *In* Validation of publication of new names and new combinations previously effectively published outside the IJSEM. List no. 92. Int. J. Syst. Evol. Microbiol. *53*: 935–937.

Cho, J.C. and S.J. Giovannoni. 2003. *Croceibacter atlanticus* gen. nov., sp. nov., a novel marine bacterium in the family *Flavobacteriaceae*. Syst. Appl. Microbiol. *26*: 76–83.

Cho, J.C. and S.J. Giovannoni. 2004. *Robiginitalea biformata* gen. nov., sp. nov., a novel marine bacterium in the family *Flavobacteriaceae* with a higher G+C content. Int. J. Syst. Evol. Microbiol. *54*: 1101–1106.

Choi, D.H. and B.C. Cho. 2006. *Lutibacter litoralis* gen. nov., sp. nov., a marine bacterium of the family *Flavobacteriaceae* isolated from tidal flat sediment. Int. J. Syst. Evol. Microbiol. *56*: 771–776.

Choi, D.H., Y.G. Kim, C.Y. Hwang, H. Yi, J. Chun and B.C. Cho. 2006. *Tenacibaculum litoreum* sp. nov., isolated from tidal flat sediment. Int. J. Syst. Evol. Microbiol. *56*: 635–640.

Chowdhury, M.B.R. and H. Wakabayashi. 1988. Effects of sodium, potassium, calcium and magnesium ions on the survival of *Flexibacter columnaris* in water. Fish Pathol. *23*: 231–235.

Christensen, W.B. 1949. Hydrogen sulfide production and citrate utilization in the differentiation of enteric pathogens and coliform bacteria. *In* Research Bulletin no. 1. Weld County Health Department, Greeley, CO.

Christensen, P.J. 1977. The history, biology, and taxonomy of the *Cytophaga* group. Can. J. Microbiol. *23*: 1599–1653.

Christensen, H., M. Bisgaard, W. Frederiksen, R. Mutters, P. Kuhnert and J.E. Olsen. 2001. Is characterization of a single isolate sufficient for valid publication of a new genus or species? Proposal to modify Recommendation 30b of the *Bacteriological Code* (1990 Revision). Int. J. Syst. Evol. Microbiol. *51*: 2221–2225.

Chung, H.-Y. 1990. On the bacterial disease of captive bullfrog. *In* Proceedings of the Republic of China-Japan Symposium on Fish Diseases, NSC Symposium Series No. 16 (edited by Kou, Wakabayashi, Liao, Chen and Lo). The National Science Council, Taipei, Taiwan, pp. 81–89.

Ciantar, M., D.A. Spratt, H.N. Newman and M. Wilson. 2001. Assessment of five culture media for the growth and isolation of *Capnocytophaga* spp. Clin. Microbiol. Infect. *7*: 158–160.

Clark, J.W. and S. Kambhampati. 2003. Phylogenetic relationships among *Blattabacterium*, endosymbiotic bacteria from the wood roach, *Cryptocercus*. Mol. Phylogenet. Evol. *26*: 82–88.

Clayton, R.A., G. Sutton, P.S. Hinkle, Jr, C. Bult and C. Fields. 1995. Intraspecific variation in small-subunit rRNA sequences in GenBank: why single sequences may not adequately represent prokaryotic taxa. Int. J. Syst. Bacteriol. *45*: 595–599.

Colding, H., J. Bangsborg, N.E. Fiehn, T. Bennekov and B. Bruun. 1994. Ribotyping for differentiating *Flavobacterium meningosepticum* isolates from clinical and environmental sources. J. Clin. Microbiol. *32*: 501–505.

Collins, M.D., H.N. Shah, A.S. McKee and R.M. Kroppenstedt. 1982. Chemotaxonomy of the genus *Capnocytophaga* (Leadbetter, Holt & Socransky). J. Appl. Bacteriol. *52*: 409–415.

Colwell, R.R., R.V. Citarella and P.K. Chen. 1966. DNA base composition of *Cytophaga marinoflava* n. sp. determined by buoyant density measurements in cesium chloride. Can. J. Microbiol. *12*: 1099–1103.

Connon, S.A. and S.J. Giovannoni. 2002. High-throughput methods for culturing microorganisms in very-low-nutrient media yield diverse new marine isolates. Appl. Environ. Microbiol. *68*: 3878–3885.

Cooper, A.J., A.P. Kalinowski, N.B. Shoemaker and A.A. Salyers. 1997. Construction and characterization of a *Bacteroides thetaiotaomicron* recA mutant: transfer of *Bacteroides* integrated conjugative elements is RecA independent. J. Bacteriol. *179*: 6221–6227.

Córdova, J.L., L. Cárdenas, L. Cárdenas and A. Yudelevich. 2002. Multiple bacterial infection of *Alexandrium catenella* (Dinophyceae). J. Plankton Res. *24*: 1–8.

Cornil, A. and A. Babes. 1890. Les bactéries et leur role dans l'étiologie, l'anatomie et l'histologie pathologiques des maladies infectieuse, Rev. 3rd edn, vol. 1. Felix Alcan, Paris.

Cottrell, M.T. and D.L. Kirchman. 2000. Community composition of marine bacterioplankton determined by 16S rRNA gene clone libraries and fluorescence in situ hybridization. Appl. Environ. Microbiol. *66*: 5116–5122.

Cousin, S., O. Päuker and E. Stackebrandt. 2007. *Flavobacterium aquidurense* sp. nov. and *Flavobacterium hercynium* sp. nov., from a hard-water creek. Int. J. Syst. Evol. Microbiol. *57*: 243–249.

Crasta, K.C., K.L. Chua, S. Subramaniam, J. Frey, H. Loh and H.M. Tan. 2002. Identification and characterization of CAMP cohemolysin as a potential virulence factor of *Riemerella anatipestifer*. J. Bacteriol. *184*: 1932–1939.

Crump, B.C., E.V. Armbrust and J.A. Baross. 1999. Phylogenetic analysis of particle-attached and free-living bacterial communities in the Columbia river, its estuary, and the adjacent coastal ocean. Appl. Environ. Microbiol. *65*: 3192–3204.

Darwish, A.M. and A.A. Ismaiel. 2005. Genetic diversity of *Flavobacterium columnare* examined by restriction fragment length polymorphism and sequencing of the 16S ribosomal RNA gene and the 16S–23S rDNA spacer. Mol. Cell. Probes *16*: 267–274.

Darwish, A.M., A.A. Ismaiel, J.C. Newton and J. Tang. 2004. Identification of *Flavobacterium columnare* by a species-specific polymerase chain reaction and renaming of ATCC43622 strain to *Flavobacterium johnsoniae*. Mol. Cell. Probes *18*: 421–427.

Daskalov, H., D.A. Austin and B. Austin. 1999. An improved growth medium for *Flavobacterium psychrophilum*. Lett. Appl. Microbiol. *28*: 297–299.

Davis, H.S. 1922. A new bacterial disease of freshwater fishes. Bull. U. S. Bur. Fish. *38*: 261–280.

Davis, J.M., M.M. Peel and J.A. Gillians. 1979. Colonization of an amputation site by *Flavobacterium odoratum* after gentamicin therapy. Med. J. Aust. *2*: 703–704.

de Beer, H. 2005. A taxonomic study of *Chryseobacterium* species in meat. PhD thesis, University of the Free State, Bloemfontein, South Africa.

de Beer, H., C.J. Hugo, P.J. Jooste, A. Willems, M. Vancanneyt, T. Coenye and P.A. Vandamme. 2005. *Chryseobacterium vrystaatense* sp. nov., isolated from raw chicken in a chicken-processing plant. Int. J. Syst. Evol. Microbiol. *55*: 2149–2153.

de Beer, H., C.J. Hugo, P.J. Jooste, M. Vancanneyt, T. Coenye and P. Vandamme. 2006. *Chryseobacterium piscium* sp. nov., isolated from fish of the South Atlantic Ocean off South Africa. Int. J. Syst. Evol. Microbiol. *56*: 1317–1322.

Decostere, A., F. Haesebrouck, E. Van Driessche, G. Charlier and R. Ducatelle. 1999. Characterization of the adhesion of *Flavobacterium columnare* (*Flexibacter columnaris*) to gill tissue. J. Fish Dis. *22*: 465–474.

Decostere, A., L.A. Devriese, R. Ducatelle and F. Haesebrouck. 2002. *Bergeyella* (*Weeksella*) *zoohelcum* associated with respiratory disease in a cat. Vet. Rec. *151*: 392.

Dees, S.B., D.E. Karr, D. Hollis and C.W. Moss. 1982. Cellular fatty acids of *Capnocytophaga* species. J. Clin. Microbiol. *16*: 779–783.

Dees, S.B., C.W. Moss, D.G. Hollis and R.E. Weaver. 1986. Chemical characterization of *Flavobacterium odoratum*, *Flavobacterium breve*, and *Flavobacterium*-like groups IIe, IIh, and IIf. J. Clin. Microbiol. *23*: 267–273.

DeRosa, M., R. Droual, R. Chin and H. Shivaprasad. 1997. Interaction of *Ornithobacterium rhinotracheale* and *Bordetella avium* in turkey poults. Proceedings of the 46th Western Poultry Disease Conference, Sacramento, pp. 52–53.

Descamps, V., S. Colin, M. Lahaye, M. Jam, C. Richard, P. Potin, T. Barbeyron, J.C. Yvin and B. Kloareg. 2006. Isolation and culture of a marine bacterium degrading the sulfated fucans from marine brown algae. Mar. Biotechnol. (NY) *8*: 27–39.

Devriese, L.A., J. Hommez, P. Vandamme, K. Kersters and F. Haesebrouck. 1995. *In vitro* antibiotic sensitivity of *Ornithobacterium rhinotracheale* strains from poultry and wild birds. Vet. Rec. *137*: 435–436.

Diarra, K., Z.G. Nong and C. Jie. 2005. Peanut milk and peanut milk based products production: a review. Crit. Rev. Food Sci. Nutr. *45*: 405–423.

Diefenthal, T. and H. Dargatz. 1995. Rapid purification of proline-specific endopeptidase from *Flavobacterium meningosepticum* heterologously expressed in *Escherichia coli*. World J. Microbiol. Biotechnol. *11*: 209–212.

Dobson, S.J., R.R. Colwell, T.A. McMeekin and P.D. Franzmann. 1993. Direct sequencing of the polymerase chain reaction: amplified 16S ribosomal RNA gene of *Flavobacterium gondwanense* sp. nov. and *Flavobacterium salegens* sp. nov., two new species from a hypersaline antarctic lake. Int. J. Syst. Bacteriol. *43*: 77–83.

Domenech, J., M.S. Reddy, J.W. Kloepper, B. Ramos and J. Gutierrez-Mañero. 2006. Combined application of the biological product LS213 with *Bacillus*, *Pseudomonas* or *Chryseobacterium* for growth promotion and biological control of soil-borne diseases in pepper and tomato. BioControl *51*: 245–258.

Donachie, S.P., J.P. Bowman and M. Alam. 2004. *Psychroflexus tropicus* sp. nov., an obligately halophilic *Cytophaga–Flavobacterium–Bacteroides* group bacterium from an Hawaiian hypersaline lake. Int. J. Syst. Evol. Microbiol. *54*: 935–940.

Dorey, M.J. 1959. Some properties of a pectolytic soil *Flavobacterium*. J. Gen. Microbiol. *20*: 91–104.

Drancourt, M., P. Berger and D. Raoult. 2004. Systematic 16S rRNA gene sequencing of atypical clinical isolates identified 27 new bacterial species associated with humans. J. Clin. Microbiol. *42*: 2197–2202.

Droual, R. and R. Chin. 1997. Interaction of *Ornithobacterium rhinotracheale* and *Escherichia coli* 78:H9 when inoculated into the air sac in turkey poults. Western Poultry Disease Conference, Sacramento. 11.

du Moulin, G.C. 1979. Airway colonization by *Flavobacterium* in an intensive care unit. J. Clin. Microbiol. *10*: 155–160.

Duchaud, E., M. Boussaha, V. Loux, J.-F. Bernardet, C. Michel, B. Kerouault, S. Mondot, P. Nicolas, R. Bossy, C. Caron, P. Bessieres, J.F. Gibrat, S. Claverol, F. Dumetz, M. Le Henaff and A. Benmansour. 2007. Complete genome sequence of the fish pathogen *Flavobacterium psychrophilum*. Nat. Biotechnol. *25*: 763–769.

Dudouyt, J., P. van Empel, Y. Gardin and D. Céline. 1995. Isolement d´un nouvel agent pathogene chez la dinde: *Ornithobacterium rhinotracheale*; conduite a tenir. Léres Journees De La Recherche Avicole. Sous Le Patronage Du Ministere De L´Agriculture Et De La Peche. Centre De Congress D´angers, 28–30 Mars, 1995. 240–243.

Dugas, J.E., L. Zurek, B.J. Paster, B.A. Keddie and E.R. Leadbetter. 2001. Isolation and characterization of a *Chryseobacterium* strain from the gut of the American cockroach, *Periplaneta americana*. Arch. Microbiol. *175*: 259–262.

Dumetz, F., E. Duchaud, S.E. LaPatra, C. Le Marrec, S. Claverol, M.C. Urdaci and M. Le Henaff. 2006. A protective immune response is generated in rainbow trout by an OmpH-like surface antigen (P18) of *Flavobacterium psychrophilum*. Appl. Environ. Microbiol. *72*: 4845–4852.

Dworkin, M. 1969. Sensitivity of gliding bacteria to actinomycin D. J. Bacteriol. *98*: 851–852.

Dyer, J.K. and R.W. Bolton. 1985. Purification and chemical characterization of an exopolysaccharide isolated from *Capnocytophaga ochracea*. Can. J. Microbiol. *31*: 1–5.

Eiler, A. and S. Bertilsson. 2007. *Flavobacteria* blooms in four eutrophic lakes: linking population dynamics of freshwater bacterioplankton to resource availability. Appl. Environ. Microbiol. *73*: 3511–3518.

Einen, J. and L. Øvreås. 2006. *Flaviramulus basaltis* gen. nov., sp. nov., a novel member of the family *Flavobacteriaceae* isolated from seafloor basalt. Int. J. Syst. Evol. Microbiol. *56*: 2455–2461.

Eisenberg, J. 1891. Bacteriologische Diagnostik. Hilfstabellen zum Gebrauche beim Praktischen Arbeiten. 3 Aufl, vol. VII–XXX. Leopold Voss, Hamburg.

El-Gohary, A.A. 1998. *Ornithobacterium rhinotracheale* (ORT) associated with hatching problems in chicken and turkey eggs. Vet. Med. J. Giza, Egypt *46*: 183–191.

El-Sukhon, S.N., A. Musa and M. Al-Attar. 2002. Studies on the bacterial etiology of airsacculitis of broilers in northern and middle Jordan with special reference to *Escherichia coli*, *Ornithobacterium rhinotracheale*, and *Bordetella avium*. Avian Dis. *46*: 605–612.

Elder, J.H. and S. Alexander. 1982. *endo*-β-*N*-Acetylglucosaminidase F: endoglycosidase from *Flavobacterium meningosepticum* that cleaves both high-mannose and complex glycoproteins. Proc. Natl. Acad. Sci. U. S. A. *79*: 4540–4544.

Elifantz, H., R.R. Malmstrom, M.T. Cottrell and D.L. Kirchman. 2005. Assimilation of polysaccharides and glucose by major bacterial groups in the Delaware Estuary. Appl. Environ. Microbiol. *71*: 7799–7805.

Engelbrecht, K., P.J. Jooste and B.A. Prior. 1996a. Quantitative and qualitative determination of the aerobic bacterial populations of Cape marine fish. S. Afr. J. Food Sci. Nutr. *8*: 60–65.

Engelbrecht, K., P.J. Jooste and B.A. Prior. 1996b. Spoilage characteristics of Gram-negative genera and species isolated from Cape marine fish. S. Afr. J. Food Sci. Nutr. *8*: 66–71.

Erganis, O., H.H. Hadimli, K. Kav, M. Corlu and D. Ozturk. 2002. A comparative study on detection of *Ornithobacterium rhinotracheale* antibodies in meat-type turkeys by dot immunobinding assay, rapid agglutination test and serum agglutination test. Avian Pathol. *31*: 201–204.

Euzéby, J.P. 1997. List of bacterial names with standing in nomenclature: a folder available on the Internet. Int. J. Syst. Bacteriol. *47*: 590–592.

Faung, S.T., L. Chiu and C.T. Wang. 1996. Platelet lysis and functional perturbation by 13-methyl myristate, the major fatty acid in *Flavobacterium ranacida*. Thromb. Res. *81*: 91–100.

Fautz, E., L. Grotjahn and H. Reichenbach. 1981. Hydroxy fatty acids as valuable chemosystematic markers in gliding bacteria and flavobacteria. *In* The *Flavobacterium-Cytophaga* Group, Gesellschaft für Biotechnologische Forschung Monograph Series No. 5. (edited by Reichenbach and Weeks). Verlag Chemie, Weinheim, pp. 17–26.

Felsenstein, J. 1985. Confidence limits on phylogenies: an approach using the bootstrap. Evolution *39*: 783–791.

Fischer, L.J., R.S. Weyant, E.H. White and F.D. Quinn. 1995. Intracellular multiplication and toxic destruction of cultured macrophages by *Capnocytophaga canimorsus*. Infect. Immun. *63*: 3484–3490.

Fitzgerald, S.L., J.M. Greyling and R.R. Bragg. 1998. Correlation between ability of *Ornithobacterium rhinotracheale* to agglutinate red blood cells and susceptibility to fosfomycin. Onderstepoort J. Vet. Res. *65*: 317–320.

Flaherty, D.K., F.H. Deck, M.A. Hood, C. Liebert, F. Singleton, P. Winzenburger, K. Bishop, L.R. Smith, L.M. Bynum and W.B. Witmer. 1984. A *Cytophaga* species endotoxin as a putative agent of occupation-related lung disease. Infect. Immun. *43*: 213–216.

Follett, E.A. and D.M. Webley. 1965. An electron microscope study of the cell surface of *Cytophaga johnsonii* and some observations on related organisms. Antonie van Leeuwenhoek *31*: 361–382.

Forlenza, S.W., M.G. Newman, A.I. Lipsey, S.E. Siegel and U. Blachman. 1980. *Capnocytophaga* sepsis: a newly recognised clinical entity in granulocytopenic patients. Lancet *1*: 567–568.

Forlenza, S.W., M.G. Newman, A.L. Horikoshi and U. Blachman. 1981. Antimicrobial susceptibility of *Capnocytophaga*. Antimicrob. Agents Chemother. *19*: 144–146.

Frankland, G.C. and P.F. Frankland. 1889. Über einige typische Mikroorganismen im Wasser und im Boden. Z. Hyg. Infektionskr. *6*: 373–400.

Fraser, S.L. and J.H. Jorgensen. 1997. Reappraisal of the antimicrobial susceptibilities of *Chryseobacterium* and *Flavobacterium* species and methods for reliable susceptibility testing. Antimicrob. Agents Chemother. *41*: 2738–2741.

Frette, L., N.O. Jorgensen, H. Irming and N. Kroer. 2004. *Tenacibaculum skagerrakense* sp. nov., a marine bacterium isolated from the pelagic zone in Skagerrak, Denmark. Int. J. Syst. Evol. Microbiol. *54*: 519–524.

Friedrich, T., S. Geibel, R. Kalmbach, I. Chizhov, K. Ataka, J. Heberle, M. Engelhard and E. Bamberg. 2002. Proteorhodopsin is a light-driven proton pump with variable vectoriality. J. Mol. Biol. *321*: 821–838.

Fung, J.C., M. Berman and T. Fiorentino. 1983. *Capnocytophaga*: a review of the literature. Am. J. Med. Technol. *49*: 589–591.

Gallego, V., M.T. García and A. Ventosa. 2006. *Chryseobacterium hispanicum* sp. nov., isolated from the drinking water distribution system of Sevilla, Spain. Int. J. Syst. Evol. Microbiol. *56*: 1589–1592.

García-López, M.L., M. Prieto and A. Otero. 1998. The physiological attributes of Gram-negative bacteria associated with spoilage of meat and meat products. *In* The Microbiology of Meat and Poultry (edited by Davies and Board). Blackie Academic & Professional, London, pp. 1–34.

Garnjobst, L. 1945. *Cytophaga columnaris* (Davis) in pure culture: a myxobacterium pathogenic to fish. J. Bacteriol. *49*: 113–128.

Gennari, M. and C. Cozzolino. 1989. Observations on *Flavobacterium*, *Cytophagaceae* and other pigmented bacteria isolated from fresh and ice stored sardines. Arch. Vet. Ital. *40*: 372–384.

Gherna, R. and C.R. Woese. 1992. A partial phylogenetic analysis of the "*Flavobacter-Bacteroides*" phylum: basis for taxonomic restructuring. Syst. Appl. Microbiol. *15*: 513–521.

Ghozlan, H., H. Deif, R.A. Kandil and S. Sabry. 2006. Biodiversity of moderately halophilic bacteria in hypersaline habitats in Egypt. J. Gen. Appl. Microbiol. *52*: 63–72.

Gich, F., K. Schubert, A. Bruns, H. Hoffelner and J. Overmann. 2005. Specific detection, isolation, and characterization of selected, previously uncultured members of the freshwater bacterioplankton community. Appl. Environ. Microbiol. *71*: 5908–5919.

Gill, V.J. 2000. Capnocytophaga. In Douglas and Benett's Principles and Practice of Infectious Diseases, 5th edn (edited by Mandell, Bennett and Dolin). Churchill Livingstone, New York, pp. 2441–2444.

Gilligan, P.H., L.R. McCarthy and B.K. Bissett. 1981. *Capnocytophaga ochracea* septicemia. J. Clin. Microbiol. *13*: 643–645.

Glaser, J. and J.L. Pate. 1973. Isolation and characterization of gliding motility mutants of *Cytophaga columnaris*. Arch. Mikrobiol. *93*: 295–309.

Godchaux, W., III and E.R. Leadbetter. 1980. *Capnocytophaga* spp. contain sulfonolipids that are novel in procaryotes. J. Bacteriol. *144*: 592–602.

Godchaux, W., III and E.R. Leadbetter. 1983. Unusual sulfonolipids are characteristic of the *Cytophaga-Flexibacter* group. J. Bacteriol. *153*: 1238–1246.

Godchaux, W.III, and E.R. Leadbetter. 1984. Sulfonolipids of gliding bacteria. Structure of the *N*-acylaminosulfonates. J. Biol. Chem. *259*: 2982–2990.

Godchaux, W.III, L. Gorski and E.R. Leadbetter. 1990. Outer membrane polysaccharide deficiency in two nongliding mutants of *Cytophaga johnsonae*. J. Bacteriol. *172*: 1250–1255.

Goldstein, E.J., D.M. Citron and C.V. Merriam. 1999. Linezolid activity compared to those of selected macrolides and other agents against aerobic and anaerobic pathogens isolated from soft tissue bite infections in humans. Antimicrob. Agents Chemother. *43*: 1469–1474.

Gómez-Consarnau, L., J.M. González, M. Coll-Lladó, P. Gourdon, T. Pascher, R. Neutze, C. Pedrós-Alió and J. Pinhassi. 2007. Light stimulates growth of proteorhodopsin-containing marine *Flavobacteria*. Nature *445*: 210–213.

González, C.J., J.A. Santos, M.L. García-López and A. Otero. 2000. Psychrobacters and related bacteria in freshwater fish. J. Food Prot. *63*: 315–321.

Gorski, L., I.W. Godchaux, E.R. Leadbetter and R.R. Wagner. 1992. Diversity in surface features of *Cytophaga johnsonae* motility mutants. J. Gen. Microbiol. *138*: 1767–1772.

Gosink, J.J., C.R. Woese and J.T. Staley. 1998. *Polaribacter* gen. nov., with three new species, *P. irgensii* sp. nov., *P. franzmannii* sp. nov. and *P. filamentus* sp. nov., gas vacuolate polar marine bacteria of the *Cytophaga–Flavobacterium–Bacteroides* group and reclassification of '*Flectobacillus glomeratus*' as *Polaribacter glomeratus* comb. nov. Int. J. Syst. Bacteriol. *48*: 223–235.

Gräf, W. 1962. Über Wassermyxobakterien. (in German with English and French abstracts). Arch. Hyg. Bakteriol. *146*: 114–125.

Graham, R., C.A. Brandly and G.L. Dunlap. 1938. Studies of duck septicemia. Cornell Vet. *28*: 1–8.

Green, S.L., D.M. Bouley, R.J. Tolwani, K.S. Waggie, B.D. Lifland, G.M. Otto and J.E. Ferrell, Jr. 1999. Identification and management of an outbreak of *Flavobacterium meningosepticum* infection in a colony of South African clawed frogs (*Xenopus laevis*). J. Am. Vet. Med. Assoc. *214*: 1833–1183.

Green, B.T., K. Green and P.E. Nolan. 2001. *Myroides odoratus* cellulitis and bacteremia: case report and review. Scand. J. Infect. Dis. *33*: 932–934.

Green, D.H., L.E. Llewellyn, A.P. Negri, S.I. Blackburn and C.J. Bolch. 2004. Phylogenetic and functional diversity of the cultivable bacterial community associated with the paralytic shellfish poisoning dinoflagellate *Gymnodinium catenatum*. FEMS Microbiol. Ecol. *47*: 345–357.

Green, S.J., E. Inbar, F.C. Michel, Jr, Y. Hadar and D. Minz. 2006. Succession of bacterial communities during early plant development: transition from seed to root and effect of compost amendment. Appl. Environ. Microbiol. *72*: 3975–3983.

Greene, C.E. 1998. Infectious diseases of the dog and cat, 2nd edn. W.B. Saunders, Philadelphia.

Greub, G., B. La Scola and D. Raoult. 2004. Amoebae-resisting bacteria isolated from human nasal swabs by amoebal coculture. Emerg. Infect. Dis. *10*: 470–477.

Grimault, E., J.C. Glerant, P. Aubry, G. Laurans, J.P. Poinsot and V. Jounieaux. 1996. Uncommon site of *Bergeyella zoohelcum*. Apropos of a case. Rev. Pneumol. Clin. *52*: 387–389.

Groben, R., G.J. Doucette, M. Kopp, M. Kodama, R. Amann and L.K. Medlin. 2000. 16S rRNA targeted probes for the identification of bacterial strains isolated from cultures of the toxic dinoflagellate *Alexandrium tamarense*. Microb. Ecol. *39*: 186–196.

Grossart, H.P., F. Levold, M. Allgaier, M. Simon and T. Brinkhoff. 2005. Marine diatom species harbour distinct bacterial communities. Environ. Microbiol. *7*: 860–873.

Groudieva, T., M. Kambourova, H. Yusef, M. Royter, R. Grote, H. Trinks and G. Antranikian. 2004. Diversity and cold-active hydrolytic enzymes of culturable bacteria associated with Arctic sea ice, Spitzbergen. Extremophiles *8*: 475–488.

Güngör, S., M. Özen, A. Akinci and R. Durmaz. 2003. A *Chryseobacterium meningosepticum* outbreak in a neonatal ward. Infect. Control. Hosp. Epidemiol. *24*: 613–617.

Gunnarsson, G., H. Baldursson and I. Hilmarsdottir. 2002. Septic arthritis caused by *Chryseobacterium meningosepticum* in an immunocompetent male. Scand. J. Infect. Dis. *34*: 299–300.

Gutell, R.R., B. Weiser, C.R. Woese and H.F. Noller. 1985. Comparative anatomy of 16-S-like ribosomal RNA. Prog. Nucleic Acid Res. Mol. Biol. *32*: 155–216.

Haburjak, J.J. and T.A. Schubert. 1997. *Flavobacterium breve* meningitis in a dog. J. Am. Anim. Hosp. Assoc. *33*: 509–512.

Hadas, E., M. Derda, J. Winiecka-Krusnell and A. Sutek. 2004. *Acanthamoeba* spp. as vehicles of pathogenic bacteria. Acta Parasitol. *49*: 276–280.

Hafez, H.M. 1998. Current status on the laboratory diagnosis of *Ornithobacterium rhinotracheale* "ORT" in poultry. Berl. Munch. Tierarztl. Wochenschr. *111*: 143–145.

Hafez, H.M. 2002. Diagnosis of *Ornithobacterium rhinotracheale*. Inter. J. Poult. Sci. *1*: 114–118.

Hafez, H.M. and W. Beyer. 1997. Preliminary investigation on *Ornithobacterium rhinotracheale* (ORT) isolates using PCR-fingerprints. Proceedings of the XIth International Congress of the World Veterinary Poultry Association Budapest, p. 51.

Hafez, H.M. and S. Friedrich. 1998. Isolierung von *Ornithobacterium rhinotracheale* „ORT" aus Mastputen in Österreich. Tierärztl. Umschau *53*: 500–504.

Hafez, H.M. and C. Popp. 2003. *Ornithobacterium rhinotracheale*: Bestimmung der Pathogenität an Hühnerembryonen. Proceedings of the 64th Semi-annual Meeting of German Poultry Diseases Group, Hannover, Germany, pp. 79–85.

Hafez, H.M. and D. Schulze. 2003. Examination on the efficacy of chemical disinfectants on *Ornithobacterium rhinotracheale in vitro*. Archiv für Geflügelkunde *67*: 153–156.

Hafez, H.M. and R. Sting. 1996. Serological surveillance on *Ornithobacterium rhinotracheale* in poultry flocks using self-made ELISA. Proceedings of the 45th Western Poultry Disease Conference, Cancun, pp. 163–164.

Hafez, H.M. and R. Sting. 1999. Investigations on different *Ornithobacterium rhinotracheale* "ORT" isolates. Avian Dis. *43*: 1–7.

Hafez, H.M., W. Kruse, J. Emele and R. Sting. 1993. Eine Atemwegsinfektion bei Mastputen durch *Pasteurella*-ähnliche Erreger: Klinik, Diagnostik und Therapie. Presented at the Proc. Int. Conf. on Poultry Diseases, Potsdam.

Hafez, H.M., A. Mazaheri and R. Sting. 2000. Efficacy of ELISA for detection of antibodies against several *Ornithobacterium rhinotracheale* serotypes. Deutsche Tierärztliche Wochenschrift *107*: 142–143.

Hamana, K. and S. Matsuzaki. 1990. Occurrence of homospermidine as a major polyamine in the authentic genus *Flavobacterium*. Can. J. Microbiol. *36*: 228–231.

Hamana, K. and Y. Nakagawa. 2001. Polyamine distribution profiles in newly validated genera and species within the *Flavobacterium-Flexibacter-Cytophaga-Sphingobacterium* complex. Microbios 106(Suppl 2): 105–116.

Hansen, G.H., O. Bergh, J. Michaelsen and D. Knappskog. 1992. *Flexibacter ovolyticus* sp. nov., a pathogen of eggs and larvae of Atlantic halibut, *Hippoglossus hippoglossus* L. Int. J. Syst. Bacteriol. *42*: 451–458.

Hantsis-Zacharov, E. and M. Halpern. 2007. *Chryseobacterium haifense* sp. nov., a psychrotolerant bacterium isolated from raw milk. Int. J. Syst. Evol. Microbiol. *57*: 2344–2348.

Hanzawa, N., S. Kanai, A. Katsuta, Y. Nakagawa and K. Yamasato. 1995. 16S rDNA-based phylogenetic analysis of marine flavobacteria. J. Mar. Biotechnol. *3*: 111–114.

Harrison, F.C. 1929. The discoloration of halibut. Can. J. Res. 1: 214–239.

Hauduroy, A., G. Ehringer, A. Urbain, G. Guillot and J. Magrou. 1937. Dictionnaire des bactéries pathogènes. Masson et Cie, Paris.

Hauduroy, P., G. Ehringer, G. Guillot, J. Magrou, A.R. Prévot, Rosset and J. Urbain. 1953. Dictionnaire des bactéries pathogènes, 2nd edn. Masson et Cie, Paris.

Hawkey, P.M., H. Malnick, S.A. Glover, N. Cook and J.A. Watts. 1984. *Capnocytophaga ochracea* infection: two cases and a review of the published work. J. Clin. Pathol. *37*: 1066–1070.

Hayes, P.R. 1977. A taxonomic study of *Flavobacteria* and related Gram-negative yellow pigmented rods. J. Appl. Bacteriol. *43*: 345–367.

Heeder, C.J., V.C. Lopes, K.V. Nagaraja, D.P. Shaw and D.A. Halvorson. 2001. Seroprevalence of *Ornithobacterium rhinotracheale* infection in commercial laying hens in the north central region of the United States. Avian Dis. *45*: 1064–1067.

Heeg, P., W. Heizmann and H. Mentzel. 1994. Infections caused by *Flavobacterium meningosepticum* in patients in a neonatal intensive care unit. Zentralbl. Hyg. Umweltmed. *195*: 282–287.

Hendrickson, J.M. and K.F. Hilbert. 1932. A new and serious septicemic disease of young ducks with a description of the causative organism. *Pfeifferella anatipestifer*. Cornell Vet. *22*: 239–252.

Herzog, P., I. Winkler, D. Wolking, P. Kampfer and A. Lipski. 2008. *Chryseobacterium ureilyticum* sp. nov., *Chryseobacterium gambrini* sp. nov., *Chryseobacterium pallidum* sp. nov. and *Chryseobacterium molle* sp. nov., isolated from beer-bottling plants. Int. J. Syst. Evol. Microbiol. *58*: 26–33.

Hikida, M., H. Wakabayashi, S. Egusa and K. Masumura. 1979. *Flexibacter* sp., a gliding bacterium pathogenic to some marine fishes in Japan. Bull. Jpn. Soc. Sci. Fish. *45*: 421–428.

Hinz, K.-H. and H.M. Hafez. 1997. Early history of *Ornithobacterium rhinotracheale* (ORT). Arch. Gefluegelk. *61*: 95–96.

Hinz, K.H., C. Blome and M. Ryll. 1994. Acute exudative pneumonia and airsacculitis associated with *Ornithobacterium rhinotracheale* in turkeys. Vet. Rec. *135*: 233–234.

Hinz, K.H., M. Ryll, B. Kohler and G. Glunder. 1998. Phenotypic characteristics of *Riemerella anatipestifer* and similar micro-organisms from various hosts. Avian Pathol. *27*: 33–42.

Holmes, B. 1987. Identification and distribution of *Flavobacterium meningosepticum* in clinical material. J. Appl. Bacteriol. *62*: 29–41.

Holmes, B. 1992. The genera *Flavobacterium, Sphingobacterium*, and *Weeksella*. *In* The Prokaryotes: A Handbook on the Biology of Bacteria: Ecophysiology, Isolation, Identification, Applications, 2nd edn, vol. 4 (edited by Balows, Trüper, Dworkin, Harder and Schleifer). Springer, New York, pp. 3620–3630.

Holmes, B. 1993. Recent developments in *Flavobacterium* taxonomy. In Advances in the taxonomy and significance of *Flavobacterium, Cytophaga* and related bacteria (edited by Jooste). University of the Orange Free State Press, Bloemfontein, South Africa, pp. 6–15.

Holmes, B. and R.J. Owen. 1979. Proposal that *Flavobacterium breve* be substituted as the type species of the genus in place of *Flavobacterium aquatile* and emended description of the genus *Flavobacterium*: status of the named species of *Flavobacterium*, Request for an Opinion. Int. J. Syst. Bacteriol. *29*: 416–426.

Holmes, B. and R.J. Owen. 1981. Emendation of the genus *Flavobacterium* and the status of the genus. Developments after the 8th edition of Bergey's Manual. In The *Flavobacterium-Cytophaga* Group, Gesellschaft für Biotechnologische Forschung Monograph Series No. 5 (edited by Reichenbach and Weeks). Verlag Chemie, Weinheim, pp. 17–26.

Holmes, B. and R.J. Owen. 1982. *Flavobacterium breve* sp. nov., nom. rev. Int. J. Syst. Bacteriol. *32*: 233–234.

Holmes, B., J.J.S. Snell and S.P. Lapage. 1977. Revised description, from clinical isolates, of *Flavobacterium odoratum* Stutzer and Kwaschnina 1929, and designation of the neotype strain. Int. J. Syst. Bacteriol. *27*: 330–336.

Holmes, B., J.J.S. Snell and S.P. Lapage. 1978. Revised description, from clinical strains, of *Flavobacterium breve* (Lustig) Bergey et al. 1923 and proposal of the neotype strain. Int. J. Syst. Bacteriol. *28*: 201–208.

Holmes, B., J.J. Snell and S.P. Lapage. 1979. *Flavobacterium odoratum*: a species resistant to a wide range of antimicrobial agents. J. Clin. Pathol. *32*: 73–77.

Holmes, B., R.J. Owen and D.G. Hollis. 1982. *Flavobacterium spiritivorum*, a new species Isolated from human clinical specimens. Int. J. Syst. Bacteriol. *32*: 157–165.

Holmes, B., R.J. Owen and T.A. McMeekin. 1984a. Genus *Flavobacterium* Bergey, Harrison, Breed, Hammer, and Huntoon 1923, 97^{AL}. In Bergey's Manual of Systematic Bacteriology, vol. 1 (edited by Krieg and Holt). Williams & Wilkins, Baltimore, pp. 353–361.

Holmes, B., R.J. Owen, A.G. Steigerwalt and D.J. Brenner. 1984b. *Flavobacterium gleum*, a new species found in human clinical specimens. Int. J. Syst. Bacteriol. *34*: 21–25.

Holmes, B., A.G. Steigerwalt, R.E. Weaver and D.J. Brenner. 1986a. *Weeksella zoohelcum* sp. nov. (formerly group-IIj), from human clinical specimens. Syst. Appl. Microbiol. *8*: 191–196.

Holmes, B., A.G. Steigerwalt, R.E. Weaver and D.J. Brenner. 1986b. *Weeksella virosa* gen. nov., sp. nov. (formerly group-IIf), found in human clinical specimens. Syst. Appl. Microbiol. *8*: 185–190.

Holmes, B., A.G. Steigerwalt, R.E. Weaver and D.J. Brenner. 1987. *In* Validation of the publication of new names and new combinations previously effectively published outside the IJSB. List no. 23. Int. J. Syst. Bacteriol. *37*: 179–180.

Holmes, B., R.E. Weaver, A.G. Steigerwalt and D.J. Brenner. 1988. A taxonomic study of *Flavobacterium spiritivorum* and *Sphingobacterium mizutae*: proposal of *Flavobacterium yabuuchiae* sp. nov. and *Flavobacterium mizutaii* comb. nov. Int. J. Syst. Bacteriol. *38*: 348–353.

Holt, R.A. 1988. *Cytophaga psychrophila*, the causative agent of bacterial cold-water disease in salmonid fish. PhD thesis, Oregon State University, Corvallis.

Holt, S.C., G. Forcier and B.J. Takacs. 1979. Fatty acid composition of gliding bacteria: oral isolates of *Capnocytophaga* compared with *Sporocytophaga*. Infect. Immun. *26*: 298–304.

Holt, R.A., A. Amandi, J.S. Rohovec and J.L. Fryer. 1989. Relation of water temperature to bacterial cold-water disease in coho salmon, chinook salmon, and rainbow trout. J. Aquat. Anim. Health *1*: 94–101.

Holt, R.A., J.S. Rohovec and J.L. Fryer. 1993. Bacterial cold-water disease. In Bacterial Diseases of Fish (edited by Inglis, Roberts and Brombage). Blackwell Scientific Publications, Oxford, pp. 3–22.

Hoque, S.N., J. Graham, M.E. Kaufmann and S. Tabaqchali. 2001. *Chryseobacterium* (*Flavobacterium*) *meningosepticum* outbreak associated with colonization of water taps in a neonatal intensive care unit. J. Hosp. Infect. *47*: 188–192.

Horn, M., M.D. Harzenetter, T. Linner, E.N. Schmid, K.D. Muller, R. Michel and M. Wagner. 2001. Members of the *Cytophaga–Flavobacterium–Bacteroides* phylum as intracellular bacteria of acanthamoebae: proposal of '*Candidatus* Amoebophilus asiaticus'. Environ. Microbiol. *3*: 440–449.

Horn, M.A., J. Ihssen, C. Matthies, A. Schramm, G. Acker and H.L. Drake. 2005. *Dechloromonas denitrificans* sp. nov., *Flavobacterium denitrificans* sp. nov., *Paenibacillus anaericanus* sp. nov. and *Paenibacillus terrae* strain MH72, N_2O-producing bacteria isolated from the gut of the earthworm *Aporrectodea caliginosa*. Int. J. Syst. Evol. Microbiol. *55*: 1255–1265.

Hosoya, R. and K. Hamana. 2004. Distribution of two triamines, spermidine and homospermidine, and an aromatic amine, 2-phenylethylamine, within the phylum *Bacteroidetes*. J. Gen. Appl. Microbiol. *50*: 255–260.

Hsueh, P.R., T.R. Hsiue and W.C. Hsieh. 1996a. Pyomyositis in intravenous drug abusers: report of a unique case and review of the literature. Clin. Infect. Dis. *22*: 858–860.

Hsueh, P.R., T.R. Hsiue, J.J. Wu, L.J. Teng, S.W. Ho, W.C. Hsieh and K.T. Luh. 1996b. *Flavobacterium indologenes* bacteremia: clinical and microbiological characteristics. Clin. Infect. Dis. *23*: 550–555.

Hsueh, P.R., L.J. Teng, S.W. Ho, W.C. Hsieh and K.T. Luh. 1996c. Clinical and microbiological characteristics of *Flavobacterium indologenes* infections associated with indwelling devices. J. Clin. Microbiol. *34*: 1908–1913.

Hsueh, P.R., L.J. Teng, P.C. Yang, S.W. Ho, W.C. Hsieh and K.T. Luh. 1997. Increasing incidence of nosocomial *Chryseobacterium indologenes* infections in Taiwan. Eur. J. Clin. Microbiol. Infect. Dis. *16*: 568–574.

Hu, C.H., Y. Xu, M.S. Xia, L. Xiong and Z.R. Xu. 2007. Effects of Cu^{2+}-exchanged montmorillonite on growth performance, microbial ecology and intestinal morphology of Nile tilapia (*Oreochromis niloticus*). Aquaculture *270*: 200–206.

Huang, B., J. Kwang, H. Loh, J. Frey, H.-M. Tan and K.-L. Chua. 2002. Development of a ELISA using a recombinant 41 kDa partial protein (P45N') for the detection of *Riemerella anatipestifer* infections in ducks. Vet. Microbiol. *88*: 339–349.

Huber, I., B. Spanggaard, K.F. Appel, L. Rossen, T. Nielsen and L. Gram. 2004. Phylogenetic analysis and *in situ* identification of the intestinal microbial community of rainbow trout (*Oncorhynchus mykiss*, Walbaum). J. Appl. Microbiol. *96*: 117–132.

Hugo, C.J. 1997. A taxonomic study of the genus *Chryseobacterium* from food and environmental sources. PhD thesis, University of the Orange Free State, Bloemfontein, South Africa.

Hugo, C.J. and P.J. Jooste. 1997. Preliminary differentiation of food strains of *Chryseobacterium* and *Empedobacter* using multilocus enzyme electrophoresis. Food Microbiol. *14*: 133–142.

Hugo, C.J. and P.J. Jooste. 2003. Culture media for genera in the family *Flavobacteriaceae*. *In* Handbook of Culture Media for Food Microbiology, 2nd edn (edited by Corry, Curtis and Baird). Elsevier Science Publishers, London, pp. 355–367.

Hugo, C.J., P.J. Jooste, P. Segers, M. Vancanneyt and K. Kersters. 1999. A polyphasic taxonomic study of *Chryseobacterium* strains isolated from dairy sources. Syst. Appl. Microbiol. *22*: 586–595.

Hugo, C.J., P. Segers, B. Hoste, M. Vancanneyt and K. Kersters. 2003. *Chryseobacterium joostei* sp. nov., isolated from the dairy environment. Int. J. Syst. Evol. Microbiol. *53*: 771–777.

Hugo, C.J., B. Bruun and P.J. Jooste. 2004a. The genera *Bergeyella* and *Weeksella*. *In* The Prokaryotes: An Evolving Electronic Resource for the Microbiological Community, 3rd edn, release 3.16 (edited by Dworkin, Falkow, Rosenberg, Schleifer and Stackebrandt). Springer, New York.

Hugo, C.J., B. Bruun and P.J. Jooste. 2004b. The genera *Empedobacter* and *Myroides*. *In* The Prokaryotes: An Evolving Electronic Resource for the Microbiological Community, 3rd edn, release 3.16 (edited by Dworkin, Falkow, Rosenberg, Schleifer and Stackebrandt). Springer, New York.

Humphrey, B.A., M.R. Dickson and K.C. Marshall. 1979. Physicochemical and in situ observations of the adhesion of gliding bacteria to surfaces. Arch. Microbiol. *120*: 231–238.

Humphry, D.R., A. George, G.W. Black and S.P. Cummings. 2001. *Flavobacterium frigidarium* sp. nov., an aerobic, psychrophilic, xylanolytic and laminarinolytic bacterium from Antarctica. Int. J. Syst. Evol. Microbiol. *51*: 1235–1243.

Hung, A.L. and A. Alvarado. 2001. Phenotypic and molecular characterization of isolates of *Ornithobacterium rhinotracheale* from Peru. Avian Dis. *45*: 999–1005.

Imshenetski, A.A. and L. Solntseva. 1945. On the imperfect forms of myxobacteria (in Russian with English summary). Mikrobiologiya *14*: 220–229.

Inoue, K. and K. Komagata. 1976. Taxonomic study on obligately psychrophilic bacteria isolated from Antarctica. J. Gen. Appl. Microbiol. *22*: 165–176.

Irgens, R.L., I. Suzuki and J.T. Staley. 1989. Gas vacuolate bacteria obtained from marine waters of Antarctica. Curr. Microbiol. *18*: 261–265.

Isaksen, M.F. and B.B. Jorgensen. 1996. Adaptation of psychrophilic and psychrotrophic sulfate-reducing bacteria to permanently cold marine environments. Appl. Environ. Microbiol. *62*: 408–414.

Isotalo, P.A., D. Edgar and B. Toye. 2000. Polymicrobial tenosynovitis with *Pasteurella multocida* and other Gram negative bacilli after a Siberian tiger bite. J. Clin. Pathol. *53*: 871–872.

Ivanova, E.P., O.I. Nedashkovskaya, J. Chun, A.M. Lysenko, G.M. Frolova, V.I. Svetashev, M.V. Vysotskii, V.V. Mikhailov, A. Huq and R.R. Colwell. 2001. *Arenibacter* gen. nov., new genus of the family *Flavobacteriaceae* and description of a new species, *Arenibacter latericius* sp. nov. Int. J. Syst. Evol. Microbiol. *51*: 1987–1995.

Ivanova, E.P., I.Y. Bakunina, T. Sawabe, K. Hayashi, Y.V. Alexeeva, N.V. Zhukova, D.V. Nicolau, T.N. Zvaygintseva and V.V. Mikhailov. 2002a. Two species of culturable bacteria associated with degradation of brown algae *Fucus evanescens*. Microb. Ecol. *43*: 242–249.

Ivanova, E.P., T. Sawabe, Y.V. Alexeeva, A.M. Lysenko, N.M. Gorshkova, K. Hayashi, N.V. Zukova, R. Christen and V.V. Mikhailov. 2002b. *Pseudoalteromonas issachenkonii* sp. nov., a bacterium that degrades the thallus of the brown alga *Fucus evanescens*. Int. J. Syst. Evol. Microbiol. *52*: 229–234.

Ivanova, E.P., Y.V. Alexeeva, S. Flavier, J.P. Wright, N.V. Zhukova, N.M. Gorshkova, V.V. Mikhailov, D.V. Nicolau and R. Christen. 2004. *Formosa algae* gen. nov., sp. nov., a novel member of the family *Flavobacteriaceae*. Int. J. Syst. Evol. Microbiol. *54*: 705–711.

Ivanova, E.P., J.P. Bowman, R. Christen, N.V. Zhukova, A.M. Lysenko, N.M. Gorshkova, N. Mitik-Dineva, A.F. Sergeev and V.V. Mikhailov. 2006. *Salegentibacter flavus* sp. nov. Int. J. Syst. Evol. Microbiol. *56*: 583–586.

Izumi, S. and F. Aranishi. 2004. Plasmid profiling of Japanese *Flavobacterium psychrophilum* isolates. J. Aquat. Anim. Health *16*: 99–103.

James, S.R., H.R. Burton, T.A. McMeekin and C.A. Mancuso. 1994. Seasonal abundance of *Halomonas meridiana, Halomonas subglaciescola, Flavobacterium gondwanense* and *Flavobacterium salegens* in 4 Antarctic lakes. Antarctic Sci. *6*: 325–332.

Janknecht, P., C.M. Schneider and T. Ness. 2002. Outbreak of *Empedobacter brevis* endophthalmitis after cataract extraction. Graefes Arch. Clin. Exp. Ophthalmol. *240*: 291–295.

Jansen, R., N. Chansiripornchai, W. Gaastra and J.P. van Putten. 2004. Characterization of plasmid pOR1 from *Ornithobacterium rhinotracheale* and construction of a shuttle plasmid. Appl. Environ. Microbiol. *70*: 5853–5858.

Jeon, Y.S., H. Chung, S. Park, I. Hur, J.H. Lee and J. Chun. 2005. jPHYDIT: a JAVA-based integrated environment for molecular phylogeny of ribosomal RNA sequences. Bioinformatics *21*: 3171–3173.

Jirjis, F.F., S.L. Noll, D.A. Halvorson, K.V. Nagaraja, F. Martin and D.P. Shaw. 2004. Effects of bacterial coinfection on the pathogenesis of avian pneumovirus infection in turkeys. Avian Dis. *48*: 34–49.

Jit, S., M. Dadhwal, O. Prakash and R. Lal. 2008. *Flavobacterium lindanitolerans* sp. nov., isolated from hexachlorocyclohexane-contaminated soil. Int. J. Syst. Evol. Microbiol. *58*: 1665–1669.

Johansen, J.E., P. Nielsen and C. Sjøholm. 1999. Description of *Cellulophaga baltica* gen. nov., sp. nov., and *Cellulophaga fucicola* gen. nov., sp. nov. and reclassification of [*Cytophaga lytica*] to *Cellulophaga lytica* gen. nov., comb. nov. Int. J. Syst. Bacteriol. *49*: 1231–1240.

Johnson, J.L. and W.S. Chilton. 1966. Galactosamine glycan of *Chondrococcus columnaris*. Science *152*: 1247–1248.

Jooste, P.J. 1985. The taxonomy and significance of *Flavobacterium-Cytophaga* strains from dairy sources. PhD thesis, University of Orange Free State, Bloemfontein, Republic of South Africa.

Jooste, P.J. and C.J. Hugo. 1999. The taxonomy, ecology and cultivation of bacterial genera belonging to the family *Flavobacteriaceae*. Int. J. Food Microbiol. *53*: 81–94.

Jooste, P.J., T.J. Britz and J. De Haast. 1985. A numerical taxonomic study of *Flavobacterium-Cytophaga* strains from dairy sources. J. Appl. Bacteriol. *59*: 311–323.

Jooste, P.J., T.J. Britz and P.M. Lategan. 1986. Screening for the presence of *Flavobacterium* strains in dairy sources. S. Afr. J. Dairy Sci. *18*: 45–50.

Joubert, P., R. Higgins, A. Laperle, I. Mikaelian, D. Venne and A. Silim. 1999. Isolation of *Ornithobacterium rhinotracheale* from turkeys in Quebec, Canada. Avian Dis. *43*: 622–626.

Jung, S.Y., S.J. Kang, M.H. Lee, S.Y. Lee, T.K. Oh and J.H. Yoon. 2005. *Gaetbulibacter saemankumensis* gen. nov., sp. nov., a novel member of the family *Flavobacteriaceae* isolated from a tidal flat sediment in Korea. Int. J. Syst. Evol. Microbiol. *55*: 1845–1849.

Jung, S.Y., T.K. Oh and J.H. Yoon. 2006. *Tenacibaculum aestuarii* sp. nov., isolated from a tidal flat sediment in Korea. Int. J. Syst. Evol. Microbiol. *56*: 1577–1581.

Junge, K., F. Imhoff, T. Staley and J.W. Deming. 2002. Phylogenetic diversity of numerically important Arctic sea-ice bacteria cultured at subzero temperature. Microb. Ecol. *43*: 315–328.

Kaiser, P. 1961. Etude de l'activité pectinolytique du sol et d'autres substrats naturels. Doctoral thesis, Université de Paris, France.

Kamma, J.J., M. Nakou and F.A. Manti. 1995. Predominant microflora of severe, moderate and minimal periodontal lesions in young adults with rapidly progressive periodontitis. J. Periodont. Res. *30*: 66–72.

Kämpfer, P., U. Dreyer, A. Neef, W. Dott and H.J. Busse. 2003. *Chryseobacterium defluvii* sp. nov., isolated from wastewater. Int. J. Syst. Evol. Microbiol. *53*: 93–97.

Kämpfer, P., V. Avesani, M. Janssens, J. Charlier, T. De Baere and M. Vaneechoutte. 2006. Description of *Wautersiella falsenii* gen. nov., sp. nov., to accommodate clinical isolates phenotypically resembling members of the genera *Chryseobacterium* and *Empedobacter*. Int. J. Syst. Evol. Microbiol. *56*: 2323–2329.

Kapke, P.A., A.T. Brown and T.T. Lillich. 1980. Carbon dioxide metabolism by *Capnocytophaga ochracea*: identification, characterization, and regulation of a phosphoenolpyruvate carboxykinase. Infect. Immun. *27*: 756–766.

Kardos, G., J. Nagy, M. Antal, A. Bistyak, M. Tenk and I. Kiss. 2007. Development of a novel PCR assay specific for *Riemerella anatipestifer*. Lett. Appl. Microbiol. *44*: 145–148.

Karpouzas, D.G., A. Fotopoulou, U. Menkissoglu-Spiroudi and B.K. Singh. 2005. Non-specific biodegradation of the organophosphorus pesticides, cadusafos and ethoprophos, by two bacterial isolates. FEMS Microbiol. Ecol. *53*: 369–378.

Kelly, K.M. and A.Y. Chistoserdov. 2001. Phylogenetic analysis of the succession of bacterial communities in the Great South Bay (Long Island). FEMS Microbiol. Ecol. *35*: 85–95.

Kempf, M.J. and M.J. McBride. 2000. Transposon insertions in the *Flavobacterium johnsoniae ftsX* gene disrupt gliding motility and cell division. J. Bacteriol. *182*: 1671–1679.

Khajamohiddin, S., P.S. Babu, D. Chakka, M. Merrick, A. Bhaduri, R. Sowdhamini and D. Siddavattam. 2006. A novel meta-cleavage product hydrolase from *Flavobacterium* sp. ATCC 27551. Biochem. Biophys. Res. Commun. *351*: 675–681.

Khan, S.T., Y. Nakagawa and S. Harayama. 2006a. *Krokinobacter* gen. nov., with three novel species, in the family *Flavobacteriaceae*. Int. J. Syst. Evol. Microbiol. *56*: 323–328.

Khan, S.T., Y. Nakagawa and S. Harayama. 2006b. *Sediminicola luteus* gen. nov., sp. nov., a novel member of the family *Flavobacteriaceae*. Int. J. Syst. Evol. Microbiol. *56*: 841–845.

Khan, S.T., Y. Nakagawa and S. Harayama. 2006c. *Sandarakinotalea sediminis* gen. nov., sp. nov., a novel member of the family *Flavobacteriaceae*. Int. J. Syst. Evol. Microbiol. *56*: 959–963.

Khan, S.T., Y. Nakagawa and S. Harayama. 2007a. *Sediminibacter furfurosus* gen. nov., sp. nov. and *Gilvibacter sediminis* gen. nov., sp. nov., novel members of the family *Flavobacteriaceae*. Int. J. Syst. Evol. Microbiol. *57*: 265–269.

Khan, S.T., Y. Nakagawa and S. Harayama. 2007b. *Galbibacter mesophilus* gen. nov., sp. nov., a novel member of the family *Flavobacteriaceae*. Int. J. Syst. Evol. Microbiol. *57*: 969–973.

Khan, S.T., Y. Nakagawa and S. Harayama. 2008. *Fulvibacter tottoriensis* gen. nov., sp. nov., a member of the family *Flavobacteriaceae* isolated from marine sediment. Int. J. Syst. Evol. Microbiol. *58*: 1670–1674.

Kim, M.K., W.T. Im, Y.K. Shin, J.H. Lim, S.H. Kim, B.C. Lee, M.Y. Park, K.Y. Lee and S.T. Lee. 2004. *Kaistella koreensis* gen. nov., sp. nov., a novel member of the *Chryseobacterium-Bergeyella-Riemerella* branch. Int. J. Syst. Evol. Microbiol. *54*: 2319–2324.

Kim, K.K., H.S. Bae, P. Schumann and S.T. Lee. 2005a. *Chryseobacterium daecheongense* sp. nov., isolated from freshwater lake sediment. Int. J. Syst. Evol. Microbiol. *55*: 133–138.

Kim, K.K., M.K. Kim, J.H. Lim, H.Y. Park and S.T. Lee. 2005b. Transfer of *Chryseobacterium meningosepticum* and *Chryseobacterium miricola* to *Elizabethkingia* gen. nov. as *Elizabethkingia meningoseptica* comb. nov. and *Elizabethkingia miricola* comb. nov. Int. J. Syst. Evol. Microbiol. *55*: 1287–1293.

Kim, B.Y., H.Y. Weon, S. Cousin, S.H. Yoo, S.W. Kwon, S.J. Go and E. Stackebrandt. 2006a. *Flavobacterium daejeonense* sp. nov. and *Flavobacterium suncheonense* sp. nov., isolated from greenhouse soils in Korea. Int. J. Syst. Evol. Microbiol. *56*: 1645–1649.

Kim, J.-S., R.S. Dungan, S.-W. Kwon and H.-Y. Weon. 2006b. The community composition of root-associated bacteria of the tomato plant. World J. Microbiol. Biotechnol. *22*: 1267–1273.

Kim, J.H., K.Y. Kim, Y.T. Hahm, B.S. Kim, J. Chun. and C.J. Cha. 2008. *Actibacter sediminis* gen. nov., sp. nov., a marine bacterium of the family *Flavobacteriaceae* isolated from tidal flat sediment. Int. J. Syst. Evol. Microbiol. *58*: 139–143.

Kimura, M. 1980. A simple method for estimating evolutionary rates of base substitutions through comparative studies of nucleotide sequences. J. Mol. Evol. *16*: 111–120.

King, E.O. 1959. Studies on a group of previously unclassified bacteria associated with meningitis in infants. Am. J. Clin. Pathol. *31*: 241–247.

Kirby, J.T., H.S. Sader, T.R. Walsh and R.N. Jones. 2004. Antimicrobial susceptibility and epidemiology of a worldwide collection of *Chryseobacterium* spp: report from the SENTRY Antimicrobial Surveillance Program (1997–2001). J. Clin. Microbiol. *42*: 445–448.

Kirchman, D.L., L. Yu and M.T. Cottrell. 2003. Diversity and abundance of uncultured *Cytophaga*-like bacteria in the Delaware estuary. Appl. Environ. Microbiol. *69*: 6587–6596.

Kisand, V., R. Cuadros and J. Wikner. 2002. Phylogeny of culturable estuarine bacteria catabolizing riverine organic matter in the northern Baltic Sea. Appl. Environ. Microbiol. *68*: 379–388.

Kisand, V., N. Andersson and J. Wikner. 2005. Bacterial freshwater species successfully immigrate to the brackish water environment in the northern Baltic. Limnol. Oceanogr. *50*: 945–956.

Kivinen, P.K., M.R. Lahtinen, E. Ruotsalainen, I.T. Harvima and M.L. Katila. 2003. *Bergeyella zoohelcum* septicaemia of a patient suffering from severe skin infection. Acta Derm. Venereol. *83*: 74–75.

Koga, Y. and A.I. Zavaleta. 2005. Intraspecies genetic variability of *Ornithobacterium rhinotracheale* in commercial birds in Peru. Avian Dis. *49*: 108–111.

Köpke, B., R. Wilms, B. Engelen, H. Cypionka and H. Sass. 2005. Microbial diversity in coastal subsurface sediments: a cultivation approach using various electron acceptors and substrate gradients. Appl. Environ. Microbiol. *71*: 7819–7830.

Krasil'nikov, N.A. 1949. Guide to the bacteria and actinomycetes. Akad. Nauk. S.S.S.R., Moscow.

Krehenbrink, M., F.B. Oppermann-Sanio and A. Steinbuchel. 2002. Evaluation of non-cyanobacterial genome sequences for occurrence of genes encoding proteins homologous to cyanophycin synthetase and cloning of an active cyanophycin synthetase from *Acinetobacter* sp. strain DSM 587. Arch. Microbiol. *177*: 371–380.

Kristiansen, J.E., A. Bremmelgaard, H.E. Busk, O. Heltberg, W. Frederiksen and T. Justesen. 1984. Rapid identification of *Capnocytophaga* isolated from septicemic patients. Eur. J. Clin. Microbiol. *3*: 236–240.

Kristensen, B., H.C. Schonheyder, N.A. Peterslund, S. Rosthoj, N. Clausen and W. Frederiksen. 1995. *Capnocytophaga* (*Capnocytophaga ochracea* group) bacteremia in hematological patients with profound granulocytopenia. Scand. J. Infect. Dis. *27*: 153–155.

Kuhnert, P. and B.M. Korczak. 2006. Prediction of whole-genome DNA–DNA similarity, determination of G+C content and phylogenetic analysis within the family *Pasteurellaceae* by multilocus sequence analysis (MLSA). Microbiology *152*: 2537–2548.

Kwon, K.K., H.S. Lee, H.B. Jung, J.H. Kang and S.J. Kim. 2006a. *Yeosuana aromativorans* gen. nov., sp. nov., a mesophilic marine bacterium belonging to the family *Flavobacteriaceae*, isolated from estuarine sediment of the South Sea, Korea. Int. J. Syst. Evol. Microbiol. *56*: 727–732.

Kwon, K.K., Y.K. Lee and H.K. Lee. 2006b. *Costertonia aggregata* gen. nov., sp. nov., a mesophilic marine bacterium of the family *Flavobacteriaceae*, isolated from a mature biofilm. Int. J. Syst. Evol. Microbiol. *56*: 1349–1353.

Laffineur, K., M. Janssens, J. Charlier, V. Avesani, G. Wauters and M. Delmée. 2002. Biochemical and susceptibility tests useful for identification of nonfermenting Gram-negative rods. J. Clin. Microbiol. *40*: 1085–1087.

Larkin, H.M. and R. Borrall. 1984. Genus III. *Flectobacillus*. *In* Bergey's Manual of Systematic Bacteriology, vol. 1 (edited by Krieg and Holt). Williams & Wilkins, Baltimore, pp. 129–132.

Lau, S.C., M.M. Tsoi, X. Li, I. Plakhotnikova, S. Dobretsov, K.W. Lau, M. Wu, P.K. Wong, J.R. Pawlik and P.Y. Qian. 2005a. *Winogradskyella poriferorum* sp. nov., a novel member of the family *Flavobacteriaceae* isolated from a sponge in the Bahamas. Int. J. Syst. Evol. Microbiol. *55*: 1589–1592.

Lau, S.C., M.M. Tsoi, X. Li, I. Plakhotnikova, S. Dobretsov, P.K. Wong, J.R. Pawlik and P.Y. Qian. 2005b. *Nonlabens tegetincola* gen. nov., sp. nov., a novel member of the family *Flavobacteriaceae* isolated from a microbial mat in a subtropical estuary. Int. J. Syst. Evol. Microbiol. *55*: 2279–2283.

Lau, S.C., M.M. Tsoi, X. Li, I. Plakhotnikova, S. Dobretsov, P.K. Wong and P.Y. Qian. 2005c. *Gramella portivictoriae* sp. nov., a novel member of the family *Flavobacteriaceae* isolated from marine sediment. Int. J. Syst. Evol. Microbiol. *55*: 2497–2500.

Lau, S.C., M.M. Tsoi, X. Li, I. Plakhotnikova, S. Dobretsov, M. Wu, P.K. Wong, J.R. Pawlik and P.Y. Qian. 2006. *Stenothermobacter spongiae* gen. nov., sp. nov., a novel member of the family *Flavobacteriaceae* isolated from a marine sponge in the Bahamas, and emended description of *Nonlabens tegetincola*. Int. J. Syst. Evol. Microbiol. *56*: 181–185.

Laughon, B.E., S.A. Syed and W.J. Loesche. 1982. API ZYM system for identification of *Bacteroides* spp., *Capnocytophaga* spp., and spirochetes of oral origin. J. Clin. Microbiol. *15*: 97–102.

Leadbetter, E.R. 1974. Order II. *Cytophagales* nomen novum. *In* Bergey's Manual of Determinative Bacteriology, 8th edn (edited by Buchanan and Gibbons). Williams & Wilkins, Baltimore, pp. 99–122.

Leadbetter, E.R., S.C. Holt and S.S. Socransky. 1979. *Capnocytophaga*: new genus of Gram-negative gliding bacteria. 1. General characteristics, taxonomic considerations and significance. Arch. Microbiol. *122*: 9–16.

Leadbetter, E.R., S.C. Holt and S.S. Socransky. 1982. *In* Validation of the publication of new names and new combinations previously effectively published outside the IJSB. List no. 8. Int. J. Syst. Bacteriol. *32*: 266–268.

Lee, S.D. 2007. *Tamlana crocina* gen. nov., sp. nov., a marine bacterium of the family *Flavobacteriaceae*, isolated from beach sediment in Korea. Int. J. Syst. Evol. Microbiol. *57*: 764–769.

Lee, C.C., M. Smith, R.E. Kibblewhite-Accinelli, T.G. Williams, K. Wagschal, G.H. Robertson and D.W. Wong. 2006a. Isolation and characterization of a cold-active xylanase enzyme from *Flavobacterium* sp. Curr. Microbiol. *52*: 112–116.

Lee, O.O., S.C. Lau, M.M. Tsoi, X. Li, I. Plakhotnikova, S. Dobretsov, M.C. Wu, P.K. Wong and P.Y. Qian. 2006b. *Gillisia myxillae* sp. nov., a novel member of the family *Flavobacteriaceae*, isolated from the marine sponge *Myxilla incrustans*. Int. J. Syst. Evol. Microbiol. *56*: 1795–1799.

Lee, H.S., K.K. Kwon, S.H. Yang, S.S. Bae, C.H. Park, S.J. Kim and J.H. Lee. 2008. Description of *Croceitalea* gen. nov. in the family *Flavobacteriaceae* with two species, *Croceitalea eckloniae* sp. nov. and *Croceitalea dokdonensis* sp. nov., isolated from the rhizosphere of the marine alga *Ecklonia kurome*. Int. J. Syst. Evol. Microbiol. *58*: 2505–2510.

Lee, D.H., H.Y. Kahng, Y.S. Lee, J.S. Jung, J.M. Kim, B.S. Chung, S.K. Park and C.O. Jeon. 2009. *Jejuia pallidilutea* gen. nov., sp. nov., a new member of the family *Flavobacteriaceae* isolated from seawater. Int. J. Syst. Evol. Microbiol. *59*: 2148–2152.

Leff, L.G., J.V. McArthur and L.J. Shimkets. 1998. Persistence and dissemination of introduced bacteria in freshwater microcosms. Microb. Ecol. *36*: 202–2211.

Leifson, E. 1963. Determination of Carbohydrate Metabolism of Marine *Bacteria*. J. Bacteriol. *85*: 1183–1184.

Lelliott, R.A. and D.E. Stead. 1987. Methods for the diagnosis of bacterial diseases in plants. Blackwell Scientific Publications, Oxford.

Leroy-Setrin, S., G. Flaujac, K. Thenaisy and E. Chaslus-Dancla. 1998. Genetic diversity of *Ornithobacterium rhinotracheale* strains isolated from poultry in France. Lett. Appl. Microbiol. *26*: 189–193.

Lewin, R.A. 1969. A classification of *Flexibacteria*. J. Gen. Microbiol. *58*: 189–206.

Lewin, R.A. and D.M. Lounsbery. 1969. Isolation, cultivation and characterization of *Flexibacteria*. J. Gen. Microbiol. *58*: 145–170.

Li, Y., Y. Kawamura, N. Fujiwara, T. Naka, H. Liu, X. Huang, K. Kobayashi and T. Ezaki. 2003a. *Chryseobacterium miricola* sp. nov., a novel species isolated from condensation water of space station Mir. Syst. Appl. Microbiol. *26*: 523–528.

Liao, C.H. and J.M. Wells. 1986. Properties of *Cytophaga johnsonae* strains causing spoilage of fresh produce at food markets. Appl. Environ. Microbiol. *52*: 1261–1265.

Liebert, C.A., M.A. Hood, F.H. Deck, K. Bishop and D.K. Flaherty. 1984. Isolation and characterization of a new *Cytophaga* species implicated in a work-related lung-disease. Appl. Environ. Microbiol. *48*: 936–943.

Lijnen, H.R., B. Van Hoef, F. Ugwu, D. Collen and I. Roelants. 2000. Specific proteolysis of human plasminogen by a 24 kDa endopeptidase from a novel *Chryseobacterium* sp. Biochemistry *39*: 479–488.

Lillich, T.T. and R. Calmes. 1979. Cytochromes and dehydrogenases in membranes of a new human periodontal bacterial pathogen, *Capnocytophaga ochracea*. Arch. Oral Biol. *24*: 699–702.

Lim, J.M., C.O. Jeon, S.S. Lee, D.J. Park, L.H. Xu, C.L. Jiang and C.J. Kim. 2008. Reclassification of *Salegentibacter catena* Ying et al. 2007 as *Salinimicrobium catena* gen. nov., comb. nov. and description of *Salinimicrobium xinjiangense* sp. nov., a halophilic bacterium isolated from Xinjiang province in China. Int. J. Syst. Evol. Microbiol. *58*: 438–442.

Lin, J.T., W.S. Wang, C.C. Yen, J.H. Liu, T.J. Chiou, M.H. Yang, T.C. Chao and P.M. Chen. 2003. *Chryseobacterium indologenes* bacteremia in a bone marrow transplant recipient with chronic graft-versus-host disease. Scand. J. Infect. Dis. *35*: 882–883.

Lin, P.Y., C. Chu, L.H. Su, C.T. Huang, W.Y. Chang and C.H. Chiu. 2004. Clinical and microbiological analysis of bloodstream infections caused by *Chryseobacterium meningosepticum* in nonneonatal patients. J. Clin. Microbiol. *42*: 3353–3355.

Lin, P.Y., C.H. Chiu, C. Chu, P. Tang and L.H. Su. 2006. Invasion of murine respiratory tract epithelial cells by *Chryseobacterium meningosepticum* and identification of genes present specifically in an invasive strain. New Microbiol. *29*: 55–62.

Lingalo, B.M. 1997. The incidence of *Flavobacteriaceae* in marine fish with special reference to *Myroides* species. BSc (Hons.) research project thesis, University of the Free State, Bloemfontein, South Africa.

Lion, C., F. Escande and J.C. Burdin. 1996. *Capnocytophaga canimorsus* infections in human: review of the literature and cases report. Eur. J. Epidemiol. *12*: 521–533.

Liu, Z.P., B.J. Wang, X. Dai, X.Y. Liu and S.J. Liu. 2006. *Zhouia amylolytica* gen. nov., sp. nov., a novel member of the family *Flavobacteriaceae* isolated from sediment of the South China Sea. Int. J. Syst. Evol. Microbiol. *56*: 2825–2829.

Liu, J., M.J. McBride and S. Subramaniam. 2007. Cell surface filaments of the gliding bacterium *Flavobacterium johnsoniae* revealed by cryo-electron tomography. J. Bacteriol. *189*: 7503–7506.

Liu, H., R. Liu, S.Y. Yang, W.K. Gao, C.X. Zhang, K.Y. Zhang and R. Lai. 2008. *Flavobacterium anhuiense* sp. nov., isolated from field soil. Int. J. Syst. Evol. Microbiol. *58*: 756–760.

Llobet-Brossa, E., R. Rosselló-Mora and R. Amann. 1998. Microbial community composition of Wadden Sea sediments as revealed by fluorescence in situ hybridization. Appl. Environ. Microbiol. *64*: 2691–2696.

Loh, H., T.P. Teo and H.C. Tan. 1992. Serotypes of '*Pasteurella*' *anatipestifer* isolates from ducks in Singapore: a proposal of new serotypes. Avian Pathol. *21*: 453–459.

London, J., R.A. Celesk, A. Kagermeier and J.L. Johnson. 1985. Emended description of *Capnocytophaga gingivalis*. Int. J. Syst. Bacteriol. *35*: 369–370.

Lopes, V., G. Rajashekara, A. Back, D.P. Shaw, D.A. Halvorson and K.V. Nagaraja. 2000. Outer membrane proteins for serologic detection of *Ornithobacterium rhinotracheale* infection in turkeys. Avian Dis. *44*: 957–962.

Lopes, V.C., A. Back, H.J. Shin, D.A. Halvorson and K.V. Nagaraja. 2002a. Development, characterization, and preliminary evaluation of a temperature-sensitive mutant of *Ornithobacterium rhinotracheale* for potential use as a live vaccine in turkeys. Avian Dis. *46*: 162–168.

Lopes, V.C., B. Velayudhan, D.A. Halvorson and K.V. Nagaraja. 2002b. Survival of *Ornithobacterium rhinotracheale* in sterilized poultry litter. Avian Dis. *46*: 1011–1014.

López, M.J., N.N. Nichols, B.S. Dien, J. Moreno and R.J. Bothast. 2004. Isolation of microorganisms for biological detoxification of lignocellulosic hydrolysates. Appl. Microbiol. Biotechnol. *64*: 125–131.

Lu, S.J., H.Q. Wang and Z.H. Yao. 2006. Isolation and characterization of gasoline-degrading bacteria from gas station leaking-contaminated soils. J. Environ. Sci. (China) *18*: 969–972.

Lustig, A. 1890. Diagnostica dei batteri delle acque con una guida alle ricerche batteriologiche e microscopiche. Rosenberg and Sellier, Turin.

Lysyk, T.J., L. Kalischuk-Tymensen, L.B. Selinger, R.C. Lancaster, L. Wever and K.J. Cheng. 1999. Rearing stable fly larvae (Diptera: Muscidae) on an egg yolk medium. J. Med. Entomol. *36*: 382–388.

Macián, M.C., M.J. Pujalte, M.C. Márquez, W. Ludwig, A. Ventosa, E. Garay and K.H. Schleifer. 2002. *Gelidibacter mesophilus* sp. nov., a novel marine bacterium in the family *Flavobacteriaceae*. Int. J. Syst. Evol. Microbiol. *52*: 1325–1329.

MacLean, L.L., M.B. Perry, E.M. Crump and W.W. Kay. 2003. Structural characterization of the lipopolysaccharide O-polysaccharide antigen produced by *Flavobacterium columnare* ATCC 43622. Eur. J. Biochem. *270*: 3440–3446.

Madetoja, J., S. Nystedt and T. Wiklund. 2003. Survival and virulence of *Flavobacterium psychrophilum* in water microcosms. FEMS Microbiol. Ecol. *43*: 217–223.

Madrid, V.M., J.Y. Aller, R.C. Aller and A.Y. Chistoserdov. 2001. High prokaryote diversity and analysis of community structure in mobile mud deposits off French Guiana: identification of two new bacterial candidate divisions. FEMS Microbiol. Ecol. *37*: 197–209.

Madsen, L. and I. Dalsgaard. 2000. Comparative studies of Danish *Flavobacterium psychrophilum* isolates: ribotypes, plasmid profiles, serotypes and virulence. J. Fish Dis. *23*: 211–218.

Malik, Y.S., K. Olsen, K. Kumar and S.M. Goyal. 2003. *In vitro* antibiotic resistance profiles of *Ornithobacterium rhinotracheale* strains from Minnesota turkeys during 1996–2002. Avian Dis. *47*: 588–593.

Manani, T.A., E.K. Collison and S. Mpuchane. 2006. Microflora of minimally processed frozen vegetables sold in Gaborone, Botswana. J. Food Prot. *69*: 2581–2586.

Manavathi, B., S.B. Pakala, P. Gorla, M. Merrick and D. Siddavattam. 2005. Influence of zinc and cobalt on expression and activity of parathion hydrolase from *Flavobacterium* sp. ATCC 27551. Pestic. Biochem. Physiol. *83*: 37–45.

Mancuso Nichols, C., J.P. Bowman and J. Guezennec. 2005. *Olleya marilimosa* gen. nov., sp. nov., an exopolysaccharide-producing marine bacterium from the family *Flavobacteriaceae*, isolated from the Southern Ocean. Int. J. Syst. Evol. Microbiol. *55*: 1557–1561.

Maneerat, S., T. Nitoda, H. Kanzaki and F. Kawai. 2005. Bile acids are new products of a marine bacterium, *Myroides* sp. strain SM1. Appl. Microbiol. Biotechnol. *67*: 679–683.

Maneerat, S., T. Bamba, K. Harada, A. Kobayashi, H. Yamada and F. Kawai. 2006. A novel crude oil emulsifier excreted in the culture supernatant of a marine bacterium, *Myroides* sp. strain SM1. Appl. Microbiol. Biotechnol. *70*: 254–259.

Manfredi, R., A. Nanetti, M. Ferri, A. Mastroianni, O.V. Coronado and F. Chiodo. 1999. *Flavobacterium* spp. organisms as opportunistic bacterial pathogens during advanced HIV disease. J. Infect. *39*: 146–152.

Männistö, M.K. and M.M. Haggblom. 2006. Characterization of psychrotolerant heterotrophic bacteria from Finnish Lapland. Syst. Appl. Microbiol. *29*: 229–243.

Manz, W., K. Wendt-Potthoff, T.R. Neu, U. Szewzyk and J.R. Lawrence. 1999. Phylogenetic composition, spatial structure, and dynamics of lotic bacterial biofilms investigated by Fluorescent In Situ Hybridization and Confocal Laser Scanning microscopy. Microb. Ecol. *37*: 225–237.

Mardy, C. and B. Holmes. 1988. Incidence of vaginal *Weeksella virosa* (formerly group IIf). J. Clin. Pathol. *41*: 211–214.

Marien, M., A. Decostere, A. Martel, K. Chiers, R. Froyman and H. Nauwynck. 2005. Synergy between avian pneumovirus and *Ornithobacterium rhinotracheale* in turkeys. Avian Pathol. *34*: 204–211.

Marmur, J. and P. Doty. 1962. Determination of the base composition of deoxyribonucleic acid from its thermal denaturation temperature. J. Mol. Biol. *5*: 109–118.

Mashimo, P.A., Y. Yamamoto, M. Nakamura and J. Slots. 1983. Selective recovery of oral *Capnocytophaga* spp. with sheep blood agar containing bacitracin and polymyxin B. J. Clin. Microbiol. *17*: 187–191.

Mauel, M.J., D.L. Miller, K.S. Frazier and M.E. Hines II. 2002. Bacterial pathogens isolated from cultured bullfrogs (*Rana catesbeina*). J. Vet. Diagn. Invest. *14*: 431–433.

McBain, A.J., R.G. Bartolo, C.E. Catrenich, D. Charbonneau, R.G. Ledder, A.H. Rickard, S.A. Symmons and P. Gilbert. 2003. Microbial characterization of biofilms in domestic drains and the establishment of stable biofilm microcosms. Appl. Environ. Microbiol. *69*: 177–185.

McBride, M.J. 2001. Bacterial gliding motility: multiple mechanisms for cell movement over surfaces. Annu. Rev. Microbiol. *55*: 49–75.

McBride, M.J. and S.A. Baker. 1996. Development of techniques to genetically manipulate members of the genera *Cytophaga, Flavobacterium, Flexibacter*, and *Sporocytophaga*. Appl. Environ. Microbiol. *62*: 3017–3022.

McBride, M.J. and M.J. Kempf. 1996. Development of techniques for the genetic manipulation of the gliding bacterium *Cytophaga johnsonae*. J. Bacteriol. *178*: 583–590.

McBride, M.J. and T.F. Braun. 2004. GldI is a lipoprotein that is required for *Flavobacterium johnsoniae* gliding motility and chitin utilization. J. Bacteriol. *186*: 2295–2302.

McBride, M.J., T.F. Braun and J.L. Brust. 2003. *Flavobacterium johnsoniae* GldH is a lipoprotein that is required for gliding motility and chitin utilization. J. Bacteriol. *185*: 6648–6657.

McCammon, S.A. and J.P. Bowman. 2000. Taxonomy of antarctic *Flavobacterium* species: description of *Flavobacterium gillisiae* sp. nov., *Flavobacterium tegetincola* sp. nov. and *Flavobacterium xanthum* sp. nov., nom. rev. and reclassification of *Flavobacterium salegens* as *Salegentibacter salegens* gen. nov., comb. nov. Int. J. Syst. Evol. Microbiol. *50*: 1055–1063.

McCammon, S.A., B.H. Innes, J.P. Bowman, P.D. Franzmann, S.J. Dobson, P.E. Holloway, J.H. Skerratt, P.D. Nichols and L.M. Rankin. 1998. *Flavobacterium hibernum* sp. nov., a lactose-utilizing bacterium from a freshwater Antarctic lake. Int. J. Syst. Bacteriol. *48*: 1405–1412.

McGuire, A.J., P.D. Franzmann and T.A. McMeekin. 1987. *Flectobacillus glomeratus* sp. nov., a curved, nonmotile, pigmented bacterium isolated from antarctic marine environments. Syst. Appl. Microbiol. *9*: 265–272.

McGuire, A.J., P.D. Franzmann and T.A. McMeekin. 1988. *In* Validation of the publication of new names and new combinations previously effectively published outside the IJSB. List no. 24. Int. J. Syst. Bacteriol. *38*: 136–137.

McSpadden Gardener, B.B. and D.M. Weller. 2001. Changes in populations of rhizosphere bacteria associated with take-all disease of wheat. Appl. Environ. Microbiol. *67*: 4414–4425.

Mesbah, M., U. Premachandran and W.B. Whitman. 1989. Precise measurement of the G+C content of deoxyribonucleic acid by high-performance liquid chromatography. Int. J. Syst. Bacteriol. *39*: 159–167.

Mez, C. 1898. Mikroskopische Wasseranalyse, Anleitung zur Untersüchung des Wassers mit besonderer Berücksichtigung von Trink- und Abwasser. J. Springer, Berlin.

Michel, C., O. Matte-Tailliez, B. Kerouault and J.-F. Bernardet. 2005. Resistance pattern and assessment of phenicol agents' minimum inhibitory concentration in multiple drug resistant *Chryseobacterium* isolates from fish and aquatic habitats. J. Appl. Microbiol. *99*: 323–332.

Miteva, V.I. and J.E. Brenchley. 2005. Detection and isolation of ultrasmall microorganisms from a 120,000-year-old Greenland glacier ice core. Appl. Environ. Microbiol. *71*: 7806–7818.

Mitra, S., Y. Matsuo, T. Haga, M. Yasumoto-Hirose, J. Yoon, H. Kasai and A. Yokota. 2009. *Leptobacterium flavescens* gen. nov., sp. nov., a marine member of the family *Flavobacteriaceae*, isolated from marine sponge and seawater. Int. J. Syst. Evol. Microbiol. *59*: 207–212.

Møller, J.D., A.C. Barnes, I. Dalsgaard and A.E. Ellis. 2005. Characterisation of surface blebbing and membrane vesicles produced by *Flavobacterium psychrophilum*. Dis. Aquat. Org. *64*: 201–209.

Mori, R. 1888. Ueber pathogene Bacterien im Canalwasser. Z. Hyg. Infektionskr. *4*: 47–54.

Morita, Y., T. Nakamura, Q. Hasan, Y. Murakami, K. Yokoyama and E. Tamiya. 1997. Cold-active enzymes from cold-adapted bacteria. J. Am. Oil Chem. Soc. *74*: 441–444.

Mossad, S.B., A.E. Lichtin, G.S. Hall and S.M. Gordon. 1997. Diagnosis: *Capnocytophaga canimorsus* septicemia. Clin. Infect. Dis. *24*: 123, 267.

Motwani, B., D. Krezolek, S. Symeonides and W. Khayr. 2004. *Myroides odoratum* cellulitis and bacteremia: a case report. Infect. Dis. Clin. Practice *12*: 343–344.

Mudarris, M. and B. Austin. 1989. Systemic disease in turbot *Scophthalmus maximus* caused by a previously unrecognised *Cytophaga*-like bacterium. Dis. Aquat. Org. *6*: 161–166.

Mudarris, M., B. Austin, P. Segers, M. Vancanneyt, B. Hoste and J.-F. Bernardet. 1994. *Flavobacterium scophthalmum* sp. nov., a pathogen of turbot (*Scophthalmus maximus* L.). Int. J. Syst. Bacteriol. *44*: 447–453.

Mulbry, W.W. and J.S. Karns. 1989. Parathion hydrolase specified by the *Flavobacterium opd* gene: relationship between the gene and protein. J. Bacteriol. *171*: 6740–6746.

Müller, K.D., E.N. Schmid and R. Michel. 1999. Intracellular bacteria of *Acanthamoebae* resembling *Legionella* spp. turned out to be *Cytophaga* sp. Zentralbl. Bakteriol. *289*: 389–397.

Müller, S., B. Kiesel and L. Berthe-Corti. 2001. *Muricauda ruestringensis* has an asymmetric cell cycle. Acta Biotechnol. *21*: 343–357.

Murayama, Y., P.A. Mashimo, L.A. Tabak, M.J. Levine and S.A. Ellison. 1982. Isolation and partial characterization of a genus common antigen and species specific antigen of *Capnocytophaga*. Jpn. J. Med. Sci. Biol. *35*: 153–170.

Naeem, K., A. Malik and A. Ullah. 2003. Seroprevalence of *Ornithobacterium rhinotracheale* in chickens in Pakistan. Vet. Rec. *153*: 533–534.

Nagaraja, K.V., A. Back, S. Sorenger, G. Rajashekara and D.A. Halvorson. 1998. Tissue distribution post-infection and antimicrobal sensitivity of *Ornithobacterium rhinotracheale*. Proceedings of the 47th Western Poultry Disease Conference, Sacramento, pp. 57–60.

Nakagawa, Y. and K. Yamasato. 1993. Phylogenetic diversity of the genus *Cytophaga* revealed by 16S ribosomal RNA sequencing and menaquinone analysis. J. Gen. Microbiol. *139*: 1155–1161.

Nakagawa, Y. and K. Yamasato. 1996. Emendation of the genus *Cytophaga* and transfer of *Cytophaga agarovorans* and *Cytophaga salmonicolor* to *Marinilabilia* gen. nov: phylogenetic analysis of the *Flavobacterium cytophaga* complex. Int. J. Syst. Bacteriol. *46*: 599–603.

Nakamura, M. and J. Slots. 1982. Aminopeptidase activity of *Capnocytophaga*. J. Periodont. Res. *17*: 597–603.

Nedashkovskaya, O.I., S.B. Kim, S.K. Han, A.M. Lysenko, M. Rohde, N.V. Zhukova, E. Falsen, G.M. Frolova, V.V. Mikhailov and K.S. Bae. 2003a. *Mesonia algae* gen. nov., sp. nov., a novel marine bacterium of the family *Flavobacteriaceae* isolated from the green alga *Acrosiphonia sonderi* (Kütz) Kornm. Int. J. Syst. Evol. Microbiol. *53*: 1967–1971.

Nedashkovskaya, O.I., M. Suzuki, M.V. Vysotskii and V.V. Mikhailov. 2003b. *Arenibacter troitsensis* sp. nov., isolated from marine bottom sediment. Int. J. Syst. Evol. Microbiol. *53*: 1287–1290.

Nedashkovskaya, O.I., M. Suzuki, M.V. Vysotskii and V.V. Mikhailov. 2003c. *Vitellibacter vladivostokensis* gen. nov., sp. nov., a new member of the phylum *Cytophaga–Flavobacterium–Bacteroides*. Int. J. Syst. Evol. Microbiol. *53*: 1281–1286.

Nedashkovskaya, O.I., S.B. Kim, S.K. Han, A.M. Lysenko, V.V. Mikhailov and K.S. Bae. 2004a. *Arenibacter certesii* sp. nov., a novel marine bacterium isolated from the green alga *Ulva fenestrata*. Int. J. Syst. Evol. Microbiol. *54*: 1173–1176.

Nedashkovskaya, O.I., S.B. Kim, S.K. Han, A.M. Lysenko, M. Rohde, M.S. Rhee, G.M. Frolova, E. Falsen, V.V. Mikhailov and K.S. Bae. 2004b. *Maribacter* gen. nov., a new member of the family *Flavobacteriaceae*, isolated from marine habitats, containing the species *Maribacter sedimenticola* sp. nov., *Maribacter aquivivus* sp. nov., *Maribacter orientalis* sp. nov. and *Maribacter ulvicola* sp. nov. Int. J. Syst. Evol. Microbiol. *54*: 1017–1023.

Nedashkovskaya, O.I., S.B. Kim, S.K. Han, M.S. Rhee, A.M. Lysenko, E. Falsen, G.M. Frolova, V.V. Mikhailov and K.S. Bae. 2004c. *Ulvibacter litoralis* gen. nov., sp. nov., a novel member of the family *Flavobacteriaceae* isolated from the green alga *Ulva fenestrata*. Int. J. Syst. Evol. Microbiol. *54*: 119–123.

Nedashkovskaya, O.I., S.B. Kim, S.K. Han, M.S. Rhee, A.M. Lysenko, M. Rohde, N.V. Zhukova, G.M. Frolova, V.V. Mikhailov and K.S. Bae. 2004d. *Algibacter lectus* gen. nov., sp. nov., a novel member of the family *Flavobacteriaceae* isolated from green algae. Int. J. Syst. Evol. Microbiol. *54*: 1257–1261.

Nedashkovskaya, O.I., M. Suzuki, A.M. Lysenko, C. Snauwaert, M. Vancanneyt, J. Swings, M.V. Vysotskii and V.V. Mikhailov. 2004e. *Cellulophaga pacifica* sp. nov. Int. J. Syst. Evol. Microbiol. *54*: 609–613.

Nedashkovskaya, O.I., M. Suzuki, M. Vancanneyt, I. Cleenwerck, A.M. Lysenko, V.V. Mikhailov and J. Swings. 2004f. *Zobellia amurskyensis* sp. nov., *Zobellia laminariae* sp. nov. and *Zobellia russellii* sp. nov., novel marine bacteria of the family *Flavobacteriaceae*. Int. J. Syst. Evol. Microbiol. *54*: 1643–1648.

Nedashkovskaya, O.I., M. Suzuki, M. Vancanneyt, I. Cleenwerck, N.V. Zhukova, M.V. Vysotskii, V.V. Mikhailov and J. Swings. 2004g. *Salegentibacter holothuriorum* sp. nov., isolated from the edible holothurian *Apostichopus japonicus*. Int. J. Syst. Evol. Microbiol. *54*: 1107–1110.

Nedashkovskaya, O.I., S.B. Kim, S.K. Han, C. Snauwaert, M. Vancanneyt, J. Swings, K.O. Kim, A.M. Lysenko, M. Rohde, G.M. Frolova, V.V. Mikhailov and K.S. Bae. 2005a. *Winogradskyella thalassocola* gen. nov., sp. nov., *Winogradskyella epiphytica* sp. nov. and *Winogradskyella eximia* sp. nov., marine bacteria of the family *Flavobacteriaceae*. Int. J. Syst. Evol. Microbiol. *55*: 49–55.

Nedashkovskaya, O.I., S.B. Kim, K.H. Lee, K.S. Bae, G.M. Frolova, V.V. Mikhailov and I.S. Kim. 2005b. *Pibocella ponti* gen. nov., sp. nov., a novel marine bacterium of the family *Flavobacteriaceae* isolated from the green alga *Acrosiphonia sonderi*. Int. J. Syst. Evol. Microbiol. *55*: 177–181.

Nedashkovskaya, O.I., S.B. Kim, K.H. Lee, V.V. Mikhailov and K.S. Bae. 2005c. *Gillisia mitskevichiae* sp. nov., a novel bacterium of the family *Flavobacteriaceae*, isolated from sea water. Int. J. Syst. Evol. Microbiol. *55*: 321–323.

Nedashkovskaya, O.I., S.B. Kim, A.M. Lysenko, G.M. Frolova, V.V. Mikhailov and K.S. Bae. 2005d. *Bizionia paragorgiae* gen. nov., sp. nov., a novel member of the family *Flavobacteriaceae* isolated from the soft coral *Paragorgia arborea*. Int. J. Syst. Evol. Microbiol. *55*: 375–378.

Nedashkovskaya, O.I., S.B. Kim, A.M. Lysenko, G.M. Frolova, V.V. Mikhailov, K.S. Bae, D.H. Lee and I.S. Kim. 2005e. *Gramella echinicola* gen. nov., sp. nov., a novel halophilic bacterium of the family *Flavobacteriaceae* isolated from the sea urchin *Strongylocentrotus intermedius*. Int. J. Syst. Evol. Microbiol. *55*: 391–394.

Nedashkovskaya, O.I., S.B. Kim, A.M. Lysenko, G.M. Frolova, V.V. Mikhailov, K.H. Lee and K.S. Bae. 2005f. Description of *Aquimarina muelleri* gen. nov., sp. nov., and proposal of the reclassification of [*Cytophaga*] *latercula* Lewin 1969 as *Stanierella latercula* gen. nov., comb. nov. Int. J. Syst. Evol. Microbiol. *55*: 225–229.

Nedashkovskaya, O.I., S.B. Kim, A.M. Lysenko, N.I. Kalinovskaya, V.V. Mikhailov, I.S. Kim and K.S. Bae. 2005g. *Polaribacter butkevichii* sp. nov., a novel marine mesophilic bacterium of the family *Flavobacteriaceae*. Curr. Microbiol. *51*: 408–412.

Nedashkovskaya, O.I., S.B. Kim, A.M. Lysenko, V.V. Mikhailov, K.S. Bae and I.S. Kim. 2005h. *Salegentibacter mishustinae* sp. nov., isolated from the sea urchin *Strongylocentrotus intermedius*. Int. J. Syst. Evol. Microbiol. *55*: 235–238.

Nedashkovskaya, O.I., M. Vancanneyt, P. Dawyndt, K. Engelbeen, K. Vandemeulebroecke, I. Cleenwerck, B. Hoste, J. Mergaert, T.L. Tan, G.M. Frolova, V.V. Mikhailov and J. Swings. 2005i. Reclassification of [*Cytophaga*] *marinoflava* Reichenbach 1989 as *Leeuwenhoekiella marinoflava* gen. nov., comb. nov. and description of *Leeuwenhoekiella aequorea* sp. nov. Int. J. Syst. Evol. Microbiol. *55*: 1033–1038.

Nedashkovskaya, O.I., S.B. Kim, J. Kwak, V.V. Mikhailov and K.S. Bae. 2006a. *Mariniflexile gromovii* gen. nov., sp. nov., a gliding bacterium isolated from the sea urchin *Strongylocentrotus intermedius*. Int. J. Syst. Evol. Microbiol. *56*: 1635–1638.

Nedashkovskaya, O.I., S.B. Kim, A.M. Lysenko, V.V. Mikhailov, I.S. Kim and K.S. Bae. 2006b. *In* List of new names and new combinations previously effectively, but not validly, published. Validation List no. 110. Int. J. Syst. Evol. Microbiol. *56*: 1459–1460.

Nedashkovskaya, O.I., S.B. Kim, M. Vancanneyt, D.S. Shin, A.M. Lysenko, L.S. Shevchenko, V.B. Krasokhin, V.V. Mikhailov, J. Swings and K.S. Bae. 2006c. *Salegentibacter agarivorans* sp. nov., a novel marine bacterium of the family *Flavobacteriaceae* isolated from the sponge *Artemisina* sp. Int. J. Syst. Evol. Microbiol. *56*: 883–887.

Nedashkovskaya, O.I., S.B. Kim, M. Vancanneyt, C. Snauwaert, A.M. Lysenko, M. Rohde, G.M. Frolova, N.V. Zhukova, V.V. Mikhailov, K.S. Bae, H.W. Oh and J. Swings. 2006d. *Formosa agariphila* sp. nov., a budding bacterium of the family *Flavobacteriaceae* isolated from marine environments, and emended description of the genus *Formosa*. Int. J. Syst. Evol. Microbiol. *56*: 161–167.

Nedashkovskaya, O.I., S.B. Kim, N.V. Zhukova, J. Kwak, V.V. Mikhailov and K.S. Bae. 2006e. *Mesonia mobilis* sp. nov., isolated from seawater, and emended description of the genus *Mesonia*. Int. J. Syst. Evol. Microbiol. *56*: 2433–2436.

Nedashkovskaya, O.I., M. Vancanneyt, L. Christiaens, N.I. Kalinovskaya, V.V. Mikhailov and J. Swings. 2006f. *Aquimarina intermedia* sp. nov., reclassification of *Stanierella latercula* (Lewin 1969) as *Aquimarina latercula* comb. nov. and *Gaetbulimicrobium brevivitae* Yoon et al. 2006 as *Aquimarina brevivitae* comb. nov. and emended description of the genus *Aquimarina*. Int. J. Syst. Evol. Microbiol. *56*: 2037–2041.

Nedashkovskaya, O.I., M. Vancanneyt, I. Cleenwerck, C. Snauwaert, S.B. Kim, A.M. Lysenko, L.S. Shevchenko, K.H. Lee, M.S. Park, G.M. Frolova, V.V. Mikhailov, K.S. Bae and J. Swings. 2006g. *Arenibacter palladensis* sp. nov., a novel marine bacterium isolated from the green alga *Ulva fenestrata*, and emended description of the genus *Arenibacter*. Int. J. Syst. Evol. Microbiol. *56*: 155–160.

Nedashkovskaya, O.I., M. Suzuki, S.B. Kim and V.V. Mikhailov. 2008. *Kriegella aquimaris* gen. nov., sp. nov., isolated from marine environments. Int. J. Syst. Evol. Microbiol. *58*: 2624–2628.

Nedashkovskaya, O.I., M. Suzuki, J.S. Lee, K.C. Lee, L.S. Shevchenko and V.V. Mikhailov. 2009. *Pseudozobellia thermophila* gen. nov., sp. nov., a bacterium of the family *Flavobacteriaceae*, isolated from the green alga *Ulva fenestrata*. Int. J. Syst. Evol. Microbiol. *59*: 806–810.

Nelson, S.S. and M.J. McBride. 2006. Mutations in *Flavobacterium johnsoniae secDF* result in defects in gliding motility and chitin utilization. J. Bacteriol. *188*: 348–351.

Nelson, S.S., P.P. Glocka, S. Agarwal, D.P. Grimm and M.J. McBride. 2007. *Flavobacterium johnsoniae* SprA is a cell surface protein involved in gliding motility. J. Bacteriol. *189*: 7145–7150.

Nematollahi, A., A. Decostere, F. Pasmans and F. Haesebrouck. 2003. *Flavobacterium psychrophilum* infections in salmonid fish. J. Fish Dis. *26*: 563–574.

Newman, M.G. and S.S. Socransky. 1977. Predominant cultivable microbiota in periodontosis. J. Periodont. Res. *12*: 120–128.

Newman, M.G. and T.N. Sims. 1979. The predominant cultivable microbiota of the periodontal abscess. J. Periodontol. *50*: 350–354.

Nichols, D.S., J.L. Brown, P.D. Nichols and T.A. McMeekin. 1997. Production of eicosapentaenoic and arachidonic acids by an Antarctic bacterium: response to growth temperature. FEMS Microbiol. Lett. *152*: 349–354.

Nichols, D., J. Bowman, K. Sanderson, C.M. Nichols, T. Lewis, T. McMeekin and P.D. Nichols. 1999. Developments with antarctic microorganisms: culture collections, bioactivity screening, taxonomy, PUFA production and cold-adapted enzymes. Curr. Opin. Biotechnol. *10*: 240–246.

Nicolas, J.L., S. Corre and J.C. Cochard. 2004. Bacterial population association with phytoplankton cultured in a bivalve hatchery. Microb. Ecol. *48*: 400–413.

Nichols, C.M., S.G. Lardiere, J.P. Bowman, P.D. Nichols, J. A E Gibson and J. Guezennec. 2005. Chemical characterization of exopolysaccharides from Antarctic marine bacteria. Microb. Ecol. *49*: 578–589.

Niswati, A., J. Murase and M. Kimura. 2005. Comparison of Bacterial Communities Associated with Microcrustaceans from the Floodwater of a Paddy Field Microcosm: Estimation Based on DGGE Pattern and Sequence Analyses. Soil Sci. Plant Nutr. *51*: 289–290.

Nogi, Y., K. Soda and T. Oikawa. 2005a. *In* Validation of publication of new names and new combinations previously effectively published outside the IJSEM. List no. 105. Int. J. Syst. Evol. Microbiol. *55*: 1743–1745.

Nogi, Y., K. Soda and T. Oikawa. 2005b. *Flavobacterium frigidimaris* sp. nov., isolated from Antarctic seawater. Syst. Appl. Microbiol. *28*: 310–315.

Nohynek, L.J., E.L. Suhonen, E.L. Nurmiaho-Lassila, J. Hantula and M. Salkinoja-Salonen. 1995. Description of four pentachlorophenol-degrading bacterial strains as *Sphingomonas chlorophenolica* sp. nov. Syst. Appl. Microbiol. *18*: 527–538.

Nulens, E., B. Bussels, A. Bols, B. Gordts and H.W. Van Landuyt. 2001. Recurrent bacteremia by *Chryseobacterium indologenes* in an oncology patient with a totally implanted intravascular device. Clin. Microbiol. Infect. *7*: 391–393.

O'Sullivan, L.A., A.J. Weightman and J.C. Fry. 2002. New degenerate *Cytophaga–Flexibacter–Bacteroides*-specific 16S ribosomal DNA-targeted oligonucleotide probes reveal high bacterial diversity in River Taff epilithon. Appl. Environ. Microbiol. *68*: 201–210.

O'Sullivan, L.A., J. Rinna, G. Humphreys, A.J. Weightman and J.C. Fry. 2006. Culturable phylogenetic diversity of the phylum '*Bacteroidetes*' from river epilithon and coastal water and description of novel members of the family *Flavobacteriaceae*: *Epilithonimonas tenax* gen. nov., sp. nov. and *Persicivirga xylanidelens* gen. nov., sp. nov. Int. J. Syst. Evol. Microbiol. *56*: 169–180.

Odor, E.M., M. Salem, C.R. Pope, B. Sample, M. Primm, K. Vance and M. Murphy. 1997. Isolation and identification of *Ornithobacterium rhinotracheale* from commercial broiler flocks on the Delmarva peninsula. Avian Dis. *41*: 257–260.

Ohishi, K., T. Yamamoto, T. Tomofuji, N. Tamaki and T. Watanabe. 2005. Isolation and characterization of aminopeptidase from *Capnocytophaga granulosa* ATCC 51502. Oral Microbiol. Immunol. *20*: 67–72.

Okabe, S., M. Odagiri, T. Ito and H. Satoh. 2007. Succession of sulfur-oxidizing bacteria in the microbial community on corroding concrete in sewer systems. Appl. Environ. Microbiol. *73*: 971–980.

Okeke, B.C. and W.T. Frankenberger, Jr. 2005. Use of starch and potato peel waste for perchlorate bioreduction in water. Sci. Total Environ. *347*: 35–45.

Olsen, H. 1966. The importance of temperature for the growth of *Flavobacterium meningosepticum*. Acta Pathol. Microbiol. Scand. *67*: 291–302.

Olsen, H. 1967. An *in vitro* study of the antibiotic sensitivity of *Flavobacterium meningosepticum*. Acta Pathol. Microbiol. Scand. *70*: 601–612.

Olsen, H. 1969. *Flavobacterium meningosepticum*. A bacteriological, epidemiological and clinical study. Dr Sci. thesis, Andelsbogtrykkeriet, Odense, Denmark.

Olson, M.E., S. Gard, M. Brown, R. Hampton and D.W. Morck. 1992. *Flavobacterium indologenes* infection in leopard frogs. J. Am. Vet. Med. Assoc. *201*: 1766–1770.

Oppong, D., V.M. King and J.A. Bowen. 2003. Isolation and characterization of filamentous bacteria from paper mill slimes. Int. Biodeter. Biodegrad. *52*: 53–62.

Ordal, E.J. and R.R. Rucker. 1944. Pathogenic myxobacteria. Proc. Soc. Exp. Biol. Med. *56*: 15–18.

Ostland, V.E., J.S. Lumsden, D.D. MacPhee and H.W. Ferguson. 1994. Characteristics of *Flavobacterium branchiophilum*, the cause of salmonid bacterial gill disease in Ontario. J. Aquat. Anim. Health *6*: 13–26.

Owen, R.J. and S.P. Lapage. 1974. A comparison of strains of King's group IIb of *Flavobacterium* with *Flavobacterium meningosepticum*. Antonie van Leeuwenhoek *40*: 255–264.

Owen, R.J. and J.J. Snell. 1976. Deoxyribonucleic acid reassociation in the classification of flavobacteria. J. Gen. Microbiol. *93*: 89–102.

Owen, R.J. and B. Holmes. 1981. Identification and classification of *Flavobacterium* species from clinical sources. In The *Flavobacterium-Cytophaga* Group. Gesellschaft für Biotechnologische Forschung Monograph Series, No. 5 (edited by Reichenbach and Weeks). Verlag Chemie, Weinheim, pp. 39–50.

Oyaizu, H. and K. Komagata. 1981. Chemotaxonomic and phenotypic characterization of the strains of species in the *Flavobacterium–Cytophaga* complex. J. Gen. Appl. Microbiol. *27*: 57–107.

Ozkalay, N., M. Anil, N. Agus, M. Helvaci and S. Sirti. 2006. Community-acquired meningitis and sepsis caused by *Chryseobacterium meningosepticum* in a patient diagnosed with thalassemia major. J. Clin. Microbiol. *44*: 3037–3039.

Parenti, D.M. and D.R. Snydman. 1985. *Capnocytophaga* species: infections in nonimmunocompromised and immunocompromised hosts. J. Infect. Dis. *151*: 140–147.

Park, M.S., S.R. Jung, M.S. Lee, K.O. Kim, J.O. Do, K.H. Lee, S.B. Kim and K.S. Bae. 2005. Isolation and characterization of bacteria associated with two sand dune plant species, *Calystegia soldanella* and *Elymus mollis*. J. Microbiol. *43*: 219–227.

Park, M., S. Lu, S.H. Ryu, B.S. Chung, W. Park, C.J. Kim and C.O. Jeon. 2006a. *Flavobacterium croceum* sp. nov., isolated from activated sludge. Int. J. Syst. Evol. Microbiol. *56*: 2443–2447.

Park, M.S., S.R. Jung, K.H. Lee, M.S. Lee, J.O. Do, S.B. Kim and K.S. Bae. 2006b. *Chryseobacterium soldanellicola* sp. nov. and *Chryseobacterium taeanense* sp. nov., isolated from roots of sand-dune plants. Int. J. Syst. Evol. Microbiol. *56*: 433–438.

Park, M., S.H. Ryu, T.H. Vu, H.S. Ro, P.Y. Yun and C.O. Jeon. 2007. *Flavobacterium defluvii* sp. nov., isolated from activated sludge. Int. J. Syst. Evol. Microbiol. *57*: 233–237.

Pate, J.L. 1988. Gliding motility in procaryotic cells. Can. J. Microbiol. *34*: 459–465.

Patel, P., M.E. Callow, I. Joint and J.A. Callow. 2003. Specificity in the settlement – modifying response of bacterial biofilms towards zoospores of the marine alga *Enteromorpha*. Environ. Microbiol. *5*: 338–349.

Pathanasophon, P., T. Tanticharoenyos and T. Sawada. 1994. Physiological characteristics, antimicrobial susceptibility and serotypes of *Pasteurella anatipestifer* isolated from ducks in Thailand. Vet. Microbiol. *39*: 179–185.

Pathanasophon, P., T. Sawada, T. Pramoolsinsap and T. Tanticharoenyos. 1996. Immunogenicity of *Riemerella anatipestifer* broth culture bacterin and cell-free culture filtrate in ducks. Avian Pathol. *25*: 705–719.

Pavlov, D., C.M. de Wet, W.O. Grabow and M.M. Ehlers. 2004. Potentially pathogenic features of heterotrophic plate count bacteria isolated from treated and untreated drinking water. Int. J. Food Microbiol. *92*: 275–287.

Pedersen, M.M., E. Marso and M.J. Pickett. 1970. Nonfermentative bacilli associated with man. 3. Pathogenicity and antibiotic susceptibility. Am. J. Clin. Pathol. *54*: 178–192.

Peng, F., M. Liu, L. Zhang, J. Dai, X. Luo, H. An and C. Fang. 2009. Planobacterium taklimakanense gen. nov., sp. nov., a member of the family *Flavobacteriaceae* that exhibits swimming motility, isolated from desert soil. Int. J. Syst. Evol. Microbiol. *59*: 1672–1678.

Perry, L.B. 1973. Gliding motility in some non-spreading *Flexibacteria*. J. Appl. Bacteriol. *36*: 227–232.

Pierce, R.L. and M.W. Vorhies. 1973. *Pasteurella anatipestifer* infection in geese. Avian Dis. *17*: 868–870.

Pinhassi, J., F. Azam, J. Hemphälä, R.A. Long, J. Martinez, U.L. Zweifel and Å. Hagström. 1999. Coupling between bacterioplankton species composition, population dynamics, and organic matter degradation. Aquat. Microb. Ecol. *17*: 13–26.

Pinhassi, J., A. Winding, S.J. Binnerup, U.L. Zweifel, B. Riemann and A. Hagstrom. 2003. Spatial variability in bacterioplankton community composition at the Skagerrak–Kattegat front. Mar. Ecol. Prog. Ser. *255*: 1–13.

Pinhassi, J., M.M. Sala, H. Havskum, F. Peters, O. Guadayol, A. Malits and C. Marrase. 2004. Changes in bacterioplankton composition under different phytoplankton regimens. Appl. Environ. Microbiol. *70*: 6753–6766.

Pinhassi, J., R. Simó, J.M. Gonzalez, M. Vila, L. Alonso-Sáez, R.P. Kiene, M.A. Moran and C. Pedrós-Alió. 2005. Dimethylsulfoniopropionate turnover is linked to the composition and dynamics of the bacterioplankton assemblage during a microcosm phytoplankton bloom. Appl. Environ. Microbiol. *71*: 7650–7660.

Pinhassi, J., J.P. Bowman, O.I. Nedashkovskaya, I. Lekunberri, L. Gomez-Consarnau and C. Pedrós-Alió. 2006. *Leeuwenhoekiella blandensis* sp. nov., a genome-sequenced marine member of the family *Flavobacteriaceae*. Int. J. Syst. Evol. Microbiol. *56*: 1489–1493.

Pitta, T.P., E.R. Leadbetter and W. GodchauxIII., 1989. Increase of ornithine amino lipid content in a sulfonolipid-deficient mutant of *Cytophaga johnsonae*. J. Bacteriol. *171*: 952–957.

Plotnikova, E.G., D.O. Rybkina, L.N. Anan'ina, O.V. Yastrebova and V.A. Demakov. 2006. Characteristics of micro-organisms isolated from technogenic soils of the Kama region. Russ. J. Ecol. *37*: 233–240.

Poirier, T.P., S.J. Tonelli and S.C. Holt. 1979. Ultrastructure of gliding bacteria: scanning electron microscopy of *Capnocytophaga sputigena*, *Capnocytophaga gingivalis*, and *Capnocytophaga ochracea*. Infect. Immun. *26*: 1146–1158.

Poirier, T.P. and S.C. Holt. 1983. Acid and alkaline phosphatases of *Capnocytophaga* species. I. Production and cytological localization of the enzymes. Can. J. Microbiol. *29*: 1350–1360.

Popp, C. and H.M. Hafez. 2002. Investigations on *Ornithobacterium rhinotracheale*. Proceedings of the 4th International Symposium on Turkey Diseases, Berlin, pp. 245–252.

Popp, C. and H.M. Hafez. 2003. *Ornithobacterium rhinotracheale*: Differenzierung Verschiedener Isolate Mittels Serotypisierung und Pulsfeld-Gelelektrophorese. Proceedings of the 64th Semi-annual Meeting of German Poultry Diseases Group, Hannover. Germany, pp. 70–78.

Prévot, A.R. 1961. Traité de Systematique Bactérienne, vol. 2, Dunod, Paris, 3–771.

Price, K.W. and M.J. Pickett. 1981. Studies of clinical isolates of flavobacteria. In The *Flavobacterium-Cytophaga* Group. Gesellschaft für Biotechnologische Forschung Monograph Series, No. 5 (edited by Reichenbach and Weeks). Verlag Chemie, Weinheim, pp. 63–77.

Pu, X. and T.J. Cutright. 2007. Degradation of pentachlorophenol by pure and mixed cultures in two different soils. Environ. Sci. Pollut. Res. Int. *14*: 244–250.

Qin, Q.L., D.L. Zhao, J. Wang, X.L. Chen, H.Y. Dang, T.G. Li, Y.Z. Zhang and P.J. Gao. 2007. *Wangia profunda* gen. nov., sp. nov., a novel marine bacterium of the family *Flavobacteriaceae* isolated from southern Okinawa Trough deep-sea sediment. FEMS Microbiol. Lett. *271*: 53–58.

Qu, J.H., H.F. Li, J.S. Yang and H.L. Yuan. 2008. *Flavobacterium cheniae* sp. nov., isolated from sediment of a eutrophic reservoir. Int. J. Syst. Evol. Microbiol. *58*: 2186–2190.

Quan, Z.X., K.K. Kim, M.K. Kim, L. Jin and S.T. Lee. 2007. *Chryseobacterium caeni* sp. nov., isolated from bioreactor sludge. Int. J. Syst. Evol. Microbiol. *57*: 141–145.

Quan, Z.X., Y.P. Xiao, S.W. Roh, Y.D. Nam, H.W. Chang, K.S. Shin, S.K. Rhee, Y.H. Park and J.W. Bae. 2008. *Joostella marina* gen. nov., sp. nov., a novel member of the family *Flavobacteriaceae* isolated from the East Sea. Int. J. Syst. Evol. Microbiol. *58*: 1388–1392.

Quatrano, R.S. and B.A. Caldwell. 1978. Isolation of a unique marine bacterium capable of growth on a wide variety of polysaccharides from macroalgae. Appl. Environ. Microbiol. *36*: 979–981.

Quilici, M.L. and C. Bizet. 1996. Ribotyping of *Chryseobacterium meningosepticum*: its use as an epidemiological tool and its correlation with serovars. Res. Microbiol. *147*: 415–425.

Radianingtyas, H., G.K. Robinson and A.T. Bull. 2003. Characterization of a soil-derived bacterial consortium degrading 4-chloroaniline. Microbiology *149*: 3279–3287.

Rappé, M.S., S.A. Connon, K.L. Vergin and S.J. Giovannoni. 2002. Cultivation of the ubiquitous SAR11 marine bacterioplankton clade. Nature *418*: 630–633.

Ratkowsky, D.A., R.K. Lowry, T.A. McMeekin, A.N. Stokes and R.E. Chandler. 1983. Model for bacterial culture growth rate throughout the entire biokinetic temperature range. J. Bacteriol. *154*: 1222–1226.

Reasoner, D.J. and E.E. Geldreich. 1985. A new medium for the enumeration and subculture of bacteria from potable water. Appl. Environ. Microbiol. *49*: 1–7.

Reichenbach, H. 1989a. Genus I. *Cytophaga* Winogradsky 1929, 577^{AL} emend. *In* Bergey's Manual of Systematic Bacteriology, vol. 3 (edited by Staley, Bryant, Pfennig and Holt). Williams & Wilkins, Baltimore, pp. 2015–2050.

Reichenbach, H. 1989b. Order I. *Cytophagales* Leadbetter 1974. *In* Bergey's Manual of Systematic Bacteriology, 8th edn, vol. 3 (edited by Staley, Bryant, Pfennig and Holt). Williams & Wilkins, Baltimore, pp. 2011–2013.

Reichenbach, H. 1989c. Genus I. *Cytophaga*. *In* Bergey's Manual of Systematic Bacteriology, vol. 3 (edited by Staley, Bryant, Pfennig and Holt). Williams & Wilkins, Baltimore, pp. 2015–2050.

Reichenbach, H. 1989d. Order *Cytophagales* Leadbetter 1974, 99^{AL}. *In* Bergey's Manual of Systematic Bacteriology, 8th edn, vol. 3 (edited by Staley, Bryant, Pfennig and Holt). Williams & Wilkins, Baltimore, pp. 2011–2073.

Reichenbach, H. 1992a. The order *Cytophagales*. *In* The Prokaryotes: A Handbook on the Biology of Bacteria: Ecophysiology, Isolation, Identification, Applications, 2nd edn, vol. 4 (edited by Balows, Trüper, Dworkin, Harder and Schleifer). Springer, New York, pp. 3631–3675.

Reichenbach, H. 1992b. *Flavobacteriaceae* fam. nov. *In* Validation of the publication of new names and new combinations previously effectively published outside the IJSB. List no. 41. Int. J. Syst. Bacteriol. *42*: 327–329.

Reina, J. and N. Borrell. 1992. Leg abscess caused by *Weeksella zoohelcum* following a dog bite. Clin. Infect. Dis. *14*: 1162–1163.

Richard, C., H. Monteil and B. Laurent. 1979a. Individualisation de six nouveaux types antigéniques de *Flavobacterium meningosepticum*. Ann. Microbiol. Inst. Pasteur *130B*: 141–144.

Richard, C., H. Monteil, A. Le Faou and B. Laurent. 1979b. Etude de souches de *Flavobacterium* isolées à Strasbourg dans un service de réanimation médicale. Individualisation d'un nouveau sérotype (G) de *Flavobacterium meningosepticum*. Méd. Mal. Infect. *9*: 124–128.

Riemer, O.V. 1904. Kurze Mitteilung über eine bei Gänsen beobachtete exsudative Septikämie und deren Erreger. Zentralbl. Bakteriol. Abt. 2 *1*: 641–648.

Rintamäki-Kinnunen, P., J.-F. Bernardet and A. Bloigu. 1997. Yellow pigmented filamentous bacteria connected with farmed salmonid fish mortality. Aquaculture *149*: 1–14.

Robertson, P.A.W. and B. Austin. 1994. Disease associated with cyprinids imported into the United Kingdom. International Symposium on Aquatic Animal Health: program and abstracts. W-17.6. Univ. of California, School of Veterinary Medicine, Davis, CA.

Romanenko, L.A., M. Uchino, G.M. Frolova and V.V. Mikhailov. 2007. *Marixanthomonas ophiurae* gen. nov., sp. nov., a marine bacterium of the family *Flavobacteriaceae* isolated from a deep-sea brittle star. Int. J. Syst. Evol. Microbiol. *57*: 457–462.

Rosado, W. and N.S. Govind. 2003. Identification of carbohydrate degrading bacteria in sub-tropical regions. Rev. Biol. Trop. 51(Suppl 4): 205–210.

Rossolini, G.M., N. Franceschini, M.L. Riccio, P.S. Mercuri, M. Perilli, M. Galleni, J.M. Frere and G. Amicosante. 1998. Characterization and sequence of the *Chryseobacterium* (*Flavobacterium*) *meningosepticum* carbapenemase: a new molecular class B beta-lactamase showing a broad substrate profile. Biochem. J. *332*: 145–152.

Rummens, J.L., J.M. Fossepre, M. De Gruyter, H. Van de Vyver, L. Neyt and H.W. Van Landuyt. 1985. Isolation of *Capnocytophaga* species with a new selective medium. J. Clin. Microbiol. *22*: 375–378.

Rummens, J.L., B. Gordts and H.W. Van Landuyt. 1986. *In vitro* susceptibility of *Capnocytophaga* species to 29 antimicrobial agents. Antimicrob. Agents Chemother. *30*: 739–742.

Ryll, M., K.H. Hinz, H. Salisch and W. Kruse. 1996. Pathogenicity of *Ornithobacterium rhinotracheale* for turkey poults under experimental conditions. Vet. Rec. *139*: 19.

Ryll, M., R. Gunther, H.M. Hafez and K.H. Hinz. 2002. Isolation and differentiation of a cytochrome oxidase-negative strain of *Ornithobacterium rhinotracheale* from turkeys. Berl. Munch. Tierarztl. Wochenschr. *115*: 274–277.

Ryu, S.H., M. Park, Y. Jeon, J.R. Lee, W. Park and C.O. Jeon. 2007. *Flavobacterium filum* sp. nov., isolated from a wastewater treatment plant in Korea. Int. J. Syst. Evol. Microbiol. *57*: 2026–2030.

Sader, H.S., R.N. Jones and M.A. Pfaller. 1995. Relapse of catheter-related *Flavobacterium meningosepticum* bacteremia demonstrated by DNA macrorestriction analysis. Clin. Infect. Dis. *21*: 997–1000.

Saha, P. and T. Chakrabarti. 2006. *Flavobacterium indicum* sp. nov., isolated from warm spring water in Assam, India. Int. J. Syst. Evol. Microbiol. *56*: 2617–2621.

Saitou, N. and M. Nei. 1987. The neighbor-joining method: a new method for reconstructing phylogenetic trees. Mol. Biol. Evol. *4*: 406–425.

Sakai, E., Y. Tokuyama, F. Nonaka, S. Ohishi, Y. Ishikawa, M. Tanaka and A. Taneno. 2000. *Ornithobacterium rhinotracheale* infection in Japan: preliminary investigations. Vet. Rec. *146*: 502–503.

Sakai, T., H. Kimura and I. Kato. 2002. A marine strain of *Flavobacteriaceae* utilizes brown seaweed fucoidan. Mar. Biotechnol. (NY) *4*: 399–405.

Sandhu, T.S. 1991. Immunogenicity and safety of a live *Pasteurella anatipestifer* vaccine in White Pekin ducklings: laboratory and field trials. Avian Pathol. *20*: 423–432.

Sandhu, T.S. and R.B. Rimlet. 1997. *Riemerella anatipestifer* infection. In Diseases of Poultry, 10th edn (edited by Calnek, Barnes, McDougald and Siazf). Iowa State University Press, Ames, IA, pp. 161–166.

Schneck, J.L. and L.F. Caslake. 2006. Genetic diversity of *Flavobacterium columnare* isolated from fish collected from warm and cold water. J. Fish. Dis. *29*: 245–248.

Schuijffel, D.F., P.C. van Empel, A.M. Pennings, J.P. van Putten and P.J. Nuijten. 2005. Successful selection of cross-protective vaccine candidates for *Ornithobacterium rhinotracheale* infection. Infect. Immun. *73*: 6812–6821.

Schuijffel, D.F., P.C. van Empel, R.P. Segers, J.P. Van Putten and P.J. Nuijten. 2006. Vaccine potential of recombinant *Ornithobacterium rhinotracheale* antigens. Vaccine *24*: 1858–1867.

Secades, P., B. Alvarez and J.A. Guijarro. 2001. Purification and characterization of a psychrophilic, calcium-induced, growth-phase-dependent metalloprotease from the fish pathogen *Flavobacterium psychrophilum*. Appl. Environ. Microbiol. *67*: 2436–2444.

Secades, P., B. Alvarez and J.A. Guijarro. 2003. Purification and properties of a new psychrophilic metalloprotease (Fpp2) in the fish pathogen *Flavobacterium psychrophilum*. FEMS Microbiol. Lett. *226*: 273–279.

Segers, P., W. Mannheim, M. Vancanneyt, K. De Brandt, K.H. Hinz, K. Kersters and P. Vandamme. 1993a. *Riemerella anatipestifer* gen. nov., comb. nov., the causative agent of septicemia anserum exsudativa, and its phylogenetic affiliation within the *Flavobacterium–Cytophaga* rRNA homology group. Int. J. Syst. Bacteriol. *43*: 768–776.

Segers, P., W. Mannheim, M. Vancanneyt, K. Debrandt, K.H. Hinz, K. Kersters and P. Vandamme. 1993b. *Riemerella anatipestifer* gen. nov., comb. nov., the causative agent of septicemia anserum exsudativa, and its phylogenetic affiliation within the *Flavobacterium cytophaga* ribosomal RNA homology group. Int. J. Syst. Bacteriol. *43*: 768–776.

Selje, N., T. Brinkhoff and M. Simon. 2005. Detection of abundant bacteria in the Weser estuary using culture-dependent and culture-independent approaches. Aquat. Microb. Ecol. *39*: 17–34.

Serdyuk, A.V. 1998. *Flavo-* and *flexibacteria* as pathogens of Atlantic salmon, rainbow, trout, cod, and king crab. Parazity i bolezni morskikh i presnovodnykh ryb Severnogo bassejna. PINRO, Murmansk, pp. 115–121.

Sethunathan, N. and T. Yoshida. 1973. A *Flavobacterium* sp. that degrades diazinon and parathion. Can. J. Microbiol. *19*: 873–875.

Sfanos, K., D. Harmody, P. Dang, A. Ledger, S. Pomponi, P. McCarthy and J. Lopez. 2005. A molecular systematic survey of cultured microbial associates of deep-water marine invertebrates. Syst. Appl. Microbiol. *28*: 242–264.

Shamsudin, M.N. and J.A. Plumb. 1996. Morphological, biochemical, and physiological characterization of *Flexibacter columnaris* isolates from four species of fish. J. Aquat. Anim. Health *8*: 335–339.

Shen, F.T., P. Kämpfer, C.C. Young, W.A. Lai and A.B. Arun. 2005. *Chryseobacterium taichungense* sp. nov., isolated from contaminated soil. Int. J. Syst. Evol. Microbiol. *55*: 1301–1304.

Shibagaki-Shimizu, T., N. Nakayama, Y. Nakajima, K. Matsuya, M. Kimura and S. Asakawa. 2006. Phylogenetic study on a bacterial community in the floodwater of a Japanese paddy field estimated by sequencing 16S rDNA fragments after denaturing gradient gel electrophoresis. Biol. Fertil. Soils *42*: 362–365.

Shimomura, K., S. Kaji and A. Hiraishi. 2005. *Chryseobacterium shigense* sp. nov., a yellow-pigmented, aerobic bacterium isolated from a lactic acid beverage. Int. J. Syst. Evol. Microbiol. *55*: 1903–1906.

Shoemaker, C.A., C.R. Arias, P.H. Klesius and T.L. Welker. 2005. Technique for identifying *Flavobacterium columnare* using whole-cell fatty acid profiles. J. Aquat. Anim. Health *17*: 267–274.

Shotts, E.B. and C.E. Starliper. 1999. Flavobacterial diseases: columnaris disease, cold-water disease and bacterial gill disease. In Fish Diseases and Disorders vol. 3, Viral, Bacterial and Fungal Infections (edited by Woo and Bruno). CABI Publishing, Wallingford, pp. 559–576.

Shukla, S.K., D.L. Paustian, P.J. Stockwell, R.E. Morey, J.G. Jordan, P.N. Levett, D.N. Frank and K.D. Reed. 2004. Isolation of a fastidious *Bergeyella* species associated with cellulitis after a cat bite and a phylogenetic comparison with *Bergeyella zoohelcum* strains. J. Clin. Microbiol. *42*: 290–293.

Shurin, S.B., S.S. Socransky, E. Sweeney and T.P. Stossel. 1979. A neutrophil disorder induced by *Capnocytophaga*, a dental micro-organism. N. Engl. J. Med. *301*: 849–854.

Siegman-Igra, Y., D. Schwartz, G. Soferman and N. Konforti. 1987. *Flavobacterium* group IIb bacteremia: report of a case and review of *Flavobacterium* infections. Med. Microbiol. Immunol. *176*: 103–111.

Silipo, A., A. Molinaro, E.L. Nazarenko, L. Sturiale, D. Garozzo, R.P. Gorshkova, O.I. Nedashkovskaya, R. Lanzetta and M. Parrilli. 2005. Structural characterization of the carbohydrate backbone of the lipooligosaccharide of the marine bacterium *Arenibacter certesii* strain KMM 3941(T). Carbohydr. Res. *340*: 2540–2549.

Singh, R., M.F. Teng, T.P. Teo and E.K. Kua. 1983. Anatipestifer disease in ducklings in Singapore. Vet. J. *7*: 53–57.

Skerman, V.B.D., V. McGowan and P.H.A. Sneath. 1980. Approved lists of bacterial names. Int. J. Syst. Bacteriol. *30*: 225–420.

Skerratt, J.H., J.P. Bowman, S. James, G. Hallegraeff and P.D. Nichols. 2002. Algicidal bacteria associated with blooms of a toxic dinoflagellate in a temperate Australian estuary. Mar. Ecol. Prog. Ser. *244*: 1–15.

Slots, J. 1981. Enzymatic characterization of some oral and nonoral gram-negative bacteria with the API ZYM system. J. Clin. Microbiol. *14*: 288–294.

Slots, J. and R.J. Genco. 1984. Black-pigmented *Bacteroides* species, *Capnocytophaga* species, and *Actinobacillus actinomycetemcomitans* in human periodontal disease: virulence factors in colonization, survival, and tissue destruction. J. Dent. Res. *63*: 412–421.

Sly, L.I., L.L. Blackall, P.C. Kraat, T. Tian-Shen and V. Sangkhobol. 1986. The use of second derivative plots for the determination of mol% guanine plus cytosine of DNA by the thermal denaturation method. J. Microbiol. Methods *5*: 139–156.

Smith, D. and D.E. Crowley. 2006. Contribution of ethylamine degrading bacteria to atrazine degradation in soils. FEMS Microbiol. Ecol. *58*: 271–277.

Smith, J.M., D.D. Frame, G. Cooper, A.A. Bickford, G.Y. Ghazikhanian and B.J. Kelly. 1987. *Pasteurella anatipestifer* infection in commercial meat-type turkeys in California. Avian Dis. *31*: 913–917.

Smith, M.C., J.P. Bowman, F.J. Scott and M.A. Line. 2000. Sublithic bacteria associated with Antarctic quartz stones. Antarctic Sci. *12*: 177–184.

Socransky, S.S., S.C. Holt, E.R. Leadbetter, A.C.R. Tanner, E. Savitt and B.F. Hammond. 1979. *Capnocytophaga*: new genus of Gram-negative gliding bacteria. 3. Physiological characterization. Arch. Microbiol. *122*: 29–33.

Sohn, J.H., J.H. Lee, H. Yi, J. Chun, K.S. Bae, T.Y. Ahn and S.J. Kim. 2004. *Kordia algicida* gen. nov., sp. nov., an algicidal bacterium isolated from red tide. Int. J. Syst. Evol. Microbiol. *54*: 675–680.

Song, Y.-L., J.L. Fryer and J.S. Rohovec. 1988. Comparison of six media for the cultivation of *Flexibacter columnaris*. Fish Pathol. *23*: 91–94.

Sørensen, H.P., T.K. Porsgaard, R.A. Kahn, P. Stougaard, K.K. Mortensen and M.G. Johnsen. 2006. Secreted beta-galactosidase from a *Flavobacterium* sp. isolated from a low-temperature environment. Appl. Microbiol. Biotechnol. *70*: 548–557.

Soriano, V.E., M.G. Longinos, P.G. Navarrete and R.P. Fernandez. 2002. Identification and characterization of *Ornithobacterium rhinotracheale* isolates from Mexico. Avian Dis. *46*: 686–690.

Soriano, V.E., N.A. Vera, C.R. Salado, R.P. Fernandez and P.J. Blackall. 2003. *In vitro* susceptibility of *Ornithobacterium rhinotracheale* to several antimicrobial drugs. Avian Dis. *47*: 476–480.

Spangler, S.K., M.A. Visalli, M.R. Jacobs and P.C. Appelbaum. 1996. Susceptibilities of non-*Pseudomonas aeruginosa* Gram-negative nonfermentative rods to ciprofloxacin, ofloxacin, levofloxacin, D-ofloxacin, sparfloxacin, ceftazidime, piperacillin, piperacillin-tazobactam, trimethoprim-sulfamethoxazole, and imipenem. Antimicrob. Agents Chemother. *40*: 772–775.

Speck, H., R.M. Kroppenstedt and W. Mannheim. 1987. Genomic relationships and species differentiation in the genus *Capnocytophaga*. Zentralbl. Bakteriol. Mikrobiol. Hyg. [A] *266*: 390–402.

Spiteller, D., K. Dettner and W. Boland. 2000. Gut bacteria may be involved in interactions between plants, herbivores and their predators: microbial biosynthesis of *N*-acylglutamine surfactants as elicitors of plant volatiles. Biol. Chem. *381*: 755–762.

Sprenger, S.J., A. Back, D.P. Shaw, K.V. Nagaraja, D.C. Roepke and D.A. Halvorson. 1998. *Ornithobacterium rhinotracheale* infection in turkeys: experimental reproduction of the disease. Avian Dis. *42*: 154–161.

Stackebrandt, E. and B.M. Goebel. 1994. Taxonomic note: a place for DNA–DNA reassociation and 16S rRNA sequence analysis in the present species definition in bacteriology. Int. J. Syst. Bacteriol. *44*: 846–849.

Stamm, W.E., J.J. Colella, M.D. Roger, R.L. Anderson and R.E. Dixon. 1975. Indwelling arterial catheters as a source of nosocomial bacteremia. An outbreak caused by *Flavobacterium* species. N. Engl. J. Med. *292*: 1099–1102.

Stanier, R.Y. 1947. Studies on non-fruiting myxobacteria. I. *Cytophaga johnsonae* n. sp., a chitin-decomposing myxobacterium. J. Bacteriol. *53*: 297–315.

Stanier, R.Y. 1957. Order VIII. *Myxobacterales* Jahn 1915. *In* Bergey's Manual of Determinative Bacteriology, 7th edn (edited by Breed, Murray and Smith). Williams & Wilkins, Baltimore, pp. 854–891.

Stevens, R.H., M.N. Sela, J. Shapira and B.F. Hammond. 1980. Detection of a fibroblast proliferation inhibitory factor from *Capnocytophaga sputigena*. Infect. Immun. *27*: 271–275.

Steyn, P.L., P. Segers, M. Vancanneyt, P. Sandra, K. Kersters and J.J. Joubert. 1998. Classification of heparinolytic bacteria into a new genus, *Pedobacter*, comprising four species: *Pedobacter heparinus* comb. nov., *Pedobacter piscium* comb. nov., *Pedobacter africanus* sp. nov. and *Pedobacter saltans* sp. nov. Proposal of the family *Sphingobacteriaceae* fam. nov. Int. J. Syst. Bacteriol. *48*: 165–177.

Strohl, W.R. and L.R. Tait. 1978. *Cytophaga aquatilis* sp. nov., a facultative anaerobe isolated from gills of freshwater fish. Int. J. Syst. Bacteriol. *28*: 293–303.

Stutzer, M. and A. Kwaschnina. 1929. Aussaaten aus den Fäzes des Menschen gelbe Kolonien bildende Bakterien (Gattung *Flavobacterium* u.a.). Zentralbl. Bakteriol. Parasitenkd Infektionskr Hyg Abt I *113*: 219–225.

Subramaniam, S., K.L. Chua, H.M. Tan, H. Loh, P. Kuhnert and J. Frey. 1997. Phylogenetic position of *Riemerella anatipestifer* based on 16S rRNA gene sequences. Int. J. Syst. Bacteriol. *47*: 562–565.

Sutter, V.L., D. Pyeatt and Y.Y. Kwok. 1981. *In vitro* susceptibility of *Capnocytophaga* strains to 18 antimicrobial agents. Antimicrob. Agents Chemother. *20*: 270–271.

Suzuki, M., Y. Nakagawa, S. Harayama and S. Yamamoto. 2001. Phylogenetic analysis and taxonomic study of marine *Cytophaga*-like bacteria: proposal for *Tenacibaculum* gen. nov. with *Tenacibaculum maritimum* comb. nov. and *Tenacibaculum ovolyticum* comb. nov., and description of *Tenacibaculum mesophilum* sp. nov. and *Tenacibaculum amylolyticum* sp. nov. Int. J. Syst. Evol. Microbiol. *51*: 1639–1652.

Syed, S.A. and W.J. Loesche. 1972. Survival of human dental plaque flora in various transport media. Appl. Microbiol. *24*: 638–644.

Szalay, D., R. Glavits, C. Nemes, A. Kosa and L. Fodor. 2002. Clinical signs and mortality caused by *Ornithobacterium rhinotracheale* in turkey flocks. Acta Vet. Hung. *50*: 297–305.

Tai, C.J., H.P. Kuo, F.L. Lee, H.K. Chen, A. Yokota and C.C. Lo. 2006. *Chryseobacterium taiwanense* sp. nov., isolated from soil in Taiwan. Int. J. Syst. Evol. Microbiol. *56*: 1771–1776.

Takeshita, T., T. Hanioka, H. Tamagawa, C. Hsieh, S. Shizukuishi and Y. Yamamoto. 1983. Neuraminidase activity of some oral anaerobic bacteria. J. Osaka Univ. Dent. Sch. *23*: 87–92.

Takeuchi, M. and A. Yokota. 1992. Proposals of *Sphingobacterium faecium* sp. nov., *Sphingobacterium piscium* sp. nov., *Sphingobacterium heparinum* comb. nov., *Sphingobacterium thalpophilum* comb. nov. and two genospecies of the genus *Sphingobacterium*, and synonymy of *Flavobacterium yabuuchiae* and *Sphingobacterium spiritivorum*. J. Gen. Appl. Microbiol. *38*: 465–482.

Tamaki, H., S. Hanada, Y. Kamagata, K. Nakamura, N. Nomura, K. Nakano and M. Matsumura. 2003. *Flavobacterium limicola* sp. nov., a psychrophilic, organic-polymer-degrading bacterium isolated from freshwater sediments. Int. J. Syst. Evol. Microbiol. *53*: 519–526.

Tanahashi, T., J. Murase, K. Matsuya, M. Hayashi, M. Kimura and S. Asakawa. 2005. Bacterial communities responsible for the decomposition of rice straw compost in a Japanese rice paddy field estimated by DGGE analysis of amplified 16S rDNA and 16S rRNA fragments. Soil Sci. Plant Nutr. *51*: 351–360.

Tanyi, J., A. Bistyak, E. Kaszanyitzky, F. Vetesi and M. Dobos-Kovacs. 1995. Isolation of *Ornithobacterium rhinotracheale* from chickens, hens and turkeys showing respiratory symptoms. Preliminary report. Magyar Allatorvosok Lapja *50*: 328–330.

Tatum, H.W., W.H. Ewing and R.E. Weaver. 1974. Miscellaneous Gram-negative bacteria. *In* Manual of Clinical Microbiology, 2nd edn (edited by Lenette, Spalding and Truant). American Society for Microbiology, Washington, DC, pp. 270–294.

Tekerekoglu, M.S., R. Durmaz, M. Ayan, Z. Cizmeci and A. Akinci. 2003. Analysis of an outbreak due to *Chryseobacterium meningosepticum* in a neonatal intensive care unit. Microbiologica *26*: 57–63.

Thompson, J.D., D.G. Higgins and T.J. Gibson. 1994. CLUSTAL W: improving the sensitivity of progressive multiple sequence alignment through sequence weighting, position-specific gap penalties and weight matrix choice. Nucleic Acids Res. *22*: 4673–4680.

Thomson, K.S., T.A. McMeekin and C.J. Thomas. 1981. Electron Microscopic Observations of *Flavobacterium aquatile* NCIB 8694 (=ATCC 11947) and *Flavobacterium meningosepticum* NCTC 10016 (=ATCC 13253). Int. J. Syst. Bacteriol. *31*: 226–231.

Tizer, K.B., J.S. Cervia, A.M. Dunn, J.J. Stavola and G.J. Noel. 1995. Successful combination vancomycin and rifampin therapy in a newborn with community-acquired *Flavobacterium meningosepticum* neonatal meningitis. Pediatr. Infect. Dis. J. *14*: 916–917.

Topley, W.W.C. and G.S. Wilson. 1929. The Principles of Bacteriology and Immunity, vol. 1. Edward Arnold, London.

Toyama, T., K. Kita-Tsukamoto and H. Wakabayashi. 1996. Identification of *Flexibacter maritimus, Flavobacterium branchiophilum* and *Cytophaga columnaris* by PCR targeted 16S ribosomal DNA. Fish Pathol. *31*: 25–31.

Travers, A.F. 1996. Concomitant *Ornithobacterium rhinotracheale* and Newcastle disease infection in broilers in South Africa. Avian Dis. *40*: 488–490.

Travers, A.F., L. Coetzee and B. Gummow. 1996. Pathogenicity differences between South African isolates of *Ornithobacterium rhinotracheale*. Onderstepoort J. Vet. Res. *63*: 197–207.

Triyanto, A.K. and H. Wakabayashi. 1999a. Genotypic diversity of strains of *Flavobacterium columnare* from diseased fishes. Fish Pathol. *34*: 65–71.

Triyanto, A.K. and H. Wakabayashi. 1999b. The use of PCR targeted 16S rDNA for identification of genomovars of *Flavobacterium columnare*. Fish Pathol. *34*: 217–218.

Tsavkelova, E.A., T.A. Cherdyntseva, S.G. Botina and A.I. Netrusov. 2007. *Bacteria* associated with orchid roots and microbial production of auxin. Microbiol. Res. *162*: 69–76.

Turan, N. and S. Ak. 2002. Investigation of the presence of *Ornithobacterium rhinotracheale* in chickens in Turkey and determination of the seroprevalance of the infection using the enzyme-linked immunosorbent assay. Avian Dis. *46*: 442–446.

Turtura, G.C., A. Perfetto and P. Lorenzelli. 2000. Microbiological investigation on black crusts from open-air stone monuments of Bologna (Italy). New Microbiol. *23*: 207–228.

Uddin, M.N. and H. Wakabayashi. 1997. Effects of temperature on growth and protease production of *Cytophaga psychrophila*. Fish Pathol. *32*: 225–226.

Ultee, A., N. Souvatzi, K. Maniadi and H. Konig. 2004. Identification of the culturable and nonculturable bacterial population in ground water of a municipal water supply in Germany. J. Appl. Microbiol. *96*: 560–568.

Ursing, J. and B. Bruun. 1987. Genetic heterogeneity of *Flavobacterium meningosepticum* demonstrated by DNA–DNA hybridization. Acta Pathol. Microbiol. Immunol. Scand. [B] *95*: 33–39.

Ursing, J. and B. Bruun. 1991. Genotypic heterogeneity of *Flavobacterium* group IIb and *Flavobacterium breve* demonstrated by DNA–DNA hybridization. APMIS *99*: 780–786.

Valtonen, M., A. Lauhio, P. Carlson, J. Multanen, A. Sivonen, M. Vaara and J. Lahdevirta. 1995. *Capnocytophaga canimorsus* septicemia: fifth report of a cat-associated infection and five other cases. Eur. J. Clin. Microbiol. Infect. Dis. *14*: 520–523.

van Beek, P.N., P.C. van Empel, G. van den Bosch, P.K. Storm, J.H. Bongers and J.H. du Preez. 1994. [Respiratory problems, growth retardation and arthritis in turkeys and broilers caused by a *Pasteurella*-like organism: *Ornithobacterium rhinotracheale* or 'Taxon 28']. Tijdschr Diergeneeskd *119*: 99–101.

Van der Meulen, H.J., W. Harder and H. Veldkamp. 1974. Isolation and characterization of *Cytophaga flevensis* sp. nov. a new agarolytic flexibacterium. Antonie van Leeuwenhoek *40*: 329–346.

van Empel, P. 1998. *Ornithobacterium rhinotracheale*. PhD thesis, University of Utrecht, Utrecht.

van Empel, P. and H. van den Bosch. 1998. Vaccination of chickens against *Ornithobacterium rhinotracheale* infection. Avian Dis. *42*: 572–578.

van Empel, P. and H.M. Hafez. 1999. *Ornithobacterium rhinotracheale*: a review. Avian Pathol. *28*: 217–227.

van Empel, P., H. van den Bosch, D. Goovaerts and P. Storm. 1996. Experimental infection in turkeys and chickens with *Ornithobacterium rhinotracheale*. Avian Dis. *40*: 858–864.

van Empel, P., H. van den Bosch, P. Loeffen and P. Storm. 1997. Identification and serotyping of *Ornithobacterium rhinotracheale*. J. Clin. Microbiol. *35*: 418–421.

van Empel, P., P. Savelkoul, R. Segers, J. Stoof, P. Loeffen and H. Van den Bosch. 1998. Molecular characterization of *Ornithobacterium rhinotracheale*. Avian Dis. *42*: 567–578.

van Empel, P., M. Vrijenhoek, D. Goovaerts and H. Van den Bosch. 1999. Immuno-histochemical and serological investigation of experimental *Ornithobacterium rhinotracheale* infection in chickens. Avian Pathol. *28*: 187–193.

Van Loock, M., T. Geens, L. De Smit, H. Nauwynck, P. van Empel, C. Naylor, H.M. Hafez, B.M. Goddeeris and D. Vanrompay. 2005. Key role of *Chlamydophila psittaci* on Belgian turkey farms in association with other respiratory pathogens. Vet. Microbiol. *107*: 91–101.

van Palenstein Helderman, W.H. 1981. Microbial etiology of periodontal disease. J. Clin. Periodontol. *8*: 261–280.

Van Trappen, S., J. Mergaert, S. Van Eygen, P. Dawyndt, M.C. Cnockaert and J. Swings. 2002. Diversity of 746 heterotrophic bacteria isolated from microbial mats from ten Antarctic lakes. Syst. Appl. Microbiol. *25*: 603–610.

Van Trappen, S., J. Mergaert and J. Swings. 2003. *Flavobacterium gelidilacus* sp. nov., isolated from microbial mats in Antarctic lakes. Int. J. Syst. Evol. Microbiol. *53*: 1241–1245.

Van Trappen, S., I. Vandecandelaere, J. Mergaert and J. Swings. 2004a. *Gillisia limnaea* gen. nov., sp. nov., a new member of the family *Flavobacteriaceae* isolated from a microbial mat in Lake Fryxell, Antarctica. Int. J. Syst. Evol. Microbiol. *54*: 445–448.

Van Trappen, S., I. Vandecandelaere, J. Mergaert and J. Swings. 2004b. *Flavobacterium degerlachei* sp. nov., *Flavobacterium frigoris* sp. nov. and *Flavobacterium micromati* sp. nov., novel psychrophilic bacteria isolated from microbial mats in Antarctic lakes. Int. J. Syst. Evol. Microbiol. *54*: 85–92.

Van Trappen, S., I. Vandecandelaere, J. Mergaert and J. Swings. 2005. *Flavobacterium fryxellicola* sp. nov. and *Flavobacterium psychrolimnae* sp. nov., novel psychrophilic bacteria isolated from microbial mats in Antarctic lakes. Int. J. Syst. Evol. Microbiol. *55*: 769–772.

van Veen, L., E. Gruys, K. Frik and P. van Empel. 2000a. Increased condemnation of broilers associated with *Ornithobacterium rhinotracheale*. Vet. Rec. *147*: 422–423.

van Veen, L., C.P. van Empel and T. Fabria. 2000b. *Ornithobacterium rhinotracheale*, a primary pathogen in broilers. Avian Dis. *44*: 896–900.

van Veen, L., E. Hartman and T. Fabri. 2001. *In vitro* antibiotic sensitivity of strains of *Ornithobacterium rhinotracheale* isolated in The Netherlands between 1996 and 1999. Vet. Rec. *149*: 611–613.

van Veen, L., M. Vrijenhoek and P. van Empel. 2004. Studies of the transmission routes of *Ornithobacterium rhinotracheale* and immunoprophylaxis to prevent infection in young meat turkeys. Avian Dis. *48*: 233–237.

Vancanneyt, M., P. Segers, L. Hauben, J. Hommez, L.A. Devriese, B. Hoste, P. Vandamme and K. Kersters. 1994. *Flavobacterium meningosepticum*, a pathogen in birds. J. Clin. Microbiol. *32*: 2398–2403.

Vancanneyt, M., P. Segers, U. Torck, B. Hoste, J.-F. Bernardet, P. Vandamme and K. Kersters. 1996. Reclassification of *Flavobacterium odoratum* (Stutzer 1929) strains to a new genus, *Myroides*, as *Myroides odoratus* comb. nov. and *Myroides odoratimimus* sp. nov. Int. J. Syst. Bacteriol. *46*: 926–932.

Vancanneyt, M., P. Vandamme, P. Segers, U. Torck, R. Coopman, K. Kersters and K.H. Hinz. 1999. *Riemerella columbina* sp. nov., a bacterium associated with respiratory disease in pigeons. Int. J. Syst. Bacteriol. *49*: 289–295.

Vandamme, P., J.-F. Bernardet, P. Segers, K. Kersters and B. Holmes. 1994a. New perspectives in the classification of the *Flavobacteria*: description of *Chryseobacterium* gen. nov., *Bergeyella* gen. nov., and *Empedobacter* nom. rev. Int. J. Syst. Bacteriol. *44*: 827–831.

Vandamme, P., P. Segers, M. Vancanneyt, K. Van Hove, R. Mutters, J. Hommez, F. Dewhirst, B. Paster, K. Kersters, E. Falsen, L.A. Devriese, M. Bisgaard, K.H. Hinz and W. Mannheim. 1994b. *Ornithobacterium rhinotracheale* gen. nov. sp. nov. isolated from the avian respiratory tract. Int. J. Syst. Bacteriol. *44*: 24–37.

Vandamme, P., M. Vancanneyt, A. van Belkum, P. Segers, W.G. Quint, K. Kersters, B.J. Paster and F.E. Dewhirst. 1996a. Polyphasic analysis of strains of the genus *Capnocytophaga* and Centers for Disease Control group DF-3. Int. J. Syst. Bacteriol. *46*: 782–791.

Vandamme, P., M. Vancanneyt, A. VanBelkum, P. Segers, W.G.V. Quint, K. Kersters, B.J. Paster and F.E. Dewhirst. 1996b. Polyphasic analysis of strains of the genus *Capnocytophaga* and centers for disease control group DF-3. Int. J. Syst. Bacteriol. *46*: 782–791.

Vandamme, P., M. Vancanneyt, P. Segers, M. Ryll, B. Köhler, W. Ludwig and K.H. Hinz. 1999. *Coenonia anatina* gen. nov., sp. nov., a novel bacterium associated with respiratory disease in ducks and geese. Int. J. Syst. Bacteriol. *49*: 867–874.

Vandekerchove, D., P. DeHerdt, H. Laevens, P. Butaye, G. Meulemans and F. Pasmans. 2004. Significance of interactions between *Escherichia coli* and respiratory pathogens in layer hen flocks suffering from colibacillosis-associated mortality. Avian Pathol. *33*: 298–302.

Vaneechoutte, M., P. Kampfer, T. De Baere, V. Avesani, M. Janssens and G. Wauters. 2007. *Chryseobacterium hominis* sp. nov., to accommodate clinical isolates biochemically similar to CDC groups II-h and II-c. Int. J. Syst. Evol. Microbiol. *57*: 2623–2628.

Varga, J., L. Fodor and L. Makrai. 2001. Characterization of some *Ornithobacterium rhinotracheale* strains and examination of their transmission via eggs. Acta Vet. Hung. *49*: 125–130.

Vela, A.I., A. Fernandez, C. Sanchez-Porro, E. Sierra, M. Mendez, M. Arbelo, A. Ventosa, L. Dominguez and J.F. Fernández-Garayzábal. 2007. *Flavobacterium ceti* sp. nov., isolated from beaked whales (*Ziphius cavirostris*). Int. J. Syst. Evol. Microbiol. *57*: 2604–2608.

Venter, H., G. Osthoff and D. Litthauer. 1999. Purification and characterization of a metalloprotease from *Chryseobacterium indologenes* Ix9a and determination of the amino acid specificity with electrospray mass spectrometry. Protein Expr. Purif. *15*: 282–295.

Vinogradov, E., M.B. Perry and W.W. Kay. 2003. The structure of the glycopeptides from the fish pathogen *Flavobacterium columnare*. Carbohydr. Res. *338*: 2653–2658.

Vishnivetskaya, T.A., M.A. Petrova, J. Urbance, M. Ponder, C.L. Moyer, D.A. Gilichinsky and J.M. Tiedje. 2006. Bacterial community in ancient Siberian permafrost as characterized by culture and culture-independent methods. Astrobiology *6*: 400–414.

Von Graevenitz, A. 1981. Clinical significance and antimicrobial susceptibility of *Flavobacteria*. In The *Flavobacterium-Cytophaga* Group. Proceedings of the International Symposium on Yellow-Pigmented Gram-Negative *Bacteria* of the *Flavobacterium-Cytophaga* Group, Braunschweig, July 8 to 11, 1980 (edited by Reichenbach and Weeks). Verlag Chemie, Weinheim, pp. 153–164.

Von Graevenitz, A. 1985. Ecology, clinical significance and antimicrobial susceptibility of infrequently encountered glucose-nonfermenting Gram-negative rods. *In* Non-fermentative Gram-negative Rods: Laboratory Identification and Clinical Aspects (edited by Gilardi). Marcel Dekker, New York, pp. 181–232.

von Graevenitz, A. 1990. Revised nomenclature of *Campylobacter laridis*, *Enterobacter intermedium*, and "*Flavobacterium branchiophila*". Int. J. Syst. Bacteriol. *40*: 211.

Vozniakovskaia Iu, M. and Z.P. Rybakova. 1969. [Some new data on the ecology and properties of a bacterium belonging to the *Promyxobacterium* genus]. Mikrobiologiya *38*: 135–142.

Wadowsky, R.M. and R.B. Yee. 1983. Satellite growth of *Legionella pneumophila* with an environmental isolate of *Flavobacterium breve*. Appl. Environ. Microbiol. *46*: 1447–1449.

Wakabayashi, H. 1991. Effects of environmental conditions on the infectivity of *Flexibacter columnaris* to fish. J. Fish Dis. *14*: 279–290.

Wakabayashi, H., M. Hikida and K. Masumura. 1986. *Flexibacter maritimus* sp. nov., a pathogen of marine fishes. Int. J. Syst. Bacteriol. *36*: 396–398.

Wakabayashi, H., G.J. Huh and N. Kimura. 1989. *Flavobacterium branchiophila* sp. nov., a causative agent of bacterial gill disease of freshwater fishes. Int. J. Syst. Bacteriol. *39*: 213–216.

Walker, V.K., G.R. Palmer and G. Voordouw. 2006. Freeze-thaw tolerance and clues to the winter survival of a soil community. Appl. Environ. Microbiol. *72*: 1784–1792.

Wang, Z.W., Y.H. Liu, X. Dai, B.J. Wang, C.Y. Jiang and S.J. Liu. 2006. *Flavobacterium saliperosum* sp. nov., isolated from freshwater lake sediment. Int. J. Syst. Evol. Microbiol. *56*: 439–442.

Wayne, L.G. 1982. Actions of the Judicial Commission of the International Committee on Systematic Bacteriology on Requests for Opinions published between July 1979 and April 1981. Int. J. Syst. Bacteriol. *32*: 464–465.

Wayne, L.G., D.J. Brenner, R.R. Colwell, P.A.D. Grimont, O. Kandler, M.I. Krichevsky, L.H. Moore, W.E.C. Moore, R.G.E. Murray, E. Stackebrandt, M.P. Starr and H.G. Trüper. 1987. Report of the ad hoc committee on the reconciliation of approaches to bacterial systematics. Int. J. Syst. Bacteriol. *37*: 463–464.

Weeks, O.B. 1955. *Flavobacterium aquatile* (Frankland and Frankland) Bergey et al., type species of the genus *Flavobacterium*. J. Bacteriol. *69*: 649–658.

Weeks, O.B. 1974. Genus *Flavobacterium* Bergey et al. 1923, 97. *In* Bergey's Manual of Determinative Bacteriology, 8th edn (edited by Buchanan and Gibbons). Williams & Wilkins, Baltimore, pp. 357–364.

Weeks, O.B. 1981. The genus *Flavobacterium*. *In* The Prokaryotes: A Handbook on the Biology of Bacteria: Ecophysiology, Isolation, Identification, Applications, vol. 1 (edited by Starr, Stolp, Trüper, Balows and Schlegel). Springer, New York, pp. 1365–1370.

Weeks, O.B. and R.S. Breed. 1957. Genus III. *Flavobacterium* Bergey et al. 1923. *In* Bergey's Manual of Determinative Bacteriology, 7th edn (edited by Breed, Murray and Smith). Williams & Wilkins, Baltimore, pp. 309–322.

Welthagen, J.J. and P.J. Jooste. 1992. Isolasie en karakterisering van gepigmenteerde psigrotrofe bakterieë uit verkoelde roumelk. S. Afr. J. Dairy Sci. *24*: 47–52.

Weon, H.Y., B.Y. Kim, S.H. Yoo, S.W. Kwon, Y.H. Cho, S.J. Go and E. Stackebrandt. 2006. *Chryseobacterium wanjuense* sp. nov., isolated from greenhouse soil in Korea. Int. J. Syst. Evol. Microbiol. *56*: 1501–1504.

Weon, H.Y., M.H. Song, J.A. Son, B.Y. Kim, S.W. Kwon, S.J. Go and E. Stackebrandt. 2007. *Flavobacterium terrae* sp. nov. and *Flavobacterium cucumis* sp. nov., isolated from greenhouse soil. Int. J. Syst. Evol. Microbiol. *57*: 1594–1598.

Weon, H.Y., B.Y. Kim, S.H. Yoo, S.W. Kwon, E. Stackebrandt and S.J. Go. 2008. *Chryseobacterium soli* sp. nov. and *Chryseobacterium jejuense* sp. nov., isolated from soil samples from Jeju, Korea. Int. J. Syst. Evol. Microbiol. *58*: 470–473.

Wery, N., U. Gerike, A. Sharman, J.B. Chaudhuri, D.W. Hough and M.J. Danson. 2003. Use of a packed-column bioreactor for isolation of diverse protease-producing bacteria from antarctic soil. Appl. Environ. Microbiol. *69*: 1457–1464.

Weyant, R.S., C.W. Moss, R.E. Weaver, D.G. Hollis, J.G. Jordan, E.C. Cook and M.I. Daneshvar. 1996. Identification of Unusual Pathogenic Gram-negative Aerobic and Facultatively Anaerobic Bacteria. Williams & Wilkins, Baltimore.

Williams, B.L. and B.F. Hammond. 1979. *Capnocytophaga*: new genus of Gram-negative gliding bacteria. 4. DNA base composition and sequence homology. Arch. Microbiol. *122*: 35–39.

Williams, B.L., D. Hollis and L.V. Holdeman. 1979. Synonymy of strains of Center for Disease Control group DF-1 with species of *Capnocytophaga*. J. Clin. Microbiol. *10*: 550–556.

Wilson, S.L., D.L. Kelley and V.K. Walker. 2006. Ice-active characteristics of soil bacteria selected by ice-affinity. Environ. Microbiol. *8*: 1816–1824.

Winn, R.E., W.F. Chase, P.W. Lauderdale and F.K. McCleskey. 1984. Septic arthritis involving *Capnocytophaga ochracea*. J. Clin. Microbiol. *19*: 538–540.

Wooley, R.E., P.S. Gibbs, T.P. Brown and J.J. Maurer. 2000. Chicken embryo lethality assay for determining the virulence of avian *Escherichia coli* isolates. Avian Dis. *44*: 318–324.

Xiang, S., T. Yao, L. An, B. Xu and J. Wang. 2005. 16S rRNA sequences and differences in bacteria isolated from the Muztag Ata glacier at increasing depths. Appl. Environ. Microbiol. *71*: 4619–4627.

Xiang, H., G.F. Wei, S. Jia, J. Huang, X.X. Miao, Z. Zhou, L.P. Zhao and Y.P. Huang. 2006. Microbial communities in the larval midgut of laboratory and field populations of cotton bollworm (Helicoverpa armigera). Can. J. Microbiol. *52*: 1085–1092.

Yabuuchi, E., E. Tanimura, A. Ohyama, I. Yano and A. Yamamoto. 1979. *Flavobacterium devorans* ATCC 10829: a strain of *Pseudomonas paucimobilis*. J. Gen. Appl. Microbiol. *25*: 95–107.

Yabuuchi, E., T. Kaneko, I. Yano, C.W. Moss and N. Miyoshi. 1983. *Sphingobacterium* gen. nov., *Sphingobacterium spiritivorum* comb. nov., *Sphingobacterium multivorum* comb. nov., *Sphingobacterium mizutae* sp. nov., and *Flavobacterium indologenes* sp. nov., glucose-nonfermenting Gram-negative rods in CDC group IiK-2 and group IIb. Int. J. Syst. Bacteriol. *33*: 580–598.

Yabuuchi, E., Y. Hashimoto, T. Ezaki, Y. Ido and N. Takeuchi. 1990. Genotypic and phenotypic differentiation of *Flavobacterium indologenes* Yabuuchi et al. 1983 from *Flavobacterium gleum* Holmes et al. 1984. Microbiol. Immunol. *34*: 73–76.

Yağci, A., N. Çerikçoğlu, M.E. Kaufmann, H. Malnick, G. Söyletir, F. Babacan and T.L. Pitt. 2000. Molecular Typing of *Myroides odoratimimus* (*Flavobacterium odoratum*) urinary tract infections in a Turkish hospital. Eur. J. Clin. Microbiol. Infect. Dis. *19*: 731–732.

Yamaguchi, S. and M. Yokoe. 2000. A novel protein-deamidating enzyme from *Chryseobacterium proteolyticum* sp. nov., a newly isolated bacterium from soil. Appl. Environ. Microbiol. *66*: 3337–3343.

Yamaguchi, S., D.J. Jeenes and D.B. Archer. 2001. Protein-glutaminase from *Chryseobacterium proteolyticum*, an enzyme that deamidates glutaminyl residues in proteins. Eur. J. Biochem. *268*: 1410–1421.

Yamamoto, T., S. Kajiura, Y. Hirai and T. Watanabe. 1994. *Capnocytophaga haemolytica* sp. nov. and *Capnocytophaga granulosa* sp. nov., from human dental plaque. Int. J. Syst. Bacteriol. *44*: 324–329.

Yang, S.J., Y.J. Choo and J.C. Cho. 2007. *Lutimonas vermicola* gen. nov., sp. nov., a member of the family *Flavobacteriaceae* isolated from the marine polychaete *Periserrula leucophryna*. Int. J. Syst. Evol. Microbiol. *57*: 1679–1684.

Yermakova, S.P., V.V. Sova and T.N. Zvyagintseva. 2002. Brown seaweed protein as an inhibitor of marine mollusk endo-(1→3)-β-D-glucanases. Carbohydr. Res. *337*: 229–237.

Yi, H. and J. Chun. 2006. *Flavobacterium weaverense* sp. nov. and *Flavobacterium segetis* sp. nov., novel psychrophiles isolated from the Antarctic. Int. J. Syst. Evol. Microbiol. *56*: 1239–1244.

Yi, H., H.M. Oh, J.H. Lee, S.J. Kim and J. Chun. 2005a. *Flavobacterium antarcticum* sp. nov., a novel psychrotolerant bacterium isolated from the Antarctic. Int. J. Syst. Evol. Microbiol. *55*: 637–641.

Yi, H., H.I. Yoon and J. Chun. 2005b. *Sejongia antarctica* gen. nov., sp. nov. and *Sejongia jeonii* sp. nov., isolated from the Antarctic. Int. J. Syst. Evol. Microbiol. *55*: 409–416.

Yin, Z.-X., W. He, W.-J. Chen, J.-H. Yan, J.-N. Yang, S.-M. Chan and J.-G. He. 2006. Cloning, expression and antimicrobial activity of an antimicrobial peptide, epinecidin-1, from the orange-spotted grouper, *Epinephelus coioides*. Aquaculture *253*: 204–211.

Ying, J.Y., Z.P. Liu, B.J. Wang, X. Dai, S.S. Yang and S.J. Liu. 2007. *Salegentibacter catena* sp. nov., isolated from sediment of the South China Sea, and emended description of the genus *Salegentibacter*. Int. J. Syst. Evol. Microbiol. *57*: 219–222.

Yoon, J.H., S.J. Kang, C.H. Lee and T.K. Oh. 2005a. *Dokdonia donghaensis* gen. nov., sp. nov., isolated from sea water. Int. J. Syst. Evol. Microbiol. *55*: 2323–2328.

Yoon, J.H., S.J. Kang, S.Y. Lee, C.H. Lee and T.K. Oh. 2005b. *Maribacter dokdonensis* sp. nov., isolated from sea water off a Korean island, Dokdo. Int. J. Syst. Evol. Microbiol. *55*: 2051–2055.

Yoon, J.H., S.J. Kang and T.K. Oh. 2005c. *Tenacibaculum lutimaris* sp. nov., isolated from a tidal flat in the Yellow Sea, Korea. Int. J. Syst. Evol. Microbiol. *55*: 793–798.

Yoon, J.H., M.H. Lee, T.K. Oh and Y.H. Park. 2005d. *Muricauda flavescens* sp. nov. and *Muricauda aquimarina* sp. nov., isolated from a salt lake near Hwajinpo Beach of the East Sea in Korea, and emended description of the genus *Muricauda*. Int. J. Syst. Evol. Microbiol. *55*: 1015–1019.

Yoon, J., S. Maneerat, F. Kawai and A. Yokota. 2006a. *Myroides pelagicus* sp. nov., isolated from seawater in Thailand. Int. J. Syst. Evol. Microbiol. *56*: 1917–1920.

Yoon, J.H., S.J. Kang, S.Y. Jung, H.W. Oh and T.K. Oh. 2006b. *Gaetbulimicrobium brevivitae* gen. nov., sp. nov., a novel member of the family *Flavobacteriaceae* isolated from a tidal flat of the Yellow Sea in Korea. Int. J. Syst. Evol. Microbiol. *56*: 115–119.

Yoon, J.H., S.J. Kang, C.H. Lee and T.K. Oh. 2006c. *Donghaeana dokdonensis* gen. nov., sp. nov., isolated from sea water. Int. J. Syst. Evol. Microbiol. *56*: 187–191.

Yoon, J.H., S.J. Kang and T.K. Oh. 2006d. *Flavobacterium soli* sp. nov., isolated from soil. Int. J. Syst. Evol. Microbiol. *56*: 997–1000.

Yoon, J.H., S.J. Kang and T.K. Oh. 2006e. *Polaribacter dokdonensis* sp. nov., isolated from seawater. Int. J. Syst. Evol. Microbiol. *56*: 1251–1255.

Yoon, J.H., S.J. Kang, J.S. Lee and T.K. Oh. 2007a. *Flavobacterium terrigena* sp. nov., isolated from soil. Int. J. Syst. Evol. Microbiol. *57*: 947–950.

Yoon, J.H., S.J. Kang and T.K. Oh. 2007b. *Chryseobacterium daeguense* sp. nov., isolated from wastewater of a textile dye works. Int. J. Syst. Evol. Microbiol. *57*: 1355–1359.

Yoon, J.H., S.J. Kang, Y.T. Jung and T.K. Oh. 2008. *Aestuariicola saemankumensis* gen. nov., sp. nov., a member of the family *Flavobacteriaceae*, isolated from tidal flat sediment. Int. J. Syst. Evol. Microbiol. *58*: 2126–2131.

Young, C.C., P. Kämpfer, F.T. Shen, W.A. Lai and A.B. Arun. 2005. *Chryseobacterium formosense* sp. nov., isolated from the rhizosphere of *Lactuca sativa* L. (garden lettuce). Int. J. Syst. Evol. Microbiol. *55*: 423–426.

Yu, J. and O. Ward. 1996. Investigation of the biodegradation of pentachlorophenol by the predominant bacterial strains in a mixed culture. Int. Biodeter. Biodegrad. *5*: 181–187.

Zdanowski, M.K., P. Weglenski, P. Golik, J.M. Sasin, P. Borsuk, M.J. Zmuda and A. Stankovic. 2004. Bacterial diversity in Adélie penguin, *Pygoscelis adeliae*, guano: molecular and morpho-physiological approaches. FEMS Microbiol. Ecol. *50*: 163–173.

Zehr, W.J. and J. Ostendorf, Jr. 1970. Case report. *Pasteurella anatipestifer* in turkeys. Avian Dis. *14*: 557–560.

Zhang, C. and Y. Rikihisa. 2004. Proposal to transfer '*Aegyptianella ranarum*', an intracellular bacterium of frog red blood cells, to the family *Flavobacteriaceae* as '*Candidatus* Hemobacterium ranarum' comb. nov. Environ. Microbiol. *6*: 568–573.

Zhang, D.C., H.X. Wang, H.C. Liu, X.Z. Dong and P.J. Zhou. 2006. *Flavobacterium glaciei* sp. nov., a novel psychrophilic bacterium isolated from the China No. 1 glacier. Int. J. Syst. Evol. Microbiol. *56*: 2921–2925.

Zhou, L., R. Srisatjaluk, D.E. Justus and R.J. Doyle. 1998. On the origin of membrane vesicles in Gram-negative bacteria. FEMS Microbiol. Lett. *163*: 223–228.

Zhou, Y., J. Dong, X. Wang, X. Huang, K.Y. Zhang, Y.Q. Zhang, Y.F. Guo, R. Lai and W.J. Li. 2007. *Chryseobacterium flavum* sp. nov., isolated from polluted soil. Int. J. Syst. Evol. Microbiol. *57*: 1765–1769.

Zhu, F., S. Wang and P. Zhou. 2003. *Flavobacterium xinjiangense* sp. nov. and *Flavobacterium omnivorum* sp. nov., novel psychrophiles from the China No. 1 glacier. Int. J. Syst. Evol. Microbiol. *53*: 853–857.

ZoBell, C.E. and H.C. Upham. 1944. A list of marine bacteria including descriptions of sixty new species. Bull. Scripps Inst. Oceanogr. Univ. Calif. *5*: 239–292.

Zorman-Rojs, O., I. Zdovc, D. Bencina and I. Mrzel. 2000. Infection of turkeys with *Ornithobacterium rhinotracheale* and *Mycoplasma synoviae*. Avian Dis. *44*: 1017–1022.

Family II. **Blattabacteriaceae** fam. nov.

Srinivas Kambhampati

Blat.ta.bac.te.ri.a.ce′a.e. N.L. n. *Blattabacterium* type genus of the family; *-aceae* ending to denote a family; N.L. fem. pl. n. *Blattabacteriaceae* the *Blattabacterium* family.

Gram-stain-negative, slightly curved and straight rods that are intracellular symbionts of cockroaches and of one termite species. These endosymbiotic bacteria are found predominantly in specialized cells of insect fat bodies. Nonmotile and nonflagellated. They play a role in the conversion of uric acid metabolites into usable nitrogenous compounds under anaerobic conditions. The endosymbionts are transmitted vertically from mother to offspring. They possess highly reduced genomes, with an estimated genome size of approximately 638 kbp.

Type genus: **Blattabacterium** Hollande and Favre 1931, 754[AL].

Further comments

Although this family is represented by a single genus and species, the clades of *Blattabacterium cuenoti* from various cockroach families should likely be elevated to the species level. Clark and Kambhampati (2003) showed that the sequence divergence (up to 2.2%) in the 16S rRNA gene of the endosymbiont from six different species of the wood-feeding cockroach, *Cryptocercus*, is comparable to that reported for congeneric free-living and endosymbiotic bacterial species (see the genus description for details). It has long been assumed that all extant cockroach species harbor endosymbiotic bacteria, however, Lo et al. (2007) could not detect *Blattabacterium* in members of the cockroach genus *Nocticola*.

Genus I. **Blattabacterium** Hollande and Favre 1931, 754[AL]

Srinivas Kambhampati

Blat.ta.bac.te′ri.um. L. fem. n. *blatta* an insect that shuns the light, the cockroach, chafer, moth, etc.; L. neut. n. *bacterium* rod; N.L. neut. n. *Blattabacterium*, an endosymbiotic bacterium harbored by cockroaches and one termite species.

Blattabacterium are present only as **intracellular symbionts (endosymbionts)** in almost all species of cockroaches (Blattaria: Dictyoptera) studied so far and one species of termite, *Mastotermes darwiniensis* (Mastotermitidae: Isoptera). Lo et al. (2007) reported that they could not detect the presence of *Blattabacterium* in members of the cockroach genus *Nocticola*. **Plump, slightly curved and straight rods**, about 1 μm in diameter and 1.6–9.0 μm in length, usually with rounded ends. Length slightly variable among cockroach species. Within a cockroach species, longer forms occur inside specialized cells of abdominal fat body called **mycetocytes**; shorter forms occur in gonads and embryos. Binary fission pairs are common in mycetocytes. **Enveloped by host cell membranes.** Gram-stain-variable to Gram-stain-positive staining but has a **Gram-negative type cell-wall** structure composed of peptidoglycan surrounded by an outer membrane. **Nonmotile. Nonflagellated. Possesses enzymes of the tricarboxylic acid cycle and respiratory cytochromes.**

DNA G+C content (mol%): 26–28 (Bd).

Type species: **Blattabacterium cuenoti** (Mercier 1906) Hollande and Favre 1931, 754[AL] (*Bacillus cuenoti* Mercier 1906, 684).

Further descriptive information

The description of *Blattabacterium* presented in this chapter is taken from Dasch et al. (1984) with updates and modifications.

Phylogenetic treatment. In the 8th edition of *Bergey's Determinative Manual*, *Blattabacterium* was placed in the tribe *Wolbachieae*, family *Rickettsiaceae*, order *Rickettsiales* (Dasch et al., 1984). Since then, molecular phylogenetic studies have been undertaken on *Blattabacterium* and their relationship to free-living bacteria. Bandi et al. (1994) sequenced the 16S rRNA gene of *Blattabacterium* from one representative each of five families of cockroaches and undertook a phylogenetic analysis that included homologous sequences from several free-living bacteria. They concluded that *Blattabacterium* belongs to the *Flavobacterium–Bacteroides* group, which, in this second edition of the *Systematics Manual*, is denoted as the phylum "*Bacteroidetes*".

Cell morphology. The light microscopic and ultrastructural appearance of *Blattabacterium* varies somewhat among different cockroach species, in different tissues of a single species, and in response to the recurrent hormonal changes of the host's incomplete metamorphosis. The endosymbionts harbored by various species of cockroaches differ primarily in length and DNA sequence (see *Taxonomic comments*, below). In *Blatta orientalis* they are 2.5–5.3 μm long; in *Cryptocercus punctulatus*, 2.5–8.1 μm; in *Heterogamia*, 5.3–9.0 μm; in *Blatta aethiopica*, 1.6 μm; and in *Blatta germanica*, 3.0 μm (Buchner, 1965). The slightly curved, longer forms occur in the mycetocytes. The shorter forms occur in the ovaries and during embryonic development (Sacchi et al., 1996, 2000). Very small dense forms 0.3 × 1.0 μm are occasionally found mixed with the larger endosymbionts in the fat body mycetocytes of *Blaberus*, *Blattella*, *Byrsotria*, *Nauphoeta*, and *Periplaneta* (Meyer and Frank, 1960; Milburn, 1966; Sacchi et al., 1996). However, there is no indication that cockroaches and *Mastotermes darwiniensis* harbor more than one species of endosymbiont in their fat body cells (Bandi et al., 1994, 1995; Clark et al., 2001; Clark and Kambhampati, 2003; Sacchi et al., 1996, 2000; Sabree et al., 2009; Sanchez-Lopez et al., 2009). The large endosymbionts may stain Gram-positive or Gram-stain-variable; the differences are possibly due to their physiologic state. Unlike most Gram-stain-positive bacteria, dilute sodium dodecyl sulfate abolishes the Gram-positive staining property of the endosymbionts (Malke and Bartsch, 1966). Cytoplasmic vacuoles and cross-walls are evident in freshly isolated endosymbionts. The Feulgen reaction of the endosymbionts is positive but diffuse (Rizki, 1954). The endosymbionts are not acid-fast, are not encapsulated, have no flagella, and stain well with Giemsa and Delafield or Heidenhain's hematoxylin. They do not stain with Sudan dyes, the Baker acid hematin test for phospholipids, or periodic acid Schiff reagent for polysaccharides. *Blattabacterium* are not fluorescent or birefringent (Richards and Brooks, 1958). Poorly fixed specimens and endosymbionts isolated in hypotonic solutions may appear swollen and yeastlike. Nucleoid areas with associated metachromatic granules that divide simultaneously with transverse fission have been observed (Buchner, 1965). Ultrastructurally, these appear to correspond to the electron-dense bodies and complex mesosome-like structures that are often associated with division septa (Anderson, 1964; Brooks, 1970; Bush and Chapman, 1961; Daniel and Brooks, 1972; Gromov and Mamkaeva, 1980).

Cell-wall composition. Muramic acid and glucosamine have been identified in the cell wall of the endosymbionts of *Periplaneta americana*, indicating the occurrence of peptidoglycan (Daniel and Brooks, 1967; Sacchi et al., 1998). This is consistent with the finding of naturally occurring endosymbiont sphaeroplasts (Milburn, 1966) and the formation of sphaeroplasts by injection of lysozyme into the host (Daniel and Brooks, 1972). However, the precise mechanism of lysozyme sphaeroplast-forming ability is not clear, since denatured lysozyme is equally effective.

Fine structure. Other than size, ultrastructural differences are negligible in *Blattabacterium* found in the nymphal fat body mycetocytes of cockroach genera representative of five divergent families of Blattaria: *Blaberus*, *Blatta*, *Byrsotria*, *Cryptocercus*, *Nauphoeta*, and *Periplaneta* (Brooks, 1970). This is in contrast to the location of the mycetocytes themselves, which may be isolated (*Cryptocercus*, and in the termite, *Mastotermes*), loosely scattered in the fat body (*Pycnoscelus*, *Blatta aethiopica*), banded in a single (*Blatta orientalis*, *Rhicnoda*), double (*Nauphoeta*), or triple row (*Blattella*), or cluster (*Ectobia*), each surrounded by a monolayer of fat body cells (Buchner, 1965; Koch, 1967; Sacchi et al., 2000). The endosymbionts are housed singly or as binary fission pairs in mycetocytes enveloped by a host-derived vacuolar membrane (Bigliardi et al., 1989). Relative to the host cytoplasm, the endosymbiont cytoplasm is electron-dense, although it is otherwise unremarkable in its content of ribosomes, loose DNA fibrils, and mesosome-like membrane invaginations. The endosymbiont cell wall is usually thin (5–10 nm) for a Gram-stain-positive bacterium and, indeed, it has an ultrastructure characteristic of Gram-stain-negative bacteria (Anderson, 1964; Brooks, 1970; Daniel and Brooks, 1972). The additional host cell membrane might be responsible for the Gram-positive staining properties reported by Gromov and Mamkaeva (1980). The encapsulating host membranes are lost from the endosymbionts only during the extracellular phases of their transfer from the fat body mycetocytes to the oocyte cytoplasm to ensure their hereditary transmission (Anderson, 1964; Daniel, 1973). At this time the endosymbionts are completely enveloped by the highly microvillous surface of the oocyte oolemma until egg maturation is complete (Sacchi et al., 1996). A similar endosymbiont transfer occurs in the rudimentary ovaries of males, but some tissue tropism is involved since the testes are not infected (Brooks and Kurtti, 1972). During embryonic development, complex extracellular and transmembrane movements of the endosymbionts occur between the time of their initial penetration of the oolemma and ultimate sequestration in the larval mycetocytes (Buchner, 1965; Sacchi et al., 1996, 2000).

By the freeze-fracture technique, alterations occur in the disposition of membrane particles on the convex surface of the plasma membrane of the endosymbionts present at the oocyte-follicle interface, following changes in the levels of host juvenile hormone (Liu, 1973, 1974). An apparent hormone-related transformation of the usual rod-shaped endosymbionts into large rounded forms filled with concentric lamellar membranes during the later instar ecdyses and oocyte development of false ovoviviparous cockroaches has been described (Milburn, 1966).

Cultural characteristics and life cycle. Cultivation of *Blattabacterium in vitro* or in tissue culture cells has not been accomplished. The endosymbionts contribute nutritional factors to the host (Sabree et al., 2009). Maintenance of endosymbionts in host insect is inhibited by treatment with the antibiotics penicillin, streptomycin, chloramphenicol, chlortetracycline, and oxytetracycline, sulfathiazole, egg white lysozyme, incubation at 37°C, and the insecticide lindane, but not aerosporin. Numerous reports of successful cultivation of the cockroach endosymbionts have appeared along with studies on the biochemistry of these cultures (Brooks, 1970). Gier (1947) and Brooks and Richards (1966) have carefully reviewed the problems inherent in preventing contamination with the diverse normal cockroach flora and in fulfilling Koch's postulates for the cultivation of endosymbionts. For none of the microorganisms so far cultivated is there a convincing relationship to the endosymbionts, as measured by definitive criteria such as common antigens, DNA base composition, common distinctive metabolic properties, and enzymes. Limited extracellular culture of endosymbionts in the presence of embryo cell cultures has been reported (Landureau, 1966) but not intracellular growth of endosymbionts in either vertebrate or invertebrate tissue culture systems.

Blattabacterium are transmitted in eggs vertically from the mother to offspring of both sexes. Using electron microscopy, Sacchi et al. (1996, 1988, 1998, 2000) and Sacchi and Grigolo (1989) studied the transovarial transmission of *Blattabacterium* in cockroaches and in the termite *Mastotermes darwiniensis*. They observed *Blattabacterium* in young follicular cells and intracellular spaces of both cockroaches and *Mastotermes darwiniensis*. In neotenic reproductives of *Mastotermes darwiniensis*, however, bacteriocytes were firmly attached to the germ layer of adjacent ovarioles. The bacterial infection occurred in the oocytes. Sacchi et al. (2000) suggested that *Blattabacterium* are discharged from the bacteriocytes by exocytosis and then cross the ovariole sheath and follicular epithelium. Giorgi and Nordin (1994), in a study of endosymbionts during embryogenesis of *Blattella germanica*, suggested that *Blattabacterium* are involved in digestion of yolk granules and migration of vitellophages to the center of the yolk. However, no experimental evidence for the production of lytic enzymes by *Blattabacterium* exists. Endosymbionts in the embryos of *Blattella germanica*, *Periplaneta americana*, and *Mastotermes darwiniensis* are located in the periplasm just beneath the vitelline membrane, and the turnover of *Blattabacterium* in the successive stages of embryo development is similar in the previously mentioned host species (Sacchi et al., 2000). The studies by Sacchi and colleagues cited above cumulatively demonstrate that the complex pattern of transovarial transmission is similar in cockroaches and *Mastotermes darwiniensis*, further confirming the close phylogenetic relationship of the endosymbionts from cockroaches and *Mastotermes darwiniensis* (Bandi et al., 1995).

Metabolism and metabolic pathways. The precise role of *Blattabacterium* in cockroach nutrition and excretion remains unknown because of the inability to culture them outside the host. However, the endosymbionts are necessary for host survival because aposymbiotic cockroaches do not survive unless fed on specialized diets containing yeast or liver (Brooks, 1963; Dasch et al., 1984). *Blattabacterium* are likely play a role in the conversion of uric acid into usable nitrogenous compounds under anaerobic conditions (Chapman, 1998; Cochran, 1985; Wren and Cochran, 1987). Whereas most terrestrial insects are uricotelic and void uric acid, cockroaches store urates internally rather than excreting them (Cochran, 1985). Under conditions of high dietary nitrogen, urates, in the form of uric acid, accumulate in cockroach fat bodies. When such individuals are fed nitrogen-poor diets, internally stored urates are mobilized. Only trace amounts of urates are present in the excreta (Nation and Patton, 1961; Srivastava and Gupta, 1961). Thus, if urates are mobilized in periods of dietary nitrogen stress and not excreted, they must be utilized for metabolic purposes (Cochran, 1985). However, insects have a limited capacity for degradation of purine molecules, and pathways for conversion of uric acid to allantoin and allantoin to allantoic acid are not known from cockroach tissues (Cochran, 1975). This suggests that another component is involved, and fat bodies are a likely site because of their involvement in uric acid synthesis and storage. Since the discovery that mycetocytes are surrounded by urocytes with contact between cell membranes, it has been suggested *Blattabacterium* might be involved in uric acid metabolism. When cockroaches are rendered aposymbiotic, there is a 20-fold increase in uric acid content of fat bodies (Malke, 1964; Malke and Schwartz, 1966). Although this suggests endosymbionts are utilizing stored uric acid for synthesis of essential amino acids (Clarebrough et al., 2000) the full genome sequence of *Blattabacterium* (see below) indicated that no uricolytic enzymes are encoded by the *Blattabacterium* genome.

Because claims for the culture of *Blattabacterium* have not been substantiated, reports of the presence of guanase, uricase, urease, xanthine oxidase, malic and lactic dehydrogenases, and synthesis of ascorbic acid based on such cultures are questionable. However, direct histochemical studies on endosymbionts present in the mycetocytes and respirometric studies on isolated endosymbionts indicate that the endosymbionts do reduce tetrazolium dyes, succinate, and cytochrome *c* while they consume oxygen. These results suggest that they have a complete set of respiratory cytochromes and an aerobic metabolism. The endosymbionts also transaminate alanine to pyruvate and glutamate to aspartate (Brooks, 1970). In many other studies, fractions enriched in endosymbionts have been compared with similarly prepared fractions from cockroaches that had been freed of their endosymbionts. Although the greater glycolytic and tricarboxylic acid cycle activities of such enriched endosymbiont fractions have been attributed to the endosymbionts (Laudani et al., 1974), the endosymbionts might merely eliminate an inhibitor that depresses such host activities. Since only the tissues containing endosymbionts exhibit significant respiration in the presence of added uric acid, it was assumed that the endosymbionts might play an important role in the uric acid catabolism of the cockroach. Recent evidence (Sabree et al., 2009; Sanchez-Lopez et al., 2009) suggested that whereas *Blattabacterium* utilize breakdown products of uric acid, they are not directly involved in uric acid catabolism.

Symbiotic insects, even when deprived of intestinal bacteria, but not aposymbiotic cockroaches (derived from chlortetracycline-treated adults), are able to synthesize labeled cysteine, glutathione, methionine, and taurine from ^{35}S-sulfate and to transfer labeled sulfur from cysteine to methionine (Block and Henry, 1961). Furthermore, only symbiotic insects can label the essential amino acids tyrosine, phenylalanine, isoleucine, valine, threonine, and arginine to a significant level from [U-^{14}C]glucose; the nonessential amino acids are labeled as well (Henry, 1962). The lighter melanization of aposymbiotic cockroaches may in part be due to their inability to synthesize tyrosine. Similarly, significant aromatic ring cleavage of phenylalanine occurs in symbiotic but not in aposymbiotic cockroaches, although this is not a major metabolic pathway even in the symbiotic cockroaches (Murdock et al., 1970).

Recently, Sabree et al. (2009) obtained the full genome sequence of *Blattabacterium* from the American cockroach, *Periplaneta americana* and Sanchez-Lopez et al. (2009) reported the full genome sequence of *Blattabacterium* from the German cockroach, *Blattella germanica*. The genome sizes of *Blattabacterium* from both host species were similar to each other at approximately 638 kbp; these small genomes are consistent with the finding that all insect endosymbionts studied to date also exhibit highly reduced genome sizes relative to free-living bacteria (Wernegreen, 2002). A major difference between *Blattabacterium* from the two host species is that whereas the one from the American cockroach contained a 3.5 kbp plasmid, no plasmid was detected in *Blattabacterium* from the German cockroach. The genome sequence confirmed that *Blattabacterium* is a member of *Flavobacteriales* (*Bacteroidetes*) and its closest living relative is *Sulcia muelleri*, an endosymbiont commonly found in sap-feeding insects (McCutcheon et al., 2009).

Sabree et al. (2009) and Sanchez-Lopez et al. (2009) reconstructed the metabolic pathways from the genomic sequence and found that although *Blattabacterium* lacks uricolytic enzymes it can recycle nitrogen from urea and ammonia, which are uric acid degradation products, into glutamate, using urease and glutamate dehydrogenase. They also suggested that *Blattabacterium* can subsequently produce all of the essential amino acids, various vitamins, and other compounds from a limited set of metabolic substrates. The metabolic pathways inferred for *Blattabacterium* from the German cockroach were highly similar to those inferred for *Blattabacterium* from the American cockroach (Sanchez-Lopez et al., 2009). Thus, the full genome sequence of *Blattabacterium* confirms its suspected role in nitrogen metabolism, including nitrogen recycling from ammonia and urea and extensive capabilities for amino acid biosynthesis.

Genetics. The *Blattabacterium* 16S rRNA gene has been sequenced from a number of cockroach species and from *Mastotermes darwiniensis* (Bandi et al., 1994, 1995; Clark et al., 2001; Clark and Kambhampati, 2003; Lo et al., 2007, 2005a; Maekawa et al., 2005b). Clark et al. (2001) and Clark and Kambhampati (2003) also sequenced the 23S rRNA gene from various *Cryptocercus* species (see *Taxonomic comments*, below).

Many insect endosymbionts, representing diverse bacterial lineages, have been shown to have a highly reduced genome size (Wernegreen, 2002). In general, insect endosymbiont genomes exhibit about 80% reduction relative to that of *Escherichia coli*. *Blattabacterium* also has a highly reduced genome that has been estimated to be about 638 kbp (Sabree et al., 2009; Sanchez-Lopez et al., 2009). The American cockroach *Blattabacterium* genome encodes 581 open reading frames (ORFs), of which 520 have a known function. More than half of these ORFs are involved in amino acid metabolism, translation and biogenesis, cell-wall and cell-membrane biogenesis or coenzyme metabolism and energy production. Thirty-three tRNA genes, corresponding to all amino acids, as well as the complete transcriptional machinery, are encoded in the genome. Relative to free-living bacteria, genes involved in transcriptional regulation, signal transduction, secondary metabolites, and defensive activities are disproportionately deleted. Those involved in translation, amino acid metabolism, and coenzyme metabolism are generally retained (Sabree et al., 2009). This pattern of gene loss and retention is similar to that noted for other insect endosymbionts (Moran and Wernegreen, 2000; Ochman and Moran, 2001). The *Blattabacterium* genome contains genes for the complete biosynthetic pathways for all amino acids with the exception of aspargine, glumatime, methionine, and serine (Sabree et al., 2009). Thus, *Blattabacterium* manufactures most amino acids from a few nitrogen-rich substrates such as glutamate, urea, and ammonia. Finally, since the *Blattabacterium* genome does not encode enzymes essential for uricolysis, it is not clear where the initial steps of uric acid degradation occur.

Antibiotic and drug sensitivity. No significant differences have been reported in the antibiotic sensitivity of endosymbionts of different cockroach species subjected to either prolonged feeding or repeated injections. Penicillin, oxytetracycline, chlortetracycline, streptomycin, sigmamycin, penetracin, chloramphenicol, and sulfathiazole, but not aerosporin, are effective (Brooks, 1970). Although significantly reduced, the endosymbiont populations are rarely completely eliminated. Aposymbiosis is most readily obtained by interruption of the hereditary transmission of the endosymbionts to the ovary. Resistant endosymbiont strains have arisen with continued antibiotic treatment.

Pathogenicity. The endosymbionts are not known to be infectious or pathogenic to insects, other invertebrates, or vertebrates. Reinfection of aposymbiotic cockroaches with isolated endosymbionts, to fulfill Koch's postulates, has not been accomplished (Brooks and Richards, 1966).

Enrichment and isolation procedures

To isolate the blattbacteria, fat bodies from frozen or fresh, surface-sterilized insects must be dissected in a buffer containing 25 mM KCl, 10 mM $MgCl_2$, 250 mM sucrose, 250 mM EDTA, 35 mM Tris/HCl, pH 7.5 (Buffer A; Charles and Ishikawa (1999). The tissue is homogenized in 10 ml of Buffer A and filtered successively through nylon mesh filters having pore sizes of 100 µm, 20 µm, and 11 µm. The filtrate is centrifuged at 1500 × *g* for 25 min at 4°C and the pellet is resuspended in 10 ml of Buffer A containing 100 mM EDTA rather than 250 mM EDTA (Buffer B). Percoll gradients (27% and 70% in 5% PEG-6000; 1% Ficoll, 1% BSA fraction V, 250 mM sucrose) are prepared. Five milliliters of 70% Percoll is placed in a 50-ml centrifuge tube followed by 5 ml of 27% Percoll and 5 ml of cell suspension. The mixture is centrifuged at 12,000 × *g* for 15 min at 4°C. The suspension is recovered from the interface of the Percoll gradients. Ten milliliters of Buffer B is added and the suspension centrifuged at 3000 × *g* for 12 min at 4°C. The pellet is redissolved in 1 ml of Buffer B and centrifuged in a microcentrifuge for 12 min at 3000 × *g*. The final pellet is redissolved in 100 µl of Buffer B for use in restriction digests and pulsed-field gel electrophoresis. Since *Blattabacterium* are intracellular symbionts, it is difficult to obtain a preparation that is not contaminated by host DNA. However, a preparation that contains 50–55% endosymbiont DNA is adequate for gel electrophoresis (S. Kambhampati, unpublished data) and shotgun cloning for subsequent use in DNA sequencing (van Ham et al., 2003). A much simpler protocol for the isolation of fractions highly enriched for *Blattabacterium* DNA was described by Sabree et al. (2009).

Procedures for testing special characteristics

Molecular diagnosis. The presence of *Blattabacterium* can be confirmed by PCR amplification using 16S rRNA and 23S rRNA gene primers shown in Table 77 (Bandi et al., 1994; Clark and Kambhampati, 2003). The primers were designed based on consensus bacterial sequences of the respective genes and will therefore likely result in the amplification of the target fragment from a variety of bacterial species. DNA sequencing is therefore required to confirm that the amplified fragment is indeed from *Blattabacterium*.

Taxonomic comments

Bandi et al. (1994) concluded that there is parallel cladogenesis between the hosts and endosymbionts (Figure 56). Clark et al. (2001) and Clark and Kambhampati (2003) concluded that, based on 16S rRNA and 23S rRNA gene sequences, there is co-cladogenesis between the wood-feeding cockroach, *Cryptocercus*, and the *Blattabacterium* harbored by them. The parallel cladogenesis between cockroaches and their endosymbionts was supported by Lo et al. (2003). Hurst et al. (1999) showed that male-killing bacteria harbored by some members of Coccinellidae are a sister group of *Blattabacterium*. Moran et al. (2003) showed that the endosymbionts in the "yellow bacteriome" of the glassy winged sharpshooter, *Homolodisca coagulata*, are closely related to *Blattabacterium* and to the bacterial species endosymbiotic in ladybeetles.

The 16S rRNA gene sequence of *Blattabacterium* has been used to infer divergence times among their hosts. The 16S

TABLE 77. Oligonucleotide primers (5′→3′) for the PCR amplification of 16S rRNA and 23S rRNA gene fragments from *Blattabacterium*

Gene	Forward primer	Reverse primer	Fragment size (bp)	References
16S rRNA	GAGAGTTTGATCCTGGCTCAG	CTACGGCTACCTTGTTACGA	~1450	Bandi et al. (1994)
16S rRNA	TGCAAGTCGAGGGGC	TCACCGCTACACCACACATTC	630	Clark et al. (2001)
16S rRNA	GAATGTGTGGTGTAGCGGTGA	GTCGAGTTGCAGACTCCAATC	640	Clark et al. (2001)
23S rRNA	GACCGATAGTGAACCAGTAC	CTGCTTCYAAGCCAACMTCC	650	Clark et al. (2001)
23S rRNA	CGGGTAAGTTCCGACCTGCA	GAGCCGACATCGAGGTGCCAA	560	Clark et al. (2001)

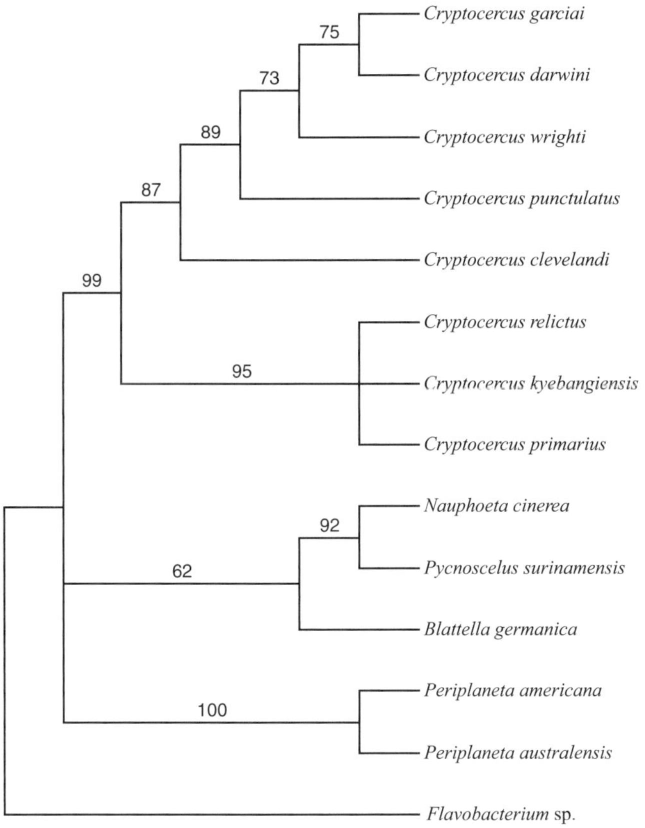

FIGURE 56. Phylogenetic relationship among *Blattabacterium* species harbored by 13 species of cockroaches based on DNA sequence of the 16S rRNA gene and parsimony analysis. The numbers above the branches are bootstrap values in per cent. The sequences for cockroach endosymbionts are from Bandi et al. (1994), Clark and Kambhampati (2003), and S. Kambhampati (unpublished). The outgroup (*Flavobacterium* sp.) sequence is from GenBank (accession no. AJ009687).

rRNA gene of free-living bacteria (Ochman and Wilson, 1987) and insect endosymbionts (Clark et al., 2001; Clark and Kambhampati, 2003; Lo et al., 2003; Maekawa et al., 2005a, 2005b; Moran et al., 1993) evolves in a clocklike manner. Therefore, it is possible to estimate divergence times among the host species harboring *Blattabacterium* by analyzing the DNA sequence of the endosymbiont 16S rRNA gene. Moran et al. (1993) estimated a rate of 0.0076–0.0232 nucleotide substitutions per site per 50 million years (MY) for the 16S rRNA gene of the aphid endosymbiont, *Buchnera aphidicola*. Lo et al. (2003) proposed a nucleotide substitution rate of 0.0087 per site per 50 MY for the 16S rRNA gene of *Blattabacterium* from various cockroach species. Maekawa et al. (2005b) and Maekawa et al. (2005a) estimated nucleotide substitution rates of 0.0084–0.0111 per site per 50 MY for the 16S rRNA gene of *Blattabacterium* harbored by *Cryptocercus* species, and 0.0087–0.0137 per site per 50 MY for the 16S rRNA gene of *Blattabacterium* harbored by two genera (*Salganea* and *Panesthia*) of Blaberidae, respectively. It must be noted, however, that all of these rates of evolution are based on a number of assumptions related to the age of termite-cockroach split, monophyly of cockroaches, endosymbiont lineage extinction, uniform rates of evolution within and among lineages, the extent to which taxon sampling is representative of all lineages, and fossil record. Unfortunately, the evidence for many of these assumptions is equivocal. Thus, the estimates of divergence times among hosts based on the *Blattabacterium* 16S rRNA gene must be interpreted with caution.

Bandi et al. (1995) sequenced the 16S rRNA gene from *Blattabacterium* of *Mastotermes darwiniensis* and compared it to that from cockroach endosymbionts. They concluded endosymbionts from the termite are monophyletic with those from cockroaches, implying monophyly of cockroaches and termites. Since *Mastotermes darwiniensis* is the only termite that is known to harbor *Blattabacterium*, all subsequent lineages of termites have evidently lost the endosymbiont (*Mastotermes darwiniensis* is the basal lineage among extant termites; Kambhampati and Eggleton, 2000).

Dasch et al. (1984) stated that the basic similarity in the DNA base composition, ultrastructure, mechanism of transmission to subsequent insect generations, and antibiotic sensitivities of the endosymbionts from different cockroaches, suggest that there is no strong basis for further speciation. However, recent evidence suggests that the endosymbiont clades in various cockroach families should probably be elevated to species status. For example, Clark and Kambhampati (2003) found that the sequence divergence (up to 2.2%) in the 16S rRNA gene of the endosymbiont from six different species of the wood-feeding cockroach, *Cryptocercus*, is comparable to that reported for congeneric free-living (Brenner et al., 1998; Fanrong et al., 1999; Wen et al., 1999; Yoon et al., 1998) and endosymbiotic (Sauer et al., 2000) bacterial species. Overall, the sequence divergence in the 16S rRNA gene of *Blattabacterium* harbored by various cockroach species shown in Figure 56 ranges from 0.6% (congeneric species) to 5.2% (taxa from different families). Thus, a thorough investigation is warranted to clarify the taxonomic status of *Blattabacterium* harbored by diverse cockroach species.

The mol% G+C of the DNA (Bd) has been determined for endosymbionts and their hosts from nine species of cockroaches in six different subfamilies. Endosymbiont DNA base compositions vary from 25.8 to 28.3 mol% G+C, except for 21.8 mol% obtained in a single determination on the endosymbionts of *Supella longipalpa*. Based on the full genome sequence, Sabree et al. (2009) reported a G+C content of 28.2 mol% for *Blattabacterium* from the American cockroach. Similarly, Sanchez-Lopez et al. (2009) estimated a G+C content of 27.1 mol% based on the full genome sequence of *Blattabacterium* from the German cockroach. Although no phylogenetic trends are obvious among the various species, the DNA G+C compositions of the endosymbionts of two *Parcoblatta* species and two *Blaberus* species are identical (27.3 and 26.5 mol%, respectively). In contrast, the cockroach DNAs differ greatly in the number (1–3), amount, and G+C compositions of the satellite DNA peaks, whereas the main band of DNA only varies from 33.5 to 36.3 mol% (Dasch, 1975).

Acknowledgements

The author's studies on cockroaches were supported by NSF grant DEB 98-06710. This is article number 06-196-B of the Kansas Agricultural Experiment Station.

List of species of the genus *Blattabacterium*

1. **Blattabacterium cuenoti** (Mercier 1906) Hollande and Favre 1931, 754[AL] (*Bacillus cuenoti* Mercier 1906, 684)

cu.en.ot'i. N.L. gen. masc. n. *cuenoti* of Cuenot; named after L. Cuenot, who studied intracellular inclusions in orthopteran insects.

The characteristics are as described for the genus. Originally studied in *Blatta orientalis*, the oriental cockroach.

DNA G+C content (mol%): 28.2 (Bd).

Type strain: none designated.

Sequence accession numbers: As of August 2009, 171 nucleotide sequences of *Blattabacterium* have been deposited in GenBank. None of these was designated a type strain. A vast majority of these are 16S rRNA gene sequences; several are 23S rRNA gene sequences from *Cryptocercus* species (Clark et al., 2001; Clark and Kambhampati, 2003). Accession numbers for *Blattabacterium* harbored by selected host species are as follows: 16S rRNA gene from *Cryptocercus* species: AF310161–310163; AF322458–AF322473 (Clark and Kambhampati, 2003); AY631413, AY631415, AY631417, AY631419, AY631421, AY631423 (S. Kambhampati, unpublished). 16S rRNA gene from various cockroach species: X75622–X75626 (Bandi et al., 1994). 16S rRNA gene from *Mastotermes darwiniensis*: Z35665 (Bandi et al., 1995). 23S rRNA gene from *Cryptocercus* species: AF322491–AF322523; AF363691–363721 (Clark and Kambhampati, 2003); AY631416, AY631418, AY631420, AY631422, AY631424, AY631414 (S. Kambhampati, unpublished). The full genome sequence of *Blattabacterium* from the American cockroach is available from GenBank under accession numbers CP00421 and CP001430.

References

Anderson, E. 1964. Oocyte differentiation and vitellogenesis in the roach *Periplaneta americana*. J. Cell. Biol. *20*: 131–155.

Bandi, C., G. Damiani, L. Magrassi, A. Grigolo, R. Fani and L. Sacchi. 1994. Flavobacteria as intracellular symbionts in cockroaches. Proc. Biol. Sci. *257*: 43–48.

Bandi, C., M. Sironi, G. Damiani, L. Magrassi, C.A. Nalepa, U. Laudani and L. Sacchi. 1995. The establishment of intracellular symbiosis in an ancestor of cockroaches and termites. Proc. Biol. Sci. *259*: 293–299.

Bigliardi, E., M. Selmi, B. Baccetti, L. Sacchi, A. Grigolo and U. Laudani. 1989. Membrane systems in endocytobiosis. I. Specializations of the vacuolar membrane in bacteriocytes of *Blattella germanica* (Dictyoptera: Blattellidae). J. Ultrastruct. Mol. Struct. Res. *102*: 66–70.

Block, R.J. and S.M. Henry. 1961. Metabolism of the sulphur amino acids and of sulphate in *Blattella germanica*. Nature *191*: 392–393.

Brenner, D.J., H.E. Muller, A.G. Steigerwalt, A.M. Whitney, C.M. O'Hara and P. Kämpfer. 1998. Two new *Rahnella* genomospecies that cannot be phenotypically differentiated from *Rahnella aquatilis*. Int. J. Syst. Bacteriol. *48*: 141–149.

Brooks, M.A. 1963. Symbiosis and aposymbiosis in arthropods. Symp. Soc. Gen. Microbiol. *13*: 200–231.

Brooks, M.A. and K. Richards. 1966. On the *in vitro* culture of intracellular symbiotes of cockroaches. J. Invertebr. Pathol. *8*: 150–157.

Brooks, M.A. 1970. Comments on the classification of intracellular symbiotes of cockroaches and a description of the species. J. Invertebr. Pathol. *16*: 249–258.

Brooks, M.A. and T.J. Kurtti. 1972. Male rudimentary ovaries: a case of cellular symbiosis in *Blattella germanica* (Dictyoptera: Blattellidae). Int. J. Insect Morphol. Embryol. *1*: 169–179.

Buchner, P. 1965. Endosymbiosis of Animals with Plant Microorganisms. Wiley, New York.

Bush, G.L. and G.B. Chapman. 1961. Electron microscopy of symbiotic bacteria in developing oocytes of the American cockroach, *Periplaneta americana*. J. Bacteriol. *81*: 267–276.

Chapman, R.F. 1998. The Insects: Structure and Function, 4th edn. Cambridge University Press, Cambridge.

Charles, H. and H. Ishikawa. 1999. Physical and genetic map of the genome of *Buchnera*, the primary endosymbiont of the pea aphid *Acyrthosiphon pisum*. J. Mol. Evol. *48*: 142–150.

Clarebrough, C., A. Mira and D. Raubenheimer. 2000. Sex-specific differences in nitrogen intake and investment by feral and laboratory-cultured cockroaches. J. Insect Physiol. *46*: 677–684.

Clark, J.W., S. Hossain, C.A. Burnside and S. Kambhampati. 2001. Coevolution between a cockroach and its bacterial endosymbiont: a biogeographical perspective. Proc. R. Soc. Lond. B. *268*: 393–398.

Clark, J.W. and S. Kambhampati. 2003. Phylogenetic relationships among *Blattabacterium*, endosymbiotic bacteria from the wood roach, *Cryptocercus*. Mol. Phylogenet. Evol. *26*: 82–88.

Cochran, D.G. 1975. Excretion in insects. *In* Insect Biochemistry and Function (edited by Candy and Kilby). Chapman & Hall, New York, pp. 177–281.

Cochran, D.G. 1985. Nitrogen excretion in cockroaches. Annu. Rev. Entomol. *30*: 29–49.

Daniel, R.S. and M.A. Brooks. 1967. Chromatographic evidence for murein from the bacteroid symbionts of *Periplaneta americana*. Experientia (Basel) *23*: 499–502.

Daniel, R.S. and M.A. Brooks. 1972. Intracellular *Bacteroides*: electron microscopy of *Periplaneta americana* injected with lysozyme. Exp. Parasitol. *31*: 232–246.

Daniel, R.S. 1973. Inheritance of intracellular bacteroids in *Periplaneta americana*. Proc. Electron Microsc. Soc. Am. *31*: 510–511.

Dasch, G.A. 1975. Morphological and molecular studies on intracellular bacterial symbionts of insects. PhD thesis, Yale University, New Haven.

Dasch, G.A., E. Weiss and K.P. Chang. 1984. Symbionts of insects. *In* Bergey's Manual of Systematic Bacteriology, vol. 1 (edited by Krieg and Holt). Williams & Wilkins, Baltimore, pp. 811–833.

Fanrong, K., G. James, M. Zhenfang, S. Gordon, B. Wang and G.L. Gilbert. 1999. Phylogenetic analysis of *Ureaplasma urealyticum*-support for the establishment of a new species, *Ureaplasma parvum*. Int. J. Syst. Bacteriol. *49*: 1879–1889.

Gier, H.T. 1947. Intracellular bacteroids in the cockroach (*Periplaneta americana* Linn.). J. Bacteriol. *53*: 173–189.

Giorgi, F. and J.H. Nordin. 1994. Structure of yolk granules in oocytes and eggs of *Blattella germanica* and their interaction with vitellophages and endosymbiotic bacteria during granule degradation. J. Insect Physiol. *40*: 1077–1092.

Gromov, B.V. and K.A. Mamkaeva. 1980. *Blattabacterium* in the fat body of the Maritime Territory relic roach, *Cryptocercus relictus*. Mikrobiologiya *49*: 1005–1007.

Henry, S.M. 1962. The significance of microorganisms in the nutrition of insects. Trans. N. Y. Acad. Sci. *24*: 676–683.

Hollande, A.C. and R. Favre. 1931. La structure cytologique de *Blattabacterium cuenoti* (Mercier) n.g. symbiote du tissue adipeux des blattides. C. R. Seances Soc. Biol. Fil. *107*: 752–754.

Hurst, G.D., C. Bandi, L. Sacchi, A.G. Cochrane, D. Bertrand, I. Karaca and M.E. Majerus. 1999. *Adonia variegata* (Coleoptera: Coccinellidae) bears maternally inherited flavobacteria that kill males only. Parasitology *118*: 125–134.

Kambhampati, S. and P. Eggleton. 2000. Taxonomy and phylogeny of termites. *In* Termites: Evolution, Sociality, Symbiosis, Ecology (edited by Abe, Bignell and Higachi). Kluwer, Dordrecht, pp. 1–23.

Koch, A. 1967. Insects and their endosymbionts. *In* Symbiosis, vol. II (edited by Henry). Academic Press, New York, pp. 1–106.

Landureau, J.C. 1966. Des cultures de cellules embryonnaires *Blattes* permettent d'obtenir la multiplication *in vitro* des bacteries symbiotiques. C. R. Hebd. Séances Acad. Sci. Ser. D, Sci. Nat. *262*: 1484–1487.

Laudani, U., G.P. Frizzi, C. Roggi and A. Montani. 1974. The function of the endosymbiotic bacteria of Blattodea. Experientia (Basel) *30*: 882–883.

Liu, T.P. 1973. The influence of juvenile hormone on the plasma membrane of symbiotic bacteria. Protoplasma *19*: 409–412.

Liu, T.P. 1974. The effect of corpora allata on the plasma membrane of the symbiotic bacteria of the oocyte surface of *Periplaneta americana* L. Gen. Comp. Endocrinol. *23*: 118–123.

Lo, N., C. Bandi, H. Watanabe, C. Nalepa and T. Beninati. 2003. Evidence for cocladogenesis between diverse dictyopteran lineages and their intracellular endosymbionts. Mol. Biol. Evol. *20*: 907–913.

Lo, N., C. Bandi, H. Watanabe, C.A. Nalepa and T. Beninati. 2007. Cockroaches that lack *Blattabacterium* endosymbionts: the phylogenetically divergent genus *Nocticola*. Biol. Lett. *3*(3): 327–330.

Maekawa, K., M. Kon, T. Matsumoto, K. Araya and N. Lo. 2005a. Phylogenetic analyses of fat body endosymbionts reveal differences in invasion times of blaberid wood-feeding cockroaches (Blaberidae: Panesthiinae) into the Japanese archipelago. Zool. Sci. *22*: 1061–1067.

Maekawa, K., Y.C. Park and N. Lo. 2005b. Phylogeny of endosymbiont bacteria harbored by the woodroach *Cryptocercus* spp. (Cryptocercidae: Blattaria): molecular clock evidence for a late Cretaceous - early Tertiary split of Asian and American lineages. Mol. Phylogenet. Evol. *36*: 728–733.

Malke, H. 1964. Production of aposymbiotic cockroaches by means of lysozyme. Nature *204*: 1223–1224.

Malke, H. and W. Schwartz. 1966. Untersuchungen uber die symbiose von tieren mit pilsen und bakterein XII. Die bedeutung der blattiden symbiose. Z. Allg. Mikrobiol. *6*: 34–68.

Malke, H.B. and G. Bartsch. 1966. Elektronenoptische unter suchung zür intracellulären bakteriensymbiose von Nauphoeta cinerea. Z. Allg. Mikrobiol. *6*: 163–176.

McCutcheon, J.P., B.R. McDonald and N.A. Moran. 2009. Convergent evolution of metabolic roles in bacterial co-symbionts of insects. Proc. Natl. Acad. Sci. U.S.A. *106*: 15394–15399.

Mercier, L. 1906. Les corps bacteroides de la blatte (*Periplaneta orientalis*): *Bacillus cuenoti* (n. sp. L. Mercier) (note preliminaire). C. R. Seances Soc. Biol. Fil. *61*: 682–684.

Meyer, G.F. and W. Frank. 1960. Elektronenmikroskopische studien über symbiontische einrichtungen bei insekten. Proc. Int. Congr. Electron Microsc. *2*: 539–542.

Milburn, N.S. 1966. Fine structure of the pleomorphic bacterioids in the mycetocytes and ovaries of several genera of cockroaches. J. Insect Physiol. *12*: 1245–1254.

Moran, N.A., M.A. Munson, P. Baumann and H. Ishikawa. 1993. A molecular clock in endosymbiotic bacteria is calibrated using insect hosts. Proc. R. Soc. Lond. B *253*: 167–171.

Moran, N.A. and J.J. Wernegreen. 2000. Lifestyle evolution in symbiotic bacteria: insights from genomics. Trends Ecol. Evol. *15*: 321–326.

Moran, N.A., C. Dale, W.A. Dunbar, H.A. Smith and H. Ochman. 2003. Intracellular symbionts of sharpshooters (Insecta: Hemiptera: Cicadellinae) form a distinct clade with a small genome. Environ. Microbiol. *5*: 116–126.

Murdock, L.L., T.L. Hopkins and R.A. Wirtz. 1970. Phenylalanine metabolism in cockroaches, *Periplaneta americana*: intracellular symbionts and aromatic ring cleavage. Comp. Biochem. Physiol. *34*: 143–146.

Nation, J.L. and R.L. Patton. 1961. A study of nitrogen excretion in insects. J. Insect Physiol. *6*: 299–308.

Ochman, H. and A.C. Wilson. 1987. Evolution in bacteria: evidence for a universal substitution rate in cellular genomes. J. Mol. Evol. *26*: 74–86.

Ochman, H. and N.A. Moran. 2001. Genes lost and genes found: evolution of bacterial pathogenesis and symbiosis. Science (Washington) *292*: 1096–1099.

Richards, A.G. and M.A. Brooks. 1958. Internal symbioses in insects. Annu. Rev. Entomol. *3*: 37–56.

Rizki, M.T.M. 1954. Deoxyribose nucleic acid in the symbiotic microorganisms of the cockroach, *Blattella germanica*. Science (Washington) *120*: 35–36.

Sabree, Z., S. Kambhampati and N.A. Moran. 2009. Nitrogen recycling and nutritional provisioning by *Blattabacterium*, the American cockroach endosymbiont. Proc. Natl. Acad. Sci. U.S.A. *106*(46): 19521–19526.

Sacchi, L., A. Grigolo, M. Mazzini, E. Bigliardi, B. Baccetti and U. Laudani. 1988. Symbionts in the oocytes of *Blattella germanica* (L.) (Dictyoptera: Blattellidae): their mode of transmission. Int. J. Insect Morphol. Embriol. *17*: 437–446.

Sacchi, L. and A. Grigolo. 1989. Endocytobiosis in *Blattella germanica* (Blattodea) recent acquisitions. Endocytobiology Cell Res. *6*: 121–147.

Sacchi, L., S. Corona, A. Grigolo, U. Laudani, M.G. Selmi and E. Bigliardi. 1996. The fate of the endocytobionts of *Blattella germanica* (Blattaria: Blattellidae) and *Periplaneta americana* (Blattaria: Blattidae) during embryo development. Ital. J. Zool. *63*: 3–12.

Sacchi, L., C.A. Nalepa, E. Bigliardi, M. Lenz, C. Bandi, S. Corona, A. Grigolo, S. Lambiase and U. Laudani. 1998. Some aspects of intracellular symbiosis during embryo development of *Mastotermes darwiniensis* (Isoptera: Mastotermitidae). Parasitologia *40*: 309–316.

Sacchi, L., C.A. Nalepa, M. Lenz, C. Bandi, S. Corona, A. Grigolo and E. Bigliardie. 2000. Transovarial transmission of symbiotic bacteria in *Mastotermes darwiniensis* (Isoptera: Mastotermitidae): Ultrastructural aspects and phylogenetic implications. Ann. Entomol. Soc. Am. *93*: 1303–1313.

Sanchez-Lopez, M.J., A. Neef, J. Pereto, R. Patino-Navarrete, M. Pignatelli, A. Latorre and A. Moya. 2009. Evolutionary convergence and nitrogen metabolism in *Blattabacterium* strain Bge, primary endosymbiont of the cockroach *Blattella germanica*. PLoS Genet. *5*: e1000721. doi: 10.1371/journal.pgen.1000721.

Sauer, C., E. Stackebrandt, J. Gadau, B. Holldobler and R. Gross. 2000. Systematic relationships and cospeciation of bacterial endosymbionts and their carpenter ant host species: proposal of the new taxon *Candidatus* Blochmannia gen. nov. Int. J. Syst. Evol. Microbiol. *50*: 1877–1886.

Srivastava, P.N. and P.D. Gupta. 1961. Excretion of uric acid in *Periplaneta americana*. J. Insect Physiol. *6*: 163–167.

van Ham, R.C.H.J., J. Kamerbeek, C. Palacios, C. Rausell, F. Abascal, U. Bastolla, J.M. Fernández, L. Jiménez, M. Postigo, F.J. Silva, J. Tamames, E. Viguera, A. Latorre, A. Valencia, F. Morán and A. Moya. 2003. Reductive genome evolution in *Buchnera aphidicola*. Proc. Natl. Acad. Sci. U.S.A. *100*: 581–586.

Wen, A., M. Fegan, C. Hayward, S. Chakraborty and L.I. Sly. 1999. Phylogenetic relationships among members of the *Comamonadaceae*, and the description of *Delftia acidovorans* (den Dooren de Jong 1926 and Tamaoka *et al.* 1987) gen. nov., comb. nov. Int. J. Syst. Bacteriol. *49*: 567–576.

Wernegreen, J.J. 2002. Genome evolution in bacterial endosymbionts of insects. Nat. Rev. Genet. *3*: 850–861.

Wren, H.N. and D.G. Cochran. 1987. Xanthine dehydrogenase activity in the cockroach endosymbiont *Blattabacterium cuenoti* (Mercier 1906) Hollande and Favre 1931 and in the cockroach fat body. Comp. Biochem. Physiol. *88*: 1023–1026.

Yoon, J.H., S.T. Lee and Y.H. Park. 1998. Inter- and intraspecific phylogenetic analysis of the genus *Nocardioides* and related taxa based on 16S rDNA sequences. Int. J. Syst. Bacteriol. *48*: 187–194.

Family III. **Cryomorphaceae** Bowman, Nichols and Gibson 2003, 1353[VP]

John P. Bowman

Cry.o.mor.pha.ce′a.e. N.L. fem. n. *Cryomorpha* the type genus of the family; - *aceae* ending to denote a family; N.L. fem. pl. n. *Cromorphaceae* the *Cryomorpha* family.

Cells are rodlike and sometimes filaments. Multiply by **binary fission.** Usually **do not contain spores or other resting cells** or gas vesicles. **Motile by gliding motility or are nonmotile. Gram-stain-negative. Strictly aerobic or facultatively anaerobic. Chemoorganotrophic** metabolism. Anaerobic growth, if it occurs, is supported by fermentation of carbohydrates. Do not fix nitrogen. **Colonies are yellow or orange due to carotenoid biosynthesis.** Most species do not form detectable flexirubin type pigments. **Most genera require seawater salts for growth. Most genera require organic compounds** present in yeast extract, peptone, and/or Casamino acids for growth. **Have a limited versatility** in regard to nutritional and catabolic activities and are **generally unreactive in most standard phenotypic tests. Most species do not utilize carbohydrates.** For fatty acid profiles, see Table 78. Cultured members of the *Cryomorphaceae* occur in a range of marine habitats including marine sediments, coastal sands, sand-filtered sea water, sea water particulates, and sea-ice algal assemblages. Also found in marine-derived refugia, i.e., quartz stone sublithic communities located in coastal ice-free regions of Antarctica. Most cultured species are able to grow at low temperatures and are isolated from low temperature ecosystems. Molecular studies, however, suggest the family is present in a wide range of non-extreme ecosystems both marine and terrestrial, across tropical to polar regions.

DNA G+C content (mol%): 35–40.

Type genus: **Cryomorpha** Bowman, Nichols and Gibson 2003, 1352[VP].

Further descriptive information

Described in 2003, the family *Cryomorphaceae* represents a broad cluster of mostly marine species of the phylum *Bacteroidetes* (Figure 57). Phylogenetically, the family is weakly monophyletic, occurring in an adjacent position to the family *Flavobacteriaceae* (Yarza et al., 2008). The family so far includes six genera, each only represented by a single species (Bowman et al., 2003, Lau et al., 2005, 2006; O'Sullivan et al., 2005). Phenotypic characteristics useful for differentiating the member genera are shown in the determinative key given below. A notable feature is the lack of activity in most standard phenotypic tests, indicative of a restricted nutritional range. Comparatively fastidious growth factor requirements possessed by these species also suggests that members of this family may be largely dependent on decaying organic biomass, and thus could have significant roles in organic remineralization processes.

The family *Cryomorphaceae* is widespread in the environment, present in a wide range of marine and terrestrial habitats as indicated by several culture-independent studies (Abell and Bowman, 2005; Acinas et al., 2004; Aguilo-Ferretjans et al., 2008; Alain et al., 2002; Bissett et al., 2006; Bowman and McCuaig, 2003; Brinkmeyer et al., 2003; Brown et al., 2005; Collins et al., 2003; Crump and Hobbie, 2005; DeLong et al., 2006; Fuchs et al., 2005; Goffredi et al., 2004; Green et al., 2004; Humayoun et al., 2003; Kirchman et al., 2003; López-Garcia et al., 2003; Nielsen et al., 2004; Pinhassi et al., 2004; Reysenbach et al., 2000; Schaefer et al., 2002; Skidmore et al., 2005; Sorensen et al., 2005; Webster et al., 2004; Zwart et al., 2002). A clade within the family *Cryomorphaceae* has been designated the AGG58 cluster, named for clones detected in marine aggregates (DeLong et al., 1993). This group has been found to be widespread in aquatic environments (O'Sullivan et al., 2004). A general consensus of phylogenetic surveys of marine-type ecosystems, suggests *Cryomorphaceae* is much less numerous than that of the marine members of the family *Flavobacteriaceae*. Whether this suggests a pattern of econiche specialization for *Cryomorphaceae* or a natural consequence of evolutionary processes among bacteria is unknown at this time.

Key to the genera of the family *Cryomorphaceae*

1. DNA G+C content is 36–37 mol%. Cells are rod shaped. Colonies are orange. Nonmotile. No growth occurs at 37°C. Oxidase-negative. Strictly aerobic. Does not utilize D-glucose. Does not reduce nitrate. Does not hydrolyze gelatin. Isolated from quartz stone sublithic communities of ice-free coastal areas of Antarctica and seawater particulates collected off Antarctic coastlines.
   → Genus I. *Cryomorpha*

2. DNA G+C content is 38–40 mol%. Cells are rod shaped. Colonies are orange. Motile by gliding. No growth occurs at 37°C. Oxidase-negative. Facultatively anaerobic. Utilizes D-glucose. Reduces nitrate to nitrate. Does not hydrolyze gelatin. Isolated from marine sediment and sea-ice algal assemblages collected off Antarctic coastlines.
   → Genus II. *Brumimicrobium*

3. DNA G+C content is 35 mol%. Cells are rod shaped or filamentous. Colonies are yellow. Motile by gliding. No growth occurs at 37°C. Oxidase-negative. Strictly aerobic. Utilizes D-glucose. Reduces nitrate to nitrate. Does not hydrolyze gelatin. Isolated from beneath frozen sand in a coastal bay of Alaska, USA.
   → Genus III. *Crocinitomix*

4. DNA G+C content is 39–40 mol%. Cells are rod shaped or filamentous. Colonies are orange. Motile by gliding. Growth occurs at 37°C. Oxidase-positive. Strictly aerobic. Utilizes D-glucose. Reduces nitrate to nitrate. Hydrolyzes gelatin. Isolated from a tank used for sand-filtering of sea water, Hong Kong, China.
   → Genus IV. *Fluviicola*

5. DNA G+C content is 37.2 mol%. Cells are rod shaped with rarer longer filaments. Colonies are yellow-orange. May form flexirubin pigments. Motile by gliding. Growth occurs at 4–25°C. Oxidase-negative. Strictly aerobic. Does not utilize glucose. Negative for nitrate and nitrite reduction. Weakly hydrolyzes gelatin. No growth occurs on seawater-containing media. Isolated from water of the River Taff, Cardiff, UK.
   → Genus V. *Lishizhenia*

6. DNA G+C content is 35.8 mol%. Cells are rod shaped. Colonies are light orange. Cells are motile by gliding. Growth occurs at 4–37°C. Oxidase-positive. Strictly aerobic. Does not utilize D-glucose. Gelatin and casein are hydrolyzed. Isolated from sand-filtered sea water collected at Port Shelter, adjacent to the Coastal Marine Laboratory, Hong Kong University of Science and Technology.
   → Genus VI. *Owenweeksia*

Genus I. **Cryomorpha** Bowman, Nichols and Gibson 2003, 1352^VP

John P. Bowman

Cry.o.mor'pha. Gr. neut. n. *kryos* icy cold, frost; Gr. fem. n. *morphe* shape or form; N.L. fem. n. *Cryomorpha* cold shape.

Cells are slender, flexible **rods with rounded ends**, 0.3–0.5 μm×0.5–3 μm. Occur singly or in pairs. **Gram-stain-negative**. Spores, resting cells, and gas vesicles not present. **Nonmotile. Colonies are orange. Flexirubin pigments are not produced. Strictly aerobic with an oxidative type of metabolism**. Catalase positive. Oxidase negative. **Chemoheterotrophic. Non-agarolytic**. The organisms have an inert metabolism in that they **do not utilize carbohydrates**, do not form acid from carbohydrates, and are unreactive in most standard biochemical tests. Strains **require sea water salts and yeast extract** for growth. Growth occurs between −2 and 22°C, with optimal growth at 15–20°C and pH 6–8. Major fatty acids are $C_{14:0}$ iso, $C_{15:0}$ iso, $C_{15:0}$ anteiso, $C_{16:0}$ iso, $C_{14:1}$ iso ω9*c*, $C_{15:1}$ iso ω10*c*, $C_{15:1}$ anteiso ω10*c*, $C_{14:0}$ iso 2-OH, and $C_{15:0}$ anteiso 2-OH. **Isolated from quartz stone sublithic communities and Southern Ocean sea water particulates.**

DNA G+C content (mol%): 36–37.

Type species: **Cryomorpha ignava** Bowman, Nichols and Gibson 2003, 1352^VP.

Further descriptive information

Cryomorpha is the type genus of the family *Cryomorphaceae* within which it forms a distinct line of descent (Figure 57).

Colonies are pinpoint to small (0.5–1 mm), orange, circular, entire edged, viscid to butyrous, and convex.

Like other members of the *Cryomorphaceae*, *Cryomorpha ignava* requires sea water and organic compounds occurring in peptone or yeast extract for growth, and it has a restricted versatility in regard to its catabolic activity. It slowly hydrolyzes Tween 80 and urea, but otherwise is unreactive in standard phenotypic tests including various commercial test strips and trays. The species has an absolute requirement for sea water and is quite stenohaline. It requires yeast extract for growth; no growth occurs in seawater media supplemented only with peptone, and growth occurs with peptone only if a vitamin solution is supplied suggesting the species requires specific vitamin growth factors that are present in yeast extract. Catalase activity in strains is very pronounced. The fatty acid profile (Table 78) is decidedly unusual, containing large amounts of branched-chain fatty acids such as $C_{14:0}$ iso 2-OH, $C_{15:0}$ anteiso 2-OH, $C_{16:1}$ iso ω11*c*, $C_{14:1}$ iso ω9*c* and $C_{14:0}$ iso that are rarely found in other Gram-stain-negative bacteria.

Enrichment and isolation procedures

Cryomorpha ignava can be isolated by direct plating of sample material onto Marine agar followed by incubation at 4–20°C. The species grows relatively slowly, with colonies taking at least 4–7 d to appear and grow to full size.

Maintenance procedures

For long-term storage *Cryomorpha ignava* can be cryopreserved at −80°C in Marine 2216 broth supplemented with 30% glycerol. Strains survive well on Marine agar plates or slants stored at 2–4°C.

Differentiation of the genus *Cryomorpha* from other genera

The most salient traits useful for distinguishing *Cryomorpha* from other members of the *Cryomorphaceae* include orange pigmentation, lack of motility, strictly aerobic oxidative metabolism, inability to use carbohydrates, lack of reactivity in most phenotypic tests, and ability to grow at low temperatures. The species also has a distinctive fatty acid profile.

List of species of the genus *Cryomorpha*

1. **Cryomorpha ignava** Bowman, Nichols and Gibson 2003, 1352^VP

ig.na'va L. fem. adj. *ignava* lazy, pertaining to the biochemically and nutritionally inert nature of the species.

The characteristics are as given for the genus, as listed in Table 78, and as given in the text.

Source: quartz stone sublithic communities and sea water particulates.

DNA G+C content (mol%): 36–37 (T_m).

Type strain: 1-22, ACAM 647, CIP 107453, LMG 21436.

Sequence accession no. (16S rRNA gene): AF170738.

Genus II. **Brumimicrobium** Bowman, Nichols and Gibson 2003, 1352^VP

John P. Bowman

Bru'mi.mi.cro'bi.um. L. fem. n. *bruma* winter; N.L. neut. n. *microbium* microbe; N.L. neut. n. *Brumimicrobium* winter microbe.

Cells are slender, flexible **rods with rounded ends**, 1–3 μm×0.3–0.5 μm, occurring singly or in pairs. **Gram-stain-negative**. Spores, resting cells, and gas vesicles are not present. **Gliding motility is present. Colonies are orange. Flexirubin pigments are not produced. Facultatively anaerobic. Possesses an oxidative metabolism under aerobic conditions. Under anaerobic growth conditions the metabolism shifts to the fermentation of sugars.** Catalase-positive. Oxidase-negative. **Chemoheterotrophic. Non-agarolytic. Utilizes D-glucose**. Reduces nitrate to nitrite. Strains **require Na^+ and yeast extract** for growth. Peptone can serve as a nitrogen source. Psychroactive. Slightly halophilic. Major fatty acids are $C_{13:0}$ iso, $C_{15:0}$ iso, and $C_{15:1}$ anteiso ω10*c*. **Isolated from polar marine sediment and sea-ice algal assemblages.**

DNA G+C content (mol%): 38–40.

Type species: **Brumimicrobium glaciale** Bowman, Nichols and Gibson 2003, 1352^VP.

Further descriptive information

The genus *Brumimicrobium* is a member of the family *Cryomorphaceae* (Bowman et al., 2003) within which it forms a distinct line of descent (Figure 57).

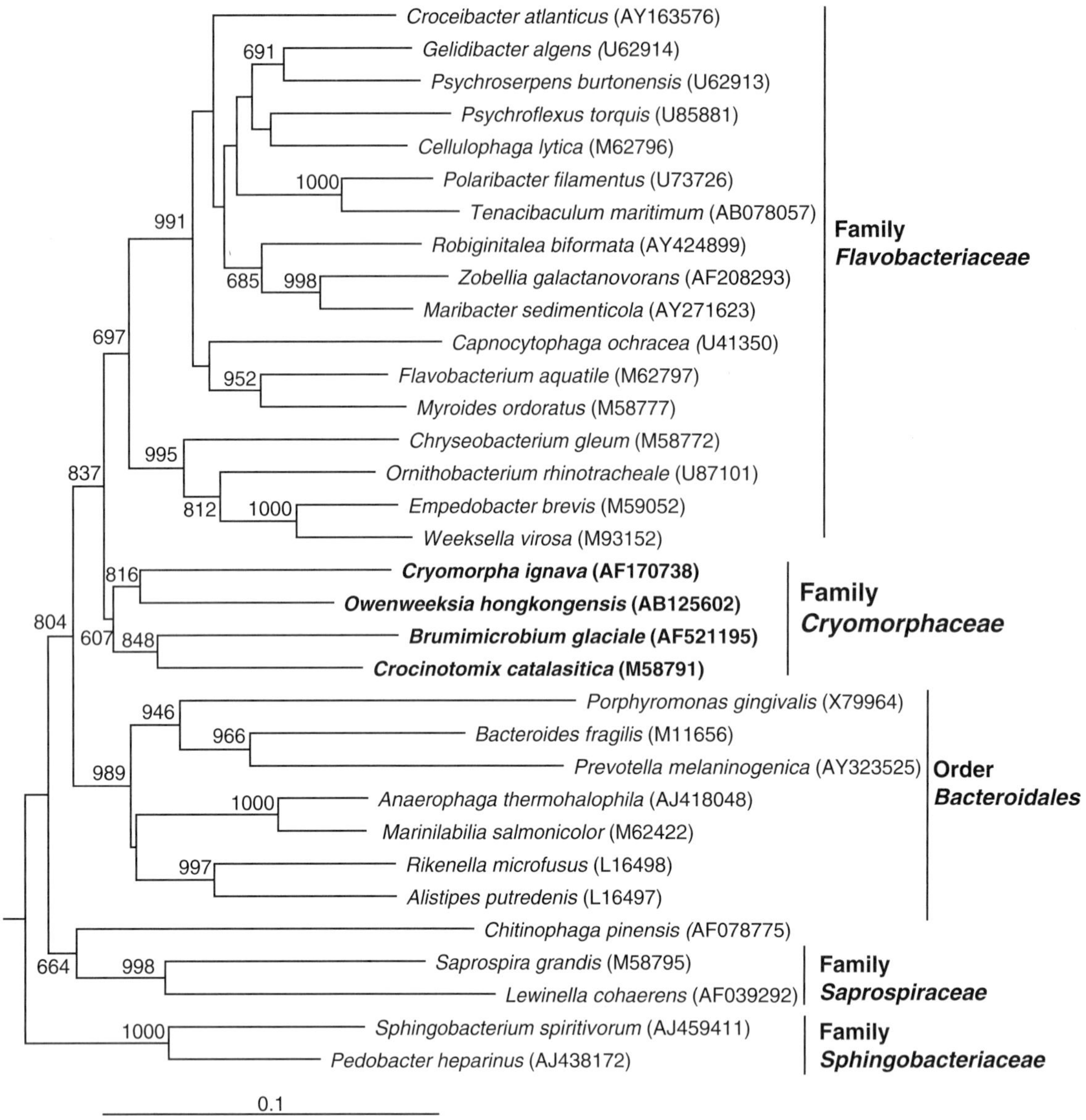

FIGURE 57. Phylogenetic tree based on 16S rRNA gene sequences showing the positions of the genera of the family *Cryomorphaceae* in relation to other related taxa of the phylum *Bacteroidetes*, including family *Flavobacteriaceae*, order *Bacteroidales*, family *Saprospiraceae* and family *Sphingobacteriaceae*. The tree is based on nearly complete 16S rRNA gene sequences clustered by maximum-likelihood and neighbor-joining methods (PHYLIP; J. Felsenstein, University of Washington, Seattle, WA, USA). Numbers at branch lengths are bootstrap values indicating the number of times the branch node was recovered following 1000 replications of the cluster analysis.

Colonies are 1–2 mm in diameter, orange, convex, circular, entire edged, and possess a gelid consistency. Gliding motility is best observed from growth on dilute media rather than Marine 2216 agar. Like other members of the *Cryomorphaceae*, the species requires factors occurring in peptone or yeast extract for growth and has a restricted catabolic versatility. It hydrolyzes Tween 80, and some strains degrade DNA. Alkaline phosphatase is produced. Some strains form β-galactosidase. Otherwise, the organisms are unreactive in standard phenotypic tests, including various commercial test strips and trays. The only known carbon and energy substrate is D-glucose, which the organism can use oxidatively under aerobic conditions, and fermentatively when under anaerobic conditions. Under aerobic conditions, no acid is formed from D-glucose. Other common substrates tested do not support growth. *Brumimicrobium glaciale* requires Na^+ for growth and is quite stenohaline. The species grows well with yeast extract; however, growth occurs in seawater media containing peptone or Casamino acids if vitamins are provided. This suggests the species has specific vitamin growth factors that are present in yeast extract.

The species is psychroactive and neutrophilic, growing between −2 and 27°C, with optimal growth at 16–19°C and pH 6–8. The fatty acid profile (Table 78) is made up mainly of C_{15} branched-chain fatty acids, a profile that differs from other members of the family *Cryomorphaceae*.

Strains of *Brumimicrobium glaciale* have been isolated from both the sea-ice and marine sediments of Antarctica (Bowman et al., 2003).

TABLE 78. Whole-cell fatty acids of type species of five genera of the family *Cryomorphaceae*[a]

Fatty acid	*Cryomorpha ignava*	*Brumimicrobium glaciale*	*Crocinotomix catalasitica*	*Fluviicola taffensis*	*Owenweeksia hongkongensis*	*Lishizhenia caseilytica*
Saturated fatty acids:						
$C_{13:0}$	–	tr	–	0.1	–	0.1
$C_{14:0}$	–	1.5	tr	3.2	–	1.6
$C_{15:0}$	tr	2.9	29.3	7.5	–	–
$C_{16:0}$	tr	1.9	1.8	3.0	tr	0.5
$C_{18:0}$	–	tr	–	–	–	–
Branched-chain fatty acids:						
$C_{11:0}$ iso	tr	–	–	–	tr	tr
$C_{13:0}$ iso	–	5.6	–	–	–	0.3
$C_{14:0}$ iso	13.8	tr	–	–	–	–
$C_{14:1}$ iso ω9*c*	6.8	tr	tr	1.0	–	–
$C_{15:1}$ iso ω10*c*	7.1	–	10.9	11.8	28.0	44.0
$C_{15:1}$ anteiso ω10*c*	8.2	45.0	–	–	–	
$C_{15:0}$ iso	7.8	36.4	36.3	44.2	18.7	34.8
$C_{15:0}$ anteiso	5.8	–	–	0.5	–	–
$C_{16:1}$ iso ω6*c*	–	tr	–		–	–
$C_{16:1}$ ω11*c* iso	2.9	–	tr	0.7	–	–
$C_{16:0}$ iso	7.8	tr	tr	0.6	tr	tr
$C_{16:0}$ anteiso	–	–	–		tr	–
$C_{17:1}$ iso ω5*c*						0.5
$C_{17:1}$ iso ω9*c*	–	–	–		7.9	–
$C_{17:1}$ iso ω12*c*	–	–	tr		–	–
$C_{17:1}$ iso ω13*c*	–	–	tr		–	–
Monounsaturated fatty acids:						
$C_{15:1}$ ω6*c*	–	–	–		1.3	–
$C_{15:1}$ ω8*c*	–	1.4	–		–	–
$C_{15:1}$ ω11*c*	–	–	8.3	1.2	–	–
$C_{16:1}$ ω7*c*	–	1.5	–		–[b]	–[b]
$C_{16:1}$ ω12*c*	–	–	4.4	4.9	–	–
$C_{17:1}$ ω13*c*	–	–	2.1		–	–
$C_{18:1}$ ω5*c*	–	–	–		tr	–
$C_{18:1}$ ω7*c*	–	tr	–		–	–
$C_{18:1}$ ω9*c*	–	tr	–		–	–
Hydroxy fatty acids:						
$C_{14:0}$ iso 2-OH	6.7	–	–			–
$C_{15:0}$ 2-OH	–	–	–	3.5	3.0	–
$C_{15:0}$ iso 2-OH	–	–	–		–[b]	–[b]
$C_{15:0}$ anteiso 2-OH	18.4	–	–			–
$C_{15:0}$ 3-OH	–	–	–	0.5	1.0	–
$C_{15:0}$ iso 3-OH	1.6	–	1.0		4.9	2.5
$C_{16:0}$ iso 3-OH	–	–	–	4.2	tr	0.4
$C_{16:0}$ 3-OH	–	–	–		–	0.5
$C_{17:0}$ 3-OH	–	–	–		1.2	–
$C_{17:0}$ iso 3-OH	–	–	–	12.3	18.1	9.3
Other	–	–	–		5.1[c]	3.3[c]

[a]Symbols: values are the percent of total fatty acids; tr, trace level detected (<1% of total fatty acids); –, not detected.

[b]Fatty acids $C_{16:1}$ ω7*c* and $C_{15:0}$ 2-OH were not resolved and make up a fatty acid summed feature comprising 10.1% of total fatty acids.

[c]Three unidentified fatty acids were also obtained with equivalent chain lengths of 11.543, 13.565 and 16.582.

Enrichment and isolation procedures

Brumimicrobium glaciale can be isolated by direct plating of sample material (sediment, sea-ice samples melted in sterile sea water at 4°C) onto Marine agar followed by incubation at 4–20°C. The species grows relatively slowly, with colonies taking several days to appear and grow to full size.

Maintenance

For long-term storage, *Brumimicrobium glaciale* can be cryopreserved at −80°C in Marine 2216 broth supplemented with 30% glycerol. Strains survive well on Marine 2216 agar plates or slants stored at 2–4°C.

Differentiation of the genus *Brumimicrobium* from other genera

The most salient traits useful for distinguishing *Brumimicrobium* from other members of the *Cryomorphaceae* include orange pigmentation, gliding motility, facultatively anaerobic metabolism, ability to use D-glucose, lack of reactivity in most phenotypic tests, and ability to grow at low temperatures. The species also has a distinctive fatty acid profile containing only a few major fatty acids.

List of species of the genus *Brumimicrobium*

1. **Brumimicrobium glaciale** Bowman, Nichols and Gibson 2003, 1352[VP]

gla'ci.al.e. L. neut. adj. *glaciale* icy, frozen.

The characteristics are as given for the genus, as listed in Table 78, and as given in the text. Inhabits sea-ice algal assemblages and polar marine sediment.

DNA G+C content (mol%): 38–40 (T_m).
Type strain: IC156, ACAM 645, CIP 107451, LMG 21434.
Sequence accession no. (16S rRNA gene): AF521195.

Genus III. **Crocinitomix** Bowman, Nichols and Gibson 2003, 1353[VP]

John P. Bowman

Cro.cin.i.to'mix L. adj. *crocinus* of or pertaining to saffron; L. fem. n. *tomix* a string or thread; N.L. fem. n. *Crocinitomix* saffron-colored thread (filament).

Cells are short **rods with rounded ends or filaments**, 0.3–0.5×0.5→20 µm. Occur singly or in pairs. **Gram-stain-negative**. Spores, resting cells, and gas vesicles not present. **Gliding motility is present. Colonies are yellow. Flexirubin pigments are not produced. Strictly aerobic, having an oxidative type of metabolism**. Catalase-positive. Oxidase-negative. **Chemoheterotrophic. Non-agarolytic. Carbohydrates are not utilized**. Acid is not formed from carbohydrates. Strains **require Na^+ and yeast extract** for growth. Growth occurs between 0 and 30°C, with optimal growth yield at 20–22°C and pH 6–8. Major fatty acids are $C_{15:0}$, $C_{15:0}$ iso, $C_{15:1}$ iso ω10*c*, $C_{15:1}$ ω11*c*, and $C_{16:1}$ ω12*c*. **Isolated from beneath frozen sand of the marine upper littoral zone (Subarctic Alaska).**

DNA G+C content (mol%): 35.

Type species: **Crocinitomix catalasitica** (Lewin 1969) Bowman, Nichols and Gibson 2003, 1353[VP] (*Microscilla aggregans* subsp. *catalatica* Lewin 1969, 197).

Further descriptive information

Crocinitomix is a member of the family *Cryomorphaceae* within which it forms a distinct line of descent (Figure 57 in the family *Cryomorphaceae* section). Colonies are saffron-colored, smooth, circular, convex, with slightly irregular edges and a butyrous consistency when grown on Marine 2216 agar. Like other members of the *Cryomorphaceae*, the species requires seawater salts and organic compounds occurring in peptone or yeast extract for growth and has a restricted versatility in regards to its catabolic activity. Growth with peptone occurs if a vitamin solution is supplied, suggesting the species has specific vitamin growth factors which are present in yeast extract.

The type strain attacks L-tyrosine and L-3,4-dihydroxyphenyl-L-alanine and produces alkaline phosphatase and arginine arylamidase activity. The type strain may also be able to form H_2S from L-cysteine (Lewin and Lounsbery, 1969), however, this activity could not be reproduced (Bowman et al., 2003). Acidification, clotting, or redigestion of curd does not occur in litmus milk. The type strain is otherwise unreactive in standard phenotypic tests including various commercial test strips and trays. The fatty acid profile (Table 78) includes monounsaturated fatty acids with unusual double bond positions including $C_{15:1}$ ω11*c*, $C_{16:1}$ ω12*c*, and $C_{17:1}$ ω13*c*. These fatty acids are apparently unique to this organism which suggests that the organism contains a novel fatty acid desaturase.

Only one isolate (ATCC 23190) of this species has been obtained to date. ATCC 23190 originated from beneath a layer of frozen coastal sand collected in the shallows of subarctic Auke Bay, Alaska.

Enrichment and isolation procedures

Crocinotomix catalasitica was isolated by direct plating of diluted sample material onto Marine agar followed by incubation at 20°C.

Maintenance procedures

For long term storage, *Crocinotomix catalasitica* can be cryopreserved at −80°C in Marine 2216 broth supplemented with 30% glycerol. Strains survive well on Marine 2216 agar plates or slants stored at 2–4°C. The organisms can also be stored by lyophilization.

Differentiation of the genus *Cryomorpha* from other genera

The most salient traits useful for distinguishing *Crocinitomix* from other members of the *Cryomorphaceae* include yellow pigmentation, gliding motility, strictly aerobic oxidative metabolism, inability to use carbohydrates, lack of reactivity in most phenotypic tests, and ability to grow at low temperatures. The species also has a distinctive fatty acid profile (Table 78).

List of species of the genus *Crocinitomix*

1. **Crocinitomix catalasitica** (Lewin 1969) Bowman, Nichols and Gibson 2003, 1353[VP] (*Microscilla aggregans* subsp. *catalatica* Lewin 1969, 197)

cat.a.la.si'ti.ca. N.L. neut. n. *catalasum* catalase; L. suff. *-icus -a -um*, suffix used in adjectives with the sense of relating to; N.L. fem. adj. *catalasitica* relating to catalase, pertaining to the ability of this species to produce catalase.

The characteristics are as given for the genus, as listed in Table 78, and as given in the text. The only known habitat is frozen sand collected from the marine upper littoral zone of Auke Bay, Alaska, USA.

DNA G+C content (mol%): 35 (T_m).
Type strain: ATCC 23190, NCIMB 1418.
Sequence accession no. (16S rRNA gene): AB078042, M58791.

Genus IV. **Fluviicola** O'Sullivan, Rinna, Humphreys, Weightman and Fry 2005, 2193[VP]

THE EDITORIAL BOARD

Flu.vi.i.co'la L. masc. n. *fluvius -ii* river; L. suff. *-cola* (from L. masc. or fem. n. *incola*) inhabitant, dweller; N.L. masc. n. *Fluviicola* river dweller.

Rods 0.4–0.5×1.5–5.7 µm, with rarer longer filaments of up to 51 µm in length. Nonflagellated. **Motile by gliding.** Gram-stain-negative. Strictly aerobic. Cell mass is pigmented yellow-orange. **Flexirubins are synthesized. Temperature range for growth, 4–25°C;** optimum, ~20°C. Catalase-positive. **Oxidase-negative. Growth does not occur in the presence of Na^+ ions. Negative for nitrate and nitrite reduction. Glucose is not utilized for growth.** Fatty acids contain a high proportion of branched-chain fatty acids (mainly $C_{15:0}$ iso and $C_{15:1}$ iso ω10*c*), as well as significant amounts of hydroxy fatty acids (mainly $C_{17:0}$ iso 3-OH). Isolated from freshwater environments.

DNA G+C content (mol%): 37.2.

Type species: **Fluviicola taffensis** O'Sullivan, Rinna, Humphreys, Weightman and Fry 2005, 2193[VP].

Enrichment and isolation

Fluviicola taffensis was isolated after incubation for 6 d on casein-yeast-tryptone (CYT) agar* at 20°C, from a 10^{-2} dilution of water from the River Taff, a river in Wales.

Differentiation of the genus *Fluviicola* from related genera

The main feature distinguishing *Fluviicola* from *Crocinitomix*, *Owenweeksia*, *Brumimicrobium*, and *Cryomorpha* is that *Fluviicola* originated from a freshwater habitat and cannot grow in the presence of Na^+ ions, whereas the other genera occur in marine habitats and require Na^+ ions or natural seawater. The aerobic metabolism of *Fluviicola* differentiates it from the fermentative metabolism of *Brumimicrobium*, and gliding motility differentiates it from *Cryomorpha*. Failure to grow at 30°C distinguishes *Fluviicola* from *Crocinitomix* and *Owenweeksia*. Unlike *Owenweeksia*, *Fluviicola* lacks oxidase activity. Unlike *Brumimicrobium*, it does not reduce nitrate and does not utilize glucose.

Taxonomic comments

O'Sullivan et al. (2005) reported that, based on 16S rRNA gene sequence analysis and on construction of a neighbor-joining phylogenetic tree, *Fluviicola* was a member of the family *Cryomorphaceae* within the phylum *Bacteroidetes*. The 16S rRNA gene sequence of *Fluviicola taffensis* showed 88% sequence similarity to that of *Crocinitomix catalasitica*.

List of species of the genus *Fluviicola*

1. **Fluviicola taffensis** O'Sullivan, Rinna, Humphreys, Weightman and Fry 2005, 2193[VP]

taf.fen'sis. N.L. masc. adj. *taffensis* pertaining to the River Taff, a river in Wales.

The characteristics are as described for the genus, with the following additional features. Colonies are 1–5 mm in diameter, circular, flat, transparent, shiny, yellow-orange, and creamy on high nutrient solid media. Growth occurs on nutrient agar, plate count agar, tryptic soy agar, and DNase agar, but not on Marine agar 2216, MacConkey agar, or Cetrimide agar. Indole is not produced. Negative for β-galactosidase, urease, and xylanase activity. Capable of DNA hydrolysis and weak gelatin hydrolysis, but not hydrolysis of agar, arginine, esculin, or starch. Acid is not produced from glucose. Does not utilize glucose, arabinose, mannose, mannitol, *N*-acetylglucosamine, maltose, gluconate, caprate, adipate, malate, citrate, or phenylacetate. Resistant to chloramphenicol, streptomycin, and kanamycin, but susceptible to penicillin G, ampicillin, rifampin, and tetracycline. The type strain was isolated from water of the River Taff, Cardiff, UK.

DNA G+C content (mol%): 37.2 (HPLC).

Type strain: RW262, DSM 16823, NCIMB 13979.

Sequence accession no. (16S rRNA gene): AF493694.

Genus V. **Lishizhenia** Lau, Ren, Wai, Qian, Wong and Wu 2006, 2321[VP]

THE EDITORIAL BOARD

Li.shi.zhe'ni.a N.L. fem. n. *Lishizhenia* named after Li Shizhen (1518–1593), the famous Chinese naturalist.

Rods 0.3–0.5 × 0.5–3.8 µm. Nonflagellated. **Cells are motile by gliding**. Nonsporeforming. Gram-stain-negative. **Halophilic**, growing between 1.0 and 7.5% (w/v) NaCl with optimum growth at 1–3%. **Organic growth factors such as yeast extract or peptone are required for growth**. Colonies are light orange. **Strict aerobes. Oxidase-positive.** Catalase- and alkaline phosphatase-positive. **Casein and gelatin are hydrolyzed. Carbohydrates are not utilized.** Major respiratory quinone is MK-6. Predominant fatty acids are $C_{15:0}$ iso, $C_{15:1}$ iso, $C_{17:0}$ iso 3-OH, and $C_{15:0}$ iso 2-OH /$C_{16:1}$ ω7*c*.

DNA G+C content (mol%): 35.8.

Type species: **Lishizhenia caseinilytica** Lau, Ren, Wai, Qian, Wong and Wu 2006, 2321[VP].

Enrichment and isolation procedures

Lishizhenia caseinilytica was isolated from the outlet of a tank storing sand-filtered sea water by direct plating of samples onto YPS-SW agar* , with incubation for 3d at 30°C. Orange

*CYT agar contains (g/l of distilled water): pancreatic digest of casein, 1.0; $CaCl_2 \cdot 2H_2O$, 0.5; $MgSO_4 \cdot 7H_2O$, 0.5; yeast extract, 0.5; and purified agar, 15.0; pH7.2.

*YPS-SW medium contains: yeast extract, 4 g; peptone, 2 g; starch, 10 g; agar, 15 g; seawater filtered through a membrane filter (0.45 µm pore diameter), 750 ml; and distilled water, 250 ml.

colonies were selected and then purified by repeated restreaking on YPS-SW agar. Cultures can be preserved at –80°C in Marine broth 2216 supplemented with 50% (w/v) glycerol.

Differentiation of the genus *Lishizhenia* from related genera

Unlike *Owenweeksia hongkongensis*, *Lishizhenia caseinilytica* has caseinase activity and a lower mol% G+C of its DNA (35.8 vs 39–40). In contrast to *Fluviicola taffensis*, it requires Na^+ ions, is oxidase positive, lacks DNase, and is susceptible to chloramphenicol. It differs from *Brumimicrobium glaciale*, *Crocinitomix catalasitica*, and *Cryomorpha ignava* by its ability to grow at 35°C, tolerance to 7.5% NaCl, and positive oxidase reaction.

Taxonomic comments

Lau et al. (2006) reported that, based on 16S rRNA gene sequence analyses, *Lishizhenia caseinilytica* formed a distinct lineage within the family *Cryomorphaceae* and was linked to *Brumimicrobium glaciale* with bootstrap support of 91% by the neighbor-joining algorithm and 80% by maximum-parsimony analysis. Sequence similarities to other members of the family *Cryomorphaceae* were less than 91%.

List of species of the genus *Lishizhenia*

1. **Lishizhenia caseinilytica** Lau, Ren, Wai, Qian, Wong and Wu 2006, 2321[VP]

ca.sei.ni.ly'ti.ca. N.L. n. *caseinum* casein; N.L. adj. *lyticus -a -um* (from Gr. adj. *lutikos -ê -on*), able to loosen, able to dissolve; N.L. fem. adj. *caseinilytica* casein-dissolving.

The characteristics are as described for the genus, with the following additional features. Enlarged cells and filamentous cells are seen occasionally in stationary phase in broth culture. Colonies are light orange, circular, convex, smooth, glistening, and translucent with an entire margin. Carotenoid pigments are present. Cells are motile by gliding but do not swarm on Marine agar. Mesophilic, growing between 4 and 37°C; optimum, 27–30°C. Growth occurs between pH 5.0 and 9.0; optimum, ~pH 7.0. No acid is formed from the following carbohydrates, nor are they utilized: *N*-acetylglucosamine, D-adonitol, amygdalin, D- and L-arabinose, D- and L-arabitol, arbutin, D-cellobiose, dextran, dulcitol, erythritol, esculin, ferric citrate, D-fructose, D- and L-fucose, D-galactose, gentiobiose, D-glucose, glycerol, glycogen, inositol, D-lactose, D-lyxose, D-maltose, D-mannose, D-melezitose, D-melibiose, methyl α-D-glucopyranoside, methyl α-D-mannopyranoside, methyl-β-D-xylopyranoside, D-mannitol, potassium gluconate, potassium 2-ketogluconate, potassium 5-ketogluconate, D-raffinose, L-rhamnose, D-ribose, salicin, D-sorbitol, L-sorbose, sucrose, D-tagatose, D-trehalose, D-turanose, xylitol, D- and L-xylose. Glucose, sucrose, and mannitol are not fermented. The cellular fatty acids (and their percentages) are as follows: $C_{13:0}$ (0.1); $C_{14:0}$ (1.6); $C_{16:0}$ (0.5); $C_{11:0}$ iso (trace); $C_{13:0}$ iso (0.3); $C_{13:1}$ at 12–13 (0.1); $C_{15:0}$ iso (44); $C_{15:1}$ iso (34.8); $C_{16:0}$ iso (trace); $C_{17:1}$ iso ω5*c* (0.5); $C_{14:0}$ iso 3-OH (trace); $C_{15:0}$ iso 3-OH (2.5); $C_{16:0}$ 3-OH (0.5); $C_{16:0}$ iso 3-OH (0.4); $C_{17:0}$ iso 3-OH (9.3); sum ($C_{15:0}$ iso 2-OH/$C_{16:1}$ ω7*c*) (2.0); and sum ($C_{16:1}$ iso/$C_{14:0}$ 3-OH) (trace). Sensitivity is exhibited toward ampicillin (10 μg), chloramphenicol (30 μg), penicillin G (2 U), rifampin (10 μg), tetracycline (30 μg), and polymyxin B (300 U). Resistance is exhibited toward kanamycin (10 μg), gentamicin sulfate (10 μg), and spectinomycin (10 μg).

Source: sand-filtered sea water collected at Port Shelter, adjacent to the Coastal Marine Laboratory, Hong Kong University of Science and Technology.

DNA G+C content (mol%): 35.8 ± 0.5 (HPLC).

Type strain: UST040201-001, JCM 13821, NRRL B-41434

Sequence accession no. (16S rRNA gene): AB176674.

Genus VI. **Owenweeksia** Lau, Ng, Ren, Lau, Qian, Wong, Lau and Wu 2005, 1055[VP]

John P. Bowman

O.wen.week'si.a N.L. fem. n. *Owenweeksia* named after Owen B. Weeks, who did a lot of work in the 1950s, 1960s and 1970s on *Flavobacterium*, *Cytophaga* and related species.

Cells are flexible short **rods**, 0.3–0.5 × 0.5–4 μm. **Filaments may be formed in media supplemented with 0.1% Tween 20.** Cells of old cultures and cells grown at low temperature are enlarged or distorted in shape. **Gram-stain-negative**. Spores and resting cells not present. **Motile by gliding**. **Strictly aerobic, having an oxidative type of metabolism**. Catalase-positive. Oxidase-positive. **Chemoheterotrophic. Non-agarolytic**. **Carbohydrates are not utilized**. Acid is not formed from carbohydrates. **Colonies are orange due to carotenoid pigments. Flexirubin pigments are not produced**. Strains **require seawater salts and either peptone or yeast extract** for growth. Growth occurs between 4–37°C and pH 5–9, with optimal growth at 25–33°C and pH 6–8. The major fatty acids are $C_{15:1}$ iso, $C_{15:0}$ iso, $C_{17:0}$ iso 3-OH, summed feature ($C_{15:0}$ iso 2-OH/$C_{16:1}$ ω7*c*) and $C_{15:0}$ iso 3-OH. The major respiratory quinone is menaquinone-6. **Isolated from a tank containing sand-filtered sea water.**

DNA G+C content (mol%): 39–40.

Type species: **Owenweeksia hongkongensis** Lau, Ng, Ren, Lau, Qian, Wong, Lau and Wu 2005, 1055[VP].

Further descriptive information

Owenweeksia hongkongensis is a member of the family *Cryomorphaceae*, forming a distinct line of descent, only distantly related to *Cryomorpha ignava* (85% 16S rRNA similarity).

Like other members of the *Cryomorphaceae*, *Owenweeksia hongkongensis* requires sea water and organic compounds occurring in peptone or yeast extract for growth and has a restricted versatility in regard to its catabolic activity. The species grows at salinities between 1–7.5%. The species produces alkaline phosphatase and esterase (substrate Tween 20), and hydrolyzes gelatin. *Owenweeksia hongkongensis* cannot degrade polysaccharides, casein, or DNA, and cannot lyse autoclaved yeast cells or hemolyze rabbit blood. H_2S and indole production are negative. The species does not utilize or produce acid from a wide range of carbohydrates which suggests that it may utilize mainly amino acids or lipids for growth. Three unidentified fatty acids have been detected in *Owenweeksia hongkongensis*, with equivalent chain lengths of 11.543, 13.565 and 16.582.

Enrichment and isolation procedures

Owenweeksia hongkongensis was isolated from a tank storing sand-filtered sea water by direct plating of samples onto YPS-SW medium and incubating at 30°C. The isolate was subsequently obtained in pure culture on YPS-SW lacking starch. *Owenweeksia hongkongensis* can be routinely cultured and stored on YP-SW (YPS-SW medium lacking starch), Marine 2216 (Difco), or in a medium containing 0.4% yeast extract in artificial sea water*.

Maintenance procedures

For long term storage, *Owenweeksia hongkongensis* can be cryopreserved at –80°C in YP-SW supplemented with 20% glycerol.

Differentiation of the genus *Owenweeksia* from other genera

The most salient traits useful for distinguishing *Owenweeksia* from other members of the *Cryomorphaceae* include orange pigmentation, gliding motility, strictly aerobic oxidative metabolism, gelatinase production, inability to use carbohydrates, and ability to grow at 30°C or more. The genus also has a substantially different fatty acid profile.

List of species of the genus *Owenweeksia*

1. **Owenweeksia hongkongensis** Lau, Ng, Ren, Lau, Qian, Wong, Lau and Wu 2005, 1055[VP]

hong.kong.en'sis. N.L. fem. adj. *hongkongensis* pertaining to Hong Kong, China.

The characteristics are as given for the genus, as listed in Table 78, and as given in the text.

Source: a tank containing sand-filtered sea water.
DNA G+C content (mol%): 39–40 (HPLC).
Type strain: UST20020801, JCM 12287, NRRL B-23963.
Sequence accession no. (16S rRNA gene): AB125062.

References

Abell, G.C. and J.P. Bowman. 2005. Ecological and biogeographic relationships of class *Flavobacteria* in the Southern Ocean. FEMS Microbiol. Ecol. *51*: 265–277.

Acinas, S.G., V. Klepac-Ceraj, D.E. Hunt, C. Pharino, I. Ceraj, D.L. Distel and M.F. Polz. 2004. Fine-scale phylogenetic architecture of a complex bacterial community. Nature *430*: 551–554.

Aguilo-Ferretjans, M.M., R. Bosch, C. Martin-Cardona, J. Lalucat and B. Nogales. 2008. Phylogenetic analysis of the composition of bacterial communities in human-exploited coastal environments from Mallorca Island (Spain). Syst. Appl. Microbiol. *31*: 231–240.

Alain, K., M. Olagnon, D. Desbruyeres, A. Page, G. Barbier, S.K. Juniper, J. Quellerou and M.A. Cambon-Bonavita. 2002. Phylogenetic characterization of the bacterial assemblage associated with mucous secretions of the hydrothermal vent polychaete *Paralvinella palmiformis*. FEMS Microbiol. Ecol. *42*: 463–476.

Bissett, A., J. Bowman and C. Burke. 2006. Bacterial diversity in organically-enriched fish farm sediments. FEMS Microbiol. Ecol. *55*: 48–56.

Bowman, J.P., C. Mancuso, C.M. Nichols and J.A.E. Gibson. 2003. *Algoriphagus ratkowskyi* gen. nov., sp. nov., *Brumimicrobium glaciale* gen. nov., sp. nov., *Cryomorpha ignava* gen. nov., sp. nov. and *Crocinitomix catalasitica* gen. nov., sp. nov., novel flavobacteria isolated from various polar habitats. Int. J. Syst. Evol.Microbiol. *53*: 1343–1355.

Bowman, J.P. and R.D. McCuaig. 2003. Diversity and biogeography of prokaryotes dwelling in Antarctic continental shelf sediment. Appl. Environ. Microbiol. *69*: 2463–2483.

Brinkmeyer, R., K. Knittel, J. Jurgens, H. Weyland, R. Amann and E. Helmke. 2003. Diversity and structure of bacterial communities in arctic versus antarctic pack ice. Appl. Environ. Microbiol. *69*: 6610–6619.

Brown, M.V., M.S. Schwalbach, I. Hewson and J.A. Fuhrman. 2005. Coupling 16S-ITS rDNA clone libraries and automated ribosomal intergenic spacer analysis to show marine microbial diversity: development and application to a time series. Environ. Microbiol. *7*: 1466–1479.

Collins, G., A. Woods, S. McHugh, M.W. Carton and V. O'Flaherty. 2003. Microbial community structure and methanogenic activity during start-up of psychrophilic anaerobic digesters treating synthetic industrial wastewaters. FEMS Microbiol. Ecol. *46*: 159–170.

Crump, B.C. and J.E. Hobbie. 2005. Synchrony and seasonality in bacterioplankton communities of two temperate rivers. Limnol. Oceanogr. *50*: 1718–1729.

DeLong, E.F., D.G. Franks and A.L. Alldredge. 1993. Phylogenetic diversity of aggregate-attached vs. free-living marine bacterial assemblages. Limnol. Oceanogr. *38*: 924–934.

DeLong, E.F., C.M. Preston, T. Mincer, V. Rich, S.J. Hallam, N.U. Frigaard, A. Martinez, M.B. Sullivan, R. Edwards, B.R. Brito, S.W. Chisholm and D.M. Karl. 2006. Community genomics among stratified microbial assemblages in the ocean's interior. Science *311*: 496–503.

Fuchs, B.M., D. Woebken, M.V. Zubkov, P. Burkill and R. Amann. 2005. Molecular identification of picoplankton populations in contrasting waters of the Arabian Sea. Aquat. Microb. Ecol. *39*: 145–157.

Goffredi, S.K., A. Waren, V.J. Orphan, C.L. Van Dover and R.C. Vrijenhoek. 2004. Novel forms of structural integration between microbes and a hydrothermal vent gastropod from the Indian Ocean. Appl. Environ. Microbiol. *70*: 3082–3090.

Green, D.H., L.E. Llewellyn, A.P. Negri, S.I. Blackburn and C.J. Bolch. 2004. Phylogenetic and functional diversity of the cultivable bacterial community associated with the paralytic shellfish poisoning dinoflagellate *Gymnodinium catenatum*. FEMS Microbiol. Ecol. *47*: 345–357.

Humayoun, S.B., N. Bano and J.T. Hollibaugh. 2003. Depth distribution of microbial diversity in Mono Lake, a meromictic soda lake in California. Appl. Environ. Microbiol. *69*: 1030–1042.

*Artificial sea water (Lewin and Lounsbery, 1969) contains (g/l of distilled water): ($CaCl_2 \cdot 2H_2O$, 1.0 g; KCl, 1.0 g; $MgSO_4 \cdot 7H_2O$, 5.0 g; and NaCl, 25 g.)

Kirchman, D.L., L. Yu and M.T. Cottrell. 2003. Diversity and abundance of uncultured cytophaga-like bacteria in the Delaware estuary. Appl. Environ. Microbiol. *69*: 6587–6596.
Lau, K.W., C.Y. Ng, J. Ren, S.C. Lau, P.Y. Qian, P.K. Wong, T.C. Lau and M. Wu. 2005. *Owenweeksia hongkongensis* gen. nov., sp. nov., a novel marine bacterium of the phylum '*Bacteroidetes*'. Int. J. Syst. Evol. Microbiol. *55*: 1051–1057.
Lau, K.W., J. Ren, N.L. Wai, P.Y. Qian, P.K. Wong and M. Wu. 2006. *Lishizhenia caseinilytica* gen. nov., sp. nov., a marine bacterium of the phylum Bacteroidetes. Int. J. Syst. Evol. Microbiol. *56*: 2317–2322.
Lewin, R.A. 1969. A classification of flexibacteria. J. Gen. Microbiol. *58*: 189–206.
Lewin, R.A. and D.M. Lounsbery. 1969. Isolation, cultivation and characterization of flexibacteria. J. Gen. Microbiol. *58*: 145–170.
López-Garcia, P., S. Duperron, P. Philippot, J. Foriel, J. Susini and D. Moreira. 2003. Bacterial diversity in hydrothermal sediment and epsilonproteobacterial dominance in experimental microcolonizers at the Mid-Atlantic Ridge. Environ. Microbiol. *5*: 961–976.
Nielsen, J.L., A. Schramm, A.E. Bernhard, G.J. van den Engh and D.A. Stahl. 2004. Flow cytometry-assisted cloning of specific sequence motifs from complex 16S rRNA gene libraries. Appl. Environ. Microbiol. *70*: 7550–7554.
O'Sullivan, L.A., K.E. Fuller, E.M. Thomas, C.M. Turley, J.C. Fry and A.J. Weightman. 2004. Distribution and culturability of the uncultivated 'AGG58 cluster' of the *Bacteroidetes* phylum in aquatic environments. FEMS Microbiol. Ecol. *47*: 359–370.
O'Sullivan, L.A., J. Rinna, G. Humphreys, A.J. Weightman and J.C. Fry. 2005. *Fluviicola taffensis* gen. nov., sp. nov., a novel freshwater bacterium of the family *Cryomorphaceae* in the phylum '*Bacteroidetes*'. Int. J. Syst. Evol. Microbiol. *55*: 2189–2194.
Pinhassi, J., M.M. Sala, H. Havskum, F. Peters, O. Guadayol, A. Malits and C. Marrase. 2004. Changes in bacterioplankton composition under different phytoplankton regimens. Appl. Environ. Microbiol. *70*: 6753–6766.
Reysenbach, A.L., K. Longnecker and J. Kirshtein. 2000. Novel bacterial and archaeal lineages from an in situ growth chamber deployed at a Mid-Atlantic Ridge hydrothermal vent. Appl. Environ. Microbiol. *66*: 3798–3806.
Schaefer, H., B. Abbas, H. Witte and G. Muyzer. 2002. Genetic diversity of 'satellite' bacteria present in cultures of marine diatoms. FEMS Microbiol. Ecol. *42*: 25–35.
Skidmore, M., S.P. Anderson, M. Sharp, J. Foght and B.D. Lanoil. 2005. Comparison of microbial community compositions of two subglacial environments reveals a possible role for microbes in chemical weathering processes. Appl. Environ. Microbiol. *71*: 6986–6997.
Sorensen, K.B., D.E. Canfield, A.P. Teske and A. Oren. 2005. Community composition of a hypersaline endoevaporitic microbial mat. Appl. Environ. Microbiol. *71*: 7352–7365.
Webster, N.S., A.P. Negri, M.M. Munro and C.N. Battershill. 2004. Diverse microbial communities inhabit Antarctic sponges. Environ. Microbiol. *6*: 288–300.
Yarza, P., M. Richter, J. Peplies, J. Euzéby, R. Amann, K.H. Schleifer, W. Ludwig, F.O. Glockner and R. Rossello-Mora. 2008. The All-Species Living Tree project: a 16S rRNA-based phylogenetic tree of all sequenced type strains. Syst. Appl. Microbiol. *31*: 241–250.
Zwart, G.J.M., B.C. Crump, M. Agterveld, F. Hagen and S.K. Han. 2002. Typical freshwater bacteria: an analysis of available 16S rRNA gene sequences from plankton of lakes and rivers. Aquat. Microb. Ecol. *28*: 141–155.

Class III. **Sphingobacteriia** class. nov.

Peter Kämpfer

Sphin.go.bac.te.ri.i′a. N.L. neut. n. *Sphingobacterium* type genus of the type order; suff. -*ia* ending proposed by Gibbons and Murray and by Stackebrandt et al. to denote a class; N.L. neut. pl. n. *Sphingobacteriia* the *Sphingobacterium* class.

The class is defined on the basis of sequence similarities of 16S rRNA gene sequences and comprises morphologically diverse nonsporeforming bacteria that stain Gram-negative. Growth is aerobic or facultatively anaerobic, and a yellow pigmentation is often formed. The class contains one order, *Sphingobacteriales*. All genera studied for this feature (mainly *Sphingobacterium*) contain high concentrations of sphingophospholipids as cellular lipid components.

DNA G+C content (mol%): 36–49.8.

Type order: **Sphingobacteriales** ord. nov.

Order I. **Sphingobacteriales** ord. nov.

Peter Kämpfer

Sphin.go.bac.te.ri′a.les. N.L. neut. n. *Sphingobacterium* type genus of the order; suff. -*ales* ending to denote an order; N.L. fem. pl. n. *Sphingobacteriales* the *Sphingobacterium* order.

Cells are rod-shaped and usually nonmotile. Aerobic or facultatively anaerobic. Limited fermentative capabilities are observed in some members. Menaquinones are of the MK-7 (rarely MK-6) type, and a fatty acid pattern containing $C_{15:0}$ iso, $C_{15:0}$ iso 2-OH, $C_{15:0}$ iso 3-OH, $C_{16:0}$, $C_{16:1}$ ω7*c*, $C_{16:0}$ 3-OH and $C_{17:0}$ iso 3-OH as the most important components are characteristic. The order contains two families, *Sphingobacteriaceae* Steyn, Segers, Vancanneyt, Sandra, Kersters and Joubert 1998 and *Chitinophagaceae* fam. nov.

Type genus: **Sphingobacterium** Yabuuchi, Kaneko, Yano, Moss and Miyoshi 1983, 592[VP].

References

Steyn, P.L., P. Segers, M. Vancanneyt, P. Sandra, K. Kersters and J.J. Joubert. 1998. Classification of heparinolytic bacteria into a new genus, *Pedobacter*, comprising four species: *Pedobacter heparinus* comb. nov., *Pedobacter piscium* comb. nov., *Pedobacter africanus* sp. nov. and *Pedobacter saltans* sp. nov. Proposal of the family *Sphingobacteriaceae* fam. nov. Int. J. Syst. Bacteriol. *48*: 165–177.
Yabuuchi, E., T. Kaneko, I. Yano, C.W. Moss and N. Miyoshi. 1983. *Sphingobacterium* gen. nov., *Sphingobacterium spiritivorum* comb. nov., *Sphingobacterium multivorum* comb. nov., *Sphingobacterium mizutae* sp. nov., and *Flavobacterium indologenes* sp. nov., glucose-nonfermenting Gram-negative rods in CDC group IIK-2 and group IIb. Int. J. Syst. Bacteriol. *33*: 580–598.

Family I. **Sphingobacteriaceae** Steyn, Segers, Vancanneyt, Sandra, Kersters and Joubert 1998, 175^{VP}

THE EDITORIAL BOARD

Sphin.go.bac.te.ri.a.ce′a.e. N.L. neut. n. *Sphingobacterium* type genus of the family; *-aceae* ending to denote a family; N.L. fem. pl. n. *Sphingobacteriaceae* family of *Sphingobacterium*

Rods, 0.3–0.6×0.5–0.6 μm. **Aerobic**, having a strictly respiratory type of metabolism. **Flagella are absent. Colonies are usually yellowish after several days. Catalase-, oxidase-, and phosphatase-positive. Indole not produced**. Homospermidine is the major polyamine. Features that differentiate the family from *Flavobacteriaceae* are the possession of **sphingophospholipids and ceramides**, the presence of **menaquinone 7**, a higher mean mol% G+C content of the DNA, and a unique fatty acid content (with $C_{15:0}$ iso, $C_{15:0}$ iso 2-OH, $C_{15:0}$ iso 3-OH, $C_{16:0}$, $C_{16:1}$ ω7*c*, $C_{16:0}$ 3-OH, and $C_{17:0}$ iso 3-OH being the most important components). Isolated from soil, activated sludge, and clinical specimens. Free-living and saprophytic, but some species can be opportunistic pathogens.

DNA G+C content (mol%): 36–45.

Type genus: **Sphingobacterium** Yabuuchi, Kaneko, Yano, Moss and Miyoshi 1983, 592^{VP}.

The family includes two genera, *Sphingobacterium* and *Pedobacter*, which are part of rRNA superfamily V and are closely related by rRNA cistron similarity (Steyn et al., 1998).

Most bacteria do not contain sphingophospholipids (SPLs), but members of *Sphingobacteriaceae* contain high concentrations of SPLs as cellular lipid components. The lipid portion of a sphingophospholipid is a sphingosine molecule in which the amino group is acylated by a fatty acid chain, forming a ceramide. The major molecular species of sphingobacterial ceramides were identified by Yano et al. (1983, 1982) as 2-*N*-2′-hydroxy-13′-methyltetradecanoyl-15-methylhexadecasphinganine, 2-*N*-13′-methyltetradecanoyl-15-methylhexadecasphinganine, and 2-*N*-13′-methyltetradecanoyl-hexadecasphinganine. Naka et al. (2003) reported that the SPLs of the type strains of *Sphingobacterium spiritivorum, Sphingobacterium antarcticum, Sphingobacterium faecium, Sphingobacterium mizutaii, Sphingobacterium multivorum*, and *Sphingobacterium thalpophilum* contained ceramide phosphorylethanolamines (CerPE-1 and CerPE-2), ceramide phosphoryl-myo-inositols (CerPI-1 and CerPI-2), and ceramide phosphorylmannose (CerPM-1). The ceramide of CerPE-1, CerPI-1, and CerPM-1 was composed of 15-methylhexadecasphinganine (isoheptadeca sphinganine, $C_{17:0}$ iso) and 13-methyltetradecanoic acid (isopentadecanoic acid, $C_{15:0}$ iso), whereas that of CerPE-2 and CerPI-2 was composed of isoheptadeca sphinganine and 2-hydroxy-13-methyltetradecanoic acid (2-hydroxy isopentadecanoic acid, $C_{15:0}$ iso 2-OH). *Sphingobacterium* and *Pedobacter* species contain homospermidine as the major polyamine (Hamana and Nakagawa, 2001a).

Genus I. **Sphingobacterium** Yabuuchi, Kaneko, Yano, Moss and Miyoshi 1983, 592^{VP}

THE EDITORIAL BOARD

Sphin.go.bac.te′ri.um. N.L. n. *sphingosinum* (from Gr. gen. n. *sphingos* of sphinx, and suff.-*ine*) sphingosine; N.L. pref. *sphingo-* pertaining to sphingosine; L. neut. n. *bacterium* a rod, and in biology a bacterium; N.L. neut. n. *Sphingobacterium* a sphingosine-containing bacterium.

Straight rods. Gram-stain-negative. Nonsporeforming. **Lack flagella** but may exhibit a flagellum-independent spreading mechanism known as "sliding motility". **Catalase-positive**. Chemoorganotrophic, without specialized growth factor requirements. **Colonies usually become yellowish after several days at room temperature. Indole and acetylmethylcarbinol are not produced. Nonproteolytic**. Gelatin is not hydrolyzed. **Heparinase negative. Acid is produced from carbohydrates oxidatively,** not fermentatively. Cellular lipids contain **sphingophospholipids** whose ceramide moieties are chiefly branched-chain dihydrosaturated $C_{17:0}$ sphingosin, and the major acid is $C_{15:0}$ iso 2-OH.

DNA G+C content (mol%): 39–42.

Type species: **Sphingobacterium spiritivorum** (Holmes et al., 1982) Yabuuchi, Kaneko, Yano, Moss and Miyoshi 1983, 594^{VP} (*Flavobacterium spiritivorum* Holmes, Owen and Hollis 1982, 161; *Flavobacterium yabuuchiae* Holmes, Weaver, Steigerwalt and Brenner 1988, 351).

Further descriptive information

Some strains of *Sphingobacterium spiritivorum* exhibit sliding motility which, as defined by Henrichsen (1972) is "a kind of surface translocation produced by the expansive forces in a growing culture in combination with special surface properties of the cells resulting in reduced friction between cell and substrate. The micromorphological pattern is that of a uniform sheet of closely packed cells in a single layer. The sheet moves slowly as a unit."

Sphingobacterium spiritivorum, Sphingobacterium multivorum, Sphingobacterium mizutaii, and *Sphingobacterium thalpophilum* are free-living and saprophytic organisms (Bernardet et al., 1996) found in environmental habitats such as soil (Fenton and Jarvis, 1994) and composted manure (Kuo et al., 2000). They have also been isolated from human clinical specimens and hospital environments and may be opportunistic pathogens (Yabuuchi et al., 1983). Blood and urine have been the most common sources for the isolation of *Sphingobacterium spiritivorum*. The species has caused cellulitis in the leg of an elderly man (Marinella, 2002); the organism was probably acquired by walking barefoot on soil. Another bacterium isolated from cellulitis of a leg of another elderly man was initially identified as *Sphingobacterium spiritivorum* but later shown by 16S rRNA gene analysis to belong to an undescribed species closely related to *Sphingobacterium mizutaii* (Tronel et al., 2003). *Sphingobacterium multivorum* has been isolated from activated sludge (Liu et al., 2004) and from various clinical specimens; it has occasionally been associated with serious infections such as septicemia in persons immunocompromised by conditions such as cystic fibrosis, kidney damage, or diabetes (Areekul et al., 1996; Freney et al., 1987; Potvliege et al., 1984; Reina et al., 1992). *Sphingobacterium mizutaii* has been isolated from clinical specimens such as ventricular fluid, synovial fluid, and urine, but its pathogenicity is unknown (Yabuuchi et al., 1983). *Sphingobacterium thalpophilum* has been isolated from blood and from leg and foot wounds (Holmes et al., 1983). Strains of *Sphingobacterium thalpophilum* have also been isolated from composted manure and are capable of converting oleic acid to 10-ketostearic acid and 10-hydroxystearic acid; these strains may be suited for developing a large-scale production of these products (Kuo et al., 2000).

Sphingobacterium antarcticum has been isolated only from Antarctic soil (Shivaji et al., 1992) and, since it cannot grow above 30°C (at least under laboratory conditions,) it seems unlikely to have clinical significance.

The type strain of *Sphingobacterium faecium* was isolated from the feces of *Bos sprunigenius Taurus*, and reference strain IFO 15337 was isolated from the feces of *Hippopotamus amphibious*. The pathogenicity, if any, of *Sphingobacterium faecium* is unknown.

An aerobic, mesophilic, Gram-stain-negative, nonmotile, rod-shaped bacterium named "*Candidatus comitans*", which is phylogenetically related to the genus *Sphingobacterium*, was reported to live in a mutualistic association with the myxobacterium *Chondromyces crocatus* (Jacobi et al., 1997, 1996). Although rod-shaped when observed in association with the myxobacteria, the cells became highly pleomorphic when grown without the myxobacteria and stopped multiplying after a few transfers. Moreover, some strains of the myxobacteria failed to grow in the absence of the companion sphingobacterium. The nature of the association between the two bacterial species is not understood, but "*Candidatus comitans*" grew well when the two bacteria were physically separated by a membrane that allowed small molecules produced by either strain to pass through.

Some sphingobacteria can utilize or degrade unusual compounds. *Sphingobacterium multivorum* isolated from aerobic sludge was reported to use the herbicide mefenacet as a carbon and energy source (Ye et al., 2004). A strain of *Sphingobacterium multivorum* that was resistant to the antimicrobial compound ε-poly-L-lysine (ε-PL), a compound that is industrially produced by *Streptomyces albulus* and used as a food additive due to its antimicrobial activities, was reported to produce an aminopeptidase that degraded the ε-PL and released L-lysine (Kito et al., 2002). A soil isolate of *Sphingobacterium multivorum* that had been grown on KDN (2-keto-3-deoxy-D-glycero-D-galacto-nononic acid)-oligosaccharide alditols as the sole carbon source synthesized an inducible, periplasmic sialidase that selectively catalyzed the hydrolysis of different types of ketosidic linkages of KDN, but not *N*-acylneuraminyl linkages (Nishino et al., 1996). Some strains of *Sphingobacterium thalpophilum* have been reported to be capable of cometabolic degradation of pentachlorophenol (PCP) when grown on glucose; the PCP-degrading enzyme was induced by the PCP itself (Liu et al., 2004).

Sphingobacterium multivorum has been used as the recipient via conjugation of a Tn*5*-marked symbiotic plasmid from *Rhizobium leguminosarum* biovar *trifolii*; root nodules were formed in white clover by the transconjugant but did not fix N_2 (Fenton and Jarvis, 1994).

Differentiation of the genus *Sphingobacterium* from other genera

The genus *Sphingobacterium* differs from the genus *Flavobacterium* by the presence of high concentrations of sphingophospholipids as cellular lipid components, by having menaquinone 7 instead of menaquinone 6, by lacking gliding motility (most strains of *Flavobacterium* are positive), and by having a higher G+C content of its DNA (39–45 vs 32–37 mol%).

Sphingobacterium differs from the genus *Pedobacter* by lacking heparinase (*Pedobacter piscium* is an exception, and *Sphingobacterium antarcticum* has not been tested), production of acid from melibiose (*Sphingobacterium antarcticum* has not been tested, and most strains of *Pedobacter* are negative), an ability to assimilate melezitose (data not reported for *Sphingobacterium antarcticum*), the presence of urease activity (however, most strains of *Sphingobacterium mizutaii* do not hydrolyze urea), and the presence of α-fucosidase activity (except *Sphingobacterium mizutaii*).

Taxonomic comments

Sphingobacterium spiritivorum and *Sphingobacterium multivorum* were formerly included in the genus *Flavobacterium* in the first edition of *Bergey's Manual of Systematic Bacteriology* (Holmes et al., 1984). Later, these species were placed in a newly created genus, *Sphingobacterium*, by Yabuuchi et al. (1983) based on DNA–DNA hybridization data and chemotaxonomic considerations, especially the presence of sphingophospholipids. These authors also included a third species, *Sphingobacterium mizutae* (name corrected to *mizutaii* by Holmes et al., 1988). *Sphingobacterium mizutaii* was renamed *Flavobacterium mizutaii* as an objective synonym by Holmes et al. (1988), however, rRNA cistron analysis (summarized by Bernardet et al., 1996) indicates that *Sphingobacterium mizutaii*, *Sphingobacterium spiritivorum*, and *Sphingobacterium multivorum*, although closely related to the genus *Flavobacterium*, are sufficiently distinct from it to deserve placement in a separate genus.

Holmes et al. (1988) created the species *Sphingobacterium yabuuchiae*, which differed from *Sphingobacterium spiritivorum*, despite a DNA–DNA hybridization value of 63% between them, by not hydrolyzing 2-naphthyl phosphate (acid phosphatase) strongly at pH 5.4. However, Takeuchi and Yokota (1992) reported that the type strains of both species did hydrolyze 2-naphthyl phosphate at pH 5.4, and that the two species show high similarity in whole-cell protein patterns in SDS–PAGE and in their physiological characteristics. Consequently, Takeuchi and Yokota concluded that the two species were synonymous, i.e., *Sphingobacterium spiritivorum*.

The species *Sphingobacterium faecium* was added to the genus by Takeuchi and Yokota (1992), based on DNA–DNA hybridization, SDS-PAGE, and chemotaxonomic features.

The heparinase-positive species *Sphingobacterium heparinum* (Payza and Korn 1956) Takeuchi and Yokota 1993 was reclassified by Steyn et al. (1998) in a new genus, *Pedobacter*, based on studies using SDS-PAGE and DNA–rRNA hybridization. Although most species of *Pedobacter* are heparinase-positive, the heparinase-negative species *Sphingobacterium piscium* Takeuchi and Yokota 1993 was also reclassified into the genus *Pedobacter* by Steyn et al. 1998 based on rRNA cistron analysis.

Sphingobacterium antarcticum was included in the genus by Shivaji et al. (1992) based on chemotaxonomic data. It differed from the other species by its low-temperature range for growth (2–30°C).

Differentiation of the species of the genus *Sphingobacterium*

Table 79 lists characteristics that differentiate the species of *Sphingobacterium.*

TABLE 79. Characteristics differentiating species of the genus *Sphingobacterium*[a,b]

Characteristic	*S. spiritivorum*	*S. anhuiense*	*S. antarcticum*	*S. canadense*	*S. composti* (Ten et al., 2006b)	*S. composti* (Yoo et al., 2007)	*S. daejeonense*	*S. faecium*	*S. kitahiroshimense*	*S. mizutaii*	*S. multivorum*	*S. siyangense*	*S. thalpophilum*
Growth at 5°C	–	+	+	–	–	–	–	+	+	–	–	+	–
Growth at 37°C	+	nr	–	nr	nr	nr	nr	+	–	+	+	nr	+
Growth at 42°C	–	–	–	–	+	+	+	–	nr	–	–	+	+
Nitrate reduction	–	nr	–	nr	nr	–	nr	–	nr	–	–	+	+
Nitrite reduction	–	nr	–	nr	nr	nr	nr	–	nr	+	–	nr	–
Hydrolysis of:													
DNA	+	w	nr	nr	+	–	–	+	nr	–	d	+	d
Esculin	nr	+	nr	+	–	+	–	+	+	nr	nr	–	nr
Gelatin	–	–	+	+	–	–	–	–	–	–	–	–	–
Starch	nr	+	nr	nr	–	–	–	+	nr	nr	nr	+	nr
Urease	+	–		+	–	–	–	+	+	nr	nr	+	nr
α-Galactosidase	+	nr	+	nr	nr	nr	–	+	nr	–	+	nr	+
Acid from ethanol (3%), mannitol	+	nr	–	nr	nr	nr	nr	–	nr	–	–	nr	–
Acid from:													
Adonitol	–	nr	nr	nr	nr	–	nr	nr	nr	d	–	nr	+
L-Arabinose	d	+	–	+	–	+	–	+	+	d	+	–	+
D-Fructose	+	nr	–	+	nr	+	nr	nr	nr	+	+	nr	+
D-Glucose	nr	+	+	+	+	+	+	+	nr	nr	nr	–	nr
Glycerol	+	nr	–	nr	nr	–	nr	nr	nr	–	+	nr	+
Inulin	d	nr	nr	nr	nr	–	nr	+	nr	–	+		+
D-Mannitol	+	–	–	–	–	–	–	–	nr	–	–	–	–
L-Rhamnose	–	–	–	–	–	–	–	–	–	+	d	+	+
Starch	–	nr	nr	+	nr	–	nr	nr	nr	–	+	nr	+

[a]Symbols: +, >85% positive; d, different strains give different reactions (16–84% positive); –, 0–15% positive; w, weak reaction; nr, not reported.

[b]Data taken from Dees et al. (1985), Holmes et al. (1988), Liu et al. (2008), Matsuyama et al. (2008), Mehnaz et al. (2007), Shivaji et al. (1992), Steyn et al. (1998), Takeuchi and Yokota (1992), Ten et al. (2006b), Yabuuchi et al. (1983), Yoo et al. (2007), and Wei et al. (2008).

List of species of the genus *Sphingobacterium*

1. **Sphingobacterium spiritivorum** (Holmes, Owen and Hollis 1982) Yabuuchi, Kaneko, Yano, Moss and Miyoshi 1983, 594[VP] (*Flavobacterium spiritivorum* Holmes, Owen and Hollis 1982, 161; *Flavobacterium yabuuchiae* Holmes, Weaver, Steigerwalt and Brenner 1988, 351)

spi.ri.ti′vo.rum. L. n. *spiritus* spirit; L. adj. suff. *vorus* devouring, eating; N.L. neut. adj. *spiritivorum* spirit-devouring, intended to refer to the ability of the organism to attack spirits (i.e., alcohol), producing acid in the process

The following data are taken from Holmes et al. (1982), Yabuuchi et al. (1983), and Takeuchi and Yokota (1992). Some strains show distinctive spreading growth on media containing 0.3% agar, but little or no spreading on media containing 0.5% agar. At the edge of the spreading growth, the cells are arranged in a monolayer, which is typical of sliding (but not gliding) translocation. Other strains show moderate spreading on semisolid agar medium. Growth occurs at 18–22°C and 37°C but not at 5°C and 42°C.

The following tests are positive: hydrolysis of DNA, esculin, tributyrin, Tween 20, Tween 80, and urea; catalase; oxidase; phosphatase; α-galactosidase; β-D-galactosidase; β-D-cellobiosidase; growth on MacConkey agar, on β-hydroxybutyrate agar, in a standard mineral base medium containing 0.2% glucose and 0.1% $(NH_4)_2SO_4$, and in the presence of 40% bile. In an ammonium salt medium, acid is produced oxidatively from D-arabinose, cellobiose, ethanol (3%), fructose, galactose, glucose, glycerol, lactose, maltose, mannitol, mannose, melezitose, raffinose, rhamnose, salicin, sucrose, trehalose, and xylose. Gelatin is hydrolyzed by the plate method but not liquefied.

The following tests are negative: acid production from adonitol, dulcitol, glycogen, inositol, D-ribose, and sorbitol; alkali production on Christensen's citrate; growth on Simmons citrate; arginine desmidase; arginine dihydrolase; casein hydrolysis; fluorescence on King medium B; gas from glucose in peptone water sugar medium; gluconate oxidation; growth on cetrimide agar; H_2S production; lysine decarboxylase; malonate utilization; nitrate and nitrite reduction; opalescence on lecithovitellin agar; ornithine decarboxylase; phenylalanine deaminase; pigment production on tyrosine agar; reduction of 0.4% selenite; and 3-ketolactose production. Starch hydrolysis was reported as negative by Holmes et al. (1982) but positive by Yabuuchi (1983).

When tested by API ZYM galleries, no activity was found for the following enzymes: esterase (C-4), lipase (C-14); L-valyl-2-napthylamide hydrolase, L-cystyl-2-naphthylamide hydrolase, *N*-glutaryl-L-phenylalanyl-2-naphthylamide hydrolase, α-D-galactosidase, β-D-glucuronidase, α-D-mannosidase, α-D-xylosidase, β-D-fucosidase, β-L-fucosidase, *N*-acetyl-α-D-glucosaminidase, lactosidase, L-tyrosyl-2-naphthylamide hydrolase, L-phenylalanyl-2-naphthylamide hydrolase, L-hydroxyprolyl-2-naphthylamide hydrolase, *N*-benzoyl-L-leucyl-2-naphthylamide hydrolase, L-isoleucyl-2-naphthylamide hydrolase, and L-prolyl-2-naphthylamide hydrolase.

The type strain is susceptible to sulfadiazine and sulfamethoxazole-trimethoprim but resistant to nalidixic acid and 14 other drugs.

Source: human clinical specimens and hospital environments.

DNA G+C content (mol%): 40 (T_m).

Type strain: CDC E7288, ATCC 33861, CCUG 13224, CDC 13224, CIP 100542, DSM 11722, GIFU 3101, JCM 1277, JCM 6897, LMG 8347, NBRC 14948, NCTC 11386.

Sequence accession no. (16S rRNA gene): D14026, EF090267, M58778.

2. **Sphingobacterium anhuiense** Wei, Zhou, Wang, Huang and Lai 2008, 2100[VP]

an.hu.i.en'se. N.L. neut. adj. *anhuiense* pertaining to Anhui, the province where the type strain was isolated.

Cells are 0.4–0.8×1.8–2.5 μm. Colonies are circular, convex, and bright yellow after 24 h cultivation at 30°C on TYB agar. Growth occurs at 4–35°C; optimum, 25–30°C. The pH range for growth is 6.0–8.0; optimum, 6.5–7.5. Growth occurs with 0–3% NaCl but not with 4% NaCl in modified TYB broth. Starch and DNA are hydrolyzed but Tween 80, casein, and urea are not. Positive for oxidase, methyl-α-D-glucosidase, β-galactosidase, and the Voges–Proskauer test, but negative for arginine dihydrolase, ornithine decarboxylase, lysine decarboxylase, tryptophan decarboxylase, urease, and H_2S production. Citrate is not utilized and indole is not produced (API 20NE system). The following substrates are utilized: *N*-acetylglucosamine, esculin, D-arabinose, L-arabinose, arbutin, cellobiose, fructose, D-fucose, L-fucose, D-galactose, glucose, inulin, D-lactose, laetrile, maltose, D-mannose, melibiose, methyl-α-D-glucoside, methyl-α-D-mannoside, raffinose, salicoside, starch, and sucrose. The following are not utilized (API 50 CHB system): D-adonitol, D-arabitol, L-arabitol, dulcitol, erythritol, D-fucose, D-gentiobiose, glycerol, glycogen, gluconate, inositol, 2-ketogluconate, 5-ketogluconate, D-lyxose, mannitol, melezitose, methyl-β-D-xyloside, L-rhamnose, D-ribose, sorbitol, L-sorbitose, D-tagatose, turanose, xylitol, D-xylose, and L-xylose. The predominant isoprenoid quinone is MK-7. Major cellular fatty acids are $C_{15:0}$ iso (32.2%), $C_{17:0}$ iso 3-OH (9.8%), and summed feature 3 ($C_{15:0}$ iso 2-OH and/or $C_{16:1}$ ω7*c*; 33.7%).

Source: forest soil in Anhui province, China.

DNA G+C content (mol%): 36.3 (HPLC).

Type strain: CW 186, CCTCC AB 207197, KCTC 22209.

Sequence accession no. (16S rRNA gene): EU364817.

3. **Sphingobacterium antarcticum** corrig. Shivaji, Ray, Rao, Saisree, Jagannadham, Kumar, Reddy and Bhargava 1992, 105[VP]

ant.arc'ti.cum. L. neut. adj. *antarcticum* southern, pertaining to the Antarctic. (*Note*: The original name, *Sphingobacterium antarcticus*, was corrected by Euzéby, 1998)

The following characteristics are taken from Shivaji et al. (1992) and Takeuchi and Yokota (1992). Rods 0.5–1.0 μm×2–3 μm. Colonies on peptone-yeast extract agar are round, pale or bright yellow, smooth, slightly convex, 1–2 mm in diameter. No growth factors are required. Optimal temperature, 25°C; range, 2–30°C, with no growth above 30°C. Optimum pH, 6.9; range for growth, 6–8. NaCl is not required but can be tolerated up to 0.5M.

The following tests are positive: catalase, oxidase, phosphatase, gelatinase, urease, and β-galactosidase; hydrolysis of Tween 20, Tween 80, esculin, and gelatin; H_2S production; growth on β-hydroxybutyrate agar and MacConkey agar; acid production from cellobiose, glucose (1% and 10%), lactose (1% and 10%), mannose, and raffinose; growth on L-arabinose, formate, D-fructose, D-galactose, glucose, glutamate, glycerol, β-hydroxybutyrate, *meso*-inositol, lactate, lactose, malate, maltose, mannitol, D-mannose, pyruvate, raffinose, L-rhamnose, D-ribose, sorbitol, sucrose, succinate, and D-xylose.

The following tests are negative: fluorescent pigment formation on King's A and B medium; gas production from carbohydrates; intracellular poly-β-hydroxybutyrate granules; starch and casein hydrolysis; reduction of nitrate and nitrite; indole production; growth on cetrimide agar; utilization of acetate, cellulose, citrate, dextrin, glycogen, inulin, malonate, melibiose, and starch as carbon sources; arginine dihydrolase; lysine decarboxylase; acid production from arabinose, ethanol, fructose, galactose, glycerol, *meso*-inositol, maltose, mannitol, melezitose, rhamnose, salicin, sorbitol, sucrose, and xylose.

The major carotenoid pigments were identified as zeaxanthin, β-cryptoxanthin, and β-carotene (Jagannadham et al., 2000). The major fatty acid is $C_{16:1}$; the long-chain bases of the sphingophospholipids are $C_{16:1}$, $C_{17:0}$, and $C_{16:0}$.

Source: soil from Schirmacher Oasis, Antarctica.

DNA G+C content: 39–40 (T_m).

Type strain: 4BY, ATCC 51969, MTCC 675.

GenBank accession numbers (16S rRNA gene): not reported.

4. **Sphingobacterium canadense** Mehnaz, Weselowski and Lazarovits 2008, 1[VP] (Effective publication: Mehnaz, Weselowski and Lazarovits 2007, 522.)

ca.na.den′se. N.L. neut. adj. *canadense* of or belonging to Canada, pertaining to its isolation from Canada.

Cells are 0.8–1.0×1.0–1.9 µm. Growth occurs on LB medium at 20–37°C but not at 41°C. Growth occurs at pH 5–10. NaCl concentrations of 0.5–3% are tolerated. After 24–48 h growth on LB, colonies are circular, entire, low convex, and smooth; they initially are off-white in color and turn yellow after a few days; the pigment is not fluorescent. Bacterial cultures are positive for indoleacetic acid production and negative for phosphate solublization. The following substrates are oxidized: dextrin, α-cyclodextrin, *N*-acetyl-D-galactosamine, *N*-acetyl-D-glucosamine, D-cellobiose, D-fructose, D-galactose, gentiobiose, D-glucose, D-lactose, lactulose, maltose, D-mannose, D-melibiose, D-raffinose, sucrose, D-trehalose, succinic acid monomethyl ester, acetic acid, formic acid, α-ketovaleric acid, propionic acid, L-asparagine, L-proline, and glycerol. The following are not oxidized: D-mannitol, L-arabinose, L-fucose, myo-inositol, maltose, L-rhamnose, D-sorbitol, xylitol, citric acid, D,L-lactic acid, malonic acid, succinic acid, D-alanine, L-alanine, potassium gluconate, trisodium citrate, capric acid, adipic acid, phenylacetic acid, L-aspartic acid, L-glutamic acid, L-histidine, L-leucine, L-ornithine, L-phenylalanine, D-serine, L-serine, and L-threonine. Acid is produced from D- and L-arabinose, D-xylose, D-glucose, D-fructose, D-mannose, methyl-α-D-glucopyranoside, esculin ferric citrate, salicin, D-cellobiose, D-maltose, D-lactose, D-melibiose, D-sucrose, inulin, D-melezitose, D-raffinose, starch, D-turanose, and L-fucose. No acid production occurs with glycerol, erythritol, D-ribose, L-xylose, D-adonitol, methyl-β-D-xylopyranoside, D-galactose, L-sorbose, L-rhamnose, dulcitol, inositol, D-mannitol, D-sorbitol, methyl-α-D-mannopyranoside, arbutin, D-trehalose, *N*-acetylglucosamine, amygdalin, glycogen, xylitol, gentiobiose, D-lyxose, D-tagatose, D-fucose, D-arabitol, L-arabitol, potassium gluconate, potassium 2-ketogluconate, and potassium 5-ketogluconate. Positive for oxidase, urease, nitrate reduction, β-glucosidase, β-galactosidase, and gelatin hydrolysis. Negative for indole production, glucose fermentation, and arginine dihydrolase. The major cellular fatty acids are $C_{16:0}$, $C_{15:0}$ iso, $C_{17:0}$ iso 3-OH and summed feature 3 ($C_{16:1}$ ω7*c*/$C_{16:1}$ ω6*c*).

Source: corn roots in London Ontario, Canada.

DNA G+C content (mol%): 40.5 (HPLC).

Type strain: CR11, LMG 23727, NCCB 100125.

Sequence accession no. (16S rRNA gene): AY787820.

5. **Sphingobacterium composti** Ten, Liu, Im, Aslam and Lee 2007, 1372[VP] (Effective publication: Ten, Liu, Im, Aslam and Lee 2007, 1732.)

com.pos′ti. N.L. gen. n. *composti* of compost.

Cells are 0.4–0.8×2.0–2.5 µm. Colonies are 1.0–1.5 mm in diameter, smooth, convex, round, glossy, and slightly yellow after 6d at 30°C on R2A medium. Growth occurs between 15 and 42°C; optimum, 30°C. Growth occurs at pH 5.5–8.5; optimum, pH 6.5–7.0. The organism grows in the presence of 4% NaCl but not 5%. Growth occurs on trypticase soy agar but not on MacConkey agar. Oxidase positive. Lipase negative. DNA is hydrolyzed but chitin, starch, cellulose, xylan, casein, collagen, and esculin are not. The following substrates are utilized for growth: D-glucose, D-galactose, D-mannose, D-lyxose, L-xylose, *N*-acetyl-D-glucosamine, salicin, D-cellobiose, D-lactose, D-maltose, D-sucrose, D-trehalose, D-mannitol, and D-sorbitol. The following substrates are not utilized: D-arabinose, L-arabinose, D-fructose, D-fucose, D-xylose, D-melibiose, D-raffinose, L-rhamnose, L-sorbose, D-ribose, pyruvate, formate, acetate, propionate, D,L-3-hydroxybutyrate, valerate, caprate, maleate, fumarate, phenylacetate, benzoate, 3-hydroxybenzoate, 4-hydroxybenzoate, citrate, lactate, malate, malonate, succinate, glutarate, tartrate, itaconate, adipate, suberate, oxalate, gluconate, ethanol, D-adonitol, dulcitol, inositol, xylitol, glycerol, amygdalin, methanol, glycogen, inulin, dextran, L-alanine, L-arginine, L-asparagine, L-aspartic acid, L-cysteine, L-glutamic acid, L-glutamine, L-histidine, glycine, L-isoleucine, L-leucine, L-lysine, L-methionine, L-phenylalanine, L-proline, L-serine, L-threonine, L-tryptophan, L-tyrosine, and L-valine. With the API 20E system, β-galactosidase activity and the Voges–Proskauer test are positive, but arginine dihydrolase, lysine decarboxylase, ornithine decarboxylase, tryptophan deaminase, urease, and gelatinase activities are not present. H_2S and indole are not produced. Acid is produced from D-glucose, D-melibiose, D-sucrose, and amygdalin but not from L-arabinose, D-mannitol, inositol, D-sorbitol, and L-rhamnose. The major fatty acids are $C_{15:0}$ iso, $C_{17:0}$ iso 3-OH, $C_{15:0}$ iso 2-OH, and/or $C_{16:1}$ ω7*c*.

Source: compost that was collected near Daejeon city in South Korea.

DNA G+C content (mol%): 36.0 (HPLC).

Type strain: T5-12, CCUG 52467, KCTC 12578, LMG 23401.

Sequence accession no. (16S rRNA gene): AB244764.

6. **Sphingobacterium composti** Yoo, Weon, Jang, Kim, Kwon, Go and Stackebrandt 2007, 1592[VP]

com.pos′ti. N.L. gen. n. *composti* of compost.

Cells are 0.5–0.6×1.0–2.0 µm. Colonies on trypticase soy agar are yellow, circular, and convex with entire margins. The temperature, pH, and NaCl ranges for growth are 10–45°C, pH 6–9, and 0–5% NaCl, respectively. Nitrate is not reduced. Indole is not produced. Glucose is not fermented. The following enzyme activities occur: oxidase, arginine dihydrolase, β-galactosidase, alkaline phosphatase, esterase lipase (C8), leucine arylamidase, valine arylamidase, acid phosphatase, naphthol-AS-BI-phosphohydrolase, α-glucosidase, β-glucosidase, and *N*-acetyl-β-glucosaminidase. No activity occurs for esterase (C4), lipase (C14), cystine arylamidase, trypsin, α-chymotrypsin, α-galactosidase, β-glucuronidase, α-mannosidase, and α-fucosidase activities. Tweens 20 and 80 are hydrolyzed but Tween 40, esculin, casein, chitin, carboxymethylcellulose, DNA, gelatin, starch, tyrosine, and urea are not. The following substrates are assimilated: D-glucose, L-arabinose, D-mannose, *N*-acetylglucosamine, D-maltose, D-sucrose, salicin, and D-melibiose. No assimilation occurs of D-mannitol, potassium gluconate, capric acid, adipic acid, malic acid, trisodium citrate, phenylacetic acid, L-rhamnose, D-ribose, inositol, itaconic acid, suberic acid, sodium malonate, sodium acetate, lactic acid, L-alanine, potassium 5-ketogluconate, glycogen, 3-hydroxybenzoic acid, L-serine, L-fucose, D-sorbitol, propionic acid, valeric acid, L-histidine, potassium 2-ketogluconate,

3-hydroxybutyric acid, 4-hydroxybenzoic acid, and L-proline. Acid is produced from D-arabinose, L-arabinose, D-galactose, D-glucose, D-fructose, D-mannose, methyl-α-D-mannopyranoside, methyl-α-D-glucopyranoside, amygdalin, arbutin, salicin, D-cellobiose, D-maltose, D-lactose, D-sucrose, D-raffinose, and L-fucose. Weak acid production occurs with D-xylose, D-melibiose, D-trehalose, and gentiobiose. No acid production occurs with glycerol, erythritol, D-ribose, L-xylose, D-adonitol, methyl-β-D-xylopyranoside, L-sorbose, L-rhamnose, dulcitol, inositol, D-mannitol, D-sorbitol, *N*-acetylglucosamine, inulin, D-melezitose, starch, glycogen, xylitol, D-turanose, D-lyxose, D-tagatose, D-fucose, D-arabitol, L-arabitol, potassium gluconate, potassium 2-ketogluconate, and potassium 5-ketogluconate. The major fatty acids are summed feature 3 ($C_{15:0}$ iso 2-OH and/or $C_{16:1}$ ω7*c*), $C_{15:0}$ iso, and $C_{17:0}$ iso 3-OH.

Source: cotton-waste composts in South Korea.

DNA G+C content (mol%): 42.3 (HPLC).

Type strain: 4M24, DSM 18850, KACC 11313.

Sequence accession no. (16S rRNA gene): EF122436.

Taxonomic note: Sphingobacterium composti Yoo et al. 2007 is a later homonym of *Sphingobacterium composti* Ten et al. 2007.

7. **Sphingobacterium daejeonense** Kim, Ten, Liu, Im and Lee 2006, 2035[VP]

dae.jeon.en'se. N.L. neut. adj. *daejeonense* pertaining to Daejeon, a city in Korea, where the type strain was isolated.

Cells are 0.5–1.0×1.2–1.8 µm. Colonies are 1–2 mm in diameter, smooth, convex, round, and slightly yellow after 3 d at 30°C on R2A medium. Growth occurs between 15 and 42°C; optimum, 30°C. The pH range for growth is pH 5.0–9.0; optimum, pH 6.5–7.0. Tolerates 5% (w/v) NaCl, but not 7%. Growth occurs on Trypticase soy agar but not on MacConkey agar. Oxidase-positive. No hydrolysis of chitin, starch, cellulose, DNA, xylan, casein, and esculin. The following substrates are utilized for growth: D-glucose, D-galactose, D-mannose, D-fructose, D-arabinose, D-lyxose, D-xylose, L-xylose, *N*-acetyl-D-glucosamine, salicin, D-cellobiose, D-lactose, D-maltose, D-melibiose, sucrose, D-trehalose, D-raffinose, D-adonitol, and amygdalin. The following substrates are not utilized for growth: D-fucose, ethanol, L-rhamnose, L-sorbose, L-arabinose, D-ribose, pyruvate, formate, acetate, propionate, D,L-3-hydroxybutyrate, valerate, caprate, maleate, fumarate, phenylacetate, benzoate, 3-hydroxybenzoate, 4-hydroxybenzoate, citrate, lactate, malate, malonate, succinate, glutarate, tartrate, itaconate, adipate, suberate, oxalate, gluconate, dulcitol, inositol, D-mannitol, D-sorbitol, xylitol, glycerol, methanol, glycogen, inulin, dextran, L-alanine, L-arginine, L-asparagine, L-aspartic acid, L-cysteine, L-glutamic acid, L-glutamine, L-histidine, glycine, L-isoleucine, L-leucine, L-lysine, L-methionine, L-phenylalanine, L-proline, L-serine, L-threonine, L-tryptophan, L-tyrosine, and L-valine. With the API 20E system, β-galactosidase and the Voges–Proskauer test are positive, but arginine dihydrolase, lysine decarboxylase, ornithine decarboxylase, tryptophan deaminase, urease, gelatinase, H_2S, and indole production are negative. Acid is produced from D-glucose, D-melibiose, and amygdalin but not from L-arabinose, D-mannitol, inositol, D-sorbitol, L-rhamnose, or sucrose. MK-7 is the predominant menaquinone. The major fatty acids are $C_{15:0}$ iso, summed feature 4 ($C_{15:0}$ iso 2-OH and/or $C_{16:1}$ ω7*c*) and $C_{17:0}$ iso 3-OH.

Source: compost collected near Daejeon city in South Korea

DNA G+C content (mol%): 38.7 (HPLC).

Type strain: TR6-04, CCUG 52468, KCTC 12579, LMG 23402.

Sequence accession no. (16S rRNA gene): AB249372.

8. **Sphingobacterium faecium** Takeuchi and Yokota 1993, 864[VP] (Effective publication: Takeuchi and Yokota 1992, 478.)

fae'ci.um. L. n. *faex faecis* dregs; L. gen. pl. n. *faecium* of feces.

The following data are taken from Takeuchi and Yokota (1993). Rods 0.4–0.5×0.5–1.0 µm. No sliding translocation occurs. Circular, entire, low convex, smooth, and opaque colonies develop on nutrient agar after 2d; on nutrient agar, a yellow or creamy white, nonfluorescent pigment is produced. Growth occurs at 5°C and 37°C but not at 42°C.

The following tests are positive: catalase, oxidase, urease, DNase, and phosphatase (alkaline and acid); esculin hydrolysis; utilization of L-arabinose, cellobiose, D-fructose, D-glucose, inulin, maltose, raffinose, starch, sucrose, and D-xylose; acid production from D- and L-arabinose, L-fucose, inulin, melibiose, melezitose, raffinose, salicin, and trehalose; production of α-galactosidase, β-glucosidase, and valine arylamidase.

The following tests are negative: trypsin; β-glucuronidase; gelatinase; indole; reduction of nitrate and nitrite; utilization of agar, arginate, cellulose, chitin, heparin, pectin, polypectate, and succinoglucan; acid production from ethanol (3%), mannitol, rhamnose, D-ribose, sorbose, xylitol, and L-xylose.

The major isoprenoid quinone is MK-7. The major nonpolar fatty acids from whole cells are $C_{15:0}$, $C_{16:1}$, $C_{15:0}$ iso 2-OH, $C_{15:0}$ iso 3-OH, and $C_{17:0}$ iso 3-OH.

Source: The type strain was isolated from the feces of *Bos sprunigenius taurus.* Reference strain NBRC 15299 was isolated form the feces of *Hippopotamus amphibious.*

DNA G+C content (mol%): 39–40 (HPLC).

Type strain: KS 0470, CIP 104193, JCM 21820, LMG 14022, NBRC 15299.

Sequence accession no. (16S rRNA gene): AJ438176.

9. **Sphingobacterium kitahiroshimense** Matsuyama, Katoh, Ohkushi, Satoh, Kawahara and Yumoto 2008, 1579[VP]

kitahi.ro.shim.en'se. N.L. neut. adj. *kitahiroshimense* pertaining to Kitahiroshima city, where the type strain was isolated).

Cells are 0.5–0.6×0.6–0.8 µm. Oxidase positive. Circular, entire, low convex, smooth colonies develop on nutrient agar after 2d, and a yellow or creamy white, nonfluorescent pigment is produced. The temperature range for growth is 4–37°C; no growth occurs at 42°C. Esculin is hydrolyzed but gelatin is not. Indole is not produced. Nitrate is not reduced. Urease, alkaline phosphatase, acid phosphatase, α-glucosidase, β-glucosidase, α-galactosidase and α-mannnosidase activities are present. The following substrates are utilized (GN2 Biolog system): dextrin, Tween 40, Tween 80, cellobiose, *N*-acetylglucosamine, D-fructose,

D-galactose, gentiobiose, α-D-glucose, α-D-lactose, maltose, D-mannose, melibiose, raffinose, L-rhamnose, sucrose, trehalose, turanose, α-ketovaleric acid, L-alanine, L-glutamic acid, L-threonine, L-proline, and L-serine. Acid is produced from trehalose, raffinose, L-arabinose, and L-xylose. The following enzyme activities do not occur: arginine dihydrolase, DNase, esterase (C4), lipase (C4), cystine arylamidase, trypsin, chymotrypsin, β-galactosidase, and fucosidase. No utilization occurs of citrate, L-arabinose, L-histidine, adonitol, erythritol, L-fucose, D-sorbitol, xylitol, formic acid, D-gluconic acid, DL-lactic acid, malonic acid, inosine, or glycerol. Acid is not produced from rhamnose. Contains branched-17:0 dihydrosphingosine as the main sphingosine. Predominant fatty acids are $C_{15:0}$ iso 2-OH and/or $C_{16:1}$ ω7*c* (summed feature 3), 40.28%; $C_{15:0}$ iso, 28.89% and $C_{17:0}$ iso 3-OH, 12.83%.

Source: soil from Kitahiroshima city, Japan.

DNA G+C content (mol%): 36.9 (HPLC).

Type strain: 10C, JCM 14970, NCIMB 14398.

Sequence accession no. (16S rRNA gene): AB361248.

10. **Sphingobacterium mizutaii** Yabuuchi, Kaneko, Yano, Moss and Miyoshi 1983, 595[VP]

mi.zu'tai.i. N.L. gen. masc. n. *mizutaii* of Mizuta, named after Shunsuke Mizuta, Japanese pediatrician, who first reported a case of meningitis in a premature baby from whose spinal fluid the type strain of the species was isolated. (*Note:* The original spelling *mitzae* was corrected by Holmes et al. (1988). Objective synonym: *Flavobacterium mizutaii* Holmes, Weaver, Steigerwalt and Brenner 1988, 353)

The following data are taken from Yabuuchi et al. (1983). The characteristics are as described for the genus, with the following additional information. No sliding or gliding translocation occurs. Bile (40%) is tolerated. Acid is produced from D-arabinose, cellobiose, fructose, glucose, lactose, maltose, mannose, melezitose, rhamnose, salicin, and sucrose in oxidation-fermentation medium. Esculin is hydrolyzed.

The following tests are negative: α-galactosidase; acid production from ethanol (3%), glycogen, inulin, and mannitol; utilization of malonate; growth at pH 5.0; fumarate respiration; hydrolysis of starch and DNA; and reaction on egg yolk agar.

Most strains do not oxidize glycogen or hydrolyze urea. Nitrate is not reduced to nitrite, but the type strain and some other strains reduce nitrite to N_2.

The major fatty acids are $C_{16:1}$, $C_{15:0}$ iso, $C_{15:0}$ iso 2-OH, and $C_{17:0}$ iso 3-OH, with small percentages of $C_{16:0}$, $C_{17:1}$ iso, and $C_{15:0}$ iso 3-OH.

The type strain is susceptible to carbenicillin, chloramphenicol, tetracycline, erythromycin, sulfadiazine, and sulfamethoxazole-trimethoprim but is resistant to penicillin, ampicillin, cephalothin, streptomycin, amikacin, kanamycin, gentamicin, clindamycin, colistin, and polymyxin B.

Pathogenicity has not been defined. The natural habitat of this organism is not known, but the organism has been isolated from clinical specimens.

DNA G+C content (mol%): 39.0–41.5 (T_m).

Type strain: ATCC 33299, CCUG 15907, CIP 101122, GIFU 1203, KC1794, LMG 8340, NBRC 14946, NCTC 12149.

Sequence accession no. (16S rRNA gene): AJ438175, D14024, M58796.

11. **Sphingobacterium multivorum** (Holmes, Owen and Weaver 1981) Yabuuchi, Kaneko, Yano, Moss and Miyoshi 1983, 594[VP] (*Flavobacterium multivorum* Holmes, Owen and Weaver 1981, 25; CDC group IIk biotype 2 of Tatum, Ewing and Weaver 1974)

mul.ti.vo'rum. L. adj. *multus* many; L. v. *vorare* to swallow; N.L. neut. adj. *multivorum* intended to mean produces acid from many carbohydrates.

The following data are taken from Holmes et al. (1981) and Yabuuchi et al. (1983). No sliding motility occurs on semisolid agar medium. A weak acid reaction is observed on Kligler iron agar slants. Growth occurs at 18–20°C and 37°C but not at 5°C and 42°C. Growth occurs in a standard mineral base medium containing 0.2% glucose and 0.1% $(NH_4)_2SO_4$.

The following tests are positive: growth on MacConkey agar; growth in heart infusion broth adjusted to pH 5.0; hydrolysis of urea, esculin, tributyrin, and Tween 20; oxidase, phosphatase, α-galactosidase, and β-galactosidase; acid production (in an ammonium salt medium) from D-arabinose, L-arabinose, cellobiose, fructose, galactose, glucose, glycerol, glycogen, inulin, lactose, maltose, mannose, melezitose, raffinose, salicin, sucrose, trehalose, and xylose.

The following tests are negative: fluorescence on King medium B; acid production from adonitol, dulcitol, ethanol (3%), inositol, mannitol, rhamnose, D-ribose, and sorbitol; alkaline reaction on Christensen citrate agar; growth on cetrimide; growth on Simmons citrate; arginine desimidase; arginine deaminase; casein hydrolysis; fluorescence on King medium B; gas from peptone–water–glucose medium; gluconate oxidation; H_2S production (by both lead acetate paper and triple sugar iron agar methods); indole production; 3-ketolactose production; lysine decarboxylase; malonate utilization; ornithine decarboxylase, phenylalanine deaminase; reduction of selenite and nitrate; reduction of nitrite to N_2; and starch hydrolysis.

Tests for the following enzyme activities are positive (using various API ZYM galleries): alkaline phosphatase, L-leucyl-2-naphthylamide hydrolase, acid phosphatase, phosphoamidase, *N*-acetyl-β-D-glucosaminidase, β-D-cellobiosidase, L-phenylalanyl-2-naphthylamide hydrolase, L-lysyl-2-naphthylamide hydrolase, L-histidyl-2-naphthylamide hydrolase, glycyl-2-naphthylamide hydrolase, α-L-aspartyl-2-naphthylamide hydrolase, L-arginyl-2-naphthylamide hydrolase, L-alanyl-2-naphthylamide hydrolase, *S*-benzyl-L-cysteyl-2-naphthylamide hydrolase, L-methionyl-2-naphthylamide hydrolase, L-glycyl-glycyl-2-naphthylamide hydrolase, glycyl-L-phenylalanyl-2-naphthylamide hydrolase, glycyl-L-prolyl-2-naphthylamide hydrolase, L-leucyl-glycyl-2-naphthylamide hydrolase, L-seryl-tyrosyl-2-naphthylamide hydrolase, L-glutaminyl-2-naphthylamide hydrolase, α-L-glutamyl-2-naphthylamide hydrolase, L-ornithyl-2-naphthylamide hydrolase, L-seryl-2-naphthylamide hydrolase, L-tryptophyl-2-naphthylamide hydrolase, and *N*-carbobenzoxy-glycyl-glycyl-L-arginyl-2-naphthylamide hydrolase.

Tests for the following enzyme activities are negative (using various API ZYM galleries): esterase (C-4), lipase (C-14),

L-cystyl-2-naphtylamide hydrolase, *N*-benzoyl-DL-arginyl-2-naphthylamide hydrolase, *N*-glutaryl-L-phenylalanyl-2-naphthylamide hydrolase, β-D-glucuronidase, α-D-mannosidase, β-D-fucosidase, β-L-fucosidase, *N*-acetyl-α-D-glucosaminidase, arylsulfatase, L-hydroxyprolyl-2-naphthylamide hydrolase, *N*-benzoyl-L-leucyl-2-naphthylamide hydrolase, *N*-carbobenzoxy-L-arginyl-4-methoxyl-2-naphthylamide hydrolase, and L-prolyl-2-naphthylamide hydrolase.

The major carotenoid pigments were identified as zeaxanthin, β-cryptoxanthin, and β-carotene (Jagannadham et al., 2000). The major fatty acids extracted from saponified whole cells of the type strain are $C_{15:0}$ iso 2-OH, $C_{15:0}$ iso, $C_{16:1}$, and $C_{17:0}$ iso 3-OH. Zeaxanthin and menaquinone-7 biosynthesis occurs via the methylerythritol phosphate pathway (Rosa-Putra et al., 2001).

The type strain is susceptible to sulfadiazine, sulfamethoxazole-trimethoprim, carbenicillin and gentamicin but resistant to penicillin, ampicillin, cephalothin, amikacin, kanamycin, colistin, and polymyxin B. Intermediate levels of susceptibility are observed with streptomycin, tetracycline, chloramphenicol, erythromycin, and nalidixic acid.

DNA G+C content (mol%): 42 (T_m).

Type strain: CDC B5533, ATCC 33613, CCUG 11736, CIP 100541, DSM 11691, GIFU 2812, JCM 21156, LMG 8342, NBRC 14947, NCTC 11343, NRRL B-14861.

Sequence accession no. (16S rRNA gene): AB100738, AY787820, D14025.

12. **Sphingobacterium siyangense** Liu, Shen, Wang and Chen 2008, 1461[VP]

si.yang.en′se. N.L. neut. adj. *siyangense* pertaining to Siyang in Jiangsu Province, China, the city where the strain was isolated).

Cells are 0.7–0.9×0.8–1.7 µm. Colonies are 1.0–2.0 mm in diameter, slightly yellowish, convex, circular and smooth with entire margins after 2 d on LB agar. Growth occurs at 4–42°C; optimum, 30–37°C. Growth occurs at pH 3.0–10.0; optimum, 6.0–8.0. Tolerant to 0–4% NaCl; optimum, 0–2%. Growth occurs on nutrient agar and cetrimide agar, but not on Simmons' citrate agar or MacConkey agar. Oxidase positive. The following tests are positive (API 20E system): β-galactosidase, arginine dihydrolase, arginine decarboxylase, urease, and nitrate reduction. Negative for ornithine decarboxylase, lysine decarboxylase, tryptophan deaminase, indole and H_2S production, and the Voges–Proskauer reaction. Starch, DNA, esculin, and Tween 20 are hydrolyzed but casein, gelatin, carboxymethylcellulose, Tween 80, and chitin are not. Acid is produced from trehalose and L-rhamnose, but not from D-glucose, cellobiose, D-galactose, sucrose, D-lactose, melezitose, melibiose, raffinose, D-mannitol, D-sorbitol, and inositol. The following substrates are utilized: D-xylose, melibiose, turanose, maltose, D-fructose, D-glucose, D-lactose, sucrose, L-sorbose, trehalose, D-galactose, raffinose, L-rhamnose, L-arabinose, cellobiose, D-ribose, D-mannose, melezitose, i-erythritol, adonitol, L-arabitol, xylitol, D-sorbitol, D-mannitol, salicin, inulin, dextrin, *N*-acetyl-D-glucosamine, gluconate, lysine, amygdalin and glycerol; inositol, acetate, malonate, and tyrosine are not utilized. The predominant fatty acids are $C_{15:0}$ iso (32.9%), $C_{16:0}$ (10.9%) and summed feature 3 ($C_{15:0}$ iso 2-OH and/or $C_{16:1}$ ω7*c*; 24.1%).

Source: soil sample from Jiangsu Province, PR China.

DNA G+C content (mol%): 38.5 (T_m).

Type strain: SY1, CGMCC 1.6855, KCTC 22131.

Sequence accession no. (16S rRNA gene): EU046272.

13. **Sphingobacterium thalpophilum** (Holmes, Hollis, Steigerwalt, Pickett and Brenner 1983) Takeuchi and Yokota 1993, 864[VP] (Effective publication: Takeuchi and Yokota 1992, 479) (*Flavobacterium thalpophilum* Holmes, Hollis, Steigerwalt, Pickett and Brenner 1983, 679)

thal.po′phi.lum. Gr. n. *thalpos* warmth; Gr. neut. adj. *philon* loving; N.L. neut. adj. *thalpophilum* warmth-loving.

The following data are taken from Holmes et al. (1983), Dees et al. (1985), and Takeuchi and Yokota (1993). No sliding motility occurs. Colonies on nutrient agar are circular, low convex, smooth, and opaque after 2 d. Eventually a yellow pigment is produced that is not fluorescent. Nonhemolytic on 5% (v/v) horse blood agar. No brown pigment is produced on tyrosine agar. Growth occurs at room temperature (18–22°C) and at 37°C and at 42°C but not at 5°C.

Does not tolerate 0.0075% KCN. Growth occurs on β-hydroxybutyrate agar (without production of lipid inclusion granules) and on MacConkey agar, but not on cetrimide agar.

The following tests are positive: acid slant and weak acid on Triple iron agar (TSI) slants; utilization of L-arabinose, cellobiose, D-fructose, D-glucose, inulin, maltose, raffinose, sucrose, and D-xylose; reduction of nitrate to nitrite; catalase, phosphatase, urease, phosphatase, and β-D-galactosidase; acid production in ammonium salt medium under aerobic conditions from glucose, adonitol, D- and L-arabinose, cellobiose, fructose, L-fucose, glycerol, inulin, lactose, maltose, D-raffinose, rhamnose, salicin, sucrose, trehalose, and xylose; acid production from 10% (w/v) glucose and lactose; hydrolysis of Tween 20, Tween 80, esculin, DNA, and tributyrin; production of α-galactosidase, β-glucosidase, esterase (C4), cystine arylamidase, valine arylamidase, α-mannosidase, and α-fucosidase.

The following tests are negative: acid production from ethanol (3%), L-lyxose, mannitol, D-ribose, sorbose, and xylitol; hydrolysis of casein; nitrite reduction; opalescence on lecithovitellin agar; citrate utilization; malonate utilization; gluconate oxidation; arginine deimidase; arginine deaminase; lysine decarboxylase; ornithine decarboxylase; phenylalanine deaminase; production of 3-ketolactose; indole production; H_2S production; selenite production; production of acid in ammonium salt medium under aerobic conditions from dulcitol, ethanol, inositol, mannitol, and sorbitol; gas from glucose in peptone water medium.

Starch hydrolysis was reported as negative by Holmes et al. (1983) but positive by Dees et al. (1985). DNase was reported negative by Takeuchi and Yokota (1992) but positive by Dees et al. (1985) and positive for six of seven strains by Holmes et al. (1983).

The following substrates are hydrolyzed (using API ZYM galleries): 2-naphthyl phosphate at pH 8.5; L-leucyl-2-naphthylamide; 2-naphthyl phosphate at pH 5.4; naphthol-AS-B-phosphodiamide; 2-naphthyl-α D-glucopyranoside; 1-naphthyl-*N*-acetyl-β-D-glucosaminide; 4-methylumbelliferyl-cellobiopyranoside; L-pyrrolidonyl-β-naphthylamide; L-lysyl-β-naphthylamide; L-histidyl-β-naphthylamide; glycyl-β-naphthylamide; L-aspartyl-β-naphthylamide; L-arginyl-β-naphthylamide; L-alanyl-β-naphthylamide; γ-L-glutamyl-β-naphthylamide; *S*-benzyl-L-cysteyl-β-naphthylamide; DL-methionyl-β-naphthylamide; glycyl-glycyl-β-naphthylamide hydrobromide; glycyl-L-phenylalanyl-β-naphthylamide; glycyl-L-prolyl-β-naphthylamide; L-leucyl-glycyl-β-naphthylamide; L-seryl-L-tyrosyl-β-naphthylamide; L-glutamine-β-naphthylamide hydrochloride; α-L-glutamyl-β-naphthylamide; L-ornithyl-β-naphthylamide; L-seryl-β-naphthylamide; L-threonyl-β-naphthylamide; L-tryptophyl-β-naphthylamide; and *N*-carbobenzoxy-glycyl-glycyl-L-arginine-β-naphthylamide.

No hydrolysis of the following substrates occurs (using API ZYM galleries): 2-naphthyl butyrate; 2-naphthyl myristate; L-cystyl-2-naphthylamide; *N*-benzoyl-DL-arginine-2-naphthylamide; *N*-glutaryl-phenylalanine-2-naphthylamide; naphthol-AS-B1-β-D-glucuronic acid; 6-bromo-2-naphthyl-α-D-mannopyranoside; 6-bromo-2-naphthyl-β-D-xylopyranoside; *p*-nitrophenyl-α-D-xylopyranoside; *p*-nitrophenyl-β-D-fucopyranoside; *p*-nitrophenyl-β-L-fucopyranoside; *o*-nitrophenyl-*N*-acetyl-α-D-glucosaminide; L-hydroxyprolyl-β-naphthylamide; *N*-benzoyl-L-leucyl-β-naphthylamide; L-isoleucyl-β-naphthylamide; and L-prolyl-β-naphthylamide hydrochloride.

The major nonpolar fatty acids from whole cells are $C_{15:0}$ iso, $C_{16:1}$, $C_{15:0}$ 2-OH, $C_{15:0}$ 3-OH, and $C_{17:0}$ iso 3-OH.

Resistant to ampicillin, carbenicillin, cephalothin, amikacin, gentamicin, kanamycin, tobramycin, chloramphenicol, and tetracycline (strain K-1232 is susceptible). Susceptible to rifampin.

DNA G+C content (mol%): is 42–45 (T_m).

Type strain: K-1173, ATCC 43320, CCUG 22397, CIP 100935, NBRC 14963, NCTC 11429, NRRL B-14902.

Sequence accession no. (16S rRNA gene): AJ438177, D14020, M58779.

Genus II. **Pedobacter** Steyn, Segers, Vancanneyt, Sandra, Kersters and Joubert 1998, 171[VP]

Rosa Margesin and Sisinthy Shivaji

Pe.do.bac′ter. Gr. neut. n. *pedon* soil; N.L. masc. n. *bacter* rod; N.L. masc. n. *Pedobacter* rod from soil.

Rod-shaped cells, 0.5–0.9×0.8–3.2 μm. **Gram-stain-negative.** Endospores are not formed. Nonmotile (*Pedobacter africanus, Pedobacter caeni, Pedobacter roseus, Pedobacter suwonensis, Pedobacter piscium, Pedobacter himalayensis*) or motile by gliding (*Pedobacter heparinus, Pedobacter cryoconitis, Pedobacter saltans*). **Aerobic.** Isolates grow at 5°C; maximum temperature for growth is 25°C (*Pedobacter cryoconitis, Pedobacter himalyensis*) or 30°C; *Pedobacter saltans* and *Pedobacter suwonensis* grow at 37°C. Colonies on solid media are round (diameter, 0.5–4 mm), convex with entire margins, pale white or pigmented (yellow to pink), sometimes mucoid/slimy (*Pedobacter caeni, Pedobacter cryoconitis, Pedobacter himalayensis, Pedobacter suwonensis*). **Catalase-, oxidase-, and phosphatase-positive; esculin hydrolysis positive. Nitrate reductase- and urease-negative.** H_2S is not produced from thiosulfate; indole is not produced from tryptophan. Chemoorganotrophic with oxidative type of metabolism. Glucose is not fermented. D-Glucose, D-lactose, D-maltose, D-mannose, L-melibiose, *N*-acetylglucosamine, salicin, sucrose, and trehalose are used as sole carbon sources. Adipate, caprate, citrate, citric acid, D-arabitol, L-arabitol, D-fucose, lysine, D-lyxose, malate, D-tagatose, dulcitol, inositol, 5-ketogluconate, phenylacetate, and L-sorbose are not utilized as sole carbon sources. Acid is not produced from D-adonitol and L-rhamnose. Major fatty acids are $C_{15:0}$ iso 2-OH/$C_{16:1}$ ω7*c*, $C_{15:0}$ iso, and $C_{17:0}$ iso 3-OH. All the species are phylogenetically closely related to one another at the 16S rRNA gene level (>95%) except *Pedobacter saltans*, which shows >10% difference.

DNA G+C content (mol%): 36–45.

Type species: **Pedobacter heparinus** (Payza and Korn 1956) Steyn, Segers, Vancanneyt, Sandra, Kersters and Joubert 1998, 175[VP]; (*Flavobacterium heparinum* Payza and Korn 1956, 854; *Cytophaga heparina* Christensen 1980, 474; *Sphingobacterium heparinum* Takeuchi and Yokota 1992, 465–482).

Further descriptive information

At the time of writing, nine species of *Pedobacter* had been described; five of them since 2003. See the *Editorial note*, below, for a listing of the 22 species since published. However, only the type species *Pedobacter heparinus* has been studied in detail. Systematic investigations do not exist. In this edition of the *Manual*, the genus *Pedobacter* is included in the phylum *Bacteroidetes*, class *Sphingobacteriia*, order *Sphingobacteriales*, and family *Sphingobacteriaceae*.

Cell morphology. Cells of *Pedobacter* are rod shaped (0.5–0.9 μm wide) with rounded or slightly tapering ends. The length of the rods varies from short (0.5–1.0 μm) to long (up to 6–10 μm) Figure 58. Filament formation has been observed with *Pedobacter caeni*, and occasionally with *Pedobacter heparinus*. No flagella or pili have been observed.

Cell-wall composition. The cell wall of *Pedobacter* species is Gram-stain-negative and there is a need to study its

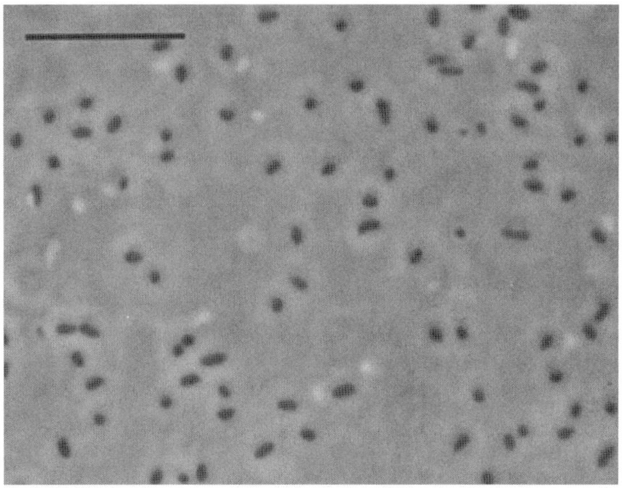

FIGURE 58. Phase-contrast photomicrograph of *Pedobacter heparinus* DSM 2366, grown in nutrient broth for 24 h at 30 °C. Bar = 10 μm.

ultrastructural characteristics. It possesses sphingolipids, but little is known with respect to other lipids that may be present. *Pedobacter cryoconitis* produces an extracellular mucous polysaccharide capsule (Margesin et al., 2003). This could be an adaptative feature, as it is generally observed in psychrophilic microorganisms (Gounot, 1999). Psychrophilic species of *Pedobacter*, namely *Pedobacter cryoconitis* and *Pedobacter himalayensis*, also produce mucus (Margesin et al., 2003; Shivaji et al., 2005), and colonies of mesophilic *Pedobacter caeni* (Vanparys et al., 2005) are slimy. In these species, the chemical nature of the mucus and the slime is yet to be determined. All species of *Pedobacter* contain MK-7 as the major menaquinone (68–90%) (Kwon et al., 2007; Shivaji et al., 2005; Steyn et al., 1998), however, the composition of the menaquinones varies from species to species. In *Pedobacter himalayensis* (Shivaji et al., 2005), the menaquinones present are MK-7 (68%), MK-7 (H_2) (5%), MK-8 (6%), MK-8 (H_2) (13%), and MK-9 (H_2) (8%). In *Pedobacter suwonensis* (Kwon et al., 2007), MK-7 constitutes 90.5% of the total, with MK-9 (5.6%), and MK-8 (3.9%) constituting less than 10% together. The predominance of MK-7 in the genus *Pedobacter* is a characteristic feature of the family *Sphingobacteriaceae* and is a discriminating chemotaxonomic feature that differentiates *Pedobacter* from *Cytophaga* (Reichenbach, 1989).

The cellular fatty acid composition of the various species of *Pedobacter* has been studied as a chemotaxonomic marker, and the predominant fatty acids are $C_{15:0}$ iso, $C_{15:0}$ iso 2-OH, $C_{15:0}$ iso 3-OH, $C_{16:0}$, $C_{16:1}$ ω7*c*, $C_{17:1}$iso ω9*c*, and $C_{17:0}$ iso 3-OH (Dees et al., 1985; Hwang et al., 2006; Kwon et al., 2007; Shivaji et al., 2005; Steyn et al., 1998; Takeuchi and Yokota, 1992; Vanparys et al., 2005, 1983; Yabuuchi and Moss, 1982). In eight of the species, namely *Pedobacter suwonensis*, *Pedobacter himalayensis*, *Pedobacter cryoconitis*, *Pedobacter heparinus*, *Pedobacter africanus*, *Pedobacter piscium*, *Pedobacter saltans*, and *Pedobacter caeni*, all of the above seven fatty acids are present, but in *Pedobacter roseus*, $C_{15:0}$ iso 3-OH, $C_{16:0}$, and $C_{17:1}$ iso ω9*c* are absent and $C_{16:0}$ iso 3-OH is unique to the species (Hwang et al., 2006). Quantitatively, the predominant fatty acid is $C_{15:0}$ iso (15–35%), followed by $C_{16:1}$ ω7*c* (20–31%) (Kwon et al., 2007; Margesin et al., 2003; Shivaji et al., 2005). As a group, the branched-chain fatty acids, ($C_{15:0}$ iso, $C_{15:0}$ iso 2-OH, $C_{15:0}$ iso 3-OH, $C_{17:1}$ iso ω9*c*, and $C_{17:0}$ iso 3-OH), constitute 70–80% of the total fatty acids. The unsaturated fatty acid $C_{16:1}$ ω7*c* is also present in substantial quantities (20–30%) and the saturated fatty acid $C_{16:0}$ constitutes 3–9% of the total cellular fatty acid content (Shivaji et al., 2005; Steyn et al., 1998). The other fatty acids that are observed in a few of the species of *Pedobacter* in trace amounts or <5% of the total fatty acid composition include $C_{14:0}$, $C_{15:0}$, $C_{15:1}$ ω6*c*, $C_{15:0}$ anteiso, $C_{16:1}$ ω5*c*, $C_{16:0}$ 10 methyl, $C_{16:0}$ 2-OH, $C_{16:0}$ 3-OH, $C_{17:1}$ anteiso ω9*c*, and $C_{17:0}$ iso 3-OH (Shivaji et al., 2005; Steyn et al., 1998; Vanparys et al., 2005). The preponderance of branched and unsaturated fatty acids in species of *Pedobacter* would imply that the membranes of these bacteria prefer increased membrane fluidity, since branched and unsaturated fatty acids are known to increase membrane fluidity and unbranched and saturated acids decrease it (Chintalapati et al., 2005; Shivaji, 2005). It is also interesting to note that a characteristic feature of all the representatives of *Sphingobacteriaciae*, including the species belonging to the genera *Sphingobacterium* and *Pedobacter*, is the presence of significant amounts of $C_{15:0}$ iso, $C_{15:0}$ iso 2-OH, $C_{16:1}$ ω7*c*, and $C_{17:0}$ iso 3-OH (Shivaji et al., 2005; Steyn et al., 1998). This feature also differentiates *Pedobacter* from other genera of the family *Flavobacteriaceae* (Bernardet et al., 1996).

Cellular polyamine analysis of *Sphingobacterium antarcticus* (Shivaji et al., 1992), *Sphingobacterium multivorum*, *Sphingobacterium spiritivorum*, *Pedobacter heparinus*, *Pedobacter piscium*, and *Flexibacter canadensis* indicated that homospermidine is the major polyamine. These three genera are phylogenetically related based on 16S rRNA gene sequence and also with respect to the presence of homospermidine. Therefore, homospermidine could serve as a chemotaxonomic marker (Hamana and Nakagawa, 2001b) of these closely related genera. *Pedobacter heparinum* and *Pedobacter piscium* contain homospermidine in amounts of 2.5 μmol/g wet weight. *Pedobacter heparinum* also contains low amounts of putrescin and spermidin (0.04 and 0.07 μmol/g wet weight, respectively). Cadaverine, spermine, and agmatine were not detected (Hamana and Nakagawa, 2001b).

Colonial characteristics. Colonies on solid media are round, convex, with entire margins. They become visible on nutrient agar after 48 h at 20–25°C, and are typically 1–4 mm in diameter. *Pedobacter caeni* colonies range from 0.5 to 8 mm in diameter (Vanparys et al., 2005). The size of the colonies varies with the temperature, age, and growth medium. Colonies are smaller when growth conditions are suboptimal and unfavorable (personal observations).

Most species of *Pedobacter* are pigmented. On nutrient agar, the pigment color is yellow in *Pedobacter cryoconitis*, *Pedobacter heparinus*, *Pedobacter africanus*, *Pedobacter piscium*, *Pedobacter saltans*, and *Pedobacter caeni*; pinkish yellow in *Pedobacter suwonensis*; and pink in *Pedobacter roseus* (Hwang et al., 2006; Kwon et al., 2007; Margesin et al., 2003; Shivaji et al., 2005; Steyn et al., 1998; Vanparys et al., 2005). Colonies of *Pedobacter himalayensis* and *Pedobacter saltans* are pale white (Shivaji et al., 2005). The intensity of pigmentation varies considerably and may be affected by the growth medium, the incubation temperature, and the incubation period. The degree of pigmentation is more pronounced with aging: colonies may be initially creamy white and then turn to yellow. The pigments are cell bound and do not diffuse into the medium, and the color may vary from translucent yellow to dirty yellow or bright yellow depending on the media composition, as observed in *Pedobacter heparinus* (Christensen, 1980; Payza and Korn, 1956; Steyn et al., 1998). The pigments are nonfluorescent in ultraviolet light. The pink pigment in *Pedobacter roseus* shows an absorption maximum at 481–482 nm and a shoulder at 499–500 nm and differs from the yellow pigment of *Pedobacter heparinus* which shows two distinct absorption peaks at 452 and 477 nm (Hwang et al., 2006). These pigments of the various pigmented species of *Pedobacter* are not of the flexirubin type and do not give a flexirubin reaction with 20% KOH (Hwang et al., 2006; Kwon et al., 2007; Steyn et al., 1998; Vanparys et al., 2005). Flexirubin pigments reversibly change color from yellow, orange, or pink to violet-red or purple-brown when exposed to KOH (Reichenbach, 1989) and are a characteristic feature of *Flexibacter*, *Sporocytophyga*, and many species of *Flavobacterium* (Hirsch, 1980; Humphrey and Marshall, 1980; Jooste, 1985; Oyaizu et al., 1982; Reichenbach et al., 1981).

Macroscopically visible formation of slime/mucus on solid media is a typical feature of *Pedobacter caeni*, *Pedobacter cryoconitis*, *Pedobacter himalayensis*, and *Pedobacter suwonensis* (Margesin et al., 2003; Shivaji et al., 2005; Vanparys et al., 2005).

Slime formation is more pronounced with aged cultures and induced by low temperature and on media with low nutrient contents (personal observations).

Cells are nonmotile (*Pedobacter africanus*, *Pedobacter caeni*, *Pedobacter roseus*, *Pedobacter suwonensis*, *Pedobacter piscium*, *Pedobacter himalayensis*) or motile by gliding (*Pedobacter heparinus*, *Pedobacter cryoconitis*, *Pedobacter saltans*) (Hwang et al., 2006; Kwon et al., 2007; Margesin et al., 2003; Shivaji et al., 2005; Steyn et al., 1998; Vanparys et al., 2005). Gliding usually results in spreading growth on solid media (Christensen and Cook, 1972). No flagella or pili have been observed (Hwang et al., 2006; Steyn et al., 1998).

Nutrition and growth conditions. All nine *Pedobacter* species are able to grow at 5°C in nutrient broth and on nutrient agar. The optimum temperature for growth (in terms of growth rate) is 20°C for species isolated from cold habitats (*Pedobacter cryoconitis*, *Pedobacter himalayensis*, *Pedobacter piscium*), whereas isolates from temperate regions grow best at 25–30°C (*Pedobacter heparinus*, *Pedobacter africanus*, *Pedobacter caeni*, *Pedobacter roseus*) or 34°C (*Pedobacter suwonensis*). The maximum growth temperature is 25°C for *Pedobacter himalayensis* and *Pedobacter cryoconitis*, and 30°C for *Pedobacter africanus* and *Pedobacter piscium*. Growth at 37°C has only been reported for *Pedobacter saltans* and *Pedobacter suwonensis* and for some strains of *Pedobacter heparinus* (Steyn et al., 1998; Takeuchi and Yokota, 1992); slow growth at 37°C has been reported for *Pedobacter caeni* (Vanparys et al., 2005).

The pH range for growth in liquid complex media is 5–8 (*Pedobacter cryoconitis*, *Pedobacter suwonensis*), 6–8 (*Pedobacter roseus*), or 6–10 (*Pedobacter himalayensis*); the optimum pH for growth is 7 (Hwang et al., 2006; Kwon et al., 2007; Margesin et al., 2003; Payza and Korn, 1956; Shivaji et al., 2005). Most *Pedobacter* species grow in the presence of 2% (w/v) NaCl, and *Pedobacter himalayensis* grows in the presence of 5% (w/v) NaCl (Shivaji et al., 2005).

Metabolism and metabolic pathways. *Pedobacter* species are chemoorganotrophic with an oxidative type of metabolism. Glucose is not fermented. The carbohydrates most frequently attacked (and utilized as sole carbon source) are D-glucose, D-lactose, D-maltose, D-mannose, melibiose, *N*-acetylglucosamine, salicin, sucrose, trehalose, and often arbutin. Adipate, caprate, citrate, citric acid, D-arabitol, L-arabitol, D-fucose, lysine, D-lyxose, malate, D-tagatose, dulcitol, inositol, 5-ketogluconate, phenylacetate, and L-sorbose are not utilized as sole carbon sources; D-xylitol and erythritol are utilized rarely. Acid is produced from a range of carbohydrates but not from D-adonitol and L-rhamnose, and rarely from D-sorbitol. Most species are not able to grow on MacConkey's agar; exceptions are *Pedobacter himalayensis* and *Pedobacter roseus* (Hwang et al., 2006; Shivaji et al., 2005).

Pedobacter strains degrade a wide range of organic compounds such as the volatile organic compounds ethyl benzene, *m*-xylene, stryrene, and *o*-xylene (Khomenkov et al., 2005); synthetic polypeptides such as high-molecular mass (5000–150,000) polymers of polyaspartic acid (Tabata et al., 2000); and diesel oil (Margesin et al., 2003).

The major respiratory isoprenoid quinone is MK-7 (91% in *Pedobacter suwonensis*, 68% in *Pedobacter himalayensis*). MK-8 (4–6%) and MK-9 (6–8%) have been found in small amounts in *Pedobacter himalayensis* (Kwon et al., 2007; Shivaji et al., 2005) (see above).

All strains show activities of β-galactosidase and phosphatase (alkaline and acidic). Most strains are proteolytic and hydrolyze gelatin. *Pedobacter cryoconitis* produces a cold-active metalloprotease (Margesin et al., 2005). Nitrate reductase, thiosulfate reductase (H_2S production from thiosulfate), urease, and tryptophan deaminase (indole production) are generally not produced. Arginine dihydrolase is only produced by *Pedobacter himalayensis*. Leucine arylamidase and naphthol phosphohydrolase is frequently produced, whereas cystine arylamidase, β-glucuronidase, and lipase (C_{14}) are rarely produced (Hwang et al., 2006; Kwon et al., 2007; Margesin et al., 2003; Shivaji et al., 2005; Steyn et al., 1998; Takeuchi and Yokota, 1992; Vanparys et al., 2005).

A characteristic property of *Pedobacter heparinus* is the production of heparinases and chondroitinases. These enzymes are polysacharide lyases that employ an elimination mechanism to cleave glycosoaminoglycans. Heparinases (EC 4.2.2.7) cleave heparin-like glycosoaminoglycans, such as (sulfated) heparin and heparan sulfate. *Pedobacter heparinus* utilizes heparin as its sole source of carbon, nitrogen, and sulfur. Heparin catabolism occurs both aerobically (higher activity at pH 6.5 than at pH 8.5) and anaerobically (higher activity at pH 8.5 than at pH 6.5) (Payza and Korn, 1956). The bacterium produces three different heparinases that differ in size, charge properties, and substrate specificities. Heparinase I primarily cleaves heparin and contains two unique calcium-binding sites, one of which is directly involved in enzymic activity (Shriver et al., 1999). Heparinase II degrades both heparin and heparan sulfate; it exhibits a stronger affinity toward heparin but its turnover rate of heparan sulfate is higher. The enzyme has been crystallized (see Shaya et al., (2004) and references therein) and differs from heparinase III in substrate specificity (Wei et al., 2005). In addition to *Pedobacter heparinus*, other *Pedobacter* species (*Pedobacter africanus*, *Pedobacter caeni*, *Pedobacter suwonensis*, *Pedobacter himalayensis*, *Pedobacter saltans*) also show heparinase activity.

Pedobacter heparinus produces two chondroitinases (Gu et al., 1995). Chondroitin is a sulfated mucopolysaccharide that consists of glucuronic acid and *N*-acetylgalactosamine units. Chondroitinase B from *Pedobacter heparinus* is the only known lyase that is strictly specific for the glycosaminoglycan dermatan sulfate. Michel et al. (2004) combined crystallographic structural data, enzyme kinetic analysis, and modeling to develop a more complete understanding of how chondroitinase B degrades dermatan sulfate and demonstrated the essential role of calcium for enzyme activity. The enzyme is an important enzymic tool for the identification and structural characterization of glycosaminoglycans, which are modified in many human diseases (Aguilar et al., 2003).

Genetics. *Flavobacterium heparinum* which is now known as *Pedobacter heparinus* is the most extensively studied species of the genus *Pedobacter* because of its ability to produce three heparinases (Lohse and Linhardt, 1992; Yang et al., 1985) and two chondroitinases (Gu et al., 1995). The genetics of this genus developed because of the commercial importance of these enzymes. Su et al. (2001) were the first to report a genetic system, pIBXF1, a mobile conjugative plasmid, for the introduction of heterologous DNA into this bacterium. pIBXF1 was constructed by assembling (a) an *Escherichia coli* mobile plasmid, a derivative of pIB21, for conjugative DNA transfer; (b) a DNA fragment from *Flavobacterium heparinum* to facilitate homologous recombination; and (c) the trimethoprim resistance gene (*Tp*^{r}) of *Flavobacterium heparinum* under the control of the *hepA* regulatory region. Blain et al. (2002) used pIBXF1 and showed that the transconjugants of *Flavobacterium heparinum* expressed high levels of the three heparinase genes (*hepA*, *hepB*, and *hepC*) encoding HepI, HepII, and HepIII, respectively, and the two chondroitinase genes (*cslA* and *cslB*) encoding ChnA and ChnB. The heparinases HepI, HepII, and HepIII and the chondroitinases ChnA and ChnB from the recombinant *Flavobacterium heparinum* strains were purified and shown to be identical to their native counterparts (Blain et al., 2002; Gu et al., 1995; Lohse and Linhardt, 1992; Yang et al., 1985). Earlier, for the want of an expression system in *Flavobacterium heparinum*, heparinases and chondroitinases were expressed in *Escherichia coli* (Sasisekharan et al., 1993; Su et al., 1996; Tkalec et al., 2000), but degradation of protein and formation of inclusion bodies made the process inefficient.

Molecular analysis of the heparinase genes, *hepA* (Sasisekharan et al., 1993), *hepB*, and *hepC* (Su et al., 1996), revealed no significant homology either at the DNA or protein levels, nor were they closely linked on the *Flavobacterium heparinum* chromosome (Su et al., 1996). Structural and functional studies employing chemical modifications and site-directed mutagenesis indicated that in HepI and HepII a histidine residue was critical for the catalytic function (Godavarti and Sasisekharan, 1998; Shriver et al., 1998), and two putative calcium-binding sites in HepI were essential for the catalytic function (Liu et al., 1999; Shriver et al., 1999). Further, crystallization of HepII revealed that it is a unit of two molecules, which is consistent with the finding that recombinant HepII functions as a dimer in solution (Shaya et al., 2004).

The two chondoritinase genes (*cslA* and *cslB*) also share no significant homology either at the DNA or peptide level. They are separated by approximately 5 kbp on the *Flavobacterium heparinum* chromosome, are translated in the same orientation (Tkalec et al., 2000), and are not linked to the *hep* genes. Crystal structures of both the enzymes (Fethiere et al., 1999, 1998; Huang et al., 1999; Li et al., 1998) suggest that the chondroitinases are very different with respect to their structures and catalytic mechanisms. Michel et al. (2004) studied the structure of chondroitinase B complexed with several dermatan sulfate and chondroitin sulfate oligosaccharides and demonstrated that chondroitinase B absolutely requires calcium for its activity, indicating that the protein-Ca^{2+}-oligosaccharide complex is functionally relevant. A feature common to the heparinases and chondroitinases from *Flavobacterium heparinum* is that these enzymes are posttranslationally modified by glycosylation (Huang et al., 1995) and that the glycosylation site(s) contains the consensus sequence Asp-Ser or Asp-Thr, similar to the sequence described for *Flavobacterium meningosepticum* (Plummer et al., 1995).

Heparinases and chondroitinase from *Flavobacterium heparinum* are presently being developed for therapeutic applications. HepI has been used to neutralize the anticoagulant properties of heparin (Baug and Zimmermann, 1993). HepI and HepIII regulate cell adhesion, differentiation, migration, and proliferation (Jackson et al., 1991; Kjellen and Lindahl, 1991; Silver, 1998). ChnA and ChnB inhibit fibroblast proliferation and tumor cell invasion, proliferation, and angiogenesis (Denholm et al., 2000).

Antibiotic or drug sensitivity. Sensitivity to sulfamethoxazole and resistance to amikacin (conferring β-lactamase activity), ampicillin, gentamicin, and kanamycin are characteristic features of most *Pedobacter* species. Resistance of the type strain, *Pedobacter heparinus*, to gentamicin (60 μg/ml) and kanamycin (100 μg/ml), and its sensitivity to chloroamphenicol (100 μg/ml), erythromycin (300 μg/ml), and tetracycline (50 μg/ml), are useful selective markers for the genetic manipulation of the strain (Su et al., 2001). Information on minimal inhibitory concentrations (MIC) of antimicrobial agents is not available.

Pedobacter species isolated from fruit bodies of *Clitocybe* species (Agaricales) is capable of detoxifying tolaasin, an extracellular toxin that induces brown blotch disease of cultivated mushrooms (Tsukamoto et al., 2002). Cells are released into water from the fungal fruit bodies without disrupting the fungal cells, which indicates that the bacteria are attached to the surface of fungal mycelia rather than residing in the fungal protoplasm.

Pathogenicity. *Pedobacter* strains are classified in Biohazard group 1, which consists of organisms that are unlikely to cause human disease. Strains should be handled under containment level 1 and can be distributed unrestrictedly to any bona fide teaching, research, or industrial institution.

Ecology. All representatives of the genus *Pedobacter* have been isolated from environmental terrestrial and aquatic habitats, such as soil (*Pedobacter heparinus*, *Pedobacter africanus*, *Pedobacter suwonensis*, *Pedobacter saltans*), activated sludge (*Pedobacter africanus*), glacier cryoconite (*Pedobacter cryoconitis*), glacial water (*Pedobacter himalayensis*), and freshwater. Other *Pedobacter* strains have been associated with fish (*Pedobacter piscium*) or fruit bodies of Agaricales (*Clitocybe* sp.), are members of the microflora in a commercial nitrifying inoculum (*Pedobacter caeni*), are found in liquid swine manure, or found in maple sap at the taphole of maple trees (Hwang et al., 2006; Kwon et al., 2007; Lagace et al., 2004; Leung and Topp, 2001; Margesin et al., 2003; Shivaji et al., 2005; Steyn et al., 1992; Tabata et al., 2000; Takeuchi and Yokota, 1992; Vanparys et al., 2005). *Pedobacter* strains have been found worldwide and under various climate conditions ranging from cold regions (Himalayan Mountains, India; Tyrolean Alps, Austria; Iceland) to temperate regions (Belgium, Russia, Canada, USA, Japan, Korea, and South Africa).

Enrichment and isolation procedures

No special procedure is required for the isolation of *Pedobacter* strains. They have been isolated either by direct isolation technique or after enrichment appropriate for aerobic bacteria. For isolation, a suitable buffered diluent is used to release microbial cells from the matrix of a representative amount of sample, and the suspension is diluted to a cell density suitable for enumeration on agar plates. Dilutions are spread onto the surface of solid media for cultivation. An enrichment procedure has been described for *Pedobacter caeni* (Vanparys et al., 2005). The time and temperature required for growth of *Pedobacter* colonies depend on the species and on the environmental conditions prevailing in the natural habitat of the isolates. Colonies appear after 2–3 d at 20–30°C, or after 4–7 d at 10°C. Recommended media for isolation of *Pedobacter* strains include commercially available trypticase soy agar (TSA; Difco), nutrient agar (NA; Difco), and R_2A agar (Difco).

Maintenance procedures

The bacteria can be maintained in the laboratory by regular transfer (subculturing) on the medium used for isolation (most commonly used media are TSA, NA, R_2A) and storage of the plates at 1–4°C. For long-term preservation, storage in liquid nitrogen (–196°C) or at –80°C in 10% skimmed milk or in R2A broth supplemented with 30% (v/v) glycerol is recommended. Culture collections distribute *Pedobacter* strains as (vacuum) dried cultures.

Procedures for testing special characters

Heparinase. The presence of heparinase can be detected by following the disappearance of heparin in inoculated growth media (Gesner and Jenkin, 1961) or by the production of acid formed as a result of the fermentation of the heparin (Salyers et al., 1977). Two plate assays for detecting heparinase activity in growing cells on heparin agar plates have been developed. The test medium described by Joubert et al. (1984) contains phytone (soybean meal digest; 10 g/l), yeast extract (1 g/l), heparin (Grade II, from porcine intestinal mucosa; Sigma; 1 mg/ml), sodium chloride (1 g/l), barium chloride (1 mg/ml), and purified agar (15 g/l). A loopful of a broth culture of the organism to be tested is spread onto the surface of the agar plate, and the inoculated plate is incubated for 4–5 d at 25°C. Sulfate is released by the breakdown of heparin and reacts with barium chloride to form an insoluble white precipitate of barium sulfate on the agar plates. The required incubation time (2–5 d) for the precipitate to form depends on the strain.

Zimmermann et al. (1990) recommends growing the bacteria (*Pedobacter heparinus*) on Luria–Bertani (LB) agar plates containing heparin (1 mg/ml) for 48 h at 30°C. The cells are washed from the plates with distilled water, and 5 ml of a protamine sulfate solution (from salmon; Sigma; 2%) is poured onto the plates. After 1–2 h at room temperature, a clear zone surrounding heparinase-producing cells is observed. Activity of concentrated/purified enzyme solutions can be detected after 1 h at 37°C. The method is based on the differential precipitation of heparin and heparinase-generated heparin fragments by protamine sulfate.

It is important to note that growth conditions must be appropriate for the strains to be tested. Purified agar or agarose must be used, because ordinary laboratory grade agar is contaminated with sulfate, which can give false-positive results. Contamination with sulfated mucopolysaccharides of animal origin must also be avoided (Joubert et al., 1984).

Chondroitinase. An agar plate assay can be used to demonstrate chondroitinase activity in growing cells. The medium and the procedure are similar to those described previously for the detection of heparinase activity, but the medium contains chondroitin sulfate (Sigma Chemical Co., grade II, from whale or shark cartilage; 5 g/l) (Joubert et al., 1984). Aguilar et al. (2003) developed an improved spectrophotometric method to measure chondroitinase activity in cell suspensions. Cells are incubated with chondroitin sulfate or dermatan sulfate, and substrate degradation is followed with the Dimethylmethylene Blue assay.

Utilization of organic compounds as sole carbon sources. This can be tested on the basis of colony formation on mineral medium agar supplemented with 0.1–0.5% of carbon source, or on the basis of growth (optical density) in liquid cultures (Hwang et al., 2006; Margesin et al., 2003; Shivaji et al., 2005).

Enzyme activities. Special care has to be given to incubation temperature when testing enzyme activity. With cold-tolerant strains, the optimum temperature for enzyme production is often considerably lower (5–10°C) than the optimum temperature for growth (Margesin et al., 2005, 2003).

Gliding motility. Gliding is the translocation of nonflagellated bacteria in contact with a solid surface and has been defined by Henrichsen (1972) as "movement which is continuous and regularly follows the long axis of the cells which are predominantly aggregated in bundles during the movement." It is his definition that is recommended here (Holmes et al., 1984). Gliding usually results in spreading growth on solid media (Christensen, 1980). Two preconditions are essential: (1) The agar plates must be humid (preferably freshly poured), because the gliding motility is dependent on humidity (amount of surface moisture) and (2) The cells must be cultured on agar with low nutrient concentration because peptone concentrations higher than 0.25–0.5% are inhibitory to gliding. Spreading may be absent when cells are grown on nutrient agar.

Gliding may be studied by microscopy of the colony edge on thinly poured agar plates after overnight incubation using a high power dry lens (Henrichsen, 1972; Holmes et al., 1984). Steyn et al. (1992) described somersaulting motility of three of four *Pedobacter saltans* strains; on a microscope slide the cells seem to somersault and then glide for one or two cell lengths. Christensen (1980) reclassified *Flavobacterium heparinum* as *Cytophaga heparina* because of its gliding motility.

Differentiation of the genus *Pedobacter* from other genera

The genera *Sphingobacterium* and *Pedobacter* constitute the family *Sphingobacteriaceae* and, although they share a number of common biochemical and chemotaxonomical features, they can be differentiated. The main differential characteristic is the concentration of fatty acids in the membrane lipids. *Sphingobacterium* species, except *Sphingobacterium mizutae*, possess greater amounts of $C_{15:0}$ iso 2-OH (16–26% in *Sphingobacterium* compared with 2–15% in *Pedobacter*). *Pedobacter* taxa, except *Pedobacter saltans*, possess significantly higher amounts of $C_{16:1}$ ω5*c* (1–5%). Significant amounts of $C_{15:0}$ iso 2-OH and/or $C_{16:1}$ ω7*c* are characteristic of all members of the family *Sphingobacteriaceae* (Steyn et al., 1998).

Other useful characteristics to differentiate the two genera pertain to the source of isolation and to selected enzyme activities and carbohydrate metabolism. *Pedobacter* strains have been isolated from various terrestrial and aquatic environmental habitats but not from clinical sources, while most representatives of *Sphingobacterium* are of clinical origin (Steyn et al., 1998) except *Sphingobacterium antarcticus*, which was isolated from soil (Shivaji et al., 1992). Both genera include facultative psychrophilic isolates. Most *Pedobacter* strains are not able to assimilate D-melezitose and to produce acid from melibiose. Activities of urease and α-fucosidase are absent in *Pedobacter* strains but are present in *Sphingobacterium*, except *Sphingobacterium mizutae*. Steyn et al. (1998) used heparinase activity to differentiate *Pedobacter* (which exhibits activity in several species) from *Sphingobacterium* (activity absent). However, Chao et al. (2003, 2004) reported heparinase production by *Sphingobacterium* species. The sphingobacterial heparinase differs from the enzyme produced by *Pedobacter heparinus* in several characteristics such as molecular mass, composition, charge properties, active site, and substrate specificities. Trypsinase activity is absent in heparinase-producing strains of *Pedobacter* (except *Pedobacter piscium*).

Taxonomic comments

The genus *Pedobacter*, represented by the type strain *Pedobacter heparinus*, was carved out from the genus *Sphingobacterium* (Steyn et al., 1998) in the family *Sphingobacteriaceae*. *Pedobacter heparinus* was first described as *Flavobacterium heparinum* (Payza and Korn, 1956) and was later reclassified as *Cytophaga heparina* (Christensen, 1980), although its position in the genus *Cytophaga* was debatable (Reichenbach, 1989). Takeuchi and Yokota (1992) proposed the transfer of *Cytophaga heparina* to the genus *Sphingobacterium* (Yabuuchi et al., 1983) as *Sphingobacterium heparinum* based on the observed homogeneity in morphological, physiological, and chemotaxonomic characteristics with *Sphingobacterium*. Subsequent studies on DNA–rRNA hybridization results using 16 heparinase-producing bacteria related to *Sphingobacterium heparinum* indicat that the genus is phylogenetically heterogeneous since $T_{m(e)}$ differences for the DNA–rRNA hybrids was 8–12°C $T_{m(e)}$ (Steyn et al., 1998). In well-characterized genera, differences in $T_{m(e)}$ values are normally in the range of 4–7°C (De Ley, 1992). Furthermore, based on the rRNA cistron similarity studies, it is clear that the 16 heparinase-producing bacteria with $T_{m(e)}$ ranging between 70 and 80°C are distinct from the *Sphingobacterium* branch, which contains strains of *Sphingobacterium spiritivorum*, *Sphingobacterium multivorum*, *Sphingobacterium mizutae*, *Sphingobacterium faecium*, and *Sphingobacterium thalpophilum* with a mean $T_{m(e)}$ of about 68°C. Prompted by these genotypic studies, the genus *Pedobacter* was created to include the heparinase-producing isolates (Steyn et al., 1998) along with the reclassified *Sphingobacterium heparinum* and *Sphingobacterium piscium* as *Pedobacter heparinus* and *Pedobacter piscium*. It is apparent that *Pedobacter* splits into two sub-branches at $T_{m(e)}$ values of 70.9°C, with *Pedobacter saltans* forming one sub-branch and all the other species including *Pedobacter heparinus* forming the second sub-branch with $T_{m(e)}$ values in the range of 74–80°C (Steyn et al., 1998). Phylogenetic studies also demonstrat that *Pedobacter heparinus* is distantly related to *Flexibacter canadensis* located at the borderline of the *Sphingobacterium* and *Pedobacter* branches (Dobson et al., 1993; Gherna and Woese, 1992; Manz et al., 1996; Nakagawa and Yamasato, 1993, 1996; Steyn et al., 1998).

The genus *Pedobacter* was described by Steyn et al. (1998) to include heparinase-producing, obligately aerobic, Gram-stain-negative rods. Even at that juncture, it was clear that the genus consisted of two distinct sub-branches, with *Pedobacter heparinus* representing one branch and *Pedobacter saltans* the other branch, based on DNA–rRNA hybrids (Steyn et al., 1998). Furthermore, DNA–DNA hybridization studies indicat that the *Pedobacter heparinus* sub-branch is further segregated into six genotypic subgroups represented by *Pedobacter heparinus* (Ia1), *Pedobacter africanus* (Ia3), three unnamed *Pedobacter* species (Ia2, Ib1 and Ib2), *Pedobacter* strain LMG 1035, and *Pedobacter piscium* (II). *Pedobacter saltans* constitutes a homogeneous group (III) by itself (Steyn et al., 1998). Between the genotypic subgroups and between the two sub-branches, the similarity at the whole genome level is <20%. It is also surprising that even within the subgroup the similarity is not always high. For instance, though the two strains of *Pedobacter heparinus* (subgroup Ia1) and the three strains of *Pedobacter saltans* (subgroup III) exhibit >90% similarity, strains of Ia2, Ia3, Ib1, and Ib2 show <60% similarity (Steyn et al., 1998). These results imply genomic heterogeneity within the genotypic groups, and there is a need to investigate the genus more thoroughly. The low DNA–DNA hybridization values indicate that the various species of the genus are not closely related. A similar low similarity has been reported earlier for species of *Cytophaga* (Behrens (1978), from Reichenbach). Between 1998 and 2006, five more new species, namely *Pedobacter cryoconitis*, *Pedobacter caeni*, *Pedobacter himalayensis*, *Pedobacter roseus*, and *Pedobacter suwonensis* have been added to the already existing four species, namely *Pedobacter heparinus*, *Pedobacter piscium*, *Pedobacter africanus*, and *Pedobacter saltans*. These new species also show very little similarity at the DNA–DNA level, with the nearest phylogenetic neighbor ranging from 19.7% between *Pedobacter cryoconitis* and *Pedobacter piscium* (Margesin et al., 2003), 14% between *Pedobacter caeni* and *Pedobacter africanus* (Vanparys et al., 2005), and 42% between *Pedobacter himalayensis* and *Pedobacter cryoconitis* (Shivaji et al., 2005). DNA–DNA hybridization values are not available for *Pedobacter roseus* vs. *Pedobacter heparinus*, and for *Pedobacter suwonensis* vs. *Pedobacter heparinus*, since the phylogenetic neighbors vary by more than 4% at the 16S rRNA gene level (Hwang et al., 2006; Kwon et al., 2007). Thus the entire genus appears to be genotypically heterogeneous, and this is further corroborated by a wide variation in the G+C content of DNA, which ranges from 36 to 45 mol% (Hwang et al., 2006; Kwon et al., 2007; Margesin et al., 2003; Shivaji et al., 2005; Steyn et al., 1998; Vanparys et al., 2005). *Pedobacter saltans* appears to be different from all the remaining eight species and this is clearly evident based on phylogenetic analysis of the 16S rRNA gene sequence. All the species of *Pedobacter* except *Pedobacter saltans* form a robust monophyletic cluster with high bootstrap values (Figure 59). *Pedobacter saltans* differs phylogenetically from all the remaining *Pedobacter* species and exhibits a difference of 9–11% at the 16S rRNA gene level with those species. The remaining species are phylogenetically closely related with one another and the similarity at the rRNA gene level ranges from 2 to 5% (Hwang et al., 2006; Kwon et al., 2007; Margesin et al., 2003; Shivaji et al., 2005; Steyn et al., 1998; Vanparys et al., 2005).

According to the criteria set by Stackebrandt and Goebel (1994) for discriminating species, it would appear that all the existing nine *Pedobacter* species at the 16S rRNA gene level differ by >2.5% and/or exhibit <70% similarity at the DNA–DNA level, and are thus different species, but a few of them, such as *Pedobacter suwonensis* (Kwon et al., 2007) and especially *Pedobacter Saltans*, differ from the nearest phylogenetic neighbor by >5% at the 16S rRNA gene level and have low DNA–DNA hybridization values; thus their species status could be reconsidered by investigating more strains and arriving at distinguishing phenotypic characteristics.

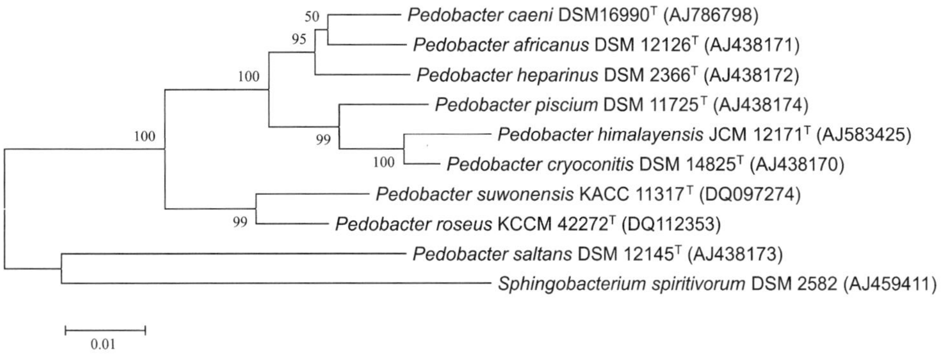

FIGURE 59. Neighbor-joining tree based on 16S rRNA gene sequences (1455 bases) showing the phylogenetic relationship between species of the genus *Pedobacter* and *Sphingobacterium spiritivorum*. Bootstrap values (expressed as percentage of 1000 replications) greater than 50% are given at the nodes.

Editorial note: An additional 22 species have been published since this chapter was written: *Pedobacter agri* (Roh et al., 2008), *Pedobacter alluvionis* (Gordon et al., 2009), *Pedobacter aquatilis* (Gallego et al., 2006), *Pedobacter borealis* (Gordon et al., 2009), *Pedobacter composti* (Lee et al., 2009), *Pedobacter daechungensis* (An et al., 2009), *Pedobacter duraquae* (Muurholm et al., 2007), *Pedobacter ginsengisoli* (Ten et al., 2006a), *Pedobacter glucosidilyticus* (Luo et al., 2010), *Pedobacter hartonius* (Muurholm et al., 2007), *Pedobacter insulae* (Yoon et al., 2007a), *Pedobacter koreensis* (Baik et al., 2007), *Pedobacter lentus* (Yoon et al., 2007c), *Pedobacter metabolipauper* (Muurholm et al., 2007), *Pedobacter nyackensis* (Gordon et al., 2009), *Pedobacter oryzae* (Jeon et al., 2009), *Pedobacter panaciterrae* (Yoon et al., 2007d), *Pedobacter sandarakinus* (Yoon et al., 2006), *Pedobacter steynii* (Muurholm et al., 2007), *Pedobacter terrae* (Yoon et al., 2007b), *Pedobacter terricola* (Yoon et al., 2007c), and *Pedobacter westerhofensis* (Muurholm et al., 2007).

Differentiation of the species of the genus *Pedobacter*

Differential characteristics of the species of *Pedobacter* are presented in Table 80.

List of species of the genus *Pedobacter*

1. **Pedobacter heparinus** (Payza and Korn 1956) Steyn, Segers, Vancanneyt, Sandra, Kersters and Joubert 1998, 175[VP] (*Flavobacterium heparinum* Payza and Korn 1956, 854; *Cytophaga heparina* Christensen 1980, 474; *Sphingobacterium heparinum* Takeuchi and Yokota 1992)

he.pa.ri'nus. Gr. neutr. n. *hepar* liver; N.L. masc. adj. *heparinus* referring to the liver and to the degradation of heparin, a sulfated mucopolysaccharide occurring in liver.

Characteristics are as described for the genus and as listed in Table 80. Cell length varies from short rods to long rods (0.5–6 μm). Motile by gliding. Colony color varies from creamy white on R2A to translucent yellow to dark orange-yellow on TSA and NA. Colonies reach a diameter of 1–3 mm after 48–72 h of cultivation, and are round, convex, with entire margins. Grows at pH 6–8 and at 5 to 30°C. Some strains grow at 37°C (Takeuchi and Yokota, 1992), others do not grow at this temperature (Payza and Korn, 1956). No growth at 42°C. No growth on MacConkey's agar. Positive for heparinase, chondroitinase, assimilation of D-mannitol, D-sorbitol, and L-fucose; acid is not produced from mannose.

DNA G+C content (mol%): 42–43 (T_m).

Type strain: ATCC 13125, CCUG 12810, CIP 104194, CIP 105498, DSM 2366, NBRC 12017, JCM 7457, LMG 4024, LMG 10339, NCIMB 9290, NRRL B-14731.

Sequence accession no. (16S rRNA gene): AJ43817, M11657.

2. **Pedobacter africanus** Steyn, Segers, Vancanneyt, Sandra, Kersters and Joubert 1998, 175[VP]

a.fri.ca'nus. L. masc. adj. *africanus* of or belonging to Africa, the source of isolation.

Characteristics are as described for the genus and as listed in Table 80. Differs from its closest phylogenetic neighbor, *Pedobacter caeni*, by its ability to assimilate L-rhamnose and D-turanose and its inability to assimilate L-xylose and to produce acid from D-lactose, D-melibiose and sucrose. Activities of α-glucosidase, *N*-acetyl-β-glucosaminidase, naphthol phosphohydrolase, and heparinase are present. Produces slimy colonies on agar plates; the colonies are translucent yellow. Grows at pH 6–8.

DNA G+C content (mol%): 43–45 (T_m).

Type strain: CCUG 39353, CIP 105499, DSM 12126, LMG 10353, NBRC 100065, NCIMB 13641.

Sequence accession no. (16S rRNA gene): AJ438171.

3. **Pedobacter caeni** Vanparys, Heylen, Lebbe and De Vos 2005, 1316[VP]

ca.e'ni. L. neut. n. *caenum* sludge; L. gen. neut. n. *caeni* of/from sludge.

Characteristics are as described for the genus and as listed in Table 80. Cell length varies from short to long (3–10 μm). Filament formation occurs. Differs from the other species of the genus by its ability to assimilate L-xylose, its inability to assimilate D-fructose, D-galactose, D-turanose, and D-xylose, its

TABLE 80. Characteristics of diagnostic value in identifying *Pedobacter* species[a,b]

Characteristic	*P. heparinus*	*P. africanus*	*P. caeni*	*P. cryoconitis*	*P. himalayensis*	*P. piscium*	*P. roseus*	*P. saltans*	*P. suwonensis*
Motile by gliding	+	–	–	+	–	–	–	+	–
Slime formation on agar plates	–	–	+	+	+	–	–	–	+
Growth temperature range (°C)	5–30	5–30	1–30	1–25	1–25	1–30	5–33	15–37	2–37
Growth at 37°C	v	–	–	–	–	–	–	+	+
Optimum temperature for growth (°C)	25	25	28	20	22	20	25–30	nd	34
Growth in presence of 5% (w/v) NaCl			nd	–	+		–	nd	
Growth on MacConkey's agar	–	–	–	–	+	–	+	–	–
Major fatty acids (%):									
$C_{17:0}$ iso 3-OH	14.0–15.2	14.7–16.1	12.1	4.2–5.8	6.0	9.2	20.9	12.7	15.8
$C_{15:0}$ iso	27.9–28.2	26.6–33.3	21.3	15.0–20.6	33.0	26.2	17.3	31.4	35.4
$C_{16:0}$ iso 3-OH	–	–	–	–	–	–	7.2	–	nd
Enzyme activities:									
N-Acetyl-β-glucosaminidase	+	+	v	+	+	v	+	+	+
Arginine dihydrolase	–	–	–	–	+	–	–	–	–
α-Chymotrypsin	v	–	–	–	v	+	+	–	–
DNase	v	–	nd	nd	nd	–	–	–	nd
Esterase lipase (C8)	+	v	(+)	–	+	+	+	v	+
α-Fucosidase	–	–	–	–	–	–	+	–	+
α-Galactosidase	v	–	–	v	–	v	+	v	+
Gelatinase	v	v	–	+	+	–	+	–	+
α-Glucosidase	+	+	–	(+)	(+)	+	+	+	+
β-Glucosidase	v	v	–	v	+	+	v	+	+
Heparinase	+	+	+	–	+	–	–	+	+
Lysine decarboxylase	+	+	nd	–	–	+	nd	+	nd
α-Mannosidase	v	v	–	–	–	+	–	–	+
Naphthol phosphohydrolase	+	+	v	+	+	+	+	+	+
Trypsin	–	–	v	v	–	v	+	–	+
Valine arylamidase	v	–	–	(+)	+	+	v	v	+
Utilization of carbon sources:									
D-Adonitol	+	–	–	–	+	–	nd	+	–
D-Arabinose	–	v	–	–	+	–	nd	v	–
D-Cellobiose	+	+	+	+	+	+	+	–	+
D-Fructose	+	v	–	v	+	+	+	v	+
L-Fucose	+	v	–	–	+	–	nd	–	–
D-Galactose	+	v	–	+	+	v	+	+	+
Gluconate	–	–	–	–	+	–	–	–	–
Glycerol	–	v	–	–	+	–	+	+	–
Glycogen	–	–	–	+	–	–	+	–	+
Inulin	+	v	–	(+)	+	–	+	–	–
2-Ketogluconate	–	–	–	+	nd	–	nd	–	–
D-Mannitol	+	–	–	–	+	–	–	–	–
D-Melezitose	v	v	–	v	–	v	nd	v	+
Methyl-α-D-mannoside	+	+	+	–	+	–	nd	–	+
L-Rhamnose	+	+	–	–	–	v	+	+	+
D-Sorbitol	+	v	–	–	v	–	–	–	–
D-Turanose	+	+	–	+	nd	+	nd	v	+
D-Xylose	+	v	–	+	+	+	nd	+	+
L-Xylose	–	–	+	–	+	–	+	–	–
Acid production from:									
D-Arabinose	v	v	+	v	–	v	+	v	+
D-Glucose	v	v	+	v	–	+	+	v	+
Inositol	–	v	+	–	v	–	–	–	–
D-Lactose	–	–	+	–	–	+	–	–	+
D-Mannitol	–	v	–	v	+	v	–	–	–
D-Mannose	–	v	v	–	v	+	+	+	+
D-Melibiose	–	–	+	–	–	v	+	–	+
Sucrose	–	–	+	–	–	+	+	–	+
Susceptibility to antibiotics:									
Penicillin (100 μg/ml)	S	S	R	R	R	S	S	S	R
Ampicillin (50 μg/ml)	R	R	R	R	R	R	R	S	R
Ampicillin (100 μg/ml)	R	S	R	R	R	R	R (weak)	S	R
Streptomycin (100 μg/ml)	R (weak)	R	R	R (weak)	S	R	S	S	S

[a]Symbols: +, 90% or more of the strains positive; –, less than 10% of the strains positive; v, strain instability; nd, data not available; S, sensitive; R, resistant.

[b]Data from Takeuchi and Yokota (1992), Steyn et al. (1998), Margesin et al. (2003), Shivaji et al. (2005), Vanparys et al. (2005), Kwon et al. (2007) and Hwang et al. (2006).

ability to produce acid from inositol, and the absence of the enzymes α-glucosidase, *N*-acetyl-β-glucosaminidase, and naphthol-phosphohydrolase. Produces slimy colonies on TSA, R2A and NA; colonies are initially creamy white and yellow when aging. The colony diameter ranges from 0.5 to 8 mm. Grows at 1–30°C and at pH 6–9. Streptomycin-sensitive.

DNA G+C content (mol%): 42.7 (T_m).

Type strain: DSM 16990, LMG 22862.

Sequence accession no. (16S rRNA gene): AJ786798.

4. **Pedobacter cryoconitis** Margesin, Sproer, Schumann and Schinner 2003, 1295[VP]

cry.o.co.ni'tis. N.L. gen. n. *cryoconitis* from cryoconite, referring to glacier cryoconite, where the strain was first found.

Characteristics are as described for the genus and as listed in Table 80. Differs from the other species of the genus by presence of DNase activity and assimilation of 2-ketogluconate. Differentiated from its closest phylogenetic neighbor, *Pedobacter himalayensis*, by its gliding motility, its ability to assimilate glycogen and to produce acid from mannitol, and its inability to assimilate D-adonitol, D-arabinose, L-fucose, glycerol, and D-mannitol. Heparinase and lysine decarboxylase are absent. Degrades diesel oil cometabolically. Temperature range for growth, 1–25°C. Produces light yellow, slimy colonies on NA and R_2A agar plates.

DNA G+C content (mol%): 43.4 (T_m).

Type strain: A37, DSM 14825, LMG 21415.

Sequence accession no. (16S rRNA gene): AJ438170.

5. **Pedobacter himalayensis** Shivaji, Chaturvedi, Reddy and Suresh 2005, 1084[VP]

him.a.lay.en'sis. N.L. masc. adj. *himalayensis* of or belonging to the Himalaya.

Characteristics are as described for the genus and as listed in Table 80. Differentiated from the other species of the genus by the presence of arginine dihydrolase activity, the ability to assimilate gluconate, and the production of acid from D-mannitol but not from D-arabinose and D-glucose. Differs from its closest phylogenetic neighbor, *Pedobacter cryoconitis*, by its ability to grow on MacConkey's agar, to grow in presence of 5% (w/v) NaCl, to produce heparinase, and to assimilate D-adonitol, D-arabinose, L-fucose, glycerol, D-mannitol, methyl-α-D-mannoside, and L-xylose. Pale-white slimy colonies are formed on R2A and NA, with absence of a yellow or pink pigment. Grows from 4 to 25°C and at pH 6–10.

DNA G+C content (mol%): 41.0 (T_m).

Type strain: HHS 22, JCM 12171, MTCC 6384.

Sequence accession no. (16S rRNA gene): AJ583425.

6. **Pedobacter piscium** (Takeuchi and Yokota 1992) Steyn, Segers, Vancanneyt, Sandra, Kersters and Joubert 1998, 175[VP] (*Sphingobacterium piscium* Takeuchi and Yokota 1992, 481)

pis'ci.um. L. masc. n. *piscis* fish; L. gen. pl. n. *piscium* of/from fishes.

Characteristics are as described for the genus and as listed in Table 80. Differs from the other species of the genus by its ability to assimilate α-mannosidase. Positive for valine arylamidase; negative for heparinase activity and assimilation of methyl-α-D-mannoside. Differs from its closest phylogenetic neighbors *Pedobacter cryoconitis* and *Pedobacter himalayensis* by its ability to grow at 30°C. Colonies are creamy white (PY) to yellow (NA, TSA).

DNA G+C content (mol%): 40–43 (T_m).

Type strain: CIP 104195, ATCC 13125, JCM 7454, LMG 14024, NBRC 14985.

Sequence accession no. (16S rRNA gene): AJ438174.

7. **Pedobacter roseus** Hwang, Choi and Cho 2006, 1834[VP]

ro.se'us. L. masc. adj. *roseus* rose-colored, pink, referring to the color of the colonies on agar plates.

Characteristics are as described for the genus and as listed in Table 80. Differs from the other species of the genus by the presence of significant amounts of the fatty acid $C_{16:0}$ iso 3-OH (7.2%) and the formation of pink colonies on R_2A agar, NA and R2A. Rods are short (length, 0.8–1.1 μm). Grows on MacConkey's agar. Glycerol, inulin, and L-xylose are assimilated. α-Fucosidase activity occurs but not β-glucosidase or heparinase activity. Grows at 5–33°C and pH 5–8.

DNA G+C content (mol%): 41.3 (T_m).

Type strain: CL-GP80, JCM 13399, KCCM 42272.

Sequence accession no. (16S rRNA gene): DQ112353.

8. **Pedobacter saltans** Steyn, Segers, Vancanneyt, Sandra, Kersters and Joubert 1998, 175[VP]

sal'tans. L. v. *saltare* to dance; L. part. adj. *saltans* dancing, referring to the gliding motility of the strain.

Characteristics are as described for the genus and as listed in Table 80. Differs from the other species of the genus by its ability to assimilate L-cellobiose. Strains of the species assimilate D-adonitol and glycerol but are unable to assimilate methyl-α-D-mannoside. Heparinase activity is present. Esterase–lipase (C8) activity is absent. Sensitive to chloroamphenicol (30 μg/ml) and tetracycline (30 μg/ml). Motile by gliding. Grows at 15–37°C and at pH 6–8. Colonies are small and creamy white on R2A, NA and TSA.

DNA G+C content (mol%): 36–38 (T_m).

Type strain: CCUG 39354, ATCC 51119, CIP 105500, DSM 12145, JCM 21818, LMG 9526, LMG 10337, NBRC 100064, NCIMB 13643.

Sequence accession no. (16S rRNA gene): AJ438173.

9. **Pedobacter suwonensis** Kwon, Kim, Lee, Jang, Seok, Kwon, Kim and Weon 2007, 481[VP]

su.won.en'sis. N.L. masc. adj. *suwonensis* of or belonging to Suwon City (Korea), where the strain was first found.

Characteristics are as described for the genus and as listed in Table 80. The cell length is 1.4–3.2 μm. Differs from the other species of the genus by its ability to assimilate D-melezitose, its wide growth temperature range (2–37°C), and its relatively high optimum temperature for growth (34°C). Activities of heparinase, α-fucosidase, and valine arylamidase are present. Pinkish-yellow colonies are formed on R2A, NA and TSA. Produces slime on R2A agar.

DNA G+C content (mol%): 44.2 (T_m).

Type strain: 15-52, DSM 18130, KACC 11317.

Sequence accession no. (16S rRNA gene): DQ097274.

Acknowledgements

R. Margesin thanks Dr F. Schinner for providing the photomicrograph (Figure 58) and Dr H.Y. Weon for providing supplementary data about *Pedobacter suwonensis*. S. Shivaji thanks Ms Preeti Chaturvedi for sequence analysis and construction of the phylogenetic tree.

References

Aguilar, J.A.K., C.R. Lima, A.G.A. Berto and Y.M. Michelacci. 2003. An improved methodology to produce *Flavobacterium heparinum* chondroitinases, important instruments for diagnosis of diseases. Biotechnol. Appl. Biochem. *37*: 115–127.

An, D.S., S.G. Kim, L.N. Ten and C.H. Cho. 2009. *Pedobacter daechungensis* sp. nov., from freshwater lake sediment in South Korea. Int. J. Syst. Evol. Microbiol. *59*: 69–72.

Areekul, S., U. Vongsthongsri, T. Mookto, S. Chettanadee and P. Wilairatana. 1996. *Sphingobacterium multivorum* septicemia: a case report. J. Med. Assoc. Thai *79*: 395–398.

Baik, K.S., Y.D. Park, M.S. Kim, S.C. Park, E.Y. Moon, M.S. Rhee, J.H. Choi and C.N. Seong. 2007. *Pedobacter koreensis* sp. nov., isolated from fresh water. Int. J. Syst. Evol. Microbiol. *57*: 2079–2083.

Baug, R.F. and J.J. Zimmermann. 1993. Heparinase in the activated clotting time assay: monitoring heparin-independent alternations in coagulation function. Perfusion Rev. *1*: 14–28.

Behrens, H. 1978. Charakterisierung der DNA gleitender Bakterien der Ordnung Cytophagales, Doctoral thesis. Technical University of Braunschweig, Germany.

Bernardet, J.-F., P. Segers, M. Vancanneyt, F. Berthe, K. Kersters and P. Vandamme. 1996. Cutting a gordian knot: emended classification and description of the genus *Flavobacterium*, emended description of the family *Flavobacteriaceae*, and proposal of *Flavobacterium hydatis* nom. nov. (basonym, *Cytophaga aquatilis* Strohl and Tait 1978). Int. J. Syst. Bacteriol. *46*: 128–148.

Blain, F., A.L. Tkalec, Z. Shao, C. Poulin, M. Pedneault, K. Gu, B. Eggimann, J. Zimmermann and H. Su. 2002. Expression system for high levels of GAG lyase gene expression and study of the hepA upstream region in *Flavobacterium heparinum*. J. Bacteriol. *184*: 3242–3252.

Chao, Y.P., N.G. Gao, X.L. Cheng, J. Yang, S.J. Qian and S.Z. Zhang. 2003. Rapid purification, characterization and substrate specificity of heparinase from a novel species of *Sphingobacterium*. J. Biochem. *134*: 365–371.

Chao, Y.P., S.X. Xiong, X. Cheng and S. Qian. 2004. Mass spectrometric evidence of heparin disaccharides for the catalytic characterization of a novel endolytic heparinase. Acta Biochim. Biophys. Sin. (Shanghai) *36*: 840–844.

Chintalapati, S., M.D. Kiran and S. Shivaji. 2005. Role of membrane lipid fatty acids in cold adaptation. Cell. Mol. Biol. *50*: 631–642.

Christensen, P. 1980. Description and taxonomic status of *Cytophaga heparina* (Payza and Korn) comb. nov. (basionym, *Flavobacterium heparinum* Payza and Korn 1956). Int. J. Syst. Bacteriol. *30*: 473–475.

Christensen, P.J. and F.D. Cook. 1972. The isolation and enumeration of cytophagas. Can J. Microbiol. *18*: 1933–1940.

De Ley, J. 1992. The *Proteobacteria*: ribosomal RNA cistron similarities and bacterial taxonomy. *In* The Prokaryotes: A Handbook on the Biology of *Bacteria*: Ecophysiology, Isolation, Identification, Applications, 2nd edn, vol. 2 (edited by Balows, Trueper, Dworkin, Harder and Schleifer). Springer, New York, pp. 2111–2140.

Dees, S.B., G.M. Carlone, D. Hollis and C.W. Moss. 1985. Chemical and phenotypic characteristics of *Flavobacterium thalpophilum* compared with those of other *Flavobacterium* and *Sphingobacterium* species. Int. J. Syst. Bacteriol. *35*: 16–22.

Dobson, S.J., R.R. Colwell, T.A. McMeekin and P.D. Franzmann. 1993. Direct sequencing of the polymerase chain reaction-amplified 16S ribosomal RNA gene of *Flavobacterium gondwanense* sp. nov. and *Flavobacterium salegens* sp. nov., two new species from a hypersaline antarctic lake. Int. J. Syst. Bacteriol. *43*: 77–83.

Euzéby, J.P. 1998. Taxonomic note: necessary correction of specific and subspecific epithets according to Rules 12c and 13b of the International Code of Nomenclature of *Bacteria* (1990 revision). Int. J. Syst. Bacteriol. *48*: 1073–1075.

Fethiere, J., B.H. Shilton, Y. Li, M. Allaire, M. Laliberte, B. Eggimann and M. Cygler. 1998. Crystallization and preliminary analysis of chondroitinase AC from *Flavobacterium heparinum*. Acta Crystallogr. D Biol. Crystallogr. *54*: 279–280.

Fethiere, J., B. Eggimann and M. Cygler. 1999. Crystal structure of chondroitinase AC: a glycosaminoglycan lyase with a new fold. J. Mol. Biol. *288*: 635–647.

Freney, J., W. Hansen, C. Ploton, H. Meugnier, S. Madier, N. Bornstein and J. Fleurette. 1987. Septicemia caused by *Sphingobacterium multivorum*. J. Clin. Microbiol. *25*: 1126–1128.

Gallego, V., M.T. Garcia and A. Ventosa. 2006. *Pedobacter aquatilis* sp. nov., isolated from drinking water, and emended description of the genus *Pedobacter*. Int. J. Syst. Evol. Microbiol. *56*: 1853–1858.

Gesner, B.M. and C.R. Jenkin. 1961. Production of heparinase by *Bacteroides*. J. Bacteriol. *81*: 595–604.

Gherna, R. and C.R. Woese. 1992. A partial phylogenetic analysis of the "*Flavobacter-Bacteroides*" phylum: basis for taxonomic restructuring. Syst. Appl. Microbiol. *15*: 513–521.

Gordon, N.S., A. Valenzuela, S.M. Adams, P.W. Ramsey, J.L. Pollock, W.E. Holben and J.E. Gannon. 2009. *Pedobacter nyackensis* sp. nov., *Pedobacter alluvionis* sp. nov. and *Pedobacter borealis* sp. nov., isolated from Montana flood-plain sediment and forest soil. Int. J. Syst. Evol. Microbiol. *59*: 1720–1726.

Gounot, A.M. 1999. Microbial life in permanently cold soils. *In* Cold-Adapted Organisms (edited by Margesin and Schinner). Springer, New York, pp. 3–15.

Gu, K., R.J. Linhardt, M. Laliberte, K. Gu and J. Zimmermann. 1995. Purification, characterization and specificity of chondroitin lyases and glycuronidase from *Flavobacterium heparinum*. Biochem. J. *312*: 569–577.

Hamana, K. and Y. Nakagawa. 2001a. Polyamine distribution profiles in the eighteen genera phylogenetically located within the *Flavobacterium-Flexibacter-Cytophaga* complex. Microbios *106*: 7–17.

Hamana, K. and Y. Nakagawa. 2001b. Polyamine distribution profiles in newly validated genera and species within the *Flavobacterium–Flexibacter–Cytophaga–Sphingobacterium* complex. Microbios *106 Suppl. 2*: 105–116.

Henrichsen, J. 1972. Bacterial surface translocation: a survey and a classification. Bacteriol. Rev. *36*: 478–503.

Hirsch, P. 1980. Distribution and pure culture studies of morphologically distinct Solar Lake microorganisms. *In* Hypersaline Brines and Evaporitic Environments (edited by Nissenbaum). Elsevier Science Publishers, Amsterdam, pp. 41–60.

Holmes, B., R.J. Owen and R.E. Weaver. 1981. *Flavobacterium multivorum*, a new species isolated from human clinical specimens and previously known as group IIk, biotype 2. Int. J. Syst. Bacteriol. *31*: 21–34.

Holmes, B., R.J. Owen and D.G. Hollis. 1982. *Flavobacterium spiritivorum*, a new species Isolated from human clinical specimens. Int. J. Syst. Bacteriol. *32*: 157–165.

Holmes, B., D.G. Hollis, A.G. Steigerwalt, M.J. Pickett and D.J. Brenner. 1983. *Flavobacterium thalpophilum*, a new species recovered from human clinical material. Int. J. Syst. Bacteriol. *33*: 677–682.

Holmes, B., R.J. Owen and T.A. McMeekin (editors). 1984. Genus *Flavobacterium*, Bergey's Manual of Systematic Bacteriology, vol. 1. Williams & Wilkins, Baltimore.

Holmes, B., R.E. Weaver, A.G. Steigerwalt and D.J. Brenner. 1988. A taxonomic study of *Flavobacterium spiritivorum* and *Sphingobacterium mizutae*: proposal of *Flavobacterium yabuuchiae* sp. nov. and *Flavobacterium mizutaii* comb. nov. Int. J. Syst. Bacteriol. *38*: 348–353.

Huang, L., H. Van Halbeek, B. Eggimann and J.J. Zimmermann. 1995. Structural characterization of the noval O-linked carbohydrate chain of heparinase I from *Flavobacterium heparinum*. Glycobiology *5*: 712.

Huang, W., A. Matte, Y. Li, Y.S. Kim, R.J. Linhardt, H. Su and M. Cygler. 1999. Crystal structure of chondroitinase B from *Flavobacterium heparinum* and its complex with a disaccharide product at 1.7 A resolution. J. Mol. Biol. *294*: 1257–1269.

Humphrey, B.A. and K.C. Marshall. 1980. Fragmentation of some gliding bacteria during the growth cycle. J. Appl. Bacteriol. *49*: 281–289.

Hwang, C.Y., D.H. Choi and B.C. Cho. 2006. *Pedobacter roseus* sp. nov., isolated from a hypertrophic pond, and emended description of the genus *Pedobacter*. Int. J. Syst. Evol. Microbiol. *56*: 1831–1836.

Jackson, R.L., S.J. Busch and A.D. Cardin. 1991. Glycosaminoglycans: molecular properties, protein interactions, and role in physiological processes. Physiol. Rev. *71*: 481–539.

Jacobi, C.A., H. Reichenbach, B.J. Tindall and E. Stackebrandt. 1996. "*Candidatus* comitans", bacterium living in coculture with *Chondromyces crocatus* (myxobacteria). Int. J. Syst. Bacteriol. *46*: 119–122.

Jacobi, C.A., B. Assmus, H. Reichenbach and E. Stackebrandt. 1997. Molecular evidence for association between the sphingobacterium-like organism "*Candidatus* comitans" and the myxobacterium *Chondromyces crocatus*. Appl. Environ. Microbiol. *63*: 719–723.

Jagannadham, M.V., M.K. Chattopadhyay, C. Subbalakshmi, M. Vairamani, K. Narayanan, C.M. Rao and S. Shivaji. 2000. Carotenoids of an Antarctic psychrotolerant bacterium, *Sphingobacterium antarcticus*, and a mesophilic bacterium, *Sphingobacterium multivorum*. Arch. Microbiol. *173*: 418–424.

Jeon, Y., J.M. Kim, J.H. Park, S.H. Lee, C.N. Seong, S.S. Lee and C.O. Jeon. 2009. *Pedobacter oryzae* sp. nov., isolated from rice paddy soil. Int. J. Syst. Evol. Microbiol. *59*: 2491–2495.

Jooste, P.J. 1985. The taxonomy and significance of *Flavobacterium–Cytophaga* strains from dairy sources, PhD thesis. University of Orange Free State, Bloemfontein, Republic of South Africa.

Joubert, J.J., E.J. van Rensburg and M.J. Pitout. 1984. A plate method for demonstrating the breakdown of heparin and chondroitin sulphate by bacteria. J. Microbiol. Methods *2*: 197–202.

Khomenkov, V.G., A.B. Shevelev, V.G. Zhukov, A.E. Kurlovich, N.A. Zagustina and V.O. Popov. 2005. [Application of molecular systematics to study of bacterial cultures consuming volatile organic compounds]. Prikl. Biokhim. Mikrobiol. *41*: 176–184.

Kim, K.H., L.N. Ten, Q.M. Liu, W.T. Im and S.T. Lee. 2006. *Sphingobacterium daejeonense* sp. nov., isolated from a compost sample. Int. J. Syst. Evol. Microbiol. *56*: 2031–2036.

Kito, M., Y. Onji, T. Yoshida and T. Nagasawa. 2002. Occurrence of epsilon-poly-L-lysine-degrading enzyme in epsilon-poly-L-lysine-tolerant *Sphingobacterium multivorum* OJ10: purification and characterization. FEMS Microbiol. Lett. *207*: 147–151.

Kjellen, L. and U. Lindahl. 1991. Proteoglycans: structures and interactions. Annu. Rev. Biochem. *60*: 443–475.

Kuo, T.M., A.C. Lanser, L.K. Nakamura and C.T. Hou. 2000. Production of 10-ketostearic acid and 10-hydroxystearic acid by strains of *Sphingobacterium thalpophilum* isolated from composted manure. Curr. Microbiol. *40*: 105–109.

Kwon, S.W., B.Y. Kim, K.H. Lee, K.Y. Jang, S.J. Seok, J.S. Kwon, W.G. Kim and H.Y. Weon. 2007. *Pedobacter suwonensis* sp. nov., isolated from the rhizosphere of Chinese cabbage (*Brassica campestris*). Int. J. Syst. Evol. Microbiol. *57*: 480–484.

Lagace, L., M. Pitre, M. Jacques and D. Roy. 2004. Identification of the bacterial community of maple sap by using amplified ribosomal DNA (rDNA) restriction analysis and rDNA sequencing. Appl. Environ. Microbiol. *70*: 2052–2060.

Lee, H.G., S.G. Kim, W.T. Im, H.M. Oh and S.T. Lee. 2009. *Pedobacter composti* sp. nov., isolated from compost. Int. J. Syst. Evol. Microbiol. *59*: 345–349.

Leung, K. and E. Topp. 2001. Bacterial community dynamics in liquid swine manure during storage: molecular analysis using DGGE/PCR of 16S rDNA. FEMS Microbiol. Ecol. *38*: 169–177.

Li, Y., A. Matte, H. Su and M. Cygler. 1998. Crystallization and preliminary X-ray analysis of chondroitinase B from *Flavobacterium heparinum*. Acta Crystallogr. Sect. D *55*: 1055–1057.

Liu, D., Z. Shriver, R. Godavarti, G. Venkataraman and R. Sasisekharan. 1999. The calcium-binding sites of heparinase I from *Flavobacterium heparinum* are essential for enzymatic activity. J. Biol. Chem. *274*: 4089–4095.

Liu, H., C.-F. Shen, K. Wang and Y.-X. Chen. 2004. Cometabolic degradation of pentachlorophenol by *Sphingobacterium multivorum*. China Environ. Sci. *24*: 294–298.

Liu, R., H. Liu, C.X. Zhang, S.Y. Yang, X.H. Liu, K.Y. Zhang and R. Lai. 2008. *Sphingobacterium siyangense* sp. nov., isolated from farm soil. Int. J. Syst. Evol. Microbiol. *58*: 1458–1462.

Lohse, D.L. and R.J. Linhardt. 1992. Purification and characterization of heparin lyases from *Flavobacterium heparinum*. J. Biol. Chem. *267*: 24347–24355.

Luo, X., Z. Wang, J. Dai, L. Zhang, J. Li, Y. Tang, Y. Wang and C. Fang. 2010. *Pedobacter glucosidilyticus* sp. nov., isolated from dry riverbed soil. Int. J. Syst. Evol. Microbiol. *60*: 229–233.

Manz, W., R. Amann, W. Ludwig, M. Vancanneyt and K.H. Schleifer. 1996. Application of a suite of 16S rRNA-specific oligonucleotide probes designed to investigate bacteria of the phylum cytophaga-flavobacter-bacteroides in the natural environment. Microbiology *142*: 1097–1106.

Margesin, R., C. Sproer, P. Schumann and F. Schinner. 2003. *Pedobacter cryoconitis* sp. nov., a facultative psychrophile from alpine glacier cryoconite. Int. J. Syst. Evol. Microbiol. *53*: 1291–1296.

Margesin, R., H. Dieplinger, J. Hofmann, B. Sarg and H. Lindner. 2005. A cold-active extracellular metalloprotease from *Pedobacter cryoconitis* - production and properties. Res. Microbiol. *156*: 499–505.

Marinella, M.A. 2002. Cellulitis and sepsis due to sphingobacterium. JAMA *288*: 1985.

Matsuyama, H., H. Katoh, T. Ohkushi, A. Satoh, K. Kawahara and I. Yumoto. 2008. *Sphingobacterium kitahiroshimense* sp. nov., isolated from soil. Int. J. Syst. Evol. Microbiol. *58*: 1576–1579.

Mehnaz, S., B. Weselowski and G. Lazarovits. 2007. *Sphingobacterium canadense* sp. nov., an isolate from corn roots. Syst. Appl. Microbiol. *30*: 519–524.

Mehnaz, S., B. Weselowski and G. Lazarovits. 2008. *In* Validation of the publication of new names and new combinations previously effectively published outside the IJSEM. List no. 119. Int. J. Syst. Evol. Microbiol. *58*: 1.

Michel, G., K. Pojasek, Y. Li, T. Sulea, R.J. Linhardt, R. Raman, V. Prabhakar, R. Sasisekharan and M. Cygler. 2004. The structure of chondroitin B lyase complexed with glycosaminoglycan oligosaccharides unravels a calcium-dependent catalytic machinery. J. Biol. Chem. *279*: 32882–32896.

Muurholm, S., S. Cousin, O. Pauker, E. Brambilla and E. Stackebrandt. 2007. *Pedobacter duraquae* sp. nov., *Pedobacter westerhofensis* sp. nov., *Pedobacter metabolipauper* sp. nov., *Pedobacter hartonius* sp. nov. and *Pedobacter steynii* sp. nov., isolated from a hard-water rivulet. Int. J. Syst. Evol. Microbiol. *57*: 2221–2227.

Naka, T., N. Fujiwara, I. Yano, S. Maeda, M. Doe, M. Minamino, N. Ikeda, Y. Kato, K. Watabe, Y. Kumazawa, I. Tomiyasu and K. Kobayashi. 2003. Structural analysis of sphingophospholipids derived from *Sphingobacterium spiritivorum*, the type species of genus *Sphingobacterium*. Biochim. Biophys. Acta *1635*: 83–92.

Nakagawa, Y. and K. Yamasato. 1993. Phylogenetic diversity of the genus *Cytophaga* revealed by 16S ribosomal RNA sequencing and menaquinone analysis. J. Gen. Microbiol. *139*: 1155–1161.

Nakagawa, Y. and K. Yamasato. 1996. Emendation of the genus *Cytophaga* and transfer of *Cytophaga agarovorans* and *Cytophaga salmonicolor* to *Marinilabilia* gen. nov: phylogenetic analysis of the *Flavobacterium cytophaga* complex. Int. J. Syst. Bacteriol. *46*: 599–603.

Nishino, S., H. Kuroyanagi, T. Terada, S. Inoue, Y. Inoue, F.A. Troy and K. Kitajima. 1996. Induction, localization, and purification of a novel sialidase, deaminoneuraminidase (KDNase), from *Sphingobacterium multivorum*. J. Biol. Chem. *271*: 2909–2913.

Oyaizu, H., K. Komagata, A. Amemura and T. Harada. 1982. A succinoglycan-decomposing bacterium, *Cytophaga arvensicola* sp. nov. J. Gen. Appl. Microbiol. *28*: 369–388.

Payza, A.N. and E.D. Korn. 1956a. Bacterial degradation of heparin. Nature *177*: 88–89.

Payza, A.N. and E.D. Korn. 1956b. The degradation of heparin by bacterial enzymes. 1. Adaption and lyophilized cells. J. Biol. Chem. *223*: 853–864.

Plummer, T.H., Jr., A.L. Tarentino and C.R. Hauer. 1995. Novel, specific O-glycosylation of secreted *Flavobacterium meningosepticum* proteins. Asp-Ser and Asp-Thr-Thr consensus sites. J. Biol. Chem. *270*: 13192–13196.

Potvliege, C., C. Dejaegher-Bauduin, W. Hansen, M. Dratwa, F. Collart, C. Tielemans and E. Yourassowsky. 1984. *Flavobacterium multivorum* septicemia in a hemodialyzed patient. J. Clin. Microbiol. *19*: 568–569.

Reichenbach, H., W. Kohl and H. Achenbach. 1981. The flexirubin-type pigments, chemosystematically useful compounds. *In* The *Flavobacterium–Cytophaga* Group. Gesellschaft für Biotechnologische

Forschung Monograph Series, No. 5 (edited by Reichenbach and Weeks). Verlag Chemie, Weinheim, pp. 101–108.

Reichenbach, H. 1989. Genus I. *Cytophaga*. *In* Bergey's Manual of Systematic Bacteriology, vol. 3 (edited by Staley, Bryant, Pfennig and Holt). Williams & Wilkins, Baltimore, pp. 2015–2050.

Reina, J., N. Borrell and J. Figuerola. 1992. *Sphingobacterium multivorum* isolated from a patient with cystic fibrosis. Eur. J. Clin. Microbiol. Infect. Dis. *11*: 81–82.

Roh, S.W., Z.X. Quan, Y.D. Nam, H.W. Chang, K.H. Kim, M.K. Kim, W.T. Im, L. Jin, S.H. Kim, S.T. Lee and J.W. Bae. 2008. *Pedobacter agri* sp. nov., from soil. Int. J. Syst. Evol. Microbiol. *58*: 1640–1643.

Rosa-Putra, S., A. Hemmerlin, J. Epperson, T.J. Bach, L.H. Guerra and M. Rohmer. 2001. Zeaxanthin and menaquinone-7 biosynthesis in *Sphingobacterium multivorum* via the methylerythritol phosphate pathway. FEMS Microbiol. Lett. *204*: 347–353.

Salyers, A.A., J.R. Vercellotti, S.E. West and T.D. Wilkins. 1977. Fermentation of mucin and plant polysaccharides by strains of *Bacteroides* from the human colon. Appl. Environ. Microbiol. *33*: 319–322.

Shaya, D., Y. Li and M. Cygler. 2004. Crystallization and preliminary X-ray analysis of heparinase II from *Pedobacter heparinus*. Acta Crystallogr. D Biol. Crystallogr. *60*: 1644–1646.

Shivaji, S., M.K. Ray, N.S. Rao, L. Saisree, M.V. Jagannadham, G.S. Kumar, G.S.N. Reddy and P.M. Bhargava. 1992. *Sphingobacterium* antarcticus sp. nov., a psychrotrophic bacterium from the soils of Schirmacher Oasis, Antarctica. Int. J. Syst. Bacteriol. *42*: 102–106.

Shivaji, S. 2005. Microbial diversity and molecular basis of cold adaptation in Antarctic bacteria. *In* Microbial Diversity: Current Perspectives and Potential Applications (edited by Satyanarayana and Johri). I.K. International, New Delhi, pp. 3–24.

Shivaji, S., P. Chaturvedi, G.S. Reddy and K. Suresh. 2005. *Pedobacter himalayensis* sp. nov., from the Hamta glacier located in the Himalayan mountain ranges of India. Int. J. Syst. Evol. Microbiol. *55*: 1083–1088.

Shriver, Z., D. Liu, Y. Hu and R. Sasisekharan. 1999. Biochemical investigations and mapping of the calcium-binding sites of heparinase I from *Flavobacterium heparinum*. J. Biol. Chem. *274*: 4082–4088.

Silver, P.J. 1998. IBT 9302 (Heparinase III): a novel enzyme for the management of reperfusion injury-related vascular damage, restenosis and wound healing. Expert Opin. Investig. Drugs *7*: 1003–1014.

Stackebrandt, E. and B.M. Goebel. 1994. Taxonomic note: a place for DNA–DNA reassociation and 16S rRNA sequence analysis in the present species definition in bacteriology. Int. J. Syst. Bacteriol. *44*: 846–849.

Steyn, P.L., B. Pot, P. Segers, K. Kersters and J.J. Joubert. 1992. Some novel aerobic heparin-degrading bacterial isolates. Syst. Appl. Microbiol. *15*: 137–143.

Steyn, P.L., P. Segers, M. Vancanneyt, P. Sandra, K. Kersters and J.J. Joubert. 1998. Classification of heparinolytic bacteria into a new genus, *Pedobacter*, comprising four species: *Pedobacter heparinus* comb. nov., *Pedobacter piscium* comb. nov., *Pedobacter africanus* sp. nov. and *Pedobacter saltans* sp. nov. Proposal of the family *Sphingobacteriaceae* fam. nov. Int. J. Syst. Bacteriol. *48*: 165–177.

Su, H., F. Blain, R.A. Musil, J.J. Zimmermann, K. Gu and D.C. Bennett. 1996. Isolation and expression in *Escherichia coli* of *hepB* and *hepC*, genes coding for the glycosaminoglycan-degrading enzymes heparinase II and heparinase III, respectively, from *Flavobacterium heparinum*. Appl. Environ. Microbiol. *62*: 2723–2734.

Su, H., Z. Shao, L. Tkalec, F. Blain and J. Zimmermann. 2001. Development of a genetic system for the transfer of DNA into *Flavobacterium heparinum*. Microbiology *147*: 581–589.

Tabata, K., H. Abe and Y. Doi. 2000. Microbial degradation of poly(aspartic acid) by two isolated strains of *Pedobacter* sp. and *Sphingomonas* sp. Biomacromolecules *1*: 157–161.

Takeuchi, M. and A. Yokota. 1992. Proposals of *Sphingobacterium faecium* sp. nov., *Sphingobacterium piscium* sp. nov., *Sphingobacterium heparinum* comb. nov., *Sphingobacterium thalpophilum* comb. nov. and two genospecies of the genus *Sphingobacterium*, and synonymy of *Flavobacterium yabuuchiae* and *Sphingobacterium spiritivorum*. J. Gen. Appl. Microbiol. *38*: 465–482.

Takeuchi, M. and A. Yokota. 1993. *In* Validation of the publication of new names and new combinations previously effectively published outside the IJSB. List no. 47. Int. J. Syst. Bacteriol. *43*: 864–865.

Tatum, H.W., W.H. Ewing and R.E. Weaver. 1974. Miscellaneous Gram-stain-negative bacteria. *In* Manual of Clinical Microbiology, 2nd edn (edited by Lennette, Spaulding and Truant). American Society for Microbiology, Washington, D.C., pp. 270–294.

Ten, L.N., Q.M. Liu, W.T. Im, M. Lee, D.C. Yang and S.T. Lee. 2006a. *Pedobacter ginsengisoli* sp. nov., a DNase-producing bacterium isolated from soil of a ginseng field in South Korea. Int. J. Syst. Evol. Microbiol. *56*: 2565–2570.

Ten, L.N., O.M Liu, W.T. Im, Z. Aslam and S.T. Lee. 2006b. *Sphingobacterium composti* sp. nov., a novel DNase-producing bacterium isolated from compost. J. Microbiol. Biotechnol. *16*: 1728–1733.

Ten, L.N., O.M Liu, W.T. Im, Z. Aslam and S.T. Lee. 2007. *In* Validation of the publication of new names and new combinations previously effectively published outside the IJSEM. List no. 116. Int. J. Syst. Evol. Microbiol. *57*: 1371–1373.

Tkalec, A.L., D. Fink, F. Blain, G. Zhang-Sun, M. Laliberte, D.C. Bennett, K. Gu, J.J. Zimmermann and H. Su. 2000. Isolation and expression in *Escherichia coli* of *cslA* and *cslB*, genes coding for the chondroitin sulfate-degrading enzymes chondroitinase AC and chondroitinase B, respectively, from *Flavobacterium heparinum*. Appl. Environ. Microbiol. *66*: 29–35.

Tronel, H., P. Plesiat, E. Ageron and P.A. Grimont. 2003. Bacteremia caused by a novel species of *Sphingobacterium*. Clin. Microbiol. Infect. *9*: 1242–1244.

Tsukamoto, T., H. Murata and A. Shirata. 2002. Identification of non-pseudomonad bacteria from fruit bodies of wild agaricales fungi that detoxify tolaasin produced by *Pseudomonas tolaasii*. Biosci Biotechnol. Biochem. *66*: 2201–2208.

Vanparys, B., K. Heylen, L. Lebbe and P. De Vos. 2005. *Pedobacter caeni* sp. nov., a novel species isolated from a nitrifying inoculum. Int. J. Syst. Evol. Microbiol. *55*: 1315–1318.

Wei, W., Y. Zhou, X. Wang, X. Huang and R. Lai. 2008. *Sphingobacterium anhuiense* sp. nov., isolated from forest soil. Int. J. Syst. Evol. Microbiol. *58*: 2098–2101.

Wei, Z., M. Lyon and J.T. Gallagher. 2005. Distinct substrate specificities of bacterial heparinases against N-unsubstituted glucosamine residues in heparan sulfate. J. Biol. Chem. *280*: 15742–15748.

Yabuuchi, E. and C.W. Moss. 1982. Cellular fatty-acid composition of strains of 3 species of *Sphingobacterium* gen nov and *Cytophaga johnsonae*. FEMS Microbiol. Lett. *13*: 87–91.

Yabuuchi, E., T. Kaneko, I. Yano, C.W. Moss and N. Miyoshi. 1983. *Sphingobacterium* gen. nov., *Sphingobacterium spiritivorum* comb. nov., *Sphingobacterium multivorum* comb. nov., *Sphingobacterium mizutae* sp. nov., and *Flavobacterium indologenes* sp. nov., glucose-nonfermenting Gram-negative rods in CDC group IiK-2 and group IIb. Int. J. Syst. Bacteriol. *33*: 580–598.

Yang, V.C., R.J. Linhardt, H. Bernstein, C.L. Cooney and R. Langer. 1985. Purification and characterization of heparinase from *Flavobacterium heparinum*. J. Biol. Chem. *260*: 1849–1857.

Yano, I., I. Tomiyasu and E. Yabuuchi. 1982. Long-chain base composition of strains of 3 species of *Sphingobacterium* gen. nov. FEMS Microbiol. Lett. *15*: 303–307.

Yano, I., S. Imaizumi, I. Tomiyasu and E. Yabuuchi. 1983. Separation and analysis of free ceramides containing 2-hydroxy fatty-acids in *Sphingobacterium* species. FEMS Microbiol. Lett. *20*: 449–453.

Ye, Y., H. Min and Y. Du. 2004. Characterization of a strain of *Sphingobacterium* sp. and its degradation to herbicide mefenacet. J. Environ. Sci. (China) *16*: 343–347.

Yoo, S.H., H.Y. Weon, H.B. Jang, B.Y. Kim, S.W. Kwon, S.J. Go and E. Stackebrandt. 2007. *Sphingobacterium composti* sp. nov., isolated from cotton-waste composts. Int. J. Syst. Evol. Microbiol. *57*: 1590–1593.

Yoon, J.H., M.H. Lee, S.J. Kang, S.Y. Park and T.K. Oh. 2006. *Pedobacter sandarakinus* sp. nov., isolated from soil. Int. J. Syst. Evol. Microbiol. *56*: 1273–1277.

Yoon, J.H., S.J. Kang, H.W. Oh and T.K. Oh. 2007a. *Pedobacter insulae* sp. nov., isolated from soil. Int. J. Syst. Evol. Microbiol. *57*: 1999–2003.

Yoon, J.H., S.J. Kang and T.K. Oh. 2007b. *Pedobacter terrae* sp. nov., isolated from soil. Int. J. Syst. Evol. Microbiol. *57*: 2462–2466.

Yoon, J.H., S.J. Kang, S. Park and T.K. Oh. 2007c. *Pedobacter lentus* sp. nov. and *Pedobacter terricola* sp. nov., isolated from soil. Int. J. Syst. Evol. Microbiol. *57*: 2089–2095.

Yoon, M.H., L.N. Ten, W.T. Im and S.T. Lee. 2007d. *Pedobacter panaciterrae* sp. nov., isolated from soil in South Korea. Int. J. Syst. Evol. Microbiol. *57*: 381–386.

Zimmermann, J.J., R. Langer and C.L. Cooney. 1990. Specific plate assay for bacterial heparinase. Appl. Environ. Microbiol. *56*: 3593–3594.

Family II. **Chitinophagaceae** fam. nov. Kämpfer, Lodders and Falsen 2010

Peter Kämpfer

Chi.ti.no.pha.ga.ce′a.e. N.L. fem. n. *Chitinophaga* type genus of the family; suff. *-aceae* ending to denote a family; N.L. fem. pl. n. *Chitinophagaceae* the *Chitinophaga* family.

This family is defined in part on the basis of sequence similarities of 16S rRNA gene sequences. The main properties are described in the description of the order *Sphingobacteriales*. Cells are often thin, rod-shaped and usually nonmotile. Swarming motility may occur. Aerobic or facultatively anaerobic. Limited fermentative capabilities are observed in some members. Menaquinones are of the MK-7 type, and a fatty acid pattern containing $C_{15:0}$ iso, $C_{17:0}$ iso 3-OH, and $C_{15:1}$ iso G are the most important components and characteristics. The family comprises two genera, *Chitinophaga* (Sangkhobol and Skerman, 1981) and *Terrimonas* (Xie and Yokota, 2006) in addition to the recently described genera *Niastella* (Weon et al., 2006), *Segetibacter* (An et al., 2007), *Niabella* (Kim et al., 2007), *Flavisolibacter* (Yoon and Im, 2007), *Sediminibacterium* (Qu and Yuan, 2008), *Parasegetibacter* (Zhang et al., 2009), *Lacibacter* (Qu et al., 2009), *Ferruginibacter* (Lim et al., 2009), *Filimonas* (Shiratori et al., 2009), *Flavihumibacter* (Zhang et al., 2010) and *Hydrotalea* (Kämpfer et al., 2010), which also share the main properties of the order *Sphingobacteriales*.

Type genus: **Chitinophaga** Sangkhobol and Skerman 1981, 288[VP].

Genus I. **Chitinophaga** Sangkhobol and Skerman 1981, 285[VP] emend. Kämpfer, Young, Sridhar, Arun, Lai, Shen and Rekha 2006, 2225[VP]

The Editorial Board

Chi.ti.no′pha.ga. N.L. n. *chitinum* chitin; Gr. v. *phagein* to devour, to eat; N.L. fem. n. *Chitinophaga* chitin eater, chitin destroyer.

Flexible rods with rounded ends, 0.5–0.8 × ~40 μm when fully developed. Occur singly. **A resting stage (microcyst), 0.8–0.9 μm in diameter, is formed but is not highly refractile.** Macroscopic fruiting bodies are not formed. **Motility by gliding is possessed by some, but not all species.** Gram-stain-negative. Aerobic. Optimum temperature, 23–24°C; maximum, 37–40°C; minimum: 10–12°C. Optimum pH, 7; maximum, 8–10; minimum, 4. Chemo-organotrophic. **Oxidative or fermentative.** Acid but no gas produced from some carbohydrates. **Some species hydrolyze chitin and some hydrolyze cellulose. Agar is not hydrolyzed.** Congo red is not absorbed. Cell masses are yellow. The major fatty acids are $C_{16:1}$ ω5*c* and $C_{15:0}$ iso. $C_{14:0}$ is not present. In those species that have been analyzed, the **quinone system is MK-7** and the **major polyamine is homospermidine**.

DNA G+C content (mol%): 43–46.

Type species: **Chitinophaga pinensis** Sangkhobol and Skerman 1981, 285[VP].

Enrichment and isolation procedures

Most strains of *Chitinophaga* can be isolated by inoculating water samples from the littoral zones of freshwater lakes and creeks onto a medium prepared by adding 1.5% of an optically clear agar to lake water which has been freed of optically visible particles by filtration through a membrane filter (pore size 0.45 μm) (Skerman, 1989). The pH is not adjusted. The medium is sterilized at 121°C for 15 min. The medium contains adequate nutrients for the development of microcolonies of most of the organisms that constitute the resident populations of freshwaters. Supplementation tends to produce overgrowth that makes isolation more difficult. Although a chitin-supplemented medium might enhance development of chitin-hydrolyzing species of *Chitinophaga*, it could also enhance the growth of other chitin-hydrolyzing bacteria at the expense of the slower-growing *Chitinophaga* cells.

The medium is inoculated by allowing ~0.05 ml to flow across the face of the medium and then removing excess surface moisture by incubating the medium for 10 min in a sterile chamber at 22°C. Petri dish cultures are examined periodically from 8 to 24 h by using a 10× phase-contract objective. Typical gliding filaments are isolated with a microloop (Skerman, 1989) before overgrowth of colonies of other microorganisms occurs. Selected cells are transferred to a lake-water agar supplemented with 0.01% peptone (Difco) and 0.01% yeast extract (Difco), which supports more luxuriant growth.

Chitinophaga skermanii was isolated from the feces of the millipede *Arthrosphaera magna* collected in India. The organism was cultured on nutrient agar (Kämpfer et al., 2006).

Strains of *Chitinophaga arvensicola* were isolated from peat samples collected from a *Sphagnum* peat bog in West Siberia (Pankratov et al., 2006). Aerobic cellulolytic communities were enriched using 120-ml serum bottles containing strips of filter paper immersed in 30 ml of 0.2× diluted liquid mineral medium ST5 (Stanier, 1942). Cell suspensions of the resulting enrichments were spread-plated onto the surface of the dilute ST5 solidified with agar and containing various supplements, such as 0.1% starch, 1% peat extract, or gellan gum.

Chitinophaga japonensis was isolated on a starch-casein agar plate from a soil sample (Fujita et al., 1996). *Chitinophaga ginsengisegetis* and *Chitinophaga ginsengisoli* were isolated from a ginseng field in South Korea (Lee et al., 2007). *Chitinophaga terrae* was isolated from soil from a field in South Korea.

Maintenance procedures

Cultures can be preserved by lyophilization or by liquid nitrogen storage. For lyophilization, cultures grown on CYEA for 48 h at 22°C are suspended in *Mist Desiccans* [which contains (per liter of distilled water) Bacto peptone (Difco), 12 g, and glucose, 30 g; Greaves (1956)], freeze-dried, and stored in ampoules under oxygen-free nitrogen at 4°C. For liquid nitrogen storage, cells from CYEA are suspended to a density of 10^9 cells/ml in filtered lake water containing 10% glycerol. This suspension, distributed and sealed in 0.5-ml amounts in 0.7-ml ampoules is precooled to −20°C for 2 h before storage in liquid nitrogen.

Differentiation of the genus *Chitinophaga* from other genera of unicellular gliding organisms

The ability to hydrolyze *p*-nitrophenyl (*p*NP)-phosphorylcholine, 2-deoxythymidine-5′-*p*NP phosphate, L-glutamate-γ-3-carboxy-*p*-nitroanilide (NA) and L-proline-*p*-NA can differentiate *Chitinophaga* species from *Terrimonas* species (Kämpfer et al., 2006); however, results for *Chitinophaga terrae*, *Chitinophaga ginsengisegetis*, and *Chitinophaga ginsengisoli* have not been reported in this regard.

Taxonomic comments

On basis of 16S rRNA gene sequence analysis and menaquinone analysis of *Cytophaga* species, Nakagawa and Yamasato (1993) concluded that the genus *Cytophaga* was so heterogeneous that it should be divided into several genera in accordance with phylogenetic relationships. Sly et al. (1999) reported that *Chitinophaga pinensis* and *Flexibacter filiformis* formed a distinct lineage based on 16S rRNA gene sequencing that also included *Flexibacter sancti* and *Cytophaga arvensicola*; later, *Flexibacter japonensis* was also added (Nakagawa et al., 2002). Kämpfer et al. (2006) proposed transferring all of these species to the genus *Chitinophaga* on the basis of phylogenetic and phenotypic data. Although the similarities of the 16S rRNA gene sequences were low (88.5–96.4%), there was no evidence for clear phenotypic differences between these organisms that would justify assignment to different genera. Kämpfer et al. (2006) also added a new species to the genus *Chitinophaga*, *Chitinophaga skermanii*.

16S rRNA gene sequence analysis by Lee et al. (2007) indicated that *Chitinophaga ginsengisegetis* had 96.5% similarity to *Chitinophaga arvensicola*, 94.5% to *Chitinophaga japonensis*, 93.4% to *Chitinophaga sancti*, 93.4% to *Chitinophaga filiformis*, 92.0% to *Chitinophaga skermanii*, and 91.9% to *Chitinophaga pinensis*. These values indicated that *Chitinophaga ginsengisegetis* was a distinct species within the genus *Chitinophaga*. *Chitinophaga ginsengisoli* exhibited 00.6% similarity to *Chitinophaga filiformis*, which suggested that they might belong in a single species; however, the two species had a DNA–DNA relatedness value of only 38%, indicating that they were separate species.

16S rRNA gene sequence analysis indicated that *Chitinophaga terrae* belonged to *Chitinophaga*, but the 16S rRNA gene sequence similarities between *Chitinophaga terrae* and established *Chitinophaga* species ranged from 90.3 to 95.7% (Kim and Jung, 2007), indicating that *Chitinophaga terrae* was a distinct species.

List of species of the genus *Chitinophaga*

1. **Chitinophaga pinensis** Sangkhobol and Skerman 1981, 285^{VP}

pi′nen.sis L. n. *pinus* a pine, pine-tree; N.L. fem. adj. *pinensis* pertaining to pines.

The characteristics are as given for the genus and as listed in Tables 81 and 82, with the following additional features. Rods 0.5–0.8 × ~40 μm, with rounded ends. Microcyst diameter, 0.8–0.9 μm. The cell mass is yellow. Motile by gliding motility. Temperature range, 10–40°C; optimum, 23–24. pH range, 4–8; some strains can grow at pH 10. NaCl tolerance, 0–1%; some strains can grow at 2%. Catalase-positive. Oxidase-negative. Chitin, gelatin, urea, casein, and Tweens 20, 40, and 80 are hydrolyzed but not cellulose, starch, alginate, or agar. Nitrate is not reduced. Phenylalanine deaminase and tryptophan deaminase activity is not present. In the O-F test for glucose catabolism, some strains are slowly oxidative while others are fermentative.

DNA G+C content (mol%): 43–46 (T_m).

Type strain: ATCC 43595, DSM 2588, IFO (now NBRC) 15968, LMG 13176, UQM 2034.

Sequence accession no. (16S rRNA gene): AF078775.

2. **Chitinophaga arvensicola** (Oyaizu, Komagata, Amemura and Harada 1983) Kämpfer, Young, Sridhar, Arun, Lai, Shen and Rekha 2006, 2225^{VP} emend. Pankratov, Kulichevskaya, Liesack and Dedysh 2006, 2764^{VP} (*Cytophaga arvensicola* Oyaizu, Komagata, Amemura and Harada 1983, 438^{VP}; effective publication: Oyaizu, Komagata, Amemura and Harada 1982, 385.)

ar.ven′si.co′la. N.L. adj. *arvensis* (from L. n. *arvum* a field) belonging to or living in the fields; L. suff. *-cola* from L. n. *incola* inhabitant; N.L. n. *arvensicola* an inhabitant of the fields.

The characteristics are as described for the genus and as listed in Tables 81 and 82, with the following additional features. Rod-shaped cells 0.4–0.5 × 10–40 μm. Motile by gliding (Pankratov et al., 2006); nonmotile by gliding (Kämpfer et al., 2006). Cells undergo a cyclic shape change in the course of culture development. Young cultures contain long and agile thread cells. On ageing, long cells divide into several nonmotile shorter (0.5–0.7 × 1.5–5 μm) cells. Old cultures consist of short rods (0.5–0.7 × 0.7–2 μm). Colonies are yellow-pigmented, irregularly shaped, with non-entire edges and flat. Produces flexirubin. Chemo-organotrophic. Aerobic or microaerobic. Growth occurs at temperatures between 4 and 37°C and at pH values between 4.5 and 8.0. NaCl inhibits growth at concentrations above 3% (w/v). Major fatty acids are $C_{15:0}$ iso, $C_{16:1}$ ω5*c*, and $C_{17:0}$ iso 3-OH. Nitrate reduction is variable. Mannitol, D-sucrose, D-cellobiose, D-rhamnose, D-galactose, inositol, D-xylose, D-glucose, and D-maltose are utilized, but not sorbitol, acetate, citrate, D-arabinose, or dulcitol. Casein, esculin, gelatin, gellan gum, and laminarin are hydrolyzed. Some strains are also capable of hydrolyzing xylan, starch, and agar. Cellulose, carboxymethylcellulose, pectin, chitin, and Tweens 20, 40, and 80 are not hydrolyzed. Acid is produced from D-glucose, D-xylose, D-cellobiose, D-maltose, D-sucrose, D-rhamnose, and D-galactose.

Source: Soils and acidic wetlands. The type strain was from soil of Osaka Prefecture, Japan.

DNA G+C content (mol%): 42.8–46.4 (Bd, T_m)

Type strain: M64, ATCC 51264, CIP 104804, DSM 3695, IAM 12650, JCM 2836, NBRC 14973.

Sequence accession no. (16S rRNA gene): AM237311.

TABLE 81. Characteristics differentiating the type strains of the species of *Chitinophaga*[a,b]

Characteristic	*C. pinensis* ACM 2034	*C. arvensicola* IAM 12650	*C. filiformis* NBRC 150656	*C. ginsengisegetis* Gsoil 040	*C. ginsengisoli* Gsoil 052	*C. japonensis* NBRC 16041	*C. sancti* NBRC 15057	*C. skermanii* CC-SG1B	*C. terrae* KP01
Cell length (μm)	<40	0.6–4	30–80	1.1–1.3	1.2–1.6	2–18	2–15	1–2	0.6–0.8
Gliding motility	+	+/–[c]	+	–	–	+	+	–	–
Filamentous shape	+	–	+	–	–	+	+	–	–
Catalase	+	+	–	–	+	+	–	+	+
Urease	+	–	nd	+	+	nd	nd	–	–
Gelatin liquefaction	+	–	+	+	+	+	+	+	–
Chitin degradation	+	–	+	–	+	–	–	nd	–
Growth at 37°C	+	–	+	+	+	+	–	+	+
Maximum NaCl concentration (%, w/v)	nd	2	0.3	2	<1	2	1	nd	nd
DNA G+C content (mol%)	45	46	45	47	48	50	43	41	46

[a]Symbols: +, positive; –, negative; nd, not determined.

[b]Data from Kämpfer et al. (2006), Xie and Yokota (2006), Takeuchi and Yokota (1992), Oyaizu et al. (1982), Kim and Jung (2007), and Lee et al. (2007).

[c]Kämpfer et al. (2006) reported no gliding motility for IAM 12650^T; Pankratov et al. (2006) reported gliding motility for DSM 3695^T.

TABLE 82. Assimilation of carbon sources by the type strains of the species of *Chitinophaga*[a,b]

Carbon source	*C. pinensis* ACM 2034	*C. arvensicola* IAM 12650	*C. filiformis* NBRC 150656	*C. ginsengisegetis* Gsoil 040	*C. ginsengisoli* Gsoil 052	*C. japonensis* NBRC 16041	*C. sancti* NBRC 15057	*C. skermanii* CC-SG1B	*C. terrae* KP01
N-Acetyl-D-glucosamine	+	+	–	+	+	+	–	+	+
Adonitol	–	–	–	nd	nd	+	–	–	nd
L-Arabinose	+	+[c]	–	+	+	+	+[c]	–	+
D-Galactose	+	–	–	nd	nd	+	–	–	nd
Gluconate	–	–	–	+[c]	+[c]	–	+[c]	–	–
Maltitol	+[d]	–	+[c]	nd	nd	+[c]	–	–	nd
D-Mannose, D-maltose	+	+	+	+	+	+	–	+	+
α-D-Melibiose	–	+	+[b]	+	+	+	+[d]	+[b]	+
L-Rhamnose	+[c]	+	–	+	+[c]	+	–	–	+
D-Ribose	–	+	–	–	–	–	–	–	–
Sucrose	+[c]	+[c]	+	+	+	+	–	–	+
D-Trehalose	+	+[c]	+	nd	nd	+	–	+[d]	nd
D-Xylose	+[c]	+	+	nd	nd	+	–	–	nd

[a]Symbols: +, positive; –, negative; nd, not determined.

[b]Data from Kämpfer et al. (2006), Kim and Jung (2007), and Lee et al. (2007).

[c]Positive after 7 d.

[d]Positive after 14 d.

3. **Chitinophaga filiformis** (Reichenbach 1989) Kämpfer, Young, Sridhar, Arun, Lai, Shen and Rekha 2006, 2225[VP] [*Flexibacter filiformis* (Solntseva 1940) Reichenbach 1989, 2067; *Myxococcus filiformis* Solntseva 1940, 221; *Flexibacter elegans* Soriano 1945, 93, non Lewin 1969, 200]

fi.li.for'mis. L. neut. n. *filum* a thread; L. suff. *-formis* like, of the shape of; N.L. fem. adj. *filiformis* thread-shaped.

The characteristics are as given for the genus and as listed in Tables 81 and 82, with the following additional features. Gliding motility is present. The organisms exhibit a cyclic shape change. Cell populations are usually highly pleomorphic. In young cultures are found long and very flexible thread cells with tapering ends, 0.4–0.5 × 30–80 μm, sometimes even longer. Later, the thread cells fragment so that even shorter cells appear when the culture ages. Finally, only short, often curved, sometimes almost coccoid rods are present which are clearly fatter and, under phase-contrast, darker than the thread cells, measuring 0.5–0.6 × 0.7–1.0 μm or less. The short cells may still continue to grow and divide as such. Usually they grow out into thread cells again when brought into fresh medium. The morphological cycle is controlled by environmental factors, particularly nutrients and temperature, and may be manipulated experimentally. On agar plates, particularly on media that allow swarming, the different morphological forms are, as a rule, found within one (usually the same) colony, with the longer cells near the edge of the swarm. The thread cells have no, or very few cross-walls, 10–30 μm apart, recognizable with the electron microscope. Each compartment appears to contain several nucleoids. The thread cells are extremely agile, gliding, bending, and twisting, but those shorter than about 6 μm are completely nonmotile. On solid media with a low nutrient content (e.g., 0.01% yeast extract) or on yeast agar (e.g., VY/2 agar; Reichenbach and Dworkin, 1981), the colonies are fast-spreading delicate swarms. On VY/2 agar the colonies have a characteristic surface pattern of circular or elongated flat mounds. On agar media with peptone concentrations above 0.3%, the colonies become more and more compact, with a convex surface and a smooth edge. Such colonies often contain only short rods. On yeast agar the cell mass is usually pale yellow, whereas on peptone media it is intensely golden yellow. This color changes quickly into purple or red-brown when alkali (e.g., 20% KOH solution) is added, and it returns to bright yellow when acid (e.g., 10% HCl) is added. The color reaction is due to the presence of flexirubin pigments, which are the main pigments of the organism. Many strains also contain substantial quantities of the bright yellow carotenoid, zeaxanthin. Very pale or totally nonpigmented strains are rarely found.

Catalase-negative. Oxidase-positive. Nitrate is not reduced. No growth occurs on seawater media. The highest NaCl concentration tolerated is 0.3%.

Peptones or a simpler organic nitrogen compound, NH_4^+ or NO_3^-, can serve as the sole source of nitrogen. Various sugars are used as carbon and energy sources and are usually metabolized with the production of acids. Acetate produced in this way is reused later; other organic acids, however, are not utilized. An acidic reaction occurs with glucose. In litmus milk, there is no acid production or coagulation; litmus is reduced and casein is hydrolyzed. Gelatin, casein, chitin, and yeast cells (in VY/2 agar) are hydrolyzed, but not starch or DNA. Indole and H_2S are not produced.

Strictly aerobic. Optimum temperature, ~35°C; this organism still grows well at 38°C. Optimum pH, ~7.

Source: Soil, decaying plant material, dung of herbivorous animals, and freshwater habitats in many different places. Its occurrence is rather common. The neotype strain was isolated in 1966 from soil collected on the island of Upolu, Samoa.

DNA G+C content (mol%): 45–47 (Bd, T_m).

Type strain: Fx e1 Reichenbach, ATCC 29495, CCUG 12809, CIP 106401, DSM 527, HAMBI 1966, NBRC 15056.

Sequence accession no. (16S rRNA gene): M58782.

4. **Chitinophaga ginsengisegetis** Lee, An, Im, Liu, Na, Cho, Jiun, Lee and Yang 2007, 1398[VP]

gin.seng.i.se.ge'tis. N.L. n. *ginsengum* ginseng; L. n. *seges, segetis* a field; N.L. gen. n. *ginsengisegetis* of a ginseng field, the source of the organism.

The characteristics are as given for the genus and as listed in Tables 81 and 82, with the following additional features. Cells are 0.4–0.6 × 1.1–1.3 μm on R2A agar after 3 d. Colonies grown on R2A agar for 3 d are smooth, circular, convex, transparent, and yellowish. Growth occurs well at 15–42°C and at pH 5.5–8.5; slow growth occurs 4°C. Growth occurs in the absence of NaCl and in the presence of 2.0% (w/v) NaCl, but not in the presence of 3% (w/v) NaCl. Nitrate is not reduced. Anaerobic growth does not occur. DNA, chitin, and xylan are not hydrolyzed. *N*-Acetyl-D-glucosamine, arabinose, L-fucose, glucose, maltose, mannose, D-melibiose, L-proline, rhamnose, salicin, D-sorbitol, and sucrose are utilized as sole carbon sources. Positive results for the utilization of L-alanine, gluconate, histidine, inositol, and L-serine are found after 7 d. Utilization of 5-ketogluconate and malate is positive after 14 d. The following compounds are not utilized: acetate, adipate, caprate, citrate, glycogen, 3-hydroxybenzoate, 4-hydroxybenzoate, 3-hydroxybutyrate, 2-ketogluconate, itaconate, L-lactate, malonate, mannitol, phenylacetate, propionate, D-ribose, D-sorbitol, suberate, and valerate. The following enzyme activities are produced (API ZYM gallery): *N*-acetyl-β-glucosaminidase, acid phosphatase, alkaline phosphatase, cystine arylamidase, esterase (C4), α-fucosidase, α-galactosidase, α-glucosidase, leucine arylamidase, trypsin, and valine arylamidase, but not chymotrypsin, esterase lipase (C8), β-galactosidase, β-glucosidase, β-glucuronidase, lipase (C14), α-mannosidase, or naphthol-AS-BI-phosphohydrolase. According to the API 20 NE gallery, the following tests are positive: arginine dihydrolase, β-galactosidase, and urease; glucose fermentation and gelatin hydrolysis are weakly positive. Negative results are obtained for nitrate reduction and tryptophan degradation. Flexirubin pigmentation is positive. $C_{15:0}$ iso and $C_{16:1}$ ω5*c* are the predominant cellular fatty acids.

Source: soil of a ginseng field in Pocheon province, South Korea.

DNA G+C content (mol%): 47.1 (HPLC).

Type strain: Gsoil 040, KCTC 12654, DSM 18108.

Sequence accession no. (16S rRNA gene): AB264798.

5. **Chitinophaga ginsengisoli** Lee, An, Im, Liu, Na, Cho, Jiun, Lee and Yang 2007, 1399[VP]

gin.sen.gi.so'li. N.L. n. *ginsengum* ginseng; L. n. *solum* soil; N.L. gen. n. *ginsengisoli* of soil of a ginseng field, the source of the organism.

The characteristics are as given for the genus and as listed in Tables 81 and 82, with the following additional features. Cells are 0.6–0.8 × 1.2–1.6 μm on R2A agar after 3 d. Colonies grown on R2A agar for 3 d are smooth, circular, convex, transparent, and yellowish. Growth occurs at 15–42°C and at pH 5.5–8.5. Growth occurs in the absence of NaCl and in the presence of <1.0% (w/v) NaCl. Nitrate is not reduced. Anaerobic growth does not occur. Chitin but not DNA or xylan is hydrolyzed. The following compounds are utilized as sole carbon sources: acetyl-D-glucosamine, arabinose, glucose, maltose, mannose, D-melibiose, L-proline, salicin, D-sorbitol, and sucrose. Positive results are also found for gluconate, inositol and rhamnose after 7 d, and for L-fucose and 5-ketogluconate after 14 d. The following are not utilized: acetate, adipate, L-alanine, caprate, citrate, glycogen, histidine, 3-hydroxybenzoate, 4-hydroxybenzoate, 3-hydroxybutyrate, itaconate, 2-ketogluconate, L-lactate, malate, malonate, mannitol, phenylacetate, propionate, D-ribose, L-serine, D-sorbitol, suberate, or valerate. The following enzymes are present (API ZYM gallery): *N*-acetyl-β-glucosaminidase, acid phosphatase, alkaline phosphatase, cystine arylamidase, esterase (C4), esterase lipase (C8), α-fucosidase, α-galactosidase, β-galactosidase, α-glucosidase, leucine arylamidase, trypsin, and valine arylamidase. Weakly positive utilization of chymotrypsin and β-glucosidase occurs. Negative for β-glucuronidase, lipase (C14), naphthol-AS-BI-phosphohydrolase, and α-mannosidase. The following tests are positive (API 20NE gallery): arginine dihydrolase, β-galactosidase, gelatin hydrolysis, glucose fermentation, and urease, but not for nitrate reduction and tryptophan degradation. Flexirubin pigmentation is positive. $C_{15:0}$ iso and $C_{16:1}$ ω5*c* are the predominant cellular fatty acids.

Source: soil of a ginseng field in Pocheon province, South Korea.

DNA G+C content (mol%): 48.4 (HPLC).

Type strain: Gsoil 052, KCTC 12592, DSM 18017.

Sequence accession no. (16S rRNA gene): AB245374.

6. **Chitinophaga japonensis** (Fujita, Okamoto, Kosako and Okuhara 1997) Kämpfer, Young, Sridhar, Arun, Lai, Shen and Rekha 2006, 2225[VP] (*Flexibacter japonensis* Fujita, Okamoto, Kosako and Okuhara 1997, 601[VP]; effective publication: Fujita, Okamoto, Kosako and Okuhara 1996, 2.)

ja.po.nen'sis. N.L. fem. adj. *japonensis* pertaining to Japan, the source of the soil from which the organism was isolated.

The characteristics are as given for the genus and as listed in Tables 81 and 82, with the following additional features. Long, slender rods or sometimes filaments. Gliding motility is present. Colonies are translucent, orange yellow, and flat with irregular edges on nutrient agar. Microcysts are not formed. No growth occurs under anaerobic condition on nutrient agar. Temperature range, 17–44°C. The following tests are positive: catalase; oxidase; nitrate reduction; hydrolysis of gelatin, esculin, and DNA; growth in the presence of 2.0% NaCl; acid production from D-glucose, D-xylose, D-galactose, sucrose, lactose, and maltose; and utilization of glycerol, D-xylose, adonitol, D-fructose, rhamnose, salicin, and inulin. The following tests are negative: Voges–Proskauer; hydrolysis of casein, chitin and starch; acid production from D-fructose, D-mannitol, D-sorbitol, inositol, and salicin; and utilization of D-mannitol, gentiobiose, and D-turanose. MK-7 is the major menaquinone, and MK-6 is a minor menaquinone.

DNA G+C content (mol%): 49.8 (HPLC).

Type strain: 758, CIP 105790, DSM 13484, JCM 9735, NBRC 16041.

Sequence accession no. (16S rRNA gene): AJ971483.

7. **Chitinophaga sancti** (Lewin 1969) Kämpfer, Young, Sridhar, Arun, Lai, Shen and Rekha 2006, 2225[VP] (*Flexibacter sancti* Lewin 1969, 199[AL])

sanc'ti. L. n. *sanctus* a saint; L. gen. n. *sancti* of a saint, perhaps named in honor of Santos Soriano, from whose laboratory the type strain was supplied (etymology is not clear).

The characteristics are as described for the genus and as listed in Table 81, with the following additional features. Rods with slightly tapering ends, 0.5 × 2–5 µm; some strains may be up to >15 µm in length. The cell mass on peptone agar is golden yellow, reportedly due to the carotenoid, zeaxanthin (Lewin, 1969), but it turns instantly purple-red when alkali is added, so that, supposedly, flexirubin type pigments are present. On yeast agar (e.g., VY/2 agar), delicate swarms develop, and, yeast cells are attacked. On solid media, a tough sticky slime is produced. Grows with peptones, Casamino acids or NO_3^- but not with glutamate as sole nitrogen source. Various sugars are metabolized, with the production of acid. Degrades carboxymethylcellulose (CMC), starch, and gelatin. Catalase-negative. Reduces NO_3^-. Not halotolerant. Maximum temperature, 35°C. Stated to resemble *Cytophaga johnsonae* in most respects, except that it does not degrade chitin (Lewin, 1969).

Source: soil (or similar material) in Argentina and the USA.

DNA G+C content (mol%): 46–47.

Type strain: ATCC 23092, DSM 784, HAMBI 1988, LMG 8377, NBRC 15057, VKM B-1428.

Sequence accession no. (16S rRNA gene): M62795.

8. **Chitinophaga skermanii** Kämpfer, Young, Sridhar, Arun, Lai, Shen and Rekha 2006, 2226[VP]

sker.ma'ni.i. N.L. gen. masc. n. *skermanii* of Skerman, in honor of Victor B.D. Skerman, an Australian microbiologist, in recognition of his numerous contributions to the taxonomy of micro-organisms.

The characteristics are as described for the genus and as listed in Tables 81 and 82, with the following additional features. No gliding motility occurs. Aerobic. Oxidase-positive. Good growth occurs after 48 h on nutrient agar, tryptic soy agar, and MacConkey agar at 30–40°C. No growth occurs at 5 or 42°C. Colonies on nutrient agar are smooth, orange, circular, translucent, and shiny with entire edges, becoming mucoid. Orange pigmentation is non-diffusible and nonfluorescent; it turns to cherry red upon the addition of 20% KOH and retains its original color on addition of HCl. Growth occurs at pH 5.5–10 and in the presence of 7% (w/v) NaCl. The detailed fatty acid profile has been given by Kämpfer et al. (2006). The following tests are positive: β-galactosidase, acetoin production, gelatinase, and oxidation of glucose, mannitol, and melibiose. The following tests are negative: arginine dihydrolase, lysine decarboxylase, citrate utilization, H_2S production, urease, tryptophan deaminase, indole production, oxidation of inositol, sorbitol, rhamnose, sucrose, amygdalin, and arabinose, and oxidase. The following compounds are utilized as sole carbon sources (tested with the Biolog GN system): α-cyclodextrin, dextrin, Tweens 40 and 80, *N*-acetyl-D-galactosamine, *N*-acetyl-D-glucosamine, cellobiose, L-fucose, gentiobiose, α-D-glucose, α-D-lactose, lactulose, maltose, D-mannose, D-melibiose, methyl β-D-glucoside, D-raffinose, sucrose, D-trehalose, turanose, monomethyl succinate, acetic acid, D-galacturonic acid, α-hydroxybutyric acid, α-ketobutyric acid, DL-lactic acid, succinic acid, DL-alanine, L-alanylglycine, L-asparagine, L-aspartic acid, L-glutamic acid, glycyl-L-aspartic acid, glycyl-L-glutamic acid, L-proline, L-serine, L-threonine, and glycerol. The following carbon sources are not utilized as sole carbon sources: D-arabitol, propionic acid, citric acid, glycogen, adonitol, L-arabinose, iso-erythritol, D-fructose, D-galactose, myo-inositol, D-mannitol, D-psicose, L-rhamnose, D-sorbitol, xylitol, pyruvic acid methyl ester, *cis*-aconitic acid, formic acid, D-galactonic acid lactone, D-glucosaminic acid, D-gluconic acid, D-glucuronic acid, β- and γ-hydroxybutyric acids, *p*-hydroxyphenylacetic acid, itaconic acid, α-ketoglutaric acid, malonic acid, D-saccharic acid, sebacic acid, bromosuccinic acid, quinic acid, succinamic acid, L-pyroglutamic acid, α-ketovaleric acid, glucuronamide, L-alaninamide, D-alanine, L-histidine, hydroxy-L-proline, D-serine, L-leucine, L-ornithine, L-phenylalanine, inosine, uridine, thymidine, DL-carnitine, γ-aminobutyric acid, urocanic acid, phenylethylamine, putrescine, 2-aminoethanol, 2,3-butanediol, DL-α-glycerol phosphate, glucose 1-phosphate and glucose 6-phosphate. The following enzyme activities are present: alkaline phosphatase, butyrate esterase, caprylate esterase, leucine arylamidase, valine arylamidase, cystine arylamidase, trypsin, acid phosphatase, naphthol-AS-BI-phosphohydrolase, α-galactosidase, α-glucosidase, β-glucosidase, and *N*-acetyl-β-glucosaminidase. No enzyme activities are present for myristate esterase, α-chymotrypsin, β-galactosidase, β-glucuronidase, α-mannosidase, and α-fucosidase.

Source: feces of the millipede *Arthrosphaera magna.*

DNA G+C content (mol%): 40.7 (HPLC).

Type strain: CC-SG1B, CCUG 52510, CIP 109140.

Sequence accession no. (16S rRNA gene): AJ971483.

9. **Chitinophaga terrae** Kim and Jung 2007, 1723[VP]

ter′rae. L. gen. n. *terrae* of the earth.

The characteristics are as described for the genus and as listed in Tables 81 and 82, with the following additional features. Nonmotile. Rods are 0.3–0.5 × 0.6–0.8 μm on R2A agar after 3 d. Colonies grown on LB agar for 3 d are smooth, circular, convex, and yellowish. Growth occurs at 15–42°C and at pH 6.0–9.0; slow growth occurs at 42°C and pH 9.0. Nitrate is reduced to nitrite but not to N_2. The following enzymes are produced: *N*-acetyl-β-glucosaminidase, acid phosphatase, alkaline phosphatase, α-chymotrypsin, cystine arylamidase, esterase (C4), esterase (C8), α-fucosidase, α-galactosidase, α-glucosidase, β-galactosidase, β-glucosidase, leucine arylamidase, α-mannosidase, naphthol-AS-BI-phosphohydrolase, trypsin, and valine arylamidase. No production occurs of arginine dihydrolase, β-glucuronidase, lipase (C14), protease (gelatin hydrolysis), and urease. The following compounds are assimilated: L-arabinose, L-fucose, D-glucose, maltose, D-mannose, D-melibiose, L-rhamnose, sucrose, *N*-acetyl-D-glucosamine, and salicin. No assimilation occurs of acetate, adipate, caprate, citrate, gluconate, 3-hydroxybenzoate, 4-hydroxybenzoate, 3-hydroxybutyrate, itaconate, 2-ketogluconate, 5-ketogluconate, lactate, L-malate, malonate, phenylacetate, propionate, suberate, n-valerate, D-ribose, *myo*-inositol, D-mannitol, D-sorbitol, L-alanine, L-histidine, L-proline, L-serine, and glycogen. $C_{15:0}$ iso, $C_{16:1}$ ω5*c*, and $C_{17:0}$ iso 3-OH are the predominant cellular fatty acids.

Source: soil from a field near Daejeon, South Korea.

DNA G+C content (mol%): 46.3 (HPLC).

Type strain: KP01, KCTC 12836, LMG 24015.

Sequence accession no. (16S rRNA gene): AB278570.

Genus II. **Terrimonas** Xie and Yokota 2006, 1120[VP]

The Editorial Board

Ter.ri.mo′nas. L. n. *terra* soil; L. fem. n. *monas* a unit, monad; N.L. fem. n. *Terrimonas* soil monad.

Rods, occurring singly. Nonmotile. No gliding motility. Gram-stain-negative. **Aerobic.** Growth is inhibited by >1.0% NaCl. Oxidase-positive. Catalase weakly positive. Positive for gelatin liquefaction, Voges–Proskauer test, and nitrate reduction. Negative for urease production and chitin degradation. Major cellular fatty acids are $C_{15:0}$ iso, $C_{15:1}$ iso, $C_{17:0}$ iso 3-OH, and summed feature 3. The predominant menaquinone is MK-7; MK-6 is a minor component. The major cellular fatty acids are $C_{15:0}$ iso, $C_{15:1}$ iso, $C_{17:0}$ iso 3-OH, and summed feature 3 (comprised of $C_{15:0}$ iso 2-OH and/or $C_{16:1}$ ω7*c*). Isolated from soil.

DNA G+C content (mol%): 47.2–48.9 (HPLC).

Type species: **Terrimonas ferruginea** Xie and Yokota 2006, 1120[VP] [*Flavobacterium ferrugineum* Sickles and Shaw 1934, 429[AL]; *Pseudobacterium ferrugineum* (Sickles and Shaw 1934) Krasil'nikov 1949, 234; *Empedobacter ferrugineum* (Sickles and Shaw 1934) Prévot 1961, 181].

Further descriptive information

Terrimonas strains grow well in nutrient broth (Difco), IAM medium 802*, and the nitrogen-free medium† of Xie and Yokota (2005) at 29°C. Although they can grow well on the latter medium, the organisms appear to be incapable of fixing N_2, based on failure to reduce acetylene and to amplify the *nifH* gene (Xie and Yokota, 2006).

Differentiation of the genus *Terrimonas* from related genera

The lack of swimming and gliding motility, and the occurrence of single cells rather than filamentous cells, and the occurrence of MK-7 as the major menaquinone, differentiates *Terrimonas* from the genus *Flavobacterium.* Lack of gliding motility and the occurrence of single cells rather than filamentous cells, differentiates *Terrimonas* from the genus *Flexithrix. Terrimonas* can be differentiated from the nongliding, nonfilamentous species *Chitinophaga arvensicola* by its ability to liquefy gelatin and to grow at 37°C. It can be distinguished from the nongliding, nonfilamentous species *Chitinophaga skermanii* by the higher mol% G+C of its DNA (47–49 vs. 41).

Taxonomic comments

Xie and Yokota (2006) isolated a bacterial strain called DY from garden soil in Japan. Sequence analysis of the 16S rRNA gene and the GyrB protein revealed that the closest relative of strain DY was the type strain of *Flavobacterium ferrugineum* Sickles and Shaw (1934), with 94.8 and 90.1% similarity, respectively. The two strains had menaquinone 7 as the major respiratory quinone, mol% G+C of the DNA (47–49), and a similar fatty acid composition.

Comparison of the 16S rDNA sequences with those of other organisms indicated that *Terrimonas* strains were most closely related to *Chitinophaga* species, sharing less than 92% sequence similarity with *Chitinophaga arvensicola, Chitinophaga japonensis, Chitinophaga sanctii, Chitinophaga filiformis,* and *Chitinophaga pinensis.* Consequently, Xie and Yokota (2006) reclassified *Flavobacterium ferrugineum* in a new genus, *Terrimonas,* as *Terrimonas ferruginea,* and they added their DY as a second species, *Terrimonas lutea.*

Differentiation of species of the genus *Terrimonas*

Table 83 lists characteristics that differentiate *Terrimonas ferruginea* from *Terrimonas lutea.*

---

*IAM medium 802 contains (g/l): polypeptone (Nihon Pharmaceutical Co.), 10; yeast extract, 2.0; and $MgSO_4 \cdot 7H_2O$, 1.0.

†The nitrogen-free medium of Xie and Yokota (2005) contains (g/l): glucose, 10.0; $CaCl_2 \cdot 2H_2O$, 0.1; $MgSO_4 \cdot 7H_2O$, 0.1; K_2HPO_4, 0.9; KH_2PO_4, 0.1; $CaCO_3$, 5·0; $FeSO_4 \cdot 7H_2O$, 0.01; $Na_2MoO_4 \cdot 2H_2O$, 0.005; pH 7·3.

TABLE 83. Characteristics differentiating the species of the genus *Terrimonas*[a,b]

Characteristic	*T. ferrug inea*	*T. lutea*
Colony pigmentation:		
Salmon-red	+	–
Yellow	–	+
Acid production from:		
Rhamnose, melibiose	+	–
Raffinose, sucrose, galactose, fructose	–	+
Assimilation of:		
Arabinose, trehalose	–	+
D-Xylose	+	–
Methyl-α-D-mannoside, N-acetylglucosamine, melibiose, sucrose, gentiobiose	–	+
Starch, glycogen	+	–
Enzymic activities (API ZYM system):		
β-Galactosidase, α-chymotrypsin	+	–
α-Fucosidase	w	–

[a]Symbols: +, 90% or more percent of strains are positive; –, 90% or more of strains are negative; w, weak reaction.

[b]Data taken from Xie and Yokota (2006).

List of species of the genus *Terrimonas*

1. **Terrimonas ferruginea** Xie and Yokota 2006, 1120. (*Flavobacterium ferrugineum* Sickles and Shaw 1934, 429[AL]; *Pseudobacterium ferrugineum* (Sickles and Shaw 1934) Krasil'nikov 1949, 234; *Empedobacter ferrugineum* (Sickles and Shaw 1934) Prévot 1961, 181)

fer.ru.gi'ne.a. L. fem. adj. *ferruginea* rust-colored.

The characteristics are as described for the genus, with the following additional features. Cells are single rods, 0.3–0.5 × 1–3 µm. Colonies are circular, nonspreading, and about 1 mm in diameter. Growth occurs at 10–37°C; optimum, 25–32°C. Colonies are salmon red on nutrient agar and IAM medium 802. No hydrolysis of chitin and chitosan occurs. Acid is produced from glucose, cellobiose, maltose, mannose, melibiose, rhamnose, and xylose, but not from sucrose, dulcitol, glycerol, inositol, mannitol, sorbitol, raffinose, fructose, or galactose. The following carbon sources are utilized: glucose, mannose, esculin, xylose, maltose, cellobiose, rhamnose, lactose, 5-ketogluconate, starch, and glycogen. Sucrose, trehalose, dulcitol, arabinose, galactose, fructose, glycerol, inositol, mannitol, methyl-α-D-mannoside, raffinose, sorbitol, and gentiobiose are not utilized. Activity of the following enzymes is strongly positive: β-galactosidase, alkaline phosphatase, leucine arylamidase, valine arylamidase, acid phosphatase, α-chymotrypsin, naphthol phosphohydrolase, C4 esterase, C8 esterase lipase, and *N*-acetyl-β-glucosaminidase. Weakly positive activity is exhibited for cystine arylamidase, trypsin, α-glucosidase, and α-fucosidase. No activity occurs for α-galactosidase, β-glucuronidase, β-glucosidase, α-mannosidase, or C14 lipase.

Source: soil from an unknown locality.

DNA G+C content (mol%): 48.9 (HPLC).

Type strain: ATCC 13524, CCUG 33443, DSM 30193, IAM 15098, JCM 21559, LMG 4021, NBRC 14992.

Sequence accession no. (16S rRNA gene): M62798.

2. **Terrimonas lutea** Xie and Yokota 2006, 1120[VP]

lu.te'a. L. fem. adj. *lutea* golden-yellow.

The characteristics are as described for the genus, with the following additional features. Cells are single rods, 0.3–0.5 × 1–3 µm. Colonies are circular, non-spreading, and about 1 mm in diameter. Growth occurs at 10–37°C; optimum 25–32°C. Colonies are yellow on nutrient agar and IAM medium 802. Chitin and chitosan are not hydrolyzed. Acid is produced from glucose, cellobiose, maltose, mannose, fructose, galactose, raffinose, and sucrose, but not from dulcitol, glycerol, inositol, mannitol, melibiose, rhamnose, xylose, or sorbitol. The following carbon sources are utilized: glucose, mannose, esculin, maltose, lactose, 5-ketogluconate, arabinose, galactose, fructose, methyl-α-D-mannoside, cellobiose, *N*-acetylglucosamine, sucrose, raffinose, gentiobiose, and melibiose. Starch, D-xylose, glycogen, glycerol, inositol, mannitol, sorbitol, rhamnose, and dulcitol are not utilized. Activity of the following enzymes is strongly positive: alkaline phosphatase, leucine arylamidase, valine arylamidase, acid phosphatase, naphthol phosphohydrolase, C4 esterase, C8 esterase lipase, and *N*-acetyl-β-glucosaminidase. Weak activity is exhibited for C14 lipase, cystine arylamidase, α- and β-glucosidases, and trypsin. No activity occurs for β-glucuronidase, α- and β-galactosidases, α-mannosidase, α-chymotrypsin, and α-fucosidase.

Source: garden soil in Japan.

DNA G+C content (mol%): 47.2 (HPLC).

Type strain: DY, CCTCC AB205006, IAM 15284, JCM 21735.

Sequence accession no. (16S rRNA gene): AB192292.

References

An, D.S., H.G. Lee, W.T. Im, Q.M. Liu and S.T. Lee. 2007. *Segetibacter koreensis* gen. nov., sp. nov., a novel member of the phylum *Bacteroidetes*, isolated from the soil of a ginseng field in South Korea. Int. J. Syst. Evol. Microbiol. *57*: 1828–1833.

Fujita, T., M. Okamoto, Y. Kosako and M. Okuhara. 1996. *Flexibacter japonensis* sp. nov., a new species that produces a novel inhibitor of human leukocyte elastase isolated from soil. Curr. Microbiol. *33*: 89–93.

Fujita, T., M. Okamoto, Y. Kosako and M. Okuhura. 1997. *In* Validation of the publication of new names and new combinations previously effectively published outside the IJSB. List no. 61. Int. J. Syst. Bacteriol. *47*: 601–602.

Greaves, R.I. 1956. The preservation of bacteria. Can. J. Microbiol. *2*: 365–371.

Kämpfer, P., C.C. Young, K.R. Sridhar, A.B. Arun, W.A. Lai, F.T. Shen and P.D. Rekha. 2006. Transfer of [*Flexibacter*] *sancti*, [*Flexibacter*] *filiformis*, [*Flexibacter*] *japonensis* and [*Cytophaga*] *arvensicola* to the genus *Chitinophaga* and description of *Chitinophaga skermanii* sp. nov. Int. J. Syst. Evol. Microbiol. *56*: 2223–2228.

Kämpfer, P., N. Lodders and E. Falsen. 2010. *Hydrotalea flava* gen. nov. sp. nov., a new species of the phylum *Bacteriodetes* and allocation of the genera *Chitinophaga, Sediminibacterium, Lacibacter, Flavihumibacter, Flavisolibacter, Niabella, Niastella, Segetibacter, Parasegetibacter, Terrimonas, Ferruginibacter, Filimonas* and *Hydrotalea* to the family *Chitinophagaceae* fam. nov. Int. J. Syst. Evol. Microbiol. *60*: (in press).

Kim, B.Y., H.Y. Weon, S.H. Yoo, S.B. Hong, S.W. Kwon, E. Stackebrandt and S.J. Go. 2007. *Niabella aurantiaca* gen. nov., sp. nov., isolated from a greenhouse soil in Korea. Int. J. Syst. Evol. Microbiol. *57*: 538–541.

Kim, M.K. and H.Y. Jung. 2007. *Chitinophaga terrae* sp. nov., isolated from soil. Int. J. Syst. Evol. Microbiol. *57*: 1721–1724.

Krasil'nikov, N.A. 1949. Guide to the bacteria and actinomycetes. *In* Akad. Nauk. S.S.S.R. Moscow, pp. 1–830.

Lee, H.G., D.S. An, W.T. Im, Q.M. Liu, J.R. Na, D.H. Cho, C.W. Jin, S.T. Lee and D.C. Yang. 2007. *Chitinophaga ginsengisegetis* sp. nov. and *Chitinophaga ginsengisoli* sp. nov., isolated from soil of a ginseng field in South Korea. Int. J. Syst. Evol. Microbiol. *57*: 1396–1401.

Lewin, R.A. 1969. A classification of *Flexibacteria*. J. Gen. Microbiol. *58*: 189–206.

Lim, J.H., S.-H. Baek and S.-T. Lee. 2009. *Ferruginibacter alkalilentus* gen. nov., sp. nov. and *Ferruginibacter lapsinanis* sp. nov., novel members of the family '*Chitinophagaceae*' in the phylum *Bacteroidetes*, isolated from freshwater sediment. Int. J. Syst. Evol. Microbiol. *59*: 2394–2399.

Nakagawa, Y. and K. Yamasato. 1993. Phylogenetic diversity of the genus *Cytophaga* revealed by 16S ribosomal RNA sequencing and menaquinone analysis. J. Gen. Microbiol. *139*: 1155–1161.

Nakagawa, Y., T. Sakane, M. Suzuki and K. Hatano. 2002. Phylogenetic structure of the genera *Flexibacter, Flexithrix*, and *Microscilla* deduced from 16S rRNA sequence analysis. J. Gen. Appl. Microbiol. *48*: 155–165.

Oyaizu, H., K. Komagata, A. Amemura and T. Harada. 1982. A succinoglycan-decomposing bacterium, *Cytophaga arvensicola* sp. nov. J. Gen. Appl. Microbiol. *28*: 369–388.

Oyaizu, H., K. Komagata, A. Amemura and T. Harada. 1983. *In* Validation of the publication of new names and new combinations previously effectively published outside the IJSB. List no. 10. Int. J. Syst. Bacteriol. *33*: 438–440.

Pankratov, T.A., I.S. Kulichevskaya, W. Liesack and S.N. Dedysh. 2006. Isolation of aerobic, gliding, xylanolytic and laminarinolytic bacteria from acidic *Sphagnum* peatlands and emended description of *Chitinophaga arvensicola* Kämpfer et al. 2006. Int. J. Syst. Evol. Microbiol. *56*: 2761–2764.

Prévot, A.R. 1961. Traité de Systematique Bactérienne, vol. 2. Dunod, Paris, 3–771.

Qu, J.H. and H.L. Yuan. 2008. *Sediminibacterium salmoneum* gen. nov., sp. nov., a member of the phylum *Bacteroidetes* isolated from sediment of a eutrophic reservoir. Int. J. Syst. Evol. Microbiol. *58*: 2191–2194.

Qu, J.-H., H.-L. Yuan, J.-S. Yang, H.-F. Li and N. Chen. 2009. *Lacibacter cauensis* gen. nov., sp. nov., a novel member of the phylum *Bacteroidetes* isolated from sediment of a eutrophic lake. Int. J. Syst. Evol. Microbiol. *59*: 1153–1157.

Reichenbach, H. and M. Dworkin. 1981. The order Cytophagales. *In* The Prokaryotes: a Handbook on Habitats, Isolation, and Identification of *Bacteria*, (edited by Starr, Stolp, Trüper, Balows and Schlegel). Springer, New York, pp. 356–379.

Reichenbach, H. 1989. Genus *Flexibacter Soriano* 1945, 92[AL] emend. *In* Bergey's Manual of Systematic Bacteriology, vol. 3 (edited by Staley, Bryant, Pfennig and Holt). Williams & Wilkins, Baltimore, pp. 2061–2071.

Sangkhobol, V. and V.B.D. Skerman. 1981. *Chitinophaga*, a new genus of chitinolytic Myxobacteria. Int. J. Syst. Bacteriol. *31*: 285–293.

Shiratori, H., Y. Tagami, T. Morishita, Y. Kamihara, T. Beppu and K. Ueda. 2009. *Filimonas lacunae* gen. nov., sp. nov., a member of the phylum *Bacteroidetes* isolated from fresh water. Int. J. Syst. Evol. Microbiol. *59*: 1137–1142.

Sickles, G.M. and M. Shaw. 1934. A systematic study of microorganisms which decompose the specific carbohydrates of the pneumococcus. J. Bacteriol. *28*: 415–431.

Skerman, V.B.D. 1989. Genus *Chitinophaga*. *In* Bergey's Manual of Systematic Bacteriology, vol. 3 (edited by Staley, Bryant, Pfennig and Holt). Williams & Wilkins, Baltimore, pp. 2074–2077.

Sly, L.I., M. Taghavi and M. Fegan. 1999. Phylogenetic position of *Chitinophaga pinensis* in the *Flexibacter–Bacteroides–Cytophaga* phylum. Int. J. Syst. Bacteriol. *49*: 479–481.

Solntseva, L.I. 1940. Biology of myxobacteria. I. *Myxococcus*. Mikrobiologiya *9*: 217–232.

Soriano, S. 1945. Un nuevo orden de bacterias: *Flexibacteriales*. Cienc. Invest. (Buenos Aires) *1*: 92–93.

Stanier, R.Y. 1942. The *Cytophaga* group: a contribution to the biology of myxobacteria. Bacteriol. Rev. *6*: 143–196.

Takeuchi, M. and A. Yokota. 1992. Proposals of *Sphingobacterium faecium* sp. nov., *Sphingobacterium piscium* sp. nov., *Sphingobacterium heparinum* comb. nov., *Sphingobacterium thalpophilum* comb. nov. and two genospecies of the genus *Sphingobacterium*, and synonymy of *Flavobacterium yabuuchiae* and *Sphingobacterium spiritivorum*. J. Gen. Appl. Microbiol. *38*: 465–482.

Weon, H.Y., B.Y. Kim, S.H. Yoo, S.Y. Lee, S.W. Kwon, S.J. Go and E. Stackebrandt. 2006. *Niastella koreensis* gen. nov., sp. nov. and *Niastella yeongjuensis* sp. nov., novel members of the phylum *Bacteroidetes*, isolated from soil cultivated with Korean ginseng. Int. J. Syst. Evol. Microbiol. *56*: 1777–1782.

Xie, C.H. and A. Yokota. 2005. Reclassification of *Alcaligenes latus* strains IAM 12599[T] and IAM 12664 and *Pseudomonas saccharophila* as *Azohydromonas lata* gen. nov., comb. nov., *Azohydromonas australica* sp. nov. and *Pelomonas saccharophila* gen. nov., comb. nov., respectively. Int. J. Syst. Evol. Microbiol. *55*: 2419–2425.

Xie, C.H. and A. Yokota. 2006. Reclassification of [*Flavobacterium*] *ferrugineum* as *Terrimonas ferruginea* gen. nov., comb. nov., and description of *Terrimonas lutea* sp. nov., isolated from soil. Int. J. Syst. Evol. Microbiol. *56*: 1117–1121; erratum: 2023.

Yoon, M.H. and W.T. Im. 2007. *Flavisolibacter ginsengiterrae* gen. nov., sp. nov. and *Flavisolibacter ginsengisoli* sp. nov., isolated from ginseng cultivating soil. Int. J. Syst. Evol. Microbiol. *57*: 1834–1839.

Zhang, K., Y. Tang, L. Zhang, J. Dai, Y. Wang, X. Luo, M. Liu, G. Luo and C. Fang. 2009. *Parasegetibacter luojiensis* gen. nov., sp. nov., a member of the phylum *Bacteroidetes* isolated from a forest soil. Int. J. Syst. Evol. Microbiol. *59*: 3058–3062.

Zhang, N.N., J.-H. Qu, H.-L. Yuan, M.-S. Sun and J.-S. Yang. 2010. *Flavihumibacter petaseus* gen. nov., sp. nov., isolated from soil of a subtropical rainforest. Int. J. Syst. Evol. Microbiol. *60*: 1609–1612.

Family III. **Saprospiraceae** fam. nov.

The Editorial Board

Sa.pro.spi.ra.ce'a.e. N.L. fem. n. *Saprospira* type genus of the family; suff. *-aceae* ending to denote family; N.L. fem. pl. n. *Saprospiraceae* the *Saprospira* family.

The family *Saprospiraceae* was circumscribed for this volume on the basis of phylogenetic analysis for the 16S rDNA sequences; the family contains the genera *Saprospira* (type genus), *Haliscomenobacter, Lewinella*, and *Aureispira*. The organisms are Gram-stain-negative rods that form long filaments; two genera form helical filaments. Flagellar motility is absent. Three genera exhibit gliding motility (*Saprospira, Lewinella*, and *Aureispira*). Two genera consist of helical gliding organisms (*Saprospira* and *Aureispira*). Nonphototrophic. Nonsporeforming. Aerobic. The color of the cell mass is pink, yellow, or orange. Found in freshwater and/or marine environments. The mol% G+C content of the DNA ranges from 33 to 53.

Further differences among the genera are listed in Table 84.

Type genus: **Saprospira** Gross 1911, 202[AL].

TABLE 84. Distinguishing characteristics for genera of the family *Saprospiraceae*[a,b]

Characteristic	*Saprospira*	*Aureispira*	*Haliscomenobacter*	*Lewinella*
Helical filaments	+	+	−	−
Sheaths	−	−	+	+
Gliding motility	+	+	−	+
Seawater required for growth	D	+	−	+
Starch hydrolysis	−	−	+	−
Trypsin activity	+	D	nr	nr
Acid phosphatase, naphthol-AS-BI-phosphohydrolase	−	+	nr	nr
Major polyamine:				
Spermidine	−	+	+	nr
Agmatine	+	−	−	nr
Habitat:				
Freshwater, wastewater, sewage	D	−	+	−
Marine environments	D	+	−	+
DNA G+C content (mol%)	33–48[c]	38–39	49	45–53[d]

[a] Symbols: +, 90% or more of strains are positive, −, 10% or less of strains are positive; D, different reactions in different species; nr, not reported or not applicable.

[b] Data from Hamana and Nakagawa (2001), Hosoya et al. (2007, 2006), Reichenbach (1989), and Sly et al. (1998).

[c] The G+C content of the type species is 46–48 mol%.

[d] The G+C content of the type species is 45 mol%.

Genus I. **Saprospira** Gross 1911, 202[AL]

Ralph A. Lewin

Sap.ro.spi′ra. Gr. adj. *sapros* rotten, putrid; L. fem. n. *spira* a spiral; N.L. fem. n. *Saprospira* spiral associated with decaying matter.

The following generic description is modified from Reichenbach (1989). **Helical filaments 10–500 µm long and 0.5–3 µm wide, multicellular**, unbranched, without sheaths. **Cells 1.5–3.5 µm in length**. Gram-stain-negative. Non-flagellate. **Move by gliding** in either longitudinal direction at speeds up to 180 µm/min, simultaneously rotating around their long axes. Resting stages not known. **Colonies on agar surfaces are spreading; they may exhibit a regular pattern of stripes. Strictly aerobic organotrophs** requiring amino acids. Pigmentation is by carotenoids – pink, yellow, orange, or red. Aquatic, marine, or freshwater, on surfaces or in bottom sediments.

DNA G+C content (mol%): 33–48 (Bd, T_m).

Type species: **Saprospira grandis** Gross 1911, 202[AL].

Further descriptive information

The saprospiras are versatile microbes. They are not rare; they are nevertheless all too little studied, perhaps because they apparently include no pathogens. Their characteristic gliding motility can be seen in the film by Reichenbach (1980). They move on solid surfaces where they can hope to find potentially nutrient substrates or prey bacteria, but how they move is still under investigation (Aizawa, 2005). The filaments are multicellular, and it is not known how the cells are coordinated to push all in one direction or the other. What induces filaments to reverse their direction, or to break is unknown.

Saprospiras are typically helical, but the function of this shape is not understood. A helical form might facilitate wriggling around particles, but experimental evidence is lacking.

All saprospiras in culture are colored, due to different kinds of carotenoids (Lewin and Fox, 1963). Since the organisms are not photosynthetic, the function of the pigments is not known. Whether saprospiras are phototactic or chemotactic is also not known.

Within the past few years the biological knowledge of *Saprospira*, and perhaps of other flexibacteria too, has taken on a new twist since the discovery that these microbes are not merely saprophytes, but that some, at least, are in effect predatory. Certain strains of marine and freshwater *Saprospira* spp., like other flexibacteria, produce extracellular toxins that can kill diatoms, cyanobacteria, and other microbes, from which essential nutrients (including predominantly amino-acids; Lewin, 1972) can be absorbed (Ashton and Robarts, 1987; Sakata et al., 1991; Sangkhobol and Skerman, 1981). Some can even catch prey organisms, by a process called ixotrophy, before killing and digesting them (Lewin, 1997). Their slime, normally used in some way for their gliding motion, serves as a kind of adhesive "fly paper" with a specific affinity for the flagella of prey bacteria, notably some strains of *Vibrio*. When these unfortunate microbes are caught, they can be moved to and fro on the *Saprospira* surface until eventually killed by exotoxins and subjected to enzymic activity, whereby nutrients are liberated to supplement solutes in the medium.

Saprospiras require some or all of the so-called "essential" amino acids that animals do, and presumably for the same reason. It is unnecessary to biosynthesize essential molecules when one can obtain them ready-made by digesting food proteins. Many saprospiras are even more like animals, living by predation on other microbes, including diatoms and dinoflagellates as well as bacteria. In some, the mucilage tracks that they employ for motility have been pre-adapted to be sticky and thereby to catch bacterial prey, notably species of *Vibrio*, by their flagella (Lewin, 1997). Then an ixotrophic saprospira can kill the prey by extracellular toxins and digest them by extracellular enzymes, thereby liberating the amino acids that are needed for growth. The natures of the sticky mucilage, and of those toxins and enzymes, need to be further studied.

The first marine lytic bacteriophage was found in a saprospira (Lewin et al., 1964), but not all saprospiras succumb to it. Some saprospiras, perhaps all, contain rhapidosomes, which look like long bacteriophage tails (Reichle and Lewin, 1968). These are hollow cylinders with protruding central axes, and are liberated in large numbers when the saprospira cells lyse. Their origin and functions are unknown.

Enrichment, isolation, and maintenance procedures*

Gross (1911) originally enriched *Saprospira grandis* and "*Saprospira nana*" by filling large dishes with bottom sand and seawater from the Mediterranean and floating cover glasses on the water surface. Overnight, the saprospiras attached themselves firmly to the glass surface. R.A. Lewin (personal communication) collected them on glass slides immersed at a depth of 1000 m off the coast of California. Other specific enrichment techniques for *Saprospira* species are not known. To isolate these organisms, samples from bottom sediments from the seacoast, rivers, or lakes are placed as streaks or spots on the surface of dry low-nutrient agar plates, e.g., seawater agar, perhaps with 0.1% NH_4Cl and 0.002% sodium acetate or with 0.03% casein peptone and 0.01% yeast extract. Corresponding media may be used for freshwater species. The crude cultures are incubated at room temperature or 30°C and checked under a dissecting microscope from time to time for spreading colonies. *Saprospira* colonies may easily be recognized by their stripes. *Saprospira grandis* grows in a very wide temperature range, and it may be favorable to incubate the plates at 6°C. The desired organism grows still reasonably well at this temperature, while the development of contaminants is much restricted. Transfers are made from the crude cultures to increasingly rich media, and finally pure cultures may be obtained by plating of cell suspensions. By that time, other gliding bacteria and agar liquefiers (a serious problem with marine samples) should be eliminated, for otherwise well-separated colonies are not easily obtained. More details about isolation procedures and composition of suitable media may be found in Lewin (1962, 1965b) and in Reichenbach and Dworkin (1981).

Cultures of *Saprospira grandis* die within a few days (30°C) or weeks (at room temperature). *Saprospira grandis*, "*Saprospira toviformis*" (Lewin and Mandel, 1970), and some strains of "*Saprospira thermalis*" (Lewin, 1965a) could be preserved in liquid nitrogen for 1 year; in most cases, addition of 10% glycerol has been essential; "*Saprospira flammula*" (Lewin, 1965a) and "*Saprospira albida*" (Lewin, 1962) did not survive (Sanfilippo and Lewin, 1970). *Saprospira grandis* strain Sa gl [ATCC 49590, DSM 2844] survives for at least 2 years when stored at –80°C in peptone-yeast extract medium (e.g., SP5 liquid medium) (Reichenbach and Dworkin, 1981). The same strain did not survive drying in skim milk.

Differentiation of the genus *Saprospira* from other genera

Helical multicellular filaments and gliding motility distinguish *Saprospira* from all other genera.† Should uncoiled, nonmotile strains, as observed in culture, also occur in nature, they would not be recognizable as saprospiras. Straight but gliding filaments from marine habitats, classified as *Microscilla marina*, are perhaps identical with *Saprospira grandis* (Lewin, 1962). From pure cultures of *Saprospira grandis*, stable variants have also been isolated that no longer form filaments but grow as 1–4-celled segments (Lewin, 1962).‡

Taxonomic comments

The problems involved in the taxonomy of *Saprospira* are like those now facing us for almost all microbes and many other kinds of organisms. We have somehow to resolve disagreements between classifications based solely on visual, biochemical, and physiological characters and those based essentially on molecular biology (MB) data. Since we as humans are primarily seeing organisms, we have no difficulty in recognizing a giraffe and assigning it to a genus. So when we look down a microscope and see helical bacteria, we naturally would like to assign them in the same way to named genera. However, whereas the shapes and sizes of giraffes are controlled by many thousands of genes, and their generations are many years long, those of microbes, with generations some ten orders of magnitude shorter, are controlled by only a handful of genes, and even these tend to be mutable. Natural selection may induce a filamentous bacterium to mutate and persist as a helix; conversely, a helical bacterium in culture may straighten spontaneously. Then what price is helicity as a key determinant for *Saprospira*? The same arguments might apply to considerations of gliding ability (Reichenbach et al., 1986). Furthermore, on the basis of present MB data, different *Saprospira*-like microbes may exhibit affinities with microbes from other diverse genera, indicative of considerable phylogenetic diversity within the named genus.

Although we cannot reconcile visual and MB classifications, we can at least make some workable arrangements. We must, if we need to name things. If we have to depend solely on what we can see under the microscope, then the broad generic name, *Saprospira*, is all we can use. But when objective MB data indicate otherwise, I think we will have to reassign – and rename – the diverse groups accordingly.

Following a proposition of Lewin (1962), nonphototrophic, helical, multicellular, filamentous bacteria with gliding motility were classified in the genus *Saprospira*. The genus had been defined earlier by Gross (1911), and his *Saprospira grandis* is very probably identical with Lewin's strains. Only the "spore" formation described by Gross and included in the species definition of *Saprospira grandis* has not been observed again and had to be discarded from the description. The present taxa are based, of course, on Lewin"s more recent and more detailed investigations and his neotype strain. Lewin later described several other species, viz., "*Saprospira thermalis*" (freshwater; Lewin, 1965a), "*Saprospira flammula*" (freshwater; Lewin, 1965a) and "*Saprospira toviformis*" (marine; Lewin and Mandel, 1970), and revived another old species, "*Spirulina albida*" Kolkwitz (1909), as "*Saprospira albida*" (freshwater;

*Taken from Reichenbach (1989).

†The genus *Aureispira* is similar to *Saprospira* but exhibits differences in trypsin activity, acid phosphatase activity, naphthol-AS-BI-phosphohydrolase activity, and also large differences in cellular fatty acid composition (Hosoya et al., 2006).

‡Further details and differential characteristics of seven putative *Saprospira* spp. are summarized in a table in Reichenbach (1989).

Lewin, 1962). Although the assumption is supported by correspondences in several morphological and physiological characteristics, it was not certain whether all these organisms are really related to one another, as no molecular taxonomic data were available. The mol% G+C range of 35–48 would at least not rule out a relatively close relationship. Unfortunately, except for *Saprospira grandis*, the type strains of all species have been lost, so that these species were not included in the Approved Lists of Bacterial Names. However, the organisms have been fairly well characterized, and there is no doubt that they really exist; their characteristics are given in detail by Reichenbach (1989).*,†

Acknowledgements

I am grateful to S. Aizawa and X. Mayali for their useful suggestions.

List of species of the genus *Saprospira*

1. **Saprospira grandis** Gross 1911, 202, emend. Lewin 1962, 560[AL]

gran′dis. L. fem. adj. *grandis* large.

The description is that given by Reichenbach (1989). Helical flexible filaments of constant width, 0.8–0.9 (to 1.2) μm depending on the strain, but variable in length; usually shorter in plate cultures (15–130 μm) than in liquid media (20–500 μm). The pitch and width of the screws may vary considerably even in one (the same) culture. The filaments may even uncoil entirely during culture. Stable variants have also been reported that grow as single cells or 2–4-celled segments (Lewin, 1962). The filaments are composed of cylindrical cells, 1–5.5 μm long depending on the strain and the culture; within one filament the longest cells are twice as long as the shortest ones. The cross-walls may be difficult to distinguish in young cultures but become well recognizable in old and drying filaments. End cells of the filaments are rounded (see photomicrographs in Reichenbach (1989).

Colonies on poor media are thin and spreading and remain relatively small, often with a pattern of parallel fine stripes (Figure 60). Colonies on rich media are round, convex, thick, and slimy, relatively small, and deep orange to red (Figures 61 and 62). In liquid media, *Saprospira grandis* grows in homogeneous suspension; upon shaking, the cultures show a silky appearance. In peptone-yeast extract medium, the generation time is 2–2.8 h (at 30°C) (Lewin, 1962, 1972). The cell mass is orange to red, due to carotenoids, mainly saproxanthin (Aasen and Liaaen-Jensen, 1966).

Grows on peptones, Casamino acids and defined amino acid mixtures; in addition, unknown growth factors, e.g., with yeast extract, have to be supplied. Glucose, galactose, sucrose, and acetate are growth stimulatory for some strains. Gelatin is liquefied. Starch, agar, alginate, and carboxymethylcellulose are not decomposed. In litmus milk, coagulation occurs, but acid is not produced, the curd is usually not digested, and litmus is not reduced. Tyrosine is degraded to colored products. H_2S is not produced from cysteine-containing media. Nitrate is not reduced to nitrite. Catalase-negative. Strictly aerobic.

DNA G+C content (mol%): 46–48 (Bd, T_m).

Type strain: ATCC 23119, JCM 21750, LMG 10407.

Sequence accession no. (16S rRNA gene): M58795.

Genus II. **Aureispira** Hosoya, Arunpairojana, Suwannachart, Kanjana-Opas and Yokota 2006, 2933[VP] emend. Hosoya, Arunpairojana, Suwannachart, Kanjana-Opas and Yokota 2006, 1950[VP]

The Editorial Board

Au.re.i.spi′ra. L. adj. *aureus* golden; L. fem. n. *spira* a spiral; N.L. fem. n. *Aureispira* golden spiral.

Cells are 0.8–1.2 × 1.5–2.5 μm, forming **flexible, unbranched, helical filaments** up to 100 μm long. Helix width is 1.5–2.0 μm; helix pitch is 4–9 μm. **Motile by gliding**. Gram-stain-negative. Colonies are yellow or yellowish-orange. Chemo-organotrophic. **Aerobic**. **Seawater is required for growth**. Positive for alkaline phosphatase and acid phosphatase. Oxidase and catalase are variable. The predominant fatty acid is $C_{20:4}$ ω6*c* (arachidonic acid); other characteristic fatty acids are $C_{16:0}$ and $C_{17:0}$ iso. The respiratory quinone is MK-7.

DNA G+C content (mol%): 38–39.

Type species: **Aureispira marina** Hosoya, Arunpairojana, Suwannachart, Kanjana-Opas and Yokota 2006, 2933[VP].

Further descriptive information

The genus belongs to the family *Saprospiraceae* in the phylum *Bacteroidetes*.

Until recently, the genus *Saprospira* was thought to be the only validly published genus of helical, nonphototrophic, gliding bacteria. However, Hosoya et al. (2006) isolated three gliding bacterial strains of helical bacteria from marine sponges and algae in Thailand and showed that these organisms represented a new genus distinct from *Saprospira*.

Enrichment and isolation procedures

Isolation can be achieved on SWG medium, consisting of seawater (filtered to remove large particles), 0.1% sodium glutamate,

---

*Modified from Reichenbach (1989).

†On the basis of analysis of 16S rRNA gene sequences, Sly et al. (1998) reported that *Saprospira grandis* is closely related to the *Haliscomenobacter* (Sly et al., 1998). Hosoya et al. (2006) reported that *Saprospira grandis* was related to *Haliscomenobacter* and *Lewinella*, and also found it closely related to their new genus *Aureispira*. Thus, the family *Saprospiraceae* would presently contain four genera - *Saprospira*, *Haliscomenobacter*, *Lewinella*, and *Aureispira*.

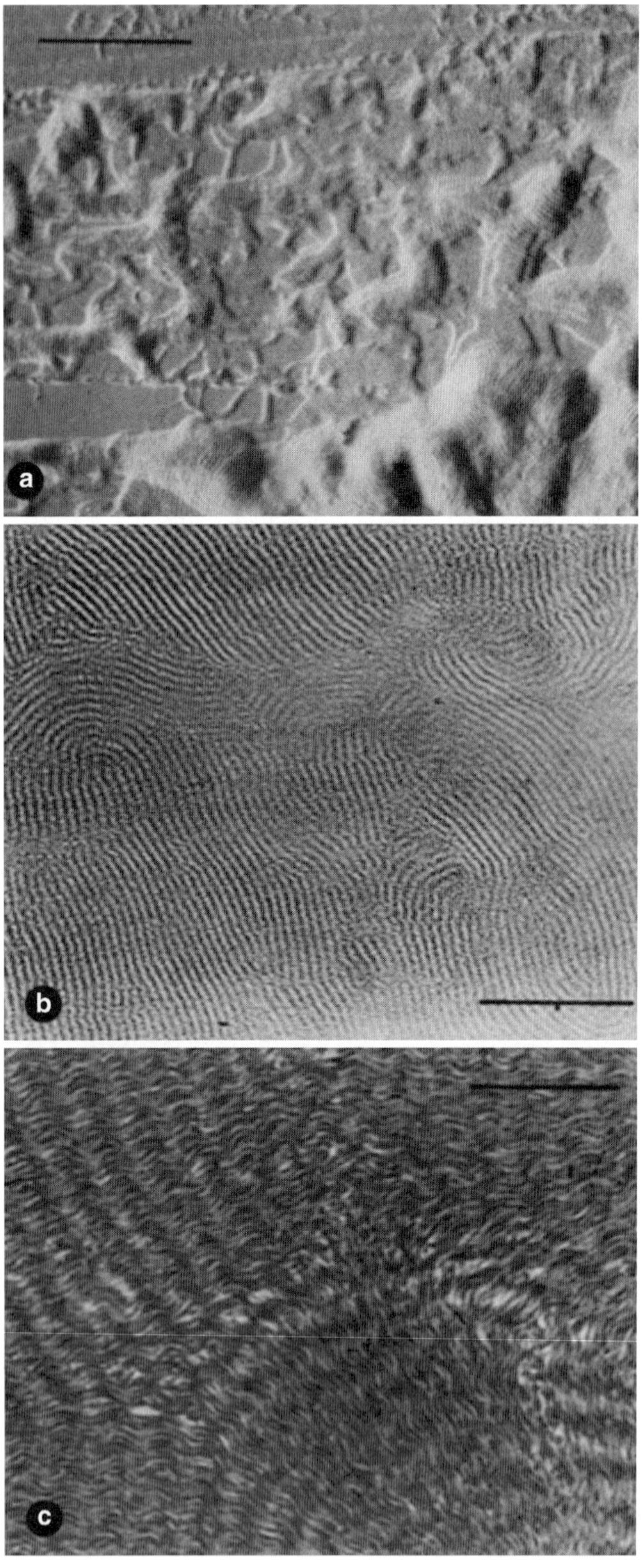

FIGURE 60. *Saprospira grandis* striped colonies. (a) Spreading colony, survey picture. The gliding filaments preferentially follow scratches in the agar surface produced by the inoculation loop. Oblique illumination. Bar = 250 μm. (b) Central part of a large colony. Brightfield illumination. Bar = 125 μm. (c) At high magnification the strict alignment of the filaments becomes recognizable. Phase-contrast. Bar = 25 μm.

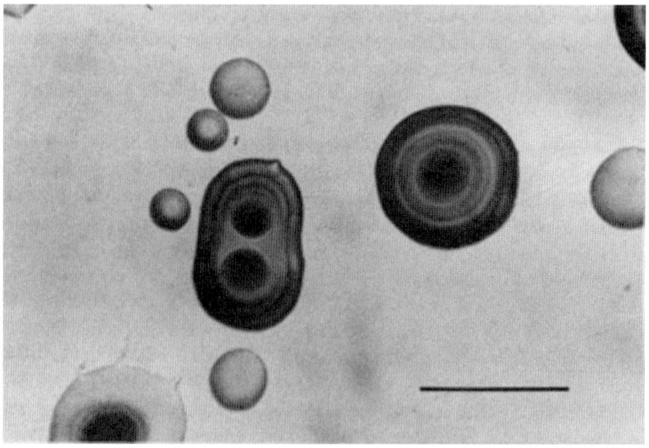

FIGURE 61. *Saprospira grandis* compact colonies on agar medium with "high" peptone content (0.1% peptone from casein + 0.02% yeast extract). Plating of an enrichment culture with *Saprospira grandis* (small smooth colonies) and a spirillum (large colonies with concentric rings). Bar = 1200 μm.

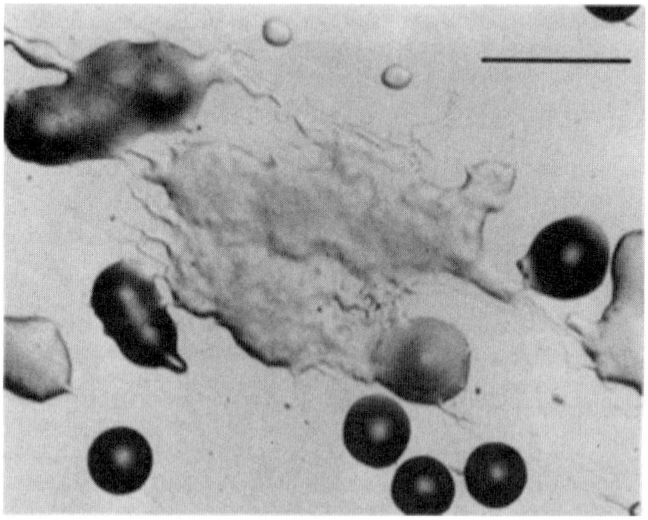

FIGURE 62. *Saprospira grandis* spreading colony on agar medium with "low" peptone content (0.03% peptone from casein + 0.01% yeast extract). Enrichment culture; the round dark colonies are contaminants. Bar = 1200 μm.

and 1.5% agar; pH 7.2. A drop from undiluted samples collected from marine sponges, algae, and coastal woody materials are allowed to flow in a narrow band (slightly off-center) across the surface of each agar plate. After drying the agar surface by incubating at 22°C for 10 min, the plates are incubated 25°C for 3–4 d and then examined by low-power phase microscopy to observe microcolony development. Purification of the gliding bacteria is performed using the micromanipulation technique of Skerman (1968) as described by Sly and Arunpairojana (1987).

Strains can be cultured and maintained on Sap2 medium, which has the following composition (g/l): tryptone, 1.0; yeast extract, 1.0; agar, 15.0. The ingredients are dissolved in 0.5× artificial seawater, having the following composition (g/l distilled water): NaCl, 15; KCl, 0.35; $MgCl_2 \cdot 6H_2O$, 5.4; $MgSO_4 \cdot 7H_2O$, 2.7; $CaCl_2 \cdot 2H_2O$, 0.5.

Differentiation of the genus *Aureispira* from other genera

Aureispira marina and *Saprospira grandis* are similar in many ways, but unlike *Saprospira grandis, Aureispira marina* is negative for trypsin, positive for acid phosphatase and naphthol-AS-BI-phosphohydrolase, has yellowish-orange colonies instead of reddish orange, and has a mol% G+C of 38–39 instead of 49.8. In addition, *Auriespira marina* contains $C_{20:4}$ ω6*c* (arachidonic acid), whereas *Saprospira grandis* does not.

Taxonomic comments

Analysis of 16S rDNA sequences of the type strain of *Aureispira marina* indicated a similarity value of 86% similarity to the type strain of *Saprospira grandis* and 78–80% to the type strains of *Lewinella cohaerens* and *Haliscomenobacter hydrossis* (Hosoya et al., 2006).

A second species, *Aureispira maritima,* was added by Hosoya et al. (2007). The DNA–DNA relatedness between the type strain of *Aureispira marina* and that of *Aureispira maritima* was less than 5%. *Aureispira maritima* can be differentiated from *Aureispira marina* by lacking C4 esterase; by producing valine arylamidase, cystine arylamidase, and trypsin; by degrading carboxymethyl-cellulose and DNA; and by having a slightly lower mol% G+C content (38.7 vs. 39.4).

List of species of the genus *Aureispira*

1. **Aureispira marina** Hosoya, Arunpairojana, Suwannachart, Kanjana-Opas and Yokota 2006, 2933[VP]

ma.ri′na. L. fem. adj. *marina* of or belonging to the sea, marine.

The characteristics are as given for the genus, with the following additional features. Optimal growth temperature, 25–30°C; no growth occurs at 8 or 37°C. pH range for growth, 6.0–8.0. Colonies are yellowish-orange. Growth occurs in artificial seawater ranging from 0.2 to 1.5× full-strength. The following tests are positive: esterase (C4); esterase lipase (C8); leucine arylamidase; naphthol-AS-BI-phosphohydrolase; degradation of casein, gelatin, and Tweens 20, 40, 60, and 80, and degradation of tyrosine to colored products. The following tests are negative: nitrate reduction; production of acetoin, H_2S, and indole; hydrolysis of agar, alginate, carboxymethylcellulose, citrate, DNA, and starch; lipase (C4); valine arylamidase; cystine arylamidase; trypsin; chymotrypsin; α-galactosidase; β-galactosidase; β-glucuronidase; α-glucosidase; β-glucosidase; *N*-acetyl-β-glucosamidase; α-mannosidase; α-fucosidase; acid production from arabinose, cellobiose, dulcitol, fructose, galactose, glucose, glycerol, inositol, lactose, maltose, mannitol, mannose, raffinose, rhamnose, sorbitol, sucrose, trehalose, and xylose.

Habitat: marine plant debris, marine sponges, and marine algae.

DNA G+C content (mol%): 38–39 (HPLC).

Type strain: 24, IAM 15389, JCM 23197, TISTR 1719.

Sequence accession no. (16S rRNA gene): AB245933

2. **Aureispira maritima** Hosoya, Arunpairojana, Suwannachart, Kanjana-Opas and Yokota 2006, 1950[VP]

ma.ri′ti.ma. L. fem. adj. *maritima* of the sea, marine, inhabiting marine environments.

The characteristics are as given for the genus, with the following additional features. Cells are 0.7–0.8 × 3–6 μm. The cells are helices 1.5–2.0 μm × 20–90 μm, with a wavelength of 4–5 μm. Optimal growth temperature, 30°C; no growth at 17 or 37°C. pH range for growth, 6.0–8.0. Growth occurs at seawater concentrations of 20–150% (w/v). Colonies are yellow. Flexirubin pigments are not detected. The following tests are positive: esterase lipase (C8), leucine arylamidase, valine arylamidase, cystine arylamidase, trypsin, and naphthol-AS-BI-phosphohydrolase, and for the degradation of casein, carboxymethylcellulose, DNA, gelatin, and Tweens 20, 40, 60 and 80. Tyrosine is degraded. Agar, alginate, citrate and starch are not attacked. Nitrate is not reduced. The following tests are negative: acetoin, H_2S and indole production; esterase (C4), lipase (C4), chymotrypsin, α-galactosidase, β-galactosidase, β-glucuronidase, α-glucosidase, β-glucosidase, *N*-acetyl-β-glucosamidase, α-mannosidase, and α-fucosidase; acid production from arabinose, cellobiose, dulcitol, fructose, galactose, glucose, glycerol, inositol, lactose, maltose, mannitol, mannose, raffinose, rhamnose, sorbitol, sucrose, trehalose, and xylose.

Habitat: marine barnacle debris.

DNA G+C content (mol%): 38.7 (HPLC).

Type strain: 59SA, IAM 15439, TISTR 1726.

Sequence accession no. (16S rRNA gene): AB278130.

Genus III. **Haliscomenobacter** van Veen, van der Kooy, Geuze and van der Vlies 1973, 213[AL] (*Streptothrix* Cohn 1875, 186, *sensu* Mulder and Van Veen 1974, 133)

Peter Kämpfer*

Ha.lis.co.me.no.bac′ter. Gr. v. *haliskomai* to fall into the hands of the enemy, to be imprisoned; N.L. masc. n. *bacter* a rod or staff; N.L. masc. n. *Haliscomenobacter* imprisoned rod.

Thin rods, 0.4–0.5 × 3–5 μm, usually **in chains, enclosed by a** narrow, hardly visible **hyaline sheath. No ferric or manganic oxides have been detected so far as depositions** in or on the sheaths. Sometimes **branching of the filaments** incidentally occurs in stationary cultures. The branching cells disrupt the sheath and form a new sheath outside the envelope. Compared with the main filaments, the lateral branches are short. Cells outside the sheaths are rarely visible; **no flagellation** or motility has been observed. Gram-stain-negative. **Aerobic**, having a strictly respiratory type of metabolism with oxygen as the

*Updated from Mulder (1989).

terminal electron acceptor. Temperature range: 8–30°C; optimum: 25–28°C. Growth is much faster at pH 7.5 than at pH 6.4.

Chemo-organotrophic. Only a few alcohols, organic acids, and sugars are used as sources of carbon and energy. Glucose, fructose, and starch are used as sources of carbon and energy; acetate, lactate, succinate, β-hydroxybutyrate, glycerol, sorbitol, and many other carbon sources are not utilized. *Haliscomenobacter* strains do not grow in the presence of high concentrations of yeast extract and peptone.

Ammonium salts and nitrates may serve as nitrogen source in the presence of vitamin B_{12} or methionine. Although, peptone, Casamino acids, and mixtures of aspartic and glutamic acids, and vitamin B_{12}, or methionine give better results, high concentrations of yeast extract and peptone inhibit growth. A strict requirement for calcium, and (to some extent) magnesium is reported. Below a concentration of 20 mg/l Ca and 5 mg/l Mg, only a few strains were able to show visible growth. Inorganic nitrogen compounds (nitrate, ammonium salts) are moderately good nitrogen sources; amino acids and peptone give better results.

Colonies. on poor agar media are hardly visible macroscopically; they **are filamentous** and <0.5 mm in diameter. On a sucrose-peptone-yeast extract medium enriched with vitamin B_{12} and thiamine, and GMBN agar, **pinkish**, smooth, slightly filamentous colonies of 1–3 mm in diameter develop. Liquid cultures turn pink, owing to the formation of carotenoid pigments.

DNA G+C content (mol%): 49.

Type species: **Haliscomenobacter hydrossis** van Veen, van der Kooy, Geuze and van der Vlies 1973, 213[AL].

Further descriptive information

On the basis of oligonucleotide cataloging, *Haliscomenobacter* was placed into the *Cytophaga–Flavobacterium* group, and in recent studies on 16S rRNA sequence analysis *Haliscomenobacter hydrossis* was grouped into the *Saprospira* group (Figure 63), which contains *Saprospira grandis*, and the *Lewinella* species as closest relatives (Gherna and Woese, 1992; Sly et al., 1998).

A detailed chemotaxonomic study (Kämpfer, 1995) revealed that the major menaquinone of *Haliscomenobacter* was type MK-7 (70–90%). The menaquinone MK-6 was also present, but only in minor amounts (10–30%). Fatty acid profiles showed four major fatty acids, listed in descending order: 20–41% 13-methyltetradecanoic acid ($C_{15:0}$ iso), 7–22% *cis*-9 hexadecenoic acid ($C_{16:1}$), 4–11% hexadecanoic acid ($C_{16:0}$), and 4–7% octadecanoic acid ($C_{18:0}$). The type strain of *Haliscomenobacter hydrossis* DSM 1100^T contained high amounts of the hydroxylated fatty acids $C_{15:0}$ iso 3-OH and $C_{15:0}$ iso 2-OH in addition to minor amounts of $C_{17:0}$ iso 3-OH. Hamana and Nakagawa (2001) reported spermidine as the major cellular polyamine in *Haliscomenobacter*.

A comprehensive study on physiological tests of ten strains of *Haliscomenobacter* showed very similar profiles. All strains were able to grow with D-fructose, D-glucose, and starch as sole carbon source in GMBN basal medium. The chromogenic substrates pNP-*N*-acetyl-β-D-glucosaminide, pNP-β-D-glucopyranoside, pNP-phosphate (pH 7.0), pNP-phosphate (pH 8.2), L-alanine-pNA, L-arginine-pNA, glycine-pNA, and L-lysine-pNA (*pNP* para-nitrophenyl; *pNA* para-nitroanilide) were hydrolyzed by all strains. Most of the physiological tests gave negative results. Very little data on the physiology of *Haliscomenobacter* has been published

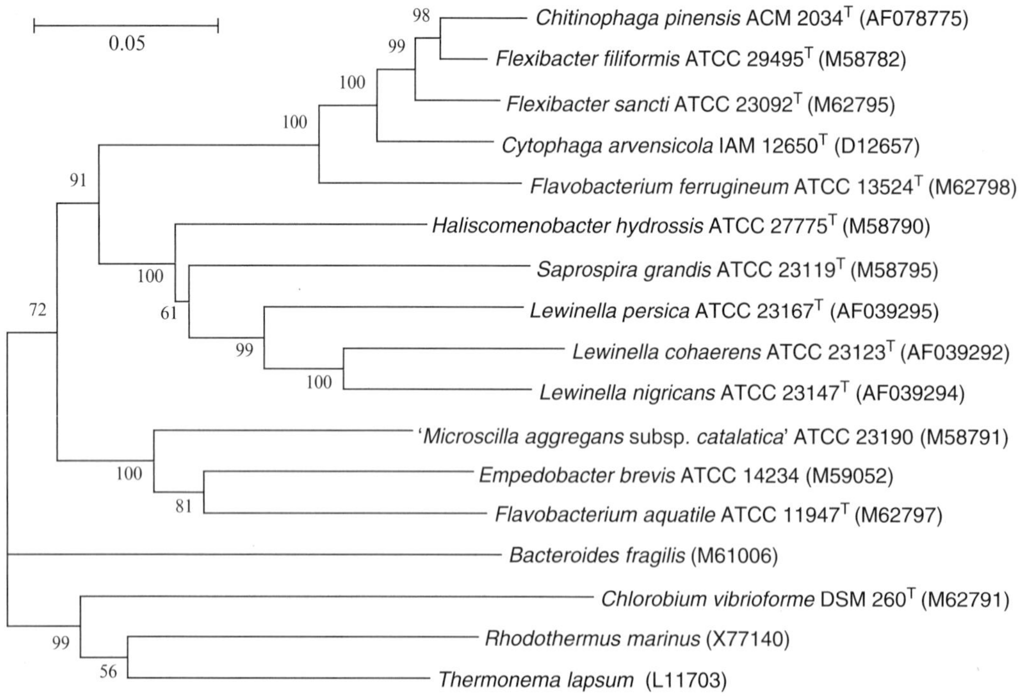

FIGURE 63. Phylogenetic analysis of *Lewinella* and related organisms based on 16S rRNA gene sequences available from the EMBL database (accession numbers are given in parentheses) constructed after multiple alignments of data. Distances (distance options according to the Kimura-2 model) and clustering with the neighbor-joining method was performed using the software package MEGA (Molecular Evolutionary Genetics Analysis) version 2.1. Bootstrap values based on 1000 replications are listed as percentages at the branching points. Bar = 0.1 nucleotide substitutions per nucleotide position.

until now (Krul, 1977; Mulder, 1989; Mulder and Deinema, 1981; van Veen et al., 1973; Ziegler et al., 1990). Krul (1977) reported that D-glucose together with some components from trypticase soy broth could be useful as energy sources. Mulder and Deinema (1981) reported that D-glucose but not glycerol and lactate can be used as carbon sources, which is in accordance with the study of Kämpfer (1995). In contrast, Krul (1977) reported sucrose utilization, but this could not be confirmed by Kämpfer (1995). In addition, Mulder (1989) listed glucosamine and lactose, and to a lesser extent mannitol, as carbon and energy sources. The utilization of lactose and D-mannitol was not observed for any isolate by Kämpfer (1995). As already mentioned above, isolates of the genus *Haliscomenobacter* are difficult to cultivate. A detailed study on the nutritional requirements of these isolates showed that D-glucose and D-fructose were good substrates for the cultivation of *Haliscomenobacter* (Kämpfer et al., 1995). In addition, the presence of calcium, magnesium, and phosphate in the media was required for growth of all isolates, and high ammonia concentrations (>2 g/1) inhibited their growth. Furthermore, *Haliscomenobacter* preferred a distinct range of phosphorus concentration (0.05–0.2 g/1), and a very low concentration of yeast extract and peptone (Kämpfer et al., 1995).

Since isolation and cultivation of these organisms is difficult and time consuming, the use of specific 16S rRNA targeted oligonucleotide probes as reported earlier (Kämpfer, 1997; Schauer and Hahn, 2005; Wagner et al., 1994) can be very useful for *in situ* detection and identification of *Haliscomenobacter*.

Haliscomenobacter hydrossis is often found in activated sludge. It is one of those filamentous bacteria that sometimes occur in large numbers in sludge (Eikelboom, 1975; Fourest et al., 2004; Pernelle et al., 2001; Schauer and Hahn, 2005). They make the sludge flocs voluminous, while many trichomes, like tiny straight needles, protrude into the surrounding water. Both phenomena are responsible for the slow settling of the flocs (bulking). There are only a few indications that the organism occurs in large masses in certain sludges. Pernelle et al. (2001) and Gaval and Pernelle (2003) reported a clear stimulation of growth of *Haliscomenobacter hydrossis* along with high amounts of easily assimilable substrates and oxygen deficiency. *Haliscomenobacter hydrossis* is not able to store polyhydroxyalkanoate under these conditions (Dionisi et al., 2002).

Under laboratory conditions, *Haliscomenobacter hydrossis* grows slowly. van Veen et al. (1982) reported maximum specific growth rates in continuous culture of approximately 0.05/h and 0.09/h for two different strains, corresponding to minimum doubling times of 14 and 9 h, respectively. These values are below those of most known bacteria occurring in wastewater and activated sludge (van Veen et al., 1982). The maintenance coefficients of these strains were low, 20 and 21 mg of glucose/g biomass/h, respectively. These values are far below those of most known bacteria and may contribute to the competitive ability of the organisms. The maximum yield coefficients (Y_G) found are 0.59 and 0.42 g of biomass/g of glucose, respectively. These observations were confirmed by Kämpfer et al. (1995), who found different carbon sources, nutrient addition, and inorganic growth factors significantly affected the growth of *Haliscomenobacter hydrossis*.

Enrichment and isolation procedures

Most often, bulking activated sludge, which upon microscopic examination is found to contain large masses of typical *Haliscomenobacter* filaments Figure 64, can be used as enrichment material. The following isolation procedure was recommended by Mulder and Deinema (1981). A sample of 0.1–0.5 ml of sludge is pipetted into tubes containing 10 ml of sterile tap water. The contents of the tubes are stirred for several minutes using a tube mixer, whereupon the flocs are allowed to settle. During moderate agitation of the floc suspensions, fragments of the threads are severed from the protruding filaments and will be seen in the supernatant. If upon microscopic observation insufficient amounts of separate filaments and free cells are present, the whole procedure is repeated. After settling of the flocs, a small amount of the supernatant is directly streaked on previously dried agar plates of the following composition (per liter of distilled water): glucose, 150 mg; $(NH_4)_2SO_4$, 50 mg; $Ca(NO_3)_2$, 10 mg; K_2HPO_4, 50 mg; $MgSO_4 \cdot 7H_2O$, 50 mg; KCl, 50 mg; $CaCO_3$, 100 mg; $FeCl_3 \cdot 6H_2O$, 5 mg; $MnSO_4 \cdot H_2O$, 2.5 mg; $CuSO_4 \cdot 5H_2O$, 0.1 mg; $ZnSO_4 \cdot 7H_2O$, 0.1 mg; $Na_2MoO_4 \cdot 2H_2O$, 0.05 mg; $CoCl_2 \cdot 6H_2O$, 0.05 mg; thiamine, 0.4 mg; vitamin B_{12}, 0.01 mg; agar (Oxoid), 10 g. After incubation for some weeks at 17–20°C, small filamentous colonies may develop that are difficult to recognize, even when viewed under a stereomicroscope. Low magnification (×150) phase-contrast microscopy facilitates detection. Sterile capillary tubes are used to transfer cells to agar plates of the following composition (in grams per liter of distilled water) (S.C.Y. agar): sucrose, 1.0; casitone (Difco), 0.75; yeast extract (Difco), 0.25; trypticase soy broth without glucose (BBL), 0.25; thiamine, 4×10^{-4}; vitamin B_{12}, 10^{-5}; agar (Oxoid), 10. Ziegler et al. (1990) investigated several additional isolation methods for filamentous bacteria and recovered most of the *Haliscomenobacter* isolates after a 1:10 dilution of the sample with MSV (Williams and Unz, 1985) and subsequent washing and centrifugation at 200 g for 5 min. I-medium

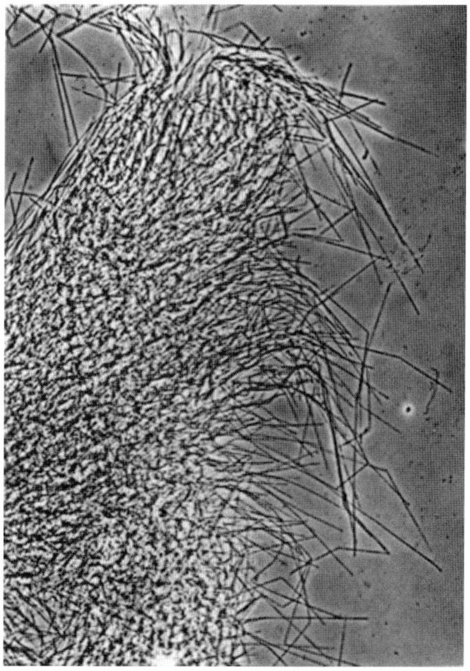

FIGURE 64. *Haliscomenobacter hydrossis* flocculent growth of pure culture showing many needle-like filaments. Phase-contrast micrograph; magnification ×1528. (Reproduced with permission from Mulder and Deinema 1981.)

TABLE 85. Characteristics that differentiate the genus *Haliscomenobacter* from *Saprospira grandis* and the genus *Lewinella*[a,b]

Characteristic	*Haliscomenobacter*	*S. grandis*	*Lewinella*
Length of cells (μm)	3–5	1–5.5	2–3
Diameter of filaments (μm)	0.4–0.5	0.8–1.2	0.5–1.5
Length of filaments (μm)	10–300	6–500	5–150
Gliding motility	–	–	+
Seawater as source of essential growth factors	–	+	+
Starch degraded	+	–	–
Predominant cellular polyamine	Spermidine	Agmatine	Spermidine
Color of cell mass	Pink	Orange-red	Orange-yellow
DNA G+C content (mol%)	49	46–48	45–53
Habitat	Wastewater, sewage	Marine	Marine

[a]Symbols: +, 90% or more of strains are positive; –, 90% or more of strains are negative.

[b]Data from Hamana and Nakagawa (2001), Reichenbach (1989), and Sly et al. (1998).

(van Veen et al., 1973) and GMBN medium (Kämpfer, 1995) were recommended for initial isolation.

Maintenance procedures

Stock cultures on the previously described S.C.Y. agar slopes, with 3 ml of sterile tap water added on the surface of the agar, are inoculated and kept at 20–25°C until turbid growth in the liquid layer. The organism remains viable during 3 months storage at 4°C and can be preserved for longer periods by common lyophilization techniques.

Differentiation of the genus *Haliscomenobacter* from other genera

A number of characteristics, summarized in Table 85, differentiate the genus *Haliscomenobacter* from the phylogenetically most closely related species *Saprospira grandis* and genus *Lewinella.*

List of species of the genus *Haliscomenobacter*

1. **Haliscomenobacter hydrossis** van Veen, van der Kooy, Geuze and van der Vlies 1973, 213[AL] (*Streptothrix hyalina* Migula 1895, 38, *sensu* Mulder and Van Veen 1974, 133)

hy.dross'is. Gr. n. *hudôr* water; *Oss* town in the Netherlands; N.L. gen. n. *hydrossis* from water of Oss.

Most of the characteristics are the same as those given for the genus. Electron micrographs revealing details of cell structure (e.g., Figure 65) have been presented by Deinema et al. (1977). Polysaccharides, not poly-β-hydroxybutyrate, serve as the reserve material of this organism. Extensive continuous-culture experiments have been carried out with *Haliscomenobacter hydrossis* growing in a complex medium (trypticase soy, yeast extract, and glucose) at different dilution rates (Krul, 1977). Owing to the pronounced proteolytic activity of the organism, complex nitrogen compounds were readily degraded. Amino acids from these compounds were used as both nitrogen and energy sources in the presence of glucose. Of these amino acids, glutamic acid, glycine, methionine, tryptophan, lysine, and arginine were relatively easily taken up. The remaining amino acids accumulated in the free state at relatively high dilution rates, apparently owing to suppression of this uptake by glucose. At dilution rates below 0.015/h, all the amino acids were taken up more or less completely. In addition to these details from continuous culture experiments, more details on the physiological properties can be obtained from Kämpfer (1995) and Kämpfer et al. (1995).

DNA G+C content (mol%): 49 (T_m).

Type strain: ATCC 27775, DSM 1100, LMG 10767.

Sequence accession no. (16S rRNA gene): M58790.

Genus IV. Lewinella Sly, Taghavi and Fegan 1998, 735[VP]

Lindsay I. Sly and Mark Fegan

Le.win.el'la. L. fem. dim. ending *-ella*; N.L. fem. dim. n. *Lewinella* named after Ralph Lewin, who first isolated these organisms.

Unbranched, flexible rods or filaments 0.5–1.5 × 5–150 μm or longer (up to several mm) consisting of individual cells 2–3 μm long. **Ensheathed**, although sheaths may be difficult to visualize. Filament breakage or cell lysis results in empty ends or spaces (necridia). Resting stages not known. **Motile by gliding**. Gram-stain-negative. Produce **orange or yellow carotenoid pigments**. Chemo-organotrophic. **Aerobic,** having a respiratory type of metabolism with oxygen as the terminal electron acceptor. Gelatin is hydrolyzed but not cellulose, chitin, or starch. **Seawater is required for growth**. Isolated from marine environments.

DNA G+C content (mol%): 44.9–53.1.

Type species: **Lewinella cohaerens** (Lewin 1970) Sly, Taghavi and Fegan 1998, 735[VP] (*Herpetosiphon cohaerens* Lewin 1970, 518[AL]).

Further descriptive information

The species assigned to the genus *Lewinella* have not been studied extensively since their description by Lewin (1970). The descriptions below are based on data and taxonomic considerations from Mandel and Lewin (1969), Lewin (1970), Holt (1989), Reichenbach (1992), Sly et al. (1998), and Hamana and Nakagawa (2001).

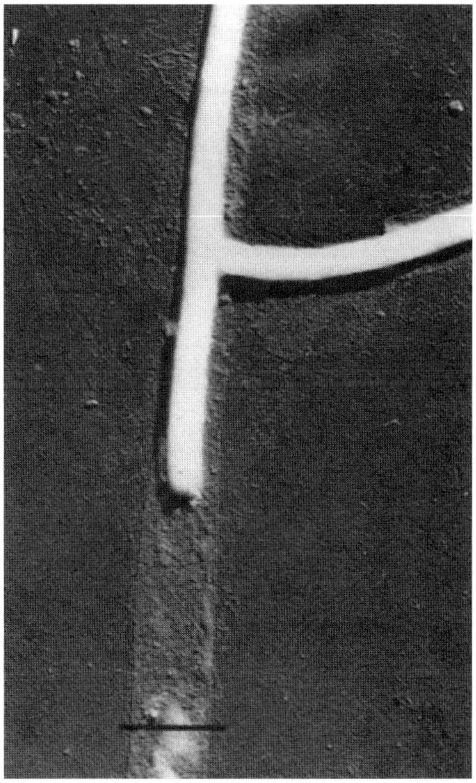

FIGURE 65. *Haliscomenobacter hydrossis* branched filament with thin hyaline sheath. Electron micrograph. Bar = 1 µm. (Reproduced with permission from Mulder and Deinema 1981.)

Enrichment and isolation procedures

According to Holt (1989), seawater agar (1.0–1.5% agar) without added nutrients can often be used. Sample material from a natural marine habitat is placed in the center or in a single streak on the plate. *Lewinella* organisms will glide away from the point of inoculation and form a characteristic rough, swirled swarm. Isolation can be achieved by selecting material from the leading edge of the swarm while viewing under low power magnification.

Differentiation of the genus *Lewinella* from other genera

The characteristics which differentiate *Lewinella* from other members of the family *Saprospiraceae* are listed in the introduction to the family Table 84. The most important features are flexible but non-helical, sheathed, unbranched filaments with gliding motility, requirement for seawater, inability to hydrolyze starch, and spermidine as the major polyamine.

Taxonomic comments

The species in this genus were originally assigned to the genus *Herpetosiphon*. However, analysis of 16S rRNA gene sequences from *Herpetosiphon* species indicated heterogeneity within that genus (Sly et al., 1998). The cluster of the three marine species of *Herpetosiphon* – *Herpetosiphon cohaerens*, *Herpetosiphon nigricans*, and *Herpetosiphon persicus* - exhibited affinities to *Saprospira* and *Haliscomenobacter* and was only distantly related to the type species of *Herpetosiphon*. Thus these three species were reassigned to a new genus, *Lewinella*, within the *Saprospiraceae* (Sly et al., 1998). Figure 66 shows the phylogenetic relationship of *Lewinella* species to other members of the family. The genus *Lewinella* forms a well-supported lineage with *Saprospira grandis* and *Haliscomenobacter hydrossis* within the *Saprospiraceae*. The 16S rRNA gene sequences of the three species have similarity values to one another of 86–88%, indicative of their separate species status (Sly et al., 1998).

Editorial note: Since the acceptance of this treatment of *Lewinella*, four new species, *Lewinella agarilytica* (Lee, 2007), *Lewinella antarctica* (Oh et al., 2009), *Lewinella lutea* (Khan et al., 2007), and *Lewinella marina* (Khan et al., 2007) have been described, and the genus and other species descriptions emended (Khan et al., 2007). Readers are referred to these papers for further information.

List of species of the genus *Lewinella*

1. **Lewinella cohaerens** (Lewin 1970) Sly, Taghavi and Fegan 1998, 735[VP] (*Herpetosiphon cohaerens* Lewin 1970, 518[AL])

co.hae′rens. L. part. adj. *cohaerens* cohering, uniting together.

The characteristics are as given for the genus, with the following additional features. Unbranched, flexible, sheathed rods or filaments 0.7 µm in diameter (1.0 µm including sheath) and 60–150 µm or more in length. Cell masses have an orange pigment believed to be the carotenoid saproxanthin. Growth is promoted by glucose and sucrose but not acetate, galactose, glycerol, or lactate. Nitrogen sources include tryptone and glutamate but not nitrate. Although no acid is produced in litmus milk, the milk is coagulated and the litmus is reduced. Tyrosine and dihydroxyphenylalanine are not degraded. Growth factors are not required. Seawater (one-half to double-strength) is required for growth.

Habitat: beach sand at Biarritz, France.

DNA G+C content (mol%): 44.9 (Bd).

Type strain: II-2, ATCC 23123, NBRC 102661, NCIMB 12855.

Sequence accession no. (16S rRNA gene): AF039292.

2. **Lewinella nigricans** (Lewin 1970) Sly, Taghavi and Fegan 1998, 736[VP] (*Herpetosiphon nigricans* Lewin 1970, 518[AL])

ni′gri.cans. L. part. adj. *nigricans* blackening.

Unbranched, flexible, sheathed rods or filaments 0.5 µm (1.0 µm including sheath) and 5–50 µm in length. Cell masses have a yellow pigment believed to be the carotenoid zeaxanthin. Glucose is a suitable carbon source. Growth is promoted by sucrose and galactose but not acetate, glycerol, or lactate. Nitrogen sources include Casamino acids, tryptone, glutamate, and nitrate. Although acid is not produced in litmus milk, the milk is coagulated and digested, and the litmus is reduced. Tyrosine is degraded with the formation of a pigmented product. Dihydroxyphenylalanine is not degraded. Growth factors are not required. Seawater (one-half to double-strength) is required for growth.

Habitat: beach sand near Lagos, Nigeria.

DNA G+C content (mol%): 53.1 (Bd).

Type strain: SS-2, ATCC 23147, NBRC 102662, NCIMB 1420.

Sequence accession no. (16S rRNA gene): AF039294.

3. **Lewinella persica** (Lewin 1970) Sly, Taghavi and Fegan 1998, 735[VP] (*Herpetosiphon persicus* Lewin 1970, 518[AL])

per′si.ca. L. fem. adj. *persica* Persian (of fruit = peach), i.e., peach-colored.

Unbranched, flexible, sheathed rods or filaments 0.7 µm (1.0 µm including sheath) and 30–150 µm or more in length.

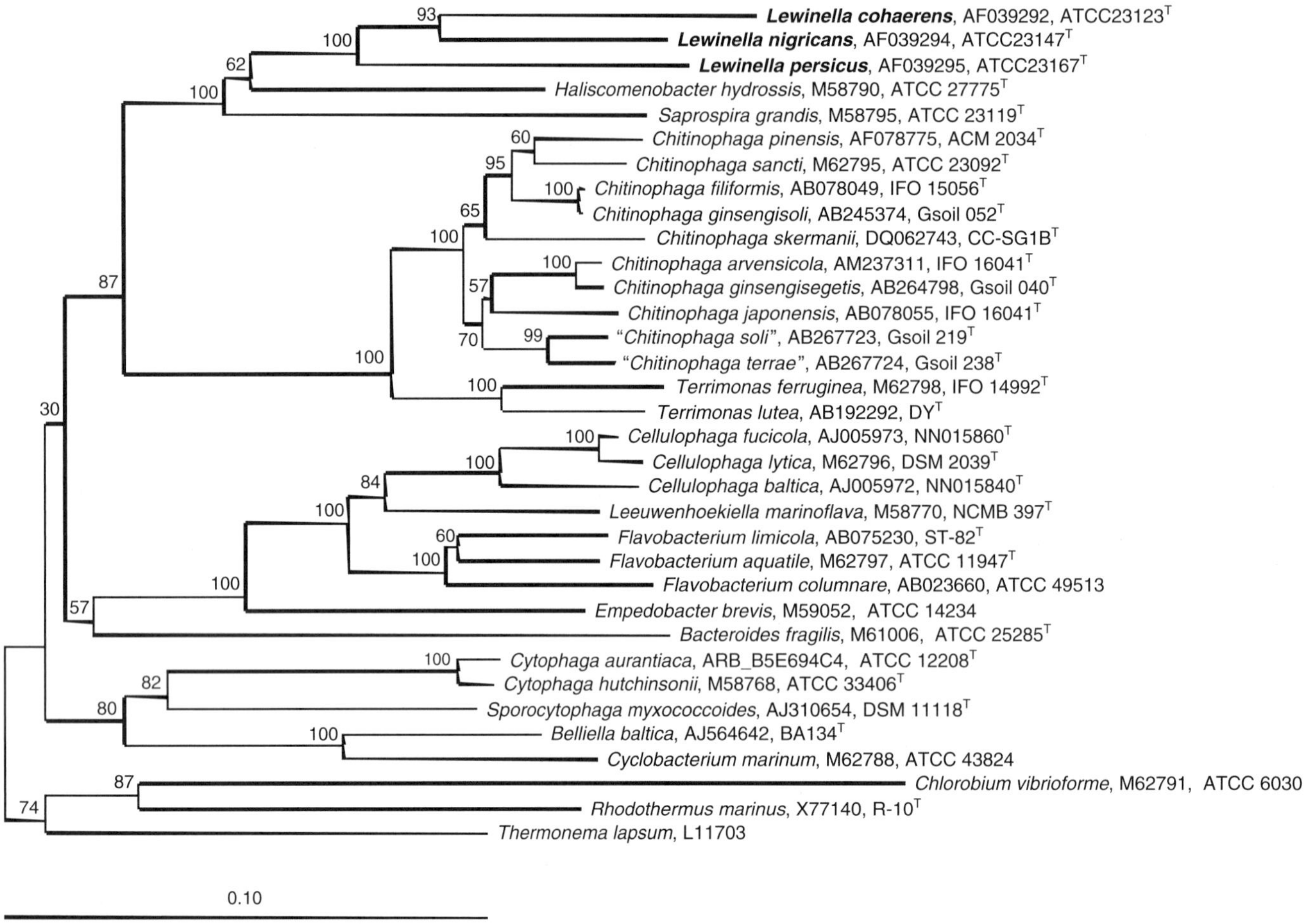

FIGURE 66. Phylogenetic tree of 16S rRNA gene sequences constructed by the neighbor-joining method showing the relationships of the species of the genus *Lewinella* to other members of the *Saprospiraceae*. Bootstrap values are shown as percentages at the branch points. The scale bar represents 10 nucleotide substitutions per 100 nucleotides.

Cell masses have an orange pigment believed to be the carotenoid saproxanthin. Glucose is a suitable carbon source. Growth is promoted by sucrose and galactose but not by acetate, glycerol, or lactate. Tryptone, glutamate, and nitrate serve as sole nitrogen sources. Although acid is not produced in litmus milk, the milk is coagulated and the litmus is reduced. Tyrosine and dihydroxyphenylalanine are not degraded. No growth factors are required. Seawater (one-half to double-strength) is required for growth.

Habitat: brown mud, Galway, Ireland.

DNA G+C content (mol%): 52.6 (Bd).

Type strain: T-3, ATCC 23167, NBRC 102663, NCIMB 1396.

Sequence accession no. (16S rRNA gene): AF039295.

Further reading

Eikelboom, D.H. 1975. Filamentous organisms observed in activated sludge. Water Res. *9*: 365–388.

Kämpfer, P. 1995. Physiological and chemotaxonomic characterization of filamentous bacteria belonging to the genus *Haliscomenobacter*. Syst. Appl. Microbiol. *18*: 363–367.

Kämpfer, P., Weltin, D., Hoffmeister, D. and Dott, W. 1995. Growth requirements of filamentous bacteria isolated from bulking and scumming sludge. Water Res. *29*: 1585–1588.

Williams, M.W. and Unz, R.F. 1985. Isolation and characterization of filamentous bacteria present in bulking activated sludge. Appl. Microbiol. Biotechnol. *22*: 273–280.

Ziegler, M., Lange, M. and Dott, W. 1990. Isolation and morphological and cytological characterization of filamentous bacteria from bulking sludge. Water Res. *24*: 1437–1451.

van Veen, W.L., D. van der Kooy, E.C.W.A. Geuze and A.W. van der Vlies. 1973 Investigations on the sheathed bacterium *Haliscomenobacter hydrossis* gen. n., sp. n., isolated from activated sludge. Antonie van Leeuwenhoek J. Microbiol. Serol. *39*: 207–216.

van Veen, W.L., J.M. Krul and C.J.E.A. Bulder. 1982. Some growth parameters of *Haliscomenobacter hydrossis* (syn. *Streptothrix hyalina*), a bacterium occurring in bulking activated sludge. Water Res. *16*: 531–534.

References

Aasen, A.J. and S.L. Liaaen-Jensen. 1966. The carotenoids of flexibacteria. II. A new xanthophyll from *Saprospira grandis*. Acta Chem. Scand. *20*: 811–819.

Aizawa, S.I. 2005. Bacterial gliding motility: visualizing invisible machinery. ASM News *71*: 71–76.

Ashton, P.J. and R.D. Robarts. 1987. Apparent predation of *Microcystis aeruginosa* Kutz. emend Elenkin by a *Saprospira*-like bacterium in a hypertrophic lake. J. Limnol. Soc. South Africa *13*: 44–47.

Cohn, F. 1875. Untersuchungen uber Bacterien. II. Beitrage z. Biol. d. Pflanzen *1*: 141–207.

Deinema, M.H., S. Henstra and E.W.v. Elgg. 1977. Structural and physiological characteristics of some sheathed bacteria. Antonie van Leeuwenhoek *43*: 19–29.

Dionisi, D., C. Levantesi, V. Renzi, V. Tandoi and M. Majone. 2002. PHA storage from several substrates by different morphological types in an anoxic/aerobic SBR. Water Sci. Technol. *46*: 337–344.

Eikelboom, D.H. 1975. Filamentous organisms observed in activated sludge. Water Res. *9*: 365–388.

Fourest, E., D. Craperi, C. Deschamps-Roupert, J.L. Pisicchio and G. Lenon. 2004. Occurrence and control of filamentous bulking in aerated wastewater treatment plants of the French paper industry. Water Sci. Technol. *50*: 29–37.

Gaval, G. and J.J. Pernelle. 2003. Impact of the repetition of oxygen deficiencies on the filamentous bacteria proliferation in activated sludge. Water Res. *37*: 1991–2000.

Gross, J. 1911. Über freilebende Spironemaceen. Mitteilungen aus der Zoologischen Station zu Neapel *20*: 188–203.

Hamana, K. and Y. Nakagawa. 2001. Polyamine distribution profiles in the eighteen genera phylogenetically located within the *Flavobacterium-Flexibacter-Cytophaga* complex. Microbios *106*: 7–17.

Holt, J.G. 1989. Genus *Herpetosiphon* Holt and Lewin 1965, 2408. *In* Bergey's Manual of Systematic Bacteriology, vol. 3 (edited by Staley, Bryant, Pfennig and Holt). Williams & Wilkins, Baltimore, pp. 2136–2138.

Hosoya, S., V. Arunpairojana, C. Suwannachart, A. Kanjana-Opas and A. Yokota. 2006. *Aureispira marina* gen. nov., sp. nov., a gliding, arachidonic acid-containing bacterium isolated from the southern coastline of Thailand. Int. J. Syst. Evol. Microbiol. *56*: 2931–2935.

Hosoya, S., V. Arunpairojana, C. Suwannachart, A. Kanjana-Opas and A. Yokota. 2007. *Aureispira maritima* sp. nov., isolated from marine barnacle debris. Int. J. Syst. Evol. Microbiol. *57*: 1948–1951.

Kämpfer, P. 1995. Physiological and chemotaxonomic characterization of filamentous bacteria belonging to the genus *Haliscomenobacter*. Syst. Appl. Microbiol. *18*: 363–367.

Kämpfer, P., D. Weltin, D. Hoffmeister and W. Dott. 1995. Growth requirements of filamentous bacteria isolated from bulking and scumming sludge. Water Res. *29*: 1585–1588.

Kämpfer, P. 1997. Detection and cultivation of filamentous bacteria from activated sludge. FEMS Microbiol. Ecol. *23*: 169–181.

Khan, S.T., Y. Fukunaga, Y. Nakagawa and S. Harayama. 2007. Emended descriptions of the genus *Lewinella* and of *Lewinella cohaerens*, *Lewinella nigricans* and *Lewinella persica*, and description of *Lewinella lutea* sp. nov. and *Lewinella marina* sp. nov. Int. J. Syst. Evol. Microbiol. *57*: 2946–2951.

Kolkwitz, R. 1909. *Schizomycetes*. Spaltpilze (*Bacteria*). Botanischer Verein der Provinz Brandenburg, Kryptogamenflora der Mark Brandenburg, vol. 5. Gerbruder Borntraeger, Leipzig *5*: 2–186.

Krul, J.M. 1977. Experiments with *Haliscomenobacter hydrossis* in continuous culture without and with *Zoogloea ramigera*. Water Res. *11*: 197–204.

Lee, S.D. 2007. *Lewinella agarilytica* sp. nov., a novel marine bacterium of the phylum *Bacteroidetes*, isolated from beach sediment. Int. J. Syst. Evol. Microbiol. *57*: 2814–2818.

Lewin, R.A. 1962. *Saprospira grandis* Gross; and suggestions for reclassifying helical aprochloroti, gliding organisms. Can. J. Microbiol. *8*: 555–563.

Lewin, R.A. and D.L. Fox. 1963. A preliminary study of the carotenoids of some flexibacteria. Can. J. Microbiol. *9*: 753–758.

Lewin, R.A., D.M. Crothers, D.L. Correll and B.E. Reimann. 1964. A phage infecting *Saprospira grandis*. Can. J. Microbiol. *10*: 75–80.

Lewin, R.A. 1965a. Isolation and some physiological features of *Saprospira thermalis*. Can. J. Microbiol. *11*: 77–86.

Lewin, R.A. 1965b. Freshwater species of *Saprospira*. Can. J. Microbiol. *11*: 135–139.

Lewin, R.A. 1970. New *Herpetosiphon* species (*Flexibacterales*). Can J. Microbiol. *16*: 517–520.

Lewin, R.A. and M. Mandel. 1970. *Saprospira* toviformis nov. spec. (*Flexibacteriales*), from a New Zealand seashore. Can. J. Microbiol. *16*: 507–510.

Lewin, R.A. 1972. Growth and nutrition of *Saprospira grandis* Gross (*Flexibacteriales*). Can. J. Microbiol. *18*: 361–365.

Lewin, R.A. 1997. *Saprospira grandis*: a flexibacterium that can catch bacterial prey by "ixotrophy". Microb. Ecol. *34*: 232–237.

Mandel, M. and R.A. Lewin. 1969. Deoxyribonucleic acid base composition of flexibacteria. J. Gen. Microbiol. *58*: 171–178.

Migula, W. 1895. *Schizomycetes* in Engler and Prantl's Die Natiirl. Pflanzen-familien, Abth. II, Theil I, la, 2–44, Leipsic, 1900.

Mulder, E.G. and W.L. Van Veen. 1974. Genus *Streptothrix*. *In* Bergey's Manual of Determinative Bacteriology, 8th edn (edited by Buchanan and Gibbons). Williams & Wilkins, Baltimore, p. 133.

Mulder, E.G. and M.H. Deinema. 1981. The sheathed bacteria. *In* The Prokaryotes: a Handbook on Habitats, Isolation, and Identification of *Bacteria* (edited by Starr, Stolp, Trüper, Balows and Schlegel). Springer, New York, pp. 425–440.

Mulder, E.G. 1989. Genus *Sphaerotilus* Kützing 1833, 386[AL]. *In* Bergey's Manual of Systematic Bacteriology, vol. 3 (edited by Staley, Bryant, Pfennig and Holt). Williams & Wilkins, Baltimore, pp. 1984–1998.

Oh, H.M., K. Lee and J.C. Cho. 2009. *Lewinella* antarctica sp. nov., a marine bacterium isolated from Antarctic seawater. Int. J. Syst. Evol. Microbiol. *59*: 65–68.

Pernelle, J.J., G. Gaval, E. Cotteux and P. Duchene. 2001. Influence of transient substrate overloads on the proliferation of filamentous bacterial populations in an activated sludge pilot plant. Water Res. *35*: 129–134.

Reichenbach, H. 1980. *Saprospira grandis* (*Leucotrichales*). Wachstum und Bewegung. Publikat. Wissensch, Filmen, Film E2424. Institut für den wissenschaftlichen Film Göttingen Ser. *13*: 1–21.

Reichenbach, H. and M. M. Dworkin. 1981. Introduction to the gliding bacteria. *In* The Prokaryotes: A Handbook on Habitats, Isolation, and Identification of *Bacteria* (edited by Starr, Stolp, Trüper, Balows and Schlegel). Springer, New York, pp. 315–327.

Reichenbach, H., W. Ludwig and E. Stackebrandt. 1986. Lack of relationship between gliding cyanobacteria and filamentous gliding heterotrophic eubacteria: comparison of 16S rRNA catalogues of *Spirulina*, *Saprospira*, *Vitreoscilla*, *Leucothrix*, and *Herpetosiphon*. Arch. Microbiol. *145*: 391–395.

Reichenbach, H. 1989. Genus *Saprospira* Gross 1911, 202. *In* Bergey's Manual of Systematic Bacteriology, vol. 3 (edited by Staley, Bryant, Pfennig and Holt). Williams & Wilkins, Baltimore, pp. 2077–2082.

Reichenbach, H. 1992. The genus *Herpetosiphon*. *In* The Prokaryotes: A Handbook on the Biology of *Bacteria*: Ecophysiology, Isolation, Identification, Applications, 2nd edn, vol. 4 (edited by Balows, Trüper, Dworkin, Harder and Schleifer). Springer, New York, pp. 3785–3805.

Reichle, R.E. and R.A. Lewin. 1968. Purification and structure of rhapidosomes. Can. J. Microbiol. *14*: 211–213.

Sakata, T., Y. Fujita and H. Yasumoto. 1991. Plaque formation by algicidal *Saprospira* sp. on a lawn of *Chaetoceros ceratosporum*. Nippon Suisan Gakkaishi *57*: 1147–1152.

Sanfilippo, A. and R.A. Lewin. 1970. Preservation of viable flexibacteria at low temperatures. Can J. Microbiol. *16*: 441–444.

Sangkhobol, V. and V.B.D. Skerman. 1981. *Saprospira* species - natural predators. Curr. Microbiol. *5*: 169–174.

Schauer, M. and M.W. Hahn. 2005. Diversity and phylogenetic affiliations of morphologically conspicuous large filamentous bacteria occurring in the pelagic zones of a broad spectrum of freshwater habitats. Appl. Environ. Microbiol. *71*: 1931–1940.

Skerman, V.B. 1968. A new type of micromanipulator and microforge. J. Gen. Microbiol. *54*: 287–297.

Sly, L.I. and A. Arunpairojana. 1987. Isolation of manganese-oxidizing *Pedomicrobium* cultures from water by micromanipulation. J. Microbiol. Methods *6*: 177–182.

Sly, L.I., M. Taghavi and M. Fegan. 1998. Phylogenetic heterogeneity within the genus *Herpetosiphon*: transfer of the marine species *Herpetosiphon cohaerens*, *Herpetosiphon nigricans* and *Herpetosiphon persicus* to the genus *Lewinella* gen. nov. in the *Flexibacter–Bacteroides–Cytophaga* phylum. Int. J. Syst. Bacteriol. *48*: 731–737.

van Veen, W.L., D. van der Kooy, E.C.W.A. Geuze and A.W.v.d. Vlies. 1973. Investigations on the sheathed bacterium *Haliscomenobacter hydrossis* gen. n., sp. n., isolated from activated sludge. Antonie van Leeuwenhoek *39*: 207–216.

van Veen, W.L., J.M. Krul and C.J.E.A. Bulder. 1982. Some growth parameters of *Haliscomenobacter hydrossis* (syn. *Streptothrix hyalina*), a bacterium occurring in bulking activated sludge. Water Res. *16*: 531–534.

Wagner, M., R. Amann, P. Kämpfer, B. Assmus, A. Hartmann, P. Hutzler, N. Springer and K.H. Schleifer. 1994. Identification and in situ detection of Gram-negative filamentous bacteria in activated sludge. Syst. Appl. Microbiol. *17*: 405–417.

Williams, M.W. and R.F. Unz. 1985. Isolation and characterization of filamentous bacteria present in bulking activated sludge. Appl. Microbiol. Biotechnol. *22*: 273–280.

Ziegler, M., M. Lange and W. Dott. 1990. Isolation and morphological and cytological characterization of filamentous bacteria from bulking sludge. Water Res. *24*: 1437–1451.

Class IV. **Cytophagia** class. nov.

Yasuyoshi Nakagawa

Cy.to.pha'gi.a. N.L. fem. n. *Cytophaga* type genus of the type order *Cytophagales*; suff. *-ia* ending proposed by Gibbons and Murray and by Stackebrandt et al. to denote a class; N.L. neut. pl. n. *Cytophagia* the class of *Cytophagales*.

Cells are short or long rods and sometimes filaments. Some genera form rings, coils, vibroids or S-shaped cells. **Spores or other resting cells are not found** except for the genus *Sporocytophaga*. **Motile by gliding or flagella, or nonmotile.** Flagella are found in only the genus *Balneola*. **Stain Gram-negative.** Growth is usually strictly aerobic, but microaerobic to anaerobic growth occurs in a few members. **Chemo-organotrophic**. **Colonies are usually pigmented**, and cell masses are yellow, orange, pink, or red owing to carotenoids, flexirubin-type pigments, or both. Widely distributed in nature. Some genera are marine organisms that require seawater salts for growth. Most species are mesophilic, but psychrophilic and thermophilic members exist.

Type order: **Cytophagales** Leadbetter 1974, 99[AL].

Taxonomic comments

The class *Cytophagia* is circumscribed for this volume on the basis of the phylogenetic analyses of the 16S rRNA sequences. The class includes the order *Cytophagales* and some other deep phylogenetic groups that are treated as orders *incertae sedis*.

Reference

Leadbetter, E.R. 1974. Order II. *Cytophagales* nomen novum. *In* Bergey's Manual of Determinative Bacteriology, 8th edn (edited by Buchanan and Gibbons). Williams & Wilkins, Baltimore, pp. 99–122.

Order I. **Cytophagales** Leadbetter 1974, 99[AL]

Yasuyoshi Nakagawa

Cy.to.pha.ga'les. N.L. fem. n. *Cytophaga* the type genus of the order; suff. *-ales* ending to denote an order; N.L. fem. pl. n. *Cytophagales* the *Cytophaga* order.

Cells are short or long rods and sometimes filaments. Some genera form rings, coils, vibroids or S-shaped cells. **Spores or other resting cells have not been found** except for the genus *Sporocytophaga*. **Motile by gliding or nonmotile.** Flagella are usually absent. **Stain Gram-negative.** Growth is usually strictly aerobic, but microaerobic to anaerobic growth occurs in a few members. **Chemo-organotrophic**. Most species are able to degrade one or several kinds of organic macromolecules such as proteins (e.g., casein, gelatin), lipids (e.g., Tweens) and esculin, starch, pectin, agar, chitin, carboxymethylcellulose or cellulose. **Colonies are usually pigmented** and cell masses are yellow, orange, pink or red owing to carotenoids, flexirubin-type pigments, or both. Widely distributed in nature. Some genera are marine organisms requiring seawater salts for growth. Most species are mesophilic, but a few species and genera are psychrophilic. The major respiratory quinone is MK-7, as far as it has been investigated. Most genera contain branched, unsaturated, or hydroxy fatty acids as predominant fatty acids.

Type genus: **Cytophaga** Winogradsky 1929, 577[AL] emend. Nakagawa and Yamasato 1996, 600.

Taxonomic comments

The order *Cytophagales* that was established by Leadbetter (1974) accommodated three families *Cytophagaceae*, *Flavobacteriaceae*, and *Bacteroidaceae*, and some genera in the 1st edition of the *Systematics* (Reichenbach, 1989). The order is circumscribed for this volume to encompass the three families *Cytophagaceae*, *Cyclobacteriaceae*, and *Flammeovirgaceae*, based on 16S rRNA sequences. The families *Flavobacteriaceae* and *Bacteroidaceae* are reassigned to the classes *Flavobacteriia* and *Bacteroidia*, respectively. The families *Cyclobacteriaceae* and *Flammeovirgaceae* were not described in previous editions of the manual. Because the three families belonging to the order *Cytophagales* share many phenotypic characteristics, they are difficult to differentiate based solely on phenotype. The major quinone is MK-7, as far as it has been investigated. The G+C contents of the DNA of the families *Cytophagaceae*, *Cyclobacteriaceae*, and *Flammeovirgaceae* are 34–65, 35–45, and 31–45 mol%, respectively.

References

Leadbetter, E.R. 1974. Order II. *Cytophagales* nomen novum. *In* Bergey's Manual of Determinative Bacteriology, 8th edn (edited by Buchanan and Gibbons). Williams & Wilkins, Baltimore, pp. 99–122.

Nakagawa, Y. and K. Yamasato. 1996. Emendation of the genus *Cytophaga* and transfer of *Cytophaga agarovorans* and *Cytophaga salmonicolor* to *Marinilabilia* gen. nov: phylogenetic analysis of the *Flavobacterium cytophaga* complex. Int. J. Syst. Bacteriol. *46*: 599–603.

Reichenbach, H. 1989. Order I. *Cytophagales* Leadbetter 1974. *In* Bergey's Manual of Systematic Bacteriology, 8th edn, vol. 3 (edited by Staley, Bryant, Pfennig, Holt). Williams & Wilkins, Baltimore, pp. 2011–2013.

Winogradsky, S. 1929. Études sur la microbiologie du sol. Sur la dégradation de la cellulose dans le sol. Ann. Inst. Pasteur (Paris) *43*: 549–633.

Family I. **Cytophagaceae** Stanier 1940, 630[AL]

Yasuyoshi Nakagawa

Cy.to.pha.ga'ce.ae. N.L. fem. n. *Cytophaga* the type genus of the family; suff. *-aceae* ending to denote a family; N.L. fem. pl. n. *Cytophagaceae* the *Cytophaga* family.

Cells are short or long rods and sometimes filaments. Some genera form rings, coils, vibroids or S-shaped cells. **Spores or other resting cells** are absent except for the genus *Sporocytophaga*. **Motile by gliding, or nonmotile.** Flagella are usually absent. **Stains Gram-negative.** Growth is usually strictly aerobic, but microaerobic to anaerobic growth occurs in members of a few genera. **Chemo-organotrophs.** Members of most species are able to degrade one or several kinds of organic macromolecules such as proteins (e.g., casein, gelatin), lipids (e.g., Tweens) and esculin, starch, pectin, agar, chitin, carboxymethylcellulose or cellulose. **Colonies are usually pigmented** and cell masses are yellow, orange, pink or red owing to carotenoids, flexirubin-type pigments, or both. Widely distributed in nature. Some genera are marine organisms requiring sea water salts for growth. Most species are mesophilic, but a few species and genera are psychrophilic. The major respiratory quinone is MK-7 as far as investigated. Predominant fatty acids in most genera are branched, unsaturated or hydroxy fatty acids such as $C_{15:0}$ iso, $C_{16:1}$, $C_{17:1}$, $C_{15:0}$ iso 2-OH and/or $C_{17:0}$ iso 3-OH.

DNA G+C content (mol%): 34–65.

Type genus: **Cytophaga** Winogradsky 1929, 577 emend. Nakagawa and Yamasato 1996, 600[AL].

Taxonomic comments

The family *Cytophagaceae* was established to accommodate one genus, *Cytophaga*, by Stanier (1940) and expanded to encompass six more genera, *Capnocytophaga*, *Chitinophaga*, *Flexibacter*, *Flexithrix*, *Microscilla*, and *Sporocytophaga* in the 1st edition of this *Manual* (Reichenbach, 1989c). Since then, many new genera have been described in the phylum *Bacteroidetes* and 16S rRNA sequence data have become available. In light of this new information, the family is delineated again in this volume. The genera *Capnocytophaga*, *Chitinophaga*, and *Flexithrix* are excluded from the family *Cytophagaceae* and reassigned to the families *Flavobacteriaceae*, *Chitinophagaceae*, and *Flammeovirgaceae*, respectively. Since 13 other genera have been added to the family *Cytophagaceae*, it now contains 17 genera. The major respiratory quinone of all genera belonging to the family is MK-7 as far as has been determined. This character is useful for differentiating the family *Cytophagaceae* from the family *Flavobacteriaceae*, which is characterized by MK-6 (Bernardet et al., 2002).

Although phenotypic differentiation of the 17 genera belonging to the family is difficult, some differential characteristics are summarized in Table 86.

Genus I. **Cytophaga** Winogradsky 1929, 577[AL] emend. Nakagawa and Yamasato 1996, 600[VP]

Yasuyoshi Nakagawa

Cy.to'pha.ga. Gr. n. *cytos* vessel, container, and in biology a cell; Gr. v. *phagein* to eat; N.L. fem. n. *Cytophaga* devourer of cell; intended to mean devourer of cell wall, cellulose digester.

Moderately long flexible rods 0.3–0.5 μm wide and 2–10 μm long with slightly tapering ends. **Motile by gliding**. Nonsporeforming. Resting stages are absent. Gram-stain-negative. **Aerobic**, having a strictly respiratory type of metabolism with oxygen as the terminal electron acceptor. Chemo-organotrophic. Cell mass is yellow to orange owing to carotenoids, flexirubin-type pigments, or both. Oxidase-positive. Catalase-negative. Terrestrial organisms, common in soil. The optimum pH for growth is 7. Crystalline cellulose (filter paper) and carboxymethyl cellulose are degraded. The major respiratory quinone is MK-7. The major polyamine is homospermidine.

DNA G+C content (mol%): 37–42.

Type species: **Cytophaga hutchinsonii** Winogradsky 1929, 578[AL].

Enrichment and isolation procedures

Standard procedures to isolate cellulose-degrading organisms can be applied. Media in which filter papers have been placed on mineral salts agar or immersed in a mineral salt solution are used. After incubation for several days to 1 month, yellow to orange spots will appear on the surface of the filter papers. Cellulose powders or carboxymethylcellulose can be used instead of filter paper. Detailed protocols have been described by Reichenbach (1989c, 1992a).

Maintenance procedures

Cultures of *Cytophaga* strains can be preserved by freezing at temperatures lower than −80°C. For freezing, cells are suspended in a suitable liquid medium containing 10% glycerol or 7% DMSO. *Cytophaga* strains are rather sensitive to drying; however, they can be preserved by the liquid drying method using a protective medium such as SM1* (Sakane et al., 1996) or by freeze drying.

Differentiation of the genus *Cytophaga* from other genera

Phylogenetic analysis shows that the genus *Cytophaga* is a phylogenetically independent genus in the phylum *Bacteroidetes* (see Figure 74 of the chapter *Flexibacter*). When the genus was emended (Nakagawa and Yamasato, 1996), the key characteristics to differentiate the genus from the related taxa were the possession of MK-7 as the major quinone and the ability to degrade cellulose. Menaquinone types can be used to differentiate the genus from the family *Flavobacteriaceae*, which is characterized by MK-6 (Bernardet et al., 2002; Nakagawa and Yamasato, 1993). Since then, a number of new genera have been proposed,

*SM1 (Sakane et al., 1996) consists of: monosodium glutamate monohydrate, 30 g; adonitol, 15 g; L(−)-cysteine-HCl, 0.5 g; and 0.1 M phosphate buffer (pH 7.0), 1000 ml.

TABLE 86. Descriptive and differential characteristics of members of the family *Cytophagaceae*

Characteristic	*Adhaeribacter*	*Arcicella*	*Cytophaga*	*Dyadobacter*	*Effluviibacter*	*Emticicia*	*Flectobacillus*	*Flexibacter*	*Hymenobacter*	*Larkinella*	*Leadbetterella*	*Meniscus*	*Microscilla*	*Pontibacter*	*Runella*	*Spirosoma*	*Sporocytophaga*
Cell shape	Rod	Rod	Rod	Rod	Rod	Rod	Rod	Rod	Rod	Rod	Rod	Rod	Rod, thread	Rod	Rod, filament	Rod, filament	Rod
Cell size (μm)	0.9–1.7 × 2.8–4.1	0.5–0.75 × 2.5–3.0	0.5–0.5 × 2–10	nd	0.3–0.5 × 1.0–3.0	0.3–0.4 × 2–5	0.3–1.0 × 2.0–10	0.2–0.6 × 10–60	0.4–1.0 × 1.3–5.0	0.5–0.9 × 1.5–3.0	0.6–0.9 × 2–7	0.7–1.0 × 2.0–3.0	0.5–0.6 × >150	0.3–0.4 × 1.2–1.9	0.5–0.9 × 2.0–14	0.5–1.0 × 1.5–50	0.3–0.5 × 5–8
Rings, coils, vibroids or S-shaped cells	–	+	–	–	+	–	+	–	–	+	–	+	–	–	+	–	–
Resting stage	nd	nd	–	nd	–	nd	nd	–	–	–	nd	–	–	–	–	–	+
Relationship to O_2	Aerobic	Aerobic	Aerobic	Aerobic	Aerobic	Aerobic	Aerobic	Aerobic	Aerobic	Aerobic	Aerobic	Aerotolerant anaerobe	Aerobic	Aerobic	Aerobic	Aerobic or facultative anaerobic	Aerobic
Presence of carotenoids	nd	nd	+	nd	nd	nd	nd	+	nd	nd	nd	nd	+	+	nd	nd	nd
Flexirubin reaction	nd	nd	D	+	nd	nd	–	+	nd	–	+	nd	nd	–	nd	–[c]	nd
Gliding motility	–	nd	+	–	–	–	–	+	–	+	–	–	+	+	–	D	+
Colony color	Pink	Light orange	Yellow-orange	Yellow-orange	Dark pink	Pink	Pink to rose	Orange	Pink to red	Pale pink	Orange	Chalky white	Orange	Pink	Pink, salmon	Yellow	Yellow
Oxidase	+	nd	+	+	+	+	+	+	D	+	+	–	nd	+	D	D	nd
Catalase	+	+	–	+	+	w	D	–	+[c]	+	+	–	nd	+	D	+	+
Alkaline phosphatase	nd	+	nd	nd	nd	nd	+	nd	+[c]	+		nd	nd	+	+[c]	D	nd
α-Galactosidase	+	w	nd	nd	nd	nd	+	nd	D	nd	w	nd	nd	–	w[c]	–[c]	nd
β-Galactosidase	+	+	nd	nd	nd	nd	D	nd	D	+	w	nd	nd	+	w[c]	+[c]	nd
NaCl requirement for growth	–	+	–	–	–	–	–	–	–	–	–	–	+	–	–	–	–
Growth at 40°C	–	+	nd	–	+	+	–	+	D	+	+	nd	–	+	D	–	nd
Hydrolysis of:																	
Esculin	nd	+	nd	D	–	nd	D	nd	D	nd	+	nd	nd	+	+[c]	+[c]	nd
Agar	–	nd	–	nd	nd	nd	nd	nd	–[c]	–	nd	nd	–	w	nd	nd	nd
Cellulose	nd	nd	+	nd	nd	–	nd	nd	–[c]	–	–	–	–	–	nd	nd	+
Starch	+	nd	–	nd	+	+	nd	+	+[c]	–	+	nd	–	–	D	nd	nd
Tween 80	nd	nd	nd	nd	nd	–	nd	nd	+[c]	–	–	nd	nd	–	+[c]	nd	nd
Major quinone	nd	nd	MK-7	nd	MK-7	nd	nd	MK-7	MK-7[c]	MK-7	MK-7	nd	MK-7	MK-7	MK-7[c]	MK-7[c]	nd
DNA G+C content (mol%)	40	34	37–42	44–50	59.5	36.9	38–41	40–43	55–65	53	33	nd	37–44	48.7	40–49	51–54	36

[a]Symbols: +, >85% positive; –, 0–15% positive; D, different reactions occur in different taxa (species of a genus); w, weak reaction; nd, not determined.

[b]Data from Baik (2006, 2007b), Buczolits et al. (2002, 2006), Chaturvedi et al. (2005), Chelius and Triplett (2000), Chelius et al. (2002), Collins et al. (2000), Dong et al. (2007), Hirsch et al. (1998b), Hwang and Cho (2006), Kim et al. (2008), Leadbetter (1989), Liu et al. (2006), Lu et al. (2007), Nakagawa and Yamasato (1993, 1996), Nedashkovskaya et al. (2005c), Nikitin et al. (2004), Raj and Maloy (1990), Reddy and Garcia-Pichel (2005), Reichenbach (1989a-c), Rickard et al. (2005), Ryu et al. (2006), Saha and Chakrabarti (2006), Suresh et al. (2006), Vancanneyt et al. (2006), Weon et al. (2005), Zhang et al. (2007), Zhang et al. (2008a, b), and Zhou et al. (2007).

[c]Not determined for all members of the genus.

making it very difficult to differentiate the genus *Cytophaga* from other related genera only by phenotypic characteristics. When additional strains are isolated, other chemotaxonomic, physiologic, and biochemical characteristics of the genus *Cytophaga* will be found.

Taxonomic comments

The genus *Cytophaga* was first described by Winogradsky (1929) for aerobic, cellulolytic, soil bacteria with probable gliding motility. Stanier (1940, 1941, 1942, 1947) investigated Gram-stain-negative bacteria that could move by gliding and degrade cellulose, agar, and/or chitin, and suggested that the gliding motility and the ability to degrade biomacromolecules were indispensable characters for the genus. These generic concepts were accepted in principle in the 1st edition of this *Manual* by Reichenbach (1989c); however, he stated that the genus is "a rather heterogeneous assembly of organisms which certainly do not belong to one single genus". In addition, several taxonomic investigations described in the reviews by Callies and Mannheim (1978), Christensen (1977), Hayes (1977), Oyaizu and Komagata (1981), and Shewan and McMeekin (1983) revealed considerable overlapping of phenotypic and chemotaxonomic characteristics between members of the genera *Cytophaga* and *Flavobacterium*. Only gliding motility could be regarded as a distinguishing feature for the two genera, but this feature is sometimes not easy to discern (Henrichsen, 1972; McMeekin and Shewan, 1978; Perry, 1973). Accordingly the two genera were called the *Flavobacterium-Cytophaga* complex. These developments are described in detail in the 1st edition of this *Manual* (Reichenbach, 1989c).

Of the 23 validly published *Cytophaga* species, 20 were recognized in the 1st edition of this *Manual*. 16S rRNA sequencing analysis showed great biological diversity within the genus *Cytophaga* (Nakagawa and Yamasato, 1993). They also showed that *Cytophaga* species were characterized by either MK-6 (*Cytophaga aquatilis, Cytophaga aurantiaca, Cytophaga columnaris, Cytophaga flevensis, Cytophaga johnsonae, Cytophaga latercula, Cytophaga lytica, Cytophaga marina, Cytophaga marinoflava, Cytophaga pectinovora, Cytophaga psychrophila, Cytophaga saccharophila, Cytophaga succinicans*) or MK-7 (*Cytophaga agarovorans, Cytophaga aprica, Cytophaga arvensicola, Cytophaga diffluens, Cytophaga fermentans, Cytophaga heparina, Cytophaga hutchinsonii, Cytophaga salmonicolor*). The types of major menaquinones correlated well with the phylogenetic relationships. Some of the species possessing MK-6 were closely related to *Flavobacterium aquatile*, the type species of the genus *Flavobacterium*. The understanding and acceptance of the diversity of the genus have led to several reclassifications and descriptions of new genera as summarized in Table 87 and described below.

Nakagawa and Yamasato (1996) emended the genus *Cytophaga* to contain only two species, *Cytophaga hutchinsonii* and *Cytophaga aurantiaca*. These are the only two species that are, based on 16S rRNA sequencing analysis, justifiably members of the genus *Cytophaga*. *Cytophaga agarovorans* and *Cytophaga salmonicolor*, which are facultatively anaerobic, usually pink-colored marine organisms, formed a distinct cluster in the class *Bacteroides* and were transferred in the new genus *Marinilabilia* (Nakagawa and Yamasato, 1996). Later, *gyrB* sequence analyses and DNA–DNA hybridization studies showed that *Marinilabilia salmonicolor* (basonym *Cytophaga salmonicolor*) and *Marinilabilia agarovorans* (basonym, *Cytophaga agarovorans*) should be unified in a single species, and the name *Marinilabilia salmonicolor* biovar. *agarovorans* was proposed for it (Suzuki et al., 1999). Eight species possessing MK-6 – *Cytophaga aquatilis, Cytophaga columnaris* (synonym, *Flexibacter columnaris*), *Cytophaga flevensis, Cytophaga johnsonae, Cytophaga pectinovora, Cytophaga psychrophila* (synonym, *Flexibacter psychrophilus*), *Cytophaga saccharophila*, and *Cytophaga succinicans* – were closely related to the genus *Flavobacterium* and therefore transferred to the genus *Flavobacterium*. The family *Flavobacteriaceae* was emended to accommodate MK-6 organisms (Bernardet et al., 1996, 2002). Accordingly, the genus *Flavobacterium* now includes both gliding-motile and non-motile organisms. This means that gliding motility is not a significant taxonomic characteristic for differentiation of genera.

TABLE 87. Other validly published *Cytophaga* species showing their proposed assignment to new or different genera

Validly published species	Revised classification	References
C. agarovorans	*Marinilabilia salmonicolor* biovar. *agarovorans*	Nakagawa and Yamasato (1996); Suzuki et al. (1999)
C. aprica	*Flammeovirga aprica*	Nakagawa et al. (1997)
C. aquatilis	*Flavobacterium hydatis*	Bernardet et al. (1996)
C. arvensicola	*Chitinophaga arvensicola*	Kämpfer et al. (2006)
C columnaris	*Flavobacterium columnare*	Bernardet et al. (1996)
C. diffluens	*Persicobacter diffluens*	Nakagawa et al. (1997)
C. fermentans	Uncertain (probably belonging to the class *Bacteroides*)	
C. flevensis	*Flavobacterium flevense*	Bernardet et al. (1996)
C. heparina	*Pedobacter heparinus*	Takeuchi and Yokota (1992); Steyn et al. (1998)
C. johnsonae	*Flavobacterium johnsoniae*	Bernardet et al. (1996)
C. latercula	*Aquimarina latercula*	Nedashkovskaya et al. (2005b); Nedashkovskaya et al. (2006)
C. lytica	*Cellulophaga lytica*	Johansen et al. (1999)
C. marina	*Tenacibaculum maritimum*	Suzuki et al. (2001)
C. marinoflava	*Leeuwenhoekiella marinoflava*	Nedashkovskaya et al. (2005d)
C. pectinovora	*Flavobacterium pectinovorum*	Bernardet et al. (1996)
C. psychrophila	*Flavobacterium psychrophilum*	Bernardet et al. (1996)
C. saccharophila	*Flavobacterium saccharophilum*	Bernardet et al. (1996)
C. salmonicolor	*Marinilabilia salmonicolor*	Nakagawa and Yamasato (1996); Suzuki et al. (1999)
C. succinicans	*Flavobacterium succinicans*	Bernardet et al. (1996)
C. uliginosa	*Zobellia uliginosa*	Bowman (2000); Barbeyron et al. (2001)

Two marine agarolytic organisms, *Cytophaga aprica* and *Cytophaga diffluens*, were transferred to the new genera *Flammeovirga* and *Persicobacter*, respectively (Nakagawa et al., 1997). A heparin-degrading species, *Cytophaga heparina* was initially transferred to the genus *Sphingobacterium*, because it contained sphingolipids (Takeuchi and Yokota, 1992), and then was reclassified in the new genus *Pedobacter* as *Pedobacter heparinum*, together with other heparinase-producing organisms (Steyn et al., 1998). *Cytophaga lytica* and *Cytophaga uliginosa* – marine species possessing MK-6 – were classified in the new genus *Cellulophaga* (Bowman, 2000; Johansen et al., 1999); later, *Cellulophaga uliginosa* was transferred to the new genus *Zobellia* (Barbeyron et al., 2001). A marine fish pathogen, *Cytophaga marina* (heterotypic synonym, *Flexibacter maritimus*) was reclassified as *Tenacibaculum maritimum* (Suzuki et al., 2001). *Cytophaga marinoflava* was assigned to the genus *Leeuwenhoekiella* (Nedashkovskaya et al., 2005d). The genus *Stanierella* was established to accommodate *Cytophaga latercula*; however, this new genus was eventually subsumed into another genus, *Aquimarina*, proposed in the same paper (Nedashkovskaya et al., 2005b, 2006). *Cytophaga arvensicola*, together with three other misclassified *Flexibacter* species – *Flexibacter filiformis*, *Flexibacter japonensis*, and *Flexibacter sancti* – was classified in the genus *Chitinophaga* (Kämpfer et al., 2006). At present, two of 23 validated species of the genus *Cytophaga* – *Cytophaga fermentans* and *Cytophaga xylanolytica* – remain to be reclassified. *Cytophaga fermentans* is a marine facultative anaerobe that possesses MK-7 and was closely related to the class *Bacteroides* based on the 16S rRNA sequence analysis (Nakagawa and Yamasato, 1993, 1996). *Cytophaga xylanolytica* is anaerobic gliding bacterium degrading xylan (Haack and Breznak, 1993). Only three partial 16S rRNA sequences (lengths, 287, 290, 350 bases, respectively) of this species are currently available; however, they show that *Cytophaga xylanolytica* is related to the class *Bacteroides*. Thus, *Cytophaga fermentans* and *Cytophaga xylanolytica* should not be included in the genus *Cytophaga*.

Differentiation of species of the genus *Cytophaga*

Table 88 lists characteristics that distinguish *Cytophaga* species from one another.

TABLE 88. Characteristics differentiating the species of the genus *Cytophaga* [a,b]

Characteristic	*C. hutchinsonii*	*C. aurantiaca*
Optimum temperature (°C)	30	20–25
Color of cell mass	Bright yellow	Bright orange
Flexirubin reaction	+	–
DNA G+C content (mol%)	39–40	37–42

[a]Symbols: +, 90% or more of strains are positive; –, 10% or less of strains are positive.

[b]Data from Reichenbach (1989c) and *Bergey's Manual of Determinative Bacteriology*, 9th edition.

List of species of the genus *Cytophaga*

1. **Cytophaga hutchinsonii** Winogradsky 1929, 578[AL]

hut.chin.so'ni.i. N.L. masc. gen. n. *hutchinsonii* of Hutchinson, named in honor of English microbiologist, H. B. Hutchinson.

The characteristics are as described for the genus with the following additional features taken from *Bergey's Manual of Determinative Bacteriology* (9th edition) and the 1st edition of this *Manual*. Cells are flexible rods, 0.3–0.5 µm wide and 2–10 µm long or longer. The cell mass is bright yellow. Optimum growth temperature is 30°C. Oxidase-positive. Catalase-negative or very weakly positive. Cellulose (filter paper) and carboxymethyl cellulose are degraded, but not agar, chitin, pectin, or starch. Peptones, Casamino acids, several amino acids including aspartic and glutamic acids, NO_3^-, and NH_4^+ serve as sole nitrogen source. Cellulose, cellobiose, and glucose are used, but arabinose, fructose, galactose, mannitol, mannose, sodium pyruvate, and xylose are not. Growth does not occur in seawater media. A flexirubin type pigment is present.

Source: unclear but assumed to be from soil (Reichenbach, 1989c).

DNA G+C content (mol%): 39 (Bd) to 40 (T_m).

Type strain: ATCC 33406, CIP 103989, DSM 1761, JCM 20678, LMG 10844, NBRC 15051, NCIMB 9469.

Sequence accession no. (complete genome): CP000383, NC008255; *(16S rRNA gene):* M58768.

2. **Cytophaga aurantiaca** (*ex* Winogradsky 1929) Reichenbach 1989c, 2036[VP] (*Cytophaga aurantiaca* Winogradsky 1929, 597)

au.ran.ti'a.ca. N.L. fem. adj. *aurantiaca* orange colored.

The characteristics are as described for the genus with the following additional features taken from *Bergey's Manual of Determinative Bacteriology* (9th edition) and the 1st edition of this *Manual*. Cells are flexible rods, 0.3–0.4 µm wide and 2–8 µm long or longer. The cell mass is bright orange. Optimum growth temperature is 20–25°C. Oxidase-positive. Catalase-negative. Cellulose (filter paper) and carboxymethyl cellulose are degraded, but not agar, casein, or starch. Peptones, NO_3^-, and NH_4^+ serve as nitrogen sources. Cellulose, cellobiose, and glucose are used. Growth does not occur in seawater media. A flexirubin type pigment is absent.

It is suggested that the type strain of *Cytophaga aurantiaca* is identical with Bortels's strain *Cytophaga aurantiaca* 51, which is isolated from soil of the swampy edge of a pond in Germany (Bortels, 1956; Reichenbach, 1989c).

DNA G+C content (mol%): 37–42 (Bd).

Type strain: ATCC 12208, DSM 3654, JCM 8511, NBRC 16043, NCIMB 8628.

Sequence accession no. (16S rRNA gene): D12658.

Other species

No species listed below should be included in the genus *Cytophaga*, because they are not closely related to the type species of the genus based on 16S rRNA sequence analysis.

1. **Cytophaga fermentans** Bachmann 1955, 549[AL]

fer.men'tans. L. part. adj. *fermentans* fermenting.

The descriptions include characteristics described in *Bergey's Manual of Determinative Bacteriology* (9th edition) and the 1st edition of this *Manual*. Cells are flexible rods, 0.3–0.7 μm wide and 8–50 μm long or longer. Motile by gliding. Nonsporeforming. Resting stages are absent. Gram-stain-negative. The cell mass from aerobic cultures is bright yellow on peptone media, and from anaerobic cultures, it is unpigmented on minimal media. Colonies are thin spreading films with yellow blobs in shallow craters. Facultatively anaerobic. Optimum growth temperature is 30°C. Agar and starch, but not alginate, cellulose (filter paper), or chitin are degraded. Peptones, asparagine, glutamine, and NH_4^+ serve as nitrogen sources, but not NO_3^- or urea. Arabinose, cellobiose, fructose, glucose, lactose, maltose, mannitol, mannose, raffinose, starch, sucrose, and xylose are used, but acetate, agar, alginate, cellulose, chitin, dulcitol, ethanol, galactose, glycerol, rhamnose, trehalose, and sorbose are not. Glucose is fermented with production of acetate, propionate, and succinate. Growth does not occur in seawater media. Nitrate is not reduced. Marine organism, requiring elevated salt concentrations.

This species is related to the class *Bacteroides* based on the 16S rRNA sequence analysis (Nakagawa and Yamasato, 1996).

DNA G+C content (mol%): 39 (Bd) to 42 (T_m).

Type strain: ATCC 19072, CIP 104805, DSM 9555, JCM 21142, NBRC 15936.

Sequence accession no. (16S rRNA gene): M58766.

2. **Cytophaga xylanolytica** Haack and Breznak 1993, 14[VP]

xy.lan.o.ly'ti.ca. N.L. n. *xylanum* xylan (a xylose-containing heteropolysaccharides in plant cell walls); N.L. fem. adj. *lytica* (from Gr. fem. adj. *lutikê*), able to loosen, able to dissolve; N.L. fem. adj. *xylanolytica* xylan-dissolving.

The characteristics are as described by Haack and Breznak (1993). Cells are flexible rods, 0.4 μm wide and 3–24 μm long or longer. Motile by gliding. Nonsporeforming. Resting stages are absent. Gram-stain-negative. The cell mass is orange to salmon due to the presence of carotenoids. Aeroduric anaerobe. No growth occurs under conventional aerobic conditions with or without CO_2 enrichment of the atmosphere. Growth occurs at 19–37°C; optimum is 30–32°C. Growth occurs at pH 6.1–8.7; optimum 7.2–8.2. Oxidase-negative. Catalase weakly positive. Sulfide is used as the sole sulfur source. Peptones, glutamine, and NH_4^+ serve as nitrogen sources, but not NO_3^-, N_2, or glutamate. Xylan, laminarin, lichenin, cellobiose, lactose, glucose, galactose, mannose, xylose, and arabinose are fermented as energy sources for growth, but not cellulose, carboxymethyl cellulose, arabinogalactan, mannan, chitin, *N*-acetylglucosamine, sorbose, ribose, or rhamnose. Acetate, propionate, and succinate are the major products of xylan, xylose, and glucose fermentation. NO_3^- and SO_4^{2-} are not used as terminal electron acceptors during anaerobic growth. O_2 is not used as terminal electron acceptor (type strain). Growth occurs in the presence of up to 3% NaCl. Marine organism, requiring elevated salt concentrations. A flexirubin type pigment and a sulfonolipid are present. The major cellular fatty acids are $C_{15:0}$ anteiso, $C_{15:0}$, and $C_{15:0}$ 3-OH.

Analysis of partial 16S rRNA sequences indicates a close relationship between this species and the class *Bacteroides*.

DNA G+C content (mol%): 45.5 (Bd).

Type strain: XM3, ATCC 51429, DSM 6779.

Sequence accession no. (*three partial 16S rRNA in lengths 287, 290, 350 bases*): AH001617.

Genus II. **Adhaeribacter** Rickard, Stead, O'May, Lindsay, Banner, Handley and Gilbert 2005, 827[VP]

Peter Gilbert* and Alexander H. Rickard

Ad.haer'i.bac.ter. L. v. *adhaereo, adhaere* to ADHERE to, to stick fast; N.L. masc. n. *bacter* from Gr. n. *baktron* rod; N.L. masc. n. *Adhaeribacter* sticky rod.

Rods, about 0.9–1.7 by 2.8–4.1 μm. In batch culture, mean size is dependent upon the phase of growth. During exponential growth, the rods occur singly, in aggregates, and in chains of up to approximately 4 cells in length. In the stationary phase, the cells occur singly. Cells are Gram-stain-negative, non-sporulating, nonflagellated, and nonmotile. Growth occurs in liquid and on solid interfaces. Strictly aerobic. **Chemoheterotrophic**. Temperature range for growth, 4–37°C; optimum, 30°C. Limited **NaCl tolerance**; growth occurs between 0–4.0% (w/v) NaCl.

Habitat: within freshwater, multi-species biofilm communities developed on a stainless-steel surface exposed to potable water at a shear rate of 305/s.

DNA G+C content (mol%): 40.0.

Type species: **Adhaeribacter aquaticus** Rickard, Stead, O'May, Lindsay, Banner, Handley and Gilbert 2005, 827[VP].

*Deceased.

Further descriptive information

Based on 16S rRNA analysis, *Adhaeribacter* represents a separate and distinct lineage within the family *Cytophagaceae* of the *Cytophaga–Flavobacterium–Bacteroides* group (Figure 67).

Colonies are pink, round, and mucoid on R2A medium (Reasoner and Geldreich, 1985) and are not easily emulsified in sterile water or phosphate buffered saline. Growth does not occur on peptone water agar or other media commonly used for the isolation of freshwater bacteria. The growth rate on agar is improved in a moist atmosphere. Cells grown in liquid R2A do not pellet tightly when centrifuged at <3500 *g*, which may be a consequence of the production of extracellular polymeric substances.

Cells that are in exponential and stationary phase are completely covered with a dense fibrillar matrix, possibly composed of extracellular polysaccharide as well as other polymeric substances (Figure 68). Bundles of fibrils exist close to the cell surface, and individual thin fibers radiate away from each cell. Furthermore, the addition of the acidic dye nigrosin reveals an extensive capsule surrounding the cells. During late exponential

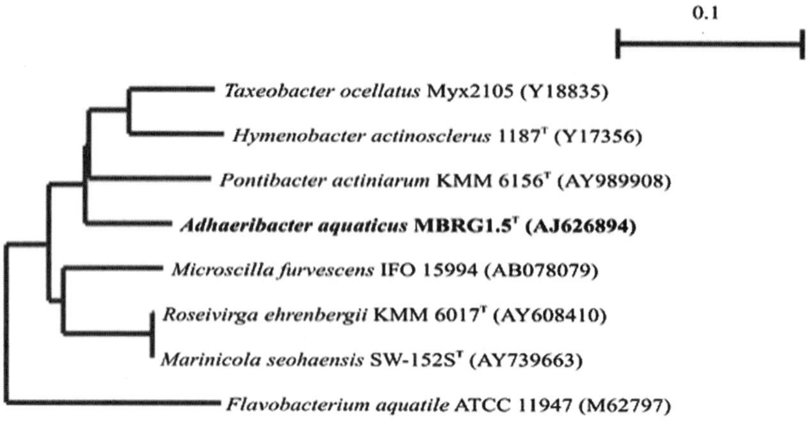

FIGURE 67. Neighbor-joining 16S rRNA phylogenetic dendrogram showing the position of *Adhaeribacter aquaticus* in relation to other closely related genera from the order *Cytophagales*. Bar represents 1 nucleotide change in every 10.

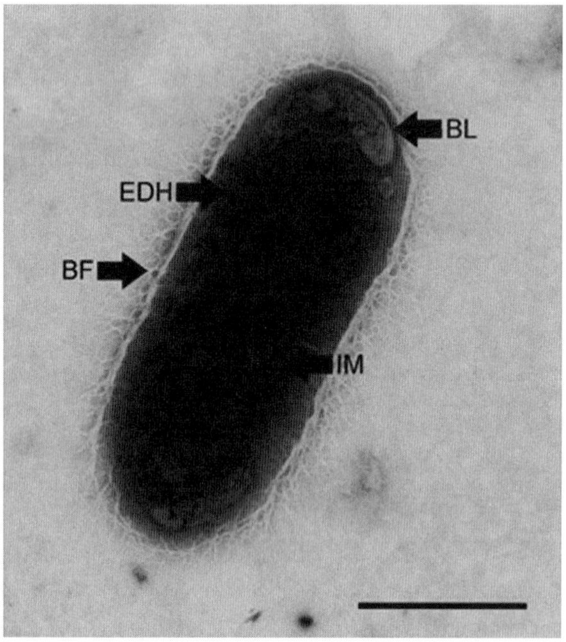

FIGURE 68. Electron micrographs of a typical negatively stained cell of *Adhaeribacter aquaticus* harvested after 72 h from batch culture. Cells were stained with 1.5% (w/v) methylamine tungstate. Cells are rod-shaped and surrounded by bundles of fibrils (BF). Each cell possesses a clearly visible electron dense "halo" (EDH). Blebbing (BL) is evident, and an inner membrane (IM) that has shrunk away from the outer membrane can be seen. Bar represents 1.0 μm.

phase and stationary phase, a large proportion of the cells plasmolyse, and blebbing and shrinkage of the inner membrane from the outer membrane occurs (Figure 68). The cause for such extensive plasmolysis is not known.

Many species of bacteria live together within biofilms (Costerton et al., 1995). *Adhaeribacter aquaticus* was isolated from a multi-species biofilm community on a stainless steel surface of a concentric cylinder reactor (Rickard et al., 2004). This device was used to demonstrate that species composition of multi-species biofilm communities is dependent upon the imposed hydrodynamic shear forces. *Adhaeribacter aquaticus* was only isolated from biofilms exposed, for 3 months, to the highest imposed shear forces (305/s). The ability of this organism to form biofilms and reside on surfaces exposed to such extreme shear rates is likely related to the copious amounts of adhesive fibrillar material that is produced by *Adhaeribacter aquaticus*.

Enrichment and isolation procedures

There is currently no known method for selective enrichment of *Adhaeribacter aquaticus* from a mixed-species freshwater community. In general, following isolation, growth of *Adhaeribacter aquaticus* is best at 30°C in aerobic liquid R2A medium or on R2A agar in a humid atmosphere. Maximum growth rates are at 30°C, and incubation at this temperature will generate visible colonies after 3 d.

Maintenance procedures

For short-term storage, *Adhaeribacter aquaticus* can be maintained on R2A agar for approximately 1 month. For long-term storage, *Adhaeribacter aquaticus* can be stored at −70°C in 50% (v/v) glycerol.

Differentiation of the genus *Adhaeribacter* from other genera

Adhaeribacter aquaticus can be phylogenetically differentiated from other genera within the order *Cytophagales* and, indeed, within the family *Cytophagaceae* by comparison of the 16S rRNA gene sequences. The validly published species *Pontibacter actiniarum* gen. nov., sp. nov. (Nedashkovskaya et al., 2005c) is the closest relative to *Adhaeribacter aquaticus*, with 89.1% 16S rRNA gene sequence similarity (Figure 67). Phenotypically, cells from no other related genera have been reported to exhibit such a high degree of plasmolysis and produce such conspicuous amounts of extracellular polymeric substances. In addition, the profile of identifiable cellular fatty acids of *Adhaeribacter aquaticus* is significantly different from that of phylogenetically related strains.

Acknowledgements

We are grateful to P.S. Handley for assistance in preparing the TEM micrograph.

List of species of the genus *Adhaeribacter*

1. **Adhaeribacter aquaticus** Rickard, Stead, O'May, Lindsay, Banner, Handley and Gilbert 2005, 827[VP]

   a.qua'ti.cus. L. masc. adj. *aquaticus* living, growing, or found in or by water, aquatic.

   The description is the same as given for the genus.

   DNA G+C content (mol%): 40.0 (T_m).

   Type strain: MBRG1.5, DSM 16391, NCIMB 14008.

   Sequence accession no. (16S rRNA gene): AJ626894.

Genus III. **Arcicella** Nikitin, Strömpl, Oranskaya and Abraham 2004, 683[VP] (*Arcocella* Nikitin, Oranskaya, Pitryuk, Chernykh and Lysenko 1994, 152)

Wolf-Rainer Abraham, Alexandre J. Macedo, Heinrich Lünsdorf and Denis I. Nikitin

Ar.ci.cel'la. L. masc. n. *arcus* the arc; L. fem. n. *cella* a store-room, and in biology a cell; N.L. fem. n. *Arcicella* arc-shaped cell.

Vibrioid, stretched spirals, or S-shaped cells 2.5–3.0 by 0.5–0.75 µm. Colonies on solid media are mucoid and light orange. Growth occurs between 4 and 40°C; optimum, 28–30°C. Optimum pH, ~7. Growth occurs in the presence of NaCl concentrations between 0.5 and 6.0% (w/v); no growth occurs above 6% or in the absence of NaCl. Aerobic. Growth occurs on a wide range of carbohydrates (monosaccharides are preferred) but not on C_1 compounds, amino acids, or salts of organic acids. Ammonium salts and nitrate can be used as nitrogen sources. Nitrate is not reduced to nitrite or N_2. Biopolymers are not hydrolyzed. The main polar lipids are 2-*N*-(2′-hydroxyisopentadecanoyl) amino-3-hydroxyisoheptadeca-4(*E*)-ene-1-(2″-aminoethyl)-phosphate and its 3-hydroxyoctadeca-4(*E*)-ene homologs, phosphatidylethylamine, 2-D-(2′-D-hydroxyisopentadecanoyl) amino-3-D-hydroxyisoheptadecane-1-sulfonic acid, and 2-*N*-(2″-hydroxyisopentadecanoyl)amino-3-hydroxyisoheptadeca-4(*E*)-ene-1-hydroxycarbonyl-6-deoxy-6-amino-mannopyranoside. The major fatty acids are $C_{16:0}$, $C_{14:0}$, $C_{18:1}$ ω6 and $C_{18:0}$. Lives in freshwater habitats.

DNA G+C content (mol%): 34.

Type species: **Arcicella aquatica** Nikitin, Strömpl, Oranskaya and Abraham 2004, 683[VP] (*Arcocella aquatica* Nikitin, Oranskaya, Pitryuk, Chernykh and Lysenko 1994, 152).

Further descriptive information

The cells are vibrioid, stretched spirals, or S-shaped, morphologically similar to the ring-forming cells of *Flectobacillus major* and, to a lesser degree, *Spirosoma linguale.*

The phospholipid and fatty acid composition of *Arcicella* and related organisms is given in Table 89.

Electron microscopic studies of *Arcicella aquatica*, in either whole mount preparations shadow-casted with Pt (Figure 69) or embedded ultrathin sections (Figure 69), revealed Gram-stain-negative cell wall architecture (Figure 69). Often the organism was found engulfed within a slimy matrix (Figure 69) and the outermost layer of the outer membrane appeared intensely stained with thorium dioxide (Figure 69). Correspondingly, protuberances of the outer membrane were coated by thorium, indicating that these surfaces are negatively charged (Lünsdorf et al., 2006). However, the outermost slime layer (Figure 69) was free from colloidal stain, which is indicative of the neutral and/or positive charge state of the slime matrix. Cells were irregular in length (2.49–6.79 µm), and width (0.54–0.76 µm), with a mean of 0.66 ± 0.05 µm (n = 33). Under actual growth conditions, intracellular inclusion bodies (ib) were often found (Figure 69), which often showed a weak hexagonal contour in cross-section, similar to that of polyhedral bodies. These cytoplasmic inclusions were 110–268 nm in diameter, with a mean of 183.4 ± 41.8 nm (n = 17). Fast Fourier Transform analysis of individual inclusions revealed a few ordered spots (Figure 69), indicative of an ordered arrangement of constituents and reflective of the regular arrangement of particles in the inclusion bodies (Figure 69). These features may be linked to the metabolic activities of diverse enzymes (see below) and thus the inclusions may represent so-called "metabolosomes" (Brinsmade et al., 2005).

Nutritional requirements were studied on a mineral medium of the following composition (g/l of distilled water): KH_2PO_4, 0.01; K_2HPO_4, 0.1; $(NH_4)_2HPO_4$, 0.2; $MgSO_4$, 0.03; NaCl, 0.01; $CaCl_2$, 0.01; $MnSO_4$, trace amounts; $FeSO_4$, trace amounts; yeast extract, 0.005%; pH, 7.2. The concentration of the carbon sources was 0.1%. Growth occurred on glucose, fructose, lactose, maltose, rhamnose, galactose, arabinose, ribose, sucrose, cellobiose, and inulin. Some growth occurred on mannitol, sorbitol, acetate, and aspartate, but no growth occurred on sorbose, dulcitol, ethanol, methanol, formate, propionate, pyruvate, monomethylamine, citrate, oxalate, succinate, and malate, or on amino acids, with the exception of aspartate.

TABLE 89. Phospholipid and fatty acid composition of *Arcicella* and related organisms[a,b]

	Individual phospholipid (% of total)			Individual fatty acids (% of total)							
Species	PE	PG	AAPG[c]	14:0	15:1	15:0	16:0	16:1	17:1	18:0	18:1
A. aquatica	35	27	38	2.5	4.4	2	2	77	0	1	0
F. major	33	25	42	1	12	4	14	44	25.5	1.5	0
S. linguale	52	8	40	nd	nd	nd	nd	nd	nd	nd	24

[a]Symbols: PE, phosphatidylethanolamine; PG, phosphatidylglycerol; AAPG, aminoacyl-phosphatidylglycerol; nd, not determined.

[b]Data from Nikitin et al. (1994).

[c]The main AAPGs of *Arcicella aquatica* are 2-*N*-(2′-hydroxyisopentadecanoyl)amino-3-hydroxyisoheptadeca-4(*E*)-ene-1-(2″-aminoethyl)-phosphate and 2-*N*-(2′-hydroxyisopentadecanoyl)amino-3-hydroxyoctadeca-4(*E*)-ene-1-(2″-aminoethyl)-phosphate.

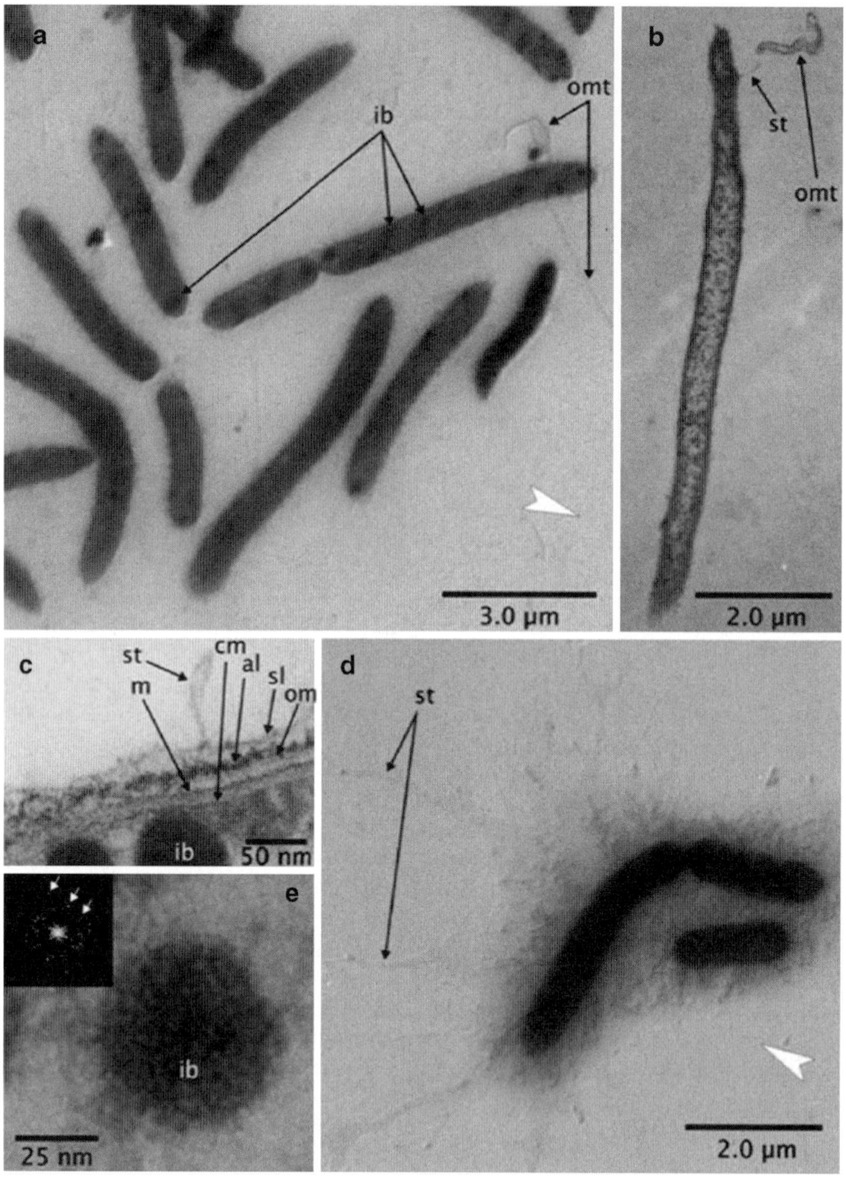

FIGURE 69. Electron micrographs of *Arcicella aquatica* (for details, see text; cm, cytoplasmic membrane; om, outer membrane; omt, outer membrane tubular extensions; m, murein. White arrows indicate direction of shadow casting.

No growth on meat infusion media is observed. Growth in liquid media at a carbon source concentration of 0.01% is characterized by a prolonged lag phase (up to 24 h) and slow growth, which proceeds for 48–72 h. Upon reaching the stationary phase, the culture retains the achieved optical density for as long as 21 d.

Strains can be cultivated on PYG medium, which has the following composition (g/l of distilled water): peptone (enzymic hydrolysate, Type I; from meat), 1.0; yeast extract ("Sigma" Y-4000), 1.0; glucose, 1.0. For agar plates, 20 g of agar is added; pH is 6.8–7.2. The medium is sterilized 120°C for 30 min. Strains grow within 4–5 d at 28°C.

Cells display high activity of alkaline phosphatase, leucine and valine arylamidase, cystine arylamidase, trypsin, acid phosphatase, naphthol-AS-BI-phosphohydrolase, β-galactosidase, α-glucosidase, catalase and *N*-acetyl-β-glucosaminidase; weak activity of esterase lipase (C_8), α-chymotrypsin, α-galactosidase, β-glucosidase and α-fucosidase; and no activity of lipase (C_{14}), α-glucuronidase, or α-mannosidase.

The genome size of *Arcicella aquatica* is 2.9×10^9 Da.

The type strain came from a neuston — a film composed of organisms found on top of or attached to the underside of the surface film of water — in the central part of Lake Trostenskoe near Zvenigorod, Moscow region.

Enrichment and isolation procedures

The type strain was isolated on a medium consisting of (g/l of distilled water): peptone, 1.0; yeast extract, 1.0; glucose, 1.0; agar (Difco), 15.0; pH 7.0–7.2. On solid media the colonies are ~3 mm in diameter, mucoid, and light orange.

Differentiation of the genus *Arcicella* from other genera

Table 90 lists some characteristics that differentiate the genus *Arcicella* from other ring-forming genera.

TABLE 90. Characteristics differentiating the genus *Arcicella* from other morphologically similar genera[a,b]

Characteristic	*Arcicella aquatica*	*Flectobacillus major*	*Runella slithyformis*	*Runella zea*	*Spirosoma linguale*	*Cyclobacterium mariensus*	*Ancyclobacter aquaticus*	*Ancyclobacter rudongensis*	*Polaribacter glomeratus*
Rings or toroids formed	+	+	+	–	+	+	Rare	+	v
Growth in absence of NaCl	–	+	+	+	+	–	nd	nd	–
Urease	–	+	–	nd	–	–	nd	+	–
Starch hydrolysis	–	+	w	–	w	–	–	–	d
Growth on:									
Acetate	+	–	–	nd	–	nd	nd	nd	nd
Dulcitol	–	+	–	nd	–	nd	nd	nd	nd
Methanol	–	–	–	–	–	–	+	+	–
Succinate, pyruvate, glutamate	–	+	–	nd	+	nd	nd	nd	nd
Propionate	–	+	–	nd	–	nd	nd	nd	nd
Tyrosine	–	+	nd	nd	nd	nd	nd	nd	nd
Aspartate	+	+	nd	nd	nd	nd	nd	nd	nd
DNA G+C content (mol%)	34	39–40	49–50	49	51–53	34–38	66–69	68	33

[a]Symbols: +, >85% positive; d, different strains give different reactions (16–84% positive); –, 0–15% positive; w, weakly positive; nd, data not reported or not applicable.

[b]Data taken from Nikitin et al. (2004), Larkin and Borrall (1984a), Gosink et al. (1998), Chelius and Triplett (2000), Chelius et al. (2002), Raj and Maloy (1990), Xin et al. (2004), and Staley and Konopka (1984).

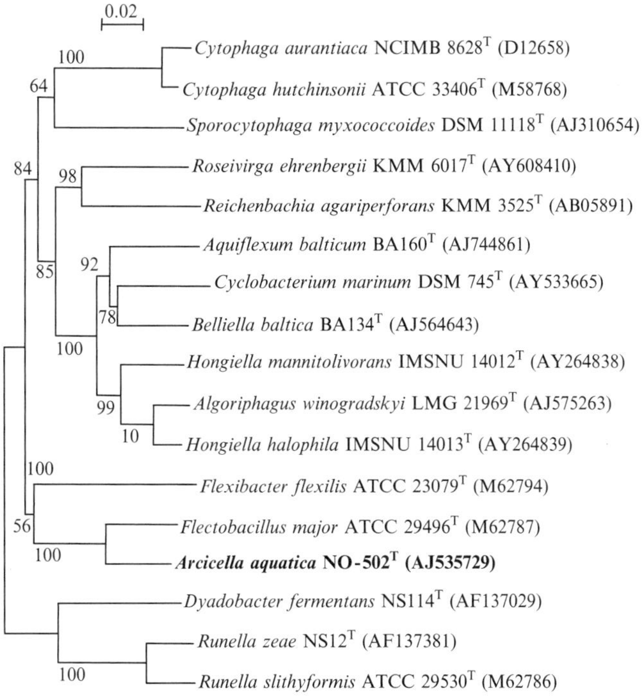

FIGURE 70. Unrooted neighbor-joining dendrogram of the phylogenetic relationships between *Arcicella aquatica* and type species of closely related genera based on a distance matrix analysis of the 16S rRNA gene sequences. Accession numbers are given in parentheses. Bootstrap percentages are indicated at tree branching points and the scale bar represents substitutions per nucleotide.

Taxonomic comments

The 16S rRNA gene sequence similarity between the type strain of *Arcicella aquatica* and the most similar published sequence of an organism with a validly published name, *Flectobacillus major*, is 93.52% (Nikitin et al., 2004) (Figure 70). This similarity value, and the differences in the polar lipids, the cellular fatty acids, and biochemical and physiological characteristics between *Flectobacillus major* and the type strain of *Arcicella aquatica*, together with the differences in the mol% G+C content of their DNA, all support the separate generic status of *Arcicella*.

List of species of the genus *Arcicella*

1. **Arcicella aquatica** Nikitin, Strömpl, Oranskaya and Abraham 2004, 683[VP] (*Arcocella aquatica* Nikitin, Oranskaya, Pitryuk, Chernykh and Lysenko 1994, 152)

a.qua.ti'ca. L. fem. adj. *aquatica* aquatic.

The characteristics are as given for the genus with the following additional features. Main phospholipids are 2-*N*-(2′-hydroxyisopentadecanoyl)amino-3-hydroxy-isoheptadeca-4(*E*)-ene-1-(2″-aminoethyl)-phosphate, 2-*N*-(2′-hydroxyisopentadecanoyl)amino-3-hydroxyoctadeca-4(*E*)-ene-1-(2″-aminoethyl)-phosphate, 1-pentadecanoyl-2-hexadecenoylphosphatidyl-2′-ethylamine, 1,2-bis-hexadecenoylphosphatidyl-2′-ethylamine and 1-hexadecanoyl-2-hexadecenoylphosphatidyl-2′-ethylamine. Cells display high activities of alkaline phosphatase, leucine, and valine arylamidase, cystine arylamidase, trypsin, acid phosphatase, naphthol-AS-BI-phosphohydrolase, β-galactosidase, α-glucosidase, catalase and *N*-acetyl-β-glucosaminidase, weak activities of esterase lipase (C8), α-chymotrypsin, α-galactosidase, β-glucosidase, and α-fucosidase, and no activities of lipase (C_{14}), α-glucuronidase, or α-mannosidase. Growth occurs on glucose, fructose, lactose, maltose, rhamnose, galactose, arabinose, ribose, sucrose, cellobiose, and inulin; some growth occurs on mannitol, sorbitol, acetate, and aspartate; no growth occurs on sorbose, dulcitol, ethanol, methanol, formate, propionate, pyruvate, monomethylamine, citrate, oxalate, succinate, malate, or amino acids, with the exception of aspartate.

DNA G+C content (mol%): 34 (T_m).

Type strain: NO-502, CIP 107990, LMG 21963.

Sequence accession no. (16S rRNA gene): AJ535729.

Genus IV. **Dyadobacter** Chelius and Triplett 2000, 755[VP], emend. Reddy and Garcia-Pichel 2005, 1298[VP]

Gundlapally S. N. Reddy and Ferran Garcia-Pichel

Dy.a.do.bac′ ter. G. n. *dyas* two in number, pair; N.L. masc. n. *bacter* from the Gr. n. *baktron* rod or staff; N.L. masc. n. *Dyadobacter* rod or staff occurring in pairs.

Rods, 0.75–2 μm in length, **in straight to curved arrangements, occurring in pairs** (Figure 71) **in young cultures and forming chains of coccoid cells in old cultures**. Nonmotile. Nonsporeforming. Gram-stain-negative. Aerobic, chemoorganoheterotrophs, unable to hydrolyze cellulose or starch or gelatin. Oxidase- and catalase-positive. **Cells produce a non-diffusible, yellow, flexirubin-like pigment. The major fatty acids are $C_{15:0}$ iso, $C_{16:1}$ ω5*c* and $C_{16:1}$ ω7*c*.**

DNA G+C content (mol%): 44–49.

Type species: **Dyadobacter fermentans** Chelius and Triplett 2000, 756[VP].

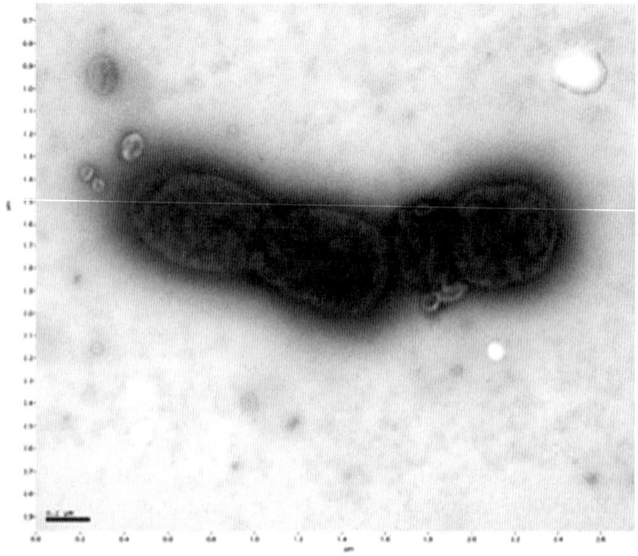

FIGURE 71. Scanning electron micrograph showing the arrangement of cells of *Dyadobacter*.

Further descriptive information

The genus belongs to the family *Cytophagaceae* in the order *Cytophagales* of the phylum *Bacteriodetes*, and presently contains six species, *Dyadobacter fermentans* (Chelius and Triplett, 2000), *Dyadobacter crusticola* (Reddy and Garcia-Pichel, 2005), *Dyadobacter hamtensis* (Chaturvedi et al., 2005), *Dyadobacter ginsengisoli* (Liu et al., 2006), *Dyadobacter beijingensis* (Dong et al., 2007), and *Dyadobacter koreensis* (Baik et al., 2007a).

Dyadobacter species grown in trypticase soy agar (TSA) or R2A medium[1] contain $C_{15:0}$ iso (19–30%), $C_{16:1}$ ω5*c* (8–20%), and $C_{16:1}$ ω7*c* (14–40%) as major fatty acids, with 20–60% of all fatty acids being the unsaturated fatty acids [$C_{16:1}$ ω5*c* and $C_{16:1}$ ω7*c*]. The minor fatty acids common to all strains are $C_{16:0}$ and $C_{16:0}$ 3-OH. Inasmuch as $C_{16:1}$ ω5*c* is present as a major fatty acid only in the species of the genus *Dyadobacter*, it can be readily used as a signature molecule to differentiate the genus *Dyadobacter* from other genera of the *Cytophagaceae*. Fatty acid profiles can also be used to differentiate among *Dyadobacter* species. $C_{17:0}$ iso 3-OH, for example, is absent from *Dyadobacter crusticola* but the most abundant fatty acid in the other five species. $C_{14:0}$ is absent from *Dyadobacter fermentans* and *Dyadobacter ginsengisoli* but present in the other four, as is $C_{15:1}$ iso. Similarly, $C_{15:0}$ iso 3-OH is absent in *Dyadobacter hamtensis* but present in all the other strains, and more significantly the fatty acid $C_{15:0}$ iso 2-OH is present only in *Dyadobacter beijingensis* and not in others.

All the species of *Dyadobacter* are yellow due to a non-diffusible flexirubin-like pigment with absorption maxima at around 428, 452, and 476 nm (Figure 72; wavelengths of peak absorption may vary slightly among species) and with a characteristic bathochromic shift (from yellow to orange and then to red) upon reaction with alkali that distinguishes them from carotenoids (Reddy and Garcia-Pichel, 2005; Weeks, 1981). These flexirubin-like pigments are not very common among bacteria and are therefore of considerable chemotaxomic value.

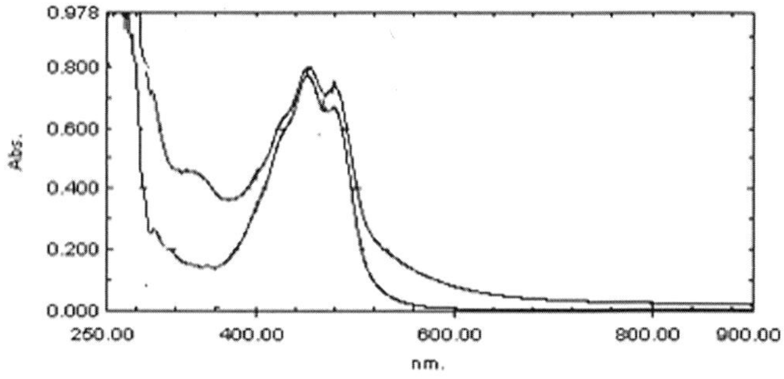

FIGURE 72. A typical absorption spectrum of flexirubin-like pigment of *Dyadobacter crusticola* in ethanol (solid line) and alkaline ethanol (dotted line).

The six species of the genus *Dyadobacter* grow well on R2A solid medium at room temperature, forming visible colonies in 24–48 h. However, their physiological adaptation to extreme conditions of temperature, pH, and salinity varies widely. For instance, the type strains of *Dyadobacter fermentans*, *Dyadobacter hamtensis*, and *Dyadobacter beijingensis* can grow from 4 to 37°C, with an optimum growth temperature of 28°C and pH 7. *Dyadobacter crusticola*, *Dyadobacter ginsengisoli*, and *Dyadobacter koreensis*, by contrast, can grow from 4 to 30°C, pH 6–8, and can tolerate a salt concentration of 1%, with a temperature optimum of 25°C and pH 7. Among the six species, only *Dyadobacter crusticola*, *Dyadobacter ginsengisoli*, and *Dyadobacter koreensis*, can grow at 5°C, albeit slowly, and thus can be deemed psychrotolerant. *Dyadobacter hamtensis* is clearly halotolerant, growing in NaCl concentration up to 11.6%, whereas *Dyadobacter koreensis* can tolerate a pH of 11, indicating that it is alkali tolerant. All the strains can grow on various rich organic media such as nutrient agar, TSA, and R2A, and type strains of *Dyadobacter fermentans* and *Dyadobacter hamtensis* can grow on Ayer's agar and peptone water as well (other species were not tested for growth on above two media). As shown in Table 91, differences in the metabolic versatility of the six species are substantial, ranging from the rather restricted capabilities of *Dyadobacter crusticola* to the wide array of carbon sources used by *Dyadobacter hamtensis*.

So far, there seems to be no unifying theme to the ecological habitat or role of *Dyadobacter* species. *Dyadobacter* could be a genus of adventitious, generalist bacteria, but, with only a few species isolated, it is probably too early to infer such a pattern. The six species originate from disparate habitats and show widely different carbon use profiles and even divergent physiological ranges for growth Table 91. The single available strain of *Dyadobacter fermentans* was isolated from internal tissue of *Zea mays* (i.e., surface-sterilized plant stems), but because the authors could demonstrate neither a beneficial nor a detrimental effect of the bacterium on the plants, its true habitat and role remain unclear. Interestingly, putative members of *Dyadobacter* have been isolated from surface-sterilized soil nematode cysts on plates of rich media (Nour et al., 2003). *Dyadobacter crusticola* was isolated from biological soil crusts (BSCs; Garcia-Pichel, 2002) where it is thought that cyanobacterial exudates serve as a source of carbon and nitrogen, and exopolysaccharide production contributes to crusting of the soil. However, extensive surveys of soil crust environments using cultivation-independent, DNA-based methods have not yet detected *Dyadobacter* (Nagy et al., 2005; Reddy and Garcia-Pichel, 2006; Smith et al., 2004), indicating that they are not a major heterotrophic component of these communities; thus, their role remains to be clarified as well. *Dyadobacter hamtensis* was isolated from melt water from a Himalayan glacier and *Dyadobacter koreensis* was isolated from fresh water. By contrast, *Dyadobacter ginsengisoli* and *Dyadobacter beijingensis* were isolated from soil and rhizosphere-associated soil, respectively. The ecological role of all the above species too was not directly studied.

Enrichment and isolation procedures

Species of *Dyadobacter* have been isolated by directly plating on agar-solidified, relatively rich media (such as R2A or BG11-PGY) and incubated at ambient temperature in the dark for up to 15 d. Their yellow pigmentation and the typical cell shape and cell aggregation patterns that are readily observable under the compound microscope in a wet or dry mount, provide an easy system for preliminary identification on enrichment cultures. Repeated streaking of single colonies usually yields axenic, clonal strains with a few iterations. Inocula used successfully include surface sterilized maize stems, surface sterilized soil nematode cysts, biological soil crusts from arid lands, glacier melt waters, soil, rhizosphere-associated soil, and fresh water.

Maintenance procedures

Stock cultures of *Dyadobacter* species can be maintained under aerobic conditions on a suitable medium such as R2A or 10× PGY or nutrient agar by monthly transfer onto fresh plates. Cultures also can be stored as glycerol stocks (18% glycerol) at −80°C. However, the best method of storing for many years is lyophilization and freezing at −80°C, which they tolerate with high recovery.

Procedures for testing special characters

Colonies of *Dyadobacter* are colored yellow due to the presence of flexirubin-like pigment, and the presence of flexirubin, a characteristic feature of the genus, can be tested directly (and differentiated from carotenoid pigments) by adding a drop of 20% KOH onto colonies of *Dyadobacter*. This will cause change in the color of the pigment from yellow to orange and then to red (Reddy and Garcia-Pichel, 2005; Weeks, 1981). This alkali-driven shift can be also carried out on ethanolic extracts and quantified spectroscopically (Figure 72). Carotenoid-like absorption maxima at 428, 452, and 478 nm broaden upon addition of 1% KOH (final concentration) and an additional peak appears around 330 nm.

Differentiation of the genus *Dyadobacter* from other genera of the family *Cytophagaceae*

The phylogenetically closest genera *Runella* and *Spirosoma* (Figure 73) can be differentiated from *Dyadobacter* on the basis of pigment, cell morphology, and its variation during the growth cycle as well as fatty acid composition. For instance, unlike *Dyadobacter*, *Runella* has salmon colored colonies, no flexirubin, no cell morphology variation during the growth cycle, and lacks the fatty acid $C_{16:1}$ ω5*c*. *Spirosoma* has no cell morphology variation during the growth cycle, degrades gelatin and starch, contains the fatty acids $C_{16:1}$ and $C_{17:0}$ iso, and lacks $C_{15:0}$ iso, $C_{16:1}$ ω5*c*, and $C_{16:1}$ ω7*c*. Within the *Cytophagaceae*, variable cell morphology is

TABLE 91. Characteristics that differentiate the species of the genus *Dyadobacter*[a]

Characteristic	*D. fermentans* (NS-114^{T})[b]	*D. crusticola* (C183-8^{T})[c]	*D. hamtensis* (HHS 11^{T})[d]	*D. ginsengisoli* (Gsoil 043)[e]	*D. beijingensis* (A54^{T})[f]	*D. koreensis* (WPCB159^{T})[g]
Source	Plant (*Zea mays*)	Soil crusts	Glacier water	Soil	Rhizosphere-associated	Freshwater
Maximum salt tolerance (as NaCl; w/v)	1.5	1.0	11.0	1.0	1.5	1.0
pH	6–8	6–8	6–8	5.5–8.5	6–8	5–9.011
Growth at 5°C	–	+	–	+	+	+
Maximum growth temperature (°C)	37	30	37	30	35	30
Lipase	–	+	–	–	nd	–
Phosphatase	–	+	+	nd	nd	+
Esculin hydrolysis	+	+	–	–	+	+
Nitrate reduction	–	–	–	+	–	–
Lysine decarboxylase	+	–	–	nd	–	–
Oxidation/fermentation:						
Glucose	+	–	+	–	+	–
Sucrose	+	–	–	–	–	–
Acid from:						
D-Glucose	+	–	+	–	+	nd
Sucrose	+	–	–	–	+	nd
D-Fructose	–	+	–	nd	+	nd
D-Maltose	+	–	–	nd	nd	nd
D-Arabinose, D-xylose	+	–	+	–	–	nd
Carbon compounds utilized:						
Sodium acetate	+	–	+	–	–	nd
D-Adonitol	–	–	+	+	+	–
D-Arabinose	+	–	–	–	–	+
L-Arabinose	+	–	+	+	+	+
Cellulose, dextran	–	–	+	–	nd	nd
Citric acid	–	–	+	–	+	nd
Dulcitol	w	+	–	–	+	–
D-Fructose	+	–	–	+	+	+
Fumarate	+	–	+	–	nd	nd
D-Galactose	+	–	+	+	+	+
Glycerol	w	–	+	–	–	–
meso-Inositol	+	+	–	–	+	–
Inulin	+	+	+	+	+	–
Lactic acid	–	+	+	–	–	nd
Malonate	+	–	+	–	nd	nd
D-Mannitol	+	+	–	–	+	–
D-Mannose	+	–	–	+	+	+
L-Melibiose	+	+	–	+	+	+
L-Rhamnose, D-rhamnose	+	–	–	+	–	w
D-Ribose	–	+	–	–	–	–
L-Sorbose	+	–	+	–	–	–
D-Sorbitol	w	+	–	–	–	–
Tartrate	+	–	+	–	–	
D-Xylose	+	+	–	+	+	+
Glycine	+	–	+	–	nd	nd
L-Alanine, L-asparagine, L-aspartic acid, L-cysteine, L-glutamic acid, L-histidine, L-isoleucine, L-leucine, L-phenylalanine, L-proline, L-serine, L-threonine. L-tyrosine, L-valine	–	–	+	nd	nd	nd
L-Arginine, L-glutamine, L-lysine, L-methionine, L-tryptophan	+	–	+	–	nd	nd

(Continued)

TABLE 91. (Continued)

Characteristic	*D. fermentans* (NS-114ᵀ)[b]	*D. crusticola* (C183-8ᵀ)[c]	*D. hamtensis* (HHS 11ᵀ)[d]	*D. ginsengisoli* (Gsoil 043)[e]	*D. beijingensis* (A54ᵀ)[f]	*D. koreensis* (WPCB159ᵀ)[g]
Sensitivity to antibiotics (μg/disc):						
Amoxycillin (30)	nd	nd	R	nd	S	R
Ampicillin (25), chloramphenicol (30), streptomycin (25)	R	R	S	R	S	S
Ciprofloxacin (30)	S	R	S	nd	nd	R
Colistin (10)	R	R	S	nd	nd	nd
Erythromycin (15)	R	R	R	nd	S	nd
Gentamicin (10)	R	S	R	nd	S	S
Kanamycin (30)	R	nd	S	R	S	nd
Lincomycin (20)	nd	nd	R	nd	S	R
Novobiocin (30)	R	S	R	nd	nd	S
Rifampin (25)	R	S	S	R	R	S
Roxithromycin (30)	R	R	S	nd	S	nd
Tetracycline (100)	S	S	S	R	R	R
Trimethoprim (25)	R	R	S	nd	nd	nd
Vancomycin (10)	S	R	R	nd	R	S
DNA G+C content (mol%)	48	48	49	48	49.2	44
16S rRNA gene sequence similarity:						
D. fermentans	100	95.9	95.2	96.4	97.4	95.2
D. crusticola	95.9	100	95.7	96.0	94.5	95.2
D. hamtensis	95.2	95.7	100	96.4	94.4	98.0
D. ginsengisoli	96.4	96.0	96.4	100	96.4	96.6
D. beijingensis	97.4	94.5	94.4	96.4	100	94.6
D. koreensis	95.2	95.2	98.0	96.6	94.6	100
Composition of cellular fatty acids:						
$C_{14:0}$	–	0.7	1.23	–	0.6	0.7
$C_{15:1}$ iso	–	0.5	3.7	–	0.5	–
$C_{15:0}$ iso	22.85	28.24	24.69	23.0	19.2	24.2
$C_{15:0}$ anteiso	–	–	–	–	1.0	–
$C_{16:1}$ ω5*c*	14.28	15.5	19.75	7.9	10.3	11.0
$C_{16:1}$ ω7*c*	28.57	39.4	14.81	44.6	17.5	34.8
$C_{16:0}$	5.71	13.6	8.64	11.0	2.9	9.4
$C_{18:0}$	–	1.9	–	–	–	–
$C_{18:1}$	–	0.34	2.46	–	–	–
$C_{15:0}$ iso 2-OH	–	–	–	–	23.4	–
$C_{15:0}$ iso 3-OH	2.9	2.4	–	3.9	2.6	2.8
$C_{16:0}$ iso 3-OH	2.85	0.5	2.46	2.3	3.1	2.5
$C_{17:0}$ iso 3-OH	22.85	–	22.22	7.6	12.4	9.5
Unknown	1.3	0.3	–	–	2.1	–

[a]Symbols: +, positive; –, negative or absent; w, weak; R, resistant; S, sensitive; nd, not done.

[b]Data from Chelius and Triplett (2000).

[c]Data from Reddy and Garcia-Pichel (2005).

[d]Data from Chaturvedi et al. (2005).

[e]Data from Liu et al. (2006).

[f]Data from Dong et al. (2007).

[g]Data from Baik et al. (2007a).

a hallmark of *Dyadobacter*, but it is also a trait of *Flectobacillus*. The latter genus, however, is pink-pigmented, not yellow, and does not contain flexirubin. The genus *Dyadobacter* can further be differentiated from *Flectobacillus* on the basis of fatty acid composition: members of *Dyadobacter* contain $C_{16:1}$ ω5*c* and $C_{16:1}$ ω7*c* as major fatty acids and lack $C_{17:0}$ iso (Table 92). Besides *Dyadobacter*, five other genera contain flexirubin-like pigments. However, compared to other genera, *Dyadobacter* is in the only one that contains the fatty acid $C_{16:1}$ ω5*c* and varies in morphology and aggregation during its growth cycle (Table 92). Other genus-specific characteristics that differentiate *Dyadobacter* from closely related genera are listed in Table 92.

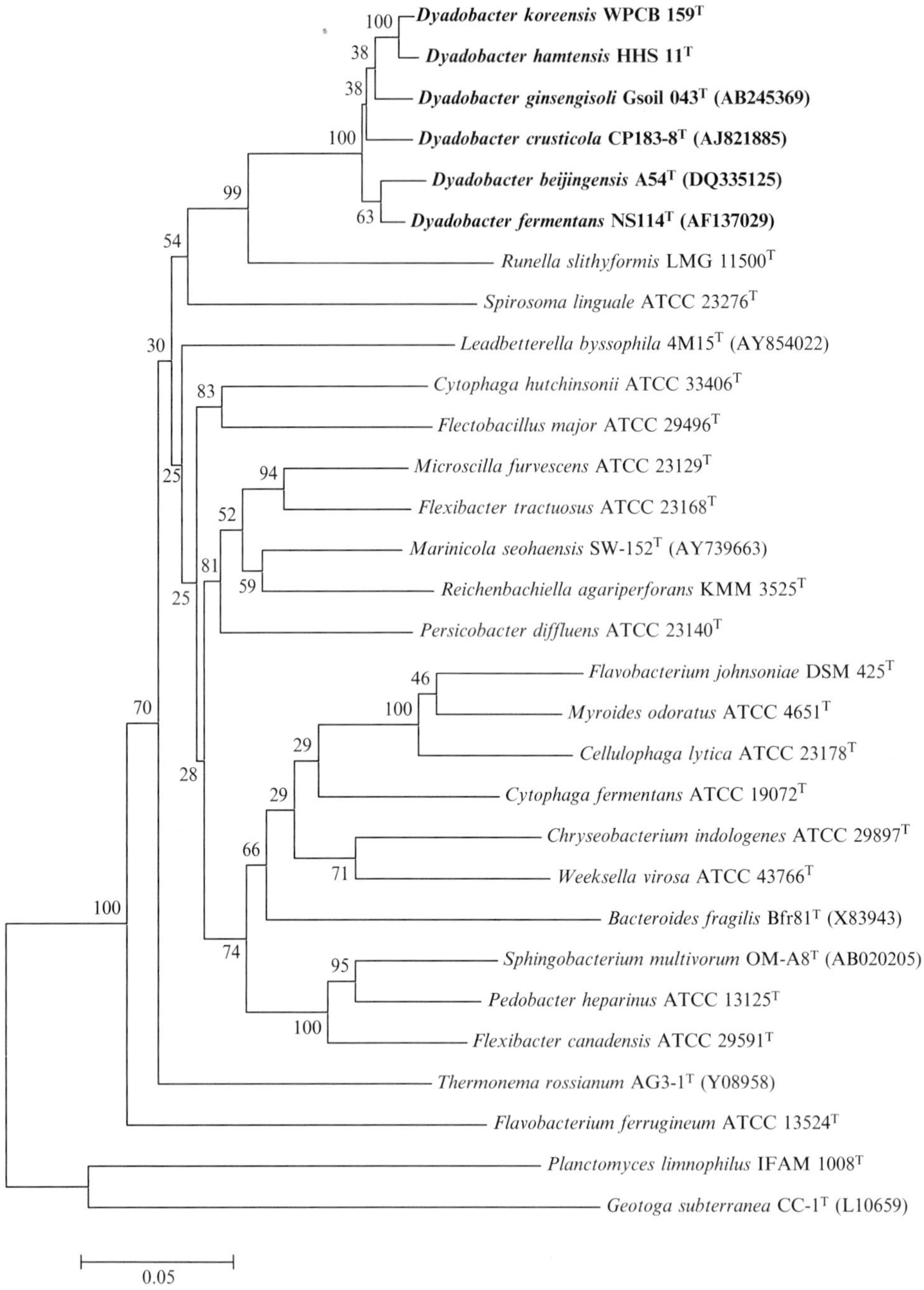

FIGURE 73. Neighbor-joining (NJ) tree based on 16S rRNA gene sequence (1337 nucleotides) showing the phylogenetic relationship between *Dyadobacter* and other related genera of the phylum *Bacteroidetes*. Bootstrap values (expressed as percentages of 1000 replicates) greater than 40% are given at nodes.

Taxonomic comments

The genus *Dyadobacter* was established to accommodate the bacteria that are straight to curved, occur in pairs in young cultures and produce a non-diffusible, yellow, flexirubin-like pigment (Chelius and Triplett, 2000). The six currently known species (*Dyadobacter fermentans*, *Dyadobacter crusticola*, *Dyadobacter hamtensis*, *Dyadobacter ginsengisoli*, *Dyadobacter beijingensis*, and *Dyadobacter koreensis*) possess the above characteristics and, therefore, can be grouped within the genus. All six species have a 16S rRNA gene sequence similarity of 94.4–98.0%. The highest similarity of 98.0% was between *Dyadobacter hamtensis* and *Dyadobacter koreensis*, followed by *Dyadobacter fermentans* and *Dyadobacter beijingensis* (a similarity of 97.4%). Since their similarity was >97.0%, DNA–DNA hybridization was performed between *Dyadobacter hamtensis* vs *Dyadobacter koreensis* and *Dyadobacter fermentans* vs *Dyadobacter beijingensis*, and showed DNA–DNA similarities of 19.4 and <25%, respectively. Thus, according to the criteria set for discriminating species (Stackebrandt and Goebel, 1994), they all are distinctly new species. Although their number is small (six), these species, interestingly, seem

TABLE 92. Characteristics that differentiate the genus *Dyadobacter* from other related genera[a]

Characteristic	*Dyadobacter*[a]	*Runella*[c]	*Spirosoma*	*Flectobacillus*[c]	*Cytophaga*[c]	*Flexibacter*[c]	*Microscilla*[c]	*Reichenbachia*[d]	*Leadbetterella*[e]	*Marincola*[f]
Colony color	Yellow	Salmon	Yellow	Pink	Bright yellow	Orange	Orange	Orange	Orange	Orange
Motility	–	–	–	–	Gliding	Gliding	Gliding	Gliding	–	Gliding
Spiral like	–	+		+	–	–	–		–	–
Filamentous	+	+		+	+	+	+		–	–
Cell morphology variation	+	–		+	–	–	–		–	–
DNA G+C content (mol%)	48	49	51–53	40	39	37	44	44.5	33	40.3
Catalase	+	w	+	w	–	na	na	+	+	+
Oxidase	+	+	+	+	+	na	–	+	+	+
Flexirubin reaction	+	–	+	–	+	na	–	+	+	+
Degradation of:										
Gelatin	–	–	+	w	–	D	+	+	+	–
Starch	–	w	w	+	–	D	+	+	+	–
Major fatty acids	$C_{15:0}$ iso, $C_{16:1}$ ω5*c*, $C_{16:1}$ ω7*c*	$C_{15:0}$ iso, $C_{16:1}$ ω7*c*/$C_{15:0}$ iso 2-OH	$C_{16:1}$, $C_{17:0}$ iso	$C_{15:0}$ iso, $C_{17:0}$ iso	na	na	na	$C_{15:0}$ iso, $C_{16:1}$	$C_{15:0}$ iso $C_{16:1}$ ω7*c*/$C_{15:0}$ iso 2-OH	$C_{15:0}$ iso, $C_{15:1}$ iso

[a]Symbols: +, 90% or more of strains are positive; –, 10% or less of strains are positive; w, weakly positive; D, different reactions occur in different taxa (species of a genus); na, not available.

[b]Data from Chelius and Triplett (2000).

[c]Data from: Yi and Chun (2004).

[d]Data from Nedashkovskaya et al. (2003).

[e]Data from Weon et al. (2005).

[f]Data from Yoon et al. (2005).

to have a kind of habitat relationship. For instance, the species *Dyadobacter fermentans* and *Dyadobacter beijingensis* have been isolated from closely related habitats – the root nodules or rhizosphere-associated soil – and exhibit a 16S rRNA gene sequence similarity of 97.4%. Similarly, *Dyadobacter hamtensis* and *Dyadobacter koreensis* were isolated from water samples and share a 16S rRNA gene sequence similarity of 98.0%.

Phylogenetic analysis based on 1337 nucleotides of the 16S rRNA gene sequence (base positions 47–1418 using *Escherichia coli* numbering, accession number J01695) indicate that species of the genus *Dyadobacter*, in accordance with their generic status, form a deeply rooted, monophyletic, and statistically very well supported clade (Figure 73). The species of *Dyadobacter* share a 16S rRNA gene sequence similarity of 94.0–98.0%, but it is less than 87% with other genera of the family *Cytophagaceae*. They share a sequence similarity of 86–87% with members of the genus *Runella* and ca. 83% with those of *Spirosoma*, these two genera being the genera most closely related to *Dyadobacter*.

According to Reddy and Garcia-Pichel (2005), three variable regions (region I: 182–207, region II: 590–649, and region III: 835–848) contribute most to the 16S rRNA gene sequence variability among known *Dyadobacter* strains. Analysis of these regions alone is sufficient to differentiate at the species level.

Differentiation of the species of the genus *Dyadobacter*

The six *Dyadobacter* species are biochemically and physiologically divergent, exhibiting very significant and easily recognizable differences, particularly in carbon utilization patterns and temperature and pH ranges for growth. Temperature, pH, and NaCl tolerance are particularly useful in differentiating *Dyadobacter crusticola*, *Dyadobacter koreensis*, and *Dyadobacter hamtensis* from other species. The pattern of sugar and amino acid utilization can also be easily used to distinguish among the species. An extensive list of differentiating tests and traits is presented in Table 91.

List of species of the genus *Dyadobacter*

1. **Dyadobacter fermentans** Chelius and Triplett 2000, 756[VP]

fe.rmen'tans. L. part. adj. *fermentans* causing fermentation.

The characteristics are as described for the genus and as listed in Tables 91 and 92, with the following additional features. Colonies of *Dyadobacter fermentans* are yellow pigmented and can grow to a diameter of 2–3 mm. Growth occurs at 15–37°C, 0–1.5% NaCl, and pH 6–8, with an optimum temperature of 28°C and optimum pH of 7. No hydrolysis of agar, gelatin, starch, and cellulose occurs. Growth occurs on starch, fumarate, 5-ketogluconate, malate, malonate, sucrose, erythritol, D- and L-xylose, β-methylxyloside, D-glucose, methyl α-D-mannoside, methyl α-D-glucoside, *N*-acetylglucosamine, amygdalin, arbutin, salicin, cellobiose, maltose, lactose, sucrose, trehalose, melezitose, D-raffinose, xylitol, gentiobiose, D-turanose, D-tagatose, and D- and L-fucose. *Dyadobacter fermentans* can ferment glucose and sucrose and produce acid from ribose. The principal fatty acids are (for cultures on

R2A medium) $C_{15:0}$ iso (22.9%), $C_{16:1}$ ω5*c* (14.3%), $C_{16:1}$ ω7*c* (28.6%), $C_{16:0}$ (5.7%), $C_{16:0}$ 3-OH (2.9%), and $C_{17:0}$ iso 3-OH (22.9%).

Source: surface-sterilized *Zea mays* (cv. Mo17) stem.

DNA G+C content (mol%): 48 (HPLC).

Type strain: NS114, ATCC 700827, CIP 107007.

Sequence accession no. (16S rRNA gene): AF137029.

2. **Dyadobacter beijingensis** Dong, Guo, Zhang, Qiu, Sun, Gong and Zhang 2007, 864[VP]

bei.jing.en.sis. N.L. masc. adj. *beijingensis* pertaining to Beijing.

The characteristics are as described for the genus and as listed in Tables 91 and 92, with the following additional features. Colonies on R2A agar are yellow, mucoid, flaky, convex, and smooth. The pigment present is of the flexirubin type, with absorption maxima at 428, 452, and 478 nm. Growth occurs at 4–35°C (optimum 28°C) but not at 37°C. Growth occurs at pH 6–8 (optimum pH 7) and in the presence of up to 1.5% NaCl. Growth occurs in peptone water, on Luria-Bertani (LB) medium, and R2A medium. No hydrolysis occurs of urea, gelatin, casein, cellulose, and starch. H_2S is not produced. The methyl red, indole, and Voges–Proskauer tests are negative. Acid is produced from D-galactose and D-rhamnose. The following substrates are utilized as sole carbon sources: erythritol, methyl β-D-xylopyranoside, D-glucose, methyl α-D-mannopyranoside, methyl α-D-glucopyranoside, arbutin, citrate, salicin, D-cellobiose, D-maltose, D-lactose, D-melibiose, sucrose, D-trehalose, D-melezitose, D-raffinose, gentiobiose, and D-fucose. No utilization occurs of glycerol, *myo*-inositol, *N*-acetylglucosamine, amygdalin, xylitol, D-turanose, D-lyxose, L-fucose, D-arabitol, L-arabitol, glycogen, D-tagatose, starch, potassium gluconate, potassium 2-ketogluconate, potassium 5-ketogluconate, and methanol. Cells are sensitive to (μg/disc) acetylspiramycin (100), carbenicillin (100), cephalothin (30), norfloxacin (10), penicillin (10), and spectinomycin (10), but are resistant to levofloxacin (100). The principal fatty acids are (culture grown on R2A medium) $C_{14:0}$ (0.6%), $C_{15:1}$ iso (0.5%), $C_{15:0}$ iso (19.2%), $C_{15:0}$ anteiso (1.0%), $C_{16:1}$ ω5*c* (10.3%), $C_{16:1}$ ω7*c* (17.5%), $C_{16:0}$ (2.9%), $C_{15:0}$ iso 2-OH (23.4), $C_{15:0}$ iso 3-OH (2.6), $C_{16:0}$ 3-OH (3.1%), $C_{17:0}$ iso 3-OH (12.4%), and unknown (2.1%).

Source: the rhizosphere of turf grasses.

DNA G+C content (mol%): 49.2 (T_m).

Type strain: A54, CGMCC 1.6375, JCM 14200.

Sequence accession no. (16S rRNA gene): DQ335125.

3. **Dyadobacter crusticola** Reddy and Garcia-Pichel 2005, 1298[VP]

crus.ti'co.la. L.n. *crusta* crust; L. suff. *-cola* (from L. n. *incola*), dweller; N.L. n. *crusticola*, a dweller of crust.

The characteristics are as described for the genus and as listed in Tables 91 and 92, with the following additional features. The colonies of *Dyadobacter crusticola* are mucoid, convex, round, and smooth and can grow to a size of 2–3 mm in diameter. Growth occurs at 5–30°C, 0–1.0% NaCl, and pH 6–8, with an optimum temperature of 25°C and pH of 7. Cells are positive for β-galactosidase but negative for urease, DNase, arginine decarboxylase, ornithine decarboxylase, phenylalanine deaminase, arginine dihydrolase, methyl red test, Voges–Proskauer test, indole production, and Simmons' citrate test. Casein and cellulose are not hydrolyzed. H_2S is not produced. Acid is produced from D-fructose but not from D-galactose, lactose, D-mannitol, D-sorbitol, and D-xylose. L-Arabinose, D-galactose, D-maltose, and D-xylose are fermented, but not D-fructose, lactose, D-mannose, and D-sorbitol. D-Cellobiose, glucose, lactose, D-fructose, D-raffinose, sucrose, and D-trehalose are used as sole carbon sources but not citrate, ethanolamine, fumaric acid, pyruvate, succinate, adenine, cytosine, guanine, thymidine, nicotinic acid, oxalate, tartaric acid, indole, and phenanthrene. Cells are sensitive to (μg/disc): carbenicillin (100), doxycycline (30), and polymyxin B (300), but are resistant to azithromycin (15), aztreonam (30), bacitracin (10), ceftriaxone (30), cephalothin (30), ethambutol (50), nitrofurantoin (150), penicillin (10), sulfisoxazole (300), and sulfathiazole (300). The pigment present is a flexirubin type with absorption maxima at 428, 452, and 478 nm. The polar lipids present are phosphatidylserine, phosphatidylglycerol and diphosphatidylglycerol (cardiolipin). The principal fatty acids are (culture grown on R2A medium) $C_{14:0}$ (0.7%), $C_{15:1}$ iso (0.5%), $C_{15:0}$ iso (28.2%), $C_{16:1}$ ω5*c* (15.5%), $C_{16:1}$ ω7*c* (39.4%), $C_{16:0}$ (13.6%), $C_{16:0}$ 3-OH (0.5%), $C_{18:0}$ (1.9%), and $C_{18:1}$ (0.3).

Source: biological soil crusts.

DNA G+C content (mol%): 48 (T_m).

Type strain: CP183–8, ATCC BAA-1036, DSM 16708.

Sequence accession no. (16S rRNA gene): AJ821885.

4. **Dyadobacter ginsengisoli** Liu, Im, Lee, Yang and Lee 2006, 1941[VP]

gin.sen.gi.so.li. N.L. n. *ginsengum* ginseng; L. n. *solum* soil; N.L. gen. n. *ginsengisoli* of the soil of a ginseng field.

The characteristics are as described for the genus and as listed in Tables 91 and 92, with the following additional features. Cells are 0.6–0.8 × 3.0–6.0 μm after 2 d on R2A agar. Colonies on R2A agar for 2 d are smooth, circular, transparent, and yellowish. Growth occurs at 4–30°C and at pH 5.5–8.5; no growth occurs at 37°C. Growth occurs in the absence of NaCl and in the presence of 1.0% NaCl, but not with 2% NaCl. Anaerobic growth does not occur. DNA, chitin, and xylan are not hydrolyzed. D-Lyxose, *N*-acetylglucosamine, salicin, D-lactose, D-maltose, and D-melibiose are utilized as sole carbon sources, but not D-fucose, ethanol, L-xylose, pyruvic acid, formic acid, propionate, 3-hydroxybutyrate, valerate, caprate, maleic acid, phenylacetate, benzoic acid, 3-hydroxybenzoate, 4-hydroxybenzoate, malate, succinic acid, glutaric acid, itaconate, adipate, suberate, oxalic acid, gluconate, xylitol, amygdalin, methanol, or glycogen. Cells are resistant to (μg/disc), carbenicillin (100) and hygromycin (100) but sensitive to streptomycin (100). The predominant menaquinone is MK-7. The principal fatty acids are (culture grown on R2A medium) $C_{15:0}$ iso (23.0%), $C_{16:1}$ ω5*c* (7.9%), $C_{16:1}$ ω7*c* (44.6%), $C_{16:0}$ (11.0%), $C_{15:0}$ iso 2-OH (3.9), $C_{15:0}$ iso 3-OH (2.3), and $C_{17:0}$ iso 3-OH (7.6%).

Source: the soil of a ginseng field in Pocheon province, South Korea.

DNA G+C content (mol%): 48 (HPLC).

Type strain: Gsoil 043, KCTC 12589, LMG 23409.

Sequence accession no. (16S rRNA gene): AB245369.

5. **Dyadobacter hamtensis** Chaturvedi, Reddy and Shivaji 2005, 2116[VP]

ham.ten'sis. N.L. masc. adj. *hamtensis* pertaining to the glacier Hamta.

The characteristics are as described for the genus and as listed in Tables 91 and 92, with the following additional features. Colonies on nutrient agar are round, 2–3 mm in diameter, and light yellow. Growth occurs at 10–37°C and pH 6–8, but not at pH 10. The optimum temperature and pH for growth are 22°C and pH 7. The organisms are positive for arginine decarboxylase and β-galactosidase, but negative for urease, arginine dihydrolase, citrate utilization, H_2S production, methyl red test, indole test, and Voges–Proskauer test. Acid is not produced from D-ribose, D-galactose, L-rhamnose, D-lactose, and D-mannose. The following compounds are utilized as sole carbon sources: sucrose, *N*-acetylglucosamine, methyl α-D-mannoside, methyl α-D-glucoside, L-fucose, melezitose, D-cellobiose, sucrose, D-trehalose, D-lactose, D-raffinose, amygdalin, dextran, glycogen, arbutin, salicin, D-erythritol, citrate, malonate, citric acid, 2-ketogluconate, inulin, sodium formate, sodium fumarate, sodium malate, sodium tartrate, polyethylene glycol, and L-creatinine, but not D-maltose, gluconate, hydroxybutyric acid, sodium succinate, sodium pyruvate, thioglycolate, starch, methanol, and agar. Cells are sensitive to (μg/disc): cefoperazone (75), cefotaxime (10), doxycycline (30), sulfamethoxazole (50), nalidixic acid (30), norfloxacin (10), lomefloxacin (30), tobramycin (15), and amikacin (30), but resistant to co-trimoxazole (25), nitrofurantoin (300), penicillin (10), cefuroxime (20), cefazolin (30), and bacitracin (10). The principal fatty acids are (culture grown on R2A medium) $C_{14:0}$ (1.2%), $C_{15:1}$ iso (3.7%), $C_{15:0}$ iso (24.7%), $C_{16:1}$ ω5*c* (19.8%), $C_{16:1}$ ω7*c* (14.8%), $C_{16:0}$ (8.6%), $C_{16:0}$ 3-OH (2.5%), $C_{18:1}$ (2.5%), and $C_{17:0}$ iso 3-OH (22.2%).

Source: a glacial water sample.

DNA G+C content (mol%): 49 (T_m).

Type strain: HHS 11, JCM 12919, MTCC 7023.

Sequence accession no. (16S rRNA gene): AJ619979.

6. **Dyadobacter koreensis**. Baik, Kim, Kim, Kim and Seong 2007a, 1228[VP]

ko.re.en'sis. N.L. masc. adj. *koreensis* pertaining to Korea.

The characteristics are as described for the genus and as listed in Tables 91 and 92, with the following additional features. Cells grow best on media such as R2A, plate count agar (PCA), and trypticase soy agar (TSA), but weakly on nutrient agar (NA). Colonies on TSA are circular, low-convex, entire margin, smooth, translucent, light yellow, and approximately 3.0 mm in diameter after 5 d at 25°C. Growth occurs at pH 5–11 (optimum pH 7) and at 4–30°C (optimum 25°C). Negative for gelatinase, arginine dihydrolase, urease, ornithine decarboxylase, and tryptophan deaminase. β-Galactosidase is present. H_2S and indole are not produced. Glucose and salicin are utilized as sole carbon and energy sources. Cells are sensitive to (μg/disc): amikacin (30), sulbactam (20), cefotaxime (30), imipenem (10), piperacillin (100), teicoplanin (30), and isepamicin (30), but resistant to aztreonam (30), clavulanic acid (20/10), bacitracin (10), cefepime (30), cefmetazole (30), ceftazidime (30), ciprofloxacin (5), moxalactam (30), oxacillin (1), piperacillin/tazobactam (100), and tobramycin (10). The major fatty acids are (culture grown on R2A medium) $C_{14:0}$ (0.7%), $C_{15:0}$ iso (24.2%), $C_{16:1}$ ω5*c* (11.0%), $C_{16:1}$ ω7*c* (34.8%), $C_{16:0}$ (9.4%), $C_{15:0}$ iso 3-OH (2.8), $C_{16:0}$ 3-OH (2.5%), and $C_{17:0}$ iso 3-OH (9.5%).

Source: freshwater of Woopo wetland, Republic of Korea.

DNA G+C content (mol%): 44 (T_m).

Type strain: WPCB159, KCTC 12534, NBRC 101116.

Sequence accession no. (16S rRNA gene): EF017660.

Genus V. **Effluviibacter** Suresh, Mayilraj and Chakrabarti 2006, 1706[VP]

The Editorial Board

Ef.flu.vi.i.bac'ter. L. neut. n. *effluvium* outflow; N.L. masc. n. *bacter* rod; N.L. masc. n. *Effluviibacter* rod from an outflow, referring to the source of isolation of the first strain.

Irregular rods 0.3–0.5 × 1.0–3.0 μm. Nonmotile. **No gliding motility**. Nonsporeforming. Gram-stain-negative. **Obligate aerobes**. Colonies are deep pink. Oxidase- and catalase-positive. **Starch and casein are hydrolyzed. Fructose and lactose are utilized. Acid is produced from glucose**. The major respiratory quinone is MK-7. The predominant whole-cell fatty acids are $C_{17:1}$ iso I/anteiso B (36.7%), $C_{15:0}$ iso (15.8%), and $C_{17:0}$ iso 3-OH (10.3%).

DNA G+C content (mol%): 59.5.

Type species: **Effluviibacter roseus** Suresh, Mayilraj and Chakrabarti 2006, 1706[VP].

Enrichment and isolation procedures

A sample of muddy water was serially diluted in physiological saline (0.9% NaCl) and plated onto nutrient agar. After prolonged incubation at 30°C, a small, pink, mucoid colony was selected and purified.

Differentiation of the genus *Effluviibacter* from other genera

Effluviibacter roseus cells are irregular rods, whereas *Pontibacter actiniarum*, *Adhaeribacter aquaticus*, and *Hymenobacter roseosalivarius* cells are uniform in shape. *Effluviibacter roseus* does not exhibit gliding motility, whereas *Pontibacter actiniarum* does. In contrast to *Pontibacter actiniarum*, *Effluviibacter roseus* hydrolyzes casein and starch, utilizes fructose and lactose, produces acid from glucose, grows at 4°C but not at 42°C, exhibits sensitivity to gentamicin and streptomycin, and has a higher mol% G+C of its DNA (59.5 vs 48.7). *Effluviibacter roseus* differs from *Adhaeribacter aquaticus* by growing in the presence of 8.0% NaCl, hydrolyzing starch, producing acid from glucose, utilizing lactose but not adonitol and *myo*-inositol, and by having a higher mol% G+C of its DNA (59.5 vs 40.0). *Effluviibacter roseus* differs from *Hymenobacter roseosalivarius* by growing in the presence of 8.0% NaCl, utilizing fructose, and having a higher mol% G+C of its DNA (59.5 vs 56.0) (Suresh et al., 2006).

Taxonomic comments

Phylogenetic analysis based on 16S rRNA gene sequences indicate that the nearest neighbor of *Effluviibacter roseus* is *Pontibacter actiniarum* (95.5% similarity), followed by *Adhaeribacter aquaticus* (89.0%) and *Hymenobacter roseosalivarius* DSM 11622[T] (88.9%) (Suresh et al., 2006).

List of species of the genus *Effluviibacter*

1. **Effluviibacter roseus** Suresh, Mayilraj and Chakrabarti 2006, 1706[VP]

ro'se.us. L. masc. adj. *roseus* rose-colored, pink.

The characteristics are as described for the genus, with the following additional features. Colonies on nutrient agar are dark pink, circular, convex, smooth, and 2–3 mm in diameter after incubation for 6–8 d. Growth occurs at 4–37°C but not at 42°C. Growth occurs at pH 6.0–10.0. NaCl is tolerated up to a concentration of 8%. The following tests are positive: methyl red, amylase, protease, lysine decarboxylase, ornithine decarboxylase, and gelatin hydrolysis. Negative tests include: indole production, citrate utilization, H_2S production, arginine dihydrolase, Voges–Proskauer, urease, hydrolysis of esculin and Tween 20, and reduction of nitrate to nitrite. D-Fructose, D-galactose, D-glucose, lactose, raffinose, and sucrose are utilized as sole carbon sources, but not acetate, adonitol, arabinose, citrate, fumarate, lactate, *myo*-inositol, salicin, sorbitol, and succinate. Histidine, L-isoleucine, L-lysine, ornithine, L-proline, and glycine are utilized as sole nitrogen sources, but not L-arginine, DL-alanine, and L-methionine. Acid is produced from D-glucose but not from lactose or sucrose. Susceptibility occurs toward vancomycin, lincomycin, streptomycin, gentamicin, and chlortetracycline, but not toward penicillin and tobramycin. The fatty acid profile is dominated by unsaturated and hydroxy fatty acids, including $C_{17:1}$ iso I/anteiso B (36.7%), $C_{15:0}$ iso (15.8%), and $C_{17:0}$ iso 3-OH (10.3%). The phospholipids present are phosphatidylglycerol, diphosphatidylglycerol, and an unknown phospholipid.

Source: muddy water from an occasional drainage system in Chandigarh, India.

DNA G+C content (mol%): 59.5 (T_m).

Type strain: SRC-1, DSM 17521, MTCC 7260.

Sequence accession no. (16S rRNA gene): AM049256.

Genus VI. **Emticicia** Saha and Chakrabarti 2006, 993[VP]

THE EDITORIAL BOARD

Em.ti.ci'ci.a. N.L. fem. n. *Emticicia* arbitrarily formed from the acronym MTCC for Microbial Type Culture Collection and GenBank, where this investigation was carried out.

Rods, usually 0.3–0.4 × 2–5 μm; in an aged culture, some cells may be longer. Cells occur singly and sometimes in pairs. Nonmotile. **No gliding motility**. Gram-stain-negative. Oxidase-positive. Catalase weakly positive. **Growth does not occur on most of the common nutritionally rich culture media but does occur on 100-fold dilutions of these media. Salt tolerance is <1% NaCl**. Colonies are light pink. **Flexirubin pigments are not produced. Nitrate is reduced to nitrite. Urease-positive**. The predominant whole-cell fatty acids are $C_{15:0}$ iso, $C_{17:0}$ iso 3-OH, $C_{15:0}$ anteiso, summed feature 3 ($C_{15:0}$ iso 2-OH and/or $C_{16:1}$ ω7*c*), $C_{15:0}$ iso 3-OH, and $C_{15:1}$ ω6*c*.

DNA G+C content (mol%): 36.9.

Type species: **Emticicia oligotrophica** Saha and Chakrabarti 2006, 993[VP].

Further descriptive information

Emticicia oligotrophica does not grow on most of the common nutritionally rich culture media such as tryptic soy broth, tryptic soy agar, yeast extract agar, plate count agar, nutrient broth, nutrient agar, Luria–Bertani agar, Mueller–Hinton agar, Zobell marine agar, etc. It does grow on modified R2A agar* and on marine agar, etc. It does grow on modified RZA Agar and on 100-fold dilutions of all the above-mentioned media. Agarose (1.5%) is used to solidify the diluted media.

Enrichment and isolation procedures

Emticicia oligotrophica was isolated from spring water on TSBA100 medium (normal strength tryptic soy broth diluted 100 times and solidified with 1.5% agarose) by dilution plating, with incubation at 37°C.

Differentiation of the genus *Emticicia* from other genera

Emticicia oligotrophica differs from *Leadbetterella byssophila* by the light pink color of its colonies (vs orange for *Leadbetterella byssophila*), reduction of nitrate, hydrolysis of urea, inability to degrade tyrosine, lack of flexirubin pigments, inability to grow on undiluted tryptic soy agar or other rich media, a relatively higher amount of $C_{15:0}$ anteiso fatty acid (12.2% of total fatty acids vs 0.4%), and a higher mol% G+C value (36.0 vs 33.0).

Taxonomic comments

Analysis of 16S rRNA gene sequences by Saha and Chakrabarti (2006) indicated that the nearest cultured neighbor of *Emticicia oligotrophica* was *Leadbetterella byssophila* (87.8% sequence similarity). Sequence similarities with other members of the phylum *Bacteroidetes* were less than 85%.

*Modified R2A (MR2A) agar contains (g/l): casamino acids, 0.025; yeast extract, 0.05; peptone, 0.075; glucose, 0.05; soluble starch, 0.05; K_2HPO_4, 0.03; $MgSO_4 \cdot 7H_2O$, 0.0024; sodium pyruvate, 0.03; and agar, 1.5; pH 7.2.

List of species of the genus *Emticicia*

1. **Emticicia oligotrophica** Saha and Chakrabarti 2006, 993[VP]

o.li.go.tro'phi.ca. Gr. adj. *oligos* few; Gr. adj. *trophikos* nursing, tending or feeding; N.L. fem. adj. *oligotrophica* eating little, referring to a bacterium living on low-nutrient media.

The characteristics are as described for the genus, with the following additional features. Colonies on 0.01 × tryptic soy agar after 2 d at 37°C are round, mucoid, light pink, and convex with almost entire margins; with prolonged incubation, the colony color deepens, it produces slime and appears glistening. Temperature range for growth, 15–42°C; pH range for growth, 5.0–11.0. Methyl red positive, Voges–Proskauer negative. Gelatin, starch, and Tween 60 are hydrolyzed, but not casein, *o*-nitrophenol-β-D-galactoside (ONPG), hypoxanthine, chitin, cellulose [carboxymethyl cellulose (CMC) or filter paper], xylan, tyrosine, or Tween 20, 40, or 80. The following compounds are utilized as sole carbon sources: D-cellobiose, D-lactose, D-melezitose, D-raffinose, and arbutin; weak utilization occurs of L-arabinose, L-arabitol, D-amygdalin, D-fructose, D-galactose, D-glucose, glycerol, *myo*-inositol, D-melibiose, and sucrose. No utilization occurs of adonitol, dulcitol, D-mannitol, D-sorbitol, L-sorbose, or xylitol. Acid is produced from arbutin, D-cellobiose, D-glucose, D-lactose, methyl α-D-glucoside, salicin, D-trehalose, and D-xylose. Weak acid production occurs from D-galactose, D-maltose, D-mannose, D-melibiose, D-melezitose, and sucrose. No acid is produced from adonitol, D-amygdalin, L-arabinose, L-arabitol, dulcitol, esculin, erythritol, D-fructose, D- or L-fucose, glycerol, glycogen, *myo*-inositol, inulin, 2-ketogluconate, D-mannitol, D-raffinose, D-ribose, D-sorbitol, L-sorbose, or xylitol. Sensitivity is exhibited toward ampicillin, bacitracin, chloramphenicol, erythromycin, gentamicin, kanamycin, lincomycin, neomycin, novobiocin, norfloxacin, penicillin G, polymyxin B, rifampin, streptomycin, sulfasomidine, and tetracycline.

Source: a warm spring water sample from Assam, India.

DNA G+C content (mol%): 36.9 (T_m).

Type strain: GPTSA100-15, DSM 17448, MTCC 6937.

Sequence accession no. (16S rRNA gene): AY904352.

Genus VII. **Flectobacillus** Larkin, Williams and Taylor 1977, 152[AL]emend. Raj and Maloy 1990, 346[VP]

THE EDITORIAL BOARD

Flec.to.ba.cil'lus. L. v. FLECTO to curve; L. masc. n. *bacillus* a little staff, rod; N.L. masc. n. *Flectobacillus* little curved rod.

Curved, nonflexible rods with variable degrees of curvature from cell to cell. **Most cells of *Flectobacillus major* are shaped like the letter C** (wide-open rings), and **sometimes cell ends touch or overlap** (closed rings). Coils are less common, and spiral forms are rare. The outer ring diameter is 5–10 μm, and the cell width is 0.6–2.0 μm. The ends of the cells are tapered or rounded or both. **Cells of *Flectobacillus lacus* are almost straight rods.** Gram-stain-negative. **Nonflagellated and nonmotile**. Colonies are 2–6 mm in diameter with a **pale pink or rose-color due to a water-insoluble pigment**. No growth occurs in media containing seawater or 3.0% NaCl. **Aerobic**, having a strictly respiratory type of metabolism with oxygen as the terminal electron acceptor. **Oxidase-positive**. Chemo-organotrophic but not methylotrophic. The natural habitats are freshwater lakes and eutrophic ponds.

DNA G+C content (mol%): 38.3–40.3.

Type species: **Flectobacillus major** (Gromov 1963) Larkin, Williams and Taylor 1977, 155[AL] (*Microcyclus major* Gromov 1963, 733).

Further descriptive information

Much of this chapter is taken from the excellent treatment by Larkin and Borrall (1984a) in the 1st edition of *Bergey's Manual of Systematic Bacteriology*. It has been updated when information that is more recent was needed.

Cells of *Flectobacillus major* typically appear as curved rods whose degree of curvature varies from nearly straight to crescent or horseshoe shaped, to rings (see Figure 3.3 in Larkin and Borrall, 1984a). An individual cell may be bent in more than one plane. Regular and irregular coils and helices as well as long sinuous filaments are rarely produced.

Colonies on MS agar (0.1% peptone, 0.1% yeast extract, 0.1% glucose, and 1.5% agar) produce a pink, water-insoluble, nonfluorescent pigment. The colonies are circular and convex with an entire margin. *Flectobacillus major* does not grow on eosin methylene blue agar, phenol red mannitol salt agar, phenylethyl alcohol agar, MacConkey agar, bismuth sulfite agar, or Salmonella-Shigella agar. Abundant growth occurs on MS agar, peptonized milk agar, yeast extract-acetate-tryptone agar, and nutrient agar (Larkin et al., 1977).

Flectobacillus is chemo-organotrophic and produces an acid reaction from a wide variety of carbohydrates including pentoses, hexoses, and disaccharides when incubated aerobically in the medium of Hugh and Leifson (1953). Acidification does not occur with any sugar alcohols. Casein, cellulose, and chitin are not hydrolyzed. Gelatin, tributyrin, and starch are hydrolyzed. Litmus milk is unchanged. *Flectobacillus* is not known to be pathogenic.

Isolation procedures

Flectobacillus major can be isolated by repeated streaking of samples onto MS agar with incubation at room temperature for up to 2 weeks. The pink pigmentation of the colonies aids detection.

Flectobacillus lacus was isolated from a surface sample from a highly eutrophic pond. The sample, at ambient temperature, was brought back to the laboratory within 15 min; 10–100 μl of sample was spread on plates containing the R2A agar medium*

*R2A agar contains (g/l of distilled water): yeast extract, 0.5; peptone, 0.5; Casamino acids, 0.5; glucose, 0.5; starch, 0.5; sodium pyruvate, 0.3; K_2HPO_4, 0.3; $MgSO_4$, 0.05; and agar, 15.0.

of Reasoner and Geldreich (1985) and incubated at 20°C for 1 week. Colonies were randomly selected and subsequently purified four times on R2A agar at 20°C.

Maintenance procedures

Strains of *Flectobacillus major* are grown on MS or nutrient agar. Incubation is at room temperature (25°C) for several days to allow abundant growth. The cultures will then survive refrigeration at 4°C for at least 3 weeks. They may also be preserved indefinitely by lyophilization.

Flectobacillus lacus can be maintained at –80°C in R2A broth supplemented with 30% (v/v) glycerol.

Procedures for testing special characters

For utilization of single carbon sources, the following medium (g/l) is used: $MgSO_4$, 0.2; $(NH_4)_2HPO_4$, 1.0; KH_2PO_4, 0.5; NaCl, 1.0; carbon source, 2.0; bromthymol blue, 0.08; agar, 20.0. The appearance of growth through four successive subcultures is considered positive even in the absence of a color change in the pH indicator (Larkin et al., 1977).

Differentiation of the genus *Flectobacillus* from other genera

Table 93 lists characteristics that differentiate the genus *Flectobacillus* from other morphologically similar organisms.

TABLE 93. Differential characteristics of the genera *Flectobacillus*, *Runella*, *Spirosoma*, *Cyclobacterium*, *Ancylobacter*, *Larkinella*, *Arcicella*, and *Polaribacter glomeratus*[a,b]

	Flectobacillus major	*Flectobacillus lacus*	*Runella slithyformis*	*Runella zea*	*Spirosoma linguale*	*Cyclobacterium marinum*	*Ancylobacter aquaticus*	*Ancylobacter rudongensis*	*Polaribacter glomeratus*	*Larkinella*	*Arcicella aquatica*
Cell shape:											
Rings formed	+	–	+	–	+	+	Rare	+	v	+	+
Coils or helices formed	Rare	–	+	–	+	+	–	nd	+	–	+
Gas vacuoles present	–	–	–	–	–	–	d	–	–	nd	nd
Growth in media containing seawater or 3% NaCl	–	–	–	–	–	+	–	+	+	–	+
Growth in the presence of 0% NaCl	+	+	+	+	+	–[c]	nd	nd	–	+	–
Pigmentation:											
White to cream	–	–	–	–	–	–	+	+	–	–	–
Yellow or tan	–	–	–	–	+	–	–	–	+	–	–
Pink or salmon	+	+	+	+	–	+	–	–	–	+	+
Urease	+	–	–	nd	–	–	nd	+	–	–	–
Methanol utilized	–	nd	–	–	–	–	+	+	–	nd	–
Acid produced oxidatively from:											
Glucose	+	+	+[d]	+	+[e,f]	+	+	+	–	–	nd
Arabinose	+	–	–	nd	+[e,f]	+	nd	+	–	–	nd
Ribose	–	–	–[g]	–	+[e]	–	+	+	–	–	nd
Glycerol	–	–	–	nd	–	–	+	+	–	–	nd
L-Rhamnose	+	–	nd	nd	nd	nd	nd	nd	nd	nd	nd
Hydrolysis of:											
Starch	+	+	w	–	w	–	–	–	d	–	–
Tributyrin	+	nd	+	nd	+	–	nd	nd	nd	nd	nd
Esculin	+[h]	+	–	nd	d	+	–	+	–	nd	nd
Habitat:											
Freshwater and/or soil	+	+	+	–	+	–	+	–	–	+	+
Marine environments	–	–	–	–	–	+	–	–	+	–	–
Plant stems or roots	–	–	–	+[i]	–	–	–	+[j]	–	–	–
DNA G+C content (mol%)	39–40	38	49–50	49	51–53	34–38	66–69	68	33	53	34

[a]Symbols: +, >85% positive; d, different strains give different reactions (16–84% positive); –, 0–15% positive; w, weak reaction; nd, data not reported or not applicable.

[b]Data taken from Larkin and Borrall (1984a, c); Gosink et al. (1998); Chelius and Triplett (2000), Chelius et al. (2002); Raj and Maloy (1990); Xin et al., 2004; Staley and Konopka (1984); Vancanneyt et al. (2006), and Hwang and Cho (2006).

[c]In contrast to the original description, NaCl is not required for growth (Nedashkovskaya et al., 2005a).

[d]Indicated as positive by Larkin and Borrall (1984a), slowly positive by Raj and Maloy (1990), and negative by Chelius and Triplett (2000).

[e]Sugar media acidification may be slow, taking up to 3 weeks (Larkin and Borrall, 1984a).

[f]Indicated as positive by Larkin and Borrall (1984a) but negative by Vancanneyt et al. (2006).

[g]Indicated as negative by Larkin and Borrall (1984a) and Raj and Maloy (1990) but positive by Chelius and Triplett (2000).

[h]Indicated as positive by Larkin and Borrall (1984a) but negative by McGuire et al. (1987).

[i]Stems of *Zea mays*.

[j]Roots of *Spartina anglica*.

Taxonomic comments

In the 1st edition of *Bergey's Manual of Systematic Bacteriology*, the genus contained two species – *Flectobacillus major* and *Flectobacillus marinus*. The latter species had been described by Raj (1976) as a species of *Microcyclus*, "*Microcyclus marinus*". In many features it resembled *Flectobacillus major*, and after studying both species Borrall and Larkin (1978) assigned it to the genus *Flectobacillus* as *Flectobacillus marinus*. Raj (1979) reported that *Flectobacillus major* and *Flectobacillus marinus* were sufficiently distinct to warrant classifying *Flectobacillus marinus* in a separate genus. The name "*Cyclobacterium marinus*" was suggested (H.D. Raj, cited by Larkin and Borrall, 1984a). Larkin and Borrall (1984a, c) believed that it would be best to retain the organism in the genus *Flectobacillus* because of the similarity in the mol% G+C content of the DNA (38–40) and because the DNA–DNA hybridization value between the type strains of *Flectobacillus marinus* and *Flectobacillus major* was 71%, using the renaturation rate method of DeLey et al. (1970). However, Raj and Maloy (1990) noted that analysis of the oligonucleotide sequences of the 16S rRNAs of these organisms (Woese et al., 1990) indicated that a relatively large evolutionary distance (ca. 20%) occurred between *Flectobacillus major* and *Flectobacillus marinus* (Woese, 1987), supporting the idea that *Flectobacillus marinus* should be separated from *Flectobacillus major* at the genus level. Consequently, Raj and Maloy (1990) reclassified *Flectobacillus marinus* as *Cyclobacterium marinus* (later corrected to *marinum* by Euzéby (1998).

Organisms similar in appearance to *Flectobacillus* have been seen in fresh waters (Larkin et al., 1977). Metcalf and Krueger (cited by Larkin and Borrall, 1984a) located two shallow lakes near Sacramento, California, that had relatively large populations of *Flectobacillus major*. Counts on MS agar ranged from zero to several thousand per ml, and more than 60 isolates were obtained from separate samplings. The isolates were strikingly similar to the type strain of *Flectobacillus major* in their morphological and physiological characteristics, but differed from it in being unable to hydrolyze urea or gelatin.

A new species of *Flectobacillus*, *Flectobacillus glomeratus*, was described by McGuire et al. (1987) from Antarctic marine environments. The cells of the strains were vibrioid to highly coiled; however, Gosink et al. (1998) found *Flectobacillus glomeratus* to be phylogenetically distinct from *Flectobacillus major* by 16S rRNA gene sequence analysis. Although the cells were curved and lacked gas vacuoles, *Flectobacillus glomeratus* did have a close phylogenetic relationship to a group of filamentous or irregularly shaped, gas-vacuolated rods isolated from sea ice and water from the Arctic and Antarctic. Consequently, Gosink et al. (1998) reclassified *Flectobacillus glomeratus* together with these rods in a new genus, *Polaribacter*.

In 2006, Hwang and Cho described a new species, *Flectobacillus lacus*, isolated from a highly eutrophic pond located within the campus of Seoul National University, Korea. From 16S rRNA gene sequence analysis, the species was closely related to *Flectobacillus major* (95.7% sequence similarity).

Differentiation of the species of the genus *Flectobacillus*

Flectobacillus major cells are shaped like the letter C, and sometimes cell ends touch or overlap, forming closed rings. Cells *of Flectobacillus lacus* are almost straight rods and do not form rings. *Flectobacillus major* also differs from *Flectobacillus lacus* by being urease positive, by using succinate as a sole carbon source, and by forming acid from arabinose and rhamnose.

Hwang and Cho (2006) reported that, in their studies, the type strain of *Flectobacillus lacus* differed from the type strain of *Flectobacillus major* in the following ways: acid was produced from D-arabitol and 5-ketogluconate but not from D- or L-arabinose, D-xylose, L-rhamnose, amygdalin, arbutin, D-lyxose, and L-fucose; adipate was assimilated but not L-arabinose, D-mannose, *N*-acetylglucosamine, D-maltose, and gluconate; DL-cysteine and succinate were not used as sole carbon sources; and no growth occurred on media containing as little as 0.5% NaCl.

List of species of the genus *Flectobacillus*

1. **Flectobacillus major** (Gromov 1963) Larkin, Williams and Taylor 1977, 155[AL] (*Microcyclus major* Gromov 1963, 733)

ma'jor. L. masc. comp. adj. *major* larger.

The characteristics are as described for the genus and as listed in Table 93, with the following additional features. Optimum growth occurs at 20–25°C. Growth occurs in the presence of 1.5% NaCl (Hwang and Cho, 2006). Catalase is weakly positive. Esculin, gelatin, starch, tributyrin, and urea are hydrolyzed, but not agar, casein, cellulose, or chitin. β-Galactosidase and phosphatase are present. The following tests are negative: acetoin production (Voges–Proskauer), H_2S production from peptone, indole production (Hwang and Cho, 2006, report the type strain as positive), lecithinase, lysine decarboxylase, nitrate reduction, ornithine decarboxylase, and phenylalanine deaminase. Acid is produced oxidatively from arabinose, cellobiose, dextrin, fructose, galactose, glucose, lactose, maltose, mannose, melibiose, α-methyl-D-glucoside, raffinose, rhamnose, salicin, sucrose, trehalose, and xylose. Acid production from inulin differs among strains. No acid is produced from dulcitol, erythritol, glycerol, mannitol, ribose, sorbitol, and sorbose. Succinate is utilized as a sole carbon source, but not acetate, benzoate, citrate, formate, malonate, propionate, or tartrate.

DNA G+C content (mol%): 39.5–40.3 (T_m).

Type strain: ATCC 29496, DSM 103, LMG 13163, VKM B-859.

Sequence accession no. (16S rRNA gene): M62787.

2. **Flectobacillus lacus** Hwang and Cho 2006, 1200[VP]

la'cus. L. gen. n. *lacus* of a lake or pond, referring to the isolation of the type strain.

The characteristics are as described for the genus and as listed in Table 93, with the following additional features. Cells are approximately 0.3–0.6 × 4.7–10.0 µm. Colonies on R2A agar are pale pink to rose in color, round, 2–6 mm in diameter, convex and smooth with entire margins. Growth occurs at 10–35°C and pH 6–9; optimal temperature and pH for growth are 25–30°C and pH 7, respectively. Growth occurs on R2A and nutrient agar without NaCl or with 0.1% (w/v) NaCl; no growth occurs in the presence of ≥0.5% (w/v)

NaCl. Grows on R2A medium and nutrient medium, but not on MacConkey medium. Flexirubin pigments are absent. Positive for catalase, gelatinase, and amylase. Negative in the Voges–Proskauer test. Positive for α-galactosidase, β-galactosidase, esculin hydrolysis, alkaline phosphatase, acid phosphatase, esterase (C4), esterase lipase (C8), leucine, valine and cystine arylamidases, trypsin, α-chymotrypsin, naphthol-phosphohydrolase, α-glucosidase, β-glucosidase, *N*-acetyl-β-glucosaminidase, α-mannosidase, and α-fucosidase. Negative for indole production, arginine dihydrolase, lipase (C14), and β-glucuronidase. The major fatty acids are $C_{16:1}$ ω5*c* (26.9 ± 10.8%), $C_{15:0}$ iso 2-OH and/or $C_{16:1}$ ω7*c* (19.2 ± 2.3%), and $C_{15:0}$ iso (12.1 ± 1.3%).

Source: a highly eutrophic pond, Gongdae Pond, located in the campus of Seoul National University, Korea.

DNA G+C content (mol%): 38.3 (HPLC).

Type strain: CL-GP79, JCM 13398, KCCM 42271.

Sequence accession no. (16S rRNA gene): DQ112352.

Genus VIII. **Flexibacter** Soriano 1945, 92[AL]

Yasuyoshi Nakagawa

Flex.i.bac'ter. L. part. adj. *flexus* bent, winding; N.L. masc. n. *bacter* from Gr. neut. n. *baktron* little stick or rod; N.L. masc. n. *Flexibacter* intended to mean flexible rod.

Slender rod-shaped cells of variable length, 0.2–0.6 µm wide and typically 10–50 µm long, flexible, and with tapering or rounded ends. **Gliding**, but nonmotile stages may occur. In some species, there is a (cyclic) change in cell morphology. The cell mass is usually pigmented **yellow to orange**, but sometimes vary pale or even colorless. The pigments are cell-bound and often are **carotenoids and/or of the flexirubin type**. **Chemo-organotrophs**. Strictly aerobic or facultatively anaerobic. Peptones and amino acid mixtures serve as nitrogen sources. Various sugars are utilized. Cellulose and agar are not attacked, but chitin and starch are often hydrolyzed.

DNA G+C content (mol%): 37–47.

Type species: **Flexibacter flexilis** Soriano 1945, 92[AL].

Further descriptive information

The major quinone of *Flexibacter canadensis, Flexibacter flexilis, Flexibacter elegans, Flexibacter filiformis, Flexibacter japonensis, Flexibacter litoralis, Flexibacter polymorphus, Flexibacter roseolus, Flexibacter ruber, Flexibacter sancti,* and both type strains of *Flexibacter aggregans* and *Flexibacter tractuosus* is MK-7, whereas MK-6 is the major quinone of *Flexibacter aurantiacus, Flexibacter columnaris* (synonym, *Cytophaga columnaris*), *Flexibacter maritimus* (heterotypic synonym, *Cytophaga marina*), *Flexibacter ovolyticus, Flexibacter psychrophilus* (synonym, *Cytophaga psychrophila*) (Kämpfer et al., 2006; Nakagawa and Yamasato, 1993, 1996); Nakagawa, unpublished). The major polyamine of *Flexibacter flexilis*, the type species of the genus is homospermidine (Hosoya and Hamana, 2004).

Enrichment and isolation procedures

No enrichment media have been designed for isolation of *Flexibacter* strains. Standard procedures to isolate soil and freshwater bacteria can be applied. Low nutrient media are preferable. Colonies of *Flexibacter* are usually orange yellow. Detailed protocols have been described by Reichenbach (1989a, b, 1992a). *Flexibacter* strains have been isolated from various environments at widely separated sites (Lewin, 1969; Lewin and Lounsbery, 1969). The type strains of *Flexibacter flexilis* came from a pond in San José, Costa Rica.

Maintenance procedures

Cultures of most *Flexibacter* strains can be preserved by freezing at temperatures lower than −80°C. For freezing, cells are suspended in the suitable liquid media containing 10% glycerol or 7% DMSO. *Flexibacter* strains are rather sensitive to drying; however, they can be preserved in a protective medium SM1 (see the chapter on *Cytophaga* for formulations) for the terrestrial organisms, or SM2 or SM3 for marine species (see the chapter on *Flammeovirga* for formulations) using a liquid drying method. The organisms can also be preserved by freeze-drying. One marine species, *Flexibacter polymorphus*, is very fastidious and is difficult to preserve by freezing −80°C or drying. Its cultures autolyze within a few days at 25°C; however, it can be successfully preserved by freezing at temperatures lower than −140°C in liquid nitrogen (LN_2) tanks.

Differentiation of the genus *Flexibacter* from other genera

Phylogenetic analysis based on 16S rRNA sequences shows that *Flexibacter flexilis* - the type species of the genus *Flexibacter* - is phylogenetically independent in the phylum *Bacteroidetes* (Figure 74). However, because few taxonomic characteristics have been investigated since the original descriptions, it is impossible to discriminate the genus *Flexibacter* from other related genera only by phenotypic characteristics. It is hoped that chemotaxonomic, physiologic, and biochemical characteristics of the genus *Flexibacter* will be investigated.

Taxonomic comments

The genus *Flexibacter* was created by Soriano (1945) to accommodate five species of gliding organisms; however, the original definition of the genus was vague and his original strains have been lost. In addition, the genus *Microscilla* proposed by Pringsheim (1951) could not be distinguished from the genus *Flexibacter*. Lewin (1969) reisolated morphologically similar organisms and redefined both genera by classifying marine organisms with longer filaments (20–100 µm) in the genus *Microscilla*. These definitions were not included in the *Bergey's Manual of Determinative Bacteriology*, 8th edition (Leadbetter, 1974). The inability to degrade polysaccharides such as agar, alginate, cellulose, and chitin was adopted in the genus definition of *Flexibacter*, and the genus *Microscilla* was grouped with the genus *Flexibacter*. However, *Flexibacter flexilis* is known to hydrolyze other polymers such as starch, and *Flexibacter. filiformis* is a potent chitin decomposer. Later, Reichenbach (1989a, b) tried to define the genus *Flexibacter* on the basis of cell morphology, DNA G+C content, and habitat. He classified marine organisms with G+C contents above 37 mol% in the genus *Microscilla*. However, he also stated that the genus *Flexibacter* is probably still heterogeneous and will

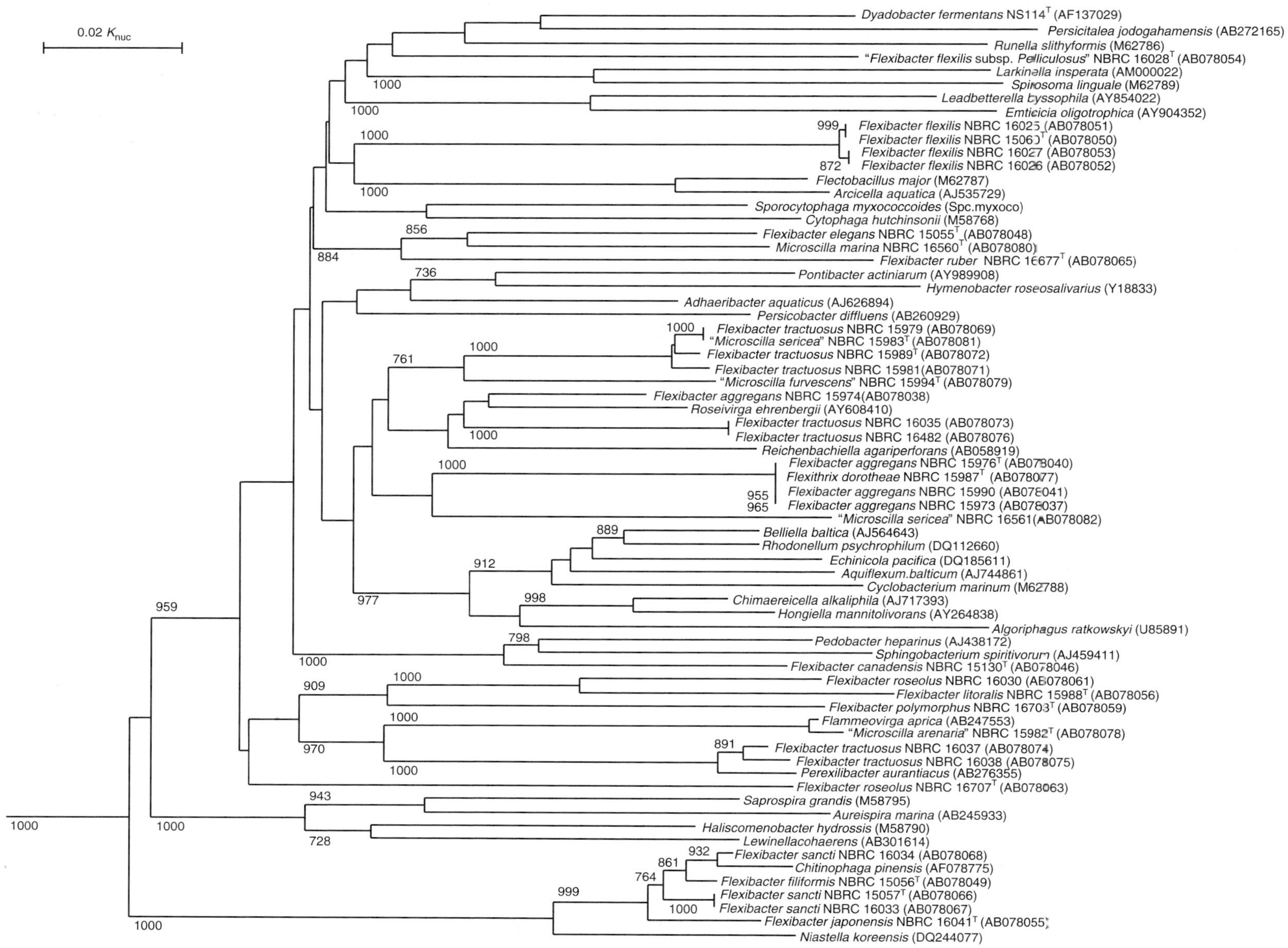

FIGURE 74. Phylogenetic tree based on 16S rRNA gene sequences of *Flexibacter* and *Microscilla* strains in the class *Sphingobacteriia*. Strains belonging to the family *Flavobacteriaceae* and *Flexibacter aggregans* subsp. *catalaticus* are not included. Their phylogenetic positions are described in the text. T indicates the type strain of the species. Bar = 0.02 K_{nuc}. Numbers at some nodes are the confidence limits estimated by 1000 bootstrap resamplings. Low values (<700) are not shown.

have to be subdivided when data on the molecular taxonomy are accumulated. Those histories are described in detail in the first edition of this *Manual* (Reichenbach, 1989a, b).

Seventeen *Flexibacter* species have been approved or validly published to date. Three species, *Flexibacter flexilis* (the type species of the genus), *Flexibacter elegans*, and *Flexibacter filiformis* were recognized as belonging to *Flexibacter* in the 1st edition of this *Manual*.

16S rRNA sequencing analysis of the genera *Flexibacter* and *Microscilla* have shown great biological diversity within the genus *Flexibacter* (Nakagawa et al., 2002). Forty strains of 17 species of *Flexibacter* diverged into 20 phylogenetic groups. Four groups were located in the family *Flavobacteriaceae*, while others – including the type species *Flexibacter flexilis* – were not. Some *Flexibacter* strains were closely related to strains of the genus *Chitinophaga*, *Microscilla*, or *Flexithrix*. A phylogenetic tree of *Flexibacter*, *Flexithrix*, and *Microscilla* strains belonging to the classes *Cytophagia* and *Sphingobacteriia* is shown in Figure 74. It can be seen that *Flexibacter flexilis*, the type species of the genus *Flexibacter*, is isolated from the other species, which suggests that the genus *Flexibacter* should be restricted to the type species. All strains in this species clustered together except for one non-validated subspecies, "*Flexibacter flexilis* subsp. *pelliculosus*", which was unrelated to the other known taxa.

Flexibacter aggregans was assigned the name "*Microscilla aggregans*" in the 1st edition of this *Manual* (Reichenbach, 1989a, b) because it is a marine organism. This species is the phylogenetically heterogeneous, diverging into four groups. The three authentic strains of *Flexibacter aggregans*, NBRC 15973, NBRC 15976^T, and NBRC 15990, constitute a tight cluster with *Flexithrix dorotheae*. This result strongly supports Reichenbach's report (1989a) of a close resemblance between the two species. Based on these results, *Flexibacter aggregans* was included in *Flexithrix dorotheae* (Hosoya and Yokota, 2007). Two strains of *Flexibacter aggregans* NBRC 15974 and NBRC 15975 remain to be reclassified. *Flexibacter aggregans* NBRC 15974 constitutes an independent lineage near *Roseivirga ehrenbergii*. The strain NBRC 15975 belongs to the family *Flavobacteriaceae*. An invalid subspecies, "*Flexibacter aggregans* subsp. *catalaticus*" was transferred to the new genus *Crocinitomix*, which belongs to the new family *Cryomorphaceae* (Bowman et al., 2003).

Flexibacter canadensis is located outside the cluster comprised of the genera *Sphingobacterium* and *Pedobacter*. Steyn et al. (1998) proposed the family *Sphingobacteriaceae* to encompass the genera *Sphingobacterium*, *Pedobacter*, and *Flexibacter canadensis*. This family is characterized by the presence of sphingolipids, but a long-chain base derived from sphingolipids was not detected in this species (Nakagawa et al., 2002). It is suggested that *Flexibacter canadensis* should be excluded from the family as a distinct genus.

Flexibacter filiformis, *Flexibacter japonensis*, and *Flexibacter sancti* are closely related to the genus *Chitinophaga* and they were subsequently transferred to the genus *Chitinophaga* (Kämpfer et al., 2006). *Flexibacter elegans*, *Flexibacter litoralis*, *Flexibacter polymorphus*, and *Flexibacter ruber* are distinct from other known species of *Flexibacter*. Thus, it is suggested that each should be excluded from the genus *Flexibacter*.

Flexibacter roseolus is a heterogeneous species: it includes two phylogenetically different strains (NBRC 16707^T and NBRC 16030), each of them is distinct from known taxa. *Flexibacter tractuosus* is also called "*Microscilla tractuosa*" because it is a marine bacterium. Eight strains assigned to the species were heterogeneous and diverged into four distinct groups. The type strain of *Flexibacter tractuosus*, NBRC 15989^T, clusters with NBRC 15979, NBRC 15981, and "*Microscilla sericea*" NBRC 15983^T. This group is considered to be the genuine *Flexibacter tractuosus* because it includes the type strain. Two strains of *Flexibacter tractuosus* NBRC 16035 and NBRC 16482 constitute one group and are distantly related to all known species. Two other strains of *Flexibacter tractuosus*, NBRC 16037 and NBRC 16038, may be reclassified in the genus *Perexilibacter* because of their close relationships to the type species of that genus. *Flexibacter tractuosus* NBRC 15980 belongs to the family *Flavobacteriaceae*.

The remaining strains belong to the family *Flavobacteriaceae*. *Flexibacter aurantiacus*, *Flexibacter columnaris*, and *Flexibacter psychrophilus* were reclassified as distinct species in the genus *Flavobacterium*, and *Flexibacter aurantiacus* is treated as the junior subjective synonym of *Flavobacterium johnsoniae* (Bernardet et al., 1996). Two non-validated subspecies exist in *Flexibacter aurantiacus*. Of these, "*Flexibacter aurantiacus* subsp. *excathedrus*" belongs to the genus *Flavobacterium*. The other subspecies, "*Flexibacter aurantiacus* subsp. *copepodarum*", is a marine organism isolated from an offshore crustacean (Lewin, 1969) and is classified in the genus *Tenacibaculum*. *Flexibacter maritimus* and *Flexibacter ovolyticus* have been transferred to the genus *Tenacibaculum* (Suzuki et al., 2001). These phylogenetic groups and their reclassification are summarized in Table 94.

In this chapter, Reichenbach's definition of the genus is generally followed; however, only the type species, *Flexibacter flexilis* can be justified on the basis of 16S rRNA sequencing analysis as belonging to the genus *Flexibacter*. Other *Flexibacter* species that should be reclassified in the future are listed as other species.

Differentiation of the species of the genus *Flexibacter*

Table 95 lists characteristics that distinguish the *Flexibacter* species from one another.

List of species of the genus *Flexibacter*

1. **Flexibacter flexilis** Soriano 1945, 92^{AL}

flex'i.lis. L. masc. adj. *flexilis* pliable, flexible.

The characteristics are as described for the genus with the following additional features taken from the 1st edition of this *Manual* and *Bergey's Manual of Determinative Bacteriology*, 9th edition.

Long flexible thread cells with tapering, occasionally crooked ends, 0.5 μm wide and 10–60 μm long (usually 10–20 μm). Motile by gliding but not particularly active. Color of the cell mass is orange, but pigmentation on many media is pale. Aerobic. The optimum growth temperature is 25°C. The optimum pH is around 7. Oxidase-positive. Catalase-negative. Peptones and Casamino acids serve as sole nitrogen source, but not glutamate, NH_4^+ or NO_3^-. Growth occurs well on peptone alone. Sugars such as glucose or sucrose may stimulate growth. Gelatin and starch are hydrolyzed, but not chitin or yeast cells. H_2S is produced. Nitrate is not reduced. Growth

TABLE 94. Phylogenetic groups of *Flexibacter* and *Microscilla* strains and their present classification

Phylogenetic group[a]	Species and strain	Present classification
1	*F. aggregans* NBRC 15975	Excluded from the genus *Flexibacter*, belongs to the family *Flavobacteriaceae*
2	*F. tractuosus* NBRC 15980	Excluded from the genus *Flexibacter*, belongs to the family *Flavobacteriaceae*
3	"*F. aurantiacus* subsp. *copepodarum*" NBRC 15978[T]	Excluded from the genus *Flexibacter*, belongs to the genus *Tenacibaculum*
	F. maritimus NBRC 15946[T]	*Tenacibaculum maritimum*[b]
	F. ovolyticus NBRC 15947[T]	*Tenacibaculum ovolyticum*[b]
4	*F. aurantiacus* NBRC 15970[T]	*Flavobacterium johnsoniae*[c]
	"*F. aurantiacus* subsp. *excathedrus*" NBRC 16024[T]	Excluded from the genus *Flexibacter*, belongs to the genus *Flavobacterium*
	F. columnaris NBRC 15943[T]	*Flavobacterium columnare*[c]
	F. psychrophilus NBRC 15942[T]	*Flavobacterium psychrophilum*[c]
5	"*F. aggregans* subsp. *catalaticus*" NBRC 15977[T]	*Crocinitomix catalasitica*[d]
6	*F. filiformis* NBRC 15056[T]	*Chitinophaga filiformis*[e]
	F. japonensis NBRC 16041[T]	*Chitinophaga japonensis*[e]
	F. sancti NBRC 15057[T], NBRC 16033, NBRC 16034	*Chitinophaga sancti*[e]
7	*F. canadensis* NBRC 15130[T]	Excluded from the genus *Flexibacter*, belongs to the family *Sphingobacteriaceae*
8	*F. elegans* NBRC 15055[T]	Excluded from the genus *Flexibacter*
9	*M. marina* NBRC 16560[T]	*Microscilla marina*
10	*F. ruber* NBRC 16677[T]	Excluded from the genus *Flexibacter*
11	*F. flexilis* NBRC 15060[T], NBRC 16025, NBRC 16026, NBRC 16027	*Flexibacter flexilis*
12	"*F. flexilis* subsp. *pelliculosus*" NBRC 16028[T]	Excluded from the genus *Flexibacter*
13	*F. litoralis* NBRC 15988[T]	Excluded from the genus *Flexibacter*
14	*F. roseolus* NBRC 16030	Excluded from the genus *Flexibacter*
15	*F. polymorphus* NBRC 16703[T]	Excluded from the genus *Flexibacter*
16	*F. roseolus* NBRC 16707[T]	Excluded from the genus *Flexibacter*
17	*F. tractuosus* NBRC 15989[T], NBRC 15979, NBRC 15981	Excluded from the genus *Flexibacter*
	"*M. sericea*" NBRC 15983[T]	Excluded from the genus *Microscilla*
18	"*M. furvescens*" NBRC 15994[T]	Excluded from the genus *Microscilla*
19	*F. tractuosus* NBRC 16035, NBRC 16482	Excluded from the genus *Flexibacter*
20	*F. aggregans* NBRC 15974	Excluded from the genus *Flexibacter*
21	"*M. sericea*" NBRC 16561	Excluded from the genus *Microscilla*
22	*F. aggregans* NBRC 15976[T], NBRC 15973, NBRC 15990	*Flexithrix dorotheae*[f]
23	"*M. arenaria*" NBRC 15982[T]	*Flammeovirga arenaria*[g]
24	*F. tractuosus* NBRC 16037, NBRC 16038	Excluded from the genus *Flexibacter*, probably belongs to the genus *Perexilibacter*

[a]Nakagawa et al. (2002).
[b]Suzuki et al. (2001).
[c]Bernardet et al. (1996).
[d]Bowman et al. (2003).
[e]Kämpfer et al. (2006).
[f]Hosoya and Yokota (2007).
[g]Takahashi et al. (2006).

does not occur on seawater media. The major carotenoid is saproxanthin. The flexirubin reaction is positive.

DNA G+C content (mol%): 40–43 (Bd).

Type strain: CR-63, ATCC 23079, CIP 103988, DSM 6793, LMG 3989, NBRC 15060.

Sequence accession no. (16S rRNA gene): AB078049.

Other species

None of the species listed below should be included in the genus *Flexibacter* because 16S rRNA sequence analysis indicates they are not closely related to the genus.

1. **Flexibacter canadensis** Christensen 1980, 431[AL]

ca.na.den′sis. N.L. masc. adj. *canadensis* Canadian.

The characteristics are as described for the genus with the following additional features taken from the 1st edition of this *Manual*, *Bergey's Manual of Determinative Bacteriology*, 9th edition, and the original description (Christensen, 1980).

Elongated rods, thin and flexible, with slightly tapering ends, 0.4–0.5 µm wide and 2–12 µm long or longer (up to 60 µm in older cultures). Motile by gliding. On solid substrates with a low nutrient content, large whitish swarms develop. On media with a higher concentration of organic compounds, growth becomes compact, pale pink, or dirty

TABLE 95. Characteristics differentiating the species of the genus *Flexibacter*[a,b]

Characteristic	*F. flexilis*	*F. canadensis*	*F. elegans*	*F. litoralis*	*F. polymorphus*	*F. roseolus*	*F. ruber*	*F. tractuosus*
Length of threads (μm)	10–60	2–60	10 to >50	>180	>200	>50	>50	5 to >50
Width of threads (μm)	0.5	0.4–0.5	0.4–0.5	0.5–0.7	0.6–1.1	nd	nd	0.5
Color of cell mass	Orange	White	Bright orange	Brick red	Orange-peach	Red	Red	Orange
Flexirubin reaction	+	nd	−	nd	nd	nd	nd	nd
Carotenoid present	Saproxanthin	nd	Saproxanthin	Flexixanthin	Saproxanthin	+	+	Saproxanthin
Relation to oxygen	Aerobe	Facultative anaerobe	Aerobe	Aerobe	Aerobe	Aerobe	Aerobe	nd
Suitable as sole nitrogen source:								
Glutamate	−	+	−	−	+	−	+	+
NH_4^+	−	+	−	nd	−	nd	nd	nd
NO_3^-	−	−	−	−	−	−	+	−
Starch hydrolysis	+	+	−	+	−	−	−	+/−
H_2S produced	+	+	−	−	−	−	−	nd
NO_3^- reduced	−	+	−	−	nd	−	+	+/−
Catalase	−	+	−	−	−	−	−	nd
Oxidase	+	+	+	nd	nd	nd	nd	nd
Growth on sea water medium	−	nd	+	+	+	+	−	+
Optimum temperature (°C)	25	18–30	30	nd	nd	nd	nd	nd
Highest temperature (°C)	40–45	40	40–45	30–35	32	40	40–45	30–45
Habitat	Fresh water	Soil	Fresh water	Marine	Marine	Hot springs	Hot springs	Marine
DNA G+C content (mol%)	40–43	37	48	31	29	34–38	37	35–40

[a]Symbols: +, >85% positive; d, different strains give different reactions (16–84% positive); −, 0–15% positive; +/−, most strains are positive; nd, not determined.

[b]Data from Reichenbach (1989a, b) and *Bergey's Manual of Determinative Bacteriology*, 9th edition.

white. Colonies often show a yellow-green-blue iridescence. Facultatively anaerobic. Oxidase- and catalase-positive. Casamino acids, gelatin, aspartate, glutamate, and NH_4^+ serve as sole nitrogen source but not NO_3^- or urea. Various sugars, glucose, and glycerol are utilized with acid production. Indole is not produced. H_2S is produced. DNA, gelatin, and starch are hydrolyzed, but not agar, chitin, alginate, cellulose, or carboxymethyl cellulose. Pectin is weakly degraded. Nitrate is reduced. H_2S is produced. Growth does not occur in the presence of 2% NaCl. The temperature range for growth is 10–40°C with optimum 18–30°C. The pH range for growth is 5–10 with an optimum of 6–8.

DNA G+C content (mol%): 37 (Bd and T_m).

Type strain: UASM 9D, ATCC 29591, CIP 104802, DSM 3403, JCM 21819, LMG 8368, NBRC 15130.

Sequence accession no. (16S rRNA gene): AB078046.

2. **Flexibacter elegans** (*ex* Lewin 1969, non Soriano 1945) Reichenbach 1989b, 2067[AL]

e'le.gans. L. masc. adj. *elegans* refined, fashionable, elegant.

The characteristics are as described for the genus with the following additional features taken from the 1st edition of this *Manual* and *Bergey's Manual of Determinative Bacteriology*, 9th edition.

Very long, fine filaments with rounded ends, 0.4–0.5 μm wide and 50 μm long, often much longer, rarely shorter than 10–20 μm. On solid substrates the filaments tend to form loops and coils. Motile by gliding. On some media the colonies are spreading. The cell mass is bright orange. Aerobic. Oxidase-positive. Catalase-negative. Peptones and Casamino acids serve as sole nitrogen source but not glutamate, NH_4^+, or NO_3. Requires threonine. Growth on media containing peptone alone is very poor or absent but may be stimulated by the addition of a sugar such as glucose. In litmus milk, there is no acid production, coagulation, or reduction of litmus. Acid is produced from glucose. Gelatin is hydrolyzed, but not chitin, starch, or yeast cells. Indole and H_2S are not produced. Nitrate is not reduced. Growth occurs on seawater media, but NaCl is not required. NaCl (2.4%) is tolerated. Optimum conditions for growth are around pH 7 and 30°C. The highest growth temperature is 40–45°C. The major carotenoid is saproxanthin. The flexirubin reaction is negative.

DNA G+C content (mol%): 48 (Bd).

Type strain: NZ-1, ATCC 23112, CIP 104801, DSM 3317, JCM 21159, LMG 10750, NBRC 15055.

Sequence accession no. (16S rRNA gene): AB078048.

3. **Flexibacter litoralis** Lewin 1969, 199[AL]

li.tora'lis. L. masc. adj. *litoralis* of or belonging to the seashore.

The characteristics are as described for the genus with the following additional features taken from the 1st edition of this *Manual* and *Bergey's Manual of Determinative Bacteriology*, 9th edition.

Threads, 0.5–0.7 μm wide and up to 180 μm long, agile, gliding, and bending, apparently without cross-walls. The cell mass is brick red. Aerobic. Catalase-negative. Peptones serve as sole nitrogen source but not glutamate, or NO_3. Requires many amino acids and thiamine. Sugars and organic acid is not utilized. Gelatin and starch are hydrolyzed, but not agar, alginate, or carboxymethyl cellulose. H_2S is not produced. Nitrate is not reduced. Growth does not occur below half-strength seawater. The highest growth temperature is 30–35°C. The major carotenoid is flexixanthin.

DNA G+C content (mol%): 31 (Bd).

Type strain: SIO-4, ATCC 23117, CIP 106402, DSM 6794, NBRC 15988.

Sequence accession no. (16S rRNA gene): AB078056.

4. **Flexibacter polymorphus** Lewin 1974, 393[AL]

po.ly.mor′phus. N.L. masc. adj. *polymorphus* (from Gr. adj. *polumorphos -on*) of many shapes, variable in form.

The characteristics are as described for the genus with the following additional features taken from the 1st edition of this *Manual* and *Bergey's Manual of Determinative Bacteriology*, 9th edition.

Long flexible filaments, 1.1 μm wide and up to several hundred μm long, with cross-walls 3.5–7 μm apart. The cross-walls are recognizable under the phase-contrast microscope. In media containing 0.1% $NaHCO_3$, finer and shorter filaments 0.6 μm wide and 10–40 μm long appear, in addition to the long ones. At a pH above 8 the cells contain optically refractile granules, presumably some lipid material. Sometimes, inflated and branched filaments also occur. The filaments are very actively gliding with speeds up to 12 μm/s (23°C). The cell mass is orange. Aerobic. Catalase-negative. Peptones, Casamino acids, and glutamate serve as sole nitrogen source but not NH_4^+ or NO_3. Requires cobalamine. Agar, alginate, cellulose, and starch are not hydrolyzed. H_2S is not produced. Growth does not occur below half-strength seawater. The highest growth temperature is 32°C. The pH range for growth is 7–8.5. The major carotenoid is saproxanthin.

DNA G+C content (mol%): 29 (Bd).

Type strain: ATCC 27820, DSM 9678, LMG 13859, NBRC 16703.

Sequence accession no. (16S rRNA gene): AB078059.

5. **Flexibacter roseolus** Lewin 1969, 199[AL]

ro.se′o.lus. L. adj. *roseus* rose-colored; L. masc. suff. *-olus* diminutive ending; N.L. masc. dim. adj. *roseolus* intended to mean with a rosy tinge.

The characteristics are as described for the genus with the following additional features taken from the 1st edition of this *Manual* and *Bergey's Manual of Determinative Bacteriology*, 9th edition.

Very long threads, more than 50 μm long. The cell mass is red. Aerobic. Catalase-negative. Peptones and Casamino acids serve as sole nitrogen source but not glutamate or NO_3^-. Gelatin is hydrolyzed but not starch. H_2S is not produced. Seawater is tolerated. The highest growth temperature is 40°C. The major carotenoid is flexixanthin, demonstrated in a strain CR-141 (=ATCC 23087, NBRC 16030).

DNA G+C content (mol%): 34–38 (Bd).

Type strain: CR-155, ATCC 23088, CIP 106406, DSM 9546, LMG 13856, NBRC 16707.

Sequence accession no. (16S rRNA gene): AB078063.

6. **Flexibacter ruber** Lewin 1969, 199[AL]

ru′ber. L. masc. adj. *ruber* red.

The characteristics are as described for the genus with the following additional features taken from the 1st edition of this *Manual* and *Bergey's Manual of Determinative Bacteriology*, 9th edition.

Very long threads, more than 50 μm long. The cell mass is red. Aerobic. Catalase-negative. Peptones, Casamino acids, glutamate and NO_3^- serve as sole nitrogen source. Various sugars are metabolized with acid production. Gelatin is hydrolyzed. H_2S is not produced. Nitrate is reduced. Growth does not occur on seawater media. The highest growth temperature is 40–45°C. The major carotenoid is flexixanthin.

DNA G+C content (mol%): 37 (Bd).

Type strain: GEY, ATCC 23103, DSM 9560, LMG 13857, NBRC 16677.

Sequence accession no. (16S rRNA gene): AB078064.

7. **Flexibacter tractuosus** Leadbetter 1974, 106[AL] ("*Microscilla tractuosa* Lewin 1969, 199)

trac.tu.o′sus. L. masc. adj. *tractuosus* that draws to itself, drawn or clumped together.

The characteristics are as described for the genus with the following additional features taken from the 1st edition of this *Manual* and *Bergey's Manual of Determinative Bacteriology*, 9th edition.

Threads, 0.5 μm wide and 5–50 μm long, or longer. The cell mass is orange. Peptones, Casamino acids, and glutamate serve as sole nitrogen source but not NO_3^-. Sugars and glycerol are usually utilized. Agar, alginate, and carboxymethyl cellulose are not degraded. Nitrate is reduced by a few strains. The highest growth temperature is 30–45°C. The major carotenoid is saproxanthin.

DNA G+C content (mol%): 35–40 (Bd).

Type strain: H-43, ATCC 23168, CIP 106410, DSM 4126, LMG 8378, NBRC 15989, VKM B-1430.

Sequence accession no. (16S rRNA gene): AB078072.

Genus IX. **Hymenobacter** Hirsch, Ludwig, Hethke, Sittig, Hoffmann and Gallikowski 1999, 1[VP] emend. Buczolits, Denner, Kämpfer and Busse 2006, 2076[VP] (Effective publication: Hirsch, Ludwig, Hethke, Sittig, Hoffmann and Gallikowski 1998a, 374.)

Sandra Buczolits and Hans-Jürgen Busse

Hy.me.no.bac′ter. Gr. masc. n. *hymen* pellicle, thin layer; N.L. masc. n. *bacter* the equivalent of Gr. neut. n. *baktron* a rod or staff; N.L. masc. n. *Hymenobacter* a rod growing in thin layers.

Rod-shaped, with polyphosphate granules near the cell poles. Cells aggregating with increasing formation of extracellular polymer and spreading in thin, red to pink layers on agar surfaces. Nonmotile. Gram-stain-negative. Colonies are flat with a small, raised center. Colony edges may show parallel arrangement of cells in the form of palisades. Aerobic. Heterotrophic, with a **preference for oligotrophic media**. Temperature range for growth, 4–37°C; optimum, 10–28°C. Some strains can grow at 42°C, but most strains do not. Cells do not grow anaerobically with or without light. Catalase- and oxidase-positive. Nonhemolytic. The carbon utilization spectrum is limited to some sugars, sugar alcohols, organic acids, and a few amino acids. Hydrolysis of gelatin, starch, xylan, Tween 80, and Tween 60 is common. DNA may also be hydrolyzed, but cellulose and pectin are not. Cells are highly sensitive towards the action of numerous antibiotics. ***meso*-Diaminopimelic acid is present in the cell-wall murein. The principal menaquinone is MK-7. The fatty acid profile consists of predominantly branched fatty acids of the iso- and anteiso-type with $C_{15:0}$ iso, $C_{15:0}$ anteiso, $C_{16:1}$ ω7*c*/$C_{15:0}$ iso 2-OH (summed feature), and $C_{17:1}$ iso I/$C_{17:1}$ anteiso B (summed feature), usually present in moderate to major amounts** (except for *Hymenobacter roseosalivarius*, which lacks $C_{15:0}$ anteiso). Unbranched fatty acids are usually present in moderate amounts, but their content is low in *Hymenobacter ocellatus*. All members of the genus contain a **polar lipid profile** consisting of **phosphatidylethanolamine, an unknown aminophospholipid (APL3), two unknown polar lipids (L3, L5)**, and a mixture of several other unknown aminophospholipids, aminolipids, phospholipids, glycolipids, and polar lipids. ***sym*-Homospermidine** is the major polyamine.

DNA G+C content (mol%): 55–65.

Type species: **Hymenobacter roseosalivarius** Hirsch, Ludwig, Hethke, Sittig, Hoffmann and Gallikowski 1999, 1[VP] (Effective publication: Hirsch, Ludwig, Hethke, Sittig, Hoffmann and Gallikowski 1998a, 382.).

Further descriptive information

The genus *Hymenobacter* was described with the single species *Hymenobacter roseosalivarius* that had been isolated from the Dry Valleys region in Antarctica (Hirsch et al., 1998a). In this study, affiliation of three strains - provisionally named "*Taxeobacter chitinovorans*", "*Taxeobacter gelupurpurascens*", and "*Taxeobacter ocellatus*" (Reichenbach, 1992b) — to this lineage was demonstrated by rRNA gene sequence similarity and phylogeny. So far, eight other species with validly published names have been effectively described as species of the genus: *Hymenobacter actinosclerus* isolated from irradiated pork (Collins et al., 2000); *Hymenobacter aerophilus* isolated during the examination of airborne bacteria in samples from the Museo Correr in Venice, Italy (Buczolits et al., 2002); *Hymenobacter norwichensis* isolated during the examination of airborne bacteria in samples from the Sainsbury Centre for Visual Arts in Norwich (UK; Buczolits et al., 2006); *Hymenobacter chitinivorans* (formerly designated "*Taxeobacter chitinovorans*"); *Hymenobacter gelipurpurascens* (formerly designated "*Taxeobacter gelupurpurascens*"); *Hymenobacter ocellatus* (formerly designated "*Taxeobacter ocellatus*") isolated from dried soil which had been stored for several years (Baik et al., 2006; Reichenbach, 1992b); *Hymenobacter rigui* isolated from wetland freshwater in Woopo, South Korea (Baik et al., 2006); *Hymenobacter xinjiangensis* isolated from soil of the Xinjing desert, China following gamma-irradiation (Zhang et al., 2007). Sources of isolation suggest that at least some *Hymenobacter* species have developed strategies to survive under unfavorable conditions such as desiccation, radiation, and cold.

Strains analyzed for their quinone systems and polyamine patterns contain menaquinone MK-7 and the predominant compound *sym*-homospermidine, respectively. The contents of certain fatty acids such as $C_{15:0}$ iso, $C_{15:0}$ anteiso, $C_{16:1}$ ω5*c*, $C_{16:1}$ ω7*c*/$C_{15:0}$ iso 2-OH (summed feature), and $C_{17:1}$ iso I/$C_{17:1}$ anteiso B (summed feature), which were detected in the range 8–37, 0–26, 2–23, 5–30, and 8–27%, respectively, suggest a high potential for differentiation of *Hymenobacter* species and identification of newly isolated strains. Even polar lipid profiles exhibit a diversity that appears to be suitable for differentiation between *Hymenobacter* species (Baik et al., 2006). Physiological and biochemical traits useful for discrimination between *Hymenobacter* species are listed in Table 96.

Nothing is known about pathogenic potential of *Hymenobacter* species.

Differentiation of the genus *Hymenobacter* from other genera

Hymenobacter species do not exhibit extraordinary phenotypic traits which allow their differentiation from closely related genera such as *Adhaeribacter*, *Pontibacter*, and *Effluviibacter*, but differentiation may be accomplished by 16S rRNA gene sequence similarities below 88.0%.

Certain quantitative differences in the fatty acid profiles might be useful for phenotypic differentiation of *Hymenobacter* species from related genera. The content of $C_{15:0}$ iso 2-OH (16.5%) in the fatty acid profile of *Adhaeribacter aquaticus* (Rickard et al., 2005) is significantly higher than in *Hymenobacter* species (Table 97) and contents of $C_{17:1}$ ω6*c* and lack of "summed feature" ($C_{15:0}$ iso 2-OH/$C_{16:1}$ ω7*c*) in the profiles of *Adhaeribacter aquaticus* and *Effluviibacter roseus* (Rickard et al., 2005; Suresh et al., 2006) might be useful for differentiation from *Hymenobacter* species. The high G+C content may distinguish *Hymenobacter* (>55 mol%) species from *Pontibacter* and *Adaeribacter* species (Nedashkovskaya et al., 2005c; Rickard et al., 2005; Zhou et al., 2007). However, the listed distinguishing traits cannot be considered to be highly reliable because each of the nearest related genera so far is represented by only one or two species, and hence the variability of the characteristics within these genera is unknown.

Taxonomic comments

Sequence similarities in the 16S rRNA genes among established species of the genus *Hymenobacter* are above 90%. This value is rather low compared to many other genera where the threshold value for genus affiliation is approximately 95% similarity. This significantly lower threshold value for delineation of the genus suggests a higher mutation rate in the 16S rRNA coding genes or the need for dissection of the genus possibly on the basis of differences in the polar lipid and fatty acid profiles and other distinguishing traits as yet unidentified. However, if the threshold value of 90% for assignment to the genus *Hymenobacter* is considered to be acceptable, deposited 16S rRNA gene sequences indicate presence in institutional culture collections of several strains that might be described as novel species of the genus, as indicated under *Other organisms*.

Phylogenetic relatedness shown in the maximum-likelihood tree after multiple alignments (Figure 75) are in good agreement with that suggested from 16S rRNA gene sequence similarities, demonstrating that *Hymenobacter ocellatus* represents the deepest branching within the genus. 16S rRNA gene based phylogenetic relatedness support placement of all unnamed strains

TABLE 96. Differential characteristics among species of the genus *Hymenobacter*[a]

Characteristic	*H. roseosalivarius*	*H. norwichensis*	*H. chitinivorans*	*H. gelipurpurascens*	*H. ocellatus*	*H. actinosclerus*	*H. aerophilus*	*H. rigui*	*H. xinjiangensis*
Growth at/in:									
1% NaCl	+[b]	−	+	+	+	nr	+	+	−
3% NaCl	nr	−	+	+	+	nr	+	−	nr
Starch	nr	+	+	+	+	+	+	+	−
Casein	nr	+	+	+	+	nr	d	nr	nr
Tyrosine	nr	+	+	nr	+	nr	+	nr	nr
4°C	+[b]	+	+	+	−	−[c]	+	+	+
37°C	−	−	−	−	+	+[d]	−	+	+
Assimilation of:									
N-Acetyl-D-glucosamine, gluconate	−	−	−	−	−	+	−	−	nr
p-Arbutin, D-ribose	−	+	−	−	−	+	−	−	nr
D-Cellobiose	d	+	−	−	−	+	−	+	nr
D-Fructose	−	+	−	−	−	−	+	+	nr
D-Galactose	−	+	−	−	−	−	−	+	nr
D-Xylose	−	+	−	−	−	−	−	−	nr
D-Glucose	−	+	−	−	−	+	+	+	nr
D-Mannose	−	+	−	−	−	+	+	−	nr
L-Rhamnose	−	−	−	−	−	−	−	−	nr
Maltitol	−	−	−	−	−	−	−	nr	nr
Sucrose	−	−	−	−	−	−	+	+	nr
Salicin	−	+	−	−	−	−	−	+	nr
D-Maltose, D-trehalose	−	+	−	−	−	−	−	+	nr
D-Mannitol	−	+	−	−	−	−	−	−	nr
Acetate, propionate	+	−	−	−	−	+	+	nr	nr
Glutarate, pyruvate	+	−	−	−	−	+	−	nr	nr
cis-Aconitate, adipate, fumarate, DL-3-hydroxybutyrate, L-malate, itaconate, DL-lactate, mesaconate, oxoglutarate, suberate, L-alanine, L-proline, L-phenylalanine	−	−	−	−	−	+	−	nr	nr
Citrate	−	−	−	−	−	+	−	−	nr
Hydrolysis of:									
pNP β-D-Glucuronide	−	−	−	−	+	−	−	nr	nr
pNP Phosphate, L-proline, *p*-nitroanilide (pNA)	nr	nr	+	+	−	+	+	+	+
pNP α-D-Glucopyranoside	+	+	−	+	+	+	+	−	nr
pNP β-Glucopyranoside	−	+	−	−	+	−	−	nr	nr
bis-pNP Phosphate	+	+	−	−	+	+	+	nr	nr
pNP Phenyl-phosphonate	−	+	−	−	−	+	+	nr	nr
L-Alanine pNA	+	+	+	+	+	+	+	nr	nr
L-Glutamate-γ-3-carboxy-pNA	−	−	−	−	−	+	+	nr	nr

[a]Symbols: +, >85% positive; d, different strains give different reactions (16–84% positive); −, 0–15% positive; nr, not reported; PNP, *para*-nitrophenyl; pNA, *p*-itroanilide.

[b]Hirsch et al. (1998a).

[c]Collins et al. (2000) reported no growth at 5°C.

[d]Collins et al. (2000) reported growth at 42°C.

within the genus *Hymenobacter*, but strain *Taxeobacter* sp. SAFR-033 appears to be a representative of a novel genus.

Maintenance procedures

For long-term storage, *Hymenobacter* cultures may be lyophilized by common procedures that are used for many bacteria; alternatively, cultures can be stored at −80°C in potassium phosphate buffer/25% glycerol (v/v).

Enrichment and isolation procedures

Temperatures between 20 and 28°C and, low nutrient media with neutral pH such as R2A* or PYE agar† appear to be most

*R2A agar has the following composition (g/l): yeast extract, 0.5; casein hydrolysate, 0.5; glucose, 0.5; starch, 0.5; K_2HPO_4, 0.3; $MgSO_4$, 0.024; Na-pyruvate, 0.3; agar, 15.0; pH 7.2 ± 0.2.

†PYE agar has the following composition (g/l): yeast extract, 3.0; peptone from casein, 3.0; agar, 15.0; pH 7.2.

TABLE 97. Fatty acid profiles of *Hymenobacter* species[a]

	H. roseosalivarius	*H. chitinivorans*	*H. gelipurpurascens*	*H. rigui*	*H. ocellatus*	*H. aerophilus*	*H. actinosclerus*	*H. norvichensis*	*H. xinjiangensis*
$C_{13:0}$ iso	–	0.6	–	–	0.6	–	–	–	–
$C_{14:0}$	–	–	–	–	–	–	–	1.3	1.3
Unknown 13.565	–	0.5	0.8	–	–	–	–	–	–
$C_{14:0}$ iso	–	0.5	–	nr	1.5	0–0.4	nr	0.8	–
$C_{15:1}$ iso G	–	0.6	0.8	nr	–	–	–	–	2.0
$C_{15:1}$ anteiso A	–	–	1.1	nr	–	0–0.5	–	–	–
Summed feature: $C_{15:1}$ iso I/$C_{13:0}$ 3-OH	–	2.1	–	–	2.7	0.8–1.4	2.3	0.8	2.2
$C_{15:0}$ iso	8.3	31.1	17.3	34.8	36.7	10.8–15.0	22.3	27.3	19.5
$C_{15:0}$ anteiso	–	3.7	23.1	5.9	3.9	18.6–22.3	25.8	10.6	3.7
$C_{15:0}$ iso 3-OH	2.7	4.1	2.3	–	4.3	1.3–1.6	1.6	2.2	2.2
$C_{15:0}$ iso 2-OH	–	–	0.9	–	–	0.5–0.9	0.8	–	–
$C_{15:1}\omega6c$	–	–	–	–	–	0.8–1.2	0.8	0.5	–
$C_{15:0}$	–	–	–	nr	–	nr	nr	0.5	–
$C_{16:1}$ iso H	2.7	2.0	3.7	–	1.3	1.4–1.5	1.5	1.4	1.8
$C_{16:0}$ iso 3-OH	1.0	–	–	–	–	–	–	–	–
$C_{16:1}$ $\omega5c$	23.3	8.9	11.3	15.0	2.0	6.6–7.9	3.7	13.6	10.6
Summed feature: $C_{16:1}$ $\omega7c$/$C_{15:0}$ iso 2-OH	29.8	13.9	17.6	13.8	5.4	21.4–22.3	13.1	23.6	20.2
$C_{16:0}$ iso	2.1	1.4	2.7	–	2.9	0.8	–	1.2	1.0
$C_{16:0}$ anteiso	–	–	–	1.9	–	0–0.7	–	–	–
$C_{16:0}$	1.1	–	–	6.4	0.6	1.6–1.7	–	2.2	6.2
$C_{16:0}$ 3-OH	1.2	0.5	–	–	–	–	–	0.6	–
Unknown 16.580	–	–	–	–	–	–	–	–	–
$C_{17:0}$ iso	1.7	2.6	1.7	5.0	2.8	2.7–4.5	1.8	1.6	1.8
$C_{17:0}$ anteiso	–	–	2.0	–	–	1.3–2.3	0.7	–	tr
$C_{17:0}$ iso 3-OH	5.8	6.2	3.7	3.1	6.8	2.6–3.5	3.1	3.0	3.4
$C_{17:0}$ 2-OH	–	–	1.7	–	–	0.8–1.3	2.0	–	tr
$C_{17:1}$ $\omega6c$	1.1	–	–	–	1.2	0–0.7	0.7	–	–
$C_{17:1}$ iso $\omega9c$	–	0.7	–	nr	0.6	0–0.5	–	–	–
Summed feature: $C_{17:1}$ iso I/$C_{17:1}$ anteiso B	18.5	20.8	9.4	14.4	26.9	17.7–18.8	19.9	8.3	8.5
$C_{18:1}$ $\omega9c$	–	–	–	–	–	–	–	–	1.1
$C_{18:1}$ $\omega7c$	–	–	–	–	–	–	–	–	4.8
$C_{18:0}$	–	–	–	–	–	–	–	–	6.5

[a]Given as relative (%) amounts; tr, traces; nr, not reported.

appropriate for isolation of hymenobacters. The ability of several *Hymenobacter* species to survive under unfavorable conditions such as exposure to increased desiccation (Buczolits et al., 2002; Hirsch et al., 1998a; Reichenbach, 1992b) or at high levels of radiation (Collins et al., 2000) may be useful for development of more specific isolation procedures.

List of species of the genus *Hymenobacter*

1. **Hymenobacter roseosalivarius** Hirsch, Ludwig, Hethke, Sittig, Hoffmann and Gallikowski 1999, 1[VP] (Effective publication: Hirsch, Ludwig, Hethke, Sittig, Hoffmann and Gallikowski 1998a, 382.)

ro'se.o.sa.li.va'ri.us. L. adj. *roseus* rose colored; L. adj. *salivarius* salivary, slimy; N.L. masc. adj. *roseosalivarius* indicating a rose-colored bacterium surrounded by much polymer.

The characteristics are as described for the genus and as listed in Tables 96 and 97, with the following additional features. Cells produce (often subpolarly) extracellular polymer and aggregate to form tight, thin, and spreading reddish layers on agar surfaces. Growth in liquid cultures is turbid. Does not grow at 37°C. L-Aspartate is used as a carbon source, but not glutamate. The maximum salt tolerance is 0.5–2.0% NaCl. Pectin and cellulose are not hydrolyzed. The fatty acid profile is given in Table 97.

Source: sandstone and soils of the Dry Valleys region of Antarctica.

DNA G+C content (mol%): 56 (HPLC).

Type strain: AA-718, CIP 106397, DSM 11622.

Sequence accession no. (16S rRNA gene): Y18833.

2. **Hymenobacter actinosclerus** Collins, Hutson, Grant and Patterson 2000, 733[VP]

ac.ti.no.scle'rus. Gr. n. *actis*, *actinos* ray, beam; Gr. adj. *scleros* hard; N.L. masc. adj. *actinosclerus* hard against rays, pertaining to the organism's radiation resistance.

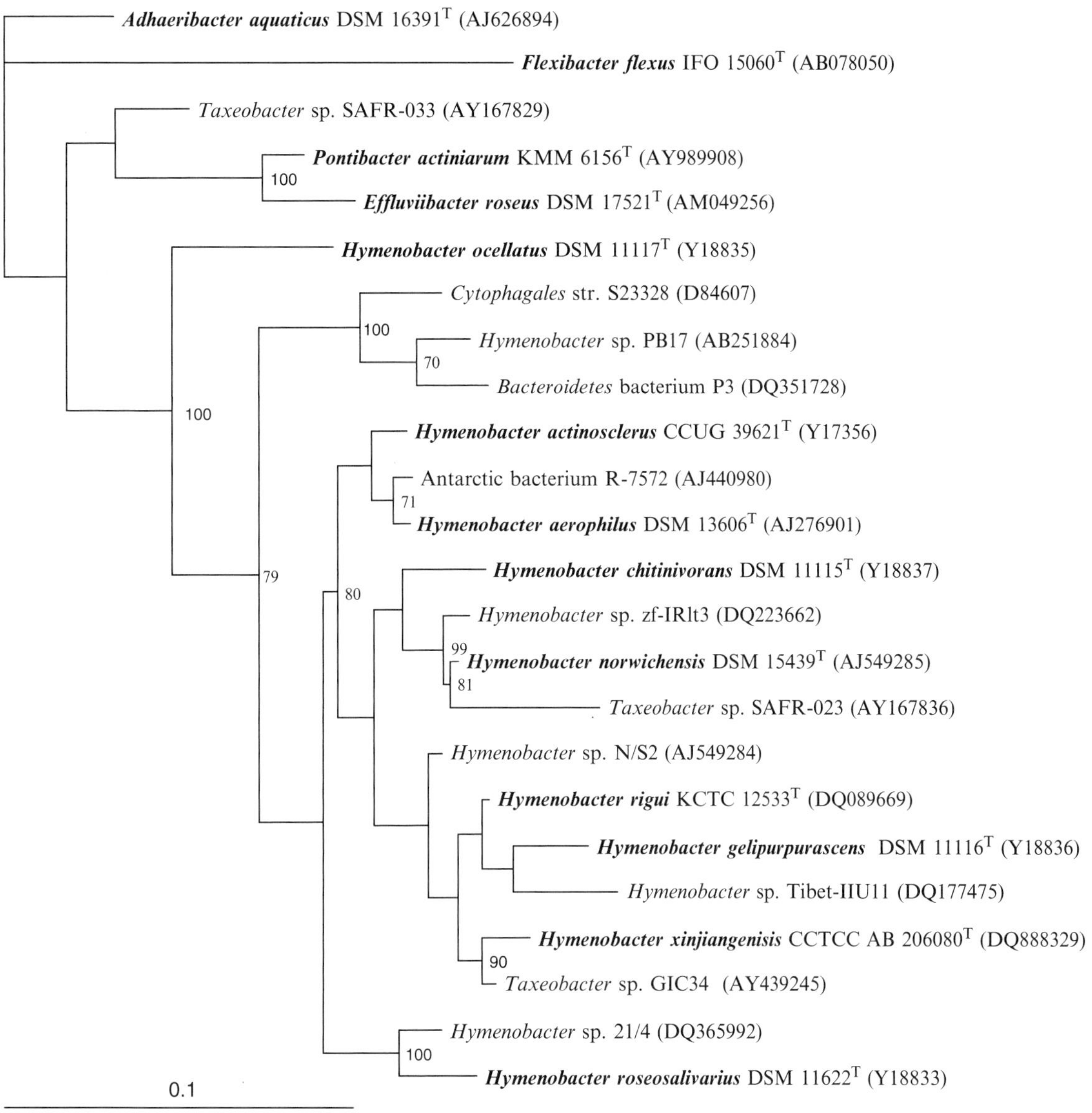

FIGURE 75. Maximum likelihood dendrogram showing phylogenetic relatedness among *Hymenobacter* species and so far unnamed isolates considered to represent novel species of the genus. Numbers at the nodes indicate bootstrap values (500 replications). The scale bar represents 10% sequence divergence.

The characteristics are as described for the genus and as listed in Tables 96 and 97, with the following additional features. Cell dimensions are 0.5–0.6 × 2.0–3.6 µm. No variation in cell morphology is observed in old cultures. Colonies on yeast extract peptone agar are circular, entire, opaque, and not easily emulsified. A water-insoluble red pigment is produced. Fluorescent pigments are not produced on King's A or B media. Chemo-organotrophic. Aerobic, having a respiratory type of metabolism, with an ability to grow under microaerobic conditions. Oxidase- and catalase-positive. Acid and gas are not produced from D-glucose. Starch is hydrolyzed, but not esculin. Alkaline phosphatase, acid phosphatase, ester lipase C8, cystine arylamidase, leucine arylamidase, valine arylamidase, *N*-acetyl-β-glucosaminidase, and phosphoamidase activity are detectable with the API ZYM system. Lipase C14, chymotrypsin, trypsin, α-galactosidase, β-galactosidase, β-glucuronidase, β-glucosidase, α-fucosidase, α-mannosidase, and urease activity are not detectable. Nitrate reduction, H_2S production, and indole production are negative. Growth occurs at 42°C but not at 5°C. Optimal growth temperature, 25–30°C. Highly radiation-resistant with a D_{10} in sodium phosphate buffer of 3.45 kGy and on minced pork of 5.05 kGy. The fatty acid profile is given in Table 97.

Source: pork chops irradiated with 1.75 kGy.

DNA G+C content (mol%): 62 (T_m).

Type strain: CCUG 39621, CIP 106628.

Sequence accession no. (16S rRNA gene): Y17356.

3. **Hymenobacter aerophilus** Buczolits, Denner, Vybiral, Wieser, Kämpfer and Busse 2002, 454[VP]

aer.o.phi'lus. Gr. n. *aer* air; N.L. masc. adj. *philus* (from Gr. masc. adj. *philos*) friend, loving; N.L. masc. adj. *aerophilus* air-loving, indicating its survival when suspended in air.

The characteristics are as described for the genus and as listed in Tables 96 and 97, with the following additional features. Cells are 0.4–0.75 × 1.3–5.0 μm. Cells grow best on nutrient-reduced media such as CasMM* and R2A agar. Colonies on CasMM and R2A agar are translucent, red, circular, entire, low-convex, smooth, and slimy; the diameter is as much as 3.0 mm after 5 d at 28°C, and the colonies spread on CasMM agar. Colonies on standard bacteriological media such as PYES† agar and TSA are opaque, red, circular, entire, convex, and smooth; diameter is up to 2.0 mm after 5 d of incubation at 28°C. No growth occurs on MacConkey or Czapek–Dox agar. The optimum growth temperature is room temperature. The temperature range for growth is 4–28°C; no growth occurs at 37°C. Tolerance maximum for growth in the presence of NaCl is 2.0%. A water-insoluble red pigment is produced; the visible absorption spectrum of the acetone–extracted pigment shows a maximum at 482 nm and two slight inflexions at 453 and 505 nm. Cells assimilate acetate, propionate, D-fructose, D-glucose, D-mannose, and sucrose and display positive reactions in the following tests: catalase, L-alanine aminopeptidase, DNase, alkaline phosphatase, esterase (C4), esterase (C8), leucine arylamidase, acid phosphatase, naphthol-AS-BI-phosphohydrolase, valine arylamidase, hydrolysis of Tween 80, *p*-nitrophenyl (pNP) α-D-glucopyranoside, pNP phenylphosphonate, L-proline *p*-nitroanilide (pNA), L-alanine pNA, L-glutamate γ-3-carboxy-pNA, bis-pNP phosphate, and 2-deoxythymidine-5′-pNP phosphate. Depending on the sensitivity of the method applied, strains show either no catalase reaction or only a weakly positive one. Cells are susceptible to bacitracin, chloramphenicol, colistin sulfate, erythromycin, fusidic acid, gentamicin, kanamycin, nitrofurantoin, penicillin G, polymyxin B, tetracycline, and vancomycin. The fatty acid profile is given in Table 97.

Source: air in the Museo Correr, Venice, Italy.

DNA G+C content (mol%): 60–63 (HPLC).

Type strain: I/26-Cor1, CCUG 49624, DSM 13606, LMG 19657.

Sequence accession no. (16S rRNA gene): AJ276901.

4. **Hymenobacter chitinivorans** Buczolits, Denner, Kämpfer and Busse 2006, 2077[VP] (*Taxeobacter chitinovorans* Reichenbach 1992b, 182)

chi.ti.ni.vo′rans. N.L. neut. n. *chitinum* chitin; L. part. adj. *vorans* devouring; N.L. part. adj. *chitinivorans* devouring chitin.

The characteristics are as described for the genus and as listed in 96 and 97, with the following additional features. Cells are approximately 0.8 × 4.0 μm. Colonies are brick red. Growth occurs on PYES agar, Czapek–Dox agar, and R2A but not on MacConkey agar. Oxidase- and catalase-positive. Nitrate reduction is weakly positive without production of N_2. Negative for production of indole from tryptophan and for arginine dihydrolase and urease. Chitin is degraded. Positive for alkaline phosphatase, esterase C4 (weakly), esterase lipase C8 (weakly), leucine arylamidase, and naphthol-AS-BI-phosphohydrolase (weakly) with the API ZYM system. Negative for lipase C14, valine arylamidase, cystine arylamidase, trypsin, chymotrypsin, acid phosphatase, α-galactosidase, β-galactosidase, β-glucuronidase, α-glucosidase, β-glucosidase, *N*-acetyl-β-glucosaminidase, α-mannosidase, and α-fucosidase. Sensitive to bacitracin (10 IU), chloramphenicol (30 μg), colistin sulfate (10 μg), erythromycin (15 μg), fusidic acid (10 μg), gentamicin (10 μg), kanamycin (30 μg), penicillin G (10 IU), vancomycin (30 μg), polymyxin B sulfate (300 IU), and tetracycline (10 μg). The fatty acid profile is given in Table 97.

Source: soil.

DNA G+C content (mol%): ~61 (analytical ultracentrifuge).

Type strain: strain Txc1, DSM 11115, LMG 21951.

Sequence accession no. (16S rRNA gene): Y18837.

5. **Hymenobacter gelipurpurascens** Buczolits, Denner, Kämpfer and Busse 2006, 2077[VP] (*Taxeobacter gelupurpurascens* Reichenbach 1992b, 182)

ge.lu.pur.pu.ras′cens. L. masc. n. *gelus -us* (or L. neut. n. *gelum -i*) the ice cold; L. part. adj. *purpurascens* turning purple; N.L. part. adj. *gelipurpurascens* becoming purple in the cold.

The characteristics are as described for the genus and as listed in Tables 96 and 97, with the following additional features. Size of cells is approximately 0.5 × 2.0 μm. Colonies are brick red, but at a growth temperature of 2–6°C, and colonies grown on a certain medium are deep blood-red (Reichenbach, 1992b). Growth occurs on PYES agar, Czapek–Dox agar, and R2A, but not on MacConkey agar. Oxidase- and catalase-positive. Nitrate reduction is weakly positive without production of N_2. Negative for production of indole from tryptophan and for arginine dihydrolase and urease. Positive for alkaline phosphatase, esterase C4 (weakly), esterase lipase C8 (weakly), leucine arylamidase, valine arylamidase (weakly), and naphthol-AS-BI-phosphohydrolase (weakly) with the API ZYM system. Negative for lipase C14, cystine arylamidase, trypsin, chymotrypsin, acid phosphatase, α-galactosidase, β-galactosidase, β-glucuronidase, α-glucosidase, β-glucosidase, *N*-acetyl-β-glucosaminidase, α-mannosidase, and α-fucosidase. Sensitive to bacitracin (10 IU), chloramphenicol (30 μg), colistin sulfate (10 μg), erythromycin (15 μg), fusidic acid (10 μg), gentamicin (10 μg), kanamycin (30 μg), penicillin G (10 IU), vancomycin (30 μg), polymyxin B sulfate (300 IU), and tetracycline (10 μg). The fatty acid profile is given in Table 97.

Source: soil.

DNA G+C content (mol%): ~57–58 (analytical ultracentrifuge).

Type strain: Txg1, DSM 11116, LMG 21873.

Sequence accession no. (16S rRNA gene): Y18836.

6. **Hymenobacter norwichensis** Buczolits, Denner, Kämpfer and Busse 2006, 2077[VP]

nor.wi.chen′sis. N.L. masc. adj. *norwichensis* of or belonging to Norwich, a city in England where the type strain was isolated.

The characteristics are as described for the genus and as listed in Tables 96 and 97, with the following additional

*CasMM agar (g/l): K_2HPO_4, 0.6; $Na_2HPO_4 \cdot 2H_2O$, 0.05; $MgSO_4 \cdot 7H_2O$, 0.05; $MgCl_2 \cdot 7H_2O$, 0.1; KNO_3, 0.2; $FeCl_3 \cdot 6H_2O$, 0.010; casein, 0.8; yeast extract, 0.4; agar, 15.0; pH 7.0.

†PYES agar (g/l): yeast extract, 3.0; peptone from casein, 3.0; sodium succinate, 2.3; agar, 15.0; pH 7.2.

features. Size of cells is approximately 2–3 × 0.8 μm. Colonies are brick red. Growth occurs on PYES agar, and R2A but not on Czapek–Dox agar or MacConkey agar. Oxidase- and catalase-positive. Nitrate reduction is positive without production of N_2. Negative for production of indole from tryptophan and for arginine dihydrolase and urease. Positive for alkaline phosphatase, esterase C4 (weakly), esterase lipase C8 (weakly), leucine arylamidase, valine arylamidase (weakly), acid phosphatase (weakly), naphthol-AS-BI-phosphohydrolase and β-glucosidase (weakly) with the API ZYM system. Negative for lipase C14, cystine arylamidase, trypsin, chymotrypsin, α-galactosidase, β-galactosidase, β-glucuronidase, α-glucosidase, *N*-acetyl-β-glucosaminidase, α-mannosidase, and α fucosidase. Sensitive to bacitracin (10 IU), chloramphenicol (30 μg), colistin sulfate (10 μg), erythromycin (15 μg), fusidic acid (10 μg), gentamicin (10 μg), kanamycin (30 μg), penicillin G (10 IU), vancomycin (30 μg), polymyxin B sulfate (300 IU), and tetracycline (10 μg). The fatty acid profile is given in Table 97.

Source: air in the Sainsbury Centre for Visual Arts in Norwich (UK).

DNA G+C content (mol%): not reported.

Type strain: NS/50, DSM 15439, LMG 21876.

Sequence accession no. (16S rRNA gene): AJ549285.

7. **Hymenobacter ocellatus** Buczolits, Denner, Kämpfer and Busse 2006, 2076[VP] (*Taxeobacter ocellatus* Reichenbach 1992b, 182)

o.cel.la'tus. L. masc. adj. *ocellatus* showing little eyes, referring to the bright granules at the cell poles.

Cells are approximately 1 × 3–6 μm. Colonies are brick red. Growth occurs on PYES agar, Czapek–Dox agar, and R2A but not on MacConkey agar. Oxidase- and catalase-positive. Nitrate reduction is weakly positive without production of N_2.Negative for production of indole from tryptophan and for arginine dihydrolase and urease. Positive for alkaline phosphatase, esterase lipase C8 (weakly), leucine arylamidase (weakly) and naphthol-AS-BI-phosphohydrolase (weakly) with the API ZYM system. Negative for esterase C4, lipase C14, valine arylamidase, cystine arylamidase, trypsin, chymotrypsin, acid phosphatase, α-galactosidase, β-galactosidase, β-glucuronidase, α-glucosidase, β-glucosidase, *N*-acetyl-β-glucosaminidase, α-mannosidase, and α-fucosidase. Sensitive to bacitracin (10 IU), chloramphenicol (30 μg), colistin sulfate (10 μg), erythromycin (15 μg), fusidic acid (10 μg), gentamicin (10 μg), kanamycin (30 μg), penicillin G (10 IU), vancomycin (30 μg), polymyxin B sulfate (300 IU), and tetracycline (10 μg). The fatty acid profile is given in Table 97.

Source: dung of an antelope.

DNA G+C content (mol%): ~65 (analytical ultracentrifuge).

Type strain: Myx 2105, Txo1, DSM 11117, LMG 21874.

Sequence accession no. (16S rRNA gene): Y18835.

8. **Hymenobacter rigui** Baik, Seong, Moon, Park, Yi and Chun 2006, 2191[VP]

ri'gui. L. gen. n. *rigui*, of a well-watered place.

Cells produce water-insoluble pinkish-red pigment. Cells grow best on R2A and TSA, and weakly on nutrient agar. Colonies on tryptic soy agar are translucent, low-convex, circular, smooth, and slimy, with a diameter up to 3.0 mm after 5 d at 25–30°C (pH 7). Growth occurs in the presence of 0–2% (w/v) NaCl (optimum 0%), at pH 5–11 (optimum pH 6), and at 4–37°C (optimum 30°C). Oxidase-negative and catalase-positive. Nitrate is not reduced to nitrite. Esculin hydrolysis is weakly positive. Citrate is not utilized. Negative for fermentation of glucose. Gelatinase is produced but not arginine dihydrolase, urease, lysine decarboxylase, ornithine decarboxylase, tryptophan deaminase, H_2S, indole, or acetoin. Positive for alkaline phosphatase, esterase (C4), esterase lipase (C8), leucine arylamidase, acid phosphatase, naphthol-AS-BI-phosphohydrolase, valine arylamidase, and α-glucosidase, but not lipase (C14), trypsin, α-chymotrypsin, β-glucuronidase, α-galactosidase, β-galactosidase, β-glucosidase, α-mannosidase, or α-fucosidase. The following substrates are utilized as sole carbon and energy sources: galactose, esculin, starch, lactose, inulin, melezitose, raffinose, glycogen, and gentiobiose. The following substrates are not utilized: glycerol, erythritol, D-arabinose, L-arabinose, ribose, D-xylose, L-xylose, adonitol, β-methyl-D-xylopyranoside, sorbose, rhamnose, dulcitol, inositol, D-mannitol, D-sorbitol, α-methyl-D-mannopyranoside, α-methyl-D-glucopyranoside, *N*-acetylglucosamine, amygdalin, arbutin, melibiose, xylitol, D-turanose, D-lyxose, D-tagatose, D-fucose, L-fucose, D-arabitol, L-arabitol, and gluconate. The fatty acid profile is given in Table 97.

Source: freshwater of Woopo wetland, Republic of Korea.

DNA G+C content (mol%): 65 (T_m).

Type strain: WPCB131, IMSNU 14116, KCTC 12533, NBRC 101118.

Sequence accession no. (16S rRNA gene): DQ089669.

9. **Hymenobacter xinjiangensis** Zhang, Liu, Tang, Zhou, Shen, Fang and Yokota 2007, 1754[VP]

xin.jiang.en'sis. N.L. masc. adj. *xinjiangensis* pertaining to Xinjiang, an autonomous region in North-West China).

Cells (~0.7 × 2–5 μm) are rod-shaped, Gram-stain-negative, aerobic, nonsporeforming bacteria. Motility is not observed. Cells grow best on nutrient-reduced media such as 0.1× TSA and PYES agar. Colonies on 0.1× TSA and PYES agar are translucent, pink, circular, entire, low-convex, and rough; diameter is up to 1.5 cm after 5 d at 28°C. The temperature range for growth is 4–37°C; optimum, 28°C. Oxidase- and catalase-positive. Esculin hydrolysis is positive. Negative for fermentation of glucose, nitrate reduction, H_2S production, citrate utilization, indole production, and urease. Using the API ZYM system, the following enzyme activities are detectable: alkaline phosphatase, esterase C4, esterase lipase C8, leucine arylamidase, valine arylamidase, cystine arylamidase, acid phosphatase, naphthol-AS-BI-phosphohydrolase, *N*-acetyl-β-glucosaminidase, and α-mannosidase. The following are not detectable: lipase C14, trypsin, chymotrypsin, α-galactosidase, β-galactosidase, β-glucuronidase, α-glucosidase, β-glucosidase, and α-fucosidase. The following are positive with the Biolog system: dextrin, D-cellobiose, i-erythritol, L-fucose, lactulose, maltose, D-mannitol, D-psicose, D-sorbitol, sucrose, D-trehalose, acetic acid, D-galacturonic acid, D-gluconic acid, D-glucosaminic, γ-hydroxybutyric acid, propionic acid, succinic acid, glucuronamide, L-alanine, glycyl-L-glutamic acid, L-serine, DL-carnitine, thymidine, and α-D-glucose-1-phosphate. Negative in the Biolog system are: α-cyclodextrin, Tween 40, Tween 80, *N*-acetyl-D-galactosamine, adonitol, L-arabinose, D-arabitol, gentiobiose, α-D-glucose, α-D-lactose, β-methyl-D-glucoside, pyruvic acid methyl ester, succinic acid methyl ester, *cis*-aconitic acid, citric acid, formic acid, D-galactonic acid lactone, α-hydroxybutyric acid, β-hydroxybutyric acid, *p*-hydroxyphenylacetic acid, itaconic acid, α-ketobutyric acid, α-ketoglutaric acid, α-ketovaleric acid, DL-lactic acid, malonic acid, quinic acid, sebacic acid, bromosuccinic acid, succinamic acid, L-alaninamide, L-alanyl-glycine, L-asparagine, glycyl-L-aspartic acid, L-histidine, hydroxy-L-proline, L-leucine, L-ornithine, L-phenylalanine, L-proline, L-pyroglutamic acid, D-serine, L-threonine, γ-aminobutyric acid, urocanic acid, inosine, uridine, phenyethylamine, putrescine, 2-aminoethanol, 2,3-butanediol, and glycerol. Sensitive to chloramphenicol, colistin sulfate, erythromycin, gentamicin, penicillin G, polymyxin B sulfate, tetracycline, and vancomycin. Tolerates high doses of gamma radiation with D_{10} of 4.8 kGy. The quinone system is menaquinone MK-7.

Source: soil of the Xinjiang Desert, China following irradiation by gamma rays.

DNA G+C content (mol%): 54 (T_m).

Type strain: X2-1g, CCTCC AB 206080, IAM 15452, JCM 23206.

Sequence accession no. (16S rRNA gene): DQ888329.

Other organisms

If a threshold value of 16S rRNA similarity is considered to be 90% for assignment to the genus *Hymenobacter*, then deposited 16S rRNA gene sequences indicate the presence of several strains in institutional culture collections that might be described as novel species of the genus. These strains, and their % rRNA gene similarity to established species of the genus *Hymenobacter*, are listed below.

1. **Antarctic bacterium** R-7572 (accession no. AJ440980) isolated from a microbial mat, Fryxell Lake, McMurdo Dry Valleys, Antarctica (Van Trappen et al., 2002), has a 16S rRNA similarity value of 91.1–96.8%.
2. **Taxeobacter** sp. GIC34 (accession no. AY439245) isolated from glacial ice core, Greenland (Miteva et al., 2004), has a 16S rRNA similarity value of 90.3–98.1%.
3. **Hymenobacter** sp. 21/4 (accession no. DQ365992) isolated from soil in Victoria Land, Antarctica (Aislabie et al., 2006) has a 16S rRNA similarity value of 90.0–97.1%.
4. **Hymenobacter** sp. zf-IRlt3 (accession no. DQ223662) isolated from ice cores, East Rongbuk Glacier, Mt. Qomolangma, Himalaya (Zhang et al., 2006), has a 16S rRNA similarity value of 90.2–98.4%.
5. **Taxeobacter** sp. SAFR-023 (accession no. AY167836) isolated from spacecraft assembly facilities, has a 16S rRNA similarity value of 87.4–96.0%.
6. **Hymenobacter** sp. Tibet-IIU11 (accession no. DQ177475) isolated from permafrost Qinghai-Tibet Plateau, China has a 16S rRNA similarity value of 90.5–96.6%.*
7. **Hymenobacter** sp. PB17 (accession no. AB251884) isolated from soil in Daejeon, South Korea has a 16S rRNA similarity value of 90.2–92.4%.†
8. **Cytophagales** str. S23328 (accession no. D84607) isolated from paddy field soil near Sendai, Japan (Mitsui et al., 1997) has a 16S rRNA similarity value of 89.4–92.3%.
9. **Bacteroidetes** bacterium P3 (accession no. DQ351728) isolated from soil of La Gorce Mountains, Antarctica has a 16S rRNA similarity value of 90.3–91.3%.
10. **Taxeobacter** sp. SAFR-033 (accession no. AY167836) isolated from spacecraft assembly facilities, shares 86.7–90.0% 16S rRNA gene sequence similarity with established species of the genus *Hymenobacter* but slightly higher values with *Effluviibacter roseus* (90.7%), *Pontibacter actiniarum* (91.6%), and *Adhaeribacter aquaticus* (91.9%).

Genus X. **Larkinella** Vancanneyt, Nedashkovskaya, Snauwaert, Mortier, Vandemeulebroecke, Hoste, Dawyndt, Frolova, Janssens and Swings 2006, 239[VP]

THE EDITORIAL BOARD

Lar.ki.nel'la. N.L. dim. fem. n. *Larkinella* named in honor of the American microbiologist John M. Larkin, who described the family *Spirosomaceae* in co-authorship with Renée Borrall.

Ring-like and horseshoe-shaped cells 0.5–0.9 μm wide and with an outer diameter of 1.5–3.0 μm. **Coils or helices are not formed. Motile by gliding**. Nonsporeforming. Gram-stain-negative. Strictly aerobic. **Produce non-diffusible pale-pink pigments. Flexirubin-type pigments are absent**. Chemo-organotrophic. Oxidase-, catalase-, and alkaline phosphatase-positive. **Acid is not produced from a large variety of carbohydrates, including cellobiose, fructose, galactose, glucose, and sucrose**. Dominant cellular fatty acids are $C_{15:0}$ iso, $C_{16:1}$ ω5*c*, $C_{17:0}$ iso 3-OH, and summed feature 3 (comprising $C_{15:0}$ iso 2-OH, $C_{16:1}$ ω7*c*, and/or $C_{16:1}$ ω7*t*). The main isoprenoid quinone is MK-7.

DNA G+C content (mol%): 53.

Type species: **Larkinella insperata** Vancanneyt, Nedashkovskaya, Snauwaert, Mortier, Vandemeulebroecke, Hoste, Dawyndt, Frolova, Janssens and Swings 2006, 239[VP].

Enrichment and isolation procedures

Larkinella insperata was isolated and purified from cooled water produced by a steam generator in a pharmaceutical company

**Editorial note:* Meanwhile, *Hymenobacter* sp. Tibet-IIU11 has been described as *Hymenobacter psychrotolerans* (Zhang et al., 2008; IJSEM 58, 1215–1220).

†*Editorial note:* Meanwhile, *Hymenobacter* sp. PB17 has been described as *Hymenobacter soli* (Kim et al., 2008; IJSEM 58, 941–945).

in Belgium in 2004. Colonies on tryptic soy agar (BBL) at 28°C under aerobic conditions produce non-diffusible pale-pink pigments under aerobic conditions.

Differentiation of the genus *Larkinella* from other genera

Characteristics differentiating the genus *Larkinella* from related or morphologically similar genera are listed in Table 93 in the chapter on the genus *Flectobacillus*.

Taxonomic comments

Vancanneyt et al. (2006) reported that, by 16S rRNA gene sequence analysis, the nearest neighbor of *Larkinella insperata* was *Spirosoma linguale*, with the type strains of the two species sharing a sequence similarity of 88.8%. Similarity values with the type strains of the type species of the genera *Arcicella*, *Dyadobacter*, *Flectobacillus*, and *Runella* were lower: 83.6–86.3%.

List of species of the genus *Larkinella*

1. **Larkinella insperata** Vancanneyt, Nedashkovskaya, Snauwaert, Mortier, Vandemeulebroecke, Hoste, Dawyndt, Frolova, Janssens and Swings 2006, 239[VP]

in.spe.ra'ta. L. fem. adj. *insperata* unexpected, referring to the unexpected source from which the bacterium was isolated.

The characteristics are as described for the genus, with the following additional features. Colonies are 1–2 mm in diameter, circular, and shiny with entire edges. Growth occurs at 10–40°C and with 0–2% NaCl. β-Galactosidase activity is present. Nitrate is not reduced. H_2S, indole, and acetoin (Voges–Proskauer reaction) are not produced. Gelatin and Tween 40 are hydrolyzed. No hydrolysis occurs of agar, casein, starch, DNA, Tweens 20 and 80, urea, cellulose (CM-cellulose or filter paper), and chitin. No acid is produced from adonitol, L-arabinose, D-cellobiose, dulcitol, D-fructose, L-fucose, D-galactose, *N*-acetylglucosamine, D-glucose, glycerol, inositol, D-lactose, D-maltose, mannitol, D-melibiose, L-raffinose, L-rhamnose, sorbitol, L-sorbose, D-sucrose, D-xylose, or L-xylose. L-Arabinose, D-glucose, D-lactose, D-mannose, and D-sucrose are utilized as sole carbon sources for growth, but not citrate, inositol, malonate, mannitol, and sorbitol. Susceptible to ampicillin, carbenicillin, and doxycycline; resistant to benzylpenicillin, chloramphenicol, erythromycin, gentamicin, kanamycin, lincomycin, neomycin, oleandomycin, polymyxin B, streptomycin, and tetracycline. Major fatty acid components (>1.0%) include $C_{14:0}$, $C_{15:0}$ iso, $C_{15:0}$ anteiso, $C_{15:0}$ iso 3-OH, $C_{16:0}$, $C_{16:1}$ ω5*c*, $C_{16:0}$ 3-OH, $C_{16:0}$ iso 3-OH, $C_{17:0}$ iso 3-OH, and summed feature 3 (comprising $C_{15:0}$ iso 2-OH, $C_{16:1}$ ω7*c* and/or $C_{16:1}$ ω7*t*).

Source: water produced by a steam generator in a pharmaceutical company in Belgium.

DNA G+C content (mol%): 53 (HPLC).

Type strain: LMG 22510, NCIMB 14103.

Sequence accession no. (16S rRNA gene): AM000022.

Genus XI. **Leadbetterella** Weon, Kim, Kwon, Park, Cha, Tindall, Stackebrandt, Trüper and Go 2005, 2299[VP]

THE EDITORIAL BOARD

Lead.bet.te.rel'la. N.L. dim. fem. n. *Leadbetterella* named in honor of Edward R. Leadbetter, who studied bacteria belonging to the CFB group.

Rods 0.6–0.9 × 2–7 μm. Nonmotile. **No gliding motility**. Strictly aerobic. Gram-stain-negative. Oxidase- and catalase-positive. Colonies are orange. **Flexirubin-type pigments are present. Growth occurs in the presence of 1% NaCl but not 3%.** Several carbohydrates are used as sole carbon sources. Esculin, gelatin, starch, and tyrosine are degraded. Major fatty acids are $C_{16:1}$ ω7*c*/$C_{15:0}$ iso 2-OH, $C_{15:0}$ iso, $C_{15:0}$ iso 2-OH/$C_{16:1}$ ω7*c*, $C_{17:0}$ iso 3-OH, and $C_{16:0}$. The respiratory quinone is MK-7. Isolated from cotton-waste composts.

DNA G+C content (mol%): 33.

Type species: **Leadbetterella byssophila** Weon, Kim, Kwon, Park, Cha, Tindall, Stackebrandt, Trüper and Go 2005, 2299[VP].

Enrichment and isolation procedures

Leadbetterella byssophila was isolated from cotton-waste composts by plating onto trypticase soy agar (TSA, pH 7.0; Difco) at 30°C.

Differentiation of the genus *Leadbetterella* from other genera

In the family *Cytophagaceae*, *Leadbetterella byssophila* is the only species that has been isolated from cotton-waste composts. Its failure to grow in the presence of 3% NaCl differentiates it from most marine genera. *Leadbetterella byssophila* differs from *Belliella*, *Cytophaga*, *Cyclobacterium*, *Hongiella*, and *Persicobacter* by its production of flexirubin-type pigments. Its lack of gliding motility distinguishes it from the members of the genus *Cytophaga* and *Reichenbachiella*. The cells of *Leadbetterella* are straight rods, unlike the curved or horseshoe-shaped cells of *Cyclobacterium*. The hydrolysis of gelatin and starch differentiates *Leadbetterella byssophila* from *Dyadobacter*, as well as the much lower mol% G+C of its DNA (33 vs 44–49). *Leadbetterella byssophila* does not hydrolyze agar, unlike *Reichenbachiella* and some species of *Cytophaga*; moreover, it exhibits a lower mol% G+C value than *Reichenbachiella* and *Cytophaga* (33 vs 44.5 and 39–46, respectively).

Taxonomic comments

On the basis on 16S rRNA analyses, Weon et al. (2005) reported that the type strain of *Leadbetterella byssophila* could be positioned in the family "*Flexibacteraceae*". Moreover, according to a CLUSTAL W alignment of the members of the phylum *Bacteroidetes*, of the named species the type strain showed highest sequence similarity (85.4%) to the type strain of *Belliella baltica*.

List of species of the genus *Leadbetterella*

1. **Leadbetterella byssophila** Weon, Kim, Kwon, Park, Cha, Tindall, Stackebrandt, Trüper and Go 2005, 2299[VP].

bys.so'phi.la. Gr. n. *byssos* cotton; N.L. fem. adj. *phila* (from Gr. fem. adj. *philê*) friend, loving; N.L. fem. adj. *byssophila* liking cotton.

The characteristics are as described for the genus, with the following additional features. Colonies on TSA are initially light orange and convex with entire margins; with prolonged incubation, they become dark orange. Growth occurs at temperatures of 15–45°C and at pH 6.0–8.0. Growth occurs on 0.5% yeast extract medium. Positive for O/F test (glucose). Positive for indole production and β-galactosidase (API 20NE). Negative for nitrate reduction and arginine dihydrolase (API 20NE). Negative for hydrolysis of casein, cellulose, chitin, DNA, Tweens 40 and 80, and urea. Growth on carbohydrates (API 20NE) occurs with *N*-acetylglucosamine, arabinose, glucose, maltose, and mannose. The following enzyme activities are present (API ZYM system): alkaline phosphatase, leucine arylamidase, valine arylamidase, trypsin, acid phosphatase, naphthol-AS-BI-phosphohydrolase, α-glucosidase, β-glucosidase, *N*-acetyl-β-glucosaminidase, and α-fucosidase (API ZYM); weak activity is detected for α-galactosidase and β-galactosidase. The following substrates are oxidized or weakly oxidized (API 50CH system): *N*-acetylglucosamine, amygdalin, D-arabinose, arbutin, D-cellobiose, esculin, D-galactose, gentiobiose, D-glucose, D-lactose, D-maltose, D-mannose, D-melibiose, methyl-α-D-glucopyranoside, methyl-α-D-mannopyranoside, L-rhamnose, salicin, starch, sucrose, and D-trehalose. The following compounds are assimilated or weakly assimilated (Biolog GN 20 system): acetic acid, *N*-acetyl-D-galactosamine, *N*-acetyl-D-glucosamine, L-alaninamide, L-alanine, L-alanylglycine, L-arabinose, D-arabitol, L-asparagine, L-aspartic acid, D-cellobiose, α-cyclodextrin, dextrin, i-erythritol, D-fructose, D-galactose, D-galacturonic acid, gentiobiose, α-D-glucose, glucose-1-phosphate, D-glucose 6-phosphate, L-glutamic acid, glycerol, α-DL-glycerol phosphate, glycogen, glycyl-L-aspartic acid, glycyl-L-glutamic acid, L-fucose, α-ketovaleric acid, DL-lactic acid, α-D-lactose, lactulose, maltose, D-mannose, D-melibiose, methyl-β-D-glucoside, methyl pyruvate, monomethylsuccinate, L-ornithine, D-raffinose, L-serine, sucrose, L-threonine, thymidine, D-trehalose, turanose, and uridine.

Source: cotton-waste composts in the Republic of Korea.

DNA G+C content (mol%): 33 (HPLC).

Type strain: 4M15, DSM 17132, KACC 11308.

Sequence accession no. (16S rRNA gene): AY854022.

Genus XII. **Meniscus** Irgens 1977, 42[AL]

Roar L. Irgens

Me.nis'cus. N.L. masc. n. *meniscus* (from Gr. masc. n. *mêniskos*) crescent moon.

Curved or straight rods, 0.7–1.0 μm in diameter and 2.0–3.0 μm in length. Cultures may contain single cells, pairs, tightly coiled spirals, S shapes (two cells, one inverted), and doughnut-shaped cells, where the ends are overlapping before division by binary fission. "Rings" have outer diameters of about 3.0 μm. Single polar flagellum. Stains **Gram-negative. Nonmotile. Resting stages not known. Encapsulated. Gas vacuoles** arranged at random within cells. **Colonies chalky white**. Chemo-organotrophic; strictly **fermentative** metabolism with no gas production. **Catalase-and oxidase-negative. Aerotolerantly anaerobic**; capable of growth under an air atmosphere when at least 1% CO_2 is present. Vitamin B_{12}, thiamin and CO_2 are required for growth. Optimum temperature, 30°C. No growth at 10 or 40°C. Isolated from anaerobic digester sludge.

DNA G+C content (mol%): 44.9 (Bd).

Type species: **Meniscus glaucopis** Irgens 1977, 42[AL].

Further descriptive information

Cells of *Meniscus* appear as curved or straight rods. The curved rods often form ringlike shapes when the ends of a cell overlap prior to cell separation. Helical, filamentous forms may also be seen. Gas vacuoles (Figures 76 and 77) may be observed by phase-contrast microscopy, and the individual gas vesicles are resolved by observing whole cells or thin sections in the transmission electron microscope.

Colonies are circular, convex in elevation, with an entire margin and smooth, glistening surface. The colonies may appear translucent or opaque, chalky white. The larger the number of gas vacuoles within the cells of the colony, the whiter the colony. The consistency of the colonies is buttery when grown anaerobically and rubbery when grown aerobically.

When grown to stationary phase in test tubes, the cells often rise to the surface where they form a white band.

The cells apparently do not respire as aerobic and anaerobic cultures have the same cell yields. Fermentation end products when grown on maltose are acetic, propionic, and succinic acids (Irgens, 1977).

Optimum growth occurs around 30°C at pH 7.0. Growth occurs at 15 and 35°C, but not at 10 or 40°C. Cobalamin (vitamin B_{12}), thiamin, and CO_2 are required for growth. Good growth in defined medium occurs with ammonium as the nitrogen source.

The following characteristics are negative: deaminase (Casamino acids), urease, acetyl methyl carbinol, indole, H_2S production and nitrate reduction.

Cells ferment agar (weakly), dextrin, melezitose, raffinose, cellobiose, sucrose, lactose, maltose, melibiose, trehalose, fructose, galactose, glucose, rhamnose (weakly), CH_3-α-D-glucoside, esculin, salicin, D-ribose, D-xylose, and arabinose. They do not ferment mannose, sorbose, glycerol, lactate, mannitol, sorbitol, adonitol, dulcitol, inositol, or amino acids. They do not hydrolyze starch, cellulose, DNA, gelatin, casein, pectin, inulin, gum arabic, tributyrin, chitin, xylan, or glycogen.

Isolated from anaerobic digester sludge but probably also present in anaerobic hypolimnion of lakes.

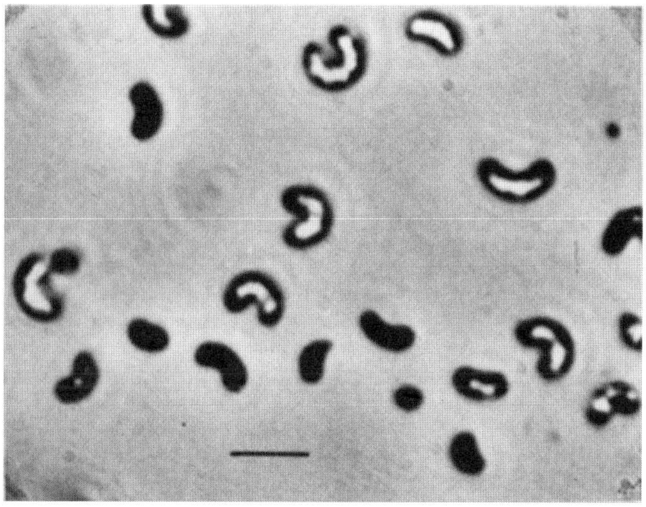

FIGURE 76. *Meniscus glaucopis* ATCC 29398. Phase-contrast. Note gas vacuoles. Bar, 2.0 μm.

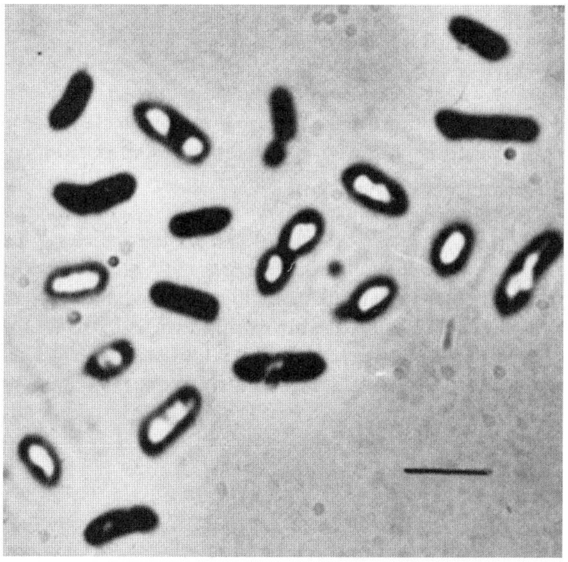

FIGURE 77. *Meniscus glaucopis* strain R. Phase-contrast. Note gas vacuoles. Bar, 2.0 μm.

Enrichment and isolation procedures

Members of this genus may be isolated on a complex medium (BGM) having the following composition (per liter): yeast extract, 1.0 g; KH_2PO_4, 0.5 g; NaCl, 0.4 g; NH_4Cl, 0.4 g; $CaCl_2 \cdot 2H_2O$, 0.01 g; sodium thioglycolate, 0.3 g; $MgSO_4 \cdot 7H_2O$, 0.2 g; $FeSO_4 \cdot 7H_2O$, 0.001 g; at least 1% CO_2; trace elements solution (TES), 1.0 ml. The TES, modified from Pfennig's formula (personal communication), contains (per liter): $ZnSO_4 \cdot 7H_2O$, 0.10 g; $MnCl \cdot 4H_2O$, 0.03 g; H_3BO_3, 0.3 g; $CoCl_2 \cdot 6H_2O$, 0.2 g; $CuCl_2 \cdot 2H_2O$, 0.01 g; $NiCl_2 \cdot 6H_2O$, 0.02 g; $Na_2MoO_4 \cdot 2H_2O$, 0.03 g; pH 3.0–4.0. The pH of the medium is adjusted to 7.3 with 10% Na_2CO_3 before autoclaving, or the medium may be autoclaved with the pH unadjusted and then adjusted to about pH 7.0 after autoclaving by the addition of 2.0 ml of 5.0 % $NaHCO_3$, sterilized by filtration, per 100 ml of medium. All solid media contain 1.0% $CaCO_3$ and 1.5% agar. Stock media contain 0.3% agar. Maltose, at a concentration of 0.2–0.5%, is used as the carbon source for growth studies and stock cultures. Pour plates are prepared using anaerobic digester sludge as the inoculum. The plates are incubated in a GasPak anaerobic jar (BBL) at 25–30°C. After 10 d chalky white colonies indicative of the presence of gas vacuoles may be observed.

Maintenance procedures

Stock cultures are maintained in test tubes rendered anaerobic by the pyrogallic acid technique, or they may be maintained on slants in cotton-stoppered test tubes in anaerobic jars. Lyophilized cultures may be stored indefinitely.

Procedures for testing special characters

The presence of gas vacuoles may be demonstrated by the disappearance of the vacuoles upon the application of a sharp blow with a hammer to a wet mount of the cells. The cover slip is protected with a 3–4-mm-thick rubber pad. Phase-contrast microscopic examination of wet mounts made with dilute India ink is used to demonstrate capsules. The hydrolysis of glycogen, inulin, pectin, gum arabic, and dextrin is tested in BGM broth without maltose. Hydrolysis is considered positive if growth occurs and the pH drops.

Differentiation of the genus *Meniscus* from other genera

Two genera of gas-vacuolated curved rods that may be confused with *Meniscus* are "*Brachyarcus*" and *Ancyclobacter* (formerly *Microcyclus*). "*Brachyarcus*" has never been isolated and differs from *Meniscus* in having cells arranged in groups (coenobia) consisting of two, four or more rings (Skuja, 1964). *Ancyclobacter* is differentiated from *Meniscus* as being a catalase-positive, obligate aerobe rather than an aerotolerant anaerobe and by having a mol% G+C of 66–69 (Van Ert and Staley, 1971).

Taxonomic comments

The assignment of *Meniscus* to the family *Cytophagaceae* is uncertain. Garrity et al. (2005) originally classified the genus within the "*Flexibacteraceae*" based upon phenotypic similarities to other members of this group. In preparation of this volume, the type genus of this family was replaced with *Cytophaga* because it has priority over *Flexibacter* (Ludwig et al., 2009), but most members of the group remained the same. In the absence of a sequence for its 16S rRNA gene, the phenotypic evidence was not sufficient to exclude *Meniscus* from the new family. However, the phenotypic evidence does not provide strong support for its assignment to this family. Most of the *Cytophagaceae* are catalase- and oxidase-positive aerobes, while *Meniscus* is catalase- and oxidase-negative and an aerotolerant anaerobe. However, within the *Cytophagaceae*, the genus *Spirosoma* contains a facultative anaerobic species, and a few other genera are either oxidase- or catalase-negative. Likewise, *Arcicella*, *Larkinella* and *Runella* form similar ring-like arrangements of cells, although all possess other morphological differences with *Meniscus*. In addition, gas vacuoles, which are abundant in *Meniscus*, have not been reported in other members of this family. Gas vesicles are also present in many morphologically and physiologically unrelated taxa and are not reliable taxonomic characteristics. At this time, it is also not possible to exclude a relationship of *Meniscus* to nonvacuolated vibrios within the *Proteobacteria*, such as *Aquaspirillum* and *Vibrio*, and possibly to other Gram-stain-negative, aerotolerant anaerobes such as *Zymomonas* and *Eikenella*. While *Meniscus* was excluded from these genera on phenotypic grounds (Irgens, 1977), a familial relationship may still exist. *Meniscus glaucopis* includes both a vibrioid strain and a straight rod strain. This is justified by the fact that all metabolic and physiological characteristics tested are identical for the two strains. These strains are merely morphovars of the same species.

List of species of the genus *Meniscus*

1. **Meniscus glaucopis** Irgens 1977, 42[AL]

glau.co'pis. N.L. masc. adj. *glaucopis* (from Gr. adj. *glaukôpis*) gleaming-eyed (an epithet of the warlike goddess *Athena*), perhaps a reference to the presence of refractile gas vacuoles.

The description of the species is the same as the genus. Isolated from anaerobic digester sludge.

DNA G+C content (mol%): 44.9 (Bd).

Type strain: ATCC 29398.

Sequence accession no. (16S rRNA gene): not available.

Genus XIII. **Microscilla** Pringsheim 1951, 140 emend. Lewin 1969, 194[AL]

Yasuyoshi Nakagawa

Mic.ro.scil'la. Gr. adj. *micros* small; L. n. *oscillum* swing; N.L. fem. n. *Microscilla* intended to mean small swinging organisms.

Long thin flexible threadlike rods usually 10–100 μm long or longer. **Motile by gliding**. Cell mass more or less intensely **orange**. **Strictly aerobic and chemo-organotrophic**. All grow on peptones as sole source of nitrogen. Chitin and cellulose are not attacked, but other polysaccharides including carboxymethyl cellulose may be decomposed. **Marine** organisms. No growth occurs in half-strength seawater.

DNA G+C content (mol%): 37–44.

Type species: **Microscilla marina** Pringsheim 1951, 140, emend. Lewin 1969, 201[AL].

Further descriptive information

The major quinone of *Microscilla marina*, "*Microscilla furvescens*", and "*Microscilla sericea*" is MK-7 (Nakagawa, unpublished). The major polyamines of *Microscilla marina* are cadaverine, spermidine, and putrescine; that of "*Microscilla furvescens*" is homospermidine. "*Microscilla sericea*" contains mainly homospermidine in addition to agmatine (Hamana and Nakagawa, 2001).

Enrichment and isolation procedures

No enrichment media have been designed for isolation of *Microscilla* strains. Standard procedures to isolate marine bacteria can be applied. Colonies of *Microscilla* are usually orange to yellow. *Microscilla* strains have been isolated from marine environments at widely separated sites (Lewin, 1969; Lewin and Lounsbery, 1969). The type strains of *Microscilla marina* and "*Microscilla sericea*" came from the marine aquarium in La Jolla, California. The type strain of "*Microscilla furvescens*" was isolated from marine sand in Samoa.

Maintenance procedures

Cultures of *Microscilla* strains can be preserved by freezing at temperatures lower than −80°C. For freezing, cells are suspended in Marine broth 2216 (Difco) containing 10% glycerol or 7% DMSO. *Microscilla* strains are rather sensitive to drying; however, they can be preserved in a protective medium SM2 or SM3 (see the chapter on *Flammeovirga* for formulations) by the liquid drying method, or by freeze drying.

Differentiation of the genus *Microscilla* from other genera

Phylogenetic analysis based on 16S rRNA sequences shows that *Microscilla marina*, the type species of the genus *Microscilla*, is phylogenetically independent in the phylum *Bacteroidetes* (see Figure 74 of the chapter *Flexibacter*). However, because few taxonomic characteristics have been investigated since the original descriptions, it is impossible to discriminate the genus *Microscilla* from other related genera only by phenotypic characteristics. It is hoped that additional strains will be isolated and useful differential chemotaxonomic, physiologic, and biochemical characteristics of the genus *Microscilla* will be investigated. All species possess MK-7 as the major quinone, which is useful for differentiating the genus from the family *Flavobacteriaceae*, which is characterized by MK-6 (Bernardet et al., 2002; Nakagawa and Yamasato, 1993).

Taxonomic comments

The genus *Microscilla* was first described by Pringsheim (1951) for actively gliding organisms including the marine type species, *Microscilla marina*, and two freshwater species. From the beginning, it was realized that the genus *Microscilla* resembles the genus *Flexibacter* established by Soriano (1945). In addition, the original Pringsheim's strains have been lost. Lewin (1969) re-isolated morphologically similar organisms and redefined both genera by classifying marine organisms with longer filaments (20–100 μm) in the genus *Microscilla*. This definition was not followed in the *Bergey's Manual of Determinative Bacteriology*, 8th edition (Leadbetter, 1974), and the genus *Microscilla* was included in the genus *Flexibacter*. Later, Reichenbach (1989a, b) tried to restrict the genus *Microscilla* to marine organisms with G+C content of the DNA of above 37 mol%. However, he also stated that too little was known about these bacteria, that these genera were still heterogeneous, and that future research including modern molecular taxonomy was a prerequisite. Those histories are described in detail in the 1st edition of this *Manual*.

Only the type species, *Microscilla marina*, has been validly published to date. In addition, four non-validated species - "*Microscilla aggregans*" (synonym, *Flexibacter aggregans*), "*Microscilla *furvescens*", "*Microscilla sericea*", and "*Microscilla tractuosa*" (synonym, *Flexibacter tractuosus*) — were included in the genus, and three other species — "*Microscilla arenaria*", *Flexibacter litoralis*, and *Flexibacter polymorphus* — were listed as *Species Incertae Sedis* in the 1st edition of the *Manual*.

16S rRNA sequencing analyses of the genera *Microscilla* and *Flexibacter* have shown a high degree of biological diversity of the genus (Nakagawa et al., 2002). The five strains in the four species *Microscilla marina*, "*Microscilla furvescens*", "*Microscilla sericea*", and "*Microscilla arenaria*" diverged into five distinct lineages, as shown in Figure 71 of the chapter on *Flexibacter* and Table 98 (see also Table 94 of the chapter on *Flexibacter*). *Microscilla marina*, the type species of the genus *Microscilla*, was a species secluded from the others, which means that other invalid *Microscilla* species should not be included in the genus. "*Microscilla arenaria*" NBRC 15982^T, which was closely related with the genus *Flammeovirga*, was reclassified as *Flammeovirga arenaria* (Takahashi et al., 2006). "*Microscilla furvescens*" is an independent lineage located next to the cluster composed of *Flexibacter tractuosus* and "*Microscilla sericea*". Two strains of "*Microscilla sericea*" were phylogenetically different. The type strain of "*Microscilla sericea*", NBRC 15983^T, clustered with *Flexibacter tractuosus* and may be classified in the same new taxon. Another strain of "*Microscilla sericea*", NBRC 16561, occupied an independent position that seemed to require a new taxon. The marine *Flexibacter* species *Flexibacter litoralis* and *Flexibacter polymorphus*, in addition to *Flexibacter aggregans* ("*Microscilla aggregans*") and *Flexibacter tractuosus* ("*Microscilla tractuosa*"), were not closely related to either the type species of the genus *Microscilla* or that of the genus *Flexibacter*. These marine *Flexibacter* species are discussed in the chapter on *Flexibacter*.

In this chapter, Reichenbach's definition of the genus is principally followed; however, based on 16S rRNA sequencing analysis, only the type species, *Microscilla marina*, can be justified as a member of the genus *Microscilla*. Other non-validated *Microscilla* species that should be reclassified in the future are listed under *Other species*.

Differentiation of species of the genus *Microscilla*

Table 99 lists characteristics that distinguish the *Microscilla* species from one another.

List of species of the genus *Microscilla*

1. **Microscilla marina** Pringsheim 1951, 140, emend. Lewin 1969, 201[AL]

ma.ri'na. L. fem. adj. *marina* of, or belonging to, the sea, marine.

The characteristics are as described for the genus with the following additional features taken from the 1st edition of this *Manual*.

Threads may become very long, more than 150 µm. The cell mass is orange. Growth occurs on media containing single- or double-strength seawater. Growth does not occur at temperatures greater than 35°C. Peptones serve as sole nitrogen source, but not glutamate or NO_3^-. Glucose is not utilized. Gelatin is degraded, but not agar, alginate, carboxymethyl cellulose, or starch. Nitrate is not reduced. The major carotenoid is saproxanthin. Only one strain — the type strain — is available.

DNA G+C content (mol%): 42 (Bd).

Type strain: S10-8, ATCC 23134, DSM 4236, LMG 18923, NBRC 16560.

Sequence accession no. (16S rRNA gene): AB078080.

Other species

The species listed below should not be included in the genus *Microscilla*, because they are not closely related to the type species of the genus, based on 16S rRNA sequence analysis.

1. **"Microscilla furvescens"** Lewin 1969, 201

fur.ves'cens. L. adj. *furvescens* becoming black.

The characteristics are as described for the genus with the following additional features taken from the 1st edition of this *Manual*.

Long threads of 10–50 µm. The cell mass is orange. Peptones, Casamino acids, glutamate, and NO_3^- serve as sole nitrogen sources. Various sugars, including glucose are utilized, but acetate is not. Agar, alginate, carboxymethyl cellulose, gelatin, and starch are degraded. Nitrate is not reduced. The major carotenoid is saproxanthin. Only one strain – the type strain – is available.

DNA G+C content (mol%): 44 (Bd).

Type strain: TV-2, ATCC 23129, LMG 13023, NBRC 15994, NCIMB 1419.

Sequence accession no. (16S rRNA gene): AB078079.

2. **"Microscilla sericea"** Lewin 1969, 201

se.ri'cea. L. fem. adj. *sericea* silk-like, silky.

TABLE 98. Phylogenetic groups of *Microscilla* strains and their present classification

Phylogenetic group	Species and strain	Present classification
1	*Microscilla marina* NBRC 16560^T	*Microscilla marina*
2	"*Microscilla arenaria*" NBRC 15982^T	*Flammeovirga arenaria*[a]
3	"*Microscilla furvescens*" NBRC 15994^T	Excluded from the genus *Microscilla*
4	"*Microscilla sericea*" NBRC 15983^T	Excluded from the genus *Microscilla*
5	"*Microscilla sericea*" NBRC 16561	Excluded from the genus *Microscilla*

[a]Takahashi et al. (2006).

TABLE 99. Characteristics differentiating the species of the genus *Microscilla*[a,b]

Characteristic	*M. marina*	"*M. furvescens*"	"*M. sericea*"
Length of threads (μm)	>150	10–50	30 to >100
Salinity range (S[c])	1–2	nd	0.5–2
Glutamate	–	+	–
NO_3^-	–	+	–
Utilization of glucose	–	+	+
Degradation of:			
Starch	–	+	+
Alginate	–	+	+
Carboxymethylcellulose	–	+	–
DNA G+C content (mol%)	42	44	38–39

[a]Symbols: +, >85% positive; –, 0–15% positive; nd, not determined.

[b]Data from Reichenbach (1989a).

[c]Expressed as strength of seawater (S); 0 = freshwater.

The characteristics are as described for the genus with the following additional features taken from the 1st edition of this *Manual*.

Long threads of 30–100 μm. The cell mass is orange. Peptones serve as the sole nitrogen source, but not Casamino acids, glutamate, or NO_3^-. Various sugars, including glucose and glycerol are utilized. Alginate, gelatin, and starch are degraded, but agar and carboxymethyl cellulose are not. Nitrate is not reduced. The major carotenoid is saproxanthin.

DNA G+C content (mol%): 38–39 (Bd).

Type strain: SIO-7, ATCC 23182, LMG 13021, NBRC 15983, NCIMB 1403.

Sequence accession no. (16S rRNA gene): AB078081.

Genus XIV. **Pontibacter** Nedashkovskaya, Kim, Suzuki, Shevchenko, Lee, Lee, Park, Frolova, Oh, Bae, Park and Mikhailov 2005c, 2585[VP]

Olga I. Nedashkovskaya and Seung Bum Kim

Pon.ti.bac′ter L. masc. n. *pontus* the sea; N.L. masc. n. *bacter* from Gr. n. *baktron* rod, N.L. masc. n. *Pontibacter* a marine bacterium.

Rods usually measuring 0.3–0.7 × 1.2–1.9 μm. **Can move by gliding**. Produce non-diffusible pink pigments. No flexirubin-type pigments are formed. **Chemo-organotrophic. Strictly aerobic**. Oxidase, catalase, alkaline phosphatase and β-galactosidase-positive. **Esculin, gelatin, and DNA are hydrolyzed**. Casein, Tween 80, chitin, and cellulose (CM-cellulose and filter paper) are not degraded, but agar, starch, and Tweens 20 and 40 may be decomposed. Carbohydrates are utilized. Can grow without seawater or sodium ions. Nitrate is not reduced to nitrite. Indole and H_2S are not produced. **The major respiratory quinone is MK-7**.

DNA G+C content (mol%): 48–52.

Type species: **Pontibacter actiniarum** Nedashkovskaya, Kim, Suzuki, Shevchenko, Lee, Lee, Park, Frolova, Oh, Bae, Park and Mikhailov 2005c, 2585[VP].

Further descriptive information

Phylogenetic analysis of almost-complete 16S rRNA gene sequences of the genus *Pontibacter* revealed that its closest relative is a single species of the genus *Adhaeribacter*, *Adhaeribacter aquaticus*, with similarity of 89.2%. The level of 16S rRNA gene sequence similarity between strains *Pontibacter actiniarum* KMM 6156[T] and *Adhaeribacter aquaticus* MBRG 1.5[T] was 95%.

The main cellular fatty acids are straight-chain unsaturated and branched-chain unsaturated fatty acids $C_{15:1}$ iso, $C_{17:0}$ iso 3-OH, summed feature 3 comprising $C_{15:0}$ iso 2-OH and $C_{16:1}\omega 7c$ or both, and summed feature 4 comprising $C_{17:1}$ iso I and $C_{17:1}$ anteiso B or both (Table 100).

On rich nutrient media strains of the genus *Pontibacter* form regular, round, shiny, with entire edges, smooth, pink colonies with diameter of 1–3 mm after cultivation of 24 h.

All strains grow at 6–36°C, at pH 7.0–10.0 and with 0–4% NaCl. According to Biolog GN2 tests, dextrin, glycogen, L-alanine, methyl β-D-glycoside, methyl pyruvate, α-ketobutyric acid, α-ketovaleric acid, DL-lactic acid, succinamic acid, alaninamide, L-alanyl-glycine, L-asparagine, L-aspartic acid, L-glutamic acid, L-proline, and threonine are utilized, but Tween 80, adonitol, i-erythritol, uridine, urocanic acid, α-D-lactose, D-mannose, citric acid, formic acid, ρ-hydroxyphenylacetic acid, malonic acid, sebacic acid, glucuronamide, D-alanine, D-serine, phenylethylamine, 2-aminoethanol, DL-α-glycerol phosphate, and glucose-1-phosphate are not utilized.

The strains of the genus *Pontibacter* are susceptible to ampicillin, benzylpenicillin, chloramphenicol, erythromycin, carbenicillin, lincomycin, and tetracycline, and resistant to polymyxin B and streptomycin.

The pontibacters were isolated from sea animals and from a desert soil sample collected in the temperate latitudes.

Enrichment and isolation procedures

The strains of the genus *Pontibacter* were isolated from an unidentified actinian (sea anemone) and a desert soil sample using the dilution plating techniques on Marine agar (Difco) and on LB agar, respectively. All isolates have been grown on media containing 0.5% of a peptone and 0.1–0.2% yeast extract (Difco). The marine representative of the genus *Pontibacter* remains viable on Marine agar (Difco) or other rich media containing natural or artificial seawater for several weeks. They have survived storage in Marine broth or artificial seawater supplemented with 20% glycerol (v/v) at –80°C for at least 5 years. A soil isolate, cultivated on LB agar, can grow well under very dry environmental conditions.

Differentiation of the genus *Pontibacter* from other genera

Strains of the genus *Pontibacter* differ from its closest phylogenetic neighbor, the freshwater bacterium *Adhaeribacter aquaticus*, by a higher DNA G+C content (48–52 mol% for *Pontibacter* strains compared with 40.0 mol% for *Adhaeribacter aquaticus*), and by the differences in fatty acid compositions (Nedashkovskaya et al., 2005c; Rickard et al., 2005) (Table 100).

TABLE 100. Cellular fatty acid content of *Pontibacter actiniarum* and its closest relative *Adhaeribacter aquaticus*[a]

Fatty acid	*P. actiniarum* KMM 6156^T	*A. aquaticus* MBRG 1.5^T
$C_{10:0}$ iso	–	1.2
$C_{15:0}$ iso	28.8	22.5
$C_{15:0}$ anteiso	0.1	4.4
$C_{16:1}\omega5c$	0.8	16.9
$C_{15:0}$ iso 2-OH	–	16.5
$C_{15:0}$ iso 3-OH	3.0	3.1
$C_{17:1}$ $\omega6c$	1.4	5.1
$C_{17:0}$ iso	2.2	–
$C_{17:0}$ iso 3-OH	6.5	12.1
Summed feature 2	2.3	–
Summed feature 3	14.7	–
Summed feature 4	31.3	11.2
Summed feature 5	1.8	–

[a]Values are percentages and values of less than 1% are not shown. Summed feature 2 consisted of one or more of the following fatty acids which could not be separated by the Microbial Identification System: $C_{15:1}$ iso I and $C_{13:0}$ 3-OH. Summed feature 3 contains one or more of the following fatty acids: $C_{15:0}$ iso 2-OH, $C_{16:1}$ $\omega7c$ and $C_{16:1}$ $\omega7t$. Summed feature 4 consisted of one or more of the following fatty acids: $C_{17:1}$ iso I and $C_{17:1}$ anteiso B. Summed feature 5 consisted of one or more of the following fatty acids: $C_{18:0}$ anteiso and $C_{18:2}$ $\omega6,9c$. Data from Nedashkovskaya et al. (2005c) and from Rickard et al. (2005).

TABLE 101. Differentiating phenotypic properties of *Pontibacter* species[a,b]

Characteristic	*P. actiniarum*	*P. akesuensis*
Hydrolysis of:		
Agar	+	–
Starch	–	+
Tweens 20 and 40	+	–
Acid production from *N*-acetylglucosamine d-fructose, L-fucose, d-glucose, inositol, D-maltose, L-raffinose, and D-sucrose	–	+
Utilization of:		
Glycyl-L-aspartic acid, Tweens 20 and 40	+	–
α-Cyclodextrin, *N*-acetyl-D-galactosamine, L-arabinose, D-arabitol, cellobiose, D-galactose, gentiobiose, α-D-glucose, D-mannitol, D-melibiose, psicose, D-raffinose, L-rhamnose, D-sorbitol, D-trehalose, turanose, lactulose, xylitol, monomethyl succinate, acetic acid, D-galactonic acid, *cis*-aconitic acid, D-galacturonic acid, D-glucosaminic acid, α-, β- and γ-hydroxybutyric acids, itaconic acid, α-ketoglutaric acid, propionic acid, quinic acid, D-saccharic acid, bromosuccinic acid, glycyl-L-glutamic acid, L-histidine, hydroxy-L-proline, L-leucine, L-ornithine, L-phenylalanine, L-pyroglutamic acid, L-serine, DL-carnitine, γ-aminobutyric acid, inosine, thymidine, putrescine, 2,3-butanediol, and glucose 6-phosphate	–	+
α-Galactosidase activity	–	+
Susceptibility to:		
Kanamycin	+	–
Gentamicin, neomycin	–	+

[a]Symbols: +, positive; –, negative.

[b]Data are taken from Nedashkovskaya et al. (2005c) and Zhou et al. (2007).

Differentiation of the species of the genus *Pontibacter*

Despite very different sources of isolation, the strains belonging to the two validly published species of the genus *Pontibacter* have many common phenotypic features. However, they can clearly be differentiated from each other by the several phenotypic traits shown in Table 101.

List of species of the genus *Pontibacter*

1. **Pontibacter actiniarum** Nedashkovskaya, Kim, Suzuki, Shevchenko, Lee, Lee, Park, Frolova, Oh, Bae, Park and Mikhailov 2005c, 2586[VP]

ac.ti.ni.a′rum. N.L. gen. pl. n. *actiniarum* of sea anemones or related animals.

Cells are 0.3–0.4 µm in width and 1.2–1.9 µm in length. On Marine agar, colonies are circular and 2–3 mm in diameter. Growth occurs at 6–43°C (optimal temperature, 25–28°C) and with 0–10% NaCl. Agar is hydrolyzed (weakly). Acid is formed from arbutin (API 50 CH gallery, bioMérieux). No acid is formed from L-arabinose, D-cellobiose, D-galactose, D-lactose, D-melibiose, L-rhamnose, L-sorbose, DL-xylose, *N*-acetylglucosamine, glycerol, adonitol, dulcitol, inositol, and mannitol. Biolog GN2 tests show that the type strain does not utilize *N*-acetyl-D-glucosamine, D-fructose, L-fucose, *myo*-inositol, α-lactose, maltose, sucrose, D-gluconic acid, D-glucuronic acid, succinic acid, and glycerol. Acetoin (Voges–Proskauer reaction) production is negative. According to the API ZYM gallery (bioMérieux), the following enzymes are produced: acid phosphatase, esterase lipase (C8), leucine- and valine-arylamidases, trypsin, naphthol-AS-BI-phosphohydrolase, α-glucosidase, and *N*-acetyl-β-glucosaminidase. The following enzymes are not produced: esterase (C4), lipase (C14), cystine arylamidase, α-chymotrypsin, β-glucosidase, β-glucuronidase, α-mannosidase, and α-fucosidase.

Source: a single strain, KMM 6156, was isolated from unidentified actinians collected in the Rudnaya Bay, Sea of Japan, Pacific Ocean.

DNA G+C content (mol%): 48.7 (T_m).

Type strain: KMM 6156, KCTC 12367, LMG 23027.

Sequence accession no. (16S rRNA gene): AY989908.

2. **Pontibacter akesuensis** Zhou, Wang, Liu, Zhang, Zhang, Lai and Li 2007, 324[VP]

a.ke.su.en'sis. N.L. masc. adj. *akesuensis* pertaining to Akesu, a city of XinJiang Province in the north-west of China from where the type strain was isolated.

According to Zhou et al. (2007), cells are 0.7 µm × 1.5–1.6 µm. On nutrient agar, colonies are round, 1–2 mm in diameter. Growth occurs at 4–36°C (optimum, 28–30°C), and with 0–4% NaCl. No acid is formed from D-cellobiose, D-galactose, D-lactose, D-melibiose, L-rhamnose, L-sorbose, and DL-xylose.

Source: a single strain, AKS 1, was isolated from the surface layer of a desert soil from Akesu, XinJiang Province, China.

DNA G+C content (mol%): 51.4 (T_m).

Type strain: AKS 1, CCTCC AB 206086, KCTC 12758.

Sequence accession no. (16S rRNA gene): DQ672723.

Genus XV. **Runella** Larkin and Williams 1978, 35[AL]

The Editorial Board

Ru.nel'la. M.E. n. *rune* an ancient alphabet; L. fem. dim. ending *-ella*; N.L. fem. n. *Runella* that which resembles figures of the runic alphabet.

Rigid straight to curved rods, the degree of curvature varying among cells within a culture. In one species (*Runella slithyformis*), the cells range from nearly straight to **crescent-shaped**, but the ends of a cell may overlap, producing a **ring-like structure** with an outside diameter of 2.0–3.0 µm; in addition, filaments up to 14 µm long may also be produced, and on rare occasions, a coil of two to three turns may be produced. In other species, the cells are either **long and filamentous** or **straight to slightly bent rods** instead of crescent-shaped or ring-like as found with *Runella slithyformis*. Gram-stain-negative. **Nonmotile.** Resting stages are not known. **Aerobic. *Runella slithyformis* has a strictly respiratory metabolism with oxygen as the terminal electron acceptor and produces acid oxidatively from a few carbohydrates, whereas *Runella zeae* can ferment sugars.** Optimum temperature, 20–35°C. **Colonies contain a pale pink or salmon-colored water-insoluble pigment.** Catalase is either positive or weakly positive. The oxidase reaction differs among species. Chemo-organotrophic. In those species so far tested, the major quinone is MK-7. Isolated from **freshwater** (*Runella slithyformis*), activated sludge (*Runella limosa* and *Runella defluvii*), and the **stems of *Zea mays*** (*Runella zeae*).

DNA G+C content (mol%): 40–49.

Type species: **Runella slithyformis** Larkin and Williams 1978, 35[AL].

Further descriptive information

Cells of *Runella slithyformis* typically appear as rods whose degree of curvature varies from nearly straight to crescent shaped (Figure 3.2 in Larkin and Borrall, 1984b). An individual cell may be bent in more than one plane. Cells of *Runella zeae* are straight to slightly bent rods that form chains of irregular shapes.

Colonies of *Runella slithyformis* on MS agar (see the chapter on *Spirosoma* for the recipe for this medium) produce a pale pink, water insoluble, nonfluorescent pigment. The colonies are circular and convex with an entire margin. Abundant growth occurs on MS agar and on nutrient agar. Scant growth occurs on chocolate agar, peptonized milk agar, and yeast extract-acetate-tryptone agar. No growth occurs on blood agar, eosin-methylene blue agar, nutrient agar containing 5% sucrose, phenol red mannitol salt agar, phenylethyl alcohol agar, trypticase soy agar (with or without 5% sucrose or 3% glucose), MacConkey agar, bismuth sulfite agar, or Salmonella-Shigella agar (Larkin and Williams, 1978). Colonies of *Runella zeae* are round, smooth, and salmon in color when grown on R2A medium* at 28°C.

Runella is chemo-organotrophic. By the technique of Hugh and Leifson (1953), acid is produced aerobically only from glucose, maltose, sucrose, and inulin. Strains differ in their ability to produce an acid reaction from rhamnose, galactose, mannose, and raffinose. Sugar alcohols are not acidified. Starch and tributyrin are hydrolyzed; esculin, cellulose, agar, chitin, and casein are not hydrolyzed. None of 11 compounds tested in the medium of Gordon and Mihm (see the chapter on *Spirosoma* for the recipe for this medium) are utilized as sole carbon sources.

Runella species are not known to be pathogenic.

Only two strains of *Runella slithyformis* have been isolated and characterized. Both were isolated from eutrophic fresh waters. Similar organisms were seen in gelatinous deposits on wet planks from a mine (Kraepelin and Passern, 1980), in marine waters (Overbeck, 1974; Sieburth, 1978) and in fresh waters (Larkin, unpublished observations).

Runella zeae was isolated from the stems of *Zea mays* (Chelius et al., 2002; Chelius and Triplett, 2000) and is not known to occur in other habitats. Only one strain has been isolated and characterized.

Runella limosa was isolated from activated sludge performing enhanced biological phosphorus removal in a sequencing batch reactor. Only one strain has been isolated and characterized.

*R2A medium of Reasoner and Geldreich (1985) contains (g/l): Yeast extract, 0.5; Proteose peptone no. 3 (Difco); Casamino acids, 0.5; glucose, 0.5; soluble starch, 0.5; K_2HPO_4, 0.3; $MgSO_4 \cdot 7H_2O$, 0.05; sodium pyruvate, 0.3; and agar, 15.0. The pH is adjusted to 7.2 with crystalline K_2HPO_4 or KH_2PO_4. The medium is sterilized by autoclaving.

Runella defluvii was isolated from the activated sludge of a domestic wastewater treatment plant. Only one strain has been isolated and characterized.

Isolation procedures

Runella slithyformis can be isolated by repeated streaking of water samples onto MS agar with incubation at room temperature for up to 2 weeks. The pale pink pigmentation of the colonies aids detection.

Runella limosa was isolated by serial dilution of a sludge sample in 1% (w/v) saline solution, with subsequent spreading onto R2A agar (Difco) and incubation at 20°C for 5 d. Subcultivation was on R2A agar at 25°C for 3 d.

For isolation of *Runella defluvii*, a sludge sample was serially diluted with 1% (w/v) saline solution, spread onto R2A agar (Difco) and incubated at 20°C for 5 d. Subcultivation was done on R2A agar at 30°C for 3 d.

Runella zeae was isolated by the procedure used by Chelius and Triplett (2000) for *Dyadobacter fermentans*. Briefly, maize seeds are surface-sterilized with sodium hypochlorite, planted in a sterilized synthetic soil, and watered with a nitrogen-free nutrient solution. After 6 weeks, the plants are harvested and stems are surface-sterilized with sodium hypochlorite, washed with sterile water, and crushed. The fluid portion is plated onto R2A agar and incubated at 28°C.

Maintenance procedures

Strains of *Runella slithyformis* are grown on MS agar or nutrient agar at room temperature for several days to allow abundant growth. They will then survive refrigeration (4°C) for at least 3 weeks. They may also be preserved indefinitely by lyophilization.

Procedures for testing special characters

Utilization of carbohydrates. For *Runella slithyformis*, the production of acid from carbohydrates by aerobic or anaerobic means is determined by the method of Hugh and Leifson (1953), in which MS agar is used but with the glucose replaced by 1% of the substrate to be tested and the agar concentration lowered to 0.3%. Incubation is continued for 8 weeks for cultures giving negative results.

Utilization of single carbon sources. For *Runella slithyformis*, the basal medium used is that of Gordon and Mihm, to which is added 0.2% of the substrate or the sodium salt of the substrate. If growth occurs through four successive subcultures, the results are considered positive even in the absence of a color change in the bromthymol blue indicator.

Differentiation of the genus *Runella* from other genera

Table 102 in the chapter on the genus *Flectobacillus* provides the primary characteristics that can be used to differentiate

TABLE 102. Characteristics differentiating the species of the genus *Runella*[a]

Characteristic	*R. slithyformis*[b]	*R. defluvii*[c]	*R. limosa*[c]	*R. zeae*[d]
Rings formed	+	–	–	–
Catalase	w	+	+	+
Oxidase	+	–	–	+
Oxidation/fermentation test with glucose and sucrose	O[e]	nr	nr	F
Acid production from:				
Glucose	+[f]	–	+	+
Ribose	–[g]	nr	nr	–
Growth at 4°C	+[e]	–	nr	–
Starch hydrolysis	w	–	–	–
Utilization of sole carbon sources:				
Acetate, starch, fumarate, D-lyxose, malate, tartrate	–[e]	nr	nr	+
D-Arabitol	w	–	–	–
Dulcitol, inositol, sorbitol	w	–	–	w
Glycerol	–	+	–	–
Glycogen	+	w	w	+
5-Ketogluconate	–	w	+	w
Maltose, trehalose	+	–	+	+
Mannitol	w	–	–	–
Methyl-β-xyloside	+	+	–	+
Growth on peptone	–[e]	nr	nr	+
Habitat:				
Freshwater	+	–	–	–
Stems of *Zea mays*	–	–	–	+
Activated sludge	–	+	+	–
Major phospholipid	nr	Phosphatidylethanolamine	Phosphatidylglycerol	nr
DNA G+C content (mol%)	49	40.1	44.5	49

[a]Symbols: +, >85% positive; d, different strains give different reactions (16–84% positive); –, 0–15% positive; w, weak reaction; nr, not reported.

[b]Results for *Runella slithyformis* are taken from Larkin and Williams (1978), Larkin and Borrall (1984b), and Lu et al. (2007).

[c]Results for *Runella defluvii* and *Runella limosa* are taken from Lu et al. (2007).

[d]Results for *Runella zeae* are taken from Chelius et al. (2000, 2002).

[e]Result from Chelius et al. (2000, 2002).

[f]Larkin and Williams (1978) and Larkin and Borrall (1984b) reported a positive reaction; Chelius et al. (2000) reported a negative reaction.

[g]Larkin and Williams (1978) and Larkin and Borrall (1984b) reported a negative reaction; Chelius et al. (2000) reported a positive reaction.

Runella slithyformis and *Runella zeae* from the morphologically similar genera.

Taxonomic comments

Analyses of the 16S rRNA gene sequences of these organisms support the classification of *Runella* as a distinct genus (Chelius and Triplett, 2000; Nikitin et al., 2004; Woese et al., 1990). Although *Runella slithyformis* and *Runella zeae* differ in morphology, fermentative ability, and other phenotypic characteristics, a 16S rRNA gene sequence similarity of 94% between the two strains suggests that they are sufficiently related to be included within a single genus (Chelius et al., 2002). The level of DNA–DNA hybridization value between *Runella zeae* and *Runella slithyformis* is 19%, indicating separate species (Chelius et al., 2002).

The addition of *Runella limosa* to the genus is supported by a 16S rRNA gene similarity of 94.8% to the type strain of *Runella slithyformis* (Ryu et al., 2006). Similarly, *Runella defluvii* exhibits a similarity value of 93.6% to *Runella slithyformis*, and a value of 97.1% to *Runella limosa*. The status of *Runella defluvii* as a species separate from *Runella limosa* is supported by a DNA–DNA hybridization value of 25% between the two species.

Differentiation of the species of the genus *Runella*

Table 102 lists characteristics that differentiate the four species of *Runella*.

List of species of the genus *Runella*

1. **Runella slithyformis** Larkin and Williams 1978, 35[AL]

slith.y.form′is. *slithy* a nonsense word from Lewis Carroll's *Jabberwocky* for a fictional organism that is "slithy" (presumably a combination of slinky and lithe); L. adj. suff. *-formis -is -e*, -like, in the shape of; N.L. fem. adj. *slithyformis* slithy in form.

The characteristics are as described for the genus and as listed in Table 102. The following additional features are taken from Larkin and Williams (1978) and Larkin and Borrall (1984b).

Good growth occurs on MS agar. Moderate growth occurs on nutrient agar. Scant growth occurs on chocolate agar, peptonized milk agar, and yeast extract-acetate-tryptone agar. No growth occurs on blood agar, eosin-methylene blue agar, nutrient agar plus 5% sucrose, phenol red-mannitol-salt agar, phenylethyl alcohol agar, trypticase soy agar, trypticase soy agar plus 3% glucose, MacConkey agar, bismuth sulfite agar, and Salmonella-Shigella agar.

A positive reaction is obtained in the following tests: β-galactosidase (ortho-nitrophenyl-β-galactoside [ONPG]); phosphatase; acid production from glucose, inulin, maltose, and sucrose; hydrolysis of tributyrin and of starch (weak).

A negative reaction is obtained in the following tests: hydrolysis of gelatin, esculin, lecithin, urea, agar, cellulose, and chitin; lysine decarboxylase; ornithine decarboxylase; phenylalanine deaminase; hemolysin production; indole production; methyl red test, Voges–Proskauer test, nitrate reduction, H_2S production from peptone; acid production from arabinose, α-methyl-D-glucoside, cellobiose, dextrin, dulcitol, erythritol, fructose, glycerol, lactose, mannitol, melibiose, ribose, salicin, sorbitol, sorbose, trehalose, and xylose; utilization of acetate, benzoate, citrate, formate, glycerol phosphate, malonate, methanol, methylamine, propionate, succinate, and tartrate as sole carbon sources. No change occurs in litmus milk.

Acid production from galactose, mannose, raffinose, and rhamnose differs between strains.

Source: freshwater.

DNA G+C content (mol%): 49–50 (T_m; absorbance ratio).

Type strain: ATCC 29530, LMG 11500.

Sequence accession no. (16S rRNA gene): M62786.

2. **Runella defluvii** Lu, Lee, Ryu, Chung, Choe and Jeon 2007, 2602[VP]

de.flu′vi.i. L. gen. n. *defluvii* of sewage.

The characteristics are as described for the genus and as listed in Table 102. The following additional features are taken from Lu et al. (2007). Rods 0.5–0.9 × 2.2–6.0 μm at 30°C on R2A agar. Colonies are slightly raised, circular and salmon pink on R2A agar. Temperature range 15–40°C, optimum, 30–35°C. pH range 6.0–9.5, optimum 7.5–8.0. Nitrate is not reduced to nitrite. Catalase-positive. Oxidase-negative. No anaerobic growth after 7 d at 30°C on R2A agar. Tyrosine, Tween 80, and esculin are hydrolyzed, but not casein, Tween 20, starch, gelatin, and urea. Acid is produced from raffinose, *myo*-inositol, lactose, L-arabinose, D-galactose, D-mannose, D-mannitol, and melibiose, but not from D-glucose, D-fructose, arbutin, or salicin. Indole, H_2S and acetoin are not produced. Citrate is not utilized (API 20E system). Enzyme activities include alkaline phosphatase, trypsin, α-chymotrypsin, *N*-acetyl-β-glucosaminidase, and naphthol-AS-BI-phosphohydrolase, but not tryptophan deaminase, esterase (C4), lipase (C14) and β-glucuronidase. Weak activities occur for esterase lipase (C8), leucine arylamidase, valine arylamidase, cystine arylamidase, acid phosphatase, α-galactosidase, β-galactosidase, α-glucosidase, β-glucosidase, α-mannosidase, and α-fucosidase (API ZYM system). Glycerol, methyl β-xyloside, methyl α-D-mannoside, and esculin are used as sole carbon sources but not erythritol, D- or L-arabinose, D-xylose, adonitol, galactose, D-glucose, D-fructose, mannose, dulcitol, inositol, mannitol, sorbitol, *N*-acetylglucosamine, salicin, cellobiose, maltose, lactose, melibiose, sucrose, trehalose, inulin, melezitose, D-raffinose, D-turanose, D- or L-arabitol, gluconate, or 2-ketogluconate. Ribose, L-xylose, sorbose, rhamnose, methyl α-D-glucoside, amygdalin, arbutin, starch, glycogen, xylitol, β-gentiobiose, D-lyxose, D-tagatose, D- and L-fucose, and 5-ketogluconate are weakly utilized (API 50CH system). Cells contain a large amount of phosphatidylethanolamine and small amounts of phosphatidylcholine and an unknown phospholipid as polar lipids. The major quinone is menaquinone-7. The cellular fatty acids are $C_{15:0}$ (29.0%), summed feature 3 ($C_{16:1}$ ω7*c* and/or $C_{15:0}$ 2-OH; 20.2%), $C_{16:1}$ ω5*c* (10.8%), $C_{17:0}$ 3-OH (9.2%),

$C_{15:0}$ 3-OH (7.4%), $C_{15:0}$ (6.5%), $C_{15:1}$ G (3.4%), $C_{16:0}$ 3-OH (3.3%), $C_{13:0}$ (2.2%), $C_{15:1}$ ω6*c* (1.8%), $C_{16:0}$ (1.4%), $C_{17:1}$ω6*c* (0.9%), $C_{14:0}$ (0.8%), $C_{15:0}$ (0.6%), and unknown ECL 14.959 (2.7%).

Source: activated sludge of a domestic wastewater treatment plant.

DNA G+C content (mol%): 40.1 mol% (HPLC).

Type strain: EMB13, DSM 17976, KCTC 12614.

Sequence accession no. (16S rRNA gene): DQ372980.

3. **Runella limosa** Ryu, Nguyen, Park, Kim and Jeon 2006, 2759[VP]

li.mo'sa. L. fem. adj. *limosa* muddy, pertaining to sludge, the natural habitat of the species.

The characteristics are as described for the genus and as listed in Table 102. The following additional features are taken from Ryu et al. (2006). Rods, 0.7–0.9 × 4.0–10.0 µm when grown at 25°C on R2A agar. Colonies are slightly raised, circular and salmon-pink on R2A agar. Optimum temperature, 25–30°C. Optimum pH, 7.5–8.0. Nitrate is not reduced to nitrite. Catalase-positive and oxidase-negative. Tyrosine, Tween 80 and esculin are hydrolyzed but not casein, Tween 20, starch, gelatin, or urea. Acid is produced from D-glucose, D-raffinose, *myo*-inositol, D-lactose, D-mannitol, and melibiose, but not from sorbitol, sucrose, rhamnose, amygdalin, D-fructose, D-galactose, D-mannose, L-arabinose, arbutin, or salicin. Alkaline phosphatase, α-chymotrypsin, *N*-acetyl-β-glucosaminidase, and naphthol-AS-BI-phosphohydrolase activities are present, but not esterase (C4), esterase lipase (C8), lipase (C14), or cystine arylamidase. Weak activity occurs for leucine arylamidase, valine arylamidase, trypsin, β-galactosidase, α-glucosidase, β-glucosidase, acid phosphatase, α-galactosidase, β-glucuronidase, α-mannosidase, and α-fucosidase. A large amount of phosphatidylglycerol is present and small amounts of two unknown phospholipids (PL1, PL2). The major quinone is MK-7. The major fatty acids are $C_{15:0}$, $C_{16:1}$ω5*c*, $C_{17:0}$ 3-OH, $C_{15:0}$ 3-OH, $C_{16:0}$ 3-OH, $C_{16:0}$, and summed feature 3 ($C_{16:1}$ω7*c* and/or $C_{15:0}$ 2-OH).

Source: sludge performing enhanced biological phosphorus removal.

DNA G+C content (mol%): 42.7 (HPLC).

Type strain: EMB111, KCTC 12615, DSM 17973.

Sequence accession no. (16S rRNA gene): DQ372985.

4. **Runella zeae** Chelius, Henn and Triplett 2002, 2062[VP]

ze'ae. L. n. *zea* a kind of grain, spelt, and also a botanical genus name (*Zea*); L. gen. n. *zeae* of *Zea*, named because the organism was isolated from maize, *Zea mays*).

The characteristics are as described for the genus and as listed in Table 102. The following additional features are taken from Chelius and Triplett (2000) and Chelius et al. (2002).

Straight to slightly bent rods that form chains of irregular shapes. Colonies are round, smooth, and salmon-colored when grown on R2A medium at 28°C. Temperature range for growth, 15–37°C.

The following reactions are positive: growth on peptone water; growth on Ayer's agar (weak); acid production from (and fermentation of) glucose and sucrose; growth on acetate, *N*-acetylglucosamine, amygdalin, D- and L-arabinose, arbutin, cellobiose, dulcitol (weak), erythritol, esculin, D-fructose, D- and L-fucose, fumarate, galactose, β-gentiobiose, D-glucose, glycogen, inositol (weak), inulin, 5-ketogluconate (weak), lactose, D-lyxose, malate, malonate, maltose, mannose, melezitose, melibiose, methyl-α-D-glucoside, methyl-α-D-mannoside, β-methylxyloside, D-raffinose, rhamnose, salicin, sorbitol (weak), sorbose (weak), sucrose, D-tagatose, tartrate, trehalose, D-turanose, and xylitol as sole carbon sources.

The following reactions are negative: nitrogenase; reduction of nitrate to nitrite; hydrolysis of agar, cellulose, and starch; growth in litmus milk; acid production from ribose; growth in the presence of 1.5% NaCl; growth on D- and L-arabitol, formate, glycerol, mannitol, and methanol as sole carbon sources.

Source: the stems of *Zea mays*.

DNA G+C content (mol%): 49 (renaturation rate).

Type strain: NS12, ATCC BAA-293, LMG 21438.

Sequence accession no. (16S rRNA gene): AF137381.

Genus XVI. **Spirosoma** Migula 1894, 237[AL]

THE EDITORIAL BOARD

Spi.ro.so'ma. Gr. n. *spira* coil; Gr. neut. n. *soma* body; N.L. neut. n. *Spirosoma* coiled body.

Rigid **straight to curved rods**, the degree of curvature varying among individual cells within a culture. The cells measure 0.5–1.0 µm × 1.5–10.8 µm. **Rings 1.5–3.0 µm in outer diameter are formed by overlapping of the ends of a cell** in some species. **Coils and helices may be present**. Long sinuous filaments up to 50 µm long may be present. Gram-stain-negative. **No swimming motility occurs. Gliding motility is present** (Vancanneyt et al., 2006). Resting stages are not known. **Obligate or facultative aerobes**. Acids are produced aerobically from a variety of carbohydrates. Optimum temperature, 20–30°C. **Colonies contain a pale to light yellow, water-insoluble carotenoid pigment. Catalase-positive** and oxidase-negative or positive depending upon the species. Chemo-organotrophic. Isolated from soil and freshwater sources.

DNA G+C content (mol%): 51–53 (T_m).

Type species: **Spirosoma linguale** (Eisenberg 1891) Migula 1894, 235[AL].

Further descriptive information

This chapter is taken largely from the previous treatment by Larkin and Borrall (1984d) in the 1st edition of *Bergey's Manual of Systematic Bacteriology*. It has been updated where more recent information was available.

Spirosoma linguale produces rings, coils, and helices and the morphology is illustrated in Figure 3.1 of Larkin and Borrall (1984d). Although a culture typically shows a ringlike morphology, it exhibits variations of shape and size under certain cultural conditions. Occasionally, cultures are composed mainly of relatively straight cells, especially after prolonged subculturing on MS agar. The curly form may be re-obtained by examining colonies from agar streaked for isolation (Larkin, unpublished observation). Maloy et al. (1978) showed that there is a relationship between morphology and the phosphate content of the medium. The rings, coils, and helices are produced when the phosphate level is below 20 mM at pH 7.2. At 20–60 mM phosphate, long nonseptate filaments develop and the ratio of diphosphatidyl glycerol to phosphatidyl glycerol is twofold higher in the filaments. Normal cells contain twice the amount of muramic acid in the peptidoglycan layer as do the filaments; furthermore, this layer in normal cells contains equal amounts of *N*-acetylglucosamine and *O*-acetylglucosamine, whereas that of filamentous cells has nearly twice as much *N*-acetylglucosamine as *O*-acetylglucosamine (Miller and Raj, 1978). The filamentous forms appear to be multinucleate (Redell et al., 1981). In contrast, the species *Spirosoma rigui* forms filaments but not rings, coils or helices (Baik et al., 2007b).

Vancanneyt et al. (2006) reported that the type strain of *Spirosoma linguale* exhibits gliding motility. *Spirosoma rigui* also exhibits gliding motility.

Colonies on MS agar* produce a yellow water-insoluble, nonfluorescent pigment, and flexirubin-type pigments are absent (Baik et al., 2007b; Vancanneyt et al., 2006). For *Spirosoma rigui*, the pigment extract had an absorption maximum of 451 nm and shoulder at 478 nm. *Spirosoma linguale* grows poorly or not at all on rich media such as chocolate agar or blood agar, or on enteric-selective media such as eosin methylene blue agar, MacConkey agar, or Salmonella-Shigella agar. Only one of the four available strains grows on Trypticase soy agar. *Spirosoma rigui* was isolated on PYGV medium (Staley, 1968) and R2A (Oxoid) agar. Best growth is obtained with R2A, MS, or PCA media. It also grows in tryptone-soy agar or TSA medium (Oxoid).

Spirosoma is chemo-organotrophic. *Spirosoma linguale* is active in the acidification of carbohydrate media. Acidification occurs with all but one (sorbose) of the 20 carbohydrates tested using the medium of Hugh and Leifson (1953) and aerobic incubation, although up to 3 weeks is sometimes required. Acidification does not occur with any sugar alcohols. Cellulose and chitin are not hydrolyzed, but some strains hydrolyze esculin and (weakly) casein. All strains hydrolyze tributyrin, and (weakly) starch. Three of the available strains of *Spirosoma linguale* produce a soft curd in litmus milk, accompanied by the reduction and then reoxidation of the litmus. A fourth strain produces only an increased alkalinity in litmus milk. Although *Spirosoma rigui* uses glucose and esculin as sole carbon sources, it does not assimilate many other carbohydrates.

Spirosoma linguale strains may be grown in the defined medium of Gordon and Mihm† (1957) with various sole carbon sources. Of 11 substrates that have been tested, only glycerol phosphate, succinate, tartrate, and malonate are utilized as sole carbon sources.

Radiorespirometric studies by Kottel and Raj (1973) revealed that *Spirosoma linguale* ATCC 23276 catabolizes glucose almost entirely (96%) by the Embden–Meyerhof pathway with a small amount (4%) being catabolized by the pentose phosphate pathway. The Entner–Doudoroff pathway is inducible, with 75% of radio-labeled gluconate being metabolized by that pathway. Gluconate recovery patterns of this strain showed evidence for the possibility of a 2–5-diketogluconate pathway or some other unorthodox pathway (Raj and Ordal, 1977). Radiorespirometric and enzymic data indicate that a functional tricarboxylic acid cycle occurs in this strain (Kottel and Raj, 1973).

Spirosoma has the MK-7 menaquinone system (Baik et al., 2007b; Urakami and Komagata, 1986). The cellular fatty acids of *Spirosoma* include large amounts of straight-chain saturated $C_{16:0}$, unsaturated $C_{16:1}$, iso $C_{15:0}$, and iso $C_{17:0}$ (Urakami and Komagata (1986). Vancanneyt et al. (2006) reported the following fatty acid profile for the type strain of *Spirosoma linguale* (% in parentheses): $C_{13:0}$ iso (3.8), $C_{15:0}$ iso (7.6), $C_{15:0}$ anteiso (trace), $C_{15:0}$ iso 3-OH (4.9), $C_{16:0}$ (7.2), $C_{16:1}$ ω5*c* (17.5), $C_{16:0}$ 3-OH (4.7), $C_{17:0}$ iso 3-OH (6.5), and summed feature 3 (47.9). Summed feature 3 consisted of one or more fatty acids that could not be separated by the Microbial Identification System): $C_{15:0}$ iso 2-OH, $C_{16:1}$ ω7*c* and/or $C_{16:1}$ ω7*t*. The composition of *Spirosoma rigui* was very similar: $C_{15:0}$ iso (9.5), $C_{15:0}$ anteiso (1.3), $C_{15:0}$ iso 3-OH (2.6), $C_{16:0}$ (8.8), $C_{16:1}$ ω5*c* (18.5), $C_{16:0}$ 3-OH (2.8), $C_{17:0}$ iso 3-OH (3.5), and summed feature 3 (45.6). (Baik et al., 2007b).

Analysis of polyamine profiles of *Spirosoma* indicate that spermidine is the major polyamine (Hamana and Nakagawa, 2001).

Spirosoma is not known to be pathogenic.

Four strains of *Spirosoma linguale* have been isolated and characterized: one from garden soil and three from freshwater. *Spirosoma linguale* has also been found in an activated sludge reactor established for the degradation of cutting fluids (Baker et al., 1983). *Spirosoma rigui* was isolated from fresh water at the Woopo wetland, Republic of Korea. *Spirosoma*-like organisms have been seen in gelatinous deposits on wet planks from a mine (Kraepelin and Passern, 1980). Although *Spirosoma* is considered noncellulolytic, cellulolytic "*Spirosoma*-like" bacteria have been isolated from the gut of the termite *Zootermopsis angusticollis* (Wenzel et al., 2002).

Isolation procedures

Spirosoma may be isolated by repeated streaking of water or diluted soil samples onto MS agar or tryptone glucose extract agar (Difco) fortified with 0.1 % yeast extract (TGEY medium; Raj, 1970) with incubation at room temperature for up to 2 weeks. The yellow pigmentation of the colonies aids detection of *Spirosoma*.

Maintenance procedures

Spirosoma linguale is grown on MS agar, TGEY or nutrient agar at room temperature until abundant growth occurs (usually 1–4 d). Cultures may then be stored at 4°C for at least 4 weeks. Preservation by lyophilization is effective for several years.

Differentiation of the genus *Spirosoma* from other genera

Table 93 in the chapter on *Flectobacillus* provides the primary characteristics that can be used to differentiate this genus from several morphologically similar genera.

---

*MS agar (g/l): peptone, 1.0; yeast extract, 1.0; glucose, 1.0; agar, 15.0.
†Gordon and Mihm's medium (g/l): $MgSO_4$, 0.2; $(NH_4)_2HPO_4$, 1.0; KH_2PO_4, 0.5; NaCl, 1.0; carbon source, 2.0; agar, 15.0; and bromthymol blue, 0.08.

Taxonomic comments

In the 8th edition of the *Manual* the genus *Microcyclus* consisted of three species which were placed together primarily because of their ability to form rings during growth. The DNA G+C values were quite different, being 39.5 mol% for "*Microcyclus major*", 51 mol% for "*Microcyclus flavus*", and 67 mol% for *Microcyclus aquaticus* (the type species). Claus (1967) and Claus et al. (1968) suggested that "*Microcyclus flavus*" and two additional isolates corresponded to the description of the forgotten genus *Spirosoma*. Moreover, Staley (1974) suggested that the grouping of the three species was unsatisfactory, as did Konopka et al. (1976), who found only a 0–14% binding of the DNA from three *Spirosoma* strains to that of *Microcyclus aquaticus*. Larkin et al. (1977) proposed the reintroduction of the genus *Spirosoma* and emended its description to include "*Microcyclus flavus*" and three other isolates.

The formation of rings or coils is not a sufficiently restrictive character to delineate a single taxonomic group. Ring formation is a characteristic that occurs in several other aerobic, chemoorganotrophic bacteria, e.g., *Flectobacillus, Runella, Cyclobacterium, Ancylobacter, Polaribacter*, and *Arcicella*. However, analyses of the 16S rRNA gene sequences of these organisms support the classification of *Spirosoma* as a distinct genus (Nikitin et al., 2004; Woese et al., 1990). Moreover, as the description of *Spirosoma rigui* proves, rings and coils are not properties of all members of this genus. This observation serves to further illustrate the unreliability of this characteristic in taxonomic assignments.

Vancanneyt et al. (2006) reported that, by 16S rRNA analyses, *Spirosoma linguale* was the nearest neighbor of *Larkinella insperata*, sharing a similarity of 88.8%.

The literature concerning the utilization of sole carbon sources by *Spirosoma linguale* is conflicting. Raj (1970) reported an inability of the species to use malonate, succinate, or tartrate, in contrast to results obtained by Larkin et al. (1977). The discrepancy may be attributable to a difference in the media employed or to the method of detecting growth. Raj (1970) used Simmons citrate agar with various organic substrates substituted for citrate and with 0.2% yeast extract added; growth responses were indicated by a color change. Larkin et al. (1977) used the agar medium of Gordon and Mihm (1957) and estimated visible growth after four successive transfers even in the absence of a color change.

The literature concerning gelatin hydrolysis also is conflicting. Larkin and Borrall (1984d) indicated that all strains hydrolyze gelatin, but Vancanneyt et al. (2006) reported that the type strain did not hydrolyze gelatin.

List of species of the genus *Spirosoma*

1. **Spirosoma linguale** (Eisenberg 1891) Migula 1894, 235[AL] (*Vibrio lingualis* Eisenberg 1891, 212; *Microcyclus flavus* Raj 1970, 62

lin.gua'le. L. n. *lingua* the tongue; L. neut. suff. *-ale* suffix denoting pertaining to; N.L. neut. adj. *linguale* of the tongue.

The characteristics are as described for the genus, with the following additional features. Morphological features are depicted in Figure 3.1 of Larkin and Borrall (1984d). Other features are as described for the genus, with the following additional characteristics. Colonies are circular and convex with an entire margin. Aerobic. Temperature range for growth of the type strain, 5–39°C (Vancanneyt et al., 2006). NaCl range for growth of the type strain, 0–1.25% (Vancanneyt et al., 2006). Starch is weakly hydrolyzed. Tweens 20, 40, and 80 are hydrolyzed by the type strain (Vancanneyt et al., 2006). Some strains hydrolyze esculin and (weakly) casein. The type strain produces β-galactosidase and alkaline phosphatase (Vancanneyt et al., 2006). The following tests are positive: acid production from arabinose, α-methyl-D-glucoside, cellobiose, dextrin, fructose, galactose, glucose, inulin, lactose, maltose, mannose, melibiose, raffinose, rhamnose, ribose, salicin, sucrose, trehalose, and xylose; utilization of glycerol phosphate, malonate, succinate, and tartrate as sole carbon sources. The type strain is reported to utilize mannose and sucrose (Vancanneyt et al., 2006). The following tests are negative: urease; lysine decarboxylase; phenylalanine deaminase; hemolysin production; indole formation; methyl red test; Voges–Proskauer test; nitrate reduction; H_2S production from peptone; acid production from sorbose, dulcitol, erythritol, glycerol, mannitol, and sorbitol; utilization of acetate, benzoate, citrate, formate, methanol, methylamine, and propionate. L-arabinose, D-glucose, and D-lactose are not utilized by the type strain (Vancanneyt et al., 2006). Found in soil and fresh water.

DNA G+C content (mol%): 51–53 (T_m).

Type strain: DSM 74, ATCC 33905, LMG 10896.

Sequence accession no. (16S rRNA gene): AM000023.

2. **Spirosoma rigui** Baik, Kim, Park, Lee, Lee, Ka, Choi and Seong 2007b, 2872[VP]

ri'gu.i. L. gen. n. *rigui* of a well-watered place, referring to the site of isolation, the Woopo wetland, Korea

The characteristics are as described for the genus, with the following additional features. Cells are rod shaped and do not form rings or coils. Colonies are circular, opaque, convex, smooth, wet and slimy. Facultatively anaerobic. Temperature range for growth, 4–37°C. NaCl range for growth, 0–1%. Uses aesculin and glucose as sole carbon and energy sources. Produces β-galactosidase but not alkaline phosphatase. Does not use the sugars cellobiose, fructose, galactose, lactose, maltose, mannose, melibiose, raffinose, rhamnose, starch, sucrose, and xylose. Does not produce H_2S or acetoin. Isolated from fresh water.

DNA G+C content (mol%): 53.3 (T_m).

Type strain: WPCB118, KCTC 12531, NBRC 101117.

Sequence accession no. (16S rRNA gene): EF507900.

Genus XVII. **Sporocytophaga** Stanier 1940, 629[AL]

EDWARD R. LEADBETTER

Spo.ro.cy.toph'a.ga. Gr. n. *spora* a seed, and in biology a spore; N.L. fem. n. *Cytophaga* genus name of a bacterium; N.L. fem. n. *Sporocytophaga* sporing *Cytophaga*.

Flexible rods with rounded ends, 0.3–0.5 × 5–8 μm, occurring singly. Sphaeroplasts and distorted cells occur in older cultures. A resting stage, the microcyst, is formed. Motile by gliding. Stains Gram-negative. Chemo-organotrophs. **Strict aerobe**. Metabolism is respiratory, with molecular oxygen used as terminal electron acceptor. **Cellobiose, cellulose, glucose** and, for some strains, **mannose are the only known sources of carbon and energy**. Agar and chitin are not known to be metabolized. Either ammonium or nitrate ions, or peptone, urea or yeast extract, can serve as sole nitrogen source. Amino acids, peptones, yeast extract or nutrient agar (Difco) cannot serve as sole carbon and energy sources. No organic growth factor requirements are known. **Catalase-positive**. Temperature optimum: ~30°C.

DNA G+C content (mol%): 36 (Bd).

Type species: **Sporocytophaga myxococcoides** (Krzemieniewska 1933) Stanier 1940, 629[AL].

Further descriptive information

Only one species of the genus has been extensively examined. *Sporocytophaga myxococcoides* was shown by Stanier (1942) to grow on glucose sterilized by filtration and by Kaars Sijpesteijn and Fåhraeus (1949) to grow on glucose autoclaved separately from other components of the medium, thus refuting the assertion that growth of the organism is obligately linked to cellulose utilization. Further studies indicate that the organism, when isolated from nature, is unable either to oxidize or to utilize glucose but putative mutants able to use glucose arise in the population (Leadbetter, unpublished observations). These "mutants" are able to metabolize immediately either cellulose or glucose, irrespective of the substrate in which they are grown. These observations thus confirm and extend those of Kaars Sijpesteijn and Fåhraeus (1949).

Recent studies of *Sporocytophaga myxococcoides* have demonstrated that the organism is able to form microcysts when either glucose or cellulose is the carbon and energy source (Gallin and Leadbetter, 1966; Leadbetter, 1963).

During growth on glucose, the Embden–Meyerhof–Parnas pathway is used (Hanstveit and Goksøyr, 1974). Cells also produce a variety of cellulases, some of which have been partially purified (Osmundsvåg and Goksøyr, 1975; Vance et al., 1980).

Major cellular lipids are sulfonolipids (Godchaux and Leadbetter, 1983) which contain a variety of fatty acyl moieties (Godchaux and Leadbetter, 1984). The sulfonolipids predominate in the outer membrane, while phospholipids are present in the inner cell membrane. Phosphatidylethanolamine and lysophosphatidylethanolamine are the major phospholipids (Holt et al., 1979). Homospermidine is the most abundant polyamine, but small amounts of putrescine and spermidine are also present (Hamana and Nakagawa, 2001). The cell-wall peptidoglycan contains diaminopimelic acid (Verma and Martin, 1967). Moreover, cells produce a slime comprised of glucose, mannose, arabinose, xylose, galactose and glucuronic acid (Martin et al., 1968).

The *Bacteroides* transposon Tn*4351* can be introduced by conjugation from *Escherichia coli* into *Sporocytophaga* and could serve as the basis for a genetic system (McBride and Baker, 1996).

List of species of the genus *Sporocytophaga*

1. **Sporocytophaga myxococcoides** (Krzemieniewska 1933) Stanier 1940, 630[AL] (*Cytophaga myxococcoides* Krzemieniewska 1933, 400)

myx.o.coc.coi'des. N.L. masc. n. *Myxococcus* genus name of a bacterium; L. suff. *-oides* (from Gr. suff. *eides* from Gr. n. *eidos* that which is seen, form, shape, figure), resembling, similar; N.L. fem. adj. *myxococcoides* resembling *Myxococcus*.

Single, often flexible rods with rounded ends, 0.3–0.5 × 5–8 μm. Sphaeroplasts and abnormally long forms may occur in old cultures. The resting stage, the microcyst, is spherical and about 1.5 μm in diameter. Both the growing (vegetative) rod and the microcyst have noticeably smaller dimensions when cultures are grown on cellulose rather than on glucose. Microcysts are notably more resistant to ultrasonic disruption than are vegetative cells.

Electron microscopic studies indicate that the vegetative cell has a fine structure typical of Gram-stain-negative bacteria, while the microcyst has a thick, fibrillar capsule exterior to a highly convoluted cell wall (Holt and Leadbetter, 1967).

Growth on cellulose (filter paper) -salts agar (or silica gel) or glucose-salts agar is gummy, and liquid cultures become viscous as a result of extracellular slime production. Cell masses are yellow, reflecting the presence of carotenoid and flexirubin-type pigments. Filter paper on the agar or silica gel surface is eventually dissolved around colonies so that translucent areas result. Colonies on glucose-salts agar medium are raised.

Other characteristics are the same as those of the genus.

Other species

Other species of *Sporocytophaga* have been described, but they are incompletely characterized:

(a) "*Sporocytophaga cauliformis*" Knorr and Gräf in Gräf 1962, 124.

(b) "*Sporocytophaga congregata*" subsp. *maroonicum* Akashi 1960, 899.

(c) "*Sporocytophaga ellipsospora*" (Imshenetski and Solntseva, 1936) Stanier 1942, 190.

(d) "*Sporocytophaga ochracea*" Ueda, Ishikawa, Itami and Asai 1952, 545.

References

Aislabie, J.M., K.-L. Chhour, D.J. Saul, S. Miyauchi, J. Ayton, R.F. Paetzold and M.R. Baulks. 2006. Dominant bacteria in soils of Marble Point and Wright Valley, Victoria Land, Antarctica. Soil Biol. Biochem. *38*: 3041–3056.

Akashi, A. 1960. Studies on the cellulose-decomposing bacteria in the rumen. J. Agri. Chem. Soc. Jpn. *34*: 895–900.

Bachmann, B.J. 1955. Studies on *Cytophaga fermentans*, n.sp., a facultatively anaerobic lower myxobacterium. J. Gen. Microbiol. *13*: 541–551.

Baik, K.S., C.N. Seong, E.Y. Moon, Y.D. Park, H. Yi and J. Chun. 2006. *Hymenobacter rigui* sp. nov., isolated from wetland freshwater. Int. J. Syst. Evol. Microbiol. *56*: 2189–2192.

Baik, K.S., M.S. Kim, E.M. Kim, H.R. Kim and C.N. Seong. 2007a. *Dyadobacter koreensis* sp. nov., isolated from fresh water. Int. J. Syst. Evol. Microbiol. *57*: 1227–1231.

Baik, K.S., M.S. Kim, S.C. Park, D.W. Lee, S.D. Lee, J.O. Ka, S.K. Choi and C.N. Seong. 2007b. *Spirosoma rigui* sp. nov., isolated from fresh water. Int. J. Syst. Evol. Microbiol. *57*: 2870–2873.

Baker, C.A., G.W. Claus and P.A. Taylor. 1983. Predominant bacteria in an activated sludge reactor for the degradation of cutting fluids. Appl. Environ. Microbiol. *46*: 1214–1223.

Barbeyron, T., S. L'Haridon, E. Corre, B. Kloareg and P. Potin. 2001. *Zobellia galactanovorans* gen. nov., sp. nov., a marine species of *Flavobacteriaceae* isolated from a red alga, and classification of *Cytophaga uliginosa* (ZoBell and Upham 1944) Reichenbach 1989 as *Zobellia uliginosa* gen. nov., comb. nov. Int. J. Syst. Evol. Microbiol. *51*: 985–997.

Bernardet, J.-F., P. Segers, M. Vancanneyt, F. Berthe, K. Kersters and P. Vandamme. 1996. Cutting a Gordian knot: emended classification and description of the genus *Flavobacterium*, emended description of the family *Flavobacteriaceae*, and proposal of *Flavobacterium hydatis* nom. nov. (basonym, *Cytophaga aquatilis* Strohl and Tait 1978). Int. J. Syst. Bacteriol. *46*: 128–148.

Bernardet, J.-F., Y. Nakagawa and B. Holmes. 2002. Proposed minimal standards for describing new taxa of the family *Flavobacteriaceae* and emended description of the family. Int. J. Syst. Evol. Microbiol. *52*: 1049–1070.

Borrall, R. and J.M. Larkin. 1978. *Flectobacillus marinus* (Raj) comb. nov., a marine bacterium previously assigned to *Microcyclus*. Int. J. Syst. Bacteriol. *28*: 341–343.

Bortels, H. 1956. [Significance of trace elements for cell vibrio and *Cytophaga* species types.]. Arch. Mikrobiol. *25*: 225–246.

Bowman, J.P. 2000. Description of *Cellulophaga algicola* sp. nov., isolated from the surfaces of Antarctic algae, and reclassification of *Cytophaga uliginosa* (ZoBell and Upham 1944) Reichenbach 1989 as *Cellulophaga uliginosa* comb. nov. Int. J. Syst. Evol. Microbiol. *50*: 1861–1868.

Bowman, J.P., C. Mancuso, C.M. Nichols and J.A.E. Gibson. 2003. *Algoriphagus ratkowskyi* gen. nov., sp. nov., *Brumimicrobium glaciale* gen. nov., sp. nov., *Cryomorpha ignava* gen. nov., sp. nov. and *Crocinitomix catalasitica* gen. nov., sp. nov., novel flavobacteria isolated from various polar habitats. Int. J. Syst. Evol. Microbiol. *53*: 1343–1355.

Brinsmade, S.R., T. Paldon and J.C. Escalante-Semerena. 2005. Minimal functions and physiological conditions required for growth of *Salmonella enterica* on ethanolamine in the absence of the metabolosome. J. Bacteriol. *187*: 8039–8046.

Buczolits, S., E.B. Denner, D. Vybiral, M. Wieser, P. Kämpfer and H.J. Busse. 2002. Classification of three airborne bacteria and proposal of *Hymenobacter aerophilus* sp. nov. Int. J. Syst. Evol. Microbiol. *52*: 445–456.

Buczolits, S., E.B. Denner, P. Kämpfer and H.J. Busse. 2006. Proposal of *Hymenobacter norwichensis* sp. nov., classification of '*Taxeobacter ocellatus*', '*Taxeobacter gelupurpurascens*' and '*Taxeobacter chitinovorans*' as *Hymenobacter ocellatus* sp. nov., *Hymenobacter gelipurpurascens* sp. nov. and *Hymenobacter chitinivorans* sp. nov., respectively, and emended description of the genus *Hymenobacter* Hirsch *et al.* 1999. Int. J. Syst. Evol. Microbiol. *56*: 2071–2078.

Callies, E. and W. Mannheim. 1978. Classification of the *Flavobacterium-Cytophaga* complex on the basis of respiratory quinones and fumarate respiration. Int. J. Syst. Bacteriol. *28*: 14–19.

Chaturvedi, P., G.S. Reddy and S. Shivaji. 2005. *Dyadobacter hamtensis* sp. nov., from Hamta glacier, located in the Himalayas, India. Int. J. Syst. Evol. Microbiol. *55*: 2113–2117.

Chelius, M.K. and E.W. Triplett. 2000. *Dyadobacter fermentans* gen. nov., sp. nov., a novel Gram-negative bacterium isolated from surface-sterilized *Zea mays* stems. Int. J. Syst. Evol. Microbiol. *50*: 751–758.

Chelius, M.K., J.A. Henn and E.W. Triplett. 2002. *Runella zeae* sp. nov., a novel Gram-negative bacterium from the stems of surface-sterilized Zea mays. Int. J. Syst. Evol. Microbiol. *52*: 2061–2063.

Christensen, P.J. 1977. The history, biology, and taxonomy of the *Cytophaga* group. Can. J. Microbiol. *23*: 1599–1653.

Christensen, P. 1980. *Flexibacter canadensis* sp. nov. Int. J. Syst. Bacteriol. *30*: 429–432.

Claus, D. 1967. Taxonomy of some highly pleomorphic bacteria Spisy Prirodoved Fak. Univ. J. E. Purkyne Brno *40*: 254–257.

Claus, D., J.E. Bergendahl and M. Mandel. 1968. DNA base composition of *Microcyclus* species and organisms of similar morphology. Arch. Mikrobiol. *63*: 26–28.

Collins, M.D., R.A. Hutson, I.R. Grant and M.F. Patterson. 2000. Phylogenetic characterization of a novel radiation-resistant bacterium from irradiated pork: description of *Hymenobacter actinosclerus* sp. nov. Int. J. Syst. Evol. Microbiol. *50*: 731–734.

Costerton, J.W., Z. Lewandowski, D.E. Caldwell, D.R. Korber and H.M. Lappin-Scott. 1995. Microbial biofilms. Annu. Rev. Microbiol. *49*: 711–745.

DeLey, J., H. Cattoir and A. Reynaerts. 1970. The quantitative measurement of DNA hybridization from renaturation rates. Eur. J. Biochem. *12*: 133–142.

Dong, Z., X. Guo, X. Zhang, F. Qiu, L. Sun, H. Gong and F. Zhang. 2007. *Dyadobacter beijingensis* sp. nov., isolated from the rhizosphere of turf grasses in China. Int. J. Syst. Evol. Microbiol. *57*: 862–865.

Eisenberg, J. 1891. Bacteriologische Diagnostik. Hilfstabellen zum Gebrauche beim Praktischen Arbeiten. 3 Aufl, vols. VII-XXX. Leopold Voss, Hamburg.

Euzéby, J.P. 1998. Taxonomic note: necessary correction of specific and subspecific epithets according to Rules 12c and 13b of the International Code of Nomenclature of *Bacteria* (1990 revision). Int. J. Syst. Bacteriol. *48*: 1073–1075.

Gallin, J.I. and E.R. Leadbetter. 1966. Morphogenesis of *Sporocytophaga*. Bacteriol. Proc. 75.

Garcia-Pichel, F. 2002. Desert environments: biological soil crusts. Encyclopedia of Environmental Microbiology. John Wiley, New York.

Garrity, G.M., J.A. Bell and T. Lilburn. 2005. The Revised Road Map to the Manual. *In* Bergey's Manual of Systematic Bacteriology, 2nd edn, vol. 2, The *Proteobacteria*, Part A, Introductory Essays (edited by Brenner, Krieg, Staley and Garrity). Springer, New York, pp. 159–220.

Godchaux, W., III and E.R. Leadbetter. 1983. Unusual sulfonolipids are characteristic of the *Cytophaga-Flexibacter* group. J. Bacteriol. *153*: 1238–1246.

Godchaux, W., III and E.R. Leadbetter. 1984. Sulfonolipids of gliding bacteria. Structure of the *N*-acylaminosulfonates. J. Biol. Chem. *259*: 2982–2990.

Gordon, R.E. and J.M. Mihm. 1957. A comparative study of some strains received as nocardiae. J. Bacteriol. *73*: 15–27.

Gosink, J.J., C.R. Woese and J.T. Staley. 1998. *Polaribacter* gen. nov., with three new species, *P. irgensii* sp. nov., *P. franzmannii* sp. nov., and *P. filamentus* sp. nov., gas vacuolate polar marine bacteria of the *Cytophaga-Flavobacterium-Bacteroides* group and reclassification of '*Flectobacillus glomeratus*' as *Polaribacter glomeratus* comb. nov. Int. J. Syst. Bacteriol. *48*: 223–235.

Gräf, W. 1962. Über Wassermyxobakterien. (in German with English and French abstracts). Arch. Hyg. Bakteriol. *146*: 114–125.

Gromov, B.V. 1963. A new bacterium of the genus *Microcyclus*. Dokl. Akad. Nauk. SSSR *152*: 733–734.

Haack, S.K. and J.A. Breznak. 1993. *Cytophaga xylanolytica* sp. nov., a xylan-degrading, anaerobic gliding bacterium. Arch. Microbiol. *159*: 6–15.

Hamana, K. and Y. Nakagawa. 2001. Polyamine distribution profiles in the eighteen genera phylogenetically located within the *Flavobacterium-Flexibacter-Cytophaga* complex. Microbios *106*: 7–17.

Hanstveit, A.O. and J. Goksøyr. 1974. The pathway of glucose catabolism in *Sporocytophaga myxococcoides*. J. Gen. Microbiol. *81*: 27–35.

Hayes, P.R. 1977. A taxonomic study of *Flavobacteria* and related Gram-negative yellow pigmented rods. J. Appl. Bacteriol. *43*: 345–367.

Henrichsen, J. 1972. Bacterial surface translocation: a survey and a classification. Bacteriol. Rev. *36*: 478–503.

Hirsch, P., W. Ludwig, C. Hethke, M. Sittig, B. Hoffmann and C.A. Gallikowski. 1998a. *Hymenobacter roseosalivarius* gen. nov., sp. nov. from continental Antartica soils and sandstone: bacteria of the *Cytophaga / Flavobacterium/ Bacteroides* line of phylogenetic descent. Syst. Appl. Microbiol. *21*: 374–383.

Hirsch, P., W. Ludwig, C. Hethke, M. Sittig, B. Hoffmann and C.A. Gallikowski. 1998b. *Hymenobacter roseosalivarius* gen. nov., sp. nov. from continental Antarctic soils and sandstone: *Bacteria* of the *Cytophaga / Flavobacterium/ Bacteroides* line of phylogenetic descent. Syst. Appl. Microbiol. *21*: 374–383.

Hirsch, P., W. Ludwig, C. Hethke, M. Sittig, B. Hoffmann and C.A. Gallikowski. 1999. *In* Validation of publication of new names and new combinations previously effectively published outside the IJSB. List no. 68. Int. J. Syst. Bacteriol. *49*: 1–3.

Holt, S.C. and E.R. Leadbetter. 1967. Fine structure of *Sporocytophaga myxococcoides*. Arch. Mikrobiol. *57*: 199–213.

Holt, S.C., J. Doundowlakis and B.J. Takas. 1979. Phospholipid composition of gliding bacteria: Oral isolates of *Capnocytophaga* compared with *Sporocytophaga*. Infect. Immun. *26*: 305–310.

Hosoya, R. and K. Hamana. 2004. Distribution of two triamines, spermidine and homospermidine, and an aromatic amine, 2-phenylethylamine, within the phylum *Bacteroidetes*. J. Gen. Appl. Microbiol. *50*: 255–260.

Hosoya, S. and A. Yokota. 2007. Reclassification of *Flexibacter aggregans* (Lewin 1969) Leadbetter 1974 as a later heterotypic synonym of *Flexithrix dorotheae* Lewin 1970. Int. J. Syst. Evol. Microbiol. *57*: 1086–1088.

Hugh, R. and E. Leifson. 1953. The taxonomic significance of fermentative versus oxidative metabolism of carbohydrates by various gram negative bacteria. J. Bacteriol. *66*: 24–26.

Hwang, C.Y. and B.C. Cho. 2006. *Flectobacillus lacus* sp. nov., isolated from a highly eutrophic pond in Korea. Int. J. Syst. Evol. Microbiol. *56*: 1197–1201.

Imshenetski, A. and L. Solntseva. 1936. On aerobic cellulose-decomposing bacteria (In Russian with En. summary). Bull. Acad. Sci. U.S.S.R. *Biol., no. 6*: 1115–1172.

Irgens, R.L. 1977. *Meniscus*, a new genus of aerotolerant, gas-vacuolated bacteria. Int. J. Syst. Bacteriol. *27*: 38–43.

Johansen, J.E., P. Nielsen and C. Sjoholm. 1999. Description of *Cellulophaga baltica* gen. nov., sp. nov., and *Cellulophaga fucicola* gen. nov., sp. nov. and reclassification of *Cytophaga lytica* to *Cellulophaga lytica* gen. nov., comb. nov. Int. J. Syst. Bacteriol. *49*: 1231–1240.

Kaars Sijpesteijn, A. and G. Fåhraeus. 1949. Adaptation of *Sporocytophaga myxococcoides* to sugars. J. Gen. Microbiol. *3*: 224–234.

Kämpfer, P., C.C. Young, K.R. Sridhar, A.B. Arun, W.A. Lai, F.T. Shen and P.D. Rekha. 2006. Transfer of [*Flexibacter*] *sancti*, [*Flexibacter*] *filiformis*, [*Flexibacter*] *japonensis* and [*Cytophaga*] *arvensicola* to the genus *Chitinophaga* and description of *Chitinophaga skermanii* sp. nov. Int. J. Syst. Evol. Microbiol. *56*: 2223–2228.

Kim, K.H., W.T. Im and S.T. Lee. 2008. *Hymenobacter soli* sp. nov., isolated from grass soil. Int. J. Syst. Evol. Microbiol. *58*: 941–945.

Konopka, A.E., R.L. Moore and J.T. Staley. 1976. Taxonomy of *Microcyclus* and other nonmotile ring-forming bacteria. Int. J. Syst. Bacteriol. *26*: 505–510.

Kottel, R.H. and H.D. Raj. 1973. Pathways of carbohydrate metabolism in *Microcyclus* species. J. Bacteriol. *113*: 341–349.

Kraepelin, G. and D. Passern. 1980. Gallertlager einer besonderen Mikroorganismengesellschaft an verbautem Grubenholz. Z. Allg. Mikrobiol. *20*: 303–314.

Krzemieniewska, H. 1933. Contribution á l'étude du genre *Cytophaga* (Winogradsky). Arch. Mikrobiol. *4*: 394–408.

Larkin, J.M. and P.M. Williams. 1978. *Runella slithyformis* gen. nov., sp. nov., a curved, nonflexible, pink bacterium. Int. J. Syst. Bacteriol. *28*: 32–36.

Larkin, J.M. and R. Borrall. 1984a. Family I. *Spirosomaceace*. *In* Bergey's Manual of Systematic Bacteriology, vol. 1 (edited by Krieg and Holt). Williams & Wilkins, Baltimore, pp. 125–132.

Larkin, J.M. and R. Borrall. 1984b. Genus *Runella*. *In* Bergey's Manual of Systematic Bacteriology, vol. 1 (edited by Krieg and Holt). Williams & Wilkins, Baltimore, pp. 128–129.

Larkin, J.M. and R. Borrall. 1984c. Deoxyribonucleic acid base composition and homology of *Microcyclus*, *Spirosoma*, and similar organisms. Int. J. Syst. Bacteriol. *34*: 211–215.

Larkin, J.M. and R. Borrall. 1984d. Genus *Spirosoma*. *In* Bergey's Manual of Systematic Bacteriology, vol. 1 (edited by Krieg and Holt). Williams & Wilkins, Baltimore, pp. 126–128.

Larkin, J.M., P.M. Williams and R. Taylor. 1977. Taxonomy of genus *Microcyclus* Ørskov 1928, reintroduction and emendation of genus *Spirosoma Migula* 1894, and proposal of a new genus, *Flectobacillus*. Int. J. Syst. Bacteriol. *27*: 147–156.

Leadbetter, E.R. 1963. Growth and morphogenesis of *Sporocytophaga myxococcoides*. Bacteriol. Proc. 42.

Leadbetter, E.R. 1974. Family I. *Cytophagaceae* Stanier 1940, 630, emend. mut. char. *In* Bergey's Manual of Determinative Bacteriology, 8th edn (edited by Buchanan and Gibbons). Williams & Wilkins, Baltimore, pp. 99–112.

Leadbetter, E.R. 1989. Genus IV. *Sporocytophaga*. *In* Bergey's Manual of Systematic Bacteriology, vol. 3 (edited by Staley, Bryant, Pfennig and Holt). Williams & Wilkins, Baltimore, p. 2061.

Lewin, R.A. 1969. A classification of *Flexibacteria*. J. Gen. Microbiol. *58*: 189–206.

Lewin, R.A. 1974. *Flexibacter polymorphus*, a new marine species. J. Gen. Microbiol. *82*: 393–403.

Lewin, R.A. and D.M. Lounsbery. 1969. Isolation, cultivation and characterization of *Flexibacteria*. J. Gen. Microbiol. *58*: 145–170.

Liu, Q.M., W.T. Im, M. Lee, D.C. Yang and S.T. Lee. 2006. *Dyadobacter ginsengisoli* sp. nov., isolated from soil of a ginseng field. Int. J. Syst. Evol. Microbiol. *56*: 1939–1944.

Lu, S., J.R. Lee, S.H. Ryu, B.S. Chung, W.S. Choe and C.O. Jeon. 2007. *Runella defluvii* sp. nov., isolated from a domestic wastewater treatment plant. Int. J. Syst. Evol. Microbiol. *57*: 2600–2603.

Ludwig, W., J. Euzéby and W.B. Whitman. 2009. Draft taxonomic outline of the *Bacteroidetes*, *Spirochaetes*, *Tenericutes* (*Mollicutes*), *Acidobacteria*, *Fibrobacteres*, *Fusobacteria*, *Dictyoglomi*, *Gemmatimonadetes*, *Lentisphaerae*, *Verrucomicrobia*, *Chlamydiae*, and *Planctomycetes*. *In* Bergey's Manual of Systematic Bacteriology, 2nd edn, vol. 4, The *Bacteroidetes*, *Spirochaetes*, *Tenericutes* (*Mollicutes*), *Acidobacteria*, *Fibrobacteres*, *Fusobacteria*, *Dictyoglomi*, *Gemmatimonadetes*, *Lentisphaerae*, *Verrucomicrobia*, *Chlamydiae*, and *Planctomycetes* (edited by Krieg, Staley, Brown, Hedlund, Paster, Ward, Ludwig and Whitman). Springer, New York.

Lünsdorf, H., I. Kristen and E. Barth. 2006. Cationic hydrous thorium dioxide colloids–a useful tool for staining negatively charged surface matrices of bacteria for use in energy-filtered transmission electron microscopy. BMC Microbiol. *6*: 59.

Maloy, S.R., L.A. Anderson and H.D. Raj. 1978. Abst. 1-136. Presented at the Annu. Meet. Am. Soc. Microbiol., Washington, DC

Martin, H.H., H.-J. Preusser and J.P. Verma. 1968. Über die Oberflächenstruktur von Myxobakterien. II. Anionische Heteropolysaccharide als Baustoffe der Schleimhülle von *Cytophaga hutchinsonii* and *Sporocytophaga myxococcoides*. Arch. Mikrobiol. *62*: 72–84.

McBride, M.J. and S.A. Baker. 1996. Development of techniques to genetically manipulate members of the genera *Cytophaga, Flavobacterium, Flexibacter,* and *Sporocytophaga.* Appl. Environ. Microbiol. *62*: 3017–3022.

McGuire, A.J., P.D. Franzmann and T.A. McMeekin. 1987. *Flectobacillus glomeratus* sp. nov., a curved, nonmotile, pigmented bacterium isolated from antarctic marine environments. Syst. Appl. Microbiol. *9*: 265–272.

McMeekin, T.A. and J.M. Shewan. 1978. A review. Taxonomic strategies for *Flavobacterium* and related genera. J. Appl. Bacteriol. *45*: 321–332.

Migula, W. 1894. Über ein neues System der Bakterien. Arb. Bakteriol. Inst. Karlsruhe. *1*: 235–238.

Miller, J.C. and H.D. Raj. 1978. Abst. K-180. Presented at the Annu. Meet. Am. Soc. Microbiol., Washington, DC

Miteva, V.I., P.P. Sheridan and J.E. Brenchley. 2004. Phylogenetic and physiological diversity of microorganisms isolated from a deep greenland glacier ice core. Appl. Environ. Microbiol. *70*: 202–213.

Mitsui, H., K. Gorlach, H.-J. Lee, R. Hattori and T. Hattori. 1997. Incubation time and media requirements of culturable bacteria from different phylogenetic groups. J. Microbiol. Methods *30*: 103–110.

Nagy, M.L., A. Perez and F. Garcia-Pichel. 2005. The prokaryotic diversity of biological soil crusts in the Sonoran Desert (Organ Pipe Cactus National Monument, AZ). FEMS Microbiol. Ecol. *54*: 233–245.

Nakagawa, Y. and K. Yamasato. 1993. Phylogenetic diversity of the genus *Cytophaga* revealed by 16S ribosomal RNA sequencing and menaquinone analysis. J. Gen. Microbiol. *139*: 1155–1161.

Nakagawa, Y. and K. Yamasato. 1996. Emendation of the genus *Cytophaga* and transfer of *Cytophaga agarovorans* and *Cytophaga salmonicolor* to *Marinilabilia* gen. nov: phylogenetic analysis of the *Flavobacterium cytophaga* complex. Int. J. Syst. Bacteriol. *46*: 599–603.

Nakagawa, Y., K. Hamana, T. Sakane and K. Yamasato. 1997. Reclassification of *Cytophaga aprica* (Lewin 1969) Reichenbach 1989 in *Flammeovirga* gen. nov. as *Flammeovirga aprica* comb. nov. and of *Cytophaga diffluens* (ex Stanier 1940; emend. Lewin 1969) Reichenbach 1989 in *Persicobacter* gen. nov. as *Persicobacter diffluens* comb. nov. Int. J. Syst. Bacteriol. *47*: 220–223.

Nakagawa, Y., T. Sakane, M. Suzuki and K. Hatano. 2002. Phylogenetic structure of the genera *Flexibacter, Flexithrix,* and *Microscilla* deduced from 16S rRNA sequence analysis. J. Gen. Appl. Microbiol. *48*: 155–165.

Nedashkovskaya, O.I., M. Suzuki, M.V. Vysotskii and V.V. Mikhailov. 2003. *Reichenbachia agariperforans* gen. nov., sp. nov., a novel marine bacterium in the phylum *Cytophaga-Flavobacterium-Bacteroides.* Int. J. Syst. Evol. Microbiol. *53*: 81–85.

Nedashkovskaya, O.I., S.B. Kim, M.S. Lee, M.S. Park, K.H. Lee, A.M. Lysenko, H.W. Oh, V.V. Mikhailov and K.S. Bae. 2005a. *Cyclobacterium amurskyense* sp. nov., a novel marine bacterium isolated from sea water. Int. J. Syst. Evol. Microbiol. *55*: 2391–2394.

Nedashkovskaya, O.I., S.B. Kim, A.M. Lysenko, G.M. Frolova, V.V. Mikhailov, K.H. Lee and K.S. Bae. 2005b. Description of *Aquimarina muelleri* gen. nov., sp. nov., and proposal of the reclassification of [*Cytophaga*] *latercula* Lewin 1969 as *Stanierella latercula* gen. nov., comb. nov. Int. J. Syst. Evol. Microbiol. *55*: 225–229.

Nedashkovskaya, O.I., S.B. Kim, M. Suzuki, L.S. Shevchenko, M.S. Lee, K.H. Lee, M.S. Park, G.M. Frolova, H.W. Oh, K.S. Bae, H.Y. Park and V.V. Mikhailov. 2005c. *Pontibacter actiniarum* gen. nov., sp. nov., a novel member of the phylum '*Bacteroidetes*', and proposal of *Reichenbachiella* gen. nov. as a replacement for the illegitimate prokaryotic generic name *Reichenbachia* Nedashkovskaya *et al.* 2003. Int. J. Syst. Evol. Microbiol. *55*: 2583–2588.

Nedashkovskaya, O.I., M. Vancanneyt, P. Dawyndt, K. Engelbeen, K. Vandemeulebroecke, I. Cleenwerck, B. Hoste, J. Mergaert, T.L. Tan, G.M. Frolova, V.V. Mikhailov and J. Swings. 2005d. Reclassification of [*Cytophaga*] *marinoflava* Reichenbach 1989 as *Leeuwenhoekiella marinoflava* gen. nov., comb. nov. and description of *Leeuwenhoekiella aequorea* sp. nov. Int. J. Syst. Evol. Microbiol. *55*: 1033–1038.

Nedashkovskaya, O.I., M. Vancanneyt, L. Christiaens, N.I. Kalinovskaya, V.V. Mikhailov and J. Swings. 2006. *Aquimarina intermedia* sp. nov., reclassification of *Stanierella latercula* (Lewin 1969) as *Aquimarina latercula* comb. nov. and *Gaetbulimicrobium brevivitae* Yoon *et al.* 2006 as *Aquimarina brevivitae* comb. nov. and emended description of the genus *Aquimarina.* Int. J. Syst. Evol. Microbiol. *56*: 2037–2041.

Nikitin, D.I., M.S. Oranskaya, I.A. Pitryuk, N.A. Chernykh and A.M. Lysenko. 1994. A new ring-forming bacterium Arcocella aquatica gen. et sp. nov. Microbiology (En. transl. from Mikrobiologiya) *63*: 87–90.

Nikitin, D.I., C. Strömpl, M.S. Oranskaya and W.R. Abraham. 2004. Phylogeny of the ring-forming bacterium *Arcicella aquatica* gen. nov., sp. nov. (*ex* Nikitin *et al.* 1994), from a freshwater neuston biofilm. Int. J. Syst. Evol. Microbiol. *54*: 681–684.

Nour, S.M., J.R. Lawrence, H. Zhu, G.D.W. Swerhone, M. Welsh, T.W. Welacky and E. Topp. 2003. *Bacteria* associated with cysts of the soybean cyst nematode (*Heterodera glycines*). Appl. Environ. Microbiol. *69*: 607–615.

Osmundsvåg, K. and J. Goksøyr. 1975. Cellulases from *Sporocytophaga myxococcoides.* Eur. J. Biochem. *57*: 405–409.

Overbeck, J. 1974. Microbiology and Biochemistry. Mitt. Int. Verein. Limnol. *20*: 198–288.

Oyaizu, H. and K. Komagata. 1981. Chemotaxonomic and phenotypic characterization of the strains of species in the *Flavobacterium-Cytophaga* complex. J. Gen. Appl. Microbiol. *27*: 57–107.

Perry, L.B. 1973. Gliding motility in some non-spreading *Flexibacteria.* J. Appl. Bacteriol. *36*: 227–232.

Pringsheim, E.G. 1951. The *Vitreoscillaceae*; a family of colourless, gliding, filamentous organisms. J. Gen. Microbiol. *5*: 124–149.

Raj, H.D. 1970. A new species: *Microcyclus flavus.* Int. J. Syst. Bacteriol. *20*: 61–81.

Raj, H.D. 1976. A new species: *Microcyclus marinus.* Int. J. Syst. Bacteriol. *26*: 528–544.

Raj, H.D. 1979. Adansonian analysis of *Microcyclus* and related bacteria. Abstract I-31. Proceedings of the Annu. Meet. Am. Soc. Microbiol.

Raj, H.D. and S.R. Maloy. 1990. Proposal of *Cyclobacterium marinus* gen. nov., comb. nov. for a marine bacterium previously assigned to the genus *Flectobacillus.* Int. J. Syst. Bacteriol. *40*: 337–347.

Raj, H.D. and E.J. Ordal. 1977. *Microcyclus* and related ring-forming bacteria. CRC Crit. Rev. Microbiol. *5*: 243–269.

Reasoner, D.J. and E.E. Geldreich. 1985. A new medium for the enumeration and subculture of bacteria from potable water. Appl. Environ. Microbiol. *49*: 1–7.

Reddy, G.S.N. and F. Garcia-Pichel. 2005. *Dyadobacter crusticola* sp. nov., from biological soil crusts in the Colorado Plateau, USA, and an emended description of the genus *Dyadobacter Chelius* and Triplett 2000. Int. J. Syst. Evol. Microbiol. *55*: 1295–1299.

Reddy, G.S.N. and F. Garcia-Pichel. 2006. The community and phylogenetic diversity of biological soil crusts in the Colorado Plateau studied by molecular fingerprinting and intensive cultivation Microb. Ecol. *52*: 345–357.

Redell, M.A., S.R. Maloy and H. D. Raj. 1981. Abst. 1–63. Proceedings of the Annu. Meet. Am. Soc. Microbiol.

Reichenbach, H. 1989a. Genus *Microscilla Pringsheim* 1951, 140, emend. Lewin 1969, 194[AL]. *In* Bergey's Manual of Systematic Bacteriology, vol. 3 (edited by Staley, Bryant, Pfennig and Holt). Williams & Wilkins, Baltimore, pp. 2071–2073.

Reichenbach, H. 1989b. Genus *Flexibacter Soriano* 1945, 92[AL] emend. *In* Bergey's Manual of Systematic Bacteriology, vol. 3 (edited by Staley, Bryant, Pfennig and Holt). Williams & Wilkins, Baltimore, pp. 2061–2071.

Reichenbach, H. 1989c. Genus I. *Cytophaga. In* Bergey's Manual of Systematic Bacteriology, vol. 3 (edited by Staley, Bryant, Pfennig and Holt). Williams & Wilkins, Baltimore, pp. 2015–2050.

Reichenbach, H. 1992a. The order *Cytophagales. In* The Prokaryotes: a Handbook on the Biology of *Bacteria*: Ecophysiology, Isolation,

Identification, Applications, 2nd edn, vol. 4 (edited by Balows, Trüper, Dworkin, Harder and Schleifer). Springer, New York, pp. 3631–3675.

Reichenbach, H. 1992b. *Taxeobacter*, a new genus of the *Cytophagales* with three new species. *In* Advances in the Taxonomy and Significance of *Flavobacterium, Cytophaga* and Related Bacteria. Proc. 2nd Int. Symposium on *Flavobacterium, Cytophaga* and related bacteria Bloemfontein, South Africa, 2–5 April 1992 (edited by Jooste). University Press, Bloemfontein, Republic of South Africa, pp. 182–185.

Rickard, A.H., A.J. McBain, A.T. Stead and P. Gilbert. 2004. Shear rate moderates community diversity in freshwater biofilms. Appl. Environ. Microbiol. *70*: 7426–7435.

Rickard, A.H., A.T. Stead, G.A. O'May, S. Lindsay, M. Banner, P.S. Handley and P. Gilbert. 2005. *Adhaeribacter aquaticus* gen. nov., sp. nov., a Gram-negative isolate from a potable water biofilm. Int. J. Syst. Evol. Microbiol. *55*: 821–829.

Ryu, S.H., T.T. Nguyen, W. Park, C.J. Kim and C.O. Jeon. 2006. *Runella limosa* sp. nov., isolated from activated sludge. Int. J. Syst. Evol. Microbiol. *56*: 2757–2760.

Saha, P. and T. Chakrabarti. 2006. *Emticicia oligotrophica* gen. nov., sp. nov., a new member of the family '*Flexibacteraceae*', phylum *Bacteroidetes*. Int. J. Syst. Evol. Microbiol. *56*: 991–995.

Sakane, T., T. Nishii, T. Itoh and K. Mikata. 1996. Protocols for long-term preservation of microorganisms by L-drying (in Japanese). Microbiol. Cult. Coll. *12*: 91–97.

Shewan, J.M. and T.A. McMeekin. 1983. Taxonomy (and ecology) of *Flavobacterium* and related genera. Annu. Rev. Microbiol. *37*: 233–252.

Sieburth, J.M. 1978. Sea Microbes. Oxford University Press, New York.

Skuja, H. 1964. Grundzuege der Algenflora und Algenvegetation der Fjeldgegenden um Abisko in Schwedisch-Lappland. Nova Acta Reg. Soc. Sci. Upsal. Ser. IV *18*: 1–139.

Smith, S.M., R.M. Abed and F. Garcia-Pichel. 2004. Biological soil crusts of sand dunes in Cape Cod National Seashore, Massachusetts, USA. Microb. Ecol. *48*: 200–208.

Soriano, S. 1945. Un nuevo orden de bacterias: *Flexibacteriales*. Cienc. Invest. (Buenos Aires) *1*: 92–93.

Stackebrandt, E. and B.M. Goebel. 1994. Taxonomic note: a place for DNA–DNA reassociation and 16S rRNA sequence analysis in the present species definition in bacteriology. Int. J. Syst. Bacteriol. *44*: 846–849.

Staley, J.T. 1968. *Prosthecomicrobium* and *Ancalomicrobium*: new prosthecate freshwater bacteria. J. Bacteriol. *95*: 1921–1942.

Staley, J.T. 1974. Genus *Microcyclus*. *In* Bergey's Manual of Determinative Bacteriology, 8th edn (edited by Buchanan and Gibbons). Williams & Wilkins, Baltimore, pp. 214–215.

Staley, J.T. and A.E. Konopka. 1984. Genus *Microcyclus* Ørskov. *In* Bergey's Manual of Systematic Bacteriology, vol. 1 (edited by Krieg and Holt). Williams & Wilkins, Baltimore, pp. 133–135.

Stanier, R.Y. 1940. Studies on the *Cytophagas*. J. Bacteriol. *40*: 619–635.

Stanier, R.Y. 1941. Studies on marine agar-digesting *Bacteria*. J. Bacteriol. *42*: 527–559.

Stanier, R.Y. 1942. The *Cytophaga* group: a contribution to the biology of myxobacteria. Bacteriol. Rev. *6*: 143–196.

Stanier, R.Y. 1947. Studies on non-fruiting myxobacteria. I. *Cytophaga johnsonae* n. sp., a chitin-decomposing myxobacterium. J. Bacteriol. *53*: 297–315.

Steyn, P.L., P. Segers, M. Vancanneyt, P. Sandra, K. Kersters and J.J. Joubert. 1998. Classification of heparinolytic bacteria into a new genus, *Pedobacter*, comprising four species: *Pedobacter heparinus* comb. nov., *Pedobacter piscium* comb. nov., *Pedobacter africanus* sp. nov. and *Pedobacter saltans* sp. nov. Proposal of the family *Sphingobacteriaceae* fam. nov. Int. J. Syst. Bacteriol. *48*: 165–177.

Suresh, K., S. Mayilraj and T. Chakrabarti. 2006. *Effluviibacter roseus* gen. nov., sp. nov., isolated from muddy water, belonging to the family "*Flexibacteraceae*". Int. J. Syst. Evol. Microbiol. *56*: 1703–1707.

Suzuki, M., Y. Nakagawa, S. Harayama and S. Yamamoto. 1999. Phylogenetic analysis of genus *Marinilabilia* and related bacteria based on the amino acid sequences of GyrB and emended description of *Marinilabilia salmonicolor* with *Marinilabilia agarovorans* as its subjective synonym. Int. J. Syst. Bacteriol. *49*: 1551–1557.

Suzuki, M., Y. Nakagawa, S. Harayama and S. Yamamoto. 2001. Phylogenetic analysis and taxonomic study of marine *Cytophaga*-like bacteria: proposal for *Tenacibaculum* gen. nov. with *Tenacibaculum maritimum* comb. nov. and *Tenacibaculum ovolyticum* comb. nov., and description of *Tenacibaculum mesophilum* sp. nov. and *Tenacibaculum amylolyticum* sp. nov. Int. J. Syst. Evol. Microbiol. *51*: 1639–1652.

Takahashi, M., K. Suzuki and Y. Nakagawa. 2006. Emendation of the genus *Flammeovirga* and *Flammeovirga aprica* with the proposal of *Flammeovirga arenaria* nom. rev., comb. nov. and *Flammeovirga yaeyamensis* sp. nov. Int. J. Syst. Evol. Microbiol. *56*: 2095–2100.

Takeuchi, M. and A. Yokota. 1992. Proposals of *Sphingobacterium faecium* sp. nov., *Sphingobacterium piscium* sp. nov., *Sphingobacterium heparinum* comb. nov., *Sphingobacterium thalpophilum* comb. nov. and two genospecies of the genus *Sphingobacterium*, and synonymy of *Flavobacterium yabuuchiae* and *Sphingobacterium spiritivorum*. J. Gen. Appl. Microbiol. *38*: 465–482.

Ueda, K., S. Ishikawa, T. Itami and T. Asai. 1952. Studies on the aerobic mesophilic cellulose-decomposing bacteria. Part 5. I. Taxonomic study. J. Agric. Chem. Soc. Jpn. *25*: 543–549.

Urakami, T. and K. Komagata. 1986. Methanol-utilizing *Ancylobacter* strains and comparison of their cellular fatty acid composition and quinone systems with those of *Spirosoma, Flectobacillus*, and *Runella* species. Int. J. Syst. Bacteriol. *36*: 415–421.

Van Ert, M. and J.T. Staley. 1971. Gas-vacuolated strains of *Microcyclus aquaticus*. J. Bacteriol. *108*: 236–240.

Van Trappen, S., J. Mergaert, S. Van Eygen, P. Dawyndt, M.C. Cnockaert and J. Swings. 2002. Diversity of 746 heterotrophic bacteria isolated from microbial mats from ten Antarctic lakes. Syst. Appl. Microbiol. *25*: 603–610.

Vancanneyt, M., O.I. Nedashkovskaya, C. Snauwaert, S. Mortier, K. Vandemeulebroecke, B. Hoste, P. Dawyndt, G.M. Frolova, D. Janssens and J. Swings. 2006. *Larkinella insperata* gen. nov., sp. nov., a bacterium of the phylum '*Bacteroidetes*' isolated from water of a steam generator. Int. J. Syst. Evol. Microbiol. *56*: 237–241.

Vance, I., C.M. Topham, S. Blayden and J. Tampion. 1980. Extracellular cellulase production by *Sporocytophaga myxococcoides*. J. Gen. Microbiol. *117*: 235–241.

Verma, J.P. and H.H. Martin. 1967. Über die Oberflächenstruktur von Myxobakterien I. Chemie und Morphologie der Zellwände von *Cytophaga hutchinsonii* und *Sporocytophaga myxococcoides*. Arch. Mikrobiol. *59*: 355–380.

Weeks, O.B. 1981. Preliminary studies of the pigments of *Flavoacterium breve* NCTC 11099 and *Flavobacterium odoratum* NCTC 11036. *In* The *Flavobacterium-Cytophaga* Group. Gesellschaft fur Biotechnologische Forschung (edited by Reichenbach and Weeks), Weinheim, pp. 108–114.

Wenzel, M., I. Schonig, M. Berchtold, P. Kämpfer and H. Konig. 2002. Aerobic and facultatively anaerobic cellulolytic bacteria from the gut of the termite *Zootermopsis angusticollis*. J. Appl. Microbiol. *92*: 32–40.

Weon, H.Y., B.Y. Kim, S.W. Kwon, I.C. Park, I.B. Cha, B.J. Tindall, E. Stackebrandt, H.G. Truper and S.J. Go. 2005. *Leadbetterella byssophila* gen. nov., sp. nov., isolated from cotton-waste composts for the cultivation of oyster mushroom. Int. J. Syst. Evol. Microbiol. *55*: 2297–2302.

Winogradsky, S. 1929. Études sur la microbiologie du sol. Sur la dégradation de la cellulose dans le sol. Ann. Inst. Pasteur (Paris) *43*: 549–633.

Woese, C.R. 1987. Bacterial evolution. Microbiol. Rev. *51*: 221–271.

Woese, C.R., S. Maloy, L. Mandelco and H.D. Raj. 1990. Phylogenetic placement of the *Spirosomaceae*. Syst. Appl. Microbiol. *13*: 19–23.

Xin, Y.H., Y.G. Zhou, H.L. Zhou and W.X. Chen. 2004. *Ancylobacter rudongensis* sp. nov., isolated from roots of *Spartina anglica*. Int. J. Syst. Evol. Microbiol. *54*: 385–388.

Yi, H. and J. Chun. 2004. *Hongiella mannitolivorans* gen. nov., sp. nov., *Hongiella halophila* sp. nov. and *Hongiella ornithinivorans* sp. nov., isolated from tidal flat sediment. Int. J. Syst. Evol. Microbiol. *54*: 157–162.

Yoon, J.H., S.J. Kang, C.H. Lee and T.K. Oh. 2005. *Marinicola seohaensis* gen. nov., sp. nov., isolated from sea water of the Yellow Sea, Korea. Int. J. Syst. Evol. Microbiol. *55*: 859–863.

Zhang, S., S. Hou, X. Ma, D. Qin and T. Chen. 2006. Culturable bacteria in Himalayan ice in response to atmospheric circulation. Biogeosci. Discuss. *3*: 765–778.

Zhang, Q., C. Liu, Y. Tang, G. Zhou, P. Shen, C. Fang and A. Yokota. 2007. *Hymenobacter xinjiangensis* sp. nov., a radiation-resistant bacterium isolated from the desert of Xinjiang, China. Int. J. Syst. Evol. Microbiol. *57*: 1752–1756.

Zhou, Y., X. Wang, H. Liu, K.Y. Zhang, Y.Q. Zhang, R. Lai and W.J. Li. 2007. *Pontibacter akesuensis* sp. nov., isolated from a desert soil in China. Int. J. Syst. Evol. Microbiol. *57*: 321–325.

Family II. **Cyclobacteriaceae** fam. nov.

Olga I. Nedashkovskaya and Wolfgang Ludwig

Cyc.lo.bac.te.ri.a'ce.ae. N.L. neut. n. *Cyclobacterium* type genus of the family; suff. *-aceae* ending to denote a family; N.L. fem. pl. n. *Cyclobacteriaceae* the *Cyclobacterium* family.

Regular and curved, ring-like or horseshoe-shaped rods that are 0.3–0.7 × 0.3–10 μm. Gram-stain-negative. Nonsporeforming. Nonflagellated and nonmotile in liquid media, but some of them can move by gliding on solid substrates.

Aerobic, with respiratory type of metabolism. Optimal temperature is 16–37°C. Chemo-organotrophs. Carbohydrates are oxidized. One species ferments glucose. Colonies are pink, red, or orange in color. For most strains, an addition of seawater or NaCl to nutrient media sufficiently increases growth rates. Oxidase-, catalase-, and alkaline phosphatase-positive. Flexirubin-type pigments are absent. Produce no indole. One taxon produces hydrogen sulfide. Nitrate may be reduced to nitrite. Most species cannot hydrolyze agar, casein, urea, or chitin.

Menaquinone 7 is a major or single respiratory quinone. Predominant fatty acids are $C_{15:0}$ iso, $C_{17:1}$ iso ω9*c* and $C_{15:0}$ iso 2-OH and/or $C_{16:1}$ ω7*c*. Most species occur in seawater, marine sediments, seaweeds, or marine animals. Some taxa inhabitant algal mates of saline lakes, marine solar salterns, soil, or nonsaline groundwater.

DNA G+C content (mol%): 35–49.

Type genus: **Cyclobacterium** Raj and Maloy 1990, 345[VP] emend. Ying, Wang, Yang and Liu 2006, 2929.

Genus I. **Cyclobacterium** Raj and Maloy 1990, 345[VP] emend. Ying, Wang, Yang and Liu 2006, 2929

The Editorial Board

Cy.clo.bac.ter'i.um. Gr. n. *cyclos* a circle L. neut. n. *bacterium* a small rod; N.L. neut. n. *Cyclobacterium* a circle-shaped bacterium.

Cells are curved, ring-like or horseshoe-shaped. Nonmotile. Colonies on Marine agar (MA) are pink and shiny. Aerobic, having a strictly respiratory type of metabolism with O_2 as the terminal electron acceptor. Neutrophilic and mesophilic. Chemo-organotrophic. Optimal growth temperature range is 25–30°C. Catalase- and oxidase-positive. The major cellular fatty acids are $C_{15:0}$ iso, summed feature 3 ($C_{15:0}$ iso 2-OH and/or $C_{16:1}$ ω7c), $C_{17:1}$ iso ω9c, $C_{17:0}$ iso 3-OH, and $C_{15:0}$ anteiso.

Habitat: marine environments.

DNA G+C content (mol%): 41–45.

Type species: **Cyclobacterium marinum** corrig. (Raj 1976) Raj and Maloy 1990, 346[VP] [*Flectobacillus marinus* (Raj 1976) Borrall and Larkin 1978, 301[AL]; *Microcyclus marinus* Raj 1976, 540].

Further descriptive information

The predominant cellular fatty acids of the type strains of *Cyclobacterium amurskyense* and *Cyclobacterium marinum* are straight-chain unsaturated, branched-chain unsaturated and saturated, namely $C_{15:0}$ iso (22.2 and 23%, respectively), $C_{15:0}$ anteiso (9.2 and 6.4%), $C_{15:1}$ iso (8.4 and 9.7%), $C_{17:1}$ iso ω9*c* (4.3 and 6.3%), $C_{17:0}$ iso 3-OH (10.7 and 12.7%), and summed feature 3 (24.3 and 23.4%), comprising $C_{16:1}$ ω7*c* and/or $C_{15:0}$ iso 2-OH (Nedashkovskaya et al., 2005).

The major cellular fatty acids of *Cyclobacterium lianum* are $C_{15:0}$ iso (28.3%), summed feature 3 ($C_{15:0}$ iso 2-OH and/or $C_{16:1}$ ω7*c*; 16.6%), $C_{17:1}$ iso ω9*c* (10.3%), $C_{17:0}$ iso 3-OH (8.0%), and $C_{15:0}$ anteiso (6.4%), similar to the profiles reported for *Cyclobacterium marinum* and *Cyclobacterium amurskyense* (Ying et al., 2006).

The cellular polyamines of *Cyclobacterium marinum* contain homospermidine, whereas those of other ring-forming genera (*Runella*, *Spirosoma*, and *Flectobacillus*) contain spermidine (Hamana and Nakagawa, 2001).

The oxidation of glucose in *Cyclobacterium marinum* occurs mainly via glycolysis, whereas gluconate is catabolized mainly via the Entner–Doudoroff pathway. These pathways act in conjunction with the tricarboxylic acid cycle and with some participation of the pentose phosphate pathway (Raj and Paveglio, 1983).

Enrichment and isolation procedures

Cyclobacterium marinum can be isolated on MS agar* or on mZ medium.† Incubation is at room temperature (25°C) for several days to allow abundant growth. The cultures will then survive refrigeration at 4°C for at least 3 weeks. They may also be preserved indefinitely by lyophilization.

Cyclobacterium amurskyense was isolated from a seawater sample collected in Amursky Bay, Gulf of Peter the Great. It can be cultured on Marine agar 2216 (Difco).

Cyclobacterium lianum was isolated from sediment of the Xijiang oilfield in the South China Sea, near Fujian Province, China. For isolation, serially diluted sediment samples were spread onto low-organic marine agar 2216 plates (differs from regular Marine agar 2216 [Difco] only by decreasing the peptone concentration from 5 to 0.5 g/l and the yeast extract from 1 to 0.1 g/l). A colony was selected after incubation at 30°C for 10 d and subcultured onto Marine agar 2216.

Differentiation of the genus *Cyclobacterium* from other genera

The genera *Runella*, *Flectobacillus*, and *Spirosoma* have a ring-like or horseshoe-like morphology that resembles that of *Cyclobacterium*. However, *Cyclobacterium* has a marine habitat and can tolerate seawater (although not necessarily requiring NaCl), whereas the other three genera have a freshwater habitat and cannot grow in the presence of seawater or 3% NaCl. Moreover, the cellular polyamines of *Cyclobacterium marinum* contain homospermidine, whereas those of *Runella*, *Spirosoma*, and *Flectobacillus* contain spermidine (Hamana and Nakagawa, 2001).

Taxonomic comments

Cyclobacterium marinum was initially classified in the *Microcyclus* by Raj (1976) as *Microcyclus marinus*. In many features, it resembled *Flectobacillus major*, the type species of the genus *Flectobacillus* Larkin, Williams and Taylor (1977), and it was assigned to that genus as *Flectobacillus marinus* by Borrall and Larkin (1978). H. D. Raj (1979) reported that *Flectobacillus major* and *Flectobacillus marinus* were sufficiently distinct as to warrant separation of *Flectobacillus marinus* into a new genus; the name *Cyclobacterium* was suggested (H. D. Raj, cited by Larkin and Borrall, 1984a). However, Larkin and Borrall (1984a, b) believed that it would be best to retain the organism in the genus *Flectobacillus* because of the similarity in the mol% G+C content of the DNA and because the DNA–DNA hybridization value between the type strains of *Flectobacillus marinus* and *Flectobacillus major* was 71%, using the renaturation rate method of DeLey et al. (1970); consequently, Larkin and Borrall recommended that *Flectobacillus marinus* should not be used to create the new genus *Cyclobacterium* and that the latter name should be discarded. However, Raj and Maloy (1990) noted that analysis of the oligonucleotide sequence catalogues of the 16S rRNA of these organisms (Woese et al., 1990) indicated a relatively large evolutionary distance (ca. 20%) between *Flectobacillus major* and *Flectobacillus marinus* (Woese, 1987), supporting the idea that *Flectobacillus marinus* should be separated from *Flectobacillus. major* at the genus level. Consequently, Raj and Maloy (1990) reclassified *Flectobacillus marinus* as *Cyclobacterium marinus* (later corrected to *marinum* by Euzéby) (1998).

Nedashkovskaya et al. (2005) reported that the level of 16S rRNA gene sequence similarity between the type strain of *Cyclobacterium amurskyense* and that of *Cyclobacterium marinum* was 96.6% (47 nucleotide differences).

Ying et al. (2006) reported that the nearest neighbors of *Cyclobacterium lianum*, based on 16S rRNA gene sequence analysis, were *Cyclobacterium marinum* (93.8% sequence similarity) and *Cyclobacterium amurskyense* (92.8% similarity). Other related genera were *Aquiflexum* (89.7%), *Belliella* (89.7%), *Hongiella* (87.8–90.3%), *Chimaereicella* (88.4%), and *Algoriphagus* (88.3–89.5%). *Hongiella* and *Chimaereicella* have since been reclassified into *Algoriphagus* by Nedashkovskaya et al. (2007b).

Differentiation of the species of the genus *Cyclobacterium*

Table 103 lists reactions differentiating the three species of *Cyclobacterium*.

List of species of the genus *Cyclobacterium*

1. **Cyclobacterium marinum** corrig. (Raj 1976) Raj and Maloy 1990, 346[VP] [*Flectobacillus marinus* (Raj 1976) Borrall and Larkin 1978, 341[AL]; *Microcyclus marinus* Raj 1976, 540]

ma.ri'num. L. neut. adj. *marinum* of the sea, marine. The original spelling *marinus* was corrected by Euzéby (1998).

The characteristics are as described for the genus, with the following additional information. In contrast to the original description, NaCl is not required for growth, and growth can occur at 42°C (Nedashkovskaya et al., 2005).

The following reactions are positive (these results also apply to *Cyclobacterium amurskyense* below; Nedashkovskaya et al., 2005): β-galactosidase, alkaline and acid phosphatase, esterase (C4), esterase lipase (C8), leucine arylamidase, valine arylamidase, cystine arylamidase, naphthol-AS-BI-phosphohydrolase, α- and β-galactosidase, α- and β-glucosidase, *N*-acetyl-β-glucosaminidase, and α-mannosidase activities; growth at 0–10% NaCl and at 4–40°C; hydrolysis of esculin; acid formation from D-cellobiose, L-fucose, D-galactose, D-lactose, L-raffinose, D-melibiose, L-rhamnose, D-trehalose, D-maltose, and D-sucrose; susceptibility to ampicillin, carbenicillin, lincomycin and oleandomycin; and utilization of glucose, D-mannose, *N*-acetyl-D-glucosamine, D-fructose, methyl α-D-mannoside, methyl α-D-glucoside, amygdalin, arbutin, salicin, inulin, melezitose, gentiobiose, D-turanose, lyxose, tagatose, D-fucose, L-fucose, ribose, sorbose, D-xylose, and methyl β-D-xyloside.

The following reactions are negative (these results also apply to *Cyclobacterium amurskyense* below; Nedashkovskaya et al., 2005): α-chymotrypsin, β-glucuronidase, α-fucosidase, arginine dihydrolase, lysine and ornithine decarboxylase activities; gliding motility; Na^+ requirement for growth; requirement for organic growth factors; nitrate reduction; flexirubin pigments; H_2S, indole, and acetoin production; degradation of agar, casein, gelatin, DNA, starch, cellulose

*MS agar for marine organisms contains (g/l): peptone, 1.0; yeast extract, 1.0; glucose, 1.0; NaCl, 30.0; agar, 15.0.

†Modified Zobell 2216 (mZ) medium contains (g/l of seawater): peptone, 5.0; yeast extract, 1.0; ferrous sulfate, 0.2; agar, 20.0 (Raj, 1976).

TABLE 103. Characteristics differentiating the species of the genus *Cyclobacterium*[a,b]

Characteristic	*C. marinum*	*C. amurskyense*	*C. lianum*
Lipase, trypsin	–	+	–
Growth at 42°C	+	–	+
Acid from:			
D-Glucose	–	–	+
L-Arabinose, DL-xylose	–	+	+
Starch	–	+	–
N-Acetyl-D-glucosamine	–	+	nd
Utilization of:			
D-Gluconate	–	+	+
L-Fucose, L-sorbose	+	+	–
Mannitol	+	–	w
Hydrolysis of:			
Tween 20	–	–	+
Tween 40	+	–	w
Susceptibility to:			
Benzylpenicillin, kanamycin	–	+	+
Streptomycin, tetracycline	+	–	–
DNA G+C content (mol%)	41.9	41.3	45.2

[a]Symbols: +, >85% positive; –, 0–15% positive; w, weak; nd, no data.

[b]Data from Nedashkovskaya et al. (2005) and Ying et al. (2006).

(CM-cellulose, filter paper), chitin, urea, and Tweens 20 and 80; acid production from D-glucose, L-sorbose, adonitol, dulcitol, glycerol, *myo*-inositol, mannitol, malate, fumarate, and citrate; utilization of glycerol, iso-erythritol, adonitol, dulcitol, *myo*-inositol, D-sorbitol, glycogen, xylitol, D-arabitol, L-arabitol, gentiobiose, 2-ketogluconate, 5-ketogluconate, caprate, adipate, malate, citrate, and phenylacetate; and susceptibility to gentamicin, neomycin, and polymyxin B.

DNA G+C content (mol%): 41.9 (T_m) (Nedashkovskaya et al., 2005).

Type strain: Raj, ATCC 25205, DSM 745, LMG 13164.

Sequence accession no. (16S rRNA gene): AJ575266, AY533665.

2. **Cyclobacterium amurskyense** Nedashkovskaya, Kim, Lee, Park, Lee, Lysenko, Oh, Mikhailov and Bae 2005, 2392[VP]

a.mur.sky.en'se. N.L. neut. adj. *amurskyense* pertaining to Amursky Bay, from which the type strain was isolated.

Cells have a width of 0.3–0.4 µm and an outer diameter of 0.9–1.2 µm. Colonies are pink, circular, low-convex, shiny with entire edges, and 1–3 mm in diameter on Marine agar 2216. Flexirubin pigments are absent. Grows in 0–10% NaCl. Growth occurs at 4–40°C. Esculin is hydrolyzed but not agar, casein, gelatin, starch, cellulose (CM-cellulose and filter paper), chitin, DNA, urea, or Tweens 20, 40, and 80. Nitrate is not reduced. H_2S, indole and acetoin (Voges–Proskauer reaction) are not produced. Positive for β-galactosidase and alkaline phosphatase. Acid is produced from the following compounds: *N*-acetylglucosamine, L-arabinose, D-cellobiose, L-fucose, D-galactose, D-lactose, D-maltose, D-melibiose, L-raffinose, L-rhamnose, D-sucrose, starch, D-trehalose, and DL-xylose. No acid is formed from adonitol, citrate, dulcitol, fumarate, D-glucose, D-glucuronic acid, glycerol, inositol, mannitol, malate, and L-sorbose. The following compounds are utilized (Biolog GN2 Microplate system): *N*-acetyl-D-glucosamine, cellobiose, dextrin, D-fructose, D-galactose, D-galacturonic acid, gentiobiose, α-D-glucose, glucose 1-phosphate, α-DL-glycerol phosphate, α-D-lactose, lactulose, maltose, D-mannose, D-melibiose, methyl β-D-glucoside, methylpyruvate, psicose, D-raffinose, L-rhamnose, sucrose, D-trehalose, and turanose.

The following compounds are oxidized: acetic acid, *N*-acetyl-D-galactosamine, L-arabinose, glucuronamide, L-glutamic acid, glycogen, L-fucose, glycerol, α-ketobutyric acid, α-ketoglutaric acid, DL-lactic acid, D-mannitol, monomethyl succinate, L-serine, and L-threonine. No oxidation occurs of *cis*-aconitic acid, adonitol, alaninamide, D-alanine, L-alanine, L-alanyl glycine, γ-aminobutyric acid, 2-aminoethanol, D-arabitol, L-asparagine, L-aspartic acid, bromosuccinic acid, 2,3-butanediol, DL-carnitine, citric acid, α-cyclodextrin, iso-erythritol, formic acid, D-galacturonic acid, D-gluconic acid, D-glucosaminic acid, glucose 6-phosphate, glycyl-L-aspartic acid, glycyl-L-glutamic acid, L-histidine, α-hydroxybutyric acid, β-hydroxybutyric acid, γ-hydroxybutyric acid, hydroxy-L-proline, *p*-hydroxyphenylacetic acid, inosine, *myo*-inositol, itaconic acid, α-ketovaleric acid, L-leucine, malonic acid, L-ornithine, L-phenylalanine, phenylethylamine, propionic acid, L-proline, putrescine, L-pyroglutamic acid, quinic acid, D-saccharic acid, sebacic acid, D-serine, D-sorbitol, succinamic acid, succinic acid, thymidine, Tweens 40 and 80, uridine, uronic acid, and xylitol.

Susceptible to ampicillin, benzylpenicillin, carbenicillin, kanamycin, oleandomycin, and lincomycin. Resistant to neomycin, streptomycin, gentamicin, polymyxin B, and tetracycline.

Source: seawater, collected in Amursky Bay, Gulf of Peter the Great, East Sea (also known as the Sea of Japan).

DNA G+C content (mol%): 41.3 (T_m).

Type strain: KMM 6143, KCTC 12363, LMG 23026.

Sequence accession no. (16S rRNA gene): AY960985.

3. **Cyclobacterium lianum** Ying, Wang, Yang and Liu 2006, 2929[VP]

li.a'num. N.L. neut. adj. *lianum* pertaining to Li, named in honor of Ji-Lun Li, who devoted himself to microbiological research and education in China.

The characteristics are as described for the genus, with the following additional features. Cells are 0.4–0.5 µm wide and the outer diameter of the rings is 1.5–1.8 µm. Colonies grown for 3 d on Marine agar are circular, 2–3 mm in diameter, light rose in color, and shiny. Growth occurs at 15–42°C; optimum, 33°C. The pH range is 6.5–9.0; optimum, 7.5–8.0. Growth occurs in 0.1–12% NaCl; optimum, 1–4%. Arginine dihydrolase, urease, and lecithinase negative. Indole and H_2S are not produced. Nitrate is not reduced. Esculin and Tween 20 are hydrolyzed, but Tweens 40 and 80 are only weakly hydrolyzed. No hydrolysis occurs of agar, casein, gelatin, starch, DNA, and carboxymethyl-cellulose. The following compounds are used as sole carbon sources: L-arabinose, cellobiose, galactose, gluconate, glucose, inulin, lactose, maltose, melezitose, D-melibiose, methyl α-D-glucoside, D-raffinose, L-rhamnose, ribose, sucrose, and trehalose. Weak utilization occurs of D-fructose, L-glutamic acid, glycerol, lactate, malate, mannitol, D-mannose, pyruvate, succinate, and D-xylose. The following are not used: adonitol, L-alanine, butyric acid, caprate, citrate, dulcitol, formate, L-fucose, *myo*-inositol, L-lysine, malonate, and L-sorbose. Acid is formed from L-arabinose, cellobiose, galactose, glucose, glycerol (weakly), inulin, lactose, maltose, melezitose, D-melibiose, methyl α-D-glucoside, D-raffinose, L-rhamnose, ribose (weakly), sucrose, trehalose, and D-xylose. The following show strong activity (API ZYM system): alkaline and acid phosphatases, leucine and valine arylamidases, naphthol-AS-BI-phosphohydrolase, β-galactosidase, α- and β-glucosidases and *N*-acetyl-β-glucosaminidase. Weak activity is exhibited for esterases C4 and C8, cystine arylamidase, α-galactosidase, and α-mannosidase. No activity is exhibited for trypsin, α-chymotrypsin, β-glucuronidase, α-fucosidase, or lipase (C14). The following compounds are oxidized (GN2 MicroPlate system): *N*-acetyl-D-glucosamine, L-alaninamide, L-alanine, L-arabinose, 2,3-butanediol, glycerol, DL-carnitine, D-cellobiose, dextrin, D-fructose, D-galactose, D-galacturonic acid, gentiobiose, α-D-glucose, glucose 1-phosphate, glucuronamide, DL-α-glycerol phosphate, DL-lactic acid, α-D-lactose, lactulose, maltose, D-mannose, D-melibiose, methyl β-D-glucoside, D-raffinose, sucrose, D-trehalose, and turanose. Weak or variable oxidation occurs with glycogen, *N*-acetyl-D-galactosamine, L-alanyl glycine, γ-aminobutyric acid, 2-aminoethanol, L-asparagine, L-aspartic acid, iso-erythritol, D-gluconic acid, glucose 6-phosphate, glutamic acid, α-ketovaleric acid, D-mannitol, monomethyl succinate, L-ornithine, L-proline, D-psicose, L-pyroglutamic acid, L-rhamnose, DL-serine, D-sorbitol, succinic acid, L-threonine, and uridine. Sensitivity is exhibited toward the following antibiotics (μg per disk): ampicillin (10), carbenicillin (100), vancomycin (30), ciprofloxacin (5), rifampicin (5), norfloxacin (10), chloramphenicol (30), benzyl penicillin (10), kanamycin (30), and erythromycin (15). Resistance is exhibited toward the following antibiotics (μg per disk): gentamicin (10), neomycin (30), polymyxin B (300), streptomycin (10), and tetracycline (30).

Source: sediment from the Xijiang oilfield in the South China Sea.

DNA G+C content (mol%): 45.2 (T_m).

Type strain: HY9, CGMCC 1.6102, JCM 14011.

Sequence accession no. (16S rRNA gene): DQ534063.

Genus II. **Algoriphagus** Bowman, Nichols and Gibson 2003, 1351^VP^, emend. Nedashkovskaya, Vancanneyt, Van Trappen, Vandemeulebroecke, Lysenko, Rohde, Falsen, Frolova, Mikhailov and Swings 2004, 1762^VP^, emend. Nedashkovskaya, Kim, Kwon, Shin, Luo, Kim and Mikhailov 2007b, 1993^VP^

Olga I. Nedashkovskaya and Marc Vancanneyt

Al.go.ri.pha′gus. L. masc. n. *algor* cold; Gr. masc. n. *phagos* glutton; N.L. masc. n. *Algoriphagus* the cold eater

Rods usually measuring 0.3–0.7 × 0.3–10.0 μm. **Gliding motility is not observed**. Produce non-diffusible carotenoid pigments. No flexirubin type of pigments are formed. **Chemo-organotrophs. Aerobic**. Oxidase, catalase, alkaline phosphatase, and β-galactosidase-positive. Arginine dihydrolase and tryptophan deaminase are not produced. **Esculin is hydrolyzed**. Cellulose (CM-cellulose and filter paper) and urea are not degraded, but agar, casein, gelatin, starch, DNA, Tweens, and chitin may be decomposed. Carbohydrates are utilized. Can grow without seawater or sodium ions. Hydrogen sulfide and indole are not produced. **The major respiratory quinone is MK-7**.

DNA G+C content (mol%): 35.0–49.0.

Type species: **Algoriphagus ratkowskyi** Bowman, Nichols and Gibson 2003, 1351^VP^.

Further descriptive information

The main cellular fatty acids are unsaturated and branched-chain unsaturated fatty acids $C_{15:0}$ iso, $C_{17:0}$ iso 3-OH, and summed feature 3 comprising $C_{15:0}$ iso 2-OH, and $C_{16:1}$ ω7*c* or both (Table 104).

On Marine agar (Difco) strains of the genus *Algoriphagus* form regular, circular, convex, shiny, smooth, with entire edges, and pink or orange colonies with diameter of 0.5–3 mm after cultivation for 48 h.

All strains grow at 6–41 °C, grow with 0–12% NaCl, and do not form acid from inositol or mannitol. Strains were isolated from sea animals and from seawater samples collected in temperate and tropic latitudes.

Enrichment and isolation procedures

Three strains of the type species *Algoriphagus ratkowskyi* were isolated from the sea-ice algal assemblages on Marine agar 2216, and one strain was recovered from cyanobacterial mat collected in a meromictic lake in Antarctica (Bowman et al., 2003) using a seawater nutrient medium (SWN), containing of 0.05 g of yeast extract, 0.05 g of tryptone, 0.05 g of bacteriological peptone, 0.05g of soluble starch, and 0.02 g of sodium pyruvate dissolved in 1000 ml of natural seawater or artificial seawater (ASW), and supplemented with 0.1 ml of a sterile vitamin solution (Bowman et al., 2003). Six strains of another species, *Algoriphagus antarcticus*, were isolated from microbial mats and cultivated on Marine agar 2216 (Van Trappen et al., 2002, 2004). Strains of *Algoriphagus chordae* and *Algoriphagus winogradskyi* were isolated from the brown alga *Chorda filum* and the green alga *Acrosiphonia sonderi*, respectively, by the standard dilution-plating technique on Marine agar 2216 (Nedashkovskaya et al., 2004). Members of the two species, *Algoriphagus mannitolivorans* and *Algoriphagus ornithinivorans*, were isolated from tidal flat sediments samples using Marine agar 2216, and one strain of *Algoriphagus halophilus* was obtained on medium R2A (Difco) supplemented with artificial sea salts (Sigma) (Yi and Chun, 2004). The representatives of *Algoriphagus aquimarinus*, *Algoriphagus locisalis*, *Algoriphagus marincola*, and *Algoriphagus yeomjeoni* were isolated from samples of seawater and cultivated on Marine agar 2216 (Nedashkovskaya et al., 2004, 2005a, b; Yoon et al., 2004). Strain *Algoriphagus vanfongensis* KMM 6241^T was isolated from 0.1 ml of tissue homogenates of the coral *Palythoa* sp. by direct plating on a medium containing [in g/l of a mixture of 30% (v/v)

TABLE 104. Fatty acid composition (%) of the *Algoriphagus* species[a,b]

Fatty acid	1. *A. alkaliphilus* AC-74^T	2. *A. aquimarinus* KMM 3958^T	3. *A. boritolerans* T-22^T	4. *A. chordae* KMM 3957^T	5. *A. halophilus* KCTC 12051^T	6. *A. locisalis* KCTC 12310^T	7. *A. mannitolivorans* KCTC 12050^T	8. *A. ornithinivorans* KCTC 12052^T	9. *A. ratkowskyi* ACAM 646^T	10. *A. terrigena* KCTC 12545^T	11. *A. vanfongensis* KMM 6241^T	12. *A. winogradskyi* KMM 3956^T	13. *A. yeomjeoni* KCTC 12309^T
$C_{15:0}$	1.6	1.6	0.6	2.5	1.0	1.5	3.0	2.7	1.6	0.3	0.7	1.2	0.6
$C_{16:0}$	0.4	1.7	3.2	1.4		0.7	0.2	0.3	2.9		0.3	1.0	0.8
$C_{11:0}$ iso	0.9		0.2			0.7				1.1	2.2		
$C_{11:0}$ anteiso	2.6	0.6	0.4	1.0	1.5	1.6	0.9		2.1	1.6	2.6	1.2	
$C_{14:0}$ iso	0.3	0.7	2.1	0.3	0.6	0.3	2.3	0.6	1.4	1.0		0.3	1.6
$C_{15:0}$ anteiso	3.4	3.2	6.4	1.9	2.8	0.8	3.7	6.1	3.6	2.4	1.6	1.6	1.2
$C_{15:0}$ iso	**32.4**	**38.9**	**16.6**	**38.1**	**28.4**	**31.2**	**32.7**	**26.5**	**30.5**	**35.3**	**21.5**	**36.6**	**28.6**
$C_{15:1}$ iso G	1.4	0.3	0.2			0.8	0.4	0.2	1.1	1.0	6.8	2.9	
$C_{16:0}$ iso	3.2	2.4	20.7	5.8	7.7	3.5	6.5	12.3	3.4	1.0	0.3	3.9	4.8
$C_{16:1}$ iso	3.8	2.4	10.9	1.5	3.5	1.6	3.5	6.4	1.7	2.5	0.6	2.1	1.9
$C_{17:1}$ anteiso ω9*c*			1.6					1.0					
$C_{17:1}$ iso ω9*c*	14.6	5.3	5.0	4.4	9.0	2.8	6.0	12.2	1.5	8.5	4.2	4.0	2.3
$C_{15:1}$ ω6*c*	1.9	2.3		1.0	1.8	1.6	4.1	1.1	2.1	1.0	0.5	1.3	0.8
$C_{16:1}$ ω5*c*	1.2	5.2	0.5	3.6	3.5	3.7	0.6	1.1	5.8	2.8		3.6	3.3
$C_{17:1}$ ω6*c*	1.7	0.7	1.3	1.2	3.4	1.7	7.0	4.5	0.5	0.7	0.7	0.8	1.0
$C_{15:0}$ iso 3-OH	3.2	1.8	0.6	1.6	2.5	2.7	2.7	1.8	2.9	3.2	5.1	2.0	3.0
$C_{16:0}$ 2-OH						0.8							1.3
$C_{16:0}$ 3-OH		0.5		1.9	0.7	1.9			1.9	0.6	0.6	0.8	1.1
$C_{16:0}$ iso 3-OH	1.3	1.0	4.5	0.9	3.1	1.7	4.7	3.4	2.8	0.6		1.9	3.3
$C_{17:0}$ 2-OH			1.1				0.4	1.8					
$C_{17:0}$ iso 3-OH	**7.4**	**5.9**	**4.9**	**6.4**	**5.9**	**7.5**	**6.4**	**6.7**	**9.2**	**6.9**	**10.7**	**6.4**	**7.8**
$C_{19:1}$ iso I											1.3		
Summed feature 3[c]	**9.2**	**20.4**	**8.9**	**22.2**	**19.0**	**29.0**	**7.4**	**6.0**	**22.3**	**24.5**	**32.6**	**24.6**	**33.7**
Summed feature 4[d]	3.9	1.8		1.7	2.7	1.5	2.5	2.0	0.9	2.5	4.9	1.6	1.3

[a]Data are taken from Ahmed et al. (2007), Nedashkovskaya et al. (2004, 2007b), Schmidt et al. (2006), Tiago et al. (2006a), Yoon et al. (2005a, b, 2006).

[b]Amount of the predominant fatty acids is shown in bold font. Values of less than 1% for all strains are not shown.

[c]Summed feature 3 consisted of one or more of the following fatty acids which could not be separated by the Microbial Identification System: $C_{15:0}$ iso 2-OH, $C_{16:1}$ ω7c, and $C_{16:1}$ ω7t.

[d]Summed feature 4 consisted of one of the following fatty acids: $C_{17:1}$ iso I and $C_{17:1}$ anteiso B.

natural seawater and 70% (v/v) distilled water]: Bacto peptone (Difco), 5.0; sucrose, 5.0; glucose, 1.0; yeast extract (Difco), 2.5; KH_2PO_4, 0.1; $MgSO_4$, 0.1 g; and Bacto agar (Difco), 15.0 (Nedashkovskaya et al., 2007a). Several members of the genus *Algoriphagus* have a terrestrial origin. Thus, strain *Algoriphagus alkaliphilus* AC-74^T was recovered from sample of the alkaline artesian water and cultivated on a modified R2A medium without NaCl (Tiago et al., 2004, 2006a). *Algoriphagus terrigena* was isolated from soil sample using the dilution-plating on 10× diluted nutrient agar (Difco) with distilled water (Yoon et al., 2006). For isolation of a single strain of *Algoriphagus boritolerans*, soil samples (5 g) were incubated in 50 ml of phosphate-buffered saline (PBS) solution at 30°C supplemented with 10 mM boron per day for several days. The supernatant was streaked on Luria-Bertani (LB) agar plates containing different levels of boron up to 200 mM (Ahmed et al., 2007). Strain T-22^T was isolated on LB agar medium containing high boron concentration and cultivated on modified R2A medium (designated R3A-V) (Tiago et al., 2004) or on marine agar 2216 at 30°C.

Maintenance procedures

Almost all of the *Algoriphagus* strains remain viable on Marine agar (Difco) or other rich medium based on natural or artificial seawater for several weeks. They have survived storage at −80°C in Marine broth or artificial seawater supplemented with 20% glycerol (v/v) for at least 5 years. Strain *Algoriphagus alkaliphilus* AC-74^T is cultivated on R3A-V medium at 30°C and maintained at −70°C in the same medium supplemented with 15% glycerol. *Algoriphagus boritolerans* is grown on R3A-V medium or on Marine agar 2216 at 30°C and maintained at −80°C in the same medium supplemented with 35% glycerol.

Differentiation of the genus *Algoriphagus* from other genera

Bacteria belonging to the genus *Algoriphagus* have many similar phenotypic features with the representatives of their closest phylogenetic relatives, the genera *Aquiflexum, Belliella, Cyclobacterium, Echinicola,* and *Rhodonellum* (Table 105). All of them are

TABLE 105. Phenotypic characteristics that differentiate the genus *Algoriphagus* from its close relatives in the family *Cyclobacteriaceae*[a,b]

Characteristics	*Algoriphagus*	*Aquiflexum*	*Belliella*	*Cyclobacterium*	*Echinicola*	*Rhodonellum*
Cell morphology:						
Regular rods	+	+	+	–	+	+
Ring-like/horseshoe-shaped	–	–	–	+	–	–
Cell size (µm)	0.3–0.7 × 0.3–10.0	0.3–0.6 × 1.1–4.8	0.3–0.5 × 0.9–3.0	0.3–0.7 × 0.8–1.5	0.3–0.5 × 1.1–2.3	0.7–1.0 × 0.8–3.0
Gliding motility	–	–	–	–	+	–
Oxidase activity	+	+	+	+	+	–
Nitrate reduction	D	+	+	–	–	–
Salinity range (%)	0–10	0–6	0–6	0–10	0–15	0–3
Growth at:						
25°C	D	+	+	+	+	–
40°C	D	+	–	+	+	–
Hydrolysis of starch	D	+	+	–	+	+
DNA G+C content (mol%)	35–49	38.4	35.4	41–45	44–46	44.2

[a]Symbols: +, >85% positive; –, 0–15% positive; D, different reactions occur in different taxa (species of a genus).

[b]Data are taken from Ahmed et al. (2007): Bowman et al. (2003), Brettar et al. (2004a, b), Nedashkovskaya et al. (2004, 2005, 2006, 2007a, b), Schmidt et al. (2006), Tiago et al. (2006a), Van Trappen et al. (2004), Yi and Chun (2004), Ying et al. (2006), Yoon et al. (2004, 2005a, b, 2006).

aerobic bacteria that form the pink-pigmented colonies on suitable solid nutrient media and grow without NaCl or seawater. However, *Algoriphagus* species may be easily differentiated from the genus *Cyclobacterium* by their inability to form ring-like or horseshoe-shaped cells on solid media (Raj and Maloy, 1990; Ying et al., 2006). Gliding motility and the absence of oxidase activity clearly distinguish representatives of the genus *Algoriphagus* from the genera *Echinicola* and *Rhodonellum*, respectively (Nedashkovskaya et al., 2006, 2007a; Schmidt et al., 2006). Notably, species of *Algoriphagus* are characterized by very diverse phenotypic features. These characteristics are suitable for the species differentiation (Table 104), but separation of members of the genus *Algoriphagus* from their close relatives, especially from members of the genera *Aquiflexum* and *Belliella*, is more difficult because their phenotypic properties are very variable. Therefore, to order discriminate *Algoriphagus* from its nearest neighbors, a polyphasic approach, including fatty acid methylester (FAME) (Table 106) and 16S rRNA-based phylogenetic analysis, should be used.

Taxonomic comments

The genus *Algoriphagus*, consisting of a single species *Algoriphagus ratkowskyi*, was established for accommodation of marine, saccharolytic, and cold-adapted *Cytophaga*-like bacteria by Bowman and co-workers in 2003. Shortly thereafter, a new genus of marine bacteria, *Hongiella*, closely related to the genus *Algoriphagus*, was described by Yi and Chun (2004). One of them, *Hongiella halophila*, was moved to the genus *Algoriphagus* because of phylogenetic relatedness (96.8–97.5%) and phenotypic similarity with the *Algoriphagus* species (Nedashkovskaya et al., 2004), while *Hongiella mannitolivorans* and *Hongiella ornithinivorans* were more distantly related (93.7–94.0 and 94.3–94.6%, respectively). The descriptions of the genera *Algoriphagus* and *Hongiella* were also emended in that study (Nedashkovskaya et al., 2004). Later, the description of the genus *Chimaereicella*, comprising a single species *Chimaereicella alkaliphila*, isolated from alkaline artesian water, was reported by Tiago and colleagues (2006a). Despite a close phylogenetic relationship between *Chimaereicella alkaliphila* and the *Algoriphagus* species (sequence similarity was 94.3–95.5%) and a resemblance in their fatty acid composition, the new bacterium was placed in a new genus based on a distinct isolation source. Later, another *Chimaereicella* species, *Chimaereicella boritolerans*, recovered from soil, was described (Ahmed et al., 2007). A level of 16S rRNA gene sequence similarity between *Chimaereicella boritolerans* and its closest relative, *Chimaereicella alkaliphila*, was 97.6%. The similarity values between *Chimaereicella boritolerans* and representative members of the genera *Algoriphagus* and *Hongiella* were 94.7–96.0 and 95.1–96.7%, respectively. DNA–DNA relatedness between a soil isolate and *Chimaereicella alkaliphila* AC-74^T was 28.3%. This fact supported the affiliation of the new isolate with the genus *Chimaereicella* as a separate species, *Chimaereicella boritolerans*. In course of studying a novel marine bacterium, designated strain KMM 6241^T, Nedashkovskaya et al. (2007b) carried out phylogenetic analysis based on sequencing of 16S rRNA gene. This analysis revealed an equidistant position of strains *Chimaereicella alkaliphila* AC-74^T relative to members of the genera *Algoriphagus* and *Hongiella*, with sequence similarity of 94.7–97.0%. Similar phenotypic features, including the presence of oxidase activity and esculin hydrolysis and the absence of gliding motility and aerobic metabolism, may argue for moving the genus *Chimaereicella* to the genus *Algoriphagus*. In addition, the species of the genus *Hongiella* possess a close relatedness with validly published *Algoriphagus* species with sequence similarities ranging from 93.7 to 96.5%. These results taken together with similarity in fatty acid compositions and phenotypic properties could be considered strong confirmation of the proposal to transfer these species of the genus *Hongiella* to the genus *Algoriphagus*. Consequently, the phylogenetic evidence and the resemblance of phenotypic characteristics (Table 105) and fatty acid composition (Table 106) support the joining of the genera *Algoriphagus*, *Chimaereicella*, and *Hongiella* into the single genus *Algoriphagus*, thereby requiring an emended description of the genus *Algoriphagus* (Nedashkovskaya et al., 2007b). Phylogenetic analysis of the almost-complete 16S rRNA gene sequences reveal that species

TABLE 106. Cellular fatty acid composition (%) of the genus *Algoriphagus* and related genera of the family *Cyclobacteriaceae*[a]

Fatty acid	*Algoriphagus*	*Aquiflexum*	*Echinicola*	*Belliella*	*Cyclobacterium*	*Rhodonellum*
$C_{11:0}$ iso	0–2.2	–	–	–	–	–
$C_{11:0}$ anteiso	0–2.6	**4.8**[b]	–	–	–	–
$C_{13:1}$ AT	–	–	0–0.2	0–1.4	–	–
$C_{14:0}$ iso	0–2.3	–	0.1–0.2	1.8–2.2	0–2.1	–
$C_{15:1}$ iso G	0–6.8	**9.4**	0–0.6	**10.0–10.3**	**3.2–8.5**	**6.3**
$C_{15:0}$ iso	**16.6–38.9**	**22.6**	**17.3–20.0**	**18.9–20.4**	**22.2–28.3**	**7.6**
$C_{15:0}$ anteiso	0.8–6.4	**18.3**	1.4–2.8	**4.2–4.8**	**6.3–9.2**	1.4
$C_{15:0}$	0.3–3.0	–	0.8–1.5	2.0–3.9	0–0.8	0.4
$C_{15:1}$ ω6*c*	0–4.1	–	1.1–1.2	1.6–2.5	0.5–1.3	2.7
$C_{16:1}$ ω5*c*	0–5.8	2.0	**4.9–7.8**	2.0–4.6	0–3.6	1.8
$C_{16:1}$ iso	0.6–10.9	**9.4**	0.3–1.0	3.2–3.8	–	**7.3**
$C_{16:0}$ iso 3-OH	–	2.0	–	–	–	–
$C_{16:0}$ iso	0.3–20.7	**4.2**	0.9–1.2	2.5–2.8	–	2.2
$C_{16:0}$	0–3.2	–	0.6–0.9	–	0.5–4.9	0.6
$C_{15:0}$ iso 3-OH	1.6–5.1	1.4	**3.4–5.0**	2.1–2.3	1.1–3.7	**3.7**
$C_{15:0}$ 3-OH	–	–	2.5–2.6	–	–	–
$C_{17:0}$ iso	0–0.8	–	0.7–1.0	0–0.5	–	0.2
$C_{17:0}$ cyclo	0–2.4	–	–	–	–	–
$C_{17:1}$ iso ω9*c*	1.5–14.6	**5.2**	**4.4–6.9**	**6.6–10.2**	**4.3–10.3**	**17.5**
$C_{17:1}$ anteiso ω9*c*	0–1.6	1.1	–	–	–	–
$C_{17:1}$ ω8*c*	0.4–0.9	–	0.4	0.9–1.5	–	0.2
$C_{17:1}$ ω6*c*	0.5–4.5	3.0	**4.3–4.8**	**4.8–9.8**	1.3–1.4	**6.8**
$C_{16:0}$ iso 3-OH	0–4.7	–	0.4–0.7	1.8–2.1	0–1.0	2.8
$C_{16:0}$ 3-OH	0–1.9	–	0.9–2.3	0–1.1	1.3–1.7	0.3
$C_{17:0}$ iso 3-OH	**4.9–10.7**	1.4	**9.4–10.0**	3.0–3.3	**8.0–10.7**	**17.5**
$C_{17:0}$ 2-OH	0–2.2	–	0.4	–	1.5–2.9	1.1
$C_{18:1}$ ω7*c*	–	–	0.7–0.8	–	0–3.0	–
$C_{18:1}$ ω5*c*	0.5–0.7	–	0.2	–	1.2–1.4	–
$C_{18:1}$ H	0–1.6	–	–	–	–	–
$C_{18:0}$	–	–	–	–	0–1.3	–
$C_{19:1}$ iso	0–1.3	1.5	0–0.7	0.8–1.6	–	–
Summed feature 3[c]	**6.0–33.7**	**6.1**	**30.7–34.5**	**7.1–11.5**	**16.2–25.1**	**12.6**
Summed feature 4[d]	0–4.9	2.5	0–5.0	**3.4–4.0**	2.5–4.4	–
Summed feature 5[e]	–	–	–	–	–	**6.5**

[a]Data are taken from Ahmed et al. (2007), Brettar et al. (2004a, b), Nedashkovskaya et al. (2004, 2005, 2006, 2007a, b), Schmidt et al. (2006), Ying et al. (2006), Yoon et al. (2005a, b, 2006).

[b]Predominant fatty acids are shown in bold. Values of less than 1% for all strains are not shown.

[c–e]Summed features consist of one or more fatty acids that could not be separated by the Microbial Identification System.

[c]Summed feature 3 is $C_{15:0}$ iso 2-OH and/or $C_{16:1}$ ω7*c*.

[d]Summed feature 4 is $C_{17:1}$ iso I and $C_{17:1}$ anteiso B.

[e]Summed feature 5 is $C_{14:0}$ 2-OH and/or $C_{15:0}$ iso 2-OH.

of the genus *Algoriphagus* have a 16S rRNA gene sequence similarity of 93.5–99.6% and that the genus *Algoriphagus* forms a cluster with the genera *Belliella, Rhodonellum, Aquiflexum, Echinicola,* and *Cyclobacterium* with sequence similarities of 91.5–93.4, 91.1–93.0, 90.8–93.2, 88.8–92.4, and 88.7–91.7%, respectively.

Differentiation of species of the genus *Algoriphagus*

The species of the genus *Algoriphagus* have many common phenotypic features. However, they can be differentiated from each other by several phenotypic traits as shown in Table 107.

List of species of the genus *Algoriphagus*

1. **Algoriphagus alkaliphilus** (Tiago, Mendes, Pires, Morais and Veríssimo 2006a) Nedashkovskaya, Kim, Kwon, Shin, Luo, Kim and Mikhailov 2007b, 1993[VP] (*Chimaereicella alkaliphila* Tiago, Mendes, Pires, Morais and Veríssimo 2006b, 925[VL]; effective publication: Tiago, Mendes, Pires, Morais and Veríssimo 2006a, 107.)

al.ka.li.phi'la. N.L. n. *alkali* (from Arabic article *al* the; Arabic n. *qaliy* ashes of saltwort, soda), alkali; Gr. adj. *philos* loving; N.L. masc. adj. *alkaliphilus* loving alkaline environments.

Cells are 0.5 × 2.1–3.3 μm. Colonies are small and red. Optimal growth occurs at about 30°C, at pH 8.0 and without NaCl. Hippurate is hydrolyzed, but elastin is not. Xylanase is not produced. D-Arabitol, L-arabitol, ribitol, α-methyl-mannoside, and 2-ketogluconate are utilized but erythritol, galactitol, β-methyl-xyloside, amygdalin, ribose, L-sorbose, glycogen, and inulin are not. Acid is formed from α-methyl-glucoside, arbutin, salicin, β-gentiobiose, D-turanose, and 5-ketogluconate. Susceptible to ceftazidin and cephalothin.

TABLE 107. Phenotypic characteristics of the *Algoriphagus* species[a,b]

Characteristic	1. *A. alkaliphilus*	2. *A. antarcticus*	3. *A. aquimarinus*	4. *A. boritolerans*	5. *A. chordae*	6. *A. halophilus*	7. *A. locisalis*	8. *A. mannitolivorans*	9. *A. marincola*	10. *A. ornithinivorans*	11. *A. ratkowskyi*	12. *A. terrigena*	13. *A. vanfongensis*	14. *A. winogradskyi*	15. *A. yeomjeoni*
Nitrate reduction	+	–	–	–	–	–	–	+	d	–	d	+	–	+	–
NaCl requirement for growth	–	–	–	–	+	–	+	–	+	–	d	+	–	–	+
Salinity range for growth (%)	0–3	0–5	0–10	0–3	1–10	0–8	1–9	0–7	1–9	0–10	0–6	1–7	0–8	0–6	1–9
Temperature range (°C)	11–39	5–25	4–34	17–37	4–32	10–41	4–35	10–42	10–45	10–40	–2–25	10–36	12–35	4–39	4–35
Hydrolysis of:															
Agar	–	–	+	–	+	–	–	–	–	–	d	–	–	+	–
Casein	–	–	+	nd	–	–	–	–	+	–	+	+	–	–	+
Gelatin	+	–	+	–	–	+	–	+	–	+	d	–	+	+	d
Starch	+	–	–	+	–	+	–	+	+	+	d	–	–	+	–
DNA	+	–	+	nd	–	–	nd	+	+	+	d	nd	–	–	–
Tween 20	nd	nd	+	nd	+	+	+	–	+	–	+	+	–	+	+
Tween 40	nd	nd	+	nd	+	+	+	–	+	+	+	+	–	+	+
Tween 80	nd	nd	+	nd	–	+	+	–	+	+	–	+	–	–	+
Acid production from:															
N-Acetylglucosamine	–	–	+	+	–	+	nd	–	–	–	+	nd	+	+	–
L-Arabinose	+	–	–	–	–	–	+	–	+	–	+	+	–	–	–
D-Cellobiose	+	–	+	+	+	–	+	+	+	–	+	+	+	+	+
D-Fructose	+	–	nd	–	nd	nd	+	–	d	–	+	+	–	nd	+
D-Galactose	+	–	–	+	+	–	+	–	+	–	+	+	–	+	+
D-Glucose	+	–	+	+	+	+	+	–	+	+	+	+	+	+	+
D-Lactose	+	–	+	+	+	–	+	–	+	–	–	+	–	+	+
D-Maltose	+	–	–	+	+	+	+	–	+	–	+	+	+	+	+
D-Mannose	+	–	+	+	+	nd	+	–	+	–	+	+	+	+	d (–)
D-Melibiose	+	–	+	–	+	–	+	–	+	–	+	+	–	+	+
D-Raffinose	–	–	–	+	+	–	+	–	+	–	+	+	–	+	d (+)
L-Rhamnose	–	–	+	+	+	–	+	–	–	–	+	–	+	+	+
D-Sucrose	+	–	+	–	+	+	+	+	+	–	–	+	–	+	+
D-Trehalose	+	–	nd	–	nd	nd	d	nd	+	–	+	+	–	nd	+
DL-Xylose	–	–	+	–	+	+	+	–	+	–	+	+	+	+	+
Utilization of:															
L-Arabinose	+	–	+	–	+	+	+	+	+	–	+	+	+	+	–
D-Galactose	+	nd	nd	+	+	+	+	–	+	–	+	+	nd	+	+
D-Glucose	+	–	+	+	+	+	+	+	+	+	+	+	+	+	+
D-Lactose	+	nd	+	+	+	+	+	+	+	+	+	+	+	+	+
D-Maltose	+	–	+	+	+	+	+	+	+	+	+	+	+	+	+
D-Mannose	+	–	+	+	+	+	+	+	+	+	+	+	+	+	+
D-Xylose	–	nd	+	–	+	+	+	+	+	+	+	+	+	+	+
D-Mannitol	–	–	+	+	–	–	–	+	–	–	+	–	–	–	–
Sorbitol	+	nd	–	–	–	–	nd	–	–	–	+	nd	–	–	nd
Inositol	–	nd	–	–	–	–	–	–	–	–	+	–	–	–	–
Glycerol	–	nd	–	nd	–	–	–	–	–	–	–	nd	–	–	–
N-Acetylglucosamine	–	–	+	–	+	+	nd	+	–	+	+	nd	+	+	–
Susceptibility to:															
Ampicillin	nd	nd	–	nd	–	–	–	+	+	+	–	–	–	–	–
Benzylpenicillin	nd	nd	–	+	–	–	–	+	+	+	–	–	–	–	–
Carbenicillin	nd	nd	+	nd	–	+	–	+	–	+	+	–	+	v	+
Gentamicin	nd	nd	–	+	–	–	–	–	–	–	–	–	+	–	–
Kanamycin	+	nd	–	–	–	–	–	–	–	+	–	–	–	–	–
Lincomycin	nd	nd	+	nd	+	+	+	+	–	+	+	–	+	v	+
Oleandomycin	nd	nd	+	nd	–	+	+	+	+	+	+	–	+	+	+
Neomycin	nd	nd	–	nd	–	–	–	–	–	–	–	–	+	–	–
Polymyxin B	nd	nd	–	nd	–	–	–	–	+	+	–	–	–	–	–
Streptomycin	nd	nd	–	+	–	–	–	+	+	–	–	–	+	–	–
Tetracycline	nd	nd	+	+	–	+	–	+	+	+	+	–	+	d	–
DNA G+C content (mol%)	43.5	39–41	41	42.5	37–40	37	42	42	43	38	35	49	43.8	39–42	41

[a]Symbols: +, >85% positive; d, different strains give different reactions (16–84% positive); –, 0–15% positive; v, variable reaction; nd, not determined.

[b]Data from Ahmed et al. (2007), Bowman et al. (2003), Nedashkovskaya et al. (2004, 2007b), Tiago et al. (2006a), Van Trappen et al. (2004), Yi and Chun (2004), Yoon et al. (2004, 2005a, b, 2006).

Source: artesian water collected at Cabeço de Vide, Southern Portugal.

DNA G+C content (mol%): 43.5 (HPLC).

Type strain: AC-74, CIP 108470, LMG 22694.

Sequence accession no. (16S rRNA gene): AJ717393.

2. **Algoriphagus antarcticus** Van Trappen, Vandecandelaere, Mergaert and Swings 2004, 1972[VP]

ant.arc'ti.cus. L. masc. adj. *antarcticus* southern, of the Antarctic, the environment from where the strains were isolated.

Cells are up to 0.5 µm in width and 2–3 µm in length. On Marine agar, colonies are 0.5–3 mm in diameter, opaque, and orange-red after 6 d incubation. Optimal growth occurs at 20°C and with 2% NaCl. Pectin and tyrosine are not hydrolyzed. No acid is produced from carbohydrates. Gluconate, caprate, adipate, and phenylacetate are not utilized.

Source: microbial mats from lakes Reid, Fryxell, and Ace, Antarctica.

DNA G+C content (mol%): 39.9–41.0 (HPLC).

Type strain: R-10710, LMG 21980, DSM 15986.

Sequence accession no. (16S rRNA gene): AJ577141.

3. **Algoriphagus aquimarinus** Nedashkovskaya, Vancanneyt, Van Trappen, Vandemeulebroecke, Lysenko, Rohde, Falsen, Frolova, Mikhailov and Swings 2004, 1762[VP]

a.qui.ma.ri'nus. L. fem. n. *aqua* water, L. masc. adj. *marinus* marine, of the sea; N.L. masc. adj. *aquimarinus*, of seawater.

Cells range from 0.5–0.7 µm in width and 1–10 µm in length. On Marine agar, colonies are 2–3 mm in diameter and pale-pink. Optimal temperature for growth is 23–25°C. No acid is produced from L-sorbose, adonitol, or glycerol.

Source: seawater from Amursky Bay, Gulf of Peter the Great, Sea of Japan.

DNA G+C content (mol%): 41.0 (T_m).

Type strain: KMM 3958, LMG 21971, CCUG 47101.

Sequence accession no. (16S rRNA gene): AJ575264.

4. **Algoriphagus boritolerans** (Ahmed, Yokota and Fujiwara 2007) Nedashkovskaya, Kim, Kwon, Shin, Luo, Kim and Mikhailov 2007b, 1993[VP] (*Chimaereicella boritolerans* Ahmed, Yokota and Fujiwara 2007, 991[VP])

bo'ri.to.le.rans. N.L. n. *borum* boron; L. part. adj. *tolerans* tolerating; N.L. part. adj. *boritolerans* boron-tolerating).

Cells are 0.3–0.4 × 1.2–3.4 µm, occurring singly and occasionally in pairs. Colonies are red and small in diameter after several days of incubation at 30°C. Growth is observed at pH 6.5–10.0. Optimal growth occurs at 28–30°C and pH 8.0–9.0. Tolerant up to 300 mM boron but grows optimally without boron supply. Acid is produced from D-turanose, gentiobiose, inulin, potassium 5-ketogluconate, methyl α-D-mannopyranoside, glycogen, and D-lyxose but not from salicin, glycerol, erythritol, D-ribose, L-sorbose, D-tagatose, D- and L-fucose, D- and L-arabitol, potassium gluconate, potassium 2-ketogluconate, methyl β-D-xylopyranoside, D-adonitol, dulcitol, D-sorbitol, amygdalyn, D-melezitose, or xylitol. Amygdalin, arbutin, and melezitose are utilized, but erythritol, D-ribose, L-xylose, methyl β-D-xylopyranoside, L-sorbose, D-melibiose, D-trehalose, D-raffinose, potassium gluconate, L-fucose or D-tagatose are not. α-Chymotrypsin, leucine arylamidase, valine arylamidase, naphthol-AS-BI-phosphohydrolase, esterase (C4), esterase lipase (C8), trypsin, acid phosphatase, and α-glucosidase are produced. Weakly susceptible to rifampin and cefoperazon but resistant to oxacillin, sulfamethizol, and metronidazole. Susceptible to amoxycycline and ofloxacin; resistant to cephalothin and chloramphenicol.

Source: naturally boron-contaminated soil of the Hisarcik area in the Kutahya province of Turkey.

DNA G+C content (mol%): 42.5 (HPLC).

Type strain: T-22, ATCC BAA-1189, DSM 17298, NBRC 101277.

Sequence accession no. (16S rRNA gene): AB197852.

5. **Algoriphagus chordae** Nedashkovskaya, Vancanneyt, Van Trappen, Vandemeulebroecke, Lysenko, Rohde, Falsen, Frolova, Mikhailov and Swings 2004, 1763[VP]

chor'dae. N.L. gen. n. *chordae* of *Chorda*, the generic name of the brown alga *Chorda filum*, from which the type strain was isolated.

Cells are 0.5–0.7 µm in width and 1–10 µm in length. On Marine agar, colonies are 2–3 mm in diameter, bright pink, and sunken into the agar. Optimal temperature for growth is 23–25°C. No acid is produced from L-sorbose, adonitol, or glycerol.

Source: brown alga *Chorda filum*, Troitsa Bay, Gulf of Peter the Great, Sea of Japan.

DNA G+C content (mol%): 37–40 (T_m).

Type strain: KMM 3957, LMG 21970, CCUG 47095.

Sequence accession no. (16S rRNA gene): AJ575265.

6. **Algoriphagus halophilus** (Yi and Chun 2004) Nedashkovskaya, Vancanneyt, Van Trappen, Vandemeulebroecke, Lysenko, Rohde, Falsen, Frolova, Mikhailov and Swings 2004, 1763[VP] (*Hongiella halophila* Yi and Chun 2004, 160[VP])

ha.lo.phi'lus. Gr. n. *hals halos* salt; Gr. adj. *philos* loving; N.L. masc.adj. *halophilus*, salt-loving.

Cells are 0.3–0.5 µm in width and 1.0–1.8 µm in length. On Marine agar, colonies are flat, translucent, and pink-orange. Optimal growth occurs at 35°C, pH 7.0, and with 1–2% NaCl or 1–2% artificial sea salts. Alginic acids and egg yolk are not decomposed. No acid is produced from L-sorbose, adonitol or glycerol. D-Cellobiose, D-fructose, and D-salicin are utilized but acetamide, acetate, benzoate, citrate, D-ribose, ethanol, glycine, inulin, 2-propanol, L-arginine, L-ascorbate, L-asparagine, L-lysine, L-ornithine, polyethylene glycol, salicylate, succinate, tartrate, or thiamin is not. Leucine arylamidase, acid phosphatase, α-chymotrypsin, naphthol-AS-BI-phosphohydrolase, β-glucuronidase and α- and β-glucosidases activities are present, but esterase (C4), esterase lipase (C8), lipase (C14), cystine arylamidase, valine arylamidase, trypsin, α-mannosidase, and α-fucosidase activities are absent. Maximum absorption of pigment occurs at 475 nm.

Source: sediment sample of getbol, the Korean tidal flat, Sea of Japan.

DNA G+C content (mol%): 37 (HPLC).

Type strain: JC 2051, KCTC 12051, DSM 15292, IMSNU 14013.

Sequence accession no. (16S rRNA gene): AY264839.

7. **Algoriphagus locisalis** Yoon, Kang and Oh 2005b, 1638[VP]

lo.ci.sa'lis. L. n. *locus* place, locality; L. gen. n. *salis*, of salt; N.L. gen. n. *locisalis*, of a place of salt.

Cells are rods 0.4–0.7 µm in width and 1.5–3.0 µm in length. On Marine agar, colonies are 1–2 mm in diameter and orange after 3 d incubation at 30°C. Optimal growth occurs at 30°C, pH 7.0–8.0, and with 2% (w/v) NaCl. Hypoxanthine, xanthine, and tyrosine are not decomposed. No acid is produced from D-melezitose or D-sorbitol. Salicin is utilized but succinate, formate, and L-glutamate are not. Utilization of D-trehalose is strain-dependent (positive for the type strain). Susceptible to chloramphenicol.

Source: seawater from a marine solar saltern of the Yellow Sea in Korea.

DNA G+C content (mol%): 42 (HPLC).

Type strain: MSS-170, KCTC 12310, JCM 12597.

Sequence accession no. (16S rRNA gene): AY835922.

8. **Algoriphagus mannitolivorans** (Yi and Chun 2004) Nedashkovskaya, Kim, Kwon, Shin, Luo, Kim and Mikhailov 2007b, 1993[VP] (*Hongiella mannitolivorans* Yi and Chun 2004, 160[VP])

man.ni.to.li.vo'rans. N.L. n. *mannitolum* mannitol; L. v. *vorare* to devour; N.L. part. adj. *mannitolivorans* mannitol-devouring, utilizing mannitol.

Cells are 0.4–0.5 × 1.1–1.7 µm. Colonies are pink-orange on Marine agar. Optimal growth is observed at 35–40°C, pH 7.0 and with 1% NaCl or 0.5–1.5% artificial sea salts. Alginic acids and egg yolk are not decomposed. Acid is not formed from D-adonitol, glycerol or D-sorbitol. D-Fructose, acetamide, acetate, benzoate, citrate, D-ribose, D-sorbitol, ethanol, glycine, inulin, 2-propanol, L-arginine, L-ascorbate, L-asparagine, L-lysine, L-ornithine, polyethylene glycol, salicylate, tartrate, and thiamine are not utilized. Leucine arylamidase, valine arylamidase, trypsin, α-chymotrypsin, acid phosphatase, naphthol-AS-BI-phosphohydrolase, and α-galactosidase are produced, but esterase lipase (C8), lipase (C14), cystine arylamidase, β-glucuronidase, α- and β-glucosidases, α-mannosidase, and α-fucosidase are not. Susceptible to chloramphenicol, doxycycline, and erythromycin. Maximum absorption of pigment occurs at 480 nm.

Source: sediment of getbol, of the Korean tidal flat.

DNA G+C content (mol%): 42 (HPLC).

Type strain: JC 2050, DSM 15301, IMNSNU 14012, KCTC 12050.

Sequence accession no. (16S rRNA gene): AY264838.

9. **Algoriphagus marincola** (Yoon, Yeo and Oh 2004) Nedashkovskaya, Kim, Kwon, Shin, Luo, Kim and Mikhailov 2007b, 1993[VP] (*Hongiella marincola* Yoon, Yeo and Oh 2004, 1848[VP])

ma.rin'co.la. L. n. *mare -is* the sea; L. n. *incola* inhabitant; N.L. n. *marincola* inhabitant of the sea.

Cells are 0.4–0.6 × 2.0–3.0 µm. Colonies are low convex, reddish-orange, and 1–2 mm in diameter after 72 h incubation at 37°C on Marine agar. Optimal growth occurs at 37°C, pH 6.5–7.5, and with 2–3% NaCl. Tyrosine is hydrolyzed weakly. Hypoxanthine, xanthine and birchwood xylan are not degraded. Acid is produced from D-melezitose but not from D-ribose, glycerol or adonitol. D-Trehalose is utilized but D-fructose, acetate, benzoate, citrate, formate, and succinate are not. Leucine arylamidase, valine arylamidase, cystine arylamidase, esterase (C4), esterase lipase (C8), acid phosphatase, trypsin, α-chymotrypsin, naphthol-AS-BI-phosphohydrolase, α-glucosidase, and β-glucosidase are produced, but lipase (C14), α-galactosidase, α-mannosidase, α-fucosidase, or β-glucuronidase is not. Susceptible to chloramphenicol, doxycycline, and erythromycin.

Source: seawater from the East Sea of Korea.

DNA G+C content (mol%): 43 (HPLC).

Type strain: SW-2, DSM 16067, JCM 12319, KCTC 12180.

Sequence accession no. (16S rRNA gene): AY533663.

Reference strain: SW-26.

Sequence accession no. (16S rRNA gene): AY533664.

10. **Algoriphagus ornithinivorans** (Yi and Chun 2004) Nedashkovskaya, Kim, Kwon, Shin, Luo, Kim and Mikhailov 2007b, 1993[VP] (*Hongiella ornithinivorans* Yi and Chun 2004, 160[VP])

or'ni.thi.ni.vo'rans. N.L. n. *ornithinum* ornithine; L. v. *vorare* to devour; N.L. part. adj. *ornithinivorans* ornithine-devouring, utilizing ornithine.

Cells are 0.3–0.4 × 0.8–2.6 µm. Colonies are pink-orange on Marine agar. Optimal growth is observed at 35–40°C, pH 7.0, and with 1% NaCl or 1.0–2.5% artificial sea salts. Maximum absorption of pigment occurs at 480 nm. Hydrolysis of alginic acids and egg yolk is not detected. Leucine arylamidase, valine arylamidase, acid phosphatase, naphthol-AS-BY-phosphohydrolase, trypsin, α-chymotrypsin, α-galactosidase, α-glucosidase, and β-glucosidase are produced, but esterase (C4), esterase lipase (C8), lipase (C14), cystine arylamidase, β-glucuronidase, α-mannosidase, or α-fucosidase are not. Acid is not produced from D-adonitol, glycerol or D-sorbitol. D-Fructose, D-trehalose, D-salicin, and L-ornithine are utilized, but acetamide, acetate, benzoate, citrate, D-ribose, D-sorbitol, ethanol, glycine, inulin, 2-propanol, L-arginine, L-ascorbate, L-asparagine, L-lysine, polyethylene glycol, salicylate, succinate, tartrate, and thiamine are not. Susceptible to chloramphenicol, doxycycline, and erythromycin.

Source: sediment of getbol, of the Korean tidal flat.

DNA G+C content (mol%): 38 (HPLC).

Type strain: JC 2052, DSM 15282, IMSNU 14014, KCTC 12052.

Sequence accession no. (16S rRNA gene): AY264840.

11. **Algoriphagus ratkowskyi** Bowman, Mancuso, Nichols and Gibson 2003, 1352[VP]

rat.kow'sky.i. N.L. gen. masc. n. *ratkowskyi* of Ratkowsky, named in honor of David A. Ratkowsky, who made significant contributions to growth modelling of bacteria, including psychrophilic bacteria.

Cells are 0.3–0.4 × 0.3–0.9 µm. On Marine agar, colonies are 1–3 mm in diameter and salmon-pink. Optimal growth occurs at 16–19°C and with 0–6% NaCl. Tributyrin is not hydrolyzed. α-Fucosidase and glutamyl glycine arylamidase activities are present; some strains produce β-glucuronidase. Acid is not formed from sugar alcohols. Salicin, β-glycerol phosphate, D-gluconate, propionate, isobutyrate, succinate, pimelate, azelate, L-proline, 2-aminobutyrate, and L-serine are utilized. Utilization of L-ornithine, glycogen, *n*-butyrate, glutarate, aconitate, and hydroxyl-L-proline is strain-dependent. The following substrates are not utilized as sole carbon sources: L-fucose, 2-ketogluconate, adonitol, D-arabitol, dulcitol, iso-erythritol, methanol, itaconate, *n*-valerate, suberate, 3-DL-hydroxybutyrate, oxaloacetate, DL-lactate, DL-tartrate, methylamine, isovalerate, heptanoate, caproate, nonanoate, adipate, 2-oxoglutarate, L-alanine, L-aspartate, L-asparagine, L-phenylalanine, L-glutamate, L-histidine, L-threonine, L-tyrosine, L-leucine, putrescine, and urate.

Source: cold marine and marine-derived habitats, including sea ice and algal mats of saline lakes.

DNA G+C content (mol%): 35–36 (T_m).

Type strain: IC025, ACAM 646, LMG 21435, CIP 107452.

Sequence accession no. (16S rRNA gene): U85891.

12. **Algoriphagus terrigena** Yoon, Lee, Kang and Oh 2006, 779[VP]

ter.ri.ge'na. L. masc. or fem. n. *terrigena* child of the earth, referring to the isolation of the type strain from soil.

Cells are 0.4–0.6 × 0.8–2.5 μm. On Marine agar, colonies are 1–2 mm in diameter and light orange after incubation for 7 d at 25°C. Optimal growth occurs at 25°C, pH 6.5–7.5, and with 2% (w/v) NaCl. Hypoxanthine, xanthine, and tyrosine are not decomposed. α-Mannosidase is present but β-glucuronidase and α-fucosidase are absent. Acid is formed from D-melezitose and D-ribose but not from D-sorbitol. Salicin is utilized as a sole carbon and energy source but not succinate or L-glutamate. Susceptible to chloramphenicol and novobiocin but resistant to cephalothin.

Source: soil of island Dokdo, Korea.

DNA G+C content (mol%): 49.0 (HPLC).

Type strain: DS-44, KCTC 12545, CIP 108837.

Sequence accession no. (16S rRNA gene): DQ178979.

13. **Algoriphagus vanfongensis** Nedashkovskaya, Kim, Kwon, Shin, Luo, Kim and Mikhailov 2007b, 1990[VP]

van.fong.en'sis. N.L. masc. adj. *vanfongensis* pertaining to the Vanfong Bay, from which the type strain was isolated.

Cells are 0.4–0.5 × 1.0–2.5 μm. Colonies are light-pink on Marine agar. Optimal growth is observed with 1–4% NaCl. Acid is not produced from L-sorbose, glycerol, adonitol or dulcitol. Gluconate, caprate, adipate, malate, citrate, or phenylacetate is not utilized. Susceptible to chloramphenicol, doxycycline, and erythromycin.

Source: coral *Palithoa* sp. collected in Vanfong Bay, South China Sea, Vietnam.

DNA G+C content (mol%): 43.8 (T_m).

Type strain: KMM 6241, DSM 17529, KCTC 12716.

Sequence accession no. (16S rRNA gene): EF392675.

14. **Algoriphagus winogradskyi** Nedashkovskaya, Vancanneyt, Van Trappen, Vandemeulebroecke, Lysenko, Rohde, Falsen, Frolova, Mikhailov and Swings 2004, 1763[VP]

wi.no.grad'sky.i. N.L. gen. masc. n. *winogradskyi*, of Winogradsky, named to honor Sergey N. Winogradsky, for his contributions to the study of *Cytophaga* like bacteria.

Cells are 0.5–0.7 × 1–10 μm. On Marine agar, colonies are 2–4 mm in diameter, bright pink, and sunken into the agar. Optimal temperature for growth is 25–28°C. No acid is produced from L-sorbose, adonitol, or glycerol.

Source: green alga *Acrosiphonia sonderi*, Troitsa Bay, Gulf of Peter The Great, Sea of Japan.

DNA G+C content (mol%): 39–42 (T_m).

Type strain: KMM 3956, LMG 21969, JCM 13505, CCUG 47094.

Sequence accession no. (16S rRNA gene): AJ575263.

15. **Algoriphagus yeomjeoni** Yoon, Kang, Jung, Lee and Oh 2005a, 869[VP]

yeom.jeo'ni. N.L. gen. n. *yeomjeoni*, of a yeomjeon, the Korean name for a marine solar saltern.

Cells are 0.4–0.7 × 1.5–2.5 μm. On Marine agar, colonies are 0.8–1.0 mm in diameter and vivid orange after 3 d incubation at 30°C. Optimal growth occurs at 25–30°C, pH 7.0–8.0, and with 2% (w/v) NaCl. Hypoxanthine, xanthine, and tyrosine are not decomposed. No acid is produced from D-melezitose, D-ribose, or D-sorbitol. Salicin is utilized but succinate and L-glutamate are not. Susceptible to chloramphenicol.

Source: seawater in a marine solar saltern of the Yellow Sea in Korea.

DNA G+C content (mol%): 41 (HPLC).

Type strain: strain MSS-160, KCTC 12309, JCM 12598.

Sequence accession no. (16S rRNA gene): AY699794.

Reference strain: MSS-161.

Genus III. **Aquiflexum** Brettar, Christen and Höfle 2004b, 2339[VP]

Ingrid Brettar, Richard Christen and Manfred G. Höfle

A.qui.fle'xum. L. fem. n. *aqua* water; L. part. adj. *flexus -a -um* bent, winding; N.L. neut. n. *Aquiflexum* to indicate the bacterium's aquatic origin and its long flexible rods.

Rods, occurring singly or in chains of up to five cells. Nonmotile (no flagella, no gliding activity), **Gram-stain-negative. Aerobic. Heterotrophic**. Oxidase- and catalase-positive. Cells contain a **high percentage of branched-chain fatty acids** (>80%), and of **$C_{15:0}$ anteiso. Cells contain carotenoids** but no flexirubin. **NaCl** is **not needed for growth**, but growth is **improved** by its presence. **Nitrate** is reduced to nitrite. **Gelatin is hydrolyzed**.

DNA G+C content (mol%): 38.4 (HPLC).

Type species: **Aquiflexum balticum** Brettar, Christen and Höfle 2004b, 2339[VP].

Further descriptive information

The genus *Aquiflexum* is so far represented by a single strain and a single species, *Aquiflexum balticum* BA160[T].

Enrichment and isolation procedures

Strain BA160[T] was isolated during a cruise onboard the research vessel (RV) *Aranda* in September 1998 from surface water (5 m, 15°C, 6% salinity, pH 8.2) from a site in the Central Baltic Sea at the entrance of the Gulf of Finland (station LL12, 59.2900° N, 22.5398° E). The strain was isolated by spreading 0.1 ml of seawater on 1/5 diluted Marine agar (Difco 2216, Marine broth or agar diluted by a factor of 5; final concentration of agar, 1.8%) and subsequently purified and cultured on this medium.

Maintenance procedures

The strain can be kept alive for months at 4°C on agar plates (dilute or regular Marine agar), mixed with glycerol and stored at −70°C, or freeze-dried. For growing the strain from old or preserved biomass, incubation at 30–35°C is recommended. At room temperature, cultivation is often not successful or shows long lag-periods.

Differentiation of the genus *Aquiflexum* from other genera

According to 16S rRNA gene sequence analysis, *Aquiflexum* is most closely related to the genera *Belliella*, *Algoriphagus*, *Hongiella*, and *Cyclobacterium* (see phylogenetic tree in the genus description of *Belliella*). The nearest relative is *Belliella baltica* (Brettar et al., 2004a), having a 16S rRNA gene sequence similarity of 92.4%. By phenotypic traits, *Aquiflexum* can be distinguished from *Belliella* by its temperature range and optimum, hydrolysis of gelatin, acid production from 13 substrates, assimilation of three substrates, and utilization of eight substrates. Compared to *Belliella baltica*, $BA160^T$ showed a higher versatility in using organic substrates, except for amino acids. Compared to *Belliella baltica*, *Aquiflexum* has a higher fraction of branched-chain fatty acids ($BA160^T$: 87%, *Belliella baltica*: 70%), lower number of detectable fatty acid compounds, and considerably different composition. The major difference is the higher abundance of *anteiso* branched fatty acids (*Aquiflexum balticum*, 22%; *Belliella baltica*, 8%). The most abundant was $C_{15:0}$ anteiso (*Aquiflexum balticum*, 19%; *Belliella baltica*, 4.5%). Additionally, $C_{17:1}$ anteiso ω9*c* was detectable for *Aquiflexum balticum*, but not for *Belliella baltica*.

A table of comparison is given in the chapter on *Belliella*.

Taxonomic comments

The genus *Aquiflexum* belongs to the family *Cyclobacteriaceae* of the class *Cytophagia*. The phylogenetic tree in the chapter on *Belliella* reflects the phylogenetic relationships within the family *Cyclobacteriaceae*.

Acknowledgements

We are grateful to H. Kuosa and the crew of the Finnish RV *Aranda* for their assistance with sampling and seawater analysis of the Baltic Sea, to J. Bötel for excellent technical assistance, and to the Deutsche Sammlung von Mikroorganismen und Zellkulturen (DSMZ) staff for analysis of fatty acids and physiological tests.

List of species of the genus *Aquiflexum*

1. **Aquiflexum balticum** Brettar, Christen and Höfle 2004b, 2339^{VP}

bal'ti.cum. N. L. neut. adj. *balticum* of or belonging to the Baltic sea, referring to the source of the type strain.

The characteristics are as given for the genus with the following additional characteristics. Cells are 0.3–0.6 μm × 1.1–4.8 μm. The dominant fatty acids are of $C_{15:0}$ iso, $C_{15:0}$ anteiso, $C_{15:1}$ iso G, and $C_{16:1}$ iso H. Colonies are circular, smooth, convex, and entire; they are red and transparent when young but become opaque with ongoing incubation (>1 week, 30°C, on ½× Marine agar). The temperature range is 4–40°C; optimum, around 30°C. Growth occurs in 0–6% NaCl; optimum, around 1.5%. The pH range is 7–9; optimum, at neutral pH. Esculin, starch, and gelatin are hydrolyzed. No degradation of cellulose or tyrosine occurs. Indole is not produced. Growth does not occur on media with 0.5% yeast, casein, DNA, chitin, or pectin. Acid is produced (using the API 50CHE system) from galactose, D-glucose, D-fructose, D-mannose, rhamnose, α-methyl-D-mannoside, α-methyl-D-glucoside, *N*-acetylglucosamine, amygdalin, arbutin, esculin salicin, cellobiose, maltose, lactose, melibiose, sucrose, trehalose, inulin, melezitose, D-raffinose, starch, glycogen, xylitol, β-gentiobiose, D-turanose, L-fucose, and 5-ketogluconate. Enzymic activities (i.e., positive results using the API 20NE and API ZYM systems) include α- and β-glucosidase; β-galactosidase; acid and alkaline phosphatase; leucine-, valine-, and cystine-arylamidase; trypsin; chymotrypsin; naphthol-phosphohydrolase; and *N*-acetyl-β-glucosaminase. The following substrates are assimilated (using the API 20NE system): glucose, arabinose, mannose, *N*-acetylglucosamine, maltose, and gluconate. The following substrates are utilized (using the Biolog GN2 system): L-arabinose, cellobiose, L-fructose, D-galactose, gentiobiose, α-D-glucose, α-D-lactose, lactulose, maltose, D-mannose, β-methyl-D-glucoside, D-psicose, D-sorbitol, sucrose, D-trehalose, turanose, mono-methyl-succinate, acetic acid, α-ketoglutaric acid, lactic acid, and propionic acid. All unmentioned tests using the API 20CHE, API ZYM, API 20NE, and Biolog GN2 test systems were negative. No utilization of amino acids was detected, but aminopeptidase was produced.

Source: marine or estuarine.

DNA G+C content (mol%): 38.4 (HPLC).

Type strain: BA160, CIP 108445, DSM 16537, LMG 22565.

Sequence accession no. (16S rRNA gene): AJ744861.

Genus IV. **Belliella** Brettar, Christen and Höfle 2004a, 69^{VP}

Ingrid Brettar, Richard Christen and Manfred G. Höfle

Bel.li.el'la. N.L. fem. dim. n. *Belliella* named in honor the aquatic microbiologist Russell Bell of the University of Uppsala.

Rods occurring as single cells or chains of up to five cells. **Nonmotile** (no flagella, no gliding activity). Gram-stain-negative. **Aerobic**. Oxidase- and catalase-positive. Chemoheterotrophic. The dominant fatty acids are $C_{15:0}$ iso, $C_{15:1}$ iso G, $C_{17:1}$ iso ω9*c*, and $C_{17:1}$ ω6*c*. Cells contain **carotenoids** but no flexirubin. **NaCl** is **not required** for growth and does not influence growth up to 3%. **Nitrate** is reduced to nitrite. **Gelatin is not hydrolyzed**.

DNA G+C content (mol%): 35.3–35.5.

Type species: **Belliella baltica** Brettar, Christen and Höfle 2004a, 69[VP].

Further descriptive information

The genus *Belliella* is so far represented by a single species comprising two strains, *Belliella baltica* BA1 and BA134^T, both isolated from the central Baltic Sea.

Enrichment and isolation procedures

The strains of *Belliella baltica* were isolated during a cruise onboard the research vessel (RV) *Aranda* in September 1998 from surface water (5 m, 15°C, 7% salinity, pH 8.4) of two stations in the Central Baltic Sea (Gotland Deep BY-15 [57.1920°N, 20.0302°E] and TEILI1 [59.2607°N, 21.3002°E]). Strains were isolated by spreading 0.1 ml of seawater on 1/5 diluted Marine agar (Difco 2216; Marine broth or agar diluted by a factor of 5; final concentration of agar, 1.8%), and subsequently purified and cultured on this medium.

Maintenance procedures

The strain can be kept alive for months at 4°C on agar plates (dilute or regular Marine agar), mixed with glycerol and stored at −70°C, or freeze-dried. Recovery from old biomass or from preserved biomass is recommended on Marine agar (half-strength) at 20–25°C.

Differentiation of the genus *Belliella* from other genera

According to 16S rRNA gene sequence analysis, *Belliella* is most closely related to the genera *Aquiflexum*, *Algoriphagus*, *Hongiella*, and *Cyclobacterium* (see Figure 78). The nearest relative is *Aquiflexum balticum* (Brettar et al., 2004b), having a 16S rRNA gene sequence similarity of 92.4%. *Belliella* can be distinguished from *Aquiflexum* on the basis of its temperature range and optimum, hydrolysis of gelatin, acid production from 13 substrates, assimilation of three substrates, and utilization of eight substrates (Table 108). Compared to *Aquiflexum balticum*, *Belliella* is less able to use organic substrates. The cellular fatty acids show significant differences between *Aquiflexum* and *Belliella*. The major

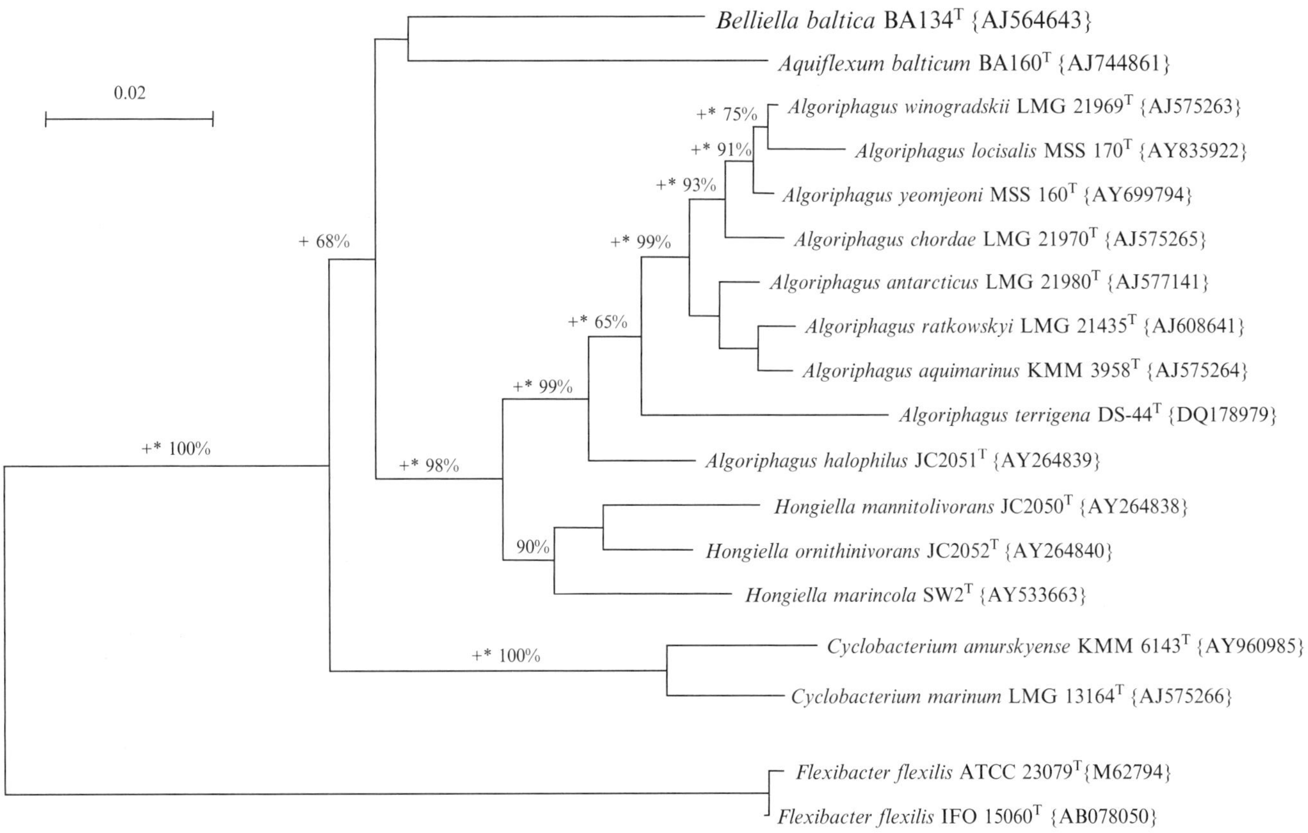

FIGURE 78. Phylogenetic position of *Belliella baltica* and *Aquiflexum balticum* within the family *Cyclobacteriaceae*. According to the phylogenetic analysis of almost complete 16S rRNA gene sequences, *Aquiflexum balticum* BA160^T and *Belliella baltica* BA134^T belong to a well defined clade (all three phylogenetic methods – see below) that also includes genera such as *Algoriphagus*, *Hongiella*, and *Cyclobacterium*. *Aquiflexum balticum* and *Belliella baltica* fail to cluster robustly with any of these genera or with each other, strongly supporting their position as independent genera. A domain (positions 85–1416 of the *Aquiflexum balticum* sequence, corresponding to a domain sequenced for every species) was chosen for phylogenetic analysis. Parsimony and maximum-likelihood (G option) were performed using programs from the PHYLIP package. For neighbor-joining, distances were calculated with DNADIST (from PHYLIP) and the Kimura 2 parameters correction. A tree was generated using the BioNJ algorithm (Gascuel, 1997). Bootstrap analysis (1000 replications) was performed using the NJ algorithm as described previously. The topology of the tree is based on NJ analysis, with the results of the other analyses reported on the figure: *, for a branch also found by maximum-likelihood; +, for a branch also found by parsimony; %, indicates % of bootstrap.

TABLE 108. Differential features of *Belliella baltica* and *Aquiflexum balticum*[a,b]

Characteristic	*Belliella baltica*	*Aquiflexum balticum*
Color	Pink/orange	Red
DNA G+C content (mol%)	35.4	38.4
Optimum NaCl concentration (%)	0–3.4	1–2
Growth at 40°C	–	+
Temperature optimum (°C)	20–30	30–35
Growth at pH 10	W	–
Gelatin hydrolysis	–	+
Acid production from:[c]		
L-Arabinose	+	–
D-Xylose	+	–
D-Mannose	–	w
Rhamnose	–	+
α-Methyl-D-mannoside	–	w
α-Methyl-D-glucoside	–	w
N-Acetylglucosamine	–	w
Amygdalin	–	w
Melezitose	–	w
Xylitol	–	w
D-Turanose	–	w
L-Fucose	–	w
5-Ketogluconate	–	w
Assimilation of:[d]		
Mannose	–	+
N-Acetylglucosamine	–	+
Gluconate	–	+
Utilization of:[e]		
D-Mannose	–	w
β-Methyl-D-glucoside	–	w
D-Psicose	–	w
D-Sorbitol	–	w
Monomethylsuccinate	–	w
α-Ketobutyric acid	+	–
α-Ketovaleric acid	+	–
L-Glutamic acid	+	–
Cellular fatty acids:[f]		
$C_{15:0}$	2.94	–
$C_{15:0}$ anteiso	4.53	18.53
$C_{15:1}$ ω6*c*	2.07	–
$C_{17:1}$ ω8*c*	1.19	–
$C_{17:1}$ anteiso ω9*c*	–	1.10

[a]Symbols: +, positive; –, negative; w, weak reaction.

[b]Data from Brettar et al. (2004a, b).

[c]Using the API50CHE system.

[d]Using the API 20NE system.

[e]Using the BIOLOG GN2 system.

[f]Values represent percent of the total fatty acids.

differences are the higher abundance of $C_{15:0}$ anteiso for *Aquiflexum balticum*, and the occurrence or absence or absence of four fatty acids (Table 108).

Taxonomic comments

The genus *Belliella* belongs to the family *Cyclobacteriaceae* of the class *Cytophagia*. The phylogenetic affiliation within the family is reflected by the phylogenetic tree (Figure 78).

Acknowledgements

We are grateful to H. Kuosa and the crew of the Finnish RV *Aranda* for their assistance with sampling and seawater analysis of the Baltic Sea.

List of species of the genus *Belliella*

1. **Belliella baltica** Brettar, Christen and Höfle 2004a, 69[VP]

bal'ti.ca. N. L. fem. adj. *baltica* of or belonging to the Baltic Sea (the source of the type strain).

The characteristics are as given for the genus with the following additional information. The cells are 0.3–0.5 × 0.9–3.0 µm. Colonies are circular, smooth, convex, and entire; they are pink and transparent when young but become orange and opaque with ongoing incubation (>2 weeks, 20°C, on ½× Marine agar). The temperature range is 4–37°C; optimum, around 25°C. NaCl is not needed for growth; growth occurs a t salinities up to 6% NaCl; optimum, 0–3%. The pH range is 7–10; optimum, around pH 7. Growth occurs on 0.5% yeast extract. Esculin, starch, and DNA are hydrolyzed. Indole is not produced. Tyrosine, cellulose, and chitin are not degraded. Growth does not occur on media with casein and pectin. Acid is produced (using the API 50CHE system) from L-arabinose, D-xylose, galactose, D-glucose, D-fructose, esculin, salicin, cellobiose, maltose, lactose, melibiose, sucrose, trehalose, D-raffinose, and starch. The following enzymic activities occur (using the API 20NE and API ZYM systems): α- and β-glucosidase; β-galactosidase; acid and alkaline phosphatase; lipase (C8); leucine-, valine-, and cystine-arylamidase; trypsin, chymotrypsin, and naphthol-phosphohydrolase. The following substrates are assimilated (using the API 20NE system): glucose, arabinose, and maltose. The following substrates are utilized (using the Biolog GN2 system): D-galactose, gentobiose, α-D-glucose, α-D-lactose, lactulose, maltose, D-trehalose, acetic acid, α-ketobutyric/glutaric/valeric-acid, and L-glutamic acid. All other tests using the API 20CHE, API ZYM, API 20NE, and Biolog GN2 test systems were negative.

Source: marine or estuarine.

DNA G+C content (mol%): 35.3–35.5 (HPLC).

Type strain: BA134, CIP 108006, DSM 15883, LMG 21964.

Sequence accession no. (16S rRNA gene): AJ564643.

Genus V. **Echinicola** Nedashkovskaya, Kim, Vancanneyt, Lysenko, Shin, Park, Lee, Jung, Kalinovskaya, Mikhailov, Bae and Swings 2006, 955[VP]

Olga I. Nedashkovskaya and Seung Bum Kim

E.chi.ni.co′la. L. masc. n. *echinus -i* a sea urchin; L. suff. *-cola* derived from L. masc. or fem. n. *incola* a dweller; N.L. fem. n. *Echinicola* a sea-urchin dweller.

Rods usually measuring 0.3–0.5 × 1.1–2.3 µm. **Motile by gliding**. Produce non-diffusible carotenoid pigments. No flexirubin-type pigments are formed. **Chemo-organotrophs. Aerobic**. Can ferment D-glucose. Oxidase-, catalase-, alkaline phosphatase-, and β-galactosidase-positive. **Starch is hydrolyzed**. Casein, urea, chitin, and cellulose (CM-cellulose and filter paper) are not decomposed, but agar, gelatin, and Tweens may be decomposed. Carbohydrates are utilized. Can grow without seawater or sodium ions. Nitrate is not reduced to nitrite. Indole is not produced. **The major respiratory quinone is MK-7**.

DNA G+C content (mol%): 44–s.

Type species: **Echinicola pacifica** Nedashkovskaya, Kim, Vancanneyt, Lysenko, Shin, Park, Lee, Jung, Kalinovskaya, Mikhailov, Bae and Swings 2006, 955[VP].

Further descriptive information

The main cellular fatty acids are straight-chain unsaturated and branched-chain unsaturated fatty acids $C_{15:0}$ iso, $C_{16:1}$ ω5*c*, $C_{15:0}$ iso 3-OH, $C_{17:1}$ iso ω9*c*, $C_{17:1}$ ω6*c*, $C_{17:0}$ iso 3-OH, summed feature 3 comprising $C_{15:0}$ iso 2-OH and $C_{16:1}$ ω7*c* or both (Table 109).

On Marine agar (Difco), strains of the genus *Echinicola* form regular, circular, convex, shiny, smooth, and pink colonies with entire edges and a diameter of 2–3 mm after cultivation for 48 h.

All strains grow at 6–41°C and with 0–12% NaCl. The *Echinicola* strains do not form acid from melibiose, raffinose, sorbose, glycerol, adonitol, dulcitol, inositol, or mannitol. They utilize arabinose, glucose, lactose, mannose, and sucrose, but not *myo*-inositol, mannitol, or sorbitol. The strains are susceptible to lincomycin and resistant to ampicillin, benzylpenicillin, gentamicin, kanamycin, neomycin, polymyxin B, streptomycin, and tetracycline.

Strains of the genus *Echinicola* were isolated from sea animals and from seawater samples collected in the temperate and tropic latitudes.

Enrichment and isolation procedures

Strains of *Echinicola pacifica* were isolated from a sea urchin, *Strongylocentrorus intermedius*, using the dilution plating technique on Marine agar 2216 (Nedashkovskaya et al., 2006). A single strain *Echinicola vietnamensis*, KMM 6221[T], was isolated from seawater by direct plating on a medium containing (in g/l of a 1:1 mixture of natural seawater and distilled water): Bacto peptone (Difco), 5.0; casein hydrolysate (Merck), 2.0; Bacto yeast extract (Difco), 2.0; glucose, 1.0; KH_2PO_4, 0.2; $MgSO_4$, 0.05; Bacto agar (Difco), 15.0. All isolates have been grown on media containing 0.5% of a peptone and 0.1–0.2% yeast extract (Difco) (Nedashkovskaya et al., 2007a).

Maintenance procedures

Echinicola strains remain viable for several weeks on Marine agar or other rich media based on natural or artificial seawater. They have survived storage −80°C for at least 5 years in Marine broth or artificial seawater supplemented with 20% glycerol (v/v).

Differentiation of the genus *Echinicola* from other genera

The genus *Echinicola* differ from its closest phylogenetic relatives, the genera *Algoriphagus*, *Belliella*, and *Cyclobacterium* by its gliding motility and by its ability to grow in the presence of 15% NaCl (Table 110). The absence of nitrate reductase and a higher mol% G+C content of its DNA (44–46 vs 35.4) separate members of the genus *Echinicola* from their nearest neighbor, *Belliella baltica*. Differences in fatty acid composition may also be helpful for discrimination of the *Echinicola* strains from their closest relatives (Table 109).

Taxonomic comments

Phylogenetic analysis of almost-complete 16S rRNA gene sequences of the genus *Echinicola* indicates that its closest

TABLE 109. Cellular fatty acid compositions (%) of the genus *Echinicola* and related genera of the phylum *Bacteroidetes*[a,b]

Fatty acid	*Echinicola*	*Algoriphagus*	*Belliella*	*Cyclobacterium*
$C_{11:0}$ iso	–[c]	0–2.2	–	–
$C_{11:0}$ anteiso	–	0–2.6	–	–
$C_{13:1}$ AT	0–0.2	–	0–1.4	–
$C_{14:0}$ iso	0.1–0.2	0–2.3	1.8–2.2	0–2.1
$C_{15:1}$ iso	0–0.6	0–6.8	**10.0–10.3**	**3.2–8.5**
$C_{15:0}$ iso	**17.3–20.0**	**16.6–38.9**	**18.9–20.4**	**22.2–28.3**
$C_{15:0}$ anteiso	1.4–2.8	0.8–6.4	**4.2–4.8**	**6.3–9.2**
$C_{15:0}$	0.8–1.5	0.3–3.0	2.0–3.9	0–0.8
$C_{15:1}$ ω6*c*	1.1–1.2	0–4.1	1.6–2.5	0.5–1.3
$C_{16:1}$ ω5*c*	**4.9–7.8**	0–5.8	2.0–4.6	0–3.6
$C_{16:1}$ iso	0.3–1.0	0.6–10.9	3.2–3.8	–
$C_{16:0}$ iso	0.9–1.2	0.3–20.7	2.5–2.8	–
$C_{16:0}$	0.6–0.9	0–3.2	–	0.5–4.9
$C_{15:0}$ iso 3-OH	**3.4–5.0**	1.6–5.1	2.1–2.3	1.1–3.7
$C_{15:0}$ 3-OH	2.5–2.6	–	–	–
$C_{17:0}$ iso	0.7–1.0	0–0.8	0–0.5	–
$C_{17:0}$ Cyclo	–	0–2.4	–	–
$C_{17:1}$ iso ω9*c*	**4.4–6.9**	1.5–14.6	**6.6–10.2**	**4.3–10.3**
$C_{17:1}$ anteiso ω9*c*	–	0–1.6	–	–
$C_{17:1}$ ω8*c*	0.4	0.4–0.9	0.9–1.5	–
$C_{17:1}$ ω6*c*	**4.3–4.8**	0.5–4.5	**4.8–9.8**	1.3–1.4
$C_{16:0}$ iso 3-OH	0.4–0.7	0–4.7	1.8–2.1	0–1.0
$C_{16:0}$ 3-OH	0.9–2.3	0–1.9	0–1.1	1.3–1.7
$C_{17:0}$ iso 3-OH	**9.4–10.0**	**4.9–10.7**	3.0–3.3	**8.0–10.7**
$C_{17:0}$ 2-OH	0.4	0–2.2	–	1.5–2.9
$C_{18:1}$ ω7*c*	0.7–0.8	–	–	0–3.0
$C_{18:1}$ ω5*c*	0.2	0.5–0.7	–	1.2–1.4
$C_{18:1}$ H	–	0–1.6	–	–
$C_{18:0}$	–	–	–	0–1.3
$C_{19:1}$ iso	0–0.7	0–1.3	0.8–1.6	–
Summed feature 3	**30.7–34.5**	**6.0–33.7**	**7.1–11.5**	**16.2–25.1**
Summed feature 4	0–5.0	0–4.9	**3.4–4.0**	2.5–4.4

[a]Values of less than 1% for all strains are not shown. The percentages of predominant fatty acids are shown in bold. Summed features consist of one or more fatty acids that could not be separated by the Microbial Identification System. Summed feature 3 is $C_{15:0}$ iso 2-OH and/or $C_{16:1}$ ω7*c*, summed feature 4 is $C_{17:1}$ iso I and $C_{17:1}$ anteiso B.

[b]Data are taken from Ahmed et al. (2007), Brettar et al. (2004a), Nedashkovskaya et al. (2004, 2005, 2006, 2007a, b), and Ying et al. (2006).

TABLE 110. Phenotypic characteristics that differentiate the genus *Echinicola* from its close relatives in the family *Cyclobacteriaceae*[a,b]

Characteristic	*Echinicola*	*Algoriphagus*	*Belliella*	*Cyclobacterium*
Cell morphology:				
Regular rods	+	+	+	–
Curved, ring-like, or horseshoe-shaped	–	–	–	+
Cell size (μm)	0.3–0.5 × 1.1–2.3	0.3–0.7 × 0.3–10.0	0.3–0.5 × 0.9–3.0	0.3–0.7 × 0.8–1.5
Gliding motility	+	–	–	–
Nitrate reduction	–	D	+	–
Salinity range (%)	0–15	0–10	0–6	0–10
Growth at 40°C	+	D	–	+
Hydrolysis of starch	+	D	+	–
DNA G+C content (mol%)	44–46	35–49	35.4	41–45

[a]Symbols: +, >85% positive; –, 0–15% positive; D, different reactions occur in different taxa (species of a genus).

[b]Data are taken from Ahmed et al. (2007), Bowman et al. (2003), Brettar et al. (2004a), Nedashkovskaya et al. (2004, 2005, 2006, 2007a, b), Raj and Maloy (1990), Tiago et al. (2006a), Van Trappen et al. (2004), Yi and Chun (2004), Ying et al. (2006), and Yoon et al. (2004, 2005a, b, 2006).

TABLE 111. Differential phenotypic characteristics of species of the genus *Echinicola*[a,b]

Characteristic	*E. vietnamensis* KMM 6221[T]	*E. pacifica* (n = 3)
Fermentation of D-glucose	–	+
Production of H_2S	–	+
Growth with 15% NaCl	+	–
Growth at 44°C	+	–
Hydrolysis of agar, gelatin, and Tween 40	–	+
Acid production from L-arabinose, D-cellobiose, D-glucose, D-lactose, D-maltose, D-mannose, L-rhamnose, and DL-xylose	–	+
Susceptibility to carbenicillin, chloramphenicol, doxycycline, erythromycin, and oleandomycin	+	–
DNA G+C content (mol%)	45.9	44–45

[a]Symbols: +, >85% positive; –, 0–15% positive; n, a number of the strains studied.
[b]Data are from Nedashkovskaya et al. (2006, 2007a).

relatives are the genera *Belliella*, *Algoriphagus*, and *Cyclobacterium*, with sequence similarity values of 91.7–92.1, 88.8–92.4, and 89.5–91.5%, respectively. A level of 16S rRNA gene sequence similarity between the two species of the genus, *Echinicola pacifica* and *Echinicola vietnamensis*, is 94.7–95.0%. The 16S rRNA gene sequence similarity between the type strain *Echinicola pacifica* KMM 6172[T] and reference strains KMM 6166 and KMM 6173 ranges from 99.4 to 99.9%. The DNA–DNA hybridization values between these strains vary from 93 to 98%.

Differentiation of the species of the genus *Echinicola*

The strains of the two species of the genus *Echinicola* have many similar phenotypic features. However, they can be differentiated from each other by the several phenotypic traits shown in Table 111.

List of species of the genus *Echinicola*

1. **Echinicola pacifica** Nedashkovskaya, Kim, Vancanneyt, Lysenko, Shin, Park, Lee, Jung, Kalinovskaya, Mikhailov, Bae and Swings 2006, 955[VP]

pa.ci'fi.ca. L. fem. adj. *pacifica* pacific, and by extension referring to the Pacific Ocean, from which the type strain was isolated.

Cells are 0.3–0.4 µm in width and 1.2–1.9 µm in length. On Marine agar, colonies are sunken into the agar. Growth occurs at 6–41°C and with 0–12% NaCl. The optimal temperature for growth 25–28°C. Decomposes gelatin (weakly) and esculin. Hydrolysis of Tweens 20 and 80 is strain-dependent. DNA is not hydrolyzed. Produces acid from *N*-acetylglucosamine. Can oxidize D-galactose and D-sucrose. Does not form acid from L-fucose. According to the API 20E gallery (bioMérieux), strain KMM 6172[T] utilizes citrate, forms acid from amygdalin, and is negative for arginine dihydrolase, lysine decarboxylase, and ornithine decarboxylase. Results of Biolog GN2 (Biolog) testing show that strain KMM 6172[T] utilizes α-cyclodextrin, dextrin, glycogen, α-D-glucose, D-fructose, L-fucose, D-galactose, gentiobiose, α-lactose, α-D-lactose, lactulose, D-melibiose, methyl β-D-glucoside, psicose, D-raffinose, D-trehalose, turanose, D-galacturonic acid, D-glucuronic acid, α-ketobutyric acid, alaninamide, L-alanine, L-alanylglycine, L-asparagine, L-aspartic acid, L-glutamic acid, hydroxy-L-proline and L-threonine. The following compounds are not utilized: Tween 80, *N*-acetyl-D-galactosamine, adonitol, L-arabitol, i-erythritol, *myo*-inositol, D-mannitol, D-sorbitol, xylitol, methyl pyruvate, monomethyl succinate, acetic acid, *cis*-aconitic acid, citric acid, formic acid, D-galactonic acid, D-gluconic acid, D-glucosaminic acid, α-, β-, and γ-hydroxybutyric acids, *p*-hydroxyphenylacetic acid, itaconic acid, α-ketoglutaric acid, α-ketovaleric acid, DL-lactic acid, malonic acid, propionic acid, quinic acid, D-saccharic acid, sebacic acid, succinic acid, bromosuccinic acid, succinamic acid, glucuronamide, D-alanine, glycyl-L-aspartic acid, glycyl-L-glutamic acid, L-histidine, L-leucine, L-ornithine, L-phenylalanine, L-proline, L-pyroglytamic acid, D-serine, L-serine, DL-carnitine, γ-aminobutyric acid, urocanic acid, inosine, uridine, thymidine, phenylethylamine, putrescine, 2-aminoethanol, 2,3-butanediol, glycerol, DL-α-glycerol phosphate, glucose 1-phosphate, and glucose-6-phosphate. H_2S is produced. Indole and acetoin (Voges–Proskauer reaction) are not produced. According to API ZYM gallery (bioMérieux), the following enzyme activities are present: α-galactosidase, acid phosphatase, esterase (C4), esterase lipase (C8), leucine arylamidase, valine arylamidase, cystine arylamidase, trypsin, α-chymotrypsin, naphthol-AS-BI-phosphohydrolase, α- and β-glucosidases, *N*-acetyl-β-glucosaminidase, α-mannosidase, and α-fucosidase. Lipase (C14) and β-glucuronidase are not present. The predominant fatty acids are $C_{15:0}$ iso (17.3–18.0%), $C_{16:1}$ ω5*c* (6.7–7.8%), $C_{17:1}$ iso ω9*c* (6.3–6.9%), $C_{17:1}$ ω6*c* (4.3–4.8%), $C_{15:0}$ iso 3-OH (3.4–5.0%), $C_{17:0}$ iso 3-OH (9.4–10.0%) and summed feature 3 (30.7–30.8%), comprising $C_{16:1}$ ω7*c* and/or $C_{15:0}$ iso 2-OH (Table 109).

Source: sea urchin *Strongylocentrotus intermedius* collected in Troitsa Bay, Gulf of Peter the Great, the East Sea (also known as the Japan Sea).

DNA G+C content (mol%): 44–45 (T_m).

Type strain: KMM 6172, KCTC 12368, LMG 23350.

Sequence accession no. (16S rRNA gene): DQ185611.

2. **Echinicola vietnamensis** Nedashkovskaya, Kim, Hoste, Shin, Beleneva, Vancanneyt and Mikhailov 2007a, 763[VP]

vi.et.nam.en'sis. N. L. fem. adj. *vietnamensis* of or belonging to Vietnam, the country of origin of the type strain.

Cells are 0.4–0.5 × 1.1–2.3 µm. Colonies are light pink on Marine agar. Growth occurs at 6–44°C and with 0–15% NaCl. The optimal temperature for growth is 30–32°C. Tweens 20 and 80 are not hydrolyzed. Acid is not produced from D-fructose, D-galactose, D-sucrose, or *N*-acetylglucosamine. The fatty acids accounting for more than 1% of the total are $C_{15:0}$ anteiso (1.4%), $C_{15:0}$ iso (20.0%), $C_{15:1}$ ω6*c* (1.2%), $C_{15:0}$ (1.5%), $C_{16:1}$ ω5*c* (4.9%), $C_{17:1}$ iso ω9*c* (4.4%), $C_{17:0}$ iso (1.0%), $C_{17:1}$ ω6*c* (4.5%), $C_{15:0}$ iso 3-OH (3.7%), $C_{16:0}$ 3-OH (2.3%), $C_{17:0}$ iso 3-OH (10.0%), summed feature 3 (34.5%), comprising $C_{16:1}$ ω7*c* and/or $C_{15:0}$ iso 2-OH and summed feature 4 (5%), comprising $C_{17:1}$ iso I and/or $C_{17:1}$ anteiso B.

Source: seawater collected in a mussel farm located in a lagoon of Nha Trang Bay, South China Sea, Vietnam.

DNA G+C content (mol%): 45.9 (T_m).

Type strain: KMM 6221, DSM 17526, LMG 23754.

Sequence accession no. (16S rRNA gene): AM406795.

Genus XV. **Rhodonellum** Schmidt, Priemé and Stougaard 2006, 2891[VP]

The Editorial Board

Rho.do.nell.um. Gr. neut. n. *rhodon* a rose; L. neut. dim. ending *-ellum*; N.L. dim. neut. n. *Rhodonellum* a small rose, referring to the red color of the colonies.

Rods 0.7–1.0 × 0.8–3.0 µm. Gram-stain-negative. Oxidase-negative. Catalase-positive. Aerobic. Chemoheterotrophic. **Colonies are pink to red due to carotenoids. Temperature range, 0–22°C; optimum, ca. 5°C. pH range, 7.5–10.7** when grown at 5–10°C. **NaCl range for growth, 0–3%;** optimum, 0.6%. Predominant fatty acids are $C_{17:1}$ iso ω9*c*, $C_{17:0}$ iso 3-OH (12.5–18.5%), and summed feature 3. Cells contain red pigment in the form of carotenoids. Optimal growth occurs above pH 9. NaCl is not required for growth, but growth is enhanced by the presence of up to 0.6% NaCl.

DNA G+C content (mol%): 44.2.

Type species: **Rhodonellum psychrophilum** Schmidt, Priemé and Stougaard 2006, 2891[VP].

Further descriptive information

Predominant fatty acids of *Rhodonellum psychrophilum* are $C_{15:1}$ iso G (6.3%), $C_{15:0}$ iso (7.6%), $C_{16:1}$ iso H (7.3%), $C_{17:1}$ iso ω9*c* (17.5%), $C_{17:1}$ ω6*c* (6.8%), $C_{17:0}$ iso 3-OH (17.5%), and summed feature 3, comprising $C_{16:1}$ ω7*c* and/or $C_{15:0}$ iso 2-OH (12.6%), and summed feature 4, comprising $C_{14:0}$ 2-OH and/or $C_{15:0}$ iso 2-OH (6.5%), which could not be distinguished by the method used (Schmidt et al., 2006).

Enrichment and isolation procedures

Rhodonellum psychrophilum was isolated from submarine ikaite tufa columns collected from the Ikka Fjord, southwest Greenland (Schmidt et al., 2006). The columns were conserved in 15% glycerol and kept at −20°C. Isolation was achieved on agar plates containing 0.1× R2A medium (Difco).

Differentiation of the genus *Rhodonellum* from other genera

Rhodonellum psychrophilum can be differentiated from *Belliellia baltica* by its pH range for growth (7.5–10.7 vs 6–10), negative oxidase reaction, temperature range of 0–22°C vs 4–37°C, an optimum temperature of 5°C vs 25–30°C, and a higher mol% G+C of the DNA (43.1 vs 35).

Taxonomic comments

Schmidt et al. (2006) reported that, based on 16S rRNA gene sequence analysis, the type strain of *Rhodonellum psychrophilum*, together with five related isolates from ikaite columns, formed a separate cluster with 86–93% gene sequence similarity to their closest relative, *Belliella baltica*.

List of species of the genus *Rhodonellum*

1. **Rhodonellum psychrophilum** Schmidt, Priemé and Stougaard 2006, 2891[VP]

psy.chro'phi.lum. Gr. adj. *psychros* cold; N.L. neut. adj. *philum* (from Gr. neut. adj. *philon*), friend, loving; N.L. neut. adj. *psychrophilum* cold-loving.

The characteristics are as described for the genus, with the following additional features. Colonies are smooth, circular, and red due to the presence of carotenoids when grown under low light intensity. Colonies are white to light red when grown at a light intensity of 20–40 $\mu E/m^2/s$. Growth occurs from pH 7.5 to above pH 10.7, with an optimum at pH 9.2–10.0. At optimal growth temperature, the range of tolerated pH is largest, whereas at, below, and above the optimal growth temperature, a narrower pH range is tolerated. NaCl is not required for growth, but up to 3% (w/v) NaCl is tolerated. Optimal growth occurs around 0.6% (w/v) NaCl. Strains can use a wide spectrum of carbon sources such as galactose, glycerol, lactose, maltose, mannose, sorbitol, and starch.

Source: the permanently alkaline and cold ikaite columns in the Ikka Fjord in southwest Greenland.

DNA G+C content (mol%): 44.2 (HPLC).

Type strain: GCM71, DSM 17998, LMG 23454.

Sequence accession no. (16S rRNA gene): DQ112660.

References

Ahmed, I., A. Yokota and T. Fujiwara. 2007. *Chimaereicella boritolerans* sp. nov., a boron-tolerant and alkaliphilic bacterium of the family *Flavobacteriaceae* isolated from soil. Int. J. Syst. Evol. Microbiol. *57*: 986–992.

Borrall, R. and J.M. Larkin. 1978. *Flectobacillus marinus* (Raj) comb. nov., a marine bacterium previously assigned to *Microcyclus*. Int. J. Syst. Bacteriol. *28*: 341–343.

Bowman, J.P., C. Mancuso, C.M. Nichols and J.A.E. Gibson. 2003. *Algoriphagus ratkowskyi* gen. nov., sp. nov., *Brumimicrobium glaciale* gen. nov., sp. nov., *Cryomorpha ignava* gen. nov., sp. nov. and *Crocinitomix catalasitica* gen. nov., sp. nov., novel *flavobacteria* isolated from various polar habitats. Int. J. Syst. Evol. Microbiol. *53*: 1343–1355.

Brettar, I., R. Christen and M.G. Höfle. 2004a. *Belliella baltica* gen. nov., sp. nov., a novel marine bacterium of the *Cytophaga-Flavobacterium-Bacteroides* group isolated from surface water of the central Baltic Sea. Int. J. Syst. Evol. Microbiol. *54*: 65–70.

Brettar, I., R. Christen and M.G. Höfle. 2004b. *Aquiflexum balticum* gen. nov., sp. nov., a novel marine bacterium of the *Cytophaga-Flavobacterium-Bacteroides* group isolated from surface water of the central Baltic Sea. Int. J. Syst. Evol. Microbiol. *54*: 2335–2341.

DeLey, J., H. Cattoir and A. Reynaerts. 1970. The quantitative measurement of DNA hybridization from renaturation rates. Eur. J. Biochem. *12*: 133–142.

Euzéby, J.P. 1998. Taxonomic note: necessary correction of specific and subspecific epithets according to Rules 12c and 13b of the International Code of Nomenclature of *Bacteria* (1990 Revision). Int. J. Syst. Bacteriol. *48*: 1073–1075.

Gascuel, O. 1997. BIONJ: an improved version of the NJ algorithm based on a simple model of sequence data. Mol. Biol. Evol. *14*: 685–695.

Hamana, K. and Y. Nakagawa. 2001. Polyamine distribution profiles in the eighteen genera phylogenetically located within the *Flavobacterium-Flexibacter-Cytophaga* complex. Microbios *106*: 7–17.

Larkin, J.M. and R. Borrall. 1984a. Family I. *Spirosomaceae*. *In* Bergey's Manual of Systematic Bacteriology, vol. 1 (edited by Krieg and Holt). Williams & Wilkins, Baltimore, pp. 125–132.

Larkin, J.M. and R. Borrall. 1984b. Deoxyribonucleic acid base composition and homology of *Microcyclus*, *Spirosoma*, and similar organisms. Int. J. Syst. Bacteriol. *34*: 211–215.

Larkin, J.M., P.M. Williams and R. Taylor. 1977. Taxonomy of genus *Microcyclus* Ørskov 1928, reintroduction and emendation of genus *Spirosoma Migula* 1894, and proposal of a new genus, *Flectobacillus*. Int. J. Syst. Bacteriol. *27*: 147–156.

Nedashkovskaya, O.I., M. Vancanneyt, S. Van Trappen, K. Vandemeulebroecke, A.M. Lysenko, M. Rohde, E. Falsen, G.M. Frolova, V.V. Mikhailov and J. Swings. 2004. Description of *Algoriphagus aquimarinus* sp. nov., *Algoriphagus chordae* sp. nov. and *Algoriphagus winogradskyi* sp. nov., from sea water and algae, transfer of *Hongiella halophila* Yi and Chun 2004 to the genus *Algoriphagus* as *Algoriphagus halophilus* comb. nov. and emended descriptions of the genera *Algoriphagus* Bowman *et al.* 2003 and *Hongiella* Yi and Chun 2004. Int. J. Syst. Evol. Microbiol. *54*: 1757–1764.

Nedashkovskaya, O.I., S.B. Kim, M.S. Lee, M.S. Park, K.H. Lee, A.M. Lysenko, H.W. Oh, V.V. Mikhailov and K.S. Bae. 2005. *Cyclobacterium amurskyense* sp. nov., a novel marine bacterium isolated from sea water. Int. J. Syst. Evol. Microbiol. *55*: 2391–2394.

Nedashkovskaya, O.I., S.B. Kim, M. Vancanneyt, A.M. Lysenko, D.S. Shin, M.S. Park, K.H. Lee, W.J. Jung, N.I. Kalinovskaya, V.V. Mikhailov, K.S. Bae and J. Swings. 2006. *Echinicola pacifica* gen. nov., sp. nov., a novel flexibacterium isolated from the sea urchin *Strongylocentrotus intermedius*. Int. J. Syst. Evol. Microbiol. *56*: 953–958.

Nedashkovskaya, O.I., S.B. Kim, B. Hoste, D.S. Shin, I.A. Beleneva, M. Vancanneyt and V.V. Mikhailov. 2007a. *Echinicola vietnamensis* sp. nov., a member of the phylum *Bacteroidetes* isolated from seawater. Int. J. Syst. Evol. Microbiol. *57*: 761–763.

Nedashkovskaya, O.I., S.B. Kim, K.K. Kwon, D.S. Shin, X. Luo, S.J. Kim and V.V. Mikhailov. 2007b. Proposal of *Algoriphagus vanfongensis* sp. nov., transfer of members of the genera *Hongiella* Yi and Chun 2004 emend. Nedashkovskaya et al., 2004 and *Chimaereicella* Tiago *et al.* 2006 to the genus *Algoriphagus*, and emended description of the genus *Algoriphagus* Bowman *et al.* 2003 emend. Nedashkovskaya *et al.* 2004. Int. J. Syst. Evol. Microbiol. *57*: 1988–1994.

Raj, H.D. 1976. A new species: *Microcyclus marinus*. Int. J. Syst. Bacteriol. *26*: 528–544.

Raj, H.D. 1979. Adansonian analysis of *Microcyclus* and related bacteria. Abstract 1-31. Proceedings of the Annu. Meet. Am. Soc. Microbiol..

Raj, H.D. and K.A. Paveglio. 1983. Contributing carbohydrate catabolic pathways in *Cyclobacterium marinus*. J. Bacteriol. *153*: 335–339.

Raj, H.D. and S.R. Maloy. 1990. Proposal of *Cyclobacterium marinus* gen. nov., comb. nov. for a marine bacterium previously assigned to the genus *Flectobacillus*. Int. J. Syst. Bacteriol. *40*: 337–347.

Schmidt, M., A. Priemé and P. Stougaard. 2006. *Rhodonellum psychrophilum* gen. nov., sp. nov., a novel psychrophilic and alkaliphilic bacterium of the phylum *Bacteroidetes* isolated from Greenland. Int. J. Syst. Evol. Microbiol. *56*: 2887–2892.

Tiago, I., A.P. Chung and A. Veríssimo. 2004. Bacterial diversity in a nonsaline alkaline environment: heterotrophic aerobic populations. Appl. Environ. Microbiol. *70*: 7378–7387.

Tiago, I., V. Mendes, C. Pires, P.V. Morais and A. Veríssimo. 2006a. *Chimaereicella alkaliphila* gen. nov., sp. nov., a Gram-negative alkaliphilic bacterium isolated from a nonsaline alkaline groundwater. Syst. Appl. Microbiol. *29*: 100–108.

Tiago, I., V. Mendes, C. Pires, P.V. Morais and A. Veríssimo. 2006b. *In* List of new names and new combinations previously effectively, but not validly, published. List no. 109. Int. J. Syst. Evol. Microbiol. *56*: 925–927.

Van Trappen, S., J. Mergaert, S. Van Eygen, P. Dawyndt, M.C. Cnockaert and J. Swings. 2002. Diversity of 746 heterotrophic bacteria isolated from microbial mats from ten Antarctic lakes. Syst. Appl. Microbiol. *25*: 603–610.

Van Trappen, S., I. Vandecandelaere, J. Mergaert and J. Swings. 2004. *Algoriphagus antarcticus* sp. nov., a novel psychrophile from microbial mats in Antarctic lakes. Int. J. Syst. Evol. Microbiol. *54*: 1969–1973.

Woese, C.R. 1987. Bacterial evolution. Microbiol. Rev. *51*: 221–271.

Woese, C.R., S. Maloy, L. Mandelco and H.D. Raj. 1990. Phylogenetic placement of the *Spirosomaceae*. Syst. Appl. Microbiol. *13*: 19–23.

Yi, H. and J. Chun. 2004. *Hongiella mannitolivorans* gen. nov., sp. nov., *Hongiella halophila* sp. nov. and *Hongiella ornithinivorans* sp. nov., isolated from tidal flat sediment. Int. J. Syst. Evol. Microbiol. *54*: 157–162.

Ying, J.Y., B.J. Wang, S.S. Yang and S.J. Liu. 2006. *Cyclobacterium lianum* sp. nov., a marine bacterium isolated from sediment of an oilfield in the South China Sea, and emended description of the genus *Cyclobacterium*. Int. J. Syst. Evol. Microbiol. *56*: 2927–2930.

Yoon, J.H., S.H. Yeo and T.K. Oh. 2004. *Hongiella marincola* sp. nov., isolated from sea water of the East Sea in Korea. Int. J. Syst. Evol. Microbiol. *54*: 1845–1848.

Yoon, J.H., S.J. Kang, S.Y. Jung, C.H. Lee and T.K. Oh. 2005a. *Algoriphagus yeomjeoni* sp. nov., isolated from a marine solar saltern in the Yellow Sea, Korea. Int. J. Syst. Evol. Microbiol. *55*: 865–870.

Yoon, J.H., S.J. Kang and T.K. Oh. 2005b. *Algoriphagus locisalis* sp. nov., isolated from a marine solar saltern. Int. J. Syst. Evol. Microbiol. *55*: 1635–1639.

Yoon, J.H., M.H. Lee, S.J. Kang and T.K. Oh. 2006. *Algoriphagus terrigena* sp. nov., isolated from soil. Int. J. Syst. Evol. Microbiol. *56*: 777–780.

Family III. **Flammeovirgaceae** fam. nov.

Olga I. Nedashkovskaya and Wolfgang Ludwig

Flam.me.o.vir.ga′ce.a.e. N.L. fem. n. *Flammeovirga* type genus of the family; suff. *-aceae* ending to denote a family; N.L. fem. pl. n. *Flammeovirgaceae* the *Flammeovirga* family.

Cells are **straight, flexible or curved rods, that** are 0.2–1.0×1.5–100 μm or longer. **Gram-stain-negative. Chemo-organotrophs.** Nonsporeforming. Nonflagellated and nonmotile in liquid media, but most of them move by gliding on solid substrates. Cells of one species can form sheathed filaments. **Colonies are pink, red, orange or apricot in color.** Colonies of the majority of strains are characterized by a spreading growth and some of them produce gelase fields and form deep craters in agar plates.

Strains of the majority species are strictly aerobic, but strains of one genus are characterized as facultatively anaerobic organisms. **Optimal grown temperature is 20–33°C. Carbohydrates are oxidized.** Two species ferment glucose. **Most strains require seawater or NaCl for growth. All strains are alkaline phosphatase-positive and arginine dihydrolase-, lysine decarboxylase-, and ornithine decarboxylase-negative.** The most strains produce oxidase and catalase. Only one species produces flexirubin type pigments. The majority of the strains cannot produce acetoin, hydrogen sulfide or indole. Nitrate may be reduced to nitrite. **Most species hydrolyze esculin, gelatin, and DNA.** Agar, casein, starch, carboxymethylcellulose and Tweens may be decomposed, but crystalline cellulose is not.

Menaquinone 7 is a major or single respiratory quinone. **Predominant fatty acids are $C_{15:0}$ iso, $C_{15:1}$ iso, $C_{17:0}$ iso 3-OH and $C_{15:0}$ iso 2-OH and/or $C_{16:1}$ ω *7c*. The main polyamines** for the majority species are **spermidine or homospermidine.** The most strains produced carotenoid pigment saproxanthin. All species **occur in different marine environments** including sea water, marine sediments, seaweeds, or marine animals, one of them was isolated from marine aquarium outflow.

DNA G+C content (mol%): 31–45.

Type genus: **Flammeovirga** Nakagawa, Hamana, Sakane and Yamasato 1997, 221[VP] emend. Takahashi, Suzuki and Nakagawa 2006, 2097[VP].

Taxonomic comments

Family *Flammeovirgaceae* comprises the recognized genera *Flammeovirga, Fabibacter, Flexithrix, Marinoscillum, Perexilibacter, Persicobacter, Rapidithrix, Reichenbachiella,* and *Roseivirga,* and generically misclassified species [*Flexibacter*] *litoralis,* [*Flexibacter*] *polymorphus,* [*Flexibacter*] *roseolus,* "*Microscilla sericea*", and "*Microscilla tractuosa*", which are phylogenetically distant from the type species of their genera. In the last two decades the strains currently affiliated with the family *Flammeovirgaceae* have been subjected to intensive taxonomic investigation by using a polyphasic approach and, especially, a phylogenetic analysis based on comparison of 16S rRNA gene sequences (Nakagawa et al., 2002). The species described as [*Cytophaga*] *aprica* and [*Cytophaga*] *diffluens* by Reichenbach (1989a) in the previous edition of *Bergey's Manual of Systematic Bacteriology,* were placed in the novel genera *Flammeovirga* and *Persicobacter,* respectively (Nakagawa et al., 1997). Later, Takahashi et al. (2006) emended the description of the genus *Flammeovirga* to include misclassified species "*Microscilla arenaria*" and the novel species *Flammeovirga yaeyamensis; Flammeovirga kamogawensis* was subsequently described by Hosoya and Yokota (2007a). A species with an uncertain taxonomic position, "[*Flexibacter*] *aggregans*" (Lewin, 1969) Leadbetter 1974, was reclassified as a later heterotypic synonym of *Flexithrix dorotheae* Lewin 1970 because of the very close phylogenetic relationship between them (Hosoya and Yokota, 2007b). In order to accommodate the single and type strain of species "*Microscilla furvescens*" and a representative of a novel species, the genus *Marinoscillum* was created (Seo et al., 2009). Recently a precise taxonomic position of the misclassified species "*Microscilla sericea*" and "*Microscilla tractuosa*" was determined, and it was proposed to reclassify them in the novel genus "*Marivirga*" as two distinct species (Nedashkovskaya et al., 2009).

In addition, the single and type strain of *Fabibacter halotolerans* forms a coherent phylogenetic cluster with species of the other member of the family, the genus *Roseivirga.* They share many phenotypic features in common and, perhaps, will be joined to the single genus.

From a view of the heterogeneity of the family *Flammeovirgaceae,* it should be noted that the representatives of all recognized and newly proposed genera such as *Flexithrix, Perexilibacter, Persicobacter, Rapidithrix, Roseivirga,* and generically misclassified species [*Flexibacter*] *litoralis,* [*Flexibacter*] *polymorphus* and [*Flexibacter*] *roseolus* can be considered as members of novel families in future.

Differential characteristics of the genera of *Flammeovirgaceae* are given in Table 112. Please note that *Marinoscillum* (Seo et al., 2009), "*Marivirga*" (Nedashkovskaya et al., 2009), *Perexilibacter* (Yoon et al., 2007), and *Rapidithrix* (Srisukchayakul et al., 2007) were proposed after the valid publication deadline for this volume and are not described in greater detail than given here.

Genus I. **Flammeovirga** Nakagawa, Hamana, Sakane and Yamasato 1997, 221[VP] emend. Takahashi, Suzuki and Nakagawa 2006, 2097[VP]

Yasuyoshi Nakagawa

Flam.me.o.vir′ga. L. adj. *flammeus* fire-colored; L. fem. n. *virga* rod; N.L. fem. n. *Flammeovirga* fire-colored rod.

Rods 0.4–0.9 μm wide and 1.7–96 μm long or longer. **Motile by gliding.** Nonsporeforming. Gram-stain-negative. **Aerobic,** having a strictly respiratory type of metabolism with oxygen as the terminal electron acceptor. Chemo-organotrophic. **Colonies spread and produce large gelase fields and deep craters in agar plates. Cell mass is orange to reddish orange.** Saproxanthin is present as the major carotenoid pigment. **Flexirubin-type pigments are absent.** Oxidase and catalase activities differ among species. **Marine organisms. Seawater is required for growth; NaCl alone can substitute.** The optimum pH for growth is 7. Nitrate is reduced. Agar, alginic acid, esculin, and starch are degraded. The major respiratory quinone is MK-7. Predominant cellular fatty acids are $C_{15:0}$ iso, $C_{20:4}$ ω6,9,12,15*c*, and $C_{16:0}$ 3-OH.

DNA G+C content (mol%): 31–36.

Type species: **Flammeovirga aprica** (Reichenbach 1989b) Nakagawa, Hamana, Sakane and Yamasato 1997, 221^VP^ emend. Takahashi, Suzuki and Nakagawa 2006, 2099^VP^ [*Cytophaga aprica* (*ex* Lewin 1969) Reichenbach 1989c, 495^VP^; *Cytophaga diffluens* var. *aprica* Lewin 1969, 197].

Further descriptive information

All species are negative for acid production from API 50CH system substrates including adonitol, D-arabinose, L-arabinose, D-arabitol, L-arabitol, dulcitol, erythritol, D-fucose, gluconate, glycerol, inositol, inulin, 2-ketogluconate, D-lyxose, methyl β-D-xyloside, D-tagatose, D-turanose, and L-xylose. All species are negative for the utilization of Biolog GN2 Microplate system substrates including *N*-acetyl-D-galactosamine, *cis*-aconitic acid, adonitol, D-alanine, γ-aminobutyric acid, 2-aminoethanol, L-arabinose, D-arabitol, bromosuccinic acid, 2,3-butanediol, DL-carnitine, citric acid, i-erythritol, formic acid, D-fructose, D-galactonic acid lactone, D-galacturonic acid, D-gluconic acid, D-glucosaminic acid, glucuronamide, D-glucuronic acid, L-glutamic acid, glycerol, DL-α-glycerol phosphate, α-hydroxybutyric acid, β-hydroxybutyric acid, γ-hydroxybutyric acid, hydroxy-L-proline, *p*-hydroxyphenylacetic acid, inosine, *myo*-inositol, itaconic acid, α-ketoglutaric acid, α-ketovaleric acid, L-leucine, malonic acid, D-mannitol, methyl-β-D-glucoside, methyl pyruvate, L-phenylalanine, phenylethylamine, D-psicose, putrescine, L-pyroglutamic acid, quinic acid, D-raffinose, D-saccharic acid, sebacic acid, D-serine, D-sorbitol, succinamic acid, succinic acid, sucrose, Tween 40, Tween 80, turanose, and xylitol.

Enrichment and isolation procedures

No enrichment media have been designed for isolation of *Flammeovirga* strains. Standard procedures to isolate marine bacteria can be applied. Colonies of *Flammeovirga* are usually orange to reddish orange and spread rapidly.

Flammeovirga strains have been isolated from marine environments at widely separated sites (Lewin, 1969; Nakagawa, 2004; Takahashi et al., 2006). The type strain of *Flammeovirga aprica* came from Kailua, Hawaii. The type strain of *Flammeovirga arenaria* was isolated from marine sand in Mexico. Strains of *Flammeovirga yaeyamensis* have been isolated from seaweeds, coastal sands, and dead leaves along the seashores of Iriomote and Ishigaki Islands (24° 20′ N 123° 45′ E and 24° 20′ N 124° 9′ E, respectively).

Maintenance procedures

Cultures of *Flammeovirga* strains can be preserved by freezing at lower than −80°C. For freezing, cells are suspended in Marine broth 2216 (Difco) containing 10% glycerol or 7% DMSO. *Flammeovirga* strains are rather sensitive to drying; however, they can be preserved by the liquid drying method using a protective medium such as SM2* or SM3† (Sakane et al., 1996), or by freeze drying.

Differentiation of the genus *Flammeovirga* from other genera

Characteristics differentiating the genus *Flammeovirga* from other mesophilic genera in the family *Flammeovirgaceae* are listed in Table 113. Presence of $C_{15:0}$ iso, $C_{16:0}$ 3-OH, and $C_{20:4}$ ω6,9,12,15*c* as major fatty acids and saproxanthin (the major carotenoid) are useful characters to discriminate the genus *Flammeovirga* from other genera.

Taxonomic comments

The genus *Flammeovirga* was created by Nakagawa et al. (1997) to accommodate the misclassified species *Cytophaga aprica*. Until 2006, the genus contained a single species, *Flammeovirga aprica*. However, Nakagawa et al. (2002) reported that "*Microscilla arenaria*" Lewin 1969 was related to the genus *Flammeovirga*, and then Takahashi et al. (2006) reclassified it as *Flammeovirga arenaria*, with the proposal of a third species, *Flammeovirga yaeyamensis* isolated from the Yaeyama Islands, Japan. These three species are clearly differentiated from each other by DNA–DNA hybridizations and phenotypic characteristics (Table 114).

Differentiation of the species of the genus *Flammeovirga*

Table 114 lists characteristics that distinguish the *Flammeovirga* species from one another.

List of species of the genus *Flammeovirga*‡

1. **Flammeovirga aprica** (Reichenbach 1989b) Nakagawa, Hamana, Sakane and Yamasato 1997, 221^VP^ emend. Takahashi, Suzuki and Nakagawa 2006, 2099^VP^ [*Cytophaga aprica* (*ex* Lewin 1969) Reichenbach 1989c, 495^VP^; *Cytophaga diffluens* var. *aprica* Lewin 1969, 197]

a′pri.ca. L. fem. adj. *aprica* sunlit, sun-loving.

The characteristics are as described for the genus and as listed in Table 113, with the following additional features. Cells are 0.5–0.9 μm wide and 1.7–96 μm long or longer. The cell mass is orange to reddish orange. Growth occurs at 15–30°C; optimum, 25°C. The pH range for growth is 6–8; optimum, 7. Growth occurs in the presence of 1–5% NaCl; optimum, 3%. Oxidase- and catalase-positive. Urease-negative. Agar, alginic acid, carboxymethylcellulose, DNA, esculin and starch are degraded, but not cellulose, chitin, gelatin, inulin, Tween 80, or tyrosine. Casein is weakly degraded. H_2S is produced. Indole-negative. The oxidation/fermentation test for catabolism of glucose is fermentative. Utilization (with the Biolog GN2 microplate system) is positive for cellobiose, α-cyclodextrin, dextrin, L-fucose, D-galactose, gentiobiose,

*SM2 consists of: monosodium glutamate monohydrate, 50 g; adonitol, 15 g; D(−)-sorbitol, 10 g; artificial seawater, 750 ml; and distilled water, 250 ml; pH 7.0.

†SM3 consists of: monosodium glutamate monohydrate, 50 g; adonitol, 15 g; D(−)-sorbitol, 10 g; L(−)-proline, 2 g; 2% methyl cellulose [4000 centipoise (cps)] solution, 100 ml; artificial seawater, 900 ml; pH 7.0.

‡*Flammeovirga kakegawaensis* was published after completion of the manuscript. Refer to Hosoya S. and A. Yokota. 2007. *Flammeovirga kamogawensis* sp. nov., isolated from coastal seawater in Japan. Int. J. Syst. Evol. Microbiol. *57*: 1327–1330.

TABLE 112. Phenotypic characteristics differentiating members of the family *Flammeovirgaceae*[a,b]

Characteristic	*Flammeovirga*	*Fabibacter*	*Flexithrix*	*Marinoscillum*
Source of isolation	Rocky sand, Hawaii, USA; brown sand, Norse Beach Puerto Penasco Sonora, Mexico; seaweed, sand, dead leaves, Yaeyama Islands, Japan sea water, Kamogawa, Japan, Pacific Ocean	Sponge *Tedania ignis*, Bahamas	Marine silt, Kerala, India; green-brown sand, Canoe Beach, Tema, Ghana	Sand, Samoa; unidentified sponge, Micronesia, Pacific Ocean
Metabolism	Fermentative/aerobic	Aerobic	Aerobic	Aerobic
Cell shape	Flexible rods	Curved rods	Flexible rods	Flexible rods
Cell size (μm)	0.4–1.0×1.7–96 or longer	0.5×1.5	2–100	0.2–0.5×10–100 or longer
Sheathed filaments formation	–	–	d	–
Colony pigmentation	Orange-reddish orange	Pink	Yellow	Orange or apricot
Gliding motility	+	+	+	+
Tryptophan deaminase	–	–	–	+
Oxidase/catalase	d/d	+/+	+/+	+/+
β-Galactosidase	d	+	+	+
Carotenoid pigments	Saproxanthin	nd	Zeaxanthin	Saproxanthin
Flexirubin type pigments production	–	–	–	–
Nitrate reduction	+	–	–	–
Indole/acetoin production	–/–	–/+	–/–	d/d
Hydrogen sulfide production	d	–	–	–
Temperature range for growth (°C) (optimum)	10–35 (25–30)	12–36 (28–30)	17–40	15–45 (33–33.5)
pH range for growth (optimum)	6.0–10.0 (7.0)	5.0–10.0	6.5–8.0	5.0–9.5 (7.5)
Salinity range for growth (% NaCl) (optimum)	1–5 (3)	0–12	2–5	0.5–12
Mg^{2+} and Ca^{2+} requirement	–	–	–	+
Hydrolysis of:				
Esculin	+	+	+	+[c]
Agar	+	–	–	–
Alginic acid	+	nd	–	+
Casein	d	–	–	d
Crystalline cellulose/carboxymethylcellulose	–/d	–	–/+	–/+[c]
Chitin	–	–	–	–
Gelatin	d	–	+	d
Starch	+	+	–	–
DNA	d	+	+	d
Tween 20	d	+	+	nd
Tween 40	d	+	+	–
Tween 80	d	+	+	–
Tyrosine	–	nd	d	nd
Urea	–	–	–	–
Acid from carbohydrates	+	+	+	+
Carbohydrates utilization	+	+	+	+
Citrate utilization	–		–	–
DNA G+C content (mol%)	31–36	42.5	35.6	41–44
Main fatty acids	$C_{15:0}$ iso, $C_{20:4}$ ω6,9,12,15*c*, $C_{16:0}$ iso 3-OH	$C_{15:0}$ iso, $C_{15:1}$ iso, $C_{15:0}$ iso 3-OH, $C_{16:0}$ iso 3-OH, $C_{17:0}$ iso 3-OH, summed feature 3	$C_{16:1}$ ω5*c*, $C_{15:0}$ iso, summed feature 3, $C_{17:0}$ iso 3-OH, $C_{16:0}$	$C_{15:0}$ iso, $C_{16:1}$ ω5*c*, $C_{17:0}$ iso 3-OH, summed feature 3
Polar lipids	nd	nd	nd	nd
Main polyamines	–	nd	Spermidine, agmatine	Homospermidine[c]

[a]Symbols: +, >90% positive; d, different strains give different reactions (11–89% positive); –, 0–0% positive; w, weak reaction; nd, not determined.

[b]Data from: Hamana and Nakagawa (2001), Hamana et al. (2008), Hosoya and Yokota (2007a, b), Lau et al. (2006), Nakagawa et al. (1997), Nedashkovskaya et al. (2003),

[c]Data for species *Marinoscillum furvescens* only.

TABLE 112. (Continued)

"*Microscilla*" *sericea*	"*Microscilla*" *tractuosa*	*Perexilibacter*	*Persicobacter*	*Rapidithrix*	*Reichenbachiella*	*Roseivirga*
Marine aquarium outflow, La Jolla, California	Beach sand, South China Sea	Sediments, Palau, Pacific Ocean	Black sandy mud, India	Seashell materials, Andaman Sea	Sea water, Sea of Japan	Green alga *Ulva fenestrata*, sea urchin *Strongylocentrotus intermedius*, Sea of Japan; sea water, Yellow Sea; sponge *Tedania ignis*, Bahamas
Aerobic	Aerobic	Aerobic	Fermentative	Aerobic	Aerobic	Aerobic
Flexible rods	Flexible rods	Straight rods	Flexible rods	Flexible rods	Flexible rods	Straight rods
0.4–0.5×10–100 or longer	0.4–0.5×10–50	0.3–0.5×10–20	0.5×4–30	0.7×20–100	0.5–0.7×5–15	0.2–0.5×2.0–4.0
–	–	–	–	–	–	–
Orange	Orange	Orange	Orange	Olive-gray	Orange	Pink-orange
+	+	+	+	+	+	d
–	–	–	–	nd	–	–
+/+	+/+	+/+	+/–	+/–	+/+	+/+
+	+	–	+	–	+	d
Saproxanthin	Saproxanthin	+	Saproxanthin	nd	nd	nd
–	–	nd	–	nd	+	–
–	–	–	+	nd	–	d
–/–	–/–	–/–	–/–	nd	–/nd	–/d
–	d	–	–	nd	–	–
10–38	10–40	4–45 (30–37)	15–40 (25–30)	nd (25–30)	4–35 (25–28)	4–44 (20–30)
nd	nd	5–10 (7)	nd	5–10	nd	5.0–10.0 (7–8)
0.5–12 (4–6)	0.5–10 (4–7)	0–3.5	1–6	nd	1–6	0–16 (2–3)
–	–	+	–	+	–	–
+	+	+	+	+	+	d
–	–	–	+	–	+	–
+	–	nd	+	nd	+	–
+	–	nd	–	–	–	–
–/–	–/–	nd	–/+	nd	–/–	–/–
–	–	nd	–	nd	–	–
+	+	+	+	+	+	+
–	–	–	+	+	+	–
+	+	–	+	nd	+	d
+	+	nd	nd	+	+	d
+	+	nd	nd	nd	–	d
+	+	nd	nd	+	–	d
nd	nd	nd	nd	+		–
–	–	nd	–	nd	–	–
+	+	+	+	+	+	d
+	+	+	+	+	+	+
–	+	–	+	+	–	d
36.1	36–37	43	40–42	40–43	44.5	40–44
$C_{15:1}$ iso, $C_{15:0}$ iso, $C_{17:0}$ iso 3-OH, $C_{15:0}$ iso 3-OH and summed feature 3	$C_{15:1}$ iso, $C_{15:0}$ iso, $C_{17:0}$ iso 3-OH and $C_{15:0}$	$C_{15:0}$ iso, $C_{16:1}$ 7*c*, $C_{16:1}$ 5*c*	$C_{15:0}$ iso, $C_{17:0}$ iso 3-OH, $C_{15:0}$, $C_{16:0}$ iso, $C_{16:0}$ iso 3-OH, $C_{20:4}$ ω6, 9,12,15c,	$C_{16:1}$ 5*c*, $C_{15:0}$ iso, $C_{17:0}$ iso 3-OH, $C_{15:0}$ iso 3-OH	Summed feature 3, $C_{16:1}$ 5*c*, $C_{15:0}$	$C_{15:1}$ iso, $C_{15:0}$ iso, $C_{17:0}$ iso 3-OH, $C_{16:0}$ iso 3-OH, $C_{15:0}$ iso 3-OH
nd	nd	nd	nd	nd	nd	Phosphatidylethanolamine, diphosphatidylglycerol, unidentified phospholipids, ninhydrin-positive lipid
Homospermidine, agmatine	nd	Spermidine	Spermidine	Spermidine	Spermidine	Spermidine

Nedashkovskaya et al. (2005a-c, 2008, 2009), Reichenbach (1989a), Seo et al. (2009), Srisukchayakul et al. (2007), Takahashi et al. (2006), and Yoon et al. (2007).

TABLE 113. Characteristics that differentiate the mesophilic genera belonging to *Flammeovirgaceae*[a]

Characteristic	*Flammeovirga*[b]	*Flexithrix*[c]	*Persicobacter*[c]
Color of cell mass	Orange to reddish orange	Golden yellow to yellow	Orange to pink
Aerobic	+	+	+ or −
Facultatively anaerobic	−	−	+
Oxidase activity	v	+	−
Catalase activity	v	+	+ or w
Urease activity	nd	w or −	−
Nitrate reduction	+	−	+
H_2S production	+	−	−
DNA G+C content (mol%)	31–36	35–37	42–44
Major fatty acids (%):[d]			
$C_{15:0}$ iso	30–54	22–28	47–56
$C_{16:0}$ 3-OH	8–12	0–1	3–7
$C_{16:1}$ ω5*c*	2–9	39–44	1–5
$C_{17:0}$ iso 3-OH	0–3	0–2	6–17
$C_{20:4}$ ω6,9,12,15*c*	8–25	0	2–7
Major carotenoids:			
Saproxanthin	+	−	+
Zeaxanthin	−	+	−

[a]Symbols: +, >85% positive; −, 0–15% positive; w, weak reaction; nd, not determined.
[b]Data from Takahashi et al. (2006).
[c]Unpublished data.
[d]Percentage of total fatty acids.

TABLE 114. Characteristics differentiating the species of the genus *Flammeovirga*[a,b]

Characteristic	*F. aprica*	*F. arenaria*	*F. yaeyamensis*
Growth at 10°C	−	+	−
Growth at 35°C	−	−	+
Optimum temperature (°C)	25	25	30
Growth at pH 10	−	−	+
Hydrolysis of:			
Carboxymethylcellulose, DNA	+	−	+
Casein	w	w	−
Gelatin, inulin, Tween 80	−	−	+
Oxidase	+	−	+ or w
Catalase	+	−	d
O-F test (glucose):			
Fermentation	+	−	+
Oxidation	−	+	−
Assimilation of:			
N-acetyl-D-glucosamine, D-melibiose	−	+	+
L-Alanyl glycine, L-ornithine, L-threonine	−	w	+ or w
Glucose 6-phosphate	w	+	−
Glucose 1-phosphate	+	+	−
Monomethyl succinate	w	−	−
L-Rhamnose, urocanic acid	−	−	+
Acid production from:			
Arbutin, melibiose	−	+	+
Mannitol, melezitose, sorbitol	−	w	−
Methyl α-D-glucoside	−	w	w
Raffinose, xylitol	−	w	+ or w
Rhamnose	−	−	+
Ribose	w	w	−
D-Xylose	+	−	+
DNA G+C content (mol%)	34.2	31.8	33.4–35.7

[a]Symbols: +, >85% positive; d, different strains give different reactions (16–84% positive); −, 0–15% positive; w, weak reaction.
[b]Data from Takahashi et al. (2006).

α-D-glucose, glucose 1-phosphate, glucose 6-phosphate, glycogen, glycyl-L-glutamic acid, DL-lactic acid, α-D-lactose, lactulose, maltose, D-mannose, monomethyl succinate, and D-trehalose, but negative for *N*-acetyl-D-glucosamine, L-alanyl glycine, D-melibiose, L-ornithine, L-rhamnose, L-threonine, and urocanic acid. Acid production (API 50CH system) is positive for *N*-acetylglucosamine, amygdalin, cellobiose, esculin, galactose, gentiobiose, glycogen, lactose, maltose, salicin, starch, and D-xylose, and weakly positive for glucose, L-fucose, mannose, and ribose, but negative for arbutin, mannitol, melibiose, melezitose, methyl α-D-glucoside, raffinose, rhamnose, sorbitol, and xylitol. The major cellular fatty acids are $C_{15:0}$ iso, $C_{20:4}$ω6, 9, 12, 15*c*, $C_{16:0}$ 3-OH, and $C_{14:0}$.

DNA G+C content (mol%): 34.2 (HPLC).

Type strain: JL-4, ATCC 23126, CIP 104807, IFO (now NBRC) 15941, JCM 21138, NCIMB 13348, NRRL B-14729.

Sequence accession no. (16S rRNA gene): AB247553.

2. **Flammeovirga arenaria** (*ex* Lewin 1969) Takahashi, Suzuki and Nakagawa 2006, 2099[VP] (*Microscilla arenaria* Lewin 1969, 197)

a.re.na'ria. L. fem. adj. *arenaria* of or pertaining to sand, (referring to the source of the organism).

The characteristics are as described for the genus and as listed in Table 113, with the following additional features. Cells are 0.5–0.9 µm wide and 2.0–40 µm long or longer. The cell mass is orange. Temperature range for growth, 10–30°C; optimum, 25°C. The pH range for growth, 6–8; optimum, 7. Growth occurs at 1–5% NaCl; optimum, 3%. Oxidase, catalase, and urease activities are negative. Agar, alginic acid, esculin, and starch are hydrolyzed, but not carboxymethylcellulose, cellulose, chitin, DNA, inulin, gelatin, Tween 80, or tyrosine. Casein is weakly degraded. H_2S is produced. Indole-negative. The O-F test of glucose catabolism is oxidative.

Utilization (with the Biolog GN2 microplate system) is positive for *N*-acetyl-D-glucosamine, cellobiose, α-cyclodextrin, dextrin, L-fucose, D-galactose, gentiobiose, α-D-glucose, glucose-1-phosphate, glucose 6-phosphate, L-glutamic acid, glycogen, glycyl-L-aspartic acid, glycyl-L-glutamic acid, α-D-lactose, lactulose, maltose, D-mannose, and D-melibiose, and weakly positive for alaninamide, L-alanine, L-alanylglycine, L-aspartic acid, DL-lactic acid, L-ornithine, and L-threonine, but negative for monomethyl succinate, L-rhamnose, and urocanic acid. Acid production (API 50CH system) is positive for *N*-acetylglucosamine, amygdalin, arbutin, cellobiose, esculin, L-fucose, galactose, gentiobiose, glucose, glycogen, lactose, maltose, mannose, melibiose, starch, and trehalose, and weakly positive for fructose, 5-ketogluconate, mannitol, methyl α-D-glucoside, methyl-α-D-mannoside, melezitose, raffinose, ribose, salicin, sorbitol, sucrose, and xylitol, but negative for rhamnose and D-xylose. The major cellular fatty acids are $C_{15:0}$ iso, $C_{14:0}$, $C_{16:0}$ 3-OH, $C_{20:4}$ ω6, 9, 12, 15c, $C_{16:1}$ ω5*c*, and $C_{15:0}$ iso 3-OH.

DNA G+C content (mol%): 31.8 (HPLC).

Type strain: HJ-1, CIP 109101, JCM 21777, LMG 18922, NBRC 15982, NCIMB 1413.

Sequence accession no. (16S rRNA gene): AB078078.

3. **Flammeovirga yaeyamensis** Takahashi, Suzuki and Nakagawa 2006, 2099[VP]

ya.e.ya.men'sis. N.L. fem. adj. *yaeyamensis* of or belonging to the Yaeyama Islands, from where the organisms were isolated.

The characteristics are as described for the genus and as listed in Table 113, with the following additional features. Cells are 0.4–0.9 µm wide and 1.7–90 µm long or longer. The cell mass is orange. Temperature range for growth, 15–35°C; optimum, 30°C. The pH range for growth is 6–10; optimum, 7. Growth occurs at 1–5% NaCl; optimum, 3%. Oxidase activity is positive. Catalase activity differs among strains. Urease-negative. Agar, alginic acid, carboxymethylcellulose, DNA, esculin, inulin, gelatin, starch and Tween 80 are degraded, but not casein, cellulose, chitin, or tyrosine. H_2S is produced. Indole-negative. The O-F test of glucose catabolism is fermentative. Utilization (with the Biolog GN2 microplate system) is positive for *N*-acetyl-D-glucosamine, cellobiose, D-galactose, gentiobiose, glycyl-L-glutamic acid, D-melibiose, L-rhamnose, and urocanic acid, and weakly positive for L-alanylglycine, α-D-glucose, maltose, L-ornithine, and L-threonine, but negative for glucose 1-phosphate, glucose 6-phosphate, and monomethyl succinate. Acid production (API 50CH system) is positive for *N*-acetylglucosamine, amygdalin, arbutin, cellobiose, esculin, galactose, gentiobiose, glucose, glycogen, lactose, maltose, mannose, melibiose, rhamnose, salicin, starch, and D-xylose, and positive or weakly positive for L-fucose, methyl α-D-glucoside, raffinose, and xylitol, but negative for mannitol, melezitose, ribose, and sorbitol. The major cellular fatty acids are $C_{15:0}$ iso, $C_{20:4}$ ω6,9,12,15*c*, $C_{16:0}$ 3-OH, $C_{15:0}$ iso 3-OH, $C_{16:0}$, and $C_{14:0}$.

DNA G+C content (mol%): 33.4–35.7 (HPLC).

Type strain: IR25-3, CIP 109099, NBRC 100898.

Sequence accession no. (16S rRNA gene): AB247554.

Genus II. **Fabibacter** Lau, Tsoi, Li, Plakhotnikova, Dobretsov, Wu, Wong, Pawlik and Qian 2006, 1062[VP]

THE EDITORIAL BOARD

Fa.bi.bac'ter. L. fem. n. *faba* bean; N.L. masc. n. *bacter* rod; N.L. masc. n. *Fabibacter* bean(-like) rod.

Curved rods 0.5 × 1.5 µm. Gram-stain-negative. Strictly aerobic. **Gliding motility is present.** Chemo-organotrophic. Colonies on marine agar are pink. The major respiratory quinone is MK-7. **Flexirubin-type pigments are not produced. Acid is produced from carbohydrates.** Oxidase- and catalase-positive. **Tween 80 is hydrolyzed but not gelatin. NaCl is not required for growth.** Isolated from the marine sponge *Tedania ignis*.

DNA G+C content (mol%): 42.5.

Type species: **Fabibacter halotolerans** Lau, Tsoi, Li, Plakhotnikova, Dobretsov, Wu, Wong, Pawlik and Qian 2006, 1062[VP].

Enrichment and isolation procedures

Fabibacter halotolerans was isolated on marine agar* from the marine sponge *Tedania ignis* in the Bahamas. The strains appeared as pink-pigmented after 48 h of cultivation at 30°C on marine agar.

Differentiation of the genus *Fabibacter* from other genera

Unlike *Roseivirga echinicomitans* and *Roseivirga ehrenbergii*, *Fabibacter halotolerans* exhibits a different cell shape (curved rod vs straight rod), gliding motility, no requirement for NaCl for growth, a higher range of tolerance to NaCl (0–12% vs 1–8%), an ability to hydrolyze Tween 80 but not gelatin, the production of acid from carbohydrates, a narrower temperature range for growth (12–36°C vs 4.0–39°C), and a slightly higher mol% G+C value for its DNA (42.5 vs 40.2–41.3). Moreover, the fatty acid profile of *Fabibacter halotolerans* differs from those described for the members of *Roseivirga* mainly by having larger quantities of $C_{15:0}$ iso 3-OH, $C_{16:0}$ iso 3-OH, and summed feature 3 (SF3) and by the additional presence of $C_{14:0}$ iso 3-OH, $C_{15:0}$ 2-OH, and $C_{15:0}$ 3-OH (Lau et al., 2006).

Lau et al. (2006) indicated that *Fabibacter* was also closely related to the *Marincola seohaensis*; however, *Marinicola seohaensis* has been formally reclassified to the genus *Roseivirga* as *Roseivirga seohaensis* (Lau et al., 2006). *Fabibacter halotolerans* differs from *Roseivirga seohaensis* by having a different cell shape (curved rod vs straight rod), orange colonies, lack of flexirubin pigments, a slightly higher mol% G+C value for its DNA (42.5 vs 40.3), no requirement for NaCl for growth, a higher range of tolerance to NaCl (0–12% vs 2–8%), the production of acid from carbohydrates, and by exhibiting activities for arginine dihydrolase, α-galactosidase, β-galactosidase, α-glucosidase, β-glucosidase, and α-mannosidase.

Taxonomic comments

Lau et al. (2006) found that the type strain of *Fabibacter* was most closely related to the members of the genera *Marinicola* (see above) and *Roseivirga*, with 93.1–93.3% 16S rRNA gene sequence similarity.

List of species of the genus *Fabibacter*

1. **Fabibacter halotolerans** Lau, Tsoi, Li, Plakhotnikova, Dobretsov, Wu, Wong, Pawlik and Qian 2006, 1062[VP]

ha.lo.to'le.rans. Gr. masc. n. *hals, halos* salt; L. part. adj. *tolerans* tolerating; N.L. part. adj. *halotolerans* salt-tolerating.

The description is as given for the genus, with the following additional features. All the characteristics described below are based on cultures grown on marine agar at 30°C for 48 h. Colonies are pink, circular, 2.0–4.0 mm in diameter, convex with a smooth surface and an entire margin. No diffusible pigment is formed. Growth occurs between 12 and 36°C (optimum, 28–30°C) and between pH 5.0 and 10.0. Sodium ions are not required for growth. The organisms tolerate up to 12% NaCl. Predominant fatty acids (>5%) are $C_{15:0}$ iso, $C_{15:1}$, $C_{15:0}$ iso 3-OH, $C_{16:0}$ iso 3-OH, $C_{17:0}$ iso 3-OH, and SF 3 (comprising $C_{15:0}$ iso 2-OH and/or 16:1 ω7*c*). These fatty acids represent 80.7% of the total. Produces acetoin, but not indole or H_2S. Nitrate is not reduced. DNA and Tweens 20, 40, and 80 are hydrolyzed, but not agar, casein, carboxymethylcellulose, chitin, or gelatin. Starch is weakly hydrolyzed. The following enzyme activities are present: *N*-Acetyl-β-glucosaminidase, acid phosphatase, alkaline phosphatase, arginine dihydrolase, α-galactosidase, β-galactosidase, α-glucosidase, β-glucosidase, α-chymotrypsin, cystine arylamidase, leucine arylamidase, valine arylamidase, esterase (C4), esterase lipase (C8), lipase (C14), α-mannosidase, trypsin, and naphthol-AS-BI-phosphohydrolase. No activities are exhibited for α-fucosidase, β-glucuronidase, lysine decarboxylase, ornithine decarboxylase, tryptophan deaminase, or urease. Growth occurs on the following sole carbon sources (API 20E, 20NE and 50CH systems): D-cellobiose, D-lactose, D-maltose, and starch. Acid is produced from the following sole carbon sources in the API 20E and 50CH systems: amygdalin, arbutin, D-cellobiose, esculin ferric citrate, D-galactose, D-glucose, gentiobiose, maltose, methyl α-D-glucopyranoside, D-raffinose, salicin, sucrose, starch, and D-trehalose. The following carbon sources are utilized (MicroLog 3 system): L-alaninamide, L-alanine, L-alanylglycine, L-aspartic acid, D-cellobiose, dextrin, D-galacturonic acid, gentiobiose, α-D-glucose, D-glucose 6-phosphate, L-glutamic acid, glycogen, glycyl-L-aspartic acid, glycyl-L-glutamic acid, α-ketobutyric acid, α-ketoglutaric acid, α-ketovaleric acid, DL-lactic acid, α-D-lactose, lactulose, maltose, D-melibiose, methyl-β-D-glucoside, L-ornithine, L-proline, L-pyroglutamic acid, D-raffinose, succinamic acid, sucrose, D-trehalose, turanose, and L-threonine. In disc-diffusion tests, susceptibility is shown toward to ampicillin (1 μg), chloramphenicol (1 μg), penicillin (1 μg), streptomycin (0.1 μg), and tetracycline (5 μg), but not to kanamycin (tested up to 100 μg).

Source: marine sponge *Tedania ignis* in the Bahamas.

DNA G+C content (mol%): 42.5±0.3 (HPLC).

Type strain: UST030701-097, JCM 13334, NRRL B-41220.

Sequence accession no. (16S rRNA gene): DQ080995.

Genus III. **Flexithrix** Lewin 1970, 513[VP] emend. Hosoya and Yokota 2007b, 1087[VP] emend.

Yasuyoshi Nakagawa

Flex'i.thrix. L. part. adj. *flexus* flexible; Gr. fem. n. *thrix* hair; N.L. fem. n. *Flexithrix* flexible hair (flexible rod).

Rods 0.4–0.9 μm wide and 1.5–70 μm long or longer. **Motile by gliding.** Nonsporeforming. Sheath is present or not. Gram-stain-negative. **Aerobic**. Chemo-organotrophic. **The cell mass is golden yellow to yellow**. Zeaxanthin is present as the major carotenoid pigment. **Flexirubin-type pigments are absent.** Oxidase- and catalase-positive. Urease activity is negative or weakly positive. **Marine organisms**. **Seawater is required for growth; NaCl alone can substitute.** The optimum pH for growth is 7. Nitrate is not reduced. H_2S and indole are not produced. Esculin, gelatin, starch, and Tween 80 are degraded. The major

*Marine agar medium (Lau et al., 2006) composed of (g/l of filter-sterilized seawater): peptone (Oxoid), 5.0; and yeast extract (Oxoid), 3.0.

respiratory quinone is MK-7. Predominant cellular fatty acids are $C_{15:0}$ iso and $C_{16:1}$ ω5*c*.

DNA G+C content (mol%): 35–37.

Type species: **Flexithrix dorotheae** Lewin 1970, 511[AL] emend. Hosoya and Yokota 2007b, 1087[VP] emend.

Further descriptive information

All strains are negative for utilization of Biolog GN2 Microplate system substrates including *cis*-aconitic acid, adonitol, γ-amino butyric acid, 2-amino ethanol, D-arabitol, bromosuccinic acid, 2,3-butanediol, DL-carnitine, citric acid, i-erythritol, formic acid, D-galactonic acid lactone, D-glucosaminic acid, glucose-6-phosphate, glycerol, DL-α-glycerol phosphate, L-histidine, α-hydroxybutyric acid, β-hydroxybutyric acid, γ-hydroxybutyric acid, *p*-hydroxy phenylacetic acid, hydroxy L-proline, inosine, *m*-inositol, itaconic acid, α-ketobutyric acid, α-ketoglutaric acid, α-ketovaleric acid, L-leucine, malonic acid, methylpyruvate, L-phenylalanine, phenyl ethylamine, propionic acid, putrescine, L-pyroglutamic acid, quinic acid, D-saccharic acid, sebacic acid, D-serine, D-sorbitol, succinamic acid, thymidine, Tween 40, Tween 80, uridine, urocanic acid, and xylitol.

All strains are negative for acid production from API 50CH system substrates including adonitol, D-arabitol, L-arabitol, dulcitol, erythritol, D-fucose, gluconate, glycerol, glycogen, inositol, 2-keto-gluconate, 5-keto-glucotate, mannitol, β-methyl-D-xyloside, sorbitol, sorbose, D-tagatose, xylitol, and L-xylose. All strains are negative for API ZYM system substrates including β-glucosidase and β-glucuronidase. All strains are sensitive to ampicillin, benzylpenicillin, chloramphenicol, erythromycin, nalidixic acid, nitrofurantoin, novobiocin, and oleandomycin, but resistant to bacitracin, carbenicillin, colistin sulfate, gentamicin, kanamycin, neomycin, polymyxin B, streptomycin, and tetracycline.

Enrichment and isolation procedures

No special enrichment media have been designed for isolation of *Flexithrix* strains. Standard procedures to isolate marine bacteria can be applied. Colonies of *Flexithrix* are usually yellow and spread on agar media. *Flexithrix* strains have been isolated from marine environments at widely separated sites (Lewin, 1969, 1970; Lewin and Lounsbery, 1969). The type strain of *Flexithrix dorotheae* came from Ernakulum, India.

Maintenance procedures

Cultures of *Flexithrix* strains can be preserved by freezing at lower than −80°C. For freezing, cells are suspended in Marine broth 2216 (Difco) containing 10% glycerol or 7% DMSO. *Flexithrix* strains are rather sensitive to drying; however, they can be preserved by the liquid drying method using a protective medium such as SM2 or SM3 (see the chapter on *Flammeovirga* for the formulations), or by freeze drying.

Differentiation of the genus *Flexithrix* from other genera

Characteristics differentiating the genus *Flexithrix* form other mesophilic genera in the family *Flammeovirgaceae* are listed in Table 113 in the chapter on *Flammeovirga*. The presence of $C_{15:0}$ iso and $C_{16:1}$ ω5*c* as major fatty acids and zeaxanthin as a major carotenoid are useful characters to discriminate the genus *Flammeovirga* from other genera.

Taxonomic comments

The genus *Flexithrix* was created by Lewin (1970) to accommodate a sheathed gliding bacteria strain QQ-3 (NBRC 15987[T]) isolated from brown silt from the coast of Kerala, India. Until the 1st edition of *Bergey's Manual of Systematic Bacteriology*, the genus contained a single strain belonging to a single species *Flexithrix dorotheae*; however, Reichenbach (1989d) mentioned that the genus *Flexithrix* closely resembles *Flexibacter aggregans* ("*Microscilla aggregans*") when it lacks sheathed filaments. Nakagawa et al. (2002) reported that three strains belonging to *Flexibacter aggregans* (NBRC 15973, NBRC 15976[T], and NBRC 15990) were closely related to the genus *Flexithrix* by 16S rRNA gene sequence analysis. Hosoya and Yokota (2007b) proposed reclassification of *Flexibacter aggregans* as *Flexithrix dorotheae* and emended the description of the genus *Flexithrix* and the species *Flexithrix dorotheae*. However, the emended descriptions did not cover characteristics of all *Flexithrix* strains, because they were based only on type strains (NBRC 15976[T] and NBRC 15987[T]). We have investigated all four strains belonging to the genus *Flexithrix*, phenotypically, chemotaxonomically, and genotypically (unpublished). The type strain of *Flexithrix dorotheae* exhibits more than 70% DNA–DNA relatedness with the three strains of *Flexibacter aggregans*, and they also share many not previously known phenotypic and chemotaxonomic characteristics. Those characteristics are described in this chapter. In addition, we found that presence of zeaxanthin (the major carotenoid) can be used to differentiate the genus *Flexithrix* from other mesophilic genera of *Flammeovirgaceae*. Therefore, we propose the emendation of the genus *Flexithrix* and the species *Flexithrix dorotheae* as indicated in this chapter.

List of species of the genus *Flexithrix*

1. **Flexithrix dorotheae** Lewin 1970, 511[AL] emend. Hosoya and Yokota 2007b, 1087[VP] emend. (*Microscilla aggregans* Lewin 1969, 197; *Flexibacter aggregans* (Lewin 1969) Leadbetter 1974, 106[AL])

do.ro.the'ae. N.L. gen. fem. n. *dorotheae* of Dorothy; named after a deceased technical assistant, Dorothy White.

The characteristics are as described for the genus, with the following additional features. Cells are 0.4–0.9 µm wide and 1.5–70 µm long or longer. The cell mass is golden yellow to yellow. Growth occurs at 10–40°C; optimum, 25–30°C. The pH range for growth is 6–11; optimum, 7. Growth occurs in the presence of 1–5% NaCl; optimum, 3%. Some strains can grow in the presence of 7% NaCl. Oxidase- and catalase-positive. Esculin, gelatin, starch, and Tween 80 are degraded, but not agar, cellulose, chitin, inulin, and yeast cells. Alginate is strongly or weakly degraded. Sole nitrogen sources include ammonium sulfate, Casamino acids, peptone, sodium glutamate, and sodium nitrate. Utilization (with the Biolog GN2 microplate system) is positive for *N*-acetyl-D-galactosamine, alaninamide, L-alanylglycine, dextrin, D-fructose, L-fucose, D-galactose, DL-lactic acid, D-melibiose, β-methyl D-glucoside, D-raffinose, L-rhamnose, D-trehalose, and turanose, and positive or weakly positive for acetic acid, *N*-acetyl-D-glucosamine, L-alanine, L-arabinose, L-asparagine, L-aspartic acid, gentiobiose, α-D-glucose, glucuronamide, D-glucuronic acid, L-glutamic acid, glycyl-L-aspartic acid, glycyl-L-glutamic acid, α-D-lactose, lactulose, maltose, L-ornithine, D-psicose, L-serine, sucrose, and L-threonine. Acid production (API 50CH system) is positive for cellobiose, esculin, fructose, galactose, lactose, maltose, melibiose, salicin, sucrose, trehalose, and D-turanose, and positive or weakly positive for *N*-acetylglucosamine, amygdalin, L-arabinose, arbutin, glucose, inulin, D-lyxose, mannose, melezitose, α-methyl-D-glucoside, α-methyl-D-mannoside, raffinose, and D-xylose. Enzyme production (API ZYM system) is positive for *N*-acetyl-β-glucosaminidase, acid phosphatase, alkaline phosphatase, β-galactosidase, leucine allyl amidase, and valine allyl amidase. The major cellular fatty acids are $C_{15:0}$ iso, $C_{16:1}$ ω5*c*, and $C_{16:0}$.

DNA G+C content (mol%): 35.9–36.1 (HPLC).

Type strain: QQ-3, ATCC 23163, DSM 6795, NBRC 15987, NBRC 102100, NCIMB 1390.

Sequence accession no. (16S rRNA gene): AB078077.

Genus IV. **Persicobacter** Nakagawa, Hamana, Sakane and Yamasato 1997, 221^VP

Yasuyoshi Nakagawa

Per.si.co.bac'ter. Gr. neut. n. *persikon* peach; N.L. masc. n. *bacter* rod; N.L. masc. n. *Persicobacter* peach rod, because the organism is a peach-colored rod.

Slender rods 0.4–0.6 μm wide and 0.9–30 μm long or longer. **Motile by gliding.** Nonsporeforming. Gram-stain-negative. **Facultatively anaerobic**. Chemo-organotrophic. **Colonies spread and produce large gelase fields and deep craters in agar plates.** The **cell mass is pink to orange**. Saproxanthin is present as the major carotenoid pigment. **Flexirubin-type pigments are absent.** Oxidase- and urease-negative. Strongly or weakly catalase-positive. **Marine organisms. Seawater is required for growth; NaCl alone can substitute.** The optimum pH for growth is 7. Nitrate is reduced. H_2S and indole are not produced. Agar, alginic acid, esculin, and gelatin are degraded. Starch is weakly degraded. The major respiratory quinone is MK-7. Predominant cellular fatty acids are $C_{15:0}$ iso and $C_{17:0}$ iso 3-OH.

DNA G+C content (mol%): 42–44.

Type species: **Persicobacter diffluens** (Reichenbach 1989b) Nakagawa, Hamana, Sakane, Yamasato 1997, 222^VP [*Cytophaga diffluens* (*ex* Stanier 1940) Reichenbach 1989c, 495^VP; *Cytophaga diffluens* Stanier 1940, 623 emend. mut. char. Lewin 1969, 197].

Further descriptive information

All species are negative for acid production from API 50CH system substrates including adonitol, L-arabinose, D-arabitol, L-arabitol, dulcitol, erythritol, D-fucose, L-fucose, gluconate, glycerol, inositol, inulin, 2-ketogluconate, 5-ketoglucomate, mannitol, melezitose, β-methyl-D-xyloside, raffinose, ribose, sorbitol, D-tagatose, D-turanose, and L-xylose. All species are negative for the utilization of Biolog GN2 Microplate system substrates including *N*-acetyl-D-galactosamine, *cis*-aconitic acid, adonitol, D-alanine, γ-amino butyric acid, 2-amino ethanol, L-arabinose, D-arabitol, bromosuccinic acid, 2,3-butanediol, DL-carnitine, citric acid, i-erythritol, formic acid, D-galactonic acid lactone, D-gluconic acid, glucuronamide, glycerol, DL-α-glycerol phosphate, L-histidine, α-hydroxybutyric acid, β-hydroxybutyric acid, γ-hydroxybutyric acid, *p*-hydroxy phenylacetic acid, hydroxy-L-proline, inosine, *m*-inositol, itaconic acid, α-ketoglutaric acid, malonic acid, D-mannitol, mono-methyl succinate, L-phenylalanine, phenyl ethylamine, propionic acid, D-psicose, putrescine, L-pyroglutamic acid, quinic acid, D-saccharic acid, sebacic acid, D-serine, D-sorbitol, succinamic acid, succinic acid, Tween 80, uridine, and xylitol. All species are sensitive to chloramphenicol, lincomycin, nitrofurantoin, novobiocin, oleandomycin, and tetracycline, but resistant to colistin sulfate, gentamicin, kanamycin, polymxin B, and streptomycin.

Enrichment and isolation procedures

No enrichment media have been designed for isolation of *Persicobacter* strains. Standard procedures to isolate marine bacteria can be applied. Colonies of *Persicobacter* are usually orange to pink, degrade agar, and spread rapidly.

Persicobacter strains have been isolated from marine environments at widely separated sites (Lewin, 1969; Lewin and Lounsbery, 1969; Nakagawa, 2004). The type strain of *Persicobacter diffluens* came from Bombay, India. The type strain of "*Persicobacter psychrovividus*" was isolated from a clam, *Ruditapes philippinarum*, collected off a seacoast in Chiba, Japan.

Maintenance procedures

Cultures of *Persicobacter* strains can be preserved by freezing at lower than −80°C. For freezing, cells are suspended in Marine broth 2216 (Difco) containing 10% glycerol or 7% DMSO. *Persicobacter* strains are rather sensitive to drying; however, they can be preserved by a liquid drying method using a protective medium SM2 or SM3 (see the chapter on *Flammeovirga* for formulations), or by freeze drying.

Differentiation of the genus *Persicobacter* from other genera

Characteristics differentiating the genus *Persicobacter* from other mesophilic genera in the family *Flammeovirgaceae* are listed in Table 115 in the chapter on *Flammeovirga*. The presence of $C_{15:0}$ iso and $C_{17:0}$ iso 3-OH as major fatty acids and saproxanthin as the major carotenoid are useful characters to discriminate the genus *Persicobacter* from other genera.

Taxonomic comments

The genus *Persicobacter* was created by Nakagawa et al. (1997) to accommodate the misclassified species *Cytophaga diffluens*. Until now, the genus contained a single species, *Persicobacter diffluens*. However, we isolated a new *Persicobacter* strain Asr22-19 from the gut of the clam (*Ruditapes philippinarum*) (Muramatsu et al., in press). This strain could grow both aerobically and anaerobically. Strain Asr22-19 with two other strains, NBRC 101035

TABLE 115. Characteristics differentiating the species of the genus *Persicobacter*[a,b]

Characteristic	*P. diffluens*	"*P. psychrovividus*"
Growth at 5–10°C	–	+
Optimum temperature (°C)	30–35	25
Optimum NaCl concentration (%)	3	5
Hydrolysis of:		
Carboxymethyl cellulose, yeast cells	+	–
Chitin	w	+
Assimilation of:		
N-Acetyl-D-glucosamine, L-glutamic acid	–	+
L-Aspartic acid, L-ornithine	–	+/w
L-Fucose, glucose 1-phosphate	w	–
DNA G+C content (mol%)	42.6–43.8	42.0–42.7

[a]Symbols: +, >85% positive; –, 0–15% positive; w, weak reaction.
[b]Data from Takahashi et al. (unpublished).

and NBRC 101041, constituted a single independent species in the genus *Persicobacter* by DNA–DNA hybridization. In addition, those three strains could be phenotypically differentiated from *Persicobacter diffluens* (Table 115). The genus *Persicobacter* was originally described as oxidase-positive, catalase-negative, and aerobic. However, we found that both species produce catalase but not oxidase. Thus, we conclude that the three strains [Asr22-19 (NBRC 101262), NBRC 101035, and NBRC 101041] should be assigned to a new species of the genus *Persicobacter* for which the name *Persicobacter psychrovividus* sp. nov. is proposed.

Differentiation of the species of the genus *Persicobacter*

Table 115 lists characteristics that distinguish the *Persicobacter* species from one another.

List of species of the genus *Persicobacter*

1. **Persicobacter diffluens** (Reichenbach 1989b) Nakagawa, Hamana, Sakane and Yamasato 1997, 222[VP] [*Cytophaga diffluens* (*ex* Stanier 1940) Reichenbach 1989c, 495[VP]; *Cytophaga diffluens* Stanier 1940, 623 emend. mut. char. Lewin 1969, 197]

dif'flu.ens. L. part. adj. *diffluens* flowing away.

The characteristics are as described for the genus and as listed in Table 115, with the following additional features. Cells are 0.4–0.5 μm wide and 0.9–30 μm long or longer. The cell mass is pink to orange. Growth occurs at 15–40°C; optimum, 30–35°C. The pH range for growth is 6–11; optimum, 7. Growth occurs in the presence of 1–5% NaCl; optimum, 3%. Some strains can grow at 45°C, pH 3, or 7% NaCl. Oxidase-negative. Catalase-positive. Agar, alginate, carboxymethyl cellulose, DNA, esculin, gelatin, and yeast cells are degraded, but not cellulose and tyrosine. Starch and chitin are weakly degraded. Utilization (with the Biolog GN2 microplate system) is positive for L-alanine, L-asparagine, cellobiose, α-cyclodextrin, dextrin, D-galactose, α-D-glucose, glycogen, glycyl-L-aspartic acid, glycyl-L-glutamic acid, DL-lactic acid, α-D-lactose, maltose, D-mannose, L-threonine, and D-trehalose, positive or weakly positive for alaninamide, D-fructose, L-fucose, gentiobiose, glucose 1-phosphate, lactulose, and L-rhamnose, but negative for *N*-acetyl-D-glucosamine, L-aspartic acid, and L-glutamic acid. Acid production (API 50CH system) is positive for *N*-acetyl glucosamine, amygdalin, arbutin, cellobiose, esculin, galactose, gentiobiose, glucose, glycogen, lactose, maltose, mannose, salicin, starch, and trehalose. The major cellular fatty acids are $C_{15:0}$ iso, $C_{17:0}$ iso 3-OH, $C_{15:0}$ anteiso, and $C_{16:0}$.

DNA G+C content (mol%): 42.6–43.8 (HPLC).

Type strain: strain B-1 Lewin, ATCC 49458, DSM 3658, JCM 8513, LMG 13036, NBRC 15940, NCIMB 1402.

Sequence accession no. (16S rRNA gene): AB260929.

Other organisms

1. **"Persicobacter psychrovividus"** Muramatsu, Takahashi, Kaneyasu, Iino, Suzuki and Nakagawa (in press)

psy.chro'viv.idus. Gr. adj. *psychros* cold; L. adj. *vividus* full of life, vigorous, active; N.L. masc. adj. *psychrovividus* active at low temperatures.

Cells are 0.4–0.6 μm wide and 0.9–6 μm long or longer. The cell mass is orange. Growth occurs at 5–45°C; optimum, 25°C. Some strains can grow at 45°C. The pH range for growth is 3–11; optimum 7. Growth occurs in the presence of 1–7% NaCl; optimum 5%. Oxidase-negative. Catalase-positive or weakly catalase-positive. Agar, alginate, esculin, gelatin, chitin, and Tween 80 are degraded, but not cellulose, carboxymethyl cellulose, yeast cells, or tyrosine. Starch is degraded weakly. Utilization (with the Biolog GN2 microplate system) is positive for *N*-acetyL-D-glucosamine, L-asparagine, α-cyclodextrin, dextrin, D-galactose, gentiobiose, α-D-glucose, L-glutamic acid, glycogen, glycyl-L-aspartic acid, glycyl-L-glutamic acid, DL-lactic acid, α-D-lactose, maltose, D-mannose, L-threonine, and D-trehalose, and positive or weakly positive for L-alanine, L-alanylglycine, L-aspartic acid, cellobiose, α-ketobutyric acid, and lactulose, but negative for L-fucose and glucose 1-phosphate. Acid production (API 50CH system) is positive for *N*-acetyl glucosamine, amygdalin, arbutin, cellobiose, esculin, galactose, gentiobiose, glucose, glycogen, lactose, maltose, mannose, rhamnose, salicin, starch, trehalose, and D-xylose. The major cellular fatty acids are $C_{15:0}$ iso, $C_{17:0}$ iso 3-OH, $C_{15:0}$ iso 3-OH, $C_{16:0}$ 3-OH, and $C_{16:0}$.

DNA G+C content (mol%): 42.0–42.7 (HPLC).

Type strain: Asr 22-19, CIP 109100, NBRC 101262.

Sequence accession no. (16S rRNA gene): AB260934.

Genus V. **Reichenbachiella** Nedashkovskaya, Kim, Suzuki, Shevchenko, Lee, Lee, Park, Frolova, Oh, Bae, Park and Mikhailov 2005c, 2587^VP (*Reichenbachia* Nedashkovskaya, Suzuki, Vysotskii and Mikhailov 2003, 82)

Olga I. Nedashkovskaya and Makoto Suzuki

Rei.chen.bach.i.el'la. N.L. fem. dim. n. *Reichenbachiella* named after Hans Reichenbach, a German microbiologist who has made a great contribution to the taxonomy of bacteria belonging to the phylum *Bacteroidetes*.

Thin rods 0.5–0.7 × 5–15 μm. **Motile by gliding.** Produce non-diffusible, orange, flexirubin pigments. **Chemo-organotrophs. Strictly aerobic.** Oxidase-, catalase-, and alkaline phosphatase-positive. Agar, alginate, gelatin, casein, starch, urea, DNA, and Tween 20 are hydrolyzed, but casein, chitin, cellulose (CM-cellulose and filter paper), and Tweens 40, 60, and 80 are not hydrolyzed. **The major respiratory quinone is MK-7. Marine**, isolated from coastal habitats in temperate latitudes. Growth does not occur without seawater or sodium ions.

DNA G+C content (mol%): 44–45.

Type species: **Reichenbachiella agariperforans** (Nedashkovskaya, Suzuki, Vysotskii and Mikhailov 2003) Nedashkovskaya, Kim, Suzuki, Shevchenko, Lee, Lee, Park, Frolova, Oh, Bae, Park and Mikhailov 2005c, 2587^VP (*Reichenbachia agariperforans* Nedashkovskaya, Suzuki, Vysotskii and Mikhailov 2003, 83).

Further descriptive information

Affiliation of the genus *Reichenbachiella* with known families of the phylum *Bacteroidetes* is not yet clear. It has no near neighbors among members of the phylum, and its closest relatives are the type strains of [*Flexibacter*] *tractuosus* and *Persicobacter diffluens*, with similarities of 16S rRNA genes of 88.1 and 86.5%, respectively. DNA–DNA hybridization values between *Reichenbachiella* and other representatives of the phylum *Bacteroidetes* are less than 86.2%.

The main cellular fatty acids are straight-chain unsaturated and branched-chain saturated fatty $C_{15:0}$ iso, $C_{16:1}$ ω5*c* and summed feature 3 comprising $C_{15:0}$ iso 2-OH and/or $C_{16:1}$ ω7*c*.

On Marine agar 2216 (Difco), *Reichenbachiella agariperforans* forms regular, round, smooth, bright orange colonies that are sunken into the agar. They have entire edges and a diameter of 3–5 mm after 48 h at 28°C. *Reichenbachiella agariperforans* grows on media containing 0.5% of a peptone and 0.1–0.2% yeast extract (Difco), prepared with natural or artificial seawater or supplemented with 2–3% NaCl. Good growth is observed on Marine agar 2216. Growth occurs at 4–35°C (optimum, 25–28°C). Growth occurs with 1–6% NaCl. The pH range for growth is 5.5–10.0 (optimum, 7.5–8.5).

The single available strain of *Reichenbachiella agariperforans* is susceptible to carbenicillin, oleandomycin, lincomycin, and tetracycline and resistant to ampicillin, benzylpenicillin, streptomycin, gentamicin, neomycin, and polymyxin B.

Enrichment and isolation procedures

The organisms were isolated from seawater by direct plating on Marine agar. Natural or artificial seawater is suitable for their cultivation. Strain KMM 3525^T has been grown on media containing 0.5% of a peptone and 0.1–0.2% yeast extract (Difco). The organisms may remain viable on Marine agar or other rich medium based on natural or artificial seawater for 1 week. They have survived storage at −80°C for at least 5 years.

Differentiation of the genus *Reichenbachiella* from other genera

Reichenbachiella differs from its closest relative [*Flexibacter*] *tractuosus* ATCC 23168^T by its production of flexirubin type pigments and agarase, NaCl requirement for growth, and the lower G+C content of its DNA. The production of catalase and flexirubin type pigments clearly distinguishes *Reichenbachiella* from its nearest neighbor, *Persicobacter diffluens*.

Taxonomic comments

Previously, the genus *Reichenbachia* and the species *Reichenbachia agariperforans* were described to accommodate strain KMM 3525^T (Nedashkovskaya et al., 2003). However, the name *Reichenbachia* is illegitimate according to the International Code of Bacteriological Nomenclature because it is a later homonym of a plant genus, and also a later homonym of an insect genus. Therefore, replacement of the names *Reichenbachia* and *Reichenbachia agariperforans* with *Reichenbachiella* and *Reichenbachiella agariperforans*, respectively, was proposed by Nedashkovskaya et al. (2005c).

Nedashkovskaya et al. (2003) reported that strain KMM 3525^T did not utilize carbohydrates but could produce acid from several of them (Nedashkovskaya et al., 2005c).

List of species of the genus *Reichenbachiella*

1. **Reichenbachiella agariperforans** (Nedashkovskaya et al., 2003) Nedashkovskaya, Kim, Suzuki, Shevchenko, Lee, Lee, Park, Frolova, Oh, Bae, Park and Mikhailov 2005c, 2587^VP (*Reichenbachia agariperforans* Nedashkovskaya, Suzuki, Vysotskii and Mikhailov 2003, 83).

a.ga.ri.per.fo'rans. N.L. n. *agarum*, agar (algal polysaccharide); L. part. adj. *perforans* perforating (making holes); N.L. part. adj. *agariperforans*, making holes in agar, i.e., bacterium making deep hollows in agar.

Cells are 0.5–0.7 × 5–15 μm. Colonies are 3–5 mm in diameter, circular, sunken into agar, shiny, with entire edges and orange-pigmented on solid media containing high nutrient levels. Growth occurs at 4–35°C, with an optimum of 25–28°C and with 1–6% NaCl. Agar, starch, alginate, gelatin, DNA, urea, and Tween 20 are hydrolyzed, but cellulose (CM-cellulose and filter paper), chitin, casein, and Tweens 40, 60, and 80 are not. Acid is formed from L-arabinose, D-cellobiose, L-fucose, D-glucose, arbutin, esculin, and *N*-acetylglucosamine, but not from D-galactose, D-lactose, D-maltose, D-melibiose, L-rhamnose, D-sucrose, DL-xylose, adonitol, dulcitol, inositol, or mannitol. Lactose, mannose, and mannitol are utilized, but citrate, fumarate, and malate are not. Leucine- and valine-arylamidases, trypsin, naphthol-AS-BI-phosphohydrolase, α- and β-galactosidases, α- and β-glucosidases, *N*-acetyl-β-glucosaminidase, and alkaline and acid phosphatases are present, but not esterase (C4), esterase lipase (C8), lipase (C14), crystine arylamidase, α-chymotrypsin, β-glucuronidase, α-mannosidase, and α-fucosidase. Flexirubin pigments are produced. Nitrate is not reduced. H_2S and indole production are negative. The predominant fatty acids are $C_{15:0}$ iso (28.6%), $C_{16:1}$ ω5*c* (21.9%), and summed feature 3 (20.7%; comprising $C_{15:0}$ iso 2-OH and/or $C_{16:1}$ ω7). The single available strain was isolated from seawater sample collected in the Amursky Bay, the Gulf of Peter the Great, the Sea of Japan.

Source: coastal marine environments.

DNA G+C content (mol%): 44.5 (T_m).

Type strain: JCM 11238, KMM 3525, NBRC 16625.

Sequence accession no. (16S rRNA gene): AB058919.

Genus VI. **Roseivirga** Nedashkovskaya, Kim, Lee, Lysenko, Shevchenko, Frolova, Mikhailov, Lee and Bae 2005a, 232[VP], emend. Nedashkovskaya, Kim, Lysenko, Park, Mikhailov, Bae and Park 2005b, 1800[VP]

Olga I. Nedashkovskaya and Seung Bum Kim

Ro.se.i.vir'ga. L. adj. *roseus* pink-colored; L. fem. n. *virga* rod; N.L. fem. n. *Roseivirga* a pink-colored and rod-shaped marine bacterium.

Thin rods usually measuring 0.2–0.5 × 2.0–4.0 µm. **Gliding motility can be observed.** Produce nondiffusible pink-orange pigments. Flexirubin type of pigments can be formed. **Chemoorganotrophs. Strictly aerobic.** Can require seawater or sodium ions for growth. Oxidase-, catalase-, and alkaline phosphatase-positive. Arginine dihydrolase, lysine, and ornithine decarboxylases, and tryptophan deaminase are absent. Agar, casein, starch, urea, cellulose (CM-cellulose and filter paper), and chitin are not attacked, but gelatin, DNA, and Tweens may be decomposed. H_2S and indole are not produced. **The major respiratory quinone is MK-7. Marine**, from coastal habitats.

DNA G+C content (mol%): 40–45.

Type species: **Roseivirga ehrenbergii** Nedashkovskaya, Kim, Lee, Lysenko, Shevchenko, Frolova, Mikhailov, Lee and Bae 2005a, 233[VP].

Further descriptive information

Phylogenetic analysis based on 16S rRNA gene sequencing indicates that the genus *Roseivirga* forms a distinct lineage within the phylum *Bacteroidetes* and the class *Cytophagia* (Figure 6). The genus *Fabibacter* is the closest relative of the *Roseivirga* species, showing sequence similarities of 93.9–95%. The range of 16S rRNA gene sequence similarities between the *Roseivirga* species is 96–99.8%.

Dominant cellular fatty acids are straight-chain unsaturated and branched-chain unsaturated fatty acids $C_{15:1}$ iso, $C_{15:0}$ anteiso, $C_{15:0}$ iso, $C_{15:0}$ iso 3-OH, and $C_{17:0}$ iso 3-OH.

On Marine agar 2216 (Difco), colonies are circular, glistening, convex, smooth, with entire edges, pink or pink-orange-pigmented, and 1–3 mm in diameter after 72 h at 25–30°C.

All isolated strains have been grown on media containing 0.5% of a peptone and 0.1–0.2% yeast extract (Difco), prepared with natural or artificial seawater or supplemented by 2–3% NaCl. Growth occurs at 4–40°C, with 1–8% NaCl and at pH 5.5–10.0. Optimal growth is observed at 21–30°C, with 2–3% NaCl and at pH 7.0–8.0. According to API ZYM testing, all studied strains of the genus *Roseivirga* produce esterase (C4), esterase lipase (C8), leucine arylamidase, valine arylamidase, α-chymotrypsin, acid phosphatase, and naphthol-AS-BI-phosphohydrolase, but not β-glucuronidase, and α-fucosidase.

Strains of the genus *Roseivirga* are susceptible to carbenicillin, lincomycin, and oleandomycin, and resistant to benzylpenicillin, gentamicin, kanamycin, neomycin, polymyxin B, streptomycin, and tetracycline. Sensitivity to ampicillin is variable.

The strains of the genus *Roseivirga* inhabit coastal marine environments. They were isolated from seaweeds, seawater, sponges, and echinoderms in the temperate latitudes.

Enrichment and isolation procedures

The roseivirgas were isolated by direct or standard dilution plating technique on marine agar (Difco). Natural or artificial seawater is suitable for cultivation of the representatives of some species. They can grow on Casamino acids, peptone, and tryptone as the sole carbon and nitrogen sources (Yoon et al., 2005). All isolated strains have been grown on media containing 0.5% peptone and 0.1–0.2% yeast extract (Difco). Strains remain viable on Marine agar (Difco) or other rich medium based on natural or artificial seawater for one or several weeks. They survive storage at −80°C for at least 5 years.

Differentiation of the genus *Roseivirga* from other genera

The genus *Roseivirga* and its closest relative, *Fabibacter halotolerans*, have many common traits (Lau et al., 2006; Nedashkovskaya et al., 2005a). However, *Roseivirga* strains can be differentiated from *Fabibacter halotolerans* by their inability to hydrolyze starch and by their distinctive fatty acid composition (Table 116).

Taxonomic comments

Since a description of the genus *Roseivirga* was published (Nedashkovskaya et al., 2005a), Yoon et al. (2005) have described a novel genus, *Marinicola*. We have found high levels of 16S rRNA gene similarity between *Marinicola seohaensis* SW-152[T] and *Roseivirga ehrenbergii* KMM 6017[T] and *Roseivirga echinicomitans* (99.8 and 99.1%, respectively) (Nedashkovskaya et al., 2005b). Consequently, we proposed placement of the genus *Marinicola* in the genus *Roseivirga*. This conclusion was supported by the results of genomic, chemotaxonomic, and phenotypic analyses, which revealed many common features between members of the genera *Marinicola* and *Roseivirga*. These data have been incorporated in the emended description of the genus *Roseivirga* (Nedashkovskaya et al., 2005b), and *Marinicola seohaensis* has been formally reclassified to the genus *Roseivirga* as *Roseivirga seohaensis* (Lau et al., 2006). Notably, DNA–DNA hybridization experiments were not conducted for genomic comparison

TABLE 116. Fatty acid content of *Roseivirga* species and *Fabibacter halotolerans*[a,b]

Fatty acid	*R. ehrenbergii* KMM 6017[T]	*R. echinicomitans* KMM 6058[T]	*R. seohaensis* LMG 1343[T]	*R. spongicola* UST 030701-084[T]	*F. halotolerans* UST 030701-097[T]
$C_{13:0}$ iso	3.2	2.9	5.2	0.7	1.6
$C_{14:0}$ iso	0.4	1.9		–	4.7
$C_{14:0}$ iso 3-OH	0.1	0.8			1.1
$C_{15:0}$ anteiso	4.5	13.1	2.4	12.5	2.5
$C_{15:0}$ iso	24.4	20.2	33.5	18.6	18.3
$C_{15:0}$	0.9	0.8	1.1		
$C_{15:0}$ 2-OH	0.2	0.3		1.9	1.9
$C_{15:0}$ 3-OH	0.2	0.4			1.3
$C_{15:1}$ anteiso	1.8	2.4		–	0.8
$C_{15:1}$ iso	34.2	20.2	20.5	12.5	14.2
$C_{15:0}$ iso 3-OH	3.0	4.1	5.6	4.9	12.5
$C_{16:0}$	0.6	0.5			
$C_{16:0}$ iso	1.1	1.8	1.2	2.0	1.2
$C_{16:1}$ iso	0.9	2.0		–	1.2
$C_{16:0}$ iso 3-OH	4.1	4.2	7.2	1.2	12.7
$C_{16:0}$ 3-OH	1.6	1.4	1.8	–	1.2
$C_{17:0}$ 2-OH	0.9	2.0		10.1	1.3
$C_{17:1}$ iso ω9*c*	–	1.1		10.8	–
$C_{17:0}$ iso 3-OH	7.7	12.1	11.2	18.3	9.3
$C_{17:0}$ iso		1.0		–	0.5
Summed feature 3[c]	1.7	1.0	4.8	5.5	13.7

[a]Values represent the percentage of the total fatty acids. Fatty acids amounting to less than 1% in all taxa are not given.

[b]Data are taken from Lau et al. (2006), Nedashkovskaya et al. (2005a, b), and Yoon et al. (2005).

[c]Summed features consist of one or more fatty acids that could not be separated by the Microbial Identification System. Summed feature 3: $C_{15:0}$ iso 2-OH and/or 16:1ω7.

of the type strains of *Roseivirga seohaensis* SW-152[T] and its nearest phylogenetic neighbors, *Roseivirga ehrenbergii* KMM 6017[T] and *Roseivirga echinicomitans* KMM 6058[T]. Therefore, an additional study on the determination of a level of DNA–DNA reassociation between the above-mentioned *Roseivirga* strains is needed to clarify the species status of strain SW-152[T].

Differentiation of the species of the genus *Roseivirga*

Although species of the genus *Roseivirga* are very close phylogenetically to each other and have similar fatty acid compositions (Table 116) and phenotypic characteristics, they can be differentiated by a set of phenotypic traits (Table 117).

List of species of the genus *Roseivirga*

1. **Roseivirga echinicomitans** Nedashkovskaya, Kim, Lysenko, Park, Mikhailov, Bae and Park 2005b, 1799[VP]

e.chi.ni.co′mi.tans. L. n. *echinus -i* sea urchin; L. pres. part. *comitans* (from L. v. *comito*) accompanying; N.L. part. adj. *echinicomitans* accompanying a sea urchin.

Cells are 0.3–0.5 µm wide and 2.1–3.2 µm long. On Marine agar, colonies are 2–3 mm in diameter, circular, shiny with entire edges, and pink-pigmented. The optimal temperature for growth is 21–23°C. Acid is not formed from L-arabinose, D-cellobiose, L-fucose, D-galactose, D-glucose, D-lactose, D-maltose, D-melibiose, L-raffinose, L-rhamnose, L-sorbose, D-sucrose, DL-xylose, *N*-acetylglucosamine, citrate, adonitol, dulcitol, glycerol, inositol, or mannitol. L-Arabinose, D-lactose, D-mannose, mannitol, inositol, malonate, gluconate, caprate, malate, and phenylacetate are not utilized. Produces (API ZYM kit) cystine arylamidase and trypsin. Utilizes (Microlog GN2 [Biolog] system) i-erythritol, D-galactose, D-sorbitol, L-leucine, L-ornithine, L-phenylalanine, L-proline, L-pyroglutamic acid, and 2,3-butanediol but not α-cyclodextrin, dextrin, glycogen, Tween 80, *N*-acetyl-D-galactosamine, *N*-acetyl-D-glucosamine, adonitol, D-arabitol, cellobiose, D-fructose, L-fucose, gentiobiose, α-D-glucose, *m*-inositol, α-lactose, α-D-lactose lactulose, D-mannitol, D-melibiose, methyl β-D-glucoside, psicose, D-raffinose, L-rhamnose, sucrose, D-trehalose, turanose, xylitol, methylpyruvate, monomethyl succinate, acetic acid, *cis*-aconitic acid, citric acid, formic acid, D-galactonic acid, D-gluconic acid, D-glucosaminic acid, D-glucuronic acid, α-hydroxybutyric acid, β-hydroxybutyric acid, γ-hydroxybutyric acid, *p*-hydroxyphenylacetic acid, itaconic acid, α-ketobutyric acid, α-ketoglutaric acid, DL-lactic acid, malonic acid, propionic acid, quinic

TABLE 117. Phenotypic characteristics differentiating *Roseivirga* species[a,b]

Characteristic	*R. echinicomitans*	*R. erhenbergii*	*R. saehaensis*	*R. spongicola*
Gliding motility	–	–	+	+
Flexirubin type pigments production	–	–	+	–
Nitrate reduction	+	–	–	–
Acetoin production	–	–	–	+
Temperature range for growth (°C)	4–31	4–39	4–40	12–44
Salinity range for growth (%)	1–8	1–8	1–8	0–16
Hydrolysis of:				
Esculin, gelatin	+	+	–	+
DNA	–	+	–	+
Tween 20	–	+	+	+
Tween 40	+	–	+	+
Tween 80	–	–	+	+
Acid from amygdalin	+	–	nd	–
Utilization of:				
Citrate	+	–	nd	+
D-Galacturonic acid, glycerol, DL-α-glycerol phosphate, inosine, thymidine	+	–	nd	–
L-Alanine	–	+	nd	–
L-Glutamic acid	–	+	–	–
α-Ketovaleric acid	–	–	nd	+
Assimilation of:				
Glucose, *N*-acetylglucosamine, maltose, adipate	+	–	–	–
Enzyme activity:				
α-Galactosidase	+	–	–	–
β-Galactosidase	+	+	–	–
α-Glucosidase	+	+	–	+
β-Glucosidase	+	–	–	+
N-Acetyl-β-glucosaminidase	+	–	+	+
Lipase (C14)	–	+	+	+
α-Mannosidase	–	+	–	–
Susceptibility to:				
Ampicillin	+	+	–	–
Benzylpenicillin, streptomycin	+	–	–	–
Tetracycline	+	–	+	–
DNA G+C content (mol%)	41.3	40.2	40.3	43–45

[a]Symbols: see standard definitions; nd, not detected.

[b]Data are taken from Lau et al. (2006), Nedashkovskaya et al. (2005a, b), and Yoon et al. (2005).

acid, D-saccharic acid, sebacic acid, succinic acid, bromosuccinic acid, succinamic acid, glucuronamide, alaninamide, D-alanine, L-alanyl glycine, L-asparagine, L-aspartic acid, glycyl L-aspartic acid, glycyl L-glutamic acid, L-histidine, hydroxy-L-proline, D-serine, L-serine, L-threonine, DL-carnitine, γ-aminobutyric acid, urocanic acid, uridine, phenylethylamine, putrescine, 2-aminoethanol, glucose 1-phosphate, and glucose 6-phosphate. Only one strain has been isolated.

Source: the sea urchin *Strongylocentrotus intermedius* in the Troitsa Bay, Sea of Japan.

DNA G+C content (mol%): 41.3 (T_m).

Type strain: KCTC 12370, KMM 6058, LMG 22587.

Sequence accession no. (16S rRNA gene): AY753206.

2. **Roseivirga ehrenbergii** Nedashkovskaya, Kim, Lee, Lysenko, Shevchenko, Frolova, Mikhailov, Lee and Bae 2005a, 233[VP]

eh.ren.ber'gi.i. N.L. gen. masc. n. *ehrenbergii* of Ehrenberg, named after the German biologist Christian Gottfried Ehrenberg (1795–1876) for his contribution to the development of microbiology.

Cells are 0.3–0.5 μm wide and 2.1–3.2 μm long. On Marine agar, colonies are 2–4 mm in diameter, circular, shiny with entire edges and pink-pigmented. Optimal growth occurs at 23–25°C. No acid is formed from L-arabinose, D-cellobiose, L-fucose, D-galactose, D-glucose, D-lactose, D-maltose, D-melibiose, L-raffinose, D-sucrose, L-rhamnose, DL-xylose, adonitol, dulcitol, glycerol, inositol, mannitol, or sorbitol. The following substrates are not utilized: D-lactose, D-mannose, D-sucrose, mannitol, inositol, sorbitol, malonate, gluconate, caprate, malate, or phenylacetate. Utilizes (Microlog GN2 [Biolog] system) glycogen, L-arabinose, methyl pyruvate, α-ketoglutaric acid, glucuronamide, alaninamide, L-alanylglycine, L-asparagine, L-aspartic acid, glycyl-L-aspartic acid, L-histidine, L-ornithine, L-proline, L-serine, L-threonine, and putrescine. API ZYM galleries indicate the presence of cystine arylamidase and trypsin.

Source: the green alga *Ulva fenestrata* collected in the Pallada Bay of the Gulf Peter the Great of the Sea of Japan.

DNA G+C content (mol%): 40.2 (T_m).

Type strain: JCM 13514, KCTC 12282, KMM 6017, LMG 22567.

Sequence accession no. (16S rRNA gene): AY608410.

3. **Roseivirga seohaensis** Lau, Tsoi, Li, Plakhotnikova, Dobretsov, Wu, Wong, Pawlik and Qian 2006, 1064[VP] (*Marinicola seohaensis* Yoon, Kang, Lee and Oh 2005, 862[VP])

seo.ha.en′sis. N.L. fem. adj. *seohaensis* of Seohae, the Korean name for the Yellow Sea in Korea, from where the organism was isolated.

Cells are 0.2–0.3 μm wide and 2.0–4.0 μm long and move by gliding. On marine agar, colonies are 1–2 mm in diameter, circular, shiny with entire edges, and pink-orange in color. Optimal growth is observed at 30°C and with 2–3% NaCl. Tween 60 is hydrolyzed but not hypoxanthine, xanthine, or L-tyrosine. No acid is formed from L-arabinose, D-cellobiose, D-fructose, L-fucose, D-galactose, D-glucose, D-lactose, D-maltose, D-mannose, D-melibiose, D-melezitose, D-raffinose, L-rhamnose, D-ribose, D-sucrose, D-trehalose, D-xylose, *N*-acetylglucosamine, adonitol, D-sorbitol, *myo*-inositol, or D-mannitol. The following substrates are not utilized: L-arabinose, D-lactose, D-mannose, D-sucrose, D-ribose, inositol, mannitol, sorbitol, DL-aspartate, L-leucine, or L-proline. Cystine arylamidase and trypsin are not produced. Susceptible to chloramphenicol, doxycycline, and erythromycin. Polar lipids are phosphatidylethanolamine, diphosphatidylglycerol, an unidentified glycolipid, an unidentified phospholipid, and a ninhydrin-positive lipid.

Source: seawater of the Yellow Sea, Korea.

DNA G+C content (mol%): 40.3 (HPLC).

Type strain: SW-152, KCTC 12312, JCM 12600.

Sequence accession no. (16S rRNA gene): AY739663.

4. **Roseivirga spongicola** Lau, Tsoi, Li, Plakhotnikova, Dobretsov, Wu, Wong, Pawlik and Qian 2006, 1063[VP]

spon.gi′co.la. L. n. *spongos -i* sponge; L. masc./fem. suff. *-cola* (from L. n. *incola*) inhabitant; N.L. n. (nominative in apposition) *spongicola* inhabitant of sponges.

Cells are 0.5 μm in width and 2.0 μm in length and move by gliding. On Marine agar, colonies are pink, 2–4 mm in diameter, and circular with entire edges. Growth occurs at pH 5.0–10.0. Optimal growth is observed at 20–30°C. Tween 60 is hydrolyzed but not hypoxanthine, xanthine, or L-tyrosine. No acid is formed from L-arabinose, D-cellobiose, D-fructose, L-fucose, D-galactose, D-glucose, D-lactose, D-maltose, D-mannose, D-melibiose, D-melezitose, D-raffinose, L-rhamnose, D-ribose, D-sucrose, D-trehalose, D-xylose, *N*-acetylglucosamine, adonitol, D-sorbitol, *myo*-inositol, or D-mannitol. The following substrates are not utilized: L-arabinose, D-lactose, D-mannose, D-sucrose, D-ribose, inositol, mannitol, sorbitol, DL-aspartate, L-leucine, or L-proline. Cystine arylamidase and trypsin are not produced. Susceptible to chloramphenicol, doxycycline, and erythromycin.

Source: the marine sponge *Tedania ignis* in the Bahamas.

DNA G+C content (mol%): 43.7 (HPLC).

Type strain: UST030701-084, JCM 13337, NRRL B-41219.

Sequence accession no. (16S rRNA gene): DQ080996.

References

Hamana, K. and Y. Nakagawa. 2001. Polyamine distribution profiles in the eighteen genera phylogenetically located within the *Flavobacterium-Flexibacter-Cytophaga* complex. Microbios *106*: 7–17.

Hamana, K., T. Itoh, Y. Benno and H. Hayashi. 2008. Polyamine distribution profiles of new members of the phylum *Bacteroidetes*. J. Gen. Appl. Microbiol. *54*: 229–236.

Hosoya, S. and A. Yokota. 2007a. *Flammeovirga kamogawensis* sp. nov., isolated from coastal seawater in Japan. Int. J. Syst. Evol. Microbiol. *57*: 1327–1330.

Hosoya, S. and A. Yokota. 2007b. Reclassification of *Flexibacter aggregans* (Lewin 1969) Leadbetter 1974 as a later heterotypic synonym of *Flexithrix dorotheae* Lewin 1970. Int. J. Syst. Evol. Microbiol. *57*: 1086–1088.

Lau, S.C., M.M. Tsoi, X. Li, I. Plakhotnikova, S. Dobretsov, M. Wu, P.K. Wong, J.R. Pawlik and P.Y. Qian. 2006. Description of *Fabibacter halotolerans* gen. nov., sp. nov. and *Roseivirga spongicola* sp. nov., and reclassification of [*Marinicola*] *seohaensis* as *Roseivirga seohaensis* comb. nov. Int. J. Syst. Evol. Microbiol. *56*: 1059–1065.

Leadbetter, E.R. 1974. Genus II *Flexibacter*. *In* Bergey's Manual of Determinative Bacteriology, 8th edn (edited by Buchanan and Gibbons). Williams & Wilkins, Baltimore, pp. 105–107.

Lewin, R.A. 1969. A classification of *Flexibacteria*. J. Gen. Microbiol. *58*: 189–206.

Lewin, R.A. 1970. *Flexithrix dorotheae* gen. et sp. nov. (*Flexibacterales*); and suggestions for reclassifying sheathed bacteria. Can. J. Microbiol. *16*: 511–515.

Lewin, R.A. and D.M. Lounsbery. 1969. Isolation, cultivation and characterization of *Flexibacteria*. J. Gen. Microbiol. *58*: 145–170.

Nakagawa, Y. 2004. Taxonomic studies of *Cytophaga*-like bacteria (in Japanese). Microbiol. Cult. Coll. *20*: 41–51.

Nakagawa, Y., K. Hamana, T. Sakane and K. Yamasato. 1997. Reclassification of *Cytophaga aprica* (Lewin 1969) Reichenbach 1989 in *Flammeovirga* gen. nov. as *Flammeovirga aprica* comb. nov. and of *Cytophaga diffluens* (*ex* Stanier 1940; emend. Lewin 1969) Reichenbach 1989 in *Persicobacter* gen. nov. as *Persicobacter diffluens* comb. nov. Int. J. Syst. Bacteriol. *47*: 220–223.

Nakagawa, Y., T. Sakane, M. Suzuki and K. Hatano. 2002. Phylogenetic structure of the genera *Flexibacter*, *Flexithrix*, and *Microscilla* deduced from 16S rRNA sequence analysis. J. Gen. Appl. Microbiol. *48*: 155–165.

Nedashkovskaya, O.I., M. Suzuki, M.V. Vysotskii and V.V. Mikhailov. 2003. *Reichenbachia agariperforans* gen. nov., sp. nov., a novel marine bacterium in the phylum *Cytophaga-Flavobacterium-Bacteroides*. Int. J. Syst. Evol. Microbiol. *53*: 81–85.

Nedashkovskaya, O.I., S.B. Kim, D.H. Lee, A.M. Lysenko, L.S. Shevchenko, G.M. Frolova, V.V. Mikhailov, K.H. Lee and K.S. Bae. 2005a. *Roseivirga ehrenbergii* gen. nov., sp. nov., a novel marine bacterium of the phylum '*Bacteroidetes*', isolated from the green alga *Ulva fenestrata*. Int. J. Syst. Evol. Microbiol. *55*: 231–234.

Nedashkovskaya, O.I., S.B. Kim, A.M. Lysenko, M.S. Park, V.V. Mikhailov, K.S. Bae and H.Y. Park. 2005b. *Roseivirga echinicomitans* sp. nov., a novel marine bacterium isolated from the sea urchin *Strongylocentrotus intermedius*, and emended description of the genus *Roseivirga*. Int. J. Syst. Evol. Microbiol. *55*: 1797–1800.

Nedashkovskaya, O.I., S.B. Kim, M. Suzuki, L.S. Shevchenko, M.S. Lee, K.H. Lee, M.S. Park, G.M. Frolova, H.W. Oh, K.S. Bae, H.Y. Park and V.V. Mikhailov. 2005c. *Pontibacter actiniarum* gen. nov., sp. nov., a novel member of the phylum '*Bacteroidetes*', and proposal of *Reichenbachiella* gen. nov. as a replacement for the illegitimate prokaryotic generic name *Reichenbachia* Nedashkovskaya *et al.* 2003. Int. J. Syst. Evol. Microbiol. *55*: 2583–2588.

Nedashkovskaya, O.I., Kim S.B., Lysenko A.M., Kalinovskaya N.I. and Mikhailov V.V. 2008. Reclassification of *Roseivirga seohaensis* (Yoon *et al.* 2005) Lau *et al.* 2006 as a later synonym of *Roseivirga ehrenbergii* Nedashkovskaya *et al.* 2005 and emendation of the species description. Int. J. Syst. Evol. Microbiol. *58*: 1194–1197.

Nedashkovskaya, O.I., M. Vancanneyt, Kim S.B. and K.S. Bae. 2009. Reclassification of '*Microscilla tractuosa*' (*ex* Lewin 1969) Reichenbach 1989 and '*Microscilla sericea*' (*ex* Lewin 1969) Reichenbach 1989 in the genus *Marivirga* gen. nov. as *Marivirga tractuosa* nom. rev., comb. nov. and *Marivirga sericea* nom. rev., comb. nov. Int. J. Syst. Evol. Microbiol. DOI: ijs.0.016121-0.

Reichenbach, H. 1989a. Order *Cytophagales* Leadbetter 1974, 99[AL]. *In* Bergey's Manual of Systematic Bacteriology, 8th edn, vol. 3 (edited by Staley, Bryant, Pfennig and Holt). Williams & Wilkins, Baltimore, pp. 2011–2073.

Reichenbach, H. 1989b. Genus I. *Cytophaga*. *In* Bergey's Manual of Systematic Bacteriology, vol. 3 (edited by Staley, Bryant, Pfennig and Holt). Williams & Wilkins, Baltimore, pp. 2015–2050.

Reichenbach, H. 1989c. *In* Validation of the publication of new names and new combinations previously effectively published outside the IJSB. List no. 31. Int. J. Syst. Bacteriol. *39*: 495–497.

Reichenbach, H. 1989d. Genus III *Flexithrix*. *In* Bergey's Manual of Systematic Bacteriology, vol. 3 (edited by Staley, Bryant, Pfennig and Holt). Williams & Wilkins, Baltimore, pp. 2058–2060.

Sakane, T., T. Nishii, T. Itoh and K. Mikata. 1996. Protocols for long-term preservation of microorganisms by L-drying (in Japanese). Microbiol. Cult. Coll. *12*: 91–97.

Seo, H.S., K.K. Kwon, S.H. Yang, H.S. Lee, S.S. Bae, J.H. Lee and S.J. Kim. 2009. *Marinoscillum* gen. nov., a member of the family '*Flexibacteraceae*', with *Marinoscillum pacificum* sp. nov. from a marine sponge and *Marinoscillum furvescens* nom. rev., comb. nov. Int. J. Syst. Evol. Microbiol. *59*: 1204–1208.

Srisukchayakul, P., C. Suwanachart, Y. Sangnoi, A. Kanjana-Opas, S. Hosoya, A. Yokota and V. Arunpairojana. 2007. *Rapidithrix thailandica* gen. nov., sp. nov., a marine gliding bacterium isolated from samples collected from the Andaman sea, along the southern coastline of Thailand. Int. J. Syst. Evol. Microbiol. *57*: 2275–2279.

Stanier, R.Y. 1940. Studies on the cytophagas. J. Bacteriol. *40*: 619–635.

Takahashi, M., K. Suzuki and Y. Nakagawa. 2006. Emendation of the genus *Flammeovirga* and *Flammeovirga aprica* with the proposal of *Flammeovirga arenaria* nom. rev., comb. nov. and *Flammeovirga yaeyamensis* sp. nov. Int. J. Syst. Evol. Microbiol. *56*: 2095–2100.

Yoon, J.H., S.J. Kang, C.H. Lee and T.K. Oh. 2005. *Marinicola seohaensis* gen. nov., sp. nov., isolated from sea water of the Yellow Sea, Korea. Int. J. Syst. Evol. Microbiol. *55*: 859–863.

Yoon, J., S. Ishikawa, H. Kasai and A. Yokota. 2007. *Perexilibacter aurantiacus* gen. nov., sp. nov., a novel member of the family 'Flammeovirgaceae' isolated from sediment. Int. J. Syst. Evol. Microbiol. *57*: 964–968.

Order II. **Incertae sedis**

The genera *Rhodothermus* and *Salinibacter* were previously assigned to the "*Crenotrichaceae*" by Garrity et al. (2005), but subsequent analyses transferred *Crenothrix* to the *Proteobacteria*. Phylogenetic analyses of the 16S rRNA suggests that these genera represent a very deep group and are only distantly related to any of the previously described orders within the *Bacteroidetes*. In view of their ambiguous status, they have been assigned to their own order *incertae sedis*.

Reference

Garrity, G.M., J.A. Bell and T. Lilburn. 2005. The Revised Road Map to the Manual. *In* Bergey's Manual of Systematic Bacteriology, 2nd edn, vol. 2, The *Proteobacteria*, Part A, Introductory Essays (edited by Brenner, Krieg, Staley and Garrity). Springer, New York, pp. 159–220.

Family I. **Rhodothermaceae** fam. nov.

Wolfgang Ludwig, Jean Euzéby and William B. Whitman

Rho.do.ther.ma.ce′a.e. N.L. masc. n. *Rhodothermus* type genus of the family; suff. *-aceae* ending to denote a family; N.L. fem. pl. n. *Rhodothermaceae* the *Rhodothermus* family.

Straight or curved rods that stain Gram-negative. Nonmotile or motile with flagella. **Chemoheterotrophic aerobes** that preferentially utilize sugars or amino acids. Catalase-positive. **Most strains form red or orange colonies**, due to a carotenoid pigment. **Moderately** (0.6 to >6% NaCl) or **extremely halophilic** (>15% NaCl). May be thermophilic, **growing at 54–77°C**, or mesophilic Common habitats include hot springs and salterns.

DNA G+C content (mol%): 64–68.

Type genus: **Rhodothermus** Alfredsson, Kristjansson, Hjörleifsdottir and Stetter 1995, 418[VP] (Effective publication: Alfredsson, Kristjansson, Hjörleifsdottir and Stetter 1988, 304.).

Genus I. **Rhodothermus** Alfredsson, Kristjansson, Hjörleifsdottir and Stetter 1995, 418[VP] (Effective publication: Alfredsson, Kristjansson, Hjörleifsdottir and Stetter 1988, 304.)

The Editorial Board

Rho.do.ther'mus. Gr. n. *rhodon* rose; Gr. masc. adj. *thermos* hot; N.L. masc. n. *Rhodothermus* the red thermophile.

Straight rods about 0.5 × 2.0–2.5 µm, with curved ends. **Occur singly, never in chains or filaments. Nonmotile.** Gram-stain-negative. A slime capsule is formed on carbohydrate-rich medium. **Most strains form red colonies**, due to a carotenoid pigment. Colonies are low convex, 3–4 mm in diameter with an entire edge. **Aerobic. Thermophilic, growing at 54–77°C.** Neutrophilic. Heterotrophic. **Growth is strictly salt-dependent, occurring in the range of 0.6 to >6% NaCl.** Catalase-positive. The oxidase reaction varies among strains. No dissimilatory nitrate reduction occurs. **Growth occurs on most common sugars.** The major cellular fatty acids are C_{15} iso, C_{15} anteiso, C_{17} iso and C_{17} anteiso. The major quinone is menaquinone 7. The habitat is submarine freshwater alkaline hot springs, marine hot springs, geothermal sites, and borehole effluents.

DNA G+C content (mol%): 64–66 (T_m).

Type species: **Rhodothermus marinus** Alfredsson, Kristjansson, Hjörleifsdottir and Stetter 1995, 418[VP] (Effective publication: Alfredsson, Kristjansson, Hjörleifsdottir and Stetter 1988, 304; *Rhodothermus obamensis* Sako, Takai, Ishida, Uchida and Katayama 1996, 1103.).

Further descriptive information

Although *Rhodothermus marinus* is considered to be nonmotile, the presence of a polar flagellum has been reported (Nunes et al., 1992a).

Red colonies are formed by most strains of *Rhodothermus marinus*, but two distinct subgroups of colorless isolates that were correlated with their geographic origin were found by Petursdottir et al. (2000).

Rhodothermus marinus accumulates osmolytes in response to increasing salinity of its growth medium. Mannosylglycerate, which occurs in some hyperthermophilic organisms and protects enzymes against inactivation by temperature and freeze-drying (Ramos et al., 1997), accumulates in *Rhodothermus marinus* in response to growth at supraoptimal temperature and salinity. The amide form, mannosylglyceramide, which has been found only in *Rhodothermus marinus* (Santos and da Costa, 2002), accumulates exclusively in response to salt stress (Nunes et al., 1995; Silva et al., 1999). Martins et al. (1999) elucidated the biosynthetic pathways of mannosylglycerate in *Rhodothermus marinus* and characterized the enzyme mannosylglycerate synthase, which catalyzes the final step in the synthesis of mannosylglycerate. *Rhodothermus marinus* is the only organism known to have two distinct pathways for the synthesis of mannosylglycerate: a two-step pathway and a single-step pathway. The level of mannosylglycerate synthase involved in the single-step pathway was selectively enhanced by heat stress, whereas mannosylglyceramide was overproduced in response to osmotic stress. The two alternative pathways for the synthesis of mannosylglycerate are regulated differently at the level of expression to play specific roles in the adaption of *Rhodothermus marinus* to two different types of stress (Borges et al., 2004).

Rhodothermus marinus is noted for its production of thermostable enzymes; a detailed review has been provided by Bjornsdottir et al. (2006). Some examples of the most heat-tolerant enzymes and their optimum temperature for activity include cellulase (100°C; Halldorsdottir et al., 1998); β-glucanase (85°C; Spilliaert et al., 1994); laminarinase (85°C; Krah et al., 1998); mannanase (85°C; Politz et al., 2000), mannosylglycerate synthase (85–90°C; Martins et al., 1999), and xylanase (80°C; Nordberg Karlsson et al. (1997, 1998).

The original isolation of *Rhodothermus marinus* was from submarine alkaline freshwater hot springs in Iceland by Alfredsson et al. (1988). Additional strains have since been isolated from various marine hot springs, geothermal sites, and borehole effluents. Nunes et al. (1992a) isolated ten strains from hot springs on a beach on the island of Sao Miguel, Azores. Moreira et al. (1996) isolated *Rhodothermus marinus* from hot springs in Naples, Italy. Petursdottir et al. (2000) isolated 81 strains from four different geothermal sites in Iceland. These locations were coastal springs at Reykjanes in Isafjardardjup in northwest Iceland, an effluent from the geothermal power plant at the Blue Lagoon in southwest Iceland, the effluent from the salt factory at Reykjanes in southwest Iceland, and coastal springs and effluent from a borehole in Oxarfjordur in northeast Iceland.

Petursdottir et al. (2000) found that even though *Rhodothermus marinus* strains were isolated at different geographic locations, they exhibited a close genetic relatedness, based on high DNA–DNA reassociation values (>68%) and almost identical 16S rRNA gene sequences. However, other measures of genetic diversity, viz., ribotyping and pulsed-field gel electrophoresis (Moreira et al., 1996) and electrophoretic analysis of allelic variation in 13 genes encoding enzymes (Pétursdóttir et al., 2000) have indicated a relatively high degree of genetic variance. Petursdottir et al. (2000) concluded that this variance is most likely the result of genetic changes occurring independently in the different geographic locations studied.

Tindall (Tindall, 1991) reported that the major cellular fatty acids of *Rhodothermus marinus* were iso-even, i.e., C_{14} iso, C_{16} iso, and C_{18} iso, in the type strain and another strain. Contrasting results were reported by Nunes et al. (1992b), who found the predominant acids to be iso-odd, i.e., C_{15} iso and C_{17} iso) and anteiso-odd (C_{15} and C_{17}) in the type strain and in the type strain and four other strains. Chung et al. (1993) found that the culture medium affected the fatty acid composition of five strains of *Rhodothermus marinus*. They cultured the organisms on four different media: basal salts medium 162 (Degryse et al., 1978) supplemented with yeast extract plus tryptone (medium A); yeast extract plus sodium glutamate (medium B); yeast extract alone (medium C) and glutamate alone (medium D). In media A and C, branched-chain C_{15} iso and C_{17} iso fatty acids were the major fatty acids, and C_{16} iso and normal-chain fatty acids were minor components. In medium B, the relative proportion of C_{16} iso increased to high levels. Glutamate had a profound effect on the fatty acid composition: in medium D, with glutamate as sole source of carbon and energy, normal-chain C_{16} constitued the major component, and normal-chain fatty acids reached about 50% of the total fatty acids. Moreira et al. (1996) reported that the fatty acid composition of 21 strains of *Rhodothermus marinus* isolated from three different locations, including a nonpigmented variant, were all highly similar. The medium used was agar-solidified basal salts medium 162 supplemented with NaCl, tryptone, and yeast extract. The major fatty acids were iso- and anteiso-branched fatty acids. Under the conditions used, $C_{17:0}$ iso was the major fatty acid, but $C_{15:0}$ anteiso was also relatively high.

Pereira et al. (2004) summarized the types of respiratory complexes from thermophilic aerobic prokaryotes. The components so far identified in the respiratory chain of *Rhodothermus* include NADH:quinone oxidoreductase type 1 (NDH-1); succinate:quinone oxidoreductasae (type B); quinol:cytochrome *c* oxidoreductase *bc*; a *caa3* oxygen reductase (type A2) and a *cbb3* oxygen reductase (type C); menaquinone 7; and HiPIP (the electron carrier between the *bc* complex and the *caa3* oxygen reductase).

The genome size of *Rhodothermus marinus* is approximately 3.3–3.6 Mbp (Moreira et al., 1996). A gene transfer system for *Rhodothermus marinus* has been established by Bjornsdottir et al. (2005).

A thorough review of the physiology and molecular biology of *Rhodothermus* has been given by Bjornsdottir et al. (2006).

Enrichment and isolation procedures

Alfredsson et al. (1988) isolated *Rhodothermus marinus* from samples collected directly from the openings of alkaline submarine hot springs in Iceland at a depth of 2–3 m. It was not possible to exclude seawater completely when sampling. The samples, which consisted of fine gravel and water, were kept refrigerated until processed further. Water samples were filtered directly, but sand and gravel samples were washed with sterile seawater, which was then filtered through membrane filters having a pore size of 0.45 µm. The filters were put on plates containing nutrient agar medium 162 of Degryse et al. (1978) containing 3% (w/v) NaCl and incubated at 72°C for 4 d. Red-pigmented colonies were selected and purified by streaking onto the same medium. The bacteria were routinely cultured on nutrient agar medium 162 containing 1% NaCl.

Sako et al. (1996) obtained strains previously classified as *Rhodothermus obamensis* from samples of effluent water and sediment from hydrothermal vents and sedimentary materials adjacent to vents. The samples were inoculated into a series of media, including Jx medium.* All of the Jx medium tubes containing sediment were turbid after 1 d of incubation at 80°C. Cultures were then streaked onto Jx medium plates and incubated at 70°C. Well-defined colonies were streaked onto another plate, and this plate was incubated at 70°C. This procedure was repeated at least five times and red colonies were selected.

Differentiation of the genus *Rhodothermus* from other genera

The combination of a high optimum temperature and a salt requirement distinguish *Rhodothermus* from other genera in the *Bacteroidetes*. The other aerobic thermophilic genus in this phylum is *Thermonema*, some strains of which require a low concentration of Na^+. *Rhodothermus* can also be differentiated from *Thermonema* by its lack of filament formation and its ability to use most common sugars and starch.

Taxonomic comments

Sako et al. (1996) isolated a thermophilic bacterium from a shallow marine hydrothermal vent environment in Tachibana Bay, Japan, that belonged to the genus *Rhodothermus* on the basis of rRNA gene sequencing. This organism was named *Rhodothermus obamensis*. However, Silva et al. (2000) reported that the type strain of *Rhodothermus obamensis*, JCM 9785, has a DNA–DNA reassociation value of 78% with the type strain of *Rhodothermus marinus*, DSM 4252. On the basis of the DNA–DNA reassociation value, 16S rRNA gene sequence comparison, and fatty acid profiles, Silva et al. concluded that *Rhodothermus obamensis* and *Rhodothermus marinus* represent the same species and that the name *Rhodothermus obamensis* should be regarded as a later synonym of *Rhodothermus marinus*.

Antón et al. (2002) reported that the *Rhodothermus marinus* was the closest cultivated relative of *Salinibacter ruber*, with a 16S rRNA gene sequence similarity of about 89%.

List of species of the genus *Rhodothermus*

1. **Rhodothermus marinus** Alfredsson, Kristjansson, Hjörleifsdottir and Stetter 1995, 418[VP] (Effective publication: Alfredsson, Kristjansson, Hjörleifsdottir and Stetter 1988, 304; *Rhodothermus obamensis* Sako, Takai, Ishida, Uchida and Katayama 1996, 1103[VP].)

ma.ri'nus, L. masc. adj. *marinus* of the sea, marine.

The characteristics are as given for the genus, with the following additional features. Optimum temperature is 65–80°C. Optimum pH is 7; optimum NaCl concentration is 2–3%. Growth occurs in the presence of 5–6% NaCl. The major cellular fatty acid components are anteiso- and C_{17} and anteiso- and C_{15} acids, and the major quinone is menaquinone 7, with smaller amounts of menaquinones 6 and 5 (Sako et al., 1996). The natural habitat is submarine alkaline hot springs.

Source: the type strain was isolated from a marine hot spring at Reykjanes, NW Iceland.

DNA G+C content (mol%): 65–67 (T_m).

Type strain: ATCC 43812, DSM 4252.

Sequence accession no. (16S rRNA gene): AF17493.

*Jx medium contains (per liter): Jamarin S synthetic seawater powder (Jamarin Laboratory, Osaka, Japan), 35 g; Jamarin S synthetic seawater solution, 5 ml; yeast extract (Difco), 1 g; and Trypticase peptone (BBL), 1 g; pH adjusted to 6.8–7.2 with H_2SO_4.

Genus II. **Salinibacter** Antón, Oren, Benlloch, Rodríguez-Valera, Amann and Rosselló-Mora 2002, 490^VP

Josefa Antón, Rudolf Amann and Ramon Rosselló-Mora

Sa.li.ni.bac′ter. L. fem. pl. n. *salinae* salterns, salt-works; N.L. masc. n. *bacter* masc. equivalent of the Gr. neut. n. *baktron* a rod; N.L. masc. n. *Salinibacter* a rod from salt-works.

Rod-shaped, often slightly curved. Gram-stain-negative. **Motile** by means of flagella. Obligately **aerobic.** Catalase- and oxidase-positive. Nitrate is not reduced. **Extremely halophilic**, requiring at least 15% NaCl (w/v) to grow; optimum, 20–30%. Chemoheterotrophic. Grows best at low substrate concentrations; high levels of organic compounds may be inhibitory. Amino acids are the preferred substrates for growth. *Salinibacter* strains form a deep branch of the phylum *Bacteroidetes* of the domain *Bacteria*. Colonies are **bright red/orange**-pigmented. Among the most salt-tolerant and salt-requiring strains within the bacterial domain.

DNA G+C content (mol%): 66.3–67.7 (HPLC).

Type species: **Salinibacter ruber** Antón, Oren, Benlloch, Rodríguez-Valera, Amann and Rosselló-Mora 2002, 490^VP.

Further descriptive information

The typical cells are slightly curved rods, but larger, round structures at the ends of the cells can also be found in some cultures (Figure 79). Maintenance of cell shape does not depend on the presence of high salt concentrations.

Salt concentrations of at least 15% (w/v) are required and all strains grow at NaCl concentrations up to saturation. *Salinibacter* apparently uses KCl to provide osmotic balance and lacks high concentrations of organic solutes (Oren et al., 2002). In this aspect, its physiology resembles that of the halophilic *Archaea* more than other aerobic bacteria. The potassium content relative to cell protein in *Salinibacter* is in the same order as for haloarchaea (in μmol/mg: 11.4–15.2 in *Salinibacter*, 13.2 in *Haloarcula marismortui*, and 12.0 in *Halobacterium salinarum*). The only other *Bacteria* known to use KCl rather than organic solutes to provide osmotic balance are the anaerobic fermentative members of the order *Halanaerobiales*. The optimal pH range for growth is 6.5–8.0 with an optimum temperature of 35–45°C. Under optimal growth conditions, the generation time is 14–18 h.

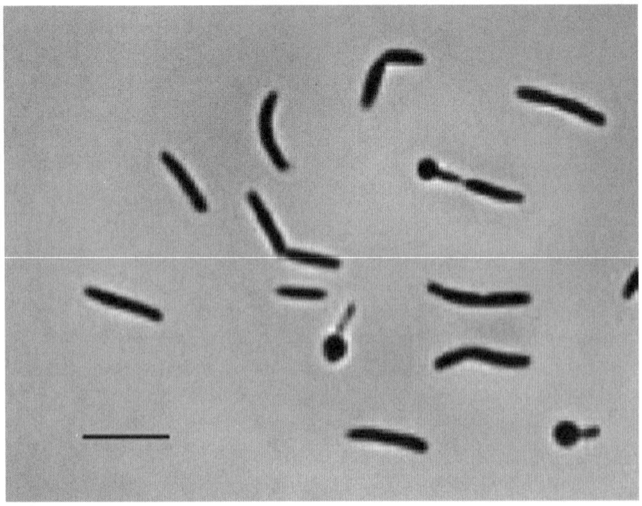

FIGURE 79. Micrograph of a *Salinibacter ruber* culture grown in 23% MGM (modified growth media with 23% total salts), showing the round structures that occasionally appear at the end of the cells. Picture courtesy of Dr. Mike Dyall-Smith, Max Planck Institute for Biochemistry, Department of Membrane Biochemistry, Am Klopferspitz 18a, Martinsried D82152, Germany. Scale bar: 5 μm.

Although the original description of *Salinibacter ruber* stated that simple sugars and organic acids (acetate, succinate) did not support growth as sole carbon and energy sources, Oren and Mana (2003) showed that the addition of glucose, maltose, and starch to a medium with yeast extract had a pronounced stimulatory effect on growth. However, sugars are not the preferred growth substrates and their consumption starts after the exhaustion of other substrates (Oren and Mana, 2003). Glycerol is probably a readily available carbon and energy source in hypersaline water bodies, for it is produced in high quantities as an osmotic solute by unicellular green algae of the genus *Dunaliella* (Oren, 2005). However, the use of radioactive substrates did not indicate an ability of *Salinibacter* to metabolize glycerol during a period of 24–48 h (Rosselló-Mora et al., 2003). On the other hand, Sher et al. (2004) showed that growth in pure culture is stimulated by glycerol, although glycerol alone was not sufficient to support growth. After 190 h of incubation, 25% of the radioactive label was incorporated into the cells, with part of the glycerol transformed into CO_2. As with haloarchaea, *Salinibacter* proteins showed a high content of acidic amino acids, a low amount of basic amino acids, a low content of hydrophobic amino acids, and a high content of serine (Oren and Mana, 2002). In addition, biochemical studies of enzymes such as NAD-dependent isocitrate dehydrogenase, NADP-dependent isocitrate dehydrogenase, NAD-dependent malate dehydrogenase, NAD-dependent glutamate dehydrogenase, and two distinct glutamate dehydrogenases showed that *Salinibacter* enzymes are adapted to function in the presence of high salt concentrations (Bonete et al., 2003; Oren and Mana, 2002). On the other hand, Madern and Zaccai (2004) found that malate dehydrogenase behaves like a non-halophilic protein, since it is completely stable in the absence of salts, its amino acid composition does not display the strong acidic character specific of halophilic proteins, and its activity is reduced by high salt concentration. As pointed out by Oren (2005), the general trend of salt dependence of *Salinibacter* proteins is clear, although there are significant differences for individual enzymes.

Salinibacter is sensitive to penicillin G, ampicillin, chloramphenicol, streptomycin, novobiocin, rifampin and ciprofloxacin. No growth inhibition was found with kanamycin, bacitracin, tetracycline and colistin. The cells are resistant to anisomycin and aphidicolin, two potent growth inhibitors of halophilic *Archaea* of the order *Halobacteriales*.

Salinibacter harbors pigment concentrations in the same order of magnitude as halophilic *Archaea* (Oren and Rodríguez-Valera, 2001). The principal pigment of *Salinibacter* is salinixanthin, a C40-carotenoid acyl glycoside [(all-*E*, 2′*S*)-2′-hydroxy-1′-[6-

O-(13-methyltetradecanoyl)-β-D-glycopyranosyloxy]-3′,4′-didehydro-1′,2′-dihydro-β,ψ-caroten-4-one], that accounts for more than 96% of the total pigments (Lutnæs et al., 2002). The remarkable structural conformity between salinixanthin and the major carotenoid acyl glucoside from *Rhodothermus marinus* is consistent with the 16S rRNA phylogeny (Lutnæs et al., 2004). Bright-red pigmentation is common in microorganisms inhabiting salt lakes and saltern ponds; for instance, the *Halobacteriaceae* possess bacterioruberins, and *Dunaliella*, synthesizes β-carotene. Thus, colony color alone is not a reliable trait to allow classification of extremely halophilic prokaryotes as members of the domains *Archaea* or *Bacteria*.

Salinibacter has a membrane lipid composition with glycerophospholipids containing ester-linked fatty acyl chains that are typical of *Bacteria*. According to Peña et al. (2005) the major lipids in *Salinibacter ruber* strains are diphosphatidylglycerol and the unknown polar lipid L7. Phosphatidylethanolamine, as well as three unknown glycolipids, several polar lipids, and one phospholipid were usually found in moderate to minor amounts. Contrary to previous observations (Oren et al., 2004), no phosphatidylglycerol was detected nor could phosphatidylcholine (PC) be unambiguously identified in any of the 17 *Salinibacter ruber* strains analyzed. In addition, a spot staining with Dragendoff reagent (specific for quaternary nitrogen found in PC but rarely in other lipids) with similar but not identical chromatographic behavior was found also by Peña et al. (2005). *Salinibacter ruber* M31 contains a novel sulfonolipid with the structure 2-carboxy-2-amino-3-*O*-(13′-methyltetradecanoyl)-4-hydroxy-18-methylnonadec-5-ene-1-sulfonic acid. This lipid accounts for 10% of total cellular lipids and appears to be a structural variant of sulfonolipids found in the cell walls of gliding *Cytophaga* and diatoms (Corcelli et al., 2004). A peak at m/z 660, corresponding to this sulfonolipid, has been proposed as the lipid signature of *Salinibacter*.

The genome of the type strain of the type species, *Salinibacter ruber* M31, has been completely sequenced by Mongodin et al. (2005). It harbors one 3,551,823 bp chromosome (62.29 mol% G+C, slightly lower, but in good agreement with earlier HPLC measurements) and one plasmid of 35,505 bp (57.9 mol% G+C) containing 2934 and 33 ORFs, respectively. The calculated isoelectric point (5.2) for the proteome of *Salinibacter* is nearer to those of haloarchaea than its closest sequenced relatives, *Bacteroides fragilis* and *Chlorobium tepidum*. M31, as do all other 16 strains characterized so far (Peña et al., 2005), has a single rRNA operon. The 16S rRNA gene sequence is almost identical for all the analyzed *Salinibacter ruber* strains, while the similarity for their 16S–23S gene spacer regions is above 97% (Peña et al., 2005).

Embden-Meyerhoff glycolytic pathway genes have been found in the chromosome, contrary to previous growth studies that suggested that *Salinibacter* uses the classic Entner-Doudoroff pathway for catabolism of glucose (Oren and Mana, 2003). Genes related to the transport and metabolism of glycerol and glycine betaine, as well as some genes related to adaptions to microoxic environments have been annotated. One so-called "hypersalinity island" encoding proteins of crucial importance to a hyperhalophilic lifestyle (such as K^+ uptake/efflux systems and cationic amino acid transporters) was also found. The M31 genome harbors genes encoding four retinal proteins: halorhodopsin (Antón et al., 2005), two sensory rhodospins and xanthorhodopsin (Balashov et al., 2005). Genome analysis suggested that the convergence on an aerobic hyperhalophilic lifestyle between haloarchaea and *Salinibacter* has arisen through convergence at the physiological level (different genes producing similar overall phenotype) and the molecular level (independent mutations yielding similar sequences or structures). Mongodin et al. (2005) hypothesize that several genes and gene clusters might have suffered lateral transfer from (or may have been laterally transferred to) haloarchaea, although the total number of apparent lateral gene transfers between *Salinibacter* and haloarchaea appears to be modest. Without a doubt, one of the most striking features of *Salinibacter* is the presence of xanthorhodopsin (XR), a proton pump with a light-harvesting carotenoid antenna XR contains two chromophores, retinal and salinixanthin, in a molar ratio of about 1:1. Light energy absorbed by the carotenoid is transferred to the retinal, extending the wavelength range of the collection of light for uphill transmembrane proton transport. Thus XR shares characteristics with the archaeal (e.g., bacteriorhodopsin and archaeorhodpsin), bacterial (proteorhodopsin), and eukaryal (letospheria rhodopsin) retinal-based light-driven proton pumps, as well as with the chlorophyll-based light-harvesting complexes and reaction centers. As pointed out by Balashov et al. (2005), "the XR complex represents the simplest electrogenic pump with an accessory antenna pigment, and it might be an early evolutionary development in using energy transfer for energy capture." Strictly speaking, heterotrophy is not the only source of energy for *Salinibacter ruber* M31.

In 2000, the high abundance and growth of a group of uncultured *Bacteria* was reported for samples from crystallizer ponds (salinity 30–37%) in a saltern in Alicante, Spain. This group, that was then named as "*Candidatus* Salinibacter" gen. nov. accounted for 5–25% in crystallizers from different locations and was affiliated with the phylum *Bacteroidetes* (Antón et al., 2000). This study followed the first evidence that *Bacteria* could be present in high numbers in crystallizer ponds that was reported in 1999 with the use of fluorescence *in situ* hybridization (FISH) with *Archaea*- and *Bacteria*-specific probes (Antón et al., 1999). This kind of approach allows for direct quantification of prokaryotes in the environment and lacks the pitfalls associated with a PCR based approach. FISH with *Bacteria*-specific probes showed that this group accounted for 11–18% in crystallizer samples from a Spanish solar saltern. The finding of abundant *Bacteria* with high cellular rRNA content in such a hypersaline environment was unexpected in light of previous reports that suggested that almost all the active biomass was of archaeal origin. In a pioneering study of the molecular microbial ecology of the salterns using a PCR-based approach, Martínez-Murcia et al. (1995) reported that the bacterial population present in crystallizers was very different (based on similarities of 16S rDNA-RFLP) from the populations inhabiting lower salinity ponds. Although these authors stated that crystallizers probably represented an extremely specialized niche for *Bacteria*, they also pointed out that only a very small proportion of the crystallizer biomass could correspond to *Bacteria*, and that the bacterial DNA detected by PCR could come from allochthonous microbiota carried over from lower salinity ponds. In 1995, Benlloch et al. analyzed 16S rRNA gene clone libraries from a crystallizer pond and obtained five bacterial clones that were partially sequenced (around 200 bp); all were related (82.6–83.6% similarities) to the α-proteobacterium *Rhodopseudomonas marina*. These sequences (accession nos: X84322, X84323 and X84324) could indeed correspond to *Salinibacter* since they have from 89 to 96% similarity with *Salinibacter ruber* in the analyzed 16S rRNA gene sequence fragment. However, the authors pointed out that "considering the salt concentration in the pond (30.8%) no known *Bacteria* could be physiologically active". Indeed, the idea that only *Archaea* could thrive in hypersaline environments was very strong. We must point out, however, that direct proof of activity of *Salinibacter* species in the highest salinity (37%) ponds has not been obtained so far. In fact, Gasol et al. (2004) found that above 32% salinity, all the prokaryotic activity was carried out by haloarchaea in the very same saltern ponds where *Salinibacter ruber* represented up to 18% of the DAPI (4′,6-diamidino-2-phenylindole) counts (total counts). This observation is based on the assumption that *Salinibacter* species are not inhibited by taurocholate, which is a potent haloarchaeal inhibitor.

The occurrence of *Salinibacter* in the environment has been studied by several methods: FISH (Antón et al., 2000; Rosselló-Mora et al., 2003), pigment analysis (Oren and Rodríguez-Valera, 2001), total DNA melting profiles and reassociation techniques (Øvreås et al., 2003), Denaturing gradient gel electrophoresis (DGGE) (Benlloch et al., 2002), and 16S rRNA gene clone library analysis, among others. These techniques show that there is a considerable degree of microdiversity among the environmental sequences related to *Salinibacter ruber*. Using the above-mentioned culture-independent approaches, *Salinibacter* has been found in crystallizer pond salterns from locations in Santa Pola (mainland Spain), Balearic (Mallorca and Ibiza) and Canary Islands, which accounted for 5–27% of the total prokaryotic community. In Santa Pola salterns, *Salinibacter* was not detected in ponds having up to 22.4% salinity, whereas it was found in increasingly high numbers (3.5–12%) in three ponds of 25, 31.6 and 37% salinity. The *Salinibacter* lipid signature peak at m/z 660 was evident in the ESI-MS profile in the lipid extract from a crystallizer pond in the Margherita di Savoia salterns (Italy) (Corcelli et al., 2004), indicating the presence of *Salinibacter* in this environment. In some instances, e.g., Andean and Eilat (Israel) salterns, *Salinibacter* was not detected by culture-independent analyses, although it could be readily isolated (Elevi-Bardavid et al., 2007; Maturrano et al., 2006). Although crystallizers are the most frequently reported habitat for *Salinibacter*, sequences with a similarity of 92–97% to those of *Salinibacter* were very abundant in 16S rRNA gene libraries constructed with DNA extracted from the different layers of an endoevaporite (crystallized gypsum-halite matrix in near-saturated salt water) from salt-works in Guerrero Negro, Mexico (Spear et al., 2003). Partial 16S rRNA gene sequences with similarities of less than 92% to *Salinibacter* have been retrieved from biofilms colonizing ancient limestone Mayan monuments in Uxmal (Mexico) (Ortega-Morales et al., 2004). *Salinibacter* 16S rRNA gene sequences were also found in clone libraries obtained from a hypersaline endoevaporite microbial mat from Eilat salterns (Sørensen et al., 2005). These sequences were most abundant in the green layer of the mat. Finally, 16S rRNA gene sequences that fall within the radiation of the genus *Salinibacter*, but represent distinct novel lineages have been recovered from evaporite crusts in brine pools at the Badwater site in Death Valley National Park, California (Elevi-Bardavid et al., 2007). Isolates having 93–94% 16S rRNA gene similarity with *Salinibacter ruber* have been obtained from these samples (Hollen et al., 2003). These authors claimed that they have isolated a new species of *Salinibacter*.

Enrichment and isolation procedures

The strains used for the genus description and for intraspecific comparative analyzes, were all isolated by plating dilutions of crystallizer samples on agar plates. They were recognized as members of the genus by 16S rRNA gene analysis or by polar lipids thin layer chromatography (TLC). The fact that *Salinibacter* is insensitive to the haloarchaeal inhibitors anisomycin and bacitracin has been used to design selective enrichment and isolation protocols (Elevi-Bardavid et al., 2007).

Strains of *Salinibacter* can be preserved by lyophilization.

Differentiation of the genus *Salinibacter* from closely related genera

No extreme halophiles are known so far among taxa closely related to *Salinibacter*, thus the differentiation from these taxa can be made on the basis of salt needed for optimum growth. However, *Salinibacter* shares its habitats with haloarchaea (i.e., the family *Halobacteriaceae*). Interestingly, there is a surprising similarity between these two groups: both are aerobic heterotrophs, maintain high intracellular K^+ concentrations, require high salt concentrations for growth with optima in the range of 12–25% total salts, are red pigmented, and have a similar mol% G+C content of their DNA. For these reasons, *Salinibacter* can be easily mistaken as haloarchaea based only on phenotypic characteristics. Therefore, we recommend phylogenetic identification as the easiest way to identify *Salinibacter* strains. Lipid and pigment analyzes (Elevi-Bardavid et al., 2007) can also be used for strain identification.

Taxonomic comments

The closest 16S rRNA sequence similarity (86.4%) relative to *Salinibacter* is *Rhodothermus*, a genus of slightly halophilic (optimum 0.5–2% NaCl), thermophilic (optimum 65–70°C) bacteria isolated from marine hot springs (Alfredsson et al., 1988). The phylogenetic branch comprising the two genera *Salinibacter* and *Rhodothermus* bifurcates close to the divergence node between the two main phyla *Bacteroidetes* and *Chlorobi*. This observation made upon 16S rRNA gene sequence analysis has been corroborated by the use of independent phylogenetic approaches based on concatenating two sets of 22 and 74, respectively, protein sequences (Soria-Carrasco et al., 2007). From the single protein phylogenies it can be observed that in nearly 30% of the cases the affiliation was with *Chlorobi*. This may be an indication of an early divergence from *Bacteroidetes*, and a future classification as a single phylum cannot be discarded (Figure 80).

Note added in proof. A novel halophilic member of the *Bacteroidetes*, *Salisaeta longa* gen. nov., sp.nov., was published after acceptance of the present chapter. It is able to grow at concentrations of 5–20% NaCl, with an optimum at 10% plus 5% $MgCl \cdot 6H_2O$ and has been described as halophilic (not extremely halophilic, as *Salinibacter ruber*). The 16S rRNA genes of *Salinibacter ruber* and *Salisaeta longa* have 88.3% similarity and thus, according to this parameter, this new species is now the closest relative of *Salinibacter ruber*, instead of *Rhodothermus marinus*. They can be differentiated based on their morphology, 16S rRNA gene sequences and salt optimum needed for growth, among other characteristics. However, a 16S rRNA primer originally designed as specific for *Salinibacter ruber* (Antón et al., 2002) is no longer specific.

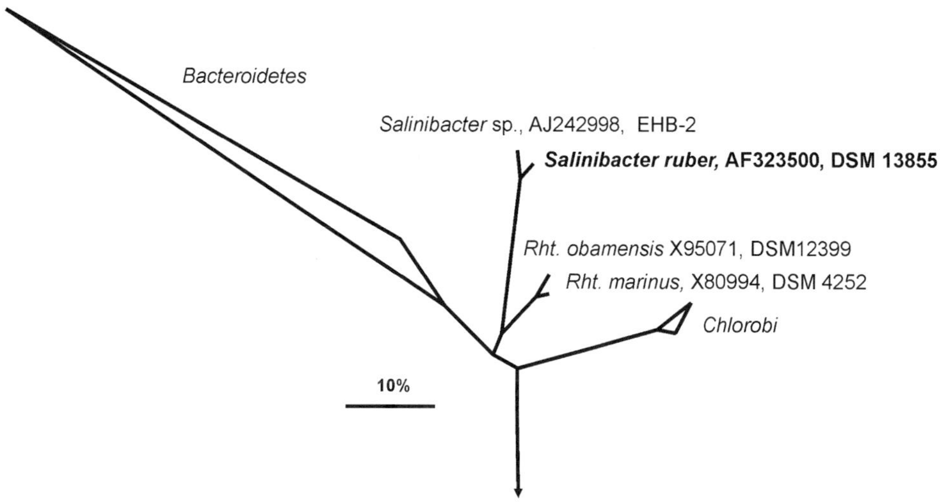

FIGURE 80. Tree reconstruction based on 16S rRNA gene sequences showing the affiliation of *Salinibacter ruber* strain M31 with its closest relative genus *Rhodothermus*, and to the hitherto uncultured *Salinibacter* species EHB-2 (Antón et al., 2000). The tree also shows the branch position relative to the rest of the members of the phyla *Bacteroidetes* and *Chlorobi*. The reconstruction was based on a dataset of more than 50,000 primary aligned 16S rRNA gene structures as implemented in the ARB software package, and corresponding to the released database available at http://www.arb-home.de (Ludwig et al., 2004). The phylogenetic analyzes were performed by using the sequences corresponding to all type strains of both phyla, and by using the neighbor-joining, maximum-parsimony, and maximum-likelihood algorithms. All treeing approaches rendered an identical topology. The bar indicates 10% of estimated sequence divergence.

Acknowledgements

We thank Dr Hans-Jürgen Busse, Institut fur Bakteriologie, Mykologie und Hygiene, Veterinärmedizinische Universität Wien, Vienna, Austria, for his help with the lipid data.

Further reading

Antón, J., A. Peña, M. Valens, F. Santos, F.O. Glöckner, M. Bauer, J. Dopazo, J. Herrero, R. Rosselló-Mora and R. Amann. 2005. *Salinibacter ruber*: genomics and biogeography. *In* Adaptation to life in high salt concentrations in *Archaea*, *Bacteria* and *Eukarya* (edited by Gunde-Cimerman, Plemenitasand Oren). Kluwer Academic Publishers, Dordrecht, The Netherlands, pp. 257–266.

Mongodin, E.F., K.E. Nelson, S. Daugherty, R.T. DeBoy, J. Wister, H. Khouri, J. Weidman, D.A. Walsh, R.T. Papke, G. Sanchez Perez, A.K. Sharma, C.L. Nesbø, D. MacLeod, E. Bapteste, W.F. Doolittle, R.L. Charlebois, B. Legault and F. Rodríguez-Valera. 2005. The genome of *Salinibacter ruber*. Convergence and gene exchange among hyperhalophilic bacteria and archaea. Proc. Natl. Acad. Sci. U. S. A. *102*: 18147–18152.

Oren, A. 2005. The genera *Rhodothermus, Thermonema, Hymenobacter* and *Salinibacter*. *In* The Prokaryotes: An Evolving Electronic Resource for the Microbiological Community, 3rd edn (edited by Dworkin, Falkow, Rosenberg, Schleifer and Stackebrandt). Springer, New York.

Peña, A., M. Valens, F. Santos, S. Buczolits, J. Antón, P. Kämpfer, H.-J. Busse, R. Amann and R. Rosselló-Mora. 2005. Intraspecific comparative analysis of the species *Salinibacter ruber*. Extremophiles *9*: 151–161.

List of species of the genus *Salinibacter*

1. **Salinibacter ruber** Antón, Oren, Benlloch, Rodríguez-Valera, Amann and Rosselló-Mora 2002, 490[VP]

ru'ber. L. masc. adj. *ruber* red.

The description is as given for the genus, with the following additional features.

Source: crystallizer pond salterns.

DNA G+C content (mol%): 66.5 (HPLC).

Type strain: M31, DSM 13855, CECT 5946.

Sequence accession no. (16S rRNA gene): AF323500.

References

Alfredsson, G.A., J.K. Kristjansson, S. Hjörleifsdottir and K.O. Stetter. 1988. *Rhodothermus marinus*, gen. nov., sp. nov., a thermophilic, halophilic bacterium from submarine hot springs in Iceland. J. Gen. Microbiol. *134*: 299–306.

Alfredsson, G.A., J.K. Kristjansson, S. Hjörleifsdottir and K.O. Stetter. 1995. *In* Validation of new names and new combinations previously effectively published outside the IJSB. List no. 60. Int. J. Syst. Bacteriol. *45*: 418–419.

Antón, J., E. Llobet-Brossa, F. Rodríguez-Valera and R. Amann. 1999. Fluorescence *in situ* hybridization analysis of the prokaryotic community inhabiting crystallizer ponds. Environ. Microbiol. *1*: 517–523.

Antón, J., R. Rosselló-Mora, F. Rodríguez-Valera and R. Amann. 2000. Extremely halophilic *Bacteria* in crystallizer ponds from solar salterns. Appl. Environ. Microbiol. *66*: 3052–3057.

Antón, J., A. Oren, S. Benlloch, F. Rodríguez-Valera, R. Amann and R. Rosselló-Mora. 2002. *Salinibacter ruber* gen. nov., sp. nov., a novel, extremely halophilic member of the *Bacteria* from saltern crystallizer ponds. Int. J. Syst. Evol. Microbiol. *52*: 485–491.

Antón, J., A. Peña, M. Valens, F. Santos, F.O. Glöckner, M. Bauer, J. Dopazo, J. Herrero, R. Rosselló-Mora and R. Amann. 2005. *Salinibacter ruber*: genomics and biogeography. *In* Adaptation to life in high salt concentrations in *Archaea*, *Bacteria* and *Eukarya* (edited by Gunde-Cimerman, Plemenitas and Oren). Kluwer Academic Publishers, Dordrecht, The Netherlands, pp. 257–266.

Balashov, S.P., E.S. Imasheva, V.A. Boichenko, J. Antón, J.M. Wang and J.K. Lanyi. 2005. Xanthorhodopsin: a proton pump with a light-harvesting carotenoid antenna. Science *309*: 2061–2064.

Benlloch, S., A.J. Martínez-Murcia and F. Rodríguez-Valera. 1995. Sequencing of bacterial and archaeal 16S rRNA genes directly amplified from a hypersaline environment. Syst. Appl. Microbiol. *18*: 574–581.

Benlloch, S., A. López-López, E.O. Casamayor, L. Øvreas, V. Goddard, F.L. Daae, G. Smerdon, R. Massana, I. Joint, F. Thingstad, C. Pedrós-Alió and F. Rodríguez-Valera. 2002. Prokaryotic diversity throughout the salinity gradiente of a coastal solar saltern. Environ. Microbiol. *4*: 349–360.

Bjornsdottir, S.H., S.H. Thorbjarnardottir and G. Eggertsson. 2005. Establishment of a gene transfer system for *Rhodothermus marinus*. Appl. Microbiol. Biotechnol. *66*: 675–682.

Bjornsdottir, S.H., T. Blondal, G.O. Hreggvidsson, G. Eggertsson, S. Petursdottir, S. Hjörleifsdottir, S.H. Thorbjarnardottir and J.K. Kristjansson. 2006. *Rhodothermus marinus*: physiology and molecular biology. Extremophiles *10*: 1–16.

Bonete, M.J., F. Perez-Pomares, S. Diaz, J. Ferrer and A. Oren. 2003. Occurrence of two different glutamate dehydrogenase activities in the halophilic bacterium *Salinibacter ruber*. FEMS Microbiol. Lett. *226*: 181–186.

Borges, N., J.D. Marugg, N. Empadinhas, M.S. da Costa and H. Santos. 2004. Specialized roles of the two pathways for the synthesis of mannosylglycerate in osmoadaptation and thermoadaptation of *Rhodothermus marinus*. J. Biol. Chem. *279*: 9892–9898.

Chung, A.P., O.C. Nunes, B.J. Tindall and M.S. da Costa. 1993. The effect of the growth medium composition on the fatty acids of *Rhodothermus marinus* and '*Thermus thermosphilus*' HB-8. FEMS Microbiol. Lett. *112*: 13–18.

Corcelli, A., V.M. Lattanzio, G. Mascolo, F. Babudri, A. Oren and M. Kates. 2004. Novel sulfonolipid in the extremely halophilic bacterium *Salinibacter ruber*. Appl. Environ. Microbiol. *70*: 6678–6685.

Degryse, W., N. Glansdorff and A. Piérard. 1978. A comparative analysis of extreme thermophilic bacteria belonging to the genus *Thermus*. Arch. Microbiol. *117*: 189–196.

Elevi-Bardavid, R., D. Ionescu, A. Oren, F.A. Rainey, B.J. Hollen, D.R. Bagaley, A.M. Small and C.M. McKay. 2007. Selective enrichment, isolation and molecular detection of *Salinibacter* and related extremely halophilic *Bacteria* from hypersaline environments. Hydrobiologia *576*: 207.

Gasol, J.M., E.O. Casamayor, I. Joint, K. Garde, K. Gustavson, S. Benlooch, B. Díez, M. Schauer, R. Massana and C. Pedrós-Alió. 2004. Control of heterotrophic prokaryotic abundance and growth rate in hypersaline planktonic environments. Aquat. Microb. Ecol. *34*: 193–206.

Halldorsdottir, S., E.T. Thorolfsdottir, R. Spilliaert, M. Johansson, S.H. Thorbjarnardottir, A. Palsdottir, G.O. Hreggvidsson, J.K. Kristjansson, O. Holst and G. Eggertsson. 1998. Cloning, sequencing and overexpression of a *Rhodothermus marinus* gene encoding a thermostable cellulase of glycosyl hydrolase family 12. Appl. Microbiol. Biotechnol. *49*: 277–284.

Hollen, B.J., D.R. Bagaley, A.M. Small, A. Oren, C.P. McKay and F.A. Rainey. 2003. Investigation of the microbial community of the salt surface layer at Badwater, Death Valley National Park. Proceedings of the American Society for Microbiology Annual Meeting, Washington, DC.

Karlsson, E.N., L. Dahlberg, N. Torto, L. Gorton and O. Holst. 1998. Enzymatic specificity and hydrolysis pattern of the catalytic domain of the xylanase Xynl from *Rhodothermus marinus*. J. Biotechnol. *60*: 23–35.

Krah, M., R. Misselwitz, O. Politz, K.K. Thomsen, H. Welfle and R. Borriss. 1998. The laminarinase from thermophilic eubacterium *Rhodothermus marinus* - conformation, stability, and identification of active site carboxylic residues by site-directed mutagenesis. Eur. J. Biochem. *257*: 101–111.

Ludwig, W., O. Strunk, R. Westram, L. Richter, H. Meier, Yadhukumar, A. Buchner, T. Lai, S. Steppi, G. Jobb, W. Forster, I. Brettske, S. Gerber, A.W. Ginhart, O. Gross, S. Grumann, S. Hermann, R. Jost, A. Konig, T. Liss, R. Lussmann, M. May, B. Nonhoff, B. Reichel, R. Strehlow, A. Stamatakis, N. Stuckmann, A. Vilbig, M. Lenke, T. Ludwig, A. Bode and K.H. Schleifer. 2004. ARB: a software environment for sequence data. Nucleic Acids Res. *32*: 1363–1371.

Lutnæs, B.F., A. Oren and S. Liaaen-Jensen. 2002. New C_{40}-carotenoid acyl glycoside as principal carotenoid in *Salinibacter ruber*, an extremely halophilic eubacterium. J. Nat. Prod. *65*: 1340–1343.

Lutnæs, B.F., A. Strand, S.K. Petursdottir and S. Liaaen-Jensen. 2004. Carotenoids of thermophilic bacteria - *Rhodothermus marinus* from submarine Icelandic hot springs. Biochem. Syst. Ecol. *32*: 455–468.

Madern, D. and G. Zaccai. 2004. Molecular adaptation: the malate dehydrogenase from the extreme halophilic bacterium *Salinibacter ruber* behaves like a non-halophilic protein. Biochimie *86*: 295–303.

Martínez-Murcia, A.J., S.G. Acinas and F. Rodríguez-Valera. 1995. Evaluation of prokaryotic diversity by restrictase digestion of 16S rDNA directly amplified from hypersaline environments. FEMS. Microb. Ecol. *17*: 247–256.

Martins, L.O., N. Empadinhas, J.D. Marugg, C. Miguel, C. Ferreira, M.S. da Costa and H. Santos. 1999. Biosynthesis of mannosylglycerate in the thermophilic bacterium *Rhodothermus marinus*. Biochemical and genetic characterization of a mannosylglycerate synthase. J. Biol. Chem. *274*: 35407–35414.

Maturrano, L., F. Santos, R. Rosselló-Mora and J. Antón. 2006. Microbial diversity in Maras salterns, a hypersaline environment in the Peruvian Andes. Appl. Environ. Microbiol. *72*: 3887–3895.

Mongodin, E.F., K.E. Nelson, S. Daugherty, R.T. DeBoy, J. Wister, H. Khouri, J. Weidman, D.A. Walsh, R.T. Papke, G. Sanchez Perez, A.K. Sharma, C.L. Nesbø, D. MacLeod, E. Bapteste, W.F. Doolittle, R.L. Charlebois, B. Legault and F. Rodríguez-Valera. 2005. The genome of *Salinibacter ruber*. Convergence and gene exchange among hyperhalophilic bacteria and archaea. Proc. Natl. Acad. Sci. U. S. A. *102*: 18147–18152.

Moreira, L., M.F. Nobre, I. Sa-correia and M.S. da Costa. 1996. Genomic typing and fatty acid composition of *Rhodothermus marinus*. Syst. Appl. Microbiol. *19*: 83–90.

Nordberg Karlsson, E., E. Bartonek-Roxa and O. Holst. 1997. Cloning and sequence of a thermostable multidomain xylanase from the bacterium *Rhodothermus marinus*. Biochim. Biophys. Acta. *1353*: 118–124.

Nunes, O.C., M.M. Donato and M.S. da Costa. 1992a. Isolation and characterization of *Rhodothermus* strains from S. Miguel Azores. Syst. Appl. Microbiol. *15*: 92–97.

Nunes, O.C., M.M. Donato, C.M. Manaia and M.S. da Costa. 1992b. The polar lipid and fatty acid composition of *Rhodothermus* strains. Syst. Appl. Microbiol. *15*: 59–62.

Nunes, O.C., C.M. Manaia, M.S. Da Costa and H. Santos. 1995. Compatible Solutes in the thermophilic bacteria *Rhodothermus marinus* and "*Thermus thermophilus*". Appl. Environ. Microbiol. *61*: 2351–2357.

Oren, A. 2005. The genera *Rhodothermus, Thermonema, Hymenobacter* and *Salinibacter*. *In* The Prokaryotes: An Evolving Electronic Resource for the Microbiological Community, 3rd edn (edited by Dworkin, Falkow, Rosenberg, Schleifer and Stackebrandt). Springer, New York.

Oren, A. and F. Rodríguez-Valera. 2001. The contribution of halophilic *Bacteria* to the red coloration of saltern crystallizer ponds. FEMS Microbiol. Ecol. *36*: 123–130.

Oren, A. and L. Mana. 2002. Amino acid composition of bulk protein and salt relationships of selected enzymes of *Salinibacter ruber*, an extremely halophilic bacterium. Extremophiles *6*: 217–223.

Oren, A. and L. Mana. 2003. Sugar metabolism in the extremely halophilic bacterium *Salinibacter ruber*. FEMS Microbiol. Lett. *223*: 83–87.

Oren, A., M. Heldal, S. Norland and E.A. Galinski. 2002. Intracellular ion and organic solute concentrations of the extremely halophilic bacterium *Salinibacter ruber*. Extremophiles *6*: 491–498.

Oren, A., F. Rodríguez-Valera, J. Antón, S. Benlloch, R. Rosselló-Mora, R. Amann, J. Coleman and N.J. Russell. 2004. Red, extremely halophilic, but not archaeal: the physiology and ecology of *Salinibacter ruber*, a bacterium isolated from saltern crystallizer ponds. *In* Halophilic Microorganisms (edited by Ventosa). Springer, New York, pp. 63–76.

Ortega-Morales, B.O., J.A. Narváez-Zapata, A. Schmalenberger, A. Sosa-López and C.C. Tebbe. 2004. Biofilms fouling ancient limestone Mayan monuments in Uxmal, Mexico: a cultivation-independent analysis. Biofilms *1*: 79–90.

Øvreas, L., F.L. Daae, V. Torsvik and F. Rodríguez-Valera. 2003. Characterization of microbial diversity in hypersaline environments by melting profiles and reassociation kinetics in combination with terminal restriction fragment length polymorphism (T-RFLP). Microb. Ecol. *46*: 291–301.

Peña, A., M. Valens, F. Santos, S. Buczolits, J. Antón, P. Kämpfer, H.-J. Busse, R. Amann and R. Rosselló-Mora. 2005. Intraspecific comparative analysis of the species *Salinibacter ruber*. Extremophiles *9*: 151–161.

Pereira, M.M., T.M. Bandeiras, A.S. Fernandes, R.S. Lemos, A.M. Melo and M. Teixeira. 2004. Respiratory chains from aerobic thermophilic prokaryotes. J. Bioenerg. Biomembr. *36*: 93–105.

Pétursdóttir, S.K., G.O. Hreggvidsson, M.S. da Costa and J.K. Kristjansson. 2000. Genetic diversity analysis of *Rhodothermus* reflects geographical origin of the isolates. Extremophiles *4*: 267–274.

Politz, O., M. Krah, K.K. Thomsen and R. Borriss. 2000. A highly thermostable endo-(1,4)-beta-mannanase from the marine bacterium *Rhodothermus marinus*. Appl. Microbiol. Biotechnol. *53*: 715–721.

Ramos, A., N. Raven, R.J. Sharp, S. Bartolucci, M. Rossi, R. Cannio, J. Lebbink, J. Van Der Oost, W.M. De Vos and H. Santos. 1997. Stabilization of enzymes against thermal stress and freeze-drying by mannosylglycerate. Appl. Environ. Microbiol. *63*: 4020–4025.

Rosselló-Mora, R., N. Lee, J. Antón and M. Wagner. 2003. Substrate uptake in extremely halophilic microbial communities revealed by microautoradiography and fluorescence *in situ* hybridization. Extremophiles *7*: 409–413.

Sako, Y., K. Takai, Y. Ishida, A. Uchida and Y. Katayama. 1996. *Rhodothermus obamensis* sp. nov., a modern lineage of extremely thermophilic marine bacteria. Int. J. Syst. Bacteriol. *46*: 1099–1104.

Santos, H. and M.S. da Costa. 2002. Compatible solutes of organisms that live in hot saline environments. Environ. Microbiol. *4*: 501–509.

Sher, J., R. Elevi, L. Mana and A. Oren. 2004. Glycerol metabolism in the extremely halophilic bacterium *Salinibacter ruber*. FEMS Microbiol. Lett. *232*: 211–215.

Silva, Z., N. Borges, L.O. Martins, R. Wait, M.S. da Costa and H. Santos. 1999. Combined effect of the growth temperature and salinity of the medium on the accumulation of compatible solutes by *Rhodothermus marinus* and *Rhodothermus obamensis*. Extremophiles *3*: 163–172.

Silva, Z., C. Horta, M.S. da Costa, A.P. Chung and F.A. Rainey. 2000. Polyphasic evidence for the reclassification of *Rhodothermus obamensis* Sako *et al.* 1996 as a member of the species *Rhodothermus marinus* Alfredsson *et al.* 1988. Int. J. Syst. Evol. Microbiol. *50*: 1457–1461.

Sørensen, K.B., D.E. Canfield, A.P. Teske and A. Oren. 2005. Community composition of a hypersaline endoevaporitic microbial mat. Appl. Environ. Microbiol. *71*: 7352–7365.

Soria-Carrasco, V., M. Valens-Vadell, A. Peña, J. Antón, R. Amann, J. Castresana and R. Rosselló-Mora. 2007. Phylogenetic position of *Salinibacter ruber* based on concatenated protein alignments. Syst. Appl. Microbiol. *30*: 171–179.

Spear, J.R., R.E. Ley, A.B. Berger and N.R. Pace. 2003. Complexity in natural microbial ecosystems: the Guerrero Negro experience. Biol. Bull. *204*: 168–173.

Spilliaert, R., G.O. Hreggvidsson, J.K. Kristjansson, G. Eggertsson and A. Palsdottir. 1994. Cloning and sequencing of a *Rhodothermus marinus* gene, *bglA*, coding for a thermostable β-glucanase and its expression in *Escherichia coli*. Eur. J. Biochem. *224*: 923–930.

Tindall, B.J. 1991. Lipid composition of *Rhodothermus marinus*. FEMS Microbiol. Lett. *80*: 65–68.

Vaisman, N. and A. Oren. 2009. *Salisaeta longa* gen. nov., sp. nov., a red, halophilic member of the *Bacteroidetes*. Int. J. Syst. Evol. Microbiol. *59*: 2571–2574.

Order III. **Incertae sedis**

Thermonema was previously assigned to the "*Flammeovirgaceae*" by Garrity et al. (2005), but subsequent analyses of the 16S rRNA suggests that it represents a very deep group and is only distantly related to any of the previously described orders within the *Bacteroidetes*. In view of its ambiguous status, it has been assigned to its own order *incertae sedis*.

Genus I. **Thermonema** Hudson, Schofield, Morgan and Daniel 1989, 487[VP]

THE EDITORIAL BOARD

Ther.mo.ne'ma. Gr. adj. *thermos* hot; Gr. neut. n. *nema* a thread; N.L. neut. n. *Thermonema* a thermophilic thread.

Apparently unicellular filaments ~0.7 μm in diameter and ~ 60 to several hundred μm long. Motile by gliding. Gram-stain-negative. **Aerobic. Optimum temperature, ~60°C; poor or no growth occurs at 70°C.** Colonies are orange (Schofield et al., 1987) or yellow (Tenreiro et al., 1997), depending on the medium. Cells possess acetone-extractable pigments with an absorbance peak at 450 nm. Flexirubin-type pigments are not produced. Oxidase- and catalase-positive. Aminopeptidase-positive. Casein, gelatin, and hippurate are hydrolyzed but not starch, xylan, or cellulose. **Growth occurs on vitamin-free Casamino acids and on a mixture of the 20 natural amino acids, but not on single amino acids. Growth does not occur on pentoses, hexoses, *N*-acetylglucosamine, disaccharides, polyols, or organic acids.** Na^+ may not be required; but if required, only low concentrations are needed. Menaquinone 7 is the major respiratory quinone. Sphingolipids are present.

Source: hot springs.

DNA G+C content (mol%): 47–51.

Type species: **Thermonema lapsum** Hudson, Schofield, Morgan and Daniel 1989, 487[VP].

Further descriptive information

The initial isolate of *Thermonema lapsum* was obtained by Schofield et al. (1987) from hot-spring water samples collected from Kuirau Park, Rotorua, New Zealand. Marteinsson et al. (2001) isolated seven marine strains belonging to the genus *Thermonema* from the concentrated fluid issuing from giant geothermal cones on the seafloor at a depth of 65 m in Eyjafjordur, Northern Iceland. The isolates were obtained only from the outer zone of the chimney, where cold seawater mixes into the vent fluid with increasing salinity and decreasing temperature. Mountain et al. (2003) obtained *Thermonema* isolates from sinters in New Zealand hot springs. Tenreiro et al. (1997) isolated *Thermonema rossianum* strains from saline hot springs along the Bay of Naples, Italy.

Schofield et al. (1987) indicated that colonies are orange on Castenholz medium D (CMD; Ramaley and Hixson, 1970). Tenreiro et al. (1997) reported colonies as bright yellow on medium 162 (Degryse et al., 1978) agar containing 1.0% NaCl.

Patel et al. (1994) determined that the major normal fatty acid components of the phospholipids and lipopolysaccharide of *Thermonema lapsum* were (in decreasing order of abundance) $C_{15:0}$ iso, $C_{15:0}$ anteiso, and $C_{15:0}$. No monounsaturated fatty acids occurred. Tenreiro et al. (1997) reported that the fatty acid composition of *Thermonema rossianum* strains was dominated by $C_{15:0}$ iso and $C_{17:0}$ iso 3-OH. $C_{15:0}$ anteiso, $C_{15:0}$ iso 2-OH, and $C_{15:0}$ iso 3-OH also occurred in moderate relative concentrations, but the concentrations of other fatty acids were minor or negligible. The total relative proportion of hydroxy fatty acids was very high, about 40% of the total fatty acids.

Homospermidine and homospermine have been detected in *Thermonema lapsum* as the major polyamines (Hamana et al., 1992).

Enrichment and isolation procedures

Schofield et al. (1987) spread hot-spring water onto Castenholz medium D solidified with 3% agar and incubated the cultures for 24 h at 70°C. Single colonies were selected, and subsequent transfers grew well at 60°C but poorly at 70°C. Spreading occurred on media containing 1.5% agar but not 3% agar. In static broth cultures, the organisms grew as a pellicle that later formed clumps. Gliding motility was observed in hanging drop preparations at room temperature.

Mountain et al. (2003) inoculated media with samples (0.1–1.0 ml) of water collected from sinters in New Zealand hot springs, and incubated the cultures at temperatures in the range 30–60°C. Microbial growth, as indicated by turbidity in liquid media (composition not specified), or by colonies on agar plates, was subcultured on solid media. Pure cultures were stored in glycerol at –80°C.

Tenreiro et al. (1997) isolated *Thermonema rossianum* strains from saline hot springs along the Bay of Naples, Italy. Water samples were filtered through membrane filters (0.45 μm pore diameter) and the filters were placed on the surfaces of medium 162 agar containing 1.0% NaCl. The plates were wrapped in plastic bags and incubated at 60°C for up to 7 d. Cultures were purified by subculturing and were maintained at –80°C in medium 162 containing 1.0% NaCl and 15.0% glycerol.

Differentiation of the genus *Thermonema* from other genera

The other aerobic thermophilic genus in the phylum *Bacteroidetes* is *Rhodothermus marinus*, which is slightly halophilic, requiring about 0.25% NaCl for growth. It can be differentiated from *Thermonema* by its lack of filament formation and its ability to use most common sugars and starch.

Taxonomic comments

Based on 1490 nucleotides constituting 97% of the 16S rRNA gene of *Thermonema lapsum*, Patel et al. (1994) concluded that *Thermonema lapsum* was the deepest member of what is now the phylum *Bacteroidetes*.

Tenreiro et al. (1997) found *Thermonema rossianum* strains to be related to other members of the phylum *Bacteroidetes* by 79–97.5% similarity in 16S rRNA gene sequences. The highest relatedness was exhibited toward *Thermonema lapsum* (97.2–97.5% sequence similarity). DNA–DNA reassociation values among *Thermonema rossianum* strains were high (>91%), but they were lower (37–41%) between these strains and *Thermonema lapsum*.

List of species of the genus *Thermonema*

1. **Thermonema lapsum** Hudson, Schofield, Morgan and Daniel 1989, 487[VP]

lap′sum. L. neut. part. adj. *lapsum* gliding, from L. v. *labor* to glide.

The characteristics are as given for the genus, with the following additional features. Optimum growth in medium 162 occurs without added NaCl (thus differentiating this species from *Thermonema rossianum*); addition of NaCl causes a decrease in growth, and no growth occurs at 4% NaCl (Tenreiro et al., 1997). Negative for α- and β-galactosidase-negative. Positive for DNase. Proteolytic. Casein, gelatin, and hippurate are hydrolyzed but not cellulose, starch, and xylan. The following basal medium supplements do not support growth: acetate, L-alanine, casein, L-cystine, galactose, gelatin, gluconate, glucose, inositol, lactose, L-malate, L-proline, propan-1-ol, pyruvate, rhamnose, ribose, skim milk, sorbitol, succinate, sucrose, and yeast extract (all at concentrations of 1 g/l) plus the glutamate amino acid family (glutamate, proline, and arginine) (all at concentrations of 3.3 g/l). The following basal medium supplements do support growth: Casamino acids, amino acid mixture 1 (aspartic acid, threonine, serine, glutamic acid, proline, glycine, alanine, valine, methionine, isoleucine, leucine, tyrosine, phenylalanine, lysine, histidine, and arginine), and amino acid mixture 2 (the same as amino acid mixture 1 but lacking methionine, phenylalanine, tyrosine, and leucine).

Source: New Zealand hot springs.

DNA G+C content (mol%): 47 (T_m).

Type strain: 23/9, ATCC 43542, DSM 5718.

Sequence accession no. (16S rRNA gene): L11703.

2. **Thermonema rossianum** Tenreiro, Nobre, Rainey, Miguel and da Costa 1997, 125[VP]

ros.si.a′num. N.L. neut. adj. *rossianum* pertaining to Rossi, in honor of Mosé Rossi, the noted Neapolitan biochemist.

The characteristics are as described for the genus, with the following additional features. Colonies on medium 162 are yellow and 2 mm in diameter after 48 h. Optimum temperature, approximately 60°C; no growth occurs at 30 and 70°C in medium 162. Optimum pH, 7.0–7.5; no growth occurs at pH 5.0 or 10.0. The optimum NaCl concentration for growth is 1.0–3.0%; no growth occurs in medium 162 without added NaCl (thus differentiating this species from *Thermonema lapsum*) or with >6.0% NaCl. The major fatty acids are $C_{15:0}$ iso and $C_{17:0}$ iso 3-OH. Nitrate is not reduced to nitrite. Casein, elastin, and gelatin are degraded; starch, xylan, and cellulose are not degraded. Casamino acids and complex amino acid mixtures are utilized for growth. Growth does not occur on single amino acids, sugars, organic acids, and polyols.

Source: thermal water tap at the Stufe di Nerone.

DNA G+C content (mol%): 50.9 (HPLC).

Type strain: NR-27, DSM 10300.

Sequence accession no. (16S rRNA gene): Y08956.

References

Degryse, W., N. Glansdorff and A. Piérard. 1978. A comparative analysis of extreme thermophilic bacteria belonging to the genus *Thermus*. Arch. Microbiol. *117*: 189–196.

Garrity, G.M., J.A. Bell and T. Lilburn. 2005. The Revised Road Map to the Manual. *In* Bergey's Manual of Systematic Bacteriology, 2nd edn, vol. 2, Part A, Introductory Essays (edited by Brenner, Krieg, Staley and Garrity). Springer, New York, pp. 159–206.

Hamana, K., H. Hamana, M. Niitsu, K. Samejima and S. Matsuzaki. 1992. Distribution of unusual long and branched polyamines in thermophilic eubacteria belonging to *Rhodothermus*, *Thermus* and *Thermonema*. J. Gen. Appl. Microbiol. *38*: 575–584.

Hudson, J.A., K.M. Schofield, H.W. Morgan and R.M. Daniel. 1989. *Thermonema lapsum* gen. nov., sp. nov., a thermophilic gliding bacterium. Int. J. Syst. Bacteriol. *39*: 485–487.

Marteinsson, V.T., J.K. Kristjánsson, H. Kristmannsdóttir, M. Dahlkvist, K. Saemundsson, M. Hannington, S.K. Petursdóttir, A. Geptner and P. Stoffers. 2001. Discovery and description of giant submarine smectite cones on the seafloor in Eyjafjordur, northern Iceland, and a novel thermal microbial habitat. Appl. Environ. Microbiol. *67*: 827–833.

Mountain, B.W., L.G. Benning and J.A. Boerema. 2003. Experimental studies on New Zealand hot spring sinters; rates of growth and textural development. Can. J. Earth Sci. *40*: 1643–1667.

Patel, B.K., D.S. Saul, R.A. Reeves, L.C. Williams, J.E. Cavanagh, P.D. Nichols and P.L. Bergquist. 1994. Phylogeny and lipid composition of *Thermonema lapsum*, a thermophilic gliding bacterium. FEMS Microbiol. Lett. *115*: 313–317.

Ramaley, R.F. and J. Hixson. 1970. Isolation of a nonpigmented, thermophilic bacterium similar to *Thermus aquaticus*. J. Bacteriol. *103*: 527–528.

Schofield, K.M., J.A. Hudson, H.W. Morgan and R.M. Daniel. 1987. A thermophilic gliding bacterium from New Zealand hot springs. FEMS Microbiol. Lett. *40*: 169–172.

Tenreiro, S., M.F. Nobre, F.A. Rainey, C. Miguel and M.S. da Costa. 1997. *Thermonema rossianum* sp. nov., a new thermophilic and slightly halophilic species from saline hot springs in Naples, Italy. Int. J. Syst. Bacteriol. *47*: 122–126.

Order IV. **Incertae sedis**

Toxothrix was previously assigned to the "*Crenotrichaceae*" by Garrity et al. (2005), but subsequent analyses transferred *Crenothrix* to the *Proteobacteria*. Phylogenetic analyses of the 16S rRNA suggests that *Toxothrix* represents a very deep group and is only distantly related to any of the previously described orders within the *Bacteroidetes*. In view of its ambiguous status, it has been assigned to its own order *incertae sedis*.

Genus I. **Toxothrix** Molisch 1925, 144[AL*]

Peter Hirsch

To.xo'thrix. Gr. n. *toxon* a bow; Gr. fem. n. *thrix* a thread; N.L. fem. n. *Toxothrix* bent thread.

Cells cylindrical, colorless, 0.5–0.75×3–6 µm, in **filaments** (trichomes) **up to 400 µm long**. A dense body (**polyphosphate?**) is often located at either end of the cell (Figure 81). Gram reaction not recorded. **Filaments often U-shaped** (Figure 81a and b) and **rotating while slowly moving forward with the rounded part in the lead**; a **mucoid substance**, excreted from several sites on the trailing ends, **is deposited as a double track** ("railroad track") **of twisted strings** each 0.2 µm wide (Figure 81). **Fan-shaped structures may be deposited laterally** along the tracks, as the arms of the U move from side to side, and between the tracks, as a result of the middle section being lifted and then touched down again (Figure 81; Krul et al., 1970). **Oxidized iron may be deposited on the mucoid threads**, rendering them yellowish brown and brittle and giving them a diameter of 2.5 µm. Pure cultures have not been obtained, but chemoorganothrophic and psychrophilic cultures have been maintained for long periods at 5 and 10°C. **Filaments are extremely fragile during laboratory examination**, and **explosive disintegration of filaments has been observed after short periods under the microscope.** Grow attached to surfaces and develop best at reduced oxygen tensions (Hässelbarth and Lüdemann, 1967) and slightly below neutrality (pH 5.1–7.7). Originally found in water reservoir near the Biological Station at the Dnjepr River in the U.S.S.R. Widely distributed in cold iron springs, brooks, forest ponds, and lakes containing ferrous iron and with reduced oxygen tension (Hirsch, 1981).

DNA G+C content (mol%): not known.

Type species: **Toxothrix trichogenes** (Cholodny 1924) Beger *in* Beger and Bringmann 1953, 332[AL].

Further descriptive information

The normal trichome does not appear to have cross-walls when viewed with the phase microscope (Figure 81). Cholodny (1924) thought the organisms had a thin, tubular sheath that split repeatedly longitudinally, thus giving rise to the "twisted thread rope." However, Krul et al. (1970) followed the formation of the double tracks and fan-shaped structures on living, undisturbed and actually growing specimens.

Toxothrix trichogenes has been reported to be cultivated by Teichmann (1935). Hirsch (1981) kept natural samples

**Editorial note*: this chapter is reproduced from Vol. 1 of the 1st edition of *Bergey's Manual of Systematic Bacteriology*.

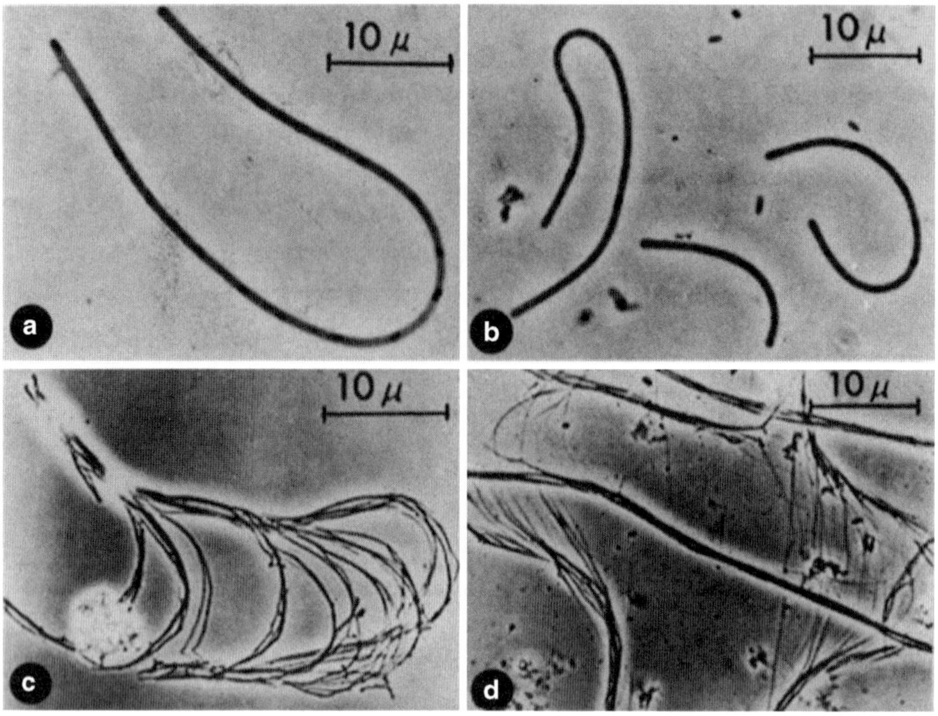

FIGURE 81. *Toxothrix trichogenes* observed in a small iron spring catch basin. (a and b) Laboratory wet mounts of living trichomes during the first minute. Phase-contrast micrographs. (c and d) Excreted polymer coated with iron oxides, through the peculiar type of motion, form fan-shaped structures (c) or double tracks (d). (Reproduced with permission from J.M. Krul et al., 1970. Antonie van Leeuwenhoek Journal of Microbiology *36:* 409–420.)

containing *Toxothrix* in the laboratory for several months; *Toxothrix* cells survived if the samples contained sediment and organic detritus and were kept cold (5°C) and dark.

The appearance of *Toxothrix* throughout the year (except for May and June) has been reported (Hirsch, 1981). Usually, it is found where *Gallionella ferruginea* grows and in waters with Fe^{2+} (1–2.7 mg/l). But *Toxothrix* cells, contrary to *Gallionella* cells, prefer habitats with a slightly higher concentration of organic compounds. An iron spring catch basin with cold, Fe-containing water and decaying leaves appears to be the optimal *Toxothrix* habitat.

Differentiation of the genus *Toxothrix* from other genera

In the absence of Fe deposition, the *Toxothrix* filaments closely resemble *Herpetosiphon*, *Haliscomenobacter*, or "*Achroonema*" filaments. But *Herpetosiphon* filaments are extremely long (300–1200 μm) and vary in their cell diameter. Also, transparent sections at the filament tips of *Herpetosiphon* (called necridia) are not present in *Toxothrix*. Strains of *Haliscomenobacter* are not known to glide or to show true branches, and their optimum pH is 7.0–8.0; also, they do not deposit iron oxides. "*Achroonema*" filaments do not glide in a U-shaped way but remain straight and fairly rigid.

Taxonomic comments

Balashova (1968) has pointed out that in some respects *Toxothrix* resembles *Gallionella*. The great differences in cell shape do not seem to support this view. Beger and Bringmann (1954) described "*Toxothrix gelatinosa*" on the basis of smaller filaments (diameter with slime threads: 1.5–1.7 μm) and the fan-shaped arrangements of individual filaments in a gelatinous matrix. However, the individual cell size (0.5×3 μm) falls within the range given for *Toxothrix trichogenes*.

List of species of the genus *Toxothrix*

1. **Toxothrix trichogenes** (Cholodny 1924) Beger *in* Beger and Bringmann 1953, 332[AL] (*Leptothrix trichogenes* Cholodny 1924, 296; *Toxothrix ferruginea* Molisch 1925, 144; *Chlamydothrix trichogenes* (Cholodny 1924) Naumann 1929, 513; *Sphaerotilus trichogenes* (Cholodny 1924) Pringsheim 1949, 234)

   tri.cho'ge. nes. Gr. n. *thrix trichos* hair; N.L. suff. *-genes* (from Gr. v. *gennaô* to produce) producing; N.L. adj. *trichogenes* hair-producing.

   Description is the same as for the genus.

   DNA G+C content (mol%): unknown.

   Type strain: no culture isolated.

References

Balashova, V.V. 1968. [On taxonomy of *Gallionella* genus]. Mikrobiologiya *37*: 715–723.

Beger, H. and G. Bringmann. 1953. Die Scheidenstruktur des Abwasserbakteriums *Sphaerotilus* und des Eisenbakteriums *Leptothrix* im elektronenmikroskopischen Bilde und ihre Bedeutung für die Systematik dieser Gattungen. Zentralbl. Bakteriol. Parasitenkd. Infektionskr. Hyg. Abt. II *107*: 318–334.

Cholodny, N. 1924. Über neue Eisenbakterienarten aus der Gattung *Leptothrix* Kütz. Zentralbl. Bakteriol. Parasitenkd. Infektionskr. Hyg. Abt. II *61*: 292–298.

Garrity, G.M., J.A. Bell and T. Lilburn. 2005. The Revised Road Map to the Manual. *In* Bergey's Manual of Systematic Bacteriology, 2nd edn, vol. 2, Part A, Introductory Essays (edited by Brenner, Krieg, Staley and Garrity). Springer, New York, pp. 159–206.

Hässelbarth, U. and D. Lüdemann. 1967. Die biologische Verockerung von Brunnen durch Massenentwicklung von Eisen- und Manganbakterien. Bohrtechnik-Brunnenbau-Rohr-leitungsbau 10/11 (Ber. DVGW Fachausschuss "Wasserfassung und Wasseranreicherung").

Hirsch, P. 1981. The genus *Toxothrix*. *In* The Prokaryotes: A Handbook on Habitats, Isolation, and Identification of Bacteria (edited by Starr, Stolp, Trüper, Balows and Schlegel). Springer, New York, pp. 409–411.

Krul, J.M., P. Hirsch and J.T. Staley. 1970. *Toxothrix trichogenes* (Chol.) Beger et Bringmann: the organism and its biology. Antonie van Leeuwenhoek *36*: 409–420.

Molisch, H. 1925. Botanische Beobachtungen in Japan. VIII. Eisenorganismen Japan. Sci. Rep. Tohoku Imp. Univ. Ser. IV Biol. *1*: 135–168.

Naumann, E. 1929. Die eisenspeichernden Bakterien. Kritische Übersicht der bisher bekannten Formen. Zentralbl. Bakteriol. Parasitenkd. Infektionskr. Hyg. Abt. II *78*: 512–515.

Pringsheim, E.G. 1949. Iron bacteria. Bacteriol. Rev. *24*: 200–245.

Teichmann, P. 1935. Vergleichende Untersuchungen über die Kultur und Morphologie einiger Eisenorganismen. PhD thesis, Prague.

Phylum XV. Spirochaetes Garrity and Holt 2001

Bruce J. Paster

Spi.ro.chae′tes. N.L. fem. pl. n. *Spirochaetales* type order of the phylum; dropping the ending to denote a phylum; N.L. fem. pl. n. *Spirochaetes* the phylum of *Spirochaetales*.

Spirochetes are Gram-stain-negative, helical or spiral-shaped, motile cells that can flex, rotate and translate through liquid and semisolid environments. Most spirochetes possess a cellular ultrastructure unique to bacteria in that they have internal organelles of motility, namely periplasmic flagella. Consequently, the spirochetes are one of the few bacterial phyla whose phenotypic characteristics, e.g., cell morphology, reflect its phylogenetic relationships as based on 16S rRNA gene sequence comparisons (Garrity and Holt, 2001), forming a distinct line of evolutionary descent among the bacteria (Garrity et al., 2005).

Spirochetes are chemo-organotrophic and, depending upon the species or phylogenetic group, may grow under anaerobic, microaerophilic, facultatively anaerobic or aerobic conditions. Some are free-living and others are host-associated, e.g., arthropods, mollusks, and mammals, including humans. Some species are known to be pathogenic.

Type order: **Spirochaetales** Buchanan 1918, 163[AL].

References

Buchanan, R.E. 1918. Studies in the nomenclature and classification of the bacteria: X. Subgroups and genera of the *Myxobacteriales* and *Spirochaetales*. J. Bacteriol. *3*: 541–545.

Garrity, G.M. and J.G. Holt. 2001. The Road Map to the Manual. *In* Bergey's Manual of Systematic Bacteriology, 2nd edn, vol. 1, The *Archaea* and the Deeply Branching and Phototrophic *Bacteria* (edited by Boone, Castenholz and Garrity). Springer, New York, pp. 119–166.

Garrity, G.M., J.A. Bell and T. Lilburn. 2005. The Revised Road Map to the Manual. *In* Bergey's Manual of Systematic Bacteriology, 2nd edn, vol. 2, The *Proteobacteria*, Part A, Introductory Essays (edited by Brenner, Krieg, Staley and Garrity). Springer, New York, pp. 159–206.

Class I. **Spirochaetia** class. nov.

Bruce J. Paster

Spi.ro.chae′ti.a. N.L. fem. n. *Spirochaeta* type genus of the type order *Spirochaetales*; suff. *-ia* ending proposed by Gibbons and Murray and by Stackebrandt et al. to denote a class; N.L. neut. pl. n. *Spirochaetia* the class of *Spirochaetales*.

Members of the class *Spirochaetia* are as described for the phylum. The class contains spirochetes in one order, *Spirochaetales*, which is presently comprised of four families, namely *Spirochaetaceae*, *Brachyspiraceae*, *Brevinemataceae*, and *Leptospiraceae*.

Type order: **Spirochaetales** Buchanan 1918, 163[AL].

Reference

Buchanan, R.E. 1918. Studies in the nomenclature and classification of the bacteria: X. Subgroups and genera of the *Myxobacteriales* and *Spirochaetales*. J. Bacteriol. *3*: 541–545.

Order I. **Spirochaetales** Buchanan 1917, 163[AL]

Bruce J. Paster

Spi.ro.chae.ta′les. N.L. fem. n. *Spirochaeta* type genus of the order; *-ales* suffix to denote an order; N.L. fem. pl. n. *Spirochaetales* the *Spirochaeta* order.

Based on 16S rRNA gene sequence comparisons, spirochetes form a coherent phylogenetic phylum (Paster et al., 1991). The order *Spirochaetales* contains the families *Spirochaetaceae*, *Brachyspiraceae*, *Brevinemataceae*, and *Leptospiraceae*, as shown in Figure 82.

Helically shaped, motile bacteria, 0.1–3 μm in diameter and 4–250 μm in length. Cells have **internal organelles of motility called periplasmic flagella** (which have been previously called axial fibrils, axial filaments, flagella, endoflagella, and periplasmic fibrils) (Canale-Parola, 1978; Paster and Canale-Parola, 1980). Periplasmic flagella are inserted subterminally at each end of the protoplasmic cylinder and extend along most of the length of the cell overlapping in the central region, but the other end of the flagella are inserted (Figure 82). This results in a n:2n:n flagellar arrangement where n ranges from 1 to 100s depending upon the species. However, the periplasmic flagella do not overlap in cells of members of the family *Leptospiraceae*. The protoplasmic cylinder and flagella are encased by an "outer sheath" which has some features analogous the outer membrane of traditional Gram-stain-negative bacteria (Figure 82). Under certain growth conditions, periplasmic flagella of some species protrudes outside of the cell (Charon et al., 1992) (Figure 82).

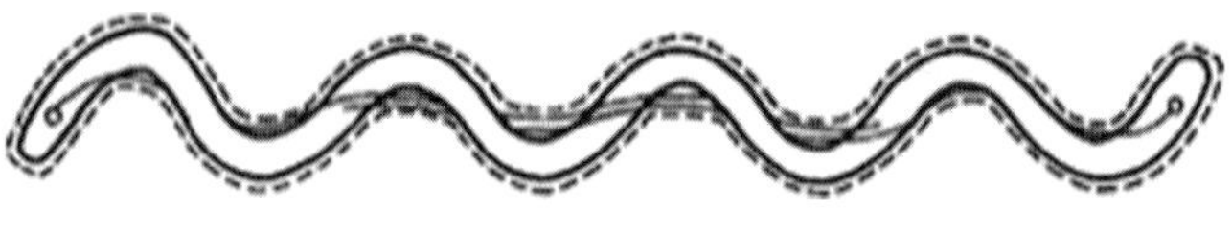

FIGURE 82. Schematic representation of a spirochete. The dotted line indicates the outer sheath encasing the helical protoplasmic cell and the periplasmic flagella which are inserted at each end of the cell.

Chemoheterotrophic. Carbohydrates, amino acids, long-chain fatty acids, or long-chain fatty alcohols serve as carbon and energy sources. **Anaerobic, facultatively anaerobic, microaerophilic, and aerobic**. Stains Gram-negative.

Free-living or in association with animal, insect, and human hosts. Some species are pathogenic.

DNA G+C content (mol%): 25–66.

Type genus: **Spirochaeta** Ehrenberg 1835, 313[AL].

Further descriptive information

Spirochetes have three main types of movements in liquids, namely locomotion, rotations about their longitudinal axis, and flexing (Canale-Parola, 1978). Cells can translocate in highly viscous environments, such as methyl cellulose or in media containing 1% agar. Cells also have been reported to creep or crawl on solid surfaces.

Some "free-living pleomorphic spirochetes", referred to as FLiPS (Ritalahti and Löffler, 2004), have been recently described. One species from the termite hindgut has been named *Spirochaeta coccoides* (Dröge et al., 2006). These species do not have the characteristic ultrastructural and behavioral features of spirochetes; namely helical, motile cells with protoplasmic cylinder and periplasmic flagella enclosed in an outer sheath. However based on 16S rRNA gene sequence analysis, these species fall within the family *Spirochaetaceae* (see chapter on *Spirochaeta*, below).

"*Spironema culicis*" was isolated recently from the mosquito and is the only species of the genus (Cechová et al., 2004). Although it has not yet been formally named, it falls phylogenetically within the family *Spirochaetaceae* (Figure 83) and likely warrants separate genus designation from other genera of the family.

Sequences of species of the genera *Treponema* and *Spirochaeta* in the family *Spirochaetaceae*, *Brevinema* in the family *Brevinemataceae*, and *Leptospira* in the family *Leptospiraceae* are unusual among bacteria in that they possess a 20–30 base 5′ extension of 16S rRNA molecule (Paster et al., 1991). The function of this 5′ extension is unknown, but the region is highly variable and was proposed to form helices of 2–12 bases. The 5′ extension has not been reported in other spirochetal species.

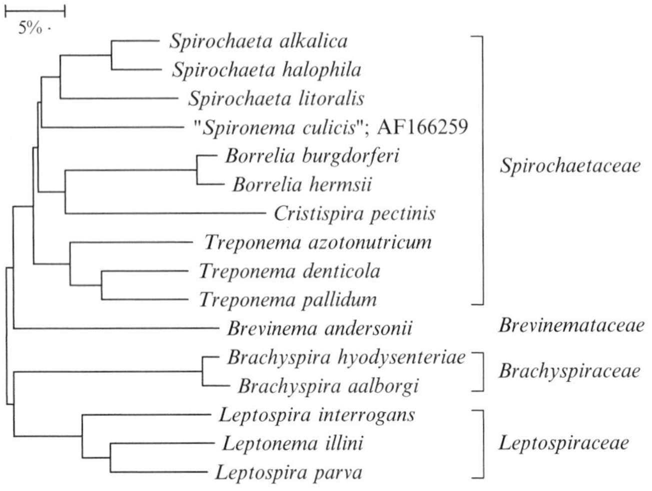

FIGURE 83. Phylogeny of the order *Spirochaetales*. The order is comprised of four families, namely *Spirochaetaceae*, *Brachyspiraceae*, *Brevinemataceae*, and *Leptospiraceae*. Bar = 5% difference in 16S rRNA gene sequences.

Key to the families of the order *Spirochaetales*

1. Cell diameter 0.1–3 µm. Ends of cells are usually not hooked. Periplasmic flagella overlap in the central region of the cell. The diamino acid in the peptidoglycan is L-ornithine. Anaerobic or facultatively anaerobic. Use carbohydrates and/or amino acids as carbon and energy sources. Free-living and host associated.
   → Family I. *Spirochaetaceae*
2. Cell diameter 0.2–0.4 µm. Cell ends may be blunt or pointed, and are not hooked. Periplasmic flagella overlap in the central region of the cell. The diamino acid in the peptidoglycan is L-ornithine. Obligately anaerobic, aerotolerant. Use monosaccharides, disaccharides, the trisaccharide trehalose, and amino sugars as carbon and energy sources. Does not use polysaccharides. Host-associated.
   → Family II. *Brachyspiraceae*
3. Cell diameter 0.2–0.3 µm. Cells are short, 4–5 µm in length with only one or two turns. Ends of cells are usually not hooked. Periplasmic flagella overlap in the central region of the cell. The diamino acid in the peptidoglycan is not known. Microaerophilic. Peptones are required for growth. Host-associated.
   → Family III. *Brevinemataceae*
4. Cell diameter 0.1–0.3 µm. Ends of cells are usually hooked. The diamino acid in the peptidoglycan is diaminopimelic acid. Periplasmic flagella do not appear to overlap in the central region of the cell. Obligately aerobic, or microaerophilic. Use long-chain fatty acids or long-chain fatty alcohols as carbon and energy sources. Do not use carbohydrates or amino acids. Free-living and host-associated.
   → Family IV. *Leptospiraceae*

References

Buchanan, R.E. 1917. Studies on the Nomenclature and Classification of the Bacteria: III. The Families of the *Eubacteriales*. J. Bacteriol. *2*: 347–350.

Canale-Parola, E. 1978. Motility and chemotaxis of spirochetes. Annu. Rev. Microbiol. *32*: 69–99.

Cechová, L., E. Durnová, S. Sikutová, J. Halouzka and M. Nemec. 2004. Characterization of spirochetal isolates from arthropods collected in South Moravia, Czech Republic, using fatty acid methyl esters analysis. J. Chromatogr. B Analyt. Technol. Biomed. Life Sci. *808*: 249–254.

Charon, N.W., S.F. Goldstein, S.M. Block, K. Curci, J.D. Ruby, J.A. Kreiling and R.J. Limberger. 1992. Morphology and dynamics of protruding spirochete periplasmic flagella. J. Bacteriol. *174*: 832–840.

Dröge, S., J. Fröhlich, R. Radek and H. König. 2006. *Spirochaeta coccoides* sp. nov., a novel coccoid spirochete from the hindgut of the termite *Neotermes castaneus*. Appl. Environ. Microbiol. *72*: 392–397.

Ehrenberg, C.G. 1835. Dritter Beitrag zur Erkemtiss grosser Organisation in der Richtung des kleinsten Raumes. Abh. Preuss. Akad. Wiss. Phys. Kl Berlin aus den Jahre 1833–1835: 143–336.

Paster, B.J. and E. Canale-Parola. 1980. Involvement of periplasmic fibrils in motility of spirochetes. J. Bacteriol. *141*: 359–364.

Paster, B.J., F.E. Dewhirst, W.G. Weisburg, L.A. Tordoff, G.J. Fraser, R.B. Hespell, T.B. Stanton, L. Zablen, L. Mandelco and C.R. Woese. 1991. Phylogenetic analysis of the spirochetes. J. Bacteriol. *173*: 6101–6109.

Ritalahti, K.M. and F.E. Löffler. 2004. Characterization of novel free-living pleomorphic spirochetes (FLiPS), Abstract 539. Presented at the 10th International Symposium on Microbial Ecology. International Society for Microbial Ecology, Geneva, Switzerland.

Family I. **Spirochaetaceae** Swellengrebel 1907, 581[AL]

Bruce J. Paster

Spi.ro.chae.ta.ce′ae. N.L. fem. n. *Spirochaeta* type genus of the family; *-aceae* ending to denote a family; N.L. fem. pl. n. *Spirochaetaceae* the *Spirochaeta* family.

Helical cells, 0.1–3.0 µm in diameter and 3.5–250 µm in length. Cells do not have hooked ends as do members of the family *Leptospiraceae*. Periplasmic flagella are inserted subterminally at each end of the cell and extend along most of the length of the cell overlapping in the central region (Figure 82). **The diamino acid in the peptidoglycan is L-ornithine.** Motile.

Anaerobic, facultatively anaerobic, or microaerophilic. Chemo-organotrophic. **Utilize carbohydrates and/or amino acids as carbon and energy sources.** Do not use long-chain fatty acids or long-chain fatty alcohols as energy sources.

Free-living or in association with animal, insect and human hosts. Some species are pathogenic. The DNA G+C content is 36–66 mol% (T_m, Bd, and genetic sequence analysis). Species examined by 16S rRNA sequence analysis are distinct from members of the families *Brachyspiraceae*, *Brevinemataceae*, and *Leptospiraceae* (Figure 83).

Type genus: **Spirochaeta** Ehrenberg 1835, 313[AL].

Key to the genera of the family *Spirochaetaceae*

1. Cells are 0.2–75 µm in diameter and 5–250 µm in length. Obligately anaerobic and facultatively anaerobic. Carbohydrates serve as energy and carbon sources. Amino acids are not used as growth substrates. Free-living in fresh water and marine environments, including mud, sediments and water of ponds, lakes, streams, and marshes. Many not-yet-cultivated species identified as based on 16S rRNA gene sequence comparisons. Not considered as pathogenic. The DNA G+C content is 45–66 mol% (T_m, Bd, and HPLC).
   → Genus I. *Spirochaeta*
2. Cells are 0.2–0.3 µm in diameter and 3–20 µm in length. Microaerophilic. Arthropod-borne pathogens of man, other mammals, and birds. The causative agents of tick-borne Lyme disease and relapsing fever and louse-borne relapsing fever in man. The DNA G+G content of a limited number of species is 27–32 mol% (T_m, HPLC, and genome sequencing).
   → Genus II. *Borrelia*
3. Cells are 0.5–3 µm in diameter and 30–180 µm in length. Hundreds of periplasmic flagella present as a bundle forming a ridge called crista. Inhabit the crystalline style of the digestive tract of aquatic mollusks. Not grown in pure culture. The DNA G+C content is not known.
   → Genus III. *Cristispira*
4. Cells are 0.1–0.4 µm in diameter and 5–20 µm in length. Obligately anaerobic. Carbohydrates and amino acids serve as energy and carbon sources. Found in the oral cavity, intestinal tract, and genital areas of humans and animals. Some species are pathogenic. Also found in the hindgut of termites. Many not-yet-cultivated species identified as based on 16S rRNA gene sequence comparisons. The DNA G+C content is 36–54 mol% (T_m, HPLC, genome sequence).
   → Genus IV. *Treponema*

Genus I. **Spirochaeta** Ehrenberg 1835, 313[AL] (*Spirochoeta* Dujardin 1841, 225, and *Spirochaete* Cohn 1872, 180 (orthographic variants of *Spirochaeta*); *Ehrenbergia* Gieszczkiewicz 1939, 24)

Susan Leschine and Bruce J. Paster

Spi.ro.chae′ta. Gr. n. *speira* (L. transliteration *spira*) a coil; Gr. fem. n. *chaitê* (L. transliteration *chaete*) hair; N.L. fem. n. *Spirochaeta* coiled hair.

Flexible helical cells 0.2–0.75 µm in diameter and 5–250 µm in length. All species have two periplasmic flagella per cell except *Spirochaeta plicatilis*, which has many periplasmic flagella. Under unfavorable conditions, spherical cells or structures 0.5–2.0 µm (occasionally up to 10 µm) in diameter are formed. Cells translocate when suspended in liquids and crawl or creep when in contact with solid surfaces. **Obligately anaerobic or facultatively anaerobic**. Under aerobic growth conditions the **facultatively anaerobic species usually produce carotenoid pigments** that give a yellow, yellow-orange, or red coloration to colonies or cells in liquid media. Thermophilic species are known. Optimum temperature range, 25–68°C. **Chemo-organotrophic, using a variety of carbohydrates as carbon and energy sources.** The main products of anaerobic carbohydrate metabolism are ethanol, acetate, CO_2, and H_2. Under aerobic conditions, facultatively anaerobic species oxidize carbohydrates yielding primarily CO_2 and acetate. Indigenous to aquatic freshwater and marine environments such as the sediments, mud and water of ponds, marshes, swamps, lakes, rivers, and hot springs. Occur commonly in H_2S-containing environments. **Free-living**. None reported to be pathogenic.

DNA G+C content (mol%): 50–65 (Bd), 45–66 (T_m), 50–58.5 (HPLC).

Type species: **Spirochaeta plicatilis** Ehrenberg 1835, 313[AL].

Further descriptive information

Cells of all species are helical in shape and possess the typical ultrastructural features of spirochetes (Canale-Parola, 1984a). The outermost structure of the cells is an "outer membrane", or "outer sheath", which encloses the coiled cell body ("protoplasmic cylinder") consisting of the cytoplasm, the nuclear region, and the peptidoglycan-cytoplasmic membrane complex. Organelles ultrastructurally similar to bacterial flagella are located in the area between the outer membrane and the protoplasmic cylinder. These organelles are essential components of the motility apparatus of spirochetes (Paster and Canale-Parola, 1980) and are usually called "periplasmic flagella". One end of each periplasmic flagellum is inserted near a pole of the protoplasmic cylinder, while the other end is not inserted. Individual periplasmic flagella extend for most of the length of *Spirochaeta* cells so that those inserted near opposite ends overlap in the central region of the organism.

The nature of the spherical structures, called "spherical bodies", that are formed under unfavorable growth conditions has not been determined. Spherical bodies occur either in physical association with helical cells or free.

When suspended in liquids, cells translocate "in straight lines or nearly straight lines, and they appear to spin rapidly about their longitudinal axis (Berg, 1976). The motility of a strain of *Spirochaeta aurantia* in liquid environments has been described (Greenberg et al., 1985, 1977c). Occasionally a cell stops momentarily and flexes, and then resumes spinning and translational motility. However, when translation resumes, the direction of movement is usually altered..." and frequently the previously leading cell end becomes the trailing end (Greenberg and Canale-Parola, 1977c). During runs, *Spirochaeta aurantia* cells have an average linear speed of approximately 16/s (Fosnaugh and Greenberg, 1988). Flexes last from a fraction of a second to several seconds. The average frequency of reversals in cell populations is approximately 0.31 reversals/ 5 s (Fosnaugh and Greenberg, 1988). Cells in motion usually retain their basic helical configuration, but they assume a variety of shapes as a result of flexing, undulating, and contracting, as well as wave propagation. Broad secondary coils or waves superimposed on the smaller primary coils are formed frequently (Canale-Parola, 1977, 1978). During creeping movements of *Spirochaeta plicatilis* on solid surfaces (Blakemore and Canale-Parola, 1973), "... the rear coils follow the tortuous path of the anterior cell end almost exactly" (Canale-Parola, 1978).

Cells retain translational motility in environments of relatively high viscosity, usually becoming immotile at viscosities of 300–1000 centipoise, depending on the strain (Greenberg and Canale-Parola, 1977a, b).

Strains of *Spirochaeta aurantia* exhibit chemotaxis toward carbohydrates, but not toward amino acids (Breznak and Canale-Parola, 1975; Greenberg and Canale-Parola, 1977c). Effective attractants for *Spirochaeta aurantia* strain M1 are as follows: D-glucose, 2-deoxy-D-glucose, α-methyl-D-glucoside, D-galactose, D-fucose, D-mannose, D-fructose, D-xylose, maltose, cellobiose, and D-glucosamine (Greenberg and Canale-Parola, 1977c). Taxis toward D-galactose and D-fucose is induced by the presence of D-galactose in the growth medium.

The helical shape of the cells is maintained by the peptidoglycan layer (Joseph and Canale-Parola, 1972). L-Ornithine is the only diaminoamino acid in the peptidoglycan of *Spirochaeta stenostrepta, Spirochaeta litoralis, Spirochaeta aurantia,* and *Spirochaeta halophila* (Joseph et al., 1973; B.J. Paster and E. Canale-Parola, unpublished data). The peptidoglycans of *Spirochaeta zuelzerae* and *Spirochaeta plicatilis* have not been tested for the presence of L-ornithine.

The peptidoglycan of *Spirochaeta stenostrepta* is composed of acylglucosamine, acylmuramic acid, L-alanine, D-glutamic acid, L-ornithine, and D-alanine (Joseph et al., 1973; Schleifer and Joseph, 1973). Peptidoglycan of similar composition is present in *Spirochaeta litoralis.* At least 50% of the peptide subunits of the peptidoglycan of *Spirochaeta stenostrepta* contain the tripeptide *N*-acyl-muramyl-L-alanyl-α-D-glutamyl-L-ornithine. Cross-linkage (30%) between the 6-amino group of L-ornithine and the carboxyl group of D-alanine occurs in the remaining peptide subunits of sequence *N*-acyl-muramyl-L-alanyl-α-D-glutamyl-L-ornithyl-D-alanine (Schleifer and Joseph, 1973).

A lipoprotein layer, adjacent and external to the peptidoglycan, has been detected in *Spirochaeta stenostrepta* (Joseph et al., 1970). This layer consists of a fine array of tightly packed, longitudinally oriented helices measuring 2.5 nm in diameter (Holt and Canale-Parola, 1968).

Colonies of *Spirochaeta* diffuse or spread through the agar medium in which they are growing. This phenomenon is especially apparent in agar media containing low substrate concentrations and 1% or less agar. Diffusion of colonies is due to migration of the growing cells through the agar medium. Migration of the cells is the result of chemotaxis toward the growth substrate and of the ability of spirochetes to locomote through agar gels (Canale-Parola, 1977, 1978).

The growth of obligately anaerobic species of *Spirochaeta* is abundant in media with 1.0 or 1.5 g of agar/100 ml, whereas the growth of some strains of facultatively anaerobic species of *Spirochaeta* is inhibited in media containing more than 1.0% agar. However, these strains grow abundantly when the agar concentration in the medium is 1% or lower. Under anaerobic conditions, *Spirochaeta stenostrepta, Spirochaeta litoralis, Spirochaeta aurantia,* and *Spirochaeta halophila* ferment carbohydrates to pyruvate via the Embden–Meyerhof pathway (Canale-Parola, 1977; Greenberg and Canale-Parola, 1976). Pyruvate is metabolized to acetyl-CoA, CO_2, and H_2 by means of a clostridial-type clastic reaction. Acetyl-CoA is converted to acetate in reactions catalyzed by phosphotransacetylase and acetate kinase, and to ethanol through a double reduction involving aldehyde and alcohol dehydrogenase activities (Canale-Parola, 1977). These pathways constitute the major anaerobic energy-yielding mechanisms utilized by the four *Spirochaeta* species mentioned above (Canale-Parola, 1977). The pathways of carbohydrate catabolism utilized by *Spirochaeta zuelzerae* have not been elucidated. This spirochete does not form ethanol, but produces succinate and larger amounts of lactate than do other species.

In addition to carbohydrates, *Spirochaeta isovalerica* ferments L-leucine, L-isoleucine, and L-valine, forming isovaleric, 2-methylbutyric, and isobutyric acids, respectively, as end products (Harwood and Canale-Parola, 1983). Fermentation of the amino acids in the absence of glucose does not support measurable growth of *Spirochaeta isovalerica,* but serves to generate ATP, which is utilized as a source of maintenance energy by the spirochete when fermentable carbohydrates are not available (Harwood and Canale-Parola, 1981a, 1983). In addition to the branched-chain fatty acids, amino acid catabolism by

Spirochaeta isovalerica yields small quantities of isobutanol and isoamyl alcohol (Harwood and Canale-Parola, 1981b, 1983).

When growing aerobically, *Spirochaeta aurantia* and *Spirochaeta halophila* derive energy by performing an incomplete oxidation of carbohydrates, with CO_2 and acetate being the main dissimilatory products. The tricarboxylic acid cycle either is not present or serves in a minor catabolic capacity in these two species. Determinations of molar growth yields and other studies indicate that, when growing aerobically, *Spirochaeta aurantia* and *Spirochaeta halophila* generate ATP via oxidative phosphorylation as well as by substrate level phosphorylation (Breznak and Canale-Parola, 1972a; Greenberg and Canale-Parola, 1976). Cytochromes b_{558} and cytochrome *o* are present in *Spirochaeta aurantia* (Breznak and Canale-Parola, 1972a).

Spirochaeta species are able to synthesize all of their cell lipids *de novo*. The chain length of cellular fatty acids varies from 12 to 18 carbons (Livermore and Johnson, 1974). *Spirochaeta aurantia* and *Spirochaeta zuelzerae* but not *Spirochaeta litoralis* and *Spirochaeta stenostrepta* synthesize unsaturated fatty acids. *Spirochaeta stenostrepta* and *Spirochaeta zuelzerae* but not *Spirochaeta litoralis* and *Spirochaeta aurantia* synthesize anteiso branched-chain fatty acids (Livermore and Johnson, 1974).

Spirochaeta species are resistant to the antibiotic rifampicin (rifampin) at concentrations of 1–50 µg rifampicin/ml. Resistance to rifampicin may be due to low affinity of the spirochetal RNA polymerase for the antibiotic.

Enrichment and isolation procedures

Anaerobic and facultatively anaerobic spirochetes occur commonly in aquatic freshwater and marine environments such as the sediments, mud and water of ponds, marshes, swamps, lakes, rivers, and hot springs, often in association with decomposing plant biomass (Harwood and Canale-Parola, 1984; Leschine, 1995). *Spirochaeta* strains are readily isolated from natural environments by means of selective procedures and usually grow abundantly in ordinary laboratory media. Anaerobic growth yields of the isolates range from 2 to 10^8 to approximately 10^9 cells/ml, but commonly are $6–8\times10^8$ cells/ml (Breznak and Canale-Parola, 1975; Canale-Parola, 1973; Greenberg and Canale-Parola, 1976; Harwood and Canale-Parola, 1984; Hespell and Canale-Parola, 1970a, b). Cell population doubling times in anaerobic cultures are 2.2–12 h, depending on the species and the growth conditions. Aerobically grown cultures yield $0.7–1.2\times10^9$ spirochetes/ml, with doubling times of 2–4 h (Breznak and Canale-Parola, 1975; Canale-Parola, 1973; Greenberg and Canale-Parola, 1976).

A procedure in which the antibiotic rifampicin (rifamycin) serves as a selective agent is quite effective for the isolation of free-living (genus *Spirochaeta*) and host-associated (genus *Treponema*) spirochetes from natural environments (Harwood et al., 1982; Patel et al., 1985; Stanton and Canale-Parola, 1979; Weber and Greenberg, 1981). This procedure (described below) is based on the observation that spirochetes in general are naturally resistant to rifampicin (Leschine and Canale-Parola, 1986; Stanton and Canale-Parola, 1979). Thus, spirochetes such as *Spirochaeta stenostrepta* and *Spirochaeta aurantia* grow in the presence of as much as 100–200 µg of rifampicin per ml of medium (Leschine and Canale-Parola, 1986), whereas the growth of many other bacteria is inhibited. The resistance of spirochetes to rifampicin is probably due to the low affinity of their RNA polymerase for the antibiotic (Allan et al., 1986; Leschine and Canale-Parola, 1986).

Enrichment procedures used in the isolation of *Spirochaeta* species are based on one or more of the following selective factors: (1) resistance to rifampicin (mentioned above), (2) the ability of spirochetes to pass through filters that retain most other bacteria, (3) the migratory movement of spirochetes through agar media (Canale-Parola, 1973, 1984b). The latter two procedures will enrich for species of *Spirochaeta* measuring less than 0.5 µm in diameter.

In the enrichment-by-filtration procedure (described below), separation of *Spirochaeta* species from most of the micro-organisms present in mud or water is achieved by techniques involving filtration through cellulose ester filter discs (e.g., Millipore filters) having a mean pore diameter of 0.3 or 0.45 µm. Spirochetes pass through these filter discs because their cell diameter is relatively small, and probably also because their motility apparatus enables them to swim freely in liquids as well as to move in contact with solid surfaces.

The enrichment-by-migration procedure uses the ability of spirochetes to move through agar gels or media containing as much as 1–2% (w/v) agar. This movement or migration occurs primarily within the agar gel, i.e., below the surface of the agar medium. In contrast, flagellated bacteria usually cannot carry out translational movement through gels or media containing agar at the above-mentioned concentrations, although several exceptions have been reported (Greenberg and Canale-Parola, 1977a). Apparently, the cell coiling of spirochetes is important for their translational motion through agar gels, inasmuch as this type of movement is impaired in mutant spirochetes lacking the cell-coiling characteristic of the parental strain (Greenberg and Canale-Parola, 1977b).

Migration of spirochetes through agar media results from the unique motility mechanism of these bacteria (Canale-Parola, 1977, 1978), as well as from chemotaxis toward the energy and carbon source (Breznak and Canale-Parola, 1975; Greenberg and Canale-Parola, 1977c; Terracciano and Canale-Parola, 1984). The role of chemotaxis in the migration of saccharolytic spirochetes through agar media has been studied (Breznak and Canale-Parola, 1975). When these spirochetes are inoculated in the center of glucose-containing agar medium plates, they grow using this sugar as their energy source. Utilization of the sugar by the spirochetes gives rise to a glucose concentration gradient that moves away from the center of the plate as more of this carbohydrate is metabolized by the spirochetes. Because the spirochetes exhibit chemotaxis toward glucose and are able to move through the agar gel, they migrate into the areas of higher glucose concentration within the gradient. Thus, the spirochetal population, by following the outward movement of the gradient, migrates toward the periphery of the plate. This behavior results in the formation of a growth "veil" or "ring" of spirochetes for which glucose serves both as the energy source for growth and as the chemoattractant (Breznak and Canale-Parola, 1975; Canale-Parola, 1973). The veil or ring increases continuously in diameter during incubation and may reach the outer edge of the plate. The migration rate of the spirochetal population is greatest in agar media containing low substrate concentrations (e.g., 0.02% glucose). In these media the substrate becomes rapidly depleted in the region where spirochetes are growing, and the spirochetal population moves toward the outer zone of higher substrate concentrations at a relatively fast rate (Breznak and Canale-Parola, 1975).

In procedures for the isolation of *Spirochaeta* species from natural environments, the chemotactic behavior and the ability of these bacteria to move through agar gels have important selective functions. In a typical isolation procedure, a small, shallow cylindrical hole is made through the surface of an agar medium containing a low concentration of carbohydrate. Rifampicin may be included in the medium as an additional selective agent for spirochetes. The medium may be in a Petri dish or a small bottle. A tiny drop of pond water, or of any other material in which spirochetes have been observed, is placed within the hole. The chemotactic, saccharolytic spirochetes in the inoculum multiply and form a growth veil that extends outwardly through the agar medium. Thus, the spirochetes in the veil move away from contaminants, which remain mainly in the vicinity of the inoculation site. Spirochetal cells from the outermost edge of the veil are used to obtain pure cultures by conventional methods, such as streaking on agar medium plates. Isolation procedures involving chemotaxis and movement through agar gels are described below.

Selective isolation techniques have not been developed for the large *Spirochaeta* species, such as *Spirochaeta plicatilis.*

Maintenance procedures

Species of *Spirochaeta* remain viable for many years when stored in the frozen state at the temperature of liquid nitrogen. Methods for liquid nitrogen storage of *Spirochaeta* species and for the preparation of other types of stock cultures of these bacteria have been described (Canale-Parola, 1973).

Differentiation of the genus *Spirochaeta* from other genera

Members of the genus *Spirochaeta* are readily differentiated from other genera of spirochetes as shown in Table 118.

Taxonomic comments

Based on 16S rRNA gene sequence comparisons (Paster et al., 1984, 1991), members of the genus *Spirochaeta* belong within the family *Spirochaetaceae* and are clearly distinct from the other genera of spirochetes (Figure 84). These data also suggest that *Spirochaeta aurantia* warrants separate genus designation, although it would still fall within the family *Spirochaetaceae* (Figure 84). However, several species of *Spirochaeta*, including *Spirochaeta stenostrepta, Spirochaeta zuelzerae,* and *Spirochaeta caldaria,* do not belong with other free-living *Spirochaeta* (shown in Figure 84), but are more closely related to members of the genus *Treponema.* The phylogenetic clustering of these species has also been confirmed by single-base signature analysis, i.e., the sequences of these species have more signature bases found in the sequences of the treponemes than in the sequences of *Spirochaeta* species. Consequently, these species should be classified as species of *Treponema,* but since there has been no formal renaming, they remain *Spirochaeta* spp. in this chapter. These "free-living"

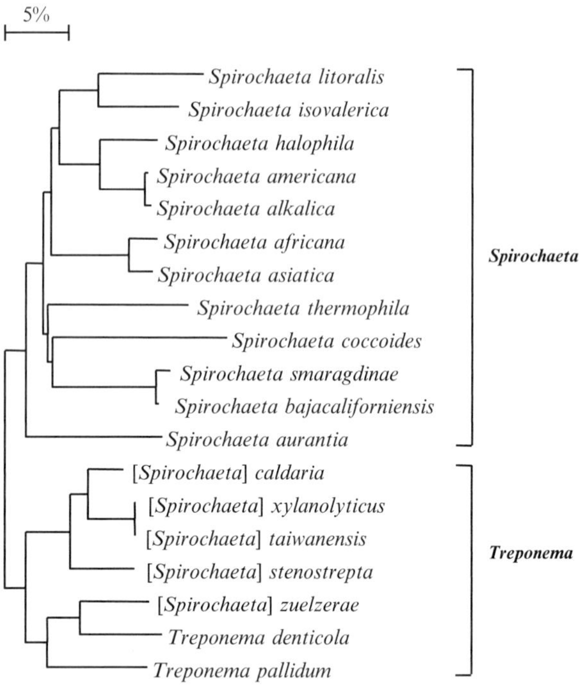

FIGURE 84. Phylogenetic tree of the genus *Spirochaeta* and related organisms, based on 16S ribosomal RNA gene (rRNA) sequences. Some of the named *Spirochaeta* species shown in brackets, namely [*Spirochaeta*] *zuelzerae,* [*Spirochaeta*] *caldaria,* [*Spirochaeta*] *stenostrepta,* [*Spirochaeta*] *xlanolyticus* and [*Spirochaeta*] *taiwanensis,* are phylogenetically more related to members of the genus *Treponema.* The scale bar represents a 5% difference in nucleotide sequence.

TABLE 118. Differentiation of the genus *Spirochaeta* from other genera of spirochetes

Characteristic	*Spirochaeta*	*Cristispira*	*Treponema*	*Borrelia*	*Brevinema*	*Leptospira* and related genera	*Brachyspira*
Free-living	+	–	–[b]	–	–	+	–
Host-associated	–	+	+	+	+	+	+
Obligate aerobes	–	–	–	–	–	+	–
Obligate anaerobes	+	–	+	–	+	–	–
Facultative anaerobes	+	–	–	–	–	–	–
Microaerophiles	–	+	–	+	–	nr	+
Energy and carbon sources:							
Carbohydrates	+	nr	+	+	nr	–	nr
Amino acids	–	nr	+	–	nr	–	nr
Long-chain fatty acids	–	nr	–	–	nr	+	nr
DNA G+C content (mol%)	41–65	nr	36–54	nr	nr	35–53	26

[a]nr, Not reported or not determined.

[b]Some species of the free-living *Spirochaeta* are more closely related to species of *Treponema* based on 16S rRNA sequence comparisons.

spirochetes might represent transitional species, i.e., they could be descendants of precursors of host-associated treponemes. Alternatively, these species may have been from a mammalian host and were disseminated via fecal contamination.

Two other purported species of *Spirochaeta* that belong more with the treponemes have been described. Two thermophilic anaerobic, xylan-degrading spirochetal strains were isolated from a hot spring in Taiwan. They have been provisionally named (unpublished) as *Spirochaeta taiwanensis* (AY735103) and *Spirochaeta xylanolyticus* (AY735097), but are likely the same species with nearly identical 16S rRNA sequences. However, these thermophilic spirochetes are also close relatives of *Spirochaeta caldaria* and consequently are likely members of the genus *Treponema*.

16S rRNA sequences of these *Spirochaeta* species possess a 20–30-base extension at the 5′ end, which is typical of 16S rRNA sequences of species of *Spirochaeta*, *Treponema*, *Leptospira*, and *Leptonema* (Defosse et al., 1995; Paster et al., 1991).

Differentiation of the species of the genus *Spirochaeta*

Sixteen species of *Spirochaeta* are presently known and listed below. Characteristics that differentiate these species are shown in Table 119. *Spirochaeta plicatilis* has not been grown in pure culture, but its ultrastructure and some of its ecological characteristics have been described (Blakemore and Canale-Parola, 1973). Ten species (*Spirochaeta stenostrepta*, *Spirochaeta litoralis*, *Spirochaeta zuelerae*, *Spirochaeta isovalerica*, *Spirochaeta bajacaliforniensis*, *Spirochaeta thermophila*, *Spirochaeta caldaria*, *Spirochaeta smargdinae*, *Spirochaeta asiatica*, and *Spirochaeta americana*) are obligate anaerobes, and two species (*Spirochaeta alkalica* and *Spirochaeta africana*) are aerotolerant anaerobes. Three other species, *Spirochaeta aurantia*, *Spirochaeta halophila*, and *Spirochaeta cellobiosiphila* are facultative anaerobes. When grown aerobically, *Spirochaeta aurantia* and *Spirochaeta halophila* but not *Spirochaeta cellobiosiphila* characteristically produce carotenoid pigments (Breznak and Warnecke, 2008; Greenberg and Canale-Parola, 1975). Most species of *Spirochaeta* are mesophilic, growing at optimum temperatures in the range 15–40°C. However, the thermophilic species, *Spirochaeta thermophila* and *Spirochaeta caldaria*, both from thermal springs, have optimum growth temperature of 66–68 and 48–52°C, respectively.

Two subspecies of *Spirochaeta aurantia* are known. One of these (*Spirochaeta aurantia* subsp. *stricta*) is characterized by significantly narrower coils than the other (*Spirochaeta aurantia* subsp. *aurantia*), and its DNA possesses a slightly lower G+C content (Breznak and Canale-Parola, 1975; Canale-Parola, 1984b).

Spirochaeta stenostrepta, *Spirochaeta zuelerae*, *Spirochaeta caldaria*, and *Spirochaeta aurantia* are freshwater species, whereas *Spirochaeta litoralis*, *Spirochaeta isovalerica*, *Spirochaeta bajacaliforniensis*, *Spirochaeta thermophila*, and *Spirochaeta cellobiosiphila* are marine species and require sodium ion (Na^+) concentrations in the range 200–480 mM for optimal growth (Aksenova et al., 1990, 1992; Breznak and Warnecke, 2008; Fracek and Stolz, 1985; Harwood and Canale-Parola, 1983; Hespell and Canale-Parola, 1970b). *Spirochaeta halophila* was isolated from a high-salinity pond and grows optimally when 750 mM NaCl, 200 mM $MgSO_4$, and 10 mM $CaCl_2$ are present in the medium (Greenberg and Canale-Parola, 1976). Other halophilic species include *Spirochaeta americana*, *Spirochaeta asiatica*, *Spirochaeta alkalica*, and *Spirochaeta africana*, which were isolated from the sediments of hypersaline lakes and require Na^+ concentrations in the range 850–1200 mM for optimal growth (Hoover et al., 2003; Zhilina et al., 1996). The latter three species are also alkaliphilic and growth does not occur below pH 8. *Spirochaeta smargdinae*, isolated from a production water sample collected from an offshore oilfield, requires at least 170 mM NaCl and grows optimally with 850 mM NaCl (Magot et al., 1997).

List of species of the genus *Spirochaeta*

1. **Spirochaeta plicatilis** Ehrenberg 1835, 313[AL]

pli.ca'ti.lis. L. fem. adj. *plicatilis* flexible.

Helical cells, 0.75 µm in diameter and usually 80–250 µm in length. Cells have regular primary coils, which are stable (persist both in the presence and absence of movement). Cells in motion may exhibit broad secondary coils superimposed on the smaller primary coils and when suspended in liquids, they display rotation about the longitudinal axis and wide waves traveling along the length of the organism. Cells creep in contact with solid surfaces (Blakemore and Canale-Parola, 1973).

Regularly spaced cross-walls or transverse septa are present (Blakemore and Canale-Parola, 1973). Long specimens may consist of chains of multicellular spirochetes. Many periplasmic flagella are present, occurring as a bundle wound around the protoplasmic cylinder. Phase-contrast photomicrographs and electron micrographs of the cells have been published (Blakemore and Canale-Parola, 1973).

Not cultivated in pure culture. Presumed to be either a microaerophile or an anaerobe that can tolerate low O_2 tensions. Present in H_2S-containing freshwater, brackish and marine mud, frequently in association with *Beggiatoa* trichomes.

DNA G+C content (mol%): not determined.

Type strain: not yet grown in pure culture.

Sequence accession no. (16S rRNA gene): none.

2. **Spirochaeta africana** Zhilina 1996, 310[VP]

a.fri.ca'na. L. fem. adj. *africana* of African continent, found in African alkaline Lake Magadi.

Motile, helical cells, 0.25–0.3 µm in diameter and 15–30 µm in length, with shorter (7.5 µm) and longer (up to 40 µm) cells occurring in culture. Outermost structure is an outer membrane enclosing periplasmic flagella and a protoplasmic cylinder. Cells have regular, stable primary coils. Cell mass is orange.

Anaerobic, aerotolerant, fermentative; utilizes carbohydrates, mainly mono- and disaccharides, as carbon and energy sources. Preferred substrates: fructose > maltose = trehalose = sucrose > cellobiose > glucose > glycogen > starch; poor growth with mannose or xylose; no growth with galactose, *N*-acetylglucosamine, or ribose (the optical isomers of the sugars were not described). Amino acids do not serve as fermentable substrates. A supplement of vitamins is required; yeast extract can be omitted from culture media. Aerotolerant; develops under a cotton plug in liquid medium.

TABLE 119. Comparison of species of the genus *Spirochaeta*[a]

Characteristic	1. *S. plicatilis*	2. *S. africana*	3. *S. alkalica*	4. *S. americana*	5. *S. asiatica*	6. *S. aurantia*	7. *S. bajacaliforniensis*	8. *S. caldaria*	9. *S. cellobiosiphila*	10. *S. halophila*	11. *S. isovalerica*	12. *S. litoralis*	13. *S. smaragdinae*	14. *S. stenostrepta*	15. *S. thermophila*	16. *S. zuelzerae*
Relationship to O_2	Unknown	O_2An	O_2An	ObAn	ObAn	FAn	ObAn	ObAn	FAn	FAn	ObAn	ObAn	ObAn	ObAn	ObAn	ObAn
DNA G+C content (mol%)	Unknown	57	57	58.5	49	61–65	50	45	41	62	64.5	51	50	60	52	56
Optimum temp. (°C)	Unknown	30–37	33–37	37	33–37	25–30	36	48–52	37	35–40	15–35	30	37	35–37	66–68	37–40
Optimum pH	Unknown	8.8–9.8	8.7–9.6	9.5	8.4–9.4	7.0–7.3	7.5	nd	7.5	7.5	7.5	7.0–7.5	7.0	7.0–7.5	7.5	7.0–8.0
Optimum NaCl (M)	Unknown	0.8–1.4	0.8–1.4	0.5	0.5–1.4	nd	0.12	nd	0.3–0.4	0.75	0.3	0.35	0.85	nd	0.25	nd
Products from glucose	Unknown	A, E, H	A, C, H	A, E, H, F	A, E, L	A, E, C, H	A, E, C, H	A, L, C, H	A, E, C, H	A, E, C, H	A, E, C, H	A, E, C, H	A, L, C, H	A, E, H, C	A, L, C, H	A, L, H, C
Environment	Fresh water, marine	Alkaline, hyper-saline lake	Alkaline, hyper-saline lake	Alkaline, hyper-saline lake	Alkaline, hyper-saline lake	Fresh water	Marine	Fresh water hot spring	Marine	High salinity pond	Marine	Marine	Oil field	Fresh water	Marine hot spring	Fresh water
Treponema 16S rRNA signature	Unknown	–	–	–	–	–	–	+	–	–	–	–	–	+	–	+

[a]Abbreviations: ObAn, obligate anaerobe; FAn, facultative anaerobe; O_2An, oxygen-tolerant anaerobe; A, acetic acid; E, ethanol; H, hydrogen; C, carbon dioxide; L, lactic acid; F, formate; nd, not determined.

Primary products of glucose fermentation are acetate, ethanol, and H_2. Lactate is a minor product in stationary phase.

Halophilic, growing in sodium carbonate medium, but not requiring it. Depends on sodium; no growth below 3% (w/v) or above 10% (w/v) NaCl. Optimal growth at pH 8.8–9.75. No growth at pH 8.0–10.8. Optimum temperature for growth, 30–37°C; range, 15–47°C; slow growth at 6°C after a long lag phase.

Source: a bacterial bloom in the brine under trona (e.g., a sedimentary deposit that results from the evaporation of seawater) from alkaline equatorial Lake Magadi.

DNA G+C content (mol%): 57.1 (T_m).

Type strain: strain Z 7692, ATCC 700263, DSM 8902.

Sequence accession no. (16S rRNA gene): X93928.

3. **Spirochaeta alkalica** Zhilina 1996, 309[VP]

al.ka.li′ca. N.L. n. *alkali* (from Arabic *al-qalyi* the ashes of saltwort), soda ash; L. fem. suff. *-ica* suffix used with the sense of pertaining to; N.L. fem. adj. *alkalica* intended to mean alkaline, developing in the alkaline medium.

Motile, helical cells, 0.4–0.5 µm in diameter and 9–18 µm in length, with shorter (6 µm) and longer (up to 35 µm) cells occurring in culture. Outermost structure is an outer membrane enclosing periplasmic flagella and a protoplasmic cylinder. Cells have regular, stable primary coils. Cell mass is orange.

Anaerobic, aerotolerant, fermentative; utilizes carbohydrates, mainly mono- and disaccharides, as carbon and energy sources. Preferred substrates: sucrose > trehalose > cellobiose > glucose = maltose > xylose > starch; poor growth with fructose, galactose, ribose, or *N*-acetylglucosamine (the optical isomers of the sugars were not described); no growth with mannose or glycogen. Amylolytic and agarolytic. Amino acids do not serve as fermentable substrates. A supplement of vitamins and yeast extract is required. Aerotolerant; growth develops under a cotton plug in liquid medium.

Primary products of glucose fermentation are acetate, H_2, and CO_2. Minor products in stationary phase are ethanol and lactate.

Alkaliphilic; growth in sodium carbonate medium optimally at pH 8.7–9.6. No growth at pH 8.3 or 10.8. Dependent on sodium; no growth below 3% (w/v) or above 10% (w/v) NaCl. Growth is possible when NaCl is substituted by equimolar Na_2CO_3+$NaHCO_3$. Requires carbonate anion. Optimum temperature for growth, 33–37°C; range, 15–44°C; slow growth at 6°C after a long lag phase.

Source: a cyanobacterial mat in a warm spring from under the horst in the equatorial alkaline Lake Magadi.

DNA G+C content (mol%): 57.1 (T_m).

Type strain: strain Z-7491, ATCC 700262, DSM 8900.

Sequence accession no. (16S rRNA gene): X93927.

4. **Spirochaeta americana** Hoover, Pikuta, Bej, Marsic, Whitman, Tang and Krader 2003, 820[VP]

a.me.ri.ca′na. N.L. fem. adj. *americana* of American continent, isolated from soda Mono Lake, California, USA.

Cells are motile and helix-shaped. Flagellum present in periplasmic space. Gram-stain-negative. Cells have regular, unstable primary coils. Sphaeroplasts are formed at the end of the growth phase.

Strictly anaerobic, catalase-negative chemoheterotroph with fermentative type of metabolism. Preferred substrates are D-glucose, fructose, maltose, sucrose, starch, and D-mannitol. Requires vitamins and yeast extract for growth. Primary end products of glucose fermentation is H_2, acetate, ethanol, and formate. Resistant to kanamycin and rifampicin, but sensitive to gentamicin, tetracycline, and chloramphenicol.

Haloalkaliphile that cannot grow at pH 7.0. Growth is dependent upon the presence of carbonate and sodium ions in the medium. No growth occurs below 2% (w/v) NaCl or above 12% (w/v) NaCl. Mesophilic. Cells can be stored frozen in a liquid medium.

Source: mud sediments of the alkaline, hypersaline, meromictic, soda Mono Lake in Northern California, USA.

DNA G+C content (mol%): 58.5 (HPLC).

Type strain: strain ASpG1, ATCC BAA-392, DSM 14872.

Sequence accession no. (16S rRNA gene): AF373921.

5. **Spirochaeta asiatica** Zhilina 1996, 311[VP]

a.si.a′ti.ca. L. fem. adj. *asiatica* from the Asian continent, in the central part of which the organism was found.

Motile, helical cells, 0.2–0.25 µm in diameter and 15–22.5 µm in length, with shorter (7.5 µm) and longer (up to 40 µm) cells occurring in culture. Outermost structure is an outer membrane enclosing periplasmic flagella and a protoplasmic cylinder. Cells have regular, nonstable primary coils. Round bodies, usually nonviable, are formed at the end of the growth period. Nonpigmented.

Strictly anaerobic, fermentative, and utilizes simple and complex carbohydrates. Preferred substrates: glucose > maltose > glycogen > mannose > trehalose > cellobiose > sucrose > starch > galactose > pectin > xylan; poor growth with xylose or arabinose; no growth with fructose, ribose, lactose, agar, or *N*-acetylglucosamine (the optical isomers of the sugars were not described). Amino acids are not fermented. A supplement of vitamins is required; yeast extract enhances growth.

Fermentation products from glucose include acetate, ethanol, and lactate; H_2 not produced.

Haloalkaliphilic; growth in soda solution at optimal pH 8.4–9.4 with limits pH 7.9–9.7. Growth is Na-dependent. No growth below 2% (w/v) NaCl or above 8% (w/v) NaCl. Optimum NaCl concentration, 3–6% (w/v). Requires carbonate anion. Optimum temperature for growth, 33–37°C; range, 20–43°C; broad thermal adaption with prolonged lag phase.

Source: mud of alkaline Lake Khatyn in Tuva, Central Asia.

DNA G+C content (mol%): 49.2 (T_m).

Type strain: strain Z-7591, ATCC 700261, DSM 8901.

Sequence accession no. (16S rRNA gene): X93926.

6. **Spirochaeta aurantia** Canale-Parola 1980, 594[VP]

au.ran′tia. N.L. fem. adj. *aurantia* orange-colored.

Helical cells, 0.3 µm in diameter and 5–50 µm in length. Most cells in cultures measure 10–20 µm in length during exponential growth. Spherical bodies 0.5–2.0 µm in diameter are present, especially in the stationary phase of growth or when the cells are incubated at temperatures unfavorable for growth (e.g., 37°C). The spherical bodies are either in association with cells or free. Each cell has two subterminally inserted periplasmic flagella in a 1:2:1 arrangement. Phase-contrast photomicrographs and electron micrographs of the cells have been published (Breznak and Canale-Parola, 1969, 1975; Canale-Parola et al., 1968).

Colonies on aerobic plates (in media containing 1 g of agar/100 ml; see Breznak and Canale-Parola, 1975) are 1–4 mm in diameter, yellow-orange to orange, round with slightly irregular edges, growing primarily within the agar medium just under the surface, sometimes with a slightly raised center. At low carbohydrate concentrations (see Breznak and Canale-Parola, 1975), the colonies are larger, and they diffuse through the agar medium in the shape of almost perfect circles. Under these growth conditions, the colonies have a lower cell density and their pigmentation is not readily apparent. Anaerobically grown colonies are white. Subsurface anaerobic colonies are spherical, fluffy, 1–3 mm in diameter.

Facultatively anaerobic, having both fermentative and respiratory types of metabolism. Carbohydrates, but not amino acids, are utilized as energy sources for growth (Breznak and Canale-Parola, 1969, 1975). Amino acids but usually not inorganic ammonium salts or nitrates serve as sole nitrogen sources. Exogenous thiamine is required by all strains tested, and riboflavin is required by most strains. Exogenous biotin is required for growth of the type strain and is stimulatory to the growth of other strains (Breznak and Canale-Parola, 1975). Nitrate is reduced to nitrite anaerobically. Oxidase-negative. Weakly catalase-positive. Superoxide dismutase (SOD) is present, and levels of SOD are higher in aerobically grown cells than anaerobically grown cells.

Optimum growth occurs between 25 and 30°C. Slow growth occurs at 15°C and usually no growth occurs at 5°C. There is poor or no growth at 37°C. Optimum growth yields result when the initial pH of the medium is 7.0–7.3.

Cells grown anaerobically, ferment glucose primarily to ethanol, acetic acid, CO_2, and H_2 (Breznak and Canale-Parola, 1969, 1972b). Under aerobic conditions, growing cells oxidize glucose mainly to CO_2 and acetic acid (Breznak and Canale-Parola, 1972a).

Cells growing aerobically produce carotenoid pigments responsible for the yellow-orange to orange color of colonies. The major carotenoid pigment is 1′,2′-dihydro-l′-hydroxytorulene (Greenberg and Canale-Parola, 1975). Nonpigmented mutants have been isolated.

Chemotactic toward carbohydrates, but not toward amino acids.

Based on 16S rRNA sequence comparisons, *Spirochaeta aurantia* branches deeply within the family *Spirochaetaceae* and may warrant separate genus designation (Figure 84).

Source: Water and mud of freshwater ponds and swamps.

DNA G+C content (mol%): 61–65 (Bd).

Type strain: strain J1, ATCC 25082, DSM 1902.

Sequence accession no. (16S rRNA gene): M57740.

6a. **Spirochaeta aurantia subsp. aurantia** Canale-Parola 1980, 594[VP]

The characteristics are as described for the species. Distinguished from the subspecies *stricta* by having a cell wavelength of 2.0–2.8 μm, a wave amplitude of 0.5 μm (the cells have loose coils), and a mol% G+C of 62–65 (Bd).

Type strain: ATCC 25082.

6b. **Spirochaeta aurantia subsp. stricta** subsp. nov.

stric′ta. L. v. *stringere* to draw tight, compress; L. fem. part. adj. *stricta* drawn tight.

The characteristics are as described for the species. Distinguished from the subspecies *aurantia* by having a cell wavelength of 1.1–1.5 μm, a wave amplitude of 0.35 μm (cells have tight coils), and a mol% G+C of 61 (Bd).

Type strain: J4T (Breznak and Canale-Parola, 1975).

7. **Spirochaeta bajacaliforniensis** Fracek and Stolz 2004, 631[VP] (Effective publication: Fracek and Stolz 1985, 324.)

ba.ja.ca.li.for.ni.en′sis. N.L. fem. adj. *bajacaliforniensis* of or belonging to Baja California for the geographical location from where it was isolated.

Helical cells, 0.2–0.3 μm by 15–45 μm. Shorter (10 μm) and longer (up to 300 μm) cells may occur. Highly motile. The amplitude is 0.5 μm and the wavelength is 1.0–1.5 μm. Two subterminally inserted periplasmic flagella are present in a 1:2:1 arrangement. The ratio of protoplasmic cylinder diameter to the cell diameter is 2:3. The surface of the protoplasmic cylinder has a characteristic polygonal pattern. Subsurface colonies are fluffy, white, and spherical.

Strictly anaerobic. Arabinose, cellobiose, galactitol, fructose, galactose, gluconate, glucose, inulin, lactose, malate, maltose, mannitol, pyruvate, rhamnose, sorbose, and trehalose are fermented (the optical isomers of the sugars were not described (Fracek and Stolz, 1985). Products of glucose fermentation are acetate, ethanol, CO_2, and H_2. No test for formate was performed. Catalase-negative.

Grow in media containing at least 20% seawater and 0.12 M NaCl. Reducing agent required in liquid medium. Growth on solid medium occurs in an 80% N_2, 17% CO_2, and 3% H_2 atmosphere.

Temperature optimum: 36°C. Do not grow below 25°C or above 44°C. Optimum growth at pH 7.5.

Source: anaerobic sulfide-rich mud underlying the laminated sediment of the microbial mats at North Pond, Laguna Figueroa, Baja California Norte, Mexico.

DNA G+C content (mol%): 50.1 (Bd).

Type strain: strain BA-2, ATCC 35968, DSM 16054.

Sequence accession no. (16S rRNA gene): M71239.

8. **Spirochaeta caldaria** Pohlschroeder 1994, 21[VP]

cal.da′ri.a. L. fem. adj. *caldaria* of warm water, inhabiting warm water.

Motile, helical cells, 0.2–0.3 μm in diameter and mostly 15–45 μm in length, the outermost structure being an outer membrane (outer sheath) enclosing the periplasmic flagella in a 1:2:1 arrangement and the protoplasmic cylinder. Cells have regular, stable primary coils. Broader secondary coils occasionally are present in cells in motion. Subsurface colonies (in media containing 1 g agar per 100 ml) are white, fluffy, cotton ball-like, approximately 2–3 mm in diameter.

Obligately anaerobic, fermentative, utilizes carbohydrates as carbon and energy sources. Amino acids do not serve as fermentable substrates for growth. Fermentable compounds include L-arabinose, D-galactose, D-glucose, D-mannose, D-fructose, D-xylose, cellobiose, cellotriose, cellotetraose, lactose, maltose, sucrose, and starch. The following substances are not fermented: D-ribose, mannitol, cellulose, xylan, glycerol, peptone, casein hydrolysate, and sodium acetate. Exogenous fatty acids, reported to be needed by *Treponema* species for cellular lipid synthesis and growth (Livermore and Johnson, 1974; Miller et al., 1991), are not required. A supplement of vitamins is required. Specific vitamin requirements have not been determined. Products of D-glucose fermentation are H_2, CO_2, acetate, and lactate.

Grows in the presence of rifampicin (100 µg/ml of medium). Growth is inhibited by penicillin G, neomycin, chloramphenicol, or tetracycline (10 µg/ml of medium each).

Thermophilic. Grows optimally between 48 and 52°C. No growth at 25 or 60°C. No growth in the presence of 0.4% NaCl or higher NaCl concentrations.

Source: freshwater hot springs.

DNA G+C content (mol%): 45 (T_m).

Type strain: strain H1, ATCC 51460, DSM 7334.

Sequence accession no. (16S rRNA gene): M71240, EU580141.

9. **Spirochaeta cellobiosiphila** Breznak and Warnecke 2008, 2762[VP]

cel.lo.bi.o.si'phi.la. N.L. neut. n. *cellobiosum* cellobiose; N.L. fem. adj. *phila* from Gr. fem. adj. *philê* friendly to, loving; N.L. fem. adj. *cellobiosiphila* loving cellobiose, isolated from a microbial mat, Little Sippewissett salt marsh, Woods Hole, MA, USA.

Cells are pale yellow, 0.3–0.4×10–12 µm in size and helical, with a body pitch of 1.4 µm. Motile by means of two (occasionally four) periplasmic flagella, of which one (or two) is inserted near each end of the cell. Facultatively anaerobic and catalase-negative. Growth occurs at 9–37°C (optimally at or near 37°C), at initial pH 5–8 (optimally at initial pH 7.5) and in media prepared with 20–100% (v/v) seawater (optimally at 60–80%) or containing 0.10–1.00 M NaCl (optimally at 0.30–0.40 M). A variety of monosaccharides and disaccharides and pectin (but not cellulose or arabinoxylan) are used as energy sources; the most rapid growth occurs on cellobiose. Neither organic acids nor amino acids are utilized as energy sources. One or more amino acids in tryptone and one or more components of yeast extract are required for growth. The products of cellobiose fermentation are acetate, ethanol, CO_2, H_2, and small amounts of formate. Aerated cultures oxidize cellobiose incompletely to acetate (and, presumably, CO_2) plus small amounts of ethanol and formate; they exhibit a $Y_{cellobiose}$ value that is 1.2-fold greater than that of cellobiose-fermenting cultures.

Source: interstitial water of a cyanobacteria-containing microbial mat collected from Little Sippewissett salt marsh, Woods Hole, MA, USA.

DNA G+C content (mol%): 41.4 (HPLC).

Type strain: strain SIP1, ATCC BAA-1285, DSM 17781.

Sequence accession no. (16S rRNA gene): EU448140.

10. **Spirochaeta halophila** Greenberg and Canale-Parola 1976, 193[AL]

ha.lo.phi'la. Gr. n. *hals, halos* salt; Gr. adj. *philus -ê -on* loving; N.L. fem. adj. *halophila* salt-loving.

Helical cells, 0.4 µm in diameter and 15–30 µm in length. Some of the cells in cultures are as short as 5 µm and as long as 60 µm. Cells have regular, stable primary coils. Spherical bodies 1–2 µm in diameter occur in cultures, especially in the stationary phase of growth or during growth at unfavorable temperatures (e.g., 45°C). Each cell has two subterminally inserted periplasmic flagella that overlap in the central region of the cell (1:2:1 arrangement). Phase-contrast micrographs and electron micrographs of the cells have been published (Greenberg and Canale-Parola, 1976).

Colonies growing aerobically on ISM plates (0.75 g of agar/100 ml of medium) are red, round, with areas of diffuse growth at their periphery, and usually 2–6 mm in diameter (after 5 d at 35°C). Each colony grows partially above and partially below the surface of the agar medium. Anaerobically grown colonies are white. When cells are streaked onto agar medium plates and incubated anaerobically, the colonies grow below the surface of the medium and are spherical, diffuse and white.

Facultatively anaerobic, having both respiratory and fermentative types of metabolism. Carbohydrates, but not amino acids, are utilized as energy sources for growth (Greenberg and Canale-Parola, 1976). Cells have specific growth requirements for relatively high concentrations of Na^+, Cl^-, Ca^{2+}, and Mg^{2+} (Greenberg and Canale-Parola, 1976). Optimum cell yields result when 0.75 M NaCl, 0.2 M $MgSO_4$, and 0.01 M $CaCl_2$ are included in growth media containing (in g/100 ml) a carbohydrate (0.5), peptone (0.2), and yeast extract (0.4). No growth occurs when any one of the three inorganic salts is omitted from the medium (e.g., ISM medium). Nitrate is reduced to nitrite anaerobically. Catalase-negative. Cells growing anaerobically ferment glucose primarily to ethanol, acetic acid, CO_2, and H_2 (Greenberg and Canale-Parola, 1976). Under aerobic conditions, growing cells oxidize glucose mainly to CO_2 and acetic acid (Greenberg and Canale-Parola, 1976).

Optimum temperature, 35–40°C. Poor growth occurs at 45°C and no growth occurs at 22°C. Cells growing aerobically produce carotenoid pigments responsible for the red color of the colonies. The major carotenoid pigment is 4-keto-l′,2′-dihydro-l′-hydroxytorulene (Greenberg and Canale-Parola, 1975). Nonpigmented mutants, occurring spontaneously in cultures, have been isolated.

Source: H_2S-containing mud of a high salinity pond (Solar Lake) located on the Sinai shore of the Gulf of Elat.

DNA G+C content (mol%): 62 (T_m, Bd).

Type strain: strain RS1, ATCC 29478, DSM 10522.

Sequence accession no. (16S rRNA gene): M88722.

11. **Spirochaeta isovalerica** Harwood and Canale-Parola 1983, 578[VP]

i.so.va.le.ri'ca. N.L. n. *acidum isovalericum* isovaleric acid; N.L. fem. adj. *isovalerica* intended to mean pertaining to isovaleric acid, the major acid formed as a fermentation product from branched-chain amino acids.

Helical cells, 0.4 µm in diameter and 10–15 µm in length. Shorter (6 µm) and longer (up to 50 µm) cells occur in cultures. Motile. Two subterminally inserted periplasmic (axial) fibrils are present in a 1:2:1 arrangement. The protoplasmic cylinder and periplasmic fibrils are enclosed in an outer sheath. Gram-stain-negative.

Chemo-organotrophic. Grows (final yield, 6–9×10^8 cells per ml) in medium containing yeast extract, trypticase, peptone, a carbohydrate, and inorganic salts under anaerobic conditions. No growth occurs aerobically. Subsurface colonies in medium containing 0.8% agar (Difco) are spherical and white, resembling cotton balls in appearance. Growth (final yield, 3–6×10^8 cells per ml) occurs in a chemically defined medium containing glucose, cysteine or sulfide, asparagine, vitamins, 0.3 M NaCl, 0.05 M $MgSO_4$, 0.01 M KCl, 0.01 M $CaCl_2$, and trace elements. Exogenously supplied vitamins are not required for growth, but a mixture of vitamins stimulates growth. Fails to grow when NaCl is omitted from the medium. Growth is stimulated by $MgSO_4$ and $CaCl_2$. Inorganic ammonium salts or amino acids serve as nitrogen sources. Carbohydrates, but not amino acids, serve as carbon and energy sources for growth.

The major products of glucose fermentation are CO_2, H_2, acetate, and ethanol. Cells have the ability to generate ATP by catabolizing L-leucine, L-isoleucine, and L-valine to form isovalerate, 2-methylbutyrate, and isobutyrate, respectively, as end products. ATP formed in this way is utilized by cells to prolong survival during periods of growth substrate starvation. Smaller amounts of isobutanol and isoamyl alcohol are also formed by cells from valine, isoleucine, and leucine. Catalase-negative. Nitrate not reduced to nitrite.

Optimum temperature range, 15–35°C. Poor growth at 5 and 39°C.

Source: anoxic marine marsh mud.

DNA G+C content (mol%): 63.6–65.6 (T_m).

Type strain: strain MA-2, ATCC 33939, DSM 2461.

Sequence accession no. (16S rRNA gene): M88720.

12. **Spirochaeta litoralis** Canale-Parola 1980, 594[VP]

li.to.ra'lis. L. fem. adj. *litoralis* of the shore.

Helical cells, 0.4–0.5 μm in diameter and 5.5–7.0 μm in length. The cells are regularly and tightly coiled during the exponential phase of growth. Spherical bodies (2.0–3.5 μm in diameter) are present in the stationary growth phase under unfavorable growth conditions (e.g., in the presence of O_2). Two subterminally inserted periplasmic flagella are present in a 1:2:1 arrangement. Phase-contrast photomicrographs and electron micrographs of the cells have been published (Hespell and Canale-Parola, 1970b; Joseph and Canale-Parola, 1972).

Subsurface colonies in agar media are spherical, fluffy, cream colored, 1–5 mm in diameter. Surface colonies (anaerobic) are round, growing partially within the agar medium, cream colored, and 2–5 mm in diameter.

Obligately anaerobic, having a fermentative type of metabolism. Various carbohydrates are fermented (Hespell and Canale-Parola, 1970b). Main products of glucose fermentation are ethanol, acetic acid, CO_2, H_2, and trace amounts of lactic, formic, and pyruvic acids (Hespell and Canale-Parola, 1970b, 1973). Nitrite is not accumulated in the medium by cells growing in the presence of nitrate. Catalase-negative.

Cells grow in media prepared with seawater, but do not grow in media prepared with freshwater unless NaCl is added (minimum concentration, 0.05 M; optimum, 0.35 M). Cells have specific requirements for Na^+ and Cl^-. Exogenous supplements of biotin, niacin, and coenzyme A are required for growth. Coenzyme A may be replaced by pantothenate, but the resulting cell yields are low. Added thiamine is stimulatory for growth. A reducing agent (e.g., sulfide or cysteine) is required for growth in laboratory media. Cells grow in chemically defined media containing glucose, $(NH_4)_2SO_4$ or amino acids, sulfide, NaCl, vitamins, coenzyme A, and inorganic salts (Hespell and Canale-Parola, 1970b).

Optimum temperature, near 30°C. Growth occurs slowly at 15°C and not at all at 5°C or 40°C. Optimum growth yields result when the initial pH of the medium is between 7.0 and 7.5.

Source: sulfide-containing marine mud.

DNA G+C content (mol%): 51 (Bd).

Type strain: strain RI, ATCC 27000, DSM 2029.

Sequence accession no. (16S rRNA gene): M88723.

13. **Spirochaeta smaragdinae** Magot, Fardeau, Arnauld, Lanau, Ollivier, Thomas and Patel 1997, 190[VP]

sma.rag.di'nae. L. masc. n. *smaragdus* emerald; N.L. gen. n. *smaragdinae* intended to mean from Emerald, the name of the oilfield in Congo, Central Africa.

Spirochetes with corkscrew-like motility. Spiral cells are 0.3–0.5 μm in diameter and 5–30 μm in length with two periplasmic flagella in a 1:2:1 arrangement. Produce translucent colonies with regular edges and a diameter of 0.5 mm after 2 weeks at 37°C.

Obligately anaerobic chemo-organotroph. Growth occurs in the presence of fructose, galactose, D-xylose, D-glucose, ribose, D-mannose, mannitol, glycerol, yeast extract, biotrypticase, and fumarate, but not with D-arabinose, rhamnose, sorbose, L-xylose, sucrose, maltose, acetate, butyrate, propionate, pyruvate, lactate, or Casamino acids. Yeast extract is required for growth and cannot be replaced by a vitamin mixture. Thiosulfate and elemental sulfur are reduced to sulfide. Glucose is oxidized to lactate, acetate, CO_2, and H_2S in the presence of thiosulfate, and to lactate, ethanol, CO_2, and H_2 in its absence. Fumarate is fermented to acetate and succinate.

Obligately halophilic. Optimum NaCl concentration for growth is 5% and the NaCl concentration range for growth is 1.0–10%. Optimum temperature for growth is 37°C with growth occurring between 20 and 40°C. Optimum pH is 7.0 with growth occurring between pH 5.5 and 8.0. Doubling time in the presence of glucose and thiosulfate under optimal conditions is about 25 h.

Source: oil-injection water from Emerald oilfield in Congo, Central Africa.

DNA G+C content (mol%): 50 (HPLC).

Type strain: strain SEBR 4228, DSM 11293, JCM 15392.

Sequence accession no. (16S rRNA gene): U80597.

14. **Spirochaeta stenostrepta** Zuelzer 1912, 17[AL]

ste.no.strep'ta. Gr. adj. *stenos* narrow; Gr. adj. *streptos -ê -on* pliant, easily bent; N.L. fem. adj. *stenostrepta* tightly coiled.

Helical cells, 0.2–0.3 μm in diameter and 15–45 μm in length. Some of the cells in cultures are shorter than 15 μm. In the late exponential and stationary phases the organisms increase in length (up to 300 μm). Long organisms occasionally pair and become entwined, or a single organism becomes partially wrapped around itself. The cells have regular, stable primary coils. Cells in motion occasionally exhibit broader secondary coils or waves superimposed on the smaller primary coils. Spherical bodies generally 1–3 μm in diameter are occasionally observed in cultures. The spherical bodies occur either free or in association with helical cells. Each cell has two subterminally inserted periplasmic flagella that overlap in the central region of the cell (1:2:1 arrangement). Phase-contrast photomicrographs and electron micrographs of the cells have been published (Canale-Parola et al., 1967, 1968; Holt and Canale-Parola, 1968).

Subsurface colonies (in GYPT medium containing 1.5 g agar/100 ml) are white, spherical, fluffy, and approximately 2–3 mm in diameter when fully developed. Smaller spherical colonies may occasionally lack the characteristic fluffiness.

Obligately anaerobic, having a fermentative type of metabolism. Various carbohydrates are fermented (Hespell and Canale-Parola, 1970a). The main products of glucose fermentation are ethanol, acetic acid, CO_2, H_2, and smaller amounts of lactic acid (Canale-Parola et al., 1967, 1968; Hespell and Canale-Parola, 1970a). Catalase-negative.

Growth reported only on complex media. Minimal growth requirements are unknown. Growth occurs between 15 and 40°C; optimum temperature is 35–37°C. Optimum growth yields result when the initial pH of the medium is between 7.0 and 7.5.

Based on 16S rRNA sequence comparisons, *Spirochaeta stenostrepta* does not belong with other free-living *Spirochaeta*, but is more closely related with members of the genus *Treponema* (Figure 84).

Source: H_2S-containing mud of a freshwater pond (Canale-Parola et al., 1967, 1968).

DNA G+C content (mol%): 60 (Bd).

Type strain: strain Zl, ATCC 25083, DSM 2028.

Sequence accession no. (16S rRNA gene): M88724.

15. **Spirochaeta thermophila** Aksenova, Rainey, Janssen, Zavarzin and Morgan 1992, 176[VP]

ther.mo'phi.la. Gr. n. *thermê* heat; Gr. fem. adj. *philê* loving; N.L. fem. adj. *thermophila* heat-loving.

Helical cells, 0.2–0.25 μm in diameter and 16–50 μm in length. An outer sheath encloses a protoplasmic cylinder; two periplasmic flagella in a 1:2:1 arrangement are subterminally anchored by an insertion disc. No cell lysis in 3% KOH.

Strictly anaerobic chemo-organotroph. Utilizes various mono-, di-, and polysaccharides but not sugar alcohols, organic acids, or amino acids. Glucose is fermented via the Embden–Meyerhof–Parnas pathway, involving a pyrophosphate-dependent phosphofructokinase. Fermentation end products from glucose are acetate, CO_2, H_2, and lactate. Ethanol and succinate not produced. No reduction of fumarate, nitrate, oxygen, sulfate, or sulfur. Indole not formed; urea not hydrolyzed. Sulfide not produced from cysteine; esculin hydrolyzed. Inhibited by penicillin, neomycin, erythromycin, tetracycline, polymyxin B, and novobiocin but resistant to rifampicin and streptomycin.

Temperature range for the type strain, 40–73°C (optimum, 66–68°C). pH range for the type strain, 5.9–7.7 (optimum, 7.5). NaCl concentration range for the type strain, 0.5–4.5% (optimum, 1.5%). Doubling time, 70 min. Temperature, pH, and salinity parameters vary for different strains, reflecting the environmental conditions prevailing at the sites of isolation.

Source: marine hot spring near the beach on Shiashkotan Island, Soviet Far East, USSR.

DNA G+C content (mol%): 52 (T_m).

Type strain: strain Z-1203, ATCC 700085, DSM 6578.

Sequence accession no. (16S rRNA gene): X62809, L09180.

16. **Spirochaeta zuelzerae** Canale-Parola 1980, 594[VP] (*Treponema zuelzerae* Veldkamp 1960, 122)

zu.el.ze'rae. N.L. fem. gen. n. *zuelzerae* of Zuelzer, named after Margarete Zuelzer, who described the occurrence of morphologically diverse spirochetes in sulfide-containing environments.

Helical cells, 0.2–0.35 μm in diameter and 8–16 μm in length. Shorter cells (as short as 2–3 μm) are occasionally observed in cultures. Long organisms (up to 80 μm) are present in old cultures. Exponentially growing cells have fairly regular, stable primary coils. Secondary coils or waves are present infrequently. Spherical bodies, generally not exceeding 3–4 μm in diameter, are formed usually at the ends of the cells in the stationary phase of growth. Two subterminally inserted periplasmic flagella are present in a 1:2:1 arrangement. Phase-contrast photomicrographs and electron micrographs of the cells have been published (Canale-Parola et al., 1968; Joseph and Canale-Parola, 1972).

The type strain was originally isolated from an enrichment culture for green photosynthetic bacteria that had been inoculated with sulfide-containing mud from a freshwater pond (Veldkamp, 1960). Subsurface colonies in agar media (for composition of medium, see Veldkamp (1960), and also SZ medium in table) are white, fluffy, spherical, with a tendency to diffuse in the agar medium. Disc-shaped colonies are present occasionally.

Obligately anaerobic, having a fermentative type of metabolism. Various carbohydrates are fermented (Veldkamp, 1960). Cells growing in media containing 0.05% $NaHCO_3$ ferment glucose mainly to acetic, lactic, and succinic acids, CO_2, and H_2 (Veldkamp, 1960). Catalase-negative.

Growth occurs at 20°C but not at 45°C. The optimum temperature range is 37–40°C. Growth is optimum when the initial pH of the medium is between 7 and 8. Inorganic ammonium salts or nitrates are not utilized as sole nitrogen sources. Added CO_2 is an absolute requirement for growth (Veldkamp, 1960). Growth has been reported only on complex media. The minimal growth requirements are unknown.

Cells have a protein antigen that gives a positive complement fixation reaction with syphilitic serum.

Based on 16S rRNA sequence comparisons, *Spirochaeta zuelzerae* does not belong with other free-living *Spirochaeta*, but is more closely related with members of the genus *Treponema* (Figure 84).

DNA G+C content (mol%): 56 (Bd).

Type strain: ATCC 19044, DSM 1903.

Sequence accession no. (16S rRNA gene): M88725.

Other organisms

1. **Spirochaeta coccoides** Dröge, Fröhlich, Radek and König 2006, 1460[VP] (Effective publication: Dröge, Fröhlich, Radek and König 2006, 396.)

coc.co'i.des. Gr. n. *kokkos* a berry; L. fem. suff. *-oides* [from Gr. suff. *-eides* (from Gr. n. *eidos* that which is seen, form, shape, figure)], resembling, similar. N.L. fem. adj. *coccoides* berry-shaped.

Placement within the genus *Spirochaeta* is questionable owing to the absence of defining morphological, ultrastructural, and behavioral features (i.e., helically shaped motile cells with protoplasmic cylinder and periplasmic fibrils enclosed in an outer sheath). In this regard, this species resembles isolates from freshwater sediments referred to as "free-living pleomorphic spirochetes" or FLiPS (Ritalahti and Löffler, 2004). 16S rRNA sequence analysis indicates that these microbes form a cluster within the *Spirochaeta–Treponema* branch of the *Spirochaetaceae* tree, and should be considered for separate taxonomic status.

Type strain: strain SPN1, DSM 17374, ATCC BAA-1237.

Sequence accession no. (16S rRNA gene): AJ698092.

It has long been known that greater diversity exists among the free-living spirochetes capable of anaerobic growth than is reflected by the species presently recognized in the genus *Spirochaeta*. For example, a free living, strictly anaerobic spirochete (strain Z4), resembling *Spirochaeta zuelzerae* morphologically, but differing in certain physiological properties, has been isolated from freshwater mud (Canale-Parola et al., 1968). The spirochete ferments glucose to acetic, lactic, succinic, and formic acids, ethanol, CO_2, and H_2. Unlike *Spirochaeta zuelzerae*, it does not require an exogenous source of CO_2 and the mol% G+C of its DNA is 59.2 (Bd).

Additionally, diverse obligate anaerobic thermophilic strains of *Spirochaeta* have been isolated from thermal springs and other high temperature environments. For example, a strain isolated from a hydrothermal spring on Raoul Island of the Kermadec archipelago of New Zealand exhibited maximum growth at 73°C (Rainey et al., 1991). This isolate fermented a range of mono-, di-, and polysaccharides, including cellulose.

Various strains of facultatively and obligately anaerobic spirochetes have been isolated from marine environments, including intertidal muds and water. An obligately anaerobic strain was isolated from water collected at a depth of 2550 m near the Galápagos hydrothermal vents in the Pacific Ocean (Harwood et al., 1982). These isolates are indigenous to marine environments inasmuch as they have Na^+ requirements typical of marine bacteria. The facultatively anaerobic marine isolates form either white or yellow colonies. Thus, they differ both in salt requirements and in pigmentation from the facultatively anaerobic species *Spirochaeta aurantia*, which is a freshwater species and forms orange colonies, and *Spirochaeta halophila*, which requires high concentrations of Ca^+ and Mg^+ and forms red colonies. However, 16S rRNA sequences are not available for these strains to determine their phylogenetic relationships with other species of *Spirochaeta*.

Diversity of not-yet-cultivated species. The breadth of diversity of species of *Spirochaeta* is impressive. By cloning and sequencing of 16S rRNA genes amplified from DNA isolated from a wide variety of environments, over 100 phylotypes of not-yet-cultivated species of *Spirochaeta* have been identified. These environments include the following: (1) microbial mats from deltas in France and Spain (Berlanga et al., 2003), Puerto Rico, and California; (2) hypersaline soda lakes in Asia, Africa, Egypt, and California; (3) volcano seep sediments in Greece and Japan; (4) marine sediments in China and Japan; (5) gray whale bone from deep-sea; (6) *Alvinella pompeiana* white tube worms and other gutless worms from deep-sea vents; (7) forested wetlands; (8) fresh water ponds or lakes in China, Japan, Greece, and Massachusetts, USA; (9) karyomastigonts (protozoa); and (10) termite hindguts. Phylogenetic analysis of representatives of these taxa and their GenBank accession numbers are shown in Figure 85.

Notably, some of these not-yet-cultivated species of *Spirochaeta* are "host-associated", i.e., found in termite hindguts (Figure 85). In contrast, most termite spirochetes fall within the genus *Treponema* (see chapter on *Treponema*). Conversely, as discussed above, there are several "free-living" species of *Spirochaeta*, e.g., *Spirochaeta zuelzerae*, *Spirochaeta stenostrepta*, and *Spirochaeta caldaria*, that are more closely related to "host-associated" members of the genus *Treponema* than to other "free-living" members of the genus *Spirochaeta*. Consequently, these data suggest that "free-living" vs "host-associated" designations may not be valid taxonomic criteria to differentiate species of *Spirochaeta* and *Treponema*.

Genus II. **Borrelia** Swellengrebel 1907, 582[AL]

Guiqing Wang and Ira Schwartz

Bor.re'li.a. N.L. fem. n. *Borrelia* named after Amedée Borrel (1867–1936).

Helical cells are 0.2–0.5 mm by 3–30 mm, composed of 3–10 loose coils. The cells are surrounded by a surface layer, an outer membrane, periplasmic flagella, and a protoplasmic cylinder (Figure 86). Typically, 15–20 periplasmic flagella (which also have been termed **endoflagella**, axial fibrils, or periplasmic fibrils) originate at each end of the cell and wind about the protoplasmic cylinder to overlap in the middle of the cell. The protoplasmic cylinder consists of a peptidoglycan layer and an inner membrane which encloses the internal components of the cells. **The cells are actively motile** with frequent reversal of the direction of translational movement. Gram-stain-negative. Stain well with Giemsa stain. Species which have been grown *in vitro* are **microaerophilic**. Nutritional requirements for *in vitro* growth are complex. **Arthropod-borne pathogens** of man, other mammals, and birds. The **causative agents of tick-borne Lyme disease and relapsing fever and louse-borne relapsing fever in man**.

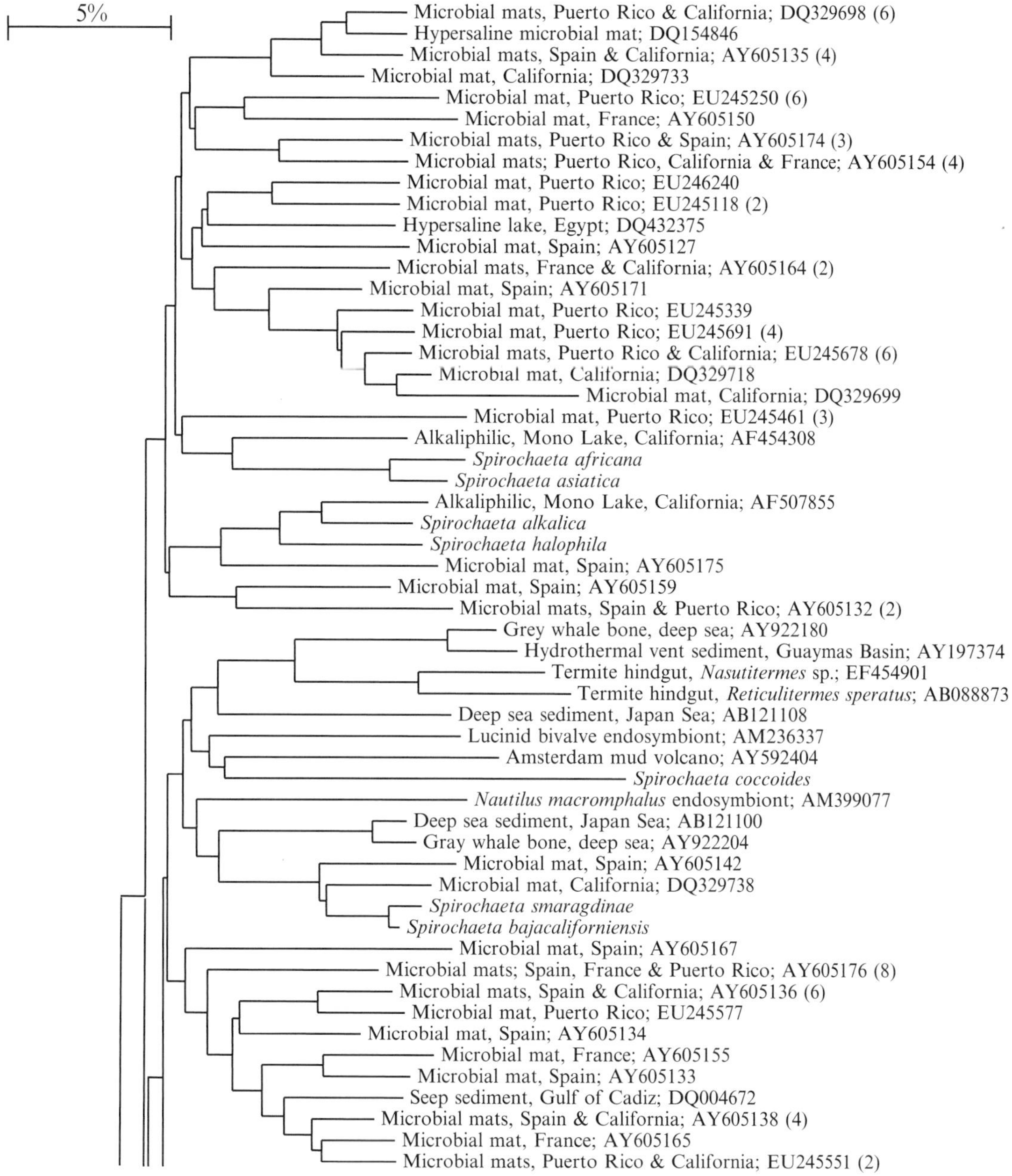

FIGURE 85. Phylogenetic tree illustrating the diversity of cultivable and not-yet-cultivated species of the genus *Spirochaeta* based on 16S rRNA sequence comparisons. Not-yet-cultivated species are noted by their environmental source, location, and GenBank accession numbers. Numbers in parentheses indicate the number of sequences available for a given phylotype. The scale bar represents a 5% difference in nucleotide sequence.

DNA G+C content (mol%): 27–32.

Type species: **Borrelia anserina** (Sakharoff 1891) Bergey, Harrison, Breed, Hammer and Huntoon 1925, 435[AL].

Further descriptive information

DNA sequence analysis of rRNA and other conserved genes (e.g., *fla*, *hbb*) has established that *Borrelia* spp. fall into two major groups (Figure 87). The first major group contains the agents of Lyme borreliosis that were first isolated from *Ixodes scapularis* ticks (Burgdorfer et al., 1982). These include the three human-pathogenic species *Borrelia burgdorferi* (Johnson et al., 1984), *Borrelia afzelii* (Canica et al., 1993), and *Borrelia garinii* (Baranton et al., 1992) and seven other species that are minimally pathogenic or nonpathogenic: *Borrelia japonica* (Kawabata et al., 1993), *Borrelia lusitaniae* (Le Fleche et al., 1997), *Borrelia tanukii*, *Borrelia turdi* (Fukunaga et al., 1996a), *Borrelia sinica* (Masuzawa et al., 2001), *Borrelia spielmanii* (Richter et al., 2006), and *Borrelia valaisiana* (Wang et al., 1997). Spirochetes in the first major group are all transmitted by hard ticks in the *Ixodes ricinus* complex, and are referred to collectively as "*Borrelia burgdorferi sensu lato*" (Baranton et al., 1992; Wang et al., 1999). The second major group includes more than 20 *Borrelia* species associated with relapsing fever that are mainly transmitted by soft-bodied or argasid ticks, with the exception

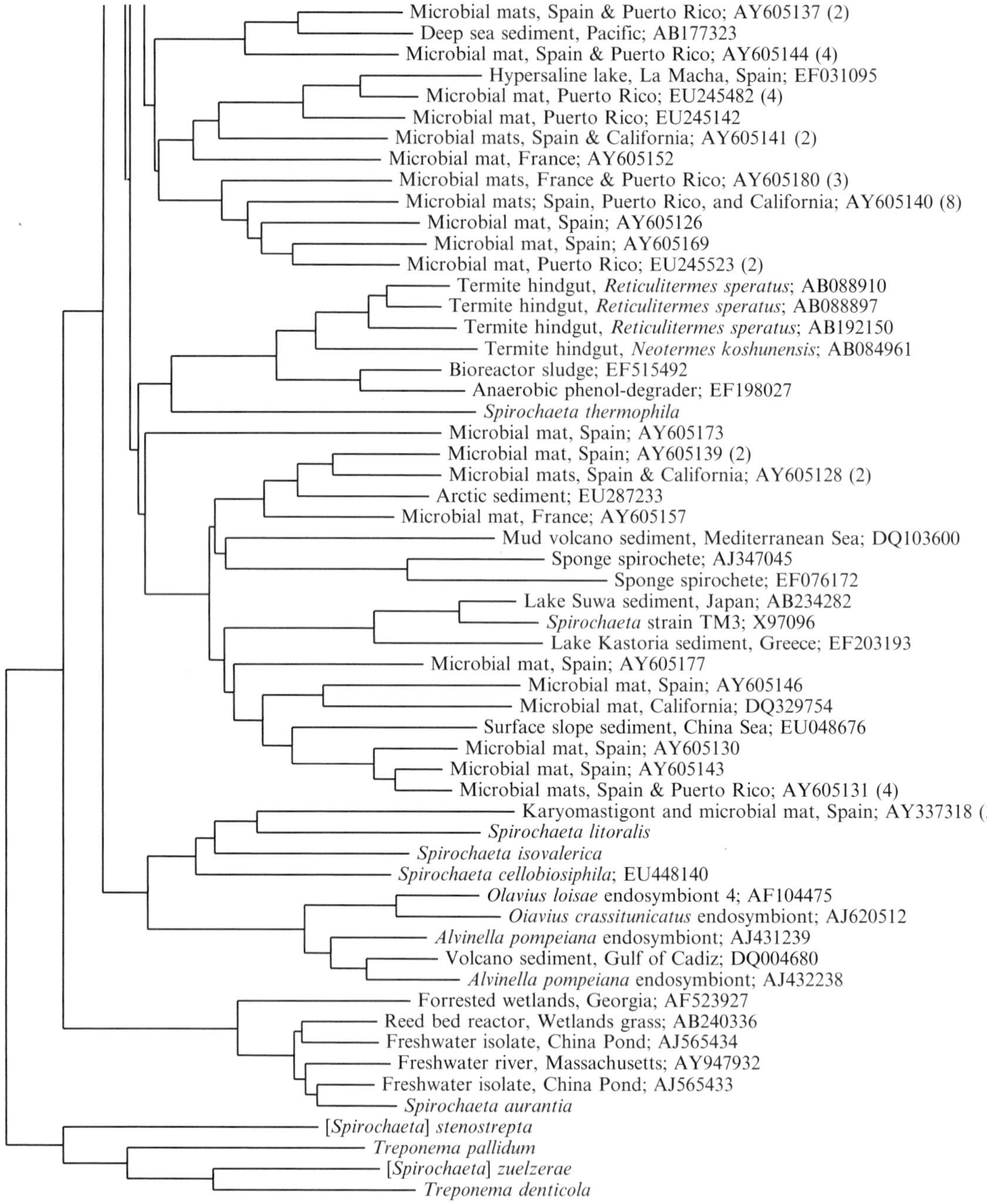

FIGURE 85. (continued)

of louse-borne *Borrelia recurrentis* (Table 121). Additional novel *Borrelia* species or strains closely related to the Lyme disease or relapsing fever borreliae have been described. Their taxonomic status and pathogenicity in humans remain to be determined.

Cells of all species are helically shaped with similar cell morphology and ultrastructure (Barbour and Hayes, 1986; Cutler, 2001; Wang et al., 2001). Among these, the Lyme disease borreliae are the longest (20–30 μm) and narrowest (0.2–0.3 μm).

Borrelia burgdorferi periplasmic flagella have both skeletal and motility functions. Inactivation of the gene encoding the major periplasmic flagellar filament protein FlaB results in nonmotile cells that are also rod-shaped rather than helical (Motaleb et al., 2000). Further, mutants may show only asymmetrical flagellar rotation in the nonchemotactic mutants (Li et al., 2002).

The lipid compositions of the outer membrane and whole cell are highly similar, suggesting that bulk transfer of lipid occurs between the cytoplasmic and outer membranes. The isolated outer membrane of *Borrelia burgdorferi* has a specific gravity of 1.12–1.19 g/cm^3, depending on the purification procedure used (Bledsoe et al., 1994; Radolf et al., 1995); this comprises approximately 16.5% of the whole spirochete by dry weight. Chemical analysis of the outer envelope of *Borrelia burgdorferi* revealed a composition of 45.9% protein, 50.8% lipid, and 3.3% carbohydrate (Coleman et al., 1986). The purified periplasmic

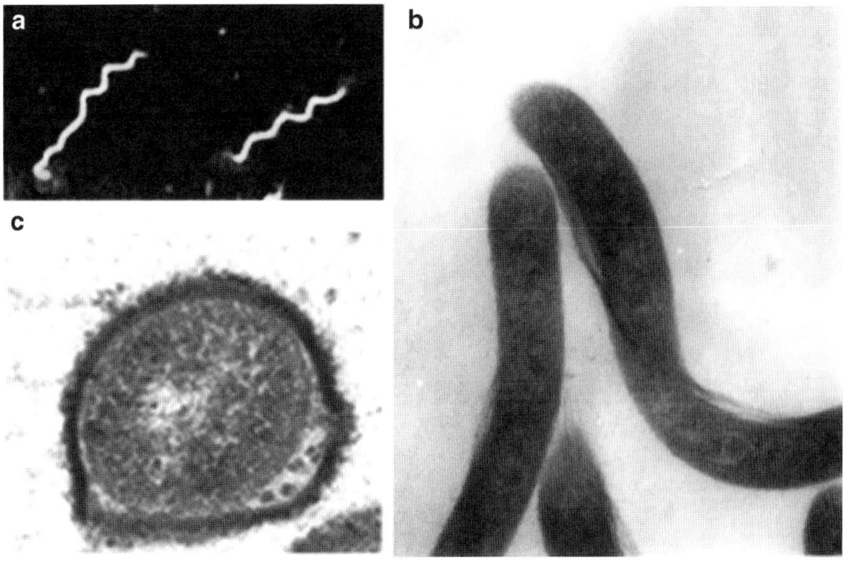

FIGURE 86. Micrographs of *Borrelia burgdorferi* by light microscopy (a) and high voltage electron microscopy (b). The cell diameter of *Borrelia burgdorferi* strain 297 shown in electron microscopy is 0.33 µm. The bundle of periplasmic flagella is clearly visible in the cell on the right (the EM micrograph was provided by K. Buttle, S.F. Goldstein and N.W. Charon). (c) Cross-section of *Borrelia burgdorferi* strain B31. The periplasmic flagella are visible in the lower right of the cell (micrograph provided by N.W. Charon, West Virginia University). Reprinted from Wang et al. (2001), with permission.

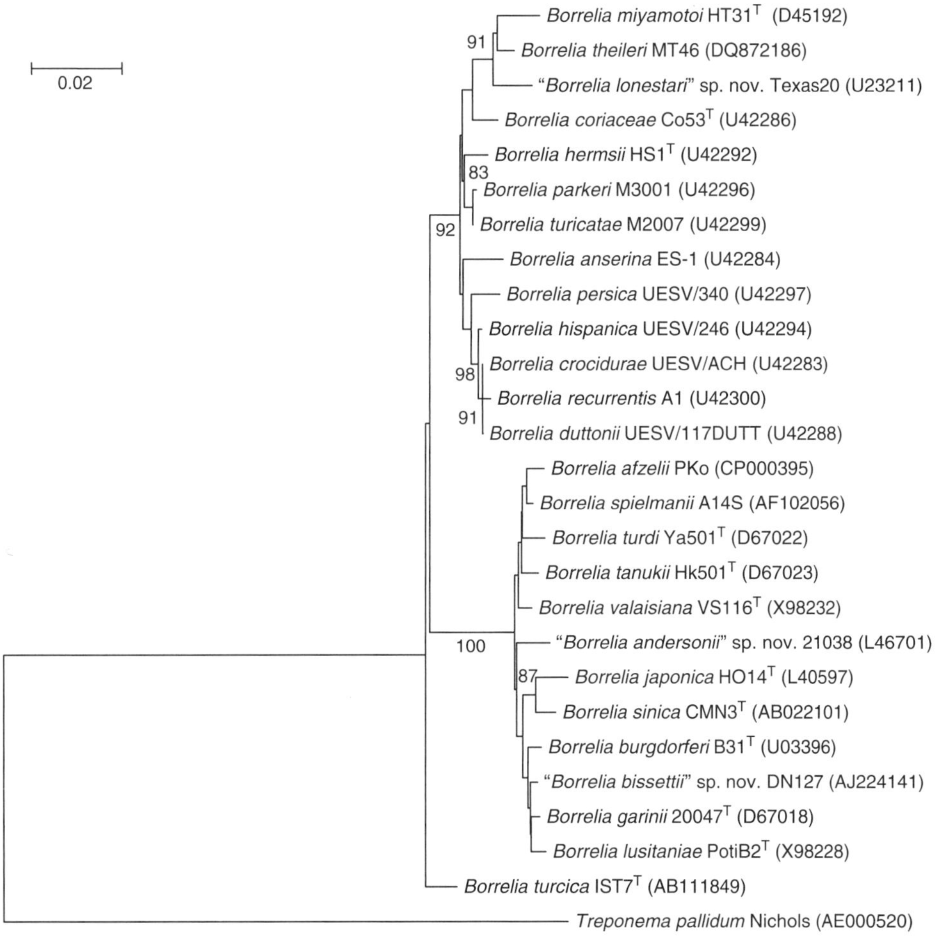

FIGURE 87. Phylogenetic tree based on 16S rRNA gene sequences of *Borrelia* species. A neighbor-joining phylogenetic tree was constructed based on Kimura's two-parameter distance estimation method using MEGA 2.1 program. Accession numbers are shown in parentheses. The numbers at the branch nodes indicate the results of the bootstrap analysis. The bar represents 2% sequence divergence. *Treponema pallidum* subsp. *pallidum* strain Nichols was used as an outgroup.

TABLE 120. Barbour–Stoenner–Kelly (BSK) II medium[a]

After detergent cleaning, all glassware is rinsed thoroughly with glass-distilled water and then autoclaved.
To 900 ml of glass-distilled water are added 100 ml of 10× concentrate of CMRL 1066 without glutamine.
Add to the 1× CMRL 1066 in the following order:
5 g Neopeptone
50 g Bovine serum albumin, Fraction V
2 g Yeastolate
6 g *N*-2-Hydroxyethylpiperazine-*N'*-2-ethanesulfonic acid
5 g Glucose
0.7 g Sodium citrate
0.8 g Sodium pyruvate
0.4 g *N*-Acetylglucosamine
2.2 g Sodium bicarbonate
Adjust the pH of medium at 20–25°C to 7.6 with 1 N NaOH.
Add 200 ml of 7% gelatin which had been dissolved in boiling water.
Sterilize by filtration with air pressure (0.2 µm nitrocellulose). Store medium at 4°C.
Before use, add unheated rabbit serum to a final concentration of 6%.
Dispense to glass or polystyrene tubes or bottles. Fill containers to 50–90% capacity and cap tightly.
Incubate at 34–37°C.

[a]Reprinted from Barbour (1984) with permission. Modifications of BSK II medium include BSK-H with removal of gelatin and different proportions of certain ingredients compared to BSK II (Pollack et al., 1993), and Kelly's medium Preac-Mursic (MKP) (Preac-Mursic et al., 1986).

membrane possesses a lower specific gravity (1.12 g/cm^2) but a higher percentage of proteins (56%) by dry weight of the membrane. Since *Borrelia burgdorferi* lacks the ability to elongate long-chain fatty acids, the fatty acid composition of the cells reflects that present in the growth medium.

Chemical analysis of *Borrelia hermsii* cells that were cultivated *in vitro* has demonstrated the presence of muramic acid and ornithine in the whole cells and in the protoplasmic cylinders, but not in the outer envelope preparations (Klaviter and Johnson, 1979). These data suggest that ornithine is a component of the cell wall. *Borrelia hermsii* contains cholesterol glucoside and its acylated derivatives (Livermore et al., 1978), and the lipid composition and metabolism of this species of *Borrelia* is remarkably similar to that of several species of mycoplasmas.

Spirochetes of the genus *Borrelia* do not have lipopolysaccharides containing lipid A (Takayama et al., 1987), and potent exotoxins are not evident.

Borrelia species are routinely cultured in liquid Barbour–Stoenner–Kelly (BSK) II medium under microaerophilic conditions. *Borrelia burgdorferi* isolates can form colonies when plated onto BSK medium solidified with 1.5% agarose (Kurtti et al., 1987). Colonies are typically observed after 2–3 weeks of incubation and may differ in morphology among different strains, e.g., compact, round colonies (mean diameter, 0.43 mm) restricted to the surface of the agarose medium, diffuse colonies (mean diameter, 1.80 mm) penetrating into the solid medium, or colonies with a raised center surrounded by a diffuse ring of spirochetes.

Borrelia burgdorferi lacks genes encoding enzymes required for the synthesis of most amino acids, fatty acids, enzyme cofactors, and nucleotides. Instead, there are 52 open reading frames (ORFs) that encode transport and binding proteins which would contribute to 16 distinct membrane transport systems for amino acids, carbohydrates, anions, and cations (Fraser et al., 1997). *Borrelia burgdorferi* also shows limited metabolic capacity. Growth of *Borrelia burgdorferi* depends largely on the availability of nutrients provided in the culture medium or from the host (mammal or tick). Analysis of the genome and reconstruction of metabolic pathways suggests that *Borrelia burgdorferi* uses glucose as a primary carbon and energy source, although other carbohydrates such as glycerol, glucosamine, fructose, and maltose may be used in glycolysis. *Borrelia burgdorferi* does not contain genes encoding enzymes of the tricarboxylic acid cycle or components of the electron transport system (Fraser et al., 1997). Thus, it is assumed that lactic acid is the main end product of glycolysis, which is consistent with the microaerophilic nature of this spirochete. Since genes for the respiratory electron transport chain were not identified, ATP production must be accomplished by substrate-level phosphorylation.

Borrelia hermsii and *Borrelia parkeri* ferment glucose, maltose, trehalose, starch, dextrin, and glycogen, but not raffinose. In contrast, *Borrelia turicatae* is able to ferment only glucose, raffinose, and dextrin.

The genome of the borreliae is composed of a small linear chromosome of approximately 1000 kb and a collection of linear and circular plasmids that are variable in number and size among species and strains. The DNA–DNA homology of Lyme disease borreliae is 76–100% among strains within species and 46–74% between different species (Baranton et al., 1992; Hyde and Johnson, 1984). For the relapsing fever spirochetes, there is >70% DNA homology between *Borrelia hermsii* and *Borrelia turicatae* (86%) or *Borrelia parkeri* (77%) that are endemic in North America, whereas 17–63% DNA relatedness was demonstrated between *Borrelia hermsii* and other relapsing fever agents (Barbour and Hayes, 1986). All *Borrelia* species studied to date have DNA G+C contents of approximately 30 mol%.

The complete genome of *Borrelia burgdorferi* has been determined (Casjens et al., 2000; Fraser et al., 1997). The genome size of the type strain *Borrelia burgdorferi sensu stricto* B31^T is 1,521,419 bp. This consists of a linear chromosome of 910,725 bp, with a DNA G+C content of 28.6 mol%, and 21 plasmids (9 circular and 12 linear) with a combined size of 610,694 bp (Casjens et al., 2000; Fraser et al., 1997). The genomes of two other pathogenic Lyme disease borreliae, *Borrelia garinii* strain PBi and *Borrelia afzelii* strain PKo, have collinear chromosomes of comparable size to *Borrelia burgdorferi* B31^T with only minor insertions and deletions. The plasmid fraction may vary significantly between different species. About 40% of *Borrelia burgdorferi* B31^T genomic DNA is of plasmid origin, whereas plasmid DNA constitutes only 29 and 36% of the analyzed genomes of *Borrelia garinii* PBi and *Borrelia afzelii* PKo, respectively (Glockner et al., 2004, 2006).

Genome analysis revealed that the Lyme disease borreliae possess some genetic structures that are unique among prokaryotes (Casjens et al., 2000; Fraser et al., 1997). These include: (1) the presence of a linear chromosome and multiple linear and circular plasmids in a single bacterium; (2) unique organization of the rRNA gene cluster, consisting of a single 16S rRNA gene and tandemly repeated 23S and 5S rRNA genes; (3) significantly higher frequency of lipoprotein-encoding genes (4.9% of the chromosomal genes and 14.5% of the plasmid genes); (4) a substantial fraction of plasmid DNA that appears to be in a state of evolutionary decay; and (5) evidence for numerous, and potentially recent, DNA rearrangements among the plasmid genes.

TABLE 121. Differential characteristics of the species of the genus *Borrelia*[a]

Species	% DNA homology with *B. hermsii*	% DNA homology with *B. burgdorferi*	Vector	Host	Distribution	Disease
Relapsing fever borreliae						
B. anserina	53–63	nd	*Argas miniatus, A. persica, A. reflexus*	Numerous birds	Worldwide	Avian borreliosis
B. baltazardii	nd	nd	Unknown	Unknown	Iran	Tick-borne relapsing fever
B. brasiliensis	nd	nd	*Ornithodoros brasiliensis*	Rodents	Brazil	
B. caucasica	nd	nd	*Ornithodoros verrucosus*	Rodents, man	Caucasus	Tick-borne relapsing fever
B. coriaceae	44–50	nd	*Ornithodoros coriaceus*	Probably deer	Western USA	Probably epizootic bovine abortion
B. crocidurae	32–35	nd	*Ornithodoros sonrai*	Rodents, man	Africa, Near East, Central Asia	Tick-borne relapsing fever
B. dugesii	nd	nd	*Ornithodoros dugesi*	Probably rodents	Mexico	
B. duttonii	17	nd	*Ornithodoros moubata*	Man	Africa	Tick-borne relapsing fever
B. graingeri	nd	nd	*Ornithodoros graingeri*	Rodents, man	East Africa	Tick-borne relapsing fever
B. harveyi	nd	nd	Unknown	Monkeys	Africa	
B. hermsii	100	30–44	*Ornithodoros hermsi*	Rodents, man	Western USA and Canada	Tick-borne relapsing fever
B. hispanica	nd	nd	*Ornithodoros erraticus*	Rodents, man	Spain, Portugal, Morocco, Algeria, Tunisia	Tick-borne relapsing fever
B. latyschewii	nd	nd	*Ornsthodoros tartakov-*	Rodents, reptiles, man	Iran, Central Asia	Tick-borne relapsing fever
B. mazzottii	nd	nd	*Ornithodoros talaje*	Rodents, armadillos, monkeys, man	Mexico and Guatemala	Tick-borne relapsing fever
B. miyamotoi	45	13–14	*Ixodes persulcatus*	Rodents	Japan	
B. parkeri	77	nd	*Ornsthodoros parkeri*	Rodents, man	Western USA	Tick-borne relapsing fever
B. persica	nd	nd	*Ornithodoros tholozani*	Rodents, bats, man	Middle East, Central Asia	Tick-borne relapsing fever
B. recurrentis	nd	nd	*Pediculus humanus,* subsp. *humanus*	Man	South America, Europe, Africa, Asia	Louse-borne relapsing fever
B. theileri	nd	nd	*Rhipicephalus decoloratus, Rhipicephalus evertsi, Boophilus micropus*	Ruminants, horses	South Africa, Australia North America, Europe	Bovine borreliosis
B. tillae	nd	nd	*Ornithodoros zumpti*	Rodents	South Africa	
B. turicatae	86	nd	*Ornithodoros turicatae*	Rodents, man	USA and Mexico	Tick-borne relapsing fever
B. venezuelensis	nd	nd	*Ornithodoros rudis*	Rodents, man	Central and South America	Tick-borne relapsing fever
B. burgdorferi sensu lato						
B. burgdorferi	30–44	100	*Ixodes scapularis* *Ixodes pacificus* *Ixodes ricinus* *Ixodes persulcatus*	Rodents	Eastern and Midwestern USA Western USA Europe Asia	Lyme borreliosis
B. afzelii	16	48	*Ixodes ricinus* *Ixodes persulcatus*	Rodents	Europe Asia	Lyme borreliosis
B. garinii	27	55	*Ixodes ricinus* *Ixodes persulcatus*	Rodents	Europe Asia	Lyme borreliosis
B. japonica	17	50–53	*Ixodes ovatus*	Rodents	Japan	
B. lusitaniae	nd	44–53	*Ixodes ricinus*	Rodents, lizards	Europe, North Africa	Probably Lyme borreliosis
B. sinica	nd	58	*Niviventer confucianus* *Ixodes ovatus*	Rodents	China	
B. spielmanii	nd	nd	*Ixodes ricinus*	Rodents	Europe	Probably Lyme borreliosis
B. tanukii	nd	50	*Ixodes tanukii* *Ixodes ovatus*	Rodents	Japan	
B. turdi	nd	58	*Ixodes turdus*	Rodents	Japan	
B. valaisiana	nd	51–65	*Ixodes ricinus* *Ixodes columnae*	Birds	EuropeAsia	Probably Lyme borreliosis
Other borreliae						
B. turcica	<20	<20	*Hyalomma aegyptium*	Unknown	Turkey	

[a]nd, Not determined.

The genomes of relapsing fever *Borrelia* species are largely linear. Each genome consists of a linear chromosome of about 1000 kb and different types of linear and circular plasmids. The size of linear plasmids is 10–200 kb and of circular plasmids is 8–40 kb. The presence of multiple copies of the chromosome and linear plasmids in some relapsing fever strains has been reported (Kitten and Barbour, 1992). The complete genomes of two relapsing fever borreliae, *Borrelia hermsii* and *Borrelia turicatae*, have been sequenced (NCBI Genome Databases, http://www.ncbi.nlm.nih.gov).

Borrelia burgdorferi is strictly a clonal organism, in spite of the observed genetic diversity and evidence for lateral gene transfer and recombination of selected plasmid genes among and within *Borrelia* species (Dykhuizen et al., 1993). Genetic manipulation of *Borrelia* species is very difficult, but basic tools are now available and their application has begun to provide information about the identities and roles of key bacterial components in both the tick vector and the mammalian host (Rosa et al., 2005). Phage particles have been observed in *Borrelia burgdorferi* culture supernatants (Eggers and Samuels, 1999).

Antigenic variation has been well documented in *Borrelia* species and extensively studied in the relapsing fever agent *Borrelia hermsii* and the Lyme disease spirochete *Borrelia burgdorferi*. About 30 variable major protein (VMP) serotypes, which are divided about equally between a variable large protein (Vlp) of about 36 kDa and variable small proteins (Vsp) of about 20 kDa, have been described in *Borrelia hermsii*. The Vlp/Vsp antigenic variation system of *Borrelia hermsii* involves an expression site that can acquire either *vlp* or *vsp* surface lipoprotein genes from up to 59 different archival copies arranged in clusters on at least five different plasmids. Gene conversion occurs through recombination events at upstream homology sequences found in each gene copy, and at downstream homology sequences found periodically among the vlp/vsp archival genes (Dai et al., 2006). Genes that are homologous to *vlp* and *vsp* of *Borrelia hermsii* are detected in other relapsing fever borreliae, including *Borrelia turicatae*, *Borrelia recurrentis*, *Borrelia crocidurae*, and *Borrelia duttonii* (Cutler, 2001, 1999). This property of relapsing fever borreliae is responsible for the numerous relapses which occur in the infected host. In *Borrelia burgdorferi*, antigenic variation can result by promiscuous recombination of plasmid-encoded *vls* sequence cassettes and may contribute to evasion of the host immune response, allowing persistence of the spirochete (Zhang et al., 1997).

More than 50 protein bands are visible in whole-cell lysates of cultivable *Borrelia* species. Serodiagnosis of Lyme disease is achieved by detection of antibodies specific to Lyme borrelial antigens (Centers for Disease Control and Prevention, 1995; Dressler et al., 1993). Since antibodies against glycerophosphodiester phosphodiesterase (GlpQ) are demonstrated only in sera from hosts infected with relapsing fever borreliae, this antigen could be used for serodiagnosis to distinguish Lyme disease from relapsing fever borrelial infections.

Borrelia species are susceptible to several classes of antimicrobial agents, including penicillins, second- and third-generation cephalosporins, tetracyclines, macrolides, and glycopeptides, but they are relatively resistant to aminoglycosides, trimethoprim, rifampicin, and quinolones (Preac-Mursic et al., 1986; Wormser et al., 2006).

Borreliae are pathogens of man, other mammals, and birds. Infections are acquired from ticks or lice which are parasitized with borreliae. Two distinct clinical syndromes are associated with *Borrelia* infections in humans: Lyme disease and relapsing fever. *Borrelia burgdorferi*, *Borrelia garinii*, and *Borrelia afzelii* are causative agents of human Lyme borreliosis. *Borrelia burgdorferi* is the sole species responsible for Lyme borreliosis in North America, whereas all three pathogenic Lyme borreliae are documented in Eurasia. Human Lyme borreliosis generally occurs in stages and may manifest as one or more symptoms including erythema migrans, carditis, neuroborreliosis, and arthritis (Steere, 2001). Association between different clinical manifestations and the infecting *Borrelia* species or genotypes has been reported (van Dam et al., 1993; Wang et al., 1999; Wormser et al., 1999).

Infections with the relapsing fever borreliae in man produce a severe septicemic illness of which two varieties are recognized: tick-borne and louse-borne relapsing fever. A single spirochete of *Borrelia duttonii*, *Borrelia hermsii*, or *Borrelia turicatae* may be sufficient to infect experimental animals.

Borrelia species exist in nature in enzootic cycles primarily involving ticks and a wide range of animal hosts. Lyme disease borreliae are transmitted among reservoirs and hosts by ticks of the family *Ixodidae*, mainly ticks within the *Ixodes ricinus* complex. *Ixodes scapularis* and *Ixodes pacificus* in the United States, and *Ixodes ricinus* and *Ixodes persulcatus* in Eurasia are the principal vectors of the spirochetes (Burgdorfer et al., 1991). All *Ixodes* species are three-host ticks; each individual tick feeds on three different host animals during its life. The white-footed mouse (*Peromyscus leucopus*) in the United States, and the wood mouse (*Apodemus sylvaticus*) and bank vole (*Clethrionomys glareolus*) in Europe are among the most important reservoirs for Lyme borreliae (Anderson, 1989; Gern et al., 1998). Migrating birds may be reservoirs of certain species such as *Borrelia garinii* and *Borrelia valaisiana*. Larval ticks acquire *Borrelia burgdorferi* by feeding on infected hosts. The spirochetes survive through the molts and remain present in all subsequent stages of the vectors. Transovarial transmission of *Borrelia burgdorferi* may occur in *Ixodes* ticks but does not represent an important factor in maintaining infected ticks in nature (Burgdorfer et al., 1991).

Tick-borne relapsing fever is transmitted by a number of different species of *Ornithodoros* ticks. Many are named after the species of *Ornithodoros* tick transmitting the infection. Numerous small animals serve as natural reservoirs for tick-borne relapsing fever borreliae. Relapsing fever borreliae are frequently vertically transmitted in infected ticks, occurring in up to 100% of offspring and in most but not all tick species (Barbour, 2005).

The causative agent of louse-borne relapsing fever, *Borrelia recurrentis*, is transmitted from human-to-human by the body louse *Pediculus humanus* subsp. *humanus*. Man acquires infection by crushing infected lice during scratching, thereby liberating borreliae which enter the body at the site of the louse bite or through skin abraded by scratching. Lice acquire *Borrelia recurrentis* by feeding on infected humans and remain infectious for the remainder of their relatively short lifespan.

The geographical distribution of the *Borrelia* species, specific arthropod vectors and hosts, are listed in Table 121.

Enrichment and isolation procedures

***In vitro* cultivation.** All Lyme disease borreliae and many species of relapsing fever borreliae (e.g., *Borrelia hermsii*, *Borrelia coriaceae*, and *Borrelia turicatae*) can be grown routinely in a cell-free liquid culture under microaerophilic conditions in Barbour–Stoenner–

Kelly II (BSK-II) medium (Barbour, 1984) or variations such as BSK-H (Pollack et al., 1993) or MKP medium (Preac-Mursic et al., 1986). These are modifications from the Kelly's medium that were used first for cultivation of certain relapsing fever borreliae (Kelly, 1971). Owing to the lack of or limited biosynthetic potential, complex nutritional requirements are necessary for cultivation of *Borrelia*. Typically, the culture medium contains serum, glucose, albumin, peptides, amino acids, vitamins, and a thickening agent such as gelatin. In addition, *N*-acetylglucosamine and long-chain saturated and unsaturated fatty acids are required. The optimum growth temperature for *Borrelia burgdorferi* is 30–34°C. The bacteria grow slowly, dividing every 8–12 h during the exponential growth phase *in vitro*. Culture-adapted isolates can usually reach cell densities of 10^7–10^8 per ml after cultivation *in vitro* for 5–7 d. *Borrelia burgdorferi* can also be grown on semi-solid BSK medium to obtain individual, pure clones (Preac-Mursic et al., 1991).

Isolation by animal inoculation. Some relapsing fever borreliae have yet to be cultivated in serial passage outside of a natural host or experimental animal. Animal inoculation remains the method of choice for isolation of these borreliae. A number of animals have been used for isolation of nonavian borreliae including mice, rats, rabbits, guinea pigs, and monkeys. Results have been variable depending on the species of *Borrelia* and the age and species of animals employed. The most consistent results have been obtained with suckling mice. Citrated blood or triturates of ticks or lice are inoculated in a volume of 0.1–0.2 ml via the subcutaneous or intraperitoneal route. At daily intervals, the mice are bled by clipping the tip of the tail with scissors to obtain a small drop of blood which is examined by darkfield microscopy or in smears stained by the Wright or Giemsa method.

Maintenance procedures

Suspensions of borreliae containing 10% glycerol retain viability for several years when stored at or below –70°C. Limited attempts to preserve borreliae by lyophilization have not been successful.

Tick-borne borreliae may also be maintained in the laboratory for several years in infected ticks with occasional feedings on experimental animals.

Differentiation of the genus *Borrelia* from other genera

Two other genera of spirochetes which may be present in biological fluids or tissues are species of *Treponema* and *Leptospira*. Borreliae stain readily and can be observed with conventional microscopy whereas leptospires and some treponemes are not visualized. The coils of borreliae are loose, coarse, and irregular. Leptospires are coiled so tightly that they are difficult to observe with darkfield microscopy. *Treponema* species are regularly and rigidly coiled and have a more uniformly helical structure than do borreliae species.

Taxonomic comments

The Lyme disease and relapsing fever borreliae are morphologically indistinguishable. However, these two major groups of *Borrelia* can be easily differentiated based on their arthropod vectors and 16S rRNA gene sequences with only a few exceptions. 16S rRNA and flagellin gene sequence analysis is reliable in discriminating members within the genus of *Borrelia* (Fukunaga et al., 1996b; Ras et al., 1996; Wang et al., 1999). In addition, several other molecular typing methods have been successfully employed for species differentiation of the Lyme disease *Borrelia*. These include rRNA restriction analysis (ribotyping), pulsed-field gel electrophoresis, randomly amplified polymorphic DNA (RAPD) fingerprinting, PCR and PCR-based restriction fragment length polymorphism (RFLP), sequence analysis of the 5S–23S intergenic spacer, and multilocus sequence analysis (Margos et al., 2008; Richter et al., 2006; Wang et al., 1999).

Traditionally, relapsing fever borreliae are classified into species according to their arthropod vectors. Those borreliae transmitted by human body lice presently belong to the species *Borrelia recurrentis*; those transmitted by ticks are differentiated on the basis of the tick-spirochete specificity theory which states that borreliae carried by a given species of tick are specific for that vector and, therefore, constitute individual species of borreliae (Barbour and Hayes, 1986; Kelly, 1984). This theory, however, is flawed by the fact that some borreliae isolated from one species of tick are capable of infecting other tick species. Moreover, many species of tick-borne borreliae have been shown to develop in experimentally infected body lice. Consequently, classification according to the traditional scheme may or may not be correct. For example, the relapsing fever borreliae *Borrelia turicatae*, *Borrelia parkeri*, and *Borrelia hermsii* are transmitted by distinct soft-bodied ticks, but they may be a single species because their DNA–DNA similarity is greater than 70%. The isolation of *Borrelia miyamotoi*, a species more closely related to relapsing fever borreliae from hard ticks (*Ixodes persulcatus* and *Ixodes ricinus*), challenges the classical concept of tick-spirochete cospeciation. Additional studies will be necessary to determine if the present classification is in fact valid.

Acknowledgements

The authors thank their many collaborators during a decade and a half of *Borrelia* research. Special thanks to Dr Gary P. Wormser for numerous insights and enlightening discussions. The research in the authors' laboratory was supported by National Institutes of Health grants AR41511 and AI45801.

Further reading

Barbour, A.G. 1986. Biology of *Borrelia* species. Microbiol. Rev. *50*: 381–400.

Cutler, S.J. 2001. Relapsing fever *Borrelia*. *In* Molecular Medical Microbiology, vol. 3 (edited by Sussman), Academic Press, London, pp. 2093–2113.

Goodman, J.L. (editor) 2005. Tick-borne Diseases of Humans. ASM Press, Washington, D.C.

Felsenfeld, O., 1971. *Borrelia*. W.H. Green, St Louis, Missouri.

Johnson, R.C. (editor). 1976. The Biology of the Parasitic Spirochetes. Academic Press, New York.

Kelly, R.T. 1984. Genus IV. *Borrelia*. *In* Bergey's Manual of Systematic Bacteriology, vol. 1 (edited by Krieg and Holt). Williams & Wilkins, Baltimore, pp. 57–62.

Wang, G., A.P. van Dam, I. Schwartz and J. Dankert. 1999. Molecular typing of *Borrelia burgdorferi sensu lato*: taxonomic, epidemiological, and clinical implications. Clin. Microbiol. Rev. *12*: 633–653.

Wang, G., G.P. Wormser and I. Schwartz. 2001. *Borrelia burgdorferi*. *In* Molecular Medical Microbiology, vol. 3 (edited by Sussman). Academic Press, London, pp. 2059–2092.

Differentiation of the species of the genus *Borrelia*

The characteristic features of species of *Borrelia* are indicated in Tables 121–123.

TABLE 122. Reactivity patterns of *Borrelia burgdorferi sensu lato* species and *Borrelia hermsii* with various species-specific monoclonal antibodies (mAbs)[a]

	Reactivity with mAb							
Species	H9724 (flagellin)	H5332 (OspA)	H3TS (OspA)	LA31 (OspA)	I17.3 (OspB)	D6 (12 kDa antigen)	O1141b (flagellin)	A116k (OspA)
B. burgdorferi	+	+	+	+	–	–	–	–
B. afzelii	+	–	–	–	+	–	–	–
B. garinii	+	+ or –	–	+	–	+	–	–
B. japonica	+	+	–	nd	–	–	+	–
B. lusitaniae	+	nd	–	nd	–	–	nd	–
B. sinica	+	–	nd	nd	nd	–	–	nd
B. spielmanii	+	nd	+	–	nd	–	nd	nd
B. tanukii	+	+	nd	nd	nd	nd	nd	nd
B. turdi	+	$+^{w}$	nd	nd	nd	nd	nd	nd
B. valaisiana	+	–	–		–	–	–	+
B. hermsii	+	–	–	–	–	–	–	–

[a]+, Positive; –, negative; $+^{w}$, weak reactive; nd, no data available;.

TABLE 123. Differentiation of Lyme disease-related *Borrelia* species based on the *Mse*I restriction polymorphism of 5S–23S rRNA (*rrfA–rrlB*) intergenic spacer amplicons[a]

Taxon	Strain	Amplicon size (bp)	RFLP pattern	MseI restriction fragments size (bp)
B. burgdorferi	B31^{T}	254	A	108, 51, 38, 29, 28
B. afzelii	VS461^{T}	246	D	108, 68, 50, 20
B. garinii	20047^{T}	253	B	108, 95, 50
	NT29	253	C	108, 57, 50, 38
B. japonica	HO14^{T}	236	E	108, 78, 50
B. lusitaniae	PotiB2^{T}	257	G	108, 81, 39, 29
	PotiB3	255	H	108, 79, 52, 16
B. sinica	CMN3^{T}	235	S	107, 48, 38, 29, 13
B. spielmanii	A14S	225	R	106, 68, 51
B. tanukii	Hk501^{T}	245	O	174, 51, 20
B. turdi	Ya501^{T}	248	P	107, 51, 38, 21, 16, 8, 7
B. vailaisiana	VS116^{T}	255	F	175, 50, 23, 7
	Am501	249	Q	169, 51, 23, 6
"*B. andersonii*" sp. nov.	19857	266	L	120, 67, 51, 28
	CA2	255	M	91, 50, 40, 28, 22, 17, 7
"*B. bissettii*" sp. nov.	DN127	257	I	108, 51, 38, 33, 27
	CA118	226	J	108, 51, 38, 29
	25015	253	K	108, 51, 34, 27, 17, 12, 4

[a]Adapted from Postic et al. (1994), Wang et al. (1999), and Masuzawa et al. (2001). For some *Borrelia* species, strain variations in the *Mse*I restriction fragment sizes are documented but are not listed in this table.

List of species of the genus *Borrelia**

1. **Borrelia anserina** (Sakharoff 1891) Bergey, Harrison, Breed, Hammer and Huntoon 1925, 435[AL] (*Spirochaeta anserina* Sakharoff 1891, 565; *Spiroschaudinnia anserina* (Sakharoff 1891) Sambon 1907, 834)

an.se'ri.na. L. fem. adj. *anserina* pertaining to geese.

The organisms are 0.2–0.3 μm wide and 8–20 μm in length, and consist of 5–8 coils.

The organisms can be grown in BSK II medium at 30°C and embryonated duck or chicken eggs. They can also be maintained in young chickens or ducks. Not infective for mice, rats, or rabbits. There are a number of antigenically different strains. The most reliable distinguishing characteristic of this species is that it is infectious for birds but not for rodents. Considered as the agent of fowl spirochetosis. Biochemical characteristics have not been determined.

Source: naturally infected *Argas miniatus* ticks and has a worldwide distribution.

DNA G+C content (mol%): not known.

*See Table 121 for listing of species categorized as relapsing fever borreliae, *Borrelia burgdorferi sensu lato*, and "other borreliae".

Type strain: not designated.

Sequence accession no. (16S rRNA gene): U42284 (strain ES-1).

2. **Borrelia afzelii** Canica, Nato, du Merle, Mazie, Baranton and Postic 1994, 182[VL]

af.ze'li.i. N.L. masc. gen. n. *afzelii* of Afzelius, named in honor of Arvid Afzelius, a Swedish physician who was the first to report the skin lesion characteristic of Lyme borreliosis in 1909.

It has been referred to as group VS461. Morphology as described for the genus. Cultural properties as described for *Borrelia burgdorferi*. rRNA gene restriction patterns after digestion by *Eco*RV contain three fragments (6.7, 3.2, and 1.6 kb). rRNA gene restriction patterns after digestion by *Hin*dIII contain five fragments (3.7, 2.1, 1.5, 0.9, and 0.6 kb).

Source: humans and *Ixodes ricinus* ticks in Europe and Asia.

DNA G+C content (mol%): 27–28 (T_m), 28 (strain VS461).

Type strain: strain VS461, ATCC 51567, CIP 103469, DSM 10508.

Sequence accession no. (16S rRNA gene): U78151 (strain VS461^T), CP000395 (strain PKo).

3. **Borrelia baltazardii** corrig. (*ex* Karimi, Hovind-Hougen, Birch-Andersen and Asmar 1979) Karimi, Hovind-Hougen, Birch-Andersen and Asmar 1983, 438[VL] ("*Borrelia baltazardi*" Karimi, Hovind-Hougen, Birch-Andersen and Asmar 1979)

bal.ta.zar'di.i. N.L. masc. gen. n. *baltazardii* of Baltazard, named after Marcel Baltazard (1908–1971), a French doctor and biologist.

Resembles *Borrelia recurrentis* in morphology.

Source: a region of Iran.

DNA G+C content (mol%): not known.

Type strain: strain Borrelia "x".

Sequence accession no. (16S rRNA gene): no sequence available.

4. **Borrelia brasiliensis** Davis 1956, 476[AL]

bra.si.li.en'sis. N.L. fem. adj. *brasiliensis* named after the specific epithet of the tick vector *Ornithodoros brasiliensis*.

Resembles *Borrelia recurrentis* in morphology. *In vitro* cultivation has not been reported. Isolated and maintained in mice, rats, and guinea pigs. Distinguishing characteristics are not well characterized. No biochemical data available.

Source: the tick *Ornithodoros brasiliensis*.

DNA G+C content (mol%): not known.

Type strain: no culture available.

Sequence accession no. (16S rRNA gene): no sequence available.

5. **Borrelia burgdorferi** Johnson 1984, 496[VP] (*Borrelia burgdorferi sensu stricto*, emend. Baranton 1992, 378)

burg.dor'fe.ri. N.L. masc. gen. n. *burgdorferi* of Burgdorfer, named in honor of Willy Burgdorfer.

The organisms are flexible helical cells with dimension of 0.2–0.3 µm by 4–30 µm. Motile with both rotational and translational movements. Consists of 3–10 coils. The coiling of the cell is regular. Seven to 11 periplasmic flagella are located at each cell end, and these flagella overlap at the central region of the cell. A multilayered outer membrane surrounds the protoplasmic cylinder, which consists of the peptidoglycan layer, cytoplasmic membrane, and the enclosed cytoplasmic contents. The diamino acid ornithine is present in the peptidoglycan. Cytoplasmic tubules are absent.

The cells are Gram-stain-negative and stain well with Giemsa and Warthin–Starry stains. Unstained cells are not visible by bright-field microscopy but are visible by dark-field or phase-contrast microscopy.

The organisms grow well in BSK II medium or its variants supplemented with *N*-acetylglucosamine and rabbit serum. The optimal growth temperature is 34–37°C, with a generation time of 8–12 h at 35°C. The cells are catalase-negative and microaerophilic. Use glucose as primary energy source, although other carbohydrates, including glycerol, glucosamine, fructose, and maltose, may be used in glycolysis. Pyruvate produced by glycolysis is converted to lactate. Chemo-organotropic.

The genome of *Borrelia burgdorferi* type strain B31 contains a linear chromosome of 910,725 bp and 21 plasmids (9 circular and 12 linear) with a combined size of 610, 694 bp. The chromosome contains 853 genes encoding a basic set of proteins for DNA replication, transcription, translation, solute transport, and energy metabolism, but no genes for cellular biosynthetic metabolic capacities. Differential expression of major surface proteins in tick vectors and mammalian hosts. It is pathogenic for humans and animals.

Source: several tick species of the genus *Ixodes* which are parasitized by the spirochete.

DNA G+C content (mol%): 28.6 (complete genome sequencing).

Type strain: strain B31, ATCC 35210, CIP 102532, DSM 4680.

Sequence accession no. (16S rRNA gene): NC_001318, U03396.

6. **Borrelia caucasica** (Kandelaki 1945) Davis 1957, 901[VL] (*Spirochaeta caucasica* Kandelaki according to Maruashvili 1945, 24)

cau.ca'si.ca. N.L. fem. adj. *caucasica* pertaining to the Caucasus.

Resembles *Borrelia recurrentis* in morphology. *In vitro* cultivation has not been reported. Isolated and maintained in mice, rats, and guinea pigs. Distinguishing characteristics are not well characterized. No biochemical data available.

Source: unknown source in Russia.

DNA G+C content (mol%): not known.

Type strain: no culture available.

Sequence accession no. (16S rRNA gene): no sequence available.

7. **Borrelia coriaceae** Johnson, Burgdorfer, Lane, Barbour, Hayes and Hyde 1987, 72[VP]

co.ri.a.ce'ae. L. fem. adj. *coriacea* of leather, made of leather; N.L. fem. gen. n. *coriaceae* named after the specific epithet of the tick vector, *Ornithodoros coriaceus*, from which the organism was isolated.

Morphology as described for the genus. Growth *in vitro* in BSK II medium but is more fastidious than other cultivable *Borrelia* species. Only three isolates (Co53, CA434, and

CA435) have been cultivated and maintained *in vitro* since 1985. Low infectivity for infant Swiss White mice, BALB/c mice, and adult New Zealand White rabbits. Probable cause of epizootic bovine abortion.

Source: the soft tick *Ornithodoros coriaceus* from Northern California.

DNA G+C content (mol%): 32.4 (T_m).

Type strain: strain Co53, ATCC 43381, CIP 104208.

Sequence accession no. (16S rRNA gene): U42286.

8. **Borrelia crocidurae** (Leger 1917) Davis 1957, 903[AL] (*Spirochaeta crocidurae* Leger 1917, 281)

cro.ci.du'rae. N.L. gen. n. *crocidurae* of *Crocidura*, a genus of Insectivora.

Resembles *Borrelia recurrentis* in morphology. Isolates from blood of patients have been grown *in vitro* in modified Kelly's medium. Isolated and maintained in young mice and rats. Four serovars have been recognized; however, the species has not been well characterized. Clusters phylogenetically with *Borrelia duttonii*, *Borrelia recurrentis*, and *Borrelia hispanica*. No biochemical data are available. Considered as one cause of tick-borne relapsing fever.

Source: isolated initially from the white-toothed shrew, *Crocidura*. The tick vector is *Alectorobius sonrai*.

DNA G+C content (mol%): not known.

Type strain: not designated.

Sequence accession no. (16S rRNA gene): U42283 (strain UESV/ACH).

9. **Borrelia dugesii** (Mazzotti 1949) Davis 1957, 902[AL] (*Spirochaeta dugesii* Mazzotti 1949, 278)

du.ge'si.i. N.L. gen. n. *dugesii* of *dugesi*, the specific epithet of the tick vector, *Ornithodoros dugesi*.

Resembles *Borrelia recurrentis* in morphology. *In vitro* cultivation has not been reported. Isolated and maintained in young mice or rats. Distinguishing characteristics: not well characterized. No biochemical data available. Considered as one cause of tick-borne relapsing fever.

Source: the argasid tick, *Ornithodoros dugesi*.

DNA G+C content (mol%): not known.

Type strain: no culture available.

Sequence accession no. (16S rRNA gene): no sequence available.

10. **Borrelia duttonii** (Novy and Knapp 1906) Bergey, Harrison, Breed, Hammer and Huntoon 1925, 434[AL] [*Spirillum duttoni* (*sic*) Novy and Knapp 1906, 296; *Spirochaeta duttoni* (Novy and Knapp 1906) Breinl 1906, 1691]

dut.to'ni.i. N.L. masc. gen. n. *duttonii* of Dutton, named after Joseph Everett Dutton (1876–1905), who died from relapsing fever during an investigative trip to the Congo.

Resembles *Borrelia recurrentis* in morphology. Cells contained 10 periplasmic flagella inserted at each end of the spirochete. Has been successfully cultivated in BSK II medium from patients with relapsing fever. Isolated and maintained in BSK II medium or young mice and rats. Genome consists of a chromosome of approximately 1 Mb, a large plasmid of approximately 200 kb, and 7–9 plasmids with sizes in the range 20–90 kb. Analysis of the SDS-PAGE profiles of *Borrelia duttonii* strains revealed a high-molecular-mass band of 33.4–34.2 kDa (variable large protein, VLP) and a low-molecular-mass band of 22.3 kDa (variable small protein, VSP). The cause of East African tick-borne relapsing fever. Man is the only reservoir.

Source: humans and the *Ornithodoros* sp. soft ticks.

DNA G+C content (mol%): 27.6 (HPLC).

Type strain: no strain designated.

Sequence accession no. (16S rRNA gene): U42288.

11. **Borrelia garinii** Baranton, Postic, Saint Girons, Boerlin, Piffaretti, Assous and Grimont 1992, 382[VP]

ga.ri'ni.i. N.L. masc. gen. n. *garinii* of Garin, in honor of Charles Garin, a French physician.

Morphology as described for the genus. Cultural properties as described for *Borrelia burgdorferi*. rRNA gene restriction patterns after digestion by *Eco*RV contain three fragments (5.3, 3.2, and 1.6 or 1.5 kb). rRNA gene restriction patterns after digestion by *Hin*dIII contain three fragments (2.1, 1.3, and 0.6 kb). Some strains produce an additional weak fragment at 2.5 kb.

Source: humans and *Ixodes* ticks in Europe and Asia.

DNA G+C content (mol%): 27–28 (T_m).

Type strain: strain 20047, ATCC 51383, CIP 103362, DSM 10534.

Sequence accession no. (16S rRNA gene): D67018.

12. **Borrelia graingeri** (Heisch 1953) Davis 1957, 903[AL] (*Spirochaeta graingeri* Heisch 1953, 133)

grain'ge.ri. N.L. gen. n. *graingeri* named after the specific epithet of the tick vector, *Ornithodoros graingeri*.

Resembles *Borrelia recurrentis* in morphology. *In vitro* cultivation has not been reported. Animal pathogenicity: present in the blood of mice and rats only transiently after inoculation. Distinguishing characteristics: not well characterized. No biochemical data available.

Source: the tick, *Ornithodoros graingeri*.

DNA G+C content (mol%): not known.

Type strain: no culture available.

Sequence accession no. (16S rRNA gene): no sequence available.

13. **Borrelia harveyi** (Garnham 1947) Davis 1948, 316[AL] (*Spirochaeta harveyi* Garnham 1947, 49)

har've.yi. N.L. masc. gen. n. *harveyi* of Harvey, named after A.E.C. Harvey.

Resembles *Borrelia recurrentis* in morphology. *In vitro* cultivation has not been reported. May be maintained in mice, rats, and monkeys. Distinguishing characteristics: not well characterized. No biochemical data available.

Source: louse-borne relapsing fever in Kenya.

DNA G+C content (mol%): not known.

Type strain: no culture available.

Sequence accession no. (16S rRNA gene): no sequence available.

14. **Borrelia hermsii** (Davis 1942) Steinhaus 1946, 453[AL] [*Spirochaeta hermsi* (*sic*) Davis 1942, 46]

herm'si.i. N.L. gen. n. *hermsii* of *hermsi*, the specific epithet of the tick vector *Ornithodoros hermsi*.

Resembles *Borrelia recurrentis* in morphology. Grows well in BSK II medium (Table 120). Readily isolated and maintained in young mice and rats. Distinguishing characteristics: ferments glucose, maltose, trehalose, starch, dextrin, and glycogen, but not raffinose.

Source: the tick *Ornithodoros hermsi.*

DNA G+C content (mol%): 29 (complete genome sequencing).

Type strain: strain HS1, ATCC 35209, CIP 104209, DSM 4682.

Sequence accession no. (16S rRNA gene): U42292.

15. **Borrelia hispanica** (de Buen 1926) Steinhaus 1946, 453[AL] (*Spirochaeta hispanica* de Buen 1926, 185)

his.pa'ni.ca. L. fem. adj. *hispanica* Spanish.

Resembles *Borrelia recurrentis* in morphology. May be cultivated *in vitro*, but the yield of organisms (7×10^6/ml) is low (Kelly, 1976). Can be maintained in guinea pigs. Causes a mild infection in young mice and rats. Can be transmitted via rat bite. Distinguishing characteristics are not well characterized. No biochemical data available.

Source: the argasid ticks *Ornithodoros nicollei* and *Ornithodoros erraticus.*

DNA G+C content (mol%): not known.

Type strain: no culture available.

Sequence accession no. (16S rRNA gene): U42294.

16. **Borrelia japonica** Kawabata, Masuzawa and Yanagihara 1994, 595[VL]

ja.po'ni.ca. N.L. fem. adj. *japonica* pertaining to Japan.

Morphology and cultural properties as described for *Borrelia burgdorferi.* rRNA gene restriction patterns after digestion by *Hin*dIII contain one fragment (2.1 kb). DNA homology between type strain HO14 and related Lyme disease borreliae: *Borrelia burgdorferi* B31^T (56%), *Borrelia garinii* 20047^T (63%), and *Borrelia afzelii* VS461^T (64%).

Source: the tick *Ixodes ovatus*, shrews, voles, and mice in Japan.

DNA G+C content (mol%): 27.9–28.3 (T_m).

Type strain: strain HO14, ATCC 51557, JCM 8951.

Sequence accession no. (16S rRNA gene): L46696, L40597.

17. **Borrelia latyschewii** (Sofiev 1941) Davis 1948, 315[AL] [*Spirochaeta latyschewi* (*sic*) Sofiev 1941, 271]

la.ty.sche'wi.i. N.L. masc. gen. n. *latyschewii* of Latyschew, named after Latyschew (Latyshev).

Resembles *Borrelia recurrentis* in morphology. *In vitro* cultivation has not been reported. Maintained in young mice. Distinguishing characteristics: not well characterized. No biochemical data available.

Source: the argasid tick, *Ornithodoros tartakovskyi.*

DNA G+C content (mol%): not known.

Type strain: no culture available.

Sequence accession no. (16S rRNA gene): no sequence available.

18. **Borrelia lusitaniae** Le Fleche, Postic, Girardet, Peter and Baranton 1997, 924[VP]

lu.si.ta.ni'ae. L. gen. n. *lusitaniae* of *Lusitania*, the Roman name for Portugal, where the organism was first isolated.

Morphology and cultural properties as described for *Borrelia burgdorferi.* Can be easily differentiated from other Lyme disease borreliae by analysis of the *Mse*I restriction polymorphism of the 5S–23S rRNA spacer. Patterns after restriction of the 16S rRNA gene by *Bfa*I contain six fragments (689, 404, 159, 100, 78, and 58 bp).

Source: Ixodes ricinus ticks, lizards, and small rodents in Europe and North Africa, and from the skin lesion of one patient in Portugal.

DNA G+C content (mol%): not known.

Type strain: strain PotiB2, CIP 105366.

Sequence accession no. (16S rRNA gene): X98228.

19. **Borrelia mazzottii** Davis 1956, 17[AL]

maz.zot'ti.i. N.L. masc. gen. n. *mazzottii* of Mazzotti, named after Luis Mazzotti, a Mexican physician who recovered a relapsing fever spirochete from *Ornithodoros talaje* in Mexico in 1953.

Resembles *Borrelia recurrentis* in morphology. *In vitro* cultivation has not been reported. Isolated and maintained in young mice and rats. Distinguishing characteristics are not well characterized. No biochemical data available.

Source: the argasid tick *Ornithodoros talaje.*

DNA G+C content (mol%): not known.

Type strain: no culture available.

Sequence accession no. (16S rRNA gene): no sequence available.

20. **Borrelia miyamotoi** Fukunaga, Takahashi, Tsuruta, Matsushita, Ralph, McClelland and Nakao 1995, 809[VP]

mi.ya.mo'to.i. N.L. masc. gen. n. *miyamotoi* of Miyamoto, in honor of Kenji Miyamoto, an entomologist who first isolated this organism from *Ixodes* ticks in Japan.

Morphology and cultural properties as described for *Borrelia burgdorferi.* The rRNA gene cluster contains only a single copy of 23S and 5S rNRA gene, which is separated from the 16S rRNA gene. Low levels of DNA relatedness (8–13%) between *Borrelia miyamotoi* HT31^T and Lyme disease borreliae: *Borrelia burgdorferi* B31^T (13%), *Borrelia garinii* 20047^T (13%), *Borrelia afzelii* VS461^T (8%), and *Borrelia japonica* HO14^T (9%). Moderate levels of DNA relatedness with *Borrelia hermsii* (44%), *Borrelia turicatae* (41%), *Borrelia parkeri* (51%), and *Borrelia coriaceae* Co53^T (41%). The pathogenicity of *Borrelia miyamotoi* is unknown.

Source: the hard tick *Ixodes persulatus*, vector of Lyme disease, and rodents in Japan. *Borrelia* strains closely related to *Borrelia miyamotoi* have been detected or cultured from *Amblyoma americanum* and *Ixodes scapularis* ticks in the United States and from *Ixodes ricinus* ticks in Sweden.

DNA G+C content (mol%): 28–29 (T_m).

Type strain: strain HT31, JCM 9579.

Sequence accession no. (16S rRNA gene): D45192.

21. **Borrelia parkeri** (Davis 1942) Steinhaus 1946, 453[AL] (*Spirochaeta parkeri* Davis 1942, 46)

par'ke.ri. N.L. gen. n. *parkeri* named after the specific epithet of the tick vector *Ornithodoros parkeri.*

Resembles *Borrelia recurrentis* in morphology. Grows well in BSK II medium (Table 120). Animal pathogenicity: isolated and maintained in young mice and rats. Distinguishing characteristics: fermentation of carbohydrates is identical to that of *Borrelia hermsii.* This suggests that the two species are in fact variants of a single species.

Source: the argasid tick *Ornithodoros parkeri*.
DNA G+C content (mol%): not known.
Type strain: no culture available.
Sequence accession no. (16S rRNA gene): U42296.

22. **Borrelia persica** (Dschunkowsky 1913) Steinhaus 1946, 453[AL] (*Spirochaeta persica* Dschunkowsky 1913, 419)

per'si.ca. L. fem. adj. *persica* Persian.

Resembles *Borrelia recurrentis* in morphology. *In vitro* cultivation has not been reported. Animal pathogenicity varies according to the strain. Most isolates can be maintained in mice or guinea pigs. Distinguishing characteristics are not well characterized. No biochemical data available.
Source: the argasid tick *Ornithodoros moubata*.
DNA G+C content (mol%): not known.
Type strain: no culture available.
Sequence accession no. (16S rRNA gene): U42297.

23. **Borrelia recurrentis** (Lebert 1874) Bergey, Harrison, Breed, Hammer and Huntoon 1925, 433[AL] [*Spirochaeta recurrentis* Lebert 1874, 273; "*Spirochaete obermeieri*" Cohn 1875; *Spiroschaudinnia recurrentis* (Lebert 1874) Sambon 1907, 833[AL]]

re.cur.ren'tis. L. part. adj. *recurrens, recurrentis* recurring; N.L. gen. n. *recurrentis* intended to mean of a recurrent fever.

The organisms are 0.3–0.6 µm wide and 8–18 µm in length, and consist of 3–8 coils. Cells with pointed ends, a mean wavelength of 1.8 µm, an amplitude of 0.8 µm, and 8–10 periplasmic flagella.

May be cultivated *in vitro* in BSK II medium (Table 120). Most strains may also be maintained in young rats. Guinea pigs are not infected. DNA–DNA hybridizations and sequencing studies of both the flagellin and 16S RNA genes revealed that the greatest similarity was between *Borrelia recurrentis* and *Borrelia duttonii*. Causes louse-borne relapsing fever.
Source: the vector of this disease is the infected human body louse (*Pediculus humanus* subsp. *humanus*).
DNA G+C content (mol%): 28.4 (HPLC).
Type strain: strain A1, ATCC 700241.
Sequence accession no. (16S rRNA gene): U42300.

24. **Borrelia sinica** Masuzawa, Takada, Kudeken, Fukui, Yano, Ishiguro, Kawamura, Imai and Ezaki 2001, 1823[VP]

si'ni.ca. M.L. fem. adj. *sinica* of China, the country from which the organism was isolated.

Morphology as described for the genus except for a maximum of four flagella at each end. Cultural properties as described for *Borrelia burgdorferi*. DNA homology of *Borrelia sinica* type strain CMN3 and other Lyme borreliae: *Borrelia afzelii* VS461^T (43%), "*Borrelia andersonii*" 21123 (39%), "*Borrelia bissettii*" DN127 (47%), *Borrelia burgdorferi* B31^T (59%), *Borrelia garinii* 20047^T (53%), *Borrelia japonica* HO14^T (52%), *Borrelia lusitaniae* PotiB2^T (47%), *Borrelia tanukii* Fi81t (51%), *Borrelia turdi* Ya501^T (46%), and *Borrelia valaisiana* VS116^T (30%).
Source: the tick *Niviventer confucianus*.
DNA G+C content (mol%): 29.6 (CMN3^T) (HPLC).
Type strain: strain CMN3^T, JCM 10505.
Sequence accession no. (16S rRNA gene): AB022101.

25. **Borrelia spielmanii** Richter, Postic, Sertour, Livey, Matuschka and Baranton 2006, 880[VP] (*Borrelia spielmani* Richter, Schlee, Allgower and Matuschka 2004)

spi.el.ma'ni.i. N.L. masc. gen. n. *spielmanii* of Spielman, in honor of Andrew Spielman, who first described the life cycle and biological relationships of *Borrelia burgdorferi*.

Morphology as described for the genus. Cultural properties as described for *Borrelia burgdorferi*. It has pathogenic potential.
Source: Ixodes ricinus ticks feeding on garden and hazel dormice, in questing ticks, and in patients in France, Germany, The Netherlands, Denmark, Slovenia, Hungary, and the Czech Republic.
DNA G+C content (mol%): 27–32 (T_m).
Type strain: strain PC-Eq17N5, CIP 108855, DSM 16813.
Sequence accession no. (16S rRNA gene): AF102056 (strain A14S).

26. **Borrelia tanukii** Fukunaga, Hamase, Okada and Nakao 1997, 1274[VL]

ta.nu'ki.i. N.L. gen. n. *tanukii* of/from *tanuki*, named after *Ixodes tanuki*, from which the organism was isolated.

Morphology as described for the genus. Cultural properties as described for *Borrelia burgdorferi*. Less than 70% DNA homology with *Borrelia burgdorferi*, *Borrelia garinii*, *Borrelia afzelii*, *Borrelia japonica*, and *Borrelia turdi*. rRNA gene restriction patterns after digestion by *Eco*RV contain two fragments (6.2 and 3.2 kb). rRNA gene restriction patterns after digestion by *Hinc*II contain three fragments (6.5, 3.2, and 1.8 kb).
Source: Ixodes tanuki ticks and several species of rodents in Japan.
DNA G+C content (mol%): 28–31 (HPLC), 29 (strain Hk501^T).
Type strain: strain Hk501, ATCC BAA-127, JCM 9662.
Sequence accession no. (16S rRNA gene): D67023.

27. **Borrelia theileri** (Laveran 1903) Bergey, Harrison, Breed. Hammer and Huntoon 1925, 435[AL] (*Spirochaeta theileri* Laveran 1903, 941)

thei'le.ri. N.L. masc. gen. n. *theileri* of Theiler, named in honor of Arnold Theiler (1867–1936), a Swiss veterinarian and regarded as the father of veterinary science of South Africa.

The organisms are 0.25–0.30 µm wide and 20–30 µm in length in cattle; they are reported to be shorter in horses. *In vitro* cultivation has not been reported. May be maintained in cattle and horses. Distinguishing characteristics: not well characterized. No biochemical data available.
Source: the cattle tick *Boophilus microplus*.
DNA G+C content (mol%): not known.
Type strain: no culture available.
Sequence accession no. (16S rRNA gene): DQ872186.

28. **Borrelia tillae** Zumpt and Organ 1961, 33[AL]

til'lae. N.L. fem. gen. n. *tillae* of Till, named after W. Till.

Resembles *Borrelia recurrentis* in morphology. *In vitro* cultivation has not been reported. Isolated and maintained in young mice and rats. Distinguishing characteristics are not well characterized. No biochemical data available.

Source: not known.
DNA G+C content (mol%): not known.
Type strain: no culture available.
Sequence accession no. (16S rRNA gene): no sequence data available.

29. **Borrelia turcica** Güner, Watanabe, Hashimoto, Kadosaka, Kawamura, Ezaki, Kawabata, Imai, Kaneda and Masuzawa 2004, 1651[VP]

tur.ci'ca. N.L. fem. adj. *turcica* of or belonging to Turkey, the country in which the organism was isolated.

Morphology is as described for the genus. Length of the cells is variable, about 10–25 μm, and the diameter is 0.2–0.28 μm, with 8–23 flagella. The mean number of flagella is 15–16. The optimal growth temperature in BSK II medium is between 34 and 37°C; cells can grow at 39°C. Fast-growing spirochete; doubling times at 34 and 37°C are 5.3 and 5.1 h, respectively.

Reactive with *Borrelia* genus-specific mAb H9326, but no reactivity with Lyme disease *Borrelia* OspA-specific mAb H5332.

The level of DNA relatedness in strain IST7^{T} was less than 20% when compared with the Lyme disease agent, *Borrelia burgdorferi*, or the relapsing fever agent, *Borrelia hermsii*. *Borrelia turcica* spirochetes formed a monophyletic cluster separated from both Lyme disease and relapsing fever-related *Borrelia* species on a phylogenetic tree based on 16S rRNA gene sequence data (Figure 87).

Source: hard tick *Hyalomma aegyptium* in Turkey.
DNA G+C content (mol%): 30 (T_m).
Type strain: strain IST7, DSM 16138, JCM 11958.
Sequence accession no. (16S rRNA gene): AB111849.

30. **Borrelia turdi** corrig. Fukunaga, Hamase, Okada and Nakao 1997, 1274[VL]

tur'di. N.L. gen. n. *turdi* of/from *turdus*, named after *Ixodes turdus*, from which the organism was isolated.

Morphology as described for the genus. Cultural properties as described for *Borrelia burgdorferi*. Less than 70% in DNA homology with *Borrelia burgdorferi*, *Borrelia garinii*, *Borrelia afzelii*, *Borrelia japonica*, and *Borrelia tanukii*. rRNA gene restriction patterns after digestion by *Eco*RV contain two fragments (6.3 and 3.2 kb). rRNA gene restriction patterns after digestion by *Hin*dIII contain three fragments (5.7, 3.2, and 2.3 kb). The pathogenicity of *Borrelia turdi* is unknown.

Source: Ixodes turdus ticks and passerine birds, *Emberiza spodocephala*, in Japan.
DNA G+C content (mol%): 28–32 (HPLC), 29 (strain Ya501^{T}).
Type strain: strain Ya501, ATCC BAA-126, JCM 9661.
Sequence accession no. (16S rRNA gene): D67022.

31. **Borrelia turicatae** (Brumpt 1933) Steinhaus 1946, 453[AL] (*Spirochaeta turicatae* Brumpt 1933, 1369)

tu.ri.ca'tae. N.L. gen. n. *turicatae* of *turicata*, the specific epithet of the tick vector, *Ornithodoros turicata*.

Resembles *Borrelia recurrentis* in morphology. Grows well in BSK II medium (Table 120). Can also be isolated and maintained in young mice and rats. Distinguishing characteristics: Ferments only glucose, raffinose, and dextrin, thus differing from *Borrelia hermsii* and *Borrelia parkeri*.

Source: the argasid tick *Ornithodoros turicata*.
DNA G+C content (mol%): 29 (complete genome sequencing).
Type strain: none designated.
Sequence accession no. (16S rRNA gene): U42299.

32. **Borrelia valaisiana** Wang, van Dam, Le Fleche, Postic, Peter, Baranton, de Boer, Spanjaard and Dankert 1997, 931[VP]

va.lai.si.a'na. N.L. fem. adj. *valaisiana* of or belonging to Valais, Switzerland, where this organism was first isolated.

Formerly *Borrelia* group VS116. Morphology as described for the genus. Cultural properties as described for *Borrelia burgdorferi*. rRNA gene restriction patterns after digestion by *Eco*RV contain three fragments (6.9, 3.2, and 1.4 kb). rRNA gene restriction patterns after digestion by *Hin*dIII contain four fragments (2.1, 1.2, 0.8, and 0.6 kb). It is probably pathogenic to humans.

Source: several species of *Ixodes* ticks in Europe and Asia.
DNA G+C content (mol%): 28–32 (T_m).
Type strain: strain VS116, CIP 105367.
Sequence accession no. (16S rRNA gene): X98232.

33. **Borrelia venezuelensis** (Brumpt 1921) Brumpt 1922b, 495[AL] [*Treponema venezuelense* Brumpt 1921, 207; "*Spirochaeta neotropicalis*" Bates and St John 1922, "*Borrelia neotropicalis*" (Bates and StJohn 1922) Steinhaus 1946]

ve.ne.zu.e.len'sis. N.L. adj. *venezuelensis* named after the specific epithet of the tick vector, *Ornithodoros rudis* (*Ornithodoros venezuelensis*).

Resembles *Borrelia recurrentis* in morphology. *In vitro* cultivation has not been reported. Isolated and maintained in young mice or rats. Distinguishing characteristics: not well characterized. No biochemical data available.

Source: the argasid tick *Ornithodoros rudis* (*Ornithodoros venezuelensis*).
DNA G+C content (mol%): not known.
Type strain: no culture available.
Sequence accession no. (16S rRNA gene): no sequence data available.

Species *Candidatus*

"*Candidatus* Borrelia texasensis": a *Borrelia* strain TXW-1, isolated from an adult male *Dermacentor variabilis* tick feeding on a coyote in Texas, USA. The levels of DNA relatedness between TXW-1 and previously described relapsing fever borreliae are extremely low (7.9–38.6%). It could not be revived from frozen cultures (Lin et al., 2005).

Other organisms belonging to the genus *Borrelia*

Three *Borrelia* species associated with Lyme disease, "*Borrelia andersonii*" sp. nov. (Marconi et al., 1995), "*Borrelia bissettii*" sp. nov. (Postic et al., 1998), and "*Borrelia californiensis*" sp. nov. (Postic et al., 2007), have been described. Their taxonomic status is not validated due to the lack of sufficient data.

"*Borrelia andersonii*" sp. nov. was chosen in honor of John Anderson, who was the first person to isolate and characterize strains in this group. It has been cultivated from *Ixodes dentatus* ticks and its wildlife host cottontail rabbits in North America. Isolates in this group possess a unique rRNA gene polymorphism (Marconi et al., 1995).

"*Borrelia bissettii*" sp. nov. was named in honor of Marjorie L. Bissett, who first described a member of this group. "*Borrelia bissettii*" sp. nov. has been isolated from *Ixodes spinipalpis* and some of its rodent hosts in the United States (Postic et al., 1998).

Several novel borrelial species or strains closely related to the relapsing fever borreliae have been described. These include an uncultivable "*Borrelia lonestari*" sp. nov. detected by PCR and 16S rRNA gene sequence analysis in the hard tick *Amblyomma americanum* in North America (Barbour et al., 1996) and possibly new *Borrelia* species in patients with relapsing fever in Spain (Anda et al., 1996) and in soft-bodied *Ornithodoros* ticks in Tanzania (Kisinza et al., 2003). Their taxonomic status and pathogenicity in humans remain to be determined.

Genus III. **Cristispira** Gross 1910, 44[AL]

Bruce J. Paster and John A. Breznak

Cris.ti.spi'ra. L. fem. n. *crista* a crest; L. fem. n. *spira* a coil; N.L. fem. n. *Cristispira* a crested coil.

Helical cells are 0.5–3.0 µm in diameter × 30–180 µm in length, displaying 2–10 helical turns. Ends of cells are blunt, rounded, or tapered. Round inclusions of unknown composition or function are seen in stained preparations or under phase microscopy. Multiple cytoplasmic vesicles bounded by a double membrane are observed in electron microscopy thin sections. **Cells divide by transverse fission. Hundreds of periplasmic flagella are bundled together and wrap around the protoplasmic cylinder forming a ridge or crest (called crista).** The cristae can be clearly seen under light microscopy when the cells stop moving. Cells flex, rotate, and swim rapidly **at more than 100 µm/s.** Under adverse environments, cells stop moving and form spherical bodies or lyse. Found in the crystalline style of the digestive tract of marine and freshwater mollusks. **Not known to be pathogenic and probably represent normal microflora.** *Cristispira*-like cells have also been observed in gastropods, e.g., snails, and in nonmollusk species. *Cristispira* has not yet been grown in pure culture.

DNA G+C content (mol%): not determined.

Type species: **Cristispira pectinis** Gross 1910, 44[AL].

Further descriptive information

Cristispira is found in the crystalline style (Figure 88), which is a mucoproteinaceous, rod-shaped organ of the digestive tract of marine and freshwater mollusks (e.g., clams, mussels, and oysters). *Cristispira* cells can be seen in great numbers from crystalline style preparations as shown in Figure 89. Cells are more numerous and motile in the inner zone rather then in the firm cortical region of the crystalline style (Perrin, 1906). Typically, mollusks with soft textured styles tend to harbor the spirochetes more frequently (Noguchi, 1921). Cristispires have been observed to protrude through, or adhere to, the outer layer of the crystalline style as shown in Figure 90 (Bernard, 1970; Tall and Nauman, 1981). Swimming motility outside of the crystalline style is rapid, more than 100 µm/s. Within the crystalline style, cristispires move slower in a creeping or crawling type of motility. Cells tend to avoid the anterior (blunt) end of the style, which usually contains food debris formed by the grinding of the style on the gastric shield region of the stomach (Breznak, 1973). It is likely that the anterior end of the style may contain substances toxic (or repellent) to cristispires (Berkeley, 1933, 1959, 1962).

Cristispires have also been observed in regions of the host other than in the crystalline style, including the style pouch, gastric shield, and fluid of the pallial cavity, stomach, anterior intestine, cecum, and rectum (Bernard, 1970; Breznak, 1973). They have been observed in aquaria water used to incubate their hosts (Bosanquet, 1911; Fantham, 1908). Not all species of mollusks examined possess *Cristispira* (Bernard, 1970; Breznak, 1973).

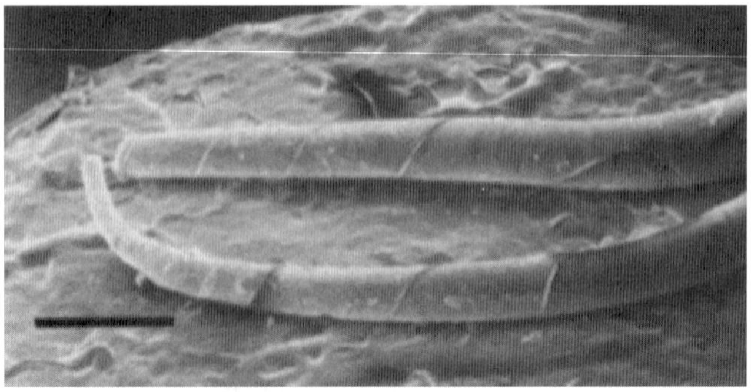

FIGURE 88. Scanning electron micrograph of the crystalline style of the oyster *Crassostrea virginica* that was cut in two pieces. Bar = 1.0 µm. (Tall and Nauman, 1981. Appl. Environ. Microbiol. *42:* 336–343; reproduced with permission from B.D. Tall.)

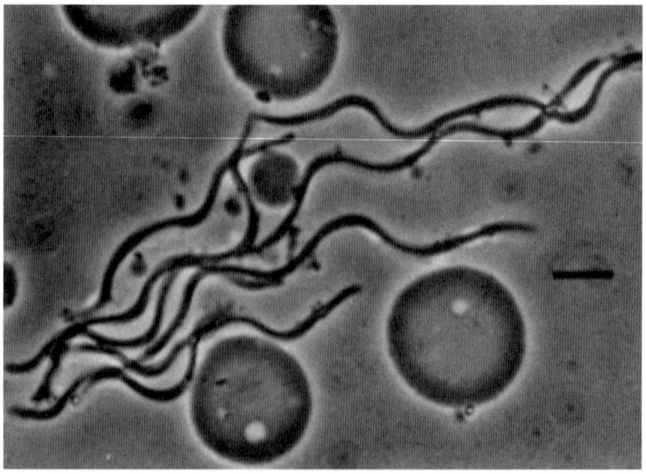

FIGURE 89. Phase-contrast micrograph of a cluster of *Cristispira* cells from the crystalline style of *Ostrea* (*Crassostrea*) *virginica*. Bar = 10 µm.

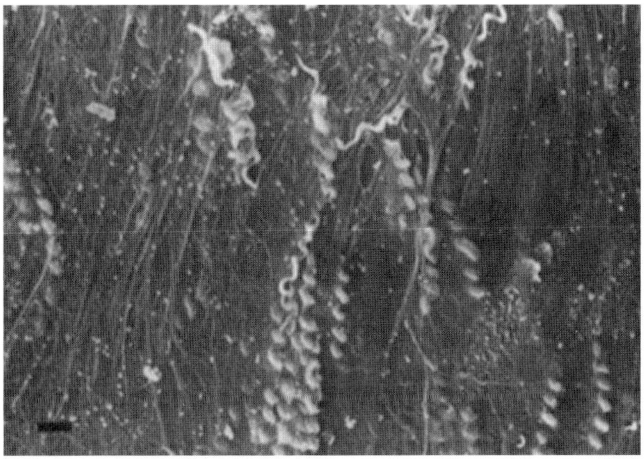

FIGURE 90. Scanning electron micrograph of *Cristispira* associated with the outer layer of the crystalline style. Differences in coiling of the spirochetes on the surface and those below the surface can be readily observed. Bar = 10 μm. (Tall and Nauman, 1981. Appl. Environ. Microbiol. *42:* 336–343; reproduced with permission from B.D. Tall.)

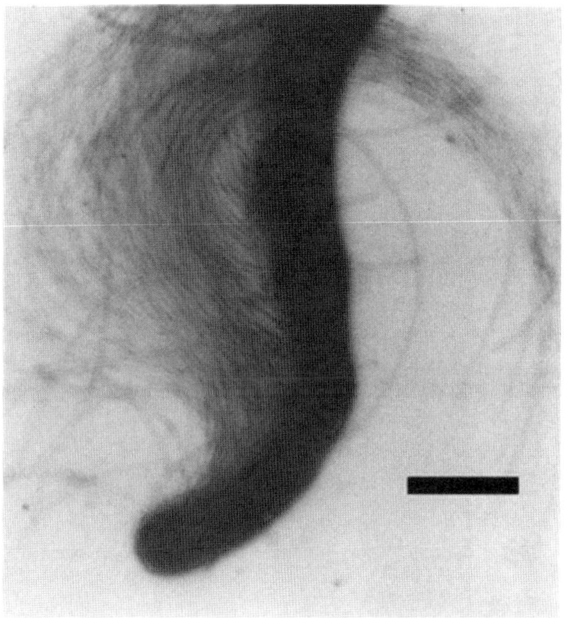

FIGURE 91. Transmission electron micrograph of a negatively stained preparation. The "crest" has been somewhat dispersed revealing the numerous periplasmic flagella. Bar = 1.0 μm. (Tall and Nauman, 1981. Appl. Environ. Microbiol. *42:* 336–343; reproduced with permission from B.D. Tall.)

In suspension, cells remain motile longer at 5–10°C than at temperatures above 20°C (Fantham, 1908; Noguchi, 1921; Perrin, 1906); glucosone inhibits motility (Berkeley, 1962); motility occurs under aerobic and putative anaerobic conditions (Berkeley, 1959). May be microaerophilic; a suspension of a high concentration of cristispires on a microscope slide will form a distinct zone, a ring, of cells near the edge of a coverslip with no cells present in the center of the slide (personal observation, BJP). Cells typically lyse in distilled water (Fantham, 1908; Perrin, 1906). Cells survive, and perhaps have limited growth, in media with proteinaceous substrates (Noguchi, 1921) and fructose (Kubomura, 1969).

Cells of *Cristispira* possess the ultrastructural characteristics typical of spirochetes, i.e., a bundle of periplasmic flagella is wrapped around the helical body of each cell. The bundle (called the crest for cells of *Cristispira*) consists of hundreds of periplasmic flagella (shown Figure 91) which are inserted near each end of the protoplasmic cylinder and appear to overlap in the central region of the cell. An outer sheath surrounds both the bundle of periplasmic flagella and the protoplasmic cylinder. *Cristispira* has the most periplasmic flagella of all the known spirochetes.

Cristispira is found usually in healthy, palatable, univalve and bivalve mollusks obtained from well aerated beds (Berkeley, 1959). It does not appear to be pathogenic to its host. In nonactively feeding mollusks or at cold temperatures, the crystalline style generally dissolves, and the numbers of *Cristispira* decrease dramatically or no *Cristispira* are found (Dimitroff, 1926; Kuhn, 1974).

The mollusk hosts that have been reported to harbor presumed *Cristispira* are listed in Table 124. *Cristispira*-like spirochetes have also been observed in starfish (Collier, 1921), tunicates (Hellmann, 1913), and termites (Hollande, 1922), however, their phylogenetic relationships to the molluscan cristispires are unknown.

Differentiation of the genus *Cristispira* from other genera

Since *Cristispira* has not yet been grown in culture, the genus can only be differentiated from other genera of spirochetes based on habitat, the presence of a crista with hundreds of periplasmic flagella, their relatively large size, and, more recently, 16S rRNA sequence comparison.

Taxonomic comments

16S rRNA genes amplified from DNA isolated from cristispire-laden crystalline styles of the common eastern oyster, *Crassotrea virginica,* were cloned into *Escherichia coli* and sequenced. Sequence comparisons indicated that *Cristispira* branches deeply among other spirochetal genera within the family *Spirochaetaceae. Cristispira* clearly represents a separate genus within this family (Figure 92). In situ hybridization experiments were used to verify that the sequence obtained was derived from the observed *Cristispira* cells (Paster et al., 1996). It is not known if there is more than one species of *Cristispira* since cells from only a single oyster were analyzed.

Consequently, the genus *Cristispira* is still considered monospecific until species can be grown or analyzed from different hosts by 16S rRNA gene sequence analysis. The early literature suggests many different specific epithets to reflect the molluscan host and subtle differences in ultrastructural traits (Breznak, 1973; Kuhn, 1974). These species have previously been referred to as species incertae sedis (Kuhn, 1974). In the 1st edition of *Bergey's Manual of Systematic Bacteriology* (Breznak, 1984), reviving such names was avoided until pure cultures could be obtained

TABLE 124. Mollusk hosts of *Cristispira*

Host	Geographical source	Reference
Amphidesma australe (Gm.)	Cheltenham Beach, Auckland, New Zealand (marine)	Judd (1979)
Anodonta cygnea Linn.	River Cam, England (freshwater)	Dobell (1912), Fantham (1908)
Anodonta grandis Say	Wisconsin, USA (freshwater)	Nelson (1918)
Anodonta mutabilis Cless.	Germany (freshwater)	Keysselitz (1906), Schellack (1909)
Anodonta (species unknown)	England (freshwater)	Bosanquet (1911)
Cardium papillosum Poli	Rovigno, Adriatic Sea (marine)	Schellack (1909)
Chama gryphoides Linn.	Rovigno, Adriatic Sea (marine)	Schellack (1909)
Chama sinistrorsa Brocchi	Rovigno, Adriatic Sea (marine)	Schellack (1909)
Clinocardium nuttallii (Con.)	British Columbia, Canada (marine)	Bernard (1970)
Crassostrea gigas Than.	British Columbia, Canada (marine)	Berkeley (1959), Bernard (1970)
Cyclas (species unknown)	Rovigno, Adriatic Sea (marine)	Schellack (1909)
Diplodonta orbella (Gould)	British Columbia, Canada (marine)	Bernard (1970)
Entodesma saxicola (Baird)	British Columbia, Canada (marine)	Bernard (1970)
Gastrochaena dubia Penn.	Rovigno, Adriatic Sea (marine)	Schellack (1909)
Lampsilis anodontoides Lea	Wisconsin, USA (freshwater)	Nelson (1918)
Lima hyans Gm.	Rovigno, Adriatic Sea (marine)	Schellack (1909)
Lima inflata Lam.	Rovigno, Adriatic Sea (marine)	Schellack (1909)
Lyonsia pugetensis Dall	British Columbia, Canada (marine)	Bernard (1970)
Macoma (species unknown)	California, USA (marine)	Berkeley (1959)
Mactra sulcataria Desh.	Fukuoka, Japan (marine)	von Prowazek (1910)
Modiola barbata Linn.	Rovigno, Adriatic Sea (marine)	Schellack (1909)
Modiola modiolus Linn.	Woods Hole, MA, USA (marine)	Noguchi (1921)
Ostrea angulata Lam.	Arcachon and LaRochelle, France (marine)	Certes (1882)
Ostrea edulis Linn.	France and Adriatic Sea (marine)	Certes (1882), Fantham (1911), Ryter and Pillot (1965), Schellack (1909), Swellengrebel (1907), von Prowazek (1910)
Ostrea lurida Carp.	British Columbia, Canada (marine)	Bernard (1970)
Ostrea talienwhaneensis Cross	Fukuoka, Japan (marine)	von Prowazek (1910)
Ostrea (*Crassostrea*) *virginica* Gm.	Tuckerton, NJ, USA; Baltimore, MD, USA; Woods Hole, MA, USA (marine)	Nelson (1918), Dimitroff (1926), Noguchi (1921), Tall and Nauman (1981)
Panope generosa Gould	British Columbia, Canada (marine)	Bernard (1970)
Paphia staminea[a]	British Columbia, Canada (marine)	Berkeley (1959)
Pecten jacobaeus Linn.	Gulf of Naples, Italy (marine)	Gross (1910)
Pinna nobilis Linn.	Rovigno, Adriatic Sea; Gulf of Naples, Italy (marine)	Gonder (1908), Schellack (1909)
Protothaca staminea (Con.)	British Columbia, Canada (marine)	Bernard (1970)
Saxicava arctica (Linn.)	Rovigno, Adriatic Sea (marine)	Schellack (1909)
Saxidomus giganteus Desh.	British Columbia, Canada (marine)	Berkeley (1959), Bernard (1970)
Semisulcospira libertina Gould (a snail)	Hiroshima, Japan (freshwater)	Terasaki (1958)
Siliqua patula Dixon	California, USA (marine)	Berkeley (1959)
Solen ensis Linn.	(?)	Fantham (1911)
Soletellina acuminata Desh.	Tamblegam Lake, Ceylon (saltwater)	Dobell (1911, 1912)
Strophitus (species unknown)	Wisconsin River, WI, USA (freshwater)	Breznak (1973)
Tapes aureus Gm.	France (marine)	Fantham (1911), Ryter and Pillot (1965)
Tapes decussatus (Linn.)	Rovigno, Adriatic Sea and France (marine)	Ryter and Pillot (1965), Schellack (1909)
Tapes laeta Wkff.	Rovigno, Adriatic Sea (marine)	Schellack (1909)
Tapes (*Venerupis*) *philippinarum* (Adams and Reeve)	Fukuoka, Urayasu, and Inage, Japan (marine)	van Prowazek (1910), Kubomura (1969)
Tapes pullastra (Moot.)	France (marine)	Ryter and Pillot (1965)
Tivela stultorum Mawe	California, USA (marine)	Berkeley (1959), Jahn and Landman (1965)
Tresus capax (Gould)	British Columbia, Canada (marine)	Bernard (1970)
Tresus nuttallii (Con.)	British Columbia, Canada (marine)	Bernard (1970)
Unio pictorum Linn.	Rovigno, Adriatic Sea (marine)	Schellack (1909)
Venerupis japonica (Desh.)	British Columbia, Canada (marine)	Bernard (1970)
Venus casta Chem.	Tamblegam Lake, Ceylon (saltwater)	Dobell (1911, 1912)
Venus mercenaria Linn.	Woods Hole, MA, USA (marine)	Noguchi (1921)
Venus verrucosa Linn.	France (marine)	Ryter and Pillot (1965)
Others:		
"Schleswig Holstein oyster"	North Germany (marine)	Möbius (1883)
"Adriatic oyster"	Adriatic Sea (marine)	Perrin (1906)
"French", "English", and "Abervrach" oysters	France and England (marine)	Fantham (1908)

[a]Authority unknown.

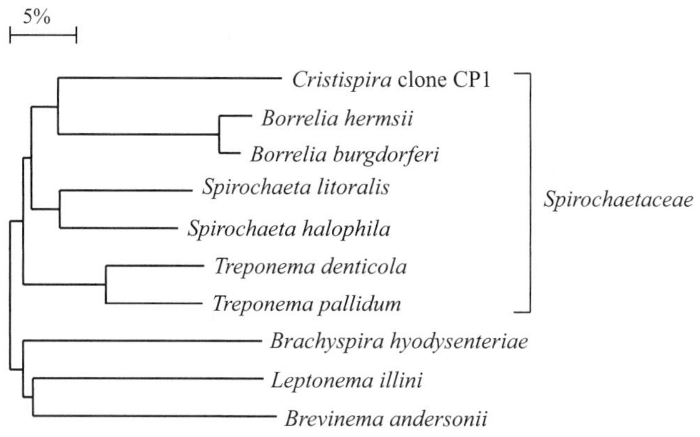

FIGURE 92. Dendrogram illustrating the phylogenetic position of *Cristispira* relative to members of other spirochetal genera. *Cristispira* clearly represents a separate genus.

for adequate characterization. Furthermore, several species of spirochetes of smaller size and with fewer periplasmic flagella than *Cristispira* have been observed in other mollusks including *Anodonta mutobilis* (Schellack, 1909), *Pachelebra* (probably *Pachylabra*) *moesta* (a snail; deMello, 1921), *Pinna squamasa* (Gonder, 1908), *Scrobicularia piperata* (Pilot and Ryter, 1965), planorbid mollusks (Richards, 1978), and *Polydora flava* (a marine polychete annelid; Mesnil and Caullery, 1916). Kuhn (1974) referred to these *Cristispira*-like species as *species inquirendae.*

Further reading

Margulis, L., L. Nault and J. Sieburth. 1991. *Cristispira* from oyster styles: complex morphology of large symbiotic spirochetes. Symbiosis *11:* 1–19.

List of species of the genus *Cristispira*

1. **Cristispira pectinis** Gross 1910, 44[AL]

pec'ti.nis. L. masc. n. *pecten* a kind of shell-fish, a scallop, and also a genus of mollusks (*Pecten*); L. gen. n. *pectinis* of a scallop, of *Pecten.*

Helical, flexible cells, 0.5–1.5 µm in diameter × 36–80 µm displaying no more than four complete helical turns. Ends of the cells are round or tapered with no terminal appendages. Stained preparations reveal cross-striations, polar granulation, and multiple inclusions. In fixed and stained preparations, a crista (i.e., a bundle of hundreds of periplasmic flagella) wraps around the length of the cell. Cells divide by transverse fission.

Source: the crystalline style and intestinal fluid of the scallop, *Pecten jacobaeus,* from the Gulf of Naples.

DNA G+C content (mol%): not determined.

Type strain: no culture isolated.

Sequence accession no. (16S rRNA gene): U42638; represents cloned 16S rRNA gene (CP1) from DNA isolated from crystalline style of oyster (Paster et al., 1996).

Genus IV. **Treponema** Schaudinn 1905, 1728[AL] (*Spironema* Vuillemin 1905, 1568; *Microspironema* Stiles and Pfender 1905, 936)

Steven J. Norris, Bruce J. Paster and Robert M. Smibert*

Trep.o.ne'ma. Gr. v. *trepô* to turn; Gr. neut. n. *nema* a thread; N.L. neut. n. *treponema* a turning thread.

Host-associated, helical cells 0.1–0.7 µm in diameter and 1–20 µm in length. Cells have tight regular or irregular spirals and one or more **periplasmic flagella** (axial fibrils or axial filaments) inserted at each end of the protoplasmic cylinder. **Cytoplasmic filaments** are seen in the protoplasmic cylinder just under the cytoplasmic membrane and running parallel with the periplasmic flagella. Under unfavorable cultural or environmental conditions, spherical cells are formed. These can also be seen in old cultures. Gram-stain-negative. Cells stain well with silver impregnation methods. Most species stain poorly, if at all, with Gram or Giemsa stain. Best observed with darkfield or phase-contrast microscopy. **Motile. Cells have rotational movement in liquid media, and translational motion in media with high viscosity [e.g., those containing 1% (w/v) methyl cellulose].** In a semisolid or solid medium, cells exhibit a serpentine type movement, sometimes referred to as creeping motility. **Strictly anaerobic or microaerophilic.** Frank pathogens (*Treponema pallidum* subspecies, *Treponema carateum,* and the rabbit pathogen *Treponema paraluiscuniculi*) represent a closely related subset within this genus and are considered microaerophiles. Limited multiplication of *Treponema pallidum* subsp. *pallidum* strains has been obtained in a tissue culture system, but none of the pathogenic *Treponema* have been cultivated continuously in artificial media or in tissue culture. **Chemo-organotrophs,** using a variety of **carbohydrates or amino acids for carbon and energy sources.** Cultivated anaerobic species are catalase- and oxidase-negative. Some require long-chain fatty acids found in serum for growth, while other cultivated species require short-chain volatile fatty acids for growth. **Host-associated. Pathogenic *Treponema pallidum* subspecies cause skin lesions, and *Treponema pallidum* (particularly subspecies *pallidum*) can cause systemic infections that, if untreated, can last for years to decades. Other species are found in the oral cavity, intestinal tract, and genital areas of humans or other mammals, and in the gut contents of wood-feeding insects.**

DNA G+C content (mol%): 37–54.

Type species: **Treponema pallidum subsp. pallidum** (Schaudinn and Hoffman 1905) Schaudinn 1905, 1728[AL].

Further descriptive information

Phylogeny and genetics. Comparison of rRNA gene sequences (Figure 93) and additional genome information indicates that

*Portions of this chapter have been adapted from the previous version in Bergey's by Smibert (1984). To assure accuracy and completeness, many of the species descriptions from the defining publications are repeated with only minor modifications. We regret that space limitations preclude the inclusion of all references relevant to this chapter.

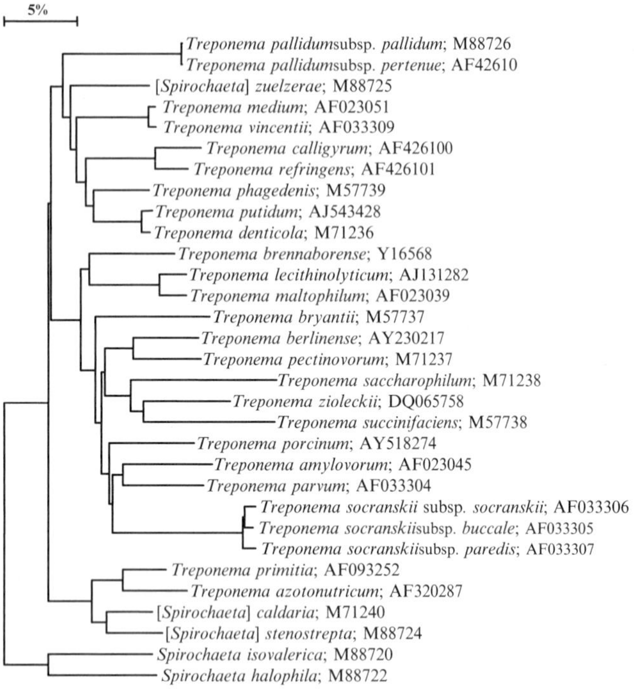

FIGURE 93. Phylogenetic tree of the genus *Treponema* and related organisms, based on 16S ribosomal RNA gene (rRNA) sequences. The distance corresponding to 5% sequence difference is indicated in the upper left corner. The GenBank accession number is provided for each organism. Approved species and candidati are not distinguished in this figure. Some of the *Spirochaeta* species shown ([*Spirochaeta*] *zuelzerae*, [*Spirochaeta*] *caldaria*, and [*Spirochaeta*] *stenostrepta*) are potential candidates for reclassification as *Treponema* species.

strains of the pathogenic species *Treponema pallidum* and *Treponema carateum* are closely related; *Treponema paraluiscuniculi*, while still a member of this group, diverges significantly from the human pathogens. In the 1984 edition of *Bergey's Manual*, Smibert (1984) reclassified the human pathogens as *Treponema pallidum* subsp. *pallidum* (venereal syphilis isolates), *Treponema pallidum* subsp. *endemicum* (endemic syphilis), and *Treponema pallidum* subsp. *pertenue* (yaws); *Treponema carateum* (pinta) has remained a separate species due to the lack of genetic information. This reclassification has withstood the test of time; indeed, the genomes of the *Treponema pallidum* subspecies differ by only a few hundred base pairs. The remaining *Treponema* species, although diverse, are clearly separated from the other genera of the family *Spirochaetaceae*.

The genome sequences of two strains (*Treponema pallidum* subsp. *pallidum* Nichols and *Treponema denticola* ATCC 35405) are available (Fraser et al., 1998; Seshadri et al., 2004). The genome size of *Treponema pallidum* is 1.138 Mb, whereas that of *Treponema denticola* is 2.843 Mb. The estimated genome sizes of other cultivatable *Treponema* species range from 2.3 to 3.9 Mb (Correia et al., 2004; Giacani et al., 2004).

The genome of *Treponema pallidum* subsp. *pallidum* Nichols does not contain any recognized extrachromosomal elements, bacteriophage sequences, or insertion elements (Fraser et al., 1998). The *Treponema denticola* chromosome contains at least one putative insertion element, a cryptic bacteriophage, and a "clustered regularly interspaced short palindromic repeat" (CRISPR) region that is thought to represent a mobile element (Seshadri et al., 2004). Some *Treponema denticola* strains contain plasmids, and one of these (pTD1) has been sequenced and shown to contain bacteriophage-related sequences (Cheetham and Katz, 1995; MacDougall et al., 1992). Myovirus-like bacteriophage particles were detected by electron microscopy in cultures of an unclassified treponeme from a bovine digital dermatitis case (Demirkan et al., 2006). Otherwise, little is known about the existence of extrachromosomal elements, bacteriophage, and mobile DNA elements in the genus *Treponema*.

Morphology. Cells of all species are helical (Stepan and Johnson, 1981) and vary in cell diameter (0.1–0.7 μm). The structure of *Treponema pallidum* subsp. *pallidum* Nichols is shown in Figure 94. The outer envelope of treponemes contains lipid, protein, and carbohydrates and is similar to the outer

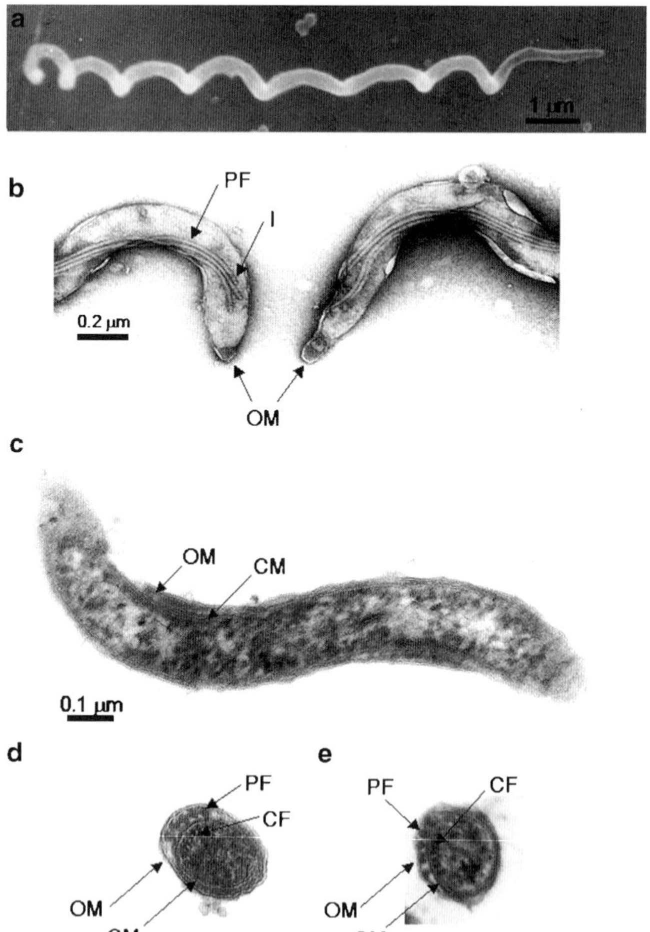

FIGURE 94. Structure of the type species *Treponema pallidum*, as exemplified by *Treponema pallidum* subspecies *pallidum* Nichols. (a) Scanning electron micrograph of an intact cell, showing the helical structure. (b) Negative stain transmission electron micrograph of the ends of two organisms. (c–e). Transmission electron micrographs of ultrathin sections. PF, periplasmic flagella; I, flagellar insertion points; OM, outer membrane; CM, cytoplasmic membrane; CF, cytoplasmic filaments (in cross-section). (Reproduced with permission from Norris et al., 2003. The Prokaryotes: an Evolving Electronic Resource for the Microbiological Community, 3rd edn. Springer, New York.)

TABLE 125. Habitats of *Treponema* species

Type	Habitat	Species
Frank pathogens	Skin, mucous membranes and (for some subspecies) internal organs of humans	*T. pallidum* subsp. *pallidum*
		T. pallidum subsp. *pertenue*
		T. pallidum subsp. *endemicum*
		T. carateum
	Genital region of rabbits with venereal spirochetosis	*T. paraluiscuniculi*
Oral spirochetes	Dental plaque in gingival crevices of humans and other animals	*T. amylovorum*
		T. denticola
		T. lecithinolyticum
		T. maltophilum
		T. medium
		T. parvum
		T. pectinovorum
		T. putidum
		T. scoliodontus
		T. socranskii
		T. vincentii
Skin-associated spirochetes	Sebaceous secretions in genital region of humans	*T. minutum*
		T. phagedenis
		T. refringens
Hoof-associated spirochetes	Hooves of cattle with digital dermatitis	*T. brennaborense*
Intestinal spirochetes	Bovine intestinal tract	*T. bryantii*
		T. saccharophilium
		T. succinifaciens
	Porcine intestinal tract	*T. berlinense*
		T. porcinum
Termite-associated spirochetes	Symbionts present in termite midguts	*T. azotonutricum*
		T. primitia

membrane of Gram-stain-negative bacteria. The lipid is mainly phospholipid and glycolipid. In general, lipopolysaccharide (LPS) appears to be absent from *Treponema* spp., and the genomes of *Treponema pallidum* subsp. *pallidum* and *Treponema denticola* lack genes required for LPS biosynthesis (Fraser et al., 1998; Seshadri et al., 2004). It has, however, been reported that *Treponema vincentii* expresses LPS (Blanco et al., 1994; Kurimoto et al., 1990). The pathogenic treponemes are further distinguished by the paucity of proteins in the outer membrane (Radolf, 1995; Radolf et al., 1989b, 1991; Walker et al., 1989); these "rare outer-membrane proteins" have not been identified definitively. The cell walls contain muramic acid, glucosamine, ornithine, glycine, and alanine (Radolf et al., 1989a). Peptidoglycan represents 1% of the dry weight of cells. Several prominent lipoproteins are associated with the cytoplasmic membrane of *Treponema pallidum* (Norris, 1993) and are implicated in the induction of inflammatory reactions in the mammalian host.

As in other spirochetes, the periplasmic flagella of *Treponema* originate at both ends of the organism and extend proximally along the length of the bacterium within the periplasmic space. The basal body and proximal hook of periplasmic flagella are similar to those of flagella found in other bacteria (Murphy et al., 2006). In the *Treponema* species examined, the flagellar filaments consist of two or more core proteins homologous to flagellins of other bacteria (e.g., FlaB1, FlaB2, and FlaB3 of *Treponema pallidum*), as well as a single sheath protein (FlaA) (Norris, 1993; Norris et al., 1988).

Treponemes have cytoplasmic filaments that extend the length of the organism along the inner layer of the cytoplasmic membrane, underlying the periplasmic flagella (Hovind-Hougen and Birch-Andersen, 1971; Izard, 2006; You et al., 1996). These ribbon-like filaments are ~5–7 nm×1 nm in cross section and occur in clusters of 4–6. The major subunit of cytoplasmic filaments is CfpA, with a molecular mass of 78–80 kDa. Most *Treponema* species examined have been found to contain cytoplasmic filaments. More recently, the cytoplasmic filaments have been shown to bridge the cytoplasmic cylinder constriction site in dividing cells, and may be anchored by a narrow plate-like structure (Izard et al., 2008).

Habitat and pathogenesis. Treponemes are primarily host-associated and are found in the microflora of humans and other animals (Table 125). The pathogenic treponemes in the *Treponema pallidum*/*Treponema carateum*/*Treponema paraluiscuniculi* group cause syphilis, yaws, pinta, and nonvenereal endemic syphilis in humans and venereal spirochetosis in rabbits; these organisms have not been cultivated continuously *in vitro*, and are usually propagated in laboratory animals, e.g., in the testes of rabbits. These bacteria are transmitted by sexual contact in the case of syphilis and venereal spirochetosis of rabbits, and by skin-to-skin contact or shared utensils for the other treponematoses. The initial infection causes lesions in the skin or mucous membranes, but *Treponema pallidum* subsp. *pallidum* and some of the other subspecies disseminate and cause systemic infection and manifestations. *Treponema pallidum* is considered a microaerophile requiring low concentrations of oxygen (1.5–5.0%) for survival and multiplication (Cox, 1994, 1980; Norris et al., 1978). Many of the other *Treponema* species identified are localized in subgingival crevices surrounding the teeth of humans and other animals and are part of the polymicrobial infection that causes periodontal disease (e.g., periodontitis) (Moore and Moore, 1994; Sela, 2001). *Treponema brennaborense* is associated with digital dermatitis in cattle, a painful ulcerative lesion occurring in the corona region of the hoof. The saprophytic *Treponema* species *Treponema minutum*, *Treponema phagedenis*, and *Treponema refringens* were isolated from material associated with the surface of the skin (smegma) in the anogenital region of humans and are nonpathogenic. *Treponema* species are also found in the intestinal tract of ruminant animals and swine, and have not as yet been implicated as pathogens. Another subset are intestinal symbionts important in the metabolism of termites.

Metabolism

The cultivated treponemes are strict anaerobes and require low redox conditions for growth. Some of these treponeme species ferment glucose and require a carbohydrate as an energy source, whereas other species gain energy from the fermentation of amino acids. Treponemes also differ in their requirement for fatty acids. Some species require long-chain fatty acids found in serum, while others require short-chain volatile fatty acids found in rumen fluid. All cultivated treponeme species require either long or short-chain fatty acids. Many of the oral organisms, including *Treponema denticola*, *Treponema vincentii*, and *Treponema scoliodontus*, require thiamine pyrophosphate in addition to serum.

Continuous culture of the pathogenic subspecies of *Treponema pallidum* has not been achieved. Limited multiplication of *Treponema pallidum* subsp. *pallidum* can be obtained in a culture system consisting of a modified tissue culture medium, Sf1Ep cottontail rabbit epithelial cell cultures, and an atmosphere containing 1.5% O_2, 5% CO_2, and 93.5% N_2 (Cox, 1994). Under these conditions, up to 100-fold increases in the number of *Treponema pallidum* over a 12-d period are observed; however, subculture does not lead to increased yields or prolonged survival (Norris and Edmondson, 1987). The genome sequence of *Treponema pallidum* subsp. *pallidum* Nichols indicates that the organism has little biosynthetic capacity (Fraser et al., 1998). The lack of continuous growth is likely due to a deficiency in required nutrients and/or irreversible damage resulting from toxic conditions in the *in vitro* environment. Cultivated treponemes (*Treponema phagedenis*, *Treponema refringens*, *Treponema denticola*, and *Treponema vincentii*) are inhibited by penicillin (100–1000 U/l), ampicillin (0.1–1.0 mg/l), oxacillin (0.1–10 mg/l), cloxacillin (0.1–1.0 mg/l), cephalothin (0.1–10 mg/l), vancomycin (0.1–10 mg/l), bacitracin (0.1–1.0 mg/l), erythromycin (0.1–1.0 mg/l), novobiocin (10–500 mg/l), tetracycline (1 mg/l), doxycycline (0.1–1.0 mg/l), chloramphenicol (100–500 mg/l), kanamycin (100–1000 mg/l), and viomycin (10–1000 mg/l). Bactericidal concentrations of antibiotics are usually much higher than inhibitory concentrations (Abramson and Smibert, 1971). All treponemes are resistant to cycloserine (500–1000 mg/l), polymyxin B (500–1000 mg/l), and nalidixic acid (500–1000 mg/l) (Abramson and Smibert, 1971). Oral and rumen treponemes are resistant to rifampicin at concentrations of 1–50 mg/l (Leschine and Canale-Parola, 1980b; Stanton and Canale-Parola, 1979), apparently due to an asparagine substitution at position 518 of the RNA polymerase β subunit RpoB (Stamm et al., 2001). Recent studies have shown that a substantial proportion of *Treponema pallidum* subsp. *pallidum* strains have a point mutation in the 16S rRNA genes that renders the organism resistant to macrolides, including azithromycin (Lukehart et al., 2004; Marra et al., 2006; Mitchell et al., 2006; Stamm and Bergen, 2000).

Enrichment and isolation procedures

The cultivable treponemes can be isolated by two general methods. The first uses membrane filters placed on the surface of agar media. The second method uses rifampicin as a selective agent (Leschine and Canale-Parola, 1980b; Stanton and Canale-Parola, 1979).

A membrane filter with a pore size of 0.2 µm is placed on an agar medium containing either 10% inactivated animal serum or 30% rumen fluid. The agar concentration should be no more than 1.3–1.4%. The sample is placed on the filter and the Petri dish quickly placed in an anaerobic jar or chamber and incubated. Treponemes and all other genera of spirochetes migrate through the filter whereas most other bacteria can not, resulting in a spirochetal enrichment. Incubation usually requires 1–2 weeks at 37°C. After incubation, the filter is removed and a white haze can be seen in the agar. Treponemes migrate through the filter and grow into the agar. A plug of agar is removed and inoculated into a pre-reduced culture medium.

In the second method, filter-sterilized rifampicin is added to either pre-reduced broth or molten agar medium in tubes to a final concentration of 1.0–2.0 mg/l. The tubes are inoculated and incubated at 37°C. Serial dilutions of samples can be made in the selective medium. The medium contains 1% agar and is allowed to solidify in the form of agar deeps. After incubation, white cottony colonies of treponemes are seen within the medium. These colonies can be removed with a Pasteur pipette and inoculated into pre-reduced medium.

The selective broth medium will contain treponemes and a few other kinds of bacteria because rifampicin does not inhibit all other bacteria. Treponemes can be isolated from the enriched selective culture by inoculating pre-reduced agar medium (1.3–1.4% agar) in 6- or 8-oz prescription bottles. The "bottle plates" can be streaked or the culture inoculated into molten agar (45°C) in the bottles, mixed, and the agar allowed to solidify on the flat side of the bottle. The gas phase in the bottle plates should be 90% N_2 and 10% CO_2.

The non-cultivatable, pathogenic *Treponema* species can be propagated through animal inoculation (Miller, 1971; Turner and Hollander, 1957). Rabbits are the most widely used host for experimental infection, although primates, guinea pigs, and hamsters can also be infected with certain *Treponema pallidum* subspecies (Table 126). Early lesions (e.g., primary or secondary lesions of syphilis patients) sampled prior to the healing stages are most likely to yield infectious organisms. Either a tissue biopsy or lesion exudate can be used for inoculation. Tissue is minced and extracted in a serum-containing medium (e.g., 0.85% NaCl with 10% heat-inactivated rabbit serum). Organisms are highly susceptible to reactive oxygen species, so ideally, the medium should be pre-equilibrated with 95% N_2: 5% CO_2 and extraction carried out under anaerobic conditions. The extract should be monitored for the presence of spirochetes by darkfield microscopy. After 10–30 min, the extract should be centrifuged at 300 × *g* for 7min to remove tissue debris, and the supernate inoculated as soon as possible by needle and syringe into the testes (1 ml per testis) or intradermally into a

TABLE 126. Differential characteristics of the noncultivable, named species of the genus *Treponema*

Characteristic[a]	1. *T. pallidum* subsp.			7. *T. carateum*	13. *T. paraluiscuniculi*
	pallidum	*pertenue*	*endemicum*		
Natural host:					
Humans	+	+	+	+	–
Rabbits	–	–	–	–	+
Nature of infection in natural host:					
Systemic; may affect most internal organs	+	–	–	–	–
Usually restricted to cutaneous lesions	–	+	+	+	+
Sexually transmitted	+	–	–	–	+
Geographical distribution:					
Worldwide	+	–	–	–	nr
Found only in tropical countries in both hemispheres	–	+	–	–	nr
Found only in tropical countries in the Western hemisphere	–	–	–	+	nr
Restricted to semi-arid regions in the Middle East, Africa, and Southeast Asia	–	–	+	–	nr
Cutaneous lesion produced in:					
Rabbits	+	+	+	–	nr
Hamsters	+[b]	+	+	–	nr
Mice	–	–		–	nr
Guinea pigs	+[c]	–	–	–	nr
Primates	+	+	+	+	nr

[a]+, Positive; –, negative; nr, not reported.

[b]Hamsters are less susceptible to subsp. *pallidum* than to subsp. *pertenue* or *endemicum*.

[c]Lesions are only slightly indurated and do not ulcerate (Turner and Hollander, 1957).

shaved area of the back (0.1 ml per site) of a rabbit. *Treponema pallidum* is highly temperature sensitive, so the rabbits should be maintained at 18–20°C and the skin inoculation sites kept free of hair. Rabbits should be examined daily for the presence of orchitis (testis swelling) or erythematous, indurated lesions at the skin inoculation sites. The infection can also be monitored by testing for Rapid Plasma Reagin (RPR) test reactivity. If lesions are not apparent at 3–4 weeks, blind passage of extracts from testicular tissue or popliteal lymph node to a naïve animal may yield positive results. Organisms can be maintained by extraction of organisms from infected tissue and inoculation of animals at 10–21-d intervals, depending on the strain. The inoculum for intratesticular inoculation should be $1–5\times10^7$ cells; yields vary greatly according to the strain but will typically be in the range of 10^9 to 10^{11} cells per animal at the time of peak orchitis.

Maintenance procedures

Cultures of all species of treponemes can be stored in the frozen state using liquid nitrogen or mechanical freezers at –80°C. Cryoprotective agents such as 10–15% (v/v) glycerol or 10% (v/v) dimethyl sulfoxide are added to cultures just prior to freezing. Lyophilization has generally not been successful for preservation of treponemes.

Taxonomic comments

Named species of *Treponema* can be subdivided into the noncultivatable, pathogenic species and the cultivatable species. Differentiation of the noncultivable species of *Treponema* is presented in Table 126. Characteristics useful for differentiating the cultivable species are given in Tables 127–129.

The uncultivated species include *Treponema pallidum* (*Treponema pallidum* subsp. *pallidum*, *Treponema pallidum* subsp. *pertenue*, and *Treponema pallidum* subsp. *endemicum*), *Treponema carateum*, and *Treponema paraluiscuniculi*. The primary distinguishing characteristics of these species and subspecies are the diseases that they cause: syphilis (subsp. *pallidum*), yaws (subsp. *pertenue*), endemic syphilis (subsp. *endemicum*), pinta (*Treponema carateum*), and venereal spirochetosis of rabbits (*Treponema paraluiscuniculi*). To the extent they have been examined, these species and subspecies are indistinguishable in terms of morphology, *in vitro* characteristics (e.g., motility and survival under different conditions), protein content as determined by SDS-PAGE, and genomic DNA–DNA hybridization. Prior infection of animals with strains of different *Treponema* species or subspecies provides incomplete cross-protection from infection with other strains, indicating that antigenic differences are present (Turner and Hollander, 1957); however few such differences have been detected at the molecular level (Baker-Zander and Lukehart, 1983, 1984; Fohn et al., 1988; Noordhoek et al., 1990).

Centurion-Lara et al. (1998) found that the 5′ and 3′ regions flanking *tpp15* contained polymorphisms that could be used to distinguish *Treponema pallidum* subsp. *pallidum* from the other pathogenic *Treponema* species and subspecies, and *Treponema paraluiscuniculi* from the human pathogens. PCR amplification followed by restriction fragment length polymorphism (RFLP) analysis revealed the presence of a unique *Eco*47III site in the 5′ region in *Treponema pallidum* subsp. *pallidum* strains, and a *Xcm*I site in the 3′ region in *Treponema paraluiscuniculi* strains. Subsequent studies by the same group (Centurion-Lara et al., 2006) showed that additional polymorphisms in the *Treponema*

TABLE 127. Available information on the characteristics of the cultivatable *Treponema* species[a]

Characteristic	2. *T. amylovorum*	3. *T. azotonutricum*	4. *T. berlinense*	5. *T. brennaborense*	6. *T. bryantii*	8. *T. denticola*	9. *T. lecithinolyticum*	10. *T. maltophilum*	11. *T. medium*	12. *T. minutum*	14. *T. parvum*	15. *T. pectinovorum*	16. *T. phagedenis*	17. *T. porcinum*	18. *T. primitia*	19. *T. putidum*	20. *T. refringens*	21. *T. saccharophilum*	22. *T. scoliodontus*	23. *T. socranskii*	24. *T. succinifaciens*	25. *T. vincentii*
Cell diameter (μm)	0.25	0.2–0.3	0.3	0.25–0.55	0.3	0.2±0.02	0.15	0.2	0.2–0.3	0.15–0.2	0.18	0.28–0.30	0.2–0.25	0.3	0.2–0.3	0.25	0.24	0.6–0.7	0.15–0.2	0.16–0.18	0.3	0.2–0.25
Cell length (μm)	7	10–12	6	5–8	3–8	7.7±0.94	5	5	5–16	9–12	1	7–15	6–12	6–8	7	10	5–8	12–20	6–16	6–15	4–8	5–16
Wavelength (μm)	1.2	1.2	nr	nr	nr	1.23±0.15	0.7	0.7	nr	1.3	0.8	nr	1.4–1.6	nr	2.3	3	1.8	nr	nr	nr	nr	1.3
Amplitude (μm)	0.3	nr	nr	nr	nr	0.5±0.05	0.3	0.3	nr	0.2	0.3	nr	0.2–0.3	nr	nr	1.5	0.2–0.3	nr	nr	nr	nr	0.2–0.3
Flagellar arrangement	3:6:3	1:2:1	1:2:1	2:4:2	1:2:1	2:4:2	1:2:1	1:2:1	5:10:5	nr	1:2:1	2:4:2	nr	1:2:1	1:2:1	2:4:2	2:4:2	16:32:16	nr	1:2:1	2:4:2	nr
DNA G+C content (mol%)	nr	50	nr	nr	36±1	37.9	nr	nr	51	37	nr	39	38–39	nr	50.9–51.0	nr	39–43	54	nr	50.5±2	36	nr
Growth requirements:																						
Optimum temperature or range	nr	30	nr	nr	37	nr	nr	nr	nr	34–40	37	37	37	nr	30	37	30–42	37–39	30–42	37	37	25–42
Serum	+	–	–	–	–	+	–	–	+	+	–	–	+	nr	–	–	+	–	+	nr	–	+
Thiamine pyrophosphate	nr	nr	nr	nr	+	+	nr	nr	nr	–	nr	nr	–	nr	nr	nr	–	nr	+	nr	–	+
N-Acetylglucosamine	nr	nr	nr	nr	nr	nr	nr	nr	nr	nr	+	nr	nr	nr	nr	nr	nr	nr	nr	nr	nr	nr
Volatile fatty acids	nr	nr	nr	nr	+	–	nr	nr	+	–	nr	+	–	nr	nr	nr	–	+	–	nr	+	–
Amino acids fermented	nr	nr	nr	nr	–	+	+	nr	nr	+	–	–	nr	–	nr	+	+	nr	nr	nr	nr	+
Esculin hydrolysis	nr	nr	–	nr	nr	+	nr	nr	+	+	nr	–	d	nr	nr	nr	+	nr	–	nr	nr	d
Indole production	nr	nr	nr	+	nr	d	nr	nr	–	+	nr	–	+	nr	nr	nr	+	nr	–	nr	nr	w
1% Glycine (growth)	nr	nr	nr	nr	nr	d	nr	nr	nr	+	nr	nr	+	nr	nr	nr	d	nr	–	nr	nr	–
Converts fumarate to succinate	nr	nr	nr	nr	nr	–	nr	nr	nr	–	nr	nr	+	nr	nr	nr	+	nr	nr	nr	nr	–

[a]Symbols: +, >85% positive; d, different strains give different reactions (16–84% positive); –, 0–15% positive; w, weak reaction; nr, not reported or not applicable.

TABLE 128. Carbohydrate utilization reported for *Treponema* species

	2. *T. amylovorum*	3. *T. azotonutricum*	4. *T. berlinense*	5. *T. brennaborense*	6. *T. bryantii*	9. *T. lecithinolyticum*	10. *T. maltophilum*	11. *T. medium*	14. *T. parvum*	15. *T. pectinovorum*	16. *T. phagedenis*	17. *T. porcinum*	18. *T. primitia*	19. *T. putidum*	21. *T. saccharophilum*	23. *T. socranskii*	24. *T. succinifaciens*
Acetate	nr	nr	nr	nr	–	nr	nr	nr	nr	nr	nr	nr	nr	nr	–	nr	–
N-Acetyl-galactosamine	nr	nr	nr	nr	nr	nr	nr	nr	+	nr	nr	nr	nr	nr	nr	nr	+
N-Acetylglucosamine	nr	nr	nr	nr	nr	+	+	nr	+	nr	nr	nr	nr	nr	nr	nr	nr
Adonitol	nr	nr	nr	nr	nr	nr	nr	–	nr	–	nr	nr	nr	nr	nr	–	nr
Amygdalin	nr	nr	nr	nr	nr	nr	nr	nr	nr	–	–	nr	nr	nr	nr	–	nr
D-Arabinose	–	–	+	nr	nr	+	d	–	–	–	–	–	d	–	nr	d	nr
D-Cellobiose	–	+	nr	nr	+	–	–	nr	–	–	–	nr	d	–	+	–	+
Dulcitol	nr	nr	nr	nr	nr	nr	nr	nr	–	–	nr	nr	nr	nr	–	nr	–
Formate	nr	nr	nr	nr	–	nr	nr	d	nr	nr	nr	nr	–	nr	–	nr	–
D-Fructose	–	+	+	nr	–	d	d	+	+	–	+	–	nr	–	+	d	–
D-Fucose	–	nr	nr	nr	nr	–	–	nr	–	nr	nr	nr	nr	–	nr	nr	nr
D-Galactose	–	nr	nr	nr	+	–	d	+	+	–	d	nr	nr	–	+	d	+
D-Galacturonic acid	–	nr	+	nr	nr	–	–	nr	–	+	nr	nr	nr	–	+	–	nr
Gluconate	nr	nr	nr	nr	–	nr	+	nr	nr	nr	nr	nr	nr	nr	–	nr	–
D-Glucose	+	+	+	nr	+	–	d	+	+	–	+	–	+	–	+	d	+
D-Glucuronic acid	–	nr	+	nr	nr	–	–	nr	–	+	nr	nr	nr	–	+	–	nr
Glycerol	nr	nr	nr	nr	–	nr	+	nr	nr	–	nr	nr	nr	nr	–	d	–
Glycogen	+	nr	nr	nr	nr	nr	nr	nr	nr	–	–	nr	nr	–	nr	d	nr
H_2+CO_2	nr	–	nr	nr	nr	nr	nr	nr	nr	nr	nr	nr	+	nr	nr	nr	nr
Inulin	nr	nr	nr	nr	–	nr	nr	+	nr	–	nr	nr	nr	nr	+	–	–
Lactate	nr	–	nr	nr	nr	nr	nr	nr	nr	nr	–	nr	–	nr	–	nr	–
D-Lactose	–	nr	nr	nr	+	–	–	–	–	–	+	nr	nr	–	+	–	+
Maltose	+	+	+	nr	–	–	+	+	–	–	–	+	+	–	+	d	+
D-Mannitol	–	–	+	nr	–	–	–	–	+	–	+	–	+	–	–	–	–
D-Mannose	–	nr	+	+	+	nr	d	+	+	–	+	–	nr	–	+	d	+
D-Melibiose	–	nr	nr	nr	nr	–	d	nr	–	–	–	nr	nr	–	nr	nr	nr
Melezitose	nr	nr	nr	nr	nr	nr	nr	nr	nr	–	–	nr	nr	nr	nr	–	nr
Pectin (polygalacturonic acid)	nr	nr	–	nr	–	nr	nr	nr	+	+	nr	–	nr	nr	+	nr	nr
Pyruvate	nr	–	nr	nr	nr	nr	nr	nr	nr	nr	+	nr	+	nr	nr	nr	–
D-Raffinose	nr	nr	nr	+	–	nr	nr	+	nr	–	nr	nr	nr	nr	+	nr	–
D-Ribose	–	+	nr	nr	–	+	d	+	+	–	d	nr	–	–	–	d	–
Salicin	nr	nr	nr	nr	nr	nr	nr	+	nr	–	–	nr	nr	nr	nr	–	nr
D-Sorbitol	nr	nr	nr	nr	–	nr	–	–	nr	–	nr	nr	nr	nr	–	–	–
Starch	+	nr	nr	nr	–	nr	+	nr	nr	–	–	nr	nr	–	+	nr	+
Succinate	nr	nr	nr	nr	–	nr	nr	nr	nr	nr	nr	nr	nr	nr	–	nr	–
D-Sucrose	–	–	+	nr	+	–	–	+	–	–	–	–	nr	–	+	d	–
D-Trehalose	–	–	+	nr	nr	–	–	+	–	–	d	–	nr	–	nr	d	–
D-Xylose	–	+	nr	nr	+	d	–	–	+	–	–	nr	+	–	–	d	+
Xylitol	nr	–	nr	nr	–	nr	nr	nr	nr	nr	nr	nr	nr	nr	–	nr	–
L-Arabinose	–	nr	nr	+	nr	–	d	nr	+	nr	nr	nr	nr	–	+	d	+
L-Fucose	–	nr	+	nr	nr	+	+	nr	–	nr	nr	–	nr	–	nr	–	nr
L-Rhamnose	–	nr	+	nr	–	–	d	nr	–	nr	nr	–	nr	–	–	d	–
L-Sorbose	–	nr	nr	nr	–	–	–	nr	–	nr	nr	nr	nr	–	–	nr	–
Uric acid	nr	–	nr	nr	nr	nr	nr	nr	nr	nr	nr	nr	nr	nr	nr	nr	nr
L-Xylose	–	nr	nr	nr	nr	–	–	nr	–	nr	nr	nr	nr	–	nr	nr	nr

Symbols: +, >85% positive; d, different strains give different reactions (16–84% positive); –, 0–15% positive; nr, not reported.

pallidum repeat (*tpr*) genes *tprC* and *tprI* could be detected by RFLP analysis and used to distinguish each of the *Treponema pallidum* subspecies and a simian isolate, *Treponema* strain Fribourg-Blanc. These are currently the only systematic means for distinguishing the pathogenic *Treponema* species and subspecies. Additional approaches have been developed for the molecular subtyping of *Treponema pallidum* subsp. *pallidum* strains (Pillay et al., 1998), and have been used in epidemiologic studies of syphilis transmission (Diclemente et al., 2004; Molepo et al., 2007; Pillay et al., 2002; Sutton et al., 2001).

Distinction of the cultivatable *Treponema* species is dependent on 16S rRNA gene sequence heterogeneity (Figure 93), phenotypic characteristics, some morphologic properties (e.g., cellular dimensions and number of flagella), and natural habitat. Little DNA–DNA reassociation data are available (Maio and Fieldsteel, 1978). 16S rRNA comparisons (Paster

TABLE 129. Enzymic activities of *Treponema* species, as determined by API ZYM analysis[a]

	2. *T. amylovorum*	4. *T. berlinense*	5. *T. brennaborense*	8. *T. denticola*[b]	9. *T. lecithinolyticum*	10. *T. maltophilum*	14. *T. parvum*	15. *T. pectinovorum*[b]	17. *T. porcinum*	19. *T. putidum*	23. *T. socranskii* subsp. *buccale*[b]	23. *T. socranskii* subsp. *paredis*[b]	23. *T. socranskii* subsp. *socranskii*[b]	25. *T. vincentii*[b]
Number of strains examined	1	6	1	2	8	3	4	1	1	7	1	1	1	2
1. Alkaline phosphatase	+	–	+	+	+	+	+	–	–	d	+	+	+	–
2. Esterase (C 4)	+	–	+	+	+	+	+	+	+	+	+	+	+	d
3. Esterase Lipase (C 8)	–	–	+	+	+	+	+	+	–	+	+	+	–	d
4. Lipase (C 14)	–	–	–	–	–	–	–	–	–	–	–	–	–	–
5. Leucine arylamidase	–	–	–	+	–	–	–	–	–	+	–	–	–	+
6. Valine arylamidase	–	–	–	–	–	–	–	–	–	–	–	–	–	–
7. Cystine arylamidase	–	–	–	d	–	–	–	–	–	d	–	–	–	–
8. Trypsin	–	–	–	+	–	–	–	–	–	+	–	–	–	–
9. α-Chymotrypsin	–	–	–	+	–	–	–	–	–	–	–	–	–	–
10. Acid phosphatase	+	+	+	+	+	+	+	+	+	+	+	+	+	+
11. Naphthol-AS-BI-phosphohydrolase	+	+	+	+	+	+	+	+	+	+	+	+	+	+
12. α-Galactosidase	–	–	–	+	–	d	–	–	–	d	–	–	–	–
13. β-Galactosidase	–	–	+	d	+	d	–	–	–	+	–	–	–	+
14. β-Glucuronidase	–	–	–	–	+	d	+	–	–	–	+	–	–	–
15. α-Glucosidase	–	–	+	–	–	+	–	–	+	–	–	–	–	–
16. β-Glucosidase	–	–	–	+	–	d	–	–	–	+	–	–	–	–
17. *N*-acetyl-β-glucosaminidase	–	–	+	–	+	–	–	–	–	–	–	–	–	+
18. α-Mannosidase	–	–	–	–	–	–	–	–	–	–	–	–	–	–
19. α-Fucosidase	+	–	–	–	+	d	–	–	–	–	–	–	–	–

[a]Symbols: +, >85% positive; d, different strains give different reactions (16–84% positive); –, 0–15% positive.

[b]Results from Wyss et al. (1996). Other results are from the defining publications.

and Dewhirst, 2000) and the available mol% G+C data indicate that the genus *Treponema* is an ancient, diverse phylogenetic group. There are instances where *Treponema* with high 16S rRNA similarity are from the same environment (e.g., *Treponema primitia* and *Treponema azotonutricum* from termite guts), and others where organisms with 16S rRNA relatedness are from different environments (e.g., *Treponema berlinense* from porcine intestinal tract and *Treponema pectinovorum* from human gingival crevices) (Figure 93). These results suggest that some *Treponema* strains may have "jumped" from one environment to another during evolution. Analyses of oral treponeme strains (Paster et al., 1998) from the Smibert collection indicate the difficulty in reconciling 16S rRNA and phenotypic data (such as fatty acid content), and the importance of using more global genotyping approaches, including DNA–DNA reassociation studies, for distinguishing phylotypes with ≥98% 16S rRNA sequence identity.

As noted in the chapter on *Spirochaeta*, three species of *Spirochaeta*, namely *Spirochaeta stenostrepta*, *Spirochaeta zuelzerae*, and *Spirochaeta caldaria*, belong with members of the genus *Treponema* rather than with other "free-living" *Spirochaeta* as indicated by 16S rRNA phylogeny (Figure 93) and single-base signature analysis (Paster et al., 1991). There has been no formal renaming, so they are described in the *Spirochaeta* chapter. As previously discussed, these "free-living" spirochetes may be descendants of precursors of host-associated treponemes or may have been disseminated via fecal contamination (Paster et al., 1991).

Intestinal spirochetes previously classified as *Treponema hyodysenteriae* and *Treponema innocens* have been reclassified as species of *Brachyspira*.

Further reading

Additional information regarding the genus *Treponema* is available in several recent books (Cabello et al., 2006; Radolf and Lukehart, 2006; Saier and Garcia-Lara, 2001), book chapters (Norris et al., 2003; Pope et al., 2006), and reviews (Antal et al., 2002; Cullen and Cameron, 2006; LaFond and Lukehart, 2006; Norris et al., 2001; Radolf et al., 1999).

Differentiation of the genus *Treponema* from other closely related genera

Characteristics useful for distinguishing *Treponema* from the other members of the family *Spirochaetaceae* are indicated in the key to the family. In terms of the pathogenic genera, *Borrelia* species are also host-dependent, but utilize arthropods (ticks or lice) as intermediate hosts and transmission agents. They are also somewhat larger in diameter and more loosely coiled than *Treponema*, and are facultative anaerobes. *Leptospira* are also associated with mammals, but are capable of surviving for prolonged periods in water or soil. They are more tightly coiled than either *Borrelia* or *Treponema*, and are obligate aerobes.

List of species of the genus *Treponema**†

1. **Treponema pallidum** (Schaudinn and Hoffman 1905) Schaudinn 1905, 1728[AL] (*Spirochaeta pallida* Schaudinn and Hoffman 1905, 528)

pal'li.dum. L. neut. adj. *pallidum* pale, pallid.

Tightly coiled, ~0.18 µm in diameter by 6–20 µm in length. The wavelength of coils is 1.1 µm and the amplitude is 0.2–0.3 µm. The ends of the cells are pointed, and a protrusion of the outer membrane at the end is often visible in well-preserved specimens by electron microscopy with negative staining (Figure 94). Two to four periplasmic flagella are inserted into each end of the cell, and overlap in the middle of the cell. Motile with graceful flexuous movements. Microaerophilic, with an optimal O_2 concentration in the range of 1–5%.

Obligate pathogens of humans. Can cause experimental infection and skin lesions in rabbits, guinea pigs, hamsters, and primates (Table 126). Sequence comparisons and DNA/DNA hybridization indicates that close to 100% DNA homology exists between the *Treponema* strains that cause syphilis, yaws, and endemic syphilis (Maio and Fieldsteel, 1980). Therefore, these organisms are considered subspecies of *Treponema pallidum* with distinctive clinical symptoms in humans and different patterns of infection in laboratory animals (Table 126). The subspecies can be distinguished by PCR and RFLP analysis of the *tpp15*, *tprC*, and *tprI* genes (Centurion-Lara et al., 1998, 2006).

Source: humans.

DNA G+C content (mol%): 52.4–53.7 (T_m).

Type strain: none designated.

1a. **Treponema pallidum subsp. pallidum** (Schaudinn and Hoffman 1905) Schaudinn 1905, 1728, subsp. nov.

The morphology and characteristics are as described for the species and as listed in Table 126. *Treponema pallidum* subsp. *pallidum* is the cause of venereal and congenital syphilis in humans. It has not been cultivated continuously in artificial media or in tissue culture. Propagated by intratesticular inoculation of rabbits.

Successful replication of *Treponema pallidum* subsp. *pallidum* (virulent Nichols strain) has been reported to occur on the surface of tissue culture cells of cottontail rabbit epithelium (Sf1EP) growing in a monolayer in an atmosphere of 1.5% O_2. A 49-fold increase (mean value) in cell numbers was reported in primary cultures of *Treponema pallidum*, but increased or prolonged multiplication with subculture has not been obtained to date (Cox, 1994; Fieldsteel et al., 1981; Norris and Edmondson, 1987). *Treponema pallidum* cells show rapid attachment to cultured mammalian cells (Fitzgerald et al., 1977a; Hayes et al., 1977). Proteins that bind to the host extracellular matrix proteins laminin and fibronectin have been identified (Cameron, 2003, 2004, 2005).

Obligate pathogen of humans. Strains such as the Nichols pathogenic strain are propagated by intratesticular inoculation of rabbits. Cutaneous inoculation of rabbits produces skin lesions. Cutaneous inoculation of hamsters, mice, and guinea pigs produces no apparent infection or visible lesions. A slight lesion is occasionally seen at the point of injection of guinea pigs.

Microaerophilic. Survives in artificial media or tissue culture longest when incubated in an atmosphere of 1–5% O_2 (Cox, 1994; Fieldsteel et al., 1977; Fitzgerald et al., 1977b, 1980; Norris et al., 1978; Sandok et al., 1978). Glucose is metabolized by way of the Embden–Meyerhof–Parnas and hexose monophosphate pathways (Schiller and Cox, 1977). Oxygen uptake by *Treponema pallidum* has been reported and is glucose-dependent (Barbieri and Cox, 1981; Cox and Barber, 1974). Oxidation of pyruvate occurs only when oxygen is present (Barbieri and Cox, 1979). Major fermentation products of glucose are acetate and CO_2 (Nichols and Baseman, 1975).

Isolated from human patients with syphilis. The reference Nichols pathogenic strain was isolated from the cerebrospinal fluid of a patient with neurosyphilis (Nichols and Hough, 1913). Shows 100% DNA/DNA homology by saturation reassociation with *Treponema pallidum* subsp. *pertenue* (Gauthier strain) but no significant DNA–DNA reassociation with *Treponema phagedenis* or *Treponema refringens* (Maio and Fieldsteel, 1978, 1980).

Source: human patients with syphilis.

DNA G+C content (mol%): 52.8 (genome sequence); 52.4–53.7 (T_m).

Type strain: none designated.

Reference strain: Nichols pathogenic.

Sequence accession no. (16S rRNA gene): M88726.

1b. **Treponema pallidum subsp. pertenue** (Castellani 1905) subsp. nov. (*Spirochaeta pertenius* Castellani 1905, 54)

per.te.nu'e. L. neut. adj. *pertenue* very thin, slender.

The morphology and characteristics as described for the species and as listed in Table 126. Pathogenic to humans. Causes yaws in humans, a contagious disease that is spread by skin-to-skin contact. *Treponema pallidum* subsp. *pertenue* has not been cultivated in artificial media or in tissue culture. Cutaneous lesions are produced at the point of inoculation in rabbits and Syrian hamsters, but not in guinea pigs. Sera from patients with yaws give positive results with serologic tests for syphilis. Attachment of *Treponema pallidum* subsp. *pertenue* to five different mammalian cell lines was compared to that of *Treponema pallidum* subsp. *pallidum* (Fieldsteel et al., 1979). *Treponema pallidum* subsp. *pertenue* attached to all five cell lines, as did *Treponema pallidum* subsp. *pallidum*. No preferential attachment was found with *Treponema pallidum* subsp. *pertenue* for nude mouse ear and cottontail rabbit epithelial (Sf1Ep) cells, but preferential attachment did occur with *Treponema pallidum* subsp. *pallidum*.

Inbred hamsters (LSH/Ss LAK) infected with *Treponema pallidum* subsp. *endemicum* (Bosnia A strain) were resistant to reinfection with both *Treponema pallidum* subsp. *pertenue* and *Treponema pallidum* subsp. *pallidum* (Schell et al., 1980).

Treponema pallidum subsp. *pertenue* (Gauthier strain) shows 100% DNA homology to *Treponema pallidum* (Nichols

---

*The 16S rRNA sequences of *Spirochaeta stenostrepta*, *Spirochaeta zuelzerae*, and *Spirochaeta caldaria* (described in the section on *Spirochaeta*) are most similar to those of members of the genus *Treponema* (Figure 93).

†*Treponema pallidum*, *Treponema carateum*, and *Treponema paraluiscuniculi* have not been cultivated continuously *in vitro*. The other species are anaerobes and have been cultivated; their characteristics are provided in Table 127, Table 128, and Table 129.

and KKJ strains) and no homology to *Treponema phagedenis* and *Treponema refringens* by DNA–DNA hybridization (Maio and Fieldsteel, 1980).

Source: Lesions from cases of yaws. Present in tropical areas of Africa, Southeast Asia, the Western Pacific Islands, and South and Central America (Antal et al., 2002).

DNA G+C content (mol%): 52–53.7 (T_m).

Type strain: none designated.

Reference strains: Gauthier or Haiti B.

Sequence accession no. (16S rRNA gene): AF42610.

1c. **Treponema pallidum subsp. endemicum** subsp. nov.

en.de'mi.cum. N.L. neut. adj *endemicum* (from Gr. adj. *endêmos -on* native, dwelling in place), endemic.

The morphology and characteristics are as described for the species and as listed in Table 126. Pathogenic to humans. The cause of nonvenereal endemic syphilis in humans, a contagious disease spread in pre-pubertal years by contact with infected individuals or shared use of contaminated utensils. *Treponema pallidum* subsp. *endemicum* has not been successfully cultivated in artificial media or tissue culture. Propagated by intratesticular inoculation of rabbits or by intradermal inoculation of hamsters. The organisms can be isolated from inguinal lymph nodes 3–4 weeks after intradermal infection. Inbred hamsters (e.g., LSH/Ss LAK) are particularly useful for study of this organism (Schell et al., 1980). Produces cutaneous lesions in rabbits, hamsters, and guinea pigs but not in mice.

Sera from patients with nonvenereal epidemic syphilis give positive results with serologic tests for syphilis.

This subspecies was created (Smibert, 1984) because the organism is considered a variant of *Treponema pallidum* and has its own clinical symptoms in human infection as well as the ability to infect and produce skin lesions in different laboratory animals, as does the organism of venereal syphilis (Table 126).

Source: lesions from patients with nonvenereal endemic syphilis. Found in semi-arid areas of Africa, the Middle East, and some areas of Southeast Asia.

DNA G+C content (mol%): not determined.

Type strain: none designated.

Reference strain: Bosnia A.

Sequence accession no. (16S rRNA gene): not determined.

2. **Treponema amylovorum** Wyss, Choi, Schüpbach, Guggenheim and Göbel 1997, 844[VP]

a.my.lo.vo'rum. Gr. n. *amylum* starch; N.L. neut. adj. *vorum* (from L. v. *voro* to devour), devouring; N.L. neut. adj. *amylovorum* starch-devouring.

Treponema amylovorum is an intermediate-sized, obligately anaerobic, helically coiled, motile treponeme (Wyss et al., 1997). The type strain HA2P^T was isolated from subgingival plaque of a deep human periodontal lesion. The cells are approximately 0.25 μm in diameter and 7 μm in length, with a wavelength of ~1.2 μm and an amplitude of ~0.3 μm. They have six periplasmic flagella (three at each end of the cell) that overlap in the center of the cell (i.e., flagellar arrangement of 3:6:3). In liquid media of low viscosity, cells exhibit active cellular rotation and jerky flexing but no directional motility. However, in media of higher viscosity or when cells creep along the surface, they exhibit slow translational movement. Cells can be stored at temperatures below −70°C in medium supplemented with 10–20% glycerol.

When streaked onto OMIZ-Pat/HuS agarose plates (Wyss et al., 1997), *Treponema amylovorum* forms dense, off-white, subsurface colonies up to 3 mm in diameter within 5 d of inoculation. *Treponema amylovorum* does not grow in chemically defined OMIZ-W1 medium and requires the addition of yeast extract and/or Neopeptone (or fractions thereof). Addition of 1% human serum is highly stimulatory, whereas fetal bovine serum or higher concentrations of human serum are inhibitory. Growth of strain HA2P^T is accompanied by acid production as detected by phenol red indicator in the medium, and strictly depends on the presence of at least one of the following carbohydrates: D-glucose, maltose, starch, or glycogen. Acid is produced from these carbohydrates. None of the other carbohydrates tested (at a concentration of 2 g/l) support growth; these carbohydrates include D-arabinose, D-cellobiose, D-fructose, D-fucose, D-galactose, D-galacturonic acid, D-glucuronic acid, D-lactose, D-mannitol, D-mannose, D-melibiose, D-ribose, D-sucrose, D-trehalose, D-xylose, L-arabinose, L-fucose, L-rhamnose, L-sorbose, and L-xylose. Following the exponential growth phase, with an estimated doubling time of less than 4 h, cells of HA2P^T rapidly lose viability and disintegrate into small vesicles. Catalase-negative. With API ZYM strips, the only (weak) enzyme activities detected in cells grown on glucose are alkaline and acid phosphatases, naphtholphosphohydrolase, C4 esterase, and α-fucosidase. In cells grown on either maltose, starch, or glycogen, weak α-glucosidase activity is also detected. Strain HA2P^T is resistant to 1 mg/l rifampicin and 100 mg/l phosphomycin.

On the basis of a phylogenetic comparison of 16S rRNA sequences, *Treponema amylovorum* is a species that is genetically distinct from previously described treponemes. Its SDS-PAGE protein and antigen profiles and its Western blot profile are readily distinguished from those of other cultivable, oral *Treponema* species. Furthermore, size and flagellation, as well as rapid flexing motility, clearly distinguish *Treponema amylovorum* from other treponeme species, such as *Treponema denticola*, *Treponema maltophilum*, *Treponema pectinovorum*, and *Treponema socranskii*. In contrast to the growth of the two asaccharolytic oral species, *Treponema denticola* and *Treponema vincentii*, growth of strain HA2P^T is strictly carbohydrate-dependent, and the range of carbohydrates utilized is clearly distinct from the range of carbohydrates utilized by *Treponema maltophilum*, *Treponema pectinovorum*, and *Treponema socranskii*.

Source: subgingival crevice of humans.

DNA G+C content (mol%): not determined.

Type strain: HA2P, ATCC 700288.

Sequence accession no. (16S rRNA gene): Y09959.

3. **Treponema azotonutricium** Graber, Leadbetter and Breznak 2004, 1319[VP]

a.zo.to.nu.tri'ci.um. N.L. neut. n. *azotum* (from Fr. *azote*), nitrogen; L. neut. adj. *nutricium* nourishing; N.L. neut. adj. *azotonutricium* nourishing with nitrogen (symbiotic dinitrogen fixation).

Cells 0.2–0.3 μm in diameter by 10–12 μm in length, with a wavelength or body pitch of 1.2 μm. Motile by two periplasmic flagella, inserted at opposite ends of the protoplasmic cylinder. Anaerobe. Catalase-negative. Yeast autolysate required for growth. Optimum temperature for growth is 30°C. Energy sources utilized for fermentative growth include D-glucose, D-fructose, D-ribose, D-xylose, D-maltose, and cellobiose. Maltose is fermented to acetate, ethanol, CO_2, and H_2 as major products. D-Mannitol, D- and L-arabinose, D-sucrose, D-trehalose, glycine, lactate, pyruvate, uric acid, and H_2 (plus CO_2) are not utilized. Exhibits nitrogenase activity and N_2-dependent growth in media low in combined N. Genome is 3901 kb and contains 50.0 mol% G+C and two *rrs* gene copies. Nucleotide sequence of the 16S rRNA places this spirochete within the "termite cluster" of the genus *Treponema*.

Source: hindgut contents of the Pacific dampwood termite *Zootermopsis angusticollis* (Hagen) (Isoptera: Termopsidae).

DNA G+C content (mol%): 50 (T_m).

Type strain: ZAS-9, ATCC BAA-888, DSM 13862.

Sequence accession no. (16S rRNA gene): AF320287.

4. **Treponema berlinense** Nordhoff, Taras, Macha, Tedin, Busse and Wieler 2005, 1678[VP]

ber.li.nen'se. N.L. neut. adj. *berlinense* pertaining to Berlin, Germany, where the type strain was isolated.

Cells show typical spirochete morphology exhibiting two to three windings with two periplasmic, subterminally inserted flagella (Nordhoff et al., 2005). Cells are approximately 0.3 µm in width and 6 µm in length. Strictly anaerobic. Good growth is observed in liquid OMIZ-Pat medium at 37°C supplemented with 10% (v/v) BHI and 10% (v/v) TSYE. On OMIZ-Pat agar plates (1–3%, w/v) supplemented with 5% (v/v) sheep blood, 10% (v/v) BHI, and 10% (v/v) TSYE, species form small, irregular, grayish swarms up to 1–2 mm in diameter, visible after 3–4 d. Addition of galacturonic or glucuronic acid promotes growth, which is enhanced further by addition of any of the following carbohydrates: D-glucose, D-fructose, maltose, D-mannitol, D-mannose, D-arabinose, L-fucose, D-trehalose, D-sucrose, and L-rhamnose. No visible growth is observed with pectin as the sole carbon source. Using the API ZYM and Rapid ID 32A systems, positive enzyme reactions are obtained only for acid phosphatase and naphthol-AS-BI-phosphohydrolase. Negative in tests for alkaline phosphatase, esterase C4, esterase lipase C8, leucine arylamidase, cystine arylamidase, trypsin, α-chymotrypsin, α-galactosidase, β-galactosidase, β-glucuronidase, α-glucosidase, β-glucosidase, *N*-acetyl-β-glucosaminidase, α-fucosidase, urease, arginine dihydrolase, α-arabinosidase, mannose, and raffinose, glutamic acid decarboxylase, α-fucosidase, arginine arylamidase, proline arylamidase, leucyl glycine arylamidase, phenylalanine arylamidase, leucine arylamidase, pyroglutamic acid arylamidase, tyrosine arylamidase, alanine arylamidase, glycine arylamidase, histidine arylamidase, glutamyl glutamic acid arylamidase, and serine arylamidase. Reduction of nitrates and indole production are not detected. The polar lipid profile contains diphosphatidylglycerol, phosphatidylethanolamine, an unknown aminophospholipid, and an unknown highly hydrophobic compound as major components. Moderate or minor amounts of phosphatidylglycerol, several unknown aminophospholipids, phospholipids, amino lipids, polar lipids, and a glycolipid are also present.

Source: swine feces in Berlin, Germany.

DNA G+C content (mol%): not determined.

Type strain: 7CPL208, ATCC BAA-909, CIP 108244, JCM 12341.

Sequence accession no. (16S rRNA gene): AY230217.

5. **Treponema brennaborense** Schrank, Choi, Grund, Moter, Heuner, Nattermann and Göbel 1999, 49[VP]

bren.na.bo.ren'se. N.L. neut. adj. *brennaborense* of or belonging to Brennabor, where the cow was raised from which the organism was first isolated.

Treponema brennaborense is an anaerobic, Gram-stain-negative, helically coiled, motile treponeme that was isolated initially from a digital dermatitis biopsy of a dairy cow. Bacterial cells are 5–8 µm long and 0.25–0.55 µm wide. One periplasmic flagellum originates subterminally at each cell pole, and the flagella overlap in the middle of the cell (i.e., have a 1:2:1 arrangement). In stationary-phase liquid cultures, the bacteria develop spherical forms. In liquid culture, the bacteria exhibit rotational movement. Growth of strain DD5/3^T is accompanied by acid production. The optimum growth temperature is 37°C and maximum cell density of approximately 8×10^8 bacteria per ml is reached after 21 h incubation. Cells can be stored frozen (–80°C) in OMIZ-Pat medium (Wyss et al., 1996) supplemented with 15% (v/v) glycerol. On semi-solid agarose plates, *Treponema brennaborense* forms diffuse, submersed white colonies up to 3 mm in diameter within 5-d incubation. Strain DD5/3^T ferments raffinose and mannose and exhibits the enzyme activities alkaline phosphatase, C_4 esterase, C_8 esterase lipase, acid phosphatase, naphtholphosphohydrolase, β-galactosidase, α-glucosidase, *N*-acetyl-β-glucosaminidase, and arginine arylamidase, as determined by the API ZYM and Rapid ID 32A systems. Catalase-negative. The addition of 2–10% (v/v) rabbit serum leads to decreased growth rate. The strain is resistant to rifampicin (1 mg/l) and phosphomycin (100 mg/l).

All previously described treponemes are genetically distinct from *Treponema brennaborense* as determined by comparative 16S rRNA sequencing. *Treponema brennaborense* is clearly distinguished by its morphology, protein pattern, and enzyme activities from the other cultivable *Treponema* species. Furthermore *Treponema brennaborense* is distinguishable from *Treponema maltophilum* by the presence of *N*-acetyl-β-glucosaminidase activity and its lack of α-galactosidase activity. *Treponema brennaborense* is clearly distinguishable from veterinary isolate 1-9185MED by its lack of trypsin and chymotrypsin activities.

Source: a digital dermatitis biopsy of a dairy cow in Brandenburg, Germany.

DNA G+C content (mol%): not determined.

Type strain: DD5/3, CIP 105900, DSM 12168.

Sequence accession no. (16S rRNA gene): Y16568.

6. **Treponema bryantii** Stanton and Canale-Parola 1981, 676[VP] (Effective publication: Stanton and Canale-Parola 1980, 145).

bry.an'ti.i. N.L. masc. gen. n. *bryantii* of Bryant, named after Marvin P. Bryant.

Helical obligate anaerobe, 3–8 µm long and 0.3 µm wide. One periplasmic flagellum is inserted at each end of the cell. No translational motility occurs at 22°C. Motile at 37°C. Requires CO_2. Grows in chemically defined reduced medium containing isobutyrate, DL-2-methyl butyrate, pyridoxal, folic acid, niacinamide, biotin, thiamine, glucose, CO_2, salts, and ammonium sulfate (Stanton and Canale-Parola, 1981). Riboflavin is stimulatory. In 0.7% Noble agar (Difco), cells form colonies in agar deeps, which are spherical and white (resembling cotton balls), and 0.5–1.0 mm in diameter after 24–36 h incubation, and eventually reach 2–3 mm in diameter. Growth in broth medium containing rumen fluid, glucose, and sodium bicarbonate yields 1.9×10^9 cells/ml. Does not utilize gluconate, succinate, acetate, formate, fumarate, sugar alcohols, or Tween 80. Grows in a medium containing cellulose and a cellulolytic bacterium such as *Bacteroides succinogenes* or *Ruminococcus albus*. No growth at 22 or 45°C.

End products of glucose fermentation (µmol/100 µmol of glucose and 84 µmol of CO_2 utilized) are acetate, 100; formate, 119; and succinate, 53. About 15% of glucose carbon is assimilated.

Growth is inhibited by penicillin (10 U disk), cephalothin (30 µg disk), tetracycline (30 µg disk), chloramphenicol (30 µg disk), erythromycin (15 µg disk), and vancomycin (30 µg disk), slightly inhibited by polymyxin B (100 U/disk), and not inhibited by rifampicin (5 µg disk or up to 10 mg/l in broth medium).

Source: bovine rumen contents.

DNA G+C content (mol%): 36±1 (T_m).

Type strain: RUS-1, ATCC 33254, DSM 1788.

Sequence accession no. (16S rRNA gene): M57737.

7. **Treponema carateum** (*ex* Brumpt 1939) sp. nov., nom. rev.

ca.ra'te.um. N.L. n. *carate* name of a South American disease, pinta; N.L. neut. adj. *carateum* of carate.

The cause of pinta or carate, a contagious disease of man transmitted by skin-to-skin contact. Morphologically similar to *Treponema pallidum*. Virulent strains have not been grown *in vitro*. Experimental transmission of the disease has been accomplished in man as well as in chimpanzees by intradermal inoculation and by direct exposure of scarified areas of skin to abraded human lesions. Has not been propagated successfully in rabbits, hamsters, or guinea pigs.

Source: exudate of cutaneous lesions of pinta. Occurs only in Mexico, Central America and parts of subtropical South America, the West Indies, and Cuba.

DNA G+C content (mol%): not determined.

Type strain: none designated.

Sequence accession no. (16S rRNA gene): not determined.

8. **Treponema denticola** (*ex* Brumpt 1922a) Chan, Siboo, Keng, Psarra, Hurley, Cheng and Iugovaz 1993, 201[VP]

den.ti'co.la. L. masc. n. *dens, dentis* tooth; L. suff. *cola* from L. n. *incola* inhabitant dweller; N.L. n. *denticola* tooth-dweller.

A small to intermediate-sized spirochete. Many characteristics are listed in Tables 127 and 129. A 2:4:2 periplasmic flagellar arrangement is common, but *Treponema denticola* strains with higher numbers of flagella have been found. Cells are motile with a jerky, but fairly rapid motion. Cells are typically 7.74±0.94 µm in length, 0.20±0.02 µm in diameter, with a wavelength and amplitude of 1.23±0.15 µm and 0.50±0.05 µm, respectively.

The organism grows well in a peptone-yeast extract-serum medium [e.g., New Oral Spirochete (NOS) medium; Leschine and Canale-Parola, 1980b) under anaerobic conditions. Surface and subsurface colonies are 0.3–1.0 mm in diameter, white, diffuse, and visible after 2 weeks incubation. *Treponema denticola* is primarily an amino acid fermenter and does not use the glycolytic pathway as a major source of energy, although it possesses genes encoding all of its enzymes (Seshadri et al., 2004). Amino acids in peptone-yeast extract-serum medium are fermented mainly to acetic acid, and to a lesser extent, lactic acid, succinic acid, and formic acid. Trace amounts of propionic acid, *n*-butyric acid, ethanol, *n*-propanol, and *n*-butanol may occasionally be found. Only 10% of the end products are from glucose. Alanine, cysteine, glycine, and serine are fermented. Arginine is metabolized to citrulline, NH_3, CO_2, proline, and small amounts of ornithine. Arginine iminohydrolase and ornithine carbamoyltransferase activity have been reported (Blakemore and Canale-Parola, 1976). Arginine can be an energy source and ornithine can be converted to putrescine and proline (Leschine and Canale-Parola, 1980a).

Treponema denticola ATCC 35405^T (Cheng et al., 1985) exhibits 76% DNA homology with *Treponema denticola* 33520 and 82% DNA homology with *Treponema denticola* ATCC 35404 (Chan et al., 1993). The genome sequence of *Treponema denticola* 35405^T has been determined (Seshadri et al., 2004). *Treponema denticola* ATCC 35405^T is susceptible to the antimicrobial agents spiramycin, metronidazole, tetracycline, penicillin G, and streptomycin but highly resistant to rifampicin (minimum inhibitory concentration, 50 mg/l).

Subdivided into biovar *denticola* (indole-positive) and biovar *comondonii* (indole-negative) (Smibert, 1984).

Methyl red-negative. Chopped meat-serum medium is neither blackened nor digested. No action on milk. Growth occurs at pH 6.5–8.0 but not at pH 6.0 or 9.6, and at 30–42°C; multiplication is minimal or absent at 25 and 45°C.

Source: oral cavity of humans and, perhaps, chimpanzees, typically from subgingival plaque.

DNA G+C content (mol%): 37.9 (genome sequence).

Type strain: ATCC 35405, CIP 103919, DSM 14222, JCM 8153.

Reference strains: ATCC 33520, ATCC 35404.

Sequence accession nos (16S rRNA gene): AE017226 (nt 610211–611726 and 1219839–1221354).

9. **Treponema lecithinolyticum** Wyss, Choi, Schüpbach, Moter, Guggenheim and Göbel 1999, 1337[VP]

le.ci.thi.no.ly'ti.cum. Gr. n. *lekithos* egg yolk; Gr. adj. *lutikos ê -on* able to loosen, dissolve; N.L. neut. adj. *lecithinolyticum* effecting the breakdown of egg yolk.

An obligately anaerobic, helically coiled, motile treponeme. Cells are approximately 5×0.15 µm, with a wavelength of 0.7 µm and an amplitude of 0.3 µm. They contain two periplasmic flagella, one originating at each end and overlapping in the center of the cell. In liquid media, the cells flex and rotate but motility is not directional. However, in media of higher viscosity, or when cells creep along a surface, motility is directional.

Since dipalmitoyl phosphatidylcholine (lecithin) inhibits growth, it is omitted from OMIZ-Pat medium (Wyss et al., 1996) (yielding OMIZ-Pat-w/oPC). When streaked onto OMIZ-Pat-w/oPC agarose, *Treponema lecithinolyticum* forms off-white, diffuse subsurface colonies up to 3 mm in diameter within 7 d incubation at 37°C. Does not grow in the chemically defined medium OMIZ-W1 (Wyss, 1992) but requires addition of yeast extract and/or Neopeptone (or fractions thereof). Cells can be stored frozen (liquid nitrogen or mechanical freezer) in OMIZ-Pat-w/oPC medium supplemented with 10–20% glycerol. Growth is strictly dependent on *N*-acetylglucosamine, strongly enhanced by further addition of D-arabinose, L-fucose, or D-ribose, D-fructose (some strains excluding OMZ 684^T), and/or D-xylose (some strains including OMZ 684^T), and not influenced by L-arabinose, D-cellobiose, D-fucose, D-galactose, D-galacturonic acid, D-glucose, D-glucuronic acid, D-lactose, maltose, D-mannitol, D-melibiose, L-rhamnose, L-sorbose, sucrose, D-trehalose, or D-xylose. Heat-inactivated human serum (1% v/v) is tolerated or stimulatory, whereas 1% fetal bovine serum is completely inhibitory. All strains are resistant to rifampicin (1 mg/l) and phosphomycin (100 mg/l).

In all eight isolates examined (Wyss et al., 1999), activities of alkaline phosphatase, acid phosphatase, β-galactosidase, β-glucuronidase, *N*-acetyl-β-glucosaminidase, phospholipase A, and phospholipase C are prominent, whereas only intermediate activities of C4-esterase, C8-esterase, naphthol phosphohydrolase, and α-fucosidase are expressed. Catalase-negative. OMZ 684^T, OMZ 685, and BL2B have strong sialidase activity (the other five strains were not tested).

Phylogenetically distinct from other cultivable treponemes on the basis of its 16S rRNA sequence. Protein and antigen patterns (SDS-PAGE) are also readily distinguished from those of other cultivable treponemes, though more conventional criteria may suffice to distinguish it from the other characterized oral spirochetes. Simultaneous expression of strong activities of phospholipase C, phospholipase A, alkaline phosphatase, acid phosphatase, β-galactosidase, β-glucuronidase, *N*-acetyl-β-glucosaminidase, and sialidase and intermediate activities of C4-esterase, C8-esterase, naphthol phosphohydrolase, and α-fucosidase distinguish *Treponema lecithinolyticum* from all other oral spirochetes. Size, flagellation, and growth characteristics additionally distinguish it from *Treponema amylovorum*, *Treponema denticola*, *Treponema medium*, and *Treponema vincentii*. Finally, *Treponema lecithinolyticum* is phenotypically distinguished from the two other lecithinolytic isolates described by Wyss et al. (1999) (i.e., OMZ 702 and BL2A, which are phylogenetically classified as *Treponema maltophilum*) by its SDS-PAGE protein profile, a ~30-kDa antigen, and activities of phospholipase A, sialidase, b-glucuronidase, and *N*-acetyl-β-glucosaminidase.

Source: Only in human subgingival plaque, with a strong association suggested for diseased versus control sites in patients with adult periodontitis and rapidly progressive periodontitis. Strains OMZ 684^T and OMZ 685 were isolated from subgingival plaque of human deep periodontal lesions.

DNA G+C content (mol%): not determined.

Type strain: OMZ 684, ATCC 700332, CIP 107075.

Reference strains: OMZ 685, ATCC 700333.

Sequence accession no. (16S rRNA gene): AJ131282.

10. **Treponema maltophilum** Wyss, Choi, Schüpbach, Guggenheim and Göbel 1996, 751^{VP}

mal.to'phi.lum. N.L. n. *maltosum* maltose; Gr. adj. *philos ê-on* loving, friendly to; N.L. neut. adj. *maltophilum* intended to mean maltose-loving.

An obligately anaerobic, helically coiled, motile treponeme (5 μm long, 0.2 μm wide with a wavelength of 0.7 μm and an amplitude of 0.3 μm) isolated from human subgingival plaque (Wyss et al., 1996). One periplasmic flagellum originates at each end, overlapping in the center of each cell in a 1:2:1 arrangement. In low-viscosity liquid media, cellular rotation produces standing waves with amplitudes of up to 2 μm, but this results in no directional motility. Translational movement, however, occurs in higher-viscosity media or when cells creep along a surface.

Cells can be stored frozen in liquid nitrogen or in a mechanical freezer in OMIZ-Pat (Wyss et al., 1996) supplemented with 10–20% glycerol.

On OMIZ-Pat agarose, off-white diffuse subsurface colonies (up to 3 mm in diameter) form within 5 d.

Treponema maltophilum does not grow in OMIZ-W1 and requires yeast extract and/or Neopeptone. Growth of most strains is strictly dependent on *N*-acetyl-β-glucosamine and at least one additional sugar. The most commonly used second sugars are D-arabinose, L-fucose, D-maltose, L-rhamnose, D-ribose, D-sucrose, and D-trehalose, but not D-glucose, which is totally ineffective. Some strains without α-fucosidase activity may not depend on *N*-acetyl-β-glucosamine. Growth is not influenced by D-cellobiose, D-fucose, D-lactose, D-mannitol, L-sorbose, or L-xylose. Fetal bovine serum at concentrations as low as 0.1% (v/v) prevents growth in OMIZ-Pat. Catalase-negative. API ZYM strips detected alkaline phosphatase, acid phosphatase, naphtholphosphohydrolase, C4 esterase, C8 esterase, and α-glucosidase activities in all strains, α- and β-galactosidase, β-glucosidase, and α-fucosidase activities in most strains, and α-fucosidase and β-glucuronidase activities in some strains. Strains with β-glucuronidase activity can grow on glucuronic acid. Immunoblotting with patient sera revealed an antigen only in strains with α-fucosidase activity. All strains examined are resistant to rifampicin (1 mg/l) and phosphomycin (100 mg/l). The type strain BR^T (α-fucosidase activity but no β-glucuronidase activity) differs markedly from reference strains HO2A (β-glucuronidase activity but no α-fucosidase activity) and PNA1 (neither α-fucosidase activity nor β-glucuronidase activity) (Wyss et al., 1996).

Phylogenetically distinct from previously described treponemes as determined by comparison of 16S rRNA sequences. Its protein and antigen patterns on SDS-PAGE differ from those of other cultivable treponemes. clearly distinguish *Treponema maltophilum* differs from *Treponema vincentii* on the basis of size, morphology, enzyme activities, and growth characteristics, from the asaccharolytic organism *Treponema denticola* by its lack of trypsin activity, from *Treponema pectinovorum* by its lack of a requirement for either glucuronic acid or galacturonic acid, and from *Treponema socranskii* by synthesis of a wide spectrum of glycosidases and utilization of a wide spectrum of carbohydrates. α-Glucosidase activity is characteristic of *Treponema maltophilum* and has never been observed in *Treponema socranskii*, although many strains of this species can ferment maltose.

Source: only in subgingival plaque samples of patients with periodontal disease.

DNA G+C content (mol%): not determined.

Type strain: BR, ATCC 51939, CIP 105146.

Reference strains: HO2A (ATCC 51940), PNA1 (ATCC 51941).

Sequence accession no. (16S rRNA gene): X87140.

11. **Treponema medium** Umemoto, Nakazawa, Hoshino, Okada, Fukunaga and Namikawa 1997, 71[VP]

me′di.um. L. neut. adj. *medium* not very great or small, medium, referring to the cell size.

A Gram-stain-negative, anaerobic, motile, helically coiled, medium-sized treponeme. The cells (5–16 µm long and 0.2–0.3 µm wide) have cytoplasmic filaments and 5–7 periplasmic flagella (axial flagella) that originate subterminally at each end and in broth cultures exhibit rotational and translational movement. The optimum growth temperature is 37°C, and colonies on agar plates are white and translucent. Ferment D-glucose, D-fructose, maltose, D-mannose, D-galactose, sucrose, D-ribose, trehalose, inulin, salicin, and D-raffinose. Produce ammonia and hydrogen sulfate and hydrolyze esculin and hippuric acid. The major acid products of strain G7201^T grown in tryptone-yeast extract-gelatin-volatile fatty acids-serum (TYGVS) medium (Ohta et al., 1986) containing 0.1% glucose are acetic acid, *n*-butyric acid, and a trace of *n*-valeric acid. Phenotypic characteristics, DNA–DNA hybridization data, G+C content of the DNA, and 16S rRNA gene sequence data indicate that human oral spirochete strain G7201^T is a member of a novel species.

Source: subgingival plaque of patients with adult periodontitis.

DNA G+C content (mol%): 51 (HPLC).

Type strain: G7201.

Sequence accession no. (16S rRNA gene): D85437.

12. **Treponema minutum** Dobell 1912, 117[AL]

mi.nu′tum. L. neut adj. *minutum* small, tiny.

Many characteristics are listed in Table 127. Two to three periplasmic flagella are inserted into each end of the cell. Motile with sluggish movement. Colonies on prereduced peptone-yeast extract-serum agar (1.4%) are visible in 9–15 d, and are 0.5–1 mm in diameter, white, and round on the agar surface. Some colonies after longer incubation are white, fluffy, and up to 1.5 mm in diameter. Colonies grow on and below the surface of the medium. Size and texture of colonies will vary with the concentration of agar in the medium. Grow well in peptone-yeast extract-serum medium under anaerobic conditions. Require animal serum (inactivated at 56–60°C for 1 h) for growth.

Amino acids in peptone-yeast extract-serum medium are fermented to a large amount of acetic acid, moderate amount of succinic acid, smaller amount of lactic acid, and trace amounts of propionic, *n*-butyric, and formic acids (most strains). Trace amounts of ethanol, *n*-propanol, and *n*-butanol are also produced by most strains. There are no additional end products in the presence of glucose.

Methyl red-negative. Skim milk is only slightly curdled. Ammonia produced by most strains. Grows at pH 6.5–8.0 but not at pH 6.0 or 9.6. Grows at 34–40°C. Chopped meat serum medium neither blackened nor digested. Slight putrid odor.

Dupouey (1963) reported that *Treponema minutum* was antigenically only slightly related to *Treponema refringens*. Not pathogenic.

Source: epidermal surfaces of male and female genitoperianal regions.

DNA G+C content (mol%): 37 (T_m).

Type strain: CIP 5162.

Sequence accession no. (16S rRNA gene): not available.

13. **Treponema paraluiscuniculi** (Jacobsthal 1920) Smibert 1974, 177[AL] (*Spirochaeta paraluis-cuniculi* Jacobsthal 1920, 571)

pa.ra.lu.is.cu.ni′cu.li. Gr. pref. *para* resembling; L. n. *lues -is* pestilences, plague, infection (here syphilis); L. n. *cuniculus -i* a rabbit; N.L. gen. n. *paraluiscuniculi* of a syphilis-like (disease) of rabbits.

Produces venereal spirochetosis (rabbit spirochetosis or rabbit syphilis) in rabbits. Morphologically similar to *Treponema pallidum*. Transmitted by sexual contact. Has not been cultivated *in vitro*. The organism can be propagated by intratesticular inoculation of rabbits. Causes a latent infection of mice, guinea pigs, and hamsters. Treponemes are found in the lymph nodes of these animals. Cutaneous lesions are found only in guinea pigs and rabbits. Nonpathogenic to humans (Graves and Downes, 1981).

The cuniculi A strain has been shown to possess homologs of *tpr* genes found in *Treponema pallidum*, but many are predicted to be nonfunctional (Giacani et al., 2004). For additional information see Smith and Persetsky (1967).

Source: lesions in the genital area of rabbits. Primarily involves the genitalia, although cutaneous lesions often occur around the face, eyes, ears, and nose.

DNA G+C content (mol%): not determined.

Type strain: none has been designated.

Reference strain: Cuniculi A.

Sequence accession no. (16S rRNA gene): not determined.

14. **Treponema parvum** Wyss, Dewhirst, Gmür, Thurnheer, Xue, Schüpbach, Guggenheim and Paster 2001, 960[VP]

par′vum. L. neut. adj. *parvum* small.

Small, obligately anaerobic, helically coiled, motile treponeme. Approximately 1 µm long and 0.18 µm wide, with a wavelength of 0.8µm and an amplitude of 0.3 µm. In rapidly growing cultures, cells may be shorter than 1 wavelength, but chains of more than 10 wavelengths are also common. Cells contain two periplasmic flagella, one originating at each pole and overlapping in the center of the cell in a 1:2:1 arrangement. Although undulation, flexing, and rotation of cells occurs in liquid medium, motility does not appear to be directional. However, in media of higher viscosity or when cells creep along a surface, movement is translational.

Cells can be stored frozen (liquid nitrogen or mechanical freezer) in OMIZ-Pat/HuS medium supplemented with

10–20% glycerol. The four isolates OMZ 832, 833^T, 842, and 843 are strictly carbohydrate-dependent; either *N*-acetyl-β-glucosamine or *N*-acetyl-β-galactosamine is sufficient for growth, though growth is strongly promoted by addition of L-arabinose, D-galactose, D-glucose, D-fructose, D-mannitol, D-mannose, pectin, D-ribose, or D-xylose but not by D-arabinose, D-cellobiose, D-fucose, L-fucose, D-galacturonic acid, D-glucuronic acid, D-lactose, D-maltose, D-melibiose, L-rhamnose, L-sorbose, D-sucrose, D-trehalose, or L-xylose. The chemically complex (undefined) components of OMIZ-Pat, i.e., YEM (fractionated yeast extract) and DANP (fractionated peptone) (Wyss and Ermert, 1996), are not strictly required for growth but are strongly stimulatory. Similarly, human serum or FBS are not required but are growth-promoting at 1% (v/v). FBS is not inhibitory even at 10% (v/v). On OMIZPat/HuS agarose, *Treponema parvum* forms off-white diffuse subsurface colonies up to 3 mm in diameter after 5 d of anaerobic incubation at 37°C. The four OMZ isolates on API ZYM tests had weak alkaline phosphatase and esterase C_4 and C_8 activities, intermediate acid phosphatase and naphthol phosphohydrolase activities, and strong β-glucuronidase activity.

Although *Treponema parvum* and *Treponema pectinovorum* are similar, the former is clearly distinguishable phenotypically by the presence of a strong β-glucuronidase activity, inability to utilize pectin as sole source of carbohydrate, and marked differences in protein and antigen patterns revealed by SDS-PAGE. Very short and thin cells, though this is not strongly diagnostic, since all cultured treponemes show variation in morphology under different growth conditions. The "Smibert-2" isolates also differ from the 1:2:1 flagellated pectinolytic treponemes isolated from nonhuman primates by Sela et al. (1987) by virtue of their inability to grow on pectin. There are 16S rRNA sequence differences between *Treponema parvum* and other species of *Treponema*, including *Treponema pectinovorum*. *Treponema parvum* is the sole representative of Group 7 of oral *Treponema* as defined by Paster et al. (1998). Its closest relatives are *Treponema pectinovorum* (88% similarity) and *Treponema amylovorum* (90% similarity).

Source: OMZ 833^T was isolated from subgingival plaque of a human deep periodontal lesion. Strain OMZ 842 was isolated from acute necrotizing ulcerative gingivitis (ANUG) lesions of a patient in China.

DNA G+C content (mol%): not determined.

Type strain: OMZ 833, ATCC 700770, DSM 16260.

Reference strain: OMZ 842, ATCC 700773.

Sequence accession no. (16S rRNA gene): AF302937.

15. **Treponema pectinovorum** Smibert and Burmeister 1983, 853^{VP}

pec.ti.no'vo.rum. N.L. n. *pectinum* pectin; N.L. neut. adj. *vorum* (from L. v. *voro* to devour) devouring; N.L. neut. adj. *pectinovorum* pectin destroying and devouring.

Obligately anaerobic, motile helically coiled treponeme. The cells are 7–15 μm long and 0.28–0.30 μm wide. They are coiled and usually have straight, slightly pointed ends. The periplasmic flagella overlap in the center of the cell in a 2:4:2 arrangement. Secondary coils are observed in motile cultures. Movement is both rotational and translational. Serpentine movement can be observed in a semisolid medium.

In Oral Treponeme Isolation (OTI) medium (Smibert and Burmeister, 1983) in bottle plates, colonies usually appear in the agar after 4–5 d. Colonies grow into the agar and are white and transluscent with slightly denser centers and entire edges. The colonies spread out and become larger after additional incubation.

These organisms grow in PY-pectin broth containing either rumen fluid or a short-chain fatty acid-heme supplement. Serum and thiamine pyrophosphate are not required. Growth only occurs in the presence of a fermentable energy source, such as pectin, polygalacturonic acid, galacturonic acid, or glucuronic acid and is greatly stimulated by the addition of a fresh filter-sterilized yeast autolysate to the medium. Growth occurs at 37°C (optimum) but not at 25 or 42°C. Broth cultures become turbid with a granular sediment that can be seen after 4–5 d of incubation. Cultures can be stored frozen in liquid nitrogen or at −85°C in a mechanical freezer.

Pectin (final pH, 5.3–5.9) is utilized. Polygalacturonic acid, galacturonic acid, and glucuronic acid are also fermented and may be substituted for pectin in PY-rumen fluid broth. Growth and acid production occur in PY-rumen fluid broth supplemented with either autoclaved or filter-sterilized pectin. Growth and acid production also occur in PY-rumen fluid broth containing 0.5% polygalacturonic acid. The pH of this medium is 5.0 after 5 d of incubation at 37°C. Adonitol, amygdalin, arabinose, cellobiose, dextrin, starch, dulcitol, erythritol, esculin, fructose, galactose, glycerol, glycogen, inositol, inulin, lactose, glycerol, glycogen, inositol, inulin, lactose, maltose, mannitol, mannose, melezitose, melibiose, mucin, raffinose, rhamnose, ribose, salicin, sorbose, sorbitol, sucrose, trehalose, xylose, and glucose are not fermented; no growth occurs in PY-rumen fluid broth containing any of these substrates.

Negative for catalase and hydrogen sulfide production. Hydrogen gas was not detected by gas chromatography of the atmospheric phase of cultures in rubber-stopper sealed tubes; gas was not detected in agar deep cultures. Gelatin, esculin, glycogen, or starch is not hydrolyzed. Indole and acetylmethylcarbinol are not produced.

The major fermentation products from PY-pectin-rumen fluid broth are acetic acid (27.9 mM) and formic acid (8.5 mM). Only traces of pyruvic and lactic acids are detected. The products from polygalacturonic acid are acetic acid (39.5 mM) and formic acid (16.7 mM) with only traces of lactic and pyruvic acids.

Distinction from other *Treponema* species is based on 16S rRNA sequence, mol% G+C content, ability to ferment pectin, polygalacturonic acid, galacturonic acid, and glucuronic acid but not other carbohydrates, cellular dimensions, and the origin of the two periplasmic flagella attached at each end of the cell (Smibert and Burmeister, 1983).

Source: human supragingival and subgingival plaque specimens but not from adults with normal, healthy gingivae and no signs of gingivitis or periodontitis.

DNA G+C content (mol%): 39 (T_m).

Type strain: ATCC 33768^T, VPI D-36DR-2^T.

Sequence accession no. (16S rRNA gene): M71237.

16. **Treponema phagedenis** (*ex* Brumpt 1922a) sp. nov., nom. rev.

pha.ge.de'nis. Gr. gen. n. *phagedenis* of a cancerous sore.

Many characteristics are listed in Tables 127 and 128. Widest cells show double contours with darkfield microscopy. Ends of the cells are blunt with no covering sheath. Three-to-eight periplasmic flagella are inserted into each end of the cell. In old cultures, the flagella may be seen trailing from the ends of the cells.

Motility in culture media is jerky with slow rotational movement. Colonies in prereduced anaerobic peptone-yeast extract-serum medium containing 1.3–1.4% agar are white, annular, 0.5–1 mm in diameter with a dense center after incubation for 2–5 d at 37°C. Colonies can grow on the surface but mainly in the agar.

Requires animal serum (heat inactivated at 56–60°C for 0.5–1 h) for growth. Bovine serum albumin supplemented with a pair of fatty acids can substitute for serum (Johnson and Eggebraten, 1971). The pair includes (a) an unsaturated fatty acid such as oleic acid and (b) a saturated fatty acid such as palmitic acid. No growth with short-chain fatty acids or α-, β-, γ-globulins.

Fermentation of glucose is by the Embden–Meyerhof–Parnas pathway. Contains ferredoxin. End products of fermentation in a serum medium without glucose are mainly acetic and *n*-butyric acids with moderate-to-small amounts of propionic and formic acids and usually small-to-trace amounts of lactic and succinic acids. Trace amounts of alcohols are also produced. In a medium containing glucose, large amounts of ethanol and *n*-butanol and smaller amounts of *n*-propanol are produced.

A very slight curd is formed in skim milk. Weakly methyl-red-positive. Does not grow at a pH of 6.0 or 9.6. Grows at 30–42°C but not or only slightly at 25 and 45°C. Reduces neutral red. Chopped meat serum medium is neither blackened nor digested. A slight fetid odor is produced in cultures.

Subdivided into reiter and kazan biovars. Biovar reiter does not hydrolyze esculin, whereas biovar kazan does. Meyer and Hunter (1967) showed that the Reiter, Kazan, and English Reiter strains are antigenically closely related. Reiter and English Reiter contained the same antigens while the Kazan strain contained an antigen not shared by the Reiter treponemes. The Nichols strain of *Treponema refringens* was antigenically unrelated. Christiansen (1964) also reported that the Reiter and Kazan 11 strains were closely related but not identical. Dupouey (1963) reported that *Treponema phagedenis* and Reiter strain were closely related antigenically, sharing at least six common antigens. The Reiter strain has a large amount of an antigen that may be shared with a number of other species including *Treponema pallidum*. More than 40 water-soluble antigens have been demonstrated in the Reiter treponeme by crossed immunoelectrophoresis. Five antigens cross-reacted with antibodies in syphilitic sera (Strandberg-Pedersen et al., 1980, 1981).

Reiter and Kazan strains have high DNA/DNA homology to each other and no homology to *Treponema refringens* (Miao and Fieldsteel, 1978; Smibert, 1974). There is no DNA/DNA homology to *Treponema denticola* (Smibert, 1974). There is no detectable DNA homology between *Treponema phagedenis* (Reiter and Kazan 5) and pathogenic Nichols strain of *Treponema pallidum* subsp. *pallidum* (Maio and Fieldsteel, 1978). Additional information on *Treponema phagedenis* Reiter can be found in an excellent review by Wallace and Harris (1967).

Source: nonpathogenic. Phagedenic ulcer on human external genitalia. Reiter treponeme from a case of primary syphilis in man and also as normal flora in the anal and genital areas of normal male and female chimpanzees.

DNA G+C content (mol%): 38–39 (T_m).

Type strain: none designated.

Reference strain: Reiter.

Sequence accession no. (16S rRNA gene): M57739.

17. **Treponema porcinum** Nordhoff, Taras, Macha, Tedin, Busse and Wieler 2005, 1678[VP]

por.ci'num. L. neut. adj. *porcinum* pertaining to swine, from which the type strain was isolated.

Cells exhibit typical spirochete morphology and are approximately 6–8 µm in length and 0.3 µm in width with 2–3 windings and 2:4:2 flagella arrangement. Strictly anaerobic. Best growth is obtained in liquid OMIZ-Pat medium at 37°C supplemented with 10% (v/v) BHI and 10% (v/v) TSYE. Growth is independent of glucuronic or galacturonic acid. D-Maltose is essential for growth, whereas any of the following carbohydrates (as the sole carbohydrate source) do not support growth: D-glucose, D-fructose, D-mannitol, D-mannose, D-arabinose, L-fucose, trehalose, sucrose, and L-rhamnose. Does not grow with pectin as a sole carbon source. On OMIZ-Pat (1–3% w/v) agar supplemented with 5% egg yolk, 10% BHI, and 10% TSYE, the species forms grayish, irregular swarms up to 2 mm in diameter, visible after 3–4 d. Reactions using the API ZYM and Rapid ID 32A system are positive for acid phosphatase, esterase C4, naphthol- AS-BI-phosphohydrolase, and α-glucosidase, and negative for alkaline phosphatase, esterase lipase C8, leucine arylamidase, cystine arylamidase, trypsin, α-chymotrypsin, α-galactosidase, β-galactosidase, β-glucuronidase, β-glucosidase, *N*-acetyl-β-glucosaminidase, α-fucosidase, urease, arginine dihydrolase, α-arabinosidase, mannose, and raffinose fermentation, glutamic acid decarboxylase, α-fucosidase, arginine arylamidase, proline arylamidase, leucyl glycine arylamidase, phenylalanine arylamidase, leucine arylamidase, pyroglutamic acid arylamidase, tyrosine arylamidase, alanine arylamidase, glycine arylamidase, histidine arylamidase, glutamyl glutamic acid arylamidase, and serine arylamidase. Reduction of nitrates and indole production are not detected. In the polar lipid profile, three unknown phospholipids and a highly hydrophobic compound predominate. Diphosphatidylglycerol, phosphatidylglycerol as well as phospholipids are present in moderate amounts. Additionally, a glycolipid and several phospholipids are present in minor amounts.

Source: swine feces in Berlin, Germany.

DNA G+C content (mol%): 38–39 (T_m).

Type strain: 14V28, ATCC BAA-908, CIP 108245, JCM 12342.

Sequence accession no. (16S rRNA gene): AY518274.

18. **Treponema primitia** Graber, Leadbetter and Breznak 2004, 1319[VP]

pri.mi'ti.a. N.L. fem. sing. n. *primitia* (nominative in apposition), the first fruit (of isolation after long work).

Cells 0.2 μm in diameter by 3–7 μm long, with a wavelength or body pitch of 2.3 μm. Motile by two periplasmic flagella, inserted at opposite ends of the protoplasmic cylinder. Anaerobe. Possesses NADH and NADPH peroxidases but neither catalase nor superoxide dismutase. Optimum temperature for growth is 30°C. Optimum pH for growth is 7.2 (range, 6.5–7.8). Homoacetogen. Energy sources used for growth include glucose, maltose, mannitol, xylose, and H_2 (plus CO_2), which are fermented to acetate as the sole product. Strain ZAS-1 also uses arabinose and cellobiose, whereas strain ZAS-2 can grow slowly by acetogenic demethylation of methoxylated benzenoids (syringate, ferulate, vanillate, and trimethoxybenzoate). Ribose, methanol, formate, CO, lactate, pyruvate, glycine, betaine, and choline are not utilized. Growth by mixotrophy (i.e., simultaneous use of H_2 and organic substrates) has been demonstrated. Laboratory-prepared yeast autolysate or certain commercial yeast extracts are required for growth. Folinate (formyltetrahydrofolate) is required for growth of strain ZAS-1, whereas folic acid or folinate is required for growth of strain ZAS-2. Cells possess homologs of the dinitrogenase reductase gene *nifH* and exhibit low levels of nitrogenase activity, but unambiguous N_2-dependent growth has not been demonstrated. Genome sizes are 3461 kb (ZAS-1) and 3835 kb (ZAS-2); G+C contents of DNA are 51.0 mol% (ZAS-1) and 50.9 mol% (ZAS-2) (by HPLC); each strain possesses 2 *rrs* gene copies. The 16S rRNA nucleotide sequences of strains ZAS-1 and ZAS-2 place them within the "termite cluster" of the genus *Treponema*.

Source: hindgut contents of the Pacific dampwood termite *Zootermopsis angusticollis* (Hagen) (Isoptera: Termopsidae).

DNA G+C content (mol%): 50.9–51.0 (HPLC).

Type strain: strain ZAS-2, ATCC BAA-887, DSM 12427.

Sequence accession no. (16S rRNA gene): AF093252 (ZAS-2^T), AF093251 (ZAS-1).

19. **Treponema putidum** Wyss, Moter, Choi, Dewhirst, Xue, Schüpbach, Göbel, Paster and Guggenheim 2004, 1121[VP]

pu'ti.dum. L. neut. adj. *putidum* stinking, fetid.

Obligately anaerobic, helically coiled, motile, asaccharolytic, and proteolytic. The human oral cavity is so far its only known habitat. Approximately 0.25 μm in diameter and approximately 10 μm long, with a wavelength of approximately 3 μm and amplitude of approximately 1.5 μm. They contain four periplasmic flagella, two originating at each cell end and overlapping in the center of the cell in an arrangement of 2:4:2. In liquid media of low viscosity, cells appear highly active with cellular rotation and jerky flexing but no directional motility. Translational movement, however, is seen in media of higher viscosity or when cells creep along a surface. Cells can be stored at temperatures below −70°C in medium supplemented with 10–20% glycerol. Within 5 d of anaerobic incubation at 37°C when streaked onto OMIZ-Pat agar (Wyss et al., 1996), dense, off-white subsurface colonies up to 3 mm in diameter are formed. Does not grow in the chemically defined OMIZ-W1 medium, but requires the addition of yeast extract and/or Neopeptone (or fractions thereof); addition of 1–10% human or fetal bovine serum is highly stimulatory. Growth is neither dependent on nor stimulated by any of the following carbohydrates, each tested at 2 g/l: D-arabinose, D-cellobiose, D-fructose, D-fucose, D-galactose, D-galacturonic acid, D-glucose, D-glucuronic acid, glycogen, D-lactose, D-maltose, D-mannitol, D-mannose, D-melibiose, D-ribose, starch, sucrose, D-trehalose, D-xylose, L-arabinose, L-fucose, L-rhamnose, L-sorbose, and L-xylose. Neuraminidase and dentilisin activities are not detected. Using API ZYM strips (Table 129), the following enzyme activities are always detected: esterase C4, esterase C8, leucyl arylamidase, trypsin, acid phosphatase, naphtholphosphohydrolase, β-galactosidase, and β-glucosidase; the following activities are never detected: lipase C14, valine arylamidase, β-glucuronidase, *N*-acetyl-β-glucosaminidase, α-mannosidase, and α-fucosidase. Other enzyme activities detectable by API ZYM are present only in some strains. Growth is resistant to rifampicin (1 mg/l), phosphomycin (100 mg/l), nalidixic acid (30 mg/l), and polymyxin (5 mg/l).

Source: subgingival plaque of a deep human periodontal lesion.

DNA G+C content (mol%): not determined.

Type strain: JZC3, OMZ 758, ATCC 700334, CIP 108088, OMZ 758.

Sequence accession no. (16S rRNA gene): AJ543428.

20. **Treponema refringens** (*ex* Castellani and Chalmers 1919) sp. nov., nom. rev.

re.frin'gens. L. part. adj. *refringens* refringent, refractive.

Many characteristics are listed in Tables 127 and 128. The average cells are 5–8 μm long and 0.24 μm wide. Some cells may appear loosely coiled. Two to four periplasmic fibrils are inserted at each end of the cell. Motile, with a slow, sluggish movement. Rotation of cells is rare, and when observed, usually slow.

Colonies on prereduced anaerobic peptone-yeast extract-serum agar (1.4%) are visible in 9–15 d. They are white, round, surface colonies 0.5–1 mm in diameter. Some colonies after longer incubation are white, fluffy, and up to 1.5 mm in diameter. Colonies grow on and below the surface of the medium. Size and texture of colonies varies with the concentration of agar in the medium.

Grows well in peptone-yeast extract-serum medium under anaerobic conditions. Requires animal serum (inactivated at 56–60°C for 1 h) for growth. Amino acids in serum medium are fermented to mostly acetic acid, moderate amounts of succinic acid, and smaller amounts of lactic acid. Some strains produce trace amounts of propionic acid, *n*-butyric acid, formic acid, ethanol, *n*-propanol, and *n*-butanol. No additional end products are produced in the presence of D-glucose.

Methyl red-negative. Skim milk is only slightly curdled. Ammonia produced by most strains. Growth occurs at pH 6.5–8.0 but not at pH 6.0 or 9.6. Grows at 30–42°C but only very slightly or not at all at 25 or 45°C. Chopped meat serum medium is neither blackened nor digested. Only a very slight putrid odor is detectable.

Dupouey (1963) reported that strains labeled *Treponema refringens* and "*Treponema calligyrum*" were closely related antigenically, sharing 4–5 common antigens, but only slightly related to *Treponema minutum*. The three strains had only one antigen in common with *Treponema pallidum*.

DNA from *Treponema refringens* shows a high homology with DNA from the avirulent Nichols strain, the Noguchi strain, and "*Treponema calligyrum*", and a very low homology with DNA from strains of *Treponema denticola* (R. M. Smibert and J. Johnson, unpublished data). No detectable DNA homology by hybridization to *Treponema phagedenis* or *Treponema pallidum* subsp. *pallidum* (Nichols) and *Treponema pallidum* subsp. *pertenue* (Maio and Fieldsteel, 1978). On this basis, "*Treponema calligyrum*" was designated a biovar of *Treponema refringens* by Smibert (1984). Biovar refringens does not grow with 1% glycine, whereas biovar calligyrum does grow in 5–6 d.

Source: condyloma acuminata lesions, occasionally from syphilitic lesions. Part of normal flora of male and female genitalia of man and animals. Not pathogenic.

DNA G+C content (mol%): 39–43 (T_m).

Type strain: none designated.

Reference strain for biovar refringens: Treponema refringens, Institut Pasteur, Paris.

Reference strain for biovar calligyrum: CIP 64.40.

Sequence accession nos (16S rRNA gene): AF426101 (*Treponema refringens* biovar *refringens*); AF426100 (*Treponema refringens* biovar *calligyrum*).

21. **Treponema saccharophilum** Paster and Canale-Parola 1985, 218[VP]

sac.cha.ro.phi'lum. Gr. n. *sacchar* sugar; Gr. adj. *philus -ê -on* loving; N.L. neut. adj. *saccharophilum* sugar-loving.

Helical cells, 0.6–0.7 µm by 12–20 µm. Cell coiling is regular except when cells are in contact with solid surfaces. A bundle of periplasmic flagella is wrapped around the cell body. At least 16 periplasmic flagella are inserted near each end of the cell. Cells swim at velocities in excess of 60 µm/s in liquid media at 37°C, but no translational motility is observed at 23°C. Translational motility ceases 1–2 min after the cells are exposed to air. Cells in contact with solid surfaces exhibit creeping motility. Obligate anaerobe. Optimum growth is at 37–39°C. At these temperatures, the final growth yield in rumen fluid-glucose-sodium bicarbonate-salts broth is 7×10^8 cells per ml and the population doubling time is 90 min. No growth at 23°C or 45°C. Subsurface colonies in agar media are spherical and opaque with diffuse edges. Utilizes as fermentable substrates for growth: L-arabinose, D-galactose, D-glucose, D-mannose, D-fructose, D-galacturonic acid, D-glucuronic acid, cellobiose, lactose, maltose, sucrose, D-raffinose, dextrin, inulin, starch, pectin, polygalacturonic acid, and arabinogalactan. Does not grow on: L-rhamnose, D-xylose, L-sorbose, D-ribose, cellulose, dextran, amino acids, D-arabitol, dulcitol, mannitol, ribitol, sorbitol, xylitol, glycerol, potassium galactonate, potassium gluconate, sodium acetate, sodium formate, sodium lactate, sodium succinate, potassium fumarate, Tween 80, glucosamine, and xylan. Exogenous isobutyric acid is required for growth, and valeric acid is stimulatory. Neither $NaHCO_3$ in media nor a CO_2-containing atmosphere is required for growth. Fermentation end products of growing cells (in micromoles per 100 pmol of glucose utilized): formate, 150; acetate, 91.2; ethanol, 79.4. Approximately 15% of the glucose carbon consumed by growing cells is assimilated in cell material. Acetate and formate are major end products of pectin or glucuronic acid fermentation. Pyruvate is metabolized via a coliform-type clastic reaction. Isolated from bovine rumen fluid using an agar medium that contained rifampicin as a selective agent and pectin as a fermentable substrate.

Source: bovine rumen contents.

DNA G+C content (mol%): 54 (T_m).

Type strain: PB, ATCC 43261, DSM 2985.

Sequence accession no. (16S rRNA gene): M71238.

22. **Treponema scoliodontus** (*ex* Noguchi 1928) sp. nov., nom. rev.

sco.li.o.don'tus. Gr. adj. *skolios* crooked, bent; Gr. n. *odous, odontos* tooth; N.L. neut. n. *scoliodontus* crooked tooth.

Many characteristics are listed in Tables 127 and 128. Very tightly coiled cells. Motile with a jerky but fairly rapid motion. Grows in peptone-yeast extract-serum medium under anaerobic conditions. Requires animal serum or ascitic fluid. Amino acids in a peptone-yeast extract-serum medium are fermented to moderate amounts of acetic acid and small amounts of formic, succinic, lactic, propionic, and *n*-butyric acids. No additional end products are produced in the presence of glucose.

Methyl red-negative. Ammonia is not produced from amino acids. No action on milk. Chopped meat serum medium is neither blackened nor digested. Produces a slight fetid odor. Grows at pH 6.5–8.0 but not at pH 6.0 or 9.6. Grows at 30–42°C.

Source: oral cavity of humans.

DNA G+C content (mol%): not known.

Type strain: none designated.

Reference strain: Treponema scoliodontus Institut Pasteur, Paris.

Sequence accession no. (16S rRNA gene): not determined.

23. **Treponema socranskii** (Noguchi 1928) Smibert, Johnson and Ranney 1984, 459[VP]

so.crans'ki.i. N.L. masc. gen. n. *socranskii* of Socransky, named for Sigmund S. Socransky, Forsyth Dental Center, Boston, USA.

Cells are 6–15 µm long and 0.16–0.18 µm wide. They have tapered ends with a slight bend or "hook" at one or both ends of the cell. Obligately anaerobic, motile, helically coiled treponeme. The species is subdivided into three distinct subspecies based on DNA homology (by hybridization) and phenotypic characteristics: subspecies *socranskii, buccale*, and *paredis* (Smibert et al., 1984). The periplasmic flagella overlap in the center of the cell, in a 1:2:1 relationship. The cells form coccoid bodies in the late stationary growth phase. Cells in broth cultures have both rotational and translational movement. Serpentine movement of cells can be seen by darkfield microscopy of cultures grown in a semisolid medium.

Colonies usually appear in Oral Treponeme Isolation (OTI) agar (Smibert and Burmeister, 1983) in 7–10 d. The colonies grow into the agar, are white and translucent, and often have slightly denser centers and edges that can be entire or irregular. T he colonies spread and become larger after extended incubation. Growth occurs only in media containing a fermentable carbohydrate and either rumen fluid (20–30%) or a mixture of short-chain fatty acids. Serum is not required. Growth is optimum at 37°C and only slight at 25 or 42°C.

The phenotypic characteristics of representative strains (10–12 of each subspecies) were examined (Smibert et al., 1984). Fermentation of glucose by a majority of the strains leads to a pH ranging from 5.1 to 5.9. A few strains belonging to *Treponema socranskii* subsp. *buccale* do not reduce the pH below 6.0 when they are grown in glucose-containing broth. These cultures have a pH range of 6.2–6.7. Inulin, lactose, Melezitose, cellobiose, salicin, D sorbitol, glycerol, amygdalin, adonitol, dulcitol, i-erythritol, inositol, and D-mannitol are not fermented, as indicated by only a slight change in the pH of the medium. Hydrogen sulfide is produced in SIM medium supplemented with rumen fluid by all but 1 of 32 strains tested.

A distinctive phenotypic trait found in all strains of all subspecies of *Treponema socranskii* (but not in any other oral spirochete) is the formation of intensely yellow colonies (or cell pellets after growth in liquid medium) in OMIZ-Pat agarose. Furthermore, live cells of *Treponema socranskii* can be recognized microscopically, since cells that are rotating around their axes have both cell tips markedly deflected, which gives the appearance of propellers on both ends of a straight helix.

None of the strains studied produces catalase, peroxidase, indole, or acetylmethylcarbinol. Esculin is not hydrolyzed. Hydrogen gas is not detected by gas chromatography of samples taken from the atmosphere above the broth medium in rubber-stoppered tubes.

The major acid fermentation products of all strains grown in PY-glucose-rumen fluid broth are acetic, lactic, and succinic acids. Trace amounts of formic acid occasionally can be found. The strains produce a mean of 8 mM lactic acid (range, 3–16 mM), 6.5 mM acetic acid (range, 2.6–10.4 mM), and 3 mM succinic acid (range, 1–6.5 mM).

Treponemes in homology group A1 are designated *Treponema socranskii* subsp. *socranskii*. The type strain of *Treponema socranskii* subsp. *socranskii* is ATCC 35536^T (=VPI DR56BRIII6). Treponemes in homology group A2 are designated *Treponema socranskii* subsp. *buccale* (buc.ca'le. L. n. *bucca* the mouth; L. neut. suff. *-ale* suffix denoting pertaining to; N.L. neut. adj. *buccale* buccal, pertaining to the mouth); the type strain is VPI D2B8^T (=ATCC 35534). The phenotypic characteristics of these subspecies are described in detail in the defining publication (Smibert et al., 1984).

Patterns of phenotypic reactions (Smibert et al., 1984) show that *Treponema socranskii* subsp. *socranskii* (homology group A1) cannot be readily differentiated from *Treponema socranskii* subsp. *buccale* (homology group A2). However, when a slide agglutination test with washed cells as the antigen was used, antisera against the type strain (strain ATCC 35536) and strain VPI D43BR1 agglutinated 11 of 12 *Treponema socranskii* subsp. *socranskii* strains (Smibert et al., 1984), and no strain of *Treponema socranskii* subsp. *buccale* or *Treponema socranskii* subsp. *paredis*. Antisera against the type strain (strain ATCC 35534) and strains VPI D11A1 and VPI D40DPEI of *Treponema socranskii* subsp. *buccale* agglutinated 7 of 10 *Treponema socranskii* subsp. *buccale* strains. Antiserum against the type strain of *Treponema socranskii* subsp. *paredis* (strain ATCC 35535) agglutinated 7 of 9 *Treponema socranskii* subsp. *paredis* strains but no *Treponema socranskii* subsp. *socranskii* or *Treponema socranskii* subsp. *buccale* strains.

Treponemes in homology group K are designated *Treponema socranskii* subsp. *paredis* (pa.re'dis. Gr. n. *pareias* cheek; N.L. gen. n. *paredis* of a cheek). L-Arabinose and rhamnose are not fermented. Other characteristics of this subspecies are the same as those of the species. The type strain of *Treponema socranskii* subsp. *paredis* is strain ATCC 35535 (= VPI D46CPE1), which was isolated from a supragingival sample from a patient with severe periodontal disease.

Treponema socranskii subsp. *paredis* can be easily separated from *Treponema socranskii* subsp. *socranskii* and *Treponema socranskii* subsp. *buccale* by the inability of *Treponema socranskii* subsp. *paredis* to ferment L-arabinose and rhamnose.

Source: subgingival sample from a patient with severe periodontal disease. It is the most frequently isolated treponeme and is usually the most numerous of the cultivable treponemes in either supragingival or subgingival samples.

DNA G+C content (mol%): 50.5±2 (T_m).

Type strains: Treponema socranskii subsp. *socranskii* ATCC 35536, JCM 8157, VPI D56BRIII6; *Treponema socranskii* subsp. *buccale* ATCC 35534, JCM 8155, VPI D2B8; *Treponema socranskii* subsp. *paredis* ATCC 35535, JCM 8156, VPI D46CPE1.

Sequence accession nos (16S rRNA gene): AF033306 (*Treponema socranskii* subsp. *socranskii*); AF033305 (*Treponema socranskii* subsp. *buccale*); AF033307 (*Treponema socranskii* subsp. *paredis*).

24. **Treponema succinifaciens** Cwyk and Canale-Parola 1981, 383^{VP} (Effective publication: Cwyk and Canale-Parola 1979, 231.)

suc.ci.ni.fa'ci.ens. N.L. n. *acidum succinicum* succinic acid; L. part. adj. *faciens* making, producing; N.L. part. adj, *succinifaciens* succinic acid-producing.

Helical, anaerobic bacterium 0.3 µm wide by 4–8 µm long. Some cells may be up to 16 µm long. May form chains. Possesses a 2:4:2 flagellar arrangement. No transitional movement at 25°C. Motile at 37°C, requires carbon dioxide. Colonies in rumen fluid agar deeps are spheroid with an opaque center and diffuse peripheral growth. Colonies are 4–8 mm in diameter after 2 d growth at 37°C. In broth, cell yields are 1.5×10^9 cells/ml with a mean generation time of 3.5 h.

The pH of a glucose culture after 48 h is about 6.0. Ferments glucose by the Embden–Meyerhof pathway. CO_2/bicarbonate and a carbohydrate are required for multiplication. Growth is supported by L-arabinose, D-xylose, D-glucose, D-mannose, D-galactose, maltose, lactose, cellobiose, dextrin, and starch, but not by D-ribose, L-sorbose, raffinose, L-rhamnose, D-fructose, sucrose, dextran, inulin, ball-milled cellulose, trehalose, glycerol, D-mannitol, D-sorbitol, dulcitol, xylitol, sodium acetate, sodium formate, sodium succinate, potassium pyruvate, sodium lactate, and potassium gluconate. Does not ferment acetate, formate, succinate, xylitol, pyruvate, lactate, gluconate, and Tween 80. End products are (in µmol/100 µmol glucose and 51 µmol of CO_2 utilized) acetate, 82; formate, 81; succinate, 58; lactate, 30; 2,3-butanediol, 5; pyruvate, 4; acetoin, 3. Catalase-negative. Poor growth occurs at 22 and 43°C.

Inhibited by penicillin G (4000 U/l), cephalothin (4 mg/l), and chloramphenicol (4 mg/l), not by erythromycin (4 mg/l), oxytetracycline (4 mg/l), polymyxin B (40,000 U/l), rifampicin (4 mg/l), streptomycin (4 mg/l), tetracycline (4 mg/l), and vancomycin (4 mg/l).

Source: colon of swine.

DNA G+C content (mol%): 36 (T_m).

Type strain: strain 6091, ATCC 33096, DSM 2489.

Sequence accession no. (16S rRNA gene): M57738.

25. **Treponema vincentii** (*ex* Brumpt 1922a) sp. nov., nom. rev.

vin.cen'ti.i. N.L. masc. gen. n. *vincentii* of Vincent, named after Jean-Hyacinthe Vincent (1862–1950), a French military physician.

Many characteristics are listed in Tables 127–129. Cells may have shallow and irregular spirals. Four to six periplasmic flagella are inserted at each end of cell. Motile with a rapid, jerky, vibratory motion. Colonies of strain N-9 are visible after incubation for 2 weeks. The colonies are white, 12–15 mm in diameter, appearing as a slight haze in the agar. Unlike other *Treponema* species, *Treponema vincentii* is reported to produce LPS (Blanco et al., 1994; Kurimoto et al., 1990). Grows in a peptone-yeast extract medium under anaerobic conditions. Requires animal serum or ascitic fluid for growth.

Amino acids are fermented to mainly acetic and *n*-butyric acids, moderate amounts of lactic acid, and smaller amounts of succinic and formic acids, and trace amounts of propionic acid, ethanol, *n*-propanol, and *n*-butanol. No additional end products are produced in the presence of glucose.

Methyl red-negative. Skim milk is not changed. Ammonia is produced in cultures. Chopped meat is neither blackened nor digested. A slight fetid odor is produced in cultures. Grows at pH 6.5–7.5 and at 25–45°C.

Meyer and Hunter (1967) reported that *Treponema vincentii* strain N-9 was antigenically distinct from *Treponema denticola* (FM) and the Nichols and Noguchi strains of *Treponema refringens*. Antigens were shared with [*Spirochaeta*] *zuelerzae* and *Treponema phagedenis* (Reiter and Kazan strains).

Source: oral cavity of humans.

DNA G+C content (mol%): not known.

Type strain: none designated.

Reference strain: N-9.

Sequence accession no. (16S rRNA gene): AF033309.

Species *Candidatus**

1. **"Treponema suis"** (Molbak et al., 2006)

su'is. L. n. *sus suis* a swine, hog, pig, boar, sow; L. gen. n. *suis* of a pig.

This *Candidatus* was identified in paraffin-embedded biopsy specimens of porcine colon. Laser capture dissection and PCR amplification were used to determine the 16S rRNA sequence, and fluorescence *in situ* hybridization (FISH) was performed to show that the spirochetes were distributed within the colonic epithelium and lamina propria. The fact that the organism was identified by FISH in nearly equal proportions in pigs with colitis (60%) and in normal controls (43%) indicates that it is not a causative agent of colitis. The 16S rRNA sequence was most closely related to *Treponema bryantii*; however, the sequence identity was only 90.1%, making it likely that this organism, provisionally called "*Treponema suis*", represents a separate species. By electron microscopy, "*Treponema suis*" is longer (6–11 µm) than other *Treponema* species identified in the porcine intestinal tract (*Treponema succinifaciens*, *Treponema berlinense*, and *Treponema porcinum*, 4–8 µm). In addition, "*Treponema suis*" has a total of 10–14 periplasmic flagella, as compared to two for the aforementioned species. The *in vitro* culture of this organism has not as yet been reported.

Source: porcine colon.

DNA G+C content (mol%): not known.

Type strain: none designated.

Sequence accession no. (16S rRNA gene): AM284386.

2. **"Treponema macrodentium"** Noguchi 1912, 82 [*Spirochaeta macrodentium* (Noguchi 1912) Pettit 1928, 182]

mac.ro.den'ti.um. Gr. adj. *makros* long; L. n. *dens dentis* tooth; N.L. gen. pl. n. *macrodentium* of large teeth.

Slender helical rods, 5–16 µm long and 0.1–0.25 µm wide. The ends of the cell are pointed. One periplasmic flagellum is inserted into each end of the cell. Motile with a fairly rapid motion. Young cells rotate rapidly on their long axis. Grows in peptone-yeast extract-medium or PPLO medium (BBL) containing 10% serum or ascitic fluid with cocarboxylase (5 mg/l), glucose (1 g/l), and cysteine (1 g/l). Requires animal serum for growth. This requirement can be replaced by isobutyric acid (20 mg/l), spermine (150 mg/l), and nicotinamide (400 mg/l). Will also grow in a medium supplemented with rumen fluid and cocarboxylase. Requires a fermentable carbohydrate as an energy source. Carbohydrates are fermented. Acid but no gas is produced. The final pH in glucose broth is 5.0–5.4. Ferments fructose, glucose, maltose, ribose, and sucrose. May ferment cellobiose, galactose, and xylose. Does not ferment mannose, rhamnose, sorbose, lactose, arabinose, trehalose, mannitol, inulin, sorbitol, or salicin. Starch is not hydrolyzed. Glucose is fermented mainly to lactic acid, moderate amounts of acetic and formic acids, and a trace of succinic acid. Gelatin is hydrolyzed. Indole-negative. Hydrogen sulfide is produced. Lactate is not used. Ammonia is not produced. Optimum temperature, 37°C. Grows at pH 7.0.

Source: subgingival crevice of humans.

DNA G+C content (mol%): 39 (T_m).

Type strain: no culture available.

Sequence accession no. (16S rRNA gene): not known.

*The following species have been described but can be found in the literature and are not yet classified formally. There are no known cultures of two of these species. They are listed so that if they are isolated again, the description can be used to aid in their identification. The names presently have no standing in nomenclature.

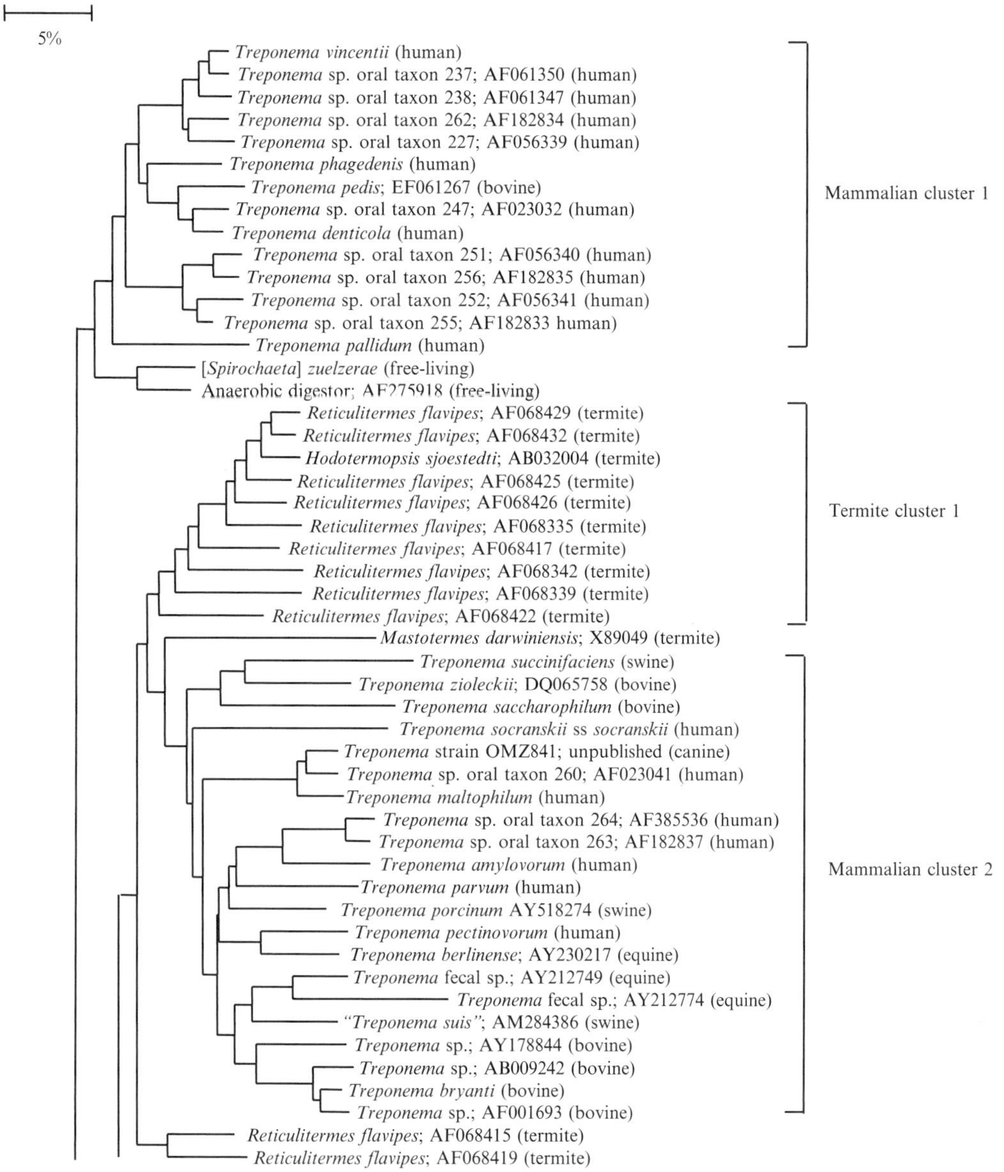

FIGURE 95. Phylogenetic tree illustrating the diversity of cultivable and not-yet-cultivable species of the genus *Treponema* based on 16S rRNA sequence comparisons. Environmental source or host for each species is noted in parentheses. Several clusters were apparent, e.g., mammalian clusters 1 and 2, termite clusters 1 and 2, and a waste water cluster 1. GenBank accession numbers for the 16S rRNA sequences of not yet cultivated species, or phylotypes, tested are shown. The scale bar represents a 5% difference in nucleotide sequence

3. **"Treponema orale"** Socransky, Listgarten, Hubersak, Cotmore and Clark 1969, 881 [*Treponema oralis* (*sic*) Socransky, Listgarten, Hubersak, Cotmore and Clark 1969, 881]

o.ra'le. L. n. *os, oris* the mouth; L. neut. suff. *-ale* suffix denoting pertaining to; N.L. neut. adj. *orale* pertaining to the mouth, of the mouth.

Slender helical cells, 6–16 µm long and 0.10–0.25 µm wide. Occasional chains are formed. One periplasmic flagellum is inserted into each end of the cell. Frequently end granules are seen in broth cultures. Motile with a jerky but fairly rapid motion. Grows in either PPLO medium without crystal violet (BBL) or peptone-yeast extract medium. Each medium contains glucose (1 g/l), cysteine (1 g/l), nicotinamide (500 mg/l), cocarboxylase (5 mg/l), spermine tetrahydrochloride (150 mg/l), and sodium isobutyrate (20 mg/l), and each is further supplemented with 10% inactivated rabbit serum or ascitic fluid, or 0.05% α-globulin. Uniform turbidity occurs in liquid media. Does not grow well on surface cultivation. Does not require carbohydrates as an energy source. Carbohydrates not fermented. Amino acids are fermented. The final pH in glucose broth is 6.8–7.2. Amino acids are fermented to acetic and propionic acids. Hydrolyzes gelatin but not starch. Indole-positive. H_2S produced. Utilizes lactate. Does not produce ammonia in cultures. Grows at pH 7.0 and at 37°C.

Paster et al. (1998) reported that a strain labeled "*Treponema oralis*" in the Smibert collection had 16S rRNA sequence and

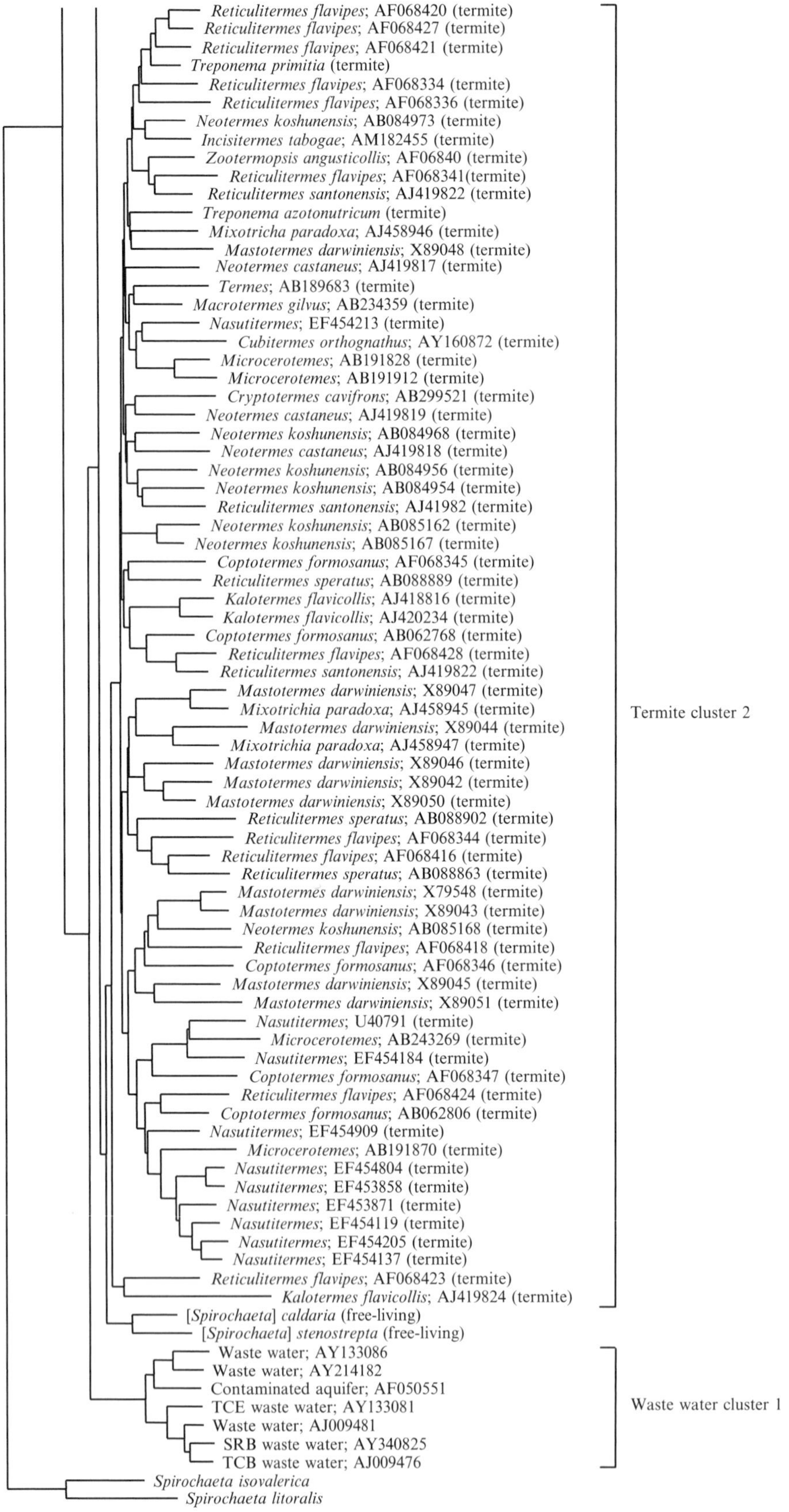

FIGURE 95. (continued)

other characteristics closely resembling those of *Treponema denticola*. The mol% G+C (37) is close to that determined for *Treponema denticola* (37.9). Therefore, this *Candidatus* species appears to be *Treponema denticola* and should be eliminated.

Source: subgingival crevice of humans.

DNA G+C content (mol%): 37.

Type strain: none.

Sequence accession no. (16S rRNA gene): none available.

4. **"Treponema zioleckii"** (Ziolecki, 1979; Ziolecki and Wojciechowicz, 1980)

zi.o.lec'ki.i. N.L. gen. masc. n. *zioleckii* of Ziolecki in honor of Alexander Ziolecki, Polish Academy of Sciences, in recognition of his contribution to the microbiology of rumen treponemes.

Cells are 0.5 µm in diameter by 5–11.5 µm in length. The number of periplasmic flagella per cell could not be estimated from electron microscopic observations, but it appears there are least four per cell (Piknova et al., 2008). Utilizes fructan, inulin, sucrose, and various plant mono- and disaccharides as fermentable substrates. Produces formate, acetate, and ethanol as endproducts of glucose fermentation. The name *Treponema zioleckii* was recently formally proposed (Piknova et al., 2008). The 16S rRNA gene sequence of *Treponema zioleckii* was > 99% similarity to strain CA, which was previously isolated from bovine rumen contents (Paster and Canale-Parola, 1982). Cells of strain CA are similar in dimensions to *Treponema zioleckii*, have 16–20 periplasmic flagella per cell, are pectinolytic, have amylolytic activity, can grow on arabinogalactan, and have a G+C content of 42 mol% (Paster and Canale-Parola, 1982).

Source: sheep rumen contents.

DNA G+C content (mol%): not known for type strain.

Type strain: kT.

Sequence accession no. (16S rRNA gene): DQ065758; strain CA: M59294.

Diversity of not-yet-cultivated species

As with species of *Spirochaeta*, species of *Treponema* are broadly diverse. In addition to known species of *Treponema*, novel phylotypes, i.e., species that have not yet been cultivated *in vitro*, have been identified by analysis of 16S rRNA genes of DNA isolated from host-associated sources, such as the human oral cavity (Choi et al., 1994; Dewhirst et al., 2000; Paster et al., 2001), termite hindguts (Lilburn et al., 1999) and the bovine rumen (Tajima et al., 1999).

Representatives of treponemal phylotypes from each of these and other environments are included in Figure 95. In one study, Paster et al. (2001) identified 49 not-yet-cultivated species of *Treponema* in human subgingival plaque (see expanded phylogenetic trees of phylotypes in Dewhirst et al., 2000, and Paster et al., 2001). The termite hindgut has an especially diverse array of treponemal phylotypes, with hundreds of potentially new species of *Treponema* identified (Berlanga et al., 2007; Lilburn et al., 1999; Ohkuma et al., 1999). The diversity of treponemes in termite hindguts is likely even more extensive considering that there are over 2,000 species of termites. Bovine and ovine digital dermatitis cases have also yielded several treponemal phylotypes (Demirkan et al., 2006; Walker et al., 1995). It is likely that other environments such as the intestinal tracts of most mammals, birds, and other insects will also contain many additional *Treponema* phylotypes.

Figure 95 illustrates the phylogenetic diversity of representatives of treponemal species and not-yet-cultivable phylotypes from human, animal, insect, and waste water environments. Interestingly, several phylogenetic clusters based on host or environment are apparent, namely mammalian, termite, and waste water.

References

Abramson, I.J. and R.M. Smibert. 1971. Inhibition of growth of treponemes by antimicrobial agents. Br. J. Vener. Dis. *47*: 407–412.

Aksenova, E.Y., V.A. Svetlichnyi and G.A. Zavazin. 1990. *Spirochaeta thermophila* sp. nov., a thermophilic marine spirochete isolated from a littoral hydrotherm of Shiashkotan Island. Microbiology (En. transl. from Mikrobiologiya) *59*: 735–741.

Aksenova, H.Y., F.A. Rainey, P.H. Janssen, G.A. Zavarzin and H.W. Morgan. 1992. *Spirochaeta thermophila* sp. nov., an obligately anaerobic, polysaccharolytic, extremely thermophilic bacterium. Int. J. Syst. Bacteriol. *42*: 175–177.

Allan, B., E.P. Greenberg and A. Kropinski. 1986. DNA-dependent RNA-polymerase from *Spirochaeta aurantia*. FEMS Microbiol. Lett. *35*: 205–210.

Anda, P., W. Sanchez-Yebra, M. del Mar Vitutia, E. Perez Pastrana, I. Rodriguez, N.S. Miller, P.B. Backenson and J.L. Benach. 1996. A new *Borrelia* species isolated from patients with relapsing fever in Spain. Lancet *348*: 162–165.

Anderson, J.F. 1989. Epizootiology of *Borrelia* in *Ixodes* tick vectors and reservoir hosts. Rev. Infect. Dis. *11 (Suppl 6)*: S1451–1459.

Antal, G.M., S.A. Lukehart and A.Z. Meheus. 2002. The endemic treponematoses. Microbes Infect. *4*: 83–94.

Baker-Zander, S.A. and S.A. Lukehart. 1983. Molecular basis of immunological cross-reactivity between *Treponema pallidum* and *Treponema pertenue*. Infect. Immun. *42*: 634–638.

Baker-Zander, S.A. and S.A. Lukehart. 1984. Antigenic cross-reactivity between *Treponema pallidum* and other pathogenic members of the family *Spirochaetaceae*. Infect. Immun. *46*: 116–121.

Baranton, G., D. Postic, I. Saint Girons, P. Boerlin, J.C. Piffaretti, M. Assous and P.A. Grimont. 1992. Delineation of *Borrelia burgdorferi sensu stricto*, *Borrelia garinii* sp. nov., and group VS461 associated with Lyme borreliosis. Int. J. Syst. Bacteriol. *42*: 378–383.

Barbieri, J.T. and C.D. Cox. 1979. Pyruvate oxidation by *Treponema pallidum*. Infect. Immun. *25*: 157–163.

Barbieri, J.T. and C.D. Cox. 1981. Influence of oxygen on respiration and glucose catabolism by *Treponema pallidum*. Infect. Immun. *31*: 992–997.

Barbour, A.G. 1984. Isolation and cultivation of Lyme disease spirochetes. Yale J. Biol. Med. *57*: 521–525.

Barbour, A.G. and S.F. Hayes. 1986. Biology of *Borrelia* species. Microbiol. Rev. *50*: 381–400.

Barbour, A.G., G.O. Maupin, G.J. Teltow, C.J. Carter and J. Piesman. 1996. Identification of an uncultivable *Borrelia* species in the hard tick Amblyomma americanum: possible agent of a Lyme disease-like illness. J. Infect. Dis. *173*: 403–409.

Barbour, A.G. 2005. Relapsing fever. *In* Tick-borne Diseases of Humans (edited by Goodman, Dennis and Sonenshine). ASM Press, Washington, D.C., pp. 268–291.

Bates, L.B. and J.H. St John. 1922. Suggestion of *Spirochaeta neotropicalis* as name for spirochaete of relapsing fever found in Panama. J. Am. Med. Assoc. *79*: 575–576.

Berg, H.C. 1976. How spirochetes may swim. J. Theor. Biol. *56*: 269–273.

Bergey, D.H., F.C. Harrison, R.S. Breed, B.W. Hammer and F.M. Huntoon. 1925. Bergey's Manual of Determinative Bacteriology, 2nd edn. Williams & Wilkins, Baltimore.

Berkeley, C. 1933. The oxidase and dehydrogenase systems of the crystalline style of *Mollusca*. Biochem. J. *27*: 1357–1365.

Berkeley, C. 1959. Some observations of *Cristispira* in the crystalline style of *Saxidomius giganteus* Deshayes and in that of some other Lamellibranchiata. Can. J. Zool. *37*: 53–58.

Berkeley, C. 1962. Toxicity of plankton to *Cristispira* inhabiting the crystalline style of a mollusk. Science *135*: 664–665.

Berlanga, M., R. Guerrero, J. A. Aas and B.J. Paster. 2003. Spirochetal diversity in microbial mats, abstract 103. Proceedings of the 103th General Meeting of the American Society for Microbiology, Washington, D.C.

Berlanga, M., B.J. Paster and R. Guerrero. 2007. Coevolution of symbiotic spirochete diversity in lower termites. Int. Microbiol. *10*: 133–139.

Bernard, F.R. 1970. Occurrence of the spirochete genus *Cristispira* in Western Canadian marine bivalves. Veliger *13*: 33–36.

Blakemore, R.P. and E. Canale-Parola. 1973. Morphological and ecological characteristics of *Spirochaeta plicatilis*. Arch. Mikrobiol. *89*: 273–289.

Blakemore, R.P. and E. Canale-Parola. 1976. Arginine catabolism by *Treponema denticola*. J. Bacteriol. *128*: 616–622.

Blanco, D.R., K. Reimann, J. Skare, C.I. Champion, D. Foley, M.M. Exner, R.E. Hancock, J.N. Miller and M.A. Lovett. 1994. Isolation of the outer membranes from *Treponema pallidum* and *Treponema vincentii*. J. Bacteriol. *176*: 6088–6099.

Bledsoe, H.A., J.A. Carroll, T.R. Whelchel, M.A. Farmer, D.W. Dorward and F.C. Gherardini. 1994. Isolation and partial characterization of *Borrelia burgdorferi* inner and outer membranes by using isopycnic centrifugation. J. Bacteriol. *176*: 7447–7455.

Bosanquet, W.C. 1911. Brief notes on the structure and development of *Spirochaeta anodontae* Keysselitz. Q. J. Microsc. Sci. *56*: 387–394.

Breinl, A. 1906. On the specific nature of the spirochaete of the African tick fever. Lancet *1*: 1690–1691.

Breznak, J.A. and E. Canale-Parola. 1969. *Spirochaeta aurantia*, a pigmented, facultatively anaerobic spirochete. J. Bacteriol. *97*: 386–395.

Breznak, J.A. and E. Canale-Parola. 1972a. Metabolism of *Spirochaeta aurantia*. II. Aerobic oxidation oxidation of carbohydrates. Arch. Mikrobiol. *83*: 278–292.

Breznak, J.A. and E. Canale-Parola. 1972b. Metabolism of *Spirochaeta aurantia*. I. Anaerobic energy-yielding pathways. Arch. Mikrobiol. *83*: 261–277.

Breznak, J.A. 1973. Biology of nonpathogenic, host-associated spirochetes. Crit. Rev. Microbiol. *2*: 457–489.

Breznak, J.A. and E. Canale-Parola. 1975. Morphology and physiology of *Spirochaeta aurantia* strains isolated from aquatic habitats. Arch. Microbiol. *105*: 1–12.

Breznak, J.A. 1984. Genus II. *Cristispira* (Gross 1910). *In* Bergey's Manual of Systematic Bacteriology, vol. 1 (edited by Krieg and Holt). Williams & Wilkins, Baltimore.

Breznak, J.A. and F. Warnecke. 2008. *Spirochaeta cellobiosiphila* sp. nov., a facultatively anaerobic, marine spirochaete. Int. J. Syst. Evol. Microbiol. *58*: 2762–2768.

Brumpt, E. 1921. Les parasites des invertébrés hématophages. *In* Thèse (edited by Lavier), Paris, p. 207.

Brumpt, E. 1922a. Les spirochetoses. *In* Nouveau Traite{acute} de Medecin (edited by Roger, Widal and Teissier). Masson et Cie, Paris, pp. 491–531.

Brumpt, E. 1922b. Les Spirochetoses. *In* Nouveau Traité de Mèdecine, 4 edn (edited by Roger, Widal and Teissier). Masson et Cie, Paris, pp. 491–531.

Brumpt, E. 1933. Étude du *Spirochaeta turicatae*, n. sp. agent de la fièvre récurrente sporadique des Etats-Unis transmis par *Ornithodorus turicata*. C. R. Soc. Biol. (Paris) *113*: 1369–1372.

Brumpt, E. 1939. Un nouveau treponeme parasite de l'homme: *Treponema carateum*, agent des carates ou "mal del Pinto". C. R. Soc. Biol. (Paris) *130*: 942–945.

Burgdorfer, W., A.G. Barbour, S.F. Hayes, J.L. Benach, E. Grunwaldt and J.P. Davis. 1982. Lyme disease-a tick-borne spirochetosis? Science *216*: 1317–1319.

Burgdorfer, W., J.F. Anderson, L. Gern, R.S. Lane, J. Piesman and A. Spielman. 1991. Relationship of *Borrelia burgdorferi* to its arthropod vectors. Scand. J. Infect. Dis. Suppl. *77*: 35–40.

Cameron, C.E. 2003. Identification of a *Treponema pallidum* laminin-binding protein. Infect. Immun. *71*: 2525–2533.

Cameron, C.E., E.L. Brown, J.M. Kuroiwa, L.M. Schnapp and N.L. Brouwer. 2004. *Treponema pallidum* fibronectin-binding proteins. J. Bacteriol. *186*: 7019–7022.

Cameron, C.E., N.L. Brouwer, L.M. Tisch and J.M. Kuroiwa. 2005. Defining the interaction of the *Treponema pallidum* adhesin Tp0751 with laminin. Infect. Immun. *73*: 7485–7494.

Canale-Parola, E., S.C. Holt and Z. Udris. 1967. Isolation of free-living, anaerobic spirochetes. Arch. Mikrobiol. *59*: 41–48.

Canale-Parola, E., Z. Udris and M. Mandel. 1968. The classification of free-living spirochetes. Arch. Mikrobiol. *63*: 385–397.

Canale-Parola, E. 1973. Isolation, growth and maintenance of anaerobic free-living spirochetes. *In* Methods in Microbiology, vol. 8 (edited by Norris and Ribbons). Academic Press, New York, pp. 61–73.

Canale-Parola, E. 1977. Physiology and evolution of spirochetes. Bacteriol. Rev. *41*: 181–204.

Canale-Parola, E. 1978. Motility and chemotaxis of spirochetes. Annu. Rev. Microbiol. *32*: 69–99.

Canale-Parola, E. 1980. Revival of the names *Spirochaeta litoralis*, *Spirochaeta zuelzerae*, and *Spirochaeta aurantia*. Int. J. Syst. Bacteriol. *30*: 594.

Canale-Parola, E. 1984a. Order I. *Spirochaetales* Buchanan 1917, 163[AL]. *In* Bergey's Manual of Systematic Bacteriology, vol. 1 (edited by Krieg and Holt). Williams & Wilkins, Baltimore, pp. 38–39.

Canale-Parola, E. 1984b. Genus I. *Spirochaeta* Ehrenberg 1835, 313[AL]. *In* Bergey's Manual of Systematic Bacteriology, vol. 1 (edited by Krieg and Holt). Williams & Wilkins, Baltimore, pp. 39–46.

Canica, M.M., F. Nato, L. du Merle, J.C. Mazie, G. Baranton and D. Postic. 1993. Monoclonal antibodies for identification of *Borrelia afzelii* sp. nov. associated with late cutaneous manifestations of Lyme borreliosis. Scand. J. Infect. Dis. *25*: 441–448.

Canica, M.M., F. Nato, L. du Merle, J.C. Mazie, G. Baranton and D. Postic. 1994. *In* Validation of the publication of new names and new combinations previously effectively published outside the IJSEM. List no. 48. Int. J. Syst. Bacteriol. *44*: 182–183.

Casjens, S., N. Palmer, R. van Vugt, W.M. Huang, B. Stevenson, P. Rosa, R. Lathigra, G. Sutton, J. Peterson, R.J. Dodson, D. Haft, E. Hickey, M. Gwinn, O. White and C.M. Fraser. 2000. A bacterial genome in flux: the twelve linear and nine circular extrachromosomal DNAs in an infectious isolate of the Lyme disease spirochete *Borrelia burgdorferi*. Mol. Microbiol. *35*: 490–516.

Castellani, A. 1905. On the presence of spirochetes in some cases of parangi (yaws, *Framboesia tropica*): Preliminary note. J. Ceylon Brit. Med. Assoc. *2*: 54.

Castellani, A. and A.J. Chalmers. 1919. Manual of Tropical Medicine, 3rd edn. Williams Wood, New York, pp. 959–960.

Centers for Disease Control and Prevention. 1995. Recommendations for test performance and interpretation from the Second National Conference on Serologic Diagnosis of Lyme disease. MMWR *44*: 590–591.

Centurion-Lara, A., C. Castro, R. Castillo, J.M. Shaffer, W.C. Van Voorhis and S.A. Lukehart. 1998. The flanking region sequences of the 15-kDa lipoprotein gene differentiate pathogenic treponemes. J. Infect. Dis. *177*: 1036–1040.

Centurion-Lara, A., B.J. Molini, C. Godornes, E. Sun, K. Hevner, W.C. Van Voorhis and S.A. Lukehart. 2006. Molecular differentiation of *Treponema pallidum* subspecies. J. Clin. Microbiol. *44*: 3377–3380.

Certes, A. 1882. Notes sur les parasites et les commensaux de l'huitre. Bull. Soc. Zool. France. *7*: 347–353.

Chan, E.C., R. Siboo, T. Keng, N. Psarra, R. Hurley, S.L. Cheng and I. Iugovaz. 1993. *Treponema denticola* (*ex* Brumpt 1925) sp. nov., nom. rev., and identification of new spirochete isolates from periodontal pockets. Int. J. Syst. Bacteriol. *43*: 196–203.

Cheetham, B.F. and M.E. Katz. 1995. A role for bacteriophages in the evolution and transfer of bacterial virulence determinants. Mol. Microbiol. *18*: 201–208.

Cheng, S.L., R. Siboo, T.C. Quee, J.L. Johnson, W.R. Mayberry and E.C. Chan. 1985. Comparative study of six random oral spirochete isolates. Serological heterogeneity of *Treponema denticola*. J. Periodont. Res. *20*: 602–612.

Choi, B.K., B.J. Paster, F.E. Dewhirst and U.B. Göbel. 1994. Diversity of cultivable and uncultivable oral spirochetes from a patient with severe destructive periodontitis. Infect. Immun. *62*: 1889–1895.

Christiansen, A.H. 1964. Studies on the antigenic structure of *T. pallidum*. 4. Comparison between the cultivable strains T. Reiter and T. Kazan Ii, applying agar gel diffusion technique and cross absorption experiments. Acta. Pathol. Microbiol. Scand. *60*: 123–130.

Cohn, F. 1875. Untersuchungen über Bakterien. Beitr. Biol. Pflanz. *1 (Heft II)*: 127–224.

Coleman, J.L., J.L. Benach, G. Beck and G.S. Habicht. 1986. Isolation of the outer envelope from *Borrelia burgdorferi*. Zentralbl. Bakteriol. Mikrobiol. Hyg. [A] *263*: 123–126.

Collier, W.A. 1921. *Cristispira helgolandica* nov. spec. und ihre Fortpflanzung. Zentralbl. Bakteriol. Parasitenkd. Infektionskr. Hyg. Abt. I. Orig. *86*: 132–134.

Correia, F.F., A.R. Plummer, B.J. Paster and F.E. Dewhirst. 2004. Genome size of human oral *Treponema* species by pulsed-field gel electrophoresis. Oral Microbiol. Immunol. *19*: 129–131.

Cox, C.D. and M.K. Barber. 1974. Oxygen uptake by *Treponema pallidum*. Infect. Immun. *10*: 123–127.

Cox, D.L. 1994. Culture of *Treponema pallidum*. *In* Methods in Enzymology, 1994/01/01 edn, vol. 236. Academic Press, pp. 390–405.

Cutler, S.J., C.O.K. Akintunde, J. Moss, M. Fukunaga, K. Kurtenbach, A. Talbert, H. Zhang, D.J.M. Wright and D.A. Warrell. 1999. Successful *in vitro* cultivation of *Borrelia duttonii* and its comparison with *Borrelia recurrentis*. Int. J. Syst. Bacteriol. *49*: 1793–1799.

Cutler, S.J. 2001. Relapsing fever *Borrelia*. *In* Molecular Medical Microbiology, vol. 3 (edited by Sussman). Academic Press, London, pp. 2093–2113.

Cwyk, W.M. and E. Canale-Parola. 1979. *Treponema succinifaciens* sp. nov., an anaerobic spirochete from the swine intestine. Arch. Microbiol. *122*: 231–239.

Cwyk, W.M. and E. Canale-Parola. 1981. *In* Validation of the publication of new names and new combinations previously effectively published outside the IJSB. List no. 7. Int. J. Syst. Bacteriol. *31*: 382–383.

Dai, Q., B.I. Restrepo, S.F. Porcella, S.J. Raffel, T.G. Schwan and A.G. Barbour. 2006. Antigenic variation by *Borrelia hermsii* occurs through recombination between extragenic repetitive elements on linear plasmids. Mol. Microbiol. *60*: 1329–1343.

Davis, G.E. 1942. Species unity or plurality of the relapsing fever spirochetes. Am. Assoc. Adv. Sci. Pub. *18*: 41–47.

Davis, G.E. 1948. The spirochetes. Annu. Rev. Microbiol. *2*: 305–334.

Davis, G.E. 1956. A relapsing fever spirochete, *Borrelia mazzottii* (sp. nov.) from *Ornithodorus talaje* from Mexico. Am. J. Hyg. *63*: 13–17.

Davis, G.E. 1957. Order IX. *Spirochaetales* Buchanan 1918. *In* Bergey's Manual of Determinative Bacteriology, 7th edn (edited by Breed, Murray and Smith). Williams & Wilkins, Baltimore, pp. 892–907.

de Buen, S. 1926. Note préliminaire sur l'épidémiologie de la fièvre récurrente espagnole. Ann. Parasitol. Hum. Comp. *4*: 182–192.

Defosse, D.L., R.C. Johnson, B.J. Paster, F.E. Dewhirst and G.J. Fraser. 1995. *Brevinema andersonii* gen. nov., sp. nov., and infectious spirochete isolated from the short-tailed shrew (*Blarina brevicauda*) and the white-footed mouse (*Peromyscus leucopus*). Int. J. Syst. Bacteriol. *45*: 78–84.

deMello, F. 1921. Protozoaires parasites du *Pachelebra moesta* Reeve. C. R. Soc. Biol. (Paris) *84*: 241–242.

Demirkan, I., H.F. Williams, A. Dhawi, S.D. Carter, C. Winstanley, K.D. Bruce and C.A. Hart. 2006. Characterization of a spirochaete isolated from a case of bovine digital dermatitis. J. Appl. Microbiol. *101*: 948–955.

Dewhirst, F.E., M.A. Tamer, R.E. Ericson, C.N. Lau, V.A. Levanos, S.K. Boches, J.L. Galvin and B.J. Paster. 2000. The diversity of periodontal spirochetes by 16S rRNA analysis. Oral Microbiol. Immunol. *15*: 196–202.

Diclemente, R.J., G.M. Wingood, R.A. Crosby, E. Rose, D. Lang, A. Pillay, J. Papp and C. Faushy. 2004. A descriptive analysis of STD prevalence among urban pregnant African-American teens: data from a pilot study. J. Adolesc. Health *34*: 376–383.

Dimitroff, V.T. 1926. Spirochetes in Baltimore market oysters. J. Bacteriol. *12*: 135–177.

Dobell, C.C. 1911. On *Cristispira veneris* nov. spec. and the affinities and classification of spirochetes. Q. J. Microsc. Sci. *56*: 507–542.

Dobell, C.C. 1912. Researches on the spirochaetes and related organisms. Arch. Protistenkd. *26*: 117–240.

Dressler, F., J.A. Whalen, B.N. Reinhardt and A.C. Steere. 1993. Western blotting in the serodiagnosis of Lyme disease. J. Infect. Dis. *167*: 392–400.

Dröge, S., J. Fröhlich, R. Radek and H. König. 2006. *Spirochaeta coccoides* sp. nov., a novel coccoid spirochete from the hindgut of the termite *Neotermes castaneus*. Appl. Environ. Microbiol. *72*: 392–397.

Dröge, S., J. Fröhlich, R. Radek and H. König. 2006. *In* Validation of publication of new names and new combinations previously effectively published outside the IJSEM. List no. 110. Int. J. Syst. Evol. Microbiol. *56*: 1459–1460.

Dschunkowsky, E. 1913. Das Rukfallfieber in Persien. Deut. Med. Wochenschr. *39*: 419–420.

Dujardin, F. 1841. Histoire naturelle des Zoophytes. Infusoires, comprenant la physiologie et la classification de ces animaux. De Roret, Paris.

Dupouey, P. 1963. Êtude immunologique de six especies de treponemes anaerobies d'origine genitale: *Treponema phagedenes*, *refringens*, *calligyra*, *minutum*, Reiter, et *pallidum*. Ann. Inst. Pasteur (Paris) *105*: 725–736; 949–970.

Dykhuizen, D.E., D.S. Polin, J.J. Dunn, B. Wilske, V. Preac-Mursic, R.J. Dattwyler and B.J. Luft. 1993. *Borrelia burgdorferi* is clonal: implications for taxonomy and vaccine development. Proc. Natl. Acad. Sci. U. S. A. *90*: 10163–10167.

Eggers, C.H. and D.S. Samuels. 1999. Molecular evidence for a new bacteriophage of *Borrelia burgdorferi*. J. Bacteriol. *181*: 7308–7313.

Ehrenberg, C.G. 1835. Dritter Beitrag zur Erkemtiss grosser Organisation in der Richtung des kleinsten Raumes. Abh. Preuss. Akad. Wiss. Phys. Kl Berlin aus den Jahre 1833–1835: 143–336.

Fantham, H.B. 1908. *Spirochaeta* (*Trypanosoma*) *balbianii* (Certes) and *Spirochaeta anodontae* (Keysselitz): their movements, structure and affinities. Q. J. Microsc. Sci. *52*: 1–73.

Fantham, H.B. 1911. Some researches on the life cycle of spirochetes. Ann. Trop. Med. Parasitol. *5*: 479–496.

Fieldsteel, A.H., F.A. Becker and J.G. Stout. 1977. Prolonged survival of virulent *Treponema pallidum* (Nichols strain) in cell-free and tissue culture systems. Infect. Immun. *18*: 173–182.

Fieldsteel, A.H., J.G. Stout and F.A. Becker. 1979. Comparative behavior of virulent strains of *Treponema pallidum* and *Treponema pertenue* in gradient cultures of various mammalian cells. Infect. Immun. *24*: 337–345.

Fieldsteel, A.H., D.L. Cox and R.A. Moeckli. 1981. Cultivation of virulent *Treponema pallidum* in tissue culture. Infect. Immun. *32*: 908–915.

Fitzgerald, T.J., P. Cleveland, R.C. Johnson, J.N. Miller and J.A. Sykes. 1977a. Scanning electron microscopy of *Treponema pallidum* (Nichols strain) attached to cultured mammalian cells. J. Bacteriol. *130*: 1333–1344.

Fitzgerald, T.J., R.C. Johnson, J.A. Sykes and J.N. Miller. 1977b. Interaction of *Treponema pallidum* (Nichols strain) with cultured mammalian cells: effects of oxygen, reducing agents, serum supplements, and different cell types. Infect. Immun. *15*: 444–452.

Fohn, M.J., F.S. Wignall, S.A. Baker-Zander and S.A. Lukehart. 1988. Specificity of antibodies from patients with pinta for antigens of *Treponema pallidum* subsp. *pallidum*. Infect. Dis. *157*: 32–37.

Fosnaugh, K. and E.P. Greenberg. 1988. Motility and chemotaxis of *Spirochaeta aurantia*: computer-assisted motion analysis. J. Bacteriol. *170*: 1768–1774.

Fracek, S.P., Jr. and J.F. Stolz. 1985. *Spirochaeta bajacaliforniensis* sp. n. from a microbial mat community at Laguna Figueroa, Baja California Norte, Mexico. Arch. Microbiol. *142*: 317–325.

Fracek, S.P.J. and J.F. Stolz. 2004. *In* Validation of publication of new names and new combinations previously effectively published outside the IJSEM. List no. 97. Int. J. Syst. Evol. Microbiol. *54*: 631–632.

Fraser, C.M., S. Casjens, W.M. Huang, G.G. Sutton, R. Clayton, R. Lathigra, O. White, K.A. Ketchum, R. Dodson, E.K. Hickey, M. Gwinn, B. Dougherty, J.F. Tomb, R.D. Fleischmann, D. Richardson, J. Peterson, A.R. Kerlavage, J. Quackenbush, S. Salzberg, M. Hanson, R. van Vugt, N. Palmer, M.D. Adams, J. Gocayne, J.C. Venter and et al. 1997. Genomic sequence of a Lyme disease spirochaete, *Borrelia burgdorferi*. Nature *390*: 580–586.

Fraser, C.M., S.J. Norris, G.M. Weinstock, O. White, G.G. Sutton, R. Dodson, M. Gwinn, E.K. Hickey, R. Clayton, K.A. Ketchum, E. Sodergren, J.M. Hardham, M.P. McLeod, S. Salzberg, J. Peterson, H. Khalak, D. Richardson, J.K. Howell, M. Chidambaram, T. Utterback, L. McDonald, P. Artiach, C. Bowman, M.D. Cotton, C. Fujii, S. Garland, B. Hatch, K. Horst, K. Roberts, M. Sandusky, J. Weidman, H.O. Smith and J.C. Venter. 1998. Complete genome sequence of *Treponema pallidum*, the syphilis spirochete. Science *281*: 375–388.

Fukunaga, M., Y. Takahashi, Y. Tsuruta, O. Matsushita, D. Ralph, M. McClelland and M. Nakao. 1995. Genetic and phenotypic analysis of *Borrelia miyamotoi* sp. nov., isolated from the ixodid tick *Ixodes persulcatus*, the vector for lyme disease in Japan. Int. J. Syst. Bacteriol. *45*: 804–810.

Fukunaga, M., A. Hamase, K. Okada and M. Nakao. 1996a. *Borrelia tanukii* sp. nov. and *Borrelia turdae* sp. nov. found from ixodid ticks in Japan: rapid species identification by 16S rRNA gene-targeted PCR analysis. Microbiol. Immunol. *40*: 877–881.

Fukunaga, M., K. Okada, M. Nakao, T. Konishi and Y. Sato. 1996b. Phylogenetic analysis of *Borrelia* species based on flagellin gene sequences and its application for molecular typing of Lyme disease borreliae. Int. J. Syst. Bacteriol. *46*: 898–905.

Fukunaga, M., A. Hamase, K. Okada and M. Nakao. 1997. *In* Validation of the publication of new names and new combinations previously effectively published outside the IJSEM. List no. 63. Int. J. Syst. Bacteriol. *47*: 1274.

Garnham, P.C.C. 1947. A new blood spirochaete in the grivet monkey, *Cercopithecus aethiops*. East Afr. Med. J. *24*: 47–51.

Gern, L., A. Estrada-Pena, F. Frandsen, J.S. Gray, T.G. Jacnson, F. Jongejan, O. Kahl, E. Korenberg, R. Mehl and P.A. Nuttall. 1998. European reservoir hosts of *Borrelia burgdorferi sensu lato*. Zentralbl. Bakteriol. *287*: 196–204.

Giacani, L., E.S. Sun, K. Hevner, B.J. Molini, W.C. Van Voorhis, S.A. Lukehart and A. Centurion-Lara. 2004. Tpr homologs in *Treponema paraluiscuniculi* Cuniculi A strain. Infect. Immun. *72*: 6561–6576.

Gieszczykiewicz, M. 1939. Zagadniene systematihki w bakteriologii – Zür Frage der Bakterien-Systematic. Bull. Acad. Polon. Sci., Ser. Sci. Biol. *1*: 9–27.

Glockner, G., R. Lehmann, A. Romualdi, S. Pradella, U. Schulte-Spechtel, M. Schilhabel, B. Wilske, J. Suhnel and M. Platzer. 2004. Comparative analysis of the *Borrelia garinii* genome. Nucleic Acids Res. *32*: 6038–6046.

Glockner, G., U. Schulte-Spechtel, M. Schilhabel, M. Felder, J. Suhnel, B. Wilske and M. Platzer. 2006. Comparative genome analysis: selection pressure on the *Borrelia* vls cassettes is essential for infectivity. BMC Genomics *7*: 211.

Gonder, R. 1908. Spirochäten aus dem Darmtraktus von Pinna: *Spirochaete pinnae* nov. spec. und *Spirochaete hartmanni* nov. spec. Zentralbl. Bakteriol. Parasitenkd. Infektionskr. Abt. I. Orig. *47*: 491–494.

Graber, J.R., J.R. Leadbetter and J.A. Breznak. 2004. Description of *Treponema azotonutricium* sp. nov. and *Treponema primitia* sp. nov., the first spirochetes isolated from termite guts. Appl. Environ. Microbiol. *70*: 1315–1320.

Graves, S. and J. Downes. 1981. Experimental infection of man with rabbit-virulent *Treponema paraluiscuniculi*. Br. J. Vener. Dis. *57*: 7–10.

Greenberg, E.P. and E. Canale-Parola. 1975. Carotenoid pigments of facultatively anaerobic spirochetes. J. Bacteriol. *123*: 1006–1012.

Greenberg, E.P. and E. Canale-Parola. 1976. *Spirochaeta halophila* sp. n., a facultative anaerobe from a high-salinity pond. Arch. Microbiol. *110*: 185–194.

Greenberg, E.P. and E. Canale-Parola. 1977a. Motility of flagellated bacteria in viscous environments. J. Bacteriol. *132*: 356–358.

Greenberg, E.P. and E. Canale-Parola. 1977b. Relationship between cell coiling and motility of spirochetes in viscous environments. J. Bacteriol. *131*: 960–969.

Greenberg, E.P. and E. Canale-Parola. 1977c. Chemotaxis in *Spirochaeta aurantia*. J. Bacteriol. *130*: 485–494.

Greenberg, E.P., B. Brahamsha and K. Fosnaugh. 1985. The motile behavior of *Spirochaeta aurantia*: a twist to chemosensory transduction in bacteria. *In* Sensing and Response in Microorganisms (edited by Eisenbach and Balaban). Elsevier, New York, pp. 107–118.

Gross, J. 1910. *Cristispira* nov. gen. Ein Beiträg zür Spirochätenfrage. Mitt. Zool. Sta. Neapel *20*: 41–93.

Güner, E.S., M. Watanabe, N. Hashimoto, T. Kadosaka, Y. Kawamura, T. Ezaki, H. Kawabata, Y. Imai, K. Kaneda and T. Masuzawa. 2004. *Borrelia turcica* sp. nov., isolated from the hard tick *Hyalomma aegyptium* in Turkey. Int. J. Syst. Evol. Microbiol. *54*: 1649–1652.

Harwood, C.S. and E. Canale-Parola. 1981a. Adenosine 5′-triphosphate-yielding pathways of branched-chain amino acid fermentation by a marine spirochete. J. Bacteriol. *148*: 117–123.

Harwood, C.S. and E. Canale-Parola. 1981b. Branched-chain amino acid fermentation by a marine spirochete: strategy for starvation survival. J. Bacteriol. *148*: 109–116.

Harwood, C.S., H.W. Jannasch and E. Canale-Parola. 1982. Anaerobic spirochete from a deep-sea hydrothermal vent. Appl. Environ. Microbiol. *44*: 234–237.

Harwood, C.S. and E. Canale-Parola. 1983. *Spirochaeta isovalerica* sp. nov., a marine anaerobe that forms branched-chain fatty acids as fermentation products. Int. J. Syst. Bacteriol. *33*: 573–579.

Harwood, C.S. and E. Canale-Parola. 1984. Ecology of spirochetes. Annu. Rev. Microbiol. *38*: 161–192.

Hayes, N.S., K.E. Muse, A.M. Collier and J.B. Baseman. 1977. Parasitism by virulent *Treponema pallidum* of host cell surfaces. Infect. Immun. *17*: 174–186.

Heisch, R.B. 1953. On a spirochaete isolated from *ornithodoros graingeri*. Parasitology *43*: 133–135.

Hellmann, G. 1913. Über die im Excretionsorgan der Ascidien der Gattung Caesira (Molgula) vorkommenden Spirochäten: *Spirochaeta caesirae* septentrionalis n. sp. und *Spirochaeta caesirae* retortiformis n. sp. Arch. Protistenkunde *29*: 22–38.

Hespell, R.B. and E. Canale-Parola. 1970a. Carbohydrate metabolism in *Spirochaeta stenostrepta*. J. Bacteriol. *103*: 216–226.

Hespell, R.B. and E. Canale-Parola. 1970b. *Spirochaeta litoralis* sp.n., a strictly anaerobic marine spirochete. Archiv. Mikrobiol. *74*: 1–18.

Hespell, R.B. and E. Canale-Parola. 1973. Glucose and pyruvate metabolism of *Spirochaeta litoralis*, an anaerobic marine spirochete. J. Bacteriol. *116*: 931–937.

Hollande, A.C. 1922. Les spirochètes de termites; processus de division: formation du schizoplaste. Arch. Zool. Espt. Gen. Notes Rev. *61*: 23–28.

Holt, S.C. and E. Canale-Parola. 1968. Fine structure of *Spirochaeta stenostrepta*, a free-living, anaerobic spirochete. J. Bacteriol. *96*: 822–835.

Hoover, R.B., E.V. Pikuta, A.K. Bej, D. Marsic, W.B. Whitman, J. Tang and P. Krader. 2003. *Spirochaeta americana* sp. nov., a new haloalkaliphilic, obligately anaerobic spirochaete isolated from soda Mono Lake in California. Int. J. Syst. Evol. Microbiol. *53*: 815–821.

Hovind-Hougen, K. and A. Birch-Andersen. 1971. Electron microscopy of endoflagella and microtubules in *Treponema* Reiter. Acta Pathol. Microbiol. Scand. [B] Microbiol. Immunol. *79*: 37–50.

Hyde, F.W. and R.C. Johnson. 1984. Genetic relationship of lyme disease spirochetes to *Borrelia*, *Treponema*, and *Leptospira* spp. J. Clin. Microbiol. *20*: 151–154.

Izard, J. 2006. Cytoskeletal cytoplasmic filament ribbon of *Treponema:* a member of an intermediate-like filament protein family. J. Mol. Microbiol. Biotechnol. *11*: 159–166.

Izard, J., C.E. Hsieh, R.J. Limberger, C.A. Mannella and M. Marko. 2008. Native cellular architecture of *Treponema denticola* revealed by cryo-electron tomography. J. Struct. Biol. *163*: 10–17.

Jacobsthal, E. 1920. Untersuchungen über eine syphilisähnliche spontanerkrankung des kaninchens (*Paralues cuniculi*). Derm. Wochenschr. *71*: 569–571.

Jahn, T.L. and M.D. Landman. 1965. Locomotion of spirochetes. Trans. Am. Microsc. Soc. *84*: 395–406.

Johnson, R.C. and L.M. Eggebraten. 1971. Fatty acid requirements of the Kazan 5 and Reiter strains of *Treponema pallidum*. Infect. Immun. *3*: 723–726.

Johnson, R.C., G.P. Schmid, F.W. Hyde, A.G. Steigerwalt and D.J. Brenner. 1984. *Borrelia burgdorferi* sp. nov., etiologic agent of lyme disease. Int. J. Syst. Bacteriol. *34*: 496–497.

Johnson, R.C., W. Burgdorfer, R.S. Lane, A.G. Barbour, S.F. Hayes and F.W. Hyde. 1987. *Borrelia coriaceae* sp. nov., putative agent of epizootic bovine abortion. Int. J. Syst. Bacteriol. *37*: 72–74.

Joseph, R., S.C. Holt and E. Canale-Parola. 1970. Ultrastructure and chemical composition of the cell wall of *Spirochaeta stenostrepta*. Bacteriol. Proc.: 57.

Joseph, R. and E. Canale-Parola. 1972. Axial fibrils of anaerobic spirochetes: ultrastructure and chemical characteristics. Arch. Mikrobiol. *81*: 146–168.

Joseph, R., S.C. Holt and E. Canale-Parola. 1973. Peptidoglycan of free-living anaerobic spirochetes. J. Bacteriol. *115*: 426–435.

Judd, W. 1979. The secretions and fine structure of bivalve crystalline style sacs. Ophelia *18*: 205–233.

Karimi, Y., K. Hovind-Hougen, A. Birch-Andersen and M. Asmar. 1979. *Borrelia persica* and *B. baltazardi* sp. nov.: experimental pathogenicity for some animals and comparison of the ultrastructure. Ann. Microbiol. (Paris) *130B*: 157–168.

Karimi, Y., K. Hovind-Hougen, A. Birch-Andersen and M. Asmar. 1983. *In* Validation of the publication of new names and new combinations previously published outside the IJSEM. List no. 10. Int. J. Syst. Bacteriol. *33*: 438–440.

Kawabata, H., T. Masuzawa and Y. Yanagihara. 1993. Genomic analysis of *Borrelia japonica* sp. nov. isolated from *Ixodes ovatus* in Japan. Microbiol. Immunol. *37*: 843–848.

Kawabata, H., T. Masuzawa and Y. Yanagihara. 1994. *In* Validation of publication of new names and new combinations previously effectively published outside the IJSEM. List no. 50. Int. J. Syst. Bacteriol. *44*: 595.

Kelly, R. 1971. Cultivation of *Borrelia hermsi*. Science *173*: 443–444.

Kelly, R. 1984. Genus IV. *Borrelia*. *In* Bergey's Manual of Systematic Bacteriology, vol. 1 (edited by Kreig and Holt). Williams & Wilkins, Baltimore, pp. 57–64.

Kelly, R.T. 1976. Cultivation and physiology of relapsing fever borreliae. *In* The Biology of Parisitic Spirochetes (edited by Johnson). Academic Press, New York, pp. 87–94.

Keysselitz, G. 1906. *Spirochaeta anodontae* nov. spec. Arb. Gesundh. Amt. Berl. *23*: 566–569.

Kisinza, W.N., P.J. McCall, H. Mitani, A. Talbert and M. Fukunaga. 2003. A newly identified tick-borne *Borrelia* species and relapsing fever in Tanzania. Lancet *362*: 1283–1284.

Kitten, T. and A.G. Barbour. 1992. The relapsing fever agent *Borrelia hermsii* has multiple copies of its chromosome and linear plasmids. Genetics *132*: 311–324.

Klaviter, E.C. and R.C. Johnson. 1979. Isolation of the outer envelope, chemical components, and ultrastructure of *Borrelia hermsi* grown *in vitro*. Acta. Trop. *36*: 123–131.

Kubomura, K. 1969. Fructose medium for the cultivation of *Cristispira* sp., a flagellate living in the crystalline style of bivalves. Sci. Rep. Saitama Univ. Ser. B. *5*: 1–5.

Kuhn, D.A. 1974. Genus II. *Cristispira* (Gross 1910). *In* Bergey's Manual of Determinative Bacteriology 8th edn (edited by Buchanan and Gibbons). Williams & Wilkins, Baltimore, pp. 171–174.

Kurimoto, T., M. Suzuki and T. Watanabe. 1990. [Chemical composition and biological activities of lipopolysaccharides extracted from *Treponema denticola* and *Treponema vincentii*]. Shigaku *78*: 208–232.

Kurtti, T.J., U.G. Munderloh, R.C. Johnson and G.G. Ahlstrand. 1987. Colony formation and morphology in *Borrelia burgdorferi*. J. Clin. Microbiol. *25*: 2054–2058.

Laveran, A. 1903. Sur la spirillose des bovidés. C. R. Acad. Sci. Paris *136*: 939–941.

Le Fleche, A., D. Postic, K. Girardet, O. Peter and G. Baranton. 1997. Characterization of *Borrelia lusitaniae* sp. nov. by 16S ribosomal DNA sequence analysis. Int. J. Syst. Bacteriol. *47*: 921–925.

Lebert, H. 1874. Rückfallstyphus und klilioses Typhoid. *In* Ziemssen's Handbuch der Speciellen Pathologie und Therapie, 2nd edn. F.C.W. Vogel, Leipzig, pp. 267–304.

Leger, A. 1917. Spirochaete de la musaraigne (*Crocidura stampfli* Tentink). Bull. Soc. Pathol. Exot. *10*: 280–281.

Leschine, S.B. and E. Canale-Parola. 1980a. Ornithine dissimilation by *Treponema denticola*. Curr. Microbiol. *3*: 305–310.

Leschine, S.B. and E. Canale-Parola. 1980b. Rifampicin as a selective agent for isolation of oral spirochetes. J. Clin. Microbiol. *12*: 792–795.

Leschine, S.B. and E. Canale-Parola. 1986. Rifampin-resistant RNA-polymerase in spirochetes. FEMS Microbiol. Lett. *35*: 199–204.

Leschine, S.B. 1995. Cellulose degradation in anaerobic environments. Annu. Rev. Microbiol. *49*: 399–426.

Li, C., R.G. Bakker, M.A. Motaleb, M.L. Sartakova, F.C. Cabello and N.W. Charon. 2002. *Asymmetrical flagellar rotation in Borrelia burgdorferi* nonchemotactic mutants. Proc. Natl. Acad. Sci. U. S. A. *99*: 6169–6174.

Lilburn, T.G., T.M. Schmidt and J.A. Breznak. 1999. Phylogenetic diversity of termite gut spirochaetes. Environ. Microbiol. *1*: 331–345.

Lin, T., L. Gao, A. Seyfang and J.H. Oliver, Jr. 2005. '*Candidatus* Borrelia texasensis', from the American dog tick *Dermacentor variabilis*. Int. J. Syst. Evol. Microbiol. *55*: 685–693.

Livermore, B.P. and R.C. Johnson. 1974. Lipids of the *Spirochaetales*: comparison of the lipids of several members of the genera *Spirochaeta*, *Treponema*, and *Leptospira*. J. Bacteriol. *120*: 1268–1273.

Livermore, B.P., R.F. Bey and R.C. Johnson. 1978. Lipid metabolism of *Borrelia hermsi*. Infect. Immun. *20*: 215–220.

Lukehart, S.A., C. Godornes, B.J. Molini, P. Sonnett, S. Hopkins, F. Mulcahy, J. Engelman, S.J. Mitchell, A.M. Rompalo, C.M. Marra and J.D. Klausner. 2004. Macrolide resistance in *Treponema pallidum* in the United States and Ireland. N. Engl. J. Med. *351*: 154–158.

MacDougall, J., D. Margarita and I. Saint Girons. 1992. Homology of a plasmid from the spirochete *Treponema denticola* with the single-stranded DNA plasmids. J. Bacteriol. *174*: 2724–2728.

Magot, M., M.L. Fardeau, O. Arnauld, C. Lanau, B. Ollivier, P. Thomas and B.K. Patel. 1997. *Spirochaeta smaragdinae* sp. nov., a new mesophilic strictly anaerobic spirochete from an oil field. FEMS Microbiol. Lett. *155*: 185–191.

Maio, R.M. and A.H. Fieldsteel. 1978. Genetics of *Treponema*: relationship between *Treponema pallidum* and five cultivable treponemes. J. Bacteriol. *133*: 101–107.

Maio, R.M. and A.H. Fieldsteel. 1980. Genetic relationship between *Treponema pallidum* and *Treponema pertenue*, two noncultivable human pathogens. J. Bacteriol. *141*: 427–429.

Marconi, R.T., D. Liveris and I. Schwartz. 1995. Identification of novel insertion elements, restriction-fragment-length-polymorphism patterns, and discontinuous 23S ribosomal RNA in lyme disease spirochetes: phylogenetic analyses of ribosomal RNA genes and their intergenic spacers in *Borrelia japonica* sp. nov. and genomic group 21038 (*Borrelia andersonii* sp. nov.) isolates. J. Clin. Microbiol. *33*: 2427–2434.

Margos, G., A.G. Gatewood, D.M. Aanensen, K. Hanincova, D. Terekhova, S.A. Vollmer, M. Cornet, J. Piesman, M. Donaghy, A. Bormane, M.A. Hurn, E.J. Feil, D. Fish, S. Casjens, G.P. Wormser, I. Schwartz and K. Kurtenbach. 2008. MLST of housekeeping genes captures geographic population structure and suggests a European origin of *Borrelia burgdorferi*. Proc. Natl. Acad. Sci. U. S. A. *105*: 8730–8735.

Marra, C.M., A.P. Colina, C. Godornes, L.C. Tantalo, M. Puray, A. Centurion-Lara and S.A. Lukehart. 2006. Antibiotic selection may contribute to increases in macrolide-resistant *Treponema pallidum*. J. Infect. Dis. *194*: 1771–1773.

Maruashvilli, G.M. 1945. On the tick borne relapsing fever. Med. Parazitol. Parazit. Bilez. *14*: 24–27.

Masuzawa, T., N. Takada, M. Kudeken, T. Fukui, Y. Yano, F. Ishiguro, Y. Kawamura, Y. Imai and T. Ezaki. 2001. *Borrelia sinica* sp. nov., a lyme disease-related *Borrelia* species isolated in China. Int. J. Syst. Evol. Microbiol. *51*: 1817–1824.

Mazzotti, L. 1949. Sobre una nueva espiroqueta de la fiebre recurrente, encontrada en Mexico. Rev. Inst. Salubr. Enferm. Trop. Mex. *10*: 277–281.

Mesnil, F. and M. Caullery. 1916. Sur un organisme spirochétoide (*Cristispira polydorae* n. sp.) de l'intestin d'une annélide polychéte. C. R. Soc. Biol. (Paris) *79*: 1118–1121.

Meyer, P.E. and E.F. Hunter. 1967. Antigenic relationships of 14 treponemes demonstrated by immunofluorescence. J. Bacteriol. *93*: 784–789.

Miller, J.N. 1971. Spirochetes in body fluids and tissues: manual of investigative methods. Charles C. Thomas, Springfield, IL.

Miller, J.N., R. M. Smibert and S.J. Norris. 1991. The genus *Treponema*. *In* The Prokaryotes: a Handbook on the Biology of *Bacteria*: Ecophysiology, Isolation, Identification, Applications, 2nd Ed edn, vol. 4 (edited by Balows, Trüper, Dworkin, Harder and Schleifer). Springer, New York, pp. 3537–3559.

Mitchell, S.J., J. Engelman, C.K. Kent, S.A. Lukehart, C. Godornes and J.D. Klausner. 2006. Azithromycin-resistant syphilis infection: San Francisco, California, 2000–2004. Clin. Infect. Dis. *42*: 337–345.

Möbius, K. 1883. *Trypanosoma balbianii* Certes im Krystallstiel schleswigholsteinischer Austern. Zool. Anz. *6*: 148.

Molbak, L., K. Klitgaard, T.K. Jensen, M. Fossi and M. Boye. 2006. Identification of a novel, invasive, not-yet-cultivated *Treponema* sp. in the large intestine of pigs by PCR amplification of the 16S rRNA gene. J. Clin. Microbiol. *44*: 4537–4540.

Molepo, J., A. Pillay, B. Weber, S. Morse and A. Hoosen. 2007. Molecular typing of *Treponema pallidum* strains from patients with neurosyphilis in Pretoria, South Africa. Sex. Transm. Infect.: sti.2006.023895.

Moore, W.E. and L.V. Moore. 1994. The bacteria of periodontal diseases. Periodontol. 2000 *5*: 66–77.

Motaleb, M.A., L. Corum, J.L. Bono, A.F. Elias, P. Rosa, D.S. Samuels and N.W. Charon. 2000. *Borrelia burgdorferi* periplasmic flagella have both skeletal and motility functions. Proc. Natl. Acad. Sci. U. S. A. *97*: 10899–10904.

Murphy, G.E., J.R. Leadbetter and G.J. Jensen. 2006. In situ structure of the complete *Treponema primitia* flagellar motor. Nature *442*: 1062–1064.

Nelson, T.C. 1918. On the origin, nature, and function of the crystalline style of lamellibranches. J. Morphol. *31*: 53–111.

Nichols, H.A. and W.H. Hough. 1913. Demonstration of *Spirochaeta pallida* in the cerebrospinal fluid from a patient with nervous relapse following the use of salvarsan. J. Am. Med. Assoc. *60*: 108–110.

Nichols, J.C. and J.B. Baseman. 1975. Carbon sources utilized by virulent *Treponema pallidum*. Infect. Immun. *12*: 1044–1050.

Noguchi, H. 1912. Cultural studies on mouth spirochetae (*Treponema microdentium* and *macrodentium*). J. Exp. Med. *15*: 81–89.

Noguchi, H. 1921. *Cristispira* in North American shellfish: a note on a *Spirillum* found in oysters. J. Exp. Med. *34*: 295–315.

Noguchi, H. 1928. The Spirochetes. *In* The New Knowledge of Bacteriology and Immunology (edited by Jordan and Falk). The University of Chicago Press, Chicago, pp. 452–497.

Noordhoek, G.T., A. Cockayne, L.M. Schouls, R.H. Meloen, E. Stolz and J.D. van Embden. 1990. A new attempt to distinguish serologically the subspecies of *Treponema pallidum* causing syphilis and yaws. J. Clin. Microbiol. *28*: 1600–1607.

Nordhoff, M., D. Taras, M. Macha, K. Tedin, H.-J. Busse and L.H. Wieler. 2005. *Treponema berlinense* sp. nov. and *Treponema porcinum* sp. nov., novel spirochaetes isolated from porcine faeces. Int. J. Syst. Evol. Microbiol. *55*: 1675–1680.

Norris, S.J., J.N. Miller, J.A. Sykes and T.J. Fitzgerald. 1978. Influence of oxygen tension, sulfhydryl compounds, and serum on the motility and virulence of *Treponema pallidum* (Nichols strain) in a cell-free system. Infect. Immun. *22*: 689–697.

Norris, S.J., J.N. Miller and J.A. Sykes. 1980. Long-term incorporation of tritiated adenine into deoxyribonucleic acid and ribonucleic acid by *Treponema pallidum* (Nichols strain). Infect. Immun. *29*: 1040–1049.

Norris, S.J. and D.G. Edmondson. 1987. Factors affecting the multiplication and subculture of *Treponema pallidum* subsp. *pallidum* in a tissue culture system. Infect. Immun. *53*: 534–539.

Norris, S.J., N.W. Charon, R.G. Cook, M.D. Fuentes and R.J. Limberger. 1988. Antigenic relatedness and N-terminal sequence homology define two classes of periplasmic flagellar proteins of *Treponema pallidum* subsp. *pallidum* and *Treponema phagedenis*. J. Bacteriol. *170*: 4072–4082.

Norris, S.J. 1993. Polypeptides of *Treponema pallidum*: progress toward understanding their structural, functional, and immunologic roles. *Treponema pallidum* Polypeptide Research Group. Microbiol. Rev. *57*: 750–779.

Norris, S.J., B.J. Paster, A. Moter and U.B. Göbel. 2003. The genus *Treponema*. *In* The Prokaryotes: an Evolving Electronic Resource for the Microbiological Community, 3rd edn (edited by Dworkin, Falkow, Rosenberg, Schleifer and Stackebrandt). Springer, New York.

Novy, F.G. and R.E. Knapp. 1906. Studies on *Spirillum obermeiri* and related organisms. J. Infect. Dis. *3*: 291–393.

Ohkuma, M., T. Iida and T. Kudo. 1999. Phylogenetic relationships of symbiotic spirochetes in the gut of diverse termites. FEMS Microbiol. Lett. *181*: 123–129.

Ohta, K., K.K. Makinen and W.J. Loesche. 1986. Purification and characterization of an enzyme produced by *Treponema denticola* capable of hydrolyzing synthetic trypsin substrates. Infect. Immun. *53*: 213–220.

Paster, B. and F. Dewhirst. 2000. Phylogenetic foundation of spirochetes. J. Mol. Microbiol. Biotechnol. *2*: 341–344.

Paster, B.J. and E. Canale-Parola. 1980. Involvement of periplasmic fibrils in motility of spirochetes. J. Bacteriol. *141*: 359–364.

Paster, B.J. and E. Canale-Parola. 1982. Physiological diversity of rumen spirochetes. Appl. Environ. Microbiol. *43*: 686–693.

Paster, B.J., E. Stackebrandt, R.B. Hespell, C.M. Hahn and C.R. Woese. 1984. The phylogeny of the spirochetes. Syst. Appl. Microbiol. *5*: 337–351.

Paster, B.J. and E. Canale-Parola. 1985. *Treponema saccharophilum* sp. nov., a large pectinolytic spirochete from the bovine rumen. Appl. Environ. Microbiol. *50*: 212–219.

Paster, B.J., F.E. Dewhirst, W.G. Weisburg, L.A. Tordoff, G.J. Fraser, R.B. Hespell, T.B. Stanton, L. Zablen, L. Mandelco and C.R. Woese. 1991. Phylogenetic analysis of the spirochetes. J. Bacteriol. *173*: 6101–6109.

Paster, B.J., D.A. Pelletier, F.E. Dewhirst, W.G. Weisburg, V. Fussing, L.K. Poulsen, S. Dannenberg and I. Schroeder. 1996. Phylogenetic position of the spirochetal genus *Cristispira*. Appl. Environ. Microbiol. *62*: 942–946.

Paster, B.J., F.E. Dewhirst, B.C. Coleman, C.N. Lau and R.L. Ericson. 1998. Phylogenetic analysis of cultivable oral treponemes from the Smibert collection. Int. J. Syst. Bacteriol. *48*: 713–722.

Paster, B.J., S.K. Boches, J.L. Galvin, R.E. Ericson, C.N. Lau, V.A. Levanos, A. Sahasrabudhe and F.E. Dewhirst. 2001. Bacterial diversity in human subgingival plaque. J. Bacteriol. *183*: 3770–3783.

Patel, B.K.C., H.W. Morgan and R.M. Daniel. 1985. Thermophilic anaerobic spirochetes in New Zealand hot springs. FEMS Microbiol. Lett. *26*: 101–106.

Perrin, W.S. 1906. Researches upon the life-history of *Trypanosoma balbianii* (Certes). Arch. Protistenkd. *7*: 131–156.

Pettit, A. 1928. Contribution à l'Étude de Spirochétidés. Seine, Vanvés.

Piknova, M., W. Guczynska, R. Miltko, P. Javorsky, A. Kasperowicz, T. Michalowski and P. Pristas. 2008. *Treponema zioleckii* sp. nov., a novel fructan-utilizing species of rumen treponemes. FEMS Microbiol. Lett. *289*: 166–172.

Pillay, A., H. Liu, C.Y. Chen, B. Holloway, W. Sturm, B. Steiner and S.A. Morse. 1998. Molecular subtyping of *Treponema pallidum* subspecies *pallidum*. Sex. Transm. Dis. *25*: 408–414.

Pillay, A., H. Liu, S. Ebrahim, C.Y. Chen, W. Lai, G. Fehler, R.C. Ballard, B. Steiner, A.W. Sturm and S.A. Morse. 2002. Molecular typing of *Treponema pallidum* in South Africa: cross-sectional studies. J. Clin. Microbiol. *40*: 256–258.

Pilot, J. and M.A. Ryter. 1965. Structure des spirochètes I. Étude des genres *Treponema*, *Borrelia* et *Leptospira* au microscope électronique. Ann. Inst. Pasteur *108*: 791–804.

Pohlschroeder, M., S.B. Leschine and E. Canale-Parola. 1994. *Spirochaeta caldaria* sp. nov., a thermophilic bacterium that enhances cellulose degradation by *Clostridium thermocellum*. Arch. Microbiol. *161*: 17–24.

Pollack, R.J., S.R. Telford, 3rd and A. Spielman. 1993. Standardization of medium for culturing Lyme disease spirochetes. J. Clin. Microbiol. *31*: 1251–1255.

Postic, D., M.V. Assous, P.A. Grimont and G. Baranton. 1994. Diversity of *Borrelia burgdorferi sensu lato* evidenced by restriction fragment length polymorphism of *rrf* (5S)–*rrl* (23S) intergenic spacer amplicons. Int. J. Syst. Bacteriol. *44*: 743–752.

Postic, D., N.M. Ras, R.S. Lane, M. Hendson and G. Baranton. 1998. Expanded diversity among Californian *Borrelia* isolates and description of *Borrelia bissettii* sp. nov. (formerly *Borrelia* group DN127). J. Clin. Microbiol. *36*: 3497–3504.

Postic, D., M. Garnier and G. Baranton. 2007. Multilocus sequence analysis of atypical *Borrelia burgdorferi sensu lato* isolates – description of *Borrelia californiensis* sp. nov., and genomospecies 1 and 2. Int. J. Med. Microbiol. *297*: 263–271.

Preac-Mursic, V., B. Wilske and G. Schierz. 1986. European *Borrelia burgdorferi* isolated from humans and ticks culture conditions and antibiotic susceptibility. Zentralbl. Bakteriol. Mikrobiol. Hyg. [A] *263*: 112–118.

Preac-Mursic, V., B. Wilske and S. Reinhardt. 1991. Culture of *Borrelia burgdorferi* on six solid media. Eur. J. Clin. Microbiol. Infect. Dis. *10*: 1076–1079.

Radolf, J.D., C. Moomaw, C.A. Slaughter and M.V. Norgard. 1989a. Penicillin-binding proteins and peptidoglycan of *Treponema pallidum* subsp. *pallidum*. Infect. Immun. *57*: 1248–1254.

Radolf, J.D., M.V. Norgard and W.W. Schulz. 1989b. Outer membrane ultrastructure explains the limited antigenicity of virulent *Treponema pallidum*. Proc. Natl. Acad. Sci. U. S. A. *86*: 2051–2055.

Radolf, J.D. 1995. *Treponema pallidum* and the quest for outer membrane proteins. Mol. Microbiol. *16*: 1067–1073.

Radolf, J.D., M.S. Goldberg, K. Bourell, S.I. Baker, J.D. Jones and M.V. Norgard. 1995. Characterization of outer membranes isolated from *Borrelia burgdorferi*, the Lyme disease spirochete. Infect. Immun. *63*: 2154–2163.

Rainey, F.A., P.H. Janssen, D.J.C. Wild and H.W. Morgan. 1991. Isolation and characterization of an obligately anaerobic, polysaccharolytic, extremely thermophilic member of the genus *Spirochaeta*. Arch. Microbiol. *155*: 396–401.

Ras, N.M., B. Lascola, D. Postic, S.J. Cutler, F. Rodhain, G. Baranton and D. Raoult. 1996. Phylogenesis of relapsing fever *Borrelia* spp. Int. J. Syst. Bacteriol. *46*: 859–865.

Richards, C.S. 1978. Spirochetes in planorbid mollusks. Trans. Am. Microsc. Soc. *97*: 191–198.

Richter, D., D.B. Schlee, R. Allgower and F.R. Matuschka. 2004. Relationships of a novel Lyme disease spirochete, *Borrelia spielmani* sp. nov., with its hosts in central Europe. Appl. Environ. Microbiol. *70*: 6414–6419.

Richter, D., D. Postic, N. Sertour, I. Livey, F.R. Matuschka and G. Baranton. 2006. Delineation of *Borrelia burgdorferi sensu lato* species by multilocus sequence analysis and confirmation of the delineation of *Borrelia spielmanii* sp. nov. Int. J. Syst. Evol. Microbiol. *56*: 873–881.

Ritalahti, K.M. and F.E. Löffler. 2004. Characterization of novel free-living pleomorphic spirochetes (FLiPS), Abstr. 539. Presented at the 10th Int. Symp. Microb. Ecol. International Society for Microbial Ecology, Geneva, Switzerland.

Rosa, P.A., K. Tilly and P.E. Stewart. 2005. The burgeoning molecular genetics of the Lyme disease spirochaete. Nat. Rev. Microbiol. *3*: 129–143.

Ryter, M.A. and J. Pilot. 1965. Structure des spirochètes. II. Étude du genre *Cristispira* au microscope optique et au microscope électronique. Ann. Inst. Pasteur *109*: 552–562.

Sakharoff, M.N. 1891. *Spirochaeta anserina* et la septicémie des oies. Ann. Inst. Pasteur (Paris) *5*: 564–566.

Sambon, L. 1907. Spiroschaudinnia. *In* Tropical Diseases, 4th edn (edited by Manson). Casseoo, London, p. 833.

Sandok, P.L., H.M. Jenkin, H.M. Matthews and M.S. Roberts. 1978. Unsustained multiplication of *Treponema pallidum* (nichols virulent strain) *in vitro* in the presence of oxygen. Infect. Immun. *19*: 421–429.

Schaudinn, F. 1905. Korrespondenzen. Deut. Med. Wochenschr. *31*: 1728.

Schaudinn, F. and E. Hoffman. 1905. Vorläufiger bericht über das Vorkommen für Spirochaeten in syphilitischen Krankheitsprodukten und bei Papillomen. Arb. Gesundh. Amt. Berlin *22*: 528–534.

Schell, R.F., J.L. LeFrock, J.K. Chan and O. Bagasra. 1980. LSH hamster model of syphilitic infection. Infect. Immun. *28*: 909–913.

Schellack, C. 1909. Studien zur Morphologie und Systematik der Spirocheten aus Muscheln. Arb. Gesundh. Amt. Berl. *30*: 379–428.

Schiller, N.L. and C.D. Cox. 1977. Catabolism of glucose and fatty acids by virulent *Treponema pallidum*. Infect. Immun. *16*: 60–68.

Schleifer, K.H. and R. Joseph. 1973. A directly cross-linked L-ornithine-containing peptidoglycan in cell walls of *Spirochaeta stenostrepta*. FEBS Lett. *36*: 83–86.

Schrank, K., B.K. Choi, S. Grund, A. Moter, K. Heuner, H. Nattermann and U.B. Gobel. 1999. *Treponema brennaborense* sp. nov., a novel spirochaete isolated from a dairy cow suffering from digital dermatitis. Int. J. Syst. Bacteriol. *49*: 43–50.

Sela, M.N., K.S. Kornman, J.L. Ebersole and S.C. Holt. 1987. Characterization of treponemes isolated from human and non-human primate periodontal pockets. Oral Microbiol. Immunol. *2*: 21–29.

Sela, M.N. 2001. Role of *Treponema denticola* in periodontal diseases. Crit. Rev. Oral. Biol. Med. *12*: 399–413.

Seshadri, R., G.S. Myers, H. Tettelin, J.A. Eisen, J.F. Heidelberg, R.J. Dodson, T.M. Davidsen, R.T. DeBoy, D.E. Fouts, D.H. Haft, J. Selengut, Q. Ren, L.M. Brinkac, R. Madupu, J. Kolonay, S.A. Durkin, S.C. Daugherty, J. Shetty, A. Shvartsbeyn, E. Gebregeorgis, K. Geer, G. Tsegaye, J. Malek, B. Ayodeji, S. Shatsman, M.P. McLeod, D. Smajs, J.K. Howell, S. Pal, A. Amin, P. Vashisth, T.Z. McNeill, Q. Xiang, E. Sodergren, E. Baca, G.M. Weinstock, S.J. Norris, C.M. Fraser and I.T. Paulsen. 2004. Comparison of the genome of the oral pathogen *Treponema denticola* with other spirochete genomes. Proc. Natl. Acad. Sci. U. S. A. *101*: 5646–5651.

Smibert, R.M. 1974. Genus *Treponema*. *In* Bergey's Manual of Determinative Bacteriology, 8th edn (edited by Buchanan and Gibbons). Williams & Wilkins, Baltimore, pp. 175–184.

Smibert, R.M. and J.A. Burmeister. 1983. *Treponema pectinovorum* sp. nov. isolated from humans with periodontitis. Int. J. Syst. Bacteriol. *33*: 852–856.

Smibert, R.M. 1984. Genus III: *Treponema* Schaudinn 1905, 1728[AL]. *In* Bergey's Manual of Systematic Bacteriology, vol. 1 (edited by Krieg and Holt). Williams & Wilkins, Baltimore, pp. 49–57.

Smibert, R.M., J.L. Johnson and R.R. Ranney. 1984. *Treponema socranskii* sp. nov., *Treponema socranskii* subsp. *socranskii* subsp. nov., *Treponema socranskii* subsp. *buccale* subsp.nov., and *Treponema socranskii* subsp. *paredis* subsp. nov. isolated from the human periodontia. Int. J. Syst. Bacteriol. *34*: 457–462.

Smith, J.L. and D.R. Persetsky. 1967. The current status of *Treponema cuniculi*. Review of the literature. Brit. J. Vener. Dis. *43*: 117–127.

Socransky, S.S., M. Listgarten, C. Hubersak, J. Cotmore and A. Clark. 1969. Morphological and biochemical differentiation of three types of small oral spirochetes. J. Bacteriol. *98*: 878–882.

Sofiev, M.S. 1941. *Spirochaeta latyschewi* n. sp. of relapsing fever type. Med. Parasitol. (Mosc.) *10*: 337–373.

Stamm, L.V. and H.L. Bergen. 2000. A point mutation associated with bacterial macrolide resistance is present in both 23S rRNA genes of an erythromycin-resistant *Treponema pallidum* clinical isolate. Antimicrob. Agents Chemother. *44*: 806–807.

Stamm, L.V., H.L. Bergen and K.A. Shangraw. 2001. Natural rifampin resistance in *Treponema* spp. correlates with presence of N531 in RpoB Rif cluster I. Antimicrob. Agents Chemother. *45*: 2973–2974.

Stanton, T.B. and E. Canale-Parola. 1979. Enumeration and selective isolation of rumen spirochetes. Appl. Environ. Microbiol. *38*: 965–973.

Stanton, T.B. and E. Canale-Parola. 1980. *Treponema bryantii* sp. nov., a rumen spirochete that interacts with cellulolytic bacteria. Arch. Microbiol. *127*: 145–156.

Stanton, T.B. and E. Canale-Parola. 1981. *In* Validation of the publication of new names and new combinations previously effectively published outside the IJSB. List No. 5. Int. J. Syst. Bacteriol. *30*: 676–677.

Steere, A.C. 2001. Lyme disease. N. Engl. J. Med. *345*: 115–125.

Steinhaus, E.A. 1946. Insect Microbiology. Comstock Publishing Co., Ithaca, New York.

Stepan, D.E. and R.C. Johnson. 1981. Helical conformation of *Treponema pallidum* (Nichols strain), *Treponema paraluis-cuniculi, Treponema denticola, Borrelia turicatae*, and unidentified oral spirochetes. Infect. Immun. *32*: 937–940.

Stiles, C.W. and C.A. Pfender. 1905. The generic name *Spironema* Vuillemin 1905 (not Meek, 1964, Mollusk) - *Microspironema* Stiles and Pfender 1905 of the parasite of syphilis. Am. Med. *10*: 936.

Strandberg-Pedersen, N., N.H. Axelsen, B.B. Jorgensen and C.S. Petersen. 1980. Antibodies in secondary syphilis against five of forty Reiter treponeme antigens. Scand. J. Immunol. *11*: 629–633.

Strandberg-Pedersen, N., N.H. Axelsen, B.B. Jorgensen and C. Sand-Peterson. 1981. Antigen analysis of *Treponema pallidum*: cross reactions between individual antigens of *Treponema pallidum* and *Treponema reiter*. Scand. J. Immunol. *11*: 629–633.

Sutton, M.Y., H. Liu, B. Steiner, A. Pillay, T. Mickey, L. Finelli, S. Morse, L.E. Markowitz and M.E. St Louis. 2001. Molecular subtyping of *Treponema pallidum* in an Arizona County with increasing syphilis morbidity: use of specimens from ulcers and blood. J. Infect. Dis. *183*: 1601–1606.

Swellengrebel, N.H. 1907. Sur la cytologie comparée des spirochètes et des spirilles. Ann. Inst. Pasteur (Paris) *21*: 562–586.

Tajima, K., R.I. Aminov, T. Nagamine, K. Ogata, M. Nakamura, H. Matsui and Y. Benno. 1999. Rumen bacterial diversity as determined by sequence analysis of 16S rDNA libraries. FEMS Microbiol. Ecol. *29*: 159–169.

Takayama, K., R.J. Rothenberg and A.G. Barbour. 1987. Absence of lipopolysaccharide in the Lyme disease spirochete, *Borrelia burgdorferi*. Infect. Immun. *55*: 2311–2313.

Tall, B.D. and R.K. Nauman. 1981. Scanning electron microscopy of *Cristispira* species in Chesapeake Bay oysters. Appl. Environ. Microbiol. *42*: 336–343.

Terasaki, Y. 1958. Studies on *Cristispira* in the crystalline style of a fresh water snail, *Semisulcospira libertine* Gould. I. The morphological characters and living condition with the style. Bull. Suzugamine Women's Coll. *5*: 7–19.

Terracciano, J.S. and E. Canale-Parola. 1984. Enhancement of chemotaxis in *Spirochaeta aurantia* grown under conditions of nutrient limitation. J. Bacteriol. *159*: 173–178.

Turner, T.B. and D.H. Hollander. 1957. Biology of the treponematoses. World Health Organization, Geneva.

Umemoto, T., F. Nakazawa, E. Hoshino, K. Okada, M. Fukunaga and I. Namikawa. 1997. *Treponema medium* sp. nov., isolated from human subgingival dental plaque. Int. J. Syst. Bacteriol. *47*: 67–72.

van Dam, A.P., H. Kuiper, K. Vos, A. Widjojokusumo, B.M. de Jongh, L. Spanjaard, A.C. Ramselaar, M.D. Kramer and J. Dankert. 1993. Different genospecies of *Borrelia burgdorferi* are associated with distinct clinical manifestations of Lyme borreliosis. Clin. Infect. Dis. *17*: 708–717.

Veldkamp, H. 1960. Isolation and characteristics of *Treponema zuelzerae* nov. spec., and anaerobic, free-living spirochete. Antonie van Leeuwenhoek *26*: 103–125.

von Prowazek, S. 1910. Parasitische Protozoen aus Japan, gesammelt von Herrn Dr. Mine in Fukuoka. Arch. Schiffs-Trop. Hyg. *14*: 297–302.

Vuillemin, P. 1905. Sur la denomination de l'agent presume de la syphilis. C. R. Acad. Sci. Paris *140*: 1567–1568.

Walker, E.M., G.A. Zampighi, D.R. Blanco, J.N. Miller and M.A. Lovett. 1989. Demonstration of rare protein in the outer membrane of *Treponema pallidum* subsp. *pallidum* by freeze-fracture analysis. J. Bacteriol. *171*: 5005–5011.

Walker, E.M., L.A. Borenstein, D.R. Blanco, J.N. Miller and M.A. Lovett. 1991. Analysis of outer membrane ultrastructure of pathogenic *Treponema* and *Borrelia* species by freeze-fracture electron microscopy. J. Bacteriol. *173*: 5585–5588.

Walker, R.L., D.H. Read, K.J. Loretz and R.W. Nordhausen. 1995. Spirochetes isolated from dairy cattle with papillomatous digital dermatitis and interdigital dermatitis. Vet. Microbiol. *47*: 343–355.

Wallace, A.L. and A. Harris. 1967. Reiter treponeme. A review of the literature. Bull. World Health Org. *36*: Suppl. 2.

Wang, G., A.P. van Dam, A. Le Fleche, D. Postic, O. Peter, G. Baranton, R. de Boer, L. Spanjaard and J. Dankert. 1997. Genetic and phenotypic analysis of *Borrelia valaisiana* sp. nov. (*Borrelia* genomic groups VS116 and M19). Int. J. Syst. Bacteriol. *47*: 926–932.

Wang, G., A.P. van Dam, I. Schwartz and J. Dankert. 1999. Molecular typing of *Borrelia burgdorferi sensu lato*: taxonomic, epidemiological, and clinical implications. Clin. Microbiol. Rev. *12*: 633–653.

Wang, G., G.P. Wormser and I. Schwartz. 2001. *Borrelia burgdorferi*. *In* Molecular Medical Microbiology, vol. 3 (edited by Sussman). Academic Press, London, pp. 2059–2092.

Weber, F.H. and E.P. Greenberg. 1981. Rifampin as a selective agent for the enumeration and isolation of spirochetes from salt marsh habitats. Curr. Microbiol. *5*: 303–306.

Wormser, G.P., D. Liveris, J. Nowakowski, R.B. Nadelman, L.F. Cavaliere, D. McKenna, D. Holmgren and I. Schwartz. 1999. Association of specific subtypes of *Borrelia burgdorferi* with hematogenous dissemination in early Lyme disease. J. Infect. Dis. *180*: 720–725.

Wormser, G.P., R.J. Dattwyler, E.D. Shapiro, J.J. Halperin, A.C. Steere, M.S. Klempner, P.J. Krause, J.S. Bakken, F. Strle, G. Stanek, L. Bockenstedt, D. Fish, J.S. Dumler and R.B. Nadelman. 2006. The clinical assessment, treatment, and prevention of lyme disease, human granulocytic anaplasmosis, and babesiosis: clinical practice guidelines by the Infectious Diseases Society of America. Clin. Infect. Dis. *43*: 1089–1134.

Wyss, C. 1992. Growth of *Porphyromonas gingivalis, Treponema denticola, T. pectinovorum, T. socranskii,* and *T. vincentii* in a chemically defined medium. J. Clin. Microbiol. *30*: 2225–2229.

Wyss, C., B.K. Choi, P. Schupbach, B. Guggenheim and U.B. Gobel. 1996. *Treponema maltophilum* sp. nov., a small oral spirochete isolated from human periodontal lesions. Int. J. Syst. Bacteriol. *46*: 745–752.

Wyss, C. and P. Ermert. 1996. *Borrelia burgdorferi* is an adenine and spermidine auxotroph. Microb. Ecol. Health Dis. *9*: 181–085.

Wyss, C., B.K. Choi, P. Schüpbach, B. Guggenheim and U.B. Göbel. 1997. *Treponema amylovorum* sp. nov., a saccharolytic spirochete of medium size isolated from an advanced human periodontal lesion. Int. J. Syst. Bacteriol. *47*: 842–845.

Wyss, C., B.K. Choi, P. Schupbach, A. Moter, B. Guggenheim and U.B. Gobel. 1999. *Treponema lecithinolyticum* sp. nov., a small saccharolytic spirochaete with phospholipase A and C activities associated with periodontal diseases. Int. J. Syst. Bacteriol. *49*: 1329–1339.

Wyss, C., F. Dewhirst, R. Gmür, T. Thurnheer, Y. Xue, P. Schüpbach, B. Guggenheim and B. Paster. 2001. *Treponema parvum* sp. nov., a small, glucuronic or galacturonic acid-dependent oral spirochaete from lesions of human periodontitis and acute necrotizing ulcerative gingivitis. Int. J. Syst. Evol. Microbiol. *51*: 955–962.

Wyss, C., A. Moter, B.K. Choi, F.E. Dewhirst, Y. Xue, P. Schüpbach, U.B. Göbel, B.J. Paster and B. Guggenheim. 2004. *Treponema putidum* sp. nov., a medium-sized proteolytic spirochaete isolated from lesions of human periodontitis and acute necrotizing ulcerative gingivitis. Int. J. Syst. Evol. Microbiol. *54*: 1117–1122.

You, Y., S. Elmore, L.L. Colton, C. Mackenzie, J.K. Stoops, G.M. Weinstock and S.J. Norris. 1996. Characterization of the cytoplasmic filament protein gene (*cfpA*) of *Treponema pallidum* subsp. *pallidum*. J. Bacteriol. *178*: 3177–3187.

Zhang, J.R., J.M. Hardham, A.G. Barbour and S.J. Norris. 1997. Antigenic variation in Lyme disease borreliae by promiscuous recombination of VMP-like sequence cassettes. Cell *89*: 275–285.

Zhilina, T.N., G.A. Zavarzin, F. Rainey, V.V. Kevbrin, N.A. Kostrikina and A.M. Lysenko. 1996. *Spirochaeta alkalica* sp. nov., *Spirochaeta africana* sp. nov., and *Spirochaeta asiatica* sp. nov., alkaliphilic anaerobes from the Continental soda lakes in Central Asia and the East African Rift. Int. J. Syst. Bacteriol. *46*: 305–312.

Ziolecki, A. 1979. Isolation and characterization of large treponemes from the bovine rumen. Appl. Environ. Microbiol. *37*: 131–135.

Ziolecki, A. and M. Wojciechowicz. 1980. Small pectinolytic spirochetes from the rumen. Appl. Environ. Microbiol. *39*: 919–922.

Zuelzer, M. 1912. Über *Spirochaeta plicatilis* Ehrbg. Und deren Verwandtschafts-beziehungen. Arch. Protistenk. *24*: 1–59.

Zumpt, F. and D. Organ. 1961. Strains of spirochaetes isolated from *Ornithodoros zumpti* Heisch & Guggisberg, and from wild rats in the Cape Province. A preliminary note. S. Afr. J. Lab. Clin. Med. *7*: 31–35.

Family II. **Brachyspiraceae**

Bruce J. Paster

Bra.chy.spi.ra.ce′ae. N.L. fem. n. *Brachyspira* type genus of the family; *-aceae* ending to denote a family; N.L. fem. pl. n. *Brachyspiraceae* the *Brachyspira* family.

The family *Brachyspiraceae* was circumscribed for this volume on the basis of phylogenetic analysis of 16S rRNA gene sequences. The family contains only one genus, *Brachyspira*. Description is the same as for the genus, *Brachyspira*.

Type genus: **Brachyspira aalborgi** Hovind-Hougen, Birch-Andersen, Henrik-Nielsen, Orholm, Pedersen, Teglbjaerg and Thaysen 1983, 896[VP] (Effective publication: Hovind-Hougen, Birch-Andersen, Henrik-Nielsen, Orholm, Pedersen, Teglbjaerg and Thaysen 1982, 1135.).

Genus I. **Brachyspira** Hovind-Hougen, Birch-Andersen, Henrik-Nielsen, Orholm, Pedersen, Teglbjaerg and Thaysen 1983, 896[VP] (Effective publication: Hovind-Hougen, Birch-Andersen, Henrik-Nielsen, Orholm, Pedersen, Teglbjaerg and Thaysen 1982, 1135.)

Thaddeus B. Stanton

Bra.chy.spi′ra. Gr. adj. *brachys* short; L. fem. n. *spira* a coil, spiral; N.L. fem. n. *Brachyspira* a short spiral, describing a bacterium that resembles a short spiral.

Brachyspira spirochetes are helical shaped bacteria with **regular coiling patterns**. Cells measure 2–11 μm by 0.2–0.4 μm. Unicellular, but dividing pairs and occasional chains of three or more cells can be observed in growing cultures. Under unfavorable growth conditions, spherical or round bodies are formed. Gram-stain negative. **Obligately anaerobic, aerotolerant**. Cell ends may be blunt or pointed. Cells have a typical spirochete cell ultrastructure, consisting of an outer sheath, helical protoplasmic cylinder, and internal flagella in the space between the protoplasmic cylinder and outer sheath. Brachyspire cells have 8–30 flagella per cell depending on the species (flagellar number usually correlates with cell size and species of smaller cells have fewer flagella). Flagella attach subterminally in equal numbers at each cell end, wrap around the protoplasmic cylinder, and their free ends overlap in the middle of the cells. Flexing and creeping motility at 22°C; translational movement in liquids at 37–42°C. Cultured anaerobically on commercially available media (trypticase soy or brain heart infusion broths) containing a carbohydrate growth substrate and supplemented with defibrinated blood or animal (calf) serum. Grows at 36–42°C, optimally at 37–39°C. Population doubling times on glucose in broth cultures are 1–5 h (not reported for *Brachyspira aalborgi*). Chemoorganotrophic, using various carbohydrates for growth. **Possess NADH oxidase for reducing molecular oxygen. Consume oxygen during growth in culture broth beneath a 1% oxygen atmosphere. Acetate, butyrate, H_2, and CO_2 are major endproducts of glucose metabolism. Higher amounts of H_2 than CO_2 are produced**. Weakly hemolytic except for *Brachyspira hyodysenteriae* which exhibits β-hemolysis (strongly hemolytic). Associated with animal and human hosts. Some species are pathogenic. **The genus *Brachyspira* is distinguished from other spirochete genera based on 16S rRNA gene sequences**. *Brachyspira* species share high 16S rRNA sequence similarity with each other. Species can be differentiated by DNA–DNA relative reassociation and MLEE (multilocus enzyme electrophoresis) analyses. Similar to other spirochete genera, *Brachyspira* is insensitive to the antibiotic rifampin.

DNA G+C content (mol%): 24.5–27.1 (T_m).

Type species: **Brachyspira aalborgi** Hovind-Hougen, Birch-Andersen, Henrik-Nielsen, Orholm, Pedersen, Teglbjaerg and Thaysen 1983, 896[VP] (Effective publication: Hovind-Hougen, Birch-Andersen, Henrik-Nielsen, Orholm, Pedersen, Teglbjaerg and Thaysen 1982, 1135[VP].).

Further descriptive information

Cell morphology and ultrastructure. By phase-contrast microscopy, *Brachyspira* species appear helical shaped with regular coils (Figure 96). They have a typical spirochete ultrastructure with protoplasmic cylinder, enclosed flagella winding around the protoplasmic cylinder, and outer sheath (membrane) (Sellwood and Bland, 1997). The species can be divided into two groups based on cell size and flagellar numbers (Table 130). *Brachyspira* species with larger cell size, such as *Brachyspira hyodysenteriae*, *Brachyspira innocens*, *Brachyspira intermedia*, *Brachyspira murdochii*, and *Brachyspira alvinipulli*, have 20–30 flagella per cell. *Brachyspira pilosicoli* and *Brachyspira aalborgi* cells are shorter in length and have 8–12 flagella per cell. As with other spirochetes, the flagella attach in roughly equal numbers at each end of the cell (Figure 97).

The end shapes of cells of different species vary (Table 130). The pointed cell ends of *Brachyspira pilosicoli* and *Brachyspira aalborgi* may serve in the attachment of these spirochetes to intestinal cells (Sellwood and Bland, 1997). Outer-membrane proteins of pathogenic *Brachyspira* species are frequently the targets of efforts to develop vaccines against intestinal spirochete diseases (Cullen et al., 2003; Cullen et al., 2004; La et al., 2004; McCaman et al., 2003; Trott et al., 2001, 2004).

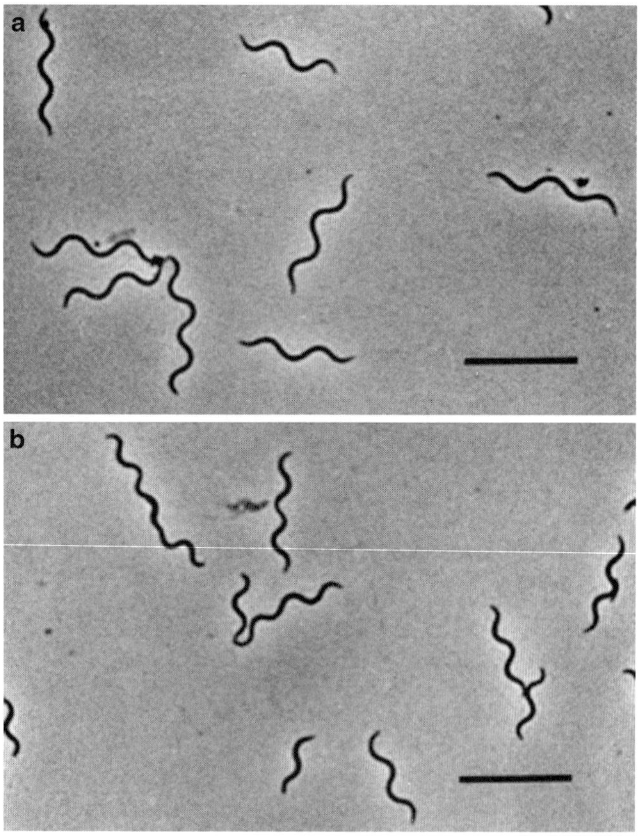

FIGURE 96. (a) *Brachyspira alvinipulli* C1[T] cells. (b) *Brachyspira hyodysenteriae* B78[T] cells. Phase-contrast photomicrographs of wet-mount preparations. Marker bars represent 10 μm. (Reproduced with permission from T.B. Stanton et al., 1998. Int. J. Syst. Bacteriol. *48*: 669–676.)

Nutrition and growth conditions. Various anaerobic, nutritionally complex broth media have been described for *Brachyspira* species (Stanton, 1997). Incubation temperatures are 38–40°C. Shaking or stirring of broth cultures is important for optimum growth. BHIS broth and Kunkle's broth are commonly used for routine growth of *Brachyspira hyodysenteriae* (Kunkle et al., 1986; Stanton and Lebo, 1988). Both media support high growth yields (1–4 × 10^9 cells/ml, direct microscope counts). In BHIS broth which contains glucose as a growth substrate, the population doubling times for most *Brachyspira* species are 1–5 h (Table 130).

HS broth (Heart infusion broth supplemented with serum) requires added carbohydrates to support optimum growth yields of *Brachyspira* species. Consequently, HS broth is useful for identifying metabolic end products (Table 130) and growth substrates (Table 131). NT broth, a serum-free medium with low protein content, supports *Brachyspira hyodysenteriae* growth (Humphrey et al., 1997).

Brachyspira species are generally aerotolerant and able to grow in sealed culture broth tubes beneath atmospheres containing 1–5% O_2. They will not initiate growth, however, in oxidized media, i.e., in media containing resazurin indicator dye that has become colored by exposure to air. A too low redox potential appears to be a controlling factor for *Brachyspira hyodysenteriae* growth (Stanton, 1997). This species can be difficult to culture in stringently prepared anaerobic broth media unless a small amount of oxygen (1%, v/v) is introduced into the culture atmosphere (Stanton and Lebo, 1988). *Brachyspira hyodysenteriae* cells can be cultured in a fermenter (12 l) beneath an air atmosphere (Stanton and Jensen, 1993).

Fermentation and growth substrates. Fermentation substrates have been identified by measuring acid production (medium pH decreases) in cultures to which the substrates have been added (Jones et al., 1986; Kinyon and Harris, 1979; Ochiai et al., 1998; Tompkins et al., 1986). Alternatively, carbon/energy sources of the spirochetes have been directly identified by measuring population density increases in HS broth, a culture medium that is growth limiting unless substrates are added (Table 131). The latter method is more sensitive for detecting substrates that support low growth yields. In HS broth, *Brachyspira hyodysenteriae*, *Brachyspira innocens*, *Brachyspira intermedia*, *Brachyspira murdochii*, *Brachyspira pilosicoli*, and *Brachyspira alvinipulli* cells use various monosaccharides, disaccharides, the trisaccharide trehalose, and amino sugars for growth (Table 131). The ability to ferment D-ribose distinguishes *Brachyspira pilosicoli* strains from other *Brachyspira* species (Fossi and Skrzypczak, 2006). *Brachyspira* strains do not use polysaccharides such as cellulose, hog gastric mucin, pectin, or glycogen. *Brachyspira hyodysenteriae* and *Brachyspira innocens* contain an inducible sucrase activity (Jensen and Stanton, 1994).

Cholesterol and phospholipids. *Brachyspira hyodysenteriae* requires cholesterol and phospholipid for growth (Lemcke and Burrows, 1980; Stanton, 1987, 1997). Cholesterol is likely required for outer membrane biosynthesis (Lemcke and

TABLE 130. *Brachyspira* species characteristics[a,b]

Characteristic	1. *B. aalborgi*	2. *B. alvinipulli*	3. *B. hyodysenteriae*	4. *B. innocens*	5. *B. intermedia*	6. *B. murdochii*	7. *B. pilosicoli*
Host	Humans	Chickens	Swine, rheas, ducks	Swine	Swine, chickens	Swine, rats	Swine, birds, dogs chickens, humans, nonhuman primates
Demonstrated pathogenicity (animal)[c]	No	Yes (chickens)	Yes (swine; rheas)	No	Yes (chickens)	No	Yes (swine)
Hemolysis type	Weak	Weak	Strong	Weak	Weak	Weak	Weak
Growth rate[d]	nr; 7–14 d	3–5 h; 3–5 d	3–5 h; 3–4 d	3–5 h; 3–5 d	2–4 h; 3–5 d	2–4 h; 3–5 d	1–2 h; 3–4 d
Cell size (µm)	2–6 × 0.2	8–11 × 0.2–0.35	7–12 × 0.3–0.4	7–12 × 0.3–0.4	8–10 × 0.35–0.45	5–8 × 0.35–0.4	5–7 × 0.23–0.3
Cell end shape	Tapered	Blunt	Blunt	Blunt	Blunt	Blunt	Pointed
Flagella per cell	8	22–30	22–28	20–26	24–28	22–26	8–12
DNA G+C content (mol%)	27.1	24.6	25.9	26	25	27	24.6
Metabolic endproducts (µmol/ml):[e]							
Acetate	nr	16.8	12.0	18.0	14	25.4	23.3
Butyrate	nr	0.8	1.7	1.8	10.3	5.2	1.5
CO_2	nr	19.4	10.3	13.1	7.2	14.8	11.3
H_2	nr	22.8	24.0	22.0	10.7	33	22.4
Ethanol	nr	2.2	nd	0.9	1.4	1	2.4
16S rRNA signature[f]	No	No	No	No	No	No	Yes
16S rRNA sequence similarity:[g]							
B. aalborgi 513A^T	100	95.7	96.5	96.5	96.0	95.9	96.4
B. hyodysenteriae B78^T	96.5	98.4	100	99.5	99.1	98.5	98.5
Biochemical reaction profile:							
D-Ribose fermentation[h]	nr	–	–	–	–	–	+
Indole production	–	–	+/–	–	+/–	–	+/–
Hippurate hydrolysis	nr	+	–	–	–	–	+/–
α-Galactosidase activity	–	–	–	+	–	–	+/–
β-Glucosidase activity	–	+	+	+	+	+	–

[a]Symbols: +, positive; –, negative; nr, not reported; nd, not detected.

[b]Only one strain of *Brachyspira aalborgii* (513A^T) and *Brachyspira alvinipulli* (strain C1^T) have been extensively studied. Data from Harris et al. (1972a, b), Kinyon and Harris (1979), Hovind-Hougen et al. (1982), Stanton and Lebo (1988), Stanton et al. (1991, 1996, 1997, 1998,), Fellstrom et al. (1995, 1997, 1999), Park et al. (1995), Trott et al. (1996a, b, 1997a–c), McLaren et al. (1997), Sellwood and Bland (1997), De Smet et al. (1998), Duhamel et al. (1998a, b), Kraaz et al. (2000), Fossi et al. (2004), and Jansson et al. (2004).

[c]Pure cultures of either the type strain or another strain have been shown to cause disease when inoculated into normal, healthy, natural host animals. *Brachyspira intermedia* infections of chickens are mild to moderate "economic loss" diseases associated with "wet litter" and "unthriftiness", in contrast to the severe pathology of swine dysentery.

[d]Values represent population doubling time in hours for broth cultures and days for colony development on agar-containing media at 37–39°C.

[e]Product yields of type strains after growth in HS broth containing glucose and beneath 1% O_2 atmosphere. All strains consumed O_2.

[f]A unique signature sequence of nucleotides was identified within the 16S rDNA of *Brachyspira pilosicoli* and has been used to design specific PCR tests for this species (Fellstrom et al., 1997; Park et al., 1995).

[g]Determined by using GenBank 16S rRNA gene sequences with accession numbers Z22781, U14930, U23033, U14920, U14927, U22838, and U23030. All sequences are for type strains except *Brachyspira murdochii* (strain 155-20).

[h]As reported by Fossi and Skrzypczak (2006).

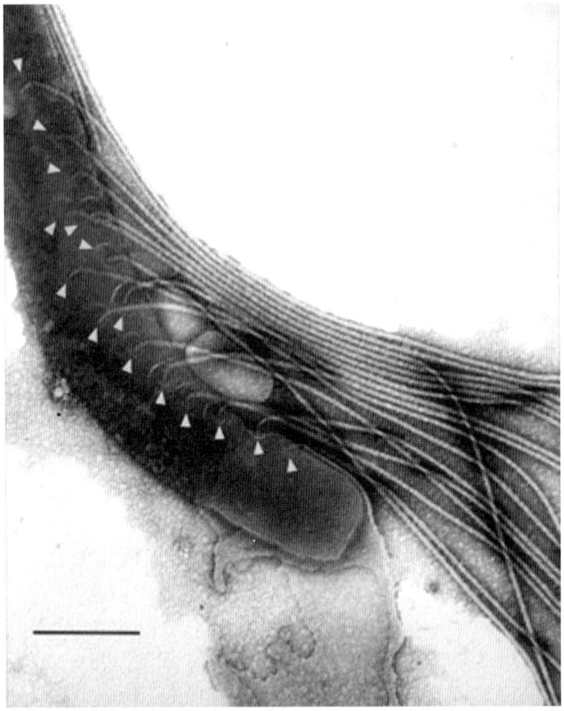

FIGURE 97. Electron micrograph of one end of *Brachyspira alvinipulli* C1^T cell negatively stained with 2% phosphotungstic acid (pH 7.0). Disrupted outer sheath enables insertion sites of 15 periplasmic flagella to be seen (white arrowheads). Marker bar = 0.25 μm. (Reproduced with permission from T.B. Stanton et al., 1998. Int. J. Syst. Bacteriol. *48:* 669–676.)

Burrows, 1980; Plaza et al., 1997; Stanton, 1987; Stanton and Cornell, 1987). The cellular fatty acids of *Brachyspira hyodysenteriae* are distinct from those of *Borrelia* and *Leptospira* species (Livesley et al., 1993). The cellular phospholipids and glycolipids of *Brachyspira hyodysenteriae* and *Brachyspira innocens* contain acyl and alkenyl side chains and the spirochetes apparently have some capacity for fatty acid and lipid biosynthesis (Matthews and Kinyon, 1984; Matthews et al., 1980a, b).

Glucose and pyruvate metabolism. *Brachyspira hyodysenteriae* uses the Embden–Meyerhof–Parnas (EMP) pathway for converting glucose to pyruvate. Pyruvate is catabolized by a clostridial-type clastic reaction to acetyl-CoA, H_2, and CO_2. Acetyl-CoA is converted to either acetate or butyrate via a branched fermentation pathway. The ATP-yielding mechanisms are substrate-level phosphorylation reactions mediated by phosphoglycerate kinase and pyruvate kinase in the EMP pathway and by acetate kinase in the conversion of acetyl phosphate to acetate (Stanton, 1989, 1997).

In *Brachyspira hyodysenteriae* cells, NADH-H^+ produced during glycolysis can be recycled or oxidized to NAD^+ by three pathways: (a) 3-hydroxybutyryl-CoA dehydrogenase and butyryl-CoA dehydrogenase (butyrate pathway); (b) NADH-ferredoxin oxidoreductase plus hydrogenase; and (c) NADH oxidase (Stanton, 1997). The *Brachyspira hyodysenteriae* NADH oxidase is a water-forming, FAD-linked enzyme (Stanton and Jensen, 1993) and the *nox* gene has been cloned (GenBank no. U19610) (Stanton and Sellwood, 1999). *Brachyspira hyodysenteriae nox*-defective mutant strains are sensitive to oxygen and are avirulent (Stanton et al., 1999).

TABLE 131. *Brachyspira* growth/fermentation substrates[a,b]

Compound	1. *B. aalborgi*	2. *B. alvinipulli*	3. *B. hyodysenteriae*	4. *B. innocens*	5. *B. intermedia*	6. *B. murdochii*	7. *B. pilosicoli*
D-Glucose	+	+	+	+	+	+	+
D-Fructose	+	+	+	+	+	+	d
Sucrose	nr	−	+	+	+	+	+
D-Trehalose	+	−	+	+	+	+	+
D-Galactose	+	−	+	+	+	−	d
D-Mannose	+	+	+	+	−	+	d
N-Acetyl-D-glucosamine	nr	+	+	+	+	+	+
D-Glucosamine	nr	+	+	+	−	−	+
D-Maltose	+	+	+	+	+	+	d
D-Cellobiose	nr	nr	−	+	−	+	+
L-Fucose	nr	−	−	+	+	+	+
D-Ribose[c]	nr	−	−	−	−	−	+
D-Xylose	nr	−	−	−	−	−	d
Lactose	+	−	+	d	−	+	nr
Pyruvate	nr	−	+	+	+	+	+

[a]Symbols: +, >85% positive; d, different strains give different reactions (16–84% positive); −, 0–15% positive; nr, not reported.

[b]Fermentation substrates for *Brachyspira aalborgi* were determined by phenol red measurements of acid production (Ochiai et al., 1997). Growth substrates for the other six *Brachyspira* species were determined by monitoring culture growth in HS broth to which the various substrates were added. Different substrates support different growth yields of cells. Data are for *Brachyspira aalborgi* NCTC 11492 (Ochiai et al., 1997), *Brachyspira hyodysenteriae* strains B78^T, B204, and B169 (Stanton and Lebo, 1988), *Brachyspira innocens* B256^T (Trott et al., 1996a), *Brachyspira pilosicoli* strains P43/6/78^T, 3295, 1648, Hrm7, Kar, WesB (Trott et al., 1996a, b), *Brachyspira intermedia* PWS/A^T (Stanton et al., 1997), *Brachyspira murdochii* 56-150^T (Stanton et al., 1997), and *Brachyspira alvinipulli* C1^T (Stanton et al., 1998). In the original studies, additional compounds were tested and were not substrates for any species.

[c]In a study of swine *Brachyspira* isolates, D-ribose fermentation was found be a unique trait of *Brachyspira pilosicoli* strains (Fossi and Skrzypczak, 2006).

NADH oxidase is universally present in *Brachyspira* species (Atyeo et al., 1999; Stanton et al., 1995). *Brachyspira* species produce higher amounts of H_2 than CO_2, indicative of the NADH-ferredoxin oxidoreductase reaction. Ethanol is produced in cultures of *Brachyspira pilosicoli*, *Brachyspira alvinipulli*, *Brachyspira murdochii*, *Brachyspira intermedia*, and *Brachyspira innocens* (Stanton, 1989; Stanton et al., 1997, 1998; Stanton and Lebo, 1988; Trott et al., 1996b, c). The major endproducts of glucose metabolism by growing cells of various *Brachyspira* species are the same as those of *Brachyspira hyodysenteriae* (Table 130). For this reason, the species are likely to have similar catabolic routes.

Iron metabolism. Several observations suggest that specific iron uptake mechanisms are present and are important for brachyspire growth in animal hosts. *Brachyspira hyodysenteriae* cells grow in broth containing an iron chelator, 2,2′-dipyridyl, and increase the expression of three unidentified high molecular mass proteins, >200, 134, and 109 kDa (Li et al., 1995). A *Brachyspira hyodysenteriae* genome locus, designated *bit* ("*Brachyspira* iron transport"), encodes six proteins that are likely to form an iron ATP-binding transport system (Dugourd et al., 1999).

Genome properties. The *Brachyspira hyodysenteriae* B78^T chromosome is circular and 3.2 Mbp (megabase pairs) in size

(Zuerner and Stanton, 1994). *Brachyspira pilosicoli* P43/6/78^T has a circular chromosome approximately 2.4 Mb (Zuerner et al., 2004). The *Brachyspira hyodysenteriae* and *Brachyspira pilosicoli* genomes are currently being sequenced under proprietary circumstances by private corporations. The genomes of the spirochetes *Brachyspira hyodysenteriae*, *Leptospira interrogans*, *Borrelia* spp., *Treponema pallidum* and *Treponema denticola* differ from one another in terms of chromosomal conformation (linear and circular), chromosomal numbers (two for *Leptospira interrogans*), size (0.95–4.9 Mb), and number and arrangement of rRNA genes (Zuerner, 1997).

Plasmids. Uncharacterized plasmids and extrachromosomal DNAs have been reported for various *Brachyspira* species (Cattani et al., 1998; Combs et al., 1989; Turner and Sellwood, 1997). Some of the extrachromosomal DNAs (Combs et al., 1989; Turner and Sellwood, 1997) are likely to be DNA from VSH-1-like bacteriophage particles spontaneously produced in brachyspire cultures. Plasmid DNA was not detected for *Brachyspira hyodysenteriae* B78^T (Zuerner and Stanton, 1994).

Prophage-like gene transfer agent (VSH-1). When treated with mitomycin C, both *Brachyspira hyodysenteriae* and *Brachyspira innocens* cells lyse and release bacteriophage-like particles (Humphrey et al., 1995). One of these prophages was purified from *Brachyspira hyodysenteriae* strain B204 cultures and was named VSH-1 (Humphrey et al., 1997). VSH-1 virions package random, 7.5 kb linear fragments of bacterial DNA. The purified virions are noninfectious, that is, they do not lyse bacteria or form plaques, and behave like generalized transducing phages (Humphrey et al., 1997; Stanton et al., 2001). The VSH-1 genome (GenBank no. AY971355) encodes head, tail, and lytic proteins (Matson et al., 2005) and is 16.5 kb (substantially larger than the virion packaged DNA). Based on MLEE analysis of *Brachyspira hyodysenteriae* strains, Trott et al. (1997c) concluded that substantial genetic recombination, likely mediated by VSH-1 gene transfer, has shaped the overall population structure of this species. VSH-1 is the first natural gene transfer mechanism to be discovered for a spirochete.

Bacteriophages have also been detected by electron microscopy in both mitomycin C-treated and untreated cultures of weakly hemolytic human intestinal spirochetes (Calderaro et al., 1998a, b). These bacteriophages resemble VSH-1 in size and morphology.

Genetic techniques – mutagenesis. *Brachyspira hyodysenteriae* strains can be mutated by allelic exchange at specific loci (Li et al., 2000; Rosey et al., 1995, 1996; Stanton et al., 1999; ter Huurne et al., 1992). By this method, *Brachyspira hyodysenteriae* genes cloned in *Escherichia coli* are inactivated *in vitro* by inserting an antimicrobial (kanamycin, chloramphenicol) resistance marker into the gene. The constructs are then introduced as plasmids into *Brachyspira hyodysenteriae* cells by electroporation. The allelic exchange technique yields mutant strains that are isogenic, except for single genetic loci, to their progenitor strains. Such strains are invaluable in determining bacterial virulence traits.

UV mutagenesis has been used to generate coumermycin A1-resistant *Brachyspira hyodysenteriae* strains with mutations in *gyrB* (DNA gyrase) genes (Stanton et al., 2001). Coumermycin resistance is an antibiotic selection marker for genetic manipulations of *Brachyspira*. Another useful selection marker is tylosin resistance. Some *Brachyspira hyodysenteriae* strains are naturally resistant to tylosin, owing to a nucleotide base change in the 23S rRNA gene (Karlsson et al., 1999).

Serotype analysis. *Brachyspira hyodysenteriae* strains have been subdivided into serotypes based on the immunological reactivities of lipo-oligosaccharides (LOS) in hot water-phenol extracts of whole cells (Baum and Joens, 1979; Li et al., 1991). A more elaborate system of immunological identification assigns *Brachyspira hyodysenteriae* strains to serogroups and subdivides the groups into serovars (Hampson et al., 1989, 1997; Lau and Hampson, 1992). This latter system has not been widely adopted because serogroups are not entirely consistent with genetic groupings based on MLEE analysis (Lee et al., 1993; Trott et al., 1997a). Enzyme-linked immunosorbent assay (ELISA) methods based on antibodies to LOS have been used to detect animal herd exposure to *Brachyspira hyodysenteriae* strains (Joens et al., 1982; Wright et al., 1989).

Antibiotic sensitivity. *Brachyspira hyodysenteriae* and other *Brachyspira* species are naturally insensitive to the antibiotics rifampin, spectinomycin, polymyxin B, and colistin at concentrations which limit growth of other intestinal bacteria (Buller and Hampson, 1994; Messier et al., 1990; Taylor and Trott, 1997; Trott et al., 1996c). Consequently, individual antibiotics or combinations of these antibiotics are useful for selectively culturing *Brachyspira hyodysenteriae* and other species (Jenkinson and Wingar, 1981; Kunkle and Kinyon, 1988).

There is evidence that *Brachyspira hyodysenteriae* and *Brachyspira pilosicoli* strains are becoming resistant to antimicrobials used to treat swine dysentery and porcine intestinal spirochetosis. These antimicrobials belong to the macrolide (i.e., tylosin, erythromycin), lincosamide (i.e., lincomycin, clindamycin), and pleuromutilin (i.e., tiamulin) classes of antibiotics (Duhamel et al., 1998a; Karlsson et al., 2003; Lobova et al., 2004; Pringle et al., 2006). Simultaneous resistance to both macrolide and lincosamide antimicrobials results from single nucleotide base changes in the 23S rRNA sequences of these spirochetes (Karlsson et al., 1999, 2004). Tiamulin resistance in a clinical isolate of *Brachyspira hyodysenteriae* has been associated with an amino acid change in ribosomal protein L3 (Pringle et al., 2004).

***Brachyspira* diseases and virulence-associated traits.** *Brachyspira hyodysenteriae*, *Brachyspira pilosicoli*, *Brachyspira intermedia*, and *Brachyspira alvinipulli* cause intestinal disease when inoculated into their healthy host animals (Table 130). Most disease-related research has focused on *Brachyspira hyodysenteriae*, the etiologic agent of swine dysentery. Several publications have described and reviewed brachyspire diseases from the viewpoints of host manifestations, clinical detection methods, therapies, and experimental models (Barrett, 1997; Galvin et al., 1997; Hampson et al., 1997; Hampson and Trott, 1999; Harris et al., 1999; Swayne, 2003; Swayne and McLaren, 1997; Taylor and Trott, 1997).

Swine dysentery (*Brachyspira hyodysenteriae*). Swine dysentery (bloody scours or black scours) is a severe intestinal disease that affects piglets, primarily in the postweaning stage of growth (8–14 weeks after birth). The disease has been reported worldwide in every major pig producing country. A typical sign of the disease is profuse bleeding into the large bowel lumen through lesions induced by *Brachyspira hyodysenteriae* cells. Afflicted animals pass loose stools containing blood and mucus and spirochetes readily seen by microscopy. These are presumptive signs of the disease. Culturing and identifying *Brachyspira* *hyodysenteriae* cells, along with histopathological observations, provide conclusive evidence of swine dysentery. Up to 90–100% of a herd can become infected, and without effective treatment, 20–30% of infected animals may die. Economic losses result from death, poor weight gain/feed efficiency, and medication expenses. Swine management strategies, including segregation by age and prophylactic administration of antibiotics (such as tylosin, lincomycin, and carbadox), are credited with reducing swine dysentery in the United States. Swine dysentery can be experimentally produced by feeding or intragastrically inoculating normal swine with *Brachyspira hyodysenteriae* cultures (Kennedy et al., 1988; Kinyon et al., 1977). However, the type strain B78^T is weakly virulent and not useful for experimental infections (Jensen and Stanton, 1993). Various approaches have been used to develop whole cell or cell subunit-based vaccines for swine dysentery (Lee et al., 2000). A commercial vaccine for swine dysentery is based on pepsin-digested *Brachyspira hyodysenteriae* cells. The immunological properties of the vaccine are now being examined (Waters et al., 2000).

Swine are the common, but not the exclusive, animal hosts for *Brachyspira hyodysenteriae*. Strains of the spirochete also have been isolated from juvenile rheas with a severe necrotizing typhlitis (Buckles et al., 1997; Jensen et al., 1996; Sagartz et al., 1992). For experimental infections, mice (Joens and Glock, 1979; Nibbelink and Wannemuehler, 1992; Rosey et al., 1996; ter Huurne et al., 1992) and 1-day-old chicks (Sueyoshi et al., 1987) have been used. Nevertheless, nuances or inconsistencies are associated with the use of these surrogate animal models (Achacha et al., 1996; Jensen and Stanton, 1993). For this reason, conclusions regarding *Brachyspira hyodysenteriae* pathogenesis based on alternative animal models should be confirmed through the use of swine infections.

Several *Brachyspira hyodysenteriae* properties are putative or demonstrated virulence-associated traits. They include lipooligosaccharide (Greer and Wannemuehler, 1989; Nibbelink and Wannemuehler, 1991; Nuessen et al., 1982, 1983), hemolysin/hemolytic activity (Hsu et al., 2001; Hutto and Wannemuehler, 1999; Kent et al., 1988; Lysons et al., 1991; Saheb et al., 1980, 1981; ter Huurne et al., 1992, 1994), chemotaxis/motility (Glock et al., 1974; Kennedy et al., 1988; Kennedy and Yancey, 1996; Milner and Sellwood, 1994), oxygen metabolism/NADH oxidase (Stanton, 1997; Stanton and Jensen, 1993), and variable surface proteins (Gabe et al., 1998). The swine host immune response also likely contributes to the pathology of swine dysentery (Hontecillas et al., 2005).

The ability to create strains with specific gene mutations has enabled direct evidence of a link between virulence and NADH oxidase activity, motility/flagella, and hemolytic activity. Mutant strains with specific deletions of *nox* (Stanton et al., 1999), *flaA* or *flaB* (Kennedy et al., 1997), or *tlyA* (Hyatt et al., 1994; Joens, 1997; ter Huurne et al., 1992) are avirulent compared to their isogenic wild-type counterparts. Although *tlyA*, *tlyB*, and *tlyC* were originally identified as hemolysin genes (Muir et al., 1992; ter Huurne et al., 1992, 1994), a subsequent study raises questions about that identification (Hsu et al., 2001).

Spirochetal colitis and spirochetal diarrhea (*Brachyspira pilosicoli*). Spirochetal colitis caused by *Brachyspira pilosicoli* is a mild to moderate diarrheal disease of swine, birds, and possibly humans (Hampson and Trott, 1999; Swayne and McLaren, 1997; Taylor and Trott, 1997). Spirochetal colitis of swine resembles a mild case or early stage of swine dysentery. Watery, mucoid diarrhea and a reduction of growth rate of affected animals are common clinical signs. The disease has been experimentally produced by inoculating pure cultures of *Brachyspira pilosicoli* into healthy swine (Taylor and Alexander, 1971; Thomson et al., 1997; Trott et al., 1996a).

Evidence that *Brachyspira pilosicoli* is a pathogen of humans is circumstantial but multifarious. *Brachyspira pilosicoli* strains have been isolated from humans, including some (homosexual males and persons living in developing countries) with intestinal disorders and who are immunocompromised (Barrett, 1997; Trott et al., 1997a, b). Human strains are virulent for healthy piglets (Trott et al., 1996a). A human volunteer became colonized after drinking cultures of *Brachyspira pilosicoli* strain Wes B (Oxberry et al., 1998). Finally, *Brachyspira pilosicoli* has been isolated from the blood of human patients (Trott et al., 1997a). The significance and the capacity of *Brachyspira pilosicoli* cells to leave the intestinal tract and circulate throughout the host's body have not been sufficiently investigated.

Virulence factors of *Brachyspira pilosicoli* are unknown. By virtue of their location, outer-membrane proteins undoubtedly mediate interactions between spirochete cells and their environment and are logically involved in host colonization and virulence (Trott et al., 2001, 2004). Cell motility, chemotaxis, and spirochete end-on attachment to host tissues are likely to be associated with colonization of the intestinal tract and therefore important for pathogenesis. In the absence of gross lesions, extensive colonization of intestinal tissues by *Brachyspira pilosicoli* cells with associated damage to microvilli could interfere with intestinal absorptive processes and lead to diarrhea (Gad et al., 1977; Taylor and Trott, 1997).

Avian intestinal spirochetosis. Many studies of avian intestinal infections associated with spirochetes were made before *Brachyspira* species had been characterized and taxonomically defined (Davelaar et al., 1986; Dwars et al., 1989; Griffiths et al., 1987). Avian diarrheal diseases can be caused by different *Brachyspira* species (Swayne, 2003; Swayne and McLaren, 1997). In addition to *Brachyspira pilosicoli* and *Brachyspira hyodysenteriae*, *Brachyspira alvinipulli* C1^T, isolated from a diarrheic chicken, is a chicken enteropathogen. The C1^T cells colonize the ceca of 1-day-old chicks and 14-month-old hens and produce mild typhlitis with discolored and watery cecal contents (Swayne et al., 1995). *Brachyspira alvinipulli* has been isolated only from poultry and resembles an uncharacterized enteropathogenic spirochete (Davelaar et al., 1986).

Brachyspira intermedia appears to be another common avian enteropathogen, inasmuch as spirochetes of that species have been isolated from birds with moderate intestinal colitis (Stephens and Hampson, 2001; Swayne, 2003; Swayne and McLaren, 1997). One-day-old chicks inoculated with pure cultures of an intestinal spirochete strain 1380, later identified by MLEE analysis as a strain of *Brachyspira intermedia* (Swayne and McLaren, 1997), shed watery feces containing spirochetes and had body weight reduction compared to control birds (Dwars et al., 1992). Avian *Brachyspira intermedia* strain HB60 isolated from a hen with diarrhea causes reduced egg production and watery feces when inoculated into healthy hens (Hampson and McLaren, 1999). A *Brachyspira intermedia* strain, freshly isolated from a swine herd with diarrhea, however, was not pathogenic for swine (Jensen et al., 2000). Virulence traits of avian enteropathogenic *Brachyspira* species have yet to be determined.

Enrichment and isolation procedures. *Brachyspira hyodysenteriae* and other *Brachyspira* species can be isolated by inoculating intestinal contents or tissues onto solid agar culture media containing antibiotics selective for the growth of those spirochetes (Calderaro et al., 2001; Hovind-Hougen et al., 1982; Jenkinson and Wingar, 1981; Kunkle and Kinyon, 1988). Achacha and Messier (1992) compared various selective media and found BJ medium containing spiramycin, rifampin, vancomycin, colistin, and spectinomycin provided the highest rate of isolation of *Brachyspira hyodysenteriae* from feces of experimentally infected swine.

Physical methods (filtration, agar migration) can be used to isolate *Brachyspira* species from intestinal contents and tissues (Harris et al., 1972b; Olson, 1996; Taylor and Alexander, 1971). These methods take advantage of the small diameter of brachyspire cells and their active motility in agar-containing media.

Maintenance procedures

Broth cultures of most *Brachyspira* species remain viable when stored at 5°C for 1–2 weeks and can be used as "working" stock cultures to inoculate fresh cultures for experiments. The refrigerated cultures should be in exponential growth phase (approx. 2×10^8 cells/ml, direct microscope counts). Oxygen exposure must be prevented for example by sealing stoppered culture tubes with plastic tape (3M Scotch #471) before storage. Long-term *in vitro* passage of strains (with possible loss of virulence) should be avoided by starting fresh working stock cultures from long-term stock cultures every several months.

Agar plate cultures with colonies of *Brachyspira hyodysenteriae* can be stored at room temperature in an anaerobic atmosphere for at least a week. The plates should be sealed (Parafilm) to prevent desiccation. A slowly growing spirochete identified as *Brachyspira aalborgi* remained viable when kept at room temperature in an anaerobic jar for over 3 months (Kraaz et al., 2000).

For long-term storage, brachyspire broth cultures in the exponential phase of growth are harvested by centrifugation (15 min, $5000 \times g$). The pelleted bacteria are resuspended at 50–100 times their original concentration in fresh sterile broth medium containing dimethylsulfoxide [DMSO; 10% (v/v), final concentration]. The cell suspension is dispensed into Nunc cryovials (0.5–1.0 ml/vial). The sealed vials are placed upright in a beaker containing enough 95% ethanol to equal the fluid level of the suspensions (and well below the tops of the cryovials). The ethanol bath provides a more uniform rate of freezing of the cells and, with the DMSO as a cryoprotectant, prevents ice crystals from damaging the bacteria. The beaker is placed in an ultra-cold freezer (−75°C). After 24 h, the frozen stock cultures are transferred to storage boxes in the freezer. After 7–10 d, a cryovial should be examined to insure recovery of contaminant-free brachyspire cells. Brachyspire cells have remained viable in frozen stocks prepared in this way for over 20 years (T.B. Stanton, unpublished observations). Completely thawed stocks should not be refrozen. However, it is often possible to subculture without thawing the frozen stocks, by scraping the surface of a frozen cell suspension with a sterile inoculation loop and inoculating fresh media with the ice scrapings. *Brachyspira hyodysenteriae* cultures also have been preserved by lyophilization (Stanton and Lebo, 1988).

Taxonomic comments

In the past 25 years, there have been several taxonomic changes for spirochetes now assigned to the genus *Brachyspira*. Initially, the designation *Treponema hyodysenteriae* was applied to both pathogenic (strongly hemolytic) and nonpathogenic (weakly hemolytic) strains of intestinal spirochetes from swine. Miao and colleagues (1978) found that these pathogenic and nonpathogenic strains share only 28% DNA homology, and the nonpathogenic strains were reclassified as a new species, *Treponema innocens* (Kinyon and Harris, 1979). In 1991, Stanton et al. proposed a reclassification of *Treponema hyodysenteriae* and *Treponema innocens* to a new genus "*Serpula*" on the basis primarily of 16S rRNA sequence and DNA homology analyses. The genus name *Serpula* was subsequently changed to *Serpulina* after it was determined that *Serpula* had prior use as a name for a genus of fungi (Stanton, 1992). More recently, Ochiai et al. (1997) proposed the unification of the genera *Serpulina* and *Brachyspira*. The genus name *Brachyspira* was first given to a human intestinal isolate, *Brachyspira aalborgi*, in 1982 (Hovind-Hougen et al., 1982). Due to the use of the genus name *Brachyspira* prior to the genus designation *Serpulina*, this proposed change is consistent with International Taxonomic rules governing bacterial nomenclature. This action raised *Brachyspira aalborgi*, currently considered a commensal member of the human intestinal microbiota, to the status of type species of the genus *Brachyspira*. *Brachyspira aalborgi* has been characterized to a limited extent (Hovind-Hougen et al., 1982; Ochiai et al., 1997) and additional biochemical, physiological, and genetic investigations of this *Brachyspira* type species are needed.

Ochiai et al. (1997) proposed the reassignment of *Serpulina hyodysenteriae* (the *Serpulina* type species), *Serpulina innocens*, and *Serpulina pilosicoli* to the genus *Brachyspira*. Unfortunately, prior to the publication of this proposal, the species *Serpulina intermedia* and *Serpulina murdochii* were described and validly named as two new species within the genus *Serpulina* based on their similarities to *Serpulina hyodysenteriae* (Stanton et al., 1997). For taxonomic and phylogenetic reasons, *Serpulina intermedia* and *Serpulina murdochii* should be reassigned to the genus *Brachyspira* and a proposal has been submitted for this reassignment (T.B. Stanton, unpublished). Thus, both species are included in this section describing the genus *Brachyspira*.

16S rRNA gene sequence comparisons indicate that members of the genus *Brachyspira* belong within the family *Spirochetaceae* and are clearly distinct from the other genera of spirochetes (Figure 98). Members of the genus *Brachyspira* cluster closely together, with several species at greater than 99% similarity. However, a phylogenetic tree with similar topology is also obtained for the phylogenetic analysis of the sequences of NADH (*nox*) genes (Mikosza et al., 2004). Other methods described below, such as MLEE, have been used to better differentiate closely related members of the genus.

Differentiation of the species of the genus *Brachyspira*

For taxonomic purposes, *Brachyspira* species can be differentiated by MLEE (multilocus enzyme electrophoresis) and DNA homology methods. In the clinical laboratory, the cultivation of strongly (β-) hemolytic spirochete colonies from the feces of dysenteric animals has been used for reliable identification of *Brachyspira hyodysenteriae* when this hemolytic type is linked to histopathology and animal disease signs. It is more difficult to differentiate "non-*hyodysenteriae*" *Brachyspira* species, all of which form weakly hemolytic (nonhemolytic) colonies for several reasons. Nonpathogenic and pathogenic species which are both nonhemolytic can be present in the same clinical samples. Diseases caused by the weakly hemolytic species are often economic diseases associated with poor animal performance

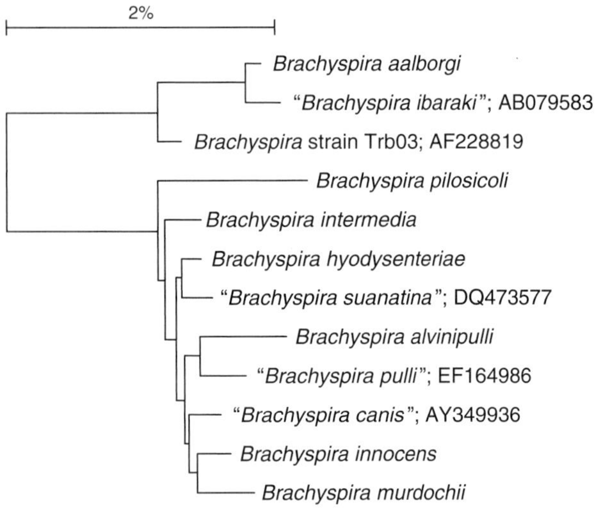

FIGURE 98. Dendrogram illustrating the phylogenetic positions of members of the genus *Brachyspira*. Species that have provisional names are indicated by quotation marks.

rather than diseases with high mortality and severe clinical signs. A further complicating factor for any clinical diagnostic test is that additional, uncharacterized *Brachyspira* species are likely to be in the intestinal tracts of humans and animals (Jansson et al., 2001; Pettersson et al., 2000).

Species differentiation based on biochemical tests (hippurate hydrolysis, indole production, and α-galactosidase and β-glucosidase) (Table 130) and 16S rRNA based RFLP-PCR methods have been proposed for differentiating swine brachyspires (Fellstrom et al., 1995, 1997). There is growing evidence, however, that *Brachyspira* species from swine at different geographical locations and from different animal species give inconsistent results in these biochemical tests (Fellstrom et al., 1999; Fossi et al., 2004; Stephens et al., 2005). Both the 16S rRNA and 23S rRNA sequences of *Brachyspira* species are highly conserved, although a 16S rRNA signature region was identified for *Brachyspira pilosicoli* and has been used effectively in a PCR assay for that species (Park et al., 1995). Various PCR techniques based on the *nox* gene are currently being tested to differentiate *Brachyspira* species in poultry and swine (La et al., 2003; Rohde et al., 2002; Townsend et al., 2005).

Further reading

Hampson, D.J. and T.B. Stanton. 1997. Intestinal Spirochetes in Domestic Animals and Humans. CAB International, Wallingford.

Hampson, D.J. and J.R. Thomson. 2004. *Brachyspira* research - special issue on colonic spirochetes of medical and veterinary significance. J. Med. Microbiol. *53*: 263–265.

Hampson, D.J. and D.J. Trott. 1999. Spirochetal diarrhea/porcine intestinal spirochetosis. *In* Diseases of Swine, 8th edn (edited by Straw, D'Allaire, Mengeling and Taylor). Iowa State University Press, Ames, pp. 553–562.

Harris, D.L., D.J. Hampson and R.D. Glock. 1999. Swine Dysentery. *In* Diseases of Swine, 8th edn (edited by Straw, D'Allaire, Mengeling and Taylor). Iowa State University Press, Ames, IA, pp. 579–600.

Stanton, T.B. 2006. Genus *Brachyspira*. *In* The Prokaryotes: A Handbook on the Biology of Bacteria, 3rd edn, vol. 7, *Proteobacteria*: Delta and Epsilon Subclasses. Deeply Rooting Bacteria (edited by Dworkin, Falkow, Rosenberg, Schleifer and Stackebrandt). Springer, New York, pp. 712–740.

Swayne, D.E. 2003. Avian intestinal spirochetosis. *In* Diseases of Poultry, 11th edn (edited by Saif, Barnes, Fadly, Glisson, Mcdougald and Swayne). Iowa State University Press, Ames, IA, pp. 826–836.

List of species of the genus *Brachyspira*

1. **Brachyspira aalborgi** Hovind-Hougen, Birch-Andersen, Henrik-Nielsen, Orholm, Pedersen, Teglbjaerg and Thaysen 1983, 896^{VP} (Effective publication: Hovind-Hougen, Birch-Andersen, Henrik-Nielsen, Orholm, Pedersen, Teglbjaerg and Thaysen 1982, 1135^{VP}.)

a.al.bor'gi. N.L. gen. n. *aalborgi* of Aalborg, named for the Danish town Aalborg in which the rectal biopsies containing the spirochete were taken from human diarrheic patients.

Exhibit characteristics common to the genus *Brachyspira*. Additional characteristics of the species are given in Tables 130 and 131. Electron microphotographs of *Brachyspira aalborgi* cells have been published (Hovind-Hougen et al., 1982). Differentiated from other *Brachyspira* species by DNA–DNA relative reassociation. *Brachyspira aalborgi* 513A exhibits 17–22% DNA–DNA relative reassociation with *Brachyspira hyodysenteriae* B78^T, *Brachyspira innocens* B256^T, and *Brachyspira pilosicoli* P43/6/78^T (filter hybridization method) (Ochiai et al., 1997).

Weakly hemolytic colonies on trypticase soy blood agar. Negative for indole production and positive for esculin hydrolysis. Ferments soluble carbohydrates (Table 131). Products of glucose metabolism have not been determined. Possess β-galactosidase and esterase, acid phosphatase, and phosphoamidase activities. Catalase-negative. *Brachyspira aalborgi* is not considered a pathogen.

Source: human intestinal tissues and contents.

DNA G+C content (mol%): 27.1 (T_m) for strain 513A.

Type strain: 513A, ATCC 43994, CIP 104603, NCTC 11492.

Sequence accession no. (16S rRNA gene): Z22781 (NCTC 11492).

2. **Brachyspira alvinipulli** Stanton, Postic and Jensen 1998, 675^{VP}

al.vi.ni.pul'li. L. adj. *alvinus -a -um* suffering from diarrhea; L. n. *pullus* a young fowl, a chicken; N.L. gen. n. *alvinipulli* of a diarrhaeic chicken, referring to the host animal from which spirochete was isolated.

Exhibit characteristics common to the genus *Brachyspira*. A phase-contrast microphotograph and an electron micrograph of *Brachyspira alvinipulli* C1 cells have been published (Stanton et al., 1998). Additional characteristics of the species are given in Tables 130 and 131. Differentiated from other *Brachyspira* species by DNA–DNA relative reassociation and MLEE analysis. Exhibits 24–39% DNA–DNA relative reassociation with *Brachyspira hyodysenteriae* B78^T, *Brachyspira innocens* B256^T, *Brachyspira pilosicoli* P43/6/78^T, *Brachyspira intermedia* PWS/A^T, and *Brachyspira murdochii* 56-150^T based on the S1 nuclease method (Stanton et al., 1998).

Weakly hemolytic colonies on trypticase soy blood agar. Grows optimally (3×10^8 to 10^9 cells/ml, direct cell counts) at 37–39°C in BHIS broth or HS broth containing carbohydrates (Table 131). Products of glucose metabolism include acetate, butyrate, CO_2, H_2, and ethanol. Positive in tests for hippurate hydrolysis, β-glucosidase, β-galactosidase, alkaline phosphatase, indoxyl acetate hydrolysis, arginine aminopeptidase, alanine aminopeptidase, and glycine aminopeptidase. Negative for indole production and catalase activity.

Shares high 16S rRNA gene sequence similarity (96–99%) with other *Brachyspira* species. Intestinal pathogen of chickens.

DNA G+C content (mol%): 24.6 (T_m) (strain C1).

Type strain: C1 (91–1207/C1), ATCC 51933, CIP 105681.

Sequence accession no. (16S rRNA gene): U23030 (strain C1).

3. **Brachyspira hyodysenteriae** (Harris, Glock, Christensen and Kinyon 1972a) emend. Ochiai, Adachi and Mori 1998, 327[VP] (*Treponema hyodysenteriae* Harris, Glock, Christensen and Kinyon 1972a, 64[AL]). Other synonyms: *Serpula hyodysenteriae* (Harris, Glock, Christensen and Kinyon 1972a) Stanton, Jensen, Casey, Tordoff, Dewhirst and Paster 1991, 56; *Serpulina hyodysenteriae* (Harris, Glock, Christensen and Kinyon 1972a) Stanton 1992, 189

hyo.dy.sen.te.ri'ae. Gr. n. *hyos* hog, pig; L. n. *dysenteria* a flux, dysentery; N.L. gen. n. *hyodysenteriae* of hog dysentery. In recognition of the species as the etiologic agent of swine dysentery.

Exhibits characteristic common to the genus *Brachyspira*. Phase-contrast photomicrographs and electron photomicrographs of cells have been published (Kinyon et al., 1977; Trott et al., 1996b). Additional characteristics of the species are given in Tables 130 and 131. Differentiated from other *Brachyspira* species by DNA–DNA relative reassociation, MLEE analysis. Strain B78^T exhibits 22% DNA–DNA relative reassociation with *Brachyspira aalborgi* NCTC 11492^T, based on the filter hybridization method (Ochiai et al., 1997). Strain B78^T exhibits 27–57% DNA–DNA relative reassociation with *Brachyspira innocens* B256^T, *Brachyspira pilosicoli* P43/6/78^T, *Brachyspira intermedia* PWS/A^T, *Brachyspira murdochii* 56-150^T, and *Brachyspira alvinipulli* C1^T, based on the S1 nuclease method (Stanton et al., 1998; Trott et al., 1996b). Unlike other *Brachyspira* species, *Brachyspira hyodysenteriae* cells form β-hemolytic (strongly hemolytic) colonies on trypticase soy blood agar and are positive for indole production. Strains have been differentiated according to serotypes, based on soluble antigens in the water phase of cells extracted with hot phenol-water.

Grows at 37–39°C in a variety of anaerobically prepared culture broth media, attaining population densities ≥10^9 cells/ml (direct cell counts), under optimum conditions. Cholesterol in nutrient amounts is required for growth, presumably for cell membrane biosynthesis. Uses carbohydrates as growth substrates (Table 131). Products of glucose metabolism include acetate, butyrate, CO_2, and H_2. Although considered an anaerobic species, growing cells under certain conditions require low (1%) oxygen tensions and consume substrate amounts of oxygen.

Cells possess NADH oxidase, superoxide dismutase, NADH peroxidase, and inducible catalase activities. Hydrogen is produced via a clostridial-type phosphoroclastic mechanism and via NADH-ferredoxin oxidoreductase. Exhibits alkaline phosphatase, C4 esterase, C8 esterase, lipase, phosphatase acid, β-galactosidase, α-glucosidase, and β-glucosidase activities. Positive for esculin hydrolysis and most strains positive for indole production.

Isolated from intestinal contents and feces of swine, rheas, and other mammals (dogs) in contact with dysenteric feces. *Brachyspira hyodysenteriae* is an intestinal pathogen of swine and rheas. *Brachyspira hyodysenteriae* cells colonize within and between intestinal epithelial cells and the overlying intestinal mucus without cell attachment. The type strain B78 has become nonpathogenic for swine likely due to long-term passage in culture. Strain B204 (ATCC 31212) has commonly been used in the United States for experimental swine infections.

DNA G+C content (mol%): 25.7–25.9% (T_m) for strains B204 and A1.

Type strain: B78, ATCC 27164, CCUG 46668, NCTC 13041.

Sequence accession no. (16S rRNA gene): M57743, U14930 (B78).

4. **Brachyspira innocens** (Kinyon and Harris 1979, 108) emend. Ochiai, Adachi and Mori 1998, 327[VP] (*Treponema innocens* Kinyon and Harris 1979, 108[AL]). Other synonyms: *Serpula innocens* (Kinyon and Harris 1979) Stanton, Jensen, Casey, Tordoff, Dewhirst and Paster 1991, 56; *Serpulina innocens* (Kinyon and Harris 1979) Stanton 1992, 189

in'no.cens. L. fem. adj. *innocens* harmless, inoffensive (referring to non-pathogenic nature of the species for swine).

Exhibit characteristics common to the genus *Brachyspira*. Additional characteristics of the species are given in Tables 130 and 131. Differentiated from other *Brachyspira* species by DNA–DNA relative reassociation and MLEE analysis. Strain B256 exhibits 19% DNA–DNA relative-reassociation with *Brachyspira aalborgi* NCTC 11492, based on the filter hybridization method (Ochiai et al., 1997). Strain B256 exhibits 26–66% relative reassociation with *Brachyspira hyodysenteriae* B78^T, *Brachyspira pilosicoli* P43/6/78^T, *Brachyspira intermedia* PWS/A^T, *Brachyspira murdochii* 56-150^T, and *Brachyspira alvinipulli* C1^T based on the S1 nuclease method (Stanton et al., 1998; Trott et al., 1996b).

Weakly hemolytic colonies on trypticase soy blood agar. Uses carbohydrates as growth substrates (Table 131). Grows optimally (9×10^8 to 1.5×10^9 cells/ml, direct counts) at 37–39°C in BHIS broth or HS broth containing carbohydrates (Table 131). Products of glucose metabolism include acetate, butyrate, CO_2, H_2, and ethanol. Population doubling times are 3–5 h. Does not produce indole, hydrolyzes esculin.

Brachyspira innocens is not considered a pathogen.

Source: intestinal contents of healthy pigs, dogs, chickens.

DNA G+C content (mol%): 25.6–25.8% (T_m) (strain B256).

Type strain: B256, ATCC 29796, CCUG 17081.

Sequence accession no. (16S rRNA gene): M57744 (strain B256).

5. **Brachyspira intermedia** (Stanton, Fournié-Amazouz, Postic, Trott, Grimont, Baranton, Hampson and Saint Girons 1997) emend. Hampson and La 2006, 1011[VP] (*Serpulina intermedia* Stanton, Fournié-Amazouz, Postic, Trott, Grimont, Baranton, Hampson and Saint Girons 1997, 1011)

in.ter.me'di.a. L. fem. adj. *intermedia* which is in the middle, referring to the fact that biochemical traits are intermediate between those possessed only by *Brachyspira hyodysenteriae* or by *Brachyspira innocens*.

Exhibit characteristics common to the genus *Brachyspira*. Additional characteristics of the species are given in Tables

130 and 131. Differentiated from other *Brachyspira* species by DNA–DNA relative reassociation and MLEE analysis. Also called "*Serpulina intermedius*". Some intestinal spirochetes referred to as "*Treponema hyodysenteriae* biotype 2" or "intermediate type" may be *Brachyspira intermedia* strains. Strain PWS/A exhibits 26–68% DNA–DNA relative reassociation with *Brachyspira hyodysenteriae* B78^T, *Brachyspira innocens* B256^T, *Brachyspira pilosicoli* P43/6/78^T, *Brachyspira murdochii* 56-150^T, and *Brachyspira alvinipulli* C1^T, based on the S1 nuclease method (Stanton et al., 1997, 1998).

Weakly to "intermediate" hemolytic colonies on trypticase soy blood agar. Uses carbohydrates as growth substrates (Table 131). Grows optimally (1.1×10^9 to 1.6×10^9 cells/ml, direct counts) at 37–39°C in BHIS broth or HS broth containing carbohydrates (Table 131). Products of glucose metabolism include acetate, butyrate, CO_2, H_2, and ethanol. Hydrolyzes esculin. Does not hydrolyze hippurate. Lacks α-galactosidase and possesses α-glucosidase and β-glucosidase activities.

Brachyspira intermedia strains have been isolated from swine, including swine with diarrhea and from commercial poultry flocks exhibiting diarrhea. Avian intestinal infections by *Brachyspira intermedia* are aptly described as "economic disease" due to production losses in flocks. Birds colonized by *Brachyspira intermedia* typically do not die and may even appear healthy. There is some thought that disease severity is either strain related or influenced by the animal diet or other environmental factors.

DNA G+C content (mol%): 25 (T_m) (strain PWS/A^T).

Type strain: PWS/A^T, ATCC 51140, CIP 105833.

Sequence accession no. (16S rRNA gene): U23033 (strain PWS/A).

6. **Brachyspira murdochii** (Stanton, Fournié-Amazouz, Postic, Trott, Grimont, Baranton, Hampson and Saint Girons 1997) emend. Hampson and La 2006, 1011^{VP} (*Serpulina intermedia* Stanton, Fournié-Amazouz, Postic, Trott, Grimont, Baranton, Hampson and Saint Girons 1997, 1011)

mur.do'chi.i. N.L. masc. gen. n. *murdochii* of Murdoch, named in recognition of work conducted at the Murdoch University in Western Australia where the type strain was identified.

Exhibit characteristics common to the genus *Brachyspira*. Additional characteristics of the species are given in Tables 130 and 131. Differentiated from other *Brachyspira* species by DNA–DNA relative reassociation and MLEE analysis. Strain 56-150^T exhibits 27–66% DNA–DNA relative reassociation with *Brachyspira hyodysenteriae* B78^T, *Brachyspira innocens* B256^T, *Brachyspira pilosicoli* P43/6/78^T, *Brachyspira intermedia* PWS/A^T, and *Brachyspira alvinipulli* C1^T, based on the S1 nuclease method (Stanton et al., 1997, 1998).

Weakly hemolytic colonies on trypticase soy blood agar. Grows optimally (9×10^8 to 1.5×10^9 cells/ml, direct counts) at 37–39°C in BHIS broth or HS broth containing carbohydrates (Table 131). Products of glucose metabolism include acetate, butyrate, CO_2, H_2, and ethanol. Does not produce indole. Does not hydrolyze hippurate. Cells lack α-galactosidase, α-glucosidase, and possess β-glucosidase activities.

Not considered a pathogen.

Source: Intestinal contents of healthy swine and rats.

DNA G+C content (mol%): 27 (T_m) (strain 56-150).

Type strain: 56-150, ATCC 51254, CIP 105832, DSM 12563.

Sequence accession no. (16S rRNA gene): AY312492 (strain 56-150).

7. **Brachyspira pilosicoli** (Trott, Stanton, Jensen, Duhamel, Johnson and Hampson 1996b) emend. Ochiai, Adachi and Mori 1998, 327^{VP} (*Serpulina pilosicoli* Trott, Stanton, Jensen, Duhamel, Johnson and Hampson 1996b, 213)

pi.lo.si'co.li. L. adj. *pilosus -a -um* hairy, napped, shaggy; L. n. *colon* or *colum* the colon; N.L. gen. n. *pilosicoli*, of a hairy colon (referring to the fact that infection and attachment by this intestinal spirochete can result in the histological appearance of a hairy covering, false brush border, on the surface of the colon).

Exhibit characteristics common to the genus *Brachyspira*. Additional characteristics of the species are given in Tables 130 and 131. Phase-contrast photomicrographs and electron microphotographs of *Brachyspira pilosicoli* cells have been published (Trott et al., 1996b). Differentiated from other *Brachyspira* species by DNA–DNA relative reassociation, MLEE analysis, and by a signature nucleotide sequence (5′-AGUUUUUUCGCUUCA-3′) in the 16S rRNA. The 16S rRNA signature sequence has been useful in clinical identification of *Brachyspira pilosicoli*. Alternative designations are *Anguillina coli* and *Serpulina pilosicoli*. Strain P43/6/78 exhibits 17% DNA–DNA relative reassociation with *Brachyspira aalborgi* NCTC 11492 based on the filter hybridization method (Ochiai et al., 1997). Strains P43/6/78 and WES-B exhibit 21–28% DNA–DNA relative reassociation with *Brachyspira hyodysenteriae* B78^T, *Brachyspira innocens* B256^T, *Brachyspira intermedia* PWS/A^T, *Brachyspira murdochii* 56-150^T, and *Brachyspira alvinipulli* C1^T, based on the S1 nuclease method (Stanton et al., 1997, 1998).

Weakly hemolytic colonies on trypticase soy blood agar. Grows optimally (1.5×10^9 to $>2.0 \times 10^9$ cells/ml, direct counts) at 37–39°C in BHIS broth or HS broth containing carbohydrates (Table 131). Products of glucose metabolism include acetate, butyrate, CO_2, H_2, and ethanol. Strains lack β-glucosidase activity.

Brachyspira pilosicoli is considered an intestinal pathogen, the etiological agent of intestinal spirochetosis. Isolated from swine with watery, mucoid diarrhea. Swine intestinal spirochetosis clinically resembles early, mild stages of swine dysentery. Typical of the disease is the attachment of *Brachyspira pilosicoli* cells by one end to colonic epithelial cells. *Brachyspira pilosicoli* has been found in intestinal contents of birds, dogs, humans, and non-human primates. *Brachyspira pilosicoli* is being investigated as a possible enteropathogen of humans, dogs, and poultry.

DNA G+C content (mol%): 25 ± 1 (T_m) (strain P43/6/78).

Type strain: P43/6/78, ATCC 51139.

Sequence accession no. (16S rRNA gene): U23032 (strain P43/6/78).

Other organisms

There are reports of intestinal spirochetes that are likely to represent new *Brachyspira* species based on their MLEE profiles and comparative analyses of 16S rRNA and NADH oxidase genes (Mikosza et al., 2004). The strains have only been partially characterized and some have been given provisional designations. For example, there are dog-associated spirochetes

"*Brachyspira canis*" (Duhamel et al., 1998b), chicken-associated spirochetes "*Brachyspira pulli*" also known as group "d" spirochetes (McLaren et al., 1997; Stephens and Hampson, 1999), duck and pig-associated spirochetes "*Brachyspira suanatina*" (Råsbäck et al., 2007), and *Brachyspira aalborgi*-like spirochetes (Mikosza et al., 2004).

References

Achacha, M. and S. Messier. 1992. Comparison of six different culture media for isolation of *Treponema hyodysenteriae*. J. Clin. Microbiol. *30*: 249–251.

Achacha, M., S. Messier and K.R. Mittal. 1996. Development of an experimental model allowing discrimination between virulent and avirulent isolates of *Serpulina* (*Treponema*) *hyodysenteriae*. Can. J. Vet. Res. *60*: 45–49.

Atyeo, R.F., T.B. Stanton, N.S. Jensen, D.S. Suriyaarachichi and D.J. Hampson. 1999. Differentiation of *Serpulina* species by NADH oxidase gene (*nox*) sequence comparisons and *nox*-based polymerase chain reaction tests. Vet. Microbiol. *67*: 47–60.

Barrett, S.P. 1997. Human intestinal spirochaetosis. *In* Intestinal *Spirochaetes* in Domestic Animals and Humans (edited by Hampson and Stanton). CAB International, Wallingford, pp. 243–265.

Baum, D.H. and L.A. Joens. 1979. Serotypes of beta-hemolytic *Treponema hyodysenteriae*. Infect. Immun. *25*: 792–796.

Buckles, E.L., K.A. Eaton and D.E. Swayne. 1997. Cases of spirochete-associated necrotizing typhlitis in captive common rheas (*Rhea americana*). Avian Dis. *41*: 144–148.

Buller, N.B. and D.J. Hampson. 1994. Antimicrobial susceptibility testing of *Serpulina hyodysenteriae*. Aust. Vet. J. *71*: 211–214.

Calderaro, A., G. Dettori, L. Collini, P. Ragni, R. Grillo, P. Cattani, G. Fadda and C. Chezzi. 1998a. Bacteriophages induced from weakly beta-haemolytic human intestinal spirochaetes by mitomycin C. J. Basic Microbiol. *38*: 323–335.

Calderaro, A., G. Dettori, R. Grillo, P. Plaisant, G. Amalfitano and C. Chezzi. 1998b. Search for bacteriophages spontaneously occurring in cultures of haemolytic intestinal spirochaetes of human and animal origin. J. Basic Microbiol. *38*: 313–322.

Calderaro, A., G. Merialdi, S. Perini, P. Ragni, R. Guegan, G. Dettori and C. Chezzi. 2001. A novel method for isolation of *Brachyspira* (*Serpulina*) *hyodysenteriae* from pigs with swine dysentery in Italy. Vet. Microbiol. *80*: 47–52.

Cattani, P., G. Dettori, A. Calderaro, R. Grillo, G. Fadda and C. Chezzi. 1998. Detection of extrachromosomal DNA in Italian isolates of weakly beta-haemolytic human intestinal spirochaetes. New Microbiol. *21*: 241–248.

Combs, B., D.J. Hampson, J.R. Mhoma and J.R. Buddle. 1989. Typing of *Treponema hyodysenteriae* by restriction endonuclease analysis. Vet. Microbiol. *19*: 351–359.

Cullen, P.A., S.A. Coutts, S.J. Cordwell, D.M. Bulach and B. Adler. 2003. Characterization of a locus encoding four paralogous outer membrane lipoproteins of *Brachyspira hyodysenteriae*. Microbes Infect. *5*: 275–283.

Cullen, P.A., D.A. Haake and B. Adler. 2004. Outer membrane proteins of pathogenic spirochetes. FEMS Microbiol. Rev. *28*: 291–318.

Davelaar, F.G., H.F. Smit, K. Hovind-Hougen, R.M. Dwars and P.C. van der Valk. 1986. Infectious typhlitis in chickens caused by spirochetes. Avian Pathol. *15*: 247–258.

De Smet, K.A., D.E. Worth and S.P. Barrett. 1998. Variation amongst human isolates of *Brachyspira* (*Serpulina*) *pilosicoli* based on biochemical characterization and 16S rRNA gene sequencing. Int. J. Syst. Bacteriol. *48*: 1257–1263.

Dugourd, D., C. Martin, C.R. Rioux, M. Jacques and J. Harel. 1999. Characterization of a periplasmic ATP-binding cassette iron import system of *Brachyspira* (*Serpulina*) *hyodysenteriae*. J. Bacteriol. *181*: 6948–6957.

Duhamel, G.E., J.M. Kinyon, M.R. Mathiesen, D.P. Murphy and D. Walter. 1998a. *In vitro* activity of four antimicrobial agents against North American isolates of porcine *Serpulina pilosicoli*. J. Vet. Diagn. Invest. *10*: 350–356.

Duhamel, G.E., D.J. Trott, N. Muniappa, M.R. Mathiesen, K. Tarasiuk, J.I. Lee and D.J. Hampson. 1998b. Canine intestinal spirochetes consist of *Serpulina pilosicoli* and a newly identified group provisionally designated "*Serpulina canis*" sp. nov. J. Clin. Microbiol. *36*: 2264–2270.

Dwars, R.M., H.F. Smit, F.G. Davelaar and V.T. Veer. 1989. Incidence of spirochaetal infections in cases of intestinal disorder in chickens. Avian Pathol. *18*: 591–595.

Dwars, R.M., F.G. Davelaar and H.F. Smit. 1992. Spirochaetosis in broilers. Avian Pathol. *21*: 261–273.

Fellstrom, C., B. Petterson, M. Uhlen, A. Gunnarsson and K.E. Johansson. 1995. Phylogeny of *Serpulina* based on sequence analyses of the 16S rRNA gene and comparison with a scheme involving biochemical classification. Res. Vet. Sci. *59*: 5–9.

Fellstrom, C., B. Pettersson, J. Thomson, A. Gunnarsson, M. Persson and K.E. Johansson. 1997. Identification of *Serpulina* species associated with porcine colitis by biochemical analysis and PCR. J. Clin. Microbiol. *35*: 462–467.

Fellstrom, C., M. Karlsson, B. Pettersson, U. Zimmerman, A. Gunnarsson and A. Aspan. 1999. Emended descriptions of indole negative and indole positive isolates of *Brachyspira* (*Serpulina*) *hyodysenteriae*. Vet. Microbiol. *70*: 225–238.

Fossi, M. and T. Skrzypczak. 2006. D-Ribose utilisation differentiates porcine *Brachyspira pilosicoli* from other porcine *Brachyspira* species. Anaerobe *12*: 110–113.

Fossi, M., T. Pohjanvirta, A. Sukura, S. Heinikainen, R. Lindecrona and S. Pelkonen. 2004. Molecular and ultrastructural characterization of porcine hippurate-negative *Brachyspira pilosicoli*. J. Clin. Microbiol. *42*: 3153–3158.

Gabe, J.D., E. Dragon, R.J. Chang and M.T. McCaman. 1998. Identification of a linked set of genes in *Serpulina hyodysenteriae* (B204) predicted to encode closely related 39-kilodalton extracytoplasmic proteins. J. Bacteriol. *180*: 444–448.

Gad, A., R. Willen, K. Furugard, R. Fors and M. Hradsky. 1977. Intestinal spirochaetosis as a cause of longstanding diarrhoea. Uppsala J. Med. Sci. *82*: 49–54.

Galvin, J.E., D.L. Harris and M.J. Wannemuehler. 1997. Prevention and control of intestinal spirochaetal disease: immunological and pharmacological mechanisms. *In* Intestinal *Spirochaetes* in Domestic Animals and Humans (edited by Hampson and Stanton). CAB International, Wallingford.

Glock, R.D., D.L. Harris and J.P. Kluge. 1974. Localization of spirochetes with the structural characteristics of *Treponema hyodysenteriae* in the lesions of swine dysentery. Infect. Immun. *9*: 167–178.

Greer, J.M. and M.J. Wannemuehler. 1989. Pathogenesis of *Treponema hyodysenteriae*: induction of interleukin-1 and tumor necrosis factor by a treponemal butanol/water extract (endotoxin). Microb. Pathog. *7*: 279–288.

Griffiths, I.B., B.W. Hunt, S.A. Lister and M.H. Lamont. 1987. Retarded growth rate and delayed onset of egg production associated with spirochaete infection in pullets. Vet. Rec. *121*: 35–37.

Hampson, D.J. and T. La. 2006. Reclassification of *Serpulina intermedia* and *Serpulina murdochii* in the genus *Brachyspira* as *Brachyspira intermedia* comb. nov. and *Brachyspira murdochii* comb. nov. Int. J. Syst. Evol. Microbiol. *56*: 1009–1012.

Hampson, D.J. and A.J. McLaren. 1999. Experimental infection of laying hens with *Serpulina intermedia* causes reduced egg production and increased faecal water content. Avian Pathol. *28*: 113–117.

Hampson, D.J. and D.J. Trott. 1999. Spirochetal diarrhea/porcine intestinal spirochetosis. *In* Diseases of Swine, 8th edn (edited by Straw, D'Allaire, Mengeling and Taylor). Iowa State University Press, Ames, pp. 553–562.

Hampson, D.J., J.R. Mhoma, B. Combs and J.R. Buddle. 1989. Proposed revisions to the serological typing system for *Treponema hyodysenteriae*. Epidemiol. Infect. *102*: 75–84.

Hampson, D.J., R.F. Atyeo and B.G. Combs. 1997. Swine dysentery. *In* Intestinal *Spirochaetes* in Domestic Animals and Humans (edited by Hampson and Stanton). CAB International, Wallingford, pp. 175–209.

Harris, D.L., R.D. Glock, C.R. Christensen and J.M. Kinyon. 1972a. Swine dysentery. I. Inoculation of pigs with the *Treponema hyodysenteriae* (new species) and reproduction of the disease. Vet. Med. *67*: 61–64.

Harris, D.L., J.M. Kinyon, M.T. Mullin and R.D. Glock. 1972b. Isolation and propagation of spirochetes from the colon of swine dysentery affected pigs. Can. J. Comp. Med. *36*: 74–76.

Harris, D.L., D.J. Hampson and R.D. Glock. 1999. Swine dysentery. *In* Diseases of Swine, 8th edn (edited by Straw, D'Allaire, Mengeling and Taylor). Iowa State University Press, Ames, IA, pp. 579–600.

Hontecillas, R., J. Bassaganya-Riera, J. Wilson, D.L. Hutto and M.J. Wannemuehler. 2005. CD4+ T-cell responses and distribution at the colonic mucosa during *Brachyspira hyodysenteriae*-induced colitis in pigs. Immunology *115*: 127–135.

Hovind-Hougen, K., A. Birch-Andersen, R. Henrik-Nielsen, M. Orholm, J.O. Pedersen, P.S. Teglbjaerg and E.H. Thaysen. 1982. Intestinal spirochetosis: morphological characterization and cultivation of the spirochete *Brachyspira aalborgi* gen. nov., sp. nov. J. Clin. Microbiol. *16*: 1127–1136.

Hovind-Hougen, K., A. Birch-Andersen, R. Henrik-Nielsen, M. Orholm, J.O. Pedersen, P.S. Teglbjaerg and E.H. Thaysen. 1983. *In* Validation of publication of new names and new combinations previously effectively published outside the IJSEM. List no. 12. Int. J. Syst. Bacteriol. *33*: 896–897.

Hsu, T., D.L. Hutto, F.C. Minion, R.L. Zuerner and M.J. Wannemuehler. 2001. Cloning of a beta-hemolysin gene of *Brachyspira* (*Serpulina*) *hyodysenteriae* and its expression in *Escherichia coli*. Infect. Immun. *69*: 706–711.

Humphrey, S.B., T.B. Stanton and N.S. Jensen. 1995. Mitomycin C induction of bacteriophages from *Serpulina hyodysenteriae* and *Serpulina innocens*. FEMS Microbiol. Lett. *134*: 97–101.

Humphrey, S.B., T.B. Stanton, N.S. Jensen and R.L. Zuerner. 1997. Purification and characterization of VSH-1, a generalized transducing bacteriophage of *Serpulina hyodysenteriae*. J. Bacteriol. *179*: 323–329.

Hutto, D.L. and M.J. Wannemuehler. 1999. A comparison of the morphologic effects of *Serpulina hyodysenteriae* or its beta-hemolysin on the murine cecal mucosa. Vet. Pathol. *36*: 412–422.

Hyatt, D.R., A.A. ter Huurne, B.A. van der Zeijst and L.A. Joens. 1994. Reduced virulence of *Serpulina hyodysenteriae* hemolysin-negative mutants in pigs and their potential to protect pigs against challenge with a virulent strain. Infect. Immun. *62*: 2244–2248.

Jansson, D.S., C. Brojer, D. Gavier-Widen, A. Gunnarsson and C. Fellstrom. 2001. *Brachyspira* spp. (*Serpulina* spp.) in birds: a review and results from a study of Swedish game birds. Anim. Health Res. Rev. *2*: 93–100.

Jansson, D.S., K.E. Johansson, T. Olofsson, T. Rasback, I. Vagsholm, B. Pettersson, A. Gunnarsson and C. Fellstrom. 2004. *Brachyspira hyodysenteriae* and other strongly beta-haemolytic and indole-positive spirochaetes isolated from mallards (*Anas platyrhynchos*). J. Med. Microbiol. *53*: 293–300.

Jenkinson, S.R. and C.R. Wingar. 1981. Selective medium for the isolation of *Treponema hyodysenteriae*. Vet. Rec. *109*: 384–385.

Jensen, N.S. and T.B. Stanton. 1993. Comparison of *Serpulina hyodysenteriae* B78, the type strain of the species, with other *S. hyodysenteriae* strains using enteropathogenicity studies and restriction fragment length polymorphism analysis. Vet. Microbiol. *36*: 221–231.

Jensen, N.S. and T.B. Stanton. 1994. Production of an inducible sucrase Activity by *Serpulina hyodysenteriae*. Appl. Environ. Microbiol. *60*: 3429–3432.

Jensen, N.S., T.B. Stanton and D.E. Swayne. 1996. Identification of the swine pathogen *Serpulina hyodysenteriae* in rheas (*Rhea americana*). Vet. Microbiol. *52*: 259–269.

Jensen, T.K., K. Moller, M. Boye, T.D. Leser and S.E. Jorsal. 2000. Scanning electron microscopy and fluorescent in situ hybridization of experimental *Brachyspira* (*Serpulina*) *pilosicoli* infection in growing pigs. Vet. Pathol. *37*: 22–32.

Joens, L.A. 1997. Virulence factors associated with *Serpulina hyodysenteriae*. *In* Intestinal *Spirochaetes* in Domestic Animals and Humans (edited by Hampson and Stanton). CAB International, Wallingford, pp. 151–172.

Joens, L.A. and R.D. Glock. 1979. Experimental infection in mice with *Treponema hyodysenteriae*. Infect. Immun. *25*: 757–760.

Joens, L.A., N.A. Nord, J.M. Kinyon and I.T. Egan. 1982. Enzyme-linked immunosorbent assay for detection of antibody to *Treponema hyodysenteriae* antigens. J. Clin. Microbiol. *15*: 249–252.

Jones, M.J., J.N. Miller and W.L. George. 1986. Microbiological and biochemical characterization of spirochetes isolated from the feces of homosexual males. J. Clin. Microbiol. *24*: 1071–1074.

Karlsson, M., C. Fellstrom, M.U. Heldtander, K.E. Johansson and A. Franklin. 1999. Genetic basis of macrolide and lincosamide resistance in *Brachyspira* (*Serpulina*) *hyodysenteriae*. FEMS Microbiol. Lett. *172*: 255–260.

Karlsson, M., C. Fellstrom, A. Gunnarsson, A. Landen and A. Franklin. 2003. Antimicrobial susceptibility testing of porcine *Brachyspira* (*Serpulina*) species isolates. J. Clin. Microbiol. *41*: 2596–2604.

Karlsson, M., C. Fellstrom, K.E. Johansson and A. Franklin. 2004. Antimicrobial resistance in *Brachyspira pilosicoli* with special reference to point mutations in the 23S rRNA gene associated with macrolide and lincosamide resistance. Microb. Drug Resist. *10*: 204–208.

Kennedy, M.J. and R.J. Yancey, Jr. 1996. Motility and chemotaxis in *Serpulina hyodysenteriae*. Vet. Microbiol. *49*: 21–30.

Kennedy, M.J., D.K. Rosnick, R.G. Ulrich and R.J. Yancey, Jr. 1988. Association of *Treponema hyodysenteriae* with porcine intestinal mucosa. J. Gen. Microbiol. *134*: 1565–1576.

Kennedy, M.J., E.L. Rosey and R.J. Yancey, Jr. 1997. Characterization of *flaA*- and *flaB*- mutants of *Serpulina hyodysenteriae:* both flagellin subunits, FlaA and FlaB, are necessary for full motility and intestinal colonization. FEMS Microbiol. Lett. *153*: 119–128.

Kent, K.A., R.M. Lemcke and R.J. Lysons. 1988. Production, purification and molecular weight determination of the haemolysin of *Treponema hyodysenteriae*. J. Med. Microbiol. *27*: 215–224.

Kinyon, J.M. and D.J. Harris. 1979. *Treponema innocens*, a new species of intestinal bacteria, and emended description of the type strain of *Treponema hyodysenteriae* Harris et al. Int. J. Syst. Bacteriol. *29*: 102–109.

Kinyon, J.M., D.L. Harris and R.D. Glock. 1977. Enteropathogenicity of various isolates of *Treponema hyodysenteriae*. Infect. Immun. *15*: 638–646.

Kraaz, W., B. Pettersson, U. Thunberg, L. Engstrand and C. Fellstrom. 2000. *Brachyspira aalborgi* infection diagnosed by culture and 16S ribosomal DNA sequencing using human colonic biopsy specimens. J. Clin. Microbiol. *38*: 3555–3560.

Kunkle, R.A. and J.M. Kinyon. 1988. Improved selective medium for the isolation of *Treponema hyodysenteriae*. J. Clin. Microbiol. *26*: 2357–2360.

Kunkle, R.A., D.L. Harris and J.M. Kinyon. 1986. Autoclaved liquid medium for propagation of *Treponema hyodysenteriae*. J. Clin. Microbiol. *24*: 669–671.

La, T., N.D. Phillips and D.J. Hampson. 2003. Development of a duplex PCR assay for detection of *Brachyspira hyodysenteriae* and *Brachyspira pilosicoli* in pig feces. J. Clin. Microbiol. *41*: 3372–3375.

La, T., N.D. Phillips, M.P. Reichel and D.J. Hampson. 2004. Protection of pigs from swine dysentery by vaccination with recombinant BmpB, a 29.7 kDa outer-membrane lipoprotein of *Brachyspira hyodysenteriae*. Vet. Microbiol. *102*: 97–109.

Lau, T.T. and D.J. Hampson. 1992. The serological grouping system for *Serpulina* (*Treponema*) *hyodysenteriae*. Epidemiol. Infect. *109*: 255–263.

Lee, J.I., D.J. Hampson, B.G. Combs and A.J. Lymbery. 1993. Genetic relationships between isolates of *Serpulina* (*Treponema*) *hyodysenteriae*, and comparison of methods for their subspecific differentiation. Vet. Microbiol. *34*: 35–46.

Lee, B.J., T. La, A.S. Mikosza and D.J. Hampson. 2000. Identification of the gene encoding BmpB, a 30 kDa outer envelope lipoprotein of

Brachyspira (*Serpulina*) *hyodysenteriae*, and immunogenicity of recombinant BmpB in mice and pigs. Vet. Microbiol. *76*: 245–257.

Lemcke, R.M. and M.R. Burrows. 1980. Sterol requirement for the growth of *Treponema hyodysenteriae*. J. Gen. Microbiol. *116*: 539–543.

Li, Z.S., M. Belanger and M. Jacques. 1991. Serotyping of Canadian isolates of *Treponema hyodysenteriae* and description of two new serotypes. J. Clin. Microbiol. *29*: 2794–2797.

Li, Z., B. Foiry and M. Jacques. 1995. Growth of *Serpulina* (*Treponema*) *hyodysenteriae* under iron-restricted conditions. Can. J. Vet. Res. *59*: 149–153.

Li, C., L. Corum, D. Morgan, E.L. Rosey, T.B. Stanton and N.W. Charon. 2000. The spirochete FlaA periplasmic flagellar sheath protein impacts flagellar helicity. J. Bacteriol. *182*: 6698–6706.

Livesley, M.A., I.P. Thompson, M.J. Bailey and P.A. Nuttall. 1993. Comparison of the fatty acid profiles of *Borrelia*, *Serpulina* and *Leptospira* species. J. Gen. Microbiol. *139*: 889–895.

Lobova, D., J. Smola and A. Cizek. 2004. Decreased susceptibility to tiamulin and valnemulin among Czech isolates of *Brachyspira hyodysenteriae*. J. Med. Microbiol. *53*: 287–291.

Lysons, R.J., K.A. Kent, A.P. Bland, R. Sellwood, W.F. Robinson and A.J. Frost. 1991. A cytotoxic haemolysin from *Treponema hyodysenteriae* - a probable virulence determinant in swine dysentery. J. Med. Microbiol. *34*: 97–102.

Matson, E.G., M.G. Thompson, S.B. Humphrey, R.L. Zuerner and T.B. Stanton. 2005. Identification of genes of VSH-1, a prophage-like gene transfer agent of *Brachyspira hyodysenteriae*. J. Bacteriol. *187*: 5885–5892.

Matthews, H.M. and J.M. Kinyon. 1984. Cellular lipid comparisons between strains of *Treponema hyodysenteriae* and *Treponema innocens*. Int. J. Syst. Bacteriol. *34*: 160–165.

Matthews, H.M., T.K. Yang and H.M. Jenkin. 1980a. *Treponema innocens* lipids and further description of an unusual galactolipid of *Treponema hyodysenteriae*. J. Bacteriol. *143*: 1151–1155.

Matthews, H.M., T.K. Yang and H.M. Jenkin. 1980b. Alk-1-enyl ether phospholipids (plasmalogens) and glycolipids of *Treponema hyodysenteriae*. Analysis of acyl and alk-1-enyl moieties. Biochim. Biophys. Acta *618*: 273–281.

McCaman, M.T., K. Auer, W. Foley and J.D. Gabe. 2003. *Brachyspira hyodysenteriae* contains eight linked gene copies related to an expressed 39-kDa surface protein. Microbes Infect. *5*: 1–6.

McLaren, A.J., D.J. Trott, D.E. Swayne, S.L. Oxberry and D.J. Hampson. 1997. Genetic and phenotypic characterization of intestinal spirochetes colonizing chickens and allocation of known pathogenic isolates to three distinct genetic groups. J. Clin. Microbiol. *35*: 412–417.

Messier, S., R. Higgins and C. Moore. 1990. Minimal inhibitory concentrations of five antimicrobials against *Treponema hyodysenteriae* and *Treponema innocens*. J. Vet. Diagn. Invest. *2*: 330–333.

Miao, R.M., A.H. Fieldsteel and D.L. Harris. 1978. Genetics of *Treponema*: characterization of *Treponema hyodysenteriae* and its relationship to *Treponema pallidum*. Infect. Immun. *22*: 736–739.

Mikosza, A.S., M.A. Munshi and D.J. Hampson. 2004. Analysis of genetic variation in *Brachyspira aalborgi* and related spirochaetes determined by partial sequencing of the 16S rRNA and NADH oxidase genes. J. Med. Microbiol. *53*: 333–339.

Milner, J.A. and R. Sellwood. 1994. Chemotactic response to mucin by *Serpulina hyodysenteriae* and other porcine spirochetes: potential role in intestinal colonization. Infect. Immun. *62*: 4095–4099.

Muir, S., M.B. Koopman, S.J. Libby, L.A. Joens, F. Heffron and J.G. Kusters. 1992. Cloning and expression of a *Serpula* (*Treponema*) *hyodysenteriae* hemolysin gene. Infect. Immun. *60*: 529–535.

Nibbelink, S.K. and M.J. Wannemuehler. 1991. Susceptibility of inbred mouse strains to infection with *Serpula* (*Treponema*) *hyodysenteriae*. Infect. Immun. *59*: 3111–3118.

Nibbelink, S.K. and M.J. Wannemuehler. 1992. An enhanced murine model for studies of *Serpulina* (*Treponema*) *hyodysenteriae* pathogenesis. Infect. Immun. *60*: 3433–3436.

Nuessen, M.E., J.R. Birmingham and L.A. Joens. 1982. Biological activity of a lipopolysaccharide extracted from *Treponema hyodysenteriae*. Infect. Immun. *37*: 138–142.

Nuessen, M.E., L.A. Joens and R.D. Glock. 1983. Involvement of lipopolysaccharide in the pathogenicity of *Treponema hyodysenteriae*. J. Immunol. *131*: 997–999.

Ochiai, S., Y. Adachi and K. Mori. 1997. Unification of the genera *Serpulina* and *Brachyspira*, and proposals of *Brachyspira hyodysenteriae* comb. nov., *Brachyspira innocens* comb. nov. and *Brachyspira pilosicoli* comb. nov. Microbiol. Immunol. *41*: 445–452.

Ochiai, S., Y. Adachi and K. Mori. 1998. *In* Validation of publication of new names and new combinations previously effectively published outside the IJSEM. List no. 64. Int. J. Syst. Bacteriol. *48*: 327–328.

Olson, L.D. 1996. Enhanced isolation of *Serpulina hyodysenteriae* by using sliced agar media. J. Clin. Microbiol. *34*: 2937–2941.

Oxberry, S.L., D.J. Trott and D.J. Hampson. 1998. *Serpulina pilosicoli*, waterbirds and water: potential sources of infection for humans and other animals. Epidemiol. Infect. *121*: 219–225.

Park, N.Y., C.Y. Chung, A.J. McLaren, R.F. Atyeo and D.J. Hampson. 1995. Polymerase chain reaction for identification of human and porcine spirochaetes recovered from cases of intestinal spirochaetosis. FEMS Microbiol. Lett. *125*: 225–229.

Pettersson, B., M. Wang, C. Fellstrom, M. Uhlen, G. Molin, B. Jeppsson and S. Ahrne. 2000. Phylogenetic evidence for novel and genetically different intestinal spirochetes resembling *Brachyspira aalborgi* in the mucosa of the human colon as revealed by 16S rDNA analysis. Syst. Appl. Microbiol. *23*: 355–363.

Plaza, H., T.R. Whelchel, S.F. Garczynski, E.W. Howerth and F.C. Gherardini. 1997. Purified outer membranes of *Serpulina hyodysenteriae* contain cholesterol. J. Bacteriol. *179*: 5414–5421.

Pringle, M., J. Poehlsgaard, B. Vester and K.S. Long. 2004. Mutations in ribosomal protein L3 and 23S ribosomal RNA at the peptidyl transferase centre are associated with reduced susceptibility to tiamulin in *Brachyspira* spp. isolates. Mol. Microbiol. *54*: 1295–1306.

Pringle, M., A. Landen and A. Franklin. 2006. Tiamulin resistance in porcine *Brachyspira pilosicoli* isolates. Res. Vet. Sci. *80*: 1–4.

Råsbäck, T., D.S. Jansson, K.E. Johansson and C. Fellstrom. 2007. A novel enteropathogenic, strongly haemolytic spirochaete isolated from pig and mallard, provisionally designated '*Brachyspira suanatina*' sp. nov. Environ. Microbiol. *9*: 983–991.

Rohde, J., A. Rothkamp and G.F. Gerlach. 2002. Differentiation of porcine *Brachyspira* species by a novel nox PCR-based restriction fragment length polymorphism analysis. J. Clin. Microbiol. *40*: 2598–2600.

Rosey, E.L., M.J. Kennedy, D.K. Petrella, R.G. Ulrich and R.J. Yancey, Jr. 1995. Inactivation of *Serpulina hyodysenteriae flaA1* and *flaB1* periplasmic flagellar genes by electroporation-mediated allelic exchange. J. Bacteriol. *177*: 5959–5970.

Rosey, E.L., M.J. Kennedy and R.J. Yancey, Jr. 1996. Dual *flaA1 flaB1* mutant of *Serpulina hyodysenteriae* expressing periplasmic flagella is severely attenuated in a murine model of swine dysentery. Infect. Immun. *64*: 4154–4162.

Sagartz, J.E., D.E. Swayne, K.A. Eaton, J.R. Hayes, K.D. Amass, R. Wack and L. Kramer. 1992. Necrotizing typhlocolitis associated with a spirochete in rheas (*Rhea americana*). Avian Dis. *36*: 282–289.

Saheb, S.A., L. Massicotte and B. Picard. 1980. Purification and characterization of *Treponema hyodysenteriae* hemolysin. Biochimie *62*: 779–785.

Saheb, S.A., N. Daigneauly-Sylvestre and B. Picard. 1981. Comparative study of the hemolysins of *Treponema hyodysenteriae* and *Treponema innocens*. Curr. Microbiol. *5*: 87–90.

Sellwood, R. and A.P. Bland. 1997. Ultrastructure of intestinal spirochaetes. *In* Intestinal *Spirochaetes* in Domestic Animals and Humans (edited by Hampson and Stanton). CAB International, Wallingford, pp. 109–149.

Stanton, T.B. 1989. Glucose metabolism and NADH recycling by *Treponema hyodysenteriae*, the agent of swine dysentery. Appl. Environ. Microbiol. *55*: 2365–2371.

Stanton, T.B. 1992. Proposal to change the genus designation *Serpula* to *Serpulina* gen. nov. containing the species *Serpulina hyodysenteriae* comb. nov. and *Serpulina innocens* comb. nov. Int. J. Syst. Bacteriol. *42*: 189–190.

Stanton, T.B. 1997. Physiology of ruminal and intestinal spirochaetes. *In* Intestinal *Spirochaetes* in Domestic Animals and Humans (edited by Hampson and Stanton). CAB International, Wallingford, pp. 7–45.

Stanton, T.B. and C.P. Cornell. 1987. Erythrocytes as a source of essential lipids for *Treponema hyodysenteriae*. Infect. Immun. *55*: 304–308.

Stanton, T.B. and N.S. Jensen. 1993. Purification and characterization of NADH oxidase from *Serpulina* (*Treponema*) *hyodysenteriae*. J. Bacteriol. *175*: 2980–2987.

Stanton, T.B. and D.F. Lebo. 1988. *Treponema hyodysenteriae* growth under various culture conditions. Vet. Microbiol. *18*: 177–190.

Stanton, T.B. and R. Sellwood. 1999. Cloning and characteristics of a gene encoding NADH oxidase, a major mechanism for oxygen metabolism by the anaerobic spirochete, *Brachyspira* (*Serpulina*) *hyodysenteriae*. Anaerobe *5*: 539–546.

Stanton, T.B., N.S. Jensen, T.A. Casey, L.A. Tordoff, F.E. Dewhirst and B.J. Paster. 1991. Reclassification of *Treponema hyodysenteriae* and *Treponema innocens* in a new genus, *Serpula* gen. nov., as *Serpula hyodysenteriae* comb. nov. and *Serpula innocens* comb. nov. Int. J. Syst. Bacteriol. *41*: 50–58.

Stanton, T.B., B.L. Hanzelka and N.S. Jensen. 1995. Survey of intestinal spirochaetes for NADH oxidase by gene probe and by enzyme assay. Microb. Ecol. Health Dis. *8*: 93–100.

Stanton, T.B., D.J. Trott, J.I. Lee, A.J. McLaren, D.J. Hampson, B.J. Paster and N.S. Jensen. 1996. Differentiation of intestinal spirochaetes by multilocus enzyme electrophoresis analysis and 16S rRNA sequence comparisons. FEMS Microbiol. Lett. *136*: 181–186.

Stanton, T.B., E. Fournié-Amazouz, D. Postic, D.J. Trott, P.A. Grimont, G. Baranton, D.J. Hampson and I. Saint Girons. 1997. Recognition of two new species of intestinal spirochetes: *Serpulina intermedia* sp. nov. and *Serpulina murdochii* sp. nov.. Int. J. Syst. Bacteriol. *47*: 1007–1012.

Stanton, T.B., D. Postic and N.S. Jensen. 1998. *Serpulina alvinipulli* sp. nov., a new *Serpulina* species that is enteropathogenic for chickens. Int. J. Syst. Bacteriol. *48*: 669–676.

Stanton, T.B., E.L. Rosey, M.J. Kennedy, N.S. Jensen and B.T. Bosworth. 1999. Isolation, oxygen sensitivity, and virulence of NADH oxidase mutants of the anaerobic spirochete *Brachyspira* (*Serpulina*) *hyodysenteriae*, etiologic agent of swine dysentery. Appl. Environ. Microbiol. *65*: 5028–5034.

Stanton, T.B., E.G. Matson and S.B. Humphrey. 2001. *Brachyspira* (*Serpulina*) *hyodysenteriae gyrB* mutants and interstrain transfer of coumermycin A(1) resistance. Appl. Environ. Microbiol. *67*: 2037–2043.

Stephens, C.P. and D.J. Hampson. 1999. Prevalence and disease association of intestinal spirochaetes in chickens in eastern Australia. Avian Pathol. *28*: 447–454.

Stephens, C.P. and D.J. Hampson. 2001. Intestinal spirochete infections of chickens: a review of disease associations, epidemiology and control. Anim. Health Res. Rev. *2*: 83–91.

Stephens, C.P., S.L. Oxberry, N.D. Phillips, T. La and D.J. Hampson. 2005. The use of multilocus enzyme electrophoresis to characterise intestinal spirochaetes (*Brachyspira* spp.) colonising hens in commercial flocks. Vet. Microbiol. *107*: 149–157.

Sueyoshi, M., Y. Adachi and S. Shoya. 1987. Enteropathogenicity of *Treponema hyodysenteriae* in young chicks. Zentralbl. Bakteriol. Mikrobiol. Hyg. [A]. *266*: 469–477.

Swayne, D.E. 2003. Avian intestinal spirochetosis. *In* Diseases of Poultry, 11th edn (edited by Saif, Barnes, Fadly, Glisson, Mcdougald and Swayne). Iowa State University Press, Ames, IA, pp. 826–836.

Swayne, D.E. and A.J. McLaren. 1997. Avian intestinal spirochaetes and avian intestinal spirochaetosis. *In* Intestinal *Spirochaetes* in Domestic Animals and Humans (edited by Hampson and Stanton). CAB International, Wallingford, pp. 267–300.

Swayne, D.E., K.A. Eaton, J. Stoutenburg, D.J. Trott, D.J. Hampson and N.S. Jensen. 1995. Identification of a new intestinal spirochete with pathogenicity for chickens. Infect. Immun. *63*: 430–436.

Taylor, D.J. and T.J. Alexander. 1971. The production of dysentery in swine by feeding cultures containing a spirochaete. Br. Vet. J. *127*: 58–61.

Taylor, D.J. and D.J. Trott. 1997. Porcine intestinal spirochaetosis and spirochaetal colitis. *In* Intestinal *Spirochaetes* in Domestic Animals and Humans (edited by Hampson and Stanton). CAB International, Wallingford, pp. 211–241.

ter Huurne, A.A., M. van Houten, S. Muir, J.G. Kusters, B.A. van der Zeijst and W. Gaastra. 1992. Inactivation of a *Serpula* (*Treponema*) *hyodysenteriae* hemolysin gene by homologous recombination: importance of this hemolysin in pathogenesis in mice. FEMS Microbiol. Lett. *92*: 109–114.

ter Huurne, A.A., S. Muir, M. van Houten, B.A. van der Zeijst, W. Gaastra and J.G. Kusters. 1994. Characterization of three putative *Serpulina hyodysenteriae* hemolysins. Microb. Pathog. *16*: 269–282.

Thomson, J.R., W.J. Smith, B.P. Murray and S. McOrist. 1997. Pathogenicity of three strains of *Serpulina pilosicoli* in pigs with a naturally acquired intestinal flora. Infect. Immun. *65*: 3693–3700.

Tompkins, D.S., S.J. Foulkes, P.G. Godwin and A.P. West. 1986. Isolation and characterisation of intestinal spirochaetes. J. Clin. Pathol. *39*: 535–541.

Townsend, K.M., V.N. Giang, C. Stephens, P.T. Scott and D.J. Trott. 2005. Application of *nox*-restriction fragment length polymorphism for the differentiation of *Brachyspira* intestinal spirochetes isolated from pigs and poultry in Australia. J. Vet. Diagn. Invest. *17*: 103–109.

Trott, D.J., C.R. Huxtable and D.J. Hampson. 1996a. Experimental infection of newly weaned pigs with human and porcine strains of *Serpulina pilosicoli*. Infect. Immun. *64*: 4648–4654.

Trott, D.J., T.B. Stanton, N.S. Jensen, G.E. Duhamel, J.L. Johnson and D.J. Hampson. 1996b. *Serpulina pilosicoli* sp. nov.: the agent of porcine intestinal spirochetosis. Int. J. Syst. Bacteriol. *46*: 206–215.

Trott, D.J., T.B. Stanton, N.S. Jensen and D.J. Hampson. 1996c. Phenotypic characteristics of *Serpulina pilosicoli* the agent of intestinal spirochaetosis. FEMS Microbiol. Lett. *142*: 209–214.

Trott, D.J., B.G. Combs, A.S. Mikosza, S.L. Oxberry, I.D. Robertson, M. Passey, J. Taime, R. Sehuko, M.P. Alpers and D.J. Hampson. 1997a. The prevalence of *Serpulina pilosicoli* in humans and domestic animals in the Eastern Highlands of Papua New Guinea. Epidemiol. Infect. *119*: 369–379.

Trott, D.J., N.S. Jensen, I. Saint Girons, S.L. Oxberry, T.B. Stanton, D. Lindquist and D.J. Hampson. 1997b. Identification and characterization of *Serpulina pilosicoli* isolates recovered from the blood of critically ill patients. J. Clin. Microbiol. *35*: 482–485.

Trott, D.J., S.L. Oxberry and D.J. Hampson. 1997c. Evidence for *Serpulina hyodysenteriae* being recombinant, with an epidemic population structure. Microbiology *143*: 3357–3365.

Trott, D.J., D.P. Alt, R.L. Zuerner, M.J. Wannemuehler and T.B. Stanton. 2001. The search for *Brachyspira* outer membrane proteins that interact with the host. Anim. Health Res. Rev. *2*: 19–30.

Trott, D.J., D.P. Alt, R.L. Zuerner, D.M. Bulach, M.J. Wannemuehler, J. Stasko, K.M. Townsend and T.B. Stanton. 2004. Identification and cloning of the gene encoding BmpC: an outer-membrane lipoprotein associated with *Brachyspira pilosicoli* membrane vesicles. Microbiology *150*: 1041–1053.

Turner, A.K. and R. Sellwood. 1997. Extracellular DNA from *Serpulina hyodysenteriae* consists of 6.5 kbp random fragments of chromosomal DNA. FEMS Microbiol. Lett. *150*: 75–80.

Waters, W.R., B.A. Pesch, R. Hontecillas, R.E. Sacco, F.A. Zuckermann and M.J. Wannemuehler. 2000. Cellular immune responses of pigs induced by vaccination with either a whole-cell sonciate or pepsin-digested *Brachyspira* (*Serpulina*) *hyodysenteriae* bacterin. Vaccine *18*: 711–719.

Wright, J.C., G.R. Wilt, R.B. Reed and T.A. Powe. 1989. Use of an enzyme-linked immunosorbent assay for detection of *Treponema hyodysenteriae* infection in swine. J. Clin. Microbiol. *27*: 411–416.

Zuerner, R.L. 1997. Genetic organization in spirochaetes. *In* Intestinal *Spirochaetes* in Domestic Animals and Humans (edited by Hampson and Stanton). CAB International, Wallingford, pp. 63–89.

Zuerner, R.L. and T.B. Stanton. 1994. Physical and genetic map of the *Serpulina hyodysenteriae* B78^T chromosome. J. Bacteriol. *176*: 1087–1092.

Zuerner, R.L., T.B. Stanton, F.C. Minion, C. Li, N.W. Charon, D.J. Trott and D.J. Hampson. 2004. Genetic variation in *Brachyspira*: chromosomal rearrangements and sequence drift distinguish *B. pilosicoli* from *B. hyodysenteriae*. Anaerobe *10*: 229–237.

Family III. **Brevinemataceae** fam. nov.

Bruce J. Paster

Bre.vi.ne.ma.ta.ce'ae. N.L. fem. n. *Brevinema -atos* type genus of the family; *-aceae* ending to denote a family; N.L. fem. pl. n. *Brevinemataceae* the *Brevinema* family.

The family *Brevinemataceae* was circumscribed for this volume on the basis of phylogenetic analysis of 16S rRNA gene sequences. The family contains only one genus, *Brevinema*.

Description is the same as for the genus *Brevinema*.

Type genus: **Brevinema** Defosse, Johnson, Paster, Dewhirst and Fraser 1995, 83[VP].

Genus I. **Brevinema** Defosse, Johnson, Paster, Dewhirst and Fraser 1995, 83[VP]

Bruce J. Paster

Bre.vi. ne'ma. L. adj. *brevis* short; Gr. neut. n. *nema* thread; N.L. neut. n. *Brevinema* a short thread.

Helical cells are 0.2–0.3 µm in diameter by 4–5 µm in length, displaying one to two helical turns. Irregular wavelengths of the cells range from 2 to 3 µm. Sheathed periplasmic flagella are in a 1:2:1 arrangement. No cytoplasmic tubules have been observed. Cells are motile by flexing, rotation, and translation. **Microaerophilic**, **host-associated**, isolated from blood and other tissues of short-tailed shrews (*Blarina brevicauda*) and white-footed mice (*Peromyscus leucopus*). Infectious for laboratory mice and Syrian hamsters.

DNA G+C content (mol%): 34–36 (T_m).

Type species: **Brevinema andersonii** Defosse, Johnson, Paster, Dewhirst and Fraser 1995, 83[VP].

Further descriptive information

Strains of *Brevinema andersonii* are homogeneous as based on enzymic, protein profile, and immunoblot data. Furthermore, there are no significant differences in fatty acid composition among the strains analyzed (Defosse et al., 1995). The major fatty acid components of *Brevinema* cells are myristic acid ($C_{14:0}$), palmitic acid ($C_{16:0}$), and oleic acid ($C_{18:1}$) with smaller amounts of stearic acid ($C_{18:0}$) and linoleic acid ($C_{18:2}$). Several fatty acids are present at low levels (less than 1%). Restriction enzyme analysis and SDS–PAGE patterns also demonstrate little or no differences among strains. Consequently, it was suggested that *Brevinema andersonii* represents a genetically homologous group, despite the diverse hosts and different geographic origins (Defosse et al., 1995).

Enrichment and isolation procedures

Brevinema andersoni has been isolated from blood and other tissues of the short-tailed shrew and the white-footed mouse using Shrew-Mouse Spirochete medium under microaerophilic conditions (Defosse et al., 1995). Fetal bovine serum, reducing agents, and peptones are required for growth. Neither supplemental bovine serum albumin, *N*-acetylglucosamine, nor pyruvate is required for growth. Optimal growth is at 30–34°C at pH 7.4 with a generation time of 11–14 h.

Differentiation of the genus *Brevinema* from other genera

Brevinema andersonii is serologically distinct from other spirochetes (Anderson et al., 1987; Defosse et al., 1995). There is little or no DNA–DNA hybridization between *Brevinema andersonii* and members of other spirochetal genera using Southern blot analysis (LeFebvre and Perng, 1989; LeFebvre et al., 1989). At the species level, *Brevinema andersonii* is differentiated using restriction enzyme analysis, SDS-PAGE, or fatty acid composition.

Taxonomic comments

Brevinema andersoni is the only named species for the genus. Based on 16S rRNA gene sequence comparisons, *Brevinema* belongs within the family *Spirochetaceae* and is clearly distinct from the other genera of spirochetes (Paster and Dewhirst, 2000). 16S rRNA sequences of *Brevinema andersonii* do not possess a 20- to 30-base extension at the 5′ end, which is typical of 16S rRNA sequences of species of *Treponema*, *Spirocheta*, *Leptospira*, and *Leptonema* (Defosse et al., 1995; Paster et al., 1991).

List of species of the genus *Brevinema*

1. **Brevinema andersonii** Defosse, Johnson, Paster, Dewhirst and Fraser 1995, 83[VP]

an.der.so'ni.i. N.L. masc. gen. n. *andersonii* of Anderson, named for John F. Anderson, who first described the organism.

The characteristics are as described for the genus. Chemoorganotrophic. Microaerophilic and catalase-negative. Growth occurs in modified BSK medium at an optimal growth temperature of 30–34°C and an optimal pH of 7.4; under these conditions the generation time is 11–14 h. Does not grow at 25°C. Exhibits the following enzymic activities: C4, C5, C6, C8, C9, and C10 esterases, C4 esterase lipase, alkaline phosphatase, acid phosphatase, and β-glucuronidase.

Source: tissues of a short-tailed shrew (*Blarina brevicauda*) captured in West Haven, Connecticut, USA (Anderson et al., 1987).

DNA G+C content (mol%): 34–36 (T_m).

Type strain: ATCC 43811, CT11616.

Sequence accession no. (16S rRNA gene): M59179.

References

Anderson, J.F., R.C. Johnson, L.A. Magnarelli, F.W. Hyde and T.G. Andreadis. 1987. New infectious spirochete isolated from short-tailed shrews and white-footed mice. J. Clin. Microbiol. *25*: 1490–1494.

Defosse, D.L., R.C. Johnson, B.J. Paster, F.E. Dewhirst and G.J. Fraser. 1995. *Brevinema andersonii* gen. nov., sp. nov., and infectious spirochete isolated from the short-tailed shrew (*Blarina brevicauda*) and the white-footed mouse (*Peromyscus leucopus*). Int. J. Syst. Bacteriol. *45*: 78–84.

LeFebvre, R.B. and G.C. Perng. 1989. Genetic and antigenic characterization of *Borrelia coriaceae*, putative agent of epizootic bovine abortion. J. Clin. Microbiol. *27*: 389–393.

LeFebvre, R.B., G.C. Perng and R.C. Johnson. 1989. Characterization of *Borrelia burgdorferi* isolates by restriction endonuclease analysis and DNA hybridization. J. Clin. Microbiol. *27*: 636–639.

Paster, B. and F. Dewhirst. 2000. Phylogenetic foundation of spirochetes. J. Mol. Microbiol. Biotechnol. *2*: 341–344.

Paster, B.J., F.E. Dewhirst, W.G. Weisburg, L.A. Tordoff, G.J. Fraser, R.B. Hespell, T.B. Stanton, L. Zablen, L. Mandelco and C.R. Woese. 1991. Phylogenetic analysis of the spirochetes. J. Bacteriol. *173*: 6101–6109.

Family IV. **Leptospiraceae** Hovind-Hougen 1979, 245[AL] emend. Levett, Morey, Galloway, Steigerwalt and Ellis 2005, 1499

Richard L. Zuerner

Lep.to.spi.ra.ce′ae. N.L. fem. n. *Leptospira* type genus of the family; *-aceae* ending to denote a family; N.L. fem. pl. n. *Leptospiraceae* the *Leptospira* family.

Helical cells, 0.1–0.3 µm in diameter and 3.5–20 µm in length. Cells have right-handed helical conformation. Cells at rest and those that are fixed have hooked ends. Actively motile cells have a spiral anterior end and a hook at the posterior end of the cell. One periplasmic flagellum (historically also referred to as axial filaments, endoflagella, or flagella), is inserted subterminally at each end of cell, but flagella rarely overlap in the center of the cell. Periplasmic flagella lie along the helix axis. **The diamino acid in peptidoglycan is α,ε-diaminopimelic acid.**

Obligate aerobes or microaerophilic. Chemoorganotrophic. **Utilize long-chain fatty acids and fatty alcohols as carbon and energy sources.** Do not use carbohydrates/amino acids as carbon or energy sources.

Free-living or in association with animal and human hosts. Some species are pathogenic.

Species examined by 16S rRNA sequence analysis are distinct from members of *Spirochetaceae.*

DNA G+C content (mol%): 33–53.

Type genus: **Leptospira** Noguchi 1917, 755[AL].

Key to the genera of the family *Leptospiraceae*

1. DNA G+C content (mol%) is 33–43% (T_m and genomic sequence analysis). Cells are 0.1 µm in diameter and 6–20 µm in length. Aerobe or microaerophilic. Long-chain fatty acids and long-chain fatty alcohols serve as carbon and energy sources. Free-living in aquatic environments, including mud, sediments, and water of ponds, lakes, and streams. May be found in fresh water and marine environments. Some species are found in association with animals. Some species are pathogenic.
   →Genus I. *Leptospira*

2. DNA G+C content (mol%) is 54% (T_m and Bd).Cells are 0.1–0.2 µm in diameter and 13–15 µm in length, with wavelength of 0.7 µm. Aerobe. Long-chain fatty acids and long-chain fatty alcohols serve as carbon and energy sources. Can grow on trypticase broth. Free-living in aquatic environments, including mud, sediments, and water of ponds, lakes, and streams. May be found in fresh water and marine environments. Type strain was found in association with animals (cattle). Nonpathogenic for hamsters.
   →Genus II. *Leptonema*

3. DNA G+C content (mol%) is 48% (T_m and Bd). Cells are 0.3 µm in diameter and 3.5–7.5 µm in length, with wavelength of 0.3–0.5 µm. Obligate aerobe. One periplasmic flagellum. Long-chain fatty acids and long-chain fatty alcohols serve as carbon and energy sources. Oxidase-positive. Found in tap water and in association with animals. Not pathogenic for hamsters.
   →Genus III. *Turneriella*

Genus I. **Leptospira** Noguchi 1917, 755[AL] emend. Faine and Stallman 1982, 461

Richard L. Zuerner

Lep.to.spi′ra. Gr. adj. *leptos* thin, narrow, fine; L. fem. n. *spira* a coil, helix; N.L. fem. n. *Leptospira* a thin helix or coil, referring to the morphology of the bacterium.

Leptospira are **long, thin, flexible rods, 0.1 µm in diameter and 6–12 µm in length**, with a regular right-handed helical coiling pattern (Carleton et al., 1979). These bacteria are unicellular but may be observed as dividing pairs or short chains of three or more cells in actively growing cultures. Resting stages are not known, but long term survival in water, with the appearance of aggregates has been described (Trueba et al., 2004). Spherically shaped cells form under unfavorable growth conditions. Bacteria stain as Gram-negative. Due to the small diameter of these bacteria, **unstained cells are not visible by bright-field microscopy.** Dark-field or phase-contrast microscopy is required for visualization of unstained cells. These are highly motile aerobic or microaerophilic bacteria. Optimum growth temperature is 28–30°C, with a generation time of 6–16 h, although **many**

primary pathogenic isolates may grow slower. Chemoorganotrophic bacteria that consume long-chain fatty acids and alcohols as primary carbon and energy sources, and carry out respiration with oxygen as the terminal electron acceptor. Optimal growth occurs in semi-solid (0.1–0.2%) agar media. Growth on 1–2% solid agar results in the formation of clear to turbid surface or subsurface colonies. Colony formation is enhanced by the addition of pyruvate. Oxidase, catalase, and/or peroxidase-positive. Some strains are β-hemolytic. Some strains are pathogenic for humans and animals, while other strains are saprophytic and found in freshwater and marine environments. **The genus *Leptospira* forms a deep unique branch of spirochetes, separate from other genera based on comparison of 16S rRNA gene sequences.** Species are differentiated by DNA–DNA relative reassociation analysis and by unique sequence polymorphisms in 16S rRNA.

DNA G+C content (mol%): 35–43.

Type species: **Leptospira interrogans** (Stimson 1907) Wenyon 1926, 1281 emend. Faine and Stallman 1982, 462 (*Spirochaeta interrogans* Stimson 1907, 541).

Further descriptive information

Taxonomic history. The taxonomy of *Leptospira* has undergone substantial revisions in the past 20 years with the use of 16S rRNA sequence comparison and DNA–DNA reassociation studies. DNA–DNA reassociation studies helped differentiate pathogenic and saprophytic species (Haapala et al., 1969). Subsequent studies by Brendle et al. (1974), Ramadass et al. (1992), and Yasuda et al. (1987) helped to revise and better define *Leptospira* species. The key paper by Brenner et al. (1999) compared 303 strains of *Leptospira*, leading to a refined definition of 12 different species and identification of five new genomospecies, one of which was given the designation *Leptospira alexanderi*. One of the species included in the Brenner et al. (1999) study was *Leptospira parva*, originally described by Hovind-Hougen et al. (1981). Due to extensive DNA sequence and 16S rRNA differences from other *Leptospira* species and to the difference from *Leptonema illini*, it was placed in a new genus of *Leptospiraceae*, *Turneriella parva* (Levett et al., 2005). Three new *Leptospira* species have been described, *Leptospira broomii* (Levett et al., 2006), *Leptospira licerasiae* (Matthias et al., 2008), and *Leptospira wolffii* (Slack et al., 2008). Thus, at the current time, the genus *Leptospira* contains 15 named species and the Subcommittee on the taxonomy of *Leptospiraceae* has recommended names for four genomospecies identified by Brenner et al., (1999) (Levett and Smythe, 2008).

Following the initial use of 16S rRNA sequence comparisons to define taxonomic relationships among spirochete genera (Paster et al., 1984, 1991); Hookey et al. (1993) expanded these analyses to help differentiate *Leptospira* species and differentiate the genera *Leptonema* and *Leptospira*. Comparison of 16S rRNA sequences and DNA–DNA reassociation studies suggest the presence of three groups of pathogenic species; Group I contains *Leptospira interrogans* and *Leptospira kirschneri*, Group II contains *Leptospira weilii*, *Leptospira borgpetersenii*, and *Leptospira santarosai*, and Group III contains *Leptospira noguchi* and *Leptospira meyeri* (Hookey, 1993). There are five species having intermediate pathogenicity: *Leptospira fainei*, *Leptospira inadai*, *Leptospira broomi*, *Leptospira licerasiae*, and *Leptospira wolffii*. The latter three species were recently recognized by the Subcommittee on the taxonomy of *Leptospiraceae* (Levett and Smythe, 2008). Saprophytic species include *Leptospira biflexa* and *Leptospira wolbachii* (Yasuda et al., 1987). The status of *Leptospira meyeri* is unclear because strains fit into both saprophytic (Yasuda et al., 1987) and pathogenic groups (Hookey et al., 1993; Victoria et al., 2008). Many of the diverse free-living strains of *Leptospira* described by Ramadass et al. (1992) and halophiles described by Cinco et al. (1975) were not included in the study by Brenner et al. (1999), so the taxonomic status of these strains is unknown. Brenner et al. (1999) discovered and described four *Leptospira* genomospecies tentatively named *Leptospira alstonii* (genomospecies 1), *Leptospira vanthielii* (genomospecies 3), *Leptospira terpstrae* (genomospecies 4), and *Leptospira yanagawae* (genomospecies 5) (Levett and Smythe, 2008).

Cell morphology, motility, and ultrastructure. *Leptospira* cells are typically visualized by dark-field light microscopy due to their long slender cell dimensions. The basic cell morphology of *Leptospira* resembles other members of the order *Spirochetales*; *Leptospira* species typically appear as long helical cells with loose coils (Figure 99). One flagellum is inserted subterminally at each end of the cell and lies along the helix axis in the periplasmic space (Goldstein et al., 1996). Scanning electron microscopic analysis shows these cells form right-handed helices, and helical handedness may affect cell motility (Carleton et al., 1979). During translational movement, the trailing end of the cell bends to form a hook thought to function as a propeller, with the leading end maintaining a spiral twist, guiding the forward movement of the bacterium (Charon et al., 1984; Charon et al., 1981). Translational movement increases with the viscosity of the medium (Greenberg and Canale-Parola, 1977a, b), a finding that suggests these bacteria are viscotaxic (Petrino and Doetsch, 1978).

Leptospira have a typical spirochete ultrastructure, having an outer membrane surrounding a protoplasmic cylinder. The cell wall contains peptidoglycan containing the diamino acid alpha, epsilon diaminopimelic acid (Azuma et al., 1975). Two flagella (also referred to as axial filaments) are inserted subterminally at each end of the cell, and wrap around the protoplasmic cylinder in the periplasmic space. The ends of the flagella rarely overlap near the midpoint of the cell. The outer membrane is

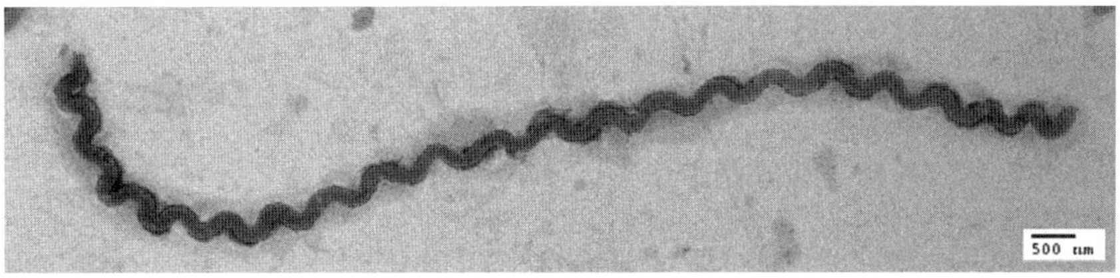

FIGURE 99. Transmission electron micrograph of *Leptospira biflexa* serovar Patoc. Bar = 500 nm.

loosely attached to the cell and easily removed by detergents (Haake et al., 1991; Zuerner et al., 1991). Rapid displacement of antibody-coated latex beads adhered to *Leptospira* cells suggests the outer membrane is quite fluid (Charon et al., 1981).

Outer membrane composition. Protein composition of the *Leptospira* outer membrane varies depending on virulence and growth conditions. For example, few proteins have been detected on the surface of virulent *Leptospira kirschneri*, while outer-membrane protein density is substantially higher in avirulent strains (Haake et al., 1991). LipL36 is expressed under normal *in vitro* growth conditions, but down-regulated during mammalian infection (Haake et al., 1998). Changes in osmolarity also alter outer-membrane protein expression. For example, expression of LigA and LigB, two bacterial immunoglobulin-like proteins increase as osmolarity rises (Choy et al., 2007; Matsunaga et al., 2007a, b; Matsunaga et al., 2005).

Under normal *in vitro* growth, the most predominant proteins on the surface of pathogenic *Leptospira* are LipL32, LipL21, and LipL41 (Cullen et al., 2002, 2003, 2005). LipL32 is the major outer-membrane protein and is unique to pathogenic *Leptospira* species. LipL32 is heat labile and, in the absence of calcium ions, is degraded by an endogenous protease to two smaller peptides (Haake et al., 2000; Zuerner et al., 1991). Although the original report describing LipL21 identified this protein as unique to pathogenic *Leptospira* species (Cullen et al., 2003), genome analysis of the *Leptospira biflexa* genome revealed the presence of a gene encoding a protein with extensive similarity to LipL21 (Picardeau et al., 2008). Protein OmpLI is an integral transmembrane porin (Haake et al., 1993). Access to genomic sequences for three species of *Leptospira*, including two pathogenic species, has resulted in cloning and analysis of many previously unknown outer-membrane proteins. Of particular interest are proteins from pathogenic strains that interact with the host. LigA and Lsa24 (also referred to as LfhA) bind extracellular matrix proteins (Barbosa et al., 2006; Choy et al., 2007), and Lsa24 binds complement factor H (Verma et al., 2006).

The leptospiral LPS is thought to be the primary antigen associated with determining serovar, and variation in LPS biosynthetic genes may occur via lateral transfer (de la Pena-Moctezuma et al., 1999) or mutation (Zuerner and Trueba, 2005), resulting in expression of antigenically distinct LPS. Leptospiral LPS is not as complex in structure as often found in typical Gram-stain-negative bacteria, and does not resolve as a ladder of variable sized products during polyacrylamide gel electrophoresis. Purified leptospiral LPS activates mouse cells via TLR2 and TLR4 (Nahori et al., 2005; Werts et al., 2001). Purified *Leptospira interrogans* Lipid A, the membrane anchor for LPS, does not induce *Limulus* amebocyte lysates to gel (Nahori et al., 2005; Que-Gewirth et al., 2004). The *Leptospira interrogans* lipid A has unusual features including a methylated 1-phosphate group (Nahori et al., 2005).

Nutrition and growth conditions. *Leptospira* are obligate aerobes but vary greatly in their ability to grow *in vitro*. Although many *Leptospira* can grow in a defined medium of relatively simple composition referred to as EMJH (Ellinghausen and McCullough, 1965; Johnson and Harris, 1967a), addition of rabbit serum is often necessary for growth of many pathogenic strains, and early attempts to grow *Leptospira* utilized media rich in rabbit serum (Fletcher, 1928; Stuart, 1946). Growth yields in EMJH media vary from 3 to 5×10^8 to $\sim 10^{10}$ cells/ml, with generation times of 6–16 h, depending on the strain. *Leptospira* consume long chain fatty acids or alcohols as primary carbon and energy sources, which are metabolized by beta-oxidation (Baseman and Cox, 1969; Henneberry and Cox, 1970). Fatty acids are commonly added to *Leptospira* media preparations through the addition of polysorbate (e.g., Tweens), nonionic detergents derived from long chain fatty acids. Tween 80, predominantly containing oleic acid, is used most commonly, but addition of Tween 40 (primarily palmitic acid) and/or Tween 60 (primarily steric acid) can be useful to support growth of fastidious strains of *Leptospira* (e.g., *Leptospira borgpetersenii* serovar Hardjo) (Ellinghausen, 1983). Leptospiral media often contains purified bovine serum albumin to bind free fatty acids and reduce their toxicity. Protein-free media for growth of some strains of *Leptospira* have been reported (Bey and Johnson, 1978; Shenberg, 1967), but these media do not support prolonged propagation of many pathogenic species. Vitamins B_1 and B_{12} are typically included in the growth medium for *Leptospira*. However, *Leptospira* synthesize heme (Guegan et al., 2003), consistent with possessing vitamin B_{12} biosynthesis genes. Furthermore, *de novo* synthesis of B_{12} may be required as genomic analysis suggests *Leptospira* may be deficient in the ability to transport exogenous B_{12} (Rodionov et al., 2003). Ammonium is provided by addition of ammonium salts to the media as amino acids do not appear to be a significant source of ammonia for *Leptospira* (Johnson and Rogers, 1964c). *Leptospira* appear to synthesize all amino acids except isoleucine by standard pathways (Charon et al., 1974). Isoleucine is synthesized either by the standard pathway, being derived from threonine, or by a pathway involving condensation of pyruvate with acetyl-S-coenzyme A (Charon et al., 1974). *Leptospira* synthesize isoleucine either by a combination of threonine and pyruvate pathways or by exclusive use of the threonine or pyruvate pathways (Westfall et al., 1983). Exogenous purines are incorporated by *Leptospira* (Johnson and Rogers, 1964a). However, *Leptospira* do not incorporate exogenous pyrimidines, thus incorporation of pyrimidine analogs such as 5-fluorouracil is a useful method for suppressing the growth of other bacteria and providing an enrichment method for isolating *Leptospira* (Johnson and Rogers, 1964b). Iron is required for growth of *Leptospira* (Faine, 1959) and is usually supplied as iron sulfate in the media. Recent analysis of iron transport in *Leptospira* revealed that iron sulfate-free media supports growth of *Leptospira interrogans*, presumably by supplying trace iron, but *Leptospira biflexa* requires addition of exogenous iron to the media for growth (Louvel et al., 2006).

Leptospira are rarely grown on agar plate media due to their slow growth. Colony formation can be observed in about 10 days for saprophytic species, but it may take 6–8 weeks for pathogenic species to form colonies. Colonies are diffuse, non-pigmented, clear to turbid, and may form below the surface. Growth, especially on solid agar media, is enhanced by the addition of pyruvate (Johnson et al., 1973). Optimal growth conditions include pH 7.2–7.4, 30°C, and media containing low salt (17–60 mM), although halophilic strains require 0.22–0.44 M for growth (Cinco and Cociancich, 1975).

Genome properties. Physical mapping of *Leptospira interrogans* serovars Pomona and Icterohemorrhagiae using pulsed-field gel electrophoresis (PFGE) and DNA hybridization revealed the presence of two chromosomal replicons in this species (Baril et al., 1992; Zuerner, 1991). Comparative mapping revealed extensive rearrangements that alter genetic organization within the same species (Zuerner et al., 1993). The genomes of four pathogenic *Leptospira* strains have been published: *Leptospira interrogans* serovar Copenhageni (Nascimento et al., 2004) and *Leptospira interrogans* serovar Lai (Ren et al., 2003), and two strains of *Leptospira borgpetersenii* serovar Hardjo (Bulach et al., 2006). Genome size ranges from 3.9 to 4.7 Mbp, consisting of one large and one small chromosome. Genome sequencing of the free-living *Leptospira biflexa* serovar Patoc shows the presence of three stable replicons, corresponding to the large and small chromosomes of pathogenic *Leptospira*, and a third replicon of unknown function (Picardeau et al., 2008). Comparative analysis of *Leptospira interrogans* serovar Lai and Copenhageni confirm previous mapping studies that suggest recombination between transposable elements contribute to changes in genetic organization by identifying an inversion flanked at both ends by insertion sequences that differentiates genetic organization of these two serovars (Nascimento et al., 2004). Extensive rearrangements differentiate the organization of *Leptospira interrogans* from *Leptospira borgpetersenii*, and strains of *Leptospira borgpetersenii* are likewise differentiated by the presence of chromosomal rearrangements. Approximately 7% of the *Leptospira borgpetersenii* serovar Hardjo genome is comprised of transposable elements, and nearly 17% of pseudogenes in *Leptospira borgpetersenii* are associated with insertion sequences (IS), leading to the conclusion that the genome of this species is undergoing IS-mediated genome erosion (Bulach et al., 2006). A primary consequence of genome erosion in *Leptospira borgpetersenii* is the apparent loss of viability in water (Bulach et al., 2006), thereby limiting dissemination of viable organisms via the environment, a common means of disease transmission for *Leptospira interrogans*. Additional variability in *Leptospira interrogans* may result from lateral transfer of genomic islands (Bourhy et al., 2007).

An unusual feature of *Leptospira* genomes is the organization of rRNA genes (Fukunaga and Mifuchi, 1989a, b). Individual rRNA genes are not linked closely to each other in *Leptospira*, but are dispersed around the large chromosome (Zuerner et al., 1993). There are two copies each of the 16S and 23S rRNA genes in *Leptospira*; there is one copy of the 5S rRNA gene in pathogenic *Leptospira*, but two copies of the 5S rRNA gene in free-living *Leptospira* (Fukunaga and Mifuchi, 1989b).

Phage, plasmid, and transposable elements and genetic manipulation in *Leptospira*. Although naturally occurring mechanisms for genetic exchange are not yet defined for *Leptospira*, several recent discoveries and advances in genetic manipulation of members of this genus suggest lateral transfer of DNA contributes to genetic variation. Bacteriophages have been detected by electron microscopy of *Leptospira* cultures. Saint Girons et al. (Saint Girons et al., 1990) described the first successful isolation and propagation of bacteriophage that infect *Leptospira biflexa*. The replication origin of bacteriophage LE1 can direct autonomous replication of a plasmid in *Leptospira biflexa*, and this discovery led to development of an *Escherichia coli*- *Leptospira biflexa* shuttle vector facilitating genetic manipulation of *Leptospira* (Girons et al., 2000). In a similar manner, discovery of genetic islands in *Leptospira interrogans* led to development of a limited-host range shuttle vector (Bourhy et al., 2007). Autonomous replication of these genetic islands may contribute to lateral transfer of DNA in nature and affect pathogenicity by spreading virulence determinants (Bourhy et al., 2007). Recent demonstration of conjugative transfer of a shuttle vector from *Escherichia coli* to *Leptospira* is consistent with evidence for lateral transfer of genetic material into *Leptospira* (Picardeau, 2008).

Intervening sequences are inserted into the 23S rRNA gene in some *Leptospira* species (Ralph and McClelland, 1993). During post-transcriptional processing, the intervening sequences are excised without ligation, therefore bacterial strains containing these elements lack an intact 23S rRNA and instead have 14S and 17S rRNA species (Hsu et al., 1990).

As noted above, genetic rearrangements occur frequently in *Leptospira* strains, and these rearrangements are often associated with transposable elements including IS-elements. There are several different IS-like elements found in *Leptospira* (Boursaux-Eude et al., 1995; Bulach et al., 2006; Nascimento et al., 2004; Ralph and McClelland, 1993; Zuerner, 1994; Zuerner and Huang, 2002; Zuerner and Trueba, 2005) that share substantial sequence similarity to elements found in diverse bacterial genera, yet lateral transfer of these elements has not been demonstrated. The presence of high numbers (~90 copies) of IS*1533* in *Leptospira borgpetersenii* serovar Hardjo is hypothesized to have a prominent role in genome degradation by generating nonfunctional pseudogenes that restrict serovar Hardjo to a host to host transmission cycle (Bulach et al., 2006). Transposition of the genetic elements derived from the eukaryotic transposon *himar*1 provides a means to generate selectable random mutants in *Leptospira* species (Bourhy et al., 2005), opening up new possibilities in genetic manipulation and analysis in this genus.

Serotype analysis. *Leptospira* are antigenically diverse and are clustered by serological analysis into serogroups. Serogroups are further subdivided into serovars, which are considered the primary serotype taxon. Serovar determination is achieved by agglutination reactions using a microscopic agglutination test (Cole et al., 1973; Ryu, 1970). Serovars are differentiated by cross-absorption agglutination reactions using high titer sera. Two strains are designated different serovars if, after sufficient cross-absorption with each sera using the heterologous strain, at least one of the sera retains 10% or more of the initial homologous agglutination titer (Stallman, 1984). Use of monoclonal antibodies can also help differentiate strains belonging to different serovars (Terpstra et al., 1985; World Health Organization, 2003). Serovar designation has historical relevance and may be essential in understanding the relationships between these bacteria and their normal maintenance hosts; maintenance hosts appear to have a stable relationship with bacteria in the same serovar but may undergo acute infection with a different serovar (Faine et al., 1999). Likewise, the same strain in its normal maintenance host rarely displays signs of acute infection, but accidental infection by that strain of a non-maintenance host can result in severe acute infection (Faine et al., 1999). Complicating the importance and relevance of serovar designation is the discovery of several serovars represented by strains belonging to different *Leptospira* species (Levett, 2001).

Antibiotic sensitivity. *Leptospira* exhibit *in vitro* sensitivity to a large number of antibiotics, including beta-lactams (penicillin, ampicillin, and amoxycillin), rifampin, tetracycline and doxycycline, cephalosporins (cefotaxime and ceftizoxime), aminoglycosides including streptomycin (Oie et al., 1983), macrolides (erythromycin and azithromycin), and fluoroquinolones including ciprofloxacin (Hospenthal and Murray, 2003). Immediate treatment of suspected infections is recommended. Traditionally, leptospirosis has been treated with intravenous penicillin for severe cases, or oral antibiotics including amoxycillin, ampicillin, doxycycline, and erythromycin (Terpstra et al., 1985; World Health Organization, 2003). There is limited clinical data on effective use of newer antibiotics to treat leptospirosis.

Environmental range. Pathogenic *Leptospira* species cause leptospirosis, one of the most widespread zoonosis known. Wild and domesticated animals are reservoirs of infection, and animal to human transmission may occur either through direct exposure to blood or urine from an infected animal, or indirectly from urine-contaminated water. After infection, the bacteria disseminate via the blood and concentrate in the kidney. *Leptospira* are voided in the urine (Figure 100), facilitating infection of new hosts (Faine et al., 1999). Leptospirosis ranges in severity from a mild influenza-like infection to a severe acute infection often resulting in death from organ failure. Chronically infected animal maintenance hosts often do not exhibit clinical signs of infection. Several *Leptospira* species exhibit an intermediate pathogenicity in humans and animal hosts, and may be opportunistic pathogens.

The normal environment for saprophytic *Leptospira* species is water or moist soil (Henry and Johnson, 1978). Cinco et al. (1975) have reported isolation of halophilic *Leptospira*, but these have not been subjected to taxonomic characterization. Recent surveys using 16S rRNA sequence analysis to assess the diversity of microbes in different environments suggest *Leptospira* and *Leptonema* are widely distributed, including in petroleum contaminated soil (Kasai et al., 2005), marine sediments (GenBank accession EU386041), and geothermal regions (GenBank accession EF205465).

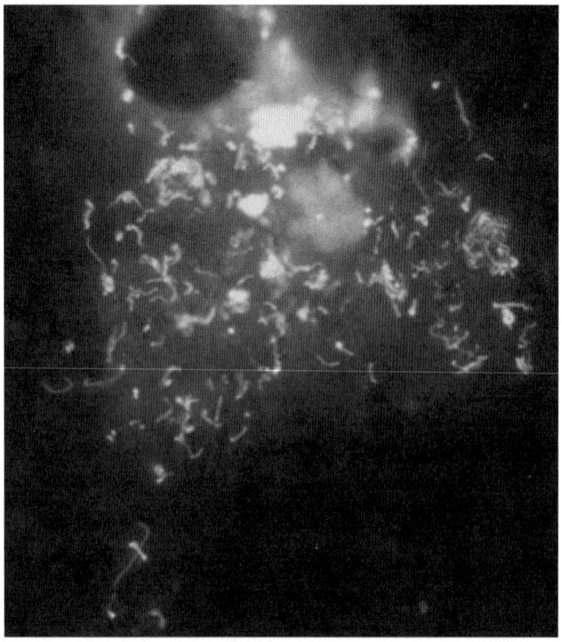

FIGURE 100. *Leptospira* in urine. A sample of urine was obtained from a dog exhibiting clinical signs of leptospirosis and stained with FITC-conjugated anti-*Leptospira* rabbit sera, counterstained with Flazo-orange, then visualized using a 60× objective on an Olympus fluorescent microscope (final magnification 600×). Under these conditions, the bacteria appear green. Individual cells and cell masses are visible.

Leptospirosis and virulence-associated traits. Virulence determinants of pathogenic *Leptospira* remain poorly characterized. This is due, in part, to the limited tools available for genetic manipulation; until quite recently, genetic manipulation of pathogenic *Leptospira* species was not possible. Thus, to date, only one protein has been clearly shown to have an affect on virulence. Mutation of the gene encoding Loa22, a protein localized in the outer membrane, resulted in attenuation of *Leptospira interrogans* (Ristow et al., 2007). Genetic complementation of a Loa22$^-$ mutant partially restored virulence near the level seen with the wild-type parental strain (Ristow et al., 2007). Hemolysins, including phospholipase C, sphingomylinase-like proteins, and pore-forming proteins, are potential virulence factors (Artiushin et al., 2004; Bernheimer and Bey, 1986; del Real et al., 1989; Lee et al., 2000, 2002; Matsunaga et al., 2007b; Segers et al., 1990, 1992). Several proteins, including bacterial immunoglobulin-like proteins LigA and LigB (Choy et al., 2007), a group of endostatin-like proteins (Barbosa et al., 2006; Stevenson et al., 2007), and the major outer-membrane protein LipL32 (Hoke et al., 2008), facilitate attachment of pathogenic *Leptospira* to mammalian extracellular matrix proteins and probably have important roles during infection. Lsa24 may also help the pathogenic *Leptospira* evade the antibacterial effects of complement by binding factor H (Verma et al., 2006), consistent with earlier findings that virulent *Leptospira* are resistant to serum (Johnson and Harris, 1967b). The unusual ability of *Leptospira* to undergo translational movement in highly viscous media (Greenberg and Canale-Parola, 1977a) may be an important factor in tissue penetration.

Enrichment and isolation procedures

Pathogenic *Leptospira* can be isolated from blood and, occasionally, spinal fluid during the early stages of infection (1–2 weeks) by growth in semi-solid (0.1–0.2% agar) or liquid growth media. Blood treated with heparin to prevent coagulation or untreated blood added directly to transport media (1% bovine serum albumin in phosphate buffer) can be used to inoculate growth media. One to two weeks after infection, and for several weeks to months following infection, *Leptospira* can be isolated from urine. *Leptospira* reside primarily in kidney and liver, therefore inoculation of media with tissue homogenates prepared from these organs collected during necropsy often results in successful isolation of bacteria. Addition of 5′-fluorouracil (100 μg/ml) to the isolation medium is useful for primary isolation of pathogenic *Leptospira*, especially from urine, by helping to suppress growth of contaminating bacteria. Primary isolation of pathogenic strains may require incubation of cultures up to 6 months with periodic microscopic evaluation. Direct intraperitoneal inoculation of weanling Golden Syrian hamsters using tissue, soil, or water samples is an alternative method to enrich for pathogenic *Leptospira* (Brendle and Alexander, 1974; Glosser et al., 1974).

Saprophytic *Leptospira* reside in streams, lakes, and moist soil (Henry and Johnson, 1978). To prepare environmental samples for culture, suspend soil in sterile distilled water, or use surface water directly, and filter through 0.45 or 0.22 μm filters before inoculating culture media. Incorporation of 5′-fluorouracil during initial isolation of environmental samples is important for restricting growth of non-*Leptospira* bacteria. Isolation of *Leptospira* from contaminated cultures can be achieved either by filtration through 0.22 µm filters or by spotting a few microliters of culture on solid culture medium. *Leptospira* will tend to migrate away from the spot, forming a halo of growth from which the bacteria can be isolated.

Leptospira growth in semi-solid media typically results in formation of a dense zone of growth (referred to as a Dinger disk) a few millimeters below the surface of the media.

Maintenance procedures

Leptospira cultures are maintained in semi-solid (0.1–0.2% agar) media at 30°C. The frequency of subculture varies greatly, dependent upon strain specific growth characteristics; some fastidious strains may take 6 months of incubation before visible growth, while more vigorous strains can be subcultured weekly. Long term cultivation of pathogenic *Leptospira* may result in loss of virulence, and therefore pathogenic strains should be frozen and stored in liquid nitrogen after initial isolation or within 1–2 *in vitro* subcultures. To prepare cultures for cryopreservation, mix a fresh Dinger disk of *Leptospira* growth in semi-solid media with an equal volume transport media and freeze at −70°C before placing in liquid nitrogen. Free-living *Leptospira* may be frozen or kept in semi-solid media for long term storage. Storage of *Leptospira* cultures by lyophilization is not recommended.

Differentiation of the genus *Leptospira* from other closely related taxa

Leptospira are obligate aerobes that consume fatty acids or alcohols as primary carbon and energy sources, distinguishing this genus from other spirochetes. Although morphologically similar to other spirochetes, *Leptospira* form tighter coils, often with at least one hooked end. These bacteria are resistant to 5′-fluorouracil, whereas other spirochetes are sensitive to this compound. The cell wall contains diaminopimelic acid vs ornithine in other spirochetes.

Further reading

Charon, N.W. and S.F. Goldstein. 2002. Genetics of motility and chemotaxis of a fascinating group of bacteria: the spirochetes. Annu. Rev. Genet. *36*: 47–73.

Faine, S. 1994. *Leptospira* and Leptospirosis. CRC Press, Boca Raton, FL.

Differentiation of species of the genus *Leptospira*

Characteristics that differentiate species of *Leptospira* are shown in Table 132. *Leptospira* species are defined by DNA–DNA reassociation analysis. In addition, 16S rRNA sequence comparison (Hookey, 1993), mapped-restriction site polymorphisms of amplified rRNA sequences (Ralph et al., 1993), and *secY* sequence comparison (Ahmed et al., 2006; Victoria et al., 2008) are also useful for differentiating *Leptospira* species. Several methods have proven helpful for strain and serovar differentiation, including restriction endonuclease analysis (Marshall et al., 1981; Robinson et al., 1982) and macrorestriction endonuclease analysis by pulsed-field gel electrophoresis (Galloway and Levett, 2008; Herrmann et al., 1991, 1992). The results of strain comparison using restriction endonuclease DNA digestion patterns usually correlate with serological analysis that separates strains based on agglutination reactions using specific sera (Ellis et al., 1991, 1988; Robinson et al., 1982, 1986; Thiermann et al., 1985). There are several examples of serovars found in more than one species (Levett, 2001), indicating the need to characterize strains by both serological and DNA-based techniques. Extensive phenotypic diversity precludes use of differential growth characteristics for differentiation of *Leptospira* species.

Taxonomic comments

The previous edition of the *Manual* described three species of *Leptospira*, namely *Leptospira interrogans*, representing pathogens, *Leptospira biflexa*, representing free-living leptospires, and *Leptospira illini species incertae sedis*. DNA–DNA renaturation studies have revealed substantial diversity within the previous classifications of *Leptospira interrogans* and *Leptospira biflexa*, and have supported the establishment of two new genera, *Leptonema*, in which *Leptonema illini species incertae sedis* now resides, and *Turneriella parva*. The taxonomic status of halophilic *Leptospira* has not been determined. Environmental surveys using 16S rRNA sequence analysis to monitor bacterial populations suggests the distribution of *Leptospira* and *Leptonema* species is diverse, and additional species are likely to be discovered.

The serovar is the primary *Leptospira* taxon, but this classification scheme is distinct from molecular taxonomy based on DNA–DNA reassociation kinetics or 16S rRNA sequence analysis. Strains representing the same serovar occur in two or more species. Likewise, serogroups also occur in different species (Table 133).

Differentiation of *Leptospira* species is difficult because these bacteria are morphologically similar and there is substantial variation in biochemical and growth characteristics within each species. Generally, free-living *Leptospira* and *Leptonema illini* share the ability to grow at 13°C. *Leptonema* is the only genus in *Leptospiraceae* that can grow on trypticase media and possesses cytoplasmic tubules. Pathogenic *Leptospira* are sensitive to 8-azaguanine, but nonpathogenic and strains with intermediate pathogenic status can vary in their capacity to grow in the presence of this compound. Likewise, 2,6-diaminopurine or copper sulfate inhibits growth of most pathogenic strains, but nonpathogenic strains and strains with intermediate pathogenic status vary in their ability to grow in the presence of these compounds. Therefore, accurate differentiation of *Leptospira* species must rely on molecular characterization comparing 16S rRNA sequences or DNA–DNA reassociation analysis (Levett and Smythe, 2006).

List of species of the genus *Leptospira*

1. **Leptospira interrogans** (Stimson 1907) Wenyon 1926, 1281[AL] emend. Faine and Stallman 1982, 462 (*Spirocheta interrogans* Stimson 1907, 541)

in.ter′ro.gans. L. part. adj. *interrogans* asking, inquiring, interrogating, here meaning shaped like a question mark.

Exhibits morphological and cultural features common to the genus *Leptospira* (Figure 101). Grows at 30°C, but does not grow at 13°C. Growth is inhibited by 8-azaguanine (225 µg/ml) or $CuSO_4$ (100 p.p.m.). Growth of some strains is inhibited by 2,6-diaminopurine (10 µg/ml). Production

TABLE 132. Differentiation of the species of the genera *Leptospira*, *Leptonema*, and *Turneriella*[a]

Characteristic	*L. interrogans*	*L. alexanderi*	*L. biflexa*	*L. borgpetersenii*	*L. broomi*	*L. fainei*	*L. inadai*	*L. kirschneri*	*L. licerasiae*	*L. meyeri*	*L. noguchi*	*L. santarosai*	*L. weilii*	*L. wolbachii*	*L. wolffii*	*"L. alstonii"*	*"L. terpstrae"*	*"L. vanthielii"*	*"L. yanagawae"*	*Leptonema illini*	*Turneriella parva*
Pathogenic in humans	+	nr	–	+	+	+	i	+	i	v	+	+	+	–	v	nr	nr	nr	nr	–	–
No. periplasmic flagella	2	2	2	2	2	2	2	2	2	2	2	2	2	2	2	2	2	2	2	2	2
Cytoplasmic tubules	–	–	–	–	–	–	–	–	–	–	–	–	–	–	–	–	–	–	–	+	–
Lipase activity	d	d	+	–	nr	nr	d	d	nr	+	d	–	–	+	nr	–	–	–	–	+	+
Growth inhibited by:																					
8-Azaguanine	+	+	d	+	nr	p	d	+	+	–	+	+	+	–	–	+	–	+	–	–	–
$CuSO_4$	+	d	d	+	nr	nr	d	d	nr	–	+	+	+	–	nr	+	–	+	–	+	–
2,6-Diaminopurine	d	+	d	–	nr	nr	d	d	nr	–	+	d	+	+	nr	+	+	+	+	+	+
Growth at 11–13°C	–	–	+	–	nr	+	v	–	nr	–	–	–	–	–	–	–	–	–	–	–	–
Growth at 30°C	+	+	+	+	+	+	+	+	+	+	+	+	+	+	+	+	+	+	+	+	+
DNA G+C content (mol%)	35	38	38.9	39.8	42	nr	42.6	nr	43.9	33.5	36.5	40.7	40.5	37.2	41.8	39.8	38.9	43.4	37.9	54.2	47–48

[a]Symbols: +, >85% positive; –, 0–15% positived, some strains; v, variation among strains; p, partial growth inhibition; i, infectious, pathogenicity not demonstrated; nr, not reported.

TABLE 133. Distribution of serogroups in different *Leptospira* species

Species	Serogroup
L. interrogans	Australis, Autumnalis, Bataviae, Canicola, Djasiman, Grippotyphosa, Hebdomadis, Icterohemorrhagiae, Louisiana, Mini, Pomona, Pyrogenes, Ranarum, Sarmin, Sehgali, Sejroe
L. alexanderi	Hebdomadis, Javanica, Manhoa, Mini
L. biflexa	Semaranga
L. borgpetersenii	Australis, Autumnalis, Ballum, Bataviae, Celledoni Hebdomadis, Javanica, Mini, Pyrogenes, Sejroe, Tarassovi
L. broomi	Undesignated
L. fainei	Hurstbridge
L. inadai	Canicola, Icterohemorrhagiae, Javanica, Lyme, Manhoa, Panama, Shermani, Tarassovi
L. kirschneri	Australis, Autumnalis, Bataviae, Canicola, Cynopteri, Djasiman, Grippotyphosa, Hebdomadis, Icterohemorrhagiae, Pomona
L. licerasiae	Iquitos
L. meyeri	Javanica, Mini, Ranarum, Sejroe, Semaranga
L. noguchi	Australis, Autumnalis, Bataviae, Djasiman, Louisiana, Panama, Pomona, Pyrogenes, Shermani, Tarassovi
L. santarosai	Autumnalis, Bataviae, Cynopteri, Grippotyphosa, Hebdomadis, Javanica, Mini, Pomona, Pyrogenes, Sarmin, Sejroe, Shermani, Tarassovi
L. weilii	Celledoni, Hebdomadis, Icterohemorrhagiae, Javanica, Manhoa, Mini, Pyrogenes, Sarmin, Sejroe, Tarassovi
L. wolbachii	Codice
L. wolffii	Undesignated
"*L. alstonii*" (genomospecies 1)	Ranarum
"*L. vanthielii*" (genomospecies 3)	Holland
"*L. terpstrae*" (genomospecies 4)	Icterohemorrhagiae
"*L. yanagawae*" (genomospecies 5)	Semaranga

of lipase varies among strains of this species. Pathogenic in mammals with disease manifestations ranging from mild influenza-like symptoms to acute, lethal infection.

The genome sequences for two strains of *Leptospira interrogans* have been determined: serovar Copenhageni strain Fiocruz L1–130 (GenBank accession nos AE016823 and AE016824) and serovar Lai strain 56601 (GenBank accession nos AE010300 and AE010301).

Source: bacteria localize in the kidneys of mammals, facilitating shedding via urine.

DNA G+C content (mol%): 35 (T_m).

Type strain: Leptospira interrogans serogroup Icterohaemorrhagiae, serovar Icterohaemorrhagiae, strain RGA, ATCC 23581, ATCC 43642, CCUG 5117, KCTC 2880.

Sequence accession no. (16S rRNA gene): AY631894, FJ154549.

2. **Leptospira alexanderi** Brenner, Kaufmann, Sulzer, Steigerwalt, Rogers and Weyant 1999, 856[VP]

a.le.xan.de'ri. N.L. masc. gen. n. *alexanderi* of Alexander, named to honor Aaron D. Alexander, an American microbiologist.

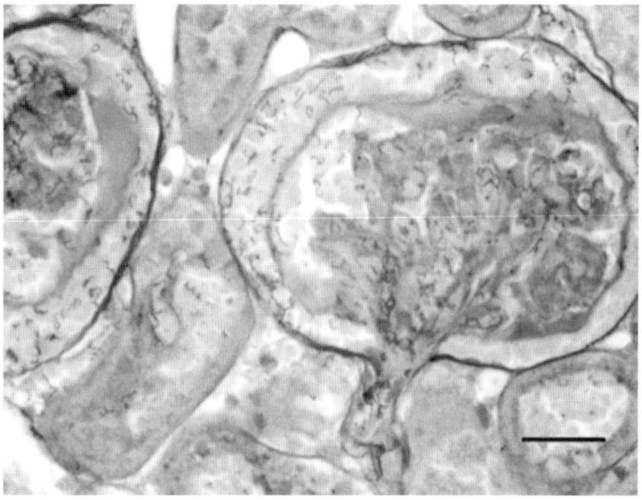

FIGURE 101. Silver-stained *Leptospira interrogans* serovar Pomona in situ. Sections of hamster kidney were isolated after infection with *Leptospira interrogans* serovar Pomona, fixed with formalin, then processed for histology. The tissue was stained with a modified PAS Steiner process. Bacteria appear black due to the precipitation of silver on the bacterial surface. A final magnification of 400× is shown. Bar = 25 µm.

Exhibits morphological and cultural features common to the genus *Leptospira*. Grows at 30°C, but does not grow at 11 or 37°C. Growth is inhibited by 8-azaguanine (225 µg/ml) or 2,6-diaminopurine (10 µg/ml). Strains vary in the ability to grow in the presence of $CuSO_4$ (100 p.p.m.), and production of lipase activity is variable. Pathogenicity for animals not reported.

Source: isolated in China from an unknown source.

DNA G+C content (mol%): 38 (T_m).

Type strain: L 60 serovar Manhao 3, ATCC 700520.

Sequence accession no. (16S rRNA gene): AY631880.

3. **Leptospira biflexa** (Wolbach and Binger 1914) Noguchi 1918, 585[AL] emend. Faine and Stallman 1982, 462 (*Spirocheta biflexa* Wolbach and Binger 1914, 23)

bi.fle'xa. L. adv. num. *bis* twice; L. part. adj. *flexus -a -um* (from. L. v. *flecto*) bent, winding; N.L. fem. adj. *biflexa* bent twice.

Exhibits morphological and cultural features common to the genus *Leptospira*. Grows at 13°C and at 30°C. Strains often grow in the presence of 8-azaguanine (225 µg/ml), 2,6-diaminopurine (10 µg/ml), and $CuSO_4$ (100 p.p.m.). All strains produce lipase. Strains are typically found in flowing or still freshwater sources, tap water, and occasionally animals. Pathogenicity not demonstrated, thought to be nonpathogenic for mammals.

The genome sequence for the type strain and one of its derivatives has been determined (GenBank accession nos CP00777–CP00779 and CP007786–CP007788).

Source: a stream in Italy.

DNA G+C content (mol%): 38.9 (genomic sequencing).

Type strain: Leptospira biflexa serovar Patoc strain Patoc I, ATCC 23582.

Sequence accession no. (16S rRNA gene): AY631876, Z12821.

4. **Leptospira borgpetersenii** Yasuda, Steigerwalt, Sulzer, Kaufmann, Rogers and Brenner 1987, 414[VP]

borg.pe'ter.sen'i.i. N.L. masc. gen. n. *borgpetersenii* of Borg-Petersen, named for C. Borg-Petersen, the Danish physician who made significant early contributions to the epidemiology and microbiology of leptospirosis in Europe.

Exhibits morphological and cultural features common to the genus *Leptospira*. Grows at 30°C, but does not grow at 11°C. Growth is inhibited by 8-azaguanine (225 µg/ml) or $CuSO_4$ (100 p.p.m.), but not by 2,6-diaminopurine (10 µg/ml). Does not produce lipase. The type strain was isolated from a Java house rat in Indonesia. Pathogenic in mammals with disease manifestations ranging from mild influenza-like symptoms to acute, lethal infection. Bacteria localize in the kidneys of their host, facilitating shedding via urine.

The genome sequences for two strains of *Leptospira borgpetersenii* serovar Hardjo (L550 and JB197) have been determined (GenBank accession nos. CP000348-CP000351).

Source: a Java house rat in Indonesia.

DNA G+C content (mol%): 39.8 (T_m).

Type strain: Veldrat Bataviae 46 serovar Javanica, ATCC 43292.

Sequence accession no. (16S rRNA gene): AY461862, AY887899, AM050572, DQ991483, FJ154600, Z21630.

5. **Leptospira broomii** Levett, Morey, Galloway and Steigerwalt 2006, 673[VP]

bro.o'mi.i. N.L. masc. gen. n. *broomii* of Broom, named for John Constable Broom (1902–1960), a Scottish bacteriologist who made substantial contributions to the study of leptospirosis.

Exhibits morphological and cultural features common to the genus *Leptospira*. Grows at 30°C, but growth at other temperatures not reported. Pathogenic in mammals with disease manifestations ranging from mild influenza-like symptoms to acute, lethal infection. Bacteria localize in the kidneys of their host, facilitating shedding via urine. However, clusters with other species (*Leptospira fainei*, *Leptospira inadai*, and *Leptospira licerasiae*) considered intermediate in pathogenicity based on 16S rRNA sequence comparison.

Source: blood of human patient with leptospirosis; additional isolates from blood, cerebrospinal fluid, and urine of human patients with leptospirosis.

DNA G+C content (mol%): 42 (T_m).

Type strain: 5399, ATCC BAA-1107, KIT 5399.

Sequence accession no. (16S rRNA gene): AY796065.

6. **Leptospira fainei** Perolat, Chappel, Adler, Baranton, Bulach, Billinghurst, Letocart, Merien and Serrano 1998, 857[VP]

fai'ne.i. N.L. masc. gen. n. *fainei* of Faine, named for Solomon Faine, an Australian medical microbiologist who made definitive contributions to the knowledge of the physiopathology and epidemiology of leptospirosis.

Exhibits morphological and cultural features common to the genus *Leptospira*. Grows at 13°C and 30°C. Growth is partially inhibited by 8-azaguanine (225 µg/ml, 30°C). May induce disease manifestations ranging from mild influenza-like symptoms to acute, potentially lethal infection. Bacteria localize in the kidneys of their host, facilitating shedding via urine. Clusters with other species (*Leptospira broomii*, *Leptospira inadai*, and *Leptospira licerasiae*) considered intermediate in pathogenicity based on 16S rRNA sequence comparison. Species contains a single serovar, Hurstbridge.

Source: the uterus of a sow in Australia; strains isolated from chronic human infections also reported.

DNA G+C content (mol%): not reported.

Type strain: BUT 6 serovar Hurstbridge.

Sequence accession no. (16S rRNA gene): AY631885, FJ154578, U60594.

7. **Leptospira inadai** Yasuda, Steigerwalt, Sulzer, Kaufmann, Rogers and Brenner 1987, 414[VP]

i.na'da.i. N.L. masc. gen. n. *inadai* of Inada, named for Ryokichi Inada, the Japanese microbiologist who is regarded by some to have first isolated leptospires from human patients.

Exhibits morphological and cultural features common to the genus *Leptospira*. Grows at 30°C, but no growth at 11°C. Growth of some strains is inhibited by 8-azaguanine (225 µg/ml), $CuSO_4$ (100 p.p.m.), or 2,6-diaminopurine (10 µg/ml). Some strains produce lipase. Strains of this species are considered to have intermediate pathogenic status, but clinical manifestations and pathogenicity have not been clearly demonstrated. Clusters with other species (*Leptospira broomii*, *Leptospira fainei*, and *Leptospira licerasiae*) considered intermediate in pathogenicity based on 16S rRNA sequence comparison.

Source: the skin of a patient with a concurrent, unrelated Lyme disease infection; strains have been isolated from a variety of rodents.

DNA G+C content (mol%): 42.6 (T_m).

Type strain: 10 serovar Lyme, ATCC 43289.

Sequence accession no. (16S rRNA gene): AY631896, Z21634.

8. **Leptospira kirschneri** Ramadass, Jarvis, Corner, Penny and Marshall 1992, 219[VP]

kirsch'ne.ri. N.L. masc. gen. n. *kirschneri* of Kirschner, named for Leopold Kirschner, a Dutch medical microbiologist who worked on leptospirosis research in Indonesia before coming to New Zealand to work at the Otago Medical School in Dunedin and whose pioneering work on leptospirosis helped focus attention on the human and animal health problem that existed at the time.

Exhibits morphological and cultural features common to the genus *Leptospira*. Grows at 30°C, but no growth at 13°C. Growth is inhibited by 8-azaguanine (225 µg/ml). Most strains are inhibited by $CuSO_4$ (100 p.p.m.) or 2,6-diaminopurine (10 µg/ml). Lipase is not produced. Strains of this species are pathogenic for mammals with disease manifestations ranging from mild influenza-like symptoms to acute, lethal infection. Bacteria localize in the kidneys of their host, facilitating shedding via urine.

Source: Indonesia,from a short-headed fruit bat.

DNA G+C content (mol%): not reported.

Type strain: 3522 C serovar Cynopteri, ATCC 49945.

Sequence accession no. (16S rRNA gene): AY631895, DQ991475, FJ154546, Z21628.

9. **Leptospira licerasiae** Matthias, Ricaldi, Cespedes, Diaz, Galloway, Saito, Steigerwalt, Patra, Ore, Gotuzzo, Gilman, Levett and Vinetz 2009, 1[VP] (Effective publication: Matthias,

Ricaldi, Cespedes, Diaz, Galloway, Saito, Steigerwalt, Patra, Ore, Gotuzzo, Gilman, Levett and Vinetz 2008, 11.)

li.ce.ra.si'ae. N.L. fem. gen. n. *licerasiae* of Liceras, named to honor Julia Liceras de Hidalgo, who obtained the first leptospiral isolates in Peru.

Exhibits morphological and cultural features common to the genus *Leptospira*. Grows at 30°C. Growth at other temperatures not reported. Growth is inhibited by 8-azaguanine (225 μg/ml). However, clusters with other species (*Leptospira broomi*, *Leptospira fainei*, and *Leptospira inadai*) considered intermediate in pathogenicity based on 16S rRNA sequence comparison. Although these strains are associated with human infections, experimental inoculation of hamsters and guinea pigs did not induce clinical signs of leptospirosis.

Source: blood obtained from human patients with undifferentiated fever.

DNA G+C content (mol%): not reported.

Type strain: VAR010 serovar Varillal, ATCC BAA-1110, KIT VAR 010, WPR VAR 010.

Sequence accession no. (16S rRNA gene): EF612284.

10. **Leptospira meyeri** Yasuda, Steigerwalt, Sulzer, Kaufmann, Rogers and Brenner 1987, 414[VP]

me.ye'ri. N.L. masc. gen. n. *meyeri* of Meyer, named to honor Karl F. Meyer, the veterinarian who established veterinary public health in the United States through his broad interests in the zoonoses, including leptospirosis.

Exhibits morphological and cultural features common to the genus *Leptospira*. Grows at 30°C, but not at 11°C. Grows in the presence 8-azaguanine (225 μg/ml), 2,6-diaminopurine (10 μg/ml), or $CuSO_4$ (100 p.p.m.). Produces lipase. Virulence has not been demonstrated, but some strains share genetic and antigenic similarity to pathogens.

Source: United States, from a leopard frog.

DNA G+C content (mol%): 33.5 (T_m).

Type strain: Iowa City Frog serovar Ranarum, ATCC 43782.

Sequence accession no. (16S rRNA gene): AY631878.

11. **Leptospira noguchii** Yasuda, Steigerwalt, Sulzer, Kaufmann, Rogers and Brenner 1987, 413[VP]

no.gu'chi.i. N.L. masc. gen. n. *noguchii* of Noguchi, named to honor Hideyo Noguchi, the Japanese microbiologist who named the genus *Leptospira*.

Exhibits morphological and cultural features common to the genus *Leptospira*. Grows at 30°C, but no growth at 11°C. Growth is inhibited by 8-azaguanine (225 μg/ml), 2,6-diaminopurine (10 μg/ml), or $CuSO_4$ (100 p.p.m.). Lipase is usually produced. Thought to be pathogenic with disease manifestations ranging from mild influenza-like symptoms to acute, lethal infection. Bacteria localize in the kidneys of their host, facilitating shedding via urine.

Source: parasitic strains isolated from mammals.

DNA G+C content (mol%): 36.5 (T_m).

Type strain: CZ 214 serovar Panama, ATCC 43288.

Sequence accession no. (16S rRNA gene): AY631886, DQ991500, Z21635.

12. **Leptospira santarosai** Yasuda, Steigerwalt, Sulzer, Kaufmann, Rogers and Brenner 1987, 413[VP]

san.ta.ro'sa.i. N.L. masc. gen. n. *santarosai* of Santa Rosa, named to honor Carlos A. Santa Rosa, the Brazilian veterinary microbiologist who was a pioneer in the study of leptospirosis as a human and animal health problem in Brazil.

Exhibits morphological and cultural features common to the genus *Leptospira*. Grows at 30°C, but does not grow at 11°C. Growth is inhibited by 8-azaguanine (225 μg/ml) or $CuSO_4$ (100 p.p.m.). Growth of some strains is inhibited by 2,6-diaminopurine (10 μg/ml). Lipase is not produced. Pathogenic in mammals with disease manifestations ranging from mild influenza-like symptoms to acute, lethal infection. Bacteria localize in the kidneys of their host, facilitating shedding via urine.

Source: the Panama Canal Zone, from a spiney rat.

DNA G+C content (mol%): 40.7 (T_m).

Type strain: LT 821 serovar Shermani, ATCC 43286.

Sequence accession no. (16S rRNA gene): AY461889, AY631883.

13. **Leptospira weilii** Yasuda, Steigerwalt, Sulzer, Kaufmann, Rogers and Brenner 1987, 413[VP]

weil'i.i. N.L. masc. gen. n. *weilii* of Weil, named to honor Adolph Weil, a German physician who was among the first to clinically differentiate leptospirosis (Weil's disease) from other types of infectious jaundice.

Exhibits morphological and cultural features common to the genus *Leptospira*. Grows at 30°C, but does not grow at 11°C. Growth is inhibited by 8-azaguanine (225 μg/ml), 2,6-diaminopurine (10 μg/ml), or $CuSO_4$ (100 p.p.m.). Lipase is not produced. Pathogenic in mammals with disease manifestations ranging from mild influenza-like symptoms to acute, lethal infection. Bacteria localize in the kidneys of their host, facilitating shedding via urine.

Source: Australia, from blood of a patient with leptospirosis.

DNA G+C content (mol%): 40.5 (T_m).

Type strain: Celledoni serovar Celledoni, ATCC 43285.

Sequence accession nos (16S rRNA gene): AY631877, DQ991486, FJ154580, U12676, Z21637.

14. **Leptospira wolbachii** Yasuda, Steigerwalt, Sulzer, Kaufmann, Rogers and Brenner 1987, 414[VP]

wol.ba'chi.i. N.L. masc. gen. n. *wolbachii* of Wolbach, named to honor Simeon Burt Wolbach (1880–1954), the American microbiologist who first identified *Leptospira* [*Spirochaeta*] *biflexa*.

Exhibits morphological and cultural features common to the genus *Leptospira*. Grows at 30°C, but does not grow at 11°C. Grows in the presence of 8-azaguanine (225 μg/ml) or 2,6-diaminopurine (10 μg/ml), but does not grow in the presence of $CuSO_4$ (100 p.p.m.). All strains produce lipase. Pathogenicity not demonstrated; thought to be nonpathogenic for mammals.

Source: water in the United States.

DNA G+C content (mol%): 37.2 (T_m).

Type strain: CDC serovar Codice, ATCC 43284.

Sequence accession no. (16S rRNA gene): AY631879.

15. **Leptospira wolffii** Slack, Kalambaheti, Symonds, Dohnt, Galloway, Steigerwalt, Chaicumpa, Bunyaraksyotin, Craig, Harrower and Smythe 2008, 2307[VP]

wolf'fi.i. N.L. masc. gen. n. *wolffii* of Wolff, named to honor Jan Wolff, a Dutch bacteriologist.

Exhibits morphological and cultural features common to the genus *Leptospira*. Grows at 30°C and 37°C, but does not grow at 13°C. Grows in the presence of 8-azaguanine (225 μg/ml). Pathogenicity not demonstrated, but thought to be of intermediate pathogenicity based on 16S rRNA and DNA:DNA reassociation similarity to *Leptospira broomii*, *Leptospira fainei*, and *Leptospira inadai*.

Source: urine of a human patient with symptoms consistent with leptospirosis.

DNA G+C content (mol%): 41.8 (T_m).

Type strain: Khorat-H2, KIT Khorat-H2, WHO LT1686.

Sequence accession no. (16S rRNA gene): EF025496.

Other organisms

1. **"Leptospira genomospecies 1"** Brenner, Kaufmann, Sulzer, Steigerwalt, Rogers and Weyant 1999, 857

Exhibits morphological and cultural features common to the genus *Leptospira*. Grows at 30°C, but does not grow at 11°C. Growth is inhibited by 8-azaguanine (225 μg/ml), 2,6-diaminopurine (10 μg/ml), or $CuSO_4$ (100 p.p.m.). Does not produce lipase. Pathogenicity not reported.

Source: a frog in China.

DNA G+C content (mol%): 39.8 (T_m).

Type strain: 79601 serovar Sichuan, ATCC 700521. Note: ATCC does not list this strain.

Sequence accession no. (16S rRNA gene): AY631881.

Taxonomic note: the Subcommittee on the taxonomy of *Leptospiraceae* proposed this species be named *Leptospira alstonii.*

2. **"Leptospira genomospecies 3"** Brenner, Kaufmann, Sulzer, Steigerwalt, Rogers and Weyant 1999, 857

Exhibits morphological and cultural features common to the genus *Leptospira*. Grows at 30°C, but does not grow at 11°C. Growth is inhibited by 8-azaguanine (225 μg/ml), 2,6-diaminopurine (10 μg/ml), or $CuSO_4$ (100 p.p.m.). Does not produce lipase. Pathogenicity not reported.

Source: water in the Netherlands.

DNA G+C content (mol%): 43.4 (T_m).

Type strain: WaZ Holland serovar Holland, ATCC 700522.

Sequence accession no. (16S rRNA gene): AY631897.

Taxonomic note: the Subcommittee on the taxonomy of *Leptospiraceae* proposed this species be named *Leptospira vanthielii.*

3. **"Leptospira genomospecies 4"** Brenner, Kaufmann, Sulzer, Steigerwalt, Rogers and Weyant 1999, 857

Exhibits morphological and cultural features common to the genus *Leptospira*. Grows at 30°C, but does not grow at 11°C. Growth is inhibited by 2,6-diaminopurine (10 μg/ml), but growth is not inhibited by 8-azaguanine (225 μg/ml) or $CuSO_4$ (100 p.p.m.). Does not produce lipase. Pathogenicity not reported.

Source: China from an unknown source.

DNA G+C content (mol%): 38.9 (T_m).

Type strain: H 2 serovar Hualin, ATCC 700522. Note ATCC lists this strain as genomospecies 3.

Sequence accession no. (16S rRNA gene): AY631888.

Taxonomic note: the Subcommittee on the taxonomy of *Leptospiraceae* proposed this species be named *Leptospira terpstrae.*

4. **"Leptospira genomospecies 5"** Brenner, Kaufmann, Sulzer, Steigerwalt, Rogers and Weyant 1999, 857

Exhibits morphological and cultural features common to the genus *Leptospira*. Grows at 30°C, but does not grow at 11°C. Growth is inhibited by 2,6-diaminopurine (10 μg/ml), but growth is not inhibited by 8-azaguanine (225 μg/ml) or $CuSO_4$ (100 p.p.m.). Does not produce lipase. Pathogenicity not reported.

Source: China from an unknown source.

DNA G+C content (mol%): 37.9 (T_m).

Type strain: Sao Paulo serovar Saopaulo, ATCC 700523.

Sequence accession no. (16S rRNA gene): AY631882.

Taxonomic note: the Subcommittee on the taxonomy of *Leptospiraceae* proposed this species be named *Leptospira yanagawae.*

Genus II. **Leptonema** Hovind-Hougen 1983, 439[VP] (Effective publication: Hovind-Hougen 1979, 250.)

Richard L. Zuerner

Lep.to.ne'ma. Gr. adj. *leptos* thin, narrow, fine; Gr. neut. n. *nema* a filament or thread; N.L. neut. n. *Leptonema* a thin filament or thread, describing a bacterium that resembles a thin filament or thread.

Long, thin, flexible rods, 0.1–0.2 μm in diameter and 13–21 μm in length, with a regular helical coiling pattern having a wavelength of 0.6–0.7 μm. Unicellular but may be observed as dividing pairs or short chains in actively growing cultures. Resting stages are not known. Gram-stain-negative. Due to small diameter, unstained cells are **not visible by bright-field microscopy**. Dark-field or phase-contrast microscopy is required for visualization of unstained cells. Highly motile. **Obligate aerobes**. Growth temperatures range between 13 and 30°C, with optimal growth at 28–30°C. **Chemoorganotrophic; consume long-chain fatty acids and fatty alcohols as primary carbon and energy sources.** Can grow on trypticase media. Uses respiration with oxygen as the terminal electron acceptor. Optimal growth occurs in semi-solid (0.2%) agar media. Growth on 1–2% solid agar results in the formation of clear to turbid subsurface colonies. Oxidase-positive. Lipase-positive. Nonpathogenic for cattle, gerbils, guinea pigs, hamsters, and mice. **Free-living in aquatic environments and soil**. Some strains found in association with animals. Does not share significant levels of sequence similarity with other *Leptospiraceae* as determined by DNA hybridization analysis. The 16S rRNA sequence is distinct from other *Leptospiraceae*.

DNA G+C content (mol%): 51–54%

Type species: **Leptonema illini** (Hanson, Tripathy, Evans and Alexander 1974) Hovind-Hougen 1983, 439 (Effective publication: Hovind-Hougen 1979, 251.) (*Leptospira illini* Hanson, Tripathy, Evans and Alexander 1974, 355).

Further descriptive information

Taxonomic history. The previous taxonomic designation for *Leptospira illini* as *Leptospira interrogans* serovar Illini and *Leptospira illini* has led to considerable confusion in the literature. In the previous edition of this manual, the taxonomic status of the genus *Leptonema* was uncertain, and the species *illini* was designated a *species incertae sedis*. The International Committee on Systematic Bacteriology Subcommittee on the Taxonomy of *Leptospira* approved the genus designation at its meeting in 1986 (Stallman, 1987).

Leptonema *illini* strain 3055, the type strain for this genus and species, was originally isolated from the urine of a healthy bull in Illinois (Hanson et al., 1974). An antigenically similar strain, A177, was isolated from a turtle in the same geographical region in the year following isolation of strain 3055^T. Early DNA hybridization studies led to the discovery that strain 3055^T lacked significant homology to either the "pathogenic" complex (*Leptospira interrogans sensu lato*), or the "saprophytic" complex (*Leptospira biflexa sensu lato*) (Brendle et al., 1974). Furthermore, the DNA G+C content (51–54 mol%) of *Leptonema illini* is substantially higher than known *Leptospira* strains (Brendle et al., 1974). Subsequent DNA hybridization studies by Ramadass et al. (1990) and Brenner et al. (1999), and analyzes of *Leptospiraceae* 16S rRNA sequences (Hookey et al., 1993; Morey et al., 2006; Paster et al., 1991) support a familial relationship between *Leptospira* and *Leptonema*, while supporting formation of a distinct genus *Leptonema* with a single known species, *Leptonema illini*.

Cell morphology, ultrastructure, and motility. Analysis of *Leptonema illini* has played a critical role in understanding cell structure and motility of the *Leptospiraceae*. *Leptonema illini* cells have a typical spirochete ultrastructure including a long slender helical morphology (Figure 102), forming right-handed helices (Carleton et al., 1979). Cells undergoing translational movement have a spiral anterior end and a hooked posterior end (Goldstein and Charon, 1990). The ability to form hooked ends is essential for translational movement and is governed by the shape of the two periplasmic flagella inserted subterminally at each end of the cell (Bromley and Charon, 1979). Nonhelical mutants are nonmotile and have periplasmic flagella with no defined shape, whereas wild-type cells and motile revertants have hooked ends and highly coiled flagella (Bromley and Charon, 1979), indicating these structures are rigid. *Leptonema* cells retain a gentle helical morphology regardless of whether the periplasmic flagella are straight or coiled (Bromley and Charon, 1979), and this is consistent with a model suggesting that the periplasmic flagella lie along the helical axis of the cell (Goldstein et al., 1996). Rotation of the periplasmic flagella is thought to cause the cytoplasmic cylinder to rotate around the axis of the flagella, and, depending on the direction of flagellar rotation, induce either a spiral or hook shape (Berg et al., 1978; Charon et al., 1984; Goldstein and Charon, 1990). A coordinated change in the cell ends from spiral to hook or from hook to spiral, enables the bacteria to rapidly change direction during translational movement (Charon et al., 1984).

A feature unique to *Leptonema* among the *Leptospiraceae* is the presence of cytoplasmic tubules that start near the insertion of the periplasmic flagella, and are about the same length as the flagella (Hovind-Hougen, 1979). These structures are also seen in species of *Treponema* and *Spirocheta*, but have not been observed in other members of *Leptospiraceae*.

Nutrition and growth conditions. Although *Leptonema* is traditionally grown on bovine serum albumin, Tween-based media used to cultivate *Leptospira*, e.g., EMJH (Ellinghausen and McCullough, 1965; Johnson and Harris, 1967a), members of this genus are distinguished by the ability to grow on trypticase media (Hanson et al., 1974). Large (8–10 mm) colonies appear on solid EMJH agar media within 7–10 d. Little information is known about the specific nutritional requirements of *Leptonema illini*. However, physiological profiles of *Leptospiraceae* using aminopeptidase substrates clustered the *Leptonema* strains together in a distinct pattern separate from either *Leptospira* or *Turneriella* (Neill et al., 1987).

Leptonema *illini* and *Leptospira biflexa* are more resistant to killing by UV or mitomycin C treatments than pathogenic *Leptospira* (Stamm and Charon, 1988).

rRNA sequence analysis. As noted above, 16S rRNA sequence analysis is useful for distinguishing *Leptonema* from *Leptospira* or *Turneriella* (Hookey et al., 1993; Morey et al., 2006; Paster et al., 1991). Unlike *Leptospira* species, which have one gene encoding 5S rRNA (*rrn*) and two genes each encoding 16S (*rrs*) and 23S (*rrl*) rRNA species, *Leptonema* has two genes

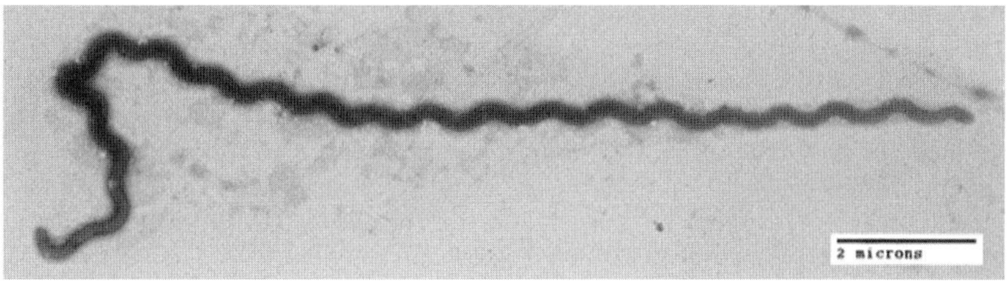

FIGURE 102. Transmission electron micrograph of *Leptonema illini* strain 3055^T. Bar = 2 μm.

encoding each rRNA species (Fukunaga et al., 1991). In addition, the two *Leptonema rrn* genes are closely linked to each other (Fukunaga and Mifuchi, 1989b), and the *rrs* and *rrl* genes are separated by a 435 bp intergenic spacer (Woo et al., 1996). In contrast, the rRNA genes in *Leptospira* species are dispersed throughout the large chromosome.

Bacteriophage. Early electron microscopy studies of *Leptonema illini* strain 3055^T cultures revealed the presence of bacteriophage-like particles with 45–50 nm heads and 60–65 × 15–20 nm tails (Hanson et al., 1974; Ritchie and Ellinghausen, 1969). Intracellular forms of these phage-like particles were also detected. Subsequent studies to characterize these particles have not been reported.

Ecology. The initial isolations of *Leptonema illini* were from cattle and turtles. Subsequent isolations of *Leptonema illini* having distinctly different antigenic profiles from the original isolates have been reported from fresh water (Bazovska et al., 1983) and a lymphocyte culture from a HIV-I infected human patient (Rocha et al., 1993). Several spirochete isolates from animal and water sources by Neill et al. (1987) have aminopeptidase substrate profiles similar to *Leptonema illini*, but these have not been characterized further by either DNA hybridization or 16S rRNA sequence analysis. Recent studies using 16S rRNA sequence analysis to profile microbial communities suggest *Leptonema illini* is broadly distributed in nature; 16S rRNA sequences matching *Leptonema illini* were detected in petroleum-contaminated soil in Japan (Kasai et al., 2005) and diseased coral communities near the Bahamas (Sekar et al., 2006).

Pathogenesis and antigenicity. Although serological surveys of cattle and swine in Illinois showed >60% positive reaction with strain 3055^T (Tripathy and Hanson, 1973a), this strain is not pathogenic for a wide variety of animal species including cattle, gerbils, guinea pigs, hamsters, and mice (Tripathy and Hanson, 1973b). Serological testing of strains 3055^T and A177 showed these strains comprised a novel antigenic group, distinct from known *Leptospira* serovars (Hanson et al., 1974). The discovery of additional serovars of *Leptonema illini* suggests this genus may also be antigenically diverse (Rocha et al., 1993).

List of species of the genus *Leptonema*

1. **Leptonema illini** (Hanson, Tripathy, Evans and Alexander 1974) Hovind-Hougen 1983, 439^{VP} (Effective publication: Hovind-Hougen 1979, 251.) (*Leptospira illini* Hanson, Tripathy, Evans and Alexander 1974, 355)

il.li'ni. N.L. gen. n. *illini* of Illinois, named after the state of Illinois, USA, where the first isolate was obtained.

Morphologically similar to *Leptospira*. Cytoplasmic tubules extend from near the insertion of periplasmic flagella and are about the same length. Aerobe. Long-chain fatty acids and long-chain fatty alcohols serve as carbon and energy sources. Can grow on trypticase media, unlike *Leptospira* or *Turneriella*. Three serovars described. Lacks significant homology to *Leptospira* or *Turneriella* as measured by DNA hybridization analysis. Has a unique 16S rRNA sequence profile. Not pathogenic for cattle, gerbils, guinea pigs, hamsters, and mice.

Source: urine of a clinically normal bull in Illinois, USA, in 1965 (Hanson et al., 1974); free living in soil or aquatic environments; some strains are found in association with animals.

DNA G+C content (mol%): 51–54% (T_m and Bd).

Type strain: 3055, NCTC 11301.

Sequence accession no. (16S rRNA gene): AY714984, Z21632.

Genus III. **Turneriella** Levett, Morey, Galloway, Steigerwalt and Ellis 2005, 1499^{VP}

Richard L. Zuerner

Tur.ne.ri'el.la. N.L. fem. dim. n. *Turneriella* named after Leslie Turner, an English microbiologist who made definitive contributions to the knowledge of leptospirosis.

Flexible helical rods, 0.3 × 3.5–7.5 µm with a wavelength of 0.3–0.5 µm. Gram-stain-negative. Unicellular. Resting stages are not known. Dark-field or phase-contrast microscopy is required for visualization of unstained cells. **Obligate aerobes**. Grows slowly at 13, 30, and 37°C, with temperature optimum of 28–30°C. **Chemoorganotrophic bacteria that consume long-chain fatty acids and fatty alcohols as primary carbon and energy sources.** Optimal growth occurs in semi-solid (0.2%) agar media. Oxidase-positive. Lipase-positive. Isolated from contaminated culture medium, tap water, and uterus of a sow. DNA hybridization and 16S rRNA sequence analyzes show this genus is distinct from other *Leptospiraceae*.

DNA G+C content (mol%): 47–48.

Type species: **Turneriella parva** (Hovind-Hougen, Ellis and Birch-Andersen 1981) Levett, Morey, Galloway, Steigerwalt and Ellis 2005, 1499^{VP} (*Leptospira parva* Hovind-Hougen, Ellis and Birch-Andersen 1981, 352).

Further descriptive information

Taxonomic history. The type strain was isolated from contaminated bovine serum albumin culture media and provisionally named *Leptospira parva* (Hovind-Hougen et al., 1981). Analysis of 16S rRNA sequence data indicated that this strain was different from other *Leptospira* and *Leptonema* species, and distinct from *Spirochetaceae* (Hookey et al., 1993). Additionally, this strain has an unusual aminopeptidase profile as compared to other *Leptospiraceae* (Neill et al., 1983, 1987). These data were consistent with DNA hybridization studies that showed no significant sequence similarity to other *Leptospiraceae* (Brenner et al., 1999). Furthermore, the G+C content of genomic DNA, at 47–48 mol% (Hovind-Hougen et al., 1981; Neill et al., 1987), is inconsistent with known *Leptospira* and *Leptonema* species. The unique position of the type strain among *Leptospiraceae* led to a proposal for a new genus, *Turneria* that was approved by the International Committee on Systematic Bacteriology Subcommittee on the Taxonomy of *Leptospiraceae* in 1990 (Marshall, 1992). However, the genus name *Turneria* was found to be illegitimate due to its use as genus names for plants and animals. A description of this bacterial genus was not published in a timely manner, leading to further confusion in the literature, with references to "*Turneria*" and *Leptospira parva incertae sedis*. The accepted name of *Turneriella* was approved by the taxonomic subcommittee in 2005 (Levett and Smythe, 2006) and a description published by Levett et al. (2005).

Enrichment and isolation procedures

The type strain and original isolate was obtained from contaminated bovine serum albumin (BSA) based media used for the cultivation of *Leptospira* (Hovind-Hougen et al., 1981). Presumably, the metabolic capabilities are similar to other *Leptospiraceae*. Cells grow slowly at 13, 30, and 37°C in liquid, solid, and semi-solid BSA-Tween based media (EMJH) capable of supporting the growth of *Leptospira*.

Maintenance procedures

These bacteria are maintained in liquid nitrogen by placing fresh growth in semi-solid BSA-Tween media, e.g., EMJH (Ellinghausen and McCullough, 1965; Johnson and Harris, 1967a), in sterile cryogenic vials. Cultures are frozen slowly to −70°C before long-term storage in liquid nitrogen.

Differentiation of *Turneriella* from other genera

The G+C content of genomic DNA is 47–48 mol%, which is distinct from *Leptospira* (33–43%) and *Leptonema* (54%). Phylogenetic analysis of 16S rRNA sequences distinguish *Turneriella* from other *Leptospiraceae* (Levett et al., 2005). Morphologically, *Turneriella* cells are shorter and have a shorter wavelength than other *Leptospiraceae*.

List of species of the genus *Turneriella*

1. **Turneriella parva** (Hovind-Hougen, Ellis and Birch-Andersen 1981) Levett, Morey, Galloway, Steigerwalt and Ellis 2005, 1499[VP] (*Leptospira parva* Hovind-Hougen, Ellis and Birch-Andersen 1981, 352)

par'va. L. fem. adj. *parva* small.

In addition to the description of the genus, this species has the following characteristics. Growth is inhibited by 200 µg 8-azaguanine/ml and 10 µg 2,6-diaminopurine/ml.

Source: contaminated bovine serum albumin (BSA)-based media.

DNA G+C content (mol%): 47–48 (T_m).

Type strain: H, ATCC BAA-1111, NCTC 11395.

Sequence accession no. (16S rRNA gene): AY293856, Z21636.

References

Ahmed, N., S.M. Devi, L. Valverde Mde, P. Vijayachari, R.S. Machang'u, W.A. Ellis and R.A. Hartskeerl. 2006. Multilocus sequence typing method for identification and genotypic classification of pathogenic *Leptospira* species. Ann. Clin. Microbiol. Antimicrob. *5*: 28.

Artiushin, S., J.F. Timoney, J. Nally and A. Verma. 2004. Host-inducible immunogenic sphingomyelinase-like protein, Lk73.5, of *Leptospira interrogans*. Infect. Immun. *72*: 742–749.

Azuma, I., T. Taniyama, Y. Yamamura, Y. Yanagihara and Y. Hattori. 1975. Chemical studies on the cell walls of *Leptospira biflexa* strain Urawa and *Treponema pallidum* strain Reiter. Jpn. J. Microbiol. *19*: 45–51.

Barbosa, A.S., P.A. Abreu, F.O. Neves, M.V. Atzingen, M.M. Watanabe, M.L. Vieira, Z.M. Morais, S.A. Vasconcellos and A.L. Nascimento. 2006. A newly identified leptospiral adhesin mediates attachment to laminin. Infect. Immun. *74*: 6356–6364.

Baril, C., J.L. Herrmann, C. Richaud, D. Margarita and I. Saint Girons. 1992. Scattering of the rRNA genes on the physical map of the circular chromosome of *Leptospira interrogans* serovar icterohaemorrhagiae. J. Bacteriol. *174*: 7566–7571.

Baseman, J.B. and C.D. Cox. 1969. Intermediate energy metabolism of *Leptospira*. J. Bacteriol. *97*: 992–1000.

Bazovska, S., K. Hovind-Hougen, A. Rudiova and E. Kmety. 1983. *Leptospira* sp. strain Dimbovitza, first isolate in Europe with characteristics of the proposed genus *Leptonema*. Int. J. Syst. Bacteriol. *33*: 325–328.

Berg, H.C., D.B. Bromley and N.W. Charon. 1978. Leptospiral motility. *In* Symp. Soc. Gen. Microbiol., vol. 28. Society for General Microbiology, Reading, pp. 285–294.

Bernheimer, A.W. and R.F. Bey. 1986. Copurification of *Leptospira interrogans* serovar pomona hemolysin and sphingomyelinase C. Infect. Immun. *54*: 262–264.

Bey, R.F. and R.C. Johnson. 1978. Protein-free and low-protein media for the cultivation of *Leptospira*. Infect. Immun. *19*: 562–569.

Bourhy, P., H. Louvel, I. Saint Girons and M. Picardeau. 2005. Random insertional mutagenesis of *Leptospira interrogans*, the agent of leptospirosis, using a *mariner* transposon. J. Bacteriol. *187*: 3255–3258.

Bourhy, P., L. Salaun, A. Lajus, C. Medigue, C. Boursaux-Eude and M. Picardeau. 2007. A genomic island of the pathogen *Leptospira interrogans* serovar Lai can excise from its chromosome. Infect. Immun. *75*: 677–683.

Boursaux-Eude, C., I. Saint Girons and R. Zuerner. 1995. IS*1500*, an IS*3*-like element from *Leptospira interrogans*. Microbiology *141*: 2165–2173.

Brendle, J.J. and A.D. Alexander. 1974. Contamination of bacteriological media by *Leptospira biflexa*. Appl. Microbiol. *28*: 505–506.

Brendle, J.J., M. Rogul and A.D. Alexander. 1974. Deoxyribonucleic acid hybridization among selected leptospiral serotypes. Int. J. Syst. Bacteriol. *24*: 205–214.

Brenner, D.J., A.F. Kaufmann, K.R. Sulzer, A.G. Steigerwalt, F.C. Rogers and R.S. Weyant. 1999. Further determination of DNA relatedness between serogroups and serovars in the family *Leptospiraceae* with a proposal for *Leptospira alexanderi* sp. nov. and four new *Leptospira* genomospecies. Int. J. Syst. Bacteriol. *49*: 839–858.

Bromley, D.B. and N.W. Charon. 1979. Axial filament involvement in the motility of *Leptospira interrogans*. J. Bacteriol. *137*: 1406–1412.

Bulach, D.M., R.L. Zuerner, P. Wilson, T. Seemann, A. McGrath, P.A. Cullen, J. Davis, M. Johnson, E. Kuczek, D.P. Alt, B. Peterson-Burch, R.L. Coppel, J.I. Rood, J.K. Davies and B. Adler. 2006. Genome reduction in *Leptospira borgpetersenii* reflects limited transmission potential. Proc. Natl. Acad. Sci. U. S. A. *103*: 14560–14565.

Carleton, O., N.W. Charon, P. Allender and S. O'Brien. 1979. Helix handedness of *Leptospira interrogans* as determined by scanning electron microscopy. J. Bacteriol. *137*: 1413–1416.

Charon, N.W., R.C. Johnson and D. Peterson. 1974. Amino acid biosynthesis in the spirochete *Leptospira*: evidence for a novel pathway of isoleucine biosynthesis. J. Bacteriol. *117*: 203–211.

Charon, N.W., C.W. Lawrence and S. O'Brien. 1981. Movement of antibody-coated latex beads attached to the spirochete *Leptospira interrogans*. Proc. Natl. Acad. Sci. U. S. A. *78*: 7166–7170.

Charon, N.W., G.R. Daughtry, R.S. McCuskey and G.N. Franz. 1984. Microcinematographic analysis of tethered *Leptospira illini*. J. Bacteriol. *160*: 1067–1073.

Choy, H.A., M.M. Kelley, T.L. Chen, A.K. Moller, J. Matsunaga and D.A. Haake. 2007. Physiological osmotic induction of *Leptospira interrogans* adhesion: LigA and LigB bind extracellular matrix proteins and fibrinogen. Infect. Immun. *75*: 2441–2450.

Cinco, M. and L. Cociancich. 1975. A suitable medium for the cultivation of halophilic leptospirae. Zentralbl. Bakteriol. Orig. A *233*: 553–555.

Cinco, M., M. Tamaro and L. Cociancich. 1975. Taxonomical, cultural and metabolic characteristics of halophilic leptospirae. Zentralbl. Bakteriol. Orig. A *233*: 400–405.

Cole, J.R., Jr., C.R. Sulzer and A.R. Pursell. 1973. Improved microtechnique for the leptospiral microscopic agglutination test. Appl. Microbiol. *25*: 976–980.

Cullen, P.A., S.J. Cordwell, D.M. Bulach, D.A. Haake and B. Adler. 2002. Global analysis of outer membrane proteins from *Leptospira interrogans* serovar Lai. Infect. Immun. *70*: 2311–2318.

Cullen, P.A., D.A. Haake, D.M. Bulach, R.L. Zuerner and B. Adler. 2003. LipL21 is a novel surface-exposed lipoprotein of pathogenic *Leptospira* species. Infect. Immun. *71*: 2414–2421.

Cullen, P.A., X. Xu, J. Matsunaga, Y. Sanchez, A.I. Ko, D.A. Haake and B. Adler. 2005. Surfaceome of *Leptospira* spp. Infect. Immun. *73*: 4853–4863.

de la Pena-Moctezuma, A., D.M. Bulach, T. Kalambaheti and B. Adler. 1999. Comparative analysis of the LPS biosynthetic loci of the genetic subtypes of serovar Hardjo: *Leptospira interrogans* subtype Hardjoprajitno and *Leptospira borgpetersenii* subtype Hardjobovis. FEMS Microbiol. Lett. *177*: 319–326.

del Real, G., R.P. Segers, B.A. van der Zeijst and W. Gaastra. 1989. Cloning of a hemolysin gene from *Leptospira interrogans* serovar hardjo. Infect. Immun. *57*: 2588–2590.

Ellinghausen, H.C., Jr. and W.G. McCullough. 1965. Nutrition of *Leptospira* Pomona and growth of 13 other serotypes: fractionation of oleic albumin complex and a medium of bovine albumin and polysorbate 80. Am. J. Vet. Res. *26*: 45–51.

Ellinghausen, H.C., Jr. 1983. Growth, cultural characteristics, and antibacterial sensitivity of *Leptospira interrogans* serovar hardjo. Cornell Vet. *73*: 225–239.

Ellis, W.A., A.B. Thiermann, J. Montgomery, A. Handsaker, P.J. Winter and R.B. Marshall. 1988. Restriction endonuclease analysis of *Leptospira interrogans* serovar hardjo isolates from cattle. Res. Vet. Sci. *44*: 375–379.

Ellis, W.A., J.M. Montgomery and A.B. Thiermann. 1991. Restriction endonuclease analysis as a taxonomic tool in the study of pig isolates belonging to the Australis serogroup of *Leptospira interrogans*. J. Clin. Microbiol. *29*: 957–961.

Faine, S. 1959. Iron as a growth requirement for pathogenic *Leptospira*. J. Gen. Microbiol. *20*: 246–251.

Faine, S. and N.D. Stallman. 1982. Amended descriptions of the genus *Leptospira* Noguchi 1917 and the species *L. interrogans* (Stimson 1907) Wenyon 1926 and *L. biflexa* (Wolbach and Binger 1914) Noguchi 1918. Int. J. Syst. Bacteriol. *32*: 461–463.

Faine, S., B. Adler, C. Bolin and P. Perolat. 1999. *Leptospira* and Leptospirosis. MediSci, Melbourne, Australia.

Fletcher, W. 1928. Recent work on leptospirosis, tsutsugamushi disease, and tropical typhus in the Federated Malay States. Trans. R. Soc. Trop. Med. Hyg. *21*: 265–282.

Fukunaga, M. and I. Mifuchi. 1989a. Unique organization of *Leptospira interrogans* rRNA genes. J. Bacteriol. *171*: 5763–5767.

Fukunaga, M. and I. Mifuchi. 1989b. The number of large ribosomal RNA genes in *Leptospira interrogans* and *Leptospira biflexa*. Microbiol. Immunol. *33*: 459–466.

Fukunaga, M., I. Horie, I. Mifuchi and M. Takemoto. 1991. Cloning, characterization and taxonomic significance of genes for the 5S ribosomal RNA of *Leptonema illini* strain 3055. J. Gen. Microbiol. *137*: 1523–1528.

Galloway, R.L. and P.N. Levett. 2008. Evaluation of a modified pulsed-field gel electrophoresis approach for the identification of *Leptospira* serovars. Am. J. Trop. Med. Hyg. *78*: 628–632.

Girons, I.S., P. Bourhy, C. Ottone, M. Picardeau, D. Yelton, R.W. Hendrix, P. Glaser and N. Charon. 2000. The LE1 bacteriophage replicates as a plasmid within *Leptospira biflexa*: construction of an *L. biflexa-Escherichia coli* shuttle vector. J. Bacteriol. *182*: 5700–5705.

Glosser, J.W., C.R. Sulzer, M. Eberhardt and W.G. Winkler. 1974. Cultural and serologic evidence of *Leptospira interrogans* serotype Tarassovi infection in turtles. J. Wildl. Dis. *10*: 429–435.

Goldstein, S.F. and N.W. Charon. 1990. Multiple-exposure photographic analysis of a motile spirochete. Proc. Natl. Acad. Sci. U. S. A. *87*: 4895–4899.

Goldstein, S.F., K.F. Buttle and N.W. Charon. 1996. Structural analysis of the *Leptospiraceae* and *Borrelia burgdorferi* by high-voltage electron microscopy. J. Bacteriol. *178*: 6539–6545.

Greenberg, E.P. and E. Canale-Parola. 1977a. Motility of flagellated bacteria in viscous environments. J. Bacteriol. *132*: 356–358.

Greenberg, E.P. and E. Canale-Parola. 1977b. Relationship between cell coiling and motility of spirochetes in viscous environments. J. Bacteriol. *131*: 960–969.

Guegan, R., J.M. Camadro, I. Saint Girons and M. Picardeau. 2003. *Leptospira* spp. possess a complete haem biosynthetic pathway and are able to use exogenous haem sources. Mol. Microbiol. *49*: 745–754.

Haake, D.A., E.M. Walker, D.R. Blanco, C.A. Bolin, M.N. Miller and M.A. Lovett. 1991. Changes in the surface of *Leptospira interrogans* serovar grippotyphosa during *in vitro* cultivation. Infect. Immun. *59*: 1131–1140.

Haake, D.A., C.I. Champion, C. Martinich, E.S. Shang, D.R. Blanco, J.N. Miller and M.A. Lovett. 1993. Molecular cloning and sequence analysis of the gene encoding OmpL1, a transmembrane outer membrane protein of pathogenic *Leptospira* spp. J. Bacteriol. *175*: 4225–4234.

Haake, D.A., C. Martinich, T.A. Summers, E.S. Shang, J.D. Pruetz, A.M. McCoy, M.K. Mazel and C.A. Bolin. 1998. Characterization of leptospiral outer membrane lipoprotein LipL36: downregulation associated with late-log-phase growth and mammalian infection. Infect. Immun. *66*: 1579–1587.

Haake, D.A., G. Chao, R.L. Zuerner, J.K. Barnett, D. Barnett, M. Mazel, J. Matsunaga, P.N. Levett and C.A. Bolin. 2000. The leptospiral major outer membrane protein LipL32 is a lipoprotein expressed during mammalian infection. Infect. Immun. *68*: 2276–2285.

Haapala, D.K., M. Rogul, L.B. Evans and A.D. Alexander. 1969. Deoxyribonucleic acid base composition and homology studies of *Leptospira*. J. Bacteriol. *98*: 421–428.

Hanson, L.E., D.N. Tripathy, L.B. Evans and A.D. Alexander. 1974. Unusual *Leptospira*, serotype illini (a new serotype). Int. J. Syst. Bacteriol. *24*: 355–357.

Henneberry, R.C. and C.D. Cox. 1970. Beta-oxidation of fatty acids by *Leptospira*. Can. J. Microbiol. *16*: 41–45.

Henry, R.A. and R.C. Johnson. 1978. Distribution of the genus *Leptospira* in soil and water. Appl. Environ. Microbiol. *35*: 492–499.

Herrmann, J.L., C. Baril, E. Bellenger, P. Perolat, G. Baranton and I. Saint Girons. 1991. Genome conservation in isolates of *Leptospira interrogans*. J. Bacteriol. *173*: 7582–7588.

Herrmann, J.L., E. Bellenger, P. Perolat, G. Baranton and I. Saint Girons. 1992. Pulsed-field gel electrophoresis of *Not*I digests of leptospiral DNA: a new rapid method of serovar identification. J. Clin. Microbiol. *30*: 1696–1702.

Hoke, D.E., S. Egan, P.A. Cullen and B. Adler. 2008. LipL32 is an extracellular matrix-interacting protein of *Leptospira spp.* and *Pseudoalteromonas tunicata*. Infect. Immun. *76*: 2063–2069.

Hookey, J.V. 1993. Characterization of *Leptospiraceae* by 16S DNA restriction fragment length polymorphisms. J. Gen. Microbiol. *139*: 1681–1689.

Hookey, J.V., J. Bryden and L. Gatehouse. 1993. The use of 16S rDNA sequence analysis to investigate the phylogeny of *Leptospiraceae* and related spirochetes. J. Gen. Microbiol. *139*: 2585–2590.

Hospenthal, D.R. and C.K. Murray. 2003. *In vitro* susceptibilities of seven *Leptospira* species to traditional and newer antibiotics. Antimicrob. Agents Chemother. *47*: 2646–2648.

Hovind-Hougen, K. 1979. *Leptospiraceae*, a new family to include *Leptospira* Noguchi 1917 and *Leptonema* gen. nov. Int. J. Syst. Bacteriol. *29*: 245–251.

Hovind-Hougen, K., W.A. Ellis and A. Birch-Andersen. 1981. *Leptospira parva sp.nov.* some morphological and biological characters. Zentralbl. Bakteriol. Mikrobiol. Hyg. A *250*: 343–354.

Hovind-Hougen, K. 1983. *In* Validation of the publication of new names and new combinations previously effectively published outside the IJSB. List no. 10. Int. J. Syst. Bacteriol. *33*: 438–440.

Hsu, D., M.J. Pan, Y.C. Zee and R.B. LeFebvre. 1990. Unique ribosome structure of *Leptospira interrogans* is composed of four rRNA components. J. Bacteriol. *172*: 3478–3480.

Johnson, R.C. and P. Rogers. 1964a. Differentiation of Pathogenic and Saprophytic Leptospires with 8-Azaguanine. J. Bacteriol. *88*: 1618–1623.

Johnson, R.C. and P. Rogers. 1964b. 5-Fluorouracil as a Selective Agent for Growth of Leptospirae. J. Bacteriol. *87*: 422–426.

Johnson, R.C. and P. Rogers. 1964c. Metabolism of leptospirae. I. Utilization of amino acids and purine, and pyrimidine bases. Arch. Biochem. Biophys. *107*: 459–470.

Johnson, R.C. and V.G. Harris. 1967a. Differentiation of pathogenic and saprophytic letospires. I. Growth at low temperatures. J. Bacteriol. *94*: 27–31.

Johnson, R.C. and V.G. Harris. 1967b. Antileptospiral activity of serum. II. Leptospiral virulence factor. J. Bacteriol. *93*: 513–519.

Johnson, R.C., J. Walby, R.A. Henry and N.E. Auran. 1973. Cultivation of parasitic leptospires: effect of pyruvate. Appl. Microbiol. *26*: 118–119.

Kasai, Y., Y. Takahata, T. Hoaki and K. Watanabe. 2005. Physiological and molecular characterization of a microbial community established in unsaturated, petroleum-contaminated soil. Environ. Microbiol. *7*: 806–818.

Lee, S.H., K.A. Kim, Y.G. Park, I.W. Seong, M.J. Kim and Y.J. Lee. 2000. Identification and partial characterization of a novel hemolysin from *Leptospira interrogans* serovar lai. Gene *254*: 19–28.

Lee, S.H., S. Kim, S.C. Park and M.J. Kim. 2002. Cytotoxic activities of *Leptospira interrogans* hemolysin SphH as a pore-forming protein on mammalian cells. Infect. Immun. *70*: 315–322.

Levett, P.N. 2001. Leptospirosis. Clin. Microbiol. Rev. *14*: 296–326.

Levett, P.N., R.E. Morey, R. Galloway, A.G. Steigerwalt and W.A. Ellis. 2005. Reclassification of *Leptospira parva* Hovind-Hougen et al. 1982 as *Turneriella parva* gen. nov., comb. nov. Int. J. Syst. Evol. Microbiol. *55*: 1497–1499.

Levett, P.N., R.E. Morey, R.L. Galloway and A.G. Steigerwalt. 2006. *Leptospira broomii* sp. nov., isolated from humans with leptospirosis. Int. J. Syst. Evol. Microbiol. *56*: 671–673.

Levett, P.N. and L. Smythe. 2006. International Committee on Systematics of Prokaryotes; Subcommittee on the taxonomy of *Leptospiraceae*: Minutes of the closed meeting, 12 and 13 November 2005, Chiang Mai, Thailand. Int. J. Syst. Evol. Microbiol. *56*: 2019–2020.

Levett, P.N. and L. Smythe. 2008. International Committee on Systematics of Prokaryotes; Subcommittee on the taxonomy of *Leptospiraceae*: Minutes of the closed meeting, 18 September 2007, Quito, Ecuador. Int. J. Syst. Evol. Microbiol. *58*: 1049–1050.

Louvel, H., S. Bommezzadri, N. Zidane, C. Boursaux-Eude, S. Creno, A. Magnier, Z. Rouy, C. Medigue, I. Saint Girons, C. Bouchier and M. Picardeau. 2006. Comparative and functional genomic analyses of iron transport and regulation in *Leptospira* spp. J. Bacteriol. *188*: 7893–7904.

Marshall, R.B., B.E. Wilton and A.J. Robinson. 1981. Identification of *Leptospira* serovars by restriction-endonuclease analysis. J. Med. Microbiol. *14*: 163–166.

Marshall, R.B. 1992. International Committee on Systematic Bacteriology: Subcommittee on the Taxonomy of *Leptospira*. Int. J. Syst. Bacteriol. *42*: 330–334.

Matsunaga, J., Y. Sanchez, X. Xu and D.A. Haake. 2005. Osmolarity, a key environmental signal controlling expression of leptospiral proteins LigA and LigB and the extracellular release of LigA. Infect. Immun. *73*: 70–78.

Matsunaga, J., M. Lo, D.M. Bulach, R.L. Zuerner, B. Adler and D.A. Haake. 2007a. Response of *Leptospira interrogans* to physiologic osmolarity: relevance in signaling the environment-to-host transition. Infect. Immun. *75*: 2864–2874.

Matsunaga, J., M.A. Medeiros, Y. Sanchez, K.F. Werneid and A.I. Ko. 2007b. Osmotic regulation of expression of two extracellular matrix-binding proteins and a haemolysin of *Leptospira interrogans*: differential effects on LigA and Sph2 extracellular release. Microbiology *153*: 3390–3398.

Matthias, M.A., J.N. Ricaldi, M. Cespedes, M.M. Diaz, R.L. Galloway, M. Saito, A.G. Steigerwalt, K.P. Patra, C.V. Ore, E. Gotuzzo, R.H. Gilman, P.N. Levett and J.M. Vinetz. 2008. Human leptospirosis caused by a new, antigenically unique *Leptospira* associated with a *Rattus* species reservoir in the peruvian Amazon. PLoS Negl. Trop. Dis. *2*: e213.

Matthias, M.A., J.N. Ricaldi, M. Cespedes, M.M. Diaz, R.L. Galloway, M. Saito, A.G. Steigerwalt, K.P. Patra, C.V. Ore, E. Gotuzzo, R.H. Gilman, P.N. Levett and J.M. Vinetz. 2009. List of new names and new combinations previously effectively, but not validly, published. List no. 125. Int. J. Syst. Evol. Microbiol. *59*: 1–2.

Morey, R.E., R.L. Galloway, S.L. Bragg, A.G. Steigerwalt, L.W. Mayer and P.N. Levett. 2006. Species-specific identification of *Leptospiraceae* by 16S rRNA gene sequencing. J. Clin. Microbiol. *44*: 3510–3516.

Nahori, M.A., E. Fournie-Amazouz, N.S. Que-Gewirth, V. Balloy, M. Chignard, C.R. Raetz, I. Saint Girons and C. Werts. 2005. Differential TLR recognition of leptospiral lipid A and lipopolysaccharide in murine and human cells. J. Immunol. *175*: 6022–6031.

Nascimento, A.L., A.I. Ko, E.A. Martins, C.B. Monteiro-Vitorello, P.L. Ho, D.A. Haake, S. Verjovski-Almeida, R.A. Hartskeerl, M.V. Marques, M.C. Oliveira, C.F. Menck, L.C. Leite, H. Carrer, L.L. Coutinho, W.M. Degrave, O.A. Dellagostin, H. El-Dorry, E.S. Ferro, M.I. Ferro, L.R. Furlan, M. Gamberini, E.A. Giglioti, A. Goes-Neto, G.H. Goldman, M.H. Goldman, R. Harakava, S.M. Jeronimo, I.L. Junqueira-de-Azevedo, E.T. Kimura, E.E. Kuramae, E.G. Lemos, M.V. Lemos, C.L. Marino, L.R. Nunes, R.C. de Oliveira, G.G. Pereira, M.S. Reis, A. Schriefer, W.J. Siqueira, P. Sommer, S.M. Tsai, A.J. Simpson, J.A. Ferro, L.E. Camargo, J.P. Kitajima, J.C. Setubal and M.A. Van Sluys. 2004. Comparative genomics of two *Leptospira interrogans* serovars reveals novel insights into physiology and pathogenesis. J. Bacteriol. *186*: 2164–2172.

Neill, S.D., L.R. Reid and W.A. Ellis. 1983. Aminopeptidase activity of *Leptospira* strains. J. Gen. Microbiol. *129*: 395–400.

Neill, S.D., R.L. Reid, S.T. Weatherup and W.A. Ellis. 1987. The use of aminopeptidase substrate specificity profiles to identify leptospires. Zentralbl. Bakteriol. Mikrobiol. Hyg. [A] *264*: 137–144.

Noguchi, H. 1917. *Spirochaeta icterohaemorrhagiae* in American wild rats and its relation to the Japanese and European strains. J. Exp. Med. *25*: 755–763.

Noguchi, H. 1918. Morphological characteristics and nomenclature of *Leptospira* (*Spirochaeta*) icterohaemorrhagiae (Inada and Ido). J. Exp. Med. *27*: 575–592.

Oie, S., K. Hironaga, A. Koshiro, H. Konishi and Z. Yoshii. 1983. *In vitro* susceptibilities of five *Leptospira* strains to 16 antimicrobial agents. Antimicrob. Agents Chemother. *24*: 905–908.

Paster, B.J., E. Stackebrandt, R.B. Hespell, C.M. Hahn and C.R. Woese. 1984. The phylogeny of the spirochetes. Syst. Appl. Microbiol. *5*: 337–351.

Paster, B.J., F.E. Dewhirst, W.G. Weisburg, L.A. Tordoff, G.J. Fraser, R.B. Hespell, T.B. Stanton, L. Zablen, L. Mandelco and C.R. Woese. 1991. Phylogenetic analysis of the spirochetes. J. Bacteriol. *173*: 6101–6109.

Perolat, P., R.J. Chappel, B. Adler, G. Baranton, D.M. Bulach, M.L. Billinghurst, M. Letocart, F. Merien and M.S. Serrano. 1998. *Leptospira fainei* sp. nov., isolated from pigs in Australia. Int. J. Syst. Bacteriol. *48*: 851–858.

Petrino, M.G. and R.N. Doetsch. 1978. 'Viscotaxis', a new behavioural response of *Leptospira interrogans* (*biflexa*) strain B16. J. Gen. Microbiol. *109*: 113–117.

Picardeau, M. 2008. Conjugative transfer between *Escherichia coli* and *Leptospira* spp. as a new genetic tool. Appl. Environ. Microbiol. *74*: 319–322.

Picardeau, M., D.M. Bulach, C. Bouchier, R.L. Zuerner, N. Zidane, P.J. Wilson, S. Creno, E.S. Kuczek, S. Bommezzadri, J.C. Davis, A. McGrath, M.J. Johnson, C. Boursaux-Eude, T. Seemann, Z. Rouy, R.L. Coppel, J.I. Rood, A. Lajus, J.K. Davies, C. Medigue and B. Adler. 2008. Genome sequence of the saprophyte *Leptospira biflexa* provides insights into the evolution of *Leptospira* and the pathogenesis of leptospirosis. PLoS ONE *3*: e1607.

Que-Gewirth, N.L., A.A. Ribeiro, S.R. Kalb, R.J. Cotter, D.M. Bulach, B. Adler, I.S. Girons, C. Werts and C.R. Raetz. 2004. A methylated phosphate group and four amide-linked acyl chains in *Leptospira interrogans* lipid A. The membrane anchor of an unusual lipopolysaccharide that activates TLR2. J. Biol. Chem. *279*: 25420–25429.

Ralph, D. and M. McClelland. 1993. Intervening sequence with conserved open reading frame in eubacterial 23S rRNA genes. Proc. Natl. Acad. Sci. U. S. A. *90*: 6864–6868.

Ralph, D., M. McClelland, J. Welsh, G. Baranton and P. Perolat. 1993. *Leptospira* species categorized by arbitrarily primed polymerase chain reaction (PCR) and by mapped restriction polymorphisms in PCR-amplified rRNA genes. J. Bacteriol. *175*: 973–981.

Ramadass, P., B.D.W. Jarvis, R.J. Corner, M. Cinco and R.B. Marshall. 1990. DNA Relatedness among Strains of *Leptospira biflexa*. Int. J. Syst. Bacteriol. *40*: 231–235.

Ramadass, P., B.D. Jarvis, R.J. Corner, D. Penny and R.B. Marshall. 1992. Genetic characterization of pathogenic *Leptospira* species by DNA hybridization. Int. J. Syst. Bacteriol. *42*: 215–219.

Ren, S.X., G. Fu, X.G. Jiang, R. Zeng, Y.G. Miao, H. Xu, Y.X. Zhang, H. Xiong, G. Lu, L.F. Lu, H.Q. Jiang, J. Jia, Y.F. Tu, J.X. Jiang, W.Y. Gu, Y.Q. Zhang, Z. Cai, H.H. Sheng, H.F. Yin, Y. Zhang, G.F. Zhu, M. Wan, H.L. Huang, Z. Qian, S.Y. Wang, W. Ma, Z.J. Yao, Y. Shen, B.Q. Qiang, Q.C. Xia, X.K. Guo, A. Danchin, I. Saint Girons, R.L. Somerville, Y.M. Wen, M.H. Shi, Z. Chen, J.G. Xu and G.P. Zhao. 2003. Unique physiological and pathogenic features of *Leptospira interrogans* revealed by whole-genome sequencing. Nature *422*: 888–893.

Ristow, P., P. Bourhy, F.W. da Cruz McBride, C.P. Figueira, M. Huerre, P. Ave, I.S. Girons, A.I. Ko and M. Picardeau. 2007. The OmpA-like protein Loa22 is essential for leptospiral virulence. PLoS Pathog. *3*: e97.

Ritchie, A.E. and H.C. Ellinghausen. 1969. Bacteriophage-like entities associated with a Leptospire. Proceedings of the Electron Microscope Society of America, Baton Rouge, pp. 228–229.

Robinson, A.J., P. Ramadass, A. Lee and R.B. Marshall. 1982. Differentiation of subtypes within *Leptospira interrogans* serovars Hardjo, Balcanica and Tarassovi, by bacterial restriction-endonuclease DNA analysis (BRENDA). J. Med. Microbiol. *15*: 331–338.

Rocha, T., E.A. Cardoso, A.M. Terrinha, J.F. Nunes, K. Hovind-Hougen and M. Cinco. 1993. Isolation of a new serovar of the genus *Leptonema* in the family *Leptospiraceae*. Zentralbl. Bakteriol. *279*: 167–172.

Rodionov, D.A., A.G. Vitreschak, A.A. Mironov and M.S. Gelfand. 2003. Comparative genomics of the vitamin B12 metabolism and regulation in prokaryotes. J. Biol. Chem. *278*: 41148–41159.

Ryu, E. 1970. Rapid microscopic agglutination test for *Leptospira* without non-specific reaction. Bull. Off. Int. Epizoot. *73*: 49–58.

Saint Girons, I., D. Margarita, P. Amouriaux and G. Baranton. 1990. First isolation of bacteriophages for a spirochaete: potential genetic tools for *Leptospira*. Res. Microbiol. *141*: 1131–1138.

Segers, R.P., A. van der Drift, A. de Nijs, P. Corcione, B.A. van der Zeijst and W. Gaastra. 1990. Molecular analysis of a sphingomyelinase C gene from *Leptospira interrogans* serovar hardjo. Infect. Immun. *58*: 2177–2185.

Segers, R.P., J.A. van Gestel, G.J. van Eys, B.A. van der Zeijst and W. Gaastra. 1992. Presence of putative sphingomyelinase genes among members of the family *Leptospiraceae*. Infect. Immun. *60*: 1707–1710.

Sekar, R., D.K. Mills, E.R. Remily, J.D. Voss and L.L. Richardson. 2006. Microbial communities in the surface mucopolysaccharide layer and the black band microbial mat of black band-diseased *Siderastrea siderea*. Appl. Environ. Microbiol. *72*: 5963–5973.

Shenberg, E. 1967. Growth of pathogenic *Leptospira* in chemically defined media. J. Bacteriol. *93*: 1598–1606.

Slack, A.T., T. Kalambaheti, M.L. Symonds, M.F. Dohnt, R.L. Galloway, A.G. Steigerwalt, W. Chaicumpa, G. Bunyaraksyotin, S. Craig, B.J. Harrower and L.D. Smythe. 2008. *Leptospira wolffii* sp. nov., isolated from a human with suspected leptospirosis in Thailand. Int. J. Syst. Evol. Microbiol. *58*: 2305–2308.

Stallman, N.D. 1984. International Committee on Systematic Bacteriology Subcommittee on the Taxonomy of *Leptospira*: Minutes of the Meeting, 6 to 10 August 1982, Boston, Massachusetts. Int. J. Syst. Bacteriol. *34*: 258–259.

Stallman, N.D. 1987. International Committee on Systematic Bacteriology Subcommittee on the Taxonomy of *Leptospira*: Minutes of the Meeting, 5 and 6 September 1986, Manchester, England. Int. J. Syst. Bacteriol. *37*: 472–473.

Stamm, L.V. and N.W. Charon. 1988. Sensitivity of pathogenic and free-living *Leptospira spp.* to UV radiation and mitomycin C. Appl. Environ. Microbiol. *54*: 728–733.

Stevenson, B., H.A. Choy, M. Pinne, M.L. Rotondi, M.C. Miller, E. Demoll, P. Kraiczy, A.E. Cooley, T.P. Creamer, M.A. Suchard, C.A. Brissette, A. Verma and D.A. Haake. 2007. *Leptospira interrogans* endostatin-like outer membrane proteins bind host fibronectin, laminin and regulators of complement. PLoS ONE *2*: e1188.

Stimson, A.M. 1907. Note on an organism found in yellow-fever tissue. Public Health Rep. *22*: 541.

Stuart, R.D. 1946. The preparation and use of a simple culture medium for leptospirae. J. Pathol. Bacteriol. *58*: 343–349.

Terpstra, W.J., H. Korver, J. van Leeuwen, P.R. Klatser and A.H. Kolk. 1985. The classification of Sejroe group serovars of *Leptospira interrogans* with monoclonal antibodies. Zentralbl. Bakteriol. Mikrobiol. Hyg. A *259*: 498–506.

Thiermann, A.B., A.L. Handsaker, S.L. Moseley and B. Kingscote. 1985. New method for classification of leptospiral isolates belonging to serogroup pomona by restriction endonuclease analysis: serovar kennewicki. J. Clin. Microbiol. *21*: 585–587.

Thiermann, A.B., A.L. Handsaker, J.W. Foley, F.H. White and B.F. Kingscote. 1986. Reclassification of North American leptospiral isolates belonging to serogroups Mini and Sejroe by restriction endonuclease analysis. Am. J. Vet. Res. *47*: 61–66.

Tripathy, D.N. and L.E. Hanson. 1973a. Studies of *Leptospira illini*, strain 3055: immunologic and serologic determinations. Am. J. Vet. Res. *34*: 563–565.

Tripathy, D.N. and L.E. Hanson. 1973b. Studies of *Leptospira illini*, strain 3055: pathogenicity for different animals. Am. J. Vet. Res. *34*: 557–562.

Trueba, G., S. Zapata, K. Madrid, P. Cullen and D. Haake. 2004. Cell aggregation: a mechanism of pathogenic *Leptospira* to survive in fresh water. Int. Microbiol. *7*: 35–40.

Verma, A., J. Hellwage, S. Artiushin, P.F. Zipfel, P. Kraiczy, J.F. Timoney and B. Stevenson. 2006. LfhA, a novel factor H-binding protein of *Leptospira interrogans*. Infect. Immun. *74*: 2659–2666.

Victoria, B., A. Ahmed, R.L. Zuerner, N. Ahmed, D.M. Bulach, J. Quinteiro and R.A. Hartskeerl. 2008. Conservation of the S10-spc-alpha locus within otherwise highly plastic genomes provides phylogenetic insight into the genus *Leptospira*. PLoS ONE *3*: e2752.

Wenyon, C.M. 1926. Protozoology: A manual for medical men, veterinarians and zoologists. Baillière, Tindall and Cox, London.

Werts, C., R.I. Tapping, J.C. Mathison, T.H. Chuang, V. Kravchenko, I. Saint Girons, D.A. Haake, P.J. Godowski, F. Hayashi, A. Ozinsky, D.M. Underhill, C.J. Kirschning, H. Wagner, A. Aderem, P.S. Tobias and R.J. Ulevitch. 2001. Leptospiral lipopolysaccharide activates cells through a TLR2-dependent mechanism. Nat. Immunol. *2*: 346–352.

Westfall, H.N., N.W. Charon and D.E. Peterson. 1983. Multiple pathways for isoleucine biosynthesis in the spirochete *Leptospira*. J. Bacteriol. *154*: 846–853.

Wolbach, S.B. and C.A.L. Binger. 1914. Notes on a filterable spirochete from fresh water. J. Med. Res. *30*: 23.

Woo, T.H., L.D. Smythe, M.L. Symonds, M.A. Norris, M.F. Dohnt and B.K. Patel. 1996. Rapid distinction between *Leptonema* and *Leptospira* by PCR amplification of 16S-23S ribosomal DNA spacer. FEMS Microbiol. Lett. *142*: 85–90.

World Health Organization. 2003. Human Leptospirosis: Guidance for Diagnosis, Surveillance and Control.

Yasuda, P.H., A.G. Steigerwalt, K.R. Sulzer, A.F. Kaufmann, F. Rogers and D.J. Brenner. 1987. Deoxyribonucleic acid relatedness between serogroups and serovars in the family *Leptospiraceae* with proposals for seven new *Leptospira* species. Int. J. Syst. Bacteriol. *37*: 407–415.

Zuerner, R.L. 1991. Physical map of chromosomal and plasmid DNA comprising the genome of *Leptospira interrogans*. Nucleic Acids Res. *19*: 4857–4860.

Zuerner, R.L., W. Knudtson, C.A. Bolin and G. Trueba. 1991. Characterization of outer membrane and secreted proteins of *Leptospira interrogans* serovar pomona. Microb. Pathog. *10*: 311–322.

Zuerner, R.L., J.L. Herrmann and I. Saint Girons. 1993. Comparison of genetic maps for two *Leptospira interrogans* serovars provides evidence for two chromosomes and intraspecies heterogeneity. J. Bacteriol. *175*: 5445–5451.

Zuerner, R.L. 1994. Nucleotide sequence analysis of IS*1533* from *Leptospira borgpetersenii*: identification and expression of two IS-encoded proteins. Plasmid *31*: 1–11.

Zuerner, R.L. and W.M. Huang. 2002. Analysis of a *Leptospira interrogans* locus containing DNA replication genes and a new IS, IS*1502*. FEMS Microbiol. Lett. *215*: 175–182.

Zuerner, R.L. and G.A. Trueba. 2005. Characterization of IS*1501* mutants of *Leptospira interrogans* serovar pomona. FEMS Microbiol. Lett. *248*: 199–205.

Hindgut spirochetes of termites and *Cryptocercus punctulatus*

Bruce J. Paster and John A. Breznak

Spirochetes are commonly observed in the hindguts of termites (Table 134) and the wood-eating cockroach *Cryptocercus punctulatus* (Grimstone, 1963). Early workers referred to these motile bacteria as spirilla (Leidy, 1877) or vibrios (Leidy, 1881), assigned them to currently recognized spirochete genera (see *Taxonomic comments*), or "spirochetes" (Damon, 1926). Later studies using electron microscopy confirmed that these bacteria were indeed true spirochetes in that they possessed ultrastructural characteristics typical of spirochetes. (Bermudes et al., 1988). Hindgut spirochetes occur free in the gut fluid as well as attached to the surfaces of hindgut protozoa. Recently, two species of spirochetes from termite hindguts have been grown in pure culture and, on the basis of 16S rRNA sequence comparisons, were determined to be species of *Treponema*, namely *Treponema azotonutricium* and *Treponema primitia* (Graber et al., 2004; see chapter on *Treponema*, in this volume). However, most of spirochetes from the termite and cockroach hindguts have not been isolated and grown in pure or mixed culture.

The size of free hindgut spirochetes ranges from about 0.2 µm in diameter × 3 µm long (Breznak and Pankratz, 1977) to as large as 1.0 µm in diameter × 100 µm long (Hollande and Gharagozlou, 1967). Likewise, the number of periplasmic flagella ranges from a few per cell to as many as 100 or more in the larger forms (Bermudes et al., 1988; To et al., 1978; Wier et al., 2000). However, the multiple periplasmic flagella in most of the large forms do not generally occur in a tight bundle as in *Cristispira*.

Bermudes et al. (1988) presented taxonomic considerations of the large uncultivable hindgut spirochetes as based on size, number of flagella, amplitude and wavelength of coils, and other ultrastructural traits. These distinctive features include the following: (a) a crenulated outer sheath (Hollande and Gharagozlou, 1967) (Figure 103); (b) a helicoidal groove or "sillon", an invagination of the outer membrane that appears to be in contact to the inner membrane (Gharagozlou, 1968; Hollande and Gharagozlou, 1967) (Figure 104); (c) the thickness of the outer and inner coat of the outer membrane (Bermudes et al., 1988); (d) a polar organelle (Bermudes et al., 1988); and (e) cytoplasmic tubules (Bermudes et al., 1988).

Some hindgut spirochetes attach by one end to the surface of certain flagellate protozoa found only in the lower termites (i.e., families Mastotermitidae, Kalotermitidae, Hodotermitidae, Serritermitidae, and Rhinotermitidae) and *Cryptocercus punctulatus*. These spirochetes may be uniformly distributed over the surface or localized to specific regions (Ball, 1969; Kirby, 1941). Some spirochetes have a structural modification of one end of the cell in their attachment to the surfaces of *Pyrsonympha* (from *Reticulitermes flavipes* and *Reticulitermes tibialis*) and *Barbulanympha* (from *Cryptocercus punctulatus*) (Bloodgood and Fitzharris, 1976; Bloodgood et al., 1974). In contrast, some protozoa have structural modifications to facilitate spirochetal attachment. For example, in *Mixotrichia paradoxa* in the termite *Mastotermes darwiniensis*, bracket-like elements in the plasma membranes serve as attachment points (Cleveland and Grimstone, 1964), whereas in polymastigotes from *Reticulitermes flavipes* the attachment points are screw-like structures (Smith et al., 1975a, b). In other termite species, both the protozoan plasma

Table 134. Distribution of spirochetes in the hindgut of termites and wood-eating cockroaches

Host genus	Source location	References
Cockroach:		
Cryptocercus punctulatus	USA	Hollande and Gharagozlou (1967)
Termite:		
Bifiditermes condonesis	Australia	To et al. (1978)
Calcaritermes (Kalotermes) nigriceps	British Guinea	Damon (1926)
Ceratokalotermes spoliator	Australia	To et al. (1978)
Coptotermes acinaciformis	Australia	To et al. (1978)
Coptotermes formosanus	Hawaii	To et al. (1978)
Coptotermes lacteus	Australia	Eutick et al. (1978)
Cryptotermes brevis	USA	Damon (1926); To et al. (1978)
Cryptotermes cavifrons	USA	To et al. (1978)
Cryptotermes gearyi	Australia	To et al. (1978)
Glyptotermes neotuberculatus	Australia	To et al. (1978)
Glyptotermes (Kalotermes) iridipennis	Australia	To et al. (1978)
Heterotermes aureus	USA	To et al. (1978)
Incisitermes rnilleri	USA	To et al. (1978)
Kalotermes (Incisitermes) minor	USA	To et al. (1978)
Kalotermes (Incisitermes) schwarzi	USA	Damon (1926); To et al. (1978)
Kalotermes (Neotermes) jouteli	USA	Margulis et al. (1981)
Kalotermes approximatus	USA	Margulis et al. (1981)
Kalotermes banksiae	Australia	Margulis et al. (1981)
Kalotermes flavicollis	France, Spain	Gharagozlou (1968); To et al. (1978)
Kalotermes snyderi	USA	Bermudes et al. (1988)
Leucopitermes lucifugus	Bastia, Corsica	Hollande (1922)
Leucopitermes tenuis	British Guinea	Damon (1926)
Marginitermes (Kalotermes) hubbardi	USA	To et al. (1978)
Mastotermes darwiniensis	Australia	Cleveland and Grimstone (1964); To et al. (1978)
Nasutitermes costalis	Puerto Rico	To et al. (1978)
Nasutitermes exitiosus	Australia	Eutick et al. (1978)
Nasutitermes morio	Puerto Rico	Damon (1926)
Neotermes castaneus	USA	To et al. (1978)
Neotermes insularis	Australia	To et al. (1978)
Paraneotermes simplicicornis	USA	To et al. (1978)
Porotermes adamsoni	Australia	To et al. (1978)
Postelectrotertnes (Kalotermes) praecox	Portugal	Hollande and Gharagozlou (1967)
Postelectrotertnes militaris		Dobell (1910, 1912)
Pterotermes occidentis	Mexico, USA	To et al. (1978)
Reticulitermes flavipes	USA	Breznak (1984); Damon (1926); Leidy (1877, 1881)
Reticulitermes hageni	USA	Damon (1926)
Reticulitermes hesperus	USA	Margulis et al. (1981)
Reticulitermes lucifugus	Japan, Italy	Ghidini and Archetti (1939); von Prowazek (1910)
Reticulitermes tibialis	USA	Bloodgood and Fitzharris (1976)
Reticulitermes virginicus	USA	Damon (1926)
Zootermopsis angusticollis	USA	Damon (1926); To et al. (1978)
Zootermopsis nevadensis	USA	Damon (1926); To et al. (1978)

membrane and the spirochetal poles are modified to form an attachment complex (Smith and Arnott, 1974).

Cleveland and Grimstone (1964) demonstrated that adherent spirochetes serve a locomotory function for *Mixotrichia paradoxa*, although not likely for propulsion. Based on these observations, Margulis et al. (1979) proposed that eukaryotic flagella and cilia evolved from ectosymbiotic spirochetes. However, antibiotic treatment of *Mixotrichia paradoxa* reduces the number of ectobionts and leads to a disintegration of the attachment systems rendering the protozoan immotile (Radek and Nitsch, 2007). Furthermore, the antibiotic treatment causes attached spirochetes to lose their helical shape and form round bodies (Radek and Nitsch, 2007).

From the data on the two cultivable hindgut spirochetes (Graber et al., 2004), *Treponema azotonutricium* ferments carbohydrates to acetate, ethanol, CO_2, and H_2 as major products and is noteworthy in that this species does have nitrogenase activity. On the other hand, *Treponema primitia* ferments carbohydrates only to acetate, with little or no nitrogenase activity. *Treponema azotonutricium* and *Treponema primitia* are obligate anaerobes, and it is likely that other hindgut spirochetes are anaerobic as they become nonmotile and begin to disintegrate when exposed to air. Hindgut spirochetes do not invade the hindgut epithelium, and the insects harboring them appear vigorous and healthy. It has been suggested that hindgut spirochetes may benefit the host. Eutick et al. (1978) observed that *Nasutitermes exitiosus* termites had a reduced life span when spirochetes were eliminated from the hindgut.

Motile spirochetes have been observed within the cytoplasm of hindgut protozoa (Kirby, 1941; Margulis et al., 1979;

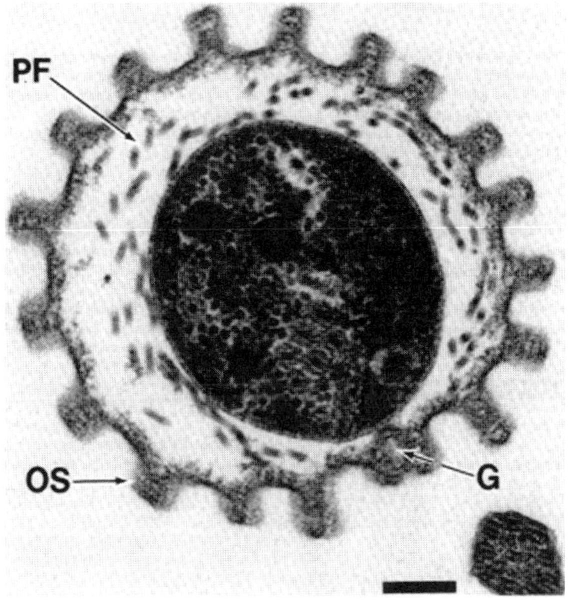

FIGURE 103. Transmission electron micrograph of transverse section of large spirochete from the hindgut of *Reticulitermes flavipes*, showing the proposed genus "*Pillotina*" with a crenulated outer sheath (OS), sillon or groove (G) and periplasmic flagella (PF). Scale bar = 0.2 μm.

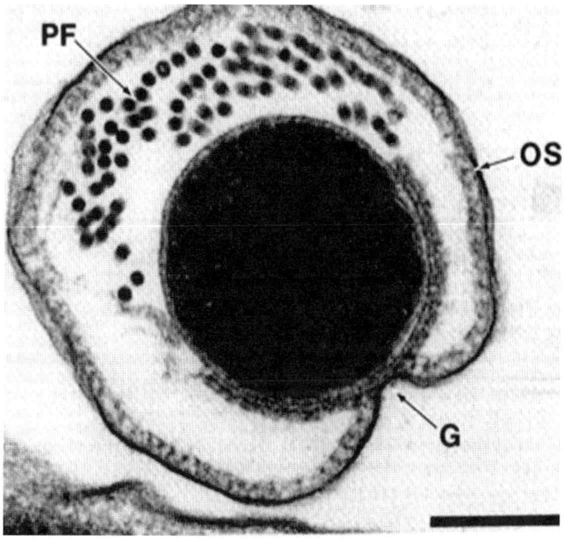

FIGURE 104. Transmission electron micrograph of transverse section of large spirochete from the hindgut of *Reticulitermes flavipes*, showing the proposed genus "*Clevelandina*" with sillon or groove (G), outer sheath (OS), and periplasmic flagella (PF). Scale bar = 0.2 μm.

To et al., 1978). However, it is not clear whether these intracellular spirochetes are truly symbiotic or were endocytosed into food vacuoles.

Taxonomic comments

Early workers classified spirochete-like organisms from termite hindguts as belonging to known spirochete genera such as "*Spirochaeta termitis*" (Dobell, 1910), "*Spirochaeta minei*" (von Prowazek, 1910), "*Spirochaeta leucotermitis*" (Hollande, 1922), and "*Spirochaeta staphylina*" (Ghidini and Archetti, 1939); "*Treponema termitis*" and "*Treponema minei*" (Dobell, 1912); and "*Cristispira termitis*" (Hollande, 1922). Electron microscopy was later used to confirm that these organisms possessed ultrastructural features characteristic of spirochetes such as a protoplasmic cylinder, periplasmic flagella (internal organelles), and an outer sheath (Margulis and Hinkle, 1992). Based on morphology and electron microscopic studies, new generic and specific epithets for true spirochetes of termite hindguts have been proposed. The framework for the morphometric analysis of large uncultivable spirochetes was formally proposed to revive four earlier-proposed species, namely *Pillotina calotermitidis*, *Diplocalyx calotermitidis*, *Hollandina pterotermitidis*, and *Clevelandina reticulitermitidis* (Bermudes et al., 1988; Margulis and Hinkle, 1992). More recently, also on the basis of morphometric analysis, two new, large pillotinaceous spirochetes, "*Canaleparolina darwiniensis*" and "*Diplocalyx cryptotermitidis*", have been proposed (Wier et al., 2000). "*Canaleparolina darwiniensis*", the large spirochete attached to the protozoan *Mixotrichia paradoxa*, is 0.5 μm × 25 μm in length and has multiple flagella in a 16:32:16 flagellar arrangement (Wier et al., 2000). "*Diplocalyx cryptotermitidis*", observed in the hindguts of the dry wood-eating termite *Cryptotermes cavifrons*, is smaller in diameter and has fewer flagella than the other large hindgut spirochetes.

Spirochetal sequences have been obtained from 16S rRNA clonal analysis of enrichments of *Mixtotricha paradoxa* and are deposited in GenBank under the accession numbers AJ458944, AJ458945, AJ458946, and AJ458947. These sequences represent four separate phylotypes that cluster deeply with the genus *Treponema* at about 90% similarity to treponemal phylotypes from termite hindguts. However, *in situ* hybridization experiments have not yet been performed to verify that the sequences obtained were derived from the spirochetes attached to *Mixotrichia paradoxa*. Consequently, these large spirochetes from the hindguts of termites and *Cryptocercus punctulatus* will be included in the family *Spirochaetaceae* until pure cultures are obtained and the phylogenetic placement among the spirochetal genera has been verified.

References

Ball, G.H. 1969. Organisms lining on and in protozoa. Res. Protozool. *3*: 567–718.

Bermudes, D., D. Chase and L. Margulis. 1988. Morphology as a basis for taxonomy of large spirochetes symbiotic in wood-eating cockroaches and termites: *Pillotina* gen. nov., nom. rev., *Pillotina calotermitidis* sp. nov., nom. rev., *Diplocalyx* gen. nov., nom. rev., *Diplocalyx calotermitidis* sp. nov., nom. rev., *Hollandina* gen. nov., nom. rev., *Hollandina pterotermitidis* sp. nov., nom. rev., and *Clevelandina reticulitermitidis* gen. nov., sp. nov. Int. J. Syst. Bacteriol. *38*: 291–302.

Bloodgood, R.A., K.R. Miller, T. Fitzharris and J.R. McIntosh. 1974. The ultrastructure of *Pyrsonympha* and its associated microorganisms. J. Morphol. *143*: 77–105.

Bloodgood, R.A. and T.P. Fitzharris. 1976. Specific associations of prokaryotes with symbiotic flagellate protozoa from hindgut of termite *Reticulitermes* and wood-eating roach *Cryptocercus*. Cytobios *17*: 103–122.

Breznak, J.A. and H.S. Pankratz. 1977. In situ morphology of the gut microbiota of wood-eating termites (*Reticulitermes flavipes* (Kollar) and *Coptotermes formosanus* Shiraki). Appl. Environ. Microbiol. *33*: 406–426.

Breznak, J.A. 1984. Hindgut spirochetes of termites and *Cryptocercus puntulatus*. *In* Bergey's Manual of Systematic Bacteriology, vol. 1 (edited by Krieg and Holt). Williams & Wilkins, Baltimore, pp. 67–70.

Cleveland, L.R. and A.V. Grimstone. 1964. The fine structure of the flagellate *Mixotrichia paradoxa* and its associated micro-organisms. Proc. R. Soc. Lond. Ser. B Biol. Sci. *159*: 668–686.

Damon, S.R. 1926. A note on the spirochaetes of termites. J. Bacteriol. *11*: 31–36.

Dobell, C.C. 1910. On some parasitic protozoa from Ceylon. Spolia Zeylanica *7*: 65–87.

Dobell, C.C. 1912. Researches on the spirochaetes and related organisms. Arch. Protistenkd. *26*: 117–240.

Eutick, H.L., P. Veivers, R.W. O'Brien and M. Slaytor. 1978. Dependence of higher termite, *Nasutitermes exitiosus* and lower termite, *Coptotermes lacteus* on their gut flora. J. Insect Physiol. *24*: 363–368.

Gharagozlou, I.D. 1968. Aspect infrastructural de *Diplocalyx calotermitidis* nov. gen., nov. sp., spirochaetale de l'intestin de *Calotermes flavicollis*. C. R. Acad. Sci. Ser. D *266*: 494–496.

Ghidini, G.M. and I. Archetti. 1939. Studi sulle termit; 2 - Le spirochete presenti in *Reticulitermes lucifugus*. Rossi. Riv. Biol. Coloniale *2*: 125–140.

Graber, J.R., J.R. Leadbetter and J.A. Breznak. 2004. Description of *Treponema azotonutricium* sp. nov. and *Treponema primitia* sp. nov., the first spirochetes isolated from termite guts. Appl. Environ. Microbiol. *70*: 1315–1320.

Grimstone, A.V. 1963. A note on the fine structure of a spirochaete. Q. J. Microsc. Sci. *104*: 145–153.

Hollande, A.C. 1922. Les spirochètes de termites; processus de division: formation du schizoplaste. Arch. Zool. Espt. Gén. Notes Rev. *61*: 23–28.

Hollande, A.C. and I.D. Gharagozlou. 1967. Morphologie infrastructurale de *Pillotina calotermitidis* nov. gen., nov. sp. spirochaetale de l'intestin de *Calotermes praecox*. C. R. Acad. Sci. *265*: 1309–1312.

Kirby, H., Jr. 1941. Organisms living on and in Protozoa. *In* Protozoa in Biological Research. Columbia University Press, New York, pp. 1009–1113.

Leidy, J. 1877. On intestinal parasites of *Termes flavipes*. Proc. Acad. Nat. Sci. *29*: 146–149.

Leidy, J. 1881. The parasites of the termites. J. Acad. Nat. Sci. *8*: 425–447.

Margulis, L., D. Chase and L.P. To. 1979. Possible evolutionary significance of spirochaetes. Proc. R. Soc. Lond. B *204*: 189–198.

Margulis, L., L.P. To and D. Chase. 1981. The genera *Pillotina, Hollandina and Diplocalyx*. *In* The Prokaryotes: a Handbook on Habitats, Isolation, and Identification of Bacteria (edited by Starr, Stolp, Trüper, Balows and Schlegel). Springer, New York, pp. 548–554.

Margulis, L. and G. Hinkle. 1992. Large symbiotic spirochetes: *Clevelandina, Cristispira, Diplocalyx, Hollandina, and Pillotina. In* The Prokaryotes: a Handbook on the Biology of Bacteria: Ecophysiology, Isolation, Identification, Applications, 2nd edn (edited by Balows, Trüper, Dworkin, Harder and Schleifer). Springer, New York, pp. 3965–3978.

Radek, R. and G. Nitsch. 2007. Ectobiotic spirochetes of flagellates from the termite *Mastotermes darwiniensis*: attachment and cyst formation. Eur. J. Protistol. *43*: 281–294.

Smith, H.E. and H.J. Arnott. 1974. Epibiotic and endobiotic bacteria associated with *Pyrsonympha vertens*-symbiotic protozoan of the termite *Reticulitermes flavipes*. Trans. Am. Microsc. Soc. *93*: 180–194.

Smith, H.E., H.E. Buhse and S.J. Stamler. 1975a. Possible formation and development of spirochaete attachment sites found on the surface of symbiotic polymastigote flagellates of the termite *Reticulitermes flavipes*. BioSystems *7*: 374–379.

Smith, H.E., S.J. Stamler and H.E. Buhse. 1975b. A scanning electron microscope survey of the surface features of polymastigote flagellates from *Reticulitermes flavipes*. Trans. Am. Microsc. Soc. *94*: 401–410.

To, L.P., L. Margulis and A.T.W. Cheung. 1978. Pillotinas and hollandinas: distribution and behavior of large spirochetes symbiotic in termites. Microbios *22*: 103–133.

von Prowazek, S. 1910. Parasitische Protozoen aus Japan, gesammelt von Herrn Dr. Mine in Fukuoka. Arch. Schiffs-Trop. Hyg. *14*: 297–302.

Wier, A., J. Ashen and L. Margulis. 2000. *Canaleparolina darwiniensis*, gen. nov., sp. nov., and other pillotinaceous spirochetes from insects. Int. Microbiol. *3*: 213–223.

Phylum XVI. Tenericutes Murray 1984a, 356[VP] (Effective publication: Murray 1984b, 33.)

Daniel R. Brown

Ten.er'i.cutes. L. adj. *tener* tender; L. fem. n. *cutis* skin; N.L. fem. n. *Tenericutes* prokaryotes of a soft pliable nature indicative of a lack of a rigid cell wall.

Members of the *Tenericutes* are **wall-less bacteria that do not synthesize precursors of peptidoglycan**.

Further descriptive information

The nomenclatural type by monotypy (Murray, 1984a) is the class *Mollicutes*, which consists of very small prokaryotes that are devoid of cell walls. Electron microscopic evidence for the absence of a cell wall was mandatory for describing novel species of mollicutes until very recently. Genes encoding the pathways for peptidoglycan biosynthesis are absent from the genomes of more than 15 species that have been annotated to date. Some species do possess an extracellular glycocalyx. The absence of a cell wall confers such mechanical plasticity that most mollicutes are readily filterable through 450 nm pores and many species have some cells in their populations that are able to pass through 220 nm or even 100 nm filters. However, they may vary in shape from coccoid to flask-shaped cells or helical filaments that reflect flexible cytoskeletal elements.

Taxonomic comments

To provide greater definition and formal nomenclature for vernacular names used in the 8th edition of *Bergey's Manual of Determinative Bacteriology* (Bergey VIII; Buchanan and Gibbons, 1974), Gibbons and Murray (1978) proposed that the higher taxa of prokaryotes be subdivided primarily according to the presence and character, or absence, of a rigid or semirigid cell wall as reflected in the determinative Gram reaction. Similar to the non-hierarchical groupings of Bergey VIII, which were based on a few readily determined criteria, the "wall-deficient" organisms grouped together in the first edition of *The Prokaryotes* included the mollicutes (Starr et al., 1981). While acknowledging the emerging 16S rRNA-based evidence that indicated a phylogenetic relationship between mollicutes and certain Gram-stain-positive bacteria in the division *Firmicutes*, Murray (1984b) proposed the separate division *Tenericutes* for the stable and distinctive group of wall-less species that are not simply an obvious subset of the *Firmicutes*.

The approved divisional rank of *Tenericutes* and the assignment of class *Mollicutes* as its nomenclatural type (Murray, 1984a) were adopted by the International Committee on Systematic Bacteriology's Subcommittee on the Taxonomy of *Mollicutes* (Tully, 1988) and subsequent valid taxonomic descriptions assigned novel species of mollicutes to the *Tenericutes*. However, the second (1992) and third (2007) editions of *The Prokaryotes* described the mollicutes instead as *Firmicutes* with low G+C DNA. The Subcommittee considered this to be an unfortunate grouping: "While the organisms are evolutionarily related to certain clostridia, the absence of a cell wall cannot be equated with Gram reaction positivity or with other members of the *Firmicutes*. It is unfortunate that workers involved in determinative bacteriology have a reference in which wall-free prokaryotes are described as Gram-positive bacteria" (Tully, 1993a). Despite numerous valid assignments of novel species of mollicutes to the *Tenericutes* during the intervening years, the class *Mollicutes* was still included in the phylum *Firmicutes* in the most recent revision of the *Taxonomic Outline of Bacteria and Archaea* (TOBA), which is based solely on the phylogeny of 16S rRNA genes (Garrity et al., 2007). The taxon *Tenericutes* is not recognized in the TOBA, although paradoxically it is the phylum consisting of the *Mollicutes* in the most current release of the Ribosomal Database Project (Cole et al., 2009). Mollicutes are specifically excluded from the most recently emended description of the *Firmicutes* in *Bergey's Manual of Systematic Bacteriology* (2nd edition, volume 3; De Vos et al., 2009) on the grounds of their lack of rigid cell walls plus analyses of strongly supported alternative universal phylogenetic markers, including RNA polymerase subunit B, the chaperonin GroEL, several different aminoacyl tRNA synthetases, and subunits of F_0F_1-ATPase (Ludwig et al., 2009; Ludwig and Schleifer, 2005). The taxonomic dignity of *Tenericutes* bestowed by its original formal validation, and upheld by a quarter of a century of valid descriptions of novel species of mollicutes, has therefore been respected in this volume of *Bergey's Manual*.

Type order: **Mycoplasmatales** Freundt 1955, 71[AL] emend. Tully, Bové, Laigret and Whitcomb 1993b, 382.

References

Buchanan, R.E. and N.E. Gibbons (editors). 1974. Bergey's Manual of Determinative Bacteriology, 8th edn. Williams & Wilkins, Baltimore.

Cole, J.R., Q. Wang, E. Cardenas, J. Fish, B. Chai, R.J. Farris, A.S. Kulam-Syed-Mohideen, D.M. McGarrell, T. Marsh, G.M. Garrity and J.M. Tiedje. 2009. The Ribosomal Database Project: improved alignments and new tools for rRNA analysis. Nucleic Acids Res. *37*: (Database issue): D141–D145.

De Vos, P., G. Garrity, D. Jones, N.R. Krieg, W. Ludwig, F.A. Rainey, K.H. Schleifer and W.B. Whitman. 2009. *In* Bergey's Manual of Systematic Bacteriology, 2nd edn, vol. 3. Springer, New York.

Freundt, E.A. 1955. The classification of the pleuropneumoniae group of organisms (*Borrelomycetales*). Int. Bull. Bacteriol. Nomencl. Taxon. *5*: 67–78.

Garrity, G.M., T.G. Lilburn, J.R. Cole, S.H. Harrison, J. Euzéby and B.J. Tindall. 2007. The Taxonomic Outline of the Bacteria and Archaea, Release 7.7, Part 11 – The Bacteria: Phyla *Planctomycetes, Chlamydiae, Spirochaetes, Fibrobacteres, Acidobacteria, Bacteroidetes, Fusobacteria,*

Verrucomicrobia, Dictyoglomi, Gemmatomonadetes, and *Lentisphaerae.* pp. 540–595. (http://www.taxonomicoutline.org/).

Gibbons, N.E. and R.G.E. Murray. 1978. Proposals concerning the higher taxa of bacteria. Int. J. Syst. Bacteriol. *28*: 1–6.

Ludwig, W. and K.H. Schleifer. 2005. Molecular phylogeny of bacteria based on comparative sequence analysis of conserved genes. *In* Microbial Phylogeny and Evolution, Concepts and Controversies, (edited by Sapp). Oxford University Press, New York, pp. 70–98.

Ludwig, W., K.H. Schleifer and W.B. Whitman. 2009. Revised road map to the phylum *Firmicutes*. *In* Bergey's Manual of Systematic Bacteriology, 2nd edn, vol. 3, The *Firmicutes* (edited by De Vos, Garrity, Jones, Krieg, Ludwig, Rainey, Schleifer and Whitman). Springer, New York, pp. 1–13.

Murray, R.G.E. 1984a. *In* Validation of the publication of new names and new combinations previously effectively published outside the IJSB. List no. 15. Int. J. Syst. Bacteriol. *34*: 355–357.

Murray, R.G.E. 1984b. The higher taxa, or, a place for everything...? *In* Bergey's Manual of Systematic Bacteriology, vol. 1 (edited by Krieg and Holt). Williams & Wilkins, Baltimore, pp. 31–34.

Starr, M.P., H. Stolp, H.G. Trüper, A. Balows and H.G. Schlegel (editors). 1981. The Prokaryotes. Springer, Berlin.

Tully, J.G. 1988. International Committee on Systematic Bacteriology, Subcommittee on the Taxonomy of Mollicutes, Minutes of the Interim Meeting, 25 and 28 August 1986, Birmingham, Alabama. Int. J. Syst. Bacteriol. *38*: 226–230.

Tully, J.G. 1993a. International Committee on Systematic Bacteriology, Subcommittee on the Taxonomy of Mollicutes, Minutes of the Interim Meetings, 1 and 2 August, 1992, Ames, Iowa. Int. J. Syst. Bacteriol. *43*: 394–397.

Tully, J.G., J.M. Bové, F. Laigret and R.F. Whitcomb. 1993b. Revised taxonomy of the class *Mollicutes* – proposed elevation of a monophyletic cluster of arthropod-associated mollicutes to ordinal rank (*Entomoplasmatales* ord. nov.), with provision for familial rank to separate species with nonhelical morphology (*Entomoplasmataceae* fam. nov.) from helical species (*Spiroplasmataceae*), and emended descriptions of the order *Mycoplasmatales*, family *Mycoplasmataceae*. Int. J. Syst. Bacteriol. *43*: 378–385.

Class I. **Mollicutes** Edward and Freundt 1967, 267[AL]

Daniel R. Brown, Meghan May, Janet M. Bradbury and Karl-Erik Johansson

Mol'li.cutes or Mol.li.cu'tes. L. adj. *mollis* soft, pliable; L. fem. n. *cutis* skin; N.L. fem. pl. n. *Mollicutes* class with pliable cell boundary.

Very small prokaryotes totally **devoid of cell walls**. Bounded by a plasma membrane only. Incapable of synthesis of peptidoglycan or its precursors. Consequently resistant to penicillin and its derivatives and sensitive to lysis by osmotic shock, detergents, alcohols, and specific antibody plus complement. **Gram-stain-negative** due to lack of cell wall, but constitute a distinct phylogenetic lineage within the Gram-stain-positive bacteria (Woese et al., 1980). **Pleomorphic**, varying from spherical or flask-shaped structures to branched or helical filaments. The coccoid and flask-shaped cells usually range from 200–500 nm in diameter, although cells as large as 2000 nm have been seen. Replicate by binary fission, but genome replication may precede cytoplasmic division, leading to the formation of multinucleated filaments. Colonies on solid media are very small, usually much less than 1 mm in diameter. The organisms tend to penetrate and grow inside the solid medium. Under suitable conditions, almost all species **form colonies that have a characteristic fried-egg appearance**. Usually nonmotile, but some species show **gliding motility**. Species that occur as helical filaments show **rotary, flexional, and translational motility**. No resting stages are known.

The species recognized so far can be grown on artificial cell-free media of varying complexity, although certain strains may be more readily isolated by cell-culture procedures. Many "*Candidatus*" species have been proposed and characterized at the molecular level, but not yet cultivated axenically. Most cultivable species **require sterols** and fatty acids for growth. However, members of some genera can grow well in either serum-free media or serum-free media supplemented with polyoxyethylene sorbitan. Most species are **facultative anaerobes**, but some are **obligate anaerobes** that are killed by exposure to minute quantities of oxygen. No tricarboxylic acid cycle enzymes, quinones, or cytochromes have been found.

All mollicutes are commensals or parasites, occurring in a wide range of vertebrate, insect, and plant hosts. Many are significant pathogens of humans, animals, insects, or plants. Genome sizes range from 580 to 2200 kbp, among the smallest recorded in prokaryotes. The genomes of more than 20 species have been completely sequenced and annotated to date (Table 135). The G+C content of the DNA is usually low, ~23–34 mol%, but in some species is as high as ~40 mol% (Bd, T_m). Can be distinguished from other bacteria in having only one or two rRNA operons (one species of *Mesoplasma* has three) and an RNA polymerase that is resistant to rifampin. The 5S rRNA contains fewer nucleotides than that of other bacteria and there are fewer tRNA genes. In some genera, instead of a stop, the UGA codon encodes tryptophan. Plasmids and viruses (phage) occur in some species.

Type order: **Mycoplasmatales** Freundt 1955, 71[AL] emend. Tully, Bové, Laigret and Whitcomb 1993, 382.

Further descriptive information

Table 136 summarizes the present classification of the *Mollicutes* into families and genera and provides the major distinguishing characteristics of these taxa. The trivial term mycoplasma has been used to denote any species included in the class *Mollicutes*, but the term mollicute(s) is now considered most appropriate as the trivial name for all members of the class, so that the trivial name mycoplasma can be retained only for members of the genus *Mycoplasma*. Hemotropic mycoplasmas are referred to by the trivial name hemoplasmas. The trivial names ureaplasma, entomoplasma, mesoplasma, spiroplasma, acholeplasma, anaeroplasma, and asteroleplasma are commonly used when reference is made to members of the corresponding genus.

Their 16S rRNA gene sequences usefully place the mollicutes into phylogenetic groups (Johansson, 2002; Weisburg et al., 1989) and an analysis of 16S rRNA gene sequences is now mandatory for characterization of novel species (Brown et al., 2007). 16S rRNA gene sequences have also shown that certain hemotropic bacteria, previously considered to be members of the *Rickettsia*, belong to the order *Mycoplasmatales*.

TABLE 135. Characteristics of sequenced mollicute genomes[a]

Species	*Mycoplasma agalactiae*	*Mycoplasma arthritidis*	*Mycoplasma capricolum* subsp. *capricolum*	*Mycoplasma conjunctivae*	*Mycoplasma gallisepticum*	*Mycoplasma genitalium*	*Mycoplasma hominis*	*Mycoplasma hyopneumoniae*[b]	*Mycoplasma mobile*	*Mycoplasma mycoides* subsp. *mycoides* SC	*Mycoplasma penetrans*	*Mycoplasma pneumoniae*	*Mycoplasma pulmonis*	*Mycoplasma synoviae*	*Mesoplasma florum*	*Ureaplasma parvum*[c]	*Ureaplasma urealyticum*[c]	*Acholeplasma laidlawii*	"*Candidatus* Phytoplasma asteris"[d]
Strain	PG2^{T}	158L3-1	ATCC 27343^{T}	HRC/581^{T}	R	G-37^{T}	PG21^{T}	J^{T}	163K^{T}	PG1^{T}	HF-2	M129	UAB CTIP	53	L1^{T}	ATCC 700970	ATCC 33699	PG8^{T}	AYWB
Size (mb)	0.88	0.82	1.0	0.85	0.99	0.58	0.67	0.90	0.78	1.21	1.36	0.82	0.96	0.80	0.79	0.75	0.87	1.5	0.71
DNA G+C content (mol%)	29	30	23	28	31	31	27	28	25	24	26	40	26	28	27	25	25	31	26
Open reading frames	742	631	812	727	726	475	537	657	633	1016	1037	689	782	659	687	653	692	1433	671
Hypothetical genes (%)	50	nd	nd	45	37	21	36	nd	27	41	41	38	38	33	nd	33	nd	nd	nd
Coding density (%)	87	80	88	90	91	90	90	87	90	80	88	87	90	91	92	91	89	90	73
References	Sirand-Pugnet et al. (2007)	Dybvig et al. (2008)	J. Glass and others, unpublished.	Calderon-Copete et al. (2009)	Papazisi et al. (2003)	Fraser et al. (1995)	Pereyre et al. (2009)	Vasconcelos et al. (2005)	Jaffe et al. (2004)	Westberg et al. (2004)	Sasaki et al. (2002)	Himmelreich et al. (1996)	Chambaud et al. (2001)	Vasconcelos et al. (2005)	Knight et al. (2004)	Glass et al. (2000)	na	na	Bai et al. (2006)

[a]nd, Not determined; na, not available.
[b]*Mycoplasma hyopneumoniae* strains 232 and 7448 were sequenced by Minion et al. (2004) and Vasconcelos et al. (2005).
[c]Data for *Ureaplasma parvum* and *Ureaplasma urealyticum* refer to serovars 3 and 10, respectively; serovars 1–14 were sequenced and deposited directly into GenBank.
[d]AYWB, Aster yellows witches' broom; the onion yellows strain was sequenced by Oshima et al. (2004).

TABLE 136. Description of the class *Mollicutes*[a]

Order	Family	Genus	Species[b]	Genome size range (kbp)	Cholesterol requirement	Habitat[c]	Defining features
I. *Mycoplasmatales*	*Mycoplasmataceae*	*Mycoplasma*	116, 9, 1, 4	580–1,350	+	H, A	
I. *Mycoplasmatales*	*Mycoplasmataceae*	*Ureaplasma*	7, 0, 0, 0	760–1,140	+	H, A	Urea hydrolysis
I. *Mycoplasmatales*[d]	*Incertae sedis*	*Eperythrozoon*	4, 0, 0, 0	Nd	nd	A	Hemotropic
I. *Mycoplasmatales*[d]	*Incertae sedis*	*Haemobartonella*	1, 0, 0, 0	Nd	nd	A	Hemotropic
II. *Entomoplasmatales*	*Entomoplasmataceae*	*Entomoplasma*	6, 0, 0, 0	870–900	+	N, P	
II. *Entomoplasmatales*	*Entomoplasmataceae*	*Mesoplasma*	11, 0, 0, 0	825–930	–	N, P	Growth with PES
II. *Entomoplasmatales*	*Spiroplasmataceae*	*Spiroplasma*	37, 0, 0, 0	780–2,220	+	N, P	Helical morphology
III. *Acholeplasmatales*	*Acholeplasmataceae*	*Acholeplasma*	18, 0, 0, 0	1,500–1,650	–	A, N, P	
III. *Acholeplasmatales*	*Incertae sedis*	"*Candidatus* Phytoplasma"	0, 27, 0, 0	530–1,350	nd	N, P	Not yet cultured *in vitro*
IV. *Anaeroplasmatales*	*Anaeroplasmataceae*	*Anaeroplasma*	4, 0, 0, 0	1,500–1,600	+	A	Strictly anaerobic
IV. *Anaeroplasmatales*	*Anaeroplasmataceae*	*Asteroleplasma*	1, 0, 0, 0	1,500	–	A	Strictly anaerobic

[a]nd, Not determined; PES, polyoxyethylene sorbitan.

[b]Numbers of species: valid, *Candidatus*, *incertae sedis*, invalid.

[c]H, human; A, vertebrate animal; N, invertebrate animal; P, plant.

[d]Affiliation of the constituent genera within the *Mycoplasmatales* has not been formalized.

The phytoplasmas, a large group of uncultivated mollicutes occurring as agents that can cycle between plant and invertebrate hosts, have been given a provisional "*Candidatus* Phytoplasma" genus designation. The 16S rRNA gene sequences from at least ten unique phylotypes, recently discovered among the human microbial flora through global 16S rRNA gene PCR (Eckburg et al., 2005), cluster distinctly enough to suggest the existence of a yet-uncircumscribed order within the class (May et al., 2009).

Non-helical mollicutes isolated from insects and plants have been placed in the order *Entomoplasmatales*, the two genera of which are distinguished by their requirement for cholesterol (Tully et al., 1993). Members of the genus *Entomoplasma* require cholesterol; those of the genus *Mesoplasma* do not. However, sterol requirements do not correlate well with phylogenetic analyses in other groups. At least four species of spiroplasmas do not require sterol for growth, but they do not form a phylogenetic group. Within the order including the obligately anaerobic mollicutes *Anaeroplasmatales*, members of the genus *Anaeroplasma* require sterols for growth, whereas members of the genus *Asteroleplasma* do not (Robinson et al., 1975; Robinson and Freundt, 1987). Thus, sterol requirement is a useful phylogenetic marker only in the *Acholeplasmatales* and *Anaeroplasmatales*.

In the past, there was some risk of confusing mollicutes with wall-less "L (Lister)-phase" variants of certain other bacteria, but simple PCR-based analyses of 16S rRNA or other gene sequences now obviate that concern. Wall-less members of the genus *Thermoplasma*, previously assigned to the *Mollicutes*, are *Archaea* and differ from all other members of this class in their 16S rRNA nucleotide sequences plus a number of important features relating to their mode of life and metabolism. Thus, they are quite unrelated to this class (Fox et al., 1980; Razin and Freundt, 1984; Woese et al., 1980). Members of the *Erysipelothrix* line of descent, also formerly assigned to the *Mollicutes*, are now assigned to the class *Erysipelotrichi* in the phylum *Firmicutes* (Stackebrandt, 2009; Verbarg et al., 2004).

Taxonomic comments

The origin of mollicutes and their relationships to other prokaryotes was controversial for many years, especially since their small genomes and comparative phenotypic simplicity suggested that they might have descended from a primitive organism. The first comparative phylogenetic analysis of the origin of mollicutes was carried out by oligonucleotide mapping of 16S rRNA gene sequences by Woese et al. (1980). The organisms then assigned to the genera *Mycoplasma*, *Spiroplasma*, and *Acholeplasma* seemed to have arisen by reductive evolution as a deep branch of the clostridial lineage leading to the genera *Bacillus* and *Lactobacillus*. This relationship had been proposed earlier (Neimark, 1979) because the low G+C mollicutes, streptococci, and lactic acid bacteria share characteristic enzymes. In particular, acholeplasma and streptococcus aldolases show high amino acid sequence similarity.

These findings were generally confirmed by studies of 5S rRNA gene sequences (Rogers et al., 1985), which included a number of acholeplasmas, anaeroplasmas, mycoplasmas, ureaplasmas, and *Clostridium innocuum*. Dendrograms constructed from evolutionary distance matrices indicated that the mollicutes form a coherent phylogenetic group that developed as a branch of the *Firmicutes*. The initial event in this evolution was proposed to be the formation of the *Acholeplasma* branch, although the position of the *Anaeroplasma* species (*Anaeroplasma bactoclasticum* and *Anaeroplasma abactoclasticum*) was not definitely established within these dendrograms. Formation of the acholeplasmas may have coincided with a reduction in genome size to about 1500–1700 kb and loss of the cell wall. With a genome size similar to the acholeplasmas, the spiroplasmas may have formed from the acholeplasmas. Later independent genome reductions to 500–1000 kb may have led to the origins of the sterol-requiring mycoplasma and ureaplasma lineages. The more extensive phylogenetic analysis of Weisburg et al. (1989) examined the 16S rRNA gene sequences of about 50 species of mollicutes and confirmed a number of these observations and provided additional insights into mollicute evolution. These results also indicated that the acholeplasmas formed upon the initial divergence of mollicutes from clostridial ancestors. Further divergence of this stem led to the sterol-requiring, anaerobic *Anaeroplasma* and the non-sterol-requiring *Asteroleplasma* branches. The *Spiroplasma* branch also appeared to originate from within the acholeplasmas, with further evolution leading to a series of repeated and independent genome reductions from nearly 2000 kb to 600–1200 kb to yield the *Mycoplasma* and *Ureaplasma* lineages.

Based on the phylogeny of 16S rRNA genes, the class *Mollicutes* was included in the phylum *Firmicutes* in the most recent revision of the *Taxonomic Outline of Bacteria and Archaea* (Garrity et al., 2007). However, the *Mollicutes* are excluded from the most recently emended description of the *Firmicutes* (De Vos et al., 2009) based on alternative phylogenetic markers, including RNA polymerase subunit B, the chaperonin GroEL, several different aminoacyl tRNA synthetases, and subunits of F_0F_1-ATPase (Ludwig and Schleifer, 2005).

The Weisburg et al. (1989) study also proposed five additional phylogenetic groupings within the mollicutes, including the anaeroplasma, asteroleplasma, spiroplasma, pneumoniae, and hominis groups (Figure 105). Phytoplasmas are similar to acholeplasmas in their 16S rRNA gene sequences and UGA codon usage (IRPCM Phytoplasma/Spiroplasma Working Team – Phytoplasma Taxonomy Group, 2004). They probably diverged from acholeplasmas at about the same time as the split of spiroplasmas into helical and non-helical lineages (Maniloff, 2002). The modern species concept for mollicutes is justified principally by DNA–DNA hybridization, serology, and 16S rRNA gene sequence similarity (Brown et al., 2007). A large number of individual species have been assigned to phylogenetic groups, clusters, and subclusters that also share other characteristics, although the cluster boundaries are sometimes subjective (Harasawa and Cassell, 1996; Johansson, 2002; Pettersson et al., 2000, 2001).

Lastly, the type order *Mycoplasmatales* is assigned to the class as this clearly appeared to be the intention of Edward and Freundt (1967) in their paper entitled "Proposal for *Mollicutes* as name of the class established for the order *Mycoplasmatales*".

Acknowledgements

The lifetime achievements in mycoplasmology and major contributions to the foundation of this material by Joseph G. Tully are gratefully acknowledged. Daniel R. Brown and Meghan May were supported by NIH grant 5R01GM076584.

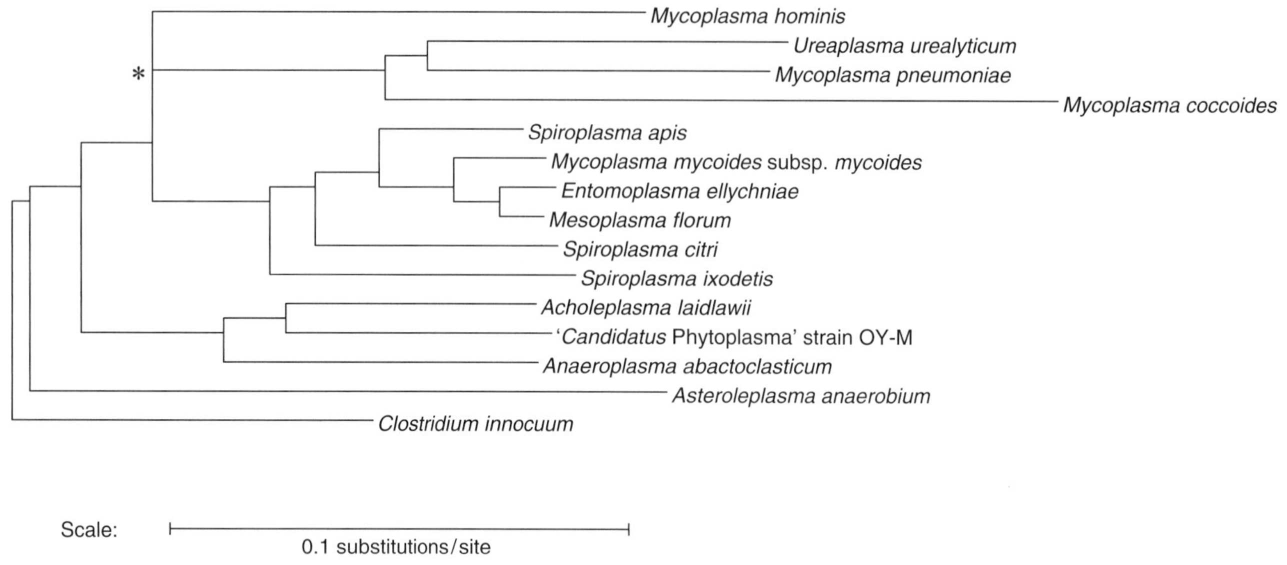

FIGURE 105. Phylogenetic grouping of the class *Mollicutes*. The phylogram was based on a Jukes–Cantor corrected distance matrix and weighted neighbor-joining analysis of the 16S rRNA gene sequences of the type genera, plus representatives of other major clusters within the *Mycoplasmatales* and *Entomoplasmatales* and a phytoplasma. *Clostridium innocuum* was the outgroup. All bootstrap values (100 replicates) are >50% except where indicated (asterisk).

Further reading

Barile, M.F., S. Razin, J.G. Tully and R.F. Whitcomb (Editors). 1979, 1985, 1989. The Mycoplasmas (five volumes). Academic Press, New York.

Maniloff, J., R.N. McElhaney, L.R. Finch and J.B. Baseman (editors). 1992. Mycoplasmas: Molecular Biology and Pathogenesis. American Society for Microbiology, Washington, D.C.

Murray, R.G.E. 1984. The higher taxa, or, a place for everything…?. *In* Bergey's Manual of Systematic Bacteriology, vol. 1 (edited by Krieg and Holt). Williams & Wilkins, Baltimore, pp. 31–34.

Razin, S. and J.G.E. Tully. 1995. Molecular and Diagnostic Procedures in Mycoplasmology, vol. 1. Academic Press, San Diego.

Taylor-Robinson, D. and J. Bradbury. 1998. Mycoplasma diseases. *In* Topley and Wilson's Principles and Practice of Microbiology, vol. 3 (edited by Hausler and Sussman). Edward Arnold, London, pp. 1013–1037.

Taylor-Robinson, D. and J.G. Tully. 1998. Mycoplasmas, ureaplasmas, spiroplasmas, and related organisms. *In* Topley and Wilson, Principles and Practice of Microbiology, 9th edn, vol. 2 (edited by Balows and Duerden). Arnold Publishers, London, pp. 799–827.

Tully, J.G. and S. Razin (editors). 1996. Molecular and Diagnostic Procedures in Mycoplasmology, vol. 2. Academic Press, San Diego, CA.

References

Bai, X., J. Zhang, A. Ewing, S.A. Miller, A. Jancso Radek, D.V. Shevchenko, K. Tsukerman, T. Walunas, A. Lapidus, J.W. Campbell and S.A. Hogenhout. 2006. Living with genome instability: the adaptation of phytoplasmas to diverse environments of their insect and plant hosts. J. Bacteriol. *188*: 3682–3696.

Brown, D., R. Whitcomb and J. Bradbury. 2007. Revised minimal standards for description of new species of the class *Mollicutes* (division *Tenericutes*). Int. J. Syst. Evol. Microbiol. *57*: 2703–2719.

Calderon-Copete, S.P., G. Wigger, C. Wunderlin, T. Schmidheini, J. Frey, M.A. Quail and L. Falquet. 2009. The *Mycoplasma conjunctivae* genome sequencing, annotation and analysis. BMC Bioinformatics *10 Suppl 6*: S7.

Chambaud, I., R. Heilig, S. Ferris, V. Barbe, D. Samson, F. Galisson, I. Moszer, K. Dybvig, H. Wroblewski, A. Viari, E.P. Rocha and A. Blanchard. 2001. The complete genome sequence of the murine respiratory pathogen *Mycoplasma pulmonis*. Nucleic Acids Res. *29*: 2145–2153.

De Vos, P., G. Garrity, D. Jones, N.R. Krieg, W. Ludwig, F.A. Rainey, K. H. Schleifer and W.B. Whitman. 2009. *In* Bergey's Manual of Systematic Bacteriology, 2nd edn, vol. 3. Springer, New York.

Dybvig, K., C. Zuhua, P. Lao, D.S. Jordan, C.T. French, A.H. Tu and A.E. Loraine. 2008. Genome of *Mycoplasma arthritidis*. Infect. Immun. *76*: 4000–4008.

Eckburg, P., E. Bik, C. Bernstein, E. Purdom, L. Dethlefsen, M. Sargent, S. Gill, K. Nelson and D. Relman. 2005. Diversity of the human intestinal microbial flora. Science *308*: 1635–1638.

Edward, D.G.ff. and E.A. Freundt. 1967. Proposal for *Mollicutes* as name of the class established for the order *Mycoplasmatales*. Int. J. Syst. Bacteriol. *17*: 267–268.

Fox, G.E., E. Stackebrandt, R.B. Hespell, J. Gibson, J. Maniloff, T.A. Dyer, R.S. Wolfe, W.E. Balch, R.S. Tanner, L.J. Magrum, L.B. Zablen, R. Blakemore, R. Gupta, L. Bonen, B.J. Lewis, D.A. Stahl, K.R. Luehrsen, K.N. Chen and C.R. Woese. 1980. The phylogeny of prokaryotes. Science *209*: 457–463.

Fraser, C.M., J.D. Gocayne, O. White, M.D. Adams, R.A. Clayton, R.D. Fleischmann, C.J. Bult, A.R. Kerlavage, G. Sutton, J.M. Kelley and et al. 1995. The minimal gene complement of Mycoplasma genitalium. Science *270*: 397–403.

Freundt, E.A. 1955. The classification of the pleuropneumoniae group of organisms (*Borrelomycetales*). Int. Bull. Bacteriol. Nomencl. Taxon. *5*: 67–78.

Garrity, G.M., T.G. Lilburn, J.R. Cole, S.H. Harrison, J. Euzéby and B.J. Tindall. 2007. The Taxonomic Outline of the Bacteria and Archaea, Release 7.7, Part 11 – The Bacteria: phyla *Planctomycetes, Chlamydiae, Spirochaetes, Fibrobacteres, Acidobacteria, Bacteroidetes, Fusobacteria, Verrucomicrobia, Dictyoglomi, Gemmatomonadetes,* and *Lentisphaerae.* pp. 540–595. (http://www.taxonomicoutline.org/).

Glass, J.I., E.J. Lefkowitz, J.S. Glass, C.R. Heiner, E.Y. Chen and G.H. Cassell. 2000. The complete sequence of the mucosal pathogen Ureaplasma urealyticum. Nature *407*: 757–762.

Harasawa, R. and G.H. Cassell. 1996. Phylogenetic analysis of genes coding for 16S rRNA in mammalian ureaplasmas. Int. J. Syst. Bacteriol. *46*: 827–829.

Himmelreich, R., H. Hilbert, H. Plagens, E. Pirkl, B.C. Li and R. Herrmann. 1996. Complete sequence analysis of the genome of the bacterium Mycoplasma pneumoniae. Nucleic Acids Res. *24*: 4420–4449.

IRPCM Phytoplasma/Spiroplasma Working Team - Phytoplasma Taxonomy Group. 2004. Description of the genus '*Candidatus* Phytoplasma', a taxon for the wall-less non-helical prokaryotes that colonize plant phloem and insects. Int. J. Syst. Evol. Microbiol. *54*: 1243–1255.

Jaffe, J.D., N. Stange-Thomann, C. Smith, D. DeCaprio, S. Fisher, J. Butler, S. Calvo, T. Elkins, M.G. Fitzgerald, N. Hafez, C.D. Kodira, J. Major, S. Wang, J. Wilkinson, R. Nicol, C. Nusbaum, B. Birren, H.C. Berg and G.M. Church. 2004. The complete genome and proteome of *Mycoplasma mobile.* Genome Res. *14*: 1447–1461.

Johansson, K.-E. 2002. Taxonomy of *Mollicutes.* In Molecular Biology and Pathogenicity of Mycoplasmas (edited by Razin and Herrmann). Kluwer Academic/Plenum Publishers, New York, pp. 1–29.

Knight, T.F., Jr. 2004. Reclassification of *Mesoplasma pleciae* as *Acholeplasma pleciae* comb. nov. on the basis of 16S rRNA and *gyrB* gene sequence data. Int. J. Syst. Evol. Microbiol. *54*: 1951–1952.

Ludwig, W. and K.H. Schleifer. 2005. Molecular phylogeny of bacteria based on comparative sequence analysis of conserved genes. *In* Microbial Phylogeny and Evolution, Concepts and Controversies (edited by Sapp). Oxford University Press, New York, pp. 70–98.

Maniloff, J. 2002. Phylogeny and Evolution. *In* Molecular Biology and Pathogenicity of Mycoplasmas. Kluwer Academic/Plenum Publishers, pp. 31–43.

May, M., R.F. Whitcomb and D.R. Brown. 2009. Mycoplasma and related organisms. *In* CRC Practical Handbook of Microbiology, 2nd edn. (edited by Goldman and Green). Taylor & Francis, pp. 456–479.

Minion, F.C., E.J. Lefkowitz, M.L. Madsen, B.J. Cleary, S.M. Swartzell and G.G. Mahairas. 2004. The genome sequence of Mycoplasma hyopneumoniae strain 232, the agent of swine mycoplasmosis. J. Bacteriol. *186*: 7123–7133.

Neimark, H.C. 1979. Phylogenetic relationships between mycoplasmas and other prokaryotes. *In* The Mycoplasmas, vol. 1 (edited by Barile and Razin). Academic Press, New York, pp. 43–61.

Oshima, K., S. Kakizawa, H. Nishigawa, H.Y. Jung, W. Wei, S. Suzuki, R. Arashida, D. Nakata, S. Miyata, M. Ugaki and S. Namba. 2004. Reductive evolution suggested from the complete genome sequence of a plant-pathogenic phytoplasma. Nat. Genet. *36*: 27–29.

Papazisi, L., T. Gorton, G. Kutish, P. Markham, G. Browning, D. Nguyen, S. Swartzell, A. Madan, G. Mahairas and S. Geary. 2003. The complete genome sequence of the avian pathogen *Mycoplasma gallisepticum* strain R(low). Microbiology *149*: 2307–2316.

Pereyre, S., P. Sirand-Pugnet, L. Beven, A. Charron, H. Renaudin, A. Barre, P. Avenaud, D. Jacob, A. Couloux, V. Barbe, A. de Daruvar, A. Blanchard and C. Bebear. 2009. Life on arginine for *Mycoplasma hominis*: clues from its minimal genome and comparison with other human urogenital mycoplasmas. PLoS. Genet. *5*: e1000677.

Pettersson, B., J.G. Tully, G. Bolske and K.E. Johansson. 2000. Updated phylogenetic description of the *Mycoplasma hominis* cluster (Weisburg *et al.* 1989) based on 16S rDNA sequences. Int. J. Syst. Evol. Microbiol. *50*: 291–301.

Pettersson, B., J.G. Tully, G. Bolske and K.E. Johansson. 2001. Re-evaluation of the classical *Mycoplasma lipophilum* cluster (Weisburg *et al.* 1989) and description of two new clusters in the hominis group based on 16S rDNA sequences. Int. J. Syst. Evol. Microbiol. *51*: 633–643.

Razin, S. and E.A. Freundt. 1984. The *Mollicutes, Mycoplasmatales,* and *Mycoplasmataceae. In* Bergey's Manual of Systematic Bacteriology, vol. 1 (edited by Krieg and Holt). Williams & Wilkins, Baltimore, pp. 740–742.

Robinson, I.M., M.J. Allison and P.A. Hartman. 1975. *Anaeroplasma abactoclasticum* gen. nov., sp. nov., obligately anaerobic mycoplasma from rumen. Int. J. Syst. Bacteriol. *25*: 173–181.

Robinson, I.M. and E.A. Freundt. 1987. Proposal for an amended classification of anaerobic mollicutes. Int. J. Syst. Bacteriol. *37*: 78–81.

Rogers, M.J., J. Simmons, R.T. Walker, W.G. Weisburg, C.R. Woese, R.S. Tanner, I.M. Robinson, D.A. Stahl, G. Olsen, R.H. Leach and J. Maniloff. 1985. Construction of the mycoplasma evolutionary tree from 5S rRNA sequence data. Proc. Natl. Acad. Sci. U.S.A. *82*: 1160–1164.

Sasaki, Y., J. Ishikawa, A. Yamashita, K. Oshima, T. Kenri, K. Furuya, C. Yoshino, A. Horino, T. Shiba, T. Sasaki and M. Hattori. 2002. The complete genomic sequence of *Mycoplasma penetrans*, an intracellular bacterial pathogen in humans. Nucleic Acids Res. *30*: 5293–5300.

Sirand-Pugnet, P., C. Lartigue, M. Marenda, D. Jacob, A. Barre, V. Barbe, C. Schenowitz, S. Mangenot, A. Couloux, B. Segurens, A. de Daruvar, A. Blanchard and C. Citti. 2007. Being pathogenic, plastic, and sexual while living with a nearly minimal bacterial genome. PLoS. Genet. *3*: e75.

Stackebrandt, E. 2009. Class III. Erysipelotrichia. *In* Bergey's Manual of Systematic Bacteriology, vol. 3 (edited by de Vos, Garrity, Jones, Krieg, Ludwig, Rainey, Schleifer and Whitman). Springer, New York, p. 1298.

Tully, J.G., J.M. Bove, F. Laigret and R.F. Whitcomb. 1993. Revised taxonomy of the class *Mollicutes* - proposed elevation of a monophyletic cluster of arthropod-associated mollicutes to ordinal rank (*Entomoplasmatales* ord. nov.), with provision for familial rank to separate species with nonhelical morphology (*Entomoplasmataceae* fam. nov.) from helical species (*Spiroplasmataceae*), and emended descriptions of the order *Mycoplasmatales,* family *Mycoplasmataceae.* Int. J. Syst. Bacteriol. *43*: 378–385.

Vasconcelos, A.T. and a. coauthors. 2005. Swine and poultry pathogens: the complete genome sequences of two strains of *Mycoplasma hyopneumoniae* and a strain of *Mycoplasma synoviae.* J. Bacteriol. *187*: 5568–5577.

Verbarg, S., H. Rheims, S. Emus, A. Fruhling, R.M. Kroppenstedt, E. Stackebrandt and P. Schumann. 2004. Erysipelothrix inopinata sp. nov., isolated in the course of sterile filtration of vegetable peptone broth, and description of *Erysipelotrichaceae* fam. nov. Int. J. Syst. Evol. Microbiol. *54*: 221–225.

Weisburg, W., J. Tully, D. Rose, J. Petzel, H. Oyaizu, D. Yang, L. Mandelco, J. Sechrest, T. Lawrence and J. Van Etten. 1989. A phylogenetic analysis of the mycoplasmas: basis for their classification. J. Bacteriol. *171*: 6455–6467.

Westberg, J., A. Persson, A. Holmberg, A. Goesmann, J. Lundeberg, K.E. Johansson, B. Pettersson and M. Uhlen. 2004. The genome sequence of *Mycoplasma mycoides* subsp. *mycoides* SC type strain PG1T, the causative agent of contagious bovine pleuropneumonia (CBPP). Genome Res. *14*: 221–227.

Woese, C.R., J. Maniloff and L.B. Zablen. 1980. Phylogenetic analysis of the mycoplasmas. Proc. Natl. Acad. Sci. U. S. A. *77*: 494–498.

Order I. **Mycoplasmatales** Freundt 1955, 71[AL] emend. Tully, Bové, Laigret and Whitcomb 1993, 382

Daniel R. Brown, Meghan May, Janet M. Bradbury, Karl-Erik Johansson and Harold Neimark

My.co.plas.ma.ta′les. N.L. neut. n. *Mycoplasma*, *-atos* type genus of the order; *-ales* ending to denote an order; N.L. fem. pl. n. *Mycoplasmatales* the *Mycoplasma* order.

The first order in the class *Mollicutes* is assigned to a group of sterol-requiring, wall-less prokaryotes that occur as commensals or pathogens in a wide range of vertebrate hosts. The description of the order is essentially the same as for the class. A single family *Mycoplasmataceae* with two genera, *Mycoplasma* and *Ureaplasma*, recognizes the prominent and distinct characteristics of the assigned organisms, based on their sterol requirements for growth, the capacity of some to hydrolyze urea, and conserved 16S rRNA gene sequences.

Type genus: **Mycoplasma** Nowak 1929, 1349 nom. cons. Jud. Comm. Opin. 22, 1958, 166.

Further descriptive information

The entire class *Mollicutes* was encompassed initially by a single order. The elevation of acholeplasmas to ordinal rank (*Acholeplasmatales* Freundt, Whitcomb, Barile, Razin and Tully 1984) recognized their major distinctions in nutritional, biochemical, physiological, and genetic characteristics from other members of the class *Mollicutes*. Subsequently, additional orders were proposed to recognize the anaerobic mollicutes and the wall-less prokaryotes from plants and insects which were phylogenetically related to the remaining *Mycoplasmatales*. Thus, the *Anaeroplasmatales* (Robinson and Freundt, 1987) recognized the strictly anaerobic, wall-less prokaryotes first isolated from the bovine and ovine rumen, and *Entomoplasmatales* (Tully et al., 1993) provided a classification for a number of the mollicutes regularly associated with plant and insect hosts. On the basis of 16S rRNA gene sequence similarities (Johansson and Pettersson, 2002), the *Mycoplasmatales* and *Entomoplasmatales* represent a clade deeply split from the *Acholeplasmatales* and *Anaeroplasmatales*.

A growth requirement for cholesterol or serum is shared by the organisms assigned to the order *Mycoplasmatales*, as well as most other organisms within the class *Mollicutes*. Therefore, tests for cholesterol requirements are essential to classification. Earlier assessments of the growth requirements for cholesterol were based upon the capacity of organisms to grow in a number of serum-free broth preparations to which various concentrations of cholesterol were added (Edward, 1971; Razin and Tully, 1970). In this test, species that do not require exogenous sterol usually show no significant growth response to increasing cholesterol concentrations. Polyoxyethylene sorbitan (Tween 80) and palmitic acid should be included in the base medium because acholeplasmas such as *Acholeplasma axanthum* and *Acholeplasma morum* require additional fatty acids for adequate growth. A modified method utilizing serial passage in selective medium has been applied successfully to a large number of mollicutes (Rose et al., 1993; Tully, 1995). The *Acholeplasmatales* grow through end-point dilutions in serum-containing medium and in serum-free preparations, or occasionally in serum-free medium supplemented with Tween 80. Mesoplasmas from the order *Entomoplasmatales* grow in serum-containing medium and in serum-free medium supplemented only with Tween 80. Most spiroplasmas, also from the *Entomoplasmatales*, and all members of the order *Mycoplasmatales* grow only in serum-containing medium.

References

Edward, D.G. 1971. Determination of sterol requirement for *Mycoplasmatales*. J. Gen. Microbiol. *69*: 205–210.

Freundt, E.A. 1955. The classification of the pleuropneumoniae group of organisms (*Borrelomycetales*). Int. Bull. Bacteriol. Nomencl. Taxon. *5*: 67–78.

Freundt, E.A., R.F. Whitcomb, M.F. Barile, S. Razin and J.G. Tully. 1984. Proposal for elevation of the family *Acholeplasmataceae* to ordinal rank: *Acholeplasmatales*. Int. J. Syst. Bacteriol. *34*: 346–349.

Johansson, K.E., Pettersson B. 2002. Taxonomy of *Mollicutes*. *In* Molecular biology and pathogenicity of mycoplasmas (edited by Razin and Herrmann). Kluwer Academic, New York, pp. 1–30.

Judicial Commission. 1958. Opinion 22. Status of the generic name Asterococcus and conservation of the generic name *Mycoplasma*. Int. Bull. Bacteriol. Nomencl. Taxon. *8*: 166–168.

Nowak, J. 1929. Morphologie, nature et cycle évolutif du microbe de la péripneumonie des bovidés. Ann. Inst. Pasteur (Paris) *43*: 1330–1352.

Razin, S. and J.G. Tully. 1970. Cholesterol requirement of mycoplasmas. J. Bacteriol. *102*: 306–310.

Robinson, I.M. and E.A. Freundt. 1987. Proposal for an amended classification of anaerobic mollicutes. Int. J. Syst. Bacteriol. *37*: 78–81.

Rose, D.L., J.G. Tully, J.M. Bove and R.F. Whitcomb. 1993. A test for measuring growth responses of *Mollicutes* to serum and polyoxyethylene sorbitan. Int. J. Syst. Bacteriol. *43*: 527–532.

Tully, J.G., J.M. Bové, F. Laigret and R.F. Whitcomb. 1993. Revised taxonomy of the class *Mollicutes* - proposed elevation of a monophyletic cluster of arthropod-associated mollicutes to ordinal rank (*Entomoplasmatales* ord. nov.), with provision for familial rank to separate species with nonhelical morphology (*Entomoplasmataceae* fam. nov.) from helical species (*Spiroplasmataceae*), and emended descriptions of the order *Mycoplasmatales*, family *Mycoplasmataceae*. Int. J. Syst. Bacteriol. *43*: 378–385.

Tully, J.G. 1995. Determination of cholesterol and polyoxyethylene sorbitan growth requirements of mollicutes. *In* Molecular and Diagnostic Procedures in Mycoplasmology, vol. 1 (edited by Razin and Tully). Academic Press, San Diego, pp. 381–389.

Family I. **Mycoplasmataceae** Freundt 1955, 71[AL] emend. Tully, Bové, Laigret and Whitcomb 1993, 382

Daniel R. Brown, Meghan May, Janet M. Bradbury, Karl-Erik Johansson and Harold Neimark

My.co.plas.ma.ta.ce′ae. N.L. neut. n. *Mycoplasma*, *-atos* type genus of the family; *-aceae* ending to denote a family; N.L. fem. pl. n. *Mycoplasmataceae* the *Mycoplasma* family.

Pleomorphic usually coccoid cells, 300–800 nm in diameter, to **slender branched filaments** of uniform diameter. Some cells have a **terminal bleb or tip structure** that mediates adhesion to certain surfaces. **Cells lack a cell wall** and are bounded only by a plasma membrane. Gram-stain-negative due to the absence of a cell wall. Usually **nonmotile**. **Facultatively anaerobic** in most instances, possessing a truncated flavin-terminated electron transport chain devoid of quinones and cytochromes. Colonies of *Mycoplasma* are usually less than 1 mm in diameter and colonies of *Ureaplasma* are much smaller than that. The typical colony has a **fried-egg or "cauliflower head" appearance**. Usually catalase-negative. **Chemo-organotrophic**, usually using either sugars or arginine, but sometimes both, or having an obligate requirement for urea as the major energy source. **Require cholesterol** or related sterols for growth. **Commensals or pathogens** of a wide range of **vertebrate hosts**. The **genome size** ranges from about **580 to 1350 kbp**, as measured by pulsed field gel electrophoresis (PFGE) or complete DNA sequencing.

DNA G+C content (mol%): about 23–40 (Bd, T_m).

Type genus: **Mycoplasma** Nowak 1929, 1349 nom. cons. Jud. Comm. Opin. 22, 1958, 166.

Further descriptive information

This family and its type genus *Mycoplasma* are polyphyletic. Two genera, *Mycoplasma* and *Ureaplasma*, are currently accepted within the family. The genus *Mycoplasma* is further divisible into phylogenetic groups on the basis of 16S rRNA gene sequence similarities (Johansson and Pettersson, 2002), including an ecologically, phenotypically, and genetically cohesive group called the mycoides cluster, which includes the type species *Mycoplasma mycoides* and other major pathogens of ruminant animals. The taxonomic position of the mycoides cluster is an important anomaly because molecular markers based upon rRNA and other gene sequences indicate that it is closely related to other genera usually associated with plant and insect hosts and currently classified within the order *Entomoplasmatales*. Members of the genus *Ureaplasma* are distinguished by their tiny colony size and capacity to hydrolyze urea.

Genus I. **Mycoplasma** Nowak 1929, 1349 nom. cons. Jud. Comm. Opin. 22, 1958, 166[AL]

Daniel R. Brown, Meghan May, Janet M. Bradbury, Mitchell F. Balish, Michael J. Calcutt, John I. Glass, Séverine Tasker, Joanne B. Messick, Karl-Erik Johansson and Harold Neimark

My.co.plas′ma. Gr. masc. n. *myces* a fungus; Gr. neut. n. *plasma* something formed or molded, a form; N.L. neut. n. *Mycoplasma* fungus form.

Pleomorphic cells, 300–800 nm in diameter, varying in shape from spherical, ovoid or flask-shaped, or twisted rods, to slender branched filaments ranging in length from 50 to 500 nm. **Cells lack a cell wall** and are bounded by a single plasma membrane. Gram-stain-negative due to the absence of a cell wall. Some have a complex internal cytoskeleton. Some have a specific **tip structure** that mediates attachment to host cells or other surfaces. **Usually nonmotile**, but gliding motility has been demonstrated in some species. **Aerobic or facultatively anaerobic**. Optimum growth at 37°C is common, but permissive growth temperatures range from 20 to 45°C. **Chemo-organotrophic**, usually using either sugars or arginine as the major energy source. **Require cholesterol** or related sterols for growth. Colonies are usually less than 1 mm in diameter. The typical colony has a **fried-egg appearance**. The **genome size** of species examined ranges from **580 kbp to about 1350 kbp**. The codon UGA encodes tryptophan in all species examined. **Commensals or pathogens** in a wide range of vertebrate hosts.

DNA G+C content (mol%): 23–40.

Type species: **Mycoplasma mycoides** (Borrel, Dujardin-Beaumetz, Jeantet and Jouan 1910) Freundt 1955, 73 (*Asterococcus mycoides* Borrel, Dujardin-Beaumetz, Jeantet and Jouan 1910, 179).

Further descriptive information

The shape of these organisms (trivial name, mycoplasmas) can depend on the osmotic pressure, nutritional quality of the culture medium, and the growth phase. Some mycoplasmas are filamentous in their early and exponential growth phases or when attached to surfaces or other cells. This form can be transitory, and the filaments may branch or fragment into chains of cocci or individual vegetative cells. Many species are typically coccoid and never develop a filamentous phase. Some species develop specialized attachment tip structures involved in colonization and virulence (Figure 106). In Romanowsky-type stained blood smears, hemotropic species (trivial name, hemoplasmas) appear as round to oval cells on the surface of erythrocytes (Figure 107). They may be found individually or, during periods of high parasitemia, in pairs or chains giving the appearance of pleomorphism. Their small size and the absence of cell wall components provide considerable plasticity to the organisms, so that cells of most species are readily filterable through 450 nm pores, and many species have some cells in the population that are able to pass through 220 nm or even 100 nm filters (Tully, 1983). Descriptions of the morphology, ultrastructure, and motility of mycoplasmas should be based on correlation of the appearance of young exponential-phase broth cultures under phase-contrast or dark-field microscopy with their appearance using negative-staining or electron microscopy (Biberfeld and Biberfeld, 1970; Boatman, 1979; Carson et al., 1992; Cole, 1983). Special attention to the osmolarity of the fixatives and buffers is required since these may alter the size and shape of the organisms. The classical isolated colony is umbonate with a fried-egg appearance, but others may have

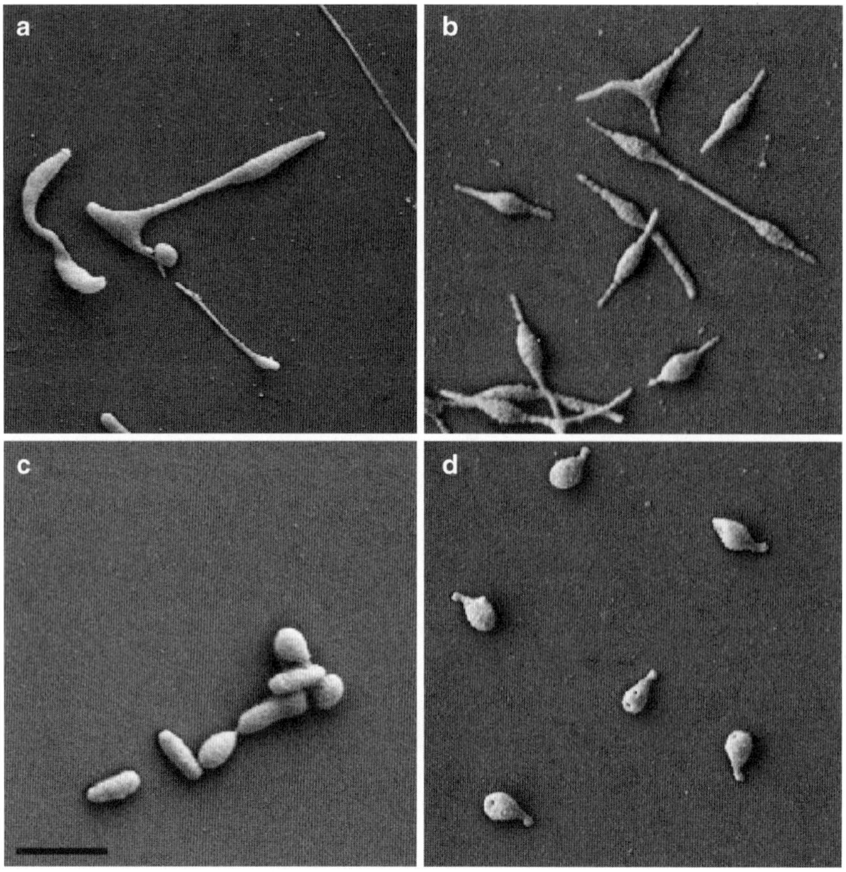

FIGURE 106. Diverse cellular morphology in the genus *Mycoplasma*. Scanning electron micrographs of cells of (a) *Mycoplasma penetrans*, (b) *Mycoplasma pneumoniae*, (c) "*Mycoplasma insons*", and (d) *Mycoplasma genitalium*. Bar = 1 μm. Images provided by Dominika Jurkovic, Jennifer Hatchel, Ryan Relich and Mitchell Balish.

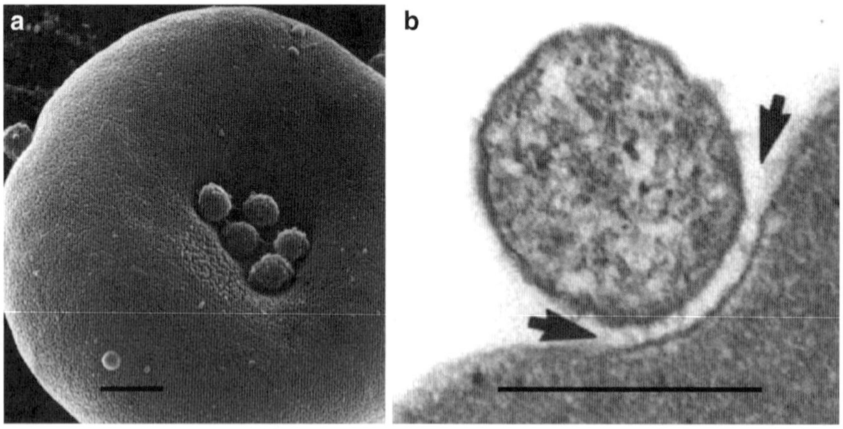

FIGURE 107. Hemotropic mycoplasmas. (a) Scanning electron micrograph of *Mycoplasma ovis* cells colonizing the surface of an erythrocyte (Neimark et al., 2004); bar = 500 nm. (b) Transmission electron micrograph showing fibrils bridging the space between a "*Candidatus* Mycoplasma kahaneii" cell and a depression in the surface of a colonized erythrocyte (Neimark et al., 2002a); bar = 250 nm. Images used with permission.

either cauliflower-like or smooth colony surfaces (Figure 108), with smooth, irregular or scalloped margins, depending on the species, agar concentration, and other growth conditions.

A significant minority of species exhibit cell polarization. This depends on Triton X-100-insoluble cytoskeletal structures involved in morphogenesis, motility, cytadherence, and cell division (Balish and Krause, 2006). In the distantly related species *Mycoplasma pneumoniae* and *Mycoplasma mobile*, the cytoskeleton underlies a terminal organelle. This prominent extension of the cytoplasm and cell membrane is the principal focus of adherence

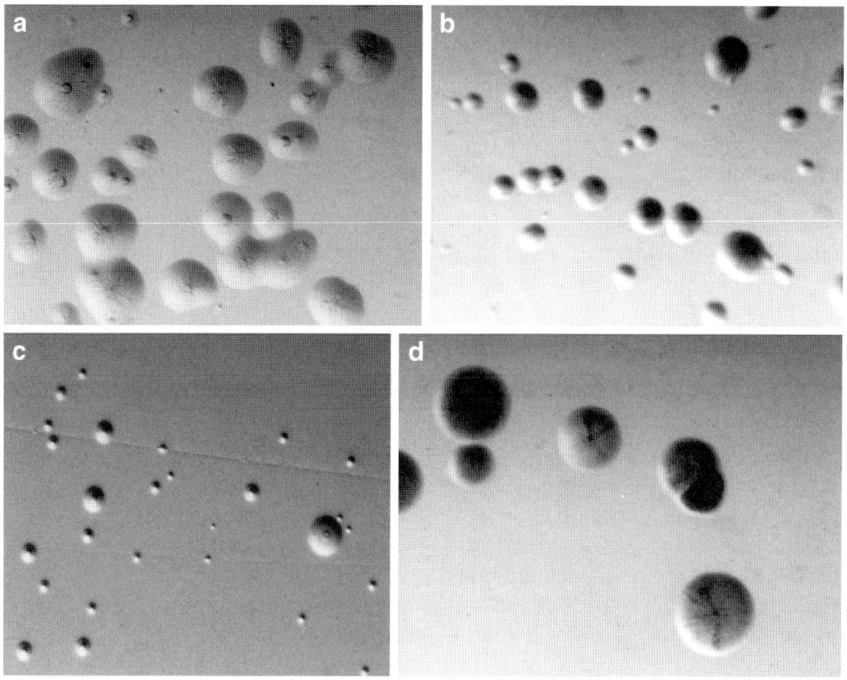

FIGURE 108. Diverse colonial morphology in the genus *Mycoplasma*. (a) *Mycoplasma mycoides* PG1^T (diameter 0.50–0.75 mm), (b) *Mycoplasma hyopneumoniae* NCTC 10110^T (diameter 0.15–0.20 mm), (c) *Mycoplasma pneumoniae* NCTC 10119^T (diameter 0.05–0.10 mm), and (d) *Mycoplasma hyorhinis* ATCC 29052 (diameter 0.25–0.30 mm) after 3, 7, 5, and 6 d growth, respectively, on Mycoplasma Experience Solid Medium at 36°C in 95% nitrogen/5% carbon dioxide. Original magnification 25×. Images provided by Helena Windsor and David Windsor.

and is the leading end of cells engaged in gliding motility. In *Mycoplasma pneumoniae*, adhesin proteins are located either all over the surface of the organelle or at its distal tip; an unrelated adhesin is concentrated at the base of the terminal organelle in *Mycoplasma mobile* (Balish, 2006). Both the formation of this attachment organelle of *Mycoplasma pneumoniae* and the localization of the adhesins depend upon cytoskeletal proteins that form an electron-dense core within its cytoplasm, which is surrounded by an electron-lucent space (Krause and Balish, 2004). The overall appearance of this core is that of two parallel, flat rods of differing thickness, with a bend near the cell-proximal end (Henderson and Jensen, 2006; Seybert et al., 2006). A bilobed button constitutes its distal end and its proximal base terminates in a bowl-like structure. Overall, both the core and the attachment organelle are 270–300 nm in length (Hatchel and Balish, 2008). Around the onset of DNA replication, a second attachment organelle is constructed (Seto et al., 2001). The motile force provided by the first organelle reorganizes the cell such that the new organelle is moved to the opposite cell pole before cell division (Hasselbring et al., 2006). These observations suggest that complex coordination exists between attachment organelle biogenesis, motor activity, the DNA replication machinery, and the cytokinetic machinery. Similar structures are present in other species of the *Mycoplasma pneumoniae* cluster, but in most cases the attachment organelle is shorter, resulting in much of the core protruding into the cell body (Hatchel and Balish, 2008). In *Mycoplasma mobile*, the terminal organelle is completely dissimilar, consisting of a cell-distal sphere with numerous tentacle-like strands extending into the cytoplasm (Nakane and Miyata, 2007). It is comprised of proteins unrelated to those found in *Mycoplasma pneumoniae*, suggesting that it has evolved independently. Further distinct cytoskeletal structures appear in *Mycoplasma penetrans* (Jurkovic and Balish, unpublished), "*Mycoplasma insons*" (Relich et al., 2009), and several species of the mycoides cluster (Peterson et al., 1973).

Attachment to eukaryotic host cells is important for the natural survival and transmission of mycoplasmas. The prominent attachment organelle of species in the *Mycoplasma pneumoniae* cluster is the most extensively characterized determinant of cytadherence. In other species, multiple adhesin proteins are involved in cytadherence. When one adhesin is blocked, cytadherence is reduced, but not completely lost. For this reason, the adhesins appear to be functionally redundant rather than synergistic in action. Numerous species possess multigene families of antigenically variable proteins, some of which have been implicated in host cell attachment or hemagglutination. While this attachment may serve as a supplemental binding mechanism in species such as *Mycoplasma gallisepticum* and *Mycoplasma hominis*, variable surface proteins are currently the only known mechanism for cytadherence and hemagglutination of *Mycoplasma synoviae* and *Mycoplasma pulmonis*. The avidity of adherence may differ among variants in *Mycoplasma pulmonis* and *Mycoplasma hominis*. Though one or more attachment mechanisms have been described for numerous species, there remains a greater number of species with no documented system for cytadherence. Strains that lose the capacity to cytadhere are almost invariably unable to survive in their hosts, but highly invasive species such as *Mycoplasma alligatoris* may not require host cell attachment for infection.

The mycoplasmas possess a typical prokaryotic plasma membrane composed of amphipathic lipids and proteins

(McElhaney, 1992a, b, c; Smith, 1992; Wieslander et al., 1992). At one time, demonstration of a single unit membrane was mandatory for defining all novel species of mollicutes (Tully, 1995a). Now, when the 16S rRNA gene sequence of a novel species is determined and the candidate is placed in one of the phylogenetic clusters of mollicutes, in the majority of cases it can be safely inferred that the organism lacks a cell wall, because the majority of others in that cluster will have been shown to be solely membrane-bound (Brown et al., 2007). The lack of a cell wall explains the resistance of the organisms to lysis by lysozyme and their susceptibility to lysis by osmotic shock and various agents causing the lysis of bacterial protoplasts (Razin, 1979, 1983). In certain species, the extracellular surface is textured with capsular material or a nap, which can be stained with ruthenium red in some cases (Rosenbusch and Minion, 1992).

These organisms represent some of the most nutritionally fastidious prokaryotes, as expected from their greatly reduced or minimalist genomes, close association with vertebrate hosts as commensals and pathogens, and total dependence upon the host to meet all nutritional requirements. They have very limited capacity for intermediary metabolism, which restricts the utility of conventional biochemical tests for identification. Detailed information on carbohydrate (Pollack, 1992, 1997, 2002; Pollack et al., 1996), lipid (McElhaney, 1992a), and amino acid (Fischer et al., 1992) metabolism is available. All species examined have truncated respiratory systems, lack a complete tricarboxylic acid cycle, and lack quinones or cytochromes, which precludes their capacity to carry out oxidative phosphorylation. Instead only low levels of ATP may be generated through glycolysis or the arginine dihydrolase pathway (Miles, 1992a, b). Fermentative species catabolize glucose or other carbohydrates to produce ATP and acid and, consequently, lower the pH of the medium. Non-fermentative species hydrolyze arginine to yield ammonia, some ATP, and carbon dioxide, and consequently raise the pH of the medium. Species such as *Mycoplasma fermentans* have both pathways. Species such as *Mycoplasma bovis* evidently lack both pathways, but are capable oxidizing pyruvate or lactate to yield ATP (Miles, 1992a; Taylor et al., 1994). Some species cause a pronounced "film and spots" reaction on media incorporating heat-inactivated horse serum or egg yolk: a wrinkled film composed of cholesterol and phospholipids forms on the surface of the medium and dark spots containing salts of fatty acids appear around the colonies.

Most mycoplasmas are aerobes or facultative anaerobes, but some species such as *Mycoplasma muris* prefer an anaerobic environment. The optimum growth of species isolated from homeothermic hosts is commonly at 37°C and the permissive temperature range of species from poikilothermic fish and reptiles is always above 20–25°C. Thus, growth of the mycoplasmas is restricted to mesophilic temperatures. Growth in liquid cultures usually produces at most light turbidity and few sedimented cells, except for the heavy turbidity and sediments usually observed with members of the *Mycoplasma mycoides* cluster. Tully (1995b) described in detail the most commonly used culture media formulations. Although colonies are occasionally first detected on blood agar, complex undefined media such as American Type Culture Collection (ATCC) medium 988 (SP-4) are usually required for primary isolation and maintenance. Cell-wall-targeting antibiotics are included to discourage growth of other bacteria. Phenol red facilitates detection of species that excrete acidic or alkaline metabolites. Growth of arginine-hydrolyzing species can be enhanced by supplementing media with arginine. Commonly used alternatives such as Frey's, Hayflick's and Friis' media differ from SP-4 mainly in the proportions of inorganic salts, amino acids, serum sources, and types of antibiotics. For species that utilize both sugars and arginine as carbon sources, the pH of the medium may initially decrease before rising later during the course of growth (Razin et al., 1998). Defined mycoplasma culture media have been described in detail (Rodwell, 1983), but provision of lipids and amino acids in the appropriate ratios is difficult technically (Miles, 1992b).

Many mobile genetic elements occur in the genus. Four plasmids have been identified in members of the mycoides cluster (Bergemann and Finch, 1988; Djordjevic et al., 2001; King and Dybvig, 1994). Each plasmid is apparently cryptic, with no discernible determinants for virulence or antibiotic resistance. DNA viruses have been isolated from *Mycoplasma bovirhinis* (Howard et al., 1980), *Mycoplasma hyorhinis* (Gourlay et al., 1983) *Mycoplasma pulmonis* (Tu et al., 2001), and *Mycoplasma arthritidis* (Voelker and Dybvig, 1999). The *Mycoplasma pulmonis* P1 virus and the lysogenic bacteriophage MAV1 of *Mycoplasma arthritidis* do not share sequence similarity (Tu et al., 2001; Voelker and Dybvig, 1999), whereas the *Mycoplasma fermentans* MFV1 prophage is strikingly similar in genetic organization to MAV1 (Röske et al., 2004). No role in pathobiology has been demonstrated for any virus or prophage.

The most abundant mobile DNAs in *Mycoplasma* are insertion sequence (IS) elements. The first identified units (IS*1138* of *Mycoplasma pulmonis*, IS*1221* of *Mycoplasma hyorhinis*, IS*1296* of *Mycoplasma mycoides* subsp. *mycoides* and IS*Mi1* of *Mycoplasma fermentans*) are members of the IS*3* family (Bhugra and Dybvig, 1993; Ferrell et al., 1989; Frey et al., 1995; Hu et al., 1990). More recently, multiple IS elements of divergent subgroups have been identified. Members of the IS*4* family include IS*1634* and IS*Mmy1* of *Mycoplasma mycoides* subsp. *mycoides* (Vilei et al., 1999; Westberg et al., 2002), IS*Mhp1* of *Mycoplasma hyopneumoniae*, IS*Mhp1*-like unit of *Mycoplasma synoviae*, and four distinct elements of *Mycoplasma bovis* (Lysnyansky et al., 2009). Among the IS*30* family members identified are IS*1630* of *Mycoplasma fermentans*, IS*Mhom1* from *Mycoplasma hominis*, IS*Mag1* of *Mycoplasma agalactiae* (Pilo et al., 2003), and two IS units of *Mycoplasma bovis*. IS-like elements have also been identified in *Mycoplasma leachii*, *Mycoplasma penetrans* (belonging to four different families), *Mycoplasma hyopneumoniae*, *Mycoplasma flocculare*, and *Mycoplasma orale*. Transposases that reside within IS units are also discernable in the genome of *Mycoplasma gallisepticum* (Papazisi et al., 2003). In select instances, almost identical IS units have been found in species from different phylogenetic clades, which strongly suggests lateral gene transfer between species. Despite their widespread distribution, IS elements are not ubiquitous in the genus. Although the type strains of *Mycoplasma bovis* (54 IS units of seven different types) and *Mycoplasma mycoides* subsp. *mycoides* (97 elements of three different types) possess large numbers of elements, the sequenced genomes of *Mycoplasma arthritidis*, *Mycoplasma genitalium*, *Mycoplasma pneumoniae*, and *Mycoplasma mobile* lack detectable IS units.

Although IS units only encode genes related to transposition, large integrating elements have also been identified in diverse *Mycoplasma* species. The Integrative Conjugal Elements (ICE) of *Mycoplasma fermentans* strain PG18^T comprise >8% of the genome and related units have been identified in *Mycoplasma agalactiae*, *Mycoplasma bovis*, *Mycoplasma capricolum*, *Mycoplasma hyopneumoniae*, and *Mycoplasma mycoides* subsp. *mycoides*. In general, such units encode 18–30 genes, can be detected in extrachromosomal forms, and are strain-variable in distribution and chromosomal insertion site. Two additional large mobile DNAs, designated Tra Islands, were identified in *Mycoplasma capricolum* California kid^T. The presence of putative conjugation genes and the variability in genomic location of Tra Islands and ICE suggest that these are agents of lateral gene transfer.

The best-studied mycoplasmas are primary pathogens of humans or domesticated animals (Baseman and Tully, 1997). About half of the listed species occur in the absence of disease, but are occasional opportunistic or secondary pathogens. The principal human pathogens are *Mycoplasma pneumoniae*, *Mycoplasma hominis*, and *Mycoplasma genitalium*, with *Mycoplasma penetrans* added to this list due to its association with HIV infections (Blanchard, 1997; Blanchard et al., 1997; Tully, 1993; Waites and Talkington, 2005). *Mycoplasma pneumoniae* is one of the main agents of community-acquired pneumonia, bronchitis, and other respiratory complications (Atkinson et al., 2008). *Mycoplasma pneumoniae* infections can also involve extra-pulmonary complications including central nervous system, cardiovascular, and dermatological manifestations. Outbreaks cause considerable morbidity and require rapid and effective therapeutic intervention (Hyde et al., 2001; Meyer and Clough, 1993). *Mycoplasma hominis* occurs more frequently in the urogenital tract of women than men and is often found in the genital tract of women with vaginitis, bacterial vaginosis, or localized intrauterine infections (Keane et al., 2000). The organism can gain access to a fetus from uterine sites and it is associated with perinatal morbidity and mortality (Gonçalves et al., 2002; Waites et al., 1988). It is also clearly associated with septicemias and respiratory infections and with transplant or joint infections in immunosuppressed persons (Brunner et al., 2000; Busch et al., 2000; Fernandez Guerrero et al., 1999; Garcia-Porrua et al., 1997; Gass et al., 1996; Hopkins et al., 2002; Mattila et al., 1999; Tully, 1993; Zheng et al., 1997). *Mycoplasma genitalium* has been associated with nongonococcal urethritis in men (Gambini et al., 2000; Jensen, 2004; Jensen et al., 2004; Taylor-Robinson et al., 2004; Taylor-Robinson and Horner, 2001; Totten et al., 2001) and urogenital disease in women (Baseman et al., 2004; Blaylock et al., 2004). *Mycoplasma genitalium* occurs more frequently in the vagina than in the cervix or urethra, but it may be involved in cervicitis (Casin et al., 2002; Manhart et al., 2001). Mycoplasmas are common agents of chronic joint inflammation in a wide variety of hosts (Cole et al., 1985). Species associated with arthritis in humans include *Mycoplasma hominis*, *Mycoplasma fermentans*, *Mycoplasma genitalium*, *Mycoplasma salivarium*, and possibly *Mycoplasma pneumoniae* in juvenile arthritis (Waites and Talkington, 2005). Humans are also susceptible to opportunistic zoonotic mycoplasmosis; immunosuppressed persons are highly susceptible. The recent molecular confirmation of a *Mycoplasma haemofelis*-like infection in an HIV-positive patient (dos Santos et al., 2008) highlights the zoonotic potential of the hemoplasmas.

Mycoplasmas colonize fish, reptiles, birds, and terrestrial and aquatic mammals. Some cause significant diseases of cattle and other ruminants, swine, poultry, or wildlife, and others are opportunistic or secondary veterinary pathogens (Simecka et al., 1992; Tully and Whitcomb, 1979). The principal bovine pathogens include the serovars historically called "Small Colony" types of *Mycoplasma mycoides* subsp. *mycoides*, and *Mycoplasma bovis*. *Mycoplasma mycoides* subsp. *mycoides* has caused major losses of livestock globally in the twentieth century due to contagious bovine pleuropneumonia and currently remains a problem in Asia and Africa (Lesnoff et al., 2004). *Mycoplasma bovis* is a widespread agent of otitis media, pneumonia, mastitis, polyarthritis, and urogenital disease in cattle and buffaloes. *Mycoplasma mycoides* subsp. *capri*, *Mycoplasma capricolum* subsp. *capricolum*, and *Mycoplasma agalactiae* are important causes of arthritis, mastitis, and agalactia in goats and sheep. *Mycoplasma mycoides* subsp. *capri* (type strain PG3^T) properly includes all of the serovars historically called "Large Colony" types of subspecies *mycoides* (Manso-Silván et al., 2009; Shahram et al., 2010). *Mycoplasma capricolum* subsp. *capripneumoniae* (type strain F38^T) causes severe contagious pleuropneumonia in goats (Leach et al., 1993; McMartin et al., 1980). Contagious bovine and caprine pleuropneumonia, and mycoplasmal agalactia of sheep or goats are subject to control through listing in the Terrestrial Animal Health Code of the Office International des Epizooties (http://oie.int) as well as strict notification and export regulations by individual countries. *Mycoplasma hyopneumoniae*, one of the most difficult species to cultivate, causes primary enzootic pneumonia in pigs and exacerbates other porcine respiratory diseases leading to substantial economic burdens. *Mycoplasma hyosynoviae* is carried in the upper respiratory tract, but causes nonsuppurative polyarthritis, usually without other serositis, especially in growing pigs.

The most important poultry pathogens are *Mycoplasma gallisepticum*, *Mycoplasma synoviae*, and *Mycoplasma meleagridis*, but more than 20 other species have been isolated from birds as diverse as ostriches, raptors, and penguins (Bradbury and Morrow, 2008). *Mycoplasma gallisepticum* can be transmitted vertically, venereally, by other direct contact, or by aerosol to cause respiratory disease and its sequelae in chickens, turkeys, and other birds. It also causes decreased egg production and egg quality in chickens. *Mycoplasma synoviae* can cause a syndrome of synovitis, tendonitis, and bursitis in addition to respiratory disease in chickens and turkeys, whereas the developmental abnormalities and airsacculitis associated with congenital or acquired *Mycoplasma meleagridis* infection seem restricted to turkeys. *Mycoplasma gallisepticum* and *Mycoplasma synoviae* are also listed in the OIE's Terrestrial Animal Health Code.

Pathogenicity of specific mycoplasmas has also been reported for companion animals (Chalker, 2005; Lemcke, 1979; Messick, 2003) and a number of wild animal hosts (Brown et al., 2005). The respiratory, reproductive, and joint diseases caused in rodents by *Mycoplasma pulmonis* and *Mycoplasma arthritidis* (Schoeb, 2000) are important models of infection and immunity in humans and other animals.

Hemoplasmas infect a variety of wild and domesticated animals and are relatively host-specific, although cross-infection of related hosts has been reported. Transmission can be achieved by ingestion of infected blood or by percutaneous inoculation. Arthropod vector transmission of some species is also supported by experimentation and by the clustered geographic distribution of hemoplasmosis in some studies (Sykes et al., 2007; Willi et al., 2006a). The pathogenicity of different hemoplasma species is variable and strain virulence also likely plays a key role in the development of disease. For example, *Mycoplasma haemofelis* can induce acute clinical disease in non-splenectomized, immunocompetent cats, whereas *Mycoplasma haemocanis* appears able to induce disease only in immunosuppressed or splenectomized dogs. Clinical syndromes range from acute fatal hemolytic anemia to chronic insidious anemia and ill-thrift. Signs may include anemia, pyrexia, anorexia, dehydration, weight loss, and infertility. The presence of erythrocyte-bound antibodies (including cold agglutinins), indicated by positive Coombs' testing, has been demonstrated in some hemoplasma-infected animals and may contribute to anemia. Animals can remain chronic asymptomatic carriers of hemoplasmas after acute infection. PCR is the diagnostic test of choice for hemoplasma infection.

Contamination of eukaryotic cell cultures with mollicutes is still a common and important yet often unrecognized problem (Tully and Razin, 1996). More than 20 species have been isolated from contaminated cell lines, but more than 90% of the contamination is thought to be caused by just five species of mycoplasma: *Mycoplasma arginini*, *Mycoplasma fermentans*, *Mycoplasma hominis*, *Mycoplasma hyorhinis*, and *Mycoplasma orale*, plus *Acholeplasma laidlawii*. *Mycoplasma pirum* and *Mycoplasma salivarium* account for most of the remainder (Drexler and Uphoff, 2002). Culture medium components of animal origin, passage of contaminated cultures, and laboratory personnel are likely to be the most significant sources of cell culture contaminants. PCR-based approaches to detection achieve sensitivity and specificity far superior to fluorescent staining methods (Masover and Becker, 1996). Another method of detection is based on mycoplasma-specific ATP synthesis activity present in contaminated culture medium (Robertson and Stemke, 1995; MycoAlert, Lonza Group). Eradication through treatment of contaminated cultures with antibiotics (Del Giudice and Gardella, 1996) is rarely successful. Strategies for prevention and control of mycoplasmal contamination of cell cultures have been described in detail (Smith and Mowles, 1996).

Several categories of potential virulence determinants are encoded in the metagenome of pathogenic mycoplasmas. Some species possess multiple types of virulence factors. Determinants such as adhesins and accessory proteins, extracellular polysaccharide structures, and pro-inflammatory or pro-apoptotic membrane lipoproteins are produced by multiple species. Several species excrete potentially toxic by-products of intermediary metabolism, including hydrogen peroxide, superoxide radicals, or ammonia. Other determinants such as extracellular endopeptidases, nucleases, and glycosidases seem irregularly distributed in the genus, whereas the ADP-ribosylating and vacuolating cytotoxin (pertussis exotoxin S1 subunit analog) of *Mycoplasma pneumoniae* and the T-lymphocyte mitogen (superantigen) of *Mycoplasma arthritidis* are evidently unique to those species. Reports of a putative exotoxin elaborated by *Mycoplasma neurolyticum* have not been substantiated by later work (Tryon and Baseman, 1992).

Candidate virulence mechanisms, such as motility, biofilm formation, or facultative intracellular invasion, are expressed by a range of pathogenic species. Several species possess systems of variable surface antigens that are thought to be important in evasion of the hosts' adaptive immune responses. In addition, a large number of species can suppress or inappropriately stimulate host immune cells and their receptors and cytokines through diverse, poorly characterized mycoplasmal components. Although candidate virulence factor discovery has accelerated significantly in recent years through whole genome annotation, the molecular basis for pathogenicity and causal relationships with disease still remain to be definitively established for most of these factors (Razin and Herrmann, 2002; Razin et al., 1998).

Because they lack lipopolysaccharide and a cell wall, and do not synthesize their own nucleotides, mycoplasmas are intrinsically resistant to polymixins, β-lactams, vancomycin, fosfomycin, sulfonamides, and trimethoprim. They are also resistant to rifampin because their RNA polymerase is not affected by that antibiotic (Bébéar and Kempf, 2005). Individual species exhibit an even broader spectrum of antibiotic resistance, such as the resistance to erythromycin and azithromycin exhibited by several species, which is apparently mediated by mutation in the 23S rRNA (Pereyre et al., 2002). Treatment of mycoplasmosis often involves the use of antibiotics that inhibit protein synthesis or DNA replication. Certain macrolides or ketolides are used when tetracyclines or fluoroquinolones are inappropriate. Fluoroquinolones, aminoglycosides, pleuromutilins, and phenicols are not widely used to treat human mycoplasmosis at present, with the exception of chloramphenicol for neonates with mycoplasmosis of the central nervous system unresponsive to other antibiotics (Waites et al., 1992), but their use in veterinary medicine is more common. The long-term antimicrobial therapy often required may be due to mycoplasmal sequestration in privileged sites, potentially including inside host cells. Mycoplasmosis in immunodeficient patients is very difficult to control with antibiotic drugs (Baseman and Tully, 1997).

Enrichment and isolation procedures

Techniques for isolation of mycoplasmas from humans, various species of animals, and from cell cultures have been described (Neimark et al., 2001; Tully and Razin, 1983). Typical steps in the isolation of mycoplasmas were outlined in the recently revised minimal standards for descriptions of new species (Brown et al., 2007). Initial isolates may contain a mixture of species, so cloning by repeated filtration through membrane filters with a pore size of 450 or 220 nm is essential. The initial filtrate and dilutions of it are cultured on solid medium and an isolated colony is subsequently picked from a plate on which only a few colonies have developed. This colony is used to found a new cultural line, which is then expanded, filtered, plated, and picked two additional times. Hemoplasmas have not yet been successfully grown in continuous culture *in vitro*, although recent work (Li et al., 2008) suggests that *in vitro* maintenance of *Mycoplasma suis* may be possible.

Maintenance procedures

Cultures of mycoplasma can be preserved by lyophilization or cryogenic storage (Leach, 1983). The serum in the culture medium provides effective cryoprotection, but addition of sucrose may enhance survival following lyophilization. Hemoplasmas can be frozen in heparin- or EDTA-anticoagulated blood cryopreserved with dimethylsulfoxide. Most species can be recovered with little loss of viability even after storage for many years.

Taxonomic comments

This polyphyletic genus is divisible on the basis of 16S rRNA and other gene sequence similarities into a large paraphyletic clade of over 100 species in two groups called hominis and pneumoniae (Johansson and Pettersson, 2002; Figure 109), plus the ecologically, phenotypically, and genetically cohesive "mycoides cluster" of five species including the type species *Mycoplasma mycoides* (Cottew et al., 1987; Manso-Silván et al., 2009; Shahram et al., 2010). The priority of *Mycoplasma mycoides* as the type species of the genus *Mycoplasma* and, hence, the family *Mycoplasmataceae* and the order *Mycoplasmatales* is, in retrospect, unfortunate.

The phylogenetic position of the mycoides cluster is eccentrically situated to the remaining species of the order *Mycoplasmatales*, amidst genera that are properly classified in the order *Entomoplasmatales*. When the order *Entomoplasmatales* was established, a century after the discovery of *Mycoplasma mycoides*, it was explicitly accepted that the taxonomic anomaly created by the phylogenetic position of the mycoides cluster will remain impractical to resolve (Tully et al., 1993). The few species in the mycoides cluster cannot simply be renamed, because confusion and peril would result, especially regarding "Small Colony" PG1^T-like strains of *Mycoplasma mycoides* subsp. *mycoides* and F38^T-like strains of *Mycoplasma capricolum* subsp. *capripneumoniae*, which are highly virulent pathogens and subject to strict international regulations.

Another controversy involves the nomenclature of uncultivated hemotropic bacteria originally assigned to the genera *Eperythrozoon* or *Haemobartonella*. It is now undisputed that, on the bases of their lack of a cell wall, small cell size, low G+C content, use of the codon UGA to encode tryptophan, regular association with vertebrate hosts, and 16S rRNA gene sequences that are most similar (80–84%) to species in the pneumoniae group of *Mycoplasma*, these organisms are properly affiliated with the *Mycoplasmatales*. However, the proposed transfers of *Eperythrozoon* and *Haemobartonella* species to the genus *Mycoplasma* (Neimark et al., 2001, 2005) were opposed on the grounds that the degree of 16S rRNA gene sequence similarity is insufficient (Uilenberg et al., 2004, 2006). The principal objection to establishing the hemoplasmas in a third genus in the *Mycoplasmataceae* (Uilenberg et al., 2006) is that this would compound the polyphyly within the pneumoniae group solely on the basis of a capacity to adhere to erythrocytes *in vivo*. In addition, the transfer of the type species *Eperythrozoon coccoides* to the genus *Mycoplasma* is complicated by priority because *Eperythrozoon* predates *Mycoplasma*. The alternative, to transfer all mycoplasmas to the genus *Eperythrozoon*, would be completely impractical and perilous in part because the epithet *Eperythrozoon* does not indicate an affiliation with *Mycoplasmatales*. The Judicial Commission of the International Committee on Systematics of Prokaryotes (ICSP) declined to rule on a request for an opinion in this matter (Neimark et al., 2005) during their 2008 meeting, but a provisional placement in the genus *Mycoplasma* has otherwise been embraced by specialists in the molecular biology and clinical pathogenicity of these and similar hemotropic organisms. At present, the designation "*Candidatus*" must still be used for new types.

Mycoplasma feliminutum was first described during a time when the only named genus of mollicutes was *Mycoplasma*. Its publication coincided with the first proposal of the genus *Acholeplasma* (Edward and Freundt, 1969, 1970), with which *Mycoplasma feliminutum* is properly affiliated through established phenotypic (Heyward et al., 1969) and 16S rRNA gene sequence (Brown et al., 1995) similarities. This explains the apparent inconsistencies with its assignment to the genus *Mycoplasma*. The name *Mycoplasma feliminutum* should therefore be revised to *Acholeplasma feliminutum* comb. nov. The type strain is Ben^T (=ATCC 25749^T; Heyward et al., 1969).

Recent work at the J. Craig Venter Institute, including complete chemical synthesis and cloning of an intact *Mycoplasma genitalium* chromosome (Gibson et al., 2008) and other work with *Mycoplasma mycoides* "Large Colony" (Lartigue et al., 2009), suggests that *de novo* synthesis of two species of mollicute is imminent. Transplantation of isolated deproteinized tet^R-selectable chromosomes from donor *Mycoplasma mycoides* "Large Colony" into recipient *Mycoplasma capricolum* cells displaced the recipient genome and conferred the genotype and phenotype of the donor (Lartigue et al., 2007). Cloning in yeast and subsequent resurrection of *Mycoplasma mycoides* "Large Colony" genomes as living bacteria demonstrate that it is possible to enliven a prokaryotic genome constructed in a eukaryotic cell (Lartigue et al., 2009). The ICSP subcommittee on the taxonomy of *Mollicutes* may be the first to accommodate a system of nomenclature and classification for species of novel prokaryotes that originate by entirely artificial speciation events (Brown and Bradbury., 2008).

Differentiation of the genus *Mycoplasma* from other genera

Properties that partially fulfill criteria for assignment to the class *Mollicutes* (Brown et al., 2007) include absence of a cell wall, filterability, and the presence of conserved 16S rRNA gene sequences. Aerobic or facultatively anaerobic growth in artificial medium and a growth requirement for sterols exclude assignment to the genera *Anaeroplasma*, *Asteroleplasma*, *Acholeplasma*, or "*Candidatus* Phytoplasma". Non-spiral cellular morphology and regular association with a vertebrate host or fluids of vertebrate origin support exclusion from the genera *Spiroplasma*, *Entomoplasma*, or *Mesoplasma*. The inability to hydrolyze urea excludes assignment to the genus *Ureaplasma*.

Acknowledgements

The lifetime achievements in mycoplasmology and substantial contributions to the preparation of this material by Joseph G. Tully are gratefully acknowledged. Daniel R. Brown and Meghan May were supported by NIH grant 5R01GM076584. Séverine Tasker was supported by Wellcome Trust grant WT077718.

Further reading

Blanchard, A. and G. Browning (editors). 2005. *Mycoplasmas*: Molecular Biology, Pathogenicity, and Strategies for Control. Horizon Press, Norwich, UK.

Maniloff, J., R.N. McElhaney, L.R. Finch and J.B. Baseman (editors). 1992. Mycoplasmas: Molecular Biology and Pathogenesis. American Society for Microbiology, Washington, D.C.

Razin, S. and J.G. Tully (editors). 1995. Molecular and Diagnostic Procedures in Mycoplasmology, vol. 1, Molecular Characterization. Academic Press, San Diego.

Tully, J.G. and S. Razin (editors). 1996. Molecular and Diagnostic Procedures in Mycoplasmology, vol. 2, Diagnostic Procedures. Academic Press, San Diego.

Differentiation of the species of the genus *Mycoplasma*

Glucose fermentation and arginine hydrolysis are discriminating phenotypic markers (Table 137), but the pleomorphism and metabolic simplicity of mycoplasmas has led to a current reliance principally on the combination of 16S rRNA gene sequencing and reciprocal serology for species differentiation. Failure to cross-react with antisera against previously recognized species provides substantial evidence for species novelty. For this reason, deposition of antiserum against a novel type strain into a recognized collection is still mandatory for novel species descriptions (Brown et al., 2007). Preliminary differentiation can be by PCR and DNA sequencing using primers specific for bacterial 16S rRNA genes or the 16S–23S intergenic region. A similarity matrix relating the candidate strain to its closest neighbors, usually species with >94% 16S rRNA gene sequence similarity, will suggest related species that should be examined for serological cross-reactivities.

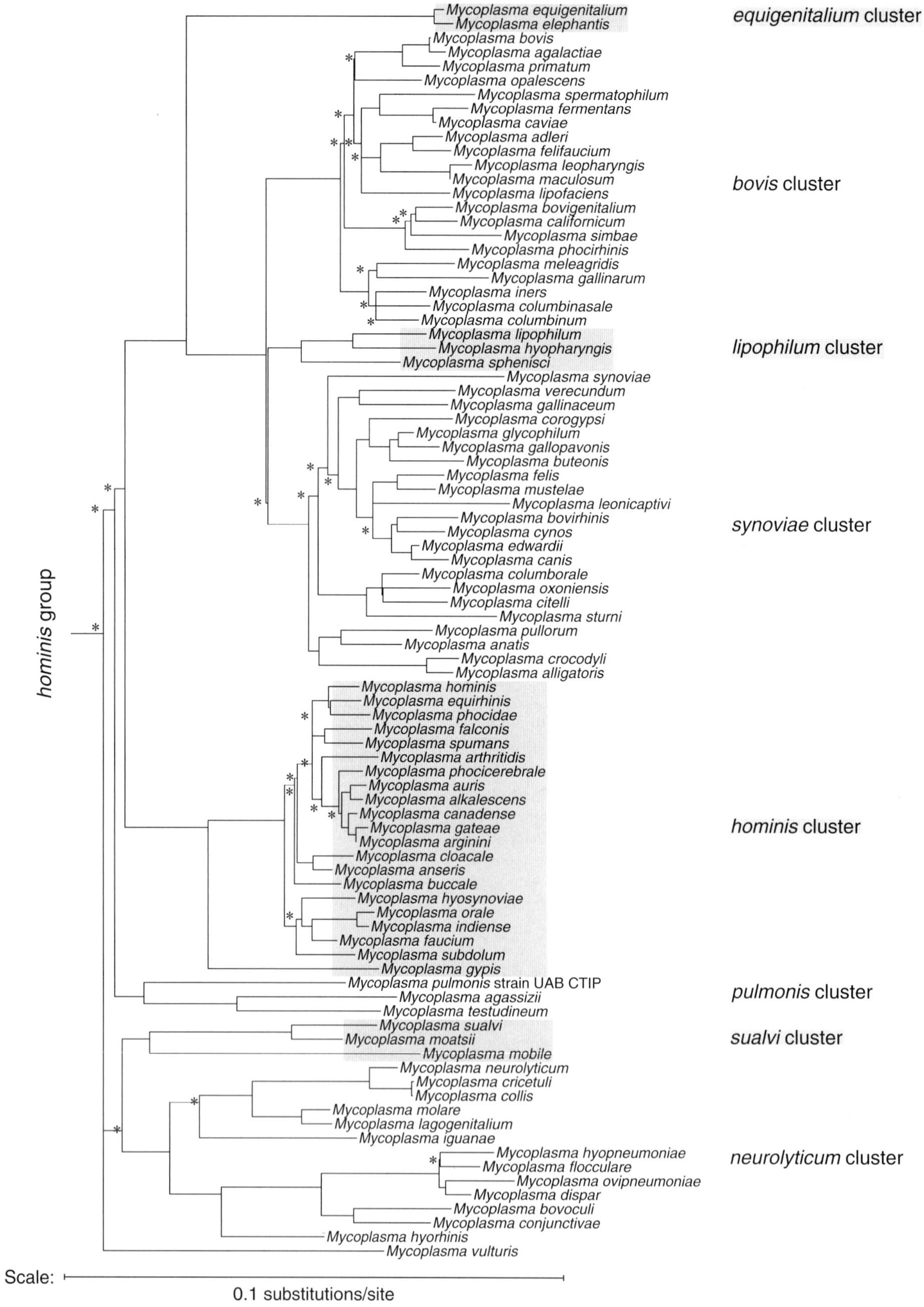

FIGURE 109. (Continued)

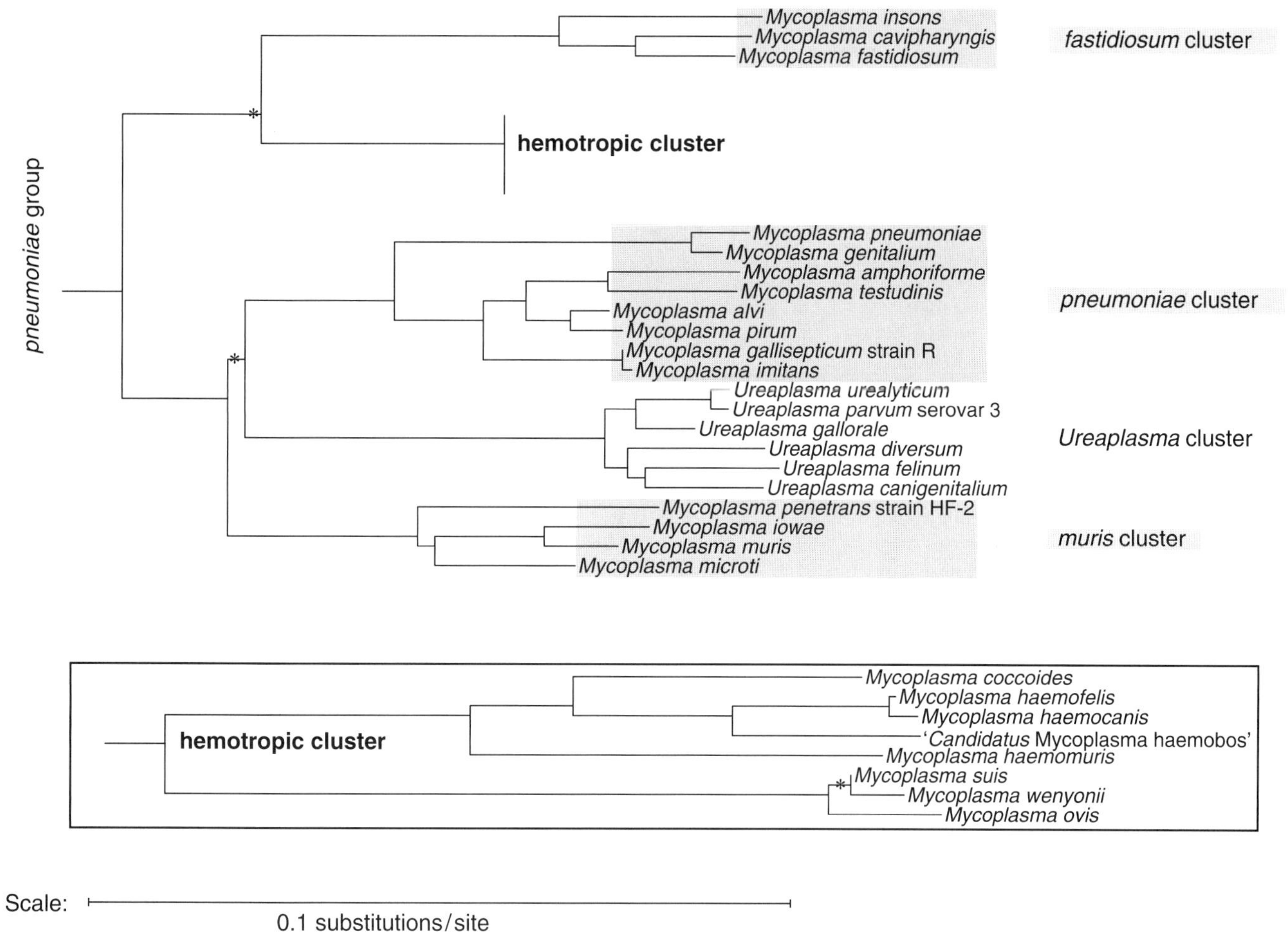

FIGURE 109. Phylogenetic relationships in the *Mycoplasma hominis* and *Mycoplasma pneumoniae* groups of the order *Mycoplasmatales*. The phylogram was based on a Jukes–Cantor corrected distance matrix and weighted neighbor-joining analysis of the 16S rRNA gene sequences of the type strains, except where noted. *Acholeplasma* (formerly *Mycoplasma*) *feliminutum* was the outgroup. The major groups and clusters are defined in terms of positions in 16S rRNA showing characteristic base composition and signature positions, plus higher-order structural synapomorphies (Johansson and Pettersson, 2002; Weisburg et al., 1989). Bootstrap values (100 replicates) <50% are indicated (*); the branching order is considered to be equivocal.

List of species of the genus *Mycoplasma*

1. **Mycoplasma mycoides** (Borrel, Dujardin-Beaumetz, Jeantet and Jouan 1910) Freundt 1955, 73[AL] (*Asterococcus mycoides* Borrel, Dujardin-Beaumetz, Jeantet and Jouan 1910, 179)

my.co.i'des. Gr. n. *mukês -êtos* mushroom or other fungus; L. suff. *-oides* (from Gr. suff. *-eides*, from Gr. n. *eidos* that which is seen, form, shape, figure) resembling, similar; N.L. neut. adj. *mycoides* fungus-like.

This is the type species of the genus. Cells are pleomorphic and capable of forming long filaments. Nonmotile. An extracellular capsule can be visualized by electron microscopy following staining with ruthenium red. Colonies on solid agar have a characteristic fried-egg appearance. Grows in modified Hayflick medium supplemented with glucose at 37°C.

The species has subsequently been divided as follows.

1a. **Mycoplasma mycoides subsp. capri** Manso-Silván, Vilei, Sachse, Djordjevic, Thiaucourt and Frey 2009, 1357[VP] (*Asterococcus mycoides* var. *capri* Edward 1953, 874; *Mycoplasma mycoides* subsp. *mycoides* var. large colony Cottew and Yeats 1978, 294)

ca'pri. L. n. *caper, -pri* goat; L. gen. n. *capri* of the goat.

Cells are pleomorphic and capable of forming long filaments and long, helical rods known as rho forms. Nonmotile. An extracellular capsule can be visualized by electron microscopy following staining with ruthenium red. Colonies on solid agar have a characteristic fried-egg appearance and are notably larger than those of *Mycoplasma mycoides* subsp. *mycoides*. Grows in modified Hayflick medium supplemented with glucose at 37°C. Formation of biofilms has been demonstrated (McAuliffe et al., 2006).

Pathogenic; causes polyarthritis, mastitis, conjunctivitis (a syndrome collectively termed contagious agalactia), pneumonia, peritonitis, and septicemia in goats; and balanitis and vulvitis in sheep. Transmission occurs via direct contact between animals or with fomites, or can be vector-borne by the common ear mite (*Psoroptes cuniculi*).

Tetracyclines are effective therapeutic agents. Eradication from herds is difficult due to the tendency of healthy animals to harbor the organism in the ear canal without seroconverting. Antigenic cross-reactivity with *Mycoplasma*

TABLE 137. Descriptive characteristics of species of *Mycoplasma*[a]

Species	Morphology	DNA G+C content (mol%)	Energy source	Medium pH shift	Serum source	Representative host	Relation to host
M. mycoides subsp. *mycoides*	Pleomorphic	24	G	A	FB	Cattle	Pathogen
M. mycoides subsp. *capri*	Pleomorphic	24	G	A	FB	Goats	Pathogen
M. adleri	Coccoidal	29.6	R	K	E	Goats	Pathogen
M. agalactiae	Coccoidal	29.7	G	A	FB, E	Goats	Pathogen
M. agassizii	Pleomorphic	nr	G	A	FB	Tortoises	Pathogen
M. alkalescens	Coccobacillary	25.9	R	K	FB	Cattle	Pathogen
M. alligatoris	Coccoidal	nr	G	A	FB	Alligators	Pathogen
M. alvi	Flask-shaped	26.4	G, R	V	FB	Cattle	Commensal
M. amphoriforme	Flask-shaped	34	G	A	FB	Humans	Opportunistic
M. anatis	Coccoidal	26.6	G	A	FB	Ducks	Opportunistic
M. anseris	Spherical	26	R	K	E	Goose	Opportunistic
M. arginini	Coccoidal	27.6	R	K	E	Mammals	Pathogen
M. arthritidis	Filamentous	30.7	R	K	FB	Rats	Pathogen
M. auris	Pleomorphic	26.9	R	K	E	Goats	Commensal
M. bovigenitalium	Pleomorphic	30.4	OH, OA	N	FB	Cattle	Pathogen
M. bovirhinis	nr	27.3	G	A	FB	Cattle	Opportunistic
M. bovis	Coccobacillary	32.9	OH, OA	N	FB, E	Cattle	Pathogen
M. bovoculi	Coccobacillary	29	G	A	E	Cattle	Pathogen
M. buccale	Coccobacillary	26.4	R	K	E	Humans	Commensal
M. buteonis	Coccoidal	27	G	A	P	Raptors	Commensal
M. californicum	Pleomorphic	31.9	OH, OA	N	E	Cattle	Pathogen
M. canadense	Coccobacillary	29	R	K	FB	Cattle	Pathogen
M. canis	Pleomorphic	28.4	G	A	FB	Dogs	Opportunistic
M. capricolum subsp. *capricolum*	Coccobacillary	23	G	A	FB, E	Goats	Pathogen
M. capricolum subsp. *capripneumoniae*	Coccobacillary	24.4	G	A	FB, E	Goats	Pathogen
M. caviae	nr	nr	G	A	FB	Guinea pigs	Commensal
M. cavipharyngis	Twisted rod	30	G	A	E	Guinea pigs	Commensal
M. citelli	Pleomorphic	27.4	G	A	FB	Squirrels	Commensal
M. cloacale	Spherical	26	R	K	E	Galliforms	Commensal
M. coccoides[b]	Coccoidal	nr	U	na	na	Mice	Pathogen
M. collis	Coccoidal	28	G	A	E	Rodents	Commensal
M. columbinasale	Coccobacillary	32	R	K	FB	Pigeons	Commensal
M. columbinum	Pleomorphic	27.3	R	K	P	Pigeons	Commensal
M. columborale	Coccoidal	29.2	G	A	P	Pigeons	Commensal
M. conjunctivae	Coccobacillary	nr	G	A	FB	Goats	Pathogen
M. corogypsi	Pleomorphic	28	G	A	P	Vultures	Pathogen
M. cottewii	Coccoid	27	G	A	E	Goats	Commensal
M. cricetuli	Pleomorphic	nr	G	A	E	Hamsters	Commensal
M. crocodyli	Coccoidal	27.6	G	A	FB	Crocodiles	Pathogen
M. cynos	Coccobacillary	25.8	G	A	FB	Dogs	Pathogen
M. dispar	Pleomorphic	29.3	G	A	FB, P	Cattle	Pathogen
M. edwardii	Coccobacillary	29.2	G	A	FB	Dogs	Opportunistic
M. elephantis	Coccoidal	24	G	A	E	Elephants	Commensal
M. equigenitalium	Pleomorphic	31.5	G	A	E	Horses	Opportunistic
M. equirhinis	Coccobacillary	nr	R	K	E	Horses	Opportunistic
M. falconis	Coccoidal	27.5	R	K	P	Falcons	Opportunistic
M. fastidiosum	Twisted rod	32.3	G	A	P	Horses	Commensal
M. faucium	Coccoidal	nr	R	K	FB	Humans	Commensal
M. felifaucium	Coccoidal	31	R	K	FB, E	Pumas	Commensal
M. feliminutum	nr	29.1	G	A	FB	Cats	Commensal
M. felis	Filamentous	25.2	G	A	FB	Cats	Pathogen
M. fermentans	Filamentous	28.7	R, G	V	FB, E	Humans	Unclear
M. flocculare	Coccobacillary	33	U	N	P	Pigs	Opportunistic
M. gallinaceum	Coccobacillary	28	G	A	P	Galliforms	Pathogen
M. gallinarum	Coccobacillary	28	R	K	P	Galliforms	Commensal
M. gallisepticum	Flask-shaped	31	G	A	FB, E	Galliforms	Pathogen
M. gallopavonis	Coccobacillary	27	G	A	P	Turkeys	Opportunistic
M. gateae	nr	28.5	U	N	FB	Cats	Opportunistic

(Continued)

TABLE 137. (Continued)

Species	Morphology	DNA G+C content (mol%)	Energy source	Medium pH shift	Serum source	Representative host	Relation to host
M. genitalium	Flask-shaped	31	G	A	FB	Humans	Pathogen
M. glycophilum	Elliptical	27.5	G	A	E	Galliforms	Commensal
M. gypis	Coccoidal	27.1	R	K	P	Vultures	Opportunistic
M. haemocanis[b]	Coccoidal	nr	U	na	na	Dogs	Pathogen
M. haemofelis[b]	Coccoidal	38.8	U	na	na	Cats	Pathogen
M. haemomuris[b]	Coccoidal	nr	U	na	na	Mice	Opportunistic
M. hominis	Coccobacillary	33.7	R	K	FB	Humans	Pathogen
M. hyopharyngis	Pleomorphic	24	R	K	P, E	Pigs	Commensal
M. hyopneumoniae	Coccobacillary	28	G	A	P, FB	Pigs	Pathogen
M. hyorhinis	Coccobacillary	27.8	G	A	FB	Pigs	Pathogen
M. hyosynoviae	Pleomorphic	28	G	A	FB	Pigs	Pathogen
M. iguanae	Coccoidal	nr	G	A	FB	Iguanas	Pathogen
M. imitans	Flask-shaped	31.9	G	A	FB	Ducks, geese	Pathogen
M. indiense	Pleomorphic	32	R	K	FB	Monkeys	Commensal
M. iners	nr	29.6	R	K	P	Galliforms	Commensal
M. insons[c]	Twisted rod	nr	G	A	FB	Iguanas	Commensal
M. iowae	Pleomorphic	25	G, R	V	FB	Turkeys	Pathogen
M. lagogenitalium	Coccoidal	23	G	A	FB	Pikas	Commensal
M. leachii	Pleomorphic	nr	G	A	E	Cattle	Pathogen
M. leonicaptivi	Pleomorphic	27	G	A	FB	Lions	Commensal
M. leopharyngis	Pleomorphic	28	G	A	FB	Lions	Commensal
M. lipofaciens	Elliptical	24.5	G, R	V	P	Galliforms	Commensal
M. lipophilum	Pleomorphic	29.7	R	K	FB, E	Humans	Unclear
M. maculosum	Coccobacillary	29.6	R	K	FB, E	Dogs	Opportunistic
M. meleagridis	Coccobacillary	28.6	R	K	P	Turkeys	Pathogen
M. microti	Coccoidal	nr	G	A	FB	Voles	Commensal
M. moatsii	Spheroidal	25.7	G, R	V	FB, E	Monkeys	Commensal
M. mobile	Flask-shaped	25	G, R	V	E	Tench	Pathogen
M. molare	Coccoidal	26	G	A	FB	Dogs	Opportunistic
M. mucosicanis[c]	Coccoidal	nr	U	nr	E	Dogs	Commensal
M. muris	Coccoidal	24.9	R	K	FB	Mice	Commensal
M. mustelae	Pleomorphic	28	G	A	E	Minks	Commensal
M. neurolyticum	Filamentous	26.2	G	A	E	Mice	Unclear
M. opalescens	nr	29.2	R	K	FB	Dogs	Commensal
M. orale	Pleomorphic	28.2	R	K	FB, E	Humans	Commensal
M. ovipneumoniae	nr	25.7	G	A	FB, P	Sheep	Pathogen
M. ovis[b]	Coccoidal	nr	U	na	na	Sheep	Pathogen
M. oxoniensis	Coccoidal	29	G	A	FB	Hamsters	Commensal
M. penetrans	Flask-shaped	25.7	G, R	V	FB	Humans	Opportunistic
M. phocicerebrale	Dumbbell	25.9	R	K	FB	Seals	Pathogen
M. phocidae	Coccoidal	27.8	G, R	V	FB, E	Seals	Opportunistic
M. phocirhinis	Coccoidal	26.5	R	K	E, P	Seals	Pathogen
M. pirum	Flask-shaped	25.5	G	A	FB	Humans	Commensal
M. pneumoniae	Flask-shaped	40	G	A	FB	Humans	Pathogen
M. primatum	Coccobacillary	28.6	R	K	FB, E	Monkeys	Opportunistic
M. pullorum	Coccobacillary	29	G	A	P, E	Galliforms	Pathogen
M. pulmonis	Flask-shaped	26.6	G	A	FB, E	Mice	Pathogen
M. putrefaciens	Coccobacillary	28.9	G	A	FB, E	Goats	Pathogen
M. salivarium	Coccoidal	27.3	R	K	E	Humans	Opportunistic
M. simbae	Pleomorphic	37	R	K	FB	Lions	Commensal
M. spermatophilum	Coccoidal	32	R	K	FB	Humans	Pathogen
M. sphenisci[c]	Pleomorphic	28	G	A	P	Penguins	Pathogen
M. spumans	Pleomorphic	28.4	R	K	FB, E	Dogs	Opportunistic
M. sturni	Pleomorphic	31	G	A	FB	Songbirds	Pathogen
M. sualvi	Coccobacillary	23.7	R, G	V	FB	Pigs	Commensal
M. subdolum	Coccoidal	28.8	R	K	FB, E, P	Horses	Opportunistic
M. suis[b]	Coccoidal	31.1	U	na	na	Pigs	Pathogen
M. synoviae	Coccoidal	34.2	G	A	P	Galliforms	Pathogen
M. testudineum	Coccoidal	nr	G	A	FB	Tortoises	Pathogen

(Continued)

TABLE 137. (Continued)

Species	Morphology	DNA G+C content (mol%)	Energy source	Medium pH shift	Serum source	Representative host	Relation to host
M. testudinis	Flask-shaped	35	G	A	FB	Tortoises	Commensal
M. verecundum	Pleomorphic	27	OA	N	FB, E	Cattle	Commensal
"*M. vulturis*"[b, c]	Coccoidal	nr	U	na	na	Vultures	Unclear
M. wenyonii[b]	Coccoidal	nr	U	na	na	Cattle	Pathogen
M. yeatsii	Coccoidal	26.6	G	na	FB	Goats	Opportunistic
M. zalophi[c]	nr	nr	G	na	FB	Sea lions	Pathogen
"*Candidatus* M. haematoparvum"[b]	Coccoidal	nr	U	na	na	Dogs	nr
"*Candidatus* M. haemobos"[b]	Coccoidal	nr	U	na	na	Cattle	nr
"*Candidatus* M. haemodidelphidis"[b]	Coccoidal	nr	U	na	na	Opossum	nr
"*Candidatus* M. haemolamae"[b]	Coccoidal	nr	U	na	na	Llamas	nr
"*Candidatus* M. haemominutum"[b]	Coccoidal	nr	U	na	na	Cats	nr
"*Candidatus* M. kahaneii"[b]	Coccoidal	nr	U	na	na	Monkeys	nr
"*Candidatus* M. ravipulmonis"[b]	Coccoidal	nr	U	na	na	Mice	Pathogen
"*Candidatus* M. haemotarandirangiferis"[b]	Coccoidal	nr	U	na	na	Reindeer	nr
"*Candidatus* M. turicensis"[b]	Coccoidal	nr	U	na	na	Cats	nr

[a]nr, Not reported; na, not applicable; G, glucose; R, arginine; OH, alcohols; OA, organic acids; U, undefined; A, acidic pH shift; K, alkaline pH shift; N, pH remains neutral; V, pH shift can be acidic or alkaline depending on the energy source provided; FB, fetal bovine serum; E, equine serum; P, porcine serum.

[b]Not yet cultivated in cell-free artificial medium. The putative organism "*Candidatus* M. haemotarandirangiferis" remains to be definitively established and the name has no standing in nomenclature.

[c]Has been grown only in co-culture with eukaryotic cells.

mycoides subsp. *mycoides* precludes the exclusive reliance on serological-based diagnostics. Experimental vaccines using formalin-inactivated *Mycoplasma mycoides* subsp. *capri* appear to protect goats from subsequent challenge (Bar-Moshe et al., 1984; de la Fe et al., 2007). This organism is under certain quarantine regulations in most non-endemic countries and is a List B pathogen in the World Organization for Animal Health (OIE) disease classification (http://oie.int).

Source: isolated from the synovial fluid, synovial membranes, udders, expelled milk, conjunctivae, lungs, blood, and ear canals of goats; and the urogenital tract of sheep (Bergonier et al., 1997; Cottew, 1979; Kidanemariam et al., 2005; Thiaucourt et al., 1996).

DNA G+C content (mol%): 24 (T_m).

Type strain: PG3, NCTC 10137, CIP 71.25.

Sequence accession nos (16S rRNA gene): U26037 (strain PG3^T), U26044 (strain Y-goat).

Further comment: Mycoplasma mycoides subsp. *capri* now refers to strains once known as *Mycoplasma mycoides* subsp. *mycoides* var. large colony as well as strains known as *Mycoplasma mycoides* subsp. *capri* (Manso-Silván et al., 2009; Shahram et al., 2010).

1b. **Mycoplasma mycoides subsp. mycoides** Manso-Silván, Vilei, Sachse, Djordjevic, Thiaucourt and Frey 2009, 1356[VP] (*Mycoplasma mycoides* subsp. *mycoides* var. small colony Cottew and Yeats 1978, 294)

my.co.i′des. Gr. n. *mukês -êtos* mushroom or other fungus; L. suff. *-oides* (from Gr. suff. *-eides* from Gr. n. *eidos* that which is seen, form, shape, figure) resembling, similar; N.L. neut. adj. *mycoides* fungus-like.

Cells are pleomorphic and capable of forming long filaments, but do not produce rho forms. Nonmotile. An extracellular capsule can be visualized by electron microscopy following staining with ruthenium red. Colonies on solid agar have a characteristic fried-egg appearance and are notably smaller than those of *Mycoplasma mycoides* subsp. *capri*. Grows in modified Hayflick medium supplemented with glucose at 37°C. Formation of biofilms has been demonstrated (McAuliffe et al., 2008).

Pathogenic; causes a characteristic, highly lethal fibrinous interstitial pneumonia and pleurisy known as contagious bovine pleuropneumonia (CBPP) in adult cattle and severe polyarthritis in calves. Transmission occurs primarily via direct contact, but can occur by droplet aerosol as well.

Tetracyclines, chloramphenicol, and fluoroquinolones are effective chemotherapeutic agents; however, treatment of endemic herds is often counterproductive as resistance can develop in carrier animals. Organisms are often sequestered in areas of coagulative necrosis in subclinically infected animals and can serve as a reservoir for reintroduction of resistant clones of *Mycoplasma mycoides* subsp. *mycoides* into a herd. Culling of infecting herds and restricting the movement of infected animals are more effective strategies for controlling spread of the disease (Windsor and Masiga, 1977). Killed and live vaccines are available for the prevention of infection, but suffer from low antigenicity, poor efficacy, and residual pathogenesis (Brown et al., 2005). Antigenic cross-reactivity with *Mycoplasma mycoides* subsp. *capri* and *Mycoplasma leachii* preclude the exclusive reliance on serological-based diagnostics. Multiple molecular diagnostics have been described (Gorton et al., 2005; Lorenzon et al., 2008; Persson et al., 1999) and many additional molecular tools such as insertion sequence typing have led to a greater understanding of the epidemiology of outbreaks (Cheng et al., 1995; Frey et al., 1995; Vilei et al., 1999). This organism is under certain quarantine regulations in most non-endemic countries and is listed in the Terrestrial Animal Health Code of the Office International des Epizooties (http://oie.int).

Source: isolated from the lungs, pleural fluid, lymph nodes, sinuses, kidneys, urine, synovial fluid, and synovial membranes of cattle and water buffalo (Gourlay and Howard, 1979; Scanziani et al., 1997; Scudamore, 1976); the respiratory tract of bison; the respiratory tract of yak; and the lungs, nasopharynx, and pleural fluid of sheep and goats (Brandao, 1995; Kusiluka et al., 2000).

DNA G+C content (mol%): 26.1 (T_m), 24.0 (strain PG1^T complete genome sequence).

Type strain: PG1, NCTC 10114, CCUG 32753.

Sequence accession nos: U26039 (16S rRNA gene), BX293980 (strain PG1^T complete genome sequence).

Further comment: *Mycoplasma mycoides* subsp. *mycoides* now refers exclusively to the agent of CBPP (Manso-Silván et al., 2009).

2. **Mycoplasma adleri** Del Giudice, Rose and Tully 1995, 31[VP]

ad'le.ri. N.L. masc. gen. n. *adleri* of Adler, referring to Henry Adler, a Californian veterinarian whose studies contributed much new information concerning the pathogenic role of caprine and avian mycoplasmas.

Cells are primarily coccoid. Nonmotile. Colonies on solid media have a typical fried-egg appearance. Grows well in Hayflick medium supplemented with arginine at 35–37°C.

Pathogenic; associated with suppurative arthritis and joint abscesses. Mode of transmission is unknown.

Source: isolated from an abscessed joint of a goat with suppurative arthritis (Del Giudice et al., 1995).

DNA G+C content (mol%): 29.6 (Bd).

Type strain: G145, ATCC 27948, CIP 105676.

Sequence accession no. (16S rRNA gene): U67943.

3. **Mycoplasma agalactiae** (Wróblewski 1931) Freundt 1955, 73[AL] (*Anulomyces agalaxiae* Wróblewski 1931, 111)

a.ga.lac.ti'ae. Gr. n. *agalactia* want of milk, agalactia; N.L. gen. n. *agalactiae* of agalactia.

Cells are primarily coccoid, but are occasionally branched and filamentous. Nonmotile. Colonies on solid media have a typical fried-egg appearance. Grows well in SP-4 or Hayflick medium supplemented with glucose at 37°C. Formation of biofilms has been demonstrated (McAuliffe et al., 2006).

Pathogenic; causes polyarthritis, mastitis, conjunctivitis (a syndrome collectively termed contagious agalactia; Bergonier et al., 1997), nonsuppurative arthritis, pneumonia, abortion, and granular vulvovaginitis (Cottew, 1983; DaMassa, 1996) in goats and sheep. Transmission occurs via direct contact, most commonly during feeding (kids and lambs) or milking (dams and ewes).

Macrolides and fluoroquinolones are effective chemotherapeutic agents; however, antimicrobial therapy is not often utilized in widespread outbreaks due to the potential for infected animals to develop carrier states with resistant strains and the tendency of antimicrobials to be excreted in milk. Control measures such as disinfection of fomites (endemic areas) and culling of infected animals (acute outbreaks) are more common practices. *Mycoplasma agalactiae* reportedly shares surface antigens with *Mycoplasma bovis* and *Mycoplasma capricolum* subsp. *capricolum* (Alberti et al., 2008; Boothby et al., 1981), potentially complicating serology-based diagnosis of infection. Molecular diagnostics that can distinguish *Mycoplasma agalactiae* from *Mycoplasma bovis* have been described (Chávez Gonzalez et al., 1995). Commercially available vaccines are widely used, but exhibit poor efficacy. Numerous experimental vaccines have been described. This organism is under certain quarantine regulations in some countries and is listed in the Terrestrial Animal Health Code of the Office International des Epizooties (http://oie.int).

Source: isolated from the joints, udders, milk, conjunctivae, lungs, vagina, liver, spleen, kidneys, and small intestine of sheep and goats.

DNA G+C content (mol%): 30.5 (T_m), 29.7 (strain PG2^T complete genome sequence).

Type strain: PG2, NCTC 10123, CIP 59.7.

Sequence accession nos: M24290 (16S rRNA gene), NC_009497 (strain PG2^T complete genome sequence).

4. **Mycoplasma agassizii** Brown, Brown, Klein, McLaughlin, Schumacher, Jacobson, Adams and Tully 2001c, 417[VP]

a.gas.si'zi.i. N.L. masc. gen. n. *agassizii* of Agassiz, referring to Louis Agassiz, a naturalist whose name was assigned to a species of desert tortoise (*Gopherus agassizii*) from which the organism was isolated.

Cells are coccoid to pleomorphic, with some strains appearing to possess a rudimentary terminal structure. Cells exhibit gliding motility. Colony forms on solid medium vary from those with a fried-egg appearance to some with mulberry characteristics. Grows well in SP-4 medium supplemented with glucose at 30°C.

Pathogenic; causes chronic upper respiratory tract disease characterized by severe rhinitis in desert tortoises, gopher tortoises, Russian tortoises, and leopard tortoises. Mode of transmission appears to be intranasal inhalation (Brown et al., 1994).

Source: isolated from the nares and choanae of desert tortoises, gopher tortoises, Russian tortoises, and leopard tortoises (Brown et al., 2001c).

DNA G+C content (mol%): not determined.

Type strain: PS6, ATCC 700616.

Sequence accession no. (16S rRNA gene): U09786.

5. **Mycoplasma alkalescens** Leach 1973, 149[AL]

al.ka.les'cens. N.L. v. *alkalesco* to make alkaline, referring to the reaction produced in arginine-containing media; N.L. part. adj. *alkalescens* alkaline-making.

Coccoid to coccobacillary cells. Motility and colony morphology have not been described for this species. Grows well in SP-4 medium supplemented with arginine at 37°C.

Pathogenic; causes febrile arthritis and sometimes mastitis, pneumonia, and otitis in cattle (Kokotovic et al., 2007; Lamm et al., 2004; Leach, 1973). Mode of transmission has not been established.

Tetracyclines and pleuromutilins are effective chemotherapeutic agents (Hirose et al., 2003). *Mycoplasma alkalescens* reportedly shares surface antigens with many arginine-fermenting *Mycoplasma* species; however, this is only likely to interfere with accurate diagnosis of infection in the case of *Mycoplasma arginini*.

Source: isolated from the synovial fluid, expelled milk, lungs, ears, prepuce, and semen of cattle.

DNA G+C content (mol%): 25.9 (T_m).

Type strain: D12, PG51, NCTC 10135, ATCC 29103.

Sequence accession no. (16S rRNA gene): U44764.

Further comment: Bovine serogroup 8 of Leach (1967).

6. **Mycoplasma alligatoris** Brown, Farley, Zacher, Carlton, Clippinger, Tully and Brown 2001a, 423[VP]

al.li.ga.to'ris. N.L. n. *alligator, -oris* an alligator; N.L. gen. n. *alligatoris* of/from an alligator.

Cells are primarily coccoid. Nonmotile. Colonies on solid medium exhibit typical fried-egg morphology. Growth is very rapid in SP-4 medium supplemented with glucose at 30°C.

Pathogenic; causes a multisystemic inflammatory illness with a lethality unprecedented among mycoplasmas. Pathologic lesions in infected animals include meningitis, interstitial pneumonia, fibrinous pleuritis, polyserositis, fibrinous pericarditis, myocarditis, endocarditis, synovitis, and splenic and hepatic necrosis. A high level of mortality of naturally and experimentally infected American alligators (*Alligator missisippiensis*) and broad-nosed caimans (*Caiman latirostris*) occurs. In contrast, *Mycoplasma alligatoris* colonizes the tonsils of experimentally infected Siamese crocodiles (*Crocodylus siamensis*) without causing overt pathology. Such findings led to the hypothesis that *Mycoplasma alligatoris* is a natural commensal of a species more closely related to crocodiles than alligators and that the extreme disease state observed in alligators results from an enzoonotic infection (Brown et al., 1996; Pye et al., 2001). The natural mode of transmission has not been established definitively; however, animals can be experimentally infected via inoculation of the glottis.

Source: isolated from the blood, synovial fluid, cerebrospinal fluid, lungs, brain, heart, liver, and spleen of naturally and experimentally infected American alligators, experimentally infected broad-nosed caimans, and from the tonsils of experimentally infected Siamese crocodiles.

DNA G+C content (mol%): not determined.

Type strain: A21JP2, ATCC 700619.

Sequence accession no. (16S rRNA gene): U56733.

Further comment: the name *Mycoplasma alligatoris* was assigned for this organism in consideration of the initial isolation from an alligator (Order: Crocodylia). The English word "alligator" is from the Spanish *el lagarto* (Latin *ille lacertu* the lizard). However, the specific epithet *lacerti*, originally proposed for this taxon, was ultimately rejected because of the modern phylogenetic distinction between lizards (Order: Lacertilia) and crocodilians.

7. **Mycoplasma alvi** Gourlay, Wyld and Leach 1977, 95[AL]

al'vi. L. n. *alvus* bowel, womb, stomach; L. gen. n. *alvi* of the bowel.

Cells are primarily coccoid; however, subsets of the population display elongated, flask-shaped cells with well-defined terminal structures. Unlike most species that exhibit polar structures, *Mycoplasma alvi* appears to be nonmotile (Bredt, 1979; Hatchel and Balish, 2008). Colonies on solid medium exhibit typical fried-egg morphology. Grows well in SP-4 medium supplemented with either glucose or arginine at 37°C.

No evidence of pathogenicity. Mode of transmission has not been established definitively.

Source: isolated from the lower alimentary tract, feces, bladder, and vagina of cows, and the intestinal tract of voles (Gourlay and Howard, 1979).

DNA G+C content (mol%): 26.4 (Bd).

Type strain: Ilsley, NCTC 10157, ATCC 29626.

Sequence accession no. (16S rRNA gene): U44765.

8. **Mycoplasma amphoriforme** Pitcher, Windsor, Windsor, Bradbury, Yavari, Jensen, Ling and Webster 2005, 2592[VP]

am.pho.ri.for'me. L. n. *amphora* amphora; L. adj. suff. *-formis -e* like, of the shape of; N.L. neut. adj. *amphoriforme* amphora-shaped, having the form of an amphora.

Cells are flask-shaped with a distinct terminal structure reminiscent of *Mycoplasma gallisepticum*. Cells exhibit low-speed gliding motility and move in the direction of the terminal structure (Hatchel et al., 2006). Colony morphology variable, from a typical fried-egg morphology to a "ground glass" appearance. Grows well in SP-4 medium supplemented with glucose at 37°C.

Pathogenicity and mode of transmission have not been established definitively.

Despite showing sensitivity to fluoroquinolones, tetracyclines, and macrolides *in vitro*, *Mycoplasma amphoriforme* appears to be successfully evasive during treatment of patients with these antibiotics. The veterinary antibiotic valnemulin is successful at controlling infection (Webster et al., 2003).

Source: isolated from the sputum of immunocompromised humans with bronchitis and related lower respiratory tract disease (Pitcher et al., 2005; Webster et al., 2003). The prevalence of *Mycoplasma amphoriforme* in the general human population is unsubstantiated.

DNA G+C content (mol%): 34.0 (fluorescent intensity).

Type strain: A39, NCTC 11740, ATCC BAA-992.

Sequence accession no. (16S rRNA gene): FJ226575.

9. **Mycoplasma anatis** Roberts 1964, 471[AL]

a.na'tis. L. n. *anas, -atis* a duck; L. gen. n. *anatis* of a duck.

Cells have been described as coccoid with ring forms, but cellular and colony morphology is generally poorly described. Motility for this species has not been assessed. Grows well in SP-4 medium supplemented with glucose 37°C.

Isolated from pathologic lesions, but attempts to reproduce disease following experimental infection have been equivocal (Amin and Jordan, 1978; Roberts, 1964). The mode of transmission has not been established definitively.

Source: isolated from pathologic lesions, the respiratory tract, hock joint, pericardium, cloaca, and meninges of ducks (Goldberg et al., 1995; Ivanics et al., 1988; Tiong, 1990).

DNA G+C content (mol%): 26.6 (Bd).
Type strain: 1340, ATCC 25524, NCTC 10156.
Sequence accession no. (16S rRNA gene): AF412970.

10. **Mycoplasma anseris** Bradbury, Jordan, Shimizu, Stipkovits and Varga 1988, 76[VP]

an′se.ris. L. gen. n. *anseris* of the goose.

Cells are primarily spherical. Nonmotile. Colonies on solid medium have a typical fried-egg appearance. Grows well in Hayflick medium supplemented with arginine at 37°C.

Opportunistic pathogen; associated with balanitis of geese, but can be isolated from clinically normal animals. Mode of transmission has not been established definitively.

Source: isolated from the phallus and cloaca of geese (Hinz et al., 1994; Stipkovits et al., 1984a).
DNA G+C content (mol%): 24.7–26.0 (Bd/T_m).
Type strain: 1219, ATCC 49234.
Sequence accession no. (16S rRNA gene): AF125584.

11. **Mycoplasma arginini** Barile, Del Giudice, Carski, Gibbs and Morris 1968, 490[AL]

ar.gi.ni′ni. N.L. n. *argininum* arginine, an amino acid; N.L. gen. n. *arginini* of arginine, referring to its hydrolysis.

Cells are primarily coccoid. Motility for this species has not been assessed. Colonies have either the typical fried-egg morphology or a granular, "berry-like" appearance. Grows well in modified Hayflick medium supplemented with arginine at 37°C.

Pathogenic; associated with pneumonia, vesiculitis, keratoconjunctivitis, and mastitis in cattle; pneumonia and keratoconjunctivitis in sheep; possibly with arthritis in goats; and septicemia in immunocompromised humans (Tully and Whitcomb, 1979; Yechouron et al., 1992). Modes of transmission have not been definitively assessed and likely vary by anatomical site.

Mycoplasma arginini is commonly associated with contamination of eukaryotic cell culture and is frequently removed by treatment of cells with antibiotics and/or maintenance of cell lines in antibiotic-containing medium. The most effective classes of antibiotics for cell culture eradication are tetracyclines, macrolides, and fluoroquinolones. Additionally, passage of eukaryotic cells in hyperimmune serum raised against *Mycoplasma arginini* has been shown to be an effective method of eradication (Jeansson and Brorson, 1985).

Source: isolated from a wide array of mammalian hosts including cattle, sheep, goats, pigs, horses, domestic dogs, domestic cats, lions, lynxes, cheetahs, chamois, camels, ibexes, humans, and mice.
DNA G+C content (mol%): 27.6 (T_m).
Type strain: G230, ATCC 23838, NCTC 10129, CIP 71.23, NBRC 14476.
Sequence accession no. (16S rRNA gene): U15794.

12. **Mycoplasma arthritidis** (Sabin 1941) Freundt 1955, 73[AL] (*Murimyces arthritidis* Sabin 1941, 57)

ar.thri′ti.dis. Gr. n. *arthritis -idos* gout, arthritis; N.L. gen. n. *arthritidis* of arthritis.

Cells are filamentous and vary in length. Motility for this species has not been assessed. Colonies on solid medium have a typical fried-egg appearance. Grows well in SP-4 medium supplemented with arginine.

Pathogenic; causes purulent polyarthritis, rhinitis, otitis media, ocular lesions, and abscesses in rats. *Mycoplasma arthritidis* is also known to superinfect lung lesions initiated by *Mycoplasma pulmonis.* Experimental inoculations via various routes can result in septicemia, acute flaccid paralysis, and pyelonephritis in rats, and chronic arthritis in mice and rabbits. *Mycoplasma arthritidis* is unique among mycoplasmas in harboring the lysogenized bacteriophage MAV1, whose contribution to virulence is equivocal, and producing the potent mitogen MAM, which appears to confer increased toxicity and lethality but to be irrelevant to arthritogenicity (Clapper et al., 2004; Luo et al., 2008; Voelker et al., 1995). The mechanism of transmission is largely dependent on the tissue infected.

Source: isolated from the synovial membranes, synovial fluid, middle ear, eye, abscessed bone, abscessed ovary, and oropharynx of wild and captive rats. Isolations have also been reported from non-human primates including rhesus monkeys and bush babies (Somerson and Cole, 1979); and from joint fluid of wild boars (Binder et al., 1990). The true origins of putative isolates from the human urethra, prostate, and cervix have been questioned (Cassell and Hill, 1979; Washburn et al., 1995).
DNA G+C content (mol%): 30.0 (T_m; strain PG6^T), 30.7 (strain 158L3-1 complete genome sequence).
Type strain: PG6, ATCC 19611, NCTC 10162, CIP 104678, NBRC 14860.
Sequence accession nos: M24580 (16S rRNA gene), CP001047 (strain 158L3-1 complete genome sequence).

13. **Mycoplasma auris** DaMassa, Tully, Rose, Pitcher, Leach and Cottew 1994, 483[VP]

au′ris. L. gen. n. *auris* of the ear, referring to the provenance of the organism, the ears of goats.

Cells are coccoid to pleomorphic. Nonmotile. Colonies on solid medium have a typical fried-egg appearance. Grows well in Hayflick medium supplemented with arginine at 37°C.

No evidence of pathogenicity. Mechanism of transmission has not been established.

Source: isolated from the external ear canals of goats (Damassa et al., 1994).
DNA G+C content (mol%): 26.9 (T_m).
Type strain: UIA, ATCC 51348, NCTC 11731, CIP 105677.
Sequence accession no. (16S rRNA gene): U67944.

14. **Mycoplasma bovigenitalium** Freundt 1955, 73[AL]

bo.vi.ge.ni.ta′li.um. L. n. *bos, bovis* the ox, bull, cow; L. pl. n. *genitalia* the genitals; N.L. pl. gen. n. *bovigenitalium* of bovine genitalia.

Cells range from coccoid to filamentous. Motility for this species has not been assessed. Colonies on solid medium have a typical fried-egg appearance. Grows well in SP-4 broth supplemented with glucose and/or arginine at 37°C and produces a "film and spots" reaction.

Pathogenic; causes vulvovaginitis, vesiculitis, epididymitis, abortion, infertility, mastitis, pneumonia, conjunctivitis, and arthritis in cattle; and pneumonia and conjunctivitis in domestic dogs. Mode of transmission is via sexual contact and/or droplet aerosol.

Control measures during outbreaks of *Mycoplasma bovigenitalium* infection include suspension of natural breeding in favor of artificial insemination with disposable instruments and, in severe cases, culling of infected animals.

Source: isolated from the udders, seminal vesicles, prepuce, semen, vagina, cervix, lungs, conjunctivae, and joint capsule of cattle; and from the lungs, prepuce, prostate, vagina, cervix, and conjunctivae of domestic dogs (Chalker, 2005; Gourlay and Howard, 1979).

DNA G+C content (mol%): 30.4 (T_m).

Type strain: PG11, ATCC 19852, NCTC 10122, NBRC 14862.

Sequence accession nos (16S rRNA gene): M24291, AY121098.

Further comment: the collection of strains formerly referred to as "*Mycoplasma* ovine/caprine serogroup 11" have been reclassified as *Mycoplasma bovigenitalium* (Nicholas et al., 2008).

15. **Mycoplasma bovirhinis** Leach 1967, 313[AL]

bo.vi.rhi'nis. L. n. *bos, bovis* the ox; Gr. n. *rhis, rhinos* nose; N.L. gen. n. *bovirhinis* of the nose of the ox.

Cell and colony morphology and motility for this species are poorly defined. Grows well in SP-4 medium supplemented with glucose at 37°C.

Pathogenic; causes pneumonia, otitis, conjunctivitis, and mastitis in cattle. *Mycoplasma bovirhinis* is often found in co-infections with other pathogens, leading to speculation that it often acts as a superinfecting agent. Mode of transmission has not been established definitively.

Source: isolated from the lungs, nasopharynx, trachea, udders, expelled milk, ears, conjunctivae, and rarely from the urogenital tract of cattle (Gourlay and Howard, 1979).

DNA G+C content (mol%): 27.3 (T_m).

Type strain: PG43, ATCC 27748, NCTC 10118, CIP 71.24, NBRC 14857.

Sequence accession no. (16S rRNA gene): U44766.

16. **Mycoplasma bovis** (Hale, Hemboldt, Plastridge and Stula 1962) Askaa and Ernø 1976, 325[AL] (*Mycoplasma agalactiae* var. *bovis* Hale, Hemboldt, Plastridge and Stula 1962, 591; *Mycoplasma bovimastitidis* Jain, Jasper and Dellinger 1967, 409)

bo'vis. L. n. *bos* the ox; L. gen. n. *bovis* of the ox.

Cells range from coccoidal to short filaments. Nonmotile. Formation of biofilms has been demonstrated (McAuliffe et al., 2006). Colonies on solid medium have the typical fried-egg appearance, with notably large centers. Grows well in SP-4 medium supplemented with glucose at 37°C and produces a "film and spots" reaction.

Pathogenic; causes mastitis, polyarthritis, keratoconjunctivitis (Gourlay and Howard, 1979), pneumonia, and otitis media (Caswell and Archambault, 2007; Maeda et al., 2003), and is rarely associated with infertility, abortion, endometritis, salpingitis, and vesiculitis (Doig, 1981; Gourlay and Howard, 1979) in cattle; pneumonia and polyarthritis in bison (Dyer et al., 2008); and is rarely associated with pneumonia, mastitis, and arthritis in goats (Egwu et al., 2001; Gourlay and Howard, 1979). Mode of transmission is via direct contact with infected animals or fomites, most commonly during feeding (suckling or trough), milking (cows), aerosol, or sexual contact.

Macrolides and fluoroquinolones are effective chemotherapeutic agents *in vitro*; however, antimicrobial therapy is not often utilized in animals with advanced disease due to poor efficacy and the tendency of antimicrobials to be excreted in milk. Control measures such as disinfection of fomites, isolation of infected animals, and euthanasia of animals showing clinical signs are more common practices. *Mycoplasma bovis* reportedly shares surface antigens with *Mycoplasma agalactiae* (Boothby et al., 1981), potentially complicating serology-based diagnosis of infection. Molecular diagnostics that can distinguish *Mycoplasma agalactiae* from *Mycoplasma bovis* have been described (Chávez Gonzalez et al., 1995). Several vaccines are commercially available, but exhibit poor efficacy in that they tend to allow for the establishment of infection while only preventing overt clinical signs.

Source: isolated from the udders, expelled milk, synovial fluid, synovial membranes, conjunctivae, lungs, ear canals, tympanic membranes, aborted calves, uterus, cervix, vagina, and semen of cattle; from the lungs and synovial fluid of bison; and from the lungs and udders of goats.

DNA G+C content (mol%): 32.9 (T_m).

Type strain: Donetta, PG45, ATCC 25523, NCTC 10131.

Sequence accession no. (16S rRNA gene): AJ419905.

Further comment: Bovine serotype 5 of Leach (1967).

17. **Mycoplasma bovoculi** Langford and Leach 1973, 1443[AL]

bo.vo'cu.li. L. n. *bos, bovis* ox, bull, cow; L. n. *oculus* the eye; N.L. gen. n. *bovoculi* of the bovine eye.

Cells are coccoid to coccobacillary. Motility for this species has not been assessed. Colonies on solid medium have a typical fried-egg appearance. Grows well in Hayflick medium supplemented with glucose at 37°C.

Pathogenic; causes conjunctivitis and keratoconjunctivitis in cattle. Face flies are the suggested mechanism of transmission, though this has yet to be established definitively. Topical application of oxytetracycline is an effective treatment for infection.

Source: isolated from the conjunctivae and semen of cattle, and from aborted calves (Langford and Leach, 1973; Singh et al., 2004).

DNA G+C content (mol%): 29.0 (T_m).

Type strain: M165/69, NCTC 10141, ATCC 29104.

Sequence accession no. (16S rRNA gene): U44768.

Further comment: Mycoplasma bovoculi was originally described as *Mycoplasma oculi* by Leach in 1973, wherein the defining publication referring to the species as *Mycoplasma bovoculi* (Langford and Leach, 1973) was cited as "in press".

18. **Mycoplasma buccale** Freundt, Taylor-Robinson, Purcell, Chanock and Black 1974, 252[AL]

buc.ca'le. L. n. *bucca* the mouth; L. neut. suff. *-ale* suffix denoting pertaining to; N.L. neut. adj. *buccale* buccal, pertaining to the mouth.

Cells range from coccoid to filamentous. Motility for this species has not been assessed. Colonies on solid medium have a typical fried-egg appearance. Grows well in Hayflick medium supplemented with arginine and herring sperm DNA at 37°C.

No evidence of pathogenicity. Mode of transmission has not been assessed definitively.

Source: isolated from the oropharynx of humans, rhesus macaques, chimpanzees, orangutans, baboons, African green monkeys, crab-eating macaques, and patas monkeys (Somerson and Cole, 1979).

DNA G+C content (mol%): 26.4 (T_m).

Type strain: CH20247, ATCC 23636, NCTC 10136, CIP 105530, NBRC 14851.

Sequence accession no. (16S rRNA gene): AF125586.

19. **Mycoplasma buteonis** Poveda, Giebel, Flossdorf, Meier and Kirchhoff 1994, 97[VP]

bu.te.o'nis. L. masc. n. *buteo, -onis* buzzard; L. masc. gen. n. *buteonis* of the buzzard.

Cells are coccoid. Motility for this species has not been assessed. Colonies on solid medium have typical fried-egg appearance. Grows well in modified Frey's medium supplemented with glucose at 37°C.

Mycoplasma buteonis may be pathogenic for saker falcons, as it was found in the respiratory tract, nervous system, and bone of a nestling with pneumonia, hepatitis, ataxia, and dyschondroplasia. No evidence of pathogenicity for buzzards. Mode of transmission has not been assessed.

Source: isolated from the trachea of buzzards; from the eggs of the lesser kestrel; and the trachea, lungs, brain, and bone marrow of the saker falcon. Has been detected in the common kestrel and the Western marsh harrier (Erdélyi et al., 1999; Lierz et al., 2008a, 2008c).

DNA G+C content (mol%): 27.0 (Bd).

Type strain: Bb/T2g, ATCC 51371.

Sequence accession no. (16S rRNA gene): AF412971.

20. **Mycoplasma californicum** Jasper, Ernø, Dellinger and Christiansen 1981, 344[VP]

ca.li.for'ni.cum. N.L. *neut.* adj. *californicum* pertaining to California.

Cells are coccoid to filamentous. Motility for this species has not been assessed. Colonies are conical in shape with distinct small centers. Grows well in modified Hayflick broth at 37°C.

Pathogenic; causes purulent mastitis in cows and rarely in sheep. Mode of transmission has not been established definitively.

Source: isolated from the udders and expelled milk of cows and ewes.

DNA G+C content (mol%): 31.9 (Bd).

Type strain: ST-6, ATCC 33461, AMRC-C 1077, NCTC 10189.

Sequence accession no. (16S rRNA gene): M24582.

21. **Mycoplasma canadense** Langford, Ruhnke and Onoviran 1976, 218[AL]

ca.na.den'se. N.L. neut. adj. *canadense* pertaining to Canada.

Cells are coccoid to coccobacillary. Motility for this species has not been assessed. Colonies on solid agar have a characteristic fried-egg appearance. Grows well in SP-4 medium supplemented with arginine at 37°C.

Pathogenic; causes mastitis and arthritis, and may be associated with infertility, abortion, and pneumonia of cattle. Mode of transmission has not been established definitively.

Source: isolated from the udders, expelled milk, synovial membranes, aborted calves, vagina, semen, and lungs (Boughton et al., 1983; Friis and Blom, 1983; Gourlay and Howard, 1979; Jackson et al., 1981).

DNA G+C content (mol%): 29.0 (T_m).

Type strain: 275C, NCTC 10152, ATCC 29418.

Sequence accession no. (16S rRNA gene): U44769.

22. **Mycoplasma canis** Edward 1955, 90[AL]

ca'nis. L. n. *canis, -is* a dog; L. gen. n. *canis* of a dog.

Cells are pleomorphic, exhibiting branched and filamentous forms. Motility for this species has not been assessed. Colonies on solid medium exhibit two stable forms: "smooth" colonies with a nongranular appearance and round edges; and "rough" colonies with a granular appearance and irregular or crenated edges. Each form maintains its characteristic appearance during repeated subculturing. Grows well in SP-4 medium supplemented with glucose at 37°C.

Opportunistic pathogen; associated with infertility and adverse pregnancy outcomes, endometritis, epididymitis, urethritis, cystitis, and pneumonia of domestic dogs (Chalker, 2005); pneumonia of cattle (ter Laak et al., 1992b); and pneumonia in immunocompromised humans (Armstrong et al., 1971). Mode of transmission is via sexual contact or aerosol. *Mycoplasma canis* appears to have a greater tendency toward upper respiratory tract commensalism and urogenital tract pathogenicity in dogs, while exhibiting pathogenicity for the respiratory tract of cattle.

Source: isolated from the cervix, vagina, prepuce, epididymis, prostate, semen, urine, bladder, oropharynx, nares, lungs, trachea, conjunctivae, kidneys, spleen, pericardium, liver, and lymph nodes of domestic dogs; from the lungs and oropharynx of cattle; from the lungs and pharynx of humans; and from the throat and rectum of baboons and African green monkeys.

DNA G+C content (mol%): 28.4 (T_m).

Type strain: PG14, ATCC 19525, NCTC 10146, NBRC 14846.

Sequence accession no. (16S rRNA gene): AF412972.

23. **Mycoplasma capricolum** Tully, Barile, Edward, Theodore and Ernø 1974, 116[AL]

ca.pri.co'lum. L. n. *caper, -pri* the male goat; N.L. -suff. *colus, -a, -um* (from L. v. *incolere* to dwell) dwelling; N.L. neut. adj. *capricolum* dwelling in a male goat.

Cells are coccobacillary. Nonmotile. Colonies on solid agar have a characteristic fried-egg appearance. Grows in SP-4 or modified Hayflick medium supplemented with glucose at 37°C.

DNA G+C content (mol%): 24.1 (T_m).

Type strain: California kid, ATCC 27343, NCTC 10154, CIP 104620.

The species has subsequently been divided as follows.

23a. **Mycoplasma capricolum subsp. capricolum** (Tully, Barile, Edward, Theodore and Ernø 1974) Leach, Ernø and MacOwan 1993, 604[VP] (*Mycoplasma capricolum* Tully, Barile, Edward, Theodore and Ernø 1974, 116)

ca.pri.co'lum. L. n. *caper, -pri* the male goat; N.L. -suff. *colus, -a, -um* (from L. v. *incolere* to dwell) dwelling; N.L. neut. adj. *capricolum* dwelling in a male goat.

Cells are coccobacillary and can produce long, helical rods known as rho forms. Nonmotile. Colonies on solid agar have a characteristic fried-egg appearance. Grows in SP-4 or modified Hayflick medium supplemented with glucose at 37°C.

Pathogenic; causes fibrinopurulent polyarthritis, mastitis, conjunctivitis (a syndrome collectively termed contagious agalactia) and septicemia in goats. Transmission occurs via direct contact between animals or with fomites.

Tetracyclines, macrolides, and tylosin are effective chemotherapeutic agents. Treatment of acutely infected animals often leads to the eradication of the organism, whereas treatment of chronically infected animals does not. Early intervention with antibiotics and improved sanitation are effective control measures (Thiaucourt et al., 1996). Antigenic cross-reactivity with *Mycoplasma capricolum* subsp. *capripneumoniae* and *Mycoplasma leachii* preclude the exclusive reliance on serological-based diagnostics. Multiple molecular diagnostics have been described (Fitzmaurice et al., 2008; Greco et al., 2001). This organism is under certain quarantine regulations in some countries and is listed in the Terrestrial Animal Health Code of the Office International des Epizooties (http://oie.int).

Source: isolated from the synovial fluid, synovial membranes, udders, expelled milk, conjunctivae, spleen, nasopharynx, oral cavity, and ear canal of goats; and the nasopharynx of sheep (Cottew, 1979).

DNA G+C content (mol%): 24.1 (T_m), 23 (strain California kid[T] complete genome).

Type strain: California kid, ATCC 27343, NCTC 10154, CIP 104620.

Sequence accession nos: U26046 (16S rRNA gene), NC_007633 (strain California kid[T] complete genome).

23b. **Mycoplasma capricolum subsp. capripneumoniae** Leach, Ernø and MacOwan 1993, 604[VP]

ca.pri.pneu.mo.ni'ae. L. n. *capra, -ae* a goat; Gr. n. *pneumonia* disease of the lungs, pneumonia; N.L. gen. n. *capripneumoniae* of a pneumonia of a goat.

Cells are coccobacillary. Nonmotile. Colonies on solid agar have a characteristic fried-egg appearance. Grows in SP-4 or modified Hayflick medium supplemented with glucose at 37°C.

Pathogenic; causes characteristic, highly lethal fibrinous pleuropneumonia known as contagious caprine pleuropneumonia (CCPP) in goats (McMartin et al., 1980). A respiratory tract disease of similar pathology found in association with *Mycoplasma capricolum* subsp. *capripneumoniae* has been reported in sheep, mouflon, and ibex (Arif et al., 2007; Shiferaw et al., 2006). Transmission occurs via droplet aerosol.

Tetracyclines and tylosin are effective chemotherapeutic agents; however, treatment of endemic herds is not often undertaken. Culling of infected herds and restricting the movement of infected animals are more common strategies for controlling spread of the disease (Thiaucourt et al., 1996). Live and killed vaccines have been described; however, each appear to afford delayed or partial protection from morbidity, and few have been completely successful at preventing infection (Browning et al., 2005). Antigenic cross-reactivity with *Mycoplasma capricolum* subsp. *capricolum* and *Mycoplasma leachii* preclude the exclusive reliance on serological-based diagnostics. Multiple molecular diagnostics have been described (Lorenzon et al., 2008; March et al., 2000; Woubit et al., 2004). This organism is under certain quarantine regulations in some countries and is listed in the Terrestrial Animal Health Code of the Office International des Epizooties (http://www.oie.int).

Source: isolated from the lower respiratory tract of goats, sheep, mouflon, and ibex.

DNA G+C content (mol%): 24.4 (Bd).

Type strain: F38, NCTC 10192.

Sequence accession no. (16S rRNA gene): U26042.

Further comment: previously known as the F38-type caprine mycoplasmas.

24. **Mycoplasma caviae** Hill 1971, 112[AL]

ca.vi'ae. N.L. n. *cavia* guinea pig (*Cavia cobaya*); N.L. gen. n. *caviae* of a guinea pig.

Cell morphology for this species has not been described and motility has not been assessed. Colonies on solid agar have a characteristic fried-egg appearance. Grows in SP-4 medium supplemented with glucose at 37°C.

No evidence of pathogenicity. Mode of transmission has not been assessed definitively.

Source: isolated from the nasopharynx and urogenital tract of guinea pigs (Hill, 1971).

DNA G+C content (mol%): not determined.

Type strain: G122, ATCC 27108, NCTC 10126.

Sequence accession no. (16S rRNA gene): AF221111.

25. **Mycoplasma cavipharyngis** Hill 1989, 371[VP] (Effective publication: Hill 1984, 3187)

ca.vi.pha.ryn'gis. N.L. n. *cavia* the guinea pig (*Cavia cobaya*); N.L. n. *pharynx -yngis* (from Gr. n. *pharugx pharuggos* throat) throat; N.L. gen. n. *cavipharyngis* of the throat of a guinea pig.

Cells are highly filamentous and filaments are twisted at intervals along their length. Regular helical forms like those of *Spiroplasma* species are not produced. Nonmotile. Growth on solid medium shows small, granular colonies with poorly defined centers. Grows in Hayflick medium supplemented with glucose at 35°C.

No evidence of pathogenicity. Mode of transmission has not been established definitively.

Source: isolated from the nasopharynx of guinea pigs (Hill, 1984).

DNA G+C content (mol%): 30 (T_m).

Type strain: 117C, NCTC 11700, ATCC 43016.

Sequence accession no. (16S rRNA gene): AF125879.

26. **Mycoplasma citelli** Rose, Tully and Langford 1978, 571[AL]

ci.tel'li. N.L. n. *Citellus* a genus of ground squirrel; N.L. gen. n. *citelli* of *Citellus*.

Cells are highly pleomorphic. Motility for this species has not been assessed. Colonies on solid agar have a characteristic fried-egg appearance. Grows well in SP-4 medium supplemented with glucose at 37°C.

No evidence of pathogenicity. Mode of transmission has not been established definitively.

Source: isolated from the trachea, lung, spleen, and liver of ground squirrels (Rose et al., 1978).

DNA G+C content (mol%): 27.4 (Bd).

Type strain: RG-2C, ATCC 29760, NCTC 10181.

Sequence accession no. (16S rRNA gene): AF412973.

27. **Mycoplasma cloacale** Bradbury and Forrest 1984, 392[VP]

clo.a.ca'le. L. neut. adj. *cloacale* pertaining to a cloaca.

Cells are primarily spherical. Nonmotile. Colonies on solid medium exhibit typical fried-egg appearance. Grows well in Hayflick medium supplemented with arginine at 37°C.

No evidence of pathogenicity. Mode of transmission has not been established definitively

Source: isolated from the cloaca of a turkey; from the lungs, trachea, ovaries, and eggs of ducks; and from chickens, pheasants, and geese (Benčina et al., 1987, 1988; Bradbury et al., 1987; Goldberg et al., 1995; Hinz et al., 1994).

DNA G+C content (mol%): 26 (Bd).

Type strain: 383, ATCC 35276, NCTC 10199.

Sequence accession no. (16S rRNA gene): AF125592.

28. **Mycoplasma collis** Hill 1983b, 849[VP]

col'lis. L. gen. n. *collis* of a hill, alluding to the author who described the species.

Cells are primarily coccoidal and are nonmotile. Growth on a solid medium shows colonies with a typical fried-egg appearance. Grows in Hayflick medium supplemented with glucose at 35–37°C.

No evidence of pathogenicity. Mode of transmission has not been assessed.

Source: isolated from the conjunctivae of captive rats and mice (Hill, 1983b). References to isolation of *Mycoplasma collis* from domestic dogs appear to have been in error (Chalker and Brownlie, 2004).

DNA G+C content (mol%): 28 (T_m).

Type strain: 58B, NCTC 10197, ATCC 35278.

Sequence accession no. (16S rRNA gene): AF538681.

29. **Mycoplasma columbinasale** Jordan, Ernø, Cottew, Hinz and Stipkovits 1982, 114[VP]

co.lum.bi.na.sa'le. L. n. *columbus* a pigeon; L. n. *nasus* nose; L. neut. suff. *-ale* suffix used with the sense of pertaining to; N.L. neut. adj. *nasale* pertaining to the nose; N.L. neut. adj. *columbinasale* pertaining to the nose of a pigeon.

Cells are coccoid to coccobacillary. Motility for this species has not been assessed. Colonies on solid medium exhibit typical fried-egg appearance. Grows well in SP-4 medium supplemented with arginine at 35–37°C. Produces a "film and spots" reaction.

No evidence of pathogenicity. Mode of transmission has not been assessed definitively.

Source: isolated from the turbinates of rock pigeons, racing pigeons, and fantail pigeons (Benčina et al., 1987; Keymer et al., 1984; Nagatomo et al., 1997; Yoder and Hofstad, 1964).

DNA G+C content (mol%): 32 (Bd).

Type strain: 694, ATCC 33549, NCTC 10184.

Sequence accession no. (16S rRNA gene): AF221112.

Further comment: previously known as avian serovar (serotype) L (Yoder and Hofstad, 1964).

30. **Mycoplasma columbinum** Shimizu, Ernø and Nagatomo 1978, 545[AL]

co.lum.bi'num. L. neut. adj. *columbinum* pertaining to a pigeon.

Cells are pleomorphic and vary from coccoid to ring forms. Motility for this species has not been assessed. Colonies on solid medium have typical fried-egg morphology. Grows in Frey's medium supplemented with arginine at 37°C. Produces a "film and spots" reaction.

No evidence of pathogenicity. Mode of transmission has not been assessed definitively.

Source: isolated from the trachea and oropharynx of feral pigeons and from the brain and lungs of racing pigeons (Benčina et al., 1987; Jordan et al., 1981; Keymer et al., 1984; Reece et al., 1986).

DNA G+C content (mol%): 27.3 (Bd).

Type strain: MMP1, ATCC 29257, NCTC 10178.

Sequence accession no. (16S rRNA gene): AF221113.

31. **Mycoplasma columborale** Shimizu, Ernø and Nagatomo 1978, 545[AL]

co.lum.bo.ra'le. L. n. *columba* pigeon; L. n. *os, oris* the mouth; L. neut. suff. *-ale* suffix used with the sense of pertaining to; N.L. neut. adj. *orale* of or pertaining to the mouth; N.L. neut. adj. *columborale* of the pigeon mouth.

Cells are pleomorphic but predominantly coccoid or exhibiting ring forms. Motility for this species has not been assessed. Growth on solid medium yields medium to large colonies with very small central zones. Grows in Frey's medium supplemented with glucose at 37°C.

Pathogenicity for pigeons is unconfirmed, but one report described airsacculitis in experimentally inoculated chickens. Mode of transmission has not been assessed definitively.

Source: isolated from the trachea and oropharynx of feral pigeons and fantail pigeons; from the oropharynx and sinuses of racing pigeons; and from corvids and house flies (Benčina et al., 1987; Bradbury et al., 2000; Jordan et al., 1981; Kempf et al., 2000; Keymer et al., 1984; MacOwan et al., 1981; Nagatomo et al., 1997; Reece et al., 1986).

DNA G+C content (mol%): 29.2 (Bd).

Type strain: MMP4, ATCC 29258, NCTC 10179.

Sequence accession no. (16S rRNA gene): AF412975.

32. **Mycoplasma conjunctivae** Barile, Del Giudice and Tully 1972, 74[AL]

con.junc.ti'va.e. N.L. n. *conjunctiva* the membrane joining the eyeball to the lids; N.L. gen. n. *conjunctivae* of conjunctiva.

Cells are coccoid to coccobacillary. Motility for this species has not been assessed. Colonies grown on solid medium may have elevated centers and a greenish, brownish, or olive color. Grows well in SP-4 medium supplemented with glucose at 37°C.

Pathogenic; causes infectious keratoconjunctivitis that can either resolve into a carrier state or result in complete or near-complete blindness in goats, sheep, chamois, and ibex (Cottew, 1979; Mayer et al., 1997). Mode of transmission is via direct contact.

Though tetracyclines are an effective antimicrobial therapy *in vivo*, treatment to eradicate *Mycoplasma conjunctivae* from herds is not often attempted as the economic burden of infection is low. Topical treatment of secondary infections is often necessary (Slatter, 2001). Several commercial diagnostic assays have been described.

Source: isolated from the conjunctivae of goats, sheep, chamois, and Alpine ibex.

DNA G+C content (mol%): not determined.

Type strain: HRC581, ATCC 25834, NCTC 10147.

Sequence accession no. (16S rRNA gene): AY816349.

33. **Mycoplasma corogypsi** Panangala, Stringfellow, Dybvig, Woodard, Sun, Rose and Gresham 1993, 589[VP]

co.ro.gyp'si. Gr. n. *korax-acos* a raven (black); Gr. n. *gyps gypos* a vulture; N.L. gen. n. *corogypsi* (*sic*) of a raven vulture.

Cells are highly pleomorphic and show small circular budding processes abutting elongated cells. Motility for this species has not been assessed. Colonies on solid medium have a fried-egg appearance. Grows in Frey's medium supplemented with glucose at 37°C.

Pathogenicity has not been established, although associated with abscess formation in a black vulture. *Mycoplasma corogypsi* has been isolated from clinically normal captive falcons, and may represent a commensal of this species. Mode of transmission has not been assessed definitively.

Source: isolated from the abscessed footpad of a black vulture, and from captive falcons (Lierz et al., 2002; Panangala et al., 1993).

DNA G+C content (mol%): 28 (Bd).

Type strain: BV1, ATCC 51148.

Sequence accession no. (16S rRNA gene): L08054.

34. **Mycoplasma cottewii** DaMassa, Tully, Rose, Pitcher, Leach and Cottew 1994, 483[VP]

cot.te'wi.i. N.L. masc. gen. n. *cottewii* of Cottew, named for Geoffrey S. Cottew, an Australian veterinarian who was a co-isolator of the organism.

Cells are primarily coccoid. Nonmotile. Formation of biofilms has been demonstrated (McAuliffe et al., 2006). Growth on solid medium shows colonies with a typical fried-egg appearance. Grows well in Hayflick medium supplemented with glucose at 37°C.

No evidence of pathogenicity. Mode of transmission has not been established.

Source: isolated from the external ear canals and rarely the sinuses of goats (Damassa et al., 1994).

DNA G+C content (mol%): 27 (T_m).

Type strain: VIS, ATCC 51347, NCTC 11732, CIP 105678.

Sequence accession no. (16S rRNA gene): U67945.

35. **Mycoplasma cricetuli** Hill 1983a, 117[VP]

cri.ce.tu'li. N.L. n. *Cricetulus* generic name of the Chinese hamster; *Cricetulus griseus*; N.L. gen. n. *cricetuli* of *Cricetulus*.

Cells are coccoid to pleomorphic. Nonmotile. Colony growth on solid medium has a fried-egg appearance with markedly small centers. Grows well in Hayflick broth supplemented with glucose at 37°C.

No evidence of pathogenicity. Mode of transmission has not been established.

Source: isolated from the conjunctivae and nasopharynx of Chinese hamsters (Hill, 1983a).

DNA G+C content (mol%): not determined.

Type strain: CH, NCTC 10190, ATCC 35279.

Sequence accession no. (16S rRNA gene): AF412976.

36. **Mycoplasma crocodyli** Kirchhoff, Mohan, Schmidt, Runge, Brown, Brown, Foggin, Muvavarirwa, Lehmann and Flossdorf 1997, 746[VP]

cro.co.dy'li. N.L. n. *Crocodylus* (from L. n. *crocodilus* crocodile) generic name of the crocodile; N.L. gen. n. *crocodyli* of *Crocodylus*.

Cells are coccoid. Nonmotile. Colonies on solid medium show a typical fried-egg appearance. Grows very rapidly in SP-4 medium supplemented with glucose at 30°C.

Pathogenic; causes exudative polyarthritis and rarely pneumonia in crocodiles. The natural mode of transmission has not been assessed definitively; however, experimental infection resulting in the reproduction of disease was achieved by intracoelomic and/or intrapulmonary inoculation.

Tetracyclines are effectively used to alleviate clinical signs in farmed crocodiles. A bacterin vaccine effective at controlling infection and preventing disease has been described (Mohan et al., 2001).

Source: isolated from the joints and lungs of Nile crocodiles (Mohan et al., 1997).

DNA G+C content (mol%): 27.6 (Bd).

Type strain: MP145, ATCC 51981.

Sequence accession no. (16S rRNA gene): AF412977.

37. **Mycoplasma cynos** Røsendal 1973, 53[AL]

cy'nos. Gr. n. *cyon, cynos* a dog; N.L. gen. n. *cynos* of a dog.

Cells are coccoid to coccobacillary. Motility for this species has not been assessed. Colonies on solid medium have defined centers and scalloped perimeters. Grows well in SP-4 medium supplemented with arginine at 37°C.

Pathogenic; causes pneumonia, bronchitis, and rarely cystitis in domestic dogs. The mode of transmission is via droplet aerosol, as demonstrated by studies housing infected and sentinel dogs (Røsendal and Vinther, 1977).

Source: isolated from the lungs, trachea, nasopharynx, urine, prepuce, prostate, cervix, vagina, and conjunctivae of domestic dogs (Chalker, 2005).

DNA G+C content (mol%): 25.8 (Bd).

Type strain: H 831, ATCC 27544, NCTC 10142.

Sequence accession no. (16S rRNA gene): AF538682.

38. **Mycoplasma dispar** Gourlay and Leach 1970, 121[AL]

dis'par. L. neut. adj. *dispar* dissimilar, different.

Cells range from coccoid to short and filamentous. Motility for this species has not been assessed. An extracellular capsule can be visualized by electron microscopy following staining with ruthenium red. Colonies on solid medium have a granular, lacy, or reticulated appearance with no or a poorly defined central area. Grows in SP-4 medium or modified Friis medium supplemented with glucose and calf thymus DNA at 37°C.

Pathogenic; causes pneumonia and rarely mastitis in cattle. Mode of transmission is by droplet aerosol.

Source: isolated from the lower respiratory tract and udders of cattle (Gourlay and Howard, 1979; Hodges et al., 1983).

DNA G+C content (mol%): 28.5–29.3 (T_m).

Type strain: 462/2, ATCC 27140, NCTC 10125.

Sequence accession no. (16S rRNA gene): AF412979.

39. **Mycoplasma edwardii** Tully, Barile, Del Giudice, Carski, Armstrong and Razin 1970, 349[AL]

ed.war'di.i. N.L. masc. gen. n. *edwardii* of Edward, named after Derrick Graham ff. Edward (1910–1978), who first isolated this organism.

Cells are coccobacillary to short and filamentous. Motility for this species has not been assessed. Colonies on solid medium show a typical fried-egg appearance. Grows well in SP-4 medium supplemented with glucose at 37°C.

Opportunistic pathogen; commonly found as a commensal of the oral and/or nasal cavities and urogenital tract of domestic dogs. *Mycoplasma edwardii* is rarely associated with pneumonia, arthritis, and septicemia of domestic dogs, often as a secondary pathogen compounding an existing lesion. Mode of transmission has not been established definitively.

Source: isolated from the oropharynx, nasopharynx, trachea, lungs, prepuce, vagina, cervix, blood, and synovial fluid of domestic dogs (Chalker, 2005; Stenske et al., 2005).

DNA G+C content (mol%): 29.2 (T_m).

Type strain: PG-24, ATCC 23462, NCTC 10132.

Sequence accession no. (16S rRNA gene): U73903.

40. **Mycoplasma elephantis** Kirchhoff, Schmidt, Lehmann, Clark and Hill 1996, 440[VP]

e.le.phan'tis. L. n. *elephas, -antis* elephant; L. gen. n. *elephantis* of the elephant.

Cells are coccoidal. Nonmotile. Colonies on solid medium show a typical fried-egg appearance. Grows well in Hayflick medium supplemented with glucose at 37°C.

Probable commensal. No pathology was observed at the site of isolation (i.e., the vagina and urethra); however, isolation was achieved almost exclusively from arthritic animals with evidence of rheumatoid factor. The possibility thus exists that the clinical status of the animals was due to sexually acquired reactive arthritis, which has been observed with other *Mycoplasma* species known to parasitize the urogenital tract (Blanchard and Bébéar, 2002). Mode of transmission has not been established definitively.

Source: isolated from the vagina and urethra of captive elephants (Clark et al., 1980, 1978).

DNA G+C content (mol%): 24 (Bd).

Type strain: E42, ATCC 51980.

Sequence accession no. (16S rRNA gene): AF221121.

41. **Mycoplasma equigenitalium** Kirchhoff 1978, 500[AL]

e.qui.ge.ni.ta'li.um. L. n. *equus, equi* the horse; L. pl. n. *genitalia* the genitals; N.L. pl. gen. n. *equigenitalium* of equine genitalia.

Cells are pleomorphic. Motility for this species has not been assessed. Colonies on solid medium show a typical fried-egg appearance. Grows in Hayflick medium supplemented with glucose at 37°C.

Opportunistic pathogen. Associated with endometritis, vulvitis, balanoposthitis, impaired fecundity, and abortion in horses; however, *Mycoplasma equigenitalium* is highly prevalent in clinically normal horses (Spergser et al., 2002). Mode of transmission is via sexual contact.

Source: isolated from the cervix, semen, and aborted foals, and rarely from the trachea, of horses (Lemcke, 1979).

DNA G+C content (mol%): 31.5 (Bd).

Type strain: T37, ATCC 29869, NCTC 10176.

Sequence accession no. (16S rRNA gene): AF221120.

42. **Mycoplasma equirhinis** Allam and Lemcke 1975, 405[AL]

e.qui.rhi'nis. L. n. *equus, equi* a horse; Gr. n. *rhis, rhinos* nose; N.L. gen. n. *equirhinis* of the nose of a horse.

Cells are coccoid to coccobacillary. Motility for this species has not been assessed. Colonies on solid medium show a typical fried-egg appearance. Grows in SP-4 or Hayflick medium supplemented with arginine at 37°C.

Opportunistic pathogen; associated with rhinitis and pneumonitis in horses, but can also be found in clinically normal animals. Mode of transmission is via droplet aerosol.

Source: isolated from the nasopharynx, nasal turbinates, trachea, tonsils, and semen of horses, and from the nasopharynx of cattle (Lemcke, 1979; Spergser et al., 2002; ter Laak et al., 1992a).

DNA G+C content (mol%): not determined.

Type strain: M432/72, ATCC 29420, NCTC 10148.

Sequence accession no. (16S rRNA gene): AF125585.

43. **Mycoplasma falconis** Poveda, Giebel, Flossdorf, Meier and Kirchhoff 1994, 97[VP]

fal.co'nis. L. gen. n. *falconis* of the falcon, the host from which the organism was first isolated.

Cells are coccoid. Motility for this species has not been assessed. Colonies on solid medium have a fried-egg appearance. Grows well in modified Frey's medium supplemented with arginine at 37°C.

Pathogenicity has not been established. Associated with respiratory tract infections of saker falcons, although can also be isolated from clinically normal birds. Mode of transmission has not been established definitively.

Source: isolated from the trachea of falcons (Lierz et al., 2002, 2008a, b).

DNA G+C content (mol%): 27.5 (Bd).

Type strain: H/T1, ATCC 51372.

Sequence accession no. (16S rRNA gene): AF125591.

44. **Mycoplasma fastidiosum** Lemcke and Poland 1980, 161[VP]

fas.ti.di.o'sum. L. neut. adj. *fastidiosum* fastidious, referring to the nutritionally fastidious nature of the organism on primary isolation.

Cells are highly filamentous and filaments are twisted at regular intervals along their length. Helical forms like those of *Spiroplasma* species are not produced. Nonmotile. Colonies on solid medium show a typical fried-egg appearance.

Grows in SP-4 or Frey's medium supplemented with glucose at 37°C.

No evidence of pathogenicity. Mode of transmission has not been assessed definitively.

Source: isolated from the nasopharynx of horses (Lemcke and Poland, 1980).

DNA G+C content (mol%): 32.3 (Bd).

Type strain: 4822, NCTC 10180, ATCC 33229.

Sequence accession no. (16S rRNA gene): AF125878.

45. **Mycoplasma faucium** Freundt, Taylor-Robinson, Purcell, Chanock and Black 1974, 253[AL]

fau'ci.um. L. pl. n. *fauces, -ium* the throat; L. gen. pl. n. *faucium* of the throat.

Cells are coccoidal. Motility for this species has not been assessed. Colonies on solid medium show a typical fried-egg appearance, but are more loosely attached to the agar surface than are the colonies of most other mycoplasmas. Grows well in SP-4 medium supplemented with arginine at 37°C. Produces a "film and spots" reaction.

Probable commensal. Most commonly found as a commensal of the human oropharynx; however, recent isolations of *Mycoplasma faucium* have been made from brain abscesses (Al Masalma et al., 2009). Mode of transmission has not been established definitively.

Source: isolated from the oropharynx and brain of humans, and from the oral cavity of numerous species of nonhuman primates (Freundt et al., 1974; Somerson and Cole, 1979).

DNA G+C content (mol%): not determined.

Type strain: DC-333, ATCC 25293, NCTC 10174.

Sequence accession no. (16S rRNA gene): AF125590.

46. **Mycoplasma felifaucium** Hill 1988, 449[VP] (Effective publication: Hill 1986, 1927)

fe.li.fau'ci.um. L. n. *felis* cat; L. pl. n. *fauces, -ium* throat; N.L. gen. pl. n. *felifaucium* of the feline throat.

Cells are primarily coccoidal. Nonmotile. Colonies on solid medium show a typical fried-egg appearance. Grows well in SP-4 or Hayflick medium supplemented with arginine at 37°C. Produces a "film and spots" reaction.

No evidence of pathogenicity. Mode of transmission has not been established definitively.

Source: isolated from the oropharynx of captive pumas (*Felis concolor*; Hill, 1986).

DNA G+C content (mol%): 31 (T_m).

Type strain: PU, NCTC 11703, ATCC 43428.

Sequence accession no. (16S rRNA gene): U15795.

47. **Mycoplasma feliminutum** Heyward, Sabry and Dowdle 1969, 621[AL]

fe.li.mi.nu'tum. L. n. *felis* a cat; L. neut. part. adj. *minutum* small; N.L. neut. adj. *feliminutum* a small colony organism isolated from cats.

Morphology is poorly defined. Motility for this species has not been assessed. Colonies are relatively small and irregular in shape. Grows well in SP-4 medium supplemented with glucose at 37°C.

No evidence of pathogenicity. Mode of transmission has not been established definitively.

Source: isolated from the oropharynx of domestic cats; from the nasopharynx, lungs, and urogenital tract of domestic dogs; and from the respiratory tract of horses (Chalker, 2005; Heyward et al., 1969; Lemcke, 1979).

DNA G+C content (mol%): 29.1 (Bd).

Type strain: Ben, ATCC 25749, NCTC 10159.

Sequence accession no. (16S rRNA gene): U16758.

Further comment: this organism was first described during a time when the only named genus of mollicutes was *Mycoplasma*. Its publication coincided with the first proposal of the genus *Acholeplasma* (Edward and Freundt, 1969, 1970), with which *Mycoplasma feliminutum* is properly affiliated through established phenotypic (Heyward et al., 1969) and 16S rRNA gene sequence (Brown et al., 1995) similarities. This explains the apparent inconsistencies with its assignment to the genus *Mycoplasma*. The name *Mycoplasma feliminutum* should therefore be revised to *Acholeplasma feliminutum* comb. nov.

48. **Mycoplasma felis** Cole, Golightly and Ward 1967, 1456[AL]

fe'lis. L. n. *felis* a cat, L. gen. n. *felis* of a cat.

Cells are coccobacillary to filamentous. Motility for this species has not been assessed. Colonies on solid media display the typical fried-egg morphology. Grows well in SP-4 medium supplemented with glucose at 37°C.

Pathogenic; associated with conjunctivitis, rhinitis, ulcerative keratitis, and polyarthritis in domestic cats, and upper and lower respiratory tract infection in horses. *Mycoplasma felis* can also be isolated from clinically normal domestic cats, domestic dogs, and horses. The mode of transmission has not been established definitively.

Source: isolated from the conjunctivae, nasopharynx, lungs, and urogenital tract of domestic cats; from the lungs, tonsils, trachea nasopharynx of horses; from the oropharynx and trachea of domestic dogs; and from the synovial fluid of an immunocompromised human (Lemcke, 1979) Røsendal, 1979; (Bonilla et al., 1997; Gray et al., 2005; Hooper et al., 1985).

DNA G+C content (mol%): 25.2 (T_m).

Type strain: CO, ATCC 23391, NCTC 10160.

Sequence accession no. (16S rRNA gene): U09787.

Further comment: the proposed species "*Mycoplasma equipharyngis*" (Kirchoff, 1974) has been reported in horses. Further characterization has demonstrated unequivocally that these isolates are *Mycoplasma felis* and all mention of "*Mycoplasma equipharyngis*" should be considered equivalent to *Mycoplasma felis* (Lemcke, 1979).

49. **Mycoplasma fermentans** Edward 1955, 90[AL]

fer.men'tans. L. part. adj. *fermentans* fermenting.

Cells are filamentous. Motility has not been established for this species. Colonies on solid media display typical fried-egg morphology. Grows well in SP-4 or Hayflick medium supplemented with either arginine or glucose at 37°C.

Pathogenicity unclear; associated with balanitis, vulvovaginitis, salpingitis, respiratory distress syndrome, pneumonia, and development of rheumatoid arthritis. *Mycoplasma fermentans* has also been tenuously linked with the progression of AIDS, chronic fatigue syndrome, Gulf War syndrome, Adamantiades-Behçet's disease, and fibromyalgia.

The connection of the preceding clinical syndromes with *Mycoplasma fermentans* is highly equivocal, as different studies have reached markedly different conclusions. Mode of transmission has not been established definitively.

Source: isolated from the urine, urethra, rectum, penis, cervix, vagina, fallopian tube, amniotic fluid, blood, synovial fluid, and throat of humans; from the cervix of an African green monkey (*Chlorocebus* sp.); and from the vagina of a sheep (Blanchard et al., 1993; Nicholas et al., 1998; Taylor-Robinson and Furr, 1997; Waites and Talkington, 2005).

DNA G+C content (mol%): 28.7 (T_m)

Type strain: PG18, ATCC 19989, NCTC 10117, NBRC 14854.

Sequence accession no. (16S rRNA gene): M24289.

50. **Mycoplasma flocculare** Meyling and Friis 1972, 289[AL]

floc.cu.la're. L. dim. n. *flocculus* a small flock or tuft of wool; L. neut. suff. *-are* suffix denoting pertaining to; N.L. neut. adj. *flocculare* resembling a small floc of wool, referring to the tendency of the organism to form clumps of flocculent material in broth culture.

Cells are coccoid to coccobacillary. Motility for this species has not been assessed. Colonies on solid media are slightly convex with a coarsely granular surface and lack a defined center. Aggregates of cells may be produced during growth in broth, appearing as small floccular elements upon gentle shaking of the culture. Grows slowly in Friis medium at 37°C.

Opportunistic pathogen; normally regarded as a commensal of the nasopharynx that can cause pneumonia in association with other pathogens, most notably *Mycoplasma hyopneumoniae.* Mode of transmission is via droplet aerosol. *Mycoplasma flocculare* reportedly shares surface antigens with *Mycoplasma hyopneumoniae,* potentially complicating serology-based diagnosis of infection (Whittlestone, 1979).

Source: isolated from the nasopharynx, lungs, pericardium, and conjunctivae of pigs.

DNA G+C content (mol%): 33 (Bd).

Type strain: Ms42, ATCC 27399, NCTC 10143.

Sequence accession no. (16S rRNA gene): L22210.

51. **Mycoplasma gallinaceum** Jordan, Ernø, Cottew, Hinz and Stipkovits 1982, 114[VP]

gal.li.na'ce.um. L. neut. adj. *gallinaceum* pertaining to a domestic fowl.

Cells are coccoid to coccobacillary. Motility for this species has not been assessed. Colonies on solid medium have typical fried-egg morphology although some are devoid of a central core. Grows well in Frey's medium supplemented with glucose at 37°C.

Opportunistic pathogen associated with tracheitis, airsacculitis, or conjunctivitis in chickens, turkeys, ducks, and pheasants. *Mycoplasma gallinaceum* has been reported to complicate cases of infectious synovitis due to *Mycoplasma synoviae* in chickens. The mode of transmission has not been assessed definitively.

Source: isolated from upper and lower respiratory tract of chickens, turkeys, pheasants, partridges, and ducks; from the conjunctivae of pheasants; and from the synovial fluid of chickens (Bradbury et al., 2001; Tiong, 1990; Welchman et al., 2002; Yagihashi et al., 1983).

DNA G+C content (mol%): 28 (Bd).

Type strain: DD, ATCC 33550, NCTC 10183.

Sequence accession no. (16S rRNA gene): L24104.

Further comment: previously known as avian serotype D (Kleckner, 1960).

52. **Mycoplasma gallinarum** Freundt 1955, 73[AL]

gal.li.na'rum. L. n. *gallina* a hen; L. gen. pl. n. *gallinarum* of hens.

Cells are coccoid to coccobacillary. Nonmotile. Colonies on solid medium have a typical fried-egg appearance. Grows well in Frey's medium supplemented with arginine at 37°C. The organism shares some antigens in immunodiffusion tests with *Mycoplasma iners, Mycoplasma columbinasale,* and *Mycoplasma meleagridis.*

Commensal of gallinaceous birds; little evidence exists for the pathogenicity of isolates in such hosts. *Mycoplasma gallinarum* may have a role in airsacculitis of geese and participate in complex infection of chickens. Mode of transmission has not been assessed definitively.

Source: isolated from the respiratory tract of chickens, turkeys, ducks, geese, red jungle fowl, bamboo partridge, sparrow, swan, and demoisella crane; and from sheep (Kisary et al., 1976; Kleven et al., 1978; Shimizu et al., 1979; Singh and Uppal, 1987).

DNA G+C content (mol%): 26.5–28.0 (T_m, Bd).

Type strain: PG16, ATCC 19708, NCTC 10120.

Sequence accession no. (16S rRNA gene): L24105.

Further comment: previously known as avian serotype B (Kleckner, 1960).

53. **Mycoplasma gallisepticum** Edward and Kanarek 1960, 699[AL]

gal.li.sep'ti.cum. L. n. *gallus* rooster, chicken; L. adj. *septicus -a -um* producing a putrefaction, putrefying, septic; N.L. neut. adj. *gallisepticum* hen-putrefying (infecting).

Cells are coccoid, ovoid, and elongated pear-shaped with a highly structured polar body, called the bleb. Cells are motile and glide in the direction of the terminal bleb. Gliding speed varies among strains. Colonies on solid medium may be small and not necessarily of typical fried-egg appearance. Grows well in SP-4 or Hayflick medium supplemented with glucose at 37°C (Balish and Krause, 2006; Hatchel et al., 2006; Nakane and Miyata, 2009).

Pathogenic. Causes a characteristic combination of pneumonia, tracheitis, and airsacculitis (collectively termed chronic respiratory disease); salpingitis and atrophy of the ovaries, isthmus, and cloaca resulting in poor egg quality and reduced hatchability; arthritis or synovitis; and keratoconjunctivitis in chickens; infectious sinusitis, coryza, airsacculitis, arthritis or synovitis, encephalitis, meningitis, ataxia, and torticollis in turkeys; conjunctivitis and coryza featuring a high mortality rate in finches and grosbeaks; and respiratory disease in additional game birds including quail, partridges, pheasants, and peafowl. Lesions established by *Mycoplasma gallisepticum* are often complicated by additional avian pathogens including *Mycoplasma synoviae,* avian strains of *Escherichia coli,* Newcastle disease virus, and infectious bronchitis virus. Established mechanisms of transmission include droplet aerosols, direct contact with infected animals or fomites, and vertical transmission.

Tetracyclines, macrolides, aminoglycosides, fluoroquinolones, and pleuromutilins are effective chemotherapeutic agents; however, treatment is typically only sought for individual birds, as medicating a commercial flock is not considered an effective control strategy. Vaccination and management strategies (i.e., single age "all in/all out" systems and culling of endemic flocks) are more commonly utilized. Multiple live and killed vaccines are commercially available, but suffer from residual pathogenicity, the need to develop a carrier state to provide protective immunity, adverse reactions, or low efficacy. Experimental vaccines have also been described. *Mycoplasma gallisepticum* shares surface antigens with *Mycoplasma imitans* and *Mycoplasma synoviae*, potentially complicating serology-based diagnosis of infection. Numerous molecular diagnostics have been described. This organism is listed in the Terrestrial Animal Health Code of the Office International des Epizooties (http://oie.int; Yogev et al., 1989; Kempf, 1998; Markham et al., 1999; Gautier-Bouchardon et al., 2002; Ferguson et al., 2004; Browning et al., 2005; Crespo and McMillan, 2008; Gates et al., 2008; Gerchman et al., 2008; Kleven, 2008).

Source: isolated from the trachea, lungs, air sacs, ovaries, oviducts, brain, arterial walls, synovial membranes, synovial fluid, conjunctivae, and eggs of chickens; from the infraorbital sinuses, air sacs, brain, meninges, conjunctivae, synovial membranes, and synovial fluid of turkeys; from the conjunctivae, infraorbital sinuses, and trachea of finches; and from the respiratory tract of quail, partridges, pheasants, peafowl, ducks, grosbeak, crows, robins, and blue jays (Benčina et al., 2003, 1988; Bradbury and Morrow, 2008; Bradbury et al., 2001; Dhondt et al., 2005, 2007; Levisohn and Kleven, 2000; Ley et al., 1996; Mikaelian et al., 2001; Murakami et al., 2002; Nolan et al., 2004; Nunoya et al., 1995; Welchman et al., 2002; Wellehan et al., 2001).

DNA G+C content (mol%): 31.8 (T_m), 31 (strain R complete genome sequence).

Type strain: PG31, X95, ATCC 19610, NCTC 10115.

Sequence accession nos: M22441 (16S rRNA gene of strain A5969), NC_004829 (strain R complete genome sequence).

Further comment: previously known as avian serotype A (Kleckner, 1960).

54. **Mycoplasma gallopavonis** Jordan, Ernø, Cottew, Hinz and Stipkovits 1982, 114[VP]

gal.lo.pa.vo'nis. N.L. n. *gallopavo, -onis* a turkey (*Meleagris gallopavo*); N.L. gen. n. *gallopavonis* of a turkey.

Cells are coccoid to coccobacillary. Motility has not been assessed for this species. Colonies on solid medium have typical fried-egg morphology. Grows well in Frey's medium supplemented with glucose at 37°C.

Opportunistic pathogen; occasionally associated with airsacculitis in turkeys, but is also isolated from clinically normal turkeys. Mode of transmission has not been assessed definitively.

Source: isolated from the choanae, trachea, and air sacs of domestic and wild turkeys (Benčina et al., 1987; Cobb et al., 1992; Hoffman et al., 1997; Luttrell et al., 1992).

DNA G+C content (mol%): 27 (Bd).

Type strain: WR1, ATCC 33551, NCTC 10186.

Sequence accession no. (16S rRNA gene): AF412980.

Further comment: previously known as avian serotype F (Kleckner, 1960).

55. **Mycoplasma gateae** Cole, Golightly and Ward 1967, 1456[AL]

ga.te'ae. N.L. gen. n. *gateae* (probably from Spanish *gato*, a cat) of a cat.

Morphology is poorly defined. Motility for this species has not been assessed. Colonies on solid medium are vacuolated and lack a well-defined central spot. Grows well in SP-4 medium at 37°C.

Opportunistic pathogen; can cause polyarthritis in domestic cats (Moise et al., 1983), but appears to be primarily a commensal species of the oral cavity. *Mycoplasma gateae* also appears to be a commensal species of domestic dogs and cattle. Mode of transmission has not been assessed definitively.

Source: isolated from the synovial membrane, oropharynx, saliva, and urogenital tract of domestic cats; from the lungs, oropharynx, trachea, and urogenital tract of domestic dogs; and from the urogenital tract of cattle (Chalker, 2005; Gourlay and Howard, 1979; Røsendal, 1979).

DNA G+C content (mol%): 28.5 (T_m).

Type strain: CS, ATCC 23392, NCTC 10161.

Sequence accession no. (16S rRNA gene): U15796.

Further comment: the original specific epithet "*gateae*", which has been perpetuated in lists of bacterial names approved by the International Committee on Systematics of Prokaryotes, the American Type Culture Collection's *Catalog of Bacteria and Bacteriophages*, and in GenBank, is illegitimate because the genitive of the medieval Latin word *gata* (female cat) would have been *gatae* and there is no word for which *gateae* would have been a legitimate genitive (Brown et al., 1995).

56. **Mycoplasma genitalium** Tully, Taylor-Robinson, Rose, Cole and Bové 1983, 395[VP]

ge.ni.ta'li.um. L. pl. n. *genitalia, -ium* the genitals; L. gen. pl. n. *genitalium* of the genitals.

Cells are predominantly flask-shaped with a terminal organelle protruding from the cell pole that is narrower than that of *Mycoplasma gallisepticum* and shorter than that of *Mycoplasma pneumoniae*. The leading end of the terminal structure is often curved. Cells exhibit gliding motility in circular patterns and glide in the direction of the terminal organelle's curvature (Hatchel and Balish, 2008). Colonies on solid media are round and possess a defined center that is somewhat less distinct than most mycoplasma species. Grows well in SP-4 medium supplemented with glucose at 37°C.

Pathogenic; causes urethritis, cervicitis, endometritis, and pelvic inflammatory disease. *Mycoplasma genitalium* is associated with infertility in humans. Mode of transmission is via sexual contact, congenitally, and possibly, in rare instances, via droplet aerosol.

Macrolides and fluoroquinolones are effective chemotherapeutic agents; however, reports indicate that treatment should be extensive, as clinical signs and detection of *Mycoplasma genitalium* tend to recur following cessation. This is potentially due to sequestration within host cells. *Mycoplasma genitalium* shares numerous surface antigens with *Mycoplasma pneumoniae*, complicating serology-based diagnosis of infection. Numerous molecular diagnostics have been described, but few have been developed commercially (Jensen, 2004; Waites and Talkington, 2005).

Source: isolated or detected in the urogenital tract, urine, rectum, synovial fluid, conjunctiva, and nasopharynx of humans (Baseman et al., 1988; de Barbeyrac et al., 1993; Jensen, 2004; Waites and Talkington, 2005).

DNA G+C content (mol%): 32.4 (Bd), 31 (strain G-37^T complete genome sequence).

Type strain: G-37, ATCC 33530, CIP 103767, NCTC 10195.

Sequence accession nos: X77334 (16S rRNA gene), NC_000908 (strain G-37^T complete genome sequence), CP000925 (strain JCVI-1.0 complete genome sequence).

57. **Mycoplasma glycophilum** Forrest and Bradbury 1984, 355^{VP} (Effective publication: Forrest and Bradbury 1984, 602)

gly.co.phi'lum. Gr. adj. *glykys* sweet (this adjective was used to coin the noun glucose); N.L. neut. adj. *philum* (from Gr. neut. adj. *philon*) friend, loving; N.L. neut. adj. *glycophilum* sweet-loving, intended to mean glucose-loving.

Cells are spherical or elliptical with an extracellular layer. Nonmotile. Growth on solid medium shows colonies with typical fried-egg appearance. Grows well in Hayflick medium supplemented with glucose at 37°C.

Pathogenicity has not been established, although there may be an association with a slight decrease in hatchability. Mode of transmission has not been assessed definitively.

Source: isolated from the respiratory tract and cloaca of chickens; and from the respiratory tract of turkeys, pheasants, partridges, ducks, and geese (Forrest and Bradbury, 1984).

DNA G+C content (mol%): 27.5 (Bd).

Type strain: 486, ATCC 35277, NCTC 10194.

Sequence accession no. (16S rRNA gene): AF412981.

58. **Mycoplasma gypis** Poveda, Giebel, Flossdorf, Meier and Kirchhoff 1994, 98^{VP}

gy'pis. Gr. n. *gyps, gypos* vulture; N.L. gen. n. *gypis* of the vulture, the host from which the organism was first isolated.

Cells are coccoid or round. Motility for this species has not been assessed. Colonies on solid medium have a fried-egg appearance. Grows in Frey's medium supplemented with arginine at 37°C. Produces a "film and spots" reaction.

Pathogenicity has not been established. Associated with respiratory tract disease of griffon vultures, but has also been isolated from healthy birds of prey. Mode of transmission has not been assessed definitively.

Source: isolated from the trachea of griffon vultures (*Griffon fulvus*), and from the trachea and air sacs of Eurasian buzzards, red kites, and Western marsh harriers (Lierz et al., 2008a, 2000; Poveda et al., 1994).

DNA G+C content (mol%): 27.1 (Bd).

Type strain: B1/T1, ATCC 51370.

Sequence accession no. (16S rRNA gene): AF125589.

59. **Mycoplasma haemocanis** (Kikuth 1928) Messick, Walker, Raphael, Berent and Shi 2002, 697^{VP} [*Bartonella canis* Kikuth 1928, 1730; *Haemobartonella* (*Bartonella*) *canis* (Kikuth 1928) Tyzzer and Weinman 1939, 151; Kreier and Ristic 1984, 726]

ha.e.mo.ca'nis Gr. neut. n. *haema* blood; L. fem. gen. n. *canis* of the dog; N.L. gen. n. *haemocanis* of dog blood.

Cells are coccoid to pleomorphic. Motility for this species has not been assessed. The morphology of infected erythrocytes is altered, demonstrating a marked depression at the site of *Mycoplasma haemocanis* attachment. This species has not been grown on any artificial medium; therefore, notable biochemical parameters are not known.

Pathogenic; causes hemolytic anemia in domestic dogs. Transmission is vector-borne and mediated by the brown dog tick (*Rhipicephalus sanguineus*).

Source: observed in association with erythrocytes of domestic dogs (Hoskins, 1991).

DNA G+C content (mol%): not determined.

Type strain: not established.

Sequence accession no. (16S rRNA gene): AF197337.

60. **Mycoplasma haemofelis** (Clark 1942) Neimark, Johansson, Rikihisa and Tully 2002b, 683^{VP} [*Eperythrozoon felis* Clark 1942, 16; *Haemobartonella felis* (Clark 1942) Flint and McKelvie 1956, 240 and Kreier and Ristic 1984, 725]

ha.e.mo.fe'lis. Gr. neut. n. *haema* blood; L. fem. gen. n. *felis* of the cat; N.L. gen. n. *haemofelis* of cat blood.

Cells are coccoid. Motility for this species has not been assessed. This species has not been grown on artificial medium; therefore, notable biochemical parameters are not known.

Pathogenic; causes hemolytic anemia in cats. The mode of transmission is percutaneous or oral; an insect vector has not been identified, although fleas have been implicated (Woods et al., 2005).

Tetracyclines and fluoroquinolones are effective therapeutic agents (Dowers et al., 2002; Tasker et al., 2006).

Source: observed in association with erythrocytes of domestic cats.

DNA G+C content (mol%): 38.5–38.8 (genome sequence survey of strain OH; Berent and Messick, 2003; J.B. Messick et al., unpublished).

Type strain: not established

Sequence accession no. (16S rRNA gene): U88563.

61. **Mycoplasma haemomuris** (Mayer 1921) Neimark, Johansson, Rikihisa and Tully 2002b, 683^{VP} (*Bartonella muris* Mayer 1921, 151; *Bartonella muris ratti* Regendanz and Kikuth 1928, 1578; *Haemobartonella muris* Tyzzer and Weinman 1939, 143)

ha.e.mo.mu'ris. Gr. neut. n. *haema* blood; L. masc. gen. n. *muris* of the mouse; N.L. gen. n. *haemomuris* of mouse blood.

Cells are coccoid and some display dense inclusion particles. Motility for this species has not been assessed. The morphology of infected erythrocytes is altered, demonstrating a marked depression at the site of *Mycoplasma haemomuris* attachment. This species has not been grown on any artificial medium; therefore, notable biochemical parameters are not known.

Opportunistic pathogen; causes anemia in splenectomized or otherwise immunosuppressed mice. Transmission is vector-borne and mediated by the rat louse (*Polypax spinulosa*).

Source: observed in association with erythrocytes of wild and captive mice, and hamsters.

DNA G+C content (mol%): not determined.

Type strain: not established.

Sequence accession no. (16S rRNA gene): U82963.

62. **Mycoplasma hominis** (Freundt 1953) Edward 1955, 90[AL] (*Micromyces hominis* Freundt 1953, 471)

ho'mi.nis. L. n. *homo, -inis* man; L. gen. n. *hominis* of man.

Cells are coccoid to filamentous. Motility for this species has not been assessed. Colonies on solid media have a typical fried-egg appearance. Grows well in SP-4 medium supplemented with arginine at 37°C.

Pathogenic; causes pyelonephritis, pelvic inflammatory disease, chorioamnionitis, and postpartum fevers in women; congenital pneumonia, meningitis, and abscesses in newborns; and rarely extragenital pathologies including bacteremia, arthritis, osteomyelitis, abscesses and wound infections, mediastinitis, pneumonia, peritonitis, prosthetic- and catheter-associated infections, and infection of hematomas. Extragenital manifestations of *Mycoplasma hominis* infection are more commonly seen in immunosuppressed individuals, but can be seen in immunocompetent patients as well. Synergism between *Mycoplasma hominis* and *Trichomonas vaginalis* infections has been reported and a recent report documents the intraprotozooal location and transmission of *Mycoplasma hominis* with *Trichomonas vaginalis* (Dessi et al., 2006; Germain et al., 1994; Vancini and Benchimol, 2008). Mode of transmission is via sexual contact, congenitally, or by artificial introduction on foreign objects (e.g., catheters) or transplanted tissues.

Macrolides and fluoroquinolones are effective chemotherapeutic agents. Combination therapy with metronidazole is required for complex infections involving *Trichomonas vaginalis*. Many commercial diagnostics are available for routine clinical use.

Source: isolated from the urogenital tract, amniotic fluid, placenta, umbilical cord blood, urine, semen, bloodstream, cerebrospinal fluid, synovial fluid, bronchoalveolar lavage fluid, peritoneal aspirates, conjunctivae, bone abscesses, and hematoma aspirates of humans; and from several species of nonhuman primates (Somerson and Cole, 1979; Taylor-Robinson and McCormack, 1979; Waites and Talkington, 2005).

DNA G+C content (mol%): 33.7 (T_m)

Type strain: PG21, ATCC 23114, NCTC 10111, CIP 103715, NBRC 14850.

Sequence accession no. (16S rRNA gene): M24473.

63. **Mycoplasma hyopharyngis** Erickson, Ross, Rose, Tully and Bové 1986, 58[VP]

hy.o.pha.ryn'gis. Gr. n. *hys, hyos* a swine; N.L. n. *pharynx -yngis* (from Gr. n. *pharugx, pharuggos* throat) throat; N.L. gen. n. *hyopharyngis* of a hog's throat.

Cells are pleomorphic. Motility for this species has not been assessed. Colonies on solid medium have a typical fried-egg appearance. Grows in medium D-TS or Hayflick medium supplemented with arginine at 37°C and produces a "film and spots" reaction.

No evidence of pathogenicity. Mode of transmission has not been established definitively.

Source: isolated from the nasopharynx of pigs (Erickson et al., 1986).

DNA G+C content (mol%): 24 (Bd, T_m).

Type strain: H3-6B F, ATCC 51909, NCTC 11705.

Sequence accession no. (16S rRNA gene): U58997.

64. **Mycoplasma hyopneumoniae** (Goodwin, Pomeroy and Whittlestone 1965) Maré and Switzer 1965, 841[AL] (*Mycoplasma suipneumoniae* Goodwin, Pomeroy and Whittlestone 1965, 1249)

hy.o.pneu.mo'ni.ae. Gr. n. *hys, hyos* a swine; Gr. n. *pneumonia* pneumonia; N.L. gen. n. *hyopneumoniae* of swine pneumonia.

Cells are coccoid to coccobacillary. Nonmotile. Colonies on solid medium are very small, lack a defined central region, and are usually convex with a granular surface. Grows very slowly in modified Friis medium, medium A26, and modified SP-4 medium supplemented with glucose at 37°C. Produces a "film and spots" reaction.

Pathogenic; causes a very characteristic chronic pneumonitis associated with ciliostasis and marked sloughing of the epithelial lining in pigs. This collection of lesions in conjunction with *Mycoplasma hyopneumoniae* is referred to as enzootic pneumonia of pigs (EPP), and is associated with high morbidity and poor feed conversion (with proportional economic loss). The mechanism of transmission is via droplet aerosol.

Tetracyclines and tylosin do not typically eradicate *Mycoplasma hyopneumoniae* from infected animals, but are effective in limiting sequelae. Maintenance of pigs on antibiotics in conjunction with management practices involving adequate nutrition, air quality, and stress reduction are commonly employed to control the effects of disease in endemic herds. Many molecular diagnostics have been described (Dubosson et al., 2004). Serological diagnostic techniques utilizing monoclonal antibodies have proven successful at distinguishing *Mycoplasma hyopneumoniae* from *Mycoplasma flocculare*, though the two share surface antigens (Armstrong et al., 1987). Numerous experimental vaccines have been described and several commercial vaccines are available. The latter appear to reduce or eliminate clinical signs rather than prevent infection (Browning et al., 2005).

Source: isolated from the lungs, nasopharynx, tonsils, trachea, and bronchiolar lavage fluid of pigs (Marois et al., 2007; Whittlestone, 1979).

DNA G+C content (mol%): 27.5 (Bd), 28 (strain J^T and strain 7448 complete genome sequence), 28.6 (strain 232 complete genome sequence).

Type strain: J, ATCC 25934, NCTC 10110.

Sequence accession nos: AY737012 (16S rRNA gene), AE017243 (strain J^T complete genome sequence), AE017244 (strain 7448 complete genome sequence), AE017332 (strain 232 complete genome sequence).

65. **Mycoplasma hyorhinis** Switzer 1955, 544[AL]

hy.o.rhi'nis. Gr. n. *hys, hyos* a swine; Gr. n. *rhis, rhinos* nose; N.L. gen. n. *hyorhinis* of a hog's nose.

Cells are coccoid to coccobacillary. Motility for this species has not been assessed. Colonies on solid media display typical fried-egg morphology. Grows well on S-4 supplemented with glucose at 37°C.

Mycoplasma hyorhinis is associated with contamination of eukaryotic cell culture and can be removed by treatment of cells with antibiotics and/or maintenance of cell lines in antibiotic-containing medium. The noted effects of *Mycoplasma hyorhinis* on cell-cycle regulation make the detection and elimination of this organism particularly pertinent (Goodison et al., 2007; Schmidhauser et al., 1990). The most effective classes of antibiotics for cell culture eradication are tetracyclines and fluoroquinolones (Borup-Christensen et al., 1988; Schmitt et al., 1988).

Pathogenic; associated with arthritis, polyserositis, and otitis media in pigs. *Mycoplasma hyorhinis* is also regarded as a commensal of the nasopharynx that can occasionally cause pneumonia, often in association with other pathogens (most notably *Mycoplasma hyopneumoniae* and *Bordetella bronchiseptica*). Mode of transmission is via droplet aerosol.

Source: isolated from the nasopharynx, lungs, ear canal, synovial fluid, serous cavity, and pericardium of pigs (Friis and Szančer, 1994; Ross, 1992; Whittlestone, 1979).

DNA G+C content (mol%): 27.8 (T_m).

Type strain: BTS-7, ATCC 17981, NCTC 10130, CIP 104968, NBRC 14858.

Sequence accession no. (16S rRNA gene): M24658.

66. **Mycoplasma hyosynoviae** Ross and Karmon 1970, 710[AL]

hy.o.sy.no.vi'ae. Gr. n. *hys, hyos* a swine; N.L. n. *synovia* fluid in the joints; N.L. gen. n. *hyosynoviae* of joint fluid of swine.

Cells are coccoid to filamentous. Motility of this species has not been assessed. Colonies display a typical fried-egg appearance at 37°C. Grows well in SP-4 medium supplemented with glucose at 37°C. A granular deposit and a waxy surface pellicle are produced during growth in broth.

Pathogenic; causes infectious synovitis, arthritis, and rarely pericarditis in pigs. Transmission occurs from sows to piglets or between adults via aerosol.

Lincosamides, fluoroquinolones, and macrolides have been used effectively for treatment in conjunction with improved disinfection and quarantine during husbandry.

Source: isolated from the synovial fluid, nasopharynx, tonsils, lymph nodes, and pericardium of pigs (Whittlestone, 1979).

DNA G+C content (mol%): 28.0 (Bd).

Type strain: S16, ATCC 25591, NCTC 10167.

Sequence accession no. (16S rRNA gene): U26730.

67. **Mycoplasma iguanae** Brown, Demcovitz, Plourdé, Potter, Hunt, Jones and Rotstein 2006, 763[VP]

i.gua'nae. N.L. gen. n. *iguanae* of the iguana lizard.

Cells are predominantly coccoid. Nonmotile. Colonies on solid medium exhibit variable (convex to umbonate) forms; mature colonies display sectored centers. Grows well in SP-4 medium supplemented with glucose between 25 and 42°C.

Associated with pathologic lesions, but unable to reproduce disease following experimental inoculation (Brown et al., 2007). Mechanism of transmission has not been established.

Source: isolated from vertebral abscesses of green iguanas.

DNA G+C content (mol%): not determined.

Type strain: 2327, ATCC BAA-1050, NCTC 11745.

Sequence accession no. (partial 16S rRNA gene sequence): AY714305.

68. **Mycoplasma imitans** Bradbury, Abdul-Wahab, Yavari, Dupiellet and Bové 1993, 726[VP]

i'mi.tans. L. part. adj. *imitans* imitating, mimicking, referring to the organism's phenotypic resemblance to *Mycoplasma gallisepticum.*

Cells are oval and flask-shaped with a short, wide attachment organelle. Cells are motile and exhibit gliding motility in the direction of the attachment organelle (Hatchel and Balish, 2008). Colonies have typical fried-egg morphology on solid medium. Grows well in SP-4 medium supplemented with glucose at 37°C.

Pathogenic; causes sinusitis in ducks, geese, and partridges. Disease has been reproduced experimentally. Mode of transmission has not been assessed definitively.

Diagnosis of infection is potentially complicated by numerous factors. Serological cross-reactions occur with *Mycoplasma gallisepticum* due to known epitopes including PvpA, the VlhA hemagglutinins, pyruvate dehydrogenase, lactate dehydrogenase, and elongation factor Tu (Jan et al., 2001; Lavrič et al., 2005; Markham et al., 1999; Rosengarten et al., 1995). In addition, the 16S rRNA genes share 99.9% identity, despite whole genome hybridization showing a relatedness of only 40–46%, potentially complicating molecular diagnostics based on the 16S rRNA gene. Two molecular methods for distinguishing these two species have been described (Harasawa et al., 2004; Marois et al., 2001).

Source: isolated from the nasal turbinates and sinuses of ducks, geese, and partridges (Dupiellet, 1984; Ganapathy and Bradbury, 1998; Kleven, 2003).

DNA G+C content (mol%): 31.9 (Bd).

Type strain: 4229, NCTC 11733, ATCC 51306.

Sequence accession no. (16S rRNA gene): L24103.

69. **Mycoplasma indiense** Hill 1993, 39[VP]

in.di.en'se. N.L. neut. adj. *indiense* pertaining to India (source of the infected primates).

Cells are pleomorphic. Nonmotile. Colonies on agar have a characteristic fried-egg appearance. Grows well in SP-4 medium supplemented with arginine at 37°C.

No evidence of pathogenicity. Mode of transmission has not been established.

Source: isolated from the throats of a rhesus monkey and a baboon (Hill, 1993).

DNA G+C content (mol%): 32 (Bd).

Type strain: 3T, NCTC 11728, ATCC 51125.

Sequence accession no. (16S rRNA gene): AF125593.

70. **Mycoplasma iners** Edward and Kanarek 1960, 699[AL]

i'ners. L. neut. adj. *iners* inactive, inert.

The cell morphology is poorly defined. Motility for this species has not been assessed. Colonies on solid medium are relatively small and of a typical fried-egg appearance. Grows in Frey's medium supplemented with arginine at 37°C.

No evidence of pathogenicity. Mode of transmission has not been assessed definitively.

Source: isolated from the respiratory tract of chickens, turkeys, geese pigeons, pheasants, and partridges; and from tissues of swine (Benčina et al., 1987; Bradbury et al., 2001; Taylor-Robinson and Dinter, 1968).

DNA G+C content (mol%): 29.1 (T_m), 29.6 (Bd).

Type strain: PG30, ATCC 19705, NCTC 10165.

Sequence accession no. (16S rRNA gene): AF221114.

Further comment: previously known as avian serotype E (Kleckner, 1960).

71. **Mycoplasma iowae** Jordan, Ernø, Cottew, Hinz and Stipkovits 1982, 114[VP]

i.o'wa.e. N.L. gen. n. *iowae* of Iowa.

Cells are pleomorphic and some display a terminal protrusion with possible attachment properties (Gallagher and Rhoades, 1983; Mirsalimi et al., 1989). Motile. Colonies on agar show typical fried-egg appearance. Grows well in SP-4 medium supplemented with either glucose or arginine at 41–43°C (Grau et al., 1991; Yoder and Hofstad, 1964).

Pathogenic; causes airsacculitis and embryo lethality resulting in reduced hatchability in turkeys. Transmission occurs vertically and by direct contact.

Tetracyclines, macrolides, and fluoroquinolones are effective chemotherapeutic agents; however, medicating a commercial flock is not considered an effective control strategy. Management tactics are more commonly utilized. Molecular diagnostic methods have been described (Ramírez et al., 2008; Raviv and Kleven, 2009). *Mycoplasma iowae* strains show considerable intra-species antigenic heterogeneity and a cross-reactive epitope with both *Mycoplasma gallisepticum* and *Mycoplasma imitans* potentially complicating serology-based diagnosis of infection (Al-Ankari and Bradbury, 1996; Dierks et al., 1967; Rosengarten et al., 1995).

Source: isolated from the air sacs, intestinal tract, and eggs of turkeys, and from the seed of an apple tree with apple proliferation disease (Bradbury and Kleven, 2008; Grau et al., 1991; Mirsalimi et al., 1989).

DNA G+C content (mol%): 25 (Bd).

Type strain: 695, ATCC 33552, NCTC 10185.

Sequence accession no. (16S rRNA gene): M24293.

Further comment: previously known as avian serotype I (Yoder and Hofstad, 1964).

72. **Mycoplasma lagogenitalium** Kobayashi, Runge, Schmidt, Kubo, Yamamoto and Kirchhoff 1997, 1211[VP]

la.go.ge.ni.ta'li.um. Gr. masc. n. *lagos* hare; L. neut. pl. gen. n. *genitalium* of genitals; N.L. gen. pl. n. *lagogenitalium* of hare's genitals.

Cells are primarily coccoid. Nonmotile. Colonies on agar have a characteristic fried-egg appearance. Grows well in SP-4 medium supplemented with glucose at 37°C.

No evidence of pathogenicity. Mechanism of transmission has not been established.

Source: isolated from the preputial smegma of Afghan pikas (*Ochotona rufescens*; Kobayashi et al., 1997).

DNA G+C content (mol%): 23 (T_m).

Type strain: 12MS, ATCC 700289, CIP 105489, DSM 22062.

Sequence accession no. (16S rRNA gene): AF412983.

73. **Mycoplasma leachii** Manso-Silván, Vilei, Sachse, Djordjevic, Thiaucourt and Frey 2009, 1356[VP]

le.a.chi'i. N.L. masc. gen. n. *leachii* of Leach, named in honor of Dr R.H. Leach, who first characterized this taxon.

Cells are pleomorphic. Nonmotile. Colonies on solid agar have a characteristic fried-egg appearance. Grows in modified Hayflick medium supplemented with glucose at 37°C.

Pathogenic; causes polyarthritis, mastitis, abortion, and pneumonia in cattle. Mode of transmission has not been established definitively.

Tetracyclines appear to control infection during acute outbreaks (Hum et al., 2000). *Mycoplasma leachii* shares surface antigens with *Mycoplasma mycoides* subsp. *mycoides*, *Mycoplasma capricolum* subsp. *capripneumoniae*, and *Mycoplasma capricolum* subsp. *capricolum*, potentially compounding serology-based diagnosis of infection.

Source: isolated from the synovial fluid, udders, expelled milk, lungs, lymph nodes, pericardium, cervix, vagina, prepuce, semen, and aborted calves of cattle (Alexander et al., 1985; Gourlay and Howard, 1979; Hum et al., 2000).

DNA G+C content (mol%): not determined.

Type strain: PG50, NCTC 10133, DSM 21131.

Sequence accession no. (16S rRNA gene): AF261730.

Further comment: the assignment of strains formerly called "*Mycoplasma* species bovine group 7 of Leach" to the species *Mycoplasma leachii* came in response to a request from the Subcommittee on the Taxonomy of *Mollicutes* of the International Committee on Systematics of Prokaryotes for a proposal for an emended taxonomy for the members of the *Mycoplasma mycoides* phylogenetic cluster (Manso-Silván et al., 2009).

74. **Mycoplasma leonicaptivi** corrig. Hill 1992, 521[VP]

le.o.ni.cap'ti.vi. L. n. *leo, -onis* the lion; L. adj. *captivus* captive; N.L. gen. n. *leonicaptivi* of the captive lion.

Cells are pleomorphic (primarily coccoid). Nonmotile. Colonies on solid medium have a typical fried-egg appearance. Growth in SP-4 broth supplemented with glucose occurs between 35 and 37°C.

No evidence of pathogenicity. Mode of transmission has not been established.

Source: isolated from the throat and respiratory tract of captive lions and leopards (Hill, 1992).

DNA G+C content (mol%): 27 (Bd).

Type strain: 3L2, NCTC 11726, ATCC 49890.

Sequence accession no. (16S rRNA gene): U16759.

Further comment: the original spelling of the specific epithet, *leocaptivus* (*sic*), has been corrected by Trüper and De'Clari (1998).

75. **Mycoplasma leopharyngis** Hill 1992, 521[VP]

le.o.pha.ryn'gis. L. masc. n. *leo, -onis* lion; N.L. n. *pharynx, -yngis* (from Gr. n. *pharugx, pharuggos* throat) throat; N.L. gen. n. *leopharyngis* (*sic*) of the throat of a lion.

Cells are coccoid to pleomorphic. Nonmotile. Colonies on solid medium have a typical fried-egg appearance under anaerobic conditions. Grows in SP-4 broth supplemented with glucose at optimum temperatures of 35–37°C. Grows well under both aerobic and anaerobic conditions.

No evidence of pathogenicity. Mechanism of transmission has not been established.

Source: isolated from the throat of lions (Hill, 1992).

DNA G+C content (mol%): 28 (Bd).

Type strain: LL2, NCTC 11725, ATCC 49889.

Sequence accession no. (16S rRNA gene): U16760.

77. **Mycoplasma lipofaciens** Bradbury, Forrest and Williams 1983, 334[VP]

li.po.fa'ci.ens. Gr. n. *lipos* animal fat, lard, tallow; L. v. *facio* to make; N.L. part. adj. *lipofaciens* fat-making, intended to refer to the production of a lipid film on solid media.

Cells are mainly spherical and elliptical. Nonmotile. Colonies on solid medium have typical fried-egg appearance. Grows in Hayflick medium supplemented with glucose or arginine at 37°C. Produces a strong "film and spots" reaction.

Commensal of birds; little evidence exists for naturally occurring pathogenicity of the isolates, although experimental inoculation of chicken or turkey eggs can result in embryo mortality. Inadvertent transmission to an investigator during experimental inoculation studies resulted in clinical signs including rhinitis and pharyngitis. Aerosol transmission has been documented in turkeys.

Source: isolated from the infraorbital sinuses of chickens; from tissues or eggs of turkeys and ducks; and from eggs of Northern goshawks (Benčina et al., 1987; Bradbury et al., 1983; Lierz et al., 2007a, b, c, 2008b).

DNA G+C content (mol%): 24.5 (Bd).

Type strain: R171, ATCC 35015, NCTC 10191.

Sequence accession no. (16S rRNA gene): AF221115.

78. **Mycoplasma lipophilum** Del Giudice, Purcell, Carski and Chanock 1974, 152[AL]

li.po.phi'lum. Gr. n. *lipos* animal fat; N.L. neut. adj. *philum* (from Gr. neut. adj. *philon*) friend, loving; N.L. neut. adj. *lipophilum* fat-loving.

Cells are pleomorphic and granular. Motility for this species has not been assessed. Colonies display typical fried-egg morphology. Growth on solid medium is associated with heavy production of film that spreads over the surface of the agar, with the development of numerous internal particles in the colonies. A film similar to that produced on agar medium develops on the surface of broth-grown cultures (Del Giudice et al., 1974). Grows in SP-4 or Hayflick medium supplemented with arginine at 37°C.

Pathogenicity for this species is unclear. This species was first isolated from a human patient with primary atypical pneumonia; however, subsequent isolations from similarly symptomatic patients have not been achieved. Mode of transmission has not been formally assessed.

Source: isolated from the upper and lower respiratory tract of a human with primary atypical pneumonia and the lower respiratory tract of rhesus monkeys (Hill, 1977).

DNA G+C content (mol%): 29.7 (Bd).

Type strain: MaBy, ATCC 27104, NCTC 10173, NBRC 14895.

Sequence accession no. (16S rRNA gene): M24581.

79. **Mycoplasma maculosum** Edward 1955, 90[AL]

ma.cu.lo'sum. L. neut. adj. *maculosum* spotted, alluding to a crinkled film covering the colonies and spreading between them, and spots appearing in the medium beneath and around the colonies.

Cells are short and filamentous, with occasional branching. Motility for this species has not been assessed. Colonies on solid medium have a typical fried-egg appearance. Grows well in Hayflick or SP-4 medium supplemented with arginine at 37°C.

Opportunistic pathogen. Cause of pneumonia in domestic dogs and rarely of meningitis in immunocompromised humans. The route of transmission is via droplet aerosol.

Source: isolated from the nasopharynx, lungs, conjunctivae, and urogenital tract of domestic dogs and the cerebrospinal fluid of an immunocompromised human (Chalker, 2005).

DNA G+C content (mol%): 26.7 (T_m), 29.6 (Bd).

Type strain: PG15, ATCC 19327, NCTC 10168, NBRC 14848.

Sequence accession no. (16S rRNA gene): AF221116.

80. **Mycoplasma meleagridis** Yamamoto, Bigland and Ortmayer 1965, 47[AL]

me.le.a'gri.dis. L. n. *meleagris, -idis* a turkey; L. gen. n. *meleagridis* of a turkey.

Cells are coccoid to coccobacillary. Motility for this species has not been assessed. An extracellular capsule can be visualized by electron microscopy following staining with ruthenium red (Green III and Hanson, 1973). Colonies on solid medium are not necessarily of typical fried-egg appearance. Grows in Frey's medium supplemented with arginine at 37–38°C (Yamamoto et al., 1965).

Pathogenic; causes airsacculitis, pneumonia, sinusitis, perosis, chondrodystrophy, bursitis, synovitis, and reduced hatchability due to embryo lethality in turkeys. Transmission is primarily vertical, but can also occur through droplet aerosol or sexual contact.

Tetracyclines, macrolides, and fluoroquinolones are effective chemotherapeutic agents; however, medicating a commercial flock is not considered an effective control strategy. The temporary use of *in ovo* antimicrobial therapy can be used to eradicate *Mycoplasma meleagridis* from a flock. Management tactics (e.g., single age "all in/all out" systems and culling of endemic flocks) are more commonly utilized (Kleven, 2008). Serological and molecular diagnostic methods have been described (Ben Abdelmoumen Mardassi et al., 2007; Ramírez et al., 2008; Raviv and Kleven, 2009).

Source: isolated from the air sacs, trachea, infraorbital sinuses, oviduct, cloaca, phallus, and eggs of turkeys, and the air sacs of buzzards, kites, and kestrels (Chin et al., 2008; Jordan, 1979; Lam et al., 2004; Lierz et al., 2000).

DNA G+C content (mol%): 27.0 (T_m), 28.6 (Bd).

Type strain: 17529, ATCC 25294, NCTC 10153.

Sequence accession no. (16S rRNA gene): L24106.

Further comment: previously known as avian serotype H (Kleckner, 1960).

81. **Mycoplasma microti** (Dillehay, Sander, Talkington, Thacker and Brown 1995) Brown, Talkington, Thacker, Brown,

Dillehay and Tully 2001b, 412[VP] (*Mycoplasma volis* Dillehay, Sander, Talkington, Thacker and Brown 1995, 633)

mi.cro'ti. N.L. n. *Microtus* a genus of field vole; N.L. gen. n. *microti* of *Microtus*.

Cells are predominantly coccoid in shape. Nonmotile. Colonies on solid medium exhibit a typical fried-egg appearance. Grows well in SP-4 supplemented with glucose in temperatures ranging from 35 to 37°C.

Opportunistic pathogen. No evidence exists for pathogenicity in the natural host; however, pneumonitis was experimentally induced in mice and rats (Evans-Davis et al., 1998).

Source: isolated from the nasopharynx and lung of prairie voles (Dillehay et al., 1995).

DNA G+C content (mol%): not determined.

Type strain: IL371, ATCC 700935.

Sequence accession no. (16S rRNA gene): AF212859.

82. **Mycoplasma moatsii** Madden, Moats, London, Matthew and Sever 1974, 464[AL]

mo.at'si.i. N.L. gen. masc. n. *moatsii* of Moats, named after Kenneth E. Moats, whose primary interest has been in the mycoplasmas of nonhuman primates.

Cells are spheroidal and some exhibit protrusions from the membrane. Motility for this species has not been assessed. Colonies exhibit typical fried-egg morphology. Grows readily in SP-4 or Hayflick broth supplemented with either arginine or glucose at an optimum temperature of 37°C.

No evidence of pathogenicity.

Source: isolated from the respiratory and reproductive tracts of grivet monkeys and from the cecum, jejunum, and colon of wild Norway rats (Giebel et al., 1990).

DNA G+C content (mol%): 25.7 (Bd).

Type strain: MK 405, ATCC 27625, NCTC 10158.

Sequence accession no. (16S rRNA gene): AF412984.

83. **Mycoplasma mobile** Kirchhoff, Beyene, Fischer, Flossdorf, Heitmann, Khattab, Lopatta, Rosengarten, Seidel and Yousef 1987, 197[VP]

mo'bi.le. L. neut. adj. *mobile* motile.

Cells are conical or flask-shaped and have a distinct terminal protrusion referred to as the "head-like structure". Cells demonstrate rapid gliding motility when adhering to charged surfaces and move in the directional of the head-like structure (Miyata et al., 2002, 2000). Colonies on solid medium have a typically fried-egg appearance. Grows well in Aluotto's medium supplemented with glucose or arginine. The temperature range for growth is 17–30°C, with optimum growth at 30°C.

Pathogenic; causes necrotic erythrodermatitis in tench. The mode of transmission has not been established.

Source: isolated from the gills of a freshwater fish (*Tinca tinca*) with "red disease" (Kirchhoff et al., 1987).

DNA G+C content (mol%): 23.5 (Bd), 24.9.

Type strain: 163K, ATCC 43663, NCTC 11711.

Sequence accession nos: M24480 (16S rRNA gene), NC_006908 (complete genome sequence of strain 163K).

84. **Mycoplasma molare** Røsendal 1974, 130[AL]

mo.la're. L. neut. adj. *molare* of or belonging to a mill, here millstone-like, referring to the heavy film reaction, which resembles the pattern on the surface of a millstone.

Cells are coccoid to pleomorphic. Nonmotile. Colonies have a typical fried-egg appearance. Grows well in SP-4 medium supplemented with glucose at 37°C. A lipid film of characteristic appearance develops on the surface and along the circumference of colonies grown on egg-yolk agar.

Opportunistic pathogen. Associated with pharyngitis and mild inflammatory lesions of the lower respiratory tract and may be associated with infertility, vaginitis, and posthitis of domestic dogs. No clear evidence for primary pathogenicity of the species. Mode of transmission has not been established definitively.

Source: isolated from the oral cavity, pharynx, cervix, vagina, and prepuce of domestic dogs (Røsendal, 1979; (Chalker, 2005).

DNA G+C content (mol%): 26.0 (Bd).

Type strain: H 542, ATCC 27746, NCTC 10144.

Sequence accession no. (16S rRNA gene): AF412985.

85. **"Mycoplasma mucosicanis"** Spergser, Langer, Muck, Macher, Szostak, Rosengarten and Busse 2010

mu.co.si.ca'nis. N.L. n. *mucosa* mucous membrane; L. n. *canis* a dog; N.L. gen. n. *mucosicanis* of mucous membranes of a dog.

Cells are pleomorphic, but primarily coccoid. Nonmotile. Colonies on solid media have a typical fried-egg morphology. Grows wells in modified Hayflick medium at 37°C and produces a "film and spots" reaction.

No evidence of pathogenicity. Mode of transmission has not been established definitively.

Source: isolated from the prepuce, semen, vagina, cervix, and oral cavity of domestic dogs (Spergser et al., 2010).

DNA G+C content (mol%): not determined.

Type strain: 1642, ATCC BAA-1895, DSM 22457.

Sequence accession no. (16S rRNA gene): AM774638.

86. **Mycoplasma muris** McGarrity, Rose, Kwiatkowski, Dion, Phillips and Tully 1983, 355[VP]

mu'ris. L. n. *mus, muris* mouse; L. gen. n. *muris* of a mouse.

Cells are primarily coccoid or coccobacillary, but exhibit a few other pleomorphic forms. Motility for this species has not been assessed. Colonies usually have a granular appearance and few colonies demonstrate the typical fried-egg appearance. Grow well in SP-4 broth supplemented with arginine at 37°C and produces a "film and spots" reaction.

No evidence of pathogenicity.

Source: isolated from the vagina of a pregnant mouse (laboratory strain RIII; McGarrity et al., 1983).

DNA G+C content (mol%): 24.9 (TLC).

Type strain: RIII-4, ATCC 33757, NCTC 10196.

Sequence accession no. (16S rRNA gene): M23939.

87. **Mycoplasma mustelae** Salih, Friis, Arseculeratne, Freundt and Christiansen 1983, 478[VP]

mu.ste'lae. N.L. n. *Mustela* (from L. n. *mustela* a weasel) the generic name of the mink *Mustela vison*; N.L. gen. n. *mustelae* of *Mustela*.

Cells are highly pleomorphic; most common morphologies include pleomorphic rings, short filamentous forms, and coccoid elements. Nonmotile. Colonies on solid medium show a typical fried-egg appearance. Growth in Hayflick medium supplemented with glucose occurs at 37°C.

No evidence of pathogenicity.

Source: isolated from the trachea and lungs of juvenile minks (*Mustela vison*; Salih et al., 1983).

DNA G+C content (mol%): 28.2 (Bd).

Type strain: MX9, ATCC 35214, NCTC 10193, AMRC-C 1486.

Sequence accession no. (16S rRNA gene): AF412986.

88. **Mycoplasma neurolyticum** (Sabin 1941) Freundt 1955, 73[AL] (*Musculomyces neurolyticus* Sabin 1941, 57)

neu.ro.ly'ti.cum. Gr. n. *neuron* nerve; N.L. adj. *lyticus -a -um* (from Gr. adj. *lutikos -ê -on*) able to loosen, able to dissolve; N.L. neut. adj. *neurolyticum* nerve-destroying.

Cells are filamentous and highly variable length. Nonmotile (Nelson and Lyons, 1965). Colonies show a typical fried-egg appearance after incubation at 37°C. Grows in Hayflick medium supplemented with glucose at 37°C (Naot et al., 1977).

Pathogenicity is currently uncertain. Potentially associated with spongiform encephalopathy and ischemic necrosis of the brain resulting in a clinical state referred to as "rolling disease" in mice and rats. Pathology may be exacerbated in the presence of additional neurotropic organisms (i.e., *Toxoplasma gondii*, *Chlamydia* spp., *Plasmodium* spp., and yellow fever virus) or during leukemic syndromes. Transmission to suckling rodents occurs shortly after birth. Treatment of *Mycoplasma neurolyticum* infections is uncommon, as pathology is typically not resolvable after the onset of clinical signs.

Source: isolated from the brain, conjunctivae, nasopharynx, and middle ear of captive mice and rats.

DNA G+C content (mol%): 22.8 (Bd), 26.2 (T_m).

Type strain: Type A, ATCC 19988, NCTC 10166, CIP 103926, NBRC 14799.

Sequence accession no. (16S rRNA gene): M23944.

Further comment: a putative exotoxin with neurological effects on rodents was formerly thought to be produced by most freshly isolated strains, although a few non-toxic strains were described (Tully and Ruchman, 1964). The findings were not substantiated by later work (Tryon and Baseman, 1992).

89. **Mycoplasma opalescens** Røsendal 1975, 469[AL]

o.pa.les'cens. L. n. *opalus* precious stone; N.L. neut. adj. *opalescens* opalescent, referring to the opalescent film produced on solid medium.

Morphology by light microscopy or ultrastructural examination is not defined. Motility for this species has not been assessed. Colonies on solid medium have a typical fried-egg appearance and possess an iridescent quality. Grows well in SP-4 medium supplemented with arginine at 37°C.

No evidence of pathogenicity.

Source: isolated from the oral cavity, prepuce, and prostate gland of domestic dogs (Røsendal, 1975).

DNA G+C content (mol%): 29.2 (Bd).

Type strain: MH5408, ATCC 27921, NCTC 10149.

Sequence accession no. (16S rRNA gene): AF538961.

90. **Mycoplasma orale** Taylor-Robinson, Canchola, Fox and Chanock 1964, 141[AL]

o.ra'le. L. n. *os, oris* the mouth; L. neut. suff. *-ale* suffix denoting pertaining to; N.L. neut. adj. *orale* pertaining to the mouth.

Cells can be either coccoid or filamentous. Motility for this species has not been assessed. Colonies on solid medium have a typical fried-egg appearance. Grows well in Hayflick or SP-4 medium supplemented with arginine at 37°C.

Mycoplasma orale is most commonly associated with contamination of eukaryotic cell culture and is frequently removed by treatment of cells with antibiotics and/or maintenance of cell lines in antibiotic-containing medium. The most effective classes of antibiotics for cell culture eradication are tetracyclines, macrolides, and fluoroquinolones. Additionally, passage of eukaryotic cells in hyperimmune serum raised against *Mycoplasma orale* has been shown to be an effective method of eradication (Vogelzang and Compeer-Dekker, 1969).

Commensal/opportunistic pathogen. Commonly found as a commensal of the human oral cavity; can cause respiratory tract infections, osteomyelitis, infectious synovitis, and abscesses in immunocompromised individuals (Paessler et al., 2002; Roifman et al., 1986).

Source: isolated from the oral cavity of subclinical humans, the sputum of an immunocompromised human with acute respiratory illness, and from synovial fluid, bone, and splenic abscesses of another immunocompromised individual.

DNA G+C content (mol%): 24.0–28.2 (T_m, Bd).

Type strain: CH19299, ATCC 23714, NCTC 10112, CIP 104969, NBRC 14477.

Sequence accession no. (16S rRNA gene): M24659.

91. **Mycoplasma ovipneumoniae** Carmichael, St George, Sullivan and Horsfall 1972, 677[AL]

o.vi.pneu.mo.ni'ae. L. fem. n. *ovis* a sheep; Gr. n. *pneumonia* pneumonia; N.L. gen. n. *ovipneumoniae* of sheep pneumonia.

Morphology and motility are poorly described. The organism produces a polysaccharide capsule with variable thickness that is dependent upon culture conditions and strain (Niang et al., 1998). Colonies grown on standard agar are convex and have a lacy or vacuolated appearance. Grows well in Friis medium or SP-4 broth supplemented with glucose at 37°C.

Pathogenic; causes chronic proliferative interstitial pneumonia, pulmonary adenomatosis, conjunctivitis (Jones et al., 1976), and mastitis under experimental conditions (Jones, 1985) of sheep and goats. Transmission occurs via droplet aerosol and can occur via intravenous inoculation in experimental infection studies.

Source: isolated from the lungs, trachea, nose, and conjunctivae of sheep and goats.

DNA G+C content (mol%): 25.7 (Bd).

Type strain: Y98, NCTC 10151, ATCC 29419.

Sequence accession no. (16S rRNA gene): U44771.

92. **Mycoplasma ovis** (Neitz, Alexander and du Toit 1934) Neimark, Hoff and Ganter 2004, 369[VP] (*Eperythrozoon ovis* Neitz, Alexander and du Toit 1934, 267)

o'vis. L. fem. n. *ovis, -is* a sheep; L. gen. n. *ovis* of a sheep.

Cells are coccoid and motility for this species has not been assessed. The morphology of infected erythrocytes is altered demonstrating a marked depression at the site of *Mycoplasma ovis* attachment. This species has not been grown on artificial medium; therefore, notable biochemical parameters are not known.

Neoarsphenamine is an effective therapeutic agent. *Mycoplasma ovis* is reported to share antigens with *Mycoplasma wenyonii* (Kreier and Ristic, 1963), potentially complicating serology-based diagnosis of infection.

Pathogenic; causes mild to severe anemia in sheep and goats that often results in poor feed conversion. Transmission occurs via blood-feeding arthropods, e.g., *Haemophysalis plumbeum, Rhipicephalus bursa, Aedes camptorhynchus,* and *Culex annulirostris* (Daddow, 1980; Howard, 1975; Nikol'skii and Slipchenko, 1969), and likely via fomites such as reused needles, shearing tools, and ear-tagging equipment (Brun-Hansen et al., 1997; Mason and Statham, 1991).

Source: observed in association with erythrocytes or unattached in suspension in the blood of sheep, goats, and rarely in eland and splenectomized deer.

DNA G+C content (mol%): not determined.

Type strain: not established.

Sequence accession no. (16S rRNA gene): AF338268.

93. **Mycoplasma oxoniensis** Hill 1991b, 24[VP]

oxo.ni.en'sis. N.L. adj. *oxoniensis* (*sic*) pertaining to Oxon, an abbreviation of Oxfordshire, where the mycoplasma was first isolated.

Cells are primarily coccoid. Nonmotile. Colonies on agar have a typical fried-egg appearance. Growth in SP-4 broth supplemented with glucose occurs at 35–37°C.

No evidence of pathogenicity. Mode of transmission is unknown.

Source: isolated from the conjunctivae of the Chinese hamster (*Cricetulus griseus*; Hill, 1991b).

DNA G+C content (mol%): 29 (Bd).

Type strain: 128, NCTC 11712, ATCC 49694.

Sequence accession no. (16S rRNA gene): AF412987.

94. **Mycoplasma penetrans** Lo, Hayes, Tully, Wang, Kotani, Pierce, Rose and Shih 1992, 363[VP]

pe.ne'trans. L. part. adj. *penetrans* penetrating, referring to the ability of the organism to penetrate into mammalian cells.

Cells are flask-shaped, with a distinct terminal structure reminiscent of the *Mycoplasma pneumoniae* attachment organelle. Cells demonstrate gliding motility when adhering to charged surfaces and move in the direction of the terminal structure. Colonies on agar plates display a typical fried-egg appearance. Grows well in SP-4 broth supplemented with either glucose or arginine at 37°C.

Opportunistic pathogen; found in the urogenital tract of immunocompromised humans, most notably HIV-positive individuals (serological detection in HIV-negative individuals is rare). Speculation regarding the ability of *Mycoplasma penetrans* to act as a cofactor in the progression of AIDS by modulation of the immune system remains intriguing, but in need of further substantiation (Blanchard, 1997). Transmission is presumed to be via sexual contact.

Source: isolated from the urine of HIV-positive humans, and from the blood, respiratory secretions, and trachea of an HIV-negative patient with multiple autoimmune syndromes (Yanez et al., 1999).

DNA G+C content (mol%): 30.5 (T_m), 25.7 (HF-2 genome sequence; Sasaki et al., 2002).

Type strain: GTU-54-6A1, ATCC 55252.

Sequence accession nos: L10839 (16S rRNA gene), NC_004432 (HF-2 complete genome sequence).

95. **Mycoplasma phocicerebrale** corrig. Giebel, Meier, Binder, Flossdorf, Poveda, Schmidt and Kirchhoff 1991, 43[VP]

pho.ci.ce.re.bra'le. L. n. *phoca* seal; N.L. neut. adj. *cerebrale* of or pertaining to the brain; N.L. neut. adj. *phocicerebrale* pertaining to the brain of a seal.

Cells are coccoid or exhibit a dumbbell shape. Motility for this species has not been assessed. Colonies on solid medium typically show a fried-egg appearance. Grows well in SP-4 medium supplemented with arginine at 37°C.

Pathogenic; associated with respiratory disease and conjunctivitis in harbor seals (Kirchhoff et al., 1989) and a distinctive ulcerative keratitis subsequent to seal bites (known as "seal finger") and secondary arthritis in humans (Baker et al., 1998; Ståby, 2004). Mode of transmission between harbor seals has not been established definitively; transmission to humans appears to be zoonotic following seal bites.

Source: isolated from the brains, noses, throats, lungs, and hearts of seals (*Phoca vitulina*) during an outbreak of respiratory disease (Kirchhoff et al., 1989), and from cutaneous lesions of humans with seal finger (Baker et al., 1998).

DNA G+C content (mol%): 25.9 (Bd).

Type strain: 1049, ATCC 49640, NCTC 11721.

Sequence accession no. (16S rRNA gene): AF304323.

Further comment: the original spelling of the specific epithet, *phocacerebrale* (*sic*), has been corrected by Königsson et al. (2001).

96. **Mycoplasma phocidae** Ruhnke and Madoff 1992, 213[VP]

pho.ci'da.e. L. n. *phoca* seal; N.L. gen. n. *phocidae* (*sic*) of a seal.

Cells are primarily coccoid. Motility for this species has not been assessed. Colonies on solid medium have a typical fried-egg appearance. Grows well in SP-4 or Hayflick medium supplemented with arginine at 37°C. Produces "film and spots" reaction.

Opportunistic pathogen; associated with secondary pneumonia of harbor seals subsequent to influenza infection. Attempts to produce disease in gray or harp seals with *Mycoplasma phocidae* in pure culture were not successful (Geraci et al., 1982). Mode of transmission has not been established definitively.

Source: isolated from the lungs, tracheae, and heart of harbor seals.

DNA G+C content (mol%): 27.8 (Bd).

Type strain: 105, ATCC 33657.

Sequence accession no. (16S rRNA gene): AF304325.

Further comment: the species designation "*Mycoplasma phocae*" was suggested by Königsson et al. (2001), but the original epithet "*phocidae*" should be retained. The suggested change is forbidden by Rule 61 (Note) of the Bacteriological Code because it would change the first syllable of the original epithet without correcting any orthographic or typographical error.

97. **Mycoplasma phocirhinis** corrig. Giebel, Meier, Binder, Flossdorf, Poveda, Schmidt and Kirchhoff 1991, 43[VP]

pho.ci.rhi'nis. L. n. *phoca* seal; Gr. n. *rhis, rhinos* nose; N.L. gen. n. *phocirhinis* of the nose of a seal.

Cells are coccoid; motility for this species has not been assessed. Colonies on solid medium usually have a fried-egg appearance. Grows well in Friis or Hayflick medium at 37°C. Produces "film and spots" reaction.

Pathogenic; associated with respiratory disease and conjunctivitis of harbor seals (Kirchhoff et al., 1989). Mode of transmission has not been established definitively.

Source: isolated from the nose, pharynx, trachea, lungs, and heart of seals (*Phoca vitulina*; Kirchhoff et al., 1989).

DNA G+C content (mol%): 26.5 (Bd).

Type strain: 852, ATCC 49639, NCTC 11722.

Sequence accession no. (16S rRNA gene): AF304324.

Further comment: the original spelling of the specific epithet, *phocarhinis* (*sic*), has been corrected by Königsson et al. (2001)

98. **Mycoplasma pirum** Del Giudice, Tully, Rose and Cole 1985, 290[VP]

pi'rum. L. neut. n. *pirum* (nominative in apposition) pear, referring to the pear-shaped morphology of the cells.

Cells are predominantly flask or pear-shaped and possess an organized terminal structure, with an outer, finely particulate nap covering the entire surface of the cell. Cells exhibit low-speed gliding motility and move in the direction of the terminal organelle (Hatchel and Balish, 2008). Colonies display a typical fried-egg appearance. Grows well in SP-4 medium supplemented with glucose at 37°C.

No evidence of pathogenicity.

Source: isolated from the rectum of immunocompetent humans and whole blood and circulating lymphocytes of HIV-positive humans (Montagnier et al., 1990). Originally isolated from cultured eukaryotic cells that were of human origin (Del Giudice et al., 1985).

DNA G+C content (mol%): 25.5 (Bd).

Type strain: HRC 70-159, ATCC 25960, NCTC 11702.

Sequence accession no. (16S rRNA gene): M23940.

99. **Mycoplasma pneumoniae** Somerson, Taylor-Robinson and Chanock 1963, 122[AL]

pneu.mo.ni'ae. Gr. n. *pneumonia* pneumonia; N.L. gen. n. *pneumoniae* of pneumonia.

Cells are highly pleomorphic; the predominant shape includes a long, thin terminal structure at one cell pole, with or without a trailing filament at the opposite pole. Cells are motile and glide in the direction of the terminal organelle when attached to cell surfaces, plastic, or glass. Colonies on solid medium usually lack the light peripheral zone, appearing rather as circular dome-shaped, granular structures. Growth is best achieved in SP-4 medium supplemented with glucose at 37°C.

Pathogenic; causes interstitial pneumonitis, tracheobronchitis, desquamative bronchitis, and pharyngitis [collectively referred to as primary atypical pneumonia (PAP); Krause and Taylor-Robinson (1992)]. Less commonly, *Mycoplasma pneumoniae* causes meningoencephalitis, otitis media, bullous myringitis, infectious synovitis, glomerulonephritis, pancreatitis, hepatitis, myocarditis, pericarditis, hemolytic anemia, and rhabdomyolysis (Waites and Talkington, 2005). The preceding can be primary lesions, but are often secondary to respiratory disease. Dysfunction of the immune system by inappropriate cytokine responses or possibly molecular mimicry following infection are associated with long-term sequelae including the development or exacerbation of asthma and chronic obstructive pulmonary disease; Stevens-Johnson syndrome and other exanthemas; and Guillain-Barre syndrome, Bell's palsy, and demyelinating neuropathies (Atkinson et al., 2008). Mode of transmission is via droplet aerosols (PAP) or sexual contact (urogential colonization).

Clinical manifestations are successfully treated with tetracyclines, fluoroquinolones, macrolides, and lincosamides (Waites and Talkington, 2005). Signs can be treated with inhaled or injected steroids. Experimental vaccinations aimed at preventing infection have been unsuccessful due to failure to elicit immune responses, retention of virulence, or invocation of immune responses that exacerbated clinical signs (Barile, 1984; Jacobs et al., 1988). *Mycoplasma pneumoniae* is reported to share antigens with *Mycoplasma genitalium* (Taylor-Robinson, 1983a), potentially complicating serology-based diagnosis of infection.

Source: isolated from the upper and lower respiratory tract, cerebrospinal fluid, synovial fluid, and urogential tract of humans.

DNA G+C content (mol%): 38.6 (T_m), 40.0 (strain M129 genome sequence).

Type strain: FH, ATCC 15531, NCTC 10119, CIP 103766, NBRC 14401.

Sequence accession nos: M29061 (16S rRNA gene), U00089 (strain M129 genome sequence).

100. **Mycoplasma primatum** Del Giudice, Carski, Barile, Lemcke and Tully 1971, 442[AL]

pri.ma'tum. L. n. *primas, primatis* chief, from which *primates*, the highest order of mammals originates; L. pl. gen. n. *primatum* of chiefs, of primates.

Cells are both spherical and coccobacillary. Motility for this species has not been assessed. Colony morphology has a fried-egg appearance. Grows well in SP-4 or Hayflick medium supplemented with arginine at 37°C.

Opportunistic pathogen; rarely associated with keratitis in humans (Ruiter and Wentholt, 1955). Mode of transmission has not been established.

Source: isolated from the oral cavity and/or urogenital tract of baboons, African green monkeys, rhesus macaques, squirrel monkeys, and humans (Hill, 1977; Somerson and Cole, 1979; Thomsen, 1974).

DNA G+C content (mol%): 28.6 (T_m).

Type strain: HRC292, ATCC 25948, NCTC 10163.

Sequence accession no. (16S rRNA gene): AF221118.

101. **Mycoplasma pullorum** Jordan, Ernø, Cottew, Hinz and Stipkovits 1982, 114[VP]

pul.lo'rum. L. n. *pullus* a young animal, especially chicken; L. gen. pl. n. *pullorum* of young chickens.

Cells are coccoid to coccobacillary. Motility for this species has not been assessed. Colonies on solid medium display typical fried-egg appearance. Grows in Frey's or Hayflick medium supplemented with glucose at 37°C.

Pathogenicity not fully established, but has been associated with tracheitis and airsacculitis in chickens, and embryo lethality resulting in reduced hatchability in chickens and turkeys. Mode of transmission has not been assessed definitively.

Source: isolated from the trachea, air sacs, and eggs of chickens; from the eggs of turkeys; and from tissues of pheasants, partridges, pigeons, and quail (Benčina et al., 1987; Bradbury et al., 2001; Kempf et al., 1991; Kleven, 2003; Lobo et al., 2004; Moalic et al., 1997; Poveda et al., 1990).

DNA G+C content (mol%): 29 (Bd).

Type strain: CKK, ATCC 33553, NCTC 10187.

Sequence accession no. (16S rRNA gene): U58504.

Further comment: previously known as avian serotype C (Adler et al., 1958).

102. **Mycoplasma pulmonis** (Sabin 1941) Freundt 1955, 73[AL] (*Murimyces pulmonis* Sabin 1941, 57)

pul.mo'nis. L. n. *pulmo, -onis* the lung; L. gen. n. *pulmonis* of the lung.

Cells are predominantly coccoid with a well-organized terminal structure. Cells are motile and glide in the direction of the terminal structure. An extracellular capsular matrix can be demonstrated by staining with ruthenium red and formation of biofilms has been demonstrated. Colonies on solid medium have a coarsely granulated and vacuolated appearance, with a lesser tendency to grow into the agar, and the central spot is consistently less well defined than in most other *Mycoplasma* species. Grows in SP-4 or modified Hayflick broth supplemented with glucose at an optimum temperature of 37°C.

Pathogenic; causes rhinitis, laryngotracheitis, bronchopneumonia (collectively described as murine respiratory mycoplasmosis in mice), otitis media, conjunctivitis, acute and chronic arthritis, oophoritis, salpingitis, epididymitis, and urethritis of rodents (chiefly mice, rats, guinea pigs, and hamsters). Transmission occurs via aerosol, fomites, sexual contact, or vertically during gestation.

Macrolides, fluoroquinolones, and tetracyclines are effective against *Mycoplasma pulmonis in vitro*; however, control measures such as decontamination of fomites, culling of infected colonies, and treatment of clinical signs with steroids are more commonly employed in clinical settings. Several candidate vaccines have been described.

Source: isolated from the respiratory and urogenital tracts, eyes, synovial fluid, and synovial membranes of (principally captive) rodents, and rarely from the nasopharynx of rabbits and horses (Allam and Lemcke, 1975; Cassell and Hill, 1979; Deeb and Kenny, 1967; Simecka et al., 1992).

DNA G+C content (mol%): 27.5–29.2 (Bd), 26.6 (strain UAB CTIP genome sequence).

Type strain: Ash, PG34, ATCC 19612, NCTC 10139, CIP 75.26, NBRC 14896.

Sequence accession nos: M23941 (16S rRNA gene sequence), NC_002771 (strain UAB CTIP genome sequence).

103. **Mycoplasma putrefaciens** Tully, Barile, Edward, Theodore and Ernø 1974, 116[AL]

pu.tre.fa'ci.ens. L. v. *putrefacio* to make rotten; L. part. adj. *putrefaciens* making rotten or putrefying, connoting the production of a putrid odor in broth and agar cultures.

Cells are predominantly coccobacillary to pleomorphic. Nonmotile. Formation of biofilms has been demonstrated (McAuliffe et al., 2006). Colony morphology has a typical fried-egg appearance. Grows well in SP-4 medium supplemented with glucose at 37°C.

Pathogenic; causes polyarthritis, mastitis, conjunctivitis (a syndrome collectively termed contagious agalactia) (Bergonier et al., 1997), abortion, salpingitis, metritis, and testicular atrophy (Gil et al., 2003) in goats.

Macrolides, fluoroquinolones, lincosamides, and tetracyclines are effective against *Mycoplasma putrefaciens*; however, control measures such as decontamination of fomites and culling of infected herds are typically recommended to discourage the development of antimicrobial-resistant strains in carrier animals (Antunes et al., 2007; Bergonier et al., 1997).

Source: isolated from the synovial fluid, udders, expelled milk, conjunctivae, ear canal, uterus, and testes of goats.

DNA G+C content (mol%): 28.9 (T_m).

Type strain: KS1, ATCC 15718, NCTC 10155.

Sequence accession no. (16S rRNA gene): M23938.

104. **Mycoplasma salivarium** Edward 1955, 90[AL]

sa.li.va'ri.um. L. neut. adj. *salivarium* slimy, saliva-like, intended to denote of saliva.

Cells are coccoid to coccobacillary. Nonmotile. Colonies are large with a typical fried-egg appearance. Grows well in Hayflick medium supplemented with arginine at 37°C and produces a "film and spots" reaction.

Mycoplasma salivarium is most frequently associated with contamination of eukaryotic cell culture and is frequently removed by treatment of cells with antibiotics and/or maintenance of cell lines in antibiotic-containing medium. The most effective classes of antibiotics for cell culture eradication are tetracyclines, macrolides, and fluoroquinolones.

Opportunistic pathogen; primarily found as a commensal of the human oral cavity, and rarely associated with arthritis, submasseteric abscesses, gingivitis, and periodontitis in immunocompromised patients (Grisold et al., 2008; Lamster et al., 1997; So et al., 1983). Mode of transmission is via direct contact with human saliva.

Source: isolated from the oral cavity, synovial fluid, dental plaque, and abscessed mandibles of humans, and the nasopharynx of pigs (Erickson et al., 1988).

DNA G+C content (mol%): 27.3 (Bd).

Type strain: PG20, H110, ATCC 23064, NCTC 10113, NBRC 14478.

Sequence accession no. (16S rRNA gene): M24661.

105. **Mycoplasma simbae** Hill 1992, 520[VP]

sim'bae. Swahili n. *simba* lion; N.L. gen. n. *simbae* of a lion.

Cells are pleomorphic and nonmotile. Colonies on solid medium have a typical fried-egg appearance. Film is produced by cultivation on egg yolk agar. Grows well in SP-4 medium at 37°C.

No evidence of pathogenicity. Mode of transmission has not been established.

Source: isolated from the throats of lions (Hill, 1992).

DNA G+C content (mol%): 37 (Bd).

Type strain: LX, NCTC 11724, ATCC 49888.

Sequence accession no. (16S rRNA gene): U16323.

106. **Mycoplasma spermatophilum** Hill 1991a, 232[VP]

sper.ma.to.phi'lum. Gr. n. *sperma, -atos* sperm or seed; N.L. neut. adj. *philum* (from Gr. neut. adj. *philon*) friend, loving; N.L. neut. adj. *spermatophilum* sperm-loving.

Cells are primarily coccoid. Nonmotile. Colonies are convex to fried-egg shaped and are of below-average size. Grows well in SP-4 medium supplemented with added arginine under anaerobic conditions at 37°C.

Pathogenic; potentially associated with infertility, as infected spermatozoa do not fertilize ova and infected fertilized ova were unable to implant following *in vitro* fertilization (Hill et al., 1987; Hill, 1991a). Mode of transmission is via sexual contact.

Source: isolated from the semen and cervix of humans with impaired fertility.

DNA G+C content (mol%): 32 (Bd).

Type strain: AH159, NCTC 11720, ATCC 49695, CIP 105549.

Sequence accession no. (16S rRNA gene): AF221119.

107. **Mycoplasma spumans** Edward 1955, 90[AL]

spu'mans. L. part. adj. *spumans* foaming, presumably alluding to thick dark markings that suggest the presence of globules inside the coarsely reticulated colonies.

Cells are coccoid to filamentous. Motility for this species has not been assessed. Colonies in early subcultures have a coarsely reticulated and vacuolated appearance. A typical fried-egg appearance of the colonies develops on repeated subculturing. Grows well in SP-4 or modified Hayflick medium supplemented with arginine.

Opportunistic pathogen; primarily found as a commensal of the nasopharynx, but has also been associated with pneumonia and arthritis of domestic dogs. Mode of transmission has not been established definitively.

Source: isolated from the lungs, nasopharynx, synovial fluid, cerebrospinal fluid, trachea, prepuce, prostate, bladder, cervix, vagina, and urine of domestic dogs (Chalker, 2005).

DNA G+C content (mol%): 28.4 (T_m).

Type strain: PG13, ATCC 19526, NCTC 10169, NBRC 14849.

Sequence accession no. (16S rRNA gene): AF125587.

108. **Mycoplasma sturni** Forsyth, Tully, Gorton, Hinckley, Frasca, van Kruiningen and Geary 1996, 719[VP]

stur'ni. N.L. n. *Sturnus* (from L. n. *sturnus* a starling or stare) a genus of birds, N.L. gen. n. *sturni* of the genus *Sturnus*, the genus of the bird from which the organism was isolated.

Cells are primarily coccoid with some irregular flask-shaped and filamentous forms seen. Motility for this species has not been assessed. Colonies on agar usually have a fried-egg appearance when grown at 37°C. Grows well in SP-4 medium supplemented with glucose at 34–37°C.

Pathogenicity has not been fully established, but it is associated with conjunctivitis in the European starling, mockingbirds, blue jays, and American crows. The organism is also found in clinically normal birds. Mode of transmission has not been established definitively.

Source: isolated from the conjunctivae of European starlings, mockingbirds, blue jays, American crows, American robins, blackbirds, rooks, carrion crows, and magpies (Frasca et al., 1997; Ley et al., 1998; Pennycott et al., 2005; Wellehan et al., 2001).

DNA G+C content (mol%): 31 (Bd).

Type strain: UCMF, ATCC 51945.

Sequence accession no. (16S rRNA gene): U22013.

109. **Mycoplasma sualvi** Gourlay, Wyld and Leach 1978, 292[AL]

su.al'vi. L. n. *sus, suis* swine; L. n. *alvus* bowel, womb, stomach; N.L. gen. n. *sualvi* of the bowel of swine.

Cells are coccobacillary and many possess organized terminal structures. Nonmotile. Colonies have a typical fried-egg appearance. Grows well in SP-4 medium supplemented with either arginine or glucose at 37°C.

No evidence of pathogenicity.

Source: isolated from the rectum, colon, small intestines, and vagina of pigs (Gourlay et al., 1978).

DNA G+C content (mol%): 23.7 (Bd).

Type strain: Mayfield B, NCTC 10170, ATCC 33004.

Sequence accession no. (16S rRNA gene): AF412988.

110. **Mycoplasma subdolum** Lemcke and Kirchhoff 1979, 49[AL]

sub.do'lum. L. neut. adj. *subdolum* somewhat deceptive, alludes to the deceptive color change that led to the original erroneous description of the strains as urea-hydrolyzing.

Cells are coccoid to coccobacilliary. Motility for this species has not been assessed. Colony growth on solid medium exhibits the typical fried-egg appearance. Grows well in SP-4, Frey's, or Hayflick medium supplemented with arginine at 37°C.

Opportunistic pathogen; equivocal evidence for virulence may represent variation among strains. Associated with impaired fecundity and abortion in horses; however, is highly prevalent in clinically normal horses (Spergser et al., 2002). Mode of transmission is via sexual contact.

Source: isolated from the cervix, semen, and aborted foals of horses (Lemcke and Kirchhoff, 1979).

DNA G+C content (mol%): 28.8 (Bd).

Type strain: TB, ATCC 29870, NCTC 10175.

Sequence accession no. (16S rRNA gene): AF125588.

111. **Mycoplasma suis** corrig. (Splitter 1950) Neimark, Johansson, Rikihisa and Tully 2002b, 683[VP] (*Eperythrozoon suis* Splitter 1950, 513)

su'is. L. gen. n. *suis* of the pig.

Cells are coccoid. Motility for this species has not been assessed. This species has not been grown on any artificial medium; therefore, notable biochemical parameters are not known.

Neoarsphenamine and tetracyclines are effective therapeutic agents. An enzyme-linked immunosorbant assay (ELISA) and PCR-based detection assays to enable diagnosis of infection have been described (Groebel et al., 2009; Gwaltney and Oberst, 1994; Hoelzle, 2008; Hsu et al., 1992).

Pathogenic; causes febrile icteroanemia in pigs. Transmission occurs via insect vectors including *Stomoxys calcitrans* and *Aedes aegypti* (Prullage et al., 1993).

Source: observed in association with the erythrocytes of pigs.

DNA G+C content (mol%): 31.1 (complete genome sequence of strain Illinois; J.B. Messick et al., unpublished).

Type strain: not established.

Sequence accession no. (16S rRNA gene): AF029394.

Further comment: the original spelling of the specific epithet, *haemosuis* (*sic*), has been corrected by the List Editor.

112. **Mycoplasma synoviae** Olson, Kerr and Campbell 1964, 209[AL]

sy.novi'ae. N.L. n. *synovia* the joint fluid; N.L. gen. n. *synoviae* of joint fluid.

Cells are coccoid and pleomorphic. Nonmotile. An amorphous extracellular layer is described. Colony appearance on solid medium is variable with some showing typical fried-egg type colonies. Grows well in Frey's medium supplemented with glucose, L-cysteine, and nicotinamide adenine dinucleotide at 37°C. Produces a "film and spots" reaction (Ajufo and Whithear, 1980; Frey et al., 1968).

Pathogenic; causes infectious synovitis, osteoarthritis, and upper respiratory disease which is often subclinical in chickens and turkeys. Also associated with a reduction in egg quality in chickens. *Mycoplasma synoviae* is often found in association with additional avian pathogens including *Mycoplasma gallisepticum*, avian strains of *Escherichia coli*, Newcastle disease virus, and infectious bronchitis virus. Direct contact with fomites and droplet aerosols are the primary mechanisms of transmission.

Tetracyclines and fluoroquinolones are effective chemotherapeutic agents; however, treatment is typically only sought for individual birds, as medicating a commercial flock is not considered an effective control strategy. Vaccination and management strategies (i.e., single age "all in/all out" systems and culling of endemic flocks) are more commonly utilized. A live vaccine is commercially available. *Mycoplasma synoviae* shares surface antigens with *Mycoplasma gallisepticum*, potentially complicating serology-based diagnosis of infection. Numerous molecular diagnostics have been described. This organism is listed in the Terrestrial Animal Health Code of the Office International des Epizooties (http://oie.int; Yogev et al., 1989; Browning et al., 2005; Kleven, 2008; Hammond et al., 2009; Raviv and Kleven, 2009)

Source: isolated from the synovial fluid, synovial membranes, and respiratory tract tissues of chickens and turkeys, and from ducks, geese, pigeons, Japanese quail, pheasants, red-legged partridges, wild turkeys, and house sparrows (Bradbury and Morrow, 2008; Feberwee et al., 2009; Jordan, 1979; Kleven, 1998).

DNA G+C content (mol%): 34.2 (Bd).

Type strain: WVU 1853, ATCC 25204, NCTC 10124.

Sequence accession nos: X52083 (16S rRNA gene), NC_007294 (strain 53 complete genome sequence).

113. **Mycoplasma testudineum** Brown, Merritt, Jacobson, Klein, Tully and Brown 2004, 1529[VP]

tes.tu.di'ne.um. L. neut. adj. *testudineum* of or pertaining to a tortoise.

Cells are predominantly coccoid in shape, though some exhibit a terminal protrusion. Cells exhibit gliding motility. Colonies on solid medium exhibit typical fried-egg forms. Grows well in SP-4 medium supplemented with glucose at 22–30°C.

Pathogenic; causes rhinitis and conjunctivitis in desert and gopher tortoises. Mode of transmission appears to be intranasal inhalation (Brown et al., 2004).

Source: isolated from the nares of desert tortoises (*Gopherus agassizii*) and gopher tortoises (*Gopherus polyphemus*).

DNA G+C content (mol%): not determined.

Type strain: BH29, ATCC 700618, MCCM 03231.

Sequence accession no. (16S rRNA gene): AY366210.

114. **Mycoplasma testudinis** Hill 1985, 491[VP]

tes.tu'di.nis. L. n. *testudo, -inis* tortoise; L. gen. n. *testudinis* of a tortoise.

Cells are pleomorphic, with many possessing a terminal organelle similar to those of *Mycoplasma gallisepticum* and *Mycoplasma amphoriforme* that periodically exhibits curvature (Hatchel et al., 2006). A subset of cells are motile and glide at high speed in the direction of the terminal structure. Colonies on solid medium have a typical fried-egg appearance. Grows well in SP-4 medium supplemented with glucose at 25–37°C, with optimum growth at 30°C.

No evidence of pathogenicity. Mode of transmission has not been established definitively.

Source: isolated from the cloaca of a Greek tortoise (Hill, 1985).

DNA G+C content (mol%): 35 (T_m).

Type strain: 01008, NCTC 11701, ATCC 43263.

Sequence accession no. (16S rRNA gene): U09788.

115. **Mycoplasma verecundum** Gourlay, Leach and Howard 1974, 483[AL]

ve.re'cun.dum. L. neut. adj. *verecundum* shy, unobtrusive, free from extravagance, alluding to the lack of obvious biochemical characteristics of the species.

Cells are highly pleomorphic, exhibiting coccoid bodies, ring forms, and branched filaments. Motility for this species has not been assessed. Growth on solid medium produces colonies with the typical fried-egg appearance. Grows well in SP-4 or Hayflick medium at 37°C.

Probable commensal; attempts to produce disease experimentally have been unsuccessful (Gourlay and Howard, 1979). Mode of transmission has not been established.

Source: isolated from the eyes of calves with conjunctivitis, the prepuce of clinically normal bulls, and from in-market kale (Gourlay and Howard, 1979; Gourlay et al., 1974; Somerson et al., 1982).

DNA G+C content (mol%): 27 (T_m).
Type strain: 107, ATCC 27862, NCTC 10145.
Sequence accession no. (16S rRNA gene): AF412989.

116. **Mycoplasma wenyonii** (Adler and Ellenbogen 1934) Neimark, Johansson, Rikihisa and Tully 2002b, 683[VP] (*Eperythrozoon wenyonii* Adler and Ellenbogen 1934, 220)

we.ny.o'ni.i. N.L. masc. gen. n. *wenyonii* of Wenyon, named after Charles Morley Wenyon (1878–1948), an investigator of these organisms.

Cells are coccoid. Motility for this species has not been assessed. This species has not been grown on any artificial medium; therefore, notable biochemical parameters are not known.

Pathogenic; causes anemia and subsequent lameness and/or infertility in cattle. Transmission is primarily vector-mediated by *Dermacentor andersoni* and reportedly can also occur vertically during gestation. Oxytetracycline is an effective therapeutic agent (Montes et al., 1994). *Mycoplasma wenyonii* is reported to share antigens with *Mycoplasma ovis* (Kreier and Ristic, 1963), potentially complicating serology-based diagnosis of infection.

Source: observed in association with the erythrocytes of cattle; Kreier and Ristic (1968) reported in addition to erythrocytes an association with platelets.

DNA G+C content (mol%): not determined.
Type strain: not established.
Sequence accession no. (16S rRNA gene): AF016546.

117. **Mycoplasma yeatsii** DaMassa, Tully, Rose, Pitcher, Leach and Cottew 1994, 483[VP]

ye.at'si.i. N.L. masc. gen. n. *yeatsii* of Yeats, named after F.R. Yeats, an Australian veterinarian who was a co-isolator of the organism.

Cells are coccoid and nonmotile. Colonies on agar have a fried-egg appearance. Grows well in SP-4 medium supplemented with glucose at 37°C. Formation of biofilms has been demonstrated (McAuliffe et al., 2006).

Opportunistic pathogen; commensal of the ear canal of goats that has rarely been found in association with mastitis and arthritis (DaMassa et al., 1991). The mode of transmission has not been established.

Source: isolated from the external ear canals, retropharyngeal lymph node, nasal cavity, udders, and milk of goats.

DNA G+C content (mol%): 26.6 (T_m).
Type strain: GIH, ATCC 51346, NCTC 11730, CIP 105675.
Sequence accession no. (16S rRNA gene): U67946.

Species *incertae sedis*

1. **Mycoplasma coccoides** (Schilling 1928) Neimark, Peters, Robinson and Stewart 2005, 1389[VP] (*Eperythrozoon coccoides* Schilling 1928, 1854)

coc.co'ides. N.L. masc. n. *coccus* (from Gr. masc. n. *kokkos* grain, seed) coccus; L. suff. *-oides* (from Gr. suff. *eides*, from Gr. n. *eidos* that which is seen, form, shape, figure), resembling, similar; N.L. neut. adj. *coccoides* coccus-shaped.

Cells are coccoid. Motility for this species has not been assessed. This species has not been grown on artificial medium; therefore, notable biochemical parameters are not known.

Pathogenic; causes anemia in wild and captive mice, and captive rats, hamsters, and rabbits. Transmission is believed to be vector-borne and mediated by the rat louse *Polyplex spinulosa* and the mouse louse *Polyplex serrata*.

Neoarsphenamine and oxophenarsine were thought to be effective chemotherapeutic agents for treatment of *Mycoplasma coccoides* infection in captive rodents, whereas tetracyclines are effective only at keeping infection at subclinical levels (Thurston, 1953).

Source: observed in association with the erythrocytes of wild and captive rodents.

DNA G+C content (mol%): not determined.
Type strain: not established.
Sequence accession no. (16S rRNA gene): AY171918.

Species *Candidatus*

1. **"*Candidatus* Mycoplasma haematoparvum"** Sykes, Ball, Bailiff and Fry 2005, 29

ha.e.ma.to.par'vum. Gr. neut. n. *haema, -atos* blood; L. neut. adj. *parvum* small; N.L. neut. adj. *haemoatoparvum* small (mycoplasma) from blood.

Source: blood of infected canines (Sykes et al., 2005).
Host habitat: circulation of infected canines.
Phylogeny: assignment to the hemoplasma cluster of the pneumoniae group of *Mollicutes* (Foley and Pedersen, 2001).
Cell morphology: wall-less; coccoid in shape.
Optimum growth temperature: not applicable.
Cultivation status: non-culturable.
Sequence accession no. (16S rRNA gene): AY854037.

2. **"*Candidatus* Mycoplasma haemobos"** Tagawa, Matsumoto and Inokuma 2008, 179

ha.e.mo'bos. Gr. neut. n. *haema* blood; L. n. *bos* an ox, a bull, a cow; N.L. n. *haemobos* (*sic*) intended to mean of cattle blood.

Source: blood of infected cattle.
Host habitat: blood of cattle.
Phylogeny: assignment to the hemoplasma cluster of the pneumoniae group of the genus *Mycoplasma*.
Cultivation status: non-culturable.
Cell morphology: wall-less; coccoid in shape.
Optimum growth temperature: not applicable.
Sequence accession no. (16S rRNA gene): EF460765.
Further comment: this organism is synonymous with "*Candidatus* Mycoplasma haemobovis" (sequence accession no. EF616468).

3. **"*Candidatus* Mycoplasma haemodidelphidis"** Messick, Walker, Raphael, Berent and Shi 2002, 697

ha.e.mo.di.del'phi.dis. Gr. neut n. *haema* blood; N.L. fem. gen. n. *didelphidis* of the opossum; N.L. gen. n. *haemodidelphidis* of opossum blood.

Source: blood of an infected opossum.
Host habitat: circulation of an infected opossum.
Phylogeny: assignment to the hemoplasma cluster of the pneumoniae group of mollicutes (Messick et al., 2002).
Cultivation status: non-culturable.
Cell morphology: wall-less; coccoid in shape.
Optimum growth temperature: not applicable.
Sequence accession no. (16S rRNA gene): AF178676.

4. **"*Candidatus* Mycoplasma haemolamae"** Messick, Walker, Raphael, Berent and Shi 2002, 697

ha.e.mo.la'ma.e. Gr. neut n. *haema* blood; N.L. gen. n. *lamae* of the alpaca; N.L. fem. gen. n. *haemolamae* of alpaca blood.

Source: blood of infected llamas.
Host habitat: circulation of infected llamas (McLaughlin et al., 1991).
Phylogeny: assignment to the hemoplasma cluster of the pneumoniae group of mollicutes (Messick et al., 2002).
Cultivation status: non-culturable.
Cell morphology: wall-less; coccoid in shape.
Optimum growth temperature: not applicable.
Sequence accession no. (16S rRNA gene): AF306346.

5. **"*Candidatus* Mycoplasma haemominutum"** Foley and Pedersen 2001, 817

ha.e.mo.mi'nu.tum. Gr. neut n. *haema* blood; L. neut. part. adj. *minutum* small in size; N.L. neut. adj. *haemominutum* small (mycoplasma) from blood.

Source: blood of infected felines (George et al., 2002; Tasker et al., 2003).
Host habitat: circulation of infected felines.
Phylogeny: assignment to the hemoplasma cluster of the pneumoniae group of mollicutes (Foley and Pedersen, 2001).
Cultivation status: non-culturable.
Cell morphology: wall-less; coccoid in shape; 300–600 nm in diameter.
Optimum growth temperature: not applicable.
Sequence accession no. (16S rRNA gene): U88564.

6. **"*Candidatus* Mycoplasma kahaneii"** Neimark, Barnaud, Gounon, Michel and Contamin 2002a, 697

ka.ha.ne'i.i. N.L. masc. gen. n. *kahaneii* of Kahane, named for I. Kahane.

Source: blood of infected monkeys (*Saimiri sciureus*) (Michel et al., 2000).
Host habitat: circulation of infected monkeys.
Phylogeny: assignment to the hemoplasma cluster of the pneumoniae group of mycoplasmas (Neimark et al., 2002a).
Cultivation status: non-culturable.
Cell morphology: wall-less; coccoid in shape.
Optimum growth temperature: not applicable.
Sequence accession no. (16S rRNA gene): AF338269.

7. **"*Candidatus* Mycoplasma ravipulmonis"** Neimark, Mitchelmore and Leach 1998, 393

ra.vi.pul.mo'nis. L. adj. *ravus* grayish; L. n. *pulmo, -onis* the lung; N.L. gen. n. *ravipulmonis* of a gray lung.

Source: lung tissue of mouse with respiratory infection (gray lung disease).
Host habitat: respiratory tissue of mice with pneumonia.
Phylogeny: forms a single species line in the hominis group of mollicutes (Neimark et al., 1998; Pettersson et al., 2000).
Cultivation status: non-culturable.
Cell morphology: wall-less; coccoid in shape; 650 nm in diameter.
Optimum growth temperature: not applicable.
Sequence accession no. (16S rRNA gene): AF001173.

8. **"*Candidatus* Mycoplasma turicensis"** Willi, Boretti, Baumgartner, Tasker, Wenger, Cattori, Meli, Reusch, Lutz and Hofmann-Lehmann 2006, 4430

tu.ri.cen'sis. L. masc. (*sic*) adj. *turicensis* pertaining to Turicum, the Latin name of Zurich, the site of the organism's initial detection.

Source: blood of infected domestic cats.
Host habitat: blood of domestic cats.
Phylogeny: assignment to the hemoplasma cluster of the pneumoniae group of the genus *Mycoplasma*.
Cultivation status: non-culturable.
Cell morphology: wall-less; coccoid in shape.
Optimum growth temperature: not applicable.
Sequence accession no. (16S rRNA gene): DQ157150.

The following proposed species has been incidentally cited, but the putative organism remains to be established definitively and the name has no standing in nomenclature.

1. **"*Candidatus* Mycoplasma haemotarandirangiferis"** Stoffregen, Alt, Palmer, Olsen, Waters and Stasko 2006, 254

ha.e.mo.ta.ran.di.ran.gi'fe.ris. Gr. neut n. *haema* blood; *Rangifer tarandus* scientific name of the reindeer; N.L. gen. n. *haemotarandirangiferis* epithet intended to indicate occurrence in blood of reindeer.

Source: blood of reindeer.
Host habitat: blood of reindeer.
Phylogeny: partial 16S rRNA gene sequences suggest possible relationships to *Mycoplasma ovis* and *Mycoplasma wenyonii* (suis cluster), and/or to *Mycoplasma haemofelis* (haemofelis cluster).
Cultivation status: non-culturable.
Cell morphology: single punctate, chaining punctate, clustering punctate, single bacillary, chaining bacillary, single rings, chaining rings, and clustering rings.
Optimum growth temperature: not applicable.
Sequence accession nos (16S rRNA gene): DQ524812–DQ524818.
Further comment: sequence accession no. DQ524819, representing clone 107LSIA (Stoffregen et al., 2006), does not support assignment of that clone to the genus *Mycoplasma* because it is most similar to 16S rRNA genes of *Fusobacterium* spp. Several different partial 16S rRNA gene sequences obtained from other clones are too variable to establish their coherence as a species.

Other organisms

1. **"Mycoplasma insons"** May, Ortiz, Wendland, Rotstein, Relich, Balish and Brown 2007, 298

   in'sons. L. neut. adj. *insons* guiltless, innocent.

   Source: trachea and choanae of a healthy green iguana (*Iguana iguana*).

   Host habitat: respiratory tract and blood of green iguanas (*Iguana iguana*).

   Phylogeny: assignment to the *Mycoplasma fastidiosum* cluster of the pneumoniae group of the genus *Mycoplasma.*

   Cultivation status: cells are culturable in SP-4 medium supplemented with glucose.

   Cell morphology: pleomorphic, but many have a highly atypical shape for a mycoplasma, often resembling a twisted rod.

   Optimum growth temperature: 30°C.

   Sequence accession no. (16S rRNA gene): DQ522159.

2. **"Mycoplasma sphenisci"** Frasca, Weber, Urquhart, Liao, Gladd, Cecchini, Hudson, May, Gast, Gorton and Geary 2005, 2979

   sphe.nis'ci. N.L. gen. n. *sphenisci* of *Spheniscus,* the genus of penguin that includes the jackass penguin (*Spheniscus demersus*) from which this mycoplasma was isolated.

   Source: choanae of a jackass penguin (*Spheniscus demersus*) with choanal discharge and halitosis.

   Host habitat: upper respiratory tract of the jackass penguin.

   Phylogeny: assignment to the *Mycoplasma lipophilum* cluster of the hominis group of the genus *Mycoplasma.*

   Cultivation status: cells are culturable in Frey's medium supplemented with glucose.

   Cell morphology: pleomorphic; some cells exhibit terminal structures.

   Optimum growth temperature: 37°C.

   Sequence accession no. (16S rRNA gene): AY756171.

3. **"Mycoplasma vulturis"** corrig. Oaks, Donahoe, Rurangirwa, Rideout, Gilbert and Virani 2004, 5911

   vul.tu'ri.i. L. gen. n. *vulturis* of a vultures, named for the host animal (Oriental white-backed vulture).

   Source: lung and spleen tissue of an Oriental white-backed vulture.

   Host habitat: upper and lower respiratory tract of Oriental white-backed vulture, where it replicates intracellularly.

   Phylogeny: assignment to the *Mycoplasma neurolyticum* cluster of the hominis group of the genus *Mycoplasma.*

   Cultivation status: cells can be grown in co-culture with chicken embryo fibroblasts, but have not been grown in pure *in vitro* culture.

   Cell morphology: coccoid; cells display intracellular vacuoles and intracellular granules of electron-dense material.

   Optimum growth temperature: 37°C.

   Sequence accession no. (16S rRNA gene): AY191226.

4. **"Mycoplasma zalophi"** Haulena, Gulland, Lawrence, Fauquier, Jang, Aldridge, Spraker, Thomas, Brown, Wendland and Davidson 2006, 43

   za.lo'phi. N.L. gen. n. *zalophi* of *Zalophus,* the genus of sea lion that includes the California sea lion (*Zalophus californianus*) from which this mycoplasma was isolated.

   Source: subdermal abscesses of captive sea lions.

   Host habitat: subdermal and intramuscular abscesses, joints, lungs, and lymph nodes of captive sea lions.

   Phylogeny: assignment to the *Mycoplasma hominis* cluster of the hominis group of the genus *Mycoplasma.*

   Cultivation status: cells are culturable in SP-4 medium supplemented with glucose.

   Cell morphology: not yet described.

   Optimum growth temperature: 37°C.

   Sequence accession no. (16S rRNA gene): AF493543.

Genus II. **Ureaplasma** Shepard, Lunceford, Ford, Purcell, Taylor-Robinson, Razin and Black 1974, 167[AL]

Janet A. Robertson and David Taylor-Robinson

U.re.a.plas'ma. N.L. fem. n. *urea* urea; Gr. neut. n. *plasma* anything formed or moulded, image, figure; N.L. neut. n. *Ureaplasma* urea form.

Coccoid cells about 500 nm in diameter; may appear as coccobacillary forms in exponential growth phase; filaments are rare. Nonmotile. Facultative anaerobes. Form **exceptionally small colonies** on solid media that are described either as **tiny (T) "fried-egg" colonies** or as **"cauliflower head" colonies** having a lobed periphery. **Unusual pH required for growth (about 6.0–6.5)**. Optimal incubation temperature for examined species is 35–37°C. Chemo-organotrophic. Like *Mycoplasma,* species of *Ureaplasma* **lack oxygen-dependent, NADH oxidase activity**. Unlike *Mycoplasma,* species of *Ureaplasma* **lack hexokinase or arginine deiminase** activities but have a **unique and obligate requirement for urea** and produce **potent ureases** that hydrolyze urea to CO_2 and NH_3 **for energy generation and growth. Genome sizes** range from 760 to 1170 kbp (PFGE). **Commensals or opportunistic pathogens** in vertebrate hosts, primarily birds and mammals (mainly primates, ungulates, and carnivores).

DNA G+C content (mol%): 25–32 (Bd, T_m).

Type species: **Ureaplasma urealyticum** Shepard, Lunceford, Ford, Purcell, Taylor-Robinson, Razin and Black 1974, 167 emend. Robertson, Stemke, Davis, Harasawa, Thirkell, Kong, Shepard and Ford 2002, 593.

Further descriptive information

Although cellular diameters as small as 100 nm and as large as 1000 nm, and minimal reproductive units of about 330 nm in diameter have been reported (Taylor-Robinson and Gourlay, 1984), published thin sections of these organisms (trivial name, ureaplasmas) show diameters only as large as 450–500 nm. The exceptions are feline ureaplasmas with thin section diameters of up to 800 nm (Harasawa et al., 1990a). Morphometric analysis of cells of the type strain of *Ureaplasma urealyticum* fixed in the exponential phase of growth showed coccoid cells with diameters of about 500 nm (Robertson et al., 1983). Similar cellular diameters have been seen in hemadsorption studies in which cells were pre-fixed during incubation before usual fixation for electron microscopy. Reports of budding and filamentous forms probably reflect the effect of cultural and handling conditions on these highly plastic cells. Although the sequence of the *Ureaplasma parvum* genome lacks any recognizable *FtsZ* genes (Glass et al., 2000), ureaplasmas appear to reproduce by binary fission. Because of their minute size, ureaplasma cells are rarely seen by light microscopy. Although they are of the Gram-stain-positive lineage, the lack of cell wall results in the organisms appearing Gram-stain-negative. They are more easily detected if stained with crystal violet alone. Electron micrographs have indicated hair-like structures, possibly pili, 5–8 nm long, radiating from the membrane (Whitescarver and Furness, 1975). An extramembranous capsule was expected from light microscopic studies. In cytochemical studies, a carbohydrate-containing, capsular structure has been demonstrated in a strain of *Ureaplasma urealyticum* (Robertson and Smook, 1976; Figure 110). The structure is a lipoglycan that has also been demonstrated on the type strain T960^T of *Ureaplasma urealyticum* and on a serovar 3 strain of *Ureaplasma parvum*; the lipoglycan composition was strain-variable (Smith, 1986). No viruses have been seen nor has viral or plasmid nucleic acid been reported in any ureaplasma.

Ureaplasma colonies are significantly smaller in diameter (≤10–175 nm) than those of *Mycoplasma* species (300–800 nm; Figure 110). For this reason, they were first described as "tiny (T) form PPLO (pleuropneumonia-like organisms) colonies" (Shepard, 1954) and later called T-mycoplasmas (Meloni et al., 1980; Shepard et al., 1974). Cultures on solid media may grow in air, but more numerous and larger colonies result in 5–15% CO_2 in N_2 or H_2 (Robertson, 1982). Colonies of many strains are detectable after overnight incubation and reach maximum dimensions within 2 d. Isolates from ungulates may require longer incubation. Temperatures of 20–40°C are permissive for growth of examined strains, but their optimal incubation temperatures are 35–37°C (Black, 1973). Ureaplasma cultures in liquid media are incubated aerobically with growth occurring in the bottom of the tube, as revealed by changes in the pH indicator. Mean generation times of 10 isolates from humans ranged from 50 to 105 min (Furness, 1975). Maximum titers of $\leq 10^8$ organisms per ml of culture produce insufficient cell mass for detectable turbidity, precluding growth measurement by turbidometric or spectrophotometric methods. Growth is best measured by broth dilution methods (Ford, 1972; Rodwell and Whitcomb, 1983; Stemke and Robertson, 1982). When immediate estimation of populations is required, ATP luminometry (Stemler et al., 1987) may be useful. An indicator system enhances colony detection and ureaplasma identification. Ammonia from urea degradation causes a rise in pH and certain cations to form a golden to deep brown precipitate on the colorless colonies, making them visible when viewed by directly transmitted light. Initially, the urease spot test used a solution of urea and 1 mM Mn^{2+} (as $MnCl_2$ or $MnSO_4$) dropped onto agar (Shepard and Howard, 1970); later, Mn^{2+} was incorporated into the agar itself to create a differential solid medium (e.g., Shepard, 1983). However, Mn^{2+} is toxic for ureaplasmas (Robertson and Chen, 1984). Equimolar $CaCl_2$ (Shepard and Robertson, 1986) gave a similar response, but allowed the recovery of live cells. Manganese susceptibility has taxonomic value (Table 138); animal isolates show differing responses (Stemke et al., 1984; Stemler et al., 1987).

Ureaplasma urealyticum has some elements of the glycolytic cycle and pentose shunt (Cocks et al., 1985), but cannot degrade glucose and lacks arginine deiminase (Woodson et al., 1965). Instead, ureaplasmas have a unique and absolute requirement for urea (0.4–1.0 mM) and a slightly acidic environment (pH 6.0–7.0); pH values outside this range can be associated with growth inhibition. The essential cytoplasmic

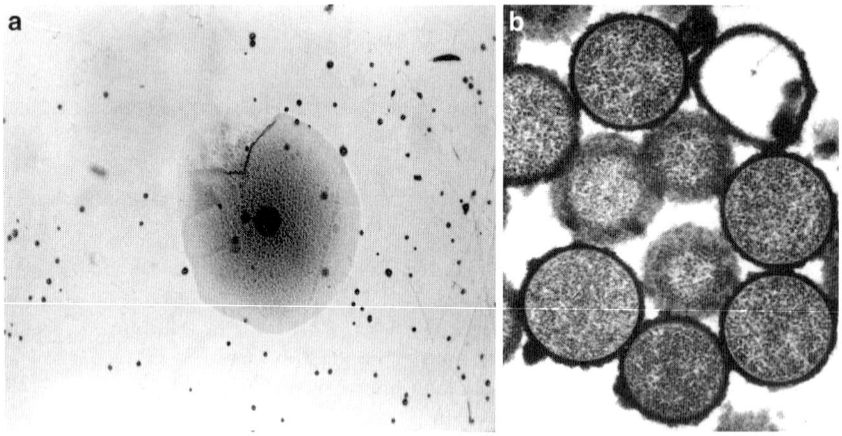

FIGURE 110. *Ureaplasma* colonial size and cellular morphology. (a) Many isolated *Ureaplasma urealyticum* colonies, accentuated by a urease spot test, surround a single large *Mycoplasma hominis* colony on a solid genital mycoplasma (GM) agar surface. Colonies of *Ureaplasma urealyticum* commonly have diameters of 15–125 µm; the diameter of the *Mycoplasma hominis* colony shown is approximately 0.9 mm. (Reproduced with permission from Robertson et al., 1983. Sexually Transmitted Diseases *10 (October-December Suppl.)*: 232–239 © Lippincott Williams & Wilkins.) (b) Transmission electron micrograph of *Ureaplasma urealyticum*, serovar 4, strain 381/74 cells, showing coccoid morphology, extramembranous capsule stained with ruthenium red, lack of a cell wall, the single limiting membrane, and apparently simple cytoplasmic contents in which only ribosomes are clearly evident. Cell diameters were 485–585 nm. (Reproduced with permission from Robertson and Smook, 1976. Journal of Bacteriology *128*: 658–660.)

TABLE 138. Phenotypic characteristics that partition the human serovar-standard strains of ureaplasmas to *Ureaplasma* species[a]

Characteristic	*U. urealyticum* serovar	*U. parvum* serovar	Reference
Isoelectric focusing and SDS-PAGE at pH 5.3 polypeptide patterns:			Sayed and Kenny (1980)
Absent		1, 3^T, 6	
Present	2, 4, 5, 7, 8^T		
1-D SDS-PAGE T960^T biovar band:[b]			Howard et al. (1981)
Absent		1, 3^T, 6	
Present	2, 4, 5, 7, 8^T		
2-D SDS-PAGE:[b]			
Biovar 1 pattern		1, 3^T, 6, 14	Mouches et al. (1981)
Biovar 2 pattern	2, 4, 5, 7, 8^T, 9, 11, 12		Swensen et al. (1983)
Growth inhibition by 1 mM Mn^{2+}:			Robertson and Chen (1984)
Temporary		1, 3^T, 6, 14	
Permanent	2, 4, 5, 7, 8^T, 9–12[c]		
Polypeptide recognized by immunoblots:			
51 and 58 kDa		1, 3^T, 6, 14	Horowitz et al. (1986)
47 kDa	2, 4, 5, 7, 8^T, 9–13		
Biovar 1 pattern		1, 3^T	Lee and Kenny (1987)
Biovar 2 pattern	2, 4, 8^T		
mAb UU8/39 recognition of membrane proteins:			Thirkell et al. (1989)
17 kDa only		1, 3^T, 6, 14	
16/17 kDa	2, 4, 5, 7, 8^T, 9–13		
mAb UU8/17 recognition of 72 kDa urease subunit:[d]			Thirkell et al. (1990)
No		1, 3^T, 6, 14	
Yes	8^T		
mAb VB10 recognition of 72 kDa urease subunit:			MacKenzie et al. (1996)
Yes		1, 3^T, 6, 14	
No	2, 4, 5, 7, 8^T, 9–13		
Urease dimorphism:			Davis et al. (1987)
Lower MW		1, 3^T, 6, 14	
Higher MW	2, 4, 5, 7, 8^T, 9–13		
Pyrophosphatase dimorphism:			Davis and Villanueva (1990)
Lower MW		1, 3^T, 6, 14	
Higher MW	2, 4, 5, 7, 8^T, 9–13		
Diaphorase bands dimorphism:			Davis and Villanueva (1990)
Inapparent		1, 3^T, 6, 14	
Apparent	2, 4, 5, 7, 8^T, 9–13		

[a]T, Type strain of species.

[b]1-D, One-dimensional; 2-D, two-dimensional.

[c]Serovar 13 gave an intermediate response and was excluded from the initial partition scheme.

[d]Hyperimmune rabbit sera and acute and convalescent sera from women with postpartum fever.

urease, comprising three subunits (Blanchard, 1990), produces a transmembrane potential, which leads to ATP synthesis (Romano et al., 1980; Smith et al., 1993). Ureaplasma dependence upon catalysis of urea for energy is the basis of growth inhibition by the urease inhibitor hydroxamic acid and its derivatives (Ford, 1972) and by fluorofamide (Kenny, 1983). The proton pump inhibitor lansoprazole and its metabolites interfere with ATP synthesis at micromolar concentrations (Nagata et al., 1995). Their small genomes make ureaplasmas dependent upon the host for amino acids, amino acid precursors, lipids, and other growth components. Ureaplasmas from humans exhibit minimal levels of acetate kinase activity (Muhlrad et al., 1981). Like other *Mycoplasmataceae*, ureaplasmas make superoxide dismutase (O'Brien et al., 1983), but, unlike the rest of this family, oxygen-dependent NADH oxidase has not been detected (Masover et al., 1977). The *Ureaplasma parvum* genome sequence includes genes for six hemin and/or Fe^{3+} transporters which are believed to be related to respiration (Glass et al., 2000). For information on additional physiological traits, see: Black (1973); Shepard et al., (1974); Shepard and Masover (1979); Taylor-Robinson and Gourlay (1984); and Pollack (1986).

Except for the obligatory requirements for supplementary urea and lower pH, ureaplasma growth requirements are similar to those of members of the genus *Mycoplasma*. The sterol requirement is met by horse, bovine, or fetal bovine serum. Because heat-sensitive pantothenic acid is a growth factor supplied by serum (Shepard and Lunceford, cited by Shepard and Masover, 1979), the serum supplement should not be "inactivated" by heating to reduce complement activity. The effect of yeast extract is variable, perhaps depending upon the particular strain requirements or the batch of extract. Other defined additives have been reported to enhance growth but, in the absence of a defined medium, evaluation is difficult. Dependence on sterols for the integrity of the cell membrane renders ureaplasmas susceptible to digitonin and certain antifungal agents, but most strains tolerate the polyene nystatin (50 U/ml) that prevents overgrowth by yeasts.

Information about ureaplasma genetics is now abundant. The genomes of 26 strains of *Ureaplasma*, including six of the seven named species and six unnamed strains, range in size from 760 to 1170 kbp (Kakulphimp et al., 1991; Robertson et al., 1990). The largest genome belongs to *Ureaplasma felinum.* The G+C content of ureaplasmal DNA is in the range 25.5–31.6 mol%, which is lower than that for other *Mollicutes* and for all other prokaryotes, greatly limiting the degeneracy in the genetic code of ureaplasmas. For the type strain of *Ureaplasma urealyticum*, UGA is a codon for tryptophan (Blanchard, 1990). The entire sequences of the genomes of two strains of *Ureaplasma parvum* serovar 3 (formerly known as *Ureaplasma urealyticum* biovar 1; Glass et al., 2000) have been determined. At 752 kbp, the organisms share with other obligate symbions, such as *Mycoplasma genitalium* (580 kbp) and *Buchnera aphidicola* (641 kbp), greatly reduced genomes. Of 641–653 total genes, 32–39 code for structural RNAs and about 610 code for proteins, about 47% of which have been classified as hypothetical genes of unknown function. Some (19%) resemble genes present in other genomes, but many (28%) appear unique. Unexpectedly absent in the *Ureaplasma parvum* genome are recognizable genes for FtsZ, for the GroEl and GroES chaperones, and for ribonucleoside-diphosphate reductase. Attempts have been made to reconcile the anomalies of gene functions assigned to this serovar of *Ureaplasma parvum* with the activities and pathways found in various ureaplasmas of humans (Glass et al., 2000; Pollack, 2001). One limitation to such analysis is that most of the physiological data that has been accumulated pertain not to *Ureaplasma parvum* but instead to the type species, *Ureaplasma urealyticum* strain T960^T, possessing a 7% larger genome. The entire sequences of the genomes of *Ureaplasma urealyticum* serovars 8 and 10 (Glass et al., 2000, 2008) have also been determined and the genomes of the type strains of all remaining *Ureaplasma urealyticum* and *Ureaplasma parvum* serovars are nearly completely annotated (Glass et al., 2008). Preliminary comparisons found that in addition to "core" and "dispensable" genomes for each species, *Ureaplasma urealyticum* had partly duplicated multiple-banded antigen (*mba*) genes (Kong et al., 1999a, 2000), and up to twice the number of genes that *Ureaplasma parvum* has for lipoproteins. Such comparisons are expected to substantially improve understanding of ureaplasmal pathogenicity and differential strain virulence.

Clinical studies based entirely on qualitative assessments of ureaplasmas are often difficult to interpret and only a few investigators have presented quantitative data (Bowie et al., 1977; De Francesco et al., 2009; Heggie et al., 2001; Taylor-Robinson et al., 1977). Factors such as hormonal levels, specific genetic attributes, and even socio-economic conditions may encourage urogenital colonization and proliferation. Nevertheless, it is clear that ureaplasmas are commensals that, on occasion, contribute to disease in susceptible human hosts. Infections attributed to ureaplasmas are often associated with an immunological component (Bowie et al., 1977), including the subset of the human population with common variable hypogammaglobulinemia (Cordtz and Jensen, 2006; Furr et al., 1994; Lehmer et al., 1991; Webster et al., 1978), in which ureaplasma-induced septic arthritis and occasionally persistent ureaplasmal urethritis is seen (Taylor-Robinson, 1985). Sexually acquired, reactive arthritis is usually linked to *Chlamydia trachomatis*, but immunological evidence of ureaplasmal involvement exists (Horowitz et al., 1994). Through urea metabolism, ureaplasmas can induce crystallization of struvite and calcium phosphates in urine *in vitro* and produce urinary calculi in animal models (Reyes et al., 2009). They are found in patients with infection stones more often than in those with metabolic stones (Grenabo et al., 1988). A statistical association with infection stones has been made (Kaya et al., 2003).

Evidence for the association of ureaplasmas with acute nongonococcal urethritis in men has been controversial, but a significant association of *Ureaplasma urealyticum* (but not *Ureaplasma parvum*) with this disease in two of three recent studies suggests a way forward to resolving this issue. Evidence for a role for ureaplasmas in acute epididymitis is, at the most, meager and it is very unlikely that they have a role in chronic prostatitis. Ureaplasmas have been associated with bacterial vaginosis and pelvic inflammatory disease, but are unlikely to be causal in either condition. It follows, therefore, that there is no convincing evidence to implicate ureaplasmas as an important cause of infertility in men or in couples. The most convincing data relating ureaplasmas to poor pregnancy outcomes have been seen when the organisms have been detected in amniotic fluid prior to membrane rupture, but there are conflicting opinions about the role of *Ureaplasma urealyticum* vs *Ureaplasma parvum.* Data since the 1980s have supported the association of neonatal ureaplasma infection with chronic lung disease and sometimes death in very low birth-weight infants. The ability of *Ureaplasma parvum* to induce chorioamnionitis and to contribute to preterm labor and fetal lung injury is supported by experimental studies in rhesus monkeys (Novy et al., 2009).

Information on the range of animal species infected with ureaplasmas and their geographic distribution is patchy, possibly because ureaplasmas are largely avirulent or, at least, not an economic threat. Isolations have been reported from squirrel, talapoin, patas, macaque and green monkeys; as well as marmosets and chimpanzees (Taylor-Robison and Gourlay, 1984). Also, there are reports of isolation from domestic dogs, raccoon dogs, cats and mink; cattle, sheep, goats, and camels; chickens and other fowl; and swine (the latter needing confirmation). Baboons, rats and mice are susceptible to experimental infection with ureaplasmas. While *Ureaplasma diversum* can inhabit all mucosal membranes of cattle, the two natural diseases it causes are subclinical respiratory infections in young calves, which occasionally develop into bronchopneumonia, and the economically important urogenital infections transmitted by bulls or their semen. The latter present as vulvovaginitis or ascend to cause infertility or abortion (Ruhnke et al., 1984; ter Laak et al., 1993). A species-specific PCR assay (Vasconcellos Cardosa et al., 2000) can circumvent culture insensitivity. As in humans, ureaplasmal diseases in animals may be influenced by the particular ureaplasma strain or many other factors. Although ureaplasmas may be present initially in certain organ and primary cell cultures, the lack of urea and higher pH in most eukaryotic cell culture systems would discourage persistence.

Although the spectrum of diseases of primary ureaplasmal etiology remains controversial, many potential virulence factors have been identified. Structural elements include a capsule, pilus-like fibrils, and the antigens of the outer membrane that constitute the serovar determinants. Erythrocytes from several animal species adhere firmly to colonies of certain strains of *Ureaplasma parvum* (Shepard and Masover, 1979), but most human isolates exhibit transient or no binding (Robertson and Sherburne, 1991). Strains of both *Ureaplasma urealyticum* and *Ureaplasma parvum* adhere to HeLa cells (Manchee and

Taylor-Robinson, 1969) and to spermatozoa (Knox et al., 2003). Demonstration of beta-hemolysis of erythrocytes by ureaplasmal products depends upon several variables; hemolysis of guinea pig erythrocytes has been most consistently observed (Black, 1973; Manchee and Taylor-Robinson, 1970; Shepard, 1967). Manchee and Taylor-Robinson (1970) described hemolysis of homologous erythrocytes by a canine ureaplasma as peroxide-associated and blocked lysis by adding catalase, except in the presence of a catalase inhibitor. Genome sequencing has revealed genes resembling those for the hemolysins HlyA and HlyC of enterohemorrhagic *Escherichia coli* (Glass et al., 2000). Ammonia and ammonium ions generated by urea hydrolysis and the alkaline environment that they create are inhibitory to the ureaplasmas (Ford and MacDonald, 1967; Shepard and Lunceford, 1967). They have well-established deleterious effects on eukaryotic cell and mammalian tissue cultures and may be the "toxin" described but never substantiated (Furness, 1973). An IgA1 protease activity has been demonstrated by ureaplasmas from both humans and canines. Human IgA1-specific protease activity, apparently similar to the type 2 serine protease of certain bacterial pathogens of humans (Spooner et al., 1992), is produced by both *Ureaplasma parvum* and *Ureaplasma urealyticum* (Kilian et al., 1984; Robertson et al., 1984), but a gene responsible for the activity has not yet been identified. Putative phopholipase A_1, A_2, and C activities (De Silva and Quinn, 1986) could not be confirmed, nor could such gene sequences be identified (J. Glass, unpublished). Biofilm production has been described recently (García-Castillo et al., 2008). Ureaplasmas have been seen intracellularly during studies in cell cultures (Mazzali and Taylor-Robinson, 1971), but may be there transiently after phagocytosis. Cells in culture (Li et al., 2000), in either experimental infection models (Moss et al., 2008) or epidemiological study subjects (Buss et al., 2003; Dammann et al., 2003; Jacobsson et al., 2003; Shobokshi and Shaarawy, 2002), have exhibited proinflammatory responses. While assessment of clinical studies is sometimes difficult because of inherent reporting bias (Klassen et al., 2002; Schelonka et al., 2005), it is anticipated that the application of bioarray technologies will lead to improved understanding of ureaplasmal mechanisms of pathogenesis.

Reviews of *in vitro* methodologies used for antimicrobial susceptibility testing for both human (Bébéar and Robertson, 1996; Waites et al., 2001) and animal (Hannan, 2000) isolates are available. Antimicrobial susceptibility patterns of *Ureaplasma urealyticum* and *Ureaplasma parvum* appear to be similar (Matlow et al., 1998). The choice of therapeutic agents active against them is limited. Tetracyclines or the macrolides (excluding lincomycin and clindamycin) are the bacteriostatic agents usually employed. Ureaplasmas from human, simian, bovine, caprine, feline, and avian sources (Koshimizu et al., 1983) withstand relatively high levels of lincomycin (10 μg/ml), to which most mycoplasmas are susceptible. Ureaplasmas also show *in vitro* resistance to rifampin and sulfonamides. Clinical resistance of ureaplasmas to tetracyclines has been long known (Ford and Smith, 1974). This high-level resistance is determined by the presence of the *tetM* determinant (Roberts and Kenny, 1986), which is now readily identified by PCR (Blanchard et al., 1997). Tetracycline-resistant strains exposed to a variety of antibiotics demonstrate a broad range of responses (Robertson et al., 1988). Clinical resistance to macrolides has also been reported (e.g., Taylor-Robinson and Furr, 1986). Certain aminoglycosides, chloramphenicol, and newer fluoroquinolones inhibit ureaplasmas, but are inappropriate for broad clinical use. Fluoroquinolone resistance is increasing (Xie and Zhang, 2006) and has been found in a previously susceptible strain (Duffy et al., 2006). Ureaplasma strains exhibiting resistance to multiple antibiotics have been found in immunocompromised hosts. In one case, the isolates were of the same serovar, but exhibited different susceptibility patterns at different anatomical sites (Lehmer et al., 1991). Because of ongoing changes in antimicrobial susceptibility patterns, the recent literature should be consulted (e.g., Beeton et al., 2009). To treat natural *Ureaplasma diversum* infections, tiamulin hydrogen fumarate, a diterpene agent in common use in veterinary medicine, may be at least as effective as the macrolide tylosin (Stipkovits et al., 1984). Others have examined the efficacy of single or combinations of antibiotics in eradicating ureaplasmas from various sites in cattle as well as from semen used for artificial insemination (ter Laak et al., 1993). The urease inhibitor, fluorofamide, has been used with varying success in eliminating ureaplasmas from animals.

When establishing antibiograms for ureaplasmas, the requirements established for common bacterial pathogens do not suffice. First, medium components may affect antibiotic activity. For instance, serum-binding reduces tetracycline activity, while the initial acidic culture reduces macrolide activity against ureaplasmas (Robertson et al., 1981). Conventional bacteria of established, low-level susceptibility can be used to measure the effect of the ureaplasmal medium on a particular antimicrobial agent. Second, the relatively slow growth of ureaplasmas as compared with many pathogenic bacteria requires that the half-life of the antibiotic be considered when test inoculum and incubation period are established. Lastly, the end points of the sensitivity tests on agar are about four-fold lower than for tests in broth (Waites et al., 1991). On consideration of the *in vivo* environment in which these organisms naturally occur, interpretation of susceptibility test end points continues to present a challenge. An international subcommittee under the aegis of the National Committee for Clinical Laboratory Standards Institute (USA) is currently finalizing a "Final Report for Development of Quality Control Reference Standards and Methods for Antimicrobial Susceptibility Testing" for *Ureaplasma* and *Mycoplasma* species infecting humans that have demonstrated variability in response to antimicrobial agents.

Enrichment and isolation procedures

Media formulations in current use for the cultivation of ureaplasmas from human sources include, in order of decreasing supplementation: the 10B broth of Shepard and Lunceford (Shepard, 1983); Taylor-Robinson's broth (Taylor-Robinson, 1983a), made without thallium acetate; and bromothymol blue broth (Robertson, 1978). Ureaplasmas are exquisitely sensitive to thallium acetate and it should not be used to inhibit other bacteria. The U4 formulation for ungulate isolates was developed by Howard et al. (1978). Consult Waites et al. (1991) for additional media formulations for isolation from humans, and Shepard (1983), Livingston and Gauer (1974), or Hannan (2000) for isolation from animals. The quality of medium components may be as important as the medium formulation. Serum supplements should be tested for their ability to support growth of the ureaplasmas of interest. Liquid medium may be stored at −20°C until required. Transport medium (e.g., 2SP) should be free of antibiotics inhibitory to ureaplasmas. Broth cultures should be diluted to ≥1:100 to reduce effects of any growth inhibitors present in the specimen. In general, *Ureaplasma parvum* is less demanding nutritionally than *Ureaplasma urealyticum* and isolates from human and other animal sources together are less demanding than those from ungulate hosts. Agar surfaces are examined at ≥40× magnification. The less fastidious strains (e.g., *Ureaplasma parvum* type strain) are more likely to produce recognizable "fried-egg" colonies than are the more fastidious strains (e.g., *Ureaplasma urealyticum* type strain), which are more likely to produce "cauliflower head" or core colonies. Cultures may be thrice cloned and the resultant culture used to initiate stocks.

Maintenance procedures

Broth cultures of ureaplasmas commonly become sterile within 12–24 h incubation at 35–37°C. "Red is dead" was the mantra of Shepard in regard to his phenol red-containing broth medium. One way to lessen this problem is to change to an indicator that changes color at a lower pH (Robertson, 1978). To Shepard's mantra we might add, "the higher the urea concentration, the steeper the death phase", an effect not countered by buffer. Incubation of diluted cultures at 30–34°C slows growth and maintains viability for up to 1 week, reducing the frequency of culture transfer and helping in transport. Short exposure of broth cultures to refrigeration reduces viability. Cells within colonies on solid medium may be recovered for approximately 1 week if the cultures are removed from incubation when growth is detectable and stored under cool, humidified conditions. For long-term storage, ultra-low temperatures (–80°C or liquid nitrogen) with a cryoprotectant [e.g., 10–20% (v/v) sterile glycerol] can maintain viability for well over a decade. For lyophilization, the pellets from broth cultures centrifuged at high speed are resuspended in a minimal volume of liquid medium or the serum used for its supplementation before processing; on reconstitution, urease activity occasionally is not immediately evident.

Differentiation of the genus *Ureaplasma* from other genera

Properties that partially fulfill criteria for assignment to the class *Mollicutes* (Brown et al., 2007) include absence of a cell wall, filterability, and the presence of conserved 16S rRNA gene sequences. Aerobic or facultatively anaerobic growth in artificial media and the necessity for sterols for growth exclude assignment to the genera *Anaeroplasma*, *Asteroleplasma*, *Acholeplasma*, and "*Candidatus* Phytoplasma". Non-spiral cellular morphology and regular association with a vertebrate host or fluids of vertebrate origin support exclusion from the genera *Spiroplasma*, *Entomoplasma*, and *Mesoplasma*. The ability to hydrolyze urea, with the inability to metabolize either glucose or arginine, excludes assignment to the genus *Mycoplasma*. For routine purposes, the colonial morphology characteristic of ureaplasmas and demonstration of the isolate's ability to catabolize urea, using the urease spot test or indicator agar, suffice for preliminary differentiation from other mollicutes. Other methods for urease detection have been largely replaced by PCRs that detect urease genes.

Taxonomic comments

The hypothetical evolutionary relationships of the *Mollicutes* have been based primarily upon 16S rRNA gene sequences (Weisburg et al., 1989). The nucleotide sequence for the 16S rRNA genes (Kong et al., 1999b; Robertson et al., 1993; Robertson et al., 1994) and the 16S–23S rRNA intergenic spacer regions of all named species plus the 14 serovars associated with humans (*Ureaplasma urealyticum* and *Ureaplasma parvum*) have been determined (Harasawa and Kanamoto, 1999; Kong et al., 1999b). The genus *Ureaplasm*a comprises two subclusters within the highly diverse pneumoniae group of the family *Mycoplasmataceae* (Johansson, 2002). One subcluster contains the human, avian, and mink isolates and the other contains feline, canine, and bovine strains (see genus *Mycoplasma* Figure 109, pneumonia group) Although the three serovars A, B, and C of *Ureaplasma diversum* are antigenically heterogeneous, the strains examined meet the 70% DNA–DNA hybridization benchmark used as an arbitrary species criterion. However, the available DNA–DNA hybridization values suggest that *Ureaplasma gallorale* and *Ureaplasma canigenitalium* might each represent more than a single species. The 16S rRNA gene sequences of ureaplasmas isolated from nonhuman primates and some less-studied vertebrates are unknown. The range of G+C contents of ureaplasmal DNA is too narrow to have much taxonomic utility. Nevertheless, the values for isolates from cattle, sheep, and goats are between 28.7 and 31.6 mol% (Howard et al., 1978), at the higher end of the range for the genus.

It is generally assumed that all ureaplasmas from avian sources belong to the species *Ureaplasma gallorale.* The seven avian isolates examined were similar to each other serologically and in SDS-PAGE, immunoblot, and RFLP profiles, and they were distinct from *Ureaplasma urealyticum* and *Ureaplasma diversum*, although resemblances between *Ureaplasma gallorale* and *Ureaplasma urealyticum* based upon one- and two-dimensional PAGE analyses have been reported (Mouches et al., 1981). However, the DNA–DNA hybridization values among *Ureaplasma gallorale* strains fall into two clusters. Homology was 70–100% for the five strains within cluster A and 96–100% for the two strains within cluster B that were studied. Between strains of cluster A and B, homology was only 51–59% and, *vice versa*, 52–69% (Harasawa et al., 1985). The taxonomic status of avian ureaplasmas, therefore, might benefit from examination of additional strains and reconsideration. The clusters appear to be more closely related than *Ureaplasma urealyticum* and *Ureaplasma parvum* (Table 139).

The taxonomy of ureaplasmas of canines is also problematic. The first report indicated G+C contents of 27.2–27.8 mol% (Bd; Howard et al., 1978). Later, the representatives of four serogroups (SI to SIV for strains DIM-C, D29M, D11N-A, and D6P-C^T, respectively) were reported to have G+C contents of 28.3–29.4 mol% (HPLC). However, the published data regarding their DNA reassociation values are confusing. Initially, Barile (1986) and Harasawa et al. (1990b) reported that the serogroup SI representative, DIM-C, had 73% homology with the serotype 2 SII representative, D29M, indicating that these two strains likely belong to the same species. However, perhaps because of the borderline value, further work was undertaken, also using [^{3}H]DNA–DNA hybridization procedures (Harasawa et al., 1993). The values obtained ranged from 41 to 63% homology among the four serogroups, i.e., each serogroup seemed to represent a distinct species. At present, the only named canine ureaplasma species is *Ureaplasma canigenitalium*; the SIV representative D6P-C^T is the type strain (Harasawa et al., 1993). Its degree of distinctiveness from the other serogroups is not emphasized in the literature.

TABLE 139. Genotypic characteristics that partition the serovar-standard strains of ureaplasmas isolated from humans to the level of species[a]

Characteristic	*U. urealyticum* serovars	*U. parvum* serovars	Reference(s)
DNA–DNA relatedness with strain 27^T DNA:			Christiansen et al. (1981)
91–102%		1, 3^T, 6	
38–60%	2, 4, 5, 7, 8^T		
DNA–DNA relatedness with strain T960^T DNA:			Christiansen et al. (1981)
49–52%		1, 3^T, 6	
69–100%	2, 4, 5, 7, 8^T		
DNA–DNA relatedness with strain 27^T DNA:			Harasawa et al. (1991)
75–100%		1, 3^T, 6, 14	
38–57%	2, 4, 5, 7, 8^T, 9–13		
DNA–DNA relatedness with strain T960^T DNA:			Harasawa et al. (1991)
48–59%		1, 3^T, 6, 14	
76–101%	2, 4, 5, 7, 8^T, 9–13		
Cleavage of DNA by Fnu4HI:			Cocks and Finch (1987)
Yes		1, 3^T, 6	
No	2, 4, 5, 7, 8^T, 9		
BamHI, HindIII, and PstI RFLP probed with RNA genes:			Razin (1983)
Biovar 1 pattern		1, 3^T, 6	
Biovar 2 pattern	2, 4, 5, 7, 8^T, 9		
EcoRI and HindIII RFLP probed with RNA genes:			Harasawa et al. (1991)
Biovar 1 pattern		1, 3^T, 6, 14	
Biovar 2 pattern	2, 4, 5, 7, 8^T, 9–13[b]		
Genome sizes determined by PFGE:			Robertson et al. (1990)
~760 kbp		1, 3^T, 6, 14	
840–1140 kbp	2, 4, 5, 7, 8^T, 9–13		
Heterogeneity of alpha polypeptide-associated urease genes:			Blanchard (1990)
Yes		1, 3^T, 6, 10[c], 12[c], 14	
No	2, 4, 5, 7, 8^T, 9, 13		
*Heterogeneity of Hind*III *site in subunit ureC of urease gene:*			Neyrolles et al. (1996)
Absent		1, 3^T, 6	
Present	2, 8^T		
*Heterogeneity of Hind*III *fragments probed with serovar 8IC61 urease probe:*			Neyrolles et al. (1996)
Biovar 1 pattern		1, 3^T, 6	
Biovar 2 pattern	2, 8^T		
Heterogeneity of urease subunit-associated genes:[d]			Kong et al. (1999b)
Biovar 1 pattern		1, 3^T, 6, 14	
Biovar 2 pattern	2, 4, 5, 7, 8^T, 9–13		
Heterogeneity of biovar-specific 16S rRNA genes determined by PCR:[d]			Robertson et al. (1993)
Strain 27^T		1, 3^T, 6, 14	
Strain T960^T	2, 4, 5, 7, 8^T, 9–13		
Biovar-specific 16S rRNA gene sequences:[d]			Robertson et al. (1994)
Strain 27^T sequence		1, 3^T, 6, 14	
Strain T960^T sequence	2, 5, 8^T		
16S rRNA, spacer regions, and urease subunit sequences:[d]			Kong et al. (1999b)
Biovar 1 pattern		1, 3^T, 6, 14	
Biovar 2 pattern	2, 4, 5, 7, 8^T, 9–13		
RFLP of 5′ region of mba genes:			Teng et al. (1994)
Biovar 1 pattern		1, 3^T, 6, 14	
Biovar 2 pattern	2, 4, 5, 7, 8^T, 9–13		
5′ end sequences of mba genes:[d,e]			Kong et al. (1999b)
Biovar 1 pattern		1, 3^T, 6, 14	
Biovar 2 pattern	2, 4, 5, 7, 8^T, 9–13		
16S–23S intergenic spacer region:[d]			Harasawa and Kanamoto (1999), Kong et al. (1999b)
Biovar 1 pattern		1, 3^T, 6, 14	
Biovar 2 pattern	2, 4, 5, 7, 8^T, 9–13		
Arbitrarily primed PCR:[d]			Grattard et al. (1995)
Biovar 1 pattern		1, 3^T, 6, 14	
Biovar 2 pattern	2, 4, 5, 7, 8^T, 9–13		

[a]T, Type strain of species.

[b]Serovar 13 response to *Eco*RI was anomalous.

[c]This anomalous pattern resulted from cultures being misidentified as serovars 10 and 12; the error was corrected by Teng et al. (1994). Kong et al. (1999b) have confirmed the efficacy of the Blanchard (1990) primers.

[d]Primers used for biovar-defining PCR(s).

[e]The multiple band (MB) antigens seen by PAGE are putative virulence markers. See text for details.

While DNA–DNA hybridization has been key to species level identification for the genus *Ureaplasma*, serology led to our knowledge of intra-species relationships (Robertson et al., 2002) and still figures importantly. Much interest has been focused specifically on the multiple-banded antigens (MBA) of the ureaplasmal cell surface. Watson et al. (1990), studying the type strain (serovar 3) of *Ureaplasma parvum*, identified the predominant ureaplasmal antigens recognized by the host during human infection. The MBA were first seen as unusual, laddered bands in immunoblots. These lipoproteins exhibited epitopes for both serovar specificity and cross-reactivity, and showed *in vitro* size variation (Teng et al., 1994; Zheng et al., 1995). Monecke et al. (2003) added evidence that the MBA are involved in a phase-switching process similar to that identified in several *Mycoplasma* species. The *mba* gene sequences have since been used to further define ureaplasma phylogeny (Knox et al., 1998; Kong et al., 1999b), as well as to characterize isolates (Knox et al., 1998; Knox and Timms, 1998; Kong et al., 1999a; Pitcher et al., 2001). The sequences of the 5′ ends of the *mba* genes of all 14 serovars have been defined (Kong et al., 1999a, 2000). Kong et al. (2000) found that for *Ureaplasma parvum*, a specific site on the gene determines serovar identity, whereas for *Ureaplasma urealyticum* at least the sequence of the 3′ end is required and, for certain other serovars, the entire sequence is involved.

The genes for the three urease subunits, *ureA*, *ureB*, and *ureC*, and adjoining regions from many isolates have been sequenced (Blanchard, 1990; Kong et al., 1999b; Neyrolles et al., 1996; Ruifu et al., 1997). Rocha and Blanchard (2002) applied bioinformatics to predict how certain gene products (e.g., the MBA, restriction and modification systems, transcription anti-termination elements, and GTP-binding proteins) might exhibit species specificity. PCRs based upon the 16S rRNA genes are the most highly conserved, followed by the 16S–23S intergenic region, the urease-associated genes, the *mba* genes, and the region upstream of them.

Acknowledgements

We thank J. Glass and collaborators at the J. Craig Venter Institute and the University of Alabama - Birmingham for the gift of the genome sequences of *Ureaplasma parvum* and *Ureaplasma urealyticum*, and J.G. Tully, K.E. Johansson, and C. Williams for their suggestions regarding the initial chapter.

Differentiation of the species of the genus *Ureaplasma*

The first paper on ureaplasma taxonomy stated that "An advantage of forming a new genus is that it confers freedom to classify new species within the genus without adhering to the principles formulated for the other genera. A numbered serovar of a *Ureaplasma* from humans is broadly equivalent to a named species within the genera *Mycoplasma* or *Acholeplasma*" (Shepard et al., 1974). The serological diversity within the type strain, *Ureaplasma urealyticum*, was regarded as no more than antigenic heterogeneity, possibly reflecting minor differences within a single epitope. To ensure that taxonomy would develop rationally, official recognition of *Ureaplasma* subspecies was avoided. Thirty-five years later, this taxonomic restraint can be appreciated.

When the first genus and species, *Ureaplasma urealyticum*, was named, it had eight known antigenic specificities (Shepard et al., 1974); 6 years later, the number had reached 14 (Robertson and Stemke, 1982) where, surprisingly, it has remained. It is surprising because the serotypes were isolated over a 20-year period representing antigens in Vancouver, BC, Canada (Ford, 1967), Camp Lejeune, NC (Shepard, 1954), and Boston, MA, USA (Lin et al., 1972), i.e., on the west and east coasts of North America. While putative untypable strains are occasionally encountered, after cloning, these have usually turned out to be serovar 3, probably the most likely to dominate in a mixed culture. However, additional serovars/genovars can be expected to emerge, especially from other parts of the world.

The named species were identified primarily by DNA–DNA hybridization and by serological tests. For taxonomic studies, the metabolism inhibition test (Purcell et al., 1966; Robertson and Stemke, 1979; Taylor-Robinson, 1983b) and either a direct or indirect immunofluorescence test (Black and Krogsgaard-Jensen, 1974; Piot, 1977; Stemke and Robertson, 1981) have been most useful. However, serological tests sometimes gave confusing cross-reactivity patterns and were not ideal for differentiating strains within a single species (Stemke and Robertson, 1985). In an attempt to circumvent problems of cross-reactions obtained with polyclonal antisera, monoclonal antibodies (mAbs) to all 14 serovars of human isolates were developed (e.g., Echahidi et al., 2001). The use of mAbs coincided with and has been largely overshadowed by the genomic revolution.

In the 1980s, the same pattern of partitioning of the serovars of human isolates was demonstrated by other phenotypic traits, traits primarily related to protein structures and functions (Table 138). When phenotypy failed to deliver a clear and convenient means of discrimination, genotypy did. The initial DNA–DNA relatedness studies of ureaplasmas from humans (Christiansen et al., 1981) confirmed the two, distinct clusters; however, the large cell biomass, special equipment, and (then) rare expertise required resulted in few strains being tested. New techniques, especially PCR, became more easily performed and less expensive so that more strains were examined and non-ambiguous results were obtained (Table 139). Supported by these strong data, the taxonomy of *Ureaplasma urealyticum* was emended and extended (Robertson et al., 2002). The ten antigenic specificities of the larger cluster (known as group 2 or as the T960 biovar) retained the *Ureaplasma urealyticum* designation with strain T960^{T} as the type strain. The remaining four antigenic specificities (known as group 1 or the parvo biovar) were renamed *Ureaplasma parvum* in recognition of that cluster's considerably smaller genome size; strain 27^{T}, the serovar 3 standard, was designated the type strain. Many PCR primers to identify species and strains of ureaplasmas have been published; commercial PCR-based kits are currently available for all named *Ureaplasma* species.

In summary, current ureaplasma taxonomy is based upon pragmatic, polyphasic criteria, i.e., a synthesis of phylogeny, phenotypy, and genotypy. For the specific requirements for taxonomic studies of *Mollicutes*, consult the most recent minimal standards document (Brown et al., 2007). Some serological testing is mandatory. Rabbit antisera to the 14 serovars of ureaplasmas of humans and certain animal species are currently available from Jerry K. Davis, Curator of the Mollicutes Collection, School of Veterinary Medicine, Purdue University, West Lafayette, IN, USA. Expertise in determining phenotypic traits specific to ureaplasmas may be accessible through collaboration with the appropriate working team of the International Research Programme for Comparative Mycoplasmology (IRPCM) of the International Organization for Mycoplasmology (IOM) at www.the-iom.org.

List of species of the genus *Ureaplasma*

1. **Ureaplasma urealyticum** Shepard, Lunceford, Ford, Purcell, Taylor-Robinson, Razin and Black 1974, 167[AL] emend. Robertson, Stemke, Davis, Harasawa, Thirkell, Kong, Shepard and Ford 2002, 593

u.re.a.ly'ti.cum. N.L. fem. n. *urea* urea; N.L. adj. *lyticus -a -um* (from Gr. adj. *lutikos -ê -on*) able to loosen, able to dissolve; N.L. neut. adj. *urealyticum* urea-dissolving or urea-digesting.

Cells are coccoid and approximately 500 nm in diameter. Coccobacillary forms are seen in exponential phase cultures. One strain has been shown to have a carbohydrate-containing capsule; it and others contain lipoglycans. Grows at temperatures between 20 and 40°C, grows better at 30–35°C and best at 36–37°C. Colonies are ≤20–50 µm in diameter with complete or partial fried-egg morphology.

Serologically distinct from all other named species in the genus, but serologically heterogeneous. Ten specific antigenic determinants are known: 2, 4, 5, and 7–13. Multiple-banded antigens are serovar-related and recognized by the host. Like *Ureaplasma parvum*, it has human IgA-specific protease activity that specifically cleaves human IgA1, but not human IgA2. Has distinctive PAGE and RFLP patterns. DNA is not restricted by endonuclease Uur9601. Genome size of the type strain is 890 kbp, whereas the sizes of the 10 known serovar standard strains range from 840 to 1140 kbp (PFGE). DNA reassociation values: within the species (serovars 2, 4, 5, and 7), 69–100%; with *Ureaplasma parvum* (serovars 1, 3, and 6), 49–52%. Opportunistic pathogen of humans; causes some cases of nongonococcal urethritis, infectious kidney stones, systemic infection in immunologically compromised hosts. Associated with a broad variety of urogenital infections for which causality remains to be established.

Source: primarily found in the genitourinary tract of female and male humans; occasionally in the oral cavity and rectum.

DNA G+C content (mol%): 25.5–27.8 (T_m; type strain) and 27.7–28.5 (Bd; serovars 2, 4, 5, and 7).

Type strain: T960, (CX8), ATCC 27618, NCTC 10177.

Sequence accession nos: M23935 and AF073450 (type strain 16S rRNA gene), AB028088 and AF059330 (type strain 16S–23S rRNA intergenic region). Complete and near-complete (>99%) genomes: serovar 2 strain ATCC 27814, NZ_ABFL00000000; serovar 4 strain ATCC 27816, NZ_AAYO00000000; serovar 5 strain ATCC 27817, NZ_AAZR00000000; serovar 7 strain ATCC 27819, NZ_AAYP00000000; serovar 8 strain ATCC 27618, NZ_AAYN00000000; serovar 9 strain ATCC 33175, NZ_AAYQ00000000; serovar 10 strain ATCC 33699, NC_011374; serovar 11 strain ATCC 33695, NZ_AAZS00000000; serovar 12 strain ATCC 33696, NZ_AAZT00000000; serovar 13 strain ATCC 33698, NZ_ABEV00000000.

2. **Ureaplasma canigenitalium** Harasawa, Imada, Kotani, Koshimizu and Barile 1993, 644[VP]

ca.ni.ge.ni.ta'li.um. L. n. *canis* dog; L. pl. n. *genitalia* the genitals; N.L. pl. gen. n. *canigenitalium* of canine genitals.

Cells are coccoid and about 500 nm in diameter; coccobacillary forms are seen. Colonies are ≤20–140 µm diameter with fried-egg morphology. Serogroup I strains represented by D6P-C[T] are serologically distinct from all other established species in the genus and from the other three serogroups of ureaplasmas isolated from dogs (represented by the strains DIM-C, D29M, and D11N-A). The species designation refers only to serogroup I strains, although strain D11N-A shows a one-way, serological cross-reaction with D6P-C[T]. It produces an IgA protease which specifically cleaves canine myeloma IgA, but not human or murine IgA. Genome size is 860 kbp (PFGE). DNA reassociation values: between D6P-C[T] and the other three canine strains (DIM-C, D29M, and D11N-A) are 41–63% versus 33% with *Ureaplasma urealyticum* (strain T960[T]).

Source: habitat is the prepuce, vagina, and oral and nasal cavities of canines.

DNA G+C content (mol%): 29.4 (HPLC).

Type strain: D6P-C, ATCC 51252, CIP 106087.

Sequence accession no. (16S rRNA gene): D78648 (type strain).

3. **Ureaplasma cati** Harasawa, Imada, Ito, Koshimizu, Cassell and Barile 1990a, 50[VP]

ca'ti. L. gen. n. *cati* of a cat.

Cells are coccoid and ≥675 nm diameter, exceeding the 450–550 nm range of most named *Ureaplasma* species. Coccobacillary forms are seen and occasionally filaments. Colonies are ≤15–140 µm in diameter with diffuse, granular appearance; some fried-egg colonies may appear after passaging. Distinct from other established species in the genus, including *Ureaplasma felinum*, antigenically and in PAGE (Harasawa et al., 1990a) and RFLP patterns (Harasawa et al., 1984). Genome size has not been determined. DNA reassociation values: 83–100% within feline serogroup SII strains (*Ureaplasma cati*) versus <10% with serogroup SI strains (*Ureaplasma felinum*).

Source: found in the oral cavity of healthy domestic cats (*Felis domestica*).

DNA G+C content (mol%): 27.9 (Bd), 28.1 (HPLC) for strain F2[T].

Type strain: F2, ATCC 49228, NCTC 11710, CIP 106088.

Sequence accession nos: D78649 (type strain 16S rRNA gene), D63685 (type strain 16S–23S rRNA intergenic spacer region).

4. **Ureaplasma diversum** Howard and Gourlay 1982, 450[VP]

di.ver'sum. L. neut. part. adj. *diversum* different, distinct, heterogeneous, referring to the difference in polypeptides and G+C content as compared to *Ureaplasma urealyticum* and to the heterogeneous antigenic structure of the species.

Cells are coccoid or coccobacillary and appear to be within the size range of other named *Ureaplasma* species although no measurements have been published. Colonies are ≤100–175 µm in diameter based on photomicrographs.

Serologically distinct from other named species but antigenically heterogeneous, comprising serogroups A, B, and C, and represented by strains A417[T], D48, and T44. These show three distinctive PAGE patterns (Howard and Gourlay, 1982), but only one RFLP pattern (Harasawa et al., 1984), and, based upon the latter criterion, were considered homogeneous.

Genome size range is 1100–1160 kbp for strains 95 TX, 1763, and 2065-B202 (PFGE). No DNA reassociation values are available.

Source: the type strain originated from a pneumonic calf lung.

DNA G+C content (mol%): 29.0 and 28.7–30.2 (Bd) for the type strain and 10 bovine isolates, respectively; thus, higher than and not overlapping the values for *Ureaplasma urealyticum* and *Ureaplasma parvum.*

Type strain: A417, ATCC 43321, NCTC 10182, CIP 106089.

Sequence accession nos: D78650 (type strain 16S rRNA gene), D63686 (type strain 16S–23S rRNA intergenic spacer region).

5. **Ureaplasma felinum** Harasawa, Imada, Ito, Koshimizu, Cassell and Barile 1990a, 50[VP]

fe.li'num. L. neut. adj. *felinum* of or belonging to a cat.

Coccoid cells of ≥800 nm diameter exceed the 450–500 nm range of most *Ureaplasma* species. Coccobacillary forms and occasional filaments are seen. Colonies are ≤15–140 μm diameter with diffuse, granular appearance; some fried-egg colonies may appear after passaging. Distinct antigenically and by PAGE and RFLP patterns (Harasawa et al., 1984) from other established species in the genus, including *Ureaplasma cati.* The genome size is 1170 kbp (PFGE). It is the largest of any ureaplasma strain examined. DNA reassociation values: 89–100% within feline serogroup SI strains (*Ureaplasma felinum*) versus <10% with serogroup II strains (*Ureaplasma cati*).

Source: found in the oral cavity of healthy domestic cats (*Felis domestica*).

DNA G+C content (mol%): 27.9 (HPLC).

Type strain: FT2-B, ATCC 49229, NCTC 11709, CIP 106090.

Sequence accession nos: D78651 (type strain 16S rRNA gene), D63687 (type strain 16S–23S rRNA intergenic spacer region).

6. **Ureaplasma gallorale** Koshimizu, Harasawa, Pan, Kotani, Ogata, Stephens and Barile 1987, 337[VP]

gal.lo.ra'le. L. n. *gallus* a barnyard fowl; L. n. *os, oris* the mouth; L. neut. suff. *-ale* suffix used with the sense of pertaining to; N.L. neut. adj. *gallorale* relating to the mouth of barnyard fowl.

Cells are coccoid and about 500 nm in diameter; coccobacillary forms are seen. Colonies are ≤15–60 μm in diameter with fried-egg morphology. Serologically distinct from all other established species in the genus. Isolates have similar SDS-PAGE, immunoblot, and RFLP patterns, but demonstrate some species heterogeneity based on reassociation values (cluster A strains D6-1^T and T9-1; cluster B strain Y8-1). Genome size is 760 kbp (PFGE). DNA reassociation values fall into two clusters: within cluster A (strains D6-1^T, D23, F2, F5, and T9-1), values are 70–100%; within cluster B (strains Y8-1 and Y4-2), values are 96–100%; between these clusters, values are 51–69%. Although these values are below expectations for a single species status, they exceed the 19–27% reassociation values with *Ureaplasma urealyticum* and *Ureaplasma diversum.*

Source: found only in oropharynx of healthy red jungle fowl (*Gallus gallus*) and chickens (*Gallus gallus* var. *domesticus*) kept as laboratory or zoo animals in Japan and in chickens and turkeys with pneumonia or airsacculitis in Hungary.

DNA G+C content (mol%): 27.6 (HPLC).

Type strain: D6-1, ATCC 43346, NCTC 11707.

Sequence accession nos: U62937 (type strain 16S rRNA gene), D63688 (type strain 16S–23S rRNA intergenic spacer region).

7. **Ureaplasma parvum** Robertson, Stemke, Davis, Harasawa, Thirkell, Kong, Shepard and Ford 2002, 593[VP]

par'vum. L. neut. adj. *parvum* small, referring to its significantly smaller genome sizes compared to *Ureaplasma urealyticum,* the other species from humans.

Cells are coccoid and about 500 nm in diameter. Coccobacillary forms are present in exponential phase cultures. Lipoglycans have been identified in a serovar 3 strain. Colonies are ≤20–140 μm in diameter with complete or partial fried-egg morphology.

Serologically distinct from all other named species in the genus, but serological heterogeneity is exhibited within the species. Four specific antigenic determinants are known: 1, 3, 6, and 14. Serovar specificities are related to the multiple-banded antigens recognized by the host. Like *Ureaplasma urealyticum, Ureaplasma parvum* has an IgA protease activity that specifically cleaves human IgA1, but not human IgA2. Distinctive PAGE and RFLP patterns. DNA restricted by endonuclease Uur9601. Genome size: 751,719 kbp for the type strain. Complete genomic sequence of the organism has been reported (Glass et al., 2000). DNA reassociation values: within the species (serovars 1, 3, and 6), 91–102%; with *Ureaplasma urealyticum* (serovars 2, 4, 5, 7–13), 38–60%.

Source: primary habitat is the genitourinary tract of female and male humans; occasionally found in the oral cavity and rectum. As yet, unclear whether an opportunistic pathogen in non-gonococcal urethritis, but likely to be so in systemic infection in immunologically compromised hosts. Associated with a broad variety of urogenital infections for which causality remains to be established.

DNA G+C content (mol%): 25.5 (from genome sequence) and 27.8–28.2 (T_m) for serovars 1 and 6.

Type strain: 27, ATCC 27815, NCTC 11736.

Sequence accession nos: L08642 and AF073456 (type strain 16S rRNA gene), AB028083 and AF059323 (type strain 16S–23S rRNA intergenic spacer region). Complete and near-complete (>99%) genomes: serovar 1 strain ATCC 27813, NZ_ABES00000000; serovar 3 strain ATCC 27815, NC_010503; serovar 3 strain ATCC 700970, NC_002162; serovar 6 strain ATCC 27818, NZ_AAZQ00000000; serovar 14 strain ATCC 33697, NZ_ABER00000000.

References

Adler, H.E., J. Fabricant, R. Yamamoto and J. Berg. 1958. Symposium on chronic respiratory diseases of poultry. I. Isolation and identification of pleuropneumonia-like organisms of avian origin. Am. J. Vet. Res. *19*: 440–447.

Adler, S. and V. Ellenbogen. 1934. A note on two new blood parasites of cattle: *Eperythrozoon* and *Bartonella.* J. Comp. Pathol. *47*: 220–221.

Ajufo, J. and K. Whithear. 1980. The surface layer of *Mycoplasma synoviae* as demonstrated by the negative staining technique. Res. Vet. Sci. *29*: 268–270.

Al-Ankari, A.R. and J.M. Bradbury. 1996. *Mycoplasma iowae*: a review. Avian Pathol. *25*: 205–229.

Al Masalma, M., F. Armougom, W. Scheld, H. Dufour, P. Roche, M. Drancourt and D. Raoult. 2009. The expansion of the microbiological spectrum of brain abscesses with use of multiple 16S ribosomal DNA sequencing. Clin. Infect. Dis. *48*: 1169–1178.

Alberti, A., P. Robino, B. Chessa, S. Rosati, M. Addis, P. Mercier, A. Mannelli, T. Cubeddu, M. Profiti, E. Bandino, R. Thiery and M. Pittau. 2008. Characterisation of *Mycoplasma capricolum* P60 surface lipoprotein and its evaluation in a recombinant ELISA. Vet. Microbiol. *128*: 81–89.

Alexander, P., K. Slee, S. McOrist, L. Ireland and P. Coloe. 1985. Mastitis in cows and polyarthritis and pneumonia in calves caused by *Mycoplasma* species bovine group 7. Aust. Vet. J. *62*: 135–136.

Allam, N.M. and R.M. Lemcke. 1975. Mycoplasmas isolated from the respiratory tract of horses. J. Hyg. (Lond.) *74*: 385–407.

Amin, M.M. and F.T. Jordan. 1978. Experimental infection of ducklings with *Mycoplasma gallisepticum* and *Mycoplasma anatis*. Res. Vet. Sci. *25*: 86–88.

Antunes, N., M. Tavío, P. Mercier, R. Ayling, W. Al-Momani, P. Assunção, R. Rosales and J. Poveda. 2007. *In vitro* susceptibilities of *Mycoplasma putrefaciens* field isolates. Antimicrob. Agents Chemother. *51*: 3452–3454.

Arif, A., J. Schulz, F. Thiaucourt, A. Taha and S. Hammer. 2007. Contagious caprine pleuropneumonia outbreak in captive wild ungulates at Al Wabra Wildlife Preservation, State of Qatar. J. Zoo Wildl. Med. *38*: 93–96.

Armstrong, C.H., M.J. Freeman and L. Sands-Freeman. 1987. Cross-reactions between *Mycoplasma hyopneumoniae* and *Mycoplasma flocculare* – practical implications for the serodiagnosis of mycoplasmal pneumonia of swine. Isr. J. Med. Sci. *23*: 654–656.

Armstrong, D., B.H. Yu, A. Yagoda and M.F. Kagnoff. 1971. Colonization of humans by *Mycoplasma canis*. J. Infect. Dis. *124*: 607–609.

Askaa, G. and H. Erno. 1976. Elevation of *Mycoplasma agalactiae* subsp. *bovis* to species rank, *Mycoplasma bovis* (Hale *et al.*) comb. nov. Int. J. Syst. Bacteriol. *26*: 323–325.

Atkinson, T.P., M.F. Balish and K.B. Waites. 2008. Epidemiology, clinical manifestations, pathogenesis and laboratory detection of *Mycoplasma pneumoniae* infections. FEMS Microbiol. Rev. *32*: 956–973.

Baker, A.S., K.L. Ruoff and S. Madoff. 1998. Isolation of *Mycoplasma* species from a patient with seal finger. Clin. Infect. Dis. *27*: 1168–1170.

Balish, M.F. 2006. Subcellular structures of mycoplasmas. Front Biosci. *11*: 2017–2027.

Balish, M.F. and D.C. Krause. 2006. Mycoplasmas: a distinct cytoskeleton for wall-less bacteria. J. Mol. Microbiol. Biotechnol. *11*: 244–255.

Bar-Moshe, B., E. Rapoport and J. Brenner. 1984. Vaccination trials against *Mycoplasma mycoides* subsp. *mycoides* (large-colony-type) infection in goats. Isr. J. Med. Sci. *20*: 972–974.

Barile, M. 1986. DNA homologies and serologic relationships among ureaplasmas from various hosts. Pediatr. Infect. Dis. *5*: S296–299.

Barile, M.F., R.A. Del Giudice, T.R. Carski, C.J. Gibbs and J.A. Morris. 1968. Isolation and characterization of *Mycoplasma arginini*: spec. nov. Proc. Soc. Exp. Biol. Med. *129*: 489–494.

Barile, M.F., R.A. Del Giudice and J.G. Tully. 1972. Isolation and characterization of *Mycoplasma conjunctivae* sp. n. from sheep and goats with keratoconjunctivitis. Infect. Immun. *5*: 70–76.

Barile, M.F. 1984. Immunization against *Mycoplasma pneumoniae* disease: a review. Isr. J. Med. Sci. *20*: 912–915.

Baseman, J., M. Cagle, J. Korte, C. Herrera, W. Rasmussen, J. Baseman, R. Shain and J. Piper. 2004. Diagnostic assessment of *Mycoplasma genitalium* in culture-positive women. J. Clin. Microbiol. *42*: 203–211.

Baseman, J.B., S.F. Dallo, J.G. Tully and D.L. Rose. 1988. Isolation and characterization of *Mycoplasma genitalium* strains from the human respiratory tract. J. Clin. Microbiol. *26*: 2266–2269.

Baseman, J.B. and J.G. Tully. 1997. Mycoplasmas: sophisticated, re-emerging, and burdened by their notoriety. Emerg. Infect. Dis. *3*: 21–32.

Bébéar, C. and J.A. Robertson. 1996. Determination of minimal inhibitory concentration. *In* Molecular and Diagnostic Procedures in Mycoplasmology, vol. 2 (edited by Tully and Razin). Academic Press, New York, pp. 189–197.

Bébéar, C.M. and I. Kempf. 2005. Antimicrobial therapy and antimicrobial resistance. *In* Mycoplasmas Molecular Biology Pathogenicity and Strategies for Control (edited by Blanchard and Browning). Horizon Bioscience, Norfolk, UK, pp. 535–568.

Beeton, M., V. Chalker, N. Maxwell, S. Kotecha and O. Spiller. 2009. Concurrent titration and determination of antibiotic resistance in *Ureaplasma* species with identification of novel point mutations in genes associated with resistance. Antimicrob. Agents Chemother. *53*: 2020–2027.

Ben Abdelmoumen Mardassi, B., A.O. Béjaoui Khiari, L., A. Landoulsi, C. Brik, B. Mlik and F. Amouna. 2007. Molecular cloning of a *Mycoplasma meleagridis*-specific antigenic domain endowed with a serodiagnostic potential. Vet. Microbiol. *119*: 31–41.

Benčina, D., D. Dorrer and T. Tadina. 1987. *Mycoplasma* species isolated from six avian species. Avian Pathol. *16*: 653–664.

Benčina, D., T. Tadina and D. Dorrer. 1988. Natural infection of ducks with *Mycoplasma synoviae* and *Mycoplasma gallisepticum* and *Mycoplasma* egg transmission. Avian Pathol. *17*: 441–449.

Benčina, D., I. Mrzel, O. Zorman Rojs, A. Bidovec and A. Dovc. 2003. Characterisation of *Mycoplasma gallisepticum* strains involved in respiratory disease in pheasants and peafowl. Vet. Rec. *152*: 230–234.

Berent, L.M. and J.B. Messick. 2003. Physical map and genome sequencing survey of *Mycoplasma haemofelis* (*Haemobartonella felis*). Infect. Immun. *71*: 3657–3662.

Bergemann, A.D. and L.R. Finch. 1988. Isolation and restriction endonuclease analysis of a *Mycoplasma* plasmid. Plasmid *19*: 68–70.

Bergonier, D., X. Berthelot and F. Poumarat. 1997. Contagious agalactia of small ruminants: current knowledge concerning epidemiology, diagnosis and control. Rev. Sci. Tech. *16*: 848–873.

Bhugra, B. and K. Dybvig. 1993. Identification and characterization of IS*1138*, a transposable element from *Mycoplasma pulmonis* that belongs to the IS*3* family. Mol. Microbiol. *7*: 577–584.

Biberfeld, G. and P. Biberfeld. 1970. Ultrastructural features of *Mycoplasma pneumoniae*. J. Bacteriol. *102*: 855–861.

Binder, A., R. Aumuller, B. Likitdecharote and H. Kirchhoff. 1990. Isolation of *Mycoplasma arthritidis* from the joint fluid of boars. Zentralbl. Veterinarmed. B *37*: 611–614.

Black, F.T. 1973. Biological and physical properties of human T-mycoplasmas. Ann. N. Y. Acad. Sci. *225*: 131–143.

Black, F.T. and A. Krogsgaard-Jensen. 1974. Application of indirect immunofluorescence, indirect haemagglutination and polyacrylamide-gel electrophoresis to human T-mycoplasmas. Acta Pathol. Microbiol. Scand. [B] Microbiol. Immunol. *82*: 345–353.

Blanchard, A. 1990. *Ureaplasma urealyticum* urease genes: use of a UGA tryptophan codon. Mol. Microbiol. *4*: 669–676.

Blanchard, A., W. Hamrick, L. Duffy, K. Baldus and G.H. Cassell. 1993. Use of the polymerase chain reaction for detection of *Mycoplasma fermentans* and *Mycoplasma genitalium* in the urogenital tract and amniotic fluid. Clin. Infect. Dis. 1*7 Suppl 1*: S272–279.

Blanchard, A. 1997. Mycoplasmas and HIV infection, a possible interaction through immune activation. Wien Klin. Wochenschr. *109*: 590–593.

Blanchard, A., L. Montagnier and M.L. Gougeon. 1997. Influence of microbial infections on the progression of HIV disease. Trends Microbiol. *5*: 326–331.

Blanchard, A. and C.M. Bébéar. 2002. Mycoplasmas of Humans. *In* Molecular Biology and Pathogenicity of Mycoplasmas (edited by Razin and Herrmann). Kluwer Academic/Plenum Publishers, New York, pp. 45–72.

Blaylock, M., O. Musatovova, J. Baseman and J. Baseman. 2004. Determination of infectious load of *Mycoplasma genitalium* in clinical samples of human vaginal cells. J. Clin. Microbiol. *42*: 746–752.

Boatman, E.S. 1979. Morphology and ultrastructure of the *Mycoplasmatales*. *In* The Mycoplasmas, vol. 1 (edited by Barile and Razin). Academic Press, New York, pp. 63–102.

Bonilla, H.F., C.E. Chenoweth, J.G. Tully, L.K. Blythe, J.A. Robertson, V.M. Ognenovski and C.A. Kauffman. 1997. *Mycoplasma felis* septic arthritis in a patient with hypogammaglobulinemia. Clin. Infect. Dis. *24*: 222–225.

Boothby, J.T., D.E. Jasper, M.H. Rollins and C.B. Thomas. 1981. Detection of *Mycoplasma bovis* specific IgG in bovine serum by enzyme-linked immunosorbent assay. Am. J. Vet. Res. *42*: 1242–1247.

Borrel, A., E. Dujardin-Beaumetz, Jeantet and C. Jouan. 1910. Le microbe de la péripneumonie. Ann. Inst. Pasteur (Paris) *24*: 168–179.

Borup-Christensen, P., K. Erb and J.C. Jensenius. 1988. Curing human hybridomas infected with *Mycoplasma hyorhinis*. J. Immunol. Methods *110*: 237–240.

Boughton, E., S.A. Hopper and P.J.R. Gayford. 1983. *Mycoplasma canadense* from bovine fetuses. Vet. Rec. *112*: 87.

Bowie, W., S. Wang, E. Alexander, J. Floyd, P. Forsyth, H. Pollock, J. Lin, T. Buchanan and K. Holmes. 1977. Etiology of nongonococcal urethritis. Evidence for *Chlamydia trachomatis* and *Ureaplasma urealyticum*. J. Clin. Invest. *59*: 735–742.

Bradbury, J.M., F. M. and A. Williams. 1983. *Mycoplasma lipofaciens*, a new species of avian origin. International Journal of Systematic and Evolutionary Microbiology *33*: 329–335.

Bradbury, J.M. and M. Forrest. 1984. *Mycoplasma cloacale*, a new species isolated from a turkey. Int. J. Syst. Bacteriol. *34*: 389–392.

Bradbury, J.M., A. Vuillaume, J.P. Dupiellet, M. Forrest, J.L. Bind and G. Gaillardperrin. 1987. Isolation of *Mycoplasma cloacale* from a number of different avian hosts in Great-Britain and France. Avian Pathol. *16*: 183–186.

Bradbury, J.M., F.T.W. Jordan, T. Shimizu, L. Stipkovits and Z. Varga. 1988. *Mycoplasma anseris* sp. nov. found in geese. Int. J. Syst. Bacteriol. *38*: 74–76.

Bradbury, J.M., O.M.S. Abdulwahab, C.A. Yavari, J.P. Dupiellet and J.M. Bove. 1993. *Mycoplasma imitans* sp. nov. is related to *Mycoplasma gallisepticum* and found in birds. Int. J. Syst. Bacteriol. *43*: 721–728.

Bradbury, J.M., C.M. Dare and C.A. Yavari. 2000. Evidence of *Mycoplasma gallisepticum* in British wild birds. Proceedings of the 13th Congress of the International Organization for Mycoplasmology Fukouka, Japan, p. 253.

Bradbury, J.M., C.A. Yavari and C.M. Dare. 2001. Mycoplasmas and respiratory disease in pheasants and partridges. Avian Pathol. *30*: 391–396.

Bradbury, J.M. and C.J. Morrow. 2008. *Mycoplasma* infections. *In* Poultry Diseases, 6th edn (edited by Pattison, Bradbury and Alexander). Elsevier, Edinburgh, pp. 220–234.

Bradbury, J.M. and S.H. Kleven. 2008. *Mycoplasma synoviae* infection. *In* Diseases of Poultry, 12th edn. (edited by Saif, Fadly, Glisson, McDougald, Nolan and Swayne). Blackwell Publishing, Ames, pp. 856–864.

Brandao, E. 1995. Isolation and identification of *Mycoplasma mycoides* subspecies *mycoides* SC strains in sheep and goats. Vet. Rec. *136*: 98–99.

Bredt, W. 1979. Motility. *In* The Mycoplasmas, vol. 1 (edited by Barile and Razin). Academic Press, New York, pp. 141–155.

Brown, D.R., G. McLaughlin and M. Brown. 1995. Taxonomy of the feline mycoplasmas *Mycoplasma felifaucium*, *Mycoplasma feliminutum*, *Mycoplasma felis*, *Mycoplasma gateae*, *Mycoplasma leocaptivus*, *Mycoplasma leopharyngis*, and *Mycoplasma simbae* by 16S rRNA gene sequence comparisons. Int. J. Syst. Bacteriol. *45*: 560–564.

Brown, D.R., R. Whitcomb and J. Bradbury. 2007. Revised minimal standards for description of new species of the class *Mollicutes* (division *Tenericutes*). Int. J. Syst. Evol. Microbiol. *57*: 2703–2719.

Brown, D.R., Clippinger T.L., Helmick K.E., Schumacher I.M., Bennett R.A., Johnson C.M., Vliet K.A., Jacobson E.R. and B. M.B. 1996. *Mycoplasma* isolation during a fatal epizootic of captive alligators (Alligator mississippiensis) in Florida. Int. Org. Mycoplasmol. Lett 4: 2.

Brown, D.R., J.M. Farley, L.A. Zacher, J.M.R. Carlton, T.L. Clippinger, J.G. Tully and M.B. Brown. 2001a. *Mycoplasma alligatoris* sp. nov., from American alligators. Int. J. Syst. Evol. Microbiol. *51*: 419–424.

Brown, D.R., D.F. Talkington, W.L. Thacker, M.B. Brown, D.L. Dillehay and J.G. Tully. 2001b. *Mycoplasma microti* sp. nov., isolated from the respiratory tract of prairie voles (*Microtus ochrogaster*). Int. J. Syst. Evol. Microbiol. *51*: 409–412.

Brown, D.R., J.L. Merritt, E.R. Jacobson, P.A. Klein, J.G. Tully and M.B. Brown. 2004. *Mycoplasma testudineum* sp. nov., from a desert tortoise (*Gopherus agassizii*) with upper respiratory tract disease. Int. J. Syst. Evol. Microbiol. *54*: 1527–1529.

Brown, D.R., L.A. Zacher, L.D. Wendland and M.B. Brown. 2005. Emerging Mycoplasmoses in Wildlife. *In* Mycoplasmas Molecular Biology Pathogenicity and Strategies for Control (edited by Blanchard and Browning). Horizon Bioscience, Norfolk, England, pp. 383–414.

Brown, D.R., D.L. Demcovitz, D.R. Plourde, S.M. Potter, M.E. Hunt, R.D. Jones and D.S. Rotstein. 2006. *Mycoplasma iguanae* sp. nov., from a green iguana (*Iguana iguana*) with vertebral disease. Int. J. Syst. Evol. Microbiol. *56*: 761–764.

Brown, D.R. and J.M. Bradbury. 2008. International Committee on Systematics of Prokaryotes of the International Union of Microbiological Societies, Subcommittee on the taxonomy of mollicutes, minutes of the meeting, 6 and 11 July 2008, Tianjin, PR China. International Journal of Systematic and Evolutionary Microbiology *58*: 2987–2990.

Brown, M.B., I.M. Schumacher, P.A. Klein, K. Harris, T. Correll and E.R. Jacobson. 1994. *Mycoplasma agassizii* causes upper respiratory tract disease in the desert tortoise. Infect. Immun. *62*: 4580–4586.

Brown, M.B., D.R. Brown, P.A. Klein, G.S. McLaughlin, I.M. Schumacher, E.R. Jacobson, H.P. Adams and J.G. Tully. 2001c. *Mycoplasma agassizii* sp. nov., isolated from the upper respiratory tract of the desert tortoise (*Gopherus agassizii*) and the gopher tortoise (*Gopherus polyphemus*). Int. J. Syst. Evol. Microbiol. *51*: 413–418.

Browning, G.F., K.G. Whithear and S.J. Geary. 2005. Vaccines to Control Mycoplasmosis. *In* Mycoplasmas: Molecular Biology, Pathogenicity, and Strategies for Control (edited by Blanchard and Browning). Horizon Bioscience, Norfolk, UK, pp. 569–599.

Brun-Hansen, H., H. Gronstol, H. Waldeland and B. Hoff. 1997. *Eperythrozoon ovis* infection in a commercial flock of sheep. Zentralbl. Veterinarmed. B *44*: 295–299.

Brunner, S., P. Frey-Rindova, M. Altwegg and R. Zbinden. 2000. Retroperitoneal abscess and bacteremia due to *Mycoplasma hominis* in a polytraumatized man. Infection *28*: 46–48.

Busch, U., H. Nitschko, F. Pfaff, B. Henrich, J. Heesemann and M. Abele-Horn. 2000. Molecular comparison of *Mycoplasma hominis* strains isolated from colonized women and women with various urogenital infections. Zentralbl. Bakteriol. *289*: 879–888.

Buss, I., R. Senthilmohan, B. Darlow, N. Mogridge, A. Kettle and C. Winterbourn. 2003. 3-Chlorotyrosine as a marker of protein damage by myeloperoxidase in tracheal aspirates from preterm infants: association with adverse respiratory outcome. Pediatr. Res. *53*: 455–462.

Carmichael, L.E., T.D. St George, N.D. Sullivan and N. Horsfall. 1972. Isolation, propagation, and characterization studies of an ovine *Mycoplasma* responsible for proliferative interstitial pneumonia. Cornell Vet. *62*: 654–679.

Carson, J.L., P.C. Hu and A.M. Collier. 1992. Cell structure and functional elements. *In* Mycoplasmas: Molecular Biology and Pathogenesis (edited by Maniloff, McElhaney, Finch, and Baseman). American Society for Microbiology, Washington, DC, pp. 63–72.

Casin, I., D. Vexiau-Robert, P. De La Salmoniere, A. Eche, B. Grandry and M. Janier. 2002. High prevalence of *Mycoplasma genitalium* in the lower genitourinary tract of women attending a sexually transmitted disease clinic in Paris, France. Sex. Transm. Dis. *29*: 353–359.

Cassell, G.H. and A. Hill. 1979. Murine and other small animal mycoplasmas. *In* The Mycoplasmas, vol. 1 (edited by Tully and Whitcomb). Academic Press, New York, pp. 235–273.

Caswell, J.L. and M. Archambault. 2007. *Mycoplasma bovis* pneumonia in cattle. Anim. Health Res. Rev. *8*: 161–186.

Chalker, V.J. and J. Brownlie. 2004. Taxonomy of the canine *Mollicutes* by 16S rRNA gene and 16S/23S rRNA intergenic spacer region sequence comparison. Int. J. Syst. Evol. Microbiol. *54*: 537–542.

Chalker, V.J. 2005. Canine mycoplasmas. Res. Vet. Sci. *79*: 1–8.

Chávez Gonzalez, Y.R., C. Ros Bascunana, G. Bolske, J.G. Mattsson, C. Fernandez Molina and K.-E. Johansson. 1995. *In vitro* amplification of the 16S rRNA genes from *Mycoplasma bovis* and *Mycoplasma agalactiae* by PCR. Vet. Microbiol. *47*: 183–190.

Cheng, X., J. Nicolet, F. Poumarat, J. Regalla, F. Thiaucourt and J. Frey. 1995. Insertion element IS1296 in *Mycoplasma mycoides* subsp. *mycoides* small colony identifies a European clonal line distinct from African and Australian strains. Microbiology *141*: 3221–3228.

Chin, R.P., G.Y. Ghazikhanian and I. Kempf. 2008. *Mycoplasma meleagridis* infection. *In* Diseases of Poultry, 12th edn. (edited by Saif, Fadly, Glisson, McDougald, Nolan and Swayne). Blackwell Publishing, Ames, pp. 834–845.

Christiansen, C., F.T. Black and E.A. Freundt. 1981. Hybridization experiments with DNA from *Ureaplasma urealyticum*, serovars I to VIII. Int. J. Syst. Bacteriol. *31*: 259–262.

Clapper, B., A.H. Tu, W.L. Simmons and K. Dybvig. 2004. Bacteriophage MAV1 is not associated with virulence of *Mycoplasma arthritidis*. Infect. Immun. *72*: 7322–7325.

Clark, H.W., J. S. Bailey, D. C. Laughlin and T.M. Brown. 1978. Isolation of *Mycoplasma* from the genital tracts of elephants. Zentralbl. Bakteriol. Parasitenkd. Infektionskr. Hyg. Abt. 1 Orig. 241:262.

Clark, H.W., D. C. Laughlin, J. S. Bailey and T.M. Brown. 1980. *Mycoplasma* Species and Arthritis in Captive Elephants. Journal of Zoo Animal Medicine *11*: 3–15.

Clark, R. 1942. *Eperythrozoon felis* (sp. nov.) in a cat. J. Afr. Vet. Med. Assoc. *13*: 15–16.

Cobb, D.T., D.H. Ley and P.D. Doerr. 1992. Isolation of *Mycoplasma gallopavonis* from free-ranging wild turkeys in coastal North Carolina seropositive and culture-negative for *Mycoplasma gallisepticum*. J. Wildl. Dis. *28*: 105–109.

Cocks, B., F. Brake, A. Mitchell and L. Finch. 1985. Enzymes of intermediary carbohydrate metabolism in *Ureaplasma urealyticum* and *Mycoplasma mycoides* subsp. *mycoides*. J. Gen. Microbiol. *131*: 2129–2135.

Cocks, B.G. and L.R. Finch. 1987. Characterization of a restriction endonuclease from *Ureaplasma urealyticum* 960 and differences in deoxyribonucleic acid modification of human ureaplasmas. Int. J. Syst. Bacteriol. *37*: 451–453.

Cole, B.C., L. Golightly and J.R. Ward. 1967. Characterization of *Mycoplasma* strains from cats. J. Bacteriol. *94*: 1451–1458.

Cole, B.C., L.R. Washburn and D. Taylor-Robinson. 1985. *Mycoplasma*-induced arthritis. *In* The Mycoplasmas, vol. 4 (edited by Razin and Barile). Academic Press, New York, pp. 108–160.

Cole, R.M. 1983. Transmission electron microscopy: Basic techniques. *In* Methods in Mycoplasmology (edited by Razin and Tully). Academic Press, New York, p. 4350.

Cordtz, J. and J. Jensen. 2006. Disseminated *Ureaplasma urealyticum* infection in a hypo-gammaglobulinaemic renal transplant patient. Scand. J. Infect. Dis. *38*: 1114–1117.

Cottew, G.S. and F.R. Yeats. 1978. Subdivision of *Mycoplasma mycoides* subsp. *mycoides* from cattle and goats into two types. Aust. Vet. J. *54*: 293–296.

Cottew, G.S. 1979. Caprine-ovine mycoplasmas. *In* The Mycoplasmas, vol. 2 (edited by Tully and Whitcomb). Academic Press, New York, pp. 103–132.

Cottew, G.S. 1983. Recovery and identification of caprine and ovine mycoplasmas. *In* Methods in Mycoplasmology, vol. 2 (edited by Tully and Razin). Academic Press, New York, pp. 91–104.

Cottew, G.S., A. Breard, A.J. DaMassa, H. Erno, R.H. Leach, P.C. Lefevre, A.W. Rodwell and G.R. Smith. 1987. Taxonomy of the *Mycoplasma mycoides* cluster. Isr. J. Med. Sci. *23*: 632–635.

Crespo, R. and R. McMillan. 2008. Facial cellulitis induced in chickens by *Mycoplasma gallisepticum* bacterin and its treatment. Avian Dis. *52*: 698–701.

Daddow, K.N. 1980. Culex annulirostris as a vector of *Eperythrozoon ovis* infection in sheep. Vet. Parasitol. *7*: 313–317.

DaMassa, A.J., E.R. Nascimento, M.I. Khan, R. Yamamoto and D.L. Brooks. 1991. Characteristics of an unusual *Mycoplasma* isolated from a case of caprine mastitis and arthritis with possible systemic manifestations. J. Vet. Diagn. Invest. *3*: 55–59.

Dammann, O., E. Allred, D. Genest, R. Kundsin and A. Leviton. 2003. Antenatal *Mycoplasma* infection, the fetal inflammatory response and cerebral white matter damage in very-low-birthweight infants. Paediatr. Perinat. Epidemiol. *17*: 49–57.

Damassa, A.J., J.G. Tully, D.L. Rose, D. Pitcher, R.H. Leach and G.S. Cottew. 1994. *Mycoplasma auris* sp. nov., *Mycoplasma cottewii* sp. nov., and *Mycoplasma yeatsii* sp. nov., new sterol-requiring mollicutes from the external ear canals of goats. Int. J. Syst. Bacteriol. *44*: 479–484.

DaMassa, A.J. 1996. *Mycoplasma* infections of goat and sheep. *In* Molecular and Diagnostic Procedures in Mycoplasmology, vol. 2 (edited by Tully and Razin). Academic Press, San Diego, pp. 265–273.

Davis, J.W., Jr., I.S. Moses, C. Ndubuka and R. Ortiz. 1987. Inorganic pyrophosphatase activity in cell-free extracts of *Ureaplasma urealyticum*. J. Gen. Microbiol. *133*: 1453–1459.

Davis, J.W., Jr. and I. Villanueva. 1990. Enzyme differences in serovar clusters of *Ureaplasma urealyticum*. *In* Recent Advances in Mycoplasmology (edited by Stanek, Cassell, Tully and Whitcomb). Gustav Fischer Verlag, New York, pp. 665–666.

de Barbeyrac, B., C. Bernet-Poggi, F. Fébrer, H. Renaudin, M. Dupon and C. Bébéar. 1993. Detection of *Mycoplasma pneumoniae* and *Mycoplasma genitalium* in clinical samples by polymerase chain reaction. Clin. Infect. Dis. *17 Suppl 1*: S83–89.

De Francesco, M., R. Negrini, G. Pinsi, L. Peroni and N. Manca. 2009. Detection of *Ureaplasma biovars* and polymerase chain reaction-based subtyping of *Ureaplasma parvum* in women with or without symptoms of genital infections. Eur. J. Clin. Microbiol. Infect. Dis. *28*: 641–646.

de la Fe, C., P. Assunção, P. Saavedra, S. Tola, C. Poveda and J. Poveda. 2007. Vaccine. Field trial of two dual vaccines against *Mycoplasma agalactiae* and *Mycoplasma mycoides* subsp. *mycoides* (large colony type) in goats *25*: 2340–2345.

De Silva, N. and P. Quinn. 1986. Endogenous activity of phospholipases A and C in *Ureaplasma urealyticum*. J. Clin. Microbiol. *23*: 354–359.

Deeb, B. and G. Kenny. 1967. Characterization of *Mycoplasma pulmonis* variants isolated from rabbits. I. Identification and properties of isolates. J. Bacteriol. *93*: 1416–1424.

Del Giudice, R.A., T.R. Carski, M.F. Barile, R.M. Lemcke and J.G. Tully. 1971. Proposal for classifying human strain Navel and related simian mycoplasmas as *Mycoplasma primatum* sp. n. J. Bacteriol. *108*: 439–445.

Del Giudice, R.A., R.H. Purcell, T.R. Carski and R.M. Chanock. 1974. *Mycoplasma lipophilum* sp. nov. Int. J. Syst. Bacteriol. *24*: 147–153.

Del Giudice, R.A., J.G. Tully, D.L. Rose and R.M. Cole. 1985. *Mycoplasma pirum* sp. nov., a terminal structured mollicute from cell cultures. Int. J. Syst. Bacteriol. *35*: 285–291.

Del Giudice, R.A., D.L. Rose and J.G. Tully. 1995. *Mycoplasma adleri* sp. nov., an isolate from a goat. Int. J. Syst. Bacteriol. *45*: 29–31.

Del Giudice, R.A. and R.S. Gardella. 1996. Antibiotic treatment of mycoplasma-infected cell cultures. *In* Molecular and Diagnostic Procedures in Mycoplasmology, vol. 2 (edited by Tully and Razin). Academic Press, San Diego, pp. 439–443.

Dessì, D., P. Rappelli, N. Diaz, P. Cappuccinelli and P.L. Fiori. 2006. *Mycoplasma hominis* and *Trichomonas vaginalis*: a unique case of symbiotic relationship between two obligate human parasites. Front. Biosci. *11*: 2028–2034.

Dhondt, A.A., S. Altizer, E.G. Cooch, A.K. Davis, A. Dobson, M.J. Driscoll, B.K. Hartup, D.M. Hawley, W.M. Hochachka, P.R. Hosseini, C.S. Jennelle, G.V. Kollias, D.H. Ley, E.C. Swarthout and K.V. Sydenstricker. 2005. Dynamics of a novel pathogen in an avian host: *Mycoplasmal conjunctivitis* in house finches. Acta Trop. *94*: 77–93.

Dhondt, A.A., K.V. Dhondt, D.M. Hawley and C.S. Jennelle. 2007. Experimental evidence for transmission of *Mycoplasma gallisepticum* in house finches by fomites. Avian Pathol. *36*: 205–208.

Dierks, R.E., J.A. Newman and B.S. Pomeroy. 1967. Characterization of avian *Mycoplasma*. Ann. N. Y. Acad. Sci. *143*: 170–189.

Dillehay, D.L., M. Sander, D.F. Talkington, W.L. Thacker and D.R. Brown. 1995. Isolation of mycoplasmas from prairie voles (Microtus ochrogaster). Lab. Anim. Sci. *45*: 631–634.

Djordjevic, S.R., W.A. Forbes, J. Forbes-Faulkner, P. Kuhnert, S. Hum, M.A. Hornitzky, E.M. Vilei and J. Frey. 2001. Genetic diversity among *Mycoplasma* species bovine group 7: clonal isolates from an outbreak of polyarthritis, mastitis, and abortion in dairy cattle. Electrophoresis *22*: 3551–3561.

Doig, P.A. 1981. Bovine genital mycoplasmosis. Can. Vet. J. *22*: 339–343.

dos Santos, A.P., R.P. dos Santos, A.W. Biondo, J.M. Dora, L.Z. Goldani, S.T. de Oliveira, A.M. de Sa Guimaraes, J. Timenetsky, H.A. de Morais, F.H. Gonzalez and J.B. Messick. 2008. Hemoplasma infection in HIV-positive patient, Brazil. Emerg Infect Dis *14*: 1922–1924.

Dowers, K.L., C. Olver, S.V. Radecki and M.R. Lappin. 2002. Use of enrofloxacin for treatment of large-form *Haemobartonella felis* in experimentally infected cats. J. Am. Vet. Med. Assoc. *221*: 250–253.

Drexler, H.G. and C.C. Uphoff. 2002. *Mycoplasma* contamination of cell cultures: Incidence, sources, effects, detection, elimination, prevention. Cytotechnology *39*: 75–90.

Dubosson, C.R., C. Conzelmann, R. Miserez, P. Boerlin, J. Frey, W. Zimmermann, H. Hani and P. Kuhnert. 2004. Development of two real-time PCR assays for the detection of *Mycoplasma hyopneumoniae* in clinical samples. Vet. Microbiol. *102*: 55–65.

Duffy, L., J. Glass, G. Hall, R. Avery, R. Rackley, S. Peterson and K. Waites. 2006. Fluoroquinolone resistance in *Ureaplasma parvum* in the United States. J. Clin. Microbiol. *44*: 1590–1591.

Dupiellet, J.-P. 1984. Mycoplasmes et acholeplasmes des palmipedes a foie gras: isolement, caracterisation, etude du role dans la pathologie. Universite de Bordeaux II.

Dyer, N., L. Hansen-Lardy, D. Krogh, L. Schaan and E. Schamber. 2008. An outbreak of chronic pneumonia and polyarthritis syndrome caused by *Mycoplasma bovis* in feedlot bison (Bison bison). J. Vet. Diagn. Invest. *20*: 369–371.

Echahidi, F., G. Muyldermans, S. Lauwers and A. Naessens. 2001. Development of an enzyme-linked immunosorbent assay for serotyping *Ureaplasma urealyticum* strains using monoclonal antibodies. Clin. Diagn. Lab. Immunol. *8*: 52–57.

Edward, D.G. and A.D. Kanarek. 1960. Organisms of the pleuropneumonia group of avian origin: their classification into species. Ann. N. Y. Acad. Sci. *79*: 696–702.

Edward, D.G. and E.A. Freundt. 1969. Proposal for classifying organisms related to *Mycoplasma* laidlawii in a family *Sapromycetaceae*, genus *Sapromyces*, within the *Mycoplasmatales*. J. Gen. Microbiol. *57*: 391–395.

Edward, D.G. and E.A. Freundt. 1970. Amended nomenclature for strains related to *Mycoplasma laidlawii*. J. Gen. Microbiol. *62*: 1–2.

Edward, D.G.f. 1953. Organisms of the pleuropneumonia group causing disease in goats. Vet. Rec. *63*: 873–874.

Edward, D.G.f. 1955. A suggested classification and nomenclature for organisms of the pleuropneumonia group. Int. Bull. Bacteriol. Nomencl. Taxon. *5*: 85–93.

Egwu, G., J. Ameh, M. Aliyu and F. Mohammed. 2001. Caprine Mycoplasmal mastitis in Nigeria. Small Rumin. Res. *39*: 87–91.

Erdélyi, K., M. Tenk and A. Dan. 1999. Mycoplasmosis associated perosis type skeletal deformity in a saker falcon nestling in Hungary. J. Wildl. Dis. *35*: 586–590.

Erickson, B.Z., R.F. Ross, D.L. Rose, J.G. Tully and J.M. Bové. 1986. *Mycoplasma hyopharyngis*, a new species from swine. Int. J. Syst. Bacteriol. *36*: 55–59.

Erickson, B.Z., R.F. Ross and J.M. Bove. 1988. Isolation of *Mycoplasma salivarium* from swine. Vet. Microbiol. *16*: 385–390.

Evans-Davis, K.D., D.L. Dillehay, D.N. Wargo, S.K. Webb, D.F. Talkington, W.L. Thacker, L.S. Small and M.B. Brown. 1998. Pathogenicity of *Mycoplasma volis* in mice and rats. Lab. Anim. Sci. *48*: 38–44.

Feberwee, A., J. de Wit and W. Landman. 2009. Induction of eggshell apex abnormalities by *Mycoplasma synoviae*: field and experimental studies. Avian Pathol. *38*: 77–85.

Ferguson, N.M., V.A. Leiting and S.H. Klevena. 2004. Safety and efficacy of the avirulent *Mycoplasma gallisepticum* strain K5054 as a live vaccine in poultry. Avian Dis. *48*: 91–99.

Fernandez Guerrero, M.L., J. Manuel Ramos and F. Soriano. 1999. *Mycoplasma hominis* bacteraemia not associated with genital infections. J. Infect. *39*: 91–94.

Ferrell, R., M. Heidari, K. Wise and M. McIntosh. 1989. A *Mycoplasma* genetic element resembling prokaryotic insertion sequences. Mol. Microbiol. *3*: 957–967.

Fischer, R.S., B.E. Fischer and R.A. Jensen. 1992. Sources of amino acids. *In* Mycoplasmas: Molecular Biology and Pathogenesis (edited by Maniloff, McElhaney, Finch and Baseman). American Society for Microbiology, Washington, D.C., pp. 201–209.

Fitzmaurice, J., M. Sewell, L. Manso-Silvan, F. Thiaucourt, W.L. McDonald and J.S. O'Keefe. 2008. Real-time polymerase chain reaction assays for the detection of members of the *Mycoplasma mycoides* cluster. N Z Vet. J. *56*: 40–47.

Flint, J.C. and McKelvie D.H. 1956. Feline infectious anemia-diagnosis and treatment. Presented at the Proc. 92nd Ann. Meet. Amer. Vet. Med. Assoc. 1955, 240–242.

Foley, J.E. and N.C. Pedersen. 2001. "*Candidatus Mycoplasma* haemominutum", a low-virulence epierythrocytic parasite of cats. Int. J. Syst. Evol. Microbiol. *51*: 815–817.

Ford, D. 1967. Relationships between *Mycoplasma* and the etiology of nongonococcal urethritis and Reiter's syndrome. Ann. N. Y. Acad. Sci. *143*: 501–504.

Ford, D. and J. MacDonald. 1967. Influence of urea on the growth of T-strain mycoplasmas. J. Bacteriol. *93*: 1509–1512.

Ford, D. 1972. Inhibition of growth of T-strain mycoplasmas by hydroxamic acids and by aurothiomalate. Antimicrob Agents Chemother *2*: 340–343.

Ford, D. and J. Smith. 1974. Non-specific urethritis associated with a tetracycline-resistant T-*mycoplasma*. Br. J. Vener. Dis. *50*: 373–374.

Forrest, M. and J.M. Bradbury. 1984. *Mycoplasma glycophilum*, a new species of avian origin. J. Gen. Microbiol. *130*: 597–603.

Forrest, M. and J. M. Bradbury. 1984. *Mycoplasma glycophilum* sp. nov. Validation List No. 15. Int. J. Syst. Bacteriol. *34*: 355–357.

Forsyth, M.H., J.G. Tully, T.S. Gorton, L. Hinckley, S. Frasca, Jr., H.J. van Kruiningen and S.J. Geary. 1996. *Mycoplasma sturni* sp. nov., from the conjunctiva of a European starling (Sturnus vulgaris). Int. J. Syst. Bacteriol. *46*: 716–719.

Frasca, S., Jr, L. Hinckley, M. Forsyth, T. Gorton, S. Geary and H. Van Kruiningen. 1997. Mycoplasmal conjunctivitis in a European starling. J. Wildl. Dis. *33*: 336–339.

Frasca, S., Jr, E. Weber, H. Urquhart, X. Liao, M. Gladd, K. Cecchini, P. Hudson, M. May, R. Gast, T. Gorton and S. Geary. 2005. Isolation and characterization of *Mycoplasma sphenisci* sp. nov. from the choana of an aquarium-reared jackass penguin (Spheniscus demersus). J. Clin. Microbiol. *43*: 2976–2979.

Freundt, E.A. 1953. The occurrence of *Micromyces* (pleuropneumonia-like organisms) in the female genito-urinary tract. Acta Pathol. Microbiol. Scand. *32*: 468–480.

Freundt, E.A. 1955. The classification of the pleuropneumoniae group of organisms (Borrelomycetales). Int. Bull. Bacteriol. Nomencl. Taxon. *5*: 67–78.

Freundt, E.A., Taylorro.D, R.H. Purcell, R.M. Chanock and F.T. Black. 1974. Proposal of *Mycoplasma buccale* nom. nov. and *Mycoplasma faucium* nom. nov. for *Mycoplasma orale* types 2 and 3, respectively. Int. J. Syst. Bacteriol. *24*: 252–255.

Frey, J., X. Cheng, P. Kuhnert and J. Nicolet. 1995. Identification and characterization of IS1296 in *Mycoplasma mycoides* subsp. *mycoides* SC and presence in related mycoplasmas. Gene *160*: 95–100.

Frey, M.L., R.P. Hanson and D.P. Anderson. 1968. A medium for the isolation of avian mycoplasmas. American Journal of Veterinary Research *29*: 2163–2171.

Friis, N.F. and E. Blom. 1983. Isolation of *Mycoplasma canadense* from bull semen. Acta Vet. Scand. *24*: 315–317.

Friis, N.F. and J. Szancer. 1994. Sensitivity of certain porcine and bovine mycoplasmas to antimicrobial agents in a liquid medium test compared to a disc assay. Acta Vet. Scand. *35*: 389–394.

Furness, G. 1973. T-mycoplasmas: their growth and production of a toxic substance in broth. J. Infect. Dis. *127*: 9–16.

Furness, G. 1975. T-mycoplasmas: Growth patterns and physical characteristics of some human strains. J. Infect. Dis. *132*: 592–596.

Furr, P.M., D. Taylor-Robinson and A.D. Webster. 1994. Mycoplasmas and ureaplasmas in patients with hypogammaglobulinaemia and their role in arthritis: microbiological observations over twenty years. Ann. Rheum. Dis. *53*: 183–187.

Gallagher, J. and K. Rhoades. 1983. Scanning electron and light microscopy of selected avian strains of *Mycoplasma iowae*. Avian Dis. *27*: 211–217.

Gambini, D., I. Decleva, L. Lupica, M. Ghislanzoni, M. Cusini and E. Alessi. 2000. *Mycoplasma genitalium* in males with nongonococcal urethritis: prevalence and clinical efficacy of eradication. Sex. Transm. Dis. *27*: 226–229.

Ganapathy, K. and J.M. Bradbury. 1998. Pathogenicity of *Mycoplasma gallisepticum* and *Mycoplasma imitans* in red-legged partridges (*Alectoris rufa*). Avian Pathol. *27*: 455–463.

García-Castillo, M., M. Morosini, M. Gálvez, F. Baquero, R. del Campo and M. Meseguer. 2008. Differences in biofilm development and antibiotic susceptibility among clinical *Ureaplasma urealyticum* and *Ureaplasma parvum* isolates. J. Antimicrob. Chemother. *62*: 1027–1030.

Garcia-Porrua, C., F.J. Blanco, A. Hernandez, A. Atanes, F. Galdo, R. Moure and A. Alonso. 1997. Septic arthritis by *Mycoplasma hominis*: a case report and review of the medical literature. Ann. Rheum. Dis. *56*: 699–700.

Gass, R., J. Fisher, D. Badesch, M. Zamora, A. Weinberg, H. Melsness, F. Grover, J.G. Tully and F.C. Fang. 1996. Donor-to-host transmission of *Mycoplasma hominis* in lung allograft recipients. Clin. Infect. Dis. *22*: 567–568.

Gates, A.E., S. Frasca, A. Nyaoke, T.S. Gorton, L.K. Silbart and S.J. Geary. 2008. Comparative assessment of a metabolically attenuated *Mycoplasma gallisepticum* mutant as a live vaccine for the prevention of avian respiratory mycoplasmosis. Vaccine *26*: 2010–2019.

Gautier-Bouchardon, A.V., A.K. Reinhardt, M. Kobisch and I. Kempf. 2002. *In vitro* development of resistance to enrofloxacin, erythromycin, tylosin, tiamulin and oxytetracycline in *Mycoplasma gallisepticum*, *Mycoplasma iowae* and *Mycoplasma synoviae*. Vet. Microbiol. *88*: 47–58.

George, J.W., B.A. Rideout, S.M. Griffey and N.C. Pedersen. 2002. Effect of preexisting FeLV infection or FeLV and feline immunodeficiency virus coinfection on pathogenicity of the small variant of *Haemobartonella felis* in cats. Am. J. Vet. Res. *63*: 1172–1178.

Geraci, J.R., D.J. Staubin, I.K. Barker, R.G. Webster, V.S. Hinshaw, W.J. Bean, H.L. Ruhnke, J.H. Prescott, G. Early, A.S. Baker, S. Madoff and R.T. Schooley. 1982. Mass mortality of harbor seals: pneumonia associated with influenza A virus. Science *215*: 1129–1131.

Gerchman, I., I. Lysnyansky, S. Perk and S. Levisohn. 2008. *In vitro* susceptibilities to fluoroquinolones in current and archived *Mycoplasma gallisepticum* and *Mycoplasma synoviae* isolates from meat-type turkeys. Vet. Microbiol. *131*: 266–276.

Germain, M., M.A. Krohn, S.L. Hillier and D.A. Eschenbach. 1994. Genital flora in pregnancy and its association with intrauterine growth retardation. J. Clin. Microbiol. *32*: 2162–2168.

Gibson, D.G., G.A. Benders, C. Andrews-Pfannkoch, E.A. Denisova, H. Baden-Tillson, J. Zaveri, T.B. Stockwell, A. Brownley, D.W. Thomas, M.A. Algire, C. Merryman, L. Young, V.N. Noskov, J.I. Glass, J.C. Venter, C.A. Hutchison, 3rd and H.O. Smith. 2008. Complete chemical synthesis, assembly, and cloning of a *Mycoplasma genitalium* genome. Science *319*: 1215–1220.

Giebel, J., A. Binder and H. Kirchhoff. 1990. Isolation of *Mycoplasma moatsii* from the intestine of wild Norway rats (Rattus norvegicus). Vet. Microbiol. *22*: 23–29.

Giebel, J., J. Meier, A. Binder, J. Flossdorf, J.B. Poveda, R. Schmidt and H. Kirchhoff. 1991. *Mycoplasma phocarhinis* sp. nov. and *Mycoplasma phocacerebrale* sp. nov., two new species from harbor seals (*Phoca vitulina* L). Int. J. Syst. Bacteriol. *41*: 39–44.

Gil, M.C., F.J. Pena, J. Hermoso De Mendoza and L. Gomez. 2003. Genital lesions in an outbreak of caprine contagious agalactia caused by *Mycoplasma agalactiae* and *Mycoplasma putrefaciens*. J. Vet. Med. B Infect. Dis. Vet. Public Health *50*: 484–487.

Glass, J., E. Lefkowitz, J. Glass, C. Heiner, E. Chen and G. Cassell. 2000. The complete sequence of the mucosal pathogen *Ureaplasma urealyticum*. Nature *407*: 757–762.

Glass, J.I., B.A. Methe, V. Paralanov, L.B. Duffy and K.B. Waites. 2008. Comparative genome analysis of all 14 *Ureaplasma parvum* and *Ureaplasma urealyticum* serovars. Presented at the International Organization for Mycoplasmology, Tianjin, China.

Goldberg, D.R., M.D. Samuel, C.B. Thomas, P. Sharp, G.L. Krapu, J.R. Robb, K.P. Kenow, C.E. Korschgen, W.H. Chipley, M.J. Conroy and et al. 1995. The occurrence of mycoplasmas in selected wild North American waterfowl. J. Wildl. Dis. *31*: 364–371.

Gonçalves, L.F., T. Chaiworapongsa and R. Romero. 2002. Intrauterine infection and prematurity. Ment. Retard. Dev. Disabil. Res. Rev. *8*: 3–13.

Goodison, S., K. Nakamura, K.A. Iczkowski, S. Anai, S.K. Boehlein and C.J. Rosser. 2007. Exogenous mycoplasmal p37 protein alters gene expression, growth and morphology of prostate cancer cells. Cytogenet Genome Res. *118*: 204–213.

Goodwin, R.F.W., A.P. Pomeroy and P. Whittlestone. 1965. Production of enzootic pneumonia in pigs with a *Mycoplasma*. Vet. Rec. *77*: 1247–1249.

Gorton, T.S., M.M. Barnett, T. Gull, R.A. French, Z. Lu, G.F. Kutish, L.G. Adams and S.J. Geary. 2005. Development of real-time diagnostic assays specific for *Mycoplasma mycoides* subspecies mycoides Small Colony. Vet. Microbiol. *111*: 51–58.

Gourlay, R.N. and R.H. Leach. 1970. A new *Mycoplasma* species isolated from pneumonic lungs of calves (*Mycoplasma dispar* sp. nov.). J. Med. Microbiol. *3*: 111–123.

Gourlay, R.N., R.H. Leach and C.J. Howard. 1974. *Mycoplasma verecundum*, a new species isolated from bovine eyes. J. Gen. Microbiol. *81*: 475–484.

Gourlay, R.N., S.G. Wyld and R.H. Leach. 1977. *Mycoplasma alvi*, a new species from bovine intestinal and urogenital tracts. Int. J. Syst. Bacteriol. *27*: 86–96.

Gourlay, R.N., S.G. Wyld and R.H. Leach. 1978. *Mycoplasma sualvi*, a new species from intestinal and urogenital tracts of pigs. Int. J. Syst. Bacteriol. *28*: 289–292.

Gourlay, R.N. and C.J. Howard. 1979. Bovine mycoplasmas. *In* The Mycoplasmas, vol. 2 (edited by Tully and Whitcomb). Academic Press, New York, pp. 49–102.

Gourlay, R.N., S.G. Wyld and M.E. Poulton. 1983. Some characteristics of *Mycoplasma* virus Hr 1, isolated from and infecting *Mycoplasma hyorhinis*. Brief report. Arch. Virol. *77*: 81–85.

Grattard, F., B. Pozzetto, B. de Barbeyrac, H. Renaudin, M. Clerc, O. Gaudin and C. Bébéar. 1995. Arbitrarily-primed PCR confirms the differentiation of strains of *Ureaplasma urealyticum* into two biovars. Mol. Cell Probes *9*: 383–389.

Grau, O., F. Laigret, P. Carle, J. Tully, D. Rose and J. Bové. 1991. Identification of a plant-derived mollicute as a strain of an avian pathogen, *Mycoplasma iowae*, and its implications for mollicute taxonomy. Int. J. Syst. Bacteriol. *41*: 473–478.

Gray, L.D., K.L. Ketring and Y.W. Tang. 2005. Clinical use of 16S rRNA gene sequencing to identify *Mycoplasma felis* and *M. gateae* associated with feline ulcerative keratitis. J. Clin. Microbiol. *43*: 3431–3434.

Greco, G., M. Corrente, V. Martella, A. Pratelli and D. Buonavoglia. 2001. A multiplex-PCR for the diagnosis of contagious agalactia of sheep and goats. Mol. Cell. Probes *15*: 21–25.

Green III, F. and R.P. Hanson. 1973. Ultrastructure and capsule of *Mycoplasma meleagridis*. Journal of Bacteriology *116*: 1011–1018.

Grenabo, L., H. Hedelin and S. Pettersson. 1988. Urinary infection stones caused by *Ureaplasma urealyticum*: a review. Scand. J. Infect. Dis. Suppl. *53*: 46–49.

Grisold, A., M. Hoenigl, E. Leitner, K. Jakse, G. Feierl, R. Raggam and E. Marth. 2008. Submasseteric abscess caused by *Mycoplasma salivarium* infection. J. Clin. Microbiol. *46*: 3860–3862.

Groebel, K., K. Hoelzle, M.M. Wittenbrink, U. Ziegler and L.E. Hoelzle. 2009. *Mycoplasma suis* invades porcine erythrocytes. Infect. Immun. *77*: 576–584.

Gwaltney, S.M. and R.D. Oberst. 1994. Comparison of an improved polymerase chain reaction protocol and the indirect hemagglutination assay in the detection of *Eperythrozoon suis* infection. J. Vet. Diagn. Invest. *6*: 321–325.

Hale, H.H., C.F. Helmboldt, W.N. Plastridge and E.F. Stula. 1962. Bovine mastitis caused by a *Mycoplasma* species. Cornell Vet. *52*: 582–591.

Hammond, P., A. Ramírez, C. Morrow and J. Bradbury. 2009. Development and evaluation of an improved diagnostic PCR for *Mycoplasma synoviae* using primers located in the haemagglutinin encoding gene vlhA and its value for strain typing. Vet. Microbiol. *136*: 61–68.

Hannan, P. 2000. Guidelines and recommendations for antimicrobial minimum inhibitory concentration (MIC) testing against veterinary *Mycoplasma* species. International Research Programme on Comparative Mycoplasmology. Vet. Res. *31*: 373–395.

Harasawa, R., K. Koshimizu, I. Pan, E. Stephens and M. Barile. 1984. Genomic analysis of avian and feline *Ureaplasmas* by restriction endonucleases. Isr. J. Med. Sci. *20*: 942–945.

Harasawa, R., K. Koshimizu, I. Pan and M. Barile. 1985. Genomic and phenotypic analyses of avian *Ureaplasma* strains. Nippon Juigaku Zasshi *47*: 901–909.

Harasawa, R., Y. Imada, M. Ito, K. Koshimizu, G.H. Cassell and M.F. Barile. 1990a. *Ureaplasma felinum* sp. nov. and *Ureaplasma cati* sp. nov. isolated from the oral cavities of cats. Int. J. Syst. Bacteriol. *40*: 45–51.

Harasawa, R., E.B. Stephens, K. Koshimizu, I.J. Pan and M.F. Barile. 1990b. DNA relatedness among established *Ureaplasma* species and unidentified feline and canine serogroups. Int. J. Syst. Bacteriol. *40*: 52–55.

Harasawa, R., K. Dybvig, H.L. Watson and G.H. Cassell. 1991. Two genomic clusters among 14 serovars of *Ureaplasma urealyticum*. Syst. Appl. Microbiol. *14*: 393–396.

Harasawa, R., Y. Imada, H. Kotani, K. Koshimizu and M.F. Barile. 1993. *Ureaplasma canigenitalium* sp. nov., isolated from dogs. Int. J. Syst. Bacteriol. *43*: 640–644.

Harasawa, R. and Y. Kanamoto. 1999. Differentiation of two biovars of *Ureaplasma urealyticum* based on the 16S-23S rRNA intergenic spacer region. J. Clin. Microbiol. *37*: 4135–4138.

Harasawa, R., D.G. Pitcher, A.S. Ramirez and J.M. Bradbury. 2004. A putative transposase gene in the 16S-23S rRNA intergenic spacer region of *Mycoplasma imitans*. Microbiology *150*: 1023–1029.

Hasselbring, B.M., J.L. Jordan, R.W. Krause and D.C. Krause. 2006. Terminal organelle development in the cell wall-less bacterium *Mycoplasma pneumoniae*. Proc. Natl. Acad. Sci. USA *103*: 16478–16483.

Hatchel, J., R. Balish, M. Duley and M. Balish. 2006. Ultrastructure and gliding motility of *Mycoplasma amphoriforme*, a possible human respiratory pathogen. Microbiology *152*: 2181–2189.

Hatchel, J.M. and M.F. Balish. 2008. Attachment organelle ultrastructure correlates with phylogeny, not gliding motility properties, in *Mycoplasma pneumoniae* relatives. Microbiology *154*: 286–295.

Haulena, M., F. Gulland, J. Lawrence, D. Fauquier, S. Jang, B. Aldridge, T. Spraker, L. Thomas, D. Brown, L. Wendland and M. Davidson. 2006. Lesions associated with a novel *Mycoplasma* sp. in California sea lions (Zalophus californianus) undergoing rehabilitation. J. Wildl. Dis. *42*: 40–45.

Heggie, A.D., D. Bar-Shain, B. Boxerbaum, A.A. Fanaroff, M.A. O'Riordan and J.A. Robertson. 2001. Identification and quantification of ureaplasmas colonizing the respiratory tract and assessment of their role in the development of chronic lung disease in preterm infants. Pediatr. Infect. Dis. J. *20*: 854–859.

Henderson, G. and G. Jensen. 2006. Three-dimensional structure of *Mycoplasma pneumoniae*'s attachment organelle and a model for its role in gliding motility. Mol. Microbiol. *60*: 376–385.

Heyward, J.T., M.Z. Sabry and W.R. Dowdle. 1969. Characterization of *Mycoplasma* species of feline origin. Am. J. Vet. Res. *30*: 615–622.

Hill, A. 1971. *Mycoplasma caviae*, a new species. J. Gen. Microbiol. *65*: 109–113.

Hill, A. 1977. The isolation of mycoplasmas from non-human primates. Vet. Rec. *101*: 117.

Hill, A.C. 1983a. *Mycoplasma cricetuli*, a new species from the conjunctivas of chinese hamsters. Int. J. Syst. Bacteriol. *33*: 113–117.

Hill, A.C. 1983b. *Mycoplasma collis*, a new species isolated from rats and mice. Int. J. Syst. Bacteriol. *33*: 847–851.

Hill, A.C. 1984. *Mycoplasma cavipharyngis*, a new species isolated from the nasopharynx of guinea pigs. J. Gen. Microbiol. *130*: 3183–3188.

Hill, A.C. 1985. *Mycoplasma testudinis*, a new species isolated from a tortoise. Int. J. Syst. Bacteriol. *35*: 489–492.

Hill, A.C. 1986. *Mycoplasma felifaucium*, a new species isolated from the respiratory tract of pumas. J. Gen. Microbiol. *132*: 1923–1928.

Hill, A.C. 1988. *In* Validation of the publication of new names and new combinations previously effectively published outside the IJSB. List no. 27. Int. J. Syst. Bacteriol. *38*: 449.

Hill, A.C. 1989. *In* Validation of the publication of new names and new combinations previously effectively published outside the IJSB. List no. 30. Int. J. Syst. Bacteriol. *39*: 371.

Hill, A.C. 1991a. *Mycoplasma spermatophilum*, a new species isolated from human spermatozoa and cervix. Int. J. Syst. Bacteriol. *41*: 229–233.

Hill, A.C. 1991b. *Mycoplasma oxoniensis*, a new species isolated from chinese hamster conjunctivas. Int. J. Syst. Bacteriol. *41*: 21–25.

Hill, A.C. 1992. *Mycoplasma simbae* sp. nov., *Mycoplasma leopharyngis* sp. nov., and *Mycoplasma leocaptivus* sp. nov., isolated from lions. Int. J. Syst. Bacteriol. *42*: 518–523.

Hill, A.C. 1993. *Mycoplasma indiense* sp. nov., isolated from the throats of nonhuman primates. Int. J. Syst. Bacteriol. *43*: 36–40.

Hill, A., M. Tucker, D. Whittingham and I. Craft. 1987. Mycoplasmas and *in vitro* fertilization. Fertil. Steril. *47*: 652–655.

Hinz, K.H., H. Pfutzner and K.P. Behr. 1994. Isolation of mycoplasmas from clinically healthy adult breeding geese in Germany. Zentralbl Veterinarmed B *41*: 145–147.

Hirose, K., H. Kobayashi, N. Ito, Y. Kawasaki, M. Zako, K. Kotani, H. Ogawa and H. Sato. 2003. Isolation of Mycoplasmas from nasal swabs of calves affected with respiratory diseases and antimicrobial susceptibility of their isolates. J. Vet. Med. B Infect. Dis. Vet. Public Health *50*: 347–351.

Hodges, R.T., M. MacPherson, R.H. Leach and S. Moller. 1983. Isolation of *Mycoplasma dispar* from mastitis in dry cows. NZ Vet. J. *31*: 60–61.

Hoelzle, L. 2008. Haemotrophic mycoplasmas: recent advances in *Mycoplasma suis*. Vet. Microbiol. *130*: 215–226.

Hoffman, R.W., M.P. Luttrell, W.R. Davidson and D.H. Ley. 1997. Mycoplasmas in wild turkeys living in association with domestic fowl. J. Wildl. Dis. *33*: 526–535.

Hooper, P.T., L.A. Ireland and A. Carter. 1985. *Mycoplasma polyarthritis* in a cat with probable severe immune deficiency. Aust. Vet. J. *62*: 352.

Hopkins, P.M., D.S. Winlaw, P.N. Chhajed, J.L. Harkness, M.D. Horton, A.M. Keogh, M.A. Malouf and A.R. Glanville. 2002. *Mycoplasma hominis* infection in heart and lung transplantation. J. Heart Lung Transplant *21*: 1225–1229.

Horowitz, S., L. Duffy, B. Garrett, J. Stephens, J. Davis and G. Cassell. 1986. Can group- and serovar-specific proteins be detected in *Ureaplasma urealyticum*? Pediatr. Infect. Dis. *5*: S325–331.

Horowitz, S., J. Horowitz, D. Taylor-Robinson, S. Sukenik, R.N. Apte, J. Bar-David, B. Thomas and C. Gilroy. 1994. *Ureaplasma urealyticum* in Reiter's syndrome. J. Rheumatol. *21*: 877–882.

Hoskins, J.D. 1991. Canine haemobartonellosis, canine hepatozoonosis, and feline cytauxzoonosis. Vet. Clin. North Am. Small Anim. Pract. *21*: 129–140.

Howard, C.J., D.H. Pocock and R.N. Gourlay. 1978. Base composition of deoxyribonucleic acid from ureaplasmas isolated from various animal species. Int. J. Syst. Bacteriol. *28*: 599–601.

Howard, C.J., R.N. Gourlay and S.G. Wyld. 1980. Isolation of a Virus, Mvbr1, from *Mycoplasma-Bovirhinis*. FEMS Microbiol. Lett. *7*: 163–165.

Howard, C.J., D.H. Pocock and R.N. Gourlay. 1981. Polyacrylamide gel electrophoretic comparison or the polypeptides from *Ureaplasma* species isolated from cattle and humans. Int. J. Syst. Bacteriol. *31*: 128–130.

Howard, C.J. and R.N. Gourlay. 1982. Proposal for a second species within the genus *Ureaplasma*, *Ureaplasma diversum* sp. nov. Int. J. Syst. Bacteriol. *32*: 446–452.

Howard, G.W. 1975. The experimental transmission of *Eperythrozoon ovis* by mosquitoes. Parasitology *71*: xxxiii.

Hsu, F.S., M.C. Liu, S.M. Chou, J.F. Zachary and A.R. Smith. 1992. Evaluation of an enzyme-linked immunosorbent assay for detection of *Eperythrozoon suis* antibodies in swine. Am. J. Vet. Res. *53*: 352–354.

Hu, W.S., R.Y. Wang, R.S. Liou, J.W. Shih and S.C. Lo. 1990. Identification of an insertion-sequence-like genetic element in the newly recognized human pathogen *Mycoplasma incognitus*. Gene *93*: 67–72.

Hum, S., A. Kessell, S. Djordjevic, R. Rheinberger, M. Hornitzky, W. Forbes and J. Gonsalves. 2000. Mastitis, polyarthritis and abortion caused by *Mycoplasma* species bovine group 7 in dairy cattle. Aust. Vet. J. *78*: 744–750.

Hyde, T., M. Gilbert, S. Schwartz, E. Zell, J. Watt, W. Thacker, D. Talkington and R. Besser. 2001. Azithromycin prophylaxis during a hospital outbreak of *Mycoplasma pneumoniae* pneumonia. J. Infect. Dis. *183*: 907–912.

Ivanics, E., R. Glávitis, G. Takacs, E. Molnár, Z. Bitay and M. Meder. 1988. An outbreak of *Mycoplasma anatis* infection associated with nervous symptoms in large-scale duck flocks. Zentralbl. Veterinarmed. B *35*: 368–378.

Jackson, G., E. Boughton and S.G. Hamer. 1981. An outbreak of bovine mastitis associated with *Mycoplasma canadense*. Vet. Rec. *108*: 31–32.

Jacobs, E., A. Stuhlert, M. Drews, K. Pumpe, H. Schaefer, M. Kist and W. Bredt. 1988. Host reactions to *Mycoplasma pneumoniae* infections in guinea-pigs preimmunized systemically with the adhesin of this pathogen. Microb. Pathog. *5*: 259–265.

Jacobsson, B., I. Mattsby-Baltzer, B. Andersch, H. Bokstrom, R.M. Holst, N. Nikolaitchouk, U.B. Wennerholm and H. Hagberg. 2003. Microbial invasion and cytokine response in amniotic fluid in a Swedish population of women with preterm prelabor rupture of membranes. Acta Obstet. Gynecol. Scand. *82*: 423–431.

Jain, N.C., D.E. Jasper and J.D. Dellinger. 1967. Cultural characteristics and serological relationships of some mycoplasmas isolated from bovine sources. J. Gen. Microbiol. *49*: 401–410.

Jan, G., M. Le Hénaff, C. Fontenelle and H. Wróblewski. 2001. Biochemical and antigenic characterisation of *Mycoplasma gallisepticum* membrane proteins P52 and P67 (pMGA). Arch. Microbiol. *177*: 81–90.

Jasper, D.E., H. Erno, J.D. Dellinger and C. Christiansen. 1981. *Mycoplasma californicum*, a new species from cows. Int. J. Syst. Bacteriol. *31*: 339–345.

Jeansson, S. and J.E. Brorson. 1985. Elimination of mycoplasmas from cell cultures utilizing hyperimmune sera. Exp. Cell. Res. *161*: 181–188.

Jensen, J.S. 2004. *Mycoplasma genitalium*: the aetiological agent of urethritis and other sexually transmitted diseases. J. Eur. Acad. Dermatol. Venereol. *18*: 1–11.

Jensen, J.S., E. Bjornelius, B. Dohn and P. Lidbrink. 2004. Comparison of first void urine and urogenital swab specimens for detection of *Mycoplasma genitalium* and *Chlamydia trachomatis* by polymerase chain reaction in patients attending a sexually transmitted disease clinic. Sex. Transm. Dis. *31*: 499–507.

Johansson, K.E. and Pettersson B. 2002. Taxonomy of *Mollicutes*. *In* Molecular biology and pathogenicity of mycoplasmas (edited by Razin and Herrmann). Kluwer Academic, New York, pp. 1–30.

Jones, G.E., A. Foggie, A. Sutherland and D.B. Harker. 1976. Mycoplasmas and ovine keratoconjunctivitis. Vet. Rec. *99*: 137–141.

Jones, G.E. 1985. The pathogenicity of some ovine or caprine mycoplasmas in the lactating mammary gland of sheep and goats. J. Comp. Pathol. *95*: 305–318.

Jordan, F.T., J.N. Howse, M.P. Adams and O.O. Fatunmbi. 1981. The isolation of *Mycoplasma columbinum* and M columborale from feral pigeons. Vet. Rec. *109*: 450.

Jordan, F.T.W. 1979. Avian mycoplasmas. *In* The Mycoplasmas, vol. 2 (edited by Tully Maniloff Whitcomb). Academic Press, New York, pp. 1–48.

Jordan, F.T.W., H. Ernø, G.S. Cottew, K.H. Kunz and L. Stipkovits. 1982. Characterization and taxonomic description of five *Mycoplasma* serovars (serotypes) of avian origin and their elevation to species rank and further evaluation of the taxonomic status of *Mycoplasma synoviae*. Int. J. Syst. Bacteriol. *32*: 108–115.

Judicial Commission. 1958. Opinion 22. Status of the generic name Asterococcus and conservation of the generic name *Mycoplasma*. Int. Bull. Bacteriol. Nomencl. Taxon. *8*: 166–168.

Kakulphimp, J., L.R. Finch and J.A. Robertson. 1991. Genome sizes of mammalian and avian Ureaplasmas. Int. J. Syst. Bacteriol. *41*: 326–327.

Kaya, S., O. Poyraz, G. Gokce, H. Kilicarslan, K. Kaya and S. Ayan. 2003. Role of genital mycoplasmata and other bacteria in urolithiasis. Scand. J. Infect. Dis. *35*: 315–317.

Keane, F.E., B.J. Thomas, C.B. Gilroy, A. Renton and D. Taylor-Robinson. 2000. The association of *Mycoplasma hominis*, *Ureaplasma urealyticum* and *Mycoplasma genitalium* with bacterial vaginosis: observations on heterosexual women and their male partners. Int. J. STD AIDS *11*: 356–360.

Kempf, I., F. Gesbert, E. Guinebert, G. M. and G. Bennejean. 1991. Isolement et caractérisation d'une souche mycoplasmique chez des faisans d'élevage. Recuil de Medecine Veterinaire *167*: 1133–1139.

Kempf, I. 1998. DNA amplification methods for diagnosis and epidemiological investigations of avian mycoplasmosis. Avian Pathol. *27*: 7–14.

Kempf, I., C. Chastel, S. Ferris, F. Dufour-Gesbert, K.E. Johansson, B. Pettersson and A. Blanchard. 2000. Isolation of *Mycoplasma columborale* from a fly (Musca domestica). Vet. Rec. *147*: 304–305.

Kenny, G. 1983. Inhibition of the growth of *Ureaplasma urealyticum* by a new urease inhibitor, flurofamide. Yale J. Biol. Med. *56*: 717–722.

Keymer, I.F., R.H. Leach, R.A. Clarke, M.E. Bardsley and R.R. McIntyre. 1984. Isolation of *Mycoplasma* spp. from racing pigeons (*Columba livia*). Avian Pathol. *13*: 65–74.

Kidanemariam, A., J. Gouws, M. van Vuuren and B. Gummow. 2005. Ulcerative balanitis and vulvitis of Dorper sheep in South Africa: a study on its aetiology and clinical features. JS Afr. Vet. Assoc. *76*: 197–203.

Kikuth, W. 1928. Über Einen neuen Anämeerreger; *Bartonella canis* nov. spec. Klin. Wochenschr. *7*: 1729–1730.

Kilian, M., M.B. Brown, T.A. Brown, E.A. Freundt and G.H. Cassell. 1984. Immunoglobulin A1 protease activity in strains of *Ureaplasma urealyticum*. Acta Pathol. Microbiol. Immunol. Scand. [B]. *92*: 61–64.

King, K.W. and K. Dybvig. 1994. Mycoplasmal cloning vectors derived from plasmid pKMK1. Plasmid *31*: 49–59.

Kirchhoff, H. 1978. *Mycoplasma equigenitalium*, a new species from cervix region of mares. Int. J. Syst. Bacteriol. *28*: 496–502.

Kirchhoff, H., P. Beyene, M. Fischer, J. Flossdorf, J. Heitmann, B. Khattab, D. Lopatta, R. Rosengarten, G. Seidel and C. Yousef. 1987. *Mycoplasma mobile* sp. nov., a new species from fish. Int. J. Syst. Bacteriol. *37*: 192–197.

Kirchhoff, H., A. Binder, B. Liess, K.T. Friedhoff, J. Pohlenz, M. Stede and T. Willhaus. 1989. Isolation of mycoplasmas from diseased seals. Vet. Rec. *124*: 513–514.

Kirchhoff, H., R. Schmidt, H. Lehmann, H.W. Clark and A.C. Hill. 1996. *Mycoplasma elephantis* sp. nov., a new species from elephants. Int. J. Syst. Bacteriol. *46*: 437–441.

Kirchhoff, H., K. Mohan, R. Schmidt, M. Runge, D.R. Brown, M.B. Brown, C.M. Foggin, P. Muvavarirwa, H. Lehmann and J. Flossdorf. 1997. *Mycoplasma crocodyli* sp. nov., a new species from crocodiles. Int. J. Syst. Bacteriol. *47*: 742–746.

Kirchoff, H. 1974. Neue spezies der Fam. *Acholeplasmataceae* und der Fam. *Mykoplasmataceae* bei Pferden. Zentralbl. Veterinarmed. B *21*: 207–210.

Kisary, J., A. El-Ebeedy and L. Stipkovits. 1976. *Mycoplasma* infection of geese. II. Studies on pathogenicity of mycoplasmas in goslings and goose and chicken embryos. Avian Pathol. *5*: 15–20.

Klassen, T.P., N. Wiebe, K. Russell, K. Stevens, L. Hartling, W.R. Craig and D. Moher. 2002. Abstracts of randomized controlled trials presented at the society for pediatric research meeting: an example of publication bias. Arch. Pediatr. Adolesc. Med. *156*: 474–479.

Kleckner, A.L. 1960. Serotypes of avian pleuropneumonia-like organisms. Am. J. Vet. Res. *21*: 274–280.

Kleven, S.H., C.S. Eidson and O.J. Fletcher. 1978. Airsacculitis induced in broilers with a combination of *Mycoplasma gallinarum* and respiratory viruses. Avian Dis. *22*: 707–716.

Kleven, S.H. 1998. Mycoplasmas in the etiology of multifactorial respiratory disease. Poult. Sci. *77*: 1146–1149.

Kleven, S.H. 2003. Mycoplasmosis: other mycoplasmal infections. *In* Diseases of Poultry, 11th edn (edited by Saif and Barnes). Wiley-Blackwell, Hoboken, NJ.

Kleven, S.H. 2008. Control of avian *Mycoplasma* infections in commercial poultry. Avian Dis. *52*: 367–374.

Knox, C., P. Giffard and P. Timms. 1998. The phylogeny of *Ureaplasma urealyticum* based on the mba gene fragment. Int. J. Syst. Bacteriol. *48 Pt 4*: 1323–1331.

Knox, C. and P. Timms. 1998. Comparison of PCR, nested PCR, and random amplified polymorphic DNA PCR for detection and typing of *Ureaplasma urealyticum* in specimens from pregnant women. J. Clin. Microbiol. *36*: 3032–3039.

Knox, C., J. Allan, J. Allan, W. Edirisinghe, D. Stenzel, F. Lawrence, D. Purdie and P. Timms. 2003. *Ureaplasma parvum* and *Ureaplasma urealyticum* are detected in semen after washing before assisted reproductive technology procedures. Fertil. Steril. *80*: 921–929.

Kobayashi, H., M. Runge, R. Schmidt, M. Kubo, K. Yamamoto and H. Kirchhoff. 1997. *Mycoplasma lagogenitalium* sp. nov., from the preputial smegma of Afghan pikas (*Ochotona rufescens rufescens*). Int. J. Syst. Bacteriol. *47*: 1208–1211.

Kokotovic, B., N.F. Friis and P. Ahrens. 2007. *Mycoplasma alkalescens* demonstrated in bronchoalveolar lavage of cattle in Denmark. Acta Vet. Scand. *49*: 2.

Kong, F., G. James, Z. Ma, S. Gordon, W. Bin and G. Gilbert. 1999a. Phylogenetic analysis of *Ureaplasma urealyticum*–support for the establishment of a new species, *Ureaplasma parvum*. Int. J. Syst. Bacteriol. *49 Pt 4*: 1879–1889.

Kong, F., X. Zhu, W. Wang, X. Zhou, S. Gordon and G.L. Gilbert. 1999b. Comparative analysis and serovar-specific identification of multiple-banded antigen genes of *Ureaplasma urealyticum* biovar 1. J. Clin. Microbiol. *37*: 538–543.

Kong, F., Z. Ma, G. James, S. Gordon and G. Gilbert. 2000. Molecular genotyping of human *Ureaplasma* species based on multiple-banded antigen (MBA) gene sequences. Int. J. Syst. Evol. Microbiol. *50 Pt 5*: 1921–1929.

Königsson, M.H., B. Pettersson and K.E. Johansson. 2001. Phylogeny of the seal mycoplasmas *Mycoplasma phocae* corrig., *Mycoplasma phocicerebrale* corrig. and *Mycoplasma phocirhinis* corrig. based on sequence analysis of 16S rDNA. Int. J. Syst. Evol. Microbiol. *51*: 1389–1393.

Koshimizu, K., M. Ito, T. Magaribuchi and H. Kotani. 1983. Selective medium for isolation of ureaplasmas from animals. Nippon Juigaku Zasshi *45*: 263–268.

Koshimizu, K., R. Harasawa, I.J. Pan, H. Kotani, M. Ogata, E.B. Stephens and M.F. Barile. 1987. *Ureaplasma gallorale* sp. nov. from the oropharynx of chickens. Int. J. Syst. Bacteriol. *37*: 333–338.

Krause, D.C. and D. Taylor-Robinson. 1992. Mycoplasmas which infect humans. *In* Mycoplasmas: Molecular Biology and Pathogenesis (edited by Maniloff, McElhaney, Finch and Baseman). American Society for Microbiology, Washington, D.C., pp. 417–444.

Krause, D.C. and M.F. Balish. 2004. Cellular engineering in a minimal microbe: structure and assembly of the terminal organelle of *Mycoplasma pneumoniae*. Mol. Microbiol. *51*: 917–924.

Kreier, J.P. and M. Ristic. 1963. Morphologic, antigenic, and pathogenic characteristics of *Eperythrozoon ovis* and *Eperythrozoon wenyoni*. Am. J. Vet. Res. *24*: 488–500.

Kreier, J.P. and M. Ristic. 1968. Haemobartonellosis, eperythrozoonosis, grahamellosis and ehrlichiosis. *In* Infectious Blood Diseases of Man and Animals (edited by Weinman and Ristic). Academic Press, New York, pp. 387–472.

Kreier, J.P. and M. Ristic. 1984. Genus III. *Haemobartonella*; Genus IV. *Eperythrozoon*. *In* Bergey's Manual of Systematic Bacteriology, vol. 1 (edited by Krieg and Holt). Williams & Wilkins, Baltimore, pp. 724–729.

Kusiluka, L.J., B. Ojeniyi, N.F. Friis, R.R. Kazwala and B. Kokotovic. 2000. Mycoplasmas isolated from the respiratory tract of cattle and goats in Tanzania. Acta Vet. Scand. *41*: 299–309.

Lam, K., A. DaMassa and G. Ghazikhanian. 2004. *Mycoplasma meleagridis*-induced lesions in the tarsometatarsal joints of turkey embryos. Avian Dis. *48*: 505–511.

Lamm, C.G., L. Munson, M.C. Thurmond, B.C. Barr and L.W. George. 2004. *Mycoplasma otitis* in California calves. J. Vet. Diagn. Invest. *16*: 397–402.

Lamster, I., J. Grbic, R. Bucklan, D. Mitchell-Lewis, H. Reynolds and J. Zambon. 1997. Epidemiology and diagnosis of HIV-associated periodontal diseases. Oral Dis. *3 Suppl 1*: S141–148.

Langford, E.V. and R.H. Leach. 1973. Characterization of a *Mycoplasma* isolated from infectious bovine keratoconjunctivitis: *M. bovoculi* sp. nov. Can. J. Microbiol. *19*: 1435–1444.

Langford, E.V., H.L. Ruhnke and O. Onoviran. 1976. *Mycoplasma canadense*, a new bovine species. Int. J. Syst. Bacteriol. *26*: 212–219.

Lartigue, C., J.I. Glass, N. Alperovich, R. Pieper, P.P. Parmar, C.A. Hutchison, 3rd, H.O. Smith and J.C. Venter. 2007. Genome transplantation in bacteria: changing one species to another. Science *317*: 632–638.

Lartigue, C., S. Vashee, M.A. Algire, R.Y. Chuang, G.A. Benders, L. Ma, V.N. Noskov, E.A. Denisova, D.G. Gibson, N. Assad-Garcia, N. Alperovich, D.W. Thomas, C. Merryman, C.A. Hutchison, 3rd, H.O. Smith, J.C. Venter and J.I. Glass. 2009. Creating bacterial strains from genomes that have been cloned and engineered in yeast. Science *325*: 1693–1696.

Lavrič, M., D. Benčina and M. Narat. 2005. *Mycoplasma gallisepticum* hemagglutinin V1hA, pyruvate dehydrogenase PdhA, lactate dehydrogenase, and elongation factor Tu share epitopes with *Mycoplasma imitans* homologues. Avian Dis. *49*: 507–513.

Leach, R.H. 1967. Comparative studies of *Mycoplasma* of bovine origin. Ann. N. Y. Acad. Sci. *143*: 305–316.

Leach, R.H. 1973. Further studies on classification of bovine strains of *Mycoplasmatales*, with proposals for new species, *Acholeplasma modicum* and *Mycoplasma alkalescens*. J. Gen. Microbiol. *75*: 135–153.

Leach, R.H. 1983. Preservation of *Mycoplasma* cultures and culture collections. *In* Methods in Mycoplasmology, vol. 1 (edited by Razin and Tully). Academic Press, New York, pp. 197–204.

Leach, R.H., H. Ernø and K.J. MacOwan. 1993. Proposal for designation of F38-type caprine mycoplasmas as *Mycoplasma capricolum* subsp. *capripneumoniae* and consequent obligatory relegation of strains currently classified as *Mycoplasma capricolum* (Tully, Barile, Edward, Theodore, and Ernø 1974) to an additional new subspecies, *M. capricolum* subsp. *capricolum* subsp. nov. Int. J. Syst. Bacteriol. *43*: 603–605.

Lee, G.Y. and G.E. Kenny. 1987. Humoral immune response to polypeptides of *Ureaplasma urealyticum* in women with postpartum fever. J. Clin. Microbiol. *25*: 1841–1844.

Lehmer, R.R., B.S. Andrews, J.A. Robertson, E.E. Stanbridge, L. de la Maza and G.J. Friow. 1991. Polyarthiritis due to *Ureaplasma urealyticum* infection in a patient with common variable immunodeficiency (CVID): Similarities to rheumatoid arthritis. Ann. Rheum. Dis. *50*: 574–576.

Lemcke, R.M. 1979. Equine Mycoplasmas. *In* The Mycoplasmas Volume 2: Human and Animal Mycoplasmas, vol. 2 (edited by Tully and Whitcomb). Academic Press, New York, pp. 177–189.

Lemcke, R.M. and H. Kirchhoff. 1979. *Mycoplasma subdolum*, a new species isolated from horses. Int. J. Syst. Bacteriol. *29*: 42–50.

Lemcke, R.M. and J. Poland. 1980. *Mycoplasma fastidiosum*: new species from horses. Int. J. Syst. Bacteriol. *30*: 151–162.

Lesnoff, M., G. Laval, P. Bonnet, S. Abdicho, A. Workalemahu, D. Kifle, A. Peyraud, R. Lancelot and F. Thiaucourt. 2004. Within-herd spread of contagious bovine pleuropneumonia in Ethiopian highlands. Prev Vet Med *64*: 27–40.

Levisohn, S. and S.H. Kleven. 2000. Avian mycoplasmosis (*Mycoplasma gallisepticum*). Rev. Sci. Tech. *19*: 425–442.

Ley, D.H., J.E. Berkhoff and J.M. McLaren. 1996. *Mycoplasma gallisepticum* isolated from house finches (Carpodacus mexicanus) with conjunctivitis. Avian Dis. *40*: 480–483.

Ley, D.H., S.J. Geary, J.E. Berkhoff, J.M. McLaren and S. Levisohn. 1998. *Mycoplasma sturni* from blue jays and northern mockingbirds with conjunctivitis in Florida. J. Wildl. Dis. *34*: 403–406.

Li, X., X. Jia, D. Shi, Y. Xiao, S. Hu, M. Liu, Z. Yuan and D. Bi. 2008. Continuous *in vitro* Cultivation of *Mycoplasma suis*. Acta Veterinaria et Zootechnica Sinica *38*: 1142–1146.

Li, Y., A. Brauner, B. Jonsson, I. van der Ploeg, O. Söder, M. Holst, J. Jensen, H. Lagercrantz and K. Tullus. 2000. *Ureaplasma urealyticum*-induced production of proinflammatory cytokines by macrophages. Pediatr. Res. *48*: 114–119.

Lierz, M., R. Schmidt, L. Brunnberg and M. Runge. 2000. Isolation of *Mycoplasma meleagridis* from free-ranging birds of prey in Germany. J. Vet. Med. B Infect. Dis. Vet. Public Health *47*: 63–67.

Lierz, M., R. Schmidt and M. Runge. 2002. *Mycoplasma* species isolated from falcons in the Middle East. Vet. Rec. *151*: 92–93.

Lierz, M., S. Deppenmeier, A. Gruber, S. Brokat and H. Hafez. 2007a. Pathogenicity of *Mycoplasma lipofaciens* strain ML64 for turkey embryos. Avian Pathol. *36*: 389–393.

Lierz, M., N. Hagen, N. Harcourt-Brown, S.J. Hernandez-Divers, D. Luschow and H.M. Hafez. 2007b. Prevalence of mycoplasmas in eggs from birds of prey using culture and a genus-specific *Mycoplasma* polymerase chain reaction. Avian Pathol. *36*: 145–150.

Lierz, M., R. Stark, S. Brokat and H. Hafez. 2007c. Pathogenicity of *Mycoplasma lipofaciens* strain ML64, isolated from an egg of a Northern Goshawk (Accipiter gentilis), for chicken embryos. Avian Pathol. *36*: 151–153.

Lierz, M., N. Hagen, S.J. Hernadez-Divers and H.M. Hafez. 2008a. Occurrence of mycoplasmas in free-ranging birds of prey in Germany. J. Wildl. Dis. *44*: 845–850.

Lierz, M., A. Jansen and H. Hafez. 2008b. Avian *Mycoplasma lipofaciens* transmission to veterinarian. Emerg. Infect. Dis. *14*: 1161–1163.

Lierz, M., E. Obon, B. Schink, F. Carbonell and H.M. Hafez. 2008c. The role of mycoplasmas in a conservation project of the lesser kestrel (*Falco naumanni*). Avian Dis. *52*: 641–645.

Lin, J., M. Kendrick and E. Kass. 1972. Serologic typing of human genital T-mycoplasmas by a complement-dependent mycoplasmacidal test. J. Infect. Dis. *126*: 658–663.

Livingston, C.J. and B. Gauer. 1974. Serologic typing of T-strain *Mycoplasma* isolated from the respiratory and reproductive tracts of cattle in the United States. Am. J. Vet. Res. *35*: 1469–1471.

Lo, S.C., M.M. Hayes, J.G. Tully, R.Y. Wang, H. Kotani, P.F. Pierce, D.L. Rose and J.W. Shih. 1992. *Mycoplasma penetrans* sp. nov., from the urogenital tract of patients with AIDS. Int. J. Syst. Bacteriol. *42*: 357–364.

Lobo, E., M.C. García, H. Moscoso, S. Martínez and S.H. Kleven. 2004. Strain heterogeneity in *Mycoplasma pullorum* isolates identified by random amplified polymorphic DNA techniques. Spanish Journal of Agricultural Research *2*: 500–503.

Lorenzon, S., L. Manso-Silvan and F. Thiaucourt. 2008. Specific real-time PCR assays for the detection and quantification of *Mycoplasma mycoides* subsp. *mycoides* SC and *Mycoplasma capricolum* subsp. *capripneumoniae*. Mol. Cell. Probes. *22*: 324–328.

Luo, W., H. Yu, Z. Cao, T.R. Schoeb, M. Marron and K. Dybvig. 2008. Association of *Mycoplasma arthritidis* mitogen with lethal toxicity but not with arthritis in mice. Infect. Immun. *76*: 4989–4998.

Luttrell, M.P., T.H. Eleazer and S.H. Kleven. 1992. *Mycoplasma gallopavonis* in eastern wild turkeys. J. Wildl. Dis. *28*: 288–291.

Lysnyansky, I., M. Calcutt, I. Ben-Barak, Y. Ron, S. Levisohn, B. Methé and D. Yogev. 2009. Molecular characterization of newly identified IS3, IS4 and IS30 insertion sequence-like elements in *Mycoplasma bovis* and their possible roles in genome plasticity. FEMS Microbiol. Lett. *294*: 172–182.

MacKenzie, C., B. Henrich and U. Hadding. 1996. Biovar-specific epitopes of the urease enzyme of *Ureaplasma urealyticum*. J. Med. Microbiol. *45*: 366–371.

MacOwan, K.J., H.G. Jones, C.J. Randall and F.T. Jordan. 1981. *Mycoplasma columborale* in a respiratory condition of pigeons and experimental air sacculitis of chickens. Vet. Rec. *109*: 562.

Madden, D.L., K.E. Moats, W.T. London, E.B. Matthew and J.L. Sever. 1974. *Mycoplasma moatsii*, a new species isolated from recently imported Grivit monkeys (*Cercopithecus aethiops*). Int. J. Syst. Bacteriol. *24*: 459–464.

Maeda, T., T. Shibahara, K. Kimura, Y. Wada, K. Sato, Y. Imada, Y. Ishikawa and K. Kadota. 2003. *Mycoplasma bovis*-associated suppurative otitis media and pneumonia in bull calves. J. Comp. Pathol. *129*: 100–110.

Manchee, R. and D. Taylor-Robinson. 1969. Enhanced growth of T-strain mycoplasmas with N-2-hydroxyethylpiperazone-N′-2-ethanesulfonic acid buffer. J. Bacteriol. *100*: 78–85.

Manchee, R. and D. Taylor-Robinson. 1970. Lysis and protection of erythrocytes by T-mycoplasmas. J. Med. Microbiol. *3*: 539–546.

Manhart, L.E., S.M. Dutro, K.K. Holmes, C.E. Stevens, C.W. Critchlow, D.A. Eschenbach and P.A. Totten. 2001. *Mycoplasma genitalium* is associated with mucopurulent cervicitis. Int. J. STD AIDS *12 (suppl. 2)*: 69.

Manso-Silván, L., E.M. Vilei, K. Sachse, S.P. Djordjevic, F. Thiaucourt and J. Frey. 2009. *Mycoplasma leachii* sp. nov. as a new species designation for *Mycoplasma* sp. bovine group 7 of Leach, and reclassification of *Mycoplasma mycoides* subsp. *mycoides* LC as a serovar of *Mycoplasma mycoides* subsp. *capri*. Int. J. Syst. Evol. Microbiol. *59*: 1353–1358.

March, J.B., C. Gammack and R. Nicholas. 2000. Rapid detection of contagious caprine pleuropneumonia using a *Mycoplasma capricolum* subsp. *capripneumoniae* capsular polysaccharide-specific antigen detection latex agglutination test. J. Clin. Microbiol. *38*: 4152–4159.

Maré, C.J. and W.P. Switzer. 1965. *Mycoplasm hyopneumoniae*, a causative agent of virus pig pneumonia. Vet. Med. *60*: 841–845.

Markham, P.F., M.F. Duffy, M.D. Glew and G.F. Browning. 1999. A gene family of *Mycoplasma imitans* closely related to the pMGA family of *Mycoplasma gallisepticum*. Microbiology *145*: 2095–2103.

Marois, C., F. Dufour-Gesbert and I. Kempf. 2001. Comparison of pulsed-field gel electrophoresis with random amplified polymorphic DNA for typing *of Mycoplasma synoviae*. Vet. Microbiol. *79*: 1–9.

Marois, C., J. Le Carrou, M. Kobisch and A.V. Gautier-Bouchardon. 2007. Isolation of *Mycoplasma hyopneumoniae* from different sampling sites in experimentally infected and contact SPF piglets. Vet. Microbiol. *120*: 96–104.

Mason, R.W. and P. Statham. 1991. The determination of the level of *Eperythrozoon ovis* parasitaemia in chronically infected sheep and its significance to the spread of infection. Aust. Vet. J. *68*: 115–116.

Masover, G., S. Razin and L. Hayflick. 1977. Localization of enzymes in *Ureaplasma urealyticum* (T-strain mycoplasma). J. Bacteriol. *130*: 297–302.

Masover, G.K. and F.A. Becker. 1996. Detection of mycoplasmas by DNA staining and fluorescent antibody methodology. *In* Molecular and Diagnostic Procedures in Mycoplasmology, vol. 2 (edited by Tully and Razin). Academic Press, San Diego, pp. 419–429.

Matlow, A., C. Th'ng, D. Kovach, P. Quinn, M. Dunn and E. Wang. 1998. Susceptibilities of neonatal respiratory isolates of *Ureaplasma urealyticum* to antimicrobial agents. Antimicrob. Agents Chemother. *42*: 1290–1292.

Mattila, P.S., P. Carlson, A. Sivonen, J. Savola, R. Luosto, J. Salo and M. Valtonen. 1999. Life-threatening *Mycoplasma hominis* mediastinitis. Clin. Infect. Dis. *29*: 1529–1537.

Mayer, D., M.P. Degiorgis, W. Meier, J. Nicolet and M. Giacometti. 1997. Lesions associated with infectious keratoconjunctivitis in alpine ibex. J. Wildl. Dis. *33*: 413–419.

May, M., G.J. Ortiz, L.D. Wendland, D.S. Rotstein, R.F. Relich, M.F. Balish and D.R. Brown. 2007. *Mycoplasma insons* sp. nov., a twisted *Mycoplasma* from green iguanas (Iguana iguana). FEMS Microbiol. Lett. *274*: 298–303.

Mayer, D., M.P. Degiorgis, W. Meier, J. Nocolet and M. Giacometti. 1997. Lesions associated with infectious keratoconjuctivitis in alpine ibex. J. Widl. Dis. *33*: 413–419.

Mayer, M. 1921. Über einige bakterienähnliche Parasiten der Erythrozyten bei Menschen und Tieren. Arch. Schiffs Trop. Hyg. *25*: 150–152.

Mazzali, R. and D. Taylor-Robinson. 1971. The behaviour of T-mycoplasmas in tissue culture. J. Med. Microbiol. *4*: 125–138.

McAuliffe, L., R. Ellis, K. Miles, R. Ayling and R. Nicholas. 2006. Biofilm formation by *Mycoplasma* species and its role in environmental persistence and survival. Microbiology *152*: 913–922.

McAuliffe, L., R.D. Ayling, R.J. Ellis and R.A. Nicholas. 2008. Biofilm-grown *Mycoplasma mycoides* subsp. *mycoides* SC exhibit both phenotypic and genotypic variation compared with planktonic cells. Vet. Microbiol. *129*: 315–324.

McElhaney, R.N. 1992a. Lipid composition, biosynthesis, and metabolism. *In* Mycoplasmas: Molecular Biology and Pathogenesis (edited by Maniloff, McElhaney, Finch and Baseman). American Society for Microbiology, Washington, D.C., pp. 231–258.

McElhaney, R.N. 1992b. Membrance function. *In* Mycoplasmas: Molecular Biology and Pathogenesis (edited by Maniloff, McElhaney, Finch and Baseman). American Society for Microbiology, Washington, D.C., pp. 259–287.

McElhaney, R.N. 1992c. Membrane structure. *In* Mycoplasmas: Molecular Biology and Pathogenesis (edited by Maniloff, McElhaney, Finch and Baseman). American Society for Microbiology, Washington, D.C., pp. 113–155.

McGarrity, G.J., D.L. Rose, V. Kwiatkowski, A.S. Dion, D.M. Phillips and J.G. Tully. 1983. *Mycoplasma muris*, a new species from laboratory mice. Int. J. Syst. Bacteriol. *33*: 350–355.

McLaughlin, B.G., P.S. McLaughlin and C.N. Evans. 1991. An *Eperythrozoon*-like parasite of llamas: attempted transmission to swine, sheep, and cats. J. Vet. Diagn. Invest. *3*: 352–353.

McMartin, D.A., K.J. MacOwan and L.L. Swift. 1980. A century of classical contagious caprine pleuropneumonia: from original description to aetiology. Br. Vet. J. *136*: 507–515.

Meloni, G.A., G. Bertoloni, F. Busolo and L. Conventi. 1980. Colony morphology, ultrastructure and morphogenesis in *Mycoplasma hominis*, *Acholeplasma laidlawii* and *Ureaplasma urealyticum*. J. Gen. Microbiol. *116*: 435–443.

Messick, J.B., P.G. Walker, W. Raphael, L. Berent and X. Shi. 2002. '*Candidatus Mycoplasma* haemodidelphidis' sp. nov., '*Candidatus Mycoplasma* haemolamae' sp. nov. and *Mycoplasma haemocanis* comb. nov., haemotrophic parasites from a naturally infected opossum (*Didelphis virginiana*), alpaca (*Lama pacos*) and dog (*Canis familiaris*): phylogenetic and secondary structural relatedness of their 16S rRNA genes to other mycoplasmas. Int. J. Syst. Evol. Microbiol. *52*: 693–698.

Messick, J.B. 2003. New perspectives about Hemotrophic *Mycoplasma* (formerly, *Haemobartonella* and *Eperythrozoon* species) infections in dogs and cats. Vet. Clin. N. Am. Small Anim. Pract. *33*: 1453–1465.

Meyer, R. and W. Clough. 1993. Extragenital *Mycoplasma hominis* infections in adults: emphasis on immunosuppression. Clin. Infect. Dis. *17 Suppl 1*: S243–249.

Meyling, A. and N.F. Friis. 1972. Serological identification of a new porcine *Mycoplasma* species, *M. flocculare*. Acta Vet. Scand. *13*: 287–289.

Michel, J.C., B. de Thoisy and H. Contamin. 2000. Chemotherapy of haemobartonellosis in squirrel monkeys (Saimiri sciureus). J. Med. Primatol. *29*: 85–87.

Mikaelian, I., D.H. Ley, R. Claveau, M. Lemieux and J.P. Berube. 2001. Mycoplasmosis in evening and pine grosbeaks with conjunctivitis in Quebec. J. Wildl. Dis. *37*: 826–830.

Miles, R.J. 1992a. Catabolism in *Mollicutes*. J. Gen. Microbiol. *138*: 1773–1783.

Miles, R.J. 1992b. Cell nutrition and growth. *In* Mycoplasmas: Molecular Biology and Pathogenesis (edited by Maniloff, McElhaney, Finch and Baseman). American Society for Microbiology, Washington, D.C., pp. 23–40.

Mirsalimi, S., S. Rosendal and R. Julian. 1989. Colonization of the intestine of turkey embryos exposed to *Mycoplasma iowae*. Avian Dis. *33*: 310–315.

Miyata, M., H. Yamamoto, T. Shimizu, A. Uenoyama, C. Citti and R. Rosengarten. 2000. Gliding mutants of *Mycoplasma mobile*: relationships between motility and cell morphology, cell adhesion and microcolony formation. Microbiology *146*: 1311–1320.

Miyata, M., W.S. Ryu and H.C. Berg. 2002. Force and velocity of *Mycoplasma mobile* gliding. J. Bacteriol. *184*: 1827–1831.

Moalic, P.Y., I. Kempf, F. Gesbert and F. Laigret. 1997. Identification of two pathogenic avian mycoplasmas as strains of *Mycoplasma pullorum*. Int. J. Syst. Bacteriol. *47*: 171–174.

Mohan, K., C.M. Foggin, P. Muvavarirwa and J. Honywill. 1997. Vaccination of farmed crocodiles (Crocodylus niloticus) against *Mycoplasma crocodyli* infection. Vet. Rec. *141*: 476.

Mohan, K., C.M. Foggin, F. Dziva and P. Muvavarirwa. 2001. Vaccination to control an outbreak of *Mycoplasma crocodyli* infection. Onderstepoort J. Vet. Res. *68*: 149–150.

Moise, N.S., J.W. Crissman, J.F. Fairbrother and C. Baldwin. 1983. *Mycoplasma gateae* arthritis and tenosynovitis in cats: case report and experimental reproduction of the disease. Am. J. Vet. Res. *44*: 16–21.

Monecke, S., J. Helbig and E. Jacobs. 2003. Phase variation of the multiple banded protein in *Ureaplasma urealyticum* and *Ureaplasma parvum*. Int. J. Med. Microbiol. *293*: 203–211.

Montagnier, L., A. Blanchard, D. Guetard, M. Lemaitre, A.M. Dirienzo, S. Chamaret, Y. Henin, E. Bahraoui, C. Dauguet, C. Axler, M. Kirstetter, R. Roue, G. pialoux and B. Dupont. 1990. A possible role of mycoplasmas as cofactors in AIDS. *In* Retroviruses of Human AIDS and Related Animal Diseases (edited by Girard and Valette). Colloque de Cent Gardes, Foundation Merieux, Lyon, pp. 9–17.

Montes, A., D. Wolfe, E. Welles, J. Tyler and E. Tepe. 1994. Infertility associated with *Eperythrozoon wenyonii* infection in a bull. J. Am. Vet. Med. Assoc. *204*: 261–263.

Moss, T., C. Knox, S. Kallapur, I. Nitsos, C. Theodoropoulos, J. Newnham, M. Ikegami and A. Jobe. 2008. Experimental amniotic fluid infection in sheep: effects of *Ureaplasma parvum* serovars 3 and 6 on preterm or term fetal sheep. Am. J. Obstet. Gynecol. *198*: 122. e121–128.

Mouches, C., D. Taylor-Robinson, L. Stipkovits and J. Bove. 1981. Comparison of human and animal Ureaplasmas by one- and two-dimensional protein analysis on polyacrylamide slab gel. Ann. Microbiol. (Paris) *132B*: 171–196.

Muhlrad, A., I. Peleg, J. Robertson, I. Robinson and I. Kahane. 1981. Acetate kinase activity in mycoplasmas. J. Bacteriol. *147*: 271–273.

Murakami, S., M. Miyama, A. Ogawa, J. Shimada and T. Nakane. 2002. Occurrence of conjunctivitis, sinusitis and upper region tracheitis in Japanese quail (Coturnix coturnix japonica), possibly caused by *Mycoplasma gallisepticum* accompanied by *Cryptosporidium* sp. infection. Avian Pathol. *31*: 363–370.

Nagata, K., E. Takagi, H. Satoh, H. Okamura and T. Tamura. 1995. Growth inhibition of *Ureaplasma urealyticum* by the proton pump inhibitor lansoprazole: direct attribution to inhibition by lansoprazole of urease activity and urea-induced ATP synthesis in *U. urealyticum*. Antimicrob. Agents Chemother. *39*: 2187–2192.

Nagatomo, H., H. Kato, T. Shimizu and B. Katayama. 1997. Isolation of mycoplasmas from fantail pigeons. J. Vet. Med. Sci. *59*: 461–462.

Nakane, D. and M. Miyata. 2007. Cytoskeletal "jellyfish" structure of *Mycoplasma mobile*. Proc. Natl. Acad. Sci. USA *104*: 19518–19523.

Nakane, D. and M. Miyata. 2009. Cytoskeletal asymmetrical dumbbell structure of a gliding mycoplasma, *Mycoplasma gallisepticum*, revealed by negative-staining electron microscopy. J. Bacteriol. *191*: 3256–3264.

Naot, Y., J.G. Tully and H. Ginsburg. 1977. Lymphocyte activation by various *Mycoplasma* strains and species. Infect. Immun. *18*: 310–317.

Neimark, H., D. Mitchelmore and R.H. Leach. 1998. An approach to characterizing uncultivated prokaryotes the Grey Lung agent and proposal of a *Candidatus* taxon for the organism, '*Candidatus Mycoplasma* ravipulmonis'. Int. J. Syst. Bacteriol. *48*: 389–394.

Neimark, H., K.E. Johansson, Y. Rikihisa and J.G. Tully. 2001. Proposal to transfer some members of the genera *Haemobartonella* and *Eperythrozoon* to the genus *Mycoplasma* with descriptions of '*Candidatus Mycoplasma haemofelis*', '*Candidatus Mycoplasma haemomuris*', '*Candidatus Mycoplasma* haemosuis' and '*Candidatus Mycoplasma wenyonii*'. Int. J. Syst. Evol. Microbiol. *51*: 891–899.

Neimark, H., A. Barnaud, P. Gounon, J.-C. Michel and H. Contamin. 2002a. The putative haemobartonella that influences *Plasmodium falciparum* parasitaemia in squirrel monkeys is a haemotrophic mycoplasma. Microbes Infect. *4*: 693–698.

Neimark, H., B. Hoff and M. Ganter. 2004. *Mycoplasma ovis* comb. nov. (formerly *Eperythrozoon ovis*), an epierythrocytic agent of haemolytic anaemia in sheep and goats. Int. J. Syst. Evol. Microbiol. *54*: 365–371.

Neimark, H., W. Peters, B.L. Robinson and L.B. Stewart. 2005. Phylogenetic analysis and description of *Eperythrozoon coccoides*, proposal to transfer to the genus *Mycoplasma* as *Mycoplasma coccoides* comb. nov. and Request for an Opinion. Int. J. Syst. Evol. Microbiol. *55*: 1385–1391.

Neimark, H.C., K.E. Johansson, Y. Rikihisa and J.G. Tully. 2002b. Revision of haemotrophic *Mycoplasma* species names. Int. J. Syst. Evol. Microbiol. *52*: 683.

Neitz, W.O., R.A. Alexander and P.J. de Toit. 1934. *Eperythrozoon ovis* (sp. nov.) infection in sheep. Onderstepoort J. Vet. Sci. *3*: 263–274.

Nelson, J.B. and M.J. Lyons. 1965. Phase-contrast and electron microscopy of murine strains of *Mycoplasma*. J. Bacteriol. *90*: 1750–1763.

Neyrolles, O., S. Ferris, N. Behbahani, L. Montagnier and A. Blanchard. 1996. Organization of *Ureaplasma urealyticum* urease gene cluster and expression in a suppressor strain of *Escherichia coli*. J. Bacteriol. *178*: 2725.

Niang, M., R.F. Rosenbusch, J.J. Andrews and M.L. Kaeberle. 1998. Demonstration of a capsule on *Mycoplasma ovipneumoniae*. Am. J. Vet. Res. *59*: 557–562.

Nicholas, R.A., A. Greig, S.E. Baker, R.D. Ayling, M. Heldtander, K.-E. Johansson, B.M. Houshaymi and R.J. Miles. 1998. Isolation of *Mycoplasma fermentans* from a sheep. Vet. Rec. *142*: 220–221.

Nicholas, R.A., Y.C. Lin, K. Sachse, H. Hotzel, K. Parham, L. McAuliffe, R.J. Miles, D.P. Kelly and A.P. Wood. 2008. Proposal that the strains of the *Mycoplasma* ovine/caprine serogroup 11 be reclassified as *Mycoplasma bovigenitalium*. Int. J. Syst. Evol. Microbiol. *58*: 308–312.

Nikol'skii, S.N. and S.N. Slipchenko. 1969. Experiments in the transmission of *Eperythrozoon ovis* by the ticks *H. plumbeum* and Rh. bursa. Veterinariia (Russian) *5*: 46.

Nolan, P.M., S.R. Roberts and G.E. Hill. 2004. Effects of *Mycoplasma gallisepticum* on reproductive success in house finches. Avian Dis. *48*: 879–885.

Novy, M.J., L. L. Duffy, M.K. Axhelm, D.W. Sadowsky, S.S. Witkin, M.G. Gravett, Cassell, G.H. and K.B. Waites. 2009. Congenital and opportunistic infections: *Ureaplasma* species and *Mycoplasma hominis*. Reproductive Science *16*: 56–70.

Nowak, J. 1929. Morphologie, nature et cycle évolutif du microbe de la péripneumonie des bovidés. Ann. Inst. Pasteur (Paris) *43*: 1330–1352.

Nunoya, T., T. Yagihashi, M. Tajima and Y. Nagasawa. 1995. Occurrence of keratoconjunctivitis apparently caused by *Mycoplasma gallisepticum* in layer chickens. Vet. Pathol. *32*: 11–18.

O'Brien, S.J., J. Simonson, S. Razin and M.F. Barile. 1983. On the distribution and characteristics of isozyme expression in *Mycoplasma*. The Yale Journal of Biology and Medicine *56*: 701–708.

Oaks, J.L., S.L. Donahoe, F.R. Rurangirwa, B.A. Rideout, M. Gilbert and M.Z. Virani. 2004. Identification of a novel *Mycoplasma* species from an Oriental white-backed vulture (*Gyps bengalensis*). J. Clin. Microbiol. *42*: 5909–5912.

Olson, N.O., K.M. Kerr and A. Campbell. 1964. Control of infectious synovitis. 13. The antigen study of three strains. Avian Dis. *8*: 209–214.

Paessler, M., A. Levinson, J.B. Patel, M. Schuster, M. Minda and I. Nachamkin. 2002. Disseminated *Mycoplasma orale* infection in a patient with common variable immunodeficiency syndrome. Diagn. Microbiol. Infect. Dis. *44*: 201–204.

Panangala, V.S., J.S. Stringfellow, K. Dybvig, A. Woodard, F. Sun, D.L. Rose and M.M. Gresham. 1993. *Mycoplasma corogypsi* sp. nov., a new species from the footpad abscess of a black vulture, *Coragyps atratus*. Int. J. Syst. Bacteriol. *43*: 585–590.

Papazisi, L., T. Gorton, G. Kutish, P. Markham, G. Browning, D. Nguyen, S. Swartzell, A. Madan, G. Mahairas and S. Geary. 2003. The complete genome sequence of the avian pathogen *Mycoplasma gallisepticum* strain R(low). Microbiology *149*: 2307–2316.

Pennycott, T., C. Dare, C. Yavari and J. Bradbury. 2005. *Mycoplasma sturni* and *Mycoplasma gallisepticum* in wild birds in Scotland. Vet. Rec. *156*: 513–515.

Pereyre, S., P. Gonzalez, B. de Barbeyrac, A. Darnige, H. Renaudin, A. Charron, S. Raherison, C. Bébéar and C.M. Bébéar. 2002. Mutations in 23S rRNA account for intrinsic resistance to macrolides in *Mycoplasma hominis* and *Mycoplasma fermentans* and for acquired resistance to macrolides in *M. hominis*. Antimicrob. Agents Chemother. *46*: 3142–3150.

Persson, A., B. Pettersson, G. Bölske and K.-E. Johansson. 1999. Diagnosis of contagious bovine pleuropneumonia by PCR-laser- induced fluorescence and PCR-restriction endonuclease analysis based on the 16S rRNA genes of *Mycoplasma mycoides* subsp. *mycoides* SC. J. Clin. Microbiol. *37*: 3815–3821.

Peterson, J.E., A.W. Rodwell and E.S. Rodwell. 1973. Occurrence and ultrastructure of a variant (rho) form of *Mycoplasma*. J. Bacteriol. *115*: 411–425.

Pettersson, B., J.G. Tully, G. Bolske and K.E. Johansson. 2000. Updated phylogenetic description of the *Mycoplasma hominis* cluster (Weisburg *et al.* 1989) based on 16S rDNA sequences. Int. J. Syst. Evol. Microbiol. *50*: 291–301.

Pilo, P., B. Fleury, M. Marenda, J. Frey and E. Vilei. 2003. Prevalence and distribution of the insertion element ISMag1 in *Mycoplasma agalactiae*. Vet. Microbiol. *92*: 37–48.

Piot, P. 1977. Comparison of growth inhibition and immunofluorescence tests in serotyping clinical isolates of *Ureaplasma urealyticum*. Br. J. Vener. Dis. *53*: 186–189.

Pitcher, D., M. Sillis and J.A. Robertson. 2001. Simple method for determining biovar and serovar types of *Ureaplasma urealyticum* clinical isolates using PCR-single-strand conformation polymorphism analysis. J. Clin. Microbiol. *39*: 1840–1844.

Pitcher, D., D. Windsor, H. Windsor, J. Brabbury, C. Yavari, J.S. Jensen, C. Ling and D. Webster. 2005. *Mycoplasma amphoriforme* sp. nov., a new species isolated from a patient with chronic brochopneumonia. Int. J. Syst. Evol. Microbiol. *55*: 2589–2594.

Pollack, J. 1986. Metabolic distinctiveness of ureaplasmas. Pediatr. Infect. Dis. *5*: S305–307.

Pollack, J.D. 1992. Carbohydrate metabolism and energy conservation. *In* Mycoplasmas: Molecular Biology and Pathogenesis (edited by Maniloff, McElhaney, Finch and Baseman). American Society for Microbiology, Washington, D.C., pp. 181–200.

Pollack, J.D., M.V. Williams, J. Banzon, M.A. Jones, L. Harvey and J.G. Tully. 1996. Comparative metabolism of *Mesoplasma, Entomoplasma, Mycoplasma*, and *Acholeplasma*. Int. J. Syst. Bacteriol. *46*: 885–890.

Pollack, J.D. 1997. *Mycoplasma* genes: a case for reflective annotation. Trends Microbiol. *5*: 413–419.

Pollack, J.D. 2001. *Ureaplasma urealyticum*: an opportunity for combinatorial genomics. Trends Microbiol. *9*: 169–175.

Pollack, J.D. 2002. The necessity of combining genomic and enzymatic data to infer metabolic function and pathways in the smallest bacteria: amino acid, purine and pyrimidine metabolism in mollicutes. Front. Biosci. *7*: d1762–1781.

Poveda, J., J. Carranza, A. Miranda, A. Garrido, M. Hermoso, A. Fernandez and J. Domenech. 1990. An epizootiological study of avian mycoplasmas in southern Spain. Avian Pathol. *19*: 627–633.

Poveda, J.B., J. Giebel, J. Flossdorf, J. Meier and H. Kirchhoff. 1994. *Mycoplasma buteonis* sp. nov. *Mycoplasma falconis* sp. nov. and *Mycoplasma gypis* sp. nov. three species from birds of prey. Int. J. Syst. Bacteriol. *44*: 94–98.

Prullage, J.B., R.E. Williams and S.M. Gaafar. 1993. On the transmissibility of *Eperythrozoon suis* by Stomoxys calcitrans and Aedes aegypti. Vet. Parasitol. *50*: 125–135.

Purcell, R., D. Taylor-Robinson, D. Wong and R. Chanock. 1966. Color test for the measurement of antibody to T-strain mycoplasmas. J. Bacteriol. *92*: 6–12.

Pye, G.W., D.R. Brown, M.F. Nogueira, K.A. Vliet, T.R. Schoeb, E.R. Jacobson and R.A. Bennett. 2001. Experimental inoculation of broad-nosed caimans (*Caiman latirostris*) and Siamese crocodiles (*Crocodylus siamensis*) with *Mycoplasma alligatoris*. J. Zoo Wildl. Med. *32*: 196–201.

Ramírez, A., C. Naylor, D. Pitcher and J. Bradbury. 2008. High interspecies and low intra-species variation in 16S-23S rDNA spacer sequences of pathogenic avian mycoplasmas offers potential use as a diagnostic tool. Vet. Microbiol. *128*: 279–287.

Raviv, Z. and S. Kleven. 2009. The development of diagnostic real-time TaqMan PCrs for the four pathogenic avian mycoplasmas. Avian Dis. *53*: 103–107.

Razin, S. 1979. Isolation and characterization of *Mycoplasma* membranes. *In* The Mycoplasmas, vol. 1 (edited by Barile and Tully). Academic Press, New York, pp. 213–229.

Razin, S. 1983. Cell lysis and isolation of membranes. *In* Methods in Mycoplasmology, vol. 1 (edited by Razin and Tully). Academic Press, New York, pp. 225–233.

Razin, S., D. Yogev and Y. Naot. 1998. Molecular biology and pathogenicity of mycoplasmas. Microbiol. Mol. Biol. Rev. *62*: 1094–1156.

Razin, S. and R. Herrmann (editors). 2002. Molecular Biology and Pathogenicity of Mycoplasmas. Academic/Plenum Press, London.

Reece, R.L., L. Ireland and P.C. Scott. 1986. Mycoplasmosis in racing pigeons. Aust. Vet. J. *63*: 166–167.

Regendanz, P. and W. Kikuth. 1928. Über Aktivierung labiler Infektionen duch Entmilzung (*Piroplasma canis*, *Nuttalia brasiliensis*, *Bartonella opossum*, *Spirochaeta didelphydis*). Arch. f. Schiffs. U. Tropenhyg. *32*: 587–593.

Relich, R.F., A.J. Friedberg and M.F. Balish. 2009. Novel cellular organization in a gliding *Mycoplasma*, *Mycoplasma insons*. J. Bacteriol. *191*: 5312–5314.

Reyes, L., M. Reinhard and M. Brown. 2009. Different inflammatory responses are associated with *Ureaplasma parvum*-induced UTI and urolith formation. BMC Infect. Dis. *9*: 9.

Roberts, D.H. 1964. The isolation of an influenza A virus and a *Mycoplasma* associated with duck sinusitis. Vet. Rec. *76*: 470–473.

Roberts, M. and G. Kenny. 1986. Dissemination of the tetM tetracycline resistance determinant to *Ureaplasma urealyticum*. Antimicrob. Agents Chemother. *29*: 350–352.

Robertson, J. and E. Smook. 1976. Cytochemical evidence of extramembranous carbohydrates on *Ureaplasma urealyticum* (T-strain *Mycoplasma*). J. Bacteriol. *128*: 658–660.

Robertson, J. 1978. Bromothymol blue broth: improved medium for detection of *Ureaplasma urealyticum* (T-strain *mycoplasma*). J. Clin. Microbiol. *7*: 127–132.

Robertson, J. and G. Stemke. 1979. Modified metabolic inhibition test for serotyping strains of *Ureaplasma urealyticum* (T-strain *Mycoplasma*). J. Clin. Microbiol. *9*: 673–676.

Robertson, J., J. Coppola and O. Heisler. 1981. Standardized method for determining antimicrobial susceptibility of strains of *Ureaplasma urealyticum* and their response to tetracycline, erythromycin, and rosaramicin. Antimicrob. Agents Chemother. *20*: 53–58.

Robertson, J. 1982. Effect of gaseous conditions on isolation and growth of *Ureaplasma urealyticum* on agar. J. Clin. Microbiol. *15*: 200–203.

Robertson, J. and G. Stemke. 1982. Expanded serotyping scheme for *Ureaplasma urealyticum* strains isolated from humans. J. Clin. Microbiol. *15*: 873–878.

Robertson, J., M. Alfa and E. Boatman. 1983. Morphology of the cells and colonies of *Mycoplasma hominis*. Sex Transm. Dis. *10*: 232–239.

Robertson, J. and M. Chen. 1984. Effects of manganese on the growth and morphology of *Ureaplasma urealyticum*. J. Clin. Microbiol. *19*: 857–864.

Robertson, J., M. Stemler and G. Stemke. 1984. Immunoglobulin A protease activity of *Ureaplasma urealyticum*. J. Clin. Microbiol. *19*: 255–258.

Robertson, J., G. Stemke, S. Maclellan and D. Taylor. 1988. Characterization of tetracycline-resistant strains of *Ureaplasma urealyticum*. J. Antimicrob. Chemother. *21*: 319–332.

Robertson, J., A. Vekris, C. Bebear and G. Stemke. 1993. Polymerase chain reaction using 16S rRNA gene sequences distinguishes the two biovars of *Ureaplasma urealyticum*. J. Clin. Microbiol. *31*: 824–830.

Robertson, J.A., G. Stemke, J.J. Davis, R. Harasawa, D. Thirkell, F. Kong, M. Shepard and D. Ford. 2002. Proposal of *Ureaplasma parvum* sp. nov. and emended description of *Ureaplasma urealyticum* (Shepard *et al.* 1974) Robertson *et al.* 2001. Int. J. Syst. Evol. Microbiol. *52*: 587–597.

Robertson, J.A., L.E. Pyle, G.W. Stemke and L.R. Finch. 1990. Human ureaplasmas show diverse genome sizes by pulsed-field electrophoresis. Nucleic Acids Res. *18*: 1451–1455.

Robertson, J.A. and R. Sherburne. 1991. Hemadsorption by colonies of *Ureaplasma urealyticum*. Infect. Immun. *59*: 2203–2206.

Robertson, J.A., L.A. Howard, C.L. Zinner and G.W. Stemke. 1994. Comparison of 16S rRNA genes within the T960 and parvo biovars of ureaplasmas isolated from humans. Int. J. Syst. Bacteriol. *44*: 836–838.

Robertson, J.A. and G.W. Stemke. 1995. Measurement of Mollicute Growth by ATP-Dependent Luminometry. *In* Molecular and Diagnostic Procedures in Mycoplasmology, vol. 1 (edited by Razin and Tully). Academic Press, San Diego, CA, pp. 65–71.

Rocha, E. and A. Blanchard. 2002. Genomic repeats, genome plasticity and the dynamics of *Mycoplasma* evolution. Nucleic Acids Res. *30*: 2031–2042.

Rodwell, A. and R.F. Whitcomb. 1983. Methods for direct and indirect measurement of *Mycoplasma* growth. *In* Methods in Mycoplasmology, vol. 1 (edited by Razin and Tully). Academic Press, New York, pp. 185–196.

Rodwell, A.W. 1983. Defined and partly defined media. *In* Methods in Mycoplasmology, vol. 1 (edited by Razin and Tully). Academic Press, New York, pp. 163–172.

Roifman, C., C. Rao, H. Lederman, S. Lavi, P. Quinn and E. Gelfand. 1986. Increased susceptibility to *Mycoplasma* infection in patients with hypogammaglobulinemia. Am. J. Med. *80*: 590–594.

Romano, N., G. Tolone, F. Ajello and R. La Licata. 1980. Adenosine 5′-triphosphate synthesis induced by urea hydrolysis in *Ureaplasma urealyticum*. J. Bacteriol. *144*: 830–832.

Rose, D.L., J.G. Tully and E.V. Langford. 1978. *Mycoplasma citelli*, a new species from ground squirrels. Int. J. Syst. Bacteriol. *28*: 567–572.

Rosenbusch, R.F. and F.C. Minion. 1992. Cell envelope: morphology and biochemistry. *In* Mycoplasmas: Molecular Biology and Pathogenesis (edited by Maniloff, McElhaney, Finch and Baseman). American Society for Microbiology, Washington, D.C., pp. 73–77.

Røsendal, S. 1974. *Mycoplasma molare*, a new canine *Mycoplasma* species. Int. J. Syst. Bacteriol. *24*: 125–130.

Røsendal, S. 1975. Canine mycoplasmas: serological studies of type and reference strains, with a proposal for the new species, *Mycoplasma opalescens*. Acta Pathol. Microbiol. Scand. Sect. B *83*: 463–470.

Røsendal, S. and O. Vinther. 1977. Experimental mycoplasmal pneumonia in dogs: electron microscopy of infected tissue. Acta Pathol. Microbiol. Scand. (B) *85B*: 462–465.

Røsendal, S. 1979. Canine and feline mycoplasmas. *In* The Mycoplasmas (edited by Tully and Whitcomb). Academic Press, New York, pp. 217–234.

Rosengarten, R., S. Levisohn and D. Yogev. 1995. A 41-kDa variable surface protein of *Mycoplasma gallisepticum* has a counterpart in *Mycoplasma imitans* and *Mycoplasma iowae*. FEMS Microbiology Letters *132*: 115–123.

Röske, K., M. Calcutt and K. Wise. 2004. The *Mycoplasma fermentans* prophage phiMFV1: genome organization, mobility and variable expression of an encoded surface protein. Mol. Microbiol. *52*: 1703–1720.

Ross, R.F. and J.A. Karmon. 1970. Heterogeneity among strains of *Mycoplasma granularum* and identification of *Mycoplasma hyosynoviae*, sp. n. J. Bacteriol. *103*: 707–713.

Ross, R.F. 1992. Mycoplasmal diseases. *In* Diseases of Swine, 7th edn (edited by Lemanske, Jr, Straw, Mengeling, D'Allaire and Taylor). Iowa State University Press, Ames, IA, pp. 537–551.

Ruhnke, H., N. Palmer, P. Doig and R. Miller. 1984. Bovine abortion and neonatal death associated with *Ureaplasma diversum*. Theriogenology *21*: 295–301.

Ruhnke, H.L. and S. Madoff. 1992. *Mycoplasma phocidae* sp. nov., isolated from harbor seals (*Phoca vitulina* L.). Int. J. Syst. Bacteriol. *42*: 211–214.

Ruifu, Y., Z. Minli, Z. Guo and X. Wang. 1997. Biovar diversity is reflected by variations of genes encoding urease of *Ureaplasma urealyticum*. Microbiol. Immunol. *41*: 625–627.

Ruiter, M. and H.M. Wentholt. 1955. Isolation of a pleuropneumonia-like organism from a skin lesion associated with a fusospirochetal flora. J. Invest. Dermatol. *24*: 31–34.

Sabin, A.B. 1941. The filterable microorganisms of the pleuropneumonia group. Bacteriol. Rev. *5*: 1–66.

Salih, M.M., N.F. Friis, S.N. Aarseculeratne, E.A. Freundt and C. Christiansen. 1983. *Mycoplasma mustelae*, a new species from mink. Int. Syst. Bacteriol. *33*: 476–479.

Sasaki, Y., J. Ishikawa, A. Yamashita, K. Oshima, T. Kenri, K. Furuya, C. Yoshino, A. Horino, T. Shiba, T. Sasaki and M. Hattori. 2002. The complete genomic sequence of *Mycoplasma penetrans*, an intracellular bacterial pathogen in humans. Nucleic Acids Res. *30*: 5293–5300.

Sayed, I. and G.E. Kenny. 1980. Comparison of the proteins and polypeptides of the eight serotypes of *Ureaplasma urealyticum* by isoelectric focussing and sodium dodecyl sulfate-polyacrylamide gel electrophoresis. Int. J. Syst. Bacteriol. *30*: 33–41.

Scanziani, E., S. Paltrinieri, M. Boldini, V. Grieco, C. Monaci, A.M. Giusti and G. Mandelli. 1997. Histological and immunohistochemical findings in thoracic lymph nodes of cattle with contagious bovine pleuropneumonia. J. Comp. Pathol. *117*: 127–136.

Schelonka, R., B. Katz, K. Waites and D.J. Benjamin. 2005. Critical appraisal of the role of *Ureaplasma* in the development of bronchopulmonary dysplasia with metaanalytic techniques. Pediatr. Infect. Dis. J. *24*: 1033–1039.

Schilling, V. 1928. *Eperythrozoon coccoides*, eine neue durch Splenektomie aktivierbare Dauerinfektion der weissen Maus. Klin. Wochenschr. *7*: 1854–1855.

Schmidhauser, C., R. Dudler, T. Schmidt and R.W. Parish. 1990. A mycoplasmal protein influences tumour cell invasiveness and contact inhibition *in vitro*. J. Cell Sci. *95*: 499–506.

Schmitt, K., W. Daubener, D. Bitter-Suermann and U. Hadding. 1988. A safe and efficient method for elimination of cell culture mycoplasmas using ciprofloxacin. J. Immunol. Methods *109*: 17–25.

Schoeb, T.R. 2000. Respiratory diseases of rodents. Vet. Clin. North Am. Exot. Anim. Pract. *3*: 481–496, vii.

Scudamore, J.M. 1976. Observations on the epidemiology of contagious bovine pleuropneumonia: *Mycoplasma mycoides* in urine. Res. Vet. Sci. *20*: 330–333.

Seto, S., G. Layh-Schmitt, T. Kenri and M. Miyata. 2001. Visualization of the attachment organelle and cytadherence proteins of *Mycoplasma pneumoniae* by immunofluorescence microscopy. J. Bacteriol. *183*: 1621–1630.

Seybert, A., R. Herrmann and A.S. Frangakis. 2006. Structural analysis of *Mycoplasma pneumoniae* by cryo-electron tomography. J. Struct. Biol. *156*: 342–354.

Shahram, M., R.A. Nicholas, A.P. Wood and D.P. Kelly. 2010. Further evidence to justify reassignment of *Mycoplasma mycoides* subspecies *mycoides* Large Colony type to *Mycoplasma mycoides* subspecies *capri*. Syst. Appl. Microbiol. *33*: 20–24.

Shepard, M. 1954. The recovery of pleuropneumonia-like organisms from Negro men with and without nongonococcal urethritis. Am. J. Syph. Gonorrhea Vener. Dis. *38*: 113–124.

Shepard, M. 1967. Cultivation and properties of T-strains of *Mycoplasma* associated with nongonococcal urethritis. Ann. N. Y. Acad. Sci. *143*: 505–514.

Shepard, M. and C. Lunceford. 1967. Occurrence of urease in T strains of *Mycoplasma*. J. Bacteriol. *93*: 1513–1520.

Shepard, M. and D. Howard. 1970. Identification of "T" mycoplasmas in primary agar cultures by means of a direct test for urease. Ann. N. Y. Acad. Sci. *174*: 809–819.

Shepard, M.C., C.D. Lunceford, D.K. Ford, R.H. Purcell, D. Taylor-Robinson, S. Razin and F.T. Black. 1974. *Ureaplasma urealyticum* gen. nov., sp. nov.: proposed nomenclature for Human-T (T-strain) mycoplasmas. Int. J. Syst. Bacteriol. *24*: 160–171.

Shepard, M.C. and G.K. Masover. 1979. Special features of ureaplasmas. *In* The Mycoplamas, vol. 1 (edited by Barile and Razin). Academic Press, New York, pp. 452–494.

Shepard, M.C. 1983. Culture media for ureaplasmas. *In* Methods in Mycoplasmology, vol. 1 (edited by Razin and Tully). Academic Press, New York, pp. 137–146.

Shepard, M.C. and J.A. Robertson. 1986. Calcium chloride as an indicator for colonies of *Ureaplasma urealyticum*. The Pediatric Infectious Diseases Journal *5*: S349.

Shiferaw, G., S. Tariku, G. Ayelet and Z. Abebe. 2006. Contagious caprine pleuropneumonia and *Mannheimia haemolytica*-associated acute respiratory disease of goats and sheep in Afar Region, Ethiopia. Rev. Sci. Tech. *25*: 1153–1163.

Shimizu, T., H. Erno and H. Nagatomo. 1978. Isolation and characterization of *Mycoplasma columbinum* and *Mycoplasma columborale*, two new species from pigeons. Int. J. Syst. Bacteriol. *28*: 538–546.

Shimizu, T., K. Numano and K. Uchida. 1979. Isolation and identification of mycoplasmas from various birds: an ecological study. Jpn. J. Vet. Sci. *41*: 273–282.

Shobokshi, A. and M. Shaarawy. 2002. Maternal serum and amniotic fluid cytokines in patients with preterm premature rupture of membranes with and without intrauterine infection. Int. J. Gynaecol. Obstet. *79*: 209–215.

Simecka, J.W., J.K. Davis, M.K. Davidson, S.R. Ross, C.T.K.-H. Städtlander and G.H. Cassell. 1992. *Mycoplasma* diseases of animals. *In* Mycoplasmas: Molecular Biology and Pathogenesis (edited by Maniloff, McElhaney, Finch, and Baseman). American Society for Microbiology, Washington, D.C., pp. 391–415.

Singh, K.C. and P.K. Uppal. 1987. Isolation of *Mycoplasma gallinarum* from sheep. Vet. Rec. *120*: 464.

Singh, Y., D.N. Garg, P.K. Kapoor and S.K. Mahajan. 2004. Isolation of *Mycoplasma bovoculi* from genitally diseased bovines and its experimental pathogenicity in pregnant guinea pigs. Indian J. Exp. Biol. *42*: 933–936.

Slatter, D.H. 2001. Fundamentals of Veterinary Ophthamology, 3rd edn. Elsevier Health Sciences, Saint Louis, MO.

Smith, A. and J. Mowles. 1996. Prevention and control of *Mycoplasma* infection of cell cultures. *In* Molecular and Diagnostic Procedures in Mycoplasmology, vol. 2 (edited by Tully and Razin). Academic Press, San Diego, pp. 445–451.

Smith, D., W. Russell, W. Ingledew and D. Thirkell. 1993. Hydrolysis of urea by *Ureaplasma urealyticum* generates a transmembrane potential with resultant ATP synthesis. J. Bacteriol. *175*: 3253–3258.

Smith, P. 1986. Mass cultivation of ureaplasmas and some applications. Pediatr. Infect. Dis. *5*: S313–315.

Smith, P.F. 1992. Membrane lipid and lipopolysaccharide structures. *In* Mycoplasmas: Molecular Biology and Pathogenesis (edited by Maniloff, McElhaney, Finch and Baseman). American Society for Microbiology, Washington, D.C., pp. 79–91.

So, A.K., P.M. Furr, D. Taylor-Robinson and A.D. Webster. 1983. Arthritis caused by *Mycoplasma salivarium* in hypogammaglobulinaemia. Br. Med. J. (Clin. Res. Ed.) *286*: 762–763.

Somerson, N.L., D. Taylor-Robinson and R.M. Chanock. 1963. Hemolyin production as an aid in the identification and quantitation of Eaton agent (*Mycoplasma pneumoniae*). Am. J. Hyg. *77*: 122–128.

Somerson, N.L. and B.C. Cole. 1979. The *Mycoplasma* flora of human and non-human primates. *In* The Mycoplasmas, vol. 1 (edited by Tully and Whitcomb). Academic Press, New York, pp. 191–216.

Somerson, N.L., J.P. Kocka, D. Rose and R.A. Del Giudice. 1982. Isolation of acholeplasmas and a *Mycoplasma* from vegetables. Appl. Environ. Microbiol. *43*: 412–417.

Spergser, J., C. Aurich, J.E. Aurich and R. Rosengarten. 2002. High prevalence of mycoplasmas in the genital tract of asymptomatic stallions in Austria. Vet. Microbiol. *87*: 119–129.

Spergser, J., S. Langer, S. Muck, K. Macher, M. Szostak, R. Rosengarten and H.-J. Busse. 2010. *Mycoplasma mucosicanis* sp. nov., isolated from the mucosa of dogs. Int. J. Syst. Evol. Microbiol.: published 23 April 2010 as doi:10.1099/ijs.0.015750-0.

Splitter, E.J. 1950. *Eperythrozoon suis*, the etiologic agent of icteroanemia–an anaplasmosis-like disease in swine. Am. J. Vet. Res. *11*: 324–329.

Spooner, R., W. Russell and D. Thirkell. 1992. Characterization of the immunoglobulin A protease of *Ureaplasma urealyticum*. Infect. Immun. *60*: 2544–2546.

Ståby, M. 2004. Seal finger-a problem among hunters once again. Lakartidningen *101*: 1910–1911.

Stemke, G. and J. Robertson. 1981. Modified colony indirect epifluorescence test for serotyping *Ureaplasma urealyticum* and an adaptation to detect common antigenic specificity. J. Clin. Microbiol. *14*: 582–584.

Stemke, G. and J. Robertson. 1982. Comparison of two methods for enumeration of mycoplasmas. J. Clin. Microbiol. *16*: 959–961.

Stemke, G.W. and J.A. Robertson. 1985. Problems associated with serotyping strains of *Ureaplasma urealyticum*. Diagn. Microbiol. Infect. Dis. *3*: 311–320.

Stemke, G., M. Stemler and J. Robertson. 1984. Growth characteristics of ureaplasmas from animal and human sources. Isr. J. Med. Sci. *20*: 935–937.

Stemler, M.E., G.W. Stemke and J.A. Robertson. 1987. ATP measurements obtained by luminometry provide rapid estimation of *Ureaplasma urealyticum* growth. J. Clin. Microbiol. *25*: 427–429.

Stenske, K.A., D.A. Bemis, K. Hill and D.J. Krahwinkel. 2005. Acute polyarthritis and septicemia from *Mycoplasma edwardii* after surgical removal of bilateral adrenal tumors in a dog. J. Vet. Intern. Med. *19*: 768–771.

Stipkovits, L., Z. Varga, M. Dobos-Kovacs and M. Santha. 1984a. Biochemical and serological examination of some *Mycoplasma* strains of goose origin. Acta Vet. Hung. *32*: 117–125.

Stipkovits, L., Z. Varga, G. Laber and J. Bockmann. 1984b. A comparison of the effect of tiamulin hydrogen fumarate and tylosin tartrate on mycoplasmas of ruminants and some animal ureaplasmas. Vet. Microbiol. *9*: 147–153.

Stoffregen, W.C., D.P. Alt, M.V. Palmer, S.C. Olsen, W.R. Waters and J.A. Stasko. 2006. Identification of a *Haemomycoplasma* species in anemic reindeer (Rangifer tarandus). J. Wildl. Dis. *42*: 249–258.

Swensen, C., J. VanHamont and B.S. Dunbar. 1983. Specific protein differences among strains of *Ureaplasma urealyticum* as determined by two-dimensional gel electrophoresis and a sensitive silver stain. Int. J. Syst. Bacteriol. *33*: 417–421.

Switzer, W.P. 1955. Studies on infectious atrophic rhinitis. IV. Characterization of a pleuropneumonia-like organism isolated from the nasal cavities of swine. Am. J. Vet. Res. *16*: 540–554.

Sykes, J.E., L.M. Ball, N.L. Bailiff and M.M. Fry. 2005. '*Candidatus Mycoplasma* haematoparvum', a novel small haemotropic mycoplasma from a dog. Int. J. Syst. Evol. Microbiol. *55*: 27–30.

Sykes, J.E., N.L. Drazenovich, L.M. Ball and C.M. Leutenegger. 2007. Use of conventional and real-time polymerase chain reaction to determine the epidemiology of hemoplasma infections in anemic and nonanemic cats. J. Vet. Intern. Med. *21*: 685–693.

Tagawa, M., K. Matsumoto and H. Inokuma. 2008. Molecular detection of *Mycoplasma wenyonii* and '*Candidatus Mycoplasma* haemobos' in cattle in Hokkaido, Japan. Vet. Microbiol. *132*: 177–180.

Tasker, S., S.H. Binns, M.J. Day, T.J. Gruffydd-Jones, D.A. Harbour, C.R. Helps, W.A. Jensen, C.S. Olver and M.R. Lappin. 2003. Use of a PCR assay to assess the prevalence and risk factors for *Mycoplasma haemofelis* and '*Candidatus Mycoplasma* haemominutum' in cats in the United Kingdom. Vet. Rec. *152*: 193–198.

Tasker, S., S.M. Caney, M.J. Day, R.S. Dean, C.R. Helps, T.G. Knowles, P.J. Lait, M.D. Pinches and T.J. Gruffydd-Jones. 2006. Effect of chronic FIV infection, and efficacy of marbofloxacin treatment, on *Mycoplasma haemofelis* infection. Vet. Microbiol. *117*: 169–179.

Taylor-Robinson, D., J. Canchola, H. Fox and R.M. Chanock. 1964. A newly identified oral *Mycoplasma* (*M. orale*) and its relationship to other human mycoplasmas. Am. J. Hyg. *80*: 135–148.

Taylor-Robinson, D. and Z. Dinter. 1968. Unexpected serotypes of mycoplasmas isolated from pigs. J. Gen. Microbiol. *53*: 221–229.

Taylor-Robinson, D., G.W. Csonka and M.J. Prentice. 1977. Human intra-urethral inoculation of ureplasmas. Q. J. Med. *46*: 309–326.

Taylor-Robinson, D. and W.M. McCormack. 1979. Mycoplasmas in human genitourinary infections. *In* The Mycoplasmas, Volume 2, Human and Animal Mycoplasmas (edited by Tully and Whitcomb). Academic Press, New York, pp. 308–357.

Taylor-Robinson, D. 1983a. Recovery of mycoplasmas from the genitourinary tract. *In* Methods in Mycoplasmology, vol. 2 (edited by Razin and Tully). Academic Press, New York, pp. 19–26.

Taylor-Robinson, D. 1983b. Metabolism inhibition tests. *In* Methods in Mycoplasmology, vol. 1 (edited by Razin and Tully). Academic Press, New York, pp. 411–421.

Taylor-Robinson, D. and R.N. Gourlay. 1984. Genus II. *Ureaplasma* Shepard, Lunceford, Ford, Purcell, Taylor-Robinson, Razin and Black 1974. *In* Bergey's Manual of Systematic Bacteriology, 8th edn, vol. 1 (edited by Kreig and Holt). Williams & Wilkins, Baltimore, pp. 770–775.

Taylor-Robinson, D. 1985. Mycoplasmal and mixed infections of the human male urogenital tract and their possible complications. *In* The Mycoplasmas, vol. 4 (edited by Razin and Barile). Academic Press, New York, pp. 27–63.

Taylor-Robinson, D. and P. Furr. 1986. Clinical antibiotic resistance of *Ureaplasma urealyticum*. Pediatr. Infect. Dis. *5*: S335–337.

Taylor-Robinson, D. and P.M. Furr. 1997. Genital *Mycoplasma* infections. Wien. Klin. Wochenschr. *109*: 578–583.

Taylor-Robinson, D. and P. Horner. 2001. The role of *Mycoplasma genitalium* in non-gonococcal urethritis. Sex. Transm. Infect. *77*: 229–231.

Taylor-Robinson, D., C. Gilroy, B. Thomas and P. Hay. 2004. *Mycoplasma genitalium* in chronic non-gonococcal urethritis. Int. J. STD AIDS *15*: 21–25.

Taylor, R.R., H. Varsani and R.J. Miles. 1994. Alternatives to arginine as energy sources for the non-fermentative *Mycoplasma gallinarum*. FEMS Microbiol. Lett. *115*: 163–167.

Teng, L., X. Zheng, J. Glass, H. Watson, J. Tsai and G. Cassell. 1994. *Ureaplasma urealyticum* biovar specificity and diversity are encoded in multiple-banded antigen gene. J. Clin. Microbiol. *32*: 1464–1469.

ter Laak, E.A., J.H. Noordergraaf and M.H. Verschure. 1993. Susceptibilities of *Mycoplasma bovis*, *Mycoplasma dispar*, and *Ureaplasma diversum* strains to antimicrobial agents *in vitro*. Antimicrob. Agents Chemother. *37*: 317–321.

ter Laak EA, Noordergraaf JH and Boomsluiter E. 1992a. The nasal mycoplasmal flora of healthy calves and cows. Zentralbl Veterinarmed B *39*: 610–616.

ter Laak EA, Noordergraaf JH and Dieltjes RP. 1992b. Prevalence of mycoplasmas in the respiratory tracts of pneumonic calves. Zentralbl Veterinarmed B *39*: 553–562.

Thiaucourt, F., G. Bolske, B. Leneguersh, D. Smith and H. Wesonga. 1996. Diagnosis and control of contagious caprine pleuropneumonia. Rev. Sci. Tech. *15*: 1415–1429.

Thirkell, D., A. Myles and W. Russell. 1989. Serotype 8- and serocluster-specific surface-expressed antigens of *Ureaplasma urealyticum*. Infect. Immun. *57*: 1697–1701.

Thirkell, D., A. Myles and D. Taylor-Robinson. 1990. A comparison of four major antigens in five human and several animal strains of ureaplasmas. J. Med. Microbiol. *32*: 163–168.

Thomsen, A.C. 1974. The isolation of *Mycoplasma primatum* during an autopsy study of the *Mycoplasma* flora of the human urinary tract. Acta Pathol. Microbiol. Scand. B Microbiol. Immunol. *82B*: 653–656.

Thurston, J.P. 1953. The chemotherapy of *Eperythrozoon coccoides* (Schilling 1928). Parasitology *43*: 170–174.

Tiong, S.K. 1990. Mycoplasmas and acholeplasmas isolated from ducks and their possible association with pasteurellas. Vet. Rec. *127*: 64–66.

Totten, P.A., M.A. Schwartz, K.E. Sjostrom, G.E. Kenny, H.H. Handsfield, J.B. Weiss and W.L. Whittington. 2001. Association of *Mycoplasma genitalium* with nongonococcal urethritis in heterosexual men. J. Infect. Dis. *183*: 269–276.

Trüper, H.G. and L. de'Clari. 1998. Taxonomic note: erratum and correction of further specific epithets formed as substantives (nouns) "in apposition". Int. J. Syst. Bacteriol. *48*: 615.

Tryon, V.V. and J.B. Baseman. 1992. Pathogenic determinants and mechanisms. *In* Mycoplasmas: Molecular Biology and Pathogenesis (edited by Maniloff, McElhaney, Finch and Baseman). American Society for Microbiology, Washington, D.C., pp. 457–471.

Tu, A.H., L.L. Voelker, X. Shen and K. Dybvig. 2001. Complete nucleotide sequence of the *Mycoplasma* virus P1 genome. Plasmid *45*: 122–126.

Tully, J.G. and I. Ruchman. 1964. Recovery, identification, and neurotoxicity of Sabin's Type A and C mouse *Mycoplasma* (PPLO) from lyophilized cultures. Proc. Soc. Exp. Biol. Med. *115*: 554–558.

Tully, J.G., M.F. Barile, R.A. Del Giudice, T.R. Carski, D. Armstrong and S. Razin. 1970. Proposal for classifying strain PG-24 and related canine mycoplasmas as *Mycoplasma edwardii* sp. n. J. Bacteriol. *101*: 346–349.

Tully, J.G., M.F. Barile, D.G.F. Edward, T.S. Theodore and H. Erno. 1974. Characterization of some caprine mycoplasmas, with proposals for new species, *Mycoplasma capricolum* and *Mycoplasma putrefaciens*. J. Gen. Microbiol. *85*: 102–120.

Tully, J.G. and R.F. Whitcomb. 1979. The Mycoplasmas Volume II: Human and Animal Mycoplasmas. Academic Press, New York.

Tully, J.G. 1983. Cloning and filtration techniques for mycoplasmas. *In* Methods in Mycoplasmology, vol. 1 (edited by Razin and Tully). Academic Press, New York, pp. 173–177.

Tully, J.G. and S.E. Razin. 1983. Methods in Mycoplasmology, vol. 2. Academic Press, New York.

Tully, J.G., D. Taylor-Robinson, D.L. Rose, R.M. Cole and J.M. Bové. 1983. *Mycoplasma genitalium*, a new species from the human urogenital tract. Int. J. Syst. Bacteriol. *33*: 387–396.

Tully, J.G. 1993. Current status of the mollicute flora of humans. Clin. Infect. Dis. *17 Suppl 1*: S2–S9.

Tully, J.G., J.M. Bove, F. Laigret and R.F. Whitcomb. 1993. Revised taxonomy of the class *Mollicutes* - proposed elevation of a monophyletic cluster of arthropod-associated *Mollicutes* to ordinal rank (*Entomoplasmatales* ord. nov.), with provision for familial rank to separate species with nonhelical morphology (*Entomoplasmataceae* fam. nov.) from helical species (*Spiroplasmataceae*), and emended descriptions of the order *Mycoplasmatales*, family *Mycoplasmataceae*. Int. J. Syst. Bacteriol. *43*: 378–385.

Tully, J.G. 1995a. International Committee on Systematic Bacteriology Subcommittee on the Taxonomy of *Mollicutes*. Revised minimal standards for description of new species of the class *Mollicutes*. Int. J. Syst. Bacteriol. *45*: 605–612.

Tully, J.G. 1995b. Culture medium formulation for primary isolation and maintenance of mollicutes. *In* Molecular and Diagnostic Procedures in Mycoplasmology, vol. 1 (edited by Razin and Tully). Academic Press, San Diego, pp. 33–39.

Tully, J.G. and S. Razin (editors). 1996. Molecular and Diagnostic Procedures in Mycoplasmology, vol. 2. Academic Press, San Diego, CA.

Tyzzer, E.E. and D. Weinman. 1939. *Haemobartonella* n.g. *(Bartonella olim pro parte) H. microti* n. sp. of the field vole, *Microtus pennsylvanicus*. Am. J. Hyg. *30*: 141–157.

Uilenberg, G., F. Thiaucourt and F. Jongejan. 2004. On molecular taxonomy: what is in a name? Exp. Appl. Acarol. *32*: 301–312.

Uilenberg, G., F. Thiaucourt and F. Jongejan. 2006. *Mycoplasma* and *Eperythrozoon* (*Mycoplasmataceae*). Comments on a recent paper. Int. J. Syst. Evol. Microbiol. *56*: 13–14.

Vancini, R.G. and M. Benchimol. 2008. Entry and intracellular location of *Mycoplasma hominis* in *Trichomonas vaginalis*. Arch. Microbiol. *189*: 7–18.

Vasconcellos Cardosa, M., A. Blanchard, S. Ferris, R. Verlengia, J. Timenetsky and R.A. Florio Da Cunha. 2000. Detection of *Ureaplasma diversum* in cattle using a newly developed PCR-based detection assay. Veterinary Microbiology *72*: 241–250.

Vilei, E.M., J. Nicolet and J. Frey. 1999. IS1634, a novel insertion element creating long, variable-length direct repeats which is specific for *Mycoplasma mycoides* subsp. *mycoides* small-colony type. J. Bacteriol. *181*: 1319–1323.

Voelker, L.L., K.E. Weaver, L.J. Ehle and L.R. Washburn. 1995. Association of lysogenic bacteriophage MAV1 with virulence of *Mycoplasma arthritidis*. Infect. Immun. *63*: 4016–4023.

Voelker, L.L. and K. Dybvig. 1999. Sequence analysis of the *Mycoplasma arthritidis* bacteriophage MAV1 genome identifies the putative virulence factor. Gene *233*: 101–107.

Vogelzang, A. and G. Compeer-Dekker. 1969. Elimination of *Mycoplasma* from various cell cultures. Antonie Van Leeuwenhoek *35*: 393–408.

Waites, K., D. Crouse and G. Cassell. 1992. Antibiotic susceptibilities and therapeutic options for *Ureaplasma urealyticum* infections in neonates. Pediatr. Infect. Dis. J. *11*: 23–29.

Waites, K., C.M. Bébéar, J.A. Robertson, D.F. Talkington and G.E. Kenny. 2001. Laboratory Diagnosis of Mycoplasmal Infections. Cumitech 34. ASM Press, Washington, D.C.

Waites, K. and D. Talkington. 2005. New developments in human disease due to mycoplasmas. *In* Mycoplasmas: Molecular Biology, Pathogenicity, and Strategies for Control (edited by Blanchard and Browning). Horizon Bioscience, Norfolk, UK, pp. 289–354.

Waites, K.B., P.T. Rudd, D.T. Crouse, K.C. Canupp, K.G. Nelson, C. Ramsey and G.H. Cassell. 1988. Chronic *Ureaplasma urealyticum* and *Mycoplasma hominis* infections of central nervous system in preterm infants. Lancet *1*: 17–21.

Waites, K.B., T.A. Figarola, T. Schmid, D.M. Crabb, L.B. Duffy and J.W. Simecka. 1991. Comparison of agar versus broth dilution techniques for determining antibiotic susceptibilities of *Ureaplasma urealyticum*. Diagn. Microbiol. Infect. Dis. *14*: 265–271.

Washburn, L.R., L.L. Voelker, L.J. Ehle, S. Hirsch, C. Dutenhofer, K. Olson and B. Beck. 1995. Comparison of *Mycoplasma arthritidis* strains by enzyme-linked immunosorbent assay, immunoblotting, and DNA restriction analysis. J. Clin. Microbiol. *33*: 2271–2279.

Watson, H.L., D.K. Blalock and G.H. Cassell. 1990. Variable antigens of *Ureaplasma urealyticum* containing both serovar-specific and serovar-cross-reactive epitopes. Infect. Immun. *58*: 3679–3688.

Webster, A.D., D. Taylor-Robinson, P.M. Furr and G.L. Asherson. 1978. Mycoplasmal (*Ureaplasma*) septic arthritis in hypogammaglobulinaemia. Br. Med. J. *1*: 478–479.

Webster, D., H. Windsor, C. Ling, D. Windsor and D. Pitcher. 2003. Chronic bronchitis in immunocompromised patients: association with a novel *Mycoplasma* species. Eur. J. Clin. Microbiol. Infect. Dis. *22*: 530–534.

Weisburg, W., J. Tully, D. Rose, J. Petzel, H. Oyaizu, D. Yang, L. Mandelco, J. Sechrest, T. Lawrence and J. Van Etten. 1989. A phylogenetic analysis of the mycoplasmas: basis for their classification. J. Bacteriol. *171*: 6455–6467.

Welchman, D.B., J. Bradbury, D. Cavanagh and N. Aebischer. 2002. Infectious agents associated with respiratory disease in pheasants. Vet. Rec. *150*: 658–664.

Wellehan, J.F., M. Calsamiglia, D.H. Ley, M.S. Zens, A. Amonsin and V. Kapur. 2001. Mycoplasmosis in captive crows and robins from Minnesota. J. Wildl. Dis. *37*: 547–555.

Westberg, J., A. Persson, B. Pettersson, M. Uhlen and K.E. Johansson. 2002. ISMmy1, a novel insertion sequence of *Mycoplasma mycoides* subsp. *mycoides* small colony type. FEMS Microbiol. Lett. *208*: 207–213.

Whitescarver, J. and G. Furness. 1975. T-mycoplasmas: a study of the morphology, ultrastructure and mode of division of some human strains. J. Med. Microbiol. *8*: 349–355.

Whittlestone, P. 1979. Porcine mycoplasmas. *In* The Mycoplasmas, vol. 2 (edited by Tully and Whitcomb). Academic Press, New York, pp. 133–176.

Wieslander, A., M.J. Boyer and H. Wróblewski. 1992. Membrane protein structure. *In* Mycoplasmas: Molecular Biology and Pathogenesis (edited by Maniloff, McElhaney, Finch and Baseman). American Society for Microbiology, Washington, D.C., pp. 93–112.

Willi, B., F.S. Boretti, C. Baumgartner, S. Tasker, B. Wenger, V. Cattori, M.L. Meli, C.E. Reusch, H. Lutz and R. Hofmann-Lehmann. 2006a. Prevalence, risk factor analysis, and follow-up of infections caused by three feline hemoplasma species in cats in Switzerland. J. Clin. Microbiol. *44*: 961–969.

Willi, B., S. Tasker, F. S. Boretti, M. G. Doherr, V. Cattori, M. L. Meli, R. G. Lobetti, R. Malik, C. E. Reusch, H. Lutz, R. Hofmann-Lehmann. 2006b. Phylogenetic analysis of "*Candidatus* Mycoplasma turicensis" isolates from pet cats in the United Kingdom, Australia, and South Africa, with analysis of risk factors for infection. J. Clin. Microbiol. *44*: 4430–4435.

Windsor, R.S. and W.N. Masiga. 1977. Investigations into the role of carrier animals in the spread of contagious bovine pleuropneumonia. Res. Vet. Sci. *23*: 224–229.

Woods, J.E., M.M. Brewer, J.R. Hawley, N. Wisnewski and M.R. Lappin. 2005. Evaluation of experimental transmission of *Candidatus Mycoplasma* haemominutum and *Mycoplasma haemofelis* by *Ctenocephalides felis* to cats. Am. J. Vet. Res. *66*: 1008–1012.

Woodson, B.A., K.S. McCarty and M.C. Shepard. 1965. Arginine metabolism in *Mycoplasma* and infected L929 fibroblasts. Arch. Biochem. *109*: 364–371.

Woubit, S., S. Lorenzon, A. Peyraud, L. Manso-Silván and F. Thiaucourt. 2004. A specific PCR for the identification of *Mycoplasma capricolum* subsp. *capripneumoniae*, the causative agent of contagious caprine pleuropneumonia (CCPP). Vet. Microbiol. *104*: 125–132.

Wroblewski, W. 1931. Morphologie et cycle évolutif des microbnes de la péripneumonie das bovides et de l'agalaxie contagieuse des chêvres et des moutons. Ann. Inst. Pasteur (Paris) *47*: 94–115.

Xie, X. and J. Zhang. 2006. Trends in the rates of resistance of *Ureaplasma urealyticum* to antibiotics and identification of the mutation site in the quinolone resistance-determining region in Chinese patients. FEMS Microbiol. Lett. *259*: 181–186.

Yagihashi, T., T. Nunoya and Y. Otaki. 1983. Effects of dual infection of chickens with *Mycoplasma synoviae* and *Mycoplasma gallinaceum* or infectious bursal disease virus on infectious synovitis. Nippon Juigaku Zasshi *45*: 529–532.

Yamamoto, R., C.H. Bigland and H.B. Ortmayer. 1965. Characteristics of *Mycoplasma meleagridis* sp. n. isolated from turkeys. J. Bacteriol. *90*: 47–49.

Yanez, A., L. Cedillo, O. Neyrolles, E. Alonso, M.C. Prevost, J. Rojas, H.L. Watson, A. Blanchard and G.H. Cassell. 1999. *Mycoplasma penetrans* bacteremia and primary antiphospholipid syndrome. Emerg. Infect. Dis. *5*: 164–167.

Yechouron, A., J. Lefebvre, H.G. Robson, D.L. Rose and J.G. Tully. 1992. Fatal septicemia due to *Mycoplasma arginini*: a new human zoonosis. Clin. Infect. Dis. *15*: 434–438.

Yoder, H.W. and M.S. Hofstad. 1964. Characterization of avian mycoplasmas. Avian Dis. *8*: 481–512.

Yogev, D., S. Levisohn and S. Razin. 1989. Genetic and antigenic relatedness between *Mycoplasma gallisepticum* and *Mycoplasma synoviae*. Vet. Microbiol. *19*: 75–84.

Zheng, X., L. Teng, H. Watson, J. Glass, A. Blanchard and G. Cassell. 1995. Small repeating units within the *Ureaplasma urealyticum* MB antigen gene encode serovar specificity and are associated with antigen size variation. Infect. Immun. *63*: 891–898.

Zheng, X., D.A. Olson, J.G. Tully, H.L. Watson, G.H. Cassell, D.R. Gustafson, K.A. Svien and T.F. Smith. 1997. Isolation of *Mycoplasma hominis* from a brain abscess. J. Clin. Microbiol. *35*: 992–994.

Family II. **Incertae sedis**

Daniel R. Brown, Séverine Tasker, Joanne B. Messick and Harold Neimark

This family accommodates the genera *Eperythrozoon* and *Haemobartonella*. These wall-less hemotropic bacteria were once placed in the family *Anaplasmataceae*, order *Rickettsiales*, because they are obligate blood parasites. None have been cultivated on artificial media, so no type strains have been established. Motility and biochemical parameters have not been definitively established for any species. These organisms are now known to be unambiguously affiliated with the order *Mycoplasmatales* on the basis of 16S rRNA similarities, plus morphology, DNA G+C contents, and evidence that they use the codon UGA to encode tryptophan (Berent and Messick, 2003), but their nomenclature remains a matter of controversy (Neimark et al., 2005; Uilenberg et al., 2006).

Genus I. **Eperythrozoon** Schilling 1928, 293[AL]

Daniel R. Brown, Séverine Tasker, Joanne B. Messick and Harold Neimark

E.pe.ry.thro.zo′on. Gr. pref. *epi* on; Gr. adj. *erythros* red; Gr. neut. n. *zoon* living being, animal; N.L. neut. n. *Eperythrozoon* (presumably intended to mean) animals on red (blood cells).

Cells adherent to host erythrocyte surfaces are coccoid and about 350 nm in diameter, but may arrange to appear as chains or deform to appear rod- or ring-shaped in stained blood smears.

Type species: **Eperythrozoon coccoides** Schilling 1928, 1854.

Further descriptive information

Hemotropic mollicutes such as the species formerly called *Eperythrozoon coccoides* (trivial name, hemoplasmas; Neimark et al., 2005) infect a variety of mammals occasionally including humans. Transmission can be through ingestion of infected blood, percutaneous inoculation, or by arthropod vectors (Sykes et al., 2007; Willi et al., 2006). The pathogenicity of different hemoplasma species is variable, and strain virulence and host immunocompetence likely play roles in the development of disease. Clinical syndromes range from acute fatal hemolytic anemia to chronic insidious anemia. Signs may include anemia, pyrexia, anorexia, dehydration, weight loss, and infertility. The presence of erythrocyte-bound antibodies has been demonstrated in some hemoplasma-infected animals and may contribute to anemia. Animals can remain chronic asymptomatic carriers of hemoplasmas after acute infection. PCR is the diagnostic test of choice for hemoplasmosis. Tetracycline treatment reduces the number of organisms in peripheral blood, but probably does not eradicate the organisms from infected animals.

Enrichment and isolation procedures

Hemoplasmas have not yet been successfully grown in continuous culture *in vitro*, although recent work (Li et al., 2008) suggests that *in vitro* maintenance of the species *Mycoplasma suis* may be possible.

Maintenance procedures

Hemoplasmas can be frozen in heparin- or EDTA-anticoagulated blood cryopreserved with dimethylsulfoxide.

Differentiation of the genus *Eperythrozoon* from other genera

A distinctive characteristic of these organisms is that they are found only in the blood of vertebrate hosts or transiently in arthropod vectors of transmission. The tenuous distinction between species of *Eperythrozoon* and those of *Haemobartonella* was based on the relatively more common visualization of eperythrozoa as ring forms (now known to be artifactual) in stained blood smears and the perception that eperythrozoa were observed with about equal frequency on erythrocytes and free in plasma, while haemobartonellae were thought to occur less often free in plasma. Properties that partially fulfill criteria for assignment of this genus to the class *Mollicutes* (Brown et al., 2007) include absence of a cell wall, filterability, and the presence of conserved 16S rRNA gene sequences. Presumptive use of the codon UGA to encode tryptophan (Berent and Messick, 2003) supports exclusion from the genera *Anaeroplasma*, *Asteroleplasma*, *Acholeplasma*, and "*Candidatus* Phytoplasma". Non-spiral cellular morphology and regular association with vertebrate hosts support exclusion from the genera *Spiroplasma*, *Entomoplasma*, and *Mesoplasma*, but sterol requirement, the degree of aerobiosis, and the capacity to hydrolyze arginine, characteristics that would help to confirm their provisional 16S rRNA-based placement in the genus *Mycoplasma*, remain unknown.

Taxonomic comments

The taxonomy and nomenclature of the uncultivated hemotropic bacteria originally assigned to the genus *Eperythrozoon* remain matters of current controversy. It is now undisputed that, on the basis of their lack of a cell wall, small cell size, low G+C content, use of the codon UGA to encode tryptophan, regular association with vertebrate hosts, and 16S rRNA gene sequences that are most similar (80–84%) to species in the pneumoniae group of genus *Mycoplasma*, these organisms are properly affiliated with the *Mycoplasmatales*. However, the proposed transfers of *Eperythrozoon* and *Haemobartonella* species to the genus *Mycoplasma* (Neimark et al., 2001, 2005) were opposed on the grounds that the degree of 16S rRNA gene sequence similarity is insufficient (Uilenberg et al., 2004, 2006). The alternative of situating the hemoplasmas in a new genus in the *Mycoplasmataceae* (Uilenberg et al., 2006) would regrettably compound the 16S rRNA gene-based polyphyly within *Mycoplasma* on no other basis than a capacity to adhere to the surface of erythrocytes *in vivo*.

The proposed transfer of the type species *Eperythrozoon coccoides* to the genus *Mycoplasma* (Neimark et al., 2005) is complicated by priority because *Eperythrozoon* predates *Mycoplasma*. However, the alternative of uniting the genera by transferring all mycoplasmas to the genus *Eperythrozoon* is completely unjustifiable considering the biological characteristics of the non-hemotropic majority of *Mycoplasma* species. The Judicial Commission of the International Committee on Systematics of Prokaryotes declined to rule on a request for an opinion in this matter during their 2008 meeting, but a provisional placement of the former *Eperythrozoon* species in the genus *Mycoplasma* has otherwise been embraced by specialists in the molecular biology and clinical pathogenicity of these and similar hemotropic organisms. At present, the designation "*Candidatus* Mycoplasma" must still be used for new types.

Further reading

Kreier, J.P. and M. Ristic. 1974. Genus IV. *Haemobartonella* Tyzzer and Weinman 1939, 143[AL]; Genus V. *Eperythrozoon* Schilling 1928, 1854[AL]. *In* Bergey's Manual of Determinative Bacteriology, 8th edn (edited by Buchanan and Gibbons). Williams & Wilkins, Baltimore, pp. 910–914.

Differentiation of the species of the genus *Eperythrozoon*

Species differentiation relies principally on 16S rRNA gene sequencing. Some species exhibit a degree of host specificity, although cross-infection of related hosts has been reported.

List of species of the genus *Eperythrozoon*

1. **Mycoplasma coccoides** (Schilling 1928) Neimark, Peters, Robinson and Stewart 2005, 1389[VP] (*Eperythrozoon coccoides* Schilling 1928, 1854)

coc.co'ides. N.L. masc. n. *coccus* (from Gr. masc. n. *kokkos* grain, seed) coccus; L. suff. *-oides* (from Gr. suff. *eides* from Gr. n. *eidos* that which is seen, form, shape, figure), resembling, similar; N.L. neut. adj. *coccoides* coccus-shaped.

Pathogenic; causes anemia in wild and captive mice, and captive rats, hamsters, and rabbits. Transmission is believed to be vector-borne and mediated by the rat louse *Polyplex spinulosa* and the mouse louse *Polyplex serrata*. Neoarsphenamine and oxophenarsine were thought to be effective chemotherapeutic agents for treatment of *Mycoplasma coccoides* infection in captive rodents, whereas tetracyclines are effective only at keeping infection at subclinical levels (Thurston, 1953).

Source: observed in association with the erythrocytes of wild and captive rodents.

DNA G+C content (mol%): not determined.

Type strain: not established.

Sequence accession no. (16S rRNA gene): AY171918.

2. **Eperythrozoon parvum** Splitter 1950, 513[AL]

par'vum. L. neut. adj. *parvum* small.

A nonpathogenic epierythrocytic parasite of pigs. Organic arsenicals are effective; tetracyclines suppress infection. Transmissible by parenteral inoculation and sometimes by massive oral inoculation. This is the only remaining species of *Eperythrozoon* whose name has standing in nomenclature that has not yet been examined by molecular genetic methods. It seems likely that, if a specimen of this organism can be found, it will prove to be a mycoplasma.

3. **Mycoplasma ovis** (Neitz, Alexander and de Toit 1934) Neimark, Hoff and Ganter 2004, 369[VP] (*Eperythrozoon ovis* Neitz, Alexander and de Toit 1934, 267)

o'vis. L. fem. gen. n. *ovis* of a sheep.

Cells are coccoid and motility for this species has not been assessed. The morphology of infected erythrocytes is altered demonstrating a marked depression at the site of *Myplasma ovis* attachment. This species has not been grown on artificial medium; therefore, notable biochemical parameters are not known.

Neoarsphenamine is an effective therapeutic agent. *Mycoplasma ovis* is reported to share antigens with *Mycoplasma wenyonii* (Kreier and Ristic, 1963), potentially complicating serology-based diagnosis of infection.

Pathogenic; causes mild to severe anemia in sheep and goats that often results in poor feed conversion. Transmission occurs via blood-feeding arthropods, e.g., *Haemophysalis plumbeum, Rhipicephalus bursa, Aedes camptorhynchus,* and *Culex annulirostris* (Daddow, 1980; Howard, 1975; Nikol'skii and Slipchenko, 1969), and likely via fomites such as reused needles, shearing tools, and ear-tagging equipment (Brun-Hansen et al., 1997; Mason and Statham, 1991).

Source: observed in association with erythrocytes or unattached in suspension in the blood of sheep, goats, and rarely in eland and splenectomized deer.

DNA G+C content (mol%): not determined.

Type strain: not established.

Sequence accession no. (16S rRNA gene): AF338268.

4. **Mycoplasma suis** corrig. (Splitter 1950) Neimark, Johansson, Rikihisa and Tully 2002, 683[VP] (*Eperythrozoon suis* Splitter 1950, 513)

su'is. L. gen. n. *suis* of the pig.

Cells are coccoid. Motility for this species has not been assessed. This species has not been grown on any artificial medium; therefore, notable biochemical parameters are not known.

Neoarsphenamine and tetracyclines are effective therapeutic agents. An enzyme-linked immunosorbant assay (ELISA) and PCR-based detection assays to enable diagnosis of infection have been described (Groebel et al., 2009; Gwaltney and Oberst, 1994; Hoelzle, 2008; Hsu et al., 1992).

Pathogenic; causes febrile icteroanemia in pigs. Transmission occurs via insect vectors including *Stomoxys calcitrans* and *Aedes aegypti* (Prullage et al., 1993).

Source: observed in association with the erythrocytes of pigs.

DNA G+C content (mol%): not determined.

Type strain: not established.

Sequence accession no. (16S rRNA gene): AF029394.

Further comment: the original spelling of the specific epithet, *haemosuis* (*sic*), has been corrected by the List Editor.

5. **Mycoplasma wenyonii** (Adler and Ellenbogen 1934) Neimark, Johansson, Rikihisa and Tully 2002, 683[VP] (*Eperythrozoon wenyonii* Adler and Ellenbogen 1934, 220)

we.ny.o'ni.i. N.L. masc. gen. n. *wenyonii* of Wenyon, named after Charles Morley Wenyon (1878–1948), an investigator of these organisms.

Cells are coccoid. Motility for this species has not been assessed. This species has not been grown on any artificial medium; therefore, notable biochemical parameters are not known.

Pathogenic; causes anemia and subsequent lameness and/or infertility in cattle. Transmission is primarily vector-mediated by *Dermacentor andersoni* and reportedly can also occur vertically during gestation. Oxytetracycline is an effective therapeutic agent (Montes et al., 1994). *Mycoplasma wenyonii* is reported to share antigens with *Mycoplasma ovis* (Kreier and Ristic, 1963), potentially complicating serology-based diagnosis of infection.

Source: observed in association with the erythrocytes and platelets of cattle (Kreier and Ristic, 1968).

DNA G+C content (mol%): not determined.

Type strain: not established.

Sequence accession no. (16S rRNA gene): AF016546.

Species of unknown phylogenetic affiliation

The phylogenetic affiliations of the following proposed organisms are unknown and their names do not have standing in nomenclature. They are listed here merely because they have been incidentally cited as species of *Eperythrozoon*.

1. **"Eperythrozoon mariboi"** Ewers 1971

The name given to uncultivated polymorphic structures observed on or in erythrocytes from flying foxes (*Pteropus macrotis*) following splenectomy (Ewers, 1971). The structures, described as fine lines, lines with rings, and rows of rings that span the diameter of the erythrocytes, differ from those of hemotropic mycoplasmas.

2. **"Eperythrozoon teganodes"** Hoyte 1962

The name given to uncultivated serially transmissible bodies observed in Giemsa-stained blood smears from cattle. The bodies only occur free in the blood plasma and do not attach to erythrocytes (Hoyte, 1962). The bodies differ from *Mycoplasma wenyonii* in morphology and include "frying-pan" shaped structures.

3. **"Eperythrozoon tuomii"** Tuomi and Von Bonsdorff 1967

Uncultivated transmissible cell wall-less bodies observed in Giemsa-stained blood smears and electron micrographs of blood from splenectomized calves. The bodies appeared in blood smears predominantly as delicate rings that did not attach to erythrocytes but were associated exclusively with thrombocytes (Tuomi and Von Bonsdorff, 1967; Uilenberg, 1967; Zwart et al., 1969).

Genus II. **Haemobartonella** Tyzzer and Weinman 1939, 305^{AL}

Daniel R. Brown, Séverine Tasker, Joanne B. Messick and Harold Neimark

Ha.e.mo.bar.to.nel'la. Gr. n. *haima* (L. transliteration *haema*) blood; N.L. fem. n. *Bartonella* a bacterial genus; N.L. fem. n. *Haemobartonella* the blood (-inhabiting) *Bartonella*.

Cells adherent to host erythrocyte surfaces are coccoid and about 350 nm in diameter, but may occur as chains or deform to appear rod- or ring-shaped in stained blood smears.

Type species: **Haemobartonella muris** (Mayer 1921) Tyzzer and Weinman 1939^{AL} (*Bartonella muris* Mayer 1921, 151; *Bartonella muris ratti* Regendanz and Kikuth 1928, 1578; *Haemobartonella muris* Tyzzer and Weinman 1939, 143).

Further descriptive information

Those organisms originally assigned to the genus *Haemobartonella* are properly affiliated with the *Mycoplasmatales*, but their transfer to the order has not yet been formalized. Any distinction between *Haemobartonella* and *Eperythrozoon* is tenuous and possibly arbitrary (Kreier and Ristic, 1974; Uilenberg et al., 2004). Enrichment, isolation and maintenance procedures, and methods of differentiation are essentially the same as those for genus *Eperythrozoon*.

List of species of the genus *Haemobartonella*

1. **Mycoplasma haemomuris** (Mayer 1921) Neimark, Johansson, Rikihisa and Tully 2002, 683^{VP} (*Bartonella muris* Mayer 1921, 151; *Bartonella muris ratti* Regendanz and Kikuth 1928, 1578; *Haemobartonella muris* Tyzzer and Weinman 1939, 143)

ha.e.mo.mu'ris. Gr. neut. n. *haema* blood; L. masc. gen. n. *muris* of the mouse; N.L. gen. n. *haemomuris* of mouse blood.

Cells are coccoid and some display dense inclusion particles. Motility for this species has not been assessed. The morphology of infected erythrocytes is altered, demonstrating a marked depression at the site of *Mycoplasma haemomuris* attachment. This species has not been grown on any artificial medium; therefore, notable biochemical parameters are not known.

Opportunistic pathogen; causes anemia in splenectomized or otherwise immunosuppressed mice. Transmission is vector-borne and mediated by the rat louse (*Polypax spinulosa*).

Source: observed in association with erythrocytes of wild and captive mice, and hamsters.

DNA G+C content (mol%): not determined.

Type strain: not established.

Sequence accession no. (16S rRNA gene): U82963.

2. **Mycoplasma haemocanis** (Kikuth 1928) Messick, Walker, Raphael, Berent and Shi 2002, 697^{VP} [*Bartonella canis* Kikuth 1928, 1730; *Haemobartonella* (*Bartonella*) *canis* (Kikuth 1928) Tyzzer and Weinman 1939, 151; Kreier and Ristic 1984, 726]

ha.e.mo.ca'nis Gr. neut. n. *haema* blood; L. fem. gen. n. *canis* of the dog; N.L. gen. n. *haemocanis* of dog blood.

Cells are coccoid to pleomorphic. Motility for this species has not been assessed. The morphology of infected erythrocytes is altered, demonstrating a marked depression at the site of *Mycoplasma haemocanis* attachment. This species has not been grown on any artificial medium; therefore, notable biochemical parameters are not known.

Pathogenic; causes hemolytic anemia in domestic dogs. Transmission is vector-borne and mediated by the brown dog tick (*Rhipicephalus sanguineus*).

Source: observed in association with erythrocytes of domestic dogs (Hoskins, 1991).

DNA G+C content (mol%): Not determined.

Type strain: not established.

Sequence accession no. (16S rRNA gene): AF197337.

3. **Mycoplasma haemofelis** (Clark 1942) Neimark, Johansson, Rikihisa and Tully 2002, 683^{VP} [*Eperythrozoon felis* Clark 1942, 16; *Haemobartonella felis* (Clark 1942) Flint and McKelvie 1956, 240 and Kreier and Ristic 1984, 725]

ha.e.mo.fe'lis. Gr. neut. n. *haema* blood; L. fem. gen. n. *felis* of the cat; N.L. gen. n. *haemofelis* of cat blood.

Cells are coccoid. Motility for this species has not been assessed. This species has not been grown on artificial medium; therefore, notable biochemical parameters are not known.

Pathogenic; causes hemolytic anemia in cats. The mode of transmission is percutaneous or oral; an insect vector has not been identified although fleas have been implicated (Woods et al., 2005).

Tetracyclines and fluoroquinolones are effective therapeutic agents (Dowers et al., 2002; Tasker et al., 2006).

Source: observed in association with erythrocytes of domestic cats.

DNA G+C content (mol%): 38.5 (genome sequence survey of strain OH; Berent and Messick, 2003).

Type strain: not established.

Sequence accession no. (16S rRNA gene): U88563.

The phylogenetic affiliations of the following proposed organism are unknown and its name does not have standing in nomenclature. It is listed here merely because it has been incidentally cited as a species of *Haemobartonella.*

1. **"Haemobartonella procyoni"** Frerichs and Holbrook 1971

Electron microscopy shows this epierythrocytic organism from a raccoon (*Procyon lotor*) is wall-less and its description indicates it probably will prove to be a hemotropic mycoplasma (Frerichs and Holbrook, 1971).

References

Adler, S. and V. Ellenbogen. 1934. A note on two new blood parasites of cattle: *Eperythrozoon* and *Bartonella.* J. Comp. Pathol. *47*: 220–221.

Berent, L.M. and J.B. Messick. 2003. Physical map and genome sequencing survey of *Mycoplasma haemofelis* (*Haemobartonella felis*). Infect. Immun. *71*: 3657–3662.

Brown, D., R. Whitcomb and J. Bradbury. 2007. Revised minimal standards for description of new species of the class *Mollicutes* (division *Tenericutes*). Int. J. Syst. Evol. Microbiol. *57*: 2703–2719.

Brun-Hansen, H., H. Gronstol, H. Waldeland and B. Hoff. 1997. *Eperythrozoon ovis* infection in a commercial flock of sheep. Zentralbl. Veterinarmed. B *44*: 295–299.

Clark, R. 1942. *Eperythrozoon felis* (sp. nov.) in a cat. J. Afr. Vet. Med. Assoc. *13*: 15–16.

Daddow, K.N. 1980. Culex annulirostris as a vector of *Eperythrozoon ovis* infection in sheep. Vet. Parasitol. *7*: 313–317.

Dowers, K.L., C. Olver, S.V. Radecki and M.R. Lappin. 2002. Use of enrofloxacin for treatment of large-form *Haemobartonella felis* in experimentally infected cats. J. Am. Vet. Med. Assoc. *221*: 250–253.

Ewers, W.H. 1971. *Eperythrozoon mariboi* sp. nov. (Protophyta: order Richettsiales) a parasite of red blood cells of the flying fox *Pteropus macrotis epularius* in New Guinea. Parasitology *63*: 261–269.

Flint, J.C. and McKelvie D.H.. 1956. Feline infectious anemia-diagnosis and treatment. Proc. 92nd Ann. Meet. Am. Vet. Med. Assoc. 1955 240–242.

Frerichs, W.M. and A.A. Holbrook. 1971. *Haemobartonella procyoni* sp. n. in the raccoon, *Procyon lotor.* J. Parasitol. *57*: 1309–1310.

Groebel, K., K. Hoelzle, M.M. Wittenbrink, U. Ziegler and L.E. Hoelzle. 2009. *Mycoplasma suis* invades porcine erythrocytes. Infect. Immun. *77*: 576–584.

Gwaltney, S.M. and R.D. Oberst. 1994. Comparison of an improved polymerase chain reaction protocol and the indirect hemagglutination assay in the detection of *Eperythrozoon suis* infection. J. Vet. Diagn. Invest. *6*: 321–325.

Hoelzle, L. 2008. Haemotrophic mycoplasmas: recent advances in *Mycoplasma suis.* Vet. Microbiol. *130*: 215–226.

Hoskins, J.D. 1991. Canine haemobartonellosis, canine hepatozoonosis, and feline cytauxzoonosis. Vet. Clin. North Am. Small Anim. Pract. *21*: 129–140.

Howard, G.W. 1975. The experimental transmission of *Eperythrozoon ovis* by mosquitoes. Parasitology *71*: xxxiii.

Hoyte, H.M.D. 1962. *Eperythrozoon teganodes* sp. nov. (*Rickettsiales*), parasitic in cattle. Parasitology *52*: 527–532.

Hsu, F.S., M.C. Liu, S.M. Chou, J.F. Zachary and A.R. Smith. 1992. Evaluation of an enzyme-linked immunosorbent assay for detection of *Eperythrozoon suis* antibodies in swine. Am. J. Vet. Res. *53*: 352–354.

Kikuth, W. 1928. Über Einen neuen Anämeerreger; *Bartonella canis* nov. spec. Klin. Wochenschr. *7*: 1729–1730.

Kreier, J.P. and M. Ristic. 1963. Morphologic, antigenic, and pathogenic characteristics of *Eperythrozoon ovis* and *Eperythrozoon wenyoni.* Am. J. Vet. Res. *24*: 488–500.

Kreier, J.P. and M. Ristic. 1968. Haemobartonellosis, eperythrozoonosis, grahamellosis and ehrlichiosis. In Infectious Blood Diseases of Man and Animals (edited by Weinman and Ristic). Academic Press, New York, pp. 387–472.

Kreier, J.P. and M. Ristic. 1974. Genus IV. *Haemobartonella* Tyzzer and Weinman 1939, 143[AL]; Genus V. *Eperythrozoon* Schilling 1928, 1854[AL]. *In* Bergey's Manual of Determinative Bacteriology, 8th edn (edited by Buchanan and Gibbons). Williams & Wilkins, Baltimore, pp. 910–914.

Kreier, J.P. and M. Ristic. 1984. Genus III. *Haemobartonella*; Genus IV. *Eperythrozoon. In* Bergey's Manual of Systematic Bacteriology, vol. 1 (edited by Krieg and Holt). Williams & Wilkins, Baltimore, pp. 724–729.

Li, X., X. Jia, D. Shi, Y. Xiao, S. Hu, M. Liu, Z. Yuan and D. Bi. 2008. Continuous *in vitro* Cultivation of *Mycoplasma suis.* Acta Vet. Zootech. Sinica *38*: 1142–1146.

Mason, R.W. and P. Statham. 1991. The determination of the level of *Eperythrozoon ovis* parasitaemia in chronically infected sheep and its significance to the spread of infection. Aust. Vet. J. *68*: 115–116.

Mayer, M. 1921. Über einige bakterienähnliche Parasiten der Erythrozyten bei Menschen und Tieren. Arch. Schiffs Trop. Hyg. *25*: 150–152.

Messick, J.B., P.G. Walker, W. Raphael, L. Berent and X. Shi. 2002. '*Candidatus Mycoplasma* haemodidelphidis' sp. nov., '*Candidatus* Mycoplasma haemolamae' sp. nov. and *Mycoplasma haemocanis* comb. nov., haemotrophic parasites from a naturally infected opossum (*Didelphis virginiana*), alpaca (*Lama pacos*) and dog (*Canis familiaris*): phylogenetic and secondary structural relatedness of their 16S rRNA genes to other mycoplasmas. Int. J. Syst. Evol. Microbiol. *52*: 693–698.

Montes, A., D. Wolfe, E. Welles, J. Tyler and E. Tepe. 1994. Infertility associated with *Eperythrozoon wenyonii* infection in a bull. J. Am. Vet. Med. Assoc. *204*: 261–263.

Neimark, H., K.E. Johansson, Y. Rikihisa and J.G. Tully. 2001. Proposal to transfer some members of the genera *Haemobartonella* and *Eperythrozoon* to the genus *Mycoplasma* with descriptions of '*Candidatus* Mycoplasma haemofelis', '*Candidatus* Mycoplasma haemomuris', '*Candidatus* Mycoplasma haemosuis' and '*Candidatus* Mycoplasma wenyonii'. Int. J. Syst. Evol. Microbiol. *51*: 891–899.

Neimark, H., B. Hoff and M. Ganter. 2004. *Mycoplasma ovis* comb. nov. (formerly *Eperythrozoon ovis*), an epierythrocytic agent of haemolytic anaemia in sheep and goats. Int. J. Syst. Evol. Microbiol. *54*: 365–371.

Neimark, H., W. Peters, B.L. Robinson and L.B. Stewart. 2005. Phylogenetic analysis and description of *Eperythrozoon coccoides,* proposal to transfer to the genus *Mycoplasma* as *Mycoplasma coccoides* comb. nov. and Request for an Opinion. Int. J. Syst. Evol. Microbiol. *55*: 1385–1391.

Neimark, H.C., K.E. Johansson, Y. Rikihisa and J.G. Tully. 2002. Revision of haemotrophic *Mycoplasma* species names. Int. J. Syst. Evol. Microbiol. *52*: 683.

Neitz, W.O., R.A. Alexander and P.J. de Toit. 1934. *Eperythrozoon ovis* (sp. nov.) infection in sheep. Onderstepoort J. Vet. Sci. *3*: 263–274.

Nikol'skii, S.N. and S.N. Slipchenko. 1969. Experiments in the transmission of *Eperythrozoon ovis* by the ticks *H. plumbeum* and *Rh. bursa.* Veterinariia (Russian) *5*: 46.

Prullage, J.B., R.E. Williams and S.M. Gaafar. 1993. On the transmissibility of *Eperythrozoon suis* by *Stomoxys calcitrans* and *Aedes aegypti.* Vet. Parasitol. *50*: 125–135.

Regendanz, P. and W. Kikuth. 1928. Über Aktivierung labiler Infektionen duch Entmilzung (*Piroplasma canis, Nuttalia brasiliensis*, Bartonella opossum, *Spirochaeta didelphydis*). Arch. f. Schiffs. U. Tropenhyg. *32*: 587–593.

Schilling, V. 1928. *Eperythrozoon coccoides*, eine neue durch Splenektomie aktivierbare Dauerinfektion der weissen Maus. Klin. Wochenschr. *7*: 1854–1855.

Splitter, E.J. 1950. *Eperythrozoon suis*, the etiologic agent of icteroanemia–an anaplasmosis-like disease in swine. Am. J. Vet. Res. *11*: 324–329.

Sykes, J.E., N.L. Drazenovich, L.M. Ball and C.M. Leutenegger. 2007. Use of conventional and real-time polymerase chain reaction to determine the epidemiology of hemoplasma infections in anemic and nonanemic cats. J. Vet. Intern. Med. *21*: 685–693.

Tasker, S., S.M. Caney, M.J. Day, R.S. Dean, C.R. Helps, T.G. Knowles, P.J. Lait, M.D. Pinches and T.J. Gruffydd-Jones. 2006. Effect of chronic FIV infection, and efficacy of marbofloxacin treatment, on *Mycoplasma haemofelis* infection. Vet. Microbiol. *117*: 169–179.

Thurston, J.P. 1953. The chemotherapy of *Eperythrozoon coccoides* (Schilling 1928). Parasitology *43*: 170–174.

Tuomi, J. and C.H. Von Bonsdorff. 1967. Ultrastructure of a microorganism associated with bovine platelets. Experientia *23*: 111–112.

Tyzzer, E.E. and D. Weinman. 1939. *Haemobartonella* n.g. *(Bartonella olim* pro parte) *H. microti* n. sp. of the field vole, *Microtus pennsylvanicus*. Am. J. Hyg. *30*: 141–157.

Uilenberg, G. 1967. [*Eperythrozoon tuomii*, n.sp. (*Rickettsiales*), the 3rd species of *Eperythrozoon* of cattle in Madagascar]. Rev. Elev. Med. Vet. Pays. Trop. *20*: 563–569.

Uilenberg, G., F. Thiaucourt and F. Jongejan. 2004. On molecular taxonomy: what is in a name? Exp. Appl. Acarol. *32*: 301–312.

Uilenberg, G., F. Thiaucourt and F. Jongejan. 2006. *Mycoplasma* and *Eperythrozoon* (*Mycoplasmataceae*). Comments on a recent paper. Int. J. Syst. Evol. Microbiol. *56*: 13–14.

Willi, B., F.S. Boretti, C. Baumgartner, S. Tasker, B. Wenger, V. Cattori, M.L. Meli, C.E. Reusch, H. Lutz and R. Hofmann-Lehmann. 2006. Prevalence, risk factor analysis, and follow-up of infections caused by three feline hemoplasma species in cats in Switzerland. J. Clin. Microbiol. *44*: 961–969.

Woods, J.E., M.M. Brewer, J.R. Hawley, N. Wisnewski and M.R. Lappin. 2005. Evaluation of experimental transmission of *Candidatus* Mycoplasma haemominutum and *Mycoplasma haemofelis* by *Ctenocephalides felis* to cats. Am. J. Vet. Res. *66*: 1008–1012.

Zwart, D., P. Leeflang and C.J. van Vorstenbosch. 1969. Studies on an *Eperythrozoon* associated with bovine thrombocytes. Zentralbl. Bakteriol. [Orig.] *210*: 82–105.

Order II. **Entomoplasmatales** Tully, Bové, Laigret and Whitcomb 1993, 381^{VP}

Daniel R. Brown, Janet M. Bradbury and Robert F. Whitcomb*

En.to.mo.plas.ma.ta′les. N.L. neut. n. *Entomoplasma* type genus of the order; *-ales* ending to denote an order: N.L. fem. pl. n. *Entomoplasmatales* the *Entomoplasma* order.

This order in the class *Mollicutes* has been assigned to a group of nonhelical and helical mollicutes that are regularly associated with arthropod or plant hosts. The description of organisms in the order is essentially the same as for the class. Two families are designated, *Entomoplasmataceae* for nonhelical mollicutes and *Spiroplasmataceae* for helical ones. The order consists of four major phylogenetic clades: the paraphyletic entomoplasmataceae clade, which consists of the genera *Entomoplasma* and *Mesoplasma*; and the Apis, Citri–Chrysopicola–Mirum, and Ixodetis clades of the genus *Spiroplasma*. All cells are chemo-organotrophic, usually fermenting glucose through the phosphoenolpyruvate-dependent sugar transferase system. Arginine may be hydrolyzed, but urea is not. Cells may require sterol for growth. Nonhelical strains that grow in serum-free media supplemented with polyoxyethylene sorbitan (PES) are currently assigned to the genus *Mesoplasma*. Temperature optimum for growth is usually 30–32°C, with a few species able to grow at 37°C. Genome sizes range from 780 to 2220 kbp by pulsed-field gel electrophoresis (PFGE), with DNA G+C contents ranging from 25 to 34 mol%. Like members of the *Mycoplasmatales*, all organisms in this order are thought to utilize the UGA codon to encode tryptophan.

Type genus: **Entomoplasma** Tully, Bové, Laigret and Whitcomb 1993, 379^{VP}.

*Deceased 21 December 2007.

Further descriptive information

The basis for the proposal for the order *Entomoplasmatales* (Tully et al., 1993) was the distinctive phylogenetic and phenotypic characteristics of culturable mollicutes regularly associated with arthropods or plants. Members of the family *Entomoplasmataceae* are nonhelical mollicutes that differ in their cholesterol or serum requirements for growth. Nonhelical organisms with a strict requirement for cholesterol were placed in the genus *Entomoplasma* (trivial name, entomoplasmas), whereas nonhelical strains able to grow in a sterol-free medium supplemented with PES were assigned to the genus *Mesoplasma* (trivial name, mesoplasmas). The proposal also included the transfer of the family *Spiroplasmataceae* from the family *Mycoplasmatales* to the family *Entomoplasmatales*. The helical organisms assigned to the genus *Spiroplasma* were within the *Spiroplasmataceae*, and genus and family descriptions of these organisms remained as proposed previously (Skripal, 1983; Whitcomb and Tully, 1984). The order *Entomoplasmatales* is a phylogenetic sister to the order *Mycoplasmatales*. These two orders together form a lineage with several unique properties, including the use of UGA as a tryptophan codon rather than a stop codon.

Taxonomic comments

The genera *Entomoplasma* and *Mesoplasma* constitute a polyphyletic sister lineage of the mycoides cluster of mycoplasmas that are eccentrically situated in the paraphyletic family *Entomoplasmataceae* (Gasparich et al., 2004). There is no current phylogenetic support for separation of *Entomoplasma* and *Mesoplasma* species based on neighbor-joining or maximum-parsimony methods of 16S rRNA gene sequence similarity analysis because they do not form coherent clusters, but are instead intermixed in one paraphyletic group (Johansson and Pettersson, 2002; Tully et al., 1998). No DNA–DNA reassociation experiments have been performed nor is there any other polyphasic taxonomic basis to support the separation. In particular, the growth requirement for sterols is not as profound a character as was initially believed and fails to justify these two species (Gasparich et al., 2004; Rose et al., 1993). For these reasons, and because *Entomoplasma* has priority (Tully et al., 1993), the species currently assigned to the genus *Mesoplasma* should most likely be transferred to the genus *Entomoplasma*. Because the transfer would include its type species, the genus *Mesoplasma* would then become illegitimate. Moreover, Knight (2004) showed that the species formerly called *Mesoplasma pleciae* (Tully et al., 1994) is properly affiliated with the genus *Acholeplasma* on undisputed grounds of 16S rRNA gene sequence similarity and preferred use of UGG rather than UGA as the codon for tryptophan. Therefore, transfer of the currently remaining members of genus *Mesoplasma* cannot be endorsed until similar analyses have been completed for all of those organisms (D.V. Volokhov, unpublished).

References

Gasparich, G.E., R.F. Whitcomb, D. Dodge, F.E. French, J. Glass and D.L. Williamson. 2004. The genus *Spiroplasma* and its non-helical descendants: phylogenetic classification, correlation with phenotype and roots of the *Mycoplasma mycoides* clade. Int. J. Syst. Evol. Microbiol. *54*: 893–918.

Johansson, K.E. and B. Pettersson. 2002. Taxonomy of *Mollicutes*. *In* Molecular Biology and Pathogenicity of Mycoplasmas (edited by Razin and Hermann). Kluwer Academic/Plenum Publishers, London, pp. 1–31.

Knight, T.F., Jr. 2004. Reclassification of *Mesoplasma pleciae* as *Acholeplasma pleciae* comb. nov. on the basis of 16S rRNA and *gyrB* gene sequence data. Int. J. Syst. Evol. Microbiol. *54*: 1951–1952.

Rose, D.L., J.G. Tully, J.M. Bove and R.F. Whitcomb. 1993. A test for measuring growth responses of *Mollicutes* to serum and polyoxyethylene sorbitan. Int. J. Syst. Bacteriol. *43*: 527–532.

Skripal, I.G. 1983. Revival of the name *Spiroplasmataceae* fam. nov., nom. rev., omitted from the 1980 Approved Lists of Bacterial Names. Int. J. Syst. Bacteriol. *33*: 408.

Tully, J.G., J.M. Bové, F. Laigret and R.F. Whitcomb. 1993. Revised taxonomy of the class *Mollicutes*–proposed elevation of a monophyletic cluster of arthropod-associated mollicutes to ordinal rank (*Entomoplasmatales* ord. nov.), with provision for familial rank to separate species with nonhelical morphology (*Entomoplasmataceae* fam. nov.) from helical species (*Spiroplasmataceae*), and emended descriptions of the order *Mycoplasmatales*, family *Mycoplasmataceae*. Int. J. Syst. Bacteriol. *43*: 378–385.

Tully, J.G., R.F. Whitcomb, K.J. Hackett, D.L. Rose, R.B. Henegar, J.M. Bove, P. Carle, D.L. Williamson and T.B. Clark. 1994. Taxonomic descriptions of eight new non-sterol-requiring *Mollicutes* assigned to the genus *Mesoplasma*. Int. J. Syst. Bacteriol. *44*: 685–693.

Tully, J.G., R.F. Whitcomb, K.J. Hackett, D.L. Williamson, F. Laigret, P. Carle, J.M. Bove, R.B. Henegar, N.M. Ellis, D.E. Dodge and J. Adams. 1998. *Entomoplasma freundtii* sp. nov., a new species from a green tiger beetle (Coleoptera: Cicindelidae). Int. J. Syst. Bacteriol. *48*: 1197–1204.

Whitcomb, R.F. and J.G. Tully. 1984. Family III. *Spiroplasmataceae* Skripal 1983, 408VP. Genus I. *Spiroplasma* Saglio, L'Hospital, Laflèche, Dupont, Bové, Tully and Freundt. *In* Bergey's Manual of Systematic Bacteriology, vol. 1 (edited by Krieg and Holt). Williams & Wilkins, Baltimore, pp. 781–787.

Family I. **Entomoplasmataceae** Tully, Bové, Laigret and Whitcomb 1993, 380VP

Daniel R. Brown, Janet M. Bradbury and Robert F. Whitcomb*

En.to.mo.plas.ma.ta.ce′ae. N.L. neut. n. *Entomoplasma*, *-atos* type genus of the family; *-aceae* ending to denote a family; N.L. fem. pl. n. *Entomoplasmataceae* the *Entomoplasma* family.

Cells are **usually coccoid** or occur as short, branched or unbranched, pleomorphic, nonhelical filaments. Filterable through membranes with a mean pore diameter of 220–450 nm. **Cells lack a cell wall** and are bounded only by a plasma membrane. **Nonmotile. Facultatively anaerobic.** The temperature range for growth varies from 10 to 37°C, with the optimum usually at 30°C. The typical colony has a **"fried-egg" appearance**. **Chemo-organotrophic**; acid is produced from glucose, with evidence of a phosphoenolpyruvate-dependent sugar transport system(s) in some members. Arginine and urea are not hydrolyzed. The organisms **may require serum or cholesterol for growth** or may grow in serum-free media plus 0.04% PES. The genome sizes range from 790 to 1140 kbp.

DNA G+C content (mol%): 26–34.

Type genus: **Entomoplasma** Tully, Bové, Laigret and Whitcomb 1993, 379VP.

Further descriptive information

All members of this paraphyletic family are nonhelical and are regularly associated with arthropod or plant hosts. They may require cholesterol or serum for growth, and most have an optimal growth temperature near 30°C. Separation of members of the genera *Entomoplasma* and *Mesoplasma* within the *Entomoplasmataceae* is based on the capacity of the *Mesoplasma* species to grow in a serum-free or cholesterol-free medium supplemented with PES (Rose et al., 1993; Tully et al., 1995), whereas *Entomoplasma* species have a growth requirement for cholesterol. The family is derived from the *Spiroplasma* lineage and is most closely related to the Apis cluster of that group. The mycoides cluster of species in the genus *Mycoplasma* is related to this family and seems to have evolved from it.

*Deceased 21 December 2007.

Genus I. **Entomoplasma** Tully, Bové, Laigret and Whitcomb 1993, 379[VP]

Daniel R. Brown, Janet M. Bradbury and Robert F. Whitcomb*

En.to.mo.plas′ma. Gr. n. *entomon* insect; Gr. neut. n. *plasma* something formed or molded, a form; N.L. neut. n. *Entomoplasma* name intended to show association with insects.

Cells are **nonhelical** and **nonmotile**, frequently pleomorphic and range in size from 200 to 1200 nm in diameter. Some cells exhibit short filamentous forms. Most species ferment glucose. Species possess the phosphoenolpyruvate-dependent sugar-phosphotransferase system. Organisms **require serum or cholesterol** for growth. The temperature range for growth ranges from 10 to 32°C, with the optimum usually at 30–32°C. The genome sizes range from 870 to 900 kbp (PFGE). All currently assigned species were isolated from insects or from plant surfaces where they were presumably deposited by insects.

DNA G+C content (mol%): 27–34.

Type species: **Entomoplasma ellychniae** Tully, Rose, Hackett, Whitcomb, Carle, Bové, Colflesh and Williamson (Tully et al., 1989) Tully, Bové, Laigret and Whitcomb 1993, 380[VP] (*Mycoplasma ellychniae* Tully, Rose, Hackett, Whitcomb, Carle, Bové, Colflesh and Williamson 1989, 288).

Further descriptive information

Cells of these organisms vary from coccoid to pleomorphic forms exhibiting short, branching, nonhelical filaments. Round cells are usually in the size range of 200–300 nm, but may be larger. Most strains were initially isolated in either M1D or SP-4 medium and all entomoplasmas grow well in SP-4 broth containing a supplement of 17% fetal bovine serum. Some strains are able to grow on media with reduced serum content. Most established species have an optimal growth temperature of 30°C, but some species grow better in broth medium maintained at 23–25°C or at 32°C. Colony growth on solid medium is best obtained on SP-4 medium incubated under anaerobic conditions at about 30°C. Under these conditions, most species produce colonies with a classic fried-egg appearance, although *Entomoplasma freundtii* is notable for its granular colony morphology.

All species show strong fermentation of glucose with production of acid and a reduction in medium pH (Table 140). Actively growing cultures in broth medium containing glucose may rapidly acidify the medium, causing partial or complete loss of viability after 7–10 d. Arginine hydrolysis and "film and spot" lipase reactions are rare among species described to date. Entomoplasmas were shown to lack some key metabolic activities found in other mollicutes, especially PP_i-dependent phosphofructokinase and dUTPase, and to possess uracil DNA glycosylase activity. Although the latter pyrimidine enzymic activity distinguished *Entomoplasma* from *Mesoplasma* species, only two *Entomoplasma* species and three *Mesoplasma* species have been tested so far for these activities (Pollack et al., 1996).

Antisera to whole cell antigens of entomoplasmas have been used extensively to provide specific identification to the species level with a variety of serologic techniques, including growth inhibition, metabolism inhibition, and agar plate immunofluorescence (Tully et al., 1989, 1990, 1998). There is no evidence for the pathogenicity of entomoplasmas to either plant or insect hosts. Like other mollicutes, the entomoplasmas are resistant to 500 U/ml penicillin G.

TABLE 140. Differential characteristics of species of the genus *Entomoplasma*[a]

Characteristic	*E. ellychniae*	*E. freundtii*	*E. lucivorax*	*E. luminosum*	*E. melaleucae*	*E. somnilux*
Glucose fermentation	+	+	+	+	+	+
Arginine hydrolysis	–	+	–	–	–	–
"Film and spots"	–	nd	+	+	–	–
Hemadsorption of guinea pig red blood cells	–	nd	–	+	–	–
DNA G+C content (mol%)	27.7	34	27.4	28.8	27	27

[a]Symbols: +, >85% positive; –, 0–15% positive; nd, not determined.

Enrichment and isolation procedures

Flowers and other plant material should be cut in the field and placed in plastic bags without touching by hand. In the laboratory, plant materials are rinsed briefly in either SP-4 or M1D media (May et al., 2008). In both of these media, fetal bovine serum is a critical component for successful growth of these organisms (Hackett and Whitcomb, 1995; Tully, 1995). The rinse medium is immediately decanted and passed through a sterile membrane filter, usually of 450 nm porosity. The filtrate is then passed through at least several tenfold dilutions in the selected culture medium. The retentate may be frozen at −70°C for later use or for retesting. The cultures are incubated at 27–30°C and monitored by dark-field microscopy and/or by observing acidification of the medium. It is important to note that several non-sterol-requiring *Acholeplasma* species have also been isolated from plant and insect material (Tully et al., 1994b).

Insect material, primarily from gut contents or hemolymph obtained by dissection or by fine-pointed glass pipettes, should be added to small volumes of SP-4 or M1D medium and filtered through a 450 nm membrane filter. Serial tenfold dilutions of the filtrate should be incubated at 27–30°C and observed for a decrease in pH of the medium. After two to three serial passages, the organisms should be purified by conventional filter-cloning techniques (Tully, 1983) and stocks of various clones and early passage isolates frozen for further identification procedures (Whitcomb and Hackett, 1996).

Maintenance procedures

Stock cultures of entomoplasmas can be maintained well in SP-4 and/or M1D broth medium containing about 17% fetal bovine serum. Most strains in the group can be adapted to grow in a broth medium containing bovine serum. Stock cultures in broth medium can be stored at −70°C for indefinite periods. For optimum preservation, the organisms should be lyophilized as broth cultures in the early exponential phase of growth and the dried cultures should be sealed under vacuum and stored at 4°C.

*Deceased 21 December 2007.

Differentiation of the genus *Entomoplasma* from other genera

Properties that partially fulfill criteria for assignment to the class *Mollicutes* (Brown et al., 2007) include absence of a cell wall, filterability, and the presence of conserved 16S rRNA gene sequences. Aerobic or facultative anaerobic growth in artificial media and the necessity for sterols for growth exclude assignment to the genera *Anaeroplasma*, *Asteroleplasma*, *Acholeplasma*, *Mesoplasma*, or "*Candidatus* Phytoplasma". Non-helical cellular morphology and regular association with arthropod or plant hosts support exclusion from the genera *Spiroplasma* or *Mycoplasma*. The inability to hydrolyze urea excludes assignment to the genus *Ureaplasma*. However, the difficulty in assigning novel species to this genus is well demonstrated by the earlier difficulties in establishing accurately the taxonomic status of these organisms (Tully et al., 1993). The availability of 16S rRNA gene sequence analyses was critical to the differentiation of these organisms from other mollicutes. Although isolates from vertebrates are very unlikely to be entomoplasmas, two bona fide *Mycoplasma* species, *Mycoplasma iowae* and *Mycoplasma equigenitalium*, have been isolated from plants [Grau et al., 1991; J.C. Vignault, J.M. Bové and J.G. Tully, unpublished (see ATCC 49192)].

Taxonomic comments

The landmark studies of Weisburg et al. (1989), using 16S rRNA gene sequences of about 50 species of mollicutes, were critical in the resolution of certain taxonomic conflicts regarding the species that became *Entomoplasma*. The first entomoplasmas to be recognized were serologically related isolates from the flowers of *Melaleuca* and *Grevillea* trees (McCoy et al., 1979). Others, found in a wide range of insect species (Tully et al., 1987), included strain ELCN-1^T from the hemolymph of the firefly beetle *Ellychnia corrusca* (Tully et al., 1989) and three serologically distinct strains isolated from gut contents of *Pyractomena* and *Photinus* beetles (Williamson et al., 1990). Although these nonhelical, sterol-requiring mollicutes were initially placed in the genus *Mycoplasma*, 16S rRNA gene sequence analysis clearly indicated that strain M1^T, previously designated *Mycoplasma melaleucae*, and strain ELCN-1^T, previously designated *Mycoplasma ellychniae*, were most closely affiliated with the *Spiroplasma* lineage of helical organisms isolated primarily from arthropods. These findings prompted a proposal to reclassify the nonhelical mollicutes from arthropods and plants in a new order, *Entomoplasmatales*, and new family, *Entomoplasmataceae*, with the genus *Entomoplasma* reserved for sterol-requiring species (Tully et al., 1993). Strains M1^T and ELCN-1^T were renamed as *Entomoplasma melaleucae* and *Entomoplasma ellychniae*, respectively. Subsequent phylogenetic analysis of *Mycoplasma freundtii*, later renamed *Entomoplasma freundtii*, confirmed the placement (Tully et al., 1998).

The paraphyletic relationship between the genera *Entomoplasma* and *Mesoplasma* is currently an unresolved problem in the systematics of this genus. It is possible that these genera, separated by the single criterion of sterol requirement, should be combined into the single genus *Entomoplasma*. However, Knight (2004) showed that *Mesoplasma pleciae* (Tully et al., 1994b) should belong to the genus *Acholeplasma* based on 16S rRNA gene sequence similarity and the preferred use of UGG rather than UGA as the codon for tryptophan. Therefore, transfer of the currently remaining members of genus *Mesoplasma* to other genera cannot be endorsed until similar analyses have been completed for all of those species (D.V. Volokhov, unpublished).

Acknowledgements

We thank Karl-Erik Johansson for helpful comments and suggestions and Gail E. Gasparich for her landmark contributions regarding the phylogenetics of the *Entomoplasmatales*. The major contributions to the foundation of this material by Joseph G. Tully are gratefully acknowledged.

Further reading

Tully, J.G. 1989. Class *Mollicutes*: new perspectives from plant and arthropod studies. *In* The Mycoplasmas, vol. 5 (edited by Whitcomb and Tully). Academic Press, San Diego, pp. 1–31.

Tully, J.G. 1996. Mollicute–host interrelationships: current concepts and diagnostic implications. *In* Molecular and Diagnostic Procedures in Mycoplasmology, vol. 2 (edited by Tully and Razin). Academic Press, San Diego, pp. 1–21.

Differentiation of the species of the genus *Entomoplasma*

The primary technique for differentiation of *Entomoplasma* species is 16S rRNA gene sequence comparisons, confirmed by serology (Brown et al., 2007). Nonhelical mollicutes that belong to a known species isolated from arthropods or plants can be readily identified serologically provided that a battery of potent antisera for classified species is available. Growth inhibition tests, performed by placing paper discs saturated with type-specific antisera on agar plates inoculated with the organism, are perhaps the most convenient and rapid serological technique to differentiate species (Clyde, 1983). The agar plate immunofluorescence test is also a convenient and rapid means of mollicute species identification. In the absence of specific conjugated antiserum, an indirect immunofluorescence test can be performed with type-specific antiserum and a fluorescein-conjugated secondary antibody. The metabolism inhibition test (Taylor-Robinson, 1983) has also been applied to differentiation of *Entomoplasma* species (Tully et al., 1998).

List of species of the genus *Entomoplasma*

1. **Entomoplasma ellychniae** (Tully, Rose, Hackett, Whitcomb, Carle, Bové, Colflesh and Williamson 1989) Tully, Bové, Laigret and Whitcomb 1993, 380^{VP} (*Mycoplasma ellychniae* Tully, Rose, Hackett, Whitcomb, Carle, Bové, Colflesh and Williamson 1989, 288)

el.lych.ni'ae. N.L. n. *Ellychnia* a genus of firefly beetles; N.L. gen. n. *ellychniae* of *Ellychnia*, from which the organism was first isolated.

This is the type species of the genus *Entomoplasma*. Cells are nonhelical, pleomorphic filaments, with some branching;

small coccoid forms, ranging in diameter from 200 to 300 nm, also occur. Passage of broth cultures through 450 and 300 nm porosity membrane filters does not reduce viable cell numbers, whereas passage through 220 nm porosity reduces cell populations by about 10%. Grows well in SP-4 medium with fetal bovine serum supplements. Does not grow well in horse serum-supplemented broth or agar media. Optimum temperature for broth growth is 30°C; can grow at 18–32°C. Colonies incubated at 30°C under anaerobic conditions have a fried-egg appearance. Does not hemadsorb guinea pig erythrocytes.

No evidence for pathogenicity for insects.

Source: isolated from the hemolymph of the firefly beetle *Ellychniae corrusca.*

DNA G+C content (mol%): 27.5 (Bd).

Type strain: ELCN-1, ATCC 43707, NCTC 11714.

Sequence accession no. (16S rRNA gene): M24292.

2. **Entomoplasma freundtii** Tully, Whitcomb, Hackett, Williamson, Laigret, Carle, Bové, Henegar, Ellis, Dodge and Adams 1998, 1203[VP]

freund'ti.i. N.L. masc. gen. n. *freundtii* of Freundt, named after Eyvind Freundt, a Danish pioneer in the taxonomy and classification of mollicutes.

Cells are predominantly coccoid in shape, ranging from 300 to 1200 nm in diameter. Organisms are readily filterable through membranes with mean pore diameters of 450, 300, and 220 nm; more than 90% of viable cells in broth culture are able to pass 220 nm porosity membranes. The temperature range for growth is 10–32°C, with an optimum at 30°C. Colonies under anaerobic conditions are granular and frequently exhibit multiple satellite forms although the organism is considered nonmotile. The organism grows well in SP-4 broth medium or other media containing horse serum supplements.

No evidence for pathogenicity for insects.

Source: isolated from the gut contents of a green tiger beetle (Coleoptera: Cicindelidae).

DNA G+C content (mol%): 34.1 (Bd).

Type strain: BARC 318, ATCC 51999.

Sequence accession no. (16S rRNA gene): AF036954.

3. **Entomoplasma lucivorax** (Williamson, Tully, Rose, Hackett, Henegar, Carle, Bové, Colflesh and Whitcomb 1990) Tully, Bové, Laigret and Whitcomb 1993, 380[VP] (*Mycoplasma lucivorax* Williamson, Tully, Rose, Hackett, Henegar, Carle, Bové, Colflesh and Whitcomb 1990, 164)

lu.ci.vo'rax. L. fem. n. *lux lucis* light; L. neut. adj. *vorax* gluttonous, devouring; N.L. neut. adj. *lucivorax* light devouring, referring to the predacious habit of the host insect, which preys on other luminescent firefly species.

Cells are either pleomorphic coccoidal or subcoccoidal, with a diameter of 200–300 nm, or are short, branched or unbranched filaments. Cells are readily filterable through membrane filters with mean pore diameters of 450, 300, and 220 nm, but do not pass 100 nm porosity membranes. Optimum temperature for growth is 30°C; can grow at 10–32°C. Nonmotile. Colonies under anaerobic conditions usually have a fried-egg appearance. Grows well in SP-4 broth medium or other media containing horse serum supplements. Colonies do not hemadsorb guinea pig erythrocytes.

No evidence of pathogenicity for insects or plants.

Source: first isolated from the gut of a firefly beetle (*Photinus pyralis*); also isolated from a flower (*Spirea ulmaria*; C. Chastel, unpublished).

DNA G+C content (mol%): 27.4 (Bd).

Type strain: PIPN-2, ATCC 49196, NCTC 11716.

Sequence accession no. (16S rRNA gene): AF547212.

4. **Entomoplasma luminosum** (Williamson, Tully, Rose, Hackett, Henegar, Carle, Bové, Colflesh and Whitcomb 1990) Tully, Bové, Laigret and Whitcomb 1993, 380[VP] (*Mycoplasma luminosum* Williamson, Tully, Rose, Hackett, Henegar, Carle, Bové, Colflesh and Whitcomb 1990, 163)

lu.mi.no'sum. L. neut. adj. *luminosum* luminous, emitting light, referring to the luminescence of the adult host from which the organism was isolated.

Cells are pleomorphic and coccoidal or subcoccoidal with a diameter of 200–300 nm. Cells also occur as short, branched or unbranched filaments. The organisms are readily filterable through membranes with mean pore diameters of 450, 300, and 220 nm, but do not pass 100 nm porosity membranes. The temperature range for growth is 10–32°C, with an optimum at 32°C. Nonmotile. Colonies under anaerobic conditions have a fried-egg appearance. The organism grows well in SP-4 broth medium or other media containing horse serum supplements. Colonies hemadsorb guinea pig erythrocytes.

No evidence of pathogenicity for insects.

Source: isolated from the gut of the firefly beetle (*Photinus marginata*).

DNA G+C content (mol%): 28.8 (Bd).

Type strain: PIMN-1, ATCC 49195, NCTC 11717.

Sequence accession no. (16S rRNA gene): AY155670.

5. **Entomoplasma melaleucae** (Tully, Rose, McCoy, Carle, Bové, Whitcomb and Weisburg 1990) Tully, Bové, Laigret and Whitcomb 1993, 380[VP] (*Mycoplasma melaleucae* Tully, Rose, McCoy, Carle, Bové, Whitcomb and Weisburg 1990, 146)

me la.leu'cae. N.L. n. *Melaleuca* a genus of tropical trees having white flowers with sweet fragrance; N.L. gen. n. *melaleucae* of *Melaleuca*, the plant from which the type strain was isolated.

Cells are pleomorphic and coccoidal or subcoccoidal, with few filamentous forms. Coccoidal forms have mean diameters of 250–300 nm. Cells are readily filterable through 450 and 300 nm porosity membrane filters, with few cells passing 220 nm porosity membranes. The temperature range for growth is 10–30°C, with an optimum at about 23°C. Nonmotile. Colonies under anaerobic conditions at 23–30°C display a fried-egg appearance. Grows well in SP-4 broth or in modified Edward medium containing fetal bovine serum. The organism does not grow well in horse serum-based broth medium. Agar colonies do not adsorb guinea pig erythrocytes.

No evidence of pathogenicity for insects or plants.

Source: isolated from flower surfaces of a subtropical plant, *Melaleuca quinquenervia*, in south Florida. Related strains have been isolated from flowers of other subtropical trees in Florida, *Melaleuca decora* and *Grevillea robusta* (silk oak), and from an anthophorine bee (*Xylocopa micans*) in the same geographic area.

DNA G+C content (mol%): 27.0 (Bd).
Type strain: M1, ATCC 49191, NCTC 11715.
Sequence accession nos (16S rRNA gene): M24478, AY345990.
Further comment: the 16S rRNA gene sequence is more similar to that of members of genus *Mesoplasma* than to others in the genus *Entomoplasma.*

6. **Entomoplasma somnilux** (Williamson, Tully, Rose, Hackett, Henegar, Carle, Bové, Colflesh and Whitcomb 1990) Tully, Bové, Laigret and Whitcomb 1993, 380[VP] (*Mycoplasma somnilux* Williamson, Tully, Rose, Hackett, Henegar, Carle, Bové, Colflesh and Whitcomb 1990, 163)

som.ni'lux. L. masc. n. *somnus* sleep; L. fem. n. *lux* light; N.L. n. *somnilux* intended to mean sleeping light, referring to the quiescent pupal stage of the host from which the organism was isolated, which precedes the luminescent adult stage.

Cells are pleomorphic and coccoidal or subcoccoidal, with a diameter of 200–300 nm; also occur as short, branched or unbranched filaments. Readily filterable through membranes with mean pore diameters of 450, 300, and 220 nm. The temperature range for growth is 10–32°C, with optimum growth at 30°C. Nonmotile. Colonies incubated under anaerobic conditions at 30°C have a fried-egg appearance. The organism grows well in SP-4 broth medium or other media containing horse serum supplements. Colonies do not adsorb guinea pig erythrocytes.

No evidence of pathogenicity for insects.
Source: isolated from a pupal gut of the firefly beetle (*Pyractomena angulata*).
DNA G+C content (mol%): 27.4 (Bd).
Type strain: PYAN-1, ATCC 49194, NCTC 11719.
Sequence accession no. (16S rRNA gene): AY157871.

Genus II. **Mesoplasma** Tully, Bové, Laigret and Whitcomb 1993, 380[VP]

Daniel R. Brown, Janet M. Bradbury and Robert F. Whitcomb*

Me.so.plas'ma. Gr. adj. *mesos* middle; Gr. neut. n. *plasma* something formed or molded, a form; N.L. neut. n. *Mesoplasma* middle form, name intended to denote a middle position with respect to sterol or cholesterol requirement.

Cells are **nonhelical** and **nonmotile**, generally **coccoid or short filamentous forms**. Coccoid cells are usually 220–300 nm in diameter, but some cells in some species can be as large as 400–500 nm. Most strains ferment glucose and most, but not all, lack the ability to hydrolyze arginine. Species possess the phosphoenolpyruvate-dependent sugar-phosphotransferase system. Neither serum nor cholesterol is required for growth, but strains show **sustained growth in a serum-free or cholesterol-free medium when the medium is supplemented with 0.04% PES**. The optimum temperature for growth is usually near 28–32°C, with some strains able to grow well at temperatures as low as 23°C or as high as 37°C. Genome sizes range from 825 to 930 kbp (PFGE).

DNA G+C content (mol%): 26–32.

Type species: **Mesoplasma florum** (McCoy, Basham, Tully, Rose, Carle and Bové 1984) Tully, Bové, Laigret and Whitcomb 1993, 380[VP] (*Acholeplasma florum* McCoy, Basham, Tully, Rose, Carle and Bové 1984, 14).

Further descriptive information

Cells are predominantly coccoid in the exponential phase of growth when examined by dark-field microscopy. Cells from broth cultures examined by transmission electron microscopy are also coccoid, with individual cells usually 220–500 nm in diameter and clearly defined by a single cytoplasmic membrane. Colony growth is best obtained on SP-4 agar medium. Plates incubated under anaerobic conditions at about 30°C usually display characteristic fried-egg type colonies after 5–7 d incubation.

Several mesoplasmas lack certain key metabolic activities found in other mollicutes, especially PP_i-dependent phosphofructokinase, dUTPase, and uracil DNA glycosylase activity (Pollack et al., 1996). Most mesoplasmas were isolated in M1D medium containing 15% fetal bovine serum (Whitcomb, 1983), but adapt well to growth in SP-4 broth containing 15–17% fetal bovine serum, or in broth medium containing a 1% bovine serum fraction supplement (Tully, 1984; Tully et al., 1994a). All species show strong fermentation of glucose with acid production (Table 141), with a rapid decline in pH of the medium and loss of viability. Arginine hydrolysis has been observed only with the type strain (PUPA-2[T]) of *Mesoplasma photuris.*

Antisera directed against whole-cell antigens of filter-cloned mesoplasmas have been used extensively to establish species and to provide species identifications. There is no evidence of pathogenicity of any currently established species in the genus for either an insect or plant host. Mesoplasmas are resistant to 500 U/ml penicillin.

Enrichment, isolation, and maintenance procedures

The culture media and procedures for isolation and maintenance of entomoplasmas from plant and insect sources can also be effectively applied for mesoplasmas.

Differentiation of the genus *Mesoplasma* from other genera

Properties that fulfill criteria for assignment to this genus are the same as those for the genus *Entomoplasma*, with the exception that the genus *Mesoplasma* is currently reserved for species that are able to grow in serum-free medium supplemented with PES (Tully et al., 1993).

Taxonomic comments

The existence of a flora of nonhelical, wall-less prokaryotes associated with arthropod or plant hosts was first documented by T.B. Clark, S. Eden-Green, and R.E. McCoy and colleagues. Some of the plant isolates were clearly related to previously described *Acholeplasma* species, such as *Acholeplasma oculi* (Eden-Green and Tully, 1979), whereas others were established as novel *Acholeplasma* species, able to grow well in broth media without any cholesterol, serum, or fatty acid supplements. However, a significant group of other similarly derived strains were able to

*Deceased 21 December 2007.

TABLE 141. Differential characteristics of species of the genus *Mesoplasma*[a]

Characteristic	*M. florum*	*M. chauliocola*	*M. coleopterae*	*M. corruscae*	*M. entomophilum*	*M. grammopterae*	*M. lactucae*	*M. photuris*	*M. seiffertii*	*M. syrphidae*	*M. tabanidae*
Glucose fermentation	+	+	+	+	+	+	+	+	+	+	+
Arginine hydrolysis	−	−	−	−	−	−	−	+	−	−	−
Hemadsorption of guinea pig red blood cells	−	+	−	+	+	−	+	−	+	+	−
DNA G+C content (mol%)	27.3	28.3	27.7	26.4	30	29.1	30	28.8	30	27.6	28.3

[a]Symbols: +, >85% positive; −, 0–15% positive.

grow in serum-free or cholesterol-free media only when small amounts of PES were added to the medium. Because these strains grew in the absence of cholesterol or serum, several of them were initially described as *Acholeplasma* species, including *Acholeplasma florum* (McCoy et al., 1984), *Acholeplasma entomophilum* (Tully et al., 1988), and *Acholeplasma seiffertii* (Bonnet et al., 1991). Although the growth response to PES in serum-free or cholesterol-free media suggested that there were fundamental differences between such mollicutes and classic acholeplasmas, conclusive taxonomic evidence was lacking. The subsequent analysis of 16S rRNA gene sequences by Weisburg et al. (1989) showed that the PES-requiring organisms were closely related to the spiroplasma group of mollicutes and were phylogenetically distant from acholeplasmas. On the basis of these findings and additional phylogenetic data, a proposal was made that the plant- and insect-derived mollicutes with growth responses to PES in serum-free or cholesterol-free media would be assigned to a new family, *Entomoplasmataceae*, and a new genus, *Mesoplasma* (Tully et al., 1993). Three of the plant-derived strains previously described as *Acholeplasma* species (*Acholeplasma florum*, *Acholeplasma entomophilum*, and *Acholeplasma seiffertii*) were transferred to the genus *Mesoplasma*, with retention of their species epithets. A single plant-derived strain that had previously been described as *Mycoplasma lactucae*, and later found to grow in serum-free or cholesterol-free media supplemented with PES, was renamed *Mesoplasma lactucae*. Later, eight novel *Mesoplasma* species were described (Tully et al., 1994a).

The paraphyletic relationship between the genera *Entomoplasma* and *Mesoplasma* is a currently unresolved problem in the systematics of this genus. It is possible that these genera, separated by the single criterion of sterol requirement, should be combined into the single genus *Entomoplasma*. However, Knight (2004) showed that *Mesoplasma pleciae* (Tully et al., 1994a) should belong to the genus *Acholeplasma* based on 16S rRNA gene sequence similarity and the preferred use of UGG rather than UGA as the codon for tryptophan. Therefore, transfer of the currently remaining members of the genus *Mesoplasma* to other genera cannot be endorsed until similar analyses have been completed for all of those species (D.V. Volokhov, unpublished).

Acknowledgements

We thank Karl-Erik Johansson for helpful comments and suggestions and Gail E. Gasparich for her landmark contributions regarding the phylogenetics of the *Entomoplasmatales*. The major contributions to the foundation of this material by Joseph G. Tully are gratefully acknowledged.

Further reading

Tully, J.G. 1989. Class *Mollicutes*: new perspectives from plant and arthropod studies. *In* The Mycoplasmas, vol. 5 (edited by Whitcomb and Tully). Academic Press, San Diego, pp. 1–31.

Tully, J.G. 1996. Mollicute-host interrelationships: current concepts and diagnostic implications. *In* Molecular and Diagnostic Procedures in Mycoplasmology, vol. 2 (edited by Tully and Razin). Academic Press, San Diego, pp. 1–21.

Differentiation of the species of the genus *Mesoplasma*

The techniques for differentiation of *Mesoplasma* species are the same as those for genus *Entomoplasma*.

List of species of the genus *Mesoplasma*

1. **Mesoplasma florum** (McCoy, Basham, Tully, Rose, Carle and Bové 1984) Tully, Bové, Laigret and Whitcomb 1993, 380[VP] (*Acholeplasma florum* McCoy, Basham, Tully, Rose, Carle and Bové 1984, 14)

flo′rum. L. gen. pl. n. *florum* of flowers, indicating the recovery site of the organism.

This is the type species of the genus. Cells are oval or coccoid. The organism is readily filterable through membranes with mean pore diameters of 450, 300, and 220 nm, but does not pass a membrane with 100 nm porosity. Temperature range for growth is 18–37°C, with an optimum at 28–30°C. Colonies on agar medium containing horse serum supplements have a typical fried-egg appearance after anaerobic incubation at 37°C. Colonies on agar do not hemadsorb guinea pig erythrocytes.

The 16S rRNA gene sequence is identical to that of *Mesoplasma entomophilum* (GenBank accession no. AF305693), but antiserum against *Mesoplasma florum* did not inhibit growth of *Mesoplasma entomophilum* or label the surfaces of *Mesoplasma entomophilum* colonies on agar (Tully et al., 1988). There are additional phenotypic distinctions between the two species.

No evidence of pathogenicity for plants or insects.

Source: first isolated from surface of flowers on a lemon tree (*Citrus limon*) in Florida, with subsequent isolations from floral surfaces of grapefruit (*Citrus *paradisi*) and powderpuff trees (*Albizia julibrissin*) in Florida (McCoy et al., 1979). Also isolated from a variety of plants and from the gut tissues of numerous species of insects (Clark et al., 1986; Tully et al., 1990; Whitcomb et al., 1982).

DNA G+C content (mol%): 27.3 (Bd, whole genome sequence).

Type strain: L1, ATCC 33453, NCTC 11704.

Sequence accession nos: AF300327 (16S rRNA gene), NC_006055 (strain L1^T genome sequence).

2. **Mesoplasma chauliocola** Tully, Whitcomb, Hackett, Rose, Henegar, Bové, Carle, Williamson and Clark 1994a, 691^{VP}

chau.li.o'co.la. N.L. n. *chaulio* first part of the genus name of goldenrod beetle (*Chauliognathus*); L. suff. *-cola* (from L. masc. or fem. n. *incola*) inhabitant; N.L. masc. n. *chauliocola* inhabitant of the goldenrod beetle.

Cells are primarily coccoid, ranging in size from 300 to 500 nm in diameter. Cells are readily filterable through membranes with mean pore diameters of 450, 300, and 220 nm, with a small number of cells able to pass through 100 nm porosity filters. Temperature range for growth is 10–37°C, with an optimum of 32–37°C. Nonmotile. Colonies incubated anaerobically at 32–37°C show fried-egg morphology. Colonies hemadsorb guinea pig erythrocytes.

No evidence of pathogenicity for plants or insects.

Source: originally isolated from gut fluid of an adult goldenrod soldier beetle (*Chauliognathus pennsylvanicus*).

DNA G+C content (mol%): 28.3 (Bd, T_m, HPLC).

Type strain: CHPA-2, ATCC 49578.

Sequence accession no. (16S rRNA gene): AY166704.

3. **Mesoplasma coleopterae** Tully, Whitcomb, Hackett, Rose, Henegar, Bové, Carle, Williamson and Clark 1994a, 692^{VP}

co.le.op.te'rae. N.L. fem. gen. n. *coleopterae* of *Coleoptera*, referring to the order of insects (Coleoptera) from which the organism was first isolated.

Cells are primarily coccoid, ranging in diameter from 300 to 500 nm. Organisms are readily filterable through membranes with mean pore diameters of 450, 300, and 220 nm. Temperature range for growth is 10–37°C, with an optimum of 30–37°C. Nonmotile. Colonies incubated anaerobically at 30°C usually have a fried-egg appearance. Agar colonies do not hemadsorb guinea pig erythrocytes.

No evidence of pathogenicity for plants or insects.

Source: original isolation was from the gut of an adult soldier beetle (*Chauliognathus* sp.).

DNA G+C content (mol%): 27.7 (Bd, T_m, HPLC).

Type strain: BARC 779, ATCC 49583.

Sequence accession no. (16S rRNA gene): DQ514605 (partial sequence).

4. **Mesoplasma corruscae** Tully, Whitcomb, Hackett, Rose, Henegar, Bové, Carle, Williamson and Clark 1994a, 691^{VP}

cor.rus'cae. N.L. fem. gen. n. *corruscae* of *corrusca*, referring to the species of firefly beetle (*Ellychnia corrusca*) from which the organism was first isolated.

Cells are primarily coccoid, ranging in diameter from 300 to 500 nm. Cells are readily filterable through membranes with mean pore diameters of 450, 300, and 220 nm. Temperature range for growth is 10–32°C, with an optimum of 30°C. Nonmotile. Colonies incubated anaerobically at 30°C usually have a fried-egg appearance. Colonies hemadsorb guinea pig erythrocytes.

No evidence of pathogenicity for plants or insects.

Source: original isolation was from the gut of an adult firefly (*Ellychnia corrusca*).

DNA G+C content (mol%): 26.4 (Bd, T_m, HPLC).

Type strain: ELCA-2, ATCC 49579.

Sequence accession no. (16S rRNA gene): AY168929.

5. **Mesoplasma entomophilum** (Tully, Rose, Carle, Bové, Hackett and Whitcomb 1988) Tully, Bové, Laigret and Whitcomb 1993, 380^{VP} (*Acholeplasma entomophilum* Tully, Rose, Carle, Bové, Hackett and Whitcomb 1988, 166)

en.to.mo.phi'lum. Gr. n. *entomon* insect; N.L. neut. adj. *philum* (from Gr. neut. adj. *philon*) friend, loving; N.L. neut. adj. *entomophilum* insect-loving.

Cells are pleomorphic, but primarily coccoid, ranging from 300 to 500 nm in diameter. Cells are readily filterable through 220 nm porosity membrane filters. The temperature range for growth is 23–32°C, with an optimum at 30°C. Nonmotile. Colonies incubated under anaerobic conditions at 30°C usually have a fried-egg appearance. Colonies hemadsorb guinea pig erythrocytes.

The 16S rRNA gene sequence is identical to that of *Mesoplasma florum* (GenBank accession no. AF300327), but antiserum against *Mesoplasma florum* did not inhibit growth of *Mesoplasma entomophilum* or label the surfaces of *Mesoplasma entomophilum* colonies on agar (Tully et al., 1988). There are additional phenotypic distinctions between the two species.

No evidence of pathogenicity for plants or insects.

Source: original isolation was from the gut contents of a tabanid fly (*Tabanus catenatus*). Also isolated from a variety of other species of insects.

DNA G+C content (mol%): 30 (Bd).

Type strain: TAC, ATCC 43706, NCTC 11713.

Sequence accession no. (16S rRNA gene): AF305693.

6. **Mesoplasma grammopterae** Tully, Whitcomb, Hackett, Rose, Henegar, Bové, Carle, Williamson and Clark 1994a, 691^{VP}

gram.mop.te'rae. N.L. fem. gen. n. *grammopterae* of *Grammoptera*, referring to the genus of beetle (*Grammoptera*) from which the organism was first isolated.

Cells are primarily coccoid, ranging in diameter from 300 to 500 nm. Cells are readily filterable through membrane filters with mean pore diameters of 450, 300, and 220 nm. Temperature range for growth is 10–37°C, with an optimum at 30°C. Nonmotile. Colonies incubated under anaerobic conditions at 30°C have a fried-egg appearance. Colonies do not hemadsorb guinea pig erythrocytes.

No evidence of pathogenicity for plants or insects.

Source: original isolation was from the gut contents of an adult long-horned beetle (*Grammoptera* sp.). Other isolations were made from adult soldier beetle (*Cantharidae* sp.) and from an adult mining bee (*Andrena* sp.).

DNA G+C content (mol%): 29.1 (Bd, T_m, HPLC).

Type strain: GRUA-1, ATCC 49580.

Sequence accession no. (16S rRNA gene): AY174170.

7. **Mesoplasma lactucae** (Rose, Kocka, Somerson, Tully, Whitcomb, Carle, Bové, Colflesh and Williamson 1990) Tully, Bové, Laigret and Whitcomb 1993, 380[VP] (*Mycoplasma lactucae* Rose, Kocka, Somerson, Tully, Whitcomb, Carle, Bové, Colflesh and Williamson 1990, 141)

lac.tu'cae. L. fem. n. *lactuca* lettuce; L. gen. n. *lactucae* of lettuce, referring to the plant from which the organism was first isolated.

Cells are primarily coccoid, ranging in size from 300 to 500 nm in diameter, with only occasional short, nonhelical, pleomorphic filaments. Cells are readily filterable through membrane filters with mean pore diameters of 450, 300, and 220 nm, and a few cells are able to pass 100 nm porosity membranes. Temperature range for growth is 18–37°C, with optimal growth at 30°C. Nonmotile. Colonies incubated under anaerobic conditions at 30°C have a fried-egg appearance. Colonies hemadsorb guinea pig erythrocytes.

No evidence of pathogenicity for plants or insects.

Source: original isolation was from lettuce (*Lactuca sativa*).

DNA G+C content (mol%): 30 (Bd).

Type strain: 831-C4, ATCC 49193, NCTC 11718.

Sequence accession no. (16S rRNA gene): AF303132. Has been reported to possess three rRNA operons (Grau, 1991).

8. **Mesoplasma photuris** Tully, Whitcomb, Hackett, Rose, Henegar, Bové, Carle, Williamson and Clark 1994a, 691[VP]

pho.tu'ris. N.L. gen. n. *photuris* of *Photuris*, referring to the genus of firefly beetle (*Photuris* sp.) from which the organism was first isolated.

Cells are primarily coccoid, ranging in diameter from 300 to 500 nm. Readily filterable through membrane filters with mean pore diameters of 450, 300, and 220 nm. Temperature range for growth is 10–32°C, with optimum at 30°C. Nonmotile. Colonies incubated under anaerobic conditions at 30°C have a fried-egg appearance. Colonies do not hemadsorb guinea pig erythrocytes.

No evidence of pathogenicity for plants or insects.

Source: original isolation was from gut fluids of larval and adult fireflies (*Photuris lucicrescens* and other *Photuris* spp.). One isolate (BARC 1976) was obtained by F.E. French from the gut of a horse fly (*Tabanus americanus*).

DNA G+C content (mol%): 28.8 (Bd, T_m, HPLC).

Type strain: PUPA-2, ATCC 49581.

Sequence accession no. (16S rRNA gene): AY177627.

9. **Mesoplasma seiffertii** (Bonnet, Saillard, Vignault, Garnier, Carle, Bové, Rose, Tully and Whitcomb 1991) Tully, Bové, Laigret and Whitcomb 1993, 380[VP] (*Acholeplasma seiffertii* Bonnet, Saillard, Vignault, Garnier, Carle, Bové, Rose, Tully and Whitcomb 1991, 48)

seif.fer'ti.i. N.L. masc. gen. n. *seiffertii* of Seiffert, in honor of Gustav Seiffert, a German microbiologist who performed pioneering studies on mollicutes that occur in soil and compost and do not require sterols for growth.

Cells are primarily coccoid, ranging in diameter from 300 to 500 nm. Cells are readily filterable through membranes with mean pore diameters of 450, 300, and 220 nm. Temperature range for growth is 20–35°C, with optimum at about 28–30°C. Nonmotile. Colonies incubated under anaerobic conditions at 30°C have a fried-egg appearance. Colonies hemadsorb guinea pig erythrocytes.

Three insect isolates of *Mesoplasma seiffertii*, two from mosquitoes and one from a horse fly, were compared to strain F7[T] of plant origin. High relatedness values of 78–98% DNA–DNA reassociation under high stringency conditions were obtained (Gros et al., 1996).

No evidence of pathogenicity for plants or insects.

Source: first isolated from floral surfaces of a sweet orange tree (*Citrus sinensis*) and from wild angelica (*Angelica sylvestris*). Also isolated from insects.

DNA G+C content (mol%): 30 (Bd).

Type strain: F7, ATCC 49495.

Sequence accession no. (16S rRNA gene): L12056.

10. **Mesoplasma syrphidae** Tully, Whitcomb, Hackett, Rose, Henegar, Bové, Carle, Williamson and Clark 1994a, 691[VP]

syr.phi'dae. N.L. fem. gen. n. *syrphidae* of a syrphid, referring to the syrphid fly family (Syrphidae), from which the organism was first isolated.

Cells are primarily coccoid, ranging in size from 300 to 500 nm in diameter. Cells readily pass membrane filters with mean pore diameters of 450, 300, and 220 nm. Temperature range for growth is 10–32°C, with optimum at 23–25°C. Nonmotile. Colonies incubated under anaerobic conditions at 23–25°C have a fried-egg appearance. Colonies hemadsorb guinea pig erythrocytes.

No evidence of pathogenicity for insects.

Source: original isolation was from the gut of an adult syrphid fly (Diptera: Syrphidae). Similar strains have been isolated from a bumblebee (*Bombus* sp.) and a skipper (Lepidoptera: Hesperiidae).

DNA G+C content (mol%): 27.6 (Bd, T_m, HPLC).

Type strain: YJS, ATCC 51578.

Sequence accession no. (16S rRNA gene): AY231458.

11. **Mesoplasma tabanidae** Tully, Whitcomb, Hackett, Rose, Henegar, Bové, Carle, Williamson and Clark 1994a, 692[VP]

ta.ba.ni.dae. N.L. fem. gen. n. *tabanidae* of a tabanid, referring to the horse fly family (Tabanidae), the host from which the organism was first isolated.

Cells are primarily coccoid, ranging in size from 300 to 500 nm in diameter. Cells readily pass membrane filters with mean pore diameters of 450, 300, and 220 nm. Temperature range for growth is 10–37°C, with optimum at 37°C. Nonmotile. Colonies incubated under anaerobic conditions at 37°C display a fried-egg appearance. Colonies do not hemadsorb guinea pig erythrocytes.

No evidence of pathogenicity for insects.

Source: original isolation was from the gut of an adult horse fly (*Tabanus abactor*).

DNA G+C content (mol%): 28.3 (Bd, T_m, HPLC).

Type strain: BARC 857, ATCC 49584.

Sequence accession no. (16S rRNA gene): AY187288.

References

Bonnet, F., C. Saillard, J.C. Vignault, M. Garnier, P. Carle, J.M. Bové, D.L. Rose, J.G. Tully and R.F. Whitcomb. 1991. *Acholeplasma seiffertii* sp. nov., a mollicute from plant surfaces. Int. J. Syst. Bacteriol. *41*: 45–49.

Brown, D.R., R.F. Whitcomb and J.M. Bradbury. 2007. Revised minimal standards for description of new species of the class *Mollicutes* (division *Tenericutes*). Int. J. Syst. Evol. Microbiol. *57*: 2703–2719.

Clark, T.B., J.G. Tully, D.L. Rose, R. Henegar and R.F. Whitcomb. 1986. Acholeplasmas and similar nonsterol-requiring mollicutes from insects: missing link in microbial ecology. Curr. Microbiol. *13*: 11–16.

Clyde, W.A., Jr. 1983. Growth inhibition tests. *In* Methods in Mycoplasmology, vol. 1 (edited by Razin and Tully). Academic Press, New York, pp. 405–410.

Eden-Green, S.J. and J.G. Tully. 1979. Isolation of *Acholeplasma* spp. from coconut palms affected by lethal yellowing disease in Jamaica. Curr. Microbiol. *2*: 311–316.

Grau, O. 1991. Analyse des gènes ribosomiques des mollicutes, application à l'identification d'un mollicute non classé et conséquences taxonomiques [thesis]. Bordeaux, France.

Grau, O., F. Laigret, P. Carle, J.G. Tully, D.L. Rose and J.M. Bové. 1991. Identification of a plant-derived mollicute as a strain of an avian pathogen, *Mycoplasma iowae*, and its implications for mollicute taxonomy. Int. J. Syst. Bacteriol. *41*: 473–478.

Gros, O., C. Saillard, C. Helias, F. LeGoff, M. Marjolet, J.M. Bové and C. Chastel. 1996. Serological and molecular characterization of *Mesoplasma seiffertii* strains isolated from hematophagous dipterans in France. Int. J. Syst. Bacteriol. *46*: 112–115.

Hackett, K.J. and R.F. Whitcomb. 1995. Cultivation of spiroplasmas in undefined and defined media. *In* Molecular and Diagnostic Procedures in Mycoplasmology, vol. 1 (edited by Razin and Tully). Academic Press, San Diego, pp. 41–53.

Knight, T.F., Jr. 2004. Reclassification of *Mesoplasma pleciae* as *Acholeplasma pleciae* comb. nov. on the basis of 16S rRNA and *gyrB* gene sequence data. Int. J. Syst. Evol. Microbiol. *54*: 1951–1952.

May, M., R.F. Whitcomb and D.R. Brown. 2008. Mycoplasma and related organisms. *In* Practical Handbook of Microbiology (edited by Goldman and Green). CRC Press, Boca Raton, pp. 467–491.

McCoy, R.E., D.S. Williams and D.L. Thomas. 1979. Isolation of mycoplasmas from flowers. Proceedings of the Republic of China-United States Cooperative Science Seminar, Symposium series 1, National Science Council, Taipei, Taiwan, pp. 75–81.

McCoy, R.E., H.G. Basham, J.G. Tully, D.L. Rose, P. Carle and J.M. Bové. 1984. *Acholeplasma florum*, a new species isolated from plants. Int. J. Syst. Bacteriol. *34*: 11–15.

Pollack, J.D., M.V. Williams, J. Banzon, M.A. Jones, L. Harvey and J.G. Tully. 1996. Comparative metabolism of *Mesoplasma, Entomoplasma, Mycoplasma,* and *Acholeplasma*. Int. J. Syst. Bacteriol. *46*: 885–890.

Rose, D.L., J.P. Kocka, N.L. Somerson, J.G. Tully, R.F. Whitcomb, P. Carle, J.M. Bové, D.E. Colflesh and D.L. Williamson. 1990. *Mycoplasma lactucae* sp. nov., a sterol-requiring mollicute from a plant surface. Int. J. Syst. Bacteriol. *40*: 138–142.

Rose, D.L., J.G. Tully, J.M. Bove and R.F. Whitcomb. 1993. A test for measuring growth responses of *Mollicutes* to serum and polyoxyethylene sorbitan. Int. J. Syst. Bacteriol. *43*: 527–532.

Taylor-Robinson, D. 1983. Metabolism inhibition tests. *In* Methods in Mycoplasmology, vol. 1 (edited by Razin and Tully). Academic Press, New York, pp. 411–421.

Tully, J.G. 1983. Cloning and filtration techniques for mycoplasmas. *In* Methods in Mycoplasmology, vol. 1 (edited by Razin and Tully). Academic Press, New York, pp. 173–177.

Tully, J.G. 1984. Genus *Acholeplasma*. *In* Bergey's Manual of Systematic Bacteriology, vol. 1 (edited by Krieg and Holt). Williams & Wilkins, Baltimore, pp. 775–781.

Tully, J.G. 1995. Determination of cholesterol and polyoxyethylene sorbitan growth requirements of mollicutes. *In* Molecular and Diagnostic Procedures in Mycoplasmology, vol. 1 (edited by Razin and Tully). Academic Press, San Diego, pp. 381–389.

Tully, J.G., D.L. Rose, R.F. Whitcomb, K.J. Hackett, T.B. Clark, R.B. Henegar, E. Clark, P. Carle and J.M. Bové. 1987. Characterization of some new insect-derived acholeplasmas. Isr. J. Med. Sci. *23*: 699–703.

Tully, J.G., D.L. Rose, P. Carle, J.M. Bové, K.J. Hackett and R.F. Whitcomb. 1988. *Acholeplasma entomophilum* sp. nov. from gut contents of a wide-range of host insects. Int. J. Syst. Bacteriol. *38*: 164–167.

Tully, J.G., D.L. Rose, K.J. Hackett, R.F. Whitcomb, P. Carle, J.M. Bové, D.E. Colflesh and D.L. Williamson. 1989. *Mycoplasma ellychniae* sp. nov., a sterol-requiring mollicute from the firefly beetle *Ellychnia corrusca*. Int. J. Syst. Bacteriol. *39*: 284–289.

Tully, J.G., D.L. Rose, R.E. McCoy, P. Carle, J.M. Bové, R.F. Whitcomb and W.G. Weisburg. 1990. *Mycoplasma melaleucae* sp. nov., a sterol-requiring mollicute from flowers of several tropical plants. Int. J. Syst. Bacteriol. *40*: 143–147.

Tully, J.G., J.M. Bové, F. Laigret and R.F. Whitcomb. 1993. Revised taxonomy of the class *Mollicutes* - proposed elevation of a monophyletic cluster of arthropod-associated mollicutes to ordinal rank (*Entomoplasmatales* ord. nov.), with provision for familial rank to separate species with nonhelical morphology (*Entomoplasmataceae* fam. nov.) from helical species (*Spiroplasmataceae*), and emended descriptions of the order *Mycoplasmatales*, family *Mycoplasmataceae*. Int. J. Syst. Bacteriol. *43*: 378–385.

Tully, J.G., R.F. Whitcomb, K.J. Hackett, D.L. Rose, R.B. Henegar, J.M. Bové, P. Carle, D.L. Williamson and T.B. Clark. 1994a. Taxonomic descriptions of eight new non-sterol-requiring *Mollicutes* assigned to the genus *Mesoplasma*. Int. J. Syst. Bacteriol. *44*: 685–693.

Tully, J.G., R.F. Whitcomb, D.L. Rose, J.M. Bové, P. Carle, N.L. Somerson, D.L. Williamson and S. Edengreen. 1994b. *Acholeplasma brassicae* sp. nov. and *Acholeplasma palmae* sp. nov., two non-sterol-requiring mollicutes from plant surfaces. Int. J. Syst. Bacteriol. *44*: 680–684.

Tully, J.G., D.L. Rose, C.E. Yunker, P. Carle, J.M. Bové, D.L. Williamson and R.F. Whitcomb. 1995. *Spiroplasma ixodetis* sp. nov., a new species from *Ixodes pacificus* ticks collected in Oregon. Int. J. Syst. Bacteriol. *45*: 23–28.

Tully, J.G., R.F. Whitcomb, K.J. Hackett, D.L. Williamson, F. Laigret, P. Carle, J.M. Bové, R.B. Henegar, N.M. Ellis, D.E. Dodge and J. Adams. 1998. *Entomoplasma freundtii* sp. nov., a new species from a green tiger beetle (Coleoptera: Cicindelidae). Int. J. Syst. Bacteriol. *48*: 1197–1204.

Weisburg, W.G., J.G. Tully, D.L. Rose, J.P. Petzel, H. Oyaizu, D. Yang, L. Mandelco, J. Sechrest, T.G. Lawrence, J. Van Etten, J. Maniloff and C.R. Woese. 1989. A phylogenetic analysis of the mycoplasmas: basis for their classification. J. Bacteriol. *171*: 6455–6467.

Whitcomb, R.F. 1983. Culture media for spiroplasmas. *In* Methods in Mycoplasmology, vol. 1 (edited by Razin and Tully). Academic Press, New York, pp. 147–158.

Whitcomb, R.F. and K.J. Hackett. 1996. Identification of mollicutes from insects. *In* Molecular and Diagnostic Procedures in Mycoplasmology, vol. 2 (edited by Tully and Razin). Academic Press, San Diego, pp. 313–322.

Whitcomb, R.F., J.G. Tully, D.L. Rose, E.B. Stephens, A. Smith, R.E. McCoy and M.F. Barile. 1982. Wall-less prokaryotes from fall flowers in central United States and Maryland. Curr. Microbiol. *7*: 285–290.

Williamson, D.L., J.G. Tully, D.L. Rose, K.J. Hackett, R. Henegar, P. Carle, J.M. Bové, D.E. Colflesh and R.F. Whitcomb. 1990. *Mycoplasma somnilux* sp. nov., *Mycoplasma luminosum* sp. nov., and *Mycoplasma lucivorax* sp. nov., new sterol-requiring mollicutes from firefly beetles (Coleoptera, Lampyridae). Int. J. Syst. Bacteriol. *40*: 160–164.

Family II. **Spiroplasmataceae** Skripal 1983, 408^{VP}

David L. Williamson, Gail E. Gasparich, Laura B. Regassa, Collette Saillard, Joël Renaudin, Joseph M. Bové and Robert F. Whitcomb*

Spi.ro.plas.ma.ta.ce'ae. N.L. neut. n. *Spiroplasma*, *-atos* type genus of the family; *-aceae* ending to denote a family; N.L. fem. pl. n. *Spiroplasmataceae* the *Spiroplasma* family.

Cells are helical during exponential growth, with rotatory, flexional, and translational motility. Genome size is variable: 780–2220 kbp. Variable sterol requirement for growth. Procedures for determining sterol requirement are as described for Family I (*Entomoplasmataceae*). Possess a phosphoenolpyruvate phosphotransferase system for glucose uptake. Reduced nicotinamide adenine dinucleotide (NADH) oxidase activity is located only in the cytoplasm. Unable to synthesize fatty acids from acetate. Other characteristics are as described for the class and order.

Type genus: **Spiroplasma** Saglio, L'Hospital, Laflèche, Dupont, Bové, Tully and Freundt 1973, 201^{AL}.

Genus I. **Spiroplasma** Saglio, L'Hospital, Laflèche, Dupont, Bové, Tully and Freundt 1973, 201^{AL}

David L. Williamson, Gail E. Gasparich, Laura B. Regassa, Collette Saillard, Joël Renaudin, Joseph M. Bové and Robert F. Whitcomb*

Spi.ro.plas'ma. Gr. n. *speira* (L. transliteration *spira*) a coil, spiral; Gr. neut. n. *plasma* something formed or molded, a form; N.L. neut. n. *Spiroplasma* spiral form.

Cells are pleomorphic, varying in size and shape from **helical** and branched nonhelical filaments to spherical or ovoid. The helical forms, usually 100–200 nm in diameter and 3–5 µm in length, generally occur during the exponential phase of growth and in some species persist during stationary phase. The cells of some species are short (1–2 µm). In certain cases, helical cells may be very tightly coiled, or the coils may show continuous variation in amplitude. Spherical cells ~300 nm in diameter and nonhelical filaments are frequently seen in the stationary phase, where they may not be viable, and in all growth phases in suboptimal growth media, where they may or may not be viable. In some species during certain phases, spherical forms may be the replicating form. **Helical filaments are motile, with flexional and twitching movements, and often show an apparent rotatory motility. Fibrils are associated with the membrane, but flagellae, periplasmic fibrils, or other organelles of locomotion are absent.** Fimbriae and pili observed on the cell surface of insect- and plant-pathogenic spiroplasmas are believed to be involved in host-cell attachment and conjugation (Ammar et al., 2004; Özbek et al., 2003), but not in locomotion. Cells divide by binary fission, with doubling times of 0.7–37 h. Facultatively anaerobic. The temperature growth range varies among species, from 5 to 41°C. **Colonies on solid media are frequently diffuse**, with irregular shapes and borders, a condition that reflects the motility of the cells during active growth (Figure 111). Colony type is strongly dependent on the agar concentration. Colony sizes vary from 0.1 to 4.0 mm in diameter. Colonies formed by nonmotile variants or mutants, or by cultures growing on inadequate media are typically umbonate with diameters of 200 µm or less. Some species, such as *Spiroplasma platyhelix*, have barely visible helicity along most of their length and display little rotatory or flexing motility. Colonies of motile, fast-growing spiroplasmas are diffuse, often with satellite colonies developing from foci adjacent to the initial site of colony development. Light turbidity may be produced in liquid cultures. Chemo-organotrophic. Acid is produced from glucose. Hydrolysis of arginine is variable. Urea, arbutin, and esculin are not hydrolyzed. Sterol requirements are variable. An optimum osmolality, usually in the range of 300–800 mOsm, has been demonstrated for some spiroplasmas. Media containing mycoplasma broth base, serum, and other supplements are required for primary growth, but after adaptation, growth often occurs in less complex media. Defined or semi-defined media are available for some species. Resistant to 10,000 U/ml penicillin. Insensitive to rifampicin, sensitive to erythromycin and tetracycline. **Isolated from the surfaces of flowers and other plant parts, from the guts and hemolymph of various insects and crustaceans, and from tick triturates. Also isolated from vascular plant fluids (phloem sap) and insects that feed on the fluids.** Specific host associations are common. The type species, *Spiroplasma citri*, is pathogenic for citrus (e.g., orange and

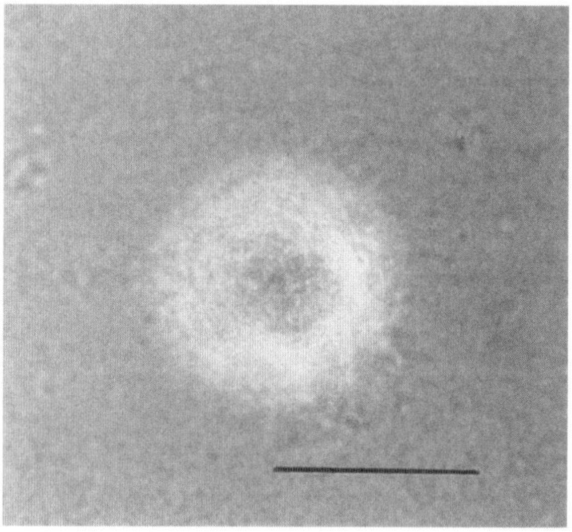

FIGURE 111. Colonial morphology of *Spiroplasma lampyridicola* strain PUP-1^{T} grown on SP-4 agar under anaerobic conditions for 4 d at 30°C. The diffuse appearance and indistinct margins reflect the motility of spiroplasmas during active growth. Bar = 50 µm. (Reprinted with permission from Stevens et al., 1997. Int. J. Syst. Bacteriol. *47*: 709–712.)

*Deceased 21 December 2007.

grapefruit), producing "stubborn" disease. Experimental or natural infections also occur in horseradish, periwinkle, radish, broad bean, carrot, and other plant species. *Spiroplasma kunkelii* is a maize pathogen. Some species are pathogenic for insects. Certain species are pathogenic, under experimental conditions, for a variety of suckling rodents (rats, mice, hamsters and rabbits) and/or chicken embryos. Genome sizes vary from 780 to 2220 kbp (PFGE).

DNA G+C content (mol%): 24–31 (T_m, Bd).

Type species: **Spiroplasma citri** Saglio, L'Hospital, Laflèche, Dupont, Bové, Tully and Freundt 1973, 202[AL].

Further descriptive information

Morphology. The morphology of spiroplasmas is most easily observed in suspensions with the light microscope under dark-field illumination (Williamson and Poulson, 1979). In the exponential phase in liquid media, most spiroplasma cells are helical filaments 90–250 nm in diameter and of variable length (Figure 112). Fixed and negatively stained cells usually show a blunt and a tapered end (Williamson, 1969; Williamson and Whitcomb, 1974). The tapered ends of the cells are a consequence of the constriction process preceding division (Garnier et al., 1981, 1984). However, they are adapted as attachment sites in some species (Ammar et al., 2004).

Motility. Helical spiroplasma cells exhibit flexing, twitching, and apparent rotation about the longitudinal axis (Cole et al., 1973; Davis and Worley, 1973). Spiroplasmas exhibit temperature-dependent chemotactic movement toward higher concentrations of nutrients, such as carbohydrates and amino acids (Daniels and Longland, 1984, 1980); but motility is random in the absence of attractants (Daniels and Longland, 1984). Both natural (Townsend et al., 1980b, 1977) and engineered (Cohen et al., 1989; Duret et al., 1999; Jacob et al., 1997) motility mutants have been described. These mutants form perfectly umbonate colonies on solid medium. Mutational analysis has highlighted the involvement of the *smc1* gene in motility. Jacob et al. (1997) demonstrated that a Tn*4001* insertion mutant with reduced flexional motility and no rotational motility could be complemented with the wild-type *scm1* gene. The *scm1* gene encodes a 409 amino acid polypeptide having ten transmembrane domains but no significant homology with known proteins. In another study, the *scm1* gene was inactivated through homologous recombination, abolishing motility (Duret et al., 1999). The disrupted *scm1*$^-$ mutant was injected into the leafhopper vector (*Circulifer haematoceps*); it multiplied actively in the insect vector and was then transmitted to periwinkle plants. The mutant induced symptoms that were indistinguishable from those caused by the motile wild-type strain showing that spiroplasma motility is not essential for phytopathogenicity and transmission to the plant host (Duret et al., 1999).

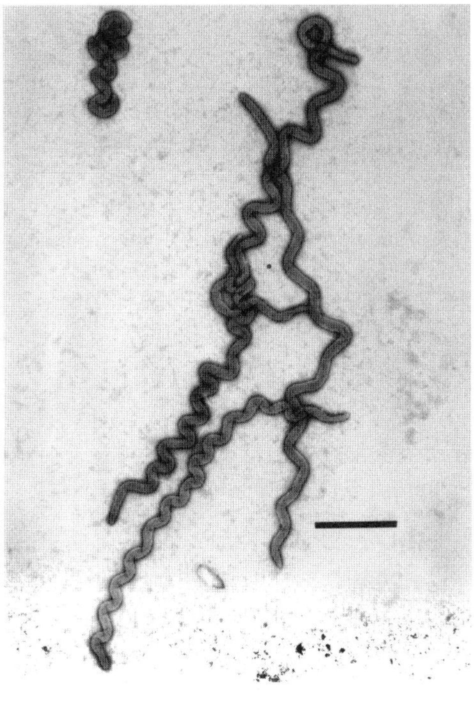

FIGURE 112. Electron micrograph of *Drosophila willistoni* strain B3SR sex-ratio spiroplasmas. Hemolymph suspension in phosphate buffered saline, glutaraldehyde vapor-fixed, and negatively stained with 1% phosphotungstic acid, pH 7.2. (Reprinted with permission from Whitcomb et al., 2007. Biodiversity and Conservation *16:* 3877–3894.)

Fibrils and motility. Microfibrils 3.6 nm in width have been envisioned in the membranes of some spiroplasmas. These structures have repeat intervals of 9 nm along their lengths (Williamson, 1974) and form a ribbon that extends the entire length of the helix (Charbonneau and Ghiorse, 1984; Williamson et al., 1984). The sequence of the fibril protein gene has been determined (Williamson et al., 1991) and the calculated mass of the fibril protein is 59 kDa. The flat, monolayered, membrane-bound ribbon composed of several well-ordered fibrils represents the internal spiroplasmal cytoskeleton. The spiroplasmal cytoskeletal ribbon follows the shortest helical line on the cellular coil. Recent studies have focused on the detailed cellular and molecular organization of the cytoskeleton in *Spiroplasma melliferum* and *Spiroplasma citri* (Gilad et al., 2003; Trachtenberg, 2004; Trachtenberg et al., 2003a, b; Trachtenberg and Gilad, 2001). Each cytoskeletal ribbon contains seven fibril pairs (or 14 fibrils) and the functional unit is a pair of aligned fibrils (Trachtenberg et al., 2003a). Paired fibrils can be viewed as chains of tetramers composed of 59 kDa monomers. Cryo-electron tomography has been used to elucidate the native state, cytoskeletal structure of *Spiroplasma melliferum* and suggested the presence of three parallel ribbons under the membrane: two appear to be composed of the fibril protein and the third is composed of the actin-like MreB protein (Kürner et al., 2005). Subsequent studies suggest the presence of a single ribbon structure (Trachtenberg et al., 2008). The subunits in the fibrils undergo conformational changes from circular to elliptical, which results in shortening of the fibrils and helix contraction, or from elliptical to circular, leading to a length increase of the fibrils and cell helix. The cytoskeleton, which is bound to the spiroplasmal membrane over its entire length, acts as a scaffold and controls the helical shape of the cell. The cell shape is therefore dynamic. Movement appears to be driven by the propagation of a pair of kinks that travel down the length of the cell along the fibril ribbons (Shaevitz et al., 2005; Wada and Netz, 2007; Wolgemuth and Charon, 2005). The contractile cytoskeleton can thus be seen as a "linear motor" in contrast to the common "rotary motor" that is part of the flagellar apparatus in bacteria (Trachtenberg, 2006).

There are several adherent proteins that copurify with the cytoskeleton, ranging in size from 26 to 170 kDa (Townsend et al., 1980a; Trachtenberg, 2006; Trachtenberg and Gilad, 2001). These proteins are apparently membrane-associated and may function as anchor proteins (Trachtenberg and Gilad, 2001). The structural organization of the cytoskeleton-associated proteins of *Spiroplasma melliferum* is beginning to be elucidated (Trachtenberg et al., 2008). The 59 kDa polypeptide is the cytoskeletal fibril protein. The 26 kDa polypeptide is probably spiralin, the major spiroplasmal membrane protein. However, the involvement of spiralin in helicity and motility is unlikely (see "Spiralin" section below), especially since spiralin is anchored on the outside surface of the cell (Bévén et al., 1996; Bové, 1993; Brenner et al., 1995; Foissac et al., 1996) and spiralin-deficient mutants maintain helicity and motility (Duret et al., 2003). The 45 kDa protein may correspond to the product of the *scm1* gene, shown to be essential for motility (Jacob et al., 1997), and the 34 kDa protein may be the product of the *mreB1* gene (W. Maccheroni and J. Renaudin, unpublished).

MreB is the bacterial homolog to eukaryotic actin (Jones et al., 2001; Van den Ent et al., 2001). Early work provided evidence for the presence of actin-like proteins in spiroplasmas. Antisera prepared against SDS-denatured invertebrate actin coupled to horseradish peroxidase specifically stained cells of *Spiroplasma citri* (Williamson et al., 1979a). Also, a protein with a molecular mass similar to that of actin (protein P25) was isolated from *Spiroplasma citri* and reacted with IgG directed against rabbit actin (Mouches et al., 1982b, c, 1983b). Monospecific antibodies raised against the P25 protein recognized not only P25 of *Spiroplasma citri*, but also a homologous protein from *Mycoplasma mycoides* PG50 and *Ureaplasma urealyticum* serotype V (Mouches et al., 1983b). More recent work has focused on the molecular organization of the genes. *mreB* genes are present in rod-shaped, filamentous, and helical bacteria, but not in coccoid, spherical bacteria, regardless of whether or not they are Gram-stain-positive or Gram-stain-negative. *mreB* genes are also absent from the pleomorphic mycoplasmas. However, *Spiroplasma citri* contains five homologs of *Bacillus subtilis mreB* genes (Maccheroni et al., 2002). Four of these (*mreB2, 3, 4,* and *5*) form a cluster on the genome and are transcribed in two separate operons. Gene *mreB1* is transcribed as a monocistronic operon and at a much higher level.

Growth characteristics. Spiroplasma cells increase in length and divide by constriction. Pulse labeling of the membrane with tritiated amino acids revealed a polar growth of the helix. Polarity was also observed by tellurium-labeling of oxido-reduction sites (Garnier et al., 1984). In the stationary or death phase, the cells are usually distorted, often forming either subovoid bodies or nonhelical filaments. Within cultured insect cells, all the spiroplasma cells were subovoid, but presumably viable (Wayadande and Fletcher, 1998). Thus, the ability of cells to grow and divide is not linked inextricably to helicity.

Growth rate. Enumerated microscopically (Rodwell and Whitcomb, 1983), spiroplasmas reach titers of 10^8–10^{11} cells/ml in medium containing horse or fetal bovine serum. Growth rates of related strains tend to be similar. Konai et al. (1996a) calculated doubling times from the time required for medium acidification. In general, spiroplasmas adapted to complex cycles or single hosts had slower growth rates than spiroplasmas known or suspected to be transmitted on plant surfaces.

Temperature. Konai et al. (1996a) determined temperature ranges and optima for a large number of spiroplasma strains. The ranges of some strains (e.g., *Spiroplasma apis*) were very wide (5–41°C), but some group I strains from leafhoppers and plants grew only at 25° and 30°C. Although some spiroplasmas grew well at 41°C, none grew at 43°C.

Biochemical reactions. All tested spiroplasmas ferment glucose with concomitant acid production, although the utilization rates may vary. Some strains of group I (e.g., members of subgroups I-4 and I-6) and all strains of *Spiroplasma mirum* ferment glucose slowly. With *Spiroplasma citri*, all strains tested grew actively on fructose and strain GII3 grew on fructose, glucose, or trehalose. The ability of spiroplasmas to utilize arginine varies (Hackett et al., 1996a). Arginine hydrolysis by some spiroplasmas can be observed only if glucose is also present in the medium. In other cases, aggressive glucose metabolism interferes with detection of arginine hydrolysis (Hackett et al., 1996a).

Regulation of the fructose and trehalose operons of Spiroplasma citri. The fructose operon of *Spiroplasma citri* (Gaurivaud et al., 2000a) became of special interest when fructose utilization was implicated in *Spiroplasma citri* phytopathogenicity (see "Mechanism of *Spiroplasma citri* phytopathogenicity" below). In particular, the role of the first gene of the operon, *fruR*, was investigated. *In vivo* transcription of the operon is greatly enhanced by the presence of fructose in the growth medium, whereas glucose has no effect. When *fruR* is not expressed (*fruR*⁻ mutants), transcription of the operon is not stimulated by fructose and the rate of fructose fermentation is decreased, indicating that FruR is an activator of the fructose operon (Gaurivaud et al., 2001). Trehalose is the major sugar in leafhoppers and other insects. The trehalose operon of *Spiroplasma citri* has a gene organization very similar to that of the fructose operon and the first gene of the trehalose operon, *treR*, also encodes a transcriptional activator of the operon (André et al., 2003).

Sterol utilization. It was originally thought that all spiroplasmas require sterol for growth. Subsequent screening by Rose et al. (1993) showed that a minority of the spiroplasmas tested were able to sustain growth in mycoplasma broth base medium without sterols. The discovery that the sterol requirement in *Mollicutes* is polyphyletic greatly diminished the significance of sterol requirements in mollicute taxonomy (Tully et al., 1993).

Metabolic pathways and enzymes. The intermediary metabolism of *Mollicutes* has been reviewed (Miles, 1992; Pollack, 2002a, b; Pollack et al., 1997). Like all mollicutes, *Spiroplasma* species apparently lack both cytochromes and, except for malate dehydrogenase, the enzymes of the tricarboxylic acid cycle. They do not have an electron-transport system and their respiration is characterized as being flavin-terminated. McElwain et al. (1988) studied *Spiroplasma citri* and Pollack et al. (1989) screened ten spiroplasma species for 67 enzyme activities. All spiroplasmas were fermentative; their 6-phosphofructokinases (6-PFKs) required ATP for substrate phosphorylation during glycolysis. This enzymic requirement is common to all mollicutes except *Acholeplasma* and *Anaeroplasma* spp. The 6-PFKs of the species in these genera require pyrophosphate and cannot use ATP. Additionally, except for *Spiroplasma floricola*, all *Spiroplasma* species have dUTPase activity. Pollack et al. (1989) also reported that all spiroplasmas except *Spiroplasma floricola* have deoxyguanosine kinase activity. They found that deoxyguanosine, but no other nucleoside, could be phosphorylated to GMP with ATP.

Spiroplasmal proteins with multiple functions have been described. The CpG-specific methylase from *Spiroplasma monobiae* appears to also have topoisomerase activity (Matsuo et al., 1994). Protein P46 of *Spiroplasma citri* is a bifunctional protein in which the N-terminal domain represents ribosomal protein L29, whereas the C-terminal domain is capable of binding a specific inverted repeat sequence. It could be involved in regulation (Le Dantec et al., 1998). Such protein multifunctionality may reflect genomic economy in the small mollicute genome (Pollack, 2002b). However, functional redundancy has also been reported; *Spiroplasma citri* apparently has two distinct membrane ATPases (Simoneau and Labarère, 1991).

Genome size, genomic maps, and chromosomal rearrangements. PFGE revealed that the genome size range for spiroplasmas varied continuously (Pyle and Finch, 1988) from 780 kbp for *Spiroplasma platyhelix* to 2220 kbp for *Spiroplasma ixodetis* (Carle et al., 1995, 1990). There is a general trend for genomic simplification in *Spiroplasma* lineages. This trend culminated in loss of helicity and motility in the *Entomoplasmataceae* and eventually to the host transfer events forming the mycoides group of mycoplasmas (Gasparich et al., 2004).

The genome size of *Spiroplasma citri* varies among strains from 1650 to 1910 kbp (Ye et al., 1995). It was found that the relative positions of mapped loci were conserved in most of the strains, but that differences in the sizes of certain fragments permitted genome size variation. Genome size can fluctuate rapidly in spiroplasma cultures after a relatively short number of *in vitro* passages (Melcher and Fletcher, 1999; Ye et al., 1996). The genome of *Spiroplasma melliferum* is 360 kbp shorter than that of *Spiroplasma citri* strain R8-A2^{T}, but DNA hybridization has shown that the two spiroplasmas share extensive DNA hybridization (65%). Comparison of their genomic maps revealed that the genome region, which is shorter in *Spiroplasma melliferum*, corresponds to a variable region in the genomes of *Spiroplasma citri* strains and that a large region of the *Spiroplasma melliferum* genome is inverted in comparison with *Spiroplasma citri*. Therefore, chromosomal rearrangements and deletions were probably major events during evolution of the genomes of *Spiroplasma citri* and *Spiroplasma melliferum*. In addition, a large amount of noncoding DNA is present as repeat sequences (McIntosh et al., 1992; Nur et al., 1986, 1987) and integrated viral DNA (Bébéar et al., 1996) may also account for differences in genome sizes of closely related species.

Base composition. The DNA G+C content for most spiroplasma groups and subgroups has been determined (Carle et al., 1995, 1990; Williamson et al., 1998). Most group I spiroplasmas and *Spiroplasma poulsonii* have a G+C content of 25–27 mol%. However, the G+C content of subgroup I-6 *Spiroplasma insolitum* is significantly higher, indicating that the base composition of spiroplasmal DNA may shift over relatively short evolutionary periods. The range of G+C content of 25–27 mol% is modal for *Spiroplasma* and is also common in the Apis clade. However, *Spiroplasma mirum* (group V), strains of *Spiroplasma apis* (group IV), and group VIII strains have a G+C content of about 29–31 mol%. Restriction sites containing only G and C nucleotides are not uniformly distributed over the genome (Ye et al., 1992).

Methylated bases. Methylated bases have been detected in spiroplasmal DNA (Nur et al., 1985). The gene encoding the CpG methylase in *Spiroplasma monobiae* has been cloned (Renbaum et al., 1990) and its mode of action studied (Renbaum and Razin, 1992).

DNA restriction patterns. Restriction patterns of spiroplasmal DNA, as determined by polyacrylamide gel electrophoresis, may be highly similar among strains of a given species (Bové et al., 1989). Variations in restriction fragment length patterns among strains of *Spiroplasma corruscae* correlated imperfectly with serological variation, so their significance was uncertain (Gasparich et al., 1998).

RNA genes. Some spiroplasmas, such as *Spiroplasma citri*, have only one rRNA operon, whereas others, such as *Spiroplasma apis*, have two (Amikam et al., 1984, 1982; Bové, 1993; Grau et al., 1988; Razin, 1985). The three rRNA genes are linked in the classical order found in bacteria: 5′-16S–23S-5S-3′. The sequence of the 16S rRNA gene (rDNA) of most spiroplasma species has been determined for phylogenetic studies (Gasparich et al., 2004; Weisburg et al., 1989). A gene cluster of ten tRNAs (Cys, Arg, Pro, Ala, Met, Ile, Ser, fMet, Asp, Phe) was identified in *Spiroplasma melliferum* (Rogers et al., 1987). Similar tRNA gene clusters have been cloned and sequenced from *Spiroplasma citri* (Citti et al., 1992).

Codon usage. In spiroplasmas, UGA is not a stop codon but encodes tryptophan. The universal tryptophan codon, UGG, is also used (Citti et al., 1992; Renaudin et al., 1986). Codon usage also reflects the A+T richness of spiroplasmal DNA (usually about 74 mol% A+T). For example, in *Spiroplasma citri*, UGA is used to code for tryptophan eight times more frequently than the universal tryptophan codon UGG (Bové, 1993; Citti et al., 1992; Navas-Castillo et al., 1992). Also, synonymous codons with U or A at the 5′ or 3′ ends are preferentially used over those with a C or G in that position.

RNA polymerase and spiroplasmal insensitivity to rifampicin. Spiroplasmas are insensitive to rifampicin. DNA-dependent RNA polymerases from *Spiroplasma melliferum* and *Spiroplasma apis* were at least 1000 times less sensitive to rifampicin than the corresponding *Escherichia coli* enzyme (Gadeau et al., 1986). Rifampicin insensitivity of *Spiroplasma citri* and all other mollicutes tested was found to be associated with the presence of an asparagine residue at position 526 in RpoB. The importance of the asparagine residue was confirmed by site-directed mutagenesis of the histidine codon (CAC) to an asparagine codon (AAC) at position 526 of *Escherichia coli* RpoB, resulting in a rifampicin-resistant mutant (Gaurivaud et al., 1996). The genetic organization surrounding the *rpoB* gene in spiroplasmas is also atypical. In many bacteria, *rpoB* is part of the β operon in which the four genes *rplK*, *rplA*, *rplJ*, and *rplL*, encoding ribosomal proteins L11, L1, L10, and L12, respectively, are located immediately upstream of *rpoB*; *rpoC* is immediately downstream of *rpoB*. In *Spiroplasma citri*, the gene organization is different in that the *hsdS* gene, encoding a component of a type I restriction-modification system, is upstream of *rpoB*. Sequences showing similarities with insertion elements are found between *hsdS* and *rpoB* (Laigret et al., 1996).

DNA polymerases and other proteins involved in DNA replication and repair. From genomic studies, it appears that *Mycoplasma* species carry the essential, multimeric enzyme for genomic DNA replication, DNA polymerase III. The subunit responsible for actual DNA biosynthesis is subunit α, encoded by *polC* (*dnaE*). The *polC* gene has been identified in all sequenced mollicute genomes, including *Spiroplasma citri*. The genes encoding the other subunits, *dnaN* (subunit β) and *dnaX* (subunits τ and γ), are also shared by the *Spiroplasma* and *Mycoplasma* species studied to date. So, it seems that spiroplasmas, like other mollicutes, possess DNA polymerase III and that it is probably the major DNA replication enzyme. However, there is also evidence for two additional DNA polymerases. A second gene for a DNA polymerase (enzyme B) was found in the *Spiroplasma citri* genome and there is evidence that the *Spiroplasma kunkelii polA* gene may encode a full-length DNA polymerase I protein (Bai and Hogenhout, 2002). DNA polymerase I is a single polypeptide that has, in addition to DNA synthesis activity, two exonuclease activities: exo-3′ to 5′ as well as exo-5′ to 3′. At this stage, it is not possible to determine the equivalence between the three spiroplasmal DNA polymerases identified by sequencing (Pol III, enzyme B, and Pol I) and those originally detected biochemically (ScA, ScB, ScC) (Charron et al., 1979, 1982). As the *Spiroplasma citri* genome sequencing project has progressed, the following *Spiroplasma citri* genes involved in DNA replication have been detected: *dnaA, dnaB, polA, dnaE, polC, dnaN, dnaX, holB, dinB* (truncated), *dnaJ, dnaK, gyrA, gyrB, parC, parE, topA, rnhB, rnhC, rnpA, rnR, rnc, yrrc, xseA, xseB,* and *ssb* (Carle et al., 2010; accession numbers AM285301–AM285339). Genes encoding DNA replication proteins have also been identified in *Spiroplasma kunkelii* (Bai and Hogenhout, 2002). *Spiroplasma citri* is highly sensitive to UV irradiation (Labarère and Barroso, 1989) and the organism has no functional *recA* gene, since a significant portion of the C-terminal part of the gene is lacking (Marais et al., 1996).

Origin of DNA replication. Even before the *Spiroplasma citri* genome project was initiated, some fragments with multiple open reading frames had been completely sequenced. For example, Ye et al. (1994b) sequenced a 5.6 kbp fragment containing genes for the replication initiation protein (*dnaA*), the beta subunit of DNA polymerase III (*dnaN*), and the DNA gyrase subunits A and B (*gyrA* and *gyrB*). Several *dnaA*-box consensus sequences were found upstream and downstream of the *dnaA* gene. From these data, it was established that the *dnaA* region was the origin of replication in *Spiroplasma citri* (Ye et al., 1994b). Zhao et al. (2004a) cloned a cell division gene cluster from *Spiroplasma kunkelii* and functionally characterized the key division gene, $ftsZ_{sk}$, and showed that it encodes a cell division protein similar to FtsZ proteins from other bacteria.

Spiralin. Spiralin, encoded by the *spi* gene, is the major membrane protein of *Spiroplasma citri* (Wróblewski et al., 1977, 1989). The deduced amino acid sequence of the protein (Bové et al., 1993; Chevalier et al., 1990; Saillard et al., 1990) corresponds well with the experimentally determined amino acid composition (Wróblewski et al., 1984). In particular, spiralin lacks tryptophan and, thus, has no UGG and/or UGA codons, which facilitates gene expression in *Escherichia coli*. Detailed analyses showed that all *Spiroplasma citri* spiralins were 241–242 amino acids long (Foissac et al., 1996). A conserved central region and an amino acid sequence repetition, including a VTKXE consensus sequence, are present in all spiralins analyzed (Foissac et al., 1997a). Spiralin confers a significant amount of the antigenic activity in group I spiroplasmas (Whitcomb et al., 1983) and has a high degree of species specificity, although minor cross-reactions have been detected (Zaaria et al., 1990). The spiralin genes of *Spiroplasma citri* and *Spiroplasma melliferum* species, which have about 65% overall DNA–DNA hybridization, shared 89% nucleotide sequence identity and 75% deduced amino acid sequence similarity (Bové et al., 1993).

Spiralin mutants were constructed through homologous recombination in *Spiroplasma citri* to examine the role of spiralin *in vivo* (Duret et al., 2003). Phenotypic characterization of mutant 9a2 showed that, in spite of a total lack of spiralin, it maintained helicity and motility similar to the wild-type strain GII3 (Duret et al., 2003). When injected into the leafhopper vector, *Circulifer haematoceps*, the mutant multiplied to a high titer, but transmission efficiency to periwinkle plants was very low compared to the wild-type strain. In the infected plants, however, the spiralin-deficient mutant multiplied well and produced the typical symptoms of the disease. In addition, preliminary results indicated that the mutant could not be acquired by insects feeding on 9a2-infected plants, suggesting that spiralin may mediate spiroplasma invasion of insect tissues (Duret et al., 2003). In order to test this possibility, *Circulifer haematoceps* leafhopper proteins were screened as putative *Spiroplasma citri*-binding molecules using Far-Western analysis (Killiny et al., 2005).These experiments showed that spiralin is a lectin capable of binding to insect 50 and 60 kDa mannose glycoproteins. Hence, spiralin could play a key role in insect transmission of *Spiroplasma citri* by mediating spiroplasma adherence to epithelial cells of the insect vector gut or salivary gland (Killiny et al., 2005). This would also explain why the spiralin-negative mutant 9a2 is poorly transmitted by the vector and is not acquired by insects feeding on 9a2-infected plants.

Viruses. Four different virus types have been found in *Spiroplasma*, SpV1-SpV4. Use of SpV1 viruses for recombinant DNA studies in *Spiroplasma citri* is described later in the section on "Tools for molecular genetics of *Spiroplasma citri*".

Cells of many spiroplasma species contain filamentous/rod-shaped viruses (SVC1 = SpV1) that are associated with nonlytic infections (Bové et al., 1989; Ranhand et al., 1980; Renaudin and Bové, 1994). They belong to the *Plectrovirus* group within the *Inoviridae*. SpV1 viruses have circular, single-stranded DNA genomes (7.5 to 8.5 kbp), some of which have been sequenced (Renaudin and Bové, 1994). SpV1 sequences also occur as prophages in the genome of the majority of *Spiroplasma citri* strains studied (Renaudin and Bové, 1994). These insertions take place at numerous sites in the chromosomes of *Spiroplasma citri* (Ye et al., 1992) and *Spiroplasma melliferum* (Ye et al., 1994a). The SpV1-ORF3 and the repeat sequences could be part of an IS-like element of chromosomal origin. Resistance of spiroplasmas to virus infection may be associated with integration of viral DNA sequences in the chromosome or extrachromosomal elements (Sha et al., 1995). The evolutionary history of these viruses is unclear, but there is some evidence for virus and plasmid co-evolution in the group I *Spiroplasma* species (Gasparich et al., 1993a) and indications of potentially widespread horizontal transmission (Vaughn and de Vos, 1995). Virus infection of spiroplasma cells can pose problems in cultures. For example, lyophilized early passages of *Spiroplasma citri* R8-A2^T proved difficult to grow and electron microscopy revealed that these cells carried large numbers of virions of virus SpV1-R8A2 (Cole et al., 1974). Likewise, SpV1 viruses have been found in *Spiroplasma poulsonii* (Cohen et al., 1987) and *Spiroplasma melliferum* (Liss and Cole., 1981); the *Spiroplasma melliferum* SpV1-KC3 virus forms plaques on various strains of *Spiroplasma melliferum*, including the type strain BC-3^T.

A second virus, reminiscent of a type B tailed bacterial virus, occurs in a small number of *Spiroplasma citri* strains (Cole et al., 1973). This SCV2 (= SpV2) virus is a polyhedron with a long, noncontractile tail. It may be associated with lytic infection. Infections in which large numbers of virions of SpV2 viruses are produced tend to be irregular and difficult to maintain under experimental conditions, so this is the least studied of the spiroplasma viruses.

A third virus (SpV3) forms polyhedral virions with short tails and has been found in many strains of *Spiroplasma citri* (Cole, 1979, 1977, 1974). The SpV3 genome is a linear double-stranded DNA molecule of 16 kbp, which can circularize to form a covalently closed molecule with single-stranded gaps, indicating that the linear molecule has cohesive ends. There is significant diversity among SpV3 viruses, extending even to major differences in genome sizes. Virus SpV3-AV9/3 was isolated from *Spiroplasma citri* strain ASP-9 (Stephens, 1980). Dickinson and Townsend (1984) isolated the SpV3 virus from plants infected with *Spiroplasma citri*. This virus, when plated on cells of *Spiroplasma citri*, had a plaque morphology typical of temperate phages. In spiroplasma cells that have been lysogenized, complete virus genomes may be integrated into the spiroplasma chromosome. These cells are then immune to superinfection by the lysogenizing virus, but susceptible to other SpV3 viruses. It is possible that lysogenization of *Spiroplasma citri* by SpV3-ai affects spiroplasma pathogenicity, particularly with respect to attenuation. *Drosophila* spiroplasmas, male-lethal or nonlethal, usually carry SpV3 viruses. Each strain of *Drosophila* spiroplasma carries an associated virus that is lytic to certain other strains (Oishi et al., 1984).

A fourth virus (SpV4), with a naked, icosahedral nucleocapsid 25 nm in diameter, was discovered (Ricard et al., 1982) in the B63 strain of *Spiroplasma melliferum*. SpV4 has a circular, single-stranded DNA genome (Renaudin and Bové, 1994; Renaudin et al., 1984a, b) and is a lytic *Spiromicrovirus* within the *Microviridae* (Chipman et al., 1998). Infection with this virus results in very clear plaques, indicating a lytic process. Host range studies (Renaudin et al., 1984a, b) have shown that only *Spiroplasma melliferum* is susceptible to SpV4. Two strains of *Spiroplasma melliferum*, including the type strain BC-3^T and B63, are not susceptible, as no plaques were formed on lawns of these spiroplasmas. These strains could be infected by transfection suggesting that resistance to the whole virus occurred at the level of adsorption or penetration of the virus (Renaudin and Bové, 1994; Renaudin et al., 1984b).

Genome sequencing. Genomic DNA sequencing efforts for two *Spiroplasma* species are in progress. For *Spiroplasma citri* GII3 (Carle et al., 2010; Saillard et al., 2008), assembly of 20,000 sequencing reads obtained from shotgun and chromosome specific libraries yielded: (1) 39 chromosomal contigs totalling 1525 kbp of the 1820 kbp *Spiroplasma citri* GII3 chromosome as well as (2) 8 circular contigs, which proved to represent seven plasmids: pSciA (7.8 kbp), pSci1 to pSci6 (12.9 to 35.3 kbp), and one viral RF DNA (SVTS2). The chromosomal contigs contained 1905 putative genes or coding sequences (CDS). Of the CDS-encoded proteins, 29% are involved in cellular processes, cell metabolism, or cell structure. CDS for viral proteins and mobile elements represented 24% of the total, whereas 47% of the CDS were for hypothetical proteins with no known function; 21% of the total CDS appeared truncated as compared to their bacterial orthologs. Families of paralogs were mainly clustered in a large region of the chromosome opposite the origin of replication. Eighty-four CDS were assigned to transport functions, including phosphoenolpyruvate phosphotransferase systems (PTS), ATP binding cassette (ABC) transporters, and ferritin. In addition to the general enzymes EI and HPr, glucose- fructose and trehalose-specific PTS permeases, and glycolytic and ATP synthesis pathways, *Spiroplasma citri* possesses a Sec-dependent protein export system and a nearly complete pathway for terpenoid biosynthesis. The sequencing of the *Spiroplasma kunkelii* CR2-3x genome (1.55 Mb) is also nearing completion (http://www.genome.ou.edu/spiro.html); the physical and genetic maps have been published (Dally et al., 2006). Several studies have begun to focus on gene content and genomic organization (Zhao et al., 2003, 2004a, b). Results show that, in addition to virus SpV1 DNA insertions, the *Spiroplasma kunkelii* genome harbors more purine and amino acid biosynthesis, transcriptional regulation, cell envelope, and DNA transport/binding genes than *Mycoplasmataceae* (e.g., *Mycoplasma genitalium* and *Mycoplasma pneumoniae*) genomes (Bai and Hogenhout, 2002).

Plasmids. Several plasmids have been discovered in spiroplasmas (Archer et al., 1981; Gasparich and Hackett, 1994; Gasparich et al., 1993a; Mouches et al., 1984a; Ranhand et al., 1980). They are especially common in spiroplasmas of group I. Eight extrachromosomal elements, including seven plasmids, were discovered during the *Spiroplasma citri* GII3 genome sequencing project. The six largest plasmids, pSci1 to pSci6, range from 12.9 to 35.3 kb (Saillard et al., 2008). *In silico* analyses of plasmid sequences revealed that they share extensive regions of homology and display a mosaic gene organization. Genes encoding proteins of the TraD-TraG, TrsE-TraE, and Soj-ParA protein families, were predicted in most of the pSci sequences. The presence of such genes, usually involved in chromosome integration, cell to cell DNA transfer, or DNA element partitioning; suggests that these molecules could be inherited vertically as well as horizontally. The largest plasmid, pSci6, encodes P32 (Killiny et al., 2006), a membrane-associated protein interestingly absent in all insect non-transmissible strains tested so far. The five remaining plasmids (pSci1 to pSci5) encode eight different *Spiroplasma citri* adhesion-related proteins. The complete sequences of plasmids pSKU146 from *Spiroplasma kunkelii* CR2-3x and pBJS-O from *Spiroplasma citri* BR3 have been reported (Davis et al., 2005; Joshi et al., 2005). These large plasmids, like the above *Spiroplasma citri* plasmids, encode an adhesin and components of a type IV translocation-related conjugation system. Characterizing the replication and stability regions of *Spiroplasma citri* plasmids resulted in the identification of a novel replication protein, suggesting that *Spiroplasma citri* plasmids belong to a new plasmid family and that the *soj* gene is involved in segregational stability of these plasmids (Breton et al., 2008a). Similar replicons were detected in various spiroplasmas of group I, such as *Spiroplasma melliferum*,

Spiroplasma kunkelii, *Spiroplasma* sp. 277F, and *Spiroplasma phoeniceum*, showing that they are not restricted to plant pathogenic spiroplasmas.

Tools for molecular genetics of *Spiroplasma citri*. Recent recombinant DNA tools are described in this section.

Several reports have been published concerning the use of SpV1 viruses as tools to introduce recombinant DNA into spiroplasmas, including optimization of transfection conditions (Gasparich et al., 1993b). The replicative form of SpV1 was used to clone and express the *Escherichia coli*-derived chloramphenicol acetyltransferase (*cat*) gene in *Spiroplasma citri*. Both the replicative form (RF) and the virion DNA produced by the transfected cells contained the *cat* gene sequences (Stamburski et al., 1991). The G fragment of the *Mycoplasma pneumoniae* cytadhesin P1 gene could also be expressed in *Spiroplasma citri* (Marais et al., 1993) using a similar method. However, the recombinant RF proved unstable, resulting in the loss of the DNA insert (Marais et al., 1996).

Recombinant plasmids have also been developed to introduce genes into *Spiroplasma citri* cells. The introduced genes include antibiotic resistance markers and wild-type genes to complement auxotrophic mutants. Most recombinant plasmids contain the origin of DNA replication (*oriC*) of the *Spiroplasma citri* chromosome (Ye et al., 1994b). One such plasmid is pBOT1 (Renaudin, 2002; Renaudin et al., 1995). This plasmid contains a 2 kbp *oriC* region, a tetracycline resistance gene (*tetM*) from Tn*916*, and the linearized *Escherichia coli* plasmid pBS with a *colE1* origin of replication. Because of its two origins of replication, *oriC* and *colE1*, pBOT1 is able to shuttle between *Spiroplasma citri* and *Escherichia coli*. When introduced into *Spiroplasma citri*, pBOT1 replicates first as a free extrachromosomal element, but later integrates into the chromosome via homologous recombination involving a single crossover event in the *oriC* region. Once integrated into the host chromosome, the whole plasmid is stably maintained. Recent studies suggest that the broad host range *Spiroplasma citri* GII3 plasmids and their shuttle derivatives may have significant advantages over *oriC* plasmids for gene transfer and expression in spiroplasmas (Breton et al., 2008a). They transform *Spiroplasma citri* (as well as *Spiroplasma kunkelii* and *Spiroplasma phoeniceum*) strains at relatively high efficiencies, the growth of the transformants is not significantly affected, they do not integrate into the chromosome, and their stability/loss can be modulated depending upon the presence/absence of the *soj* gene.

Spiroplasma citri mutants have been produced by random and targeted approaches. The transposon Tn*4001* has been used successfully for random mutagenesis of *Spiroplasma citri* (Foissac et al., 1997c). For targeted gene inactivation, plasmids derived from pBOT1 have been used to disrupt genes (e.g., fructose operon, motility gene *scm1*) through homologous recombination involving a single crossover event (Duret et al., 1999; Gaurivaud et al., 2000c). More recently, Lartigue et al. (2002) developed vector pC2, in which the *oriC* fragment was reduced to the minimal sequence needed to promote plasmid replication; this vector increases recombination frequency at the target gene. To avoid the extensive passaging that was required for recombination prior to transformant screening, vector pC55 was designed using a selective tetracycline resistance marker that is only expressed after the plasmid has integrated into the chromosome at the target gene. This approach was used to inactivate the spiralin gene (*spi*) and the gene encoding the IICB component of the glucose phosphotransferase system permease (*ptsG*) (André et al., 2005; Duret et al., 2003; Lartigue et al., 2002). Another series of recombinant plasmids, the pGOT vectors, allow for selection of rare recombination events by using two distinct selective markers. First, transformants are screened for their resistance to gentamicin and next, site-specific recombinants are selected for based on their resistance to tetracycline, which can only be expressed through recombination at the target gene. In this way, inactivation of the *crr* gene, encoding the glucose phosphotransferase permease IIA component, was obtained (Duret et al., 2005). The use of the transposon $\gamma\delta$ TnpR/res recombination system to produce unmarked mutations (i.e., without insertion of antibiotic markers) in *Spiroplasma citri* was demonstrated by the production of a disrupted *arcA* mutant (Duret et al., 2005); *arcA* encodes arginine deiminase. In this system, the target gene is disrupted by integration of a plasmid containing target gene sequences along with the *tetM* gene flanked by binding-specific recombination (res) sites. After integration of the plasmid, a second plasmid is introduced that encodes the resolvase TnpR. TnpR mediates the resolution of the cointegrate at the res sites, thereby removing *tetM* but leaving behind a mutated version of the target gene. The TnpR-encoding plasmid is lost spontaneously when selective pressure is removed.

Antigenic structure. Growth inhibition tests (Whitcomb et al., 1982) were used in the early years to identify spiroplasma species or groups, but metabolism inhibition (Williamson et al., 1979b; Williamson and Whitcomb, 1983) and deformation tests (Williamson et al., 1978) are now used almost exclusively (see below).

Antigenic variability, which has been described for some *Mycoplasma* species (Rosengarten and Wise, 1990; Yogev et al., 1991), has not been demonstrated in spiroplasmas (R. Rosengarten, personal communication).

Group classification. The classification of spiroplasmas was first proposed by Junca et al. (1980) and has been revised periodically (Tully et al., 1987; Williamson et al., 1998). These classifications are based on serological reactions of the organisms in growth inhibition, deformation and metabolism inhibition tests and/or characteristics of their genomes. Development of a classification scheme has resulted in the delineation of spiroplasma groups and subgroups (Table 142). In the scheme, "groups" have been defined as clusters of similar organisms, all of which possess negligible DNA–DNA hybridization with representatives of other groups, but moderate to high levels of hybridization (20–100%) with each other. Groups are, therefore, putative species. This level of genomic differentiation correlates well with substantial differences in serology. Thirty-four groups were presented in a revised classification of spiroplasmas in 1998 (Williamson et al., 1998). Four additional groups (XXXV–XXXVIII) were proposed recently as the result of a global spiroplasma environmental survey (Whitcomb et al., 2007) and more are anticipated (Jandhyam et al., 2008). Subgroups have been defined by the International Committee on Systematics of Bacteria (ICSB) Subcommittee on the Taxonomy of *Mollicutes* (ICSB, 1984) as clusters of spiroplasma strains showing intermediate levels of intragroup DNA–DNA hybridization (10–70%) and possessing corollary serological relationships. Three spiroplasma groups [group I (Junca et al., 1980; Saillard et al., 1987), group VIII (Gasparich et al., 1993c), and

TABLE 142. Biological properties of spiroplasmas[a]

Group	Spiroplasma	Strain	ATCC no.	Morphology[b]	Genome[c]	G+C[d]	Arg[e]	Dt[f]	OptT[g]	Host
I-1	*S. citri*	R8-A2^{T}	27556^{T}	Long helix	1820	26	+	4.1	32	Phloem/leafhopper
I-2	*S. melliferum*	BC-3^{T}	33219^{T}	Long helix	1460	26	+	1.5	37	Honey bee
I-3	*S. kunkelii*	E275^{T}	29320^{T}	Long helix	1610	26	+	27.3	30	Phloem/leafhopper
I-4	*Spiroplasma* sp.	277F	29761	Long helix	1620	26	+	2.3	32	Rabbit tick
I-5	*Spiroplasma* sp.	LB-12	33649	Long helix	1020	26	–	26.3	30	Plant bug
I-6	*S. insolitum*	M55^{T}	33502^{T}	Long helix	1810	28	–	7.2	30	Flower surface
I-7	*Spiroplasma* sp.	N525	33287	Long helix	1780	26	+	4.7	32	Green June beetle
I-8	*S. phoeniceum*	P40^{T}	43115^{T}	Long helix	1860	26	+	16.8	30	Phloem/vector
I-9	*S. penaei*	SHRIMP^{T}	BAA-1082^{T} (CAIM 1252^{T})	Helix	nd	29	+	nd	28	Pacific white shrimp
II	*S. poulsonii*	DW-1^{T}	43153^{T}	Long helix	1040	26	nd	15.8	30	*Drosophila* hemolymph
III	*S. floricola*	OMBG	29989^{T}	Helix	1270	26	–	0.9	37	Plant surface
IV	*S. apis*	B31^{T}	33834^{T}	Helix	1300	30	+	1.1	34.5	Honey bee
V	*S. mirum*	SMCA^{T}	29335^{T}	Helix	1300	30	+	7.8	37	Rabbit tick
VI	*S. ixodetis*	Y32^{T}	33835^{T}	Tight coil	2220	25	–	9.2	30	Ixodid tick
VII	*S. monobiae*	MQ-1^{T}	33825^{T}	Helix	940	28	–	1.9	32	*Monobia* wasp
VIII-1	*S. syrphidicola*	EA-1^{T}	33826^{T}	Minute helix	1230	30	+	1.0	32	Syrphid fly
VIII-3	*Spiroplasma* sp.	TAAS-1	51123	Minute helix	1170	31	+	1.4	37	Horse fly
VIII-2	*S. chrysopicola*	DF-1^{T}	43209^{T}	Minute helix	1270	29	+	6.4	30	Deer fly
IX	*S. clarkii*	CN-5^{T}	33827^{T}	Helix	1720	29	+	4.3	30	Green June beetle
X	*S. culicicola*	AES-1^{T}	35112^{T}	Short helix	1350	26	–	1.0	37	Mosquito
XI	*S. velocicrescens*	MQ-4^{T}	35262^{T}	Short helix	1480	26	–	0.6	37	*Monobia* wasp
XII	*S. diabroticae*	DU-1^{T}	43210^{T}	Helix	1350	25	+	0.9	32	Beetle
XIII	*S. sabaudiense*	Ar-1343^{T}	43303^{T}	Helix	1175	29	+	4.1	30	Mosquito
XIV	*S. corruscae*	EC-1^{T}	43212^{T}	Helix	nd	26	–	1.5	32	Horse fly/beetle
XV	*Spiroplasma* sp.	I-25	43262	Wave-coil	1380	26	–	3.4	30	Leafhopper
XVI-1	*S. cantharicola*	CC-1^{T}	43207^{T}	Helix	nd	26	–	2.6	32	Cantharid beetle
XVI-2	*Spiroplasma* sp.	CB-1	43208	Helix	1320	26	–	2.6	32	Cantharid beetle
XVI-3	*Spiroplasma* sp.	Ar-1357	51126	Helix	nd	26	–	3.4	30	Mosquito
XVII	*S. turonicum*	Tab4c^{T}	700271^{T}	Helix	1305	25	–	nd	30	Horse fly
XVIII	*S. litorale*	TN-1^{T}	43211^{T}	Helix	1370	25	–	1.7	32	Horse fly
XIX	*S. lampyridicola*	PUP-1^{T}	43206^{T}	Unstable helix	1375	25	–	9.8	30	Firefly
XX	*S. leptinotarsae*	LD-1^{T}	43213^{T}	Motile funnel	1085	25	+	7.2	30	Colorado potato beetle
XXI	*Spiroplasma* sp.	W115	43260	Helix	980	24	–	4.0	30	Flower surface
XXII	*S. taiwanense*	CT-1^{T}	43302^{T}	Helix	1195	26	–	4.8	30	Mosquito
XXIII	*S. gladiatoris*	TG-1^{T}	43525^{T}	Helix	nd	26	–	4.1	31	Horse fly
XXIV	*S. chinense*	CCH^{T}	43960^{T}	Helix	1530	29	–	0.8	37	Flower surface
XXV	*S. diminutum*	CUAS-1^{T}	49235^{T}	Short helix	1080	26	–	1.0	32	Mosquito
XXVI	*S. alleghenense*	PLHS-1^{T}	51752^{T}	Helix	1465	31	+	6.4	30	Scorpion fly
XXVII	*S. lineolae*	TALS-2^{T}	51749^{T}	Helix	1390	25	–	5.6	30	Horse fly
XXVIII	*S. platyhelix*	PALS-1^{T}	51748^{T}	Wave-coil	780	29	+	6.4	30	Dragonfly
XXIX	*Spiroplasma* sp.	TIUS-1	51751	Rare helices	840	28	–	3.6	30	Tiphiid wasp
XXX	*Spiroplasma* sp.	BIUS-1	51750	Late helices	nd	28	–	0.9	37	Flower surface
XXXI	*S. montanense*	HYOS-1^{T}	51745^{T}	Helix	1225	28	+	0.7	32	Horse fly
XXXII	*S. helicoides*	TABS-2^{T}	51746^{T}	Helix	nd	27	–	3.0	32	Horse fly
XXXIII	*S. tabanidicola*	TAUS-1^{T}	51747^{T}	Helix	1375	26	–	3.7	30	Horse fly
XXXIV	*Spiroplasma* sp.	B1901	700283	Helix	1295	25	–	nd	nd	Horse fly
XXXV	*Spiroplasma* sp.	BARC 4886	BAA-1183	Helix	nd	nd	–	0.6	32	Horse fly
XXXVI	*Spiroplasma* sp.	BARC 4900	BAA-1184	Helix	nd	nd	–	1.0	30	Horse fly
XXXVII	*Spiroplasma* sp.	BARC 4908	BAA-1187	Helix	nd	nd	–	1.2	32	Horse fly
XXXVIII	*Spiroplasma* sp.	GSU5450	BAA-1188	Helix	nd	nd	–	1.5	32	Horse fly
Nd	*S. atrichopogonis*	GNAT3597^{T}	BAA-520^{T} (NBRC 100390^{T})	Helix	nd	28	+	nd	30	Biting midge
Nd	*S. leucomae*	SMA^{T}	BAA-521^{T} (NBRC 100392^{T})	Helix	nd	24	+	nd	30	Satin moth

[a]nd, Not determined.

[b]For descriptions of morphotypes, see text.

[c]Genome size (kbp).

[d]DNA G+C content (mol%).

[e]+, Catabolizes arginine.

[f]Doubling time (h) (Konai et al., 1996a).

[g]Optimum growth temperature (°C).

group XVI (Abalain-Colloc et al., 1993)] have been divided into a total of 15 subgroups. "Serovars" have been defined as genotypic clusters varying substantially in metabolism inhibition and deformation serology, but that are insufficiently differentiated from members of existing groups or subgroups to warrant separation. However, with the discovery of a large number of strains for some groups (e.g., group VIII), the serovar/subgroup picture has become very confused (Regassa et al., 2004; see *Phylogeny*, below).

Procedures for species descriptions and minimal standards. Species descriptions of spiroplasmas have been in accord with recommendations of minimum standards proposed by the ICSP (International Committee on Systematics of Prokaryotes) Subcommittee on the Taxonomy of *Mollicutes* (Brown et al., 2007).

Cloning. Production of spiroplasma lineages produced from a single cell or clonings are performed largely by serial dilution of filtered cultures using 96-well microtiter plates (Whitcomb et al., 1986; Whitcomb and Hackett, 1987). At a certain dilution, which varies from plating to plating, the mean number of cells per well decreases so that fewer than about 8 of the 96-wells support growth of a spiroplasma clone. Very probably, such clones arise from a single spiroplasmal cell.

Cellular morphology. Using dark-field microscopy, cultures should appear helical and motile during at least one growth phase (see "Morphology" and "Motility" above). However, morphological exceptions do occur (see "Differentiating characters" below and reviewed by Gasparich et al., 2004).

16S rRNA gene sequence analysis. Preliminary identification is performed by PCR amplification using universal 16S rRNA (Gasparich et al., 2004) or other described primers (e.g., Fukatsu and Nikoh, 1998; Jandhyam, 2008). DNA sequence analysis using a BLAST search provides preliminary placement within the genus *Spiroplasma*. Those strains showing close phylogenetic relationships based on 16S rRNA gene sequence analyses should then be screened using serological tests.

Serological tests. The deformation test (Williamson et al., 1978) is used routinely for serological analyses. Reciprocal titers of ≥320 are generally required for definitive group placement. Deformation is defined as entire or partial loss of helicity. At the end point, cells are often seen in which an unaffected part of the helical filament exhibits flexing motility despite the presence of a bleb on another part of the cell. The deformation titer is the reciprocal of the final antiserum dilution that exhibits deformation of ≥50% of the cells. Antiserum should be produced for any strain thought to represent a novel serogroup and any positive test against characterized groups requires a recriprocal test using the newly prepared antiserum.

The high levels of specificity and sensitivity of the metabolism inhibition test make it especially useful for defining groups and subgroups (Williamson et al., 1979b; Williamson and Whitcomb, 1983). Other serological tests have also been employed for characterization of spiroplasmas. Growth inhibition tests were used for delineation of spiroplasma groups I through XI (Whitcomb et al., 1982), but were not used thereafter. Growth inhibition tests are problematic for spiroplasmas because they require development of procedures for obtaining colonies. The spiroplasma motility inhibition test (Hackett et al., 1997) has proved useful for determination of intraspecific variation in *Spiroplasma leptinotarsae*. ELISA has been used for detection of *Spiroplasma kunkelii* (Gordon et al., 1985) and *Spiroplasma citri* (Saillard and Bové, 1983).

Optimum growth temperature. Optimal growth temperatures between 10 and 41°C have been determined (Konai et al., 1996a).

Substrate metabolism. The ability to ferment glucose and produce acid must be examined (Aluotto et al., 1970). The ability to hydrolyze arginine and produce ammonia should be assessed (Barile, 1983). See the section on "Biochemical reactions" above for more details.

Ecology. The species description must include ecological information such as isolation site within the host and cultivation conditions, common and binomial host name, geographical location of host (with GPS), any known interaction between the spiroplasma and its host, and, in the case of a pathogen, disease symptoms observed.

Antibiotic sensitivities. In early studies (Bowyer and Calavan, 1974; Liao and Chen, 1981b), spiroplasmas proved to be especially sensitive *in vitro* to tetracycline, erythromycin, tylosin, tobramycin, and lincomycin. Strains have been isolated that are permanently resistant to kanamycin, neomycin, gentamicin, erythromycin, and several tetracycline antibiotics (Liao and Chen, 1981b). Insensitivity to rifampicin has been studied in relation to its inhibition of transcription (see "RNA polymerase and spiroplasmal insensitivity to rifampicin" above) and penicillin insensitivity is seen for all spiroplasmas due to the lack of a cell wall. Natural amphipathic peptides such as Gramicidin S alter the membrane potential of spiroplasma cells and induce the loss of cell motility and helicity (Bévén and Wróblewski, 1997). The toxicity of the lipopeptide antibiotic globomycin was found to be correlated with an inhibition of spiralin processing (Bévén et al., 1996). As with Gramicidin S, the antibiotic was effective against spiroplasmas, but not *Mycoplasma mycoides*. Natural 18-residue peptaibols (trichorzins PA) are bacteriocidal to spiroplasmas (Bévén et al., 1998). The mode of action appears to be permeabilization of the host cell membrane.

Hosts, ecology, and pathogenicity

Hosts. Almost all spiroplasmas have been found to be associated with arthropods or an arthropod connection is strongly suspected. Hackett et al. (1990) searched for mollicutes in a wide variety of insect orders. Isolates were obtained from six orders and 14 insect families. Only one of these orders, Odonata (dragonflies), was primitive (heterometabolous) and it was speculated that the spiroplasma from a dragonfly host might have been acquired via predation. Hackett et al. (1990) suggested that the *Spiroplasma/Entomoplasma* clade may have arisen in a paraneopteran-holometabolan ancestor, coevolved with these orders, and never adapted to more primitive insect orders. Some insect families have an especially rich spiroplasma, entomoplasma, and mesoplasma flora.

Insect gut. The majority of spiroplasmas appear to be maintained in an insect gut/plant surface cycle. Clark (1984) hypothesized several types of gut infection in which persistence in the gut and the ability to invade hemolymph varied among spiroplasma species. It has been hypothesized (Hackett and Clark, 1989) that the gut cycle was primitive and that other cycles were derived from it. Spiroplasmas have been isolated from guts of tabanids (Diptera: Tabanidae) worldwide (French et al., 1997, 1990, 1996; Jandhyam et al., 2008; Le Goff et al., 1991, 1993; Regassa and Gasparich, 2006; Vazeille-Falcoz et al., 1997; Whitcomb et al., 1997a). Examination of diversity trends among the tabanid isolates suggests that spiroplasma diversity increases with temperature, resulting in more diversity in southern climes in the Northern Hemisphere (Whitcomb et al., 2007). Although evidence points strongly to multiple cycles of horizontal transmission, the sites where such transmission occurs remain unknown. However, some tabanids utilize honeydew (excreta of sucking insects) deposited on leaf surfaces, suggesting a possible transmission mechanism. Mosquitoes (Chastel and Humphery-Smith, 1991) are also common spiroplasma hosts (Lindh et al., 2005). Additionally, spiroplasmas inhabit the gut of ground beetles (*Harpalus pensylvanicus* and *Anisodactylus sanctaecrucis*) as evidenced by 16S rRNA gene sequence analysis of the digestive tract bacterial flora (Lundgren et al., 2007).

Plant surfaces. Flowers and other plant surfaces represent a major site where spiroplasmas and other microbes are transmitted from insect to insect (Clark, 1978; Davis, 1978; McCoy et al., 1979). Members of several spiroplasma groups have been isolated only from flowers and strains of several other spiroplasmas have been isolated from both insects and flowers. Biological evidence suggests that mosquito spiroplasmas are transmitted from insect to insect on flowers (Chastel et al., 1990; Le Goff et al., 1990). It is not known whether any of the so-called "flower spiroplasmas" can exist as true epiphytes. Isolations of spiroplasmas from a variety of insects (Clark, 1982; Hackett et al., 1990) suggest that it is likely that many or most of these flower isolates are deposited passively by visiting arthropods.

Plant phloem and sucking insects. Several spiroplasmas possess a life cycle that involves infection of plant phloem and homopterous insects (Bové, 1997; Fletcher et al., 1998; Garnier et al., 2001; Saglio and Whitcomb, 1979). In the course of passage through the insect, spiroplasmas pass through, accumulate, or multiply in gut epithelial cells and salivary cells. They also accumulate in the insect neurolemma. Large accumulations of spiroplasma cells occur frequently in the hemolymph, where they undoubtedly multiply (Whitcomb and Williamson, 1979). Spiroplasmas may multiply in a number of sucking insect species that have been exposed to diseased plants, but often only a single vector or several vector species transmit spiroplasmal pathogens from plant to plant (summarized in Calavan and Bové, 1989; Whitcomb, 1989; Kersting and Sengonca, 1992).

Sex ratio organisms. Once thought to be a genetic factor, the sex ratio trait in *Drosophila* was shown by Poulson and Sakaguchi (1961) to be induced by a micro-organism, *Spiroplasma poulsonii* (Williamson et al., 1999). A number of other spiroplasmas in a variety of insect hosts have been identified that also cause sex ratio distortions, including isolates from the chrysomelid beetle *Adalia bipunctata* (Hurst and Jiggins, 2000; Hurst et al., 1999) and the butterfly *Danaus chrysippus* (Jiggins et al., 2000). In addition, 16S rRNA gene sequence analysis identified spiroplasmas as the causative agent for male-killing: in a population of *Harmonia axyridis* (ladybird beetle) in Japan (Nakamura et al., 2005); in populations of *Drosophila neocardini*, *Drosophila ornatifrons* and *Drosophila paraguayensis* from Brazil (Montenegro et al., 2006, 2005); in populations of *Anisosticta novemdecimpunctata* (ladybird beetle) in Britain (Tinsley and Majerus, 2006); in a population of *Adalia bipunctata* (Sokolova et al., 2002); in several strains from the Tucson *Drosophila* stock culture collection (Mateos et al., 2006); and in *Drosophila melanogaster* populations from Uganda and Brazil (Pool et al., 2006). Other organisms closely associated with their insect hosts were discovered inferentially by PCR studies (Fukatsu and Nikoh, 2000, 2001) and also appear to be related to *Spiroplasma mirum*. They also cause preferential male killing in an infected *Drosophila* population (Anbutsu and Fukatsu, 2003). Natural infection rates of male-killing spiroplasmas in *Drosophila melanogaster* are about 2.3%, as determined for a Brazilian population (Montenegro et al., 2005), and vary between 0.1 and 3% for Japanese populations of *Drosophila hydei* (Kageyama et al., 2006). The male-killing spiroplasma strain isolated from *Adalia bipunctata* was used to artificially infect eight different coccinellid beetle species. The data suggest that host range could serve to limit horizontal transfer to closely related host species (Tinsley and Majerus, 2007). Supporting this hypothesis was the study that showed the interspecific lateral transmission of spiroplasmas from *Drosophila nebulosa* to *Drosophila willistoni* via ectoparasitic mites (Jaenike et al., 2007). A recent multilocus analysis by Haselkorn et al. (2009) showed that *Drosophila* species are infected with at least four distinct spiroplasma haplotypes.

Studies on *Drosophila* infections by the sex-ratio organism showed that it did not induce the innate immunity of the insect (Hurst et al., 2003). The sex-ratio spiroplasmas have been shown to be vertically transmitted through female hosts, with spiroplasmas present during oogenesis (Anbutsu and Fukatsu, 2003). Although the exact mechanism of male-killing has not been determined, studies have shown that male killing occurs shortly after formation of the host dosage compensation complex (Bentley et al., 2007) and that male *Drosophila melanogaster* mutants lacking any of the five genes involved in the dosage compensation complex are not killed (Veneti et al., 2005). In the Kenyan butterfly *Danaus chrysippus*, a correlation between male killing and a recessive allele for a gene controlling infection susceptibility has been reported. Moreover, infections seemed to have a negative effect on body size (Herren et al., 2007).

Ticks. Three *Spiroplasma* species have been isolated from ticks. Two of these, *Spiroplasma mirum* and *Spiroplasma* sp. 277F, are from the rabbit tick *Haemaphysalis leporispalustris* (Tully et al., 1982; Williamson et al., 1989). The third species was isolated from *Ixodes pacificus* ticks and named *Spiroplasma ixodetis* (Tully et al., 1995). 16S rRNA gene sequence analysis of spiroplasmas originally isolated from *Ixodes* ticks and growing in a Buffalo Green Monkey mammalian cell culture line showed a high degree of identity with the *Spiroplasma ixodetis* 16S rRNA gene (Henning et al., 2006). Analysis of the 16S rRNA gene sequence from DNA extracted from unfed *Ixodes ovatus* from Japan indicated the presence of spiroplasmas that were also closely related to *Spiroplasma ixodetis* (Taroura et al., 2005). The ability of tick spiroplasmas, including *Spiroplasma ixodetis*, to multiply at 37°C reflects the role of vertebrates as tick hosts. The ability of *Spiroplasma ixodetis* to grow at 32°C as well as 37°C (Tully et al., 1982) may reflect the ecology of some of the cold-blooded vertebrate hosts of these ticks. There is no evidence that any of these spiroplasmas are transmitted to vertebrate hosts of the ticks.

Crustaceans. *Spiroplasma* sp. have recently been isolated in both freshwater and salt-water crustaceans.

Spiroplasma penaei (strain SHRIMP[T]) was isolated from the hemolymph of Pacific white shrimp (*Penaeus vannamei*) after high mortalities were observed in an aquaculture pond in Columbia, South America (Nunan et al., 2004). The pathogenic agent was the spiroplasma (Nunan et al., 2005). Although not cultivated, 16S rRNA gene sequence analysis also revealed the presence of spiroplasmas in the gut of the hydrothermal shrimp *Rimicaris exoculata* (Zbinden and Cambon-Bonavita, 2003). In another outbreak, Chinese mitten crab (*Eriocheir sinensis*) reared in aquaculture ponds in China became infected with tremor disease. The causative agent was determined to be a spiroplasma with 99% 16S rRNA gene sequence identity to *Spiroplasma mirum* (Wang et al., 2004a, b). However, recent studies suggest that the infective agent may be a species similar to, but distinct from, *Spiroplasma mirum* (Bi et al., 2008). The same organism also infects red swamp crayfish (*Procambarus clarkii*) that are co-reared with the Chinese mitten crab (Bi et al., 2008; Wang et al., 2005) as well as the shrimp *Penaeus vannamei* (Bi et al., 2008).

Other hosts. Spiroplasmas have been identified in a variety of other hosts, although not necessarily linked to the gut habitat. The first spiroplasma isolated from a lepidopteran came from the hemolymph of white satin moth larvae (*Leucoma salicis* L.) from Poland (designated strain SMA[T]) and was serologically distinct from any previously described spiroplasma group (Oduori et al., 2005). Another novel spiroplasma (designated strain GNAT3597[T]) was isolated from biting midges from the genus *Atrichopogon* (Koerber et al., 2005). Spiroplasmas that are closely related to the male-killing spiroplasmas in ladybird beetles (Majerus et al., 1999; Tinsley and Majerus, 2006) have also been identified in the predatory mite *Neoseiulus californicus* using 16S rRNA gene sequence analysis (Enigl and Schausberger, 2007). A broad survey of 16 spider families for the presence of endosymbionts using 16S rRNA gene sequence analysis revealed that six families contained spiroplasmas, including *Agelenidae*, *Araneidae*, *Gnaphosidae*, *Linyphiidae*, *Lycosidae*, and *Tetragnathidae* (Goodacre et al., 2006).

Biogeography. Spiroplasmas have been identified from hosts in Africa, Asia, Australia, Europe, South America, and North America. While they are worldwide in distribution, studies suggest that biodiversity may be greatest in warm climates (Whitcomb et al., 2007). Because spiroplasmas are host-associated, it seems reasonable that *Spiroplasma* species distribution would be limited by host biogeography. Early studies indicated that some spiroplasmas have discrete geographic distributions (Whitcomb et al., 1990). As the diversity of sampling sites increases, the view of spiroplasma biogeography is likely to shift (Regassa and Gasparich, 2006). Distinct distributions may exist, but probably on a larger geographic scale. While it is not clear what factors account for spiroplasma ranges, the level of host specificity and host overwintering ranges may contribute to the biogeography of *Spiroplasma* species (Whitcomb et al., 2007).

Pathogenicity. Symptoms of infection and confirmation of Koch's postulates have been reported for the etiologic roles of: *Spiroplasma citri* in "stubborn" disease of citrus (Calavan and Bové, 1989; Markham et al., 1974); corn stunt spiroplasma (Chen and Liao, 1975; Nault and Bradfute, 1979; Williamson and Whitcomb, 1975); *Spiroplasma phoeniceum* in aster, an experimental host (Saillard et al., 1987); *Spiroplasma poulsonii* in *Drosophila pseudoobscura* (Williamson et al., 1989); *Spiroplasma penaei* in *Penaeus vannamei* (Nunan et al., 2005); and *Spiroplasma eriocheiris* (Wang et al., 2010) in the Chinese mitten crab, *Eriocheir sinensis* (Wang et al., 2004b). Recent studies have focused on spiroplasma infection and replication in the midgut and Malpighian tubules of leafhoppers (Özbek et al., 2003). The use of immunofluorescence confocal laser scanning microscopy has revealed the presence of *Spiroplasma kunkelii* in the midgut, filter chamber, Malpighian tubules, hindgut, fat tissues, hemocytes, muscle, trachea, and salivary glands of leafhopper hosts, but not in the nerve cells of the brain or nerve ganglia (Ammar and Hogenhout., 2005). Plant spiroplasmas may also be pathogenic for unusual vectors (Whitcomb and Williamson, 1979), but are much less so for their usual host (Madden and Nault, 1983; Nault et al., 1984). In fact, some spiroplasmas are beneficial to their leafhopper hosts (Ebbert and Nault, 1994) and it has been hypothesized that infection plays an important role in the host's overwintering strategies (Moya-Raygoza et al., 2007a, b; Summers et al., 2004).

Spiroplasma mirum is experimentally pathogenic for a variety of suckling animals, causing cataract and other ocular symptoms, neural pathology (Clark and Rorke, 1979), and malignant transformation in cultured cells (Kotani et al., 1990). *Spiroplasma melliferum* also persists and causes pathology in suckling mice (Chastel et al., 1990, 1991). *Spiroplasma eriocheiris* is neurotropic to brain tissue in experimentally injected chicken embryos (Wang et al., 2003). There are two recent reports of spiroplasmas in aquatic invertebrates. Nunan et al. (2005) characterized a spiroplasma in commercially raised shrimp that led to a lethal disease. *Spiroplasma melliferum* and *Spiroplasma apis* cause disease in honey bees (Clark, 1977; Mouches et al., 1982a, 1983a). Intrathoracic inoculation of *Spiroplasma taiwanense* reduced the survival and impaired the flight capacity of inoculated mosquitoes (Humphery-Smith et al., 1991a), and inoculation of *Spiroplasma taiwanense* per os decreased the survival of mosquito larvae in laboratory trials (Humphery-Smith et al., 1991b). *Spiroplasma poulsonii* causes sex ratio abnormalities (male-killing) in *Drosophila* (Williamson and Poulson, 1979). Male-killing spiroplasma strains related to *Spiroplasma poulsonii* cause necrosis in neuroblastic and fibroblastic cells (Kuroda et al., 1992). The significance of some biological properties of spiroplasmas is incompletely understood. For example, membranes of *Spiroplasma monobiae* are potent inducers of tumor necrosis factor alpha secretion and of blast transformation (Sher et al., 1990a, b) in insect cell culture.

Spiroplasmas are implicated by circumstantial evidence, in the view of some workers, to be associated with human disease. Bastian first claimed in 1979 that spiroplasmas were associated with Creutzfeldt–Jakob Disease (CJD), an extremely rare scrapie-like disease of humans (Bastian, 1979). Bastian and Foster (2001) reported finding spiroplasma 16S rRNA genes in CJD- and scrapie-infected brains that were not observed in controls. More recent studies (Bastian et al., 2004) presented evidence to show that spiroplasma 16S rRNA genes were found in brain tissue samples from scrapie-infected sheep, chronic wasting disease-infected cervids, and CJD-infected humans. All the brain tissues from non-infected controls were negative for spiroplasmal DNA. These authors further showed that the sequence of the PCR products from the infected brains was 96% identical to the *Spiroplasma mirum* 16S rRNA gene. However, these results could not be replicated in an independent blind study of uninfected and Scrapie-infected hamster brains using the same primers (Alexeeva et al., 2006). A recent study to fulfill Koch's postulate reported the transfer of spiroplasma from transmissible spongiform encephalopathy (TSE) brains and *Spiroplasma mirum* to induce spongiform encephalopathy in ruminants (Bastian et al., 2007). The current status of the involvement of spiroplasmas in TSE is the subject of recent reviews (Bastian, 2005; Bastian and Fermin, 2005). Other proposed connections between mollicutes and human disease have been evaluated by Baseman and Tully (1997).

Mechanism of Spiroplasma citri phytopathogenicity. Transposon (Tn*4001*) mutants have been examined extensively to elucidate the molecular mechanisms associated with *Spiroplasma citri* phytopathogenicity. One of these mutants, GMT553, highlighted the involvement of selective carbohydrate utilization in *Spiroplasma citri* pathogenicity (see review by Bové et al., 2003). When introduced into periwinkle plants via injected leafhoppers (*Circulifer haematoceps*), GMT553 multiplied in the plants as actively as wild-type *Spiroplasma citri* strain GII3, but did not induce symptoms (Foissac et al., 1997b, c; Gaurivaud et al., 2000b). In this mutant, the transposon was found to be inserted in *fruR*, a transcriptional activator of the fructose operon (*fruRAK*; Gaurivaud et al., 2000a). The second gene of the operon, *fruA*, encodes fructose permease, which enables uptake of fructose; and the third gene, *fruK*, encodes 1-phosphofructokinase. In mutant GMT553, transcription of the fructose operon is abolished and, hence, the mutant cannot utilize fructose as a carbon or energy source (Gaurivaud et al., 2000a). Mutant GMT553 was functionally complemented for fructose utilization and phytopathogenicity in *trans* by a recombinant *fruR–fruA–fruK* operon, *fruA–fruK* partial operon, or *fruA* alone, but not *fruR* or *fruR–fruA* (Gaurivaud et al., 2000a, b). It should be pointed out that both fructose^{+} and fructose^{-} spiroplasmas are able to utilize glucose.

Further insight into *Spiroplasma citri* phytopathogenicity in relation to sugar metabolism comes from the production of a spiroplasma mutant unable to use glucose (André et al., 2005). The import of glucose into *Spiroplasma citri* cells involves a phosphotransferase (PTS) system composed of two distinct polypeptides encoded by (1) *crr* (glucose PTS permease IIA^{Glc} component) and (2) *ptsG* (glucose PTS permease IICB^{Glc} component). A *ptsG* mutant (GII3-glc1) proved unable to import glucose. When introduced into periwinkle (*Catharanthus roseus*) plants through leafhopper transmission, the mutant induced severe symptoms similar to those obtained with wild-type GII3, in strong contrast to the fructose operon mutant, GMT553, which was virtually non-pathogenic. These results indicated that fructose and glucose utilization were not equally involved in pathogenicity and are consistent with biochemical data showing that, in the presence of both sugars, *Spiroplasma citri* preferentially used fructose. NMR analyses of carbohydrates in plant extracts revealed the accumulation of soluble sugars, particularly glucose, in plants infected by wild-type *Spiroplasma citri* GII3 or GII3-glc1, but not in those infected by GMT553. In the infected plant, *Spiroplasma citri* cells are restricted to the sieve tubes. In the companion cell, sucrose is cleaved by invertase to fructose and glucose. In the sieve tube, wild-type *Spiroplasma citri* cells will use fructose preferentially over glucose leading to a decreased fructose concentration and, consequently, to an increase of invertase activity, which in turn results in glucose accumulation. Glucose accumulation is known to induce stunting and repression of photosynthesis genes in *Arabidopsis thaliana*. Such symptoms are precisely those observed in periwinkle plants infected by wild-type *Spiroplasma citri* (André et al., 2005).

Genes that are up- or down-regulated in plants following infection with *Spiroplasma citri* have been studied by differential display analysis of mRNAs in healthy and symptomatic periwinkle plants (Jagoueix-Eveillard et al., 2001). Expression of the transketolase gene was inhibited in plants infected by the wild-type spiroplasma, but not by the non-phytopathogenic mutant GMT553, further indicating that sugar metabolism and transport are important factors in pathogenicity. Sugar PTS system permeases have been shown to be important in rapid adaptation to sugar differences between plant host and insect vector (André et al., 2003).

Leafhopper transmission of Spiroplasma citri. Spiroplasmas are acquired by leafhopper vectors that imbibe sap from the sieve tubes of infected plants. However, in order to be transmitted to a plant, the mollicutes need first to multiply in the insect vector after crossing the gut barrier (Wayadande and Fletcher, 1995). They multiply to high titers (10^6–10^7/ml) in the insect hemolymph, but only when they have reached the salivary glands can they be inoculated into a plant. One gene required for efficient transmission, *sc76*, was inactivated in a transposon mutant (G76) with reduced transmissibility (Boutareaud et al., 2004); *sc76* encodes a putative lipoprotein. Plants infected with the G76 mutant showed symptoms 4–5 weeks later than those infected with wild-type GII3, but when they appeared, the symptoms induced were severe. Mutant G76 multiplied in plants and leafhoppers as efficiently as the wild-type strain. However, leafhoppers injected with the wild-type spiroplasma transmitted the spiroplasma to 100% of exposed plants. In contrast, those injected with mutant G76 infected only 50% of the plants. This inefficiency was shown to be associated with a numerical decrease in spiroplasma cells in the salivary glands that correlated with reduced output from the stylets of transmitting leafhoppers; the number of mutant cells transmitted through Parafilm membranes was less than 5% of numbers of wild-type cells transmitted based on colony-forming units. Functional complementation of the G76 mutant with the *sc76* gene restored the wild-type phenotype. Because both wild-type and mutant cells multiplied to equally high titers in the hemolymph, the results suggest that the mutant is inefficiently passed from the hemolymph into the salivary glands or that it may multiply to a lower titer in the glands.

Transmission of *Spiroplasma citri* by leafhopper vectors must involve adherence to and invasion of insect host cells. Electron microscopic studies of leafhopper midgut by Ammar et al. (2004) have demonstrated the attachment of *Spiroplasma kunkelii* cells by a tip structure to the cell membrane between microvilli of epithelial cells. *Spiroplasma citri* surface protein P89 was shown to mediate adhesion of the spiroplasma to cells of the vector *Circulifer tenellus* and was designated SARP1 (Berg et al., 2001; Yu et al., 2000). The gene encoding SARP1, *arp1*, was cloned and characterized from *Spiroplasma citri* BR3-T. The putative gene product SARP1 contains a novel domain at the N terminus, called "sarpin" (Berg et al., 2001). The *arp*1 gene is located on plasmid pBJS-O in *Spiroplasma citri* (Joshi et al., 2005). The *Spiroplasma kunkelii* plasmid pSKU146 encodes an adhesin that is a homolog of SARP1 (Davis et al., 2005). Other spiroplasma plasmids encode additional adhesin-related proteins. As indicated above (see *Plasmids*), *Spiroplasma citri* GII3 contains six large plasmids, pSci1 to pSci6 (Saillard et al., 2008). Although plasmids pSci1 to pSci5 encode eight different *Spiroplasma citri* adhesin-related proteins (ScARPs), they are not required for insect transmission (Berho et al., 2006b). One of the ScARPs, protein P80, shared 63% similarity and 45% identity with SARP1. Protein P80 is carried by plasmid pSci4 and has been named ScARP4a. The ScARP-encoding genes could not be detected in DNA from non-transmissible strains (Berho et al., 2006b). Sequence alignments of ScARP proteins revealed that they share common features including a conserved signal peptide followed by six to eight repeats of 39–42 amino acids each, a central conserved region of 330 amino acids, and a transmembrane domain at the C terminus (Saillard et al., 2008).

Plasmid pSci6 carries the gene for protein P32, which is present in all *Spiroplasma citri* strains capable of being transmitted by the leafhopper vector *Circulifer haematoceps*, but absent from all non-transmissible strains (Killiny et al., 2006). Complementation studies with P32 alone did not restore transmissibility (Killiny et al., 2006). However, if the pSci6 plasmid was transferred to an insect-non-transmissible *Spiroplasma citri* strain, then the phenotype could be converted to insect-transmissible, indicating the likely presence of additional transmissibility factors on pSci6 (Berho et al., 2006a). Indeed, recent data indicates that factors essential for transmissibility are encoded by a 10 kbp fragment of pSci6 (Breton et al., 2010). The finding that the insect-transmissible strain *Spiroplasma citri* Alc254 contains only a single plasmid, pSci6 (S. Richard and J. Renaudin, unpublished) also reinforces the hypothesis that pSci6-encoded determinants play a key role in insect transmission of *Spiroplasma citri* by its leafhopper vector.

Enrichment and isolation procedures

Isolation. Success in the isolation of fastidious spiroplasmas is influenced strongly by the titer of the inoculum. Spiroplasmas have been isolated from salivary glands, gut, and nerve tissues of their insect hosts. Many spiroplasmas envisioned by dark-field microscopy have proved to be noncultivable (Hackett and Clark, 1989). Initial insect extracts in growth media are passed through a 0.45 µm filter. The filtrate is then observed daily for pH indicator change. An alternative to filtration involves the use of antibiotics or other inhibitors (Grulet et al., 1993; Markham et al., 1983; Whitcomb et al., 1973). Spiroplasma isolations from infected plants are best obtained from sap expressed from vascular bundles of hosts showing early disease symptoms. Plant sap often contains spiroplasmal substances (Liao et al., 1979) whose presence in primary cultures may necessitate blind passage or serial dilution.

Isolation media. M1D medium (Whitcomb, 1983) has been used for primary isolations of the large proportion of spiroplasma species. SP-4 medium, a rich formulation derived from experiments with M1D, is necessary for isolation of *Spiroplasma mirum* from fluids of the embryonated egg (Tully et al., 1982). SP-4 medium is also required for isolation of *Spiroplasma ixodetis* (Tully et al., 1981). Some very fastidious spiroplasmas such as *Spiroplasma poulsonii* (Hackett et al., 1986) and *Spiroplasma leptinotarsae* (Hackett and Lynn, 1985) were isolated by cocultivation with insect cells. However, the requirement for cocultivation of *Spiroplasma leptinotarsae* can be circumvented by placing the primary cultures in BBL anaerobic GasPak jar systems with low redox potential and enhanced CO_2 atmosphere (Konai et al., 1996b). By lowering the pH of the growth medium from 7.4 to 6.2 and using bromocresol purple as a pH indicator (pH 5.2 yellow to pH 6.8 purple), it was possible to perform metabolism inhibition tests involving *Spiroplasma leptinotarsae* as the antigen. The same low-pH medium containing 2.0% Noble agar permitted the growth of colonies (Williamson, unpublished data). Cohen and Williamson (1988) reported that a fortuitous contamination of H-2 medium by a slow-growing, pink-colored yeast (*Rhodotorula rubra*) permitted primary isolation of the non-male-lethal variant of the *Dorsophila willistoni* spiroplasma. After 10–12 passages with yeast, the spiroplasmas were able to grow in yeast-free H-2 medium.

Maintenance procedures. Adaptation. Most spiroplasmas can be adapted to a wide variety of media formulations. Spiroplasmas commonly grow more slowly upon transfer to new media. Initial reduction in growth rate is probably related to a combination of differences in nutrients, pH, osmolality, etc. Isolates may grow at only slightly reduced rates during the first 1–5 passages in a new medium. However, if the new medium is markedly deficient, the growth rate may decrease precipitously after 5–10 passages. Continuous careful passaging may result in growth rate recovery to levels similar to that in the initial medium. For such adaptations, best results are achieved by starting with a 1:1 ratio of old and new media and gradually withdrawing the old formulation. *Spiroplasma clarkii*, after continuous passage for hundreds of generations, finally adapted to extremely simple media (Hackett et al., 1994). Adaptation may involve mutation and/or activation of adaptive enzymes, or, possibly, other mechanisms. Growth rates in such simple media were much slower than those in rich media.

Maintenance media. *Spiroplasma citri* can be cultivated in a relatively simple medium that utilizes sorbitol to maintain osmolality (Saglio et al., 1971). A modification of this medium (BSR) has been used extensively for *Spiroplasma citri* (Bové and Saillard, 1979), in which the horse serum content was lowered to 10% and the fresh yeast extract was omitted. Other simple media, such as C-3G (Liao and Chen, 1977), are suitable for maintenance or large-batch cultivation of fast-growing spiroplasmas. This medium is also adequate for primary isolation of *Spiroplasma kunkelii* (Alivizatos, 1988). However, cultivation of more fastidious spiroplasmas is best achieved in M1D medium (Hackett and Whitcomb, 1995; Whitcomb, 1983; Williamson and Whitcomb, 1975) if they derive from plant or insect habitats. SP-4 medium (Tully et al., 1977) is very suitable if spiroplasmas derive from tick habitats. SM-1 medium (Clark, 1982) has also been successfully employed for many insect spiroplasmas.

Defined media. *Spiroplasma floricola* and some strains of *Spiroplasma apis* have been cultivated in chemically defined media (Chang, 1989, 1982).

Preservation. Spiroplasmas are routinely preserved by lyophilization (FAO/WHO, 1974). Most spiroplasmas can be maintained at −70°C indefinitely. Preservation success at −20°C is irregular and uncertain.

Differentiation of the genus *Spiroplasma* from other closely related taxa

Spiroplasmas can be clearly differentiated from all other microorganisms by their unique properties of helicity and motility, combined with the complete absence of periplasmic fibrils, cell walls, or cell wall precursors. However, spiroplasmas may be nonhelical under some environmental conditions or when cultures are in the stationary phase of growth. Morphological study of the organisms in the exponential phase of growth usually reveals characteristic helical forms. However, the existence of spiroplasmas that appear entirely or largely as nonhelical forms (e.g., *Spiroplasma ixodetis* and group XXIII strain TIUS-1) raises the theoretical possibility that an organism situated at an apomorphic (advanced) position on the spiroplasma phylogenetic tree could totally lack helicity or motility. In fact, the clade containing *Mycoplasma mycoides* and the *Entomoplasmataceae* has apparently done exactly that. *Spiroplasma floricola* produces nonhelical, but viable, cells early in stationary phase, which can begin within 24 h of medium inoculation. For reasons such as this, it is necessary to examine cultures throughout the growth cycle to ensure that an adequate search for helical cells has been made.

Taxonomic comments

Early history. The term "spiroplasma" was first coined as a trivial term to describe helical organisms shown to be associated with corn stunt disease (Davis et al., 1972a, b) that could not, at that time, be cultivated (Davis and Worley, 1973). Shortly thereafter, when similar organisms associated with citrus stubborn disease were characterized (Saglio et al., 1973), the trivial term was adopted as the generic name and the stubborn organism was named *Spiroplasma citri*. This species was the first cultured spiroplasma and the first cultured mollicute of plant origin. Shortly after the stubborn agent was named, the genus *Spiroplasma* was elevated to the status of a family (Skripal, 1974) and added to the Approved Lists of Bacterial Names (Skripal, 1983). The organism that was eventually named *Spiroplasma mirum* (Tully et al., 1982) was isolated by Clark (1964) in embryonated chicken eggs soon after the discovery of the organism later named *Spiroplasma poulsonii*. Because *Spiroplasma mirum* readily passed through filters, it was first mistaken for a virus. The subgroup I-4 277F spiroplasma was cultivated in 1968, but was mistaken for a spirochete (Pickens et al., 1968). The first organism to be initially recognized as a spiroplasma was *Spiroplasma kunkelii*, which was envisioned by dark-field and electron microscopy in 1971–1972 and cultivated in 1975 (Liao and Chen, 1977; Williamson and Whitcomb, 1975). More than a decade passed before Clark (1982) showed that spiroplasmas, many of them fast-growing, occurred principally in insects.

Species concept. The species concept in spiroplasmas, as in all bacteria, was based on DNA–DNA reassociation (ICSB Subcommittee on the Taxonomy of *Mollicutes*, 1995; Johnson, 1994; Rosselló-Mora and Amann, 2001; Stackebrandt et al., 2002; Wayne et al., 1987). In practice, DNA–DNA reassociation results with spiroplasmas have proven difficult to standardize. Estimates of reassociation between *Spiroplasma citri* (subgroup I-1) and *Spiroplasma kunkelii* (subgroup I-3) varied between 30 and 70%, depending on the method employed and the degree of stringency (Bové and Saillard, 1979; Christiansen et al., 1979; Lee and Davis, 1980; Liao and Chen, 1981a; Rahimian and Gumpf, 1980). Given these challenges, an alternative method was identified in serology. Surface serology of spiroplasmas has proven to be a robust surrogate for DNA–DNA hybridization assays.

Phylogeny. Phylogenetic studies of *Spiroplasma* became possible when Carl Woese and colleagues, searching for a molecular chronometer by which microbial evolution could be reconstructed, found that rRNA met most or all of the desired criteria (reviewed by Woese, 1987). Today, sequencing of rRNA genes has become a universal tool for phylogenetic reconstruction. Early phylogenetic analyses involved distance estimates (DeSoete, 1983). Later, neighbor-joining (Saitou and Nei, 1987) was introduced into mollicute phylogeny (Maniloff, 1992) and several mollicute workers have used maximum-likelihood (Felsenstein, 1993). The extensive and classical studies of K.-E. Johansson's group (Johansson et al., 1998; Pettersson et al., 2000) were completed using neighbor-joining, but selectively confirmed by maximum-likelihood and maximum-parsimony (Swofford, 1998). Gasparich et al. (2004) studied the phylogeny of *Spiroplasma* and its nonhelical descendants using parsimony, maximum-likelihood, distance, and neighbor-joining analyses, which generated 24 phylogenetic inferences that were common to all, or almost all, of the trees. More recently, Bayesian analysis [MrBayes (http://mrbayes.csit.fsu.edu/index.php)] was used to examine an expanded *Spiroplasma* Apis clade based on 16S rRNA and 16S–23S ITS sequences; the analyses showed congruency between Bayesian and maximum-parsimony trees (Jandhyam et al., 2008).

Woese et al. (1980) presented a 16S rRNA gene-based phylogenetic tree for *Mollicutes*, including *Spiroplasma*, indicating that these wall-less bacteria were related to members of the phylum *Firmicutes* such as *Lactobacillus* spp. and *Clostridium innocuum*. The tree suggested that *Mollicutes* might be monophyletic. However, a later study by Weisburg et al. (1989) with 40 additional species of *Mollicutes* including ten spiroplasmas, failed to confirm the monophyly of *Mollicutes* at the deepest branching orders. The Woese et al. (1980) model also suggested that the genus *Mycoplasma* might not be monophyletic, in that the type species, *Mycoplasma mycoides*, and two related species, *Mycoplasma capricolum* and *Mycoplasma putrefaciens*, appeared to be more closely related to the Apis clade of *Spiroplasma* than to the other *Mycoplasma* species. This conclusion was supported by analyses of the 5S rRNA genes (Rogers et al., 1985). All trees so far obtained indicate that the acholeplasma-anaeroplasma (*Acholeplasmatales–Anaeroplasmatales*) and spiroplasma-mycoplasma (*Mycoplasmatales–Entomoplasmatales*) lineages are monophyletic, but are separated by an ancient divergence.

In-depth analysis of characterized spiroplasmas and their nonhelical descendants indicates the existence of four major clades within the monophyletic spiroplasma-mycoplasma lineage (Gasparich et al., 2004; Figure 113). One of the four clades consists of the nonhelical species of the mycoides group (as defined by Johansson, 2002) as well as the six species of *Entomoplasma* and twelve species of *Mesoplasma* (the *Entomoplasmataceae*); this assemblage was designated the Mycoides-Entomoplasmataceae clade. The analyses indicated that the remaining three clades represented *Spiroplasma* species. One of these clades, the Apis clade, was found to be a sister to the Mycoides-Entomoplasmataceae clade. The Apis clade contains a large number of species from diverse insect hosts, many of which possess life cycles

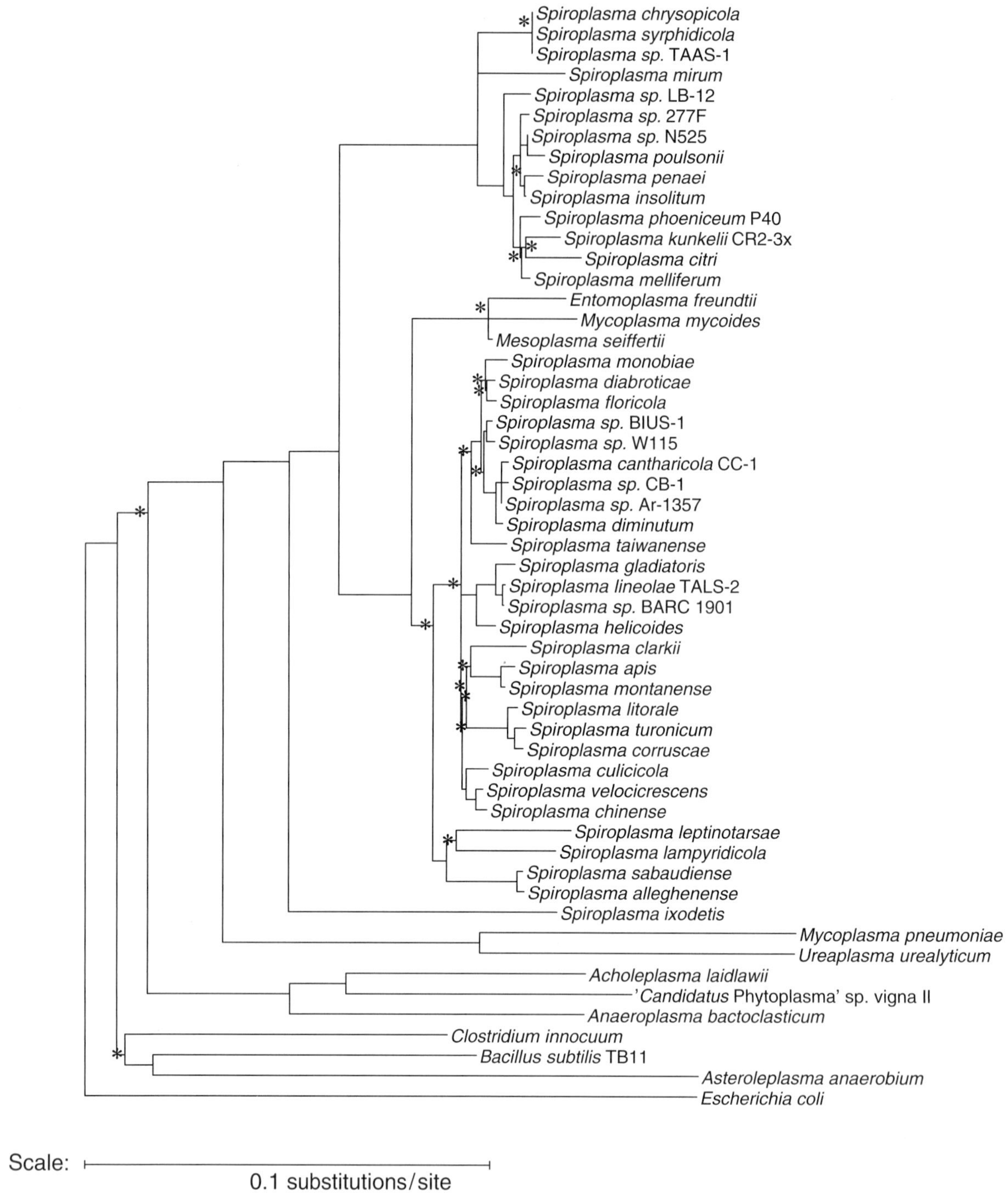

FIGURE 113. Phylogenetic relationships of members of the class *Mollicutes* and selected members of the phylum *Firmicutes*. The phylogram was based on a Jukes-Cantor corrected distance matrix and weighted neighbor-joining analysis of the 16S rRNA gene sequences of the type strains, except where noted. *Escherichia coli* was the outgroup. Bootstrap values (100 replicates) <50% are indicated (*). The GenBank accession numbers for 16S rRNA gene sequences used are: *Mycoplasma mycoides* (U26039); *Mycoplasma pneumoniae* (M29061); *Entomoplasma freundtii* (AF036954); *Mesoplasma seiffertii* (AY351331); *Spiroplasma apis* (M23937); *Spiroplasma clarkii* (M 24474); *Spiroplasma gladiatoris* (M24475); *Spiroplasma taiwanense* (M24476); *Spiroplasma monobiae* (M24481); *Spiroplasma diabroticae* (M24482); *Spiroplasma melliferum* (AY325304); *Spiroplasma citri* (M23942); *Spiroplasma mirum* (M24662); *Spiroplasma ixodetis* (M24477); *Spiroplasma* sp. strain N525 (DQ186642); *Spiroplasma poulsonii* (M24483); *Spiroplasma penaei* (AY771927); *Spiroplasma phoeniceum* (AY772395); *Spiroplasma kunkelii* (DQ319068); *Spiroplasma cantharicola* (DQ861914); *Spiroplasma lineolae* (DQ860100); *Spiroplasma* sp. strain 277F (AY189312); *Spiroplasma* sp. strain LB-12 (AY189313); *Spiroplasma insolitum* (AY189133); *Spiroplasma floricola* (AY189131); *Spiroplasma syrphidicola* (AY189309); *Spiroplasma chrysopicola* (AY189127); *Spiroplasma* sp. strain TAAS-1 (AY189314); *Spiroplasma culicicola* (AY189129); *Spiroplasma velocicrescens* (AY189311); *Spiroplasma sabaudiense* (AY189308); *Spiroplasma corruscae* (AY189128); *Spiroplasma* sp. strain CB-1 (AY189315); *Spiroplasma* sp. strain Ar-1357 (AY189316); *Spiroplasma turonicum* (AY189310); *Spiroplasma litorale* (AY189306); *Spiroplasma lampyridicola* (AY189134); *Spiroplasma leptinotarsae* (AY189305); *Spiroplasma* sp. strain W115 (AY189317); *Spiroplasma chinense* (AY189126); *Spiroplasma diminutum* (AY189130); *Spiroplasma alleghenense* (AY189125); *Spiroplasma* sp. strain BIUS-1 (AY189319); *Spiroplasma montanense* (AY189307); *Spiroplasma helicoides* (AY189132); *Spiroplasma* sp. strain BARC 1901 (AY189320); *Ureaplasma urealyticum* (M23935); "*Candidatus* Phytoplasma" sp. Vigna II (AJ289195); *Acholeplasma laidlawii* (M23932); *Anaeroplasma bactoclasticum* (M25049); *Clostridium innocuum* (M23732); *Asteroleplasma anaerobium* (M22351); *Bacillus subtilis* (AF058766); *Escherichia coli* (J01859).

involving transmission between the guts of insects and plant surfaces. One of these species, *Spiroplasma* sp. TIUS-1 (group XXVIII) has very poor helicity and a genome size of 840 kbp, smaller than that of most other spiroplasmas. This species diverged from the spiroplasma lineage close to the node of entomoplasmal divergence and can be envisioned as a "missing link" in the evolutionary development of the Mycoides-Entomoplasmataceae clade. The other two *Spiroplasma* clades are the monospecific Ixodetis clade (group VI) and the Citri-Chrysopicola-Mirum clade (with representatives from groups I, II, V, and VIII). The Citri-Chrysopicola-Mirum clade contains *Spiroplasma mirum*, *Spiroplasma poulsonii*, the three subgroups of the Chrysopicola (group VIII) clade, and the nine subgroups of the Citri (group I) clade. Members of group I and group VIII show close intragroup relationships, as indicated by the similarities of their 16S rRNA gene sequences (Gasparich et al., 2004). DNA–DNA reassociation studies for group I (Bové et al., 1983, 1982; Junca et al., 1980) spiroplasmas supported the subgroup cluster. The Chrysopicola clade (group VIII) subgroups have met a different fate. Although their DNA–DNA similarities in reassociation procedures were slightly less than 70%, their 16S rRNA gene sequence similarities were >99% (Gasparich et al., 1993c). The strains of this group, including not only the subgroups, but a plethora of isolates from the same ecological context, appear to form a matrix of interrelated strains. Boundaries that seemed clear when the subgroups were initially described, eventually eroded beyond recognition. The 16S rRNA gene sequence similarities are too high to permit cladistic analysis and even 16S–23S rRNA spacer region sequence analysis failed to resolve the existing subgroups (Regassa et al., 2004). Over time, the concept of the microbial species has undergone a subtle change. It is now recognized (Rosselló-Mora and Amann, 2001; Stackebrandt et al., 2002) that microbial species must at times consist of strain clusters that may contain species with <70% similarity as determined by DNA–DNA reassociation. Group VIII spiroplasmas may comprise such a cluster and efforts to subdivide this cluster may have been inadvisable.

Character mapping of non-genetic features has been completed in conjunction with phylogenetic analyses (Gasparich et al., 2004). Serological classifications of spiroplasmas are generally supported by the trees, but the resolution of genetic analyses appears to be much greater than that of serology. Genome size and G+C content were moderately conserved among closely related strains. Apparent conservation of slower growth rates in some clades was most likely attributable to host affiliation; spiroplasmas of all groups that were well adapted to a specific host had slower growth rates. Sterol requirements were polyphyletic, as was the ability to grow in the presence of PES, but not serum.

Acknowledgements

We gratefully acknowledge J. Dennis Pollack for assistance on sections concerning intermediary metabolism.

Further reading

Whitcomb, R.F. and J.G. Tully (editors). 1989. The Mycoplasmas, vol. 5, Spiroplasmas, acholeplasmas, and mycoplasmas of plants and arthropods. Academic Press, New York.

List of species of the genus *Spiroplasma*

1. **Spiroplasma citri** Saglio, L'Hospital, Laflèche, Dupont, Bové, Tully and Freundt 1973, 202[AL]

cit'ri. L. masc. n. *citrus* the citrus; N.L. masc. n. *Citrus* generic name; N.L. gen. n. *citri* of *Citrus*, to denote the plant host.

Cells are helices that divide in mid-exponential phase when they have four turns. Helical filaments are usually 100–200 nm in diameter and 2–4 µm in length. Cells are longer in late exponential phase and early stationary phase. Nonviable cells in late exponential phase are nonhelical.

Colonies on solid media containing 20% horse serum and 0.8% Noble agar (Difco) are umbonate, 60–150 µm in diameter. Moderate turbidity is produced in liquid cultures. Biological properties are listed in Table 142.

Serologically distinct from other *Spiroplasma* species, groups, and subgroups, but shares some cross-reactivity with members of other group I subgroups. Has close phylogenetic affinities with other group I members, and with *Spiroplasma poulsonii* in trees constructed using 16S rRNA gene sequences.

Pathogenic for citrus plants and a variety of plant hosts (aster, periwinkle, broad bean) following transmission by infected insects (leafhoppers).

DNA–DNA renaturation experiments confirm serological data that indicate that the differences between the type strain (subgroup I-1) and other subgroups of group I are great enough to warrant its designation as a distinct species. The genome size is 1820 kbp (PFGE).

Source: isolated from leaves, seed coats, and fruits of citrus plants (orange and grapefruit) infected with stubborn disease, and from other naturally infected plants (e.g., periwinkle, horseradish or brassicaceous weeds) or insects. Known from Mediterranean and other warm climates of Europe, North Africa, Near and Middle East, and the Western United States (California and Arizona).

DNA G+C content (mol%): 25–27 (T_m, Bd).

Type strain: ATCC 27556, Morocco strain, R8-A2.

Sequence accession no. (16S rRNA gene): M23942.

2. **Spiroplasma alleghenense** Adams, Whitcomb, Tully, Clark, Rose, Carle, Konai, Bové, Henegar and Williamson 1997, 762[VP]

al.le.ghen.en'se. N.L. neut. adj. *alleghenense* of the Allegheny Mountains, referring to the geographic origin of the type strain, the range of the Appalachian Mountains from which it was derived.

Cells are motile helical filaments, 100–300 nm in diameter. Under many growth conditions, cells in medium are deformed. Colonies on solid medium containing 3.0% Noble agar are small and granular and never have a fried-egg appearance. Biological properties are listed in Table 142.

Serologically distinct from other *Spiroplasma* species, groups, and subgroups. When tested as an antigen, cross-reacts broadly with many nonspecific sera (one-way reaction). Has close phylogenetic relationship to *Spiroplasma sabaudiense* (group XIII) and strain TIUS-1 (group XXVIII) in trees constructed using 16S rRNA gene sequences. The genome size is 1,465 kbp (PFGE).

Source: isolated from the hemolymph of a common scorpion fly, *Panorpa helena* in West Virginia, USA.

DNA G+C content (mol%): 31 ± 1 (T_m, Bd).

Type strain: ATCC 51752, PLHS-1.

Sequence accession no. (16S rRNA gene): AY189125.

3. **Spiroplasma apis** Mouches, Bové, Tully, Rose, McCoy, Carle-Junca, Garnier and Saillard 1984b, 91[VP] (Effective publication: Mouches, Bové, Tully, Rose, McCoy, Carle-Junca, Garnier and Saillard 1983a, 383.)

a'pis. L. fem. n. *apis*, *-is* a bee, and also the genus name of the honey bee, *Apis mellifera*; L. gen. n. *apis* of a bee, of *Apis mellifera*, the insect host for this species.

The morphology is as described for the genus. Helical filaments are usually 100–150 nm in diameter and 3–10 µm in length. Colonies on solid medium containing 20% fetal bovine serum and 0.8% Noble agar (Difco) are usually diffuse, rarely exhibiting central zones of growth into the agar. Colonies on solid medium with 2.25% Noble agar and 1–5% bovine serum fraction are smaller, but exhibit central zones of growth into the agar and some peripheral growth on the agar surface around the central zones. Marked turbidity is produced during growth in most spiroplasma media (BSR, M1A, SP-4). Biological properties are listed in Table 142.

Serologically distinct from other *Spiroplasma* species, groups, and subgroups. Many strains show partial cross-reactions when tested against sera to strain B31[T] (Tully et al., 1980). These strains show more than 80% DNA–DNA reassociation with strain B31[T], but their exact taxonomic status is unclear. Some strains show a very low level reciprocal cross-reaction with *Spiroplasma montanense* in deformation serology. In accordance with serology, *Spiroplasma apis* and *Spiroplasma montanense* are sister species in phylogenetic trees constructed using 16S rRNA gene sequences. The genome size is 1300 kbp (PFGE).

Etiologic agent of May disease of honey bees in southwestern France. Various strains of the organism exhibit experimental pathogenicity for young honey bees in feeding experiments.

Source: isolated from honey bees (*Apis mellifera*) and from flower surfaces in widely separated geographic regions (France, Corsica, Morocco, USA).

DNA G+C content (mol%): 29–31 (T_m, Bd).

Type strain: ATCC 33834, B31.

Sequence accession no. (16S rRNA gene): AY736030.

4. **Spiroplasma atrichopogonis** Koerber, Gasparich, Frana and Grogan 2005, 291[VP]

a.tri.cho.po.go'nis. N.L. gen. n. *atrichopogonis* of *Atrichopogon*, systematic genus name of a biting midge (Diptera: Ceratopogonidae).

The morphology is as described for the genus. Cells are helical and motile. Biological properties are listed in Table 142.

Serologically distinct from previously established *Spiroplasma* species, groups, and subgroups. The genome size has not been determined.

Source: isolated from a pooled sample of two nearly identical species of biting midges (*Atrichopogon geminus* and *Atrichopogon levis*).

DNA G+C content (mol%): 28.8 ± 1 (T_m).

Type strain: ATTC BAA-520, NBRC 100390, GNAT3597.

Sequence accession no. (16S rRNA gene): not available.

5. **Spiroplasma cantharicola** Whitcomb, Chastel, Abalain-Colloc, Stevens, Tully, Rose, Carle, Bové, Henegar, Hackett, Clark, Konai and Williamson 1993a, 423[VP]

can.thar.i'co.la. Gr. *kantharos* scarab beetle; L. suff. *-cola* (from L. n. *incola*) inhabitant, dweller; N.L. n. *cantharicola* an inhabitant of a family of beetles.

The morphology is as described for the genus. Cells are helical and motile. Colonies on solid medium containing 0.8% Noble agar are diffuse, without fried-egg morphology. Biological properties are listed in Table 142.

Serologically distinct from other *Spiroplasma* species, groups, and subgroups, but shares some reciprocal cross-reactivity with members of other group XVI subgroups.

Not yet classified phylogenetically, but no doubt closely related to subgroups XVI-2 and XVI-3, which are sisters forming a clade related to *Spiroplasma diminutum* in phylogenetic trees constructed using 16S rRNA gene sequences. Moreover, DNA–DNA renaturation experiments confirm that the differences between the type strain and other subgroups of group XVI are great enough to warrant its designation as a distinct species. The genome size is 1320 kbp (PFGE).

Source: isolated from the gut of an adult cantharid beetle (*Cantharis carolinus*) in Maryland, USA. Based on its residence in the gut of a flower-visiting insect, this species is thought to be transmitted on flowers.

DNA G+C content (mol%): 26 ± 1 (T_m, Bd, HPLC).

Type strain: ATCC 43207, CC-1.

Sequence accession no. (16S rRNA gene): DQ861914.

6. **Spiroplasma chinense** Guo, Chen, Whitcomb, Rose, Tully, Williamson, Ye and Chen 1990, 424[VP]

chi.nen'se. N.L. neut. adj. *chinense* of China, the location where the organism was first isolated.

The morphology is as described for the genus. Cells are motile helical filaments ~160 nm in diameter. Colonies on solid medium containing 0.8–1.0% Noble agar are diffuse with many small satellite colonies; growth on 2.25% agar produces smaller rough or granular colonies and fewer satellite forms. Biological properties are listed in Table 142.

Serologically distinct from other *Spiroplasma* species, groups, and subgroups. Phylogenetically, this species is closely related to *Spiroplasma velocicrescens* in phylogenetic trees constructed using 16S rRNA gene sequences. The genome size is 1530 kbp (PFGE).

Source: isolated from flower surfaces of bindweed (*Calystegia hederacea*) in Jiangsu, People's Republic of China.

DNA G+C content (mol%): 29 ± 1 (T_m).

Type strain: ATCC 43960, CCH.

Sequence accession no. (16S rRNA gene): AY189126.

7. **Spiroplasma chrysopicola** Whitcomb, French, Tully, Gasparich, Rose, Carle, Bové, Henegar, Konai, Hackett, Adams, Clark and Williamson 1997b, 718[VP]

chry.so.pi′co.la. N.L. n. *Chrysops* a genus of deer flies in the Tabanidae; L. suff. *-cola* (from L. n. *incola*) inhabitant, dweller; N.L. n. *chrysopicola* inhabiting *Chrysops* spp.

Helical motile filaments are short and thin, passing a 220 nm filter quantitatively. Grows to titers as high as 10^{11}/ml. Colonies on solid medium containing 2.25% Noble agar have dense centers and smooth edges (a fried-egg appearance) and do not have satellites. Biological properties are listed in Table 142.

Serologically distinct from other *Spiroplasma* species, groups, and subgroups, but exhibits some reciprocal or one-way cross-reactivity with members of other group VIII subgroups. Some strains of group VIII spiroplasmas may be difficult to identify to subgroup. Shares less than 70% DNA–DNA reassociation with *Spiroplasma syrphidicola* and strain TAAS-1 (subgroup VIII-3). Phylogenetically, this species is closely related to other group VIII strains in trees constructed using 16S rRNA gene sequences. The 16S rRNA gene similarity coefficients of group VIII spiroplasmas are >0.99, so this gene is insufficient for distinguishing species in group VIII. The genome size is 1270 kbp (PFGE). Pathogenicity for insects has not been determined.

Source: isolated from the gut of a deer fly (*Chrysops* sp.) in Maryland, USA. Other strains from deer flies have been collected from as far west as Wyoming, from New England, and very rarely, as far south as Georgia, USA.

DNA G+C content (mol%): 30 ± 1 (Bd).

Type strain: ATCC 43209, DF-1.

Sequence accession no. (16S rRNA gene): AY189127.

8. **Spiroplasma clarkii** Whitcomb, Vignault, Tully, Rose, Carle, Bové, Hackett, Henegar, Konai and Williamson 1993c, 264[VP]

clar′ki.i. N.L. masc. gen. n. *clarkii* of Clark, in honor of Truman B. Clark, a pioneer spiroplasma ecologist.

The morphology is as described for the genus. The helical motile filaments remain stable throughout exponential growth. Colonies on solid medium containing 0.8% Noble agar are diffuse, without fried-egg morphology. Biological properties are listed in Table 142.

Serologically distinct from other *Spiroplasma* species, groups, and subgroups. Phylogenetically, this species is placed in the classical Apis cluster of spiroplasmas, but it does not have an especially close neighbor in trees constructed using 16S rRNA gene sequences. The genome size is 1720 kbp (PFGE). Pathogenicity for insects has not been determined.

Source: isolated from the gut of a larval scarabaeid beetle (*Cotinus nitida*) in Maryland, USA.

DNA G+C content (mol%): 29 ± 1 (T_m, Bd, HPLC).

Type strain: ATCC 33827, CN-5.

Sequence accession no. (16S rRNA gene): M24474.

9. **Spiroplasma corruscae** Hackett, Whitcomb, French, Tully, Gasparich, Rose, Carle, Bové, Henegar, Clark, Konai, Clark and Williamson 1996c, 949[VP]

cor.rus′cae. N.L. gen. n. *corruscae* of *corrusca*, referring to the species of firefly beetle (*Ellychnia corrusca*) from which the organism was first isolated.

The morphology is as described for the genus. Cells are helical and motile. Colonies on solid medium containing 2.25% Noble agar are slightly diffuse to discrete and generally without the characteristic fried-egg morphology. Biological properties are listed in Table 142.

Serologically distinct from previously established *Spiroplasma* species, groups, and subgroups. Phylogenetically, closely related to *Spiroplasma turonicum* and *Spiroplasma litorale* in trees constructed using 16S rRNA gene sequences. The genome size has not been determined.

Source: isolated from the gut of an adult lampyrid beetle (*Ellychnia corrusca*) in Maryland in early spring, but found much more frequently in horse flies in summer months. Other strains have been collected from Canada and Georgia, Connecticut, South Dakota, and Texas, USA.

DNA G+C content (mol%): 26 ± 1 (T_m, Bd).

Type strain: ATCC 43212, EC-1.

Sequence accession no. (16S rRNA gene): AY189128.

10. **Spiroplasma culicicola** Hung, Chen, Whitcomb, Tully and Chen 1987, 368[VP]

cu.li.ci′co.la. L. n. *culex*, *-icis* a gnat, midge, and also a genus of mosquitoes (*Culex*, family Culicidae); L. suffix *-cola* (from L. n. *incola*) inhabitant, dweller; N.L. n. *culicicola* intended to mean an inhabitant of the Culicidae.

Cells are pleomorphic, but are commonly very short motile helices, 1–2 μm in length. Colonies on solid medium containing 1% Noble agar have a fried-egg appearance with satellites. Biological properties are listed in Table 142.

Serologically distinct from other *Spiroplasma* species, groups, and subgroups. Phylogenetically, this species is placed in the classical Apis cluster of spiroplasmas, but does not have an especially close neighbor in trees constructed using 16S rRNA gene sequences. The genome size is 1350 kbp (PFGE).

Source: isolated from a triturate of a salt marsh mosquito (*Aedes sollicitans*) collected in New Jersey, USA.

DNA G+C content (mol%): 26 ± 1 (T_m, Bd).

Type strain: ATCC 35112, AES-1.

Sequence accession no. (16S rRNA gene): AY189129.

11. **Spiroplasma diabroticae** Carle, Whitcomb, Hackett, Tully, Rose, Bové, Henegar, Konai and Williamson 1997, 80[VP]

di.a.bro.ti′cae. N.L. gen. n. *diabroticae* of *Diabrotica*, referring to *Diabrotica undecimpunctata*, the chrysomelid beetle from which the organism was isolated.

The morphology is as described for the genus. Cells are helical, motile filaments, 200–300 nm in diameter. Colonies on solid medium containing 0.8% Noble agar are diffuse, without fried-egg morphology. Biological properties are listed in Table 142.

Serologically distinct from other established *Spiroplasma* species, groups, and subgroups. Phylogenetically, closely related to *Spiroplasma floricola* in trees constructed using 16S rRNA gene sequences. The genome size is 1350 kbp (PFGE).

Source: isolated from the hemolymph of an adult chrysomelid beetle, *Diabroticae undecimpunctata howardi*.

DNA G+C content (mol%): 25 ± 1 (T_m, Bd, HPLC).

Type strain: ATCC 43210, DU-1.

Sequence accession no. (16S rRNA gene): M24482.

12. **Spiroplasma diminutum** Williamson, Tully, Rosen, Rose, Whitcomb, Abalain-Colloc, Carle, Bové and Smyth 1996, 232[VP]

di.min.u'tum. L. v. *deminuere* to break into small pieces, make smaller; L. neut. part. adj. *diminutum* made smaller, reflecting a smaller size.

The morphology is as described for the genus. Cells are short (1–2 µm), helical filaments, 100–200 nm in diameter that appear to be rapidly moving, irregularly spherical bodies when exponential phase broth cultures are examined under dark-field illumination. Colonies on solid medium containing 1.6% Noble agar have dense centers, granular perimeters, and nondistinct edges with satellite colonies. Biological properties are listed in Table 142.

Serologically distinct from other *Spiroplasma* species, groups, and subgroups. Phylogenetically, closely related to group XVI spiroplasmas in trees constructed using 16S rRNA gene sequences. The genome size is 1080 kbp (PFGE).

Source: isolated from a frozen triturate of adult female *Culex annulus* mosquitoes collected in Taishan, Taiwan.

DNA G+C content (mol%): 26 ± 1 (T_m, Bd, HPLC).

Type strain: ATCC 49235, CUAS-1.

Sequence accession no. (16S rRNA gene): AY189130.

13. **Spiroplasma floricola** Davis, Lee and Worley 1981, 462[VP]

flor.i'co.la. L. n. *flos, -oris* a flower; L. suff. *-cola* (from L. n. *incola*) inhabitant, dweller; N.L. n. *floricola* flower-dweller.

The morphology is as described for the genus. Helical cells are 150–200 nm in diameter and 2–5 µm in length. Colonies on solid media have granular central regions surrounded by satellite colonies that probably form after migration of cells from the central focus. Biological properties are listed in Table 142.

Serologically distinct from other *Spiroplasma* species, groups, and subgroups. Phylogenetically, closely related to *Spiroplasma diabroticae* and various flower spiroplasmas in trees constructed using 16S rRNA gene sequences. The genome size of strain OBMG is 1270 kbp. Experimentally pathogenic for insects and embryonated chicken eggs.

Source: isolated from flowers of tulip tree and magnolia trees in Maryland, USA. Other strains have been collected from coleopterous insects.

DNA G+C content (mol%): 25 (T_m).

Type strain: ATCC 29989, 23-6.

Sequence accession no. (16S rRNA gene): AY189131.

14. **Spiroplasma gladiatoris** Whitcomb, French, Tully, Gasparich, Rose, Carle, Bové, Henegar, Konai, Hackett, Adams, Clark and Williamson 1997b, 718[VP]

gla.di.a'to.ris. L. gen. n. *gladiatoris* of a gladiator, reflecting the initial isolation of the organism from the horse fly *Tabanus gladiator.*

Morphology is as described for the genus. Cells are motile helical filaments. Colonies on solid medium containing 3% Noble agar are granular with dense centers and diffuse edges, do not have satellites, and never have a fried-egg appearance. Biological properties are listed in Table 142.

Serologically distinct from other *Spiroplasma* species, groups, and subgroups. This species has a specific antigen, common to several spiroplasmal inhabitants of horse flies, that confers a high level of one way cross-reactivity when it is used as an antigen. Phylogenetically, closely related to two other tabanid spiroplasmas, *Spiroplasma helicoides* and group XXXIV strain B1901, in phylogenetic trees constructed using 16S rRNA gene sequences. The genome size has not been determined.

Source: isolated from the gut of a horse fly (*Tabanus gladiator*) in Maryland, USA. Other strains have been collected at various locations in the southeastern United States.

DNA G+C content (mol%): 26 ± 1 (Bd).

Type strain: ATCC 43525, TG-1.

Sequence accession no. (16S rRNA gene): M24475.

15. **Spiroplasma helicoides** Whitcomb, French, Tully, Gasparich, Rose, Carle, Bové, Henegar, Konai, Hackett, Adams, Clark and Williamson 1997b, 718[VP]

he.li.co.i'des. Gr. n. *helix* spiral; Gr. suff. *-oides* like, resembling, similar; N.L. neut. adj. *helicoides* spiral-like.

The morphology is as described for the genus. Cells are motile helical filaments that lack a cell wall. Colonies on solid medium containing 2.25% Noble agar have dense centers and smooth edges, do not have satellites, and have a perfect fried-egg appearance. Biological properties are listed in Table 142.

Serologically distinct from other *Spiroplasma* species, groups, and subgroups. This species has a specific antigen, common to several spiroplasmal inhabitants of horse flies, that confers a high level one-way cross-reaction when it is used as antigen. Phylogenetically, closely related to two other tabanid spiroplasmas, *Spiroplasma gladiatoris* and *Spiroplasma* sp. BARC 1901, in trees constructed using 16S rRNA gene sequences. Genome size has not been determined.

Source: isolated from the gut of a horse fly *Tabanus abactor* collected in Oklahoma, USA. Other strains have been collected in Georgia, USA.

DNA G+C content (mol%): 26 ± 1 (Bd).

Type strain: ATCC 51746, TABS-2.

Sequence accession no. (16S rRNA gene): AY189132.

16. **Spiroplasma insolitum** Hackett, Whitcomb, Tully, Rose, Carle, Bové, Henegar, Clark, Clark, Konai, Adams and Williamson 1993, 276[VP]

in.so'li.tum. L. neut. adj. *insolitum* unusual or uncommon, to denote unusual base composition.

Cells in exponential phase are long, motile, helical cells that lack true cell walls and periplasmic fibrils. Colonies on solid SP-4 medium containing 0.8 or 2.25% Noble agar are diffuse, with small central zones of growth surrounded by small satellite colonies. Colonies on solid SP-4 medium containing horse serum and 0.8% Noble agar show fried-egg morphology. Biological properties are listed in Table 142.

Serologically distinct from other *Spiroplasma* species and groups, but cross-reacts reciprocally in complex patterns of relatedness with group I subgroups and *Spiroplasma poulsonii.* DNA–DNA renaturation experiments confirm that the differences between the type strain and other subgroups of group I are great enough to warrant its designation as a distinct species. Has close phylogenetic affinities with other group I members and with *Spiroplasma poulsonii* in trees constructed using 16S rRNA gene sequences. The genome size is 1850 kbp (PFGE). Pathogenicity for insects has not been determined.

Source: the type strain was isolated from a fall flower (Asteraceae: *Bidens* sp.) collected in Maryland, USA. Similar isolates have been found in the hemocoel of click beetles. Also isolated from other composite and onagracead flowers and from the guts of many insects visiting these flowers, including cantharid and meloid beetles; syrphid flies; andrenid and megachilid bees; and four families of butterflies.

DNA G+C content (mol%): 28 ± 1 (T_m, Bd).

Type strain: ATCC 33502, M55.

Sequence accession no. (16S rRNA gene): AY189133.

17. **Spiroplasma ixodetis** Tully, Rose, Yunker, Carle, Bové, Williamson and Whitcomb 1995, 27[VP]

ix.o.de'tis. N.L. gen. n. *ixodetis* of *Ixodes*, the genus name of *Ixodes pacificus* ticks, from which the organism was first isolated.

Cells are coccoid forms, 300–500 nm in diameter, straight and branched filaments, or tightly coiled helical organisms. Motility is flexional, but not translational. Colonies on solid medium containing 2.25% Noble agar usually have the appearance of fried eggs. Biological properties are listed in Table 142.

Serologically distinct from other *Spiroplasma* species, groups, and subgroups. Phylogenetically unique; occurs at base of spiroplasma lineage in trees constructed using 16S rRNA gene sequences. The genome size is 2220 kbp (PFGE).

Source: isolated from macerated tissue suspensions prepared from pooled adult *Ixodes pacificus* ticks (Ixodidae) collected in Oregon, USA.

DNA G+C content (mol%): 25 ± 1 (T_m, Bd, HPLC).

Type strain: ATCC 33835, Y32.

Sequence accession no. (16S rRNA gene): M24477.

18. **Spiroplasma kunkelii** Whitcomb, Chen, Williamson, Liao, Tully, Bové, Mouches, Rose, Coan and Clark 1986, 175[VP]

kun.kel'i.i. N.L. masc. gen. n. *kunkelii* of Kunkel, named after Louis Otto Kunkel (1884–1960), to honor his major and fundamental contributions to the study of plant mollicutes.

Cells in exponential phase are helical, motile filaments, 100–150 nm in diameter and 3–10 μm long to nonhelical filaments or spherical cells, 300–800 nm in diameter. Colonies on solid medium containing 0.8% Noble agar are usually diffuse, rarely exhibiting central zones of growth into agar. Colonies on solid C-3G medium containing 5% horse serum or on media containing 2.25% Noble agar frequently have a fried-egg morphology. Biological properties are listed in Table 142.

Serologically distinct from other *Spiroplasma* species, groups, and subgroups, but shares complex patterns of reciprocal cross-reactivity with members of other group I subgroups and *Spiroplasma poulsonii*. DNA–DNA renaturation experiments confirm that the serological differences between the type strain and other subgroups of group I are great enough to warrant its designation as a distinct species. Has close phylogenetic affinities with other group I members and with *Spiroplasma poulsonii* in trees constructed using 16S rRNA gene sequences. The genome size is 1610 kbp (PFGE). Pathogenicity for plants and insects has been experimentally verified.

Source: isolated from maize displaying symptoms of corn stunt disease and from leafhoppers associated with diseased maize, largely in the neotropics.

DNA G+C content (mol%): 26 ± 1 (T_m, Bd).

Type strain: ATCC 29320, E275.

Sequence accession no. (16S rRNA gene): DQ319068 (strain CR2-3x).

19. **Spiroplasma lampyridicola** Stevens, Tang, Jenkins, Goins, Tully, Rose, Konai, Williamson, Carle, Bové, Hackett, French, Wedincamp, Henegar and Whitcomb 1997, 711[VP]

lam.py.ri.di'co.la. N.L. n. *Lampyridae* the firefly beetle family; L. suff. *-cola* (from L. n. *incola*) inhabitant, dweller; N.L. n. *lampyridicola* an inhabitant of members of the Lampyridae.

The morphology is as described for the genus. Cells are motile helical filaments. Colonies on solid medium containing 3.0% Noble agar are small and granular with dense centers, but do not have a true fried-egg appearance. Biological properties are listed in Table 142.

Serologically distinct from other *Spiroplasma* species, groups, and subgroups. When tested as antigen, cross-reacts (one-way) with many specific spiroplasma antisera. Phylogenetically, a sister to *Spiroplasma leptinotarsae* in trees constructed using 16S rRNA gene sequences. The genome size is 1375 kbp (PFGE).

Source: isolated from the gut fluids of a firefly beetle (*Photuris pennsylvanicus*) collected in Maryland, USA. Also known from Georgia and New Jersey, USA.

DNA G+C content (mol%): 26 ± 1 (T_m, Bd).

Type strain: ATCC 43206, PUP-1.

Sequence accession no. (16S rRNA gene): AY189134.

20. **Spiroplasma leptinotarsae** Hackett, Whitcomb, Clark, Henegar, Lynn, Wagner, Tully, Gasparich, Rose, Carle, Bové, Konai, Clark, Adams and Williamson 1996b, 910[VP]

lep.ti.no.tar'sae. N.L. gen. n. *leptinotarsae* of *Leptinotarsa*, referring to *Leptinotarsa decemlineata*, the Colorado potato beetle.

Cells *in vivo* are usually seen in the resting stage, in which they consist of coin-like compressed coils. When placed in fresh medium, these bodies turn immediately into "spring"- or "funnel"-shaped spirals, which are capable of very rapid translational motility. After a relatively small number of passes *in vitro*, this spectacular morphology is lost and the cells return to the modal morphology as described for the genus. Colonies on solid medium containing 2.0% Noble agar are slightly diffuse to discrete and produce numerous satellites. Biological properties are listed in Table 142.

Serologically distinct from other *Spiroplasma* species, groups, and subgroups. When tested as antigen, cross-reacts with many spiroplasma antisera (one-way). Phylogenetically, a sister to *Spiroplasma lampyridicola* in trees constructed using 16S rRNA gene sequences. The genome size is 1,085 kbp (PFGE).

Source: isolated from the gut of Colorado potato beetle (*Leptinotarsa decemlineata*) larvae in Maryland, USA. Also isolated from beetles collected in Maryland, Michigan, New Mexico, North Carolina, Texas, Canada, and Poland.

DNA G+C content (mol%): 25 ± 1 (T_m, Bd, HPLC).

Type strain: ATCC 43213, LD-1.

Sequence accession no. (16S rRNA gene): AY189305.

21. **Spiroplasma leucomae** Oduori, Lipa and Gasparich 2005, 2449[VP]

leu.co'mae. N.L. gen. n. *leucomae* of *Leucoma*, systematic genus name of the white satin moth (Lepidoptera: Lymantriidae), the source of the type strain.

Morphology is as described for the genus. Cells are filamentous, helical, motile, and approximately 150 nm in diameter. They freely pass through filters with pores of 450 and 220 nm, but do not pass through filters with 100 nm pores. Biological properties are listed in Table 142.

Serologically distinct from previously established *Spiroplasma* species, groups, and subgroups. The genome size has not been determined. Pathogenicity for the moth larvae is not known.

Source: isolated from fifth instar satin moth larvae (*Leucoma salicis*).

DNA G+C content (mol%): 24 ± 1 (T_m).

Type strain: ATCC BAA-521, NBRC 100392, SMA.

Sequence accession no. (16S rRNA gene): DQ101278.

22. **Spiroplasma lineolae** French, Whitcomb, Tully, Carle, Bové, Henegar, Adams, Gasparich and Williamson 1997, 1080[VP]

lin.e.o'lae. N.L. n. *lineola* a species of tabanid fly; N.L. gen. n. *lineolae* of *Tabanus lineola*, from which the organism was isolated.

The morphology is as described for the genus. Cells are motile, helical filaments, 200–300 nm in diameter. Colonies on solid medium containing 3% Noble agar are small, granular, and never have a fried-egg appearance. Biological properties are listed in Table 142.

Serologically distinct from other *Spiroplasma* species, groups, and subgroups. Phylogenetic position has not been determined, but its other taxonomic properties suggest that it may be related to other tabanid spiroplasmas of the Apis cluster. The genome size is 1390 kbp (PFGE).

Source: type strain isolated from the viscera of the tabanid fly *Tabanus lineola* collected in coastal Georgia. A strain from *Tabanus lineola* has been collected in Costa Rica (Whitcomb et al., 2007).

DNA G+C content (mol%): 25 ± 1 (T_m, Bd).

Type strain: ATCC 51749, TALS-2.

Sequence accession no. (16S rRNA gene): DQ860100.

23. **Spiroplasma litorale** Konai, Whitcomb, French, Tully, Rose, Carle, Bové, Hackett, Henegar, Clark and Williamson 1997, 361[VP]

li.to.ra'le. L. neut. adj. *litorale* of the shore or coastal area.

The morphology is as described for the genus. Cells are motile, helical filaments. Colonies on solid medium containing 2.25% Noble agar are granular with dense centers, uneven margins, and multiple satellites, and never have fried-egg appearance. Biological properties are listed in Table 142.

Serologically distinct from other *Spiroplasma* species, groups, and subgroups. Phylogenetically, closely related to two other tabanid spiroplasmas, *Spiroplasma turonicum* and *Spiroplasma litorale*, in trees constructed using 16S rRNA gene sequences. The genome size is 1370 kbp (PFGE).

Source: isolated from the gut of a female green-eyed horse fly (*Tabanus nigrovittatus*) from the Outer Banks of North Carolina. Also collected from coastal Georgia and both Atlantic and Pacific coasts of Costa Rica.

DNA G+C content (mol%): 25 ± 1 (Bd).

Type strain: ATCC 34211, TN-1.

Sequence accession no. (16S rRNA gene): AY189306.

24. **Spiroplasma melliferum** Clark, Whitcomb, Tully, Mouches, Saillard, Bové, Wróblewski, Carle, Rose, Henegar and Williamson 1985, 305[VP]

mel.li'fe.rum. L. adj. *mellifer*, *-fera*, *-ferum* honey-bearing, honey-producing; L. neut. adj. *melliferum* intended to mean isolated from the honey bee (*Apis mellifera*).

Morphology is as described for the genus. Cells are pleomorphic, varying from helical filaments that are 100–150 nm in diameter and 3–10 µm in length to nonhelical filaments or spherical cells that are 300–800 nm in diameter. The motile cells lack true cell wells and periplasmic fibrils. Colonies on solid medium supplemented with 0.8% Noble agar are usually diffuse, rarely exhibiting central zones of growth into agar. Colonies on solid medium containing 2.25% Noble agar are smaller, but frequently have a fried-egg morphology. Physiological and genomic properties are listed in Table 142.

Serologically distinct from other *Spiroplasma* species, groups, and subgroups, but shares complex patterns of reciprocal cross-reactivity with members of other group I subgroups and *Spiroplasma poulsonii*. Has close phylogenetic affinities with other group I members and with *Spiroplasma poulsonii* in trees constructed using 16S rRNA gene sequences. DNA–DNA renaturation experiments confirm that the serological differences between the type strain and other subgroups of group I are great enough to warrant its designation as a distinct species. The genome size is 1460 kbp (PFGE). Pathogenic for honey bees in natural and experimental oral infections.

Source: isolated from hemolymph and gut of honey bees (*Apis mellifera*) in widely separated geographic regions. Also recovered from hemolymph of bumble bees, leafcutter bees, and a robber fly, and the intestinal contents of sweat bees, digger bees, bumble bees, and a butterfly. Also recovered from a variety of plant surfaces (flowers) in widely separated geographic regions.

DNA G+C content (mol%): 26–28 (T_m, Bd).

Type strain: ATCC 33219, BC-3.

Sequence accession no. (16S rRNA gene): AY325304.

25. **Spiroplasma mirum** Tully, Whitcomb, Rose and Bové 1982, 99[VP]

mi'rum. L. neut. adj. *mirum* extraordinary.

The morphology is as described for the genus. Helical filaments measure 100–200 nm in diameter and 3–8 µm in length. Colonies on solid media containing fetal bovine serum and 0.8–2.25% Noble agar (Difco) are diffuse and without central zones of growth into the agar. Solid media prepared with 1.25% agar and in which fetal bovine serum has been replaced with bovine serum fraction yield colonies with central zones of growth into the agar and no peripheral growth on the surface of the medium. Moderate turbidity is produced during growth in liquid media. Biological properties are listed in Table 142. This species has been cultivated in a defined medium.

Serologically distinct from other *Spiroplasma* species, groups, and subgroups. Phylogenetically, in trees constructed using 16S rRNA gene sequences, this species is basal to group I and group VIII spiroplasmas on the one hand, and to the Apis cluster and *Entomoplasmataceae* on the other. It is the most primitive (plesiomorphic) spiroplasma with modal helicity. The genome size is 1300 kbp (PFGE). Produces experimental ocular and nervous system disease and death in intracerebrally inoculated suckling animals (rats, mice, hamsters, and rabbits). Pathogenic for chicken embryos via yolk sac inoculation. Experimentally pathogenic for the wax moth (*Galleria mellonella*).

Source: the type strain was isolated from rabbit ticks (*Haemaphysalis leporispalustris*) collected in Georgia, USA. Other strains have been collected in Georgia, Maryland, and New York, USA.

DNA G+C content (mol%): 30–31 (T_m).

Type strain: ATCC 29335, SMCA.

Sequence accession no. (16S rRNA gene): M24662.

26. **Spiroplasma monobiae** Whitcomb, Tully, Rose, Carle, Bové, Henegar, Hackett, Clark, Konai, Adams and Williamson 1993b, 259[VP]

mo.no.bi′ae. N.L. n. *Monobia* a genus of vespid wasps; N.L. gen. n. *monobiae* of the genus *Monobia*, from which the organism was isolated.

The morphology is as described for the genus, with motile helical filaments. Colonies on solid medium containing 2.25% Noble agar are diffuse and never have a fried-egg appearance. Biological properties are listed in Table 142.

Serologically distinct from other *Spiroplasma* species, groups, and subgroups. Phylogenetically, a member of the Apis clade, but with no especially close neighbors in trees constructed using 16S rRNA gene sequences. The genome size is 940 kbp (PFGE).

Source: isolated from the hemolymph of an adult vespid wasp (*Monobia quadridens*) collected in Maryland, USA. Based on its residence in the gut of a flower-visiting insect, this species is thought to be transmitted on flowers.

DNA G+C content (mol%): 28 ± 1 (T_m, Bd, HPLC).

Type strain: ATCC 33825, MQ-1.

Sequence accession no. (16S rRNA gene): M24481.

27. **Spiroplasma montanense** Whitcomb, French, Tully, Rose, Carle, Bové, Clark, Henegar, Konai, Hackett, Adams and Williamson 1997c, 722[VP]

mon.ta.nen′se. N.L. neut. adj. *montanense* pertaining to Montana, where the species was first isolated.

The morphology is as described for the genus. Cells are motile, helical filaments that lack a cell wall. Colonies on solid medium containing 2.25% Noble agar are granular and have dense centers, irregular margins, and numerous small satellites. Biological properties are listed in Table 142.

Serologically distinct from other *Spiroplasma* species, groups, and subgroups. Reacts reciprocally in deformation serology at very low levels in deformation tests with *Spiroplasma apis*. "Bridge strains" have been isolated in Georgia with substantial cross-reactivity with both *Spiroplasma montanense* and *Spiroplasma apis*. Sister to *Spiroplasma apis* in trees constructed using 16S rRNA gene sequences. The genome size is 1225 kbp (PFGE).

Source: isolated from the gut of the tabanid fly *Hybomitra opaca*, in southwestern Montana. Other isolates have been obtained from New England, Connecticut, and southeastern Canada.

DNA G+C content (mol%): 28 ± 1 (Bd).

Type strain: ATCC 51745, HYOS-1.

Sequence accession no. (16S rRNA gene): AY189307.

28. **Spiroplasma penaei** Nunan, Lightner, Oduori and Gasparich 2005, 2320[VP]

pe.na′e.i. N.L. n. *Penaeus* a species of shrimp; N.L. gen. *penaei* of *Penaeus*, referring to *Penaeus vannamei*, from which the organism was isolated.

The morphology is as described for the genus. Cells are helical and motile. Biological properties are listed in Table 142.

Serologically distinct from previously characterized *Spiroplasma* species, groups, and subgroups, but shares some cross-reactivity with members of other group I subgroups. Has close phylogenetic affinities with other group I members and with *Spiroplasma poulsonii* in trees constructed using 16S rRNA gene sequences. The genome size has not been determined. Pathogenicity has been indicated by injection into *Penaeus vannamei*.

Source: isolated from the hemolymph of the Pacific white shrimp, *Penaeus vannamei*.

DNA G+C content (mol%): 29 ± 1 (T_m).

Type strain: CAIM 1252, SHRIMP, ATCC BAA-1082.

Sequence accession no. (16S rRNA gene): AY771927.

29. **Spiroplasma phoeniceum** Saillard, Vignault, Bové, Raie, Tully, Williamson, Fos, Garnier, Gadeau, Carle and Whitcomb 1987, 113[VP]

phoe.ni′ce.um. N.L. neut. adj. *phoeniceum* (from L. neut. adj. *phonicium*) of Phoenice, an ancient country that was located on today's Syrian coast, referring to the geographical origin of the isolates.

Morphology is as described for the genus. Colonies on solid medium containing 0.8% Noble agar show fried-egg morphology. Physiological and genomic properties are listed in Table 142.

Serologically distinct from other *Spiroplasma* species, groups, and subgroups, but shares some cross-reactivity with members of other group I subgroups and *Spiroplasma poulsonii*. Has close phylogenetic affinities with other group I members and with *Spiroplasma poulsonii* in trees constructed using 16S rRNA gene sequences. Has been shown to be transmissible to leafhoppers by injection and experimentally pathogenic to aster inoculated by the injected leafhoppers. DNA–DNA renaturation experiments confirm that the differences between the type strain and other subgroups of group I are great enough to warrant its designation as a distinct species. The genome size is 1860 kbp (PFGE).

Source: isolated from periwinkles that were naturally infected in various locations along the Syrian coastal area.

DNA G+C content (mol%): 26 ± 1 (T_m, Bd).

Type strain: ATCC 43115, P40.

Sequence accession no. (16S rRNA gene): AY772395.

30. **Spiroplasma platyhelix** Williamson, Adams, Whitcomb, Tully, Carle, Konai, Bové and Henegar 1997, 766[VP]

pla.ty.he'lix. Gr. adj. *platys* flat; Gr. fem. n. *helix* a coil or spiral; N.L. fem. n. *platyhelix* flat coil, referring to the flattened nature of the helical filament.

Cells are flattened, helical filaments, 200–300 nm in diameter. They show no rotatory or translational motility, but exhibit contractile movements in which tightness of coiling moves along the axis of the filament. Colonies on solid medium containing 2.25% Noble agar form perfect fried-egg colonies with dense centers, smooth edges, and without satellites. Biological properties are listed in Table 142.

Serologically distinct from other *Spiroplasma* species, groups, and subgroups. The genome size is 780 kbp (PFGE).

Source: isolated from the gut of a dragonfly, *Pachydiplax longipennis*, collected in Maryland, USA.

DNA G+C content (mol%): 29 ± 1 (Bd).

Type strain: ATCC 51748, PALS-1.

Sequence accession no. (16S rRNA gene): AY800347.

31. **Spiroplasma poulsonii** Williamson, Sakaguchi, Hackett, Whitcomb, Tully, Carle, Bové, Adams, Konai and Henegar 1999, 616[VP]

poul.so'ni.i. N.L. masc. gen. n. poulsonii of Poulson, named in memory of Donald F. Poulson, in whose laboratory at Yale University this spiroplasma was discovered and studied intensively.

Morphology is as described for the genus. Long, motile, helical filaments, 200–250 nm in diameter occur *in vivo* in *Drosophila* hemolymph and *in vitro*. Colonies on solid medium containing 1.8% Noble agar are small (60–70 µm in diameter), have dense centers and uneven edges, and are without satellites. Biological properties are listed in Table 142.

Serologically distinct from other *Spiroplasma* species, groups, and subgroups, but shares some reciprocal cross-reactivity with members of group I subgroups. Phylogenetically related to group I spiroplasmas in trees constructed using 16S rRNA gene sequences. The genome size is 2040 kbp (PFGE). Spiroplasmas causing sex-ratio abnormalities occur naturally in *Drosophila* spp. collected in Brazil, Colombia, Dominican Republic, Haiti, Jamaica, and the West Indies. Non-male-lethal spiroplasmas also occur in natural populations of *Drosophila hydei* in Japan. Pathogenicity (lethality to male progeny) has been confirmed by injection into *Drosophila pseudoobscura* female flies. Vertical transmissibility is lost after cultivation and cloning.

Source: isolated from the hemolymph of *Drosophila pseudoobscura* females infected by hemolymph transfer of the Barbados-3 strain of *Drosophila willistoni* SR organism.

DNA G+C content (mol%): 26 ± 1 (T_m, Bd).

Type strain: ATCC 43153, DW-1.

Sequence accession no. (16S rRNA gene): M24483.

32. **Spiroplasma sabaudiense** Abalain-Colloc, Chastel, Tully, Bové, Whitcomb, Gilot and Williamson 1987, 264[VP]

sa.bau.di.en'se. L. neut. adj. *sabaudiense* of *Sabaudia*, an ancient country of Gaul, corresponding to present day Savoy, referring to the geographic origin of the isolate.

The morphology is as described for the genus. Cells are helical filaments, 100–160 nm in diameter and 3.1–3.8 µm long. Motile. Colonies on solid medium containing 1.6% Noble agar are diffuse, rarely exhibiting fried-egg morphology, with numerous satellite colonies. Physiological and genomic properties are listed in Table 142.

Serologically distinct from other *Spiroplasma* species, groups, and subgroups. Phylogenetically, related to *Spiroplasma alleghenense* and *Spiroplasma* sp. TIUS-1 in trees constructed using 16S rRNA gene sequences. The genome size is 1175 kbp (PFGE).

Source: isolated from a triturate of female *Aedes* spp. mosquitoes in Savoy, France.

DNA G+C content (mol%): 30 ± 1 (T_m, Bd).

Type strain: ATCC 43303, Ar-1343.

Sequence accession no. (16S rRNA gene): AY189308.

33. **Spiroplasma syrphidicola** Whitcomb, Gasparich, French, Tully, Rose, Carle, Bové, Henegar, Konai, Hackett, Adams, Clark and Williamson 1996, 799[VP]

syr.phi.di'co.la. N.L. pl. n. *Syrphidae* a family of flies; L. suff. *-cola* (from L. masc. or fem. n. *incola*) inhabitant, dweller; N.L. masc. n. *syrphidicola* inhabitant of syrphid flies, the insects from which the organism was isolated.

Helical motile filaments are short and thin, passing a 220 nm filter quantitatively. Grows to titers as high as 10^{11}/ml. These short, thin, abundant cells are provisionally diagnostic for group VIII. Colonies on solid medium containing 2.25% Noble agar are irregular with satellites, diffuse, and never have a fried-egg appearance. Growth on solid medium containing 1.6% Noble agar is diffuse. Biological properties are listed in Table 142.

Serologically distinct from other *Spiroplasma* species, groups, and subgroups, but shares some reciprocal cross-reactivity with members of other group VIII subgroups. Placement of group VIII strains into subgroups has become increasingly difficult as more strains have accumulated. Phylogenetically, this species is closely related to other group VIII strains in trees constructed using 16S rRNA gene sequences. The 16S rRNA gene sequence similarity coefficients of group VIII spiroplasmas are >0.99, so this gene is insufficient for species separations in group VIII. DNA–DNA renaturation experiments confirm that the differences between the type strain and other subgroups of group VIII are great enough to warrant its designation as a distinct species. Genome size is 1230 kbp (PFGE).

Source: isolated from the hemolymph of the syrphid fly *Eristalis arbustorum* in Maryland, USA. Strains that are provisionally identified as *Spiroplasma syrphidicola* have been obtained from horse flies collected from several locations in the southeastern United States.

DNA G+C content (mol%): 30 ± 1 (Bd).

Type strain: ATCC 33826, EA-1.

Sequence accession no. (16S rRNA gene): AY189309.

34. **Spiroplasma tabanidicola** Whitcomb, French, Tully, Gasparich, Rose, Carle, Bové, Henegar, Konai, Hackett, Adams, Clark and Williamson 1997b, 718[VP]

ta.ba.ni.di'co.la. N.L. n. *Tabanidae* family name for horse flies; L. suff. *-cola* (from L. n. *incola*) inhabitant, dweller; N.L. n. *tabanidicola* an inhabitant of horse flies.

The morphology is as described for the genus. Cells are motile, helical filaments that lack a cell wall. Colonies on solid medium containing 3% Noble agar are uneven and granular with dense centers and irregular edges, do not have satellites, and never have a fried-egg appearance. Physiological and genomic properties are listed in Table 142.

Serologically distinct from other *Spiroplasma* species, groups, and subgroups. However, some strains may show a very low level reciprocal serological cross-reaction in deformation serology with *Spiroplasma gladiatoris*. This species has a specific antigen, common to several spiroplasmal inhabitants of horse flies, that confers a high level one-way cross-reaction when it is used as antigen. The genome size is 1375 kbp (PFGE).

Source: isolated from the gut of a horse fly belonging to the *Tabanus abdominalis-limbatinevris* complex.

DNA G+C content (mol%): 26 ± 1 (Bd).

Type strain: ATCC 51747, TAUS-1.

Sequence accession no. (16S rRNA gene): DQ004931.

35. **Spiroplasma taiwanense** Abalain-Colloc, Rosen, Tully, Bové, Chastel and Williamson 1988, 105[VP]

tai.wan.en'se. N.L. neut. adj. *taiwanense* of or belonging to Taiwan, referring to the geographic origin of the isolate.

The morphology is as described for the genus. Cells are motile, helical filaments, 100–160 nm in diameter and 3.1–3.8 µm long. Colonies on solid medium containing 1.6% Noble agar have fried-egg morphology. Biological properties are listed in Table 142.

Serologically distinct from other *Spiroplasma* species, groups, and subgroups. Phylogenetically, this species is in the classical Apis cluster of spiroplasmas, but does not have an especially close neighbor in trees constructed using 16S rRNA gene sequences. The genome size is 1195 kbp (PFGE).

Source: isolated from a triturate of female mosquitoes (*Culex tritaeniorhynchus*) at Taishan, Taiwan, Republic of China.

DNA G+C content (mol%): 25 ± 1 (T_m, Bd).

Type strain: ATCC 43302, CT-1.

Sequence accession no. (16S rRNA gene): M24476.

36. **Spiroplasma turonicum** Hélias, Vazeille-Falcoz, Le Goff, Abalain-Colloc, Rodhain, Carle, Whitcomb, Williamson, Tully, Bové and Chastel 1998, 460[VP]

tu.ro'ni.cum. L. neut. adj. *turonicum* of Touraine, the province in France from which the organism was first isolated.

The morphology is as described for the genus. Cells are motile, helical filaments. Colonies on solid medium containing 3% Noble agar exhibit a "cauliflower-like" appearance and do not have a fried-egg morphology. Biological properties are listed in Table 142.

Serologically distinct from previously established *Spiroplasma* species. Phylogenetically, related to two other tabanid spiroplasmas, *Spiroplasma corruscae* and *Spiroplasma litorale*, in trees constructed using 16S rRNA gene sequences. The genome size is 1305 kbp (PFGE).

Source: isolated from a triturate of a single horse fly (*Haematopota pluvialis*) collected in France.

DNA G+C content (mol%): 25 ± 1 (Bd).

Type strain: ATCC 700271, Tab4c.

Sequence accession no. (16S rRNA gene): AY189310.

37. **Spiroplasma velocicrescens** Konai, Whitcomb, Tully, Rose, Carle, Bové, Henegar, Hackett, Clark and Williamson 1995, 205[VP]

ve.lo.ci.cres'cens. L. adj. *velox, -ocis* fast, quick; L. part. adj. *crescens* growing; N.L. n. part. adj. *velocicrescens* fast-growing.

The morphology is as described for the genus. Cells are helical, motile filaments, 200–300 nm in diameter. Colonies on solid medium containing 0.8% Noble agar are diffuse and never have a fried-egg appearance. Biological properties are listed in Table 142.

Serologically distinct from other *Spiroplasma* species, groups, and subgroups. Phylogenetically, this species is sister to *Spiroplasma chinense* in trees constructed using 16S rRNA gene sequences. The genome size is 1480 kbp (PFGE).

Source: isolated from the gut of a vespid wasp, *Monobia quadridens*, collected in Maryland, USA. Based on its residence in the gut of a flower-visiting insect, this species is thought to be transmitted on flowers.

DNA G+C content (mol%): 27 ± 1 (T_m, Bd, HPLC).

Type strain: ATCC 35262, MQ-4.

Sequence accession no. (16S rRNA gene): AY189311.

References

Abalain-Colloc, M.L., C. Chastel, J.G. Tully, J.M. Bové, R.F. Whitcomb, B. Gilot and D.L. Williamson. 1987. *Spiroplasma sabaudiense* sp. nov. from mosquitos collected in France. Int. J. Syst. Bacteriol. *37*: 260–265.

Abalain-Colloc, M.L., L. Rosen, J.G. Tully, J.M. Bové, C. Chastel and D.L. Williamson. 1988. *Spiroplasma taiwanense* sp. nov. from *Culex tritaeniorhynchus* mosquitos collected in Taiwan. Int. J. Syst. Bacteriol. *38*: 103–107.

Abalain-Colloc, M.L., D.L. Williamson, P. Carle, J.H. Abalain, F. Bonnet, J.G. Tully, M. Konai, R.F. Whitcomb, J.M. Bové and C. Chastel. 1993. Division of group XVI spiroplasmas into subgroups. Int. J. Syst. Bacteriol. *43*: 342–346.

Adams, J.R., R.F. Whitcomb, J.G. Tully, E.A. Clark, D.L. Rose, P. Carle, M. Konai, J.M. Bové, R.B. Henegar and D.L. Williamson. 1997. *Spiroplasma alleghenense* sp. nov., a new species from the scorpion fly *Panorpa helena* (Mecoptera: Panorpidae). Int. J. Syst. Bacteriol. *47*: 759–762.

Alexeeva, I., E.J. Elliott, S. Rollins, G.E. Gasparich, J. Lazar and R.G. Rohwer. 2006. Absence of *Spiroplasma* or other bacterial 16S rRNA genes in brain tissue of hamsters with scrapie. J. Clin. Microbiol. *44*: 91–97.

Alivizatos, A.S. 1988. Isolation and culture of corn stunt *Spiroplasma* in serum-free medium. J. Phytopathol. *122*: 68–75.

Aluotto, B.B., R. G. Wittler, C.O.Williams and J. E. Faber. 1970. Standardized bacteriologic techniques for characterization of *Mycoplasma* species. Int. J. Syst. Bacteriol. *20*: 35–58.

Amikam, D., S. Razin and G. Glaser. 1982. Ribosomal RNA genes in *Mycoplasma*. Nucleic Acids Res. *10*: 4215–4222.

Amikam, D., G. Glaser and S. Razin. 1984. Mycoplasmas (*Mollicutes*) have a low number of rRNA genes. J. Bacteriol. *158*: 376–378.

Ammar, E. and S.A. Hogenhout. 2005. Use of immunofluorescence confocal laser scanning microscopy to study distribution of the bacterium corn stunt spiroplasma in vector leafhoppers (Hemiptera: Cicadellidae) and in host plants. Ann. Entomol. Soc. Am. *98*: 820–826.

Ammar, E.D., D. Fulton, X. Bai, T. Meulia and S.A. Hogenhout. 2004. An attachment tip and pili-like structures in insect- and plant-pathogenic spiroplasmas of the class *Mollicutes*. Arch. Microbiol. *181*: 97–105.

Anbutsu, H. and T. Fukatsu. 2003. Population dynamics of male-killing and non-male-killing spiroplasmas in *Drosophila melanogaster*. Appl. Environ. Microbiol. *69*: 1428–1434.

André, A., W. Maccheroni, F. Doignon, M. Garnier and J. Renaudin. 2003. Glucose and trehalose PTS permeases of *Spiroplasma citri* probably share a single IIA domain, enabling the spiroplasma to adapt quickly to carbohydrate changes in its environment. Microbiology *149*: 2687–2696.

André, A., M. Maucourt, A. Moing, D. Rolin and J. Renaudin. 2005. Sugar import and phytopathogenicity of *Spiroplasma citri*: glucose and fructose play distinct roles. Mol. Plant. Microbe. Interact. *18*: 33–42.

Archer, D.B., J. Best and C. Barber. 1981. Isolation and restriction mapping of a spiroplasma plasmid. J. Gen. Microbiol. *126*: 511–514.

Bai, X. and S.A. Hogenhout. 2002. A genome sequence survey of the mollicute corn stunt spiroplasma *Spiroplasma kunkelii*. FEMS Microbiol. Lett. *210*: 7–17.

Barile, M.F. 1983. Arginine hydrolysis. *In* Methods in Mycoplasmology, vol. 1 (edited by Razin and Tully). Academic Press, New York, pp. 345–349.

Baseman, J.B. and J.G. Tully. 1997. Mycoplasmas: sophisticated, re-emerging, and burdened by their notoriety. Emerg. Infect. Dis. *3*: 21–32.

Bastian, F.O. 1979. *Spiroplasma*-like inclusions in Creutzfeldt-Jakob disease. Arch. Pathol. Lab. Med. *103*: 665–669.

Bastian, F.O. 2005. *Spiroplasma* as a candidate agent for the transmissible spongiform encephalopathies. J. Neuropathol. Exp. Neurol. *64*: 833–838.

Bastian, F.O. and J.W. Foster. 2001. *Spiroplasma* sp. 16S rDNA in Creutzfeldt-Jakob disease and scrapie as shown by PCR and DNA sequence analysis. J. Neuropathol. Exp. Neurol. *60*: 613–620.

Bastian, F.O., S. Dash and R.F. Garry. 2004. Linking chronic wasting disease to scrapie by comparison of *Spiroplasma mirum* ribosomal DNA sequences. Exp. Mol. Pathol. *77*: 49–56.

Bastian, F.O. and C.D. Fermin. 2005. Slow virus disease: deciphering conflicting data on the transmissible spongiform encephalopathies (TSE) also called prion diseases. Microsc. Res. Tech. *68*: 239–246.

Bastian, F.O., D.E. Sanders, W.A. Forbes, S.D. Hagius, J.V. Walker, W.G. Henk, F.M. Enright and P.H. Elzer. 2007. *Spiroplasma* spp. from transmissible spongiform encephalopathy brains or ticks induce spongiform encephalopathy in ruminants. J. Med. Microbiol. *56*: 1235–1242.

Bébéar, C.M., P. Aullo, J.M. Bove and J. Renaudin. 1996. *Spiroplasma citri* virus SpV1: characterization of viral sequences present in the spiroplasma host chromosome. Curr. Microbiol. *32*: 134–140.

Bentley, J.K., Z. Veneti, J. Heraty and G.D. Hurst. 2007. The pathology of embryo death caused by the male-killing *Spiroplasma* bacterium in *Drosophila nebulosa*. BMC Biol. *5*: 9.

Berg, M., U. Melcher and J. Fletcher. 2001. Characterization of *Spiroplasma citri* adhesion related protein SARP1, which contains a domain of a novel family designated sarpin. Gene *275*: 57–64.

Berho, N., S. Duret, J.L. Danet and J. Renaudin. 2006a. Plasmid pSci6 from *Spiroplasma citri* GII-3 confers insect transmissibility to the non-transmissible strain *S. citri* 44. Microbiology *152*: 2703–2716.

Berho, N., S. Duret and J. Renaudin. 2006b. Absence of plasmids encoding adhesion-related proteins in non-insect-transmissible strains of *Spiroplasma citri*. Microbiology *152*: 873–886.

Bévén, L., M. LeHenaff, C. Fontenelle and H. Wróblewski. 1996. Inhibition of spiralin processing by the lipopeptide antibiotic globomycin. Curr. Microbiol. *33*: 317–322.

Bévén, L. and H. Wróblewski. 1997. Effect of natural amphipathic peptides on viability, membrane potential, cell shape and motility of mollicutes. Res. Microbiol. *148*: 163–175.

Bévén, L., D. Duval, S. Rebuffat, F.G. Riddell, B. Bodo and H. Wróblewski. 1998. Membrane permeabilisation and antimycoplasmic activity of the 18-residue peptaibols, trichorzins PA. Biochim. Biophys. Acta *1372*: 78–90.

Bi, K., H. Huang, W. Gu, J. Wang and W. Wang. 2008. Phylogenetic analysis of Spiroplasmas from three freshwater crustaceans (*Eriocheir sinensis, Procambarus clarkia and Penaeus vannamei*) in China. J. Invertebr. Pathol. *99*: 57–65.

Boutareaud, A., J.L. Danet, M. Garnier and C. Saillard. 2004. Disruption of a gene predicted to encode a solute binding protein of an ABC transporter reduces transmission of *Spiroplasma citri* by the leafhopper *Circulifer haematoceps*. Appl. Environ. Microbiol. *70*: 3960–3967.

Bové, J.M. and C. Saillard. 1979. Cell biology of spiroplasmas. *In* The Mycoplasmas, vol. 3 (edited by Whitcomb and Tully). Academic Press, New York, pp. 83–153.

Bové, J.M., C. Saillard, P. Junca, J.R. DeGorce-Dumas, B. Ricard, A. Nhami, R.F. Whitcomb, D. Williamson and J.G. Tully. 1982. Guanine-plus-cytosine content, hybridization percentages, and *Eco*RI restriction enzyme profiles of spiroplasmal DNA. Rev. Infect. Dis. *4 Suppl*: S129–136.

Bové, J.M., C. Mouches, P. Carle-Junca, J.R. Degorce-Dumas, J.G. Tully and R.F. Whitcomb. 1983. Spiroplasmas of Group I: the *Spiroplasma citri* cluster. Yale J. Biol. Med. *56*: 573–582.

Bové, J.M., P. Carle, M. Garnier, F. Laigret, J. Renaudin and C. Saillard. 1989. Molecular and cellular biology of spiroplasmas. *In* The Mycoplasmas, vol. 5 (edited by Whitcomb and Tully). Academic Press, New York, pp. 243–364.

Bové, J.M. 1993. Molecular features of mollicutes. Clin. Infect. Dis. *17 Suppl 1*: S10–31.

Bové, J.M., X. Foissac and C. Saillard. 1993. Spiralins. *In* Subcellular Biochemistry. *Mycoplasma* Cell Membranes (edited by Rottem and Kahane). Plenum Press, New York, pp. 203–223.

Bové, J.M. 1997. Spiroplasmas: infectious agents of plants, arthropods and vertebrates. Wien. Klin. Wochenschr. *109*: 604–612.

Bové, J.M., J. Renaudin, C. Saillard, X. Foissac and M. Garnier. 2003. *Spiroplasma citri*, a plant pathogenic mollicute: relationships with its two hosts, the plant and the leafhopper vector. Annu. Rev. Phytopathol. *41*: 483–500.

Bowyer, J.W. and E.C. Calavan. 1974. Antibiotic sensitivity *in vitro* of the mycoplasmalike organism associated with citrus stubborn disease. Phytopathology *64*: 346–349.

Brenner, C., H. Duclohier, V. Krchnak and H. Wroblewski. 1995. Conformation, pore-forming activity, and antigenicity of synthetic peptide analogues of a spiralin putative amphipathic alpha helix. Biochim. Biophys. Acta *1235*: 161–168.

Breton, M., S. Duret, N. Arricau-Bouvery, L. Beven and J. Renaudin. 2008a. Characterizing the replication and stability regions of *Spiroplasma citri* plasmids identifies a novel replication protein and expands the genetic toolbox for plant-pathogenic spiroplasmas. Microbiology *154*: 3232–3244.

Breton, M., S. Duret, J.L. Danet, M.P. Dubrana and J. Renaudin. 2010. Sequences essential for transmission of *Spiroplasma citri* leafhopper vector, *Circulifer haematoceps*, revealed by plasmid curing and replacement based on incompatibility. Appl. Environ. Microbiol. *76*: 3198–3205.

Brown, D.R., R.F. Whitcomb and J.M. Bradbury. 2007. Revised minimal standards for description of new species of the class *Mollicutes* (division *Tenericutes*). Int. J. Syst. Evol. Microbiol. *57*: 2703–2719.

Calavan, E.C. and J.M. Bové. 1989. Molecular and cellular biology of spiroplasmas. *In* The Mycoplasmas, vol. 5 (edited by Whitcomb and Tully). Academic Press, New York, pp. 425–485.

Carle, P., J.G. Tully, R.F. Whitcomb and J.M. Bové. 1990. Size of the spiroplasmal genome and guanosine plus cytosine content of spiroplasmal DNA. Zentralbl. Bakteriol. Suppl. *20*: 926–931.

Carle, P., F. Laigret, J.G. Tully and J.M. Bové. 1995. Heterogeneity of genome sizes within the genus *Spiroplasma*. Int. J. Syst. Bacteriol. *45*: 178–181.

Carle, P., C. Saillard, N. Carrere, S. Carrere, S. Duret, S. Eveillard, P. Gaurivaud, G. Gourgues, J. Gouzy, P. Salar, E. Verdin, M. Breton, A. Blanchard, F. Laigret, J.M. Bové, J. Renaudin and X. Foissac. 2010. Partial chromosome sequence of *Spiroplasma citri* reveals extensive viral invasion and important gene decay. Appl. Environ. Microbiol. *76*: 3420–3446.

Carle, P., R.F. Whitcomb, K.J. Hackett, J.G. Tully, D.L. Rose, J.M. Bové, R.B. Henegar, M. Konai and D.L. Williamson. 1997. *Spiroplasma diabroticae* sp. nov., from the southern corn rootworm beetle, *Diabrotica undecimpunctata* (Coleoptera: Chrysomelidae). Int. J. Syst. Bacteriol. *47*: 78–80.

Chang, C.J. and T.A. Chen. 1982. *Spiroplasmas*: cultivation in chemically defined medium. Science *215*: 1121–1122.

Chang, C.J. 1989. Nutrition and cultivation of spiroplasmas. *In* The Mycoplasmas, vol. 5 (edited by Whitcomb and Tully). Academic Press, New York, pp. 201–241.

Charbonneau, D.L. and W.C. Ghiorse. 1984. Ultrastructure and location of cytoplasmic fibrils in *Spiroplasma-Floricola* OBMG. Curr. Microbiol. *10*: 65–71.

Charron, A., C. Bébéar, G. Brun, P. Yot, J. Latrille and J.M. Bové. 1979. Separation and partial characterization of two deoxyribonucleic acid polymerases from *Spiroplasma citri*. J. Bacteriol. *140*: 763–768.

Charron, A., M. Castroviejo, C. Bébéar, J. Latrille and J.M. Bové. 1982. A third DNA polymerase from *Spiroplasma citri* and two other spiroplasmas. J. Bacteriol. *149*: 1138–1141.

Chastel, C., B. Gilot, F. Le Goff, B. Divau, G. Kerdraon, I. Humphery-Smith, R. Gruffax and A.M. Simitzis-Le Flohic. 1990. New developments in the ecology of mosquito spiroplasmas. Zentralbl. Bakteriol. Suppl. *20*: 445–460.

Chastel, C. and I. Humphery-Smith. 1991. Mosquito spiroplasmas. Adv. Dis. Vector Res. *7*: 149–205.

Chastel, C., F. Le Goff and I. Humphery-Smith. 1991. Multiplication and persistence of *Spiroplasma melliferum* strain A56 in experimentally infected suckling mice. Res. Microbiol. *142*: 411–417.

Chen, T.A. and C.H. Liao. 1975. Corn stunt spiroplasma: Isolation, cultivation and proof of pathogenesis. Science *188*: 1015–1017.

Chevalier, C., C. Saillard and J.M. Bové. 1990. Organization and nucleotide sequences of the *Spiroplasma citri* genes for ribosomal protein S2, elongation factor Ts, spiralin, phosphofructokinase, pyruvate kinase, and an unidentified protein. J. Bacteriol. *172*: 2693–2703.

Chipman, P.R., M. Agbandje-McKenna, J. Renaudin, T.S. Baker and R. McKenna. 1998. Structural analysis of the *Spiroplasma* virus, SpV4: implications for evolutionary variation to obtain host diversity among the *Microviridae*. Structure *6*: 135–145.

Christiansen, C., G. Askaa, E.A. Freundt and R.F. Whitcomb. 1979. Nucleic-acid hybridization experiments with *Spiroplasma citri* and the corn stunt and suckling mouse cataract spiroplasmas. Curr. Microbiol. *2*: 323–326.

Citti, C., L. Marechal-Drouard, C. Saillard, J.H. Weil and J.M. Bové. 1992. *Spiroplasma citri* UGG and UGA tryptophan codons: sequence of the two tryptophanyl-tRNAs and organization of the corresponding genes. J. Bacteriol. *174*: 6471–6478.

Clark, HF. 1964. Suckling mouse cataract agent. J. Infect. Dis. *114*: 476–487.

Clark, HF. and L.B. Rorke. 1979. Spiroplasmas of tick origin and their pathogenicity. *In* The Mycoplasmas, vol. 3 (edited by Whitcomb and Tully). Academic Press, New York, pp. 155–174.

Clark, T.B. 1977. *Spiroplasma* sp., a new pathogen in honey bees. J. Invertebr. Pathol. *29*: 112–113.

Clark, T.B. 1978. Honey bee spiroplasmosis, a new problem for beekeepers. Am. Bee J. *118*: 18–19.

Clark, T.B. 1982. Spiroplasmas: diversity of arthropod reservoirs and host-parasite relationships. Science *217*: 57–59.

Clark, T.B. 1984. Diversity of spiroplasma host-parasite relationships. Isr. J. Med. Sci. *20*: 995–997.

Clark, T.B., R.F. Whitcomb, J.G. Tully, C. Mouches, C. Saillard, J.M. Bové, H. Wroblewski, P. Carle, D.L. Rose, R.B. Henegar and D.L. Williamson. 1985. *Spiroplasma melliferum*, a new species from the honeybee (*Apis mellifera*). Int. J. Syst. Bacteriol. *35*: 296–308.

Cohen, A.J., D.L. Williamson and K. Oishi. 1987. SpV3 viruses of *Drosophila* spiroplasmas. Isr. J. Med. Sci. *23*: 429–433.

Cohen, A.J. and D.L. Williamson. 1988. Yeast supported growth of *Drosophila* species spiroplasmas. Proceedings of the 7th International Congress of the International Organization for Mycoplasmology, Vienna, Austria.

Cohen, A.J., D.L. Williamson and P.R. Brink. 1989. A motility mutant of *Spiroplasma melliferum* induced with nitrous acid. Curr. Microbiol. *18*: 219–222.

Cole, R.M., J.G. Tully, T.J. Popkin and J.M. Bové. 1973. Morphology, ultrastructure, and bacteriophage infection of the helical mycoplasma-like organism (*Spiroplasma citri* gen. nov., sp. nov.) cultured from "stubborn" disease of citrus. J. Bacteriol. *115*: 367–384.

Cole, R.M., J.G. Tully and T.J. Popkin. 1974. Virus-like particles in *Spiroplasma citri*. Colloq. Inst. Natl. Santé Rech. Med. *33*: 125–132.

Cole, R.M., W.O. Mitchell and C.F. Garon. 1977. *Spiroplasma citri* 3: propagation, purification, proteins, and nucleic acid. Science *198*: 1262–1263.

Cole, R.M. 1979. *Mycoplasma* and *Spiroplasma* viruses: ultrastructure. *In* The Mycoplasmas, vol. 1 (edited by Barile and Razin). Academic Press, New York, pp. 385–410.

Dally, E.L., T.S. Barros, Y. Zhao, S. Lin, B.A. Roe and R.E. Davis. 2006. Physical and genetic map of the *Spiroplasma kunkelii* CR2–3x chromosome. Can. J. Microbiol. *52*: 857–867.

Daniels, M.J., J.M. Longland and J. Gilbart. 1980. Aspects of motility and chemotaxis in spiroplasmas. J. Gen. Microbiol. *118*: 429–436.

Daniels, M.J. and J.M. Longland. 1984. Chemotactic behavior of spiroplasmas. Curr. Microbiol. *10*: 191–193.

Davis, R.E., J.F. Worley, R.F. Whitcomb, T. Ishijima and R.L. Steere. 1972a. Helical filaments produced by a mycoplasma-like organism associated with corn stunt disease. Science *176*: 521–523.

Davis, R.E., R.F. Whitcomb, T.A. Chen and R.R. Granados. 1972b. Current status of the aetiology of corn stunt disease. *In* Pathogenic Mycoplasmas (edited by Elliott and Birch). Elsevier-Excerpta Medica-North-Holland, Amsterdam, pp. 205–214.

Davis, R.E. and J.F. Worley. 1973. *Spiroplasma*: Motile, helical microorganism associated with corn stunt disease. Phytopathology *63*: 403–408.

Davis, R.E. 1978. *Spiroplasma* associated with flowers of tulip tree (*Liriodendron tulipifera* L). Can. J. Microbiol. *24*: 954–959.

Davis, R.E., I.M. Lee and J.F. Worley. 1981. *Spiroplasma floricola*, a new species isolated from surfaces of flowers of the tulip tree, *Liriodendron tulipifera* L. Int. J. Syst. Bacteriol. *31*: 456–464.

Davis, R.E., E.L. Dally, R. Jomantiene, Y. Zhao, B. Roe, S. Lin and J. Shao. 2005. Cryptic plasmid pSKU146 from the wall-less plant pathogen *Spiroplasma kunkelii* encodes an adhesin and components of a type IV translocation-related conjugation system. Plasmid *53*: 179–190.

DeSoete, G. 1983. A least square algorithm for fitting additive trees to proximity data. Psychometrika *48*: 621–626.

Dickinson, M.J. and R. Townsend. 1984. Characterization of the genome of a rod-shaped virus Infecting *Spiroplasma citri*. J. Gen. Virol. *65*: 1607–1610.

Duret, S., J.L. Danet, M. Garnier and J. Renaudin. 1999. Gene disruption through homologous recombination in *Spiroplasma citri*: an *scm1*-disrupted motility mutant is pathogenic. J. Bacteriol. *181*: 7449–7456.

Duret, S., N. Berho, J.L. Danet, M. Garnier and J. Renaudin. 2003. Spiralin is not essential for helicity, motility, or pathogenicity but is required for efficient transmission of *Spiroplasma citri* by its leafhopper vector *Circulifer haematoceps*. Appl. Environ. Microbiol. *69*: 6225–6234.

Duret, S., A. Andre and J. Renaudin. 2005. Specific gene targeting in *Spiroplasma citri*: improved vectors and production of unmarked mutations using site-specific recombination. Microbiology *151*: 2793–2803.

Ebbert, M. and L.R. Nault. 1994. Improved overwintering ability in *Dalbulus maidis* (Homoptera: Cicadellidae) vectors infected with *Spiroplasma kunkelii* (*Mycoplasmatales: Spiroplasmataceae*). Environ. Entomol. *23*: 634–644.

Enigl, M. and P. Schausberger. 2007. Incidence of the endosymbionts *Wolbachia, Cardinium* and *Spiroplasma* in phytoseiid mites and associated prey. Exp. Appl. Acarol. *42*: 75–85.

FAO/WHO. 1974. Preservation of mycoplasmas by lyophilization. World Health Organization working document VPH/MIC/741. FAO/WHO Programme on Comparative Mycoplasmology Working Group. World Health Organization, Geneva.

Felsenstein, J. 1993. PHYLIP (Phylogeny Inference Package) 3.57 edn. Department of Genetics, University of Washington, Seattle.

Fletcher, J., A. Wayadande, U. Melcher and F.C. Ye. 1998. The phytopathogenic mollicute–insect vector interface: a closer look. Phytopathology *88*: 1351–1358.

Foissac, X., C. Saillard, J. Gandar, L. Zreik and J.M. Bové. 1996. Spiralin polymorphism in strains of *Spiroplasma citri* is not due to differences in posttranslational palmitoylation. J. Bacteriol. *178*: 2934–2940.

Foissac, X., J.M. Bové and C. Saillard. 1997a. Sequence analysis of *Spiroplasma phoeniceum* and *Spiroplasma kunkelii* spiralin genes and comparison with other spiralin genes. Curr. Microbiol. *35*: 240–243.

Foissac, X., J.L. Danet, C. Saillard, P. Gaurivaud, F. Laigret, C. Paré and J.M. Bové. 1997b. Mutagenesis by insertion of Tn4001 into the genome of *Spiroplasma citri*: Characterization of mutants affected in plant pathogenicity and transmission to the plant by the leafhopper vector *Circulifer haematoceps*. Mol. Plant Microbe Interact. *10*: 454–461.

Foissac, X., C. Saillard and J.M. Bové. 1997c. Random insertion of transposon Tn4001 in the genome of *Spiroplasma citri* strain GII3. Plasmid *37*: 80–86.

French, F.E., R.F. Whitcomb, J.G. Tully, K.J. Hackett, E.A. Clark, R.B. Henegar, A.G. Wagner and D.L. Rose. 1990. Tabanid spiroplasmas of the southeast USA: new groups and correlation with host life history strategy. Zentralbl. Bakteriol. Suppl. *20*: 919–922.

French, F.E., R.F. Whitcomb, J.G. Tully, D.L. Williamson and R.B. Henegar. 1996. Spiroplasmas of *Tabanus lineola*. IOM Lett. *4*: 211–212.

French, F.E., R.F. Whitcomb, J.G. Tully, P. Carle, J.M. Bové, R.B. Henegar, J.R. Adams, G.E. Gasparich and D.L. Williamson. 1997. *Spiroplasma lineolae* sp. nov., from the horsefly *Tabanus lineola* (Diptera: Tabanidae). Int. J. Syst. Bacteriol. *47*: 1078–1081.

Fukatsu, T. and N. Nikoh. 1998. Two intracellular symbiotic bacteria from the mulberry psyllid *Anomoneura mori* (Insecta, Homoptera). Appl. Environ. Microbiol. *64*: 3599–3606.

Fukatsu, T. and N. Nikoh. 2000. Endosymbiotic microbiota of the bamboo pseudococcid *Antonina crawii* (Insecta, Homoptera). Appl. Environ. Microbiol. *66*: 643–650.

Fukatsu, T., T. Tsuchida, N. Nikoh and R. Koga. 2001. *Spiroplasma* symbiont of the pea aphid, *Acyrthosiphon pisum* (Insecta: Homoptera). Appl. Environ. Microbiol. *67*: 1284–1291.

Gadeau, A.P., C. Mouches and J.M. Bové. 1986. Probable insensitivity of mollicutes to rifampin and characterization of spiroplasmal DNA-dependent RNA polymerase. J. Bacteriol. *166*: 824–828.

Garnier, M., M. Clerc and J.M. Bové. 1981. Growth and division of spiroplasmas: morphology of *Spiroplasma citri* during growth in liquid medium. J. Bacteriol. *147*: 642–652.

Garnier, M., M. Clerc and J.M. Bové. 1984. Growth and division of *Spiroplasma citri*: elongation of elementary helices. J. Bacteriol. *158*: 23–28.

Garnier, M., X. Foissac, P. Gaurivaud, F. Laigret, J. Renaudin, C. Saillard and J.M. Bové. 2001. Mycoplasmas, plants, insect vectors: a matrimonial triangle. C. R. Acad. Sci. III *324*: 923–928.

Gasparich, G.E., K.J. Hackett, E.A. Clark, J. Renaudin and R.F. Whitcomb. 1993a. Occurrence of extrachromosomal deoxyribonucleic acids in spiroplasmas associated with plants, insects, and ticks. Plasmid *29*: 81–93.

Gasparich, G.E., K.J. Hackett, C. Stamburski, J. Renaudin and J.M. Bové. 1993b. Optimization of methods for transfecting *Spiroplasma citri* strain R8A2 HP with the spiroplasma virus SpV1 replicative form. Plasmid *29*: 193–205.

Gasparich, G.E., C. Saillard, E.A. Clark, M. Konai, F.E. French, J.G. Tully, K.J. Hackett and R.F. Whitcomb. 1993c. Serologic and genomic relatedness of group-VIII and group-XVII spiroplasmas and subdivision of spiroplasma group-VIII into subgroups. Int. J. Syst. Bacteriol. *43*: 338–341.

Gasparich, G.E. and K.J. Hackett. 1994. Characterization of a cryptic extrachromosomal element isolated from the mollicute *Spiroplasma taiwanense*. Plasmid *32*: 342–343.

Gasparich, G.E., K.J. Hackett, F.E. French and R.F. Whitcomb. 1998. Serologic and genomic relatedness of group XIV spiroplasma isolates from a lampyrid beetle and tabanid flies: an ecologic paradox. Int. J. Syst. Bacteriol. *48*: 321–324.

Gasparich, G.E., R.F. Whitcomb, D. Dodge, F.E. French, J. Glass and D.L. Williamson. 2004. The genus *Spiroplasma* and its non-helical descendants: phylogenetic classification, correlation with phenotype and roots of the *Mycoplasma mycoides* clade. Int. J. Syst. Evol. Microbiol. *54*: 893–918.

Gaurivaud, P., F. Laigret and J.M. Bové. 1996. Insusceptibility of members of the class *Mollicutes* to rifampin: studies of the *Spiroplasma citri* RNA polymerase β-subunit gene. Antimicrob. Agents Chemother. *40*: 858–862.

Gaurivaud, P., J.L. Danet, F. Laigret, M. Garnier and J.M. Bové. 2000a. Fructose utilization and phytopathogenicity of *Spiroplasma citri*. Mol. Plant Microbe Interact. *13*: 1145–1155.

Gaurivaud, P., F. Laigret, M. Garnier and J.M. Bové. 2000b. Fructose utilization and pathogenicity of *Spiroplasma citri*: characterization of the fructose operon. Gene *252*: 61–69.

Gaurivaud, P., F. Laigret, E. Verdin, M. Garnier and J.M. Bové. 2000c. Fructose operon mutants of *Spiroplasma citri*. Microbiology *146*: 2229–2236.

Gaurivaud, P., F. Laigret, M. Garnier and J.M. Bove. 2001. Characterization of FruR as a putative activator of the fructose operon of *Spiroplasma citri*. FEMS Microbiol. Lett. *198*: 73–78.

Gilad, R., A. Porat and S. Trachtenberg. 2003. Motility modes of *Spiroplasma melliferum* BC3: a helical, wall-less bacterium driven by a linear motor. Mol. Microbiol. *47*: 657–669.

Goodacre, S.L., O.Y. Martin, C.F. Thomas and G.M. Hewitt. 2006. *Wolbachia* and other endosymbiont infections in spiders. Mol. Ecol. *15*: 517–527.

Gordon, D.T., L.R. Nault, N.H. Gordon and S.E. Heady. 1985. Serological detection of corn stunt spiroplasma and maize rayado fino virus in field-collected *Dalbulus* spp. from Mexico. Plant Dis. *69*: 108–111.

Grau, O., F. Laigret and J.M. Bové. 1988. Analysis of ribosomal RNA genes in two spiroplasmas, one acholeplasma and one unclassified mollicute. Zentralbl. Bakteriol. Suppl. *20*: 895–897.

Grulet, O., I. Humphery-Smith, C. Sunyach, F. Le Goff and C. Chastel. 1993. Spiromed: a rapid and inexpensive spiroplasma isolation technique. J. Microbiol. Methods *17*: 123–128.

Guo, Y.H., T.A. Chen, R.F. Whitcomb, D.L. Rose, J.G. Tully, D.L. Williamson, X.D. Ye and Y.X. Chen. 1990. *Spiroplasma chinense* sp. nov. from flowers of *Calystegia hederacea* in China. Int. J. Syst. Bacteriol. *40*: 421–425.

Hackett, K.J. and D.E. Lynn. 1985. Cell-assisted growth of a fastidious spiroplasma. Science *230*: 825–827.

Hackett, K.J., D.E. Lynn, D.L. Williamson, A.S. Ginsberg and R.F. Whitcomb. 1986. Cultivation of the *Drosophila* sex-ratio *Spiroplasma*. Science *232*: 1253–1255.

Hackett, K.J. and T.B. Clark. 1989. Ecology of Spiroplasmas. *In* The Mycoplasmas, vol. 5 (edited by Whitcomb and Tully). Academic Press, New York, pp. 113–200.

Hackett, K.J., R.F. Whitcomb, R.B. Henegar, A.G. Wagner, E.A. Clark, J.G. Tully, F. Green, W.H. McKay, P. Santini, D.L. Rose, J.J. Anderson and D.E. Lynn. 1990. Mollicute diversity in arthropod hosts. Zentralbl. Bakteriol. Suppl. *20*: 441–454.

Hackett, K.J., R.F. Whitcomb, J.G. Tully, D.L. Rose, P. Carle, J.M. Bové, R.B. Henegar, T.B. Clark, E.A. Clark, M. Konai, J.R. Adams and D.L. Williamson. 1993. *Spiroplasma insolitum* sp. nov., a new species of group-I spiroplasma with an unusual DNA base composition. Int. J. Syst. Bacteriol. *43*: 272–277.

Hackett, K.J., R.H. Hackett, E.A. Clark, G.E. Gasparich, J.D. Pollack and R.F. Whitcomb. 1994. Development of the first completely defined medium for a spiroplasma, *Spiroplasma clarkii* strain CN-5. IOM Lett. *3*: 446–447.

Hackett, K.J. and R.F. Whitcomb. 1995. Cultivation of spiroplasmas in undefined and defined media. *In* Molecular and Diagnostic Procedures in Mycoplasmology, vol. 1 (edited by Razin and Tully). Academic Press, San Diego, pp. 41–53.

Hackett, K.J., E.A. Clark, R.F. Whitcomb, M. Camp and J.G. Tully. 1996a. Amended data on arginine utilization by *Spiroplasma* species. Int. J. Syst. Bacteriol. *46*: 912–915.

Hackett, K.J., R.F. Whitcomb, T.B. Clark, R.B. Henegar, D.E. Lynn, A.G. Wagner, J.G. Tully, G.E. Gasparich, D.L. Rose, P. Carle, J.M. Bové, M. Konai, E.A. Clark, J.R. Adams and D.L. Williamson. 1996b. *Spiroplasma leptinotarsae* sp. nov., a mollicute uniquely adapted to its host, the Colorado potato beetle, *Leptinotarsa decemlineata* (Coleoptera: Chrysomelidae). Int. J. Syst. Bacteriol. *46*: 906–911.

Hackett, K.J., R.F. Whitcomb, F.E. French, J.G. Tully, G.E. Gasparich, D.L. Rose, P. Carle, J.M. Bové, R.B. Henegar, T.B. Clark, M. Konai, E.A. Clark and D.L. Williamson. 1996c. *Spiroplasma corruscae* sp. nov., from a firefly beetle (Coleoptera: Lampyridae) and tabanid flies (Diptera: Tabanidae). Int. J. Syst. Bacteriol. *46*: 947–950.

Hackett, K.J., J.J. Lipa, G.E. Gasparich, D.E. Lynn, M. Konai, M. Camp and R.F. Whitcomb. 1997. The spiroplasma motility inhibition test, a new method for determining intraspecific variation among Colorado potato beetle spiroplasmas. Int. J. Syst. Bacteriol. *47*: 33–37.

Haselkorn, T.S., T.A. Markow and N.A. Moran. 2009. Multiple introductions of the *Spiroplasma* bacterial endosymbiont into *Drosophila*. Mol Ecol *18*: 1294–1305.

Hélias, C., M. Vazeille-Falcoz, F. Le Goff, M.L. Abalain-Colloc, F. Rodhain, P. Carle, R.F. Whitcomb, D.L. Williamson, J.G. Tully, J.M. Bové and C. Chastel. 1998. *Spiroplasma turonicum* sp. nov. from *Haematopota* horse flies (Diptera: Tabanidae) in France. Int. J. Syst. Bacteriol. *48*: 457–461.

Henning, K., S. Greiner-Fischer, H. Hotzel, M. Ebsen and D. Theegarten. 2006. Isolation of *Spiroplasma* sp. from an *Ixodes* tick. Int. J. Med. Microbiol. *296 Suppl 40*: 157–161.

Herren, J.K., I. Gordon, P. W. H. Holland and D. Smith. 2007. The butterfly *Danaus chrysippus* (Lepidoptera: Nymphalidae) in Kenya is variably infected with respect to genotype and body size by a maternally transmitted male-killing endosymbiont (*Spiroplasma*). Int. J. Trop. Ins. Sc.: 62–69.

Humphery-Smith, I., O. Grulet and C. Chastel. 1991a. Pathogenicity of *Spiroplasma taiwanense* for larval *Aedes aegypti* mosquitoes. Med. Vet. Entomol. *5*: 229–232.

Humphery-Smith, I., O. Grulet, F. Le Goff and C. Chastel. 1991b. *Spiroplasma* (*Mollicutes*: *Spiroplasmataceae*) pathogenic for *Aedes aegypti* and *Anopheles stephensi* (Diptera: Culicidae). J. Med. Entomol. *28*: 219–222.

Hung, S.H.Y., T.A. Chen, R.F. Whitcomb, J.G. Tully and Y.X. Chen. 1987. *Spiroplasma culicicola* sp. nov. from the salt-marsh mosquito *Aedes sollicitans*. Int. J. Syst. Bacteriol. *37*: 365–370.

Hurst, G.D.D., H. Anbutsu, M. Kutsukake and T. Fukatsu. 2003. Hidden from the host: *Spiroplasma* bacteria infecting *Drosophila* do not cause an immune response, but are suppressed by ectopic immune activation. Insect Mol. Biol. *12*: 93–97.

Hurst, G.D.D., J.H.G. von der Schulenburg, T.M.O. Majerus, D. Bertrand, I.A. Zakharov, J. Baungaard, W. Volkl, R. Stouthamer and M.E.N. Majerus. 1999. Invasion of one insect species, *Adalia bipunctata*, by two different male-killing bacteria. Insect Mol. Biol. *8*: 133–139.

Hurst, G.D.D. and F.M. Jiggins. 2000. Male-killing bacteria in insects: mechanisms, incidence, and implications. Emerg. Infect. Dis. *6*: 329–336.

International Committee on Systematics of Bacteria. 1984. Minutes of the interim meeting. 30 August and 6 September 1982, Tokyo, Japan Int. J. Syst. Bacteriol. *34*: 361–365.

International Committee on Systematics of Bacteria Subcommittee on the Taxonomy of *Mollicutes*. 1995. Revised minimum standards for description of new species of the class *Mollicutes* (division *Tenericutes*). Int J. Syst. Bacteriol *45*: 605–612.

Jacob, C., F. Nouzieres, S. Duret, J.M. Bové and J. Renaudin. 1997. Isolation, characterization, and complementation of a motility mutant of *Spiroplasma citri*. J. Bacteriol. *179*: 4802–4810.

Jaenike, J., M. Polak, A. Fiskin, M. Helou and M. Minhas. 2007. Interspecific transmission of endosymbiotic *Spiroplasma* by mites. Biol. Lett. *3*: 23–25.

Jagoueix-Eveillard, S., F. Tarendeau, K. Guolter, J.L. Danet, J.M. Bové and M. Garnier. 2001. *Catharanthus roseus* genes regulated differentially by mollicute infections. Mol. Plant Microbe Interact. *14*: 225–233.

Jandhyam, H., C. R. Bates, T. E. Young, L. Beatti, G. E. Gasparich, F. E. French and L. B. Regassa. 2008. Global spiroplasma biodiversity in a single host. Presented at the 17th Congress of International Organization for Mycoplasmology, Beijing, China. Abstract no. 206, p. 130.

Jiggins, F.M., G.D. Hurst, C.D. Jiggins, J.H. von der Schulenburg and M.E. Majerus. 2000. The butterfly *Danaus chrysippus* is infected by a male-killing *Spiroplasma* bacterium. Parasitology *120*: 439–446.

Johansson, K.-E., M.U.K. Heldtander and B. Pettersson. 1998. Characterization of mycoplasmas by PCR and sequence analysis with universal 16S rDNA primers. *In* Methods in Molecular Biology: *Mycoplasma* protocols, vol. 104 (edited by Miles and Nicholas). Humana Press, Totawa, NJ, pp. 145–165.

Johansson, K.-E. and B. Pettersson. 2002. Taxonomy of *Mollicutes*. *In* Molecular Biology and Pathogenicity of Mycoplasmas (edited by Razin and Herrmann). Kluwer Academic/Plenum Publishers, London, pp. 1–31.

Johnson, J.L. 1994. Similarity analysis of DNAs. *In* Methods for General and Molecular Bacteriology (edited by Gerhardt, Murray, Wood and Krieg). ASM Press, Washington, D.C., pp. 656–682.

Jones, L.J., R. Carballido-Lopez and J. Errington. 2001. Control of cell shape in bacteria: helical, actin-like filaments in *Bacillus subtilis*. Cell *104*: 913–922.

Joshi, B.D., M. Berg, J. Rogers, J. Fletcher and U. Melcher. 2005. Sequence comparisons of plasmids pBJS-O of *Spiroplasma citri* and pSKU146 of *S. kunkelii*: implications for plasmid evolution. BMC Genomics *6*: 175.

Junca, P., C. Saillard, J. Tully, O. Garcia-Jurado, J.R. Degorce-Dumas, C. Mouches, J.C. Vignault, R. Vogel, R. McCoy, R. Whitcomb, D. Williamson, J. Latrille and J.M. Bové. 1980. Characterization of spiroplasmas isolated from insects and flowers in continental France, Corsica and Morocco. Proposals for a taxonomical classification of spiroplasmas. [transl. from. Fr.] C. R. Hebd. Des Seances Acad. Sci. Ser. D Sci. Nat. *290*: 1209–1211.

Kageyama, D., H. Anbutsu, M. Watada, T. Hosokawa, M. Shimada and T. Fukatsu. 2006. Prevalence of a non-male-killing spiroplasma in natural populations of *Drosophila hydei*. Appl Environ Microbiol *72*: 6667–6673.

Kersting, U. and C. Sengonca. 1992. Detection of insect vectors of the citrus stubborn disease pathogen, *Spiroplasma citri* Saglio *et al.*, in the citrus growing area of south Turkey. J. Appl. Entomol. *113*: 356–364.

Killiny, N., M. Castroviejo and C. Saillard. 2005. *Spiroplasma citri* spiralin acts *in vitro* as a lectin binding to glycoproteins from its insect vector *Circulifer haematoceps*. Phytopathology *95*: 541–548.

Killiny, N., B. Batailler, X. Foissac and C. Saillard. 2006. Identification of a *Spiroplasma citri* hydrophilic protein associated with insect transmissibility. Microbiology *152*: 1221–1230.

Koerber, R.T., G.E. Gasparich, M.F. Frana and W.L. Grogan, Jr. 2005. *Spiroplasma atrichopogonis* sp. nov., from a ceratopogonid biting midge. Int. J. Syst. Evol. Microbiol. *55*: 289–292.

Konai, M., R.F. Whitcomb, J.G. Tully, D.L. Rose, P. Carle, J.M. Bové, R.B. Henegar, K.J. Hackett, T.B. Clark and D.L. Williamson. 1995. *Spiroplasma velocicrescens* sp. nov., from the vespid wasp *Monobia quadridens*. Int. J. Syst. Bacteriol. *45*: 203–206.

Konai, M., E.A. Clark, M. Camp, A.L. Koeh and R.F. Whitcomb. 1996a. Temperature ranges, growth optima, and growth rates of *Spiroplasma* (*Spiroplasmataceae*, class *Mollicutes*) species. Curr. Microbiol. *32*: 314–319.

Konai, M., K.J. Hackett, D.L. Williamson, J.J. Lipa, J.D. Pollack, G.E. Gasparich, E.A. Clark, D.C. Vacek and R.F. Whitcomb. 1996b. Improved cultivation systems for isolation of the Colorado potato beetle spiroplasma. Appl. Environ. Microbiol. *62*: 3453–3458.

Konai, M., R.F. Whitcomb, F.E. French, J.G. Tully, D.L. Rose, P. Carle, J.M. Bové, K.J. Hackett, R.B. Henegar, T.B. Clark and D.L. Williamson. 1997. *Spiroplasma litorale* sp. nov., from tabanid flies (Tabanidae: Diptera) in the southeastern United States. Int. J. Syst. Bacteriol. *47*: 359–362.

Kotani, H., G.H. Butler and G.J. McGarrity. 1990. Malignant transformation by *Spiroplasma mirum*. Zentralbl. Bakteriol. Suppl. *20*: 145–152.

Kürner, J., A.S. Frangakis and W. Baumeister. 2005. Cyro-electron tomography reveals the cytoskeletal structure of *Spiroplasma melliferum*. Science *307*: 436–438.

Kuroda, Y., Y. Shimada, B. Sakaguchi and K. Oishi. 1992. Effects of sex-ratio (SR)-spiroplasma infection on *Drosophila* primary embryonic cultured cells and on embryogenesis. Zool. Sci. *9*: 283–291.

Labarère, J. and G. Barroso. 1989. Lethal and mutation frequency responses of *Spiroplasma citri* cells to UV irradiation. Mutat. Res. *210*: 135–141.

Laigret, F., P. Gaurivaud and J.M. Bové. 1996. The unique organization of the *rpo*B region of *Spiroplasma citri*: a restriction and modification system gene is adjacent to *rpo*B. Gene *171*: 95–98.

Lartigue, C., S. Duret, M. Garnier and J. Renaudin. 2002. New plasmid vectors for specific gene targeting in *Spiroplasma citri*. Plasmid *48*: 149–159.

Le Dantec, L., M. Castroviejo, J.M. Bové and C. Saillard. 1998. Purification, cloning, and preliminary characterization of a *Spiroplasma citri* ribosomal protein with DNA binding capacity. J. Biol. Chem. *273*: 24379–24386.

Le Goff, F., M. Marjolet, J. Guilloteau, I. Humphery-Smith and C. Chastel. 1990. Characterization and ecology of mosquito spiroplasmas from Atlantic biotopes in France. Ann. Parasitol. Hum. Comp. *65*: 107–110.

Le Goff, F., I. Humphery-Smith, M. Leclercq and C. Chastel. 1991. Spiroplasmas from European Tabanidae. Med. Vet. Entomol. *5*: 143–144.

Le Goff, F., M. Marjolet, I. Humphery-Smith, M. Leclercq, C. Hélias, F. Suplisson and C. Chastel. 1993. Tabanid spiroplasmas from France: characterization, ecology and experimental study. Ann. Parasitol. Hum. Comp. *68*: 150–153.

Lee, I.M. and R.E. Davis. 1980. DNA homology among diverse spiroplasma strains representing several serological groups. Can. J. Microbiol. *26*: 1356–1363.

Liao, C.H. and T.A. Chen. 1977. Culture of corn stunt spiroplasma in a simple medium. Phytopathology *67*: 802–807.

Liao, C.H., C.J. Chang and T.A. Chen. 1979. Spiroplasmastatic action of plant tissue extracts. Proceedings of the R. O. C. U. S. Coop. Science Seminar *Mycoplasma* Diseases of Plants, Taipei, pp. 99–103.

Liao, C.H. and T.A. Chen. 1981a. Deoxyribonucleic acid hybridization between *Spiroplasma citri* and the corn stunt spiroplasma. Curr. Microbiol. *5*: 83–86.

Liao, C.H. and T.A. Chen. 1981b. *In vitro* susceptibility and resistance of two spiroplasmas to antibiotics. Phytopathology *71*: 442–445.

Lindh, J.M., O. Terenius and I. Faye. 2005. 16S rRNA gene-based identification of midgut bacteria from field-caught *Anopheles gambiae* sensu lato and *A. funestus* mosquitoes reveals new species related to known insect symbionts. Appl. Environ. Microbiol. *71*: 7217–7223.

Liss, A. and R.M. Cole. 1981. Spiroplasmavirus group I: isolation, growth, and properties. Curr. Microbiol. *5*: 357–362.

Lundgren, J.G., R. M. Lehman and J. Chee-Sanford. 2007. Bacterial communities within digestive tracts of ground beetles (Coleoptera: Carabidae). Ann. Entomol. Soc. Am. *100*: 275–282.

Maccheroni, W., J. L. Danet, S. Duret-Nurbel, J. M. Bové, M. Garnier and J. Renaudin. 2002. Cell shape determination in *Spiroplasma citri*: organization of *mreB* genes and effect of *mreB1* disruption on insect transmission and pathogenicity. Proceedings of the 15th Conference of the International Organization of Citrus Virologists (edited by Duran-Vila, Milne and da Graça), Riverside, California, p. 443.

Madden, L.V. and L.R. Nault. 1983. Differential pathogenicity of corn stunting mollicutes to leafhopper vectors in *Dalbulus* and *Baldulus* species. Phytopathology *73*: 1608–1614.

Majerus, T.M., J.H. Graf von der Schulenburg, M.E. Majerus and G.D. Hurst. 1999. Molecular identification of a male-killing agent in the ladybird *Harmonia axyridis* (Pallas) (Coleoptera: Coccinellidae). Insect. Mol. Biol. *8*: 551–555.

Maniloff, J. 1992. Phylogeny of mycoplasmas. *In* Mycoplasmas: Molecular Biology and Pathogenesis (edited by Maniloff, McElhaney, Finch and Baseman). American Society for Microbiology, Washington, D.C., pp. 549–559.

Marais, A., J.M. Bové, S.F. Dallo, J.B. Baseman and J. Renaudin. 1993. Expression in *Spiroplasma citri* of an epitope carried on the G fragment of the cytadhesin P1 gene from *Mycoplasma pneumoniae*. J. Bacteriol. *175*: 2783–2787.

Marais, A., J.M. Bové and J. Renaudin. 1996. *Spiroplasma citri* virus SpV1-derived cloning vector: deletion formation by illegitimate and homologous recombination in a spiroplasmal host strain which probably lacks a functional recA gene. J. Bacteriol. *178*: 862–870.

Markham, P.G., R. Townsend, M. Bar Joseph, M.J. Daniels, A. Plaskitt and B.M. Meddins. 1974. Spiroplasmas are causal agents of citrus little-leaf disease. Ann. Appl. Biol. *78*: 49–57.

Markham, P.G., T.B. Clark and R.F. Whitcomb. 1983. Culture techniques for spiroplasmas from arthropods. *In* Methods in Mycoplasmology, vol. 2 (edited by Tully and Razin). Academic Press, New York, pp. 217–223.

Mateos, M., S.J. Castrezana, B.J. Nankivell, A.M. Estes, T.A. Markow and N.A. Moran. 2006. Heritable endosymbionts of *Drosophila*. Genetics *174*: 363–376.

Matsuo, K., J. Silke, K. Gramatikoff and W. Schaffner. 1994. The CpG-specific methylase SssI has topoisomerase activity in the presence of Mg^{2+}. Nucleic Acids Res. *22*: 5354–5359.

McCoy, R.E., D.S. Williams and D.L. Thomas. 1979. Isolation of mycoplasmas from flowers. Proceedings of the Republic of China-United States Cooperative Science Seminar, Symposium series 1, National Science Council, Taipei, Taiwan, pp. 75–81.

McElwain, M.C., D.K.F. Chandler, M.F. Barile, T.F. Young, V.V. Tryon, J.W. Davis, J.P. Petzel, C.J. Chang, M.V. Williams and J.D. Pollack. 1988. Purine and pyrimidine metabolism in *Mollicutes* species. Int. J. Syst. Bacteriol. *38*: 417–423.

McIntosh, M.A., G. Deng, J. Zheng and R.V. Ferrell. 1992. Repetitive DNA sequences. *In* Mycoplasmas: Molecular Biology and Pathogenesis (edited by Maniloff, McElhaney, Finch and Baseman). American Society for Microbiology, Washington, D.C., pp. 363–376.

Melcher, U. and J. Fletcher. 1999. Genetic variation in *Spiroplasma citri*. Eur. J. Plant Pathol. *105*: 519–533.

Miles, R.J. 1992. Catabolism in *Mollicutes*. J. Gen. Microbiol. *138*: 1773–1783.

Montenegro, H., V.N. Solferini, L.B. Klaczko and G.D. Hurst. 2005. Male-killing *Spiroplasma* naturally infecting *Drosophila melanogaster*. Insect. Mol. Biol. *14*: 281–287.

Montenegro, H., L. M. Hatadani, H. F. Medeiros and L.B. Klaczko. 2006. Male killing in three species of the tripunctata radiation of *Drosophila* (Diptera: Drosophilidae). J. Zoo. Syst. Evol. Res. *44*: 130–135.

Mouches, C., J.M. Bové, J. Albisetti, T.B. Clark and J.G. Tully. 1982a. A spiroplasma of serogroup IV causes a May-disease-like disorder of honeybees in southwestern France. Microb. Ecol. *8*: 387–399.

Mouches, C., A. Menara, B. Geny, D. Charlemagne and J.M. Bové. 1982b. Synthesis of *Spiroplasma citri* protein specifically recognized by rabbit immunoglobulin to rabbit actin. Rev. Infect. Dis. *4*: S277.

Mouches, C., A. Menara, J.G. Tully and J.M. Bové. 1982c. Polyacrylamide gel analysis of spiroplasmas proteins and its contribution to the taxonomy of spiroplasmas. Rev. Infect. Dis. *4 Suppl*: S141–147.

Mouches, C., J.M. Bové, J.G. Tully, D.L. Rose, R.E. McCoy, P. Carle-Junca, M. Garnier and C. Saillard. 1983a. *Spiroplasma apis*, a new species from the honey bee *Apis mellifera*. Ann. Microbiol. (Paris) *134A*: 383–397.

Mouches, C., T. Candresse, G.J. McGarrity and J.M. Bové. 1983b. Analysis of spiroplasma proteins: contribution to the taxonomy of group IV spiroplasmas and the characterization of spiroplasma protein antigens. Yale J. Biol. Med. *56*: 431–437.

Mouches, C., G. Barroso, A. Gadeau and J.M. Bové. 1984a. Characterization of two cryptic plasmids from *Spiroplasma citri* and occurrence of their DNA sequences among various spiroplasmas. Ann. Microbiol. (Paris) *135A*: 17–24.

Mouches, C., J. M. Bové, J. G. Tully, D. L. Rose, R. E. McCoy, P. Carle-Junca, M. Garnier and C. Saillard. 1984b. *In* Validation of the publication of new names and new combinations previously effectively published outside the IJSB. List no. 13. Int. J. Syst. Bacteriol. *34*: 91–92.

Moya-Raygoza, G., S.A. Hogenhout and L.R. Nault. 2007a. Habitat of the corn leafhopper (Hemiptera: Cicadellidae) during the dry (winter) season in Mexico. Environ Entomol *36*: 1066–1072.

Moya-Raygoza, G., V. Palomera-Avalos and C. Galaviz-Mejia. 2007b. Field overwintering biology of *Spiroplasma kunkelii* (*Mycoplasmatales*: *Spiroplasmataceae*) and its vector *Dalbulus maidis* (Hemiptera: Cicadellidae). Ann. Appl. Biol. *151*: 373–379.

Nakamura, K., H. Ueno and K. Miura. 2005. Prevalence of inherited male-killing microorganisms in Japanese populations of ladybird beetle *Harmonia axyridis* (Coleoptera: Coccinellidae). Ann. Ent. Soc. Am. *98*: 96–99.

Nault, L.R. and O.E. Bradfute. 1979. Corn stunt: involvement of a complex of leafhopper-borne pathogens. *In* Leafhopper Vectors and Plant Disease Agents (edited by Maramorosch and Harris). Academic Press, New York, pp. 561–586.

Nault, L.R., L.V. Madden, W.E. Styer, B.W. Triplehorn, G.F. Shambaugh and S.E. Heady. 1984. Pathogenicity of corn stunt spiroplasma and maize bushy stunt mycoplasma to their vector, *Dalbulus longulus*. Phytopathology *74*: 977–979.

Navas-Castillo, J., F. Laigret, J.G. Tully and J.M. Bové. 1992. The mollicute *Acholeplasma florum* possesses a gene of phosphoenolpyruvate-sugar phosphotransferase system and it uses UGA as tryptophan codon. C. R. Acad. Sci. Ser. III Life Sci. *315*: 43–48.

Nunan, L.M., C.R. Pantoja, M. Salazar, F. Aranguren and D.V. Lightner. 2004. Characterization and molecular methods for detection of a novel spiroplasma pathogenic to *Penaeus vannamei*. Dis. Aquat. Organ. *62*: 255–264.

Nunan, L.M., D.V. Lightner, M.A. Oduori and G.E. Gasparich. 2005. *Spiroplasma penaei* sp. nov., associated with mortalities in *Penaeus vannamei*, Pacific white shrimp. Int. J. Syst. Evol. Microbiol. *55*: 2317–2322.

Nur, I., M. Szyf, A. Razin, G. Glaser, S. Rottem and S. Razin. 1985. Procaryotic and eucaryotic traits of DNA methylation in spiroplasmas (mycoplasmas). J. Bacteriol. *164*: 19–24.

Nur, I., G. Glaser and S. Razin. 1986. Free and integrated plasmid DNA in spiroplasmas. Curr. Microbiol. *14*: 169–176.

Nur, I., D.J. LeBlanc and J.G. Tully. 1987. Short, interspersed, and repetitive DNA sequences in *Spiroplasma* species. Plasmid *17*: 110–116.

Oduori, M.A., J.J. Lipa and G.E. Gasparich. 2005. *Spiroplasma leucomae* sp. nov., isolated in Poland from white satin moth (*Leucoma salicis* L.) larvae. Int. J. Syst. Evol. Microbiol. *55*: 2447–2450.

Oishi, K., D.F. Poulson and D.L. Williamson. 1984. Virus-mediated change in clumping properties of *Drosophila* SR spiroplasmas. Curr. Microbiol. *10*: 153–158.

Özbek, E., S.A. Miller, T. Meulia and S.A. Hogenhout. 2003. Infection and replication sites of *Spiroplasma kunkelii* (Class: Mollicutes) in midgut and Malpighian tubules of the leafhopper *Dalbulus maidis*. J. Invertebr. Pathol. *82*: 167–175.

Pettersson, B., J.G. Tully, G. Bolske and K.E. Johansson. 2000. Updated phylogenetic description of the *Mycoplasma hominis* cluster (Weisburg *et al.* 1989) based on 16S rDNA sequences. Int. J. Syst. Evol. Microbiol. *50*: 291–301.

Pickens, E.G., R.K. Gerloff and W. Burgdorfer. 1968. Spirochete from the rabbit tick, *Haemaphysalis leporispalustris* (Packard). I. Isolation and preliminary characterization. J. Bacteriol. *95*: 291–299.

Pollack, J.D., M.C. McElwain, D. Desantis, J.T. Manolukas, J.G. Tully, C.J. Chang, R.F. Whitcomb, K.J. Hackett and M.V. Williams. 1989. Metabolism of members of the *Spiroplasmataceae*. Int. J. Syst. Bacteriol. *39*: 406–412.

Pollack, J.D., M.V. Williams and R.N. McElhaney. 1997. The comparative metabolism of the mollicutes (mycoplasmas): the utility for taxonomic classification and the relationship of putative gene annotation and phylogeny to enzymatic function in the smallest free-living cells. Crit. Rev. Microbiol. *23*: 269–354.

Pollack, J.D. 2002a. The necessity of combining genomic and enzymatic data to infer metabolic function and pathways in the smallest bacteria: amino acid, purine and pyrimidine metabolism in mollicutes. Front. Biosci. *7*: d1762–1781.

Pollack, J.D. 2002b. Central carbohydrate pathways: Metabolic flexibility and the extra role of some "housekeeping enzymes". *In* Molecular Biology and Pathogenicity of Mycoplasmas (edited by Razin and Herrmann). Kluwer Academic/Plenum Publishers, New York, pp. 163–199.

Pool, J.E., A. Wong and C.F. Aquadro. 2006. Finding of male-killing *Spiroplasma* infecting *Drosophila melanogaster* in Africa implies transatlantic migration of this endosymbiont. Heredity *97*: 27–32.

Poulson, D.F. and B. Sakaguchi. 1961. Nature of "sex-ratio" agent in *Drosophila*. Science *133*: 1489–1490.

Pyle, L.E. and L.R. Finch. 1988. A physical map of the genome of *Mycoplasma mycoides* subspecies *mycoides* Y with some functional loci. Nucleic Acids Res. *16*: 6027–6039.

Rahimian, H. and D.J. Gumpf. 1980. Deoxyribonucleic acid relationship between *Spiroplasma citri* and the corn stunt spiroplasma. Int. J. Syst. Bacteriol. *30*: 605–608.

Ranhand, J.M., W.O. Mitchell, T.J. Popkin and R.M. Cole. 1980. Covalently closed circular deoxyribonucleic acids in spiroplasmas. J. Bacteriol. *143*: 1194–1199.

Razin, S. 1985. Molecular biology and genetics of mycoplasmas (*Mollicutes*). Microbiol. Rev. *49*: 419–455.

Regassa, L.B., K.M. Stewart, A.C. Murphy, F.E. French, T. Lin and R.F. Whitcomb. 2004. Differentiation of group VIII *Spiroplasma* strains with sequences of the 16S–23S rDNA intergenic spacer region. Can. J. Microbiol. *50*: 1061–1067.

Regassa, L.B. and G.E. Gasparich. 2006. Spiroplasmas: evolutionary relationships and biodiversity. Front. Biosci. *11*: 2983–3002.

Renaudin, J., M.C. Pascarel, M. Garnier, P. Carle-Junca and J.M. Bové. 1984a. SpV4, a new *Spiroplasma* virus with circular, single-stranded DNA. Ann. Virol. *135E*: 163–168.

Renaudin, J., M.C. Pascarel, M. Garnier, P. Carle and J.M. Bové. 1984b. Characterization of spiroplasma virus group 4 (SV4). Isr. J. Med. Sci. *20*: 797–799.

Renaudin, J., M.C. Pascarel, C. Saillard, C. Chevalier and J.M. Bové. 1986. Chez les spiroplasmes le codon UGA n'est pas non-sens et semble coder pour le tryptophane. C. R. Acad. Sci. Ser. III *303*: 539–540.

Renaudin, J. and J.M. Bové. 1994. SpV1 and SpV4, spiroplasma viruses with circular, single-stranded DNA genomes, and their contribution to the molecular biology of spiroplasmas. Adv. Virus Res. *44*: 429–463.

Renaudin, J., A. Marais, E. Verdin, S. Duret, X. Foissac, F. Laigret and J.M. Bové. 1995. Integrative and free *Spiroplasma citri* oriC plasmids: Expression of the *Spiroplasma phoeniceum* spiralin in *Spiroplasma citri*. J. Bacteriol. *177*: 2870–2877.

Renaudin, J. 2002. Extrachromosomal elements and gene transfer. *In* Molecular Biology and Pathogenicity of Mycoplasmas (edited by Razin and Herrmann). Academic/Plenum Press, New York, pp. 347–370.

Renbaum, P., D. Abrahamove, A. Fainsod, G.G. Wilson, S. Rottem and A. Razin. 1990. Cloning, characterization, and expression in *Escherichia coli* of the gene coding for the CpG DNA methylase from *Spiroplasma* sp. strain MQ1 (M-SssI). Nucleic Acids Res. *18*: 1145–1152.

Renbaum, P. and A. Razin. 1992. Mode of action of the *Spiroplasma* CpG methylase M-SssI. FEBS Lett. *313*: 243–247.

Ricard, B., M. Garnier and J.M. Bové. 1982. Characterization of spiroplasmal virus 3 from spiroplasmas and discovery of a new spiroplasmal virus (SpV4). Rev. Infect. Dis. *4*: S275.

Rodwell, A.W. and R.F. Whitcomb. 1983. Methods of direct and indirect measurement of mycoplasma growth. *In* Methods in Mycoplasmology, vol. 1 (edited by Razin and Tully). Academic Press, New York, pp. 185–196.

Rogers, M.J., J. Simmons, R.T. Walker, W.G. Weisburg, C.R. Woese, R.S. Tanner, I.M. Robinson, D.A. Stahl, G. Olsen, R.H. Leach and J. Maniloff. 1985. Construction of the mycoplasma evolutionary tree from 5S rRNA sequence data. Proc. Natl. Acad. Sci. U. S. A. *82*: 1160–1164.

Rogers, M.J., A.A. Steinmetz and R.T. Walker. 1987. Organization and structure of tRNA genes in *Spiroplasma melliferum*. Isr. J. Med. Sci. *23*: 357–360.

Rose, D.L., J.G. Tully, J.M. Bové and R.F. Whitcomb. 1993. A test for measuring growth responses of *Mollicutes* to serum and polyoxyethylene sorbitan. Int. J. Syst. Bacteriol. *43*: 527–532.

Rosengarten, R. and K.S. Wise. 1990. Phenotypic switching in mycoplasmas: phase variation of diverse surface lipoproteins. Science *247*: 315–318.

Rosselló-Mora, R. and R. Amann. 2001. The species concept for prokaryotes. FEMS Microbiol. Rev *25*: 39–67.

Saglio, P., D. Laflèche, C. Bonissol and J.M. Bové. 1971. Isolation, culture and electronmicroscopy of mycoplasma-like structures associated with stubborn disease of citrus and their comparison with structures observed in citrus plants affected by greening disease. [transl. from Fr.] Physiol. Vég. *9*: 569–582.

Saglio, P., M. L'Hospital, D. Laflèche, G. Dupont, J.M. Bové, J.G. Tully and E.A. Freundt. 1973. *Spiroplasma citri* gen. and sp. nov.: a mycoplasma-like organism associated with stubborn disease of citrus. Int. J. Syst. Bacteriol. *23*: 191–204.

Saglio, P.H.M. and R.F. Whitcomb. 1979. Diversity of wall-less prokaryotes in plant vascular tissue, fungi and invertebrate animals. *In* The Mycoplasmas, vol. 3 (edited by Whitcomb and Tully). Academic Press, New York, pp. 1–36.

Saillard, C. and J.M. Bové. 1983. Application of ELISA to spiroplasma detection and classification. *In* Methods in Mycoplasmology, vol. 1 (edited by Razin and Tully). Academic Press, New York, pp. 471–476.

Saillard, C., J.C. Vignault, J.M. Bové, A. Raie, J.G. Tully, D.L. Williamson, A. Fos, M. Garnier, A. Gadeau, P. Carle and R.F. Whitcomb. 1987. *Spiroplasma phoeniceum* sp. nov., a new plant-pathogenic species from Syria. Int. J. Syst. Bacteriol. *37*: 106–115.

Saillard, C., C. Chevalier and J.M. Bové. 1990. Structure and organization of the spiralin gene. Zentralbl. Bakteriol. Suppl. *20*: 897–901.

Saillard, C., P. Carle, S. Duret-Nurbel, R. Henri, N. Killiny, S. Carrere, J. Gouzy, J.M. Bové, J. Renaudin and X. Foissac. 2008. The abundant extrachromosomal DNA content of the *Spiroplasma citri* GII3-3X genome. BMC Genomics *9*: 195.

Saitou, N. and M. Nei. 1987. The neighbor-joining method: a new method for reconstructing phylogenetic trees. Mol. Biol. Evol. *4*: 406–425.

Sha, Y., U. Melcher, R.E. Davis and J. Fletcher. 1995. Resistance of *Spiroplasma citri* lines to the virus SVTS2 is associated with integration of viral DNA sequences into host chromosomal and extrachromosomal DNA. Appl. Environ. Microbiol. *61*: 3950–3959.

Shaevitz, J.W., J.Y. Lee and D.A. Fletcher. 2005. *Spiroplasma* swim by a processive change in body helicity. Cell *122*: 941–945.

Sher, T., A. Yamin, M. Matzliach, S. Rottem and R. Gallily. 1990a. Partial biochemical characterization of spiroplasma membrane component inducing tumor necrosis factor alpha. Anticancer Drugs *1*: 83–87.

Sher, T., A. Yamin, S. Rottem and R. Gallily. 1990b. *In vitro* induction of tumor necrosis factor alpha, tumor cytolysis, and blast transformation by *Spiroplasma* membranes. J. Natl. Cancer Inst. *82*: 1142–1145.

Simoneau, P. and J. Labarère. 1991. Evidence for the presence of two distinct membrane ATPases in *Spiroplasma citri*. J. Gen. Microbiol. *137*: 179–185.

Skripal, I.G. 1974. On improvement of taxonomy of the class *Mollicutes* and establishment in the order *Mycoplasmatales* of the new family *Spiroplasmataceae* fam. nov. Mikrobiol. Zh. (Kiev). *36*: 462–467.

Skripal, I.G. 1983. Revival of the name *Spiroplasmataceae* fam. nov., nom. rev., omitted from the 1980 Approved Lists of Bacterial Names. Int. J. Syst. Bacteriol. *33*: 408.

Sokolova, M.I., N.S. Zinkevich and I.A. Zakharov. 2002. *Bacteria* in ovarioles of females from maleless families of ladybird beetles *Adalia bipunctata* L. (Coleoptera: Coccinellidae) naturally infected with *Rickettsia*, *Wolbachia*, and *Spiroplasma*. J. Invertebr. Pathol. *79*: 72–79.

Stackebrandt, E., W. Frederiksen, G.M. Garrity, P.A. Grimont, P. Kämpfer, M.C. Maiden, X. Nesme, R. Rosselló-Mora, J. Swings, H.G. Trüper, L. Vauterin, A.C. Ward and W.B. Whitman. 2002. Report of the ad hoc committee for the re-evaluation of the species definition in bacteriology. Int. J. Syst. Evol. Microbiol. *52*: 1043–1047.

Stamburski, C., J. Renaudin and J.M. Bove. 1991. First step toward a virus-derived vector for gene cloning and expression in spiroplasmas, organisms which read UGA as a tryptophan codon: synthesis of chloramphenicol acetyltransferase in *Spiroplasma citri*. J. Bacteriol. *173*: 2225–2230.

Stephens, M.A. 1980. Studies on *Spiroplasma* viruses. PhD thesis, University of East Anglia, Norwich, UK.

Stevens, C., A.Y. Tang, E. Jenkins, R.L. Goins, J.G. Tully, D.L. Rose, M. Konai, D.L. Williamson, P. Carle, J. Bove, K.J. Hackett, F.E. French, J. Wedincamp, R.B. Henegar and R.F. Whitcomb. 1997. *Spiroplasma lampyridicola* sp. nov., from the firefly beetle *Photuris pennsylvanicus*. Int. J. Syst. Bacteriol. *47*: 709–712.

Summers, C.G., A. S. Newton and D.C. Opgenorth. 2004. Overwintering of corn leafhopper, *Dalbulus maidis* (Homoptera: Cicadellidae), and *Spiroplasma kunkelii* (*Mycoplasmatales*: *Spiroplasmataceae*) in California's San Joaquin Valley. Environ. Entomol. *33*: 1644–1651.

Swofford, D.L. 1998. PAUP: Phylogenetic analysis using parsimony and other methods, 4 edn. Sinauer Associates, Sunderland, MA.

Taroura, S., Y. Shimada, Y. Sakata, T. Miyama, H. Hiraoka, M. Watanabe, K. Itamoto, M. Okuda and H. Inokuma. 2005. Detection of DNA of '*Candidatus* Mycoplasma haemominutum' and *Spiroplasma* sp. in unfed ticks collected from vegetation in Japan. J. Vet. Med. Sci. *67*: 1277–1279.

Tinsley, M.C. and M.E. Majerus. 2006. A new male-killing parasitism: *Spiroplasma* bacteria infect the ladybird beetle *Anisosticta novemdecimpunctata* (Coleoptera: Coccinellidae). Parasitology *132*: 757–765.

Tinsley, M.C. and M.E. Majerus. 2007. Small steps or giant leaps for male-killers? Phylogenetic constraints to male-killer host shifts. BMC Evol. Biol. *7*: 238.

Townsend, R., P.G. Markham, K.A. Plaskitt and M.J. Daniels. 1977. Isolation and characterization of a nonhelical strain of *Spiroplasma citri*. J. Gen. Microbiol. *100*: 15–21.

Townsend, R., D.B. Archer and K.A. Plaskitt. 1980a. Purification and preliminary characterization of *Spiroplasma* fibrils. J. Bacteriol. *142*: 694–700.

Townsend, R., J. Burgess and K.A. Plaskitt. 1980b. Morphology and ultrastructure of helical an nonhelical strains of *Spiroplasma citri*. J. Bacteriol. *142*: 973–981.

Trachtenberg, S. and R. Gilad. 2001. A bacterial linear motor: cellular and molecular organization of the contractile cytoskeleton of the helical bacterium *Spiroplasma melliferum* BC3. Mol. Microbiol. *41*: 827–848.

Trachtenberg, S., S.B. Andrews and R.D. Leapman. 2003a. Mass distribution and spatial organization of the linear bacterial motor of *Spiroplasma citri* R8A2. J. Bacteriol. *185*: 1987–1994.

Trachtenberg, S., R. Gilad and N. Geffen. 2003b. The bacterial linear motor of *Spiroplasma melliferum* BC3: from single molecules to swimming cells. Mol. Microbiol. *47*: 671–697.

Trachtenberg, S. 2004. Shaping and moving a *Spiroplasma*. J. Mol. Microbiol. Biotechnol. *7*: 78–87.

Trachtenberg, S. 2006. The cytoskeleton of *Spiroplasma*: a complex linear motor. J. Mol. Microbiol. Biotechnol. *11*: 265–283.

Trachtenberg, S., L.M. Dorward, V.V. Speransky, H. Jaffe, S.B. Andrews and R.D. Leapman. 2008. Structure of the cytoskeleton of *Spiroplasma melliferum* BC3 and its interactions with the cell membrane. J. Mol. Biol. *378*: 778–789.

Tully, J.G., R.F. Whitcomb, H.F. Clark and D.L. Williamson. 1977. Pathogenic mycoplasmas: cultivation and vertebrate pathogenicity of a new *Spiroplasma*. Science *195*: 892–894.

Tully, J.G., D.L. Rose, O. Garciajurado, J.C. Vignault, C. Saillard, J.M. Bové, R.E. McCoy and D.L. Williamson. 1980. Serological analysis of a new group of spiroplasmas. Curr. Microbiol. *3*: 369–372.

Tully, J.G., D.L. Rose, C.E. Yunker, J. Cory, R.F. Whitcomb and D.L. Williamson. 1981. Helical mycoplasmas (spiroplasmas) from *Ixodes* ticks. Science *212*: 1043–1045.

Tully, J.G., R.F. Whitcomb, D.L. Rose and J.M. Bové. 1982. *Spiroplasma mirum*, a new species from the rabbit tick (*Haemaphysalis leporispalustris*). Int. J. Syst. Bacteriol. *32*: 92–100.

Tully, J.G., D.L. Rose, E. Clark, P. Carle, J.M. Bové, R.B. Henegar, R.F. Whitcomb, D.E. Colflesh and D.L. Williamson. 1987. Revised group classification of the genus *Spiroplasma* (class *Mollicutes*), with proposed new groups XII to XXIII. Int. J. Syst. Bacteriol. *37*: 357–364.

Tully, J.G., J.M. Bové, F. Laigret and R.F. Whitcomb. 1993. Revised taxonomy of the class *Mollicutes* - proposed elevation of a monophyletic cluster of arthropod-associated *mollicutes* to ordinal rank (*Entomoplasmatales* ord. nov.), with provision for familial rank to separate species with nonhelical morphology (*Entomoplasmataceae* fam. nov.) from helical species (*Spiroplasmataceae*), and emended descriptions of the order *Mycoplasmatales*, family *Mycoplasmataceae*. Int. J. Syst. Bacteriol. *43*: 378–385.

Tully, J.G., D.L. Rose, C.E. Yunker, P. Carle, J.M. Bové, D.L. Williamson and R.F. Whitcomb. 1995. *Spiroplasma ixodetis* sp. nov., a new species from *Ixodes pacificus* ticks collected in Oregon. Int. J. Syst. Bacteriol. *45*: 23–28.

Van den Ent, F., L. A. Amos and J. Löwe. 2001. Prokaryotic origin of the actin cytoskeleton. Nature: 39–44.

Vaughn, E.E. and W.M. de Vos. 1995. Identification and characterization of the insertion element IS*1070* from *Leuconostoc lactis* NZ6009. Gene *155*: 95–100.

Vazeille-Falcoz, M., C. Hélias, F. Le Goff, F. Rodhain and C. Chastel. 1997. Three spiroplasmas isolated from *Haematopota* sp. (Diptera: Tabanidae) in France. J. Med. Entomol. *34*: 238–241.

Veneti, Z., J.K. Bentley, T. Koana, H.R. Braig and G.D.D. Hurst. 2005. A functional dosage compensation complex required for male killing in *Drosophila*. Science *307*: 1461–1463.

Wada, H. and R.R. Netz. 2007. Model for self-propulsive helical filaments: kink-pair propagation. Phys. Rev. Lett. *99*: 108102.

Wang, W., L. Rong, W. Gu, K. Du and J. Chen. 2003. Study on experimental infections of *Spiroplasma* from the Chinese mitten crab in crayfish, mice and embryonated chickens. Res. Microbiol. *154*: 677–680.

Wang, W., J. Chen, K. Du and Z. Xu. 2004a. Morphology of spiroplasmas in the Chinese mitten crab *Eriocheir sinensis* associated with tremor disease. Res Microbiol *155*: 630–635.

Wang, W., B. Wen, G.E. Gasparich, N. Zhu, L. Rong, J. Chen and Z. Xu. 2004b. A *Spiroplasma* associated with tremor disease in the Chinese mitten crab (*Eriocheir sinensis*). Microbiology *150*: 3035–3040.

Wang, W., W. Gu, Z. Ding, Y. Ren, J. Chen and Y. Hou. 2005. A novel *Spiroplasma* pathogen causing systemic infection in the crayfish *Procambarus clarkii* (Crustacea: Decapod), in China. FEMS Microbiol. Lett. *249*: 131–137.

Wang, W., W. Gu, G.E. Gasparich, K. Bi, J. Ou, Q. Meng, T. Liang, Q. Feng, J. Zhang and Y. Zhang. 2010. *Spiroplasma eriocheiris* sp. nov., a novel species associated with mortalities in *Eriocheiris sinensis*, Chinese mitten crab. Int. J. Syst. Evol. Microbiol. ijs.0.020529-Ov1-ijs.0.020529-0.

Wayadande, A.C. and J. Fletcher. 1995. Transmission of *Spiroplasma citri* lines and their ability to cross gut and salivary gland barriers within the leafhopper vector *Circulifer tenellus*. Phytopathology *85*: 1256–1259.

Wayadande, A.C. and J. Fletcher. 1998. Development and use of an established cell line of the leafhopper *Circulifer tenellus* to characterize *Spiroplasma citri*-vector interactions. J. Invertebr. Pathol. *72*: 126–131.

Wayne, L.G., D.J. Brenner, R.R. Colwell, P.A.D. Grimont, O. Kandler, M.I. Krichevsky, L.H. Moore, W.E.C. Moore, R.G.E. Murray, E. Stackebrandt, M.P. Starr and H.G. Trüper. 1987. Report of the ad hoc committee on the reconciliation of approaches to bacterial systematics. Int. J. Syst. Bacteriol. *37*: 463–464.

Weisburg, W.G., J.G. Tully, D.L. Rose, J.P. Petzel, H. Oyaizu, D. Yang, L. Mandelco, J. Sechrest, T.G. Lawrence, J. Van Etten, J. Maniloff and C.R. Woese. 1989. A phylogenetic analysis of the mycoplasmas: basis for their classification. J. Bacteriol. *171*: 6455–6467.

Whitcomb, R.F., J.G. Tully, J.M. Bové and P. Saglio. 1973. Spiroplasmas and acholeplasmas: multiplication in insects. Science *182*: 1251–1253.

Whitcomb, R.F. and D.L. Williamson. 1979. Pathogenicity of mycoplasmas for arthropods. Zentralbl. Bakteriol. Orig. A *245*: 200–221.

Whitcomb, R.F., J. G. Tully, P. McCawley and D.L. Rose. 1982. Application of the growth-inhibition test to *Spiroplasma* taxonomy. Int. J. Syst. Bacteriol. *32*: 387–394.

Whitcomb, R.F. 1983. Culture media for spiroplasmas. *In* Methods in Mycoplasmology, vol. 1 (edited by Razin and Tully). Academic Press, New York, pp. 147–158.

Whitcomb, R.F., T.A. Chen, D.L. Williamson, C. Liao, J.G. Tully, J.M. Bové, C. Mouches, D.L. Rose, M.E. Coan and T.B. Clark. 1986. *Spiroplasma kunkelii* sp. nov.: characterization of the etiologic agent of corn stunt disease. Int. J. Syst. Bacteriol. *36*: 170–178.

Whitcomb, R.F. and K.J. Hackett. 1987. Cloning by limiting dilution in liquid media: an improved alternative for cloning mollicute species. Isr. J. Med. Sci. *23*: 517.

Whitcomb, R.F. 1989. The biology of *Spiroplasma kunkelii*. *In* The Mycoplasmas, vol. 5 (edited by Whitcomb and Tully). Academic Press, New York, pp. 487–544.

Whitcomb, R.F., K. J. Hackett, J. G. Tully, E. A. Clark, F. E. French, R. B. Henegar, D. L. Rose and A.G. Wagner. 1990. Tabanid spiroplasmas as a model for mollicute biogeography. Zentrabl. Bakteriol. Suppl.: 931–934.

Whitcomb, R.F., C. Chastel, M. Abalain-Colloc, C. Stevens, J.G. Tully, D.L. Rose, P. Carle, J.M. Bové, R.B. Henegar, K.J. Hackett, T.B. Clark, M. Konai and D.L. Williamson. 1993a. *Spiroplasma cantharicola* sp. nov., from cantharid beetles (Coleoptera: Cantharidae). Int. J. Syst. Bacteriol. *43*: 421–424.

Whitcomb, R.F., J.G. Tully, D.L. Rose, P. Carle, J.M. Bové, R.B. Henegar, K.J. Hackett, T.B. Clark, M. Konai, J. Adams and D.L. Williamson. 1993b. *Spiroplasma monobiae* sp. nov. from the vespid wasp *Monobia quadridens* (Hymenoptera, Vespidae). Int. J. Syst. Bacteriol. *43*: 256–260.

Whitcomb, R.F., J.C. Vignault, J.G. Tully, D.L. Rose, P. Carle, J.M. Bové, K.J. Hackett, R.B. Henegar, M. Konai and D.L. Williamson. 1993c. *Spiroplasma clarkii* sp. nov. from the green June beetle (Coleoptera, Scarabaeidae). Int. J. Syst. Bacteriol. *43*: 261–265.

Whitcomb, R.F., G.E. Gasparich, F.E. French, J.G. Tully, D.L. Rose, P. Carle, J.M. Bové, R.B. Henegar, M. Konai, K.J. Hackett, J.R. Adams, T.B. Clark and D.L. Williamson. 1996. *Spiroplasma syrphidicola* sp.

nov., from a syrphid fly (Diptera: Syrphidae). Int. J. Syst. Bacteriol. *46*: 797–801.

Whitcomb, R.F., F. E. French, J. G. Tully, P. Carle, R. Henegar, K. J. Hackett, G. E. Gasparich and D.L. Williamson. 1997a. *Spiroplasma* species, groups, and subgroups from North American Tabanidae. Curr. Microbiol. *35*: 287–293.

Whitcomb, R.F., F.E. French, J.G. Tully, G.E. Gasparich, D.L. Rose, P. Carle, J. Bové, R.B. Henegar, M. Konai, K.J. Hackett, J.R. Adams, T.B. Clark and D.L. Williamson. 1997b. *Spiroplasma chrysopicola* sp. nov., *Spiroplasma gladiatoris* sp. nov., *Spiroplasma helicoides* sp. nov., and *Spiroplasma tabanidicola* sp. nov., from tabanid (Diptera: Tabanidae) flies. Int. J. Syst. Bacteriol. *47*: 713–719.

Whitcomb, R.F., F.E. French, J.G. Tully, D.L. Rose, P.M. Carle, J.M. Bove, E.A. Clark, R.B. Henegar, M. Konai, K.J. Hackett, J.R. Adams and D.L. Williamson. 1997c. *Spiroplasma montanense* sp. nov., from *Hybomitra* horseflies at northern latitudes in north America. Int. J. Syst. Bacteriol. *47*: 720–723.

Whitcomb, R.F., J. G. Tully, G. E. Gasparich, L. B. Regassa, D. L. Williamson and F.E. French. 2007. *Spiroplasma* species in the Costa Rican highlands: implications for biogeography and biodiversity. Biodivers. Conserv. *16*: 3877–3894.

Williamson, D.L. 1969. The sex ratio spirochete in *Drosophila robusta*. Jpn. J. Genet. *44*: 36–41.

Williamson, D.L. 1974. Unusual fibrils from the spirochete-like sex ratio organism. J. Bacteriol. *117*: 904–906.

Williamson, D.L. and R.F. Whitcomb. 1974. Helical wall-free prokaryotes in *Drosophila*, leafhoppers and plants. Colloq. Inst. Natl. Santé Rech. Med. *33*: 283–290.

Williamson, D.L. and R.F. Whitcomb. 1975. Plant mycoplasmas: a cultivable spiroplasma causes corn stunt disease. Science *188*: 1018–1020.

Williamson, D.L., R. F. Whitcomb and J.G. Tully. 1978. The *Spiroplasma* deformation test, a new serological method. Curr. Microbiol. *1*: 203–207.

Williamson, D.L., D.I. Blaustein, R.J.C. Levine and M.J. Elfvin. 1979a. Anti-actin-peroxidase staining of the helical wall-free prokaryote *Spiroplasma citri*. Curr. Microbiol. *2*: 143–145.

Williamson, D.L., J. G. Tully and R.F. Whitcomb. 1979b. Serological relationships of spiroplasmas as shown by combined deformation and metabolism inhibition tests. Int. J. Syst. Bacteriol. *29*: 345–351.

Williamson, D.L. and D.F. Poulson. 1979. Sex ratio organisms (spiroplasmas) of *Drosophila*. *In* The Mycoplasmas, vol. 3 (edited by Whitcomb and Tully). Academic Press, New York, pp. 175–208.

Williamson, D.L. and R.F. Whitcomb. 1983. Special serological tests for spiroplasma identification. *In* Methods in Mycoplasmology, vol. 2 (edited by Razin and Tully). Academic Press, New York, pp. 249–259.

Williamson, D.L., P.R. Brink and G.W. Zieve. 1984. *Spiroplasma fibrils*. Isr. J. Med. Sci. *20*: 830–835.

Williamson, D.L., J. G. Tully and R.F. Whitcomb. 1989. The genus *Spiroplasma*. *In* The Mycoplasmas, vol. 5 (edited by Whitcomb and Tully). Academic Press, San Diego, pp. 71–111.

Williamson, D.L., J. Renaudin and J.M. Bové. 1991. Nucleotide sequence of the *Spiroplasma citri* fibril protein gene. J. Bacteriol. *173*: 4353–4362.

Williamson, D.L., J.G. Tully, L. Rosen, D.L. Rose, R.F. Whitcomb, M.L. Abalain-Colloc, P. Carle, J.M. Bové and J. Smyth. 1996. *Spiroplasma diminutum* sp. nov., from *Culex annulus* mosquitoes collected in Taiwan. Int. J. Syst. Bacteriol. *46*: 229–233.

Williamson, D.L., J.R. Adams, R.F. Whitcomb, J.G. Tully, P. Carle, M. Konai, J.M. Bove and R.B. Henegar. 1997. *Spiroplasma platyhelix* sp. nov., a new mollicute with unusual morphology and genome size from the dragonfly *Pachydiplax longipennis*. Int. J. Syst. Bacteriol. *47*: 763–766.

Williamson, D.L., R.F. Whitcomb, J.G. Tully, G.E. Gasparich, D.L. Rose, P. Carle, J.M. Bové, K.J. Hackett, J.R. Adams, R.B. Henegar, M. Konai, C. Chastel and F.E. French. 1998. Revised group classification of the genus *Spiroplasma*. Int. J. Syst. Bacteriol. *48*: 1–12.

Williamson, D.L., B. Sakaguchi, K.J. Hackett, R.F. Whitcomb, J.G. Tully, P. Carle, J.M. Bové, J.R. Adams, M. Konai and R.B. Henegar. 1999. *Spiroplasma poulsonii* sp. nov., a new species associated with male-lethality in *Drosophila willistoni*, a neotropical species of fruit fly. Int. J. Syst. Bacteriol. *49*: 611–618.

Woese, C.R., J. Maniloff and L.B. Zablen. 1980. Phylogenetic analysis of the mycoplasmas. Proc. Natl. Acad. Sci. U. S. A. *77*: 494–498.

Woese, C.R. 1987. Bacterial evolution. Microbiol. Rev. *51*: 221–271.

Wolgemuth, C.W. and N.W. Charon. 2005. The kinky propulsion of *Spiroplasma*. Cell *122*: 827–828.

Wróblewski, H., K.E. Johansson and S. Hjérten. 1977. Purification and characterization of spiralin, the main protein of the *Spiroplasma citri* membrane. Biochim. Biophys. Acta *465*: 275–289.

Wróblewski, H., S. Nyström, A. Blanchard and A. Wieslander. 1989. Topology and acylation of spiralin. J. Bacteriol. *171*: 5039–5047.

Wróblewski, H., D. Robic, D. Thomas and A. Blanchard. 1984. Comparison of the amino acid compositions and antigenic properties of spiralins purified from the plasma membranes of different spiroplasmas. Ann. Microbiol. (Paris) *135A*: 73–82.

Ye, F., F. Laigret, J.C. Whitley, C. Citti, L.R. Finch, P. Carle, J. Renaudin and J.M. Bové. 1992. A physical and genetic map of the *Spiroplasma citri* genome. Nucleic Acids Res. *20*: 1559–1565.

Ye, F., F. Laigret and J.M. Bové. 1994a. A physical and genomic map of the prokaryote *Spiroplasma melliferum* and its comparison with the *Spiroplasma citri* map. C. R. Acad. Sci. Ser. III *317*: 392–398.

Ye, F., J. Renaudin, J.M. Bové and F. Laigret. 1994b. Cloning and sequencing of the replication origin (*oriC*) of the *Spiroplasma citri* chromosome and construction of autonomously replicating artificial plasmids. Curr. Microbiol. *29*: 23–29.

Ye, F., F. Laigret, P. Carle and J.M. Bové. 1995. Chromosomal heterogeneity among various strains of *Spiroplasma citri*. Int. J. Syst. Bacteriol. *45*: 729–734.

Ye, F., U. Melcher, J.E. Rascoe and J. Fletcher. 1996. Extensive chromosome aberrations in *Spiroplasma citri* strain BR3. Biochem. Genet. *34*: 269–286.

Yogev, D., R. Rosengarten, R. Watson-McKown and K.S. Wise. 1991. Molecular basis of *Mycoplasma* surface antigenic variation: a novel set of divergent genes undergo spontaneous mutation of periodic coding regions and 5′ regulatory sequences. EMBO J. *10*: 4069–4079.

Yu, J., A.C. Wayadande and J. Fletcher. 2000. *Spiroplasma citri* surface protein P89 implicated in adhesion to cells of the vector *Circulifer tenellus*. Phytopathology *90*: 716–722.

Zaaria, A., C. Fontenelle, M. Le Henaff and H. Wróblewski. 1990. Antigenic relatedness between the spiralins of *Spiroplasma citri* and *Spiroplasma melliferum*. J. Bacteriol. *172*: 5494–5496.

Zbinden, M. and M.A. Cambon-Bonavita. 2003. Occurrence of *Deferribacterales* and *Entomoplasmatales* in the deep-sea Alvinocarid shrimp *Rimicaris exoculata* gut. FEMS Microbiol. Ecol. *46*: 23–30.

Zhao, Y., R.W. Hammond, R. Jomantiene, E.L. Dally, I.M. Lee, H. Jia, H. Wu, S. Lin, P. Zhang, S. Kenton, F.Z. Najar, A. Hua, B.A. Roe, J. Fletcher and R.E. Davis. 2003. Gene content and organization of an 85-kb DNA segment from the genome of the phytopathogenic mollicute *Spiroplasma kunkelii*. Mol. Genet. Genomics *269*: 592–602.

Zhao, Y., R.W. Hammond, I.M. Lee, B.A. Roe, S. Lin and R.E. Davis. 2004a. Cell division gene cluster in *Spiroplasma kunkelii*: functional characterization of *ftsZ* and the first report of *ftsA* in mollicutes. DNA Cell Biol. *23*: 127–134.

Zhao, Y., H. Wang, R.W. Hammond, R. Jomantiene, Q. Liu, S. Lin, B.A. Roe and R.E. Davis. 2004b. Predicted ATP-binding cassette systems in the phytopathogenic mollicute *Spiroplasma kunkelii*. Mol. Genet. Genomics *271*: 325–338.

Order III. **Acholeplasmatales** Freundt, Whitcomb, Barile, Razin and Tully 1984, 348[VP]

Daniel R. Brown, Janet M. Bradbury and Karl-Erik Johansson

A.cho.le.plas.ma.ta'les. N.L. neut. n. *Acholeplasma* type genus of the order; *-ales* ending to denote an order; N.L. fem. pl. n. *Acholeplasmatales* the *Acholeplasma* order.

This order in the class *Mollicutes* is assigned to a group of wall-less prokaryotes that do not require sterol for growth and occur in a wide variety of habitats, including many vertebrate hosts, insects, and plants. A single family, *Acholeplasmataceae*, and a single genus, *Acholeplasma*, recognize the prominent and distinct characteristics of the assigned organisms.

Type genus: **Acholeplasma** Edward and Freundt 1970, 1[AL].

Further descriptive information

The trivial name acholeplasma(s) is commonly used when reference is made to species of this order. The initial proposal for elevation of the acholeplasmas to ordinal rank (Freundt et al., 1984) was based primarily on the universal lack of a sterol requirement for growth of *Acholeplasma* species, in addition to other major genetic, nutritional, biochemical, and physiological characteristics that distinguish them from other members of the class *Mollicutes*. A subsequent proposal for an additional order, *Entomoplasmatales* (Tully et al., 1993), within the class to distinguish a group of mollicutes that are phylogenetically more closely related to the *Mycoplasmatales* than to acholeplasmas necessitated further revisions within the class.

Although most mollicutes require exogenous cholesterol or serum for growth, all species within the genus *Acholeplasma* and some assigned to the genera *Asteroleplasma*, *Spiroplasma*, and *Mesoplasma* do not have that requirement. The species that do not have a sterol requirement can easily be excluded from the sterol-requiring taxa by tests that measure growth responses to cholesterol or to a number of serum-free broth preparations (Edward, 1971; Razin and Tully, 1970; Rose et al., 1993; Tully, 1995). For instance, the *Acholeplasmatales* grow through end-point dilutions in serum-containing medium and in serum-free preparations, indicating the absence of a growth requirement for cholesterol.

Analyses of rRNA and other genes have shown that a large group of uncultured, plant-pathogenic organisms referred to by the trivial name phytoplasmas (Sears and Kirkpatrick, 1994) are closely related to acholeplasmas (Lim and Sears, 1992; Toth et al., 1994). The 16S rRNA gene sequences for members of the genus *Acholeplasma* that have been determined so far show that the acholeplasmas form two clades, one of which is a sister lineage to the phytoplasmas, although the formal taxonomic assignment of "*Candidatus* Phytoplasma" proposed gen. nov. (IRPCM Phytoplasma/Spiroplasma Working Team – Phytoplasma Taxonomy Group, 2004) currently remains *incertae sedis*.

References

Edward, D.G. and E.A. Freundt. 1970. Amended nomenclature for strains related to *Mycoplasma laidlawii*. J. Gen. Microbiol. *62*: 1–2.

Edward, D.G. 1971. Determination of sterol requirement for *Mycoplasmatales*. J. Gen. Microbiol. *69*: 205–210.

Freundt, E.A., R.F. Whitcomb, M.F. Barile, S. Razin and J.G. Tully. 1984. Proposal for elevation of the family *Acholeplasmataceae* to ordinal rank: *Acholeplasmatales*. Int. J. Syst. Bacteriol. *34*: 346–349.

IRPCM Phytoplasma/Spiroplasma Working Team – Phytoplasma Taxonomy Group. 2004. Description of the genus '*Candidatus* Phytoplasma', a taxon for the wall-less non-helical prokaryotes that colonize plant phloem and insects. Int. J. Syst. Evol. Microbiol. *54*: 1243–1255.

Lim, P.O. and B.B. Sears. 1992. Evolutionary relationships of a plant-pathogenic mycoplasmalike organism and *Acholeplasma laidlawii* deduced from two ribosomal protein gene sequences. J. Bacteriol. *174*: 2606–2611.

Razin, S. and J.G. Tully. 1970. Cholesterol requirement of mycoplasmas. J. Bacteriol. *102*: 306–310.

Rose, D.L., J.G. Tully, J.M. Bové and R.F. Whitcomb. 1993. A test for measuring growth responses of *Mollicutes* to serum and polyoxyethylene sorbitan. Int. J. Syst. Bacteriol. *43*: 527–532.

Sears, B.B. and B.C. Kirkpatrick. 1994. Unveiling the evolutionary relationships of plant pathogenic mycoplasmalike organisms. ASM News *60*: 307–312.

Toth, K.F., N. Harrison and B.B. Sears. 1994. Phylogenetic relationships among members of the class *Mollicutes* deduced from rps3 gene sequences. Int. J. Syst. Bacteriol. *44*: 119–124.

Tully, J.G., J.M. Bové, F. Laigret and R.F. Whitcomb. 1993. Revised taxonomy of the class *Mollicutes* - proposed elevation of a monophyletic cluster of arthropod-associated mollicutes to ordinal rank (*Entomoplasmatales* ord. nov.), with provision for familial rank to separate species with nonhelical morphology (*Entomoplasmataceae* fam. nov.) from helical species (*Spiroplasmataceae*), and emended descriptions of the order *Mycoplasmatales*, family *Mycoplasmataceae*. Int. J. Syst. Bacteriol. *43*: 378–385.

Tully, J.G. 1995. Determination of cholesterol and polyoxyethylene sorbitan growth requirements of mollicutes. *In* Molecular and Diagnostic Procedures in Mycoplasmology, vol. 1 (edited by Razin and Tully). Academic Press, San Diego, pp. 381–389.

Family I. **Acholeplasmataceae** Edward and Freundt 1970, 1[AL]

Daniel R. Brown, Janet M. Bradbury and Karl-Erik Johansson

A.cho.le.plas.ma.ta.ce'ae. N.L. neut. n. *Acholeplasma*, *-atos* type genus of the family; *-aceae* ending to denote a family; N.L. fem. pl. n. *Acholeplasmataceae* the *Acholeplasma* family.

Type genus: **Acholeplasma** Edward and Freundt 1970, 1[AL].

Further descriptive information

This family is monotypic, so its properties are essentially those of the genus *Acholeplasma*.

Genus I. **Acholeplasma** Edward and Freundt 1970, 1[AL]

Daniel R. Brown, Janet M. Bradbury and Karl-Erik Johansson

A.cho.le.plas′ma. Gr. pref. *a* not; Gr. n. *chole* bile; Gr. neut. n. *plasma* something formed or molded, a form; N.L. neut. n. *Acholeplasma* name intended to indicate that cholesterol, a constituent of bile, is not required.

Cells are **spherical**, with a diameter of about 300 nm, or **filamentous**, 2–5 μm long. **Nonmotile.** Colonies have a **"fried-egg" appearance** and may reach 2–3 mm in diameter. Facultatively anaerobic; most strains grow readily in simple media. All members **lack a sterol requirement for growth. Chemo-organotrophic**, most species utilizing glucose and other sugars as the major energy sources. Many strains are capable of fatty acid biosynthesis from acetate. Arginine and urea are not hydrolyzed. Pigmented carotenoids occur in some species. All species are resistant, or only slightly susceptible, to 1.5% digitonin. Saprophytes found in soil, compost, wastewaters, or commensals of vertebrates, insects, or plants. None are known to be a primary pathogen, but they may cause cytopathic effects in tissue cultures. The **genome sizes** range from about **1500 to 2100 kbp**. All species examined utilize the universal genetic code in which UGA is a stop codon.

DNA G+C content (mol%): 27–38.

Type species: **Acholeplasma laidlawii** (Sabin 1941) Edward and Freundt 1970, 1[AL] (*Sapromyces laidlawi* Sabin 1941, 334).

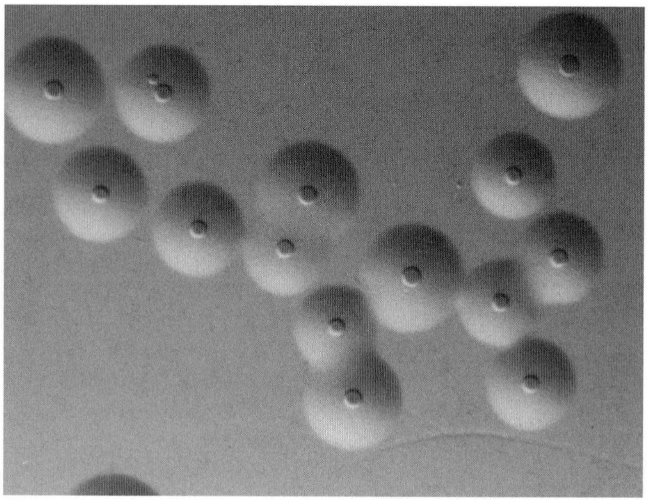

FIGURE 114. Colonies of *Acholeplasma laidlawii* PG8[T] (=NCTC 10116[T]; diameter 0.15–0.25 mm) after 3 d growth on Mycoplasma Experience Solid Medium at 36°C in 95% nitrogen/5% carbon dioxide. Original magnification = 25×. Image provided by Helena Windsor and David Windsor.

Further descriptive information

Cells of acholeplasmas typically appear as pleomorphic coccoid, coccobacillary, or short filamentous forms when grown in mycoplasma broth containing 20% horse serum or 1% bovine serum fraction. Viable spherical cells usually have a minimum diameter of about 300 nm. Filaments may be as much as 500 nm in length, but some longer filaments and branching filaments occur in some strains. Filaments often show beading with eventual development of coccoid forms. Cellular morphology may also depend upon the ratio of unsaturated to saturated fatty acids in the medium. Adjustment of preparative materials to the osmolarity of the culture medium is necessary for proper morphological examination.

Most acholeplasmas exhibit heavy turbidity when grown aerobically in broth containing 5–20% serum, usually of horse or fetal bovine origin, or when grown in 1% bovine serum fraction broth at 37°C. Less turbidity is evident when most acholeplasmas are cultured in serum-free broth and some species may be inhibited in media containing 20% horse serum. Strains of some acholeplasmas (*Acholeplasma morum*, *Acholeplasma modicum*, and *Acholeplasma axanthum*) may not grow well in serum-free medium unless glucose and some fatty acids (Tween 80 and palmitic acid) are included. Colonies on solid medium containing serum or bovine serum fraction are usually large (100–200 nm in diameter) with the classical "fried-egg" appearance after 24–72 h at 37°C (Figure 114). Colonies of *Acholeplasma axanthum* and several other acholeplasmas may show only central zones of growth into the agar or other unusual colony forms, such as mulberry-like colonies. Most acholeplasmas display optimum growth at 37°C. Growth is much slower at 25–27°C and strains may require 7–10 d to reach the turbidity observed after 24 h at 37°C. Species of plant origin (*Acholeplasma brassicae* and *Acholeplasma palmae*) have an optimum growth temperature of 30°C.

Most species in the genus are strong fermenters and produce acid from glucose metabolism, although a few species such as *Acholeplasma parvum* may not ferment glucose or other carbohydrates (Table 143). Fermentation of mannose is usually negative, although several species do catabolize this carbohydrate. All *Acholeplasma* species examined possess a fructose 1,6-diphosphate-activated lactate dehydrogenase, which is a property shared with certain streptococci.

Gourlay (1970) found that a fresh isolate of *Acholeplasma laidlawii* from a bovine source was infected with a filamentous, single-stranded DNA virus designated L1 (Bruce et al., 1972; Maniloff, 1992). Later, L2 and L3 viruses were also isolated from *Acholeplasma laidlawii* (Gourlay, 1971, 1972, 1973; Gourlay et al., 1973). L2 virus is a quasi-spherical, double-stranded DNA virus (Maniloff et al., 1977), and L3 is a short-tailed phage with double-stranded DNA (Garwes et al., 1975; Gourlay, 1974; Haberer et al., 1979; Maniloff et al., 1977). Another virus isolated from *Acholeplasma laidlawii* is L172, a single-stranded DNA, quasi-spherical virus that is different from L1 (Liska, 1972). Two viruses have been isolated from other acholeplasmas, including one from *Acholeplasma modicum*, designated M1 (Congdon et al., 1979), and from *Acholeplasma oculi* strain PG49 (designated O1) (Ichimaru and Nakamura, 1983). The nucleic acid structure of the last two viruses has not been defined.

Antisera to filter-cloned whole-cell antigens are utilized in several serological techniques to assess the antigenic structure of acholeplasmas and to provide identification of the organism to the species level (Tully, 1979). The three most useful techniques are growth inhibition (Clyde, 1983), plate immunofluorescence

TABLE 143. Differential characteristics of the species of the genus *Acholeplasma*[a]

Characteristic	*A. laidlawii*	*A. axanthum*	*A. brassicae*	*A. cavigenitalium*	*A. equifetale*	*A. granularum*	*A. hippikon*	*A. modicum*	*A. morum*	*A. multilocale*	*A. oculi*	*A. palmae*	*A. parvum*	*A. pleciae*	*A. vituli*
Glucose fermentation	+	+	+	+	+	+	+	+	+	+	+	+	–	+	+
Mannose fermentation	–	–	–	–	+	–	+	–	–	+	–	–	nd	nd	+
Arbutin hydrolysis	+	+	–	–	nd	–	nd	–	+	nd	+	–	nd	nd	–
Esculin hydrolysis	+	+	nd	nd	nd	–	nd	–	+	nd	+	nd	–	nd	–
Film and spots	+	–	nd	–	+	–	+	–	–	+	–	nd	nd	nd	–
Benzyl viologen reduction	+	+	+	+	+	+	+	+	+	–	+	+	+	nd	nd
DNA G+C content (mol%)	31–36	31	35.5	36	30.5	30–32	33	29	34	31	27	30	29	31.6	37.6–38.3

[a]Symbols: +, >85% positive; –, 0–15% positive; nd, not determined.

(Gardella et al., 1983; Tully, 1973), and metabolism inhibition (Taylor-Robinson, 1983).

Acholeplasmas may be the most common mollicutes in vertebrate animals and they are found frequently in the upper respiratory tract and urogenital tract of such hosts (Tully, 1979, 1996). Eukaryotic cells in continuous culture are frequently contaminated with acholeplasmas, primarily from the occurrence of acholeplasmas in animal serum used in tissue culture media. At least five *Acholeplasma* species have been identified on plant surfaces (*Acholeplasma axanthum*, *Acholeplasma brassicae*, *Acholeplasma laidlawii*, *Acholeplasma oculi*, and *Acholeplasma palmae*), possibly representing contamination from insects. However, with the exception of *Acholeplasma pleciae* (Knight, 2004), the only acholeplasmas identified from insects have been from mosquitoes. *Acholeplasma laidlawii* was identified in a pool of *Anopheles sinensis*, and a strain of *Acholeplasma morum* was present in a pool of *Armigeres subalbatus* (D.L. Williamson and J.G. Tully, unpublished).

Little evidence exists for a pathogenic role of acholeplasmas in natural diseases. The widespread distribution of acholeplasmas in both healthy and diseased animal tissues and of antibodies against acholeplasmas in most animal sera complicates experimental pathogenicity studies. However, *Acholeplasma axanthum* was pathogenic for goslings and young goose embryos (Kisary et al., 1975, 1976). Inoculation into leafhoppers, including those known to be vectors of plant mycoplasma diseases, shows multiplication and prolonged persistence of acholeplasmas in host tissues (Eden-Green and Markham, 1987; Whitcomb et al., 1973; Whitcomb and Williamson, 1975), but there is no evidence that the few *Acholeplasma* species found on plant surfaces play any role in plant or insect disease.

A few recent reports are available on the antibiotic sensitivity of acholeplasmas and whether the actions of these drugs are inhibitory to growth or kill cells. Acholeplasmas are sensitive to the following antibiotics (minimum inhibitory concentration range in µg/ml): tetracycline, 0.5–25.0; erythromycin, 0.03–1.0; lincomycin, 0.25–1.0; tylosin tartartate, 0.1–12.5; and kanamycin, 20–200 (Kato et al., 1972; Lewis and Poland, 1978; Ogata et al., 1971).

Enrichment and isolation procedures

Typical steps in isolation of all mollicutes were outlined in the recently revised minimal standards for descriptions of novel species (Brown et al., 2007). Techniques for isolation of acholeplasmas from animal tissues and from cell cultures have been described (Tully, 1983). Although colonies are occasionally first detected on blood agar, complex undefined media such as American Type Culture Collection medium 988 (SP-4) are usually required for primary isolation and maintenance. Cell wall-targeting antibiotics are included to discourage growth of other bacteria. Phenol red facilitates detection of species that excrete acidic or alkaline metabolites. Commonly used alternatives such as Frey's, Hayflick's and Friis' media differ from SP-4 mainly in the proportions of inorganic salts, amino acids, serum sources, and types of antibiotics included. Broths are incubated aerobically at 37°C for 14 d and examined periodically for turbidity or pH changes, either to acid or alkaline levels. Tubes showing turbidity are plated to agar prepared from the same medium formulation, and the plates are incubated at 37°C in an atmosphere of 95% N_2, 5% CO_2, as in the GasPak system. Tubes without obvious turbidity should be plated at the end of the 14-d incubation period. Initial isolates may contain a mixture of species, so cloning by repeated filtration through membrane filters with a pore size of 450 or 220 nm is essential. The initial filtrate and dilutions of it are cultured on solid medium and an isolated colony is subsequently picked from a plate on which only a few colonies have developed. This colony is used to found a new cultural line, which is then expanded, filtered, plated, and picked two additional times. Identification is confirmed by additional biochemical and serological tests.

Maintenance procedures

Stock acholeplasma cultures can be maintained in either mycoplasma broth medium containing 5–20% serum or in the serum-fraction broth formulation at room temperature (25–30°C) with only weekly transfer (Tully, 1995). Maintenance is best in broth medium devoid of glucose, since excess acid production reduces viability. Stock cultures can also be maintained indefinitely when frozen at –70°C. Agar colonies can also be maintained for 1–2 weeks at 25°C if plates are sealed to prevent drying. For optimum preservation, acholeplasmas should be lyophilized directly in the culture medium when the broth cultures reach a mid-exponential phase, usually 1–2 d at 37°C. Lyophilized cultures should be sealed under vacuum and stored at 4°C (Leach, 1983).

Differentiation of the genus *Acholeplasma* from other genera

Properties that partially fulfill criteria for assignment to the class *Mollicutes* (Brown et al., 2007) include absence of a cell wall, filterability, and the presence of conserved 16S rRNA gene sequences. They usually possess two 16S rRNA operons. Aerobic or facultative anaerobic growth in artificial media and the absence of a requirement for sterols or cholesterol for growth exclude assignment to the genera *Anaeroplasma*, *Asteroleplasma*, "*Candidatus* Phytoplasma", *Mycoplasma*, or *Ureaplasma*. Absence of a spiral cellular morphology, regular association with a vertebrate host or fluids of vertebrate origin, and regular use of the codon UGG to encode tryptophan (Knight, 2004) and UGA as a stop codon (Tanaka et al., 1989, 1991) support exclusion from the genera *Spiroplasma*, *Entomoplasma*, or *Mesoplasma*. Reduction of the redox indicator benzyl viologen has been reported to be fairly specific for differentiation of the genus *Acholeplasma* from other mollicutes (Pollack et al., 1996a). Only *Acholeplasma multilocale* failed to give a positive reaction, although several *Mesoplasma* and *Entomoplasma* species yielded variable responses to the test (Pollack et al., 1996a). Most acholeplasmas have membrane-localized NADH oxidase activity, in comparison to the NADH oxidase activity located in the cytoplasm of other genera within the class. Another special characteristic is the occurrence in most acholeplasmas of unique pyrimidine enzymic activities, especially a dUTPase enzyme, with the possible exception again of *Acholeplasma multilocale* (Pollack et al., 1996b). Acholeplasmas may possess a number of other biological characteristics that may distinguish them from other genera within the class *Mollicutes*, including polyterpenol synthesis (Smith and Langworthy, 1979), positional distribution of fatty acids (Rottem and Markowitz, 1979), the presence of superoxide dismutase (Kirby et al., 1980; Lee and Kenny, 1984; Lynch and Cole, 1980; O'Brien et al., 1981), and the presence of spacer tRNA (Nakagawa et al., 1992). However, most of these features have not been established for even a majority of *Acholeplasma* species.

Taxonomic comments

Acholeplasma genome sizes range from 1215 to 2095 kbp by pulsed-field gel electrophoresis or complete DNA sequencing, but most are in a more narrow range of 1215–1610 kbp (Carle et al., 1993; Neimark et al., 1992) that overlaps with genome sizes of many *Spiroplasma* species. Tests of eight *Acholeplasma* species showed less than 8% DNA–DNA hybridization between type strains and surprisingly extensive genomic heterogeneity within species (Aulakh et al., 1983; Stephens et al., 1983a, b). The highest level of relatedness, 21% DNA–DNA hybridization, was between the type strains of *Acholeplasma laidlawii* and *Acholeplasma granularum*. Some strain pairs, such as within *Acholeplasma laidlawii*, shared as little as 40% DNA–DNA hybridization, differences that in other genera would have justified subdivision of an apparently diverse strain complex into component species. However, no polyphasic taxonomic basis was found to support such designations. Restriction endonuclease digest patterns also reflect heterogeneity within some species (Razin et al., 1983). The DNA–DNA hybridization and restriction digest patterns of eight *Acholeplasma axanthum* strains isolated from a variety of hosts and habitats differed markedly from each other and some heterogeneity occurred among six different *Acholeplasma oculi* strains.

Mesoplasma pleciae was first isolated from the hemolymph of a larva of a *Plecia* corn root maggot and assigned to the genus *Mesoplasma* because sustained growth occurred in serum-free mycoplasma broth only when the medium contained 0.04% Tween 80 fatty acid mixture (Tully et al., 1994a). However, 16S rRNA gene sequence similarities and its preferred use of UGG rather than UGA to encode tryptophan support proper reclassification as *Acholeplasma pleciae* comb. nov. (Knight, 2004); the type strain is PS-1^T (Tully et al., 1994a).

Mycoplasma feliminutum was first described during a time when the only named genus of mollicutes was *Mycoplasma*. Its publication coincided with the first proposal of the genus *Acholeplasma* (Edward and Freundt, 1969, 1970), with which *Mycoplasma feliminutum* is properly affiliated through established phenotypic (Heyward et al., 1969) and 16S rRNA gene sequence (Brown et al., 1995; Johansson and Pettersson, 2002) similarities. This explains the apparent inconsistencies with its assignment to the genus *Mycoplasma*. The name *Mycoplasma feliminutum* should therefore be revised to *Acholeplasma feliminutum* comb. nov.; the type strain is Ben^T (=ATCC 25749^T; Heyward et al., 1969).

The lack of signature enzymic activities cast serious doubt on the status of *Acholeplasma multilocale* PN525^T as an authentic member of the genus *Acholeplasma* (Pollack et al., 1996b). It may be affiliated with an unrecognized metabolic subgroup, but it seems more likely to be a strain of *Mycoplasma* or *Entomoplasma*.

Acknowledgements

The major contributions to the foundation of this material by Joseph G. Tully are gratefully acknowledged.

Further reading

Taylor-Robinson, D. and J.G. Tully. 1998. Mycoplasmas, ureaplasmas, spiroplasmas, and related organisms. *In* Topley and Wilson, Principles and Practice of Microbiology, 9th edn, vol. 2 (edited by Balows and Duerden). Arnold Publishers, London, pp. 799–827.

Tully, J.G. 1989. Class *Mollicutes*: new perspectives from plant and arthropod studies. *In* The Mycoplasmas (edited by Whitcomb and Tully). Academic Press, San Diego, pp. 1–31.

Differentiation of the species of the genus *Acholeplasma*

Esculin hydrolysis by a β-D-glucosidase and arbutin hydrolysis are sometimes useful diagnostic tests for differentiation of some acholeplasmas (Bradbury, 1977; Rose and Tully, 1983). The production of carotenoid pigments, principally neurosporene, has been used to differentiate some acholeplasmas, especially *Acholeplasma axanthum* and *Acholeplasma modicum* (Mayberry et al., 1974; Smith and Langworthy, 1979; Tully and Razin, 1970). Carotenoids are also synthesized in some strains of *Acholeplasma laidlawii* under certain growth conditions (Johansson, 1974). The "film and spots" reaction, which occurs in a number of *Mycoplasma* and several *Acholeplasma* species, relates to the production of crystallized calcium soaps of fatty acids on the surface of agar plates (Edward, 1954; Fabricant and Freundt, 1967). Fatty acids are liberated from the serum or supplemental egg yolk (Fabricant and Freundt, 1967; Thorns and Boughton, 1978) in the agar medium by the lipolytic activity of the organisms. Failure to cross-react with antisera against previously recognized species provides evidence for species novelty. For this reason, deposition of antiserum against a novel type strain into a recognized collection is still mandatory for new species descriptions (Brown et al., 2007). Preliminary differentiation can be by PCR and DNA sequencing using primers specific for bacterial 16S rRNA genes or the 16S–23S intergenic region. A similarity matrix relating the candidate strain to its closest neighbors, usually species with >0.94 16S rRNA gene sequence similarity, will suggest an assemblage of related species that should be examined for serological cross-reactivities.

List of species of the genus *Acholeplasma*

1. **Acholeplasma laidlawii** (Sabin 1941) Edward and Freundt 1970, 1[AL] (*Sapromyces laidlawi* Sabin 1941, 334)

laid.law'i.i. N.L. gen. masc. n. *laidlawii* of Laidlaw, named after Patrick P. Laidlaw, one of the microbiologists who first isolated this species.

This is the type species of the genus *Acholeplasma*. Filaments, usually relatively short, although much longer branched filaments may develop in media with certain ratios of saturated to unsaturated fatty acids. Coccoid forms may predominate in certain cultures including co-culture with eukaryotic cells. Agar colonies are large for a mollicute and exhibit well-developed central zones and peripheral growth on horse serum agar. On serum-free agar, colonies are smaller and may show only the central zone of growth into the agar. Relatively strong turbidity is produced during growth in broth containing serum. Temperature range for growth is 20–41°C with optimum at 37°C, even for strains recovered from plant or non-animal sources. Usually produces large amounts of carotenoids when cultivated in the presence of PPLO serum fraction (Difco).

Serologically distinct from most established species in the genus, but partial cross-reactions may occur with *Acholeplasma granularum* strains. DNA–DNA hybridization between strains of this species range from 40 to >80%. *Acholeplasma granularum* strain BTS-39^T showed 20% hybridization with *Acholeplasma laidlawii* strain PG8^T. Pathogenicity has not been established.

Source: isolated from sewage, manure, humus, soil, and many animal hosts and their tissues, including some isolates from the human oral cavity, vagina, and wounds. Has been recovered from the surfaces of some plants, although few isolations have been reported from insect hosts. Frequent contaminant of eukaryotic cell cultures.

DNA G+C content (mol%): 31.7–35.7 (Bd, T_m).

Type strain: ATCC 23206, PG8, NCTC 10116, CIP 75.27, NBRC 14400.

Sequence accession no. (16S rRNA gene): U14905.

Further comment: on the Approved Lists of Bacterial Names and on the Approved Lists of Bacterial Names (Amended Edition), this taxon is incorrectly cited as *Acholeplasma laidlawii* [Freundt 1955 (*sic*)] Edward and Freundt (1970).

2. **Acholeplasma axanthum** Tully and Razin 1970, 754[AL]

a.xan'thum. Gr. pref. *a* not, without; Gr. adj. *xanthos -ê -on* yellow; N.L. neut. adj. *axanthum* without yellow (pigment).

Predominantly coccobacillary and coccoid with a few short myceloid elements. Large colonies with clearly marked centers form on horse serum agar; colonies on serum-free agar are smaller and usually lack the peripheral growth around their center. Agar colonies produce zones of β-hemolysis by the overlay technique. Growth in media devoid of serum or serum fraction is much poorer than for other acholeplasmas. Minimal nutritional requirements are poorly defined, but marked stimulation of growth with polyoxyethylene sorbitan (Tween 80) suggests a requirement for fatty acids. Temperature range for growth is 22–37°C with optimum growth at 37°C. Synthesis of carotenoid pigments can be demonstrated only when large volume cultures are tested. Produces sphingolipids. No evidence for pathogenicity.

Source: originally isolated from murine leukemia tissue culture cells, but numerous subsequent isolations of the organism from bovine serum and a variety of bovine tissue sites (nasal cavity, lymph nodes, kidney) suggest cell-culture contamination was of bovine serum origin. Also isolated from variety of other animals and surfaces of some plants.

DNA G+C content (mol%): 31 (Bd).

Type strain: S-743, ATCC 25176, NCTC 10138.

Sequence accession no. (16S rRNA gene): AF412968.

3. **Acholeplasma brassicae** Tully, Whitcomb, Rose, Bové, Carle, Somerson, Williamson and Eden-Green 1994b, 683[VP]

bras.si'cae. L. fem. gen. n. *brassicae* of cabbage, referring to the plant origin of the organism.

Cells are primarily coccoid. Temperature range for growth is 18–37°C. Optimal growth occurs at 30°C. No evidence for pathogenicity.

Source: isolated as a surface contaminant from broccoli (*Brassica oleracea* var. *italica*).

DNA G+C content (mol%): 35.5 (Bd, T_m, HPLC).

Type strain: 0502, ATCC 49388.

Sequence accession no. (16S rRNA gene): AY538163.

4. **Acholeplasma cavigenitalium** Hill 1992, 591[VP]

ca.vi.ge.ni.ta'li.um. N.L. n. *cavia* guinea pig (*Cavia cobaya*); L. pl. n. *genitalia -ium* the genitals; N.L. pl. gen. n. *cavigenitalium* of guinea pig genitals.

Pleomorphic cells, mostly coccoid. Grows on broth or agar medium under aerobic conditions, with optimum temperature between 35 and 37°C. Colonies on agar medium have typical fried-egg appearance. Originally described as a non-fermenter, but the type strain ferments glucose. Does not grow well on SP-4 broth or in horse serum broth, but grows well on simple base medium with additions of 10–15% fetal bovine serum. No evidence for pathogenicity.

Source: isolated from the vagina of guinea pigs.

DNA G+C content (mol%): 36 (Bd).

Type strain: GP3, NCTC 11727, ATCC 49901.

Sequence accession no. (16S rRNA gene): AY538164.

5. **Acholeplasma entomophilum** Tully, Rose, Carle, Bové, Hackett and Whitcomb 1988, 166[VP]

en.to.mo.phi'lum. Gr. n. *entomon* insect; N.L. neut. adj. *philum* (from Gr. neut. adj. *philon*) friend, loving; N.L. neut. adj. *entomophilum* insect-loving.

Cells are pleomorphic, but primarily coccoid. Colonies on solid medium usually have a fried-egg appearance. Acid is produced from glucose, but not mannose. Carotenoids are not produced. "Film and spot" reaction is negative. Agar colonies hemadsorb guinea pig erythrocytes. Strains require 0.4% Tween 80 or fatty acid supplements for growth in serum-free media. Temperature range for growth is 23–32°C, with optimum growth at about 30°C. Pathogenicity has not been established.

Source: isolated from gut contents of tabanid flies, beetles, butterflies, honey bees, and moths, and from flowers.

DNA G+C content (mol%): 30 (Bd).

Type strain: TAC, ATCC 43706.

Sequence accession no. (16S rRNA gene): M23931.

Further comment: with the proposal of the order *Entomoplasmatales* (Tully et al., 1993), *Acholeplasma entomophilum* was transferred to the family *Entomoplasmataceae.* The name *Acholeplasma entomophilum* was therefore revised to *Mesoplasma entomophilum* comb. nov. The type strain is TAC^T (=ATCC 43706^T; Tully et al., 1988).

6. **Acholeplasma equifetale** Kirchhoff 1978, 81[AL]

eq.ui.fe.ta'le. L. n. *equus* horse; N.L. adj. *fetalis -is -e* pertaining to the fetus; N.L. neut. adj. *equifetale* pertaining to the horse fetus.

Cells are pleomorphic, but predominantly coccoid. Colonies on solid medium containing serum usually have a fried-egg appearance; on serum-free medium, colonies are similar, but usually smaller. Growth temperature range is 22–37°C. Pathogenicity has not been established.

Source: isolated from the lung and liver of aborted horse fetuses. Also recovered from the respiratory tract of apparently normal horses and the respiratory tract and cloacae of broiler chickens (Bradbury, 1978).

DNA G+C content (mol%): 30.5 (Bd).

Type strain: C112, ATCC 29724, NCTC 10171.

Sequence accession no. (16S rRNA gene): AY538165.

Further comment: Kirchhoff is incorrectly cited as "Kirchoff" on the Approved Lists of Bacterial Names.

7. **Acholeplasma florum** McCoy, Basham, Tully, Rose, Carle and Bové 1984, 14[VP]

flo'rum. L. gen. pl. n. *florum* of flowers, indicating the recovery site of the organism.

Cells are ovoid. Colonies on agar are umbonate. Films and spots are produced on serum-containing media. Glucose is utilized, but mannose is not. Carotenes are not produced, nor is β-D-glucosidase. Pathogenicity has not been established.

Source: the known strains were isolated from flower surfaces.

DNA G+C content (mol%): 27.3 (Bd).

Type strain: L1, ATCC 33453.

Sequence accession nos: AF300327 (16S rRNA gene), NC_006055 (strain L1^T complete genome).

Further comment: with the proposal of the order *Entomoplasmatales* (Tully et al., 1993), *Acholeplasma florum* was transferred to the family *Entomoplasmataceae.* The name *Acholeplasma florum* was therefore revised to *Mesoplasma florum* comb. nov. The type strain is L1^T (=ATCC 33453^T; McCoy et al., 1984).

8. **Acholeplasma granularum** (Switzer 1964) Edward and Freundt 1970, 2[AL] (*Mycoplasma granularum* Switzer 1964, 504)

gra.nu.la'rum. N.L. fem. n. *granula* (from L. neut. n. *granulum*) a small grain, a granule; N.L. gen. pl. n. *granularum* of small grains, made up of granules, granular.

Cells are pleomorphic, with short filaments and coccoid cells. Colonies on solid medium are large with clearly marked central zones and a fried-egg appearance. Colonies on serum-free medium are smaller and may lack the peripheral zone of growth around central core. Temperature range for growth is 22–37°C, with optimum around 37°C. Agar colonies produce a zone of β-hemolysis by the overlay technique using sheep erythrocytes. DNA–DNA hybridization studies showed 20–22% hybridization with *Acholeplasma laidlawii*, but none with other acholeplasmas. Pathogenicity has not been established. Aerosol challenge of specific pathogen-free pigs did not induce clinical or histological evidence of disease.

Source: isolated frequently from the nasal cavity of swine, with occasional isolates from swine lung and feces. Also isolated from the conjunctivae and nasopharynx of horses, and the genital tract of guinea pigs. Occasional contaminant of eukaryotic cell cultures.

DNA G+C content (mol%): 30.5–32.4 (T_m, Bd).

Type strain: BTS-39, ATCC 19168, NCTC 10128.

Sequence accession no. (16S rRNA gene): AY538166.

9. **Acholeplasma hippikon** Kirchhoff 1978, 81[AL]

hip.pi'kon. Gr. neut. adj. *hippikon* pertaining to the horse.

Cells are pleomorphic with predominantly coccoid forms. Colonies on solid medium containing horse serum typically have a fried-egg appearance, with smaller colonies on serum-free agar medium. Growth occurs over a temperature range of 22–37°C, with optimal growth at 35–37°C. Agar colonies produce β-hemolysis with the overlay technique, using a variety of animal red blood cells. Pathogenicity has not been established.

Source: isolated from the lung of aborted horse fetuses.

DNA G+C content (mol%): 33.1 (Bd).

Type strain: C1, ATCC 29725, NCTC 10172.

Sequence accession no. (16S rRNA gene): AY538167.

Further comment: Kirchhoff is incorrectly cited as "Kirchoff" on the Approved Lists of Bacterial Names.

10. **Acholeplasma modicum** Leach 1973, 147[AL]

mo'di.cum. L. neut. adj. *modicum* moderate, referring to moderate growth.

Cells are pleomorphic, with spherical, ring-shaped, and coccobacillary forms. Colonies on solid medium are distinctly smaller than those of most other acholeplasmas. Very small colonies without peripheral zones of growth are noted on serum-free solid medium. Very light turbidity is observed in serum-free broth, but more turbidity is found in broth containing serum. Growth temperature range is 22–37°C, with optimum growth around 35–37°C. Can be shown to produce carotenoids when large volumes of cells are examined. Agar colonies produce α- or β-hemolysis by the overlay technique using sheep, ox, or guinea pig red blood cells. Pathogenicity has not been established.

Source: isolated from various tissues of cattle, including blood, bronchial lymph nodes, thoracic fluids, lungs, and semen. Also isolated from nasal secretions of pigs, and occasionally from chickens, turkeys, and ducks.

DNA G+C content (mol%): 29.3 (T_m).

Type strain: PG49, ATCC 29102, NCTC 10134.

Sequence accession no. (16S rRNA gene): M23933.

11. **Acholeplasma morum** Rose, Tully and Del Giudice 1980, 653[VP]

mor'um. L. n. *morum* a mulberry, denoting the mulberry-like appearance of agar colonies of the organism.

Cells are pleomorphic, predominantly coccoid or coccobacillary forms, but with some beaded filaments. Colonies on solid medium without serum supplements are very small in size and have only central zones without any peripheral growth. Optimal growth on solid medium occurs with a 10% serum concentration and colony growth appears to be suppressed in a medium with 20% serum. Optimal growth in broth is apparent when 5–10% serum is added or when 1% bovine serum fraction supplements are added, but poor growth occurs in broth containing 20% horse serum. Growth in serum-free broth usually requires some fatty acid supplements, such as palmitic acid or polyoxyethylene sorbitan (Tween 80). Temperature range for growth is 23–37°C, with optimum growth at about 35–37°C.

Pathogenicity has not been established. Calf kidney cell cultures containing the organism show cytopathogenic effects.

Source: originally recovered from commercial fetal bovine serum and from calf kidney cultures containing fetal bovine serum. One isolation, in broth containing horse serum, was from a pool of *Armigeres subalbatus* mosquitoes collected by Leon Rosen in Taiwan in 1978 (strain SP7; D.L. Williamson and J.G. Tully, unpublished).

DNA G+C content (mol%): 34.0 (T_m).

Type strain: 72-043, ATCC 33211, NCTC 10188.

Sequence accession no. (16S rRNA gene): AY538168.

12. **Acholeplasma multilocale** Hill, Polak-Vogelzang and Angulo 1992, 516[VP]

mul.ti.lo.ca'le. L. adj. *multus* many, numerous; L. adj. *localis -is -e* of or belonging to a place, local; N.L. neut. adj. *multilocale* referring to more than one location.

Pleomorphic cells. Colonies on agar medium have a typical fried-egg appearance. Organisms grow well in broth medium at 35–37°C. No evidence for pathogenicity.

Source: isolated from the nasopharynx of a horse and the feces of a rabbit.

DNA G+C content (mol%): 31 (Bd).

Type strain: PN525, NCTC 11723, ATCC 49900.

Sequence accession no. (16S rRNA gene): AY538169.

13. **Acholeplasma oculi** corrig. al-Aubaidi, Dardiri, Muscoplatt and McCauley 1973, 126[AL]

o'cu.li. L. n. *oculus* the eye; L. gen. n. *oculi* of the eye.

Cells are pleomorphic, including spherical, ring-shaped, and coccobacillary forms. Medium-sized colonies with typical fried-egg appearance are formed on horse serum agar. Colonies on serum-free agar are smaller and may lack the peripheral growth around the central core. Growth occurs at temperatures of 25–37°C. Agar colonies produce zones of hemolysis by the overlay technique using sheep red blood cells.

Pathogenicity is not well established. Intravenous inoculation of goats produced signs of pneumonia and death within 6 d. Conjunctival inoculation of goats produced mild conjunctivitis.

Source: isolated from the conjunctiva of goats with keratoconjunctivitis; porcine nasal secretions; equine nasopharynx, lung, spinal fluid, joint, and semen; the urogenital tract of cattle; and the external genitalia of guinea pigs. Present in amniotic fluid of pregnant women (Waites et al., 1987). Occasionally isolated from ducks and turkeys, with unreported isolations from an ostrich. Also several isolations from palm trees and other plants (Eden-Green and Tully, 1979; Somerson et al., 1982). Isolations from eukaryotic cell cultures may represent contamination of bovine origin.

DNA G+C content (mol%): 27 (T_m).

Type strain: 19-L, ATCC 27350, NCTC 10150.

Sequence accession no. (16S rRNA gene): U14904.

Further comment: originally named *Acholeplasma oculusi* by al-Aubaidi et al. (1973); the orthographic error was corrected by al-Aubaidi (1975).

14. **Acholeplasma palmae** Tully, Whitcomb, Rose, Bové, Carle, Somerson, Williamson and Eden-Green 1994b, 683[VP]

pal'mae. L. fem. gen. n. *palmae* of a palm tree, referring to the plant from which the organism was isolated.

Cells are primarily coccoid. Colonies on solid medium usually have a fried-egg appearance. The temperature range for growth is 18–37°C, with optimal growth occurring at 30°C. No evidence for pathogenicity. It is one of the closest phylogenetic relatives of the phytoplasmas.

Source: isolated from the crown tissues of a palm tree (*Cocos nucifera*) with lethal yellowing disease.

DNA G+C content (mol%): 30 (Bd, T_m, HPLC).

Type strain: J233, ATCC 49389.

Sequence accession no. (16S rRNA gene): L33734.

15. **Acholeplasma parvum** Atobe, Watabe and Ogata 1983, 348[VP]

par'vum. L. neut. adj. *parvum* small, intended to refer to the poor biochemical activities and tiny agar colonies of the organism.

Pleomorphic coccobacillary cells. Colonies on agar medium present a typical fried-egg appearance under both aerobic and anaerobic conditions. Initial reports of growth in the absence of cholesterol or serum have been made, but growth on serum-free medium is not well confirmed. The organism does not grow in most standard media for acholeplasmas or in most other medium formulations for sterol-requiring mycoplasmas. Needs special growth factor of 1% phytone or soytone peptone supplements; growth is sometimes better with the addition of 15% fetal bovine serum. Organisms grow on agar better than in broth; growth is better under aerobic conditions than under anaerobic conditions and better at 22–30°C than at 37°C. No evidence of fermentation of any carbohydrate, including glucose, salicin, and esculin. No evidence for pathogenicity.

Source: isolated from the oral cavities and vagina of healthy horses.

DNA G+C content (mol%): 29.1 (T_m).

Type strain: H23M, ATCC 29892, NCTC 10198.

Sequence accession no. (16S rRNA gene): AY538170.

16. **Acholeplasma pleciae** (Tully, Whitcomb, Hackett, Rose, Henegar, Bové, Carle, Williamson and Clark 1994a) Knight 2004, 1952[VP] (*Mesoplasma pleciae* Tully, Whitcomb, Hackett, Rose, Henegar, Bové, Carle, Williamson and Clark 1994a, 690)

ple.ci′ae. N.L. gen. n. *pleciae* of *Plecia*, referring to the genus of corn maggot (*Plecia* sp.) from which the organism was first isolated.

Cells are primarily coccoid. Colonies on solid media incubated under anaerobic conditions at 30°C have a fried-egg appearance. Supplements of 0.04% polyoxyethylene sorbitan (Tween 80) are required for growth in serum-free media. Temperature range for growth is 18–32°C, with optimal growth at 30°C. Agar colonies do not hemadsorb guinea pig erythrocytes. No evidence for pathogenicity.

Source: originally isolated from the hemolymph of a larva of the corn root maggot (*Plecia* sp.).

DNA G+C content (mol%): 31.6 (Bd, T_m, HPLC).

Type strain: PS-1, ATCC 49582.

Sequence accession no. (16S rRNA gene): AY257485.

17. **Acholeplasma seiffertii** Bonnet, Saillard, Vignault, Garnier, Carle, Bové, Rose, Tully and Whitcomb 1991, 48[VP]

seif.fer′ti.i. N.L. gen. masc. n. *seiffertii* of Seiffert, in honor of Gustav Seiffert, a German microbiologist who performed pioneering studies of sterol-nonrequiring mollicutes that occur in soil and compost.

Cells are primarily coccoid. Colonies on solid medium usually have the appearance of fried-eggs. Acid produced from glucose and mannose. Colonies on agar hemadsorb guinea pig erythrocytes. Temperature range for growth is 20–35°C; optimum growth occurs at 28°C. No evidence for pathogenicity.

Source: isolated from floral surfaces of a sweet orange (*Citrus sinensis*) and wild angelica (*Angelica sylvestris*).

DNA G+C content (mol%): 30 (Bd).

Type strain: F7, ATCC 49495.

Sequence accession no. (16S rRNA gene): AY351331.

Further comment: with the proposal of the order *Entomoplasmatales* (Tully et al., 1993), *Acholeplasma seiffertii* was transferred to the family *Entomoplasmataceae*. The name *Acholeplasma seiffertii* was therefore revised to *Mesoplasma seiffertii* comb. nov. The type strain is F7^T (=ATCC 49495^T; Bonnet et al., 1991).

18. **Acholeplasma vituli** Angulo, Reijgers, Brugman, Kroesen, Hekkens, Carle, Bové, Tully, Hill, Schouls, Schot, Roholl and Polak-Vogelzang 2000, 1130[VP]

vi.tu′li. L. n. *vitulus* calf; L. gen. n. *vituli* of calf, referring to the provenance or occurrence of the organism in fetal calf serum.

Cells are predominantly coccoid in shape. Colonies on solid media demonstrate a fried-egg appearance under both aerobic and anaerobic conditions. Temperature range for growth is 25–37°C. No evidence for pathogenicity.

Source: isolated from fetal bovine serum or contaminated eukaryotic cell cultures containing serum.

DNA G+C content (mol%): 38.3 (Bd), 37.6 (T_m).

Type strain: FC 097-2, ATCC 700667, CIP 107001.

Sequence accession no. (16S rRNA gene): AF031479.

References

al-Aubaidi, J.M. 1975. Orthographic errors in the name *Acholeplasma oculusi*. Int. J. Syst. Bacteriol. *25*: 221.

al-Aubaidi, J.M., A.H. Dardiri, C.C. Muscoplatt and E.H. McCauley. 1973. Identification and characterization of *Acholeplasma oculusi* spec. nov. from the eyes of goats with keratoconjunctivitis. Cornell Vet. *63*: 117–129.

Angulo, A.F., R. Reijgers, J. Brugman, I. Kroesen, F.E.N. Hekkens, P. Carle, J.M. Bové, J.G. Tully, A.C. Hill, L.M. Schouls, C.S. Schot, P.J.M. Roholl and A.A. Polak-Vogelzang. 2000. *Acholeplasma vituli* sp. nov., from bovine serum and cell cultures. Int. J. Syst. Evol. Microbiol. *50*: 1125–1131.

Atobe, H., J. Watabe and M. Ogata. 1983. *Acholeplasma parvum*, a new species from horses. Int. J. Syst. Bacteriol. *33*: 344–349.

Aulakh, G.S., E.B. Stephens, D.L. Rose, J.G. Tully and M.F. Barile. 1983. Nucleic acid relationships among *Acholeplasma* species. J. Bacteriol. *153*: 1338–1341.

Bonnet, F., C. Saillard, J.C. Vignault, M. Garnier, P. Carle, J.M. Bové, D.L. Rose, J.G. Tully and R.F. Whitcomb. 1991. *Acholeplasma seiffertii* sp. nov., a mollicute from plant surfaces. Int. J. Syst. Bacteriol. *41*: 45–49.

Bradbury, J. 1977. Rapid biochemical tests for characterization of the *Mycoplasmatales*. J. Clin. Microbiol. *5*: 531–534.

Bradbury, J.M. 1978. *Acholeplasma equifetale* in broiler chickens. Vet. Rec. *102*: 516.

Brown, D., G. McLaughlin and M. Brown. 1995. Taxonomy of the feline mycoplasmas *Mycoplasma felifaucium*, *Mycoplasma feliminutum*, *Mycoplasma felis*, *Mycoplasma gateae*, *Mycoplasma leocaptivus*, *Mycoplasma leopharyngis*, and *Mycoplasma simbae* by 16S rRNA gene sequence comparisons. Int. J. Syst. Bacteriol. *45*: 560–564.

Brown, D., R. Whitcomb and J. Bradbury. 2007. Revised minimal standards for description of new species of the class *Mollicutes* (division *Tenericutes*). Int. J. Syst. Evol. Microbiol. *57*: 2703–2719.

Bruce, J., R.N. Gourlay, R. Hull and D.J. Garwes. 1972. Ultrastructure of *Mycoplasmatales* virus laidlawii I. J. Gen. Virol. *16*: 215–221.

Carle, P., D.L. Rose, J.G. Tully and J.M. Bové. 1993. The genome size of spiroplasmas and other mollicutes. Int. Org. Mycoplasmol. Lett. *2*: 263.

Clyde, W.A., Jr. 1983. Growth inhibition tests. *In* Methods in Mycoplasmology, vol. 1 (edited by Razin and Tully). Academic Press, New York, pp. 405–410.

Congdon, A.L., E.S. Boatman and G.E. Kenny. 1979. *Mycoplasmatales* virus MV-M1: discovery in *Acholeplasma modicum* and preliminary characterization. Curr. Microbiol. *3*: 111–115.

Eden-Green, S.J. and P.G. Markham. 1987. Multiplication and persistence of *Acholeplasma* spp. in leafhoppers. J. Invertebr. Pathol. *49*: 235–241.

Eden-Green, S.J. and J.G. Tully. 1979. Isolation of *Acholeplasma* spp. from coconut palms affected by lethal yellowing disease in Jamaica. Curr. Microbiol. *2*: 311–316.

Edward, D.G. 1954. The pleuropneumonia group of organisms: a review, together with some new observations. J. Gen. Microbiol. *10*: 27–64.

Edward, D.G. and E.A. Freundt. 1969. Proposal for classifying organisms related to *Mycoplasma laidlawii* in a family *Sapromycetaceae*, genus *Sapromyces*, within the *Mycoplasmatales*. J. Gen. Microbiol. *57*: 391–395.

Edward, D.G. and E.A. Freundt. 1970. Amended nomenclature for strains related to *Mycoplasma laidlawii*. J. Gen. Microbiol. *62*: 1–2.

Fabricant, J. and E.A. Freundt. 1967. Importance of extension and standardization of laboratory tests for the identification and classification of mycoplasma. Ann. N. Y. Acad. Sci. *143*: 50–58.

Gardella, R.S., R.A. Del Giudice and J.G. Tully. 1983. Immunofluorescence. *In* Methods in Mycoplasmology, vol. 1 (edited by Razin and Tully). Academic Press, New York, pp. 431–439.

Garwes, D., B. Pike, S. Wyld, D. Pocock and R. Gourlay. 1975. Character ization of *Mycoplasmatales* virus-laidlawii 3. J. Gen. Virol. *29*: 11–24.

Gourlay, R.N. 1970. Isolation of a virus infecting a strain of *Mycoplasma laidlawii*. Nature *225*: 1165.

Gourlay, R.N. 1971. *Mycoplasmatales* virus-laidlawii 2, a new virus isolated from *Acholeplasma laidlawii*. J. Gen. Virol. *12*: 65–67.

Gourlay, R.N. 1972. Isolation and characterization of mycoplasma viruses. Proceedings of the CIBA Found. Symp., pp. 145–156.

Gourlay, R.N. 1973. *Mycoplasma* viruses: isolation, physicohemical, and biological properties. Ann. N. Y. Acad. Sci. *225*: 144–148.

Gourlay, R.N. 1974. *Mycoplasma* viruses: isolation, physicochemical, and biological properties. CRC Crit. Rev. Microbiol. *3*: 315–331.

Gourlay, R.N., J. Brownlie and C.J. Howard. 1973. Isolation of T-mycoplasmas from goats, and the production of subclinical mastitis in goats by the intramammary inoculation of human T-mycoplasmas. J. Gen. Microbiol. *76*: 251–254.

Haberer, K., G. Klotz, J. Maniloff and A. Kleinschmidt. 1979. Structural and biological properties of mycoplasmavirus MVL3: an unusual virus-procaryote interaction. J. Virol. *32*: 268–275.

Heyward, J.T., M.Z. Sabry and W.R. Dowdle. 1969. Characterization of *Mycoplasma* species of feline origin. Am. J. Vet. Res. *30*: 615–622.

Hill, A. 1992. *Acholeplasma cavigenitalium* sp. nov., isolated from the vagina of guinea pigs. Int. J. Syst. Bacteriol. *42*: 589–592.

Hill, A., A. Polak-Vogelzang and A. Angulo. 1992. *Acholeplasma multilocale* sp. nov., isolated from a horse and a rabbit. Int. J. Syst. Bacteriol. *42*: 513–517.

Ichimaru, H. and M. Nakamura. 1983. Biological properties of a plaque-inducing agent obtained from *Acholeplasma oculi*. Yale J. Biol. Med. 56: 761–763.

Johansson, K-E. 1974. Fractionation of membrane proteins from Acholeplasma laidlawii by preparative agarose suspension electrophoresis. *In* Protides of the Biological Fluids – 21st Colloquium (edited by Peeters). Pergamon Press, Oxford, pp. 151–156.

Johansson, K.E., Pettersson B. 2002. Taxonomy of *Mollicutes*. *In* Molecular Biology and Pathogenicity of Mycoplasmas (edited by Razin and Herrmann). Kluwer Academic/Plenum Press, New York, pp. 1–30.

Kato, H., T. Murakami, S. Takase and K. Ono. 1972. Sensitivities *in vitro* to antibiotics of *Mycoplasma* isolated from canine sources. Jpn. J. Vet. Sci. *34*: 197–206.

Kirby, T., J. Blum, I. Kahane and I. Fridovich. 1980. Distinguishing between manganese-containing and iron-containing superoxide dismutases in crude extracts of cells. Arch. Biochem. Biophys. *20*: 551–555.

Kirchhoff, H. 1978. *Acholeplasma equifetale* and *Acholeplasma hippikon*, two new species from aborted horse fetuses. Int. J. Syst. Bacteriol. *28*: 76–81.

Kisary, J. and L. Stipkovits. 1975. Effect of *Mycoplasma gallinarum* on the replication *in vitro* of goose parvovirus strain "B". Acta Microbiol. Acad. Sci. Hung. *22*: 305–307.

Kisary, J., A. El-Ebeedy and L. Stipkovits. 1976. *Mycoplasma* infection of geese. II. Studies on pathogenicity of mycoplasmas in goslings and goose and chicken embryos. Avian Pathol. *5*: 15–20.

Knight, T.F., Jr. 2004. Reclassification of *Mesoplasma pleciae* as *Acholeplasma pleciae* comb. nov. on the basis of 16S rRNA and *gyrB* gene sequence data. Int. J. Syst. Evol. Microbiol. *54*: 1951–1952.

Leach, R.H. 1973. Further studies on classification of bovine strains of *Mycoplasmatales*, with proposals for new species, *Acholeplasma modicum* and *Mycoplasma alkalescens*. J. Gen. Microbiol. *75*: 135–153.

Leach, R.H. 1983. Preservation of *Mycoplasma* cultures and culture collections. *In* Methods in Mycoplasmology, vol. 1 (edited by Razin and Tully). Academic Press, New York, pp. 197–204.

Lee, G.Y. and G.E. Kenny. 1984. Immunological heterogeneity of superoxide dismutases in the *Acholeplasmataceae*. Int. J. Syst. Bacteriol. *34*: 74–76.

Lewis, J. and J. Poland. 1978. Sensitivity of mycoplasmas of the respiratory tract of pigs and horses to erythromycin and its use in selective media. Res. Vet. Sci. *24*: 121–123.

Liska, B. 1972. Isolation of a new *Mycoplasmatales* virus. Stud. Biophys. *34*: 151–155.

Lynch, R.E. and B.C. Cole. 1980. *Mycoplasma pneumoniae*: a prokaryote which consumes oxygen and generates superoxide but which lacks superoxide dismutase. Biochem. Biophys. Res. Commun. *96*: 98–105.

Maniloff, J. 1992. *Mycoplasma* viruses. *In* Mycoplasmas: Molecular Biology and Pathogenesis (edited by Maniloff, McElhaney, Finch and Baseman). American Society for Microbiology, Washington, DC, pp. 41–59.

Maniloff, J., J. Das and J.R. Christensen. 1977. Viruses of mycoplasmas and spiroplasmas. Adv. Virus Res. *21*: 343–380.

Mayberry, W.R., P.F. Smith and T.A. Langworthy. 1974. Heptose-containing pentaglycosyl diglyceride among the lipids of *Acholeplasma modicum*. J. Bacteriol. *118*: 898–904.

McCoy, R.E., H.G. Basham, J.G. Tully, D.L. Rose, P. Carle and J.M. Bové. 1984. *Acholeplasma florum*, a new species isolated from plants. Int. J. Syst. Bacteriol. *34*: 11–15.

Nakagawa, T., T. Uemori, K. Asada, I. Kato and R. Harasawa. 1992. *Acholeplasma laidlawii* has tRNA genes in the 16S–23S spacer of the rRNA operon. J. Bacteriol. *174*: 8163–8165.

Neimark, H.C., J.G. Tully, D. Rose and C. Lange. 1992. Chromosome size polymorphism among mollicutes. Int. Org. Mycoplasmol. Lett. *2*: 261.

O'Brien, S.J., J.M. Simonson, M.W. Grabowski and M.F. Barile. 1981. Analysis of multiple isoenzyme expression among twenty-two species of *Mycoplasma* and *Acholeplasma*. J. Bacteriol. *146*: 222–232.

Ogata, M., H. Atobe, H. Kushida and K. Yamamoto. 1971. *In vitro* sensitivity of mycoplasmas isolated from various animals and sewage to antibiotics and nitrofurans. J. Antibiot. (Tokyo) *24*: 443–451.

Pollack, J.D., J. Banzon, K. Donelson, J.G. Tully, Jr, J.W. Davis, K.J. Hackett, C. Agbanyim and R.J. Miles. 1996a. Reduction of benzyl viologen distinguishes genera of the class *Mollicutes*. Int. J. Syst. Bacteriol. *46*: 881–884.

Pollack, J.D., M.V. Williams, J. Banzon, M.A. Jones, L. Harvey and J.G. Tully. 1996b. Comparative metabolism of *Mesoplasma*, *Entomoplasma*, *Mycoplasma*, and *Acholeplasma*. Int. J. Syst. Bacteriol. *46*: 885–890.

Razin, S., J. Tully, D. Rose and M. Barile. 1983. DNA cleavage patterns as indicators of genotypic heterogeneity among strains of *Acholeplasma* and *Mycoplasma* species. J. Gen. Microbiol. *129*: 1935–1944.

Rose, D.L. and J.G. Tully. 1983. Detection of β-D-glucosidase: hydrolysis of esculin and arbutin. *In* Methods in Mycoplasmology (edited by Tully). Academic Press, New York, pp. 385–389.

Rose, D.L., J.G. Tully and R.A. Del Giudice. 1980. *Acholeplasma morum*, a new non-sterol-requiring species. Int. J. Syst. Bacteriol. *30*: 647–654.

Rottem, S. and O. Markowitz. 1979. Unusual positional distribution of fatty acids in phosphatidylglycerol of sterol-requiring mycoplasmas. FEBS Lett. *107*: 379–382.

Sabin, A.B. 1941. The filterable microorganisms of the pleuropneumonia group. Bacteriol. Rev. *5*: 1–66.

Smith, P. and T. Langworthy. 1979. Existence of carotenoids in *Acholeplasma axanthum*. J. Bacteriol. *137*: 185–188.
Somerson, N., J. Kocka, D. Rose and R. Del Giudice. 1982. Isolation of acholeplasmas and a mycoplasma from vegetables. Appl. Environ. Microbiol. *43*: 412–417.
Stephens, E.B., G.S. Aulakh, D.L. Rose, J.G. Tully and M.F. Barile. 1983a. Intraspecies genetic relatedness among strains of *Acholeplasma laidlawii* and of *Acholeplasma axanthum* by nucleic acid hybridization. J. Gen. Microbiol. *129*: 1929–1934.
Stephens, E.B., G.S. Aulakh, D.L. Rose, J.G. Tully and M.F. Barile. 1983b. Interspecies and intraspecies DNA homology among established species of *Acholeplasma*: a review. Yale J. Biol. Med. *56*: 729–735.
Switzer, W.P. 1964. Mycoplasmosis. *In* Diseases of Swine, 2nd edn (edited by Dunne). Iowa State University Press, Ames, IA, pp. 498–507.
Tanaka, R., A. Muto and S. Osawa. 1989. Nucleotide sequence of tryptophan tRNA gene in *Acholeplasma laidlawii*. Nucleic Acids Res. *17*: 5842.
Tanaka, R., Y. Andachi and A. Muto. 1991. Evolution of tRNAs and tRNA genes in *Acholeplasma laidlawii*. Nucleic Acids Res. *19*: 6787–6792.
Taylor-Robinson, D. 1983. Metabolism inhibition tests. *In* Methods in Mycoplasmology, vol. 1 (edited by Razin and Tully). Academic Press, New York, pp. 411–421.
Thorns, C. and E. Boughton. 1978. Studies on film production and its specific inhibition, with special reference to *Mycoplasma bovis* (*M. agalactiae* var. bovis). Zentralbl. Veterinarmed. B *25*: 657–667.
Tully, J.G. 1973. Biological and serological characteristics of the acholeplasmas. N. Y. Acad. Sci. *225*: 74–93.
Tully, J.G. 1979. Special features of the acholeplasmas. *In* The Mycoplasmas, vol. 1 (edited by Barile and Razin). Academic Press, New York, pp. 431–449.
Tully, J.G. 1983. Methods in mycoplasmology, vol. 2, Diagnostic Mycoplasmology. Academic Press, New York.
Tully, J.G. 1995. Determination of cholesterol and polyoxyethylene sorbitan growth requirements of mollicutes. *In* Molecular and Diagnostic Procedures in Mycoplasmology, vol. 1 (edited by Razin and Tully). Academic Press, San Diego, pp. 381–389.
Tully, J.G. 1996. Mollicute-host interrelationships: current concepts and diagnostic implications. *In* Molecular and Diagnostic Procedures in Mycoplasmology, vol. 2 (edited by Tully and Razin). Academic Press, San Diego, pp. 1–21.
Tully, J. and S. Razin. 1970. *Acholeplasma axanthum*, sp. n.: a new sterol-nonrequiring member of the *Mycoplasmatales*. J. Bacteriol. *103*: 751–754.
Tully, J.G., D.L. Rose, P. Carle, J.M. Bové, K.J. Hackett and R.F. Whitcomb. 1988. *Acholeplasma entomophilum* sp. nov. from gut contents of a wide-range of host insects. Int. J. Syst. Bacteriol. *38*: 164–167.
Tully, J.G., J.M. Bove, F. Laigret and R.F. Whitcomb. 1993. Revised taxonomy of the class *Mollicutes* – proposed elevation of a monophyletic cluster of arthropod-associated mollicutes to ordinal rank (*Entomoplasmatales* ord. nov.), with provision for familial rank to separate species with nonhelical morphology (*Entomoplasmataceae* fam. nov.) from helical species (*Spiroplasmataceae*), and emended descriptions of the order *Mycoplasmatales*, family *Mycoplasmataceae*. Int. J. Syst. Bacteriol. *43*: 378–385.
Tully, J.G., R.F. Whitcomb, K.J. Hackett, D.L. Rose, R.B. Henegar, J.M. Bové, P. Carle, D.L. Williamson and T.B. Clark. 1994a. Taxonomic descriptions of eight new non-sterol-requiring *Mollicutes* assigned to the genus *Mesoplasma*. Int. J. Syst. Bacteriol. *44*: 685–693.
Tully, J.G., R.F. Whitcomb, D.L. Rose, J.M. Bové, P. Carle, N.L. Somerson, D.L. Williamson and S. Eden-Green. 1994b. *Acholeplasma brassicae* sp. nov. and *Acholeplasma palmae* sp. nov., two non-sterol-requiring mollicutes from plant surfaces. Int. J. Syst. Bacteriol. *44*: 680–684.
Waites, K.B., J.G. Tully, D.L. Rose, P.A. Marriott, R.O. Davis and G.H. Cassell. 1987. Isolation of *Acholeplasma oculi* from human amniotic fluid in early pregnancy. Curr. Microbiol. *15*: 325–327.
Whitcomb, R.F. and D.L. Williamson. 1975. Helical wall-free prokaryotes in insects: multiplication and pathogenicity. Ann. N. Y. Acad. Sci. *266*: 260–275.
Whitcomb, R.F., J.G. Tully, J.M. Bové and P. Saglio. 1973. Spiroplasmas and acholeplasmas: multiplication in insects. Science *182*: 1251–1253.

Family II. **Incertae sedis**

This family includes the phytoplasma strains of the order *Acholeplasmatales*. Although never cultured in cell-free media, these plant pathogens and symbionts have been well studied by culture-independent methods.

Genus I. "***Candidatus*** **Phytoplasma**" gen. nov. IRPCM Phytoplasma/Spiroplasma Working Team 2004, 1244

Nigel A. Harrison, Dawn Gundersen-Rindal and Robert E. Davis

Phy.to.plas′ma. Gr. masc. n. *phytos* a plant; Gr. neut. n. *plasma* something formed or molded, a form.

Phytoplasmas (Sears and Kirkpatrick, 1994) are wall-less, nutritionally fastidious, phytopathogenic prokaryotes 0.2–0.8 μm in diameter that morphologically resemble members of the *Mollicutes*. Sequencing of nearly full-length PCR-amplified 16S rRNA genes (Gundersen et al., 1994; Namba et al., 1993; Seemüller et al., 1994), combined with earlier studies (Kuske and Kirkpatrick, 1992b; Lim and Sears, 1989), provided the first comprehensive phylogeny of the organisms and showed that they constitute a unique, monophyletic clade within the *Mollicutes*. These organisms are most closely related to members of the genus *Acholeplasma* within the *Anaeroplasma* clade as defined by Weisburg et al. (1989). Sustained culture in cell-free media has not yet been demonstrated for any phytoplasma. Their genome sizes have been estimated to range from 530 to 1350 kb, and the G+C content of phytoplasma DNA is about 23–30 mol%. The presence of a characteristic oligonucleotide sequence in the 16S rRNA gene, CAA GAY BAT KAT GTK TAG CYG GDC T, and standard codon usage indicate that phytoplasmas represent a distinct taxon for which the name "*Candidatus* Phytoplasma" has been adopted by specialists in the molecular biology and pathogenicity of these and similar phytopathogenic organisms (IRPCM Phytoplasma/Spiroplasma Working Team – Phytoplasma Taxonomy Group, 2004). At present, the designation "*Candidatus*" must still be used for new types.

Further descriptive information

Phytoplasma cells typically have a diameter less than 1 µm and are polymorphic. Viewed in ultra-thin section by electron microscopy Figure 115, they appear ovoid, oblong, or filamentous in plant and insect hosts (Doi et al., 1967; Hearon et al., 1976). Transmission electron microscopy of semi-thick (0.3 µm) sections (Thomas, 1979) and serial sections (Chen and Hiruki, 1978; Florance and Cameron, 1978; Waters and P. Hunt., 1980), and scanning electron microscopy studies (Bertaccini et al., 1999; Haggis and Sinha, 1978; Marcone et al., 1996) have done much to clarify the gross cellular morphology of phytoplasmas. They range from spherical to filamentous, often with extensive branching reminiscent of that seen in *Mycoplasma mycoides*. Small dense rounded forms ~0.1 µm in diameter, formerly considered to be "elementary bodies" when seen in thin section, were shown to represent constrictions in filamentous forms. Dumbbell-shaped forms once thought to be "dividing cells" are actually branch points of filamentous forms, whereas forms thought to have internal vesicles have been shown to have involuted membranes oriented such that the plane of the sections cut through the cell membrane twice (McCoy, 1979). Phytoplasma cell membranes are resistant to digitonin and sensitive to hypotonic salt solutions, and, as such, are similar to those of non-sterol requiring mollicutes (Lim et al., 1992).

Phytoplasmas are consistently observed within phloem sieve elements (Christensen et al., 2004; McCoy et al., 1989; Oshima et al., 2001b; Webb et al., 1999) and occasionally have been reported in both companion cells (Rudzinska-Langwald and Kaminska., 1999; Sears and Klomparens, 1989) and parenchyma cells (Esau et al., 1976; Siller et al., 1987) of infected plants. Sieve elements are specialized living cells that lack nuclei when mature and transport photosynthate from leaves not only to growing tissues, but also to other tissues unable to photosynthesize (Oparka and Turgeon, 1999; Sjölund, 1997). This applies particularly to roots that require considerable energy for the uptake of water and nutrients (Flores et al., 1999). Phloem sap is unique in that it contains from 12 to 30% sucrose and is under high hydrostatic (turgor) pressure (Evert, 1977). Sieve elements have pores in their end plates and lateral walls, allowing passage of photosynthate to adjacent sieve tube elements. The sieve pores, which have an average diameter of ~0.2 µm, are of sufficient size to allow passage of spherical and filamentous phytoplasma cells from one sieve element to another (McCoy, 1979). The chemical composition of sieve sap is complex, containing sugars, minerals, free amino acids, proteins, and ATP (Van Helden et al., 1994). This rich milieu, with its high osmotic and hydrostatic pressures, serves to support extensive multiplication of phytoplasmas *in planta*. Phytoplasmas also multiply in the internal tissues and organs of their insect vectors (Kirkpatrick et al., 1987; Lefol et al., 1994; Marzachi et al., 2004; Nasu et al., 1970), which are primarily leafhoppers, planthoppers, and psyllids (D'Arcy and Nault, 1982; Jones, 2002; Weintraub and Beanland, 2006). In many respects, the composition of insect hemolymph is similar to that of plant phloem sap, as both contain high levels of complex and simple organic compounds (Moriwaki et al., 2003; Saglio and Whitcomb, 1979).

Physical maps of several phytoplasma genomes have been constructed (Firrao et al., 1996; Lauer and Seemüller, 2000; Marcone and Seemüller, 2001; Padovan et al., 2000). The presence of extrachromosomal DNAs or plasmids in numerous phytoplasmas has also been reported (Davis et al., 1988; Denes and Sinha, 1991; Kuboyama et al., 1998; Kuske and Kirkpatrick, 1990; Liefting et al., 2004; Lin et al., 2009; Nakashima and Hayashi, 1997; Nishigawa et al., 2003; Oshima et al., 2001a; Tran-Nguyen and Gibb, 2006) and suggested as a potential means of intermolecular recombination (Nishigawa et al., 2002b). Phytoplasma-associated extrachromosomal DNAs have been shown to contain genes encoding a putative geminivirus-related replication (Rep) protein (Liefting et al., 2006; Nishigawa et al., 2001; Rekab et al., 1999) and a single-stranded DNA-binding

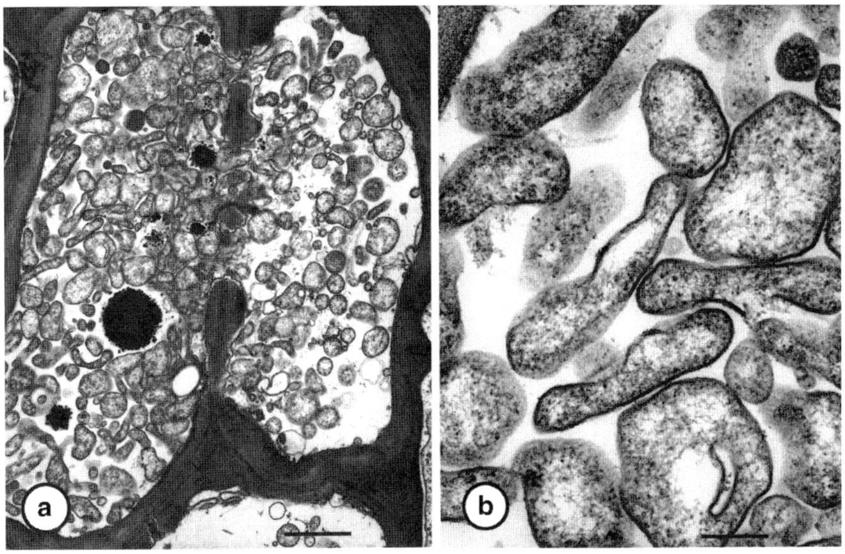

FIGURE 115. Electron micrographs of ultrathin sections of leaf petiole from a sunnhemp (*Crotalaria juncea* L.) plant displaying Crotalaria phyllody disease symptoms. (a) Polymorphic phytoplasma cells occluding the lumen of adjacent leaf phloem sieve tube elements. Bar = 2 µm. (b) Ultratructural morphology indicates phytoplasma cells are bounded by a unit membrane and contain DNA fibrils and ribosomes. Bar = 200 nm. Images provided by Phil Jones.

protein (Nishigawa et al., 2002a), as well as a putative gene similar to DNA primase of other bacterial chromosomes (Liefting et al., 2004) and still other genes of as yet unknown identity. Moreover, heterogeneity in extrachromosomal DNAs has been associated with reduced pathogenicity and loss of insect vector transmissibility (Denes and Sinha, 1992; Nishigawa et al., 2002a, 2003).

Onion yellows mild strain (OY-M) was the first phytoplasma genome to be completely sequenced. The genome of this aster yellows group strain consists of a circular chromosome of 860,632 bp. It also contains two extrachromosomal DNAs, EcOYM (5025 bp) and plasmid pOYM (3932 bp) (Nishigawa et al., 2003; Oshima et al., 2002), representing two different classes based on the type of replication protein encoded. While EcOYM contains a *rep* gene homologous to that of the geminiviruses, pOYM has a *rep* gene that encodes a unique protein with characteristics of both viral-*rep* and plasmid-*rep* (Namba, 2002). The chromosome is a circular DNA molecule with a G+C content of 28 mol% and contains 754 open reading frames (ORFs), comprising 73% of the chromosome. Of these, 66% of ORFs exhibit significant homology to gene sequences currently archived in the GenBank database. Putative proteins encoded by ORFs could be assigned to one of six different functional categories: (1) information storage and processing (260 ORFs); (2) metabolism (107 ORFs); (3) cellular processes (77 ORFs); (4) poorly characterized, i.e., with homology to uncharacterized proteins of other organisms (50 ORFs); or (5) others, i.e., without homology to any known proteins (260 ORFs). Like mycoplasmal genomes, the OY-M phytoplasma genome lacks many genes related to amino acid and fatty acid biosynthesis, the tricarboxylic acid cycle, and oxidative phosphorylation. However, OY-M phytoplasma differs from mycoplasma in that it lacks genes for the phosphotransferase system and for metabolizing UDP-galactose to glucose 1-phosphate, suggesting that it possesses a unique sugar intake and metabolic system. Furthermore, OY-M phytoplasma lacks most of the genes needed to synthesize nucleotides and ATP suggesting that it probably assimilates these and other necessary metabolites from host cytoplasm. Many genes, such as those for glycolysis, are present as multiple redundant copies representing 18% of the total genome. Twenty-seven genes encoding transporter systems such as malate, metal-ion and amino acid transporters, some of which have multiple copies, were identified, suggesting that phytoplasmas aggressively import many metabolites from the host cell. Other than genes encoding glucanase and hemolysin-like proteins, no other genes presently known to be related to bacterial pathogenicity were evident in the OY-M phytoplasma genome, suggesting novel mechanisms for virulence.

Annotation of the OY-M phytoplasma genome has been followed by three other phytoplasma genome annotations. Aster yellows witches'-broom phytoplasma ("*Candidatus* Phytoplasma asteris"-related strain AY-WB) possesses a circular 706,569 nucleotide chromosome and plasmids AYWB-pI (3872 bp), -pII (4009 bp), -pIII (5104 bp), and -pIV (4316 bp) (Bai et al., 2006). Australian tomato big bud phytoplasma ("*Candidatus* Phytoplasma australiense"-related strain TBB) has a circular 879,324 bp chromosome and a 3700 bp plasmid (Tran-Nguyen et al., 2008), whereas apple proliferation phytoplasma ("*Candidatus* Phytoplasma mali"-related strain AT) has a linear 601,943 bp chromosome (Kube et al., 2008). The chromosome of "*Candidatus* Phytoplasma mali" is characterized by large terminal inverted repeats and covalently closed hairpin ends. Analysis of protein-coding genes revealed that glycolysis, the major energy-yielding pathway supposed for OY-M phytoplasma, is incomplete in AT phytoplasma. It also differs from OY-M and AY-WB phytoplasmas by a lower G+C content (21.4 mol%), fewer paralogous genes, a strongly reduced number of ABC transporters for amino acids, and an extended set of genes for homologous recombination, excision repair, and SOS response.

Comparative genomics have also recently identified ORFs shared by AY-WB phytoplasma and the distantly-related corn stunt pathogen *Spiroplasma kunkelii* that are absent from obligate animal and human pathogenic mollicutes. These proteins were identified as polynucleotide phosphorylase (PNPase), cmp-binding factor (CBF), cytosine deaminase, and YlxR protein and could be important for insect transmission or plant pathogenicity. Also identified were four additional proteins, ppGpp synthetase, HAD hydrolase, AtA (AAA type ATPase), and P-type Mg^{2+} transport ATPase, that seemed to be more closely related between AY-WB and *Spiroplasma kunkelii* than to their mycoplasmal counterparts (Bai et al., 2004).

Phytoplasmas possess a unique genome architecture that is characterized by multiple, nonrandomly distributed sequence-variable mosaics (SVMs) of clustered genes, originally recognized in a study of closely related "*Candidatus* Phytoplasma asteris"-related strains CPh and OY-M (Jomantiene and Davis, 2006). Targeted genome sequencing and comparative genomics indicated that this genome architecture is a common characteristic among phytoplasmas, leading to the proposal that the origin of SVMs was an ancient event in the evolution of the phytoplasma clade (Jomantiene et al., 2007), perhaps as a result of recurrent targeted attacks by mobile elements such as phages (Wei et al., 2008a). Jomantiene and Davis (2006) proposed that sizes and numbers of SVMs could account in part for the known variation in genome size among phytoplasma strains; this concept was independently suggested by Bai et al. (2006) on the basis of results from a comparative study of two completely sequenced phytoplasma genomes. Nucleotide sequences within SVMs included full length or pseudogene forms of *fli*A, an ATP-dependent Zn protease gene, *tra5*, *smc*, *himA*, *tmk*, and *ssb* (encoding single-stranded DNA-binding protein), genes potentially encoding hypothetical proteins of unknown function, genes exhibiting similarities to transposase, and a phage-related gene (Jomantiene et al., 2007). A similar set of nucleotide sequences occurs in AY-WB genomic regions termed potential mobile units (PMUs) by Bai et al. (2006). The presence of sequences encoding putatively secreted and/or transmembrane, cell surface-interacting proteins indicates that these genomic features are likely to be significant for phytoplasma/host interactions (Bai et al., 2006; Jomantiene and Davis, 2006; Jomantiene et al., 2007).

Short (17–35 bases) conserved, imperfect palindromic DNA sequences (PhREPs) that are present in SVMs possibly play a role in phytoplasma genome plasticity and targeting of mobile genetic elements. SVMs can be viewed as composites formed by the acquisition of genes through horizontal transfer, recombination, and rearrangement, and capture of mobile elements recurrently targeted to SVMs, leading Jomantiene et al. (2007) to suggest that SVMs provide loci for acquisition of new genes and targeting of mobile genetic elements to specific regions in phytoplasma chromosomes.

The chromosomes of avirulent, mildly, moderately, and highly virulent strains of "*Candidatus* Phytoplasma mali" (Seemüller and Schneider, 2007) differ from one another in size and exhibit distinct restriction endonuclease patterns when cleaved with rare cutting enzymes. PCR-based DNA amplifications, primed separately by eight primer pairs, revealed target sequence heterogeneity among all "*Candidatus* Phytoplasma mali"-related strains tested, but no correlations linked molecular markers with strain virulence or the maximum titer obtained upon infection of apple trees. In a separate study, a comparison of mild (OY-M) and severe (OY-W) strains of onion yellows (OY) phytoplasma indicated that severe symptoms were associated with higher populations of OY-W in infected host plants (Oshima et al., 2001b). A cluster of eight genes, considered essential for glycolysis, were subsequently identified within a similar 30 kb genomic region of both strains (Oshima et al., 2007). Of these, five genes (*smtA*, *greA*, *osmC*, *eno*, and *pfkA*) were randomly duplicated in OY-W, possibly influencing glycolytic activitiy. A higher consumption of metabolites such as sugars in the intracellular environment of the phloem may explain differences between OY-W and OY-M in growth rate, which in turn may be linked, directly or indirectly, to symptom severity.

Cloned fragments of phytoplasma DNA have been widely employed as probes in dot and Southern blot hybridization assays to identify and characterize phytoplasmas (reviewed by Lee and Davis, 1992; Lee et al. (2000). Southern blot restriction fragment length polymorphism (RFLP) analysis has enabled investigations of genetic relationships among phytoplasmas associated with similar hosts or with symptomatologically similar diseases (Kison et al., 1997, 1994; Kuske et al., 1991; Schneider and Seemüller, 1994b). Several discrete phytoplasma groups, each comprising strains that shared extensive sequence homology and little or no apparent homology with other phytoplasmas, were identified by this type of analysis. Lee and co-workers (1992) coined the term "genomic strain cluster" to denote each of seven discrete genotypic groups resolved by employing a selection of phytoplasma genomic DNA probes (reviewed by Lee and Davis, 1992). Of these, aster yellows (AY) was the largest group, represented by 15 genetically variable strains that were further delineated into three distinct genomic types (types I, II, and III) or subclusters (Lee et al., 1992). Significantly, major groupings later revealed by RFLP analysis of 16S rRNA genes were consistent with those defined by monoclonal antibody typing (Lee et al., 1993a) and other molecular methods (Lee et al., 1998b), but differed from distinctions made in traditional classification based solely on biological properties such as plant host range, symptomatology, and insect vector specificity (Chiykowski and Sinha, 1990).

Polyclonal antibodies (PAbs) have been produced against phytoplasma-enriched extracts (intact organisms or membrane fractions) partially purified from plants (reviewed by Chen et al., 1989) and against vector leafhopper-derived immunogens (Errampelli and Fletcher, 1993; Kirkpatrick et al., 1987). Most PAbs exhibit relatively high background reactions with healthy host antigens; thus, generation of useful polyclonal antisera has been limited so far to a few phytoplasmas maintained at high titer in host tissues. Phytoplasmas can be differentiated on the basis of their antigenic properties through the use of PAbs in enzyme-linked immunosorbent (ELISA), immunofluorescence, or Western blot assays. Antigenic similarity revealed among phytoplasmas by these assays is often in agreement with relationships demonstrated by vector transmission studies. Detection of antigenically distinct phytoplasmas in plants exhibiting very similar disease symptoms attests to the unreliability of symptom expression alone as a means of differentiating phytoplasmas.

Improvements in phytoplasma extraction methods have provided a source of immunogens for monoclonal antibody (MAb) production (Chang et al., 1995; Hsu et al., 1990; Jiang et al., 1989; Loi et al., 2002, 1998; Shen and Lin, 1993; Tanne et al., 2001). Used in ELISA, dot or tissue blot immunoassay, immunocapture PCR (Rajan et al., 1995), immunofluorescence microscopy, or immunosorbent electron microscopy (ISEM) (Clark, 1992; Shen and Lin, 1994), MAbs have demonstrated considerable promise for detection and differentiation of phytoplasmas infecting a broad range of host plants, including woody perennials (Guo et al., 1998). Due to their high degree of specificity, monoclonals seem most suited for differentiating very closely related strains (Lee et al., 1993a).

Isolation, cloning, and expression of immunodominant protein genes have identified putative proteins that account for a major portion of the membrane proteins of several phytoplasmas (Arashida et al., 2008; Barbara et al., 2002; Berg et al., 1999; Blomquist et al., 2001; Galetto et al., 2008; Kakizawa et al., 2004, 2009; Morton et al., 2003; Suzuki et al., 2006; Yu et al., 1998). When these purified proteins were used as immunogens, the resulting polyclonal antisera exhibited high specific titers and low background reactions in ELISA and Western blot analyses that were designed to detect phytoplasma proteins in infected hosts. Similarly, the *secA* gene was cloned from an onion yellows (OY-M) strain of aster yellows phytoplasma (Kakizawa et al., 2001) and used to raise an anti-SecA rabbit antibody against a purified partial SecA protein expressed in *Escherichia coli*. Light microscopy of thin sections of garland chrysanthemum (*Chrysanthemum coronarium*) treated by immunohistochemical straining revealed that the SecA protein was present in phloem of OY-M-infected but not healthy host plants. In addition, antisera against both OY-M phytoplasma SecA protein and GyrA protein of *Acholeplasma laidlawii* reacted with proteins of several unrelated phytoplasmas extracted from plant tissues (Koui et al., 2002; Wei et al., 2004a).

Phytoplasmas are the apparent etiological agents of diseases of at least 1000 plant species worldwide (McCoy et al., 1989; Seemüller et al., 1998). Although they can be transmitted from infected to healthy plants by scion or root grafts, most plant to plant spread occurs naturally via phloem-feeding insect vectors primarily of the family *Cicadellidae* (leafhoppers) and, less commonly, by planthoppers (*Fulgoroidea*) of the family *Ciixidae* and psyllids (*Psylloidea*) (D'Arcy and Nault, 1982; Tsai, 1979; Weintraub and Beanland, 2006). Phytoplasmas are transmitted in a circulative-propagative manner that typically involves a transmission latent period from 2 to 8 weeks (Carraro et al., 2001; Webb et al., 1999). The insect vector becomes infected upon ingesting phytoplasmas in phloem sap of infected plants. After an incubation period of one to several weeks, the phytoplasma multiplies to high titer in the salivary glands and the insect becomes capable of infecting the phloem of the healthy plants on which it feeds (Kunkel, 1926; Lee et al., 1998a; Nasu et al., 1970). Generally, phytoplasma infection does not appear to significantly affect the activity, weight, longevity, or fecundity of vector insects (Garnier et al., 2001). Some phytoplasmas can be vectored by many species of leafhoppers (McCoy et al., 1989; Nielson, 1979) and different insect species may serve as vectors in different geographic regions. Several vectors also have the ability to transmit more than one type of phytoplasma, whereas other phytoplasmas are transmitted by one or a few vector species to a narrow range of plant species (Lee et al., 1998a). There is mounting evidence also for transovarial transmission of some phytoplasmas (Alma et al., 1997; Hanboonsong et al., 2002; Kawakita et al., 2000; Tedeschi et al., 2006).

Plants may serve as both natural and experimental hosts to several different phytoplasmas. Dual or mixed infections involving related or unrelated phytoplasmas are known to occur naturally in plants and appear to be more common in perennial than annual plants (Bianco et al., 1993; Lee et al., 1995). Also, closely related phytoplasma strains are capable of inducing dissimilar symptoms on the same plant species (White et al., 1998), whereas similar symptoms on the same host plant may be induced by unrelated phytoplasmas (Harrison et al., 2003). The ability to accurately identify phytoplasmas by using DNA-based methods has shown that these organisms are more genetically diverse than was once thought (Davis and Sinclair, 1998). The geographic occurrence of phytoplasmas is determined largely by geographic ranges and feeding behavior (mono-, oligo-, or polyphagous) of the vector species, the relative susceptibility of the preferred host plant species, and the native host ranges of plant and insect hosts (Lee et al., 1998a). Phytoplasmas can be introduced into new geographic regions by long-distance dispersal of infectious vectors (Lee et al., 2003) and by movement of infected plants or vegetative plant parts. Most recently, phytoplasma DNA has been detected in embryos of aborted seed from diseased plants (Cordova et al., 2003; Nipah et al., 2007) and seed transmission of phytoplasmas infecting alfalfa (*Medicago sativa* L.) has been demonstrated (Khan et al., 2002).

An array of characteristic symptoms is associated with phytoplasma infection of several hundred plant species worldwide. Symptoms vary according to the particular host species, stage of host infection and the associated phytoplasma strain (reviewed by Davis and Lee, 1992; Hogenhout et al., 2008; Kirkpatrick, 1989, 1992; Lee et al., 2000; McCoy et al., 1989; Seemüller et al., 2002; Sinclair et al., 1994). Some symptoms indicative of profound disturbances in the normal balance of growth regulators in plants include virescence (greening of petals), phyllody (conversion of floral organs into leafy structures), big bud, floral proliferation, sterility of flowers, proliferation of adventitious or axillary shoots, internode elongation and etiolation, generalized stunting (small flowers, leaves and fruits or shortened internodes), unseasonal discoloration of leaves or shoots (yellow to purple discoloration), leaf curling, cupping or crinkling, witches'-brooms (bunchy growth at stem apices), vein clearing, vein enlargement, phloem discoloration, and general plant decline such as die-back of twigs, branches and trunks (Lee and Davis, 1992; Lee et al., 2000; McCoy et al., 1989).

Infection of herbaceous host plants is followed by rapid intraphloemic spread of phytoplasma from leaves to roots, often accompanied by six-fold increases in phytoplasma populations in these tissues between 14 and 28 d post-inoculation (Kuske and Kirkpatrick, 1992a; Wei et al., 2004b). Phytoplasma concentrations ranging from 2.2×10^8 to 1.5×10^9 cells per gram of tissue have been measured in high titer herbaceous hosts such as periwinkle (*Catharanthus roseus*) and in certain woody perennial hosts such as alder (*Alnus*) and most poplar (*Populus*) species. Lowest phytoplasma concentrations, from 370 to 34,000 cells per gram of tissue, were detected in apple trees that were grafted on resistant rootstocks and in oak (*Quercus robur*) or hornbeam (*Carpinus betulus*) trees exhibiting nonspecific leaf yellowing symptoms (Berg and Seemüller, 1999).

Colonization is usually marked by phloem dysfunction and a reduction in photosynthetic capacity. Alterations in phloem function have been correlated with structural degeneration of sieve elements due possibly to physical blockage by colonizing phytoplasma or the action of a phytotoxin (Guthrie et al., 2001; Siddique et al., 1998). The onset of symptoms may be accompanied by substantial impairment of the photosynthetic rate of mature leaves and by fluctuations in carbohydrate and amino acid levels in source versus sink leaves (Lepka et al., 1999; Tan and Whitlow, 2001). Leaf yellowing is associated with: decreases in chlorophyll content, carotenoids, and soluble proteins (Bertamini and Nedunchezhian, 2001); abnormal stomatal function (Martinez et al., 2000); histopathological changes such the amount of total polyphenols; loss of cellular integrity (Musetti et al., 2000); fluctuations in hydrogen peroxide; peroxidase activity and glutathione content in diseased versus healthy plant tissues (Musetti et al., 2004); and increases in calcium (Ca^{2+}) ions in cells (Musetti and Favali, 2003; Rudzinska-Langwald and Kaminska, 2003). Such adverse changes are accompanied by differential regulation of genes encoding proteins involved in floral development (Pracros et al., 2006), photosynthesis, sugar transport, and response to stress or in pathways of lipid and phenylpropanoid or phytosterol synthesis (Albertazzi et al., 2009; Carginale et al., 2004; Hren et al., 2009; Jagoueix-Eveillard et al., 2001).

The organisms degenerate and lose their cellular contents following treatment of infected plants with tetracycline antibiotics (Kamińska and Śliwa, 2003; Sinha and Peterson, 1972). Tetracycline sensitivity and the lack of sensitivity to cell wall-inhibiting antibiotics such as penicillin (Davis and Whitcomb, 1970; Ishii et al., 1967) also support their inclusion in the *Mollicutes.* Protective or therapeutic treatments with tetracycline antibiotics for phytoplasma disease control have been extended to a few high-value crop plants such as coconut for control of palm lethal yellowing, and to cherries and peaches for control of X-disease (McCoy, 1982; Nyland, 1971; Raju and Nyland, 1988). Administered by trunk injection, treatment of each tree with 1.0 g (protective dose) or 3.0 g (therapeutic dose) three times per year was sufficient for control of coconut lethal yellowing disease (McCoy, 1982).

Enrichment and isolation procedures

Isolation of phytoplasma-enriched fractions from plant and insect host tissues is possible by differential centrifugation and filtration after first disrupting tissues in osmotically-augmented buffers (Kirkpatrick et al., 1995; Lee et al., 1988; Sinha, 1979; Thomas and Balasundaran, 2001). Further purification of phytoplasmas is possible by centrifugation of enriched preparations in discontinuous Percoll density gradients (Davis et al., 1988; Gomez et al., 1996; Jiang and Chen, 1987) or by affinity chromatography using phytoplasma-specific antibodies coupled to Protein A-Sepharose columns (Jiang et al., 1988; Seddas et al., 1995). Viability of these enriched preparations may be assessed by infectivity tests in which aliquots of the phytoplasma preparations are micro-injected into vector insects, which are then fed on healthy indicator plants (Nasu et al., 1974; Sinha, 1979; Whitcomb et al., 1966a, b). Separation of enriched phytoplasma DNA from mixtures with host DNA is also possible by use of cesium chloride-bisbenzimide buoyant density gradient centrifugation (Kollar and Seemüller, 1989). Present as an uppermost band in final gradients, phytoplasma DNA fractionated by this means was suitable for endonuclease digestion and cloning for DNA probe development (Harrison et al., 1992, 1991; Kollar and Seemüller, 1990).

Maintenance procedures

Viable phytoplasmas have been maintained for at least 6 years in intact vector insects frozen at −70°C (Chiykowski, 1983). Viable X-disease phytoplasmas have been maintained for 2 weeks in salivary glands suspended in a tissue culture medium (Nasu et al., 1974). Extracts of phytoplasma-infected insects prepared in a $MgCl_2$/glycine buffer, osmotically adjusted to 800 milliosmoles/kg with sucrose, retained their infectivity for up to 3 d (Smith et al., 1981). Phytoplasma strains have been routinely maintained in diseased plants kept in an insect-proof greenhouse or in plantlets grown in tissue culture (Bertaccini et al., 1992; Davis and Lee, 1992; Jarausch et al., 1996; Sears and Klomparens, 1989; Wongkaew and Fletcher, 2004). While plant to plant transmission is accomplished naturally by vector insects and, in some cases, through grafts, experimental transmissions commonly include the use of plant parasitic dodders (*Cuscuta* sp.) (Marcone et al., 1999a). Although phytoplasma strains are commonly maintained in plants by periodic graft inoculation, maintenance of phytoplasmas exclusively in plants can result in strain attenuation over time and an associated loss of transmissibility by vector insects (Chiykowski, 1988; Denes and Sinha, 1992).

Differentiation of the genus "*Candidatus* Phytoplasma" from other genera

Phytoplasma-specific nucleic acid probes and PCR technology have largely supplanted traditional methods of electron microscopy and biological criteria for sensitive detection, identification, and genetic characterization of phytoplasmas. Molecular-based analyses have shown phytoplasma genomes to be A+T rich (Kollar and Seemuller, 1989; Oshima et al., 2004) and to range from 530 to 1350 kbp in size (Marcone et al., 1999b, 2001; Neimark and Kirkpatrick, 1993). Before any phytoplasma genomes were sequenced, phytoplasmas were shown to contain two rRNA operons (Davis, 2003a; Harrison et al., 2002; Ho et al., 2001; Jomantiene et al., 2002; Jung et al., 2003a; Lee et al., 1998b; Liefting et al., 1996; Marcone et al., 2000; Schneider and Seemüller, 1994a). Other genes that have been identified include ribosomal protein genes (Gundersen et al., 1994; Lee et al., 1998b; Lim and Sears, 1992; Martini et al., 2007; Miyata et al., 2002a; Toth et al., 1994) of the S10-*spc* operon (Miyata et al., 2002a), a nitroreductase gene (Jarausch et al., 1994), DNA gyrase genes (Chuang and Lin, 2000), genes encoding elongation factors G and *Tu* (An et al., 2006; Berg and Seemüller, 1999; Koui et al., 2003; Marcone et al., 2000; Miyata et al., 2002b; Schneider et al., 1997), *secA*, *secY*, and *secE* genes of a functional Sec protein translocation system (Kakizawa et al., 2001, 2004), *gidA*, *potB*, *potC*, and *potD* (Mounsey et al., 2006), a gene encoding an RNase P ribozyme (Wagner et al., 2001), *recA* (Chu et al., 2006), *rpoC* (Lin et al. 2006), *polC* (Chi and Lin., 2005), and insertion sequence (IS)-like elements (Lee et al., 2005). Numerous other putative genes or pseudogenes have been identified recently after partially or fully sequencing random fragments of genomic DNA cloned from phytoplasmas by various methods (Bai et al., 2004; Cimerman et al., 2006, 2009; Davis et al., 2003b, 2005; Garcia-Chapa et al., 2004; Liefting and Kirkpatrick, 2003; Melamed et al., 2003; Miyata et al., 2003; Streten and Gibb, 2003).

Development of phytoplasma-specific rRNA gene primers has permitted PCR-mediated amplification of various regions of the rRNA operons (Ahrens and Seemüller, 1992; Baric and Dalla Via, 2004; Davis and Lee, 1993; Deng and Hiruki, 1991; Gundersen and Lee, 1996; Lee et al., 1993b) (Namba et al., 1993; Smart et al., 1996). RFLP analysis of PCR-amplified rDNA provided a practical solution to the problem of phytoplasma identification and classification (Lee et al., 2000, 1998b). Pairwise comparisons of disparate strains were marked by considerable differences in RFLP patterns, whereas strains that were considered closely related on the basis of similar biological properties were often, although not always, indistinguishable on the basis of RFLP patterns. Alternatively, heteroduplex mobility analysis has demonstrated greater sensitivity than RFLP analysis for detecting minor variability in 16S rRNA genes of closely related phytoplasma strains (Cousin et al., 1998; Wang and Hiruki, 2000), since RFLP analysis is limited to detection of recognition sites for restriction endonucleases. Cluster analysis of rDNA RFLP patterns provided the first means to differentiate between known and unknown phytoplasmas from a wide range of plant hosts and geographic locations, and to resolve phytoplasmas into well-defined phylogenetic groups and subgroups (Ahrens and Seemüller, 1992; Lee et al., 1993b; Schneider et al., 1993, 1995).

Taxonomic comments

The inability to cultivate phytoplasmas outside of their plant and insect hosts has thus far rendered traditional methods impractical as aids for taxonomy of these organisms. Unlike their culturable *Mollicute* relatives, which were originally classified based only upon biological and phenotypic properties in pure culture, phytoplasmas cannot be classified by these criteria. Through application of DNA-based methods, it is now possible to accurately identify and characterize phytoplasmas and to assess their genetic interrelationships. These capabilities have assisted development of classification systems, first based on hybridization data, later based on 16S rDNA RFLPs, and ultimately on phylogenetic analysis of 16S rRNA genes and other conserved gene sequences. Classification schemes founded upon these molecular criteria have been refined and expanded upon over time, with the goal of defining a taxonomy for these unique organisms. In a phytoplasma classification scheme proposed by Lee et al. (1993b), based on analyses of rDNA RFLPs, a total of nine primary 16S rDNA groups (termed 16Sr groups) and 14 subgroups were initially recognized. Phytoplasma groups delineated by these analyses were consistent with genomic strain clusters previously identified by DNA hybridization analysis (Lee et al., 1992), although a greater diversity among strains comprising group 16SrI (aster yellows and related strains) was indicated by the earlier hybridization data. Subgroups within a given 16Sr group were distinguished by the presence of one or more restriction sites in a phytoplasma strain that differed from those in all existing members of a given subgroup. For those strains in which intra-rRNA operon heterogeneity was detected, subgroup designations were assigned according to the combined patterns of both 16S rRNA genes.

RFLP analysis of more variable ribosomal protein genes (Gundersen et al., 1994; Lee et al., 2004a, b) or *tuf* genes (Marcone et al., 2000; Schneider et al., 1997) has provided a means for more detailed subdivision of phytoplasma primary groups delineated by 16S rDNA RFLP data. This strategy for finer subgroup differentiation has been used to modify and expand upon earlier classifications and to incorporate many newly identified phytoplasma strains. Based on RFLP analysis of nearly full-length 16S rRNAs, at least 15 primary 16Sr groups have been recognized (Lee et al., 1998b; Montano et al., 2001): 16SrI, Aster yellows; 16SrII, Peanut witches'-broom; 16SrIII, X-disease; 16SrIV, Coconut lethal yellows; 16SrV, Elm yellows; 16SrVI, Clover proliferation; 16SrVII, Ash yellows; 16SrVIII, Loofah witches'-broom; 16SrIX, Pigeonpea witches'-broom; 16SrX, Apple proliferation; 16SrXI, Rice yellow dwarf; 16SrXII, Stolbur; 16SrXIII, Mexican periwinkle virescence; 16SrXIV, Bermuda grass white leaf; and 16SrXV Hibiscus witches'-broom. A total of 45 subgroups were identified when ribosomal protein gene RFLP data was also considered in the analyses.

Sequencing of 30 nearly full-length amplified 16S rRNA genes was undertaken by Namba et al. (1993), Gundersen et al. (1994), and Seemüller et al. (1994) from a diversity of strains previously characterized by rDNA RFLP analysis. These collective efforts, combined with earlier studies (Kuske and Kirkpatrick, 1992b; Lim and Sears, 1989), provided the first comprehensive phytoplasma phylogeny. In recognition of their unique phylogenetic status, the trivial name "phytoplasma" was initially proposed (Sears and Kirkpatrick, 1994) and has since been adopted formally (IRPCM Phytoplasma/Spiroplasma Working Team – Phytoplasma Taxonomy Group, 2004) to collectively name these fastidious, phytopathogenic mollicutes previously known as mycoplasma-like organisms. Within the phytoplasma clade, major subclades (primary groups representing "*Candidatus*" species) include: (1) Stolbur; (2) Aster yellows; (3) Apple proliferation; (4) Coconut lethal yellowing; (5) Pigeonpea witches'-broom; (6) X-disease; (7) Rice yellow dwarf; (8) Elm yellows; (9) Ash yellows; (10) Sunnhemp witches'-broom; (11) Loofah witches'-broom; (12) Clover proliferation; and (13) Peanut witches'-broom (Kirkpatrick et al., 1995; Schneider et al., 1995; White et al., 1998). Primary phytoplasma groups including 19 novel groups, namely Australian grapevine yellows (AUSGY), Italian bindweed stolbur (IBS), Buckthorn witches'-broom (BWB), *Spartium* witches'-broom (SpaWB), *Galactia* little leaf (GaLL), *Vigna* little leaf (ViLL), Clover yellow edge (CYE), Hibiscus witches'-broom (HibWB), Pear decline (PD), European stone fruit yellows (ESFY), Japanese hydrangea phyllody (JHP), *Psammotettix cephalotes*-borne (BVK), Italian alfalfa witches'-broom (IAWB), *Cirsium* phyllody (CirP), Bermuda grass white leaf (BGWL), Sugarcane white leaf (SCWL), Tanzanian lethal decline (TLD), *Stylosanthes* little leaf (StLL), and *Pinus sylvestris* yellows (PinP), that were absent from previous classification schemes have been subsequently defined. These new taxonomic entities were delineated on the basis of phylogenetic tree branching patterns, differences in 16S rRNA gene sequence similarities that were 1.2–2.3% or greater and, in some instances, by additional considerations such as plant host and vector specificity, primer specificity, and RFLP comparisons of ribosomal and nonribosomal DNA, as well as serological comparisons (Seemüller et al., 2002, 1998).

Most recently, Wei et al. (2007) applied computer-simulated RFLP analysis for classification of phytoplasma strains. Through comparisons of virtual RFLP patterns of 16S rRNA genes and calculations of coefficients of RFLP similarity, the authors classified all available 16S rRNA gene sequences, including sequences from 250 previously unclassified phytoplasma strains, into a total of 28 16Sr RFLP groups. These included ten new groups and dozens of new subgroup lineages (Cai et al., 2008; Wei et al., 2008b). Each new group represents a potential "*Candidatus* Phytoplasma" species level taxon. This information was used to augment the 16Sr RFLP classification system (Lee et al., 2000, 1998b, 1993b) with the following additional groups: 16SrXVI, Sugarcane yellow leaf syndrome; 16SrXVII, Papaya bunchy top group; 16SrXVIII, American potato purple top wilt group; 16SrXIX, Japanese chestnut witches'-broom group; 16SrXX, Buckthorn witches'-broom group; 16SrXXI, Pine shoot proliferation group; 16SrXXI, Nigerian coconut lethal decline (LDN) group; 16SrXXIII, Buckland valley grapevine yellows group; 16SrXXIV, Sorghum bunchy shoot group; 16SrXXV, Weeping tea witches'-broom group; 16SrXXVI, Mauritius sugarcane yellow D3T1 group; 16SrXXVII, Mauritius sugarcane yellow D3T2 group; and 16SrXXVIII, Havana derbid phytoplasma group. The virtual RFLP patterns are available for online use as reference patterns at http://www.ba.ars.usda.gov/data/mppl/virtualgel.html.

The spacer region (SR) separating the 16S from the 23S rRNA gene of phytoplasmas was also shown to be a reliable phylogenetic marker. Phylogenetic trees derived from the entire 16S–23S SR (Gibb et al., 1998; Kenyon et al., 1998) or variable regions flanking the $tRNA^{ile}$ gene (Kirkpatrick et al., 1995; Schneider et al., 1995) differentiated phytoplasmas into groups that were concordant with the major groups established previously from analyses of 16S rRNA genes. Phytoplasmas collectively differ in their 16S rRNA gene sequence by no more than 14%, whereas their respective 16S–23S SR sequences differ by as much as 22%. This added variation has contributed to improved accuracy of phytoplasma classification at the subgroup level. Similarly, phylogenetic analysis of ribosomal protein genes, *secY*, *secA*, or 23S rRNA genes has been employed to differentiate closely related phytoplasma strains, as well as to aid the group and subgroup classification of diverse phytoplasmas (Daire, 1993; Hodgetts et al., 2008; Lee et al., 1998b, 2004a, 2006b; Martini et al., 2007; Reinert, 1999). Such studies have led to finer differentiation among phytoplasma subgroups and to enriched descriptions of "*Candidatus* Phytoplasma" species (Lee et al., 2004a, 2006a, b).

A polyphasic system for taxonomy based on integration of genotypic, phenotypic, and phylogenetic information employed for bacterial classification (Murray et al., 1990; Stackebrandt and Goebel, 1994) has proved problematic for nonculturable phytoplasmas. In response to a rapidly growing database of phylogenetic markers, even in the absence of species-defining biological or phenotypic characters, the Working Team on Phytoplasmas of the International Research Programme of Comparative Mycoplasmology (IRPCM Phytoplasma/Spiroplasma

Working Team – Phytoplasma Taxonomy Group, 2004) proposed that taxonomy of phytoplasmas be based primarily upon phylogenetic analyses. This proposal was agreed to and adopted as policy by the ICSB Subcommittee on the Taxonomy of *Mollicutes* (1993, 1997), which also recommended that the provisional taxonomic status of "*Candidatus*", originally proposed by Murray and Schleifer (1994), be used for assigning genera names as follows: "*Candidatus* Phytoplasma" (from *phytos*, Greek for plant; *plasma*, Greek for thing molded) [(Mollicutes) NC; NA; O; NAS (GenBank no. M30790); oligonucleotide sequence of unique region of the 16S rRNA gene is CAA GAY BAT KAT GTK TAG CYG GDC T; P (Plant, phloem; Insect, salivary gland); M]. (IRPCM Phytoplasma/Spiroplasma Working Team – Phytoplasma Taxonomy Group, 2004). By this same approach, major groups within the genus also delineated by phylogenetic analysis of near full-length 16S rRNA gene sequences were considered to represent one or more distinct species.

Current guidelines for "*Candidatus* Phytoplasma" species descriptions (Anonymous, 2000; Firrao et al., 2005; IRPCM Phytoplasma/Spiroplasma Working Team – Phytoplasma Taxonomy Group, 2004) are based upon identification of a significantly unique 16S rRNA gene sequence >1200 bp in length. The strain from which the sequence is obtained should be designated as the reference strain. Strains with minimal differences in the 16S rRNA sequence, relative to the reference strain, should be referred to as related strains. In general, a strain can be described as a new "*Candidatus* Phytoplasma" species if its 16S rRNA gene sequence has less than 97% identity to any previously described "*Candidatus* Phytoplasma" species (ICSB Subcommittee on the Taxonomy of *Mollicutes*, 2001). There are cases in which phytoplasmas may share more than 97% of their 16S rRNA gene sequence, but clearly represent ecologically distinct populations and, thus, they may warrant description as separate species. In such cases, the description of two different species is recommended when all of the following conditions apply: (1) the two phytoplasmas are transmitted by different vector species; (2) the two phytoplasmas have a different natural plant host, or at least their symptomatology is significantly different in the same plant host; (3) there is evidence of significant molecular diversity between phytoplasmas as determined by DNA hybridization assays with cloned nonribosomal DNA markers, serological reactions, or by PCR-based assays. The taxonomic rank of subspecies should not be used. Reference strains should be available to the scientific community in graft-inoculated or *in vitro* micropropagated host plants or as DNA if perpetuation of strains in infected host plants is not feasible. Descriptions of "*Candidatus* Phytoplasma" species should be preferably submitted to the International Journal of Systematic and Evolutionary Microbiology (http://ijs.sgmjournals.org/).

Recent phylogenetic investigations, including the present analyses (Figure 116), suggest 97.5% 16S rRNA gene sequence similarity may represent a more suitable upper threshold for "*Candidatus* Phytoplasma" species separation, in that taxonomic subgroups designated based on 16S rRNA gene sequence similarities of ≤97.5% more consistently define species that are phylogenetically distinct from nearest related species. Regardless of homology criteria, a taxonomy is emerging for the phytoplasmas in the absence of cultivability where species and related strains of a species are clearly recognized with due consideration of the genetic, ecological, and environmental constraints unique to this group of plant- and insect-associated *Mollicutes*. To a large extent, the present taxonomy employs vernacular names based on associated diseases, but is constantly shifting towards a traditional taxonomy as more and more "*Candidatus* Phytoplasma" species continue to be recognized and proposed.

List of species of the genus "*Candidatus* Phytoplasma"

In accordance with the current guidelines for "*Candidatus* Phytoplasma" species descriptions, the following species have been designated. Proposed assignments to the class *Mollicutes* are based on nucleic acid sequences. None of these species have been cultivated independently of their host, and their metabolism and growth temperatures are unknown.

1. **"*Candidatus* Phytoplasma allocasuarinae"** Marcone, Gibb, Streten and Schneider 2004a, 1028

   Vernacular epithet: Allocasuarina yellows phytoplasma, strain AlloY[R].

   Gram reaction: not applicable.

   Morphology: other.

   Sequence accession no. (16S rRNA gene): AY135523.

   Unique region of 16S rRNA gene: 5′-TTTATTCGAGAGGGCG-3′.

   Habitat, association, or host: phloem of *Allocasuarina muelleriana* (Slaty she-oak).

2. **"*Candidatus* Phytoplasma americanum"** Lee, Bottner, Secor and Rivera-Varas 2006a, 1596

   Vernacular epithet: Potato purple top, strain APPTW12-NE[R].

   Gram reaction: not applicable.

   Morphology: other.

   Sequence accession no. (16S rRNA gene): DQ174122.

   Unique regions of 16S rRNA gene: 5′-GTTTCTTCGGAAA-3′ (68–80), 5′-GTTAGAAATGACT-3′ (142–153), 5′-GCTGGTGGCTT-3′ (1438–1448).

   Habitat, association, or host: Solanum tuberosum phloem.

3. **"*Candidatus* Phytoplasma asteris"** Lee, Gundersen-Rindal, Davis, Bottner, Marcone and Seemüller 2004a, 1046

   Vernacular epithet: Aster yellows (AY) phytoplasma, strain OAY[R].

   Gram reaction: not applicable.

   Morphology: other.

   Sequence accession no. (16S rRNA gene): M30790.

   Unique regions of 16S rRNA gene: 5′-GGGAGGA-3′, 5′-CTGACGGTACC-3′, and 5′-CACAGTGGAGGTTATCAGTTG-3′.

   Habitat, association, or host: phloem of *Oenothera hookeri* (Evening primrose).

4. **"*Candidatus* Phytoplasma aurantifolia"** Zreik, Carle, Bové and Garnier 1995, 452

   Vernacular epithet: Witches'-broom disease of lime phytoplasma, strain WBDL[R].

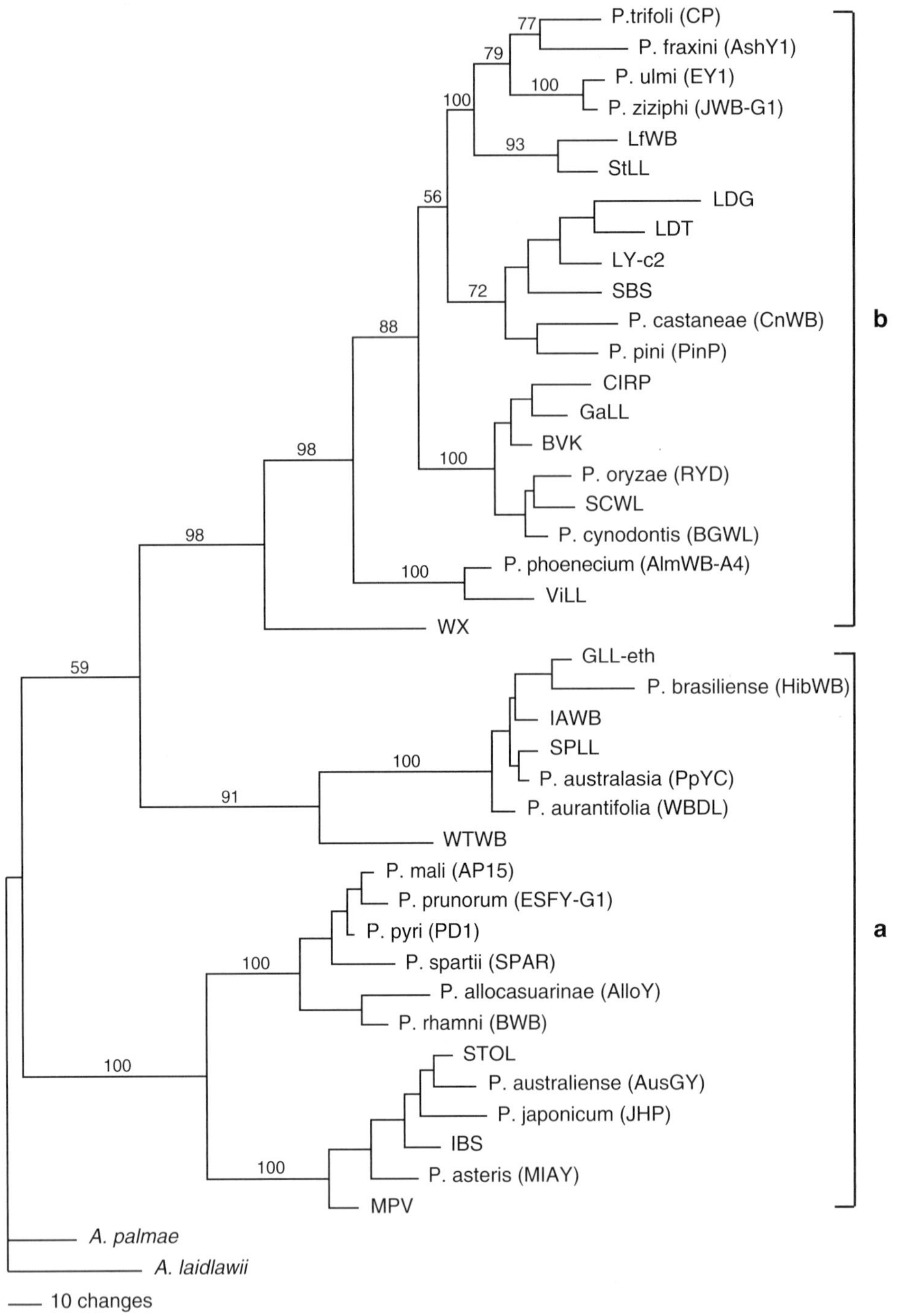

FIGURE 116. Phylogenetic analysis of the phytoplasmas. Phylogenetic trees were constructed by parsimony analyses of phytoplasma 16S rRNA gene sequences using the computer program PAUP (Swofford, 1998). The closely related culturable *Acholeplasma palmae* was employed as the outgroup. Because phytoplasma taxa are too numerous to present in a single inclusive tree, a global phylogeny of representative phytoplasmas is first presented. The global tree is divided into lower (a) and upper (b) regions. Each region of the global phylogeny is then expanded into inclusive trees, a and b, which collectively include 145 phytoplasmas from diverse geographic origins. Taxonomic subgroups, representing phytoplasmas sharing at least 97.5% 16S rRNA gene sequence similarity, are identified on each inclusive tree. Each phylogenetically distinct subgroup is equivalent to a subclade (or putative species) within the genus "*Candidatus* Phytoplasma". In all trees, branch lengths are proportional to the number of inferred character state transformations. Bootstrap (confidence) values greater than or equal to 50 are shown on the branches. Phytoplasmas for which 16S rRNA gene sequences of at least 1200 bp in length have been determined (312 total) are listed by subgroup in Table 144 along with their sequence accession numbers.

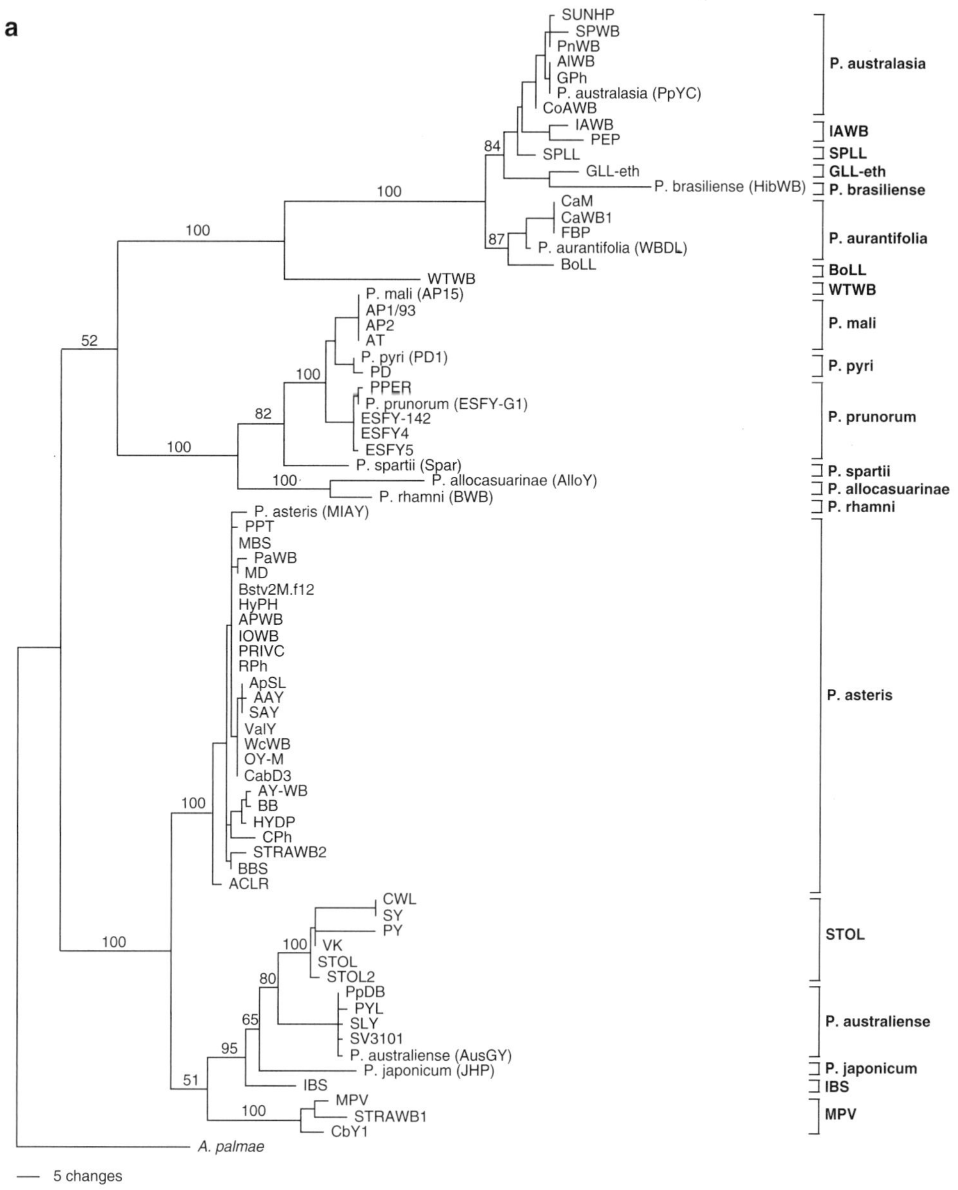

FIGURE 116. (Continued)

Gram reaction: not applicable.

Morphology: other.

Sequence accession no. (16S rRNA gene): U15442.

Unique region of 16S rRNA gene: 5′-GCAAGTGGTGAACCATTTGTTT-3′.

Habitat, association, or host: phloem of *Citrus*; hemolymph and salivary glands of *Hishimonus phycitis* (Cicadellidae).

5. **"*Candidatus* Phytoplasma australasia"** White, Blackall, Scott and Walsh 1998, 949

Vernacular epithet: Papaya yellow crinkle phytoplasma, strain PpYC^{R}.

Gram reaction: not applicable.

Morphology: other.

Sequence accession no. (16S rRNA gene): Y10097.

Unique regions of 16S rRNA gene: 5′-TAAAAGGCATCTTTTATC-3′ and 5′-CAAGGAAGAAAAGCAAATGGCGAACCATTTGTTT-3′.

Habitat, association, or host: phloem of *Carica papaya* and *Lycopersicon esculentum*.

6. **"*Candidatus* Phytoplasma australiense"** Davis, Dally, Gundersen, Lee and Habili 1997, 268

Vernacular epithet: Australian grapevine yellows phytoplasma, strain AUSGY^{R}.

Gram reaction: not applicable.

Morphology: other.

Sequence accession no. (16S rRNA gene): L76865.

Unique regions of 16S rRNA gene: 5′-CGGTAGAAATATCGT-3′ and 5′-TTTATCTTTAAAAGACCTCGCAAGA-3′.

Habitat, association, or host: Vitis phloem.

7. **"*Candidatus* Phytoplasma brasiliense"** Montano, Davis, Dally, Hogenhout, Pimentel and Brioso 2001, 1117

Vernacular epithet: Hibiscus witches'-broom (HibWB) phytoplasma, strain HibWB26^{R}.

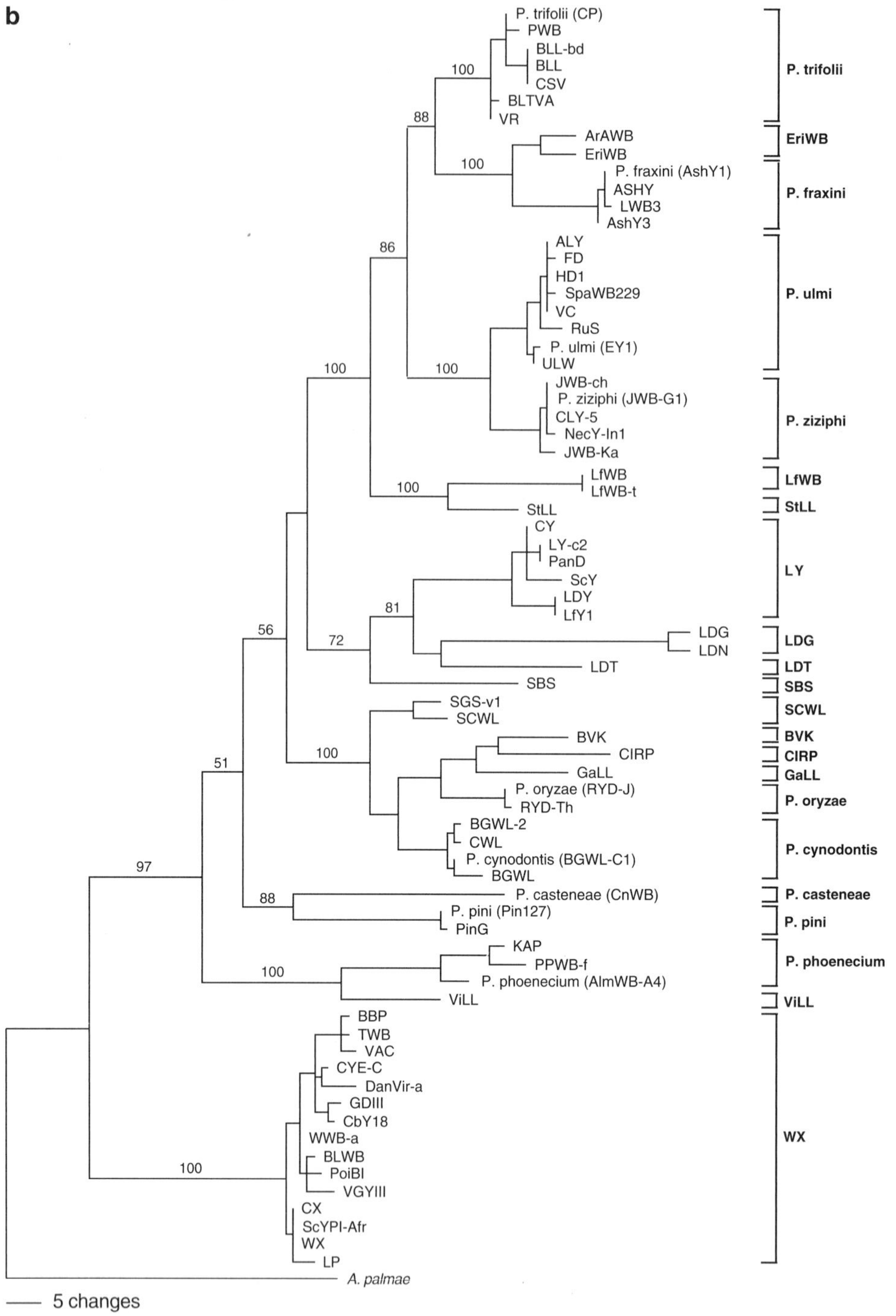

FIGURE 116. (Continued)

Gram reaction: not applicable.

Morphology: other.

Sequence accession no. (16S rRNA gene): AF147708.

Unique regions of 16S rRNA gene: 5′-GAAAAAGAAAG-3′, 5′-TCTTTCTTT-3′, 5′-CAG-3′, 5′-ACTTTG-3′, and 5′-GTCA AAAC-3′.

Habitat, association, or host: Hibiscus phloem.

8. **"*Candidatus* Phytoplasma caricae"** Arocha, López, Piñol, Fernández, Picornell, Almeida, Palenzuela, Wilson and Jones 2005, 2462

Vernacular epithet: Cuban papaya phytoplasma, strain PAY^{R}.

Gram reaction: not applicable.

Morphology: other.

Sequence accession no. (16S rRNA gene): AY725234.

TABLE 144. Provisional groupings, strain designations, associated plant disease, geographic origin and accession numbers of 16S rRNA gene sequences derived from phytoplasmas[a]

Subgroup	Strain	Associated plant disease	"*Candidatus* Phytoplasma species"	Geographic origin	16S accession no.[a]
AlloY	AlloY	Allocasuarina yellows	P. allocasuarinae	Australia	AY135523*
AP	AP15R	Apple proliferation	P. mali	Italy	AJ542541*
	AT	Apple proliferation		Germany	X68375*
	AP2	Apple proliferation		Germany	AF248958*
	AP1/93	Apple proliferation		France	AJ542542*
	D365/04	Apple proliferation		Slovenia	EF025917
	APSb	Apple proliferation		Italy	EF193361
ESFY	ESFY-G1R	European stone fruit yellows	P. prunorum	Germany	AJ542544*
	ESFY5	European stone fruit yellows		Austria	AY029540*
	ESFY4	European stone fruit yellows		Czech Republic	Y11933*
	PPER	European stone fruit yellows		Germany	X68374
	ESFY-142	European stone fruit yellows		Spain	AJ575108*
	ESFY-173	European stone fruit yellows		Spain	AJ575106
	ESFY-215	European stone fruit yellows		Spain	AJ575105
AshY	AshY1	Ash yellows	P. fraxini	USA, New York	AF092209*
	AshY3	Ash yellows		USA, Utah	AF105315*
	ASHY	Ash yellows		Germany	X68339*
	LWB3	Lilac witches'-broom		USA, Massachusetts	AF105317*
EriWB	EriWB	*Erigeron* witches'-broom		Brazil	AY034608
	ArAWB	Argentinian alfalfa witches'-broom		Argentina	AY147038
AusGY	AusGY	Australian grapevine yellows	P. australiense	Australia	L76865
	PpDB	Papaya die-back		Australia	Y10095*
	PYL	*Phormium* yellow leaf, rrnA		New Zealand	U43569
	PYLb	*Phormium* yellow leaf		New Zealand	U43570*
	SLY	Strawberry lethal yellows		Australia	AJ243045*
	SV3101	Strawberry virescence		Tonga	AY377868*
AY	MIAY	*Oenothera* virescence	P. asteris	USA, Michigan	M30790*
	OY-M	Onion yellows		Japan	NC005303*
	MBS	Maize bushy stunt		Mexico	AY265208*
	APWB	*Aphanamixis polystachya* witches'-broom		Bangladesh	AY495702*
	IOWB	*Ipomoea obscura* witches'-broom		Taiwan	AY265205*
	HyPH	Hydrangea phyllody		Italy	AY265207*
	HYPh	Hydrangea phyllody		France	AY265219
	RPh	Oilseed rape phyllody		Czech Republic	U89378*
	MD	Mulberry dwarf		South Korea	AY075038*
	GDS				DQ112021
	AYWB_ro4	Aster yellows witches'-broom		Ohio, USA	NC007716
	PaWB	Paulownia witches'-broom		Korea	AF279271*
	PY1	Periwinkle yellows		China	AF453328
	BVGY				AY083605
	AAY	American aster yellows		Southern USA	X68373*
	CabD4	Cabbage proliferation		USA, Texas	AY180932
	AY-BW	Aster yellows		USA, Ohio	AY389820
	ApSL	Apple sessile leaf		Lithuania	AY734454
	SAY	Severe aster yellows		USA, California	M86340*
	ValY	Valeriana yellows, rrnA		Lithuania	AY102274*
	PRIVC	Primrose virescence		Germany	AY265210*
	WcWB	Watercress witches'-broom		USA, Hawaii	AY665676*
	CabD3	Cabbage proliferation		USA, Texas	AY180947*
	AY-sb	Sugar beet aster yellows		Hungary	AF245439
	Bstv2Mf12				AY180951
	ACLR-AY	Apricot chlorotic leafroll		Spain	AY265211
	ACLR	Apricot chlorotic leafroll		Europe	X68338*
	BBS3	Blueberry stunt		USA, Michigan	AY265213
	STRAWB2	Strawberry green petal		USA, Florida	U96616*
	PoY	Populus yellows		Croatia	AF503568
	KVG	Clover phyllody		Germany	X83870

(continued)

TABLE 144. (continued)

Subgroup	Strain	Associated plant disease	"*Candidatus* Phytoplasma species"	Geographic origin	16S accession no.[a]
	CPh	Clover phyllody		Canada	AF222066*
	HYDP	Hydrangea phyllody		Belgium	AY265215*
	AY-WB	Aster yellows		USA, Ohio	AY389827*
	BB	Tomato big bud		USA, Arkansas	AY180955*
	PPT	Potato purple top		Mexico	AF217247*
THP	THP	Tomato 'hoja de perejil'	P. lycopersici	Bolivia	AY787136
	Derbid	Derbid phytoplasma		Cuba	AY744945
BWB	BWB	Buckthorn witches'-broom	P. rhamni	Germany	X76431*
BGWL	BGWL-C1	Bermudagrass white leaf	P. cynodontis	Italy	AJ550984*
	BGWL	Bermudagrass white leaf		Italy	Y16388*
	BGWL-2	Bermudagrass white leaf		Thailand	AF248961*
	CWL	Cynodon white leaf		Australia	AF509321*
BVK	BVK	*Psammotettic cephalotes*-borne		Germany	X76429*
CIRP	CIRP	*Cirsium* phyllody		Germany	X83438*
CnWB	CnWB	Chestnut witches'- broom	P. castaneae	Korea	AB054986*
CP	CPR	Clover proliferation	P. trifolii	Canada	AY390261*
	BLL	Brinjal little leaf		India	X83431*
	BLTVA	Columbia basin potato purple top		USA, Washington	AY692280*
	VR	Vinca virescence		USA, California	AY500817*
	PWB	Potato witches'-broom		Canada	AY500818*
	CSV	*Centauria stolstitialis* virescence		Italy	AY270156*
EY	EY1	Elm yellows	P. ulmi	USA, New York	AY197655*
	ULW	Ulmus witches'-broom		Italy	X68376*
	FD	Flavescence doree		Italy	X76560*
	RuS	Rubus stunt		Italy	AY197648*
	HD1	Hemp dogbane yellows		USA, New York	AY197654*
	VC	Asymptomatic Virginia creeper		USA, Florida	AF305198*
SpaWB	SpaWB229	Spartium witches'-broom	P. spartii	Italy	AY197652*
JWB	JWB-G1T	Jujube witches'-broom Gifu isolate 1	P. ziziphi	Japan	AB052876*
	JWB-Ka	Jujube witches'-broom Korea isolate 1		Korea	AB052879*
	JWB-ch	*Ziziphus jujube* witches'-broom		China	AF305240
	NecY-In1	Nectarine yellows		India	AY332659*
	CLY-5	Cherry lethal yellows		China	AY197659*
FBP	WBDL	Witches'-broom disease of lime	P. aurantifolia	Oman	U15442*
	CaWB-YNO1	Cactus witches'-broom		China	AJ293216
	FBP	Faba bean phyllody		Sudan	X83432*
	PPLL	Pigeon pea little leaf		Australia	AJ289191
BoLL	BoLL	Bonamia little leaf		Australia	Y15863*
GaLL	GaLL	Galactia little leaf		Australia	Y15865*
GLL-eth	GLL-eth	Gliricidia little leaf		Ethiopia	AF361018*
HibWB	HibWB	Hibiscus witches'-broom	P. brasiliense	Brazil	AF147708*
IAWB	IAWB	Alfalfa witches'-broom		Italy	Y16390*
	PEP	Pichris echioides phyllody		Italy	Y16393*
IBS	IBS	Italian bindweed yellows		Southern Italy	Y16391*
StrawY	StrawY	Strawberry lethal yellows	P. fragariae	Lithuania	DQ086423
JHP	JHP	Japanese hydrangea phyllody	P. japonicum	Japan	AB010425
LDG	LDG	Cape St Paul wilt		Ghana	Y13912*
	LDN	Awka disease of coconut		Nigeria	Y14175*
LDT	LDT	Coconut lethal disease		Tanzania	X80177*
LfWB	LfWB	Loofah WB		Taiwan	L33764*
	LfWB-t	Loofah WB		Taiwan	AF086621*
LY	CPY	*Carludovica palmata* yellows		Mexico	AF237615
	LDY	Yucatan coconut decline		Mexico	U18753*

(continued)

TABLE 144. (continued)

Subgroup	Strain	Associated plant disease	"*Candidatus* Phytoplasma species"	Geographic origin	16S accession no.[a]
	LfY1	Coconut leaf yellowing		Mexico	AF500329*
	LfY5(PE65)	Coconut leaf yellowing		Mexico	AF500334
	LY-c2	Coconut lethal yellows		USA, Florida	AF498309*
	LY-JC8	Coconut lethal yellows		Jamaica	AF498307
	PanD	Pandanus decline		USA, Florida	AF361020*
	ScY	Sugarcane yellows, group 4		Mauritius	AJ539178*
ScY	SCD3T2	Sugarcane yellows, group 3		Mauritius	AJ539180
	SCD3T1	Sugarcane yellows, group 3		Mauritius	AJ539179
MPV	MPV	Mexican periwinkle virescence		Mexico	AF248960*
	PerWB-FL	Periwinkle witches'-broom		USA, Florida	AY204549
	CbY1	Chinaberry yellows		Bolivia	AF495882*
	STRAWB1	Strawberry green petal		USA, Florida	U96614*
PD	PD1	Pear decline	P. pyri	Italy	AJ542543*
	PD	Pear decline		Germany	X76425*
	PYLR	Peach yellow leafroll		USA, California	Y16394
	EPC	Pear decline		Iran	DQ471321
PinP	Pin127R	*Pinus halepensis* yellows	P. pini	Spain	AJ632155*
	PinG	*Pinus sylvestris* yellows		Germany	AJ310849*
PPWB	AlmWB-A4	Almond witches'-broom	P. phoenecium	Lebanon	AF515636*
	KAP	*Knautia arvensis* phyllody		Italy	Y18052*
	PPWB-f	Pigeonpea witches'-broom		USA, Florida	AF248957*
RYD	RYD-J	Rice yellow dwarf	P. oryzae	Japan	D12581
	RYD-Th	Rice yellow dwarf		Thailand	AB052873*
SBS	SBS	Sorghum bunchy shoot		Australia	AF509322*
SCWL	SCWL	Sugarcane white leaf		Thailand	X76432*
	SGS-v1	Sorghum grassy shoot, variant 1		Australia	AF509324*
SpaWB	Spar	Spartium witches'-broom	P. spartii	Italy	X92869*
SPLL	SPLL	Sweet potato little leaf		Australia	X90591*
SPWB	PpYC	Papaya yellow crinkle	P. australasia	Australia	Y10097*
	GPh	Gerbera phyllody		Japan?	AB026155*
	PnWB	Peanut witches' broom		Taiwan	L33765*
	CoAWB	Cocky apple witches'-broom		Australia	AJ295330*
	SPLL	Sweet potato little leaf		Australia	AJ289193
	SUNHP	Sunnhemp phyllody		Thailand	X76433*
	AlWB	Alfalfa witches'-broom		Oman	AY169322*
	TBB	Australian tomato big bud		Australia	Y08173
StLL	StLL	Stylosanthes little leaf		Australia	AJ289192*
STOL	STOL	Stolbur of Capsicum annum		Europe	X76427*
	VK	Grapevine yellows		Europe	X76428*
	2642BN	Grapevine yellows	P. solani	France	AJ964960
ViLL	ViLL	Vigna little leaf		Australia	Y15866*
CIWB	IM-3	*Cassia italica* witches'-broom	P. omaniense	Oman	EF666051
WTWB	WTWB	Weeping tea witches'-broom		Australia	AF521672*
WX	BBP	Blueberry proliferation		Lithuania	AY034090*
	BLWB	Black locust witches'-broom		USA, Maryland	AF244363*
	CbY18	Chinaberry yellows		Bolivia	AF495657*
	CX	Canadian peach X		Canada	L33733*
	CYE	Clover yellow edge		Canada	AF175304*
	DanVir-a	Dandelion virescence, rrnA		Lithuania	AF370119*
	LP	Little peach		USA, S. Carolina	AF236122*
	PoiBI	Poinsettia branch-inducing		Southern USA	AF190223*
	ScYP I-Afr	Sugarcane yellows		Africa	AF056095*
	TWB	Tsuwabuki WB		Japan	D12580*
	VAC	Vaccinium witches'-broom		Germany	X76430*
	VGYIII	Virginia grapevine yellows		USA, Virginia	AF060875*
	WWB-a	Walnut witches'-broom, rrnA		USA, Georgia	AF190226*
	WX	Western X		USA, California	L04682*

[a]Accession numbers denoted by an asterisk were used as sources of 16S rRNA gene sequences for comprehensive phylogenetic analysis of subgroup phytoplasmas from diverse geographic origins.

Unique regions of 16S rRNA gene: 5′-AAA-3′ (196–198), 5′-ATT-3′ (600–603), 5′-AGGCGCC-3′ (1089–1095), 5′-GCG-GATTTAGTCACTTTTCAGGC-3′ (1379–1401).

Habitat, association, or host: Carica papaya phloem.

9. **"*Candidatus* Phytoplasma castaneae"** Jung, Sawayanagi, Kakizawa, Nishigawa, Miyata, Oshima, Ugaki, Lee, Hibi and Namba 2002, 1548

Vernacular epithet: Chestnut witches' broom phytoplasma, strain CnWB[R].

Gram reaction: not applicable.

Morphology: other.

Sequence accession no. (16S rRNA gene): AB054986.

Unique regions of 16S rRNA gene: 5′-CTAGTTTAAAAA-CAATGCTC-3′ and 5′-CTCATCTTCCTCCAATTC-3′.

Habitat, association, or host: Castanea crenata phloem.

10. **"*Candidatus* Phytoplasma cynodontis"** Marcone, Schneider and Seemüller 2004b, 1081

Vernacular epithet: Bermuda grass white leaf (BGWL) phytoplasma, strain BGWL-C1[R].

Gram reaction: not applicable.

Morphology: other.

Sequence accession no. (16S rRNA gene): AJ550984.

Unique region of 16S rRNA gene: 5′-AATTAGAAGGCAT-CTTTTAAT-3′.

Habitat, association, or host: phloem of *Cynodon dactylon* (Bermuda grass).

11. **"*Candidatus* Phytoplasma fragariae"** Valiunas, Staniulis and Davis 2006, 280

Vernacular epithet: Strawberry yellows phytoplasma, strain StrawY[R].

Gram reaction: not applicable.

Morphology: other.

Sequence accession no. (16S rRNA gene): DQ086423.

Unique regions of 16S rRNA gene: 5′-GTGCAATGCT-CAACGTTGTGAT-3′, 5′-AATTGCA-3′, and 5′-TGAGTAAT-CAAGAGGGAG-3′.

Habitat, association, or host: phloem of *Fragaria* x *ananassa.*

12. **"*Candidatus* Phytoplasma fraxini"** Griffiths, Sinclair, Smart and Davis 1999, 1613

Vernacular epithet: Ash yellows phytoplasma, strain AshY[R] and lilac witches'-broom (LWB) phytoplasma.

Gram reaction: not applicable.

Morphology: other.

Sequence accession no. (16S rRNA gene): AF092209.

Unique regions of 16S rRNA gene: 5′-CGGAAACCCCT-CAAAAGGTTT-3′ and 5′-AGGAAAGTC-3′.

Habitat, association, or host: phloem of *Fraxinus* and *Syringa.*

13. **"*Candidatus* Phytoplasma graminis"** Arocha, López, Piñol, Fernández, Picornell, Almeida, Palenzuela, Wilson and Jones 2005, 2462

Vernacular epithet: Sugarcane yellow leaf phytoplasma, strain SCYP[R].

Gram reaction: not applicable.

Morphology: other.

Sequence accession no. (16S rRNA gene): AY725228.

Unique regions of 16S rRNA gene: 5′-TTTG-3′ (465–468), 5′-TTG-3′ (478–480), 5′-GGG-3′ (1552–1554), 5′-TAA-3′ (1381–1383), and 5′-ATTTACGTTTCTG-3′ (1392–1404).

Habitat, association, or host: Saccharum officinarum phloem.

14. **"*Candidatus* Phytoplasma japonicum"** Sawayanagi, Horikoshi, Kanehira, Shinohara, Bertaccini, Cousin, Hiruki and Namba 1999, 1284

Vernacular epithet: Japanese *Hydrangea* phyllody phytoplasma, strain JHP[R].

Gram reaction: not applicable.

Morphology: other.

Sequence accession no. (16S rRNA gene): AB010425.

Unique regions of 16S rRNA gene: 5′-GTGTAGCCG-GGCTGAGAGGTCA-3′ and 5′-TCCAACTCTAGCTAAA-CAGTTTCTG-3′.

Habitat, association, or host: Hydrangea phloem.

15. **"*Candidatus* Phytoplasma lycopersici"** Arocha, Antesana, Montellano, Franco, Plata and Jones 2007, 1709

Vernacular epithet: Tomato "hoja de perejil" phytoplasma, strain THP[R].

Gram reaction: not applicable.

Morphology: other.

Sequence accession no. (16S rRNA gene): AY787136.

Unique regions of 16S rRNA gene: 5′-CTTA-3′ (positions 175–178), 5′-AATGGT-3′ (198–203), 5′-ATA-3′ (229–231), 5′-TGGAGGAA-3′ (234–242), 5′-CACG-3′ (302–305), 5′-TCT-3′ (315–317), 5′-GCT-3′ (334–336), 5′-TAT-3′ (336–338), 5′-TAC-3′ (413–415), and 5′-AGC-3′ (434–436).

Habitat, association, or host: Lycopersicon esculentum phloem.

16. **"*Candidatus* Phytoplasma mali"** Seemüller and Schneider 2004, 1224

Vernacular epithet: Apple proliferation (AP) phytoplasma, strain AP15[R].

Gram reaction: not applicable.

Morphology: other.

Sequence accession no. (16S rRNA gene): AJ542541.

Unique region of 16S rRNA gene: 5′-AATACTCGAAACCA-GTA-3′.

Habitat, association, or host: Malus phloem.

17. **"*Candidatus* Phytoplasma omanense"** Al-Saady, Khan, Calari, Al-Subhi and Bertaccini 2008, 464

Vernacular epithet: Cassia witches'-broom (CWB) phytoplasma, strain IM-1[R].

Gram reaction: not applicable.

Morphology: other.

Sequence accession no. (16S rRNA gene): EF666051.

Unique regions of 16S rRNA gene: 5′-AAAAAACAGT-3′ (467–474), 5′-TTGC-3′ (642–645), 5′-GTTAAAG-3′ (853–861), 5′-TAATT-3′ (1010–1014), and 5′-AAATT-3′ (1052–1056).

Habitat, association, or host: Cassia italica phloem.

18. **"*Candidatus* Phytoplasma oryzae"** Jung, Sawayanagi, Wongkaew, Kakizawa, Nishigawa, Wei, Oshima, Miyata, Ugaki, Hibi and Namba 2003c, 1928

Vernacular epithet: Rice yellow dwarf (RYD) phytoplasma, strain RYD-Th[R].

Gram reaction: not applicable.

Morphology: other.

Sequence accession nos (16S rRNA gene): D12581, AB052873 (RYD-Th).

Unique regions of 16S rRNA gene: 5′-AACTGGATAGGAAAT-TAAAAGGT-3′ and 5′-ATGAGACTGCCAATA-3′.

Habitat, association, or host: Oryza sativa phloem.

19. **"*Candidatus* Phytoplasma phoenicium"** Verdin, Salar, Danet, Choueiri, Jreijiri, El Zammar, Gélie, Bové and Garnier 2003, 837

Vernacular epithet: Almond witches'-broom (AlmWB) phytoplasma, strain AlmWB-A4[R].

Gram reaction: not applicable.

Morphology: other.

Sequence accession no. (16S rRNA gene): AF515636.

Unique region of 16S rRNA gene: 5′-CCTTTTTCGGAAGG-TATG-3′.

Habitat, association, or host: Prunus amygdalus phloem.

20. **"*Candidatus* Phytoplasma pini"** Schneider, Torres, Martín, Schröder, Behnke and Seemüller 2005, 306

Vernacular epithet: Pinus halepensis yellows (Pin) phytoplasma, strain Pin127S[R].

Gram reaction: not applicable.

Morphology: other.

Sequence accession no. (16S rRNA gene): AJ632155.

Unique regions of 16S rRNA gene: 5′-GGAAATCTTTCG-GGATTTTAGT-3′ and 5′-TCTCAGTGCTTAACGCTGT-TCT-3′.

Habitat, association, or host: Pinus phloem.

21. **"*Candidatus* Phytoplasma prunorum"** Seemüller and Schneider 2004, 1224

Vernacular epithet: European stone fruit yellows (ESFY) phytoplasma, strain ESFY-G1[R].

Gram reaction: not applicable.

Morphology: other.

Sequence accession no. (16S rRNA gene): AJ542544.

Unique regions of 16S rRNA gene: 5′-AATACCCGAAACCA-GTA-3′ and 5′-TGAAGTTTTGAGGCATCTCGAA-3′.

Habitat, association, or host: Prunus phloem.

22. **"*Candidatus* Phytoplasma pyri"** Seemüller and Schneider 2004, 1224

Vernacular epithet: Pear decline (PD) phytoplasma, strain PD1[R].

Gram reaction: not applicable.

Morphology: other.

Sequence accession no. (16S rRNA gene): AJ542543.

Unique regions of 16S rRNA gene: 5′-AATACTCAAAACCA-GTA-3′ and 5′-ATACGGCCCAAACTCATACGGA-3′.

Habitat, association, or host: Pyrus phloem.

23. **"*Candidatus* Phytoplasma rhamni"** Marcone, Gibb, Streten and Schneider 2004a, 1028

Vernacular epithet: Buckthorn witches'-broom phytoplasma, strain BWB[R].

Gram reaction: not applicable.

Morphology: other.

Sequence accession nos (16S rRNA gene): X76431, AJ583009.

Unique regions of 16S rRNA gene: 5′-CGAAGTATTTCGA-TAC-3′.

Habitat, association, or host: phloem of *Rhamnus catharticus* (buckthorn).

24. **"*Candidatus* Phytoplasma solani"** Firrao, Gibb and Streton 2005, 251

Vernacular epithet: Stolbur phytoplasma; subgroup A reference type of the stolbur phytoplasma taxonomic group 16SrXII (Lee et al., 2000).

Gram reaction: not applicable.

Morphology: other.

Sequence accession no. (16S rRNA gene): AJ970609 (strain PO; Cimerman et al., 2006).

Unique region of 16S rRNA gene: not reported.

Habitat, association, or host: many species of *Solanaceae* plus several species in other plant families, and *Fulguromorpha* spp. planthopper vectors.

25. **"*Candidatus* Phytoplasma spartii"** Marcone, Gibb, Streten and Schneider 2004a, 1028

Vernacular epithet: Spartium witches'-broom phytoplasma, strain SpaWB[R].

Gram reaction: not applicable.

Morphology: other.

Sequence accession no. (16S rRNA gene): X92869.

Unique region of 16S rRNA gene: 5′-TTATCCGCGTTAC-3′.

Habitat, association, or host: phloem of *Spartium junceum* (Spanish broom).

26. **"*Candidatus* Phytoplasma tamaricis"** Zhao, Sun, Wei, Davis, Wu and Liu 2009, 2496

Vernacular epithet: Salt cedar witches'-broom phytoplasma, strain SCWB1[R].

Gram reaction: not applicable.

Morphology: other.

Sequence accession no. (16S rRNA gene): FJ432664.

Unique regions of 16S rRNA gene: 5′-ATTAGGCATCTAG-TAACTTTG-3′, 5′-TGCTCAACATTGTTGC-3′, 5′-AGCTTT-GCAAAGTTG-3′, and 5′-TAACAGAGGTTATCAGAGTT-3′.

Habitat, association, or host: phloem of *Tamarix chinensis* (salt cedar).

27. **"*Candidatus* Phytoplasma trifolii"** Hiruki and Wang 2004, 1352

Vernacular epithet: Clover proliferation phytoplasma, strain CP[R].

Gram reaction: not applicable.

Morphology: other.

Sequence accession no. (16S rRNA gene): AY390261.

Unique regions of 16S rRNA gene: 5′-TTCTTACGA-3′ and 5′-TAGAGTTAAAAGCC-3′.

Habitat, association, or host: Trifolium phloem.

28. **"*Candidatus* Phytoplasma ulmi"** Lee, Martini, Marcone and Zhu 2004b, 345

Vernacular epithet: Elm yellows phytoplasma (EY) phytoplasma, strain EY1[R].

Gram reaction: not applicable.

Morphology: other.

Sequence accession nos (16S rRNA gene): AY197655, AY197675, and AY197690.

Unique regions of 16S rRNA gene: 5′-GGAAA-3′ and 5′-CGT-TAGTTGCC-3′.

Habitat, association, or host: Ulmus americana phloem.

29. **"*Candidatus* Phytoplasma vitis"** Firrao, Gibb and Streton 2005, 251

Vernacular epithet: Flavescence dorée phytoplasma; strains are genetically heterogenous and vary in degree of virulence, but all are referable to subgroups C or D of the elm yellows phytoplasma taxonomic group 16SrV (Lee et al., 2000).

Gram reaction: not applicable.

Morphology: other.

Sequence accession nos (16S rRNA gene): AY197645 (16SrV subgroup C), AY197644 (16SrV subgroup D1).

Unique regions of 16S rRNA gene: not reported.

Habitat, association, or host: grapevines (*Vitis vinifera*) and the leafhopper vector *Scaphoideus titanus.*

30. **"*Candidatus* Phytoplasma ziziphi"** Jung, Sawayanagi, Kakizawa, Nishigawa, Wei, Oshima, Miyata, Ugaki, Hibi and Namba 2003b, 1041

Vernacular epithet: Jujube witches'-broom phytoplasma, strain JWB[R].

Gram reaction: not applicable.

Morphology: other.

*Sequence accession nos (16S rRNA gene):*AB052875–AB052879.

Unique regions of 16S rRNA gene: 5′-TAAAAAGGCATCTTTTTGTT-3′ and 5′-AATCCGGACTAAGACTGT-3′.

Habitat, association, or host: Ziziphyus jujube phloem.

References

Ahrens, U. and E. Seemüller. 1992. Detection of DNA of plant pathogenic mycoplasma-like organisms by a polymerase chain reaction that amplifies a sequence of the 16S rRNA gene. Phytopathology *82*: 828–832.

Al-Saady, N.A., A.J. Khan, A. Calari, A.M. Al-Subhi and A. Bertaccini. 2008. '*Candidatus* Phytoplasma omanense', associated with witches'-broom of *Cassia italica* (Mill.) Spreng. in Oman. Int. J. Syst. Evol. Microbiol. *58*: 461–466.

Albertazzi, G., A.C.J. Milc, E. Francia, E. Roncaglia, F. Ferrari, E. Tagliafico, E. Stefani and N. Pecchioni. 2009. Gene expression in grapevine cultivars in response to Bois Noir phytoplasma infection. Plant Science *176*: 792–804.

Alma, A., D. Bosco, A. Danielli, A. Bertaccini, M. Vibio and A. Arzone. 1997. Identification of phytoplasmas in eggs, nymphs and adults of *Scaphoideus titanus* Ball reared on healthy plants. Insect. Mol. Biol. *6*: 115–121.

An, F.-Q., Y.-F. Wu, X.-Q. Sun, P.-W. Gu and Y. Yang. 2006. Homologic analysis of tuf gene for elongation factor Tu of phytoplasma from wheat blue dwarf. Sci. Agric. Sinica *39*: 74–80.

Anonymous. 2000. Report of consultations: Phytoplasmas, spiroplasmas, mesoplasmas and entomoplasmas working team. Presented at the International Research Programme on Comparative Mycoplasmology (IRPCM) of the International Organization for Mycoplasmology (IOM) Fukuoka, Japan.

Arashida, R., S. Kakizawa, Y. Ishii, A. Hoshi, H.Y. Jung, S. Kagiwada, Y. Yamaji, K. Oshima and S. Namba. 2008. Cloning and characterization of the antigenic membrane protein (Amp) gene and *in situ* detection of Amp from malformed flowers infected with Japanese hydrangea phyllody phytoplasma. Phytopathology *98*: 769–775.

Arocha, Y., M. Lopez, B. Pinol, M. Fernandez, B. Picornell, R. Almeida, I. Palenzuela, M.R. Wilson and P. Jones. 2005. '*Candidatus* Phytoplasma graminis' and '*Candidatus* Phytoplasma caricae', two novel phytoplasmas associated with diseases of sugarcane, weeds and papaya in Cuba. Int. J. Syst. Evol. Microbiol. *55*: 2451–2463.

Arocha, Y., O. Antesana, E. Montellano, P. Franco, G. Plata and P. Jones. 2007. '*Candidatus* Phytoplasma lycopersici', a phytoplasma associated with 'hoja de perejil' disease in Bolivia. Int. J. Syst. Evol. Microbiol. *57*: 1704–1710.

Bai, X., J. Zhang, I.R. Holford and S.A. Hogenhout. 2004. Comparative genomics identifies genes shared by distantly related insect-transmitted plant pathogenic mollicutes. FEMS Microbiol. Lett. *235*: 249–258.

Bai, X., J. Zhang, A. Ewing, S.A. Miller, A. Jancso Radek, D.V. Shevchenko, K. Tsukerman, T. Walunas, A. Lapidus, J.W. Campbell and S.A. Hogenhout. 2006. Living with genome instability: the adaptation of phytoplasmas to diverse environments of their insect and plant hosts. J. Bacteriol. *188*: 3682–3696.

Barbara, D.J., A. Morton, M.F. Clark and D.L. Davies. 2002. Immunodominant membrane proteins from two phytoplasmas in the aster yellows clade (chlorante aster yellows and clover phyllody) are highly divergent in the major hydrophilic region. Microbiology *148*: 157–167.

Baric, S. and J. Dalla-Via. 2004. A new approach to apple proliferation detection: a highly sensitive real-time PCR assay. J. Microbiol. Methods *57*: 135–145.

Berg, M., D.L. Davies, M.F. Clark, H.J. Vetten, G. Maier, C. Marcone and E. Seemüller. 1999. Isolation of the gene encoding an immunodominant membrane protein of the apple proliferation phytoplasma and expression and characterization of the gene product. Microbiology *145*: 1937–1943.

Berg, M. and E. Seemüller. 1999. Chromosomal organization and nucleotide sequence of the genes coding for the elongation factors G and Tu of the apple proliferation phytoplasma. Gene *226*: 103–109.

Bertaccini, A., R.E. Davis, R.W. Hammond, M. Vibio, M.G. Bellardi and I.-M. Lee. 1992. Sensitive detection of mycoplasmalike organisms in field-collected and *in vitro* propagated plants of *Brassica, Hydrangea* and *Chrysanthemum* by polymerase chain reaction. Ann. Appl. Biol. *121*: 593–599.

Bertaccini, A., J. Fráova, S. Paltrinieri, M. Martini, M. Navrátil, C. Lugaresi, Nebesárová and M. Simkova. 1999. Leek proliferation: a new phytoplasma disease in the Czech Republic and Italy. Eur. J. Plant Pathol. *105*: 487–493.

Bertamini, M. and N. Nedunchezhian. 2001. Effects of phytoplasma [stolbur-subgroup (Bois noir-BN)] on photosynthetic pigments, saccharides, ribulose 1,5-biphosphate carboxylase, nitrate and nitrite reductases, and photosynthetic activities in field-grown grapevine (*Vitis vinifera* L. cv. Chardonnay) leaves. Photosynthetica *39*: 119–122.

Bianco, P.A., R.E. Davis, J.P. Prince, I.-M. Lee, D.E. Gundersen, A. Fortusini and G. Belli. 1993. Double and single infections by aster yellows and elm yellows MLOs in grapevines and symptoms characteristic of Flavescence dorée. Rev. Patol. Veg. *3*: 69–82.

Blomquist, C.L., D.J. Barbara, D.L. Davies, M.F. Clark and B.C. Kirkpatrick. 2001. An immunodominant membrane protein gene from the Western X-disease phytoplasma is distinct from those of other phytoplasmas. Microbiology *147*: 571–580.

Cai, H., W. Wei, R.E. Davis, H. Chen and Y. Zhao. 2008. Genetic diversity among phytoplasmas infecting *Opuntia* species: virtual RFLP analysis identifies new subgroups in the peanut witches'-broom phytoplasma group. Int. J. Syst. Evol. Microbiol. *58*: 1448–1457.

Carginale, V., G. Maria, C. Capasso, E. Ionata, F. La Cara, M. Pastore, A. Bertaccini and A. Capasso. 2004. Identification of genes expressed in response to phytoplasma infection in leaves of *Prunus armeniaca* by messenger RNA differential display. Gene *332*: 29–34.

Carraro, L., N. Loi and P. Ermacora. 2001. Transmission characteristics of the European stone fruit yellows phytoplasma and its vector *Cacopsylla pruni.* Eur. J. Plant Pathol. *107*: 695–700.

Chang, F.L., C.C. Chen and C.P. Lin. 1995. Monoclonal antibody for the detection and identification of a phytoplasma associated with rice yellow dwarf. Eur. J. Plant Pathol. *101*: 511–518.

Chen, M.H. and C. Hiruki. 1978. The preservation of membranes of tubular bodies associated with mycoplasma-like organisms by tannic acid. Can. J. Bot. *56*: 2878–2882.

Chen, T.A., D.A. Lei and C.P. Lin. 1989. Detection and identification of plant and insect mollicutes. *In* The Mycoplasmas, vol. 5 (edited by Whitcomb and Tully). Academic Press, New York, pp. 393–424.

Chi, K.L. and C.P. Lin. 2005. Cloning and analysis of *polC* gene of phytoplasma associated with peanut witches' broom. Plant Pathol. Bull. *14*: 51–58.

Chiykowski, L.N. 1983. Frozen leafhoppers as a vehicle for long-term storage of different isolates of the aster yellows agents. Can. J. Plant Pathol. *5*: 101–106.

Chiykowski, L.N. 1988. Maintenance of yellows-type mycoplasmalike organisms. *In* Tree Mycoplasmas and *Mycoplasma* Diseases (edited by Hiruki). The University of Alberta Press, Edmonton, Alberta, Canada, pp. 123–134.

Chiykowski, L.N. and R.C. Sinha. 1990. Differentiation of MLO diseases by means of symptomatology and vector transmission. Rec. Adv. Mycoplasmol. *Suppl. 20*: 280–287.

Christensen, N.M., M. Nicolaisen, M. Hansen and A. Schulz. 2004. Distribution of phytoplasmas in infected plants as revealed by real-time PCR and bioimaging. Mol. Plant Microbe. Interact. *17*: 1175–1184.

Chu, Y.R., W. Y. Chen and C.P. Lin. 2006. Cloning and sequence analses of recA gene of phytoplasma associated with peanut witches' broom. Plant Pathol. Bull. *15*: 211–218.

Chuang, J.G. and C.P. Lin. 2000. Cloning of *gyrB* and *gyrA* genes of phytoplasma associated with peanut witches' broom. Plant Pathol. Bull. *9*: 157–166.

Cimerman, A., G. Arnaud and X. Foissac. 2006. Stolbur phytoplasma genome survey achieved using a suppression subtractive hybridization approach with high specificity. Appl. Environ. Microbiol. *72*: 3274–3283.

Cimerman, A., D. Pacifico, P. Salar, C. Marzachi and X. Foissac. 2009. Striking diversity of *vmp1*, a variable gene encoding a putative membrane protein of the stolbur phytoplasma. Appl. Environ. Microbiol. *75*: 2951–2957.

Clark, M.F. 1992. Immunodiagnostic techniques for plant mycoplasma-like organisms. *In* Techniques for the Rapid Detection of Plant Pathogens (edited by Duncan and Torrance). Blackwell Scientific Publications, Oxford, pp. 34–45.

Cordova, I., P. Jones, N.A. Harrison and C. Oropeza. 2003. *In situ* PCR detection of phytoplasma DNA in embryos from coconut palms with lethal yellowing disease. Mol. Plant Pathol. *4*: 99–108.

Cousin, M.T., J. Roux, E. Boudon-Padieu, R. Berges, E. Seemüller and C. Hiruki. 1998. Use of heteroduplex mobility analysis (HMA) for differentiating phytoplasma isolates causing witches' broom disease of *Populus nigra* vc Italica and stolbur or big bud symptoms on tomato. J. Phytopathol. *146*: 97–102.

D'Arcy, C.J. and L.R. Nault. 1982. Insect transmission of plant viruses and mycoplasmalike and rickettsialike organisms. Plant Dis. *66*: 99–104.

Daire, X., D. Clair, J. Larrue, E. Boudon-Padieu and A. Caudwell. 1993. Diversity among mycoplasma-like organisms inducing grapevine yellows in France. Vitis *32*: 159–163.

Davis, M.J., J.H. Tsai, R.L. Cox, L.L. McDaniel and N.A. Harrison. 1988. Cloning of chromosomal and extrachromosomal DNA of the mycoplasma-like organism that causes maize bushy stunt disease. Mol. Plant Microbe Interact. *1*: 295–302.

Davis, R.E. and R.F. Whitcomb. 1970. Evidence on possible mycoplasma etiology of aster yellows disease. I. Suppression of symptom development in plants by antibiotics. Infect. Immun. *2*: 201–208.

Davis, R.E. and I.-M. Lee. 1992. Mycoplasmalike organisms as plant disease agents. ATCC Quart. Newsl. *4*: 8–11.

Davis, R.E. and I.-M. Lee. 1993. Cluster-specific polymerase chain reaction amplification of 16S rDNA sequences for detection and identification of mycoplasmalike organisms. Phytopathology *63*: 1008–1011.

Davis, R.E., E.L. Dally, D.E. Gundersen, I.M. Lee and N. Habili. 1997. "*Candidatus* Phytoplasma australiense," a new phytoplasma taxon associated with Australian grapevine yellows. Int. J. Syst. Bacteriol. *47*: 262–269.

Davis, R.E. and W.A. Sinclair. 1998. Phytoplasma identity and disease etiology. Phytopathology *88*: 1372–1376.

Davis, R.E., R. Jomantiene, A. Kalvelyte and E.L. Dally. 2003a. Differential amplification of sequence heterogeneous ribosomal RNA genes and classification of the '*Fragaria multicipita*' phytoplasma. Microbiol. Res. *158*: 229–236.

Davis, R.E., R. Jomantiene, Y. Zhao and E.L. Dally. 2003b. Folate biosynthesis pseudogenes, *PsifolP* and *PsifolK*, and an O-sialoglycoprotein endopeptidase gene homolog in the phytoplasma genome. DNA Cell Biol. *22*: 697–706.

Davis, R.E., R. Jomantiene and Y. Zhao. 2005. Lineage-specific decay of folate biosynthesis genes suggests ongoing host adaptation in phytoplasmas. DNA Cell Biol. *24*: 832–840.

Davis, R.E., R. Jomantiene, E. L. Dally and T. K. Wolf. 1998. Phytoplasmas associated with grapevine yellows in Virginia belong to group 16SrI, subgroup A (tomato big bud phytoplasma subgroup), and group 16SrIII, new subgroup I. Vitis *37*: 131–137.

Denes, A.S. and R.C. Sinha. 1991. Extrachromosomal DNA elements of plant-pathogenic mycoplasma-like organisms. Can. J. Plant Pathol. *13*: 26–32.

Denes, A.S. and R.C. Sinha. 1992. Alteration of clover phyllody mycoplasma DNA after *in vitro* culturing of phyllody-diseased clover. Can. J. Plant Pathol. *14*: 189–196.

Deng, S. and C. Hiruki. 1991. Amplification of 16S rRNA genes from culturable and non-culturable mollicutes. J. Microbiol. Meth. *14*: 53–61.

Doi, Y., M. Teranaka, K. Yora and H. Asuyama. 1967. *Mycoplasma* or PLT-group-like organisms found in the phloem elements of plants infected with mulberry dwarf, potato witches-broom, aster yellows, or Paulownia witches broom. Ann. Phytopathol. Soc. Jpn. *33*: 256–266.

Errampelli, D. and J. Fletcher. 1993. Production of monospecific polyclonal antibodies made against aster yellows MLO-associated antigen. Phytopathology *83*: 1279–1282.

Esau, K., A.C. Magyarosy and V. Breazeale. 1976. Studies of the mycoplasma-like organism (MLO) in spinach leaves affected by the aster yellows disease. Protoplasma *90*: 189–203.

Evert, R.F. 1977. Phloem structure and histochemistry. Annu. Rev. Plant Physiol. *28*: 199–222.

Firrao, G., C.D. Smart and B.C. Kirkpatrick. 1996. Physical map of the western X-disease phytoplasma chromosome. J. Bacteriol. *178*: 3985–3988.

Firrao, G., K. Gibb and C. Streten. 2005. Short taxonomic guide to the genus '*Candidatus* Phytoplasma'. J. Plant Pathol. *87*: 249–263.

Florance, E.R. and H.T. Cameron, 1978. Three-dimensional structure and morphology of mycoplasma-like bodies associated with albino disease of *Prunus avium*. Phytopathology *68*: 75–80.

Flores, H.E., J. M. Vivanco and V.M. Loyola-Vargas. 1999. 'Radicle' biochemistry: the biology of root-specific metabolism. Trends Plant Sci. *4*: 220–226.

Galetto, L., J. Fletcher, D. Bosco, M. Turina, A. Wayadande and C. Marzachi. 2008. Characterization of putative membrane protein genes of the '*Candidatus* Phytoplasma asteris', chrysanthemum yellows isolate. Can. J. Microbiol. *54*: 341–351.

Garcia-Chapa, M., A. Batlle, D. Rekab, M.R. Rosquete and G. Firrao. 2004. PCR-mediated whole genome amplification of phytoplasmas. J. Microbiol. Methods *56*: 231–242.

Garnier, M., X. Foissac, P. Gaurivaud, F. Laigret, J. Renaudin, C. Saillard and J.M. Bové. 2001. Mycoplasmas, plants, insect vectors: a matrimonial triangle. C. R. Acad. Sci. III *324*: 923–928.

Gibb, K.S., B. Schneider and A.C. Padovan. 1998. Differential detection and genetic relatedness of phytoplasmas in papaya. Plant Pathol. *47*: 325–332.

Gomez, G.G., L.R. Conci, D.A. Ducasse and S.F. Nome. 1996. Purification of the phytoplasma associated with China-tree (*Melia azedarach* L.) decline and the production of a polyclonal antoserum for its detection. J. Phytopathol. *144*: 473–477.

Griffiths, H.M., W.A. Sinclair, C.D. Smart and R.E. Davis. 1999. The phytoplasma associated with ash yellows and lilac witches'-broom: '*Candidatus* Phytoplasma fraxini'. Int. J. Syst. Bacteriol. *49*: 1605–1614.

Gundersen, D.E., I.M. Lee, S.A. Rehner, R.E. Davis and D.T. Kingsbury. 1994. Phylogeny of mycoplasmalike organisms (phytoplasmas): a basis for their classification. J. Bacteriol. *176*: 5244–5254.

Gundersen, D.E. and I.M. Lee. 1996. Ultrasensitive detection of phytoplasmas by nested-PCR assays using two universal primer pairs. Phytopathol. Mediterr. *35*: 144–151.

Guo, Y.H., Z.M. Cheng, J.A. Walla and Z. Zhang. 1998. Diagnosis of X-disease phytoplasma in stone fruits by a monoclonal antibody developed directly from a woody plant. J. Environ. Hortic. *16*: 33–37.

Guthrie, J.N., K.B. Walsh, P.T. Scott and T.S. Rasmussen. 2001. The phytopathology of Australian papaya dieback: a proposed role for the phytoplasma. Physiol. Mol. Plant Pathol. *58*: 23–30.

Haggis, G.H. and R.C. Sinha. 1978. Scanning electron microscopy of mycoplasmalike organisms after freeze fracture of plant tissues affected with clover phyllody and aster yellows. Phytopathology *68*: 677–680.

Hanboonsong, Y., C. Choosai, S. Panyim and S. Damak. 2002. Transovarial transmission of sugarcane white leaf phytoplasma in the insect vector *Matsumuratettix hiroglyphicus* (Matsumura). Insect. Mol. Biol. *11*: 97–103.

Harrison, N.A., J.H. Tsai, C.M. Bourne and P.A. Richardson. 1991. Molecular cloning and detection of chromosomal and extrachromosomal DNA of mycoplasma-like organisms associated with witches' broom disease of pigeon pea in Florida. Mol. Plant Microbe Interact. *4*: 300–307.

Harrison, N.A., C.M. Bourne, R.L. Cox, J.H. Tsai and P.A. Richardson. 1992. DNA probes for detection of mycoplasma-like organisms associated with lethal yellowing disease of palms in Florida. Phytopathology *82*: 216–224.

Harrison, N.A., W. Myrie, P. Jones, M.L. Carpio, M. Castillo, M.M. Doyle and C. Oropeza. 2002. 16S rRNA interoperon sequence heterogeneity distinguishes strain populations of palm lethal yellowing phytoplasma in the Caribbean region. Ann. Appl. Biol. *141*: 183–193.

Harrison, N.A., E. Boa and M.L. Carpio. 2003. Characterization of phytoplasmas detected in Chinaberry trees with symptoms of leaf yellowing and decline in Bolivia. Plant Pathol. *52*: 147–157.

Hearon, S.S., R.H. Lawson, F.F. Smith, J.T. Mckenzie and J. Rosen. 1976. Morphology of filamentous forms of a mycoplasmalike organism associated with hydrangea virescence. Phytopathology *66*: 608–616.

Hiruki, C. and K. Wang. 2004. Clover proliferation phytoplasma: '*Candidatus* Phytoplasma trifolii'. Int. J. Syst. Evol. Microbiol. *54*: 1349–1353.

Ho, K.C., C.C. Tsai and T.L. Chung. 2001. Organization of ribosomal RNA genes from a Loofah witches' broom phytoplasma. DNA Cell Biol. *20*: 115–122.

Hodgetts, J., N. Boonham, R. Mumford, N. Harrison and M. Dickinson. 2008. Phytoplasma phylogenetics based on analysis of *secA* and 23S rRNA gene sequences for improved resolution of candidate species of '*Candidatus* Phytoplasma'. Int. J. Syst. Evol. Microbiol. *58*: 1826–1837.

Hogenhout, S.A., K. Oshima, D. Ammar el, S. Kakizawa, H.N. Kingdom and S. Namba. 2008. Phytoplasmas: bacteria that manipulate plants and insects. Mol. Plant. Pathol. *9*: 403–423.

Hren, M., M. Ravnikar, J. Brzin, P. Ermacora, L. Carraro, P. A. Bianco, P. Casati, M. Borgo, E. Angelini, A. Rotter and K. Gruden. 2009. Induced expression of sucrose synthase and alcohol dehydrogenase I genes in phytoplasma-infected grapevine plants grown in the field. Plant Pathol. *58*: 170–180.

Hsu, H.T., I.M. Lee, R.E. Davis and Y.C. Wang. 1990. Immunization for generation of hybridoma antibodies specifically reacting with plants infected with a mycoplasmalike organism (MLO) and their use in detection of MLO antigens. Phytopathology *80*: 946–950.

ICSB Subcommittee on the Taxonomy of *Mollicutes*. 1993. Minutes of the Interim meetings, 1 and 2 August, 1992, Ames, Iowa. Int. J. Syst. Bacteriol. *43*: 394–397.

ICSB Subcommittee on the Taxonomy of *Mollicutes*. 1997. Minutes of the interim meetings, 12 and 18 August, 1996, Orlando, Florida, USA Int. J. Syst. Bacteriol. *47*: 911–914.

ICSB Subcommittee on the Taxonomy of *Mollicutes*. 2001. Minutes of the interim meetings, 13 and 19 July 2000, Fukuoka, Japan. Int. J. Syst. Evol. Microbiol. *51*: 2227–2230.

IRPCM Phytoplasma/*Spiroplasma* Working Team – Phytoplasma Taxonomy Group. 2004. Description of the genus '*Candidatus* Phytoplasma', a taxon for the wall-less non-helical prokaryotes that colonize plant phloem and insects. Int. J. Syst. Evol. Microbiol. *54*: 1243–1255.

Ishii, T., Y. Doi, K. Yora and H. Asuyama. 1967. Suppressive effects of antibiotics of tetracycline group on symptom development of mulberry dwarf disease. Ann. Phytopathol. Soc. Jpn. *33*: 267–275.

Jagoueix-Eveillard, S., F. Tarendeau, K. Guolter, J.L. Danet, J.M. Bové and M. Garnier. 2001. *Catharanthus roseus* genes regulated differentially by mollicute infections. Mol. Plant Microbe Interact. *14*: 225–233.

Jarausch, W., C. Saillard, F. Dosba and J.M. Bové. 1994. Differentiation of mycoplasmalike organisms (MLOs) in European fruit trees by PCR using specific primers derived from the sequence of a chromosomal fragment of the apple proliferation MLO. Appl. Environ. Microbiol. *60*: 2916–2923.

Jarausch, W., C. Saillard and F. Dosba. 1996. Long-term maintenance of nonculturable apple proliferation phytoplasmas in their micropropagated natural host plant. Plant Pathol. *45*: 778–786.

Jiang, Y.P. and T.A. Chen. 1987. Purification of mycoplasma-like organisms from lettuce with aster yellows disease. Phytopathology *77*: 949–953.

Jiang, Y.P., J.D. Lei and T.A. Chen. 1988. Purification of aster yellows agent from diseased lettuce using affinity chromatography. Phytopathology *78*: 828–831.

Jiang, Y.P., T.A. Chen, L.N. Chiykowski and R.C. Sinha. 1989. Production of monoclonal antibodies to peach eastern-X disease and their use in disease detection. Can. J. Plant Pathol. *11*: 325–331.

Jomantiene, R., R.E. Davis, D. Valiunas and A. Alminaite. 2002. New group 16SrIII phytoplasma lineages in Lithuania exhibit rRNA interoperon sequence heterogeneity. Eur. J. Plant Pathol. *108*: 507–517.

Jomantiene, R. and R.E. Davis. 2006. Clusters of diverse genes existing as multiple, sequence-variable mosaics in a phytoplasma genome. FEMS Microbiol. Lett. *255*: 59–65.

Jomantiene, R., Y. Zhao and R.E. Davis. 2007. Sequence-variable mosaics: composites of recurrent transposition characterizing the genomes of phylogenetically diverse phytoplasmas. DNA Cell Biol. *26*: 557–564.

Jones, P. 2002. Phytoplasma plant pathogens. *In* Plant Pathologists Pocketbook (edited by Waller). CAB International, Wallingford, UK, pp. 126–139.

Jung, H.Y., T. Sawayanagi, S. Kakizawa, H. Nishigawa, S. Miyata, K. Oshima, M. Ugaki, J.T. Lee, T. Hibi and S. Namba. 2002. '*Candidatus* Phytoplasma castaneae', a novel phytoplasma taxon associated with chestnut witches' broom disease. Int. J. Syst. Evol. Microbiol. *52*: 1543–1549.

Jung, H.Y., S. Miyata, K. Oshima, S. Kakizawa, H. Nishigawa, W. Wei, S. Suzuki, M. Ugaki, T. Hibi and S. Namba. 2003a. First complete nucleotide sequence and heterologous gene organization of the two rRNA operons in the phytoplasma genome. DNA Cell Biol. *22*: 209–215.

Jung, H.Y., T. Sawayanagi, S. Kakizawa, H. Nishigawa, W. Wei, K. Oshima, S. Miyata, M. Ugaki, T. Hibi and S. Namba. 2003b. '*Candidatus* Phytoplasma ziziphi', a novel phytoplasma taxon associated with jujube witches'-broom disease. Int. J. Syst. Evol. Microbiol. *53*: 1037–1041.

Jung, H.Y., T. Sawayanagi, P. Wongkaew, S. Kakizawa, H. Nishigawa, W. Wei, K. Oshima, S. Miyata, M. Ugaki, T. Hibi and S. Namba. 2003c. '*Candidatus* Phytoplasma oryzae', a novel phytoplasma taxon associated with rice yellow dwarf disease. Int. J. Syst. Evol. Microbiol. *53*: 1925–1929.

Kakizawa, S., K. Oshima, T. Kuboyama, H. Nishigawa, H. Jung, T. Sawayanagi, T. Tsuchizaki, S. Miyata, M. Ugaki and S. Namba. 2001. Cloning and expression analysis of *Phytoplasma* protein translocation genes. Mol. Plant Microbe Interact. *14*: 1043–1050.

Kakizawa, S., K. Oshima, H. Nighigawa, H.Y. Jung, W. Wei, S. Suzuki, M. Tanaka, S. Miyata, M. Ugaki and S. Namba. 2004. Secretion of immunodominant membrane protein from onion yellows phytoplasma through the Sec protein-translocation system in *Escherichia coli*. Microbiology *150*: 135–142.

Kakizawa, S., K. Oshima, Y. Ishii, A. Hoshi, K. Maejima, H.Y. Jung, Y. Yamaji and S. Namba. 2009. Cloning of immunodominant membrane protein genes of phytoplasmas and their *in planta* expression. FEMS Microbiol. Lett. *293*: 92–101.

Kamińska, M. and H. Śliwa. 2003. Effect of antibiotics on symptoms of stunting disease of *Magnolia lilliflora* plants. J. Phytopathol. *151*: 59–63.

Kawakita, H., T. Saiki, W. Wei, W. Mitsuhashi, K. Watanabe and M. Sato. 2000. Identification of mulberry dwarf phytoplasmas in the genital organs and eggs of leafhopper *Hishimonoides sellatiformis*. Phytopathology *90*: 909–914.

Kenyon, L., N.A. Harrison, G.R. Ashburner, E.R. Boa and P.A. Richardson. 1998. Detection of a pigeon pea witches' broom-related phytoplasma in trees of *Gliricidia sepium* affected by little-leaf disease in Central America. Plant Pathol. *47*: 671–680.

Khan, A.J., S. Botti, S. Paltrinieri, A.M. Al-Subhi and A.F. Bertaccina. 2002. Phytoplasmas in alfalfa seedlings: infected or contaminated seed? Proceedings of the 14th International Congress of the International Organization for Mycoplasmology, Vienna, Austria, p. 148.

Kirkpatrick, B.C., D.C. Stenger, T.J. Morris and A.H. Purcell. 1987. Cloning and detection of DNA from a nonculturable plant pathogenic *Mycoplasma*-like organism. Science *238*: 197–200.

Kirkpatrick, B.C. 1989. Strategies for characterizing plant pathogenic mycoplasma-like organisms and their effects on plants. *In* Plant-Microbe Interactions, Molecular and Genetic Perspectives (edited by Kosuge and Nester). McGraw-Hill, New York, pp. 241–293.

Kirkpatrick, B.C. 1992. *Mycoplasma*-like organisms: plant and invertebrate pathogens. *In* The Prokaryotes: a Handbook on the Biology of *Bacteria*: Ecophysiology, Isolation, Identification, Applications, 2nd edn, vol. 4 (edited by Balows, Trüper, Dworkin, Harder and Schleifer). Springer, New York, pp. 4050–4067.

Kirkpatrick, B.C., N.A. Harrison, I.-M. Lee, H. Neimark and B.B. Sears. 1995. Isolation of *Mycoplasma*-like organism DNA from plant and insect hosts. *In* Molecular and Diagnostic Procedures in Mycoplasmology, vol. 2 (edited by Razin and Tully). Academic Press, New York, pp. 105–117.

Kison, H., B. Schneider and E. Seemüller. 1994. Restriction fragment length polymorphisms within the apple proliferation mycoplasmalike organism. J. Phytopathol. *141*: 395–401.

Kison, H., B.C. Kirkpatrick and E. Seemüller. 1997. Genetic comparison of the peach yellow leaf roll agent with European fruit tree phytoplasmas of the apple proliferation group. Plant Pathol. Bull. *46*: 538–544.

Kollar, A. and E. Seemüller. 1989. Base composition of the DNA of mycoplasmalike organisms associated with various plant diseases. J. Phytopathol. *127*: 177–186.

Kollar, A. and E. Seemüller. 1990. Chemical composition of the phloem exudate of *Mycoplasma*-infected trees. J. Phytopathol. *128*: 99–111.

Koui, T., T. Natsuaki and S. Okuda. 2002. Antiserum raised against gyrase A of *Acholeplasma laidlawii* reacts with phytoplasma proteins. FEMS Microbiol. Lett. *206*: 169–174.

Koui, T., N. Tomohide and S. Okuda. 2003. Phylogenetic analysis of elongation factor Tu gene of phytoplasmas from Japan. J. Gen. Plant Pathol. *69*: 316–319.

Kube, M., B. Schneider, H. Kuhl, T. Dandekar, K. Heitmann, A. M. Migdoll, R. Reinhardt and E. Seemüller. 2008. The linear chromosome of the plant-pathogenic *Mycoplasma* '*Candidatus* Phytoplasma mali'. BMC Genomics *9*: 306.

Kuboyama, T., C.C. Huang, X. Lu, T. Sawayanagi, T. Kanazawa, T. Kagami, I. Matsuda, T. Tsuchizaki and S. Namba. 1998. A plasmid isolated from phytopathogenic onion yellows phytoplasma and its heterogeneity in the pathogenic phytoplasma mutant. Mol. Plant Microbe Interact. *11*: 1031–1037.

Kunkel, L.O. 1926. Studies on aster yellows. Am. J. Bot. *13*: 646–705.

Kuske, C.R. and B.C. Kirkpatrick. 1990. Identification and characterization of plasmids from the western aster yellows mycoplasmalike organism. J. Bacteriol. *172*: 1628–1633.

Kuske, C.R., B.C. Kirkpatrick and E. Seemüller. 1991. Differentiation of virescence MLOS using western aster yellows mycoplasma-like organism chromosomal DNA probes and restriction fragment length polymorphism analysis. J. Gen. Microbiol. *137*: 153–159.

Kuske, C.R. and B.C. Kirkpatrick. 1992a. Distribution and multiplication of western aster yellows mycoplasmalike organisms in *Catharanthus roseus* as determined by DNA hybridization analysis. Phytopathology *82*: 457–462.

Kuske, C.R. and B.C. Kirkpatrick. 1992b. Phylogenetic relationships between the western aster yellows mycoplasmalike organism and other prokaryotes established by 16S rRNA gene sequence. Int. J. Syst. Bacteriol. *42*: 226–233.

Lauer, U. and E. Seemüller. 2000. Physical map of the chromosome of the apple proliferation phytoplasma. J. Bacteriol. *182*: 1415–1418.

Lee, I.-M. and R.E. Davis. 1992. Mycoplasmas which infect plants and insects. *In* Mycoplasmas: Molecular Biology and Pathogenesis (edited by Maniloff, McElhaney, Finch and Baseman). American Society for Microbiology, Washington, D.C., pp. 379–390.

Lee, I.-M., A. Bertaccini, M. Vibio and D.E. Gundersen. 1988. Detection and investigation of genetic relatedness among aster yellows and other mycoplasmalike organisms by using cloned DNA and RNA probes. Mol. Plant Microbe Interact. *1*: 303–310.

Lee, I.-M., R.E. Davis, T.A. Chen, L.N. Chiykowski, J. Fletcher, C. Hiruki and D.A. Schaff. 1992. A genotype-based system for identification and classification of mycoplasmalike organisms (MLOs) in the aster yellows MLO strain cluster. Phytopathology *82*: 977–986.

Lee, I.-M., R.E. Davis and H.T. Hsu. 1993a. Differentiation of strains in the aster yellows mycoplasmalike organism strain cluster by serological assay with monoclonal antibodies. Plant Dis. *77*: 815–817.

Lee, I.-M., R.W. Hammond, R.E. Davis and D.E. Gundersen. 1993b. Universal amplification and analysis of pathogen 16S rDNA for classification and identification of mycoplasmalike organisms. Phytopathology *83*: 834–842.

Lee, I.-M., A. Bertaccini, M. Vibio and D.E. Gundersen. 1995. Detection of multiple phytoplasmas in perennial fruit trees with decline symptoms in Italy. Phytopathology *85*: 728–735.

Lee, I.-M., D.E. Gundersen-Rindal and A. Bertaccini. 1998a. Phytoplasma: ecology and genomic diversity. Phytopathology *88*: 1359–1366.

Lee, I.-M., D.E. Gundersen-Rindal, R.E. Davis and I.M. Bartoszyk. 1998b. Revised classification scheme of phytoplasmas based an RFLP analyses of 16S rRNA and ribosomal protein gene sequences. Int. J. Syst. Bacteriol. *48*: 1153–1169.

Lee, I.-M., R.E. Davis and D.E. Gundersen-Rindal. 2000. Phytoplasma: phytopathogenic *Mollicutes*. Annu. Rev. Microbiol. *54*: 221–255.

Lee, I.-M., M. Martini, K.D. Bottner, R.A. Dane, M.C. Black and N. Troxclair. 2003. Ecological implications from a molecular analysis of phytoplasmas involved in an aster yellows epidemic in various crops in Texas. Phytopathology *93*: 1368–1377.

Lee, I.-M., D.E. Gundersen-Rindal, R.E. Davis, K.D. Bottner, C. Marcone and E. Seemüller. 2004a. '*Candidatus* Phytoplasma asteris', a novel phytoplasma taxon associated with aster yellows and related diseases. Int. J. Syst. Evol. Microbiol. *54*: 1037–1048.

Lee, I.-M., M. Martini, C. Marcone and S.F. Zhu. 2004b. Classification of phytoplasma strains in the elm yellows group (16SrV) and proposal of '*Candidatus* Phytoplasma ulmi' for the phytoplasma associated with elm yellows. Int. J. Syst. Evol. Microbiol. *54*: 337–347.

Lee, I.-M., Y. Zhao and K.D. Bottner. 2005. Novel insertion sequence-like elements in phytoplasma strains of the aster yellows group are putative new members of the IS*3* family. FEMS Microbiol. Lett. *242*: 353–360.

Lee, I.-M., K.D. Bottner, G. Secor and V. Rivera-Varas. 2006a. "*Candidatus* Phytoplasma americanum", a phytoplasma associated with a potato purple top wilt disease complex. Int. J. Syst. Evol. Microbiol. *56*: 1593–1597.

Lee, I.-M., Y. Zhao and K.D. Bottner. 2006b. SecY gene sequence analysis for finer differentiation of diverse strains in the aster yellows phytoplasma group. Mol. Cell. Probes *20*: 87–91.

Lefol, C., J. Lherminier, E. Boudon-Padieu, J. Larrue, C. Louis and A. Caudwell. 1994. Propagation of Flavescence Dorèe MLO (mycoplasma-like organisms) in the leafhopper vector *Euscelidius variegatus*. Kbm. J. Invertebr. Pathol. *63*: 285–293.

Lepka, P., M. Stitt, E. Moll and E. Seemüller. 1999. Effect of phytoplasmal infection on concentration and translocation of carbohydrates and amino acids in periwinkle and tobacco. Physiol. Mol. Plant Pathol. *55*: 59–68.

Liefting, L.W., M.T. Andersen, R.E. Beever, R.C. Gardner and R.L.S. Forster. 1996. Sequence heterogeneity in the two 16S rRNA genes of *Phormium* yellow leaf phytoplasma. Appl. Environ. Microbiol. *62*: 3133–3139.

Liefting, L.W. and B.C. Kirkpatrick. 2003. Cosmid cloning and sample sequencing of the genome of the uncultivable mollicute, western X-disease phytoplasma, using DNA purified by pulsed-field gel electrophoresis. FEMS Microbiol. Lett. *221*: 203–211.

Liefting, L.W., M.E. Shaw and B.C. Kirkpatrick. 2004. Sequence analysis of two plasmids from the phytoplasma beet leafhopper-transmitted virescence agent. Microbiology *150*: 1809–1817.

Liefting, L.W., M.T. Andersen, T.J. Lough and R.E. Beever. 2006. Comparative analysis of the plasmids from two isolates of "*Candidatus* Phytoplasma australiense". Plasmid *56*: 138–144.

Lim, P.O. and B.B. Sears. 1989. 16S rRNA sequence indicates that plant-pathogenic mycoplasmalike organisms are evolutionarily distinct from animal mycoplasmas. J. Bacteriol. *171*: 5901–5906.

Lim, P.O. and B.B. Sears. 1992. Evolutionary relationships of a plant-pathogenic mycoplasmalike organism and *Acholeplasma laidlawii* deduced from two ribosomal protein gene sequences. J. Bacteriol. *174*: 2606–2611.

Lim, P.O., B.B. Sears and K.L. Klomparens. 1992. Membrane properties of a plant-pathogenic mycoplasmalike organism. J. Bacteriol. *174*: 682–686.

Lin, C.-L., T. Zhou, H.-F. Li, Z.-F. Fan, Y. Li, C.-G. Piao and G.-Z. Tian. 2009. Molecular characterisation of two plasmids from paulownia witches'-broom phytoplasma and detection of a plasmid encoded protein in infected plants. Eur. J. Plant Pathol. *123*: 321–330.

Lin, C.-Y., Chen, W.-Y., and C.P. Lin. 2006. Cloning and analysis of *rpoC* gene of phytoplasma associated with peanut witches' broom. Plant Pathol. Bull. *15*: 129–138.

Loi, N., P. Ermacora, T.A. Chen, L. Carraro and R. Osler. 1998. Monoclonal antibodies for the detection of tagetes witches' broom agent. J. Plant Pathol. *80*: 171–174.

Loi, N., P. Ermacora, L. Carraro, R. Osler and T.A. Chen. 2002. Production of monoclonal antibodies against apple proliferation phytoplasma and their use in serological detection. Eur. J. Plant Pathol. *108*: 81–86.

Marcone, C., A. Ragozzino, B. Schneider, U. Lauer, C.D. Smart and E. Seemüller. 1996. Genetic characterization and classification of two phytoplasmas associated with spartium witches' broom disease. Plant Dis. *80*: 365–371.

Marcone, C., F. Hergenhahn, A. Ragozzino and E. Seemüller. 1999a. Dodder transmission of pear decline, European stone fruit yellows, rubus stunt, *Picris echioides* yellows and cotton phyllody phytoplasmas to periwinkle. J. Phytopathol. *147*: 187–192.

Marcone, C., H. Neimark, A. Ragozzino, U. Lauer and E. Seemüller. 1999b. Chromosome sizes of phytoplasmas composing major phylogenetic groups and subgroups. Phytopathology *89*: 805–810.

Marcone, C., I.-M. Lee, R.E. Davis, A. Ragozzino and E. Seemüller. 2000. Classification of aster yellows-group phytoplasmas based on combined analyses of rRNA and tuf gene sequences. Int. J. Syst. Evol. Microbiol. *50*: 1703–1713.

Marcone, C., A. Ragozzino, I. Camele, G.L. Rana and E. Seemüller. 2001. Updating and extending genetic characterization and classification of phytoplasmas from wild and cultivated plants in southern Italy. J. Plant Pathol. *83*: 133–138.

Marcone, C. and E. Seemüller. 2001. A chromosome map of the European stone fruit yellows phytoplasma. Microbiology *147*: 1213–1221.

Marcone, C., K.S. Gibb, C. Streten and B. Schneider. 2004a. '*Candidatus* Phytoplasma spartii', '*Candidatus* Phytoplasma rhamni' and '*Candidatus* Phytoplasma allocasuarinae', respectively associated with spartium witches'-broom, buckthorn witches'-broom and allocasuarina yellows diseases. Int. J. Syst. Evol. Microbiol. *54*: 1025–1029.

Marcone, C., B. Schneider and E. Seemüller. 2004b. '*Candidatus* Phytoplasma cynodontis', the phytoplasma associated with Bermuda grass white leaf disease. Int. J. Syst. Evol. Microbiol. *54:* 1077–1082.

Martinez, S., I. Cordova, B.E. Maust, C. Oropeza and J.M. Santamaria. 2000. Is abscisic acid responsible for abnormal stomatal closure in coconut palms showing lethal yellowing? J. Plant Physiol. *156*: 319–322.

Martini, M., I.M. Lee, K.D. Bottner, Y. Zhao, S. Botti, A. Bertaccini, N.A. Harrison, L. Carraro, C. Marcone, A.J. Khan and R. Osler. 2007. Ribosomal protein gene-based phylogeny for finer differentiation and classification of phytoplasmas. Int. J. Syst. Evol. Microbiol. *57*: 2037–2051.

Marzachi, C., R.G. Milne and D. Bosco. 2004. Phytoplasma-plant-vector relationships. *In* Recent Research Developments in Plant Pathology, vol. 3 (edited by Pandalai). Research Signpost, Trivandrum, India.

McCoy, R.E. 1979. Mycoplasmas and yellows diseases. *In* The Mycoplasmas, vol. III, Plant and Insect Mycoplasmas (edited by Whitcomb and Tully). Academic Press, New York, pp. 229–264.

McCoy, R.E. 1982. Use of tetracycline antibiotics to control yellows diseases. Plant Dis. *66*: 539–542.

McCoy, R.E., A. Caudwell, C.J. Chang, T.A. Chen, L.N. Chiykowski, M.T. Cousin, J.L. Dale, G.T.N. deLeeuw, D.A. Golino, K.J. Hackett, B.C. Kirkpatrick, R. Marwitz, H. Petzold, R.C. Sinha, M. Suguira, R.F. Whitcomb, I.L. Yang, B.M. Zhu and E. Seemüller. 1989. Plant diseases associated with mycoplasma-like organisms. *In* The Mycoplasmas, vol. V (edited by Whitcomb and Tully). Academic Press, San Diego, pp. 545–640.

Melamed, S., E. Tanne, R. Ben-Haim, O. Edelbaum, D. Yogev and I. Sela. 2003. Identification and characterization of phytoplasmal genes, employing a novel method of isolating phytoplasmal genomic DNA. J. Bacteriol. *185*: 6513–6521.

Miyata, S., K. Furuki, K. Oshima, T. Sawayanagi, H. Nishigawa, S. Kakizawa, H.Y. Jung, M. Ugaki and S. Namba. 2002a. Complete nucleotide sequence of the S10-*spc* operon of phytoplasma: Gene organization and genetic code resemble those of *Bacillus subtilis*. DNA Cell Biol. *21*: 527–534.

Miyata, S., K. Furuki, T. Sawayanagi, K. Oshima, T. Kuboyama, T. Tsuchizaki, M. Ugaki and S. Namba. 2002b. The gene arrangement and sequence of *str* operon of phytoplasma resemble those of *Bacillus* more than those of *Mycoplasma*. J. Gen. Plant Pathol. *68*: 62–67.

Miyata, S., K. Oshima, S. Kakizawa, H. Nishigawa, H.Y. Jung, T. Kuboyama, M. Ugaki and S. Namba. 2003. Two different thymidylate kinase gene homologues, including one that has catalytic activity, are encoded in the onion yellows phytoplasma genome. Microbiology *149*: 2243–2250.

Montano, H.G., R.E. Davis, E.L. Dally, S. Hogenhout, J.P. Pimentel and P.S. Brioso. 2001. '*Candidatus* Phytoplasma brasiliense', a new phytoplasma taxon associated with hibiscus witches' broom disease. Int. J. Syst. Evol. Microbiol. *51*: 1109–1118.

Moriwaki, N., K. Matsuchita, M. Nishina and Y. Kono. 2003. High concentrations of trehalose in aphid hemolymph Appl. Entomol. Zool. *38*: 241–248.

Morton, A., D.L. Davies, C.L. Blomquist and D.J. Barbara. 2003. Characterization of homologues of the apple proliferation immunodominant membrane protein gene from three related phytoplasmas. Mol. Plant Pathol. *4*: 109–114.

Mounsey, K.E., C. Streten and K.S. Gibb. 2006. Sequence characterization of four putative membrane-associated proteins from sweet potato little strain V4 phytoplasma. Plant Pathol. *55*: 29–35.

Murray, R.G.E., D.J. Brenner, R.R. Colwell, P. de Vos, P. Goodfellow, P.A.D. Grimont, N. Pfennig, E. Stackebrandt and G.A. Zavarin. 1990. Report of the ad hoc committee on approaches to taxonomy within the proteobacteria. Int. J. Syst. Bacteriol. *40*: 213–215.

Murray, R.G.E. and K.H. Schleifer. 1994. Taxonomic notes: a proposal for recording the properties of putative taxa of procaryotes. Int. J. Syst. Bacteriol. *44*: 174–176.

Musetti, R., M.A. Favali and L. Pressacco. 2000. Histopathology and polyphenol content in plants infected by phytoplasmas. Cytobios *102*: 133–147.

Musetti, R. and M.A. Favali. 2003. Cytochemical localization of calcium and X-ray microanalysis of *Catharanthus roseus* L. infected with phytoplasmas. Micron *34*: 387–393.

Musetti, R., L.S. Di Toppi, P. Ermacora and M.A. Favali. 2004. Recovery in apple trees infected with the apple proliferation phytoplasma: an ultrastructural and biochemical study. Phytopathology *94*: 203–208.

Nakashima, K. and T. Hayashi. 1997. Sequence analysis of extrachromosomal DNA of sugarcane white leaf phytoplasma. Ann. Phytopathol. Soc. Jpn. *63*: 21–25.

Namba, S., H. Oyaizu, S. Kato, S. Iwanami and T. Tsuchizaki. 1993. Phylogenetic diversity of phytopathogenic mycoplasmalike organisms. Int. J. Syst. Bacteriol. *43*: 461–467.

Namba, S. 2002. Molecular biological studies on phytoplasmas. J. Gen. Plant Pathol. *68*: 257–259.

Nasu, S., D.D. Jensen and J. Richardson. 1970. Electron microscopy of mycoplasma-like bodies associated with insect and plant hosts of peach western X-disease. Virology *41*: 583–595.

Nasu, S., D. D. Jensen and J. Richardson. 1974. Primary culture of the western X-disease mycoplasma-like organism from *Colladonus montanus* leafhopper vectors. Appl. Entomol. Zool. *9*: 115–126.

Neimark, H. and B.C. Kirkpatrick. 1993. Isolation and characterization of full-length chromosomes from non-culturable plant-pathogenic mycoplasma-like organisms. Mol. Microbiol. *7*: 21–28.

Nielson, M.W. 1979. Taxonomic relationships of leafhopper vectors of plant pathogens. *In* Leafhopper Vectors and Plant Disease Agents (edited by Maromorosch and Harris). Academic Press, New York, pp. 3–27.

Nipah, J.O., P. Jones and M.J. Dickinson. 2007. Detection of lethal yellowing phytoplasmas in embryos from coconuts infected with Cape St. Paul wilt disease in Ghana. Plant Pathol. *56*: 777–784.

Nishigawa, H., S. Miyata, K. Oshima, T. Sawayanagi, A. Komoto, T. Kuboyama, I. Matsuda, T. Tsuchizaki and S. Namba. 2001. *In planta* expression of a protein encoded by the extrachromosomal DNA of a phytoplasma and related to geminivirus replication proteins. Microbiology *147*: 507–513.

Nishigawa, H., K. Oshima, S. Kakizawa, H.Y. Jung, T. Kuboyama, S. Miyata, M. Ugaki and S. Namba. 2002a. A plasmid from a non-insect-transmissible line of a phytoplasma lacks two open reading frames that exist in the plasmid from the wild-type line. Gene *298*: 195–201.

Nishigawa, H., K. Oshima, S. Kakizawa, H.Y. Jung, T. Kuboyama, S. Miyata, M. Ugaki and S. Namba. 2002b. Evidence of intermolecular recombination between extrachromosomal DNAs in phytoplasma: a trigger for the biological diversity of phytoplasma? Microbiology *148*: 1389–1396.

Nishigawa, H., K. Oshima, S. Miyata, M. Ugaki and S. Namba. 2003. Complete set of extrachromosomal DNAs from three pathogenic lines of onion yellows phytoplasma and use of PCR to differentiate each line. J. Gen. Plant Pathol. *69*: 194–198.

Nyland, G. 1971. Remission of symptoms of pear decline in pear and peach X-disease in peach after treatment with a tetracycline. Phytopathology *61*: 904–905.

Oparka, K.J. and R. Turgeon. 1999. Sieve elements and companion cells-traffic control centers of the phloem. Plant Cell. *11*: 739–750.

Oshima, K., S. Kakizawa, H. Nishigawa, T. Kuboyama, S. Miyata, M. Ugaki and S. Namba. 2001a. A plasmid of phytoplasma encodes a unique replication protein having both plasmid- and virus-like domains: clue to viral ancestry or result of virus/plasmid recombination? Virology *285*: 270–277.

Oshima, K., T. Shiomi, T. Kuboyama, T. Sawayanagi, H. Nishigawa, S. Kakizawa, S. Miyata, M. Ugaki and S. Namba. 2001b. Isolation and characterization of derivative lines of the onion yellows phytoplasma that do not cause stunting or phloem hyperplasia. Phytopathology *91*: 1024–1029.

Oshima, K., S. Miyata, T. Sawayanagi, S. Kakizawa, H. Nishigawa, H. Jung, K. Furuki, M. Yanazaki, S. Suzuki, W. Wei, T. Kuboyama, M. Ugaki and S. Namba. 2002. Minimal set of metabolic pathways suggested from the genome of onion yellow phytoplasma. J. Plant Pathol. *68*: 225–236.

Oshima, K., S. Kakizawa, H. Nishigawa, H.Y. Jung, W. Wei, S. Suzuki, R. Arashida, D. Nakata, S. Miyata, M. Ugaki and S. Namba. 2004. Reductive evolution suggested from the complete genome sequence of a plant-pathogenic phytoplasma. Nat. Genet. *36*: 27–29.

Oshima, K., S. Kakizawa, R. Arashida, Y. Ishii, A. Hoshi, Y. Hayashi, S. Kakiwada and S. Namba. 2007. Presence of two glycolytic gene clusters in a severe pathogenic line of *Candidatus* Phytoplasma asteris. Mol. Plant Pathol. *8*: 481–489.

Padovan, A.C., G. Firrao, B. Schneider and K.S. Gibb. 2000. Chromosome mapping of the sweet potato little leaf phytoplasma reveals genome heterogeneity within the phytoplasmas. Microbiology *146*: 893–902.

Pracros, P., J. Renaudin, S. Eveillard, A. Mouras and M. Hernould. 2006. Tomato flower abnormalities induced by stolbur infection are associated with changes of expression of floral development genes. Mol. Plant Microbe Interact. *19*: 62–68.

Rajan, J., M.F. Clark, M. Barba and A. Hadidi. 1995. Detection of apple proliferation and other MLOs by immuno-capture PCR (IC-PCR). Acta Hortic. *386*: 511–514.

Raju, B.C. and G. Nyland. 1988. Chemotherapy of mycoplasma diseases of fruit trees. *In* Tree Mycoplasmas and *Mycoplasma* Diseases (edited by Hiruki). University of Alberta Press, Edmonton, Alberta, Canada, pp. 207–216.

Reinert, W. 1999. Detection and further differentiation of plant pathogenic phytoplasmas (*Mollicutes*, Eubacteria) in Germany regarding phytopathological aspects. PhD Dissertation. Dem Fachbereich Biologie der Technischen Universität Darmstadt (in German), p. 148.

Rekab, D., L. Carraro, B. Schneider, E. Seemüller, J. Chen, C.J. Chang, R. Locci and G. Firrao. 1999. Geminivirus-related extrachromosomal DNAs of the X-clade phytoplasmas share high sequence similarity. Microbiology *145*: 1453–1459.

Rudzinska-Langwald, A. and M. Kaminska. 1999. Cytopathological evidence for transport of phytoplasma in infected plants. Bot. Pol. *68*: 261–266.

Rudzinska-Langwald, A. and M. Kaminska. 2003. Changes in the ultrastructure and cytoplasmic free calcium in *Gladiolus* x *hybridus* Van Houtte roots infected by aster yellows phytoplasma. Acta Soc. Bot. Pol. *72*: 269–282.

Saglio, P.H.M. and R.F. Whitcomb. 1979. Diversity of wall-less prokaryotes in plant vascular tissue, fungi and invertebrate animals. *In* The Mycoplasmas, vol. 3 (edited by Whitcomb and Tully). Academic Press, New York, pp. 1–36.

Sawayanagi, T., N. Horikoshi, T. Kanehira, M. Shinohara, A. Bertaccini, M.T. Cousin, C. Hiruki and S. Namba. 1999. '*Candidatus* Phytoplasma japonicum', a new phytoplasma taxon associated with Japanese *Hydrangea* phyllody. Int. J. Syst. Bacteriol. *49*: 1275–1285.

Schneider, B., U. Ahrens, B.C. Kirkpatrick and E. Seemüller. 1993. Classification of plant-pathogenic mycoplasma-like organisms using restriction-site analysis of PCR-amplified 16S rDNA. J. Gen. Microbiol. *139*: 519–527.

Schneider, B. and E. Seemüller. 1994a. Presence of two sets of ribosomal genes in phytopathogenic mollicutes. Appl. Environ. Microbiol. *60*: 3409–3412.

Schneider, B. and E. Seemüller. 1994b. Studies on taxonomic relationships of mycoplasma-like organisms by Southern blot analysis. J. Phytopathol. *141*: 173–185.

Schneider, B., E. Seemüller, C.D. Smart and B.C. Kirkpatrick. 1995. Phylogenetic classification of plant pathogenic mycoplasmalike organisms or phytoplasmas. *In* Molecular and Diagnostic Procedures in Mycoplasmology: Molecular Characterization, vol. 1 (edited by Razin and Tully). Academic Press, San Diego, pp. 369–380.

Schneider, B., K.S. Gibb and E. Seemüller. 1997. Sequence and RFLP analysis of the elongation factor Tu gene used in differentiation and classification of phytoplasmas. Microbiology *143*: 3381–3389.

Schneider, B., E. Torres, M.P. Martin, M. Schroder, H.D. Behnke and E. Seemuller. 2005. '*Candidatus* Phytoplasma pini', a novel taxon from *Pinus silvestris* and *Pinus halepensis*. Int. J. Syst. Evol. Microbiol. *55*: 303–307.

Sears, B.B. and K.L. Klomparens. 1989. Leaf tip cultures of the evening primrose allow stable, aspectic culture of mycoplasma-like organism. Can. J. Plant Pathol. *11*: 343–348.

Sears, B.B. and B.C. Kirkpatrick. 1994. Unveiling the evolutionary relationships of plant pathogenic mycoplasmalike organisms. ASM News *60*: 307–312.

Seddas, A., R. Meignoz, C. Kuszala and E. Boudon-Padieu. 1995. Evidence for the physical integrity of flavescence dorée phytoplasmas purified by affinity chromatography by immunoaffinity from infected plants or leafhoppers and the plant pathogenicity of phytoplasmas from leafhoppers. Plant Pathol. *44*: 971–978.

Seemüller, E. and B. Schneider. 2007. Differences in virulence and genomic features of strains of '*Candidatus* Phytoplasma mali', the Apple Proliferation agent. Phytopathology *97*: 964–970.

Seemüller, E., B. Schneider, R. Mäurer, U. Ahrens, X. Daire, H. Kison, K.H. Lorenz, G. Firrao, L. Avinent, B.B. Sears and E. Stackebrandt. 1994. Phylogenetic classification of phytopathogenic mollicutes by sequence analysis of 16S ribosomal DNA. Int. J. Syst. Bacteriol. *44*: 440–446.

Seemüller, E., C. Marcone, U. Lauer, A. Ragozzino and M. Göschl. 1998. Current status of molecuar classification of the phytoplasmas. J. Plant Pathol. *80*: 3–26.

Seemüller, E., M. Garnier and B. Schneider. 2002. Mycoplasmas of plants and insects. *In* Molecular Biology and Pathogenicity of Mycoplasmas (edited by Razin and Hermann). Kluwer Academic/Plenum Publishers, Dordrecht, The Netherlands, pp. 91–116.

Seemüller, E. and B. Schneider. 2004. '*Candidatus* Phytoplasma mali', '*Candidatus* Phytoplasma pyri' and '*Candidatus* Phytoplasma prunorum', the causal agents of apple proliferation, pear decline and European stone fruit yellows, respectively. Int. J. Syst. Evol. Microbiol. *54*: 1217–1226.

Shen, W.C. and C.P. Lin. 1993. Production of monoclonal antibodies against a mycoplasmalike organism associated with sweetpotato witches' broom. Phytopathology *83*: 671–675.

Shen, W.C. and C.P. Lin. 1994. Application of immunofluorescent staining, tissue blotting techniques against a mycoplasmalike organism assoicated with sweetpotato witches' broom. Plant Pathol. Bull. *3*: 79–83.

Siddique, A.B.M., J.N. Giuthrie, K.B. Walsh, D.T. White and P.T. Scott. 1998. Histopathology and within-plant distribution of the phytoplasma associated with Australian papaya dieback. Plant Dis. *82*: 1112–1120.

Siller, W., B. Kuhbandner, R. Marwitz, H. Petzold and E. Seemüller. 1987. Occurrence of mycoplasma-like organisms in parenchyma cells of *Cuscuta odorata* (Ruiz et Pav.). J. Phytopathol. *119*: 147–159.

Sinclair, W.A., H.M. Griffiths and I.M. Lee. 1994. Mycoplasmalike organisms as causes of slow growth and decline of trees and shrubs. J. Arboric. *20*: 176–189.

Sinha, R.C. and E.A. Peterson. 1972. Uptake and persistence of oxytetracycline in aster plants and vector leafhoppers in relation to inhibition of clover phyllody agent. Phytopathology *62*: 377–383.

Sinha, R.C. 1979. Purification and serology of mycoplasma-like organisms from aster yellows-infected plants. Can. J. Plant Pathol. *1*: 65–70.

Sjölund, R.D. 1997. The phloem sieve element: a river runs through it. Plant Cell *9*: 1137–1146.

Smart, C.D., B. Schneider, C.L. Blomquist, L.J. Guerra, N.A. Harrison, U. Ahrens, K.H. Lorenz, E. Seemuller and B.C. Kirkpatrick. 1996. Phytoplasma-specific PCR primers based on sequences of the 16S–23S rRNA spacer region. Appl. Environ. Microbiol. *62*: 2988–2993.

Smith, A.J., R.E. McCoy and J.H. Tsai. 1981. Maintenance *in vitro* of the aster yellows mycoplasmalike organism. Phytopathology *71*: 819–822.

Stackebrandt, E. and B.M. Goebel. 1994. Taxonomic note: a place for DNA–DNA reassociation and 16S rRNA sequence analysis in the present species definition in bacteriology. Int. J. Syst. Bacteriol. *44*: 846–849.

Streten, C. and K.S. Gibb. 2003. Identification of genes in the tomato big bud phytoplasma and comparison to those in sweet potato little leaf-V4 phytoplasma. Microbiology *149*: 1797–1805.

Suzuki, S., K. Oshima, S. Kakizawa, R. Arashida, H.Y. Jung, Y. Yamaji, H. Nishigawa, M. Ugaki and S. Namba. 2006. Interaction between the membrane protein of a pathogen and insect microfilament complex determines insect-vector specificity. Proc. Natl. Acad. Sci. U. S. A. *103*: 4252–4257.

Swofford, D.L. 1998. PAUP: Phylogenetic analysis using parsimony and other methods, 4th edn. Sinauer Associates, Sunderland, MA.

Tan, P.K. and T. Whitlow. 2001. Physiological responses of *Catharanthus roseus* (periwinkle) to ash yellows phytoplasmal infection. New Phytol. *150*: 759–769.

Tanne, E., E. Boudon-Padieu, D. Clair, M. Davidovich, S. Melamed and M. Klein. 2001. Detection of phytoplasma by polymerase chain reaction of insect feeding medium and its use in determining vectoring ability. Phytopathology *91*: 741–746.

Tedeschi, R., V. Ferrato, J. Rossi and A. Alma. 2006. Possible phytoplasma transovarial transmission in the psyllids *Cacopsylla melanoneura* and *Cacopsylla pruni*. Plant Pathol. *55*: 18–24.

Thomas, D.L. 1979. Mycoplasmalike bodies associated with lethal declines of palms in Florida. Phytopathology *69*: 928–934.

Thomas, S. and M. Balasundaran. 2001. Purification of sandal spike phytoplasma for the production of polyclonal antibody. Curr. Sci. *80*: 1489–1494.

Toth, K.F., N. Harrison and B.B. Sears. 1994. Phylogenetic relationships among members of the class *Mollicutes* deduced from *rps3* gene sequences. Int. J. Syst. Bacteriol. *44*: 119–124.

Tran-Nguyen, L.T. and K.S. Gibb. 2006. Extrachromosomal DNA isolated from tomato big bud and *Candidatus* Phytoplasma australiense phytoplasma strains. Plasmid *56*: 153–166.

Tran-Nguyen, L.T., M. Kube, B. Schneider, R. Reinhardt and K.S. Gibb. 2008. Comparative genome analysis of "*Candidatus* Phytoplasma australiense" (subgroup tuf-Australia I; rp-A) and "*Ca.* Phytoplasma asteris" strains OY-M and AY-WB. J. Bacteriol. *190*: 3979–3991.

Tsai, J.H. 1979. Vector transmission of mycoplasmal agents of plant diseases. *In* The Mycoplasmas, vol. III, Plant and Insect Mycoplasmas (edited by Whitcomb and Tully). Academic Press, New York, pp. 266–307.

Valiunas, D., J. Staniulis and R.E. Davis. 2006. '*Candidatus* Phytoplasma fragariae', a novel phytoplasma taxon discovered in yellows diseased strawberry, *Fragaria* x *ananassa*. Int. J. Syst. Evol. Microbiol. *56*: 277–281.

Van Helden, M., W.F. Tjallinghii and T.A. Van Beek. 1994. Phloem collection from lettuce (*Lactuca sativa* L.): Chemical comparison among collection methods. J. Chem. Ecol. *20*: 3191–3206.

Verdin, E., P. Salar, J.L. Danet, E. Choueiri, F. Jreijiri, S. El Zammar, B. Gelie, J.M. Bové and M. Garnier. 2003. '*Candidatus* Phytoplasma phoenicium' sp. nov., a novel phytoplasma associated with an emerging lethal disease of almond trees in Lebanon and Iran. Int. J. Syst. Evol. Microbiol. *53*: 833–838.

Wagner, M., C. Fingerhut, H.J. Gross and A. Schön. 2001. The first phytoplasma RNase P RNA provides new insights into the sequence requirements of this ribozyme. Nucleic Acids Res. *29*: 2661–2665.

Wang, K. and C. Hiruki. 2000. Heteroduplex mobility assay detects DNA mutations for differentiation of closely related phytoplasma strains. J. Microbiol. Methods *41*: 59–68.

Waters, H. and P. Hunt. 1980. The *in vivo* three-dimensional form of a plant mycoplasma-like organism revealed by the analysis of serial ultra-thin sections. J. Gen. Microbiol. *116*: 111–131.

Webb, D.R., R.G. Bonfiglioli, L. Carraro, R. Osler and R.H. Symons. 1999. Oligonucleotides as hybridization probes to localize phytoplasmas in host plants and insect vectors. Phytopathology *89*: 894–901.

Wei, W., S. Kakizawa, H.Y. Jung, S. Suzuki, M. Tanaka, H. Nishigawa, S. Miyata, K. Oshima, M. Ugaki, T. Hibi and S. Namba. 2004a. An antibody against the SecA membrane protein of one phytoplasma reacts with those of phylogenetically different phytoplasmas. Phytopathology *94*: 683–686.

Wei, W., S. Kakizawa, S. Suzuki, H.Y. Jung, H. Nishigawa, S. Miyata, K. Oshima, M. Ugaki, T. Hibi and S. Namba. 2004b. *In planta* dynamic analysis of onion yellows phytoplasma using localized inoculation by insect transmission. Phytopathology *94*: 244–250.

Wei, W., R.E. Davis, I.M. Lee and Y. Zhao. 2007. Computer-simulated RFLP analysis of 16S rRNA genes: identification of ten new phytoplasma groups. Int. J. Syst. Evol. Microbiol. *57*: 1855–1867.

Wei, W., R.E. Davis, R. Jomantiene and Y. Zhao. 2008a. Ancient, recurrent phage attacks and recombination shaped dynamic sequence-variable mosaics at the root of phytoplasma genome evolution. Proc. Natl. Acad. Sci. U. S. A. *105*: 11827–11832.

Wei, W., I.M. Lee, R.E. Davis, X. Suo and Y. Zhao. 2008b. Automated RFLP pattern comparison and similarity coefficient calculation for rapid delineation of new and distinct phytoplasma 16Sr subgroup lineages. Int. J. Syst. Evol. Microbiol. *58*: 2368–2377.

Weintraub, P.G. and L. Beanland. 2006. Insect vectors of phytoplasmas. Annu. Rev. Entomol. *51*: 91–111.

Weisburg, W., J. Tully, D. Rose, J. Petzel, H. Oyaizu, D. Yang, L. Mandelco, J. Sechrest, T. Lawrence and J. Van Etten. 1989. A phylogenetic analysis of the mycoplasmas: basis for their classification. J. Bacteriol. *171*: 6455–6467.

Whitcomb, R.F., D.D. Jensen and J. Richardson. 1966a. The infection of leafhoppers by western X-disease. virus: II. Fluctuation of virus concentration in the hemolymph after injection. Virology *28*: 454–458.

Whitcomb, R.F., D.D. Jensen and J. Richardson. 1966b. The infection of leafhoppers by the western X-disease virus: I. Frequency of transmission after injection or acquisition feeding. Virology *28*: 448–453.

White, D.T., L.L. Blackall, P.T. Scott and K.B. Walsh. 1998. Phylogenetic positions of phytoplasmas associated with dieback, yellow crinkle and mosaic diseases of papaya, and their proposed inclusion in '*Candidatus* Phytoplasma australiense' and a new taxon, '*Candidatus* Phytoplasma australasia'. Int. J. Syst. Bacteriol. *48*: 941–951.

Wongkaew, P. and J. Fletcher. 2004. Sugarcane white leaf phytoplasma in tissue culture: long-term maintenance, transmission, and oxytetracycline remission. Plant Cell Rep. *23*: 426–434.

Yu, Y.L., K.W. Yeh and C.P. Lin. 1998. An antigenic protein gene of a phytoplasma associated with sweet potato witches' broom. Microbiology *144*: 1257–1262.

Zhao, Y., Q. Sun, W. Wei, R.E. Davis, W. Wu and Q. Liu. 2009. '*Candidatus* Phytoplasma tamaricis', a novel taxon discovered in witches'-broom-diseased salt cedar (*Tamarix chinensis* Lour.). Int. J. Syst. Evol. Microbiol. *59*: 2496–2504.

Zreik, L., P. Carle, J.M. Bové and M. Garnier. 1995. Characterization of the mycoplasmalike organism associated with witches' broom disease of lime and proposition of a *Candidatus* taxon for the organism, "*Candidatus* Phytoplasma aurantifolia". Int. J. Syst. Bacteriol. *45*: 449–453.

Order IV. **Anaeroplasmatales** Robinson and Freundt 1987, 81[VP]

Daniel R. Brown, Janet M. Bradbury and Karl-Erik Johansson

A.na.e.ro.plas.ma.ta′les. N.L. neut. n. *Anaeroplasma*, *-atos* type genus of the order; *-ales* ending to denote an order; N.L. fem. pl. n. *Anaeroplasmatales* the *Anaeroplasma* order.

This order in the class *Mollicutes* represents a unique group of strictly anaerobic, wall-less prokaryotes (trivial name, anaeroplasmas) first isolated from the bovine and ovine rumen. Other than their anaerobiosis, the description of organisms in the order is essentially the same as for the class. A single family, *Anaeroplasmataceae*, with two genera, was proposed to recognize the two most prominent characteristics of the organisms: a requirement of sterol supplements for growth by those strictly anaerobic organisms now assigned to the genus *Anaeroplasma*; and strictly anaerobic growth in the absence of sterol supplements by those now assigned to the genus *Asteroleplasma*. Genome sizes range from 1542 to 1794 kbp as estimated by renaturation kinetics. The DNA G+C content ranges from 29 to 40 mol%. All species examined utilize the universal genetic code in which UGA is a stop codon. Phylogenetic studies indicate that members of the *Anaeroplasmatales* are much more closely related to the *Acholeplasmatales* than to the *Mycoplasmatales* or *Entomoplasmatales* (Weisburg et al., 1989).

Type genus: **Anaeroplasma** Robinson, Allison and Hartman 1975, 179[AL].

Further descriptive information

The initial proposal for elevation of the anaeroplasmas to an order of the class *Mollicutes* (Robinson and Freundt, 1987) was based upon the description of three novel species and the observation that some anaeroplasmas did not have a sterol requirement for growth.

The obligate requirement for anaerobic growth conditions is the single most important property in distinguishing members of the *Anaeroplasmatales* from other mollicutes. The anaeroplasmas exist in a natural environment where the oxidation potential is maintained at a low level by the metabolism of associated micro-organisms. Anaerobic methods for preparing media and culture techniques for the organisms are essentially those described by Hungate (1969), with media and inocula maintained in closed vessels and exposure to air avoided during inoculation and incubation. A primary isolation medium and clarified rumen fluid broth have been described (Bryant and Robinson, 1961; Robinson, 1983; Robinson et al., 1975).

References

Bryant, M.P. and I.M. Robinson. 1961. An improved nonselective culture medium for ruminal bacteria and its use in determining diurnal variation in numbers of bacteria in the rumen. J. Dairy Sci. *44*: 1446–1456.

Hungate, R.E. 1969. A roll tube method for cultivation of strict anaerobes. *In* Methods in Microbiology, vol. 3B (edited by Norris and Ribbons). Academic Press, London, pp. 117–132.

Robinson, I.M., M.J. Allison and P.A. Hartman. 1975. *Anaeroplasma abactoclasticum* gen. nov., sp. nov., obligately anaerobic mycoplasma from rumen. Int. J. Syst. Bacteriol. *25*: 173–181.

Robinson, I.M. 1983. Culture media for anaeroplasmas. *In* Methods in Mycoplasmology, vol. 1, (edited by Razin and Tully). Academic Press, New York, pp. 159–162.

Robinson, I.M. and E.A. Freundt. 1987. Proposal for an amended classification of anaerobic mollicutes. Int. J. Syst. Bacteriol. *37*: 78–81.

Weisburg, W., J. Tully, D. Rose, J. Petzel, H. Oyaizu, D. Yang, L. Mandelco, J. Sechrest, T. Lawrence and J. Van Etten. 1989. A phylogenetic analysis of the mycoplasmas: basis for their classification. J. Bacteriol. *171*: 6455–6467.

Family I. **Anaeroplasmataceae** Robinson and Freundt 1987, 80[VP]

Daniel R. Brown, Janet M. Bradbury and Karl-Erik Johansson

A.na.e.ro.plas.ma.ta.ce'ae. N.L. neut. n. *Anaeroplasma*, *-atos* type genus of the family; *-aceae* ending to denote a family; N.L. fem. pl. n. *Anaeroplasmataceae* the *Anaeroplasma* family.

All members have an obligate requirement for anaerobiosis. Organisms assigned to the genus *Anaeroplasma* require sterol supplements for growth. Organisms assigned to the genus *Asteroleplasma* grow in the absence of sterol supplements. Other characteristics are as described for the type genus.

Type genus: **Anaeroplasma** Robinson, Allison and Hartman 1975, 179[AL].

Further descriptive information

The obligate requirement for anaerobic growth conditions and for growth only in media containing cholesterol is established with the methods described by Hungate (1969), with media and inocula maintained in closed vessels and exposure to air avoided during inoculation and incubation. The primary isolation medium and the clarified rumen fluid broth supplemented with cholesterol have been described (Robinson, 1983).

Genus I. **Anaeroplasma** Robinson, Allison and Hartman 1975, 179[AL]

Daniel R. Brown, Janet M. Bradbury and Karl-Erik Johansson

A.na.e.ro.plas'ma. Gr. prefix *an* without; Gr. masc. n. *aer* air; Gr. neut. n. *plasma* a form; N.L. neut. n. *Anaeroplasma* intended to denote "anaerobic mycoplasma".

Cells are predominantly **coccoid**, about 500 nm in diameter; clusters of up to ten coccoid cells may be joined by short filaments. Older cells have a variety of pleomorphic forms. **Cells lack a cell wall** and are bound by a single plasma membrane. Gram-stain-negative due to absence of cell wall. **Obligately anaerobic**; the inhibitory effect of oxygen on growth is not alleviated during repeated subcultures. **Require sterol** supplements for growth. **Nonmotile**. Optimal temperature, 37°C; no growth at 26 or 47°C. Optimal pH, 6.5–7.0. Surface colonies have a dense center with a translucent periphery, or "fried-egg" appearance. Subsurface colonies are golden, irregular, and often multilobed. Strains **vary in their ability to ferment various carbohydrates**. The products of carbohydrate fermentation include acids (generally acetic, formic, propionic, lactic, and succinic), ethanol, and gases (primarily CO_2, but some strains also produce H_2). Bacteriolytic and nonbacteriolytic strains have been described. **Commensals** in the bovine and ovine rumen.

DNA G+C content (mol%): 29–34 (T_m, Bd).

Type species: **Anaeroplasma abactoclasticum** Robinson, Allison and Hartman 1975, 179[AL].

Further descriptive information

Cells of *Anaeroplasma* examined by phase-contrast microscopy appear as single cells, clumps, dumbbell forms, and clusters of coccoid forms joined by short filaments. In electron micrographs of negatively stained preparations, pleomorphic forms are observed; these include filamentous cells, budding cells, and cells with bleb-like structures.

All species examined have similar fermentation products of acetate, formate, lactate, ethanol, and carbon dioxide (Robinson et al., 1975). *Anaeroplasma abactoclasticum* is the only species known not to digest casein. *Anaeroplasma abactoclasticum* strains are the only ones known to produce succinate through fermentation. *Anaeroplasma bactoclasticum*, *Anaeroplasma intermedium*, and *Anaeroplasma varium* are the only species known to produce hydrogen and propionate during their fermentation.

The roll-tube anaerobic culture technique (Hungate, 1969), with pre-reduced medium maintained in a system for exclusion of oxygen, is used to culture the organisms (Robinson, 1983; Robinson and Allison, 1975; Robinson et al., 1975; Robinson and Hungate, 1973). Anaerobic mollicutes in a sewage sludge digester were cultured in an anaerobic cabinet (Rose and Pirt, 1981). Although it is possible that other types of anaerobic culture techniques might be acceptable (anaerobic culture jar or GasPak system), the effective use of such equipment has not been demonstrated (Robinson, 1983). Strains with bacteriolytic activity are detected with the addition of autoclaved *Escherichia coli* cells to the Primary Isolation Medium (PIM) described below. Clear zones around colonies of anaeroplasmas, when viewed by a stereoscopic microscope, are suggestive of bacteriolytic anaeroplasmas. Colonies can be subcultured to clarified rumen fluid broth (CRFB) medium described below.

A slide agglutination test was first used to show that the antigens of anaerobic mollicutes were not related to established *Mycoplasma* or *Acholeplasma* species found in cattle (Robinson and Hungate, 1973). Later, the agglutination test was adapted to either a plate or tube test and combined with an agar gel diffusion test and a modified growth inhibition procedure to examine the antigenic interrelationships among the anaerobic mollicutes (Robinson and Rhoades, 1977). On the basis of these tests, a serological grouping of anaerobic mollicutes appeared compatible with the group separations based upon cultural, biochemical, and biophysical properties of the organisms (Robinson, 1979; Robinson and Rhoades, 1977).

There is no current evidence for the pathogenicity of any of the *Anaeroplasma* species described so far. Obligately anaerobic mollicutes appear to be a heterogeneous group that has been found so far only in the rumen of cattle and sheep (Robinson, 1979; Robinson et al., 1975). Each new isolated group of these organisms seems to have different properties, suggesting that additional undescribed species are likely to exist. The ecological role of these organisms in the rumen has not been determined. Although the titer of these organisms in the rumen appears to be low when compared to titers of other rumen organisms, the mollicutes probably contribute to the pool of microbial fermentation products at that site. Growth of anaeroplasmas is inhibited by thallium acetate (0.2%), bacitracin (1000 mg/ml), streptomycin (200 mg/ml), and D-cycloserine (500 mg/ml), but not by benzylpenicillic acid (1000 U/ml).

Enrichment and isolation procedures

The PIM medium used to grow and detect anaerobic mycoplasmas (Robinson, 1983; Robinson et al., 1975; Robinson and Hungate, 1973) contains: 40% (v/v) rumen fluid strained through cheesecloth, autoclaved, and clarified by centrifugation; 0.05% (w/v) glucose; 0.05% (w/v) cellobiose; 0.05% (w/v) starch; 3.75% (v/v) of a mineral solution consisting of 1.7×10^{-3} M K_2HPO_4, 1.3×10^{-3} M KH_2PO_4, 7.6×10^{-4} M NaCl, 3.4×10^{-3} M $(NH_4)_2SO_4$, 4.1×10^{-4} M $CaCl_2$, and 3.8×10^{-4} M $MgSO_4 \cdot 7H_2O$; 0.2% (w/v) trypticase; 0.1% (w/v) yeast extract; 0.0001% (w/v) resazurin; 0.5% (w/v) autoclaved *Escherichia coli* cells; 0.4% (w/v) Na_2CO_3; 0.05% (w/v) cysteine hydrochloride; 1.5% (w/v) agar; and 0.0006% (w/v) benzylpenicillic acid. Pure cultures are established by picking individual colonies from PIM roll tubes and subculturing into CRFB medium. CRFB medium contains the same ingredients and concentrations as PIM, except: glucose, cellobiose, and starch concentrations are 0.2% (w/v), and autoclaved *Escherichia coli* cells, agar, and benzylpenicillic acid are omitted. Growth also occurs in a rumen fluid-free medium (Medium D) in which growth factors supplied in rumen fluid are replaced by lipopolysaccharide (Boivin; Difco) and cholesterol (Robinson, 1983; Robinson et al., 1975), or in a completely defined medium in which the trypticase, yeast extract, and lipopolysaccharide of Medium D are replaced by amino acids, vitamins, and phosphatidylcholine esterified with unsaturated fatty acids (Robinson, 1979, 1983). An agar-overlay plating technique carried out in an anaerobic hood has also been reported to be an effective isolation procedure (Robinson, 1979). Anaerobic mollicutes grow only in a prereduced medium maintained in a system for exclusion of oxygen. When resazurin is used in the test medium and becomes oxidized, mollicutes will fail to grow.

Maintenance procedures

Cultures are viable after storage for as long as 5 years at −40°C in CRFB medium. They may also be preserved by lyophilization using standard techniques for other mollicutes (Leach, 1983). However, the type strains of several species of *Anaeroplasma* are no longer available from the American Type Culture Collection because it was impossible to revive the cultures sent by the depositors.

Differentiation of the genus *Anaeroplasma* from other genera

Properties that partially fulfill criteria for assignment to the class *Mollicutes* (Brown et al., 2007) include absence of a cell wall, filterability, and the presence of conserved 16S rRNA gene sequences. The obligately anaerobic nature of *Anaeroplasma* species is a distinctive and stable characteristic among these organisms. Strictly anaerobic growth plus the requirement for sterol supplements for growth exclude assignment to any other taxon in the class. Moreover, the bacteriolytic capability possessed by some of the anaeroplasmas has not been reported for other mycoplasmas. Plasmalogens (alkenyl-glycerol ethers), which are found in various anaerobic bacteria but not in aerobic bacteria, are major components of polar lipids from anaeroplasmas (Langworthy et al., 1975); this further supports the contention that anaeroplasmas are distinct from other mollicutes.

Taxonomic comments

The first organism in the group to be described was referred to as *Acholeplasma bactoclasticum* (type strain JR^T = ATCC 27112^T; Robinson and Hungate, 1973) because the organism was thought to lack a sterol requirement for growth. Later, when other obligately anaerobic mollicutes were isolated, these and the JR^T strain were found to require sterol for growth. A proposal was then made to form the new genus *Anaeroplasma* to accommodate strain JR^T (as *Anaeroplasma bactoclasticum*; Robinson and Allison, 1975) and a second anaerobic mollicute designated *Anaeroplasma abactoclasticum* (Robinson et al., 1975). These developments prompted a proposal for an amended classification of anaerobic mollicutes, which included descriptions of *Anaeroplasma varium* and *Anaeroplasma intermedium*, the family *Anaeroplasmataceae*, and the order *Anaeroplasmatales* within the class *Mollicutes* (Robinson and Freundt, 1987).

Early serological studies suggested the existence of several distinct species of anaeroplasmas. Subsequent reports on DNA–DNA hybridization, DNA base composition, and genome size comparisons of organisms in the group also indicated the existence of a number of species in two distinct genera of anaerobic mollicutes (Christiansen et al., 1986; Stephens et al., 1985). Strains initially assigned to *Acholeplasma abactoclasticum* [serovar 3, type strain $6\text{-}1^T$ (=ATCC 27879^T)] were found to be a single species with about 80% interstrain DNA–DNA hybridization. However, strains $A\text{-}2^T$ (serovar 1) and $7LA^T$ (serovar 2), previously included in the description of *Acholeplasma bactoclasticum*, are the type strains of separate species designated *Anaeroplasma varium* and *Anaeroplasma intermedium*, respectively (Robinson and Freundt, 1987). Strains of *Anaeroplasma* all have DNA G+C contents in the range 29.3–33.7 mol%, whereas the base composition of serovar 4 strains 161^T, 162, and 163 clustered above 40 mol%. These were assigned to the new genus *Asteroleplasma* [type strain 161^T (=ATCC 27880^T)] whose members are anaerobic, but do not require sterol for growth. Genome sizes ranged from 1542 to 1715 kbp for *Anaeroplasma* species, as determined by renaturation kinetics (Christiansen et al., 1986). Although the genome sizes reported were in the expected range for members of the class *Mollicutes*, no data are currently available on genome sizes estimated by the more accurate pulsed-field gel electrophoresis technique. A phylogenetic analysis of members of the *Anaeroplasmatales*, based upon 16S rRNA gene sequence comparison, was carried out by Weisburg et al. (1989). *Anaeroplasma* and *Acholeplasma* are sister genera basal on the mollicute tree.

Acknowledgements

The major contributions to the foundation of this material by Joseph G. Tully are gratefully acknowledged.

Further reading

Johansson, K.-E. 2002. Taxonomy of *Mollicutes*. *In* Molecular Biology and Pathogenicity of Mycoplasmas (edited by Razin and Herrmann). Kluwer Academic/Plenum Publishers, New York, pp. 1–29.

Differentiation of the species of the genus *Anaeroplasma*

The technical challenges of cultivating these anaerobic mollicutes have led to a current reliance principally on the combination of 16S rRNA gene sequencing and reciprocal serology for species differentiation. Serological characterization of anaeroplasmas has been performed with agglutination, modified metabolism inhibition, and immunodiffusion tests (Robinson and Rhoades, 1977). Failure to cross-react with antisera against previously recognized species provides substantial evidence for species novelty. DNA–DNA hybridization values between species examined are less than 5%. Bacteriolytic and nonbacteriolytic organisms occur within the genus. When grown on agar media containing a suspension of autoclaved *Escherichia coli* cells, bacteriolytic strains form colonies surrounded by a clear zone due to lysis of the suspended cells by a diffusible enzyme(s). On media lacking suspended cells, bacteriolytic and nonbacteriolytic strains of *Anaeroplasma* cannot be distinguished from each other on the basis of colonial or cellular morphology.

List of species of the genus *Anaeroplasma*

1. **Anaeroplasma abactoclasticum** Robinson, Allison and Hartman 1975, 179[AL]

a.bac.to.clas'ti.cum. Gr. pref. *a* without; Gr. *bakt-* (L. transliteration *bact-*) part of the stem of the Gr. dim. n. *bakterion* (L. transliteration *bacterium*) a small rod; N.L. adj. *clasticus, -a, um* (from Gr. adj. *klastos, -ê, -on* broken in pieces) breaking; N.L. neut. adj. *abactoclasticum* intended to denote "not bacteriolytic".

This is the type species of the genus *Anaeroplasma.* Cells are coccoid, about 500 nm in diameter, sometimes joined into short chains by filaments. Colonies on solid medium are subsurface, but nevertheless present a typical fried-egg appearance. Growth is inhibited by 20 mg/ml digitonin. A major distinguishing factor is the lack of the extracellular bacterioclastic and proteolytic enzymes that characterize the lytic species.

No evidence of a role in pathogenicity.

Source: occurs primarily in the bovine and ovine rumen.

DNA G+C content (mol%): 29.3 (Bd).

Type strain: 6-1, ATCC 27879.

Sequence accession no. (16S rRNA gene): M25050.

2. **Anaeroplasma bactoclasticum** (Robinson and Hungate 1973) Robinson and Allison 1975, 186[AL] (*Acholeplasma bactoclasticum* Robinson and Hungate 1973, 180)

bac.to.clas'ti.cum. Gr. *bakt-* (L. transliteration *bact-*) part of the stem of the Gr. dim. n. *bakterion* (L. transliteration *bacterium*) a small rod; N.L. adj. *clasticus, -a, um* (from Gr. adj. *klastos, -ê, -on* broken in pieces) breaking; N.L. neut. adj. *bactoclasticum* bacteria-breaking.

Pleomorphic and coccoid cells ranging in size from 550 to 2000 nm in diameter. Cells cluster and sometimes form short chains. Colonies on solid medium have a typical fried-egg appearance. Optimal temperature for growth is between 30 and 47°C. Growth is inhibited by 20 mg/ml digitonin. Skim milk is cleared by a proteolytic, extracellular enzyme and certain bacteria are lysed by an extracellular enzyme that attacks the peptidoglycan layer of the cell wall. Shares some serological relationship to other established species in the genus, but can be distinguished by agglutination, modified metabolism inhibition, and agar gel immunodiffusion precipitation tests.

No evidence of pathogenicity.

Source: occurs in the bovine and ovine rumen.

DNA G+C content (mol%): 32.5 to 33.7 (T_m, Bd).

Type strain: JR, ATCC 27112.

Sequence accession no. (16S rRNA gene): M25049.

3. **Anaeroplasma intermedium** Robinson and Freundt 1987, 79[VP]

in.ter.me'di.um. L. neut. adj. *intermedium* intermediate.

Cellular morphology and colonial features are similar to those of *Acholeplasma bactoclasticum.* Serologically distinct from other species in the genus by agglutination, metabolism inhibition, and agar gel immunodiffusion precipitation tests (Robinson and Rhoades, 1977).

No evidence of pathogenicity.

Source: occurs in the bovine and ovine rumen.

DNA G+C content (mol%): 32.5 (Bd).

Type strain: 7LA, ATCC 43166.

Sequence accession no. (16S rRNA gene): not available.

4. **Anaeroplasma varium** Robinson and Freundt 1987, 79[VP]

va'ri.um. L. neut. adj. *varium* diverse, varied, intended to mean different from *Anaeroplasma bactoclasticum.*

Cellular morphology and colonial features are similar to those of *Acholeplasma bactoclasticum.* Serologically distinct from other species in the genus by agglutination, metabolism inhibition, and agar gel immunodiffusion precipitation tests (Robinson and Rhoades, 1977).

No evidence of pathogenicity.

Source: occurs in the bovine and ovine rumen.

DNA G+C content (mol%): 33.4 (T_m).

Type strain: A-2, ATCC 43167.

Sequence accession no. (16S rRNA gene): M23934.

Genus II. **Asteroleplasma** Robinson and Freundt 1987, 79[VP]

Daniel R. Brown, Janet M. Bradbury and Karl-Erik Johansson

A.ste.rol.e.plas'ma. Gr. pref. *a* not; N.L. neut. n. *sterolum* sterol; *e* combining vowel; Gr. neut. n. *plasma* something formed or molded, a form; N.L. neut. n. *Asteroleplasma* name intended to indicate that sterol is not required for growth.

Cellular and colonial morphology **similar to species of the genus** *Anaeroplasma.* Nonmotile. The three strains that form the new genus and species are **obligately anaerobic** and **capable of growth in the absence of cholesterol or serum supplements**. Temperature optimum for growth is 37°C. No evidence of bacteriolytic activity. The organisms are serologically distinct from other members in the family *Anaeroplasmataceae.* Occur in the ovine rumen.

DNA G+C content (mol%): about 40 (T_m, Bd).

Type species: **Asteroleplasma anaerobium** Robinson and Freundt 1987, 79[VP].

Further descriptive information

The most prominent characteristics of the organisms are strictly anaerobic growth and growth in the absence of sterol supplements. The G+C contents of strains analyzed to date are higher than the values for *Anaeroplasma* species (Stephens et al., 1985). DNA–DNA reassociation values clearly show that strains 161^T,

162, and 163 of *Asteroleplasma anaerobium* are genetically related and distinct from established species in the genera *Acholeplasma* or *Anaeroplasma* (Stephens et al., 1985). Tube agglutination tests and gel diffusion precipitation tests showed that strains assigned to *Asteroleplasma anaerobium* are serologically distinct from *Anaeroplasma* species (Robinson and Rhoades, 1977). No data have been reported on antibiotic sensitivity or pathogenicity of asteroleplasmas. Strains have been isolated only from sheep rumen. Isolation and maintenance techniques are similar to those reported for *Anaeroplasma* species.

Differentiation of the genus *Asteroleplasma* from other genera

Properties that partially fulfill criteria for assignment to the class *Mollicutes* (Brown et al., 2007) include absence of a cell wall, filterability, and the presence of conserved 16S rRNA gene sequences. The obligately anaerobic nature of *Asteroleplasma* is a distinctive and stable characteristic. Strictly anaerobic growth plus growth in the absence of sterol supplements exclude assignment to any other taxon in the class. Extracellular bacteriolytic and proteolytic enzymes are absent. Growth is not inhibited by 20 mg/ml digitonin.

Taxonomic comments

The taxonomic position of strains 161^T, 162, and 163 of obligately anaerobic mollicute serovar 4 was delineated through observations that they did not require sterol for growth (Robinson et al., 1975), were serologically distinct (Robinson and Rhoades, 1977), and had G+C contents that were much higher than those of *Anaeroplasma* species (Christiansen et al., 1986). Less than 5% DNA–DNA relatedness existed between these strains and species assigned to the genus *Anaeroplasma* (Stephens et al., 1985). Lastly, the significance of the group and the need to clarify its taxonomic status was emphasized when it was demonstrated that a significant proportion of the anaerobic mollicute population in the bovine and ovine rumen does not require sterol for growth (Robinson and Rhoades, 1982). The phylogenetic analysis of Weisburg et al. (1989) indicated that *Asteroleplasma anaerobium* had branched from the *Firmicutes* lineage independently of *Acholeplasma* and *Anaeroplasma*. Further, *Asteroleplasma* shared two of three important synapomorphies that united the *Mycoplasma* and *Spiroplasma* lineages. Thus, the question of possible monophyly and the true phylogenetic position of *Asteroleplasma* with respect to other mollicutes remains open.

Acknowledgements

The major contributions to the foundation of this material by Joseph G. Tully are gratefully acknowledged.

Further reading

Johansson, K.-E. 2002. Taxonomy of *Mollicutes*. *In* Molecular Biology and Pathogenicity of Mycoplasmas (edited by Razin and Herrmann). Kluwer Academic/Plenum Publishers, New York, pp. 1–29.

List of species of the genus *Asteroleplasma*

1. **Asteroleplasma anaerobium** Robinson and Freundt 1987, 79^{VP}

a.na.e.ro'bi.um. Gr. pref. *an* not; Gr. n. *aer* air; Gr. n. *bios* life; N.L. neut. adj. *anaerobium* not living in air.

This is the type species of the genus *Asteroleplasma*. Cell morphology and colonial characteristics are similar to those of other members of the order *Anaeroplasmatales*. Strains 161^T, 162, and 163 form a homogeneous and distinct serological group, as judged by about 80% DNA–DNA hybridization and serological agglutination, metabolism inhibition, and agar gel immunodiffusion precipitation tests.

No evidence of pathogenicity.

Source: all isolates have been identified from the bovine or ovine rumen.

DNA G+C content (mol%): 40.2–40.5 (Bd, T_m).

Type strain: 161 (the type strain ATCC 27880 no longer exists).

Sequence accession no. (16S rRNA gene): M22351.

References

Brown, D., R. Whitcomb and J. Bradbury. 2007. Revised minimal standards for description of new species of the class *Mollicutes* (division *Tenericutes*). Int. J. Syst. Evol. Microbiol. *57*: 2703–2719.

Christiansen, C., E.A. Freundt and I.M. Robinson. 1986. Genome size and deoxyribonucleic acid base composition of *Anaeroplasma abactoclasticum*, *Anaeroplasma bactoclasticum*, and a sterol-nonrequiring anaerobic mollicute. Int. J. Syst. Bacteriol. *36*: 483–485.

Hungate, R.E. 1969. A roll tube method for cultivation of strict anaerobes. *In* Methods in Microbiology, vol. 3B (edited by Norris and Ribbons). Academic Press, London, pp. 117–132.

Langworthy, T., W. Mayberry, P. Smith and I. Robinson. 1975. Plasmalogen composition of *Anaeroplasma*. J. Bacteriol. *122*: 785–787.

Leach, R.H. 1983. Preservation of *Mycoplasma* cultures and culture collections. *In* Methods in Mycoplasmology, vol. 1 (edited by Razin and Tully). Academic Press, New York, pp. 197–204.

Robinson, I.M. and M.J. Allison. 1975. Transfer of *Acholeplasma bactoclasticum* Robinson and Hungate to genus *Anaeroplasma* (*Anaeroplasma bactoclasticum* Robinson and Hungate comb. nov.), emended description of species. Int. J. Syst. Bacteriol. *25*: 182–186.

Robinson, I.M., M.J. Allison and P.A. Hartman. 1975. *Anaeroplasma abactoclasticum* gen. nov., sp. nov., obligately anaerobic mycoplasma from rumen. Int. J. Syst. Bacteriol. *25*: 173–181.

Robinson, I.M. and K.R. Rhoades. 1977. Serological relationships between strains of anaerobic mycoplasmas. Int. J. Syst. Bacteriol. *27*: 200–203.

Robinson, I.M. 1979. Special features of anaeroplasmas. *In* The Mycoplasmas, vol. 1 (edited by Barile and Razin). Academic Press, New York, pp. 515–528.

Robinson, I.M. and K.R. Rhoades. 1982. Serologic relationships between strains of anaerobic mycoplasmas. Rev. Infect. Dis. *4*: S271.

Robinson, I.M. 1983. Culture media for anaeroplasmas. *In* Methods in Mycoplasmology, vol. 1 (edited by Razin and Tully). Academic Press, New York, pp. 159–162.

Robinson, I.M. and E.A. Freundt. 1987. Proposal for an amended classification of anaerobic mollicutes. Int. J. Syst. Bacteriol. *37*: 78–81.

Robinson, J.P. and R.E. Hungate. 1973. *Acholeplasma bactoclasticum* sp. n., an anaerobic mycoplasma from the bovine rumen. Int. J. Syst. Bacteriol. *23*: 171–181.

Rose, C. and S. Pirt. 1981. Conversion of glucose to fatty acids and methane: roles of two mycoplasmal agents. J. Bacteriol. *147*: 248–254.

Stephens, E., I. Robinson and M. Barile. 1985. Nucleic acid relationships among the anaerobic mycoplasmas. J. Gen. Microbiol. *131*: 1223–1227.

Weisburg, W., J. Tully, D. Rose, J. Petzel, H. Oyaizu, D. Yang, L. Mandelco, J. Sechrest, T. Lawrence and J. Van Etten. 1989. A phylogenetic analysis of the mycoplasmas: basis for their classification. J. Bacteriol. *171*: 6455–6467.

Phylum XVII. Acidobacteria phyl. nov.

J. Cameron Thrash and John D. Coates

A.ci'do.bac.ter'i.a. *Acidobacteriales* type order of the phylum; removing the *-ales* ending and inserting *-a* to denote phylum; N.L. neut. n. *Acidobacteria* the phylum of *Acidobacteriales*.

The phylum currently contains two classes, three orders, three families, and six described genera. The phylum is identified on the basis of phylogenetic analysis of 16S rRNA gene sequences (Figure 117).

Culture-independent methods have identified significant diversity within the phylum based on 16S rRNA gene sequences (Barns et al., 1999; Hugenholtz et al., 1998; Kuske et al., 1997; Ludwig et al., 1997; Meisinger et al., 2007; Quaiser et al., 2003; Zimmermann et al., 2005). Currently, more than 3000 *Acidobacteria* sequences exist in public databases making up 26 coherent groups within the phylum (Barns et al., 2007). *Acidobacteria* sequences have been detected in a wide variety of habitats including soils, sediments, hot springs, marine snow, feces (Barns et al., 1999), a variety of cave environments (Meisinger et al., (2007), and references therein), and metal-contaminated soils (Barns et al., 2007). Although few of these representatives have been cultivated, many new isolates have recently been reported (Janssen et al., 2002; Joseph et al., 2003; McCaig et al., 2001; Sait et al., 2006, 2002; Stevenson et al., 2004). Future analysis of these additional isolates should improve understanding of the various functional roles played by members of this phylum, supplementing that for the six genera already described.

Table 145 provides characteristics that can be used to distinguish the six genera.

Type order: **Acidobacteriales** ord. nov.

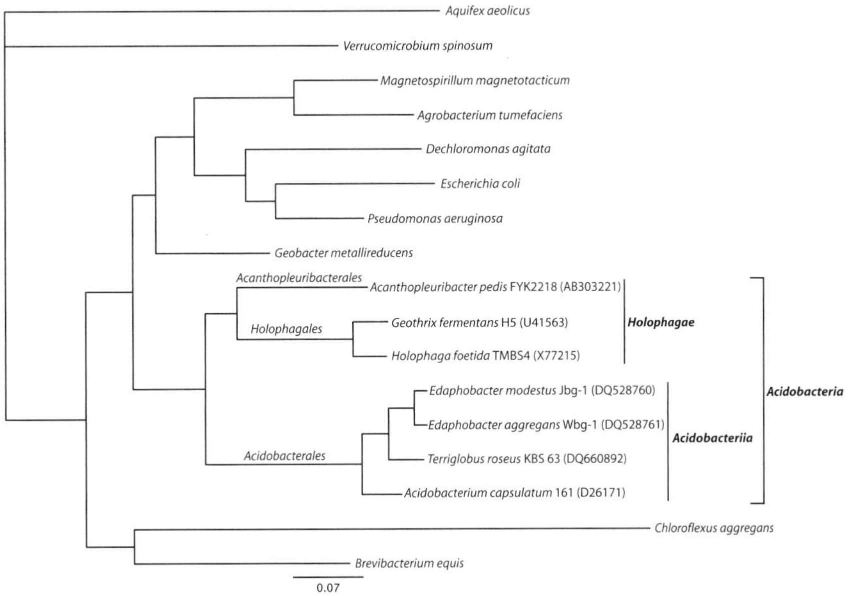

FIGURE 117. Phylogenetic tree of the *Acidobacteria* based on the 16S rRNA gene. The phylum contains six described genera in two classes, three orders, and three families. The class *Acidobacteriia* contains the genera *Acidobacterium, Edaphobacter*, and *Terriglobus* in a single family, *Acidobacteriaceae*, in the order *Acidobacteriales*. The class *Holophagae* contains the genera *Holophaga*, and *Geothrix* in the family *Holophagaceae* in the order *Holophagales*, and the genus *Acanthopleuribacter* in the family *Acanthopleuribacteraceae* in the order *Acanthopleuribacterales*. Bar = 0.07 substitutions per site. Tree was constructed using Bayesian analysis (Huelsenbeck and Ronquist, 2001; Ronquist and Huelsenbeck, 2003).

TABLE 145. Differential characteristics of the genera of the phylum *Acidobacteria*[a]

	Acidobacterium	*Edaphobacter*	*Terriglobus*	*Holophaga*	*Geothrix*	*Acanthopleuribacter*
Size (μm)	0.3–0.8 × 1.1–2.3	0.5–0.9 × 1–2.1	0.5–0.7 × 0.9–1.4	0.5–0.7 × 1–3	0.1 × 1–2	0.7–1.0 × 2.4–4.7
Colony color	Pale orange to orange	Beige	White or pink	Beige	White on iron(III), red on fumarate	Yellow
Motility	+	+/–	–	–	–	+
Flagella	+	na	–	–	–	+
Sporeforming	–	–	na	–	–	na
Respiratory	+	+	+	–	+	+
Electron acceptors utilized	Oxygen	Oxygen	Oxygen	–	Nitrate, Mn(IV), Fe(III), AQDS, fumarate	Oxygen
Fermentative	–	–	–	+	+	–
Compounds fermented	–	–	–	3,4,5-Trimethoxybenzoate, syringate, 5-hydroxyvanillate, phloroglucinolmonomethylether, sinapate, ferulate, caffeate, gallate, 2,4,6-trihydroxybenzoate, pyrogallol, phloroglucinol, and pyruvate	Citrate, fumarate	–
Electron donors for respiration	D-Glucose, L-arabinose, D-xylose, D-galactose, D-mannose, maltose, arbutin, cellobiose, starch, β-gentiobiose, lactose, trehalose, amygdalin	D-Glucose, L-aspartate, L-glutamate, L-glutamine, L-ornithine, L-arabinose, D-fructose, D-glucosamine, D-lactose, D-lyxose, trehalose, D-xylose, D-glucuronic acid, L-rhamnose, D-sorbitol, L-lyxitol, D-mannitol, *myo*-inositol, xylitol, Casamino acids, yeast extract, peptone	D-Glucose, D-fructose, D-galactose, D-mannose, D-xylose, D-arabinose, sucrose, D-maltose, D-raffinose, D-cellobiose, succinate, D-glucuronate, D-gluconic acid	–	Acetate, propionate, palmitate, succinate, fumarate, lactate, yeast extract	α-D-Glucose, L-alanine, hydroxy- L-proline, L-serine, L-threonine, inosine, uridine, thymidine
Optimum temp (°C)	na	30	12–23	28–32	35	30
Temp range (°C)	25–37	15–37	12–23	10–35	25–40	15–30
Optimum pH	na	5.5	5–6	6.8–7.5	na	7.0–8.0
pH range	3–6	4.0–7.0	4.5–7.0	5.5–8.0	na	5.0–9.0
NaCl tolerance (%)	<3.5	na	na	0.1–1.5	na	50–150% ASW[b]
DNA G+C content (mol%)	60.8	55.8–56.9	59.8	62.5	na	56.7

[a]Symbols: +, present or confirmed capability; –, absent/none or confirmed incapability; na, not available; AQDS, anthraquinone-2,6-disulfonate.

[b]ASW, Artificial Seawater, Nagai Chemical Products.

References

Barns, S.M., S.L. Takala and C.R. Kuske. 1999. Wide distribution and diversity of members of the bacterial kingdom *Acidobacterium* in the envrionment. Appl. Environ. Microbiol. *65*: 1731–1737.

Barns, S.M., E.C. Cain, L. Sommerville and C.R. Kuske. 2007. *Acidobacteria* phylum sequences in uranium-contaminated subsurface sediments greatly expand the known diversity within the phylum. Appl. Environ. Microbiol. *73*: 3113–3116.

Huelsenbeck, J.P. and F. Ronquist. 2001. MRBAYES: Bayesian inference of phylogenetic trees. Bioinformatics *17*: 754–755.

Hugenholtz, P., B.M. Goebel and N.R. Pace. 1998. Impact of culture-independent studies on the emerging phylogenetic view of bacterial diversity. J. Bacteriol. *180*: 4765–4774.

Janssen, P.H., P.S. Yates, B.E. Grinton, P.M. Taylor and M. Sait. 2002. Improved culturability of soil bacteria and isolation in pure culture of novel members of the divisions *Acidobacteria, Actinobacteria, Proteobacteria,* and *Verrucomicrobia*. Appl. Environ. Microbiol. *68*: 2391–2396.

Joseph, S.J., P. Hugenholtz, P. Sangwan, C.A. Osborne and P.H. Janssen. 2003. Laboratory cultivation of widespread and previously uncultured soil bacteria. Appl. Environ. Microbiol. *69*: 7210–7215.

Kuske, C.R., S.M. Barns and J.D. Busch. 1997. Diverse uncultivated bacterial groups from soils of the arid southwestern United States that are present in many geographic regions. Appl. Environ. Microbiol. *63*: 3614–3621.

Ludwig, W., S.H. Bauer, M. Bauer, I. Held, G. Kirchhof, R. Schulze, I. Huber, S. Spring, A. Hartmann and K.H. Schleifer. 1997. Detection and in situ identification of representatives of a widely distributed new bacterial phylum. FEMS Microbiol. Lett. *153*: 181–190.

McCaig, A.E., S.J. Grayston, J.I. Prosser and L.A. Glover. 2001. Impact of cultivation on characterisation of species composition of soil bacterial communities. FEMS Microbiol. Ecol. *35*: 37–48.

Meisinger, D.B., J. Zimmermann, W. Ludwig, K.-H. Schleifer, G. Wanner, M. Schmid, P.C. Bennett, A.S. Engel and N.M. Lee. 2007. *In situ* detection of novel *Acidobacteria* in microbial mats from a chemolithoautotrophically based cave ecosystem (Lower Kane Cave, WY, USA). Environ. Microbiol. *9*: 1523–1534.

Quaiser, A., T. Ochsenreiter, C. Lanz, S.C. Schuster, A.H. Treusch, J. Eck and C. Schleper. 2003. Acidobacteria form a coherent but highly diverse group within the bacterial domain: evidence from environmental genomics. Mol. Microbiol. *50*: 563–575.

Ronquist, F. and J.P. Huelsenbeck. 2003. MrBayes 3: Bayesian phylogenetic inference under mixed models. Bioinformatics *19*: 1572–1574.

Sait, M., P. Hugenholtz and P.H. Janssen. 2002. Cultivation of globally distributed soil bacteria from phylogenetic lineages previously only detected in cultivation-independent surveys. Environ. Microbiol. *4*: 654–666.

Sait, M., K.E.R. Davis and P.H. Janssen. 2006. Effect of pH on isolation and distribution of members of subdivision 1 of the phylum *Acidobacteria* occurring in soil. Appl. Environ. Microbiol. *72*: 1852–1857.

Stevenson, B.S., S.A. Eichorst, J.T. Wertz, T.M. Schmidt and J.A. Breznak. 2004. New strategies for cultivation and detection of previously uncultured microbes. Appl. Environ. Microbiol. *70*: 4748–4755.

Zimmermann, J., J.M. Gonzalez and C. Saiz-Jimenez. 2005. Detection and phylogenetic relationships of highly diverse uncultured acidobacterial communities in Altamira Cave using 23S rRNA sequence analysis. Geomicrobiol. J. *22*: 379–388.

Class I. **Acidobacteriia** Cavalier-Smith 2002, 12[VP]

J. Cameron Thrash and John D. Coates

A.ci.do.bac.te.ri′i.a. N.L. neut. n. *Acidobacterium* type genus of the type order *Acidobacteriales*; N.L. suff. *-ia*, ending proposed by Gibbons and Murray and by Stackebrandt et al. to denote a class; N.L. neut. pl. n. *Acidobacteriia*, the class of *Acidobacteriales*.

The class *Acidobacteriia* was circumscribed for this volume on the basis of phylogenetic analysis of 16S rRNA gene sequences; the class contains the sole order *Acidobacteriales*.

Type order: **Acidobacteriales** ord. nov.

Reference

Cavalier-Smith, T. 2002. The neomuran origin of archaebacteria, the negibacterial root of the universal tree and bacterial megaclassification. Int. J. Syst. Evol. Microbiol. *52*: 7–76.

Order I. **Acidobacteriales** Cavalier-Smith 2002, 12[VP]

J. Cameron Thrash and John D. Coates

A.ci′do.bac.ter.i.a′les. N.L. neut. n. *Acidobacterium* type genus of the order; *-ales* ending to denote order; N.L. fem. pl. n. *Acidobacteriales* the *Acidobacterium* order.

The order *Acidobacteriales* was circumscribed for this volume on the basis of phylogenetic analysis of 16S rRNA gene sequences; the order contains the family *Acidobacteriaceae*.

Type genus: **Acidobacterium** Kishimoto, Kosako and Tano 1991a, 456 (Effective publication: Kishimoto, Kosako and Tano 1991b, 6.).

References

Cavalier-Smith, T. 2002. The neomuran origin of archaebacteria, the negibacterial root of the universal tree and bacterial megaclassification. Int. J. Syst. Evol. Microbiol. *52*: 7–76.

Kishimoto, N., K. Inagaki, T. Sugio and T. Tano. 1991a. *In* Validation of the publication of new names and combinations previously effectively published outside the IJSB. List no. 38. Int. J. Syst. Bacteriol. *41*: 456–457.

Kishimoto, N., Y. Kosako and T. Tano. 1991b. *Acidobacterium capsulatum* gen. nov., sp. nov.: an acidophilic chemoorganotrophic bacterium containing menaquinone from acidic mineral environment. Curr. Microbiol. *22*: 1–7.

Family I. **Acidobacteriaceae** fam. nov.

J. Cameron Thrash and John D. Coates

A.ci'do.bac.ter.i.a'ce.ae. N.L. neut. n. *Acidobacterium* type genus of the family; suff. *-aceae* denoting family; N.L. fem. pl. n. *Acidobacteriaceae* the *Acidobacterium* family.

The family *Acidobacteriaceae* was circumscribed for this volume on the basis of phylogenetic analysis of 16S rRNA gene sequences; the family contains the genera *Acidobacterium* (type genus), *Edaphobacter*, and *Terriglobus*. All known species grow as **chemoorganotrophs**.

Type genus: **Acidobacterium** Kishimoto, Kosako and Tano 1991b, 456 (Effective publication: Kishimoto, Kosako and Tano 1991c, 6.).

Genus I. **Acidobacterium** Kishimoto, Kosako and Tano 1991b, 456[VP] (Effective publication: Kishimoto, Kosako and Tano 1991c, 6)

J. Cameron Thrash and John D. Coates

A.ci'do.bac.ter'i.um. N.L. n. *acidum* an acid; L. neut. n. *bacterium* a small rod; N.L. neut. n. *Acidobacterium* an acid-loving rod.

Rod-shaped cells, 0.3–0.8 μm × 1.1–2.3 μm. Gram-stain-negative. Nonphotosynthetic. Nonsporeforming. **Cells occur singly, in pairs, or in short chains**. **Capsule** produced. **Aerobic**, having an obligately aerobic metabolism. **Acidophilic**. Growth at pH 3.0–6.0, but not at 6.5. No growth at 3.5% NaCl. Do not denitrify. Catalase-positive. Chemoorganotrophic. Growth on a variety of sugars including glucose as the sole carbon source. No growth on alcohols including methanol. Growth is inhibited by 0.25 mM acetate, 2 mM lactate, and 4 mM succinate. Growth at 25–37°C but not at 42°C. **Motile by peritrichous flagella**. The major quinone is menaquinone with eight isoprene units (MK-8). The major fatty acid is $C_{15:0}$. 2- and 3-Hydroxy fatty acids are absent except for trace amounts of $C_{12:0}$.

Source: acid mine drainage in the Yanahaya mine, Okayama, Japan.

DNA G+C content (mol%): 59.7–60.8.

Type species: **Acidobacterium capsulatum** Kishimoto, Kosako and Tano 1991b, 456[VP] (Effective publication: Kishimoto, Kosako and Tano 1991c, 6.).

Further descriptive information

As seen by electron microscopy, cells have peritrichous flagella. A capsule can be visualized by India ink staining.

The only species in the genus was isolated from a mine in Okayama, Japan by plating acid mine drainage samples on glucose-yeast extract (GYE) plates (Kishimoto and Tano, 1987). Many closely related organisms identified by both culture and culture-independent techniques have been reported from a variety of ecosystems including acid mine drainage (Kishimoto et al., 1991c), mine mud and soil (Kishimoto et al., 1991c), and other acidic soils (Sait et al., 2006). *Acidobacterium* contains the following major fatty acids: $C_{14:0}$, $C_{15:1}$, $C_{16:1}$, $C_{16:0}$, $C_{17:1}$, $C_{17:0}$, $C_{18:1}$, $C_{18:0}$, $C_{15:0}$ iso, $C_{17:1}$ iso, $C_{17:0}$ iso, and $C_{19:0}$ cyclo.

Enrichment and isolation procedures

The type strain was isolated on GYE plates from an acid mine drainage community obtained from the Yanahaya mine in Okayama, Japan. Samples were collected and plated within 8 h (Kishimoto et al., 1991c; Kishimoto and Tano, 1987). GYE media is composed of (g/l distilled water): $(NH_4)_2SO_4$, 2.0; KCl, 0.1; K_2HPO_4, 0.1; $MgSO_4 \cdot 7H_2O$, 0.5; glucose, 1.0, yeast extract (Difco Laboratories), 0.1. pH adjusted to 3.0 with H_2SO_4. Cells are grown at 30°C.

Maintenance procedures

The type strain is maintained as a lyophilized culture at the DSMZ.

Differentiation of the genus *Acidobacterium* from other genera

Acidobacterium is strictly aerobic whereas *Geothrix* and *Holophaga* are strictly anaerobic. *Acidobacterium* is acidophilic, but *Geothrix* and *Holophaga* grow preferentially at about neutral pH. *Acidobacterium* has 79% 16S rRNA gene sequence identity with *Geothrix* and *Holophaga*, 91% identity with *Edaphobacter*, and 92% identity with *Terriglobus*. *Terriglobus* does not contain the fatty acid $C_{18:1}$ ω9*c*.

List of species of the genus *Acidobacterium*

1. **Acidobacterium capsulatum** Kishimoto, Kosako and Tano 1991b, 456[VP] (Effective publication: Kishimoto, Kosako and Tano 1991c, 6.)

cap.su.la'tum. N.L. neut. adj. *capsulatum* capsuled.

Colonies on GYE gellan gum plates are pale orange to orange.

The characteristics are as described for the genus with the following additional information. The cells are positive for β-galactosidase, β-glucuronidase, α-fucosidase, *N*-acetyl-β-glucosaminidase, and α-glucosidase activity (Kishimoto et al., 1991a). Additionally, cells produce β-glucosidase and secrete it extracellularly (Kishimoto et al., 1991a, c). Cells also have cellulase, xylanase, and trehalase activity (Inagaki et al., 1998, 2001).

DNA G+C content (mol%): 60.8 (HPLC).

Type strain: 161, ATCC 51196, DSM 11244, JCM 7670, NCIMB 13165, NRBC 15755.

Sequence accession no. (16S rRNA gene): D26171.

Genus II. **Edaphobacter** Koch, Gich, Dunfield and Overmann 2008, 1119[VP]

J. Cameron Thrash and John D. Coates

E.da'pho.bac'ter. Gr. neut. n. *edaphos* soil; N.L. masc. n. *bacter* a short rod; N.L. masc. n. *Edaphobacter* rod of the soil.

Rod-shaped cells, 1.0–2.1 µm × 0.5–0.9 µm. Gram-stain-negative. Nonsporeforming. **Cells occur singly or in clumps**. Do not form capsule. **Aerobic**, having an obligately aerobic metabolism with oxygen as the terminal electron acceptor. **Acidophilic**. Growth at pH 4.0–7.0, with a pH optimum of 5.5. Catalase-positive. Chemoorganotrophic. Growth on a variety of sugars, some amino acids, yeast extract, and peptone as the sole carbon source. Growth between 15 and 37°C, with a temperature optimum of 30°C. May be motile or nonmotile. Major fatty acids include $C_{16:1}$ ω7*c*/$C_{15:0}$ iso 2-OH and $C_{17:1}$ ω8*c*. Isolated from mollisol on Jochberg in southern Germany and leptosol in a forest near Würzburg, Germany.

DNA G+C content (mol%): 55.8–56.9.

Type species: **Edaphobacter modestus** Koch, Gich, Dunfield and Overmann 2008, 1121[VP].

Further descriptive information

Cells form round beige colonies when grown on agar or gellan gum. *Edaphobacter* are positive for the following enzyme activities: alkaline phosphatase, acid phosphatase, naphthol-AS-BI-phosphohydrolase, esterase, esterase lipase, leucyl arylamidase, α-chymotrypsin, α-galactosidase, β-galactosidase, β-glucuronidase, α-glucosidase, β-glucosidase, and α-fucosidase.

Edaphobacter contain the following major fatty acids in common between both species: $C_{16:0}$, $C_{15:0}$ iso, $C_{17:0}$ iso, and may contain the following additional fatty acids: $C_{14:0}$, $C_{15:0}$, $C_{17:0}$, $C_{18:0}$, $C_{19:0}$, $C_{14:1}$ ω5*c*, $C_{15:1}$ ω6*c*, $C_{16:1}$ iso H, $C_{17:1}$ ω6*c*, $C_{17:1}$ ω8*c*, $C_{18:1}$ ω9*c*, $C_{18:1}$ iso H, $C_{14:0}$ iso, $C_{16:0}$ iso, $C_{17:1}$ iso ω9*c*, $C_{17:0}$ 10-methyl, $C_{18:0}$ iso, $C_{16:1}$ ω7*c*/$C_{15:0}$ iso 2-OH, and $C_{19:1}$ ω11*c*/$C_{19:1}$ ω9*c*.

Species in the genus were isolated from mollisol on Jochberg in southern Germany and leptosol in forest near Würzburg, Germany.

Enrichment and isolation procedures

Edaphobacter modestus Jbg-1 was isolated from mollisol at 1400 m on Jochberg, Germany. Enrichments were inoculated from soil sampled within the top 3 cm into pH 6.3 HEPES-buffered (10 mM) media containing pectin, chitin, starch, cellulose, xylan, and curdlan (1% each), and included cAMP, *N*-(oxohexanoyl)-DL-homoserine lactone, *N*-(butyryl)-DL-homoserine lactone at 10 µM each. Incubations were carried out at 15°C. An enrichment positive for acidobacteria-specific PCR products was streaked on agar plates containing casein peptone (0.05%), glucose (0.01%), and yeast extract (0.025%) (Koch et al., 2008).

Edaphobacter aggregans Wbg-1 was isolated from forest leptosol in a deciduous forest near Würzburg, Germany. Agar plates containing defined pH 7.0 phosphate-buffered (100 mM) media, as described by Koch et al. (2008), were sprinkled with soil sampled within the top 8–13 cm and incubated at 20°C in a sealed, 20% methane (in air) headspace. After successive transfers on this media, colonies were transferred to R2A agar plates.

Maintenance procedures

Both strains are maintained as lyophilized culture at the DSMZ.

Differentiation of the genus *Edaphobacter* from other genera

While both *Acidobacterium* and *Terriglobus* can utilize cellobiose, D-galactose, maltose, and D-mannose, *Edaphobacter* species cannot. Likewise, *Edaphobacter* can make use of amino acids, such as L-glutamate and L-glutamine, which the other genera cannot. Additionally, 16S rRNA gene sequence identity between the type species and the other genera distinguishes *Edaphobacter* as a unique genus: 95% identity with *Terriglobus*, 91% identity with *Acidobacterium*.

Differentiation of species of the genus *Edaphobacter*

The two species of the genus *Edaphobacter* are distinguished physiologically by differential substrate utilization. Additionally, *Edaphobacter modestus* and *Edaphobacter aggregans* contain very different fatty acid profiles. The dominant fatty acid for *Edaphobacter modestus* is $C_{16:1}$ ω7*c*/$C_{15:0}$ iso 2-OH, with $C_{15:0}$ iso and $C_{16:0}$ as secondary fatty acids. In contrast, the dominant fatty acid for *Edaphobacter aggregans* is $C_{17:1}$ ω8*c*, with $C_{16:0}$ iso and $C_{17:0}$ as secondary fatty acids. *Edaphobacter modestus* can be motile, *Edaphobacter aggregans* is nonmotile, and each have slightly different growth ranges for temperature and pH. Also, by DNA–DNA hybridization, these two species share only 11.5–13.6% similarity (Koch et al., 2008).

List of species of the genus *Edaphobacter*

1. **Edaphobacter modestus** Koch, Gich, Dunfield and Overmann 2008, 1121[VP]

mo.des'tus. L. masc. adj. *modestus* moderate, referring to the adaptation of the type strain to low substrate concentrations.

The type strain is adapted to low substrate concentrations.

The characteristics are as described for the genus with the following additional information. Rods are 1.0–1.8 µm × 0.5–0.7 µm. Motile. Positive for cytochrome oxidase. Growth at pH 4.5–7.0 with an optimum of pH 5.5. Growth at 15–30°C with an optimum of 30°C. Differential to Wbg-1, Jbg-1 can utilize L-arabinose, D-fructose, D-glucosamine, D-lyxose, trehalose, D-xylose, L-rhamnose, D-sorbitol, L-lyxitol, D-mannitol, *myo*-inositol, and xylitol. Contains the following fatty acids: $C_{16:0}$, $C_{15:0}$ iso, $C_{17:0}$ iso, $C_{14:0}$, $C_{18:0}$, $C_{14:1}$ ω5*c*, $C_{17:1}$ iso ω9*c*, and $C_{16:1}$ ω7*c*/$C_{15:0}$ iso 2-OH.

DNA G+C content (mol%): 55.8 (HPLC).

Type strain: Jbg-1, ATCC BAA-1329, DSM 18101.

Sequence accession no. (16S rRNA gene): DQ528760.

2. **Edaphobacter aggregans** Koch, Gich, Dunfield and Overmann 2008, 1121[VP]

ag'gre.gans. L. part. adj. *aggregans* assembling, aggregating.

The type strain may grow in clumps.

The characteristics are as described for the genus with the following additional information. Rods are 1.5–2.1 µm × 0.7–0.9 µm. Nonmotile. Growth at pH 4.0–7.0 with an optimum of pH 5.5. Growth at 15–37°C with an optimum of 30°C. Differential to Jbg-1, Wbg-1 can utilize L-aspartate, L-ornithine, and D-glucuronic acid. Contains the following fatty acids: $C_{16:0}$, $C_{15:0}$ iso, $C_{17:0}$ iso, $C_{15:0}$, $C_{17:0}$, $C_{19:0}$, $C_{15:1}$ ω6*c*, $C_{16:1}$ iso H, $C_{17:1}$ ω6*c*, $C_{17:1}$ ω8*c*, $C_{18:1}$ ω9*c*, $C_{18:1}$ iso H, $C_{14:0}$ iso, $C_{16:0}$ iso, $C_{17:0}$ 10-methyl, $C_{18:0}$ iso, $C_{16:1}$ ω7*c*/$C_{15:0}$ iso 2-OH, and $C_{19:1}$ ω11*c*/$C_{19:1}$ ω9*c*.

DNA G+C content (mol%): 56.9 (HPLC).

Type strain: Wbg-1, ATCC BAA-1497, DSM 19364.

Sequence accession no. (16S rRNA gene): DQ528761.

Genus III. **Terriglobus** Eichorst, Breznak and Schmidt 2007b, 1933[VP] (Effective publication: Eichorst, Breznak and Schmidt 2007a, 2715.)

J. Cameron Thrash and John D. Coates

Ter.ri.glo'bus. L. fem. n. *terra* earth; L. masc. n. *globus* ball, clump; N.L. masc. n. *Terriglobus* clump of earth.

Rod-shaped cells, 0.5–0.7 µm × 0.9–1.4 µm. Gram-stain-negative. **Cells occur singly or in clumps**. Extracellular matrix produced. **Aerobic**, having an obligately aerobic metabolism. **Acidophilic**. Growth at pH 4.5–7.0, with optimum between pH 5–6. Catalase-positive. Negative for oxidase activity. Chemoorganotrophic. Growth on a variety of sugars as well as succinate as the sole carbon source. Growth at 12–23°C, but not at 4 or 37°C. Nonmotile. The major fatty acid is $C_{16:1}$ ω7*c*/$C_{15:0}$ iso 2-OH.

Source: never-tilled soils at the Michigan State University W. K. Kellogg Biological Station Long-Term Ecological Research (KBS LTER) site in Michigan, USA.

DNA G+C content (mol%): 59.8.

Type species: **Terriglobus roseus** Eichorst, Breznak and Schmidt 2007b, 1933[VP] (Effective publication: Eichorst, Breznak and Schmidt 2007a, 2716.).

Further descriptive information

Cells contain carotenoids with absorption maxima in chloroform at 473, 505, and 539 nm. *Terriglobus* can grow at oxygen content from 2 to 20%.

Terriglobus contains the following major fatty acids: $C_{14:0}$, $C_{15:0}$, $C_{16:0}$, $C_{16:0}$ N OH, $C_{17:0}$, $C_{18:0}$, $C_{20:0}$, $C_{14:1}$ ω5*c*, $C_{15:1}$ ω6*c*, $C_{16:1}$ ω5*c*, $C_{18:1}$ ω7*c*, $C_{18:1}$ ω5*c*, $C_{20:2}$ ω6,9*c*, $C_{13:0}$ iso, $C_{15:0}$ iso, $C_{15:0}$ anteiso, $C_{17:1}$ iso ω5*c*, $C_{17:0}$ iso, $C_{17:0}$ anteiso, $C_{19:0}$ anteiso, $C_{15:1}$ iso H/$C_{13:0}$ 3-OH, and $C_{16:1}$ ω7*c*/$C_{15:0}$ iso 2-OH.

The only species in the genus was isolated from never-tilled soils at the Michigan State University KBS LTER site in Michigan, USA.

Enrichment and isolation procedures

Approximately 30 g from 2.5 cm × 10.0 cm cores of never-tilled soil from the Michigan State University Kellogg Biological Station LTER site was added to 100 ml phosphate-buffered solution containing 137 mM NaCl, 2.7 mM KCl, 10 mM Na_2HPO_4, and 2 mM KH_2PO_4, adjusted to pH 7.0, supplemented with 2.24 mM $Na_4P_2O_7$ decahydrate and 1 mM dithiothreitol, and stirred vigorously for 30 min. The slurry was allowed to settle for 30 min. Aliquots of supernatant were then used to inoculate serial dilutions in isolation media and plated as described in Eichorst et al. (2007a).

Maintenance procedures

The type strain is maintained as lyophilized culture at the DSMZ.

Differentiation of the genus *Terriglobus* from other genera

Terriglobus can be distinguished from *Acidobacterium* and *Edaphobacter* by 16S rRNA gene sequence identity (92 and 95%, respectively) and fatty acid composition, as *Terriglobus* does not contain the fatty acid $C_{18:1}$ ω9*c*, which is present in *Acidobacterium* and *Edaphobacter*.

List of species of the genus *Terriglobus*

1. **Terriglobus roseus** Eichorst, Breznak and Schmidt 2007b, 1933[VP] (Effective publication: Eichorst, Breznak and Schmidt 2007a, 2716.)

ro'se.us. L. masc. adj. *roseus* rose-colored, pink.

Colonies on R2A medium are round and pink.

The characteristics are as described for the genus with the following additional information. Carotenoid concentration increases with increasing oxygen partial pressures.

DNA G+C content (mol%): 59.8 (HPLC).

Type strain: KBS 63, DSM 18391, NRRL B-41598.

Sequence accession no. (16S rRNA gene): DQ660892.

References

Eichorst, S.A., J.A. Breznak and T.M. Schmidt. 2007a. Isolation and characterization of soil bacteria that define *Terriglobus* gen. nov., in the phylum *Acidobacteria*. Appl. Environ. Microbiol. *73*: 2708–2717.

Eichorst, S.A., J.A. Breznak and T.M. Schmidt. 2007b. *In* Validation of the publication of new names and combinations previously effectively published outside the IJSEM. List no. 117. Int. J. Syst. Evol. Microbiol. *57*: 1933–1934.

Inagaki, K., K. Nakahira, K. Mukai, T. Tamura and H. Tanaka. 1998. Gene cloning and characterization of an acidic xylanase from *Acidobacterium capsulatum*. Biosci. Biotechnol. Biochem. *62*: 1061–1067.

Inagaki, K., N. Ueno, T. Tamura and H. Tanaka. 2001. Purification and characterization of an acid trehalase from *Acidobacterium capsulatum*. J. Biosci. Bioeng. *91*. 141–146.

Kishimoto, N. and T. Tano. 1987. Acidophilic heterotrophic bacteria isolated from acidic mine drainage, sewage, and soils. J. Gen. Appl. Microbiol. *33*: 11–25.

Kishimoto, N., K. Inagaki, T. Sugio and T. Tano. 1991a. Purification and properties of an acidic β-glucosidase from *Acidobacterium capsulatum*. J. Ferment. Bioeng. *71*: 318–321.

Kishimoto, N., K. Inagaki, T. Sugio and T. Tano. 1991b. *In* Validation of the publication of new names and combinations previously effectively published outside the IJSB. List no. 38. Int. J. Syst. Bacteriol. *41*: 456–457.

Kishimoto, N., Y. Kosako and T. Tano. 1991c. *Acidobacterium capsulatum* gen. nov., sp. nov.: an acidophilic chemoorganotrophic bacterium containing menaquinone from acidic mineral environment. Curr. Microbiol. *22*: 1–7.

Koch, I.H., F. Gich, P.F. Dunfield and J. Overmann. 2008. *Edaphobacter modestus* gen. nov., sp. nov., and *Edaphobacter aggregans* sp. nov., acidobacteria isolated from alpine and forest soils. Int. J. Syst. Evol. Microbiol. *58*: 1114–1122.

Sait, M., K.E.R. Davis and P.H. Janssen. 2006. Effect of pH on isolation and distribution of members of subdivision 1 of the phylum *Acidobacteria* occurring in soil. Appl. Environ. Microbiol. *72*: 1852–1857.

Class II. **Holophagae** Fukunaga, Kurahashi, Yanagi, Yokota and Harayama 2008, 2601^VP^

J. Cameron Thrash and John D. Coates

Ho.lo.pha′gae. N.L. fem. n. *Holophaga* type genus of the type order *Holophagales*; *-ae* ending to designate a class; N.L. fem. pl. n. *Holophagae* the class of *Holophagales*.

The class *Holophagae* was designated by Fukunaga et al. (2008) on the basis of phylogenetic analysis of 16S rRNA gene sequences. It is equivalent to subdivision 8 previously defined by Hugenholtz et al. (1998) and contains the orders *Holophagales* and *Acanthopleuribacterales* (Fukunaga et al., 2008).

Type order: **Holophagales** Fukunaga, Kurahashi, Yanagi, Yokota and Harayama 2008, 2600^VP^.

References

Fukunaga, Y., M. Kurahashi, K. Yanagi, A. Yokota and S. Harayama. 2008. *Acanthopleuribacter pedis* gen. nov., sp. nov., a marine bacterium isolated from a chiton, and description of *Acanthopleuribacteraceae* fam. nov., *Acanthopleuribacterales* ord. nov., *Holophagaceae* fam. nov., *Holophagales* ord. nov. and *Holophagae* classis nov. in the phylum '*Acidobacteria*'. Int. J. Syst. Evol. Microbiol. *58*: 2597–2601.

Hugenholtz, P., C. Pitulle, K.L. Hershberger and N.R. Pace. 1998. Novel division level bacterial diversity in a Yellowstone hot spring. J. Bacteriol. *180*: 366–376.

Order I. **Holophagales** Fukunaga, Kurahashi, Yanagi, Yokota and Harayama 2008, 2600^VP^

J. Cameron Thrash and John D. Coates

Ho.lo.pha.ga′les. N.L. fem. n. *Holophaga* type genus of the order; *-ales* ending to denote order; N.L. fem. pl. n. *Holophagales* the *Holophaga* order.

The order *Holophagales* was designated by Fukunaga et al. (2008) on the basis of phylogenetic analysis of 16S rRNA gene sequences; the order contains the family *Holophagaceae*.

Type genus: **Holophaga** Liesack, Bak, Kreft and Stackebrandt 1995, 197^VP^ (Effective publication: Liesack, Bak, Kreft and Stackebrandt 1994, 88).

References

Fukunaga, Y., M. Kurahashi, K. Yanagi, A. Yokota and S. Harayama. 2008. *Acanthopleuribacter pedis* gen. nov., sp. nov., a marine bacterium isolated from a chiton, and description of *Acanthopleuribacteraceae* fam. nov., *Acanthopleuribacterales* ord. nov., *Holophagaceae* fam. nov., *Holophagales* ord. nov. and *Holophagae* classis nov. in the phylum '*Acidobacteria*'. Int. J. Syst. Evol. Microbiol. *58*: 2597–2601.

Liesack, W., F. Bak, J.U. Kreft and E. Stackebrandt. 1994. *Holophaga foetida* gen. nov., sp. nov., a new, homoacetogenic bacterium degrading methoxylated aromatic compounds. Arch. Microbiol. *162*: 85–90.

Liesack, W., F. Bak, J.U. Kreft and E. Stackebrandt. 1995. *In* Validation of the publication of new names and combinations previously effectively published outside the IJSB. List no. 52. Int. J. Syst. Bacteriol. *45*: 197–198.

Family I. **Holophagaceae** Fukunaga, Kurahashi, Yanagi, Yokota and Harayama 2008, 2600^{VP}

J. Cameron Thrash and John D. Coates

Ho.lo.pha.ga′ce.ae. N.L. fem. n. *Holophaga* type genus of the family; suff. *-aceae* denoting family; N.L. fem. pl. n. *Holophagaceae* the *Holophaga* family.

The family *Holophagaceae* was designated by Fukunaga et al. (2008) on the basis of phylogenetic analysis of 16S rRNA gene sequences; the family contains the genera *Holophaga* (type genus) and *Geothrix*. All known species grow as **chemoorganotrophs**.

Type genus: **Holophaga** Liesack, Bak, Kreft and Stackebrandt 1995, 197^{VP} (Effective publication: Liesack, Bak, Kreft and Stackebrandt 1994, 85–90.).

Genus I. **Holophaga** Liesack, Bak, Kreft and Stackebrandt 1995, 197^{VP} (Effective publication: Liesack, Bak, Kreft and Stackebrandt 1994, 88.)

J. Cameron Thrash and John D. Coates

Ho.lo′pha.ga. Gr. adj. *holos* entire; Gr. v. *phagein* to eat; N.L. fem. n. *Holophaga* eating it all.

Rod-shaped cells, 0.5–0.7 μm × 1–3 μm. Gram-stain-negative. Nonphotosynthetic. Nonsporeforming. Anaerobic, having **strictly anaerobic** metabolism that is **obligately fermentative**. Ferments trimethoxybenzoate and syringate to dimethylsulfide and methanediol. Ferments pyruvate and trihydroxybenzenes such as pyrogallol and gallate, as well as methylated derivatives of hydroxybenzene including 3,4,5-trimethoxybenzoate (TMB). Aromatics are fermented to acetate. Optimal growth is at 28–32°C. No growth below 10°C or above 35°C. Optimum growth from pH 6.8 to 7.5. No growth below pH 5.5 or above pH 8.0. Growth with NaCl from 0.1 to 1.5%. Nonmotile. Contains *c*-type cytochromes.

Source: anoxic mud near Konstanz, Germany.

DNA G+C content (mol%): 62.5.

Type species: **Holophaga foetida** Liesack, Bak, Kreft and Stackebrandt 1995, 197^{VP} (Effective publication: Liesack, Bak, Kreft and Stackebrandt 1994, 89.).

Further descriptive information

Only the following compounds are known to support growth: TMB, syringate, 5-hydroxyvanillate, phloroglucinolmonomethylether, sinapate, ferulate, caffeate, gallate, 2,4,6-trihydroxybenzoate, pyrogallol, phloroglucinol, and pyruvate. Cells also utilize the methyl substituents of methylated aromatic compounds. Methyl substituents are subject to simultaneous or alternative homoacetogenic metabolism. If sulfide is present in the media, methlylated sulfides are formed. Trihydroxybenzenes are degraded through the phloroglucinol pathway.

The only species in the genus was isolated from black anoxic mud near Konstanz, Germany.

Enrichment and isolation procedures

The type strain was isolated with deep-agar dilutions from black anoxic ditch mud near Konstanz, Germany. Mud was added to liquid media dilution series supplemented with 3 mM TMB or syringate (Bak et al., 1992). Dilutions of 10^{-4} and higher showed positive growth and were successfully subcultured and transferred to deep-agar tubes for isolation. All enrichment procedures were carried out under anoxic conditions at 28°C in the dark.

Maintenance procedures

The type strain is maintained as lyophilized culture at the DSMZ.

Differentiation of the genus *Holophaga* from other genera

Holophaga is anaerobic and obligately fermentative, distinguishing it from the respiratory *Acanthopleuribacter*, and *Geothrix*. *Holophaga* can complete methyl-transfer reactions to sulfide generating methanethiol and dimethylsulfide. 16S rRNA gene sequence divergence also distinguishes *Holophaga* from *Geothrix* (94%), and *Acanthopleuribacter* (84%).

List of species of the genus *Holophaga*

1. **Holophaga foetida** Liesack, Bak, Kreft and Stackebrandt 1995, 197^{VP} (Effective publication: Liesack, Bak, Kreft and Stackebrandt 1994, 89.)

   foe.ti′da. L. fem. adj. *foetida* stinking, here referring to the production of malodorous methanethiol and dimethlysulfide.

   Colonies on agar are lens-shaped, translucent, beige. The characteristics are as described for the genus with the following additional information.

   DNA G+C content (mol%): 62.5 (HPLC).

   Type strain: TMBS4, DSM 6591.

   Sequence accession no. (16S rRNA gene): X77215.

Genus II. **Geothrix** Coates, Ellis, Gaw and Lovley 1999, 1620^{VP}

J. Cameron Thrash and John D. Coates

Ge′o.thrix. Gr. n. *gê* earth; Gr. fem. n. *thrix* hair; N.L. fem. n. *Geothrix* hair of earth, refers to the cell morphology under fumarate-reducing conditions.

Rod-shaped cells 0.1 μm × 1–2 μm. Nonphotosynthetic. Nonsporeforming. **Cells occur singly and in chains. Anaerobic**, having an **obligately anaerobic metabolism** that can be **either fermentative or respiratory using nitrate, Mn(IV),**

anthraquinone-2,6-disulfonate (AQDS), poorly crystalline iron(III) oxide, and iron(III) chelated with nitrilotriacetic acid [Fe(III)-NTA] or citrate as alternative electron acceptors. Chemoorganotrophic. Growth on yeast extract and organic acids including acetate and lactate, but not on alcohols including ethanol. Capable of fermenting organic acids including citrate and fumarate. End products of citrate fermentation are acetate and succinate. Optimal growth is 35–40°C. No growth at or below 25°C. Nonmotile. Contains *c*-type cytochromes.

Source: petroleum-contaminated aquifer in Hanahan, South Carolina, USA.

DNA G+C content (mol%): unknown.

Type species: **Geothrix fermentans** Coates, Ellis, Gaw and Lovley 1999, 1620[VP].

Further descriptive information

When growing with fumarate as the sole electron acceptor in liquid medium, cells form long intertwined chains of hundreds of cells. In contrast, when growing on iron(III) as the sole electron acceptor, cells grow singly or in short chains of two to three cells.

The only species in the genus was isolated from petroleum-contaminated aquifer sediment in South Carolina, USA, by enrichment first with toluene and iron(III) NTA and then with acetate and iron(III) NTA.

Enrichment and isolation procedures

The type strain was isolated on iron(III) pyrophosphate/acetate plates from a contaminated aquifer enrichment culture growing on toluene and iron(III) NTA. The original inoculum was from petroleum-contaminated aquifer sediment from the Defense Fuel Supply Center in Hanahan, South Carolina, USA. The enrichment sediments were amended with 2 mmol kg NTA and 10 μM toluene. After sediments were adapted to consistent oxidation of toluene coupled to iron(III) reduction, 1 g of this enrichment sediment was used to inoculate liquid enrichment cultures growing on iron(III) NTA and acetate (Coates et al., 1999). After enrichment, the organism was isolated on iron(III) pyrophosphate/acetate plates where growth resulted in clearing zones in the green-colored iron(III) pyrophosphate around the colonies. All incubations were carried out at 30°C under strictly anaerobic conditions with a N_2/CO_2 (80:20) headspace.

Maintenance procedures

Cultures of *Geothrix* can be maintained on plates or liquid culture with weekly transfers. Long term storage can be done with lyophilization or freezing at −80°C. Cryoprotective agents like 50% glycerol should be added before freezing and must be made anaerobic prior to addition of cells.

Differentiation of the genus *Geothrix* from other genera

Terminal electron-accepting processes distinguish *Geothrix* from *Holophaga* and *Acanthopleuribacter*. *Geothrix* is an obligate anaerobe whereas *Acanthopleuribacter* is an obligate aerobe and *Holophaga* is non-respiratory. Additionally, *Geothrix* can also ferment, *Acanthopleuribacter* does not. Low (94% and 84%) 16S rRNA gene sequence homology distinguishes *Geothrix* from *Holophaga* and *Acanthopleuribacter*, respectively. *Holophaga* is obligately fermentative, whereas *Geothrix* is capable of fermentation and anaerobic respiration of a variety of alternative electron acceptors including iron(III) and nitrate.

List of species of the genus *Geothrix*

1. **Geothrix fermentans** Coates, Ellis, Gaw and Lovley 1999, 1620[VP]

fer.men′tans. L. part. adj. *fermentans* fermenting.

Colonies on iron(III) pyrophosphate are white, less than 1mm in diameter. Colonies are red when grown on fumarate medium.

The characteristics are as described for the genus with the following additional information. Optimum growth temperature is 35°C. Contact is not required for reduction of insoluble iron(III) oxide (Nevin and Lovley, 2002). Both chelating and electron shuttling properties are identified in soluble fractions of insoluble iron(III) oxide-reducing cultures of *Geothrix fermentans* (Nevin and Lovley, 2002).

DNA G+C content (mol%): unknown.

Type strain: H-5, ATCC 700665, DSM 14018.

Sequence accession no. (16S rRNA gene): U41563.

References

Bak, F., K. Finster and F. Rothfuß. 1992. Formation of dimethylsulfide and methanethiol from methoxylated aromatic compounds and inorganic sulfide by newly isolated anaerobic bacteria. Arch. Microbiol. *157*: 529–534.

Coates, J.D., D.J. Ellis, C.V. Gaw and D.R. Lovley. 1999. *Geothrix fermentans* gen. nov., sp. nov., a novel Fe(III)-reducing bacterium from a hydrocarbon-contaminated aquifer. Int. J. Syst. Bacteriol. *49*: 1615–1622.

Fukunaga, Y., M. Kurahashi, K. Yanagi, A. Yokota and S. Harayama. 2008. *Acanthopleuribacter pedis* gen. nov., sp. nov., a marine bacterium isolated from a chiton, and description of *Acanthopleuribacteraceae* fam. nov., *Acanthopleuribacterales* ord. nov., *Holophagaceae* fam. nov., *Holophagales* ord. nov. and *Holophagae* classis nov. in the phylum '*Acidobacteria*'. Int. J. Syst. Evol. Microbiol. *58*: 2597–2601.

Liesack, W., F. Bak, J.U. Kreft and E. Stackebrandt. 1994. *Holophaga foetida* gen. nov., sp. nov., a new, homoacetogenic bacterium degrading methoxylated aromatic compounds. Arch. Microbiol. *162*: 85–90.

Liesack, W., F. Bak, J. U. Kreft and E. Stackebrandt. 1995. *In* Validation of the publication of new names and combinations previously effectively published outside the IJSB. List no. 52. Int. J. Syst. Bacteriol. *45*: 197–198.

Nevin, K.P. and D.R. Lovley. 2002. Mechanisms for accessing insoluble Fe(III) oxide during dissimilatory Fe(III) reduction by *Geothrix fermentans*. Appl. Environ. Microbiol. *68*: 2294–2299.

Order II. **Acanthopleuribacterales** Fukunaga, Kurahashi, Yanagi, Yokota and Harayama 2008, 2600[VP]

J. Cameron Thrash and John D. Coates

A.can'tho.pleu.ri.bac'te.ra'les. N.L. masc. n. *Acanthopleuribacter* type genus of the order; *-ales* ending to denote order; N.L. fem. pl. n. *Acanthopleuribacterales* the *Acanthopleuribacter* order.

The order *Acanthopleuribacterales* was designated by Fukunaga et al. (2008) on the basis of phylogenetic analysis of 16S rRNA gene sequences; the order contains the family *Acanthopleuribacteraceae*.

Type genus: **Acanthopleuribacter** Fukunaga, Kurahashi, Yanagi, Yokota and Harayama 2008, 2600[VP].

Reference

Fukunaga, Y., M. Kurahashi, K. Yanagi, A. Yokota and S. Harayama. 2008. *Acanthopleuribacter pedis* gen. nov., sp. nov., a marine bacterium isolated from a chiton, and description of *Acanthopleuribacteraceae* fam. nov., *Acanthopleuribacterales* ord. nov., *Holophagaceae* fam. nov., *Holophagales* ord. nov. and *Holophagae* classis nov. in the phylum '*Acidobacteria*'. Int. J. Syst. Evol. Microbiol. *58*: 2597–2601.

Family I. **Acanthopleuribacteraceae** Fukunaga, Kurahashi, Yanagi, Yokota and Harayama 2008, 2600[VP]

J. Cameron Thrash and John D. Coates

A.can'tho.pleu.ri.bac'te.ra'ce.ae. N.L. masc. n. *Acanthopleuribacter* type genus of the family; suff. *-aceae* denoting family; N.L. fem. pl. n. *Acanthopleuribacteraceae* the *Acanthopleuribacter* family.

The family *Acanthopleuribacteraceae* was designated by Fukunaga et al. (2008) on the basis of phylogenetic analysis of 16S rRNA gene sequences; the family contains the sole genus *Acanthopleuribacter*.

Type genus: **Acanthopleuribacter** Fukunaga, Kurahashi, Yanagi, Yokota and Harayama 2008, 2600[VP].

Genus I. **Acanthopleuribacter** Fukunaga, Kurahashi, Yanagi, Yokota and Harayama 2008, 2600[VP]

J. Cameron Thrash and John D. Coates

A.can'tho.pleu.ri.bac'ter. N.L. n. *Acanthopleura* mollusk genus designation; N.L. masc. n. *bacter* a rod; N.L. masc. n. *Acanthopleuribacter* a rod from *Acanthopleura*, in reference to the source of the type strain, the chiton *Acanthopleura japonica*.

Rod-shaped cells, 0.7–1.0 μm × 2.4–4.7 μm. Gram-negative by TEM inspection. **Aerobic**, having an obligately aerobic metabolism. **Mesophilic**. Growth at pH 5.0–9.0, optimally at pH 7.0–8.0. Do not reduce nitrate or nitrite. Chemoorganotrophic. Growth on a variety of amino acids including L-serine, as well as α-D-glucose as the sole carbon source. Growth on marine agar from 15 to 30°C with optimum growth at 30°C. **Motile by peritrichous flagella**. The major quinones are menaquinone with six and seven isoprene units (MK-6, MK-8). The major fatty acids are $C_{15:0}$ iso, $C_{17:0}$ iso, $C_{16:0}$; major hydroxy fatty acids are $C_{13:0}$ iso 3-OH and $C_{17:0}$ iso 3-OH.

Source: Acanthopleura japonica chiton from the Boso peninsula, Chiba, Japan.

DNA G+C content (mol%): 56.7.

Type species: **Acanthopleuribacter pedis** Fukunaga, Kurahashi, Yanagi, Yokota and Harayama 2008, 2600[VP].

Further descriptive information

As seen by electron microscopy, cells have peritrichous flagella and gram-negative cell wall structure.

The only species in the genus was isolated from foot tissue of a *Acanthopleura japonica* chiton taken from a beach near the Natural History Museum and Institute at Katsuura, Boso peninsula, Chiba, Japan.

Enrichment and isolation procedures

The type strain was isolated from a *Acanthopleura japonica* chiton. The chiton was washed in sterile artificial seawater (ASW, Naigai Chemical Products), dissected, and tissues homogenized to serve as inoculum for serial dilutions in sterile ASW. Dilutions yielding cells were plated on 1:5 dilution marine agar 2216 (Difco) and incubated at 20°C for 8 d.

Maintenance procedures

The type strain is maintained as a lyophilized culture at the NBRC.

Differentiation of the genus *Acanthopleuribacter* from other genera

Acanthopleuribacter is obligately aerobic whereas *Geothrix* and *Holophaga* are obligately anaerobic. *Acanthopleuribacter* is strictly respiratory, whereas *Geothrix* is respiratory and fermentative and *Holophaga* is strictly fermentative. *Acanthopleuribacter* has 84% 16S rRNA gene sequence identity with *Geothrix* and *Holophaga*.

List of species of the genus *Acanthopleuribacter*

1. **Acanthopleuribacter pedis** Fukunaga, Kurahashi, Yanagi, Yokota and Harayama 2008, 2600[VP]

pe′dis. L. gen. n. *pedis* of the foot, referring to the chiton foot from which the type strain was isolated.

Colonies on marine agar plates are smooth circular, and yellow.

The characteristics are as described for the genus with the following additional information. Cells contain the following major fatty acids: $C_{11:0}$ iso, $C_{13:0}$ iso, $C_{14:0}$, $C_{15:0}$ iso, $C_{16:0}$, $C_{16:0}$ N alcohol, $C_{17:0}$ iso, $C_{16:1}$ ω9*c*, $C_{17:1}$ ω10*c* iso, $C_{20:5}$ ω3*c*, $C_{13:0}$ iso 3-OH and $C_{17:0}$ iso 3-OH. Cells also have catalase, oxidase, protease, β-glucosidase, alkaline phosphatase, leucine arylamidase, valine arylamidase, trypsin, acid phosphatase, naphthol-AS-BI-phosphohydrolase, *N*-acetyl-β-glucosaminidase, esterase C4, ester lipase C8, and cystine arylamidase activity.

DNA G+C content (mol%): 56.7 (HPLC).

Type strain: FYK2218, NRBC 101209, KCTC = 12899.

Sequence accession no. (16S rRNA gene): AB303221.

Reference

Fukunaga, Y., M. Kurahashi, K. Yanagi, A. Yokota and S. Harayama. 2008. *Acanthopleuribacter pedis* gen. nov., sp. nov., a marine bacterium isolated from a chiton, and description of *Acanthopleuribacteraceae* fam. nov., *Acanthopleuribacterales* ord. nov., *Holophagaceae* fam. nov., *Holophagales* ord. nov. and *Holophagae* classis nov. in the phylum '*Acidobacteria*'. Int. J. Syst. Evol. Microbiol. *58*: 2597–2601.

Phylum XVIII. Fibrobacteres Garrity and Holt 2001

Anne M. Spain, Cecil W. Forsberg and Lee R. Krumholz

Fi'bro.bac.ter'es. N.L. masc. n. *Fibrobacter* type genus of the type order; *-es* ending to denote phylum N.L. fem. pl. n. *Fibrobacteres* the *Fibrobacter* phylum.

The phylum *Fibrobacteres* currently consists of three classes circumscribed on the basis of phylogenetic analysis of 16S rRNA gene sequences, including one cultivated class, *Fibrobacteria* class. nov. *Fibrobacterales* is the type order, and contains a single family and genus.

The phylum *Fibrobacteres* is most closely related to the *Bacteroidetes* [*Cytophaga-Flavobacterium-Bacteroides* (CFB)] phylum based on signature sequences of proteins (Griffiths and Gupta, 2001; Gupta, 2004; Gupta and Lorenzini, 2007). Phylogenic analyses based on 16S rRNA gene sequences support the relatedness of *Fibrobacteres* to this phylum, but indicate it is even more closely related to a candidate division TG3 (Hongoh et al., 2005, 2006a), and also shares a common ancestor with *Gemmatimonadetes* (Figure 118). The phylum contains three classes (Table 146, Figure 118), only one of which contains cultivated isolates and is formally named and described herein.

Although all of the currently extant species were isolated from the rumen or other locations of the gastrointestinal tract, 16S rRNA genes related to the *Fibrobacteres* have been observed in a number of environments (Table 146, Figure 119). It is therefore likely that this phylum is much more broadly distributed in the environment than indicated on the basis of the habitat of the pure culture representatives. Clones within the class *Fibrobacteria* (but not belonging to *Fibrobacterales*) were detected in an acid-impacted lake and a sulfidic cave stream biofilm (Figure 119), as well as within the termite species *Macrotermes gilvus* (Hongoh et al., 2006b). The class-level lineage *Fibrobacteria*-2 was originally named a subphylum of *Fibrobacteres* and is solely composed of clone sequences derived from the gut of different species of termites (Hongoh et al., 2006a). The third class-level lineage of *Fibrobacteria*, denoted here as "Environmental clones", consists of clones detected in both soil and water downstream of manure (Figure 119). Although only three sequences are shown in Figure 119, unpublished sequences of <1000 bp belonging to this class-level lineage have been deposited in GenBank and also come from soil and water sources.

Type order: **Fibrobacterales** ord. nov.

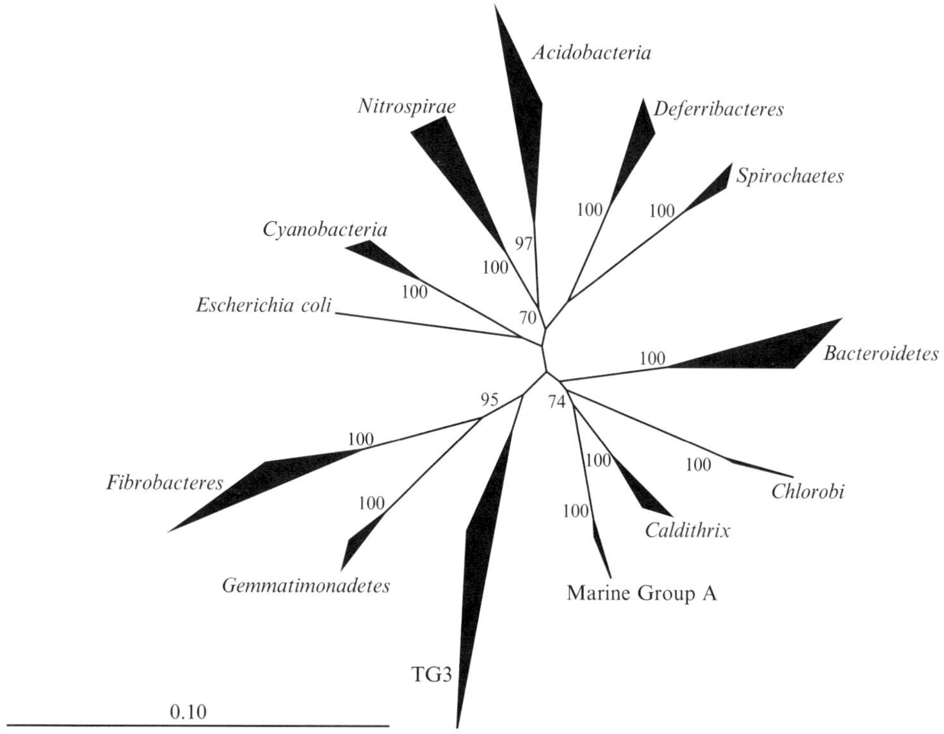

FIGURE 118. Maximum-likelihood tree of the phylum *Fibrobacteres* in relation to closely related phyla. The tree was constructed from two or more aligned nearly full-length (>1300 bp) 16S rRNA gene sequences from each phylum-level lineage using the fastDNAml method. Bootstrap values >50 are shown on the left or above corresponding branches and are based on 1000 replicates.

TABLE 146. Taxonomic divisions within the phylum *Fibrobacteres*

Lineage	Delineation [Max. (mean ± SD)][a] Within lineage	To *Fibrobacter succinogenes* str. S85[c]	Isolate sources[b] Environmental clones	Cultivated isolates
Phylum. *Fibrobacteres*	0.29 (0.18 ± 0.06)	0.25 (0.15 ± 0.07)		
Class I. *Fibrobacteria*	0.18 (0.10 ± 0.06)	0.16 (0.08 ± 0.06)		
Order I. *Fibrobacterales*	0.11 (0.07 ± 0.03)	0.09 (0.05 ± 0.03)		
Family I. *Fibrobacteraceae*				
Genus I. *Fibrobacter*				
F. succinogenes	0.06 (0.05 ± 0.01)	0.06 (0.04 ± 0.02)	Equine cecum Bovine rumen	Ovine rumen Bovine rumen
F. intestinalis	0.04	0.09 (0.09 ± 0.00)	na	Ovine rumen Rat cecum Porcine cecum
Unclassified *Fibrobacteria*	0.18 (0.14 ± 0.03)	0.16 (0.13 ± 0.02)	Termite gut Sulfidic cave stream biofilm Acid-impacted lake	na
Class II. *Fibrobacteria-2*	0.14 (0.09 ± 0.02)	0.22 (0.20 ± 0.01)	Termite gut	na
Class III. Environmental clones	0.19 (0.14 ± 0.04)	0.25 (0.22 ± 0.02)	Soil and water[d]	na

[a]Delineation values are derived from a distance matrix of all sequences in Figure 118, based on Jukes–Cantor-corrected distances. Maximum delineation values are the maximum pairwise distances between any two sequences within a given lineage; mean delineation values are the mean distances between any two sequences from Figure 118 within a given lineage.

[b]na, Not applicable.

[c]Pairwise distances between sequences from all lineages were compared to that from *Fibrobacter succinogenes* str. S85, used as a reference.

[d]Water source was 10 m downstream from manure application site.

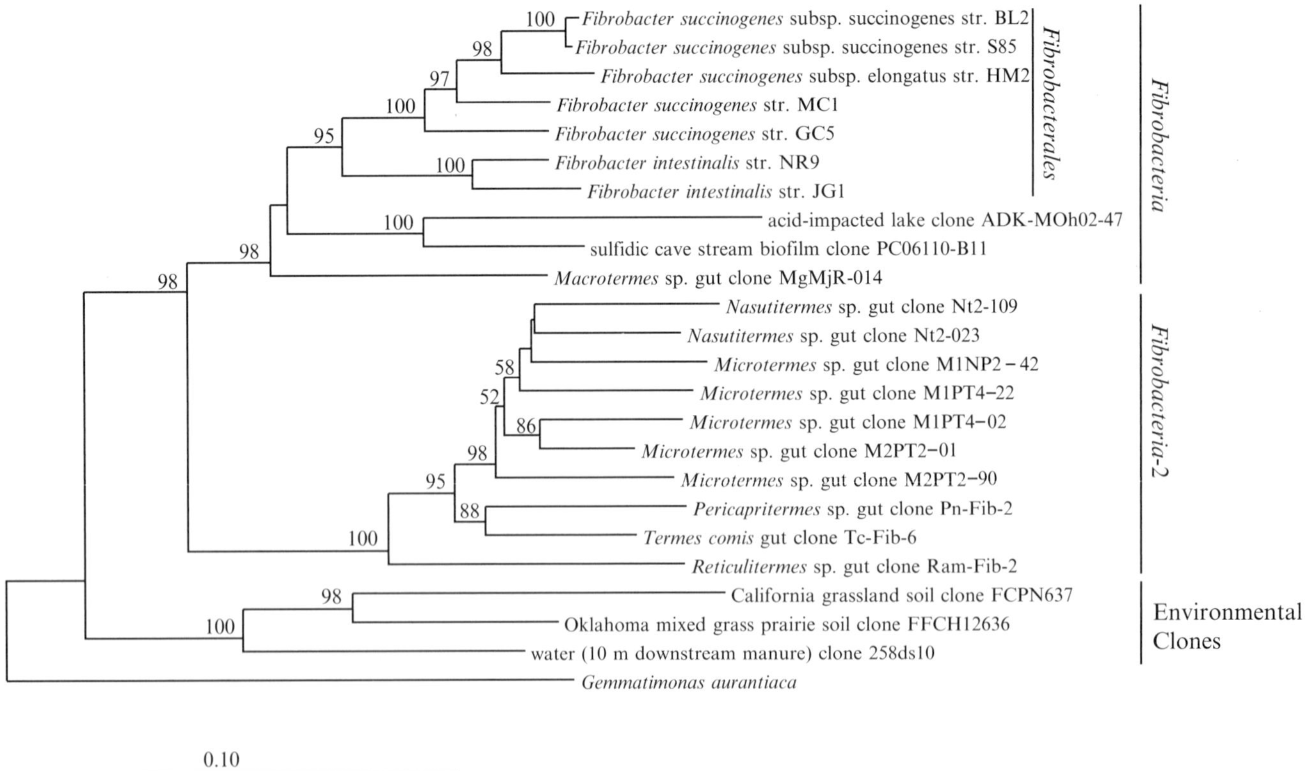

FIGURE 119. Distance phylogram of species within the three classes of the phylum *Fibrobacteres*. Environmental clone sequences are only included in lineages with no cultivated representatives. The tree was constructed from nearly full-length (>1300 bp) 16S rRNA gene sequences using the neighbor-joining algorithm with Jukes–Cantor corrected distances. Bootstrap values >50 are shown above corresponding branches and are based on 1000 replicates.

References

Garrity, G.M. and J.G. Holt. 2001. The Road Map to the Manual. *In* Bergey's Manual of Systematic Bacteriology, 2nd edn, vol. 1, The *Archaea* and the Deeply Branching and Phototrophic *Bacteria* (edited by Boone, Castenholz and Garrity). Springer, New York, pp. 119–166.

Griffiths, E. and R.S. Gupta. 2001. The use of signature sequences in different proteins to determine the relative branching order of bacterial divisions: evidence that *Fibrobacter* diverged at a similar time to *Chlamydia* and the *Cytophaga–Flavobacterium–Bacteroides* division. Microbiology *147*: 2611–2622.

Gupta, R.S. 2004. The phylogeny and signature sequences characteristics of *Fibrobacteres*, *Chlorobi*, and *Bacteroidetes*. Crit. Rev. Microbiol. *30*: 123–143.

Gupta, R.S. and E. Lorenzini. 2007. Phylogeny and molecular signatures (conserved proteins and indels) that are specific for the *Bacteroidetes* and *Chlorobi* species. BMC Evol. Biol. *7*: 71.

Hongoh, Y., P. Deevong, T. Inoue, S. Moriya, S. Trakulnaleamsai, M. Ohkuma, C. Vongkaluang, N. Noparatnaraporn and T. Kudo. 2005. Intra- and interspecific comparisons of bacterial diversity and community structure support coevolution of gut microbiota and termite host. Appl. Environ. Microbiol. *71*: 6590–6599.

Hongoh, Y., P. Deevong, S. Hattori, T. Inoue, S. Noda, N. Noparatnaraporn, T. Kudo and M. Ohkuma. 2006a. Phylogenetic diversity, localization, and cell morphologies of members of the candidate phylum TG3 and a subphylum in the phylum *Fibrobacteres*, recently discovered bacterial groups dominant in termite guts. Appl. Environ. Microbiol. *72*: 6780–6788.

Hongoh, Y., L. Ekpornprasit, T. Inoue, S. Moriya, S. Trakulnaleamsai, M. Ohkuma, N. Noparatnaraporn and T. Kudo. 2006b. Intracolony variation of bacterial gut microbiota among castes and ages in the fungus-growing termite *Macrotermes gilvus*. Mol. Ecol. *15*: 505–516.

Class I. **Fibrobacteria** class. nov.

Anne M. Spain, Cecil W. Forsberg and Lee R. Krumholz

Fi.bro.bac.te′ri.a. N.L. masc. n. *Fibrobacter* type genus of the type order of the class; suff. *-ia* ending proposed by Gibbons and Murray and by Stackebrandt et al. to denote a class; N.L. neut. pl. n. *Fibrobacteria* the *Fibrobacter* class.

The class *Fibrobacteria* is circumscribed here on the basis of 16S rRNA sequences. The class contains the single type order *Fibrobacterales* ord. nov.

Order I. **Fibrobacterales** ord. nov.

Anne M. Spain, Cecil W. Forsberg and Lee R. Krumholz

Fi.bro.bac.ter.a′les. N.L. masc. n. *Fibrobacter* type genus of the order; *-ales* ending to denote order; N.L. fem. pl. n. *Fibrobacterales* the *Fibrobacter* order.

The order *Fibrobacterales* is circumscribed herein on the basis of phylogenetic sequences. The order contains the sole family, *Fibrobacteraceae* fam.nov. Cells are rod-shaped of varying lengths and are capable of growth on cellulose. Members of the genus *Fibrobacter* are anaerobic.

Type genus: **Fibrobacter** Montgomery, Flesher and Stahl 1988, 434[VP].

Reference

Montgomery, L., B. Flesher and D. Stahl. 1988. Transfer of *Bacteroides succinogenes* (Hungate) to *Fibrobacter* gen. nov. as *Fibrobacter succinogenes* comb. nov. and description of *Fibrobacter intestinalis* sp. nov. Int. J. Syst. Bacteriol. *38*: 430–435.

Family I. **Fibrobacteraceae** fam. nov.

Anne M. Spain, Cecil W. Forsberg and Lee R. Krumholz

Fi.bro.bac.ter.a′ceae. N.L. masc. n. *Fibrobacter* type genus of the family, *-aceae* ending to denote family; N.L. fem. pl. n. *Fibrobacteraceae* the *Fibrobacter* family.

Cells are Gram-stain-negative obligately anaerobic and rod-to-ovoid in shape depending on the strain and the culture conditions. Older cultures form sphaeroplasts. Motility appears to occur by gliding, but cannot be observed microscopically. Flagella have not been detected. Membranes are composed of straight-chain fatty acids and phospholipids are predominantly ethanolamine plasmalogens.

All of the strains grow by degrading cellulose and cellulose containing plant compounds. Cells will also grow on the soluble sugars, glucose, or cellobiose. The major fermentation products are succinate and acetate. Although bacteria will grow on soluble forms of cellulose, they are thought to exist primarily as cells that are directly attached to decomposing plant material.

Type genus: **Fibrobacter** Montgomery, Flesher and Stahl 1988, 434[VP].

Genus I. **Fibrobacter** Montgomery, Flesher and Stahl 1988, 434[VP]

Anne M. Spain, Cecil W. Forsberg and Lee R. Krumholz

Fi'bro.bac'ter. L. fem. n. *fibra* fiber or filament in plants or animals; N.L. masc. n. *bacter* rod or staff; N.L. masc n. *Fibrobacter* bacterial rod that subsists on fiber.

Cells are rod-shaped (0.3–0.5 μm × 0.8–1.6 μm) or ovoid (0.8–1.6 μm × 0.8–1.6 μm). Obligately anaerobic non-sporing and not detectably motile by microscopy. They are able to migrate through agar when growing on cellulose, suggesting a gliding form of motility. Ferment a narrow range of carbohydrates including glucose, cellobiose, and cellulose. Other sugars including lactose or maltose are fermented by a few species. The fermentation products are acetate and succinate, and sometimes formate at low levels. Cells require CO_2, straight-chain and branched-chain fatty acids in the media as well as ammonia as the N source. Membranes are composed of straight-chain fatty acids, and phospholipids are predominantly ethanolamine plasmalogens. Habitat is the mammalian gastrointestinal tract.

DNA G+C content (mol%): 45–51.

Type species: **Fibrobacter succinogenes** Montgomery, Flesher and Stahl 1988, 434[VP].

Taxonomic comments and further descriptive information

The first pure culture of *Fibrobacter* was described as *Bacteroides succinogenes* by (Hungate, 1950). Prévot (1966) considered *Bacteroides succinogenes* to belong in the genus *Ruminobacter*, but the name *Ruminobacter succinogenes* was not validly published. More recently, phylogenetic analysis showed that strains of *Bacteroides succinogenes* were phylogenetically distinct from other species of *Bacteroides* necessitating the formation of a new genus (Montgomery et al., 1988). Sequences of small-subunit rRNA of several strains were shown to have less than 72% similarity with *Bacteroides fragilis* providing evidence that these organisms constituted a distinct evolutionary line of descent at the phylum level.

There are currently no phenotypic characteristics that are useful for distinguishing the two species *Fibrobacter intestinalis* and *Fibrobacter succinogenes*. Small-subunit rRNA analysis must be used. *Fibrobacter succinogenes* subsp. *succinogenes* can be distinguished from *Fibrobacter succinogenes* subsp. *elongatus* based on cell morphology, the former being ovoid and the latter more slender and rod-shaped.

Within the genus, strains of *Fibrobacter succinogenes* subsp. *succinogenes* and *Fibrobacter succinogenes* subsp. *elongatus* have a 16S rRNA sequence similarity of 95.3–98.1% (Amann et al., 1992) and DNA hybridization of less than 20%. Between the two species, the 16S rRNA similarity is 91.8–92.9%, with a DNA hybridization of less than 10%.

Enrichment and isolation procedures

All of the isolates to date have been obtained from the gastrointestinal (GI) tract and have been described as two species within *Fibrobacter*. These organisms were isolated by directly diluting GI tract contents into roll tubes containing media with milled filter paper as substrate. Hungate originally isolated *Fibrobacter succinogenes* using an agar mineral medium containing clarified rumen fluid (a source of nutrients) and acid treated milled cotton (Hungate, 1950) in a mineral solution containing bicarbonate and a reducing agent. This original medium contained 1–2% agar and involved shake or roll tubes. Bryant and Doetsch (1954) later isolated a variety of strains from the rumen, most likely using similar techniques. Much more recently, an effective technique was described by Macy et al. (1982), which involves diluting the contents of the rat cecum using a mineral solution buffered with 0.5% $NaHCO_3$ and reduced with cysteine and sulfide. Dilutions were then used as inoculum for a similar agar medium containing 0.5% agar and 0.5% pebble-milled Whatman no. 1 filter paper. Each serum tube (1.6 × 15 cm) contained a low volume (3 ml) of agar medium. Tubes were then cooled on a tube roller producing an extremely thin agar layer. Tubes were then incubated at 39°C for several days. Cells do not form colonies under these conditions, but the clearings due to cellulose hydrolysis can be easily visualized in this soft thin agar. Cells were picked from the cleared areas and transferred to a rich sugar based medium.

Alternatively, *Fibrobacter succinogenes* can be isolated from enrichment cultures (Stewart et al., 1981). The enrichment media contained minerals, vitamins, fatty acids, and dewaxed cotton (0.25%). After approximately five transfers in the enrichment media, colonies of *Fibrobacter succinogenes* are isolated by diluting into roll tubes containing a cellobiose-based medium.

The above isolation techniques are tedious and also technically challenging. As such, they have not been used commonly outside of the rumen microbiology field. It is likely that this is the main reason why there are no isolates from any environment other than the GI tract.

Miscellaneous comments

Accessibility to cellulose digestion sites in the cell-wall matrix has been suggested as the rate-limiting factor in cellulose digestion (Dehority and Tirabasso, 1998). It has been proposed that access to cell-wall polymers is limited by the small pore size between polymers, which is on the order of 2–4 nm. This size is not sufficient to allow free diffusion of simple globular enzymes with masses greater than 20 kDa into the wall matrix (Gardner et al., 1999). Since the plant cell wall is a matrix of different polymers, a combination of cellulase and hemicellulase enzymes acting simultaneously is therefore essential. It was originally assumed that *Fibrobacter succinogenes* cells contained an array of the key enzymes for plant cell-wall biodegradation because it can burrow into the plant cell-wall matrix (Cheng et al., 1983). Insight into this conclusion was clearly documented (Matulova et al., 2005) by growth of *Fibrobacter succinogenes* strain S85 on ^{13}C enriched wheat straw and by assessment of the products of hydrolysis using a combination of nuclear magnetic resonance spectroscopy, and sugar linkage and compositional analysis. They observed the absence of acetylated xylooligosaccharides among the hydrolysis products despite the highly acetylated state of wheat straw cell-wall materials (Bourquin and Fahey, 1994), thereby documenting extensive enzymic deacetylation. Free sugars and polymers that accumulated in the culture fluid were xylose, arabinose, and arabinoglucuronoxylan oligosaccharides indicating extensive hemicellulase action. They also observed simultaneous degradation of hemicellulose and cellulose, and furthermore, amorphous and crystalline regions of cellulose were degraded at the same rate, which supported the concept of the concerted action of numerous enzymes in the surface degradation of the cell walls. Glucose did not accumulate in the medium, indicating rapid utilization of plant cell oligosaccharides with minimal cellodextrin export and recycling as previously suggested (Wells et al., 1995). The accumulation of xylose and arabinose occurred as was expected since they are not used as a carbon source (Matte and Forsberg, 1992; Matte et al., 1992). However, 4-OMe-α-glucuronic acid was not detected despite the presence of α-glucuronidase (Smith and Forsberg, 1991), which may be explained by low activity of the enzyme. This research clearly documented that digestion of the cell-wall matrix by *Fibrobacter succinogenes* involves the interaction of a complex array of fibrolytic enzymes.

Enzymology and cloning studies prior to sequencing the genome of *Fibrobacter succinogenes* S85 documented the presence of seven endoglucanases, as well as a cellodextrinase, a chloride-stimulated cellobiosidase, a lichenase, and a α-glucuronidase (Forsberg et al., 2000). Added to this array of enzymes were at least three xylanases (Jun et al., 2003), an arabinose debranching xylanase (Matte and Forsberg, 1992), three acetyl xylan esterases (Kam et al., 2005; McDermid et al., 1990a), an arabinofuranosidase, and a ferulic acid esterase (McDermid et al., 1990b). The debranching nature of both the acetylxylan esterase and the xylanase clearly have very important roles in plant cell-wall biodegradation because they improve access of other enzymes to previously unavailable substrate.

Bera-Maillet et al. (2004) conducted experiments to determine whether ten glycosyl hydrolase genes, previously cloned from *Fibrobacter succinogenes* S85, were present in other strains of *Fibrobacter succinogenes* and in *Fibrobacter intestinalis* strain NR9. Almost all of the glycosyl hydrolase genes were detected in strains of *Fibrobacter succinogenes* closely related to strain S85, and a few were present in *Fibrobacter intestinalis* NR9. Only *celF*, a member of glycosyl hydrolase family 51, a 118-kDa endoglucanase of carbohydrate binding module families 11 and 30, was detected in all strains of *Fibrobacter succinogenes* and in *Fibrobacter intestinalis* NR9. A concluding remark from this study was that *Fibrobacter succinogenes* strain S85 is a representative model of the species for studying the mechanism of cellulose and plant cell-wall digestion.

To elucidate the role of adhesion of *Fibrobacter* cells to microcrystalline cellulose in the digestion of cellulose, adhesion defective mutants were isolated from both *Fibrobacter succinogenes* S85 (Gong and Forsberg, 1989) and *Fibrobacter intestinalis* DR7 (Miron and Forsberg, 1998). In both studies it was observed that the mutants either grew more slowly on cellulose or not at all, but their growth on glucose was unaffected, which suggested that adhesion was essential for cellulose digestion.

Sequencing of the genome of *Fibrobacter succinogenes* strain S85 has dramatically modified the approach to studies on the mechanism of plant cell-wall digestion. The genome of *Fibrobacter succinogenes* S85 is 3.8 Mbp and contains 3252 open reading frames putatively encoding proteins. Of these, 120 appear to encode enzymes involved in plant cell-wall biodegradation on the basis of amino acid sequence similarity to plant cell-wall-degrading enzymes of other bacteria and fungi. The degradative enzymes include 91 glycosyl hydrolases, 14 carbohydrate esterases, and nine pectate hydrolases (Morrison et al., 2003). The glycosyl hydrolases include 33 cellulases (in families 5, 8, 9, 10, 44, 45, 51, and 74), and 27 xylanases (mainly in families 8, 10, 11, 30, 39, and 43). Only five of these cellulases have cellulose binding modules (CBMs) while 13 of the xylanases have CBMs. Hemicellulose debranching enzymes include 14 cinnamoyl esterases (ten containing CBMs) and nine pectate lyases (four containing CBMs). From a count of putative plant cell-wall-degrading enzymes, the number present in *Fibrobacter succinogenes* exceeds those present in *Thermobifida fusca*, *Clostridium thermocellum*, and *Cytophaga hutchinsonii*. However, no proteins were found with similarity to scaffoldin, cohesin, or dockerin proteins characteristic of cellulosomal cellulase complexes present in the ruminal bacteria *Ruminococcus albus* and *Ruminococcus flavefaciens* and the cellulolytic ruminal fungi (Doi and Kosugi, 2004; Lynd et al., 2002). Furthermore, there were no genes encoding cellulases from families 6 and 48, which typically contain exoglucanases found in both cellulosomal and non-cellulosomal cellulase systems that degrade crystalline cellulose, for example, those of *Thermobifida fusca* and *Clostridium thermocellum*. Furthermore, in contrast to the cellulase systems of *Thermobifida fusca* and *Clostridium thermocellum*, which once synthesized remain active, that of *Fibrobacter succinogenes* is inactivated upon suspension of growth (Maglione et al., 1997). This further supports the conclusion that the mechanism of cellulose digestion by *Fibrobacter succinogenes* is very different from that of other organisms in nature.

The availability of the genome has not led to an immediate solution of the mechanism of cellulose digestion by *Fibrobacter succinogenes*. Nonetheless, it has greatly enhanced our ability to dissect the cellulase system through the application of one- and two-dimensional electrophoresis and mass spectroscopy of tryptic digests of nanogram quantities of proteins. A total of 11 novel cellulose binding proteins were identified as important candidates for roles in adhesion of cells to cellulose as all were absent from the outer membrane of a mutant strain of *Fibrobacter succinogenes* S85 incapable of binding to cellulose. None of these proteins contained a classic CBM (Jun et al., 2007). Some of these proteins have affinity specifically for amorphous cellulose, as documented, with a second non-adherent mutant that bound to amorphous cellulose, but not to crystalline cellulose, while others are specific for crystalline cellulose. The binding properties of the proteins were determined by mixing detergent-solubilized outer membranes from the two mutants and the wild-type bacterium with amorphous and crystalline cellulose, and analysis of the bound proteins by SDS-PAGE and mass spectrometry. One of these proteins was a pilin, and a role for pili in binding of *Ruminococcus albus* to cellulose has been documented (Pegden et al., 1998). In the same study, 16 proteins induced by growth of *Fibrobacter succinogenes* on cellulose were also identified, and this included Cel10A (FSU0257) a previously characterized chloride-stimulated cellobiosidase (Huang et al., 1988; Jun et al., 2007). A critical question is whether all of the putative glycosyl hydrolases are expressed at a level sufficiently high to have a role in plant cell-wall digestion, or whether only selected hydrolase enzymes are responsible for this phenotype. Several published studies (Jun et al., 2007; Qi et al., 2007, 2008a) and unpublished data have led to the identification of a short list of glycoside hydrolases that are expressed by *Fibrobacter succinogenes* S85 (Table 147). However, this is not the complete list since the nucleotide sequences of several biochemically characterized xylanases (Matte and Forsberg, 1992), acetyl xylan esterases (McDermid et al., 1990a), feruloyl

TABLE 147. Cellulase and hemicellulase genes known to be expressed in *Fibrobacter succinogenes* S85[a,b]

				Modular structure							
FSU#	Name	Mw	IP	SP	CD	CBM	Fn	Ig	BTD	Unknown	Reference(s)
Exoglucanase-like:											
257	Cel10A (ClCBase)	60.2	9.0	+	+	4	–	–	–	–	Qi et al. (2007)
2070	Cel5C (CedA)	40.6	5.3	–	+	–	–	–	–	–	Iyo and Forsberg (1994)
Endoglucanases:											
1947	Cel45C	52.8	5.2	+	+	–	–	–	–	+	Qi et al. (2007)
2005	Cel5B	40.2	4.8	+	+	–	–	–	–		Unpublished
2914	Cel5H	96.8	5.2	+	+	11, 11	–	–	10.8		Qi et al. (2007)
2303	Cel8B	76.5	5.4	+	+	–	–	–	10.3	+	Qi et al. (2007)
809	Cel9H	231.0	4.9	+	+	–	+	–	9.6	+	Jun et al. (2007)
810	Cel9I	213.6	5.0	+	+	–	+	+	–	+	Unpublished
382	Cel51A (CelF)	116.3	7.3	+	+	11, 30, 30	–	–	9.8	–	Malburg et al. (1997)
2772	Cel5G (Cel-3)	70.8	4.6	+	+	11	–	–	–		McGavin et al. (1989)
2361	Cel9B (CelE)	64.7	5.8	+	+	–	–	+	11.4 (46 r)		Qi et al. (2007)
2362	Cel9C (CelD)	69.1	5.9	+	+	–	–	+	11.6 (46 r)		Malburg et al. (1996)
1346	Cel5K (CelG)	60.3									Iyo and Forsberg (1996)
226	Lic16C (mlg)	35.2m		+	+					+	Teather and Erfle (1990), Wen et al. (2005)
Glucan glucohydrolase:											
2558	Cel9D	77m	5.27	+	+	–	–	+	10.5 (69 r)		Qi et al. (2008a)
Xylanases:											
777	Xyn11C (XynC)	66.4	6.2	+	+	–	–	–		+	Paradis et al. (1993), Zhu et al. (1994)
2294	Xyn10C (XynB)	64.9	5.34	+	+	6					Jun et al. (2003)
2292	Xyn10A (XynD)	68.8	5.12	+	+	6					Jun et al. (2003)
2292	Xyn10B (XynE)	66.7	5.25	+	+	6					Jun et al. (2003)
Uncertain level of expression:											
1346	Cel5K	54.7	7.3	+	+	–	–	–	–	+	Unpublished
451	Cel9G (EGB)	62.5	5.0	–	+	–	–	+	–	–	Bera-Maillet et al. (2000), Broussolle et al. (1994)

[a]Symbols: +, present; –, absent.

[b]Abbreviations: FSU, *Fibrobacter succinogenes*; Mw, molecular weight; IP, isoelectric point; SP, signal peptide; CD, catalytic domain; CBM, carbohydrate-binding module; Fd, fibronection-like domain; Ig, immunoglobin-like domain; BTD, basic terminal domain, number of residues (r) in the domain is in parentheses.

esterase(s) (McDermid et al., 1990b), and an α-glucuronidase (Smith and Forsberg, 1991) have not been determined. It will be important to obtain the sequences of these genes, because of their potentially important roles in hemicellulose digestion.

Beginning with highly expressed cellulases, the synergistic interaction of the enzymes was studied (Qi et al., 2007). These experiments led to the conclusion that the two predominant endoglucanases Cel51A (FSU0382) and Cel9B (FSU2361) formerly characterized as endoglucanases 1 and 2, respectively (McGavin and Forsberg, 1988), probably play central roles in cellulose digestion. However, one of the most interesting enzymes recently identified is a glucan glucohydrolase Cel9D (FSU2558) that preferentially cleaves glucose residues from the non-reducing ends of long chains of glucose residues, but does not readily hydrolyze cellobiose (Qi et al., 2008a). It exhibits a strong synergistic interaction with Cel51A (FSU0382) and Cel8B (FSU2303), both of which produce a mixture of cellooligosaccharides as their cellulose hydrolysis products.

Based on our knowledge of the *Fibrobacter succinogenes* cellulase system, a schematic diagram illustrating the cellulases and CBMs involved in cellulose digestion is presented in Figure 120. Although information about the proteins involved in cellulose digestion is extensive, precise knowledge of the mechanism of decrystallization of highly crystalline cellulose, and the proteins involved in the process, is still lacking. *Fibrobacter succinogenes* S85 obviously has a very efficient mechanism for the initial amorphization of crystalline cellulose, since it grows as rapidly as any other cellulolytic bacterium with crystalline cellulose as the sole source of carbon and energy.

Fibrobacter intestinalis strains are known for their capacity to degrade and utilize cellulose as a source of carbon. The availability of the *Fibrobacter succinogenes* genome sequence has been valuable in the identification of numerous *Fibrobacter intestinalis* glycosyl hydrolases with similarity to those in *Fibrobacter succinogenes*. Application of both forward and reverse suppressive subtractive hybridization (Qi et al., 2005, 2008b) found 37 glycosyl hydrolases in *Fibrobacter intestinalis* strain DR7 that, on the basis of low stringency hybridization, were similar to those in *Fibrobacter succinogenes*. In addition, five other cellulase similarities were already known. This genome sequence provides information that will enable the cloning and subsequent enzymic characterization of many *Fibrobacter intestinalis* cellulases.

List of species of the genus *Fibrobacter*

1. **Fibrobacter succinogenes** (*Bacteroides succinogenes* Hungate 1950, 13[AL]) Montgomery, Flesher and Stahl 1988, 434[VP]

suc.ci.no'ge.nes. N.L. n. *acidum succinum* succinic acid; N.L. suff. *-genes* (from Gr. v. *gennaô* to produce), producing; N.L. adj. *succinogenes* succinic-acid-producing.

Cellulose is also used by all strains. Soluble substrates include cellobiose, D-glucose, and sometimes lactose. Other compounds such as starch, pectin, maltose, and trehalose have been used, but consistent results have not been obtained. Habitats include the mammalian GI tract. Several studies have addressed the question of distribution of *Fibrobacter* within the intestine of cattle, sheep, goats, and horses (Koike et al., 2004; Lin et al., 1994; Lin and Stahl, 1995). *Fibrobacter succinogenes* subsp. *succinogenes* was present in roughly equal levels in comparison to *Fibrobacter succinogenes* subsp. *elongatus* in the rumina of sheep and hay fed cattle. *Fibrobacter succinogenes* subsp. *succinogenes* was the dominant *Fibrobacter* subspecies in rumina of goats. However, this subspecies was only a minor component if observed at all in the lower GI tract of horses and goats and in the rumina of commercial (non-hay fed) steers. These data in conjunction with earlier results support the conjecture that *Fibrobacter succinogenes* subsp. *succinogenes* is a critical component of the rumen microbial community during feeding of hay.

Source: bovine rumen.
DNA G+C content (mol%): 48–49 (Bd).
Type strain: S85, ATCC 19169.
Sequence accession no. (16S rRNA gene): AJ496032.

1a. **Fibrobacter succinogenes subsp. succinogenes** (*Bacteroides succinogenes* Hungate 1950, 13[AL]) Montgomery, Flesher and Stahl 1988, 434[VP]

suc.ci.no'ge.nes. N.L. n. *acidum succinum* succinic acid; N.L. suff. *-genes* (from Gr. v. *gennaô* to produce), producing; N.L. adj. *succinogenes* succinic-acid-producing.

Cells are ovoid to lemon shaped (0.8–1.6 µm × 0.9–1.6 µm). Sphaeroplasts form after growth ceases. Colonies on cellobiose agar are lenticular, translucent, and light brown when examined by transmitted light. Individual colonies poorly clear cellulose contained in agar unless they are cultured with *Treponema bryanti* (Kudo et al., 1987). Optimal clearing occurs in roll tubes containing 0.5% each of milled cellulose and purified agar (Montgomery and Macy, 1982). Broth cultures are evenly turbid and produce a smooth sediment. All of the strains tested require biotin, and para-aminobenzoic acid (PABA) is stimulatory.

Source: bovine rumen.
DNA G+C content (mol%): 48–49 (Bd).
Type strain: S85, ATCC 19169
Sequence accession no. (16S rRNA gene): AJ496032.

1b. **Fibrobacter succinogenes subsp. elongatus** corrig. Montgomery, Flesher and Stahl 1988, 434[VP]

e'lon.ga'tus. L. masc. part. adj. *elongatus* elongated, stretched out.

Cells are rod-shaped (0.3–0.5 µm × 0.8–2.0 µm). Sphaeroplasts form after growth ceases. Colonies on cellobiose agar are lenticular, translucent, and light brown when examined by transmitted light. Some strains produce a yellow pigment. Clear zones are formed in cellulose agar, but colonies are indistinct and transient in the absence of fermentable sugars. Clearing is best at 0.5% each of agar and cellulose. Broth cultures are evenly turbid and produce a smooth sediment; some strains produce a yellow pigment. Vitamin requirements are biotin and sometimes cyanocobalamine (B_{12}) and PABA. Habitats include the mammalian intestines, rumina, and ceca. In studies characterizing the distribution of *Fibrobacter* species in the GI tracts of cattle, sheep, goats, and horses (Koike et al., 2004; Lin et al., 1994; Lin and Stahl, 1995), *Fibrobacter succinogenes* subsp. *elongatus* was shown to make up the majority of *Fibrobacter* cells in commercial (grain fed) steers and in the lower GI tract and cecum of one of three goats and in the rumina of the other two goats. It was present at roughly equal levels relative to

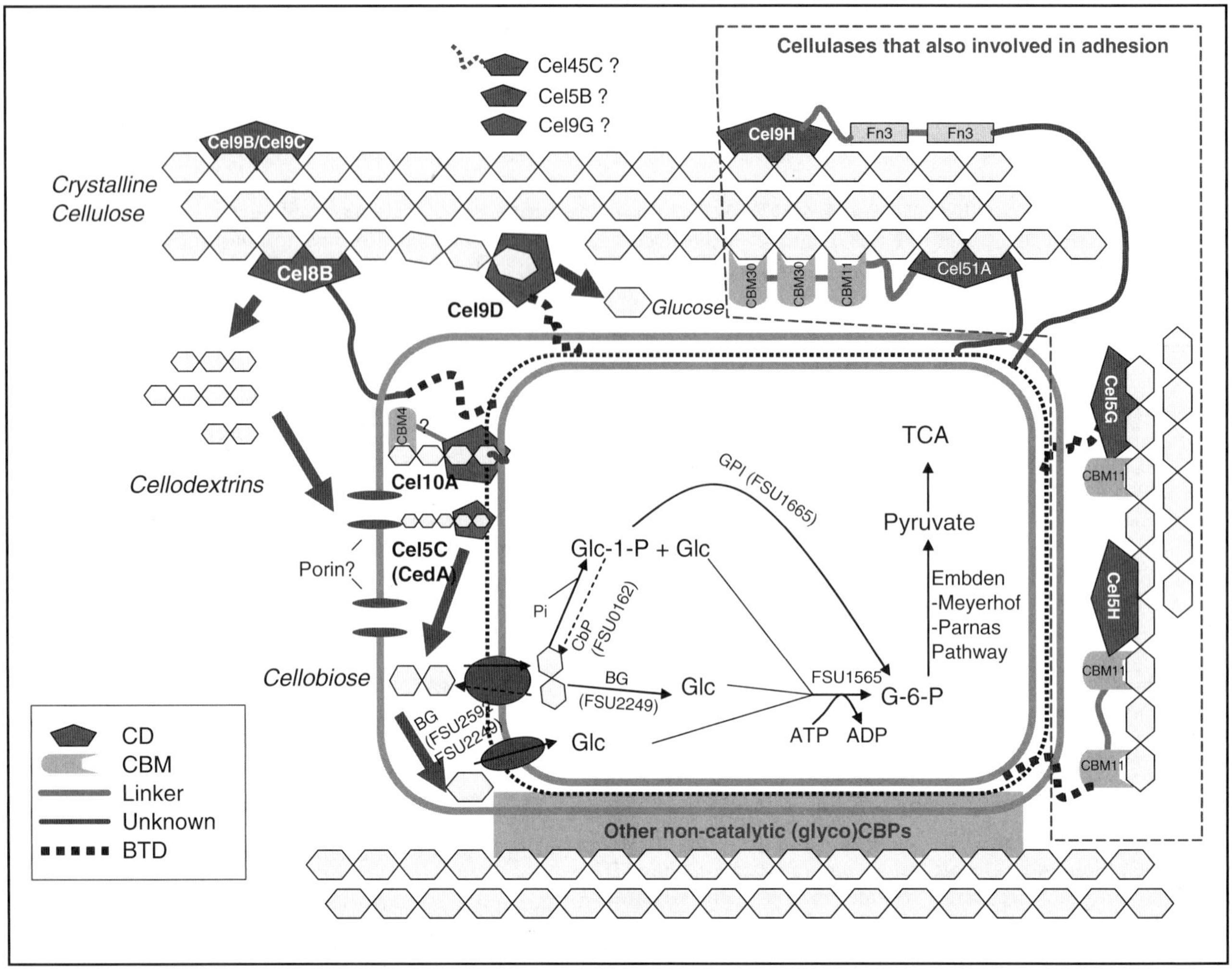

FIGURE 120. Putative strategy for cellulose degradation by *Fibrobacter succinogenes*. This figure shows the cellulases produced by *Fibrobacter succinogenes*. Degradation of cellulose could be divided into three steps. In STEP 1, cellulases Cel9B, Cel9C, Cel51A, Cel8B, and Cel5G cleave the cellulose chains probably at amorphous regions and degrade the strand(s) into cellodextrins; STEP 2, cellodextrinases Cel5C and Cl-stimulated cellobiosidase Cel10A degrade the cellodextrins produced in STEP 1 to cellobiose; STEP 2′, cellodextrins and single glycan chains on cellulose are degraded to glucose by Cel9D; STEP 3, cellobiose is degraded by cell associated β-glucosidase to glucose intra- or extracellularly. Glucose and cellobiose produced from STEP 2 and 3 can be transported into cells by unknown permeases (Maas and Glass, 1991) and metabolized. Cellulases Cel5G, Cel5H, Cel8B, and Cel9D, which contain the BTDII (amino acid Basic C-Terminal Domain), are cell associated. Cel9H, Cel10A, and Cel51A were also known to be cell associated. Adhesion of *Fibrobacter succinogenes* was shown to be mediated by both catalytic and non-catalytic proteins. Cel5H, Cel51A, Cel10A, and Cel5G contain CBMs (cellulose-binding modules), which are cell associated, and may mediate the attachment of cells to cellulose. Cel9H was known as a cellulose-binding cellulase, but its catalytic properties remain to be studied. Three other cellulases, Cel45C, Cel5B, and Cel9G are also produced by this bacterium. However, since the distribution of these enzymes has not been studied and no CBM or other noncatalytic domains (CDs) could be predicted, the role of these enzymes is unknown (contributed by Dr Meng Qi, University of Guelph).

Fibrobacter succinogenes subsp. *succinogenes* in the rumina of hay fed cattle.

Source: ovine rumen.

DNA G+C content (mol%): 51 (Bd) (for strain REH9-1), not determined (for type strain).

Type strain: HM2, ATCC 43856.

Sequence accession no. (16S rRNA gene): AJ496186, M62689.

2. **Fibrobacter intestinalis** Montgomery, Flesher and Stahl 1988, 434[VP]

in.tes.tin.a′lis. L. n. *intestinum* intestines; L. masc. suff. *-alis* suffix denoting pertaining to; N.L. *masc.* adj. *intestinalis* pertaining to the intestines, referring to the original site of isolation.

Cells are rods (0.3–0.4 μm × 0.8–2.0 μm) and form sphaeroplasts after growth ceases. Colonies on cellobiose agar are lenticular, translucent, and light brown when examined by transmitted light. Clear zones are formed in cellulose agar, but colonies are indistinct and transient in the absence of fermentable sugars. Broth cultures are evenly turbid and produce a smooth sediment; some strains produce a yellow pigment. Vitamin requirements are cyanocobalamin (vitamin B_{12}), PABA, and sometimes thiamine and biotin. Habitats include mammalian intestines, rumina, and ceca.

Source: rat cecum.

DNA G+C content (mol%): 45 (Bd).

Type strain: NR9, ATCC 43854.

Sequence accession no. (16S rRNA gene): AJ496284, M62695.

References

Amann, R.I., C. Lin, R. Key, L. Montgomery and D.A. Stahl. 1992. Diversity among *Fibrobacter* isolates: towards a phylogenetic classification. Syst. Appl. Microbiol. *15*: 23–31.

Bera-Maillet, C., V. Broussolle, P. Pristas, J.P. Girardeau, G. Gaudet and E. Forano. 2000. Characterisation of endoglucanases EGB and EGC from *Fibrobacter succinogenes*. Biochim. Biophys. Acta *1476*: 191–202.

Bera-Maillet, C., Y. Ribot and E. Forano. 2004. Fiber-degrading systems of different strains of the genus *Fibrobacter*. Appl. Environ. Microbiol. *70*: 2172–2179.

Bourquin, L.D. and G.C. Fahey, Jr. 1994. Ruminal digestion and glycosyl linkage patterns of cell wall components from leaf and stem fractions of alfalfa, orchardgrass, and wheat straw. J. Anim. Sci. *72*: 1362–1374.

Broussolle, V., E. Forano, G. Gaudet and Y. Ribot. 1994. Gene sequence and analysis of protein domains of EGB, a novel family E endoglucanase from *Fibrobacter succinogenes* S85. FEMS Microbiol. Lett. *124*: 439–447.

Bryant, M.P. and R.N. Doetsch. 1954. A study of actively cellulolytic rod-shaped bacteria of the bovine rumen. J. Dairy Sci. *37*: 1176–1183.

Cheng, K.J., C.S. Stewart, D. Dinsdale and J.W. Costerton. 1983. Electron microscopy of bacteria involved in the digestion of plant cell walls. Anim. Feed Sci. Technol. *10*: 93–120.

Dehority, B.A. and P.A. Tirabasso. 1998. Effect of ruminal cellulolytic bacterial concentrations on *in situ* digestion of forage cellulose. J. Anim. Sci. *76*: 2905–2911.

Doi, R.H. and A. Kosugi. 2004. Cellulosomes: plant-cell-wall-degrading enzyme complexes. Nat. Rev. Microbiol. *2*: 541–551.

Forsberg, C.W., E. Forano and A. Chesson. 2000. Microbial adherence to plant cell wall and enzymatic hydrolysis. *In* Ruminant Physiology Digestion, Metabolism, Growth and Reproduction (edited by Cronje). CABI Publishing, Wallingford, pp. 79–98.

Gardner, P.T., T.J. Wood, A. Chesson and T. Stuchbury. 1999. Effect of degradation on the porosity and surface area of forage cell walls of differing lignin content. J. Sci. Food Agric. *79* 11–18.

Gong, J. and C.W. Forsberg. 1989. Factors affecting adhesion of *Fibrobacter succinogenes* subsp. *succinogenes* S85 and adherence-defective mutants to cellulose. Appl. Environ. Microbiol. *55*: 3039–3044.

Huang, L., C.W. Forsberg and D.Y. Thomas. 1988. Purification and characterization of a chloride-stimulated cellobiosidase from *Bacteroides succinogenes* S85. J. Bacteriol. *170*: 2923–2932.

Hungate, R.E. 1950. The anaerobic mesophilic cellulolytic bacteria. Bacteriol. Rev. *14*: 1–49.

Iyo, A.H. and C.W. Forsberg. 1994. Features of the cellodextrinase gene from *Fibrobacter succinogenes* S85. Can. J. Microbiol. *40*: 592–596.

Iyo, A.H. and C.W. Forsberg. 1996. Endoglucanase G from *Fibrobacter succinogenes* S85 belongs to a class of enzymes characterized by a basic C-terminal domain. Can. J. Microbiol. *42*: 934–943.

Jun, H.S., J.K. Ha, L.M. Malburg, Jr, G.A. Verrinder and C.W. Forsberg. 2003. Characteristics of a cluster of xylanase genes in *Fibrobacter succinogenes* S85. Can. J. Microbiol. *49*: 171–180.

Jun, H.S., M. Qi, J. Gong, E.E. Egbosimba and C.W. Forsberg. 2007. Outer membrane proteins of *Fibrobacter succinogenes* with potential roles in adhesion to cellulose and in cellulose digestion. J. Bacteriol. *189* 6806–6815.

Kam, D.K., H.S. Jun, J.K. Ha, G.D. Inglis and C.W. Forsberg. 2005. Characteristics of adjacent family 6 acetylxylan esterases from *Fibrobacter succinogenes* and the interaction with the Xyn10E xylanase in hydrolysis of acetylated xylan. Can. J. Microbiol. *51*: 821–832.

Koike, S., J. Pan, T. Suzuki, T. Takano, C. Oshima, Y. Kobayashi and K. Tanaka. 2004. Ruminal distribution of the cellulolytic bacterium *Fibrobacter succinogenes* in relation to its phylogenetic grouping. Anim. Sci. J. *75*: 417–422.

Kudo, H., K.J. Cheng and J.W. Costerton. 1987. Interactions between *Treponema bryantii* and cellulolytic bacteria in the *in vitro* degradation of straw cellulose. Can. J. Microbiol. *33*: 244–248.

Lin, C.Z. and D.A. Stahl. 1995. Taxon-specific probes for the cellulolytic genus *Fibrobacter* reveal abundant and novel equine-associated populations. Appl. Environ. Microbiol. *61*: 1348–1351.

Lin, C.Z., B. Flesher, W.C. Capman, R.I. Amann and D.A. Stahl. 1994. Taxon specific hybridization probes for fiber-digesting bacteria suggest novel gut-associated *Fibrobacter*. Syst. Appl. Microbiol. *17*: 418–424.

Lynd, L.R., P.J. Weimer, W.H. van Zyl and I.S. Pretorius. 2002. Microbial cellulose utilization: fundamentals and biotechnology. Microbiol. Mol. Biol. Rev. *66*: 506–577.

Maas, L.K. and T.L. Glass. 1991. Cellobiose uptake by the cellulolytic ruminal anaerobe *Fibrobacter* (*Bacteroides*) *succinogenes*. Can. J. Microbiol. *37*: 141–147.

Macy, J.M., J.R. Farrand and L. Montgomery. 1982. Cellulolytic and non-cellulolytic bacteria in rat gastrointestinal tracts Appl. Environ. Microbiol. *44*: 1428–1434.

Maglione, G., J.B. Russell and D.B. Wilson. 1997. Kinetics of cellulose digestion by *Fibrobacter succinogenes* S85. Appl. Environ. Microbiol. *63*: 665–669.

Malburg, L.M.J., A.H. Iyo and C.W. Forsberg. 1996. A novel family 9 endoglucanase gene (*celD*), whose product cleaves substrates mainly to glucose, and its adjacent upstream homolog (*celE*) from *Fibrobacter succinogenes* S85. Appl. Environ. Microbiol. *62*: 898–906.

Malburg, S.R., L.M. Malburg, T. Liu, A.Y. Iyo, C. Forsberg. 1997. Catalytic properties of the cellulose-binding endoglucanase F from *Fibrobacter succinogenes* S85. Appl. Environ. Microbiol. *63*: 2449–2453.

Matte, A. and C.W. Forsberg. 1992. Purification, characterization, and mode of action of endoxylanases 1 and 2 from *Fibrobacter succinogenes* S85. Appl. Environ. Microbiol. *58*: 157–168.

Matte, A., C.W. Forsberg and A.M. Verrinder Gibbins. 1992. Enzymes associated with metabolism of xylose and other pentoses by *Prevotella* (*Bacteroides*) *ruminicola* strains, *Selenomonas ruminantium* D, and *Fibrobacter succinogenes* S85. Can. J. Microbiol. *38*: 370–376.

Matulova, M., R. Nouaille, P. Capek, M. Pean, E. Forano and A.M. Delort. 2005. Degradation of wheat straw by *Fibrobacter succinogenes* S85: a liquid- and solid-state nuclear magnetic resonance study. Appl. Environ. Microbiol. *71*: 1247–1253.

McDermid, K.P., C.W. Forsberg and C.R. MacKenzie. 1990a. Purification and properties of an acetylxylan esterase from *Fibrobacter succinogenes* S85. Appl. Environ. Microbiol. *56*: 3805–3810.

McDermid, K.P., C.R. MacKenzie and C.W. Forsberg. 1990b. Esterase activities of *Fibrobacter succinogenes* subsp. *succinogenes* S85. Appl. Environ. Microbiol. *56*: 127–132.

McGavin, M. and C.W. Forsberg. 1988. Isolation and characterization of endoglucanases 1 and 2 from *Bacteroides succinogenes* S85. J. Bacteriol. *170*: 2914–2922.

McGavin, M.J., C.W. Forsberg, B. Crosby, A.W. Bell, D. Dignard and D.Y. Thomas. 1989. Structure of the cel-3 gene from *Fibrobacter succinogenes* S85 and characteristics of the encoded gene product, endoglucanase 3. J. Bacteriol. *171*: 5587–5595.

Miron, J. and C.W. Forsberg. 1998. Features of *Fibrobacter intestinalis* DR7 mutant which is impaired with its ability to adhere to cellulose. Anaerobe *4*: 35–43.

Montgomery, L. and J.M. Macy. 1982. Characterization of rat cecum cellulolytic bacteria. Appl. Environ. Microbiol. *44*: 1435–1443.

Montgomery, L., B. Flesher and D. Stahl. 1988. Transfer of *Bacteroides succinogenes* (Hungate) to *Fibrobacter* gen. nov. as *Fibrobacter succinogenes* comb. nov. and description of *Fibrobacter intestinalis* sp. nov. Int. J. Syst. Bacteriol. *38*: 430–435.

Morrison, M., K. Nelson, I. Cann, C. Forsberg, R.I. Mackie, J.B. Russell, B.A. White, D.B. Wilson, K. Amya, B. Cheng, S. Qi, H.S. Jun, S. Mulligan, K. Tran, H. Carty, H. Khouri, W. Nelson, S. Daugherty and C. Fraser. 2003. The *Fibrobacter succinogenes* strain S85 sequencing project. 3rd ASM-TIGR, Microbial Genome Meeting, New Orleans.

Paradis, F.W., H. Zhu, P.J. Krell, J.P. Phillips and C.W. Forsberg. 1993. The *xynC* gene from *Fibrobacter succinogenes* S85 codes for a xylanase with two similar catalytic domains. J. Bacteriol. *175*: 7666–7672.

Pegden, R.S., M.A. Larson, R.J. Grant and M. Morrison. 1998. Adherence of the Gram-positive bacterium *Ruminococcus albus* to cellulose and identification of a novel form of cellulose-binding protein which belongs to the pil family of proteins. J. Bacteriol. *180*: 5921–5927.

Prévot, A.R. 1966. Manual for the Classification and Determination of the Anaerobic Bacteria. Lea & Febiger, Philadelphia.

Qi, M., K.E. Nelson, S.C. Daugherty, W.C. Nelson, I.R. Hance, M. Morrison and C.W. Forsberg. 2005. Novel molecular features of the fibrolytic intestinal bacterium *Fibrobacter intestinalis* not shared with *Fibrobacter succinogenes* as determined by suppressive subtractive hybridization. J. Bacteriol. *187*: 3739–3751.

Qi, M., H.S. Jun and C.W. Forsberg. 2007. Characterization and synergistic interactions of *Fibrobacter succinogenes* glycoside hydrolases. Appl. Environ. Microbiol. *73*: 6098–6105.

Qi, M., H.S. Jun and C.W. Forsberg. 2008a. Cel9D, an atypical 1,4-β-D-glucan glucohydrolase from *Fibrobacter succinogenes*: characteristics, catalytic residues, and synergistic interactions with other cellulases. J. Bacteriol. *190*: 1976–1984.

Qi, M., K.E. Nelson, S.C. Daugherty, W.C. Nelson, I.R. Hance, M. Morrison and C.W. Forsberg. 2008b. Genomic differences between *Fibrobacter succinogenes* S85 and *Fibrobacter intestinalis* DR7 identified by suppression subtractive hybridization. Appl. Environ. Microbiol. *74*: 987–993.

Smith, D.C. and C.W. Forsberg. 1991. α-Glucuronidase and other hemicellulase activities of *Fibrobacter succinogenes* S85 grown on crystalline cellulose or ball-milled barley straw. Appl. Environ. Microbiol. *57*: 3552–3557.

Stewart, C.S., C. Paniagua, D. Dinsdale, K-J. Cheng and S.H. Garrow. 1981. Selective isolation and characteristics of *Bacteriodes succinogenes* from the rumen of a cow. Appl. Environ. Microbiol. *41*: 504–510.

Teather, R.M. and J.D. Erfle. 1990. DNA sequence of a *Fibrobacter succinogenes* mixed-linkage beta-glucanase (1,3–1,4-β-D-glucan 4-glucanohydrolase) gene. J. Bacteriol. *172*: 3837–3841.

Wells, J.E., J.B. Russell, Y. Shi and P.J. Weimer. 1995. Cellodextrin efflux by the cellulolytic ruminal bacterium *Fibrobacter succinogenes* and its potential role in the growth of nonadherent bacteria. Appl. Environ. Microbiol. *61*: 1757–1762.

Wen, T.N., J.L. Chen, S.H. Lee, N.S. Yang and L.F. Shyur. 2005. A truncated *Fibrobacter succinogenes* 1,3–1,4-β-D-glucanase with improved enzymatic activity and thermotolerance. Biochemistry *44*: 9197–9205.

Zhu, H., F.W. Paradis, P.J. Krell, J.P. Phillips and C.W. Forsberg. 1994. Enzymatic specificities and modes of action of the two catalytic domains of the XynC xylanase from *Fibrobacter succinogenes* S85. J. Bacteriol. *176*: 3885–3894.

Phylum XIX. Fusobacteria Garrity and Holt 2001, 140

James T. Staley and William B. Whitman

Fu.so.bac.te'ri.a. N.L. neut. n. *Fusobacterium* genus of the phylum with ending *-ia* to denote phylum; N.L. neut. pl. n. *Fusobacteriia* the phylum of *Fusobacterium*.

The phylum *Fusobacteria* is described in part on the basis of phylogenetic analyses of the 16S rRNA gene sequences of its members. The phylum contains rod-shaped bacteria that stain Gram-negative. Described species are fermentative and produce a variety of organic acids when grown on carbohydrates, amino acids or peptides. Some species are pathogenic to humans.

Type order: **Fusobacteriales** Staley and Whitman, this volume.

Reference

Garrity, G.M. and J.G. Holt. 2001. The Road Map to the *Manual*. *In* Bergey's Manual of Systematic Bacteriology, 2nd edn, vol. 1, The *Archaea* and the Deeply Branching and Phototrophic *Bacteria* (edited by Boone, Castenholz and Garrity). Springer, New York, pp. 119–166.

Class I. **Fusobacteriia** class. nov.

James T. Staley and William B. Whitman

Fu.so.bac.te'ri.ia. N L. neut. n. *Fusobacterium* type genus of the order *Fusobacteriales*; suff. *-ia* ending proposed by Gibbons and Murray and by Stackebrandt et al. to denote class; N.L. neut. pl. n. *Fusobacteriia* the *Fusobacterium* class.

The class *Fusobacteria* is described in part on the basis of phylogenetic analyses of the 16S rRNA gene sequences of its members. The class contains facultative aerobic to obligately anaerobic organisms that stain as Gram-negative rods and ferment carbohydrates or amino acids and peptides to produce various organic acids including acetic, propionic, butyric, formic or succinic, depending on the substrate and species. Species occur in sediments as well as the oral or intestinal habitats of animals. Some species are human pathogens.

Type order: **Fusobacteriales** Staley and Whitman, this volume.

Order I. **Fusobacteriales** ord. nov.

James T. Staley and William B. Whitman

Fu.so.bac.te.ri.a'les. N.L. neut. n. *Fusobacterium* type genus of the order; suff. *-ales* ending denoting order; N.L. fem. pl. n. *Fusobacteriales* the *Fusobacterium* order.

The order *Fusobacteriales* is described in part on the basis of phylogenetic analyses of the 16S rRNA gene sequences of its members. The order contains facultative aerobic to obligately anaerobic organisms that stain as Gram-negative rods. All described species are nonmotile and fermentative. Organisms ferment carbohydrates or amino acids and peptides to produce various organic acids including acetic, propionic, butyric, formic or succinic depending on the substrate and species. Species occur in anoxic environments including sediments as well as the oral or intestinal habitats of animals, including mammals. Some species are pathogenic to humans.

Type genus: **Fusobacterium** Knorr 1922, 4[AL].

Reference

Knorr, M. 1922. Über die fusospirilläre Symbiose, die Gattung *Fusobacterium* (K.B. Lehmann) und *Spirillum sputigenum*. Zugleich ein Beiträg zür Bakteriologie der Mundhohle. II. Mitteilung. Die. Gattung *Fusobacterium*. I Abt. Orig. Zentralbl. Bakteriol. Parasitenkd. Infektionskr. Hyg. *89*: 4–22.

Family I. **Fusobacteriaceae** fam. nov.

James T. Staley and William B. Whitman

Fu.so.bac.te.ri.a.ce'a.e. N. L. neut. n. *Fusobacterium* type genus of the family; suff. *-aceae* ending denoting family; N.L. fem. pl. n. *Fusobacteriaceae* the *Fusobacterium* family.

The family *Fusobacteriaceae* is described in part on the basis of phylogenetic analyses of the 16S rRNA gene sequencess of its members. Micro-aerotolerant to obligately anaerobic organisms that stain as Gram-stain-negative rods. All named species are nonmotile and fermentative. Ferment carbohydrates or amino acids and peptides to produce various organic acids including acetic, propionic, butyric, formic or succinic depending on the substrate and species. Occur in anoxic environments including sediments as well as the oral and intestinal habitats of animals including mammals.

Comprises the genera *Cetobacterium, Fusobacterium, Ilyobacter*, and *Propionigenium*.

Type genus: **Fusobacterium** Knorr 1922, 4[AL].

Genus I. **Fusobacterium** Knorr 1922, 4[AL]

Saheer E. Gharbia, Haroun N. Shah and Kirstin J. Edwards

Fu.so.bac.te'ri.um. L. n. *fusus* a spindle; L. neut. n. *bacterium* a small rod; N.L. neut. n. *Fusobacterium* a small spindle-shaped rod.

Nonsporeforming rods that are **Gram-stain-negative** and **obligately anaerobic**. Metabolize peptone or carbohydrates in PY-glucose to **produce butyrate**, often with acetate and lower levels of lactate, propionate, succinate, and formate.

DNA G+C content (mol%): 26–34.

Type species: **Fusobacterium nucleatum** Knorr 1922, 17[AL] [*Bacillus fusiformis* Veillon and Zuber 1898, 540 and other combinations using "*Fusiformis*" except the organism described as *Fusobacterium fusiforme* by Hoffman in the 7th edition of the *Manual*; Group I, Spaulding and Rettger 1937, 535; Group III (and probably *Fusobacterium polymorphum*) Baird-Parker 1960, 458; *not Fusobacterium plauti-vincentii* Knorr 1922, 5.].

Further descriptive information

The genus *Fusobacterium* includes species that do not ferment adonitol, arabinose, dulcitol, glycerol, glycogen, inocitol, inulin, mannitol, melezitose, rhamnose, ribose, sorbitol, or sorbose. Cellobiose is not fermented and esculin is not hydrolyzed except by *Fusobacterium mortiferum* and *Fusobacterium necrogenes*; does not reduce nitrate (except for *Fusobacterium ulcerans*); and does not produce catalase, lecithinase, or acetylmethycarbinol. In addition to butyric, propionic, and acetic acids, species produce variable amounts of butanol from PY medium. Small amounts of formate, lactate, succinate, and ethanol may be produced. Some species convert threonine or lactate to propionate. Pyruvate is converted to acetate and butyrate and sometimes also to formate, succinate, and lactate. H_2S is produced. All species, except *Fusobacterium naviforme* and *Fusobacterium russii*, produce propionate from threonine. Lactate is converted to propionate by *Fusobacterium necroforum* and *Fusobacterium equinum*. Fusobacteria form volatile sulfur compounds from cysteine and methionine (Claesson et al., 1990; Pianotti et al., 1986). All species produce indole except for *Fusobacterium mortiferum, Fusobacterium necrogenes, Fusobacterium russii, Fusobacterium ulcerans*, and some strains of *Fusobacterium varium*. Apart from a weak reaction by *Fusobacterium necrophorum*, only *Fusobacterium equinum* produces esterases, while *Fusobacterium canifelinum* is the only fluoroquinolone resistant species of the genus. Additional features are described in Table 148.

Cell morphology. The cells are pleomorphic, some are arranged into filaments, and are spindle-shaped with pointed ends (fusiform) in a few species, while others are coccobacilli. Width is variable. The cells may be single, in pairs end-to-end, or form long coiled filaments. Staining may be irregular and spheroplasts are common in some species. The cells of *Fusobacterium nucleatum* are slender, spindle-shaped with tapered or pointed ends 0.4–0.7 μm thick and 4–10 μm long, and appear singly, in tandem pairs, or in bundles of roughly parallel bacilli. Filaments are often seen in old cultures of *Fusobacterium periodonticum* and *Fusobacterium simiae* which have similar cellular morphology to *Fusobacterium nucleatum*. The cells of *Fusobacterium necrophorum* are pleomorphic, often curved, with rounded and sometimes tapered ends, and they may have spherical enlargements. Free coccoid bodies and especially filaments are common. Cell length may vary from coccobacilli to long threads in clinical samples. *Fusobacterium naviforme* strains have boat-shaped cells. Gonidial forms may be seen in old cultures of *Fusobacterium gonidiaformans. Fusobacterium varium* is a small bacillus that does not form filaments. Strains of *Fusobacterium mortiferum* are extremely pleomorphic with globular forms, swellings, and threads. Short rods are predominant in cultures of *Fusobacterium equinum*. The other fusobacteria are pleomorphic filamentous organisms.

Cell-wall composition. Fusobacteria have a cell-wall structure based on two membranes separated by a periplasmic space containing a peptidoglycan layer. *Meso*-lanthioine replaces diaminopimelic acid in *Fusobacterium nucleatum, Fusobacterium gonidiaformans, Fusobacterium necrophorum, Fusobacterium russii, Fusobacterium necrogenes, Fusobacterium simiae*, and *Fusobacterium periodonticum* (Gharbia and Shah, 1990; Kato et al., 1981; Vasstrand et al., 1982), while *Fusobacterium varium, Fusobacterium naviforme*, and *Fusobacterium ulcerans* retain *meso*-diaminopimelic acid and *Fusobacterium mortiferum* contain both versions. *Fusobacterium equinium* and *Fusobacterium canifelinum* have not been tested. Information on the fatty acid content and cell-wall composition of various *Fusobacterium* species can be found in Hofstad (1979), Kato et al. (1981, 1979), Miyagawa et al. (1979), Hofstad and Skaug (1980), Jantzen and Hofstad

TABLE 148. Characteristics differentiating the species and subspecies of *Fusobacterium*[a]

Characteristic	*F. nucleatum* subsp. *nucleatum*	*F. nucleatum* subsp. *fusiforme*	*F. nucleatum* subsp. *polymorphum*	*F. nucleatum* subsp. *vincentii*	*F. nucleatum* subsp. *animalis*	*F. canifelinum*	*F. equinum*	*F. gonidiaformans*	*F. mortiferum*	*F. naviforme*	*F. necrogenes*	*F. necrophorum* subsp. *necrophorum*	*F. necroforum* subsp. *funduleforme*	*F. perfoetans*	*F. periodonticum*	*F. russii*	*F. simae*	*F. ulcerans*	*F. varium*
Cellobiose	–	–	–	–	–	–	–	–	w^-	–	w^-	–	–	–	–	–	–	–	–
Esculin hydrolysis	–	–	–	–	–	–	–	–	$+^-$	–	+	–	–	–	–	–	–	–	–
Utilization of:																			
Fructose	w^-	w^-	w^-	w^-	w^-	w^-	–	–	w^a	–	w^a	$-^w$	$-^w$	w	w^-	–	w	–	w^a
Gelatin	$-^w$	$-^w$	$-^w$	$-^w$	$-^w$	$-^w$	$-^w$	–	–	–	–	v	v	–	–	–	–	–	–
Glucose	$-^w$	$-^w$	$-^w$	$-^w$	$-^w$	$-^w$	–	w^-	a^w	$-^w$	w^a	$-^w$	$-^w$	w	w^-	–	w	a^w	w^a
Lactose	–	–	–	–	–	–	–	–	w^a	–	–	–	–	–	–	–	–	–	–
Maltose	–	–	–	–	–	–	–	–	w^-	–	–	–	–	–	–	–	–	–	–
Mannose	–	–	–	–	–	–	–	–	w^a	–	w^a	–	–	–	–	–	–	a^w	w
Raffinose	–	–	–	–	–	–	–	–	v	–	$-^w$	–	–	–	–	–	–	–	–
Sucrose	–	–	–	–	–	–	–	–	v	–	$-^w$	–	–	w	–	–	–	–	–
Indole	+	+	+	+	+	+	+	+	–	+	–	+	+	–	+	–	+	–	$+^-$
Nitrate	–	–	–	–	–	–		–	–	–	–	–	–	–	–	–	–	+	–
Bile growth	–	–	–	–	+	–	+	–	+	–	+	v	v	–	–	–	+	+	+
Lipase	–	–	–	–	–	–	+	–	–	–	–	$+^-$		–	–	–	w	–	$-^w$
β-Hemolysis	–	–	–	–	–	–	–	–	–	–	–	+	–	–	–	–	–	–	–
Gas produced in agar	–	–	–	–	–	–	nt	+	+	–	+	+	+	+	–	–	–	+	+
Lactate→propionate	–	–	–	–	–	–	+	–	–	–	–	+	+	–	–	–	–	–	–
Threonine→propionate	+	+	+	+	+	+	+	+	+	–	+	+	+	+	+	–	+	+	+
Activity of:																			
N-Acetyl-glucosaminidase	–	–	–	–	–	–	–	–	–	–	–	–	–	+	–	–	–	–	–
Alkaline phosphatase	–	–	–	–	–	–	–	–	+	–	+	+	–	–	–	w	–	–	w
Acid-phosphatase	–	–	–	–	–	–	+	w	+	–	+	+	+	w	–	+	–	+	w
α-Galactosidase	–	–	–	–	–	–	–	–	+	–	+	–	–	–	–	–	–	–	–
β-Galactosidase	–	–	–	–	–	–	–	–	+	–	+	–	–	+	–	–	–	–	–
β-Glucoronidase	–	–	–	–	–	–	–	–	–	+	–	–	–	–	–	–	–	–	–
α-Glucosidase	–	–	–	–	–	–	–	–	–	–	–	–	–	–	–	–	–	–	–
β-Glucosidase	–	–	–	–	–	–	–	–	$+^-$	–	–	–	–	–	–	–	–	–	–
Esterase (C4)	–	–	–	–	–	–	+	–	–	–	–	w	w	–	–	–	w	–	–
Esterase (C8)	–	–	–	–	–	–	+	–	–	–	–	w	w	–	–	–	–	–	–

[a]Symbols: w, weakly positive; w^-: most strains weakly positive, some negative; $-^w$: most strains negative, some weakly positive; a^w: strongly acid with some weakly acid reactions; w^a: usually weakly acid with occasional strong acid reactions; nt, not tested. (Adapted from Conrads et al., 2004a.)

(1981), and Vasstrand (1981). LPS exhibits O-antigenic specificity linked to lipid A through 2-keto-3-deoxyoctonate. The lipid A of *Fusobacterium nucleatum* is structurally similar to that of *Enterobactereaceae* and cross-reacts serologically with antibodies to *Escherichia coli* lipid A (Dahlen and Mattsby-Baltzer, 1983). *Fusobacterium necrophorum* isolates have a rough-type LPS, whereas that of *Fusobacterium necrophorum* subsp. *necrophorum* is of a smooth type (Brown et al., 1997). The repeat unit of the O-antigenic polysaccharide of *Fusobacterium necrophorum* LPS is an acid identified as a 2-amino-2-deoxy-2-C-methyl-pentonic acid (2-amino-2-methyl-3,4,5-trihydroxypentanoic acid), a novel structure not found before in nature (Hermansson et al., 1993). The main protein of the outer membrane of *Fusobacterium nucleatum*, designated FomA, has a molecular mass of 40 kDa (Bakken et al., 1989b). It is a nonspecific porin present in the outer membrane as a trimer (Kleivdal et al., 1995). The gene encoding the FomA porin has been sequenced (Bolstad et al., 1994). The deduced topology is similar to that of porin structures from other bacteria. *fomA* has been cloned and expressed in *Escherichia coli* (Haake and Wang, 1997; Jensen et al., 1996). Several other outer-membrane proteins have been predicted from the genome sequence, among them some very-high-molecular-mass-proteins (Kapatral et al., 2002). FomA and outer-membrane proteins with molecular masses of 55, 60 and 70 kDa are the major antigens in *Fusobacterium nucleatum* (Bakken et al., 1989a).

The cellular fatty acids in the *Fusobacterium* species examined are straight-chain saturated and monoenoic acids of chain lengths C_{12}-C_{18} and *O*-3-hydroxy-tetradecanoate is distinctive to the oral species *Fusobacterium nucleatum*, *Fusobacterium simiae*, and *Fusobacterium periodonticum* (Jantzen and Hofstad, 1981).

Comparison of small-subunit rRNA sequences has revealed levels of sequence similarity that are consistent with the single genus, but intrageneric heterogeneity is evident (Lawson et al., 1991; Tanner et al., 1994). *Fusobacterium nucleatum*, the species isolated most frequently from humans, is divided into five subspecies; *Fusobacterium nucleatum* subsp. *nucleatum*, subsp. *polymorphum*, subsp. *vincetii*, subsp. *fusiforme*, and subsp. *animalis*, described on the basis of electrophoretic patterns of whole-cell proteins and DNA homology (Dzink et al., 1990) and electrophoretic mobility of two enzymes and DNA homology (Gharbia and Shah, 1992). Recent evidence to support the heterogeneity within *Fusobacterium nucleatum* and the existence of distinct subspecies within *Fusobacterium nucleatum* was obtained by comparison of their 16S–23S internal transcribed spacer regions (Conrads et al., 2002).

Fusobacterium periodonticum is phylogenetically similar to *Fusobacterium nucleatum* (Jousimies-Somer and Summanen, 2002). *Fusobacterium nucleatum*-like isolates from cats and dogs were found to be distinct from *Fusobacterium nucleatum* both in their 16S rRNA and DNA–DNA hybridization. These have been reclassified as *Fusobacterium canifelinum* which are resistant to levofloxacin (MIC >4 μg/ml) and other fluoroquinolones. The resistance is due to two substitutions in the quinolone resistance determining region of *gyrA* relative to *Fusobacterium nucleatum*. The first replacement is of Ser79 with leucine and the second is Gly83 is replaced with arginine (Conrads et al., 2005).

Animal isolates of *Fusobacterium necrophorum* form two subspecies: *Fusobacterium necrophorum* subsp. *necrophorum* and *Fusobacterium necrophorum* subsp. *funduliforme* (Shinjo et al., 1991) corresponding to biovars A and B. A distinctive feature separating the subspecies genetically is the presence of isoleucine and alanine tRNA gene in *Fusobacterium necrophorum* subsp. *necroforum*, while *Fusobacterium necrophorum* subsp. *funduliforme* contains the isoleucine tRNA gene (Jin et al., 2002).

Genetic differences between the species have been reported in leukotoxin A (*lktA*) between both subspecies (Narayanan et al., 2001). Subspecies *necrophorum* is more frequently isolated, often in pure culture, from liver abscesses than subsp. *funduliforme*. Leukotoxin, an exotoxin, is a major virulence factor. In *Fusobacterium necrophorum* subsp. *necrophorum*, Lkt is a high-molecular-mass protein that is encoded by a tricistronic leukotoxin operon (*lktBAC*) and induces apoptosis and necrosis of bovine leukocytes in a dose-dependent manner. The subsp. *funduliforme* produces lower concentration of leukotoxin and hence is less virulent than subsp. *necrophorum*. The low toxicity associated with subsp. *funduliforme* leukotoxin, a less virulent subspecies, may in part be due to the differences in the *lktA* gene and reduced transcription (Tadepalli et al., 2008).

Fusobacterium mortiferum, *Fusobacterium varium*, and *Fusobacterium ulcerans* share several phenotypic properties similar to *Fusobacterium varium*, but *Fusobacterium ulcerans* reduces nitrate to nitrite (Citron, 2002). It also contains unique fragments of 1000 and 550 bp not present in *Fusobacterium varium* and *Fusobacterium mortiferum* (Claros et al., 1999). *Fusobacterium necrogenes* is distantly related to the other *Fusobacterium* species. Based on 16S–23S rDNA spacer region sequences (Conrads et al., 2002) *Fusobacterium gonidiaformans* is genealogically related to *Fusobacterium necrophorum* (Nicholson et al., 1994).

Metabolism. The core metabolism of *Fusobacterium nucleatum* is similar to that of *Clostridium*, *Lactococcus*, and *Enterococcus* species (Kapatral et al., 2002). More than 137 transporters for the uptake of substrates such as peptides, sugars, metal ions, and cofactors have been detected. Amino acids and small peptides comprise the major sources of energy for all *Fusobacterium* species (Gharbia and H.N. Shah, 1989), however, peptides influence the uptake of amino acids enhancing the utilization of histidine and glutamate while threonine, methionine, and asparagine were repressed (Gharbia et al., 1989). Acidic and cationic amino acids are the main acids incorporated. Glutamate, histidine, lysine, and serine appear to be key amino acids in *Fusobacterium nucleatum* (Gharbia and Shah, 1991a; Rogers et al., 1992). Biosynthetic pathways exist for glutamate, aspartate, and glutamylglutamate to be used as growth substrates for *Fusobacterium nucleatum* (Takahashi and Sato, 2002). Glutamate is a key catabolic substrate in *Fusobacterium* species (Gharbia and Shah, 1991b). It is catabolized via the 2-oxoglutarate pathway with the production of acetate and butyrate as end products (Gharbia and Shah, 1991b). Glutamate may also be degraded by the methylaspartate pathway in *Fusobacterium varium* (Gharbia and Shah, 1991b). Enzymes representative for the mesaconate pathway for catabolism of glutamate have been detected in *Fusobacterium varium*, *Fusobacterium mortiferum*, and *Fusobacterium ulcerans* (Gharbia and Shah, 1991b). *Fusobacterium varium* and *Fusobacterium mortiferum* also possess enzymes representative of the 4-aminobutyrate pathway.

Amino acids are imported as monomers, di-, or oligopeptides, and an active transport of the dipeptide L-cysteinglycine has been detected in *Fusobacterium nucleatum* (Carlsson et al., 1994).

Fusobacterium species differ in their ability to use fermentable carbohydrates as energy sources for growth (Robrish et al., 1991). *Fusobacterium nucleatum* and other species utilize glucose for biosynthesis of intracellular glycopolymers that can be degraded to produce energy under conditions of amino-acid deprivation (Robrish et al., 1987). The accumulation of glucose is dependent on energy generated by the fermentation of amino acids (Robrish et al., 1987). *Fusobacterium mortiferum* is an exception in that accumulation of sugars is independent of a fermentable amino acid. Significantly, *Fusobacterium mortiferum* has the ability to metabolize various sugars as energy sources for growth (Robrish et al., 1991). Sugars utilized include α- and β-glycosides, which are transported by the phosphoenolpyruvate-dependent sugar: phosphotransferase system. Thus, it utilizes sucrose and its isomeric α-D-glucosyl-D-fructoses as energy sources for growth (Pikis et al., 2002).

The genes encoding phospho-β-glucosidase (P-β-glucosidase; EC 3.22.1.86) and 6-phospho-α-D-glucosidase (maltose-6-phosphate hydrolase; EC 3.2.1.122) known as *pbgA* and *malH*, respectively, have been expressed in *Escherichia coli* (Bouma et al., 1997; Thompson et al., 1997).

Genetics. The genome of *Fusobacterium nucleatum* subsp. *nucleatum* (strain ATCC 25586ᵀ) contains 2.17 Mb encoding 2067 open reading frames (ORFs) organized on a single circular chromosome (Kapatral et al., 2002). About 2.3% of the ORFs are unique to *Fusobacterium nucleatum*. The genome analysis has revealed several key aspects of the pathways of organic acid, amino acid, carbohydrate, and lipid metabolism. Nine very-high-molecular-mass outer-membrane proteins are predicted from the sequence, none of which has been reported in the literature. More than 137 transporters for the uptake of a variety of substrates such as peptides, sugars, metal ions, and cofactors have been identified. Biosynthetic pathways exist for only three amino acids: glutamate, aspartate, and asparagine. The remaining amino acids are imported as such or as di- or oligopeptides that are subsequently degraded in the cytoplasm. A principal source of energy appears to be the fermentation of glutamate to butyrate. Additionally, desulfuration of cysteine and methionine yields ammonia, H_2S, methyl mercaptan, and butyrate, which are capable of arresting fibroblast growth, thus preventing wound healing and aiding penetration of the gingival epithelium.

Analysis of the draft genome sequence of *Fusobacterium nucleatum* subsp. *vincentii* (ATCC 49256), and comparison of this sequence to the genome ATCC 25586 sequence resolved that 441 ORFs have no orthologs in strain ATCC 25586. Of these, 118 ORFs have no known function and are unique, whereas 323 ORFs have functional orthologs in other organisms. Genes for eukaryotic serine/threonine kinase and phosphatase, transpeptidase E-transglycosylase Pbp1A were also identified among the ATCC 49256-specific ORFs. Unique ABC transporters, cryptic phages, and three types of restriction-modification systems have been identified in ATCC 49256. ORFs for ethanolamine utilization, thermostable carboxypeptidase, glutamyl-transpeptidase, and deblocking aminopeptidases are absent from ATCC 49256. Both strains lack a catalase-peroxidase system, but possess thioredoxin/glutaredoxin enzymes. Genes for resistance to antibiotics such as acriflavin, bacitracin, bleomycin, daunorubicin, florfenicol, and other general multidrug resistance are present (Kapatral et al., 2003)

The genome of *Fusobacterium nucleatum* subsp. *polymorphum* ATCC 10953 contains a chromosome of approximately 2.4 Mbp and a plasmid (pFN3) of 11.9 kbp, and there are 2361 proteins encoded in the genome. Plasmid pFN3 from the strain was also sequenced and analyzed. When compared to the other two available fusobacterial genomes (*Fusobacterium nucleatum* subsp. *nucleatum* and *Fusobacterium nucleatum* subsp. *vincentii*) 627 ORFs unique to *Fusobacterium nucleatum* subsp. *polymorphum* ATCC 10953 were identified. A large percentage of these mapped within one of 28 regions or islands containing five or more genes. Seventeen percent of the clustered proteins that demonstrated similarity were most similar to proteins from the clostridia, with others being most similar to proteins from other Gram stain positive organisms such as *Bacillus* and *Streptococcus*. A 10 kb region homologous to the *Salmonella typhimurium* propanediol utilization locus was identified, as was a prophage and integrated conjugal plasmid. The genome contains five composite ribozyme/transposons similar to the Cd*ISt* IStrons described in *Clostridium difficile*.

Additionally, three plasmids isolated from *Fusobacterium nucleatum* strains were sequenced. These are designated pFN1, pPA52, and pKH9 (Bachrach et al., 2004; Haake et al., 2000; McKay et al., 1995).

Ecology. *Fusobacterium* are normal inhabitants of the mucous membrane of humans and animals. The habitat of *Fusobacterium nucleatum* and *Fusobacterium periodonticum* is the human oral cavity; the gingival crevice is the niche of both. Strains are isolated from the oral microflora of adults and children and are also found in edentulous infants. The vagina is likely to be the primary endogenous site for *Fusobacterium naviforme* and *Fusobacterium gonidiaformans* (Hill, 1993). The habitat of *Fusobacterium mortiforum* and *Fusobacterium varium* is the gastrointestinal tract, where they are present in small numbers. *Fusobacterium necrogenes*, which was originally isolated from a chicken abscess and from cecal contents of ducks, is very rarely isolated from humans. The habitat of *Fusobacterium ulcerans*, isolated from tropical ulcers, is unknown.

Fusobacterium necrophorum is a normal inhabitant of the alimentary tract of cattle, horses, sheep, and pigs and is frequently isolated from cats and dogs. Reports, mainly from the first part of the twentieth century, indicate its presence in a range of wild animals including reptiles. However, the relatively frequent isolation of *Fusobacterium necroforum* from soft-tissue infections of the oral cavity and the upper respiratory tract compared to elsewhere in the body, suggest that these sites are the principal human habitat of this organism.

Fusobacterium russii is normally found in the canine and feline oral flora (Love et al., 1987), but it has also been isolated from human feces. The oral cavity of horses and the stump-tailed macaque (*Macaca arctoides*) are the habitat of *Fusobacterium equinum* and *Fusobacterium simiae*, respectively. *Fusobacterium canifelinum* was isolated from the microflora of infected cat and dog bite wounds in humans (Conrads et al., 2005).

Maintenance procedures

Fusobacteria are not particularly demanding with regard to a low oxidation-reduction potential. They are, however, killed fairly readily by exposure to oxygen. This is possibly due to their susceptibility to peroxides. Growth is best at 35–37°C and at a pH near 7.

All *Fusobacterium* species grow on blood agar based on proteose-peptone, tryptone, or trypticase. Good growth is usually obtained in a rich semifluid medium such as brain heart medium supplemented with yeast extract. Batch cultivation is best performed in a fluid medium containing a reducing agent.

Laboratory isolation and identification

Fusobacterium species can be separated from other related taxa based on their phylogenetic relatedness and distinctive phenotypic characteristics (Table 149).

The isolation of fusobacteria from an aerobic infection requires proper collection of the specimen, the use of an anaerobic transport vial, optimal anaerobiosis (an anaerobic chamber is not necessary), and a rich medium for optimal growth. Fastidious Anaerobe Agar Base (FAA, Lab M) supplemented with 5% sheep or horse blood is recommended for culture.

The use of a selective medium is necessary for the enumeration of viable fusobacteria in specimens of normal flora. The addition of josamycin, vancomycin, and norfloxacin plus 5% defibrinated horse blood to FAA supports the growth of *Fusobacterium* species while completely inhibiting most other obligate anaerobes and facultative bacteria (Brazier et al., 1991). Rifampin blood agar (Sutter et al., 1971) is useful for the selective isolation of *Fusobacterium varium* and *Fusobacterium mortiferum*. Diagnostic tables are convenient for the examination of bile resistance, production of alkaline phosphatase, and orthonitrophenol-β-D-galactopyranoside (ONPG)-test. Commercially available kits based on the detection of preformed enzymes and designed for identification of anaerobes may be of help in supporting a suspected identity of a *Fusobacterium* isolate.

Surface colonies of *Fusobacterium nucleatum* are 1–2 mm in diameter after incubation for 2 d. They are white to yellow-gray in color, speckled, smooth or breadcrumb-like. The colonies are usually nonhemolytic. Cultures of *Fusobacterium nucleatum* produce an unpleasant, but not fetid, smell. After incubation for 2 d, colonies of animal isolates of *Fusobacterium necrophorum* produce a foul, putrid odor. Fluid or semifluid cultures are characterized by abundant production of gas, and colonies are about 2 mm in diameter, flat, circular with scalloped to erose edge, opaque, and white to gray (subsp. *necrophorum*); or 1 mm wide, circular with entire margins, gray, translucent, and with smooth surfaces (subsp. *funduliforme*) (Shinjo et al., 1991).

The colonial morphology of *Fusobacterium necrophorum* isolated from human infections is similar to that of *Fusobacterium necrophorum* subsp. *funduliforme*. The colonies are small and have been described as creamy, opaque, smooth, umbonate or raised, and with entire edges (Hall et al., 1997). The colonial morphology of most other *Fusobacterium* species is similar to that of *Fusobacterium necrophorum* subsp. *funduliforme*.

All *Fusobacterium* species produce butyric acid. They are catalase and nitrate negative, sensitive to kanamycin (1000 µg disk) and colistin (10 µg disk), resistant to vancomycin (5 µg disk), and produce a rancid odor.

Determination of the electrophoretic migration of glutamate dehydrogenase and 2-oxoglutarate reductase provides a rapid method for the identification of most *Fusobacterium* species (Gharbia and Shah, 1990). Identification based on gas-liquid chromatography of cellular fatty acids may be of value in reference laboratories, provided a database is used as a basis for comparison. A DNA probe has been used successfully to detect

TABLE 149. Characteristics differentiating *Fusobacterium* from genetically related taxa[a]

Reaction	*Fusobacterium nucleatum*	*Cetobacterium ceti*	*Leptotrichia buccalis*	*Sebaldella termitidis*	*Sneathia sanguinegens*	*Streptobacillus moniliformis*
Catalase	–	–	–	–	–	–
Oxidase	nd	–	nd	nd	–	–
Indole	+	+	–	–	–	–
Arginine dihydroylase	nd	nd	nd	nd	+	+
Nitrate to nitrite	–	v	–	–	–	–
Esculin hydrolysis	–	–	+	nd	+	v
Gelatinase	–	–	–	–	nd	–
Starch hydrolysis	nd	–	v	–	–	nd
ONPG	nd	+	nd	nd	nd	–
Acid from:						
Arabinose	nd	nd	–	–	–	–
Cellobiose	–	nd	+	nd	nd	–
Fructose	–w	nd	+	+	nd	+
Galactose	nd	nd	d	nd	nd	+
Glucose	–w	nd	+	+	+	+
Glycogen	nd	nd	nd	nd	–	+
Lactose	–	nd	+	–	–	–
Maltose	–	nd	+	+	–	+
Mannose	–	nd	+	nd	v	+
Mannitol	–	nd	–	+	–	–
Starch	–	nd	d	–	–	+
Sucrose	–	nd	+	+	–	–
Trehalose	–	nd	+	+	–	–
Xylose	–	nd	–	+	nd	–

[a]Symbols: +, 90% or more strains are positive; –, 90% or more strains are negative; d, 11–89% of the strains are positive; ONPG, *O*-nitrophenyl-β-D-galactopyranoside; v, variable; w, weak; nd, no data.

Fusobacterium nucleatum directly in samples of subgingival plaque (Lippke et al., 1991), and checkerboard DNA–DNA hybridization has been used successfully to detect *Fusobacterium nucleatum* and subspecies in periodontal and endodontic infections (Socransky et al., 1998; Sunde et al., 2000). A PCR-detection of part of the major 40 kDa outer-membrane protein, the FomA porin of *Fusobacterium nucleatum*, is specific for *Fusobacterium nucleatum* when used under condition of high stringency (Bolstad and Jensen, 1993). Sequencing of 16S rRNA from clinical isolates provides a discriminatory tool for separating all species (Figure 121).

List of species of the genus *Fusobacterium*

1. **Fusobacterium nucleatum** Knorr 1922, 17[AL] [*Bacillus fusiformis* Veillon and Zuber 1898, 540 and other combinations using "*Fusiformis*" except the organism described as *Fusobacterium fusiforme* by Hoffman in the 7th edition of the *Manual*; Group I, Spaulding and Rettger 1937, 535; Group III (and probably *Fusobacterium polymorphum*) Baird-Parker 1960, 458; *not Fusobacterium plauti-vincentii* Knorr 1922, 5.]

nu.cle.a'tum. L. neut. adj. *nucleatum* having a kernel, intended to mean nucleated.

Cells from glucose broth cultures are 0.4–0.7 × 3 µm, have tapered to pointed ends, and often have central swellings and intracellular granules. Cell length is variable but is usually fairly uniform within actively growing cultures. Cells do not possess pili or flagella (Dahlen et al., 1978; Falkler and Hawley, 1977). Surface colonies on blood agar are 1–2 mm in diameter, circular to slightly irregular, convex to pulvinate, translucent often with a "flecked" appearance, usually nonhemolytic (horse or rabbit blood), but may be slightly hemolytic under the area of confluent growth or may produce greenish discoloration of the blood agar upon exposure to oxygen. Glucose broth cultures have a flocculent or granular sediment with or without turbidity, a final pH of 5.6–6.2, and a foul "bad breath" odor. Produces DNase (Porschen and Sonntag, 1974). No phosphatase detected (Porschen and Sonntag, 1974). Most strains produce H_2S. Capable of hemagglutinating human and animal erythrocytes. Contains the diamino acid lanthionine in cell-wall peptidoglycan as a major component; no lysine, diaminopimelic acid, or ornithine (Gharbia and Shah, 1990; Kato et al., 1979; Vasstrand, 1981). Heptose and KDO are present in the lipopolysaccharide (Hofstad, 1974). Grows in the presence of up to 6% oxygen; survives exposure to air for 100 min (Loesche, 1969).

Strains are resistant to 3 µg/ml of erythromycin (broth disc test). Some *Fusobacterium nucleatum* isolates produce β-lactamase, an activity not reported for other *Fusobacterium* species. A penicillin-hydrolyzing β-lactamase inhibited by clavulanic acid has been isolated from *Fusobacterium nucleatum* (Turner et al., 1985). *Fusobacterium nucleatum*, similar to other species of the genus, are susceptible to amoxycillin/clavulanate, carbapenems, chloramphenicol, quinolone clinafloxalin, linezolide, and nitroimidazoles (Goldstein et al., 1999). Resistance to cefoxitin and clindamycin is very low. Resistance to tetracyclines is common in *Fusobacterium nucleatum* since a high percentage of strains harbor a chromosomal *tetM* locus. Alternatively, a tetM determinant on a conjugal transposon (Roberts and Lansciardi, 1990) was also detected. Resistance to glycylglycines has not been observed (Downes et al., 1999). *Fusobacterium nucleatum* is sensitive to the bactericidal action of protegrins and other antibacterial peptides (Miyasaki et al., 1998). Other characteristics of the species are given in Table 148.

DNA G+C content (mol%): 26–28 (T_m).

Type strain: ATCC 25586, CCUG 32989, CCUG 33059, CIP 101130, JCM 8532, LMG 13131.

Sequence accession no. (16S rRNA gene): AJ133496.

Further comments: *Fusobacterium nucleatum* is divided into

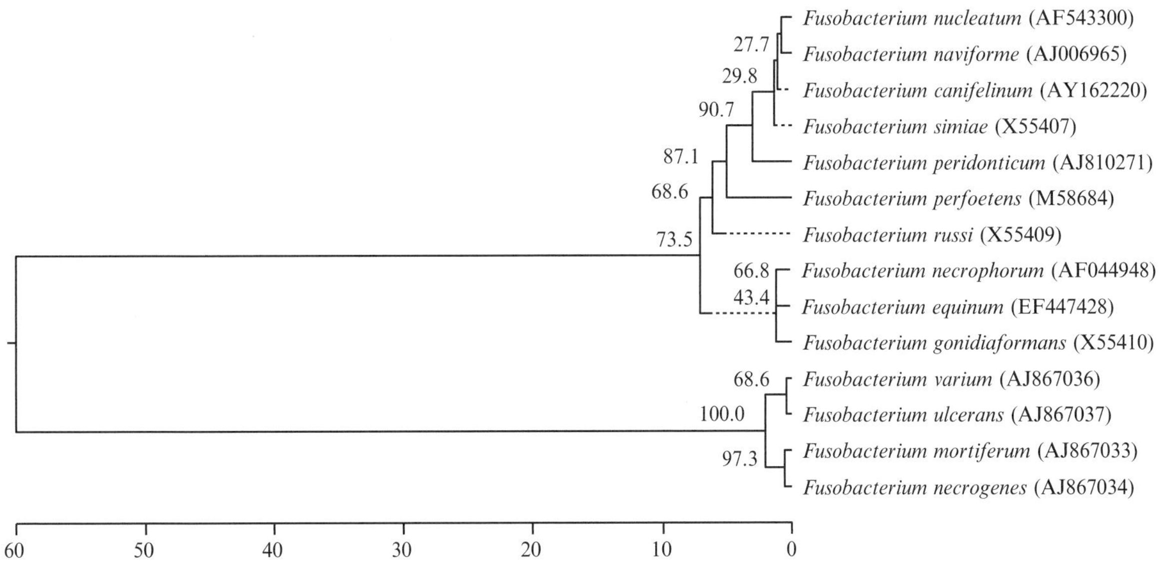

FIGURE 121. Unrooted neighbor-joining tree of the 16S rRNA gene sequences of *Fusobacterium*. GenBank accession numbers are given after species names. The numbers above branches are bootstrap percentages from 1000 resampled datasets. Bar = 0.1 difference per nucleotide.

five subspecies: *Fusobacterium nucleatum* subsp. *nucleatum*, subsp. *animalis*, subsp. *fusiforme*, subsp. *polymorphum*, and subsp. *vincentii*, described on the basis of electrophoretic patterns of whole-cell proteins and DNA homology (Dzink et al., 1990) and electrophoretic mobility of glutamate dehydrogenase and 2-oxoglutarate dehydrogenases and DNA homology (Gharbia and Shah, 1992). The five subspecies differed from each other and from other closely related species by comparisons of their 16S–23S internal transcribed spacer regions (Conrads et al., 2002).

1a. **Fusobacterium nucleatum subsp. nucleatum** (Knorr 1922) Dzink, Sheenan and Socransky 1990, 77[VP] (*Fusobacterium nucleatum* Knorr 1922, 17)

nu.cle.a'tum. L. neut. adj. *nucleatum* having a kernel, intended to mean nucleated.

DNA G+C content (mol%): 26–28 (T_m).

Type strain: ATCC 25586, CCUG 32989, CCUG 33059, CIP 101130, JCM 8532, LMG 13131.

Sequence accession no. (16S rRNA gene): AJ133496.

Further information: the subspecies name *Fusobacterium nucleatum* subsp. *nucleatum* Knorr (1922) is automatically created by the valid publication of *Fusobacterium nucleatum* subsp. *polymorphum* (*ex* Knorr (1922) Dzink (1990), and the valid publication of *Fusobacterium nucleatum* subsp. *vincentii* (*ex* Knorr (1922) Dzink (1990) [Rule 40d (formerly Rule 46)].

1b. **Fusobacterium nucleatum subsp. animalis** Gharbia and Shah 1992, 297[VP]

a.ni.ma'lis. L. n. *animal, -alis* an animal; L. gen. n. *animalis* of an animal.

Type strain: ATCC 51191, CCUG 32879, CIP 104879, JCM 11025, NCTC 12276.

Sequence accession no. (16S rRNA gene): X55404.

1c. **Fusobacterium nucleatum subsp. fusiforme** (*ex* Veillon and Zuber 1898) Gharbia and Shah 1992, 297[VP] ("*Sphaerophorus fusiformis*" Veillon and Zuber 1898; Sebald 1962)

fu.si.for'me. L.n. *fusus* a spindle; L. neut. suff. *forme* in shape of; N.L. neut. adj. *fusiforme* spindle-shaped.

Type strain: ATCC 51190, CCUG 32880, CIP 104878, DSM 19508, JCM 11024, NCTC 11326.

Sequence accession no. (16S rRNA gene): AM849219.

1d. **Fusobacterium nucleatum subsp. polymorphum** (Knorr 1922) Dzink, Sheenan and Socransky 1990, 77[VP] ("*Fusobacterium polymorphum*" Knorr 1922)

po.ly.mor'phum. N.L. neut. adj. *polymorphum* (from Gr. adj. *polumorphos -on*) multiform, polymorphic.

Type strain: ATCC 10953, CCUG 9126, DSM 20482, JCM 12990, NCTC 10562.

Sequence accession no. (16S rRNA gene): AF287812.

1e. **Fusobacterium nucleatum subsp. vincentii** (Knorr 1922) Dzink, Sheenan and Socransky 1990, 77[VP] ("*Fusobacterium plauti-vincentii*" Knorr 1922)

vin.cen'.ti.i. N.L. masc. gen. n. *vincentii* of Vincent, referring to Henri Vincent who studied the organism originally isolated from Vincent's angina and necrotizing ulcerative gingivitis.

Type strain: ATCC 49256, CCUG 37843, CIP 104988, DSM 19507, JCM 11023.

Sequence accession no. (16S rRNA gene): AJ006964, AM887529.

2. **Fusobacterium canifelinum** Conrads, Citron, Mutters, Jang and Goldstein 2004b, 1909[VP] (Effective publication: Conrads, Citron, Mutters, Jang and Goldstein 2004a, 412.)

ca. ni.felí num. L. gen. pl. n. *canum* of dogs; L. neut. adj. *felinum* of or belonging to a cat; N.L. neut. adj. *canifelinum* of dogs and cats.

Strains grown on supplemented Brucella blood-agar for 2 d; colonies are convex, 1–2 mm in diameter with a slightly lobate margin, white, opaque with a granular internal appearance. Cells are slender Gram-stain-negative rods with pointed ends. Glucose and fructose are fermented weakly with a terminal pH 5.7–6.0. End product analysis of PY-glucose by gas-liquid chromatography reveals acetic and butyric acids. Threonine, but not lactate, is converted to propionate. The isolates produce indole, fail to grow in bile, do not hydrolyze esculin, and do not produce acid from lactose, maltose, mannose, raffinose, and sucrose. On agar dilution sensitivity tests, all isolates tested were susceptible to penicillin G, and metronidazole. All strains were resistant (MIC >4 µg/ml) to levofloxacin, moxifloxacin, gemifloxacin, and other fluoroquinolones. *Fusobacterium canifelinum* can be distinguished phenotypically from other species of the genus *Fusobacterium* by the resistance to fluoroquinolones.

Source: a purulent dog-bite wound in a human patient.

DNA G+C content (mol%): 26–28 (T_m).

Type strain: RMA 1036, ATCC BAA-689, CCUG 49733, DSM 15542.

Sequence accession no. (16S rRNA gene): AY162221.

3. **Fusobacterium equinum** Dorsch, Love and Bailey 2001, 1962[VP]

e.quin'um. L neut. adj. *equinum* of horses.

On sheep-blood agar after 2 d, colonies are circular with an entire undulate margin, convex to umbonate, 1–2 mm in diameter, and creamish in color. Gram staining of single colonies on primary plates reveals pleomorphic Gram-stain-negative rods (coccobacilli to rods, with rounded ends, and curved rods often with irregular staining), but on subculture, coccobacilli and short rods predominate. The terminal pH in glucose liquid medium ranges from pH 6.8–7.0. On GLC analysis, acetic, propionic, and butyric acids are produced and lactate and threonine are converted to propionate. The isolates produce indole and grow in bile and are nonmotile and fail to hydrolyze esculin and starch. The isolates do not ferment esculin, fructose, glucose, lactose, maltose, mannose, starch, and sucrose. They do not hemagglutinate chicken erythrocytes. They are lipase and lecithinase positive. On broth disc sensitivity tests, all isolates are sensitive to penicillin G, amoxycillin, chloramphenicol, doxycycline, and metronidazole; all strains except VPB 4076 and VPB 4014 are sensitive to erythromycin.

Source: pus obtained from a discharged sinus associated with the pirulent para-oral lesion in a horse.

DNA G+C content (mol%): 29–31 (T_m).

Type strain: JCM 11174, NCTC 13176, VPB 4027.

Sequence accession no. (16S rRNA gene): AJ295750.

4. **Fusobacterium gonidiaformans** (Tunnicliff and Jackson 1925) Moore and Holdeman 1970, 45[AL] [*Bacillus gonidiaformans* Tunnicliff and Jackson 1925, 430; *Sphaerophorus gonidiaformans* (Tunnicliff and Jackson 1925) Prévot 1938, 299]

go.ni.di.a.for'mans. Gr. n. *gone* offspring, seed; N.L. n. *gonidium* gonidium; N.L. pl. n. *gonidia* gonidia; L. part. adj. *formans* forming; N.L. part. adj. *gonidiaformans* gonidia forming.

Cells from glucose broth cultures are pleomorphic and vacuolated, 0.4–0.7 × 0.7–3.0 μm, often with degenerate filaments or long strands. The spheroid or gonidial forms implied by the name of this organism are seen most often in old cultures or in media that are not highly reduced. Surface colonies on horse blood agar plates are punctiform to 1 mm in diameter, circular entire, low convex, translucent, and smooth. Glucose broth cultures are turbid with smooth sediment and a final pH of 5.6–6.2. Produces DNase (Porschen and Sonntag, 1974). No phosphatase is detected (Porschen and Spaulding, 1974). Hippurate is hydrolyzed. Other characteristics of the species are given in Table 148.

Source: the intestinal and urogenital tracts of humans, various types of human infections and from a lamb with pneumonia.

DNA G+C content (mol%): 33 (T_m) (Gharbia and Shah, 1990).

Type strain: ATCC 25563, CCUG 16790.

Sequence accession no. (16S rRNA gene): X55410.

5. **Fusobacterium mortiferum** (Harris 1901) Moore and Holdeman 1970, 45[AL] [*Bacillus mortiferus* Harris 1901, 546; *Sphaerophorus mortiferus* (*sic*) (Harris 1901) Prévot 1938, 299]

mor.ti'fer.um. L. neut. adj. *mortiferum* death-bringing, death-bearing.

Cells from glucose broth cultures are 0.8–1.0 × 1.5–10 μm occurring singly and in pairs and short chains. Cells stain irregularly and may be extremely pleomorphic. Surface colonies on horse blood agar are 1–2 mm in diameter, circular with entire, diffuse, or slightly scalloped edge; convex; translucent; smooth. Glucose broth cultures are uniformly turbid with smooth or semiviscous sediment. No superoxide dismutase is detected (Gregory et al., 1978). Lysine decarboxylase is not produced (Werner, 1974). Produces DNase and phosphatase (Porschen and Sonntag, 1974; Porschen and Spaulding, 1974). Heptose and KDO are present in the lipopolysaccharide (Hofstad, 1974). *Fusobacterium mortiferum* is the only member of the genus *Fusobacterium* to have a mixture of mesolanthionine and diaminopimelic acids at an equimolar ration in its cell wall. Some strains are susceptible to cephalothin, cefazolin, cefoxitin, and lincomycin (Finegold, 1977). Resistant to 3 μg/ml erythromycin; susceptible to 12 μg/ml chloramphenicol, 2 U/ml penicillin G, and 6 μg/mltetracycline (broth disk method).

Source: blood and various human clinical specimens, intestinal tract and feces; and one from irradiated mice.

DNA G+C content (mol%): 26–28 (T_m).

Type strain: ATCC 25557, CCUG 14475.

Sequence accession no. (16S rRNA gene): AJ867032.

6. **Fusobacterium naviforme** (Jungano 1909) Moore and Holdeman 1970, 45[AL] [*Bacillus naviformis* Jungano 1909, 123; *Ristella naviformis* (Jungano 1909) Prévot 1938, 291]

na.vi.for'me. L. n. *navis* ship; L. neut. suff. *forme* in shape of; N.L. neut. adj. *naviforme* in the shape of a ship.

Cells from glucose broth cultures are 4–10 μm that are often concave. Cells in old cultures have a beaded appearance. Surface colonies are punctiform to 2.0 mm in diameter, circular entire, low convex, gray-white, translucent with mottled appearance when viewed by obliquely transmitted light. Glucose broth cultures are lightly turbid with smooth to clumpy sediment and final pH of 5.5–6.4. Produces DNase (Porschen and Sonntag, 1974). No phosphatase is detected (Porschen and Spaulding, 1974). Heptose and KDO are present in the lipopolysaccharide (Hofstad, 1974). All strains produce indole from glucose-peptone medium (Gharbia and H.N. Shah, 1989) The cell walls contain *meso*-lanthionine as the major diaminopimelic acid. Some strains are resistant to 3 μg/ml erythromycin. Susceptible to 12 μg/ml chloramphenicol, 1.6 μg/ml clindamycin, 2 U/ml penicillin G, and 6 μg/ml tetracycline.

Source: the large intestine of a laboratory rat. Other strains have been isolated from the human gingival sulcus, from various human clinical specimens, and the bovine rumen.

DNA G+C content (mol%): 32–33 (T_m).

Type strain: ATCC 25832, CCUG 50052, NCTC 13121.

Sequence accession no. (16S rRNA gene): not available.

7. **Fusobacterium necrogenes** (Weinberg, Nativelle and Prévot 1937) Moore and Holdeman 1970, 45[AL] [*Bacillus necrogenes* Weinberg, Nativelle and Prévot 1937, 681; *Spherophorus necrogenes* (*sic*) (Weinberg, Nativelle and Prévot 1937) Prévot 1938, 298]

ne.cro'ge.nes. Gr. n. *necros* the dead; N.L. suff. *-genes* (from Gr. v. *gennaô* to produce) producing; N.L. adj. *necrogenes* dead-producing, here necrosis-producing.

This description is based on our study of the type and three phenotypically similar strains. Cells from 1-d-old glucose broth cultures are extremely pleomorphic with coccoid cells about 0.3–0.8 μm and thin filamentous forms 0.2–0.8 μm in diameter and up to 20 μm in length. Cells in older cultures are somewhat more uniform, irregularly staining rods 0.7–0.8 × 1.5–4.0 μm. Surface colonies on horse blood agar are minute to 0.5 mm in diameter, circular, flat to low convex, entire, translucent, white, smooth, and shiny, and colonies are surrounded by zones of β-hemolysis. Glucose broth cultures are moderately turbid with pH of 5.7–6.0 Glutamine decarboxylase negative (Terada et al., 1976). Heptose and KDO present in the lipopolysaccharide (Hofstad, 1974).

Source: isolated by Kawamura (1926) from necrotic abscess of a chicken. Barnes strain EB/D/1/4a was isolated from cecal contents of a duck (Barnes and Impey, 1968). Other strains have been isolated from human feces.

DNA G+C content (mol%): 27–28 (T_m).

Type strain: ATCC 25556, CCUG 4949, NCTC 10723.

Sequence accession no. (16S rRNA gene): AJ867034.

8. **Fusobacterium necrophorum** (Flügge 1886) Moore and Holdeman 1969, 12[AL] [*Bacillus necrophorus* Flügge 1886, 273;

Fusiformis necrophorus (Flügge) Topley and Wilson 1929, 299; *Sphaerophorus necrophorus* (Flügge) Prévot 1938, 298]

ne.cro'pho.rum. Gr. n. *necros* the dead; Gr. v. *phoreô* to bear; N.L. neut. adj. *necrophorum* dead producing, here necrosis producing.

Cells in glucose broth cultures are 0.5–0.7 µm in diameter with swellings up to 1.8 µm. The ends of the cells may be round or tapered. Cell length ranges from coccoid bodies to filaments over 1.00 µm. Filamentous forms with granular inclusions are more common in broth, while bacilli are more common in older cultures and growth on agar. Surface colonies on blood agar are 1–2 mm in diameter; circular with scalloped to erose edges, convex to umbonate, often with bumpy, ridged, or uneven surface; translucent to opaque, often with mosaic internal structure when viewed by transmitted light. Most strains produce either α- or β-hemolysis on rabbit blood agar. In general, the β-hemolytic strains are lipase positive (on egg yolk agar), and the α-hemolytic or nonhemolytic strains are lipase negative. No lecithinase is produced. Glucose broth cultures have a smooth, flocculent, granular, or stringy sediment and usually are turbid. The final pH of fructose and glucose cultures is 5.6–6.3. A few strains produce a pH of 5.8–5.9 in maltose medium.

Human rabbit and guinea pig red blood cells are agglutinated; bovine and ovine red blood cells are not (Simon, 1975). Dextran is not hydrolyzed (Holbrook and McMillan., 1977). No superoxide dismutase (Gregory et al., 1978) or lysine decarboxylase (Werner, 1974) is detected. Produce DNase (Porschen and Sonntag, 1974). No phosphatase is detected (Porschen and Spaulding, 1974). Heptose and KDO are present in the lipopolysaccharide (Hofstad, 1974).

Source: the natural cavities of humans and other animals and from clinical specimens (necrotic lesions, abscesses, and blood) of humans and other animals particularly liver abscesses and foot rot of cattle. For a review of natural and experimental pathogenicity, see Prévot et al. (1967) and Langworth (1977).

Source: bovine liver abscess; Fievez strain 2358.

DNA G+C content (mol%): 31–34 (T_m) [chromatographic separation (Sebald, 1962)].

Type strain: ATCC 25286, CCUG 9994, CIP 104559, JCM 3718, VPI 2891.

Sequence accession no. (16S rRNA gene): AJ867039.

Further comments: two subspecies have been described among isolates of *Fusobacterium necrophorum*. These correspond to Biovar A and B.

8a. **Fusobacterium necrophorum subsp. necrophorum** (Flügge 1886) Shinjo, Fujisawa and Mitsuoka 1991, 396[VP] (*Fusobacterium necrophorum* Moore and Holdeman 1969, 12)

ne.cro'pho.rum. Gr. n. *necros* the dead; Gr. v. *phoreô* to bear; N.L. neut.adj. *necrophorum* dead producing, here necrosis producing.

Previously known as *Fusobacterium necrophorum* biovar A. The subspecies name *Fusobacterium necrophorum* subsp. *necrophorum* (Flügge, 1886) Moore and Holdeman (1969) is automatically created by the valid publication of *Fusobacterium necrophorum* subsp. *funduliforme* (*ex* Hallé (1898) Shinjo et al. (1991) [Rule 40d (formerly Rule 46)] and was previously known as biovar B.

Type strain: ATCC 25286, CCUG 9994, CIP 104559, JCM 3718, VPI 2891.

Sequence accession no. (16S rRNA gene): AJ867039.

8b. **Fusobacterium necrophorum subsp. funduliforme** (*ex* Hallé 1898) Shinjo, Fujisawa and Mitsuoka 1991, 396[VP] ("*Sphaerophorus funduliformis*" Hallé 1898)

fun.du.li.for'me. L. n. *fundulus* sausage; L. neut. suff. *forme* in shape of; N.L. neut. adj. *funduliforme* sausage-shaped.

Type strain: Fn524, ATCC 51357, CCUG 42162, CIP 104859, DSM 19678, JCM 3724.

Sequence accession no. (16S rRNA gene): AM905356.

9. **Fusobacterium perfoetens** (Tissier 1905) Moore and Holdeman 1973, 72[AL] [*Coccobacillus perfoetens* Tissier 1905, 110; *Ristella perfoetens* (Tissier 1905) Prévot 1938, 291; *Sphaerophorus perfoetens* (Tissier, 1905) Sebald 1962, 149]

per.foe'tens. L. prep. *per* very; L. part. adj. *foetens* stinking; N.L. part. adj. *perfoetens* very stinking.

Description is from Prévot et al. (1967), Weinberg et al. (1937), van Assche and Wilssens (1977). Cells from glucose cultures are 0.6–0.8 × 0.8–1.0 µm, oval, never elongated, occurring singly, in pairs, in chains of no more than three cells, or in masses. No flagella or capsule. Colonies in deep agar (2 d) are 1 mm in diameter and lenticular. Surface colonies on blood agar are 1–2 mm in diameter, circular with an entire edge, convex to raised, grayish white, translucent, smooth, and nonhemolytic on horse blood. Colonies of some strains are slightly umbonate with diffuse edges and a slightly mottled or granular appearance.

Growth in glucose broth is rapid. Cultures are turbid with a fine to ropy sediment and a pH of 5.6. Gas and a fetid odor are produced. Galactose is weakly fermented (Van Assche and Wilssens, 1977). Growth is enhanced in media containing fructose, glucose, mannose, sucrose, and trehalose. The pH of cultures in these media ranges from 5.6 to 5.95. Produces CO_2 and NH_3. Major amounts of lactic acid produced from PYG cultures (Van Assche and Wilssens, 1977) no lactic acid detected in PYG cultures of the type strain.

Inhibited by 0.001% polymyxin. Resistant to 0.001% brilliant green. Optimum temperature is 37°C; good growth occurs at 45°C, poor growth at 25–30°C. Survives up to 24 h of exposure to air.

Source: isolated by Tissier in 1900 from an infant with diarrhea and in 1905 from nursing infants. Strain CC1 isolated in 1947 by Prévot from the cecum of a horse and studied by Sebald (1962) has been lost. Van Assche and Wilssens (1977) studied 6 isolates from the feces of a 2-week-old pig, one of which has been designated the neotype strain.

DNA G+C content (mol%): 28–30 (T_m) (Sebald, 1962; Van Assche and Wilssens, 1977).

Type strain: ATCC 29250.

Sequence accession no. (16S rRNA gene): M58684.

10. **Fusobacterium peridonticum** Slots, Potts and Mashimo 1984, 270[VP] (Effective publication: Slots, Potts and Mashimo 1983, 963.)

pe.ri.o'don.ti.cum. Gr. prep. *peri* around; Gr. n. *odous, ontos* tooth; L. neut. suff. *-icum* suffix used with the sense of pertaining to; N.L. neutr. adj. *periodonticum* pertaining to periodonte.

Description is from Slots et al. (1982). Obligately anaerobic, nonmotile, nonsporeforming, Gram-stain-negative rod. Mean cell size on blood agar is 0.5–1.0 × 4.0–7.0 μm, but filaments longer than 100 μm are often present. Cells had pointed to slightly rounded ends and generally occurred singly or in pairs lying end to end. After anerobic incubation for 2–3 d on blood agar, the colonies measured from 2.0 to 3.0 mm in diameter, were circular, convex, entire or slightly scalloped at the edge, slightly rough, and granular and opaque. Glucose broths were turbid, with flocculent or stringy sediment. Indole is produced, hydrogen sulfide formed from cysteine, litmus milk reduced, hippurate hydrolyzed, and fructose, galactose, and glucose fermented. Growth is inhibited by 1 μg/ml chloramphenicol, penicillin G, and tetracycline and was unaffected by 5 μg/ml erthryomycin and vancomycin. Addition of 2% oxgall inhibited the growth.

DNA G+C content (mol%): 28 (T_m).

Type strain: EK1-15, ATCC 33693, CCUG 14345, JCM 12991.

Sequence accession no. (16S rRNA gene): X55405.

11. **Fusobacterium russii** (Hauduroy, Ehringer, Urbain, Guillot and Magrou 1937) Moore and Holdeman 1970, 45[AL] (*Bacteroides russii* Hauduroy, Ehringer, Urbain, Guillot and Magrou 1937, 73)

rus'si.i. N.L. masc. gen. n. *russii* of Russ, named after V. Russ, the bacteriologist who first cultured this organism.

Cells in glucose broth are 1.5–4.0 μm in length and may form thin filaments 10–15 μm in length. Surface colonies on horse blood agar are 0.5–1 mm in diameter, circular, smooth, shiny, entire, convex, and translucent. The type strain is β-hemolytic on horse blood agar. Glucose-peptone broth cultures are turbid, often with stringy sediment and have a final pH of 5.9–6.1. Produces DNase and phosphatase (Porschen and Sonntag, 1974; Porschen and Spaulding, 1974). Heptose and KDO are present in the lipopolysaccharide (Hofstad, 1974). The major dibasic amino acid is *meso*-lanthionine. Strains do not ferment any carbohydrates nor produce lactate threonine, but phosphatase activities were reported (Table 148). Susceptible to 12 μg/ml chloramphenicol, 1.6 μg/ml clindamycin, and 2 U/ml penicillin G. Some strains are resistant to 6 μg/ml tetracycline.

Source: isolated by Russ (1905) from perianal abscess. Also isolated from infections of cats, including actinomycosis of cats and from human and animal feces.

DNA G+C content (mol%): 31 (T_m).

Type strain: ATCC 25533.

Sequence accession no. (16S rRNA gene): X55409.

12. **Fusobacterium simiae** Slots and Potts 1982, 193[VP]

sim'i.ae. L. fem. n. *simia* monkey; L. gen. n. *simiae* of the monkey.

Cellular morphology resembles that of *Fusobacterium nucleatum* (*q.v.*, species 1). Fructose and glucose are fermented (pH of 5.5–5.6). Twenty-seven other substrates tested are not fermented. Indole and lipase are produced; hippurate is hydrolyzed. Grows in media containing 2% oxgall. From glucose, butyrate is the major fermentation product; major amounts of acetate and small amounts or propionate, lactate, and succinate also are produced. Neither hydrogen nor gas is detected. Lactate and threonine are converted to propionate. The type strain has 48% DNA homology with the type and one other strain of *Fusobacterium nucleatum* and 9% homology with the type strain of *Fusobacterium necrophorum.*

Source: the mouth of the stump-tailed macaque (*Macaca arctoides*).

DNA G+C content (mol%): 27–28 (T_m).

Type strain: 7511 R2-13, ATCC 33568, CCUG 16798 (Slots and Potts 7511 R2–13).

Sequence accession no. (16S rRNA gene): X55407.

13. **Fusobacterium ulcerans** Adriaans and Shah 1988, 447[VP]

ul'ce. rans. L. part. adj. *ulcerans* making sore, causing to ulcerate, referring to the source of isolation.

Gram-stain-negative, nonsporeforming, obligately anaerobic, rod-shaped 0.5–4.5 μm long. Most strains consist of long cells with pointed ends. Surface colonies on anaerobic blood agar plates are 2–3 mm in diameter, circular, entire, domed to low convex, cream colored, and nonhemolytic. A second morphological type consists of rod-shaped organisms that are 0.8–4.5 × 0.2 μm and have a large round swelling (diameter 0.5 μm) in the center of the cell. Surface colonies on blood agar plates are 1–2 mm in diameter, circular, entire, and low convex. They are translucent or white in color, butyrous, and nonhemolytic. The optimal temperature for growth is 37°C with colonies appearing in 36–48 h. Culture in brain heart infusion broth produce generalized turbidity after incubation for 3–4 d.

Fermentation of peptone-yeast-glucose broth produces large amounts of butyric acid and small to moderate amounts of acetic, propionic, lactic, and succinic acids. All strains convert threonine to propionic acid. Both colonial types produce gas when grown in deep culture. No H_2S is detected. Acid is produced from glucose, and some strains ferment glucose and mannose. No acid is produced from fructose, lactose, sucrose, maltose, arabinose, raffinose, or rhamnose. Indole, catalase, lecithinase, urease, lipase, and oxidase are not produced. Esculin and starch are not hydrolyzed. Nitrate is reduced by all strains.

All strains are susceptible to penicillin and phosphamycin and are resistant to rifampin.

DNA G+C content (mol%): 29.2–29.5 (T_m).

Type strain: ATCC 49185, CCUG 50053, NCTC 12111.

Sequence accession no. (16S rRNA gene): AJ867037.

14. **Fusobacterium varium** (Eggerth and Gagnon 1933) Moore and Holdeman 1969, 12[AL] [*Bacteroides varius* Eggerth and Gagnon 1933, 409; *Sphaerophorus varius* (Eggerth and Gagnon 1933) Prévot 1938, 299]

va'ri.um. L. neut. adj. *varium* diverse, different, various.

Cells from glucose broth cultures are pleomorphic, coccoid and rod-shaped, and stain unevenly. Cells are 0.3–0.7 × 0.7–2.0 μm and occur singly and in pairs. Surface colonies on blood agar are punctiform to 1 mm in diameter, circular with entire edges, flat to low convex, translucent, usually with gray-white centers, and colorless edges. Glucose broth cultures are turbid with smooth sediment and a final pH of 5.3–5.7. Dextran is not hydrolyzed (Holbrook and

McMillan., 1977). Produces lysine decarboxylase (Werner, 1974). Produces DNase (Porschen and Sonntag, 1974). No phosphate detected (Porschen and Spaulding, 1974). Heptose and KDO are present in the lipopolysaccharide (Hofstad, 1974).

A temperate lysogenic bacteriophage is harbored by the reference ATCC 27725 which causes a lytic response in other strains of *Fusobacterium varium* including the reference strain ATCC 8501. The phage is activated by incubation at 45°C and UV exposure (Gharbia, personal observation)

Source: human feces, purulent infections of humans (upper respiratory tract, surgical wounds, peritonitis), cecal contents of mice, intestinal contents of *Blatta orientalis* (roach), posterior intestinal tract of *Recticulitermes lucifugus* (termite) and vaginal swab of chinchilla.

DNA G+C content (mol%): 29 (T_m) (Sebald, 1962).

Type strain: ATCC 8501, CCUG 4858, NCTC 10560.

Sequence accession no. (16S rRNA gene): AJ867036.

Genus II. **Cetobacterium** Foster, Ross, Naylor, Collins, Ramos, Fernández-Garayzábal and Reid 1996, 362[VP] (Effective publication: Foster, Ross, Naylor, Collins, Ramos, Fernández-Garayzábal and Reid 1995, 206.)

Kirstin J. Edwards, Julie M. J. Logan and Saheer E. Gharbia

Ce.to.bac.te'ri.um. Gr. n. *kêtos* whale; L. neut. n. *bacterium* a rod; N.L. neut. n. *Cetobacterium* a bacterium found in association with whales.

Short pleomorphic, nonsporeforming, rod-shaped cells. Central swelling and filaments may be present. Gram-stain-negative. Nonmotile. Microaerotolerant. Catalase negative. **Fermentative. Acetic acid is the major end product from peptones or carbohydrates**; butyric, propionic, lactic, and succinic acids may or may not be formed in small amounts. Indole produced and ONPG hydrolyzed. Alkaline phosphatase, and acid phosphatase positive. Produces small to moderate amounts of phosphohydrolase. May or may not produce α- or β-galactosidase and α-glucosidase. Urease may or may not be produced. Lecithinase and lipase negative. Gelatin is not hydrolyzed. Nitrate may or may not be reduced to nitrite. Resistant to 20% bile. Resistant to vancomycin. Sensitive to kanamycin and colistin sulfate discs. Susceptible to cefoxitin, clindamycin, imipenem, and metronidazole. Isolated from mammalian intestinal tract and oral cavity.

DNA G+C content (mol%): 29–31.

Type species: **Cetobacterium ceti** Foster, Ross, Naylor, Collins, Ramos, Fernández-Garayzábal and Reid 1996, 362[VP] (Effective publication: Foster, Ross, Naylor, Collins, Ramos, Fernández-Garayzábal and Reid 1995, 206.).

Further descriptive information

Cetobacterium was first isolated from the intestinal contents of a porpoise and from a mouth lesion of a minke whale (Foster et al., 1995). *Cetobacterium somerae* has subsequently been isolated from human feces (Finegold et al., 2003a). Colonies are gray, waxy, circular with scalloped to erose edges, slightly-raised, smooth, dull, and opaque with a diameter of 2–4 mm. Weak hemolysis was observed with both sheep and horse blood. No growth was observed for *Cetobacterium ceti* following subculture in an atmosphere of 10% CO_2 or in air, whereas *Cetobacterium somerae* grows in 2% but not 6% oxygen (Finegold et al., 2003a).

Differentiation of the genus *Cetobacterium* from other genera

Cetobacterium differs from *Fusobacterium* species in producing acetic and propionic acids, whereas members of the genus *Fusobacterium* produce butyric acid (Foster et al., 1995). Sequencing the 16S rRNA gene demonstrates the highest sequence similarities with *Fusobacterium* (91–94%) and 92% similarity with *Propionigenium modestum.* Significant sequence similarity was also observed with *Leptotrichia* and *Sebedella* (86%) (Foster et al., 1995). A large amount of acetic acid with lesser amounts of propionic, lactic, and succinic acids permits differentiation from *Propionigenium modestum* which does not ferment carbohydrates, but produces large amounts propionic and lesser amounts of acetic acid only from succinate and other substrates (Schink and Pfennig, 1982) A positive indole reaction allows further separation from *Propionigenium modestum.*

List of species of the genus *Cetobacterium*

1. **Cetobacterium ceti** Foster, Ross, Naylor, Collins, Ramos, Fernández-Garayzábal and Reid 1996, 362[VP] (Effective publication: Foster, Ross, Naylor, Collins, Ramos, Fernández-Garayzábal and Reid 1995, 206.)

ce.ti. L. gen. n. *ceti* of a whale.

Description is from Foster et al. (1995). Surface colonies on blood agar are 2–4 mm in diameter after 48 h at 37°C, gray, waxy, circular, with scalloped to erose edges, slightly raised, smooth, dull, opaque, and weakly hemolytic on sheep and horse blood. No growth at 25°C or 45°C. Catalase-negative. Indole, ONPG and phosphatase positive. Lecithinase, lipase, DNase, nitrate, urea, esculin, gelatin, and starch negative. Resistant to 20% bile. Resistant to vancomycin. Sensitive to colistin sulfate and kanamycin. Sensitivity to penicillin varies. The major volatile fatty acids produced are acetic, propionic, lactic, and succinic. Butyric acid is not produced.

DNA G+C content (mol%): 29 (T_m).

Type strain: M-3333, ATCC 700028, NCIMB 703026.

Sequence accession no. (16S rRNA gene): X78419.

2. **Cetobacterium somerae** Finegold, Vaisanen, Molitoris, Tomzynski, Song, Liu, Collins and Lawson 2003b, 1219[VP] (Effective publication: Finegold, Vaisanen, Molitoris, Tomzynski, Song, Liu, Collins and Lawson 2003a, 180.)

so'me.rae. N.L. gen. fem. n. *somerae* of Somer, to honor Hannele Jousimies-Somer, a contemporary Finish microbiologist, in recognition of her important contributions to anaerobic microbiology.

Description is from Finegold et al. (2003a). Rod-shaped. Gram-stain-negative. Microaerotolerant. After 48 h incubation anaerobically at 37°C on brucella blood agar, colonies are 2–3 mm in diameter, smooth, circular, entire, and gray in color. Colonies do not fluoresce under UV light. Catalase negative. Indole positive after 48 h incubation. Acetic acid is the major end product in peptone yeast broth; small amounts of propionic and butyric acids may be formed; a trace of succinic acid was produced by all strains. Nitrate is reduced to nitrite. Resistant to 20% ox bile. Esculin may or may not be hydrolyzed. Gelatin is not hydrolyzed. By traditional tests, ONPG is hydrolyzed, but lecithinase, lipase, and β-lactamase are not produced. Urease may or may not be detected. Using the API ZYM system, α- and β-galactosidase and alkaline phosphatase are produced; phosphohydrolase is produced in lesser amounts, and α-glucosidase may or may not be produced. Susceptible to kanamycin, colistin sulfate, cefoxitin, clindamycin, imipenem, and metronidazole. Resistant to ampicillin, penicillin G, ramoplanin, trimethoprim/sulfamethoxazole, and vancomycin. Predominant long-chain cellular fatty acids are $C_{14:0}$, $C_{16:0}$, and $C_{16:1}$ ω9*c*.

Source: human fecal material.

DNA G+C content (mol%): 31 (HPLC).

Type strain: WAL 14325, ATCC BAA-474, CCUG 46254.

Sequence accession no. (16S rRNA gene): AJ438155.

Genus III. **Ilyobacter** Stieb and Schink 1985, 375^VP (Effective publication: Stieb and Schink 1984, 145.)

Bernhard Schink, Peter H. Janssen and Andreas Brune

I.ly.o.bac′ter. Gr. fem. n. *ilys* mud; N.L. masc. n. *bacter* rod; N.L. masc. n. *Ilyobacter* a mud-inhabiting rod.

Strictly anaerobic chemoorganotrophic bacteria with fermentative metabolism, nonphotosynthetic, inorganic electron acceptors not used. Nonsporeforming.

Chemoorganotrophic, fermentative type of metabolism. Media containing a reductant are necessary for growth. Catalase negative. Isolated from anoxic environments.

DNA G+C content (mol%): 31.7–36.7.

Type species: **Ilyobacter polytropus** Stieb and Schink 1985, 375^VP (Effective publication: Stieb and Schink 1984, 145.).

Further descriptive information

The genus *Ilyobacter* consists so far of four species, *Ilyobacter polytropus*, *Ilyobacter delafieldii*, *Ilyobacter insuetus*, and *Ilyobacter tartaricus*. The genus was created to house Gram-stain-negative, obligately anaerobic, nonsporeforming bacteria that do not contain cytochromes and use unusual substrates for growth. Its members differ from those of most other genera of strictly anaerobic bacteria by their unusual patterns of substrate utilization and product formation. Fermentation products include acetate, butyrate, and (on some substrates) also formate and ethanol. Malate and fumarate are fermented to acetate, formate, and propionate. The DNA base ratio of *Ilyobacter* species ranges from 32 to 36 mol% G+C, and so is clearly lower that that of any *Bacteroides*, *Selenomonas*, or *Pelobacter* species. With the exception of *Ilyobacter delafieldii* (see below), all *Ilyobacter* species are short to coccoid rods, often in pairs or short chains.

Ilyobacter polytropus was enriched and isolated from marine sediment with 3-hydroxybutyrate, which was fermented to acetate and butyrate. Glycerol was fermented to 1,3-propanediol and 3-hydroxypropionate. Acetate and formate were the only products of pyruvate or citrate fermentation. Glucose and fructose were fermented to acetate, formate, and ethanol. Malate and fumarate were fermented to acetate, formate, and propionate.

Ilyobacter tartaricus was enriched and isolated from marine sediment with L-tartrate as sole source of carbon and energy. Tartrate, citrate, pyruvate, and oxaloacetate are fermented to acetate, formate, and CO_2. In addition, ethanol is formed from fructose and glucose.

Ilyobacter insuetus was isolated from marine sediment with quinic acid (1,3,4,5-tetrahydroxy-cyclohexane-1-carboxylic acid, sodium salt) as the sole source of carbon and energy. This bacterium is restricted to the fermentation of hydroaromatic substrates. Of more than 30 different substrates tested, only quinic acid and shikimic acid (3,4,5-trihydroxy-1-cyclohexene-1-carboxylic acid) are utilized. Neither sugars, alcohols, other carboxylic acids, amino acids, nor aromatic compounds are fermented. Thus, this species represents an extreme case of specialization in substrate utilization.

Ilyobacter delafieldii (Janssen and Harfoot., 1991) was enriched and isolated from estuarine sediment with crotonate as substrate. It ferments crotonate, 3-hydroxybutyrate, lactate, pyruvate, and poly-β-hydroxybutyrate to acetate, propionate, butyrate, CO_2, and H_2. Poly-β-hydroxybutyrate is hydrolyzed outside the cell without cell contact, by a PHB depolymerase that is excreted into the growth medium (Janssen and Harfoot, 1990). So far, this bacterium is unique in its capacity to degrade extracellular PHB anaerobically (Schink et al., 1992).

The taxonomic status of *Ilyobacter delafieldii* is unclear. Since it stains Gram-negative and resembles *Ilyobacter polytropus* in many of its metabolic capacities, it was originally assigned to the genus *Ilyobacter*. However, sequence analysis of its 16S rRNA gene later revealed that it should be grouped within the genus *Clostridium* (Janssen, unpublished), even though spore formation could not be demonstrated (Janssen and Harfoot, 1990). The cell-wall architecture of *Ilyobacter delafieldii* strain 10cr1 (Janssen and Harfoot, 1990) is not typical of Gram-stain-negative bacteria but resembles that of a Gram-stain-positive bacterium with a complex cell-wall structure. For this reason, we do not include *Ilyobacter delafieldii* any further in this genus.

Ilyobacter tartaricus has generated major interest because of its Na^+-translocating ATP-synthase system. Tartrate and oxaloacetate are metabolized via pyruvate; the oxaloacetate decarboxylase is a Na^+-translocating, membrane-bound enzyme. The Na^+-gradient established this way is used for ATP-synthesis via a membrane-bound Na^+-translocating ATP-synthase enzyme, and contributes to the overall energy balance of the cell. Together with a similar enzyme in *Propionigenium modestum* (Hilpert et al., 1984), this Na^+-ATPase has become one of the model systems to study the architecture of this type of F_1F_0-ATPases and especially the linkage between Na^+-ion transport and ATP synthesis (Neumann et al., 1998). The three-dimensional structure of this Na^+-ATPase was recently resolved in detail (Meier et al., 2005).

The study of Na^+-translocating ATPases has also contributed significantly to a better understanding of proton transport in the more common H^+-translocating F_1F_0-ATPases (Gemperli et al., 2003).

It appears that anoxic marine sediments are the typical habitats of these bacteria. At least with *Ilyobacter tartaricus*, the energy metabolism is based on sodium ions as coupling ions in energy conservation.

Enrichment and isolation procedures

A strictly anoxic, sulfide-reduced mineral medium with 10 mM of either 3-hydroxybutyrate, shikimate, or L-tartrate as the sole organic carbon and energy source, and incubation at 27–30°C has proven to be highly selective for the enrichment of *Ilyobacter polytropus*, *Ilyobacter insuetus*, or *Ilyobacter tartaricus*, respectively. The carbonate-buffered standard medium used for enrichment and isolation has been described in detail (Schink and Pfennig, 1982; Widdel and Pfennig, 1981).

After two to three transfers, the bacteria can be isolated in anoxic agar deep dilution series (Pfennig, 1978) or in roll tubes (Balch et al., 1979). Streaking on Petri dishes in an anoxic glove box has not yet been tried with these bacteria.

Maintenance procedures

Cultures are maintained either by repeated transfer at intervals of 2–3 months or by freezing in liquid nitrogen using techniques common for strictly anaerobic bacteria.

Differentiation of the genus *Ilyobacter* from other genera

The three species remaining in the genus *Ilyobacter* differ from most other strictly anaerobic bacteria in their unusual patterns of substrate utilization and product formation and their low G+C content. With the exception of the genus *Propionigenium*, they are clearly separated from all other genera by their 16S rRNA gene sequences.

Comparative 16S rRNA gene sequence analysis places the members of the genera *Propionigenium* and *Ilyobacter* (with the exception of *Ilyobacter delafieldii*) into the *Fusobacteria* phylum (Brune et al., 2002). Both genera form a distinct cluster, clearly separated from the *Sebaldella–Streptobacillus–Leptotrichia* lineage and the *Fusobacterium* branch. While the 16S rRNA gene sequences did not allow resolving the branching order within the *Ilyobacter–Propionigenium* cluster, the 23S rRNA gene sequences supported a monophyletic status at least for the genus *Ilyobacter* (Brune et al., 2002). Although the metabolic properties of *Propionigenium* and *Ilyobacter* species are sufficiently different to justify maintenance of two separate genera, the situation is unsatisfying and asks for a future taxonomic revision of this group.

Further reading

Dimroth, P., C. von Ballmoos and T. Meier. 2006. Catalytic and mechanical cycles in F-ATP synthases. Fourth in the Cycles Review Series. EMBO Rep. *7*: 276–282.

Brune, A. and B. Schink. 1992. Anaerobic degradation of hydroaromatic compounds by newly isolated fermenting bacteria. Arch. Microbiol. *158*: 320–327.

List of species of the genus *Ilyobacter*

1. **Ilyobacter polytropus** Stieb and Schink 1985, 375[VP] (Effective publication: Stieb and Schink 1984, 145.)

po.ly'tro.pus. N.L. masc. adj. *polytropus* (from Gr. masc. adj. *polytropos*) turning many ways, versatile, referring to metabolic versatility.

Rod-shaped cells, 0.7 × 1.5–3.0 µm in size with rounded ends, single or in pairs. Nonmotile, Gram-stain-negative, nonsporeforming.

Strictly anaerobic chemoorganotroph. 3-Hydroxybutyrate and crotonate fermented to acetate and butyrate. Glycerol fermented to 1,3-propanediol and 3-hydroxypropionate. Malate and fumarate fermented to acetate, formate, and propionate. Glucose and fructose fermented to acetate, formate, and ethanol. No other organic acids, sugars, or alcohols metabolized. Sulfate, sulfur, thiosulfate, and nitrate not reduced. Growth occurs in mineral media with a reductant. Indole not formed; gelatin and urea not hydrolyzed. No catalase activity. No cytochromes detectable.

Growth requires mineral media with a reductant and at least 1% sodium chloride.

Selective enrichment in NaCl-containing mineral media with 3-hydroxybutyrate as substrate.

pH Range: 6.5–8.5, optimum at 7.0–7.5. Temperature range: 10–35°C, optimum growth temperature 30°C. Habitats: anoxic marine or brackish water sediment.

DNA G+C content (mol%): 32.2 ± 0.5 (T_m).

Type strain: CuHbu1, ATCC 51220, DSM 2926.

Sequence accession no. (16S rRNA gene): AJ307981.

2. **Ilyobacter insuetus** Brune, Evers, Kalm, Ludwig and Schink 2002, 431[VP]

in.su.e'tus. L. masc. part. adj. *insuetus* unusual, extraordinary, referring to the organism's metabolism.

Rod-shaped to coccoid cells, 0.8–1.0 µm in diameter and 1.0–1.5 µm long, with rounded ends. Nonmotile, Gram-stain-negative, nonsporeforming.

Strictly anaerobic chemoorganotroph. Quinic acid and shikimic acid utilized for growth and fermented to acetate, propionate, butyrate, H_2, and CO_2. No growth with sugars (cellobiose, fructose, glucose, erythrose, lactose, ribose, xylose), alcohols (*meso*-erythritol, ethanol, glycerol, mannitol), carboxylic acids (citrate, crotonate, fumarate, glycolate, 2-hydroxybutyrate, 3-hydroxybutyrate, 4-hydroxybutyrate, lactate, malate, 2-oxobutyrate, pyruvate, sorbate, tartrate), amino acids (alanine, aspartate, glycine, threonine), or aromatic compounds (gallate, phloroglucinol, protocatechuate, resorcinol, 3,4,5-trimethoxybenzoate, 3,4,5-trimethoxycinnamate). Sulfate, sulfur, thiosulfate, nitrate, and ferric iron not reduced. Strict anaerobe. No catalase activity; no superoxide dismutase activity; no cytochromes. Growth requires mineral media with a reductant and at least 0.7% sodium chloride.

Selective enrichment in NaCl-containing mineral media with quinic acid as the sole source of carbon and energy.

pH Range: 6.0–9.0, optimum at 7.0–8.0. Temperature range: 15–40°C, optimum growth temperature 30°C. Habitats: anoxic marine sediment.

DNA G+C content (mol%): 35.7 ± 1.0 (HPLC).

Type strain: VenChi2, ATCC BAA-291, DSM 6831.

Sequence accession no. (16S rRNA gene): AJ307980.

3. **Ilyobacter tartaricus** Schink 1985, 375[VP] (Effective publication: Schink 1984, 413.)

tar.ta'ri.cus. N.L. n. *acidum tartaricum* tartaric acid; N.L. masc. adj. *tartaricus* referring to tartaric acid as isolation substrate.

Rod-shaped cells, 1.0–1.2 × 1.2–2.5 µm in size, often in chains. Surrounded by slime capsules, nonmotile, Gram-stain-negative, nonsporeforming.

Strictly anaerobic chemoorganotroph. Growth on L-tartrate, citrate, pyruvate, oxaloacetate, glucose, fructose, raffinose, glycerol. Fermentation products include acetate, formate, and ethanol. No growth on formate, acetate, lactate, methanol, ethanol, ethylene glycol, 2,3 butanediol, glycerate, malate, fumarate, glyoxylate, glycolate, mannose, maltose, lactose, sucrose, cellobiose, sorbose, rhamnose, trehalose, xylose, arabinose, peptone, yeast extract. Sulfate, sulfur, thiosulfate or nitrate not reduced. Indole not formed; gelatin and urea not hydrolyzed. No catalase activity, no cytochromes.

Growth requires mineral media with a reductant and at least 1% sodium chloride.

Selective enrichment from marine sediments with L-tartrate as sole carbon and energy source.

pH Range: 5.5–8.0, optimum at 6.5–7.2. Temperature range: 10–40°C, optimum growth temperature 32°C. Habitats: anoxic marine sediment.

DNA G+C content (mol%): 33.1 ± 1.0 (T_m).

Type strain: CraTa2, ATCC 35898, DSM 2382.

Sequence accession no. (16S rRNA gene): AJ307982.

Genus IV. **Propionigenium** Schink and Pfennig 1983, 896[VP] (Effective publication: Schink and Pfennig 1982, 215.)

Bernhard Schink and Peter H. Janssen

Pro.pi.o.ni.ge'ni.um. N.L. n. *acidum propionicum* propionic acid; L. v. *genere* to make, produce; N.L. neut. n. *Propionigenium* propionic acid maker.

Strictly anaerobic chemoorganotrophic bacteria with fermentative metabolism, nonphotosynthetic, inorganic electron acceptors not used. Nonsporeforming.

Chemoorganotrophic, fermentative type of metabolism, preferentially using dicarboxylic acids as substrates. Media containing a reductant are necessary for growth. Catalase-negative. Isolated from anoxic marine or freshwater sediments.

DNA G+C content (mol%): 32.9–41.

Type species: **Propionigenium modestum** Schink and Pfennig 1983, 896[VP] (Effective publication: Schink and Pfennig 1982, 215.).

Further descriptive information

The genus *Propionigenium* consists so far of two species, *Propionigenium modestum* and *Propionigenium maris*. *Propionigenium modestum* comprises four strains of physiologically and morphologically similar isolates from various sources (Schink and Pfennig, 1982). This genus was created to house strictly anaerobic bacteria that are able to grow by decarboxylation of succinate to propionate. Pure cultures could be obtained only with enrichment cultures from marine sources; freshwater enrichments grew much slower, and pure cultures were finally isolated when the sodium chloride concentration of the medium was increased to 100–150 mM. A further species, *Propionigenium maris*, was created later to comprise bacteria similar to *Propionigenium modestum* but which are metabolically much more versatile and are able to ferment, carbohydrates, amino acids, and other organic acids in addition to C4 dicarboxylic acids (Janssen and Liesack, 1995).

Propionigenium modestum was originally isolated from a black, anoxic, marine sediment sample taken from the Canal Grande in Venice, Italy (Schink and Pfennig, 1982). Similar strains were isolated later from many other marine habitats. Enrichments from freshwater sediments sometimes produced cells of similar morphology to *Propionigenium modestum*, and these could be cultivated only in media with increased (100–150 mM) sodium chloride concentrations. Also, *Propionigenium maris* was isolated from marine sediments (Janssen and Liesack, 1995).

It has to be assumed that anoxic marine sediments are the typical habitats of these bacteria. Their energy metabolism is based on sodium ions as coupling ions in energy conservation. With this ability, they are well adapted to a marine environment. Several marine bacteria have been found to use sodium ions as energy couplers in various functions, e.g., respiration.

Propionigenium maris-like bacteria were isolated also from burrows of bromophenol-producing marine infauna, where they apparently are involved in reductive debromination of bromophenols (Watson et al., 2000). They probably use organic excretions of the infauna as electron donors for this reductive reaction.

Enrichment and isolation procedures

A strictly anoxic, sulfide-reduced mineral medium with 20 mM succinate as the sole organic carbon and energy source, and incubation at 27–30°C has proven to be highly selective for the enrichment of *Propionigenium modestum* if marine sediment samples of about 5-ml volume are used as the inoculum. The carbonate-buffered standard medium used for enrichment and isolation has been described in detail (Schink and Pfennig, 1982; Widdel and Pfennig, 1981).

After two to three transfers in liquid medium, gas should no longer be formed by the enrichment cultures, and a dominant population of short, coccoid rods should be established. These bacteria can be isolated in anoxic agar deep dilution series (Pfennig, 1978) or in roll tubes (Balch et al., 1979). Streaking Petri dishes in an anoxic glove box has not yet been tried with these bacteria. Preparation of pure cultures requires two subsequent dilution series; purity should be checked after growth in selective mineral medium and in complex medium.

Propionigenium maris requires yeast extract (0.1% w/v) for growth in pure culture. Although it is not recommended to add yeast extract in the liquid enrichment cultures, it is needed in the purification step.

Maintenance procedures

Cultures are maintained either by repeated transfer at intervals of 2–3 months or by freezing in liquid nitrogen using techniques common for strictly anaerobic bacteria. No information exists about survival upon lyophilization.

Differentiation of the genus *Propionigenium* from other genera

In phase-contrast microscopy, cells of *Propionigenium modestum* appear as short, coccoid rods with a diameter of 0.5–0.6 μm and a length of 0.5–2.0 μm, often in short chains. They are Gram-stain-negative and nonsporeforming. *Propionigenium maris* also forms coccoid to ovoid cells or short rod-like cells similar in size to *Propionigenium modestum*.

Both *Propionigenium* species are strictly anaerobic and do not tolerate increased oxygen tensions. They are specialists for the utilization of C4-dicarboxylic acids. Whereas *Propionigenium modestum* is restricted to use of only few such compounds, *Propionigenium maris* uses also other substrates such as sugars and amino acids, and depends on yeast extract as a medium additive. No cytochromes have been detected, which is consistent with the absence of electron transport phosphorylation.

Early 16S rRNA gene sequence analyses revealed that *Propionigenium maris* and *Propionigenium modestum* are closely related and form a distinct lineage within a phylogenetically coherent group characterized by *Fusobacterium nucleatum* and other *Fusobacterium* species, together with *Clostridium rectum*, *Leptotrichia buccalis*, and *Sebaldella termitidis* (Both et al., 1991; Janssen and Liesack, 1995). This group has been elevated to the rank of phylum, *Fusobacteria*. In addition to *Propionigenium*, this phylum embraces the genera *Fusobacterium*, *Ilyobacter*, *Leptotrichia*, *Sebaldella*, *Streptobacillus*, and *Sneathia* (Garrity et al., 2002).

List of species of the genus *Propionigenium*

1. **Propionigenium modestum** Schink and Pfennig 1983, 896[VP] (Effective publication: Schink and Pfennig 1982, 215.)

mo.de'stum. L. neut. adj. *modestum* modest, referring to an extremely modest type of metabolism.

Rod-shaped to coccoid cells, 0.5–0.6 in diameter × 0.5–2.0 μm long, with rounded ends, single, in pairs, or in chains. Nonmotile, Gram-stain-negative, nonsporeforming.

Strictly anaerobic chemoorganotroph. Succinate, fumarate, malate, aspartate, oxaloacetate, and pyruvate utilized for growth and fermented to propionate (acetate), and CO_2. No other organic acids and no sugars or alcohols metabolized. Sulfate, sulfur, thiosulfate, and nitrate not reduced. Indole not formed; gelatin and urea not hydrolyzed. No catalase activity.

Growth requires mineral media with a reductant and at least 1% sodium chloride. Selective enrichment in NaCl-containing mineral media with succinate as substrate.

pH range 6.5–8.4, optimum at 7.1–7.7. Temperature range 15–40°C, optimum 33°C. No cytochromes detectable. Habitats: anoxic marine or brackish water sediment.

DNA G+C content (mol%): 33.9 ± 1.0 (T_m).

Type strain: GraSucc2, ATCC 35614, DSM 2376.

Sequence accession no. (16S rRNA gene): X54275.

2. **Propionigenium maris** Janssen and Liesack 1996, 362[VP] emend. Watson, Matsui, Leaphart, Wiegel, Rainey and Lovell 2000, 1040 (Effective publication: Janssen and Liesack 1995, 33.)

ma'ris. L. neut. n. *mare* the sea; L. gen. n. *maris* of the sea, referring to the tidal mat flats from which this organism was isolated.

Coccoid to oval short rods with rounded ends, 1.0 μm in diameter × 1.2–2.5 μm long; under some culture conditions up to 50 μm long. Gram-stain-negative, nonsporeforming.

Strictly anaerobic chemoorganotroph. Fermentative metabolism, external electron acceptors not used, however, bromophenols can be reductively dehalogenated. No cytochromes formed. In the presence of yeast extract, several carbohydrates and amino and organic acids are fermented. These substrates include succinate, fumarate, pyruvate, citrate, 3-hydroxybutyrate, glucose, fructose, maltose, aspartate, lysine, threonine, glutamate, and cysteine. Typical products of fermentation are propionate, acetate, and formate, depending on the substrate. Carbohydrates are fermented to formate, acetate, ethanol, and lactate. 3-Hydroxbutyrate is fermented to acetate and butyrate. Hydrogen is produced from carbohydrates and yeast extract, ammonia from amino acids, and sulfide from cysteine. Sulfate, sulfur, thiosulfate, and nitrate not reduced. Indole formed from L-tryptophan. Esculin and urea not hydrolyzed. No catalase activity.

Growth in salt-water media with at least 5 and up to 55 g NaCl per liter. Anoxic conditions required for growth.

pH range 5.3–8.8, optimum at 6.9–7.7. Temperature range 15–40°C, optimum 34–37°C. Habitat: anoxic marine or brackish water sediment.

DNA G+C content (mol%): 40.0 ± 1.0 (T_m)

Type strain: 10succ1, DSM 9537.

Sequence accession no. (16S rRNA gene): X84049.

References

Adriaans, B. and H. Shah. 1988. *Fusobacterium ulcerans* sp. nov. from tropical ulcers. Int. J. Syst. Bacteriol. *38*: 447–448.

Bachrach, G., S.K. Haake, A. Glick, R. Hazan, R. Naor, R.N. Andersen and P.E. Kolenbrander. 2004. Characterization of the novel *Fusobacterium nucleatum* plasmid pKH9 and evidence of an addiction system. Appl. Environ. Microbiol. *70*: 6957–6962.

Baird-Parker, A.C. 1960. The classification of *Fusobacteria* from the human mouth. J. Gen. Microbiol. *22*: 458–469.

Bakken, V., S. Aaro, T. Hofstad and E.N. Vasstrand. 1989a. Outer membrane proteins as major antigens of *Fusobacterium nucleatum*. FEMS Microbiol. Immunol. *1*: 473–483.

Bakken, V., S. Aaro and H.B. Jensen. 1989b. Purification and partial characterization of a major outer-membrane protein of *Fusobacterium nucleatum*. J. Gen. Microbiol. *135*: 3253–3262.

Balch, W.E., G.E. Fox, L.J. Magrum, C.R. Woese and R.S. Wolfe. 1979. Methanogens: reevaluation of a unique biological group. Microbiol. Rev. *43*: 260–296.

Barnes, E.M. and C.S. Impey. 1968. Anaerobic gram negative nonsporing bacteria from the caeca of poultry. J. Appl. Bacteriol. *31*: 530–541.

Bolstad, A.I. and H.B. Jensen. 1993. Polymerase chain reaction-amplified nonradioactive probes for identification of *Fusobacterium nucleatum*. J. Clin. Microbiol. *31*: 528–532.

Bolstad, A.I., J. Tommassen and H.B. Jensen. 1994. Sequence variability of the 40-kDa outer membrane proteins of *Fusobacterium nucleatum* strains and a model for the topology of the proteins. Mol. Gen. Genet. *244*: 104–110.

Both, B., G. Kaim, J. Wolters, K.H. Schleifer, E. Stackebrandt and W. Ludwig. 1991. *Propionigenium modestum*: a separate line of descent within the eubacteria. FEMS Microbiol. Lett. *62*: 53–58.

Bouma, C.L., J. Reizer, A. Reizer, S.A. Robrish and J. Thompson. 1997. 6-phospho-α-D-glucosidase from *Fusobacterium mortiferum*: cloning, expression, and assignment to family 4 of the glycosylhydrolases. J. Bacteriol. *179*: 4129–4137.

Brazier, J.S., D.M. Citron and E.J. Goldstein. 1991. A selective medium for *Fusobacterium* spp. J. Appl. Bacteriol. *71*: 343–346.

Brown, R., H.G. Lough and I.R. Poxton. 1997. Phenotypic characteristics and lipopolysaccharides of human and animal isolates of *Fusobacterium necrophorum*. J. Med. Microbiol. *46*: 873–878.

Brune, A., S. Evers, G. Kaim, W. Ludwig and B. Schink. 2002. *Ilyobacter insuetus* sp. nov., a fermentative bacterium specialized in the degradation of hydroaromatic compounds. Int. J. Syst. Evol. Microbiol. *52*: 429–432.

Carlsson, J., J.T. Larsen and M.B. Edlund. 1994. Utilization of glutathione (L-γ-glutamyl-L-cysteinylglycine) by *Fusobacterium nucleatum* subspecies nucleatum. Oral Microbiol. Immunol. *9*: 297–300.

Citron, D.M. 2002. Update on the taxonomy and clinical aspects of the genus *Fusobacterium*. Clin. Infect. Dis. *35*: S22–27.

Claesson, R., M.B. Edlund, S. Persson and J. Carlsson. 1990. Production of volatile sulfur compounds by various *Fusobacterium* species. Oral Microbiol. Immunol. *5*: 137–142.

Claros, M.C., Y. Papke, N. Kleinkauf, D. Adler, D. M. Citron, S. Hunt-Gerardo, Th. Montag, E.J.C. Goldstein and A.C. Rodloff. 1999. Characteristics of *Fusobacterium ulcerans*, a new and unusual species compared with *Fusobacterium varium* and *Fusobacterium mortiferum*. Anaerobe *5*: 137–140.

Conrads, G., M.C. Claros, D.M. Citron, K.L. Tyrrell, V. Merriam and E.J. Goldstein. 2002. 16S-23S rDNA internal transcribed spacer sequences for analysis of the phylogenetic relationships among species of the genus *Fusobacterium*. Int. J. Syst. Evol. Microbiol. *52:* 493–499.

Conrads, G., D.M. Citron, R. Mutters, S. Jang and E.J.C. Goldstein. 2004a. *Fusobacterium canifelinum* sp. nov., from the oral cavity of cats and dogs. Syst. Appl. Microbiol. *27:* 407–413.

Conrads, G., D.M. Citron, R. Mutters, S. Jang and E.J.C Goldstein. 2004b. *In* Validation of publication of new names and new combinations previously effectively published outside the IJSEM. List no. 100. Int. J. Syst. Evol. Microbiol. *54*: 1909–1910.

Conrads, G., D.M. Citron and E.J. Goldstein. 2005. Genetic determinant of intrinsic quinolone resistance in *Fusobacterium canifelinum*. Antimicrob. Agents Chemother. *49*: 434–437.

Dahlen, G., H. Nygren and H.A. Hannsson. 1978. Immunoelectron microscopic localization of lipopolysaccharides in the cell wall of *Bacteroides oralis* and *Fusobacterium nucleatum*. Infect. Immun. *19*: 265–271.

Dahlen, G. and I. Mattsby-Baltzer. 1983. Lipid A in anaerobic bacteria. Infect. Immun. *39*: 466–468.

Dorsch, M., D.N. Love and G.D. Bailey. 2001. *Fusobacterium equinum* sp. nov., from the oral cavity of horses. Int. J. Syst. Evol. Microbiol. *51:* 1959–1963.

Downes, J., A. King, J. Hardie and I. Phillips. 1999. Evaluation of the Rapid ID 32A system for identification of anaerobic Gram-negative bacilli, excluding the *Bacteroides fragilis* group. Clin. Microbiol. Infect. *5*: 319–326.

Dzink, J.L., M.T. Sheenan and S.S. Socransky. 1990. Proposal of three subspecies of *Fusobacterium nucleatum* Knorr 1922: *Fusobacterium nucleatum* subsp. *nucleatum* subsp. nov., comb. nov., *Fusobacterium nucleatum* subsp. *polymorphum* subsp. nov., nom. rev., comb. nov., and *Fusobacterium nucleatum* subsp. *vincentii* subsp. nov., nom. rev., comb. nov. Int. J. Syst. Bacteriol. *40*: 74–78.

Eggerth, A.H. and B.H. Gagnon. 1933. The *Bacteroides* of human feces. J. Bacteriol. *25*: 389–413.

Falkler, W.A., Jr. and C.E. Hawley. 1977. Hemagglutinating activity of *Fusobacterium nucleatum*. Infect. Immun. *15*: 230–238.

Finegold, S.M. 1977. Anaerobic bacteria in human disease. Academic Press, New York.

Finegold, S.M., M.L. Vaisanen, D.R. Molitoris, T.J. Tomzynski, Y. Song, C. Liu, M.D. Collins and P.A. Lawson. 2003a. *Cetobacterium somerae* sp. nov. from human feces and emended description of the genus *Cetobacterium*. Syst. Appl. Microbiol. *26*: 177–181.

Finegold, S.M., M.L. Vaisanen, D.R. Molitoris, T.J. Tomzynski, Y. Song, C. Liu, M.D. Collins and P.A. Lawson. 2003b. *In* Valid publication of new names and new contributions previously effectively published outside the IJSEM. List no. 93. Int. J. Syst. Evol. Microbiol. *53*: 1219–1220.

Flügge, C. 1886. Die Mikrooganismen. F.C.W. Vogel, Leipzig.

Foster, G., H.M. Ross, R.D. Naylor, M.D. Collins, C.P. Ramos, F. Fernández-Garayzábal and R.J. Reid. 1995. *Cetobacterium ceti* gen. nov., sp. nov., a new Gram-negative obligate anaerobe from sea mammals. Lett. Appl. Microbiol. *21*: 202–206.

Foster, G., H.M. Ross, R.D. Naylor, M.D. Collins, C.R. Pascual, F. Fernández-Garayzábal and R.J. Reid. 1996. *In* Validation of the publication of new names and new contributions previously effectively published outside the IJSEM. List no. 56. Int. J. Syst. Evol. Microbiol. *46*: 362–363.

Garrity, G.M., K.L. Johnson, J.A. Bell and D.B. Searles. 2002. Taxonomic outline of the Prokaryotes. *In* Bergey's Manual of Systematic Bacteriology, 2nd edn. Springer, New York.

Gemperli, A.C., P. Dimroth and J. Steuber. 2003. Sodium ion cycling mediates energy coupling between complex I and ATP synthase. Proc. Natl. Acad. Sci. U.S.A. *100*: 839–844.

Gharbia, S.E. and H.N. Shah. 1989. The uptake of amino acids from a chemically defined medium by *Fusobacterium* species. Curr. Microbiol. *18*: 189–193.

Gharbia, S.E., H.N. Shah and S.G. Welch. 1989. The influence of peptides on the uptake of amino acids in *Fusobacterium* species; predicted interactions with *Porphyromonas gingivalis* Curr. Microbiol. *19*: 231–235.

Gharbia, S.E. and H.N. Shah. 1990. Identification of *Fusobacterium* species by the electrophoretic migration of glutamate dehydrogenase and 2-oxoglutarate reductase in relation to their DNA base composition and peptidoglycan dibasic amino acids. J. Med. Microbiol. *33*: 183–188.

Gharbia, S.E. and H.N. Shah. 1991a. Comparison of the amino acid uptake profile of reference and clinical isolates of *Fusobacterium nucleatum* subspecies. Oral Microbiol. Immunol. *6*: 264–269.

Gharbia, S.E. and H.N. Shah. 1991b. Pathways of glutamate catabolism among *Fusobacterium* species. J. Gen. Microbiol. *137*: 1201–1206.

Gharbia, S.E. and H.N. Shah. 1992. *Fusobacterium nucleatum* subsp. *fusiforme* subsp. nov. and *Fusobacterium nucleatum* subsp. *animalis* subsp. nov. as additional subspecies within *Fusobacterium nucleatum*. Int. J. Syst. Bacteriol. *42*: 296–298.

Goldstein, E.J., D.M. Citron and C.V. Merriam. 1999. Linezolid activity compared to those of selected macrolides and other agents against aerobic and anaerobic pathogens isolated from soft tissue bite infections in humans. Antimicrob. Agents Chemother. *43*: 1469–1474.

Gregory, E.M., W.E. Moore and L.V. Holdeman. 1978. Superoxide dismutase in anaerobes: survey. Appl. Environ. Microbiol. *35*: 988–991.

Haake, S.K. and X. Wang. 1997. Cloning and expression of *fomA*, the major outer-membrane protein gene from *Fusobacterium nucleatum* T18. Arch. Oral Biol. *42*: 19–24.

Haake, S.K., S.C. Yoder, G. Attarian and K. Podkaminer. 2000. Native plasmids of *Fusobacterium nucleatum*: characterization and use in development of genetic systems. J. Bacteriol. *182*: 1176–1180.

Hall, V., B.I. Duerden, J.T. Magee, H.C. Ryley and J.S. Brazier. 1997. A comparative study of *Fusobacterium necrophorum* strains from human and animal sources by phenotypic reactions, pyrolysis mass spectrometry and SDS-PAGE. J. Med. Microbiol. *46*: 865–871.

Hallé, J. 1898. Recherches sur la bactériologie du canal génital de la femme (état normal et pathologique). Thesis, Paris.

Harris, N.M. 1901. *Bacillus mortiferus* (nov. spec.). J. Exp. Med. *6*: 519–547.

Hauduroy, A., G. Ehringer, A. Urbain, G. Guillot and J. Magrou. 1937. Dictionnaire des bactéries pathogènes. Masson et Cie, Paris.

Hermansson, K., M.B. Perry, E. Altman, J.R. Brisson and M.M. Garcia. 1993. Structural studies of the O-antigenic polysaccharide of *Fusobacterium necrophorum*. Eur. J. Biochem. *212*: 801–809.

Hill, G.B. 1993. Investigating the source of amniotic fluid isolates of fusobacteria. Clin. Infect. Dis. *16 Suppl. 4*: S423–424.

Hilpert, W., B. Schink and P. Dimroth. 1984. Life by a new decarboxylation-dependent energy conservation mechanism with Na^+ as coupling ion. EMBO J. *3*: 1665–1670.

Hofstad, T. 1974. The distribution of heptose and 2-keto-3-deoxy-octonate in *Bacteroidaceae*. J. Gen. Microbiol. *85*: 314–320.

Hofstad, T. 1979. Serological responses to antigens of *Bacteroidaceae*. Microbiol. Rev. *43*: 103–115.

Hofstad, T. and N. Skaug. 1980. Fatty acids and neutral sugars present in lipopolysaccharides isolated from *Fusobacterium* species. Acta Pathol. Microbiol. Scand. B *88*: 115–120.

Holbrook, W.P. and C. McMillan. 1977. The hydrolysis of dextran by gram negative non-sporeforming anaerobic bacilli. J. Appl. Bacteriol. *42*: 259–273.

Janssen, P.H. and C.G. Harfoot. 1990. *Ilyobacter delafieldii* sp. nov., a metabolically restricted anaerobic bacterium fermenting PHB. Arch. Microbiol. *154*: 253–259.

Janssen, P.H. and C.G. Harfoot. 1991. *In* Validation of the publication of new names and new combinations previously effectively published outside the IJSB. List no. 37. Int. J. Syst. Bacteriol. *41*: 331.

Janssen, P.H. and W. Liesack. 1995. Succinate decarboxylation by *Propionigenium maris* sp. nov., a new anaerobic bacterium from an estuarine sediment. Arch. Microbiol. *164*: 29–35.

Janssen, P.H. and W. Liesack. 1996. *In* Validation of the publication of new names and new combinations previously effectively published outside the IJSB. List no. 56. Int. J. Syst. Bacteriol. *46*: 362–363.

Jantzen, E. and T. Hofstad. 1981. Fatty acids of *Fusobacterium* species: taxonomic implications. J. Gen. Microbiol. *123*: 163–171.

Jensen, H.B., J. Skeidsvoll, A. Fjellbirkeland, B. Hogh, P. Puntervoll, H. Kleivdal and J. Tommassen. 1996. Cloning of the *fomA* gene, encoding the major outer membrane porin of *Fusobacterium nucleatum* ATCC10953. Microb. Pathog. *21*: 331–342.

Jin, J., D. Xu, W. Narongwanichgarn, Y. Goto, T. Haga and T. Shinjo. 2002. Characterization of the 16S-23S rRNA intergenic spacer regions among strains of the *Fusobacterium necrophorum* cluster. J. Vet. Med. Sci. *64*: 273–276.

Jousimies-Somer, H. and P. Summanen. 2002. Recent taxonomic changes and terminology update of clinically significant anaerobic gram-negative bacteria (excluding spirochetes). Clin. Infect. Dis. *35*: S17–21.

Jungano, M. 1909. Sur la flore anaérobie du rat. C.R. Soc. Biol. Paris *66*: 112–114; 122–124.

Kapatral, V., I. Anderson, N. Ivanova, G. Reznik, T. Los, A. Lykidis, A. Bhattacharyya, A. Bartman, W. Gardner, G. Grechkin, L. Zhu, O. Vasieva, L. Chu, Y. Kogan, O. Chaga, E. Goltsman, A. Bernal, N. Larsen, M. D'Souza, T. Walunas, G. Pusch, R. Haselkorn, M. Fonstein, N. Kyrpides and R. Overbeek. 2002. Genome sequence and analysis of the oral bacterium *Fusobacterium nucleatum* strain ATCC 25586. J. Bacteriol. *184*: 2005–2018.

Kapatral, V., N. Ivanova, I. Anderson, G. Reznik, A. Bhattacharyya, W.L. Gardner, N. Mikhailova, A. Lapidus, N. Larsen, M. D'Souza, T. Walunas, R. Haselkorn, R. Overbeek and N. Kyrpides. 2003. Genome analysis of *F. nucleatum* sub spp vincentii and its comparison with the genome of *F. nucleatum* ATCC 25586. Genome Res. *13*: 1180–1189.

Kato, K., T. Umemoto, H. Sagawa and S. Kotani. 1979. Lanthionine as an essential constituent of cell wall peptidoglycan of *Fusobacterium nucleatum*. Curr. Microbiol. *3*: 147–151.

Kato, K., T. Umemoto, H. Fukuhara, H. Sagawa and S. Kotani. 1981. Variation of dibasic amino acid in the cell wall peptidoglycan of bacteria of genus *Fusobacterium*. FEMS Microbiol. Lett. *10*: 81–85.

Kawamura, Y. 1926. A coryne-bacillus as a cause of abscess in the feet of hens. J. Japan. Soc. Vet. Sci. *5*: 22.

Kleivdal, H., R. Benz and H.B. Jensen. 1995. The *Fusobacterium nucleatum* major outer-membrane protein (FomA) forms trimeric, water-filled channels in lipid bilayer membranes. Eur. J. Biochem. *233*: 310–316.

Knorr, M. 1922. Über die fusospirilläre Symbiose, die Gattung *Fusobacterium* (K.B. Lehmann) und *Spirillum sputigenum*. Zugleich ein Beiträg zűr Bakteriologie der Mundhohle. II. Mitteilung. Die. Gattung *Fusobacterium*. I Abt. Orig. Zentralbl. Bakteriol. Parasitenkd. Infektionskr. Hyg. *89*: 4–22.

Langworth, B.F. 1977. *Fusobacterium necrophorum*: its characteristics and role as an animal pathogen. Bacteriol. Rev. *41*: 373–390.

Lawson, P.A., S.E. Gharbia, H.N. Shah, D.R. Clark and M.D. Collins. 1991. Intrageneric relationships of members of the genus *Fusobacterium* as determined by reverse transcriptase sequencing of small-subunit rRNA. Int. J. Syst. Bacteriol. *41*: 347–354.

Lippke, J.A., W.J. Peros, M.W. Keville, E.D. Savitt and C.K. French. 1991. DNA probe detection of *Eikenella corrodens*, *Wolinella recta* and *Fusobacterium nucleatum* in subgingival plaque. Oral Microbiol. Immunol. *6*: 81–87.

Loesche, W.J. 1969. Oxygen sensitivity of various anaerobic bacteria. Appl. Microbiol. *18*: 723–727.

Love, D.N., E.P. Cato, J.L. Johnson, R.F. Jones and M. Bailey. 1987. Deoxyribonucleic acid hybridization among strains of fusobacteria isolated from soft tissue infections of cats: Comparison with human and animal type strains from oral and other sites. Int. J. Syst. Bacteriol. *37*: 23–26.

McKay, T.L., J. Ko, Y. Bilalis and J.M. DiRienzo. 1995. Mobile genetic elements of *Fusobacterium nucleatum*. Plasmid *33*: 15–25.

Meier, T., P. Polzer, K. Diederichs, W. Welte and P. Dimroth. 2005. Structure of the rotor ring of F-Type Na^+-ATPase from *Ilyobacter tartaricus*. Science *308*: 659–662.

Miyagawa, E., R. Azuma and T. Suto. 1979. Cellular fatty acid composition in Gram-negative obligately anaerobic rods. J. Gen. Microbiol. *25*: 41–51.

Miyasaki, K.T., R. Iofel, A. Oren, T. Huynh and R.I. Lehrer. 1998. Killing of *Fusobacterium nucleatum*, *Porphyromonas gingivalis* and *Prevotella intermedia* by protegrins. J. Periodontal. Res. *33*: 91–98.

Moore, W.E.C. and L.V. Holdeman. 1969. Anaerobic Gram-negative non-sporeforming rods. *In* Outline of Clinical Methods in Anaerobic Bacteriology, 1st revn (edited by Cato, Cummins, Holdeman, Johnson, Moore, Smibert and Smith). Virginia Polytechnic Institute Anaerobe Laboratory, Blacksburg, Virginia.

Moore, W.E.C. and L.V. Holdeman. 1970. *Fusobacterium*. *In* Outline of Clinical Methods in Anaerobic Bacteriology, 2nd revn (edited by Cato, Cummins, Holdeman, Johnson, Moore, Smibert and Smith). Virginia Polytechnic Institute Anaerobe Laboratory, Blacksburg, Virginia.

Moore, W.E.C. and L.V. Holdeman. 1973. New names and combinations in genera *Bacteroides* Castellani and Chalmers, *Fusobacterium* Knorr, *Eubacterium* Prevot, *Propionibacterium* Delwich, and *Lactobacillus* Orla-Jensen. Int. J. Syst. Bacteriol. *23*: 69–74.

Narayanan, S.K., T.G. Nagaraja, M.M. Chengappa and G.C. Stewart. 2001. Cloning, sequencing, and expression of the leukotoxin gene from *Fusobacterium necrophorum*. Infect. Immun. *69*: 5447–5455.

Neumann, S., U. Matthey, G. Kaim, and P. Dimroth. 1998. Purification and properties of the F_1F_0 ATPase of *Ilyobacter tartaricus*, a sodium ion pump. J. Bacteriol. *180*: 3312–3316.

Nicholson, L.A., C.J. Morrow, L.A. Corner and A.L. Hodgson. 1994. Phylogenetic relationship of *Fusobacterium necrophorum* A, AB, and B biotypes based upon 16S rRNA gene sequence analysis. Int. J. Syst. Bacteriol. *44*: 315–319.

Pfennig, N. 1978. *Rhodocyclus purpureus* gen. nov. and sp. nov. a ring-shaped, vitamin-B_{12}-requiring member of family *Rhodospirillaceae*. Int. J. Syst. Bacteriol. *28*: 283–288.

Pianotti, R., S. Lachette and S. Dills. 1986. Desulfuration of cysteine and methionine by *Fusobacterium nucleatum*. J. Dent. Res. *65*: 913–917.

Pikis, A., S. Immel, S.A. Robrish and J. Thompson. 2002. Metabolism of sucrose and its five isomers by *Fusobacterium mortiferum*. Microbiology *148*: 843–852.

Porschen, R.K. and S. Sonntag. 1974. Extracellular deoxyribonuclease production by anaerobic bacteria. Appl. Microbiol. *27*: 1031–1033.

Porschen, R.K. and E.H. Spaulding. 1974. Phosphatase activity of anaerobic organisms. Appl. Microbiol. *27*: 744–747.

Prévot, A.R. 1938. Etudes de systematique bacterienne. III. Invalidite du genre *Bacteroides* Castellani et Chalmers demembrement et reclassification. Ann. Inst. Pasteur *20*: 285–307.

Prévot, A.R., A. Turpin and P. Kaiser. 1967. Les bactéries anaérobies. Dunod, Paris.

Roberts, M.C. and J. Lansciardi. 1990. Transferable TetM in *Fusobacterium nucleatum*. Antimicrob. Agents Chemother. *34*: 1836–1838.

Robrish, S.A., C. Oliver and J. Thompson. 1987. Amino acid-dependent transport of sugars by *Fusobacterium nucleatum* ATCC 10953. J. Bacteriol. *169*: 3891–3897.

Robrish, S.A., C. Oliver and J. Thompson. 1991. Sugar metabolism by fusobacteria: regulation of transport, phosphorylation, and polymer formation by *Fusobacterium mortiferum* ATCC 25557. Infect. Immun. *59*: 4547–4554.

Rogers, A.H., N.J. Gully, A.L. Pfennig and P.S. Zilm. 1992. The breakdown and utilization of peptides by strains of *Fusobacterium nucleatum*. Oral Microbiol. Immunol. *7*: 299–303.

Russ, V.R. 1905. Über ein Inflenzabacillenahnliches anaerobes. Stabchen. Zentralbl. Bakteriol. Parasitenkd. Infektionskr. Hyg., I Abt. Orig. *39*: 357.

Schink, B. and N. Pfennig. 1982. *Propionigenium modestum* gen.nov. sp. nov. a new strictly anaerobic, non-sporing bacterium growing on succinate. Arch. Microbiol. *133*: 209–216.

Schink, B. and N. Pfennig. 1983. *In* Validation of the publication of new names and new combinations previously effectively published outside the IJSB. List no. 12. Int. J. Syst. Bacteriol. *33*: 896–897.

Schink, B. 1984. Fermentation of tartrate enantiomers by anaerobic bacteria, and description of two new species of strict anaerobes, *Ruminococcus pasteurii* and *Ilyobacter tartaricus*. Arch. Microbiol. *139*: 409–414.

Schink, B. 1985. *In* Validation of the publication of new names and new combinations previously effectively published outside the IJSB. List no. 18. Int. J. Syst. Bacteriol. *35*: 375–376.

Schink, B., P.H. Janssen and J. Frings. 1992. Microbial degradation of natural and of new synthetic polymers. FEMS Microbiol. Rev. *9*: 311–316.

Sebald, M. 1962. Étude sur les bactéries anaérobies gram-négatives asporulées. Thèses de l'Université Paris, Imprimerie Barnéoud S.A., Laval, France.

Shinjo, T., T. Fujisawa and T. Mitsuoka. 1991. Proposal of two subspecies of *Fusobacterium necrophorum* (Flugge) Moore and Holdeman: Fusobacteriumnecrophorum subsp. necrophorum subsp. nov., nom. rev (*ex* Flugge 1886), and *Fusobacterium necrophorum* subsp. *funduliforme* subsp. nov., nom. rev. (*ex* Halle 1898). Int. J. Syst. Bacteriol. *41*: 395–397.

Simon, P.C. 1975. A simple method for rapid identification of *Sphaerophorus necrophorus* isolates. Can. J. Comp. Med. *39*: 349–353.

Slots, J. and T.V. Potts. 1982. *Fusobacterium simiae*, a new species from monkey dental plaque. Int. J. Syst. Bacteriol. *32*: 191–194.

Slots, J., T.V. Potts and P.A. Mashimo. 1983. *Fusobacterium periodonticum*, a new species from the human oral cavity. J. Dent. Res. *62*: 960–963.

Slots, J., T.V. Potts and P.A. Mashimo. 1984. *In* Validation of the publication of new names and new combinations previously effectively published outside the IJSB. List no. 14. Int. J. Syst. Bacteriol. *34*: 349–353.

Socransky, S.S., A.D. Haffajee, M.A. Cugini, C. Smith and R.L. Kent, Jr. 1998. Microbial complexes in subgingival plaque. J. Clin. Periodontol. *25*: 134–144.

Spaulding, E.H. and L.F. Rettger. 1937. *Fusobacterium* genus I. Biochemical and serological classification. J. Bacteriol. *34*: 535–548.

Stieb, M. and B. Schink. 1984. A new 3-hydroxybutyrate fermenting anaerobe, *Ilyobacter polytropus*, gen. nov., sp. nov., possessing various fermentation pathways. Arch. Microbiol. *140*: 139–146.

Stieb, M. and B. Schink. 1985. *In* Validation of the publication of new names and new combinations previously effectively published outside the IJSB. List no. 18. Int. J. Syst. Bacteriol. *35*: 375–376.

Sunde, P.T., L. Tronstad, E.R. Eribe, P.O. Lind and I. Olsen. 2000. Assessment of periradicular microbiota by DNA–DNA hybridization. Endod. Dent. Traumatol. *16*: 191–196.

Sutter, V.L., P.T. Sugihara and S.M. Finegold. 1971. Rifampin-blood-agar as a selective medium for the isolation of certain anaerobic bacteria. Appl. Microbiol. *22*: 777–780.

Tadepalli, S., G.C. Stewart, T.G. Nagaraja and S.K. Narayanan. 2008. Leukotoxin operon and differential expressions of the leukotoxin gene in bovine *Fusobacterium necrophorum* subspecies. Anaerobe *14*: 13–18.

Takahashi, N. and T. Sato. 2002. Dipeptide utilization by the periodontal pathogens *Porphyromonas gingivalis*, *Prevotella intermedia*, *Prevotella nigrescens* and *Fusobacterium nucleatum*. Oral Microbiol. Immunol. *17*: 50–54.

Tanner, A., M.F. Maiden, B.J. Paster and F.E. Dewhirst. 1994. The impact of 16S ribosomal RNA-based phylogeny on the taxonomy of oral bacteria. Periodontol. 2000 *5*: 26–51.

Terada, A., K. Uchida and T. Mitsuoka. 1976. Die Bacteroidaceenflora in den faeces von schweinen. Zentralbl. Bakteriol. Parasitenkd. Infektionskr. Hyg. Abt. Orig. *234*: 362–370.

Thompson, J., S.A. Robrish, C.L. Bouma, D.I. Freedberg and J.E. Folk. 1997. Phospho-β-glucosidase from *Fusobacterium mortiferum*: purification, cloning, and inactivation by 6-phosphoglucono-delta-lactone. J. Bacteriol. *179*: 1636–1645.

Tissier, H. 1905. Répartition des microbes dans l'intestin du nourrisson. Ann. Inst. Pasteur (Paris) *19*: 109–123.

Topley, W.W.C. and G.S. Wilson. 1929. The Principles of Bacteriology and Immunity, vol. 1. Edward Arnold, London.

Tunnicliff, R. and L. Jackson. 1925. *Bacillus gonidiaformans* (n. sp.) - an hitherto undescribed organism. J. Infect. Dis. *36*: 430–438.

Turner, K., L. Lindqvist and C.E. Nord. 1985. Purification and properties of a novel β–lactamase from *Fusobacterium nucleatum*. Antimicrob. Agents Chemother. *27*: 943–947.

Van Assche, P.F. and A.T. Wilssens. 1977. *Fusobacterium perfoetens* (Tissier) Moore and Holdeman 1973: description and proposed neotype strain. Int. J. Syst. Bacteriol. *27*: 1–5.

Vasstrand, E.N. 1981. Lysozyme digestion and chemical characterization of the peptidoglycan of *Fusobacterium nucleatum* Fev 1. Infect. Immun. *33*: 75–82.

Vasstrand, E.N., H.B. Jensen, T. Miron and T. Hofstad. 1982. Composition of peptidoglycans in *Bacteroidaceae*: determination and distribution of lanthionine. Infect. Immun. *36*: 114–122.

Veillon, A. and A. Zuber. 1898. Recherches sur quelques microbes strictement anaérobies et leur rôle en pathologie. Arch. Med. Exp. *10*: 517–545.

Watson, J., G.Y. Matsui, A. Leaphart, J. Wiegel, F.A. Rainey and C.R. Lovell. 2000. Reductively debrominating strains of *Propionigenium maris* from burrows of bromophenol-producing marine infauna. Int. J. Syst. Evol. Microbiol. *50*: 1035–1042.

Weinberg, M., R. Nativelle and A.R. Prévot. 1937. Les microbes anaérobies. Masson et Cie, Paris.

Werner, H. 1974. Demonstration of lysine decarboxylase activity in the obligately anaerobic bacterium *Sphaerophorus varius*. Zentrabl. Bakteriol. Parasitendk. Infektionskr. Hyg. I. Abt. Orig. A *226*: 364–368.

Widdel, F. and N. Pfennig. 1981. Studies on dissimilatory sulfate-reducing bacteria that decompose fatty acids. 1. Isolation of new sulfate-reducing bacteria enriched with acetate from saline environments: description of *Desulfobacter postgatei* gen. nov., sp. nov. Arch. Microbiol. *129*: 395–400.

Family II. **Leptotrichiaceae** fam. nov.

James T. Staley and William B. Whitman

Lep.to.tri.chi.a.ce′a.e. N.L. fem. n. *Leptotrichia* type genus of the family; suff. *-aceae* ending to denote a family; N.L. fem. pl. n. *Leptotrichiaceae* the *Leptotrichia* family.

The family *Leptotrichiaceae* is described in part on the basis of phylogenetic analyses of the 16S rRNA gene sequences of its members. Facultative to obligately anaerobic organisms that stain as Gram-negative rods. All described species are nonmotile and fermentative. Ferment carbohydrates to produce various organic acids including lactic, acetic, formic or succinic depending on the substrate and species. Some species are fastidious and require serum or blood for growth. Some species have been isolated from human clinical specimens and are pathogenic to humans. Some species occur in the human oral cavity and others in the hindgut of termites. Comprises the genera *Leptotrichia, Sneathia, Streptobacillus,* and *Sebaldella.*

Type genus: **Leptotrichia** Trevisan 1879, 138[AL].

Genus I. **Leptotrichia** Trevisan 1879, 138[AL]

Kirstin J. Edwards and Saheer E. Gharbia

Lep.to.tri.chi′a. Gr. adj. *leptos* fine, small; Gr. fem. n. *thrix, thricos* hair; N.L. fem. n. *Leptotrichia* fine hair.

Straight or slightly curved rods, 0.5–3.0 × 5–15 µm, **with one or both ends pointed or rounded**. Frequently arranged in pairs, separate filaments, or chains, often with flattened ends. No club formation or branching. Nonmotile. Gram-stain-negative. **Anaerobic on first isolation; many strains subsequently grow aerobically in the presence of CO_2**. Optimum temperature 35–37°C. Good growth occurs at pH 7.0–7.4. Chemoorganotrophic. **Metabolize carbohydrates with formation of acid without gas**. **The major product of glucose fermentation is lactic acid**. Acetic and succinic acids may be produced in trace amounts. Hydrogen sulfide and indole are not produced. Nitrate is not reduced. The primary habitat is the oral cavity of humans, though also found in the female periurethral region and present in the oral cavity of some animals.

DNA G+C content (mol%): 25–29.7.

Type species: **Leptotrichia buccalis** (Robin 1853) Trevisan 1879, 147[AL] (*Leptothrix buccalis* Robin 1853, 345).

Further descriptive information

Colonies of *Leptotrichia* grown anaerobically on blood agar or tryptone-yeast extract medium are distinctive. After 2–3 d of incubation they are smooth, colorless, 2–3 mm in diameter, convex and with a convoluted surface. The colonies are sometimes raised with a filamentous edge, particularly after incubation for 24 h or less (Hamilton and Zahler, 1957; Kasai, 1961). The colonies are nonhemolytic and nonadherent to the medium. Pleomorphism in colony morphology may be seen (Kasai, 1961). An anaerobic atmosphere with 5–10% CO_2 is essential for good growth of *Leptotrichia* on solid medium. Many strains become aerotolerant upon transfer. *Leptotrichia* is highly saccharolastic. Several mono- and disaccharides are fermented with the production of D- and L-lactic acid alone, or accompanied by trace amounts of acetic and succinic acids. CO_2 is not produced. A few strains hydrolyze starch within 2–4 d. Pyruvate is fermented with production of CO_2 and acetic and formic acids (Jackins and Barker, 1951). Within the *Leptotrichia* there is significant heterogeneity in enzymic/biochemical reactions (Eribe et al., 2002) and in cellular fatty acid content (Eribe et al., 2002; Hofstad and Jantzen, 1982). Significant variation between *Leptotrichia* is also observed in SDS-PAGE profiles of whole-cell proteins and RAPD patterns of DNA (Eribe and Olsen, 2002).

The primary habitat is the human oral cavity, where they are typically found in plaque (Eribe et al., 2004; Hofstad, 1984), but they have also been isolated from the normal flora of the periurethral region of healthy girls and the genitalia of women (Evaldson et al., 1980; Moore et al., 1976; Söderberg et al., 1979) and occasionally been recovered from blood, mainly in immunocompromised patients with neutropenia and from endocarditis (Eribe et al., 2004; Hammann et al., 1993; Messiaen et al., 1996; Patel et al., 1999; Reig et al., 1985; Tee et al., 2001; Vernelen et al., 1996; Weinberger et al., 1991). *Leptotrichia buccalis* is normally considered to play a role in tooth decay and periodontal disease (Hofstad, 1984; Krywolap and Page, 1977) especially in immunocompromised patients. Though it is not solely dependent on tooth eruption since the organism has been isolated from the mouth of pre-dentate infants (McCarthy et al., 1965). It has also been isolated from animals which have been fed commercial pellets (Schwartz et al., 1995).

Differentiation of the genus *Leptotrichia* from other genera

The first full description of *Leptotrichia buccalis* described the organism as a Gram-negative bacterium related to *Fusobacterium nucleatum* (Böe and Thjotta, 1944; Thjötta et al., 1939). Hamilton and Zahler (1957), Gilmour et al. (1961), and Kasai (1965) concluded that *Leptotrichia buccalis* was a Gram-positive organism related to *Lactobacillus*. But an electron microscopy study and the isolation of a potent endotoxin from the organism definitely established that the organism is a Gram-negative bacterium (Hofstad and Selvig, 1969). Its ability to produce lactic acid as the only major acid from glucose fermentation distinguishes it from other closely related genera such as *Fusobacterium*, and analysis of 16S rRNA sequences clearly distinguishes *Leptotrichia* from other related genera (Figure 122). Until 1995, there was only one species defined, *Leptotrichia buccalis*. The closely related species "*Leptotrichia sanguinegens*" (Hanff et al., 1995b), also known as "*Leptotrichia microbii*" (Hanff et al., 1995a) was recently transferred to the genus *Sneathia* as *Sneathia sanguinegens* (Collins et al., 2001). The recently defined species, *Leptotrichia trevisanii* (Tee et al., 2001), *Leptotrichia goodfellowii*, *Leptotrichia hofstadii*,

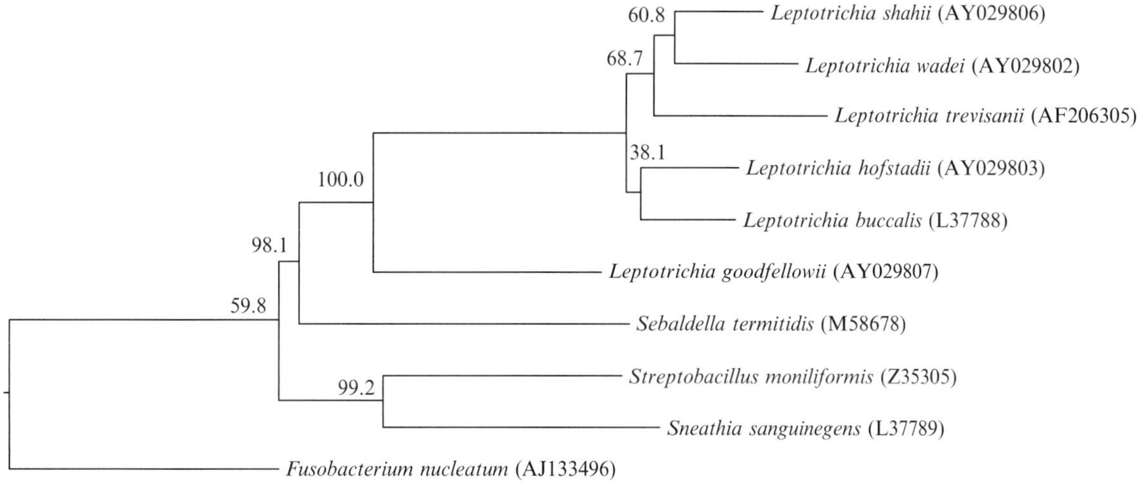

FIGURE 122. Phylogenetic tree of full-length 16S rRNA gene sequences of the genus *Leptotrichia* and related species. GenBank accession numbers are given after species names. The tree was constructed using the neighbor-joining method, and the numbers at branching points are bootstrap percentages.

Leptotrichia shahii, and *Leptotrichia wadei* (Eribe et al., 2004) have been added. The main characteristics of all species are described in Table 150. A further species has been described and named "*Leptotrichia amnionii*" (Shukla et al., 2002), however, 16S rRNA sequencing suggests that "*Leptotrichia amnionii*" would be better assigned to the genus *Sneathia* (Eribe et al., 2004), and a full description of the strain has not been validated.

List of species of the genus *Leptotrichia*

1. **Leptotrichia buccalis** (Robin 1853) Trevisan 1879, 147[AL] (*Leptothrix buccalis* Robin 1853, 345)

buc.ca'lis. L. n. *bucca* the mouth; L. fem. suff. *-alis* suffix denoting pertaining to; N.L. fem. adj. *buccalis* buccal, pertaining to the mouth.

Straight or slightly curved rods, 0.8–1.5 × 5–15 µm, with one or both ends pointed or rounded. Frequently arranged in pairs, chains, or separate filaments. No club formation or branching. Nonmotile. Gram-stain-negative, often with Gram-stain-positive granules distributed evenly along the long axis. May be Gram-stain-positive in very young cultures. Anaerobic on first isolation; many strains subsequently grow aerobically in the presence of CO_2. Optimum growth temperature, 35–37°C; little or no growth occurs at 25°C. Good growth occurs at pH 7.0–7.4. Chemoorganotrophic. Metabolize carbohydrates with formation of acid without gas. The major product of glucose fermentation is lactic acid. Acetic and succinic acids may be produced in trace amounts. Catalase, hydrogen sulfide, and indole are not produced. Nitrate is not reduced. The main habitat is the oral cavity of humans, though also found in the female periurethral region, and can be present in the oral cavity of animals fed with commercial pellets.

DNA G+C content (mol%): 25 (T_m).

Type strain: ATCC 14201, CCUG 34316, CIP 105792, DSM 1135, JCM 12969, NCTC 10249.

Sequence accession no. (16S rRNA gene): L37788.

2. **Leptotrichia goodfellowii** Eribe, Paster, Caugant, Dewhirst, Stomberg, Lacy and Olsen 2004, 589[VP]

good.fel'low.i.i. N.L. masc. gen. n. *goodfellowii* of Goodfellow, named in honor of Michael Goodfellow, for his contributions to microbial systematics.

The description is from Eribe et al. (2004). After 2–6 d of anaerobic incubation at 37°C, colonies on Columbia or BHI agar plates supplemented with 5% human blood, hemin, and menadione are 0.8–2.0 mm in diameter, speckled, convex, irregular, pink in the periphery, and grayish light brown in the rest of the colony. They have a glistening surface, are opaque, dry, and β-hemolytic. Catalase positive and esculin weakly positive. Oxidase and indole are not produced. Colonies grow best anaerobically and sparsely aerobically. Growth occurs at 37°C but not at 25 or 42°C; optimal temperature for growth is 37°C. Gram-stain-negative, nonsporeforming, nonmotile rods. Cells are arranged in pairs, some slightly curved, others in chains joined by flattened ends. Arginine dihydrolase, β-galactosidase, β -glucosidase, *N*-acetyl-β-glucosaminidase, alkaline phosphatase, arginine arylamidase, leucine arylamidase, and histidine arylamidase are produced. Mannose is fermented. Isolated from human blood.

DNA G+C content (mol%): 25 (HPLC).

Type strain: LB 57, CCUG 32286, CIP 107915.

Sequence accession no. (16S rRNA gene): AY029807.

3. **Leptotrichia hofstadii** Eribe, Paster, Caugant, Dewhirst, Stomberg, Lacy and Olsen 2004, 589[VP]

hof.stad'i.i. N.L. masc gen. n. *hofstadii* of Hofstad, named in honor of Tor Hofstad, for his contributions to *Leptotrichia* taxonomy.

TABLE 150. Characteristics differentiating species of the genus *Leptotrichia*[a]

Characteristic	*L. buccalis*	*L. goodfellowii*	*L. hofstadii*	*L. shahii*	*L. trevisanii*	*L. wadei*
Growth at:						
25°C	+	–	–	+	nt	–
42°C	+	–	–	–	nt	–
Hemolysis of human blood	–	+	+	–	nt	+
Production of:						
N-Acetyl-β-glucosaminidase	–	+	–	–	+	–
Alkaline phosphatase	+	+	+	–	+	–
α-Arabinosidase	–	–	–	+	–	–
Arginine arylamidase	–	+	–	–	+	–
Arginine dihydrolase	–	+	–	–	nt	–
α-Galactosidase	+	–	–	–	–	–
β-Galactosidase	–	+	–	–	–	–
β-Galactosidase-6-phosphate	+	–	+	–	nt	–
α-Glucosidase	+	–	+	+	+	+
β-Glucosidase	+	+	+	–	+	+
Histidine arylamidase	–	+	–	–	nt	–
Leucine arylamidase	–	+	–	–	–	–
Tyrosine arylamidase	+	–	–	–	nt	–
Fermentation of:						
Mannose	+	–	+	–	nt	–
Raffinose	+	–	–	–	nt	–
Saturated fatty acids:						
$C_{14:0}$	10	14	6	9	nt	7
$C_{16:0}$	39	41	48	32	nt	45
Unsaturated straight-chain fatty acids:						
$C_{18:1}$ *c*11/*t*9/*t*6 or unknown with ECL of 17.834	42	28	27	36	nt	24
Hydroxy fatty acids:						
$C_{14:0}$ 3-OH or $C_{15:0}$ DMA	7	9	9	5	nt	8
Catalase	–	+	+	+	+	+
Esculin	+	+	+	+	nt	+

[a]Symbols: +, >85% positive; d, different strains give different reactions (16–84% positive);–, 0–15% positive; w, weak reaction; nt, not tested; ECL, Equivalent chain-length; DMA, dimethyl acetal.

The description is from Eribe et al. (2004). After 2–6 d of anaerobic incubation at 37°C, colonies on Columbia or BHI agar plates supplemented with 5% human blood, hemin, and menadione are 0.5–1.8 mm in diameter. They have a glistening and granular surface, are opaque, dry, and β-hemolytic. Older colonies can be up to 4.0–6.5 mm, circular, convex, entire (some are irregular and lobate), and grayish in color with a dark central spot. Catalase positive and esculin weakly positive. Oxidase and indole are not produced. Colonies grow best anaerobically and sparsely aerobically. Growth occurs at 37°C but not at 25 or 42°C; optimal temperature is 37°C. Gram-stain-negative, nonsporeforming, nonmotile rods. Cells are arranged in pairs, some slightly curved, others in chains joined by flattened ends. β-Galactosidase-6-phosphate, α-glucosidase, β-glucosidase, and alkaline phosphatase are produced. Mannose is fermented.

Source: saliva of a healthy person.

DNA G+C content (mol%): 25 (HPLC).

Type strain: LB 23, CCUG 47504, CIP 107917.

Sequence accession no. (16S rRNA gene): AY029803.

4. **Leptotrichia shahii** Eribe, Paster, Caugant, Dewhirst, Stomberg, Lacy and Olsen 2004, 589[VP]

sha'hi.i. N.L. masc. gen. n. *shahii* of Shah, named in honor of Haroun N. Shah, a Trinidad-born microbiologist, for his contributions to microbiology.

The description is from Eribe et al. (2004). After 2–6 days of anaerobic incubation at 37°C, colonies on Columbia or BHI agar plates supplemented with 5% human blood, hemin, and menadione are 1.0–1.5 mm in diameter, very filamentous to rhizoid or convoluted, pale-speckled, and grayish in color, with a dark central spot in old colonies. These are opaque, semi-dry in consistency, and nonhemolytic. Catalase-positive and esculin weakly positive. Oxidase and indole are not produced. Colonies grow best anaerobically and sparsely aerobically. Growth occurs at 25 and 37°C but not at 42°C; optimal temperature is 37°C. Gram-stain-negative, nonsporeforming, nonmotile rods. Cells are arranged in pairs, some slightly curved, others in chains joined by flattened ends. α-Glucosidase and α-arabinosidase are produced.

Source: a patient with gingivitis.

DNA G+C content (mol%): 25 (HPLC).

Type strain: LB 37, CCUG 47503, CIP 107916, VPI N06A-34.
Sequence accession no. (16S rRNA gene): AY029806.

5. **Leptotrichia trevisanii** Tee, Midolo, Janssen, Kerr and Dyall-Smith 2002, 686[VP] (Effective publication: Tee, Midolo, Janssen, Kerr and Dyall-Smith 2001,768.)

tre.vi.sa'ni.i. N.L. masc. gen. n. *trevisanii* of V. Trevisan, who proposed the genus *Leptotrichia* in 1879.

The description is from Tee et al. (2001).The organism is characterized by long, fusiform, nonmotile, Gram-stain-negative bacilli that are 0.8–0.9 × 6–13 µm with tapered ends. Many cells are slightly curved. Cells are arranged in pairs and chains and, where they join, the ends are flattened. Isolates are nonsporeforming, catalase positive, indole negative, and oxidase negative. They are negative for urease, negative for hydrolysis of the *p*-NP-sugars α-D-arabinoside, α-D-galactoside, and α-L-fucoside, and are unable to hydrolyze *O*-NP-β-D-galactoside. They are positive for hydrolysis of *p*-NP derivatives of α-D-glucoside, β-D-glucoside, *N*-acetyl-β-D-glucosamine, and *p*-NP-phosphate. The organism is unable to hydrolyze the naphthylamide derivatives of leucyl-glycine, glycine, proline, serine, and pyrrollidone, but is able to hydrolyze arginine and phenylalanine derivatives. *Leptotrichia trevisanii* grows both anaerobically and under 5% CO_2 (in air) on subsequent culture on solid media. It ferments glucose to produce lactic acid as the major organic end product. On horse blood agar plates, colonies appeared smooth, grayish in color, low convex, and erose edged; they showed a dark central spot with transillumination and grew to about 2 mm diameter after 5 d of incubation. Older colonies showed a slightly convoluted surface.

DNA G+C content (mol%): 29.7 (HPLC).
Type strain: "Wee Tee" 1999, ATCC 700907, DSM 22070.
Sequence accession no. (16S rRNA gene): AF206305.

6. **Leptotrichia wadei** Eribe, Paster, Caugant, Dewhirst, Stomberg, Lacy and Olsen 2004, 591[VP]

wade'i. N.L. masc. gen. n. *wadei* of Wade, named in honor of William G. Wade, for his contributions to microbiology.

The description is from Eribe et al. (2004). After 2–6 d of anaerobic incubation at 37°C, colonies on Columbia or BHI agar plates supplemented with 5% human blood, hemin, and menadione are 0.5–3.0 mm in diameter, convex, sparsely filamentous to irregular, and grayish brown in color, with a dark central spot in old colonies. The surface appearance is glistening and smooth with a rough edge. Colonies are opaque, dry in consistency, and β-hemolytic. Esculin and catalase are positive. Oxidase and indole are not produced. Growth occurs best anaerobically and sparsely aerobically at 37°C but not at 25 or 42°C. Gram-stain-negative, nonsporeforming, nonmotile rods. Cells are arranged in pairs, some slightly curved, others in chains joined by flattened ends. α-Glucosidase and β-glucosidase are produced.

Source: saliva of a healthy person.
DNA G+C content (mol%): 25 (HPLC).
Type strain: LB 16, CCUG 47505, CIP 107918.
Sequence accession no. (16S rRNA gene): AY029802.

Genus II. **Sebaldella** Collins and Shah 1986, 349[VP]

Saheer E. Gharbia and Kirstin J. Edwards

Se.bal.del'la. N.L. dim. ending *-ella*; N.L. fem. dim. n. *Sebaldella* named after the French microbiologist Madeleine Sebald, who first described the organism.

Rods. Nonsporeforming. Nonmotile. Gram-stain-negative. Anaerobic. Acid produced from glucose and some other sugars. The **major end products of glucose fermentation are acetic and lactic acids**; formic acid may also be produced. Hexose monophosphate shunt enzymes, glucose-6-phosphate dehydrogenase, and 6-phosphogluconate dehydrogenase are absent. Glutamate dehydrogenase and malate dehydrogenase are absent. Nonhydroxylated and 3-hydroxylated long-chain fatty acids are present. The fatty acids are primarily of the straight-chain saturated and monounsaturated types. Menaquinones are absent.

DNA G+C content (mol%): 32–36.

Type species: **Sebaldella termitidis** (Sebald 1962) Collins and Shah 1986, 349[VP] (*Sphaerphorus siccus* var. *termitidis* Sebald 1962, 124; *Bacteroides termitidis* Holdeman and Moore 1970, 33).

Enrichment and isolation procedures

Surface colonies are 1–2 mm in diameter, circular, and transparent to opaque. Colonies in deep agar are lenticular and nonpigmented.

Differentiation of the genus *Sebaldella* from other genera

Sebaldella differs from *Bacteroides fragilis* and related species by exhibiting a lower G+C content (32–36 mol%), by producing acetic and lactic acids as major end products of glucose fermentation, and by the absence of glutamate and malate dehydrogenases (Collins and Shah, 1986). *Sebaldella* also differs from the *Bacteroides fragilis* group in lipid composition. The long chain fatty acids of *Bacteroides* are predominantly of the straight-chain saturated, anteiso- and iso-methyl branched chain types, with monosaturated acids either absent or present in only trace amounts. In contrast, *Sebaldella* primarily synthesizes acids of the straight chain saturated and monosaturated types and methyl branched acids are absent (Miyagawa et al., 1979; Shah and Collins, 1983). *Sebaldella* and *Bacteroides* also differ in isoprenoid quinone composition, with *Bacteroides* possessing menaquinones and *Sebaldella* lacking respiratory quinones (Collins and Jones, 1981; Shah and Collins, 1980). Based on 16S rRNA studies, *Sebaldella* is phylogentically distinct from *Bacteroides* and all other described eubacterial phyla (Paster et al., 1985).

List of species of the genus *Sebaldella*

1. **Sebaldella termitidis** (Sebald 1962) Collins and Shah 1986, 349[VP] (*Sphaerphorus siccus* var. *termitidis* Sebald 1962, 124; *Bacteroides termitidis* Holdeman and Moore 1970, 33.)

ter.mi'ti.dis. L. n. *termes, -itis* wood-eating worm; N.L. fem. adj. *termitidis* pertaining to the termite.

The description is from Collins and Shah (1986). The Gram-stain-negative, obligately anaerobic, nonmotile, rod-shaped cells are 0.3–0.5 × 2–12 μm with central swellings and occur singly, in pairs, and in filaments. Surface colonies are 1–2 mm in diameter, circular, and transparent to opaque. Colonies in deep agar are lenticular and nonpigmented. Acetic and lactic acids are the major end products of glucose metabolism; formic acid may also be produced. Acid is produced from glucose, fructose, maltose, mannitol, mannose, rhamnose, sucrose, trehalose, and xylose. Acid is not produced from arabinose, melazitose, or starch; low levels of acid may be produced from lactose (delayed reaction). Most strains produce H_2S. Gelatin is not liquefied; coagulated proteins are not attacked. Urease, chitinase, and indole are not produced. Nitrate is not reduced. Uric acid is degraded to CO_2, acetate, and ammonia. Malate dehydrogenase and glutamate dehdrogenase are not produced.

Nonhydroxylated and 3-hydroxylated long-chain fatty acids are present. The fatty acids are of the straight-chain saturated and monounsaturated types, with hexadecanoic and octadecenoic acids predominating. Menaquinones are not produced. Isolated from posterior intestinal contents of termites, where these organisms are part of the predominant bacterial flora.

DNA G+C content (mol%): 32–36 (Bd).

Type strain: ATCC 33386, NCTC 11300.

Sequence accession no. (16S rRNA gene): M58678.

Genus III. **Sneathia** Collins, Hoyles, Törnqvist, von Essen and Falsen 2002, 687[VP] (Effective publication: Collins, Hoyles, Törnqvist, von Essen and Falsen 2001, 360.)

Julie M.J. Logan, Kirstin J. Edwards and Saheer E. Gharbia

Sneath'i.a. N.L. fem. n. *Sneathia* named after the British microbiologist Peter H.A. Sneath, in recognition of his outstanding contributions to microbial systematics.

Gram-stain-negative, asporogenous, **rod-shaped bacteria**; nonmotile. Cells may display pleomorphism and filaments may be observed. **Anaerobic**, although some strains may show poor growth in CO_2. **Fermentative metabolism. Acid but no gas is produced from glucose**. Acid is not produced from ribose or maltose. Lactic acid, formic acid, and minor amounts of acetic acid are the end products of glucose metabolism; succinic acid may be produced. **Fastidious**; require serum or blood for growth. Optimum temperature for growth 35–37°C. Catalase and oxidase negative. Esculin and hippurate are hydrolyzed but starch is not. β-Glucuronidase is produced. Indole is not produced. Voges-Proskauer negative. Nitrate is not reduced to nitrite.

DNA G+C content (mol%): 22–25.

Type species: **Sneathia sanguinegens** Collins, Hoyles, Törnqvist, von Essen and Falsen 2002, 687 (Effective publication: Collins, Hoyles, Törnqvist, von Essen and Falsen 2001, 360.).

Further descriptive information

Originally, Hanff et al. (1995b) reported the isolation from blood cultures of an unusual Gram-stain-negative anaerobic rod-shaped organism from four obstetric patients with post-partum fever, two neonates, and a 100-year-old woman. This fastidious, serum-requiring bacterium was considered by Hanff et al. (1995b) to be a member of the genus *Leptotrichia* and was designated "*Leptotrichia sanguinegens*". The species was, however, not validly published, and no type strain was designated. In 2001, Collins et al. isolated three strains from amniotic fluid and blood from non-obstetric patients that resembled the bacterium described by Hanff et al. (1995b). However, based on both phenotypic and phylogenetic evidence, Collins et al. concluded that these isolates corresponded to "*Leptotrichia sanguinegens*" and proposed the new genus and species *Sneathia sanguinegens*.

Differentiation of the genus *Sneathia* from other genera

Sneathia sanguinegens can be readily identified in the clinical laboratory on the basis of its cellular morphology and fastidious growth requirements combined with its API Rapid ID32A and ZYM biochemical profiles. In particular, using these systems it can be readily distinguished from *Streptobacillus moniliformis* by its positive β-glucuronidase reaction and by failing to produce chymotrypsin and proline arylamidase. Similarly, *Sneathia sanguinegens* can be easily distinguished from *Leptotrichia buccalis* in requiring serum or blood for growth, by producing β-glucuronidase, and by its negative α-glucosidase and β-glucosidase reactions. On the basis of 16S rDNA sequencing, *Sneathia* isolates demonstrated high sequence similarity of >99.8% and are distinct from their closest named relatives such as *Streptobacillus moniliformis*, *Sebaldella termiditis*, and *Leptotrichia buccalis* (Collins et al., 2001).

List of species of the genus *Sneathia*

1. **Sneathia sanguinegens** Collins, Hoyles, Törnqvist, von Essen and Falsen 2002, 687[VP] (Effective publication: Collins, Hoyles, Törnqvist, von Essen and Falsen 2001, 360.)

san.gui.ne'gens. L. n. *sanguis, -inis* blood; L. part adj. *egens* needing; N.L part. adj *sanguinegens* needing blood; because the organism requires blood or serum.

Description is from Collins et al. (2001). Gram-stain-negative anaerobic or facultatively anaerobic, nonspore-forming, nonmotile, rod-shaped cells. Colonies on chocolate or blood agar are pin-point and convex after 72 h. Fastidious, requiring blood or serum for growth. Catalase and oxidase negative. Lactic acid, formic acid, and minor amounts of acetic acid are the end products of glucose metabolism; succinic acid may be produced. Acid but no gas is produced from glucose. Acid may or may not be produced from mannose and raffinose. Acid is not produced from L-arabinose, D-arabitol, cyclodextrin, glycogen, lactose, mannitol, maltose, melebiose, melezitose, methyl-β-D-glucopyranoside, pullulan, D-ribose, sorbitol, sucrose, tagatose, or trehalose. Alkaline phosphatase, acid phosphatase, arginine arylamidase, phosphoamidase, and β-glucuronidase are detected. Arginine dihydrolase, alanine arylamidase, alanine phenylalanine proline arylamidase, α-arabinosidase, chymotrypsin, cystine arylamidase, α-frucosidase, α-galactosidase, β-galactosidase, β-galactosidase-6-phosphate, α-glucosidase, β-glucosidase, glutamic acid decarboxylase, glycyl tryptophan arylamidase, pyroglutamic acid arylamidase, leucine glycin arylamidase, lipase C14, α-mannosidase, β-mannosidase, *N*-acetyl-β-glucosaminidase, proline arylamidase, trypsin, valine arylamidase, and urease are not detected. Activity may or may not be detected for ester lipase C8, esterase C4, glutamyl glutamic acid arylamidase, glycine arylamidase, histidine arylamidase, phenyl alanine arylamidase, leucine arylamidase, serine arylamidase, and tyrosine arylamidase. Indole-negative and Voges–Proskauer-negative. Esculin and hippurate are hydrolyzed but starch is not. Nitrate is not reduced to nitrite. Habitat is not known.

Source: human clinical specimens (blood, amniotic fluid).

DNA G+C content (mol%): 22–25 (HPLC).

Type strain: CCUG 41628, CIP 106906.

Sequence accession no. (16S rRNA gene): AJ344093.

Genus IV. **Streptobacillus** Levaditi, Nicolau and Poincloux 1925, 1188[AL]

Saheer E. Gharbia and Kirstin J. Edwards

Strep.to.ba.cil'lus. Gr.adj. *streptos* twisted, curved; L. masc. n. *bacillus* a small rod; N.L. masc. n. *Streptobacillus* a twisted or curved small rod.

Rods with rounded or pointed ends. Occur singly or form long, wavy chains. Nonsporeforming. Nonmotile. Gram-stain-negative. Conversion to L-phase or transitional-phase variant may occur spontaneously during cultivation. **Capable of growth anaerobically or aerobically. Ferments glucose to produce acid but not gas**. Optimum temperature 35–37°C. Catalase and oxidase negative. Indole not produced. Nitrate not reduced to nitrite. Require serum, ascitic fluid, or blood for growth. **Isolated from the throat and nasopharynx of wild and laboratory rats. Causes rat-bite fever in man.**

DNA G+C content (mol%): 24–26.

Type species: **Streptobacillus moniliformis** Levaditi, Nicolau and Poincloux 1925, 1188[AL].

Further descriptive information

The sole species forms small (1–2 mm), smooth, convex, grayish, nonhemolytic colonies on 5% horse blood Columbia agar after 72 h of incubation in an atmosphere of carbon dioxide at an optimum temperature of 35–37°C. Direct examinations reveal pleomorphic, Gram-stain-negative filamentous rods with lateral bulbous swellings. In serum-supplemented liquid media, bacterial growth shows a typical cottonball like appearance. The organisms exist in two variant types, the "normal" bacillary form and the inducible or spontaneously occurring L-form that exhibits the typical "fried-egg" colony morphology (Wittler and Cary, 1974). The latter is regarded as an apathogenic *Streptobacillus moniliformis* variant (Freundt, 1956). *Streptobacillus moniliformis* can be distinguished from other biochemically related bacteria by its negative reactions for catalase, oxidase, indole production and reduction of nitrate to nitrite (Savage, 1984).

Streptobacillus moniliformis is a pathogen for humans that was first isolated as a cause of rat-bite fever by Hugo Schottmüller in 1914. He named the organism *Streptothrix muris ratti*. Clinical pictures of the disease are similar despite two different modes of transmission. Oral uptake of *Streptobacillus moniliformis* via contaminated food leads to a disease known as Haverhill fever, named after the place (Haverhill, MA, USA) where the first well-documented epidemic was observed (Place and Sutton, 1934). Affected individuals had consumed unpasteurized milk or milk products to which rats had access. In another well-documented epidemic (Chelmsford, UK), boarding school pupils became infected after using water from a spring in the vicinity of which rats were observed (McEvoy et al., 1987). As in Haverhill, *Streptobacillus moniliformis* was not isolated from captured rats. However, in both cases, epidemiological data suggested it was most probable that foodstuffs or water contaminated by rats were responsible for the epidemics.

Rats are also the source of the second type of human streptobacillosis, the so-called rat-bite fever. (Another form of rat-bite fever called sodoku is caused by *Spirillum minus* which is not the subject of this review.) *Streptobacillus moniliformis* can be transmitted by rat bite, but recent reports suggest that not only the rat bite, but also simple contact with (pet) rats (Clausen, 1987; Rygg and Bruun, 1992), may result in rat-bite fever. The disease is characterized by an acute onset with chills, vomiting, malaise, headache, irregularly relapsing fever, erythematous rash (especially of the extremities), and arthralgia. Untreated, it often leads to a severe septic polyarthritis and lymphadenopathy. If untreated, mortality is estimated to be about 13% (Roughgarden, 1965; Simon and Wilson, 1986). Complications of streptobacillary rat-bite fever are endocarditis (Rey et al., 1987; Rupp, 1992), pericarditis (Carbeck et al., 1967) brain abscess (Oeding and Pedersen, 1950), amnionitis (Faro et al., 1980), septicemia (Brown and Nunemaker, 1942; Dellamonica et al., 1979; Renaut et al., 1982; Rygg and Bruun, 1992), interstitial pneumonia, prostatitis, and pancreatitis (Delannoy et al., 1991).

So far as is known, the rat is the natural reservoir of *Streptobacillus moniliformis* and therefore plays the dominant role in harboring and transmitting the infectious agent. Most probably the microorganism is a member of the commensal flora of the upper respiratory tract. Hence, the main isolation sites in healthy rats are the nasopharynx (Strangeways, 1933), larynx, upper trachea (Peagle et al., 1976), and the middle ear (Koopman et al., 1991). Although of only low pathogenicity for the rat, *Streptobacillus moniliformis* may act as a secondary invader (Weisbroth, 1979) in conjunction with presumptive pathogens such as *Pasteurella pneumotropica*, *Mycoplasma pulmonis* causing otitis media (Olson and McCune, 1968; Wullenweber et al., 1992), conjunctivitis (Young and Hill, 1974), bronchopneumonia (Bell and Elmes, 1969), and chronic pneumonia (Gay et al., 1972).

Taxonomic comments

This organism has been given various names over the years. They are found in original research papers as well as in textbooks on pathogenic bacteria. The list of references is: *Streptothrix muris ratti* (Schottmüller, 1914), *Nocardia muris* (de Mello and Pais, 1918), *Actinomyces muris ratti* (Lieske, 1921; Schottmüller, 1914), *Haverhillia multiformis* (Parker and Hudson, 1926), *Actinomyces muris* (de Mello and Pais, 1918; Topley and Wilson, 1936), *Asterococcus muris* (de Mello and Pais, 1918; Heilman, 1941), *Proactinomyces muris* (de Mello and Pais, 1918; Krasil'nikov, 1941), *Haverhillia moniliformis* (Levaditi et al., 1925; Prévot, 1948), *Actinobacillus muris* (de Mello and Pais, 1918; Wilson and Miles, 1955).

Differentiation of the genus *Streptobacillus* from other genera

Streptobacillus can be differentiated from genera which are found in the same habitat, including *Cardiobacterium*, *Actinobacillus*, and *Haemophilus*, on the basis of its serum requirement, flocculent growth in broth, small butyrous colony on agar, characteristic microscopic appearance, absence of catalase and oxidase activity, and failure to reduce nitrate to nitrite or produce indole. Related genera will be positive for one or more of the catalase, oxidase, nitrate reduction or indole production characteristics while *Streptobacillus* is negative for all (Lapage, 1974; Midgley et al., 1970).

Gas-liquid chromatography analysis of the fatty acid pattern shows palmitic, stearic, oleic, and linoleic acid as major components (Edwards and Finch, 1986). On the basis of numerical analysis of SDS-PAGE protein profiles, seven subgroups were identified among 31 *Streptobacillus moniliformis* cultures representing 22 different strains of human, murine, and avian origin isolated in Europe, USA, and Australia. As these groups show a close similarity, it was concluded that *Streptobacillus moniliformis* is a very homologous species (Costas and Owen, 1987).

List of species of the genus *Streptobacillus*

1. **Streptobacillus moniliformis** Levaditi, Nicolau and Poincloux 1925, 1188[AL]

mo.ni.li.for'mis. L. n. *monile* necklace; L. masc. suff. *formis* in the shape of; N.L. masc. adj. *moniliformis* necklace-shaped.

The description and characteristics are as described for the genus.

DNA G+C content (mol%): 24–26 (T_m).

Type strain: ATCC 14647, CCUG 2469, CCUG 13453, DSM 12112, NCTC 10651.

Sequence accession no. (16S rRNA gene): Z35305.

Reference

Bell, D.P. and P.C. Elmes. 1969. Effects of certain organisms associated with chronic respiratory disease on SPF and conventional rats. J. Med. Microbiol. *2*: 511–519.

Böe, J. and T. Thjotta. 1944. The position of *Fusobacterium* and *Leptotrichia* in the bacteriological system. Acta Pathol. Microbiol. Scand. *21*: 441–450.

Brown, T.M. and J.C. Nunemaker. 1942. Rat-bite fever. a review of the American cases with reevaluation of etiology; report of cases. Bull. Johns Hopkins Hosp. *70*: 201–236.

Carbeck, R.B., J.F. Murphy and E.M. Britt. 1967. Streptobacillary rat-bite fever with massive pericardial effusion. J.A.M.A. *201*: 133–134.

Clausen, C. 1987. Septic arthritis due to *Streptobacillus moniliformis*. Clin. Microbiol. Newsl. *9*: 123–124.

Collins, M.D. and D. Jones. 1981. Distribution of isoprenoid quinone structural types in bacteria and their taxonomic implication. Microbiol. Rev. *45*: 316–354.

Collins, M.D. and H.N. Shah. 1986. Reclassification of *Bacteroides termitidis* Sebald (Holdeman and Moore) in a new genus *Sebaldella termitidis*, as *Sebaldella termitidis* comb. nov. Int. J. Syst. Bacteriol. *36*: 349–350.

Collins, M.D., L. Hoyles, E. Törnqvist, R. von Essen and E. Falsen. 2001. Characterization of some strains from human clinical sources which resemble "*Leptotrichia sanguinegens*": description of *Sneathia sanguinegens* sp. nov., gen. nov. Syst. Appl. Microbiol. *24*: 358–361.

Collins, M.D., L. Hoyles, E. Törnqvist, R. von Essen and E. Falsen. 2002. *In* Validation of publication of new names and new combinations previously effectively published outside the IJSEM. Validation List no. 85. Int. J. Syst. Evol. Microbiol. *52*: 685–690.

Costas, M. and R.J. Owen. 1987. Numerical analysis of electrophoretic protein patterns of *Streptobacillus moniliformis* strains from human, murine and avian infections. J. Med. Microbiol. *23*: 303–311.

de Mello, F. and A.S.A. Pais. 1918. Um caso de nocardiose pulmonar simulando a tísica. Arq. Hig. Pat. Exot. Lisboa *6*: 133–206.

Delannoy, D., P. Savinel, M.H. Balquet, J.P. Canonne, J. Amourette and P.Y. Bugnon. 1991. Manifestations digestives et pulmonaires rélévant une septicémie à *Streptobacillus moniliformis*: présentation atypique d' une pathologie rare et méconnue. La Revue de Médicine Interne. *3*: 5158.

Dellamonica, P., E. Delbeke, D. Giraud and G. Illy. 1979. Septicémies à *Streptobacillus moniliformis*: à propos d' un cas-revue de la littérature. Méd. Mal. Infect. *9*: 226–229.

Edwards, R. and R.G. Finch. 1986. Characterisation and antibiotic susceptibilities of *Streptobacillus moniliformis*. J. Med. Microbiol. *21*: 39–42.

Eribe, E.R., B.J. Paster, D.A. Caugant, F.E. Dewhirst, V.K. Stromberg, G.H. Lacy and I. Olsen. 2004. Genetic diversity of *Leptotrichia* and description of *Leptotrichia goodfellowii* sp. nov., *Leptotrichia hofstadii* sp. nov., *Leptotrichia shahii* sp. nov. and *Leptotrichia wadei* sp. nov. Int. J. Syst. Evol. Microbiol. *54*: 583–592.

Eribe, E.R.K., T. Hofstad and I. Olsen. 2002. Enzymatic/biochemical and cellular fatty acid analyses of *Leptotrichia* isolates. Microb. Ecol. Health Dis. *14*: 137–148.

Eribe, E.R.K. and I. Olsen. 2002. SDS-PAGE of whole-cell proteins and random amplified polymorphic DNA (RAPD) analyses of *Leptotrichia* isolates. Microb. Ecol. Health Dis. *14*: 193–202.

Evaldson, G., G. Carlstrom, A. Lagrelius, A.S. Malmborg and C.E. Nord. 1980. Microbiological findings in pregnant women with premature rupture of the membranes. Med. Microbiol. Immunol. *168*: 283–297.

Faro, S., C. Walker and R.L. Pierson. 1980. Amnionitis with intact amniotic membranes involving *Streptobacillus moniliformis*. Obstet. Gynecol. *55*: 9S-11S.

Freundt, E.A. 1956. Experimental investigations into the pathogenicity of the L-phase variant of *Streptobacillus moniliformis*. Acta Pathol. Microbiol. Scand. *38*: 248–256.

Gay, F.W., M.E. Maguire and A. Baskerville. 1972. Etiology of chronic pneumonia in rats and a study of the experimental disease in mice. Infect. Immun. *6*: 83–91.

Gilmour, M.N., J.A.H. Howell and B.G. Bibby. 1961. The classification of organisms termed *Leptotrichia (Leptrotrix) buccalis*. I. Review of the literature and proposed separation into *Leptotrichia buccalis* Trevisan 1879 and *Bacterionema* gen. nov. *B. matruchotii* (Mendel 1919) comb. nov. Bacteriol. Rev. *25*: 131–141.

Hamilton, R.D. and S.A. Zahler. 1957. A study of *Leptotrichia buccalis*. J. Bacteriol. *73*: 386–393.

Hanff, P.A., J.A. Rosol-Donoghue, C.A. Spiegel, K.A. Wilson and L.A. Moore 1995a, posting date. *Sneathia sanguinegens*. NCBI Taxonomy Browser. [Online.]

Hanff, P.A., J.A. Rosol-Donoghue, C.A. Spiegel, K.H. Wilson and L.H. Moore. 1995b. *Leptotrichia sanguinegens* sp. nov., a new agent of postpartum and neonatal bacteremia. Clin. Infect. Dis. *20 Suppl. 2*: S237–239.

Hammann, R., A. Iwand, J. Brachmann, K. Keller and A. Werner. 1993. Endocarditis caused by a *Leptotrichia buccalis*-like bacterium in a patient with a prosthetic aortic valve. Eur. J. Clin. Microbiol. Infect. Dis. *12*: 280–282.

Heilman, F.R. 1941. A study of *Asterococcus muris* (*Streptobacillus moniliformis*). II. Cultivation and biochemical activities. J. Infect. Dis. *69*: 45–51.

Hofstad, T. and K.A. Selvig. 1969. Ultrastructure of *Leptotrichia buccalis*. J. Gen. Microbiol. *56*: 23–26.

Hofstad, T. and E. Jantzen. 1982. Fatty acids of *Leptotrichia buccalis*: taxonomic implications. J. Gen. Microbiol. *128*: 151–153.

Hofstad, T. 1984. Genus III. *Leptotrichia*. *In* Bergey's Manual of Systematic Bacteriology, vol. 1 (edited by Krieg). Williams & Wilkins, Baltimore, pp. 637–641.

Holdeman, L.V. and W.E.C. Moore (editors). 1970. *Bacteroides*, Outline of Clinical Methods in Anaerobic Bacteriology, 2nd revn. Virginia Polytechnic Institute Anaerobe Laboratory, Blacksburg, VA.

Jackins, H.C. and H.A. Barker. 1951. Fermentative processes of the fusiform bacteria. J. Bacteriol. *61*: 101–114.

Kasai, G.J. 1961. A study of *Leptotrichia buccalis*. I. Morphology and preliminary observations. J. Dent. Res. *40*: 800–811.

Kasai, G.J. 1965. A study of *Leptotrichia buccalis*. II. Biochemical and physiological observations. J. Dent. Res. *44*: 1015–1022.

Koopman, J.P., M.E. van den Brink, P.P.C.A. Vennix, W. Kuypers, R. Boot and R.H. Bakker. 1991. Isolation of *Streptobacillus moniliformis* from the middle ear of rats. Lab. Anim. *25*: 35–39.

Krasil'nikov. 1941. Proactinomyces. *In* Guide to the *Actinomycetes*. Izd. Akad. Nauk., U.S.S.R, Moskau, p. 76.

Krywolap, G.N. and L.R. Page. 1977. Oral *Fusobacterium*, *Leptotrichia* and *Bacterionema*: II. Pathogenicity: a review of the literature. J. Baltimore Coll. Dent. Surg. *32*: 26–32.

Lapage, S.P. 1974. Genus *Cardiobacterium* Slotnick and Daugherty. *In* Bergey's Manual of Determinative Bacteriology, 8th edn (edited by Buchanan and Gibbons). Williams & Wilkins, Baltimore, pp. 377–378.

Levaditi, C., S. Nicolau and P. Poincloux. 1925. Sur le role étiologique de *Streptobacillus moniliformis* (nov. spec.) dans l'erythème polymorph aigu septicémique. C. R. Hebd. Séances Acad. Sci. (Paris) *180*: 1188–1190.

Lieske, R. 1921. Morphologie und Biologie der Strahlenpilze (Actinomyceten). Borntraeger Bros., Leipzig.

McCarthy, C., M.L. Snyder and R.B. Parker. 1965. The indigenous oral flora of man. I. The newborn to the 1-year-old infant. Arch. Oral Biol. *10*: 61–70.

McEvoy, M.B., N.D. Noah and R. Pilsworth. 1987. Outbreak of fever caused by *Streptobacillus moniliformis*. Lancet. *2*: 1361–1363.

Messiaen, T., C. Lefebvre and A. Geubel. 1996. Hepatic abscess likely related to *Leptotrichia buccalis* in an immunocompetent patient. Liver *16*: 342–343.

Midgley, J., S.P. LaPage, B.A. Jenkins, G.I. Barrow, M.E. Roberts and A.G. Buck. 1970. *Cardiobacterium hominis* endocarditis. J. Med. Microbiol. *3*: 91–98.

Miyagawa, E., R. Azuma and T. Suto. 1979. Cellular fatty acid composition in Gram-negative obligately anaerobic rods. J. Gen. Microbiol. *25*: 41–51.

Moore, W.E.C., J.L. Johnson and L.V. Holdeman. 1976. Emendation of *Bacteroidaceae* and *Butyrivibrio* and descriptions of *Desulfomonas* gen. nov. and ten new species in genera *Desulfomonas*, *Butyrivibrio*, *Eubacterium*, *Clostridium*, and *Ruminococcus*. Int. J. Syst. Bacteriol. *26*: 238–252.

Neumann, S., U. Matthey, G. Kaim, and P. Dimroth. 1998. Purification and properties of the F_1F_0 ATPase of *Ilyobacter tartaricus*, a sodium ion pump. J. Bacteriol. *180*: 3312–3316.

Oeding, P. and H. Pedersen. 1950. *Streptothrix muris ratti* (*Streptobacillus moniliformis*) isolated from a brain abscess. Acta Pathol. Microbiol. Scand. *27*: 436–442.

Olson, L.D. and E.L. McCune. 1968. Histopathology of chronic otitis media in the rat. Lab. Anim. Care *18*: 478–485.

Parker, J.F. and N.P. Hudson. 1926. The etiology of Haverhill fever (erythema arthriticum epidemicum). Am. J. Pathol. *2*: 375–379.

Paster, B.J., W. Ludwig, W.G. Weisburg, E. Stackebrant, R.B. Hespel, C.M. Hahn, H. Reichenback, K.O. Stetter and C.R. Woese. 1985. A phylogenic grouping of the bacteroides, cytophagas, and certain flavobacteria. Syst. Appl. Microbiol. *6*: 34–42.

Patel, J.B., J. Clarridge, M.S. Schuster, M. Waddington, J. Osborne and I. Nachamkin. 1999. Bacteremia caused by a novel isolate resembling *Leptotrichia* species in a neutropenic patient. J. Clin. Microbiol. *37*: 2064–2067.

Peagle, R.D., R. P. Tewari, W.N. Berhard and E. Peters. 1976. Microbial flora of the larynx, trachea and large intestine of the rat after long-term inhalation of 100 per cent oxygen. Anesthesiology *44*: 287–290.

Place, E.H. and L.E. Sutton. 1934. Erythema arthriticum epidemicum (Haverhill fever). Arch. Intern. Med. *54*: 659–684.

Prévot, A.R. 1948. Manuel de classification et de determination des bactéries anaérobies, 2nd edn. Masson et Cie, Paris.

Reig, M., F. Baquero, M. Garcia-Campello and E. Loza. 1985. *Leptotrichia buccalis* bacteremia in neutropenic children. J. Clin. Microbiol. *22*: 320–321.

Renaut, J.J., C. Pecquet, C. Verlingue, H. Barriere, M. Deriennic and A.L. Courticu. 1982. Septicémie à *Streptobacillus moniliformis*. Nouv. Presse. Med. *11*: 1143.

Rey, J.L., G. Laurans, A. Pleskof, M. Guerlin, J. Orfila, C. Tribouillov, P. Bernasconi and J.P. Lesbre. 1987. Les endocardites à *Streptobacillus moniliformis*. A propos de deux cas. Ann. Cardiol. Angéiol. *36*: 297–300.

Robin, C. 1853. Histoire naturelle des végétaux parasites qui crossent sur l'homme et sur les animaux vivants. J.-B. Baillière, Paris.

Roughgarden, J.W. 1965. Antimicrobial Therapy of ratbite fever. A review. Arch. Intern. Med. *116*: 39–54.

Rupp, M.E. 1992. *Streptobacillus moniliformis* endocarditis: case report and review. Clin. Infect. Dis. *14*: 769–772.

Rygg, M. and C.F. Bruun. 1992. Rat bite fever (*Streptobacillus moniliformis*) with septicemia in a child. Scand. J. Infect. Dis. *24*: 535–540.

Savage, N.L. 1984. Genus *Streptobacillus* Levaditi, Nicolau and Poincloux 1925. *In* Bergey's Manual of Systematic Bacteriology, vol. 1 (edited by Krieg). Williams & Wilkins, Baltimore.

Schottmüller, H. 1914. Zur Atiologie und Klinik der Bisskrankheit (Ratten-, Katzen-, Eichhornchen-Bisskrankheit). Dermatol. Wochenschr. Erganzungsh. *58*: 77.

Schwartz, D.N., B. Schable, F.C. Tenover and R.A. Miller. 1995. *Leptotrichia buccalis* bacteremia in patients treated in a single bone marrow transplant unit. Clin. Infect. Dis. *20*: 762–767.

Sebald, M. 1962. Étude sur les bactéries anaérobies gram-négatives asporulées. Thèses de l'Université Paris, Imprimerie Barnéoud S.A., Laval, France.

Shah, H.N. and M.D. Collins. 1980. Fatty acid and isoprenoid quinone composition in the classification of *Bacteroides melaninogenicus* and related taxa. J. Appl. Bacteriol. *48*: 75–87.

Shah, H.N. and M.D. Collins. 1983. Genus *Bacteroides.* A chemotaxonomical perspective. J. Appl. Bacteriol. *55*: 403–416.

Shukla, S.K., P.R. Meier, P.D. Mitchell, D.N. Frank and K.D. Reed. 2002. *Leptotrichia amnionii* sp. nov., a novel bacterium isolated from the amniotic fluid of a woman after intrauterine fetal demise. J. Clin. Microbiol. *40*: 3346–3349.

Simon, M.W. and H.D. Wilson. 1986. *Streptobacillus moniliformis* endocarditis. A case report. Clin. Pediatr. (Phila) *25*: 110–111.

Söderberg, G., A.A. Lindberg and C.E. Nord. 1979. *Bacteroides fragilis* in acute salpingitis. Infection *7*: 226–230.

Strangeways, W.I. 1933. Rats as carriers of *Streptobacillus moniliformis.* J. Pathol. Bacteriol. *37*: 45–51.

Tee, W., P. Midolo, P.H. Janssen, T. Kerr and M.L. Dyall-Smith. 2001. Bacteremia due to *Leptotrichia trevisanii* sp. nov. Eur. J. Clin. Microbiol. Infect. Dis. *20*: 765–769.

Tee, W., P. Midolo, P.H. Janssen, T. Kerr and M.L. Dyall-Smith. 2002. *In* Validation of publication of new names and new combinations previously effectively published outside the IJSEM. List no. 85. Int. J. Syst. Evol. Microbiol. *52*: 685–690.

Thjötta, T., O. Hartmann and J. Böe. 1939. A study of the *Leptotrichia* Trevisan. History, morphology, biological and serological characterisitics. Skr. Nor. Vidensk.-Akad. Oslo I. Mat.-Naturvidensk. Kl. *5*: 1–199.

Topley, W.W.C. and G.S. Wilson. 1936. The Principles of Bacteriology and Immunity, 2nd edn. Edward Arnold, London, p. 274.

Trevisan, V. 1879. Prime linee d'introduzione allo studio dei *Batterj italiani.* Rendiconti dell'Istituto Lombardo di Scienze Series 2 *12*: 133–151.

Vernelen, K., I. Mertens, J. Thomas, J. Vandeven, J. Verhaegen and L. Verbist. 1996. Bacteremia with *Leptotrichia buccalis*: report of a case and review of the literature. Acta Clin. Belg. *51*: 265–270.

Weinberger, M., T. Wu, M. Rubin, V.J. Gill and P.A. Pizzo. 1991. *Leptotrichia buccalis* bacteremia in patients with cancer: report of four cases and review. Rev. Infect. Dis. *13*: 201–206.

Weisbroth, S.H. 1979. Bacterial and mycotic diseases. *In* The Laboratory Rat, vol. 1, Biology and Diseases (edited by Baker, Lindsey and Weisbroth). Academic Press, New York, pp. 193–241.

Wilson, G.S. and A.A. Miles. 1955. Topley and Wilson's Principles of Bacteriology and Immunology, 3rd edn, vol. 1. Williams & Wilkins, Baltimore.

Wittler, R.G. and S.G. Cary. 1974. Genus *Streptobacillus Levaditi,* Nicolau and Poincloux. 1925,1188. *In* Bergey's Manual of Determinative Bacteriology, 8th ed. (edited by Buchanan and Gibbons). Williams & Wilkins, Baltimore, pp. 378–381.

Wullenweber, M., C. Jonas and I. Kunstyr. 1992. *Streptobacillus moniliformis* isolated from otitis media of conventionally kept laboratory rats. J. Exp. Anim. Sci. *35*: 49–57.

Young, C. and A. Hill. 1974. Conjunctivitis in a colony of rats. Lab. Anim. *8*: 301–304.

Phylum XX. Dictyoglomi phyl. nov.

Bharat K. C. Patel

Dic′ty.o.glo′mi. N.L. n. *Dictyoglomus* type genus of the type order of the phylum; *-i* ending to denote phylum; N.L. neut. pl. n. *Dictyoglomi* the phylum of the order *Dictyoglomales*.

The phylum is currently represented by a single class, order, family, and genus. The phylum forms a deep line of descent with its related phyla *Thermomicrobia* and *Deinococcus–Thermus* (Figure 123). Gram-stain-negative, strictly anaerobic, thermophilic, and chemoorganotrophic rod-shaped to filamentous cells that form spherical balls known as rotund bodies. Members produce a range of thermostable enzymes of significance to the biotechnology industries.

Type order: **Dictyoglomales** ord. nov.

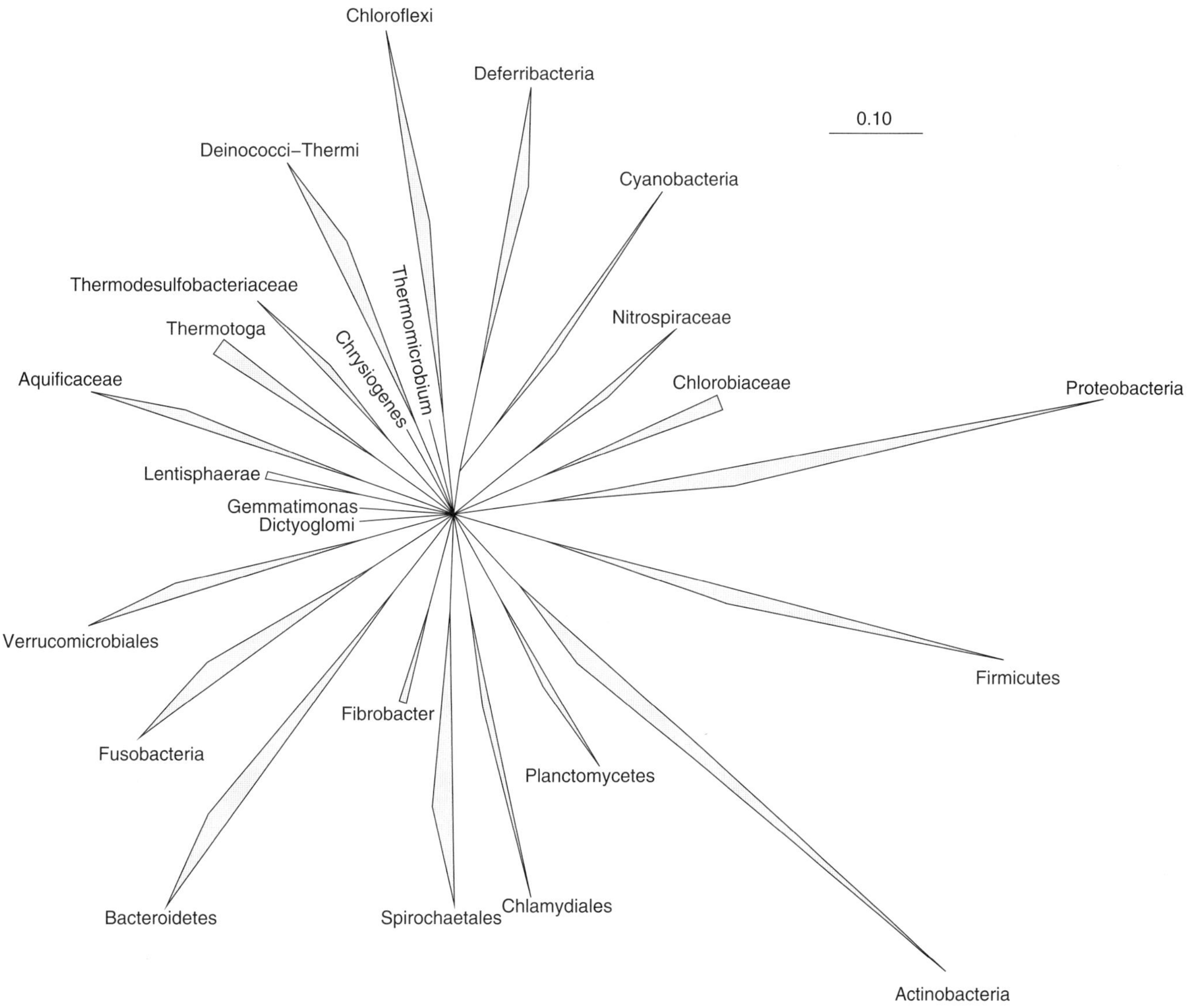

FIGURE 123. Phylogenetic position of phylum *Dictyoglomi*. The scale bar represents 10 changes per 100 nucleotides.

Class I. **Dictyoglomia** class. nov.

BHARAT K. C. PATEL

Dic.ty.o.glo′mi.a. N.L. n. *Dictyoglomus* type genus of the type order of the class; suff. *-ia* ending proposed by Gibbons and Murray and by Stackebrandt et al. to denote a class; N.L. neut. pl. n. *Dictyoglomia* the class of the order *Dictyoglomales*.

Slender (thin) rod to filamentous cells, 5–30 × 0.35–0.45 μm, occurring singly or in pairs. Spherical balls of cells, known as rotund bodies, are commonly observed. Stain Gram-negative, and cell walls possess an outer and an inner cell wall layer. Spores are not formed; cells are nonmotile and no flagella have been observed. **Thermophilic**. Neutrophilic. Chemoorganotrophic, **strict anaerobe, ferments glucose. The principal fermentation end products are acetate, lactate, ethanol, CO_2, and H_2**. Occur in natural thermal hot springs and man-made thermal environments.

Type order: **Dictyoglomales** ord. nov.

Order I. **Dictyoglomales** ord. nov.

BHARAT K. C. PATEL

Dic′ty.o.glo.ma′les. N.L. neut. n. *Dictyoglomus* type genus of the order; *-ales* ending to denote an order; N.L. fem. pl. n. *Dictyoglomales* the order of *Dictyoglomus*.

Only one order, *Dictyoglomales*, currently exists, therefore the description of the order is the same as the class *Dictyoglomi*.

Type genus: **Dictyoglomus** Saiki, Kobayashi, Kawagoe and Beppu 1985, 253[VP].

Reference

Saiki, T., Y. Kobayashi, K. Kawagoe and T. Beppu. 1985. *Dictyoglomus thermophilum* gen. nov., sp. nov., a chemoorganotrophic, anaerobic, thermophilic bacterium. Int. J. Syst. Bacteriol. *35*: 253–259.

Family I. **Dictyoglomaceae** fam. nov.

BHARAT K. C. PATEL

Dic′ty.o.glo.ma′ce.ae. N.L. neut. n. *Dictyoglomus* type genus of the family; *-aceae* ending to denote a family; N.L. fem. pl. n. *Dictyoglomaceae* the family of *Dictyoglomus*.

Only one family, *Dictyoglomaceae*, is accepted in the order *Dictyoglomales*. The description of the family is the same as for the class. Only one genus is accepted in the family *Dictyoglomaceae*.

Type genus: **Dictyoglomus** Saiki, Kobayashi, Kawagoe and Beppu 1985, 256[VP].

Genus I. **Dictyoglomus** Saiki, Kobayashi, Kawagoe and Beppu 1985, 256[VP]

BHARAT K. C. PATEL

Dic′ty.o.glo′mus. Gr. n. *dictyon* net; L. neut. n. *glomus* ball; N.L. neut. n. *Dictyoglomus* net ball.

Slender rod to filamentous cells (5–30 × 0.35–0.45 μm), which **occur singly or in pairs**. Nonsporeforming and nonmotile; **no flagella** have been observed. Stain Gram-negative, and cell walls possess outer and an inner cell wall layers. **Spherical balls of cells**, termed rotund bodies, formed due to either cell aggregation or by cell division, are commonly observed. The walls of the cells in the rotund bodies are portioned with the rotund bodies acquiring the outer wall layer with the cells attached to the inner wall layer. **Thermophilic, neutrophilic, chemoorganotrophs** which are strict **anaerobes. Ferment carbohydrates to acetate, lactate, ethanol, CO_2, and H_2. Occur in natural thermal hot springs** and man-made thermal environments (Table 151).

DNA G+C content (mol%): 29–34.

Type species: **Dictyoglomus thermophilum** Saiki, Kobayashi, Kawagoe and Beppu 1985, 256[VP].

Further descriptive information

Dictyoglomus species can easily be identified as they all form thin filaments and rotund bodies (Figure 124) similar to that observed for *Thermus* species. Only three out of 95 positive enrichment cultures initiated from 370 volcanic hot spring samples yielded *Dictyoglomus* isolates (Patel et al., 1987). None was isolated from the 300 positive enrichment cultures initiated from non-volcanic geothermal waters of the Great Artesian Basin of Australia (Patel, unpublished results), suggesting that, unlike other thermoanaerobes, *Dictyoglomus* species are restricted in nature.

Members of *Dictyoglomus* have so far been isolated only from natural and man-made thermal habitats. To date, two species (*Dictyoglomus thermophilum* H-6-12[T] (Saiki et al., 1985) and *Dictyoglomus turgidum* Z-1310[T]) (Svetlichnii and Svetlichnaya, 1988), both of which have been isolated from volcanic hot springs,

TABLE 151. Characteristics differentiating members of the genus *Dictyoglomus*[a]

Characteristics	*D. thermophiilum* strain H-6-12^T	*D. turgidum* strain Z-1310^T	*Dictyoglomus* strain Rt46-B1	*Dictyoglomus* strain B1
References	Saiki et al. (1985)	Svetlichnii and Svetlichnaya (1988)	Patel et al. (1987)	Mathrani and Ahring (1991)
Habitat	Hot spring, Kumamoto Prefecture, Japan	Hot spring, Kamchatka, Russia	Hot spring, Kuirau Park, New Zealand	Pulp mill cooling tower, Kirkniemi, Finland
Cell morphology and size (μm)	Rods to filaments (5–20 × 0.4–0.6)	Rods to filaments (10–30 × 0.3–0.4)	Rods to filaments (5–25 × 0.35–0.45)	Rods to filaments (up to 5–20 × 0.3)
Size of rotund bodies (μm)	15–100	15–100	3–20	5–25
DNA G+C content (mol%)	29	32.5	29.5	34
Temperature range for growth (°C) (optimum)	51–80 (73–78)	48–86 (72)	40–82 (70)	>50–<80 (72)
pH range for growth (optimum)	5.4–8.9 (7.0)	5.2–9.0 (7.0–7.1)	6.0–8.5 (7.5)	5.0–9.0 (7.2)
Generation time (min)	258	240	462	300
Substrates utilized:				
Arabinose	+	+	+	nd
Carboxymethylcellulose	nd	+	nd	nd
Casein hydrolysate	nd	+	nd	nd
Cellobiose	+	+	+	nd
Cellulose	–	nd	nd	nd
Fructose	+	nd	nd	nd
Fucose	+	nd	nd	nd
Galactose	+	+	+	nd
Glucose	+	+	+	nd
Glycerol	–	nd	nd	nd
Glycogen	nd	nd	nd	nd
Inositol	–	nd	nd	nd
Lactose	+	+	+	nd
Maltose	+	+	+	nd
Mannitol	–	nd	nd	nd
Mannose	+	+	+	nd
Pectin	+	nd	nd	nd
Pyruvate	–	nd	nd	nd
Raffinose	+	nd	nd	nd
Rhamnose	–	nd	nd	nd
Sorbitol	–	nd	nd	nd
Starch	+	+	+	nd
Sucrose	+	+	+	nd
Xylose	+	nd	nd	nd
Enzymes produced:				
Amylase	+	+	+	nd
Cellulase	–	–	–	nd
Xylanase	+	+	+	+
End-products from carbohydrates	Acetate, H_2, lactate, CO_2 (from starch)	Acetate, H_2, lactate, ethanol (from glucose)	Acetate, H_2, lactate, ethanol (from glucose)	Acetate, H_2, (from beech xylan)
Resistance to vancomycin (100 μg/ml)	–	nd	nd	–

[a]Symbols: +, >85% positive; –, 0–15% positive; w, weak reaction; nd, not determined.

have been validly published. Several other strains have also been described. These include *Dictyoglomus* strains Z-1311 to Z-1317 and *Dictyoglomus* strain Rt46-B1 (formerly named *Fervidothrix*), Rt8N2, and Tos-80-la-d. All of these were isolated from volcanic hot springs (pH ranging from 2.8 to 9.0 and temperatures ranging from 55 to 90°C) from Japan, Italy, and Russia (Patel et al., 1986, 1987; Plant et al., 1987; Svetlichnii and Svetlichnaya, 1988). *Dictyoglomus* strain B1 was isolated from a man-made pulp mill cooling tower which had a temperature range of 70–80°C and a pH range of 5.5–6.5 (Mathrani and Ahring, 1991, 1992).

With the exception of *Dictyoglomus* strain B isolated from a cooling tower of a pulp mill, all other *Dictyoglomus* isolates are reported to be nutritionally versatile and use a wide range of carbohydrates. However, all strains studied to date are slow growers with a generation time reported to vary between 260 and 600 min depending on the substrate used. A variety of extracellular enzymes have been identified for their potential in

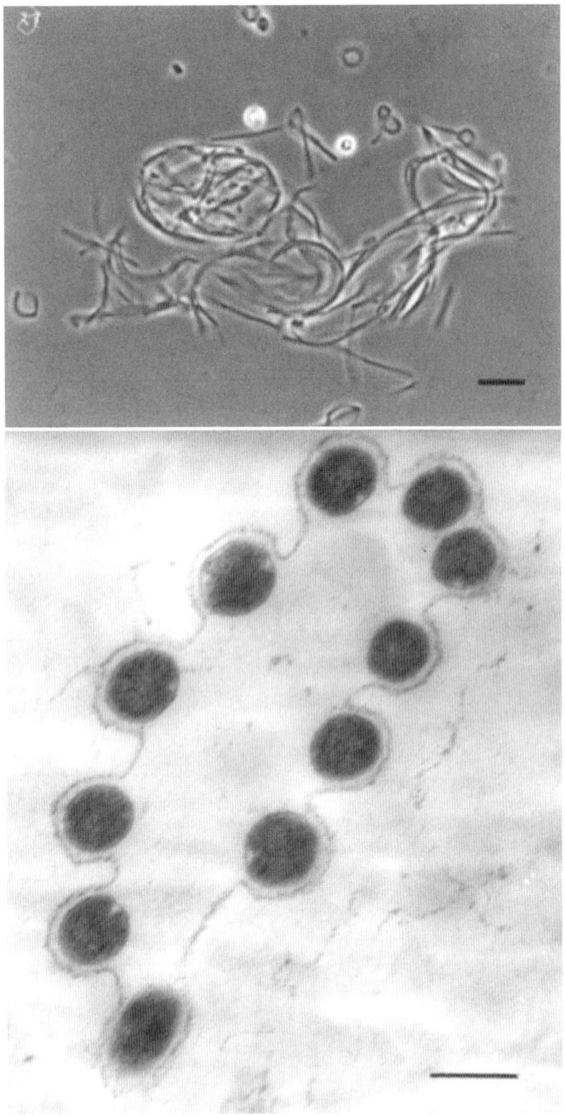

FIGURE 124. Large spherical rotund bodies of *Dictyoglomus thermophilum* (strain H-6-12^T). (top) Phase-contrast microphotograph; bar = 10 µm. (bottom) Thin section of cells showing that they are connected by the outer cell wall layer to form the rotund bodies; bar = 0.5 µm.

biotechnology. A number of reports have shown that xylanases from *Dictyoglomus* strains have the potential to pretreat wood pulp in paper manufacturing processing and achieve comparable levels of whiteness with much less bleach (Kenealy and Jeffries, 2003; Morris et al., 1998; Ratto et al., 1994). This has led to marked improvements in methods for the production of xylanases, both by conventional culture methods (Adamsen et al., 1995) and by gene recombinant techniques (Gibbs et al., 1995; Te'o et al., 2000).

The structure of a xylanase has also been elucidated (Sunna and Bergquist, 2003) and metagenomic approaches have been used to clone and study *Dictyoglomus*-like xylanase genes directly from natural volcanic thermal samples. Amylases from *Dictyglomus* are reported to have a high degree of homology to amylases from a number of hyperthermophilic members of domain *Archaea*. This has stimulated an interest in the origin, evolution, and functioning of amylolytic enzymes in thermophilic and hyperthermophilic microbes (Fukusumi et al., 1988; Horinouchi et al., 1988; Janecek, 1998; Jeon et al., 1997; Kobayashi et al., 1988). Studies on other enzymes such as mannases (Gibbs et al., 1999) and phosphofructokinase (Ding et al., 1999, 2000) have also been reported. Nielsen et al. (2007) provided evidence that the bioaugmentation of xylanase-producing *Dictyoglomus* species and cellulase/xylanase-producing *Caldicellusiruptor* species in cattle manure, led to a 93% increase in methane production. Their physiological processes in pure and/or mixed cultures are not well understood but if this significant gap in our knowledge were to be bridged, then we may be able to significantly improve the microbial energy-yielding processes. At the time of writing, the genomes of *Dictyoglomus thermophilum* and *Caldicellulosiruptor saccharolyticus* are being sequenced, and it will be interesting to await the analysis of the data in a bid to unravel the mysteries of their metabolic pathways.

Enrichment and isolation procedures

Members of the genus *Dictyoglomus* can be enriched and isolated on complex anaerobic media containing yeast extract and/or peptones and a variety of fermentable polysaccharides. All media are prepared and dispensed anaerobically using Hungate techniques in an atmosphere of 100% nitrogen. *Dictyoglomus thermophilum* are grown in a medium containing per liter (pH 7.2): 1.5 g KH_2PO_4, 4.2 g $Na_2HPO_4 \cdot 12H_2O$, 0.5 g NH_4Cl, 0.38 g $MgCl_2 \cdot 6H_2O$, 0.05 g $CaCl_2$, 0.039 g $Fe(NH_4)_2(SO_4)_2 \cdot 6H_2O$, 1.0 g Na_2CO_3, 2.0 mg resazurin, 1.0 g L-cysteine. HCl, 10.0 ml Trace Metal Solution (from a ×100 stock containing per liter 290.0 mg $CoCl_2 \cdot 6H_2O$, 240.0 mg $Na_2MoO_4 \cdot 2H_2O$, 17.0 mg Na_2SeO_3, 200.0 mg $MnCl_2 \cdot 4H_2O$, 280.0 mg $ZnSO_4 \cdot 7H_2O$), 2.0 g yeast extract, and 2.0 g peptone) and 10.0 ml filter-sterilized Wolfe's vitamin solution (from a ×100 stock solution containing 2.0 mg biotin, 2.0 mg folic acid, 10.0 mg pyridoxine hydrochloride, 5.0 mg thiamine HCl, 5.0 mg riboflavin, 5.0 mg nicotinic acid, 5.0 mg calcium D-(+)-pantothenate, 0.1 mg vitamin B12, 5.0 mg *p*-aminobenzoic acid, and 5.0 mg thioctic acid), added aseptically to the sterile medium with soluble starch (5 g/l) as the fermentable substrate.

Dictyoglomus turgidum and *Dictyoglomus* strain B are cultured on a complex medium containing xylan as the fermentable substrate, whereas *Dictyoglomus* strain Rt46-B1 and a number of other volcanic hot spring isolates are enriched on the complex medium Trypticase Yeast Extract Medium (TYE). TYEG contains glucose as the fermentable substrate (Patel et al., 1985a, b). Enrichments are usually initiated by adding sediment and/or water samples into anaerobic medium followed by incubation for up to 7 d at temperatures 70–80°C. In general, the use of glucose should be avoided if the enrichments are to be incubated for more than a few days at temperatures higher than 70°C as furfurals, a toxic by-product are produced under these conditions. Polysaccharides such as xylan and starch are preferable, as they do not form furfurals.

Maintenance procedures

Dictyoglomus strains grown in Trypticase Yeast Extract Starch (TYES) or Trypticase Yeast Extract Glucose (TYEG) Medium remain viable for at least up to 3 months on the bench at room temperature and can be successfully subcultured. For long-term storage, mid-exponential-phase grown cultures are preserved in a 50:50 glycerol:TYEG medium to which 0.1 ml of iron sulfide (Brock and Od'ea, 1977) is added, and cultures are stored at −20 or −80°C.

Differentiation of the genus *Dictyoglomus* from other genera

Though members of *Dictyoglomus* are thermophilic (optimum growth temperature >70°C), strictly anaerobic, carbohydrate-fermenting bacteria, they can be differentiated from all other thermoanaerobes due to their unique slender filamentous-shaped cells and their phylogenetic position in the tree of life. *Dictyoglomus* strains possess a number of phenotypic traits, such as the production of rotund bodies, that are in common with certain members of the order *Thermotogales* and *Thermales*. Members of *Dictyoglomus* and of the order *Thermotogales* can be readily differentiated from *Thermales* as the former contains large amounts of C_{16} and C_{18} fatty acids whereas the latter contain C_{15} iso and C_{17} iso in their cell envelope phospholipid fatty acids (Patel et al., 1991). There are numerous phenotypic features which differentiate members of the *Thermotogales* order and *Dictyoglomus* species, e.g., the presence of terminal spheroids or blebs (ballooning of the ends of the cell) in members of the order *Thermotogales* but not in *Dictyoglomus* species. Morphologically, members of *Dictyoglomus* species are slender filamentous cells. Additionally, members of *Thermotogales* appear to be restricted to volcanic springs and oilfields with low salinity, but this is not the case with *Dictyoglomus* species. Furthermore, *Dictyoglomus* strains are slow growers (e.g., 4.3–10 h generation times in batch cultures in a complex glucose medium), but members of the order *Thermotogales* are fast growers (1.5 h generation time in the same medium).

Taxonomic comments

Early 16S rRNA gene based phylogeny placed *Dictyoglomus thermophilum* H-6-12^T and *Dictyoglomus* strain Rt46-B1 as a deep line of descent in the vicinity of the phylum *Thermotogales* (Love et al., 1993). The high concentrations of saturated phospholipid fatty acids in the cell membranes is another trait that is shared between members of *Dictyoglomus* and the phylum *Thermotogales* (Patel et al., 1991). However, the phylogenetic relationship was not confidently predicted by bootstrap analysis, and the inclusion of different sequences changed the tree topology suggesting that the placement of members of *Dictyglomus* as a deep line of descent was tenuous. The re-evaluation of the phylogeny, based on the selection of new sequences that have been added to the 16S rRNA database, suggests that the members of *Dictyoglomus* do form a separate line of descent, but that it is closer to the members of the phylum *Firmicutes* (Figure 1).

Acknowledgements

This research was supported by the Australian Research Council and by Griffith University Grants Scheme.

List of species of the genus *Dictyoglomus*

1. **Dictyoglomus thermophilum** Saiki, Kobayashi, Kawagoe and Beppu 1985, 256^{VP}

ther.mo'phil.um Gr. n. *thermê* heat; N.L. neut. adj. *philum* (from Gr. neut. adj. *philon*) loving; N.L. neut. adj. *thermophilum* heat-loving.

Rods to filaments measuring 5–20 × 0.4–0.6 µm **that occur singly, in pairs, and in bundles**. Several to several dozen **cells form large spherical bodies** (50–100 µm in diameter) known as rotund bodies (Figure 124). Stain Gram-negative. Nonsporeforming, the cells are nonmotile, and flagella are not observed. Pigment is not formed. **Strict fermentative anaerobe** which grows optimally at 73°C (temperature growth range of >45 and <80°C) and a pH of 7 (pH range of >5.4 and <8.9). Amygdalin, arabinose, cellobiose, fructose, fucose, galactose, glucose, lactose, maltose, raffinose, xylose, mannose, melezitose, melibiose, pectin, starch, ribose, sucrose, trehalose, and insoluble potato starch can be utilized but not rhamnose, glycerol, inositol, mannitol, sorbitol, pyruvate, cellulose, xylitol, insoluble corn, and wheat starch. **Acetate, lactate, H_2, and CO_2 are produced from insoluble potato starch fermentation**. Tests for methyl red, indole production, and nitrate reduction are negative. Voges-Proskauer test is positive. Catalase is produced. Growth is inhibited by antibacterial agents such as streptomycin, tetracycline, neomycin, and tunicamycin at 10 µg/ml, and chloramphenicol, vancomycin, and actinomycin D at 100 µg/ml. Lysozyme (78 µg/ml) and sodium dodecyl sulfate (1%) independently do not lyse the cells, but lysozyme sensitizes cells to subsequent lysis by sodium dodecyl sulfate (0.1%).

Source: slightly alkaline hot springs (Tsuetate Hot Spring) in Kumamoto Prefecture, Japan.

DNA G+C content (mol%): 29 (T_m).

Type strain: H-6-12, ATCC 35947, DSM 3960.

Sequence accession no. (16S rRNA gene): X69194.

2. **Dictyoglomus turgidum** corrig. Svetlichnii and Svetlichnaya 1995, 879^{VP} (Effective publication: Svetlichnii and Svetlichnaya 1988, 369.)

tur'gi.dum L. neut. adj. *turgidum* swollen, inflated, because the cells form spherical aggregates.

Rods to filaments measuring 10–30 × 0.3–0.4 µm which occur singly, in pairs, and in bundles. Several to several dozen cells form large spherical bodies (50–100 µm in diameter) known as rotund bodies. Stain Gram-negative. Nonsporeforming; the cells are nonmotile and flagella are not observed. Pigment is not formed. Strict fermentative anaerobe which grows optimally at 72°C (temperature growth range of 48–86°C) and a pH of 7 (pH range of 5.2–9.0). Cellobiose, fructose, glucose, lactose, maltose, raffinose, sucrose, rhamnose, inositol, mannitol, sorbitol, pyruvate, starch, pectin, glycogen, microcrystalline cellulose, carboxymethylcellulose, lignin, humic acids, yeast extract, peptone, casein hydrolysate, and Casmino acids are fermented, but not rabinose, galactose, xylose, mannose, and glycerol. Ethanol, acetate, H_2, and CO_2 are produced as end products of fermentation. Growth is inhibited by antibacterial agents such as 100 µg/ml streptomycin, 50 µg/ml chloramphenicol and penicillin, but it is resistant to 100 µg/ml vancomycin and rifampin.

Source: hot springs of Uzon volcano crater, Kamchatka, Russia.

DNA G+C content (mol%): 32.5 (T_m).

Type strain: Z-1310, DSM 6724.

Sequence accession no. (16S rRNA gene): not determined.

References

Adamsen, A.K., J. Lindhagen and B.K. Ahring. 1995. Optimization of extracellular xylanase production by *Dictyoglomus* sp. B1 in continuous culture. Appl. Microbiol. Biotechnol. *44*: 327–332.

Brock, T.D. and K. Od'ea. 1977. Amorphous ferrous sulfide as a reducing agent for culture of anaerobes. Appl. Environ. Microbiol. *33*: 254–256.

Ding, Y.H., R.S. Ronimus and H.W. Morgan. 1999. Purification and properties of the pyrophosphate-dependent phosphofructokinase from *Dictyoglomus thermophilum* Rt46 B.1. Extremophiles *3*: 131–137.

Ding, Y.H.R., R.S. Ronimus and H.W. Morgan. 2000. Sequencing, cloning, and high-level expression of the *pfp* gene, encoding a PPi-dependent phosphofructokinase from the extremely thermophilic eubacterium *Dictyoglomus thermophilum*. J. Bacteriol. *182*: 4661–4666.

Fukusumi, S., A. Kamizono, S. Horinouchi and T. Beppu. 1988. Cloning and nucleotide sequence of a heat-stable amylase gene from an anaerobic thermophile, *Dictyoglomus thermophilum*. Eur. J. Biochem. *174*: 15–21.

Gibbs, M.D., R.A. Reeves and P.L. Bergquist. 1995. Cloning, sequencing, and expression of a xylanase gene from the extreme thermophile *Dictyoglomus thermophilum* Rt46B.1 and activity of the enzyme on fiber-bound substrate. Appl. Environ. Microbiol. *61*: 4403–4408.

Gibbs, M.D., R.A. Reeves, A. Sunna and P.L. Bergquist. 1999. Sequencing and expression of a beta-mannanase gene from the extreme thermophile *Dictyoglomus thermophilum* Rt46B.1, and characteristics of the recombinant enzyme. Curr. Microbiol. *39*: 351–0357.

Horinouchi, S., S. Fukusumi, T. Ohshima and T. Beppu. 1988. Cloning and expression in *Escherichia coli* of two additional amylase genes of a strictly anaerobic thermophile, *Dictyoglomus thermophilum*, and their nucleotide sequences with extremely low guanine-plus-cytosine contents. Eur. J. Biochem. *176*: 243–253.

Janecek, S. 1998. Sequence of archaeal *Methanococcus jannaschii* alpha-amylase contains features of families 13 and 57 of glycosyl hydrolases: a trace of their common ancestor? Folia Microbiol. *43*: 123–128.

Jeon, B.S., H. Taguchi, H. Sakai, T. Ohshima, T. Wakagi and H. Matsuzawa. 1997. 4-a-Glucanotransferase from the hyperthermophilic archaeon *Thermococcus litoralis* – enzyme purification and characterization, and gene cloning, sequencing and expression in *Escherichia coli*. Eur. J. Biochem. *248*: 171–178.

Kenealy, W.R. and T.W. Jeffries. 2003. Enzyme processes for pulp and paper: a review of recent developments. Wood Deterior. Preserv. *845*: 210–239.

Kobayashi, Y., M. Motoike, S. Fukuzumi, T. Ohshima, T. Saiki and T. Beppu. 1988. Heat-stable amylase complex produced by a strictly anaerobic and extremely thermophilic bacterium, *Dictyoglomus thermophilum*. Agric. Biol. Chem. (Tokyo) *52*: 615–616.

Love, C.A., B.K.C. Patel, W. Ludwig and E. Stackebrandt. 1993. The phylogenetic position of *Dictyoglomus thermophilum* based on 16S rRNA sequence analysis. FEMS Microbiol. Lett. *107*: 317–320.

Mathrani, I.M. and B.K. Ahring. 1991. Isolation and characterization of a strictly xylan-degrading *Dictyoglomus* from a man-made, thermophilic anaerobic environment. Arch. Microbiol. *157*: 13–17.

Mathrani, I.M. and B.K. Ahring. 1992. Thermophilic and alkalophilic xylanases from several *Dictyoglomus* isolates. Appl. Microbiol. Biotechnol. *38*: 23–27.

Morris, D.D., M.D. Gibbs, C.W. Chin, M.H. Koh, K.K. Wong, R.W. Allison, P.J. Nelson and P.L. Bergquist. 1998. Cloning of the *xynB* gene from *Dictyoglomus thermophilum* Rt46B.1 and action of the gene product on kraft pulp. Appl. Environ. Microbiol. *64*: 1759–1765.

Nielsen, H.B., Z. Mladenovska and B.K. Ahring. 2007. Bioaugmentation of a two-stage thermophilic (68°C/55°C) anaerobic digestion concept for improvement of the methane yield from cattle manure. Biotechnol. Bioeng. *97*: 1638–1643.

Patel, B.K.C., H.W. Morgan and R.M. Daniel. 1985a. *Fervidobacterium nodosum* gen. nov. and spec. nov., a new chemoorganotrophic, caldoactive, anaerobic bacterium. Arch. Microbiol. *141*: 63–69.

Patel, B.K.C., H.W. Morgan and R.M. Daniel. 1985b. A simple and efficient method for preparing anaerobic media. Biotechnol. Lett. *7*: 227–228.

Patel, B.K.C., H.W. Morgan and R.M. Daniel. 1986. Studies on some thermophilic glycolytic anaerobic bacteria from New Zealand hot springs. Syst. Appl. Microbiol. *8*: 128–136.

Patel, B.K.C., H.W. Morgan, J. Wiegel and R.M. Daniel. 1987. Isolation of an extremely thermophilic chemo-organotrophic anaerobe similar to *Dictyoglomus thermophilum* from New-Zealand hot springs. Arch. Microbiol. *147*: 21–24.

Patel, B.K.C., J.H. Skerratt and P.D. Nichols. 1991. The phospholipid ester-Linked fatty acid composition of thermophilic bacteria. Syst. Appl. Microbiol. *14*: 311–316.

Plant, A.R., B.K.C. Patel, H.W. Morgan and R.M. Daniel. 1987. Starch degradation by thermophilic anaerobic bacteria. Syst. Appl. Microbiol. *9*: 158–162.

Ratto, M., I.M. Mathrani, B. Ahring and L. Viikari. 1994. Application of thermostable xylanase of *Dictyoglomus* sp. in enzymic treatment of Kraft Pulps. Appl. Microbiol. Biotechnol. *41*: 130–133.

Saiki, T., Y. Kobayashi, K. Kawagoe and T. Beppu. 1985. *Dictyoglomus thermophilum* gen. nov., sp. nov., a chemoorganotrophic, anaerobic, thermophilic bacterium. Int. J. Syst. Bacteriol. *35*: 253–259.

Sunna, A. and P.L. Bergquist. 2003. A gene encoding a novel extremely thermostable 1,4-β-xylanase isolated directly from an environmental DNA sample. Extremophiles *7*: 63–70.

Svetlichnii, V.A. and T.P. Svetlichnaya. 1988. *Dictyoglomus turgidus* sp. nov., a new extreme thermophilic eubacterium isolated from hot springs in the Uzon Volcano Crater. Microbiology (En. transl. from Mikrobiologiya) *57*: 364–370.

Svetlichnii, V.A. and T.P. Svetlichnaya. 1995. *In* Validation of the publication of new names and new combinations previously effectively published outside the IJSB. List no. 55. Int. J. Syst. Bacteriol. *45*: 879–880.

Te'o, V.S., A.E. Cziferszky, P.L. Bergquist and K.M. Nevalainen. 2000. Codon optimization of xylanase gene *xynB* from the thermophilic bacterium *Dictyoglomus thermophilum* for expression in the filamentous fungus *Trichoderma reesei*. FEMS Microbiol. Lett. *190*: 13–19.

Phylum XXI. Gemmatimonadetes Zhang, Sekiguchi, Hanada, Hugenholtz, Kim, Kamagata and Nakamura 2003, 1161VP

Yoichi Kamagata

Gem.ma.ti.mo.na.de′tes. N.L. fem. pl. n. *Gemmatimonadales* type order of the phylum; N.L. fem. pl. n. *Gemmatimonadetes* the phylum of the order *Gemmatimonadales*.

The phylum *Gemmatimonadetes* is defined on a phylogenetic basis by comparative 16S rRNA gene sequence analysis of one isolated strain and uncultured representatives from multiple terrestrial and aquatic habitats. Heterotrophic Gram-stain-negative bacteria lacking diaminopimelic acid (DAP) in their cell envelopes.

Type order: **Gemmatimonadales** Zhang, Sekiguchi, Hanada, Hugenholtz, Kim, Kamagata and Nakamura 2003, 1161VP.

Reference

Zhang, H., Y. Sekiguchi, S. Hanada, P. Hugenholtz, H. Kim, Y. Kamagata and K. Nakamura. 2003. *Gemmatimonas aurantiaca* gen. nov., sp. nov., a Gram-negative, aerobic, polyphosphate-accumulating micro-organism, the first cultured representative of the new bacterial phylum *Gemmatimonadetes* phyl. nov. Int. J. Syst. Evol. Microbiol. *53*: 1155–1163.

Class I. **Gemmatimonadetes** Zhang, Sekiguchi, Hanada, Hugenholtz, Kim, Kamagata and Nakamura 2003, 1161VP

Yoichi Kamagata

Gem.ma.ti.mo.na.de′tes. N.L. fem. pl. n. *Gemmatimonadales* type order of the class; N.L. fem. pl. n. *Gemmatimonadetes* the class of the order *Gemmatimonadales*.

The description is the same as for the phylum *Gemmatimonadetes.*

Type order: **Gemmatimonadales** Zhang, Sekiguchi, Hanada, Hugenholtz, Kim, Kamagata and Nakamura 2003, 1161VP.

Reference

Zhang, H., Y. Sekiguchi, S. Hanada, P. Hugenholtz, H. Kim, Y. Kamagata and K. Nakamura. 2003. *Gemmatimonas aurantiaca* gen. nov., sp. nov., a Gram-negative, aerobic, polyphosphate-accumulating micro-organism, the first cultured representative of the new bacterial phylum *Gemmatimonadetes* phyl. nov. Int. J. Syst. Evol. Microbiol. *53*: 1155–1163.

Order I. **Gemmatimonadales** Zhang, Sekiguchi, Hanada, Hugenholtz, Kim, Kamagata and Nakamura 2003, 1161VP

Yoichi Kamagata

Gem.ma.ti.mo.na.da′les. N.L. fem. n. *Gemmatimonas* type genus of the order; *-ales* ending to denote an order; N.L. fem. pl. n. *Gemmatimonadales* the order of the genus *Gemmatimonas.*

The description is the same as for the phylum *Gemmatimonadetes.*

Type genus: **Gemmatimonas** Zhang, Sekiguchi, Hanada, Hugenholtz, Kim, Kamagata and Nakamura 2003, 1161VP.

Reference

Zhang, H., Y. Sekiguchi, S. Hanada, P. Hugenholtz, H. Kim, Y. Kamagata and K. Nakamura. 2003. *Gemmatimonas aurantiaca* gen. nov., sp. nov., a Gram-negative, aerobic, polyphosphate-accumulating micro-organism, the first cultured representative of the new bacterial phylum *Gemmatimonadetes* phyl. nov. Int. J. Syst. Evol. Microbiol. *53*: 1155–1163.

Family I. **Gemmatimonadaceae** Zhang, Sekiguchi, Hanada, Hugenholtz, Kamagata and Nakamura 2003, 1161[VP]

Yoichi Kamagata

Gem.mati.mo.na.da.ce′ae. N.L. fem. n. *Gemmatimonas* type genus of the family; *-aceae* ending to denote a family; N.L. fem. pl. n. *Gemmatimonadaceae* the family of the genus *Gemmatimonas*.

The description is the same as for the genus *Gemmatimonas*.

Type genus: **Gemmatimonas** Zhang, Sekiguchi, Hanada, Hugenholtz, Kamagata and Nakamura 2003, 1161[VP].

Genus I. **Gemmatimonas** Zhang, Sekiguchi, Hanada, Hugenholtz, Kamagata and Nakamura 2003, 1161[VP]

Yoichi Kamagata

Gem.ma′ti.mo′nas. L. adj. *gemmatus* provided with buds; L. fem. n. *monas* a unit; N.L. fem. n. *Gemmatimonas* a budding unit.

Gram-stain-negative. Cells are **motile, rod-shaped**. **Spores are not formed**. Cells divide by binary fission but often show budding forms. Mesophilic. Heterotrophic. Cells grow under aerobic conditions. **The major respiratory quinone is MK-9**. The main fatty acids are $C_{15:0}$ iso, $C_{16:1}$ and $C_{14:0}$. The cell wall contains no diaminopimelic acid (DAP) isomers.

DNA G+C content (mol%): 66.0 (HPLC).

Type species: **Gemmatimonas aurantiaca** Zhang, Sekiguchi, Hanada, Hugenholtz, Kamagata and Nakamura 2003, 1161[VP].

Further descriptive information

The genus *Gemmatimonas* contains only one species, *Gemmatimonas aurantiaca*. The type strain, T-27^T, is a Gram-stain-negative, rod-shaped aerobe. Cells often appear to divide by budding division. Strain T-27^T grows at 25–35°C with an optimum growth temperature of 30°C, while no growth is observed below 20°C or above 37°C within 20 d of incubation. The pH range for growth is 6.5–9.5, with an optimum at pH 7.0. Cells are rod-shaped, Gram-stain-negative, motile, 0.7 µm in width and 2.5–3.2 µm in length (mean size 2.8 µm) (Figure 125). Spore formation is not observed. Diffusible pigment production is not observed. Electron microscopy reveals that cells of strain T-27^T possess a Gram-stain-negative cell envelope, with the cytoplasmic and outer membranes readily visible (Figure 126). Electron-dense and transparent inclusion bodies are present in transmission electron micrograph (TEM) images (Figure 126). Cells reproduce by binary fission and often show budding morphology, suggesting asymmetrical cell division (Figures 125 and 126).

Strain T-27^T is able to utilize a limited range of substrates, such as yeast extract, peptone, succinate, acetate, gelatin and benzoate. Neisser staining is positive and 4,6-diamidino-2-phenylindole (DAPI)-stained cells display a yellow fluorescence, indicative of polyphosphate inclusions. Menaquinone-9 is the major respiratory quinone. The cellular fatty acids of the strain are mainly composed of $C_{15:0}$ iso, $C_{16:1}$ and $C_{14:0}$. The G+C content of the genomic DNA is 66 mol%.

Comparative analyses of 16S rRNA gene sequences indicate that strain T-27^T belongs to the candidate division BD (also called KS-B), a phylum-level lineage in the bacterial domain, to date comprised exclusively of environmental 16S rDNA clone sequences. Hence, a new genus and species were proposed, *Gemmatimonas aurantiaca*, as the first cultivated representative of the *Gemmatimonadetes*. Environmental sequence data indicate that this phylum is widespread in nature and has a phylogenetic breadth (19% 16S rDNA sequence divergence) that is greater than well-known phyla such as the *Actinobacteria* (18% divergence).

Based on the standard set of phenotypic tests, strain T-27^T is not physiologically conspicuous. Many phylogenetically diverse bacteria are Gram-stain-negative aerobic heterotrophs and a number of bacteria can accumulate polyphosphate, including "*Candidatus* Accumulibacter phosphatis" (a member of the *Proteobacteria*; Hesselmann et al., 1999), *Microlunatus phosphovorus* and *Tetrasphaera* species (members of the *Actinobacteria*) (Hanada et al., 2002; Maszenan et al., 2000; Nakamura et al., 1995). Strain T-27^T also has a pedestrian, if somewhat limited, carbon substrate utilization profile. Therefore, if isolates are screened using only phenotypic tests, phylogenetically novel microorganisms may be overlooked. The chemotaxonomic and ultrastructural data obtained for strain T-27^T, however, hint at a

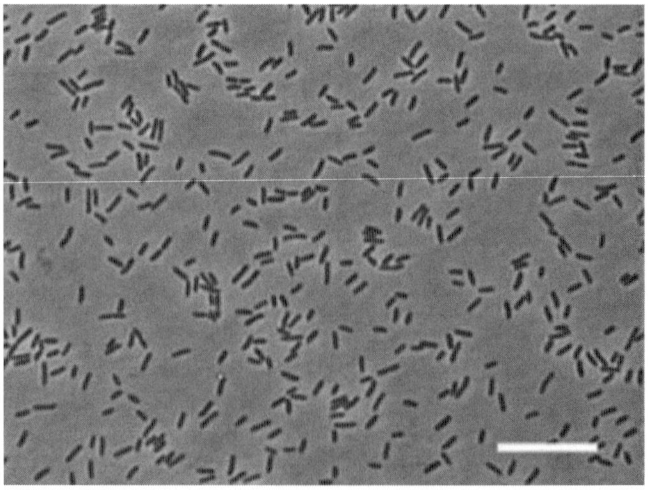

FIGURE 125. Phase-contrast photomicrograph of cells of *Gemmatimonas aurantiaca* strain T-27^T grown aerobically in NM-1 liquid medium at 30°C, often showing asymmetrical cell division. Bar, 10 µm.

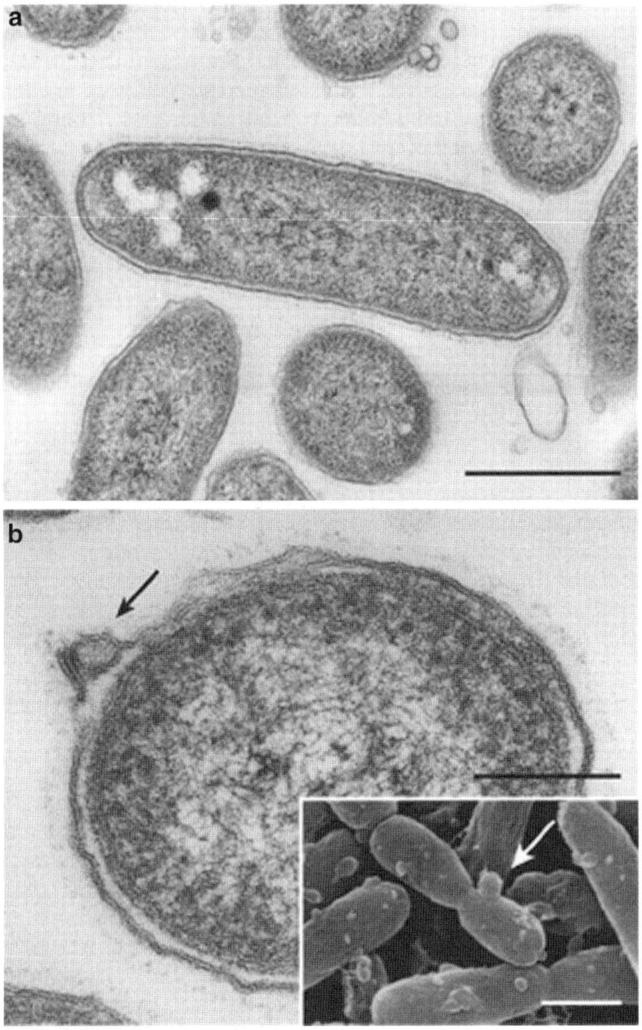

FIGURE 126. Electron micrographs of cells of *Gemmatimonas aurantiaca* strain T-27^T. (a) Transmission electron micrograph showing a Gram-negative cell envelope structure. Bar, 0.5 µm. (b) Transmission and scanning (inset) electron micrographs of cells of strain T-27^T, suggesting asymmetric cell division. Budding structures are shown by arrows. Bar, 0.25 µm (inset, 0.5 µm).

more exotic ancestry. The strain has an unusual combination of dominant fatty acids: 45% $C_{15:0}$ iso (usually characteristic of Gram-stain-positive bacteria) and 27% $C_{16:1}$ (more often found in Gram-stain-negative heterotrophs). Similarly, it is unusual for Gram-stain-negative bacteria to lack DAP in their cell-wall peptidoglycan and to grow aerobically using MK-9 as the major respiratory quinone. Also, the cell envelope is unusual in that the space between the membranes (periplasmic space) is atypically wide for Gram-stain-negative organisms. Genome sequencing now under way will give further insight into this novel lineage within the domain bacteria.

Enrichment and isolation procedures

Strain T-27^T was isolated from a laboratory-scale sequential batch wastewater treatment process operated under enhanced biological phosphorous removal (EBPR) conditions. A strategy was taken to isolate slowly growing, polyphosphate-accumulating bacteria that may be important but uncharacterized members of EBPR communities. Low-speed centrifugations were applied initially to sonicated sludge samples to enrich for heavy and/or large cells that may contain various accumulated materials, such as polyphosphate, poly-β-hydroxyalkanoate and glycogen. Following enrichment, cells were plated onto a heterotrophic agar medium, NM-1, and incubated for up to 12 weeks. Pinpoint and slowly appearing colonies were preferentially selected from the isolation plates and purified through repeated isolation of single colonies on NM-1 media. Twenty isolates were obtained in this manner and checked for polyphosphate inclusions by Neisser staining. Among the Neisser-positive isolates (13/20), one strain, designated T-27^T, warranted further investigation based upon its novel 16S rRNA gene sequence. After 2 weeks incubation, colonies of strain T-27^T were circular, smooth, faintly orange to pink and only 1–2 mm in diameter. This slow growth, resulting in pinpoint colonies, may explain why no closely related isolates have been obtained in pure culture previously.

Taxonomic comments

Comparative sequence analysis indicates that strain T-27^T is not closely related to any known cultured microorganisms in recognized bacterial phyla (85% sequence identity). Instead, it is a member of candidate division BD, a previously known phylum-level lineage in the domain *Bacteria* containing only environmental clones obtained from a number of diverse habitats (Hugenholtz et al., 2001). The environmental clones most closely related to strain T-27^T (Ebpr21 and SBRH63, 99 and 97% sequence identity, respectively) were obtained from two geographically remote activated sludges operated under EBPR conditions (Hugenholtz et al., 2001; Liu et al., 2001). This suggests that strain T-27^T and related organisms may be ubiquitous polyphosphate-accumulating bacteria in EBPR processes. Based on the phylogenetic analysis, the phylum *Gemmatimonadetes* is comprised of four subclasses (subclasses 1–4), and the strain T-27^T belongs to subclass 1 (Zhang et al., 2003). Refer to Zhang et al. (2003) for the evolutionary-distance dendrogram of the type strain, associated representatives of this phylum, and other bacterial phyla.

Over 2000 environmental clone sequences belonging to the *Gemmatimonadetes* are identified in the public databases (as of July 2010) by similarity searches and comparative analysis. They are retrieved not only from terrestrial (soils, activated sludge, saline cave waters and clinical samples) but marine habitats (deep-sea sediments, gas hydrates, arctic bacterioplankton, coastal mobile mud, and marine sponge symbionts). Recently, it was found that members of the phylum *Gemmatimonadetes* make up an average of 2% of soil bacterial communities (Janssen, 2006). Recently, four other isolates within this phylum have been isolated from soil (Davis et al., 2005; Joseph et al., 2003), but they remain unnamed. Although they belong to the same subclass as *Gemmatimonas aurantiaca*, their 16S rRNA genes are 7–8% divergent from those of *Gemmatimonas aurantiaca* and 5–7% divergent from each other. These four isolates therefore could extend the phylogenetic coverage of this phylum.

Differentiation of the species of the genus *Gemmatimonas*

The type and only species is *Gemmatimonas aurantiaca*.

List of species of the genus *Gemmatimonas*

1. **Gemmatimonas aurantiaca** Zhang, Sekiguchi, Hanada, Hugenholtz, Kamagata and Nakamura 2003, 1161[VP]

au.ran'ti.a'ca. N.L. fem. adj. *aurantiaca* orange-colored.

Cells are **rod-shaped** (0.7 μm in width and 2.5–3.2 μm in length: mean size 2.8 μm). Cells possess a **Gram-stain-negative** cell envelope. Cells reproduce by **binary fission and** often show **budding** morphology. Growth occurs between 25 and 35°C with the optimum at 30°C. The pH range is 6.5–9.5. Optimum growth occurs at pH 7.0. The optimum doubling time of growth is 12 h. Catalase and oxidase are produced. **Aerobic**. No fermentative growth is observed. Nitrate reduction is negative.

Neisser staining is positive. Cells are stained yellow with DAPI. Menaquinone-9 is the major respiratory quinone. The cellular fatty acids of the strain are mainly composed of $C_{15:0}$ iso, $C_{16:1}$ and $C_{14:0}$. The following can be utilized as sole carbon sources: yeast extract, peptone, succinate, acetate, gelatin and benzoate. The following can be utilized weakly as sole carbon sources: glucose, sucrose, galactose, melibiose, maltose, formate and hydroxybutyrate. The following carbon sources are not utilized: fructose, mannose, lactose, trehalose, raffinose, arabinose, xylose, rhamnose, glycerol, ethanol, propanol, erythritol, mannitol, sorbitol, lactate, citrate, pyruvate, propionate, malate, butyrate, glutamate, aspartate, alanine, starch, glycogen, gentiobiose, turanose, methyl pyruvate, 2-oxoglutarate, α-ketovalerate, serine, histidine, glycine, leucine and Casamino acids.

Source: the type strain was isolated from a laboratory-scale anaerobic-aerobic sequential batch reactor operated under EBPR conditions.

DNA G+C content (mol%): 66.0 (HPLC).

Type strain: T-27, JCM 11422, DSM 14586.

Sequence accession no. (16S rRNA gene): AB072735.

References

Davis, K.E., S.J. Joseph and P.H. Janssen. 2005. Effects of growth medium, inoculum size, and incubation time on culturability and isolation of soil bacteria. Appl. Environ. Microbiol. *71*: 826–834.

Hanada, S., W.T. Liu, T. Shintani, Y. Kamagata and K. Nakamura. 2002. *Tetrasphaera elongata* sp. nov., a polyphosphate-accumulating bacterium isolated from activated sludge. Int. J. Syst. Evol. Microbiol. *52*: 883–887.

Hugenholtz, P., G.W. Tyson, R.I. Webb, A.M. Wagner and L.L. Blackall. 2001. Investigation of candidate division TM7, a recently recognized major lineage of the domain *Bacteria* with no known pure-culture representatives. Appl. Environ. Microbiol. *67*: 411–419.

Janssen, P.H. 2006. Identifying the dominant soil bacterial taxa in libraries of 16S rRNA and 16S rRNA genes. Appl. Environ. Microbiol. *72*: 1719–1728.

Joseph, S.J., P. Hugenholtz, P. Sangwan, C.A. Osborne and P.H. Janssen. 2003. Laboratory cultivation of widespread and previously uncultured soil bacteria. Appl. Environ. Microbiol. *69*: 7210–7215.

Liu, W.T., A.T. Nielsen, J.H. Wu, C.S. Tsai, Y. Matsuo and S. Molin. 2001. *In situ* identification of polyphosphate- and polyhydroxyalkanoate-accumulating traits for microbial populations in a biological phosphorus removal process. Environ. Microbiol. *3*: 110–122.

Maszenan, A.M., R.J. Seviour, B.K.C. Patel, P. Schumann, J. Burghardt, Y. Tokiwa and H.M. Stratton. 2000. Three isolates of novel polyphosphate- accumulating Gram-positive cocci, obtained from activated sludge, belong to a new genus, *Tetrasphaera* gen. nov., and description of two new species, *Tetrasphaera japonica* sp. nov. and *Tetrasphaera australiensis* sp. nov. Int. J. Syst. Evol. Microbiol. *50*: 593–603.

Nakamura, K., A. Hiraishi, Y. Yoshimi, M. Kawaharasaki, K. Masuda and Y. Kamagata. 1995. *Microlunatus phosphovorus* gen. nov., sp. nov., a new Gram-positive polyphosphate-accumulating bacterium isolated from activated-sludge. Int. J. Syst. Bacteriol. *45*: 17–22.

Zhang, H., Y. Sekiguchi, S. Hanada, P. Hugenholtz, H. Kim, Y. Kamagata and K. Nakamura. 2003. *Gemmatimonas aurantiaca* gen. nov., sp. nov., a Gram-negative, aerobic, polyphosphate-accumulating micro-organism, the first cultured representative of the new bacterial phylum *Gemmatimonadetes* phyl. nov. Int. J. Syst. Evol. Microbiol. *53*: 1155–1163.

Phylum XXII. Lentisphaerae Cho, Vergin, Morris and Giovannoni 2004a, 1005^{VP} (Effective publication: Cho, Vergin, Morris and Giovannoni 2004b, 617.)

Brian P. Hedlund, Jang-Cheon Cho, Muriel Derrien and Kyle C. Costa

Len.ti.spha.e′rae. N.L. fem. n. *Lentisphaera* type genus of the type order of the phylum; N.L. fem. pl. n. *Lentisphaerae* phylum of the genus *Lentisphaera*. *Lentisphaerae* the phylum of bacteria having 16S rRNA gene sequences related to those of members of the *Lentisphaerales*.

The phylum *Lentisphaerae* is defined by phylogenetic analysis based on 16S rRNA gene sequences of cultured strains from seawater and human feces and environmental clone sequences retrieved mainly from **marine habitats, freshwater habitats, anaerobic digesters, and vertebrate feces**. The phylum includes two bacterial genera, *Lentisphaera* and *Victivallis*, both of which are **chemo-organotrophic cocci** with a **Gram-negative** cell structure. **Saccharolytic**, only able to use mono- and di-saccharides, sugar alcohols, or sugar acids. Both produce **extracellular slime material**.

Type order: **Lentisphaerales** Cho, Vergin, Morris and Giovannoni 2004a, 1005^{VP} (Effective publication: Cho, Vergin, Morris and Giovannoni 2004b, 617.).

Further descriptive information

The phylum *Lentisphaerae* has been shown to be related to the phyla *Planctomycetes*, *Chlamydiae*, and *Verrucomicrobia* along with candidate phyla "*Poribacteria*" and OP3 within the so-called PVC superphylum (Cho et al., 2004b; Wagner and Horn, 2006). Within the PVC superphylum, *Lentisphaerae* are most closely related to the *Verrucomicrobia*, with which they share an obligately chemo-organotrophic metabolism focused on carbohydrate metabolism and the general characteristic of being difficult to isolate in pure culture. Possibly, these two phyla are difficult to isolate because they are both K-strategists (Noll et al., 2005). As such, they may be outgrown in enrichment cultures owing to relatively slow growth rates, or inhibited by high concentrations of organic substrates in common isolation media. In addition, these phyla, particularly *Lentisphaerae*, are generally outnumbered by other bacteria in nature (Cho et al., 2004b; Lee et al., 1996; Zoetendal et al., 2003) and several isolates cannot grow on 1.5% (w/v) agar plates, which are used in most microbiology studies (Janssen et al., 2002, 2004; Sangwan et al., 2005; Zoetendal et al., 2003). Attempts to grow these organisms should probably involve serial dilutions of natural samples into media containing low concentrations of monosaccharides, disaccharides, sugar alcohols, or sugar acids as growth substrates. Since this isolation strategy is minimally selective, high throughput isolation and screening approaches are recommended. It may be productive to avoid standard agar plates (1.5% w/v) in favor of broth culture, gellan gum, or semisolid agar.

Prior to the isolation of *Lentisphaera* (Cho et al., 2004b) and *Victivallis* (Zoetendal et al., 2003), sequences in this lineage were grouped within candidate phylum VadinBE97, known only from cultivation-independent microbial censuses (Rappe and Giovannoni, 2003). Currently, 16S rRNA genes representing uncultured bacteria make up >90% of the sequences in the phylum. These environmental clones comprise several unique order- to family-level lineages (Figure 127, Table 152). The two cultivated orders plus three uncultivated lineages were termed subgroups L1–L5 by Cho et al. (2004b); however, to account for the paraphyly of subgroup L4, accommodate new sequences, and avoid confusion, we use the term subphylum to denote well-supported deep monophyletic groups of three or more environmental clone sequences. This term has no taxonomic value, therefore the taxonomy of these subphyla should be re-evaluated and formally addressed as new organisms are isolated and characterized.

The two cultivated genera belong to two well-supported clades, formally designated the orders *Lentisphaerales* and *Victivallales*. The order *Lentisphaerales* occupies a deep phylogenetic position within the *Lentisphaerae* and is comprised of *Lentisphaera* along with two other 16S rRNA gene sequences from marine environments. This clade is well supported and phylogenetically distant from other members of the phylum.

The order *Victivallales* is comprised of *Victivallis*, indigenous to the human colon, along with 16S rRNA gene sequences from a turkey cecum (Scupham, 2007), dugong feces, termite gut contents, and anaerobic digesters fed with waste from a wine distillery or municipal sludge (Chouari et al., 2005; Delbes et al., 2000; Gordon et al., 1997). The order *Victivallales* groups with subphylum 3 (formerly subgroup L2 (Cho et al., 2004b), which is represented by 16S rRNA gene sequences from the cattle rumen, an anaerobic wine distillery waste digester (Gordon et al., 1997), and freshwater lake and river sediments (Crump and Hobbie, 2005; Nercessian et al., 2005). Subphylum 4 (formerly subgroup L3 (Cho et al., 2004b) and subphylum 5 are each comprised of sequences from mucosal secretions of marine invertebrates or deep-sea marine sediments (Alain et al., 2002; Li et al., 1999; Zhu et al., 2008).

Subphyla 6 and 7 form a well-supported clade. Subphylum 6 16S rRNA gene sequences have been recovered from a variety

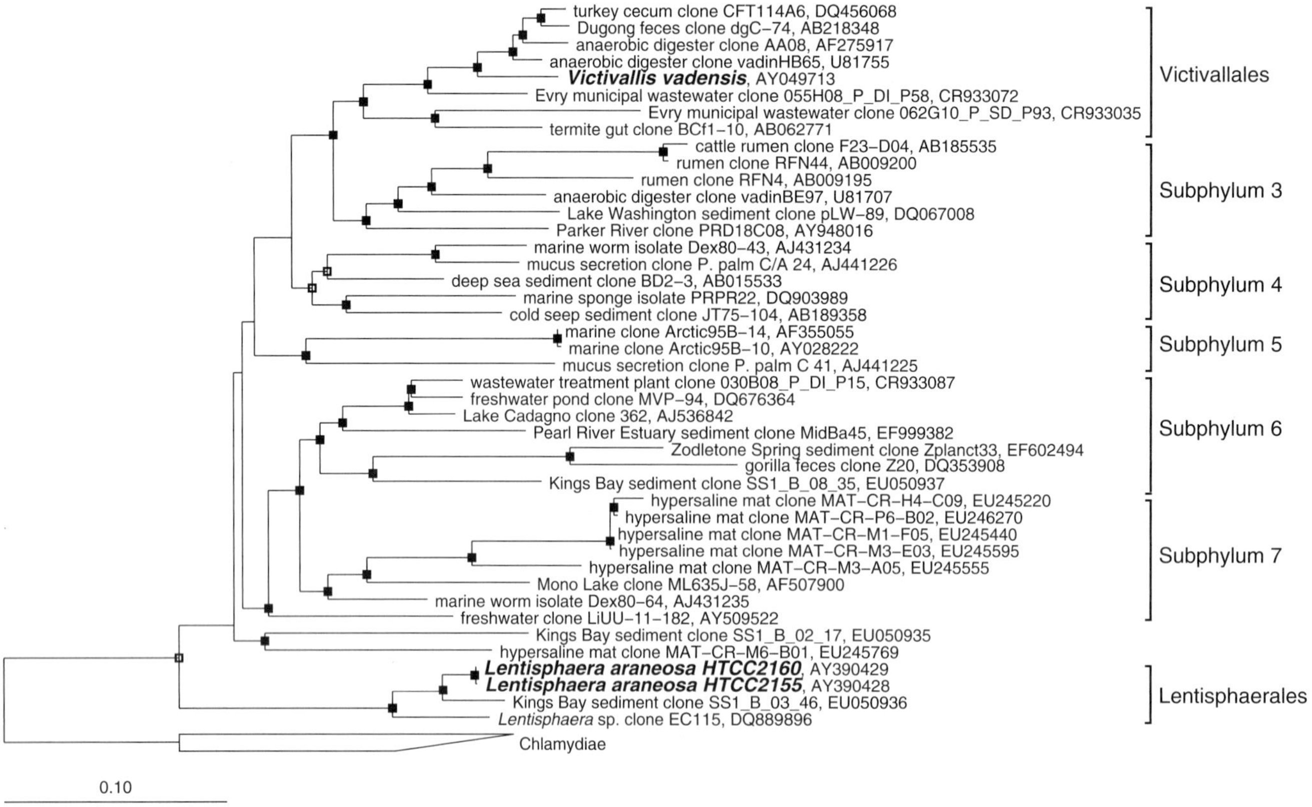

FIGURE 127. Maximum likelihood tree with heuristic correction produced by using ARB (Ludwig et al., 2004). Solid squares represent nodes supported by maximum likelihood, parsimony, and distance (neighbor-joining, Kimura 2-parameter correction). Open nodes were supported by two of the three methods. Parsimony and neighbor joining trees were made with 1000 bootstraps and heuristic corrections. Alignments were unmasked. The dataset was purged of sequences suggested to be anomalous by Mallard (Ashelford et al., 2006).

TABLE 152. Habitat distribution of the orders of *Lentisphaera* including major uncultivated groups

	Freshwater		Marine					
	Sediment	Water	Sediment	Water	Invertebrate mucous	Vertebrate digestive tract	Anaerobic digesters	Hypersaline
Order *Lentisphaerales*				•				
Order *Victivallales*						•	•	
Subphylum 3, uncultivated	○					•	○	
Subphylum 4, uncultivated			○		•			
Subphylum 5, uncultivated			○		•			
Subphylum 6, uncultivated	•	○						
Subphylum 7, uncultivated	○							•

Symbols: •, strains isolated and/or 16S rRNA gene clones from several different cultivation-independent studies of habitat type; ○, more than one 16S rRNA gene clone from a cultivation-independent study of habitat type. No symbol, little or no evidence of colonization of the habitat.

of anaerobic habitats, including municipal sewage (Chouari et al., 2005), anoxic freshwater sediments (Briee et al., 2007; Elshahed et al., 2007), and feces of a wild gorilla (Frey et al., 2006). Subphylum 7 is composed of clone sequences from aquatic habitats varying in salinity, including a basal sequence from a freshwater lake (Eiler and Bertilsson, 2004), a marine clone from the marine tube worm *Alvinella pompejana*, and several derived sequences isolated from hypersaline environments (Humayoun et al., 2003; Isenbarger et al., 2008). The branching pattern within subphylum 7 suggests a freshwater ancestor with later niche invasion into saline and hypersaline environments.

References

Alain, K., M. Olagnon, D. Desbruyeres, A. Page, G. Barbier, S.K. Juniper, J. Quellerou and M.A. Cambon-Bonavita. 2002. Phylogenetic characterization of the bacterial assemblage associated with mucous secretions of the hydrothermal vent polychaete *Paralvinella palmiformis*. FEMS Microbiol. Ecol. *42*: 463–476.

Ashelford, K.E., N.A. Chuzhanova, J.C. Fry, A.J. Jones and A.J. Weightman. 2006. New screening software shows that most recent large 16S rRNA gene clone libraries contain chimeras. Appl. Environ. Microbiol. *72*: 5734–5741.

Briee, C., D. Moreira and P. Lopez-Garcia. 2007. Archaeal and bacterial community composition of sediment and plankton from a suboxic freshwater pond. Res. Microbiol. *158*: 213–227.

Cho, J.C., K.L. Vergin, R.M. Morris and S.J. Giovannoni. 2004a. *In* Validation of publication of new names and new combinations previously effectively published outside the IJSEM. List no. 98. Int. J. Syst. Evol. Microbiol. *54*: 1005–1006.

Cho, J.C., K.L. Vergin, R.M. Morris and S.J. Giovannoni. 2004b. *Lentisphaera araneosa* gen. nov., sp. nov., a transparent exopolymer producing marine bacterium, and the description of a novel bacterial phylum, *Lentisphaerae*. Environ. Microbiol. *6*: 611–621.

Chouari, R., D. Le Paslier, C. Dauga, P. Daegelen, J. Weissenbach and A. Sghir. 2005. Novel major bacterial candidate division within a municipal anaerobic sludge digester. Appl. Environ. Microbiol. *71*: 2145–2153.

Crump, B.C. and J.E. Hobbie. 2005. Synchrony and seasonality in bacterioplankton communities of two temperate rivers. Limnol. Oceanogr. *50*: 1718–1729.

Delbes, C., R. Moletta and J.J. Godon. 2000. Monitoring of activity dynamics of an anaerobic digester bacterial community using 16S rRNA polymerase chain reaction – single-strand conformation polymorphism analysis. Environ. Microbiol. *2*: 506–515.

Eiler, A. and S. Bertilsson. 2004. Composition of freshwater bacterial communities associated with cyanobacterial blooms in four Swedish lakes. Environ. Microbiol. *6*: 1228–1243.

Elshahed, M.S., N.H. Youssef, Q. Luo, F.Z. Najar, B.A. Roe, T.M. Sisk, S.I. Buhring, K.U. Hinrichs and L.R. Krumholz. 2007. Phylogenetic and metabolic diversity of *Planctomycetes* from anaerobic, sulfide- and sulfur-rich Zodletone Spring, Oklahoma. Appl. Environ. Microbiol. *73*: 4707–4716.

Frey, J.C., J.M. Rothman, A.N. Pell, J.B. Nizeyi, M.R. Cranfield and E.R. Angert. 2006. Fecal bacterial diversity in a wild gorilla. Appl. Environ. Microbiol. *72*: 3788–3792.

Gordon, J.J., E. Zumstein, P. Dabert, F. Habouzit and R. Moletta. 1997. Molecular microbial diversity of an anaerobic digestor as determined by small-subunit rDNA sequence analysis. Appl. Environ. Microbiol. *63*: 2802–2813.

Humayoun, S.B., N. Bano and J.T. Hollibaugh. 2003. Depth distribution of microbial diversity in Mono Lake, a meromictic soda lake in California. Appl. Environ. Microbiol. *69*: 1030–1042.

Isenbarger, T.A., M. Finney, C. Rios-Velazquez, J. Handelsman and G. Ruvkun. 2008. Miniprimer PCR, a new lens for viewing the microbial world. Appl. Environ. Microbiol. *74*: 840–849.

Janssen, P.H., P.S. Yates, B.E. Grinton, P.M. Taylor and M. Sait. 2002. Improved culturability of soil bacteria and isolation in pure culture of novel members of the divisions *Acidobacteria*, *Actinobacteria*, *Proteobacteria*, and *Verrucomicrobia*. Appl. Environ. Microbiol. *68*: 2391–2396.

Lee, S.Y., J. Bollinger, D. Bezdicek and A. Ogram. 1996. Estimation of the abundance of an uncultured soil bacterial strain by a competitive quantitative PCR method. Appl. Environ. Microbiol. *62*: 3787–3793.

Li, L., C. Kato and K. Horikoshi. 1999. Bacterial diversity in deep-sea sediments from different depths. Biodivers. Conserv. *8*: 659–677.

Ludwig, W., O. Strunk, R. Westram, L. Richter, H. Meier, Yadhukumar, A. Buchner, T. Lai, S. Steppi, G. Jobb, W. Forster, I. Brettske, S. Gerber, A.W. Ginhart, O. Gross, S. Grumann, S. Hermann, R. Jost, A. Konig, T. Liss, R. Lussmann, M. May, B. Nonhoff, B. Reichel, R. Strehlow, A. Stamatakis, N. Stuckmann, A. Vilbig, M. Lenke, T. Ludwig, A. Bode and K.H. Schleifer. 2004. ARB: a software environment for sequence data. Nucleic Acids Res. *32*: 1363–1371.

Nercessian, O., E. Noyes, M.G. Kalyuzhnaya, M.E. Lidstrom and L. Chistoserdova. 2005. Bacterial populations active in metabolism of C1 compounds in the sediment of Lake Washington, a freshwater lake. Appl. Environ. Microbiol. *71*: 6885–6899.

Noll, M., D. Matthies, P. Frenzel, M. Derakshani and W. Liesack. 2005. Succession of bacterial community structure and diversity in a paddy soil oxygen gradient. Environ. Microbiol. *7*: 382–395.

Rappe, M.S. and S.J. Giovannoni. 2003. The uncultured microbial majority. Annu. Rev. Microbiol. *57*: 369–394.

Sangwan, P., X. Chen, P. Hugenholtz and P.H. Janssen. 2004. *Chthoniobacter flavus* gen. nov., sp. nov., the first pure-culture representative of subdivision two, *Spartobacteria* classis nov., of the phylum *Verrucomicrobia*. Appl. Environ. Microbiol. *70*: 5875–5881.

Sangwan, P., S. Kovac, K.E. Davis, M. Sait and P.H. Janssen. 2005. Detection and cultivation of soil *Verrucomicrobia*. Appl. Environ. Microbiol. *71*: 8402–8410.

Scupham, A.J. 2007. Succession in the intestinal microbiota of preadolescent turkeys. FEMS Microbiol. Ecol. *60*: 136–147.

Wagner, M. and M. Horn. 2006. The *Planctomycetes*, *Verrucomicrobia*, *Chlamydiae* and sister phyla comprise a superphylum with biotechnological and medical relevance. Curr. Opin. Biotechnol. *17*: 241–249.

Zhu, P., Q. Li and G. Wang. 2008. Unique microbial signatures of the alien Hawaiian marine sponge *Suberites zeteki*. Microb. Ecol. *55*: 406–414.

Zoetendal, E.G., C.M. Plugge, A.D. Akkermans and W.M. de Vos. 2003. *Victivallis vadensis* gen. nov., sp. nov., a sugar-fermenting anaerobe from human faeces. Int. J. Syst. Evol. Microbiol. *53*: 211–215.

Class I. **Lentisphaeria** class. nov.

Jang-Cheon Cho, Muriel Derrien and Brian P. Hedlund

Len.ti.spha.e′ria. N.L. fem. n. *Lentisphaera* type genus of the type order; N.L. neut. pl. n. *Lentisphaeria* class of the order *Lentisphaerales. Lentisphaeria* the class of bacteria having 16S rRNA gene sequences related to those of the members of the *Lentisphaerales*.

The class *Lentisphaeria* encompasses bacteria and related 16S rRNA gene sequences mainly from marine habitats, freshwater habitats, anaerobic digesters, and vertebrate feces. It contains the orders *Lentisphaerales* and *Victivallales* in the phylum *Lentisphaerae* and is currently represented only by the genera *Lentisphaera* and *Victivallis*. The delineation of the class is primarily determined based on phylogenetic information of 16S rRNA sequences.

Type order: **Lentisphaerales** Cho, Vergin, Morris and Giovannoni 2004a, 1005[VP] (Effective publication: Cho, Vergin, Morris and Giovannoni 2004b, 617.).

References

Cho, J.C., K.L. Vergin, R.M. Morris and S.J. Giovannoni. 2004a. *In* Validation of publication of new names and new combinations previously effectively published outside the IJSEM. List no. 98. Int. J. Syst. Evol. Microbiol. *54*: 1005–1006.

Cho, J.C., K.L. Vergin, R.M. Morris and S.J. Giovannoni. 2004b. *Lentisphaera araneosa* gen. nov., sp. nov., a transparent exopolymer producing marine bacterium, and the description of a novel bacterial phylum, *Lentisphaerae.* Environ. Microbiol. *6*: 611–621.

Order I. **Lentisphaerales** Cho, Vergin, Morris and Giovannoni 2004a, 1005^{VP} (Effective publication: Cho, Vergin, Morris and Giovannoni 2004b, 617.)

Jang-Cheon Cho and Brian P. Hedlund

Len.ti.spha.e.ra′les. N.L. fem. n. *Lentisphaera* type genus of the order; *-ales* ending to denote an order; N.L. fem. pl. n. *Lentisphaerales* the order of the genus *Lentisphaera*

The order *Lentisphaerales* encompasses bacteria and related 16S rRNA gene sequences (detected by cultivation-independent methods and retrieved from pelagic marine habitats) within the phylum *Lentisphaerae.* The order comprises two strains of the genus *Lentisphaera.* The delineation of the order is primarily determined based on phylogenetic information of 16S rRNA sequences.

Type genus: **Lentisphaera** Cho, Vergin, Morris and Giovannoni 2004a, 1005^{VP} (Effective publication: Cho, Vergin, Morris and Giovannoni 2004b, 618.).

References

Cho, J.C., K.L. Vergin, R.M. Morris and S.J. Giovannoni. 2004a. *In* Validation of publication of new names and new combinations previously effectively published outside the IJSEM. List no. 98. Int. J. Syst. Evol. Microbiol. *54*: 1005–1006.

Cho, J.C., K.L. Vergin, R.M. Morris and S.J. Giovannoni. 2004b. *Lentisphaera araneosa* gen. nov., sp. nov., a transparent exopolymer producing marine bacterium, and the description of a novel bacterial phylum, *Lentisphaerae.* Environ. Microbiol. *6*: 611–621.

Family I. **Lentisphaeraceae** fam. nov.

Jang-Cheon Cho and Brian P. Hedlund

Len.ti.spha.e.ra.ce′ae. N.L. fem. n. *Lentisphaera* type genus of the family; *-aceae* ending to denote a family; N.L. fem. pl. n. *Lentisphaeraceae* the family of the genus *Lentisphaera.*

The family *Lentisphaeraceae* encompasses bacteria and related 16S rRNA gene sequences (detected by cultivation-independent methods and retrieved from pelagic marine habitats) within the phylum *Lentisphaerae.* The family comprises two strains of the genus *Lentisphaera.* The delineation of the family is primarily determined based on phylogenetic information of 16S rRNA sequences.

Type genus: **Lentisphaera** Cho, Vergin, Morris and Giovannoni 2004a, 1005^{VP} (Effective publication: Cho, Vergin, Morris and Giovannoni 2004b, 618.).

Genus I. **Lentisphaera** Cho, Vergin, Morris and Giovannoni 2004a, 1005^{VP} (Effective publication: Cho, Vergin, Morris and Giovannoni 2004b, 618.)

Jang-Cheon Cho and Stephen J. Giovannoni

Len.ti.spha.e′ra. L. adj. *lentus* sticky; L. fem. n. *sphaera* sphere; N.L. fem. n. *Lentisphaera* a sticky sphere.

Cells are spherical (coccus). Gram-stain negative. Gram-negative cell structure. Cells have thin layer of extracellular slime and buds or appendages around the cells. Nonmotile. Strictly aerobic, chemoheterotrophic, and facultatively oligotrophic bacteria. Do not produce pigments. Require NaCl for growth. **Do not utilize amino acids as sole carbon sources. Produce transparent exopolymers.** The major cellular fatty acids are $C_{16:1}$ ω9*c* (50.8%) and $C_{14:0}$ (25.9%). **The genome size is 6.02 Mb, based on full-genome sequencing.** Phylogenetically, the genus belongs to the order *Lentisphaerales* in the phylum *Lentisphaerae.*

DNA G+C content (mol%): 48–49.

Type species: **Lentisphaera araneosa** Cho, Vergin, Morris and Giovannoni 2004a, 1005^{VP} (Effective publication: Cho, Vergin, Morris and Giovannoni 2004b, 618.).

Further descriptive information

The genus *Lentisphaera* currently contains only one species with the type strain, *Lentisphaera araneosa* HTCC2155^{T}. Phylogenetic analyses based on almost complete 16S rRNA gene sequences of strains HTCC2155^{T} and HTCC2160 show that the genus forms a distinct clade together with *Victivallis vadensis* and several environmental sequences of the candidate phylum VadinBE97. The phylogenetic clade containing the genera *Lentisphaera* and *Victivallis* is clearly separated from the nearest phylum, *Verrucomicrobia* (Figure 128). Based on this phylogenetic analysis, the genus *Lentisphaera* is classified as the type genus in the order *Lentisphaerales* of the phylum *Lentisphaerae.* The comparative analyses of 16S rRNA gene sequence show that *Lentisphaera araneosa* has only 65.1–82.1% similarity with the sequences of representatives of all other phyla in the domain *Bacteria*.

Cells are nonmotile, spherical forms with 0.8 μm in diameter, dividing by binary fission. Cells have a Gram-stain-negative cell structure. Flagella, endospore, and intracellular granules are not found. Colonies are 0.6–1.1 mm in diameter, milkish, uniformly circular, convex, and opaque, with smooth surfaces and entire margins, after incubation on marine agar 2216 at 20°C for 3 weeks. The genus *Lentisphaera* is an obligately aerobic, NaCl-requiring, and psychrotolerant marine chemoheterotroph. The temperature range, pH range, and salinity range for growth are 4–25°C (optimum at 16–20°C), pH 7.0–9.0 (optimum at pH 8.0), and 0.75–15% (w/v) of NaCl (optimum at 3.0%), respectively. β Galactosidase activity is positive. Catalase, oxidase, den itrification, indole production, glucose acidification, arginine dihydrolase, urease, and hydrolysis of esculin and gelatin are negative. Cells utilize the following carbon sources (0.02%) as sole carbon sources; D-glucose, D-galactose, D-fructose, β-lactose, D-trehalose D-cellobiose, D-maltose, D-mannose, D-glucosamine, D-mannitol, D-sorbitol, pyruvic acid, succinic acid, and D-malic acid. However, the cells do not utilize amino acids as sole carbon sources.

The most important physiological property of the genus *Lentisphaera* is transparent exopolymer (TEP) production in oligotrophic conditions. The genus produces TEP in artificial seawater medium (ASW; 30 g of NaCl, 1.0 g of $MgCl_2 \cdot 6H_2O$, 4.0 g of Na_2SO_4, 0.7 g of KCl, 0.15 g of $CaCl_2 \cdot 2H_2O$, 0.5 g of NH_4Cl, 0.2 g of $NaHCO_3$, 0.1 g of KBr, 0.27 g of KH_2PO_4, 0.04 g of $SrCl_2 \cdot 6H_2O$, 0.025 g of H_3BO_3, 0.001 g of NaF, 10 ml of Tris–Cl [pH 8.0]), and 1 ml of SN trace metal solution (Waterbury et al., 1986) per 1 l deionized water) and LNHM medium (0.2 μm-filtered and autoclaved seawater supplemented with 1.0 μM NH_4Cl and 0.1 μM KH_2PO_4) amended with 1× mixed carbons (MC; LNHM plus 1× MC); 1× MC was composed of 0.001% (w/v) of each of the following carbon compounds: D-glucose, D-ribose, succinic acid, pyruvic acid, glycerol, *N*-acetyl D-glucosamine, and 0.002% (v/v) of ethanol. The culture medium increases in viscosity at the end of exponential growth phase and reaches a maximum in stationary phase. The TEP stains bright to deep blue with the dye Alcian Blue. The composition of the TEP is rhamnose (62.2%), galactose (14.2%), mannose (12.2%), and glucose (11.1%).

Lentisphaera araneosa was isolated from the surface seawater of the Pacific Ocean; thus its major habit is considered to be marine environments. The members of the phylum *Lentisphaerae* including cultured and uncultured members seem to inhabit mainly marine environments and animal (or human)-related anaerobic terrestrial environments. Marine habitats include seawaters of the Arctic Ocean, the Pacific Ocean, deep-sea sediments, and mucous of the hydrothermal vent polychaete *Paralvinella palmiformis*. Bulk nucleic acid hybridization values determined by using *Lentisphaera araneosa*-specific probe (5′-TTAGCAAGTAAGGATATGGGT-3′) from the Pacific Ocean and the oligotrophic Atlantic seawater samples indicate that *Lentisphaera araneosa* is a common marine bacterium which accounts for less than 1% of total bacterial 16S rRNA.

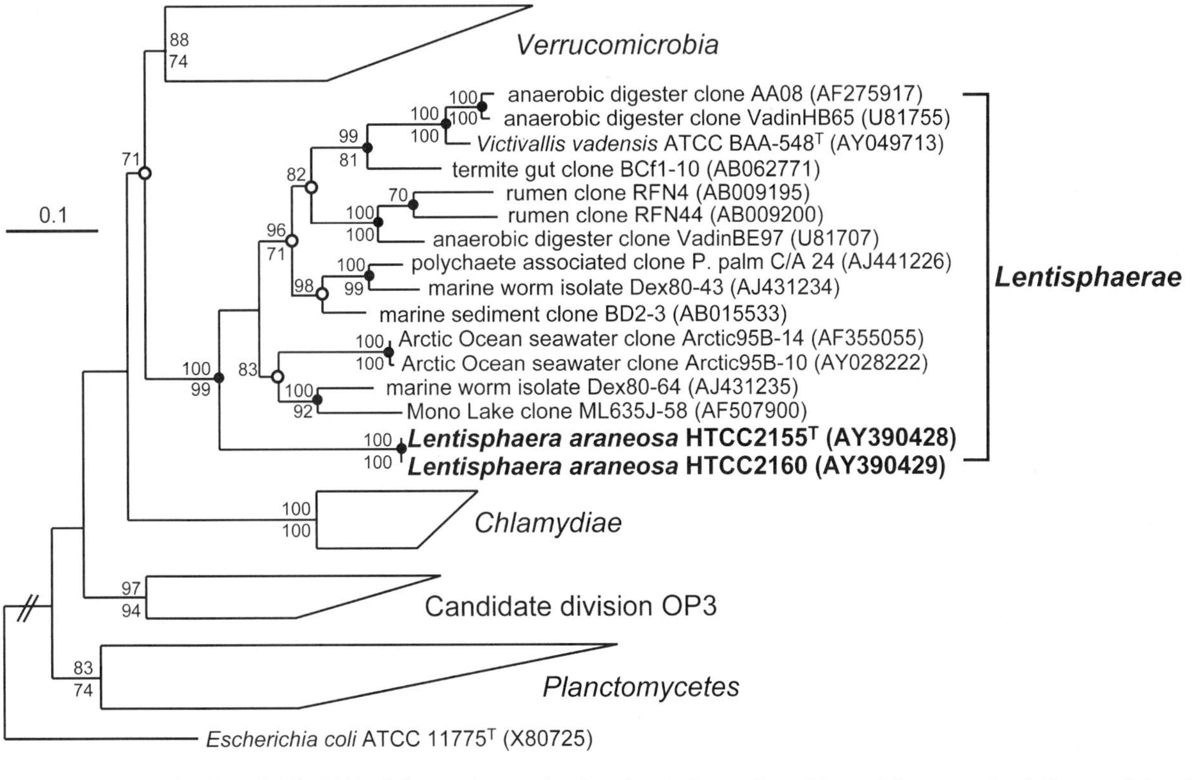

FIGURE 128. Maximum-likelihood 16S rRNA phylogenetic tree showing the phylogenetic positions of the genus *Lentisphaera* and the phylum *Lentisphaerae*. Bootstrap proportions over 70% from both neighbor-joining (above nodes) and maximum parsimony (below nodes) are shown. The closed circles and open circles at each node in the *Lentisphaerae* indicate recovered nodes in all three treeing methods and recovered nodes in two treeing methods respectively. Scale bar, 0.1 substitutions per nucleotide position.

Enrichment and isolation procedures

The original liquid cultures of *Lentisphaera araneosa* were cultivated from the Oregon coast of the Pacific Ocean surface by a dilution-to-extinction method, using LNHM amended with 1× MC after incubation at 16°C for 21 d. The liquid cultures were spread on LNHM plus 1× MC agar medium and colonies were isolated after incubation for 20 d at 16°C. The colonies were transferred on marine agar 2216 and stored as glycerol suspension in liquid nitrogen. Currently there is no special method to enrich *Lentisphaera* from the mixed marine microbial community. Because the growth rate of *Lentisphaera araneosa* is much slower than that of standard colony forming marine bacteria, use of an oligotrophic medium such as LNHM plus 1× MC medium or 1/10 marine R2A is recommended rather than Zobell's medium or marine agar 2216 for the isolation of *Lentisphaera araneosa* from the complex marine microbial community. Recommended incubation temperature to isolate the genus is 16°C.

Maintenance procedures

Frozen stocks as a glycerol suspension (10–30%) stored in liquid nitrogen or at −70°C are routinely used to start a new culture. Either colonies scraped from the surface of agar medium or broth cultures can be used for preparing glycerol stocks. Neither marine broth 2216 nor 1/10 marine R2A broth is suitable for maintaining or preserving the cultures. Incubating the cultures in LNHM plus 1× MC medium or ASW plus 10× MC is recommended for the maintenance and storage of the cultures. Working cultures can be maintained on marine agar 2216 at 16°C for 2 months. Working cultures (slants) can be maintained at 4°C for 3 months.

Procedures for testing special characters

Colonies of *Lentisphaera araneosa* can be visualized as small rounded colonies after 10 d of incubation. To check for culture purity, DAPI stained cells are viewed by epifluorescent microscopy. When *Lentisphaera araneosa* is cultured in ASW plus 10× MC medium, the viscosity of the medium can be easily checked by swirling the culture flasks and measuring viscosity. *Lentisphaera araneosa* can be specifically identified by using *Lentisphaera araneosa*-specific PCR primers HTCC2155-195F (5′-AAGGTTACGCTTA GGGATGA-3′) and HTCC2155-1345R (5′-GTAGCTGATGC-CCATTTACT-3′), and a standard PCR protocol using an annealing temperature of 55°C.

Differentiation of the genus *Lentisphaera* from other related genera

Phylogenetically, the most closely related genus is *Victivallis* within the order *Victivallales* in the phylum *Lentisphaera*. The genus *Lentisphaera* is differentiated from the genus *Victivallis* by 16S rDNA sequence similarity (84.4%), habitat (seawater vs human), oxygen requirement (obligately aerobic vs obligately anaerobic), mode of glucose utilization (oxidation vs fermentation), DNA G+C composition (48.3 mol% vs 59.2 mol%), and optimum growth temperature (20°C vs. 37°C).

Taxonomic comments

The genus *Lentisphaera* is the type genus of the order *Lentisphaerales*, which encompasses bacteria retrieved from marine habitats, within the phylum *Lentisphaerae*. The phylum *Lentisphaerae* was named after the genus *Lentisphaera*, although *Victivallis* is the first validly published genus in the phylum. Because the category phylum is not covered by the Rules of the Bacteriological Code, *Lentisphaerae* as nomenclature of the phylum is legitimate. As the phylum contains some cultured-but-undescribed species and many uncultured organisms, culturing endeavors and more formal description of novel taxa in the phylum would be very helpful to elucidate the taxonomy of the phylum *Lentisphaerae*.

List of species of the genus *Lentisphaera*

1. **Lentisphaera araneosa** Cho, Vergin, Morris and Giovannoni 2004a, 1005^{VP} (Effective publication: Cho, Vergin, Morris and Giovannoni 2004b, 618.)

   a.ra.ne.o′sa. L. fem. adj. *araneosa* similar to cobwebs, pertaining to the morphology of transparent exopolymer particles produced by the species.

   The characteristics of the species are as described for the genus.

   DNA G+C content (mol%): 48–49 (HPLC).

   Type strain: HTCC2155, ATCC BAA-859, KCTC 12141.

   Sequence accession no. (16S rRNA gene): AY390428.

References

Cho, J.C., K.L. Vergin, R.M. Morris and S.J. Giovannoni. 2004a. *In* Validation of publication of new names and new combinations previously effectively published outside the IJSEM. List no. 98. Int. J. Syst. Evol. Microbiol. *54*: 1005–1006.

Cho, J.C., K.L. Vergin, R.M. Morris and S.J. Giovannoni. 2004b. *Lentisphaera araneosa* gen. nov., sp. nov., a transparent exopolymer producing marine bacterium, and the description of a novel bacterial phylum, *Lentisphaerae*. Environ. Microbiol. *6*: 611–621.

Waterbury, J.B., S.W. Watson, F.W. Valois and D.G. Franks. 1986. Biological and ecological characterization of the marine unicellular cyanobacterium *Synechococcus*. *In* Canadian Bulletin Fisheries and Aquatic Sciences, vol. 214 (edited by Platt and Li). Department of Fisheries and Oceans, Ottawa, pp. 71–120.

Order II. **Victivallales** Cho, Vergin, Morris and Giovannoni 2004a, 1005[VP] (Effective publication: Cho, Vergin, Morris and Giovannoni 2004b, 618.)

Muriel Derrien, Jang-Cheon Cho and Brian P. Hedlund

Vic.ti.val.lal'les. N.L. fem. n. *Victivallis* type genus of the order; *-ales* ending to denote an order; N.L. fem. pl. n. *Victivallales* the order of the genus *Victivallis*

The order comprises the genus *Victivallis* and uncultured 16S rRNA genes from mainly anaerobic environments that branch within the phylum *Lentisphaerae*. Clones have been retrieved from the intestinal contents of humans, dugongs, turkeys, and termites, as well as anaerobic digesters and wastewater. The delineation of the order is primarily based on phylogenetic information from 16S rRNA sequences.

Type genus: **Victivallis** Zoetendal, Plugge, Akkermans and de Vos 2003, 214[VP] (Effective publication: Cho, Vergin, Morris and Giovannoni 2004b, 618.).

References

Cho, J.C., K.L. Vergin, R.M. Morris and S.J. Giovannoni. 2004a. *In* Validation of publication of new names and new combinations previously effectively published outside the IJSEM. List no. 98. Int. J. Syst. Evol. Microbiol. *54:* 1005–1006.

Cho, J.C., K.L. Vergin, R.M. Morris and S.J. Giovannoni. 2004b. *Lentisphaera araneosa* gen. nov., sp. nov., a transparent exopolymer producing marine bacterium, and the description of a novel bacterial phylum, *Lentisphaerae*. Environ. Microbiol. *6:* 611–621.

Zoetendal, E.G., C.M. Plugge, A.D. Akkermans and W.M. de Vos. 2003. *Victivallis vadensis* gen. nov., sp. nov., a sugar-fermenting anaerobe from human faeces. Int. J. Syst. Evol. Microbiol. *53:* 211–215.

Family I. **Victivallaceae** fam. nov.

Muriel Derrien, Jang-Cheon Cho and Brian P. Hedlund

Vic.ti.val.la.ce'ae. N.L. fem. n. *Victivallis* type genus of the family; *-aceae* ending to denote a family; N.L. fem. pl. n. *Victivallaceae* the family of the genus *Victivallis*

The family comprises the genus *Victivallis* and uncultured 16S rRNA genes from mainly anaerobic environments that branch within the phylum *Lentisphaerae*. Clones have been retrieved from the intestinal contents of humans, dugongs, turkeys, and termites, as well as anaerobic digesters and wastewater. The delineation of the family is primarily based on phylogenetic information from 16S rRNA sequences.

Type genus: **Victivallis** Zoetendal, Plugge, Akkermans and de Vos 2003, 214[VP].

Genus I. **Victivallis** Zoetendal, Plugge, Akkermans and de Vos 2003, 214[VP]

Muriel Derrien, Caroline M. Plugge, Willem M. de Vos and Erwin G. Zoetendal

Vic.ti.val'lis. L. masc. n. *victus* food; L. fem. n. *vallis* valley; N.L. fem. n. *victivallis* food valley which refers to the Wageningen "Food Valley". This is an area in The Netherlands in which food science is a major research topic.

Coccoid cells, occurring singly and in pairs. The diameters of the cells vary between 0.5 and 1.3 µm. **Gram-negative cell structure. Nonmotile. Strictly anaerobic**. *Victivallis* does not grow on standard agar plates, but beige, shiny, lens-shaped colonies are formed on agar plates with 0.75% (w/v) agar after 10 d of incubation. Cells grow optimally at 37°C and pH 6.5. Growth of *Victivallis* is **chemoorganotrophic** and restricted to a variety of **sugars. Acetate, ethanol, H_2** and **bicarbonate** are the main fermentation products from glucose.

16S rRNA gene sequence analysis indicated that *Victivallis* belongs to the recently described order *Victivallales* of the phylum *Lentisphaerae*. The closest cultured relative is *Lentisphaera araneosa* (84.4% 16S rRNA gene similarity).

DNA G+C content (mol%): 59.2 (HPLC).

Type species: **Victivallis vadensis** Zoetendal, Plugge, Akkermans and de Vos 2003, 214[VP].

Further descriptive information

Light microscopic and transmission electronic microscopic (TEM) analyses revealed that *Victivallis vadensis* is a coccoid cell with Gram-negative cell structure (Figure 129). The cells are variable in diameter (0.5–1.3 µm) and surrounded by haloes, which consist of an extracellular slime layer. In addition, many intracellular electron-dense structures are observed (Figure 129).

Victivallis vadensis was initially isolated in a bicarbonate-buffered anaerobic mineral salts medium (for details see Plugge (2005) supplemented with 10 mM cellobiose and 0.7% (v/v) clarified sterile rumen fluid. However, rumen fluid is not required for growth. In this basal medium *Victivallis* can be cultivated using cellobiose, fructose, galactose, glucose, lactose, lactulose, maltose, maltotriose, mannitol, melibiose, *myo*-inositol, raftilose, rhamnose, ribose, sucrose and xylose as sole carbon and energy source. All sugars are utilized fermentatively. Besides growth on the above-mentioned defined media, *Victivallis* grows well on Wilkens–Chalgren broth (Oxoid; 16 g/l) and the medium described by Kamlage et al. (1999) (designated KA medium) with minor modifications [no hemin, bacteriological peptone (Oxoid) instead of tryptic peptone from meat], both supplemented with 0.7% clarified sterile rumen fluid. Optimal growth conditions include strict anoxic conditions, pH 6.5,

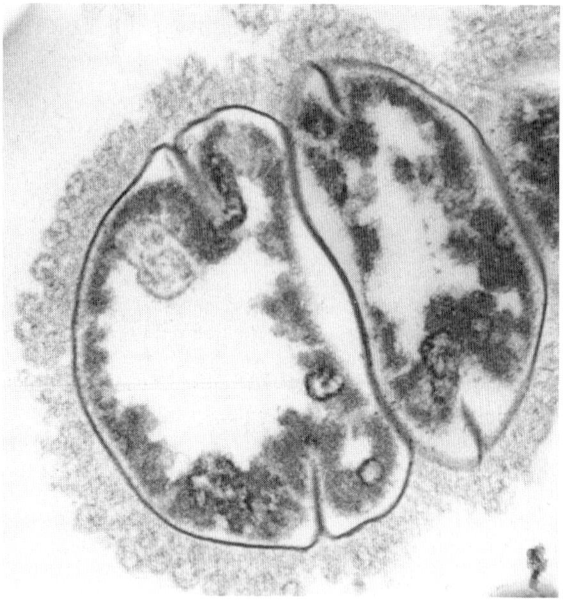

FIGURE 129. Transmission electronic micrograph of the type strain, *Victivallis vadensis*. Scale bar, approximately 0.5 μm.

temperature of 37°C and addition of 0.2% yeast extract. Under these conditions, the doubling time may reach 0.5 h. *Victivallis* can grow syntrophically with *Methanospirillum hungatei* JF-1^T (DSM 864) in co-culture; glucose is converted exclusively to acetate and methane.

Little is known about the habitat of *Victivallis* since it has only been isolated from a single fecal sample. Denaturating gradient gel electrophoretic analysis (DGGE) of 16S rRNA genes indicated that it is not a dominant member from this fecal sample. Two related uncultured *Victivallales* clones which share only 94% 16S rRNA sequence similarity with *Victivallis vadensis* were detected in an anaerobic digester that was fed with wastewater containing plant material (Godon et al., 1997). This indicates that *Victivallis* may also be present in these types of habitats.

Enrichment and isolation procedures

Victivallis can be enriched and eventually isolated from human feces in liquid anaerobic basal medium, with cellobiose as sole carbon and energy source, by repeated transfers in serial dilution. For a detailed protocol concerning the preparation of basal medium see Plugge (2005). Incubations are performed at 37°C under anaerobic conditions under a gas phase of 182 kPa (1.8 atm) N_2/CO_2 (80:20, v/v). The antibiotics streptomycin and polymyxin B may be used to inhibit growth of potential contaminants. Triplicate incubations in soft agar (0.75%) with subsequent transfer of single colonies into liquid medium will result in the isolation of *Victivallis*. Growth in liquid cultures will take 1 d; formation of colonies on soft agar will take approximately 10 d.

Maintenance procedures

Victivallis can be stored in basal medium containing cellobiose or any other sugar for several months at 4°C under anaerobic conditions. For longer periods up to several years, cultures can be maintained in basal medium containing 25% glycerol and 0.1% of a titanium-citrate solution (100 mM) at −80°C. *Victivallis* can also be freeze-dried.

Procedures for testing special characters

Victivallis cells are characterized as coccoid cells with Gram-negative cell structure varying in size between 0.5 and 1.3 μm, and surrounded by a slime layer. This slime layer increases the viscosity of TE or 1× PBS when cells are washed with these solutions for DNA isolation. A special characteristic of *Victivallis* is its inability to grow on standard agar plates.

Differentiation of the genus *Victivallis* from other genera

Phylogenetic analysis of 16S rRNA genes revealed that *Victivallis* belongs to the order *Victivallales* of the phylum *Lentisphaerae* (Figure 130). The closest cultured relative of the type strain *Victivallis vadensis* is *Lentisphaera araneosa*, another isolate from the phylum *Lentisphaerae*, although it only shares 84.4% similarity at the16S rRNA gene level (Cho et al., 2004). While *Victivallis* and *Lentisphaera* both have limited metabolic capacities, the two genera differ by: habitat (human intestine vs sea water); oxygen requirement (strict anaerobe vs strict aerobe), glucose utilization (fermentation vs oxidation), DNA G+C composition (59.2 vs 48.3 mol%), and optimum growth temperature (37°C vs 20°C).

Taxonomic comments

To date, *Victivallis vadensis* is the only isolate representing the genus *Victivallis* and order *Victivallales*. This limits our ability to comprehensively describe this group at the order, genus and species level. Fortunately, culture-independent data shows the detection of bacteria that are more closely related (94% 16S rRNA gene sequence similarity). This suggests that additional species of *Victivallis* or of a closely related genus are waiting to be isolated.

Differentiation of the species of the genus *Victivallis*

Only one species is described, the type species *Victivallis vadensis*.

List of the species of the genus *Victivallis*

1. **Victivallis vadensis** Zoetendal, Plugge, Akkermans and de Vos 2003, 214^{VP}

va.den'sis. N.L. fem. adj. *vadensis* of or belonging to Vada, referring to Wageningen. *Victivallis vadensis* refers to the Wageningen "Food Valley".

Cells are coccoid-shaped, 0.5–1.3 μm, nonmotile, with Gram-stain-negative cell structure and surrounded by an extracellular slime layer. Cells occur singly or in pairs. In pure culture, the cells can grow on a variety of sugars, only under strictly anaerobic conditions. Glucose is converted to acetate, ethanol, H_2 and bicarbonate. Growth on solid agar media is possible below 0.75% (w/v) agar. Cells grow optimally on bicarbonate-buffered mineral salts medium with cellobiose at 37°C and pH 6.5.

Source: human feces.

DNA G+C content (mol%): 59.2 (HPLC).

Type strain: Cello, DSM 14823, ATCC BAA-548.

Sequence accession no. (16S rRNA gene): AY049713.

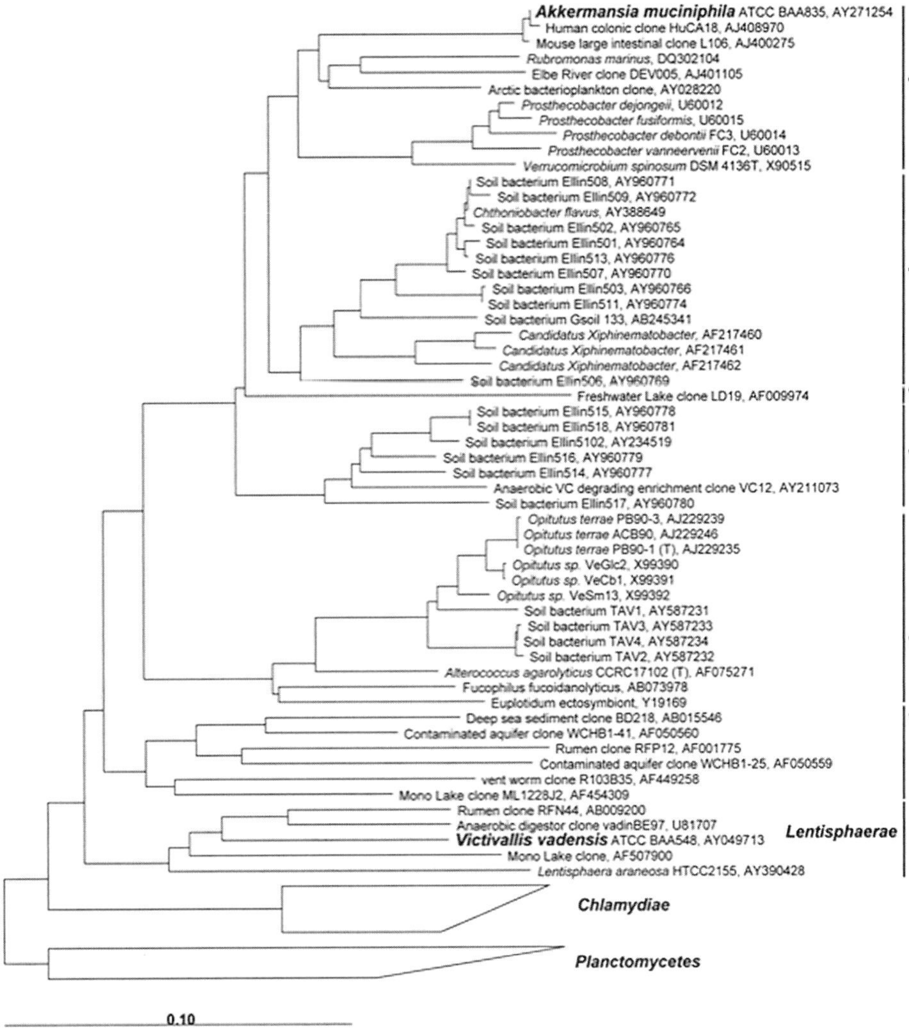

FIGURE 130. Phylogenetic neighbor-joining dendogram showing the relationships of *Victivallis vadensis* with representatives of the phyla *Verrucomicrobia* and *Lentisphaerae*. Bar = 0.1% sequence difference.

References

Cho, J.C., K.L. Vergin, R.M. Morris and S.J. Giovannoni. 2004. *Lentisphaera araneosa* gen. nov., sp. nov., a transparent exopolymer producing marine bacterium, and the description of a novel bacterial phylum, *Lentisphaerae*. Environ. Microbiol. *6*: 611–621.

Godon, J.J., E. Zumstein, P. Dabert, F. Habouzit and R. Moletta. 1997. Molecular microbial diversity of an anaerobic digestor as determined by small-subunit rDNA sequence analysis. Appl. Environ. Microbiol. *63*: 2802–2813.

Kamlage, B., L. Hartmann, B. Gruhl and M. Blaut. 1999. Intestinal microorganisms do not supply associated gnotobiotic rats with conjugated linoleic acid. J. Nutr. *129*: 2212–2217.

Plugge, C.M. 2005. Anoxic media design, preparation, and considerations. *In* Methods in Enzymology, vol. 397. Academic Press, New York, NY, pp. 3–16.

Zoetendal, E.G., C.M. Plugge, A.D. Akkermans and W.M. de Vos. 2003. *Victivallis vadensis* gen. nov., sp. nov., a sugar-fermenting anaerobe from human faeces. Int. J. Syst. Evol. Microbiol. *53*: 211–215.

Phylum XXIII. Verrucomicrobia phyl. nov.

Brian P. Hedlund

Ver.ru'co.mi.cro'bi.a. N.L. fem. pl. n. *Verrucomicobiales* type order of the phylum; N.L. neut. pl. n. *Verrucomicrobia* the *Verrucomicrobiales* phylum.

The phylum *Verrucomicrobia* is defined by phylogenetic analysis of 16S rRNA gene sequences of cultured strains and environmental clone sequences retrieved from a wide variety of environments. All cultivated members of the phylum stain Gram-negative; many have intracellular compartments bounded by internal membranes. Menaquinones are the dominant respiratory quinones; ubiquinones have not been detected. Most members are chemoheterotrophs, preferring carbohydrates including complex natural polysaccharides. Recently isolated members are thermoacidophilic methylotrophs.

Type order: **Verrucomicrobiales** Ward-Rainey, Rainey, Schlesner and Stackebrandt 1996, 625[VP] (Effective publication: Ward-Rainey, Rainey, Schlesner and Stackebrandt 1995, 3249.) emend. Yoon, Matsuo, Adachi, Nozawa, Matsuda, Kasai and Yokota 2008, 1002.

Further descriptive information

To date, there are seven class-level groups in the phylum including three classes that have been formally defined, *Verrucomicrobiae*, *Spartobacteria*, and *Opitutae*. Two additional subphyla groups include recently cultivated organisms that have not yet been formally defined. Lastly, two subphyla groups are represented solely by 16S rRNA gene sequences from cultivation-independent studies (Figure 131). Each class-level group exhibits distinct habitat preferences that are not yet fully elucidated (Table 153). Classes *Verrucomicrobiae*, *Spartobacteria*, and *Opitutae* are each described in their own class chapters. Here, traits of the four remaining class-level groups are described.

Subphylum 3 is known mostly from 16S rRNA gene sequences from cultivation-independent studies of soils and marine environments. Recently, six strains were isolated from soils by prolonged incubation on gellan gum-solidified medium with xylan as the sole carbon and energy source, including Ellin 514, Ellin 516, and Ellin 518 (Sangwan et al., 2005; Figure 132). The genus name "*Pedosphaera*" will be proposed to encompass these Gram-stain-negative, saccharolytic cocci (Sangwan and Janssen, pers. Comm.). Ellin 514 has been shown to have a compartmentalized cell including a membrane-enclosed compartment containing a condensed nucleoid and ribosomes (termed the pirellulosome) and a ribosome-free compartment between the pirellulosome and the cytoplasmic membrane (termed the paryphoplasm) (Lee et al., 2009). This cellular architecture has also been described in *Verrucomicrobium*, *Prosthecobacter*, and *Chthoniobacter* species (Lee et al., 2009). The phylogenetic structure of subphylum 3 includes two primary lineages (Figure 132). The lineage including "*Pedosphaera*" is comprised almost entirely of 16S rRNA genes recovered from soil habitats (Cruz-Martinez et al., 2009; Janssen, 2006; Sangwan et al., 2005) where it is typically the second most abundant class-level group of *Verrucomicrobia* in clone libraries at 0–4.7% of total 16S rRNA gene clones (Janssen, 2006). The second lineage is comprised of 16S rRNA genes recovered from marine habitats, particularly in the deep sea (Lau and Armbrust, 2006; Penn et al., 2006).

Subphylum 6 has been predominantly recovered from methane-emitting mud volcanoes showing evidence of methane consumption (Alain et al., 2006; Niemann et al., 2006). Subsequently, three separate groups reported isolation of methanotrophic *Verrucomicrobia* from thermoacidic mud volcanoes, provisionally named "*Methylacidiphilum infernorum*" (Dunfield et al., 2007), "*Methylacidiphilum kamchatkensis*" (Islam et al., 2008), and "*Acidimethylosilex fumarolicum*" (Pol et al., 2007). "*Methylacidiphilum infernorum*" and "*Methylacidiphilum kamchatkensis*" were isolated from Hell's Gate, New Zealand, and Uzon Caldera, Kamchatka, Russia, respectively. These two isolates have similar pH growth optima at pH 3.5 and ranges from pH 0.8 to 5.8. The temperature ranges for growth were 40–65°C with optima at 55°C. However, only "*Methylacidiphilum kamchatkensis*" appears to be able to fix atmospheric N_2. "*Acidimethylosilex fumarolicum*" was isolated from the Pozzouli Solfatara, Italy with a pH range of 1.0–6.0 (optimum 2.5) at 60°C. All three isolates have three divergent operons encoding the components of the particulate (membrane-bound) methane monooxygenase (pMMO) and possess intracellular membrane structures described as tubular or polyhedral which are likely to increase membrane surface area for pMMO in the absence of stacked membrane structures typical of methylotrophic *Proteobacteria*. Analysis of pyrosequencing data from "*Acidimethylosilex fumarolicum*" (Pol et al., 2007) and a complete genome from "*Methylacidiphilum infernorum*" (Hou et al., 2008) show simple, modified methanotrophic pathways with many homologs from methylotrophic *Proteobacteria*. The genome size of "*Methylacidiphilum infernorum*" is much smaller than *Proteobacteria* methylotrophs at 2.3 Mb; it was suggested that genes for C1 metabolism were obtained by horizontal gene transfer from *Proteobacteria*.

Subphylum 5 is known only by cultivation-independent studies (Figure 133). Two major clades exist. One is found primarily in anaerobic sediments such as those showing anaerobic ammonia oxidation activity (Freitag and Prosser, 2003), contaminated aquifers (Dojka et al., 1998), geothermal mats (both marine and freshwater) (Hirayama et al., 2007), RNA from freshwater sediments (Nercessian et al., 2005), and hypersaline terminal-basin lakes (Humayoun et al., 2003). The second clade is also found in anaerobic habitats such as those listed above, however, this clade has also been detected in digestive tracts of vertebrates

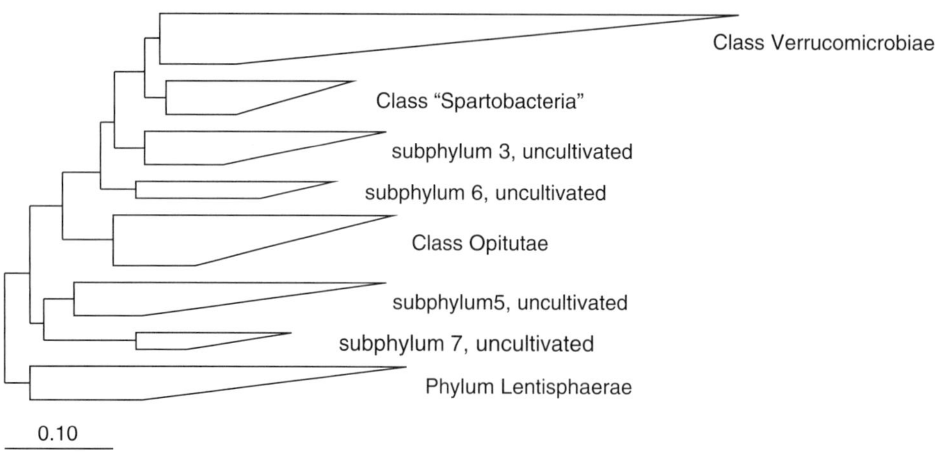

FIGURE 131. Phylogeny of *Verrucomicrobia* calculated with maximum likelihood methods with heuristic correction showing nearly complete 16S rRNA gene sequences present in version 91 of the SILVA database (Pruesse et al., 2007). All class-level groups were included and wedged to show the relationship between existing classes and class-level groups or "subphyla" not yet formally described. Subphylum 3 includes strains recently isolated by Janssen's group but not yet formally described. The genus name "*Pedosphaera*" will be proposed (Sangwan et al., 2005). Recent thermoacidophilic methylotrophs provisionally named "*Methylacidiphilum infernorum*" (Dunfield et al., 2007), "*Methylacidiphilum kamchatkensis*" (Islam et al., 2008), "*Acidimethylosilex fumarolicum*" (Pol et al., 2007), and related 16S rRNA gene sequences belong to subphylum 6. The phylogeny was produced by using ARB (Ludwig et al., 2004) using *Escherichia coli* 16S rRNA gene nucleotides 110–1274. Alignments were unmasked.

TABLE 153. Habitat distribution of the classes of *Verrucomicrobia* including uncultivated class-level groups from studies published prior to June 2008[a,b]

Habitat	Class *Verrucomicrobiae*	Class *Opitutae*	Class *Spartobacteria*	Class 3, uncultivated[c]	Class 5, uncultivated	Class 6, uncultivated[c]	Class 7, uncultivated
Soil	○	●	●	●		○	
Sediment:							
Fresh	○		○		●		
Marine	●			○	●		○
Water:							
Fresh	●	○	○		○		
Marine	●	●		●	○		
Symbionts or commensals:							
Vertebrate	●				●		
Invertebrate	●	●	●	○	○	○	○
Geothermal		○		○	○	●	○

[a]Symbols: ●, multiple strains isolated from different samples and/or >25% of 16S rRNA gene clones in the class from cultivation-independent studies come from habitat type; ○, single strain isolated or multiple strains from the same sample and/or <25% of 16S rRNA gene clones from cultivation-independent studies come from habitat type. No symbol, little or no evidence of colonization of the habitat.

[b]Here, class-level groups of *Verrucomicrobia* are used synonymously with original subphylum designations (Hugenholtz et al., 1998).

[c]Although listed as "uncultivated" here, classes 3 and 6 are now represented in culture (Dunfield et al., 2007; Islam et al., 2008; Pol et al., 2007; Sangwan et al., 2005).

(Frey et al., 2006). Subphylum 7 is also represented only by 16S rRNA gene sequences from a variety of environments, although it is currently only represented by a small number of sequences. Subphylum 7 appears to be monophlyletic with subphylum 5, although it is separated by a relatively large phylogenetic distance and is treated separately here.

The relationship between the *Verrucomicrobia* and other phyla has been the subject of some research. It was originally suggested by phylogenetic analyses of a relatively small number of 16S rRNA gene sequences that the *Verrucomicrobia* were most closely related to the *Planctomycetes* and the *Chlamydia* (Hedlund et al., 1996, 1997; Ward-Rainey et al., 1995). Recently, this has been supported by more robust analyses based on alignments of large numbers of 16S rRNA gene sequences (Wagner and Horn, 2006), concatenated ribosomal proteins (Hou et al., 2008), and BLAST analysis of putative proteins from whole-genome sequences (Hou et al., 2008). It has been proposed that the *Verrucomicrobia* belong to the so-called PVC superphylum, which is comprised of the well-characterized phyla *Planctomycetes*, *Verrucomicrobia*, and *Chlamydia* along with the *Lentisphaera*, *Poribacteria*, and candidate phylum OP3. The relationships among these phyla and others will likely be further clarified through comparisons of whole genomes of a number of *Verrucomicrobia* (Galperin, 2008). The relationship between the *Verrucomicrobia* and *Planctomycetes* is also bolstered by the report of very similar cell compartmentalization (Lee et al., 2009).

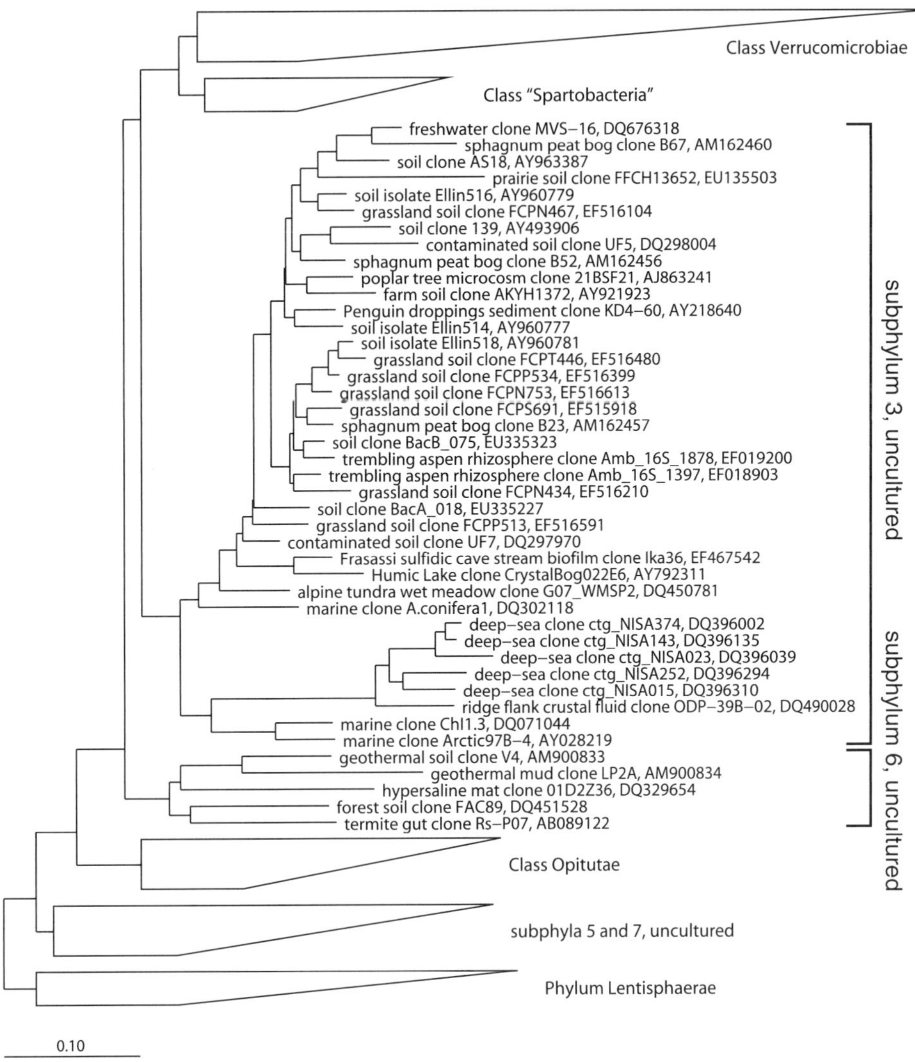

FIGURE 132. Phylogeny of *Verrucomicrobia* subphyla 3 and 6. Calculated as in Figure 131. Subphylum 3 includes strains Ellin 514, Ellin 516, and Ellin 518 recently isolated by Janssen's group but not yet formally described. The genus name "*Pedosphaera*" will be proposed (Sangwan et al., 2005). Recent thermoacidophilic methylotrophs provisionally named "*Methylacidiphilum infernorum*" (Dunfield et al., 2007), "*Methylacidiphilum kamchatkensis*" (Islam et al., 2008), "*Acidimethylosilex fumarolicum*" (Pol et al., 2007), and related 16S rRNA gene sequences belong to subphylum 6. The phylogeny was produced by using ARB (Ludwig et al., 2004) using *Escherichia coli* 16S rRNA gene nucleotides 110–1274. Alignments were unmasked.

Four members of the *Verrucomicrobia*, representing three class-level lineages, were used for detailed ultrastructural characterization using cryosubstituted cross sections revealing complex cell structure. All four *Verrucomicrobia* were characterized by the presences of an intracytoplasmic membrane that bounds a condensed nucleoid and ribosomes and a ribosome-free paryphoplasm between the intracytoplasmic membrane and the cytoplasmic membrane. As in the *Planctomycetes*, the function of the complex cell structure is not currently understood.

References

Alain, K., T. Holler, F. Musat, M. Elvert, T. Treude and M. Kruger. 2006. Microbiological investigation of methane- and hydrocarbon-discharging mud volcanoes in the Carpathian Mountains, Romania. Environ. Microbiol. *8*: 574–590.

Cruz-Martinez, K., K.B. Suttle, E.L. Brodie, M.E. Power, G.L. Andersen and J.F. Banfield. 2009. Despite strong seasonal responses, soil microbial consortia are more resilient to long-term changes in rainfall than overlying grassland. ISME J. *3*: 738–744.

Dojka, M.A., P. Hugenholtz, S.K. Haack and N.R. Pace. 1998. Microbial diversity in a hydrocarbon- and chlorinated-solvent-contaminated aquifer undergoing intrinsic bioremediation. Appl. Environ. Microbiol. *64*: 3869–3877.

Dunfield, P.F., A. Yuryev, P. Senin, A.V. Smirnova, M.B. Stott, S. Hou, B. Ly, J.H. Saw, Z. Zhou, Y. Ren, J. Wang, B.W. Mountain, M.A. Crowe, T.M. Weatherby, P.L. Bodelier, W. Liesack, L. Feng, L. Wang and M. Alam. 2007. Methane oxidation by an extremely acidophilic bacterium of the phylum *Verrucomicrobia*. Nature *450*: 879–882.

Freitag, T.E. and J.I. Prosser. 2003. Community structure of ammonia-oxidizing bacteria within anoxic marine sediments. Appl. Environ. Microbiol. *69*: 1359–1371.

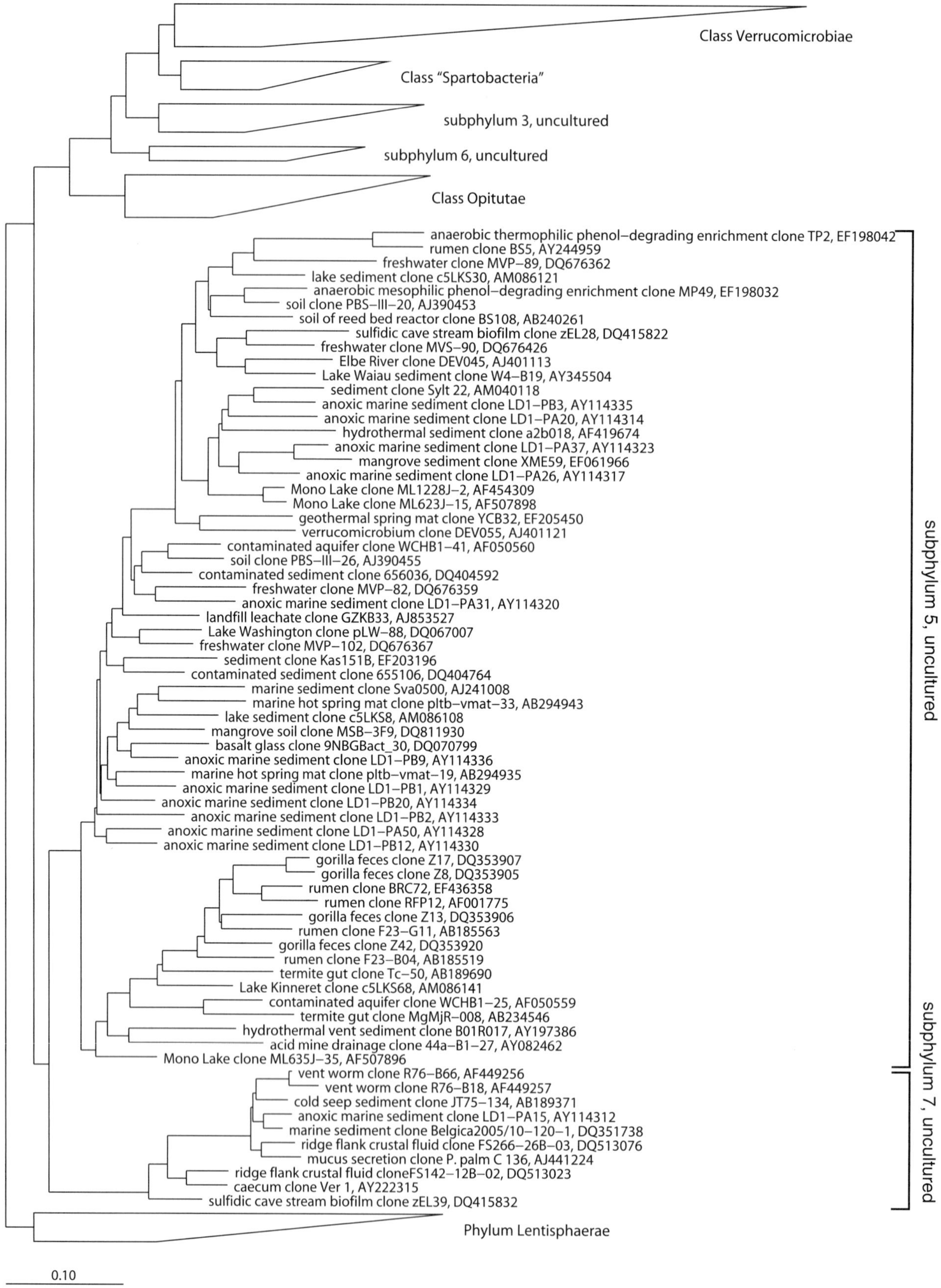

FIGURE 133. Phylogeny of *Verrucomicrobia* subphyla 5 and 7. Calculated as in Figure 131. Subphyla 5 and 7 are, to date, represented only by 16S rRNA gene sequences from cultivation-independent studies.

Frey, J.C., J.M. Rothman, A.N. Pell, J.B. Nizeyi, M.R. Cranfield and E.R. Angert. 2006. Fecal bacterial diversity in a wild gorilla. Appl. Environ. Microbiol. *72*: 3788–3792.

Galperin, M.Y. 2008. New feel for new phyla. Environ. Microbiol. *10*: 1927–1933.

Hedlund, B.P., J.J. Gosink and J.T. Staley. 1996. Phylogeny of *Prosthecobacter*, the fusiform caulobacters: members of a recently discovered division of the *Bacteria*. Int. J. Syst. Bacteriol. *46*: 960–966.

Hedlund, B.P., J.J. Gosink and J.T. Staley. 1997. *Verrucomicrobia* div. nov., a new division of the bacteria containing three new species of *Prosthecobacter*. Antonie van Leeuwenhoek *72*: 29–38.

Hirayama, H., M. Sunamura, K. Takai, T. Nunoura, T. Noguchi, H. Oida, Y. Furushima, H. Yamamoto, T. Oomori and K. Horikoshi. 2007. Culture-dependent and -independent characterization of microbial communities associated with a shallow submarine hydrothermal system occurring within a coral reef off Taketomi Island, Japan. Appl. Environ. Microbiol. *73*: 7642–7656.

Hou, S., K.S. Makarova, J.H. Saw, P. Senin, B.V. Ly, Z. Zhou, Y. Ren, J. Wang, M.Y. Galperin, M.V. Omelchenko, Y.I. Wolf, N. Yutin, E.V. Koonin, M.B. Stott, B.W. Mountain, M.A. Crowe, A.V. Smirnova, P.F. Dunfield, L. Feng, L. Wang and M. Alam. 2008. Complete genome sequence of the extremely acidophilic methanotroph isolate V4, *Methylacidiphilum infernorum*, a representative of the bacterial phylum *Verrucomicrobia*. Biol. Direct *3*: 26.

Hugenholtz, P., B.M. Goebel and N.R. Pace. 1998. Impact of culture-independent studies on the emerging phylogenetic view of bacterial diversity. J. Bacteriol. *180*: 4765–4774.

Humayoun, S.B., N. Bano and J.T. Hollibaugh. 2003. Depth distribution of microbial diversity in Mono Lake, a meromictic soda lake in California. Appl. Environ. Microbiol. *69*: 1030–1042.

Islam, T., S. Jensen, L.J. Reigstad, Ø. Larsen and N.K. Birkeland. 2008. Methane oxidation at 55°C and pH 2 by a thermoacidophilic bacterium belonging to the *Verrucomicrobia* phylum. Proc. Natl. Acad. Sci. U. S. A. *105*: 300–304.

Janssen, P.H. 2006. Identifying the dominant soil bacterial taxa in libraries of 16S rRNA and 16S rRNA genes. Appl. Environ. Microbiol. *72*: 1719–1728.

Lau, W.W. and E.V. Armbrust. 2006. Detection of glycolate oxidase gene *glcD* diversity among cultured and environmental marine bacteria. Environ. Microbiol. *8*: 1688–1702.

Lee, K.C., R.I. Webb, P.H. Janssen, P. Sangwan, T. Romeo, J.T. Staley and J.A. Fuerst. 2009. Phylum *Verrucomicrobia* representatives share a compartmentalized cell plan with members of bacterial phylum *Planctomycetes*. BMC Microbiol. *9*: 5.

Ludwig, W., O. Strunk, R. Westram, L. Richter, H. Meier, Yadhukumar, A. Buchner, T. Lai, S. Steppi, G. Jobb, W. Forster, I. Brettske, S. Gerber, A.W. Ginhart, O. Gross, S. Grumann, S. Hermann, R. Jost, A. Konig, T. Liss, R. Lussmann, M. May, B. Nonhoff, B. Reichel, R. Strehlow, A. Stamatakis, N. Stuckmann, A. Vilbig, M. Lenke, T. Ludwig, A. Bode and K.H. Schleifer. 2004. ARB: a software environment for sequence data. Nucleic Acids Res. *32*: 1363–1371.

Nercessian, O., E. Noyes, M.G. Kalyuzhnaya, M.E. Lidstrom and L. Chistoserdova. 2005. Bacterial populations active in metabolism of C_1 compounds in the sediment of Lake Washington, a freshwater lake. Appl. Environ. Microbiol. *71*: 6885–6899.

Niemann, H., T. Losekann, D. de Beer, M. Elvert, T. Nadalig, K. Knittel, R. Amann, E.J. Sauter, M. Schluter, M. Klages, J.P. Foucher and A. Boetius. 2006. Novel microbial communities of the Haakon Mosby mud volcano and their role as a methane sink. Nature *443*: 854–858.

Penn, K., D. Wu, J.A. Eisen and N. Ward. 2006. Characterization of bacterial communities associated with deep-sea corals on Gulf of Alaska seamounts. Appl. Environ. Microbiol. *72*: 1680–1683.

Pol, A., K. Heijmans, H.R. Harhangi, D. Tedesco, M.S.M. Jetten and H.J.M. Op den Camp. 2007. Methanotrophy below pH 1 by a new *Verrucomicrobia* species. Nature *450*: 874–878.

Pruesse, E., C. Quast, K. Knittel, B. Fuchs, W. Ludwig, J. Peplies and F.O. Glöckner. 2007. SILVA: a comprehensive online resource for quality checked and aligned rRNA sequence data compatible with ARB. Nucleic Acids Res. *35*: 7188–7196

Sangwan, P., S. Kovac, K.E. Davis, M. Sait and P.H. Janssen. 2005. Detection and cultivation of soil *Verrucomicrobia*. Appl. Environ. Microbiol. *71*: 8402–8410.

Wagner, M. and M. Horn. 2006. The *Planctomycetes*, *Verrucomicrobia*, *Chlamydiae* and sister phyla comprise a superphylum with biotechnological and medical relevance. Curr. Opin. Biotechnol. *17*: 241–249.

Ward-Rainey, N., F.A. Rainey, H. Schlesner and E. Stackebrandt. 1995. Assignment of a hitherto unidentified 16S rDNA species to a main line of descent within the domain *Bacteria*. Microbiology *141*: 3247–3250.

Ward-Rainey, N., F.A. Rainey, H. Schlesner and E. Stackebrandt. 1996. *In* Validation of the publication of new names and new combinations previously effectively published outside the IJSB. List no. 57. Int. J. Syst. Bacteriol. *46*: 625–626.

Yoon, J., Y. Matsuo, K. Adachi, M. Nozawa, S. Matsuda, H. Kasai and A. Yokota. 2008. Description of *Persicirhabdus sediminis* gen. nov., sp. nov., *Roseibacillus ishigakijimensis* gen. nov., sp. nov., *Roseibacillus ponti* sp. nov., *Roseibacillus persicicus* sp. nov., *Luteolibacter pohnpeiensis* gen. nov., sp. nov. and *Luteolibacter algae* sp. nov., six marine members of the phylum '*Verrucomicrobia*', and emended descriptions of the class *Verrucomicrobiae*, the order *Verrucomicrobiales* and the family *Verrucomicrobiaceae*. Int. J. Syst. Evol. Microbiol. *58*: 998–1007.

Class I. **Verrucomicrobiae** Hedlund, Gosink and Staley 1998, 328^{VP} (Effective publication: Hedlund, Gosink and Staley 1997, 35.) emend. Yoon, Matsuo, Adachi, Nozawa, Matsuda, Kasai and Yokota 2008, 1002

Brian P. Hedlund

Ver.ru'co.mi.cro.bi'a.e. N.L. fem. pl. n. *Verrucomicrobiales* type order of the class; *-ae* ending to denote a class; N.L. fem. pl. n. *Verrucomicrobiae* the *Verrucomicrobiales* class.

On the basis of 16S rRNA gene sequence analyses, this group comprises one of the primary lineages in the phylum *Verrucomicrobia* and is equivalent to subdivision 1 (Hugenholtz et al., 1998) of the phylum. The majority of its members inhabit marine and freshwater habitats or vertebrate digestive tracts, including humans, however, some have been recovered from soils. The *Verrucomicrobia* currently comprise one order, at present represented by three families, *Verrucomicrobiaceae*, *Akkermansiaceae*, and *Rubritaleaceae*. Each family includes one or more cultivated and taxonomically described genus and is found primarily in a single habitat type, namely, freshwater habitats, vertebrate digestive tracts, and marine habitats, respectively. All members of this group stain Gram-negative.

Type order: **Verrucomicrobiales** Ward-Rainey, Rainey, Schlesner and Stackebrandt 1996, 625^{VP} (Effective publication: Ward-Rainey, Rainey, Schlesner and Stackebrandt 1995, 3249.)

emend. Yoon, Matsuo, Adachi, Nozawa, Matsuda, Kasai and Yokota 2008, 1002.

Further descriptive information

There are at least three well-supported family-level groups within the class *Verrucomicrobia*, each of which have been formally described, *Verrucomicrobiaceae*, *Akkermansiaceae*, and *Rubritaleaceae* (Figure 134).

The family *Verrucomicrobiaceae* is represented by the genera *Verrucomicrobium* and *Prosthecobacter* as well as 16S rRNA genes from cultivation-independent studies primarily from freshwater habitats. Both genera possess prosthecae. Although prosthecae are phylogenetically widespread, particularly among members of the *Alphaproteobacteria*, the *Verrucomicrobiaceae* are distinguished by the possession of fimbriae, menaquinones, and specialization in carbohydrate degradation (Hedlund et al., 1997). *Prosthecobacter* has become the focus of investigation since the discovery of homologs of tubulin genes, *btubA* and *btubB*, in *Prosthecobacter vanneerveni*, *Prosthecobacter dejongeii*, *Prosthecobacter debontii*, and *Prosthecobacter fusiformis* (Jenkins et al., 2002). A number of studies have elucidated the distribution of these tubulin genes and those encoding other structural components among *Verrucomicrobia* as well as the biochemical properties of their gene products. Although initial analysis of the partial-genome sequence of *Prosthecobacter dejongeii* failed to reveal genes of the *fts* operon (Jenkins et al., 2002) which is normally involved in contractile ring formation and function during bacterial cell division, the genome sequence of *Verrucomicrobium spinosum* did contain *ftsZ* and *ftsA* (Yee et al., 2007). Subsequently, degenerate-primed PCR eventually yielded both genes from four *Prosthecobacter* species (Pilhofer et al., 2007b) and genomic studies revealed ancestral division and cell wall (*dcw*) gene clusters (*ddl*, *ftsQ*, *ftsA*, and *ftsZ*) in all three formally described classes of *Verrucomicrobia* as well as a taxonomically uncharacterized strain in subphylum 3, strain Ellin514 (Pilhofer et al., 2008). Bacterial tubulins appear to have a narrow phylogenetic distribution in the *Verrucomicrobia* since genome projects of members of five class-level groups (subphyla) have failed to reveal tubulin homologs. Furthermore, a new species of *Prosthecobacter*, *Prosthecobacter fluviatilis*, was recently described, which, based on negative PCR results, may lack bacterial tubulin genes (Takeda et al., 2008). However, it should be noted that ectosymbionts of the marine ciliate *Euplotidium* that belong to the *Opitutales*, class *Opitutae*, also appear to have tubulins based on biochemical and structural studies, although no gene sequences exist therefore it is currently not possible to determine whether the genes from *Prosthecobacter* are evolutionarily related (Petroni et al., 2000). The patchy distribution of tubulin genes in the *Verrucomicrobiaceae* justifies relegation of these genes to the pan-genome, rather than the core genome, which strongly suggests a non-essential role in these organisms. Consistent with this patchy phylogenetic distribution is the variable genomic context of tubulin genes and a kinesin-like gene among *Prosthecobacter* genomes (Pilhofer et al., 2007a). Thus, all available data suggest that cell division in *Verrucomicrobia* is likely to be dependent on prototypical *dcw* components and that bacterial tubulins and the universally conserved kinesin-like gene serve an alternate, non-essential function. The presence of the bacterial actin homolog that is responsible for maintaining shapes other than a sphere (*mreB*) in *Verrucomicrobiales*, and the apparent absence of tubulins in *Prosthecobacter fluviatilis*, seem to rule out a role for tubulins in prosthecae formation or maintenance. Biochemical studies of *Prosthecobacter* tubulin gene products show that BtubA and BtubB form dimers that polymerize to form protofilaments *in vitro* with concomitant GTP hydrolysis in the absence of a chaperone, and the protofilaments aggregate to form pairs and, subsequently, bundles that are dozens of protofilaments thick (Schlieper et al., 2005; Sontag et al., 2005, 2009). Ultimately, the biological role of *Prosthecobacter* tubulins may best be addressed genetically; no genetic system to study *Verrucomicrobia* has yet been described.

The families *Akkermansiaceae* and *Rubritaleaceae* are each represented by one cultivated genus, *Akkermansia* and *Rubritalea*, respectively. *Akkermansia* and closely related 16S rRNA gene sequences have been recovered exclusively from intestinal contents of vertebrates, including humans, mice, rabbits, and dugong, suggesting a wide distribution in phylogenetically, ecologically, and physiologically distinct mammals (Hayashi et al., 2002; Hold et al., 2002; Salzman et al., 2002; Nelson et al., 2003; Mangin et al., 2004; Eckburg et al., 2005; R.A. Hutson and M.D. Collins, not published). Some related 16S rRNA gene sequences have been reported from soils or freshwater habitats, but it is currently unclear whether these represent free-living organisms or fecal contaminants. Several species of *Rubritalea* have been isolated from diverse marine invertebrates and sediments, and many related 16S rRNA gene sequences have been recovered from cultivation-independent censuses of marine habitats, particularly pelagic samples (Scheuermayer et al., 2006). However, a few 16S rRNA gene sequences have been recovered from freshwater samples and from activated sludge. The lack of habitat- or host-specific phylogenetic structure among *Rubritaleaceae* environmental clones may suggests a marine clade with salinity broad tolerance.

Since the chapters for this edition of *Bergey's Manual* were prepared, three additional genera in the *Verrucomicrobiaceae*, collectively represented by ten strains and six new species, were isolated from marine sediment, water, and red alga: *Persicirhabdus*, *Roseobacillus*, and *Luteolibacter* (Yoon et al., 2008). The descriptions of these organisms are generally consistent with the class description provided here. It was suggested that these isolates be placed in the family *Verrucomicrobiaceae*, the only existing family of the *Verrucomicrobiales* at the time; however, with the updated taxonomy proposed here, all three new genera should be placed in the *Rubritaleaceae*.

References

Eckburg, P.B., E.M. Bik, C.N. Bernstein, E. Purdom, L. Dethlefsen, M. Sargent, S.R. Gill, K.E. Nelson and D.A. Relman. 2005. Diversity of the human intestinal microbial flora. Science *308*: 1635–1638.

Hayashi, H., M. Sakamoto and Y. Benno. 2002. Fecal microbial diversity in a strict vegetarian as determined by molecular analysis and cultivation. Microbiol. Immunol. *46*: 819–831.

Hedlund, B.P., J.J. Gosink and J.T. Staley. 1997. *Verrucomicrobia* div. nov., a new division of the bacteria containing three new species of *Prosthecobacter*. Antonie van Leeuwenhoek *72*: 29–38.

Hedlund, B.P., J.J. Gosink and J.T. Staley. 1998. *In* Validation of the publication of new names and new combinations previously effectively published outside the IJSB. List no. 64. Int. J. Syst. Bacteriol. *48*: 327–328.

Hold, G.L., S.E. Pryde, V.J. Russell, E. Furnie and H.J. Flint. 2002. Assessment of microbial diversity in human colonic samples by 16S rDNA sequence analysis. FEMS Microbiol. Ecol. *39*: 33–39.

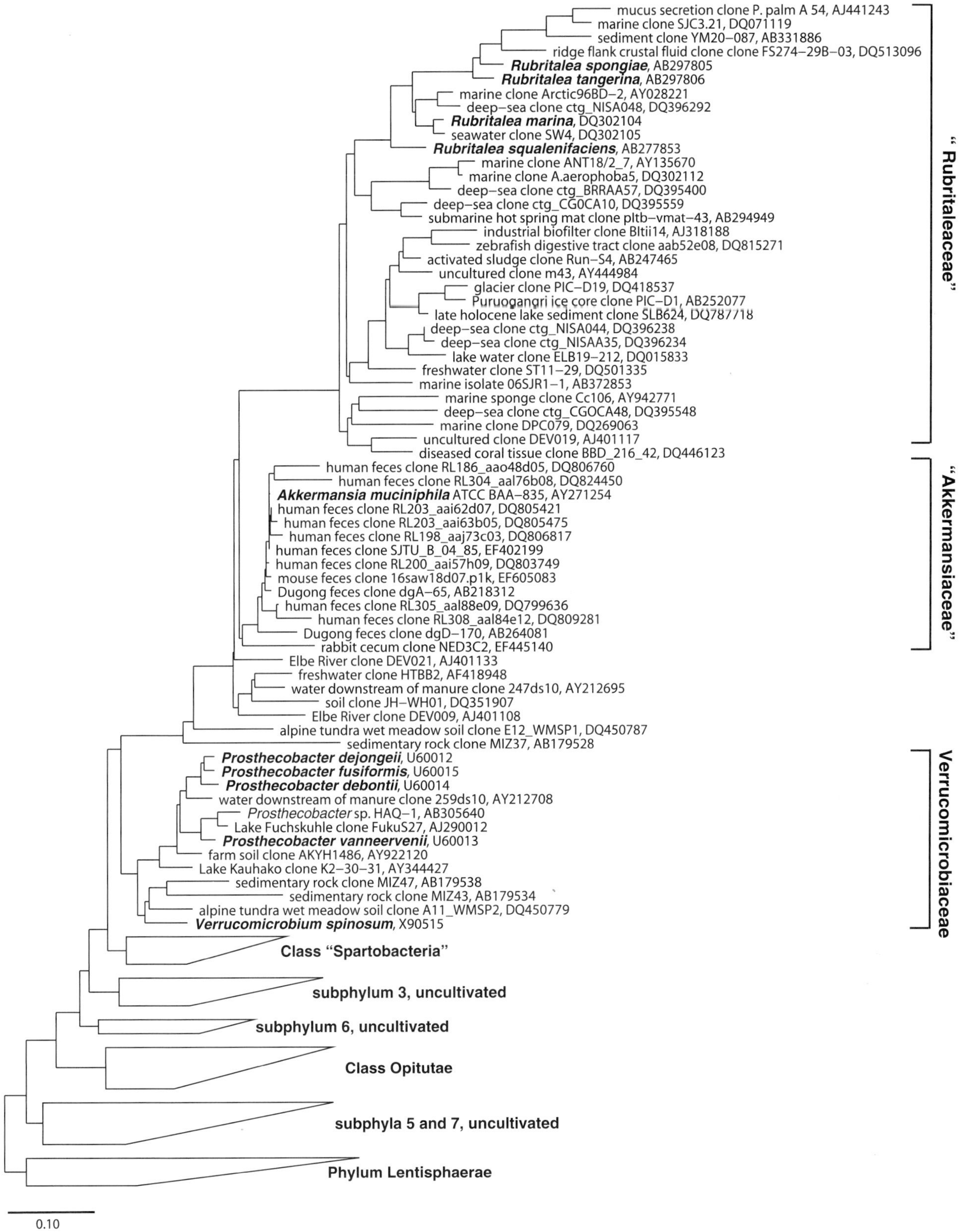

FIGURE 134. Maximum likelihood phylogeny of *Verrucomicrobia* with heuristic correction showing nearly complete 16S rRNA gene sequences present in version 91 of the SILVA database (Pruesse et al., 2007). All class-level groups were included, but all those except the *Verrucomicrobiae* were wedged. The phylogeny was produced by using ARB (Ludwig et al., 2004) using *Escherichia coli* 16S rRNA gene nucleotides 110–1274. Alignments were unmasked.

Hugenholtz, P., B.M. Goebel and N.R. Pace. 1998. Impact of culture-independent studies on the emerging phylogenetic view of bacterial diversity. J. Bacteriol. *180*: 4765–4774.

Jenkins, C., R. Samudrala, I. Anderson, B.P. Hedlund, G. Petroni, N. Michailova, N. Pinel, R. Overbeek, G. Rosati and J.T. Staley. 2002. Genes for the cytoskeletal protein tubulin in the bacterial genus *Prosthecobacter*. Proc. Natl. Acad. Sci. U.S.A. *99*: 17049–17054.

Ludwig, W., O. Strunk, R. Westram, L. Richter, H. Meier, Yadhukumar, A. Buchner, T. Lai, S. Steppi, G. Jobb, W. Forster, I. Brettske, S. Gerber, A.W. Ginhart, O. Gross, S. Grumann, S. Hermann, R. Jost, A. Konig, T. Liss, R. Lussmann, M. May, B. Nonhoff, B. Reichel, R. Strehlow, A. Stamatakis, N. Stuckmann, A. Vilbig, M. Lenke, T. Ludwig, A. Bode and K.H. Schleifer. 2004. ARB: a software environment for sequence data. Nucleic Acids Res. *32*: 1363–1371.

Mangin, I., R. Bonnet, L. Seksik, L. Rigottier-Gois, M. Sutren, Y. Bouhnik, C. Neut, M.D. Collins, J.-F. Colombel, P. Marteau and J. Dore. 2004. Molecular inventory of faecal microflora in patients with Crohn's disease. FEMS Microbiol. Ecol. *50*: 25–36.

Nelson, K.E., S.H. Zinder, I. Hance, P. Burr, D. Odongo, D. Wasawo, A. Odenyo and R. Bishop. 2003. Phylogenetic analysis of the microbial populations in the wild herbivore gastrointestinal tract: insights into an unexplored niche. Environ. Microbiol. *5*: 1212–1220.

Petroni, G., S. Spring, K.H. Schleifer, F. Verni and G. Rosati. 2000. Defensive extrusive ectosymbionts of *Euplotidium* (Ciliophora) that contain microtubule-like structures are bacteria related to *Verrucomicrobia*. Proc. Natl. Acad. Sci. U.S.A. *97*: 1813–1817.

Pilhofer, M., A.P. Bauer, M. Schrallhammer, L. Richter, W. Ludwig, K.H. Schleifer and G. Petroni. 2007a. Characterization of bacterial operons consisting of two tubulins and a kinesin-like gene by the novel Two-Step Gene Walking method. Nucleic Acids Res. *35*: e135.

Pilhofer, M., G. Rosati, W. Ludwig, K.H. Schleifer and G. Petroni. 2007b. Coexistence of tubulins and *ftsZ* in different *Prosthecobacter* species. Mol. Biol. Evol. *24*: 1439–1442.

Pilhofer, M., K. Rappl, C. Eckl, A.P. Bauer, W. Ludwig, K.H. Schleifer and G. Petroni. 2008. Characterization and evolution of cell division and cell wall synthesis genes in the bacterial phyla *Verrucomicrobia*, *Lentisphaerae*, *Chlamydiae*, and *Planctomycetes* and phylogenetic comparison with rRNA genes. J. Bacteriol. *190*: 3192–3202.

Pruesse, E., C. Quast, K. Knittel, B. Fuchs, W. Ludwig, J. Peplies and F.O. Glöckner. 2007. SILVA: a comprehensive online resource for quality checked and aligned rRNA sequence data compatible with ARB. Nucleic Acids Res. *35*: 7188–7196.

Salzman, N.H., H. de Jong, Y. Paterson, H.J. Harmsen, G.W. Welling and N.A. Bos. 2002. Analysis of 16S libraries of mouse gastrointestinal microflora reveals a large new group of mouse intestinal bacteria. Microbiology *148*: 3651–3660.

Scheuermayer, M., T.A. Gulder, G. Bringmann and U. Hentschel. 2006. *Rubritalea marina* gen. nov., sp. nov., a marine representative of the phylum '*Verrucomicrobia*', isolated from a sponge (*Porifera*). Int. J. Syst. Evol. Microbiol. *56*: 2119–2124.

Schlieper, D., M.A. Oliva, J.M. Andreu and J. Lowe. 2005. Structure of bacterial tubulin BtubA/B: evidence for horizontal gene transfer. Proc. Natl. Acad. Sci. U.S.A. *102*: 9170–9175.

Sontag, C.A., J.T. Staley and H.P. Erickson. 2005. *In vitro* assembly and GTP hydrolysis by bacterial tubulins BtubA and BtubB. J. Cell. Biol. *169*: 233–238.

Sontag, C.A., H. Sage and H.P. Erickson. 2009. BtubA–BtubB heterodimer is an essential intermediate in protofilament assembly. PLoS One *4*: e7253.

Takeda, M., A. Yoneya, Y. Miyazaki, K. Kondo, H. Makita, M. Kondoh, I. Suzuki and J. Koizumi. 2008. *Prosthecobacter fluviatilis* sp. nov., which lacks the bacterial tubulin *btubA* and *btubB* genes. Int. J. Syst. Evol. Microbiol. *58*: 1561–1565.

Ward-Rainey, N., F.A. Rainey, H. Schlesner and E. Stackebrandt. 1995. Assignment of a hitherto unidentified 16S rDNA species to a main line of descent within the domain *Bacteria*. Microbiology *141*: 3247–3250.

Ward-Rainey, N., F.A. Rainey, H. Schlesner and E. Stackebrandt. 1996. *In* Validation of the publication of new names and new combinations previously effectively published outside the IJSB. List no. 57. Int. J. Syst. Bacteriol. *46*: 625–626.

Yee, B., F.F. Lafi, B. Oakley, J.T. Staley and J.A. Fuerst. 2007. A canonical FtsZ protein in *Verrucomicrobium spinosum*, a member of the bacterial phylum *Verrucomicrobia* that also includes tubulin-producing *Prosthecobacter* species. BMC Evol. Biol. *7*: 37.

Yoon, J., Y. Matsuo, K. Adachi, M. Nozawa, S. Matsuda, H. Kasai and A. Yokota. 2008. Description of *Persicirhabdus sediminis* gen. nov., sp. nov., *Roseibacillus ishigakijimensis* gen. nov., sp. nov., *Roseibacillus ponti* sp. nov., *Roseibacillus persicicus* sp. nov., *Luteolibacter pohnpeiensis* gen. nov., sp. nov. and *Luteolibacter algae* sp. nov., six marine members of the phylum '*Verrucomicrobia*', and emended descriptions of the class *Verrucomicrobiae*, the order *Verrucomicrobiales* and the family *Verrucomicrobiaceae*. Int. J. Syst. Evol. Microbiol. *58*: 998–1007.

Order I. **Verrucomicrobiales** Ward-Rainey, Rainey, Schlesner and Stackebrandt 1996, 625[VP] (Effective publication: Ward-Rainey, Rainey, Schlesner and Stackebrandt 1995, 3249.) emend. Yoon, Matsuo, Adachi, Nozawa, Matsuda, Kasai and Yokota 2008, 1002

Brian P. Hedlund and Muriel Derrien

Ver.ru.co.mi.cro.bi.a'les. N.L. neut. n. *Verrucomicrobium* type genus of the order; *-ales* ending to denote an order; N.L. fem. pl. n. *Verrucomicrobiales* the order of the genus *Verrucomicrobium*

Encompasses bacteria isolated from freshwater or marine aquatic ecosystems and from animal feces, within the class *Verrucomicrobiae*. The order contains the families *Verrucomicrobiaceae*, *Akkermansiaceae*, and *Rubritaleaceae*. Delineation of the order is determined primarily by phylogenetic information from 16S rRNA gene sequences. All stain Gram-negative and have a Gram-negative cell structure. All are chemoorganotrophic. Includes obligate aerobes, facultative anaerobes, and strict anaerobes.

Type genus: **Verrucomicrobium** Schlesner 1988, 221[VP] (Effective publication: Schlesner 1987, 56.).

References

Schlesner, H. 1987. *Verrucomicrobium spinosum* gen. nov., sp. nov., a fimbriated prosthecate bacterium. Syst. Appl. Microbiol. *10*: 54–56.

Schlesner, H. 1988. *In* Validation of the publication of new names and new combinations previously effectively published outside the IJSB. List no. 25. Int. J. Syst. Bacteriol. *38*: 220–222.

Ward-Rainey, N., F.A. Rainey, H. Schlesner and E. Stackebrandt. 1995. Assignment of a hitherto unidentified 16S rDNA species to a main line of descent within the domain *Bacteria*. Microbiology *141*: 3247–3250.

Ward-Rainey, N., F.A. Rainey, H. Schlesner and E. Stackebrandt. 1996. *In* Validation of the publication of new names and new combinations previously effectively published outside the IJSB. List no. 57. Int. J. Syst. Bacteriol. *46*: 625–626.

Yoon, J., Y. Matsuo, K. Adachi, M. Nozawa, S. Matsuda, H. Kasai and A. Yokota. 2008. Description of *Persicirhabdus sediminis* gen. nov., sp. nov., *Roseibacillus ishigakijimensis* gen. nov., sp. nov., *Roseibacillus ponti* sp. nov., *Roseibacillus persicicus* sp. nov., *Luteolibacter pohnpeiensis* gen. nov., sp. nov. and *Luteolibacter algae* sp. nov., six marine members of the phylum '*Verrucomicrobia*', and emended descriptions of the class *Verrucomicrobiae*, the order *Verrucomicrobiales* and the family *Verrucomicrobiaceae*. Int. J. Syst. Evol. Microbiol. *58*: 998–1007.

Family I. **Verrucomicrobiaceae** Ward-Rainey, Rainey, Schlesner and Stackebrandt 1996, 625[VP] (Effective publication: Ward-Rainey, Rainey, Schlesner and Stackebrandt 1995, 3249.) emend. Yoon, Matsuo, Adachi, Nozawa, Matsuda, Kasai and Yokota 2008, 1002

Brian P. Hedlund

Ver.ru.co.mi.cro.bi.a.ce'ae. N.L. neut. n. *Verrucomicrobium* type genus of the family; L. suff. *-aceae* ending to denote a family; N.L. fem. pl. n. *Verrucomicrobiaceae* the *Verrucomicrobium* family

Prosthecate cells with fimbriae. Rod- or vibrioid-shaped; cells may be fusiform. Nonsporeforming. Nonmotile. **Cells stain Gram-negative and have a Gram-negative cell structure. Aerobic or facultatively anaerobic via fermentation.** Mesophilic. **Chemoorganotrophic. Saccharolytic.** Use predominantly mono- and disaccharides and their derivatives. Menaquinones are the only respiratory quinones that have been detected.

DNA G+C content (mol%): 54–60.

Type genus: **Verrucomicrobium** Schlesner 1988, 221[VP] (Effective publication: Schlesner 1987, 56.).

Further descriptive information

Encompasses bacteria isolated from freshwater lakes and associated 16S rRNA gene sequences recovered from freshwater environments and soils within the order *Verrucomicrobiales*. Delineation of the family is determined primarily by phylogenetic information from 16S rRNA gene sequences.

Key to the genera of the family *Verrucomicrobiaceae*

1. The DNA G+C content is 58–59 mol%. Cells are rod-shaped with many short (<2 μm) prosthecae. Fimbriae extend from tips of prosthecae. Growth on monosaccharides or disaccharides as carbon and energy sources aerobically or via fermentation without gas production. Isolated from fresh water or brackish environments.
   → Genus I. *Verrucomicrobium*

2. The DNA G+C content is 54–60 mol%. Fusiform cells are rod- or vibrioid-shaped with a single prostheca extending from the old pole of the cell. Growth on monosaccharides or disaccharides under aerobic conditions. Unable to grow anaerobically. Isolated from freshwater environments.
   → Genus II. *Prosthecobacter*

Genus I. **Verrucomicrobium** Schlesner 1988, 221[VP] (Effective publication: Schlesner 1987, 55.)

Heinz Schlesner

Ver.ru.co.mi.cro'bi.um. L. fem. n. *verruca* wart; N.L. neut. n. *microbium* (from Gr. adj. *mikros* small and Gr. n. *bios* life), a microbe; N.L. neut. n. *Verrucomicrobium* warty microbe.

Unicellular bacterium, rod shaped with numerous prosthecae extending from all locations on the cell surface. **Fimbriae varying in length and number extrude from the tips of the prosthecae.** Gram-stain-negative; nonmotile. Nonsporeforming. PHB is not stored. No gas vesicles. Mesophilic. Chemoorganotrophic; facultatively anaerobic. Glucose is fermented; nitrate is not reduced under anaerobic conditions. Oxidase and catalase positive. The cell wall contains *m*-diaminopimelic acid. The main respiratory quinone is MK10. MK9, MK10(H_2), and MK11 are minor quinones. Major fatty acids are $C_{14:0}$, $C_{14:0}$ iso, $C_{15:0}$ anteiso, $C_{15:0}$, $C_{16:1}$ ω5, $C_{16:0}$, $C_{17:1}$ ω12. The major phospholipids are phosphatidylglycerol and phosphatidylmethylethanolamine.

DNA G+C content (mol%): 58–59.

Type species: **Verrucomicrobium spinosum** Schlesner 1988, 221[VP] (Effective publication: Schlesner 1987, 55.).

Further descriptive information

Verrucomicrobium spinosum is the only described species of *Verrucomicrobium*. Bacteria with multiple prosthecae are grouped in different phyla: the phototrophic *Ancalochloris* (Gorlenko and Lebedeva, 1971) and *Prosthecochloris* (Gorlenko, 1970) are members of the phylum *Chlorobi* (Garrity and Holt, 2001), while the heterotrophic bacteria *Ancalomicrobium*, *Prosthecomicrobium* (Staley, 1968), and *Stella* (Vasilyeva, 1985) are *Proteobacteria* (Garrity and Holt, 2001). None of the above, however, has fimbriate prosthecae as does *Verrucomicrobium spinosum* (Figure 135; Schlesner, 1992).

Prosthecate bacteria with fimbriae extending from the tips of prosthecae have rarely been observed. Such bacteria occurred in an enrichment culture of Lake Plußsee water to which vitamin solution no. 6 (see below) was added (P. Hirsch, Kiel, personal communication). Lake Plußsee is a small eutrophic lake near Plön, Holstein, Germany. The type strain of *Verrucomicrobium spinosum* was isolated from Lake Vollstedter See (Holstein, Germany), a shallow (max. depth 2 m) eutrophic lake. Another strain came from the Schrevenparkteich, a pond in a public park in Kiel, Germany. This pond was very eutrophic because flocks of water fowl lived there and were fed by visitors to the park.

Isolates of *Verrucomicrobium* utilize the following for carbon and energy: esculin, amygdalin, cellobiose, fructose, galactose,

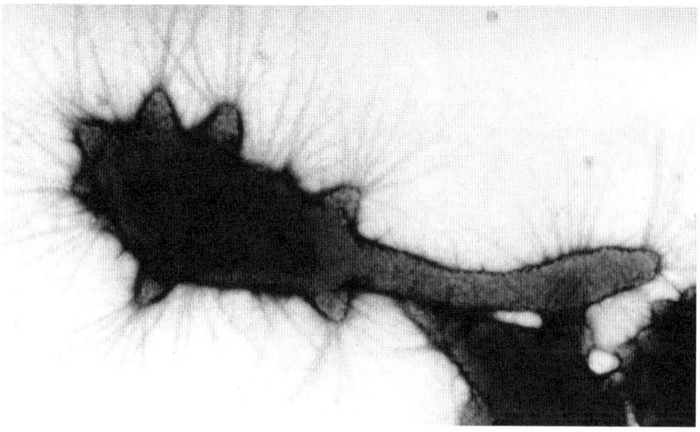

FIGURE 135. Morphology of *Verrucomicrobium spinosum*. Numerous fimbriae are excreted from the tips of short conical and tube-like prosthecae. Negatively stained with phosphotungstic acid.

glucose, glucuronate, glycerol, lactose, maltose, mannose, melibiose, melizitose, *N*-acetylglucosamine, raffinose, rhamnose, salicin, sucrose, and trehalose. Amino acids and sugar alcohols are not utilized. The following are fermented with acid but no gas production: cellobiose, fructose, galactose, glucose, lactose, maltose, mannose, melibiose, *N*-acetylglucosamine, rhamnose, sucrose, and trehalose. No growth occurs on acetate, adipate, D-arabinose, L-arabinose, aspartic acid, butyrate, caproate, ethanol, formate, fucose, glutamic acid, glutamine, histidine, β-hydroxybutyrate, inositol, inulin, lactate, lyxose, malate, mannitol, methanol, 2-oxoglutarate, proline, propionate, pyruvate, ribose, serine, sorbitol, sorbose, succinate, threonine, and valerate. Ammonia, *N*-acetylglucosamine, nitrate, and urea are used as sole nitrogen sources. Gelatin and starch are hydrolyzed, but casein and Tween 80 are not. Phosphatase and urease are produced. However, thiosulfate is not reduced to H_2S, and ammonia is not produced from peptone.

Oligonucleotide cataloging and reverse transcriptase sequencing of 16S rRNA (Albrecht et al., 1987) as well as sequencing of the 5S rRNA (Bomar and Stackebrandt, 1987) indicates that *Verrucomicrobium spinosum* is a representative of a novel phylum. This finding is supported by the sequence analyzes of 16S rRNA (Ward-Rainey et al., 1995) and 23S rRNA (Ward et al., 2000) genes. The presence of *dnaK* (HSP70) multigene family was reported (Ward-Rainey et al., 1997).

Enrichment and isolation procedures

Verrucomicrobium species can be enriched in Erlenmeyer flasks with 50 ml of enrichment medium over a sediment of $CaCO_3$.

Enrichment medium: *N*-acetylglucosamine, 1.0 g; Hutner's basal salts (Cohen-Bazire et al., 1957), 20 ml; vitamin solution no. 6 (Staley, 1968), 10 ml. Combine ingredients, dilute to 1 l with distilled water, and adjust pH to 9.7. Autoclave, cool to room temperature, then add $NaH_2PO_4 \cdot H_2O$ aseptically to give a concentration of 0.65 mmol/liter.

Upcoming colonies have to be examined microscopically for prosthecate cells. This is conveniently done by applying the time-saving toothpick-procedure (Hirsch et al., 1977) which even allows the examination of very small colonies: A sterile wooden toothpick is dipped onto a colony and then on the agar surface of a Petri dish containing the appropriate medium, thus inoculating the agar. To allow inoculation of a single Petri dish with bacteria from a large number of colonies, a grid can be drawn with a marker on the backside of the Petri dish with areas of about 5 × 5 mm. After inoculation of the agar medium, the toothpick generally contains enough bacteria to prepare a smear for microscopic examination. Three specimens can be placed on one slide. Prosthecate bacteria from positive colonies will be isolated by subsequent streaks on agar-solidified medium M13 (for recipe see chapter "Genus *Blastopirellula*").

Maintenance procedures

The strains can be kept on agar slants with M13 at 4–6°C for several months. For long-term preservation, lyophilization or storage in liquid nitrogen is recommended. Deep freezing at −70°C with 50% glycerol is also possible.

Differentiation of the genus *Verrucomicrobium* from other closely related taxa

Verrucomicrobium is easily distinguished from the anaerobic phototrophic genera *Ancalochloris* and *Prosthecochloris* and also from the heterotrophic genus *Stella* that normally has six appendages lying in one plane. The most morphologically similar genera are *Ancalomicrobium* and *Prosthecomicrobium*. The major phenotypic features for differentiation of *Verrucomicrobium* are the fimbriae extruding from the tips of the prosthecae, which are not found in the genera *Ancalomicrobium* and *Prosthecomicrobium*. The respiratory quinones of *Verrucomicrobium spinosum* are menaquinones while in the genera *Ancalomicrobium*, *Prosthecomicrobium*, and *Stella* only ubiquinone Q10 was found (Sittig and Schlesner, 1993). Another important feature is the DNA base ratio which is considerably lower in *Verrucomicrobium* (Table 154).

Further reading

Schlesner, H., C. Jenkins and J.T. Staley. 2005. The phylum *Verrucomicrobia*: a phylogenetically heterogeneous bacterial group. *In* The Prokaryotes: an Evolving Electronic Resource for the Microbiological Community, 3rd edn (edited by Dworkin). Springer, New York.

TABLE 154. Characteristics differentiating *Verrucomicrobium* from *Ancalomicrobium* and *Prosthecomicrobium*[a]

Characteristic	*Verrucomicrobium*	*Ancalomicrobium*	*Prosthecomicrobium*
Short prosthecae (<2 μm)	+	–	D
Fimbriae from tips of prosthecae	+	–	–
Colony color	Light yellow	Chalky white	Colorless, yellow or pink
Fermentation of glucose	+	+	–
G+C content of DNA	58–59	70.4	64–70

[a]Symbols: +, yes; –, no; D, varies depending on species.

List of species of the genus *Verrucomicrobium*

1. **Verrucomicrobium spinosum** Schlesner 1988, 221[VP] (Effective publication: Schlesner 1987, 55.)

spi'no.sum. L. neut. adj. *spinosum* thorny, spiny.

Cells measure 0.8–1.0 × 1.0–3.8 μm. Conical prosthecae with a length of about 0.5 μm extend in all directions from the cell surface. Occasionally, one or two longer prosthecae (up to 2 μm) may occur, and often one of them is polarly inserted. These prosthecae are more or less tube-like. Colonies on medium M 13 are light yellow. Growth is optimal between 26 and 33°C, and the maximum growth temperature is 34°C. The organism is stenohaline (i.e., able to tolerate artificial sea water up to a salinity of 17‰ or 1% (w/v) NaCl). Only a limited number of substrates can be utilized as the sole carbon and energy source, mainly hexoses, di- or trisaccharides, or derivatives of glucose. C_1- or C_2-compounds are not utilized, nor are fatty acids or amino acids. Various sugars are fermented without gas formation. Nitrate is not reduced under anaerobic conditions.

Other characters are listed in Table 154. The chemical composition is identical to that in the genus description.

DNA G+C content (mol%): 59 (T_m).

Type strain: ATCC 43997, DSM 4136, IFAM 1439.

Sequence accession no. (16S rRNA gene): X90515.

Genus II. **Prosthecobacter** Staley, de Bont and de Jonge 1980, 595[VP] (Effective publication: Staley, de Bont and de Jonge 1976, 341.)

Brian P. Hedlund

Pros.the.co.bac'ter. Gr. fem. n. *prosthece* appendage; N.L. masc. n. *bacter* masc. form of Gr. neut. n. *baktron* rod; N.L. masc. n. *Prosthecobacter* appendage (-producing) rod.

Unicellular, Gram-stain-negative, fusiform rod shaped to vibrioid bacteria with a single polar prostheca. Nonmotile. Prosthecae do not branch and do not bear buds. Cells attach to various substrata by a holdfast structure located at the distal tip of the prostheca. **Reproduction by binary fission results in the formation of a partially to fully differentiated prosthecate daughter cell which is a mirror image of the mother cell.** Rosettes are rare in pure culture, and, when formed, are comprised of few cells. **Chemoorganotrophic.** Growth occurs in defined media using ammonium salts as nitrogen source and glucose as carbon source. Obligately **aerobic. Catalase-positive.** Cell wall contains *m*-diaminopimelic acid. Menaquinones are dominant respiratory quinones.

DNA G+C content (mol%): 54–60.

Type species: **Prosthecobacter fusiformis** Staley, de Bont and de Jonge 1980, 595[VP] (Effective publication: Staley, de Bont and de Jonge 1976, 341.).

Further descriptive information

Prosthecobacter was originally described in the literature by Henrici and Johnson (1935) and isolated into pure culture many years later (de Bont et al., 1970). To date, only four strains have been isolated, and each represents a unique species within the genus. *Prosthecobacter* is obligately aerobic and saccharolytic, exhibiting strong growth on a variety of mono- and disaccharides (de Bont et al., 1970; Hedlund et al., 1997) (Table 155). During growth on the defined medium MMB (Staley et al., 1976) with glucose, menaquinone MK-6(H_2) was the major quinone detected (Hedlund et al., 1996).

Electron microscopy shows that cells of *Prosthecobacter* has a cell structure that is typical of Gram-stain-negative prokaryotes and that *Prosthecobacter* appendages contain cytoplasmic material (Figure 136). The cells are heavily fibriated except for the distal tips of the prosthecae.

Study of the 95% complete genome of *Prosthecobacter dejongeii* revealed genes highly homologous to alpha and beta tubulins in eukaryotes, denoted *btuba* (for bacterial tubulin a) and *btubb* (for bacterial tubulin B) (Jenkins et al., 2002), and homologs of these are found in other *Prosthecobacter* species, including two homologs of each tubulin in *Prosthecobacter debontii* (Pilhofer et al., 2007). These genes are genuine tubulin homologs since they share higher primary sequence homology to eukaryotic tubulins than

TABLE 155. Characteristics differentiating species of the genus *Prosthecobacter*[a]

Characteristic	*P. fusiformis*	*P. debontii*	*P. dejongeii*	*P. vanneervenii*
Utilization of:				
N-Acetylglucosamine	–	+	+	+
D-Arabinose	+	+	–	+
D-Fructose	–	+	–	+
D-Glucosamine	–	–	–	+
Glycogen	–	–	+	+
D-Raffinose	–	–	–	+
D-Ribose	–	+	+	+
DNA G+C content (mol%)	56.1	57.1	54.6	60.1
Growth temperature range (°C)	4–40	8–38	1–35	10–38
Generation time in MMB (h)[b]	nd	nd	6	24

[a]Symbols: +, >85% positive; –, 0–15% positive; nd, not determined.
[b]Staley et al. (1976).

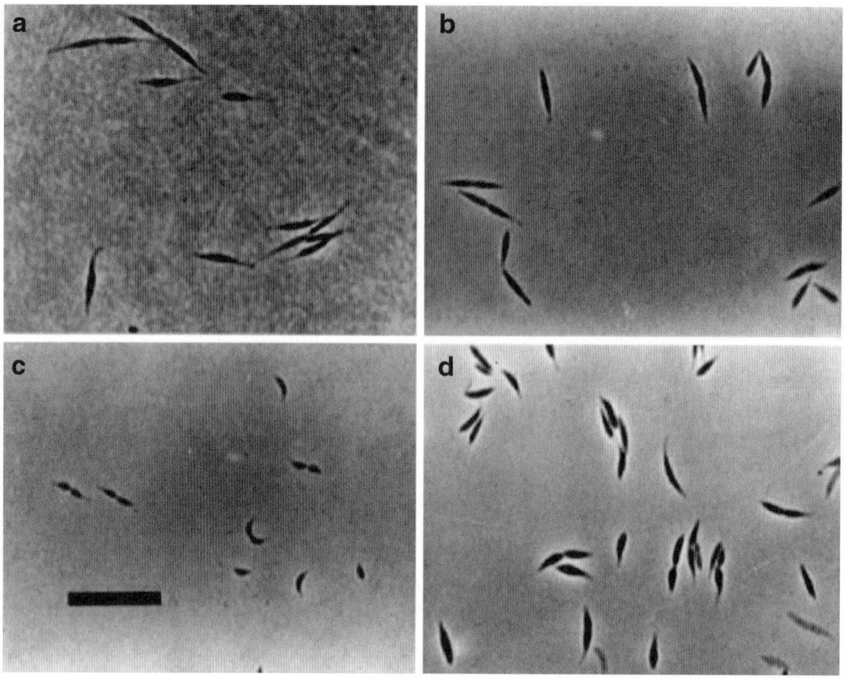

FIGURE 136. Phase-contrast micrographs of the four species of *Prosthecobacter*. (a) *Prosthecobacter fusiformis* (FC4); (b) *Prosthecobacter vanneervenii* (FC2); (c) *Prosthecobacter debontii* (FC3); (d) *Prosthecobacter dejongeii* (FC1). Bar = 10 µm. (Reproduced with permission from Staley et al., 1976. Antonie van Leeuwenhoek *42:* 333–342.)

they do to the bacterial tubulin homolog *ftsZ*. In addition, the putative amino acid sequences for BtubA and BtubB possess 9 of 9 motifs specific for tubulin and only 2–3 of the 6 FtsZ signature motifs. The presence of tubulin genes in the *Prosthecobacter* genome indicates a shared ancestry between verrucomicrobia and eukaryotes either through lateral gene transfer or larger genome fusion as may occur through an ancient endosymbiosis. The finding that *Prosthecobacter* contains few eukaryotic signature proteins suggests the former (Staley et al., 2005). The observation that microtubule-like structures in the verrucomicrobial ectosymbiont of the ciliate *Euplotidium* cross-react with anti-tubulin antibodies, and the demonstration that polymerization of these microtubule-like structures is sensitive to nocodazole and low temperature (4°C) provides a clue that tubulins may be found elsewhere within the verrucomicrobia (Petroni et al., 2000). However, the putative tubulin homologs in *Euplotidium* symbionts have not been sequenced, and tubulin homologs are absent from the genome of *Verrucomicrobium* (Pilhofer et al., 2007).

Some work has been done to try to elucidate the function of *Prosthecobacter* tubulins. Jenkins et al. (2002) showed that the tubulins are cotranscribed along with a third gene that is highly homologous to a kinesin light chain; kinesin is the the motor protein for microtubule motor protein in eukaryotes. However, *Prosthecobacter* tubulins overexpressed in *Escherichia coli* do not cross-react with antibovine tubulin antibodies, and no microtubule-like structures are visible in electron micrographs (Jenkins et al., 2002). Comparative modeling suggested *Prosthecobacter* tubulins are more stable as monomers than dimers, suggesting polymerization does not occur (Jenkins et al., 2002); however, they were subsequently shown to form protofilaments in *Escherichia coli* with equimolar amounts of BtubA/B (Schlieper et al., 2005; Sontag et al., 2005). Like eukaryotic microtubules, BtubA/B protofilaments are GTP and magnesium-dependent; however, unlike eukaryotic tubulins BtubA/B, protofilaments form *in vitro* without chaperones (Schlieper et al., 2005). Crystal structures of both BtubA and a BtubA/B have been solved confirming they are highly similar to eukaryotic tubulins, including surface loops (Schlieper et al., 2005). Unfortunately, the role of BtubA/B in *Prosthecobacter* remains unknown.

Phylogeny and classification. Only four strains of *Prosthecobacter* have been isolated, each representing a distinct species according to DNA–DNA hybridization, 16S rDNA homologies of 94.6–97.9%, and morphological and phenotypic differences (Hedlund et al., 1997). The genus is clearly distinct from the most closely related cultivated organism, *Verrucomicrobium spinosum*, since *Prosthecobacter* forms a monophyletic clade and is morphologically and physiologically distinct.

Enrichment and isolation procedures

Isolation of *Prosthecobacter* into pure culture requires persistence and patience. Only four isolates have been described in the literature. Three of the four *Prosthecobacter* strains were isolated by enriching heterotrophs in freshwater samples using the enrichment procedure of Houwink (1951) by adding peptone to 0.01% to the water sample and incubating 1–3 weeks. When *Prosthecobacter* cells were visible in the enrichment, the sample was streaked onto the same medium solidified with 1.5% agar. Colonies were screened by phase-contrast microscopy.

Maintenance procedures

Lyophilized cultures stored at −20°C have been revived after at least 20 years of storage. In addition, cultures containing MMB supplemented with 15% glycerol (v/v) and stored at −80°C have been successfully revived after several years.

Procedures for testing special characteristics

For determination of carbon substrates, Hedlund et al. (1997) added filter-sterilized substrates to 0.2% (w/v) to the defined medium MMB (Staley et al., 1976) in microtiter wells and measured growth at 3-, 5-, and 10-d incubations at room temperature.

Differentiation of the genus *Prosthecobacter* from other closely related taxa

Prosthecobacter can be distinguished from other heterotrophic prosthecate bacteria based on morphology and motility. *Prosthecomicrobium, Ancalomicrobium,* and *Verrucomicrobium* possess many prosthecae per cell, allowing differentiation from singly prosthecate *Prosthecobacter. Caulobacter* and *Asticcacaulus* are similar in possessing a single polar or subpolar prostheca yet they divide asymmetrically, yielding a motile non-stalked daughter cell and a sessile stalked mother cell. *Prosthecobacter,* like its relative *Verrucomicrobium,* is heavily fibriate and *Prosthecobacter* prosthecae are distinct from *Caulobacter* and *Asticcacaulus* because they lack crossbands and possess a bulbous tip (Figure 137). *Prosthecobacter* and its close relative *Verrucomicrobium* can also be distinguished from the prosthecate *Alphaprotoebacteria* based on the predominance of menaquinones for respiration, rather than ubiquinones. Due to the morphological diversity within the genus, presumptive *Prosthecobacter* isolates should be identified by 16S rDNA sequencing or other molecular method.

Taxonomic comments

Prosthecobacter was observed in freshwater samples by Henrici and Johnson (1935), who described them as the "fusiform type" of *Caulobacter,* and, prior to phylogenetic studies, the organisms were referred to in the literature as "fusiform caulobacters". Given their phylogenetic position in the *Verrucomicrobia,* it is clear *Prosthecobacter* is phylogenetically unrelated to the *Alphaproteobacteria Caulobacter* and *Asticcacaulus.*

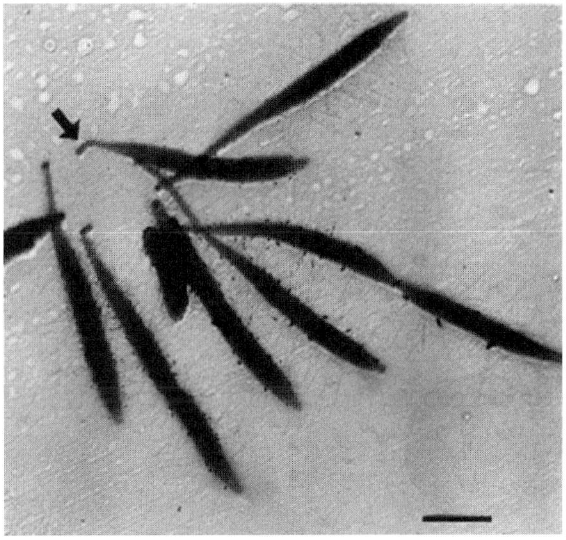

FIGURE 137. Shadowed electron micrograph of *Prosthecobacter vanneervenii* showing numerous radiating fimbriae and distinctive bulbs at prostheca tips (arrow). Bar = 2 μm. (Reproduced with permission from Hedlund et al., 1996. Int. J. Syst. Bacteriol. *46:* 960–966.)

Further reading

Schlieper, D., M.A. Oliva, J.M. Andreu and J. Lowe. 2005. Structure of bacterial tubulin BtubA/B: evidence for horizontal gene transfer. Proc. Natl. Acad. Sci. U. S. A. *102*: 9170–9175.

List of species of the genus *Prosthecobacter*

1. **Prosthecobacter fusiformis** Staley, de Bont and de Jonge 1980, 595[VP] (Effective publication: Staley, de Bont and de Jonge 1976, 341.)

fu.si.for'mis. L. n. *fusus* spindle; L. masc. adj. suff. *-formis* (from L. n. *forma* shape, form) -like, in the shape of; N.L. masc. adj. *fusiformis* spindle-shaped.

Fusiform cells are usually straight, long (3–10 μM), and thin (0.5 μM). Carbon sources include D-glucose, D-galactose, D-mannose, D-sucrose, D-lactose, D-melibiose, D-maltose, D-xylose, L-rhamnose, D-trehalose, cellobiose, and D-arabinose. Vitamins not required for growth. Temperature range for growth 4–40°C. Colonies of type strain are yellow, opaque, circular, convex, and with entire margins.

DNA G+C content (mol%): 56.1 (Bd).

Type strain: FC4, ATCC 25309.

Sequence accession no. (16S rRNA gene): U60015.

2. **Prosthecobacter debontii** Hedlund, Gosink and Staley 1998, 327[VP] (Effective publication: Hedlund, Gosink and Staley, 1997, 35.)

de.bon'ti.i. N.L. masc. gen. n. *debontii* of de Bont, named in honor of J.A.M. de Bont, a Dutch microbiologist who isolated and described the first member of the genus, *Prosthecobacter fusiformis* (FC4).

Fusiform cells are usually vibrioid, short (2–8 μM) and thick (0.5 μM). Carbon sources include D-glucose, D-galactose, D-mannose, D-sucrose, D-lactose, D-melibiose, D-maltose, D-xylose, L-rhamnose, D-trehalose, cellobiose, D-ribose, D-arabinose, D-fructose, and *N*-acetylglucosamine. Vitamins not required for growth. Temperature range for growth is 4–40°C. Colonies of type strain are yellow, opaque, circular, convex, and with entire margins.

DNA G+C content (mol%): 57.1 (Bd).

Type strain: FC3, ATCC 700200.

Sequence accession no. (16S rRNA gene): U60014.

3. **Prosthecobacter dejongeii** Hedlund, Gosink and Staley 1998, 327[VP] (Effective publication: Hedlund, Gosink and Staley, 1997, 36.)

de.jon.ge'i.i. N.L. masc. gen. n. *dejongeii* of de Jonge, named in honor of Klaaske de Jonge, a Dutch microbiologist who studied *Prosthecobacter.*

Fusiform cells are usually straight, long (3–10 μM), and thick (1.0 μM). Carbon sources include D-glucose, D-galactose, D-mannose, D-sucrose, D-lactose, D-melibiose, D-maltose, D-xylose, L-rhamnose, D-trehalose, cellobiose, D-ribose, glycogen, and *N*-acetylglucosamine. Vitamins not required for growth. Temperature range for growth is 1–35°C. Colonies of type strain are pale yellow, opaque, circular, convex, and with entire margins.

DNA G+C content (mol%): 54.6 (Bd).

Type strain: FC1, ATCC 27091.

Sequence accession no. (16S rRNA gene): U60012.

4. **Prosthecobacter vanneervenii** Hedlund, Gosink and Staley 1998, 327[VP] (Effective publication: Hedlund, Gosink and Staley, 1997, 37.)

van.ne.er.ve'ni.i. N.L. masc. gen. n. *vanneervenii* of van Neerven, named in honor of Alex van Neerven, a Dutch microbiologist who studied *Prosthecobacter* and other prosthecate bacteria.

Fusiform cells are usually straight, short (2–8 μM), and thin (0.5 μM). Carbon sources include D-glucose, D-galactose, D-mannose, D-sucrose, D-lactose, D-melibiose, D-maltose, D-xylose, L-rhamnose, D-trehalose, cellobiose, D-ribose, glycogen, D-fructose, D-raffinose, D-glucosamine, and *N*-acetylglucosamine. Vitamins not required for growth. Temperature range for growth is 10–38°C. Colonies of type strain are pale yellow, opaque, circular, convex, and with entire margins.

DNA G+C content (mol%): 60.1 (Bd).

Type strain: FC2, ATCC 700199.

Sequence accession no. (16S rRNA gene): U60013.

References

Albrecht, W., A. Fischer, J. Smida and E. Stackebrandt. 1987. *Verrucomicrobium spinosum*, a eubacterium representing an ancient line of descent. Syst. Appl. Microbiol. *10*: 57–62.

Bomar, D. and E. Stackebrandt. 1987. 5S rRNA sequences from *Nitrobacter winogradskyi, Caulobacter crescentus, Stella humosa* and *Verrucomicrobium spinosum*. Nucleic Acids Res. *15*: 9597.

Cohen-Bazire, G., W.R. Sistrom and R.Y. Stanier. 1957. Kinetic studies of pigment synthesis by non-sulfur purple bacteria. J. Cell Phys. *49*: 25–68.

de Bont, J.A.M., J.T. Staley and H.S. Pankratz. 1970. Isolation and description of a non-motile, fusiform, stalked bacterium, a representative of a new genus. Antonie van Leeuwenhoek *36*: 397–407.

Garrity, G.M. and J.G. Holt. 2001. The Road Map to the *Manual*. *In* Bergey's Manual of Systematic Bacteriology, 2nd edn, vol. 1, The *Archaea* and the Deeply Branching and Phototrophic *Bacteria* (edited by Boone, Castenholz and Garrity). Springer, New York, pp. 119–166.

Gorlenko, V.M. 1970. A new phototrophic green sulphur bacterium: *Prosthecochloris aestuarii* nov. gen. nov. spec. Z. Allg. Mikrobiol. *10*: 147–149.

Gorlenko, V.M. and E.V. Lebedeva. 1971. New green sulphur bacteria with appendages [in Russian, with English summary]. Mikrobiologya *40*: 1035–1039.

Hedlund, B.P., J.J. Gosink and J.T. Staley. 1996. Phylogeny of *Prosthecobacter*, the fusiform caulobacters: members of a recently discovered division of the *Bacteria*. Int. J. Syst. Bacteriol. *46*: 960–966.

Hedlund, B.P., J.J. Gosink and J.T. Staley. 1997. *Verrucomicrobia* div. nov., a new division of the bacteria containing three new species of *Prosthecobacter*. Antonie van Leeuwenhoek *72*: 29–38.

Hedlund, B.P., J.J. Gosink and J.T. Staley. 1998. *In* Validation of the publication of new names and new combinations previously effectively published outside the IJSB. List no. 64. Int. J. Syst. Bacteriol. *48*: 327–328.

Henrici, A.T. and D.E. Johnson. 1935. Studies of freshwater bacteria: II. Stalked bacteria, a new order of *Schizomycetes*. J. Bacteriol. *30*: 61–93.

Hirsch, P., M. Müller and H. Schlesner. 1977. New aquatic budding and prosthecate bacteria and their taxonomic position. *In* Symposium on Aquatic Microbiology, Lancaster, UK. Academic Press, London, pp. 107–133.

Houwink, A.L. 1951. *Caulobacter* versus *Bacillus* spec. div. Nature *168*: 654–655.

Jenkins, C., R. Samudrala, I. Anderson, B.P. Hedlund, G. Petroni, N. Michailova, N. Pinel, R. Overbeek, G. Rosati and J.T. Staley. 2002. Genes for the cytoskeletal protein tubulin in the bacterial genus *Prosthecobacter*. Proc. Natl. Acad. Sci. U. S. A. *99*: 17049–17054.

Petroni, G., S. Spring, K.H. Schleifer, F. Verni and G. Rosati. 2000. Defensive extrusive ectosymbionts of *Euplotidium* (Ciliophora) that contain microtubule-like structures are bacteria related to *Verrucomicrobia*. Proc. Natl. Acad. Sci. U. S. A. *97*: 1813–1817.

Pilhofer, M., G. Rosati, W. Ludwig, K.H. Schleifer and G. Petroni. 2007. Coexistence of tubulins and *ftsZ* in different *Prosthecobacter* species. Mol. Biol. Evol. *24*: 1439–1442.

Schlesner, H. 1987. *Verrucomicrobium spinosum* gen. nov., sp. nov., a fimbriated prosthecate bacterium. Syst. Appl. Microbiol. *10*: 54–56.

Schlesner, H. 1988. *In* Validation of the publication of new names and new combinations previously effectively published outside the IJSB. List no. 25. Int. J. Syst. Bacteriol. *38*: 220–222.

Schlesner, H. 1992. The genus *Verrucomicrobium*. *In* The Prokaryotes: a Handbook on the Biology of Bacteria: Ecophysiology, Isolation, Identification, Applications, 2nd edn, vol. 4 (edited by Balows, Dworkin, Harder, Schleifer and Trüper). Springer, New York, pp. 3806–3808.

Schlieper, D., M.A. Oliva, J.M. Andreu and J. Lowe. 2005. Structure of bacterial tubulin BtubA/B: evidence for horizontal gene transfer. Proc. Natl. Acad. Sci. U. S. A. *102*: 9170–9175.

Sittig, M. and H. Schlesner. 1993. Chemotaxonomic investigation of various prosthecate and or budding bacteria. Syst. Appl. Microbiol. *16*: 92–103.

Sontag, C.A., J.T. Staley and H.P. Erickson. 2005. *In vitro* assembly and GTP hydrolysis by bacterial tubulins BtubA and BtubB. J. Cell. Biol. *169*: 233–238.

Staley, J.T. 1968. *Prosthecomicrobium* and *Ancalomicrobium*: new prosthecate freshwater bacteria. J. Bacteriol. *95*: 1921–1942.

Staley, J.T., J.A.M. de Bont and K. de Jonge. 1976. *Prosthecobacter fusiformis* nov. gen. et sp., the fusiform caulobacter. Antonie van Leeuwenhoek *42*: 333–342.

Staley, J.T., J.A.M. de Bont and K. de Jonge. 1980. *Prosthecobacter fusiformis* gen. and sp. nov., nom. rev. Int. J. Syst. Bacteriol. *30*: 595.

Staley, J.T., H. Bouzek and C. Jenkins. 2005. Eukaryotic signature proteins of *Prosthecobacter dejongeii* and *Gemmata* sp. Wa-1 as revealed by *in silico* analysis. FEMS Microbiol. Lett. *243*: 9–14.

Vasilyeva, L.V. 1985. *Stella*, a new genus of soil prosthecobacteria, with proposals for *Stella humosa* sp. nov. and *Stella vacuolata* sp. nov. Int. J. Syst. Bacteriol. *35*: 518–521.

Ward-Rainey, N., F.A. Rainey, H. Schlesner and E. Stackebrandt. 1995. Assignment of a hitherto unidentified 16S rDNA species to a main line of descent within the domain *Bacteria*. Microbiology *141*: 3247–3250.

Ward-Rainey, N., F.A. Rainey, H. Schlesner and E. Stackebrandt. 1996. *In* Validation of the publication of new names and new combinations previously effectively published outside the IJSB. List no. 57. *In* Int. J. Syst. Bacteriol. *46*: 625–626.

Ward-Rainey, N., F.A. Rainey and E. Stackebrandt. 1997. The presence of a *dnaK* (HSP70) multigene family in members of the orders *Planctomycetales* and *Verrucomicrobiales*. J. Bacteriol. *179*: 6360–6366.

Ward, N.L., F.A. Rainey, B.P. Hedlund, J.T. Staley, W. Ludwig and E. Stackebrandt. 2000. Comparative phylogenetic analyses of members of the order *Planctomycetales* and the division *Verrucomicrobia*: 23S rRNA gene sequence analysis supports the 16S rRNA gene sequence-derived phylogeny. Int. J. Syst. Evol. Microbiol. *50*: 1965–1972.

Yoon, J., Y. Matsuo, K. Adachi, M. Nozawa, S. Matsuda, H. Kasai and A. Yokota. 2008. Description of *Persicirhabdus sediminis* gen. nov., sp. nov., *Roseibacillus ishigakijimensis* gen. nov., sp. nov., *Roseibacillus ponti* sp. nov., *Roseibacillus persicicus* sp. nov., *Luteolibacter pohnpeiensis* gen. nov., sp. nov. and *Luteolibacter algae* sp. nov., six marine members of the phylum '*Verrucomicrobia*', and emended descriptions of the class *Verrucomicrobiae*, the order *Verrucomicrobiales* and the family *Verrucomicrobiaceae*. Int. J. Syst. Evol. Microbiol. *58*: 998–1007.

Family II. **Akkermansiaceae** fam. nov.

Brian P. Hedlund and Muriel Derrien

Ak.ker.man.si.a.ce'a.e. N.L. fem. n. *Akkermansia* type genus of the family; L. suff. *-aceae* ending to denote a family; N.L. fem. pl. n. *Akkermansiaceae* the *Akkermansia* family.

Ovoid cells. Nonsporeforming. Nonmotile. **Cells stain Gram-negative and have a Gram-negative cell structure. Strictly anaerobic.** Mesophilic. **Chemoorganotrophic and obligately fermentative.** Able to ferment mucin.

DNA G+C content (mol%): 47.6.

Type genus: **Akkermansia** Derrien, Vaughan, Plugge and de Vos 2004, 1474[VP].

Encompasses bacteria isolated from the human intestine and associated 16S rRNA gene sequences recovered from intestines or feces of humans and other mammals within the order *Verrucomicrobiales*. Delineation of the family is determined primarily by phylogenetic information from 16S rRNA gene sequences. Currently, the family includes only the genus *Akkermansia*.

Genus I. **Akkermansia** Derrien, Vaughan, Plugge and de Vos 2004, 1474[VP]

Muriel Derrien, Caroline M. Plugge, Willem M. de Vos and Erwin G. Zoetandal

Ak.ker.man'si.a. N.L. fem. n. *Akkermansia* named after Antoon Akkermans, Dutch microbiologist recognized for his contribution to microbial ecology.

Oval-shaped cells, occurring singly, in pairs, and rarely in chains, about 0.6 µm × 0.7 µm. **Gram-stain-negative. Nonmotile. Strictly anaerobic.** Colonies are distinct, whitish in color, and reach maximum size on 0.75% agar plates after 6 d of incubation. Cells grow optimally at 37°C and pH 6.5. Growth is **chemoorganotrophic** and restricted to a small number of sugars. **Acetate, propionate, and ethanol** are the major fermentation products from mucin. Mucolytic in pure culture.

The 16S rRNA gene sequence indicates that the genus *Akkermansia* belongs to the phylum *Verrucomicrobia* and, together with the genera *Verrucomicrobium* and *Prosthecobacter*, belongs to the order *Verrucomicrobiales*.

DNA G+C content (mol%): 47.6.

Type species: **Akkermansia muciniphila** Derrien, Vaughan, Plugge and de Vos 2004, 1474[VP].

Further descriptive information

For the growth of *Akkermansia*, a mucin based medium is recommended. In active cultures, *Akkermansia* appears oval-shaped (Figure 138), with diameter of 0.7 µm in light microscopy. It is usually detected in pairs or aggregates and rarely in chains. The doubling time of the strain is approximately 1.5 h in mucin medium. *Akkermansia* forms round and white colonies that are about 0.7 mm in diameter after 6 d of anaerobic incubation in 0.75% agar. No motility is observed under the microscope. Filaments have been observed by electronic microscopy and can be attributed to extracellular polymers. During mucin fermentation, acetate, propionate, and ethanol are produced as major end products. No growth is observed on glucose, cellobiose, lactose, galactose, xylose, fucose, rhamnose, maltose, succinate, acetate, fumarate, butyrate, lactate, casitone, Casamino acids, tryptone, peptone, yeast extract, proline, glycine, aspartate, serine, threonine, glutamate, alanine, *N*-acetylglucosamine, or *N*-acetylgalactosamine. The only mucin-free medium known to support growth contains peptone, yeast extract, tryptone, casitone, *N*-acetylglucosamine, *N*-acetylgalactosamine, and glucose.

The type species, *Akkermansia muciniphila*, was originally isolated from a fecal sample from a healthy volunteer.

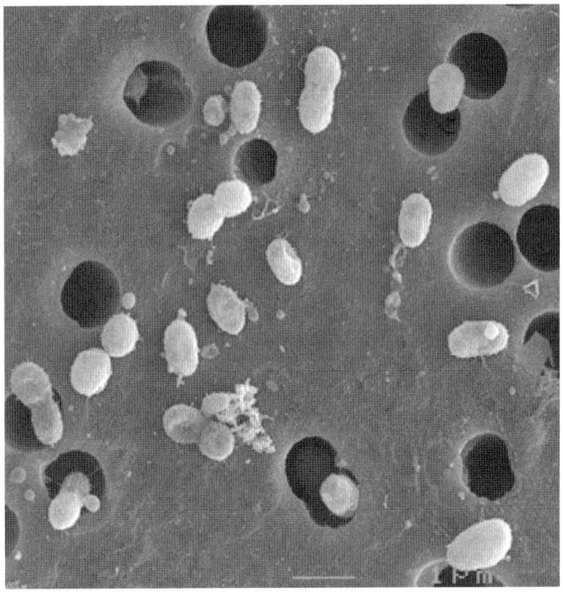

FIGURE 138. Scanning electronic micrograph of the type strain, *Akkermansia muciniphila*, grown in mucin medium. Bar = 1 µm.

Consistent with other studies in which *Akkermansia* was found in different 16S rRNA clone libraries, the bacterium accounted for around 1% of the total fecal cells (Derrien et al., submitted). *Akkermansia* was detected in many fecal and intestinal samples (Hayashi et al., 2002; Hold et al., 2002; Salzman et al., 2002; Nelson et al., 2003; Mangin et al., 2004; Eckburg et al., 2005; R.A. Hutson and M.D. Collins, not published). *Akkermansia* has never been detected in environmental samples such as water, soil, or anaerobic digesters from which most of the *Verrucomicrobia* members are isolated.

Enrichment and isolation procedures

Akkermansia can be enriched using a bicarbonate buffered mineral salts medium prepared according to Plugge (2005). This basal medium is supplemented with 0.25% (v/v) commercial hog gastric mucin purified with ethanol precipitation (Hoskins and Boulding, 1981). Growth is optimal at 37°C under strict anaerobic conditions provided by a gas phase of 182 kPa

(1.8 atm.) N_2/CO_2 (80:20, v/v). Growth in liquid cultures is apparent after 1 d. Isolated colonies can be obtained by inoculating a liquid culture into the same medium containing 0.75% agar and incubating the bottles for about 6 d.

Maintenance procedures

Akkermansia can be stored in basal medium containing mucin for several months at 4°C under anaerobic conditions. For long-term storage, cultures can be maintained at −80°C in mucin medium containing 25% glycerol as a cryoprotectant, or they can be freeze-dried.

Procedures for testing special characteristics

Based on the *Akkermansia* 16S rRNA gene sequence, a specific 20-base-long oligonucleotide probe for fluorescent *in situ* hybridization (FISH) has been developed and validated against a panel of intestinal pure cultures (Table 156). The validation was done by increasing the stringency using formamide. The specific conditions of hybridization are 20% formamide at 50°C (Derrien et al., submitted). In addition, two specific primers for PCR were also designed based on the 16S rRNA gene sequence and validated: AM1 and AM2 (Table 156). They allow the amplification of a 300-bp fragment and were validated against intestinal clones and pure cultures. The optimum annealing temperature is 56°C (Derrien et al., submitted).

Differentiation of the genus *Akkermansia* from other genera

Based on the differences in the 16S rRNA sequences, *Akkermansia* can be clearly distinguished from other genera of the order *Verrucomicrobiales* (Figure 139), which includes *Prosthecobacter*

TABLE 156. Nucleic acid composition of the primers and probe specific to the type species *Akkermansia muciniphila* 16S rRNA gene sequence

Designation	Sequence (5′ → 3′)	Direction
AM1 (PCR primer)	CAG CAC GTG AAG GTG GGG AC	Forward
AM2 (PCR primer)	CCT TGC GGT TGG CTT CAG AT	Reverse
MUC 1437 (FISH probe)	CCTTGCGGTTGG CTTCAGAT	

(Hedlund et al., 1997; Staley et al., 1976) and *Verrucomicrobium* (Schlesner, 1987). Although *Akkermansia* shares some common features with *Prosthecobacter* and *Verrucomicrobium* (Gram-stain-negative, growth without vitamins), *Akkermansia* is clearly differentiated from the genera *Prosthecobacter* and *Verrucomicrobium* with respect to its habitat (human versus environmental habitats), oxygen requirement (strict anaerobic versus strict aerobic), shape (oval-shaped versus rod-shaped), and DNA G+C composition (47.6 mol% vs. 54.6–60.1 mol% for *Prosthecobacter* and 57.9–59.3 mol% for *Verrucomicrobium*).

Taxonomic comments

The nearly complete 16S rRNA gene sequence (1433 nt) of the type species of *Akkermansia* showed 99% identity to human colonic clones HuCA18 and HuCC13 (Hold et al., 2002) and mouse colonic clone L10-6 (Salzman et al., 2002). The most closely related cultivated species was *Verrucomicrobium spinosum*, which is only distantly related (92% based on 16S rRNA sequence similarity).

List of species of the genus *Akkermansia*

1. **Akkermansia muciniphila** Derrien, Vaughan, Plugge and de Vos 2004, 1474[VP]

mu.ci.ni′phi.la. N.L. neut. n. *mucinum* mucin; N.L. adj. *philus -a -um* (from Gr. adj. *philos -ê -on*) friend, loving; N.L. fem. adj. *muciniphila* mucin-loving.

Cells are oval-shaped, nonmotile, and Gram-stain-negative. The long axis of single cells is 0.6–0.7 µm, depending on the growth substrate used. Cells occur singly, in pairs, in short chains, and in aggregates. Growth occurs at 20–40°C and pH 5.5–8.0, with optimum growth at 37°C and pH 6.5. Strictly anaerobic. Able to grow on gastric mucin, brain heart infusion (Difco), and Columbia broth (Difco), and on *N*-acetylglucosamine, *N*-acetylgalactosamine, and glucose when these three sugars are in the presence of peptone, yeast extract, casitone, and tryptone. Cellobiose, lactose, galactose, xylose, fucose, rhamnose, maltose, succinate, acetate, fumarate, butyrate, lactate, casitone, Casamino acids, tryptone, peptone, yeast extract, proline, glycine, aspartate, serine, threonine, and glutamate do not support growth. Capable of using mucin as carbon, energy, and nitrogen source. Able to release sulfate in a free form from mucin fermentation. In mucin medium, cells are covered with filaments. Growth occurs without vitamins. Colonies appear white with a diameter of 0.7 mm in soft agar mucin medium.

DNA G+C content (mol%): 47.6 (HPLC).

Type strain: Muc, ATCC BAA-835, CIP 107961.

Sequence accession no. (16S rRNA gene): AY271254.

References

Derrien, M., E.E. Vaughan, C.M. Plugge and W.M. de Vos. 2004. *Akkermansia muciniphila* gen. nov., sp. nov., a human intestinal mucin-degrading bacterium. Int. J. Syst. Evol. Microbiol. *54*: 1469–1476.

Eckburg, P.B., E.M. Bik, C.N. Bernstein, E. Purdom, L. Dethlefsen, M. Sargent, S.R. Gill, K.E. Nelson and D.A. Relman. 2005. Diversity of the human intestinal microbial flora. Science *308*: 1635–1638.

Hayashi, H., M. Sakamoto and Y. Benno. 2002. Fecal microbial diversity in a strict vegetarian as determined by molecular analysis and cultivation. Microbiol. Immunol. *46*: 819–831.

Hedlund, B.P., J.J. Gosink and J.T. Staley. 1997. *Verrucomicrobia* div. nov., a new division of the bacteria containing three new species of *Prosthecobacter*. Antonie van Leeuwenhoek *72*: 29–38.

Hold, G.L., S.E. Pryde, V.J. Russell, E. Furnie and H.J. Flint. 2002. Assessment of microbial diversity in human colonic samples by 16S rDNA sequence analysis. FEMS Microbiol. Ecol. *39*: 33–39.

Hoskins, L.C. and E.T. Boulding. 1981. Mucin degradation in human colon ecosystems. Evidence for the existence and role of bacterial subpopulations producing glycosidases as extracellular enzymes. J. Clin. Invest. *67*: 163–172.

Mangin, I., R. Bonnet, L. Seksik, L. Rigottier-Gois, M. Sutren, Y. Bouhnik, C. Neut, M.D. Collins, J-F. Colombel, P. Marteau and J. Dore. 2004. Molecular inventory of faecal microflora in patients with Crohn's disease. FEMS Microbiol. Ecol. *50*: 25–36.

Nelson, K.E., S.H. Zinder, I. Hance, P. Burr, D. Odongo, D. Wasawo, A. Odenyo and R. Bishop. 2003. Phylogenetic analysis of the microbial

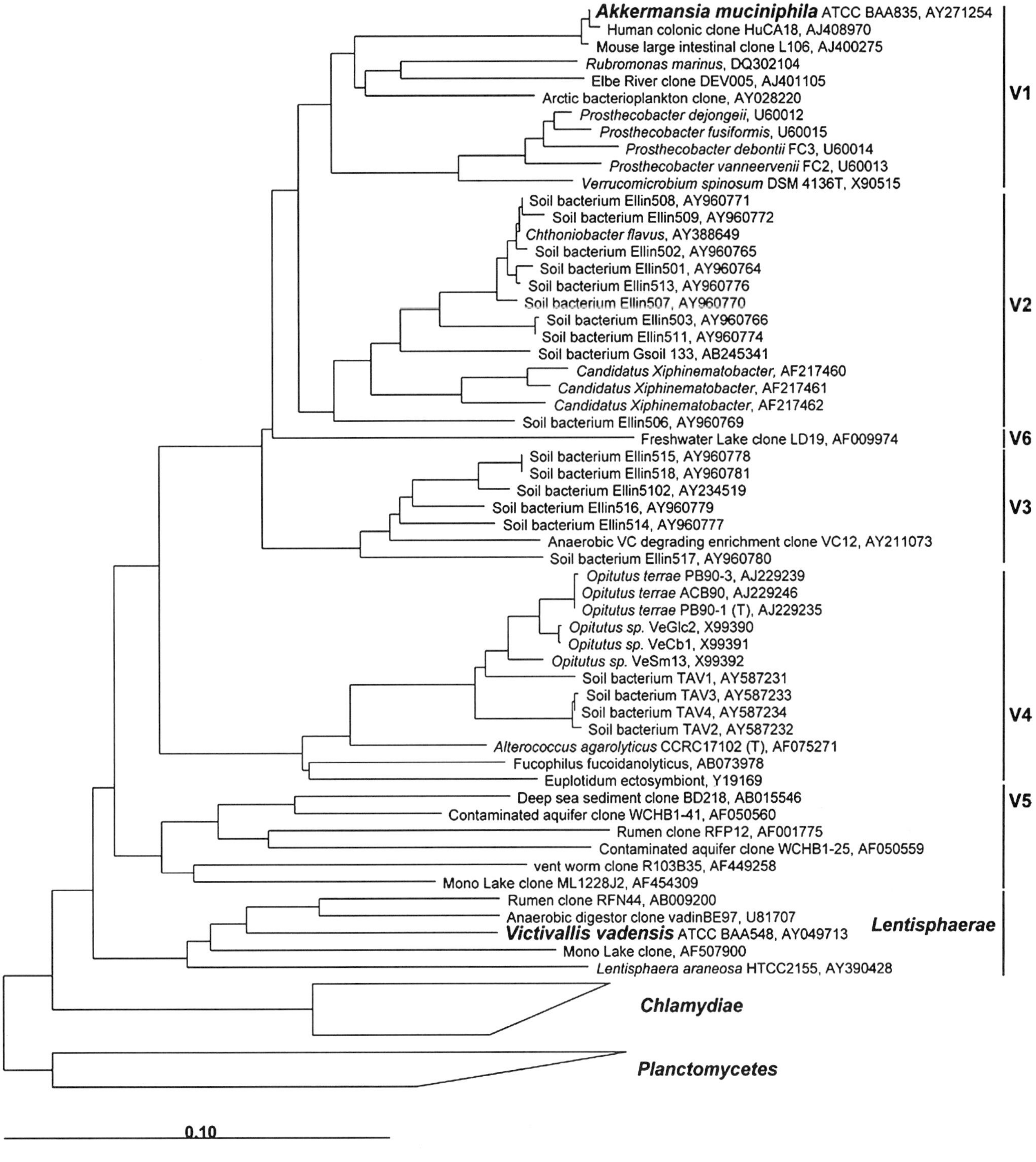

FIGURE 139. Phylogenetic neighbor-joining dendrogram showing the relationships of *Akkermansia muciniphila* with representatives of the phylum *Verrucomicrobia*. Bar = 0.1% sequence difference.

populations in the wild herbivore gastrointestinal tract: insights into an unexplored niche. Environ. Microbiol. *5*: 1212–1220.

Plugge, C.M. 2005. Anoxic media design, preparation, and considerations. *In* Methods in Enzymology, vol. 397. Academic Press, New York, pp. 3–16.

Salzman, N.H., H. de Jong, Y. Paterson, H.J. Harmsen, G.W. Welling and N.A. Bos. 2002. Analysis of 16S libraries of mouse gastrointestinal microflora reveals a large new group of mouse intestinal bacteria. Microbiology *148*: 3651–3660.

Schlesner, H. 1987. *Verrucomicrobium spinosum* gen. nov., sp. nov., a fimbriated prosthecate bacterium. Syst. Appl. Microbiol. *10*: 54–56.

Staley, J.T., J.A.M. de Bont and K. de Jonge. 1976. *Prosthecobacter fusiformis* nov. gen. et sp., the fusiform caulobacter. Antonie van Leeuwenhoek *42*: 333–342.

Family III. **Rubritaleaceae** fam. nov.

Brian P. Hedlund

Ru.bri.ta.le.a.ce'a.e. N.L. fem. n. *Rubritalea* type genus of the family; L. suff. *-aceae* ending to denote a family; N.L. fem. pl. n. *Rubritaleaceae* the *Rubritalea* family.

Coccoid or rod-shaped bacteria from **marine habitats**. Cells stain **Gram-negative** and have a **Gram-negative cell structure. Nonmotile.** Obligately aerobic or facultatively anaerobic, reducing nitrate to nitrite. Oxidize a wide variety of organic molecules for growth. **Menaquinones** are the only respiratory quinones detected.

DNA G+C content (mol%): 47.7–52.4.

Type genus: **Rubritalea** Scheuermayer, Gulder, Bringmann and Hentschel 2006, 2123[VP].

Further descriptive information

Encompasses bacteria isolated from marine habitats and associated 16S rRNA gene sequences recovered from marine environments within the order *Verrucomicrobiales.* Delineation of the family is determined primarily by phylogenetic information from 16S rRNA gene sequences. Currently, the family is represented only by the genus *Rubritalea.*

Genus I. **Rubritalea** Scheuermayer, Gulder, Bringmann and Hentschel 2006, 2123[VP]

Brian P. Hedlund, Jaewoo Yoon and Hiroaki Kasai

Ru.bri.ta'le.a. L. adj. *ruber -bra -brum* red; L. fem. n. *talea* a rod, staff; N.L. fem. n. *Rubritalea* a red-colored rod.

Coccoid or rod-shaped bacteria from **marine habitats**. **Gram-stain-negative;** have a **Gram-negative cell structure**. **Nonmotile.** Obligately aerobic or facultatively anaerobic, reducing nitrate to nitrite. Colonies are **red or pink** in color due to carotenoids. Some produce squalene. Cell walls contain ***meso-diaminopimelic acid.*** **Chemoorganotrophic**. Oxidize a wide variety of organic molecules for growth. Most abundant cellular fatty acids are **$C_{14:0}$ iso, $C_{16:0}$ iso, and $C_{16:1}$ ω*7c*** (Table 157). **Menaquinones** are the only respiratory quinones detected, primarily MK-8 and MK-9. Phylogenetically, the genus belongs to the order *Verrucomicrobiales* in the phylum "*Verrucomicrobia*".

DNA G+C content (mol%): 47.7–52.4.

Type species: **Rubritalea marina** Scheuermayer, Gulder, Bringmann and Hentschel 2006, 2123[VP].

Further descriptive information

The genus *Rubritalea* currently includes five species, including the type species *Rubritalea* marinus (Figure 140, Table 158). Each of the species of *Rubritalea* were isolated from marine environments, and all 16S rRNA gene sequences closely related to *Rubritalea* that have been recovered from cultivation independent studies originate from marine samples. Therefore, all data suggest that this genus is an exclusive and widely distributed inhabitant of the marine environment, however, it is noteworthy that not all species of *Rubritalea* require sodium ions. The natural habitat of *Rubritalea* species within the marine habitat is less clear. Four of the five species were isolated from marine invertebrates, including *Rubritalea marina* from homogenized tissue of the Mediterranean sponge *Axinella polypoides* (Scheuermayer et al., 2006), *Rubritalea spongiae* from an unidentified sponge off the coast of Japan (Yoon et al., 2007), and *Rubritalea squalenifaciens* from *Halichondria okadai* off the coast of Japan (Kasai et al., 2007). In addition, *Rubritalea tangerina* was isolated from homogenized visceral tissue from a sea hare (Yoon et al., 2007). In contrast, *Rubritalea sabuli* was isolated from marine sediment off the island of Pohnpei, part of the Federated States of Micronesia (Yoon et al., 2008). Scheuermayer et al. (2006) pointed out that sponges may not be the natural habitat of *Rubritalea marina.* Although sponges host specific and robust intracellular commensal populations, sponges are filter-feeders and therefore also concentrate planktonic microorganisms for phagocytocis. The authors recovered related 16S rRNA gene sequences from both bulk sea water and sponges from the location from which *Rubritalea marina* was isolated. There was no particular phylogenetic structure suggesting a specific association, suggesting *Rubritalea marina* may derive from the planktonic environment. An alternative hypothesis is that *Rubritalea* is a symbiont of sponges with a planktonic dispersal form, however, the isolation of species from visceral tissue of a sea hare and from marine sediments, and the lack of environment-specific phylogenetic clustering within the genus, suggest this genus may be more widely distributed in marine environments.

Each of the species of *Rubritalea* produces red or pink nondiffusible pigments that are extractable in methanol, ethanol, and/or acetone. The acetone-extracted pigments of *Rubritalea squalenifaciens* have been the best characterized (Kasai et al., 2007; Shindo et al., 2008). Carotenoid pigments were separated from squalene in acetone extracts by an HPLC/PAD (photodiode array detection)/APCI (atmospheric pressure chemical ionization)-MS (mass spectrometry) system (TermoFinnigan). The carotenoid absorption maxima were at 312, 470, 490, and 518 nm (Kasai et al., 2007). Further mass spectrometry work identified three novel acyl glyco-carotenoic acids, diapolycopenedioic acid xylosyl esters, one of which was shown to have potent antioxidant activity in a singlet oxygen suppression model (Shindo et al., 2008). *Rubritalea spongiae* and *Rubritalea tangerina* contain the same pigments as shown by the HPLC/PAD/APCI-MS system (Yoon et al., 2007). *Rubritalea sabuli* contained carotenoid with the same UV-visible absorption spectrum, however, the HPLC retention time was different, and the molecule could not be identified by MS due to the presence of several different molecules in the extract (Yoon et al., 2008). Less work was done on the pigment of *Rubritalea marina,* however, the absorption maximum at 495 nm suggests a similar carotenoid (Scheuermayer et al., 2006).

TABLE 157. Predominant membrane fatty acids in species of the genus *Rubritalea*[a]

Fatty acid	*R. marina*	*R. sabuli*	*R. spongiae*	*R. squalenifaciens*	*R. tangerina*
Saturated:					
$C_{14:0}$	nd	<1	<1	<1	2.9
$C_{14:0}$	4.5	1	2.6	1.9	1.4
$C_{14:0}$	23	2.5	7.3	4.4	23.1
$C_{14:0}$	nd	–	2.8	<1	1.1
$C_{14:0}$	nd	–	<1	<1	2.4
Unsaturated:					
$C_{15:1}$ ω6*c*	nd	2.1	2	1	<1
$C_{16:1}$ ω7*c*	21.5	7	11.8	7	12.7
$C_{18:1}$ ω7*c*	nd	–	–	–	1.5
Branched:					
$C_{14:0}$ iso	22	49.4	40.6	43.1	35.5
$C_{16:0}$ iso	nd	29.1	16.1	20.6	12.2
$C_{15:0}$ anteiso	6	5.1	12.9	18.1	4.7
$C_{17:0}$ anteiso	nd	1.3	2	1.6	–

[a]Symbols: +, >85% positive; –, 0–15% positive; nd, not described; –, not detected. Values are percentages of total fatty acid.

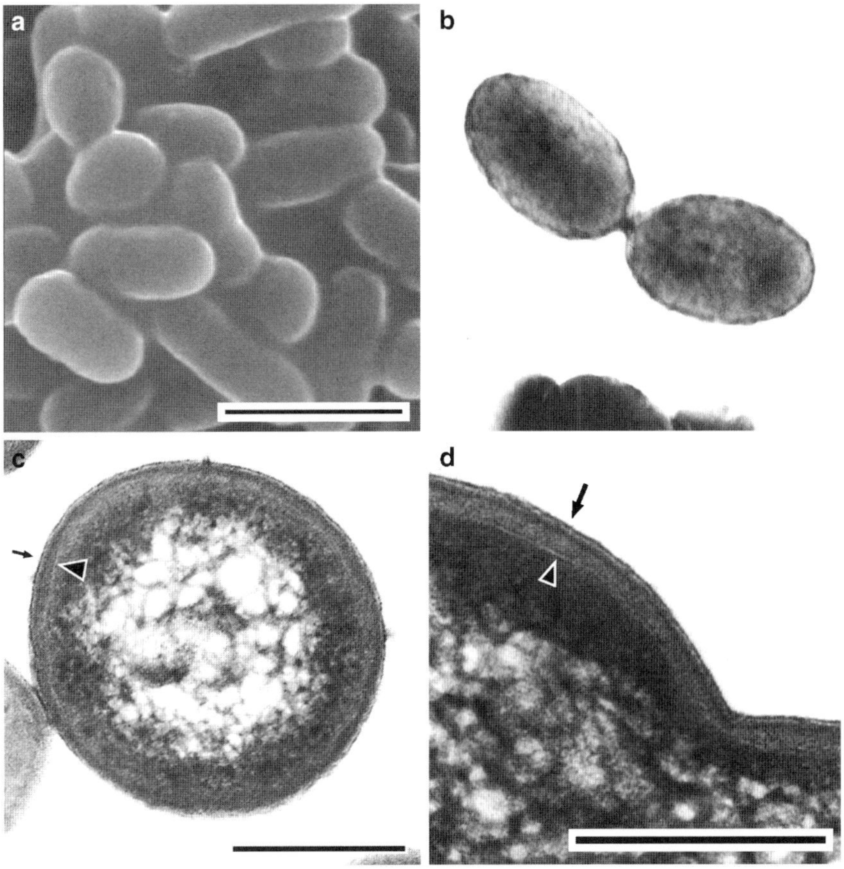

FIGURE 140. Electron micrographs of cells of *Rubritalea squalenifaciens* strain HOact23[T]. (a) Scanning electron micrograph of cells growing on an agar medium. (b) Transmission electron micrograph of negatively stained whole cells. (c, d) Transmission electron micrographs of thin sections of cells; triangles and arrow heads point to cytoplasmic and outer membranes, respectively. Bars = 1 μm (a–c) and 500 nm (d). (Reproduced with permission from Kasai et al., 2007. Int. J. Syst. Evol. Microbiol. *57:* 1630–1634.)

Electron microscopy has identified possible buds in *Rubritalea marina.* However, no such structures were identified in other species of *Rubritalea,* and it was suggested that the other species divide by binary fission. Electron microscopy has shown that *Rubritalea* has a Gram-negative cell structure (Figure 140).

Enrichment and isolation procedures

Each of the species of *Rubritalea* was isolated by directly plating mechanically homogenized samples (tissues of marine invertebrates or sand) onto an unselective, complex, marine medium solidified with 1.5% agar. At present, no specific enrichment and isolation strategies are known for the genus.

TABLE 158. Characteristics differentiating species of the genus *Rubritalea*

Characteristic	*R. marina*	*R. sabuli*	*R. spongiae*	*R. squalenifaciens*	*R. tangerina*
Isolation source	Marine sponge	Marine sediment	Marine sponge	Marine sponge	Sea hare
Cell shape	Coccoid or rod	Coccoid or rod	Coccoid or rod	Rod	Coccoid or rod
Colony color	Red	Pink	Reddish pink	Reddish pink	Reddish orange
pH range	6.8–8.2	6.5–8.5	6.5–8	7.5–8.5	6.5–8.5
Temperature range (°C)	8–30	10–37	4–37	15–37	15–37
NaCl range (%)	1.4–3.8	0–8	1–7	1–4	0–9
Production of:					
Catalase	−	+	+	+	−
Oxidase	+	−	+	+	+
Nitrate reduction	+	−	+	+	−
Oxidation of:					
Acetic acid	+	+	+	−	−
N-Acetyl-D-galactosamine	−	+	−	+	−
N-Acetyl-D-glucosamine	−	+	−	+	−
cis-Aconitic acid	−	−	+	−	+
Cellobiose	+	+	−	−	−
Citric acid	−	−	−	−	+
Dextrin	+	+	−	−	−
Formic acid	+	−	−	−	−
L-Fructose	−	+	+	−	−
L-Fucose	−	−	+	−	+
D-Glucuronic acid	−	+	−	+	−
D-Glucose	+	+	+	+	−
Glucose 1-phosphate	−	−	+	−	+
Glucose 6-phosphate	−	−	−	−	+
Inosine	−	−	+	−	−
α-Ketobutyric acid	−	−	+	−	−
α-Ketoglutaric acid	−	−	+	−	−
D-Mannitol	−	−	+	−	−
D-Mannose	+	+	−	−	+
Melibiose	+	+	+	+	−
D-Sorbitol	−	−	+	−	−
Succinic acid	−	+	−	−	−
Trehalose	−	+	−	+	−
Enzyme activity of:					
N-Acetyl-β-glucosaminidase	−	+	−	+	+
Chymotrypsin	−	−	−	+	−
Esterase (C4)	+	−	−	+	−
Esterase lipase (C8)	+	−	−	+	−
α-Fucosidase	−	−	+	−	−
Leucine arylamidase	+	+	−	+	+
Trypsin	+	−	−	−	−

Maintenance procedures

For long-term storage, *Rubritalea* can be maintained at −80°C in 75% artificial sea water medium containing 30% glycerol as a cryoprotectant or they can be freeze-dried.

Differentiation of the genus *Rubritalea* from other genera

Phylogenetically, the most closely related genera are *Prosthecobacter*, *Verrucomicrobium*, and *Akkermansia* within the order *Verrucomicrobiales* in the phylum "*Verrucomicrobia*" (Figure 141). The genus *Rubritalea* is differentiated from *Prosthecobacter* and *Verrucomicrobium* by 16S rRNA gene sequence similarity, habitat (marine vs freshwater), and by morphology. Both *Prosthecobacter* and *Verrucomicrobium* are prosthecate and have fimbriae. *Rubritalea* is differentiated from *Akkermansia* by 16S rRNA gene sequence similarity, habitat (marine vs freshwater), and relation to oxygen; *Akkermansia* is a fermentative obligate anaerobe.

List of species of the genus *Rubritalea*

1. **Rubritalea marina** Scheuermayer, Gulder, Bringmann and Hentschel 2006, 2123[VP]

ma.ri'na. L. fem. adj. *marina* of or belonging to the sea, marine.

Description as for the genus. Coccoid or rod-shaped, and red in color. Able to reduce nitrate to nitrite. Able to grow in medium containing 60–160% ASW, at pH values 6.8–8.2 and at temperatures 8–30°C. Menaquinones MK-8 and MK-9 are present. Catalase-negative. Oxidase-positive. Able to grow with glucose, xylose, melibiose, cellobiose, lactose, pyruvate, or pectin as the sole energy source under aerobic conditions. Positive for alkaline phosphatase, esterase (C4),

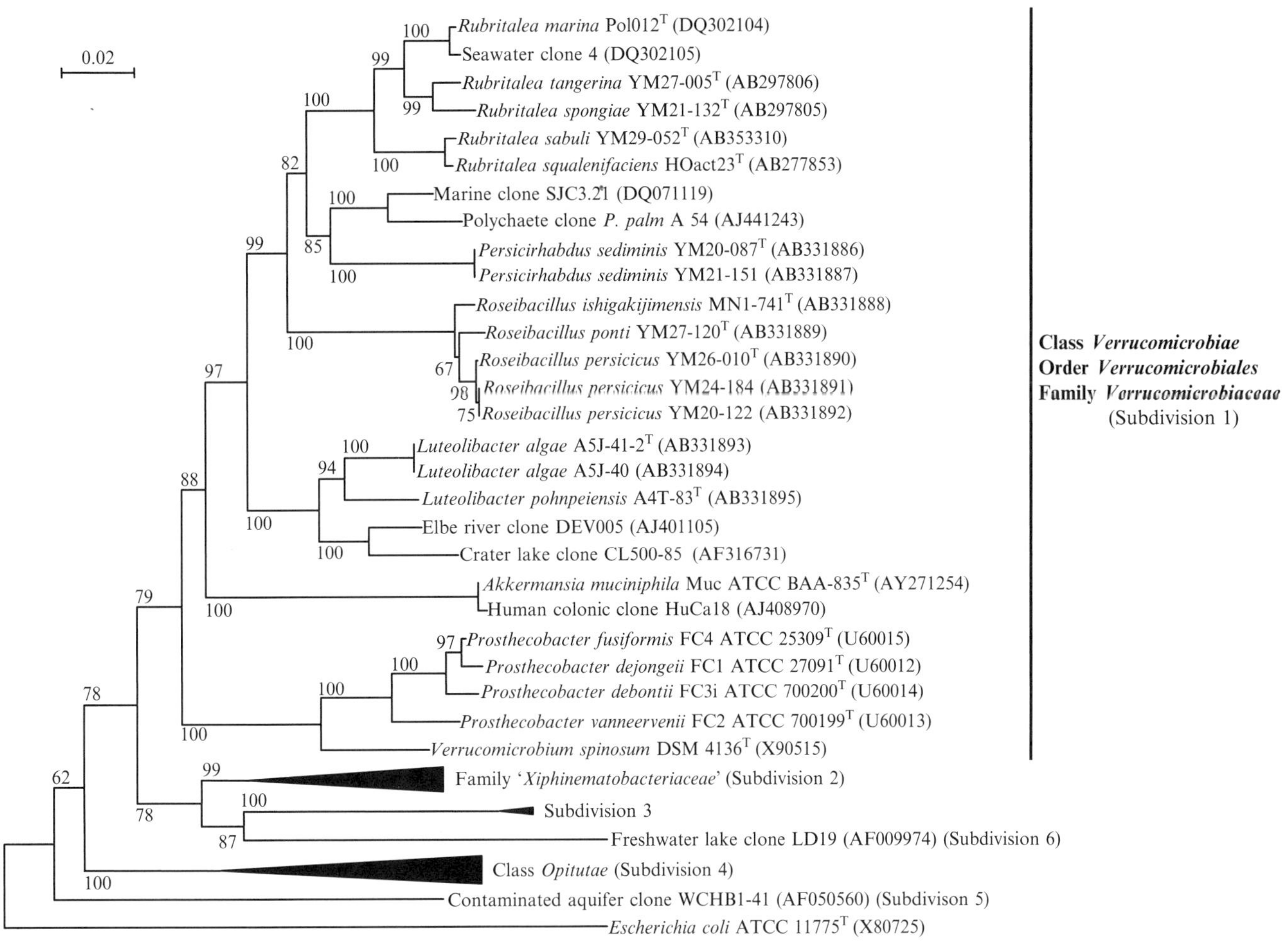

FIGURE 141. Neighbor-joining tree showing the phylogenetic relationship of genus *Rubritalea* and related species of the family *Verrucomicrobiaceae* within the phylum "*Verrucomicrobia*" based on 16S rRNA gene sequences. Numbers at nodes are percentage bootstrap values derived from 1000 replications. Sequence determined in this study is shown in bold. The sequence of *Escherichia coli* ATCC 11775^T was used as an outgroup. Bar = 2% estimated difference in nucleotide sequence position. (Modified with permission from Yoon et al., 2008. Int. J. Syst. Evol. Microbiol. *58:* 992–997.)

esterase lipase (C8), leucine arylamidase, trypsin, acid phosphatase, naphthol-AS-BI-phosphohydrolase, β-glucosidase, and β-galactosidase. Positive for oxidation of dextrin, D-cellobiose, D-fructose, D-galactose, α-D-glucose, D-mannose, D-melibiose, acetic acid, and formic acid.

Source: the Mediterranean sponge *Axinella polypoides.*

DNA G+C content (mol%): 50.9 (HPLC).

Type strain: Pol012, CIP 108984, DSM 177716.

Sequence accession no. (16S rRNA gene): DQ302104.

2. **Rubritalea sabuli** Yoon, Matsuo, Matsuda, Adachi, Kasai and Yokota 2008, 994^{VP}

sa'bu.li. L. gen. n. *sabuli* of sand.

Description as for the genus. Obligately aerobic, not able to reduce nitrate to nitrite. Coccoid (0.6–0.8 µm in diameter) or rod-shaped (0.5–1.0 × 1.2–1.5 µm). Colonies grown on MA are circular, convex, and pink in color. The temperature range for growth is 10–37°C, with optimal growth at 30–37°C and no growth at 4 or 45°C. The pH range for growth is 6.5–8.5. NaCl is not required for growth, but is tolerated up to 8% (w/v). The major respiratory menaquinones are MK-8 and MK-9. Catalase-positive. Oxidase-negative. Positive for alkaline phosphatase, leucine arylamidase, acid phosphatase, naphthol-AS-BI-phosphohydrolase, and *N*-acetyl-β-glucosaminidase tests. Negative for esterase (C4), esterase lipase (C8), lipase (C4), valine arylamidase, cystine arylamidase, trypsin, chymotrypsin, α-galactosidase, β-galactosidase, β-glucuronidase, α-glucosidase, β-glucosidase, α-mannosidase and α-fucosidase tests. Dextrin, *N*-acetyl-D-galactosamine, *N*-acetyl-D-glucosamine, L-arabinose, cellobiose, D-fructose, D-galactose, gentiobiose, D-glucose, D-lactose, lactulose, maltose, D-mannose, melibiose, methyl β-D-glucoside, trehalose, furanose, methyl pyruvate, acetic acid, D-glucuronic acid, succinic acid, alaninamide, L-alanine, L-alanyl glycine, L-glutamic acid, L-serine, and L-threonine are oxidized.

Source: marine sediment on the island of Pohnpei, Micronesia.

DNA G+C content (mol%): 47.7 (HPLC).

Type strain: YM29-052, KCTC 22127, MBIC08323.

Sequence accession no. (16S rRNA gene): AB353310.

3. **Rubritalea spongiae** Yoon, Matsuo, Matsuda, Adachi, Kasai and Yokota 2007, 2339[VP]

spon.gi'ae. L. gen. n. *spongiae* of a sponge, referring to the isolation source of the microorganism.

Description as for the genus. Cells are coccoid (0.6–1.0 μm in diameter) or rod-shaped (0.5–1.0 × 0.8–1.2 μm). Aerobic; not able to reduce nitrate to nitrite. Colonies on 1/2 strength R2A agar with 75% artificial sea water are circular, convex, and reddish pink in color. The temperature range for growth is 4–37°C, with optimum growth at 30–37°C. No growth occurs at 45°C. The pH range for growth is 6.5–8.0. NaCl is required for growth; tolerates up to 7% (w/v) NaCl. Catalase- and oxidase-positive. Major respiratory menaquinones are MK-8 and MK-9. Positive for alkaline phosphatase, acid phosphatase, naphthol-AS-BI-phosphohydrolase, and α-fucosidase. Negative for esterase (C4), esterase lipase (C8), lipase (C4), leucine arylamidase, valine arylamidase, cystine arylamidase, trypsin, chymotrypsin, α-galactosidase, β-galactosidase, β-glucuronidase, α-glucosidase, β-glucosidase, *N*-acetyl-β-glucosaminidase and α-mannosidase. D-Fructose, L-fucose, D-glucose, D-mannitol, D-melibiose, D-sorbitol, acetic acid, *cis*-aconitic acid, α-ketobutyric acid, α-ketoglutaric acid, inosine, and α-D-glucose 1-phosphate are oxidized, but cyclodextrin, dextrin, glycogen, Tween 40, Tween 80, *N*-acetyl-D-galactosamine, *N*-acetyl-D-glucosamine, adonitol, L-arabinose, D-arabitol, cellobiose, iso-erythritol, D-galactose, gentiobiose, myo-inositol, α-D-lactose, lactulose, maltose, D-mannose, methyl β-D-glucoside, D-psicose, D-raffinose, L-rhamnose, sucrose, trehalose, furanose, xylitol, pyruvic acid methyl ester, succinic acid monomethyl ester, citric acid, formic acid, D-galactonic acid lactone, D-galacturonic acid, D-gluconic acid, D-glucosaminic acid, D-glucuronic acid, α-hydroxybutyric acid, β-hydroxybutyric acid, γ-hydroxybutyric acid, *p*-hydroxyphenylacetic acid, itaconic acid, α-ketovaleric acid, DL-lactic acid, malonic acid, propionic acid, quinic acid, D-saccharic acid, sebacic acid, succinic acid, bromosuccinic acid, succinamic acid, glucuronamide, alaninamide, D-alanine, L-alanine, L-alanyl glycine, L-asparagine, L-aspartic acid, L-glutamic acid, glycyl L-aspartic acid, glycyl L-glutamic acid, L-histidine, hydroxy-L-proline, L-leucine, L-ornithine, L-phenylalanine, L-proline, L-pyroglutamic acid, D-serine, L-serine, L-threonine, DL-carnitine, γ-aminobutyric acid, urocanic acid, uridine, thymidine, phenylethylamine, putrescine, 2-aminoethanol, 2,3-butanediol, glycerol, DL-α-glycerol phosphate, and D-glucose 6-phosphate are not oxidized.

Source: an unidentified marine sponge.

DNA G+C content (mol%): 48.0 (HPLC).

Type strain: YM21–132, KCTC 12906, MBIC08281.

Sequence accession no. (16S rRNA gene): AB297805.

4. **Rubritalea squalenifaciens** Kasai, Katsuta, Sekiguchi, Matsuda, Adachi, Shindo, Yoon, Yokota and Shizuri 2007, 1633[VP]

squa.le.ni.fa'ci.ens. N.L. n. *squalenum* squalene; L. part. adj. *faciens* producing, from L. v. *facio* to produce; N.L. part. adj. *squalenifaciens* squalene-producing.

Description as for the genus. Cells are rod-shaped and red-pink in color. Able to reduce nitrate to nitrite. Grows in sea water medium containing 1–4% NaCl, at pH 7.5–8.5, and temperatures of 15–37°C. Menaquinones MK-8, MK-9 and MK-10 are present. The peptidoglycan in the cell wall contains *meso*-diaminopimelic acid, glutamic acid, and alanine. Able to grow with *N*-acetyl-D-glucosamine, D-galactose, D-glucose, lactose, melibiose, sucrose, or xylose as the sole carbon source under aerobic conditions. Positive for catalase, oxidase, alkaline phosphatase, esterase (C4), esterase lipase (C8), leucine arylamidase, chymotrypsin, acid phosphatase, naphthol-AS-BI-phosphohydrolase, and *N*-acetyl-β-glucosaminidase. Produces acid from *N*-acetyl-D-galactosamine, *N*-acetyl-D-glucosamine, D-glucose, D-glucuronic acid, D-galactose, D-melibiose, and D-trehalose.

Source: the marine sponge *Halichondria okadai.*

DNA G+C content (mol%): 52.4 (HPLC).

Type strain: HOact23, DSM18772, MBIC08254, NBRC 103619.

Sequence accession no. (16S rRNA gene): AB277853.

5. **Rubritalea tangerina** Yoon, Matsuo, Matsuda, Adachi, Kasai and Yokota 2007, 2340[VP]

tan.ge'ri.na. N.L. fem. adj. *tangerina* tangerine, referring to the reddish-orange color of colonies.

Description as for the genus. Coccoid (0.5–0.8 μm in diameter) or rod-shaped (0.5–0.8 × 1.0–1.5 μm). Facultatively anaerobic; able to grow on half strength R2A agar with 75% artificial sea water in an AnaeroPack (Mitsubishi Gas Chemical Co., Inc.). Able to reduce nitrate to nitrite. Colonies grown on half-strength R2A agar with 75% artificial sea water are circular, convex, and reddish orange in color. Temperature range for growth is 15–37°C, with optimum growth at 30–37°C. No growth occurs at 4 or 45°C. pH Range for growth is 6.5–8.5. NaCl is not required for growth, but can tolerate up to 9% (w/v) NaCl. Catalase-negative but oxidase-positive. Major respiratory menaquinones are MK-8 and MK-9.

Positive for alkaline phosphatase, leucine arylamidase, acid phosphatase, naphthol-AS-BI-phosphohydrolase, and *N*-acetyl-β-glucosaminidase. Negative for α-galactosidase, β-galactosidase, α-glucosidase, valine arylamidase, trypsin, esterase (C4), esterase lipase (C8), lipase (C4), cystine arylamidase, chymotrypsin, β-glucuronidase, β-glucosidase, α-mannosidase, and α-fucosidase. L-Fucose, D-mannose, *cis*-aconitic acid, citric acid, succinic acid, α-D-glucose 1-phosphate and D-glucose 6-phosphate are oxidized, but cyclodextrin, dextrin, glycogen, Tween 40, Tween 80, *N*-acetyl-D-galactosamine, *N*-acetyl-D-glucosamine, adonitol, L-arabinose, D-arabitol, cellobiose, iso-erythritol, D-fructose, D-galactose, gentiobiose, D-glucose, myo-inositol, α-D-lactose, lactulose, maltose, D-mannitol, D-melibiose, methyl-β-D-glucoside, D-psicose, D-raffinose, L-rhamnose, D-sorbitol, sucrose, trehalose, furanose, xylitol, pyruvic acid methyl ester, succinic acid monomethyl ester, acetic acid, formic acid, D-galactonic acid lactone, D-galacturonic acid, D-gluconic acid, D-glucosaminic acid, D-glucuronic acid, α-hydroxybutyric acid, β-hydroxybutyric acid, chydroxybutyric acid, *p*-hydroxyphenylacetic acid, itaconic acid, α-ketobutyric acid, α-ketoglutaric acid, α-ketovaleric acid, DL-lactic acid, malonic acid, propionic acid, quinic acid, D-saccharic acid, sebacic acid, bromosuccinic acid, succinamic acid, glucuronamide, alaninamide, D-alanine, L-alanine, L-alanyl glycine, L-asparagine, L-aspartic acid, L-glutamic acid, glycyl L-aspartic acid, glycyl L-glutamic acid, L-histidine, hydroxy-L-proline, L-leucine, L-ornithine,

L-phenylalanine, L-proline, L-pyroglutamic acid, D-serine, L-serine, L-threonine, DL-carnitine, caminobutyric acid, urocanic acid, inosine, uridine, thymidine, phenylethylamine, putrescine, 2-aminoethanol, 2,3-butanediol, glycerol, and DL-α-glycerol phosphate are not oxidized.

Source: the visceral specimen of an unidentified sea hare.
DNA G+C content (mol%): 50.3 (HPLC).
Type strain: YM27-005, KCTC 12907, MBIC08282.
Sequence accession no. (16S rRNA gene): AB297806.

References

Kasai, H., A. Katsuta, H. Sekiguchi, S. Matsuda, K. Adachi, K. Shindo, J. Yoon, A. Yokota and Y. Shizuri. 2007. *Rubritalea squalenifaciens* sp. nov., a squalene-producing marine bacterium belonging to subdivision 1 of the phylum '*Verrucomicrobia*'. Int. J. Syst. Evol. Microbiol. *57*: 1630–1634.

Scheuermayer, M., T.A. Gulder, G. Bringmann and U. Hentschel. 2006. *Rubritalea marina* gen. nov., sp. nov., a marine representative of the phylum '*Verrucomicrobia*', isolated from a sponge (*Porifera*). Int. J. Syst. Evol. Microbiol. *56*: 2119–2124.

Shindo, K., E. Asagi, A. Sano, E. Hotta, N. Minemura, K. Mikami, E. Tamesada, N. Misawa and T. Maoka. 2008. Diapolycopenedioic acid xylosyl esters A, B, and C, novel antioxidative glyco-C_{30}-carotenoic acids produced by a new marine bacterium *Rubritalea squalenifaciens*. J. Antibiot. (Tokyo) *61*: 185–191.

Yoon, J., Y. Matsuo, S. Matsuda, K. Adachi, H. Kasai and A. Yokota. 2007. *Rubritalea spongiae* sp. nov. and *Rubritalea tangerina* sp. nov., two carotenoid- and squalene-producing marine bacteria of the family *Verrucomicrobiaceae* within the phylum '*Verrucomicrobia*', isolated from marine animals. Int. J. Syst. Evol. Microbiol. *57*: 2337–2343.

Yoon, J., Y. Matsuo, S. Matsuda, K. Adachi, H. Kasai and A. Yokota. 2008. *Rubritalea sabuli* sp. nov., a carotenoid- and squalene-producing member of the family *Verrucomicrobiaceae*, isolated from marine sediment. Int. J. Syst. Evol. Microbiol. *58*: 992–997.

Class II. **Opitutae** Choo, Lee, Song and Cho 2007, 535[VP]

JANG-CHEON CHO, PETER H. JANSSEN, KYLE C. COSTA AND BRIAN P. HEDLUND

O.pi.tu'ta.e. N.L. fem. pl. n. *Opitutales* type order of the class; *-ae* ending to denote a class; N.L. fem. pl. n. *Opitutae* class of the order *Opitutales*.

On the basis of 16S rRNA gene sequence analyses, this group comprises one of the primary lineages in the phylum *Verrucomicrobia* and is equivalent to subdivision 4 (Hugenholtz et al., 1998) of the phylum. The majority of its members inhabit soils, marine environments, or invertebrate digestive tracts; however, some have been recovered from freshwater lakes, hypersaline microbial mats, or coastal hot springs. Others are obligate ectosymbionts of marine cilia. The *Opitutae* currently comprise two orders, the type order *Opitutales*, at present represented by the genera *Opitutus* and *Alterococus*, and the order *Puniceicoccales*, represented by the genera *Puniceicoccus*, *Cerasicoccus*, *Coraliomargarita*, and *Pelagicoccus*. All members of this group stain Gram-negative. All tested members are resistant to β-lactam antibiotics and apparently lack a peptidoglycan cell wall.

Type order: **Opitutales** Choo, Lee, Song and Cho 2007, 536[VP].

Further descriptive information

To date, there appear to be two well-supported family level groups within the class *Opitutae*, both of which have been formally described, *Puniceicoccaceae* and *Opitutaceae* (Figure 142).

The family *Puniceicoccaceae* is represented by the genera *Puniceicoccus*, *Cerasicoccus*, *Coraliomargarita*, and *Pelagicoccus* all of which were isolated from marine or estuarine habitats. *Puniceicoccus*, *Cerasicoccus*, and *Coraliomargarita* form a phylogenetic cluster along with 16S rRNA gene sequences recovered primarily from marine environments (e.g., Schafer et al., 2000; Sekar et al., 2006). In addition to those formally characterized genera and sequences from uncultivated organisms, closely related but taxonomically uncharacterized isolates also exist. "*Fucophilus fucoidanolyticus*" was isolated from gut contents of the sea cucumber *Stichopus japonicus* based on its ability to degrade fucoidan obtained from macrooalgae on which the sea cucumber grazes (Sakai et al., 2003a). Subsequently, enzymes responsible for fucoidan hydrolysis were used to digest fucose and the resulting oligosaccharide products were identified (Sakai et al., 2003b). Additionally, 16S rRNA gene sequences representing "*Lentimonas marisflavi*" and isolate YN31-114 exist in GenBank. A separate phylogenetic cluster of *Puniceicoccaceae* is comprised entirely of sequences obtained through cultivation-independent studies. This group includes two subclades, one represented by sequences retrieved from marine environments (e.g., Suzuki et al., 2001; DeLong et al., 2006) and the other retrieved from termite gut contents, freshwater lakes, and other continental habitats (Hongoh et al., 2005; Nakajima et al., 2005; Tajima et al., 2000). The genus *Pelagicoccus* is somewhat phylogenetically distinct from other members of the family and may warrant a separate family designation in the future. *Verrucomicrobia* ectosymbionts of the marine ciliate *Euplotidium* also group with the *Puniceicoccaceae* (Petroni et al., 2000). The *Euplotidium* ectosymbionts, termed epixenosomes, divide by binary fission, forming rows of structures on the surface of the ciliate host. As they mature, they form intracellular bundles of tubules that form a basket and an apical zone of condensed DNA similar to chromatin. When the ciliate is disturbed by predators, the epixenosomes are ejected as harpoon-like structures which offer protection from grazing. Several lines of evidence suggest that epixenosome tubules are homologous to eukaryotic microtubules, including tubule diameter, 22 ± 3 nm, and sensitivity to several microtubule inhibitors. The discovery of tubulins in the genus *Prosthecobacter* (Jenkins et al., 2002) raises the possibility that tubulins are widespread in the *Verrucomicrobia* and are evolutionarily ancient in the lineage (Jenkins et al., 2002; Löwe et al., 2004). However, it has been argued that tubulin genes in some *Verrucomicrobia* reflect recent horizontal transfer based on the patchy distribution of tubulin genes in the phylum, the clear relationship between *Verrucomicrobia* tubulins and those in contemporary eukaryotes, and the presence of canonical division and cell-wall gene content in *Verrucomicrobia* (Jenkins et al.,

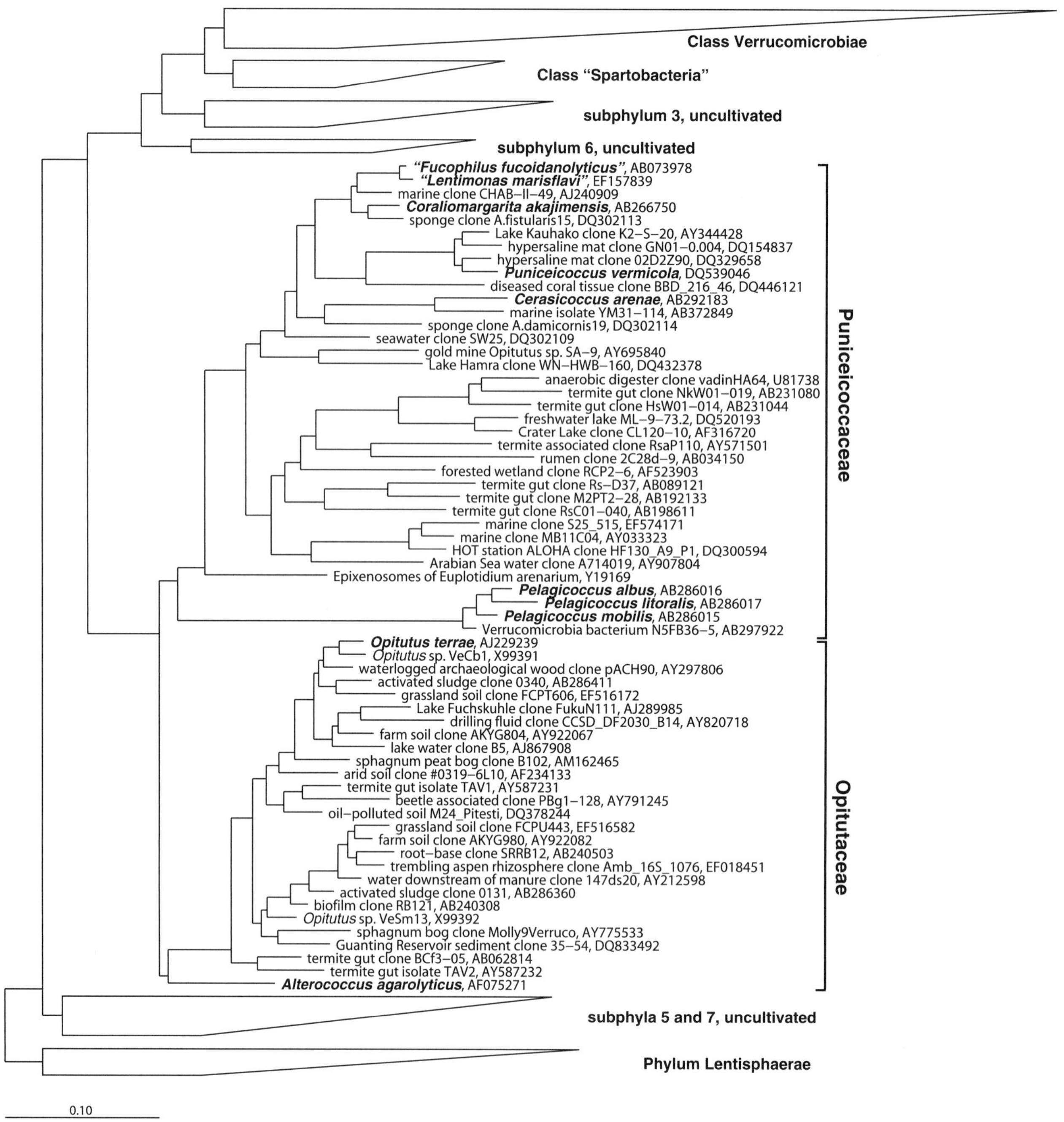

FIGURE 142. Maximum likelihood phylogeny of *Verrucomicrobia* with heuristic correction showing nearly complete 16S rRNA gene sequences present in version 91 of the SILVA database (Pruesse et al., 2007). All class-level groups were included, but all those except the *Opitutae* were wedged. The phylogeny was produced by using ARB (Ludwig et al., 2004) using *Escherichia coli* 16S rRNA gene nucleotides 110–1274. Alignments were unmasked.

2002; Pilhofer et al.,2007, 2008; Schlieper et al., 2005; Sontag et al., 2005). These epixenosomes have been called "*Candidatus* Epixenosoma ejectans", though no formal proposal for this nomenclature exists.

The family *Opitutaceae* is represented by two phylogenetically, physiologically, and ecologically distinct genera, *Opitutus* and *Alterococcus*. However, *Alterococcus* is quite phylogenetically distinct from *Opitutus* and 16S rRNA genes obtained in cultivation-independent studies, so it is possible that a separate family may be proposed in the future to accommodate *Alterococcus* as the phylogenetic depth and breadth of the *Opitutae* is elucidated. The phylogenetic breadth of the *Opitutae* is well represented taxonomically when taxonomically uncharacterized isolates from larval termite guts are considered (Stevenson et al., 2004). These isolates include four strains which, based on their phylogenetic distance from *Opitutus* and *Alterococcus* (<93% 16S rRNA gene identity), likely represent two different genera in the *Opitutaceae*, one represented by strain TAV1 and the other represented by TAV2, 3, and 4. The isolates were obtained by plating homogenized material from freshly collected larval termite guts onto agar plates with dilute yeast extract and peptone as the growth substrate and incubating under hypoxic conditions [2% (v/v) O_2] with 5% (v/v) CO_2. Catalase was added to the medium to protect cells from oxidative damage. Although they were not characterized in detail, Stevenson suggested that all four strains were facultative or aerotolerant anaerobes and all were small cocci, similar to *Opitutus*. Environmental 16S rRNA gene sequences related to *Opitutus* have been described from a variety of soils, freshwater habitats, activated sludge, and termite gut contents (e.g., Glöckner et al., 2000; Holmes et al., 2000; Shinzato et al., 2005; Morales et al., 2006).

References

Choo, Y.J., K. Lee, J. Song and J.C. Cho. 2007. *Puniceicoccus vermicola* gen. nov., sp. nov., a novel marine bacterium, and description of *Puniceicoccaceae* fam. nov., *Puniceicoccales* ord. nov., *Opitutaceae* fam. nov., *Opitutales* ord. nov. and *Opitutae* classis nov. in the phylum '*Verrucomicrobia*'. Int. J. Syst. Evol. Microbiol. *57*: 532–537.

DeLong, E.F., C.M. Preston, T. Mincer, V. Rich, S.J. Hallam, N.U. Frigaard, A. Martinez, M.B. Sullivan, R. Edwards, B.R. Brito, S.W. Chisholm and D.M. Karl. 2006. Community genomics among stratified microbial assemblages in the ocean's interior. Science *311*: 496–503.

Glöckner, F.O., E. Zaichikov, N. Belkova, L. Denissova, J. Pernthaler, A. Pernthaler and R. Amann. 2000. Comparative 16S rRNA analysis of lake bacterioplankton reveals globally distributed phylogenetic clusters including an abundant group of actinobacteria. Appl. Environ. Microbiol. *66*: 5053–5065.

Holmes, A.J., J. Bowyer, M.P. Holley, M. O'Donoghue, M. Montgomery and M.R. Gillings. 2000. Diverse, yet-to-be-cultured members of the *Rubrobacter* subdivision of the *Actinobacteria* are widespread in Australian arid soils. FEMS Microbiol. Ecol. *33*: 111–120.

Hongoh, Y., P. Deevong, T. Inoue, S. Moriya, S. Trakulnaleamsai, M. Ohkuma, C. Vongkaluang, N. Noparatnaraporn and T. Kudo. 2005. Intra- and interspecific comparisons of bacterial diversity and community structure support coevolution of gut microbiota and termite host. Appl. Environ. Microbiol. *71*: 6590–6599.

Hugenholtz, P., B.M. Goebel and N.R. Pace. 1998. Impact of culture-independent studies on the emerging phylogenetic view of bacterial diversity. J. Bacteriol. *180*: 4765–4774.

Jenkins, C., R. Samudrala, I. Anderson, B.P. Hedlund, G. Petroni, N. Michailova, N. Pinel, R. Overbeek, G. Rosati and J.T. Staley. 2002. Genes for the cytoskeletal protein tubulin in the bacterial genus *Prosthecobacter*. Proc. Natl. Acad. Sci. U. S. A. *99*: 17049–17054.

Löwe, J., F. van den Ent and L.A. Amos. 2004. Molecules of the bacterial cytoskeleton. Annu. Rev. Biophys. Biomol. Struct. *33*: 177–198.

Ludwig, W., O. Strunk, R. Westram, L. Richter, H. Meier, Yadhukumar, A. Buchner, T. Lai, S. Steppi, G. Jobb, W. Forster, I. Brettske, S. Gerber, A.W. Ginhart, O. Gross, S. Grumann, S. Hermann, R. Jost, A. Konig, T. Liss, R. Lussmann, M. May, B. Nonhoff, B. Reichel, R. Strehlow, A. Stamatakis, N. Stuckmann, A. Vilbig, M. Lenke, T. Ludwig, A. Bode and K.H. Schleifer. 2004. ARB: a software environment for sequence data. Nucleic Acids Res. *32*: 1363–1371.

Morales, S.E., P.J. Mouser, N. Ward, S.P. Hudman, N.J. Gotelli, D.S. Ross and T.A. Lewis. 2006. Comparison of bacterial communities in New England *Sphagnum* bogs using terminal restriction fragment length polymorphism (T-RFLP). Microb. Ecol. *52*: 34–44.

Nakajima, H., Y. Hongoh, R. Usami, T. Kudo and M. Ohkuma. 2005. Spatial distribution of bacterial phylotypes in the gut of the termite *Reticulitermes speratus* and the bacterial community colonizing the gut epithelium. FEMS Microbiol. Ecol. *54*: 247–255.

Petroni, G., S. Spring, K.H. Schleifer, F. Verni and G. Rosati. 2000. Defensive extrusive ectosymbionts of *Euplotidium* (Ciliophora) that contain microtubule-like structures are bacteria related to *Verrucomicrobia*. Proc. Natl. Acad. Sci. U. S. A. *97*: 1813–1817.

Pilhofer, M., G. Rosati, W. Ludwig, K.H. Schleifer and G. Petroni. 2007. Coexistence of tubulins and *ftsZ* in different *Prosthecobacter* species. Mol. Biol. Evol. *24*: 1439–1442.

Pilhofer, M., K. Rappl, C. Eckl, A.P. Bauer, W. Ludwig, K.H. Schleifer and G. Petroni. 2008. Characterization and evolution of cell division and cell wall synthesis genes in the bacterial phyla *Verrucomicrobia*, *Lentisphaerae*, *Chlamydiae*, and *Planctomycetes* and phylogenetic comparison with rRNA genes. J. Bacteriol. *190*: 3192–3202.

Pruesse, E., C. Quast, K. Knittel, B. Fuchs, W. Ludwig, J. Peplies and F.O. Glöckner. 2007. SILVA: a comprehensive online resource for quality checked and aligned rRNA sequence data compatible with ARB. Nucleic Acids Res. *35*: 7188–7196.

Sakai, T., K. Ishizuka and I. Kato. 2003a. Isolation and characterization of a fucoidan-degrading marine bacterium. Mar. Biotechnol. (N.Y.) *5*: 409–416.

Sakai, T., K. Ishizuka, K. Shimanaka, K. Ikai and I. Kato. 2003b. Structures of oligosaccharides derived from *Cladosiphon okamuranus* fucoidan by digestion with marine bacterial enzymes. Mar. Biotechnol. (N.Y.) *5*: 536–544.

Schafer, H., P. Servais and G. Muyzer. 2000. Successional changes in the genetic diversity of a marine bacterial assemblage during confinement. Arch. Microbiol. *173*: 138–145.

Schlieper, D., M.A. Oliva, J.M. Andreu and J. Lowe. 2005. Structure of bacterial tubulin BtubA/B: evidence for horizontal gene transfer. Proc. Natl. Acad. Sci. U. S. A. *102*: 9170–9175.

Sekar, R., D.K. Mills, E.R. Remily, J.D. Voss and L.L. Richardson. 2006. Microbial communities in the surface mucopolysaccharide layer and the black band microbial mat of black band-diseased *Siderastrea siderea*. Appl. Environ. Microbiol. *72*: 5963–5973.

Shinzato, N., M. Muramatsu, T. Matsui and Y. Watanabe. 2005. Molecular phylogenetic diversity of the bacterial community in the gut of the termite *Coptotermes formosanus*. Biosci. Biotechnol. Biochem. *69*: 1145–1155.

Sontag, C.A., J.T. Staley and H.P. Erickson. 2005. In vitro assembly and GTP hydrolysis by bacterial tubulins BtubA and BtubB. J. Cell. Biol. *169*: 233–238.

Stevenson, B.S., S.A. Eichorst, J.T. Wertz, T.M. Schmidt and J.A. Breznak. 2004. New strategies for cultivation and detection of previously uncultured microbes. Appl. Environ. Microbiol. *70*: 4748–4755.

Suzuki, M.T., O. Beja, L.T. Taylor and E.F. Delong. 2001. Phylogenetic analysis of ribosomal RNA operons from uncultivated coastal marine bacterioplankton. Environ. Microbiol. *3*: 323–331.

Tajima, K., S. Arai, K. Ogata, T. Nagamine, H. Matsui, M. Namakura, R.I. Aminov and Y. Benno. 2000. Rumen bacterial community transition during adaptation to high-grain diet. Anaerobe *6*: 273–284.

Order I. **Opitutales** Choo, Lee, Song and Cho 2007, 536[VP]

Jang-Cheon Cho, Peter H. Janssen, Wung Yang Shieh and Brian P. Hedlund

O.pi.tu.ta′les. N.L. masc. n. *Opitutus* type genus of the order; *-ales* ending to denote an order; N.L. fem. pl. n. *Opitutales* the order of the genus *Opitutus*.

Encompasses bacteria isolated from soil environments and a coastal marine hot spring within the class *Opitutae.* The order contains the family *Opitutaceae.* Delineation of the order is determined primarily by phylogenetic information from 16S rRNA gene sequences. All known members of the order are cocci that are Gram-stain-negative. All are chemoorganotrophic facultative or obligate anaerobes capable of fermenting carbohydrates.

Type genus: **Opitutus** Chin, Liesack and Janssen 2001, 1967[VP].

References

Chin, K.J., W. Liesack and P.H. Janssen. 2001. *Opitutus terrae* gen. nov., sp. nov., to accommodate novel strains of the division '*Verrucomicrobia*' isolated from rice paddy soil. Int. J. Syst. Evol. Microbiol. *51*: 1965–1968.

Choo, Y.J., K. Lee, J. Song and J.C. Cho. 2007. *Puniceicoccus vermicola* gen. nov., sp. nov., a novel marine bacterium, and description of *Puniceicoccaceae* fam. nov., *Puniceicoccales* ord. nov., *Opitutaceae* fam. nov., *Opitutales* ord. nov. and *Opitutae* classis nov. in the phylum '*Verrucomicrobia*'. Int. J. Syst. Evol. Microbiol. *57*: 532–537.

Family I. **Opitutaceae** Choo, Lee, Song and Cho 2007, 536[VP]

Jang-Cheon Cho, Peter H. Janssen, Wung Yang Shieh and Brian P. Hedlund

O.pi.tu.ta.ce′a.e. N.L. masc. n. *Opitutus* type genus of the family; *-aceae* ending to denote a family; N.L. fem. pl. n. *Opitutaceae* the family of the genus *Opitutus*.

Cocci or coccobacilli that are 0.4–0.9 µm in diameter. Nonsporeforming. **Motile** via a single flagellum. **Gram-stain-negative.** Colonies are nonpigmented. **Facultatively or obligately anaerobic.** Mesophilic or moderately thermophilic. **Chemoorganotrophic. Ferment mono- and disaccharides to organic acids** with or without production of H_2 and CO_2. Peptides and amino acids are not fermented.

DNA G+C content (mol%): 65–67.

Type genus: **Opitutus** Chin, Liesack and Janssen 2001, 1967[VP].

Further descriptive information

Encompasses all bacteria within the order *Opitutales.* Currently, the family comprises the genera *Opitutus* and *Alterococcus* together with strains isolated from termite guts that have not been taxonomically characterized. Related 16S rRNA genes representing uncultured bacteria have been retrieved mainly from soil environments, arthropod guts (termites and beetles), and freshwater. Delineation of the family is determined primarily by phylogenetic information from 16S rRNA gene sequences.

Key to the genera of the family *Opitutaceae*

1. The DNA G+C content is 65 mol%. Cocci less than 0.7 µm in diameter. Growth occurs at 10–37°C. Strictly anaerobic. Ferments mono-, di-, and polysaccharides. Reduces nitrate to nitrite. Catalase-negative. Isolated from flooded soils.
   → Genus I. *Opitutus*
2. The DNA G+C content is 65.5–67.0 mol%. Cocci more than 0.7 µm in diameter. Growth occurs at 38–58°C. Facultatively anaerobic. Ferments mono- and disaccharides. Grows aerobically on peptides. Catalase-positive. Isolated from coastal hot springs.
   → Genus II. *Alterococcus*

Genus I. **Opitutus** Chin, Liesack and Janssen 2001, 1967[VP]

Peter H. Janssen

O.pi.tu′tus. L. fem. n. *Ops, Opis* a Roman Earth and harvest goddess; L. part. adj. *tutus* protected; N.L. masc. n. *Opitutus* the one protected by Ops.

Cocci or coccobacilli. Stain **Gram-negative**. **Motile** with flagellum. No endospores. **Anaerobe**; media containing a suitable reductant shorten the lag phase. Chemo-organotrophic metabolism. **Monosaccharides, disaccharides, and polysaccharides are fermented**, but alcohols, amino acids, and organic acids are not. Acetate, propionate, CO_2, and H_2 are the fermentation end products; the ratios of these are dependent on the partial pressure of H_2. **Nitrate is reduced to nitrite**. Sulfate, sulfur, thiosulfate, and fumarate are not used as terminal electron acceptors.

DNA G+C content (mol%): 65.

Type species: **Opitutus terrae** Chin, Liesack and Janssen 2001, 1968[VP].

Further descriptive information

Cell pellets are pink in color, and cells contain cytochrome *b.* Glucose is fermented by a modified Embden-Meyerhof–Parnas pathway, with a pyrophosphate-dependent 6-phosphofructokinase (Janssen, 1998). The ratio of the major organic end products, acetate and propionate, is dependent on the partial pressure of hydrogen, with a shift to increased propionate production occurring at hydrogen partial pressures of greater than 20 Pa (Chin and Janssen, 2002).

The type strain of *Opitutus terrae* appears to be an obligate anaerobe, but related strains appear to be at least aerotolerant anaerobes (Janssen et al., 1997; Stevenson et al., 2004). Similarly, while the type strain reduces nitrate, other isolates do not appear able to do so (Chin et al., 1999; Janssen et al., 1997).

Enrichment and isolation procedures

Janssen et al. (1997) enriched three strains closely related to *Opitutus terrae* from small inocula prepared by diluting slurries of anaerobic soils in anaerobic media containing mono- and disaccharides. Later, Chin et al. (1999) used similar methods to isolate the type strain of *Opitutus terrae* and two closely related strains, also from anaerobic soil. The growth substrates in those experiments were polysaccharides. Stevenson et al. (2004) isolated four strains closely related to members of the genus *Opitutus* from the digestive tract of worker larvae of the termite *Reticulitermes flavipes* by inoculating gut homogenates onto agar plates with yeast extract as the growth substrate. These plates were incubated under hypoxic conditions [2% (v/v) O_2] with 5% (v/v) CO_2. These isolates are therefore at least aerotolerant, and Stevenson et al. (2004) suggest that some strains may be facultative anaerobes. At this time, there is no known selective medium for *Opitutus* strains. The strategies used by Stevenson et al. (2004) and Sangwan et al. (2005), monitoring cultures using a PCR-based assay or with group-specific oligonucleotide probes targeting 16S rRNA to detect the presence of members of the phylum *Verrucomicrobia*, are the only approaches published to obtain isolates other than by chance isolation.

Maintenance procedures

Cultures of *Opitutus terrae* and its close relatives can be stored for at least 1 year at 4°C in anaerobic media in sealed bottles or tubes either completely filled or under a nitrogen plus carbon dioxide headspace. Maintenance is by subculture at 3- to 6-month intervals.

Taxonomic comments

The genus is represented by only one isolate assigned to a named species, *Opitutus terrae*. This means that the assignment of characteristics to the generic and specific descriptions is somewhat arbitrary. Other related isolates (Chin et al., 1999; Janssen et al., 1997) have not been assigned to named species, but appear to be related to *Opitutus terrae*, and are members of that species or of other species of the genus *Opitutus*. These unassigned isolates appear to be phenotypically similar, and their 16S rRNA genes share significant 16S rRNA gene sequence identity (95.3–100%) with *Opitutus terrae*. In the absence of more detailed characterization of other isolates, and the delineation of further species, the separation of genus- and species-specific characteristics is somewhat arbitrary. The four isolates obtained from termite gut (Stevenson et al., 2004) are less closely related to *Opitutus terrae* (<93% 16S rRNA gene sequence identity), and their taxonomic status is less certain. All of these isolates are taxonomically distinct from *Alterococcus agarlyticus* and "*Fucophilus fucoidanolyticus*". with which they share <90% 16S rRNA gene sequence identity.

List of species of the genus *Opitutus*

1. **Opitutus terrae** Chin, Liesack and Janssen 2001, 1968[VP]
ter'ra.e. L. gen. n. *terrae* of the earth.

Cocci, 0.4–0.6 µm in diameter. Gram-stain-negative. Motile by means of a flagellum. No spores are formed. The colonies in agar deeps are unpigmented and granular in appearance.

Grows with glucose, fructose, galactose, mannose, galacturonic acid, mannitol, arabinose, cellobiose, maltose, sucrose, lactose, melibiose, xylan, pectin, and starch. Xylose, ribose, sorbose, methyl-α-glucopyranoside, cellulose, chitin, arabinogalactan, pyruvate, lactate, fumarate, malate, tartrate, citrate, crotonate, glycerol (± acetate), aspartate, alanine, serine, leucine, isoleucine, glutamate, proline, lysine, and H_2 (± acetate) do not support growth of the type strain. Propionate and acetate are the major end products of fermentation. Succinate, lactate, formate, ethanol, and H_2 are also produced. Methanol is also formed from pectin. Nitrate is reduced to nitrite. Sulfur, sulfate, sulfite, thiosulfate, and fumarate are not used as electron acceptors. Esculin is hydrolyzed, but gelatin and urea are not.

Anaerobe. The type strain grows at pH 5.5–9.0, with maximum growth rates at pH 7.5–8.0. Growth of the type strain is possible at temperatures of 10–37°C, but not at 4°C or at 40°C.

The type strain was isolated from artificial rice paddy systems constructed using soil from rice paddies in Vercelli, Italy.

DNA G+C content (mol%): 65 (draft genome sequence).

Type strain: Pb90-1, DSM 11246.

Sequence accession no. (16S rRNA gene): AJ229235.

Genus II. **Alterococcus** Shieh and Jean 1998, 644[VP]

Brian P. Hedlund and Wung Yang Shieh

Al.te.ro.coc'cus. L. adj. *alter*, *-tera*, *-terum* another, different; N.L. masc. n. *coccus* (from Gr. masc. n. *kokkos*) a grain or berry; N.L. masc. n. *Alterococcus* another coccus.

Spherical cells with **Gram-negative cell structure**. Gram-stain-negative. **Motile** by means of a single flagellum. **Facultative anaerobes** capable of aerobic growth on peptone and yeast extract. **Ferment monosaccharides or disaccharides for carbon and energy.** Ferment glucose primarily to butyrate, together with propionate or formate. Lactate and acetate also produced. No gas produced. Do not ferment peptides. **Agarolytic.** Oxidase- and catalase-positive. **Moderately thermophilic**, growing at 40–56°C but not 30 or 60°C. **Slightly halophilic**, growing in media containing 1–3% NaCl but not in those containing 0 or 5% NaCl. **Require sodium.** $C_{15:0}$ anteiso is the most abundant cellular fatty acid. Phylogenetically, the genus belongs to the order *Opitutales* in the phylum *Verrucomicrobia*. Isolated from two hot springs in the intertidal zone of Lutao, Taiwan, based on their ability to hydrolyze agar in an agar-solidified marine medium.

DNA G+C content (mol%): 65.5–67.0.

Type species: **Alterococcus agarolyticus** Shieh and Jean 1998, 644[VP].

Further descriptive information

The genus *Alterococcus* currently contains only one species and five strains, with the type strain, *Alterococcus agarolyticus* BCRC 17102^T (original designation, CCRC 19135^T), and no closely related isolates or 16S rRNA gene sequences from cultivation-independent studies. The only other named genus in the Opitutales is *Opitutus*. *Opitutus terrae* was one of several strains of ultramicrobacteria belonging to the *Verrucomicrobia* that were isolated from rice paddy soils (Janssen et al., 1997). In addition, capnophilic, facultatively anaerobic isolates from termite guts (Stevenson et al., 2004) are as yet undescribed taxonomically. Related 16S rRNA genes representing a number of uncultivated organisms from soils, freshwater, activated sludge, and arthropod guts share less than 90% 16S rRNA gene sequence percentage identity.

Alterococcus was isolated from two different coastal hot springs on the island of Lutao, off Taiwan, which were fed seawater heated to 39–51°C. The temperature (48°C) and salinity (2.0–2.5% Na^+ w/v) optima of *Alterococcus* suggest it is indigenous to that habitat. As such, *Alterococcus* is the only known thermophilic member of the *Verrucomicrobia*, though 16S rRNA gene sequences belonging to the *Verrucomicrobia* have been recovered from ~70°C sediment taken from Obsidian Pool, Yellowstone National Park (Hugenholtz et al., 1998).

Cells stain Gram-negative and have a Gram-negative cell structure (Figures 143 and 144). Cells are spherical, 0.8–0.9 µm in diameter and divide by binary fission. Cells are motile via a single flagellum. The preferred medium for aerobic growth is PY broth, which can be solidified with 1.5% (w/v) Bacto agar (Difco) or partially solidified with 0.5% (w/v) Bacto agar. PY broth contains (per l) 4.0 g of Bacto Peptone (Difco), 2.0 g of Bacto yeast extract (Difco), 20 g of NaCl, 0.5 g of $MgSO4{\cdot}7H_2O$, 0.01 g of $CaCl_2$, and 4.5 g of 3-(*N*-morpholino)-2-hydroxypropanesulfonic acid (MOPSO; Sigma), pH 7.0. The medium can be supplemented with carbohydrates (0.5% w/v) for fermentative growth. The genus *Alterococcus* is capable of fermenting the following substrates at 0.5%: cellobiose, galactose, glucose, lactose, sucrose, trehalose, and xylose. Cells grown in PYG (PY containing 0.5% glucose) or semisolid PY agar produce butyrate together with propionate or formate as the primary fermentation products, with lesser amounts of lactate, and acetate. No gas is formed. When grown for 2 days at 50°C on PY plates containing 1.5% agar (w/v), colonies are white, circular, opaque, and surrounded by shallow depressions in the agar, indicative of agarolytic activity. The temperature range, pH range, and salinity range for growth are 38–58°C (optimum, 48°C), pH 7.0–8.5 (optimum, pH 8.0), and 1.0–3.0% (w/v) of NaCl (optimum, 2.0–2.5%), respectively. Agarase, catalase, and oxidase are positive. Arginine dihydrolase, lysine and ornithine decarboxylases, DNase, lipase (Tween 80), gelatinase, and casein hydrolysis are negative. Amylase is weak.

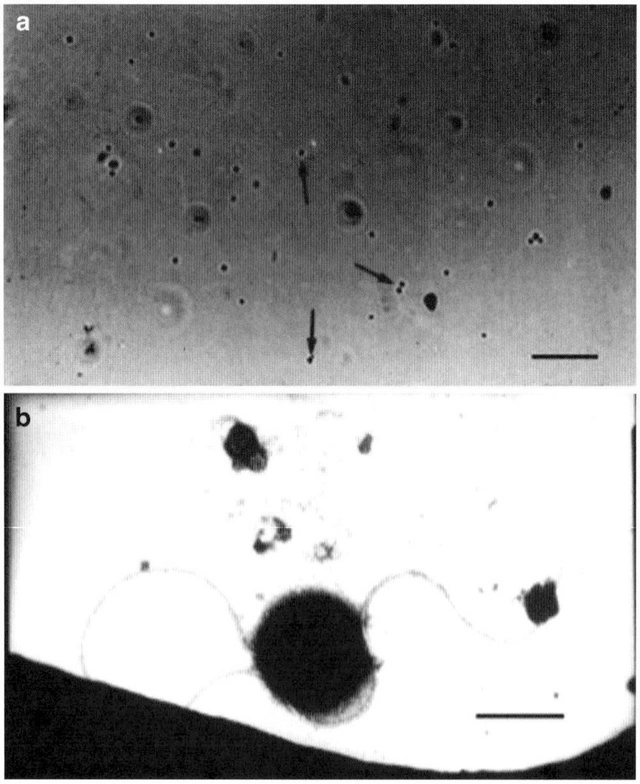

FIGURE 143. (a) Phase-contrast micrograph showing coccoid cells of *Alterococcus agarolyticus* strain $ADT3^T$. (b) One coccoid cell with a single flagellum (partially covered by the grid) as viewed by transmission electron microscopy. (Reprinted with permission from Shieh and Jean, 1998. Can. J. Microbiol. *44:* 637–648.)

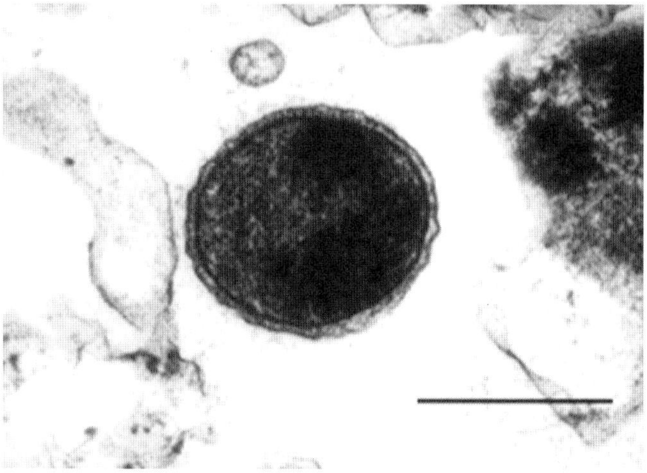

FIGURE 144. Ultrathin section electron micrograph of *Alterococcus agarolyticus* strain $ADT3^T$, showing a spherical cell with Gram-negative cell structure. (Reprinted with permission from Shieh and Jean, 1998. Can. J. Microbiol. *44:* 637–648.)

When grown to early stationary phase in PY broth, the most abundant cellular fatty acid was $C_{15:0}$ anteiso with lesser amounts of $C_{16:0}$, $C_{16:0}$ iso, $C_{17:0}$, $C_{14:0}$, $C_{16:0}$ 2-OH, and $C_{18:0}$.

Enrichment and isolation procedures

Samples from coastal 39–51°C hot springs are serially diluted and spread directly onto PY agar. Agarolytic colonies appearing after 3–7 d incubation in the dark at 50°C are streaked on the same medium for purity.

Maintenance procedures

Cultures can be maintained for at least 3 months at 45°C. Long-term storage can be achieved by suspending the cells in NaCl/

glycerol solution (2 g of NaCl and 15 g of glycerol in 100 ml of deionized water) and freezing the cell suspension in liquid nitrogen or at −80°C after pre-cooling to −4°C for 1 h.

Differentiation of the genus *Alterococcus* from other related genera

Phylogenetically, the most closely related genus *Opitutus* is within the order *Opitutales*. *Alterococcus* is differentiated from the genus *Opitutus* by 16S rDNA sequence dissimilarity (<90% identity), habitat (coastal hot springs vs soils), sodium requirement, maximum growth temperature (58°C vs 37°C), cell size (0.8–0.9 µm vs 0.4–0.6 µm diameter), relationship to oxygen (facultative anaerobe vs anaerobe), carbohydrate fermentation products (mixed acid vs organic acids with H_2 and CO_2) and DNA G+C composition (65.5–67.0 mol% vs 74 mol%). Since few *Verrucomicrobia* isolates exist, suspected relatives of *Alterococcus* should not be identified without 16S rRNA gene sequencing.

Taxonomic comments

This genus was originally placed in the *Enterobacteriaceae* as a sister taxon to *Escherichia* and *Shigella* (Shieh and Jean, 1998); however, subsequent phylogenetic analyses of the almost complete 16S rRNA gene from the type strain indicated that the genus is a deeply branching member of the order *Opitutales* in the *Verrucomicrobia*.

List of species of the genus *Alterococcus*

1. **Alterococcus agarolyticus** Shieh and Jean 1998, 644[VP]

a.ga.ro.ly′ti.cus. N.L. n. *agarum* (from Malayan n. *agar*) agar, a complex gelling polysaccharide from marine red algae; N.L. masc. adj. *lyticus* (from Gr. masc. adj. *lutikos*), able to loosen, able to dissolve; N.L. masc. adj. *agarolyticus* agar-dissolving.

Young colonies are white, circular, and opaque and form shallow depressions on PY agar media after 1–2 d of incubation and deep depressions after 1 week. The characteristics are as described for the genus. Cells spherical, normally 0.8–0.9 µm in diameter and occurring mostly singly or in pairs. Despite the presence of a buffer, cultures in PYG broth reduce the pH by 1.5 and 2.1 units under aerobic and anaerobic conditions, respectively. Cells lyse rapidly after growth. Liquefy agar and ferment glucose, cellobiose, galactose, lactose, sucrose, trehalose, and xylose. Cannot ferment amino acids. Grow only aerobically on peptone and yeast extract. Growth occurs between 38 and 58°C, optimum 48°C; 1.0–3.5% NaCl, optimum 2.0–2.5%; pH 7.0–8.5. Cells are susceptible to ampicillin, chloramphenicol, erythromycin, penicillin G, and tetracycline.

DNA G+C content (mol%): 65.5–67.0 (HPLC).

Type strain: ADT3, BCRC 17102 (original designation, CCRC 19135), CIP 106113.

Sequence accession no. (16S rRNA gene): AF075271.

References

Chin, K.J. and P.H. Janssen. 2002. Propionate formation by *Opitutus terrae* in pure culture and in mixed culture with a hydrogenotrophic methanogen and implications for carbon fluxes in anoxic rice paddy soil. Appl. Environ. Microbiol. *68*: 2089–2092.

Chin, K.J., D. Hahn, U. Hengstmann, W. Liesack and P.H. Janssen. 1999. Characterization and identification of numerically abundant culturable bacteria from the anoxic bulk soil of rice paddy microcosms. Appl. Environ. Microbiol. *65*: 5042–5049.

Chin, K.J., W. Liesack and P.H. Janssen. 2001. *Opitutus terrae* gen. nov., sp. nov., to accommodate novel strains of the division '*Verrucomicrobia*' isolated from rice paddy soil. Int. J. Syst. Evol. Microbiol. *51*: 1965–1968.

Hugenholtz, P., C. Pitulle, K.L. Hershberger and N.R. Pace. 1998. Novel division level bacterial diversity in a Yellowstone hot spring. J. Bacteriol. *180*: 366–376.

Janssen, P.H. 1998. Pathway of glucose catabolism by strain VeGlc2, an anaerobe belonging to the *Verrucomicrobiales* lineage of bacterial descent. Appl. Environ. Microbiol. *64*: 4830–4833.

Janssen, P.H., A. Schuhmann, E. Mörschel and F.A. Rainey. 1997. Novel anaerobic ultramicrobacteria belonging to the *Verrucomicrobiales* lineage of bacterial descent isolated by dilution culture from anoxic rice paddy soil. Appl. Environ. Microbiol. *63*: 1382–1388.

Sangwan, P., S. Kovac, K.E. Davis, M. Sait and P.H. Janssen. 2005. Detection and cultivation of soil *Verrucomicrobia*. Appl. Environ. Microbiol. *71*: 8402–8410.

Shieh, W.Y. and W.D. Jean. 1998. *Alterococcus agarolyticus*, gen.nov., sp.nov., a halophilic thermophilic bacterium capable of agar degradation. Can. J. Microbiol. *44*: 637–645.

Stevenson, B.S., S.A. Eichorst, J.T. Wertz, T.M. Schmidt and J.A. Breznak. 2004. New strategies for cultivation and detection of previously uncultured microbes. Appl. Environ. Microbiol. *70*: 4748–4475.

Order II. **Puniceicoccales** Choo, Lee, Song and Cho 2007, 536[VP]

Jang-Cheon Cho, Jaewoo Yoon and Brian P. Hedlund

Pu.ni.ce.i.coc.ca′les. N.L. masc. n. *Puniceicoccus* type genus of the order; *-ales* ending to denote an order; N.L. fem. pl. n. *Puniceicoccales* the order of the genus *Puniceicoccus*.

Encompasses bacteria isolated from marine environments within the class *Opitutae*. The order contains the family *Puniceicoccaceae*. Delineation of the order is determined primarily by phylogenetic information from 16S rRNA gene sequences. All known members of the order are cocci that are Gram-stain-negative and lack muramic acid and diaminopimelic acid. Cells are resistant to β-lactam antibiotics. All are chemoorganotrophic obligate aerobes or facultative anaerobes. Menaquinone-7 is the major respiratory quinone.

Type genus: **Puniceicoccus** Choo, Lee, Song and Cho 2007, 536[VP].

Reference

Choo, Y.J., K. Lee, J. Song and J.C. Cho. 2007. *Puniceicoccus vermicola* gen. nov., sp. nov., a novel marine bacterium, and description of *Puniceicoccaceae* fam. nov., *Puniceicoccales* ord. nov., *Opitutaceae* fam. nov., *Opitutales* ord. nov. and *Opitutae* classis nov. in the phylum '*Verrucomicrobia*'. Int. J. Syst. Evol. Microbiol. *57*: 532–537.

Family I. **Puniceicoccaceae** Choo, Lee, Song and Cho 2007, 536^{VP}

Jang-Cheon Cho, Jaewoo Yoon and Brian P. Hedlund

Pu.ni.ce.i.coc.ca.ce′a.e. N.L. masc. n. *Puniceicoccus* type genus of the family; *-aceae* ending to denote a family; N.L. fem. pl. n. *Puniceicoccaceae* the family of *Puniceicoccus*.

Cocci. Gram-stain-negative. Aerobic or facultatively anaerobic via fermentation. Mesophilic. **Chemoorganotrophic. Saccharolytic.** Use predominantly mono-, di-, and polysaccharides and their derivatives. **Neither muramic acid nor diaminopimelic acid identified** in cell-wall extracts. **Resistant to β-lactam antibiotics. The major respiratory quinone is menaquinone MK-7.**

DNA G+C content (mol%): 52.1–57.4.

Type genus: **Puniceicoccus** Choo, Lee, Song and Cho 2007, 536^{VP}.

Further descriptive information

Encompasses Gram-stain-negative bacteria retrieved mainly from marine and aquatic environments, within the order *Puniceicoccales.* Currently, the family comprises the genera *Puniceicoccus, Pelagicoccus, Coraliomargarita, Cerasicoccus,* and "*Fucophilus*", together with several uncultured marine and lacustrine bacteria. Delineation of the family is determined primarily by phylogenetic information from 16S rRNA gene sequences.

Key to the genera of the family *Puniceicoccaceae*

1. The DNA G+C content is 52.1 mol%. Cells are nonmotile and colonies are pale red. Capable of anaerobic growth. Catalase- and oxidase-negative. Produces acid from cellobiose, glucose, and lactose. Requires Na^+ ions. Isolated from marine environments.
   →Genus I. *Puniceicoccus*
2. The DNA G+C content is 54 mol%. Cells are nonmotile and colonies are pale pink. Obligately aerobic. Catalase- and oxidase-positive. Produces acid from glucose, lactose, and mannose. Hydrolyzes starch. Does not require Na^+ ions. Isolated from marine environments.
   →Genus II. *Cerasicoccus*
3. The DNA G+C content is 53.9 mol%. Cells are nonmotile and colonies are white. Obligately aerobic. Catalase-negative and oxidase-positive. Produces acid from mannitol and mannose. Requires Na^+ ions. Isolated from marine environments.
   →Genus III. *Coraliomargarita*
4. The DNA G+C content is 51.6–57.2 mol%. Cells of some species are motile and colonies are white. Obligately aerobic or facultatively anaerobic. Oxidase-positive. Some species produce acid from cellobiose, lactose, and melibiose. Requires Na^+ ions. Isolated from marine environments.
   →Genus IV. *Pelagicoccus*

Genus I. **Puniceicoccus** Choo, Lee, Song and Cho 2007, 536^{VP}

Jang-Cheon Cho

Pu.ni.ce.i.coc′cus. L. adj. *puniceus* pinkish red; N.L. masc. n. *coccus* from Gr. masc. n. *kokkos* a berry; N.L. masc. n. *Puniceicoccus* a pinkish-red-colored coccus.

Facultatively anaerobic **nonmotile cocci** (0.6–1.0 μm in diameter). **Gram-stain-negative.** Chemoheterotrophic. **Produces carotenoid pigments**. Requires NaCl for growth. Oxidase- and catalase-negative. Utilizes a variety of carbon compounds as sole carbon sources. The predominant fatty acids are $C_{15:0}$ anteiso and $C_{18:0}$. The major quinone is **MK-7**.

DNA G+C content (mol%): 52.1.

Type species: **Puniceicoccus vermicola** Choo, Lee, Song and Cho 2007, 536^{VP}.

Further descriptive information

Colonies on marine agar 2216 grown at 30°C for 5 d are 0.3–0.5 mm in diameter, uniformly circular, smooth, convex, opaque, and pale-reddish colored. Colony size increases up to 3 mm in diameter after 3 weeks of incubation. While the organism is facultatively anaerobic, anaerobic cultures grow much slower than aerobic cultures.

The temperature, pH, and NaCl concentration ranges for growth are 8–37°C (optimum, 25–30°C), pH 5.0–12.0 (optimum, pH 9.0), and 1.0–7.5% (w/v) NaCl (optimum, 3.0–3.5%). The cellular fatty acid constituents are $C_{15:0}$ anteiso (30.9%), $C_{18:0}$ (24.7%), $C_{16:0}$ (7.9%), $C_{17:0}$ (7.0%), $C_{14:0}$ iso (5.3%), $C_{14:0}$ (4.9%), $C_{17:0}$ anteiso (3.6%), $C_{18:0}$ 3-OH (2.5%), and $C_{12:0}$ 3-OH (2.1%). Tests for nitrate reduction, indole production, glucose fermentation, arginine dihydrolase, urease, esculin hydrolysis, gelatinase, and PNPG (β-galactosidase) are negative.

Enzyme activities (API ZYM test) and carbon source oxidation results (Biolog test) are shown in Choo et al. (2007). The type strain of the genus is resistant to ampicillin, chloramphenicol, gentamicin, kanamycin, penicillin G, rifampin, streptomycin, and vancomycin but susceptible to erythromycin and tetracycline.

Enrichment and isolation procedures

The organism was isolated from the digestive tract of a marine clamworm, *Periserrula leucophryna*, inhabiting a tidal flat of the Yellow Sea. The organism can be isolated by spreading animal-tissue homogenates on marine agar 2216 after incubation for at least 1 week. Currently, there is no special method to enrich *Puniceicoccus* from the mixed marine microbial community.

Maintenance procedures

Frozen stocks as a glycerol suspension (10–30%) stored in liquid nitrogen or at −70°C are routinely used to start a new culture. Lyophilization of liquid cultures is also recommended for long-term storage. Working stock cultures can be maintained on slants at 4°C for 3 months.

Differentiation of the genus *Puniceicoccus* from other genera

A recent report by Yoon et al. (2007b) provides a table of characteristics that differentiate *Puniceicoccus* from the other related genera *Cerasicoccus*, *Coraliomargarita*, and *Pelagicoccus*.

Taxonomic comments

The genus *Puniceicoccus* is the type genus of the family *Puniceicoccaceae* and the order *Puniceicoccales*. Classification of *Puniceicoccus* as a separate genus of the class *Opitutae* is based on 16S rDNA sequence-based phylogeny as well as phenotypic characteristics. According to 16S rDNA sequence analyzes, the most closely related genera to the genus *Puniceicoccus* are *Cerasicoccus*, *Coraliomargarita*, and *Pelagicoccus*. The only species currently included in the genus is *Puniceicoccus vermicola*.

List of species of the genus *Puniceicoccus*

1. **Puniceicoccus vermicola** Choo, Lee, Song and Cho 2007, 536[VP]

ver.mi'co.la. L. n. *vermis* worm; L. suff. *-cola* from L. n. *incola* inhabitant; N.L. n. *vermicola* inhabitant of worms.

The characteristics of the species are as described for the genus.

DNA G+C content (mol%): 52.1 (HPLC).

Type strain: IMCC 1545, JCM 14086, KCCM 42343, NBRC 101964.

Sequence accession no. (16S rRNA gene): DQ539046.

Genus II. **Cerasicoccus** Yoon, Matsuo, Matsuda, Adachi, Kasai and Yokota 2007a, 2070[VP]

Jaewoo Yoon and Brian P. Hedlund

Ce.ra.si.coc'cus. L. neut. n. *cerasum* a cherry; Gr. masc. n. *kokkos* berry; N.L. masc. n. *Cerasicoccus* pale-pink-colored coccus, referring to the pale-pink colour of the bacterium.

Cocci, about 0.8 μm × 1.0 μm in diameter. **Gram-stain-negative. Nonmotile. Obligately aerobic;** nitrate not reduced. Growth is **chemoorganotrophic** and restricted to a small number of sugars and organic acids. Colonies are pale pink due to carotenoids. **Neither muramic acid nor diaminopimelic acid identified in cell-wall extracts. Resistant to β-lactam antibiotics**. Predominant cellular fatty acids are $C_{14:0}$ and $C_{18:1}$ ω9*c*. The major respiratory quinone is **MK-7**.

DNA G+C content (mol%): 54.0.

Type species: **Cerasicoccus arenae** Yoon, Matsuo, Matsuda, Adachi, Kasai and Yokota 2007a, 2070[VP].

Further descriptive information

Only a single strain of *Cerasicoccus* has ever been isolated. Like other members of the order *Puniceicoccales*, *Cerasicoccus* was isolated from the marine environment. Sequences closely related to *Cerasicoccus* have not been described in cultivation-independent censuses, so their ecological function and distribution in nature remain unclear.

Cells of Cerasicoccus are uniform cocci about 0.8 μm × 1.0 μm in diameter (Figure 145). Amino acid analysis of the cell-wall hydrolysate indicates the absence of muramic acid and diaminopimelic acid in the cell wall, which suggests that the strain does not contain peptidoglycan. Consistent with this, *Cerasicoccus* is highly resistant to β-lactam antibiotics. However, like other members of the order *Puniceicoccales*, no detailed studies of the cell structure of *Cerasicoccus* have been conducted, so it is currently unknown whether these cells possess a structure similar to peptidoglycan or whether they have two membranes or one.

Enrichment and isolation procedures

Currently no procedures to enrich *Cerasicoccus* are known. Strain YM26-026[T] was isolated from marine sand collected from the shore of the Gulf of Touni, Touni-cho, Kamaishi, Iwate, Japan. The sample was washed gently with 0.1 N HCl for 5 min, neutralized by addition of 5 ml of sterile seawater, and then homogenized with a glass rod. The homogenate (50 μl) was

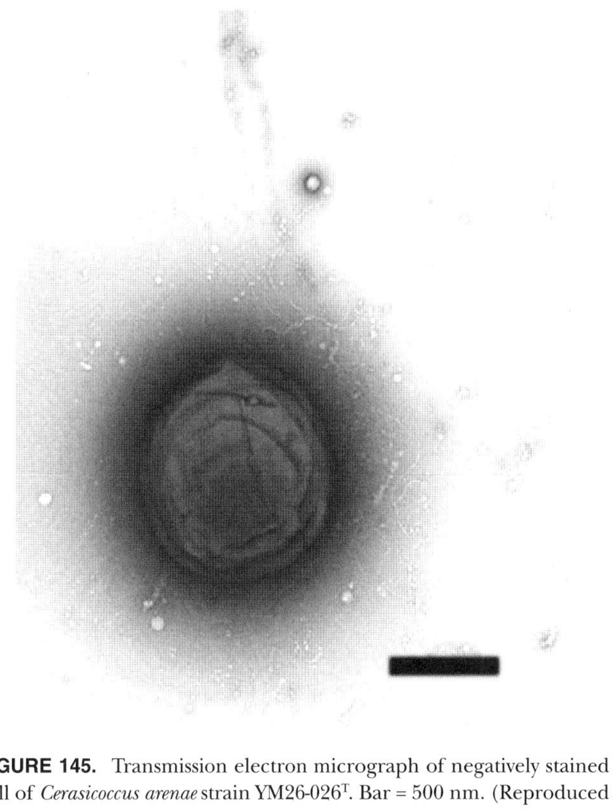

FIGURE 145. Transmission electron micrograph of negatively stained cell of *Cerasicoccus arenae* strain YM26-026[T]. Bar = 500 nm. (Reproduced with permission from Yoon et al., 2007a. Int. J. Syst. Evol. Microbiol. *57:* 2067–2072.)

plated onto medium "P" (Yoon et al., 2007b) and incubated at 25°C for 30 d. A pale-pink pigmented colony was subsequently purified on marine broth 2216 (Difco) containing 1.5% agar.

Maintenance procedures

For long-term storage, *Cerasicoccus* can be maintained at −80°C in 75% artificial seawater medium containing 30% glycerol as a cryoprotectant or they can be freeze-dried.

Procedures for testing special characteristics

API 20E, API 50CH, and API ZYM strips (bioMérieux) were used to determine physiological and biochemical characteristics. API 20E, API 50CH, and API ZYM were read after 72 h incubation at 30°C and 4 h incubation at 37°C, respectively. β-Lactam antibiotic susceptibility was determined on 1/5 marine agar 2216, using 8 mm paper discs (Advantec) at the following antibiotic concentrations: ampicillin (1, 10, 100, 500, or 1000 µg/ml) and penicillin G (1, 10, 100, 500, or 1000 µg/ml). Cell walls were prepared by the methods described by Schleifer and Kandler (1972) and amino acids in an acid hydrolysate of the cell walls were identified by TLC (Harper and Davis, 1979) and by HPLC, as their phenylthiocarbamoyl derivatives, with a model LC-10AD HPLC apparatus (Shimadzu) equipped with a Wakopak WS-PTC column (Wako Pure Chemical Industries, 1989) (Yokota et al., 1993).

Differentiation of the genus *Cerasicoccus* from other genera

Cerasicoccus shares many phenotypic characteristics with other members of the order *Puniceicoccales*, which includes *Puniceicoccus* (Choo et al., 2007), *Pelagicoccus* (Yoon et al., 2007b, d), and *Coraliomargarita* (Yoon et al., 2007c). All are marine cocci that are Gram-stain-negative. All are extremely resistant to β-lactam antibiotics and lack muramic acid and *meso*-diaminopimelic acid. All have similar DNA G+C content (53.9–57.4 mol%) and possess menaquinones MK-7 as the predominant respiratory quinone. Due to the relatively small number of isolates representing each of these genera, differential phenotypic characteristics are difficult to identify at present, therefore, 16S rRNA gene sequencing and phylogenetic analysis is recommended for differentiation (Figure 146).

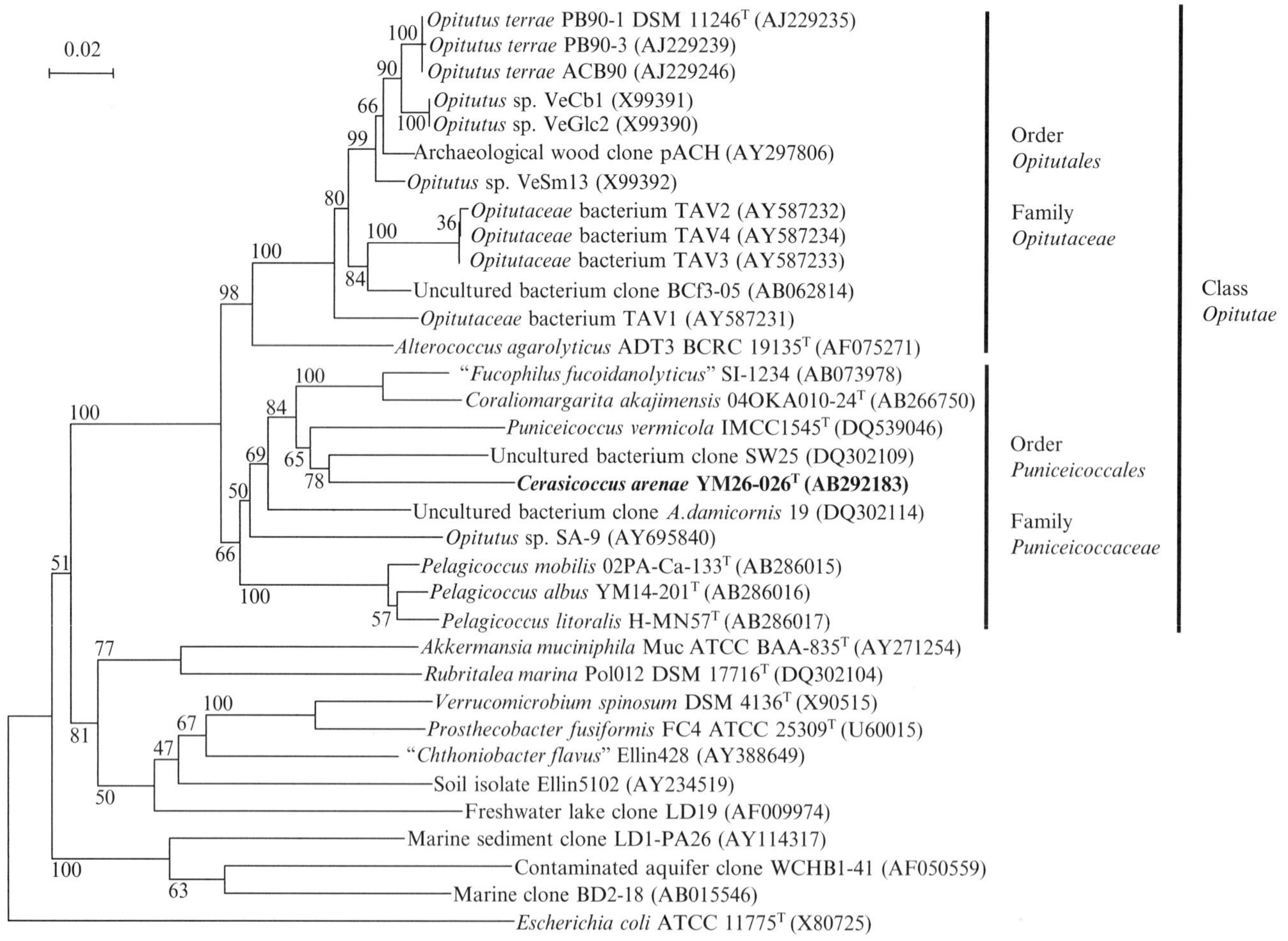

FIGURE 146. Neighbor-joining phylogenetic tree based on 16S rRNA gene sequences analysis showing the position of *Cerasicoccus* in the class *Opitutae*. Numbers at the nodes indicate bootstrap value derived from 1000 replications. Bar = 2% sequence divergence. (Modified with permission from Yoon et al., 2007a. Int. J. Syst. Evol. Microbiol. *57:* 2067–2072.)

List of species of the genus *Cerasicoccus*

1. **Cerasicoccus arenae** Yoon, Matsuo, Matsuda, Adachi, Kasai and Yokota 2007a, 2070[VP]

a.re'na.e. L. gen. n. *arenae* of sand.

Main characteristics are the same as those given for the genus. Neither cellular gliding movement nor swarming growth is observed. Colonies grown on 1/5 marine agar medium are circular, convex, and pale-pink pigmented. The temperature range for growth is 10–30°C, optimally at 25–30°C. No growth occurs at 4 or 45°C. The pH range for growth is 6–9. Seawater is not required for growth. Growth occurs from 0 to 8% (w/v) NaCl. Growth occurs in the presence of ampicillin (1–1000 µg/ml) and penicillin G (1–1000 µg/ml). Esculin and starch are hydrolyzed but agar, DNA, gelatin, and urea are not hydrolyzed. The reactions for *o*-nitrophenyl-β-D-galactosidase (ONPG) and tryptophan deaminase are positive, but acetoin, citrate utilization, arginine dihydrolase, lysine decarboxylase, ornithine decarboxylase, hydrogen sulfide, and indole production are negative. Acid is produced from L-arabinose, D-lyxose, galactose, glucose, fructose, mannose, methyl-α-D-glucopyranoside, esculine ferric citrate, lactose, gentiobiose, D-turanose, and 5-keto-gluconate, but not from trehalose, D-tagatose, D-fucose, L-fucose, D-arabitol, L-arabitol and erythritol, mannitol, sorbitol, glycerol, D-arabinose, ribose, D-xylose, L-xylose, adonitol, methyl-β-D-xylopyranoside, sorbose, rhamnose, dulcitol, inositol, methyl-α-D-mannnopyranoside, *N*-acetylglucosamine, amygdalin, arbutin, salicin, cellobiose, maltose, melibiose, sucrose, inulin, melezitose, rafinose, starch, glycogen, xylitol, gluconate, and 2-keto-gluconate. Alkaline phosphatase, naphthol-AS-BI-phosphohydrolase, and β-galactosidase are positive, but acid phosphatase, α-galactosidase, α-glucosidase, leucine arylamidase, valine arylamidase, trypsine, esterase (C4), esterase lipase (C8), lipase (C4), cystein arylamidase, chymotrypsin, β-glucuronidase, β-glucosidase, *N*-acetyl-β-glucosaminidase, α-mannosidase, and α-fucosidase are negative. The usual components of bacterial cell walls such as muramic acid and diaminopimelic acid could not be detected. Major fatty acid components (>1.0%) include $C_{14:0}$ iso (3.5%), $C_{14:0}$ (38.7%), $C_{14:0}$ 2-OH (4.4%), $C_{16:0}$ (2.3%), $C_{16:0}$ 3-OH (1.3%), $C_{18:1}$ ω9*c* (43.3%), $C_{18:0}$ (1.5%), and $C_{20:0}$ (1.5%).

DNA G+C content (mol%): 54.0 (HPLC).

Type strain: YM26-026, KCTC 12870, MBIC 08280, NBRC 103621.

Sequence accession no. (16S rRNA gene): AB292183.

Genus III. **Coraliomargarita** Yoon, Yasumoto-Hirose, Katsuta, Sekiguchi, Matsuda, Kasai and Yokota 2007c, 962[VP]

Jaewoo Yoon and Brian P. Hedlund

Co.ra.li.o.mar.ga.ri'ta. Gr. n. *koralion* coral; L. fem. n. *margarita* a pearl; N.L. fem. n. *Coraliomargarita* coral pearl, referring to a white-colony-forming coccoid microorganism isolated from seawater in a sample bottle of hard coral.

Cocci, about 0.5 µm × 1.2 µm in diameter. **Gram-stain-negative. Nonmotile. Obligately aerobic;** nitrate not reduced. Growth is **chemoorganotrophic** and restricted to a small number of mono- and disaccharides, sugar alcohols, and organic acids. Colonies are white. **Neither muramic acid nor diaminopimelic acid identified in cell-wall extracts. Resistant to β-lactam antibiotics**. Predominant cellular fatty acids are $C_{14:0}$, $C_{18:1}$ ω9*c*, and $C_{18:0}$. The major respiratory quinone is **MK-7**.

DNA G+C content (mol%): 53.9.

Type species: **Coraliomargarita akajimensis** Yoon, Yasumoto-Hirose, Katsuta, Sekiguchi, Matsuda, Kasai and Yokota 2007c, 962[VP].

Further descriptive information

Only a single strain of *Coraliomargarita* has ever been isolated. Like other members of the order *Puniceicoccales*, *Coraliomargarita* was isolated from the marine environment. Few sequences closely related to *Coraliomargarita* have been described in cultivation independent censuses, so their ecological function and distribution in nature remain unclear.

Cells of *Coraliomargarita* are uniform nonmotile cocci about 0.5 µm × 1.2 µm in diameter (Figure 147). Amino acid analysis of the cell-wall hydrolysate indicates the absence of muramic acid and diaminopimelic acid in the cell wall, suggesting that the strain does not contain peptidoglycan. Consistent with this, *Coraliomargarita* is highly resistant to β-lactam antibiotics. However, like other members of the order *Puniceicoccales*, no detailed studies of the cell structure of *Coraliomargarita* have been conducted, so it is currently unknown whether these cells possess a structure similar to peptidoglycan.

Enrichment and isolation procedures

Currently no procedures to enrich *Coraliomargarita* are known, however, in the future it may be possible to devise strategies to enrich for or isolate *Coraliomargarita* based on its resistance to β-lactam antibiotics. *Coraliomargarita* was isolated from a sample containing the hard coral *Galaxea fascicularis* Linnaeus 1767 and surrounding seawater, which was diluted 1/10 in autoclaved, filtered seawater (v/v) and plated onto MA [3.74 g of marine broth 2216 (Difco), 750 ml of filtered seawater, 250 ml of distilled water, and 15 g of agar].

Maintenance procedures

For long-term storage, *Coraliomargarita* can be maintained at −80°C in 75% artificial seawater medium containing 30% glycerol as a cryoprotectant, or they can be freeze-dried.

Procedures for testing special characteristics

API 20E and API 50CH strips (bioMérieux) were used to determine physiological and biochemical characteristics. API 20E, API 50CH, and API ZYM were read after 48 h incubation at 30°C and 48 h incubation at 30°C, respectively. β-Lactam antibiotic susceptibility test against novel isolate checked on 1/5 marine agar 2216, using 8 mm paper disc (Advantec)

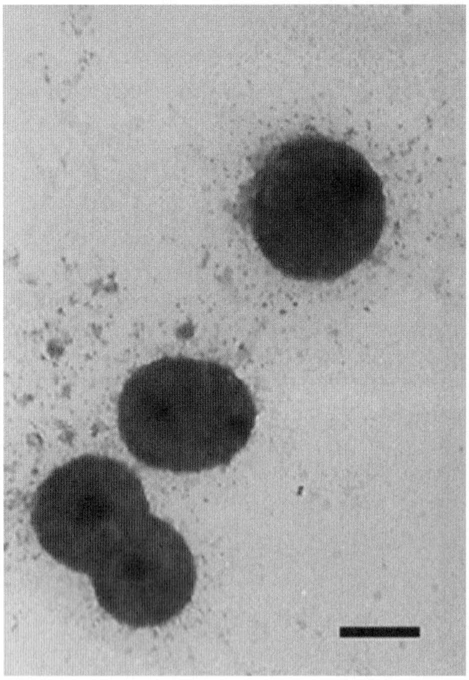

FIGURE 147. Transmission electron micrograph of negatively stained cell of *Coraliomargarita akajimensis* strain 04OKA010-24^{T}. Bar = 500 nm. (Reproduced with permission from Yoon et al., 2007c. Int. J. Syst. Evol. Microbiol. *57:* 959–963.)

at the following antibiotic concentrations: ampicillin (1, 10, 100, 500, or 1000 μg/ml) and penicillin G (1, 10, 100, 500, or 1000 μg/ml) (Yoon et al., 2007b). Cell walls were prepared by the methods described by Schleifer and Kandler (1972) and amino acids in an acid hydrolysate of the cell walls were identified by TLC (Harper and Davis, 1979) and by HPLC, as their phenylthiocarbamoyl derivatives, with a model LC-10AD HPLC apparatus (Shimadzu) equipped with a Wakopak WS-PTC column (Wako Pure Chemical Industries) (Yokota et al., 1993).

Differentiation of the genus *Coraliomargarita* from other genera

Coraliomargarita shares many phenotypic characteristics with other members of the order *Puniceicoccales*, which includes *Puniceicoccus* (Choo et al., 2007), *Pelagicoccus* (Yoon et al., 2007b, d), and *Cerasicoccus* (Yoon et al., 2007a). All are marine cocci that are Gram-stain-negative. All are extremely resistant to β-lactam antibiotics and lack muramic acid and *meso*-diaminopimelic acid. All have similar DNA G+C content (51.6–57.4 mol%) and possess menaquinones MK-7 as the predominant respiratory quinone. Due to the relatively small number of isolates representing each of these genera, differential phenotypic characteristics are difficult to identify at present, therefore, 16S rRNA gene sequencing and phylogenetic analysis is recommended for differentiation (Figure 148).

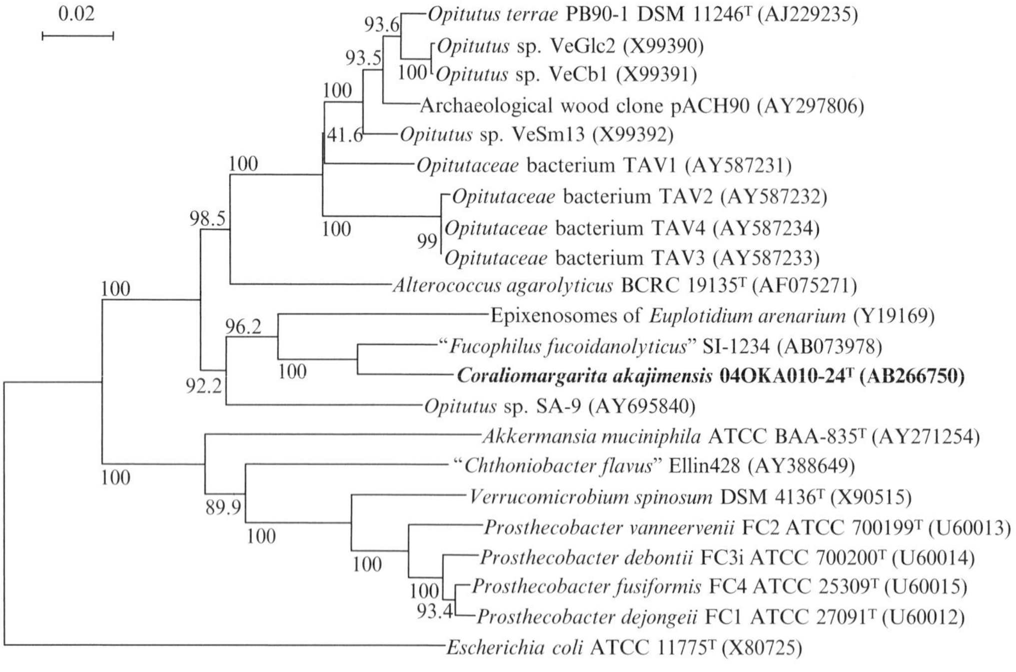

FIGURE 148. Neighbor-joining phylogenetic tree based on 16S rRNA gene sequences analysis showing the position of *Coraliomargarita* in the class *Opitutae*. Numbers at the nodes indicate bootstrap value derived from 1000 replications. Bar = 2% sequence divergence. (Modified with permission from Yoon et al., 2007c. Int. J. Syst. Evol. Microbiol. *57:* 959–963.)

List of species of the genus *Coraliomargarita*

1. **Coraliomargarita akajimensis** Yoon, Yasumoto-Hirose, Katsuta, Sekiguchi, Matsuda, Kasai and Yokota 2007c, 962[VP]

a.ka.ji.men'sis. N.L. fem. adj. *akajimensis* pertaining to Akajima, an island in Okinawa, from which the type strain was isolated.

Main characteristics are the same as those given for the genus. In addition, cells are cocci, 0.5–1.2 μm in diameter. Neither cellular gliding movement nor swarming growth is observed. Colonies grown on 1/2 R2A agar medium with 75% Artificial seawater are circular, convex, and white. The optimum temperature range for growth is 20–30°C, and no growth occurs at 4 or 45°C. The pH range for growth is 7.0–9.0. NaCl is required for growth and can be tolerated in an up to 5% solution (w/v). Urea and DNA are hydrolyzed but agar, casein, esculin, and gelatin are not hydrolyzed. Acetoin, *o*-nitrophenyl-β-D-galactosidase (ONPG), and tryptophan deaminase are positive, but the reactions for citrate utilization, arginine dihydrolase, lysine decarboxylase, ornithine decarboxylase, hydrogen sulfide, and indole production are negative. Acid is produced from glycerol, galactose, fructose, mannose, mannitol, sorbitol, trehalose, D-turanose, D-lyxose, D-tagatose, D-fucosue, L-fucose, D-arabitol, L-arabitol, and 5-keto-gluconate but not from erythritol, D-arabinose, L-arabinose, ribose, D-xylose, L-xylose, adonitol, methyl-β-D-xylopyranoside, glucose, sorbose, rhamnose, dulcitol, inositol, methyl-α-D-mannnopyranoside, methyl-α-D-glucopyranoside, *N*-acetylglucosamine, amygdalin, arbutin, salicin, cellobiose, maltose, lactose, melibiose, sucrose, inulin, melezitose, rafinose, starch, glycogen, xylitol, gentiobiose, gluconate, and 2-keto-gluconate. The usual components of bacterial cell walls such as muramic acid and diaminopimelic acid could not be detected. Major fatty acid components (>2.0%) include $C_{14:0}$ iso(8.2%), $C_{14:0}$ (24.2%), $C_{15:0}$ anteiso (2.9%), $C_{16:0}$ (3.3%), $C_{18:1}$ ω9*c* (23.5%), $C_{18:0}$ (15.5%), $C_{19:0}$ (2.8%), and $C_{21:0}$ (6.9%).

Source: seawater in the sampling bottle of the hard coral, *Galaxea fascicularis* Linnaeus 1767, collected at Majanohama, Akajima in Japan.

DNA G+C content (mol%): 53.9 (HPLC).

Type strain: 04OKA010-24, IAM 15411, JCM 23193, KCTC 12865, MBIC 06463, NBRC 103620.

Sequence accession no. (16S rRNA gene): AB266750.

Genus IV. **Pelagicoccus** Yoon, Yasumoto-Hiorse, Matsuo, Nozawa, Matsuda, Kasai and Yokota 2007d, 1381[VP]

Jaewoo Yoon and Brian P. Hedlund

Pe.la.gi.coc'cus. L. n. *pelagus* the open sea, the ocean; N.L. masc. n. *coccus* (from Gr. masc. n. *kokkos*), berry; N.L. masc. n. *Pelagicoccus* referring to a coccoid-shaped bacterium from the sea.

Cocci, 0.5 μm × 1.2 μm in diameter. **Gram-stain-negative.** Some species motile via one or more flagella. Obligately aerobic or facultatively anaerobic; nitrate not reduced. Growth is **chemoorganotrophic** and restricted to a small number of mono- and disaccharides, polysaccharides, and organic acids. Colonies are white or pale yellow. **Neither muramic acid nor diaminopimelic acid identified in cell-wall extracts. Resistant to β-lactam antibiotics**. Predominant cellular fatty acids are $C_{16:0}$, $C_{16:1}$ ω7*c*, and $C_{15:0}$ anteiso. The major respiratory quinone is **MK-7**.

DNA G+C content (mol%): 51.6–57.4.

Type species: **Pelagicoccus mobilis** Yoon, Yasumoto-Hiorse, Matsuo, Nozawa, Matsuda, Kasai and Yokota 2007d, 1381[VP].

Further descriptive information

All strains of the genus *Pelagicoccus* have been isolated from shallow coastal marine or estuarine waters. All of the isolates were associated with natural or unnatural solid substrates, suggesting that these organisms are normally epibionts rather than free-living in their environment. Since *Pelagicoccus croceus* was isolated from a marine plant and *Pelagicoccus albus* and *Pelagicoccus litoralis* were enriched by using an in situ cultivation approach using plant exudates as enrichment substrates, the outer surface of marine plants is likely a common habitat of members of this genus. However, these organisms have never been enumerated in their environment, and they have not been detected or studied by cultivation-independent approaches so their ecological function and full distribution in nature remain unclear.

Cells of *Pelagicoccus* are uniform cocci, with different size, depending on the species (Figure 149). *Pelagicoccus mobilis* is motile via one or more flagellum; neither flagella nor mobility has been observed in the other species of *Pelagicoccus*. Some cells of *Pelagicoccus mobilis* have cell appendages. Amino acid analysis of the cell-wall hydrolysate indicates the absence of muramic acid and diaminopimelic acid in the cell wall, which suggests that the isolates do not contain peptidoglycan. Consistent with this, *Pelagicoccus* is highly resistant to β-lactam antibiotics. However, like other members of the order *Puniceicoccales*, no detailed studies of the cell structure of *Pelagicoccus* have been conducted, so it is currently unknown whether these cells possess a structure similar to peptidoglycan.

Enrichment and isolation procedures

Currently no procedures to specifically enrich *Pelagicoccus* are known. However, with the knowledge that neither muramic acid nor diaminopimelic acid were identified in cell-wall extracts, and that *Pelagicoccus* isolates are resistant to high concentrations of β-lactam antibiotics, a future strategy that relies on resistance to these antibiotics may be devised.

Pelagicoccus mobilis was isolated by plating a seawater sample from a coral reef near Palau directly onto marine broth 2216 (Difco) supplemented with 1% (w/v) $CaCO_3$ and 1.5% (w/v) agar (Yoon et al., 2007d). *Pelagicoccus croceus* was isolated from a homogenate of a leaf of the marine plant *Enhalus acoroides* (L.f.) Royale from the Kuira River mangrove estuary, Iriomotejima, Okinawa, Japan. The leaf was homogenized in sterile distilled water, diluted, and plated directly onto diluted, solidified marine broth 2216 (Difco) supplemented with 10 mg jack bean lectin concanavalin A (Wako)/ml. To prepare this medium, 3.74 g marine broth 2216, 15 g agar, 250 ml artificial seawater (Lyman and Fleming, 1940) and 750 ml distilled water were

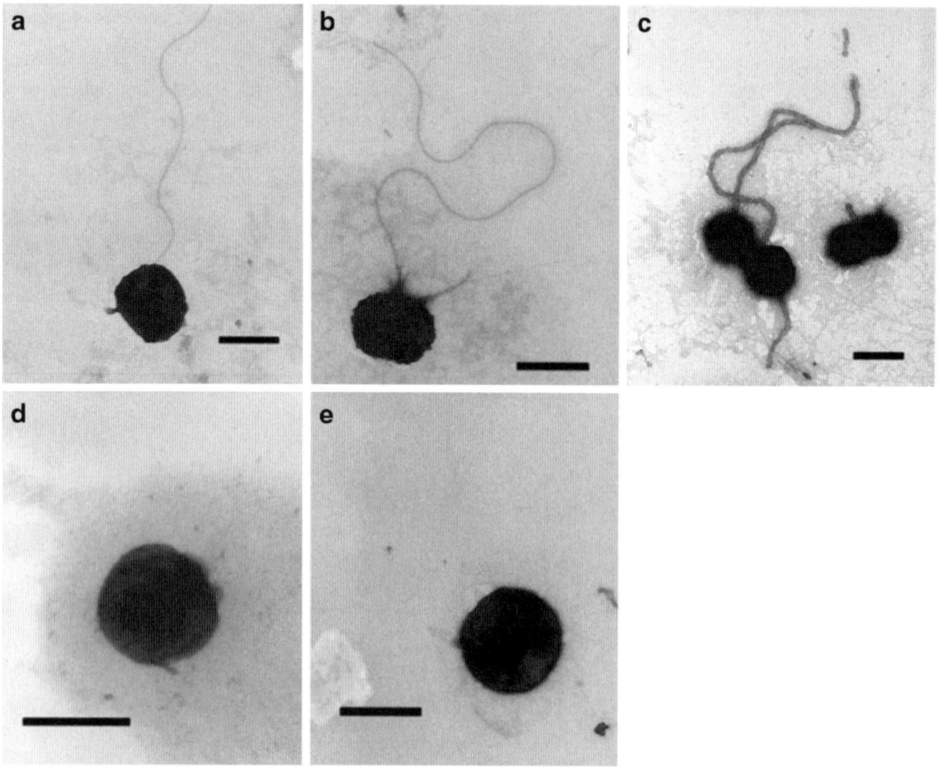

FIGURE 149. Transmission electron micrographs of negatively stained cells of *Pelagicoccus mobilis* (a–c), *Pelagicoccus albus* (d), and *Pelagicoccus litoralis* (e). Bar = 500 nm. (Reproduced with permission from Yoon et al., 2007d. Int. J. Syst. Evol. Microbiol. *57:* 1377–1385.)

mixed and autoclaved, after which a solution of 2.5 ml concanavalin A prepared at a concentration of 4 mg/ml in sterile distilled water was added (Yoon et al., 2007b).

Pelagicoccus albus and *Pelagicoccus litoralis* were enriched *in situ* by using polyurethane foam (PUF) blocks impregnated with agar containing dilute organic compounds (Yasumoto-Hirose et al., 2006). Strains of *Pelagicoccus albus* were enriched by a 3-d incubation of a PUF block off the coast of Palau. The PUF block was impregnated with sterile seawater containing 0.1% (w/v) 7-hydroxyflavone and 1.5% (w/v) agar. Strains of *Pelagicoccus litoralis* were enriched by a 2-week incubation of PUF blocks supplemented with HV agar (Hayakawa and Nonomura, 1987) in 90% seawater containing 0.1% (w/v) collidine or 0.1% (w/v) vanillin. PUF blocks were homogenized with a glass rod in 5 ml of sterile seawater. A 50–100 ml sample of the homogenate was plated onto "P" medium (*Pelagicoccus albus*; Yoon et al., 2007d) or onto HV medium containing 90% seawater with 40 mg/ml of nalidixic acid and 100 mg/ml of cycloheximide (*Pelagicoccus litoralis*; Yoon et al., 2007d).

Maintenance procedures

For long-term storage, *Pelagicoccus* can be maintained at −80°C in 75% artificial seawater medium containing 30% glycerol as a cryoprotectant or they can be freeze-dried.

Procedures for testing special characteristics

API 20E, API 50CH, and API ZYM strips (bioMérieux) were used to determine physiological and biochemical characteristics. API 20E and API 50CH tests were read after 72 h incubation at 30°C and API ZYM tests were read after 4 h incubation at 37°C, respectively. β-Lactam antibiotic susceptibility test against novel isolate checked on marine agar 2216 and half-strength R2A agar, using 8 mm paper disc (Advantec) at the following antibiotic concentrations: ampicillin (1, 10, 100, 500, or 1000 µg/ml) and penicillin G (1, 10, 100, 500, or 1000 µg/ml) (Yoon et al., 2007b). Cell walls were prepared by the methods described by Schleifer and Kandler (1972) and amino acids in an acid hydrolysate of the cell walls were identified by TLC (Harper and Davis, 1979) and by HPLC, as their phenylthiocarbamoyl derivatives, with a model LC-10AD HPLC apparatus (Shimadzu) equipped with a Wakopak WS-PTC column (Wako Pure Chemical Industries, 1989) (Yokota et al., 1993).

Differentiation of the genus *Pelagicoccus* from other genera

Pelagicoccus shares many phenotypic characteristics with other members of the order *Puniceicoccales*, which includes *Puniceicoccus* (Choo et al., 2007), *Coraliomargarita* (Yoon et al., 2007c), and *Cerasicoccus* (Yoon et al., 2007a). All are marine cocci that are Gram-stain-negative. All are extremely resistant to β-lactam antibiotics and lack muramic acid and *meso*-diaminopimelic acid. All have similar DNA G+C content (51.6–57.4 mol%) and possess menaquinone MK-7 as the predominant respiratory quinone. Due to the relatively small number of isolates representing each of these genera, differential phenotypic characteristics are difficult to identify at present, therefore, 16S rRNA gene sequencing and phylogenetic analysis are recommended for differentiation (Figure 150).

Differentiation of the species of the genus *Pelagicoccus*

The species of *Pelagicoccus* can be differentiated by 16S rRNA gene sequencing and phylogenetic analysis (Figure 150). Characteristics that may be useful in the differentiation of the species of the genus *Pelagicoccus* are listed in Table 159.

List of species of the genus *Pelagicoccus*

1. **Pelagicoccus mobilis** Yoon, Yasumoto-Hiorse, Matsuo, Nozawa, Matsuda, Kasai and Yokota 2007d, 1381[VP]

mo.bi'lis. L. masc. adj. *mobilis* movable, mobile, referring to the ability to move by flagella.

Main characteristics are the same as those given for the genus. In addition, cells are obligately aerobic cocci, 0.5–0.7 µm in diameter. Motile by means of single or multiple flagella. Appendages are found on some cells. Neither cellular gliding movement nor swarming growth is observed. Colonies grown on 1/2 R2A agar medium with 75% artificial seawater are circular, convex, and white. The temperature range for growth is 20–37°C, optimally at 28–30°C, but no growth occurs at 4 or 45°C. The pH range for growth is 7.0–9.0. NaCl is required for growth; can tolerate up to 4% (w/v) NaCl. Growth occurs in the presence of ampicillin (1–1000 µg/ml) and penicillin G (1–1000 µg/ml). Catalase reaction is negative, whereas oxidase is positive. Esculin and urea are hydrolyzed, but agar, DNA, starch, and gelatin are

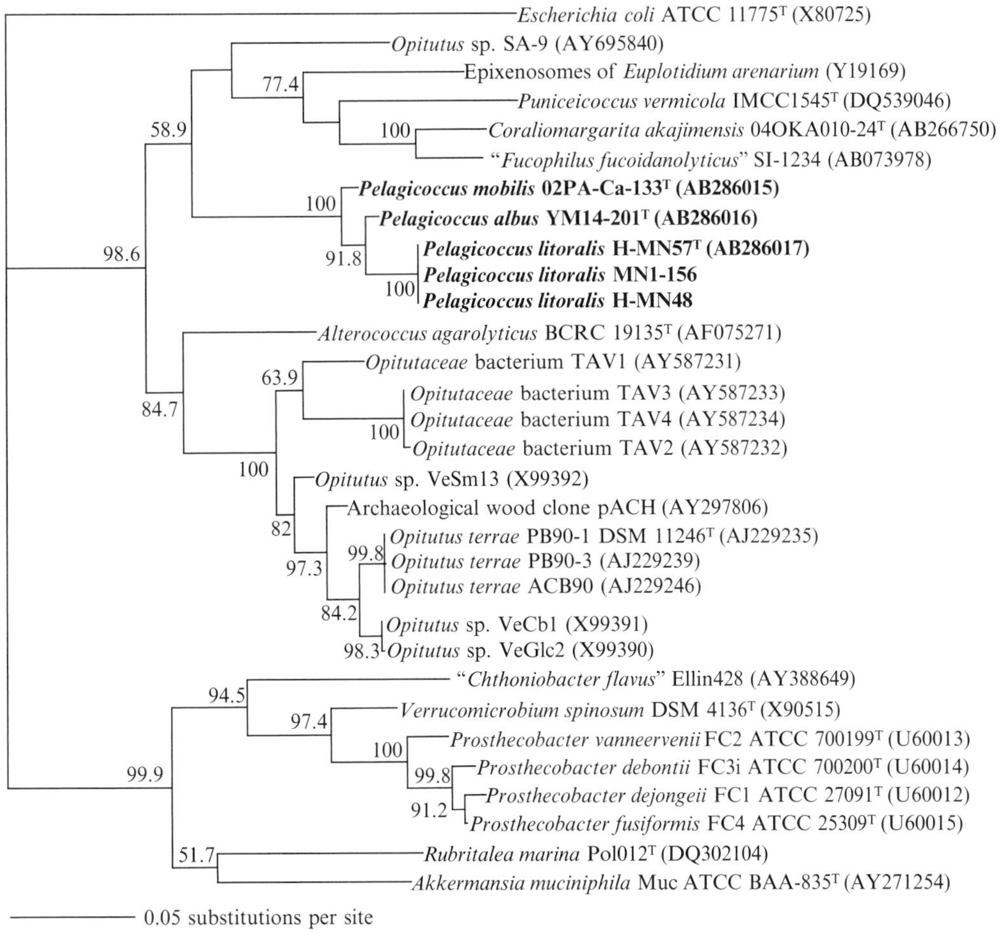

FIGURE 150. Maximum-likelihood phylogenetic tree showing the positions of *Pelagicoccus* strains. The numbers at the branch nodes are bootstrap values based on 1000 replications. The sequence of *Escherichia coli* ATCC 11775[T] was used as the outgroup. Bar = 0.05 substitutions per site. (Modified with permission from Yoon et al., 2007d. Int. J. Syst. Evol. Microbiol. *57:* 1377–1385.)

TABLE 159. Characteristics differentiating species of the genus *Pelagicoccus*[a]

Characteristic	*P. mobilis*	*P. albus*	*P. litoralis*	*P. croceus*
Cell diameter (μm)	0.5–0.7	0.8–1.2	1.0–1.2	0.5–1.0
Flagella	+	–	–	–
Appendage	+	–	–	–
Motility	+	–	–	–
O_2 requirement	Obligately aerobic	Facultatively anaerobic	Obligately aerobic	Obligately aerobic
Production of:				
Catalase	–	+	+	+
ONPG	–	+	d	+
Growth at 4°C	–	–	+	–
Growth at 37°C	+	+	–	–
NaCl range (%)	1–4	3–7	1–4	1–5
Hydrolysis of:				
DNA	–	+	–	–
Urea	+	–	–	–
Acid production from:				
Cellobiose	+	–	–	–
Glucose	+	–	–	–
Lactose	+	–	–	w
Melibiose	–	–	+	w
Enzyme activity:				
Acid phosphatase	+	+	–	+
α-Galactosidase	–	+	–	–
α-Glucosidase	–	+	–	–
N-Acetyl-β-glucosaminidase	+	–	–	+

[a]Symbols: +, >85% positive; d, different strains give different reactions (16–84% positive); –, 0–15% positive; w, weak reaction.

not hydrolyzed. The reactions for acetoin, *o*-nitrophenyl-β-D-galactosidase (ONPG), tryptophan deaminase, citrate utilization, arginine dihydrolase, lysine decarboxylase, ornithine decarboxylase, hydrogen sulfide, and indole production are negative. Acid is produced from esculin ferric citrate, salicin, cellobiose, lactose, sucrose, trehalose, starch, glycogen, and gentiobiose, but not from glycerol, galactose, fructose, mannose, mannitol, sorbitol, D-turanose, D-lyxose, D-tagatose, D-fucose, L-fucose, D-arabitol, L-arabitol, 5-keto-gluconate, erythritol, D-arabinose, L-arabinose, ribose, D-xylose, L-xylose, adonitol, methyl-β-D-xylopyranoside, glucose, sorbose, rhamnose, dulcitol, inositol, methyl-α-D-mannnopyranoside, methyl-α-D-glucopyranoside, *N*-acetylglucosamine, amygdalin, arbutin, maltose, melibiose, inulin, melezitose, rafinose, xylitol, gluconate, and 2-keto-gluconate. Alkaline phosphatase, acid phosphatase, naphthol-AS-BI-phosphohydrolase, and *N*-acetyl-β-glucosaminidase are positive, but esterase (C4), esterase lipase (C8), lipase (C4), leucine arylamidase, valine arylamidase, cystein arylamidase, trypsin, chymotrypsin, α-galactosidase, β-galactosidase, β-glucuronidase, α-glucosidase, β-glucosidase, α-mannosidase, and α-fucosidase are negative. The usual components of bacterial cell walls such as muramic acid and diaminopimelic acid could not be detected. Major fatty acid components (>1.0%) include $C_{14:0}$ (5.8%), $C_{15:0}$ (5.9%), $C_{16:0}$ (23.3%), $C_{17:0}$ (2.1%), $C_{16:1}$ ω7*c* (15.1%), $C_{14:0}$ iso (2.1%), $C_{16:0}$ iso (2.3%), and $C_{15:0}$ anteiso (29.8%).

Source: seawater from Palau.

DNA G+C content (mol%): 57.4 (HPLC).

Type strain: 02PA-Ca-133, IAM 15422, JCM 23203, KCTC 13126, MBIC 08004.

Sequence accession no. (16S rRNA gene): AB286015.

2. **Pelagicoccus albus** Yoon, Yasumoto-Hiorse, Matsuo, Nozawa, Matsuda, Kasai and Yokota 2007d, 1381[VP]

al'bus. L. masc. adj. *albus* white, referring to the dull-white color of colonies.

Main characteristics are the same as those given for the genus. In addition, cells are facultatively anaerobic cocci, 0.8–1.2 μm in diameter. Neither cellular gliding movement nor swarming growth is observed. Colonies grown on 1/2 R2A agar medium with 75% artificial seawater are circular, convex, and white. The temperature range for growth is 15–37°C, optimally at 25–27°C, but no growth occurs at 4 or 45°C. The pH range for growth is 6.5–9.0. NaCl is required for growth; can tolerate up to 7% (w/v) NaCl. Growth occurs in the presence of ampicillin (1–1000 μg/ml) and penicillin G (1–1000 μg/ml). Catalase- and oxidase-positive. DNA and esculin are hydrolyzed but agar, starch, gelatin, and urea are not. The reactions for acetoin, *o*-nitrophenyl-β-D-galactosidase (ONPG), and tryptophan deaminase are positive, but citrate utilization, arginine dihydrolase, lysine decarboxylase, ornithine decarboxylase, hydrogen sulfide, and indole production are negative. Acid is produced from esculin ferric citrate and 5-keto-gluconate, but not from glycerol, galactose, fructose, mannose, mannitol, sorbitol, trehalose, D-turanose, D-lyxose, D-tagatose, D-fucose, L-fucose, D-arabitol, L-arabitol, erythritol, D-arabinose, L-arabinose, ribose, D-xylose, L-xylose, adonitol, methyl-β-D-xylopyranoside, glucose, sorbose, rhamnose, dulcitol, inositol, methyl-α-D-mannnopyranoside, methyl-α-D-glucopyranoside, *N*-acetylglucosamine, amygdalin, arbutin, salicin, cellobiose, maltose, lactose, melibiose, sucrose, inulin, melezitose, rafinose, starch, glycogen, xylitol, gentiobiose, gluconate, and 2-keto-gluconate. Alkaline phosphatase, acid phosphatase, naphthol-AS-BI-phosphohydrolase, α-galactosidase, β-galactosidase, and α-glucosidase are positive, but leucine arylamidase, valine arylamidase, trypsine, esterase (C4), esterase lipase (C8), lipase (C4), cystein arylamidase, chymotrypsine, β-glucuronidase, β-glucosidase, *N*-acetyl-β-glucosaminidase, α-mannosidase, and α-fucosidase are negative. The usual components of bacterial cell walls such as muramic acid and diaminopimelic acid could not be detected. Major fatty acid components (>1.0%) include $C_{14:0}$ (4%), $C_{15:0}$ (1.7%), $C_{16:0}$ (23.8%), $C_{18:0}$ (1.2%), $C_{16:1}$ ω7*c* (14.5%), $C_{16:0}$ iso(2.3%), $C_{15:0}$ anteiso (37.5%), $C_{13:0}$ 2-OH (5.5%), and $C_{16:0}$ 3-OH (1.4%).

Source: seawater from Palau using an *in situ* cultivation technique.

DNA G+C content (mol%): 57.2 (HPLC).

Type strain: YM14-201, IAM 15421, JCM 23202, KCTC 13124, MBIC 08272.

Sequence accession no. (16S rRNA gene): AB286016.

3. **Pelagicoccus croceus** Yoon, Oku, Matsuo, Matsuda, Kasai and Yokota 2007b, 2877[VP]

cro.ce'us. L. masc. adj. *croceus* saffron-colored, yellow, golden, referring to the pale yellow color of colonies.

Main characteristics are the same as those given for the genus. In addition, cells are obligately aerobic, nonmotile cocci, 0.5–1.0 µm in diameter. Neither cellular gliding movement nor swarming growth is observed. Colonies grown on 1/5-strength marine agar 2216 are circular, convex, and pale yellow. The temperature range for growth is 20–30°C, optimally at 25–30°C, but no growth occurs at 4 or 45°C. The pH range for growth is 6.5–9.0. NaCl is required for growth; tolerates up to 5% (w/v). Growth occurs in the presence of ampicillin (1–1000 mg/ml) and penicillin G (1–1000 mg/ml). Catalase- and oxidase-positive. Nitrate is not reduced. Esculin is hydrolyzed but agar, DNA, starch, gelatin, and urea are not. The reaction for ONPG is positive, but acetoin, tryptophan deaminase, citrate utilization, arginine dihydrolase, lysine decarboxylase, ornithine decarboxylase, hydrogen sulfide, and indole production are negative. Acid is produced from methyl β-D-xylopyranoside, methyl α-D-mannnopyranoside, esculin ferric citrate, lactose, melibiose, D-turanose, D-lyxose, D-tagatose, and 5-ketogluconate, but not from sucrose, glycerol, galactose, fructose, mannose, mannitol, sorbitol, trehalose, D-fucose, L-fucose, D-arabitol, L-arabitol, erythritol, D-arabinose, L-arabinose, ribose, D-xylose, L-xylose, adonitol, glucose, sorbose, rhamnose, dulcitol, inositol, methyl α-D-glucopyranoside, *N*-acetyl-D-glucosamine, amygdalin, arbutin, salicin, cellobiose, maltose, inulin, melezitose, raffinose, starch, glycogen, xylitol, gentiobiose, gluconate, or 2-ketogluconate. Alkaline phosphatase, leucine arylamidase, and acid phosphatase are positive, but β-galactosidase, naphthol-AS-BI-phosphohydrolase, agalactosidase, α-glucosidase, valine arylamidase, trypsin, esterase (C4), esterase lipase (C8), lipase (C4), cystine arylamidase, chymotrypsin, β-glucuronidase, β-glucosidase, *N*-acetyl-β-glucosaminidase, α-mannosidase, and afucosidase are negative. The usual components of bacterial cell walls such as muramic acid and diaminopimelic acid could not be detected. Major fatty acid components (>1.0 %) include $C_{14:0}$ (1.3%), $C_{15:0}$ (21.2%), $C_{16:0}$ (20.7%), $C_{17:0}$ (6.8%), $C_{15:1}$ ω6*c* (2.6%), $C_{16:1}$ ω7*c* (12.7%), $C_{18:1}$ ω9*c* (1.5%), $C_{16:0}$ iso (1.8%), $C_{15:0}$ anteiso (25.4%), and $C_{17:0}$ anteiso (1.5%).

Source: leaf surface of seagrass *Enhalus acoroides* (L.f.) Royle.

DNA G+C content (mol%): 51.6 (HPLC).

Type strain: N5FB36-5, KCTC 12903, MBIC 08282.

Sequence accession no. (16S rRNA gene): AB297922.

4. **Pelagicoccus litoralis** Yoon, Yasumoto-Hiorse, Matsuo, Nozawa, Matsuda, Kasai and Yokota 2007d, 1383[VP]

li.to.ra'lis. L. masc. adj. *litoralis* pertaining to the coast.

Main characteristics are the same as those given for the genus. In addition, cells are obligately aerobic cocci, 1.0–1.2 µm in diameter. Neither cellular gliding movement nor swarming growth is observed. Colonies grown on 1/2 R2A agar medium with 75% artificial seawater are circular, convex, and white. The temperature range for growth is 4–30°C, optimally at 25–27°C, but no growth occurs at 45°C. The pH range for growth is 6.5–9.0. NaCl is required for growth and can tolerate up to 4% (w/v) NaCl. Growth occurs in the presence of ampicillin (1–1000 µg/ml) and penicillin G (1–1000 µg/ml). Catalase- and oxidase-positive. Esculin is hydrolyzed but agar, DNA, starch, gelatin, and urea are not. The reactions for acetoin, *o*-nitrophenyl-β-D-galactosidase (ONPG), and tryptophan deaminase are positive, but citrate utilization, arginine dihydrolase, lysine decarboxylase, ornithine decarboxylase, hydrogen sulfide, and indole production are negative. Acid is produced from esculin ferric citrate, melibiose, sucrose, and 5-keto-gluconate, but not from glycerol, galactose, fructose, mannose, mannitol, sorbitol, trehalose, D-turanose, D-lyxose, D-tagatose, D-fucose, L-fucose, D-arabitol, L-arabitol, erythritol, D-arabinose, L-arabinose, ribose, D-xylose, L-xylose, adonitol, methyl-β-D-xylopyranoside, glucose, sorbose, rhamnose, dulcitol, inositol, methyl-α-D-mannnopyranoside, methyl-α-D-glucopyranoside, *N*-acetylglucosamine, amygdalin, arbutin, salicin, cellobiose, maltose, lactose, inulin, melezitose, rafinose, starch, glycogen, xylitol, gentiobiose, gluconate, and 2-keto-gluconate. Alkaline phosphatase and β-galactosidase are positive, but acid phosphatase, naphthol-AS-BI-phosphohydrolase, α-galactosidase, α-glucosidase, leucine arylamidase, valine arylamidase, trypsin, esterase (C4), esterase lipase (C8), lipase (C4), cystein arylamidase, chymotrypsin, β-glucuronidase, β-glucosidase, *N*-acetyl-β-glucosaminidase, α-mannosidase, and α-fucosidase are negative. The usual components of bacterial cell walls such as muramic acid and diaminopimelic acid could not be detected. Major fatty acid components (>1.0%) include $C_{14:0}$ (4.4%), $C_{15:0}$ (4.5%), $C_{16:0}$ (14.3%), $C_{17:0}$ (1.3%), $C_{18:0}$ (1.4%), $C_{16:1}$ ω7*c* (20.7%), $C_{14:0}$ iso (1.1%), $C_{16:0}$ iso (1.7%), $C_{15:0}$ anteiso (38.1%), and $C_{13:0}$ 2-OH (5.7%).

Source: seawater from Heita Bay, Japan, using an *in situ* cultivation technique.

DNA G+C content (mol%): 56.4 (HPLC).

Type strain: H-MN57, IAM 15423, JCM 23204, KCTC 13125, MBIC 08273, NBRC 103622.

Sequence accession no. (16S rRNA gene): AB286017.

References

Choo, Y.J., K. Lee, J. Song and J.C. Cho. 2007. *Puniceicoccus vermicola* gen. nov., sp. nov., a novel marine bacterium, and description of *Puniceicoccaceae* fam. nov., *Puniceicoccales* ord. nov., *Opitutaceae* fam. nov., *Opitutales* ord. nov. and *Opitutae* classis nov. in the phylum '*Verrucomicrobia*'. Int. J. Syst. Evol. Microbiol. *57*: 532–537.

Harper, J.J. and G.H.G. Davis. 1979. Two-dimensional thin-layer chromatography for amino acid analysis of bacterial cell walls. Int. J. Syst. Bacteriol. *29*: 56–58.

Hayakawa, M. and H. Nonomura. 1987. Humic acid-vitamine agar, a new medium for the selective isolation of soil actinomycetes. J. Ferment. Technol. *65*: 501–509.

Lyman, J. and R.H. Fleming. 1940. Composition of sea water. J. Mar. Res. *3*: 134–146.

Schleifer, K.H. and O. Kandler. 1972. Peptidoglycan types of bacterial cell walls and their taxonomic implications. Bacteriol. Rev. *36*: 407–477.

Yasumoto-Hirose, M., M. Nishijima, M.K. Ngirchechol, K. Kanoh, Y. Shizuri and W. Miki. 2006. Isolation of marine bacteria by *in situ* culture on media-supplemented polyurethane foam. Mar. Biotechnol. (N.Y.) *8*: 227–237.

Yokota, A., T. Tamura, T. Nishii and T. Hasegawa. 1993. *Kineococcus aurantiacus* gen. nov., sp. nov., a new aerobic, Gram-positive, motile coccus with *meso*-diaminopimelic acid and arabinogalactan in the cell wall. Int. J. Syst. Bacteriol. *43*: 52–57.

Yoon, J., Y. Matsuo, S. Matsuda, K. Adachi, H. Kasai and A. Yokota. 2007a. *Cerasicoccus arenae* gen. nov., sp. nov., a carotenoid-producing marine representative of the family *Puniceicoccaceae* within the phylum '*Verrucomicrobia*', isolated from marine sand. Int. J. Syst. Evol. Microbiol. *57*: 2067–2072.

Yoon, J., N. Oku, S. Matsuda, H. Kasai and A. Yokota. 2007b. *Pelagicoccus croceus* sp. nov., a novel marine member of the family *Puniceicoccaceae* within the phylum '*Verrucomicrobia*' isolated from seagrass. Int. J. Syst. Evol. Microbiol. *57*: 2874–2880.

Yoon, J., M. Yasumoto-Hirose, A. Katsuta, H. Sekiguchi, S. Matsuda, H. Kasai and A. Yokota. 2007c. *Coraliomargarita akajimensis* gen. nov., sp. nov., a novel member of the phylum '*Verrucomicrobia*' isolated from seawater in Japan. Int. J. Syst. Evol. Microbiol. *57*: 959–963.

Yoon, J., M. Yasumoto-Hirose, Y. Matsuo, M. Nozawa, S. Matsuda, H. Kasai and A. Yokota. 2007d. *Pelagicoccus mobilis* gen. nov., sp. nov., *Pelagicoccus albus* sp. nov. and *Pelagicoccus litoralis* sp. nov., three novel members of subdivision 4 within the phylum '*Verrucomicrobia*', isolated from seawater by *in situ* cultivation. Int. J. Syst. Evol. Microbiol. *57*: 1377–1385.

Class III. **Spartobacteria** class. nov.

Peter H. Janssen, Kyle C. Costa and Brian P. Hedlund

Spar.to.bac.te′ri.a. Gr. adj *spartos* sown, in relation to the Spartoi, the sown men of the Cadmus myth, who sprung from the soil; L. neut. n. *bacterium* (from Gr. dim. neut. n. *bakterion* a small rod) rod; N.L. neut. pl. n. *Spartobacteria* sown small rods.

On the basis of 16S rRNA gene sequence analyses, this group comprises one of the primary lineages in the phylum *Verrucomicrobia* and is equivalent to subdivision 2 (Hugenholtz et al., 1998) of the phylum. The majority of its members appear to be soil-inhabiting bacteria, while some are endosymbionts of nematodes. A few 16S rRNA gene sequences have been recovered from freshwater habitats. The *Spartobacteria* currently comprise one order, the *Chthoniobacterales*, and a single family, *Chthoniobacteriaceae*. All members of this group are Gram-stain-negative.

Type order: **Chthoniobacterales** ord. nov.

Further descriptive information

To date, there appear to be three groups within the class *Spartobacteria*; however, since they are only separated by shallow branch length and only one is represented by pure cultures, they are currently all grouped within the family *Chthoniobacteriaceae* (Figure 151). Each of these groups is found predominantly in soils, where they are the dominant verrucomicrobia in PCR-based censuses of soil bacteria (Axelrood et al., 2002; Borneman and Triplett, 1997; Chow et al., 2002; Dunbar et al., 2002; Furlong et al., 2002; Holmes et al., 2000; Janssen, 2006; Liles et al., 2003; Macrae et al., 2000; Ochsenreiter et al., 2003; Ward-Rainey et al., 1995). *Spartobacteria* comprise up to 21.1% of 16S rRNA gene clones in clone libraries derived from soil DNA and are generally highly represented. Janssen (2006) reported a mean of 6.3% spartobacteria clones in a compilation of 16S rRNA gene clone library data derived from 21 studies of diverse soils using bacterial and universal PCR primers and Sanger DNA sequencing. More recent studies based on 454 pyrosequencing of hundreds of thousands of 16S rRNA gene fragments from geographically disparate soil samples showed somewhat lower percentages of sequences representing the phylum *Verrucomicrobia* (0.11–5%) (Elshahed et al., 2008; Fulthorpe et al., 2008; Lauber et al., 2009; Roesch et al., 2007). One phylotype, EA25, was shown by quantitative PCR to comprise up to 2×10^8 cells per gram of soil (Lee et al., 1996). *Spartobacteria* appear to be highly active in some soils, comprising from ~1 to ~10% of bacterial 16S rRNA in soils (Buckley and Schmidt, 2001, 2003; Felske and Akkermans, 1998; Felske et al., 1998, 2000).

The family *Chthoniobacteriaceae* is represented by ten isolates of which the species *Chthoniobacter flavus* is the only that has been taxonomically characterized (Sangwan et al., 2005). *Chthoniobacter flavus* is an aerobic saccharolytic heterotroph isolated from Australian soil (Sangwan et al., 2004). Relatives of this group have been detected in phylogenetic censuses of diverse soils including an alpine meadow (Costello and Schmidt, 2006), a broad-leaf evergreen forest (Chan et al., 2006), farm soil (Tringe et al., 2005), and grasslands. A somewhat separate clade is represented by three candidate species that appear to be obligate intracellular symbionts of different species of nematode (Vandekerckhove et al., 2002, 2000). Related 16S rRNA gene sequences have been recovered from geographically and geochemically distinct soils (Chan et al., 2006; Graff and Conrad, 2005; Kim et al., 2007), a contaminated aquifer (Dojka et al., 1998), and a sulfidic cave stream (Macalady et al., 2006). These uncultivated phylotypes are distant from "*Candidatus* Xiphinematobacter", so it is unclear whether they have a symbiotic or free-living lifestyle. A more distantly related phylotype from a humic lake may or may not be related to this group (Newton et al., 2006). The third group in the class is represented only by 16S rRNA gene sequences retrieved from soils and has no cultivated relatives.

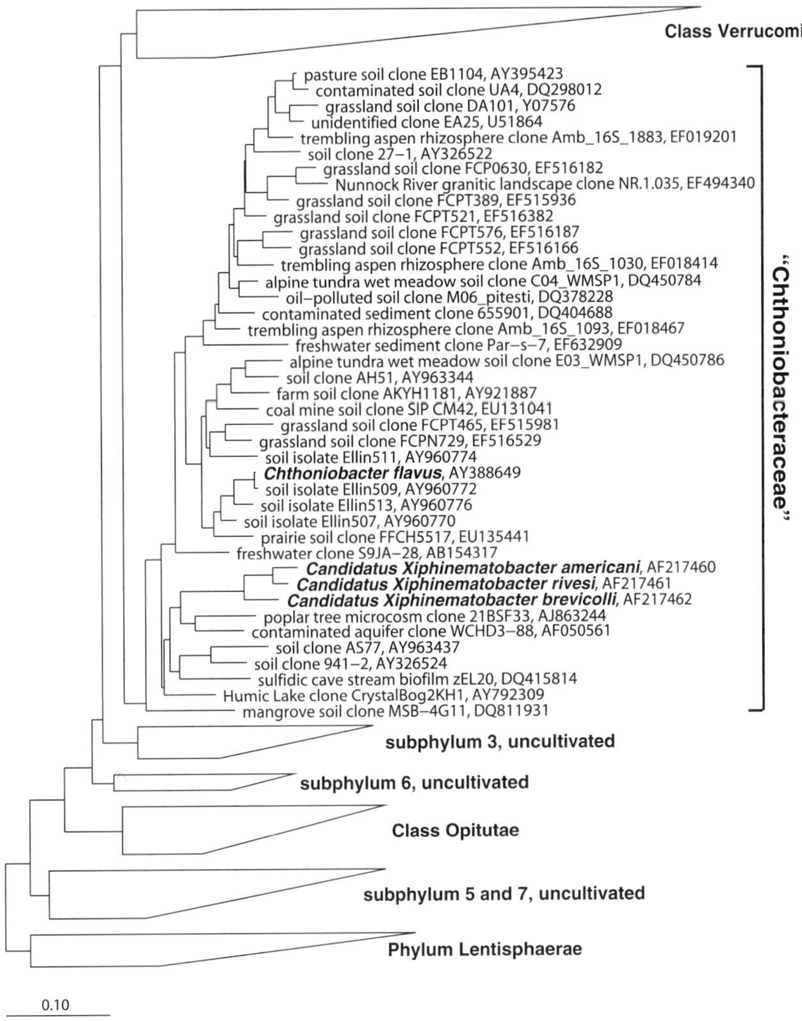

FIGURE 151. Maximum likelihood phylogeny of *Verrucomicrobia* with heuristic correction showing nearly complete 16S rRNA gene sequences present in version 91 of the SILVA database (Pruesse et al., 2007). All class-level groups were included, but all those except the *Spartobacteria* were wedged. The phylogeny was produced by using ARB (Ludwig et al., 2004) using *Escherichia coli* 16S rRNA gene nucleotides 110–1274. Alignments were unmasked.

References

Axelrood, P.E., M.L. Chow, C.C. Radomski, J.M. McDermott and J. Davies. 2002. Molecular characterization of bacterial diversity from British Columbia forest soils subjected to disturbance. Can. J. Microbiol. *48*: 655–674.

Borneman, J. and E.W. Triplett. 1997. Molecular microbial diversity in soils from eastern Amazonia: evidence for unusual microorganisms and microbial population shifts associated with deforestation. Appl. Environ. Microbiol. *63*: 2647–2653.

Buckley, D.H. and T.M. Schmidt. 2001. Environmental factors influencing the distribution of rRNA from *Verrucomicrobia* in soil. FEMS Microbiol. Ecol. *35*: 105–112.

Buckley, D.H. and T.M. Schmidt. 2003. Diversity and dynamics of microbial communities in soils from agro-ecosystems. Environ. Microbiol. *5*: 441–452.

Chan, O.C., X. Yang, Y. Fu, Z. Feng, L. Sha, P. Casper and X. Zou. 2006. 16S rRNA gene analyses of bacterial community structures in the soils of evergreen broad-leaved forests in south-west China. FEMS Microbiol. Ecol. *58*: 247–259.

Chow, M.L., C.C. Radomski, J.M. McDermott, J. Davies and P.E. Axelrood. 2002. Molecular characterization of bacterial diversity in Lodgepole pine (*Pinus contorta*) rhizosphere soils from British Columbia forest soils differing in disturbance and geographic source. FEMS Microbiol. Ecol. *42*: 347–357.

Dojka, M.A., P. Hugenholtz, S.K. Haack and N.R. Pace. 1998. Microbial diversity in a hydrocarbon- and chlorinated-solvent- contaminated aquifer undergoing intrinsic bioremediation. Appl. Environ. Microbiol. *64*: 3869–3877.

Dunbar, J., S.M. Barns, L.O. Ticknor and C.R. Kuske. 2002. Empirical and theoretical bacterial diversity in four Arizona soils. Appl. Environ. Microbiol. *68*: 3035–3045

Elshahed, M.S., N.H. Youssef, A.M. Spain, C. Sheik, F.Z. Najar, L.O. Sukharnikov, B.A. Roe, J.P. Davis, P.D. Schloss, V.L. Bailey and L.R. Krumholz. 2008. Novelty and uniqueness patterns of rare members of the soil biosphere. Appl. Environ. Microbiol. *74*: 5422–5428.

Felske, A. and A.D. Akkermans. 1998. Prominent occurrence of ribosomes from an uncultured bacterium of the *Verrucomicrobiales* cluster in grassland soils. Lett. Appl. Microbiol. *26*: 219–223.

Felske, A., A.D. Akkermans and W.M. de Vos. 1998. Quantification of 16S rRNAs in complex bacterial communities by multiple competitive reverse transcription-PCR in temperature gradient gel electrophoresis fingerprints. Appl. Environ. Microbiol. *64*: 4581–4587.

Felske, A., A. Wolterink, R. Van Lis, W.M. de Vos and A.D. Akkermans. 2000. Response of a soil bacterial community to grassland succession as monitored by 16S rRNA levels of the predominant ribotypes. Appl. Environ. Microbiol. *66*: 3998–4003.

Fulthorpe, R.R., L.F. Roesch, A. Riva and E.W. Triplett. 2008. Distantly sampled soils carry few species in common. ISME J. *2*: 901–910.

Furlong, M.A., D.R. Singleton, D.C. Coleman and W.B. Whitman. 2002. Molecular and culture-based analyses of prokaryotic communities from an agricultural soil and the burrows and casts of the earthworm *Lumbricus rubellus*. Appl. Environ. Microbiol. *68*: 1265–1279.

Graff, A. and R. Conrad. 2005. Impact of flooding on soil bacterial communities associated with poplar (*Populus* sp.) trees. FEMS Microbiol. Ecol. *53*: 401–415.

Holmes, A.J., J. Bowyer, M.P. Holley, M. O'Donoghue, M. Montgomery and M.R. Gillings. 2000. Diverse, yet-to-be-cultured members of the *Rubrobacter* subdivision of the *Actinobacteria* are widespread in Australian arid soils. FEMS Microbiol. Ecol. *33*: 111–120.

Hugenholtz, P., B.M. Goebel and N.R. Pace. 1998. Impact of culture-independent studies on the emerging phylogenetic view of bacterial diversity. J. Bacteriol. *180*: 4765–4774.

Janssen, P.H., P.S. Yates, B.E. Grinton, P.M. Taylor and M. Sait. 2002. Improved culturability of soil bacteria and isolation in pure culture of novel members of the divisions *Acidobacteria, Actinobacteria, Proteobacteria,* and *Verrucomicrobia*. Appl. Environ. Microbiol. *68*: 2391–2396.

Janssen, P.H. 2006. Identifying the dominant soil bacterial taxa in libraries of 16S rRNA and 16S rRNA genes. Appl. Environ. Microbiol. *72*: 1719–1728.

Kim, J.-S., Sparovek, G., Longo, R.M., De Melo, W.J. and D. Crowley. 2007. Bacterial diversity of terra preta and pristine forest soil from the Western Amazon. Soil Biol. Biochem. *39*: 684–690.

Lauber, C.L., M. Hamady, R. Knight and N. Fierer. 2009. Pyrosequencing-based assessment of soil pH as a predictor of soil bacterial community structure at the continental scale. Appl. Environ. Microbiol. *75*: 5111–5120.

Lee, S.Y., J. Bollinger, D. Bezdicek and A. Ogram. 1996. Estimation of the abundance of an uncultured soil bacterial strain by a competitive quantitative PCR method. Appl. Environ. Microbiol. *62*: 3787–3793.

Liles, M.R., B.F. Manske, S.B. Bintrim, J. Handelsman and R.M. Goodman. 2003. A census of rRNA genes and linked genomic sequences within a soil metagenomic library. Appl. Environ. Microbiol. *69*: 2684–2691.

Ludwig, W., O. Strunk, R. Westram, L. Richter, H. Meier, Yadhukumar, A. Buchner, T. Lai, S. Steppi, G. Jobb, W. Forster, I. Brettske, S. Gerber, A.W. Ginhart, O. Gross, S. Grumann, S. Hermann, R. Jost, A. Konig, T. Liss, R. Lussmann, M. May, B. Nonhoff, B. Reichel, R. Strehlow, A. Stamatakis, N. Stuckmann, A. Vilbig, M. Lenke, T. Ludwig, A. Bode and K.H. Schleifer. 2004. ARB: a software environment for sequence data. Nucleic Acids Res. *32*: 1363–1371.

Macalady, J.L., E.H. Lyon, B. Koffman, L.K. Albertson, K. Meyer, S. Galdenzi and S. Mariani. 2006. Dominant microbial populations in limestone-corroding stream biofilms, Frasassi cave system, Italy. Appl. Environ. Microbiol. *72*: 5596–5609.

Macrae, A., D.L. Rimmer and A.G. O'Donnell. 2000. Novel bacterial diversity recovered from the rhizosphere of oilseed rape (*Brassica napus*) determined by the analysis of 16S ribosomal DNA. Antonie van Leeuwenhoek *78*: 13–21.

Newton, R.J., A.D. Kent, E.W. Triplett and K.D. McMahon. 2006. Microbial community dynamics in a humic lake: differential persistence of common freshwater phylotypes. Environ. Microbiol. *8*: 956–970.

Ochsenreiter, T., D. Selezi, A. Quaiser, L. Bonch-Osmolovskaya and C. Schleper. 2003. Diversity and abundance of *Crenarchaeota* in terrestrial habitats studied by 16S RNA surveys and real time PCR. Environ. Microbiol. *5*: 787–797.

Pruesse, E., C. Quast, K. Knittel, B. Fuchs, W. Ludwig, J. Peplies and F.O. Glöckner. 2007. SILVA: a comprehensive online resource for quality checked and aligned rRNA sequence data compatible with ARB. Nucleic Acids Res. *35*: 7188–7196.

Roesch, L.F., R.R. Fulthorpe, A. Riva, G. Casella, A.K. Hadwin, A.D. Kent, S.H. Daroub, F.A. Camargo, W.G. Farmerie and E.W. Triplett. 2007. Pyrosequencing enumerates and contrasts soil microbial diversity. ISME J. *1*: 283–290.

Sangwan, P., X. Chen, P. Hugenholtz and P.H. Janssen. 2004. *Chthoniobacter flavus* gen. nov., sp. nov., the first pure-culture representative of subdivision two, *Spartobacteria* classis nov., of the phylum *Verrucomicrobia*. Appl. Environ. Microbiol. *70*: 5875–5881.

Sangwan, P., S. Kovac, K.E. Davis, M. Sait and P.H. Janssen. 2005. Detection and cultivation of soil *Verrucomicrobia*. Appl. Environ. Microbiol. *71*: 8402–8410.

Tringe, S.G., C. von Mering, A. Kobayashi, A.A. Salamov, K. Chen, H.W. Chang, M. Podar, J.M. Short, E.J. Mathur, J.C. Detter, P. Bork, P. Hugenholtz and E.M. Rubin. 2005. Comparative metagenomics of microbial communities. Science *308*: 554–557.

Vandekerckhove, T.T., A. Willems, M. Gillis and A. Coomans. 2000. Occurrence of novel verrucomicrobial species, endosymbiotic and associated with parthenogenesis in *Xiphinema americanum*-group species (Nematoda, Longidoridae). Int. J. Syst. Evol. Microbiol. *50*: 2197–2205.

Vandekerckhove, T.T., A. Coomans, K. Cornelis, P. Baert and M. Gillis. 2002. Use of the *Verrucomicrobia*-specific probe EUB338-III and fluorescent in situ hybridization for detection of "*Candidatus* Xiphinematobacter" cells in nematode hosts. Appl. Environ. Microbiol. *68*: 3121–3125.

Ward-Rainey, N., F.A. Rainey, H. Schlesner and E. Stackebrandt. 1995. Assignment of a hitherto unidentified 16S rDNA species to a main line of descent within the domain *Bacteria*. Microbiology *141*: 3247–3250.

Order I. **Chthoniobacterales** ord. nov.

Peter H. Janssen and Brian P. Hedlund

Chtho.ni.o.bac.ter.a'les. N.L. masc. n. *Chthoniobacter* type genus of the order; *-ales* ending to denote an order; N.L. fem. pl. n. *Chthoniobacterales* the *Chthoniobacter* order.

Encompasses bacteria isolated from soil environments and maternally transmitted endosymbionts of nematodes within the class *Spartobacteria* (Sangwan et al., 2004). The order contains the family *Chthoniobacteraceae*. Delineation of the order is determined primarily by phylogenetic information from 16S rRNA gene sequences. All known members of the order are rod-shaped or pleomorphic cells that are Gram-stain-negative and have a Gram-stain-negative cell structure.

Type genus: **Chthoniobacter** gen. nov.

Reference

Sangwan, P., X. Chen, P. Hugenholtz and P.H. Janssen. 2004. *Chthoniobacter flavus* gen. nov., sp. nov., the first pure-culture representative of subdivision two, *Spartobacteria* classis nov., of the phylum *Verrucomicrobia*. Appl. Environ. Microbiol. *70*: 5875–5881.

Family I. **Chthoniobacteraceae** fam. nov.

Peter H. Janssen and Brian P. Hedlund

Chtho.ni.o.bac.te.ra.ce'ae. N.L. masc. n. *Chthoniobacter* type genus of the family; *-aceae* ending to denote a family; N.L. fem. pl. n. *Chthoniobacteraceae* the *Chthoniobacter* family.

Rod-shaped or pleomorphic cells that stain **Gram-negative**. Nonmotile. Nonsporeforming. Includes free-living micro-organisms in soil and symbionts of nematodes. Free-living organisms are aerobic and saccharolytic and use predominantly mono-, di-, and polysaccharides and their derivatives. Menaquinones are the only respiratory quinones detected, primarily MK-10 and MK-11.

DNA G+C content (mol%): 61.

Type genus: **Chthoniobacter** Sangwan, Chen, Hugenholtz and Janssen 2010[VP] (Effective publication: Sangwan, Chen, Hugenholtz and Janssen 2004, 5880.)

Further descriptive information

Encompasses bacteria isolated from soil and associated 16S rRNA gene sequences recovered from soils and obligate cytoplasmic symbionts of nematodes within the order *Chthonobacteriales*. Delineation of the family is determined primarily by phylogenetic information from 16S rRNA gene sequences. Currently the family is represented by the genera *Chthoniobacter* and "*Candidatus* Xiphinematobacter".

Genus I. **Chthoniobacter** Sangwan, Chen, Hugenholtz and Janssen 2010[VP] (Effective publication: Sangwan, Chen, Hugenholtz and Janssen 2004, 5880.)

Peter H. Janssen

Chtho.ni.o.bac'ter. Gr. adj. *chthonios* born from the soil, also the name of one of the Spartoi of the Gr. Cadmus myth; N.L. masc. n. *bacter* the equivalent of Gr. neut. n. *baktron* rod or staff; N.L. masc. n. *Chthoniobacter* rod from the soil.

Rods. Gram-stain-negative, with Gram-negative cell-wall structure. Nonmotile. No endospores. **Aerobe.** Chemo-organotrophic metabolism. **Monosaccharides, disaccharides, and polysaccharides support growth. Cells contain menaquinones** and yellow pigment.

DNA G+C content (mol%): 61.

Type species: **Chthoniobacter flavus** Sangwan, Chen, Hugenholtz and Janssen 2010[VP] (Effective publication: Sangwan, Chen, Hugenholtz and Janssen 2004, 5880).

Enrichment and isolation procedures

Janssen et al. (2002) isolated the type strain of *Chthoniobacter flavus* from soil using plates of diluted nutrient agar at pH 6.0. Later, Sangwan et al. (2005) screened colonies appearing on a range of media with a group-specific oligonucleotide probe to detect the presence of members of the phylum *Verrucomicrobia*, and obtained another nine isolates closely related to *Chthoniobacter flavus*. These were all isolated from soil. Some of these isolates may belong to *Chthoniobacter flavus*, but others are likely to be members of new species of the genus. At this time, there is no known selective medium for strains of *Chthoniobacter* spp. The strategy used by Sangwan et al. (2005) of screening with group-specific oligonucleotide probes to detect the presence of members of the phylum *Verrucomicrobia* is the only published approach to obtain isolates other than by chance isolation.

Maintenance procedures

Cultures of *Chthoniobacter flavus* can be stored at −70°C in Difco nutrient broth containing 15% (v/v) glycerol. Cultures on plates can be maintained by 6-monthly subculture, and the plates stored at 4°C in sealed polyethylene bags.

Taxonomic comments

Only one strain, the type strain of *Chthoniobacter flavus*, has been formally assigned to this genus. Of nine strains of verrucomicrobia isolated by Sangwan et al. (2005), six were closely related to *Chthoniobacter flavus*, and shared >96% 16S rRNA gene sequence identity with the type strain. These are probably members of the genus *Chthoniobacter*. The other three other isolates appear to be more distantly related (<93% 16S rRNA gene sequence identity with *Chthoniobacter flavus*), but all nine strains were phenotypically very similar to each other and to *Chthoniobacter flavus*. In the absence of more detailed characterization of other isolates, and the delineation of further species, the separation of genus- and species-specific characteristics is somewhat arbitrary.

List of species of the genus *Chthoniobacter*

1. **Chthoniobacter flavus** Sangwan, Chen, Hugenholtz and Janssen 2004, 5880

fla'vus L. masc. adj. *flavus* yellow.

Cells are oval in shape, singly or in pairs, each 1.4 μm long and 0.9 μm in diameter. Cells are not motile and have no flagella. Cells contain poly-β-hydroxybutyrate and polyphosphate inclusions. Gram-stain negative, and Gram-negative cell-envelope structure, with peptidoglycan with direct cross-linkages of the A1γ *meso*-Dpm type. The major cellular fatty acids are tetradecanoic acid, 12-methyl-tetradecanoic acid, *cis*-9-hexadecenoic acid and/or 2-hydroxy-13-methyltetradecanoic acid, and hexadecanoic acid. Yellow pigmentation. Cells contain menaquinones MK-10 and MK-11. Form convex and entire colonies on agar surfaces.

Aerobic growth with sugars, sugar polymers, and pyruvate, but not with other organic acids, amino acids, or alcohols. No growth under anaerobic conditions, or with nitrate as electron acceptor. Catalase- and oxidase-negative. No *nifH* gene.

Grows at pH values of 4–7, and at temperatures between 25 and 34°C, but not at 37°C.

The type strain was isolated from pasture soil in Victoria, Australia.

DNA G+C content (mol%): 53 (draft genome sequence).

Type strain: Ellin428, ATCC BAA-1822.

Sequence accession no. (16S rRNA gene): AY388649.

Genus II. "***Candidatus* Xiphinematobacter**" Vandekerckhove, Willems, Gillis and Coomans 2000, 2203

Tom T. M. Vandekerckhove, Jason B. Navarro, August Coomans and Brian P. Hedlund

Xi.phi.ne.ma.to.bac'ter. Gr. neut. n. *Xiphinema, -atos* the genus name of the host organism; N.L. masc. n. *bacter* the equivalent of Gr. neut. n. *baktron* a rod; N.L. masc. n. *Xiphinematobacter* the rod-shaped microbe associated with *Xiphinema*.

Full-grown cells are **rod-shaped with rounded ends, 0.7–1.0 × 2.1–3.2 µm;** however, cells in the J_1 (first juvenile) stage of nematode development have a wrinkled, pleomorphic shape. The longer entities usually consist of a mother cell from which a daughter cell is budding, giving rise to serial pairs typical of this bacterial genus. In thin sections **cells have two or three membranes** consisting of, from inside to outside, a cytoplasmic membrane, an electron-dense outer membrane, and, in many individuals, a vacuolar membrane which is probably derived from the host cell membrane and which often shows discontinuities. **No peptidoglycan layer is evident**; however, a periplasmic hexagonally arrayed monolayer of 10 nm protein units is sometimes present. **Gram-stain-negative, nonmotile, and nonsporulating. DNA is often condensed at cell poles.** Bacteria live as **obligate cytoplasmic symbionts with maternal transmission in nematodes of the *Xiphinema americanum* group (Nematoda, Longidoridae), in which they are presumed to induce thelytokous (mother-to-daughter) parthenogenesis**.

DNA G+C content (mol%): not determined.

Type species: **"*Candidatus* Xiphinematobacter brevicolli"** Vandekerckhove, Willems, Gillis and Coomans 2000, 2203.

Further descriptive information

Intracellular bacterial symbionts in the female reproductive tissue of the dagger nematode *Xiphinema* (Figure 152) were identified by Vandekerckhove et al. (2000), who recovered 16S rRNA gene sequences representing the phylum *Verrucomicrobia* in three closely related species of nematode. Phylogenetic analysis showed that the symbionts make up a distinct monophyletic lineage within the class *Spartobacteria*. Subsequently the endosymbionts were confirmed as members of the phylum by fluorescent *in situ* hybridization (FISH) using a *Verrucomicrobia*-specific probe (Vandekerckhove et al., 2002) and the symbionts were grouped within the genus "*Candidatus* Xiphinematobacter".

DAPI (4′,6-diamidino-2-phenylindole) staining confirms transmission electron microscopic observations that "*Candidatus* Xiphinematobacter" DNA is normally condensed at cell poles (Figure 153), similar to cells of ectosymbionts of the ciliate *Euplotidium* (Petroni et al., 2000). Unravelling of the vertical (maternal) transmission pathway by conventional epifluorescence microscopy upon DAPI staining (Coomans et al., 2000) is corroborated by confocal laser scanning fluorescence microscopy using the FISH approach (Vandekerckhove et al., 2002).

"*Candidatus* Xiphinematobacter" appears to be the only bacterium living in association with dagger nematodes. Even the guts of *Xiphinema* specimens are devoid of other bacteria since these nematodes feed on root hair cells through a 400–450 nm diameter stylet, which shears incoming cells or blocks their entry. In juveniles, the endosymbionts occur in the gut epithelial cells. However, to be transmitted vertically "*Candidatus* Xiphinematobacter" must access the developing oocytes as host nematodes molt from the final juvenile stage (J_4) to adult. It appears that the primary route for the endosymbionts to travel from the gut epithelium to the oocytes occurs indirectly through yolk transfer (Coomans et al., 2000). Yolk is synthesized by the gut epithelia and secreted into the adjacent ovarian wall (or nurse) cells, in which endosymbionts proliferate. From there endosymbionts appear to be transferred to developing oocytes again with the yolk. "*Candidatus* Xiphinematobacter" undergoes a second proliferation in the egg; bacteria move to surround the nucleus of the intestinal stem cell of the early embryo (Figure 152), poising them to grow within the gut epithelia of the developing juvenile.

"*Candidatus* Xiphinematobacter" has been demonstrated in the male gut epithelium of rare males of *Xiphinema*; however they are less numerous and more often coccus-like than in an average female. Endosymbionts are not found in the ripe testes (Figure 152) and vertical transmission through sperm is highly unlikely since sperm contain an extremely reduced cytoplasm. There is no evidence of an extracellular or free-living phase of "*Candidatus* Xiphinematobacter" and horizontal transmission seems extremely unlikely.

"*Candidatus* Xiphinematobacter" has been observed in each of the hundreds of female specimens of the *Xiphinema americanum* group that have been examined, though occasionally individuals with few symbionts are discovered. It has been suggested that "*Candidatus* Xiphinematobacter" symbionts induce thelytokous (mother-to-daughter) parthenogenesis in their nematode hosts since males of these species are rare (Coomans and Willems, 1998). This activity was previously only ascribed to the alphaproteobacterium *Wolbachia*, which is an obligate intracellular symbiont of arthropods and filarial nematodes. Indeed, the only known parthenogenic species of *Xiphinema*, members of the *Xiphinema americanum* group, are

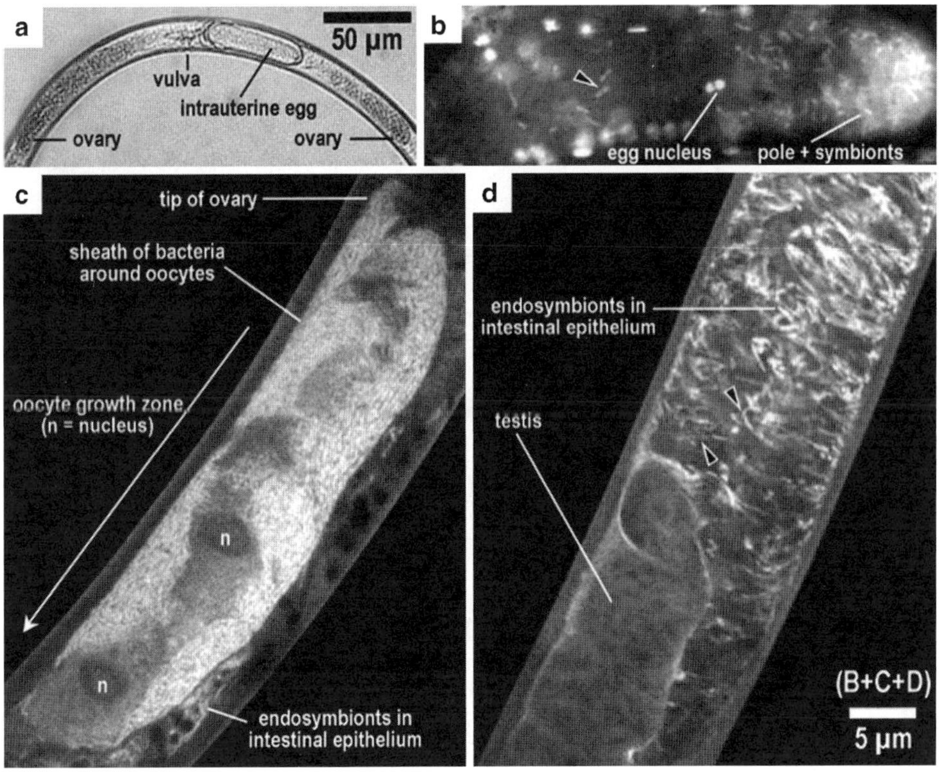

FIGURE 152. Light microscopy of *Xiphinema brevicollum*. Arrowheads point to DNA lumps (often two, sometimes more when budding) at bacterial cell poles. (a) Transmitted light. Adult female with relevant parts of reproductive system. The morphology is described in more detail in Coomans and Claeys (1998). (b) DAPI epifluorescence (Coomans et al., 2000). Intrauterine egg with dividing nucleus and endosymbiont accumulation at the egg pole where the gut stem cell will arise. A few very bright somatic nematode nuclei are also visible at the periphery. (c, d) Confocal fluorescence using FISH with the verrucomicrobial 16S rRNA probe EUB338-III (Vandekerckhove et al., 2002). Adult female (c) with densely packed symbionts in an ovary and underlying gut, and adult male (d) with testis lacking symbionts.

also the only members of the genus known to harbor bacterial endosymbionts. It is unknown whether other sex ratio altering strategies employed by *Wolbachia* are used by "*Candidatus* Xiphinematobacter", such as induction of feminization, killing of male embryos, or cytoplasmic incompatibility (reviewed by Stouthamer et al., 1999).

Phylogeny and classification

The three candidate species of "*Candidatus* Xiphinematobacter" are symbionts of different species of host nematode. The symbionts are phylogenetically well defined; they form a monophyletic group and share a mean 93% 16S rRNA gene homology. Related nucleic acid sequences from soil may or may not represent endosymbionts. The congruent branching order of "*Candidatus* Xiphinematobacter" and host nematodes phylogenies is consistent with the vertical, rather than horizontal, transmission of the symbionts.

Differentiation of the genus "*Candidatus* Xiphinematobacter" from closely related taxa

"*Candidatus* Xiphinematobacter" can be presumptively identified by its location within intestinal cells or ovarian cells in juveniles or adults, respectively, of "dagger" nematodes of the *Xiphinema americanum* group (Nematoda, Longidoridae). The genus *Wolbachia* is not found in this group of nematodes.

Furthermore, "*Candidatus* Xiphinematobacter" displays a possibly unique combination of morphological peculiarities such as the occurrence of numerous serial pairs of cells (e.g., Figures 152 and 153), polar lumps of DNA (e.g., Figures 152 and 153), and an aberrant Gram-stain-negative cell wall where the peptidoglycan layer is replaced by a paracrystalline protein monolayer (Figure 153).

Confirmation of the phylogenetic identification can be obtained by 16S rRNA gene sequencing and phylogenetic staining (e.g., FISH), as done by Vandekerckhove et al. (2000).

List of species of the genus "*Candidatus* Xiphinematobacter"

1. **"*Candidatus* Xiphinematobacter brevicolli"** Vandekerckhove, Willems, Gillis and Coomans 2000, 2203

bre.vi.col'li. N.L. gen. n. *brevicolli* of the *brevicollum* species, i.e., lives in the species *Xiphinema brevicollum*.

Characteristics as for the genus, of which it is the type species. Specific host is the nematode *Xiphinema brevicollum*, with which is has established the oldest of the three symbioses. It shares a mean corrected 16S rRNA gene identity with

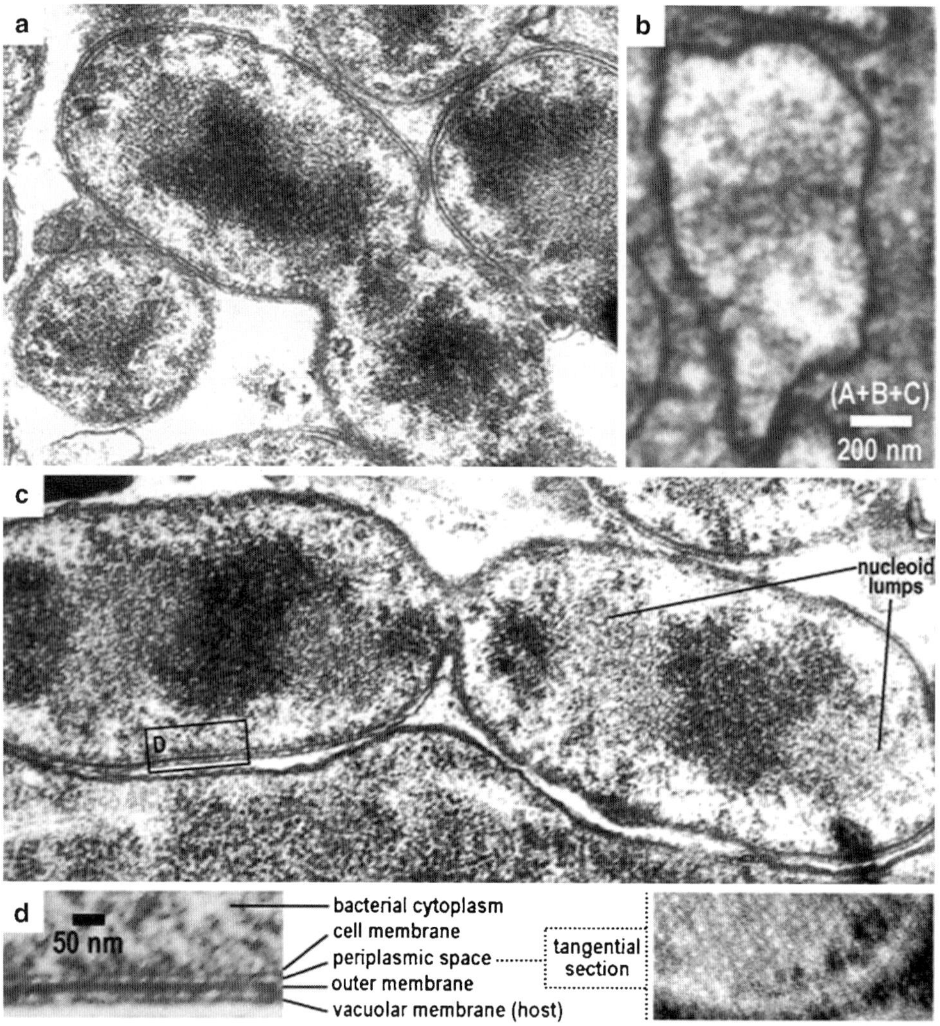

FIGURE 153. Transmission electron microscopy of "*Candidatus* Xiphinematobacter" endosymbionts of dagger nematodes (Vandekerckhove et al., 2000). (a) Typical appearance of an actively budding cell. (b) Pleomorphic cell variant observed in the J_1 stage of nematode development. (c) Serial pair of bacterial mother and daughter cells just before budding off. Note the polar DNA lumps in both cells and the trilaminar envelope around them (as detailed in part d). (d) High magnification of the trilaminar structure surrounding "*Candidatus* Xiphinematobacter". Note the absence of a peptidoglycan layer in the periplasm (left, cross section), which is replaced by a paracrystalline monolayer of hexagonally arrayed globular proteins (right, tangential section).

"*Candidatus* Xiphinematobacter americani" sp. nov. and "*Candidatus* Xiphinematobacter rivesi" sp. nov. of about 90%.

DNA G+C content (mol%): not determined.

Type strain: no type strain.

Sequence accession no. (16S rRNA gene): AF217462.

2. **"*Candidatus* Xiphinematobacter americani"** Vandekerckhove, Willems, Gillis and Coomans 2000, 2203

a.me.ri.ca'ni. N.L. gen. n. *americani* of the *americanum* species, i.e., lives in the species *Xiphinema americanum.*

Characteristics as for the genus. Specific host is the nematode *Xiphinema americanum.* Closest relative is "*Candidatus* Xiphinematobacter rivesi" sp. nov., having a 16S rRNA gene sequence 3.15% distant.

DNA G+C content (mol%): not determined.

Type strain: no type strain.

Sequence accession no. (16S rRNA gene): AF217460.

3. **"*Candidatus* Xiphinematobacter rivesi"** Vandekerckhove, Willems, Gillis and Coomans 2000, 2203

ri.ve'si. N.L. gen. n. *rivesi* of Rives, i.e., lives in the species *Xiphinema rivesi.*

Characteristics as for the genus. Specific host is the nematode *Xiphinema rivesi.* Closest relative is "*Candidatus* Xiphinematobacter americani" sp. nov., having a 16S rRNA gene sequence 3.15% distant.

DNA G+C content (mol%): not determined.

Type strain: no type strain.

Sequence accession no. (16S rRNA gene): AF217461.

References

Coomans, A. and M. Claeys. 1998. Structure of the female reproductive system of *Xiphinema americanum* (Nematoda, Longidoridae). Fundam. Appl. Nematol. *21*: 569–580.

Coomans, A. and A. Willems. 1998. What are symbiotic bacteria doing in the ovaria of *Xiphinema americanum*-group subspecies? Nematologica *44*: 323–326.

Coomans, A., T.T.M. Vandekerckhove and M. Claeys. 2000. Transovarial transmission of symbionts in *Xiphinema brevicollum* (Nematoda: Longidoridae). Nematology *2*: 455–461.

Petroni, G., S. Spring, K.H. Schleifer, F. Verni and G. Rosati. 2000. Defensive extrusive ectosymbionts of *Euplotidium* (Ciliophora) that contain microtubule-like structures are bacteria related to *Verrucomicrobia*. Proc. Natl. Acad. Sci. U. S. A. *97*: 1813–1817.

Sangwan, P., X. Chen, P. Hugenholtz and P.H. Janssen. 2004. *Chthoniobacter flavus* gen. nov., sp. nov., the first pure-culture representative of subdivision two, *Spartobacteria* classis nov., of the phylum *Verrucomicrobia*. Appl. Environ. Microbiol. *70*: 5875–5881.

Sangwan, P., S. Kovac, K.E. Davis, M. Sait and P.H. Janssen. 2005. Detection and cultivation of soil *Verrucomicrobia*. Appl. Environ. Microbiol. *71*: 8402–8410.

Stouthamer, R., J.A. Breeuwer and G.D. Hurst. 1999. *Wolbachia pipientis*: microbial manipulator of arthropod reproduction. Annu. Rev. Microbiol. *53*: 71–102.

Vandekerckhove, T.T., A. Willems, M. Gillis and A. Coomans. 2000. Occurrence of novel verrucomicrobial species, endosymbiotic and associated with parthenogenesis in *Xiphinema americanum*-group species (Nematoda, Longidoridae). Int. J. Syst. Evol. Microbiol. *50*: 2197–2205.

Vandekerckhove, T.T., A. Coomans, K. Cornelis, P. Baert and M. Gillis. 2002. Use of the *Verrucomicrobia*-specific probe EUB338-III and fluorescent in situ hybridization for detection of "*Candidatus* Xiphinematobacter" cells in nematode hosts. Appl. Environ. Microbiol. *68*: 3121–3125.

Phylum XXIV. Chlamydiae Garrity and Holt 2001

Matthias Horn

Chla.my.di′ae. N.L. fem. pl. n. *Chlamydiales* type order of the phylum; N.L. fem. pl. n. *Chlamydiae* the phylum of the order *Chlamydiales.*

The phylum *Chlamydiae* is based on phylogenetic analysis of 16S rRNA sequences, clearly separating its members (showing ≥80% 16S rRNA sequence similarity with each other) from all other *Bacteria* by a long, deep branch in phylogenetic trees (Figure 154). All known members of this phylum are nonmotile, obligate intracellular bacteria and multiply in eukaryotic hosts (animals including humans, or protozoa). Unlike most other bacteria, they show a developmental cycle characterized by morphologically and physiologically distinct stages. All chlamydiae characterized so far possess small genomes (1–2.4 Mb); they are metabolically impaired and need to import essential building blocks like nucleotides, amino acids, and cofactors from their host cells. The intracellular life style of chlamydiae is thus thought to be an ancient trait of this phylum. Chlamydiae show a Gram-negative type cell wall, but they lack detectable amounts of peptidoglycan. They do not encode the cell division protein FtsZ, which is otherwise considered essential for bacteria. Chlamydiae possess several unique proteins found in no other microorganisms (n = 59) (Griffiths et al., 2006) and conserved insertions and deletions in widely distributed proteins (Griffiths and Gupta, 2007; Griffiths et al., 2005; Gupta and Griffiths, 2006).

Members of the phyla *Lentisphaera* and *Verrucomicrobia* seem to be the closest free-living relatives of *Chlamydiae* based on phylogenetic analysis of the 16S rRNA gene, concatenated protein datasets, and two unique inserts in LysRS and RpoB proteins (Griffiths and Gupta, 2007; Strous et al., 2006; Wagner and

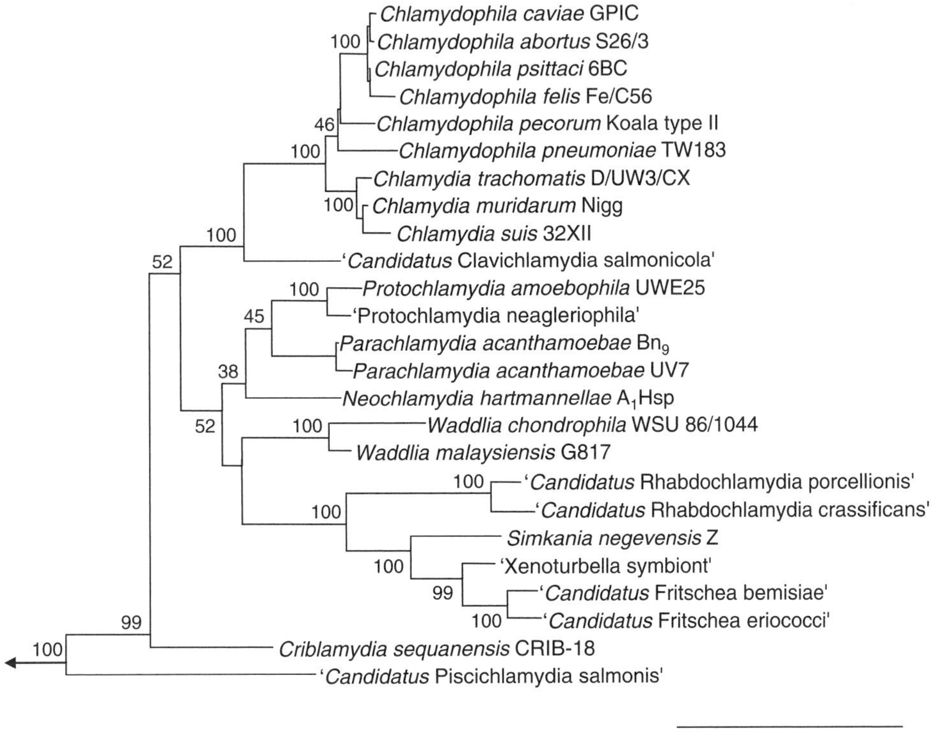

FIGURE 154. Phylogenetic tree showing the relationships of representative members of the *Chlamydiae.* The tree has been calculated with the ARB software, using a manually curated version of the SILVA database (SSU Ref, version 92) and the PhyML treeing algorithm implemented in ARB (Guindon et al., 2005; Ludwig et al., 2004; Pruesse et al., 2007). The HKY substitution model, empirical base frequency estimates, and a fixed transition/transversion ratio of 4.00 were used; non-parametric bootstrap analysis (100 resamplings) and a filter excluding highly variable positions were applied. Bar = 10% estimated evolutionary distance; arrow, to outgroup.

Horn, 2006). In addition, a phylogenetic relationship of the *Chlamydiae* with the phyla *Planctomycetes* and *Poribacteria* has been proposed (Cavalier-Smith, 2002; Wagner and Horn, 2006).

The phylum *Chlamydiae* currently comprises a single class, *Chlamydiia*, containing only one order, the *Chlamydiales*.

Type order: **Chlamydiales** Storz and Page 1971, 334[AL].

References

Cavalier-Smith, T. 2002. The phagotrophic origin of eukaryotes and phylogenetic classification of Protozoa. Int. J. Syst. Evol. Microbiol. *52*: 297–354.

Garrity, G.M. and J.G. Holt. 2001. The Road Map to the *Manual. In* Bergey's Manual of Systematic Bacteriology, 2nd edn, vol. 1, The *Archaea* and the Deeply Branching and Phototrophic *Bacteria* (edited by Boone, Castenholz and Garrity). Springer, New York, pp. 119–166.

Griffiths, E., A.K. Petrich and R.S. Gupta. 2005. Conserved indels in essential proteins that are distinctive characteristics of *Chlamydiales* and provide novel means for their identification. Microbiology *151*: 2647–2657.

Griffiths, E., M.S. Ventresca and R.S. Gupta. 2006. BLAST screening of chlamydial genomes to identify signature proteins that are unique for the *Chlamydiales*, *Chlamydiaceae*, *Chlamydophila* and *Chlamydia* groups of species. BMC Genomics *7*: 14.

Griffiths, E. and R.S. Gupta. 2007. Phylogeny and shared conserved inserts in proteins provide evidence that *Verrucomicrobia* are the closest known free-living relatives of chlamydiae. Microbiology *153*: 2648–2654.

Guindon, S., F. Lethiec, P. Duroux and O. Gascuel. 2005. PHYML Online – a web server for fast maximum likelihood-based phylogenetic inference. Nucleic Acids Res. *33*: W557–W559.

Gupta, R.S. and E. Griffiths. 2006. Chlamydiae-specific proteins and indels: novel tools for studies. Trends Microbiol. *14*: 527–535.

Ludwig, W., O. Strunk, R. Westram, L. Richter, H. Meier, Yadhukumar, A. Buchner, T. Lai, S. Steppi, G. Jobb, W. Forster, I. Brettske, S. Gerber, A.W. Ginhart, O. Gross, S. Grumann, S. Hermann, R. Jost, A. Konig, T. Liss, R. Lussmann, M. May, B. Nonhoff, B. Reichel, R. Strehlow, A. Stamatakis, N. Stuckmann, A. Vilbig, M. Lenke, T. Ludwig, A. Bode and K.H. Schleifer. 2004. ARB: a software environment for sequence data. Nucleic Acids Res. *32*: 1363–1371.

Pruesse, E., C. Quast, K. Knittel, B. Fuchs, W. Ludwig, J. Peplies and F.O. Glöckner. 2007. SILVA: a comprehensive online resource for quality checked and aligned rRNA sequence data compatible with ARB. Nucleic Acids Res. *35*: 7188–7196.

Storz, J. and L.A. Page. 1971. Taxonomy of the chlamydiae: reasons for classifying organisms of the genus *Chlamydia*, family *Chlamydiaceae*, in a separate order *Chlamydiales* ord. nov. Int. J. Syst. Bacteriol. *21*: 332–334.

Strous, M., E. Pelletier, S. Mangenot, T. Rattei, A. Lehner, M.W. Taylor, M. Horn, H. Daims, D. Bartol-Mavel, P. Wincker, V. Barbe, N. Fonknechten, D. Vallenet, B. Segurens, C. Schenowitz-Truong, C. Medigue, A. Collingro, B. Snel, B.E. Dutilh, H.J. Op den Camp, C. van der Drift, I. Cirpus, K.T. van de Pas-Schoonen, H.R. Harhangi, L. van Niftrik, M. Schmid, J. Keltjens, J. van de Vossenberg, B. Kartal, H. Meier, D. Frishman, M.A. Huynen, H.W. Mewes, J. Weissenbach, M.S. Jetten, M. Wagner and D. Le Paslier. 2006. Deciphering the evolution and metabolism of an anammox bacterium from a community genome. Nature *440*: 790–794.

Wagner, M. and M. Horn. 2006. The *Planctomycetes*, *Verrucomicrobia*, *Chlamydiae* and sister phyla comprise a superphylum with biotechnological and medical relevance. Curr. Opin. Biotechnol. *17*: 241–249.

Class I. **Chlamydiia** class. nov.

Matthias Horn

Chla.my.di'ia. N.L. fem. pl. n. *Chlamydiales* type order of the class; N.L. fem. pl. n. *Chlamydiia* the *Chlamydiales* class.

Members of the class *Chlamydiia* have the following common features: an obligate intracellular life style, the absence of flagella and peptidoglycan, and a proteinaceous cell wall. The class *Chlamydiia* contains but a single order, the *Chlamydiales*.

Type order: **Chlamydiales** Storz and Page 1971, 334[AL].

Reference

Storz, J. and L.A. Page. 1971. Taxonomy of the chlamydiae: reasons for classifying organisms of the genus *Chlamydia*, family *Chlamydiaceae*, in a separate order *Chlamydiales* ord. nov. Int. J. Syst. Bacteriol. *21*: 332–334.

Order I. **Chlamydiales** Storz and Page 1971, 334[AL]

Cho-chou Kuo, Matthias Horn and Richard S. Stephens

Chla.my.di.a'les. N.L. fem. n. *Chlamydia* type genus of the order; -*ales* ending to denote an order; N.L. fem. pl. n. *Chlamydiales* the *Chlamydia* order.

Gram-stain-negative, obligately intracellular microorganisms that live in eukaryotes and may cause various diseases. The mode of multiplication within the cytoplasmic vacuoles is characterized by reorganization of small, rigid-walled, metabolically inactive infectious forms (elementary bodies) into large, flexible-walled, metabolically active forms (reticulate bodies) that are generally noninfectious and divide by binary fission. The developmental cycle is complete when the reticulate bodies reorganize into elementary bodies that survive extracellulary to infect new host cells upon their release from the host cell. The order *Chlamydiales* traditionally contained a single family, the *Chlamydiaceae*, which included the type genus *Chlamydia*. The recent discovery of novel chlamydia-related bacteria expanded the order *Chlamydiales* to include four additional families, having 80–90% 16S rRNA sequence similarity with the *Chlamydiaceae* (Everett et al., 1999; Rurangirwa et al., 1999; Thomas et al., 2006). Three further families are added based on the descriptions of "*Candidatus* Piscichlamydia salmonis", "*Candidatus* Clavichlamydia salmonicola", and "*Candidatus* Rhabdochlamydia porcellionis" (Draghi et al., 2004; Karlsen et al., 2008; Kostanjsek et al., 2004), respectively, all of which show 80–90% 16S rRNA sequence similarity with all other *Chlamydiales* families (Figure 154).

For classification of members of the *Chlamydiales* at the family and genus level, the 16S rRNA sequence similarity thresholds proposed by Everett et al. (1999) are currently accepted. Members of a *Chlamydiales* family generally share ≥90% 16S rRNA similarity with each other; members of a *Chlamydiales* genus generally share ≥95% 16S rRNA sequence similarity, with the *Chlamydiaceae* being a notable exception. As the proposed separation of the *Chlamydiaceae* into the two genera *Chlamydia* and *Chlamydophila* (Everett et al., 1999) is not being used consistently in the scientific literature, it was recently proposed to abandon the genus name *Chlamydophila* and to transfer all *Chlamydophila* species to the genus *Chlamydia* (Bavoil and Wyrick, 2006; Kalayoglu and Byrne, 2006).

The order *Chlamydiales* consists currently of eight families, *Chlamydiaceae, Clavichlamydiaceae, Criblamydiaceae, Parachlamydiaceae, Piscichlamydiaceae, Rhabdochlamydiaceae, Simkaniaceae,* and *Waddliaceae.*

Further evidence for an even greater diversity of yet uncultured chlamydiae exists (Corsaro et al., 2003; Horn and Wagner, 2001; Ossewaarde and Meijer, 1999). Several oligonucleotide probes for the identification of chlamydiae by fluorescence *in situ* hybridization at different taxonomic levels are available (Loy et al., 2007; Poppert et al., 2002).

DNA G+C content (mol%): 35.8–41.3.

Type genus: **Chlamydia** Jones, Rake and Stearns 1945, 55[AL] emend. Everett, Bush and Andersen 1999, 430.

References

Bavoil, P.M. and P.B. Wyrick. 2006. Chlamydia: genomics and pathogenesis. Horizon Bioscience, Norwich, UK.

Corsaro, D., M. Valassina and D. Venditti. 2003. Increasing diversity within Chlamydiae. Crit. Rev. Microbiol. *29*: 37–78.

Draghi, A., 2nd, V.L. Popov, M.M. Kahl, J.B. Stanton, C.C. Brown, G.J. Tsongalis, A.B. West and S. Frasca, Jr. 2004. Characterization of "*Candidatus* piscichlamydia salmonis" (order *Chlamydiales*), a chlamydia-like bacterium associated with epitheliocystis in farmed Atlantic salmon (*Salmo salar*). J. Clin. Microbiol. *42*: 5286–5297.

Everett, K.D.E., R.M. Bush and A.A. Andersen. 1999. Emended description of the order *Chlamydiales*, proposal of *Parachlamydiaceae* fam. nov. and *Simkaniaceae* fam. nov., each containing one monotypic genus, revised taxonomy of the family *Chlamydiaceae*, including a new genus and five new species, and standards for the identification of organisms. Int. J. Syst. Bacteriol. *49*: 415–440.

Horn, M. and M. Wagner. 2001. Evidence for additional genus-level diversity of *Chlamydiales* in the environment. FEMS Microbiol. Lett. *204*: 71–74.

Jones, H., G. Rake and B. Stearns. 1945. Studies on lymphogranuloma venereum. III. The action of the sulfonamides on the agent of lymphogranuloma venereum. J. Infect. Dis. *76*: 55–69.

Kalayoglu, M.V. and G.I. Byrne. 2006. The genus *Chlamydia* - medical. *In* The Prokaryotes: a Handbook on the Biology of Bacteria, 3rd edn, vol. 7, *Proteobacteria*: Delta and Epsilon Subclasses. Deeply Rooting Bacteria (edited by Dworkin, Falkow, Rosenberg, Schleifer and Stackebrandt). Springer, New York, pp. 741–754.

Karlsen, M., A. Nylund, K. Watanabe, J.V. Helvik, S. Nylund and H. Plarre. 2008. Characterization of '*Candidatus* Clavochlamydia salmonicola': an intracellular bacterium infecting salmonid fish. Environ. Microbiol. *10*: 208–218.

Kostanjsek, R., J. Strus, D. Drobne and G. Avgustin. 2004. '*Candidatus* Rhabdochlamydia porcellionis', an intracellular bacterium from the hepatopancreas of the terrestrial isopod *Porcellio scaber* (Crustacea: Isopoda). Int. J. Syst. Evol. Microbiol. *54*: 543–549.

Loy, A., F. Maixner, M. Wagner and M. Horn. 2007. probeBase – an online resource for rRNA-targeted oligonucleotide probes: new features 2007. Nucleic Acids Res. *35*: D800–D804.

Ossewaarde, J.M. and A. Meijer. 1999. Molecular evidence for the existence of additional members of the order *Chlamydiales*. Microbiology *145*: 411–417.

Poppert, S., A. Essig, R. Marre, M. Wagner and M. Horn. 2002. Detection and differentiation of chlamydiae by fluorescence *in situ* hybridization. Appl. Environ. Microbiol. *68*: 4081–4089.

Rurangirwa, F.R., P.M. Dilbeck, T.B. Crawford, T.C. McGuire and T.F. McElwain. 1999. Analysis of the 16S rRNA gene of micro-organism WSU 86-1044 from an aborted bovine foetus reveals that it is a member of the order *Chlamydiales*: proposal of *Waddliaceae* fam. nov., *Waddlia chondrophila* gen. nov., sp. nov. Int. J. Syst. Bacteriol. *49*: 577–581.

Storz, J. and L.A. Page. 1971. Taxonomy of the chlamydiae: reasons for classifying organisms of the genus *Chlamydia*, family *Chlamydiaceae*, in a separate order *Chlamydiales* ord. nov. Int. J. Syst. Bacteriol. *21*: 332–334.

Thomas, V., N. Casson and G. Greub. 2006. *Criblamydia sequanensis*, a new intracellular *Chlamydiales* isolated from Seine river water using amoebal co-culture. Environ. Microbiol. *8*: 2125–2135.

Family I. **Chlamydiaceae** Rake 1957, 957[AL]

Cho-Chou Kuo and Richard S. Stephens

Chla.my.di.a.ce′ae. N.L. n. *Chlamydia* type genus of the family; *-aceae* ending to denote family; N.L. fem. pl. n. *Chlamydiaceae* the *Chlamydia* family.

In the previous edition of *Bergey's Manual of Systematic Bacteriology* (Volume 1, 1984), the family contained a single genus, *Chlamydia.* The family was emended to include an additional genus, *Chlamydophila*, by Everett et al. in 1999. The new genus and five species, *Chlamydophila abortus, Chlamydophila caviae, Chlamydophila felis, Chlamydophila pecorum, Chlamydophila pneumoniae,* and *Chlamydophila psittaci* were validly published, but the scientific community has not uniformly adopted the use of the new genus name. For this reason, in this edition of the *Manual* the original genus *Chlamydia* is retained as the sole genus in the family and the five additional species are classified in the genus.

DNA G+C content (mol%): 39–45.

Type genus: **Chlamydia** Jones, Rake and Stearns 1945, 55[AL].

Genus I. Chlamydia Jones, Rake and Stearns 1945, 55[AL]

Cho-Chou Kuo, Richard S. Stephens, Patrik M. Bavoil and Bernhard Kaltenboeck

Chla.my'di.a. Gr. fem. n. *chlamys, chlamydis* a cloak; N.L. fem. n. *Chlamydia* a cloak.

Coccoid, nonmotile, obligate intracellular organisms, 0.2–1.5 µm in diameter that parasitize and multiply in the cytoplasm of eukaryotic cells within membrane bound vacuoles, termed inclusions, by a unique developmental cycle. The cycle is characterized by physical changes in the outer membrane and nucleoid structure of metabolically inert elementary bodies that change into larger metabolically active reticulate bodies that divide by binary fission. The cycle is complete when reticulate bodies reorganize into intermediate bodies then to elementary bodies that exit the host cells and start a new generation of the developmental cycle. In a young inclusion, only reticulate bodies are present. However in a mature inclusion, all three forms (reticulate, intermediate, and elementary bodies) are present because the developmental cycle is not synchronized. The intracellular survival of the organism is facilitated by its ability to prevent fusion of lysosomes to chlamydia-containing inclusions by an unknown mechanism.

Elementary bodies are 0.2–0.4 µm in diameter, have electron-dense DNA condensed with protein and a few ribosomes, and are surrounded by rigid trilaminar cell walls resistant to sonic lysis. Elementary bodies are infectious. **Reticulate bodies** are 0.6–1.5 µm in diameter, have less dense, fibrillar nuclear material, more ribosomes, and plastic trilaminar walls that are sensitive to sonic lysis. Reticulate bodies are not infectious. **Intermediate bodies** are organisms in transition from reticulate bodies to elementary bodies that are variable in size and have a small plaque of dense nuclear material in the center of the cytoplasm with a characteristic target-like appearance. Infectivity of intermediate bodies has not been demonstrated.

Cell walls of elementary and reticulate bodies exhibit regular hexagonal arrays of subunits on the inner surface and single patches of hexagonal ordered hemispheric projections on their outer surface that resemble in structure the type III secretion apparatus of Gram-stain-negative bacteria. Cell walls contain lipopolysaccharide and are analogous in structure and composition to walls of Gram-stain-negative bacteria, however, muramic acid is either absent or is present only in trace amounts in elementary bodies.

Chlamydiae cause a variety of diseases in humans, other mammals, and birds. Cultivation of *Chlamydia* in cell-free medium has not been achieved. They may be propagated in laboratory animals, the yolk sac of chicken embryos, or in cell culture. Chlamydiae require their hosts for low-molecular-weight synthetic intermediates that they use to synthesize DNA, RNA, proteins, and lipids. The genome size is among the smallest of prokaryotes, 1–1.3 × 10^6 bp.

DNA G+C content (mol%): 39–45.

Type species: **Chlamydia trachomatis** (Busacca 1935) Rake 1957, 958[AL] [*Rickettsia trachomae* (*sic*) Busacca 1935, 567].

Further descriptive information

The voluminous *Chlamydia* literature has been exhaustively reviewed elsewhere. Some classical reviews, but mainly recent publications in books and articles of broad coverage on the subject treated in this chapter have been cited individually under

TABLE 160. Characteristics of chlamydial elementary and reticulate bodies[a,b]

Characteristic	Elementary	Reticulate
General:		
Diameter (µm)	0.2–0.4	0.5–1.5
Density (g/cm^3)	1.21	1.18
Time of appearance in developmental cycle	Late	Early
Infectivity for new host cells	+	–
Intravenous lethality for mice	+	–
Immediate cytotoxicity in cell culture	+	–
Cell wall:		
Rigidity	Rigid	Flexible
Cross-linking of outer membrane	+	–
Susceptibility to:		
Mechanical stress	–	+
Osmotic stress	–	+
Lysis by trypsin	–	+
Trilamminar structure	+	+
Projections on surface	+	+
Hexagonal arrays on inner surface	+	+
Muramic acid	–	nd
Synthesis inhibited by penicillin	+	–
Genus-specific antigen	+	+
Hemagglutinin	+	–
Nucleic acids:		
DNA	Compact	Disperse
RNA/DNA ratio	1	3–4
Ribosomes	Scanty	Abundant
Metabolism:		
Net generation of ATP	–	–
ATP/ADP transport system	–	–
ATP-dependent liberation of CO_2 from glucose 6-P, pyruvate, glutamate, and aspartate	+[c]	+[c]
ATP-dependent host-free protein synthesis	–	+

[a]Symbols: +, >85% positive; –, 0–15% positive; nd, not determined.

[b]Adapted and modified from Table 9.13 in the chapter on the Genus *Chlamydia* in *Bergey's Manual of Systematic Bacteriology*, 1984, Vol. 1, p. 730.

[c]These experiments were conducted with mixtures of the two chlamydial cell types, so it is not certain whether only one or both types were responsible for the observed metabolic activity.

Further Reading. Articles on the specific subjects discussed in this chapter are cited in the Bibliography Section in this book.

Cell morphology and developmental cycle. The structural forms of chlamydiae as well as their composition and biology are highly dependent on the developmental cycle. Therefore, the description of these characteristics can only be described in the context of the developmental cycle. The properties of the two principal cell forms are summarized in Table 160 and Figure 155. The developmental cycle may be divided into three phases: (1) attachment and entry of elementary bodies into host cells and their reorganization into reticulate bodies,

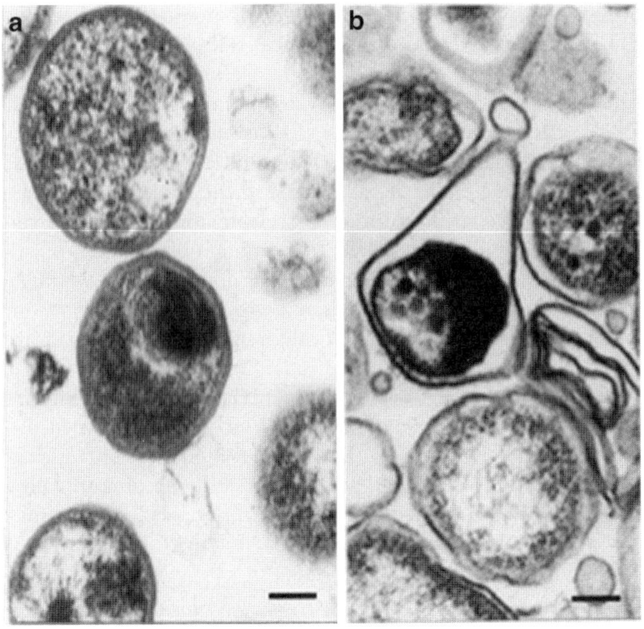

FIGURE 155. Elementary body form of chlamydiae is round, except for *Chlamydia pneumoniae* which is pear-shaped. (a) *Chlamydia trachomatis* serovar B, strain TW-5, (b). *Chlamydia pneumoniae* strain TW-183. The reticulate body is round for all species. Bars = 0.1 µm. (Reproduced with permission from Grayston et al., 1989. Int. J. Syst. Bacteriol. *39:* 88–90.)

(2) multiplication of reticulate bodies, and (3) conversion of a large fraction of the reticulate body population into a new generation of elementary bodies that are released from the host cells. The cycle is not synchronized, and each of the first two phases overlaps into the next succeeding one (Figure 156).

The initial stage of interaction between chlamydiae and host cells involves the electrostatic interaction of elementary bodies with host cells, which requires divalent cations (Kuo and Grayston, 1976). This interaction is enhanced by polycations and inhibited by polyanions (Kuo et al., 1973). The binding is further secured by other ligands, such as the hsp70 molecular chaperone (Raulston et al., 1993), the OmcB protein (Stephens et al., 2001; Ting et al., 1995), and the major outer-membrane protein (Su et al., 1990). Ligands, such as the high mannose oligosaccharide, may be involved (Kuo et al., 1996). These interactions lead to the entry of elementary bodies into host cells by microfilament independent and perhaps receptor-mediated endocytosis (Wyrick et al., 1989), although the mechanism remains uncharacterized. Candidate receptors, mannose receptor (Kuo et al., 2002), mannose 6-phosphate/insulin-like growth factor-2 receptor (Puolakkainen et al., 2005), and estrogen receptor (Davis et al., 2002), have been suggested, however, use of these receptors may vary in different cell types and species. Uptake of chlamydia appears to be an actin-mediated process that may be directed by *Chlamydia*-specific proteins (Clifton et al., 2005).

Ingestion of chlamydial elementary bodies requires expenditure of energy by the host but not by the parasite. Elementary bodies contain intrinsic ATPase activity. Disulfide bonds of the major outer-membrane protein of chlamydiae are reduced perhaps by the reducing agents of the host cell, such as glutathione,

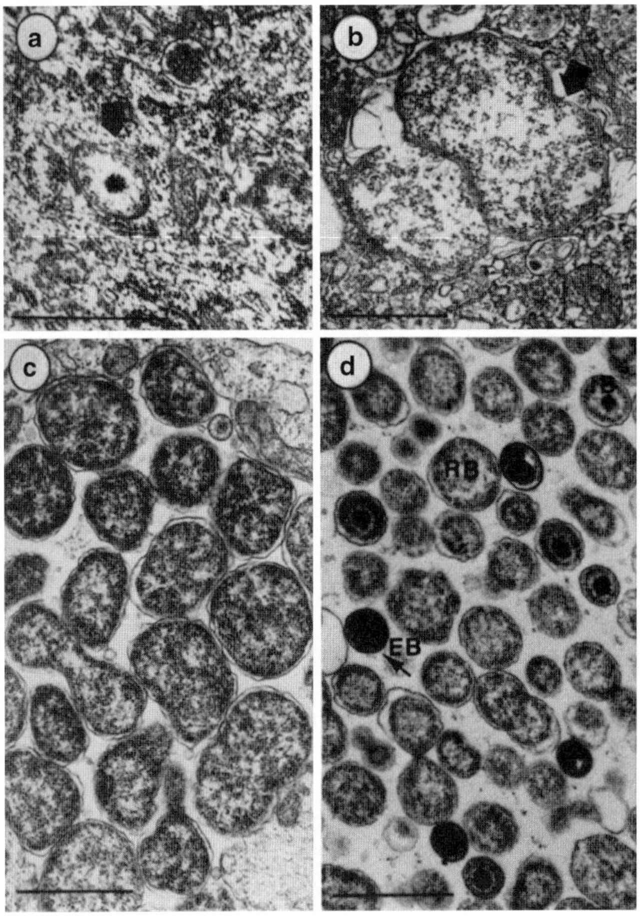

FIGURE 156. Transmission electron micrograph of thin sections of mouse fibroblasts (L cells) infected with *Chlamydia psittaci* strain meningopneumonitis. Bars = 1.0 µm. (a) At 2.5 h after infection. The arrow points to an elementary body that has just begun to differentiate into a reticulate body. (b) at 12 h after infection. The arrow points to a dividing reticulate body. (c) at 20 h after infection. The chlamydial population is almost entirely composed of dividing reticulate bodies. (d) at 30 h after infection. Some reticulate bodies are still dividing, but others have begun to reorganize into elementary bodies. EB, elementary body; RB, reticulate body; IB, intermediate body – a chlamydial cell intermediate in appearance between an elementary body and a reticulate body. (Reproduced with permission from Tribby et al., 1973. J. Infect. Dis. *127:* 155–163.)

to render the outer membrane permeable to host ATP which is then hydrolyzed by ATPase to generate energy for chlamydial metabolism immediately after entry into the host cells (Peeling et al., 1989). Ingested elementary bodies enter host cells inside phagosomes, which are cytoplasmic vacuoles surrounded by membranes derived from the plasma membranes of the host cell. Following ingestion of the phagocytized microorganisms and throughout the developmental cycle, some unidentified chlamydial activity(ies) prevents the fusion of lysosomes with phagosomes and destruction of the phagocytized organism. This fundamental property is seen only with live, but not with dead, elementary bodies. The *Chlamydia*-bearing phagosomes are transported rapidly (within minutes) to a perinuclear region within the host cell (Majeed and Kihlstrom, 1991). In a monkey model of subcutaneous autotransplantation of salpingeal fimbrial tissues, this process happens within 10 min (Patton et al., 1998). Within the inclusion vacuole, the elementary body transforms quickly into the initial reticulate body which then grows and divides. The reticulate body is devoid of infectivity; its initially dense nucleoid is dispersed into more evenly distributed fibrillar DNA, ribosomes increase in number, and the cell wall becomes thinner, more flexible, and more fragile. Completion of the reorganization is signaled by division of the reticulate body, which is first seen 8–10 h after infection (Chi et al., 1987).

Multiplication of chlamydiae is by binary fission of reticulate bodies without apparent septation. Budding of daughter cells that are much smaller in size, contain no nucleoids, and are not infectious has been observed. Nuclear segregation is accomplished by separation of two low electron-dense zones filled with fine fibrillar material and indentation of walls at the plane of separation. From approximately 10–16 h after infection, a large fraction of the reticulate bodies is at some stage of division. The proportion of dividing forms decreases thereafter, but division occurs throughout the developmental cycle. Generally, within 24 h after infection, the inclusion contains only reticulate bodies (Figure 156), and the doubling time during exponential-phase growth is 2–3 h. Multiplication of reticulate bodies takes place within an expanding membrane-bound vacuole, or inclusion, that is an extension in time and space of the phagocytic vacuole in which the parent elementary body was brought into the host cell. Chlamydiae acquire endogenously synthesized macromolecules produced by the Golgi apparatus, including sphingolipids (Scidmore et al., 1996). Thus, the biogenesis of the inclusion membrane, including the acquisition of lipids, occurs by the intersection of the lipid distribution pathway. In addition, proteins produced by the reticulate body remodel the inclusion membrane (Rockey and Rosquist, 1994). By 12–15 h after infection, chlamydial inclusions are large enough to be seen by ordinary light microscopy of stained cells or by phase-contrast microscopy of live cells. Near the end of the developmental cycle, an inclusion may contain hundreds of chlamydial cells and fill nearly the entire cytoplasm of the host cell.

At 20–25 h after infection, host cell-associated infectivity begins to appear, and elementary bodies are seen again in electron micrographs (Figure 156). The beginning of the maturation process is signaled by the up-regulation of nuclear-binding protein genes that produce eukaryotic-like histone proteins that condense nuclear material into an electron-dense mass located at the center of the cytoplasm. These condense within reticulate bodies to form intermediate bodies (Hackstadt et al., 1991; Wagar and Stephens, 1988). Mature elementary bodies do not divide. Reticulate bodies continue to divide and reorganize into elementary bodies until the host cell can no longer support the multiplication of chlamydiae. Therefore, there is no clear-cut termination to the developmental cycle, although the end is usually considered to come 48–72 h after infection, at which time the maximum yields of infectious elementary bodies are obtained depending on the chlamydial strain. At this stage, all three developmental forms (reticulate, intermediate, and elementary bodies) coexist in an inclusion (Figure 156). The mechanism by which chlamydiae are released from host cells is poorly understood. Under cinematographic observation, the reticulate bodies are stationary and localized at the periphery of the inclusion, while the elementary bodies are in Brownian movement (Neeper et al., 1990). At the point of contact with the inclusion membrane the outer envelope of the reticulate body is rearranged into a rigid, well-developed trilaminar structure (Figure 157) (Peterson and de la Maza, 1988; Yang et al., 1994). Prior to the release of the inclusion contents, inclusions appear to rotate within the cell, extend and retract large blunt protrusions at the plasma membrane of the host cell which last 1–2 h, and eventually rupture to release the contents of the inclusion (Neeper et al., 1990). Two independent modes of cellular exit that occur with approximately equal frequency *in vitro* have been characterized. One is extrusion of the inclusion from the infected cell (Hybiske and Stephens, 2007; Neeper et al., 1990), and the second mechanism is lysis of the intracellular inclusion followed by lysis of the cell (Hybiske and Stephens, 2007). Terminal release of lysosomal enzymes may speed dissolution of moribund cells. The yield of infectious chlamydiae per host cell varies from 10 to nearly 1000 depending on many factors such as the chlamydial strain, the host cell, the culture conditions, and the method by which infectivity is measured.

Cell walls and fine structure. The chlamydial cell exhibits a double-layer outer membrane (cell wall), a cytoplasmic membrane that is separated from the outer membrane by a narrow periplasmic space (Figure 155), hexagonally arrayed subunits lining the inner surface, and spatially related patches of hexagonal projections on its outer surface (Figure 158) (Matsumoto, 1981, 1982). The hexagonally arranged projections are likely type III secretion system-like structures (Bavoil and Hsia, 1998; Hsia et al., 1997). It appears that these structures are also found in the reticulate body (Matsumoto, 1981). The morphology of chlamydial organisms is typically round except for the human biovar of *Chlamydia pneumoniae* which is pleiomorphic but typically pear-shaped and exhibits a large periplasmic space (Figure 155). *Chlamydia pneumoniae* with round morphology have been isolated, though these isolates also have a wider periplasmic space than the typical round elementary bodies

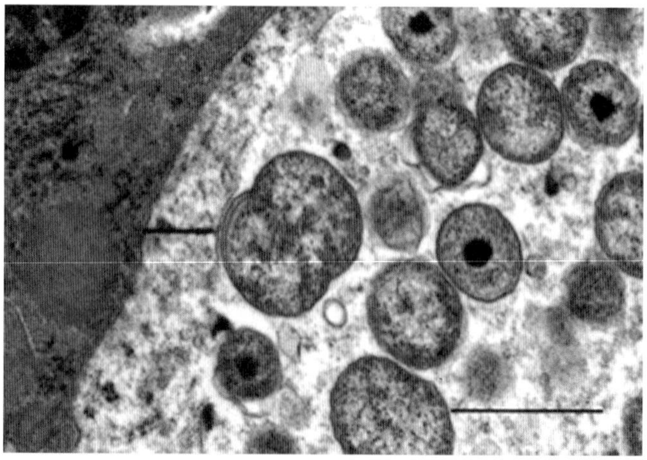

FIGURE 157. Transmission electron-micrograph of a portion of a *Chlamydia pneumoniae* inclusion in a ciliated bronchial epithelial cell in the mouse lungs demonstrating the outer envelope of reticulate body rearranged into a rigid well-developed trilaminar structure at the point of contact with the inclusion membrane (arrow). The reticulate body is undergoing division. Bar = 1.0 μm. (Reproduced with permission from Yang et al., 1994. J. Infect. Dis. *170:* 464–467.)

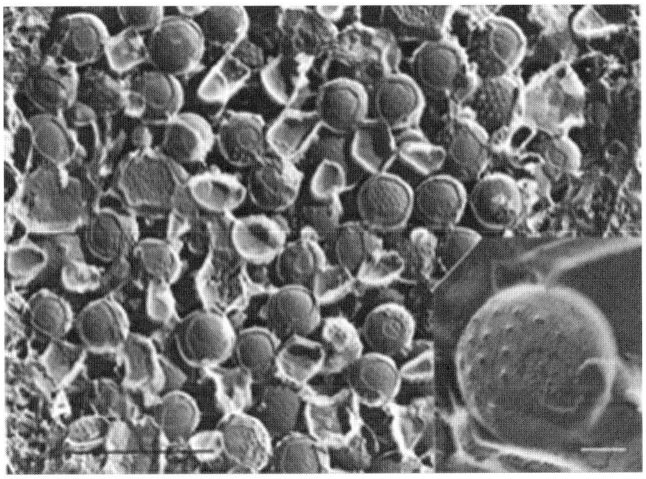

FIGURE 158. Hexagonally arrayed subunits lining the inner surface and spatially related patches of hexagonal projections on the outer surface revealed by freeze-etching of elementary bodies in *Chlamydia psittaci.* Bar = 1.0 µm. Inset: individual elementary body illustrating the arrangement of projections. Bar = 100 nm. (Reproduced with permission from Matsumoto, 1982. J. Bacteriol. *151:* 1040–1042.)

(Miyashita et al., 1997). One half of the protein content of the outer membrane is in the form of one protein with a molecular weight of approximately 40,000 (Caldwell et al., 1981) and having species and type-specific immunoreactivity (Kuo and Chi, 1987).

Nutrition and growth. Chlamydiae competitively acquire biosynthetic intermediates from the metabolic pools of their hosts, a process augmented by treatment of the host cell with cycloheximide or emetine (Alexander, 1968; Plaunt and Hatch, 1988; Reed et al., 1981). Chlamydiae have free access to some constituents of these pools but not to exogenous substrates for ATP synthesis. Therefore, it is essential that chlamydiae transport and use exogenous ATP for their own biosynthetic needs. Nutritional requirements of chlamydiae cannot be strictly defined because cell-free growth has not been achieved. Therefore, determinations of the nutritional requirements of chlamydiae have been performed in embryonated eggs or in cultured cells. The observed requirements for amino acids determined in cell culture (Allan and Pearce, 1983a, b; Kuo and Grayston, 1990a) indicate that the number of required amino acids varies among the species, the biovars within the species, and the cell types (HeLa or McCoy) used for determination (Table 161).

Metabolism. Investigations of the metabolism of host-free chlamydiae must be interpreted with two cautions in mind: (1) that the metabolic activities detected using host-free preparations are not due to contamination with host enzymes; and (2) that the correct cell type, i.e., the reticulate body or elementary body, is examined. Metabolic activity is traditionally analyzed at 20 h post-infection when the metabolic activity of chlamydiae is at its peak and the organisms are all in the reticulate body form, or at 48 h post-infection when there are mixed populations of developmental forms and the majority are elementary bodies. It has long been apparent that the reticulate body is the metabolically active cell type, while the elementary body is metabolically inactive. In recent years, the cloning of mutant cells deficient in specific enzymes of energy and nucleotide metabolic pathways has provided additional means for analyzing chlamydial metabolism (Iliffe-Lee and McClarty, 1999; Tjaden et al., 1999).

When separated from host cells, chlamydiae have few metabolic activities. They do not catabolize glucose, but they produce carbon dioxide from glucose 6-phosphate, glutamate, pyruvate, and aspartate in the presence of ATP, NADP, and other cofactors (Iliffe-Lee and McClarty, 1999). However, chlamydiae do not metabolize glutamate beyond succinate, and glutamate does not serve as an energy source.

Cytochromes, flavins, and reactions that result in a net gain of ATP have not been demonstrated. Because chlamydiae have appeared in the past to be unable to synthesize their own high energy compounds, they have been described as "energy parasites" (Hatch et al., 1982). Genomic analysis has, however, revealed that *Chlamydia* species encode enzymes of energy producing pathways and substrate-level phosphorylation, indicating that the energy burden of the chlamydiae is at least partially fulfilled by the *Chlamydia* themselves (Iliffe-Lee and McClarty, 1999).

Chlamydiae synthesize protein, lipid, and glycogen. Glycogen is produced and deposited within cytoplasmic inclusions of *Chlamydia trachomatis*, *Chlamydia muridarum*, and *Chlamydia suis*, but not in amounts detectable by iodine staining in inclusions of other chlamydial species (Gordon and Quan, 1965). Folic acid and its derivatives are synthesized by most strains of *Chlamydia trachomatis*, *Chlamydia muridarum*, and *Chlamydia suis* but not by other chlamydial species (Lin and Moulder, 1966). Chlamydiae also synthesize lipids such as phosphatidyl glycerol and branched-chain fatty acids that are characteristic of prokaryotes and are not found in their host cells (Reed et al., 1981).

Chlamydiae are auxotrophic for ATP, GTP, and UTP. CTP is obtained from the host cell or synthesized from UTP (Tipples and McClarty, 1993). Chlamydiae contain two structural genes encoding nucleotide transport proteins (Npt), $Npt1_{Ct}$ and $Npt2_{Ct}$, respectively, for the uptake of ribonucleoside triphosphate and transport of energy. $Npt1_{Ct}$ mediates uptake of ATP by chlamydiae and $Npt2_{Ct}$ mediates the exchange of ADP for ATP after ATP is catabolized to ADP by chlamydiae (Tjaden et al., 1999). Paradoxically, chlamydiae have the functional capacity to produce their own ATP and reducing power because they contain the required functional enzymes. These enzymes are pyruvate kinase, phosphoglycerate kinase, and glyceraldehde-3-phophate dehydrogenase of glycolysis and glucose-6-phosphate dehydrogenase of the pentose phosphate pathway. The existence of these enzymes has been demonstrated by complementation in *Eschericia coli* mutants deficient in these specific genes (Iliffe-Lee and McClarty, 1999).

Genetics. The DNA G+C content ranges from 39 to 45 mol% (T_m) (Table 162). The percentage DNA homology between the species and between the biovars within the species is presented in Table 162 (Cox et al., 1988; Fukushi and Hirai, 1992; Gerloff et al., 1970; Kingsbury and Weiss, 1968). Genome sequences for ten strains representing five chlamydial species have been reported with many more in progress. *Chlamydia* species have small genomes relative to most free-living organisms. The genome size varies from 1.03×10^6 bp for *Chlamydia trachomatis* to 1.23×10^6 bp for *Chlamydia pneumoniae* (Table 163). One of the remarkable outcomes from comparing genomes is how similar they are despite being separated

TABLE 161. Amino acid requirements of *Chlamydia* species determined in cell culture[a,b]

Amino acid	*C. trachomatis* Ocular	*C. trachomatis* Genital	*C. trachomatis* LGV	*C. abortus*	*C. caviae*	*C. felis*	*C. pneumoniae* (human)	Meningopneumonitis
Arginine	–	–	–	–	–	–	(d)	–
Cysteine	–	–	–	–	–	–	+	–
Glutamine	d	d	+	+	–	+	d	–
Histidine	+	+[c]	+	–	–	–	+	–
Isoleucine	–	–	–	–	–	–	(d)	–
Leucine	d	d	+	+	+	+	(+)	+
Lysine	–	–	–	–	–	–	–	–
Methionine	–	d	+	–	–	–	+	–
Phenylalanine	+	d	+	+	+	–	+	+
Threonine	–	–	–	–	–	–	(d)	–
Tryptophan	+	–	–	–	–	–	+	d
Tyrosine	–	–	–	+	–	–	+	–
Valine	+	+	+	+	+	+	(+)	+

[a]Symbols: +, >85% positive; d, different strains give different reactions (16–84% positive); –, 0–15% positive; ( ) indicates values from 90% amino acid depletion when 100% depletion of amino acid resulted in cell detachment.

[b]Determined in McCoy cells (*Chlamydia pneumoniae* in HeLa cells).

[c]All serovars except H.

TABLE 162. DNA G+C content and DNA relatedness based on DNA–DNA hybridization within and between species of the genus *Chlamydia*[a,b]

DNA from species	G+C (mol %)	Percent relatedness to DNA from: *C. trachomatis*	*C. psittaci*	*C. pneumoniae*
C. trachomatis:				
Trachoma	42–45	96–100	≤5	≤5
Genital	42–45	96–100	≤5	≤5
LGV	42–45	95–97	12	na
C. muridarum	42–44	63	na	na
C. psittaci	39–42	≤5	100	≤5
Meningopneumonitis[c]	39–43	11–12	93–96	na
C. pneumoniae	40	≤5–6	≤5–7	94–100
C. pecorum	39–42	1–10	na	na
C. abortus	na	6	85	na
C. caviae	na	≤5	32	≤5
C. felis	na	11	37	6–8

[a]Strains tested: trachoma and genital (serovars A–E); *Chlamydia psittaci* (6BC, parakeet, ornithosis, chicken); *Chlamydia pneumoniae* (TW183, AR39, AR458, LR65); Data for *Chlamydia suis* are not available.

[b]na, Data not available.

[c]Ferret, unclassified.

by a 100 million years of evolution. Gene content identity of any chlamydial genome to that for the prototypical *Chlamydia trachomatis* (D/UW-3/Cx) is greater than 90%. Given the similarity of the intracellular environment (niche) and the small genome size for all *Chlamydia* species, it could be expected that they would share a largely identical complement of genes. However, even more remarkable is the high level of conservation of the order of their genes, called genome synteny. Gene order is one of the least-conserved characteristics of bacteria as synteny is the first genomic property to be lost during evolution. This has not been the case for the *Chlamydiaceae*. For example, alignment of the genomes of *Chlamydia trachomatis* and *Chlamydia muridarum* results in precise conservation of gene order throughout the entire genome despite the fact that they are separated from a common ancestor over 50 million years ago as evidenced by sharing only 56% nucleotide identity between the genomes. Measurement of synteny between all available genomes reveals analogous outcomes (Table 163). The conclusion is that species of *Chlamydia* have been separated for a long geological time as reflected, for example, in 16S rRNA phylogenetic analyzes; nevertheless, they have changed very incrementally in the framework of biological evolution.

Mutants. Targeted mutant strains have not been generated due to the lack of a genetic tool for *Chlamydiaceae*. A nitrosoguanidine-generated, temperature-sensitive attenuated mutant of *Chlamydia abortus* has been isolated and is used as a live vaccine against abortion of ewes in Europe (Rodolakis, 1983).

Plasmids. Most *Chlamydia* species have a conserved 7500 bp plasmid (Joseph et al., 1986; Palmer and Falkow, 1986) that is not required for growth *in vitro* (Peterson et al., 1990). The plasmid has been detected in nearly all isolates of *Chlamydia trachomatis*, *Chlamydia psittaci* and in one strain of *Chlamydia pneumoniae* (Thomas et al., 1997). There are four plasmid copies per elementary body chromosome for *Chlamydia trachomatis* (Pickett et al., 2005). Clinical isolates representing the

TABLE 163. Characteristics differentiating species of the genus *Chlamydia*[a]

Characteristic	*C. trachomatis* genital biovar	*C. trachomatis* LGV biovar	*C. abortus*	*C. caviae*	*C. felis*	*C. muridarum*	*C. pecorum*	*C. pneumoniae*	*C. psittaci*	*C. suis*
Natural host	Human	Human	Ruminants, pig	Guinea pig	Cat	Rodent	Ruminants, pig, koala	Human, horse, koala	Bird	Pig
Site of infection	Eye, genital	Genital lymph node	Genital, respiratory	Eye, genital	Eye, respiratory	Respiratory?	Intestinal, urogenital, respiratory	Respiratory	Intestinal, respiratory	Genital, respiratory
16S rRNA difference (%)[b]	0	0	4.8	4.7	4.8	1.6	4.8	6.1	4.8	2.5
Genome size (bp)	1,042,519	1,038,841	1,144,377	1,173,390	nd	1,069,412	nd	1,230,230	nd	nd
Gene content identity[a]	100	nd	92.4	92.5	nd	96	nd	91.2	nd	nd
Genome synteny	100	nd	87.6	85.4	nd	87.4	nd	89.8	nd	nd
Plasmid	+	+	–	+	+	+	+	–	+	+
Bacteriophage	–	–	+	+	–	–	+	"+ (rare)"	+	–
pmp gene paralogs	9	9	18	17	nd	9	nd	21	nd	nd

[a]Symbols: +, >85% positive; –, 0–15% positive; nd, not determined.

[b]Compared to *Chlamydia trachomatis* (D/UW-3/Cx).

lymphogranuloma venereum (LGV) biovar and the genital biovar that lack the common plasmid have been characterized (Peterson et al., 1990; Stothard et al., 1998). This suggests that the plasmid is not essential for growth and persistence *in vivo* (i.e., pathogenesis). Plasmids from each species encode eight analogous proteins whose functions in the biology of *Chlamydia* are unknown, further mystifying the role of plasmid in chlamydial biology.

Phages. Several phages that infect *Chlamydia* species termed chlamydiaphages have been isolated and characterized. The first chlamydiaphage, named Chp1, was isolated from a *Chlamydia psittaci* outbreak in a duck farm in England in 1982 (Richmond et al., 1982; Storey et al., 1989a; Storey et al., 1989b). Chp1 forms large characteristic paracrystalline virion arrays that allow its identification by transmission electron microscopy. Several other chlamydiaphages were subsequently isolated from *Chlamydia caviae* (PhiCPG1) (Hsia et al., 2000), *Chlamydia pneumoniae* (PhiAR39) (Read et al., 2000), *Chlamydia abortus* (Chp2) (Liu et al., 2000), and *Chlamydia pecorum* (Chp3) (Garner et al., 2004). However, phages have not been isolated from *Chlamydia trachomatis* or *Chlamydia muridarum.* All chlamydiaphages isolated to date are members of the microviridae and are genetically and structurally related to phages of *Spiroplasma melliferum* and *Bdellovibrio bacteriovorus* and more distantly to the coliphage phiX174. The phage host range is specified by a peptidic loop of the major capsid protein, VP1 (Read et al., 2000), and sequence variation within the loop is consistent with the observed overlapping host ranges of the phages. For instance, PhiCPG1 and PhiAR39, which have identical VP1 sequences, are both able to infect *Chlamydia caviae* and *Chlamydia pneumoniae.* However, Chp2, whose VP1 sequence differs, may also infect *Chlamydia caviae* and other *Chlamydia* species (Everson et al., 2002), suggesting that distinct receptors to multiple phages exist at the surface of chlamydiae. Receptors for these phages have not been identified. The phage infectious cycle was characterized by transmission electron microscopy for phage PhiCPG1 infection of *Chlamydia caviae* (Hsia et al., 2000). Briefly, PhiCP1 virions gain access to intracellular chlamydiae by first attaching to and cointernalizing with extracellular elementary bodies. Once inside, the nascent inclusion virions infect the newly formed reticulate body. Abundant phage progeny are generated by rolling circle replication of the single-stranded DNA genome and packaging into phage procapsids. Phage-infected reticulate bodies appear as aberrantly enlarged cells similar to *in vitro* persistent forms of chlamydiae. Infectious phage progeny are released by phage-specified lysis of the inclusion membrane approximately 24 h post phage infection.

Antigenic structure. All chlamydiae share a similar antigenic framework consisting of common and specific antigenic determinants (Table 164). Monoclonal antibodies to the most important chlamydial antigens are commercially available. The *Chlamydiaceae* are characterized by a common lipopolysaccharide (LPS) antigen. This antigen is heat-stable and is the antigen for complement fixation tests. The common antigenic determinant is a LPS core terminating in a 2-keto-3-deoxyoctanoic acid (Brade et al., 1986). Antibodies to LPS react strongly to antigen present within inclusions and in the reticulate body membranes, however, reactivity to elementary bodies by immunofluorescence is weak. All chlamydiae encode a major 60,000 molecular weight heat-shock protein (HSP-60) family member (Bavoil et al., 1990) that is associated with seroreactivity in individuals at risk for pathogenic sequelae of infection (Wagar et al., 1990). This should not be confused with another 60,000 molecular weight protein (OmcB) that is cysteine-rich and present in elementary-body outer membranes but not reticulate-body outer membranes (Batteiger et al., 1985; Hatch et al., 1986). This protein is a potent immunogen and useful marker for serological studies of chlamydial infection.

TABLE 164. Structure and function of the major immunoreactive antigens identified in *Chlamydia*[a]

Molecular weight	Name	Function
≥ 98,000	Pmp	Unknown. Six Pmp subtypes: A, B, C/D, E/F, G/I, and H
75,000	HSP	Chaperone HSP-70 family
60,000	OmcB	Structural, cysteine-rich OMP
60,000	HSP	Chaperone HSP-60 family
40,000	OmpA (MOMP)	Outer-membrane porin. Contains serological determinants of species-, subspecies-, and serovar-specificity
40,000	PorB	Outer-membrane porin
12,500	OmcA	Structural lipoprotein, cysteine-rich OMP
~12,000	LPS	Endotoxin, genus antigen

[a]Abbreviations: Pmp, polymorphic membrane protein; HSP, heat shock protein; Crp, cysteine-rich protein (cross linking of disulfide bonding of cysteine-rich proteins give rigidity to the outer membrane of the EB); Omc, outer membrane complex protein; MOMP, major outer membrane protein; LPS, lipopolysaccharide.

A major antigen is the major outer-membrane protein (OmpA or MOMP) that is the quantitatively predominant protein in *Chlamydia trachomatis* (Caldwell et al., 1981) and other *Chlamydia* species (Campbell et al., 1990). This protein is an outer membrane porin (Bavoil et al., 1984) and is antigenically complex (Stephens et al., 1982). For *Chlamydia trachomatis* and *Chlamydia suis,* four variable sequence (VS) loops are exposed on the surface and accessible to binding by antibodies (Stephens et al., 1988). *Chlamydia trachomatis* serovar-specific antigens have been mapped and are associated with vs1 and vs2; more broadly reactive determinants, including a species-specific antigen, are localized with vs4. For *Chlamydia trachomatis* (Fan and Stephens, 1997) and *Chlamydia pneumoniae* (Wolf et al., 2001), some of the OmpA antigenic determinants are conformation dependent. Sequence based antigenic variations of this protein singularly account for the antigenic specificities originally characterized by the micro-immunofluorescence test in which 15 prototype serovars were defined (Figure 159) (Wang and Grayston, 1970). Serovars A through K, including Ba, are associated with ocular and urogenital tract infection and serovars L_1, L_2, and L_3 are serovars of the LGV biovar. Antigenic variation is thought to derive from selection by the host immune response to infection (Stephens, 1989). While the other *Chlamydia* species genomes each have *ompA*, only *Chlamydia trachomatis, Chlamydia pecorum,* and *Chlamydia suis* have alleles of *ompA* that significantly differ among isolates. A new porin, PorB, was discovered following genome sequencing (Kubo and Stephens, 2000). Like OmpA, PorB is surface-exposed but, unlike OmpA, PorB is not quantitatively predominant and does not display significant sequence variation among isolates within a species.

An entire multigene family of antigens [polymorphic membrane proteins (Pmps)] was also discovered from genome

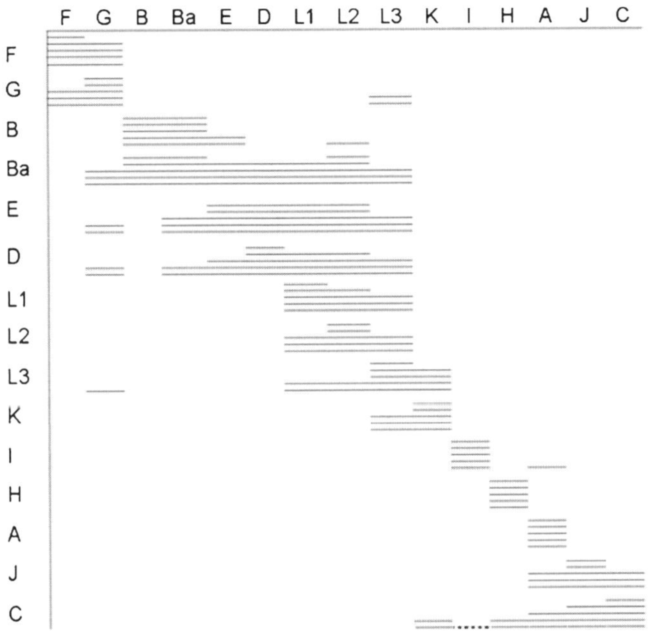

FIGURE 159. Microimmunofluorescence pattern of *Chlamydia trachomatis* antisera prepared in mice. A difference of one line expresses a twofold difference from the homologous titer which is shown with five lines. (Reproduced with permission from Wang et al., 1973. Infect. Immun. *7:* 356–360; adapted from the 1984 edition of the *Bergey's Manual of Systematic Bacteriology*, vol. 1; Chapter Genus I *Chlamydia*, Figure 9.14.)

sequencing to include nine representatives in *Chlamydia trachomatis* (Stephens et al., 1998) and 21 representatives in *Chlamydia pneumoniae* (Kalman et al., 1999) and intermediate numbers of paralogs for other species. The protein sequences are very different, but they are unambiguously related (Grimwood et al., 2001). The functions of these proteins are unknown, but some have been shown to be surface-exposed outer-membrane proteins, and their expression is dependent upon phase variation.

Antibiotic sensitivity. The susceptibility of chlamydiae to antibiotics has been tested in yolk sac of chicken embryos (Gordon et al., 1957; Molder et al., 1955) or cell culture (Kuo and Grayston, 1988; Kuo et al., 1977b). The growth of chlamydiae in culture is inhibited by the tetracyclines, macrolides, azalides, chloramphenicol, rifampin, and fluoroquinolones. Chlamydial multiplication is not blocked by aminoglycosides, bacitracin, or vancomycin. *Chlamydia trachomatis*, *Chlamydia muridarum*, and *Chlamydia suis* are sensitive to sulfonamides, whereas other species, except the 6BC strain of *Chlamydia psittaci*, are resistant. Beta-lactam antibiotics (penicillin), which inhibit the synthesis of peptidoglycan, interrupt the developmental cycle by preventing the maturation of reticulate bodies into elementary bodies. However, if these antibiotics are removed, development proceeds normally. The inhibitory effect of penicillin on chlamydiae which contain no or only trace amounts of muramic acid has been attributed to the presence of penicillin-binding proteins in their outer membrane (Barbour et al., 1982). D-Cycloserine acts similarly to penicillin, but *Chlamydia trachomatis* is much more susceptible to growth inhibition by this antimicrobial than is *Chlamydia psittaci*. Antagonism is not observed among sulfa drugs, tetracycline, erythromycin, chloramphenicol, penicillin, and ciprofloxacine. An additive inhibition is observed between tetracycline and penicillin, tetracycline and erythromycin, tetracycline and chloramphenicol, erythromycin and penicillin, erythromycin and chloramphenicol (How et al., 1985) and azalides and rifampin derivatives (Kuo, C.C., unpublished data). Synergistic activity is observed between trimethoprim and sulfamethoxazole (How et al., 1985). The drugs of choice for treatment of chlamydial infections are tetracyclines, macrolide, and azalides. Long-acting antibiotics (doxycycline) and those antibiotics that achieve high intracellular concentrations (azithromycin) are preferred.

Chlamydial strains resistant to sulfonamides, penicillin, chlorotetracycline, and rifampin have been produced by *in vitro* passage in the presence of these drugs. However, stable drug-resistant mutants of *Chlamydia trachomatis* or *Chlamydia pneumoniae* have not been isolated. *Chlamydia suis* in American pigs is resistant to tetracycline by virtue of the integration in its genome of a Tet(R) transposon-like genetic element from another bacterium (Dugan et al., 2004).

Pathogenesis. Chlamydiae are among the most widely distributed pathogens of humans and animals and they produce variable clinical syndromes. Humans are the natural hosts of *Chlamydia trachomatis* (Grayston and Wang, 1975) and the human biovar of *Chlamydia pneumoniae* (Kuo et al., 1986). *Chlamydia trachomatis* causes ocular infection, which may lead to blindness (trachoma), a disease still found in tropical and subtropical endemic areas (Grayston et al., 1985), and it causes genitourinary tract infections, which may cause tubal factor infertility (fallopian tube obstruction), which is a serious complication of chronic salpingitis (Grayston and Wang, 1975) (Table 165). The human biovar of *Chlamydia pneumoniae* is a common respiratory pathogen with manifestations of pharyngitis, bronchitis, and pneumonia (Grayston et al., 1989) and is associated with atherosclerosis and coronary heart disease (Kuo et al., 1993) (Table 165). The organism has been isolated from the atherosclerotic lesions of the artery. A unique characteristic of human infection with these chlamydial agents is their ability to establish persistent or chronic infection, as exemplified by the difficulty in reisolating the organism from lesion sites, such as from the occluded fallopian tube in *Chlamydia trachomatis*

TABLE 165. Diseases caused by human chlamydial agents

Species	Biovar	Disease or syndrome
Chlamydia trachomatis	Ocular[a]	Inclusion conjunctivitis of newborns, children, and adults; trachoma and blindness
	Genital	Infantile pneumonia; urethritis in men; epididymitis; urethral syndrome in women; cervicitis; endometritis; salpingitis, tubal factor infertility; proctitis; perihepatitis and peritonitis (Fitz–Hugh–Curtis syndrome)?; bartholinitis?; endocarditis?
	LGV	Lymphogranuloma venereum
Chlamydia pneumoniae	Human	Pharyngitis; bronchitis; pneumonia; sinusitis?; otitis media?; atherosclerosis (a co-risk factor with hyperlipidemia for atherosclerosis)?

[a]Trachoma biovar in old nomenclature

salpingitis (Campbell et al., 1993) and from the lungs and atherosclerotic arteries in *Chlamydia pneumoniae* infections (Kuo et al., 1993). However, the presence of organisms (chlamydial DNA, RNA, and antigens) can still be detected by PCR, RT-PCR, or immunohistochemistry. This observation has been attributed to the arrest of the developmental cycle at the reticulate body stage by depletion of intracellular tryptophan or other required amino acids (Allan and Pearce, 1983b, a). Tryptophan may become limiting as a result of gamma-interferon produced in the context of the host's innate immune response (Beatty et al., 1994). Gamma-interferon induces indoleamine 2,3 dioxygenase (IDO) that catabolizes oxidative decyclization of tryptophan (Kane et al., 1999).

Chlamydia psittaci causes systemic infection in birds and the infection is often asymptomatic (Table 166). *Chlamydia psittaci* is highly infectious to humans. Humans may contract *Chlamydia psittaci* infection by the airborne route, either by direct contact with infected birds or indirectly by inhalation of dust contaminated with excreta of infected birds. The major manifestation of *Chlamydia psittaci* infection in humans is atypical pneumonia, and the infection in humans is often systemic. Psittacosis is a term used to describe the infections acquired from a psittacine bird, and ornithosis is the infection contracted from birds other than psittacine. *Chlamydia psittaci* is widespread in wild and domestic birds including parrots, parakeets, pigeons, ducks, and turkeys. Outbreaks of *Chlamydia psittaci* infection may occur in poultry, and transmissions of infection to poultry farmers occur sporadically. Quarantine of trapped wild birds in combination with preventive antibiotic (tetracycline) treatment has been effective in controlling spread of infection from wild to domesticated birds.

Mammalian chlamydial species cause various diseases unique to their hosts (Table 166). The major syndromes caused by mammalian chlamydial agents are: pneumonitis in mice caused by *Chlamydia muridarum*; conjunctivitis and respiratory infection in koalas caused by the biovar koala of *Chlamydia pneumoniae*; encephalitis, polyarthritis, pneumonia, enteritis, and genital infection caused by *Chlamydia pecorum* in ruminants, and urogenital infection and infertility in koala; conjunctivitis, pneumonia, and enteritis in swine caused by *Chlamydia suis*; abortion in sheep, goats, and cattle caused by *Chlamydia abortus*; follicular conjunctivitis and interstitial keratitis in guinea pigs caused by *Chlamydia caviae*; and conjunctivitis in cats caused by *Chlamydia felis*. Human infection due to other mammalian chlamydial infections is rare, because mammalian chlamydial agents are much less virulent to humans. Sporadic cases of severe flu-like infection and abortion occur after exposure of farmers in the third trimester of pregnancy to *Chlamydia abortus* by contact with aborting sheep or goats (Herring et al., 1987; Johnson et al., 1985). Pet owners or laboratory personnel may develop conjunctivitis acquired from cats with conjunctivitis due to *Chlamydia felis* (Schachter et al., 1969).

Phylogenetic relationships based on 16S rRNA. Several investigators have evaluated the relationships among many of the *Chlamydia* and *Chlamydia*-related organisms by tree-methodologies using 16S rRNA (Corsaro et al., 2002; Everett et al., 1999; Petersson et al., 1997; Pudjiatmoko et al., 1997; Weisburg et al., 1986). Complete 16S rRNA sequences of most of

TABLE 166. Disease caused by avian and mammalian chlamydial agents

Species	Host	Disease
Chlamydia abortus	Sheep, goats, cattle, swine	Abortion, vaginitis, endometritis, seminal vesiculitis, mastitis; often latent infection
Chlamydia caviae	Guinea pigs	Follicular conjunctivitis, interstitial keratitis
Chlamydia felis	Cats	Conjunctivitis with or without rhinitis
Chlamydia muridarum	Mice, hamsters	Pneumonitis; often latent ileitis
Chlamydia pecorum	Sheep, cattle, swine, koala	Encephalomyelitis, polyarthritis, pneumonia, enteritis, vaginitis, endometritis in sheep and cattle; polyarthritis, serocitis, enteritis, pneumonia in swine; keratoconjunctivitis, vaginitis, ovarian cyst, infertility in koala; often latent infection
Chlamydia pneumoniae:		
Biovar koala	Koala	Rhinitis, pneumonia
Biovar equine	Horse	Rhinitis[a]
Chlamydia psittaci	Birds	Systemic, often latent respiratory and enteric infection
Chlamydia suis	Swine	Conjunctivitis, pneumonia, enteritis, polyarthritis
Chlamydia abortus	Sheep, goats, cattle, swine	Abortion, vaginitis, endometritis, seminal vesiculitis, mastitis; often latent infection
Chlamydia caviae	Guinea pigs	Follicular conjunctivitis, interstitial keratitis
Chlamydia felis	Cats	Conjunctivitis with or without rhinitis
Chlamydia muridarum	Mice, hamsters	Pneumonitis; often latent ileitis
Chlamydia pecorum	Sheep, cattle, swine, koala	Encephalomyelitis, polyarthritis, pneumonia, enteritis, vaginitis, endometritis in sheep and cattle; polyarthritis, serocitis, enteritis, pneumonia in swine; keratoconjunctivitis, vaginitis, ovarian cyst, infertility in koala; often latent infection
Chlamydia pneumoniae:		
Biovar koala	Koala	Rhinitis, pneumonia
Biovar equine	Horse	Rhinitis[a]
Chlamydia psittaci	Birds	Systemic, often latent respiratory and enteric infection
Chlamydia suis	Swine	Conjunctivitis, pneumonia, enteritis, polyarthritis

[a]Only one isolate from a horse with serous nasal discharge has been reported.

the prototypical strains are available, and the resulting trees from different reports using the same or different methods are markedly similar. All *Chlamydia*-related organisms are deeply separated from other bacteria and form their own bacterial division as they lack a significant relationship to other bacterial divisions (Pace, 1997). Distantly related organisms such as the *Piscichlamydia* (Draghi et al., 2004), *Protochlamydia* (Collingro et al., 2005a, b), and *Parachlamydia* (Amann et al., 1997) are distinct from chlamydial organisms that establish the strongly monophyletic *Chlamydiaceae* family (Figure 160). Three groups diverged from a common ancestor that contain species related to the *Chlamydia trachomatis*, *Chlamydia psittaci*, and *Chlamydia pneumoniae* groups. New species names have also been proposed (Everett et al., 1999), which are often representative of the relationship with the natural host and are adopted herein. The *Chlamydia trachomatis* grouping contains the murine biovar mouse pneumonitis (MoPn) (Nigg and Eaton, 1944) *Chlamydia muridarum* and related organisms isolated from pigs named *Chlamydia suis* (Hoelzle et al., 2000). The *Chlamydia psittaci* grouping contains strains from birds (*Chlamydia psittaci*), cats (*Chlamydia felis*), guinea pigs (*Chlamydia caviae*), and ruminants and pigs (*Chlamydia abortus*). The third group contains *Chlamydia pneumoniae* (Grayston et al., 1989), originally isolated from humans, and *Chlamydia pecorum* isolated from cattle (Fukushi and Hirai, 1992).

Analyzes of a variety of other single components such as protein and 23S rRNA sequences support these relationships. The application of simple DNA amplification procedures for the detection of novel 16S or 23S sequences related to *Chlamydia* has resulted in the detection of *Chlamydia*-like organisms from many animals, fish, reptiles, and environmental samples. These activities will likely continue, but many do not result in full-length gene sequences and are not supported by microbiological isolation.

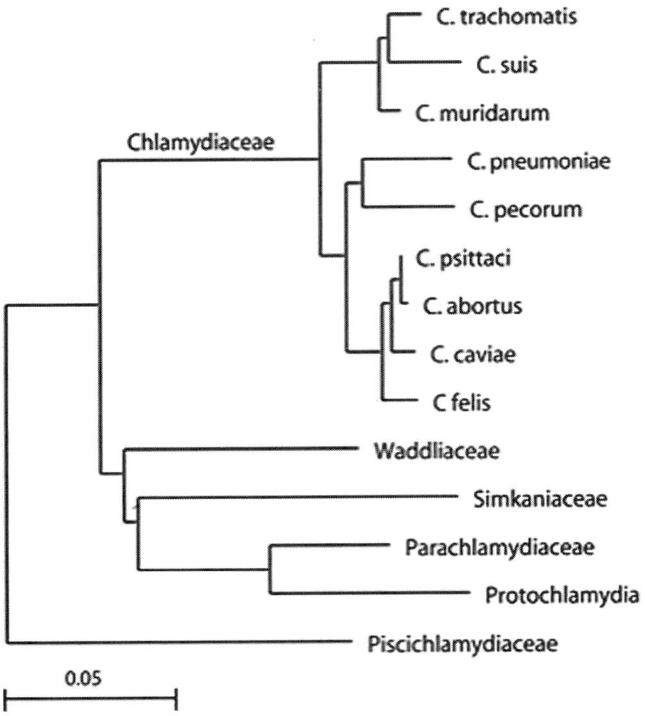

FIGURE 160. 16S rRNA gene phylogeny of *Chlamydiaceae* and *Chlamydiaceae*-related organisms using neighbor joining with Jukes–Cantor distance corrections. Bar = 0.05 substitutions/site.

Enrichment and isolation procedures

Laboratory handling of *Chlamydia*-containing specimens and growth of chlamydiae should be conducted by employing biosafety level 2 (BSL-2) or higher containment facilities and practice (Chosewood and Wilson 2007). Strains from all chlamydial species grow in the yolk sacs of chicken embryos. All chlamydial strains multiply in a variety of cell cultures with the assistance of infection-promoting procedures such as pretreatment of host cells with polycations (Kuo et al., 1972b), centrifugation of the inoculum onto host cell monolayers (Gordon et al., 1969), and inhibition of the synthesis of host macromolecules with cycloheximide (Ripa and Mardh, 1977). The efficiency of growth in cell culture varies among the species and the biovars within the species. For example, *Chlamydia psittaci* grows more efficiently than other mammalian chlamydial species in both chicken embryos and cell cultures, while the LGV biovar multiplies more readily than the trachoma and genital biovars of *Chlamydia trachomatis* in both systems. Chlamydiae also infect and are lethal to mice following intracerebral inoculation. This test of virulence varies among chlamydial strains and is similar to their behavior in chicken embryos and cell cultures. *Chlamydia psittaci* strains are particularly virulent to humans. The risk is so great that attempts to isolate *Chlamydia psittaci* from birds should be made only by experienced workers in specially equipped laboratories.

Collection and processing of specimens. Collection of an appropriate specimen for isolation of chlamydiae depends on the disease, the host, and the chlamydial agent. Clinical samples, especially swab samples from the bodily openings and intestinal tissues, are contaminated with other bacteria and yeasts, therefore antibacterial and antifungal agents that are not inhibitory to chlamydiae are added to the transport medium. Frequently used isolation mixtures contain gentamicin, or streptomycin and vancomycin, and amphotericin B. Clinical samples are usually suspended in phosphate buffer containing sucrose and glutamate or cell culture medium, supplemented with suitable antimicrobial agents. A suitable medium contains 0.2 M sucrose in phosphate buffer and 10 μg/ml gentamicin and 3.75 μg/ml amphotericin B. Intracellular chlamydiae are released by shaking with glass beads or by sonication. Specimens may be kept at 4°C if inoculated within 24 h. If they must be held for longer times, they should be frozen at −75°C or lower.

Isolation in cell culture. The cell lines most commonly used for isolation of chlamydiae are McCoy cells (a heterodiploid mouse line) (Gordon et al., 1969) and HeLa 229 cells (derived from a human cervical carcinoma) (Kuo et al., 1977a; Kuo et al., 1972b). Other cell lines are used for certain species. For example, HL (human line) (Kuo and Grayston, 1990b) and Hep-2 (human nasopharyngeal carcinoma) (Roblin et al., 1992) cells for *Chlamydia pneumoniae* and L-929 (mouse fibroblast) or BGMK (buffalo green monkey kidney) cells for *Chlamydia psittaci*, *Chlamydia pecorum*, *Chlamydia felis*, and *Chlamydia suis*. Enhancement of infection is achieved by pretreating cell monolayers with 20 μg/ml di-ethyl-amino-ethyl (DEAE)-dextran, centrifuging the inoculum onto the monolayer at 1000–3000× g for 1 h at 20–35°C, and incubating the inoculated cells with medium containing 0.5–1.0 μg/ml cycloheximide. After 2 or 3 d of incubation at 35 or 37°C, the presence of chlamydial inclusions is determined by fluorescent antibody stain of the cell monolayer.

Isolation in chicken embryos. Six-to-seven-day-old fertile hen eggs from a flock maintained on antibiotic-free feed are candled for viability, inoculated into their yolk sacs and incubated at 35–37°C for 14 d (T'ang et al., 1957). On the third day of incubation, eggs containing dead embryos are discarded. The eggs are candled every day thereafter. All dead embryos are examined for the presence of chlamydiae by staining impress smears of the yolk sac with Gimenez or modified Macchiavello. Negative primary passage may be passed again blind. Criteria for positive isolation are the presence of elementary bodies and serially transmissible embryo mortality in the absence of contaminating bacteria or viruses. For isolation of human strains of chlamydiae, the yolk sac is less sensitive than cell culture.

Isolation in mice. Most chlamydial strains tested were found to infect mice after intranasal instillation. BALB/c or A/J inbred mice are preferred for isolation. Under light metofane or isoflurane anesthesia, 20–40 μl of the inoculum is deposited on the nostrils of 3 week-old mice that readily inhale the inoculum due to anesthesia-induced hyperventilation. This method has been used for isolation of chlamydiae from avian hosts and of virulent strains from other hosts, but is less sensitive than cell culture for isolation of low-virulence strains. Within one week after inoculation, mice develop typical lung lesions consisting of grayish foci of consolidation, sometimes involving a major portion of the lung. For verification of the infection, mice are sacrificed after 1 week, and impression smears from lesions are stained with Gimenez or FITC-labeled monoclonal antibody against chlamydial LPS, or lung tissue is examined by PCR. Standard bacteriological culture is used to confirm the absence of contaminating bacteria, and cell culture is used for exclusion of viral infections. In negative isolation attempts, subculture in BGMK cells is preferred over blind passages.

Maintenance procedures

Chlamydiae are maintained and propagated in chicken embryos and cell cultures by the procedure just described. Chlamydiae are best preserved by freezing at −75°C or lower in buffers containing sucrose and glutamic acid, such as SPG (sucrose, 75 g; KH_2PO_4, 0.52 g; Na_2HPO_4, 1.22 g; glutamic acid, 0.72 g; H_2O to 1 l; pH 7.4–7.6).

Chlamydiae can be propagated to a high titer by serial passages in chicken embryo yolk sac cultures or cell monolayer cultures in a large flask. For example, it is possible to achieve 100% infection of HeLa 229 cells by serial passage without the assistance of centrifugation or inhibitors of host-cell macromolecular synthesis with *Chlamydia trachomatis*, *Chlamydia psittaci*, and *Chlamydia muridarum* (Kuo et al., 1977a) and of HL (human line) (Kuo and Grayston, 1990b) and Hep-2 (Roblin et al., 1992) cells with *Chlamydia pneumoniae*. After amplification of chlamydial titer to grow to 100% infectivity in stationary cell cultures, the LGV biovar of *Chlamydia trachomatis* and *Chlamydia psittaci* can be propagated in a suspension culture of L-cells. Alternatively, BGMK monolayer cells can be used for effective propagation of chlamydiae, and elementary bodies released into the culture medium can be concentrated into stocks of high purity and infectivity (Li et al., 2005).

Procedures for testing special characters

Several methods are used for quantification of infectivity titer and enumeration of bacterial particles. For example, *Chlamydia psittaci*, the LGV biovar *Chlamydia trachomatis*, and *Chlamydia muridarum* readily form plaques on monolayers of L-cells (a heterodiploid mouse line) after extended incubation. Ocular and genital biovar strains of *Chlamydia trachomatis* can also be coaxed to form plaques (Matsumoto et al., 1998). The infectivity titer is expressed as plaque forming units. A simple method is to determine the inclusion-forming units by infecting a monolayer by centrifugation and culturing infected cells with medium containing inhibitors of host macromolecular synthesis, incubating for 48–72 h, and staining with Giemsa, iodine, or fluorescent antibodies and counting the inclusions. The number of chlamydial inclusions in each of a number of microscopic fields is counted. The inclusion counts are standardized, and the titer is expressed in term of inclusion forming units per milliliter (Furness et al., 1960). The burst size is defined in term of the number of infectious elementary bodies produced from a single inclusion. for example, the burst size of *Chlamydia pneumoniae* in HL cells is 73 with the assistance of centrifugation and inhibition of host cell macromolecular synthesis by cycloheximide. The m.o.i. is the number of infectious organisms inoculated per cell in a monolayer culture. Chlamydial infectivity in cell culture, chicken embryos, and mice may be expressed in term of the 50% infectious dose or the 50% lethal dose. Infectivity for chicken embryos inoculated via the yolk sac may also be estimated from the nearly linear relation between the logarithm of the inoculum dilution and the survival time of the inoculated embryos.

Total bacterial counts may be obtained by electron microscopic examination of grids of known geometry onto which purified chlamydial particles have been deposited. The particle counts are greater than inclusion forming units. This may be as much as a factor of 30 with *Chlamydia trachomatis* cultured in HeLa cells and purified by sucrose gradient centrifugation (C.- C. Kuo, unpublished data).

The ocular and genital biovars of *Chlamydia trachomatis* may be differentiated by testing their ability to grow in tryptophan-free medium supplemented with indole. Because the genital biovar is able to utilize exogenous indole for the synthesis of tryptophan while the ocular biovar is not, growth is observed with the genital but not with the ocular biovar (Fehlner-Gardiner et al., 2002).

Applications of real-time quantitative amplification of chlamydial and host DNA and gene sequence techniques provide robust methods for the quantification of chlamydiae and determination of growth rates *in vitro*, and for measuring chlamydial infectious burdens *in vivo*.

Differentiation of the family *Chlamydiaceae* from closely related families

The common properties and 16S rRNA relatedness of five families of the order *Chlamydiales* have been described in the preceding section under the Order *Chlamydiales*. Therefore, only some other biological and pathogenic characteristics useful for differentiation of the family *Chlamydiaceae* from the families *Parachlamydiaceae*, *Piscichlamydiaceae*, *Waddliaceae*, and *Simkaniaceae* are presented below.

The natural hosts for *Parachlamydiae* are free-living amoebae, although they have been recovered from soil (Collingro et al., 2005a, b); for *Piscichlamydia*, Atlantic salmon (Draghi et al., 2004); for *Waddlia*, bovines (Dilbeck et al., 1990); and for *Simkania*, soil (Kahane et al., 1995). Isolations of *Simkaniaceae* from humans with respiratory infection have been reported. However, the pathogenesis of *Parachlamydia* in humans has not been described. These intracellular microorganisms form inclusions and exhibit similar developmental cycles. However, the developmental forms vary in morphology. Specifically, *Piscichlamydia* forms variable elongated reticulate bodies with distinctive head and tail cells and the intermediate form which does not evolve to the elementary form (Draghi et al., 2004). Antigenically, cross-reactivity to chlamydial LPS has been demonstrated with *Piscichlamydia*, but not with *Parachlamydia*.

Differentiation of species, biovars, and serovars of the genus *Chlamydia*

Nine species of Chlamydia are currently recognized including *Chlamydia trachomatis, Chlamydia muridarum, Chlamydia pecorum, Chlamydia pneumoniae, Chlamydia psittaci, Chlamydia suis, Chlamydia abortus, Chlamydia caviae,* and *Chlamydia felis. Chlamydia trachomatis* is further subdivided into three biovars (Table 167) and *Chlamydia pneumoniae* into three biovars (Table 168). In the following sections, if the chlamydial property being described is not common to all members of the genus, the species or biovars to which the description applies will be stated.

Characteristics useful for differentiation of the species of *Chlamydia* are listed in Tables 161, 162, and 163. Nucleic acid-based techniques have emerged as technically easier for use in species differentiation than direct determination of biological differences. Initially, whole-genome hybridization and restriction fragment length polymorphism (RFLP) methods required large quantities of pure DNA of all chlamydial species to be differentiated. These methods have been largely replaced by PCR amplification of genes containing species-specific nucleotide signature sequences. These species signatures are then determined by RFLP, oligonucleotide probing, or, increasingly, by DNA sequencing of the amplification product. Direct sequencing has the advantage that the nucleotide sequences can be immediately aligned to sequences deposited in GenBank, and comparison with amplification products from standards of the chlamydial species is unnecessary. The availability in the near future of genome sequences of all chlamydial species, even for multiple strains of *Chlamydia trachomatis* and *Chlamydia pneumoniae*, allows the use of any gene with species signatures. Nevertheless, it will be preferable to rely for routine differentiation on the *ompA* and rRNA genes. The *ompA* gene contains highly polymorphic serovar determinants interspersed among conserved domains that allow genus-specific priming and species as well as serovar differentiation (Kaltenboeck et al., 1993, 1997). The 16S and 23S rRNA genes (but not the intergenic rRNA gene spacer sequence) are less polymorphic but allow design of primers with higher GC content than *ompA* primers, and thus more robust and sensitive PCR methods for direct amplification and species differentiation in disease specimens (DeGraves et al., 2003).

TABLE 167. Differentiation of *Chlamydia trachomatis* biovars[a]

Characteristic	Ocular	Genital	LGV
Tryptophan synthase:			
Functional	–	+	+
Utilization of indole	–	+	+
Preferred site of infection:			
Squamocolumnar epithelial cells	+	+	–
Lymph nodes	–	–	+
Behavior in laboratory animals:			
Intracerebral lethality in mice	–	–	+
Follicular conjunctivitis in primates	+	+	–
Behavior in cell culture:			
Plaque in L cells	–	–	+
Infection markedly enhanced by:			
Centrifugation onto cell monolayer	+	+	–
Treatment of host cells with diethyl- aminoethyl-dextran	+	+	–

[a]Symbols. +, >85% positive; , 0 15% positive.

TABLE 168. Differentiation of *Chlamydia pneumoniae* biovars

Human	Koala	Equine
Pharyngitis	Rhinitis	Serous nasal discharge
Bronchitis	Pneumonia	
Pneumonia		
One strain	na[a]	Only one isolate[b]
No plasmid in EB	na	Plasmid in EB
EB pear-shape[c]	na	EB oval
Large periplasmic space	na	Tight periplasmic space

[a]na, No information is available.

[b]No second isolate to confirm.

[c]Isolates with oval elementary bodies have been reported for the human biovar. However, unlike the elementary body of other chlamydial species which demonstrates a tight periplasmic space, these isolates still manifest a discernible periplasmic space, although less wide than that described originally for pear-shaped elementary bodies of *Chlamydia pneumoniae*.

Differentiation of the three biovars of *Chlamydia trachomatis* is listed in Table 167. Differentiation of the three biovars of *Chlamydia pneumoniae* is provided in Table 168.

Only a single serovar or serotype has been identified by the micro-immunofluorescence test in all chlamydial species, except for *Chlamydia trachomatis* which has 15 serovars, and for *Chlamydia psittaci* which has 6 serovars. Cross-reaction patterns of the micro-immunofluorescence antibody reaction in *Chlamydia trachomatis* are depicted in Figure 159 (Wang et al., 1973). No cross-reactivities are observed when *Chlamydia psittaci* serovar-specific monoclonal antibodies are used in the micro-immunofluorescence test for differentiation of *Chlamydia psittaci* serovars (Andersen, 1991).

Taxonomic considerations

The *Chlamydiales* represent a large clade of the bacteria that were pathogens of single eukaryotic cells nearly one billion years ago (Greub and Raoult, 2003; Stephens, 2002). Since that time, they have become obligate intracellular pathogens that have had very limited opportunity to acquire genes from other organisms. The ability to sample the microbial gene pool has played a decisive role in the adaptation to changing environments, evolution, and speciation of bacteria. From the perspective of chlamydiae, the intracellular vacuole is a largely invariant environment whether it is within a single-celled organism, or within a cell of a bird, koala, or human. During this billion years of history, all bacteria acquired and collected mutations in their 16S rRNA genes while rapidly evolving and adapting to novel environmental niches (i.e., speciation). The chlamydiae, too, collected 16S rRNA mutations, however, chlamydiae lacked the opportunity or the necessity to evolve in significantly new directions. While there has been considerable gene loss, even the amount of intragenomic recombination has been limited as evidenced by the high level of genome synteny among *Chlamydia* species.

From 1971, chlamydiae have been classified within one order (*Chlamydiales*) and one family (*Chlamydiaceae*) and with one genus (*Chlamydia*) (Molder et al., 1984). With the advent of PCR, *Chlamydia*-like DNA has been found in many types of animals and environmental samples; the extent of diversity of *Chlamydia* began to emerge, and this continues today, unfortunately without microbiological isolation. During this period, there was a growing need for more accommodating and accurate polyphasic taxonomic characterization, and this proceeded through a deliberate process highlighted by the detailed and comprehensive characterization of *Chlamydia pneumoniae* (Grayston et al., 1989) and *Chlamydia pecorum* (Fukushi and Hirai, 1992). In 1999, the taxonomy of *Chlamydia* was thrown in disarray and created confusion in the field (Schachter et al., 2001) when Everett et al. (1999) proposed a new second genus for the order *Chlamydiales* and family *Chlamydiaceae* in 1999. The genus proposal was based upon an arbitrary difference of greater than 4% in 16S rRNA gene sequences. This resulted in the separation of *Chlamydia pneumoniae*, *Chlamydia pecorum*, and *Chlamydia psittaci* into a new genus *Chlamydophila* leaving the *Chlamydia trachomatis* biovars in the original *Chlamydia* genus. This separation was inconsistent with the 4% genus threshold, since the 16S rRNA genes of several species within the proposed *Chlamydophila* genus differ by 4% or more.

One of the goals of taxonomy is to maintain stability in the nomenclature and to avoid unnecessary confusion according to Principle 1, Subprinciples 1 and 2 of the International Code of Nomenclature of Bacteria (Sneath, 1992). Indeed, for bacterial pathogens, renaming them even though it may be justified scientifically, may not be advisable. For example, the Judicial Commission of the International Committee on Nomenclature of Bacteria rejected the proposal for renaming the pathogen, *Yersinia pestis*, as a subspecies *Yersinia pseudotuberculosis* thereby giving rise to the name *Yersinia pseudotuberculosis* subspecies *pestis* even though this was justified based on DNA–DNA hybridization (Becovier et al., 1980; Judicial Commission, 1985). Because of the ruling of the Judicial Commission, the validated name *Yersinia pseudotuberculosis* subsp. *pestis* was rejected and the name *Yersinia pestis* was conserved on the approved list, due to the inherent confusion its renaming would cause to the field as well as the potential hazard it posed as a *nomen periculosum* for the general public welfare (Judicial Commission, 1985; Williams, 1984).

These principles can be applied to the pathogens of the family *Chlamydiaceae*. Therefore, for this edition of *Bergey's Manual of Systematic Bacteriology*, the sole genus name *Chlamydia* is being retained for all of the species of the family *Chlamydiaceae*, all of which are human or animal pathogens.

Until now there has been a single *Chlamydia trachomatis* biovar designation called the trachoma biovar that included strains responsible for natural ocular infection and trachoma and strains that naturally caused urogenital infection and disease. Although the trachoma-causing strains were typically represented by serovars A-C and the urogenital strains by serovars D-K, no biological markers were available to differentiate strains that caused these different diseases. It now appears that trachoma-causing strains lack a functional tryptophan synthase, whereas urogenital strains contain a functional enzyme (Fehlner-Gardiner et al., 2002). The enzymic function can easily be tested *in vitro* or by gene sequence techniques for non-isolated samples. It is proposed that trachoma strains lacking tryptophan synthase function compose the "trachoma" biovar and those from the urogenital tract that have a functional enzyme compose a new "genital" biovar. Thus, *Chlamydia trachomatis* consists of three biovars including ocular, genital, and LGV.

Further reading

Allegra, L and F. Blasi (editors). 1999. *Chlamydia pneumoniae.* Springer-Verlag Italiana, Milano.

Barron A.L. (Editor). 1988. Microbiology of *Chlamydia.* CRC Press, Boca Raton.

Bavoil, P. and P. Wyrick (editors). 2006. *Chlamydia*: Genomics, Pathogenesis and Implications for Control. Horizon Press, Norwich, UK.

Campbell, L.A. and C.-C. Kuo. 2004. *Chlamydia pneumoniae* - an infectious risk factor for atherosclerosis? Nature Rev. Microbiol. *2*: 23–32.

Campbell, L.A., C.-C. Kuo and C. Gaydos. 2006. Chlamydial Infections. *In* Detrick, B., R. G. Hamilton and J. D. Folds. (editors) Manual of Molecular and Clinical Laboratory Immunology. 7th edn. American Society for Microbiology Press, Washington, D.C., pp. 518–525.

Campbell, L.A., J.M. Marrazzo, W.E. Stamm and C.-C. Kuo. 2001. Chapter 27. Chlamydiae. *In* Laboratory Diagnosis of Bacterial Infections (edited by Cimolai). Marcel Dekker, New York, pp. 795–821.

Gordon, F.B. (editor). 1962. The Biology of the Trachoma Agent. New York Academy of Science, New York.

Hanna, L, J. Schachter, C.R. Dawson and P. Thygeson (editors). 1967. Trachoma and Allied Diseases. Am. J. Ophthal. *63*: Part II.

Kuo, C.-C., L.A. Jackson, L.A. Campbell and J.T. Grayston. 1995. *Chlamydia pneumoniae* (TWAR). Clin. Microbiol. Rev. *8*: 451–461.

Molder, J.W. (editor). 1989. Intracellular Parasitism. CRC Press, Boca Raton.

Nichols, R.L. (editor) 1971. Trachoma and Related Diseases Caused by Chlamydial Agents. Excerpta Medica, Amsterdam.

Stephens, R.S. (editor). 1999. *Chlamydia*: *Intracellular Biology*, *Pathogenesis*, and *Immunity*. American Society for Microbiology Press, Washington, D.C.

Storz, J. 1971. *Chlamydia* and *Chlamydia*-induced Diseases. Charles C. Thomas, Springfield, IL.

Storz, J. and B. Kaltenboeck, 1993. Disease diversity of Chlamydial Infections. *In* Rickettsial and Chlamydial Diseases of Domestic Animals (edited by Woldehiwet and Ristic). Pergamon Press, Oxford, pp. 363–392.

List of species of the genus *Chlamydia*

1. **Chlamydia trachomatis** (Busacca 1935) Rake 1957, 958[AL] [*Rickettsia trachomae* (*sic*) Busacca 1935, 567]

tra.cho'ma.tis. Gr. n. *trachusma -atos* roughness; N.L. n. *trachoma, -atis* the disease trachoma; N.L. gen. n. *trachomatis* of trachoma.

The characteristics are as described for the genus and as listed in Tables 161, 162, 163, and 167. The inclusion morphology is depicted in Figure 161.

All *Chlamydia trachomatis* isolates have come from humans. *Chlamydia trachomatis* isolates are classified into three biovars, ocular, genital, and lymphogranuloma venereum (LGV) on the basis of tissue tropism, pathogenesis in humans and laboratory animals, and their biological characteristics in cell cultures (Kuo et al., 1972a). The molecular basis of the separation of ocular from genital biovar is the ability to utilize exogenous indole for tryptophan biosynthesis by the genital, but not by the ocular biovar. The LGV biovar also utilizes indole for tryptophan biosynthesis. Three biovars are genetically (Table 162) and serologically (Figure 159) closely related. Serologically distinguishable from other chlamydial species by the species-specific monoclonal antibody in the micro-immunofluorescence test. Isolates are serotyped by the same test, however, a broad-reactivity among BED serovars and LGV are noted. Strains of ocular biovars parasitize mucous membranes of the conjunctiva, while strains of the genital biovar parasitize the squamous and columnar epithelium of the urogenital mucosa. The LGV biovar has tropism for lymphoid organs and is more invasive than the other two biovars. This behavior parallels its behavior in animals and cultured cells.

DNA G+C content (mol%): 43–44.2 (T_m).

Type strain: A/Har-13 (Trachoma type A strain HAR-13), ATCC VR-571-B.

Sequence accession no. (16S rRNA gene): D89067, E17344, NR 025888.

Further information concerning the three biovars of *Chlamydia trachomatis* is given below:

(i) Biovar ocular. Chlamydiae of this biovar exclusively parasitize the conjunctival mucosal cells. Transmission is by discharge from the inflamed eyes commonly in family or school. The primary infection is in the form of follicular conjunctivitis, which may heal spontaneously. However, in endemic areas, repeated infections sustain the chronicity of infection and cause scarring of the conjunctiva and cornea. The end stage of the disease is blindness. Table 165 lists the diseases caused by the ocular biovar of *Chlamydia trachomatis*. Serovars of the ocular biovar are A, B, Ba, and C.

DNA G+C content (mol%) of five strains: 44.0 (T_m).

Type strain: PK-2, serovar C, ATCC VR-576.

Sequence accession no. (16S rRNA gene) for strain C/TW-3/OT: D85720.

(ii) Biovar genital. Strains of this biovar are sexually transmitted. The primary site of infection is the mucosal membranes of the urogenital tract. Diseases caused by the genital biovar are listed in Table 165. The infection may be auto-transferred to the conjunctiva or to contacts to cause acute follicular conjunctivitis. Chronic infection due to the genital biovar of *Chlamydia trachomatis* is a common cause of infertility in women. Serovars of genital isolates are D through K. No difference in the virulence among the serovars has been observed.

DNA G+C content (mol%) of two strains is: 44.2 (T_m).

Type strain: UW-3/Cx, serovar D, ATCC VR-885.

Sequence accession no. (16S rRNA gene) for strain D/UW-3Cx: D85721.

(iii) Biovar lymphogranuloma venereum (LGV). Strains of this biovar are sexually transmitted and cause systemic infection with the major manifestation of lymphadenopathy or swelling of lymph nodes draining the extragenitalia. Three serovars are recognized which are L_1, L_2, and L_3.

DNA G+C content (mol%): 43.0 (T_m).

Type strain: 434, serovar L_2, ATCC VR-902B.

Sequence accession no. (16S and 23S rRNA genes): U68443.

2. **Chlamydia abortus** comb. nov. Everett, Bush and Andersen 1999 (*Chlamydophila abortus* Everett, Bush and Andersen 1999, 434)

   a.bor'tus. L. n. *abortus, -us* an abortion, a miscarriage; L. gen. n. *abortus* of an abortion, a miscarriage.

The characteristics are as described for the genus and as listed in Tables 162, 163, and 166. This species was separated from *Chlamydia psittaci* with the establishment of five new

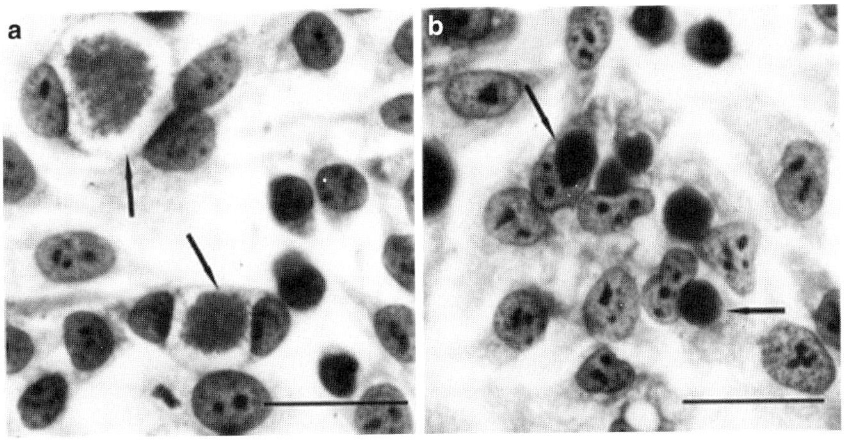

FIGURE 161. Two different morphologies of chlamydial inclusions. Infected host cell monolayers were fixed with methanol and stained with May-Greenwald Giemsa. Bars = 20 µm. Arrows point to a typical inclusion. (a) Vacuolar, glycogen-positive (*Chlamydia trachomatis, Chlamydia muridarum* and *Chlamydia suis*). (b) Solid, glycogen-negative (*Chlamydia psittaci, Chlamydia pneumoniae, Chlamydia pecorum, Chlamydia abortus, Chlamydia caviae* and *Chlamydia felis*). (Courtesy of author, C.-C. Kuo.)

species in addition to *Chlamydia trachomatis*, *Chlamydia pecorum*, *Chlamydia pneumoniae*, and *Chlamydia psittaci*.

Chlamydia abortus is endemic in ruminants, and it is a frequent cause of late-term abortion in flocks of sheep or goats (enzootic abortion of ewes, EAE), but less frequently in cattle. It has also been associated with seminal vesiculitis in bulls and with bovine infertility. *Chlamydia abortus* has also infrequently been identified as the cause of abortion in other mammals such as horse and pig, and experimentally causes abortion in laboratory animals such as rabbits, guinea pigs, and mice. Infection of pregnant women with *Chlamydia abortus* is typically contracted during contact with small ruminants and can cause flu-like symptoms of systemic infection followed by abortion (Herring et al., 1987; Johnson et al., 1985). *Chlamydia abortus* isolates worldwide represent a single serovar and have virtually identical *ompA* and rRNA genes.

DNA G+C content (mol%): not determined.

Type strain: B577, Ovine Chlamydial Abortion strain B-577, ATCC VR-656.

Sequence accession no. (16S rRNA gene): AB001783, D85709.

3. **Chlamydia caviae** comb. nov. Everett, Bush and Andersen 1999 (*Chlamydophila caviae* Everett, Bush and Andersen 1999, 434.)

ca.vi'ae. N.L. gen. n. *caviae* of *Cavia*, of the guinea pig (*Cavia cobaya*) because the type strain was isolated from *Cavia cobaya*.

The characteristics are as described for the genus and as listed in Tables 162, 163, and 166. This species was separated from *Chlamydia psittaci* with the establishment of five new species in addition to *Chlamydia trachomatis*, *Chlamydia pecorum*, *Chlamydia pneumoniae*, and *Chlamydia psittaci*.

Chlamydia caviae is a chlamydial pathogen that has exclusively been isolated from guinea pigs. The *ompA* sequences of five *Chlamydia caviae* isolates are identical. The primary isolation site is the conjunctiva (guinea pig inclusion conjunctivitis, GPIC), but inoculation of the guinea pig genital tract results in disease and infertility that are remarkably similar to those caused by human genital infection with *Chlamydia trachomatis*.

DNA G+C content (mol%): not determined.

Type strain: GPIC, ATCC VR 813.

Sequence accession no. (16S rRNA gene): D85708.

4. **Chlamydia felis** comb. nov. Everett, Bush and Andersen 1999 (*Chlamydophila felis* Everett, Bush and Andersen 1999, 434.)

fe'lis. L. n. *felis, -is* a cat; L. gen. n. *felis* of a cat.

The characteristics are as described for the genus and as listed in Table 162, 163, and 166. This species was separated from *Chlamydia psittaci* with the establishment of five new species in addition to *Chlamydia trachomatis*, *Chlamydia pecorum*, *Chlamydia pneumoniae*, and *Chlamydia psittaci*.

Chlamydia felis is frequently isolated from conjunctivitis and rhinitis in cats. All strains represent a single serovar, have highly conserved *ompA* genes, and rRNA genes differ by less than 0.6%. Human conjunctivitis caused by infection with *Chlamydia felis* contracted from pet cats appears to be common (Schachter et al., 1969).

DNA G+C content (mol%): not determined.

Type strain: FP Baker, ATCC VR-120.

Sequence accession no. (16S rRNA gene): D85701.

5. **Chlamydia muridarum** Everett, Bush and Andersen 1999, 431[VP]

mu.ri.da'rum. N.L. pl. gen. n. *muridarum* of the *Muridae* (i.e., the mouse/hamster family).

The characteristics are as described for the genus and as listed in Tables 162, 163, and 166. This species was separated from *Chlamydia trachomatis* with the establishment of five new species in addition to *Chlamydia trachomatis*, *Chlamydia pecorum*, *Chlamydia pneumoniae*, and *Chlamydia psittaci*.

Two virtually identical *Chlamydia muridarum* strains are currently known, isolated from the lungs of clinically healthy mice (strain MoPn Nigg II) and from the intestinal tract of a hamster with proliferative ileitis (strain SFPD). Strain MoPn has been found to be highly virulent for mice and can cause severe interstitial pneumonia. It is also frequently used in mouse models of chlamydial genital tract infection and infertility.

DNA G+C content (mol%): 42–44 (T_m).

Type strain: Nigg II, ATCC VR123.

Sequence accession no. (16S rRNA gene): D85718.

6. **Chlamydia pecorum** Fukushi and Hirai 1992, 307[VP]

pe.co'rum. L. gen. pl. n. *pecorum* of flocks of sheep or herds of cattle.

The characteristics are as described for the genus and as listed in Tables 162, 163, and 166. This was the fourth species established for the genus, after *Chlamydia trachomatis*, *Chlamydia psittaci*, and *Chlamydia pneumoniae*.

Chlamydia pecorum has been isolated from ruminants, swine, and koalas, and is frequently found in low numbers on the mucosal membranes of healthy host animals. *Chlamydia pecorum* causes severe diseases in these animals such as sporadic bovine encephalomyelitis (SBE) in feedlot cattle, and enteritis, pneumonia, polyarthritis, and urogenital infection in cattle and swine. In koalas, *Chlamydia pecorum* causes keratoconjunctivitis, vaginitis, cystitis, and infertility, and is strongly associated with the decline of the koala population in combination with stress resulting from destruction of the koala habitat. *Chlamydia pecorum* isolates show polymorphic *ompA* genes, suggesting the existence of numerous as yet uncharacterized serovars. Human infection with *Chlamydia pecorum* has not been observed.

DNA G+C content (mol%): 39.3 (T_m).

Type strain: Bo/E58, ATCC VR-628.

Sequence accession no. (16S rRNA gene): D88371.

7. **Chlamydia pneumoniae** Grayston, Kuo, Campbell and Wang 1989, 88[VP]

pneu.mo'ni.ae. Gr. n. *pneumonia* pneumonia, inflammation of the lungs; N.L. gen. n. *pneumoniae* of pneumonia.

The characteristics are as described for the genus and as listed in Tables 162, 163, 165, 166, and 168. The inclusion morphology is depicted in Figure 155.

The proposal for establishing this species, which was the third species in the genus, was based on the ultrastructural morphology of the elementary bodies that is typically pear-shaped (Figure 155), presence of a species-specific antigen identifiable with a species-specific monoclonal antibody, and its clear separation from the two other species by the DNA–DNA homology (Table 162). The species includes three biovars: human, koala, and horse (Table 168). Since 1983, no other isolate has been obtained to confirm the original isolate from horse (Spencer and Johnson, 1983; Willis et al., 1990). The inclusions contain no glycogen, hence stain negative with iodine. Diseases caused by *Chlamydia pneumoniae* in humans are listed in Table 165 and diseases in animals in Table 166.

DNA G+C content (mol%): 40 (T_m).

Type strain: TW-183, ATCC VR-2282.

Sequence accession no. (16S rRNA gene): L06108, NR 026527, Z49873.

8. **Chlamydia psittaci** (Lillie 1930) Page 1968, 60[AL] (*Rickettsia psittaci* Lillie 1930, 778)

psit'ta.ci. L. n. *psittacus* a parrot; L. gen. n. *psittaci* of a parrot.

The characteristics are as described for the genus and as listed in Tables 162 and 163. This was the type species and strain of the first two species, *Chlamydia trachomatis* and *Chlamydia psittaci*, established for the genus.

Chlamydia psittaci is comprised of six serovars, A–F, of avian chlamydial isolates (Andersen, 1991, 1997) that cause latent and clinically overt infections of the intestinal and respiratory tracts in many species of birds. Upon exposure of birds to crowding and stress, these infections often become severe, with diarrhea, air sacculitis, pneumonia, and fatal systemic spread with pericarditis and splenomegaly. Human *Chlamydia psittaci*-induced disease, typically contracted by inhalation of dried bird feces, is called psittacosis or ornithosis if contracted from psittacine or from non-psittacine birds, respectively. Psittacosis is characterized by early flu-like symptoms followed by severe, sometimes fatal atypical interstitial pneumonia that requires intensive antibiotic and symptomatic therapy and resolves only slowly after several weeks. *Chlamydia psittaci* isolates should be handled only in a biosafety level 3 laboratory.

DNA G+C content (mol%): 41.3 (T_m).

Type strain: 6BC, Psittacosis strain 6 BC, ATCC VR-125.

Sequence accession no. (16S rRNA gene): AB001778, U68447.

9. **Chlamydia suis** Everett, Bush and Andersen 1999, 431[VP]

su'is. L. n. *sus suis* a swine; L. gen. n. *suis* of a swine.

The characteristics are as described for the genus and as listed in Tables 163 and 166. This species was separated from *Chlamydia trachomatis* with the establishment of five new species in addition to *Chlamydia trachomatis*, *Chlamydia pecorum*, *Chlamydia pneumoniae*, and *Chlamydia psittaci*.

The S45 type strain of *Chlamydia suis* was isolated by Koelbl in the 1960s in Austria from a pig with inapparent intestinal infection (Koelbl, 1969; Koelbl et al., 1970). Since 1995, high incidences and tetracycline-resistant strains of *Chlamydia suis* have been found in enteric and conjunctival porcine specimens, suggesting that *Chlamydia suis* infections are endemic in swine (Dugan et al., 2004). *Chlamydia suis* strains show marked *ompA* gene polymorphism, particularly of the variable domains, but less than 1.1% 16S rRNA gene diversity. This suggests the existence of numerous serovars, similar to *Chlamydia trachomatis*. Human infection with *Chlamydia suis* has not been observed.

DNA G+C content (mol%): not determined.

Type strain: S45, ATCC VR-1474.

Sequence accession no. (16S rRNA gene): U73110.

Other organisms

Chlamydia-like organisms have been described in reptiles (turtle, tortoise, and boar) (Homer et al., 1994; Vanrompay et al., 1994), amphibians (frog) (Bodetti et al., 2002), bivalves (mussel and clam) (Cajaraville and Angulo, 1991; Harshbarger and Chang, 1977; Page and Cutlip, 1982), and crab (Sparks et al., 1985). These *Chlamydia*-like organisms have ultra-structural morphology and developmental forms resembling those of chlamydiae in that the organisms grow within a cytoplasmic vacuole (inclusion) and undergo a developmental cycle from the reticulate body-like form to an elementary body-like form. The relatedness of these organisms to other *Chlamydia* species has been demonstrated by electron microscopy, PCR analysis of OMP and rRNA genes, and immunoreactivity with *Chlamydia*-specific antibodies. Isolation of the organisms by inoculation of chicken eggs has been reported from Moorish tortoise and green sea turtle (Homer et al., 1994; Vanrompay et al., 1994). However, taxonomic relatedness of these microorganisms to other chlamydial organisms has yet to be determined.

References

Alexander, J.J. 1968. Separation of protein synthesis in meningopneumonitisgent from that in L cells by differential susceptibility to cycloheximide. J. Bacteriol. *95*: 327–332.

Allan, I. and J.H. Pearce. 1983a. Differential amino acid utilization by *Chlamydia psittaci* (strain guinea pig inclusion conjunctivitis) and its regulatory effect on chlamydial growth. J. Gen. Microbiol. *129*: 1991–2000.

Allan, I. and J.H. Pearce. 1983b. Amino acid requirements of strains of *Chlamydia trachomatis* and *C. psittaci* growing in McCoy cells: relationship with clinical syndrome and host origin. J. Gen. Microbiol. *129*: 2001–2007.

Amann, R., N. Springer, W. Schonhuber, W. Ludwig, E.N. Schmid, K.D. Muller and R. Michel. 1997. Obligate intracellular bacterial parasites of acanthamoebae related to *Chlamydia* spp. Appl. Environ. Microbiol. *63*: 115–121.

Andersen, A.A. 1991. Serotyping of *Chlamydia psittaci* isolates using serovar-specific monoclonal antibodies with the microimmunofluorescence test. J. Clin. Microbiol. *29*: 707–711.

Andersen, A.A. 1997. Two new serovars of *Chlamydia psittaci* from North American birds. J. Vet. Diagn. Invest. *9*: 159–164.

Barbour, A.G., K. Amano, T. Hackstadt, L. Perry and H.D. Caldwell. 1982. *Chlamydia trachomatis* has penicillin-binding proteins but not detectable muramic acid. J. Bacteriol. *151*: 420–428.

Batteiger, B.E., W.J.t. Newhall and R.B. Jones. 1985. Differences in outer membrane proteins of the lymphogranuloma venereum and trachoma biovars of *Chlamydia trachomatis*. Infect. Immun. *50*: 488–494.

Bavoil, P., A. Ohlin and J. Schachter. 1984. Role of disulfide bonding in outer membrane structure and permeability in *Chlamydia trachomatis*. Infect. Immun. *44*: 479–485.

Bavoil, P., R.S. Stephens and S. Falkow. 1990. A soluble 60 kiloDalton antigen of *Chlamydia* spp. is a homologue of *Escherichia coli* GroEL. Mol. Microbiol. *4*: 461–469.

Bavoil, P.M. and R.C. Hsia. 1998. Type III secretion in *Chlamydia*: a case of deja vu? Mol. Microbiol. *28*: 860–862.

Beatty, W.L., T.A. Belanger, A.A. Desai, R.P. Morrison and G.I. Byrne. 1994. Tryptophan depletion as a mechanism of gamma interferon-mediated chlamydial persistence. Infect. Immun. *62*: 3705–3711.

Becovier, H., H.H. Mollaret, J.M. Alonso, J. Brault, G.R. Fanning, A.G. Steigerwalt and D.J. Brenner. 1980. Intra- and interspecies relatedness of *Yersinia pestis* by DNA hybridization and its relationship to *Yersinia pseudotuberculosis*. Curr. Microbiol. *4*: 225–229.

Bodetti, T.J., E. Jacobson, C. Wan, L. Hafner, A. Pospischil, K. Rose and P. Timms. 2002. Molecular evidence to support the expansion of the hostrange of *Chlamydophila pneumoniae* to include reptiles as well as humans, horses, koalas and amphibians. Syst. Appl. Microbiol. *25*: 146–152.

Brade, L., S. Schramek, U. Schade and H. Brade. 1986. Chemical, biological, and immunochemical properties of the *Chlamydia psittaci* lipopolysaccharide. Infect. Immun. *54*: 568–574.

Busacca, A. 1935. Un germe caractères de rickettsies (*Rickettsia trachomæ*) dans tissues trachomateux. Arch. Ophthalmol. *52*: 567–572.

Cajaraville, M.P. and E. Angulo. 1991. *Chlamydia*-like organisms in digestive and duct cells of mussels from the Basque coast. J. Invertebr. Pathol. *58*: 381–386.

Caldwell, H.D., J. Kromhout and J. Schachter. 1981. Purification and partial characterization of the major outer membrane protein of *Chlamydia trachomatis*. Infect. Immun. *31*: 1161–1176.

Campbell, L.A., C.C. Kuo, S.P. Wang and J.T. Grayston. 1990. Serological response to *Chlamydia pneumoniae* infection. J. Clin. Microbiol. *28*: 1261–1264.

Campbell, L.A., D.L. Patton, D.E. Moore, A.L. Cappuccio, B.A. Mueller and S.P. Wang. 1993. Detection of *Chlamydia trachomatis* deoxyribonucleic acid in women with tubal infertility. Fertil Steril. *59*: 45–50.

Chi, E.Y., C.C. Kuo and J.T. Grayston. 1987. Unique ultrastructure in the elementary body of *Chlamydia* sp. strain TWAR. J. Bacteriol. *169*: 3757–3763.

Chosewood, L.C. and D.E. Wilson 2007. Biosafety in Microbiological and Biomedical Laboratories, 5th edn, U.S. Department of Health and Human Services, Public Health Service Centers for Disease Control and Prevention and National Institutes of Health. U.S. Government Printing Office, Washington, D.C.

Clifton, D.R., C.A. Dooley, S.S. Grieshaber, R.A. Carabeo, K.A. Fields and T. Hackstadt. 2005. Tyrosine phosphorylation of the chlamydial effector protein Tarp is species specific and not required for recruitment of actin. Infect. Immun. *73*: 3860–3868.

Collingro, A., S. Poppert, E. Heinz, S. Schmitz-Esser, A. Essig, M. Schweikert, M. Wagner and M. Horn. 2005a. Recovery of an environmental *Chlamydia* strain from activated sludge by co-cultivation with *Acanthamoeba* sp. Microbiology *151*: 301–309.

Collingro, A., E.R. Toenshoff, M.W. Taylor, T.R. Fritsche, M. Wagner and M. Horn. 2005b. '*Candidatus* Protochlamydia amoebophila', an endosymbiont of *Acanthamoeba* spp. Int. J. Syst. Evol. Microbiol. *55*: 1863–1866.

Corsaro, D., D. Venditti and M. Valassina. 2002. New parachlamydial 16S rDNA phylotypes detected in human clinical samples. Res. Microbiol. *153*: 563–567.

Cox, R.L., C.-C. Kuo, J.T. Grayston and L.A. Campbell. 1988. Deoxyribonucleic acid relatedness of *Chlamydia* sp. strain TWAR to *Chlamydia trachomatis* and *Chlamydia* psitaci. Int. J. Syst. Bacteriol. *38*: 265–267.

Davis, C.H., J.E. Raulston and P.B. Wyrick. 2002. Protein disulfide isomerase, a component of the estrogen receptor complex, is associated with *Chlamydia trachomatis* serovar E attached to human endometrial epithelial cells. Infect. Immun. *70*: 3413–3418.

DeGraves, F.J., D. Gao and B. Kaltenboeck. 2003. High-sensitivity quantitative PCR platform. BioTechniques *34*: 106–110, 112–105.

Dilbeck, P.M., J.F. Evermann, T.B. Crawford, A.C. Ward, C.W. Leathers, C.J. Holland, C.A. Mebus, L.L. Logan, F.R. Rurangirwa and T.C. McGuire. 1990. Isolation of a previously undescribed rickettsia from an aborted bovine fetus. J. Clin. Microbiol. *28*: 814–816.

Draghi, A., 2nd, V.L. Popov, M.M. Kahl, J.B. Stanton, C.C. Brown, G.J. Tsongalis, A.B. West and S. Frasca, Jr. 2004. Characterization of "*Candidatus* Piscichlamydia salmonis" (order *Chlamydiales*), a chlamydia-like bacterium associated with epitheliocystis in farmed Atlantic salmon (*Salmo salar*). J. Clin. Microbiol. *42*: 5286–5297.

Dugan, J., D.D. Rockey, L. Jones and A.A. Andersen. 2004. Tetracycline resistance in *Chlamydia suis* mediated by genomic islands inserted into the chlamydial *inv*-like gene. Antimicrob. Agents Chemother. *48*: 3989–3995.

Everett, K.D.E., R.M. Bush and A.A. Andersen. 1999. Emended description of the order *Chlamydiales*, proposal of *Parachlamydiaceae* fam. nov. and *Simkaniaceae* fam. nov., each containing one monotypic genus, revised taxonomy of the family *Chlamydiaceae*, including a new genus and five new species, and standards for the identification of organisms. Int. J. Syst. Bacteriol. *49*: 415–440.

Everson, J.S., S.A. Garner, B. Fane, B.L. Liu, P.R. Lambden and I.N. Clarke. 2002. Biological properties and cell tropism of Chp2, a bacteriophage of the obligate intracellular bacterium *Chlamydophila abortus*. J. Bacteriol. *184*: 2748–2754.

Fan, J. and R.S. Stephens. 1997. Antigen conformation dependence of *Chlamydia trachomatis* infectivity neutralization. J. Infect. Dis. *176*: 713–721.

Fehlner-Gardiner, C., C. Roshick, J.H. Carlson, S. Hughes, R.J. Belland, H.D. Caldwell and G. McClarty. 2002. Molecular basis defining human *Chlamydia trachomatis* tissue tropism. A possible role for tryptophan synthase. J. Biol. Chem. *277*: 26893–26903.

Fukushi, H. and K. Hirai. 1992. Proposal of *Chlamydia pecorum* sp. nov. for *Chlamydia* strains derived from ruminants. Int. J. Syst. Bacteriol. *42*: 306–308.

Furness, G., D.M. Graham and P. Reeve. 1960. The titration of trachoma and inclusion blennorrhoea viruses in cell cultures. J. Gen. Microbiol. *23*: 613–619.

Garner, S.A., J.S. Everson, P.R. Lambden, B.A. Fane and I.N. Clarke. 2004. Isolation, molecular characterisation and genome sequence of a bacteriophage (Chp3) from *Chlamydophila pecorum*. Virus Genes *28*: 207–214.

Gerloff, R.K., D.B. Ritter and R.O. Watson. 1970. Studies on thermal denaturation of DNA from various chlamydiae. J. Infect. Dis. *121*: 65–69.

Gordon, F.B., V.W. Andrew and J.C. Wagner. 1957. Development of resistance to penicillin and to chlortetracycline in psittacosis virus. Virology *4*: 156–171.

Gordon, F.B. and A.L. Quan. 1965. Occurence of glycogen in inclusions of the psittacosis-lymphogranuloma venereum-trachoma agents. J. Infect. Dis. *115*: 186–196.

Gordon, F.B., I.A. Harper, A.L. Quan, J.D. Treharne, R.S. Dwyer and J.A. Garland. 1969. Detection of *Chlamydia* (Bedsonia) in certain infections of man. I. Laboratory procedures: comparison of yolk sac and cell culture for detection and isolation. J. Infect. Dis. *120*: 451–462.

Grayston, J.T. and S. Wang. 1975. New knowledge of chlamydiae and the diseases they cause. J. Infect. Dis. *132*: 87–105.

Grayston, J.T., S.P. Wang, L.J. Yeh and C.C. Kuo. 1985. Importance of reinfection in the pathogenesis of trachoma. Rev. Infect. Dis. *7*: 717–725.

Grayston, J.T., C.C. Kuo, L.A. Campbell and S.P. Wang. 1989. *Chlamydia pneumoniae* sp. nov. for *Chlamydia* sp. strain twar. Int. J. Syst. Bacteriol. *39*: 88–90.

Greub, G. and D. Raoult. 2003. History of the ADP/ATP-translocase-encoding gene, a parasitism gene transferred from a *Chlamydiales* ancestor to plants 1 billion years ago. Appl. Environ. Microbiol. *69*: 5530–5535.

Grimwood, J., L. Olinger and R.S. Stephens. 2001. Expression of *Chlamydia pneumoniae* polymorphic membrane protein family genes. Infect. Immun. *69*: 2383–2389.

Hackstadt, T., W. Baehr and Y. Ying. 1991. *Chlamydia trachomatis* developmentally regulated protein is homologous to eukaryotic histone H1. Proc. Natl. Acad. Sci. U. S. A. *88*: 3937–3941.

Harshbarger, J.C. and S.C. Chang. 1977. *Chlamydiae* (with phages), mycoplasmas, and richettsiae in Chesapeake Bay bivalves. Science *196*: 666–668.

Hatch, T.P., E. Al-Hossainy and J.A. Silverman. 1982. Adenine nucleotide and lysine transport in *Chlamydia psittaci*. J. Bacteriol. *150*: 662–670.

Hatch, T.P., M. Miceli and J.E. Sublett. 1986. Synthesis of disulfide-bonded outer membrane proteins during the developmental cycle of *Chlamydia psittaci* and *Chlamydia trachomatis*. J. Bacteriol. *165*: 379–385.

Herring, A.J., I.E. Anderson, M. McClenaghan, N.F. Inglis, H. Williams, B.A. Matheson, C.P. West, M. Rodger and P.P. Brettle. 1987. Restriction endonuclease analysis of DNA from two isolates of *Chlamydia psittaci* obtained from human abortions. Br. Med. J. (Clin. Res. Ed.) *295*: 1239.

Hoelzle, L.E., G. Steinhausen and M.M. Wittenbrink. 2000. PCR-based detection of chlamydial infection in swine and subsequent PCR-coupled genotyping of chlamydial *omp1*-gene amplicons by DNA-hybridization, RFLP-analysis, and nucleotide sequence analysis. Epidemiol. Infect. *125*: 427–439.

Homer, B.L., E.R. Jacobson, J. Schumacher and G. Scherba. 1994. Chlamydiosis in mariculture-reared green sea turtles (*Chelonia mydas*). Vet. Pathol. *31*: 1–7.

How, S.J., D. Hobson, C.A. Hart and R.E. Webster. 1985. An *in-vitro* investigation of synergy and antagonism between antimicrobials against *Chlamydia trachomatis*. J. Antimicrob. Chemother. *15*: 533–538.

Hsia, R.C., Y. Pannekoek, E. Ingerowski and P.M. Bavoil. 1997. Type III secretion genes identify a putative virulence locus of *Chlamydia*. Mol. Microbiol. *25*: 351–359.

Hsia, R.C., L.M. Ting and P.M. Bavoil. 2000. Microvirus of chlamydia psittaci strain guinea pig inclusion conjunctivitis: isolation and molecular characterization. Microbiology *146*: 1651–1660.

Hybiske, K. and R.S. Stephens. 2007. Mechanisms of host cell exit by the intracellular bacterium *Chlamydia*. Proc. Natl. Acad. Sci. U. S. A. *104*: 11430–11435.

Iliffe-Lee, E.R. and G. McClarty. 1999. Glucose metabolism in *Chlamydia trachomatis*: the 'energy parasite' hypothesis revisited. Mol. Microbiol. *33*: 177–187.

Johnson, F.W., B.A. Matheson, H. Williams, A.G. Laing, V. Jandial, R. Davidson-Lamb, G.J. Halliday, D. Hobson, S.Y. Wong, K.M. Hadley and et al. 1985. Abortion due to infection with *Chlamydia psittaci* in a sheep farmer's wife. Br. Med. J. (Clin. Res. Ed.) *290*: 592–594.

Jones, H., G. Rake and B. Stearns. 1945. Studies on lymphogranuloma venereum. III. The action of the sulfonamides on the agent of lymphogranuloma venereum. J. Infect. Dis. *76*: 55–69.

Joseph, T., F.E. Nano, C.F. Garon and H.D. Caldwell. 1986. Molecular characterization of *Chlamydia trachomatis* and *Chlamydia psittaci* plasmids. Infect. Immun. *51*: 699–703.

Judicial Commission. 1985. Opinion 60. Rejection of the name *Yersinia pseudotuberculosis* subsp. *pestis* (van Loghem) Bechovier et al. 1991 and conservation of the name *Yersinia pestis* (Lehman and Neumann) van Loghem 1944 for the plague bacillus. Int. J. Syst. Bacteriol. *35*: 540–541.

Kahane, S., E. Metzer and M.G. Friedman. 1995. Evidence that the novel microorganism 'Z' may belong to a new genus in the family *Chlamydiaceae*. FEMS Microbiol. Lett. *126*: 203–207.

Kalman, S., W. Mitchell, R. Marathe, C. Lammel, J. Fan, R.W. Hyman, L. Olinger, J. Grimwood, R.W. Davis and R.S. Stephens. 1999. Comparative genomes of *Chlamydia pneumoniae* and *C. trachomatis*. Nat. Genet. *21*: 385–389.

Kaltenboeck, B., K.G. Kousoulas and J. Storz. 1993. Structures of and allelic diversity and relationships among the major outer membrane protein (*ompA*) genes of the four chlamydial species. J. Bacteriol. *175*: 487–502.

Kaltenboeck, B., N. Schmeer and R. Schneider. 1997. Evidence for numerous *omp1* alleles of porcine *Chlamydia trachomatis* and novel chlamydial species obtained by PCR. J. Clin. Microbiol. *35*: 1835–1841.

Kane, C.D., R.M. Vena, S.P. Ouellette and G.I. Byrne. 1999. Intracellular tryptophan pool sizes may account for differences in gamma interferon-mediated inhibition and persistence of chlamydial growth in polarized and nonpolarized cells. Infect. Immun. *67*: 1666–1671.

Kingsbury, D.T. and E. Weiss. 1968. Lack of deoxyribonucleic acid homology between species of the genus *Chlamydia*. J. Bacteriol. *96*: 1421–1423.

Koelbl, O. 1969. Untersuchungen uebe das Vorkommen von Miyagawanellen beim Schwein. Wien Tierarztl Mschr *56*: 332–335.

Koelbl, O., H. Burtscher and J Hebensreit. 1970. Polyarthritis bei Schlachtschweinen. Mikrobiologische, histologische und fleischhygienische Aspekte. Wien Tierarztl. Mschr *57*: 355–361.

Kubo, A. and R.S. Stephens. 2000. Characterization and functional analysis of PorB, a *Chlamydia* porin and neutralizing target. Mol. Microbiol. *38*: 772–780.

Kuo, C.-C., S.-P. Wang and J.T. Grayston. 1977a. Growth of trachoma organisms in HeLa 229 cell culture. *In* Nongonococal Urethritis and Related Infection (edited by Hobson and Holmes). American Society for Microbiology, Washington, D.C., pp. 328–336.

Kuo, C.-C., H.-H. Chen, S.-P. Wang and J.T. Grayston. 1986. Identification of a new group of *Chlamydia psittaci* strains called TWAR. J. Clin. Microbiol. *24*: 1034–1037.

Kuo, C., S. Wang and J.T. Grayston. 1972a. Differentiation of TRIC and LGV organisms based on enhancement of infectivity by DEAE-dextran in cell culture. J. Infect. Dis. *125*: 313–317.

Kuo, C., S. Wang, B.B. Wentworth and J.T. Grayston. 1972b. Primary isolation of TRIC organisms in HeLa 229 cells treated with DEAE-dextran. J. Infect. Dis. *125*: 665–668.

Kuo, C., N. Takahashi, A.F. Swanson, Y. Ozeki and S. Hakomori. 1996. An N-linked high-mannose type oligosaccharide, expressed at the major outer membrane protein of *Chlamydia trachomatis*, mediates attachment and infectivity of the microorganism to HeLa cells. J. Clin. Invest *98*: 2813–2818.

Kuo, C.C., S.P. Wang and J.T. Grayston. 1973. Effect of polycations, polyanions and neuraminidase on the infectivity of trachoma-inclusin conjunctivitis and lymphogranuloma venereum organisms HeLa cells: sialic acid residues as possible receptors for trachoma-inclusion conjunction. Infect. Immun. *8*: 74–79.

Kuo, C.C. and T. Grayston. 1976. Interaction of *Chlamydia trachomatis* organisms and HeLa 229 cells. Infect. Immun. *13*: 1103–1109.

Kuo, C.C., S.P. Wang and J.T. Grayston. 1977b. Antimicrobial activity of several antibiotics and a sulfonamide against *Chlamydia trachomatis* organisms in cell culture. Antimicrob. Agents Chemother. *12*: 80–83.

Kuo, C.C. and E.Y. Chi. 1987. Ultrastructural study of *Chlamydia trachomatis* surface antigens by immunogold staining with monoclonal antibodies. Infect. Immun. *55*: 1324–1328.

Kuo, C.C. and J.T. Grayston. 1988. *In vitro* drug susceptibility of *Chlamydia* sp. strain TWAR. Antimicrob. Agents Chemother. *32*: 257–258.

Kuo, C.C. and J.T. Grayston. 1990a. Amino acid requirements for growth of *Chlamydia pneumoniae* in cell cultures: growth enhancement by lysine or methionine depletion. J. Clin. Microbiol. *28*: 1098–1100.

Kuo, C.C. and J.T. Grayston. 1990b. A sensitive cell line, HL cells, for isolation and propagation of *Chlamydia pneumoniae* strain TWAR. J. Infect. Dis. *162*: 755–758.

Kuo, C.C., A. Shor, L.A. Campbell, H. Fukushi, D.L. Patton and J.T. Grayston. 1993. Demonstration of *Chlamydia pneumoniae* in atherosclerotic lesions of coronary arteries. J. Infect. Dis. *167*: 841–849.

Kuo, C.C., M. Puolakkainen, T.M. Lin, M. Witte and L.A. Campbell. 2002. Mannose-receptor positive and negative mouse macrophages differ in their susceptibility to infection by *Chlamydia* species. Microb. Pathog. *32*: 43–48.

Li, D., A. Vaglenov, T. Kim, C. Wang, D. Gao and B. Kaltenboeck. 2005. High-yield culture and purification of *Chlamydiaceae* bacteria. J. Microbiol. Methods *61*: 17–24.

Lillie, R.D. 1930. Psittacosis-rickettsia-like inclusions in man and in experimental animals. Pub. Health Rep *45*: 773–778.

Lin, H.S. and J.W. Moulder. 1966. Patterns of response to sulfadiazine, D-cycloserine and D-alanine in members of the psittacosis group. J. Infect. Dis. *116*: 372–376.

Liu, B.L., J.S. Everson, B. Fane, P. Giannikopoulou, E. Vretou, P.R. Lambden and I.N. Clarke. 2000. Molecular characterization of a bacteriophage (Chp2) from *Chlamydia psittaci*. J. Virol. *74*: 3464–3469.

Majeed, M. and E. Kihlstrom. 1991. Mobilization of F-actin and clathrin during redistribution of *Chlamydia trachomatis* to an intracellular site in eucaryotic cells. Infect. Immun. *59*: 4465–4472.

Matsumoto, A. 1981. Electron microscopic observations of surface projections and related intracellular structures of *Chlamydia* organisms. J. Electron Microsc. (Tokyo) *30*: 315–320.

Matsumoto, A. 1982. Surface projections of *Chlamydia psittaci* elementary bodies as revealed by freeze-deep-etching. J. Bacteriol. *151*: 1040–1042.

Matsumoto, A., H. Izutsu, N. Miyashita and M. Ohuchi. 1998. Plaque formation by and plaque cloning of *Chlamydia trachomatis* biovar trachoma. J. Clin. Microbiol. *36*: 3013–3019.

Miyashita, N., A. Matsumoto, R. Soejima, T. Kishimoto, M. Nakajima, Y. Niki and T. Matsushima. 1997. Morphological analysis of *Chlamydia pneumoniae*. J. Jpn. Chemother. *45*: 255–264.

Molder, J.W., B.R.S. McCormack, F.M. Gogolak, M.M. Zebovitz and M.K. Itatani. 1955. Production and properties of a penicillin-resistant strain of feline pneumonitis virus. J. Infect. Dis. *96*: 57–74.

Molder, J.W., T.P. Hatch, C.-C. Kuo, J. Schachter and J. Storz. 1984. Genus *Chlamydia*. *In* Bergey's Manual of Systematic Bacteriology, vol. 1 (edited by Krieg). Williams & Wilkins, Baltimore, pp. 729–739.

Neeper, I.D., D.L. Patton and C.C. Kuo. 1990. Cinematographic observations of growth cycles of *Chlamydia trachomatis* in primary cultures of human amniotic cells. Infect. Immun. *58*: 2042–2047.

Nigg, C. and M.D. Eaton. 1944. Isolation from normal mice of a pneumotropic virus which forms elementary bodies. J. Exp. Med. *79*: 497–510.

Pace, N.R. 1997. A molecular view of microbial diversity and the biosphere. Science *276*: 734–740.

Page, L.A. 1968. Proposal for the recognition of two species in the genus *Chlamydia* Jones, Rake, and Stearns 1945. Int. J. Syst. Bacteriol. *18*: 51–66.

Page, L.A. and R.C. Cutlip. 1982. Morphological and immunological confirmation of the presence of *Chlamydiae* in the gut tissues of the Chesapeake Bay clam, *Mercenaria mercenaria*. Curr. Microbiol. *7*: 297–300.

Palmer, L. and S. Falkow. 1986. A common plasmid of *Chlamydia trachomatis*. Plasmid *16*: 52–62.

Patton, D.L., P.K. Cummings, C. Sweeney, T. Yvonne and C. C. Kuo. 1998. *In vivo* uptake of *Chlamydia trachomatis* by fallopian tube epithelial cells is rapid. *In* Chlamydial Infections. Proceedings of the Ninth International Symposium on Human Chlamydial Infection (edited by Stephens, Byrne, Christiansen, Clarke, Grayston, D. Rank, Ridgway, Saikku, Schachter and Stamm), San Francisco, pp. 87–90.

Peeling, R.W., J. Peeling and R.C. Brunham. 1989. High-resolution 31P nuclear magnetic resonance study of *Chlamydia trachomatis*: induction of ATPase activity in elementary bodies. Infect. Immun. *57*: 3338–3344.

Peterson, E.M. and L.M. de la Maza. 1988. *Chlamydia* parasitism: ultrastructural characterization of the interaction between the chlamydial cell envelope and the host cell. J. Bacteriol. *170*: 1389–1392.

Peterson, E.M., B.A. Markoff, J. Schachter and L.M. de la Maza. 1990. The 7.5-kb plasmid present in *Chlamydia trachomatis* is not essential for the growth of this microorganism. Plasmid *23*: 144–148.

Petersson, B., A. Andersson, T. Leitiner, O. Olsvik, M. Uhlen, C. Storey and C. Black. 1997. Evolutionary relationships among members of the genus *Chlamydia* based on 16S rDNA analysis. J. Bacteriol. *179*: 4195–4205.

Pickett, M.A., J.S. Everson, P.J. Pead and I.N. Clarke. 2005. The plasmids of *Chlamydia trachomatis* and *Chlamydophila pneumoniae* (N16): accurate determination of copy number and the paradoxical effect of plasmid-curing agents. Microbiology *151*: 893–903.

Plaunt, M.R. and T.P. Hatch. 1988. Protein synthesis early in the developmental cycle of *Chlamydia psittaci*. Infect. Immun. *56*: 3021–3025.

Pudjiatmoko, H. Fukushi, Y. Ochiai, T. Yamaguchi and K. Hirai. 1997. Phylogenetic analysis of the genus *Chlamydia* based on 16S rRNA gene sequences. Int. J. Syst. Bacteriol. *47*: 425–431.

Puolakkainen, M., C.C. Kuo and L.A. Campbell. 2005. *Chlamydia pneumoniae* uses the mannose 6-phosphate/insulin-like growth factor 2 receptor for infection of endothelial cells. Infect. Immun. *73*: 4620–4625.

Rake, E.G. 1957. Family II. *Chlamydiaceae* fam. nov. *In* Bergey's Manual of Determinative Bacteriology, 7th edn (edited by Breed, Murray and Smith). Williams & Wilkins, Baltimore, pp. 957–968.

Raulston, J.E., C.H. Davis, D.H. Schmiel, M.W. Morgan and P.B. Wyrick. 1993. Molecular characterization and outer membrane association of a *Chlamydia trachomatis* protein related to the Hsp70 family of proteins. J. Biol. Chem. *268*: 23139–23147.

Read, T.D., C.M. Fraser, R.C. Hsia and P.M. Bavoil. 2000. Comparative analysis of *Chlamydia* bacteriophages reveals variation localized to a putative receptor binding domain. Microb. Comp. Genomics *5*: 223–231.

Reed, S.I., L.E. Anderson and H.M. Jenkin. 1981. Use of cycloheximide to study independent lipid metabolism of *Chlamydia trachomatis* cultivated in mouse L cells grown in serum-free medium. Infect. Immun. *31*: 668–673.

Richmond, S.J., P. Stirling and C.R. Ashley. 1982. Virus infecting the reticulate bodies of an avian strain of *Chlamydia psittaci*. FEMS Microbiol. Lett. *14*: 31–36.

Ripa, K.R. and P. A. Mardh. 1977. Cultivation of *Chlamydia trachomatis* in cycloheximide-treated McCoy cells. J. Clin. Microbiol. *6*: 328–331.

Roblin, P.M., W. Dumornay and M.R. Hammerschlag. 1992. Use of HEp-2 cells for improved isolation and passage of *Chlamydia pneumoniae*. J. Clin. Microbiol. *30*: 1968–1971.

Rockey, D.D. and J.L. Rosquist. 1994. Protein antigens of *Chlamydia psittaci* present in infected cells but not detected in the infectious elementary body. Infect. Immun. *62*: 106–112.

Rodolakis, A. 1983. *In vitro* and *in vivo* properties of chemically induced temperature-sensitive mutants of *Chlamydia psittaci* var. ovis: screening in a murine model. Infect. Immun. *42*: 525–530.

Schachter, J., H.B. Ostler and K.F. Meyer. 1969. Human infection with the agent of feline pneumonitis. Lancet *1*: 1063–1065.

Schachter, J., R.S. Stephens, P. Timms, C. Kuo, P.M. Bavoil, S. Birkelund, J. Boman, H. Caldwell, L.A. Campbell, M. Chernesky, G. Christiansen, I.N. Clarke, C. Gaydos, J.T. Grayston, T. Hackstadt, R. Hsia, B. Kaltenboeck, M. Leinonnen, D. Ojcius, G. McClarty, J. Orfila, R. Peeling, M. Puolakkainen, T.C. Quinn, R.G. Rank, J. Raulston, G.L. Ridgeway, P. Saikku, W.E. Stamm, D. Taylor-Robinson, S.P. Wang and P.B. Wyrick. 2001. Radical changes to chlamydial taxonomy are not necessary just yet. Int. J. Syst. Evol. Microbiol. *51*: 249; author reply 251–253.

Scidmore, M.A., E.R. Fischer and T. Hackstadt. 1996. Sphingolipids and glycoproteins are differentially trafficked to the *Chlamydia trachomatis* inclusion. J. Cell Biol. *134*: 363–374.

Sneath, P.H.A. 1992. International Code of Nomenclature of Bacteria (1990 Revision). American Society for Microbiology, Washington, D.C.

Sparks, A.K., J.F. Morado and J.W. Hawkes. 1985. A systemic microbial disease in the Dungeness crab, *Cancer magister*, caused by a *Chlamydia*-like organism. J. Invertebr. Pathol. *45*: 204–217.

Spencer, W.N. and F.W. Johnson. 1983. Simple transport medium for the isolation of *Chlamydia psittaci* from clinical material. Vet. Rec. *113*: 535–536.

Stephens, R.S., M.R. Tam, C.C. Kuo and R.C. Nowinski. 1982. Monoclonal antibodies to *Chlamydia trachomatis*: antibody specificities and antigen characterization. J. Immunol. *128*: 1083–1089.

Stephens, R.S., E.A. Wagar and G.K. Schoolnik. 1988. High-resolution mapping of serovar-specific and common antigenic determinants of the major outer membrane protein of *Chlamydia trachomatis*. J. Exp Med. *167*: 817–831.

Stephens, R.S. 1989. Antigenic variation of *Chlamydia trachomatis*. *In* Intracellular Parasitism (edited by Molder). CRC Press, Boca Raton, FL, pp. 51–62.

Stephens, R.S., S. Kalman, C. Lammel, J. Fan, R. Marathe, L. Aravind, W. Mitchell, L. Olinger, R.L. Tatusov, Q. Zhao, E.V. Koonin and R.W. Davis. 1998. Genome sequence of an obligate intracellular pathogen of humans: *Chlamydia trachomatis*. Science *282*: 754–759.

Stephens, R.S., K. Koshiyama, E. Lewis and A. Kubo. 2001. Heparin-binding outer membrane protein of chlamydiae. Mol. Microbiol. *40*: 691–699.

Stephens, R.S. 2002. *Chlamydiae* and evolution: a billion years and counting. *In* Chlamydial Infections. Proceedings of the Tenth International Symposium on Human Chlamydial Infections (edited by Schachter, Christiansen, Clarke et al.), Basim Yeri, Turkey, pp. 3–12.

Storey, C.C., M. Lusher and S.J. Richmond. 1989a. Analysis of the complete nucleotide sequence of Chp1, a phage which infects avian *Chlamydia psittaci*. J. Gen. Virol. *70*: 3381–3390.

Storey, C.C., M. Lusher, S.J. Richmond and J. Bacon. 1989b. Further characterization of a bacteriophage recovered from an avian strain of *Chlamydia psittaci*. J. Gen. Virol. *70*: 1321–1327.

Stothard, D.R., J.A. Williams, B. Van Der Pol and R.B. Jones. 1998. Identification of a *Chlamydia trachomatis* serovar E urogenital isolate which lacks the cryptic plasmid. Infect. Immun. *66*: 6010–6013.

Su, H., N.G. Watkins, Y.X. Zhang and H.D. Caldwell. 1990. *Chlamydia trachomatis*-host cell interactions: role of the chlamydial major outer membrane protein as an adhesin. Infect. Immun. *58*: 1017–1025.

T'ang, F., J. Chang, Y. Huang and K. Wang. 1957. Studies on the etiology of trachoma with special reference to isolation of the virus in chick embryo. Chin. Med. J. *75*: 429–447.

Thomas, N.S., M. Lusher, C.C. Storey and I.N. Clarke. 1997. Plasmid diversity in *Chlamydia*. Microbiology *143*: 1847–1854.

Ting, L.M., R.C. Hsia, C.G. Haidaris and P.M. Bavoil. 1995. Interaction of outer envelope proteins of *Chlamydia psittaci* GPIC with the HeLa cell surface. Infect. Immun. *63*: 3600–3608.

Tipples, G. and G. McClarty. 1993. The obligate intracellular bacterium *Chlamydia trachomatis* is auxotrophic for three of the four ribonucleoside triphosphates. Mol. Microbiol. *8*: 1105–1114.

Tjaden, J., H.H. Winkler, C. Schwoppe, M. Van Der Laan, T. Mohlmann and H.E. Neuhaus. 1999. Two nucleotide transport proteins in *Chlamydia trachomatis*, one for net nucleoside triphosphate uptake and the other for transport of energy. J. Bacteriol. *181*: 1196–1202.

Tribby, II, R.R. Friis and J.W. Moulder. 1973. Effect of chloramphenicol, rifampicin, and nalidixic acid on *Chlamydia psittaci* growing in L cells. J. Infect. Dis. *127*: 155–163.

Vanrompay, D., W. De Meurichy, R. Ducatelle and F. Haesebrouck. 1994. Pneumonia in Moorish tortoises (*Testudo graeca*) associated with avian serovar A *Chlamydia psittaci*. Vet. Rec. *135*: 284–285.

Wagar, E.A. and R.S. Stephens. 1988. Developmental-form-specific DNA-binding proteins in *Chlamydia* spp. Infect. Immun. *56*: 1678–1684.

Wagar, E.A., J. Schachter, P. Bavoil and R.S. Stephens. 1990. Differential human serologic response to two 60,000 molecular weight *Chlamydia trachomatis* antigens. J. Infect. Dis. *162*: 922–927.

Wang, S.P. and J.T. Grayston. 1970. Immunologic relationship between genital TRIC, lymphogranuloma venereum, and related organisms in a new microtiter indirect immunofluorescence test. Am. J. Ophthalmol. *70*: 367–374.

Wang, S.P., C.C. Kuo and J.T. Grayston. 1973. A simplified method for immunological typing of trachoma-inclusion conjunctivitis-lymphogranuloma venereum organisms. Infect. Immun. *7*: 356–360.

Weisburg, W.G., T.P. Hatch and C.R. Woese. 1986. Eubacterial origin of *Chlamydiae*. J. Bacteriol. *167*: 570–574.

Williams, J.E. 1984. Proposal to reject the new combination *Yersinia pseudotuberculosis* subsp. *pestis* for violation of the First Principle of the International Code of Nomenclature of Bacteria. Int. J. Syst. Bacteriol. *34*: 268–269.

Willis, J.M., G. Watson, M. Lusher, T.S. Mair, D. Wood and S.J. Richmond. 1990. Characterization of *Chlamydia psittaci* isolated from a horse. Vet. Microbiol. *24*: 11–19.

Wolf, K., E. Fischer, D. Mead, G. Zhong, R. Peeling, B. Whitmire and H.D. Caldwell. 2001. *Chlamydia pneumoniae* major outer membrane protein is a surface-exposed antigen that elicits antibodies primarily directed against conformation-dependent determinants. Infect. Immun. *69*: 3082–3091.

Wyrick, P.B., J. Choong, C.H. Davis, S.T. Knight, M.O. Royal, A.S. Maslow and C.R. Bagnell. 1989. Entry of genital *Chlamydia trachomatis* into polarized human epithelial cells. Infect. Immun. *57*: 2378–2389.

Yang, Z.P., P.K. Cummings, D.L. Patton and C.C. Kuo. 1994. Ultrastructural lung pathology of experimental *Chlamydia pneumoniae* pneumonitis in mice. J. Infect. Dis. *170*: 464–467.

Family II. "*Candidatus* Clavichlamydiaceae"

Matthias Horn

Cla.vi.chla.my.di.a.ce'ae. N.L. fem. n. *Clavichlamydia* type genus of the family; *-aceae* ending to denote a family, N.L. fem. pl. n. *Clavichlamydiaceae* the *Clavichlamydia* family.

The family "*Candidatus* Clavichlamydiaceae" is currently solely based on 16S rRNA sequences moderately related to members of the *Chlamydiaceae* (around 90% sequence similarity) (Figure 154). The family "*Candidatus* Clavichlamydiaceae" is a sister family of the *Chlamydiaceae* and contains a single genus, "*Candidatus* Clavichlamydia".

Type genus: **"*Candidatus* Clavichlamydia"** corrig. Karlsen, Nylund, Watanabe, Helvik, Nylund and Plarre 2008.

Genus I. "*Candidatus* Clavichlamydia" corrig. Karlsen, Nylund, Watanabe, Helvik, Nylund and Plarre 2008

Matthias Horn

Cla.vi.chla.my'di.a. L. n. *clava* a knotty branch, rough stick, cudgel, club; N.L. fem. n. *Chlamydia* taxonomic name of a bacterial genus; N.L. fem. n. *Clavichlamydia* a club *Chlamydia* (the morphology of the bacteria includes the characteristic head-and-tail cells that have been described earlier from salmonid fish suffering from epitheliocystis).

Pleomorphic or elongated, nonmotile, obligate intracellular bacteria, up to 2 μm in length. Cells show a **developmental cycle** with morphologically distinct stages and thrive **inside a host vacuole.** Organisms were found within gill lesions of fish and are differentiated from all other chlamydiae by the morphology of the proposed **elementary bodies, which show a characteristic head-and-tail form** (Figure 162). Members have not yet been obtained in cell culture.

Type species: **"*Candidatus* Clavichlamydia salmonicola"** corrig. Karlsen, Nylund, Watanabe, Helvik, Nylund and Plarre 2008.

Further descriptive information

Three distinct morphological forms, representing different developmental stages, have been observed inside the host-derived vacuole that surrounded the bacteria. The proposed reticulate bodies are large pleomorphic cells (up to 2 μm) with

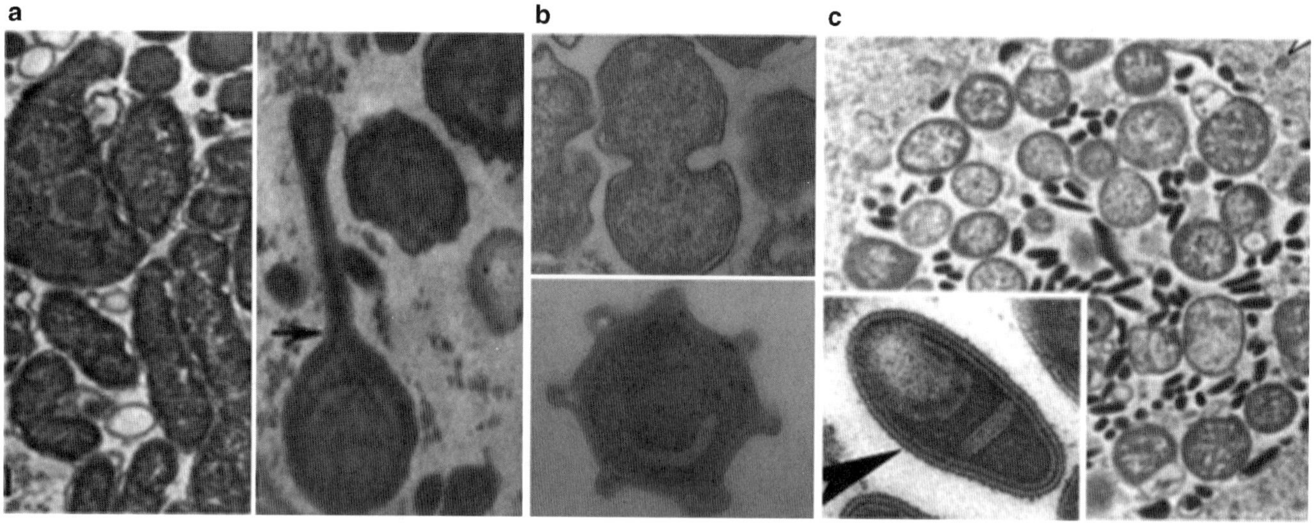

FIGURE 162. Morphological diversity of developmental stages of the *Chlamyidae*. (a) "*Candidatus* Clavichlamydia salmonicola", (b) *Criblamydia sequanensis* and (c) "*Candidatus* Rhabdochlamydia porcellionis". Images are reproduced from Kostanjsek et al. (2004), Thomas et al. (2006) and Karlsen et al. (2008). See text for details.

an inner structure resembling the reticulate bodies of *Chlamydiaceae* (Figure 162). The proposed intermediate bodies were slightly smaller with a condensed cytoplasm. The proposed elementary bodies were about 1 μm long and consisted of a head and a tail region (Figure 162). Transitional forms between these three developmental stages have been observed (Karlsen et al., 2008). Bacteria with a similar morphology have previously been observed in the gills of steelhead trout and lake trout, and in nonproliferative gill lesions of Atlantic salmon (Bradley et al., 1988; Draghi et al., 2004; Rourke et al., 1984), but these bacteria were never identified on a molecular level.

Pathogenesis. "*Candidatus* Clavichlamydia salmonicola" has been found infecting the gills of salmonid fish in freshwater (Atlantic salmon, *Salmo salar*, and wild brown trout, *Salmo trutta*), forming large cysts (up to 56 μm) consisting of single hypertrophied epithelial cells. Although a direct causal relationship has not yet been demonstrated, "*Candidatus* Clavichlamydia salmonicola" has been suggested to represent one possible etiological agent of epitheliocystis in fish from freshwater and marine environments (Karlsen et al., 2008).

Taxonomic note

The two available 16S rRNA sequences from the genus *Clavichlamydia* share slightly greater than 90% similarity with the 16S rRNA sequences of *Chlamydia abortus*, *Chlamydia felis*, *Chlamydia pecorum*, *Chlamydia pneumoniae*, and *Chlamydia psittaci*, but slightly less than 90% with *Chlamydia muridarum*, *Chlamydia suis*, and *Chlamydia trachomatis*. While the unique morphology of the developmental stages of members of the genus "*Candidatus* Clavichlamydia" might favor the placement of this genus into a new family separate from the *Chlamydiaceae*, this might be subject to change when more sequence information and additional descriptions become available.

The names *Clavochlamydiaceae*, *Clavochlamydia*, and "*Candidatus* Clavochlamydia salmonicola" have been changed by the editors to "*Candidatus* Clavichlamydiaceae", "*Candidatus* Clavichlamydia", and "*Candidatus* Clavichlamydia salmonicola" during preparation of this chapter. The connecting vowel must be "i" because the first part of the name is derived from Latin.

List of tentative species of the genus "*Candidatus* Clavichlamydia"

1. **"*Candidatus* Clavichlamydia salmonicola"** corrig. Karlsen, Nylund, Watanabe, Helvik, Nylund and Plarre 2008.

   sal.mo.ni.co'la L. n. *salmo -onis* salmon; L. suff. *-cola* (from L. n. *incola*), inhabitant, dweller; N.L. n. *salmonicola* a salmon-dweller (infecting fish of the genus *Salmo*).

   The genus "*Candidatus* Clavichlamydia" corrig. is based on the description of "*Candidatus* Clavochlamydia salmonicola" (Karlsen et al., 2008) (Figure 154).

   Sequence accession no. (16S rRNA gene): EF577391, EF577392.

References

Bradley, T.M., C.E. Newcomer and K.O. Maxwell. 1988. Epitheliocystis associated with massive mortalities of cultured lake trout *Salvelinus namaycush*. Dis. Aquat. Org. *4*: 9–17.

Draghi, A., 2nd, V.L. Popov, M.M. Kahl, J.B. Stanton, C.C. Brown, G.J. Tsongalis, A.B. West and S. Frasca, Jr. 2004. Characterization of "*Candidatus* piscichlamydia salmonis" (order *Chlamydiales*), a chlamydia-like bacterium associated with epitheliocystis in farmed Atlantic salmon (*Salmo salar*). J. Clin. Microbiol. *42*: 5286–5297.

Karlsen, M., A. Nylund, K. Watanabe, J.V. Helvik, S. Nylund and H. Plarre. 2008. Characterization of '*Candidatus* Clavochlamydia salmonicola': an intracellular bacterium infecting salmonid fish. Environ. Microbiol. *10*: 208–218.

Rourke, A.W., R.W. Davis and T.M. Bradley. 1984. A light and electron microscope study of epitheliocystis in juvenile steelhead trout, *Salmo gairdneri* Richardson. J. Fish Dis. *7*: 301–309.

Thomas, V., N. Casson and G. Greub. 2006. *Criblamydia sequanensis*, a new intracellular *Chlamydiales* isolated from Seine river water using amoebal co-culture. Environ. Microbiol. *8*: 2125–2135.

Family III. **Criblamydiaceae** Thomas, Casson and Greub 2006, 2131

MATTHIAS HORN

Cri.bla.my.di.a.ce′ae. N.L. fem. n. *Criblamydia* type genus of the family; *-aceae* ending to denote a family; N.L. fem. pl. n. *Criblamydiaceae* the *Criblamydia* family.

The family *Criblamydiaceae* currently contains a single genus, *Criblamydia*, comprising a single species, which grows in amoebae of the genus *Acanthamoeba* and shows highest 16S rRNA sequence similarity to members of the *Parachlamydiaceae* (89%, Figure 154) (Thomas et al., 2006).

Type genus: **Criblamydia** Thomas, Casson and Greub 2006, 2131.

Genus I. **Criblamydia** Thomas, Casson and Greub 2006, 2131

MATTHIAS HORN

Crib.la.my′di.a. N.L. fem. n. *Cribia* arbitrary name derived from the abbreviation CRIB (Center for Research on Intracellular Bacteria); N.L. fem. n. *Chlamydia* taxonomic name of a bacterial genus; N.L. fem. n. *Criblamydia* arbitrary name intended to mean a *Chlamydia* named after CRIB.

Coccoid to pleomorphic, nonmotile, obligately intracellular bacteria, 0.5–1 µm in diameter. Cells show a **developmental cycle** with morphologically distinct stages and **grow within host-derived vacuoles**. Organisms are differentiated from all other chlamydiae by the **star-shaped elementary body** seen with transmission electron microscopy (Figure 162). The genus *Criblamydia* contains a single species, *Criblamydia sequanensis* (Figure 154), which has been recovered from a water sample from the river Seine (France) by co-cultivation with *Acanthamoeba castellanii* ATCC 30010 (Thomas et al., 2006).

Type species: **Criblamydia sequanensis** Thomas, Casson and Greub 2006, 2131.

Further descriptive information

The natural host of the sole *Criblamydia* strain known to date (CRIB-18) is not known. Like many chlamydiae, it grew well in the free-living amoebae *Acanthamoeba castellanii* ATCC 30010 and *Acanthamoeba polyphaga* Linc-AP1, but could not propagate in *Hartmannella vermiformis* ATCC 50237 and *Dictyostelium discoideum* DH1-10, or mammalian cells (Thomas et al., 2006). In *Acanthamoeba castellanii, Criblamydia sequanensis* formed 1–10 large vacuoles within 72–96 h, which led to lysis of the amoeba host cell after 4–6 d. No bacteria were observed in amoeba cysts. The reticulate bodies (0.8–1.0 µm in diameter) possess a granular cytoplasm and a three-layer cell wall. The elementary bodies (0.5 µm in diameter) frequently contain electron-translucent oblong laminar structures, possess a five-layer cell wall, and show a star-like morphology with 3–7 branches in transmission electron micrographs, which possibly represents a fixation artifact due to some unique cell-wall properties (Thomas et al., 2006).

List of species of the genus *Criblamydia*

1. **Criblamydia sequanensis** Thomas, Casson and Greub 2006, 2131.

sequ.a.nen′sis.L. n. *Sequana* Latin name of the river Seine (France); N.L. fem. adj. *sequanensis* of or belonging to *Sequana* from which the type strain was recovered.

The characteristics of the sole member of the genus and species are the same as that provided for the genus description. Its isolation by co-culture with amoebae in the presence of ampicillin and vancomycin suggests resistance to these antibiotics. The 16S rRNA-targeted oligonucleotide probe S-S-Crib-0064-a-A-20 (5′-GTTACCCAAATACTTCGTTC-3′) can be used for specific detection using fluorescence in situ hybridization (Thomas et al., 2006).

Type strain: CRIB-18, not yet deposited in a public culture collection.

Sequence accession no. (16S rRNA gene): DQ124300.

Reference

Thomas, V., N. Casson and G. Greub. 2006. *Criblamydia sequanensis*, a new intracellular *Chlamydiales* isolated from Seine river water using amoebal co-culture. Environ. Microbiol. *8*: 2125–2135.

Family IV. **Parachlamydiaceae** Everett, Bush and Andersen 1999, 434^{VP}

MATTHIAS HORN

Pa.ra.chla.my.di.a.ce′ae. N.L. fem. n. *Parachlamydia* type genus of the family; *-aceae* ending to denote a family; N.L. fem. pl. n. *Parachlamydiaceae* the *Parachlamydia* family.

The *Parachlamydiaceae* naturally thrive in free-living amoebae (Everett et al., 1999). The family currently contains three genera, *Parachlamydia, Neochlamydia,* and *Protochlamydia* (Figure 154). Further symbionts of acanthamoebae, which were identified as being related to the *Parachlamydiaceae* but which were not described in greater detail, might represent additional genera within this family (Fritsche et al., 2000).

Type genus: **Parachlamydia** Everett, Busch and Andersen 1999, 435^{VP}.

Genus I. **Parachlamydia** Everett, Bush and Andersen 1999, 435[VP]

Matthias Horn

Pa.ra.chla.my'di.a. Gr. prep. *para* like, alongside of; N.L. fem. n. *Chlamydia* taxonomic name of a bacterial genus; N.L. fem. n. *Parachlamydia* resembling the genus *Chlamydia*.

Coccoid, nonmotile, obligate intracellular bacteria, 0.4–0.6 µm in diameter. Cells show a **developmental cycle** with morphologically distinct stages and **grow within host-derived vacuoles.** Organisms naturally infect free-living amoebae of the genus *Acanthamoeba*. The genus *Parachlamydia* currently contains but a single species, *Parachlamydia acanthamoebae* (Amann et al., 1997; Everett et al., 1999) (Figure 154). A number of additional symbionts of *Acanthamoeba* species closely related to. *Parachlamydia acanthamoebae* have been identified but not described in greater detail (Amann et al., 1997; Collingro et al., 2005a; Heinz et al., 2007; Thomas et al., 2006).

Type species: **Parachlamydia acanthamoebae** Everett, Bush and Andersen 1999, 435[VP].

Further descriptive information

Two distinct morphological forms, almost similar in size (0.4–0.6 µm) and resembling the reticulate and elementary bodies of the *Chlamydiaceae*, have originally been observed inside the host-derived vacuoles in which the bacteria live (Michel et al., 1994). An inclusion typically contains many bacteria. In addition, a third developmental stage, the crescent body, has been proposed and was associated with prolonged co-incubation time of 6 d (Greub and Raoult, 2002a). The crescent body of *Parachlamydia acanthamoebae* has a crescent shape, shows a thickened cell wall similar to the elementary body, and is considered to represent an additional infective stage. Elementary and crescent bodies enter the amoeba host by phagocytosis and differentiate within inclusions in less than 8 h to reticulate bodies (Greub and Raoult, 2002a). All developmental stages except for crescent bodies have also been observed outside of inclusions (directly in the cytoplasm), although the inclusion seemed to be the preferred intracellular niche of *Parachlamydia acanthamoebae*. Bacteria are released from their host cells within vesicles or by lysis of the amoebae. The lytic activity of *Parachlamydia acanthamoebae* was dependent on temperature, showing low levels at temperatures below 30°C and reaching a maximum at temperatures of 32–37°C (Greub et al., 2003c). Infection of acanthamoebae largely inhibited cyst formation (Greub and Raoult, 2002a; Michel et al., 1994).

The *Parachlamydia* species strain OEW1, most closely related to *Parachlamydia acanthamoebae*, has been found naturally sharing its amoeba host cells with a second, betaproteobacterial symbiont related to "*Candidatus* Procabacter" (Heinz et al., 2007). Co-cultivation with amoebae has been a valuable and straightforward method to directly recover *Parachlamydia* species (and other chlamydiae) from complex environmental samples without the need to isolate the natural host (Collingro et al., 2005a; Thomas et al., 2006).

In addition to being found in free-living amoebae, in the laboratory *Parachlamydia acanthamoebae* is able to enter and (to a limited degree to) multiply in Vero cells, in human macrophages, pneumocytes, and lung fibroblasts (Casson et al., 2006; Collingro et al., 2005a; Greub et al., 2003d). In monocyte-derived macrophages, *Parachlamydia acanthamoebae* trafficked through the endocytic pathway and replicated in a modified vacuole, which was acidic, Lamp-1 positive, and cathepsin negative (Greub et al., 2005). This is fundamentally different from the *Chlamydiaceae*, which bypass the endocytic pathway (Dautry-Varsat et al., 2004).

In human macrophages, *Parachlamydia acanthamoebae* did not induce an oxidative burst or pro-inflammatory cytokine production, but showed a cytopathic effect, most likely due to apoptosis (Greub et al., 2003d, 2005a). In contrast, no cytopathic effect was observed on pneumocytes and lung fibroblasts, indicating that rather than macrophages, these or similar cell types might represent replicative niches for *Parachlamydia acanthamoebae* in humans (Casson et al., 2006).

Parachlamydia acanthamoebae, like other chlamydiae, uses an ATP/ADP translocase to thrive as an energy parasite within its host cell (Greub and Raoult, 2003; Schmitz-Esser et al., 2004).

Pathogenesis. Free-living amoebae are considered the natural hosts of *Parachlamydia* species, but there is also accumulating evidence for a potential pathogenicity of these bacteria for humans (extensively reviewed by Greub and Raoult, 2002b; Corsaro and Venditti, 2004; Corsaro and Greub, 2006). "Hall's coccus", later identified as a *Parachlamydia acanthamoebae* strain, was one of the first chlamydia-like organisms that has been implicated in human disease. It was originally found in an *Acanthamoeba* species isolated from the source of an outbreak of humidifier fever in Vermont, USA. In a separate study, approximately 1% of patients with pneumonia of undetermined cause showed elevated antibody titers against these bacteria (Birtles et al., 1997). To date, *Parachlamydia* species have primarily been associated with respiratory disease including community acquired pneumonia, bronchitis, and aspiration pneumonia, and also with atherosclerosis (Corsaro et al., 2002; Greub et al., 2003a, b, 2006; Marrie et al., 2001). These studies are purely based on serological (immunofluorescence assays) or molecular evidence (PCR, nested PCR, real-time PCR). *Parachlamydia acanthamoebae* has not yet been isolated from a patient, and Frederick's and Relman's revisions of Koch's postulates for sequence-based identification of microbial pathogens have not yet been fulfilled (Fredericks and Relman, 1996). Considering the slow growth of *Parachlamydia acanthamoebae* in human macrophages, pneumocytes, and lung fibroblasts compared to the *Chlamydiaceae* or the intracellular pathogen *Legionella pneumophila* (a difference of three orders of magnitude), and taking into account evidence for *Parachlamydia acanthamoebae* infections in immunocompromised patients (Greub et al., 2003a; Marrie et al., 2001), *Parachlamydia acanthamoebae* might be an opportunistic respiratory pathogen.

All *Parachlamydia* strains tested were resistant to beta-lactam antibiotics and fluoroquinolones, but were susceptible to macrolides, tetracycline, and rifampicin (Casson and Greub, 2006; Maurin et al., 2002; Michel et al., 1994).

List of species of the genus *Parachlamydia*

1. **Parachlamydia acanthamoebae** Everett, Bush and Andersen 1999, 435[VP]

   a.can.tha.mo.e'bae. N.L. n. *Acanthamoeba* taxonomic name of a genus of *Acanthamoebidae*; N.L. gen. n. *acanthamoebae* of (living in) members of the genus *Acanthamoeba*.

   The characteristics of the sole member of the genus and species are the same as that provided for the genus description. The original host amoeba *Acanthamoeba* sp. Bn_9 was isolated from human nasal mucosa of a healthy volunteer in Germany (Michel et al., 1994). *Parachlamydia acanthamoebae* can also use other *Acanthamoeba* species as hosts.

   The 16S rRNA-targeted oligonucleotide probe Bn_9658 (5′-TCCGTTTTCTCCGCCTAC-3′) can be used for the detection of *Parachlamydia acanthamoebae* by fluorescence *in situ* hybridization (Amann et al., 1997), but this probe targets also members of the genus *Protochlamydia*.

   Several *Parachlamydia acanthamoebae* strains have been described (Berg17, Bn9, "Hall's coccus", OEW1, Seine, UV7) showing nearly identical 16S rRNA gene sequences. The genome sequence of *Parachlamydia acanthamoebae* UV7 is currently being determined.

   DNA G+C content (mol%): not available.

   Type strain: Bn_9, ATCC VR 1476.

   Sequence accession no. (16S rRNA gene): not available.

Genus II. **Neochlamydia** Horn, Wagner, Müller, Schmid, Fritsche, Schleifer and Michel 2001, 1229[VP]
(Effective publication: Horn, Wagner, Müller, Schmid, Fritsche, Schleifer and Michel 2000, 1236.)

Matthias Horn

Ne.o.chla.my′di.a. Gr. pref. *neo-* (from. Gr. adj. *neos*) new; N.L. fem. n. *Chlamydia* name of a bacterial genus; N.L. fem. n. *Neochlamydia* a new *Chlamydia*, referring to the modest phylogenetic relationship to the *Chlamydiaceae*.

Coccoid, nonmotile, obligate intracellular bacteria, 0.4–0.6 µm in diameter. Bacteria show a **developmental cycle** with morphologically distinct stages and are **directly located in the host cell cytoplasm**. Organisms naturally **infect the amoeba *Hartmannella vermiformis***, but no other free-living amoebae including the *Acanthamoeba* species tested; the only exception was the social amoeba *Dictyostelium discoideum* $Berg_{25}$. The genus *Neochlamydia* currently contains a single species, *Neochlamydia hartmannellae* (Figure 154) (Horn et al., 2000).

Type species: **Neochlamydia hartmannellae** Horn, Wagner, Müller, Schmid, Fritsche, Schleifer and Michel 2001, 1229[VP] (Effective publication: Horn, Wagner, Müller, Schmid, Fritsche, Schleifer and Michel 2000, 1237.).

Further descriptive information

Two distinct morphological forms, representing different developmental stages, have been observed inside the amoeba *Hartmannella vermiformis* host. Reticulate bodies were irregularly spherical in shape (0.4–0.6 µm in diameter) and showed a granular cytoplasm similar to the reticulate bodies of members of the *Chlamydiaceae*. The coccoid elementary bodies were of similar size (0.5–0.6 µm in diameter). In contrast to all other chlamydiae known so far, *Neochlamydia hartmannellae* did not seem to be located within inclusions, but directly in the cytoplasm of its host cell, indicating that these organisms possess an escape mechanism from the phagosome (Horn et al., 2000).

Infection of the amoeba *Hartmannella vermiformis* with *Neochlamydia hartmannellae* suppressed cyst formation and led to rapid lysis of the amoeba host within 5 d. This suggests a limited adaption of host and parasite, possibly due to a relatively short evolutionary relationship, and might indicate the existence of alternative protist hosts for *Neochlamydia hartmannellae* in the environment (Horn et al., 2000).

Bacteriophage-like particles have been observed by electron microscopy in abnormal, enlarged *Neochlamydia hartmannellae* cells (up to 1.3 µm in diameter). This putative bacteriophage (Neo-Ph/1) is 68 nm in diameter, much larger than known chlamydia phages of the genus *Chlamydiamicrovirus* (Schmid et al., 2001).

Pathogenesis. *Neochlamydia hartmannellae* is primarily a parasite of amoebae. *Neochlamydia hartmannellae* has also been identified as possible causative agent of epitheliocystis in fish (Draghi et al., 2007; Meijer et al., 2006). In addition, there is preliminary molecular evidence for an association with feline ocular disease (von Bomhard et al., 2003). To date, there is no evidence for an association of *Neochlamydia hartmannellae* with disease in humans. *Neochlamydia hartmannellae* is resistant to the quinolone ciprofloxacin (Casson and Greub, 2006).

List of species of the genus *Neochlamydia*

1. **Neochlamydia hartmannellae** Horn, Wagner, Müller, Schmid, Fritsche, Schleifer and Michel 2001, 1229[VP] (Effective publication: Horn, Wagner, Müller, Schmid, Fritsche, Schleifer and Michel 2000, 1237.)

   hart.man.nel′lae. N.L. gen. n. *hartmannellae*, of *Hartmannella* (taxonomic name of a genus of *Hartmannellidae*), referring to the name of the host amoeba, *Hartmannella vermiformis* strain A_1Hsp, in which the organism was first discovered.

   The characteristics of the sole member of the genus and species are the same as that provided for the genus description. The original host amoeba (*Hartmannella vermiformis* A_1Hsp) of the type strain has been isolated from a water conduit system of a dental unit in Germany.

   The 16S rRNA-targeted oligonucleotide probe S-S-ParaC-0658-a-A-18 (5′- TCCATTTTCTCCGTCTAC -3′) can be used for the detection of *Neochlamydia hartmannellae* by fluorescence *in situ* hybridization (Horn et al., 2000). This probe also targets two closely related symbionts of *Acanthamoeba species* (Fritsche et al., 2000).

DNA G+C content (mol%): not available.
Type strain: A_1Hsp, ATCC 50802.
Sequence accession no. (16S rRNA gene): AF177275.

Genus III. **Protochlamydia** Horn, gen. nov. (previously known as "*Candidatus* Protochlamydia" Collingro et al. 2005b)

Matthias Horn

Pro.to.chla.my'di.a. Gr. adj. *protos* first, foremost; N.L. fem. n. *Chlamydia* name of a bacterial genus; N.L. fem. n. *Protochlamydia* first *Chlamydia*, referring to the similarity of these bacteria to the chlamydial ancestor.

Coccoid, nonmotile, obligately intracellular bacteria, 0.5–1 µm. Cells show a **developmental cycle** with morphologically distinct stages and **grow within host-derived vacuoles.** Organisms naturally **infect amoebae of the genus *Acanthamoeba*.** The genus *Protochlamydia* currently contains a single species, *Protochlamydia amoebophila* (Figure 154) (Collingro et al., 2005b; Fritsche et al., 2000). A 16S rRNA gene sequence showing 97% similarity with the 16S rRNA gene of *Protochlamydia amoebophila* has been deposited at GenBank/EMBL/DDBJ (DQ632609). This sequence represents a second species within the genus *Protochlamydia*, tentatively named *Protochlamydia naegleriophila* (Casson, Michel, Goy, Müller and Greub, unpublished), but no further details have yet been published.

DNA G+C content (mol%): 35.8.

Type species: **Protochlamydia amoebophila** Horn, sp. nov. (previously known as "*Candidatus* Protochlamydia amoebophila" Collingro et al. 2005b).

Further descriptive information

The developmental stages of *Protochlamydia* species resemble those of the *Chlamydiaceae.* Elementary bodies are 0.5–0.8 µm in diameter; reticulate bodies are 0.7–1.0 µm in diameter. Bacteria are located inside inclusions, which, in contrast to *Parachlamydia acanthamoebae*, typically comprise only one or very few bacteria. No stages similar to persistent forms of the *Chlamydiaceae* have been observed (Collingro et al., 2005b; Fritsche et al., 2000).

Parachlamydia amoebophila had a cytopathic effect on its original host *Acanthamoeba* sp. UWE25, but also grows in various other *Acanthamoeba* strains, where its life cycle seems to be well coordinated with growth of the amoeba host. *Parachlamydia amoebophila* shows only marginal cytopathic effect on these host cells at temperatures around 20°C, but hampers their multiplication (Collingro et al., 2004). In contrast to *Neochlamydia hartmannellae* and *Parachlamydia acanthamoebae*, infection of *Acanthamoeba* species does not seem to inhibit cyst formation, and *Parachlamydia amoebophila* is occasionally also seen in *Acanthamoeba* cysts (Collingro et al., 2005a; Fritsche et al., 2000). *Parachlamydia amoebophila* is also able to use the social amoeba *Dictyostelium discoideum* as host (Skriwan et al., 2002).

The circular genome of the type strain, *Parachlamydia amoebophila* UWE25, is the largest chlamydial genome available to date (2.4 Mb), contains a low G+C content of 36 mol%, and encodes three rRNA operons (Horn et al., 2004). Based on comparative genome analysis, several metabolic pathways are truncated in *Parachlamydia amoebophila*, and therefore essential building blocks like nucleotides, many amino acids (except for glycine, alanine, serine, aspartic acid, glutamine, glutamic acid, and proline), and several cofactors (except for riboflavin, heme, folate, and menaquinone) need to be acquired from the host cell. Compared to the *Chlamydiaceae*, however, *Parachlamydia amoebophila* has generally greater metabolic capabilities. The *Parachlamydia amoebophila* genome, for example, encodes a complete tricarboxylic acid cycle and possesses a more versatile respiratory chain (Horn et al., 2006).

Like other intracellular bacteria, *Parachlamydia amoebophila* uses a variety of transport proteins to compensate for the lack of essential biosynthetic pathways. As an example, five nucleotide transporters are encoded in the *Parachlamydia amoebophila* genome, whose joint action catalyzes the import of RNA nucleotides and the cofactor nicotinamide adenine dinucleotide (Haferkamp et al., 2004, 2006; Schmitz-Esser et al., 2004). In addition, one of these transporters functions as an ATP/ADP translocase importing ATP from the amoeba host in exchange for bacterial ADP and thus serving to parasitize the host's ATP pool (Schmitz-Esser et al., 2004; Trentmann et al., 2007).

Parachlamydia amoebophila encodes a complete gene set for peptidoglycan biosynthesis and the cysteine-rich proteins OmcA and OmcB, but lacks a gene for the major outer-membrane protein OmpA. *Parachlamydia amoebophila* does not cross-react with antibodies against OmpA of members of the *Chlamydiaceae* or the *Chlamydia*-specific lipopolysaccharide epitope (Collingro et al., 2005b; Horn et al., 2004; Horn et al., 2006).

Several proteins that have been implicated in pathogenicity of the *Chlamydiaceae* are encoded in the *Parachlamydia amoebophila* genome. For example, *Parachlamydia amoebophila* possesses proteins targeted to the inclusion membrane (Inc) and expresses the *Chlamydia*-specific protease-like activity factor CPAF. *Parachlamydia amoebophila* also possesses a type three secretions system, distributed among several loci across the genome, which seems to be characteristic for the *Chlamydiae* in contrast to proteobacterial type three secretion systems, which are generally encoded on a plasmid or a single region on the chromosome (Horn et al., 2004; Hueck, 1998; Peters et al., 2007). On the other hand, many proteins considered important for host cell interaction of the *Chlamydiaceae* are missing in *Parachlamydia amoebophila*, e.g., polymorphic membrane proteins (Pmp). *Parachlamydia amoebophila* instead encodes several proteins potentially involved in host cell interaction but absent in the *Chlamydiaceae*, such as a number of leucine-rich repeat proteins and a type four secretion system encoded on a genomic island which has most likely been acquired by lateral gene transfer and might be involved in protein export or conjugative DNA transfer (Greub et al., 2004; Horn et al., 2004).

Phylogenetic analysis suggests that the last common ancestor of *Parachlamydia amoebophila* and the *Chlamydiaceae* lived at least 700 million years ago. This ancient microorganism had, at that time, most likely already lived within a eukaryotic host cell and used several basic mechanisms for host cell interaction which are conserved to date, and which most likely developed during interplay with ancient unicellular eukaryotes (Horn et al., 2004).

Pathogenesis. Free-living amoebae are considered the natural hosts of *Protochlamydia* species. The presence of *Parachlamydia amoebophila* increased the *in vitro* cytopathogenicity of *Acanthamoeba* host cells on human fibroblasts (Fritsche et al., 1998). To date, there is no evidence for a direct pathogenic potential of *Parachlamydia amoebophila* for humans.

List of species of the genus *Protochlamydia*

1. **Protochlamydia amoebophila** Horn, sp. nov. (previously known as "*Candidatus* Protochlamydia amoebophila" Collingro, Toenshoff, Taylor, Fritsche, Wagner and Horn 2005b).

a.moe'bo.phi.la. N.L. n. *amoeba* an amoeba; N.L. adj. *philus -a -um* (from Gr. adj. *philos -ê -on*) friend, loving; N.L. fem. adj. *amoebophila* amoeba-loving, referring to the intracellular life style within amoebae.

The characteristics of the sole member of the genus and species are the same as that provided for the genus description. The original host amoeba *Acanthamoeba* species UWE25 was isolated from a soil sample in western Washington State, USA; the current host is *Acanthamoeba* species UWC1 (Fritsche et al., 1993, 1998).

The 16S rRNA-targeted oligonucleotide probe Bn_9658 (5′-TCCGTTTTCTCCGCCTAC-3′) can be used for the detection of *Protochlamydia amoebophila* by fluorescence *in situ* hybridization (Amann et al., 1997), but this probe also targets members of the genus *Parachlamydia*.

A complete genome sequence is available for the type strain *Protochlamydia amoebophila* UWE25 (Horn et al., 2004).

DNA G+C content (mol%): 35.8.

Type strain: UWE25, ATCC PRA-7.

Sequence accession no. (16S rRNA gene): AF083615.

Taxonomic note: the correct specific epithet should be *amoebiphila*.

References

Amann, R., N. Springer, W. Schonhuber, W. Ludwig, E.N. Schmid, K.D. Muller and R. Michel. 1997. Obligate intracellular bacterial parasites of acanthamoebae related to *Chlamydia* spp. Appl. Environ. Microbiol. *63*: 115–121.

Birtles, R.J., T.J. Rowbotham, C. Storey, T.J. Marrie and D. Raoult. 1997. *Chlamydia*-like obligate parasite of free-living amoebae. Lancet *349*: 925–926.

Casson, N. and G. Greub. 2006. Resistance of different *Chlamydia*-like organisms to quinolones and mutations in the quinoline resistance-determining region of the DNA gyrase A- and topoisomerase-encoding genes. Int. J. Antimicrob. Agents *27*: 541–544.

Casson, N., N. Medico, J. Bille and G. Greub. 2006. Parachlamydia acanthamoebae enters and multiplies within pneumocytes and lung fibroblasts. Microbes Infect. *8*: 1294–1300.

Collingro, A., C. Baranyi, R. Michel, M. Wagner, H. Aspöck, Horn, M., and J. Walochnik. 2004. Chlamydial endocytobionts of free-living amoebae differentially affect the growth rate of their hosts. Eur. J. Protistol. *40*: 57–60.

Collingro, A., S. Poppert, E. Heinz, S. Schmitz-Esser, A. Essig, M. Schweikert, M. Wagner and M. Horn. 2005a. Recovery of an environmental *Chlamydia* strain from activated sludge by co-cultivation with *Acanthamoeba* sp. Microbiology *151*: 301–309.

Collingro, A., E.R. Toenshoff, M.W. Taylor, T.R. Fritsche, M. Wagner and M. Horn. 2005b. '*Candidatus* Protochlamydia amoebophila', an endosymbiont of *Acanthamoeba* spp. Int. J. Syst. Evol. Microbiol. *55*: 1863–1866.

Corsaro, D. and D. Venditti. 2004. Emerging chlamydial infections. Crit. Rev. Microbiol. *30*: 75–106.

Corsaro, D., D. Venditti and M. Valassina. 2002. New parachlamydial 16S rDNA phylotypes detected in human clinical samples. Res. Microbiol. *153*: 563–567.

Dautry-Varsat, A., M.E. Balana and B. Wyplosz. 2004. Chlamydia-host cell interactions: recent advances on bacterial entry and intracellular development. Traffic *5*: 561–570.

Draghi, A., II, J. Bebak, V.L. Popov, A.C. Noble, S.J. Geary, A.B. West, P. Byrne and S. Frasca, Jr. 2007. Characterization of a *Neochlamydia*-like bacterium associated with epitheliocystis in cultured Arctic charr Salvelinus alpinus. Dis. Aquat. Org. *76*: 27–38.

Everett, K.D.E., R.M. Bush and A.A. Andersen. 1999. Emended description of the order *Chlamydiales*, proposal of *Parachlamydiaceae* fam. nov. and *Simkaniaceae* fam. nov., each containing one monotypic genus, revised taxonomy of the family *Chlamydiaceae*, including a new genus and five new species, and standards for the identification of organisms. Int. J. Syst. Bacteriol. *49*: 415–440.

Fredericks, D.N. and D.A. Relman. 1996. Sequence-based identification of microbial pathogens: a reconsideration of Koch's postulates. Clin. Microbiol. Rev *9*: 18–33.

Fritsche, T.R., D. Sobek and R.K. Gautom. 1998. Enhancement of in vitro cytopathogenicity by *Acanthamoeba* spp. following acquisition of bacterial endosymbionts. FEMS Microbiol. Lett. *166*: 231–236.

Fritsche, T.R., M. Horn, M. Wagner, R.P. Herwig, K.H. Schleifer and R.K. Gautom. 2000. Phylogenetic diversity among geographically dispersed *Chlamydiales* endosymbionts recovered from clinical and environmental isolates of *Acanthamoeba* spp. Appl. Environ. Microbiol. *66*: 2613–2619.

Greub, G. and D. Raoult. 2002a. Crescent bodies of *Parachlamydia acanthamoeba* and its life cycle within *Acanthamoeba polyphaga*: an electron micrograph study. Appl. Environ. Microbiol. *68*: 3076–3084.

Greub, G. and D. Raoult. 2002b. *Parachlamydiaceae*: potential emerging pathogens. Emerg. Infect. Dis. *8*: 625–630.

Greub, G., P. Berger, L. Papazian and D. Raoult. 2003a. *Parachlamydiaceae* as rare agents of pneumonia. Emerg. Infect. Dis. *9*: 755–756.

Greub, G., I. Boyadjiev, B. La Scola, D. Raoult and C. Martin. 2003b. Serological hint suggesting that *Parachlamydiaceae* are agents of pneumonia in polytraumatized intensive care patients. Ann. N. Y. Acad. Sci. *990*: 311–319.

Greub, G., B. La Scola and D. Raoult. 2003c. *Parachlamydia acanthamoeba* is endosymbiotic or lytic for *Acanthamoeba polyphaga* depending on the incubation temperature. Ann. N. Y. Acad. Sci. *990*: 628–634.

Greub, G., J.L. Mege and D. Raoult. 2003d. *Parachlamydia acanthamoebae* enters and multiplies within human macrophages and induces their apoptosis [corrected]. Infect. Immun. *71*: 5979–5985.

Greub, G., F. Collyn, L. Guy and C.A. Roten. 2004. A genomic island present along the bacterial chromosome of the *Parachlamydiaceae* UWE25, an obligate amoebal endosymbiont, encodes a potentially functional F-like conjugative DNA transfer system. BMC Microbiol. *4*: 48.

Greub, G., J.L. Mege, J.P. Gorvel, D. Raoult and S. Meresse. 2005. Intracellular trafficking of *Parachlamydia acanthamoebae*. Cell. Microbiol. *7*: 581–589.

Greub, G., O. Hartung, T. Adekambi, Y.S. Alimi and D. Raoult. 2006. Chlamydialike organisms and atherosclerosis. Emerg. Infect. Dis. *12*: 705–706.

Heinz, E., I. Kolarov, C. Kastner, E.R. Toenshoff, M. Wagner and M. Horn. 2007. An *Acanthamoeba* sp. containing two phylogenetically different bacterial endosymbionts. Environ. Microbiol. *9*: 1604–1609.

Horn, M., M. Wagner, K.D. Muller, E.N. Schmid, T.R. Fritsche, K.H. Schleifer and R. Michel. 2000. *Neochlamydia hartmannellae* gen. nov., sp. nov. (*Parachlamydiaceae*), an endoparasite of the amoeba *Hartmannella vermiformis*. Microbiology *146*: 1231–1239.

Horn, M., M. Wagner, K.D. Müller, E.N. Schmid, T.R. Fritsche, K.H. Schleifer and R. Michel. 2001. *In* Validation of the publication of new names and new combinations previously effectively published outside the IJSEM. List no. 81. Int. J. Syst. Evol. Microbiol. *51*: 1229.

Horn, M., A. Collingro, S. Schmitz-Esser, C.L. Beier, U. Purkhold, B. Fartmann, P. Brandt, G.J. Nyakatura, M. Droege, D. Frishman, T. Rattei, H.W. Mewes and M. Wagner. 2004. Illuminating the evolutionary history of chlamydiae. Science *304*: 728–730.

Horn, M., A. Collingro, S. Schmitz-Esser and M. Wagner. 2006. Environmental chlamydia genomics. *In* Chlamydia: genomics and pathogenesis (edited by Bavoil and Wyrick). Horizon Bioscience, Norwich, UK, pp. 25–44.

Marrie, T.J., D. Raoult, B. La Scola, R.J. Birtles and E. de Carolis. 2001. *Legionella*-like and other amoebal pathogens as agents of community-acquired pneumonia. Emerg. Infect. Dis. *7*: 1026–1029.

Meijer, A., P.J. Roholl, J.M. Ossewaarde, B. Jones and B.F. Nowak. 2006. Molecular evidence for association of chlamydiales bacteria with epitheliocystis in leafy seadragon (*Phycodurus eques*), silver perch (*Bidyanus bidyanus*), and barramundi (*Lates calcarifer*). Appl. Environ. Microbiol. *72*: 284–290.

Michel, R., B. Hauröder-Philippczyk, K.-D. Müller and I. Weishaar. 1994. *Acanthamoeba* from human nasal mucosa infected with an obligate intracellular parasite. Eur. J. Protistol. *30*: 104–110.

Schmid, E.N., K.D. Mueller and R. Michel. 2001. Evidence for bacteriophages within *Neochlamydia hartmannellae*, an obligate endoparasitic bacterium of the free-living amoeba *Hartmannella vermiformis*. Endocytobiol. Cell Res. *14*: 115–119.

Skriwan, C., M. Fajardo, S. Hagele, M. Horn, M. Wagner, R. Michel, G. Krohne, M. Schleicher, J. Hacker and M. Steinert. 2002. Various bacterial pathogens and symbionts infect the amoeba *Dictyostelium discoideum*. Int. J. Med. Microbiol. *291*: 615–624.

Thomas, V., N. Casson and G. Greub. 2006. *Criblamydia sequanensis*, a new intracellular *Chlamydiales* isolated from Seine river water using amoebal co-culture. Environ. Microbiol. *8*: 2125–2135.

von Bomhard, W., A. Polkinghorne, Z. Huat Lu, L. Vaughan, A. Vogtlin, D.R. Zimmermann, B. Spiess and A. Pospischil. 2003. Detection of novel chlamydiae in cats with ocular disease. Am. J. Vet. Res. *64*: 1421–1428.

Family V. "*Candidatus* Piscichlamydiaceae"

Matthias Horn

Pi.sci.chla.my.di.a.ce'ae. N.L. fem. n. *Piscichlamydia* type genus of the family; *-aceae* ending to denote a family, N.L. fem. pl. n. *Piscichlamydiaceae* the *Piscichlamydia* family.

The family "*Candidatus* Piscichlamydiaceae" is to date solely based on 16S rRNA sequences distantly related to all other members of the *Chlamydiae* (80–83% sequence similarity; Figure 154). The family "*Candidatus* Piscichlamydiaceae" contains a single genus, "*Candidatus* Piscichlamydia" (Draghi et al., 2004).

Type genus: **"*Candidatus* Piscichlamydia"** Draghi, Popov, Kahl, Stanton, Brown, Tsongalis, West and Frasca 2004, 5286.

Genus I. "*Candidatus* Piscichlamydia" Draghi, Popov, Kahl, Stanton, Brown, Tsongalis, West and Frasca 2004, 5286

Matthias Horn

Pi.sci.chla.my'di.a. L. n. *piscis, -is* fish; N.L. fem. n. *Chlamydia* name of a bacterial genus; N.L. fem. n. *Piscichlamydia Chlamydia*-like organism affecting fish.

Coccoid to rod shaped, nonmotile, obligately intracellular bacteria, up to 1.8 μm in length.

Cells show a developmental cycle with morphologically distinct stages similar to those of the *Chlamydiaceae* and thrive inside a host vacuole. Organisms were found within gill lesions of fish and have not yet been obtained in cell culture.

Type species: **"*Candidatus* Piscichlamydia salmonis"** Draghi, Popov, Kahl, Stanton, Brown, Tsongalis, West and Frasca 2004, 5286.

Further descriptive information

Two distinct morphological forms, representing different developmental stages, have been observed inside the host-derived vacuole that surrounded the bacteria. The proposed reticulate bodies are elongate, oblong, or spherical (0.7–1.8 μm) with even, granular cytoplasm resembling the reticulate bodies of *Chlamydiaceae*. The second morphological stage was round to oval (0.6–0.8 μm) with a central, condensed cytoplasm. Although these forms were proposed to represent intermediate bodies, they might in fact correspond to elementary bodies as no other stage resembling the elementary bodies of other chlamydiae has been observed (Draghi et al., 2004). Immunogold labeling experiments using antibodies against the *Chlamydia* genus-specific LPS epitope KDOp-(2–8)-KDOp-(2–4)-KDO suggested the presence of a similar trisaccharide in the LPS of "*Candidatus* Piscichlamydia salmonis" (Draghi et al., 2004).

Pathogenesis. "*Candidatus* Piscichlamydia salmonis" has been found infecting the gills of salmonid fish (Atlantic salmon, *Salmo salar*) in marine water, forming proliferative lesions and inclusions in hypertrophied epithelial cells at the tips of lamellae. "*Candidatus* Piscichlamydia salmonis" has been suggested to represent one possible etiological agent of epitheliocystis in fish from marine environments (Draghi et al., 2004).

List of tentative species of the genus "*Candidatus* Piscichlamydia"

1. **"*Candidatus* Piscichlamydia salmonis"** Draghi, Popov, Kahl, Stanton, Brown, Tsongalis, West and Frasca 2004, 5286

sal.mo'nis. L. n. *salmo, -onis* salmon; L. gen. n. *salmonis* of salmon (infecting fish of the genus *Salmo*).

The genus "*Candidatus* Piscichlamydia" is based on the description of "*Candidatus* Piscichlamydia salmonis" (Draghi et al., 2004) (Figure 154).

Sequence accession no. (16S rRNA gene): AY462244.

Reference

Draghi, A., 2nd, V.L. Popov, M.M. Kahl, J.B. Stanton, C.C. Brown, G.J. Tsongalis, A.B. West and S. Frasca, Jr. 2004. Characterization of "*Candidatus* Piscichlamydia salmonis" (order *Chlamydiales*), a chlamydia-like bacterium associated with epitheliocystis in farmed Atlantic salmon (*Salmo salar*). J. Clin. Microbiol. *42*: 5286–5297.

Family VI. **Rhabdochlamydiaceae** fam. nov.

Matthias Horn

Rhab.do.chla.my.di.a.ce'ae. N.L. fem. n. *Rhabdochlamydia* typo genus of the family; *-aceae* ending to denote a family; N.L. fem. pl. n. *Rhabdochlamydiaceae* the *Rhabdochlamydia* family.

Members of the family *Rhabdochlamydiaceae* have been found infecting arthropods, but can also thrive in amoebae and are moderately related to the *Simkaniaceae* (around 86% 16S rRNA sequence similarity; Figure 154). The family *Rhabdochlamydiaceae* is a sister family of the *Simkaniaceae* and contains a single genus, *Rhabdochlamydia*.

Type genus: **Rhabdochlamydia** gen. nov.

Genus I. **Rhabdochlamydia** Horn, gen. nov. (previously known as "*Candidatus* Rhabdochlamydia" Kostanjšek, Štrus, Drobne and Avguštin 2004)

Matthias Horn

Rhab.do.chla.my'di.a. Gr. fem. n. *rhabdos* stick, rod; N.L. fem. n. *Chlamydia* taxonomic name of a bacterial genus; N.L. fem. n. *Rhabdochlamydia* referring to the rod-like morphology of the elementary bodies of this genus.

Coccoid to rod shaped, nonmotile, obligately intracellular bacteria, 0.35–4 µm. Cells show a **developmental cycle** with morphologically distinct stages and **grow within host-derived vacuoles.** Organisms naturally **infect arthropods** and are differentiated from all other chlamydiae by the **rod-shaped elementary body**, which has a **five-layer cell wall** and shows one or two **electron-translucent oblong structures in the cytoplasm** (Figure 162). The genus *Rhabdochlamydia* is based on the description of "***Candidatus* Rhabdochlamydia porcellionis**" and ***Rhabdochlamydia crassificans*** (Corsaro et al., 2007; Kostanjšek et al., 2004) (Figure 154).

Type species: **Rhabdochlamydia crassificans** Horn, gen. nov. (previously known as "*Candidatus* Rhabdochlamydia" Kostanjšek, Štrus, Drobne and Avguštin 2004).

Further descriptive information

Three distinct morphological forms, representing different developmental stages, have been observed inside the host-derived vacuoles in which the bacteria live. Reticulate bodies are spherical in shape (0.9–4 µm in diameter), show a granular cytoplasm, resemble the reticulate bodies of members of the *Chlamydiaceae*, and have a three-layer cell wall. The elementary bodies are rod-shaped (0.25–0.7 µm long, with a diameter of 0.1–0.25 µm), electron-dense, possess a five-layer cell wall, and show one to two electron-translucent, elongate inclusions in the cytoplasm, generally at one pole (Corsaro et al., 2007; Kostanjšek et al., 2004). Transitional stages between reticulate and elementary bodies, intermediate bodies, have been observed, which largely resemble reticulate bodies, but are smaller in size (0.35–0.65 µm) and show an electron-dense central area (Kostanjšek et al., 2004). Elementary bodies with a similar morphology have previously been described for organisms classified as *Rickettsiella* species in spiders and scorpions (Morel, 1976; Weiss et al., 1984).

Pathogenesis. The *Rhabdochlamydia* species identified to date have been found in the hepatopancreas of the terrestrial isopod *Porcellio scaber* (Kostanjšek et al., 2004) and the fat body of the cockroach *Blatta orientalis*, respectively (Corsaro et al., 2007). These arthropods might represent the natural hosts for the respective *Rhabdochlamydia* species, which can co-exist with their hosts in an asymptomatic state, but which may also cause disease, e.g., abdominal swelling (Corsaro et al., 2007; Kostanjšek et al., 2004).

List of species of the genus *Rhabdochlamydia*

1. **Rhabdochlamydia crassificans** Corsaro, Thomas, Goy, Venditti, Radek and Greub 2005, 226

cras'si.fi.cans. L. adj. *crassus* thick; L. ending *-ficans* (from L. v. *facio facere* to make) making; N.L. adj. *crassificans* making thick, referring to the body swelling of cockroaches upon infection.

The characteristics are the same as for the genus, except noted. Reticulate bodies are about 0.9 µm in diameter; elementary bodies are about 0.5 µm × 0.25 µm (Corsaro et al., 2007); two intermediate stages have been described resembling the elementary and the reticulate bodies, respectively (Radek, 2000).

Rhabdochlamydia crassificans was found infecting the cockroach *Blatta orientalis*. Infection occurred mainly in the fat body but also systemically, affecting gut epithelium,

Malpighian tubules, blood cells, and ovarioles (Radek, 2000). Cells infected with *Rhabdochlamydia crassificans* contained several small to large inclusions (7–50 μm in diameter with single vacuoles as large as 100 μm). Mycetocytes containing the primary cockroach symbiont *Blattabacterium* species were also observed to be co-infected with *Rhabdochlamydia crassificans* (Radek, 2000). Infection of *Blattabacterium orientalis* with *Rhabdochlamydia crassificans* caused severe abdominal swelling resulting in individuals unable to move and with a decreased life expectancy (Radek, 2000), but asymptomatic cockroaches were also found (Corsaro et al., 2007). On the population level, infection with *Rhabdochlamydia crassificans* led to stagnation of reproduction of *Blattabacterium orientalis* colonies. Transmission of *Rhabdochlamydia crassificans* may be transovarial or via feces.

Rhabdochlamydia crassificans can be grown in the amoeba *Acanthamoeba castellanii* ATCC 30010 (Casson et al., 2007).

Rhabdochlamydia crassificans has previously been classified as *Rickettsiella crassificans* (Radek, 2000).

Type strain: CRIB-01 (not yet deposited in a public culture collection).

Sequence accession no. (16S rRNA gene): AY928092.

2. **"*Candidatus* Rhabdochlamydia porcellionis"** Kostanjšek, Štrus, Drobne and Avguštin 2004, 548

por.cel.li.o'nis. L. n. *porcellio, -onis* a cheslip, woodlouse, sowbug, and also a zoological genus name (*Porcellio*); L. gen. n. *porcellionis* of *Porcellio* (infecting terrestrial isopods of the genus *Porcellio*).

Not yet obtained in cell culture. The characteristics are the same as for the genus, except as noted.

Reticulate bodies are 1–4 μm in diameter; intermediate bodies are 0.35–0.65 μm. Mature elementary bodies are 0.25–0.7 μm in length and 0.1–0.15 μm in diameter (Kostanjšek et al., 2004).

"*Candidatus* Rhabdochlamydia porcellionis" was found infecting the hepatopancreas of the terrestrial isopod ("woodlouse") *Porcellio scaber.* Infection can be noted by characteristic, white spots (100–200 μm in diameter) along the glands, which represent enlarged epithelial cells with multiple smaller or a single large vacuole (up to 30 μm in diameter) containing different developmental stages of the bacteria, which are eventually released into the gland lumina. Infection of *Porcellio scaber* with "*Candidatus* Rhabdochlamydia porcellionis" leads to a reorganization of the structure and function of the digestive glands and possibly to death of the infected individuals although no differences in feeding rate has been observed between uninfected and infected individuals during a period of 5 d. Between 0 and 10% of individuals in natural *Porcellio scaber* populations are generally infected (Drobne et al., 1999; Kostanjšek et al., 2004).

"*Candidatus* Rhabdochlamydia porcellionis" has previously been classified as *Chlamydia isopodii* (Shay et al., 1985) or placed in the polyphyletic genus *Rickettsiella* (Drobne et al., 1999).

The 16S rRNA-targeted oligonucleotide probe S-S-RhaC-0992-a-A-20 (5′-GATGCTGTCCTTTGCATTTC-3′) can be used for specific detection using fluorescence *in situ* hybridization (Kostanjšek et al., 2004).

Sequence accession no. (16S rRNA gene): AY223862.

References

Casson, N., J.M. Entenza and G. Greub. 2007. Serological cross-reactivity between different *Chlamydia*-like organisms. J. Clin. Microbiol. *45*: 234–236.

Corsaro, D., V. Thomas, G. Goy, D. Venditti, R. Radek and G. Greub. 2005. '*Candidatus* Rhabdochlamydia crassificans', an intracellular bacterial pathogen of the cockroach *Blatta orientalis* (Insecta: Blattodea). Syst. Appl. Microbiol. *30*: 221–228.

Corsaro, D., V. Thomas, G. Goy, D. Venditti, R. Radek and G. Greub. 2007. '*Candidatus* Rhabdochlamydia crassificans', an intracellular bacterial pathogen of the cockroach *Blatta orientalis* (Insecta: Blattodea). Syst. Appl. Microbiol. *30*: 221–228.

Drobne, D., J. Strus, N. Znidarsic and P. Zidar. 1999. Morphological description of bacterial infection of digestive glands in the terrestrial isopod *Porcellio scaber* (Isopoda, crustacea). J. Invertebr. Pathol. *73*: 113–119.

Kostanjšek, R., J. Štrus, D. Drobne and G. Avguštin. 2004. '*Candidatus* Rhabdochlamydia porcellionis', an intracellular bacterium from the hepatopancreas of the terrestrial isopod *Porcellio scaber* (Crustacea: Isopoda). Int. J. Syst. Evol. Microbiol. *54*: 543–549.

Morel, G. 1976. Studies on *Porochlamydia buthi* g. n., sp. n., an intracellular pathogen of the scorpion *Buthus occitanus*. J. Invertebr. Pathol. *28*: 167–175.

Radek, R. 2000. Light and electron microscopic study of a *Rickettsiella* species from the cockroach *Blatta orientalis*. J. Invertebr. Pathol. *76*: 249–256.

Shay, M.T., A. Bettica, G.M. Vernon and E.R. Witkus. 1985. *Chlamydia isopodii* sp. n., an obligate intracellular parasite of *Porcellio scaber*. Exp. Cell Biol. *53*: 115–120.

Weiss, E., G.A. Dasch and K.P. Chang. 1984. Genus VIII. *Rickettsiella* Philip 1956, 267^{AL}. *In* Bergey's Manual of Systematic Bacteriology, vol. 1 (edited by Krieg and Holt). Williams & Wilkins, Baltimore, pp. 713–717.

Family VII. **Simkaniaceae** Everett, Bush and Andersen 1999, 435^{VP}

Matthias Horn

Sim.ka.ni.a.ce'ae. N.L. fem. n. *Simkania* type genus of the family; *-aceae* ending to denote a family. N.L. fem. pl. n. *Simkaniaceae* the *Simkania* family.

The family *Simkaniaceae* originally comprised but a single genus, the type genus *Simkania* (Everett et al., 1999). A second genus, *Fritschea*, based on the description of "*Candidatus* Fritschea bemisiae" and "*Candidatus* Fritschea eriococci" is added (Everett et al., 2005; Thao et al., 2003). The recently identified endosymbiont of the enigmatic marine animal *Xenoturbella* constitutes a third genus within the family *Simkaniaceae*, but awaits formal description (Israelsson, 2007) (Figure 154). The 23S rRNA genes of all known members of the *Simkaniaceae* (including the *Xenoturbella* symbiont, but with the exception of "*Candidatus* Fritschea eriococci") contain group I introns encoding a DNA endonuclease most similar to group I introns in the 23S rRNA of chloroplasts and mitochondria of algae. The family *Simkaniaceae* is a sister family of the *Rhabdochlamydiaceae*.

Type genus: **Simkania** Everett, Bush and Andersen 1999, 435[VP].

Genus I. **Simkania** Everett, Bush and Andersen 1999, 435[VP]

Patrik M. Bavoil, Richard S. Stephens, Bernhard Kaltenboeck and Cho-Chou Kuo

Sim.ka'ni.a. N.L. fem. n. *Simkania* arbitrary name formed from the personal name Simona Kahane, the discoverer of the type species.

Coccoid or slightly elongated, nonmotile, obligately intracellular organisms, 0.2–0.7 μm in diameter. Cells grow in the cytosolic compartment of eukaryotic host cells within membrane-bound vacuoles termed inclusions. Organisms are differentiated from members of the *Chlamydiaceae* by their resistance to penicillin, the infectivity of both replicating and proposed differentiated forms toward eukaryotic cells, and their infectivity of amoebae.

DNA G+C content (mol%): 42.5.

Type species: **Simkania negevensis** Everett, Bush and Andersen 1999, 435[VP].

Further descriptive information

Isolation, growth, cell morphology, and other characteristics. The expanding scientific literature is reviewed in which *Simkania* illustrates significant structural and biological differences from the *Chlamydiaceae*. First isolated as a contaminant of cell culture (Kahane et al., 1993), *Simkania negevensis* also grows in Vero cells, HeLa cells, and trophozoites of acanthamoebae, and survives in amoebal cysts (Kahane et al., 2001).

Rapid growth occurs in Vero cells for a period of 2–3 d followed by a "stationary" phase of 8 or more days (Kahane et al., 1999). Growing simkaniae are characterized by replicating and proposed differentiated forms that are 0.2–0.3 and 0.3–0.7 μm in diameter, respectively; both forms are infectious (Kahane et al., 2002). Differentiated forms differ from chlamydial elementary bodies structurally in that they appear either as coccoid or slightly elongated with an electron-dense, eccentrically located nucleoid (Kahane et al., 2002). Replicating forms are more pleomorphic than their chlamydial counterparts and often appear elongated. Multiple *Simkania* inclusions are observed in each infected cell. The genome size of *Simkania negevensis* has been determined to be 1.7 Mbp.

Pathogenesis. Although a direct causal relationship has not been demonstrated, *Simkania negevensis* has been serologically associated with community-acquired pneumonia in adults (Lieberman et al., 1997), bronchiolitis in infants (Kahane et al., 1998), and acute exacerbation of chronic obstructive pulmonary disease (Lieberman et al., 2002). The species is resistant to penicillin (Kahane et al., 1993).

List of species of the genus *Simkania*

1. **Simkania negevensis** Everett, Bush and Andersen 1999, 435[VP]

ne.ge.ven'sis N.L. fem. adj. *negevensis* of or pertaining to the Negev Desert in Israel.

The characteristics of the sole member of the genus and species are the same as that provided for the genus description.

DNA G+C content (mol%): 42.5 (T_m).

Type strain: Z, ATCC VR-1471.

Sequence accession no. (16S rRNA gene): U68460.

Genus II. "*Candidatus* Fritschea" Everett, Thao, Horn, Dyszynski and Baumann 2005

Matthias Horn

Fri'tsche.a. N.L. fem. n. *Fritschea* named after Thomas R. Fritsche, an American physician and parasitologist, in honor of his contributions to our current knowledge on chlamydial diversity.

Coccoid to rod-shaped, nonmotile, obligately intracellular bacteria, up to 2.5 μm in length. Cells show a **developmental cycle** with morphologically distinct stages similar to those of the *Chlamydiaceae*. Organisms were found in whitefly and scale insects, but have not yet been obtained in cell culture. The genus is based on the description of "*Candidatus* Fritschea bemisiae" and "*Candidatus* Fritschea eriococci" (Everett et al., 2005; Thao et al., 2003) (Figure 154).

Type species: **"Candidatus Fritschea bemisiae"** Everett, Thao, Horn, Dyszynski and Baumann 2005, 1585.

Further descriptive information

Two distinct morphological forms, representing different developmental stages of "*Candidatus* Fritschea bemisiae", have been observed inside bacteriocytes of the whitefly *Bemisia tabaci*, which also contained gammaproteobacterial primary and secondary symbionts (Costa et al., 1995). The intracellular niche of "*Candidatus* Fritschea bemisiae" has not yet been characterized in much detail; it is currently unclear whether "*Candidatus* Fritschea bemisiae" thrives directly in the cytoplasm or within a host-derived vacuole. The proposed reticulate bodies are elongate (0.7–2.5 μm in length and 0.7–0.8 μm in width) with even, granular cytoplasm resembling the reticulate bodies of *Chlamydiaceae*. Occasionally, filamentous structures can be seen in the cytoplasm of reticulate bodies. The second morphological stage is spherical, much smaller (0.2–0.25 μm in diameter), and very electron-dense (Costa et al., 1995). No morphological data are available for "*Candidatus* Fritschea eriococci".

List of species of the genus *Fritschea*

1. **"*Candidatus* Fritschea bemisiae"** Everett, Thao, Horn, Dyszynski and Baumann 2005, 1585.

be.mi'si.ae. N.L. gen. fem. n. *bemisiae* of *Bemisia*, discovered in whitefly insects of the genus *Bemisia*.

Not yet obtained in cell culture. The characteristics are the same as for the genus. "*Candidatus* Fritschea bemisiae" has so far only been found in the whitefly insect *Bemisia tabaci* (biotype A). "*Candidatus* Fritschea bemisiae" could not be detected in *Bemisia tabaci* biotypes B and Q (Chiel et al., 2007). A 16.6 kb genome fragment of "*Candidatus* Fritschea bemisiae" has been sequenced, indicating a genomic G+C content of 40 mol% and encoding six proteins most similar to proteins of *Protochlamydia amoebophila* and other chlamydial species (Thao et al., 2003). The 23S rRNA gene of "*Candidatus* Fritschea bemisiae" contains a group I intron similar to the group I intron found in the 23S rRNA of *Simkania negevensis*. The oligonucleotide 5′-GATGCTGTCCTTTGCATTTC-3′ should be suited for specific detection of "*Candidatus* Fritschea bemisiae" by fluorescence *in situ* hybridization (Thao et al., 2003).

Sequence accession no. (16.6 kb genome fragment including the rRNA operon): AY140910.

2. **"*Candidatus* Fritschea eriococci"** Everett, Thao, Horn, Dyszynski and Baumann 2005, 1586

e.ri.o.cocci. N.L. gen. masc. n. *eriococci* of *Eriococcus*, discovered in scale insects of the genus *Eriococcus*.

Not yet obtained in cell culture; no morphological data available. The rRNA genes of "*Candidatus* Fritschea eriococci" have been identified in the scale insect *Eriococcus spurious* during a screening of a collection of insects for *Fritschea* species (Thao et al., 2003). The 16S rRNA shows 97% sequence similarity with the 16S rRNA of "*Candidatus* Fritschea bemisiae". "*Candidatus* Fritschea eriococci" is so far the only member of the family *Simkaniaceae* that does not contain a group I intron in the 23S rRNA gene.

Sequence accession no. (16S and 23S rRNA genes): AY140911.

References

Chiel, E., Y. Gottlieb, E. Zchori-Fein, N. Mozes-Daube, N. Katzir, M. Inbar and M. Ghanim. 2007. Biotype-dependent secondary symbiont communities in sympatric populations of *Bemisia tabaci*. Bull. Entomol. Res. *97*: 407–413.

Costa, H.S., D.M. Westcot, D.E. Ullman, R. Rosell, J.K. Brown and M.W. Johnson. 1995. Morphological variation in *Bemisia* endosymbionts. Protoplasma *189*: 194–202.

Everett, K.D.E., R.M. Bush and A.A. Andersen. 1999. Emended description of the order *Chlamydiales*, proposal of *Parachlamydiaceae* fam. nov. and *Simkaniaceae* fam. nov., each containing one monotypic genus, revised taxonomy of the family *Chlamydiaceae*, including a new genus and five new species, and standards for the identification of organisms. Int. J. Syst. Bacteriol. *49*: 415–440.

Everett, K.D.E., M. Thao, M. Horn, G.E. Dyszynski and P. Baumann. 2005. Novel chlamydiae in whiteflies and scale insects: endosymbionts '*Candidatus* Fritschea bemisiae' strain Falk and '*Candidatus* Fritschea eriococci' strain Elm. Int. J. Syst. Evol. Microbiol. *55*: 1581–1587.

Israelsson, O. 2007. Chlamydial symbionts in the enigmatic *Xenoturbella* (Deuterostomia). J. Invertebr. Pathol. *96*: 213–220.

Kahane, S., R. Gonen, C. Sayada, J. Elion and M.G. Friedman. 1993. Description and partial characterization of a new *Chlamydia*-like microorganism. FEMS Microbiol. Lett. *109*: 329–333.

Kahane, S., D. Greenberg, M.G. Friedman, H. Haikin and R. Dagan. 1998. High prevalence of "*Simkania*" Z, a novel *Chlamydia*-like bacterium, in infants with acute bronchiolitis. J. Infect. Dis. *177*: 1425–1429.

Kahane, S., K.D.E. Everett, N. Kimmel and M.G. Friedman. 1999. *Simkania negevensis* strain Z^T: growth, antigenic and genome characteristics. Int. J. Syst. Bacteriol. *49*: 815–820.

Kahane, S., B. Dvoskin, M. Mathias and M.G. Friedman. 2001. Infection of *Acanthamoeba polyphaga* with *Simkania negevensis* and *S. negevensis* survival within amoebal cysts. Appl. Environ. Microbiol. *67*: 4789–4795.

Kahane, S., N. Kimmel and M.G. Friedman. 2002. The growth cycle of *Simkania negevensis*. Microbiology *148*: 735–742.

Lieberman, D., S. Kahane, D. Lieberman and M.G. Friedman. 1997. Pneumonia with serological evidence of acute infection with the *Chlamydia*-like microorganism "Z". Am. J. Respir Crit. Care Med. *156*: 578–582.

Lieberman, D., B. Dvoskin, D.V. Lieberman, S. Kahane and M.G.. Friedman. 2002. Serological evidence of acute infection with the *Chlamydia*-like microorganism *Simkania negevensis* (Z) in acute exacerbation of chronic obstructive pulmonary disease. Eur. J. Clin. Microbiol. Infect. Dis. *21*: 307–309.

Thao, M.L., L. Baumann, J.M. Hess, B.W. Falk, J.C.K. Ng, P.J. Gullan and P. Baumann. 2003. Phylogenetic evidence for two new insect-associated chlamydia of the family *Simkaniaceae*. Curr. Microbiol. *47*: 46–50.

Family VIII. **Waddliaceae** Rurangirwa, Dilbeck, Crawford, McGuire and McElwain 1999, 580^{VP}

The Editorial Board

Wadd'li.a'ce.ae. N.L fem. n. *Waddlia* the type genus of the family; *-aceae* ending to denote family; N.L. fem. pl. *Waddliaceae* the *Waddlia* family.

The family *Waddliaceae* includes the sole genus, *Waddlia* (Rurangirwa et al., 1999) which are **obligately intracellular bacteria of cattle and possibly other animals**. The organisms are classified based upon 16S rRNA gene phylogeny.

DNA G+C content (mol%): not available.

Type genus: **Waddlia** Rurangirwa, Dilbeck, Crawford, McGuire and McElwain 1999, 580^{VP}.

Genus I. **Waddlia** Rurangirwa, Dilbeck, Crawford, McGuire and McElwain 1999, 580^{VP}

The Editorial Board

Wadd'li.a. N.L. fem. n. *Waddlia* arbitrary name derived from the abbreviation WADDL (Washington Animal Disease Diagnostic Laboratory).

Obligate intracellular bacteria in animals. Cells of the type strain were isolated from aborted fetuses of cattle.

DNA G+C content (mol%): not available.

Type species: **Waddlia chondrophila** Rurangirwa, Dilbeck, Crawford, McGuire and McElwain 1999, 580^{VP}.

Further descriptive information

Strains do not react with antisera from other species used for typing the chlamydiae or rickettsiae. The 16S rRNA genes are most closely related to that of the *Chlamydiaceae* because they exhibit 80–90% similarity to strains of that family.

List of species of the genus *Waddlia*

1. **Waddlia chondrophila** Rurangirwa, Dilbeck, Crawford, McGuire and McElwain 1999, 580^{VP}

chon.dro'phi.la. Gr. n. *chondros* clump; N.L. fem. adj. *phila* (from Gr. fem. adj. *philê*), friend, loving; N.L. fem. adj. *chondrophila* liking clumps, in reference to the association of the bacterium with mitochondria (the noun mitochondria was coined in 1898 by the microbiologist Carl Benda, from Gr. *mitos* thread and Gr. *chondrion* little granule).

Gram-stain-negative, coccoid bacteria 0.2–0.5 μm in diameter which are **obligately intracellular parasites of cattle.** *Waddlia chondrophila* was isolated from the tissues of a bovine fetus, which was aborted in the first trimester. This organism's description is identical to that of "WSU 86-1044" that was reported by Dilbeck et al. (1990) and Kocan et al. (1990). The species includes only the type strain, WSU 86-1044^{T} (=ATCC VR 1470^{T}). The full-length 16S rRNA gene sequence is 15% different from that of *Chlamydiaceae* species. Furthermore, it is 12.8% different from *Parachlamydia acanthamoebae* and 15% different from *Simkania negevensis* thereby justifying its placement in a separate family of the *Chlamydiales* (Everett et al., 1999).

Cells can be grown in bovine turbinate cells (BT) where they produce multiple cytoplasmic vacuoles. BT infection is terminated with tetracycline or chloroform treatment. Cells divide by binary transverse fission. Two developmental cell forms are produced. The dense cell form is infective. The reticulate cell type, usually associated with mitochondria, divides by binary transverse fission.

DNA G+C content (mol%): not available.

Type strain: WSU 86-1044, ATCC VR-1470.

Sequence accession no. (16S rRNA gene): AF042496.

Other organism

1. **"Waddlia malaysiensis"**

ma.lay.si.en'sis. N.L. fem. adj. *malaysiensis* of or pertaining to Malaysia

Although there are no other validly published species in the genus *Waddlia*, a strain obtained from bats has been reported. Based upon 16S rRNA gene sequence analyses it is most closely related to *Waddlia chondrophila*. A new species was proposed for the bat parasites, "*Waddlia malaysiensis*", but its name has not been validated (Chua et al., 2005).

References

Chua, P.K., J.E. Corkill, P.S. Hooi, S.C. Cheng, C. Winstanley and C.A. Hart. 2005. Isolation of *Waddlia malaysiensis*, a novel intracellular bacterium, from fruit bat (*Eonycteris spelaea*). Emerg. Infect. Dis. *11*: 271–277.

Dilbeck, P.M., J.F. Evermann, T.B. Crawford, A.C. Ward, C.W. Leathers, C.J. Holland, C.A. Mebus, L.L. Logan, F.R. Rurangirwa and T.C. McGuire. 1990. Isolation of a previously undescribed rickettsia from an aborted bovine fetus. J. Clin. Microbiol. *28*: 814–816.

Everett, K.D.E, R.M. Bush and A.A. Andersen. 1999. Emended description of the order *Chlamydiales*, proposal of *Parachlamydiaceae* fam. nov. and *Simkaniaceae* fam. nov., each containing one monotypic genus, revised taxonomy of the family *Chlamydiaceae*, including a new genus and five new species, and standards for the identification of organisms. Int. J. Syst. Bacteriol. *49*: 415–440.

Kocan, K.M., T.B. Crawford, P.M. Dilbeck, J.F. Evermann and T.C. McGuire. 1990. Development of a rickettsia isolated from an aborted bovine fetus. J. Bacteriol. *172*: 5949–5955.

Rurangirwa, F.R., P.M. Dilbeck, T.B. Crawford, T.C. McGuire and T.F. McElwain. 1999. Analysis of the 16S rRNA gene of micro-organism WSU 86-1044 from an aborted bovine foetus reveals that it is a member of the order *Chlamydiales*: proposal of *Waddliaceae* fam. nov., *Waddlia chondrophila* gen. nov., sp. nov. Int. J. Syst. Bacteriol. *49*: 577–581.

Phylum XXV. Planctomycetes Garrity and Holt 2001, 137 emend. Ward (this volume)

Naomi L. Ward

Planc.to.my.ce'tes. N.L. fem. pl. n. *Planctomycetales* type order of the phylum; *-etes* ending to denote a phylum; N.L. fem. pl. n. *Planctomycetes* the phylum of *Planctomycetales*.

The phylum *Planctomycetes* was circumscribed for this volume on the basis of phylogenetic analysis of 16S rRNA gene sequences; the phylum contains the class *Planctomycetia*.

Type order: **Planctomycetales** Schlesner and Stackebrandt 1987, 179^{VP} (Effective publication: Schlesner and Stackebrandt 1986, 175.).

References

Garrity, G.M. and J.G. Holt. 2001. The Road Map to the *Manual*. *In* Bergey's Manual of Systematic Bacteriology, 2nd edn, vol. 1, The *Archaea* and the Deeply Branching and Phototrophic *Bacteria* (edited by Boone, Castenholz and Garrity). Springer, New York, pp. 119–166.

Schlesner, H. and E. Stackebrandt. 1986. Assignment of the genera *Planctomyces* and *Pirella* to a new family *Planctomycetaceae* fam. nov. and description of the order *Planctomycetales* ord. nov. Syst. Appl. Microbiol. *8*: 174–176.

Schlesner, H. and E. Stackebrandt. 1987. *In* Validation of the publication of new names and new combinations previously effectively published outside the IJSB. List no. 23. Int. J. Syst. Bacteriol. *37*: 179–180.

Class I. **Planctomycetia** class. nov.

Naomi L. Ward

Planc.to.my.ce.t'i.a. N.L. pl. n. *Planctomycetales* type order of the class; *-ia* ending to denote a class; N.L. pl. n. *Planctomycetia* the class of *Planctomycetales*.

The class *Planctomycetia* was circumscribed for this volume on the basis of phylogenetic analysis of 16S rRNA gene sequences; the class contains the orders *Planctomycetales* and *Brocadiales*.

Type order: **Planctomycetales** Schlesner and Stackebrandt 1987, 179^{VP} (Effective publication: Schlesner and Stackebrandt 1986, 175.) emend. Ward (this volume).

References

Schlesner, H. and E. Stackebrandt. 1986. Assignment of the genera *Planctomyces* and *Pirella* to a new family *Planctomycetaceae* fam. nov. and description of the order *Planctomycetales* ord. nov. Syst. Appl. Microbiol. *8*: 174–176.

Schlesner, H. and E. Stackebrandt. 1987. *In* Validation of the publication of new names and new combinations previously effectively published outside the IJSB. List no. 23. Int. J. Syst. Bacteriol. *37*: 179–180.

Order I. **Planctomycetales** Schlesner and Stackebrandt 1987, 179^{VP} (Effective publication: Schlesner and Stackebrandt 1986, 175) emend. Ward (this volume)

Naomi L. Ward

Planc.to.my.ce.ta'les. N.L. masc. n. *Planctomyces, -etis* type genus of the order; *-ales* ending to denote an order; N.L. fem. pl. n. *Planctomycetales* the order of *Planctomyces*.

The description is the same as for the family *Planctomycetaceae*.

Type genus: **Planctomyces** Gimesi 1924, 4^{AL}.

Taxonomic note: the original description gave the family *Planctomycetaceae* as the nomenclatural type. The type is here corrected to the genus *Planctomyces*, in accordance with Opinion 79 of the Judicial Commission of the International Committee on Systematics of Prokaryotes (2005).

References

Gimesi, N. 1924. Hydrobiologiai tanulmanyok (Hydrobiologische Studien). I. *Planctomyces bekefii* Gim. nov. gen. et sp. (in Hungarian, with German transl.). Kiadja a Magyar Ciszterci Rend, Budapest, pp. 1–8.

Judicial Commission of the International Committee on Systematics of Prokaryotes. 2005. The nomenclatural types of the orders *Acholeplasmatales, Halanaerobiales, Halobacteriales, Methanobacteriales, Methanococcales, Methanomicrobiales, Planctomycetales, Prochlorales, Sulfolobales, Thermococcales, Thermoproteales* and *Verrucomicrobiales* are the genera *Acholeplasma, Halanaerobium, Halobacterium, Methanobacterium, Methanococcus, Methanomicrobium, Planctomyces, Prochloron, Sulfolobus, Thermococcus, Thermoproteus* and *Verrucomicrobium*, respectively. Opinion 79. Int. J. Syst. Evol. Microbiol. *55*: 517–518.

Schlesner, H. and E. Stackebrandt. 1986. Assignment of the genera *Planctomyces* and *Pirella* to a new family *Planctomycetaceae* fam. nov. and description of the order *Planctomycetales* ord. nov. Syst. Appl. Microbiol. *8*: 174–176.

Schlesner, H. and E. Stackebrandt. 1987. *In* Validation of the publication of new names and new combinations previously effectively published outside the IJSB. List no. 23. Int. J. Syst. Bacteriol. *37*: 179–180.

Family I. **Planctomycetaceae** Schlesner and Stackebrandt 1987, 179[VP] (Effective publication: Schlesner and Stackebrandt 1986, 175) emend. Ward (this volume)

Naomi L. Ward

Planc.to.my.ce.ta.ce'ae. N.L. masc. n. *Planctomyces*, *-etis* type genus of the family; *-aceae* ending to denote a family; N.L. fem. pl. n. *Planctomycetaceae* the *Planctomyces* family.

Gram-stain-negative bacteria observed in, or isolated from, marine, brackish, freshwater and groundwater habitats, as well as soil, peat bog, compost, manure, sewage sludge, and animal (prawn, sponge, coral, lice, termite, human colon) tissues. The occurrence of planctomycetes in high temperature, hypersaline, acidic and alkaline environments has been reported. Some species reported to be associated with the occurrence of algal blooms. The use of 16S rRNA as a molecular marker has suggested that planctomycetes, like other previously obscure taxa such as the acidobacteria and verrucomicrobia, have a cosmopolitan distribution.

Mature cell shape may be tear-drop to pear-shaped, spherical to ovoid, or bulbiform. One species (*Isosphaera pallida*) is multicellular and filamentous. Mature cells may have major multifibrillar (i.e., nonprosthecate) appendages described as stalks, spikes, spines, fimbriae, or bristles. Stalks may be encrusted with iron or manganese oxide deposits; the mechanism (active oxidation or passive deposition) has not been determined. Holdfasts may be present either at the distal end of the stalk or at the narrow or pointed cell pole of mature cell. Formation of rosettes may occur. Crateriform structures and pili present either on reproductive pole of mature cell only, or scattered all over cell surface.

Reproduction, where observed, is by budding. Buds may be motile by means of a monotrichous flagellum inserted either close to the reproductive pole or opposite at the subsequent attachment pole. With regard to shape and distribution of crateriform structures, buds resemble the mother cell.

Relatively few strains have been obtained in pure culture, and the type species of the type genus has never been isolated. Most axenic cultures are aerobic or facultatively anaerobic chemoheterotrophs, with carbohydrates serving as prime sources of carbon for growth. Growth temperature optimum 20–41°C.

Cell envelope (in axenic cultures where analysis has been performed) lacks murein (peptidoglycan). Proteins, often rich in glutamate and cystine/cysteine, are the main cell envelope component. Main isoprenoid quinone is MK-6, a menaquinone, in species which have been analyzed.

Cells may exhibit different types of compartmentalization, unusual in the prokaryotic domains, where the nucleoid and RNA are separated from the remainder of the cell inside a compartment designated the pirellulosome. This cell plan has recently been shown to be shared with members of the verrucomicrobia.

Phylogenetic analysis by means of 16S rRNA and other molecular chronometers suggests that the planctomycetes form a monophyletic group only distantly related to other members of the domain *Bacteria*. This distance is supported by the absence of certain otherwise universal oligonucleotides in planctomycete 16S rRNA genes. There is considerable evidence for a superphylum relationship between the planctomycetes, chlamydiae, and verrucomicrobia; deeper relationships to other phyla are still unclear. Planctomycete 5S rRNA is shorter (109–111 nt) than the "minimal" length of 118 nt found in other prokaryotes, and uniquely lacks an insertion at position 66. Some species display unlinked organization of the *rrn* operon.

DNA G+C content (mol%): 50–64.

Type genus: **Planctomyces** Gimesi 1924, 8[AL].

Key to the genera of the family *Planctomycetaceae* available in pure culture

1. Single ovoid to spherical cells or rosettes; stalks very common; single polar flagellum; crateriform structures distributed over entire cell surface or reproductive pole; pink to red pigmentation; no capsule; no growth at pH 5; growth at pH 8; no growth at 4°C; growth at 35°C.
   → Genus *Planctomyces*

2. Single ellipsoid cells; stalks short and rarely observed; nonpigmented; capsule present; growth at pH 5; no growth at pH 8; growth at 4°C; no growth at 35°C.
   → Genus *Schlesneria*

3. Single cells or rosettes; stalk absent; single polar flagellum; crateriform structures only on reproductive pole; single large pirellulosome; nonpigmented; limited tolerance for saline conditions; cells contain spermidine.
   → Genus *Pirellula*

4. Single cells or rosettes; stalk absent; single large pirellulosome; nonpigmented; requires high concentration of sodium chloride and calcium.
   → Genus *Blastopirellula*

5. Single cells or rosettes; stalk absent; single subpolar flagellum; multiple smaller pirellulosome structures; pigmented; requires high concentration of sodium chloride and calcium; cells contain putrescine and cadaverine; possesses phosphatidylcholine.
   → Genus *Rhodopirellula*

6. Multicellular, filamentous; stalk absent; no flagellum; motile by gliding; phototactic; forms gas vesicles; crateriform structures distributed over entire cell surface; strictly aerobic; no growth at pH 4.5–5.5; no growth at 4°C; growth above 35°C; no $C_{18:2}$ fatty acids; pink pigmented.
   → Genus *Isosphaera*

7. Single, paired, or shapeless aggregates of spherical cells; stalk absent; no gliding motility; not phototactic; aerobic or microaerophilic; growth at pH 4.5–5.5; growth at 4°C; no growth above 35°C; presence of $C_{18:2}$ fatty acids.
   → Genus *Singulisphaera*

8. Single spherical to ovoid cells; no rosettes; stalk absent; flagellum is polar bundle; crateriform structures distributed

over entire cell surface; evidence for a membrane-bounded nuclear body; sterol synthesis; no growth below pH 5.0.
→ Genus *Gemmata*

9. Single ellipsoidal cells or rosettes; stalk present; single polar flagellum; crateriform structures distributed over entire cell surface; no sterol synthesis; growth below pH 5.0.
→ Genus *Zavarzinella*

Genus I. **Planctomyces** Gimesi 1924, 4[AL]

Naomi L. Ward, James T. Staley and Jean M. Schmidt

Planc.to.my'ces. Gr. adj. *planktos* wandering, floating; Gr. masc. n. *mukês* fungus; N.L. masc. n. *Planctomyces* floating fungus.

Cells are spherical, ovoid, ellipsoidal, teardrop-shaped, or bulbiform. Often relatively **large** (ignoring appendages and aggregations, individual vegetative cells range up to 3.5 µm in greatest dimension; immature buds smaller). Have at least one **major multifibrillar (nonprosthecate) appendage** (called a **spike, spire, fascicle, bristle, or stalk**) which does not always have the true stalk function of connecting the cell to a substratum. **Divide by budding.** A **holdfast**-which is not always an easily visualized, discrete structure-is often present at the distal end of an appendage or at one end of the cell. Often form **homologous aggregations, rosettes or bouquets,** by attachment at the holdfasts in natural habitats. Produce **crateriform surface structures** (surface pits 12 nm in diameter, circumscribed by a grommet with a 30–36 nm outside diameter) and **pili** in characteristic patterns. **Gram-stain-negative** bacteria. **Lack peptidoglycan. Resistant to β-lactam antibiotics. Divide by budding.** Some species have a **dimorphic life cycle**: **a sessile mother cell** buds; the bud develops into a **swarmer** that is **motile** by means of a **flagellum** (often **ensheathed**); after maturation, the swarmer loses its flagellum and becomes a sessile, budding mother cell. All cultivated species are aerobes or facultative anaerobes. Occur worldwide in both eutrophic and oligotrophic freshwaters, as well as in estuarine and marine habitats. Sometimes become **encrusted with iron and manganese oxides.** Often **associated** in nature with **algae** and **cyanobacteria**. Although several species have now been **isolated in pure culture,** the type species and various other rosette-forming species **have not been cultivated axenically. Carbohydrates** are reported to be the main carbon source for cultivated members of the genus, which are relatively **slow-growing**.

DNA G+C content (mol%): 50.5–57.7.

Type species: **Planctomyces bekefii** Gimesi 1924, 4[AL] (*Blastocaulis sphaerica* Henrici and Johnson 1935, 84 (Hirsch 1972); morphotype la of the *Blastocaulis–Planctomyces* group (Schmidt and Starr 1980b, 1981).

Further descriptive information

Phylogenetic treatment. The three described *Planctomyces* species for which 16S rRNA gene sequences are available (*Planctomyces brasiliensis, Planctomyces limnophilus*, and *Planctomyces maris*), as well as several undescribed species isolated by Heinz Schlesner (Schlesner, 1994) form a phylogenetically coherent group that appears to share its most recent common ancestor with members of the genus *Pirellula* (Griepenburg et al., 1999; Ward et al., 1995). This relationship between the two genera is also reflected in analysis of the 23S rRNA gene (Ward et al., 2000), as well as that encoding the 70-kDa heat-shock protein HSP70 (Ward-Rainey et al., 1996). In contrast, a recent analysis of ribonuclease P RNA (Butler and Fuerst, 2004) does not appear to support a closer relationship between *Planctomyces* and *Pirellula* than between either genus and the remaining cultivated planctomycete genera, *Isosphaera, Rhodopirellula, Blastopirellula*, and *Gemmata*.

Cell morphology. Members of the genus *Planctomyces* rank among the most morphologically distinctive of the *Bacteria*, with relatively large cells (up to 3.5 µm in greatest dimension) that assume a spherical, ovoid, ellipsoidal, teardrop-shaped, or bulbiform shape (Tables 169, 170). They exhibit a diversity of major multifibrillar non-prosthecate appendages, variously named spikes, spires, fascicles, bristles, and stalks. In some species, the appendage (stalk) connects the mature cell to its substratum via a holdfast structure, while in others the holdfast is located at the opposite pole of the cell. There are no reports of the stalk serving a reproductive function as found in, for example, the genus *Hyphomonas*. Several species, including the uncultivated *Planctomyces bekefii, Planctomyces guttaeformis*, and *Planctomyces stranskae*, form rosettes by adhering at the holdfasts, yielding a distinctive and visually appealing structure. Like other members of their family, *Planctomyces* species exhibit crateriform structures (surface pits 12 nm in diameter, circumscribed by a grommet with a 30–36-nm outside diameter) and pili on their cell surface; the distribution of these features can be diagnostic for the species.

Cell-wall composition. The absence of peptidoglycan as the principal cell-wall structural compound is a distinguishing hallmark of the planctomycetes (Giovannoni et al., 1987a; König et al., 1984; Liesack et al., 1986; Stackebrandt et al., 1986) and shared by all *Planctomyces* species for which cell-wall analysis has been performed. As is the case for other planctomycetes, the cell walls of *Planctomyces* species are proteinaceous, presumably strengthened by the disulfide bonds formed by large quantities of glutamate and cystine/cysteine residues (Liesack et al., 1986).

Lipid composition. Like other planctomycetes, members of the genus *Planctomyces* contain the ester-linked polar lipids characteristic of the *Bacteria* (Kerger et al., 1988). Cellular fatty acids include palmitic, palmitoleic, and oleic acids, as well as 18-carbon monounsaturated $C_{18:1}$ ω9*c* and $C_{18:1}$ ω11*c* fatty acids. *Planctomyces* strains tested possess hydroxy fatty acids of the 3-hydroxyeicosanoic acid ($C_{20:0}$ 3-OH) type, but lack 3-hydroxyoctadecanoic acid ($C_{18:0}$ 3-OH) (Sittig and Schlesner, 1993), a fatty acid profile which can be used to distinguish them from members of the genus *Pirellula*. Phospholipid composition has been found to vary between groups of related *Planctomyces* strains (Sittig and Schlesner, 1993).

Polyamine distribution. Griepenburg and coworkers (Griepenburg et al., 1999) demonstrated that polyamine distribution is a useful chemotaxonomic marker at and below the

TABLE 169. Characteristics differentiating species of the genus *Planctomyces*[a]

Characteristic	*P. bekefii*	*P. brasiliensis*	*P. guttaeformis*	*P. limnophilus*[b]	*P. maris*	*P. stranskae*
Cell diameter (μm)	1.4–1.7	0.7–1.8	0.6–1.3	1.1–1.5	0.4–1.5	1.3–1.7
Cell shape	Spherical	Spherical to ovoid	Bulbiform	Spherical to ovoid	Spherical to ovoid	Bulbiform
Characteristic cellular appendage(s)	Tubular stalk, tapering multifibrillar spires	Stalk of loosely twisted fibrils	No stalk, multifibrillar spike at globose end	Braided or twisted, multifibrillar stalk (fascicle)	Braided, multifibrillar stalk (fascicle)	No stalk, multifibrillar bristles at globose end
Appendage length (μm)	<0.2–several	nr	5–10	nr	Variable, up to 5	1–2
Appendage width (μm)	0.25–0.35	nr	nr	nr	0.05–0.10	nr
Appendage fibril diameter (nm)	12–13	nr	nr	nr	5	nr
Distribution of crateriform structures	Uniform	Uniform	Globose end of cell	Uniform	Uniform	Globose end of cell
Distribution of pili/fimbrae	Uniform	Uniform (adult cell)	Globose end of cell	Uniform	Uniform	Uncertain
Morphotype *sensu* Schmidt and Starr	Ia	IIIa	Va	IIIa	IIIb	Vb
Rosette formation observed	+	nr	+	nr	nr	+
Motility observed	–	nr	–	+	+	–
Flagellum observed	nr	+	–	+	+	–
Acid-fast	nr	nr	nr	–	nr	nr
Endospore formation	nr	nr	nr	–	nr	nr
Source/habitat	Eutrophic, aquatic, slightly alkaline	Salt pit	Eutrophic, aquatic	Freshwater lake	Marine	Eutrophic, aquatic
Associated with algae/cyanobacteria	+	nr	+	nr	nr	+
Mn and or Fe oxide deposition	+	nr	nr	nr	nr	+
Cultivated in axenic culture	–	+	–	+	+	–
Colony surface	nr	Rough, dry	nr	Smooth, glistening	Smooth, glistening	nr
Colony color	nr	Yellow to ochre	nr	Red	White/cream/colorless	nr
ASW[c] tolerance (%)	nr	20–300	nr	<52	25–150	nr
ASW optimum (%)	nr	40–180	nr	nr	nr	nr
Salinity growth range (%, w/v)	nr	0.7–10.0	nr	<0.1	1.5–4.0	nr
NaCl growth range (mM)	nr	100–170	nr	nr	100 to >300	nr
pH growth range	nr	nr	nr	6.2–8.0	nr	nr
pH growth optimum	nr	nr	nr	6.2–7.0	nr	nr
Temperature growth range (°C)	nr	nr to 38	nr	17–39	6–38	nr
Temperature growth optimum (°C)	nr	27–35	nr	30–32	30–33	nr
Resistance to β-lactam antibiotics	+	nr	+	nr	nr	+
DNA G+C content (mol%)	nr	55.5–57.7	nr	53.2–54.4	50.5–52.0	nr
Genome size (× 10^9 Da)	nr	2.81	nr	2.67	3.62	nr
D-Cellobiose	nr	+	nr	+	+	nr
D(–)Fructose	nr	–	nr	–	–	nr
D-Fucose	nr	–	nr	nr	–	nr
Furanose	nr	nr	nr	nr	+	nr
D(+)Galactose	nr	+	nr	An[e]	+	nr
D(+)Glucose	nr	+	nr	An	+	nr
Lactose	nr	nr	nr	nr	–	nr
D(–)Lyxose	nr	–	nr	nr	nr	nr
Maltose	nr	+	nr	An	+	nr
D(+)Mannose	nr	+	nr	nr	+	nr

α-D-Melezitose	nr	–	nr	nr	–	nr
Melibiose	nr	+	nr	nr	+	nr
Raffinose	nr	–	nr	nr	nr	nr
Rhamnose	nr	+	nr	nr	+	nr
D-Ribose	nr	+	nr	–	–	nr
Sucrose	nr	nr	nr	An	nr	nr
L(–)Sorbose	nr	–	nr	nr	–	nr
Sucrose	nr	nr	nr	nr	–	nr
Trehalose	nr	+	nr	nr	+	nr
D(+)Xylose	nr	–	nr	nr	+	nr
Anaerobic gas Hugh–Leifson	nr	nr	nr	–	nr	nr
Adonitol	nr	nr	nr	nr	–	nr
Arabitol	nr	nr	nr	nr	–	nr
Dulcitol	nr	nr	nr	nr	–	nr
Erythritol	nr	nr	nr	nr	–	nr
Ethanol	nr	–	nr	+	–	nr
Glycerol	nr	–	nr	+	nr	nr
D(–)Mannitol	nr	–	nr	–	–	nr
Methanol	nr	–	nr	+	–	nr
Naphthol	nr	nr	nr	nr	–	nr
D(–)Sorbitol	nr	–	nr	nr	–	nr
Acetate	nr	–	nr	–	–	nr
Benzoate	nr	nr	nr	nr	–	nr
Caproate	nr	nr	nr	–	–	nr
Citrate	nr	nr	nr	–	–	nr
Formate	nr	nr	nr	+	–	nr
Fumarate	nr	–	nr	–	–	nr
Gluconate	nr	–	nr	nr	nr	nr
Glucuronate	nr	+	nr	–	+	nr
Glutarate	nr	nr	nr	–	nr	nr
Lactate	nr	–	nr	–	+	nr
Malate	nr	–	nr	–	–	nr
Maleic acid	nr	nr	nr	nr	–	nr
Malonate	nr	nr	nr	nr	–	nr
α-Oxoglutarate	nr	nr	nr	–	–	nr
Phthalate	nr	nr	nr	–	–	nr
Propionate	nr	nr	nr	–	–	nr
Pyruvate	nr	–	nr	–	–	nr
Succinate	nr	–	nr	–	–	nr
Tartrate	nr	nr	nr	–	–	nr
Uric acid	nr	nr	nr	nr	–	nr
Valerate	nr	nr	nr	nr	–	nr
DL-Alanine	nr	nr	nr	–	nr	nr
L-Arginine	nr	nr	nr	–	–	nr
L-Arginine-HCl	nr	nr	nr	nr	–	nr
L-Asparagine	nr	nr	nr	nr	–	nr
L-Aspartate	nr	nr	nr	–	nr	nr
DL-Aspartate	nr	nr	nr	nr	–	nr
DL-Citrulline	nr	nr	nr	nr	–	nr

(continued)

TABLE 169. (continued)

Characteristic	*P. bekefii*	*P. brasiliensis*	*P. guttaeformis*	*P. limnophilus*[b]	*P. maris*	*P. stranskae*
L-Cystine	nr	nr	nr	nr	–	nr
L-Cysteine-HCl	nr	nr	nr	nr	–	nr
L-Glutamic acid	nr	–	nr	–	–	nr
L-Glutamine	nr	–	nr	nr	–	nr
L-Glycine	nr	nr	nr	–	–	nr
L-Histidine	nr	nr	nr	–	nr	nr
L-Histidine-HCl	nr	nr	nr	nr	–	nr
L-Hydroxyproline	nr	nr	nr	nr	–	nr
L-Leucine	nr	nr	nr	–	–	nr
L-Lysine-HCl	nr	nr	nr	nr	–	nr
DL-Methionine	nr	nr	nr	nr	–	nr
DL-Norleucine	nr	nr	nr	nr	–	nr
DL-Ornithine-HCl	nr	nr	nr	nr	–	nr
DL-Phenylalanine	nr	nr	nr	–	nr	nr
L-Phenylalanine	nr	nr	nr	nr	–	nr
L-Proline	nr	nr	nr	–	–	nr
L-Serine	nr	–	nr	–	–	nr
L-Threonine	nr	nr	nr	nr	–	nr
L-Tryptophan	nr	nr	nr	nr	–	nr
DL-Tyrosine	nr	nr	nr	nr	–	nr
DL-Valine	nr	nr	nr	nr	–	nr
Creatine	nr	nr	nr	nr	–	nr
Creatinine	nr	–	nr	nr	–	nr
N-Acetylglucosamine	nr	+	nr	+	+	nr
DL-Aminobutyric acid	nr	nr	nr	nr	–	nr
Amygdalin	nr	–	nr	nr	–	nr
Dextrin	nr	nr	nr	–	nr	nr
Inulin	nr	nr	nr	–	nr	nr
Salicin	nr	nr	nr	–	–	nr
Pectin	nr	nr	nr	nr	+	nr
Esculin	nr	nr	nr	nr	+	nr
Formamide	nr	nr	nr	–	nr	nr
Methane	nr	nr	nr	+	nr	nr
Methylamine hydrochloride, as a carbon source	nr	nr	nr	+	nr	nr
Urea, as a carbon source	nr	nr	nr	+	nr	nr
Gelatin hydrolysis	nr	+	nr	+	nr	nr
Starch hydrolysis	nr	+	nr	–	–	nr
Nitrate reduction	nr	+	nr	+ (dissimilatory), – (assimilatory)	nr	nr
Anaerobic gas formation with nitrate	nr	nr	nr	–	nr	nr
Nitrification	nr	nr	nr	–	nr	nr
Ammonia, as a nitrogen source	nr	+	nr	nr	nr	nr
Nitrate, as a nitrogen source	nr	+	nr	nr	nr	nr
N-Acetylglucosamine, as a nitrogen source	nr	+	nr	nr	nr	nr
Urea, as a nitrogen source	nr	–	nr	–	nr	nr

$(NH_4)_2SO_4$, as a nitrogen source	nr	nr	nr	+	nr	nr
$NaNO_2$, as a nitrogen source	nr	nr	nr	–	nr	nr
$NaNO_3$, as a nitrogen source	nr	nr	nr	–	nr	nr
Methylamine hydrochloride, as a nitrogen source	nr	nr	nr	–	nr	nr
Growth in or change of litmus milk	nr	nr	nr	–	nr	nr
Extracellular DNase	nr	nr	nr	–	nr	nr
Catalase	nr	+	nr	nr	nr	nr
Cytochrome oxidase	nr	+	nr	nr	nr	nr
H_2S formation	nr	+	nr	+	nr	nr
Tolerance to 30% (v/v) CO	nr	nr	nr	+	nr	nr
Tolerance to 50% (v/v) CO	nr	nr	nr	–	nr	nr
Decarboxylation of lysine or arginine	nr	nr	nr	–	nr	nr
Deamination of phenylalanine or lysine	nr	nr	nr	–	nr	nr
Oligocarbophilic growth	nr	nr	nr	–	nr	nr
Urease	nr	nr	nr	–	nr	nr
Formation of acetoin	nr	nr	nr	–	nr	nr
Formation of indole	nr	nr	nr	–	–	nr
Requirement for vitamins	nr	–	nr	–	–	nr
Presence of fatty acid $C_{20:1}$	nr	+	nr	–	–	nr

[a]Symbols: +, >85% positive; –, 0–15% positive; nr, not reported.

[b]Some carbon sources listed as positive tested positive at concentrations below 0.1%.

[c]Artificial seawater (Lyman and Fleming, 1940).

[d]Data from Kölbel-Boelke et al. (1985).

[e]An, Anaerobic acid formation.

TABLE 170. Diagnostic characteristics of the species of the genus *Planctomyces*

Characteristic	*P. bekefii*	*P. brasiliensis*	*P. guttaeformis*	*P. limnophilus*	*P. maris*	*P. stranskae*
Cell diameter (μm)	1.4–1.7	0.7–1.8	0.6–1.3	1.1–1.5	0.4–1.5	1.3–1.7
Cell shape	Spherical	Spherical to ovoid	Bulbiform	Spherical to ovoid	Spherical to ovoid	Bulbiform
Characteristic cellular appendage(s)	Tubular stalk	Stalk of loosely twisted fibrils	No stalk, multifibrillar spike at globose end	Braided or twisted, multifibrillar stalk (fascicle)	Braided, multifibrillar stalk (fascicle)	No stalk, multifibrillar bristles at globose end
Distribution of crateriform structures	Uniform	Uniform	Globose end of cell	Uniform	Uniform	Globose end of cell
Distribution of pili/fimbrae	Uniform	Uniform (adult cell)	Globose end of cell	Uniform	Uniform	Uncertain
Morphotype *sensu* Schmidt and Starr	Ia	IIIa	Va	IIIa	IIIb	Vb
Source/habitat	Eutrophic, aquatic, slightly alkaline	Salt pit	Eutrophic, aquatic	Freshwater lake	Marine	Eutrophic, aquatic

species level. *Planctomyces limnophilus* and related strains differ from other planctomycetes tested in containing large amounts of putrescine, rather than spermidine and *sym*-homospermidine, as the dominant polyamine component.

Fine structure. *Planctomyces maris* exhibits the cellular compartmentalization described as characteristic for members of the genus *Pirellula,* namely the presence of a single membrane which divides the cell into the pirellulosome (containing the nucleoid and most of the RNA) and the paryphoplasm or polar cap region (Lindsay et al., 1997).

Life cycle. All species available in pure culture have a dimorphic life cycle, characterized by a budding mode of reproduction in which the cell envelope of the daughter bud is formed *de novo* (Staley, 1973; Tekniepe et al., 1981). The resulting daughter cells are motile, by means of a polar to subpolar flagellum, and initially lack stalks. Stalks eventually form during maturation of the daughter cell into its sessile adult form.

Nutrition and growth conditions. Carbohydrates are reported to be the main carbon source for cultivated members of the genus (Tables 169, 170), but *Planctomyces limnophilus* can also use gelatin, while *Planctomyces brasiliensis* utilizes both gelatin and starch (Hirsch and Müller, 1985; Schlesner, 1989). The use of selective carbon and nitrogen sources, e.g., *N*-acetylglucosamine, proved very successful for the isolation and maintenance of a large group of novel isolates (Schlesner, 1994).

Some species, e.g., *Planctomyces maris,* are obligate aerobes, while *Planctomyces limnophilus* and *Planctomyces brasiliensis* are facultative anaerobes, in the sense that they can ferment carbohydrates. As a group, planctomycete species are relatively slow-growing; the shortest recorded generation time is 13 h for *Planctomyces maris* (Bauld and Staley, 1976).

Metabolism and metabolic pathways. There have been virtually no studies of *Planctomyces* metabolism beyond the physiological information garnered for the purpose of species description.

Genetics. There is a limited amount of available information on genome organization in *Planctomyces limnophilus* (Ward-Rainey, 1996) which suggests a single circular molecule of approximately 5.2 Megabases. The same study confirmed the unlinked *rrn* (rRNA) operon organization reported by Menke and coworkers (Menke et al., 1991) and revealed duplicate copies of several housekeeping genes. Other genetic studies of *Planctomyces* species are few and far between: Leary and coworkers (Leary et al., 1998) cloned the *rpoN* gene, encoding alternative sigma factor δ^{54}, from *Planctomyces limnophilus* and demonstrated complementation in a *Salmonella typhimurium* mutant. The presence of this alternative sigma factor, involved in diverse metabolic functions such as nitrogen fixation, hydrogen metabolism, and degradation of aromatic compounds (Kustu et al., 1989), could catalyze future studies of genetic control mechanisms in planctomycetes. The introduction of foreign genes into *Planctomyces maris* from *Pseudomonas putida* has been accomplished (Dahlberg et al., 1998) but establishment of genetic systems for other planctomycetes lies in the future.

Bacteriophages and plasmids. There are only two reports of the presence of bacteriophages in *Planctomyces* species (Gliesche, 1980; Majewski, 1985) which describe the tailed bacteriophage PI-89 from the type strain of *Planctomyces limnophilus* as a member of the family Styloviridae morphogroup B1. The presence of plasmids in the type strain of *Planctomyces limnophilus* is suggested by the observation of a discrete band of high mobility when undigested chromosomal DNA from these strains is subjected to electrophoresis (Ward-Rainey, 1996).

Antibiotic sensitivity. As might be expected from the composition of the cell wall, there is a corresponding resistance of cultivated *Planctomyces* species (and species not yet available in pure culture for which enrichments have been obtained) to antibiotics targeting the synthesis of peptidoglycan, including both β-lactams and non-β-lactams such as D-cycloserine (König et al., 1984; Schmidt and Starr, 1982). β-Lactamase activity was tested and found to be absent in *Planctomyces maris* (as well as *Pirellula marina*) (Claus et al., 2000).

Ecology. *Planctomyces* species, like planctomycetes in general, are now known to be remarkably ubiquitous in their distribution, being found in freshwater and marine habitats, as well as non-aquatic environments such as soil and sewage sludge. Reported geographic locations include European, Asian, North American, South American, and Australian sites. Much of our knowledge of this wide distribution has arisen from molecular microbial ecology studies utilizing 16S rRNA as a marker gene, but also from a handful of studies (Fuerst et al., 1991; Schlesner, 1994) in which novel isolation methodologies were applied. *Planctomyces* species have been observed in environments in all trophic states, with some reports of higher numbers occurring in association with eutrophic and polluted waters; other reports (Staley et al., 1980) indicated that the ratio of planctomycetes to total heterotrophs is independent of trophic state. In any case, algal or cyanobacterial blooms (often a response to eutrophication) are sometimes followed by increases in numbers of the rosette-forming planctomycetes (*Planctomyces bekefii, Planctomyces guttaeformis, Planctomyces stranskae*) (Granberg, 1969), perhaps due to the hydrogen sulfide and increased iron and manganese concentrations resulting from phytoplankton decomposition (Kristiansen, 1971). Many rosette-forming species have been reported to accumulate iron and/or manganese on the stalk surface (Schmidt and Starr, 1981, 1982); the mechanism by which these deposits are acquired (passive accumulation vs. active oxidation) is not known. Associations with invertebrates have also been reported; some isolates obtained from the giant tiger prawn *Penaeus monodon* are related to members of the genus *Planctomyces* (Fuerst et al., 1997), with the closest phylogenetic relationship to *Planctomyces brasiliensis.*

The ecological role of *Planctomyces* species is poorly understood, despite their ubiquity, in large part due to the paucity of isolates available in pure culture and the limited scope of physiological research into these strains. Most of the available physiological data were acquired for the purpose of taxonomic description. All available cultured members of the genus are chemoheterotrophs, with carbohydrates serving as the major carbon source.

Taxonomic comments

The genus *Planctomyces* Gimesi 1924 was originally established to accommodate the peculiar aquatic "fungus" observed by Gimesi (1924) in a pond (Lake Lágymányos) in Budapest, Hungary. Gimesi's organism, named by him *Planctomyces bekefii*, has been reported many times from all over the world (Fott and Komarek, 1960; Heynig, 1961; Hortobágyi, 1965, 1968; Kristiansen, 1971; Olah and Hajdu, 1973; Parra, 1972; Schmidt and Starr, 1978; Teiling, 1942; Tell, 1975; Wawrik, 1952; Zavarzin, 1961).

The notion that this organism was a fungus persisted during the next half century; in fact, several additional species were assigned to this "fungal" genus *Planctomyces* (Hortobágyi, 1965; Skuja, 1964; Wawrik, 1952). This situation was not altered until Hirsch (1972) reported that a freshwater bacterium, described under the name "*Blastocaulis sphaerica*" (Henrici and Johnson, 1935), was identical with *Planctomyces bekefii.* Examination of material from the type localities of both organisms (Schmidt and Starr, 1979b) has provided conclusive evidence that the two organisms are indeed indistinguishable and that both are bacterial rather than fungal. Thus, a bacterial genus ("*Blastocaulis*" Henrici and Johnson 1935) has the dubious distinction of being an exact later synonym of a group which had, for some 50 years, been considered to be a "fungal" genus (*Planctomyces* Gimesi 1924).

Neither the type species (*Planctomyces bekefii*) nor several other species have ever been cultivated axenically, making it difficult to establish the limits of the genus *Planctomyces* solely on the basis of the cultivated members. Because of this taxonomic uncertainty, this rather heterogeneous assemblage has been treated in the past as morphotypes of the *Blastocaulis–Planctomyces* group (Schmidt, 1978; Schmidt et al., 1981). It has been pointed out (Starr and Schmidt, 1989) that some organisms comprising the genus *Planctomyces* most likely do not fit within the boundaries of a single well-conceived bacterial genus. However, the paucity and fragmentary nature of existing knowledge about these bacteria, including unavailability of axenic cultures of the type species of the genus *Planctomyces* (as well as of other key organisms), presently precludes precise delineation of alternative genera. Hence, as in the previous edition of the *Manual* (Starr and Schmidt, 1989), all validly published members of this group are formally classified here as species of the genus *Planctomyces*, within a framework of the vernacular morphotype designations. Where later synonyms have been published, they are referenced here, although good definitive comparative studies are often lacking. Due to the fact that the available phenotypic information differs greatly for cultivated and non-cultivated species, we have presented comprehensive information on all species in Table 169, diagnostic information pertinent to all species in Table 170, and diagnostic information relevant only to cultivated species in Table 171.

Enrichment and isolation procedures

Of the five morphotypes originally described by Schmidt and Starr (1978, 1979a, b, 1981), only representatives of morphotypes III and IV have been isolated in axenic culture. Although they have not yet been isolated axenically, morphotypes I, II, and V have been substantially enriched in the laboratory.

Enrichment and axenic isolation of morphotypes III and IV. Morphotypes III and IV can be enriched (Schmidt, 1978; Schmidt and Starr, 1978, 1981) by adding a small amount of peptone (0.001–0.005%) to freshwater samples placed in beakers covered with clear plastic wrap and held for several days or weeks at room temperature (24–28°C). These budding and appendaged bacteria occur in the surface film, often closely associated with green algae and cyanobacteria. Exposure of the enrichment beakers to ambient light is usually accompanied by development of algae and cyanobacteria. Enrichment of morphotypes III and IV is improved by the presence of these phototrophs and, consequently, by incubation of the samples in the light rather than in the dark. Microscopic examination of the samples is done on a weekly basis by brief contact of a small coverslip with the surface film in the beaker; these preparations are then observed by phase-contrast microscopy, looking for ellipsoidal, ovoid, or spherical budding bacteria possessing a very thin, nonprosthecate major appendage. The appendage may be difficult to visualize by light microscopy, and it may be desirable to touch electron microscope grids to promising surface films and, after negative staining, examine these in the transmission electron microscope. Enrichment cultures containing such organisms in their surface films are suitable candidates for streaking on solid media as described below.

The water agar-coverslip technique (Hirsch et al., 1977) is another useful method for enrichment and isolation of morphotypes III and IV. Relatively nutrient-poor water samples are solidified with 1.8–2.0% agar without any additions ("water agar"); a layer 0.5–0.75 cm deep is placed in the bottom of a Petri dish 20 cm in diameter. In addition, coverslips are coated with the same water agar. Several such coated coverslips, as well as several uncoated ones, are inserted perpendicularly into the solidified layer of agar in the Petri dish. The water sample providing the inoculum is then added in a volume that almost fills

TABLE 171. Diagnostic characteristics of the cultivated species of the genus *Planctomyces*[a]

Characteristic	*P. brasiliensis*	*P. limnophilus*[b]	*P. maris*
Colony surface	Rough, dry	Smooth, glistening	Smooth, glistening
Colony color	Yellow to ochre	Red	White/cream/colorless
Salinity growth range (%, w/v)	0.7–10.0	<1.0	1.5–4.0
Temperature growth optimum (°C)	27–35	30–32	30–33
DNA G+C content (mol%)	55.5–57.7	53.2–54.4	50.5–52.0
Genome size ($\times 10^9$ Da)[c]	2.81	2.67	3.62
D-Ribose	+	–	–
Ethanol	–	+	–
Methanol	–	+	–
Glucuronate	+	–	+
Lactate	–	–	+
Starch hydrolysis	+	–	–
Presence of fatty acid $C_{20:1}$	+	–	–

[a]Symbols: +, >85% positive; –, 0–15% positive.

[b]Some carbon sources listed as positive tested positive at concentrations below 0.1%.

[c]Data from Kölbel-Boelke et al. (1985).

the Petri dish, leaving only the top 2–3 mm of the coverslips not submerged. The dishes are incubated at room temperatures under ambient light to encourage development of phototrophs which, in turn, seem to enhance development of these budding bacteria. The coverslips that were not coated with water agar are monitored weekly or even more often, by examining them with a phase-contrast microscope for the occurrence of attached budding bacteria. When such bacteria are present, the agar-coated coverslips provide the inoculum for preparation of streak or dilution-spread plates. The locations at which the budding bacteria were observed microscopically to occur on the uncoated coverslips is used as a guide in selecting the regions of the agar-coated coverslips to be plated.

Isolation media. The media used for isolation of morphotypes III and IV are typically low in nutrients. Often the natural water sample itself, with 1.5–1.8% agar added, is suitable. Staley's (Staley, 1981a) MMB agar (0.15% peptone, 0.015% yeast extract, 0.1% glucose, 0.025% ammonium sulfate, 1.0% vitamin solution, 1–2% Hutner's vitamin-free mineral base, 1.5% agar, and distilled water) also is useful. A number of strains have been isolated by using Schmidt's (Schmidt, 1978) medium I (0.02% peptone, 0.01% yeast extract, 0.1% filter-sterilized glucose, 1.0% Hutner's vitamin-free mineral base, 1.0–1.5% agar, and distilled water) and Staley's (Staley, 1981b) PYG medium (0.025% peptone, 0.025% yeast extract, 0.025% glucose, 2.0% Hutner's vitamin-free mineral base, 1.0% vitamin solution, 1.5% agar, and distilled water or artificial sea water). Media with higher concentrations of organic nutrients may also be employed. Sometimes, isolations are facilitated by additions of 5 mM $MgSO_4$, 0.005 M Trizma buffer (pH 7.5–7.9; Sigma), and/or NaCl or artificial sea water (the latter is particularly desirable for isolations from brackish or marine sources). Use of solidifying agents other than agar may improve the chances of securing axenic cultures.

Direct streaking and spreading of diluted material are the most effective of the various routines used for inoculating the primary isolation plates. Morphotypes III and IV grow rather slowly in culture. Plates must be incubated for 2–5 weeks at room temperature; hence, precautions must be taken to avoid desiccation of the medium, such as sealing the Petri dishes with Parafilm or using plastic bags or boxes. Colonies are very small, rather inconspicuous, usually quite adherent or cohesive, and white or pigmented (pink or light brown). Patch plates, prepared with the same medium used for the primary isolation, may then be used to produce enough material for microscopic examination and restreaking.

Enrichment of morphotype Ia. The enrichment of morphotype Ia, *Planctomyces bekefii*, the type species of the genus *Planctomyces*, has been facilitated by taking advantage of the considerable resistance of all members of this group to antibiotics which, in other bacteria, affect peptidoglycan integrity. Addition of a β-lactam antibiotic such as ampicillin (about 250 μg/ml) to pond water samples containing detectable but low amounts ($1–5 \times 10^4$/ml) of morphotype Ia rosettes results, within 3 d, in a 5–6-fold increase in the number of rosettes as determined by direct microscopic counts. This increase may have resulted from (a) release of nutrients or growth factors from other bacteria lysed by action of the antibiotic or (b) inhibition by the antibiotic of organisms antagonistic to the development of *Planctomyces bekefii* rosettes. The effect of β-lactam antibiotics appears to be dependent upon the types and numbers of other microbes present in the water samples. Many workers, including phycologists and limnologists, have reported that occurrence of planktonic morphotype Ia often coincides with occurrence of various algae and cyanobacteria. The dominant algae and cyanobacteria present during and immediately preceding *Planctomyces bekefii* blooms in natural aquatic environments may be useful as "helpers" to enhance growth of *Planctomyces bekefii* in co-cultivation trials in the laboratory. Axenic cultivation of *Planctomyces bekefii* (morphotype Ia), the type species of the genus *Planctomyces*, and the closely related "*Planctomyces crassus*" (morphotype Ib) remains a challenge. Some success has been achieved in enriching co-cultures of *Planctomyces bekefii* using dilute peptone media. The accompanying Gram-stain-negative bacterium could be purified but purification of *Planctomyces bekefii* using spent medium of the accompanying bacterium was unsuccessful (J. T. Staley, personal communication).

Enrichment of morphotype II. Morphotype II, an ovoid budding organism with a long multifibrillar stalk, occurs individually and in clusters ("bouquets") in pond water enrichments to which 0.001–0.005% peptone has been added. The pond waters in which this relatively uncommon organism occurs are typically eutrophic. About 1 or 2 weeks after peptone is added to pond waters of this sort, these morphotype II organisms are found in the surface films. They have not been isolated axenically, nor have they been cultivated on any solid medium.

Attempted enrichment of morphotype V. *Planctomyces guttaeformis* (morphotype Va) and *Planctomyces stranskae* (morphotype Vb) are planktonic bacteria bulbiform in shape and characteristically occurring in multicellular rosettes in pond or lake waters where they accompany algal and/or cyanobacterial blooms. They seem not to be inhibited by addition of ampicillin (250 μg/ml); neither do they seem to be enriched by the presence of this antibiotic, as is *Planctomyces bekefii*. Additions of various organic nutrients (e.g., peptones, amino acid mixtures, individual amino acids, vitamin mixtures, and/or individual vitamins) to pond water samples containing easily detectable numbers of morphotype V have not provided any clue to specific enrichment conditions for these bacteria. Their natural aquatic habitats usually have a distinctly eutrophic character, and the number of rosettes of these bacteria often increases significantly (4–5-fold) in such water samples held in the collection bottle for a week or so in the laboratory without any additions (e.g., antibiotics, preservatives, or possible nutrients). No representative of morphotype V has been observed to grow on any solid medium.

Maintenance procedures

The existing axenic cultures, all belonging to morphotypes III and IV, have successfully been preserved by lyophilization of cell suspensions in skim milk. Addition of 10% (v/v) glycerol to liquid cultures of morphotypes III and IV, followed by freezing and storage at −70°C, has satisfactorily preserved these strains for several months.

Acknowledgements

We gratefully acknowledge the use of text written by M.P. Starr and J.M. Schmidt for the chapter on the genus *Planctomyces* (Starr and Schmidt, 1989) contained in the previous edition of the *Manual*. The sections *Taxonomic comments*, *Enrichment and isolation procedures*, and *Maintenance procedures*, as well as the genus summary and all but one of the species descriptions, are used here virtually unaltered from the previous edition.

Further reading

Starr, M.P., R.M. Sayre and J.M. Schmidt. 1983. Assignment of ATCC 27377 to *Planctomyces staleyi* sp. nov. and conservation of *Pasteuria ramosa* Metchnikoff 1888 on the basis of type descriptive material. Request for an Opinion. Int. J. Syst. Bacteriol. *33*: 666–671.

Ward, N., J.T. Staley, J.A. Fuerst, S.J. Giovannoni, H. Schlesner and E. Stackebrandt. 2004. The order *Planctomycetales*, including the genera *Planctomyces*, *Pirellula*, *Gemmata*, and *Isosphaera*, and the *Candidatus* genera Brocadia, Kuenenia, and Scalindua. *In* The Prokaryotes: an Evolving Electronic Resource for the Microbiological Community, 3rd edn, release 3.18 (edited by Dworkin, Falkow, Rosenberg, Schleifer and Stackebrandt). Springer, New York.

Differentiation of species of the genus *Planctomyces*

The species of *Planctomyces* can be differentiated from one another by the features listed in Table 169.

List of species of the genus *Planctomyces*

1. **Planctomyces bekefii** Gimesi 1924, 4[AL] (*Blastocaulis sphaerica* Henrici and Johnson 1935, 84 Hirsch 1972); morphotype la of the *Blastocaulis–Planctomyces* group (Schmidt and Starr 1980b, 1981)

be.ke'fi.i. N.L. gen. masc. n. *bekefii* of Békefi, named for Remigius Békefi (1858–1924), cultural historian, university professor, and abbot of the Hungarian Cistercian Order.

Gram-stain-negative, budding bacteria. Cellular shape is spherical. Mature cells are 1.4–1.7 μm in diameter. Each mature cell has a tubular, nonprosthecate stalk (Figure 163) 0.25–0.35 μm wide and varying in length from very short (<0.2 μm) to several micrometers. The tubular stalk is comprised of a regular array (Figure 163) of longitudinally aligned microtubules (12–13 nm in diameter) enmeshed in an osmophilic matrix (Figure 163). The distal end of each stalk is characteristically attached to several other distal stalk tips to form multicellular homologous rosettes (Figure 163), however, no distinctive or obvious holdfast structure can be discerned at the junction. Crateriform surface structures (Figure 163) are uniformly distributed over the entire cell. Numerous cephalotrichous, tapering, multifibrillar spires (Figure 163) emanate from the hemisphere of the cell opposite the site of attachment of the stalk. Pili are present over the entire cell.

Nonmotile. Divide by budding from a site precisely opposite to the pole from which the stalk originates. The nonstalked and nonmotile daughter cells, after they are released from the mother cell, may associate with the center of the originating or another rosette and subsequently form a stalk, thus resulting in an increased number of stalked cells in a rosette. Stalks may be encrusted with manganese and/or iron oxides (Schmidt et al., 1982). Resistant to β-lactam antibiotics. Rosettes are planktonic, occurring in eutrophic freshwater lakes, ponds, and reservoirs, typically with waters of a slightly alkaline pH. Occurrence usually is concomitant with or subsequent to blooms of various algae and cyanobacteria, especially during the late summer in temperate climates. Mesophilic. Worldwide in distribution. Complete life cycle is unknown, however, nonstalked budding bacteria with morphological and ultrastructural features (shape, size, occurrence and distribution of crateriform surface structures and spires, distinctive cell envelopes) identical to those of *Planctomyces bekefii* occur in clusters in the same aquatic samples. Although the relationship of these nonstalked forms to *Planctomyces bekefii* has not yet been demonstrated conclusively, the circumstantial evidence is strong that they may be immature (i.e., nonstalked) forms of *Planctomyces bekefii*. *Planctomyces bekefii* has not been cultivated axenically; hence, archival (Gimesi, 1924) and modern (Schmidt and Starr, 1980a, b) descriptive type material from the type locality, a pond in the Lágymányos district of Budapest, Hungary, must suffice in place of a type culture.

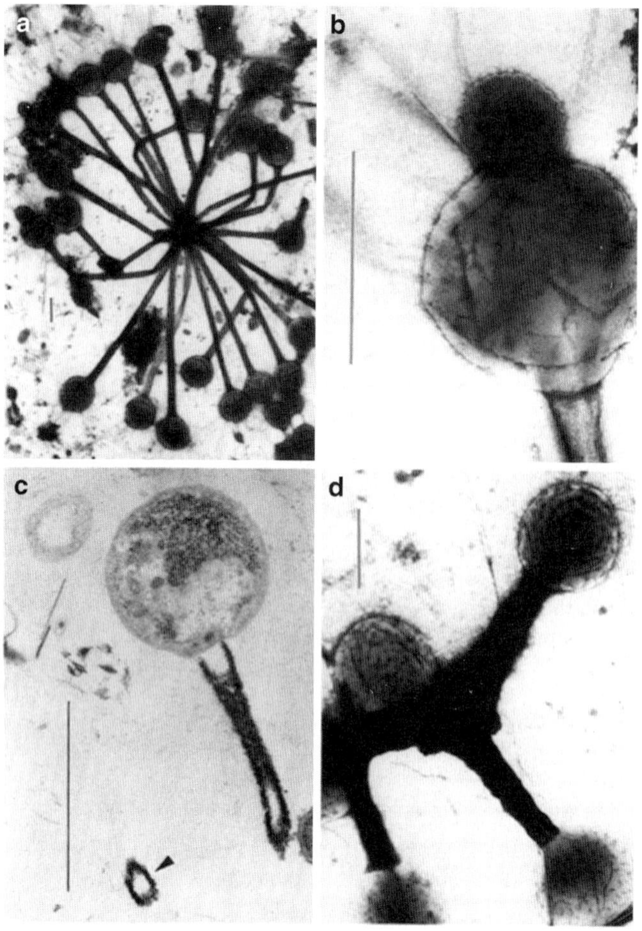

FIGURE 163. Morphotype I of the *Blastocaulis–Planctomyces* group. (a) Morphotype Ia (*Planctomyces bekefii*) from the eutrophic man-made pond at The Lakes, Tempe, Arizona, USA; rosette with several budding cells. 1.0% uranyl acetate; ×3960. (b) Budding cell of morphotype Ia (*Planctomyces bekefii*) from Lake Mendota, Madison, Wisconsin, USA. Note bud, spires, crateriform surface structures, and stalk substructure showing parallel fibrils. 1.0% uranyl acetate; ×40,135. (c) Thin section of cell of morphotype Ia (*Planctomyces bekefii*) from the pond at The Lakes, Tempe, Arizona, showing a tangential section of the tubular stalk as well as a separate transverse section (arrow) of another stalk. Fixed by the acroleinglutaraldehyde procedure of Burdett and Murray (1974) (see also Schmidt and Starr, 1980a); ×36,000. (d) Rosette of morphotype Ib "*Planctomyces crassus*" in a sample from the railroad-bridge pond, Budapest, Hungary, showing heavy metallic oxide encrustation of its stalks (Schmidt and Starr, 1981). 1.0% uranyl acetate; ×15,000. Bars = 1.0 µm.

No species belonging to morphotype II has been validly published, nor has the axenic cultivation of any strain been reported, although substantial laboratory enrichment has been achieved (Schmidt and Starr, 1979a). The unmistakable arrangement into a bouquet of several stalked cells of a typical representative of morphotype II, as sketched by Henrici and Johnson (1935), is reproduced here in Figure 164. This organism was referred to as "*Blastocaulis* sp." by these authors, with the remark that "specific names are withheld until further studies have been completed" (Henrici and Johnson, 1935). Unfortunately, no further publication about this organism by these authors seems to have appeared.

DNA G+C content (mol%): not determined.

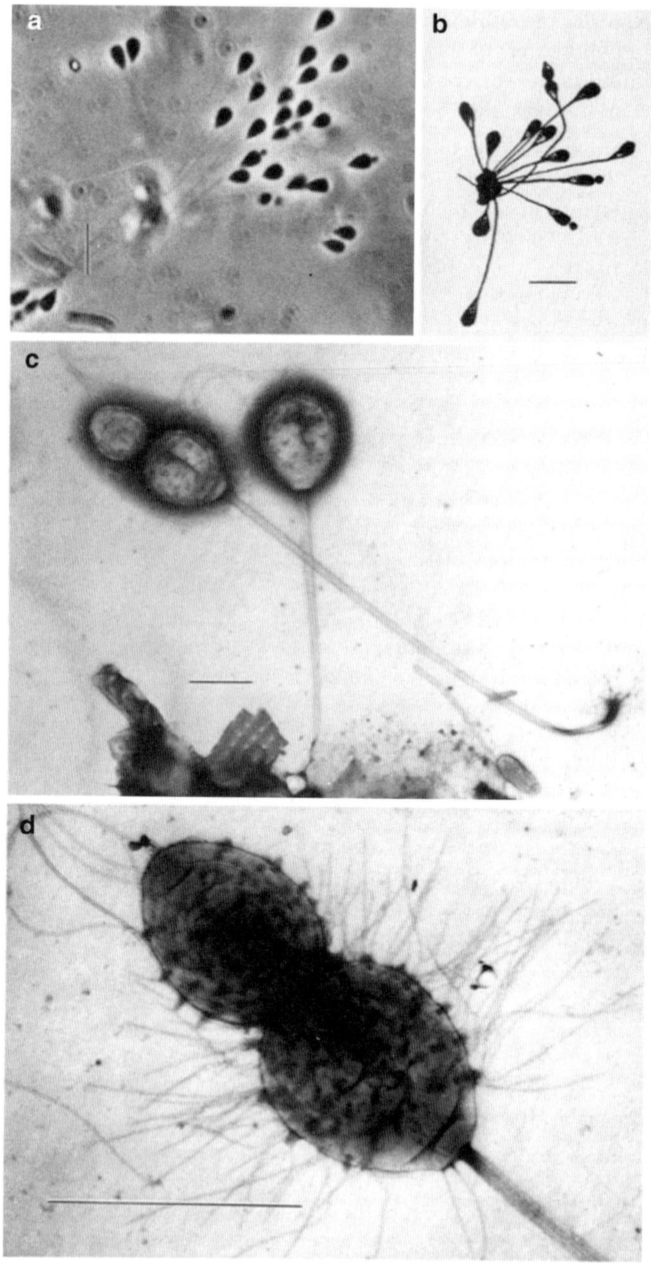

FIGURE 164. Morphotype II of the *Blastocaulis–Planctomyces* group. (a) Several morphotype II cells, some with buds, arranged in a typical bouquet, from a dilute peptone enrichment of water from a lake near Hesteskodam, Hillerød, Denmark. These organisms can be found in the surface film or pellicle of such enrichments. ×1900. (b) Drawing of "*Blastocaulis* sp." (Henrici and Johnson, 1935). ×1900. (c) Two morphotype II organisms, one with a bud and each with a long major appendage (a stalk) terminating in a holdfast (one holdfast is attached to detritus). Several cephalotrichous flagella and numerous crateriform surface structures are found on the budding (nonstalked) hemisphere. Dilute peptone enrichment of water sample from a highly eutrophic pond at the sewage works in Strandfontein, Republic of South Africa. 1.0% uranyl acetate; ×11,200. (d) A budding cell of morphotype II with crateriform surface structures everted as cornicula, as they often do when stained with 1.0% uranyl acetate (Schmidt and Starr, 1979b); pili, several flagella, and the major appendage (a stalk) are evident. Dilute peptone enrichment of water sample from an ornamental fountain, University of the Witwatersrand, Johannesburg, Republic of South Africa. 1.0% uranyl acetate; ×51,300. Bars = 5.0 µm for (a) and (b) and 1.0 µm for (c) and (d).

Type strain: description, not isolated in axenic culture.

Sequence accession no. (16S rRNA gene): not determined.

2. **Planctomyces brasiliensis** Schlesner 1990, 105[VP] (Effective publication: Schlesner 1989, 161.)

bra.si.li.en′sis. N.L. masc. adj. *brasiliensis* pertaining to the country of Brazil.

Cells are spherical to ovoid, with a diameter of 0.7–1.8 µm. Colonies have a dry and rough surface and a yellow to ochre pigmentation. Growth is optimal between 27 and 35°C, the maximum temperature is less than 38°C. Nitrate is reduced to nitrite under anaerobic conditions. Carbon sources utilized are: D-cellobiose, D(+)-galactose, D(+)-glucose, maltose, D(+)-mannose, melibiose, rhamnose, ribose, trehalose, *N*-acetylglucosamine, and glucuronate. Carbon sources not utilized: D(−)-fructose, D-fucose, D(−)-lyxose, α-D-melezitose, raffinose, L(−)-sorbose, D(+)-xylose, methanol, ethanol, glycerol, D(−)-mannitol, D(−)-sorbitol, acetate, fumarate, lactate, malate, pyruvate, succinate, L-glutamic acid, L-glutamine, L-serine, amygdalin, gluconate, and creatinine. Ammonia, nitrate, and *N*-acetylglucosamine are utilized as a nitrogen source; urea is not utilized. Gelatin and starch are hydrolyzed; DNA and Tween 80 are not. Catalase, cytochrome oxidase, and H_2S are produced. The molecular mass of genomic DNA (strain 1448[T]) is 2.81×10^9 Da.

DNA G+C content (mol%): 55.1–57.7 (T_m).

Type strain: ATCC 49424, DSM 5305, IFAM 1448, JCM 21570, NBRC 103401.

Sequence accession no. (16S rRNA gene): AJ231190, X81949, X85247.

3. **Planctomyces guttaeformis** (*ex* Hortobágyi 1965) Starr and Schmidt 1984, 473[VP] [*Planctomyces guttaeformis* Hortobágyi 1965, 111 emend. Hajdú (Hortobágyi and Hajdú 1984)]; morphotype Va of the *Blastocaulis–Planctomyces* group (Schmidt and Starr 1979a; Starr and Schmidt 1984)

gut.ta.e.for′mis. L. n. *gutta, -ae* a drop; L. gen. n. *guttae* of a drop; L. suff. *-formis* in the form or shape of; N.L. masc. adj. *guttaeformis* drop-shaped.

Gram-stain-negative, budding bacteria. Mature cells are relatively large (2.9–3.0 µm in length, 1.15–1.3 µm for the diameter of the globose end, and 0.6–0.7 µm for the diameter of the cylindrical end). Bulbiform (bulb-shaped; in this case, a cylinder blending into a spherical terminal dilatation; shaped like a common screw-in incandescent lamp). A prominent multifibrillar appendage, a spike (Figure 165), extends from the globose end of the cell. Division is by budding, with a mirror-image bud (Figure 165) formed at the globose end of the cell; the bud is often slightly off center with respect to the longitudinal axis of the mother cell, whereas the spike is usually exactly polar. Numerous crateriform surface structures and fine pili occur on the globose end of the cell (Figure 165). Fine structure is that of a typical bacterium except for the rather thin cell envelope, the spike and its basal attachment site, and the crateriform surface structures. Cells are resistant to β-lactam antibiotics. Occur in rosettes (Figure 165) consisting of as many as a dozen cells attached via holdfasts located at the narrower ends of the several cells. Neither motility nor flagella have been observed. Common in eutrophic aquatic habitats, usually accompanying phototrophs (algae and cyanobacteria). Species has not been cultivated axenically, though blooms occur in laboratory enrichments. Descriptive type material in Starr and Schmidt (1984). The type locality is a eutrophic man-made pond in a residential district (The Lakes) in Tempe, Arizona, USA. Frequently confused with *Planctomyces stranskae*, from which it can readily be distinguished on the basis of the characteristic cellular sizes and shapes, the presence of a multifibrillar spike in *Planctomyces guttaeformis* and its absence in *Planctomyces stranskae*, and the presence of numerous multifibrillar bristles in *Planctomyces stranskae* and their absence in *Planctomyces guttaeformis*.

DNA G+C content (mol%): not determined.

Type strain: description, not isolated in axenic culture.

Sequence accession no. (16S rRNA gene): not determined.

4. **Planctomyces limnophilus** Hirsch and Müller 1986, 355[VP] (Effective publication: Hirsch and Müller 1985, 278.) (This taxon is based on two red-pigmented strains belonging to morphotype IIIA of the *Blastocaulis–Planctomyces* group; Schmidt and Starr, 1981. Synonymy is unclear, since direct comparisons of *Planctomyces limnophilus* with nonpigmented morphotype IIIa strains or with the earlier published *Gemmata obscuriglobus* Franzmann and Skerman (1984), a representative of morphotype IIIc of this group, seem not to have been made.)

lim.no′phi.lus. Gr. n. *limnos* lake; Gr. masc. adj. *philos* loving; N.L. masc. adj. *limnophilus* lake-loving.

Gram-stain-negative, budding bacteria. Mature cells are spherical to ellipsoidal, 1.1–1.5 µm in diameter. Numerous crateriform structures and pili occur on the cell surface. Mature cells have a braided or twisted, multifibrillar, nonprosthecate stalk (a fascicle) extending from a pole exactly opposite the budding pole; the noncellular end of the fascicle serves as a "holdfast" device. Multiplication is by budding on the pole of the cell distal to the fascicle. Daughter cells are monotrichously and polarly or subpolarly flagellated; the flagellum is attached at or near the pole from which the stalk originates. Nonacidfast; colonies red; do not form endospores.

Temperature range: 17–39°C; optimum: 30–32°C. NaCl tolerance: <1% (w/v). pH tolerance: 6.2–8.0; pH optimum: 6.2–7.0.

The following substances are utilized as carbon sources (0.1% (w/v)): D-glucose, D-galactose, maltose, cellobiose, and *N*-acetylglucosamine. The following substances are not utilized as carbon sources (0.1% (w/v) except where different concentrations are noted): glucuronic acid, D-fructose, D-ribose, mannitol, starch, dextrin, inulin, salicin, pyruvate, citrate, α-oxoglutarate, succinate, fumarate, malate, formamide, methylamine hydrochloride (0.136%), formate (0.136%), urea (0.09%), methane (0.5%), methanol (0.4%), ethanol (0.4%), lactate, acetate, propionate, tartrate, glutarate, caproate, phthalate, glycerol (0.186%), L-arginine, L-aspartate, DL-alanine, L-glutamate, L-glycine, L-histidine, L-leucine, DL-phenylalanine, L-proline and L-serine. No aerobic acid formation from D-glucose, sucrose, D-fructose, maltose, D-galactose, or mannitol. Anaerobic acid formation from D-glucose, sucrose, maltose, or galactose; no anaerobic acid formation from D-fructose or mannitol. No anaerobic gas formation in the Hugh–Leifson tests.

The following substances are utilized as nitrogen sources at 0.1% (w/v) except where a different concentration is noted:

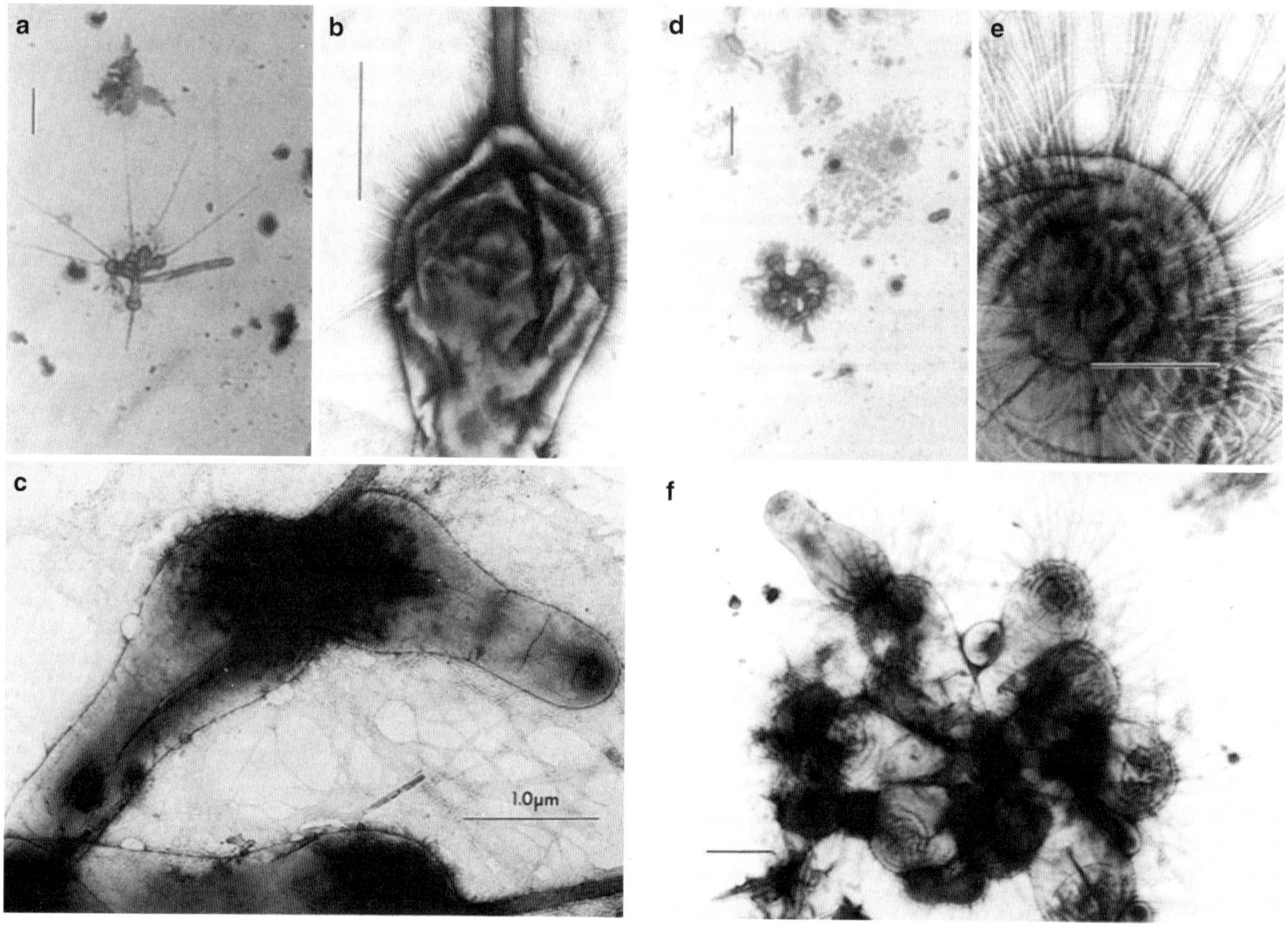

FIGURE 165. Morphotype V of the *Blastocaulis–Planctomyces* group. (a) Stained (Kodaka et al., 1982) preparation of a rosette of morphotype Va (*Planctomyces guttaeformis*), homologous as to cell type, showing the typical bulbiform cellular shape and the prominent spikes. ×1800. (b) Bulbous end of morphotype Va (*Planctomyces guttaeformis*) cell, showing pili, crateriform surface structures, and basal portion of the spike. 1.0% uranyl acetate; ×25,200. (c) Budding cell of morphotype Va (*Planctomyces guttaeformis*) from the pond at The Lakes, Tempe, Arizona, showing pili, crateriform surface structures, and base of the spike. The bud is a mirror image of the bulbiform mother cell. 1.0% uranyl acetate; ×27,333. (d) Stained (Kodaka et al., 1982) preparation showing a rosette, homologous as to cell type, of morphotype Vb (*Planctomyces stranskae*). Note that this organism lacks the spikes typical of morphotype Va but does have a prominent halo of bristles with entrapped bacteria. ×1800. (e) Globose hemisphere of a morphotype Vb (*Planctomyces stranskae*) cell from the pond at The Lakes, Tempe, Arizona, showing the typical multifibrillar bristles. Neither a spike nor the "socket" from which the spike emanates in morphotype Va is present. 1.0% uranyl acetate; ×45,000. (f) Rosette of cells of morphotype Vb (*Planctomyces stranskae*) cells from the pond at The Lakes, Tempe, Arizona, one with a mirror-image bud. Note the numerous multifibrillar bristles on the globose hemispheres of each cell. 1.0% uranyl acetate; ×11,560. Bars = 5.0 µm for (a) and (d); 1.0 µm for (b), (c) and (f); and 0.5 µm for (e).

$(NH_4)_2SO_4$ (0.05%), Bacto peptone, Bacto yeast extract, or Casamino acids (vitamin-free). The following substances are not utilized as nitrogen sources: $NaNO_2$ (0.2–0.69%), $NaNO_3$ (0.2–0.85%), methylamine hydrochloride (0.675%), and urea (0.46%).

Vitamins not required. The following reactions are positive: nitrate reduction (dissimilatory), gelatin liquefaction, H_2S formation, and tolerance to 30 vol% carbon monoxide. The following reactions are negative: decarboxylation of lysine and arginine, deamination of phenylalanine and lysine, oligocarbophilic growth, urease, nitrification, nitrate reduction (assimilatory), anaerobic gas formation with nitrate, formation of acetoin and indole, growth in or changes of litmus milk, tolerance to 50 vol% carbon monoxide, and extracellular DNase.

Genome size (strain 1008): $2.67 \pm 0.05 \times 10^9$ daltons.

Source: surface water of the freshwater lake Plußee, Holstein, Germany.

DNA G+C content (mol%): 53.24 ± 0.59.

Type strain: Mü 290, ATCC 43296, DSM 3776, IFAM 1008.

Sequence accession no. (16S rRNA gene): X62911.

5. **Planctomyces maris** (*ex* Bauld and Staley 1976) Bauld and Staley 1980, 657[VP] (morphotype IIIb of the *Blastocaulis–Planctomyces* group (Schmidt and Starr, 1981). Other names are sometimes encountered for organisms belonging to morphotype III; most of these names have no taxonomic standing. Two red-pigmented strains belonging to morphotype IIIA provided the basis for the description of the species *Planctomyces limnophilus* by Hirsch and Müller (1985).

A single morphotype IIIC strain is the basis of a description by Franzmann and Skerman (1984) of the genus *Gemmata* and its sole species *Gemmata obscuriglobus*. However, rigorous comparisons of these taxa with *Planctomyces maris* and/or with other presently unnamed representatives of morphotype III (Schmidt and Starr, 1978, 1981) remain to be made.

ma'ris. L. gen. n. *maris* of the sea.

Gram-stain-negative, budding bacteria. Mature cells are spherical or ellipsoidal, 1.0–1.5 µm in major dimension. Numerous crateriform surface structures and pili (5–6 nm in diameter) are distributed randomly over the entire cell (Figure 166). In the budding process, the newly released daughter cells (0.4 µm in minimum diameter) are usually smaller than mature cells. Mature cells generally have a braided, multifibrillar, nonprosthecate stalk extending from a pole exactly opposite to the budding pole (Figure 166). Individual fibrils (5 nm in diameter) of the multifibrillar stalk, emanating from separate pores at the basal structure when anchored to the cell, are joined into a bundle (a fascicle) to form the stalk, which may be up to 5 µm long. The many finer fibrils, which extend from the distal tip of the stalk, may serve as a "holdfast" device (Figure166); some stalked cells, joined by the distal tips of their stalks, are depicted here in Figure 166. The daughter cells are motile by a single, subpolar, sheathed flagellum which is attached to the cell at or near the pole from which the stalk originates.

Of marine origin, with an absolute requirement for 1.5–4.0% (w/v) NaCl. Heterotrophic, obligately aerobic, mesophilic (optimum temperature: 30–33°C). Colonies on marine agar (Difco) are white or slightly cream-colored, 2–4 mm in diameter, slightly mucoid, and glistening. Carbon sources utilized: *N*-acetylglucosamine, cellobiose, esculin, furanose, galactose, glucose, glucuronic acid, lactic acid, maltose, mannose, melibiose, pectin, rhamnose, trehalose, and xylose. Vitamins are not required, but colonies grow more rapidly in the presence of peptone and yeast extract than in the glucose-salt medium. Growth is relatively slow, with a minimum doubling time of 13 h at 30°C.

The DNA G+C content of ATCC 29201 is 50.5 mol% (Bauld and Staley, 1976) or 52.0 mol% (M. Mandel, personal communication, 1979) as determined by buoyant density.

Strains essentially identical to *Planctomyces maris* in morphological and ultrastructural features, and probably also related in other ways, have been isolated from freshwater samples originating in many parts of the world (Franzmann and Skerman, 1984; Gebers et al., 1985; Hirsch and Müller, 1985; Hirsch et al., 1977; Schmidt, 1978; Schmidt and Starr, 1978, 1981). With the exceptions of *Gemmata obscuriglobus* Franzmann and Skerman (1984), a taxon based upon a single morphotype IIIc strain not yet compared rigorously with other members of morphotype III, and *Planctomyces limnophilus* Hirsch and Müller (1985), a taxon based solely upon two red pigmented morphotype IIIa strains, none of these strains has as yet been validly named. They do not require 1.5% NaCl.

Genome sizes (renaturation kinetics) of the few strains examined ranged from 2.67–3.65 × 10^9 Da (Kölbel-Boelke et al., 1985); to the extent examined, no peptidoglycan can be detected by chemical analysis (O. Kandler, personal communication, 1979–1980; König et al. (1984) or electron microscopy (König et al., 1984; Schmidt and Starr, 1978, 1981), thus correlating with the demonstrated resistance to β-lactam antibiotics (König et al., 1984; Schmidt and Starr, 1980b), existence of other characteristics atypical of ordinary eubacteria (Schmidt and Starr, 1981), and their postulated evolutionary origin (Stackebrandt et al., 1984). Some of these freshwater strains form pink or red colonies, though their colonies are more typically cream colored or white; some develop a brown (melanin-like) pigmentation after several days of growth. The single strain tested was not pathogenic to mice (Famurewa et al., 1983).

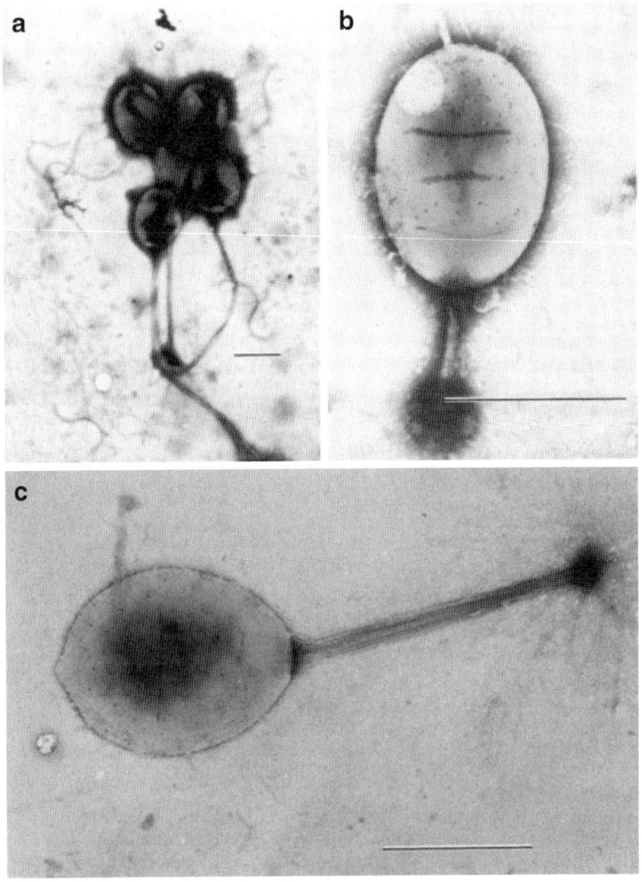

FIGURE 166. Morphotype III of the *Blastocaulis–Planctomyces* group. (a) Several cells of *Planctomyces maris*, ATCC 29201, morphotype IIIb of the *Blastocaulis–Planctomyces* group. Stained with 1.0% potassium phosphotungstate following fixation with 0.5% glutaraldehyde. A braided, multifibrillar fascicle (a stalk) extends from a pole of each cell, and the distal ends of the fascicles are attached to each other. ×8960. (b) Cell of a morphotype III organism from a freshwater enrichment (water sample from Old Main Fountain, Arizona State University, Tempe, Arizona, USA). The crateriform structures are scattered over most of the cell surface. Note the ellipsoidal cellular shape. The prominent, braided, multifibrillar fascicle with its electron-dense (i.e., heavily stained) holdfast is typical of morphotype III. 1.0% potassium phosphotungstate; ×35,100. (c) Cell of morphotype III from an enrichment (water sample from Willow Springs Lake, northern Arizona), showing a very early state of budding (the tiny protuberance at the pole opposite to the stalk) as well as the typical braided stalk with electron-dense holdfast. 0.5% potassium phosphotungstate; ×28,700. Bars = 1.0 µm for (a)–(c).

DNA G+C content (mol%): 50–52 (Bd).
Type strain: 534-30, ATCC 29201, DSM 8797.
Sequence accession no. (16S rRNA gene): AJ231184, X62910.

6. **Planctomyces stranskae** (*ex* Wawrik 1952) Starr and Schmidt 1984, 473[VP] (*Planctomyces stranskae* Wawrik 1952, 448; morphotype Vb of the *Blastocaulis–Planctomyces* group (Starr and Schmidt, 1984)

stran'skae. N.L. fem. gen. n. *stranskae* of Stransky; named originally by F. Wawrik in honor of her biology teacher, M.L. Stransky.

Gram-stain-negative, budding bacteria. Mature cells are relatively large (3.0–3.5 µm in length, 1.5–1.7 µm for the diameter of the globose end, about 1.3–1.4 µm for the initial diameter of the conical end, and 1.0–1.1 µm for the diameter of the terminus of the cell). Bulbiform (i.e., bulb shaped; in this case, a truncated cone blending into a spherical terminal dilatation; shaped like a common screw-in incandescent lamp). Numerous multifibrillar appendages, termed bristles (Figure 165), extend from the globose end of the cell; many smaller bacteria are associated with the bristles, especially in preparations made directly from natural samples. Division is by budding, with a mirror-image bud (Figure 165) formed at the globose end of the cell. Numerous crateriform surface structures (Figure 165) occur on the globose end of the cell; presence of pili is uncertain. Fine structure is that of a typical bacterium, except for a rather thin cell envelope, the crateriform surface structures, and the many multifibrillar bristles. Cells are resistant to β-lactam antibiotics. Occur in rosettes consisting of as many as a dozen cells attached via holdfasts at the narrower ends of the several cells. Neither motility nor flagella have been observed. Common in eutrophic freshwater habitats, usually accompanying phototrophs (algae and cyanobacteria). Metallic oxide encrustations of the cells have been observed in specimens from Austria and Arizona. Species has not been cultivated axenically, though blooms occur in laboratory enrichments. Descriptive type material in Starr and Schmidt (1984). The type locality is a eutrophic man-made pond in a residential district (The Lakes) in Tempe, Arizona, USA. Frequently confused with *Planctomyces guttaeformis*, from which it can readily be distinguished on the basis of the characteristic cellular sizes and shapes, the absence of a multifibrillar spike in *Planctomyces stranskae* and its presence in *Planctomyces guttaeformis*, and the presence of numerous multifibrillar bristles in *Planctomyces stranskae* and their absence in *Planctomyces guttaeformis*.
DNA G+C content (mol%): not determined.
Type strain: description, not isolated in axenic culture.
Sequence accession no. (16S rRNA gene): not determined.

Species *incertae sedis*

1. **"Planctomyces condensatus"** Skuja 1964, 16

Skuja (1964) stated that this organism differs from *Planctomyces bekefii* Gimesi "durch die viel kürzeren und sich zentralwärts nicht merklich verdickten Zellstielchen, die kleineren Kolonien, durch das gewöhnlich solitäre Vorkommen von terminalen Kugelzellen sowie die verhältnissmässig mächtige und anscheinend immer auftretende zentrale Eisenhydroxykonkretion." ("in its much shorter cell stalks which are not markedly thicker towards the center of the rosette; the smaller colonies; the typically solitary occurrence of terminal spherical cells, as well as the relatively heavy and always apparent concretion of the central ferric hydroxide or as well as the relatively heavy and always apparent concretion of iron hydroxide at the center of the rosette.")

2. **"Planctomyces crassus"** Hortobágyi 1965, 111 (Morphotype Ib of the *Blastocaulis–Planctomyces* group; Schmidt and Starr, 1981)

In the German résumé of his paper, Hortobágyi (1965), using the mycological terminology then in vogue, stated that "*Planctomyces crassus*" was distinguished from all other species of the genus *Planctomyces* "durch die Ausbildung der Sporenträger und die aus den Armen hevorschiebenden Sporen" ["by the development of the conidiophores and the conidia which emerge from it"] -a view based on comparisons of the remarkably broad and tapering stalks (in the mycological terminology used by Hortobágyi, these are called "Sporenträger" or "conidiophores") of "*Planctomyces crassus*" with the rather thin and uniformly dimensioned stalks of *Planctomyces bekefii*, and of the spherical cells ("Sporen" or "conidia") of "*Planctomyces crassus*" with the bulbiform cells of *Planctomyces guttaeformis*. Our examination (Schmidt and Starr, 1981) of this relatively uncommon bacterial form (Figure 163) supports the distinctions made by Hortobágyi (1965) and adds others concerning its bacterial nature, the fine structure of the stalk, and deposition of metallic oxides.

3. **"Planctomyces crassus subsp. maximus"** Hortobágyi 1980

Its author stated: "The subspecies differs from *Planctomyces crassus* Hortobágyi in its robust thallus similar to that of *Planctomyces condensatus* Skuja, but its shape is different[,] as is the development of reproductive cells."

4. **"Planctomyces ferrimorula"** Wawrik 1956, 296

The entire description of this rare organism by Wawrik (1956) follows: "In einer zarten Gallerthülle lagern locker um eine zentrale Eisenoxyhydratausfällung von 1–8 µm diam., 1–100 und mehr farblose kugelige Zellen von 0.5–1.5 µm diam. Das Endstadium der Entwicklung ist ein morulaartiges Gebilde von 12–20 µm diam., in dem sich die Zellen nach allen Richtungen des Raumes dicht an die dunkelbraune Eisen-konkretion angelagert haben. Es wurden auch zusammengesetzte Kolonien beobachtet." ("1–100 or more colorless spherical cells of 0.5–1.5 µm diameter are stored in a delicate slime layer around a central iron hydroxide deposit/precipitate of 1–8 µm diameter. The final stage of development is a morula-like object of 12–20 µm diameter, in which the cells have attached closely in all directions around the dark-brown iron accretion. Compound colonies were also observed.")

5. **"Planctomyces gracilis"** Hortobágyi 1965, 112

   This unusual rosette-forming and filamentous organism (Figure 167) is not a member of the *Blastocaulis–Planctomyces* group (Starr and Schmidt, 1984).

6. **"Planctomyces hajdui"** Hortobágyi 1980

   Its author stated "The species is nearest to *Planctomyces crassus* Hortobágyi but differs from it in its broad thallus and the projecting thinner stalks."

7. **"Planctomyces kljasmensis"** (Razumov 1949) Hirsch 1972, 110

   An accessible description of this organism has been presented by Zavarzin (1961) under the names "*Blastocaulis kljasmensis*" Razumov (1949) and "*Gallionella planctonica*" Krasil'nikov (1949). From Zavarzin's description, this organism – except for its smaller cells (reported by Zavarzin as spherical and 0.3–0.5 µm in diameter, although he stated that "sometimes larger cells are encountered") – seems to be similar to iron-encrusted *Planctomyces bekefii*.

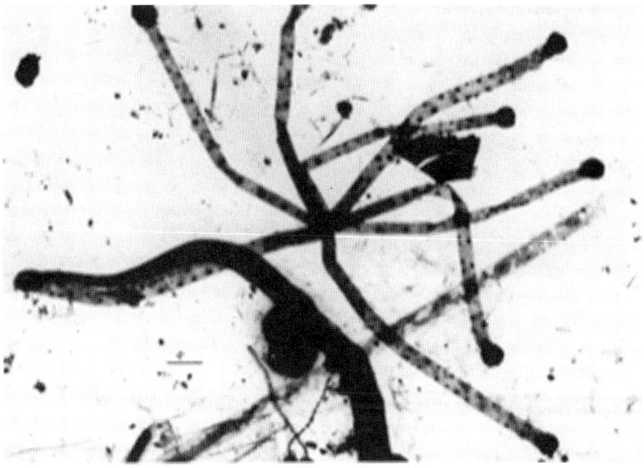

FIGURE 167. Rosette of "*Planctomyces gracilis*" Hortobágyi (1965). Water sample from the pond at The Lakes, Tempe, Arizona, USA. This unusual rosette-forming and filamentous bacterium, which lacks crateriform surface structures and multifibrillar appendages, does not belong to the *Blastocaulis–Planctomyces* group (Starr and Schmidt, 1984). 1.0% uranyl acetate; ×6710; bar = 1.0 µm.

Genus II. **Blastopirellula** Schlesner, Rensmann, Tindall, Gade, Rabus, Pfeiffer and Hirsch 2004, 1578[VP]

Heinz Schlesner

Blas.to.pi.rel'lu.la. Gr. masc. n. *blastos* bud, shoot; N.L. fem. n. *Pirellula* name of a bacterial genus; N.L. fem. n. *Blastopirellula* a budding *Pirellula*.

Cells are ovoid, ellipsoidal or pear-shaped, occurring singly or in rosettes by attachment at the smaller cell pole. Cells divide by budding. Buds are produced directly from the broader cell pole of the mother cell; **they may appear bean-shaped. Crateriform structures and fimbriae are found in the upper cell region. Cells have an intracytoplasmic membranous structure (pirellulosome).** Gram-stain-variable. Buds may have a single flagellum inserted subpolarly at the proximal pole. Adult cells are immobile. Strictly aerobic. **Colonies are grayish to brownish white**. Mesophilic. Requires sea water for growth. Nonsporeforming. PHB is not stored. Chemoheterotrophic. Carbon and energy sources are mainly carbohydrates. C1-compounds are not used. *N*-Acetylglucosamine serves as carbon and nitrogen source. Catalase- and cytochrome oxidase-positive. **The proteinaceous cell wall lacks peptidoglycan.** The major polyamine is *sym*-homospermidine. **The major menaquinone is MK-6.** The major phospholipid is phosphatidylglycerol. The major fatty acids present are $C_{15:0}$, $C_{16:0}$ iso, $C_{16:1}$ ω7, $C_{16:0}$, $C_{17:1}$ ω8, $C_{17:0}$, $C_{18:1}$ ω9, $C_{18:1}$ ω7, $C_{18:0}$, $C_{19:1}$ ω8, and $C_{20:1}$ ω9.

DNA G+C content (mol%): 54–57 (HPLC).

Type species: **Blastopirellula marina** (Schlesner 1986) Schlesner, Rensmann, Tindall, Gade, Rabus, Pfeiffer and Hirsch 2004, 1578 (*Pirella marina* Schlesner 1986, 180; *Pirellula marina* Schlesner and Hirsch 1987, 441).

Further descriptive information

Blastopirellula marina is the only described species of *Blastopirellula*. Morphologically, it is characterized by a polar organization of the cells. The small cell pole (holdfast pole) excretes a sticky substance of still unknown composition which enables the cell to attach to surfaces or to form rosettes. The reproductive pole produces the bean-shaped bud. Crateriform structures are distributed over the entire cell surface of the buds, however, they are only found around the reproductive pole of adult cells. In the same area, numerous fimbriae are found (Figures 168, 169).

Blastopirellula marina possesses a characteristic type of cell compartmentalization, in which a single membrane (the intracytoplasmic membrane) divides the cell into two separate regions, the pirellulosome and a polar cap region, later termed "the paryphoplasm" (Lindsay et al., 1997; Figure 170). The pirellulosome contains the nucleoid and most of the ribosomes. A structure analogous to the pirellulosome has also been reported in other genera of the *Planctomycetales* (Lindsay et al., 2001).

The cell wall lacks peptidoglycan and possesses a proteinaceous sacculus instead, as was found in all strains of *Planctomycetales* examined so far. This was first reported for several strains of *Planctomyces* and *Pirellula* (then *Pasteuria*; König et al., 1984).

The Gram reaction of untreated adult cells is positive; buds and swarmer cells react negatively. Adult cells also give the negative Gram reaction after washing. The reason might be a glycocalix as was shown for *Rhodopirellula* and which may retain the dyes (refer to Figure 185).

Blastopirellula is a member of the attached living microflora and, like several other representatives of such immobile bacteria, it produces motile daughter cells, thus exhibiting a biphasic life cycle. After ripening, the bud separates from the mother cell and moves away till it attaches to a substratum. Then the flagellum is released (see Figure 186, below). When there is a high cell density, swarmer cells may form rosettes by contact with their holdfast poles (Figure 171).

The main carbon and energy sources are carbohydrates. Interestingly, *N*-acetylglucosamine, the monomer of chitin, serves as both carbon and nitrogen source. Chitinase activity, however, was not detected.

Though *Blastopirellula* is quite similar to *Pirellula* and *Rhodopirellula* in morphology and physiology, the 16S rRNA sequence

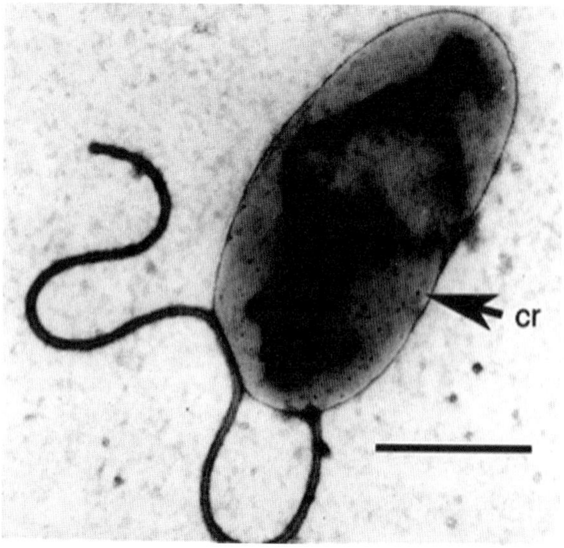

FIGURE 168. Bean-shaped swarmer cell of *Blastopirellula marina* strain SH106^T with single flagellum shows crateriform structures (cr) on the cell surface. Negatively stained with uranyl acetate. Bar = 0.5 µm.

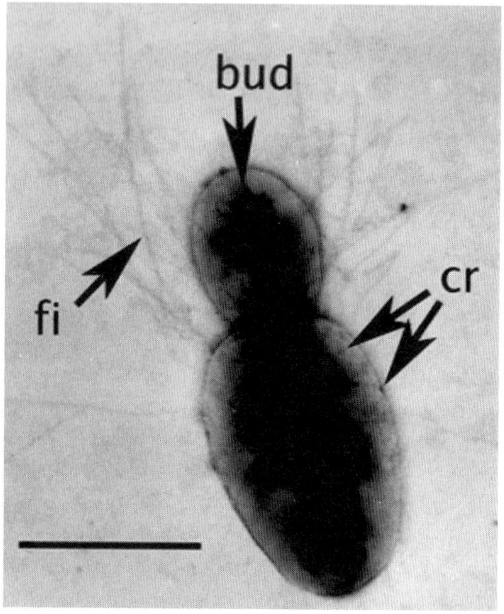

FIGURE 169. Mature cell of *Blastopirellula marina* strain SH106^T with bud. Crateriform structures (cr) and fimbriae (fi) are seen in the region of the reproductive pole. Negatively stained with phosphotungstic acid. Bar = 1 µm.

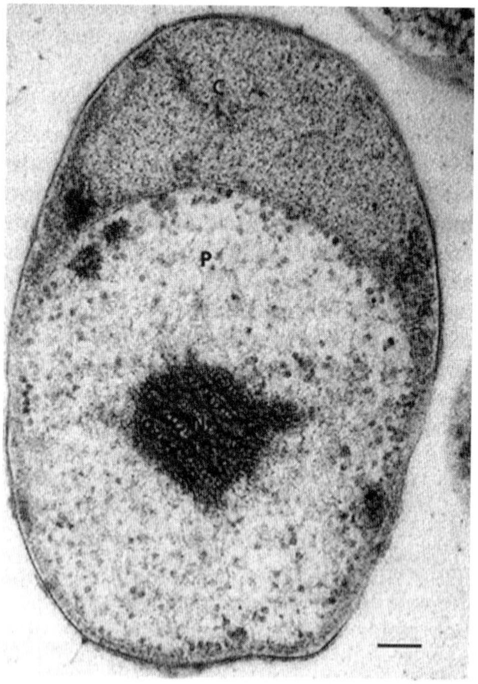

FIGURE 170. Thin section of cryosubstituted cell of *Blastopirellula marina* displaying cell compartmentalization into pirellulosome (P) and polar cap region (paryphoplasm, C) by a single intracytoplasmic membrane (ICM). Bar = 0.1 µm. (Reproduced with permission from Fuerst, 1995. Microbiology *143:* 739–748.)

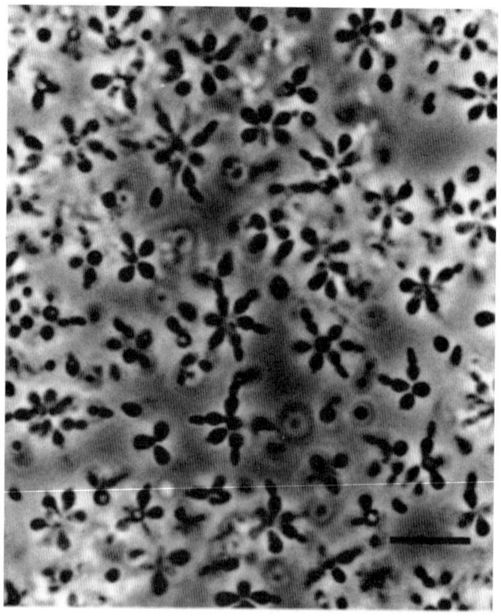

FIGURE 171. Rosette-formation of *Blastopirellula marina* strain SH116. Note the bean-shaped buds. Phase-contrast. Bar = 10 µm.

analysis showed less than 90% similarity between the type strains of these genera (refer to Table 174, below). The length of the 5S rRNA of *Rhodopirellula baltica* strain SH1^T is in the range of that for other planctomycetae, i.e. 109–111 bases rather than 118 bases as for the majority of *Bacteria* and *Archaea* (Bomar et al., 1988). The analysis of sequences encoding 23S rRNA also separates the genera *Pirellula* and *Blastopirellula* (Ward et al., 2000). The sequences encoding the 70-kDa heat-shock protein (Ward-Rainey et al., 1997) and the β-subunit of ATP synthase (Rönner et al., 1991) were analyzed and compared to sequences of other bacteria and supported the findings of 16S rRNA analyzes that the planctomycetae were a phylogentically diverse group within the *Bacteria* (Griepenburg et al., 1999; Ward et al., 1995). The comparison of elongation factor-Tu proteins from *Blastopirellula marina* and *Isosphaera pallida* with numerous other bacteria (Jenkins and Fuerst, 2001) did not support a close relationship between planctomycetes and chlamydiae (the only other group bacteria with a cell wall lacking peptidoglycan) that had been proposed earlier as sister taxa (Weisburg et al., 1986).

The presence of an an extrachromosomal element was detected by pulsed field gel electrophoresis (PFGE) of undigested chromosomal DNA (Ward-Rainey, 1996). Physical mapping of the genome of *Blastopirellula marina* using probes for housekeeping genes indicated that the genes encoding 16S rRNA and 23S rRNA are separated in the genome and that more than one chromosomal locus exists for several of the genetic markers (Ward-Rainey, 1996). Unlinked rrn organization had been previously reported by Liesack and Stackebrandt (1989).

Blastopirellula marina can be considered as a marine bacterium as is does not grow in freshwater media. Essential components of artificial sea water were Ca^{2+}, Na^+, and Cl^-. Seven strains were isolated from the brackish water of Kiel Fjord and Schlei Fjord (both part of the Baltic Sea) over a period of 15 years, suggesting that these bacteria belong to the autochthonous microflora of this habitat (Schlesner et al., 2004).

Enrichment and isolation procedures

Adult cells are normally attached to surfaces and so mainly swarmer cells are sampled from the free water column. For this reason, an enrichment of cells is necessary before starting isolation attempts. Three methods of enrichment have been successfully used (Schlesner, 1986): (a) 100 ml samples with 0.05% glucose in 250 ml Erlenmeyer flasks with cotton plugs were kept aerobically at room temperature (20–23°C) in daylight, (b) as in (a) but the substrate was 0.1% *N*-acetylglucosamine + 100 µg/ml ampicillin, (c) the Petri-dish method (Hirsch et al., 1977; Schlesner, 1994): The bottom of a 20 cm diameter glass Petri dish was covered with a 2 cm layer of sterilized water agar before adding the water sample. Sterilized cover glasses with or without water agar coat were placed vertically and partially into the agar layer and monitored microscopically from time to time. As soon as the desired bacteria had appeared on such sample cover glasses, other cover glasses were taken out for agar streak isolations.

Streaks can be performed on media M13, M13a or M30. The addition of ampicillin and cycloheximide (0.2 mg/l each) reduces growth of ampicillin-sensitive bacteria and of fungi.

Medium 13 (Schlesner, 1986) (g/l or ml/l) contains: peptone, 0.25; yeast extract, 0.25; glucose, 0.25; artificial sea water (ASW; Lyman and Fleming, 1940, 250); vitamin solution no. 6 (Staley, 1968), 10; Hutner's basal salts medium (HBM; Cohen-Bazire et al., 1957) 20; 0.1 m Tris/HCl pH 7.5, 50 ml; distilled water, 670. The recipes of the other media are given in the Genus *Rhodopirellula* chapter, below.

Maintenance procedures

Besides the media mentioned above, M13f is a suitable culture medium. Addition of ampicillin and cycloheximide is not necessary for pure cultures. The optimum incubation temperature is 30°C.

The strains grown on slants can be kept at 4–5°C for at least 3 months. They are easily revived from lyophilized cultures and can be stored at −70°C in a solution of 50% glycerol in the culture medium.

Differentiation of the genus *Blastopirellula* from other closely related taxa

Properties of the type strains of *Rhodopirellula baltica*, *Blastopirellula marina*, and *Pirellula staleyi* that allow differentiation of the three genera are summarized in Table 174.

Further reading

Fuerst, J.A. 1995. The planctomycetes: emerging models for microbial ecology, evolution and cell biology. Microbiology *141*: 1493–1506.

Neef, A., R. Amann, H. Schlesner and K.-H. Schleifer. 1998. Monitoring a widespread bacterial group: *in situ* detection of planctomycetes with 16S rRNA-targeted probes. Microbiology *144*: 3257–3266.

Sittig, M. and H. Schlesner. 1993. Chemotaxonomic investigation of various prosthecate and/or budding bacteria. Syst. Appl. Microbiol. *16*: 92–103.

Ward, N., J.T. Staley, J.A. Fuerst, S.J. Giovannoni, H. Schlesner and E. Stackebrandt. 2004. The order *Planctomycetales*, including the genera *Planctomyces*, *Pirellula*, *Gemmata*, and *Isosphaera*, and the *Candidatus* genera Brocadia, Kuenenia, and Scalindua. *In* The Prokaryotes: an Evolving Electronic Resource for the Microbiological Community, 3rd edn, release 3.18 (edited by Dworkin, Falkow, Rosenberg, Schleifer and Stackebrandt). Springer, New York.

List of species of the genus *Blastopirellula*

1. **Blastopirellula marina** (Schlesner 1986) Schlesner, Rensmann, Tindall, Gade, Rabus, Pfeiffer and Hirsch 2004, 1578^VP^ (*Pirella marina* Schlesner 1986, 180; *Pirellula marina* Schlesner and Hirsch 1987, 441)

ma.ri'na. L. fem. adj. *marina* of, or belonging to, the sea, marine.

Cells are 0.7–1.5 × 1.0–2.0 µm in size. A single flagellum is subpolarly inserted at the proximal pole. Colonies are round, smooth and grayish to brownish white. Growth is optimal between 27 and 33°C. No growth at 38°C. Ca^{2+}, Na^+, and Cl^- are required for growth; vitamins are not required. Strictly aerobic. Glucose is not fermented; nitrate cannot serve as an electron acceptor. Other characteristics are listed in Table 172.

The chemical composition is identical to that in the genus description.

Source: brackish water of Kiel Fjord (Baltic Sea).

DNA G+C content (mol%): 56 (HPLC).

Type strain: SH 106, ATCC 49069, DSM 3645, IFAM 1313.

Sequence accession no. (16S rRNA gene): X62912.

Genus III. **Gemmata** Franzmann and Skerman 1985, 375^VP^ (Effective publication: Franzmann and Skerman 1984, 266.)

John A. Fuerst, Kuo-Chang Lee and Margaret K. Butler

Gem.ma'ta. L. v. *gemmare* to put forth buds, to bud; N.L. fem. n. *Gemmata* (from L. fem. part. adj. *gemmata* put forth buds, budded) budded (bacteria), referring to the cell division mode of the bacterium.

Spherical to pear-shaped cells, 1.4–3.0 × 1.4–3.0 µm. Gram-stain-negative. **Cells reproduce by budding**, with the single daughter cell a mirror image of the mother cell, and there may be a narrow bud attachment site. Cells can bud repeatedly. **Cells possess a multitrichous bundle of flagella.** Aerobic, having a strictly aerobic type of metabolism with

TABLE 172. Characteristics of *Blastopirellula marina*[a]

Characteristic	
Carbon source utilization:	
N-Acetyglucosamine, cellobiose, chondroitin sulfate, dextrin, fructose, fucose, galactose, gluconate, glucuronate, glucose, glutamic acid, glycerol, lactose, lyxose, maltose, mannitol, mannose, melibiose, melizitose, pyruvate, raffinose, rhamnose, salicin, sucrose, trehalose, xylose	+
Acetate, adonitol, arabitol, citrate, dulcitol, erythritol, esculin, ethanol, formate, fumarate, glutamine, inulin, lactate, malate, methanol, 2-oxoglutarate, propionate, succinate	–
Nitrogen source utilization:	
N-Acetylglucosamine, ammonia, Casamino acids, gelatin, nitrate, peptone, yeast extract	+
Urea	–
Hydrolysis of:	
DNA, esculin, gelatin, glycerol, starch	+
Casein, cellulose, chitin	–
Activity of:	
Catalase, cytochrome oxidase, lipase	+
Urease	–
Production of:	
H_2S	+
Acetoin, indole	–

[a]Symbols: +, >85% positive; –, 0–15% positive.

oxygen as the terminal electron acceptor. **Colonies have pink pigmentation.**

Cell walls of the type species are proteinaceous and do not possess detectable peptidoglycan. **Crateriform structures on cell surface are uniformly distributed over the cell. In thin sections of chemically fixed or cryosubstituted cells and freeze-fracture replicas, nucleoids are surrounded by an envelope consisting of two closely apposed membranes** which are contained within a region of cell cytoplasm bounded in turn by an intracytoplasmic membrane. Via phase-contrast light microscopy, this nuclear region may appear as an inclusion of different phase refractility. Cells possess the shared planctomycete cell plan, i.e., cell is compartmentalized into a ribosome-free paryphoplasm and ribosome- and nucleoid-containing pirellulosome separated via a single intracytoplasmic membrane. **16S rRNA sequence has a characteristic insertion absent in other planctomycetes.**

DNA G+C content (mol%): 64.4 ± 1.0 (T_m).

Type species: **Gemmata obscuriglobus** Franzmann and Skerman 1985, 375[VP] (Effective publication: Franzmann and Skerman 1984, 266.).

Further descriptive information

Cells reproduce by budding from one pole to produce a smaller daughter cell of similar shape attached via a narrow region to the mother cell (Figure 172).

Cells possess crateriform structures distributed uniformly over cell surface (Figure 173), multitrichous flagella (emerging as a tuft from one side of the coccus in *Gemmata obscuriglobus*), and may also possess fimbriae (Figure 174). Crateriform structures are retained in cell envelopes isolated after treatment with 10% sodium dodecyl sufate, and the proteinaceous cell wall lacks peptidoglycan as indicated by the absence of muramic acid and diaminopimelic acid (Stackebrandt et al., 1986). Alignments of the 16S rRNA sequence of *Gemmata obscuriglobus* and related strains display a characteristic insertion of at least ten bases between *Escherichia coli* position 998 and 999 relative to sequences of other strains due to a unique secondary structure bulge loop (Liesack and Stackebrandt, 1992; Wang et al., 2002). At least two distinct *Gemmata* strains, *Gemmata obscuriglobus* and *Gemmata* Wa1-1 (a strain with 93% 16S rRNA similarity to *Gemmata obscuriglobus* and therefore possibly a distinct species) are capable of synthesizing sterols including lanosterol and parkeol, via squalene monooxygenase and oxidosqualene cyclase (Pearson et al., 2003). Draft genome sequences exist for *Gemmata obscuriglobus* and for *Gemmata strain* Wa1-1. Genes involved in C_1 transfer mediated by methanopterin and methanofuran are present in *Gemmata obscuriglobus* and in another member of the clade of organisms clustering with *Gemmata obscuriglobus* in phylogenetic trees, *Gemmata* Wa1-1 (Chistoserdova et al., 2004).

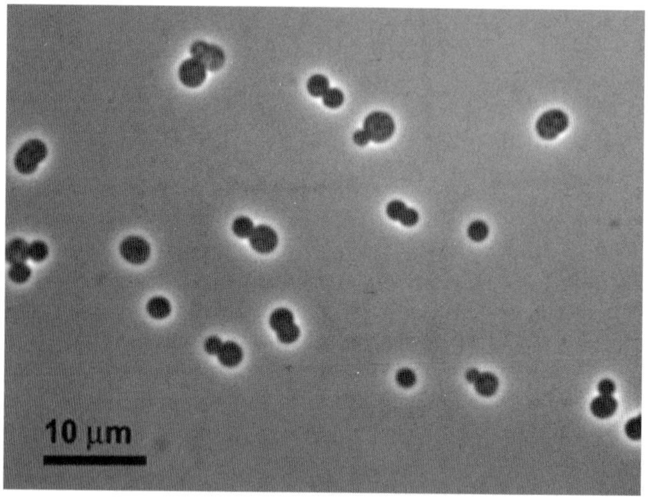

FIGURE 172. Phase-contrast micrograph of budding *Gemmata obscuriglobus* cells grown for 5 d in PYGV broth. Bar = 10 µm.

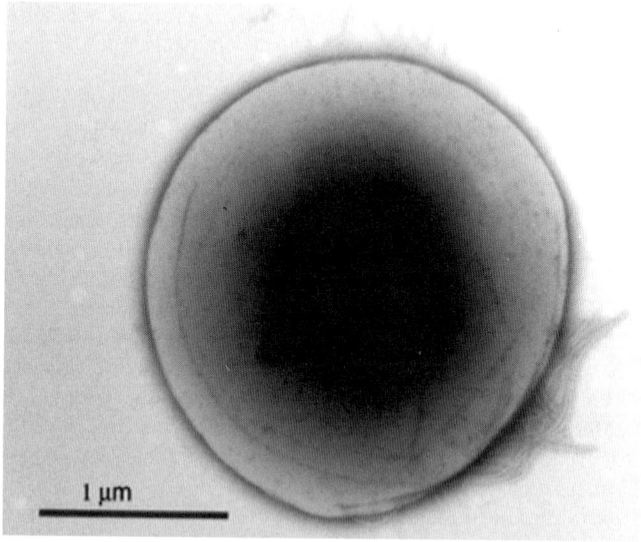

FIGURE 173. Transmission electron micrograph of cell of *Gemmata obscuriglobus* negatively stained with ammonium molybdate, showing crateriform structures with uniform distribution over the cell surface. Bar = 1 µm.

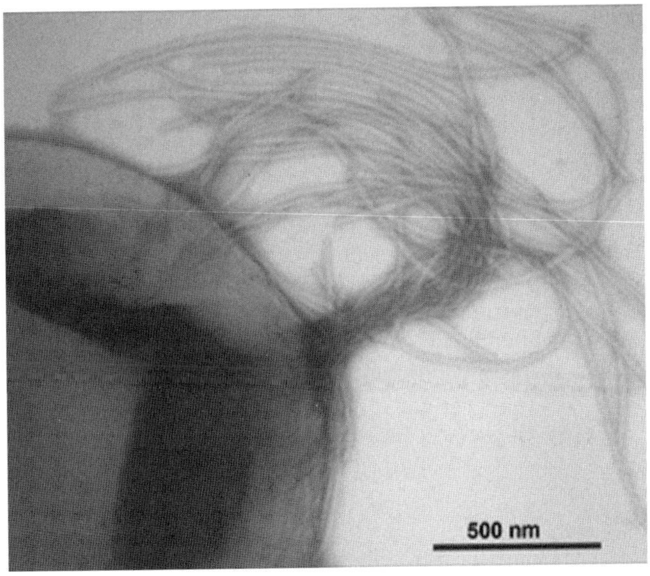

FIGURE 174. Transmission electron micrograph of cell of *Gemmata obscuriglobus* negatively stained with ammonium molybdate, showing a multitrichous tuft of flagella emerging from a restricted region. Bar = 500 nm.

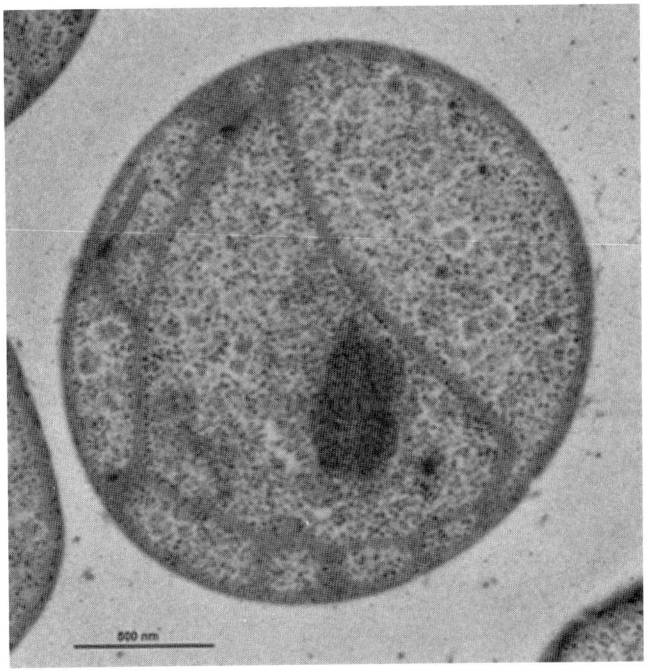

FIGURE 175. Transmission electron micrograph of thin sectioned cell of *Gemmata obscuriglobus* prepared via high-pressure freezing and cryosubstitution, showing the double-membrane envelope surrounding a consensed fibrillar nucleoid, forming the nuclear body characteristic of *Gemmata obscuriglobus* and close relatives. Bar = 500 nm.

Enrichment and isolation procedures

The type strain was isolated from a freshwater inoculum (derived from Maroon Dam, southeast Queensland, Australia) using micromanipulation methods of Skerman (1968) using lake water agar (Franzmann and Skerman, 1981). Other strains have been isolated using a soil inoculum (Wang et al., 2002). Antibiotics inhibiting peptidoglycan synthesis and protein synthesis in other bacteria have been used effectively for *Gemmata* strain isolation, e.g., M1 isolation medium of Schlesner (1994) supplemented with the antifungals amphotericin B (0.25–0.5 μg/ml)or cycloheximide (20–100 μg/ml), ampicillin at 200 μg/ml or penicillin G at 500 μg/ml and streptomycin at 1000 μg/ml (Wang et al., 2002). Resistance to peptidoglycan-synthesis-inhibiting antibiotics forms the basis for selection, but resistance of at least some strains of *Gemmata* to the protein synthesis-inhibitor streptomycin also aids selectivity of this medium. For experimental studies or broth enrichment cultures, growth in PYGV broth (Staley, 1968) can avoid clumping and allow determination of exponential phase.

Maintenance procedures

Staley's complex medium (pH 7.2) (Staley, 1973) was used for maintenance of *Gemmata obscuriglobus* after initial isolation using lake water agar. Half-strength soil substitution equivalent agar (½ SSE) (Lindsay et al., 1995, 2001) and M1 agar (Schlesner, 1994) have also been used successfully for maintenance.

Frequent subculture (every 2 weeks) may be needed to prevent loss of viability. Cultures can be successfully preserved via freeze-drying and via 20% glycerol suspensions at −70 to −80°C.

Procedures for testing special characters

For the study of cell ultrastructure and the presence of double membrane envelopes enclosing the nucleoids, cryosubstitution is recommended as described by Lindsay et al. (2001).

Differentiation of the genus *Gemmata* from other genera

Cells of *Gemmata obscuriglobus* and other *Gemmata*-like strains clustering with *Gemmata obscurigobus* tend to be coccoid rather than ovoid or teardrop-shaped, so are similar to cells of *Planctomyces* and *Isosphaera* but differ from *Pirellula*. Unlike *Isosphaera pallida* which forms filaments exhibiting intercalary budding, *Gemmata* strains are usually single cells or occur with one bud. With respect to internal cell structure, *Gemmata* strains so far examined all display a double-membrane-bounded nucleoid (Figure 175) forming a nuclear body compartment surrounded by an envelope consisting of two membranes closely apposed to each other; this nuclear body is found within the riboplasm of the pirellulosome (Fuerst and Webb, 1991; Lindsay et al., 2001; Wang et al., 2002). This double-membrane-bounded nuclear body has not been observed in the genera *Pirellula, Blastopirellula, Rhodopirellula*, or *Isosphaera* and also does not occur in at least *Planctomyces maris* within the genus *Planctomyces* (Lindsay et al., 2001). *Gemmata obscuriglobus*, at least, does not form a holdfast structure or, unlike *Planctomyces* species, a stalk. *Gemmata* forms a tuft of several flagella (Figure 174), unlike other planctomycetes such as *Pirellula* and *Blastopirellula* with single polar flagella, or *Isosphaera* which is a gliding bacterium not known to possess flagella. Unlike planctomycetes producing only motile swarmer buds, mother cells of *Gemmata* possess flagella and most cells in a young culture are motile whether budding or not.

Gemmata differs from *Pirellula* and *Blastopirellula* in the distribution of the dense circular pits called crateriform structures on its cell surface, which are uniformly distributed in the case of *Gemmata* unlike the polar distribution in *Pirellula* and *Blastopirellula*, but similar to the uniform distribution in *Planctomyces*. *Gemmata obscuriglobus* possesses a higher DNA G+C content than species of *Pirellula* (56–59 mol%), *Blastopirellula* (53.6–57.4 mol%), *Rhodopirellula* (53–57 mol%), *Planctomyces* (50–58 mol%), and *Isosphaera* (62 mol%).

Taxonomic comments

The characteristic cell compartmentation of *Gemmata* and strains clustering phylogenetically in the *Gemmata* group, involving a double membrane-bounded region containing a fibrillar nucleoid, is found in all cultured strains of this group where this property has been examined via thin sections of cells prepared via cryosubstitution following cryofixation (e.g., via high pressure freezing); see Figure 175. This membrane-bounded structure has been termed a "nuclear body" (though some confusion with a nonanalogous structure within the eukaryote nucleus with that name is possible if that term is used). It thus appears to be phylogenetically conserved within *Gemmata*-like strains, and it is diagnostic for the genus as it now stands. Double-membrane bounded nucleoids are not present in other planctomycete genera such as *Pirellula*, *Blastopirellula*, *Rhodopirellula*, or *Isosphaera*, or in anammox genera "*Candidatus* Kuenenia", "*Candidatus* Brocadia", or "*Candidatus* Scalindua", and within the genus *Planctomyces* (at least not in *Planctomyces maris*, though the structures of *Planctomyces limnophilus* and *Planctomyces brasiliensis* are too complex to exclude such occurrence). The nucleoid in the other planctomycete genera appears to be enveloped only by the single intracytoplasmic membrane and lies within the ribosome-containing pirellulosome compartment bounded by that membrane.

List of species of the genus *Gemmata*

1. **Gemmata obscuriglobus** Franzmann and Skerman 1985, 375[VP] (Effective publication: Franzmann and Skerman 1984, 266.)

ob.scu.ri.glo′bus. L. adj. *obscurus* dark; L. n. *globus* a sphere; N.L. n. *obscuriglobus* a dark sphere.

Spherical to pear-shaped cells, 1.4–3.0 × 1.4–3.0 μm, which form new cells by budding through a narrow attachment site. Cells contain an inclusion of different phase density correlated with a DAPI stain- positive DNA-containing compartment. In thin sections of cryosubstituted cells via transmission electron microscopy, the condensed nucleoids appear surrounded by an envelope consisting of two closely apposed membranes. Gram-stain-negative. Cell wall is proteinaceous in composition and does not contain detectable markers for peptidoglycan such as muramic acid or for Gram-stain-negative peptidoglycan such as diaminopimelic acid. A discrete holdfast is not produced, and attachement to glass surfaces appears to be effected by secreted slime from a broad area on the cell. Secreted stalks are not produced. Cells are motile by multitrichous flagella, and all cells in a motile population appear motile, not only buds. Colony color is pink or rose red. Generation time on Staley maintenance medium (Staley, 1973) at 24°C is about 13 h. Catalase-positive. Oxidase-negative. Glucose, galactose, mannose, xylose, ribose, rhamnose, melezitose, salicin, and starch are utilized as carbon sources but fructose, pyruvate, lactate, acetate, tartrate, propionate, phthalate, glycerol, mannitol, ethanol, methanol, adonitol, or a large number of amino acids are not. The temperature range for growth is 16–35°C.

DNA G+C content (mol%): 64.4 ± 1.0 (T_m).

Type strain: DSM 5831, ACM 2246 (originally designated UQM 2246 before reassignment of abbreviation for the relevant culture collection).

Sequence accession no. (16S rRNA gene): AJ231191, X54522, X56305, X85248.

Genus IV. **Isosphaera** Giovannoni, Schabtach and Castenholz 1995, 619[VP] (Effective publication: Giovannoni, Schabtach and Castenholz 1987b, 283.)

Richard W. Castenholz

[*Isocystis* Borzi 1878 (cyanobacteria)]

I.so.spha.e′ra. Gr. adj. *isos* equal; L. fem. n. *sphaera* ball or sphere; N.L. fem. n. *Isosphaera* sphere of equal size.

Spherical cells 2.0–2.5 μm in diameter **forming chains** of indefinite length, often over 100 cells. Cell division by **budding,** intercalary or terminal. No branching occurs. Capsules are not present. Gas vesicles are usually present in all cells, mostly in one cluster (gas vacuole or "pseudovacuole"). Resting cells are not known. Gram variable. **Motility by gliding;** pili present, no flagella. **Phototactic** on solid substrate. **Strictly aerobic.** Temperature range ~40–55°C in culture. Optimal pH range 7.8–8.8.

Chemoheterotrophic, aerobic respiratory metabolism. Glucose, ribose, and lactate utilized as carbon sources. Oligotrophic, i.e D-glucose above 0.05% is inhibitory, as is ribose over 0.25%. Casamino acids tend to enhance growth, but not adequate as the sole carbon source. Carotenoids present; no chlorophyll or bacteriochlorophyll present. Cell envelopes are proteinaceous with phospholipids containing beta-hydroxylated fatty acids in the lipopolysacharide lipid A outer layer. Lipids are ester-linked. **Peptidoglycan is absent**.

DNA G+C content (mol%): 62.2.

Type species: **Isosphaera pallida** (Woronichin 1927) Giovannoni, Schabtach and Castenholz 1995, 619 corrig. Euzéby and Kudo 2001, 1938 (Effective publication: Giovannoni, Schabtach and Castenholz 1987b, 283.) (*Isocystis pallida* Woronichin 1927 – as a cyanobacterium; *Torulopsidosira pallida* (Woronochin) Geitler 1963 – as a yeast).

Further descriptive information

Cell size is variable in actively growing cultures. Negatively stained specimens show no sign of capsules, although cells sometimes appear capsulated under phase-contrast microscopy. No resting cells are known, but some cells appear refractile. Small refractile spheres may be seen in many cells. These are clusters of gas vesicles (Figure 176). Intercalary budding of cells may also been seen (Figure 176). Numerous pili are visualized by transmission electron microscopy (TEM) (Giovannoni et al., 1987b). Gliding motility is relatively slow (~ 0.05 μm/s). Chains of cells and entire colonies are phototactic, the latter looking like pink "comets" (see Giovannoni et al., 1987b). Motility and phototaxis has been demonstrated on plates of 0.4% agarose, Gelrite, and 1.5% Bacto-agar.

Ultrastructure visualized by TEM has shown periodically spaced ring-shaped structures on the outer surface of the cell wall in negatively stained cells. TEM of thin sections of the cell envelope shows two electron-dense layers separated by an electron-transparent layer (Figure 177). Cytoplasmic staining with 4′,6-diamidino-2-phenylindole has shown that the DNA is restricted to a distinct region of the cell, surrounded by an envelope-like structure resembling a nuclear membrane (Figure 178). Cell-wall and lipid composition has been characterized by Giovannoni et al. (1987a) and the polyamine pattern by others (Griepenburg et al., 1999). The DNA G+C content is 62.2 mol% (Giovannoni et al., 1987b). At this point, only *Isosphaera* and *Gemmata* strains have contents of over 60 mol%; other members of the *Planctomycetales* have values lower than 60 mol% (Griepenburg et al., 1999).

As with other members of the order *Planctomycetales*, growth of *Isosphaera pallida* is not inhibited by penicillin G (2338 units/ml) or related antibiotics.

Ecological comments. Although the initial observations and recovery of *Isosphaera pallida* were from hot springs with relatively mild temperatures (~37–55°C) and TDS of less than 2 ppt, mixed *Isosphaera*/cyanobacterial cultures were also collected

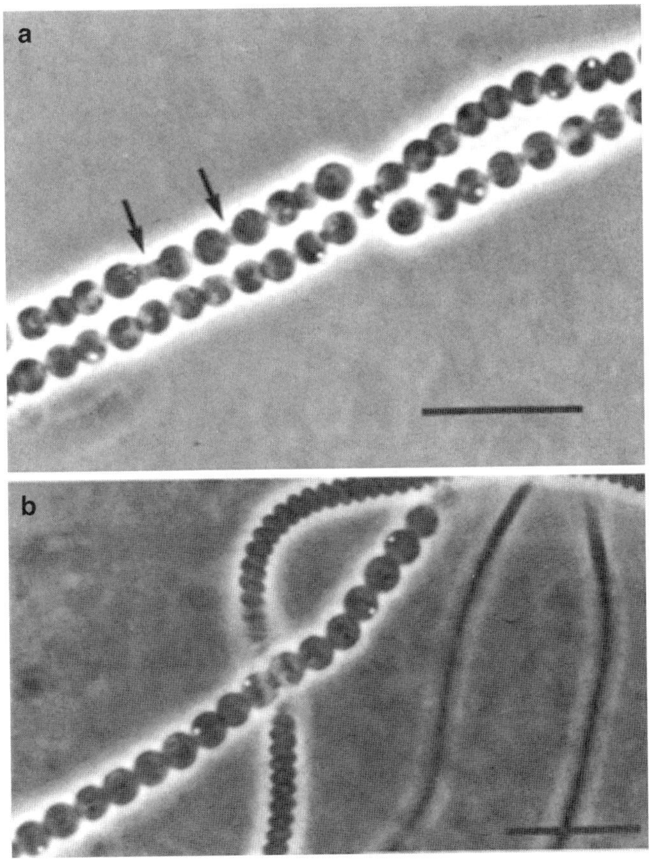

FIGURE 176. Phase-contrast photomicrographs of *Isosphaera pallida*. (a) Exponential-phase cells of strain IS1B isolated from Kah-nee-ta Hot springs, Oregon. Arrows indicate pair of buds forming between adjacent cells. Spherical white bodies are gas vacuoles (clusters of gas vesicles). (b). Field specimen from Mammoth Hot Springs, Yellowstone National Park. The tightly coiled filament (trichome) is a species of the cyanobacterium, *Spirulina*. Again, the light spots showing in some cells of *Isosphaera* are gas vacuoles. Bars = 10 μm.

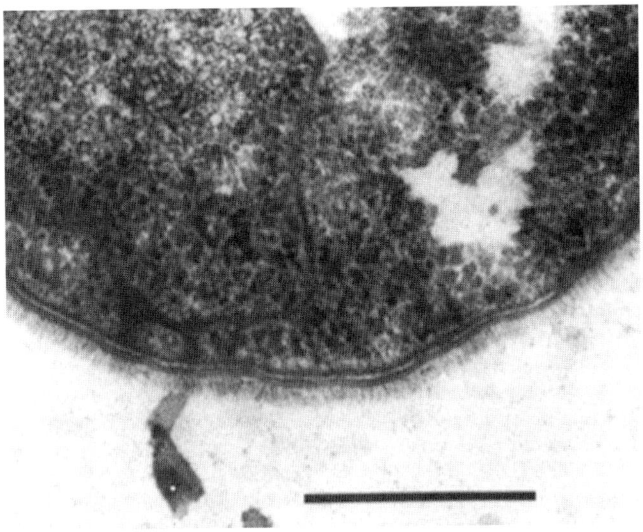

FIGURE 177. Transmission electron micrograph of a thin section of *Isosphaera pallida* showing wall ultrastructure. Bar = 0.5 μm.

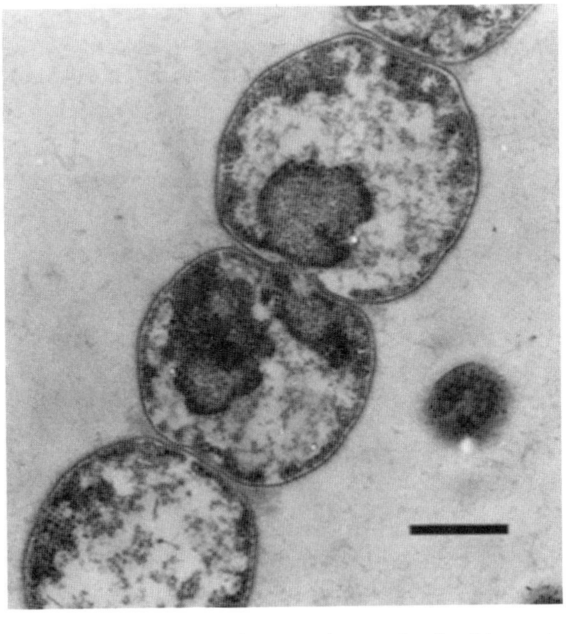

FIGURE 178. Transmission electron micrograph of a thin section of *Isosphaera pallida* showing a late stage bud (center cell). Also note dark boundary around nuclear area in two of the cells. Bar = ~0.8 μm.

from a slightly saline hot spring in Utah (~13 ppt), USA, but isolation attempts failed. More recently, *Isosphaera*-like isolates and environmental sequences that also cluster with *Isosphaera* have been identified from nonthermal environments such as leakage water from a compost heap (Wang et al., 2002) and also from a municipal wastewater treatment plant (Chouari et al., 2003). In the former case, isolates could be grown at 28°C without supplemental CO_2. The original six isolates from various hot springs and pools in Wyoming (including Yellowstone National Park) produced gas vesicles in discreet clusters (Giovannoni et al., 1987b). Since O_2 is in lower concentrations in warm waters, the gas vesicles of this aerobic bacterium may be an adaption that allows a buoyant ascent from bottom mats toward surface waters where O_2 is higher. Little more is known of the ecology of *Isosphaera* species.

Enrichment and isolation procedures

Isosphaera pallida has been seen microscopically and isolated from numerous North American hot springs including Mammoth Hot Springs and Lower Geyser Basin, both in Yellowstone National Park, (Wyoming), Thermopolis hot springs, Wyoming, and Kah-nee-ta Hot Springs, Oregon, as well as other hot springs in North America and Europe (Anagnostidis and Rathsack-Künzenbach, 1967; Giovannoni et al., 1987b).

It invariably occurs in association with cyanobacteria in neutral to alkaline hot springs within the temperature range of 35–55°C. It is generally a minor component, but can be identified easily by light microscopy. Since it is a chemoheterotroph, it is presumably dependent on products released from the cyanobacterial primary producers.

Isosphaera pallida may be isolated by streaking dilute samples onto plates of IMC medium (Table 173) and incubated in the dark at 45°C under a filter-sterilized mixture of 5% CO_2 and 95% air. Under these conditions, the bicarbonate/CO_2 buffer system maintains a pH of ~7.9. Small (1–2 mm) pink circular colonies appear after 1–2 weeks. Colonies are firm and nonviscous.

Maintenance in culture

Isosphaera cultures have been maintained by monthly transfer on agar slants of IMC medium at 45°C in an atmosphere of 5% CO_2 (Table 173). These slants may also be stored (after substantial growth has occurred) at room temperature in air (cotton-plugged) for 1 month without losing viability. Viability may also be conserved in sealed tubes at −80°C.

Differentiation of the genus *Isosphaera* from other genera

Size and morphological shape are adequate for preliminary identification. If in growth phase, the budding of cells may be seen in addition to the small glistening spheres that are the clusters of gas vesicles (Figure 176). A possible exception are some cyanobacteria that form chains of spherical cells (e.g., *Nostoc* spp.), but normally can be distinguished by blue-green to greenish color that reflects their content of chlorophyll *a* and blue or red phycobilin pigments. The presence of these pigments can also be confirmed by autofluorescence under an epifluorescent microscope. There are still some biologists who believe that the original "*Isocystis pallida*" (described as a cyanobacterium), from which *Isosphaera pallida* was renamed, still exists as a cyanobacterium with the identical morphotypic characteristics of *Isosphaera* (e.g., Hindak, 2001), but the cyanobiologists Komárek and Anagnostidis (2005) have left this possibility open. It is obvious from physiological data, lack of chlorophyll, and the 16S rRNA sequences, that *Isosphaera* is definitely related to the Planctomycetales and not to the cyanobacteria. Colonies of *Isosphaera* are pale pink, but filaments show no color under a light microscope unless en masse. *Gemmata, Pirellula,* and *Planctomyces* comprise the other cultured members of this order. None of these three genera bear a resemblance to *Isosphaera* morphologically and none are filamentous, but they do have in common a membrane-like boundary around the nucleoid.

Taxonomic comments

The 16S rRNA gene sequence (1460 bp) of the original strain IS1B of *Isosphaera pallida* from Kah-nee-ta Hot Springs is available

TABLE 173. Composition of *Isosphaera* medium IMC

Solution/medium	Ingredient	Concentration/quantity
Solution A[a]	$CaCl_2 \cdot 2H_2O$	0.32 g/l
	$MgSO_4 \cdot 7H_2O$	0.40 g/l
	KCl	0.50 g/l
	NaCl	1.00 g/l
	$(NH_4)_2SO_4$	0.50 g/l
	KH_2PO_4	0.30 g/l
	$FeCl_3$	0.292 mg/l
	Micronutrient solution[b]	10 ml/l (make up with 1 l distilled H_2O), then autoclave
Solution B	$NaHCO_3$	42 g/l (sterilize by autoclaving, followed by sparging with sterile CO_2 for 1 h)
IMC medium	Solution A (sterile)	250 ml
	Distilled H_2O (sterile)	650 ml, then add:
	Sterile solution B	100 ml, plus
	D-Glucose (final conc.)	0.025% (filter-sterilized)
	Casamino acids (final conc.)	0.025% (filter-sterilized)
	Vitamin mix[c]	0.5 ml/l (filter-sterilized)

[a]Adjust solution A to pH 7.6 with NaOH (store at 4°C until autoclaved).

[b]Use any commonly used micronutrient solution (see Castenholz, 1988).

[c]The vitamin solution contains (per ml) nicotinic acid, 2 mg; thiamine HCl, 1 mg; *p*-aminobenzoic acid, 0.2 mg; biotin, 0.02 mg; and B_{12}, 0.02 mg.

from GenBank (accession no. X64372). A partial sequence (2779 bp) of the 23S rDNA of another culture strain is also available, as well as 16S rDNA sequences of other strains and of the translational elongation factor-Tu (*tuf*) gene. Phylogenetic trees place *Isosphaera pallida* and other *Isosphaera*-like relatives distinctly within the order *Planctomycetales* but only distantly related to other members of the order such as *Gemmata*, *Pirellula*, and *Planctomyces* (Chouari et al., 2003; Wang et al., 2002). A comparative analysis of ribonuclease P RNA of the cultured members of the *Planctomycetes* was completed by Butler and Fuerst (2004). The tRNA processing endonuclease ribonuclease P contains a highly conserved and required RNA molecule (RNase P RNA) that is the catalytic subunit of the enzyme. They were able to show how the secondary structures of RNase P RNA differed in *Isosphaera pallida* (and a related *Isosphaera*-like isolate) from isolates of *Pirellula*, *Planctomyces*, *Gemmata*, and *Gemmata*-like strains.

"*Isocystis pallida*" Woronichin 1927 was described as pale blue-green in color, but otherwise morphologically the same as *Isosphaera pallida*. The original collection was from warm springs in the Caucasus. Because of the budding nature of the cells, Geitler (1963) reclassified "*Isocystis pallida*" and other species of "*Isocystis*" as yeasts (thereby conceding they were nonphotosynthetic). Anagnostidis and Rathsack-Kunzenbach (1967), who were very familiar with these same morphotypes from hot springs in Greece, concluded that "*Isocystis pallida*" was a cyanobacterium. A weak red epifluorescence was seen in some of their collections. This fluorescence was not seen in the North American collections and cultures. If their collections were mixed with highly fluorescent cyanobacteria, there is commonly a "spillover" that appears to give fluorescence to nonfluorescent cells.

Further reading

Giovannoni, S.J. and R.W. Castenholz. 1989. Genus "*Isosphaera*" Giovannoni, Schabtach and Castenholz. *In* Bergey's Manual of Systematic Bacteriology, vol. 3 (edited by Staley, Bryant, Pfennig and Holt). Williams & Wilkins, Baltimore, pp. 1959–1961.

Staley, J.T., J.A.F. Fuerst, S. Giovannoni and H. Schlesner. 1992. The order *Planctomycetales* and the genera *Planctomyces*, *Pirellula*, *Gemmata* and *Isosphaera*. *In* The Prokaryotes: a Handbook on the Biology of Bacteria: Ecophysiology, Isolation, Identification, Applications, 2nd edn, vol. 4 (edited by Balows, Trüper, Dworkin, Harder and Schleifer). Springer, New York, pp. 3710–3731.

List of species of the genus *Isosphaera*

1. **Isosphaera pallida** (Woronichin 1927) Giovannoni, Schabtach and Castenholz 1995, 619[VP] corrig. Euzéby and Kudo 2001, 1938 (Effective publication: Giovannoni, Schabtach and Castenholz 1987b, 283.) (*Isocystis pallida* Woronichin 1927 – as a cyanobacterium; *Torulopsidosira pallida* (Woronochin) Geitler 1963 – as a yeast)

pal'li.da L. fem. adj. *pallida* pale.

Only one species has been isolated and described in detail. Therefore, the characteristics are those described for the genus above. Other strains have been isolated but not systematically. Some probably represent other species of this genus (Griepenburg et al., 1999; Wang et al., 2002).

DNA G+C content (mol%): 62.2 (Bd).

Type strain: IS1B, ATCC 43644, DSM 9630.

Sequence accession no. (16S rRNA gene): AJ231195, X64372.

Genus V. **Pirellula** Schlesner and Hirsch 1987, 441[VP] emend. Schlesner, Rensmann, Tindall, Gade, Rabus, Pfeiffer and Hirsch 2004, 1577

John A. Fuerst and Margaret K. Butler

Pi.rel'lu.la. L. n. *pirum* pear; L. fem. dim. ending *-ella*; L. fem. dim. ending *-ula*; N.L. fem. n. *Pirellula* very small pear, referring to the shape of the bacterium.

Pear or teardrop-shaped cells with one pointed attachment pole, 0.5–3.0 × 1.0–5.0 µm. Gram-stain-negative. **Cells do not possess a stalk**, but may form a fibrillar holdfast. **Frequently form rosettes** by attachment with the pointed cell pole. The holdfast occurs on one pole of the cell (the narrow pole of ovoid cells) and facilitates attachment to other cells to form rosettes. **Crateriform structures are distributed at one pole only. Reproduction is solely by bud formation**. Buds are formed at or near the opposite, wider pole which thus constitutes the reproductive pole. Newly formed buds are motile by **polar to subpolar monotrichous flagella**. When thin sectioned cells prepared via cryosubstitution are examined via electron microscopy, they display a **characteristic organization including a major cell compartment, the pirellulosome, in which ribosomes and a condensed nucleoid are enclosed by an intracytoplasmic membrane, and a ribosome-free paryphoplasm region between this intracytoplasmic membrane and the closely apposed cytoplasmic membrane and cell wall**. Chemoorganotrophic; obligately aerobic. The original description was emended following the description of *Rhodopirellula* and *Blastopirellula* to include more recent data (Schlesner et al., 2004). **The major polyamine is** *sym*-homospermidine. The major respiratory lipoquinone present is MK-6. The major phospholipid present is phosphatidylglycerol. A number of other lipids are present that have characteristic R_f values, but whose structures are not currently known. **The lipid pattern is characteristic. The major fatty acids present are $C_{14:0}$, $C_{16:1}$ ω7, $C_{16:0}$, $C_{18:1}$ ω9, $C_{18:1}$ ω7, $C_{18:0}$, and $C_{20:1}$ ω11.**

DNA G+C content (mol%): 56.4 ± 0.4 to 59.0 (T_m, Bd).

Type species: **Pirellula staleyi** Schlesner and Hirsch 1987, 441[VP] emend. Schlesner, Rensmann, Tindall, Gade, Rabus, Pfeiffer and Hirsch 2004, 1577.

Further descriptive information

Cells of *Pirellula staleyi* share with *Rhodopirellula baltica* and *Blastopirellula marina* a similar compartmentalized cell structure when thin sections of cells prepared via cryosubstitution are viewed by transmission electron microscopy (Lindsay et al., 2001). The condensed nucleoid and ribosome-like particles are contained within a single intracytoplasmic membrane forming a pirellulosome organelle. A ribosome-free paryphoplasm region lies between the intracytoplasmic membrane and the cytoplasmic membrane closely apposed to the cell wall, often forming a ribosome-free polar cap at one end of the cell (Figure 179). Crateriform structures, electron-dense pits on the

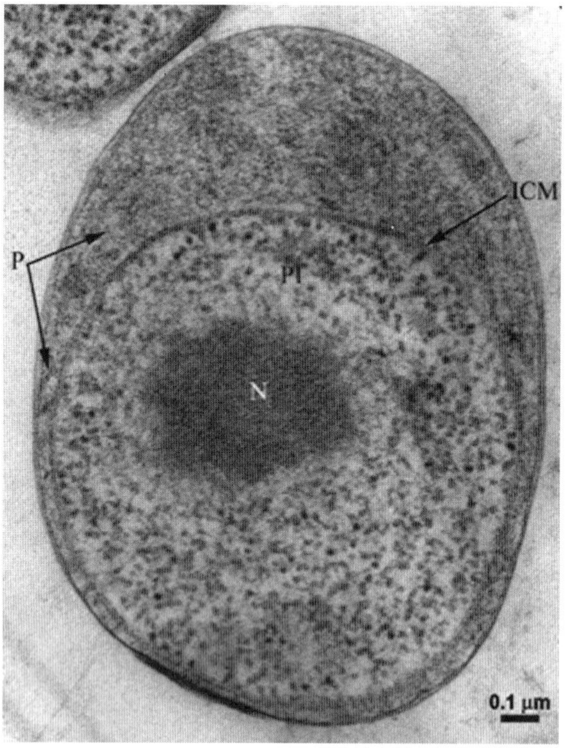

FIGURE 179. Transmission electron micrograph of thin sectioned cryosubstituted cell of *Pirellula staleyi* ATCC 27377 showing pirellulosome (PI) containing condensed nucleoid (N) and ribosome-like particles, surrounded by a single intracytoplasmic membrane (ICM), and the ribosome-free paryphoplasm (P) surrounding the pirellulosome and forming a polar cap at one end of the cell. Bar = 0.1 µm.

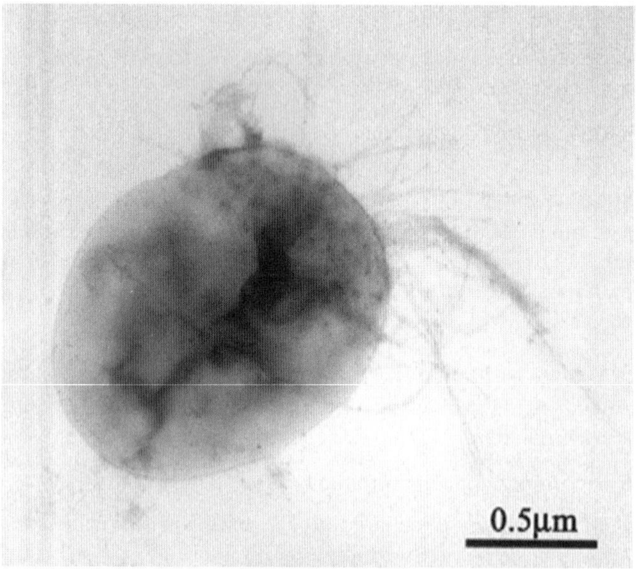

FIGURE 180. Transmission electron micrograph of negatively stained *Pirellula* strain ATCC 35122 showing polar distribution of crateriform structures on same pole as fimbriae appendages. Bar = 0.5 µm.

cell surface of planctomycetes, are distributed over one pole only (Figure 180), and in *Pirellula staleyi* appear to extend over only less than half the cell surface in contrast to *Blastopirellula*. The cell wall, as in other planctomycetes which have been examined, is proteinaceous, with amino acids accounting for 82%

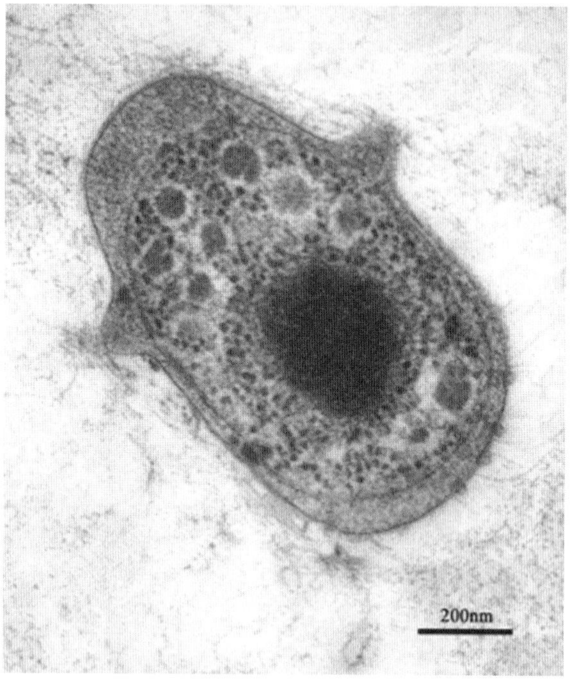

FIGURE 181. Transmission electron micrograph of thin section of cryosubstituted cell of *Pirellula* strain ATCC 35122, showing internal pirellulosome containing a condensed nucleoid surrounded by single intracytoplasmic membrane and polar cap, and hump-like prosthecate protrusions of wall and paryphoplasm on both sides of the cell. Bar = 200 nm.

of isolated cell-wall dry weight in the type strain ATCC 27377 (Liesack et al., 1986). Unlike *Rhodopirellula baltica*, the polar cap region in *Pirellula staleyi* does not appear to possess further small membrane-bounded ribosome containing regions. Both the type strain ATCC 27377 and the strain ATCC 35122 (which shares 100% 16S rRNA identity with the type strain) display one or two pointed-humplike protrusions that are opposite each other on the cell in the case where two are visible, constituting prosthecae visible at the electron microscope level (Figure 181). The type strain ATCC 27377 and another closely related strain ATCC 35122 attach to glass surfaces when grown in broth, consistent with potential habitat as periphyton.

Of the type strains of the three "*Pirellula*-group" genera, *Pirellula staleyi*, *Rhodopirellula baltica*, and *Blastopirellula marina*, only *Rhodopirellula baltica* is pigmented pink to red. However, it is possible that pigmented strains of *Pirellula* occur, as ICPB 4232 (a so-called "morphotype IV" strain of freshwater planctomycete cultures) from which ATCC 35122 with 100% identity in 16S rRNA sequence to ATCC 27377 was derived as a nonpigmented subclone, was reported to produce brown colonies (Tekniepe et al., 1981).

Enrichment and isolation procedures

Pirellula staleyi was originally isolated from freshwater (Lake Lansing, Michigan, USA) by adding 100 ml of freshwater sample to a sterile 150 ml beaker containing 10 mg peptone and incubating at room temperature (Staley, 1973). After approximately a month incubation, 0.1 ml of a 1:100 dilution of this enrichment was spread onto a peptone yeast extract medium (0.025% each of peptone, yeast extract, and glucose in a vitamin mineral salts medium) and incubated for 3 weeks, after which light yellow pigmented colonies containing budding bacteria appeared and were restreaked on the same medium. *Pirellula staleyi*-like organisms have also been isolated from freshwater sources such as Campus Lake, Baton Rouge, Louisiana (Tekniepe et al., 1981), the source of ATCC 35122 which shares 100% 16S rRNA sequence identity with ATCC 27377. These organisms were isolated by streaking and patch-plating of colonies appearing only after 5–7 d on Medium II agar of Schmidt and Starr containing 0.04% peptone, 0.02% yeast extract, 5–10 mM $MgCl_2$, and Hutner's vitamin-free mineral base (Schmidt and Starr, 1981).

These approaches can be combined with a selective one employing peptidoglycan synthesis-inhibiting antibiotics such as ampicillin and/or protein-synthesis inhibiting antibiotics such as streptomycin as used for planctomycete isolation by Schlesner (1994) and Wang et al. (2002). *Pirellula*-like strains including nonpigmented ones were isolated from freshwater sources by Schlesner (1994) using *N*-acetylglucosamine (2.0 g/l) as sole carbon and nitrogen source and ampicillin and cycloheximide (0.002% each) as selective agents against other bacteria and fungi, respectively.

Maintenance procedures

The type strain, ATCC 27377, can be maintained for at least 6 months at ca. 4°C on a medium (Staley, 1968; Van Ert and Staley, 1971) containing 0.5% Bacto-peptone, 0.05% yeast extract, vitamin and mineral salts solutions, and 1.5% Difco agar (Staley, 1973). It has been preserved via freeze-drying and at −70°C in Protect (Technical Service Consultants, Ltd) commercial vials. It grows well on Caulobacter tap water agar (ATCC Medium 36 – Caulobacter medium) at 26–28°C. ATCC 35122 can be grown on Caulobacter tap water agar or M1 agar (Schlesner, 1994) at 26–28°C. Cells are weakly inhibited by artificial light at 2400 lx.

Differentiation of the genus *Pirellula* from other genera

Features distinguishing *Pirellula* from the morphologically similar *Blastopirellula* and *Rhodopirellula* are summarized in Table 174. For example, *Pirellula staleyi* has only limited tolerance to artificial sea water and hydrolyzes casein, unlike *Blastopirellula* and *Rhodopirellula*, and the major fatty acid composition is characteristic (e.g., relatively high percentage of $C_{20:1}$ ω11). *Pirellula* can be distinguished from *Planctomyces* and *Gemmata* on the basis of its mature cell shape as teardrop to pear-shaped (rather than spherical to ovoid) with the attachment pole slightly pointed, the absence of a stalk (relative to *Planctomyces*), and the presence of crateriform structures on one cell pole only (claimed to be the reproductive cell pole) (Schlesner and Hirsch, 1984) relative to *Planctomyces* and *Gemmata* where these are distributed uniformly over the cell surface. *Pirellula* can be distinguished from *Gemmata* on the basis of the absence in thin sections of cryosubstituted cells of any double membrane-bounded nuclear body within the pirellulosome of *Pirellula*, but it shares this simple type of pirellulosome with *Blastopirellula* and *Rhodopirellula*. *Pirellula* is distinguished from *Blastopirellula* and *Rhodopirellula* on the basis of the high sodium chloride and calcium requirements of the latter (12–175% and 12–200% artificial seawater, respectively) relative to the limited tolerance (0–50% artificial sea water) of *Pirellula staleyi* to artificial sea water (Schlesner et al., 2004). Colonies of *Rhodopirellula* are pink to red, but those of *Pirellula* are unpigmented DNA base composition is of limited value in distinguishing *Pirellula* from *Blastopirellula* and *Rhodopirellula*. *Pirellula staleyi* type strain has a DNA G+C content varying with author from 56 to 59 mol% (Schlesner and Hirsch, 1984), and this overlaps with those of *Rhodopirellula baltica* (53–57 mol% (Schlesner et al., 2004) and *Blastopirellula marina* (53.6–57.4 mol% (Schlesner, 1986).

16S rRNA gene sequence similarity values are of significance in delineating this genus, but the extent cannot be defined at present, since strains with less than 95% sequence similarity to members of this genus should probably be placed in separate genera (Schlesner et al., 2004).

Taxonomic comments

The type strain ATCC 27377, originally isolated from Lake Lansing, Michigan, had been assigned to *Pasteuria ramosa* on the basis of a morphological resemblance to a bacterial parasite of cladoceran invertebrates in the genus *Daphnia* isolated by Metchnikoff in the nineteenth century (Staley, 1973). Again on the basis largely of morphological characteristics, ATCC 27377 was reassigned (Starr et al., 1983) to the genus *Planctomyces* as *Planctomyces staleyi*, and then, on the basis of 16S rRNA oligonucleotide analysis and DNA base composition, to the genus *Pirella* (Schlesner and Hirsch, 1984). *Pirella* was later rejected as a name (Schlesner and Hirsch, 1987) due to homonymy with the name of a fungus *Pirella* Bainier 1883, therefore ATCC 27377 was proposed as the type strain of the new genus *Pirellula*, initially including *Pirellula staleyi* and *Pirellula marina*. Finally, the genus *Pirellula* was emended by Schlesner et al. (2004) following the description of the two new genera *Blastopirellula* to include the former *Pirellula marina* and *Rhodopirellula* to include the new species *Rhodopirellula baltica*. *Pirellula staleyi* is thus now the only described species of the genus.

Strain ATCC 35122 was isolated (designated ICPB 4362) as a "white" nonpigmented subclone of ICPB 4232, originally isolated from freshwater Campus lake, Baton Rouge, Louisiana, which tended to form brown colonies with age (Tekniepe et al., 1981). It displays 100% 16S rRNA sequence identity with ATCC 27377 (Butler et al., 2002) and shares crateriform structure distribution, ultrastructural compartmentalization features, and hump-like protrusions with ATCC 27377 (Butler et al., 2002), forming a second culture collection reference culture of a member of genus *Pirellula*.

List of species of the genus *Pirellula*

1. **Pirellula staleyi** Schlesner and Hirsch 1987, 441[VP] emend. Schlesner, Rensmann, Tindall, Gade, Rabus, Pfeiffer and Hirsch 2004, 1577

sta'le.yi. N.L. gen. masc. n. *staleyi* of Staley, named after James T. Staley, who isolated strain ATCC 27377, under the name *Pasteuria ramosa.*

The description is the same as the genus. Glucose, galactose, mannose, xylose, ribose, fucose, rhamnose, cellobiose, melitzitose, inulin, salicin, starch, dextrin, glucuronic acid, *N*-acetylglucosamine, pyruvate, pectin, lactose, maltose, melibiose, raffinose, sucrose, and trehalose are utilized as carbon sources. The maximum salt tolerance is 50% artificial seawater.

DNA G+C content (mol%): 56.4 ± 0.4–59.0 (T_m, Bd).

Type strain: ATCC 27377, DSM 6068.

Sequence accession no. (16S rRNA gene): AJ231183, X81946.

Genus VI. **Rhodopirellula** Schlesner, Rensmann, Tindall, Gade, Rabus, Pfeiffer and Hirsch 2004, 1577[VP]

HEINZ SCHLESNER

Rho.do.pi.rel'lu.la. Gr. neut. n. *rhodon* a rose; N.L. fem. n. *Pirellula* name of a bacterial genus; N.L. fem. n. *Rhodopirellula* a red *Pirellula*.

Cells are ovoid, ellipsoidal, or pear-shaped, occurring singly or in rosettes by attachment at the smaller cell pole. Cells divide by budding. Buds are produced directly from the broader cell pole of the mother cell. **Crateriform structures and fimbriae are found in the upper cell region. Cells have intracytoplasmic membranous structures.** Gram-stain-variable. Buds may have a single flagellum inserted subpolarly at the proximal pole. Adult cells are immobile. Strictly aerobic. **Colonies are pink to red in color**. Mesophilic. Requires seawater for growth. Nonspore-forming. PHB is not stored. Chemoheterotrophic. Carbon and energy sources are mainly carbohydrates. C1-compounds are not used. *N*-acetylglucosamine serves as carbon and nitrogen source. Catalase- and cytochrome oxidase-positive. **The proteinaceous cell wall lacks peptidoglycan.** The major polyamines are putrescine, cadaverine, and *sym*-homospermidine. **The major menaquinone is MK-6.** The major fatty acids are $C_{16:1}$ ω7, $C_{16:0}$, $C_{17:1}$ ω8, $C_{17:0}$, $C_{18:1}$ ω7, $C_{18:1}$ ω9, and $C_{18:0}$. The major phospholipids are phosphatidylcholine and phosphatidylglycerol.

DNA G+C content (mol%): 53–57 (HPLC).

Type species: **Rhodopirellula baltica** Schlesner, Rensmann, Tindall, Gade, Rabus, Pfeiffer and Hirsch 2004, 1577[VP].

Further descriptive information

Rhodopirellula baltica is the only described species of *Rhodopirellula.* Morphologically, it is characterized by a polar organization of the cells. The small cell pole ("holdfast pole") excretes a sticky substance of still unknown composition (Figure 182) to enable the cell to attach to surfaces or to form rosettes (see Figure 171 in the chapter Genus *Blastopirellula*). The reproductive pole produces the bud which grows to a smaller mirror image of the mother cell. Crateriform structures are distributed over the entire cell surface of the buds, however, they are only found around the reproductive pole of adult cells. In the same area, numerous fimbriae are found. Studies of electron microscopical images suggest, that the fimbriae originate from the crateriform structures (Figure 183).

Thin cryosubstituted sections of cells showed membranous structures surrounding the nuclear material and the majority of the ribosomes. Analysis of cross-sections of strain SH 1[T] as well as SH 796 (Gade et al., 2004) revealed several small structures in addition to a large central one (Figure 184; Schlesner et al., 2004). This microscopic appearance differs from that of the pirellulosome structure described for *Pirellula staleyi* and *Blastopirellula marina* (Lindsay et al., 1997, 2001).

Adult cells excrete a glycocalix (Figure 185) which might be responsible for the positive Gram reaction of adult cells as it may retain the dyes. Buds and swarmer cells react negatively, and even adult cells give the negative Gram reaction after washing.

The cell wall lacks peptidoglycan and possesses a proteinaceous sacculus instead as has been found in all strains of *Planctomycetales* examined so far. This was first reported for several strains of *Planctomyces* and *Pirellula* (then *Pasteuria*; König et al., 1984). A more detailed study of the cell-wall composition

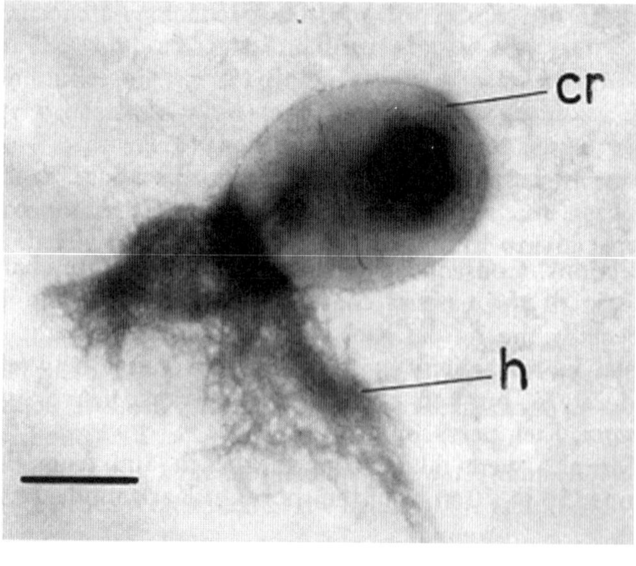

FIGURE 182. *Rhodopirellula baltica* strain SH1[T]. Cell with extensive holdfast (h) production. Crateriform structures (cr) are restricted to the reproductive cell pole. Bar = 0.5 µm. Negatively stained with phosphotungstic acid.

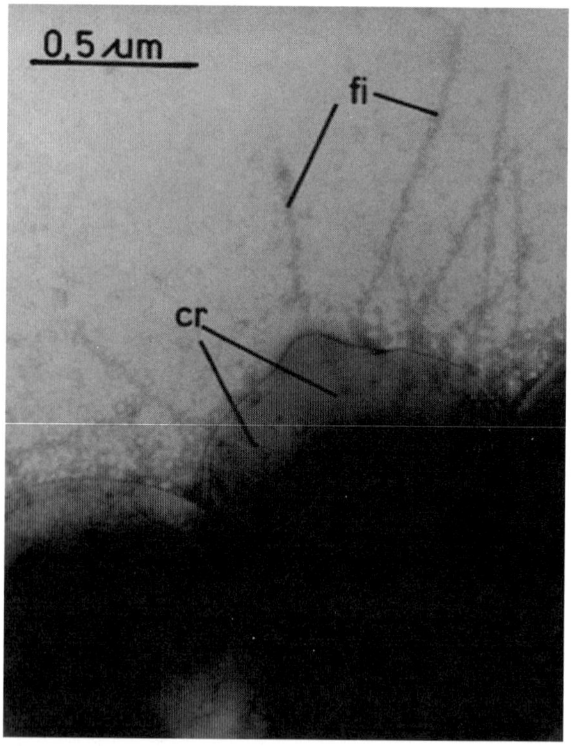

FIGURE 183. *Rhodopirellula baltica* strain SH143. Fimbriae (fi) originate from crateriform structures (cr). Bar = 0.5 µm. Negatively stained with phosphotungstic acid.

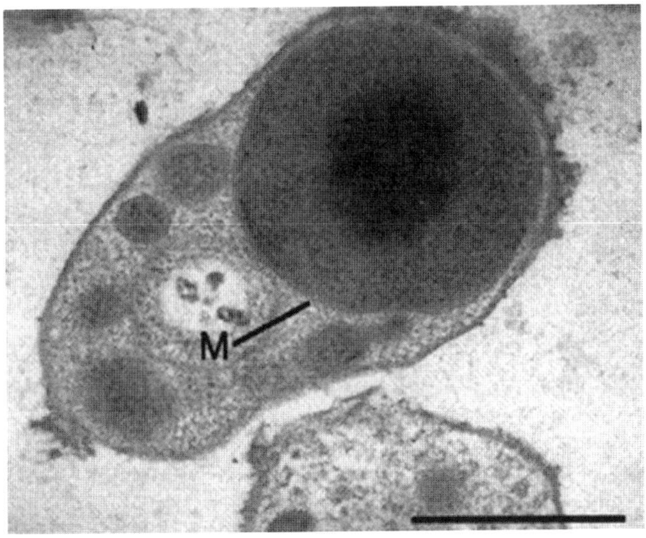

FIGURE 184. Thin section of cryosubstituted cell of *Rhodopirellula baltica* strain SH1^T displaying intracellular compartmentalization. Membrane (M) engulfing the pirellulosome like structures. Bar = 0.5 µm.

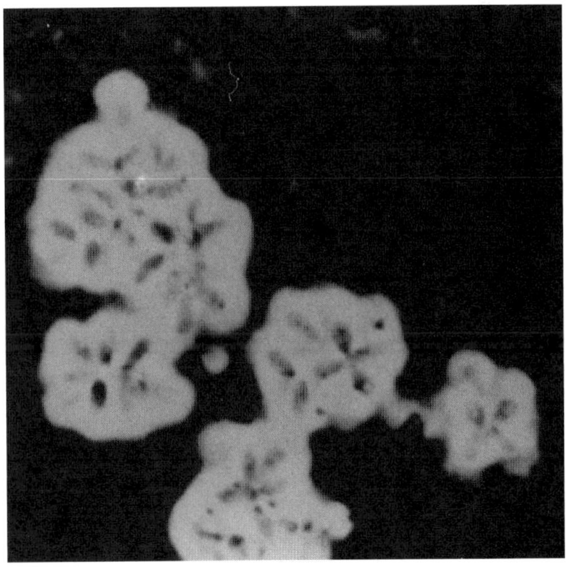

FIGURE 185. *Rhodopirellula baltica* strain SH165 negatively contrasted with India ink demonstrates slime formation.

included the type strain of *Rhodopirellula baltica* (Liesack et al., 1986).

Rhodopirellula is a member of the attached living microflora and, in common with several other representatives of such immobile bacteria, it produces motile daughter cells, thus exhibiting a biphasic life cycle. After ripening, the bud separates from the mother cell and moves away till it attaches to a substratum. Then the flagellum is released (Figure 186). In case of high cell densities, swarmer cells may form rosettes by contact with their holdfast poles.

Carbohydrates are the main substrates as carbon and energy sources. Interestingly, *N*-acetylglucosamine, the monomer of chitin, serves as carbon and nitrogen source; chitinase activity, however, was not detected.

Though *Rhodopirellula* is quite similar to *Pirellula staleyi* and *Blastopirellula marina* in morphology and physiology, the 16S rRNA sequence analysis showed less than 90% similarity between the type strains of these genera (Table 174). The length of the 5S rRNA of *Rhodopirellula baltica* strain SH1^T is in the range of that for other planctomyceta, i.e. 109–111 bases rather than 118 bases as for the majority of *Bacteria* and *Archaea* (Bomar et al., 1988).

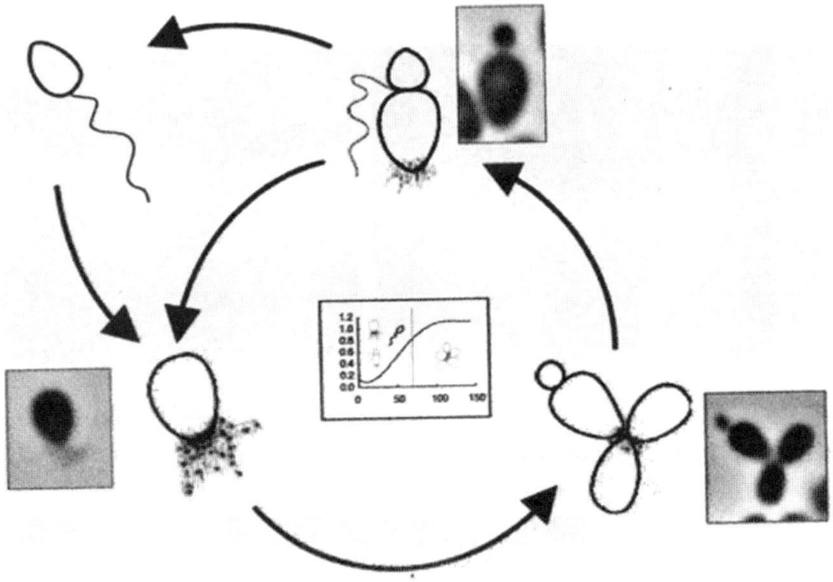

FIGURE 186. Schematic drawing of the life cycle of *Rhodopirellula baltica*. For clarity, typical cell structures, such as fimbriae and crateriform structures are not shown. (Reproduced with permission from Gade et al., 2005. Environ. Microbiol. 7: 1074–1084.)

TABLE 174. Characteristics differentiating *Pirellula staleyi*, *Blastopirellula marina*, and *Rhodopirellula baltica*[a,b]

Characteristic	*Pirellula staleyi* ATCC 27377[T]	*Blastopirellula marina* DSM 3645[T]	*Rhodopirellula baltica* SH 1[T]
Cell size (µm)	0.9–1.0 × 1.0–1.5	0.7–1.5 × 1.0–2.0	1.0–2.5 × 1.2–2.3
Pigmentation	None	None	Pink to red
Salinity tolerance (% ASW)	0–50	12–175	12–200
Vitamin requirement	–	–	B_{12}
Carbon source utilization:			
Chondroitin sulfate	–	+	+
Fucose	+	+	–
Glutamic acid	–	+	–
Glycerol	–	+	+
Hydrolysis of casein	+	–	–
Lipase	–	+	–
Polyamines (µmol/g dry weight):[c]			
Cadaverine	–	–	15.3
sym-Homospermidine	50.2	11.3	19.6
Phosphatidylcholine	–	–	+
Putrescine	–	–	19.5
Spermidine	0.8	–	–
Cell-wall amino acids (molar ratios):			
Cysteine	3.6[d]	1.2[e]	9.2[d]
Glutamate	9.0[d]	11.3[e]	36.3[d]
Threonine	3.0[d]	3.8[e]	9.0[d]
Valine	1.7[d]	2.4[e]	8.2[d]
Fatty acids (relative %):			
$C_{14:0}$	4.9	–	0.5
$C_{15:0}$	–	5.9	0.5
$C_{16:0}$ iso	–	4.9	–
$C_{16:1}$ ω7	3.5	4.1	8
$C_{16:0}$	33.8	27.5	39.2
$C_{17:1}$ ω8	14.4	–	4
$C_{17:0}$	5.3	–	1.2
$C_{18:1}$ ω9	26.6	26.6	40.8
$C_{18:1}$ ω7	2	2.3	1.6
$C_{18:0}$	3.3	2.5	4.3
$C_{19:1}$ ω8	–	2.6	–
$C_{19:0}$	–	2.7	–
$C_{20:1}$ ω11	15.7	1.2	–
16S rRNA gene sequence similarity (%) to ATCC 27377[T]/SH 1[Tf]	100/85.0	87.5/87.1	85.0/100

[a]Symbols: +, positive, main component; –, negative, not present; (+), present in small amounts.

[b]Data from Table 4 of Schlesner et al. (2004).

[c]Griepenburg et al. (1999).

[d]Liesack et al. (1986).

[e]König et al. (1984).

[f]Ward et al. (1995).

An unusual characteristic seems to be the large genome size. Initial investigations of genome sizes of budding bacteria were performed using DNA renaturation kinetics and indicated that the type strain of *Rhodopirellula baltica* had a considerably larger genome than most of the other strains of budding bacteria (belonging to the genera *Hyphomicrobium*, *Hyphomonas*, *Filomicrobium*, and *Pedomicrobium*) (Kölbel-Boelke et al., 1985). Recently, the complete genome sequence of the type strain has been determined (Glöckner et al., 2003) (GenBank accession no. BX119912). With a size of 7.145 Mb, it is one of the largest circular bacterial genomes known to date. The genome contains unexpected genes, e.g., 110 genes possibly encoding sulfatases (Glöckner et al., 2003) or archaea-like genes for C1-transfer enzymes (Bauer et al., 2004). It could be speculated that growth with chondroitin sulfate, as was recently found (Schlesner et al., 2004), may require the activity of a specific sulfatase that liberates the carbohydrate moiety. Although the organism has a proteinaceous cell-wall, remnants of genes for peptidoglycan synthesis were found. Genes for lipid A biosynthesis and homologs to the flagellar L- and P-ring protein indicate a former Gram-negative type of cell wall. Genomic and proteomic studies have also identified genes involved in *N*-acetylglucosamine metabolism (Rabus et al., 2002) which may be a key process in marine systems where *Rhodopirellula*, *Blastopirellula*, and *Planctomyces* species are abundant (Glöckner et al., 1999; Llobet-Brossa et al., 1998; Schlesner, 1994).

In proteomic analysis of carbohydrate catabolism and regulation, almost all enzymes of glycolysis and TCA cycle were identified, and almost all enzymes of the oxidative branch of the pentose phosphate cycle were detected (Gade et al., 2005a, c).

Growth phase dependent changes of protein composition were quantitatively monitored. Results indicated an opposing regulation of TCA and oxidative pentose phosphate cycle, down regulation of several enzymes involved in amino acid biosyntheses, and an up regulation of the alternative sigma factor σ^H in stationary phase (Gade et al., 2005b).

Rhodopirellula baltica can be considered a marine bacterium as it does not grow in freshwater media. Essential components of artificial seawater were Ca^{2+}, Na^+, and Cl^-. The majority of the 25 strains of the species were isolated from the brackish water of the Kiel Fjord (Baltic Sea) over a period of 25 years, suggesting that these bacteria belong to the autochthonous microbial community of the habitat (Schlesner et al., 2004). Interestingly, two strains were isolated from the Mediterranean sponge *Aplysina aerophoba* (Gade et al., 2004), indicating the widespread occurrence of the species. In addition, DeLong et al. (1993) obtained molecular clones of 16S rRNA gene sequences with high similarity to *Rhodopirellula baltica* from marine snow (Pacific Ocean).

Enrichment and isolation procedures

Adult cells are normally attached to surfaces and so mainly swarmer cells are sampled from the free water column. For this reason, an enrichment of cells is necessary before starting isolation attempts. A method successfully applied is to keep 100 ml samples in cotton-plugged 250 ml Erlenmeyer flask over a sediment of 10 g chitin in the laboratory at room temperature (20–23°C) for 3 weeks before streaking over agar-solidified media. The use of *N*-acetylglucosamine as sole carbon and nitrogen source (M30) will reduce the number of upcoming colonies considerably. Further reduction can be achieved by adding filter-sterilized antibiotics affecting the development of peptidoglycan, e.g., ampicillin (0.002%, w/v), to the medium after autoclaving. Addition of 0.002% cycloheximide reduces the growth of fungi. Incubation temperature should not exceed 28°C. Upcoming pigmented colonies are examined microscopically. This is conveniently done by the tooth-pick procedure (Hirsch et al., 1977; Schlesner, 1994). A colony is touched with a sterile tooth-pick and transferred to a master plate (Petri dish with a grid on the backside and filled with the same solidified medium) and then to a labeled glass slide. If microscopic observation of the slide reveals cells morphologically similar to *Rhodopirellula* species, the inoculum on the master plate is labeled, cultivated, and purified by subsequent streaks on solidified medium. Suitable media are M13a*, M13f†, or M30‡.

Maintenance procedures

The strains grown on slants can be kept at 4–5°C for at least 3 months. They are easily revived from lyophilized cultures and can be stored at −70°C in a solution of 50% glycerol in the culture medium.

Differentiation of the genus *Rhodopirellula* from other genera

Major differentiating properties are shown in Table 174.

Further reading

Fuerst, J.A. 1995. The planctomycetes: emerging models for microbial ecology, evolution and cell biology. Microbiology *141*: 1493–1506.

Gade, D., J. Thiermann, D. Markowsky and R. Rabus. 2003. Evaluation of two-dimensional difference gel electrophoresis for protein profiling. Soluable proteins of the marine bacterium *Pirellula* sp strain 1. J. Mol. Microbiol. Biotec. *5*: 240–251.

Neef, A., R. Amann, H. Schlesner and K.-H. Schleifer. 1998. Monitoring a widespread bacterial group: *in situ* detection of planctomycetes with 16S rRNA-targeted probes. Microbiology *144*: 3257–3266.

Sittig, M. and H. Schlesner. 1993. Chemotaxonomic investigation of various prosthecate and/or budding bacteria. Syst. Appl. Microbiol. *16*: 92–103.

Ward, N., J.T. Staley, J.A. Fuerst, S.J. Giovannoni, H. Schlesner and E. Stackebrandt. 2004. The order *Planctomycetales*, including the genera *Planctomyces*, *Pirellula*, *Gemmata*, and *Isosphaera*, and the *Candidatus* genera Brocadia, Kuenenia, and Scalindua. *In* The Prokaryotes: an Evolving Electronic Resource for the Microbiological Community, 3rd edn, release 3.18 (edited by Dworkin, Falkow, Rosenberg, Schleifer and Stackebrandt). Springer, New York.

List of species of the genus *Rhodopirellula*

1. **Rhodopirellula baltica** Schlesner, Rensmann, Tindall, Gade, Rabus, Pfeiffer and Hirsch 2004, 1577[VP]

bal'ti.ca. N.L. fem. adj. *baltica* pertaining to the Baltic Sea, the place of isolation.

Cells are 1.0–2.5 × 1.2–2.3 μm in size. A single flagellum is subpolarly inserted at the proximal pole. Colonies are round, smooth, and pink to red in color. Growth is optimal between 28 and 30°C. No growth is observed above 32°C. Vitamin B_{12}, Ca^{2+}, Na^+, and Cl^- are required for growth. Strictly aerobic. Glucose is not fermented, nitrate cannot serve as an electron acceptor. Other characters are listed in Table 175.

The circular genome contains 7.145 MB. The chemical composition is identical to that in the genus description.

Source: brackish water of Kiel Fjord (Baltic Sea).

DNA G+C content (mol%): 55 (HPLC).

Type strain: SH 1, DSM 10527, IFAM 1310, NCIMB 13988.

Sequence accession no. (16S rRNA gene): X81938.

---

*Medium 13a (Schlesner, 1994) (g/l or ml/l)): peptone, 0.25; yeast extract, 0.25; glucose, 0.25; artificial seawater (ASW; Lyman and Fleming, (1940)), 500; vitamin solution no. 6 (Staley, 1968), 10; Hutner's basal salts medium (HBM; Cohen-Bazire et al., (1957)) 20; 0.1 M Tris/HCl pH 7.5, 50 ml; distilled water, 420.

†Medium 13f (Schlesner et al., 2004): (g/l or ml/l)): peptone, 0.75; yeast extract, 0.75; glucose, 5.0; artificial seawater (ASW; Lyman and Fleming, (1940)), 250; vitamin solution no. 6 (Staley, 1968), 10; Hutner's basal salts medium (HBM; Cohen-Bazire et al., (1957)) 20; 0.1 M Tris/HCl pH 7.5, 50 ml; distilled water, 670.

‡Medium 30 (Staley et al., 1992): (g/l or ml/l)): *N*-acetylglucosamine, 2.0; artificial seawater (ASW; Lyman and Fleming, (1940)), 250; vitamin solution no. 6 (Staley, 1968), 10; Hutner's basal salts medium (HBM; Cohen-Bazire et al., (1957)) 20; 0.1 m Tris/HCl pH 7.5, 50 ml; distilled water, 670.

TABLE 175. Characteristics of *Rhodopirellula baltica*[a]

Characteristic	
Carbon source utilization:	
N-Acetylglucosamine, amygdalin, cellobiose, chondroitin sulfate, dextrin, esculin, fructose, galactose, gluconate, glucose, glucuronate, glycerol, lactose, lyxose, maltose, mannose, melibiose, melezitose, raffinose, rhamnose, ribose, salicin, sucrose, trehalose, xylose	+
Acetate, adipate, adonitol, alanine, arabitol, arginine, asparagine, aspartate, benzoate, caproate, citrate, cysteine, cystine, dulcitol, erythritol, ethanol, formate, fucose, fumarate, glutamate, glutamine, glutarate, glycine, histidine, indole, inositol, inulin, isoleucine, lactate, leucine, lysine, malate, mannitol, methanol, methionine, methylamine, methylsulfonate, norleucine, ornithine, 2-oxoglutarate, pectin, phenylalanine, phthalate, proline, propionate, pyruvate, serine, sorbitol, sorbose, succinate, tartrate, threonine, tryptophan, tyrosine, urea, valine	–
Nitrogen source utilization:	
N-Acetylglucosamine, ammonium, Casamino acids, gelatin, nitrate, peptone, yeast extract	+
Nicotinate, urea	–
Hydrolysis of:	
Esculin, gelatin, starch	+
Alginate, casein, cellulose, chitin Tween 80	–
Hemolytic activity with blood of:	
Calf, horse, sheep	–
Activity of:	
Catalase, cytochrome oxidase	+
Urease	–
Production of:	
H_2S	+
Acetoin, indole	–
Sensitivity to:	
Tetracycline	+
Ampicillin, penicillin, streptomycin	–

[a]Symbols: +, >85% positive; –, 0–15% positive.

Genus VII. **Schlesneria** Kulichevskaya, Ivanova, Belova, Baulina, Bodelier, Rijpstra, Sinninghe Damsté, Zavarzin and Dedysh 2007, 2685[VP]

Svetlana N. Dedysh, Irina S. Kulichevskaya and George A. Zavarzin

Schles.ne'ri.a. N.L. fem. n. *Schlesneria* named after the German microbiologist Heinz Schlesner in recognition of his work on planctomycete diversity and ecology.

Ellipsoid-shaped cells, occurring singly, in pairs, or **in rosettes. Free cell poles are covered with crateriform surface structures** and possess **an unusual spur-like projection.** Encapsulated. **Reproduce by budding.** Gram-stain-negative. Adult cells are immobile. Daughter cells are motile by means of two subpolar flagella. No resting stages known. **Stalk-like structures are short and rarely observed.** Chemoorganotrophic aerobes. Capable of growth in microaerobic conditions. Colonies are small and nonpigmented. Various carbohydrates or *N*-acetylglucosamine are the preferred substrates. No growth factors are required. Organic nitrogen compounds, ammonium, nitrate, and some amino acids serve as nitrogen sources. **Possess hydrolytic capabilities. Oxidase- and catalase-positive.** Urease-negative. Dissimilatory nitrate reduction is positive. Mesophilic and **moderately acidophilic bacteria** growing in the pH range of 4.2–7.5, with an optimum at pH 5.0–6.2. **Sensitive to NaCl.** This genus is a member of the phylum *Planctomycetes*, order *Planctomycetales*, family *Planctomycetaceae*. Acidic wetlands are the main habitat.

DNA G+C content (mol%): 54.4–56.5.

Type species: **Schlesneria paludicola** Kulichevskaya, Ivanova, Belova, Baulina, Bodelier, Rijpstra, Sinninghe Damsté, Zavarzin and Dedysh 2007, 2686[VP].

Further descriptive information

A single species, *Schlesneria paludicola*, has been described based on characterization of three independent isolates (Kulichevskaya et al., 2007). These strains possess ellipsoidal cells that occur singly or in pairs or are arranged in rosettes (Figure 187). Mature cells are 0.6–1.5 µm × 1.3–2.1 µm; they are immobile and reproduce by budding. After separation from the mother cells, the buds display high motility due to possession of subpolar flagella. In most cases, two flagella per cell are observed. Thin sections reveal a characteristic type of cell organization in which a single membrane separates the cytoplasm into two major compartments. The nucleoid and most ribosomes are located in one of these compartments. Examination of negatively stained cells using electron microscopy shows the presence of crateriform pits which display a polar distribution and cover approximately one-third of the cell surface (Figure 187). An unusual spur-like projection is located on one pole of the cell which gives a cell shape somewhat similar to a lemon. In addition, the cell surface is covered with numerous, 4–8 nm thick, fimbria-like fibrillar structures and 10–15 nm thick fibrillar appendages. Some cells in cultures are connected to each other by means of stalk-like

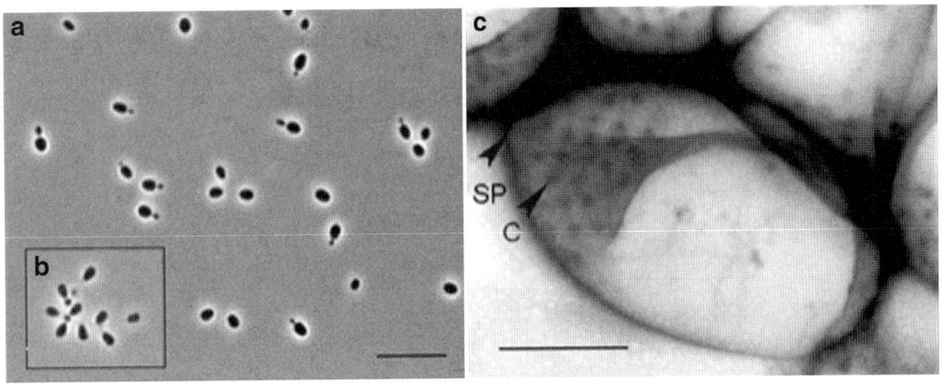

FIGURE 187. (a, b) Phase-contrast micrographs of mature and budding single cells (a) and rosettes (b) of *Schlesneria paludicola*. (c) Electron micrograph of negatively stained cells displaying crateriform surface structures with polar distribution and a polar lemon-shaped cell projection. C, Crateriform structures; SP, spur-like projection. Bars = 10 μm (a and b) and 0.5 μm (c). [(c) Printed with permission of Olga I. Baulina.]

structures, which resemble the bundles of twisted fibrils. However, these structures are short, rarely observed, and cannot be discerned by observation with the phase-contrast microscope.

On agar media, visible growth of *Schlesneria paludicola* appears after 2–3 weeks of incubation at 24°C. Colonies are small and uncolored. Members of the genus *Schlesneria* are aerobic chemoorganotrophs. Best growth is observed on media with carbohydrates or *N*-acetylglucosamine. All strains are also capable of slow growth on media with various biopolymers. Representatives of *Schlesneria paludicola* are able to reduce nitrate to nitrite. They are moderately acidophilic (pH range of 4.2–7.5) and mesophilic (temperature range of 4–32°C) organisms. High sensitivity to NaCl reflects adaptation of these bacteria to dilute environments such as ombrotrophic peat bogs.

Cells of *Schlesneria paludicola* contain menaquinone-6 (MK-6) as the predominant isoprenoid quinone. The major fatty acids are $C_{16:0}$ (43.5–49.6%) and $C_{16:1}$ ω7*c* (40.3–46.8%); the major neutral lipids are *n*-C_{31} polyunsaturated alkenes, *n*-C_{26}–2,25-diols, and *n*-C_{24}–2,23-diols.

Enrichment and isolation procedures

Microbial biofilms enriched with cells of *Schlesneria paludicola* and other peat-inhabiting planctomycetes can be obtained by using the water agar-coverslip technique developed by Hirsch et al. (1977). Sterile glass coverslips are inserted vertically into a sterile water agar layer in Petri dishes. The plates are then inoculated with 10–15 ml of peat water and incubated in the dark at 15–24°C for 1–2 months. Periodically, a few coverslips are taken out and subjected to examination by means of fluorescence *in situ* hybridization (FISH) with oligonucleotide probes specific for the *Planctomycetes* (Neef et al., 1998) to monitor development of the target bacteria as described by Dedysh et al. (2006) and Kulichevskaya et al. (2006). When good growth of planctomycetes is observed on these coverslips, replicates are taken out for agar streak isolation. The latter is performed using agar medium M31 (modification of medium 31 described by Staley et al., 1992) containing (g per liter of distilled water) KH_2PO_4, 0.1; Hutner's basal salts, 20 ml; *N*-acetylglucosamine, 1.0; Na-ampicillin, 0.2; peptone, 0.1; yeast extract, 0.1; agar-agar (Difco), 18, pH 5.5–5.8. Incubation is at 20–24°C for 1–2 months. The colonies appearing on the plates are randomly picked for FISH-based identification and examination by phase microscopy for the presence of ellipsoidal cells arranged in pairs or in rosettes. This procedure is continued until individual colonies of target bacteria are ultimately identified and obtained in pure culture.

Maintenance procedures

Strains can be maintained on agar medium M31 by subculturing once in 2–3 months. Alternatively, they can be kept in liquid medium M31 with 10% glycerol at −70°C or stored by lyophilization.

Differentiation of the genus *Schlesneria* from other genera

Table 176 lists characteristics of *Schlesneria* that differentiate it from other aerobic budding bacteria that lack peptidoglycan. Cell shape and formation of rosettes makes *Schlesneria* different from *Isosphaera*, *Singulisphaera*, and *Gemmata*. High sensitivity to NaCl distinguishes it from *Pirellula*, *Blastopirellula*, and *Rhodopirellula*. With the only exception of *Singulisphaera*, the capability to grow at pH below 5.0 distinguishes *Schlesneria* from all other currently known planctomycetes. Finally, the absence of long stalks and minor amounts of $C_{18:1}$ ω9*c* in the fatty acid profile helps to differentiate *Schlesneria* from *Planctomyces*.

Taxonomic comments

Comparative sequence analysis of the 16S rRNA gene shows that *Schlesneria paludicola* is a member of the phylum *Planctomycetes* and belongs to a phylogenetic lineage defined by the genus *Planctomyces*. Highest 16S rRNA gene sequence relatedness (87%) is displayed with *Planctomyces limnophilus* (Hirsch and Müller, 1985). Other validly described cultured species of the genus *Planctomyces*, i.e. *Planctomyces maris* (Bauld and Staley, 1976) and *Planctomyces brasilensis* (Schlesner, 1989), show lower levels of 16S rRNA gene relatedness (84–86%).

Three taxonomically described strains of *Schlesneria paludicola* were isolated from acidic *Sphagnum*-dominated boreal wetlands of northern Russia. Two other phylogenetically related (97.5–97.8% 16S rRNA gene sequence similarity) but taxonomically uncharacterized planctomycetes, strains Schlesner 638 and 642, were isolated from compost leakage water (Schlesner, 1994).

TABLE 176. Characteristics differentiating *Schlesneria* from other aerobic budding bacteria lacking peptidoglycan[a]

Characteristic	*Schlesneria*	*Blastopirellula*	*Gemmata*	*Isosphaera*	*Pirellula*	*Planctomyces*	*Rhodopirellula*	*Singulisphaera*
Cell shape	Ellipsoidal	Ovoid or ellipsoidal	Spherical to ovoid	Spherical	Ovoid or ellipsoidal	Ovoid or spherical	Ovoid or ellipsoidal	Spherical
Cells arranged in rosettes	+	+	–	–	+	+	+	–
Cells arranged in filaments	–	–	–	+	–	–	–	–
Presence of long stalks	–	–	–	–	–	+	–	–
Pigmentation	–	–	+	+	–	v	+	–
Growth below pH 5	+	–	–	–	–	–	–	+
Growth above 40°C	–	–	–	+	–	–	–	–
Tolerance of >0.5% NaCl	–	+	nd	nd	+	+	+	–
Presence of $C_{18:2}$ fatty acids	–	–	–	–	–	–	–	+
High content of $C_{18:1}\omega9c$	–	+	–	nd	+	+	+	+
DNA G+C content (mol%)	54–57	54–57	64	62	54–57	51–58	53–57	57–60

[a]Symbols: +, >85% positive; –, 0–15% positive; nd, not determined; v, variable (colorless, light rose, red, and yellow).

List of species of the genus *Schlesneria*

1. **Schlesneria paludicola** Kulichevskaya, Ivanova, Belova, Baulina, Bodelier, Rijpstra, Sinninghe Damsté, Zavarzin and Dedysh 2007, 2686[VP]

pa.lu.di.co'la. L. n. *palus, -udis* a marsh, bog; L. suff. *cola* (from L. n. *incola*) inhabitant, dweller; N.L. n. *paludicola* a bog-dweller.

Description as for the genus with the following additional information. Colonies are circular, smooth, raised, opaque, uncolored, 1–3 mm in diameter. Carbon sources (0.05%, w/v) include glucose, *N*-acetylglucosamine, cellobiose, maltose, sucrose, trehalose, fucose, and salicin. Capable of hydrolyzing fucoidan, laminarin, esculin, pectin, chondroitin sulfate, pullulan, gelatin, and xylan. Nitrogen sources (0.05%, w/v) are ammonia, nitrate, *N*-acetylglucosamine, Bacto peptone, Bacto Yeast Extract, alanine, aspartate, glutamine, and threonine. Growth factors are not required, but yeast extract slightly increases the growth rate. Members of this species are resistant to ampicillin, streptomycin, chloramphenicol, lincomycin, and novobiocin, but are sensitive to neomycin, kanamycin, and gentamicin. Optimal growth occurs at pH 5.0–6.2 and at temperatures 15 – 26°C. NaCl inhibits growth at concentrations above 0.5% (w/v).

DNA G+C content (mol%): 54.4–56.5 (T_m).

Type strain: MPL7, ATCC BAA-1393, VKM 2452.

Sequence accession no. (16S rRNA gene): AM162407.

Genus VIII. **Singulisphaera** Kulichevskaya, Ivanova, Baulina, Bodelier, Sinninghe Damsté and Dedysh 2008, 1191[VP]

Svetlana N. Dedysh and Irina S. Kulichevskaya

Sin.gu.li.spha.e'ra. L. adj. *singuli* single, separate; L. fem. n. *sphaera* sphere; N.L. fem. n. *Singulisphaera* a single spherical cell.

Spherical cells, up to 3–4 µm in diameter, occurring **singly**, in pairs, or in shapeless aggregates. **Crateriform structures** are scattered **over the whole cell surface**. Encapsulated. Immobile. Reproduce by budding. **Stalk-like structures are absent.** Attach to surfaces by means of **amorphous holdfast material**. Gram-stain-negative. Chemoorganotrophic aerobes. Capable of growth in microaerobic conditions. Colonies are opaque and nonpigmented. Various carbohydrates or *N*-acetylglucosamine are the preferred substrates. No growth factors are required. **Possess hydrolytic capabilities.** Organic nitrogen compounds, ammonium, and some amino acids serve as nitrogen sources. **Nitrate is not utilized.** Catalase, oxidase, and urease positive. Dissimilatory nitrate reduction is negative. Mesophilic and **moderately acidophilic bacteria** growing in the pH range of 4.2–7.5, with an optimum at pH 5.0–6.2. **Sensitive to NaCl.** This genus is a member of the phylum *Planctomycetes*, order *Planctomycetales*, family *Planctomycetaceae*. Acidic wetlands are the main habitat.

DNA G+C content (mol%): 57.8–59.9.

Type species: **Singulisphaera acidiphila** Kulichevskaya, Ivanova, Baulina, Bodelier, Sinninghe Damsté and Dedysh 2008, 1191[VP].

Further descriptive information

A single species, *Singulisphaera acidiphila*, has been described based on characterization of four independent isolates (Kulichevskaya et al., 2008). Mature cells of this species vary in size from 1.6 to 2.6 µm (Figure 188), but some cells in old cultures are up to 3–4 µm in diameter (Figure 188). No stable cell chains or filaments are present in cultures of *Singulisphaera acidiphila*. Examination of old cultures grown on agar plates shows the presence of an amorphous holdfast substance excreted from the cell poles (Figure 188). Thin sections show the pattern of cell compartmentalization typical for *Isosphaera*-like planctomycetes (Figure 188) (Fuerst, 2005; Giovannoni et al., 1987b; Lindsay et al., 2001). The cell wall consists of two electron-dense layers separated by an electron-transparent layer; the total wall thickness is about 12 nm. Crateriform structures with a diameter of 25 nm are scattered over the whole cell surface. An enlarged view of the cross-section of one of these pit-like structures shows that it represents a cell-wall invagination, which is connected to underlying paryphoplasm (Figure 188). The intracytoplasmic membrane compartmentalizes the cell to produce irregularly shaped membrane-bounded regions. The paryphoplasm in cells of *Singulisphaera acidiphila* is invaginated in a way to form a central region and numerous peripheral regions which are filled with an electron-dense substance. Nucleoid and electron-dense ribosome-like particles are located in the central part of the cell.

On agar media, *Singulisphaera acidiphila* forms raised, opaque, uncolored, circular colonies with an entire edge and a smooth surface. One-month-old colonies are 1–4 mm in diameter. Members of the genus *Singulisphaera* are obligately aerobic chemoheterotrophs, however, they grow well in microaerobic conditions. Most sugars and *N*-acetylglucosamine are the preferred growth substrates, but good growth occurs also on media with various biopolymers. Organic acids are either not utilized or poorly utilized by some of the strains. Representatives of *Singulisphaera acidiphila* are moderately acidophilic (pH range of 4.2–7.5) and mesophilic (temperature range of 4–33°C). High sensitivity to NaCl reflects adaption of these bacteria to dilute environments such as ombrotrophic peat bogs. Growth inhibition of 50–80% occurs in the presence of NaCl in the medium at concentrations of 0.2–0.5% (w/v), and NaCl at concentrations above 0.5% (w/v) completely inhibits growth.

Cells of *Singulisphaera acidiphila* contain menaquinone-6 (MK-6) as the predominant isoprenoid quinone. The major fatty acids are $C_{16:0}$ (23.9–36.6%), $C_{18:1}$ ω9*c* (23.5–46.7%), and $C_{18:2}$ ω6,12*c* (13.8–25.6%); the latter is genus-characteristic. The neutral lipids are dominated by an *n*-$C_{31:9}$ hydrocarbon; squalene, diplopterol, and 3-methyl-diplopterol are also present.

Enrichment and isolation procedures

Enrichment strategy and the medium used for isolation procedure (M31) are the same as for members of the genus *Schlesneria* (above). The colonies appearing on the plates are subjected to examination by phase microscopy for the presence of spherical cells that occur singly or in shapeless aggregates.

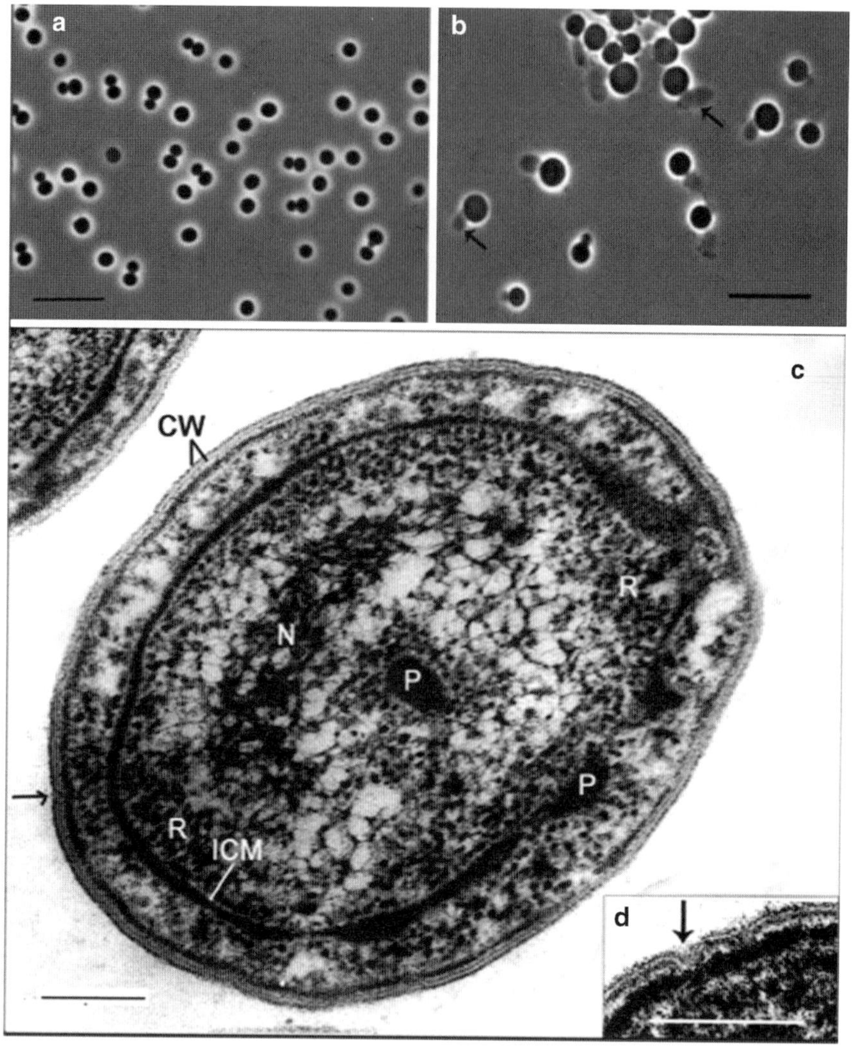

FIGURE 188. (a) Phase-contrast micrographs of cells of *Singulisphaera acidiphila* grown in liquid culture and (b) on agar medium; black arrows show amorphous holdfast substance excreted by the cells grown on agar medium. (c) Electron micrograph of an ultrathin section of a cell of *Singulisphaera acidiphila*; CW, cell wall; ICM, intracytoplasmic membrane; P, paryphoplasm; R, ribosome-like particles; N, nucleoid. Black arrow indicates pit-like invagination of a cell wall [enlarged view is shown in (d)]. Bars = 10 μm [(a) and (b)] and 0.2 μm [(c) and (d)]. [(c) and (d), Printed with permission of Olga I. Baulina.]

Maintenance procedures

Strains can be maintained on agar medium M31 by subculturing once in 2–3 months. Alternatively, they can be kept in liquid medium M31 with 10% glycerol at −70°C or stored by lyophilization.

Differentiation of the genus *Singulisphaera* from other genera

Table 176 gives characteristics of *Singulisphaera* that differentiate it from other aerobic budding bacteria that lack peptidoglycan. Spherical cell shape makes *Singulisphaera* different from *Schlesneria, Planctomyces, Pirellula, Blastopirellula,* and *Rhodopirellula.* Absence of cell filaments, pigmentation, and gliding motility differentiate *Singulisphaera* from *Isosphaera.* High sensitivity to NaCl distinguishes it from *Pirellula, Blastopirellula,* and *Rhodopirellula.* The ability to grow at pH below 5.0 distinguishes it from all other currently known planctomycetes with the only exception being *Schlesneria.* Finally, the presence of $C_{18:2}$ fatty acids is a genus-characteristic feature of *Singulisphaera* that is unique among other planctomycetes.

Taxonomic comments

Comparative sequence analysis of the 16S rRNA gene shows that *Singulisphaera acidiphila* is a member of the order *Planctomycetales* and belongs to a phylogenetic lineage defined by the genus *Isosphaera.* The 16S rRNA gene sequence identity between the *Isosphaera pallida* (Giovannoni et al., 1987b) and representatives of the *Singulisphaera acidiphila* is about 90%.

Four taxonomically described strains of *Singulisphaera acidiphila* were obtained from acidic *Sphagnum*-dominated boreal wetlands of northern Russia.

List of species of the genus *Singulisphaera*

1. **Singulisphaera acidiphila** Kulichevskaya, Ivanova, Baulina, Bodelier, Sinninghe Damsté and Dedysh 2008, 1191[VP]

a.ci.di'phi.la. N.L. n. *acidum* acid from L. adj. *acidus* sour; N.L. adj. *philus -a -um* (from Gr. adj. *philos -ê -on*), friend, loving; N.L. fem. adj. *acidiphila* acid-loving.

Description as for the genus with the following additional information. Carbon sources (0.05%, w/v) include glucose, fructose, galactose, lactose, cellobiose, maltose, mannose, melibiose, rhamnose, ribose, trehalose, saccharose, xylose, leucrose, *N*-acetylglucosamine, and salicin. Ability to utilize fucose, lactate, and pyruvate is variable. Capable of hydrolyzing laminarin, pectin, chondroitin sulfate, esculin, gelatin, pullulan, lichenan, and xylan. Shows the following enzyme activities: alkaline and acid phosphatase, esterase, esterase lipase, leucine arylamidase, cystine arylamidase, valine arylamidase, phosphohydrolase, *N*-acetyl-β-glucosaminidase, and β-galactosidase. Nitrogen sources (0.05%, w/v) are ammonia, *N*-acetylglucosamine, Bacto Peptone, Bacto Yeast Extract, alanine, aspartate, arginine, glutamine, threonine, tryptophan, and glycine. Some strains can also utilize asparagine, isoleucine, lysine, phenylalanine, proline, and valine. Nitrate or nitrite are not utilized. Vitamins are not required. Resistant to ampicillin, streptomycin, chloramphenicol, lincomycin, kanamycin, and novobiocin, but sensitive to neomycin and gentamicin. Optimal growth occurs at pH 5.0–6.2 and at temperatures 20–26°C. NaCl inhibits growth at concentrations above 0.5% (w/v).

DNA G+C content (mol%): 57.8–59.9 (T_m).

Type strain: MOB10, ATCC BAA-1392, DSM 18658, VKM B-2454.

Sequence accession no. (16S rRNA gene): AM850678.

References

Anagnostidis, K. and R. Rathsack-Künzenbach. 1967. *Isocystis pallida* - Blaualge oder hefeartiger Pilz. 29: 191–198.

Bauer, M., T. Lombardot, H. Teeling, N.L. Ward, R.I. Amann and F.O. Glöckner. 2004. Archaea-like genes for C_1-transfer enzymes in *Planctomycetes*: phylogenetic implications of their unexpected presence in this phylum. J. Mol. Evol. *59*: 571–586.

Bauld, J. and J.T. Staley. 1976. *Planctomyces maris* sp. nov., marine isolate of *Planctomyces blastocaulis* group of budding bacteria. J. Gen. Microbiol. *97*: 45–55.

Bauld, J. and J.T. Staley. 1980. *Planctomyces maris* sp. nov., nom. rev. Int. J. Syst. Bacteriol. *30*: 657–657.

Bomar, D., S. Giovannoni and E. Stackebrandt. 1988. A unique type of eubacterial 5S rRNA in members of the order *Planctomycetales*. J. Mol. Evol. *27*: 121–125.

Burdett, I.D. and R.G. Murray. 1974. Septum formation in *Escherichia coli*: characterization of septal structure and the effects of antibiotics on cell division. J. Bacteriol. *119*: 303–324.

Butler, M.K., J. Wang, R.I. Webb and J.A. Fuerst. 2002. Molecular and ultrastructural confirmation of classification of ATCC 35122 as a strain of *Pirellula staleyi*. Int. J. Syst. Evol. Microbiol. *52*: 1663–1667.

Butler, M.K. and J.A. Fuerst. 2004. Comparative analysis of ribonuclease P RNA of the *Planctomycetes*. Int. J. Syst. Evol. Microbiol. *54*: 1333–1344.

Castenholz, R.W. 1988. Culturing methods (cyanobacteria). *In* Cyanobacteria: Methods in Enzymology, vol. 167 (edited by Packer and Glazer), pp. 68–93.

Chistoserdova, L., C. Jenkins, M.G. Kalyuzhnaya, C.J. Marx, A. Lapidus, J.A. Vorholt, J.T. Staley and M.E. Lidstrom. 2004. The enigmatic *planctomycetes* may hold a key to the origins of methanogenesis and methylotrophy. Mol. Biol. Evol. *21*: 1234–1241.

Chouari, R., D. Le Paslier, P. Daegelen, P. Ginestet, J. Weissenbach and A. Sghir. 2003. Molecular evidence for novel planctomycete diversity in a municipal wastewater treatment plant. Appl. Environ. Microbiol. *69*: 7354–7363.

Claus, H., H.H. Martin, C.A. Jantos and H. Konig. 2000. A search for beta-lactamase in chlamydiae, mycoplasmas, planctomycetes, and cyanelles: bacteria and bacterial descendants at different phylogenetic positions and stages of cell wall development. Microbiol. Res. *155*: 1–6.

Cohen-Bazire, G., W.R. Sistrom and R.Y. Stanier. 1957. Kinetic studies of pigment synthesis by non-sulfur purple bacteria. J. Cell Phys. *49*: 25–68.

Dahlberg, C., M. Bergstrom and M. Hermansson. 1998. *In situ* detection of high levels of horizontal plasmid transfer in marine bacterial communities. Appl. Environ. Microbiol. *64*: 2670–2675.

Dedysh, S.N., T.A. Pankratov, S.E. Belova, I.S. Kulichevskaya and W. Liesack. 2006. Phylogenetic analysis and in situ identification of bacteria community composition in an acidic Sphagnum peat bog. Appl. Environ. Microbiol. *72*: 2110–2117.

DeLong, E.F., D.G. Franks and A.I. Alldredge. 1993. Phylogenetic diversity of aggregate-attached vs. free-living marine bacterial assemblages. Limnol. Oceanogr. *38*: 924–934.

Euzéby, J.P. and T. Kudo. 2001. Corrigenda to the Validation Lists. Int. J. Syst. Evol. Microbiol. *51*: 1933–1938.

Famurewa, O., H.G. Sonntag and P. Hirsch. 1983. Avirulence of 27 bacteria that are budding, prosthecate, or both. Int. J. Syst. Bacteriol. *33*: 565–572.

Fott, B. and J. Komarek. 1960. Das Phytoplankton der Teiche im Teschner Schlesien. Preslia *32*: 113–141.

Franzmann, P.D. and V.B.D. Skerman. 1981. *Agitococcus lubricus* gen. nov. sp. nov., a lipolytic, twitching coccus from fresh water. Int. J. Syst. Bacteriol. *31*: 177–183.

Franzmann, P.D. and V.B. Skerman. 1984. *Gemmata obscuriglobus*, a new genus and species of the budding bacteria. Antonie van Leeuwenhoek *50*: 261–268.

Franzmann, P.D. and V.B. Skerman. 1985. *In* Validation of the publication of new names and new combinations previously effectively published outside the IJSB. List no. 18. Int. J. Syst. Bacteriol. *35*: 375–376.

Fuerst, J.A., S.K. Sambhi, J.L. Paynter, J.A. Hawkins and J.G. Atherton. 1991. Isolation of a bacterium resembling *Pirellula* species from primary tissue culture of the giant tiger prawn (*Penaeus monodon*). Appl. Environ. Microbiol. *57*: 3127–3134.

Fuerst, J.A. and R.I. Webb. 1991. Membrane-bounded nucleoid in the eubacterium *Gemmata obscuriglobus*. Proc. Natl. Acad. Sci. U. S. A. *88*: 8184–8188.

Fuerst, J.A., H.G. Gwilliam, M. Lindsay, A. Lichanska, C. Belcher, J.E. Vickers and P. Hugenholtz. 1997. Isolation and molecular identification of planctomycete bacteria from postlarvae of the giant tiger prawn, *Penaeus monodon*. Appl. Environ. Microbiol. *63*: 254–262.

Fuerst, J.A. 2005. Intracellular compartmentation in *Planctomycetes*. Annu. Rev. Microbiol. *59*: 299–328.

Gade, D., H. Schlesner, F.O. Glöckner, R. Amann, S. Pfeiffer and M. Thomm. 2004. Identification of *Planctomycetes* with order-, genus-, and strain-specific 16S rRNA-targeted probes. Microb. Ecol. *47*: 243–251.

Gade, D., J. Gobom and R. Rabus. 2005a. Proteomic analysis of carbohydrate catabolism and regulation in the marine bacterium *Rhodopirellula baltica*. Proteomics *5*: 3672–3683.

Gade, D., T. Stührmann, R. Reinhardt and R. Rabus. 2005b. Growth phase dependent regulation of protein composition in *Rhodopirellula baltica*. Environ. Microbiol. 7: 1074–1084.

Gade, D., D. Theiss, D. Lange, E. Mirgorodskaya, T. Lombardot, F.O. Glockner, M. Kube, R. Reinhardt, R. Amann, H. Lehrach, R. Rabus and J. Gobom. 2005c. Towards the proteome of the marine bacterium *Rhodopirellula baltica*: mapping the soluble proteins. Proteomics *5*: 3654–3671.

Gebers, R., U. Wehmeyer, T. Roggentin, H. Schlesner, J. Kölbel-Boelke and P. Hirsch. 1985. Deoxyribonucleic acid base compositions and nucleotide distributions of 65 strains of budding bacteria. Int. J. Syst. Bacteriol. *35*: 260–269.

Geitler, L. 1963. Die angebliche Cyanophyceae *Isosphaera pallida* is ein hefeartiger Pilz. Arch. Mikrobiol. *46*: 238–242.

Gimesi, N. 1924. Hydrobiologiai tanulmanyok (Hydrobiologische Studien). I. *Planctomyces bekefii* Gim. nov. gen. et sp. (in Hungarian, with German transl.). Kiadja a Magyar Ciszterci Rend, Budapest, pp. 1–8.

Giovannoni, S.J., W. Godchaux, 3rd, E. Schabtach and R.W. Castenholz. 1987a. Cell wall and lipid composition of *Isosphaera pallida*, a budding *Eubacterium* from hot springs. J. Bacteriol. *169*: 2702–2707.

Giovannoni, S.J., E. Schabtach and R.W. Castenholz. 1987b. *Isosphaera pallida*, gen. and comb. nov., a gliding, budding *Eubacterium* from hot springs. Arch. Microbiol. *147*: 276–284.

Giovannoni, S.J., E. Shabtach and R.W. Castenholz. 1995. *In* Validation of the publication of new names and new combinations previously effectively published outside the IJSB. List no. 54. Int. J. Syst. Bacteriol. *45*: 619–620.

Gliesche, C. 1980. Isolierung und Charakterisierung von Bakteriophagen knospender Bakterien (Isolation and characterization of bacteriophages of budding bacteria). Diploma thesis, Christian-Albrechts-Universitaet, Kiel.

Glöckner, F.O., B.M. Fuchs and R. Amann. 1999. Bacterioplankton compositions of lakes and oceans: a first comparison based on fluorescence in situ hybridization. Appl. Environ. Microbiol. *65*: 3721–3726.

Glöckner, F.O., M. Kube, M. Bauer, H. Teeling, T. Lombardot, W. Ludwig, D. Gade, A. Beck, K. Borzym, K. Heitmann, R. Rabus, H. Schlesner, R. Amann and R. Reinhardt. 2003. Complete genome sequence of the marine planctomycete *Pirellula* sp. strain 1. Proc. Natl. Acad. Sci. U. S. A. *100*: 8298–8303.

Granberg, K. 1969. Kasviplanktonin merkityksesta vesilaitoksen raakvaveden tarkkailussa. Limnol. Foren. I. Finland Limnol. Symp. *1968*: 34–43.

Griepenburg, U., N. Ward-Rainey, S. Mohamed, H. Schlesner, H. Marxsen, F.A. Rainey, E. Stackebrandt and G. Auling. 1999. Phylogenetic diversity, polyamine pattern and DNA base composition of members of the order *Planctomycetales*. Int. J. Syst. Bacteriol. *49*: 689–696.

Henrici, A.T. and D.E. Johnson. 1935. Studies of freshwater bacteria: II. Stalked bacteria, a new order of *Schizomycetes*. J. Bacteriol. *30*: 61–93.

Heynig, H. 1961. Zur Kenntnis des Planktons mitteldeutscher Gewaesser. Arch. Protistenkd. *105*: 407–416.

Hindák, F. 2001. Thermal micro-organisms from a hot spring on the coast of Lake Bogoria, Kenya. Nova Hedwigia, Beiheft *123*: 77–93.

Hirsch, P. 1972. Two identical genera of budding and stalked bacteria: *Planctomyces* Gimesi 1924 and Blastocaulis Henrici and Johnson 1935. Int. J. Syst. Bacteriol. *22*: 107–111.

Hirsch, P., M. Müller and H. Schlesner. 1977. New aquatic budding and prosthecate bacteria and their taxonomic position. *In* Symposium on Aquatic Microbiology, Lancaster, UK. Academic Press, London, pp. 107–133.

Hirsch, P. and M. Müller. 1985. *Planctomyces limnophilus* sp. nov., a stalked and budding bacterium from freshwater. Syst. Appl. Microbiol. *6*: 276–280.

Hirsch, P. and M. Müller. 1986. *In* Validation of the publication of new names and new combinations previously effectively published outside the IJSB. List no. 20. Int. J. Syst. Bacteriol. *36*: 354–356.

Hortobágyi, T. 1965. Neue *Planctomyces* Arten. Tot. Koezlem. *52*: 111–119.

Hortobágyi, T. 1968. *Planctomyces* from Vietnam. Acta Phytopathol. Acad. Sci. Hung *3*: 271–273.

Hortobágyi, T. 1980. Aquatic bacteria and fungi in Danube River and in the water producing systems of the Budapest Waterworks. Acta Microbiol. Acad. Sci. Hung. *27*: 259–268.

Hortobágyi, T. and L. Hajdú. 1984. A critical survey of *Planctomyces* research. Acta Bot. Hung. *30*: 3–9.

Jenkins, C. and J.A. Fuerst. 2001. Phylogenetic analysis of evolutionary relationships of the planctomycete division of the domain bacteria based on amino acid sequences of elongation factor Tu. J. Mol. Evol. *52*: 405–418.

Kerger, B.D., C.A. Mancuso, P.D. Nichols, D.C. White, T. Langworthy, M. Sittig, H. Schlesner and P. Hirsch. 1988. The budding bacteria, *Pirellula* and *Planctomyces*, with atypical 16S rRNA and absence of peptidoglycan, show eubacterial phospholipids and uniquely high proportions of long chain beta-hydroxy fatty acids in the lipopolysaccharide lipid A. Arch. Microbiol. *149*: 255–260.

Kodaka, H., A.Y. Armfield, G.L. Lombard and V.R. Dowell, Jr. 1982. Practical procedure for demonstrating bacterial flagella. J. Clin. Microbiol. *16*: 948–952.

Kölbel-Boelke, J., R. Gebers and P. Hirsch. 1985. Genome size determination for 33 strains of budding bacteria. Int. J. Syst. Bacteriol. *35*: 270–273.

Komárek, J. and K. Anagnostidis. 2005. Cyanoprokaryota 2. *In Oscillatoriales*. Elsevier/Spektrum, Munich, pp. 336–339.

König, H., H. Schlesner and P. Hirsch. 1984. Cell wall studies on budding bacteria of the *Planctomyces - Pasteuria* group and on a *Prosthecomicrobium* sp. Arch. Microbiol. *138*: 200–205.

Krasil'nikov, N.A. 1949. Guide to the bacteria and actinomycetes. *In* Akad. Nauk. S.S.S.R. Moscow, pp. 1–830.

Kristiansen, J. 1971. On *Planctomyces bekefii* and its occurrence in Danish lakes and ponds. Bot. Tidsskr. *66*: 293–392.

Kulichevskaya, I.S., T.A. Pankratov and S.N. Dedysh. 2006. Detection of representatives of the *Planctomycetes* in *Sphagnum* peat bogs by molecular and cultivation approaches. Microbiology (En. transl. from Mikrobiologiya) *75*: 329–335.

Kulichevskaya, I.S., A.O. Ivanova, S.E. Belova, O.I. Baulina, P.L.E. Bodelier, W.I.C. Rijpstra, J.S. Sinninghe Damsté, G.A. Zavarzin and S.N. Dedysh. 2007. *Schlesneria paludicola* gen. nov., sp. nov., the first acidophilic member of the order *Planctomycetales*, from *Sphagnum*-dominated boreal wetlands. Int. J. Syst. Evol. Microbiol. *57*: 2680–2687.

Kulichevskaya, I.S., A.O. Ivanova, O.I. Baulina, P.L.E. Bodelier, J.S. Sinninghe Damsté and S.N. Dedysh. 2008. *Singulisphaera acidiphila* gen. nov., sp. nov., a non-filamentous, *Isosphaera*-like planctomycete from acidic northern wetlands. Int. J. Syst. Evol. Microbiol. *58*: 1186–1193.

Kustu, S., E. Santero, J. Keener, D. Popham and D. Weiss. 1989. Expression of sigma 54 (*ntrA*)-dependent genes is probably united by a common mechanism. Microbiol. Rev. *53*: 367–376.

Leary, B.A., N. Ward-Rainey and T.R. Hoover. 1998. Cloning and characterization of *Planctomyces limnophilus rpoN*: complementation of a *Salmonella typhimurium rpoN* mutant strain. Gene *221*: 151–157.

Liesack, W., H. König, H. Schlesner and P. Hirsch. 1986. Chemical composition of the peptidoglycan-free cell envelopes of budding bacteria of the *Pirella/Planctomyces* group. Arch. Microbiol. *145*: 361–366.

Liesack, W. and E. Stackebrandt. 1989. Evidence for unlinked rrn operons in the planctomycete *Pirellula marina*. J. Bacteriol. *171*: 5025–5030.

Lindsay, M., R.I. Webb, H.M. Hosmer and J.A. Fuerst. 1995. Effects of fixative and buffer on morphology and ultrastrcuture of a freshwater planctomycete, *Gemmata obscuriglobus*. J. Microbiol. Methods *21*: 45–54.

Lindsay, M.R., R.I. Webb and J.A. Fuerst. 1997. Pirellulosomes: a new type of membrane-bounded cell compartment in planctomycete bacteria of the genus *Pirellula*. Microbiology *143*: 739–748.

Lindsay, M.R., R.I. Webb, M. Strous, M.S. Jetten, M.K. Butler, R.J. Forde and J.A. Fuerst. 2001. Cell compartmentalisation in *Planctomycetes*: novel types of structural organisation for the bacterial cell. Arch. Microbiol. *175*: 413–429.

Llobet-Brossa, E., R. Rossello-Mora and R. Amann. 1998. Microbial community composition of wadden sea sediments as revealed by fluorescence *in situ* hybridization. Appl. Environ. Microbiol. *64*: 2691–2696.

Lyman, J. and R.H. Fleming. 1940. Composition of sea water. J. Mar. Res. *3*: 134–146.

Majewski, D.M. 1985. Molekularbiologische Charakterisierung von Bakteriophagen knospender Bakterien (Molecular biological characterisation of bacteriophages of budding bacteria). Diploma thesis, Christian-Albrechts-Universitaet, Kiel.

Menke, M.A.O.H., W. Liesack and E. Stackebrandt. 1991. Ribotyping of 16S and 23S rRNA genes and organization of rrn operonsin members of the bacterial genera *Gemmata, Planctomyces, Thermotoga, Thermus* and *Verrucomicrobium*. Arch. Microbiol. *155*: 263–271.

Neef, A., R. Amann, H. Schlesner and K.H. Schleifer. 1998. Monitoring a widespread bacterial group: *in situ* detection of *Planctomycetes* with 16S rRNA-targeted probes. Microbiology *144*: 3257–3266.

Olah, J. and L. Hajdu. 1973. Electron microscopic morphology of *Planctomyces bekefii* (*sic*) Gimesi. Arch. Hydrobiol. *71*: 271–275.

Parra, O. 1972. Presencia del genero *Planctomyces* (Fungi Imperfecti - Moniliales) en Chile. Bol. Soc. Arg. Bot. *14*: 282–284.

Pearson, A., M. Budin and J.J. Brocks. 2003. Phylogenetic and biochemical evidence for sterol synthesis in the bacterium *Gemmata obscuriglobus*. Proc. Natl. Acad. Sci. U. S. A. *100*: 15352–15357.

Rabus, R., D. Gade, R. Hellwig, M. Bauer, F.O. Glöckner, M. Kube, H. Schlesner, R. Reinhardt and R. Amann. 2002. Analysis of *N*-acetylglucosamine metabolism in the marine bacterium *Pirellula* sp. strain 1 by a proteomic approach. Proteomics *2*: 649–655.

Razumov, A.S. 1949. *Gallionella kljasmensis* sp. n. a bacterial component of the plankton (in Russian). Mikrobiologiya *18*: 442–446.

Rönner, S., W. Liesack, J. Wolters and E. Stackebrandt. 1991. Cloning and sequencing of a large fragment of the *atpD*-gene of *Pirellula marina*: a contribution to the phylogeny of *Planctomycetales*. Endocytobios Cell Res. *7*: 219–229.

Schlesner, H. and P. Hirsch. 1984. Assignment of ATCC 27377 to *Pirella* gen. nov. as *Pirella staleyi* comb. nov. Int. J. Syst. Bacteriol. *34*: 492–495.

Schlesner, H. 1986. *Pirella marina* sp. nov., a budding, peptidoglycanless bacterium from brackish water. Syst. Appl. Microbiol. *8*: 177–180.

Schlesner, H. and E. Stackebrandt. 1986. Assignment of the genera *Planctomyces* and *Pirella* to a new family *Planctomycetaceae* fam. nov. and description of the order *Planctomycetales* ord. nov. Syst. Appl. Microbiol. *8*: 174–176.

Schlesner, H. and P. Hirsch. 1987. Rejection of the genus name *Pirella* for pear-shaped budding bacteria and proposal to create the genus *Pirellula* gen. nov. Int. J. Syst. Bacteriol. *37*: 441–441.

Schlesner, H. and E. Stackebrandt. 1987. *In* Validation of the publication of new names and new combinations previously effectively published outside the IJSB. List no. 23. Int. J. Syst. Bacteriol. *37*: 179–180.

Schlesner, H. 1989. *Planctomyces brasiliensis* sp. nov., a halotolerant bacterium from a salt pit. Syst. Appl. Microbiol. *12*: 159–161.

Schlesner, H. 1990. *In* Validation of the publication of new names and new combinations previously effectively published outside the IJSB. List no. 32. Int. J. Syst. Bacteriol. *40*: 105–106.

Schlesner, H. 1994. The development of media suitable for the microorganisms morphologically resembling *Planctomyces* spp., *Pirellula* spp., and other *Planctomycetales* from various aquatic habitats using dilute media. Syst. Appl. Microbiol. *17*: 135–145.

Schlesner, H., C. Rensmann, B.J. Tindall, D. Gade, R. Rabus, S. Pfeiffer and P. Hirsch. 2004. Taxonomic heterogeneity within the *Planctomycetales* as derived by DNA–DNA hybridization, description of *Rhodopirellula baltica* gen. nov., sp. nov., transfer of *Pirellula marina* to the genus *Blastopirellula* gen. nov. as *Blastopirellula marina* comb. nov. and emended description of the genus *Pirellula*. Int. J. Syst. Evol. Microbiol. *54*: 1567–1580.

Schmidt, J.M. 1978. Isolation and ultrastructure of freshwater strains of *Planctomyces*. Curr. Microbiol. *1*: 65–70.

Schmidt, J.M. and M.P. Starr. 1978. Morphological diversity of freshwater bacteria belonging to the *Blastocaulis–Planctomyces* group as observed in natural populations and enrichments. Curr. Microbiol. *1*: 325–330.

Schmidt, J.M. and M.P. Starr. 1979a. Morphotype V of the *Blastocaulis–Planctomyces* group of budding and appendaged bacteria: *Planctomyces guttaeformis* Hortobágyi (*sensu* Hajdu). Curr. Microbiol. *2*: 195–200.

Schmidt, J.M. and M.P. Starr. 1979b. Corniculate cell surface protrusions in morphotype II of the *Blastocaulis–Planctomyces* group of budding and appendaged bacteria. Curr. Microbiol. *3*: 187–190.

Schmidt, J.M. and M.P. Starr. 1980a. Some ultrastructural features of *Planctomyces bekefii*, morphotype I of the *Blastocaulis–Planctomyces* group of budding and appendaged bacteria. Curr. Microbiol. *4*: 189–194.

Schmidt, J.M. and M.P. Starr. 1980b. Current sightings, at the respective type localities and elsewhere, of *Planctomyces bekefii* Gimesi 1924 and *Blastocaulis sphaerica* Henrici and Johnson 1935. Curr. Microbiol. *4*: 183–188.

Schmidt, J.M., W.P. Sharp and M.P. Starr. 1981. Manganese and iron encrustations and other features of *Planctomyces crassus* Hortobágyi 1965, morphotype Ib of the *Blastocaulis–Planctomyces* group of budding and appendaged bacteria, examined by electron microscopy and X-ray micro-analysis. Curr. Microbiol. *5*: 241–246.

Schmidt, J.M. and M.P. Starr. 1981. The *Blastocaulis–Planctomyces* group of budding and appendaged bacteria. *In* The Prokaryotes: a Handbook on Habitats, Isolation, and Identification of *Bacteria*, vol. 1 (edited by Starr, Stolp, Trüper, Balows and Schlegel). Springer, New York, pp. 496–504.

Schmidt, J.M., W.P. Sharp and M.P. Starr. 1982. Metallic-oxide encrustations of the nonprosthecate stalks of naturally occurring populations of *Planctomyces bekefii*. Curr. Microbiol. *7*: 389–394.

Schmidt, J.M. and M.P. Starr. 1982. Ultrastructural features of budding cells in a prokaryote belonging to morphotype IV of the *Blastocaulis–Planctomyces* group. Curr. Microbiol. *7*: 7–11.

Sittig, M. and H. Schlesner. 1993. Chemotaxonomic investigation of various prosthecate and or budding bacteria. Syst. Appl. Microbiol. *16*: 92–103.

Skerman, V.B. 1968. A new type of micromanipulator and microforge. J. Gen. Microbiol. *54*: 287–297.

Skuja, H. 1964. Grundzuege der Algenflora und Algenvegetation der Fjeldgegenden um Abisko in Schwedisch-Lappland. Nova Acta Reg. Soc. Sci. Upsal. Ser. IV *18*: 1–139.

Stackebrandt, E., W. Ludwig, W. Schubert, F. Klink, H. Schlesner, T. Roggentin and P. Hirsch. 1984. Molecular genetic evidence for early evolutionary origin of budding peptidoglycan-less eubacteria. Nature *307*: 735–737.

Stackebrandt, E., U. Wehmeyer and W. Liesack. 1986. 16S Ribosomal RNA and cell wall analysis of *Gemmata obscuriglobus*, a new member of the order *Planctomycetales*. FEMS Microbiol. Lett. *37*: 289–292.

Staley, J.T. 1968. *Prosthecomicrobium* and *Ancalomicrobium*: new prosthecate freshwater bacteria. J. Bacteriol. *95*: 1921–1942.

Staley, J.T. 1973. Budding bacteria of the *Pasteuria–Blastobacter* group. Can. J. Microbiol. *19*: 609–614.

Staley, J.T., K.C. Marshall and V.B.D. Skerman. 1980. Budding and prosthecate bacteria from freshwater habitats of various trophic states. Microb. Ecol. *5*: 245–251.

Staley, J.T. 1981a. The genera *Prosthecomicrobium* and *Ancalomicrobium*. *In* The Prokaryotes: a Handbook on Habitats, Isolation, and Identification of *Bacteria* (edited by Starr, Stolp, Trüper, Balows and Schlegel). Springer, New York, pp. 456–460.

Staley, J.T. 1981b. The genus *Pasteuria*. *In* The Prokaryotes: a Handbook on Habitats, Isolation, and Identification of *Bacteria* (edited by Starr, Stolp, Trüper, Balows and Schlegel). Springer, New York, pp. 490–492.

Staley, J.T., J.A. Fuerst, S. Giovannoni and H. Schlesner. 1992. The order *Planctomycetales* and the genera *Planctomyces, Pirellula, Gemmata* and *Isosphaera*. *In* The Prokaryotes: a Handbook on the Biology of *Bacteria*: Ecophysiology, Isolation, Identification, Applications, 2nd edn, vol. 4 (edited by Balows, Trüper, Dworkin, Harder and Schleifer). Springer, New York, pp. 3710–3731.

Starr, M.P., R.M. Sayre and J.M. Schmidt. 1983. Assignment of ATCC 27377 to *Planctomyces staleyi* sp. nov. and conservation of *Pasteuria ramosa* Metchnikoff 1888 on the basis of type descriptive material: Request for an Opinion. Int. J. Syst. Bacteriol. *33*: 666–671.

Starr, M.P. and J.M. Schmidt. 1984. *Planctomyces stranskae* (*ex* Wawrik 1952) sp. nov., nom. rev. and *Planctomyces guttaeformis* (*ex* Hortobágyi 1965) sp. nov., nom. rev. Int. J. Syst. Bacteriol. *34*: 470–477.

Starr, M.P. and J.M. Schmidt. 1989. Genus *Planctomyces* Gimesi 1924. *In* Bergey's Manual of Systematic Bacteriology, vol. 3 (edited by Staley, Bryant, Pfennig and Holt). Williams & Wilkins, Baltimore, pp. 1946–1958.

Teiling, E. 1942. Schwedische Planktonalgen. 3. Neue oder wenig bekannte Formen. Bot. Not. *1942*: 63–68.

Tekniepe, B.L., J.M. Schmidt and M.P. Starr. 1981. Life cycle of a budding and appendaged bacterium belonging to morphotype IV of the *Blastocaulis–Planctomyces* group. Curr. Microbiol. *5*: 1–6.

Tell, G. 1975. Presencia de *Planctomyces bekefii* (Fungi Imperfecti, Moniliales) en la Argentina. Physis (B. Aires) *34*: 71.

Van Ert, M. and J.T. Staley. 1971. Gas-vacuolated strains of *Microcyclus aquaticus*. J. Bacteriol. *108*: 236–240.

Wang, J., C. Jenkins, R.I. Webb and J.A. Fuerst. 2002. Isolation of *Gemmata*-like and *Isosphaera*-like planctomycete bacteria from soil and freshwater. Appl. Environ. Microbiol. *68*: 417–422.

Ward-Rainey, N. 1996. Genetic diversity in members of the order *Planctomycetales*. PhD thesis, University of Warwick, Coventry.

Ward-Rainey, N., F.A. Rainey, E.M. Wellington and E. Stackebrandt. 1996. Physical map of the genome of *Planctomyces limnophilus*, a representative of the phylogenetically distinct planctomycete lineage. J. Bacteriol. *178*: 1908–1913.

Ward-Rainey, N., F.A. Rainey and E. Stackebrandt. 1997. The presence of a *dnaK* (HSP70) multigene family in members of the orders *Planctomycetales* and *Verrucomicrobiales*. J. Bacteriol. *179*: 6360–6366.

Ward, N., F.A. Rainey, E. Stackebrandt and H. Schlesner. 1995. Unraveling the extent of diversity within the order *Planctomycetales*. Appl. Environ. Microbiol. *61*: 2270–2275.

Ward, N.L., F.A. Rainey, B.P. Hedlund, J.T. Staley, W. Ludwig and E. Stackebrandt. 2000. Comparative phylogenetic analyses of members of the order *Planctomycetales* and the division *Verrucomicrobia*: 23S rRNA gene sequence analysis supports the 16S rRNA gene sequence-derived phylogeny. Int. J. Syst. Evol. Microbiol. *50*: 1965–1972.

Wawrik, F. 1952. *Planctomyces*-Studien. Sydowia Ann. Mycol. Ser. II *6*: 443–451.

Wawrik, F. 1956. Neue Planktonorganismen aus Waldviertler Fischteichen. Oesterr. Bot. Z. *103*: 291–299.

Weisburg, W.G., T.P. Hatch and C.R. Woese. 1986. Eubacterial origin of *Chlamydiae*. J. Bacteriol. *167*: 570–574.

Woronichin, N.N. 1927. Materiali k agologitscheskoj flore i rastitjelnosti mineralnich istotchnikov gruppie Kaukaskich mineralnich wod. Trav. Inst. Balneol. aux Eaux Miner du Caucase *5*: 90–91.

Zavarzin, G.A. 1961. Budding bacteria. Mikrobiologiya *30*: 774–791.

Order II. "*Candidatus* Brocadiales" ord. nov.

Mike S.M. Jetten, Huub J.M. Op den Camp, J. Gijs Kuenen and Marc Strous

Bro.ca.di.a′les. N.L. fem. n. "*Candidatus* Brocadia" type genus of the order; *-ales* ending to denote an order; N.L. fem. pl. n. *Brocadiales* the order of "*Candidatus* Brocadia".

The description is the same as for the family "*Candidatus* Brocadiaceae".

Type genus: "***Candidatus* Brocadia**" Strous et al. 1999; Kuenen and Jetten 2001.

References

Kuenen, J.G. and M.S.M. Jetten. 2001. Extraordinary anaerobic ammonium-oxidizing bacteria. ASM News *67*: 456–463.

Strous, M., J.A. Fuerst, E.H.M. Kramer, S. Logemann, G. Muyzer, K.T. van de Pas-Schoonen, R. Webb, J.G. Kuenen and M.S.M. Jetten. 1999. Missing lithotroph identified as new planctomycete. Nature *400*: 446–449.

Family I. "*Candidatus* Brocadiaceae" fam. nov.

Mike S.M. Jetten, Huub J.M. Op den Camp, J. Gijs Kuenen and Marc Strous

Bro.ca.di.a.ce′ae. N.L. fem. n. [*Candidatus*] Brocadia type genus of the family; *-aceae* ending to denote a family; N.L. fem. pl. n. *Brocadiaceae* the family of "*Candidatus* Brocadia".

Coccoid cells, generally 0.7–1.1 × 1.1–1.3 μm. Depending on the growth conditions, anammox cells occur singly or in aggregates. Phenomenologically "Gram-stain-negative". No endospores are formed. Ultrastructure similar to that of the Planctomycetales, with typical membrane-bound riboplasm and paryphoplasm. An internal organelle (the anammoxosome), the locus of anammox metabolism, is present in the cytoplasm. Nonmotile. Obligately anaerobic. **Facultatively chemolithoautotrophic on ammonium and nitrite.** Nitrite (which is converted to nitrate) is the electron donor for autotrophic CO_2-fixation. Electron transport is cytochrome-based with nitrate, nitrite, Mn(IV), or Fe(III) as terminal electron acceptor. **Hydrazine is produced from ammonium and hydroxylamine.** Nitrate: nitrite oxidoreductase activity is present. Catalase- and hydrogen peroxidase-positive. **Media supporting an autotrophic lifestyle, containing ammonium, nitrite, and bicarbonate, are required.** Optimum growth temperature for enriched species 15–40°C. Very slow growth with doubling times of 2–3 weeks.

Type genus: **"Candidatus Brocadia"** Strous et al. 1999a; Kuenen and Jetten 2001.

Taxonomic comments

The family "*Candidatus* Brocadiaceae" contains the bacteria responsible for anaerobic oxidation of ammonium (anammox). So far all anammox bacteria identified on the basis of 16S rRNA gene sequencing comprise a monophyletic cluster within the *Planctomycetes* lineage of descent (Jetten et al., 2005b; Quan et al., 2008; Schmid et al., 2005; Woebken et al., 2008). Although knowledge on the microbiology of the anammox bacteria is steadily increasing, at this point the information is still too limited to describe the five recognized genera of the family "*Candidatus* Brocadiaceae" separately (Strous and Jetten, 2004). As isolation of pure cultures using traditional methods has so far not been successful, all type species have the *Candidatus* status (Kartal et al., 2007b, 2008; Kuenen and Jetten, 2001; Quan et al., 2008; Schmid et al., 2001, 2003; Strous et al., 1999a; Woebken et al., 2008). Highly purified cultures have been obtained by physical separation. The genome of one of the anammox bacteria "*Candidatus* Kuenenia stuttgartiensis" has been sequenced (Strous et al., 2006).

Further descriptive information

Phylogeny and detection. Comparative 16S rRNA gene sequence analysis shows more than 15% sequence difference between the anammox bacteria and cultivated members of the *Planctomycetales* (Figure 189). The similarity of the five anammox genera is well below the 97% threshold value typically used to separate species. The deep monophyletic branching of the anammox bacteria has been confirmed by analysis of large datasets of concatenated ribosomal proteins and genes (Strous et al., 2006). Since the anammox bacteria have several features in common with members of the order *Planctomycetales*, notably its compartmentalized ultrastructure, it is not surprising that the oligonucleotide probe S-P-Planc-0046-a-A-18 (Table 177; Neef et al., 1998) also hybridizes with all five genera of anammox bacteria. Most of the initial probes designed for anammox bacteria targeted only "*Candidatus* Brocadia anammoxidans" (Table 177; Schmid et al., 2005; Strous et al., 1999a). However, probe S * Amx 0820 a A 22 also hybridized with "*Candidatus* Kuenenia stuttgartiensis". "*Candidatus* Kuenenia" and "*Candidatus* Brocadia" can be distinguished by fluorescence *in situ* hybridization (FISH) microscopy using the probes S-S-Kst-0157-a-A-18 and S-S-Ban-0162-a-A-18 (Table 177; Schmid et al., 2005). Probe S-*-Kst-1275-a-A-20 (Table 177) is specific for "*Candidatus* Kuenenia stuttgartiensis". The genus "*Candidatus* Scalindua" can be detected by probes S-G-Sca-1309-a-A-21, S-*-Scabr-1114-a-A-22 and S-*-BS-820-a-A-22 (Table 177). Specific probes for the genera "*Candidatus* Anammoxoglobus" and "*Candidatus* Jettenia" have recently become available (Kartal et al., 2007b; Quan et al., 2008). The 23S rRNA targeting probe L-*-Amx-1900-a-A-21 was designed to detect "*Candidatus* Brocadia" and "*Candidatus* Kuenenia" (Schmid et al., 2001). Recently the intergenic spacer and large parts of the 23S ribosomal gene of the five different anammox genera have become available to design new probes (Quan et al., 2008; Woebken et al., 2008).

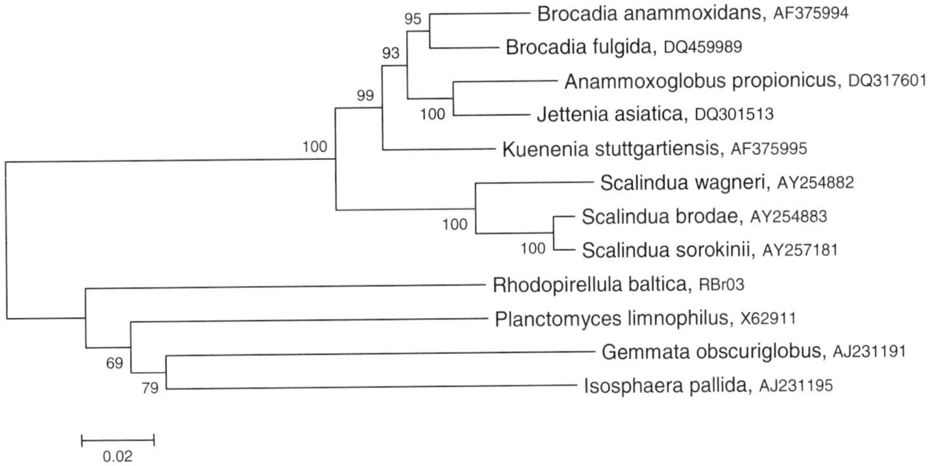

FIGURE 189. Phylogenetic affiliation of the *Brocadiales* to the *Planctomycetales*, based on 16S rRNA gene analysis. The evolutionary history was inferred using the neighbor-joining method (Saitou and Nei, 1987). The optimal tree with the sum of branch length = 0.93 is shown. The percentage of replicate trees in which the associated taxa clustered together in the bootstrap test (500 replicates) is shown next to the branches (Felsenstein, 1985). The tree is drawn to scale, with branch lengths in the same units as those of the evolutionary distances used to infer the phylogenetic tree. The evolutionary distances were computed using the maximum composite likelihood method (Tamura et al., 2004) and are expressed as the number of base substitutions per site. All positions containing alignment gaps and missing data were eliminated only in pairwise sequence comparisons (pairwise deletion option). There were a total of 1633 positions in the final dataset. Phylogenetic analyses were conducted in MEGA4 (Tamura et al., 2007). Accession numbers are given after the species names.

TABLE 177. Phenotypic characteristics of members of the order "*Candidatus* Brocadiales"

Characteristic	*Brocadia*	*Anammoxoglobus*	*Jettenia*	*Kuenenia*	*Scalindua*
Gram stain	–	–	–	–	–
Motility	–	–	–	–	–
Catalase	+	+	nd	+	+
Oxidase	+	+	nd	+	+
Nitrate reduction	+	+	nd	+	+
Nitrite, gas	+	+	+	+	+
N_2H_4 production	+	+	nd	+	+
HZO gene	+	+	+	+	+
Anammoxosome	+	+	+	+	+
Ladderane lipids	+	+	nd	+	+
Salt tolerance	–	nd	nd	+	+
Phosphate tolerance	–	nd	nd	+	nd
Oxidation of organic acids:					
Formate	+	++	nd	+	+
Acetate	++	++	nd	+	+
Propionate	+	++	nd	+	–
Oxidation of methylamine	+	nd	nd	nd	nd
Oxidation of hydrogen	–	nd	nd	–	–
Hybridization in FISH:					
Probe AMX368	+	+	+	+	+
Probe AMX820	+	–	–	+	–
Probe KST1275	–	–	–	+	–
Probe BS820	–	–	–	–	+

Symbols: +, greater than 90% of strains positive; ++, high activity; –, less than 10% positive; nd, not determined.

Cell-wall composition. Peptidoglycan appears to be absent and electron microscopy shows a visible proteinaceous S-layer (van Niftrik et al., 2008a, b). The absence of a complete peptidoglycan biosynthesis route was confirmed by genomic analysis of *Kuenenia stuttgartiensis* (Strous et al., 2006).

Ultrastructure. Like the cells of other planctomycetes, anammox bacteria possess the shared planctomycete cell plan involving a single-membrane-bounded compartment containing the nucleoid and ribosomes, as well as a paryphoplasm region surrounding the outer rim of the cell (Fuerst, 2005). Within the riboplasm, however, members of the family "*Candidatus* Brocadiaceae" possess another single membrane-bounded compartment, the anammoxosome, unique to these bacteria. The anammoxosome harbors at least one specialized enzyme, a hydrazine/hydroxylamine oxidoreductase functional in the mechanism of anaerobic ammonium oxidation (Lindsay et al., 2001; Schalk et al., 2000; Shimamura et al., 2007, 2008; van Niftrik et al., 2008a, b). The anammoxosome is wrapped in a single-membrane envelope possessing unique cyclobutane-containing ladderane lipids which may confer relatively high density and reduced permeability to the anammoxosome membrane.

Ladderane lipids, fatty acids, and sterols. The main components in the membrane lipids of anammox bacteria were shown to be so-called "ladderane lipids" (Damste et al., 2002, 2005). These lipids are comprised of three-to-five linearly concatenated cyclobutane moieties with *cis* ring junctions, which occur as fatty acids, fatty alcohols, alkyl glycerol monoethers, dialkyl glycerol diethers, and alkyl acyl glycerol ether/esters (Damste et al., 2005). The internal membrane surrounding the anammoxosome organelle is nearly exclusively composed of ladderane lipids as a barrier against diffusion of protons and intermediates (van Niftrik et al., 2004). The most abundant fatty acids of anammox bacteria are 14-methylpentadecanoic acid (C_{16}), 10-methylhexadecanoic acid, and 9,14-dimethylpentadecanoic acid. In addition, the anammox bacteria also contain squalene, and a number of hopanoids (hop-22[29]-ene, diplopterol, 17β,21β(H)-bis-homohopanoic acid, 17β,21β(H)-32-hydroxy-trishomohopanoic acid, 22,29,30-trisnor-21-oxo-hopane; Damste et al., 2004). Recently the intact phospholipids of anammox bacteria containing phosphoethanolamine and phosphocholine headgroups have been investigated and compared to each other (Boumann et al., 2006; Rattray, 2008).

Nutrition and growth conditions. A medium sustaining a chemolithoautotrophic lifestyle on ammonium, nitrite, and carbon dioxide in the absence of oxygen is required for growth of anammox bacteria (Jetten et al., 2005a; van de Graaf et al., 1995, 1996). Furthermore, a continuous flow reactor set up with very efficient biomass retention is required to maintain optimal conditions for (exponential) growth (Strous et al., 1998, 2002; van der Star, 2008a). A sequencing fed-batch reactor has been proven to be very suitable for the enrichment and maintenance of a growing anammox culture. Recently, also a membrane reactor was described to produce high quality single cells (van der Star, 2008b).

The basic mineral medium contains (in g/l): $KHCO_3$, 1.25; NaH_2PO_4, 0.05; $CaCl_2 \cdot 2H_2O$, 0.3; $MgSO_4 \cdot 7H_2O$, 0.2; $FeSO_4$, 0.00625; EDTA, 0.00625; trace element solution, 1.25 ml/l. The trace element solution contains (g/l): EDTA, 15; $ZnSO_4 \cdot 7H_2O$, 0.43; $CoCl_2 \cdot 6H_2O$, 0.24; $MnCl_2 \cdot 4H_2O$, 1.0; $CuSO_4 \cdot H_2O$, 0.25; $NaMoO_4 \cdot 2H_2O$, 0.22; $NiCl_2 \cdot 6H_2O$, 0.19; $NaSeO_4 \cdot 10H_2O$, 0.21; H_3BO_4, 0.014; $NaWO_4 \cdot 2H_2O$, 0.050. For the enrichment of marine "*Candidatus* Scalindua" species an appropriate seawater medium should be composed (Kartal et al., 2006; Nakajima et al., 2008; van de Vossenberg, 2008; Windey et al., 2005). The genus "*Candidatus* Anammoxoglobus" and "*Candidatus* Brocadia fulgida" obtain a competitive advantage when using propionate and acetate, as an additional energy source, respectively. Thus propionate or acetate can be added to the ammonium-nitrite enrichment medium in a C/N ratio of 1/6 (Guven et al., 2005; Kartal et al., 2007b, 2008).

The medium concentrations of $NaNO_2$, $(NH_4)_2SO_4$, and $NaNO_3$ are initially set to 5, 5, and 10 mM each. Suitable starting material for the enrichment of anammox bacteria is activated sludge from a treatment plant with a long sludge age, or anoxic sediments which receive a continuous supply of ammonium and nitrate (Dapena-Mora et al., 2004; Fujii et al., 2002; Kartal et al., 2007b, 2008; Schmid et al., 2000, 2003; Toh et al., 2002; van de Vossenberg, 2008; van der Star et al., 2007; Zhang et al., 2007). The biomass concentration in the enrichment cultures should stay above 1 g dry wt per liter to prevent loss of activity. The inflowing ammonium and nitrite concentrations can be increased gradually (to a maximum of 45 mM each) as long as the nitrite is completely consumed. Successful enrichments often have a reddish, pink, or brownish appearance due to the high cytochrome content of the anammox bacteria. In these enrichments anammox bacteria comprise about 80–85% of the community as determined by specific oligonucleotide probes using FISH. Purified cells can be obtained via physical separation using density centrifugation (Kartal et al., 2007a, 2008; Strous et al., 1999a, 2002).

Antibiotic sensitivity. Penicillin G can be added to the media at 0.1 mg/ml to suppress growth of undesirable contaminants (van de Graaf et al., 1995). Whether the insensitivity of anammox bacteria to Penicillin G is caused by lack of peptidoglycan or the growth in aggregates remains to be established.

Ecology. Anammox bacteria were first discovered in wastewater treatment systems and most successful enrichments have used inocula from such environments (Jetten et al., 2005a, b; Schmid et al., 2000, 2003). However, it is now clear that the occurrence of anammox bacteria is practically ubiquitous. Many studies have detected the presence and activity of anammox bacteria in more than 30 natural freshwater and marine ecosystems all over the world (Francis et al., 2007; Op den Camp et al., 2006; Penton et al., 2006; Tsushima et al., 2007). The detection and quantification of the anammox bacteria in these ecosystems was based on a combination of different techniques: enrichment cultures, nutrient profiles, ^{15}N labeling incubations with sediments or water samples, ladderane lipid analysis, FISH microscopy, or 16S rRNA gene sequencing (Dalsgaard et al., 2005; Egli et al., 2001; Kuypers et al., 2003, 2005; Pynaert et al., 2003; Rich et al., 2008; Risgaard-Petersen et al., 2004; Rysgaard and Glud, 2004; Schubert et al., 2006; Tal et al., 2005; Thamdrup and Dalsgaard, 2002). As the anammox bacteria require the simultaneous presence of ammonium and nitrite for growth, they are typical interface organisms, thriving in sediments, biofilms, and stratified waterbodies. In anoxic sediments with low organic carbon content, the anammox bacteria can account for 20–79% of total N_2 production. Studies in the anoxic water columns of the Black Sea and the Golfe Dulce showed that anammox bacteria were responsible for 20–50% of the total N_2 production (Dalsgaard et al., 2003; Kuypers et al., 2003; Thamdrup et al., 2004). Anammox bacteria are also mainly responsible for nitrogen loss in the oxygen minimum zone (OMZ) that are most productive regions of the world oceans (Hamersley et al., 2007; Kuypers et al., 2005; Thamdrup et al., 2006). The strong N-deficit in the OMZ was until now attributed to denitrification, but recent studies showed unequivocally that the anammox bacteria are responsible for the majority of nitrogen transformation in the OMZs. Based on these observations, it is likely that anammox also plays an important role in other OMZ waters and sediments of the ocean.

In various ecosystems, anammox bacteria will be dependent on the activity of aerobic ammonium-oxidizing bacteria especially when the oxygen supply is limited for example in biofilms or aggregates (Schmidt et al., 2002). In these oxygen-limited environments, the ammonium-oxidizing bacteria would oxidize ammonium to nitrite and keep the oxygen concentration low, while anammox bacteria would convert the produced nitrite and the remaining ammonium to dinitrogen gas (Strous et al., 1997; Third et al., 2001, 2005). Such conditions have been established in many different man-made ecosystems (Schmidt et al., 2002). FISH analysis and activity measurements showed that aerobic as well as anammox bacteria were present and active, but aerobic nitrite oxidizers (i.e., *Nitrobacter* or *Nitrospira*) were not detected (Third et al., 2001, 2005). It seems likely that under these conditions anaerobic and aerobic ammonium oxidizers form a quite stable community. The cooperation of ammonium-oxidizing bacteria is not only relevant for wastewater treatment, but might play an important role in natural environments at the oxic/anoxic interface (Kindaichi et al., 2007; Kuypers et al., 2005; Lam et al., 2007; Nielsen et al., 2005; Woebken et al., 2007).

Ecophysiology. The anammox bacteria have a relatively well-studied ecophysiology (Strous et al., 1999b). The anammox bacteria grow very slowly. This is caused by a low maximum substrate conversion rate, rather than by a low biomass yield. The K_s values of anammox bacteria for ammonium and nitrite are below the detection level (<5 μM). Anammox bacteria are reversibly inhibited by very low levels (<1 μM) of oxygen and irreversibly inhibited by high nitrite (>10 mM), low methanol (<0.5 mM), and acetylene concentrations (Guven et al., 2005; Jensen et al., 2007; Strous et al., 1999b; Third et al., 2005). All anammox bacteria investigated so far produce hydrazine from hydroxylamine and ammonium (Kartal et al., 2007b; van de Graaf et al., 1997; van der Star, 2008a). Anammox bacteria may be able to use organic acids as additional electron donors (Table 177) for nitrate reduction to nitrite and nitrite reduction to ammonium (Guven et al., 2005; Kartal et al., 2007a, b, 2008; van de Vossenberg, 2008). The genome sequence (4.2 Mb) of "*Candidatus* Kuenenia stuttgartiensis" has been assembled from community bacterial artificial chromosome (BAC), fosmid, and shotgun libraries (Strous et al., 2006). Genomic evidence, confirmed by CO dehydrogenase enzyme assays, points to the acetyl-CoA pathway for CO_2 fixation. This is consistent with the observed carbon discrimination and the presence of very ^{13}C-depleted (–70%) lipids in anammox bacteria (Schouten et al., 2004). The presence of cytochrome *c*-based respiration of nitrate and nitrite could also be confirmed by analysis of the genome assembly.

Isolation procedures. Owing to the long generation time of 2–3 weeks and the cell density-dependent activity, anammox organisms are most likely incompatible with any currently existing isolation protocol. However, anammox cells can be purified and separated from other organisms present in the community using a density-gradient centrifugation (Strous et al., 1999a, 2002). For this procedure, anammox aggregates are washed in HEPES buffer pH 7.8 containing 5 mM bicarbonate. Subsequently, the anammox cells are mildly sonicated. After centrifugation, the larger aggregates form a light red pellet on top of the dark red pellet of the single cells. The aggregates are removed, the single cells are resuspended in the HEPES/bicarbonate buffer, and Percoll is added. This mixture is centrifuged and anammox cells form a red band at the lower part of the gradient (density 1.107–1.138 g/l). After extraction of the band, the Percoll is removed by washing with HEPES/bicarbonate buffer. The resulting cell suspension contains usually only one contaminating organism for every 200–800 target organisms. The anammox activity of purified cells can be tested after addition of hydrazine, ammonium, and nitrite (Strous et al., 1999a).

Differentiation of the anammox bacteria

As stated in the introduction, information is too limited to separately describe the five recognized genera of the family "*Candidatus* Brocadiaceae". In Table 177, the most prominent features of the five different anammox genera are listed. After enrichment of the anammox bacteria on ammonium and nitrite media, sequencing of the 16S rRNA gene, and FISH analysis with specific probes are most indicative to determine which anammox species has been obtained. Higher concentrations of phosphate may favor enrichment of "*Candidatus* Kuenenia" species, whereas inclusion of higher concentrations of salt might favor the marine anammox species. The addition of propionate or acetate results in the enrichment of "*Candidatus* Anammoxoglobus propionicus" or "*Candidatus* Brocadia fulgida", respectively.

Further reading

Kuenen, J.G. 2008. Anammox bacteria: from discovery to application. Nat. Rev. Microbiol. *6*: 320–326.

Pilcher, H. 2005. Microbiology: pipe dreams. Nature *437*: 1227–1228.

List of genera of the family "*Candidatus* Brocadiaceae"

1. **"*Candidatus* Brocadia"** Strous et al. 1999a; Kuenen and Jetten 2001

Bro.ca′di.a. N.L. fem. n. *Brocadia* named after place of discovery at the Gist-Brocades factory in Delft, the Netherlands, and the gaudy brocade color of the bacteria.

Currently the genus includes "*Candidatus* Brocadia anammoxidans" and "*Candidatus* Brocadia fulgida".

2. **"*Candidatus* Anammoxoglobus"** Kartal et al. 2007b

A.nam.mo.xo.glo′bus. N.L. n. *anammox* abbreviation of anaerobic ammonium oxidation (arbitrary name); L. masc. n. *globus* sphere referring to the spherical shape; N.L. masc. n. *Anammoxoglobus.*

Currently the genus includes "*Candidatus* Anammoxoglobus propionicus".

3. **"*Candidatus* Jettenia"** Quan et al. 2008

Jet.te′ni.a. N.L. fem. n. *Jettenia* named after Mike S.M. Jetten.

Currently the genus includes "*Candidatus* Jettenia asiatica".

4. **"*Candidatus* Kuenenia"** Schmid et al. 2001

Ku.e.ne′ni.a. N.L. fem. n. *Kuenenia* named in honor after Johannes Gijsbrecht Kuenen for his contribution as founding father of anammox research.

Currently the genus includes "*Candidatus* Kuenenia stuttgartiensis".

5. **"*Candidatus* Scalindua"** Schmid et al. 2003; Kuypers et al. 2003; Woebken et al. 2008

Sca.lin′du.a. L. fem. n. *scala* ladder; L. v. *induo* to dress out or fitt with; N.L. fem. n. *scalindua* intended to mean dressed out with ladders, referring to the presence of unique ladderane lipids.

Currently the genus includes "*Candidatus* Scalindua brodae", "*Candidatus* Scalindua sorokinii", "*Candidatus* Scalindua wagneri", and "*Candidatus* Scalindua arabica".

References

Boumann, H.A., E.C. Hopmans, I. van de Leemput, H.J. Op den Camp, J. van de Vossenberg, M. Strous, M.S.M. Jetten, J.S. Sinninghe Damste and S. Schouten. 2006. Ladderane phospholipids in anammox bacteria comprise phosphocholine and phosphoethanolamine headgroups. FEMS Microbiol. Lett. *258*: 297–304.

Dalsgaard, T., D.E. Canfield, J. Petersen, B. Thamdrup and J. Acuna-Gonzalez. 2003. N_2 production by the anammox reaction in the anoxic water column of Golfo Dulce, Costa Rica. Nature *422*: 606–608.

Dalsgaard, T., B. Thamdrup and D.E. Canfield. 2005. Anaerobic ammonium oxidation (anammox) in the marine environment. Res. Microbiol. *156*: 457–464.

Damste, J.S.S., M. Strous, W.I.C. Rijpstra, E.C. Hopmans, J.A.J. Geenevasen, A.C.T. van Duin, L.A. van Niftrik and M.S.M. Jetten. 2002. Linearly concatenated cyclobutane lipids form a dense bacterial membrane. Nature *419*: 708–712.

Damste, J.S.S., W.I.C. Rijpstra, S. Schouten, J.A. Fuerst, M.S.M. Jetten and M. Strous. 2004. The occurrence of hopanoids in planctomycetes: implications for the sedimentary biomarker record. Org. Geochem. *35*: 561–566.

Damste, J.S.S., W.I.C. Rijpstra, J.A.J. Geenevasen, M. Strous and M.S.M. Jetten. 2005. Structural identification of ladderane and other membrane lipids of planctomycetes capable of anaerobic ammonium oxidation (anammox). FEBS J. *272*: 4270–4283.

Dapena-Mora, A., S.W.H. Van Hulle, J.L. Campos, R. Mendez, P.A. Vanrolleghem and M.S.M. Jetten. 2004. Enrichment of anammox biomass from municipal activated sludge: experimental and modelling results. J. Chem. Technol. Biotechnol. *79*: 1421–1428.

Egli, K., U. Fanger, P.J.J. Alvarez, H. Siegrist, J.R. van der Meer and A.J.B. Zehnder. 2001. Enrichment and characterization of an anammox bacterium from a rotating biological contactor treating ammonium-rich leachate. Arch. Microbiol. *175*: 198–207.

Felsenstein, J. 1985. Confidence limits on phylogenies: an approach using the bootstrap. Evolution *39*: 783–791.

Francis, C.A., J.M. Beman and M.M.M. Kuypers. 2007. New processes and players in the nitrogen cycle: the microbial ecology of anaerobic and archaeal ammonia oxidation. ISME J. *1*: 19–27.

Fuerst, J.A. 2005. Intracellular compartmentation in *Planctomycetes*. Annu. Rev. Microbiol. *59*: 299–328.

Fujii, T., H. Sugino, J.D. Rouse and K. Furukawa. 2002. Characterization of the microbial community in an anaerobic ammonium-oxidizing biofilm cultured on a nonwoven biomass carrier. J. Biosci. Bioeng. *94*: 412–418.

Guven, D., A. Dapena, B. Kartal, M.C. Schmid, B. Maas, K. van de Pas-Schoonen, S. Sozen, R. Mendez, H.J.M. Op den Camp, M.S.M. Jetten, M. Strous and I. Schmidt. 2005. Propionate oxidation by and methanol inhibition of anaerobic ammonium-oxidizing bacteria. Appl. Environ. Microbiol. *71*: 1066–1071.

Hamersley, M.R., G. Lavik, D. Woebken, J.E. Rattray, P. Lam, E.C. Hopmans, J.S.S. Damste, S. Kruger, M. Graco, D. Gutierrez and M.M.M. Kuypers. 2007. Anaerobic ammonium oxidation in the Peruvian oxygen minimum zone. Limnol. Oceanogr. *52*: 923–933.

Jensen, M.M., B. Thamdrup and T. Dalsgaard. 2007. Effects of specific inhibitors on anammox and denitrification in marine sedirnents. Appl. Environ. Microbiol. *73*: 3151–3158.

Jetten, M.S.M., I. Cirpus, B. Kartal, L. van Niftrik, K.T. van de Pas-Schoonen, O. Sliekers, S. Haaijer, W. van der Star, M. Schmid, J. van de Vossenberg, I. Schmidt, H. Harhangi, M. van Loosdrecht, J.G. Kuenen, H.O. den Camp and M. Strous. 2005a. 1994–2004: 10 years of research on the anaerobic oxidation of ammonium. Biochem. Soc. Trans. *33*: 119–123.

Jetten, M.S.M., M. Schmid, K. van de Pas-Schoonen, J.S.S. Damste and M. Strous. 2005b. Anammox organisms: enrichment, cultivation, and environmental analysis. Environ. Microbiol. *397:* 34–57.

Kartal, B., M. Koleva, R. Arsov, W. van der Star, M.S.M. Jetten and M. Strous. 2006. Adaptation of a freshwater anammox population to high salinity wastewater. J. Biotechnol. *126*: 546–553.

Kartal, B., M.M. Kuypers, G. Lavik, J. Schalk, H.J. Op den Camp, M.S.M. Jetten and M. Strous. 2007a. Anammox bacteria disguised as denitrifiers: nitrate reduction to dinitrogen gas via nitrite and ammonium. Environ. Microbiol. *9*: 635–642.

Kartal, B., J. Rattray, L.A. van Niftrik, J. van de Vossenberg, M.C. Schmid, R.I. Webb, S. Schouten, J.A. Fuerst, J.S.S. Damste, M.S.M. Jetten and M. Strous. 2007b. *Candidatus* "Anammoxoglobus propionicus" a new propionate oxidizing species of anaerobic ammonium oxidizing bacteria. Syst. Appl. Microbiol. *30*: 39–49.

Kartal, B., L. van Niftrik, J. Rattray, J. van de Vossenberg, M.C. Schmid, J.S.S. Damste, M.S.M. Jetten and M. Strous. 2008. *Candidatus* '*Brocadia fulgida*': an autofluorescent anaerobic ammonium oxidizing bacterium. FEMS Microbiol. Ecol. *63*: 46–55.

Kindaichi, T., I. Tsushima, Y. Ogasawara, M. Shimokawa, N. Ozaki, H. Satoh and S. Okabe. 2007. In situ activity and spatial organization of anaerobic ammonium-oxidizing (anammox) bacteria in biofilms. Appl. Environ. Microbiol. *73*: 4931–4939.

Kuenen, J.G. and M.S.M. Jetten. 2001. Extraordinary anaerobic ammonium-oxidizing bacteria. ASM News *67*: 456–463.

Kuypers, M.M.M., A.O. Sliekers, G. Lavik, M. Schmid, B.B. Jorgensen, J.G. Kuenen, J.S.S. Damste, M. Strous and M.S.M. Jetten. 2003. Anaerobic ammonium oxidation by anammox bacteria in the Black Sea. Nature *422*: 608–611.

Kuypers, M.M.M., G. Lavik, D. Woebken, M. Schmid, B.M. Fuchs, R. Amann, B.B. Jorgensen and M.S.M. Jetten. 2005. Massive nitrogen loss from the Benguela upwelling system through anaerobic ammonium oxidation. Proc. Natl. Acad. Sci. U. S. A. *102*: 6478–6483.

Lam, P., M.M. Jensen, G. Lavik, D.F. McGinnis, B. Muller, C.J. Schubert, R. Amann, B. Thamdrup and M.M.M. Kuypers. 2007. Linking crenarchaeal and bacterial nitrification to anammox in the Black Sea. Proc. Natl. Acad. Sci. U. S. A. *104*: 7104–7109.

Lindsay, M.R., R.I. Webb, M. Strous, M.S.M. Jetten, M.K. Butler, R.J. Forde and J.A. Fuerst. 2001. Cell compartmentalisation in *Planctomycetes*: novel types of structural organisation for the bacterial cell. Arch. Microbiol. *175*: 413–429.

Nakajima, J., M. Sakka, T. Kimura, K. Furukawa and K. Sakka. 2008. Enrichment of anammox bacteria from marine environment for the construction of a bioremediation reactor. Appl. Mycol. Biotechnol. *77*: 1159–1166.

Neef, A., R. Amann, H. Schlesner and K.H. Schleifer. 1998. Monitoring a widespread bacterial group: *in situ* detection of *Planctomycetes* with 16S rRNA-targeted probes. Microbiology *144*: 3257–3266.

Nielsen, M., A. Bollmann, O. Sliekers, M.S.M. Jetten, M. Schmid, M. Strous, I. Schmidt, L.H. Larsen, L.P. Nielsen and N.P. Revsbech. 2005. Kinetics, diffusional limitation and microscale distribution of chemistry and organisms in a CANON reactor. FEMS Microbiol. Ecol. *51*: 247–256.

Op den Camp H.J.M., B. Kartal, D. Guven, L.A.M.P. van Niftrik, S.C.M. Haaijer, W.R.L. van der Star, K.T. van de Pas-Schoonen, A. Cabezas, Z. Ying, M.C. Schmid, M.M.M. Kuypers, J. van de Vossenberg, H.R. Harhangi, C. Picioreanu, M.C.M. van Loosdrecht, J.G. Kuenen, M. Strous and M.S.M. Jetten. 2006. Global impact and application of the anaerobic ammonium-oxidizing (anammox) bacteria. Biochem. Soc. Trans. *34*: 174–178.

Penton, C.R., A.H. Devol and J.M. Tiedje. 2006. Molecular evidence for the broad distribution of anaerobic ammonium-oxidizing bacteria in freshwater and marine sediments. Appl. Environ. Microbiol. *72*: 6829–6832.

Pynaert, K., B.F. Smets, S. Wyffels, D. Beheydt, S.D. Siciliano and W. Verstraete. 2003. Characterization of an autotrophic nitrogen-removing biofilm from a highly loaded lab-scale rotating biological contactor. Appl. Environ. Microbiol. *69*: 3626–3635.

Quan, Z.-X., S.K. Rhee, J.E. Zuo, Y. Yang, J.W. Bae, J.R. Park, S.T. Lee and Y.H. Park. 2008. Diversity of ammonium-oxidizing bacteria in anaerobic ammonium-oxidizing (anammox) reactor. Environ. Microbiol. *10*: 3130–3139.

Rattray, J.E., J. van de Vossenberg, E.C. Hopmans, B. Kartal, L. van Niftrik, W.I. Rijpstra, M. Strous, M.S.M. Jetten, S. Schouten and J.S. Damste. 2008. Ladderane lipid distribution in four genera of anammox bacteria. Arch. Microbiol. *190*: 51–66.

Rich, J.J., O.R. Dale, B. Song and B.B. Ward. 2008. Anaerobic ammonium oxidation (Anammox) in Chesapeake Bay sediments. Microb. Ecol. *55*: 311–320.

Risgaard-Petersen, N., R.L. Meyer, M. Schmid, M.S.M. Jetten, A. Enrich-Prast, S. Rysgaard and N.P. Revsbech. 2004. Anaerobic ammonium oxidation in an estuarine sediment. Aquat. Microb. Ecol. *36*: 293–304.

Rysgaard, S. and R.N. Glud. 2004. Anaerobic N_2 production in Arctic sea ice. Limnol. Oceanogr. *49*: 86–94.

Saitou, N. and M. Nei. 1987. The neighbor-joining method: a new method for reconstructing phylogenetic trees. Mol. Biol. Evol. *4*: 406–425.

Schalk, J., S. de Vries, J.G. Kuenen and M.S.M. Jetten. 2000. Involvement of a novel hydroxylamine oxidoreductase in anaerobic ammonium oxidation. Biochemistry *39*: 5405–5412.

Schmid, M., U. Twachtmann, M. Klein, M. Strous, S. Juretschko, M.S.M. Jetten, J.W. Metzger, K.H. Schleifer and M. Wagner. 2000. Molecular evidence for genus level diversity of bacteria capable of catalyzing anaerobic ammonium oxidation. Syst. Appl. Microbiol. *23*: 93–106.

Schmid, M., S. Schmitz-Esser, M.S.M. Jetten and M. Wagner. 2001. 16S–23S rDNA intergenic spacer and 23S rDNA of anaerobic ammonium-oxidizing bacteria: implications for phylogeny and in situ detection. Environ. Microbiol. *3*: 450–459.

Schmid, M., K. Walsh, R. Webb, W.I.C. Rijpstra, K. van de Pas-Schoonen, M.J. Verbruggen, T. Hill, B. Moffett, J. Fuerst, S. Schouten, J.S.S.

Damste, J. Harris, P. Shaw, M.S.M. Jetten and M. Strous. 2003. *Candidatus* "Scalindua brodae", spec. nov., *Candidatus* "Scalindua wagneri", spec. nov., two new species of anaerobic ammonium oxidizing bacteria. Syst. Appl. Microbiol. *26*: 529–538.

Schmid, M.C., B. Maas, A. Dapena, K.V. de Pas-Schoonen, J. van de Vossenberg, B. Kartal, L. van Niftrik, I. Schmidt, I. Cirpus, J.G. Kuenen, M. Wagner, J.S.S. Damste, M. Kuypers, N.P. Revsbech, R. Mendez, M.S.M. Jetten and M. Strous. 2005. Biomarkers for *in situ* detection of anaerobic ammonium-oxidizing (anammox) bacteria. Appl. Environ. Microbiol. *71*: 1677–1684.

Schmidt, I., O. Sliekers, M. Schmid, I. Cirpus, M. Strous, E. Bock, J.G. Kuenen and M.S.M. Jetten. 2002. Aerobic and anaerobic ammonia oxidizing bacteria competitors or natural partners? FEMS Microbiol. Ecol. *39*: 175–181.

Schouten, S., M. Strous, M.M.M. Kuypers, W.I.C. Rijpstra, M. Baas, C.J. Schubert, M.S.M. Jetten and J.S.S. Damste. 2004. Stable carbon isotopic fractionations associated with inorganic carbon fixation by anaerobic ammonium-oxidizing bacteria. Appl. Environ. Microbiol. *70*: 3785–3788.

Schubert, C.J., E. Durisch-Kaiser, B. Wehrli, B. Thamdrup, P. Lam and M.M.M. Kuypers. 2006. Anaerobic ammonium oxidation in a tropical freshwater system (Lake Tanganyika). Environ. Microbiol. *8*: 1857–1863.

Shimamura, M., T. Nishiyama, H. Shigetomo, T. Toyomoto, Y. Kawahara, K. Furukawa and T. Fujii. 2007. Isolation of a multiheme protein with features of a hydrazine-oxidizing enzyme from an anaerobic ammonium-oxidizing enrichment culture. Appl. Environ. Microbiol. *73*: 1065–1072.

Shimamura, M., T. Nishiyama, K. Shinya, Y. Kawahara, K. Furukawa and T. Fujii. 2008. Another multiheme protein, hydroxylamine oxidoreductase, abundantly produced in an anammox bacterium besides the hydrazine-oxidizing enzyme. J. Biosci. Bioeng. *105*: 243–248.

Strous, M. and M.S.M. Jetten. 2004. Anaerobic oxidation of methane and ammonium. Annu. Rev. Microbiol. *58*: 99–117.

Strous, M., E. van Gerven, J.G. Kuenen and M.S.M. Jetten. 1997. Effects of aerobic and microaerobic conditions on anaerobic ammonium-oxidizing (Anammox) sludge. Appl. Environ. Microbiol. *63*: 2446–2448.

Strous, M., J.J. Heijnen, J.G. Kuenen and M.S.M. Jetten. 1998. The sequencing batch reactor as a powerful tool for the study of slowly growing anaerobic ammonium-oxidizing microorganisms. Appl. Mycol. Biotechnol. *50*: 589–596.

Strous, M., J.A. Fuerst, E.H.M. Kramer, S. Logemann, G. Muyzer, K.T. van de Pas-Schoonen, R. Webb, J.G. Kuenen and M.S.M. Jetten. 1999a. Missing lithotroph identified as new planctomycete. Nature *400*: 446–449.

Strous, M., J.G. Kuenen and M.S.M. Jetten. 1999b. Key physiology of anaerobic ammonium oxidation. Appl. Environ. Microbiol. *65*: 3248–3250.

Strous, M., J.G. Kuenen, J.A. Fuerst, M. Wagner and M.S.M. Jetten. 2002. The anammox case – a new experimental manifesto for microbiological eco-physiology. Antonie van Leeuwenhoek *81*: 693–702.

Strous, M., E. Pelletier, S. Mangenot, T. Rattei, A. Lehner, M.W. Taylor, M. Horn, H. Daims, D. Bartol-Mavel, P. Wincker, V. Barbe, N. Fonknechten, D. Vallenet, B. Segurens, C. Schenowitz-Truong, C. Medigue, A. Collingro, B. Snel, B.E. Dutilh, H.J.M. Op den Camp, C. van der Drift, I. Cirpus, K.T. van de Pas-Schoonen, H.R. Harhangi, L. van Niftrik, M. Schmid, J. Keltjens, J. van de Vossenberg, B. Kartal, H. Meier, D. Frishman, M.A. Huynen, H.W. Mewes, J. Weissenbach, M.S.M. Jetten, M. Wagner and D. Le Paslier. 2006. Deciphering the evolution and metabolism of an anammox bacterium from a community genome. Nature *440*: 790–794.

Tal, Y., J.E.M. Watts and H.J. Schreier. 2005. Anaerobic ammonia-oxidizing bacteria and related activity in Baltimore Inner Harbor sediment. Appl. Environ. Microbiol. *71*: 1816–1821.

Tamura, K., M. Nei and S. Kumar. 2004. Prospects for inferring very large phylogenies by using the neighbor-joining method. Proc. Natl. Acad. Sci. U. S. A. *101*: 11030–11035.

Tamura, K., J. Dudley, M. Nei and S. Kumar. 2007. MEGA4: Molecular Evolutionary Genetics Analysis (MEGA) software version 4.0. Mol. Biol. Evol. *24*: 1596–1599.

Thamdrup, B. and T. Dalsgaard. 2002. Production of N_2 through anaerobic ammonium oxidation coupled to nitrate reduction in marine sediments. Appl. Environ. Microbiol. *68*: 1312–1318.

Thamdrup, B., T. Dalsgaard, M.M. Jensen and J. Petersen. 2004. Anammox and the marine N cycle. Geochim. Cosmochim. Acta Microbiol. *68*: A325–A325.

Thamdrup, B., T. Dalsgaard, M.M. Jensen, O. Ulloa, L. Farias and R. Escribano. 2006. Anaerobic ammonium oxidation in the oxygen-deficient waters off northern Chile. Limnol. Oceanogr. *51*: 2145–2156.

Third, K.A., A.O. Sliekers, J.G. Kuenen and M.S.M. Jetten. 2001. The CANON system (completely autotrophic nitrogen-removal over nitrite) under ammonium limitation: interaction and competition between three groups of bacteria. Syst. Appl. Microbiol. *24*: 588–596.

Third, K.A., J. Paxman, M. Schmid, M. Strous, M.S.M. Jetten and R. Cord-Ruwisch. 2005. Enrichment of anammox from activated sludge and its application in the CANON process. Microb. Ecol. *49*: 236–244.

Toh, S.K., R.I. Webb and N.J. Ashbolt. 2002. Enrichment of autotrophic anaerobic ammonium-oxidizing consortia from various wastewaters. Microb. Ecol. *43*: 154–167.

Tsushima, I., Y. Ogasawara, T. Kindaichi, H. Satoh and S. Okabe. 2007. Development of high-rate anaerobic ammonium-oxidizing (Anammox) biofilm reactors. Water Res. *41*: 1623–1634.

van de Graaf, A.A., A. Mulder, P. de Bruijn, M.S.M. Jetten, L.A. Robertson and J.G. Kuenen. 1995. Anaerobic oxidation of ammonium is a biologically mediated process. Appl. Environ. Microbiol. *61*: 1246–1251.

van de Graaf, A.A., P. deBruijn, L.A. Robertson, M.S.M. Jetten and J.G. Kuenen. 1996. Autotrophic growth of anaerobic ammonium-oxidizing micro-organisms in a fluidized bed reactor. Microbiology *142*: 2187–2196.

van de Graaf, A.A., P. de Bruijn, L.A. Robertson, M.S.M. Jetten and J.G. Kuenen. 1997. Metabolic pathway of anaerobic ammonium oxidation on the basis of N-15 studies in a fluidized bed reactor. Microbiology *143*: 2415–2421.

van de Vossenberg, J., J.E. Rattray, W. Geerts, B. Kartal, L. van Niftrik, E.G. van Donselaar, J.S. Sinninghe Damste, M. Strous and M.S.M. Jetten. 2008. Enrichment and characterization of marine anammox bacteria associated with global nitrogen gas production. Environ. Microbiol. *10*: 3120–3129.

van der Star, W.R.L., W.R. Abma, D. Blommers, J.W. Mulder, T. Tokutomi, M. Strous, C. Picioreanu and M.C.M. van Loosdrecht. 2007. Startup of reactors for anoxic ammonium oxidation: experiences from the first full-scale anammox reactor in Rotterdam. Water Res. *41*: 4149–4163.

van der Star, W.R., A.I. Miclea, U.G. van Dongen, G. Muyzer, C. Picioreanu and M.C. van Loosdrecht. 2008a. The membrane bioreactor: a novel tool to grow anammox bacteria as free cells. Biotechnol. Bioeng. *101*: 286–294.

van der Star, W.R., M.J. van de Graaf, B. Kartal, C. Picioreanu, M.S.M. Jetten and M.C. van Loosdrecht. 2008b. Response of anammox bacteria to hydroxylamine. Appl. Environ. Microbiol. *74*: 4417–4426.

van Niftrik, L.A., J.A. Fuerst, J.S.S. Damste, J.G. Kuenen, M.S.M. Jetten and M. Strous. 2004. The anammoxosome: an intracytoplasmic compartment in anammox bacteria. FEMS Microbiol. Lett. *233*: 7–13.

van Niftrik, L., W.J.C. Geerts, E.G. van Donselaar, B.M. Humbel, R.I. Webb, J.A. Fuerst, A.J. Verkleij, M.S.M. Jetten and M. Strous. 2008a. Linking ultrastructure and function in four genera of anaerobic ammonium-oxidizing bacteria: cell plan, glycogen storage, and localization of cytochrome *c* proteins. J. Bacteriol. *190*: 708–717.

van Niftrik, L., W.J.C. Geerts, E.G. van Donselaar, B.M. Humbel, A. Yakushevska, A.J. Verkleij, M.S.M. Jetten and M. Strous. 2008b. Combined structural and chemical analysis of the anammoxosome: a membrane-bounded intracytoplasmic compartment in anammox bacteria. J. Struct. Biol. *161*: 401–410.

Windey, K., I. De Bo and W. Verstraete. 2005. Oxygen-limited autotrophic nitrification-denitrification (OLAND) in a rotating biological contactor treating high-salinity wastewater. Water Res. *39*: 4512–4520.

Woebken, D., B.A. Fuchs, M.A.A. Kuypers and R. Amann. 2007. Potential interactions of particle-associated anammox bacteria with bacterial and archaeal partners in the Namibian upwelling system. Appl. Environ. Microbiol. *73*: 4648–4657.

Woebken, D., P. Lam, M.M. Kuypers, S.W. Naqvi, B. Kartal, M. Strous, M.S.M. Jetten, B.M. Fuchs and R. Amann. 2008. A microdiversity study of anammox bacteria reveals a novel *Candidatus* Scalindua phylotype in marine oxygen minimum zones. Environ. Microbiol. *10*: 3106–3119.

Zhang, Y., X.H. Ruan, H. den Camp, T.J.M. Smits, M.S.M. Jetten and M.C. Schmid. 2007. Diversity and abundance of aerobic and anaerobic ammonium-oxidizing bacteria in freshwater sediments of the Xinyi River (China). Environ. Microbiol. *9*: 2375–2382.

Author index

Abraham, Wolf-Rainer, 377
Amann, Rudolf, 460
Antón, Josefa, 460
Balish, Mitchell F., 575
Bavoil, Patrik M., 846, 875
Benno, Yoshimi, 70, 78
Bernardet, Jean-François, 105, 106, 112, 180, 202
Berthe-Corti, Luise, 240
Bové, Joseph M., 654
Bowman, John P., 112, 155, 176, 219, 231, 258, 261, 266, 275, 277, 322, 323, 326, 328
Bradbury, Janet M., 568, 574, 575, 644–646, 649, 687, 688, 719, 720, 722
Brettar, Ingrid, 433, 434
Breznak, John A., 498, 563
Brown, Daniel R., 567, 568, 574, 575, 639, 640, 642, 644–646, 649, 687, 688, 719, 720, 722
Brune, Andreas, 759
Bruns, Alke, 240
Bruun, Brita, 180, 202
Buczolits, Sandra, 397
Busse, Hans-Jürgen, 397
Butler, Margaret K., 897, 903
Calcutt, Michael J., 575
Castenholz, Richard W., 900
Chattaway, Marie Ann, 86
Cho, Jang-Cheon, 199, 234, 785, 787, 788, 791, 817, 820, 823, 824
Christen, Richard, 214, 433, 434
Chun, Jongsik, 161, 228, 271
Coates, John D., 725, 727–734
Coomans, August, 838
Costa, Kyle C., 785, 817, 834
Davis, Robert E., 696
Dedysh, Svetlana N., 910, 913
Derrien, Muriel, 785, 787, 791, 802, 809
de Vos, Willem M., 791, 809
Edwards, Kirstin J., 748, 758, 766, 769–771
Euzéby, Jean, 1, 21, 25, 49, 457
Fegan, Mark, 366
Finegold, Sydney M., 27, 56, 62
Forsberg, Cecil W., 737, 739
Fuerst, John A., 897, 903
Garcia-Pichel, Ferran, 380
Garrity, George M., xxv
Gasparich, Gail E., 654
Gharbia, Saheer E., 86, 748, 758, 766, 769–771
Gilbert, Peter, 375
Giovannoni, Stephen J., 199, 264, 785, 788
Glass, John I., 575
Gundersen-Rindal, Dawn, 696
Hafez, Hafez M., 250
Harrison, Nigel A., 696
Hedlund, Brian P., 785, 787, 788, 791, 795, 799, 802, 803, 805, 809, 812, 817, 820, 821, 823–825, 827, 829, 834, 836–838
Hirsch, Peter, 467
Höfle, Manfred G., 433, 434
Holt, Stanley C., 168
Horn, Matthias, 843, 844, 865, 867–870, 872–875
Hugo, Celia J., 180
Irgens, Roar L., 406
Ivanova, Elena P., 214
Janssen, Peter H., 759, 761, 817, 820, 834, 836, 837
Jetten, Mike S.M., 918
Johansson, Karl-Erik, 568, 574, 575, 687, 719, 720, 722
Kaltenboeck, Bernhard, 846, 875
Kamagata, Yoichi, 781, 782
Kambhampati, Srinivas, 315
Kämpfer, Peter, 285, 330, 351, 363
Kasai, Hiroaki, 812
Kim, Sang-Jin, 228
Kim, Seung Bum, 157, 158, 166, 226, 235, 239, 283, 288, 410, 437, 453
Könönen, Eija, 56
Krieg, Noel R., xxv, 25, 61, 85
Krumholz, Lee R., 737, 739
Kuenen, J. Gijs, 918
Kulichevskaya, Irina S., 910, 913
Kuo, Cho-chou, 844–846, 875
Leadbetter, Edward R., 418
Lee, Kuo-Chang, 897
Leschine, Susan, 473
Lewin, Ralph A., 359
Liu, Chengxu, 27
Logan, Julie M. J., 758, 770
Ludwig, Wolfgang, 1, 21, 25, 49, 457
Lünsdorf, Heinrich, 377
Macedo, Alexandre J., 377
Margesin, Rosa, 339
May, Meghan, 568, 574, 575
Messick, Joanne B., 575, 639, 640, 642
Mikhailov, Valery V., 166, 226, 232, 284, 288
Nakagawa, Yasuyoshi, 370, 371, 392, 408, 442, 448, 450
Navarro, Jason B., 838
Nedashkovskaya, Olga I., 157, 158, 161, 166, 226, 232, 235, 239, 283, 284, 288, 292, 410, 423, 426, 437, 442, 452, 453
Neimark, Harold, 574, 575, 639, 640, 642
Nikitin, Denis I., 377
Norris, Steven J., 501
Olsen, Ingar, 71
Op den Camp, Huub J.M., 918
Paster, Bruce J., 471, 473, 498, 501, 531, 545, 563
Patel, Bharat K. C, 775, 776
Plugge, Caroline M, 791, 809
Rajakurana, Lakshani, 86
Rautio, Merja, 56
Reddy, Gundlapally S.N., 380
Regassa, Laura B., 654
Renaudin, Joël, 654
Rickard, Alexander H., 375
Robertson, Janet A., 613
Rosselló-Mora, Ramon, 460
Saillard, Collette, 654
Sakamoto, Mitsuo, 78
Schink, Bernhard, 51, 759, 761
Schlesner, Heinz, 803, 895, 906
Schmidt, Jean M., 881
Schwartz, Ira, 484
Shah, Haroun N., 27, 45, 55, 62, 86, 748, 769
Shieh, Wung Yang, 820
Shivaji, Sisinthy, 339
Sly, Lindsay I., 366
Smibert, Robert M., 501
Song, Yuli, 27, 56, 817, 820, 823, 824
Spain, Anne M., 737, 739
Staley, James T., 255, 747, 748, 766, 799, 805, 881
Stanton, Thaddeus B., 531, 872
Stephens, Richard S., 844–846, 875
Strous, Marc, 918
Summanen, Paula, 62
Suzuki, Makoto, 49, 76, 78, 102, 279, 284, 292, 410, 442, 452
Tanner, Anne R.C., 78, 197
Tasker, Séverine, 575, 639, 640, 642
Taylor-Robinson, David, 613
Thrash, Cameron, 725–729, 731, 732, 734
Vandekerckhove, Tom T.M., 838
Van Trappen, Stefanie, 221
Wang, Guiqing, 484
Ward, Naomi L., 879–881
Whitcomb, Robert F., 644–646, 649, 654
Whitman, William B., 1, 21, 25, 49, 457, 747, 748, 766
Williamson, David L., 654
Yi, Hana, 271
Yoon, Jaewoo, 812, 823–825, 827, 829
Zoetendal, Erwin G., 791
Zuerner, Richard L., 546, 556, 558

Index of scientific names of *Archaea* and *Bacteria*

Key to the fonts and symbols used in this index:

Nomenclature	
Lower case, Roman	Genera, species, and subspecies of bacteria. Every bacterial name mentioned in the text is listed in the index. Specific epithets are listed individually and also under the genus.*
CAPITALS, ROMAN:	Names of taxa higher than genus (tribes, families, orders, classes, divisions, kingdoms).
Pagination	
Roman:	Pages on which taxa are mentioned.
Boldface:	Indicates page on which the description of a taxon is given.†

* Infrasubspecific names, such as serovars, biovars, and pathovars, are not listed in the index.

† A description may not necessarily be given in the *Manual* for a taxon that is considered as *incertae sedis* or that is listed in an addendum or note added in proof; however, the page on which the complete citation of such a taxon is given is indicated in boldface type.

Index of scientific names of *Archaea* and *Bacteria*

aalborgi (Brachyspira), 10, 472, 531–534, 537, **538**, 539–541
abactoclasticum (Anaeroplasma), 12, 571, 572, 720–**722**
abortus (Chlamydophila), 13, 845, 859
acanthamoebae (Parachlamydia), 13, **868**, 869, 870, 877
Acanthopleuribacter, 12, 725, 726, 732, 733, **734**
ACANTHOPLEURIBACTERACEAE, 12, 725, **734**
ACANTHOPLEURIBACTERALES, 12, 725, 731, **734**
Acanthopleuribacter pedis, 734, **735**
acetatigenes (Proteiniphilum), 65, 77, **78**, 82
Acetivibrio, 42–44
Acetivibrio cellulolyticus, 39
Acetofilamentum, 3, 25, 26, **41**, 42, 43, 44
Acetofilamentum rigidum, 41, **42**
Acetomicrobium, 3, 25, 26, **42**, 43, 44
Acetomicrobium faecale, **43**, 44
Acetomicrobium flavidum, 42, **43**, 44
Acetothermus, 3, 26, 42, 43, **44**
Acetothermus paucivorans, **44**
Acholeplasma, 11, 568, 570, 571, 581, 583, 596, 620, 640, 645–647, 649, 650, 656, 667, 687, **688**, 689–694, 696, 720–723
Acholeplasma axanthum, 11, 574, 688–690, **691**
Acholeplasma brassicae, 11, 688, 689, **691**
Acholeplasma cavigenitalium, 11, 689, **691**
Acholeplasma entomophilum, 650, 651, **692**
Acholeplasma equifetale, 689, **692**
Acholeplasma florum, 649, 650, **692**
Acholeplasma granularum, 11, 689–691, **692**
Acholeplasma hippikon, 689, **692**
Acholeplasma laidlawii, 11, 569, 580, 668, 688–690, **691**, 692, 699
Acholeplasma modicum, 11, 688–690, **692**
Acholeplasma morum, 11, 574, 688, 689, **693**
Acholeplasma multilocale, 11, 689, 690, **693**
Acholeplasma oculi, 11, 649, 688–690, **693**
Acholeplasma palmae, 11, 688, 689, **693**, 704
Acholeplasma parvum, 11, 688, 689, **693**
Acholeplasma pleciae, 11, 689, 690, **694**
Acholeplasma seiffertii, 650, 652, **694**
ACHOLEPLASMATACEAE, 11, 570, **687**, 688, 690, 692, 694
ACHOLEPLASMATALES, 10, 11, 12, 570, 571, 574, 667, **687**, 696, 719
Acholeplasma vituli, 11, 689, **694**
acidifaciens (Bacteroides), 3
acidificum (Flavobacterium), 5
Acidimethylosilex fumarolicum, 795–797
acidiphila (Singulisphaera), 16, 913–914, **915**
ACIDOBACTERIA, 1, 2, 12, 21, **725**, 726, 729, 880
ACIDOBACTERIACEAE, 12, 23, 725, 727, **728**
ACIDOBACTERIALES, 2, 12, 725, **727**
ACIDOBACTERIIA, 12, 725, **727**
Acidobacterium, 12, 725–727, **728**, 729, 730
Acidobacterium capsulatum, 12, **728**
acidurans (Flavobacterium), 5, 138
Actibacter, 110
Actibacter sediminis, 5
actiniarum (Pontibacter), 9, 387, 388, **401**, 404, 410, 411
Actinobacillus muris, 772
ACTINOBACTERIA, 2, 782
actinosclerus (Hymenobacter), 9, 398, 399, **400**, 401
Adhaeribacter, 372, **375**, 376, 377, 398, 410
Adhaeribacter aquaticus, 9, 375, 376, **377**, 387, 388, 398, 404, 410, 411
adleri (Mycoplasma), 11, 582, 584, **587**
Aegyptianella pullorum, 110
Aegyptianella ranarum, 110, 207
aequorea (Leeuwenhoekiella), 7, 233, **234**
Aequorivita, 107, **155**–157, 200, 232, 245, 265, 284, 285
Aequorivita antarctica, 7, 155–**157**
Aequorivita crocea, 7, 155–**157**
Aequorivita lipolytica, 7, 155–**157**
Aequorivita sublithincola, 7, 156, **157**
Aeromonas hydrophila, 211
Aeromonas salmonicida, 211
aerophilus (Hymenobacter), 9, 398, 399, **401**, 402
aeruginosa (Pseudomonas), 725
Aestuariicola, 110
Aestuariicola saemankumensis, 5
Afipia, 196
africana (Spirochaeta), 10, 476, **477**, 478, 485
africanus (Pedobacter), 330, 339–342, 344, **345**
afzelii (Borrelia), 10, 485, 488, 490, **493**, 495, 496
agalactiae (Mycoplasma), 11, 569, 578, 579, 582, 584, **587**, 590
agarilytica (Lewinella), 8, 367
agariperforans (Reichenbachia), 452
agariperforans (Reichenbachiella), 9, **452**
agariphila (Formosa), 106, 215, 216, **217**
agarivorans (Salegentibacter), 7, 266, **267**, 268
agarovorans (Cytophaga), 50, 373
agarovorans (Marinilabilia), 4, 50–52
agassizii (Mycoplasma), 11, 582, 584, **587**
aggregans (Edaphobacter), 729, **730**
aggregans (Flexibacter), 9, 392, **394**, 408, 409, 449
aggregata (Costertonia), 5, **199**
agitata (Dechloromonas), 725
agri (Pedobacter), 8, 345
Agrobacterium radiobacter, 183
Agrobacterium tumefaciens, 111
akajimensis (Coraliomargarita), 13, 818, 826, 827, 828, **829**, 831
akesuensis (Pontibacter), **412**
Akkermansia, 13, 800, **809**, 810, 811, 814
AKKERMANSIACEAE, 13, 799, 800, 802, **809**, 810, 811, 814
Akkermansia muciniphila, 809, **810**, 811
albensis (Prevotella), 5, 65, **91**, 103
albulus (Streptomyces), 332
albus (Pelagicoccus), 818, 829, 830, **831**, 832
albus (Ruminococcus), 512, 741
alexanderi (Leptospira), 10, 547, **553**
algae (Formosa), 7, 214, 215, **217**, 218
algae (Luteolibacter), 799, 802, 803
algae (Mesonia), 7, 111, 160, 223, 226, 233, 239, **240**, 242
algens (Gelidibacter), 6, 215, 219, 220, **221**, 291
Algibacter, 107, 110, **157**, 158, 159, 178, 214–216, 218, 221–223, 230, 249, 269, 288, 289, 291
Algibacter lectus, 6, 157, **158**, 214, 218, 238, 249, 269
Algibacter mikhailovii, 6
algicida (Kordia), 159, 228–**230**, 242
algicola (Cellulophaga), 5, 6, 176, 177, 178, **179**
algicola (Lacinutrix), 6
Algoriphagus, 9, 424, **426**, 427–435, 437–439
Algoriphagus alkaliphilus, 9, 427, **429**, 430
Algoriphagus antarcticus, 9, 430, **431**
Algoriphagus aquimarinus, 9, 426, 427, 430, **431**
Algoriphagus boritolerans, 9, 427, 430, **431**
Algoriphagus chordae, 9, 427, 430, **431**
Algoriphagus halophilus, 9, 427, 430, **431**
Algoriphagus locisalis, 9, 426, 427, 430, **432**
Algoriphagus mannitolivorans, 9, 426, 427, 430, **432**
Algoriphagus marincola, 9, 426, 430, **432**
Algoriphagus ornithinivorans, 9, 426, 427, 430, **432**
Algoriphagus ratkowskyi, 9, 426–428, 430, **432**
Algoriphagus terrigena, 9, 427, 430, **433**
Algoriphagus vanfongensis, 9, 426, 427, **433**
Algoriphagus winogradskyi, 9, 426, 427, 430, **433**
Algoriphagus yeomjeoni, 9, 426, 427, 430, **433**
algoritergicola (Bizionia), 7, **167**
Alistipes, 4, 31, 32, 50, 54, **56**–59, 61, 63
Alistipes finegoldii, 58, 59, **60**
Alistipes onderdonkii, 4, 58, 59, **60**, 61
Alistipes putredinis, 4, 31, 32, 56, **58**, 59
Alistipes shahii, 4, 58, 59, **60**
alkalescens (Mycoplasma), 11, 582, 584, **587**, 588
alkalica (Spirochaeta), 10, 472, 476–478, **479**, 485
Alkaliflexus, 4, 49, 50, **53**
Alkaliflexus imshenetskii, 4, **53**

alkalilentus (Ferruginibacter), 358
alkaliphila (Balneola), 9
alkaliphila (Chimaereicella), 428, 429
Alkaliphilus, 427, 429
alkaliphilus (Algoriphagus), 9, 427, **429**, 430
alkalitolerans (Dyadobacter), 9
alleghenense (Spiroplasma), 11, 661, 668, **669**–670
allerginae (Cytophaga), 132, 136, **153**
alligatoris (Mycoplasma), 11, 577, 582, 584, **588**
allocasuarinae (Candidatus Phytoplasma), **703**
alluvionis (Pedobacter), 345
ALPHAPROTEOBACTERIA, 800, 807
alvi (Mycoplasma), 11, 582, 583, **588**
alvinipulli (Brachyspira), 532, 534, 535, 536, **538**, 539, 540
americana (Spirochaeta), 10, 476–478, **479**, 485
americani (Candidatus Xiphinematobacter), 13, **840**
americanum (Candidatus Phytoplasma), **703**
amnii (Prevotella), 4
amnionii (Leptotrichia), 767
amoebophila (Candidatus Protochlamydia), 870, **870**, 871, **871**
amoebophila (Protochlamydia), 870, **871**, 876
amphoriforme (Mycoplasma), 11, 583, 584, **588**, 610
amurskyense (Cyclobacterium), **423**, 424, 425, 435
amurskyensis (Zobellia), 6, **294**
amylolytica (Zhouia), 5, **292**
amylophilus (Bacteroides), 32
amylophilus (Ruminobacter), 32
amylovorum (Treponema), 10, 502, **510**, 513, 521
anaerobium (Asteroleplasma), 12, 722, **723**
Anaerophaga, 3, 26, 49, 50, **51**, 52, 53
Anaerophaga thermohalophila, 4, 51, **52**
Anaeroplasma, 12, 56, 571, 581, 618, 640, 647, 656, 667, 690, 696, 719, **720**, 721, 722, 723
Anaeroplasma abactoclasticum, 12, 571, 572, 720–**722**
Anaeroplasma bactoclasticum, 12, 571, 668, 720–**722**
Anaeroplasma intermedium, 12, 720–**722**
ANAEROPLASMATACEAE, 12, 570, 719, **720**, 721–723
ANAEROPLASMATALES, 10, 12, 570, 571, 574, 667, **719**, 721, 723
Anaeroplasma varium, 720–**722**
Anaerorhabdus, 3, 25, 26, 32, **45**
Anaerorhabdus furcosa, 32, 45
ANAPLASMATACEAE, 639
anatina (Coenonia), 5, 65, 111, **198**, 250, 253, 254, 292
anatipestifer (Riemerella), 5, 182, 185, 197, 198, 250, 253, 254, 262, **263**, 264
anatis (Mycoplasma), 11, 582, 584, **588**, 589
Ancalochloris, 803, 804
Ancalomicrobium, 803, 804, 807
Ancylobacter, 390, 417
Ancylobacter aquaticus, 390
Ancylobacter rudongensis, 379, 390
andersonii (Borrelia), 10, 492, 496, 497, **545**
andersonii (Brevinema), 10, 492, 496, 497, 545
anhuiense (Flavobacterium), 138
anhuiense (Sphingobacterium), 8, 333–334
anserina (Borrelia), 10, 485, 489, **492**, 493
anseris (Mycoplasma), 11, 109, 584, **589**
antarctica (Aequorivita), 7, 155–**157**
antarctica (Lewinella), 8, 367
antarctica (Sejongia), 5, 111, 182, 185, 271–273, **274**
antarcticum (Flavobacterium), 5, 112, 127, 129, 130, 137, **139**
antarcticus (Algoriphagus), 9, 430, **431**
antarcticus (Ulvibacter), 7
aphidicola (Buchnera), 319, 616
apis (Spiroplasma), 572, 656–657, 661, 664, 666–668, **670**, 674–675, 677
aprica (Cytophaga), 52, 373, 374, 443
aprica (Flammeovirga), 9, 52, 373, 393, **443**
Aquaspirillum, 408
aquatica (Arcicella), 9, 377–379, **380**, 390
aquaticum (Chromobacterium), 138
aquaticum (Chryseobacterium), 5
aquaticus (Adhaeribacter), 9, 375, 376, **377**, 387, 388, 398, 404, 410, 411
aquaticus (Ancylobacter), 390
aquatile (Empedobacter), 138
aquatile (Flavobacterium), 5, 112, 127, 128, 129, 135–137, **138**, 139, 145, 181, 189, 241, 274, 373
aquatilis (Bacillus), 138
aquatilis (Bacterium), 138
aquatilis (Chromobacterium), 138
aquatilis (Cytophaga), 145, 373
aquatilis (Flavobacterium), 138
aquatilis (Pedobacter), 345
aquidurense (Flavobacterium), 112, 128, 133, 136, **139**
Aquiflexum, 424, 427, 428, 429, **433**, 434, 435
Aquiflexum balticum, 9, 433, **434**, 435, 436
Aquimarina, 107, 110, **158**, 159–161, 221–223, 229, 269, 292, 374
aquimarina (Muricauda), 6, 240–**244**
Aquimarina brevivitae, 7, 159, 160, **161**
Aquimarina intermedia, 7, 159, 160, **161**
Aquimarina latercula, 7, 159, 160, **161**, 373
Aquimarina muelleri, 7, 158, 159, **160**
aquimarinus (Algoriphagus), 9, 426, 427, 430, **431**
aquivivus (Maribacter), 5–6, 235–236, **237**
araneosa (Lentisphaera), 13, 788, 789, **790**, 792
Arcicella, 372, **377**, 378–380, 390, 405, 407, 417
Arcicella aquatica, 9, 377–379, **380**, 390
Arcicella rosea, 9
arenae (Cerasicoccus), 13, 825, **827**
arenaria (Flammeovirga), 395, 409, 443, **447**
Arenibacter, 107, **161**, 162–164, 232, 233, 241, 245, 265, 270, 284, 293
Arenibacter certesii, 5, 162, 163, **164**
Arenibacter echinorum, 5
Arenibacter latericius, 5–6, 162, **163**, 164
Arenibacter palladensis, 162, 163, **164**
Arenibacter troitsensis, 5, 162, 163, **164**
arginini (Mycoplasma), 11, 580, 582, 584, 588, **589**
armeniacum (Limibacter), 9
aromativorans (Yeosuana), 6, 238, **291**
arothri (Chryseobacterium), 5
arthritidis (Mycoplasma), 11, 569, 578–580, 582, 584, **589**
arvensicola (Chitinophaga), 8, 351, **352**, 353, 356, 373
arvensicola (Cytophaga), 352, 364, 373, 374
asacharolytica (Porphyromonas), 4, 62, **64**, 69, 89
asaccharolyticus (Bacteroides), 32
asiatica (Spirochaeta), 10, 476–478, **479**, 485
asteris (Candidatus Phytoplasma), 569, 698, **703**
Asteroleplasma, 12, 568, 570, 571, 581, 618, 640, 647, 668, 687, 690, 721, **722**, 723
Asteroleplasma anaerobium, 12, 722, **723**
atlanticus (Croceibacter), 6, 111, 199, **200**, 223, 242, 259, 324
atlanticus (Psychrilyobacter), 12
atrichopogonis (Spiroplasma), 661, **670**
aurantia (Spirochaeta), 10, 474–478, **479**, 480, 484, 486
aurantiaca (Cytophaga), 154, 368, 373, **374**, 379
aurantiaca (Gemmatimonas), 13, 782–**784**
aurantiaca (Niabella), 8
aurantiacus (Flexibacter), **153**, 154, 392, 394
aurantiacus var. excathedrus (Flexibacter), 113, 136, **153**, 154, 392
aurantiacus (Perexilibacter), 393
aurantia subsp. aurantia (Spirochaeta), 477, **480**
aurantia subsp. stricta (Spirochaeta), 477, **480**
aurantifolia (Candidatus Phytoplasma), **703**
Aureispira, 7, 8, 22, 358–**361**, 363
Aureispira marina, 8, 361, **363**, 393
Aureispira maritima, 8, **363**
aureus (Staphylococcus), 72, 211
auris (Mycoplasma), 11, 582, 584, **589**
australasia (Candidatus Phytoplasma), **705**
australiense (Candidatus Phytoplasma), 698, **705**
axanthum (Acholeplasma), 11, 574, 688–690, **691**
azotonutricium (Treponema), 10, 472, 508, **510**, 521, 563, 564

Bacillus, 571, 748, 751
Bacillus aquatilis, 138
Bacillus brevis, 210, 211
Bacillus fusiformis, 748, 753
Bacillus subtilis, 111, 656, 668
BACTERIA, 1, 13, 460, 461, 462, 783, 789, 843, 880, 881, 896, 907
bacteriovorus (Bdellovibrio), 852
Bacterium aquatilis, 138
Bacterium breve, 210, 211
Bacterium canale, 210, 211
Bacterium canalis parvis, 210, 211
Bacterium melaninogenicum, 86, 88, 101
Bacterium zoogleoformans, 101
BACTEROIDACEAE, 3, 4, **25**, 26, 28, 30, 32, 34, 36, 38, 40–44, 88, 370
BACTEROIDALES, 3, **25**, 324
Bacteroides, 3, 4, 25, 26, **27**, 30, 35, 37, 38, 39, 52, 55, 58, 63, 65, 72–74, 77, 79, 81, 87, 89, 93, 112, 373–375, 769
Bacteroides acidifaciens, 3
Bacteroides amylophilus, 32
Bacteroides asaccharolyticus, 32
Bacteroides barnesiae, 3
Bacteroides bivius, 32, 91
Bacteroides buccae, 32, 93
Bacteroides buccalis, 32, 93
Bacteroides caccae, 3, 28, **34**
Bacteroides capillosus, 32, **38**, 39
Bacteroides capillus, 3, 32, 93
Bacteroides cellulosilyticus, 3
Bacteroides cellulosolvens, 3, 38, **39**
Bacteroides coagulans, 39, **40**
Bacteroides coprocola, 3, 28, **34**
Bacteroides coprophilus, 3
Bacteroides coprosuis, 3, 28, **34**
Bacteroides corporis, 32, 94
Bacteroides denticola, 95
Bacteroides disiens, 32, 95

Bacteroides distasonis, 33, 34, 35, 74
Bacteroides dorei, 3, 28, **35**
Bacteroides eggerthii, 3, 28, 32, **35**
Bacteroides endodontalis, 32, 67
Bacteroides finegoldii, 3, **35**
Bacteroides forsythus, 3, 32, 33, 74, 79–81
Bacteroides fragilis, 3, 27, 30, 32, **33**–34, 36, 38, 45, 55, 62, 79, 87, 89, 132, 242, 461, 769
Bacteroides furcosus, 3, 32, 45
Bacteroides galacturonicus, 39, **40**
Bacteroides gallinarum, 3
Bacteroides gingivalis, 32, 67
Bacteroides goldsteinii, 28, **35**
Bacteroides gracilis, 32
Bacteroides helcogenes, 3, 30, **36**
Bacteroides heparinolyticus, 32
Bacteroides hypermegas, 32
Bacteroides intermedius, 32, 96
Bacteroides intestinalis, 3, **36**
Bacteroides levii, 32, 68
Bacteroides loescheii, 32, 96
Bacteroides macacae, 68
Bacteroides massiliensis, 3, 29, **36**
Bacteroides melaninogenicus, 32, 62, 64, 68, 88
Bacteroides melaninogenicus subsp. macacae, 68
Bacteroides merdae, 29, 33, **36**
Bacteroides microfusus, 32, 56
Bacteroides multiacidus, 32
Bacteroides nodosus, 32
Bacteroides nordii, 3, 29, **36**
Bacteroides ochraceus, 32, 88, 174
Bacteroides oralis, 32, 93, 95, 174
Bacteroides oris, 98
Bacteroides ovatus, 3, 29, **36**, 37
Bacteroides pectinophilus, 39, **40**
Bacteroides pentosaceus, 32, 93
Bacteroides plebeius, 3, 29, **37**
Bacteroides pneumosintes, 32
Bacteroides polypragmatus, 3, 39, **40**
Bacteroides praeacutus, 32
Bacteroides putredinis, 55, 56, 58
Bacteroides pyogenes, 3, 30, **37**
Bacteroides ruminicola, 32, 92, 93, 95, 99
Bacteroides ruminicola subsp. brevis, 92
Bacteroides salanitronis, 3
Bacteroides salivosus, 32, 69
Bacteroides salyersiae, 3, 29, **37**
Bacteroides splanchnicus, 3, 4, 32, 39, **40**, 41, 74
Bacteroides stercoris, 3, 29, 30, **37**
Bacteroides succinogenes, 32, 740, 743
Bacteroides suis, 3, 29, **37**
Bacteroides tectus, 3, 29, **37**,38
Bacteroides termitidis, 769, 770
Bacteroides thetaiotaomicron, 29, 30, **38**, 62, 131
Bacteroides uniformis, 3, 29, 30, **38**, 73
Bacteroides ureolyticus, 32
Bacteroides veroralis, 32, 101
Bacteroides vulgatus, 3, 29, 30, **38**
Bacteroides xylanisolvens, 3
Bacteroides xylanolyticus, 39, **41**
Bacteroides zoogleoformans, 32, 33
BACTEROIDETES, 1–16, 21–24, **25**, 27, 31, 52, 53, 56, 58, 62, 65, 71, 77, 79, 106, 108, 110, 112, 128, 131, 132, 180, 188, 198, 200, 203, 204, 255, 259, 315, 317, 322, 324, 327, 361, 371, 384, 388, 392, 404, 405, 408, 438, 452, 453, 457, 459–462, 466, 467
BACTEROIDIA, 2, 4, **25**, 370
bactoclasticum (Anaeroplasma), 12, 571, 668, 720–**722**
bajacaliforniensis (Spirochaeta), 10, 476–478, **480**, 485
Balneola, 3, 7, 370
Balneola alkaliphila, 9
Balneola vulgaris, 9
baltazardii (Borrelia), 10, 489, **493**
baltica (Belliella), 9, 405, 434, 435, **436**, 437, 440
baltica (Cellulophaga), 5, 6, 176–178, **179**, 180
baltica (Rhodopirellula), 16, 896, 897, 903–905, **906**, 907–910
balticum (Aquiflexum), 9, 433–436
balustinum (Chryseobacterium), 5, 182–184, 186–190, **191**, 193, 194, 212, 228
balustinum (Flavobacterium), 191
barnesiae (Bacteroides), 3
Barnesiella, 61, **70**, 71
Barnesiella intestinihominis, 4
Barnesiella viscericola, 4, 70, **71**
baroniae (Prevotella), 4, 65, **91**
Bartonella, 642
basaltis (Flaviramulus), 6, 213, **214**
Bdellovibrio, 852
Bdellovibrio bacteriovorus, 852
Beggiatoa, 477
beijingensis (Dyadobacter), 380, 381, 384, 385, **386**
bekefii (Planctomyces), 16, **881**, 887, 888, 889, 890, 894, 895
bellariivorans (Prolixibacter), 3
Belliella, 405, 424, 427–429, **434**, 435–439
Belliella baltica, 9, 405, 434, 435, **436**, 437, 440
bemisiae (Candidatus Fritschea), 874, **875**, 876
bemisiae (Fritschea), 874, 875, **876**
bennonis (Porphyromonas), 69
bergensis (Prevotella), 65, **91**
Bergeyella, 136, **165**, 180, 181, 197, 203, 205, 208, 212, 227, 228, 250, 263, 271, 273, 274, 287, 288
Bergeyella zoohelcum, 5, 110, **165**, 166, 182, 189, 197, 288
berlinense (Treponema), 502, 508, **511**, 520
BETAPROTEOBACTERIA, 868
Bifidobacterium, 102
Bifidobacterium bifidum, 102
bifidum (Bifidobacterium), 102
biflexa (Leptospira), 10, 547–549, 551, **553**, 555, 557
biformata (Robiginitalea), 6, 259, **264**, 265
Bilophila, 30
bissettii (Borrelia), 496–498
bivia (Prevotella), 65, **91**, 92, 103
bivius (Bacteroides), 32, 91
Bizionia, 7, 135, 157, 158, **166**, 167, 218, 221, 222, 249, 291
Bizionia algoritergicola, 7, **167**
Bizionia gelidisalsuginis, 7, **167**, 168
Bizionia myxarmorum, 7, 167, **168**
Bizionia paragorgiae, 7, **166**, 167, 218, 249
Bizionia saleffrena, 7, 167, **168**
blandensis (Leeuwenhoekiella), 7, 132, 233, **234**
Blastopirellula, 804, 880, 881, **895**, 896, 897, 899, 900, 903–906, 908, 911, 912, 914
Blastopirellula marina, 16, 895, 896. **897**, 898, 904, 905, 907, 908
BLATTABACTERIACEAE, 5, 7, 105, **315**, 316–320
Blattabacterium, 105, **315**, 316–320
Blattabacterium cuenoti, 7, 315, **320**
Blatta orientalis, 316, 320, 758, 873
Bordetella, 601
Bordetella bronchiseptica, 601
borealis (Pedobacter), 345
borgpetersenii (Leptospira), 10, 547–549, **553**, 554
boritolerans (Algoriphagus), 9, 427, 430, **431**
boritolerans (Chimaereicella), 428, 431
Borrelia, 10, 473, 476, **484**, 485–488, 490, 491, 492–498, 508
Borrelia afzelii, 10, 485, 488, 490, **493**, 495, 496
Borrelia andersonii, 10, 492, 496, 497, **545**
Borrelia anserina, 10, 485, 489, **492**, 493
Borrelia baltazardii, 10, 489, **493**
Borrelia bissettii, 496–498
Borrelia brasiliensis, 10, 489, **493**
Borrelia burgdorferi, 10, 485, 487, 488, 490, 491, **493**, 494–497
Borrelia californiensis, 497
Borrelia caucasica, 10, 489, **493**
Borrelia coriaceae, 10, 489, 490, **493**, 494, 495
Borrelia crocidurae, 10, **494**
Borrelia dugesii, 489, **494**
Borrelia duttonii, 10, 489, 490, **494**, 496
Borrelia garinii, 485, 488, 490, **494**, 495–497
Borrelia graingeri, 10, 489, **494**
Borrelia harveyi, 10, 489, **494**
Borrelia hermsii, 10, 488–492, **494**, 495, 497
Borrelia hispanica, 10, 494, **495**
Borrelia japonica, 10, 485, 489, 492, **495**, 496, 497
Borrelia latyschewii, 10, 489, **495**
Borrelia lonestari, 498
Borrelia lusitaniae, 485, 489, 492, **495**, 496
Borrelia mazzottii, 489, **495**
Borrelia miyamotoi, 487, 489, 491, **495**
Borrelia parkeri, 488, 489, 491, **495**, 497
Borrelia persica, 489, 495, **496**
Borrelia recurrentis, 486, 489–491, 493–495, **496**, 497
Borrelia sinica, 485, 489, 492, **496**
Borrelia spielmanii, 485, 489, **496**
Borrelia tanukii, 10, 485, 489, 492, **496**, 497
Borrelia theileri, 489, **496**
Borrelia tillae, 489, **496**
Borrelia turdi, 10, 485, 487, 489, 496, **497**
Borrelia turcica, 10, 489, **497**
Borrelia turicatae, 10, 488–491, 495, **497**
Borrelia valaisiana, 485, 489, 490, 492, 496, **497**
botulinum (Clostridium), 102
bovigenitalium (Mycoplasma), 11, 578, 582, 584, **589**, 590
bovirhinis (Mycoplasma), 11, 578, 582, 584, **590**
bovis (Chryseobacterium), 5
bovis (Mycoplasma), 11, 578, 579, 582, 584, 587, **590**
bovoculi (Mycoplasma), 11, 582, 584, **590**
Brachyspira, 9, 10, 476, 508, **531**, 537–540
Brachyspira aalborgi, 10, 472, 531–534, 537, **538**, 539–541
Brachyspira alvinipulli, 532, 534, 535, 536, **538**, 539, 540
BRACHYSPIRACEAE, 9, 10, 471–473, **531**, 532–540
Brachyspira hyodysenteriae, 508, 531, 532, 534–538, **539**, 540
Brachyspira innocens, 532, 534, 535, 537–538, **539**, 540
Brachyspira intermedia, 532–534, 536–**539**, 540, 550

Brachyspira murdochii, 532–534, 538–**540**
Brachyspira pilosicoli, 532–**540**
Brachyspira suanatina, 541
branchiophilum (Flavobacterium), 112, 129, 130, 132, 133, 134, 135, 136, **139**
brasiliense (Candidatus Phytoplasma), **705**
brasiliensis (Borrelia), 10, 489, **493**
brasiliensis (Planctomyces), 881, 886, 887, **891**
brassicae (Acholeplasma), 11, 688, 689, **691**
brennaborense (Treponema), 502, 504, **511**
breve (Bacterium), 210, 211
breve (Empedobacter), 210, 212
breve (Flavobacterium), 210, 211
Brevibacterium, 725
brevicolli (Candidatus Xiphinematobacter), 13, **839**
Brevinema, 9, 10, 476, 501, 545
Brevinema andersonii, 10, 492, 496, 497, 545
BREVINEMATACEAE, 10, 471–473, **545**
brevis (Bacillus), 210, 211
brevis (Empedobacter), 5, 137, 189, 206, 208, **210**, 211, 274, 286, 287
brevis (Flavobacterium), 210, 211
brevis (Prevotella), 4, 65, **92**
brevis (Pseudobacterium), 210, 212
brevivitae (Aquimarina), 7, 159, 160, **161**
brevivitae (Gaetbulimicrobium), 159
broadyi (Subsaxibacter), 6, 111, 223, 242, **276**, 277
BROCADIACEAE, 918, **919**, 922
BROCADIACEAE (CANDIDATUS), 918, **919**, 920–922
BROCADIALES, 16, 879, **918**, 920
BROCADIALES (CANDIDATUS), **918**, 919–922
bronchiseptica (Bordetella), 601
broomii (Leptospira), 10, 547, **554**, 556
Brucella, 30, 35–37, 56, 57, 60, 95, 99, 754, 759
Brumimicrobium, 7, 16, **323**, 324–326
Brumimicrobium glaciale, 7, 16, 323–**326**, 328, 329
bryantii (Prevotella), 4, 65, **92**
bryantii (Treponema), 502, **511**, 743
buccae (Bacteroides), 32, 93
buccae (Prevotella), 65, **93**
buccale (Mycoplasma), 582, 584, **590**
buccalis (Bacteroides), 32, 93
buccalis (Leptotrichia), 12, 762, **766**, 770
buccalis (Prevotella), 4, 65, **93**
Buchnera aphidicola, 319, 616
burgdorferi (Borrelia), 10, 485, 487, 488, 490, 491, **493**, 494–497
burtonensis (Psychroserpens), 7, 111, 214, 215, 223, 242, **261**, 262, 288, 289, 324
buteonis (Mycoplasma), 582, 584, **591**
butkevichii (Polaribacter), **256**
byssophila (Leadbetterella), 9, 388, 405, **406**

caccae (Bacteroides), 3, 28, **34**
caeni (Chryseobacterium), 5, 190
caeni (Pedobacter), 339–342, 344, **345**
caldaria (Spirochaeta), 10, 476–478, **480**, 484, 502, 508, 522
Caldicellulosiruptor, 778
Caldicellulosiruptor saccharolyticus, 778
californicum (Mycoplasma), 11, 582, 584, **591**
californiensis (Borrelia), 497
calotermitidis (Diplocalyx), 16, 565
calotermitidis (Pillotina), 10, 16, 565, 585
Campylobacter, 39
Campylobacter gracilis, 32
Campylobacter rectus, 79
canadense (Mycoplasma), 11, 582, 584, **591**
canadense (Sphingobacterium), 8, 333, **334**, 335
canadensis (Flexibacter), 8, 340, 344, 392, 394, **395**, 396
canale (Bacterium), 210, 211
Canaleparolina darwiniensis, 565
canalis parvis (Bacterium), 210, 211
Candidatus Borrelia texasensis, **497**
CANDIDATUS BROCADIACEAE, **918**, 919–922
CANDIDATUS BROCADIALES, **918**, 919–922
Candidatus Cardinium hertigii, 9
Candidatus Clavichlamydia, **865**, 866
CANDIDATUS CLAVICHLAMYDIACEAE, 845, **865**
Candidatus Clavichlamydia salmonicola, 13, 843, 845, 865, **866**
Candidatus Clavochlamydia salmonicola, 866
Candidatus comitans, 332
Candidatus Fritschea, 13, 874, **875**, 876
Candidatus Fritschea bemisiae, 874, **875**, 876
Candidatus Fritschea eriococci, 16, 874, 875, **876**
Candidatus Mycoplasma haematoparvum, 586, **611**
Candidatus Mycoplasma haemobos, **611**
Candidatus Mycoplasma haemodidelphidis, 586, **612**
Candidatus Mycoplasma haemolamae, 586, **612**
Candidatus Mycoplasma haemominutum, 586, **612**
Candidatus Mycoplasma haemotarandirangiferis, **612**
Candidatus Mycoplasma kahaneii, **612**
Candidatus Mycoplasma ravipulmonis, 586, **612**
Candidatus Mycoplasma turicensis, **612**
Candidatus Phytoplasma, 11, 570, 581, 618, 640, 647, 668, 687, 690, 696, 701, 702, 703, 704 , 707–709
Candidatus Phytoplasma allocasuarinae, **703**
Candidatus Phytoplasma americanum, **703**
Candidatus Phytoplasma asteris, 569, 698, **703**
Candidatus Phytoplasma aurantifolia, **703**
Candidatus Phytoplasma australasia, **705**
Candidatus Phytoplasma australiense, 698, **705**
Candidatus Phytoplasma brasiliense, **705**
Candidatus Phytoplasma caricae, **706**
Candidatus Phytoplasma castaneae, **710**
Candidatus Phytoplasma cynodontis, **710**
Candidatus Phytoplasma fragariae, **710**
Candidatus Phytoplasma fraxini, **710**
Candidatus Phytoplasma graminis, **710**
Candidatus Phytoplasma japonicum, **710**
Candidatus Phytoplasma lycopersici, **710**
Candidatus Phytoplasma mali, 698, 699, **710**
Candidatus Phytoplasma omanense, **710**
Candidatus Phytoplasma oryzae, **710**
Candidatus Phytoplasma phoenicium, **711**
Candidatus Phytoplasma pini, **711**
Candidatus Phytoplasma prunorum, **711**
Candidatus Phytoplasma pyri, **711**
Candidatus Phytoplasma rhamni, **711**
Candidatus Phytoplasma spartii, **711**
Candidatus Phytoplasma trifolii, **711**
Candidatus Phytoplasma ulmi, **711**
Candidatus Phytoplasma ziziphi, **712**
Candidatus Piscichlamydia, **872**, 873
CANDIDATUS PISCICHLAMYDIACEAE, 13, **872**
Candidatus Piscichlamydia salmonis, 13, 845, 872, **873**
Candidatus Procabacter, 868
Candidatus Protochlamydia, **870**
Candidatus Protochlamydia amoebophila, 870, **870**, 871, **871**
Candidatus Rhabdochlamydia porcellionis, 13, 845, 866, 873, **874**
Candidatus Xiphinematobacter, 837, **838**, 839, 840
Candidatus Xiphinematobacter americani, 13, **840**
Candidatus Xiphinematobacter brevicolli, 13, **839**
Candidatus Xiphinematobacter rivesi, 13, **840**
cangingivalis (Porphyromonas), **64**
canifelinum (Fusobacterium), 748, 750, **754**
canigenitalium (Ureaplasma), 618, **621**
canimorsus (Capnocytophaga), 5, 168, 172, **174**, 175
canis (Haemobartonella), 11, 599, 642
canis (Mycoplasma), 11, 582, 584, **591**
canoris (Porphyromonas), **64**
cansulci (Porphyromonas), 4, **64**, 66, 67
cantharicola (Spiroplasma), 11, 661, 668, **670**
capillosus (Bacteroides), 32, **38**, 39
capillus (Bacteroides), 3, 32, 93
Capnocytophaga, 52, 72–75, 88, 107, 135, **168**, 169–173, 174, 175, 176, 198, 250
Capnocytophaga canimorsus, 5, 168, 172, **174**, 175
Capnocytophaga cynodegmi, 5, 168, 172, 174, **175**
Capnocytophaga gingivalis, 5, 168, 169, 172, 174, **175**
Capnocytophaga granulosa, 5, 168–170, 174, **175**, 176
Capnocytophaga haemolytica, 5, 168, 172, 174, 175, **176**
Capnocytophaga ochracea, 5, 32, 168, 169, 171, 172, **174**, 175
Capnocytophaga sputigena, 5, 168, 169, 171, 172, 174, **176**
capnocytophagoides (Dysgonomonas), 71–74, **75**, 76
capricolum (Mycoplasma), 11, 579, 581, **591**, 592
capricolum subsp. capricolum (Mycoplasma), 569, 579, 584, 587, **592**, 602
capricolum subsp. capripneumoniae (Mycoplasma), 579, 581, 584, **592**, 602
Capsularis zoogleoformans, 101
capsulatum (Acidobacterium), 12, **728**
capsulatum (Novosphingobium), 138
carateum (Treponema), 501, 502, 503, 505, **512**
Cardiobacterium, 772
caricae (Candidatus Phytoplasma), **706**
caseinilytica (Lishizhenia), 7, 327, **328**
castaneae (Candidatus Phytoplasma), **710**
catalasitica (Crocinitomix), 7, 16, **326**, 327, 328, 395
catena (Salegentibacter), 7, 266, 267, **268**
catena (Salinimicrobium), 7
cati (Ureaplasma), 11, **621**, 622
catoniae (Oribaculum), 66–67
catoniae (Porphyromonas), 4, 62, 64, **66**
caucasica (Borrelia), 10, 489, **493**
cauliformis (Sporocytophaga), 113, 136, 153, **154**, 418
Caulobacter, 807, 905
caviae (Chlamydophila), 13, 845, 860
caviae (Mycoplasma), 11, 582, 584, **592**
cavigenitalium (Acholeplasma), 11, 689, **691**
cavipharyngis (Mycoplasma), 11, 583, 584, **592**

cellobiosiphila (Spirochaeta), 477–478, **481**, 486
cellulolyticus (Acetivibrio), 39
Cellulophaga, 107, 176, 177, **178**, 179, 201, 230, 255, 284, 292, 293, 374
Cellulophaga algicola, 5, 6, 176, 177, 178, **179**
Cellulophaga baltica, 5, 6, 176–178, **179**, 180
Cellulophaga fucicola, 6, 176–178, **179**, 229
Cellulophaga lytica, 6, 176–178, **179**, 229, 230, 373
Cellulophaga pacifica, 5, 6, 176–179, **180**
Cellulophaga uliginosa, 179, 295
cellulosilyticus (Bacteroides), 3
cellulosolvens (Bacteroides), 3, 38, **39**
Cerasicoccus, 817, 824–826, **827**, 828, 830
Cerasicoccus arenae, 13, 825, **827**
certesii (Arenibacter), 5, 162, 163, **164**
ceti (Cetobacterium), 12, 752, **758**
ceti (Flavobacterium), 138
Cetobacterium, 12, 748, **758**
Cetobacterium ceti, 12, 752, **758**
Cetobacterium somerae, **758**
chauliocola (Mesoplasma), 11, 650, **651**
cheniae (Flavobacterium), 138
Chimaereicella, 9, 424, 428
Chimaereicella alkaliphila, 428, 429
Chimaereicella boritolerans, 428, 431
chinense (Spiroplasma), 11, 661, 668, **670**, 677
chitinivorans (Hymenobacter), 9, 398–400, **402**
Chitinophaga, 7, 8, **351**, 352, 353, 356–358, 371, 374, 394
Chitinophaga arvensicola, 8, 351, **352**, 353, 356, 373
CHITINOPHAGACEAE, 7, 8, 330, **351**, 352–357, 371
Chitinophaga filiformis, 8, 352, **353**, 356, 395
Chitinophaga ginsengisegetis, 8, 352, 353, **354**
Chitinophaga ginsengisoli, 8, 351, 352, 353, **354**
Chitinophaga japonensis, 8, 351, 352, 353, **355**, 395
Chitinophaga pinensis, 8, 351, 352, 353, 356
Chitinophaga sancti, 8, 352, 353, **355**, 356, 395
Chitinophaga skermanii, 8, 351, 352, 353, **355**, 356
Chitinophaga soli, 368
Chitinophaga terrae, 351, 352, 353, **356**
Chlamydia, 845, **846**, 847, 849–855, 857, 858, 861, 865, 866–870, 872, 873
CHLAMYDIACEAE, 13, 844, **845**, 850, 852, 855, 856, 858, 865, 866, 868–870, 872, 873, 875, 877
CHLAMYDIAE, 1, 2, 13, 15, 21, 24, 785, 786, **843**, 844, 852, 870, 872
Chlamydia isopodii, 874
CHLAMYDIALES, 13, 842, **844**, 845, 856–858, 877
Chlamydia muridarum, 13, 849, 850, 851–857, 859, **860**, 866
Chlamydia pecorum, 850–852, 854, 855, 857–859, **860**, 861, 866
Chlamydia pneumoniae, 847–859, **860**, 861, 866
Chlamydia psittaci, 847, 849–860, **861**, 866
Chlamydia suis, 849–855, 857, 859, **861**, 866
Chlamydia trachomatis, 13, 616, 846, 849–853, 855–857, **858**, 859–861, 866
CHLAMYDIIA, **844**
Chlamydophila, 13, 845, 858
Chlamydophila abortus, 13, 845, 859
Chlamydophila caviae, 13, 845, 860
Chlamydophila felis, 13, 845, 860
Chlamydophila pecorum, 13, 845
Chlamydophila pneumoniae, 13
Chlamydophila psittaci, 13, 845
Chlorobi, 462, 463, 803
CHLOROBIACEAE, 775
Chlorobium tepidum, 461
Chlorobium vibrioforme, 364, 368
Chloroflexi, 775
Chloroflexus aggregans, 725
Chondromyces crocatus, 332
chondrophila (Waddlia), 16, **877**
chordae (Algoriphagus), 9, 427, 430, **431**
Chromobacterium aquaticum, 138
Chromobacterium aquatilis, 138
Chryseobacterium, 5, 107, 108, 138, 165, **180**, 181–196, 203–206, 208–212, 227, 250, 263, 271, 273, 274
Chryseobacterium aquaticum, 5
Chryseobacterium arothri, 5
Chryseobacterium balustinum, 5, 182–184, 186–190, **191**, 193, 194, 212, 228
Chryseobacterium bovis, 5
Chryseobacterium caeni, 5, 190
Chryseobacterium daecheongense, 5, 181–184, 190, **191**, 195, 196, 212
Chryseobacterium daeguense, 190
Chryseobacterium defluvii, 5, 182, 183, 184, 190
Chryseobacterium flavum, 5, 190
Chryseobacterium formosense, 5, 182, 183, 184, 189, 190, **192**, 273, 274
Chryseobacterium gambrini, 5, 190
Chryseobacterium gleum, 5, 137, 180, 183, 186, 188, 189, **190**, 191, 192, 195, 196, 206, 208
Chryseobacterium gregarium, 5
Chryseobacterium haifense, 5, 190
Chryseobacterium hispanicum, 5, 180, 182, 183, 184, **192**
Chryseobacterium hominis, 5, 190
Chryseobacterium hungaricum, 5
Chryseobacterium indologenes, 5, 182, 183, 184, 186–191, **192**, 196, 206, 208
Chryseobacterium indoltheticum, 5, 137, 182–184, 188, 189, 191, **192**, 193–195
Chryseobacterium jejuense, 5, 190
Chryseobacterium jeonii, 5
Chryseobacterium joostei 5, 181–184, 187–189, **193**, 194, 196
Chryseobacterium luteum, 5, 190
Chryseobacterium massiliae, 180, 187, 188, 189, **196**
Chryseobacterium meningosepticum, 110, 186, 189, 203, 205, 209, 253, 254
Chryseobacterium miricola, 5, 186, 189, 203, 205, 209, 210
Chryseobacterium molle, 5, 190
Chryseobacterium oranimense, 5
Chryseobacterium pallidum, 5, 190
Chryseobacterium piscium, 182–184, 188, **193**, 194
Chryseobacterium proteolyticum, 180, 181, 184, 186, 188, 189, **196**
Chryseobacterium scophthalmum, 5, 181–184, 186–189, 192, 193, **194**, 212
Chryseobacterium shigense, 5, 181–184, 188, 190, **194**
Chryseobacterium soldanellicola, 5, 182, 183, 185–187, 190, **194**, 195
Chryseobacterium soli, 5, 190
Chryseobacterium taeanense, 5, 182–184, 186, 190, **195**
Chryseobacterium taichungense, 182, 183, 185, 187, 189, 190, **195**
Chryseobacterium taiwanense, 5, 182, 183, 185, 190, **195**
Chryseobacterium ureilyticum, 5, 190
Chryseobacterium vrystaatense, 5, 182, 183, 185, 188, **195**, 196
Chryseobacterium wanjuense, 5, 182, 183, 185, 190, **196**
Chrysiogenes, 775
chrysopicola (Spiroplasma), 11, 661, 668–669, **670**, 671
Chthoniobacter, 836, **837**
Chthoniobacter flavus, 13, 834, **837**
circumdentaria (Porphyromonas), 4, **67**
citelli (Mycoplasma), 11, 582, 584, **592**, 593
citri (Spiroplasma), 11, 572, 654–**669**
clarkii (Spiroplasma), 661, 666, 668, **671**
Clavichlamydia (Candidatus), **865**, 866
CLAVICHLAMYDIACEAE (CANDIDATUS), 845, **865**, 866
Clavichlamydia salmonicola (Candidatus), 13, 843, 845, 865, **866**
Clavochlamydia, 866
CLAVOCHLAMYDIACEAE, 866
Clavochlamydia salmonicola (Candidatus), 866
Clevelandina, 9, 565
Clevelandina reticulitermitidis, 10, 16, 565
cloacale (Mycoplasma), 11, 582, 584, **593**
Cloacibacterium, 110, **197**
Cloacibacterium normanense, **197**
Clostridium, 38, 39, 750, 773
Clostridium botulinum, 102
Clostridium difficile, 751
Clostridium innocuum, 571, 572, 667, 668
Clostridium orbiscindens, 38
Clostridium rectum, 762
Clostridium thermocellum, 741
coagulans (Bacteroides), **39**, 40
coccoides (Eperythrozoon), 11, 581, 611, **640**, 641
coccoides (Mycoplasma), 11, 572, 583, 584, **611**, 641
coccoides (Spirochaeta), 10, 472–473, 476, **484**, 485
Coenonia, **198**, 292
Coenonia anatina, 5, 65, 111, **198**, 250, 253, 254, 292
cohaerens (Herpetosiphon), 366, 367
cohaerens (Lewinella), 8, 363, 366–367
coleopterae (Mesoplasma), 11, 650, **651**
coli (Escherichia), 1, 72, 89, 111, 205, 211, 253, 263, 281, 318, 341, 342, 385, 418, 499, 535, 549, 597, 610, 617, 657, 658, 660, 668, 699, 720–722, 750, 751, 796, 797, 801, 806, 815, 818, 831, 835, 898
collis (Mycoplasma), 11, 582, 584, **593**
columbina (Riemerella), 182, 185, 197, 253, 254, 263, **264**
columbinasale (Mycoplasma), 11, 582, 584, **593**, 597
columbinum (Mycoplasma), 11, 582, 584, **593**
columborale (Mycoplasma), 11, 582, 584, **593**
columnare (Flavobacterium), 112, 120, 127–134, 136, 137, **140**, 154, 373, 395
columnaris (Cytophaga), 140, 373, 392
columnaris (Flexibacter), 140, 373, 392, 394
comitans (Candidatus), 332
composti (Pedobacter), 345
composti (Sphingobacterium), 8, 333, **335**, 336
conjunctivae (Mycoplasma), 11, 569, 582, 584, **593**, 594
copepodicola (Lacinutrix), 6, 231, **232**, 249
copri (Prevotella), **93**

coprocola (Bacteroides), 3, 28, **34**
coprophilus (Bacteroides), 3
coprosuis (Bacteroides), 3, 28, **34**
Coraliomargarita, 14, 24, 817, 824–826, **827**–830
Coraliomargarita akajimensis, 13, 818, 826, **827**, 828, 829, 831
coriaceae (Borrelia), 10, 489, 490, **493**, 494, 495
corogypsi (Mycoplasma), 11, 582, 584, **594**
corporis (Bacteroides), 32, 94
corporis (Prevotella), 4, 65, **94**
corruscae (Mesoplasma), 11, 650, **651**
corruscae (Spiroplasma), 11, 657, 661, 668, **671**, 677
corynebacterioides (Nocardia), 138
corynebacteroides (Rhodococcus), 138
Costertonia, 5, 110, **199**
Costertonia aggregata, 5, **199**
cottewii (Mycoplasma), 11, 584, **594**
Crenothrix, 7, 457, 467
CRENOTRICHACEAE, 3, 7, 8, 457, 467
crevioricanis (Porphyromonas), **67**
Criblamydia, 24, 843, 866, **867**
CRIBLAMYDIACEAE, 13, 15, 24, 845, **867**
Criblamydia sequanensis, 13, 843, 866, **867**
cricetuli (Mycoplasma), 11, 582, 584, **594**
Cristispira, 10, 23, 473, 476, **498**, 499, 500, 501, 563, 565
Cristispira pectinis, 10, 472, **498**, 501
crocatus (Chondromyces), 332
crocea (Aequorivita), 7, 155–**157**
Croceibacter, 6, 107, **199**, 200, 265
Croceibacter atlanticus, 6, 111, 199, **200**, 223, 242, 259, 324
Croceitalea, 110
croceum (Flavobacterium), 112, 128, 129, 130, 133, 135–137, **141**
croceus (Pelagicoccus), 829, **833**
crocidurae (Borrelia), 10, **494**
Crocinitomix, 6, 22, 322, **326**, 327, 328, 394
Crocinitomix catalasitica, 7, 16, **326**, 327, 328, 395
crocodyli (Mycoplasma), 11, 582, 584, **594**
crusticola (Dyadobacter), 380, 381, 384, 385, **386**
cryoconitis (Pedobacter), 339–342, 344, **347**
Cryomorpha, 3, 5, 6, 7, 22, 105, 106, 110, **322**, 323, 324, 325, 326, 327, 328, 329, 394
CRYOMORPHACEAE, 3, 5, 6, 7, 22, 105, 106, 110, **322**, 323–329, 394
Cryomorpha ignava, 7, 16, **323**, 324, 325, 328
cryptotermitidis (Diplocalyx), 565
cucumis (Flavobacterium), 138
cuenoti (Blattabacterium), 7, 315, **320**
culicicola (Spiroplasma), 11, 661, 668, **671**
CYANOBACTERIA, 737, 775
CYCLOBACTERIACEAE, 9, 370, **423**, 424–440
Cyclobacterium, 8, 9, 390, 405, 417, **423**, 424–429, 434, 435, 437–439
Cyclobacterium amurskyense, **423**, 424, 425, 435
Cyclobacterium lianum, 423, 424, 425
Cyclobacterium marinum, 9, 368, 379, 390, 393, **423**, 424, 435
cynodegmi (Capnocytophaga), 5, 168, 172, 174, **175**
cynodontis (Candidatus Phytoplasma), **710**
cynos (Mycoplasma), 11, 582, 584, **594**
Cytophaga, 8, 31, 49, 51, 52, 53, 55, 106, 110, 135, 137, 168, 179, 189, 190, 209, 210, 229, 244, 256, 257, 288, 293, 324, 328, 340, 344, 352, 364, 370, **371**, 373, 374, 375, 418, 428, 433, 442, 456, 457, 461, 737
Cytophaga agarovorans, 50, 373
Cytophaga allerginae, 132, 136, **153**
Cytophaga aprica, 52, 373, 374, 442, 443
Cytophaga aquatilis, 145, 373
Cytophaga arvensicola, 352, 364, 373, 374
Cytophaga aurantiaca, 154, 368, 373, **374**, 379
CYTOPHAGACEAE, 3, 8, 9, 22, 370, **371**, 372, 374, 375, 376, 378, 380–382, 384, 385, 386, 388, 390, 392, 394, 396, 398, 400, 402, 404–408, 410, 412, 414, 416, 418
Cytophaga columnaris, 140, 373, 392
Cytophaga diffluens, 52, 373, 374, 442, 443, 450, 451
Cytophaga fermentans, 4, 50, 52, 53, 57, 373, **374**, 384
Cytophaga flevensis, 373
Cytophaga heparina, 339, 343–345, 373, 374
Cytophaga hutchinsonii, 9, 179, 368, 371, 373, **374**, 379, 384, 393, 741
Cytophaga johnsonae, 146, 241, 355, 373
Cytophaga latercula, 161, 229, 373, 374
CYTOPHAGALES, 8, 22, 105, **370**, 376, 380, 401, 404
Cytophaga lytica, 52, 176, 179, 373, 374
Cytophaga marina, 279, 281, 373, 374
Cytophaga marinoflava, 57, 179, 234, 373, 374
Cytophaga pectinovora, 148, 373
Cytophaga psychrophila, 149, 154, 373, 392
Cytophaga saccharophila, 150, 373
Cytophaga salmonicolor, 49–51, 370, 373
Cytophaga succinicans, 151, 373
Cytophaga uliginosa, 179, 293, 295, 374
Cytophaga xylanolytica, 49–52, 374, **375**
Cytophagia, 3, 7, 8, 9, **370**, 394, 436, 453

daecheongense (Chryseobacterium), 5, 181–184, 190, **191**, 195, 196, 212
daechungensis (Pedobacter), 8, 345
daeguense (Chryseobacterium), 190
daejeonense (Flavobacterium), 129, 133, 136, 137, **141**
daejeonense (Sphingobacterium), 8, 333, **336**
darwiniensis (Canaleparolina), 565
Dechloromonas agitata, 725
Deferribacter, 737, 775
DEFERRIBACTERES, 737
defluvii (Chryseobacterium), 5, 182–184, 190
defluvii (Flavobacterium), 112, 128, 130, 133, 136, 137
defluvii (Runella), 412, 413, **414**
degerlachei (Flavobacterium), 127, 128, 132, 137, **141**, 144, 149
Deinococci, 775
Deinococcus, 775
delafieldii (Ilyobacter), 759, **760**
denitrificans (Flavobacterium), 106, 112, 127, 129, 130, 133, 136, 137
dentalis (Prevotella), 5, 94, **105**
denticola (Bacteroides), 95
denticola (Prevotella), 4, 65, **95**
denticola (Treponema), 472, 476, 486, 501, 502, 503, 504, 510, **512**, 513, 516, 523, 535
deserti (Hymenobacter), 9
diabroticae (Spiroplasma), 11, 661, 668, **671**, 672
diaphorus (Krokinobacter), 7, 230, **231**
Dichelobacter nodosus, 32
DICTYOGLOMALES, 12, 775, **776**
DICTYOGLOMACEAE, 12, 23, **776**
DICTYOGLOMI, 1, 2, 12, 21, 23, **775**, 776
DICTYOGLOMIA, 12, **776**
Dictyoglomus, 13, 23, **776**, 777–779
Dictyoglomus thermophilum, 776, 778, **779**
Dictyoglomus turgidum, 778, **779**
difficile (Clostridium), 751
diffluens (Cytophaga), 52, 373, 374, 443, **450**, 451
diffluens (Persicobacter), 9, 54, 450, **451**, 452
diminutum (Spiroplasma), 11, 661, 668, 670, **672**
Diplocalyx, 9, 16, 23, 565
Diplocalyx calotermitidis, 16, 565
Diplocalyx cryptotermitidis, 565
disiens (Bacteroides), 32, 95
disiens (Prevotella), 65, **95**
dispar (Mycoplasma), 11, 582, 584, **594**, 595
distasonis (Bacteroides), 33, **34**, 35, 74
distasonis (Parabacteroides), 4, 33, 35, 70, 77, 78
diversum (Ureaplasma), 11, 616–618, **621**, 622
dokdonensis (Donghaeana), **202**
dokdonensis (Maribacter), 5–6, 236, **237**
dokdonensis (Polaribacter), **256**
Dokdonia, 7, 107, **201**, 229
Dokdonia donghaensis, 7, **201**
domesticum (Pseudosphingobacterium), 8
Donghaeana, **201**, 202
Donghaeana dokdonensis, **202**
donghaensis (Dokdonia), 7, **201**
dorei (Bacteroides), 3, 28, **35**
dorotheae (Flexithrix), 9, 394, 396, 442, **449**
dugesii (Borrelia), 489, **494**
duraquae (Pedobacter), 345
duttonii (Borrelia), 10, 489, 490, **494**, 496
Dyadobacter, **380**, 381, 382–386
Dyadobacter alkalitolerans, 9
Dyadobacter beijingensis, 380, 381, 384, 385, **386**
Dyadobacter crusticola, 380, 381, 384, 385, **386**
Dyadobacter fermentans, 380, 381, 384, **385**, 386, 413
Dyadobacter ginsengisoli, 380, 381, 384, **386**
Dyadobacter hamtensis, 380, 381, 384, 385, **387**
Dyadobacter koreensis, 380, 381, 384, 385, **387**
Dysgonomonas, 4, 61, **71**, 72–76
Dysgonomonas capnocytophagoides, 71–74, **75**, 76
Dysgonomonas gadei, 4, **71**, 72–74
Dysgonomonas mossii, 4, 71–74, **75**, 76

Echinicola, 427–429, **437**, 438, 439
echinicola (Gramella), 7, **226**
Echinicola pacifica, 9, **437**, 439
Echinicola vietnamensis, 437, 439, **440**
echinicomitans (Roseivirga), 448, 453, **454**
echinorum (Arenibacter), 5
eckloniae (Flagellimonas), 5, 16
Edaphobacter, 725, 726, 728, **729**, 730
Edaphobacter aggregans, 729, **730**
Edaphobacter modestus, 12, **729**
edwardii (Mycoplasma), 11, 582, 584, **595**
Effluviibacter, 372, **387**, 388, 398
Effluviibacter roseus, 9, **387**, 388, 398
eggerthii (Bacteroides), 3, 28, 32, **35**
ehrenbergii (Roseivirga), 9, 394, 448, 453, 454, **455**
EHRLICHIACEAE, 110
eikastus (Krokinobacter), 7, 230, **231**
Eikenella, 408
elegans (Flexibacter), 353, 392, 394, **396**
elephantis (Mycoplasma), 11, 582, 584, **595**
Elizabethkingia, 107, 108, 180, 181, 186, 189, **202**, 203–206, 208, 209, 212, 273, 274

Elizabethkingia meningoseptica, 5, 110, 137, 138, 182, 184–186, 189, 202–208, **209**
Elizabethkingia miricola, 182, 185, 203–206, 208–209, **210**
ellychniae (Entomoplasma), 11, **646**, 647–648
ellychniae (Mycoplasma), 646, 647
Empedobacter, 107, 108, 136, 180, 181, 203, 206, **210**, 211, 285, 286
Empedobacter aquatile, 138
Empedobacter breve, 210, 212
Empedobacter brevis, 5, 137, 189, 206, 208, **210**, 211, 274, 286, 287
Emticicia, 372, **388**, 389
Emticicia ginsengisoli, 9
Emticicia oligotrophica, **388**, 389
endodontalis (Bacteroides), 32, 67
endodontalis (Porphyromonas), **67**
enoeca (Prevotella), 4, 65, **95**
ENTEROBACTERIACEAE, 823
Enterococcus, 183, 750
entomophilum (Acholeplasma), 650, 651, **692**
entomophilum (Mesoplasma), 11, 650, **651**, 692
Entomoplasma, 11, 568, 570, 571, 581, 618, 640, **644**, 645, 646, 647, 649, 650, 662, 667, 690
Entomoplasma ellychniae, 11, **646**, 647–648
Entomoplasma freundtii, 646, 647, **648**, 668
Entomoplasma lucivorax, 11, 646, **648**
Entomoplasma luminosum, 646, **648**
Entomoplasma somnilux, 646, **649**
ENTOMOPLASMATACEAE, 11, 570, 574, 644, **645**, 647, 650, 654, 657, 667, 675, 692, 694
ENTOMOPLASMATALES, 10, 11, 570–572, 574, 575, 581, **644**, 647, 650, 687, 692, 694, 719
Eperythrozoon, 11, 570, 581, 639, **640**, 641, 642
Eperythrozoon coccoides, 11, 581, 611, **640**, 641
Eperythrozoon mariboi, **642**
Eperythrozoon ovis, 11, 606, 641
Eperythrozoon parvum, **641**
Eperythrozoon suis, 609, 641
Eperythrozoon teganodes, **642**
Eperythrozoon tuomii, **642**
Eperythrozoon wenyonii, 611, 641
epidermidis (Staphylococcus), 67, 68, 211
Epilithonimonas tenax, 5, 181, 182, 185, **212**
epiphytica (Winogradskyella), 7, **289**
equifetale (Acholeplasma), 689, **692**
equigenitalium (Mycoplasma), 11, 582, 584, **595**, 647
equinum (Fusobacterium), 748, 751, **754**
equirhinis (Mycoplasma), 11, 582, 584, **595**
eriococci (Fritschea), 16, 874, 875, **876**
eriococci (Candidatus Fritschea), 16, 874, 875, **876**
Erysipelothrix, 571
ERYSIPELOTRICHI, 571
Escherichia, 823
Escherichia coli, 1, 72, 89, 111, 205, 211, 253, 263, 281, 318, 341, 342, 385, 418, 499, 535, 549, 597, 610, 617, 657, 658, 660, 668, 699, 720–722, 750, 751, 796, 797, 801, 806, 815, 818, 831, 835, 898
esteraromaticum (Microbacterium), 138
Eudoraea, 110
eximia (Winogradskyella), 7, **290**

Fabibacter, 442, 444, **447**, 448, 453
Fabibacter halotolerans, 9, 442, 447, **448**, 453, 454
faecale (Acetomicrobium), **43**, 44
faecium (Sphingobacterium), 8, 331–333, **336**, 344
fainei (Leptospira), 10, 547, **554**, 555, 556
falconis (Mycoplasma), 11, 582, 584, **595**
falsenii (Wautersiella), 5, **285**, 286, 287
farinofermentans (Flavobacterium), 138
fastidiosum (Mycoplasma), 11, 583, 584, **595**, 596, 613
faucium (Mycoplasma), 11, 582, 584, **596**
felifaucium (Mycoplasma), 11, 582, 584, **596**
feliminutum (Mycoplasma), 11, 581, 584, **596**, 690
felinum (Ureaplasma), 11, 616, 621, **622**
felis (Chlamydophila), 13, 845, 860
felis (Haemobartonella), 599, 612
felis (Mycoplasma), 11, 582, 584, **596**
fermentans (Cytophaga), 4, 50, 52, 53, 57, 373, **374**, 384
fermentans (Dyadobacter), 380, 381, 384, **385**, 386, 413
fermentans (Geothrix), 12, **733**
fermentans (Mycoplasma), 11, 578–580, 582, 584, **596**, 597
ferruginea (Gallionella), 468
ferrugineum (Flavobacterium), 356, 357
Ferruginibacter, 35
Fibrobacter, 12, 32, **739**, 740, 741, 743
FIBROBACTERES, 1, 2, 12, 21, 23, **737**, 738
FIBROBACTERIA, 12, 737, **739**
Fibrobacter intestinalis, 739–741, 743, **744**
Fibrobacter succinogenes, 12, 32, 738, 739, **740**, 741, 742, 743, 744
Fibrobacter succinogenes subsp. elongatus, 740, **743**
Fibrobacter succinogenes subsp. succinogenes, 740, **743**
filamentus (Polaribacter), 5, **255**, 256
filiformis (Chitinophaga), 8, 352, **353**, 356, 395
filiformis (Flexibacter), 352, 353, 374, 392, 394
Filimonas, 351
Filomicrobium, 908
filum (Flavobacterium), 138
finegoldii (Alistipes), 58, 59, **60**
finegoldii (Bacteroides), 3, **35**
FIRMICUTES, 3, 10, 25, 38, 39, 567, 571, 667, 668, 723, 779
Flagellimonas, 5, 110
Flagellimonas eckloniae, 5, 16
Flammeovirga, 8, 9, 374, 392, 408, 409, 442, **443**, 444, 446, 449, 450
Flammeovirga aprica, 9, 52, 373, 393, **443**
Flammeovirga arenaria, 395, 409, 443, **447**
FLAMMEOVIRGACEAE, 3, 8, 9, 370, **442**, 443–456, 465
Flammeovirga kamogawensis, 442
Flammeovirga yaeyamensis, 442, 443, **447**, 457
flava (Sediminitomix), 9
flavefaciens (Ruminococcus), 741
flavescens (Leptobacterium), 7
flavescens (Muricauda), 6, 241–**244**
flavidum (Acetomicrobium), 42, **43**, 44
Flavihumibacter, 351
Flavihumibacter petaseus, 358
Flaviramulus basaltis, 6, 213, **214**
Flavisolibacter, 8, 351
Flavisolibacter ginsengisoli, 8
FLAVOBACTERIA, 2, 16, 27, 58, 62, 109, 127, 128, 132, 135, 188, 211, 214, 215, 241, 243
FLAVOBACTERIACEAE, 5, 6, 105, **106**, 107–112, 128, 132, 134–137, 139, 154, 168, 176–181, 187–189, 200–203–206, 209–216, 219–223, 227, 229–231, 234, 238, 239, 241, 242, 244, 245, 250, 253–255, 258–266, 269, 271, 273, 275–279, 284, 285, 288, 292, 322, 324, 331, 340, 370, 371, 373, 393–395, 408
Flavobacteriales, 5, 7, **105**, 106, 112, 317
FLAVOBACTERIIA, 2, 5, 6, 7, **105**, 106, 112, 133, 317, 370
Flavobacterium, 5, 31, 52, 105, 106, 107, **112**, 115, 117, 119, 120, 121, 123, 125, 127, 128, 129, 130–134, 135, 136, 137, 138, 140, 143, 145–147, 148, 153, 154, 181, 186, 189, 190, 202, 203, 205, 207–209, 211, 246, 250, 286, 319, 328, 332, 340, 373, 394
Flavobacterium acidificum, 5
Flavobacterium acidurans, 5, 138
Flavobacterium anhuiense, 138
Flavobacterium antarcticum, 5, 112, 127, 129, 130, 137, **139**
Flavobacterium aquatile, 5, 112, 127, 128, 129, 135–**138**, 145, 181, 189, 241, 274, 373
Flavobacterium aquatilis, 138
Flavobacterium aquidurense, 112, 128, 133, 136, **139**
Flavobacterium balustinum, 191
Flavobacterium branchiophilum, 112, 129, 130, 132, 133, 134, 135, 136, **139**
Flavobacterium breve, 210, 211
Flavobacterium brevis, 210, 211
Flavobacterium ceti, 138
Flavobacterium cheniae, 138
Flavobacterium columnare, 112, 120, 127–134, 136, 137, **140**, 154, 373, 395
Flavobacterium croceum, 112, 128, 129, 130, 133, 135–137, **141**
Flavobacterium cucumis, 138
Flavobacterium daejeonense, 129, 133, 136, 137, **141**
Flavobacterium defluvii, 112, 128, 130, 133, 136, 137
Flavobacterium degerlachei, 127, 128, 132, 137, **141**, 144, 149
Flavobacterium denitrificans, 106, 112, 127, 129, 130, 133, 136, 137
Flavobacterium farinofermentans, 138
Flavobacterium ferrugineum, 356, 357
Flavobacterium filum, 138
Flavobacterium flevense, 129, 130, 136, 137, 373
Flavobacterium frigidarium, 129, 130, 137, **142**
Flavobacterium frigidimaris, 130, 131, 133, 136, 137, **143**
Flavobacterium frigoris, 127, 128, 132, 137, **143**, 144, 149
Flavobacterium fryxellicola, 137, **143**, 149
Flavobacterium gelidilacus, 112, 130, 137, **144**
Flavobacterium gillisiae, 137, **144**
Flavobacterium glaciei, 112, 129, 130, 133, 136, 137, **144**
Flavobacterium gleum, 180, 190
Flavobacterium gondwanense, 260, 261
Flavobacterium granuli, 112, 134, 137, **144**
Flavobacterium heparinum, 339, 341, 342, 343, 344, 345
Flavobacterium hercynium, 112, 128, 133, 136, **145**
Flavobacterium hibernum, 129, 130, 131, **145**
Flavobacterium hydatis, 112, 128, 130, 132, 134, 136, **145**, 153, 373

Flavobacterium indicum, 112, 127, 130, 133, 135–137, **146**
Flavobacterium indologenes, 192
Flavobacterium indoltheticum, 192
Flavobacterium johnsoniae, 112, 127–134, 136, 140, **146**, 147, 153, 154, 206, 259, 373, 394, 395
Flavobacterium limicola, 128–130, **147**
Flavobacterium lindanitolerans, 138
Flavobacterium meningosepticum, 203, 209, 342
Flavobacterium micromati, 128, 130, 137, 143, 144, **148**, 149
Flavobacterium mizutaii, 138, 332, 337
Flavobacterium multivorum, 337
Flavobacterium oceanosedimentum, 5, 138
Flavobacterium odoratum, 135, 136, 245, 246
Flavobacterium omnivorum, 5, 112, 117, 129, 133, 137, **148**
Flavobacterium pectinovorum, 129, 130, 136, **148**, 154, 373
Flavobacterium psychrolimnae, 5, 112, 125, 137, 143, 144, **149**
Flavobacterium psychrophilum, 5, 112, 125, 127–134, 136, 140, **149**, 373, 395
Flavobacterium resinovorum, 138
Flavobacterium saccharophilum, 5, 125, 129, 130, 136, **149**, 150, 154, 373
Flavobacterium salegens, 266
Flavobacterium saliperosum, 5, 107, 112, 125, 130, 133, 136, 137, **150**
Flavobacterium scophthalmum, 194
Flavobacterium segetis, 5, 112, 125, 129, 130, 133, 136, 137, **150**
Flavobacterium soli, 129, 130, 133, 136, 137, **151**
Flavobacterium spiritivorum, 331, 333
Flavobacterium succinicans, 5, 112, 125, 130, 132, 134, 135, 136, **151**, 373
Flavobacterium suncheonense, 5, 119, 129, 130, 133, 136, 137, **151**
Flavobacterium tegetincola, 5, 119, 129, 130, **151**
Flavobacterium terrae, 5, 138
Flavobacterium terrigena, 5, 138
Flavobacterium thalpophilum, 338
Flavobacterium thermophilum, 5, 138
Flavobacterium uliginosum, 293, 295
Flavobacterium weaverense, 5, 112, 119, 129, 130, 133, 136, 137, **152**
Flavobacterium xanthum, 5, 119, 128–130, 136, 137, 144, 149, **152**
Flavobacterium xinjiangense, 119, 129, 130, 133, 137, **152**
Flavobacterium yabuuchiae, 137, 331, 333
flavum (Chryseobacterium), 5, 190
flavum (Promyxobacterium), 136, **154**
flavus (Chthoniobacter), 13, 834, **837**
flavus (Salegentibacter), 7, 266, 267, **268**
Flectobacillus, 258, 383, 385, **389**, 391, 405, 413, 416, 417, 423, 424
Flectobacillus glomeratus, 255, 257, 391
Flectobacillus lacus, 389, **391**
Flectobacillus major, 9, 258, 377, 379, **389**, 390, 391, 424
Flectobacillus marinus, **391**, 423, 424
flevense (Flavobacterium), 129, 130, 136, 137, 373
flevensis (Cytophaga), 373
Flexibacter, 51, 135, 136, 147, 241, 340, 372, 374, 385, **392**, 394, 395, 396, 407, 408, 409, 442
FLEXIBACTERACEAE, 3, 8, 9, 405, 407
Flexibacter aggregans, 9, 392, **394**, 408, 409, 449
Flexibacter aurantiacus, **153**, 154, 392, 394
Flexibacter aurantiacus var. excathedrus, 113, 136, **153**, 154, 392
Flexibacter canadensis, 8, 340, 344, 392, **394**, 395
Flexibacter columnaris, 140, 373, 392, 394
Flexibacter elegans, 353, 392, 394, **396**
Flexibacter filiformis, 352, 353, 374, 392, 394
Flexibacter flexilis, 9, **392**, 394, 395
Flexibacter japonensis, 352, 355, 374, 392, 394
Flexibacter litoralis, 392, 394, **396**, 409
Flexibacter maritimus, 279, 281, 374, 392, 394
Flexibacter ovolyticus, 283, 392, 394
Flexibacter polymorphus, 392, 394, 409
Flexibacter psychrophilus, 149, 373, 392, 394
Flexibacter roseolus, 9, 392, 394, 397
Flexibacter ruber, 9, 392, 394, **397**
Flexibacter sancti, 352, 355, 374, 392, 394
Flexibacter tractuosus, 9, 229, 392, 394, **397**, 409
flexilis (Flexibacter), 9, **392**, 394, 395
Flexithrix, 371, 394, 442, 444, 446, **448**, 449
Flexithrix dorotheae, 9, 394, 396, 442, **449**
flocculare (Mycoplasma), 11, 578, 582, 584, **597**
floricola (Spiroplasma), 656–657, 661, 666–668, **672**
florum (Acholeplasma), 649, 650, **692**
florum (Mesoplasma), 11, 569, 572, 649, **650**
fluorescens (Pseudomonas), 211
Fluviicola, 322, **327**
Fluviicola taffensis, 7, 325, **327**, 328
foetida (Holophaga), 12, **732**
Formosa, 16, 107, 110, 157, 158, 166, 178, 192, **214**, 215–218, 221–223, 291
formosa (Zavarzinella), 16
Formosa agariphila, 106, 215, 216, **217**
Formosa algae, 7, 214, 215, **217**, 218
formosense (Chryseobacterium), 5, 182–184, 189, 190, **192**, 273, 274
forsythus (Bacteroides), 3, 32, 33, 74, 79–81
fragariae (Candidatus Phytoplasma), **710**
fragilis (Bacteroides), 3, 27, 30, 32, **33–34**, 36, 38, 45, 55, 62, 79, 87, 89, 132, 242, 461, 769
franzmannii (Polaribacter), 229, **256**
fraxini (Candidatus Phytoplasma), **710**
freundtii (Entomoplasma), 646, 647, **648**, 668
frigidarium (Flavobacterium), 129, 130, 137, **142**
frigidimaris (Flavobacterium), 130, 131, 133, 136, 137, **143**
frigoris (Flavobacterium), 127, 128, 132, 137, **143**, 144, 149
Fritschea bemisiae, 874, 875, **876**
Fritschea bemisiae (Candidatus), 874, **875**, 876
Fritschea eriococci, 16, 874, 875, **876**
Fritschea eriococci (Candidatus), 16, 874, 875, **876**
fryxellicola (Flavobacterium), 137, **143**, 149
fucanivorans (Mariniflexile), 6
fucicola (Cellulophaga), 6, 176–178, **179**, 229
Fulvibacter, 110
Fulvibacter tottoriensis, 7
Fulvivirga kasyanovii, 9
fumarolicum (Acidimethylosilex), 795–797
furcosa (Anaerorhabdus), 32, 45
furcosus (Bacteroides), 3, 32, 45
furfurosus (Sediminibacter), 6
furvescens (Marinoscillum), 444
furvescens (Microscilla), 384, 393, 408, **409**, 442
fusiforme (Fusobacterium), 753
fusiformis (Bacillus), 748, 753
fusiformis (Sphaerophorus), 754
FUSOBACTERIA, 2, 12, **747**, 748–772
FUSOBACTERIACEAE, 12, 23, **748**
FUSOBACTERIALES, 12, **747**
FUSOBACTERIIA, 12, **747**
Fusobacterium, 12, 612, **747**, 748–755, 757, 758, 760, 762, 766
Fusobacterium canifelinum, 748, 750, **754**
Fusobacterium equinum, 748, 751, **754**
Fusobacterium fusiforme, 753
Fusobacterium gonidiaformans, 748, 750, 751, **755**
Fusobacterium mortiferum, 748, 750–52, **755**
Fusobacterium naviforme, 748, 751, **755**
Fusobacterium necrogenes, 748, 750, 751, **755**
Fusobacterium necrophorum, 748, 750, 751, 752, **755–757**
Fusobacterium necrophorum subsp. funduliforme, 750, 752, **756**
Fusobacterium necrophorum subsp. necrophorum, 750, **756**
Fusobacterium nucleatum, 12, 79, 748b, 750, 751, 752, **753**, 754, 757, 762, 766
Fusobacterium nucleatum subsp. animalis, 750, 753, **754**
Fusobacterium nucleatum subsp. fusiforme, 750, 753, **754**
Fusobacterium nucleatum subsp. nucleatum, 750, 751, 753, **754**
Fusobacterium nucleatum subsp. polymorphum, 750, 751, 753, **754**
Fusobacterium nucleatum subsp. vincentii, 750, 751, 753, **754**
Fusobacterium perfoetens, **756**
Fusobacterium periodonticum, 748, 750, 751, **756**
Fusobacterium plautii, 753, 754
Fusobacterium plauti-vincentii, 753, 754
Fusobacterium polymorphum, 753, 754
Fusobacterium russii, 748, 751, **757**
Fusobacterium simiae, 748, 750, 751, **757**
Fusobacterium ulcerans, 748, 750, 751, **757**
Fusobacterium varium, 748, 750, 751, 752, **757**, 758

gadei (Dysgonomonas), 4, **71**, 72–74
Gaetbulibacter, 108, 110, 158, 214–216, **218**, 288, 289, 291
Gaetbulibacter marinus, 6
Gaetbulibacter saemankumensis, 6, 214, **218**, 291
Gaetbulimicrobium, 108, 110, 159, 292
Gaetbulimicrobium brevivitae, 159
galactanivorans (Zobellia), 6, 236, **293**
galacturonicus (Bacteroides), 39, **40**
Galbibacter, 110
Galbibacter mesophilus, 5
gallinaceum (Mycoplasma), 11, 582, 584, **597**
gallinarum (Bacteroides), 3
gallinarum (Mycoplasma), 11, 582, 584, **597**
Gallionella, 468
Gallionella ferruginea, 468
gallisepticum (Mycoplasma), 11, 569, 577–579, 583, 584, 588, **597**, 598, 601, 602, 610
gallopavonis (Mycoplasma), 11, 582, 584, **598**
gallorale (Ureaplasma), 11, 618, **622**
gambrini (Chryseobacterium), 5, 190
GAMMAPROTEOBACTERIA, 875
garinii (Borrelia), 485, 488, 490, **494**, 495–497
gateae (Mycoplasma), 11, 582, 584, **598**

Gelidibacter, 108, 159, 181, 203, 214, 216, **219**–221, 276–278, 291
Gelidibacter algens, 6, 215, 219, 220, **221**, 291
Gelidibacter gilvus, 6, 219, 220, **221**, 291
Gelidibacter mesophilus, 5, 6, 219, 220, **221**
Gelidibacter salicanalis, 6, 219, 220, **221**, 278
gelidilacus (Flavobacterium), 112, 130, 137, **144**
gelidisalsuginis (Bizionia), 7, **167**, 168
gelipurpurascens (Hymenobacter), 9, 398–400, **402**
Gemmata, 881, 889, 893, **897**–903, 905, 909, 911, 912
Gemmata obscuriglobus, 16, 891, 893, **898**, 899, 900
GEMMATIMONADACEAE, 13, **782**, 784
GEMMATIMONADALES, 13, **781**
GEMMATIMONADETES, 1, 2, 13, 21, 23, 737, **781**–783
Gemmatimonas, 13, 781, **782**–784
Gemmatimonas aurantiaca, 13, 782–**784**
genikus (Krokinobacter), 7, **230**, 233
genitalium (Mycoplasma), 11, 569, 576, 578, 579, 581, 583, 585, **598**, 599, 607, 616, 659
Geothrix, 725, 726, 728, **732**, 733, 734
Geothrix fermentans, 12, **733**
Gillisia, 5, 108, 144, **221**, 222, 223–225, 255
gillisiae (Flavobacterium), 137, **144**
Gillisia hiemivivida, 7, 221, 222, **224**, 225
Gillisia illustrilutea, 7, 221, 222, 224, **225**
Gillisia limnaea, 7, 221, 222, **224**
Gillisia mitskevichiae, 7, 221, 222, 224, **225**
Gillisia myxillae, 7, 221, 222, 224, **225**
Gillisia sandarakina, 7, 221, 222, 224, **225**, 226
Gilvibacter sediminis, 7
gilvus (Gelidibacter), 6, 219, 220, **221**, 291
gingivalis (Bacteroides), 32, 67
gingivalis (Capnocytophaga), 5, 168, 169, 172, 174, **175**
gingivalis (Porphyromonas), 62–64, **67**, 68, 79
gingivicanis (Porphyromonas), **68**, 69, 70
ginsengisegetis (Chitinophaga), 8, 352, 353, **354**
ginsengisoli (Chitinophaga), 8, 351, 352, 353, **354**
ginsengisoli (Dyadobacter), 380, 381, 384, **386**
ginsengisoli (Emticicia), 9
ginsengisoli (Flavisolibacter), 8
ginsengisoli (Olivibacter), 8
glaciale (Brumimicrobium), 7, 16, 323–**326**, 328, 329
glaciei (Flavobacterium), 112, 129, 130, 133, 136, 137, **144**
gladiatoris (Spiroplasma), 11, 661, 668, **672**, 677
glaucopis (Meniscus), 9, 406–**408**
glaucopis (Meniscus), 9, **406**, 407, 408
gleum (Chryseobacterium), 5, 137, 180, 183, 186, 188, 189, **190**, 191, 192, 195, 196, 206, 208
gleum (Flavobacterium), 180, 190
glomeratus (Flectobacillus), 255, 257, 391
glomeratus (Polaribacter), **257**, 379, 390
glucosidilyticus (Pedobacter), 245
glycophilum (Mycoplasma), 11, 582, 585, **599**
goldsteinii (Bacteroides), 28, **35**
goldsteinii (Parabacteroides), 4, 33, 35
gondwanense (Flavobacterium), 260, 261
gondwanensis (Psychroflexus), 138, 260, **261**
gonidiaformans (Fusobacterium), 748, 750, 751, **755**
goodfellowii (Leptotrichia), 12, 766, **767**
gracilis (Bacteroides), 32
gracilis (Campylobacter), 32
gracilis (Mucilaginibacter), 8
graingeri (Borrelia), 10, 489, **494**
Gramella, 108, **226**, 239
Gramella echinicola, 7, **226**
Gramella portivictoriae, 7, 226, **227**
graminis (Candidatus Phytoplasma), **710**
grammopterae (Mesoplasma), 11, 650, **651**
grandis (Saprospira), 8, 324, 359, 360, **361**, 362–364, 366–368, 393
granularum (Acholeplasma), 11, 689–691, **692**
granuli (Flavobacterium), 112, 134, 137, **144**
granulosa (Capnocytophaga), 5, 168–170, 174, **175**, 176
gregarium (Chryseobacterium), 5
gromovii (Mariniflexile), 6, 238–239
gulae (Porphyromonas), **68**, 70
guttaeformis (Planctomyces), 16, 881, 887, **891**, 892, 894
gypis (Mycoplasma), 11, 582, 585, **599**

haematoparvum (Candidatus Mycoplasma), 586, **611**
Haemobartonella, 11, 23, 570, 581, 599, 639–641, **642**, 643
Haemobartonella canis, 11, 599, 642
Haemobartonella felis, 599, 642
Haemobartonella muris, 11, 599, 642
Haemobartonella procyoni, **643**
haemobos (Candidatus Mycoplasma), **611**
haemocanis (Mycoplasma), 11, 579–580, 583, 585, **599**, **642**, 643
haemodidelphidis (Candidatus Mycoplasma), 586, **612**
haemofelis (Mycoplasma), 11, 579, 583, 585, **599**, 612, **642**, 643
haemolamae (Candidatus Mycoplasma), 586, **612**
haemolytica (Capnocytophaga), 5, 168, 172, 174–**176**
haemominutum (Candidatus Mycoplasma), 586, **612**
haemomuris (Mycoplasma), 11, 583, 585, **599**, 600, **642**
haemotarandirangiferis (Candidatus Mycoplasma), **612**
Haemophilus, 772
haifense (Chryseobacterium), 5, 190
HALANAEROBIALES, 460, 879
Halanaerobium, 879
Haliscomenobacter hydrossis, 8, 363–365, **366**–368
Hallella, 5, 94, 105
Hallella seregens, 5, 94, 105
halmophila (Halomonas), 138
Haloarcula marismortui, 460
HALOBACTERIACEAE, 461, 462
HALOBACTERIALES, 460,
Halobacterium, 460
Halobacterium salinarum, 460
Haloferula, 13
Haloferula harenae, 13
Haloferula helveola, 13
Haloferula phyci, 13
Haloferula rosea, 13
Haloferula sargassicola, 13
Halomonas, 138
Halomonas halmophila, 138
halophila (Hongiella), 428, 431
halophila (Spirochaeta), 10, 472, 474–478, **481**, 484–485, 501–502
halophilus (Algoriphagus), 9, 427, 430, **431**
halotolerans (Fabibacter), 9, 442, 447–448, 453–454
hamtensis (Dyadobacter), 380, 381, 384, 385, **387**
harenae (Haloferula), 13
hartmannellae (Neochlamydia), 843, **869**
hartonius (Pedobacter), 345
harveyi (Borrelia), 10, 489, **494**
helcogenes (Bacteroides), 3, 30, **36**
helicoides (Spiroplasma), 11, 661, 668, **672**, 677
helveola (Haloferula), 13
heparina (Cytophaga), 339, 343–345, 373, 374
heparinolytica (Prevotella), 5
heparinolyticus (Bacteroides), 32
heparinum (Flavobacterium), 339, 341, 342, 343, 344, 345
heparinum (Sphingobacterium), 332, 339, 344–345
heparinus (Pedobacter), 8, 138, 330, **339**, 340–342, 344, 345
hercynium (Flavobacterium), 112, 128, 133, 136, **145**
hermsii (Borrelia), 10, 488–492, 494–495, 497
Herpetosiphon, 366, 367, 468
Herpetosiphon cohaerens, 366, 367
Herpetosiphon nigricans, 367
Herpetosiphon persicus, 367
hibernum (Flavobacterium), 129–131, **145**
hiemivivida (Gillisia), 7, 221, 222, **224**, 225
himalayensis (Pedobacter), 339–342, 344, **347**
hippikon (Acholeplasma), 689, **692**
hispanica (Borrelia), 10, 494, **495**
hispanicum (Chryseobacterium), 5, 180, 182–184, **192**
hofstadii (Leptotrichia), 12, 766, **767**, 768
Hollandina, 9, 10, 23, 565
Hollandina pterotermitidis, 10, 565
Holophaga, 12, 23, 725, 726, 728, **731**, 732, 733, 734
HOLOPHAGACEAE, 12, 23, 725, 731, **732**
HOLOPHAGAE, 12, 12, 23, 725, **731**
Holophaga foetida, 12, **732**
HOLOPHAGALES, 12, 23, 725, **731**
holothuriorum (Salegentibacter), 7, 266, 267, **269**
hominis (Chryseobacterium), 5, 190
hominis (Mycoplasma), 11, 569, 572, 577–580, 582, 583, 585, **600**, 613, 614
Hongiella, 9, 405, 424, **428**, 431, 432, 434, 435
Hongiella halophila, 428, 431
Hongiella mannitolivorans, 428, 432
Hongiella marincola, 432
Hongiella ornithinivorans, 428, 432
hongkongensis (Owenweeksia), 7, 324, 325, 328, **329**
Humibacter, 351
hungaricum (Chryseobacterium), 5
hutchinsonii (Cytophaga), 9, 179, 368, 371, 373, **374**, 379, 384, 393, 741
hydatis (Flavobacterium), 112, 128, 130, 132, 134, 136, **145**, 153, 373
hydrophila (Aeromonas), 211
hydrossis (Haliscomenobacter), 8, 363–365, **366**–368
Hymenobacter, 9, 22, 372, 387, 388, **397**, 404
Hymenobacter actinosclerus, 9, 398, 399, **400**, 401
Hymenobacter aerophilus, 9, 398, 399, **401**, 402
Hymenobacter chitinivorans, 9, 398–400, **402**
Hymenobacter deserti, 9

Hymenobacter gelipurpurascens, 9, 398–400, **402**
Hymenobacter norwichensis, 9, 398–400, **402**, 403
Hymenobacter ocellatus, 9, 398–400, **403**
Hymenobacter psychrotolerans, 9, 404
Hymenobacter rigui, 9, 398–400, **403**
Hymenobacter roseosalivarius, 9, 387, 388, 398, 399, **400**
Hymenobacter soli, 9, 404
Hymenobacter xinjiangensis, 9, 398, 399, **403**, 404
hyodysenteriae (Brachyspira), 508, 531, 532, 534–538, **539**, 540
hyodysenteriae (Serpula), 539
hyodysenteriae (Serpulina), 537, 539
hyodysenteriae (Treponema), 508, 535, 540
hyopharyngis (Mycoplasma), 11, 582, 585, **600**
hyopneumoniae (Mycoplasma), 11, 569, 577–579, 582, 585, 597, **600**, 601
hyorhinis (Mycoplasma), 11, 577–578, 580, 582, 585, **600–601**
hyosynoviae (Mycoplasma), 11, 579, 582, 585, **601**
hypermegale (Megamonas), 32
hypermegale (Megamonas), 32
hypermegas (Bacteroides), 32
Hyphomicrobium, 908
Hyphomonas, 881, 908

ignava (Cryomorpha), 7, 16, **323**, 324, 325, 328
ignava (Johnsonella), 47, 84, 104
iguanae (Mycoplasma), 11, 582, 585, **601**
illini (Leptonema), 10, 547, 551, 557, **558**
illustrilutea (Gillisia), 7, 221, 222, 224, **225**
Ilyobacter, 12, **759**, 760, 761
Ilyobacter delafieldii, 759, 760
Ilyobacter insuetus, 759, **760**
Ilyobacter polytropus, 12, 759, **760**
Ilyobacter tartaricus, 12, 759, 760, **761**
imitans (Mycoplasma), 11, 583, 585, 598, **601**, 602
imshenetskii (Alkaliflexus), 4, **53**
inadai (Leptospira), 10, 547, **554**, 555, 556
indicum (Flavobacterium), 112, 127, 130, 133, 135–137, **146**
indiense (Mycoplasma), 11, 582, 585, **601**
indologenes (Chryseobacterium), 5, 182–184, 186–191, **192**, 196, 206, 208
indologenes (Flavobacterium), 192
indoltheticum (Chryseobacterium), 5, 137, 182–184, 188, 189, 191, **192**, 193–195
indoltheticum (Flavobacterium), 192
iners (Mycoplasma), 11, 582, 585, 597, **601**, 602
innocens (Brachyspira), 532, 534, 535, 537–538, **539**, 540
innocens (Serpula), 539
innocens (Serpulina), 537, 539
innocens (Treponema), 508, 537
innocuum (Clostridium), 571, 572, 667, 668
insolitum (Spiroplasma), 11, 657, 661, 668, **672**, 673
insons (Mycoplasma), 576, 577, 583, 585, **613**
insperata (Larkinella), 9, 404, **405**, 417
insuetus (Ilyobacter), 759, **760**
insulae (Pedobacter), 8
intermedia (Aquimarina), 7, 159, 160, **161**
intermedia (Brachyspira), 532–534, 536–**539**, 540, 550
intermedia (Prevotella), 65, 79, 82, 87, **96**, 99
intermedia (Serpulina), 537, 539, 540
intermedium (Anaeroplasma), 12, 720–**722**
intermedius (Bacteroides), 32, 96
interrogans (Leptospira), 10, 535, **547**, 553, 557
intestinalis (Bacteroides), 3, **36**
intestinalis (Fibrobacter), 738–741, 743, **744**
intestinihominis (Barnesiella), 4
iowae (Mycoplasma), 11, 583, 585, **602**, 647
irgensii (Polaribacter), 106, 241, **258**, 259
ishigakijimensis (Roseibacillus), 13
isopodii (Chlamydia), 874
isoptericolens (Treponema), 502
Isosphaera, 880, 881, 889, 897, 899, **900**, 901–903, 909, 911, 913, 914
Isosphaera pallida, 16, 880, 896, 899, **900**, 901–903, 914, 919
isovalerica (Spirochaeta), 10, 474–478, **481**, 486, 502, 522
ixodetis (Spiroplasma), 11, 572, 657, 661, 663, 666–668, **673**

japonensis (Chitinophaga), 8, 351, 352, 353, **355**, 395
japonensis (Flexibacter), 352, 355, 374, 392, 394
japonica (Borrelia), 10, 485, 489, 492, **495**, 496, 497
japonicum (Candidatus Phytoplasma), **710**
jejuense (Chryseobacterium), 5, 190
Jejuia, 110
Jejuia pallidilutea, 305
jeonii (Chryseobacterium), 5
jeonii (Sejongia), 5, 182, 185, 271–273, **274**
jodogahamensis (Persicitalea), 9
johnsonae (Cytophaga), 146, 241, 355, 373
Johnsonella ignava, 47, 84, 104
johnsoniae (Flavobacterium), 112, 127–134, 136, 140, **146**, 147, 153, 154, 206, 259, 373, 394, 395
joostei (Chryseobacterium), 5, 181–184, 187–189, **193**, 194, 196
Joostella, 5, 110
Joostella marina, 5

kahaneii (Candidatus Mycoplasma), **612**
Kaistella, 108, 180, 208, 212, **227**, 228, 271, 273, 274
Kaistella koreensis, 5, 182, 184, 203, 227, **228**, 274
kamogawensis (Flammeovirga), 442
kasyanovii (Fulvivirga), 9
Kingella, 250
kirschneri (Leptospira), 10, 547, 548, **554**
kitahiroshimense (Sphingobacterium), 8, 333, **336**
Kordia, 108, 159, **228–230**, 269
Kordia algicida, 159, 228–**230**, 242
koreensis (Dyadobacter), 380, 381, 384, 385, **387**
koreensis (Kaistella), 5, 182, 184, 203, 227, **228**, 274
koreensis (Niastella), 8, 351, 393
koreensis (Pedobacter), 345
koreensis (Segetibacter), 8
Kriegella, 110
Krokinobacter, 7, **230**, 231
Krokinobacter diaphorus, 7, 230, **231**
Krokinobacter eikastus, 7, 230, **231**
Krokinobacter genikus, 7, **230**, 233
kunkelii (Spiroplasma), 655, 658–662, 664–668, **673**, 698

Lacibacter, 351
Lacinutrix, 158, **231**, 249, 288, 289
Lacinutrix algicola, 6
Lacinutrix copepodicola, 6, 231, **232**, 249
Lacinutrix mariniflava, 6
Lactobacillus, 571, 667, 766
Lactococcus, 750
lactucae (Mesoplasma), 11, 650, **652**
lactucae (Mycoplasmai), 650, 652
lacus (Flectobacillus), 389, **391**
lagogenitalium (Mycoplasma), 11, 582, 585, **602**
laidlawii (Acholeplasma), 11, 569, 580, 668, 688–690, **691**, 692, 699
laminariae (Zobellia), 6, **294**
lampyridicola (Spiroplasma), 11, 654, 661, 668, **673**
Larkinella, 390, **404**, 405, 407
Larkinella insperata, 9, 404, **405**, 417
latercula (Aquimarina), 7, 159, 160, **161**, 373
latercula (Cytophaga), 161, 229, 373, 374
latercula (Stanierella), 107, 111, 159, 161, 223, 242
latericius (Arenibacter), 5–6, 162, **163**, 164
latyschewii (Borrelia), 10, 489, **495**
leachii (Mycoplasma), 578, 585–587, 592, **602**
Leadbetterella, **405**, 406
Leadbetterella byssophila, 9, 388, 405, **406**
lecithinolyticum (Treponema), 502, **512**, 513
lectus (Algibacter), 6, 157, **158**, 214, 218, 238, 249, 269
Leeuwenhoekiella, 200, **232**, 233, 234, 284, 374
Leeuwenhoekiella aequorea, 7, 233, **234**
Leeuwenhoekiella blandensis, 7, 132, 233, **234**
Leeuwenhoekiella marinoflava, 7, 232, 233, **234**, 284
Legionella, 196
Legionella pneumophila, 211, 868
leguminosarum (Rhizobium), 332
Lentisphaera, 13, 785–790, 796, 843
Lentisphaera araneosa, 13, 788, 789, **790**, 792
LENTISPHAERACEAE, 13, 23, **788**
LENTISPHAERAE, 2, 13, 21, **785**, 786–793
LENTISPHAERALES, 785, 787, **788**
LENTISPHAERIA, 13, **787**
lentus (Pedobacter), 345
leonicaptivi (Mycoplasma), 11, 582, 585, **602**
leopharyngis (Mycoplasma), 11, 582, 585, **602**, 603
leptinotarsae (Spiroplasma), 11, 661–662, 666, 668, **673**, 674
Leptobacterium, 5
Leptobacterium flavescens, 7
Leptonema, 477, 545–547, 550–552, **556**, 557, 558
Leptonema illini, 10, 547, 551, **557**, 558
Leptospira, 10, 23, 472, 476, 477, 491, 508, 534, 545, **546**–559
Leptospira alexanderi, 10, 547, **553**
Leptospira biflexa, 10, 547–549, 551, **553**, 555, 557
Leptospira borgpetersenii, 10, 547–549, **553**, 554
Leptospira broomii, 10, 547, **554**, 556
LEPTOSPIRACEAE, 9, 10, 471–473, **546**, 547–559
Leptospira fainei, 10, 547, **554**, 555, 556
Leptospira inadai, 10, 547, **554**, 555, 556
Leptospira interrogans, 10, 535, **547**, 548–553, **557**
Leptospira kirschneri, 10, 547, 548, **554**
Leptospira licerasiae, 10, 547, **554**, 555
Leptospira meyeri, 547, **555**
Leptospira noguchii, 10, **555**

Leptospira parva, 547, 558, 559
Leptospira santarosai, 10, **555**
Leptospira weilii, 10, 547, **555**
Leptospira wolbachii, 10, 547, **555**
Leptospira wolffii, 10, 547, **555**, 556
Leptotrichia, 758, **766**–769
Leptotrichia amnionii, 767
Leptotrichia buccalis, 12, 762, **766**, 767, 770
Leptotrichia goodfellowii, 12, 766, **767**
Leptotrichia hofstadii, 12, 766, **767**, 768
Leptotrichia sanguinegens, 766, 770
Leptotrichia shahii, 767, **768**
Leptotrichia trevisanii, 766, **769**
Leptotrichia wadei, 12, **769**
leucomae (Spiroplasma), 11, **674**
levii (Bacteroides), 32, 68
levii (Porphyromonas), 64, **68**
Lewinella, 358, 364, **366**, 367, 368
Lewinella agarilytica, 8, 367
Lewinella antarctica, 8, 367
Lewinella cohaerens, 8, 363, 366–367
Lewinella lutea, 8, 367
Lewinella marina, 8, 367
Lewinella nigricans, 8, **367**
Lewinella persica, 8, **367**, 368
lianum (Cyclobacterium), 423, 424, **425**
licerasiae (Leptospira), 10, 547, **554**, 555
Limibacter armeniacum, 9
limicola (Flavobacterium), 128–130, **147**
limnaea (Gillisia), 7, 221, 222, **224**
limnophilus (Planctomyces), 881, 886, **891**, 892, 893, 900, 911
limosa (Runella), 412–414, **415**
lindanitolerans (Flavobacterium), 138
lineolae (Spiroplasma), 11, 661, 668, **674**
linguale (Spirosoma), 9, 377, 379, 384, 390, 393, 405, **415**, 416, 417
lipofaciens (Mycoplasma), 11, 582, 585, **603**
lipolytica (Aequorivita), 7, 155–**157**
lipophilum (Mycoplasma), 11, 582, 585, **603**, 613
Lishizhenia, **327**, 328
Lishizhenia caseinilytica, 7, 327, **328**
litorale (Spiroplasma), 11, 661, 668, 671, **674**, 677
litoralis (Flexibacter), 392, 394, **396**, 409
litoralis (Lutibacter), 5, 234, **235**
litoralis (Pelagicoccus), 818, 826, 829, 830, **831**, 833
litoralis (Spirochaeta), 10, 472, 474–478, **482**, 486, 501, 522
litoralis (Ulvibacter), 7, 283, **284**
locisalis (Algoriphagus), 9, 426, 427, 430, **432**
loescheii (Bacteroides), 32, 96
loescheii (Prevotella), 4, 65, 82, **96**
lonestari (Borrelia), 498
longa (Salisaeta), 462
lucivorax (Entomoplasma), 11, 646, **648**
lucivorax (Mycoplasma), 648
luminosum (Entomoplasma), 646, **648**
luminosum (Mycoplasma), 648
lusitaniae (Borrelia), 485, 489, 492, **495**, 496
Lutaonella, 110
lutea (Lewinella), 8, 367
lutea (Rudanella), 9
Luteolibacter, 800
Luteolibacter algae, 799, 802, 803
Luteolibacter pohnpeiensis, 13, 799, 802, 803
luteum (Chryseobacterium), 5, 190
luteus (Sediminicola), 6, 242, **270**
Lutibacter, 110, 229, **234**, 235, 280
Lutibacter litoralis, 5, 234, **235**
lutimaris (Muricauda), 6
Lutimonas vermicola, 5
lycopersici (Candidatus Phytoplasma), **710**
lytica (Cellulophaga), 6, 176–178, **179**, 229, 230, 373
lytica (Cytophaga), 52, 176, 179, 373, 374

macacae (Bacteroides), 68
macacae (Porphyromonas), 64, **68**, 69
macrodentium (Treponema), **520**
maculosa (Prevotella), 4
maculosum (Mycoplasma), 11, 582, 585, **603**
major (Flectobacillus), 9, 258, 377, 379, **389**, 390, 391, 424
mali (Candidatus Phytoplasma), 698, 699, **710**
maltophilia (Xanthomonas), 211
maltophilum (Treponema), 502, 510, 511, **513**, 514
mannitolivorans (Algoriphagus), 9, 426, 427, 430, **432**
mannitolivorans (Hongiella), 428, 432
Maribacter, 6, 108, 162–163, 178, 199, 232–234, **235**, 236–238, 241, 245, 270, 284, 293
Maribacter aquivivus, 5–6, 235–236, **237**
Maribacter dokdonensis, 5–6, 236, **237**
Maribacter orientalis, 5–6, 236, **237**, 238
Maribacter sedimenticola, 5–6, 111, 162, 235–237, 242, 324
Maribacter ulvicola, 5–6, 235–237, **238**
mariboi (Eperythrozoon), **642**
marilimosa (Olleya), 6, 111, 242, **249**
marina (Aureispira), 8, 361, **363**, 393
marina (Blastopirellula), 16, 895, 896. **897**, 898, 904, 905, 907, 908
marina (Cytophaga), 279, 281, 373, 374
marina (Joostella), 5
marina (Lewinella), 8, 367
marina (Microscilla), 9, 360, 393, 395, 408, **409**, 410
marina (Pirella), 895, 897
marina (Pirellula), 886, 895, 905
marina (Rubritalea), 13, 801, 812, **814**, 826, 831
marincola (Algoriphagus), 9, 426, 430, **432**
marincola (Hongiella), 432
Marinicola, 448, 453
Marinicola seohaensis, 384, 448, 453, 456
mariniflava (Lacinutrix), 6
Mariniflexile, 5, 110, **238**, 239
Mariniflexile fucanivorans, 6
Mariniflexile gromovii, 6, 238–239
Marinilabilia, 4, **49**, 50–53, 373
Marinilabilia agarovorans, 4, 50–52
MARINILABILIACEAE, 3, 4, 25, **49**
Marinilabilia salmonicolor, 4, 49, 50, **51**–53, 324, 373
marinoflava (Cytophaga), 57, 179, 234, 373, 374
marinoflava (Leeuwenhoekiella), 7, 232, 233, **234**, 284
Marinoscillum, 442, 444
Marinoscillum furvescens, 444
marinum (Cyclobacterium), 9, 368, 379, 390, 393, **423**, 424, 435
marinus (Flectobacillus), **391**, 423, 424
marinus (Gaetbulibacter), 6
marinus (Rhodothermus), 9, **458**, 459, 461, 462, 466
maris (Planctomyces), 881, 886, **892**, 893, 899, 900, 911
maris (Propionigenium), 761, **762**
marismortui (Haloarcula), 460
maritima (Aureispira), 8, **363**
maritimus (Flexibacter), 279, 281, 374, 392, 394
maritypicum (Microbacterium), 138
Marixanthomonas ophiurae, 7
marshii (Prevotella), 65, **97**
Massilia, 36, 102, 196
massiliae (Chryseobacterium), 180, 187–189, **196**
massiliensis (Bacteroides), 3, 29, **36**
massiliensis (Prevotella), 87, **102**
mazzottii (Borrelia), 489, **495**
medium (Treponema), 502, 513, **514**
Megamonas, 3, 17, 25, 32
Megamonas hypermegale, 32
melaleucae (Entomoplasma), 647, **648**
melaleucae (Mycoplasma), 647, 648
melaninogenica (Prevotella), 4, 65, 82, **86**, 88
melaninogenicum (Bacterium), 86, 88
melaninogenicus (Bacteroides), 32, 62, 64, 68, 88
melaninogenicus subsp. macacae (Bacteroides), 68
meleagridis (Mycoplasma), 11, 579, 582, 585, 597, **603**
melliferum (Spiroplasma), 11, 655–659, 661, 664, 668, **674**, 852
meningoseptica (Elizabethkingia), 5, 110, 137, 138, 182, 184–186, 189, 202–208, **209**
meningosepticum (Chryseobacterium), 110, 186, 189, 203, 205, 209, 253, 254
meningosepticum (Flavobacterium), 203, 209, 342
Meniscus, 372, **406**, 407–408
Meniscus glaucopis, 9, 406–**408**
merdae (Bacteroides), 29, 33, **36**
merdae (Parabacteroides), 33
Mesoflavibacter, 5, 22, 22, 110
Mesoflavibacter zeaxanthinifaciens, 7
Mesonia, 5, 22, 108, 221–223, 226, **239**, 240, 255, 269
Mesonia algae, 7, 111, 160, 223, 226, 233, 239, **239**, 240, 242
Mesonia mobilis, 7, 240, **240**
mesophilus (Galbibacter), 5
mesophilus (Gelidibacter), 5, 6, 219, 220, **221**
Mesoplasma, 11, 23, 568, 570–571, 574, 581, 618, 640, 644–647, **649**, 650, 667, 687, 690
Mesoplasma chauliocola, 11, 650, **651**
Mesoplasma coleopterae, 11, 650, **651**
Mesoplasma corruscae, 11, 650, **651**
Mesoplasma entomophilum, 11, 650, **651**, 692
Mesoplasma florum, 11, 569, 572, 649, **650**
Mesoplasma grammopterae, 11, 650, **651**
Mesoplasma lactucae, 11, 650, **652**
Mesoplasma photuris, 11, 649–650, **652**
Mesoplasma pleciae, 573, 645, 647, 650, 690, 694
Mesoplasma seiffertii, 11, 650, **652**, 668, 694
Mesoplasma syrphidae, 11, 650, **652**
Mesoplasma tabanidae, 11, 650, **652**
metabolipauper (Pedobacter), 345
METHANOBACTERIALES, 879
Methanobacterium, 879
Methanobacterium formicicum, 78
METHANOCOCCALES, 879
Methanococcus, 879
METHANOMICROBIALES, 879
Methanomicrobium, 879
Methanospirillum hungatei, 792
meyeri (Leptospira), 547, **555**
Microbacterium esteraromaticum, 138
Microbacterium maritypicum, 138

Microcyclus, 389, 391, 407, 417, 424
Microcyclus aquaticus, 471
microfusus (Bacteroides), 32, 56
microfusus (Rikenella), 4, **55**, 56
Microlunatus phosphovorus, 782
micromati (Flavobacterium), 128, 130, 137, 143, 144, **148**, 149
Microscilla, 8, 22, 135, 371, 372, 385, 392–395, **408**, 409, 410
Microscilla furvescens, 384, 393, 408, **409**, 442
Microscilla marina, 9, 360, 393, 395, 408, **409**, 410
Microscilla sericea, 393, 394, 408, **409**, 442
microti (Mycoplasma), 11, 583, 585, **603**, 604
mikhailovii (Algibacter), 6
minutum (Treponema), 10, 504, **514**, 518
miricola (Chryseobacterium), 5, 186, 189, 203, 205, 209, 210
miricola (Elizabethkingia), 182, 185, 203–206, 208–209, **210**
mirum (Spiroplasma), 656–657, 661, 663–669, **674**, 675
mishustinae (Salegentibacter), 7, 266–268, **269**
mitskevichiae (Gillisia), 7, 221, 222, 224, **225**
Mitsuaria chitosanitabida, 186
Mitsuokella, 32, 94
Mitsuokella dentalis, 94, 105
Mitsuokella multacida, 32
miyamotoi (Borrelia), 487, 489, 491, **495**
mizutaii (Flavobacterium), 138, 332, 337
mizutaii (Sphingobacterium), 8, 331–333, **337**
moatsii (Mycoplasma), 11, 582, 585, **604**
mobile (Mycoplasma), 11, 569, 576–578, 582, 585, **604**
mobilis (Mesonia), 7, 240
mobilis (Mesonia), **240**
mobilis (Pelagicoccus), 13, **818**, 829, 830, 831
modestum (Propionigenium), 12, 758, 759, 761, **762**
modestus (Edaphobacter), 12, **729**
modicum (Acholeplasma), 11, 688–690, **692**
molare (Mycoplasma), 11, 582, 587, 604
molle (Chryseobacterium), 5, 190
MOLLICUTES, 1, 2, 10, 16, 21, 23, 567, **568**, 570–572, 574, 581, 602, 611, 616, 618, 620, 640, 644, 647, 656, 660, 662, 667, 668, 687, 690, 696, 700, 703, 719, 721, 723
moniliformis (Streptobacillus), 12, 752, 767, 770–**772**
monobiae (Spiroplasma), 657, 661, 664, 668, **675**
montanense (Spiroplasma), 11, 661, 668, 670, **675**
Moraxella anatipestifer, 263
mortiferum (Fusobacterium), 748, 750–752, **755**
morum (Acholeplasma), 11, 574, 688, 689, **693**
mossii (Dysgonomonas), 4, 71–74, **75**, 76
Mucilaginibacter, 7, 8, 22
Mucilaginibacter gracilis, 8
Mucilaginibacter paludis, 8
muciniphila (Akkermansia), 809, **810**, 811
muelleri (Aquimarina), 7, 158, 159, **160**
multiacidus (Bacteroides), 32
multiformis (Prevotella), 65, **97**
multilocale (Acholeplasma), 11, 689, 690, **693**
multisaccharivorax (Prevotella), 65, **97**
multivorum (Flavobacterium), 337
multivorum (Sphingobacterium), 8, 137, 330–332, **337**, 340, 344, 384
murdochii (Brachyspira), 532–534, 538–**540**
murdochii (Serpulina), 537
Muricauda, 5, 22, 106, 108, 110, 135, 162, 163, 199, 232, 233, **240**, 241–245, 265, 270, 284, 293
Muricauda aquimarina, 6, 240–**244**
Muricauda flavescens, 6, 241–**244**
Muricauda lutimaris, 6
Muricauda ruestringensis, 6, 111, 162, 223, 240–**244**, 265
muridarum (Chlamydia), 13, 849, 850, 851–857, 859, **860**, 866
muris (Actinobacillus), 772
muris (Haemobartonella), 11, 599, 642
muris (Mycoplasma), 11, 240, 578, 583, 585, 604
mustelae (Mycoplasma), 11, 582, 585, 604
mycoides (Mycoplasma), 11, 569, 572, 575, 577–581, **583**, 586, 587, 602, 644, 645, 656, 657, 662, 667–669, 697
mycoides subsp. capri (Mycoplasma), 579, **583**, 586
mycoides subsp. mycoides (Mycoplasma), 569, 572, 578, 579, 581, 583, **586**, 587, 602
Mycoplasma mycoides subsp. capri, 579, **583**, 586
Mycoplasma mycoides subsp. mycoides, 569, 572, 578, 579, 581, 583, **586**, 587, 602
Mycoplasma, 10, 11, 23, 488, 568, 571, 574, **575**, 576–581, 583–585, 587, 640, 641, 644, 645, 647, 650, 654, 656, 658, 660, 667, 669, 688–690, 693, 698, 702, 719–721, 723
Mycoplasma adleri, 11, 582, 584, **587**
Mycoplasma agalactiae, 11, 569, 578, 579, 582, 584, **587**, 590
Mycoplasma agassizii, 11, 582, 584, **587**
Mycoplasma alkalescens, 11, 582, 584, **587**, 588
Mycoplasma alligatoris, 11, 577, 582, 584, **588**
Mycoplasma amphoriforme, 11, 583, 584, **588**, 610
Mycoplasma anatis, 11, 582, 584, **588**, 589
Mycoplasma anseris, 11, 109, 584, **589**
Mycoplasma arginini, 11, 580, 582, 584, 588, **589**
Mycoplasma arthritidis, 11, 569, 578–580, 582, 584, **589**
Mycoplasma auris, 11, 582, 584, **589**
Mycoplasma bovigenitalium, 11, 582, 584, **589**, 590
Mycoplasma bovirhinis, 11, 578, 582, 584, **590**
Mycoplasma bovis, 11, 578, 579, 582, 584, 587, **590**
Mycoplasma bovoculi, 11, 582, 584, **590**
Mycoplasma buccale, 582, 584, **590**
Mycoplasma buteonis, 582, 584, **591**
Mycoplasma californicum, 11, 582, 584, **591**
Mycoplasma canadense, 11, 582, 584, **591**
Mycoplasma canis, 11, 582, 584, **591**
Mycoplasma capricolum, 11, 579, 581, **591**, 592
Mycoplasma capricolum subsp. capricolum, 569, 579, 584, 587, **592**, 602
Mycoplasma capricolum subsp. capripneumoniae, 579, 581, 584, **592**, 602
Mycoplasma caviae, 11, 582, 584, **592**
Mycoplasma cavipharyngis, 11, 583, 584, **592**
Mycoplasma citelli, 11, 582, 584, **592**, 593
Mycoplasma cloacale, 11, 582, 584, **593**
Mycoplasma coccoides, 11, 572, 583, 584, **611**, 641
Mycoplasma collis, 11, 582, 584, **593**
Mycoplasma columbinasale, 11, 582, 584, **593**, 597
Mycoplasma columbinum, 11, 582, 584, **593**
Mycoplasma columborale, 11, 582, 584, **593**
Mycoplasma conjunctivae, 11, 569, 582, 584, **593**, 594
Mycoplasma corogypsi, 11, 582, 584, **594**
Mycoplasma cottewii, 11, 584, **594**
Mycoplasma cricetuli, 11, 582, 584, **594**
Mycoplasma crocodyli, 11, 582, 584, **594**
Mycoplasma cynos, 11, 582, 584, **594**
Mycoplasma dispar, 11, 582, 584, **594**, 595
Mycoplasma edwardii, 11, 582, 584, **595**
Mycoplasma elephantis, 11, 582, 584, **595**
Mycoplasma ellychniae, 646, 647
Mycoplasma equigenitalium, 11, 582, 584, **595**, 647
Mycoplasma equirhinis, 11, 582, 584, **595**
Mycoplasma falconis, 11, 582, 584, **595**
Mycoplasma fastidiosum, 11, 583, 584, **595**, 596, 613
Mycoplasma faucium, 11, 582, 584, **596**
Mycoplasma felifaucium, 11, 582, 584, **596**
Mycoplasma feliminutum, 11, 581, 584, **596**, 690
Mycoplasma felis, 11, 582, 584, **596**
Mycoplasma fermentans, 11, 578–580, 582, 584, **596**, 597
Mycoplasma flocculare, 11, 578, 582, 584, **597**
Mycoplasma gallinaceum, 11, 582, 584, **597**
Mycoplasma gallinarum, 11, 582, 584, **597**
Mycoplasma gallisepticum, 11, 569, 577–579, 583, 584, 588, **597**, 598, 601, 602, 610
Mycoplasma gallopavonis, 11, 582, 584, **598**
Mycoplasma gateae, 11, 582, 584, **598**
Mycoplasma genitalium, 11, 569, 576, 578, 579, 581, 583, 585, **598**, 599, 607, 616, 659
Mycoplasma glycophilum, 11, 582, 585, **599**
Mycoplasma gypis, 11, 582, 585, **599**
Mycoplasma haematoparvum (Candidatus), 586, **611**
Mycoplasma haemobos (Candidatus), **611**
Mycoplasma haemocanis, 11, 579–580, 583, 585, **599**, 642, 643
Mycoplasma haemodidelphidis (Candidatus), 586, **612**
Mycoplasma haemofelis, 11, 579, 583, 585, **599**, 612, 642–643
Mycoplasma haemolamae (Candidatus), 586, **612**
Mycoplasma haemominutum (Candidatus), 586, **612**
Mycoplasma haemomuris, 11, 583, 585, **599**, 600, **642**
Mycoplasma haemotarandirangiferis (Candidatus), **612**
Mycoplasma hominis, 11, 569, 572, 577–580, 582, 583, 585, **600**, 613, 614
Mycoplasma hyopharyngis, 11, 582, 585, **600**
Mycoplasma hyopneumoniae, 11, 569, 577–579, 582, 585, 597, **600**, 601
Mycoplasma hyorhinis, 11, 577–578, 580, 582, 585, **600**, 601
Mycoplasma hyosynoviae, 11, 579, 582, 585, **601**
Mycoplasma iguanae, 11, 582, 585, **601**
Mycoplasma imitans, 11, 583, 585, 598, **601**, 602
Mycoplasma indiense, 11, 582, 585, **601**
Mycoplasma iners, 11, 582, 585, 597, **601**, 602
Mycoplasma insons, 576, 577, 583, 585, **613**
Mycoplasma iowae, 11, 583, 585, **602**, 647
Mycoplasma kahaneii (Candidatus), **612**
Mycoplasma lactucae, 650, 652
Mycoplasma lagogenitalium, 11, 582, 585, **602**

Mycoplasma leachii, 578, 585–587, 592, **602**
Mycoplasma leonicaptivi, 11, 582, 585, **602**
Mycoplasma leopharyngis, 11, 582, 585, **602**, 603
Mycoplasma lipofaciens, 11, 582, 585, **603**
Mycoplasma lipophilum, 11, 582, 585, **603**, 613
Mycoplasma lucivorax, 648
Mycoplasma luminosum, 648
Mycoplasma maculosum, 11, 582, 585, **603**
Mycoplasma melaleucae, 647, 648
Mycoplasma meleagridis, 11, 579, 582, 585, 597, **603**
Mycoplasma microti, 11, 583, 585, **603**, 604
Mycoplasma moatsii, 11, 582, 585, **604**
Mycoplasma mobile, 11, 569, 576–578, 582, 585, **604**
Mycoplasma molare, 11, 582, 587, **604**
Mycoplasma mucosicanis, **604**
Mycoplasma muris, 11, 240, 578, 583, 585, **604**
Mycoplasma mustelae, 11, 582, 585, **604**
Mycoplasma mycoides, 11, 569, 572, 575, 577–581, **583**, 586, 587, 602, 644, 645, 656, 657, 662, 667–669, 697
Mycoplasma neurolyticum, 580, 582, **605**, 613
Mycoplasma opalescens, 582, **605**
Mycoplasma orale, 578, 580, 582, **605**
Mycoplasma ovipneumoniae, 582, **605**
Mycoplasma ovis, 576, 583, **606**, 611, 612, **641**
Mycoplasma oxoniensis, 582, **606**
Mycoplasma penetrans, 569, 576–579, 583, **606**
Mycoplasma phocicerebrale, 582, **606**
Mycoplasma phocidae, 582, **606**
Mycoplasma phocirhinis, 582, **607**
Mycoplasma pirum, 580, 583, **607**
Mycoplasma pneumoniae, 569, 572, 576–580, 583, 598, 599, 606, **607**, 659, 660, 668
Mycoplasma primatum, 582, **607**
Mycoplasma pullorum, 582, **608**
Mycoplasma pulmonis, 569, 577–579, 582, 589, **608**, 772
Mycoplasma putrefaciens, **608**, 667
Mycoplasma ravipulmonis (Candidatus), 586, **612**
Mycoplasma salivarium, 579, 580, **608**
Mycoplasma simbae, 609, **609**
Mycoplasma somnilux, 649
Mycoplasma spermatophilum, 582, **609**
Mycoplasma sphenisci, **613**
Mycoplasma spumans, 582, **609**
Mycoplasma sturni, 582, **609**
Mycoplasma sualvi, 582, **609**
Mycoplasma subdolum, 582, **609**
Mycoplasma suis, 580, 583, 640, **641**, **609**
Mycoplasma synoviae, 569, 577–579, 582, 597, 598, **610**
Mycoplasma testudineum, 582, **610**
Mycoplasma testudinis, 583, **610**
Mycoplasma turicensis (Candidatus), **612**
Mycoplasma verecundum, 582, **610**
Mycoplasma volis, 604
Mycoplasma vulturis, 582, **613**
Mycoplasma wenyonii, 583, 606, **611**, 612, 641, 642
Mycoplasma yeatsii, **611**
Mycoplasma zalophi, **613**
MYCOPLASMATACEAE, 11, 23, 570, 574, **575,** 580, 581, 615, 619, 640, 659
MYCOPLASMATALES, 10, 11, 23, 567, 568, 570–573, **574**, 580, 581, 583, 639, 640, 642, 644, 667, 687, 719
MYROIDACEAE, 5
myxarmorum (Bizionia), 7, 167, **168**
myxillae (Gillisia), 7, 221, 222, 224, **225**
myxococcoides (Sporocytophaga), 9, 368, 379, 393, **418**
myxolifaciens (Robiginitalea), 6

naegleriophila (Protochlamydia), 870
naviforme (Fusobacterium), 748, 751, **755**
necrogenes (Fusobacterium), 748, 750, 751, **755**
necrophorum (Fusobacterium), 748, 750, 751, 752, **755**, 757
necrophorum subsp. funduliforme (Fusobacterium), 750, 752, **756**
necrophorum subsp. necrophorum (Fusobacterium), 750, **756**
negevensis (Simkania), 13, 843, **875–877**
Neochlamydia, 15, 867, **869**
Neochlamydia hartmannellae, 849, **869**
neurolyticum (Mycoplasma), 580, 582, **605**, 613
Niabella aurantiaca, 8
Niastella, 8, 351
Niastella koreensis, 8, 351, 393
Niastella yeongjuensis, 8
nigrescens (Prevotella), 65, 82, 87, **98**
nigricans (Herpetosiphon), 367
nigricans (Lewinella), 8, **367**
Nitrobacter, 921
Nitrospira, 921
NITROSPIRACEAE, 775
Nocardia corynebacterioides, 138
nodosus (Bacteroides), 32
noguchii (Leptospira), 10, **555**
Nonlabens, 5, 7, 202, **248**
Nonlabens tegetincola, 7, 111, 202, **248**, 275
nordii (Bacteroides), 3, 29, **36**
normanense (Cloacibacterium), **197**
norwichensis (Hymenobacter), 9, 398–400, **402**, 403
Novosphingobium capsulatum, 138
Novosphingobium resinovorum, 138
Nubsella, 7, 8
Nubsella zeaxanthinifaciens, 8
nucleatum (Fusobacterium), 2, 79, 748, 750, 751, 752, **753**, 754, 757, 762, 766
nucleatum subsp. animalis (Fusobacterium), 750, 753, **754**
nucleatum subsp. fusiforme (Fusobacterium), 750, 753, **754**
nucleatum subsp. nucleatum (Fusobacterium), 750, 751, 753, **754**
nucleatum subsp. polymorphum (Fusobacterium), 750, 751, 753, **754**
nucleatum subsp. vincentii (Fusobacterium), 750, 751, 753, **754**
nyackensis (Pedobacter), 345

obamensis (Rhodothermus), 458, 459
obscuriglobus (Gemmata), 16, 891, 893, **898**, 899, 900
oceanosedimentum (Flavobacterium), 5, 138
ocellatus (Hymenobacter), 9, 398–400, **403**
ochracea (Capnocytophaga), 5, 32, 168, 169, 171, 172, **174**, 175
ochraceus (Bacteroides), 32, 88, 174
oculi (Acholeplasma), 11, 649, 688–690, **693**
odoratum (Flavobacterium), 135, 136, 245, 246
Odoribacter, 3, 4, 82
Odoribacter splanchnicus, 3, 4, 27, 32, 39, 40, 57, 65, 74, 82
oligotrophica (Emticicia), **388**, 389
Olivibacter ginsengisoli, 8
Olivibacter sitiensis, 8
Olivibacter soli, 8
Olivibacter terrae, 8
Olleya, 5, 6, 22, 108, 111, 135, 158, 216, 242, **249**, 288, 289
Olleya marilimosa, 6, 111, 242, **249**
omanense (Candidatus Phytoplasma), **710**
omnivorum (Flavobacterium), 5, 112, 117, 129, 133, 137, **148**
onderdonkii (Alistipes), 4, 58, 59, **60**, 61
opalescens (Mycoplasma), 582, **605**
ophiurae (Marixanthomonas), 7
OPITUTACEAE, 13, 14, 23, 817–**820**, 826, 828, 831
OPITUTAE, 13, 14, 23, 795, 796, 800, 801, 815, **817**, 818–820, 823, 825, 826, 828, 835
OPITUTALES, 13, 23, 800, 817, **820**, 821–823, 826
Opitutus, 13, 14, 23, 817–**820**, 821–823, 826, 831
Opitutus terrae, 13, 818, **820**, 821, 822, 826, 828, 831
orale (Mycoplasma), 578, 580, 582, **605**
orale (Treponema), **521**
oralis (Bacteroides), 32, 93, 95, 174
oralis (Prevotella), 65, **98**
oranimense (Chryseobacterium), 5
Orbiscindens (Clostridium), 38
Oribaculum, 66
Oribaculum catoniae, 66–67
orientalis (Blatta), 316, 320, 758, 873
orientalis (Maribacter), 5–6, 236, **237**, 238
oris (Bacteroides), 98
oris (Prevotella), 65, 93, **98**
ornithinivorans (Algoriphagus), 9, 426, 427, 430, **432**
ornithinivorans (Hongiella), 428, 432
Ornithobacterium, 5, 6, 22, 108, 111, 180, 203, 205, 234, **250**, 251–254, 274
Ornithobacterium rhinotracheale, 2, 5, 111, 250–252, **253**, 254, 274, 324
oryzae (Candidatus Phytoplasma), **710**
oryzae (Pedobacter), 345
oryzae (Xylanibacter), 5, **102**, 103
oulorum (Prevotella), **99**, 103
ovatus (Bacteroides), 3, 29, **36**, 37
ovipneumoniae (Mycoplasma), 582, **605**
ovis (Eperythrozoon), 11, 606, 641
ovis (Mycoplasma), 576, 583, **606**, 611, 612, **641**
ovolyticus (Flexibacter), 283, 392, 394
Owenweeksia, 6, 7, 22, 322, 324, 325, 327, **328**, 329
Owenweeksia hongkongensis, 7, 324, 325, 328, **329**
oxoniensis (Mycoplasma), 582, **606**

pacifica (Cellulophaga), 5, 6, 176–179, **180**
pacifica (Echinicola), 9, **437**, 439
palladensis (Arenibacter), 162, 163, **164**
pallens (Prevotella), 65, **99**
pallida (Isosphaera), 16, 880, 896, 899, **900**, 901–903, 914, 919
pallidilutea (Jejuia), 305
pallidum (Chryseobacterium), 5, 190
pallidum (Treponema), 10, 472, 476, 486, 487, 501, 502, 504, 505, 507, **509**, 514, 521, 535
pallidum subsp. endemicum (Treponema), 502, 505, **510**
pallidum subsp. pallidum (Treponema), 487, 501, 502, 503, 504, 505, 507, 509, **510**, 516
pallidum subsp. pertenue (Treponema), 502, 505, **509**

palmae (Acholeplasma), 11, 688, 689, **693**, 704
Paludibacter, 61, 70, **76**, 81
Paludibacter propionicigenes, 4, **76**, 82
paludicola (Schlesneria), 16, 910, 911, **913**
paludis (Mucilaginibacter), 8
panaciterrae (Pedobacter), 345
Parabacteroides, 4, 33, 61, 70, 78–81
Parabacteroides distasonis, 4, 33, 35, 70, 77, 78
Parabacteroides goldsteinii, 4, 33, 35
Parabacteroides merdae, 33
Parachlamydia, 855, 857, **867**, 868
Parachlamydia acanthamoebae, 13, **868**, 869, 870, 877
PARACHLAMYDIACEAE, 856, **867**
paragorgiae (Bizionia), 7, **166**, 167, 218, 249
paraluiscuniculi (Treponema), 501, 502, 503, 505, **514**
Parapedobacter, 8
Parasegetibacter, 351
parkeri (Borrelia), 488, 489, 491, **495**, 497
parva (Leptospira), 547, 558, 559
parva (Turneriella), 10, 547, 551, **558**, 559
parvum (Acholeplasma), 11, 688, 689, **693**
parvum (Eperythrozoon), **641**
parvum (Treponema), 502, **514**, 515, 521
parvum (Ureaplasma), 11, 614–621, **622**
Pasteurella pneumotropica, 772
Pasteuria, 895, 906
Pasteuria ramosa, 889, 905
paucimobilis (Pseudomonas), 211
paucimobilis (Sphingomonas), 138
paucivorans (Acetothermus), **44**
pecorum (Chlamydia), 850–852, 854, 855, 857–859, **860**, 861, 866
pecorum (Chlamydophila), 13, 845
Pectinis (Cristispira), 10, 472, **498**, 501
pectinophilus (Bacteroides), 39, **40**
pectinovora (Cytophaga), 148, 373
pectinovorum (Flavobacterium), 129, 130, 136, **148**, 154, 373
pectinovorum (Treponema), 502, 508, 510, 513, **515**, 521
pedis (Acanthopleuribacter), 734, **735**
pedis (Treponema), 502
Pedobacter, 8, 330–332, **339**, 340–345
Pedobacter africanus, 330, 339–342, 344, **345**
Pedobacter agri, 8, 345
Pedobacter alluvionis, 345
Pedobacter aquatilis, 345
Pedobacter borealis, 345
Pedobacter caeni, 339–342, 344, **345**
Pedobacter composti, 345
Pedobacter cryoconitis, 339–342, 344, **347**
Pedobacter daechungensis, 8, 345
Pedobacter duraquae, 345
Pedobacter glucosidilyticus, 245
Pedobacter hartonius, 345
Pedobacter heparinus, 8, 138, 330, **339**, 340–342, 344, 345
Pedobacter himalayensis, 339–342, 344, **347**
Pedobacter insulae, 8
Pedobacter koreensis, 345
Pedobacter lentus, 345
Pedobacter metabolipauper, 345
Pedobacter nyackensis, 345
Pedobacter oryzae, 345
Pedobacter panaciterrae, 345
Pedobacter piscium, **347**
Pedobacter roseus, 339–341, 344, **347**
Pedobacter saltans, 330, 339, 340–344, **347**
Pedobacter sandarakinus, 345
Pedobacter steynii, 345
Pedobacter suwonensis, 339, 340–342, 344, **347**
Pedobacter terrae, 345
Pedobacter terricola, 345
Pedomicrobium, 908
Pelagicoccus, 817, 824–826, 828, **829**
Pelagicoccus albus, 818, 829, 830, 831, **832**
Pelagicoccus croceus, 829, **833**
Pelagicoccus litoralis, 818, 826, 829, 830, 831, **833**
Pelagicoccus mobilis, 13, 818, 829, 830, **831**
Pelobacter, 759
penaei (Spiroplasma), 11, 661, 664, 668, **675**
penetrans (Mycoplasma), 569, 576–579, 583, **606**
pentosaceus (Bacteroides), 32, 93
Perexilibacter, 394, 442, 445
Perexilibacter aurantiacus, 393
perfoetens (Fusobacterium), **756**
periodonticum (Fusobacterium), 748, 750, 751, **756**
persica (Borrelia), 489, 495, **496**
persica (Lewinella), 8, **367**, 368
persicicus (Roseibacillus), 13
Persicirhabdus, 800
Persicirhabdus sediminis, 13
Persicitalea, 8
Persicitalea jodogahamensis, 9
Persicivirga, 109, **254**, 255
Persicivirga xylanidelens, 7, 254, **255**
Persicobacter, 54, 374, 405, 442, 445, 446, **450**, 451
Persicobacter diffluens, 9, 54, 450, **451**, 452
persicus (Herpetosiphon), 367
pertenue (Treponema), 502
pestis (Yersinia), 858
petaseus (Flavihumibacter), 358
Petrimonas, 61, **77**
Petrimonas sulfuriphila, 4, **77**
phagedenis (Treponema), 502, 504, 509, 510, **516**, 521
phocicerebrale (Mycoplasma), 582, **606**
phocidae (Mycoplasma), 582, **606**
phocirhinis (Mycoplasma), 582, **607**
phoenicium (Candidatus Phytoplasma), **711**
phoeniceum (Spiroplasma), 11, 660–661, 664, 668, **675**
photuris (Mesoplasma), 11, 649–650, **652**
phyci (Haloferula), 13
Phytoplasma (Candidatus), 11, 570, 581, 618, 640, 647, 668, 687, 690, **696**, 701, 702, 703, 704, 707–709
Phytoplasma allocasuarinae (Candidatus), **703**
Phytoplasma americanum (Candidatus), **703**
Phytoplasma asteris (Candidatus), 569, 698, **703**
Phytoplasma aurantifolia (Candidatus), **703**
Phytoplasma australasia (Candidatus), **705**
Phytoplasma australiense (Candidatus), 698, **705**
Phytoplasma brasiliense (Candidatus), **705**
Phytoplasma caricae (Candidatus), **706**
Phytoplasma castaneae (Candidatus), **710**
Phytoplasma cynodontis (Candidatus), **710**
Phytoplasma fragariae (Candidatus), **710**
Phytoplasma fraxini (Candidatus), **710**
Phytoplasma graminis (Candidatus), **710**
Phytoplasma japonicum (Candidatus), **710**
Phytoplasma lycopersici (Candidatus), **710**
Phytoplasma mali (Candidatus), 698, 699, **710**
Phytoplasma omanense (Candidatus), **710**
Phytoplasma oryzae (Candidatus), **710**
Phytoplasma phoenicium (Candidatus), **711**
Phytoplasma pini (Candidatus), **711**
Phytoplasma prunorum (Candidatus), **711**
Phytoplasma pyri (Candidatus), **711**
Phytoplasma rhamni (Candidatus), **711**
Phytoplasma spartii (Candidatus), **711**
Phytoplasma trifolii (Candidatus), **711**
Phytoplasma ulmi (Candidatus), **711**
Phytoplasma ziziphi (Candidatus), **712**
Pibocella, 241, 245
Pibocella ponti, 6, 106
Pillotina, 9, 16, 565, 585
Pillotina calotermitidis, 10, 16, 565, 585
pilosicoli (Brachyspira), 532–**540**
pilosicoli (Serpulina), 537, 540
pinensis (Chitinophaga), 8, 351, 352, 353, 356
pini (Candidatus Phytoplasma), **711**
Pirella, 879, 905
Pirella marina, 895, 897
Pirellula, 880, 881, 886, 889, 895–897, 899, 900, 902, **903**, 904–906
Pirellula marina, 886, 895, 905
Pirellula staleyi, 16, 897, 903, 904, **905**, 906–908
pirum (Mycoplasma), 580, 583, **607**
Piscichlamydia salmonis (Candidatus), 872, **873**
piscium (Chryseobacterium), 182–184, 188, **193**, 194
piscium (Pedobacter), **347**
piscium (Sphingobacterium), 332, 344, 347
PLANCTOBACTERIA, 2
PLANCTOMYCEA, 2
Planctomyces, 16, **879**, 880, 881, 886–889
Planctomyces bekefii, 16, **881**, 887, 888, 889, 890, 894, 895
Planctomyces brasiliensis, 881, 886, 887, **891**
Planctomyces guttaeformis, 16, 881, 887, **891**, 892, 894
Planctomyces limnophilus, 881, 886, **891**, 892, 893, 900, 911
Planctomyces maris, 881, 886, **892**, 893, 899, 900, 911
Planctomyces stranskae, 881, 887, 889, 891, 892, **894**
PLANCTOMYCETACEAE, 16, 879, **880**, 910, 913
PLANCTOMYCETALES, 16, **879**, 889, 895, 897, 901–903, 909, 910, 913, 918, 919
PLANCTOMYCETES, 2, 13, 16, 785, 796, 797, 844, **879**, 880–915
PLANCTOMYCETIA, 16, **879**
Planobacterium, 110
Planomicrobium okeanokoites, 138
platyhelix (Spiroplasma), 11, 654, 657, 661, **676**
plautii (Fusobacterium), 753, 754
plauti-vincentii (Fusobacterium), 753, 754
plebeius (Bacteroides), 3, 29, **37**
pleciae (Acholeplasma, 11, 689, 690, **694**
pleciae (Mesoplasma), 573, 645, 647, 650, 690, 694
plicatilis (Spirochaeta), 9–10, 473, 474, 476, **477**, 478
pneumoniae (Chlamydia), 847–859, **860**, 861, 866
pneumoniae (Chlamydophila), 13
pneumoniae (Mycoplasma), 569, 572, 576–580, 583, 598, 599, 606, **607**, 659, 660, 668
pneumophila (Legionella), 211, 868
pneumosintes (Bacteroides), 32
pneumotropica (Pasteurella), 772
pohnpeiensis (Luteolibacter), 13, 799, 802, 803

Polaribacter, 109, 181, 203, 229, 235, **255**, 256, 258, 391, 417
Polaribacter butkevichii, **256**
Polaribacter dokdonensis, **256**
Polaribacter filamentus, 5, **255**, 256
Polaribacter franzmannii, 229, **256**
Polaribacter glomeratus, **257**, 379, 390
Polaribacter irgensii, 106, 241, **258**, 259
polymorphum (Fusobacterium), 753, 754
polymorphus (Flexibacter), 392, 394, 409
polypragmatus (Bacteroides), 3, 39, **40**
polytropus (Ilyobacter), 12, 759, **760**
ponti (Pibocella), 6, 106
ponti (Roseibacillus), 13
Pontibacter, 372, 398, **410**, 411
Pontibacter actiniarum, 9, 387, 388, **401**, 404, 410, 411
Pontibacter akesuensis, **412**
porcellionis (Candidatus Rhabdochlamydia), 13, 845, 866, 873, **874**
porcinum (Treponema), 502, **516**, 520
poriferorum (Winogradskyella), 7, **290**
Porphyromonas, 4, 31, 32, 58, 61, **62**, 63, 64, 70, 81, 87
Porphyromonas asaccharolytica, 4, 62, **64**, 69, 89
Porphyromonas bennonis, 69
Porphyromonas cangingivalis, **64**
Porphyromonas canoris, **64**
Porphyromonas cansulci, 4, **64**, 66, 67
Porphyromonas catoniae, 4, 62, 64, **66**
Porphyromonas circumdentaria, 4, **67**
Porphyromonas crevioricanis, **67**
Porphyromonas endodontalis, **67**
Porphyromonas gingivalis, 62–64, **67**, 68, 79
Porphyromonas gingivicanis, **68**, 69, 70
Porphyromonas gulae, **68**, 70
Porphyromonas levii, 64, **68**
Porphyromonas macacae, 64, **68**, 69
Porphyromonas salivosa, 64, 68, **69**, 70
Porphyromonas somerae, 62, **69**
Porphyromonas uenonis, 62, 64, **69**
portivictoriae (Gramella), 7, 226, **227**
poulsonii (Spiroplasma), 11, 657, 659, 661, 663–664, 666–669, 672–673, **676**
praeacuta (Tissierella), 32
praeacutus (Bacteroides), 32
Prevotella, 4, 5, 31, 32, 55, 58, 63, 64, 73, 78, 81, **85**, 86, 87–89, 90, 92, 93, 96, 102, 105
Prevotella albensis, 5, 65, **91**, 103
Prevotella amnii, 4
Prevotella baroniae, 4, 65, **91**
Prevotella bergensis, 65, **91**
Prevotella bivia, 65, **91**, 92, 103
Prevotella brevis, 4, 65, **92**
Prevotella bryantii, 4, 65, **92**
Prevotella buccae, 65, **93**
Prevotella buccalis, 4, 65, **93**
PREVOTELLACEAE, 3, 4, 25, **85**–103
Prevotella copri, **93**
Prevotella corporis, 4, 65, **94**
Prevotella dentalis, 5, **94**, 105
Prevotella denticola, 4, 65, **95**
Prevotella disiens, 65, **95**
Prevotella enoeca, 4, 65, **95**
Prevotella heparinolytica, 5
Prevotella intermedia, 65, 79, 82, 87, **96**, 99
Prevotella loescheii, 4, 65, 82, **96**
Prevotella maculosa, 4
Prevotella marshii, 65, **97**
Prevotella massiliensis, 87, **102**
Prevotella melaninogenica, 4, 65, 82, **86**, 88
Prevotella multiformis, 65, **97**
Prevotella multisaccharivorax, 65, **97**
Prevotella nigrescens, 65, 82, 87, **98**
Prevotella oralis, 65, **98**
Prevotella oris, 65, 93, **98**
Prevotella oulorum, **99**, 103
Prevotella pallens, 65, **99**
Prevotella ruminicola, 65, 91, 92, **99**
Prevotella salivae, 65, **100**
Prevotella shahii, 65, **100**
Prevotella stercorea, **100**
Prevotella tannerae, 5, 65, **100**
Prevotella timonensis, **101**
Prevotella veroralis, 65, **101**, 105
Prevotella zoogleoformans, 27, 33, **101**
primatum (Mycoplasma), 582, **607**
primitia (Treponema), 10, 508, **517**, 521, 563, 564
Procabacter (Candidatus), 868
Prochlorococcus, 879
Prochloron, 879
procyoni (Haemobartonella), **643**
profunda (Zunongwangia), 7
Prolixibacter bellariivorans, 3
Promyxobacterium flavum, 136, **154**
propionicigenes (Paludibacter), 4, **76**, 82
Propionigenium, 12, 748, 760, **761**, 762
Propionigenium maris, 761, **762**
Propionigenium modestum, 12, 758, 759, 761, **762**
Prosthecobacter, 795, 800, 803, **805**, 806–810, 814, 817
Prosthecochloris, 803, 804
Prosthecomicrobium, 803, 804
Proteiniphilum, 4, 61, **77**, 78
Proteiniphilum acetatigenes, 65, 77, **78**, 82
PROTEOBACTERIA, 7, 408, 457, 467, 775, 782, 795, 803, 870
proteolyticum (Chryseobacterium), 180, 181, 184, 186, 188, 189, **196**
Protochlamydia, 855, 867, 869, **870**
Protochlamydia amoebophila, 870, **871**, 876
Protochlamydia amoebophila (Candidatus), 870, **870**, 871, 871
Protochlamydia naegleriophila, 870
prunorum (Candidatus Phytoplasma), **711**
Pseudobacterium brevis, 210, 212
Pseudomonas, 183, 187
Pseudomonas aeruginosa, 725
Pseudomonas fluorescens, 211
Pseudomonas paucimobilis, 211
Pseudomonas putida, 886
Pseudosphingobacterium, 7, 8
Pseudosphingobacterium domesticum, 8
pseudotuberculosis (Yersinia), 858
pseudotuberculosis subsp. pestis (Yersinia), 858
Pseudozobellia, 110
psittaci (Chlamydia), 847, 849–860, **861**, 866
psittaci (Chlamydophila), 13, 845
Psychrilyobacter, 12
Psychrilyobacter atlanticus, 12
Psychroflexus, 5, 221, 222, 230, 248, 255, **258**, 259, 260, 261, 269, 292
Psychroflexus gondwanensis, 138, 260, **261**
Psychroflexus salinarum, 7
Psychroflexus torques, 7, 111, 160, 223, 242, **258**, 259, 260, 269, 324
Psychroflexus tropicus, 258, **261**
psychrolimnae (Flavobacterium), 5, 112, 125, 137, 143, 144, **149**
psychrophila (Cytophaga), 149, 154, 373, 392
psychrophilum (Flavobacterium), 5, 112, 125, 127–134, 136, 140, **149**, 373, 395
psychrophilum (Rhodonellum), 9, **440**
psychrophilus (Flexibacter), 149, 373, 392, 394
Psychroserpens, 6, 135, 158, 214, 230, 231, **261**, 262, 288, 289
Psychroserpens burtonensis, 7, 111, 214, 215, 223, 242, **261**, 262, 288, 289, 324
psychrotolerans (Hymenobacter), 9, 404
pterotermitidis (Hollandina), 10, 565
pullorum (Aegyptianella), 110
pullorum (Mycoplasma), 582, **608**
pulmonis (Mycoplasma), 569, 577–579, 582, 589, **608**, 772
PUNICEICOCCACEAE, 13, 14, 24, 817, 818, **824**–826
PUNICEICOCCALES, 817, **823**, 824, 825, 826, 827, 828, 829, 830
Puniceicoccus, 14, 817, **824**, 825
Puniceicoccus vermicola, 13, 818, **824**, 825, 826, 831
putida (Pseudomonas), 886
putidum (Treponema), 502, **517**
putredinis (Alistipes), 4, 31, 32, 56, **58**, 59
putredinis (Bacteroides), 55, 56, 58
putrefaciens (Mycoplasma), **608**, 667
pyogenes (Bacteroides), 3, 30, **37**
pyri (Candidatus Phytoplasma), **711**

radiobacter (Agrobacterium), 183
ramosa (Pasteuria), 889, 905
ranarum (Aegyptianella), 110, 207
Rapidithrix, 442
Rapidithrix thailandica, 9
ratkowskyi (Algoriphagus), 9, 426–428, 430, **432**
ravipulmonis (Candidatus Mycoplasma), 586, **612**
rectum (Clostridium), 762
rectus (Campylobacter), 79
recurrentis (Borrelia), 486, 489–491, 493–495, **496**, 497
refringens (Treponema), 502, 504, 509, 510, 516, **517**, 518
Reichenbachia, 385, 452
Reichenbachia agariperforans, 452
Reichenbachiella, 9, 22, 405, 442, 445, **452**
Reichenbachiella agariperforans, 9, **452**
resinovorum (Flavobacterium), 138
resinovorum (Novosphingobium), 138
reticulitermitidis (Clevelandina), 10, 16, 565
Rhabdochlamydia, 13, 24, 845, 866, **873**, 874
RHABDOCHLAMYDIACEAE, 13, 24, 845, **873**, 875
Rhabdochlamydia crassificans, 13, **873**, 874
Rhabdochlamydia porcellionis (Candidatus), 13, 845, 866, 873, **874**
rhamni (Candidatus Phytoplasma), **711**
rhinotracheale (Ornithobacterium), 2, 5, 111, 250–252, **253**, 254, 274, 324
Rhizobium leguminosarum, 332
Rhodococcus corynebacteroides, 138
Rhodonellum psychrophilum, 9, **440**
Rhodopirellula, 24, 880, 881, 895, 897, 899, 900, 903, 905, **906**, 907–909, 911, 912, 914
Rhodopirellula baltica, 16, 896, 897, 903–905, **906**, 907–910
RHODOTHERMACEAE, 3, 9, **457**, 458–468
Rhodothermus, 7, 8, 22, 25, **457**, 458, 459, 462, 463
Rhodothermus marinus, 9, **458**, 459, 461, 462, 466
Rhodothermus obamensis, 458, 459

Rickettsia, 568
RICKETTSIACEAE, 315
RICKETTSIALES, 110, 315, 639
Rickettsiella, 873, 874
Riemerella, 22, 108, 109, 165, 180, 181, 189, 197, 203, 205, 208, 212, 227, 228, 250, **262**, 263, 271, 273, 274
Riemerella anatipestifer, 5, 182, 185, 197, 198, 250, 253, 254, 262, **263**, 264
Riemerella columbina, 182, 185, 197, 253, 254, 263, **264**
rigidum (Acetofilamentum), 41, **42**
rigui (Hymenobacter), 9, 398–400, **403**
rigui (Spirosoma), 9, 416, **417**
Rikenella, 21, 32, 50, **54**, 55, 56, 58, 63
RIKENELLACEAE, 4, 25, **54**, 55–60
Rikenella microfusus, 4, **55**, 56
rivesi (Candidatus Xiphinematobacter), 13, **840**
Robiginitalea, 22, 109, 135, 162, 163, 241, 245, **264**, 265, 270, 284, 293
Robiginitalea biformata, 6, 259, **264**, 265
Robiginitalea myxolifaciens, 6
rosea (Arcicella), 9
rosea (Haloferula), 13
Roseibacillus, 23
Roseibacillus ishigakijimensis, 13
Roseibacillus persicicus, 13
Roseibacillus ponti, 13
Roseivirga, 22, 442, 445, 448, **453**, 454, 455
Roseivirga echinicomitans, 448, 453, **454**
Roseivirga ehrenbergii, 9, 394, 448, 453, 454, **455**
Roseivirga seohaensis, 448, 453, 454, **456**
Roseivirga spongicola, **456**
roseolus (Flexibacter), 9, 392, 394, 397
roseosalivarius (Hymenobacter), 9, 387, 388, 398, 399, **400**
roseus (Effluviibacter), 9, **387**, 388, 398
roseus (Pedobacter), 339–341, 344, **347**
ruber (Flexibacter), 9, 392, 394, **397**
ruber (Salinibacter), 9, 459–462, **463**
Rubritalea, 13, 23, 800, **812**, 813–815
Rubritalea marina, 13, 801, 812, **814**, 826, 831
Rubritalea sabuli, 13, 812, **815**
Rubritalea spongiae, 13, 812, **816**
Rubritalea squalenifaciens, 13, 801, 812, 813, **816**
Rubritalea tangerina, 13, 812, **816**
Rudanella, 22
Rudanella lutea, 9
rudongensis (Ancylobacter), 379, 390
ruestringensis (Muricauda), 6, 111, 162, 223, 240–**244**, 265
ruminicola (Bacteroides), 32, 92, 93, 95, 99
ruminicola (Prevotella), 65, 91, 92, **99**
ruminicola subsp. brevis (Bacteroides), 92
Ruminobacter, 740
Ruminobacter amylophilus, 32
Ruminococcus albus, 512, 741
Ruminococcus flavefaciens, 741
Runella, 22, 372, 381, 385, 390, 405, 407, **412**–414, 417, 423, 424
Runella defluvii, 412, 413, **414**
Runella limosa, 412–414, **415**
Runella slithyformis, 9, 379, 390, 412, 413, **414**
Runella zeae, 412–414, **415**
russellii (Zobellia), 6, **294**
russii (Fusobacterium), 748, 751, **757**

sabaudiense (Spiroplasma), 11, 661, 668–670, **676**
sabuli (Rubritalea), 13, 812, **815**
saccharolyticus (Caldicellulosiruptor), 778
Saccharophila (Cytophaga), 150, 373
saccharophilum (Flavobacterium), 5, 125, 129, 130, 136, **149**, 150, 154, 373
saccharophilum (Treponema), 502, **518**
saemankumensis (Aestuariicola), 5
saemankumensis (Gaetbulibacter), 6, 214, **218**, 291
salanitronis (Bacteroides), 3
saleffrena (Bizionia), 7, 167, **168**
salegens (Flavobacterium), 266
salegens (Salegentibacter), 7, 111, 138, 160, 223, 229, 233, 242, **266**, 267
Salegentibacter, 5, 22, 109, 221–223, 226, 230, 239, 248, 255, **266**, 267–269
Salegentibacter agarivorans, 7, 266, **267**, 268
Salegentibacter catena, 7, 266, 267, **268**
Salegentibacter flavus, 7, 266, 267, **268**
Salegentibacter holothuriorum, 7, 266, 267, **269**
Salegentibacter mishustinae, 7, 266–268, **269**
Salegentibacter salegens, 7, 111, 138, 160, 223, 229, 233, 242, **266**, 267
Salegentibacter salinarum, 7
salicanalis (Gelidibacter), 6, 219, 220, **221**, 278
salinarum (Halobacterium), 460
salinarum (Psychroflexus), 7
salinarum (Salegentibacter), 7
Salinibacter, 7, 8, 22, 25, 457, **460**–463
Salinibacter ruber, 9, 459–462, **463**
Salinimicrobium, 5, 22, 110
Salinimicrobium catena, 7
Salinimicrobium xinjiangense, 7
saliperosum (Flavobacterium), 5, 107, 112, 125, 130, 133, 136, 137, **150**
Salisaeta longa, 462
salivae (Prevotella), 65, **100**
salivarium (Mycoplasma), 579, 580, **608**
salivosa (Porphyromonas), 64, 68, **69**, 70
salivosus (Bacteroides), 32, 69
Salmonella, 389, 412, 414, 416
Salmonella typhimurium, 751, 886
salmoneum (Sediminibacterium), 8
salmonicida (Aeromonas), 211
salmonicola (Candidatus Clavichlamydia), 13, 843, 845, 865, **866**
salmonicola (Candidatus Clavochlamydia), 866
salmonicolor (Cytophaga), 49–51, 370, 373
salmonicolor (Marinilabilia), 4, 49, 50, **51**–53, 324, 373
salmonis (Candidatus Piscichlamydia), 13, 845, 872, **873**
saltans (Pedobacter), 330, 339, 340–344, **347**
salyersiae (Bacteroides), 3, 29, **37**
sancti (Chitinophaga), 8, 352, 353, **355**, 356, 395
sancti (Flexibacter), 352, 355, 374, 392, 394
sandarakina (Gillisia), 7, 221, 222, 224, **225**, 226
Sandarakinotalea, 5, 7, 22, 110, **269**, 270
Sandarakinotalea sediminis, 7, 269, **270**
sandarakinus (Pedobacter), 345
sanguinegens (Leptotrichia), 766, 770
sanguinegens (Sneathia), 12, 752, 766, 767, **770**
sanguinis (Streptococcus), 79
santarosai (Leptospira), 10, **555**
Saprospira, 7, 22, 358, **359**–361, 364, 367
SAPROSPIRACEAE, 7, 8, 324, **358**, 359–368
Saprospira grandis, 8, 324, 359, 360, **361**–364, 366–368, 393
sargassicola (Haloferula), 13
saxinquilinus (Subsaximicrobium), 6, 277, **278**, 279
Schlesneria, 16, 24, 880, **910**, 911–913, 914
Schlesneria paludicola, 16, 910, 911, **913**
scoliodontus (Treponema), 503, 504, 506, **518**
scophthalmum (Chryseobacterium), 5, 181–184, 186–189, 192, 193, **194**, 212
scophthalmum (Flavobacterium), 194
Sebaldella, 12, 23, 32, 760, 762, 766, **769**, 770
Sebaldella termitidis, 12, 32, 752, 762, 767, 769, **770**
sedimenticola (Maribacter), 5–6, 111, 162, 235–237, 242, 324
Sediminibacter furfurosus, 6
Sediminibacterium, 7, 8, 22, 351
Sediminibacterium salmoneum, 8
Sediminicola, 5, 22, 110, **270**, 271
Sediminicola luteus, 6, 242, **270**
sediminis (Actibacter), 5
sediminis (Gilvibacter), 7
sediminis (Persicirhabdus), 13
sediminis (Sandarakinotalea), 7, 269, **270**
Sediminitomix, 8, 22
Sediminitomix flava, 9
Segetibacter, 7, 8, 22, 351
Segetibacter koreensis, 8
segetis (Flavobacterium), 5, 112, 125, 129, 130, 133, 136, 137, **150**
seiffertii (Acholeplasma), 650, 652, **694**
seiffertii (Mesoplasma), 650, **652**, 668, 694
Sejongia, 6, 22, 108, 109, 180, 182, 184, 189, 203, 208, 212, **271**–274
Sejongia antarctica, 5, 111, 182, 185, 271–273, **274**
Sejongia jeonii, 5, 182, 185, 271–273, **274**
Selenomonas, 759
seohaensis (Marinicola), 384, 448, 453, 456
seohaensis (Roseivirga), 448, 453, 454, **456**
seregens (Hallella), 5, 94, 105
sericea (Microscilla), 393, 394, 408, **409**, 442
Serpula, 537
Serpula hyodysenteriae, 539
Serpula innocens, 539
Serpulina, 9, 10, 537
SERPULINACEAE, 9
Serpulina hyodysenteriae, 537, 539
Serpulina innocens, 537, 539
Serpulina intermedia, 537, 539, 540
Serpulina murdochii, 537
Serpulina pilosicoli, 537, 540
shahii (Alistipes), 4, 58, 59, **60**
shahii (Leptotrichia), 767, **768**
shahii (Prevotella), 65, **100**
Shigella, 389, 412, 414, 416, 823
shigense (Chryseobacterium), 5, 181–184, 188, 190, **194**
simbae (Mycoplasma), 609, **609**
simiae (Fusobacterium), 748, 750, 751, **757**
Simkania, 15, 24, 857, 874, **875**
SIMKANIACEAE, 13, 15, 24, 845, 856, 857, 873, **874**, 875, 876
Simkania negevensis, 13, 843, **875**–**877**
Singulisphaera, 16, 24, 880, 911, 912, **913**, 914, 915
Singulisphaera acidiphila, 16, 913, 914, **915**
sinica (Borrelia), 485, 489, 492, **496**
sitiensis (Olivibacter), 8
siyangense (Sphingobacterium), 8, 333, **338**
skermanii (Chitinophaga), 8, 351, 352, 353, **355**, 356
slithyformis (Runella), 9, 379, 390, 412, 413, **414**

smaragdinae (Spirochaeta), 10, 476, 478, **482**, 485
Sneathia, 12, 23, 752, 762, 766, 767, **770**
Sneathia sanguinegens, 12, 752, 766, 767, **770**
socranskii (Treponema), 502, 510, 513, **518**, 519
socranskii subsp. buccale (Treponema), 502, 519
socranskii subsp. paredis (Treponema), 502, 519
socranskii subsp. socranskii (Treponema), 502, 519, 521
soldanellicola (Chryseobacterium), 5, 182, 183, 185, 186, 187, 190, **194**, 195
soli (Chitinophaga), 368
soli (Chryseobacterium), 5, 190
soli (Flavobacterium), 129, 130, 133, 136, 137, **151**
soli (Hymenobacter), 9, 404
soli (Olivibacter), 8
somerae (Cetobacterium), **758**
somerae (Porphyromonas), 62, **69**
somnilux (Entomoplasma), 646, **649**
somnilux (Mycoplasma), 649
spartii (Candidatus Phytoplasma), **711**
SPARTOBACTERIA, 13, 14, 24, 795, 796, 801, 818, **834**, 835, 836, 838
spermatophilum (Mycoplasma), 582, **609**
Sphaerophorus fusiformis, 754
sphenisci (Mycoplasma), **613**
SPHINGOBACTERIA, 2, 25
SPHINGOBACTERIACEAE, 3, 7, 8, 22, 110, 324, 330, **331**, 339, 340, 343, 344, 394, 395
SPHINGOBACTERIIA, 2, 3, 7, 8, 22, **330**, 331–347, 393, 394
SPHINGOBACTERIALES, 7, 22, **330**, 339, 351
Sphingobacterium, 7, 8, 22, 135, 137–138, 210, 330, **331**–339, 340, 343–344, 374, 394
Sphingobacterium anhuiense, 8, 333, **334**
Sphingobacterium antarcticum, 8, 331–333, **334**
Sphingobacterium canadense, 8, 333, **334**, 335
Sphingobacterium composti, 8, 333, **335**, 336
Sphingobacterium daejeonense, 8, 333, **336**
Sphingobacterium faecium, 8, 331–333, **336**, 344
Sphingobacterium heparinum, 332, 339, 344–345
Sphingobacterium kitahiroshimense, 8, 333, **336**
Sphingobacterium mizutaii, 8, 331–333, **337**
Sphingobacterium multivorum, 8, 137, 330–332, **337**, 340, 344, 384
Sphingobacterium piscium, 332, 344, 347
Sphingobacterium siyangense, 8, 333, **338**
Sphingobacterium spiritivorum, 8, 137, 324, 330–**333**, 340, 344–345, 393
Sphingobacterium thalpophilum, 137, 331, 332, **338**, 344
Sphingomonas, 138
Sphingomonas paucimobilis, 138
spielmanii (Borrelia), 485, 489, **496**
spinosum (Verrucomicrobium), 13, 800, 803, 804, **805**, 806
Spirillum, 362
spiritivorum (Flavobacterium), 331, 333
spiritivorum (Sphingobacterium), 8, 137, 324, 330, **331**, 332, 333, 340, 344–345, 393
Spirochaeta, 9–10, 23, 471, **472**, 473–477, 483–486, 502, 508, 523
Spirochaeta africana, 10, 476, **477**, 478, 485
Spirochaeta alkalica, 10, 472, 476–478, **479**, 485
Spirochaeta americana, 10, 476–478, **479**, 485
Spirochaeta asiatica, 10, 476–478, **479**, 485
Spirochaeta aurantia, 10, 474–478, **479**, 480, 484, 486
Spirochaeta aurantia subsp. aurantia, 477, **480**
Spirochaeta aurantia subsp. stricta, 477, **480**
Spirochaeta bajacaliforniensis, 10, 476–478, **480**, 485
Spirochaeta caldaria, 10, 476–478, **480**, 484, 502, 508, 522
SPIROCHAETACEAE, 9, 10, 23, 471, 472, **473**–485, 499, 501–502, 508, 565
Spirochaeta cellobiosiphila, 477–478, **481**, 486
Spirochaeta coccoides, 10, 472–473, 476, **484**, 485
SPIROCHAETAE, 2
Spirochaeta halophila, 10, 472, 474–478, **481**, 484–485, 501–502
Spirochaeta isovalerica, 10, 474–478, **481**, 486, 502, 522
SPIROCHAETALES, 9–10, 23, **471**–472, 775
Spirochaeta litoralis, 10, 472, 474–478, **482**, 486, 501, 522
Spirochaeta plicatilis, 9–10, 473, 474, 476, **477**, 478
Spirochaeta smaragdinae, 10, 476, 478, **482**, 485
Spirochaeta stenostrepta, 10, 474–478, **482**, 483, 484, 486, 502, 508, 522
Spirochaeta thermophila, 10, 476–478, **483**, 486
Spirochaeta zuelzerae, 10, 474–476, 478, **483**, 484, 486, 502, 508, 521
SPIROCHAETES, 2, 9, 9–10, 10, 16, 23, **471**, 472–559, 737
SPIROCHAETIA, 9, **471**
Spiroplasma, 11, 23, 570, 571, 581, 592, 595, 618, 640, 644, 645, 647, **654**, 656, 657, 658, 659, 661, 662, 663, 664, 667–677, 687, 690, 723
Spiroplasma alleghenense, 11, 661, 668, **669**, 670
Spiroplasma apis, 572, 656, 657, 661, 664, 666–668, **670**, 674–675, 677
Spiroplasma atrichopogonis, 661, **670**
Spiroplasma cantharicola, 11, 661, 668, **670**
Spiroplasma chinense, 11, 661, 668, **670**, 677
Spiroplasma chrysopicola, 11, 661, 668–669, **670**, 671
Spiroplasma citri, 11, 572, 654–**669**
Spiroplasma clarkii, 661, 666, 668, **671**
Spiroplasma corruscae, 11, 657, 661, 668, **671**, 677
Spiroplasma culicicola, 11, 661, 668, **671**
Spiroplasma diabroticae, 11, 661, 668, **671**, 672
Spiroplasma diminutum, 11, 661, 668, 670, **672**
Spiroplasma floricola, 656–657, 661, 666–668, **672**
Spiroplasma gladiatoris, 11, 661, 668, **672**, 677
Spiroplasma helicoides, 11, 661, 668, **672**, 677
Spiroplasma insolitum, 11, 657, 661, 668, **672**, 673
Spiroplasma ixodetis, 11, 572, 657, 661, 663, 666–668, **673**
Spiroplasma kunkelii, 655, 658–662, 664–668, **673**, 698
Spiroplasma lampyridicola, 11, 654, 661, 668, **673**
Spiroplasma leptinotarsae, 11, 661–662, 666, 668, **673**, 674
Spiroplasma leucomae, 11, **674**
Spiroplasma lineolae, 11, 661, 668, **674**
Spiroplasma litorale, 11, 661, 668, 671, **674**, 677
Spiroplasma melliferum, 11, 655–659, 661, 664, 668, **674**, 852
Spiroplasma mirum, 656–657, 661, 663–669, **674**, 675
Spiroplasma monobiae, 657, 661, 664, 668, **675**
Spiroplasma montanense, 11, 661, 668, 670, **675**
Spiroplasma penaei, 11, 661, 664, 668, **675**
Spiroplasma phoeniceum, 11, 660–661, 664, 668, **675**
Spiroplasma platyhelix, 11, 654, 657, 661, **676**
Spiroplasma poulsonii, 11, 657, 659, 661, 663–664, 666–669, 672–673, **676**
Spiroplasma sabaudiense, 11, 661, 668–670, **676**
Spiroplasma syrphidicola, 11, 661, 668–670, **676**
Spiroplasma tabanidicola, 661, **677**
SPIROPLASMATACEAE, 11, 23, 570, 644, **654**, 656–677
Spiroplasma taiwanense, 661, 664, 668, **677**
Spiroplasma turonicum, 11, 661, 668, 671, 674, **677**
Spiroplasma velocicrescens, 11, 661, 668, 670, **677**
Spirosoma, 22, 372, 381, 385, 390, 407, 412, **415**–417, 423, 424
SPIROSOMACEAE, 8, 404
Spirosoma linguale, 9, 377, 379, 384, 390, 393, 405, 415, 416, **417**
Spirosoma panaciterrae, 9
Spirosoma rigui, 9, 416, **417**
splanchnicus (Bacteroides), 3, 4, 32, 39, **40**, 41, 74
splanchnicus (Odoribacter), 3, 4, 27, 32, 39, 40, 57, 65, 74, 82
spongiae (Rubritalea), 13, 812, **816**
spongiae (Stenothermobacter), 7, 111, 202, **275**
spongicola (Roseivirga), **456**
Sporocytophaga, 8, 51, 154, 370–372, **418**
Sporocytophaga cauliformis, 113, 136, 153, **154**, 418
Sporocytophaga myxococcoides, 9, 368, 379, 393, **418**
spumans (Mycoplasma), 582, **609**
sputigena (Capnocytophaga), 5, 168, 169, 171, 172, 174, **176**
squalenifaciens (Rubritalea), 13, 801, 812, 813, **816**
staleyi (Pirellula), 16, 897, 903, 904, **905**, 906–908
Stanierella, 109–110, 221–223, 374
Stanierella latercula, 107, 111, 159, 161, 223, 242
Staphylococcus aureus, 72, 211
Staphylococcus epidermidis, 67, 68, 211
Stella, 803, 804
stenostrepta (Spirochaeta), 10, 474–478, **482**, 483, 484, 486, 502, 508, 522
Stenothermobacter, 5, 22, 109, 202, **275**
Stenothermobacter spongiae, 7, 111, 202, **275**
stercorea (Prevotella), **100**
stercoris (Bacteroides), 3, 29, 30, **37**
steynii (Pedobacter), 345
stranskae (Planctomyces), 881, 887, 889, 891, 892, **894**
Streptobacillus, 12, 23, 760, 762, 766, **771**, 772

Streptobacillus moniliformis, 12, 752, 767, 770–**772**
Streptococcus, 571, 751
Streptococcus sanguinis, 79
Streptomyces albulus, 332
sturni (Mycoplasma), 582, **609**
sualvi (Mycoplasma), 582, **609**
suanatina (Brachyspira), 541
subdolum (Mycoplasma), 582, **609**
sublithincola (Aequorivita), 7, 156, **157**
Subsaxibacter, 6, 109, 216, 219, **275**, 276, 277, 278, 291
Subsaxibacter broadyi, 6, 111, 223, 242, **276**, 277
Subsaximicrobium, 6, 109, 216, 219, 276, **277**, 278, 279, 291
Subsaximicrobium saxinquilinus, 6, 277, **278**, 279
Subsaximicrobium wynnwilliamsii, 6, 111, 223, 242, 277, **278**
subtilis (Bacillus), 111, 656, 668
succinicans (Cytophaga), 151, 373
succinicans (Flavobacterium), 5, 112, 125, 130, 132, 134, 135, 136, **151**, 373
succinifaciens (Treponema), 502, **519**, 520, 521
succinogenes (Bacteroides), 32, 740, 743
succinogenes subsp. elongatus (Fibrobacter), 740, **743**
succinogenes subsp. succinogenes (Fibrobacter), 740, **743**
suis (Bacteroides), 3, 29, **37**
suis (Chlamydia), 849–855, 857, 859, **861**, 866
suis (Eperythrozoon), 609, 641
suis (Mycoplasma), 580, 583, 640, **641**, **609**
suis (Treponema), 521, **520**
sulfuriphila (Petrimonas), 4, **77**
suncheonense (Flavobacterium), 5, 119, 129, 130, 133, 136, 137, **151**
suwonensis (Pedobacter), 339, 340–342, 344, **347**
synoviae (Mycoplasma), 569, 577–579, 582, 597, 598, **610**
syrphidae (Mesoplasma), 11, 650, **652**
syrphidicola (Spiroplasma), 11, 661, 668, 671, **676**

tabanidae (Mesoplasma), 11, 650, **652**
tabanidicola (Spiroplasma), 661, **677**
taeanense (Chryseobacterium), 5, 182–184, 186, 190, **195**
taffensis (Fluviicola), 7, 325, **327**, 328
taichungense (Chryseobacterium), 182, 183, 185, 187, 189, 190, **195**
taiwanense (Chryseobacterium), 5, 182, 183, 185, 190, **195**
taiwanense (Spiroplasma), 661, 664, 668, **677**
tangerina (Rubritalea), 13, 812, **816**
tannerae (Prevotella), 5, 65, **100**
tanukii (Borrelia), 10, 485, 489, 492, **496**, 497
tartaricus (Ilyobacter), 12, 759, 760, **761**
tectus (Bacteroides), 3, 29, **37**, 38
teganodes (Eperythrozoon), **642**
tegetincola (Flavobacterium), 5, 119, 129, 130, **151**
tegetincola (Nonlabens), 7, 111, 202, **248**, 275
tenax (Epilithonimonas), 5, 181, 182, 185, **212**
TENERICUTES, 1, 2, 10, 21, 23, **567**
tepidum (Chlorobium), 461
termitidis (Bacteroides), 769, 770
termitidis (Sebaldella), 12, 32, 752, 762, 767, 769, **770**
terrae (Chitinophaga), 351, 352, 353, **356**
terrae (Flavobacterium), 5, 138
terrae (Olivibacter), 8
terrae (Opitutus), 13, 818, **820**, 821, 822, 826, 828, 831
terrae (Pedobacter), 345
terricola (Pedobacter), 345
terrigena (Algoriphagus), 9, 427, 430, **433**
terrigena (Flavobacterium), 5, 138
testudineum (Mycoplasma), 582, **610**
testudinis (Mycoplasma), 583, **610**
texasensis (Candidatus Borrelia), **497**
thailandica (Rapidithrix), 9
thalassocola (Winogradskyella), 7, 241, 288, **289**, 290
thalpophilum (Flavobacterium), 138
thalpophilum (Sphingobacterium), 8, 137, 331–332, **338**, 344
theileri (Borrelia), 489, **496**
thermocellum (Clostridium), 741
thermohalophila (Anaerophaga), 4, 51, **52**
thermophila (Spirochaeta), 10, 476–478, **483**, 486
thermophilum (Dictyoglomus), 776, 778, **779**
thermophilum (Flavobacterium), 5, 338
thetaiotaomicron (Bacteroides), 29, 30, **38**, 62, 131
tillae (Borrelia), 489, **496**
timonensis (Prevotella), **101**
Tissierella, 32
Tissierella praeacuta, 32
torquis (Psychroflexus), 7, 111, 160, 223, 242, **258**, 259, 260, 269, 324
tottoriensis (Fulvibacter), 7
Toxothrix, 23, **467**, 468
Toxothrix trichogenes, 8, 9, 467, **468**
trachomatis (Chlamydia), 13, 616, 846, 849–853, 855–857, **858**, 859–861, 866
Treponema, 10, 23, 472, 473, 475, 476, 477, 481, 483, 484, 491, **501**–**511**, 515, 517, 523, 545, 557, 563, 565
Treponema amylovorum, 10, 502, **510**, 513, 521
Treponema azotonutricium, 10, 472, 508, **510**, 521, 563, 564
Treponema berlinense, 502, 508, **511**, 520
Treponema brennaborense, 502, 504, **511**
Treponema bryantii, 502, **511**, 743
Treponema carateum, 501, 502, 503, 505, **512**
Treponema denticola, 472, 476, 486, 501, 502, 503, 504, 510, **512**, 513, 516, 523, 535
Treponema hyodysenteriae, 508, 535, 540
Treponema innocens, 508, 537
Treponema isoptericolens, 502
Treponema lecithinolyticum, 502, **512**, 513
Treponema macrodentium, **520**
Treponema maltophilum, 502, 510, 511, **513**, 514
Treponema medium, 502, 513, **514**
Treponema minutum, 10, 504, **514**, 518
Treponema orale, **521**
Treponema pallidum, 10, 472, 476, 486, 487, 501, 502, 504, 505, 507, **509**, 514, 521, 535
Treponema pallidum subsp. endemicum, 502, 505, **510**
Treponema pallidum subsp. pallidum, 487, 501, 502, 503, 504, 505, 507, 509, **510**, 516
Treponema pallidum subsp. pertenue, 502, 505, **509**
Treponema paraluiscuniculi, 501, 502, 503, 505, **514**
Treponema parvum, 502, **514**, 515, 521
Treponema pectinovorum, 502, 508, 510, 513, **515**, 521
Treponema pedis, 502
Treponema pertenue, 502
Treponema phagedenis, 502, 504, 509, 510, **516**, 521
Treponema porcinum, 502, **516**, 520
Treponema primitia, 10, 508, **517**, 521, 563, 564
Treponema putidum, 502, **517**
Treponema refringens, 502, 504, 509, 510, 516, **517**, 518
Treponema saccharophilum, 502, **518**
Treponema scoliodontus, 503, 504, 506, **518**
Treponema socranskii, 502, 510, 513, **518**, 519
Treponema socranskii subsp. buccale, 502, 519
Treponema socranskii subsp. paredis, 502, 519
Treponema socranskii subsp. socranskii, 502, 519, 521
Treponema succinifaciens, 502, **519**, 520, 521
Treponema suis, 521, **520**
Treponema vincentii, 502, 503, 504, 510, 513, **520**
Treponema zioleckii, 502, **523**
trevisanii (Leptotrichia), 766, **769**
trichogenes (Toxothrix), 8, 9, **467**, 468
trifolii (Candidatus Phytoplasma), **711**
troitsensis (Arenibacter), 5, 162, 163, **164**
tropicus (Psychroflexus), 258, **261**
tumefaciens (Agrobacterium), 111
tuomii (Eperythrozoon), **642**
turcica (Borrelia), 10, 489, **497**
turdi (Borrelia), 10, 485, 487, 489, 496, **497**
turgidum (Dictyoglomus), 778, **779**
turicatae (Borrelia), 10, 488–491, 495, **497**
turicensis (Candidatus Mycoplasma), **612**
Turneria, 559
Turneriella, 10, 552, 557, **558**
Turneriella parva, 10, 547, 551, 558, **559**
turonicum (Spiroplasma), 11, 661, 668, 671, 674, **677**
typhimurium (Salmonella), 751, 886

uenonis (Porphyromonas), 62, 64, **69**
ulcerans (Fusobacterium), 748, 750, 751, **757**
uliginosa (Cellulophaga), 179, 295
uliginosa (Cytophaga), 179, 293, 295, 374
uliginosa (Zobellia), 6, 138, 179, 235, 293, **295**
uliginosum (Flavobacterium), 293, 295
ulmi (Candidatus Phytoplasma), **711**
Ulvibacter, 22, 109, 221–223, 270, **283**, 284
Ulvibacter antarcticus, 7
Ulvibacter litoralis, 7, 283, **284**
ulvicola (Maribacter), 5–6, 235–237, **238**
uniformis (Bacteroides), 3, 29, 30, **38**, 73
urealyticum (Ureaplasma), 11, 613, 614–620, **621**, 622, 656, 668
Ureaplasma, 11, 23, 568, 571, 575, 581, **613**, 614–622, 647
Ureaplasma canigenitalium, 618, **621**
Ureaplasma cati, 11, **621**, 622
Ureaplasma diversum, 11, 616–618, **621**, 622
Ureaplasma felinum, 11, 616, 621, **622**
Ureaplasma gallorale, 11, 618, **622**
Ureaplasma parvum, 11, 614–621, **622**
Ureaplasma urealyticum, 11, 613, 614–620, **621**, 622, 656, 668
ureilyticum (Chryseobacterium), 5, 190
ureolyticus (Bacteroides), 32

vadensis (Victivallis), 13, 788, 791, **792**, 793
vaeyamensis (Flammeovirga), 442, 443, **447**, 457
valaisiana (Borrelia), 485, 489, 490, 492, 496, **497**
vanfongensis (Algoriphagus), 9, 426, 427, **433**

varium (Anaeroplasma), 720–**722**
varium (Fusobacterium), 748, 750, 751, 752, **757**, 758
Veillonella parvula, 79
velocicrescens (Spiroplasma), 11, 661, 668, 670, **677**
verecundum (Mycoplasma), 582, **610**
vermicola (Lutimonas), 5
vermicola (Puniceicoccus), 13, 818, **824**, 825, 826, 831
veroralis (Bacteroides), 32, 101
veroralis (Prevotella), 65, **101**, 105
VERRUCOMICROBIA, 1, 2, 13, 14, 16, 21, 23, 785, 788, 793, **795**–801, 806, 807, 809, 811, 812, 814, 815, 817, 818, 821–823, 834, 835, 837, 838, 843
VERRUCOMICROBIACEAE, 13, 23, 799, 800, 802, **803**, 815, 817
VERRUCOMICROBIAE, 13, 14, 23, 795, 796, **799**, 801, 802, 815, 818, 835
VERRUCOMICROBIALES, 13, 23, 795, 799, 800, **802**, 803, 809, 810, 812, 814
Verrucomicrobium, 23, 795, 800, 802, **803**, 804–807, 809, 810, 814
Verrucomicrobium spinosum, 13, 800, 803, 804, **805**, 806
Vibrio, 359, 408
vibrioforme (Chlorobium), 364, 368
VICTIVALLACEAE, 13, 23, **791**
VICTIVALLALES, 13, 23, 785, 787, 790, **791**, 792
Victivallis, 13, 23, 785, 787, 790, **791**, 792
Victivallis vadensis, 13, 788, 791, **792**, 793
vietnamensis (Echinicola), 437, 439, **440**
vincentii (Treponema), 502, 503, 504, 510, 513, **520**
virosa (Weeksella), 5, 110, 189, 211, 274, **286**, 287, 288
viscericola (Barnesiella), 4, 70, **71**
Vitellibacter, 22, 109, 155, 232, 245, 265, 270, **284**, 285
Vitellibacter vladivostokensis, 7, 155, 156, 284, **285**
vituli (Acholeplasma), 11, 689, **694**
vladivostokensis (Vitellibacter), 7, 155, 156, 284, **285**
volis (Mycoplasma), 604
vrystaatense (Chryseobacterium), 5, 182, 183, 185, 188, **195**, 196
vulgaris (Balneola), 9
vulgatus (Bacteroides), 3, 29, 30, **38**
vulturis (Mycoplasma), 582, **613**

Waddlia, 24, 857, 876, **877**
WADDLIACEAE, 16, 24, 845, 856, **876**
Waddlia chondrophila, 16, **877**
wadei (Leptotrichia), 12, **769**
wanjuense (Chryseobacterium), 5, 182, 183, 185, 190, **196**
Wautersiella, 22, **285**, 286
Wautersiella falsenii, 5, **285**, 286, 287
weaverense (Flavobacterium), 5, 112, 119, 129, 130, 133, 136, 137, **152**
Weeksella, 22, 108, 109, 165, 180, 203, 205, 211, 250, 285, **286**, 287, 288
Weeksella virosa, 5, 110, 189, 211, 274, **286**, 287, 288
Weeksella zoohelcum, 165, 288
weilii (Leptospira), 10, 547, **555**
wenyonii (Eperythrozoon), 611, 641
wenyonii (Mycoplasma), 583, 606, **611**, 612, 641, 642
Winogradskyella, 22, 109, 158, 215, **288**, 289, 290, 292
Winogradskyella epiphytica, **289**
Winogradskyella eximia, **290**
Winogradskyella poriferorum, 7, **290**
Winogradskyella thalassocola, 7, 241, 288, **289**, 290
winogradskyi (Algoriphagus), 9, 426, 427, 430, **433**
Wolbachia, 838, 839
wolbachii (Leptospira), 10, 547, **555**
wolffii (Leptospira), 10, 547, **555**, 556
wynnwilliamsii (Subsaximicrobium), 6, 111, 223, 242, 277, **278**

Xanthomonas maltophilia, 211
xanthum (Flavobacterium), 5, 119, 128–130, 136, 137, 144, 149, **152**
xinjiangense (Flavobacterium), 119, 129, 130, 133, 137, **152**
xinjiangense (Salinimicrobium), 7
xinjiangensis (Hymenobacter), 9, 398, 399, **403**, 404
Xiphinematobacter (Candidatus), 837, **838**, 839, 840
Xiphinematobacter americani (Candidatus), 13, **840**
Xiphinematobacter brevicolli (Candidatus), 13, **839**
Xiphinematobacter rivesi (Candidatus), 13, **840**
Xylanibacter, 21, 85, **102**, 103
Xylanibacter oryzae, 5, **102**, 103
xylanidelens (Persicivirga), 7, 254, **255**
xylanisolvens (Bacteroides), 3
xylanolytica (Cytophaga), 49–52, 374, **375**
xylanolyticus (Bacteroides), 39, **41**

yabuuchiae (Flavobacterium), 137, 331, 333
yaeyamensis (Flammeovirga), 442, 443, **447**, 457
yeatsii (Mycoplasma), **611**
yeomjeoni (Algoriphagus), 9, 426, 427, 430, **433**
yeongjuensis (Niastella), 8
Yeosuana, 22, 110, **291**
Yeosuana aromativorans, 6, 238, **291**
Yersinia pestis, 858
Yersinia pseudotuberculosis, 858
Yersinia pseudotuberculosis subsp. pestis, 858

zalophi (Mycoplasma), **613**
Zavarzinella, 881
Zavarzinella formosa, 16
zeae (Runella), 412–414, **415**
Zeaxanthinibacter, 22
zeaxanthinifaciens (Mesoflavibacter), 7
zeaxanthinifaciens (Mesoflavibacter), 7
zeaxanthinifaciens (Nubsella), 8
Zhouia, 5, 22, 110, **292**
Zhouia amylolytica, 5, **292**
zioleckii (Treponema), 502, **523**
ziziphi (Candidatus Phytoplasma), **712**
Zobellia, 22, 109, 135, 162, 163, 178, 199, 232, 233, 235, 236, 241, 244, 245, 265, 270, 284, **292**, 293, 294, 374
Zobellia amurskyensis, 6, **294**
Zobellia galactanivorans, 6, 236, **293**
Zobellia laminariae, 6, **294**
Zobellia russellii, 6, **294**
Zobellia uliginosa, 6, 138, 179, 235, 293, **295**
zoogleoformans (Bacterium), 101
zoogleoformans (Bacteroides), 32, 33
zoogleoformans (Capsularis), 101
zoogleoformans (Prevotella), 27, 33, **101**
Zoogloea, 101
zoohelcum (Bergeyella), 5, 110, **165**, 166, 182, 189, 197, 288
zoohelcum (Weeksella), 165, 288
zuelzerae (Spirochaeta), 10, 474–476, 478, **483**, 484, 486, 502, 508, 521
Zunongwangia, 22
Zunongwangia profunda, 7
Zymomonas, 408

Printed in the United States of America